Plastics Technology Handbook

Plastics Technology Handbook: Volume 1: Introduction-Properties-Fabrication-Processes

First published in 2010 by
Momentum Press®, LLC
222 East 46th Street, New York, N.Y. 10017
www.momentumpress.net

ISBN-13: 978-1-60650-079-8 (hard back, case bound)
ISBN-10: 1-60650-079-1 (hard back, case bound)

ISBN-13: 978-1-60650-081-1 (e-book)
ISBN-10: 1-60650-081-3 (e-book)

DOI forthcoming

Cover Design by Jonathan Pennell
Interior Design by Scribe, Inc. (www.scribenet.com)

First Edition October 2010

10 9 8 7 6 5 4 3 2 1

Printed in Taiwan

Plastics Technology Handbook

Volume 1

Introduction • Properties Fabrication • Processes

Edited by

Donald V. Rosato, PhD, MBA, MS, BS, PE

Plastisource Inc.

Society of Plastics Engineers

Plastics Pioneers Association

University of Massachusetts–Lowell Plastics Advisory Board

Marlene G. Rosato, BASc (ChE) P Eng

Gander International Inc.

Canadian Society of Chemical Engineers

Association of Professional Engineers of Ontario

Product Development and Management Association

Nick R. Schott, PhD, MS, BS (ChE), PE

University of Massachusetts–Lowell Professor of Plastics Engineering & Plastics Department Head (Retired)

Plastics Institute of America

Secretary & Director for Educational and Research Programs

MOMENTUM PRESS

Momentum Press, LLC, New York

Contents

Figures

Tables

PREFACE

This book, as a four-volume set, offers a simplified, practical, and innovative approach to understanding the design and manufacture of products in the world of plastics. Its unique review will expand and enhance your knowledge of plastic technology by defining and focusing on past, current, and future technical trends. Plastics behavior is presented to enhance one's capability when fabricating products to meet performance requirements, reduce costs, and generally be profitable. Important aspects are also presented for example to gain understanding of the advantages of different materials and product shapes. Information provided is concise and comprehensive.

Prepared with the plastics technologist in mind, this book will be useful to many others. The practical and scientific information contained in this book is of value to both the novice including trainees and students, and the most experienced fabricators, designers, and engineering personnel wishing to extend their knowledge and capability in plastics manufacturing including related parameters that influence the behavior and characteristics of plastics. The tool maker (mold, die, etc.), fabricator, designer, plant manager, material supplier, equipment supplier, testing and quality control personnel, cost estimator, accountant, sales and marketing personnel, new venture type, buyer, vendor, educator/trainer, workshop leader, librarian, industry information provider, lawyer, and consultant can all benefit from this book. The intent is to provide a review of the many aspects of plastics that range from the elementary to practical to the advanced and more theoretical approaches. People with different interests can focus on and interrelate across subjects in order to expand their knowledge within the world of plastics.

Over 20000 subjects covering useful pertinent information are reviewed in different chapters contained in the four volumes of this book, as summarized in the expanded table of contents and index. Subjects include reviews on materials, processes, product designs, and so on. From a pragmatic standpoint, any theoretical aspect that is presented has been prepared so that the practical person will understand it and put it to use. The theorist, in turn will gain an insight into the practical limitations that exist in plastics as they exist in other materials such as steel, wood, and so on. There is no material that is "perfect." The four volumes of this book together contain 1800 plus figures and 1400 plus tables providing extensive details to supplement the different subjects.

In working with any material (plastics, metal, wood, etc.), it is important to know its behavior in order to maximize product performance relative to cost/efficiency. Examples of different plastic materials and associated products are reviewed with their behavior patterns. Applications span toys, medical devices, cars, boats, underwater devices, containers, springs, pipes, buildings, aircraft, and spacecraft. The reader's product to be designed and/or fabricated can directly or indirectly be related to products reviewed in this book. Important are behaviors associated with and interrelated with the many different plastics materials (thermoplastics, thermosets, elastomers, reinforced plastics) and the many fabricating processes (extrusion, injection molding, blow molding, forming, foaming, reaction injection molding, and rotational molding). They are presented so that the technical or nontechnical reader can readily understand the interrelationships of materials to processes.

This book has been prepared with the awareness that its usefulness will depend on its simplicity and its ability to provide essential information. An endless amount of data exists worldwide for the many plastic materials that total about 35000 different types. Unfortunately, as with other materials, a single plastic material does not exist that will meet all performance requirements. However, more so than with any other materials, there is a plastic that can be used to meet practically any product requirement(s). Examples are provided of different plastic products relative to critical factors ranging from meeting performance requirements in different environments to reducing costs and targeting for zero defects. These reviews span small to large and simple to complex shaped products. The data included provide examples that span what is commercially available. For instance, static physical properties (tensile, flexural, etc.), dynamic physical properties (creep, fatigue, impact, etc.), chemical properties, and so on, can range from near zero to extremely high values, with some having the highest of any material. These plastics can be applied in different environments ranging from below and on the earth's surface, to outer space.

Pitfalls to be avoided are reviewed in this book. When qualified people recognize the potential problems that can exist, these problems can be designed around or eliminated so that they do not affect the product's performance. In this way, costly pitfalls that result in poor product performance or failure can be reduced or eliminated. Potential problems or failures are reviewed with solutions also presented. This failure/solution review will enhance the intuitive skills of people new to plastics as well as those who are already working in plastics. Plastic materials have been produced worldwide over many years for use in the design and fabrication of all kinds of plastic products that profitably and successfully meet high quality, consistency, and long-life standards. All that is needed is to understand the behavior of plastics and properly apply these behaviors.

Patents or trademarks may cover certain of the materials, products, or processes presented. They are discussed for information purposes only and no authorization to use these patents or trademarks is given or implied. Likewise, the use of general descriptive names, proprietary names, trade names, commercial designations, and so on does not in any way imply that they may be used freely. While the information presented represents useful information that can be studied or

analyzed and is believed to be true and accurate, neither the authors, contributors, reviewers, nor the publisher can accept any legal responsibility for any errors, omissions, inaccuracies, or other factors. Information is provided without warranty of any kind. No representation as to accuracy, usability, or results should be inferred.

Preparation for this book drew on information from participating industry personnel, global industry and trade associations, and the authors' worldwide personal, industrial, and teaching experiences.

DON & MARLENE ROSATO AND NICK SCHOTT, 2010

About the Editors

Dr. Donald V. Rosato, president of PlastiSource, Inc., a prototype manufacturing, technology development, and marketing advisory firm in Massachusetts, United States, is internationally recognized as a leader in plastics technology, business, and marketing. He has extensive technical, marketing, and plastics industry business experience ranging from laboratory testing to production to marketing, having worked for Northrop Grumman, Owens-Illinois, DuPont/Conoco, Hoechst Celanese/Ticona, and Borg Warner/G.E. Plastics. He has developed numerous polymer-related patents and is a participating member of many trade and industry groups. Relying on his unrivaled knowledge of the industry plus high-level international contacts, Dr. Rosato is also uniquely positioned to provide an expert, inside view of a range of advanced plastics materials, processes, and applications through a series of seminars and webinars. Among his many accolades, Dr. Rosato has been named Engineer of the Year by the Society of Plastics Engineers. Dr. Rosato has written extensively, authoring or editing numerous papers, including articles published in the *Encyclopedia of Polymer Science and Engineering*, and major books, including the *Concise Encyclopedia of Plastics*, *Injection Molding Handbook 3rd ed.*, *Plastic Product Material and Process Selection Handbook*, *Designing with Plastics and Advanced Composites*, and *Plastics Institute of America Plastics Engineering, Manufacturing and Data Handbook*. Dr. Rosato holds a BS in chemistry from Boston College, MBA at Northeastern University, MS in plastics engineering from University of Massachusetts Lowell, and PhD in business administration at University of California, Berkeley.

Marlene G. Rosato, with stints in France, China, and South Korea, has very comprehensive international plastics and elastomer business experience in technical support, plant start-up and troubleshooting, manufacturing and engineering management, business development and strategic planning with Bayer/Polysar and DuPont and does extensive international technical, manufacturing, and management consulting as president of Gander International Inc. She also has an extensive writing background authoring or editing numerous papers and major books, including the *Concise Encyclopedia of Plastics*, *Injection Molding Handbook 3rd ed.*, and the *Plastics Institute of America Plastics Engineering, Manufacturing and Data Handbook*. A senior member of the Canadian Society of Chemical Engineering and the Association of Professional Engineers of Canada, Ms. Rosato is a licensed professional engineer of Ontario, Canada. She received a Bachelor of Applied Science in chemical engineering from the University of British Columbia with continuing education at McGill University in Quebec, Queens University and the University of Western Ontario both in Ontario, Canada, and also has extensive executive management training.

Professor Nick Schott, a long-time member of the world-renowned University of Massachusetts Lowell Plastics Engineering Department faculty, served as its department head for a quarter of a century. Additionally, he founded the Institute for Plastics Innovation, a research consortium affiliated with the university that conducts research related to plastics manufacturing, with a current emphasis on bioplastics, and served as its director from 1989 to 1994. Dr. Schott has received numerous plastics industry accolades from the SPE, SPI, PPA, PIA, as well as other global industry associations and is renowned for the depth of his plastics technology experience, particularly in processing-related areas. Moreover, he is a quite prolific and requested industry presenter, author, patent holder, and product/process developer, in addition to his quite extensive and continuing academic responsibilities at the undergraduate to postdoctoral level. Among America's internationally recognized plastics professors, Dr. Nick R. Schott most certainly heads everyone's list not only within the 2500 plus global UMASS Lowell Plastics Engineering alumni family, which he has helped grow, but also in broad global plastics and industrial circles. Professor Schott holds a BS in ChE from UC Berkeley, and an MS and PhD from the University of Arizona.

CHAPTER 1

INTRODUCTION TO PLASTICS

WORLDWIDE IMPORTANCE

It would be difficult to imagine the modern world without plastics. Practically all markets worldwide use plastics. Today they are an integral part of everyone's lifestyle, with products varying from commonplace domestic to sophisticated scientific products. Nowadays designers readily turn to plastics. Exceptional progress has been made worldwide in all markets over the past century. As a matter of fact, many of the technical wonders we take for granted would be impossible without versatile, economical plastics.

The information in this book reviews the world of plastics: plastic materials, processes, product designs, and markets that continue to generate the worldwide growth of plastics (Figs. 1.1 to 1.7). Topics from material and product performance to cost analysis are reviewed. Advancing plastic technologies continues to be the top priority in the creation of expanding worldwide markets. In the past, fabricators focused on economies of scale: large plants and mass production. Going forward, fabricators will also concentrate on economies of scope: flexible plants with mass customization. Innovation and responsiveness will replace low rates of change and stability (141).

There have been a number of paradigm shifts in the plastics business model, owing to market changes. Gone are the days of just buying plastic and fabricating. Now industries want design collaboration, numerical analysis and virtual prototyping, global specifications, shorter technology life cycles, quick market introduction windows, and product stewardship such as dematerialization and multiple life cycles. Expectations are higher for plastic materials as well. Metals-to-plastic conversions, micromolded parts, reinforced structural parts, shielded housings, thermoplastic elastomer applications, and parts for harsh environments are making use of a variety of recently developed engineering plastics and filler systems. Machinery builders have kept up with the numerous innovations in processes and materials.

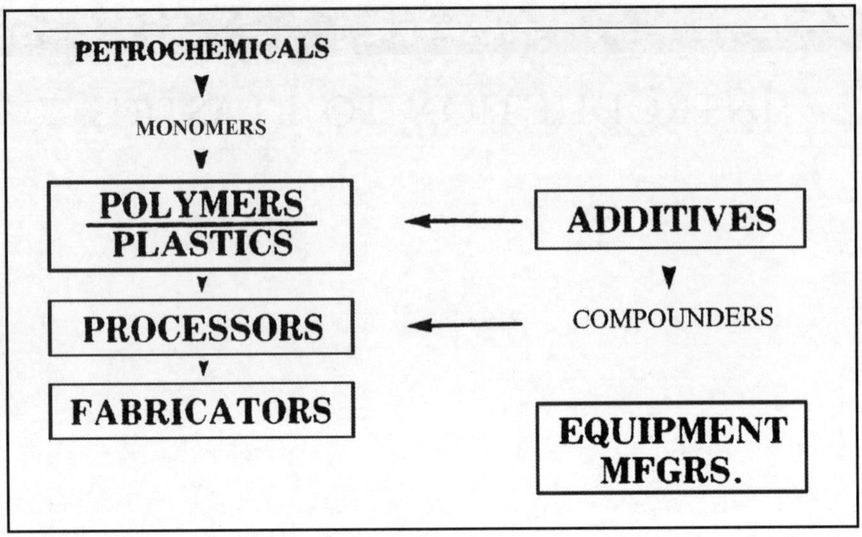

Figure 1.1 Overview chart of petrochemicals to monomers to polymers to plastics to processors to fabricators

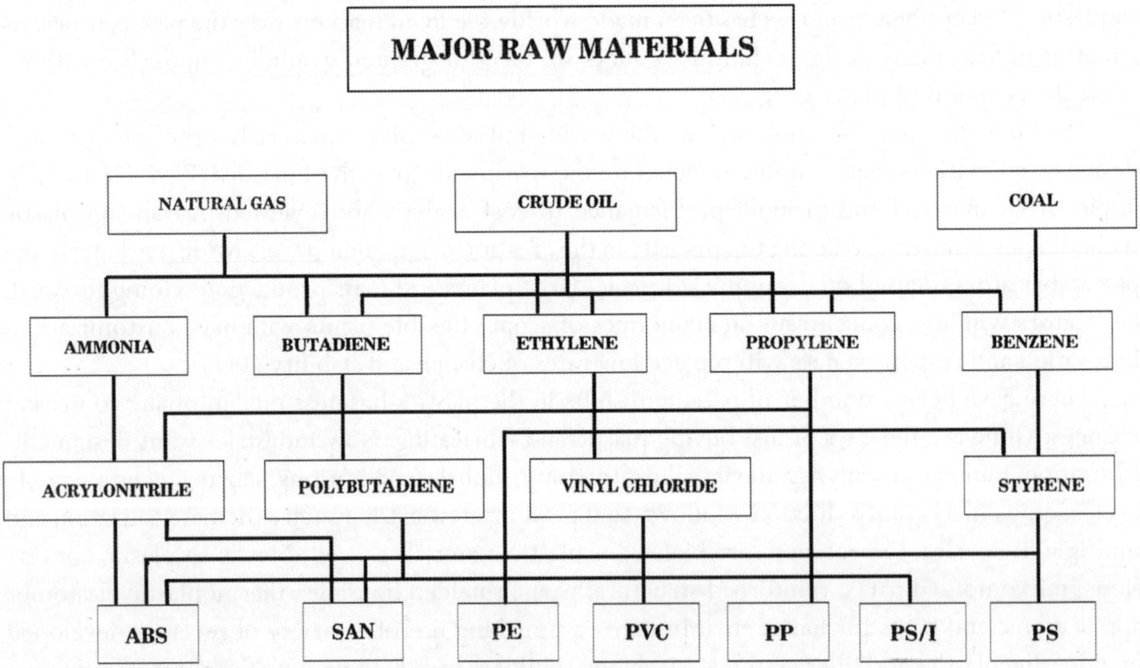

Figure 1.2 Simplified flowchart from major raw material to plastic materials

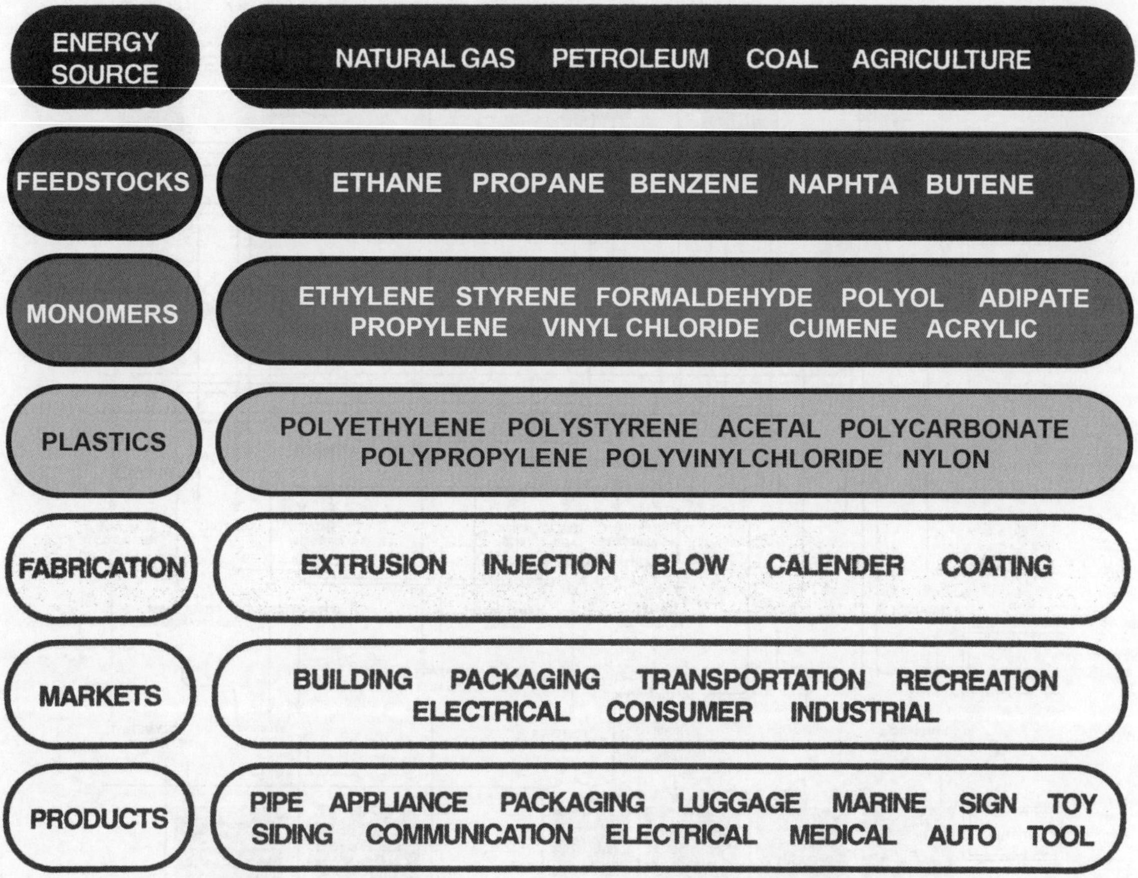

Figure 1.3 Flowchart from energy sources via fabricators to plastic products

Plastics are a worldwide, multibillion-dollar industry in which a steady flow of new plastic materials, new fabrication processes, new design concepts, and new market demands has caused rapid and tremendous growth. The profound impact of plastics to people worldwide and in all industries worldwide is built upon the plastics industry's intelligent practical application of technologies that range from chemistry to engineering. Materials utilize the versatility and vast array of inherent plastic properties as well as high-speed/low-energy processing techniques. The result has been the development of cost-effective products that in turn continue to have exceptional benefits for people and industries worldwide.

Plastic plays an important role in the development of our society worldwide. With properties ranges that can be widely adjusted and ease of processing, plastics can be used to produce highly integrated conventional and customized product solutions. The plastics sector is far from having exhausted the innovation potential that exists. What the worldwide plastics industry offers is

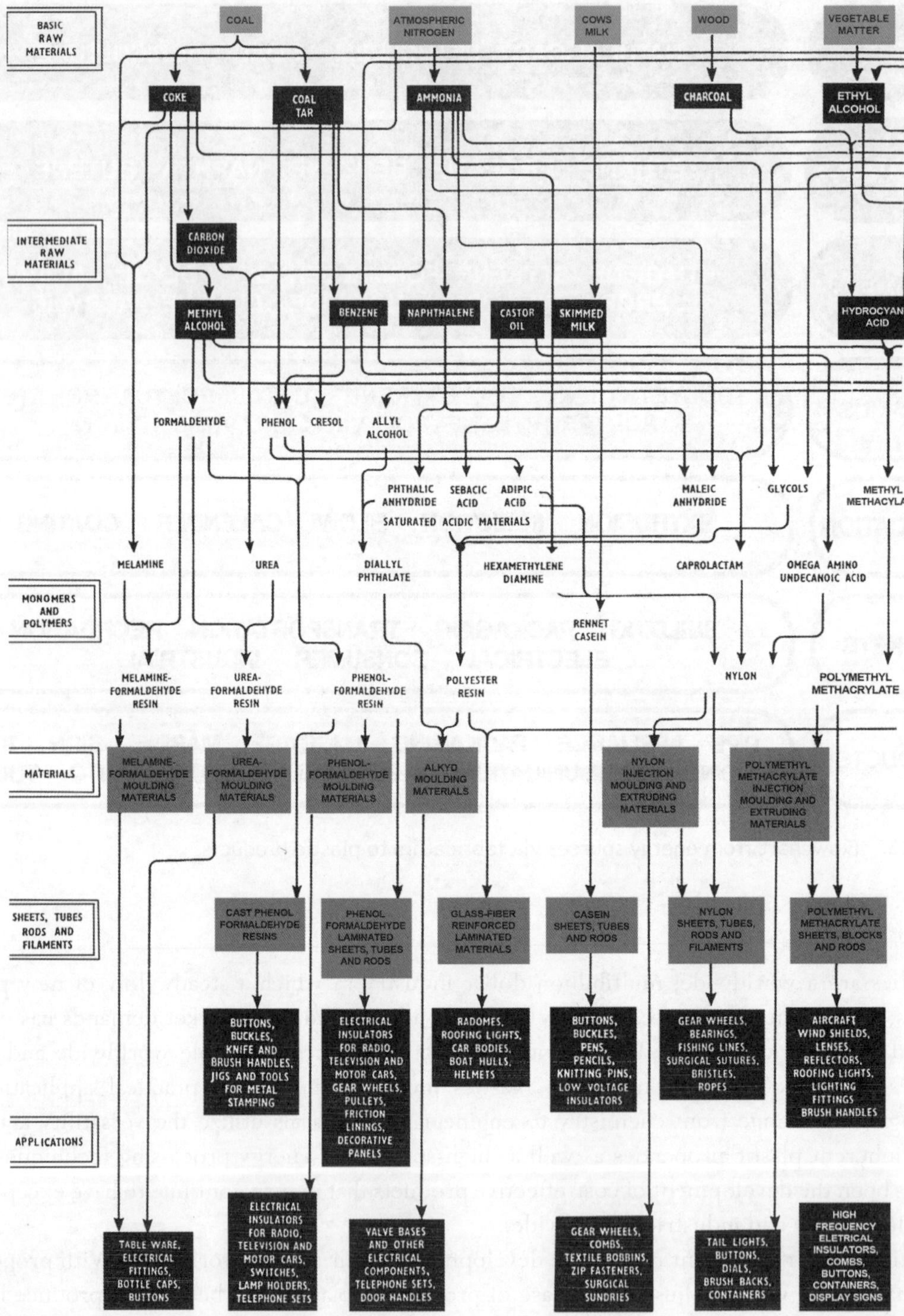

Figure 1.4 Detailed flowchart from raw material to plastic products

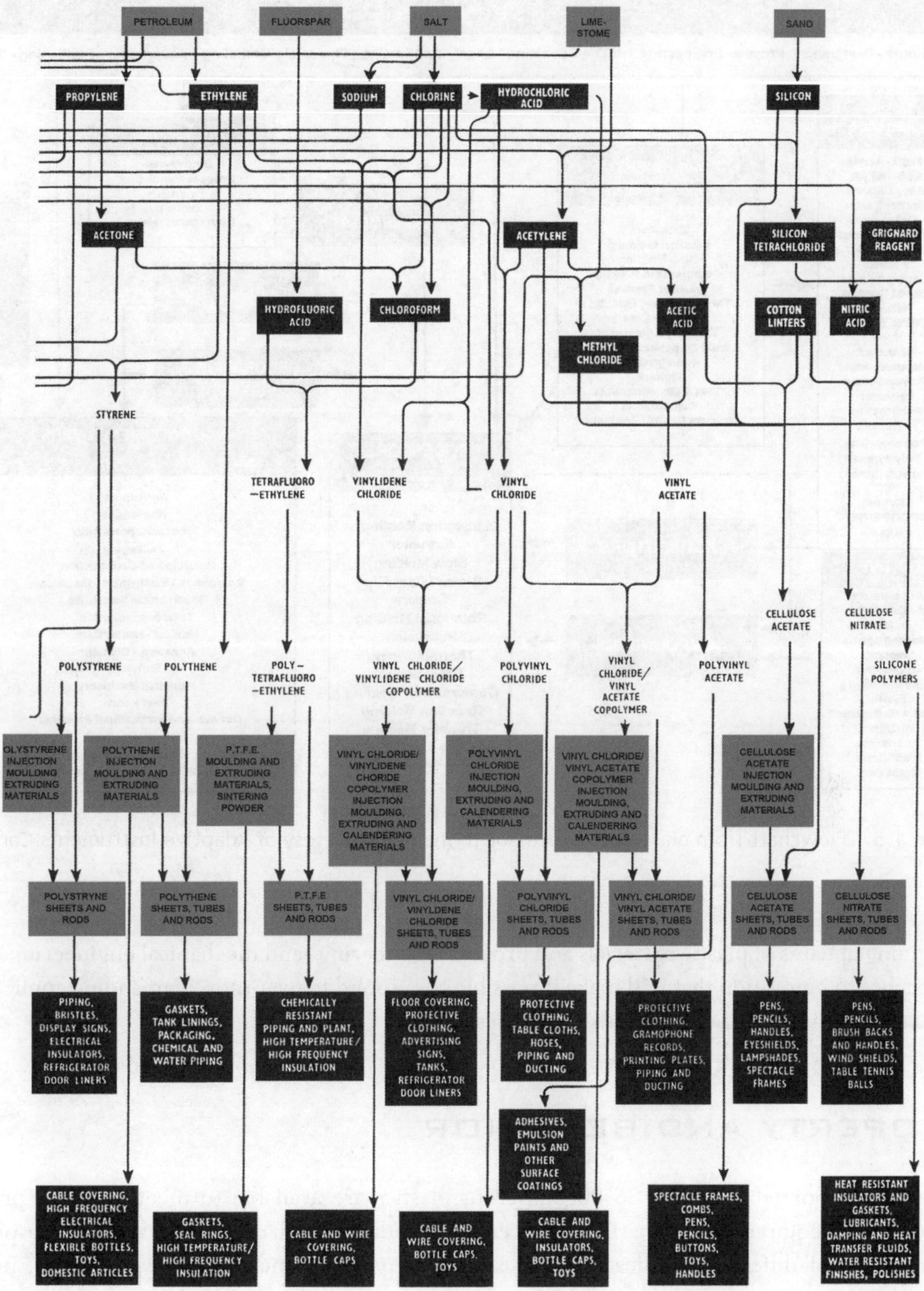

Figure 1.4 *(continued)*

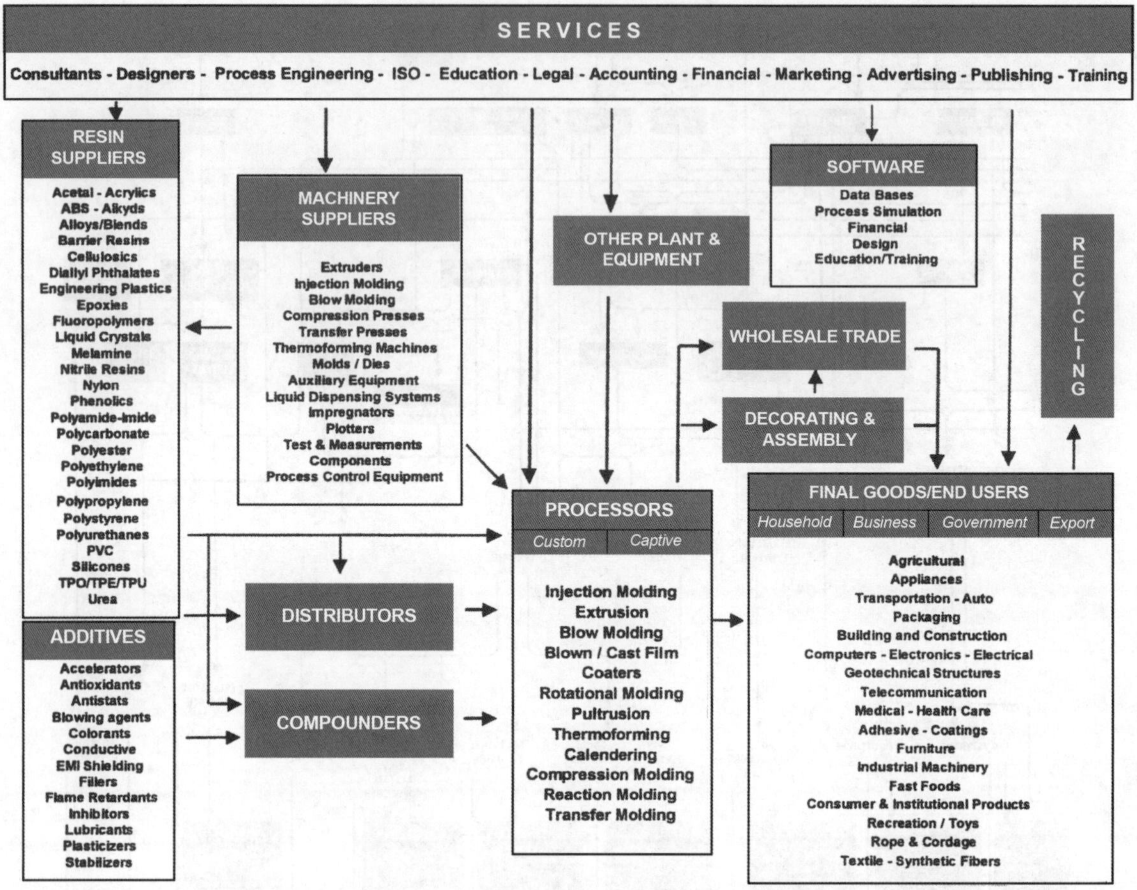

Figure 1.5 Flowchart from plastics to processor to market (courtesy of Adaptive Instruments Corp.)

continuing updates of plastic materials and process engineering- and mechanical engineering-based approaches to innovation that will make it possible to respond to ever more demanding applications or the substitution of other materials by plastics.

PROPERTY AND BEHAVIOR

It has been reported that over 35,000 different plastics are available to meet different product performance requirements (Fig. 1.8), processing standards, and/or cost factors. These plastics are made up of different families of plastics such as polyethylenes, polyvinyl chloride, nylons, fluoroplastics, epoxies, and neoprenes (chapter 2). In turn these families of plastics are compounded into hundreds to thousands of materials meeting different product requirements.

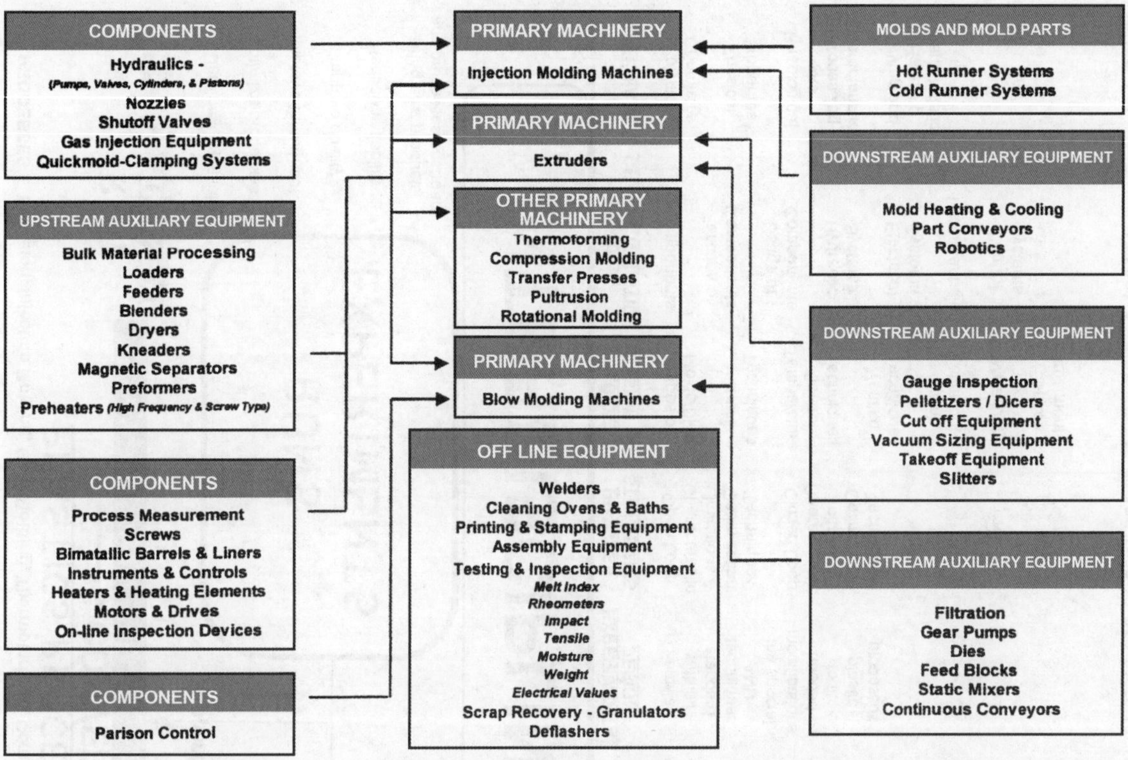

Figure 1.6 Flowchart from equipment to fabricating processes (courtesy of Adaptive Instruments Corp.)

The usefulness of the different plastic materials results from the fact that they include properties such as light weight, resistance in different environments (corrosion resistance, weather resistance, etc.), excellent chemical resistance, and/or a wide range of colors/appearances (chapter 22). Tables 1.1 to 1.7 provide an introduction to a few plastics and some of their properties. The remainder of this book will provide detailed information on many different plastics regarding their diverse properties, fabricating processes, design behaviors, and markets that they serve worldwide.

When designing and/or fabricating a product, a specific plastic is used. It is identified as a type from a plastic producer and/or requirements for a plastic material. Data throughout this book that identifies a plastic such as polyethylene (PE) may differ, since literally thousands of PEs are available. These data are presented to provide examples in their use for a specific plastic. Data for a specific plastic are available from plastic producers and various databases (chapter 25).

As shown in Figures 1.9 and 1.10, plastics are now among the most widely used materials both in the United States and globally, having surpassed steel on a volume basis in 1983. At the start of this century (year 2000), plastics surpassed steel on a weight basis. These figures do not include the two major materials consumed, namely, wood and nonmetallic materials (stone, clay, concrete, glass, etc.). Each represents about 45% by volume of all materials consumed. The remaining 10% consists of plastic, steel, and other materials.

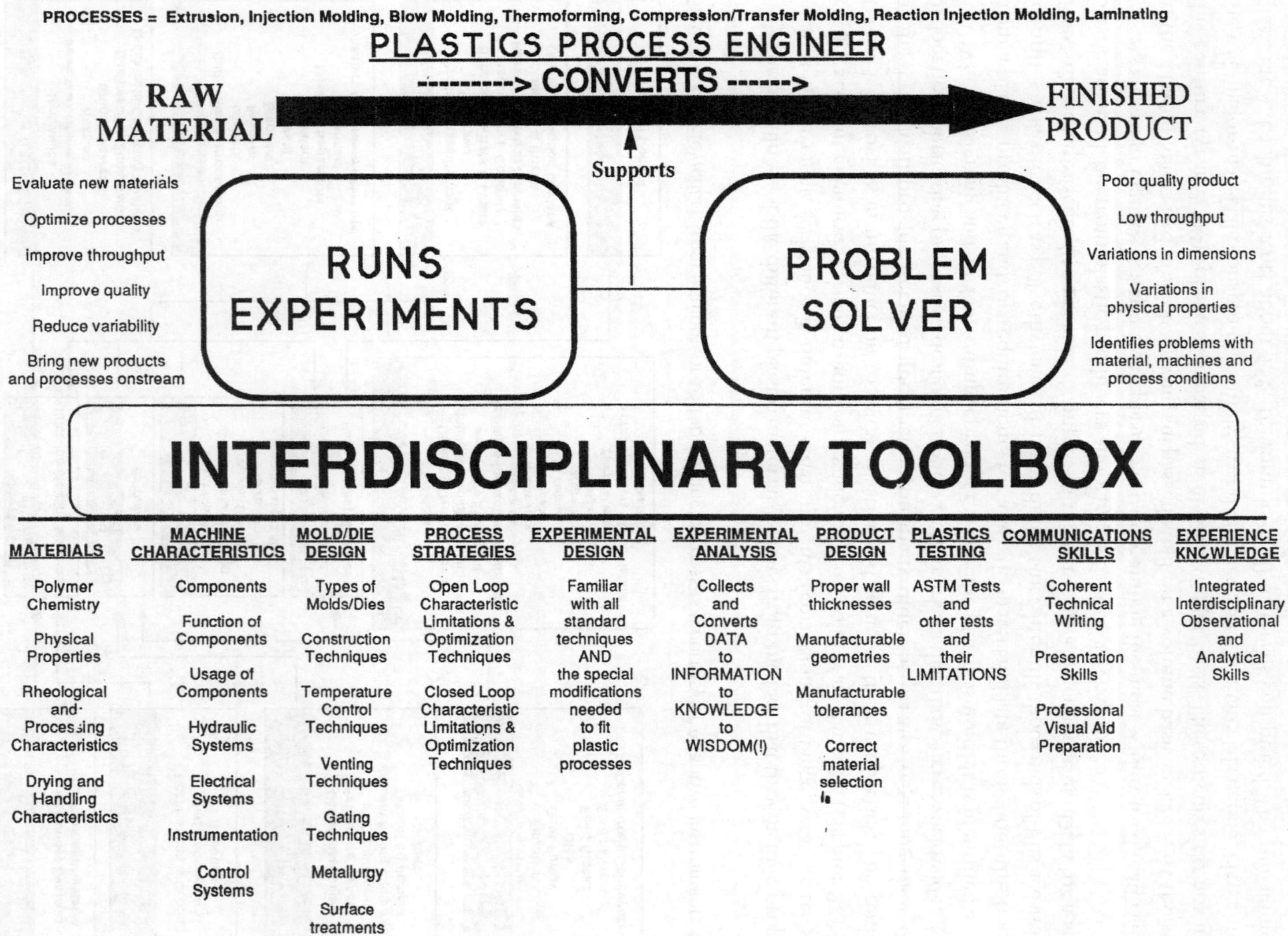

Figure 1.7 Flowchart that converts plastics to finished products (courtesy of Allerlei Consultants)

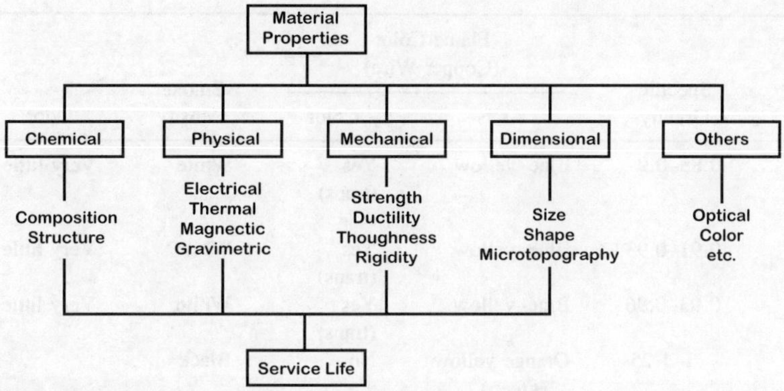

Figure 1.8 Introduction to properties

Material	Spec. gravity	lb/cu in.
ABS	1.04–1.06	0.0375–0.0383
Acetal	1.42	0.0514
Cellulose acetate	1.27	0.0458
Ethyl cellulose	1.10	0.0397
Nylon	1.14	0.0412
Polycarbonate	1.20	0.0433
Polethylene		
Low density	0.914–0.940	0.330–0.339
High density and copolymers	0.941 and up	0.0340–0.0346
Polpropylene	0.905	0.0327
Polystyrene		
General purpose	1.06	0.0383
High impact	1.03–1.04	0.0433–0.0494
Poly(vinyl chloride) compound	1.20–1.37	0.0433–0.0494
Styrene acrylonitrile copolymer	1.07	0.0386
Vinyl acetate copolymer	0.94–0.95	0.0339–0.0343
Aluminum SAE–306	2.77	0.100
Aluminum SAE–309	2.64	0.0953
Brass, yellow	8.5	0.307
Brass, 85/5/5/5	8.75	0.316
Iron, pig, basic	7.1	0.256
Magnesium AZ–91B	1.81	0.283
Steel, CR alloy	7.85	0.283
Steel, tool, standard 0.95C	7.82	0.282
Steel, Stainless 304	7.92	0.286
Zinc SAE–903	6.6	0.238

Table 1.1 Comparison of plastic and other materials weightwise

Material	Specific Gravity	Flame Color (Copper Wire) As is	Color	Smoke Density	Odor	Solvents
Polypropylene	0.85–0.9	Blue–yellow	Yes (trans)	White	Very little	Heavy
LDPE	0.91–0.93	Blue–yellow	Yes (trans)	White	Very little	Candle wax
HDPE	0.93–0.96	Blue–yellow	Yes (trans)	White	Very little	Candle wax
Epoxy	1–1.25	Orange yellow (green)	No	Black		Phenolic
Chlorinated PE	1–2.4	Green	Yes			
Polystyrene	1.05–1.08	Orange yellow	Yes	Black	Dense	Sweet marigolds
Polyvinyl butyral	1.07–1.08	Blue mantle Yellow	Yes (trans)			Rancid butter
Nylon	1.09–1.14	Blue mantle Yellow	Yes			Burnt hair
Ethyl cellulose	1.1–1.16	Blue white	Yes			Sweet
Polyester	1.12–1.46	Yellow	No	Black	Dense	Sweet (resinous)
Vinyl chloride	1.15–1.65	(Green) Yellow–orange	Yes Softening	White to Green	Little	Acrid chlorine
Acrylic	1.18–1.19	Blue mantle Yellow–orange	Yes (trans)	Some Black		Floral burnt fat

Table 1.2 Examples of plastic properties

Plastics success is illustrated by the many millions of plastic products manufactured worldwide; during the start of the twenty-first century, over 350,100 million lb (159 million tons) (Table 1.8) were consumed. The United States consumed over 100,000 million lb; about 90% are thermoplastic (TP) and 10% thermoset (TS) plastics. U.S. and European consumption compose about one-third of the world total. Even though there are worldwide about 35,000 different types of plastic materials, most are not used in large quantities; they have specific performance and/or cost capabilities geared generally for specific products and specific processes that include many thousands of end uses.

CHEMISTRY OF POLYMERS

The materials reviewed in this book, as in the industry, are identified by different terms such as *polymer*, *plastic*, *resin*, *elastomer*, *reinforced plastic* (RP), and *composite unreinforced* or *reinforced plastic*. They are somewhat synonymous. Polymers, the basic ingredients in plastics, can be defined as high molecular weight organic chemical compounds, synthetic or natural substances consisting of molecules. Practically all of these polymers are compounded with other products (additives, fillers, reinforcements, etc.) to

Resin material	Impact strength notched izod, ft-lb/in., ½ in. bar	Tensile strength, psi × 10^3	Tensile modulus, psi × 10^3	Elongation, (%)	Flexural strength, psi × 10^3	Compressive strength, psi × 10^3
1. Acetal	1.2–2.3	8.8–10	400–410	12–75	13–14	18
2. ABS	1.0–9.5	3.5–10.5	200–450	10–100	5–15	5–11
3. Acrylic	0.4	8.7–10.5	380–430	3–6	14–16	14–17
4. Acrylic high impact	0.5–2.3	5.5–8	225–330	23–38	8.5–12	7–12
5. Cellulose acetate	.5–5.6	2.3–8.1	—	10–70	2.2–11.5	2.2–10.9
6. Cellulose acetate butyrate	0.8–6.3	2.6–6.9	—	40–88	1.8–9.3	2.1–9.4
7. Cellulose propionate	0.9–10.2	1.8–7.3	—	30–100	2.8–11	3–9.6
8. Chlorinated polyether	0.4	6	160	60–160	5	—
9. Chlorotrifluoroethylene	3.5	6	150–190	60–190	8–10	6–12
10. Ethyl cellulose	1.7–6	2.3–6.5	—	10–40	3–6.7	—
11. Ethyl vinyl acetate	No break	20–40	3.0–15	500–1500	—	—
12. FEP	No break	2–3.2	60–80	350	—	—
13. Nylon 6	0.9–4	9.5–12.4	200–450	25–300	9–16.6	4–11
14. Nylon 6/6	0.9–2	11.2–13.1	410–480	60–300	14.6	5–13
15. Nylon 6/10	—	7–8.5	160–280	50–300	10.5	4–6
16. Polyallomer	1.5–12	2.9–4.2	100–170	400–650	4–5	—
17. Polycarbonate	2–3	8–9	345	75	11–13.5	12.5
18. Polyethylene low density	No break	1–2.4	14–38	20–800	—	—
19 Polyethylene medium density	No break	1.7–2.8	50–80	80–600	—	—
20. Polyethylene high density	0.5–23	2.8–5	75–160	10–800	1–4	0.8–3.6
21. Polyethylene high molecular weight	>20	5.4	102	525	3.5	2.4
22. Polypropylene	0.5–15	2.3–5	150–250	10–700	4.5–6	6–8
23. Polystyrene	0.25–0.65	5–9	400–500	1–2.5	7–15	11.5–16
24. Polystyrene high impact	0.7–1.5	3.5–8	300–400	10–40	5.5–12.5	8–16
25. Polyurethane	No break	4.5–8	—	100–600	0.7–1	20
26. Poly(vinyl chloride) (flexible)	Varied	1–4	—	100–450	—	—
27. Poly(vinyl chloride) (rigid)	0.4–22	6–9	200–600	5–40	8–15	10–11
28. Poly(vinyl dichloride) (rigid)	0.8–6.3	7.5–8.8	348–450	65	14.2–17	—
29. Styrene acrylonitrile (SAN)	0.3–55	10–12	500	1–3.2	17	15–17.5
30. TFE fluorocarbon	No break	2–5	50–100	75–400	—	4–1.2
31. Ionomer	5.7–14	3.5–5.5	28–40	300–450	—	—
32. Phenoxy	1.5–12	8–9.5	350–410	50–100	12–14.5	—
33. Polyphenylene oxide	—	—	—	50–80	—	—
34. Polysulfone	1.3	10.2	360	50–100	15.4	15.4

Table 1.3 Thermoplastic properties

Resin material	Compressive modulus, psi $\times 10^3$	Heat distortion temperature, °F, 264 psi	Heat resistance, continuous °F	Coefficient thermal expansion, in./in.-°C $\times 10^{-5}$	Thermal conductivity, cal/cm²-sec-°C-cm $\times 10^{-4}$	Volume resistivity, ohm-cm
1. Acetal	410	255	185	8.1–8.5	1.6–5.5	$1–10^{13}$
2. ABS	200–450	185–230	160–235	5.7–10	5–8	$10^{15}–10^{17}$
3. Acrylic	350–430	167–198	130–200	5–8.5	4.4	$>10^{14}$
4. Acrylic high impact	250–360	169–190	140–195	6.5–10.5	4.0	$10^{16}–10^{17}$
5. Cellulose acetate	—	111–195	140–175	8–16	4–8	$10^{10}–10^{12}$
6. Cellulose acetate butyrate	—	113–202	140–175	11–17	4–8	$10^{10}–10^{12}$
7. Cellulose propionate	—	121–228	140–175	11–16	4–8	$10^{12}–10^{16}$
8. Chlorinated polyether	130	185–210	250–275	8	3.13	1.5×10^{16}
9. Chlorotrifluoroethylene	180	160–170	390	5–7	4–6	10^{18}
10. Ethyl cellulose	—	150–200	140–180	10–20	3.8–7	—
11. Ethyl vinyl acetate	—	—	120–170	10–20	8	—
12. FEP	70	124	400	8.3–10.5	5.9	$>10^{18}$
13. Nylon 6	347	150–175	250	8.3	5.9	$3.3^{13}–4.5^{13}$
14. Nylon 6/6	400	200	250	10	5.8	$10^{14}–10^{15}$
15. Nylon 6/10	—	145	220	10	5.5	4.5×10^{12}
16. Polyallomer	—	115–140	250	8–11	2–4	$>10^{16}$
17. Polycarbonate	345	270	250–270	7	4.6	2.1×10^{16}
18. Polyethylene low density	—	—	140–175	10–20	8	$10^{15}–10^{18}$
19 Polyethylene medium density	—	—	150–180	10–20	8	$10^{15}–10^{18}$
20. Polyethylene high density	50–110	110–125	180–250	10–20	8	$6\times10^{15}–10^{18}$
21. Polyethylene high molecular weight	110	120	250	13	8	$>10^{16}$
22. Polypropylene	—	140–200	250	6–8.5	2.8–4	6.5×10^{16}
23. Polystyrene	300–560	167–203	150–180	6–8	1.9–3.3	$10^{17}–10^{21}$
24. Polystyrene high impact	—	150–200	120–170	6.5–8.5	1–3	$10^{13}–10^{17}$
25. Polyurethane	—	—	150–180	10–20	5	$2–10^{11}$
26. Poly(vinyl chloride) (flexible)	—	—	—	7–25	3–4	$10^{11}–10^{15}$
27. Poly(vinyl chloride) (rigid)	300–400	140–175	160–165	5–10	3–5	$10^{12}–10^{16}$
28. Poly(vinyl dichloride) (rigid)	—	212–220	185–210	7–8	3–4	10^{15}
29. Styrene acrylonitrile (SAN)	650	200–208	—	7	3	10^{15}
30. TFE fluorocarbon	70–90	132	550	5.5(25–60°C)	6	$>10^{18}$
31. Ionomer	—	—	140	12–13	5.8	$>10^{16}$
32. Phenoxy	325	175–188	—	3.2–3.8	—	$2.75–5\times10^{-5}$
33. Polyphenylene oxide	—	375	—	—	—	10^{17}
34. Polysulfone	370	345	300	$3.1–10^{-5}$ in./in. °F	1.8 BTU/hr-sq ft-ft lb	5×10^{16}

Table 1.3 *(continued)*

Resin material	Dielectric constant, 60 cycles	Delectric strength, ST $\frac{1}{8}$-in. thickness, volts/mil	Power factor, 60 cycles	Arc resistance, sec	Water absorption, 24 hr, %	Rockwell hardness
1. Acetal	3.7–3.8	500	.004–.005	129	0.12–0.25	M94, R120
2. ABS	2.6–3.5	300–450	.003–.007	45–90	0.2–0.4	R50–120
3. Acrylic	3.7	450–500	.04–.05	No tracking	0.3	M84–97
4. Acrylic high impact	3.5–3.7	450–480	.04–.05	No tracking	0.2–0.3	M20–67
5. Cellulose acetate	3.5–7.5	290–600	.01–.06	50–130	2.1–4.2	R35–118
6. Cellulose acetate butyrate	3.5–6.4	250–400	.01–.04	—	0.9–2.2	R31–116
7. Cellulose propionate	3.4–4.2	300–450	.01–.04	170–190	1.2–2.8	R15–120
8. Chlorinated polyether	3	400	.01	—	.01	R100
9. Chlorotrifluoroethylene	2.65	450	.015	>360	Nil	R85–112
10. Ethyl cellulose	—	—	—	—	1.3–1.5	R50–110
11. Ethyl vinyl acetate	2.3	—	—	—	<.01	R3–7
12. FEP	2.1	500	.0002	>165	<.01	—
13. Nylon 6	6.1	300–400	0.4–0.6	140	1.5	R107–119
14. Nylon 6/6	3.6–4.0	300–400	.014	140	1.3	R118–123
15. Nylon 6/10	4.0–7.6	300–400	.04–.05	140	0.4	R111
16. Polyallomer	3.2	500–1000	.0001–.0005	—	<.01	R50–85
17. Polycarbonate	3.17	400	.0009	120	15	M80, R118
18. Polyethylene low density	2.28	450–1000	.0001–.0005	Melts	<.02	R10
19 Polyethylene medium density	2.3	450–1000	.0001–.0005	Melts	<.02	R15
20. Polyethylene high density	2.3	450–1000	.0001–.0005	Melts	<.01	R30–60
21. Polyethylene high molecular weight	2.3	710	.0003	—	<.01	R55
22. Polypropylene	2.1–2.27	450–1000	.0001–.0005	13–185	<.01	R30–99
23. Polystyrene	2.5–2.65	500–700	.0001–.0005	60–100	.03–.05	M60–80
24. Polystyrene high impact	2.5–3.5	500	.003–.005	60–90	.05–.10	M25–69
25. Polyurethane	6.7–7.5	450–550	.015–.017	—	0.60–0.80	M28, R60
26. Poly(vinyl chloride) (flexible)	5–9	300–1000	.08–.15	—	0.15–0.75	—
27. Poly(vinyl chloride) (rigid)	3.4	425–1040	.01–.02	—	.07–.40	R100–120
28. Poly(vinyl dichloride) (rigid)	—	1200	—	—	0.11	R118
29. Styrene acrylonitrile (SAN)	2.8–3	400–500	—	—	0.23–0.28	M30–83
30. TFE fluorocarbon	2.1	400	<.0001	No tracking	.01	R58
31. Ionomer	2.4–2.5	1000	0.1	—	0.1–1.4	D60–65
32. Phenoxy	4.1	404–520	.0012–.0009	70	0.13	R113–118
33. Polyphenylene oxide	—	400–500	—	—	—	—
34. Polysulfone	2.82	425	.0008–.0056	122	0.22	M69, R120

Table 1.3 *(continued)*

Resin material	Flammability, in./min	Specific gravity	Mold shrinkage, in./in.	Clarity
1. Acetal	1.1	1.410–1.425	.022	Translucent to opaque
2. ABS	1.0–2	1.01–1.07	.003–.007	Opaque
3. Acrylic	0.6–0.7	1.18–1.18	.002–.006	Transparent
4. Acrylic high impact	1.1–1.2	1.11–1.18	.004–.008	Translucent to opaque
5. Cellulose acetate	0.5–2	1.23–1.34	.001–.007	Transparent
6. Cellulose acetate butyrate	0.5–1.5	1.15–1.22	.003–.006	Transparent
7. Cellulose propionate	0.5–1.5	1.16–1.23	.001–.006	Transparent
8. Chlorinated polyether	Self-ext.	1.4	.004–.006	Semitranslucent to opaque
9. Chlorotrifluoroethylene	Nil	2.90–2.14	.010–.015	Transparent to opaque
10. Ethyl cellulose	—	1.11–1.13	—	Transparent to opaque
11. Ethyl vinyl acetate	Slow burning	0.93–0.95	.01–.02	Transparent
12. FEP	Nonflam.	2.14	.03	Transparent to opaque
13. Nylon 6	Self-ext.	1.13–1.14	.007–.011	Transparent to opaque
14. Nylon 6/6	Self-ext.	1.13–1.15	.007–.015	Transparent to opaque
15. Nylon 6/10	Self-ext.	1.07–1.09	.015	—
16. Polyallomer	Slow burning	0.90–.906	.01–.02	Transparent to opaque
17. Polycarbonate	Self-ext.	1.2	.005–.007	Transparent
18. Polyethylene low density	Slow burning	.910–.925	.01–.03	Transparent to opaque
19 Polyethylene medium density	Slow burning	.926–.940	.01–.035	Transparent to opaque
20. Polyethylene high density	Slow burning	0.94–0.98	.01–.04	Translucent to opaque
21. Polyethylene high molecular weight	Very slow	0.94	.03	Translucent to opaque
22. Polypropylene	Slow burning	0.90–.0908	.008–.025	Translucent to opaque
23. Polystyrene	0.5–2.5	1.05–1.06	.002–.006	Transparent
24. Polystyrene high impact	0.5–2.5	1.04–1.06	.003–.005	Translucent to opaque
25. Polyurethane	Slow burning	1.20–1.26	.008–.012	Translucent to opaque
26. Poly(vinyl chloride) (flexible)	Self-ext.	1.15–1.80	.002–.004	Transparent to opaque
27. Poly(vinyl chloride) (rigid)	Self-ext.	1.33–1.58	—	Transparent to opaque
28. Poly(vinyl dichloride) (rigid)	Self-ext.	1.50–1.54	.007–.008	Translucent to opaque
29. Styrene acrylonitrile (SAN)	0.47–0.7	1.07–1.08	.003–.004	Transparent
30. TFE fluorocarbon	Nonflam.	2.13–2.18	.02–.06	Transparent to opaque
31. Ionomer	9–1.1	0.94–0.96	.001–.005	Transparent
32. Phenoxy	Slow burning, Self-ext.	1.17–1.34	.003–.004	Transparent to opaque
33. Polyphenylene oxide	Self-ext.	1.06	—	—
34. Polysulfone	Self-ext.	1.24–1.25	.0076	Transparent to opaque

Table 1.3 *(continued)*

Resin material	Impact strength, izod ft-lb/in.	Tensile strength, psi×10³	Flexural strength, psi×10³	Flexural modulus, psi×10³	Compressive strength, psi×10³	Heat distortion °F@264 psi
1. Alkyd glass-filled general purpose	1.5–4	5–9	12–18	1700–2000	21–29	>400
2. Alkyd-glass MAI 30	3–4	7–9	14–19	2000	22–24	>400
3. Alkyd-glass MAI 60	8–10	5–9	12–20	2000–2500	24–36	>400
4. Alkyd mineral	.30–.35	3–8	6–15	2200–3000	16–20	350–400
5. DAP glass MIL–M–19833 GDI–30	6–15	7–9.5	17–19	1250–1600	24–45	390–500
6. DAP mineral (MDG) MIL–M–14F MIL–P–4389	.28–.40	5.5–6.5	9.6–10	1200	22–25	300–320
7. DAP orlon MIL–F–14F SD 1–5	.55–4.5	6.8	9–10.5	710	20–30	240–266
8. DAP glass MIL–M–14F SDG	0.4–1.2	6.7–9.2	10–19	1300	25–30	350–500
9. DAP unfilled	0.3	4	7–9	600	22–24	310
10. Epoxy glass	5.5	11.5	22	3000–3200	30–38	500
11. Epoxy mineral	0.35–0.5	6–11	19.5	900–1200	37	260–360
12. Melamine cellulose	.25–.35	5–10	10–16	1300–1600	25–24	350–410
13. Melamine cloth	.55–.90	7–10	12–15	1600	30–35	310
14. Melamine glass MMI 30	4–16	6–10	15–24	2400	20–32	400
15. Phenolic asbestos MFA 30	3.7	8	16	2500	24	400
16. Phenolic cloth CFI–10	1.05–2.2	6.5–7	0.5–10	1000–1200	20–25	300–360
17. Phenolic cloth CFI–20	2.3	6.5–7.5	10–12.5	1000	22–25	300
18. Phenolic flock/flour	.34–.50	6–9	7–12	1000	20–30	270–325
19. Phenolic glass	9–17	7–11	13.5–22	3000	14–35	600
20. Phenolic GP	.26–.38	6–9	7–12	1000	25–35	275–370
21. Phenolic GPI–100	13	16.8	24.9	2865	39	>550
22. Phenolic MFE	0.5	9.2	17.5	2600	32	415
23. Phenolic mineral MFG	0.64–1.5	4–7	8.5–11.5	1400	17.5–20	320–500
24. Phenolic mineral MFH	0.26	5–8	6–9	1500	25–35	300–450
25. Phenolic mineral MFI–20	5	8.7	18.6	2500	34.5	>500
26. Phenolic nylon	0.5	8	12	600	24	290
27. Phenolic rubber/flour	0.5	4	5.5–7	400	16–18	250
28. Urea cellulose	.24–.35	5–10	10–18	1300–1600	25–38	266–380
29. Silicone	8–20	4–8	13–19	2100–2500	10–13	>900
30. Silicone mineral	.25–.39	2.5–4.4	6.8–8	1250–2270	11–18	340–900

Table 1.4 Thermoset plastic properties

Resin material	Heat resistance, continuous °F	Thermal expansion in./°C$\times 10^{-5}$	Thermal conductivity, cal/sec-cm^2-°C-cm$\times 10^{-4}$	Volume resistivity, ohm/cm	Dielectric constant, 60 cycles	Dielectric strength, STI/8V.P.M.
1. Alkyd glass-filled general purpose	300	2–5	10–15	1×10^{15}	6.7	275–375
2. Alkyd-glass MAI 30	350	2–4	10–15	—	6	350
3. Alkyd-glass MAI 60	300	2–5	8–12	10^{12}–10^{14}	5–5.6	375
4. Alkyd mineral	275–400	2–5	15–25	10^{12}–10^{14}	5.7–6.3	375
5. DAP glass MIL–M-19833 GDI–30	365–465	2.7–3.5	6–8	10^{12}	4.5	350–400
6. DAP mineral (MDG) MIL–M–14F MIL–P–4389	350–440	3.5–4.2	13.7	10^{13}+	5.2	350–450
7. DAP orlon MIL–F–14F SD 1–5	300–350	4–5.4	6.3–6.9	1×10^{10}	3.8	366–400
8. DAP glass MIL–M–14F SDG	365–465	1.2–3.5	6.7	10^{13}	4.25	420
9. DAP unfilled	350	—	—	2×10^{16}	3.6	450
10. Epoxy glass	300-500	1.08–1.2	8.5	3.8×10^{15}	5.5(@100)	360 volts/mil
11. Epoxy mineral	550	3.7	18.1	10^{14}–10^{16}	4.4	400
12. Melamine cellulose	210	2–5.7	7–10.1	0.8–2.0$\times 10^{12}$	7.9–9.5	300–400
13. Melamine cloth	250	2.5–3	10.6	1–3$\times 10^{11}$	8.1–12.6	250–340
14. Melamine glass MMI 30	300–400	1.2–2	11.5	2×10^{11}	9.7–11.1	170–240
15. Phenolic asbestos MFA 30	450	.5–1	14	—	—	85
16. Phenolic cloth CFI–10	275	2–3	9.3	5–8$\times 10^{11}$	6.1–21.2	250–400
17. Phenolic cloth CFI–20	340	1.5–3	7	6×10^{10}	5.2–21	300
18. Phenolic flock/flour	300	2–3.9	7–8	2×10^{11}	6–10	300–375
19. Phenolic glass	325–450	.8–1.6	9.7	1×10^{12}	5.6–7.2	350–400
20. Phenolic GP	300	3–4.1	5–8.1	2×10^{11}	5–10	300–425
21. Phenolic GPI–100	—	—	32	—	5.4(1MC)	382
22. Phenolic MFE	—	1.1	15	—	4.5(1MC)	360
23. Phenolic mineral MFG	325–425	1.5–2.5	15.87	1×10^{12}	40–60	150–250
24. Phenolic mineral MFH	400–450	1.5	10	1×10^{12}	9–15	250–350
25. Phenolic mineral MFI–20	475	1.5	14	1.6×10^{10}	45	85
26. Phenolic nylon	275	7.5	7.50–18	1×10^{12}	4	450
27. Phenolic rubber/flour	275	1.5–4	5	3.4×10^{9}	9	300
28. Urea cellulose	170	2.7	7–10.1	.5–5$\times 10^{11}$	7–9.5	300–400
29. Silicone	450–700	.8	7.5	3×10^{14}	4.35	250–280
30. Silicone mineral	400–700+	2.5–6	4–10	1×10^{4}	3.6–4.5	280–400

Table 1.4 *(continued)*

Resin material	Power factor, 60 cycles	Arc resistance, sec	Water absorption %-24 hr	Rockwell hardness	Specific gravity	Mold shrinkage, in./in.
1. Alkyd glass-filled general purpose	.01–.02	180+	.07–.2	—	1.93–2.32	.003–.006
2. Alkyd-glass MAI 30	.005–.01	180+	.05	—	2.3	.001–.005
3. Alkyd-glass MAI 60	.020–.030	180+	.07–.10	—	2.02–2.13	.001–.004
4. Alkyd mineral	.030–.045	180+	.05–.12	E108	2.17–2.24	.004–.007
5. DAP glass MIL–M–19833 GDI–30	.010–.017	130–180	.05–.25	M108	1.57–186	.0009–.004
6. DAP mineral (MDG) MIL–M–14F MIL–P–4389	.06–.40	140	.2–.5	M100	1.58–1.74	.004–.008
7. DAP orlon MIL–F–14F SD 1–5	.025–.035	115	.1–.2	M108	1.31–1.34	.009–.010
8. DAP glass MIL–M–14F SDG	.010–.017	125–180	.05	E80	1.6–1.75	.002–.004
9. DAP unfilled	.008	120	.09	M114–116	1.27	.007
10. Epoxy glass	.015	125–187	.03–.05	M105–115	1.8	.0005
11. Epoxy mineral	.011	180–190	.06–.15	M110	1.85	.002
12. Melamine cellulose	.030–.080	120–140	.1–.6	M118–124	1.47–1.52	.006–.015
13. Melamine cloth	.1–.34	125–150	.3–.6	—	1.4–1.5	.003–.005
14. Melamine glass MMI 30	.14–.23	180–186	.09–.6	—	1.9–2	.001–.003
15. Phenolic asbestos MFA 30	—	190	.35	M112	1.95	.0005–.0015
16. Phenolic cloth CFI–10	.16–.64	—	.8–1	M95–107	1.36–1.40	.002–.005
17. Phenolic cloth CFI–20	.64	4	.9–1	E79	1.37–1.40	.002–.005
18. Phenolic flock/flour	.25-.3	80(15518)	.3–1	M95–116	1.34–1.38	.004–.008
19. Phenolic glass	.02–.05	10–60	.5–1	M90–99	1.70–1.90	.0009
20. Phenolic GP	.05–.5	26–102	.2–1	M95–117	1.34–1.46	.004–.008
21. Phenolic GPI–100	.026(1MC)	125	.03	—	1.69	.0001–.001
22. Phenolic MFE	.013(1MC)	180+	.03	—	1.85	.002
23. Phenolic mineral MFG	.25–.50	100–190	.12–.4	M90–115	1.72–1.86	.001–.002
24. Phenolic mineral MFH	.20	100–180	.2	M100	1.6–2	.002–.005
25. Phenolic mineral MFI–20	.28	165	.04	M105	1.76	.0001–.001
26. Phenolic nylon	.02	80	.20	M108	1.22	.012–.016
27. Phenolic rubber/flour	.14	8–10	1.4–2	M50	1.29–1.32	.007–.010
28. Urea cellulose	.035–.040	80–130	.5–.7	M116–120	1.47–1.52	.006–.14
29. Silicone	.003–.02	175–240	.10–.30	M87	1.88	.0005
30. Silicone mineral	.002–.01	220–240+	.05–.22	M75–90	1.85–2.82	.005–.009

Table 1.4 *(continued)*

Resin material	Impact strength, notched izod ft lb/in., ½ in. bar	Tensile strength psi×10^3	Tensile modulus psi×10^3	Elongation, %	Flexural strength psi×10^3	Compressive strength psi×10^3	Compressive modulus psi×10^3
1. Acetal glass filled	2–3	10.5–12.5	600–1000	2–3	15–16	10.5–13	—
2. Acrylonitrile butadiene stryrene glass filled	2.0–2.4	15–18	800–1030	2.5–3	23–27	15–18	3.75–490
3. Nylon 6 glass filled	3–4.3	21–30	—	1–3	27–37	19–22	—
4. Nylon 6/6 glass filled	2.5–3.7	20–30	—	1–3	28–37	18–24	—
5. Polycarbonate glass filled	3–4	18.5–20	1200–1700	1.4–2.2	25–27	18–19	—
6. Polyethylene glass filled	4.5	11	—	3.5	12	6	—
7. Polypropylene glass filled	3.5–7	8–9	450–900	2–3	10–11	6–7	—
8. Polystyrene glass filled	1.4–4.0	11–14.3	990–1290	1.1–1.3	15–20	14.5–17	—
9. Polyurethane glass filled	6–7	7–9	70	3–4	5.5–8.4	3.9	—
10. Styrene acrylonitrile glass filled	2.5–3	18	—	1–3	23–26	22	—

Resin material	Heat distortion, temp. °F, 264 psi	Coefficient thermal expansion, in./in.-°C×10^{-4}	Thermal conductivity, cal/cm²-sec-°C/cm×10^{-4}	Volume resistivity, ohm-cm	Dielectric constant 60 cycles	Dielectric strength, ST ⅛″ thickness, volts/mil	Arc resistance sec
1. Acetal glass filled	325–335	3.4–6.3	—	4×10^{16}	—	—	—
2. Acrylonitrile butadiene stryrene glass filled	230–240	2.9–3.6	—	10^{14}–10^{15}	—	—	—
3. Nylon 6 glass filled	420–423	1.4–2.3	5.9	1.5–3.0×10^{14}–10^{15}	4.5	400–580	92
4. Nylon 6/6 glass filled	498–502	1.6–3.1	5.2	2.6–5.5×10^{15}	4.0–4.5	480–500	100–148
5. Polycarbonate glass filled	295–300	1.6–2.0	1.29–1.80	1.5×10^{15}	3.7–3.8	475–482	7
6. Polyethylene glass filled	260	3	—	2.9×10^{16}	—	—	—
7. Polypropylene glass filled	280–300	2.9–4.3	—	1.7×10^{16}	2.26–2.54	317–475	73–77
8. Polystyrene glass filled	195–220	3.9–4.5	0.4	3.2–3.7×10^{16}	2.8–3.5	350–450	23–75
9. Polyurethane glass filled	180–200	—	—	—	—	—	—
10. Styrene acrylonitrile glass filled	225–230	2.7–3.3	—	4.4×10^{16}	3.5	510	16–65

Resin material	Water absorption 24 hr (%)	Rockwell hardness	Specific gravity	Mold shrinkage, in./in.	Clarity	Average minimum price per lb	Trade names
1. Acetal glass filled	0.45–1.1	M75–90	1.55–1.7	0.002	Opaque	\$1.28	Formaldefil;[a] Delrin.[b]
2. Acrylonitrile butadiene stryrene glass filled	.18–.21	M75–100	1.23–1.36	.001–.003	Opaque	\$0.86	Absafil.[a]
3. Nylon 6 glass filled	0.9–1.1	E55–70	1.37–1.52	.003–.006	Translucent	\$1.40	Nylafil.[a]
4. Nylon 6/6 glass filled	.06–0.7	E60–80	1.37–1.52	.004–.007	Translucent	\$1.40	Nylafil.[a]
5. Polycarbonate glass filled	.08–.11	E55–90	1.35–1.52	.0025–.0010	Translucent	\$1.45	Polycarbafil;[a] Lexan.[c]
6. Polyethylene glass filled	0.04	R60	1.30	—	Translucent	\$0.63	Ethofil.[a]
7. Polypropylene glass filled	0.01–.05	R-90–105	1.06–1.24	.003–.008	Translucent	\$0.86	Profil;[a] Pro-Fax.[d]
8. Polystyrene glass filled	.05–.07	E35–55	1.2–1.3	.0005–.004	Translucent	\$0.35	Styrafil.[a]
9. Polyurethane glass filled	0.2–0.3	R70–85	1.4–1.65	—	Opaque	\$2.22	Urafil.[a]
10. Styrene acrylonitrile glass filled	.08–.15	E60–70	1.36–1.46	.0005–.0025	Translucent	\$0.47	Acrylafil.[a]

[a] Fiberfil, Inc.
[b] DuPont
[c] Sabic Innevative Plastics
[d] Hercules Powder Co.

Table 1.5 Reinforced thermoplastic properties

Resin	Properties	Uses	Limitations
Diallyl phthalate polymers	Good electricals, dimensional stability, chemical and heat resistance	Prepregs, ducting, radomes, aircraft, and missiles	
Epoxies	Good electricals, chemical resistance, high strength	Printed circuit board tooling, filament winding	Require heat curing for maximum performance.
Melamines	Good electricals, chemical and heat resistance	Decorative, electrical (arc and track resistance), circuit breakers	
Phenolics	Low cost, chemical resistance, good electricals, heat resistance; nonflammable; can be used to 350–400°F	Work horse—diverse mechanical and electrical applications	Dissolve in caustic unless specially treated
Polyesters	Good all-around properties, ease of fabrication, low cost, versatile	Corrugated sheeting, seating, boats, automotive, tanks and piping, aircraft, tote boxes	Not for strong oxidizers, aromatic solvents, concentrated caustic
Silicones	Heat resistance, good electricals	Electrical, aerospace	

Table 1.6 Reinforced thermoset plastic properties

Property	Thermoplastics	Thermosets
Elasticity	EVA, PVC, TPR	Silicone
Moisture Absorption	Chlorotrifluorethylene	Alkyd-glass fiber
Steam Resistance	Polysulfone	DAP
Flame Resistance	TFE, P1	Melamine
Water Immersion	Chlorinated polyether	DAP
Stress Craze Resistance	Polypropylene	All
High Temperature	TFE, PPS, P1, PAS	Silicones
Gasoline Resistance	Acetal	Phenolic
Impact	UHMW PE	Epoxy-glass fiber
Cold Flow	Polysulfone	Melamine-glass fiberglass
Chemical Resistance	TFE, FEP, PE, PP	Epoxy
Scratch Resistance	Acrylic	Allyl diglycol carbonate (C-39)
Abrasive Wear	Polyurethane	Phenolic-canvas
Colors	Acetate, PS	Urea, melamine
Low Temperature	TFE	DAP
Low Cost	PP, PE, PVC, PS	Phenolic
Low Gravity	Polypropylene methylpentene	Phenolic/nylon
Thermal Expansion	Phenoxy glass	Epoxy-glass fiber
Volume Resistivity	TFE	DAP
Dielectric Strength	PVC	DAP, polyester

Table 1.7 Brief summary of thermoplastic and thermoset properties

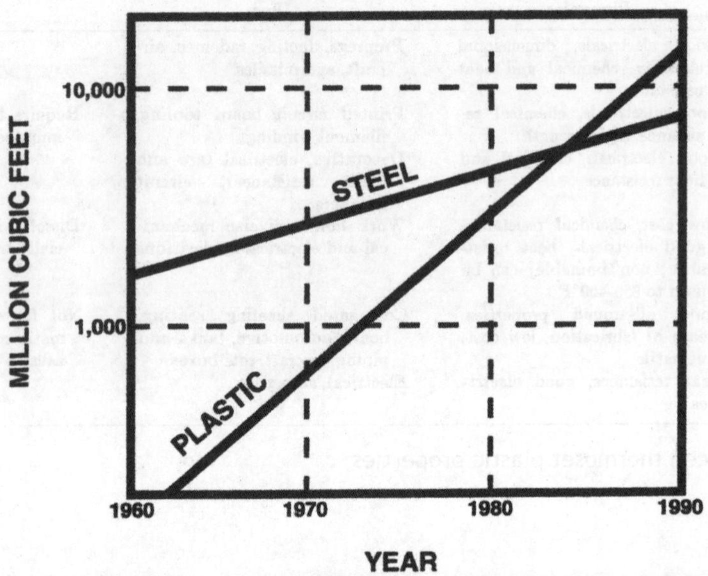

Figure 1.9 Volume of plastic and steel worldwide crossed about 1983 (courtesy of PlastiSource)

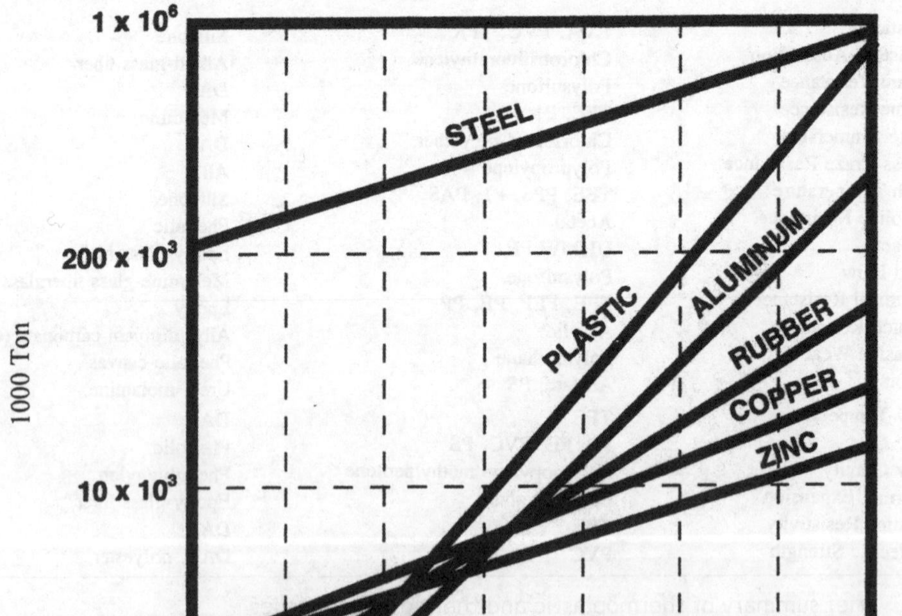

Figure 1.10 Weight of plastic and steel worldwide crossed about 2000 (courtesy of PlastiSource)

Plastic \ Region	United States	Canada	Mexico	Brazil	Other Latin America[1]	Western Europe[2]	Eastern Europe[3]	Japan	China	Other Asia-Pacific[4]	Africa & Middle East	Rest of World	Grand Total
LLDPE	8518	660	480	653	950	4193	1043	1539	2761	6141	580	417	27,935
LDPE	7798	1381	1216	1590	619	10,754	2613	1904	1938	3391	1210	457	34,871
HDPE	16,065	1286	1002	1388	417	9278	2344	2258	1218	4967	1325	532	42,080
Urethane	5215	475	380	410	390	5681	1828	1562	1825	4850	390	310	23,316
PVC	16,698	1500	605	1456	1867	13188	2492	3815	5338	11,673	1250	773	60,655
Polystyrene	6589	825	405	280	495	6180	1236	2205	2299	5275	190	352	26,331
Polypropylene	15,739	810	709	1366	1307	13716	2713	5001	2667	11,176	1275	738	57,217
ABS	1509	145	175	365	290	1410	282	948	460	950	305	92	6,931
Acrylic	613	75	80	165	140	576	115	302	155	325	140	37	2,723
Unsaturated Polyester	1731	180	125	250	210	1036	207	1285	645	1175	215	95	7,154
Nylon	1267	110	90	170	145	1210	242	339	170	425	155	59	4,382
PET	4430	410	192	482	386	2464	493	1178	715	5441	390	224	16,805
Poly-carbonate	857	90	70	155	140	607	121	443	215	475	145	45	3,363
Thermoplastic Polyester	346	45	30	65	50	243	48	164	85	322	42	20	1,460
Acetal	389	40	30	70	45	317	63	212	106	235	63	21	1,591
Recycle Plastics	1800	166	121	195	162	1625	220	502	453	1254	165	88	6,751
Other Plastics[5]	8257	763	412	875	101	7283	1773	348	294	5326	740	363	26,535
Total	97,821	8,961	6,122	9,935	7,714	79,761	17,833	24,005	21,344	63,401	8,580	4,623	350,100

[1] Argentina, Chile, Columbia, Venezuela & all other. [2] European Union plus Norway &Switzerland. [3] Includes Russia and Balkans [4] Australia, India, Indonesia, Malaysia, North Korea, Pakistan, South Korea, Taiwan, Thailand, Philippines, Singapore, Vietnam. [5] High Performance, other thermosets, specialty elastomers, tailored blends and alloys.

Table 1.8 Estimated worldwide consumption of different plastics in million lb (courtesy of PlastiSource)

provide many different properties and/or processing capabilities. Thus, *plastics* is the correct term to use except in very few applications in which only the polymer is used to fabricate products.

The term *plastic* is not a definitive one. Metals, for instance, are also permanently deformable and therefore have a plastic behavior. How else could roll aluminum be made into foil for kitchen use, or tungsten wire be drawn into a filament for an incandescent light bulb, or a 90-ton ingot of steel be forged into a rotor for a generator? Likewise, the different glasses, which contain compounds of metals and nonmetals, can be permanently shaped at high temperatures. These cousins to polymers and plastics are not considered plastics within the plastics industry.

Various stages in the manufacture of plastics exist (Table 1.9). An elementary understanding of the chemical activity taking place on a molecular level provides the basis for a grasp of the relationships between plastics technology and manufacturing and the rapidly changing competitive situation in the plastics industry. The discovery of new ways to force molecules to combine gives rise to new plastics (312).

Natural gas, crude oil, and coal can be starting points for a variety of plastics (Figs. 1.1 to 1.6). They undergo some primary processing such as distillation, cracking, or solvent extraction to produce ethylene (C_2H_4), propylene (C_3H_6), or benzene (C_6H_6), precursors to plastics. The chemical composition of plastics is basically organic polymers that are very large molecules composed of connecting chains of carbon (C) items generally linked to hydrogen atom elements (H) and often also oxygen (O), nitrogen (N), chlorine (Cl), fluorine (F), and sulfur (S).

A polymer is a large molecule built up by a repetition of small simple chemical units. These large molecules are formed by the reaction of monomers. For example, the monomer for the plastic polyvinyl chloride (PVC) is vinyl chloride. When the vinyl chloride monomer is subjected to heat and pressure it undergoes a process called polymerization (Table 1.10): the joining together of many small molecules in repeat units to make a very large molecule. Structural representations of the monomer repeat unit and polymer are shown later on in this chapter.

The number of repeat units in PVC may range from 800 to 1,600, which in turn produce different polymers. In some cases a polymer molecule will have a linear configuration, much as a chain is built up from its links. In other cases the molecules are branched or interconnected to form three dimensional networks. The particular configuration, which is a function of the plastic materials and manufacturing process involved, largely determines the properties of the finished plastic article.

Even though monomers are generally quite reactive (polymerizable), they usually require the addition of catalysts, initiators, pH control, heat, and/or a vacuum to speed and control the polymerization reaction that will result in optimizing the manufacturing process and final product. When pure monomers can be converted directly to pure polymers, it is called the bulk polymerization process, but often it is more convenient to run the polymerization reaction in an organic solvent (solution polymerization), in a water emulsion (emulsion polymerization), or as organic droplets dispersed in water (suspension polymerization). Often the catalyst system chosen exerts precise control over the structure of the polymers formed. These are referred to as stereospecific systems. Examples of the structures of the common polymers and chemical characteristics versus polymer properties are presented in Tables 1.11 and 1.12.

Process:	Chemical				Physical and mechanical		
Stages	Basic chemical	Monomers	Polymerization		Processing	Fabricating	Finishing
	Petroleum is converted to *petrochemicals* such as: ethylene, benzene, propylene, acetylene	Petrochemicals plus other chemicals are converted into *monomers* such as: ethylene, vinyl chloride, acrylonitrile, styrene, propylene	One or more monomers are polymerized to form *copolymers* or polymers such as: polyethylene, poly(vinyl-chloride), styrene-acrylonitrile, butadiene, copolyme (ABS), polystyrene polypropylene	Additives fillers reinforcements, plasticizers stabilizers, color pigments, antiexidants, inhibitors, and other chemicals are sometimes *added* to the base polymers to form *compounds* suitable for use by processors or as coatings for paper, wood etc. or in paints and adhesives	The plastics *compounds* are *formed* into a variety of solid shapes such as sheets, tubes, rods, film, and other shapes, by the heat and/or pressure of casting, molding, extrusion, or other means of processing. This step may provide a finished product such as plastic pipe, etc.	These solid shapes may be *fabricated by* thermoforming, machining, etc., to create usable plastics articles such as toys or appliances	In some cases there is a *finishing* step such as the printing of surface designs on vinyl film, welding parts, machining, polishing, etc.

Table 1.9 Flow pattern from basic materials to products

Polymer	Bulk	Solution	Suspension	Emulsion
PVC		B	A	B
Polystyrene	A	B	B	B
Polyethylene	A	A	B	B
Polypropylene		A		
Acrylic polymers	A	B	B	B
AES Polymers				
Poly(vinyl alcohol)		A		A
Poly(vinyl acetate)	B	B	B	A
Nylon	A	B		

A = Method of polymerization most commonly used commercially.
B = Method of polymerization sometimes used commercially.

Method	Advantages	Disadvantages
Bulk (batch)	Homogeneous Minimum contamination; simple equipment for making casting	Strong exothermic; broadened molecular-weight distribution at high conversion; complex if small particles required
Bulk (continuous)	Lower conversion per pass leading to better heat control and narrower molecular-weight distribution	Requires agitation, material tranfer, separation, and recycling
Emulsion	Rapid polymerization to high molecular weight and narrow distribution, with ready heat control; emulsion may be directly usable	Contamination with emulsifier, etc, almost inevitably leading to poor color and color stability; washing, drying, and compacting may be required
Solution	Ready control of heat of polymerization; polymer solution may be directly usable	Not useful for dry polymer because of difficulty of complete solvent removal
Suspension	Heterogeneous Ready control of heat of polymerization; suspension of resulting granular polymer may be directly usable	Continuous agitation required; contamination by stabilizer possible; washing, drying, possibly compacting required

Table 1.10 Examples of polymerization methods

Polymer	Structure
High-density polyethylene	$-(CH_2CH_2)-$
Low-density polyethylene	$-(CH_2CH_2)-(CH_2\underset{\underset{C_2H_5}{\vert}}{C}H)-(CH_2\underset{\underset{C_4H_9}{\vert}}{C}H)-(CH_2\underset{\underset{C_nH_{2n+1}}{\vert}}{C}H)-$
Ionomer	$-(CH_2CH_2)-(CH_2\overset{\overset{CH_3}{\vert}}{\underset{\underset{O=C-O^- Na^+}{\vert}}{C}})-$
Ethylene/vinyl acetate copolymer	$-(CH_2CH_2)-(CH_2\underset{\underset{OCCH_3(=O)}{\vert}}{C}H)-$
Polypropylene	$-(CH_2\underset{\underset{CH_3}{\vert}}{C}H)-$
Polystyrene	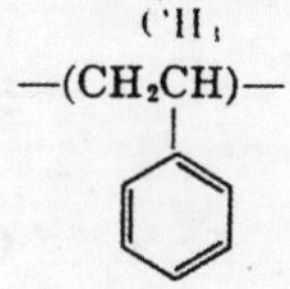
Styrene/acrylonitrile copolymer	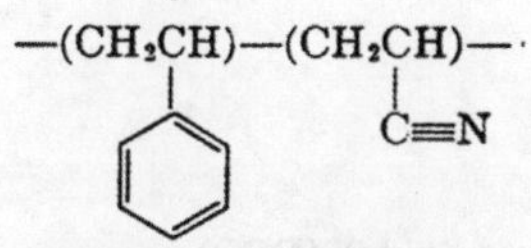
Impact styrene	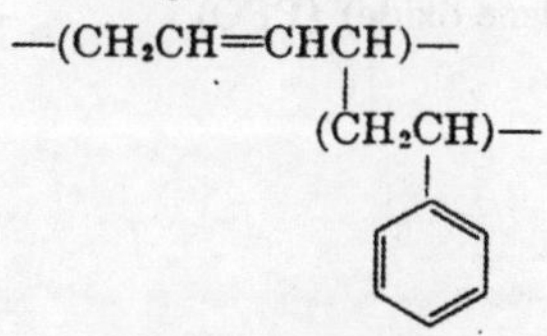
Acrylonitrile/butadiene/styrene terpolymer (ABS)	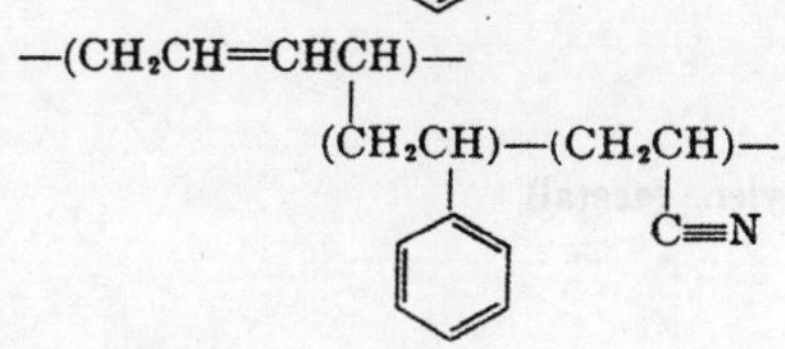
Polyvinyl chloride	$-(CH_2\underset{\underset{Cl}{\vert}}{C}H)-$
Vinyl chloride/vinyl acetate copolymer	$-(CH_2\underset{\underset{Cl}{\vert}}{C}H)-(CH_2\underset{\underset{OCCH_3(=O)}{\vert}}{C}H)-$
Polyvinylidene chloride	$-(CH_2\overset{\overset{Cl}{\vert}}{\underset{\underset{Cl}{\vert}}{C}})-$
Polytetrafluoroethylene	$-(CF_2CF_2)-$
Fluorinated ethylene/propylene copolymer	$-(CF_2CF_2)-(CF_2\underset{\underset{CF_3}{\vert}}{C}F)-$

Table 1.11 Examples of polymer structures

Polymer	Structure
Polychlorotrifluoroethylene	$-(CF_2CF(Cl))-$
Polyvinylidene fluoride	$-(CH_2CF_2)-$
Polymethyl methacrylate	$-(CH_2C(CH_3)(C(=O)OCH_3))-$
Polyacrylonitrile	$-(CH_2CH(C{\equiv}N))-$
Phenoxy resin	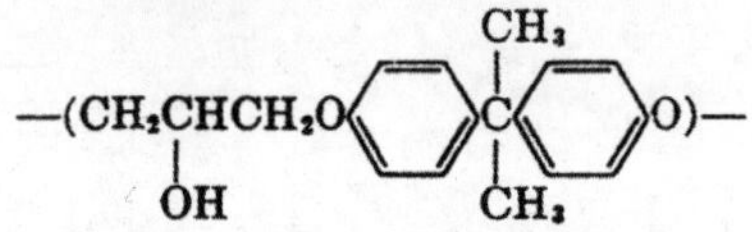
Epoxy resin	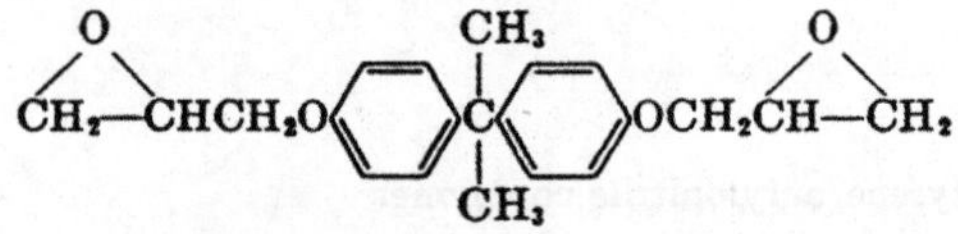
Poly(2,6-dimethylphenylene oxide) (PPO)	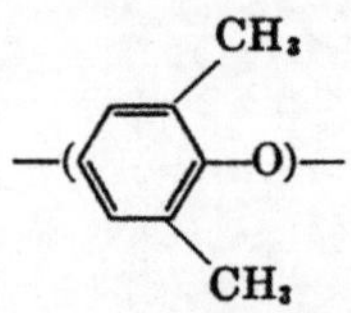
Polysulfone	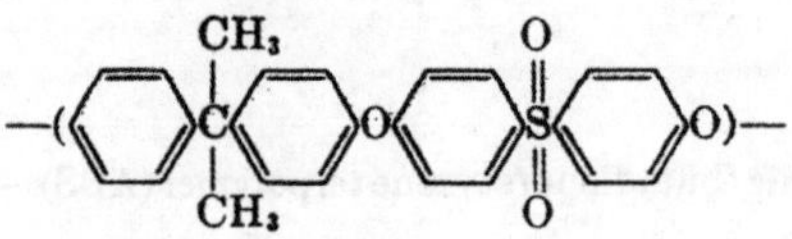
Polyoxymethylene (acetal)	$-(CH_2O)-$
Cellulose triacetate	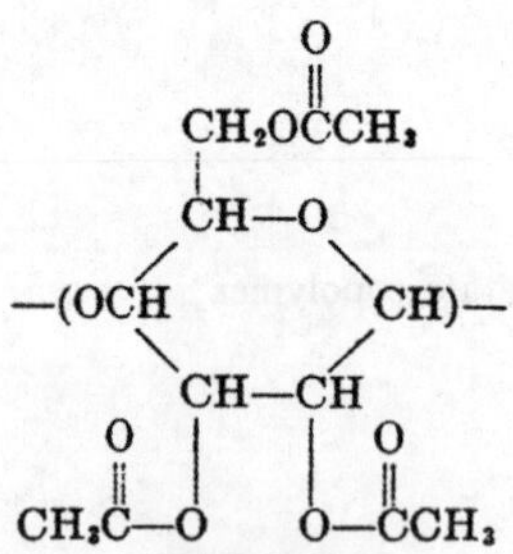
Ethyl cellulose	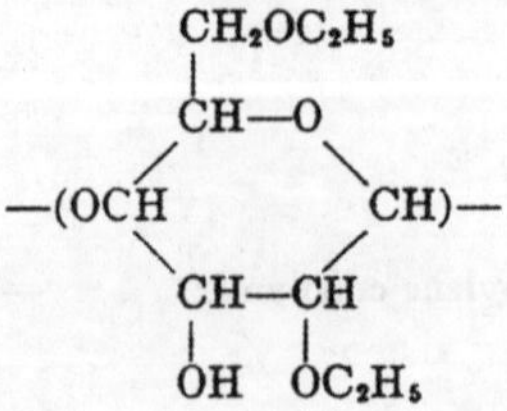

Table 1.11 *(continued)*

Polycarbonate.	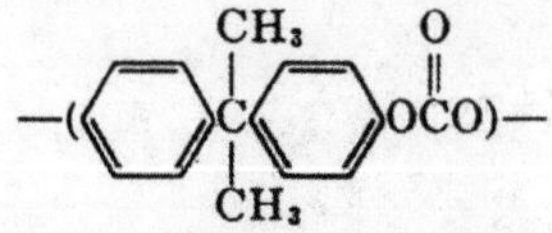
Poly(ethylene terephthalate)	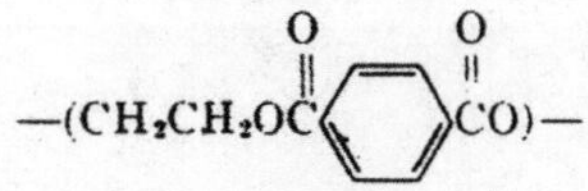
Unsaturated polyester (cured)	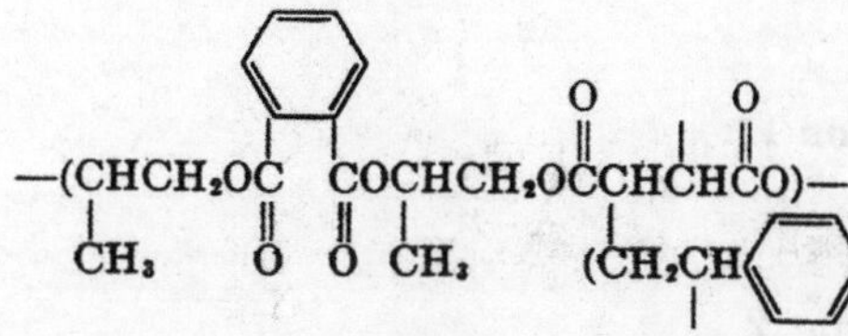
Diallyl phthalate	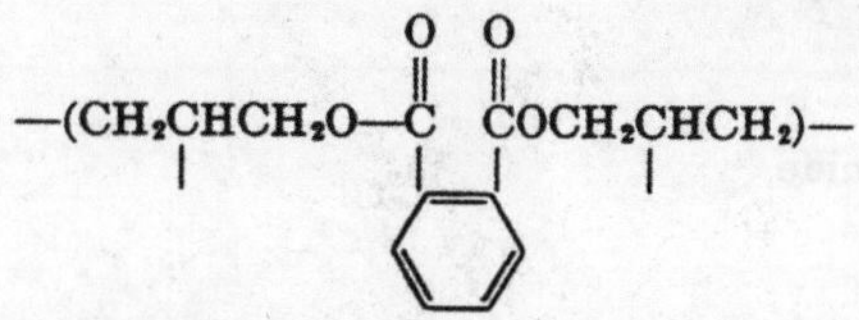
Alkyd	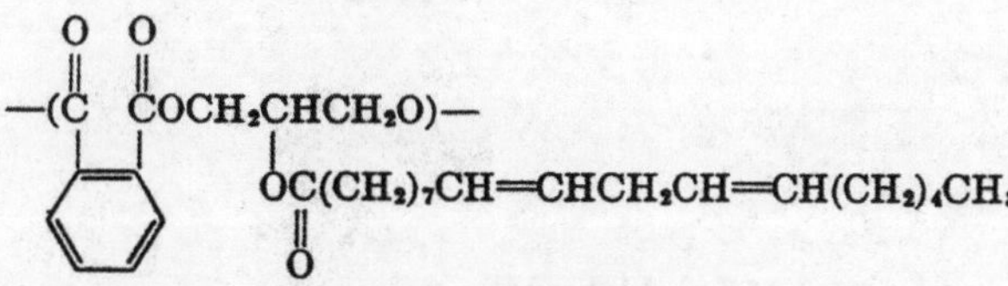
Polyether polyurethane	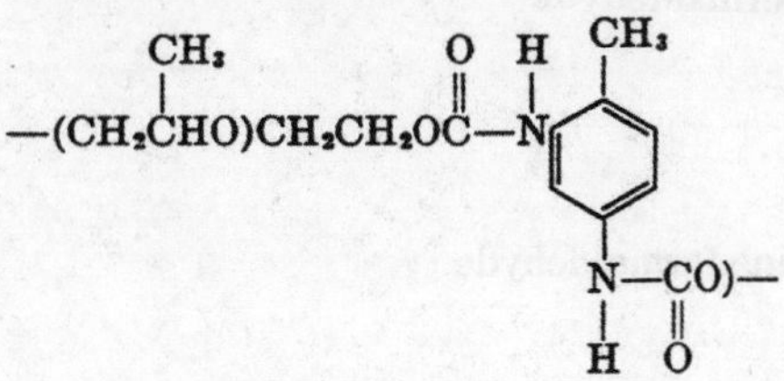
Polyester polyurethane	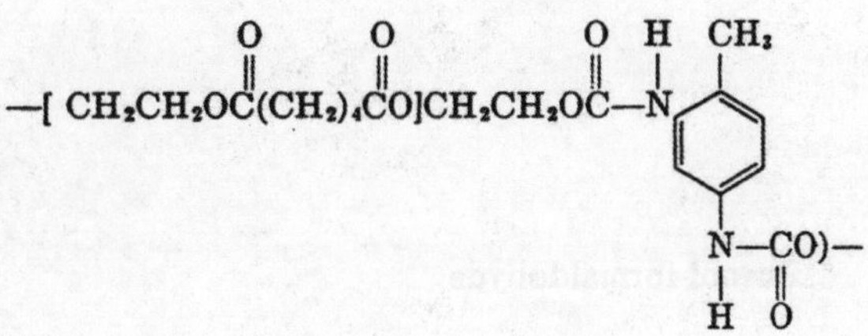

Table 1.11 *(continued)*

Polymer	Structure
Nylon 6	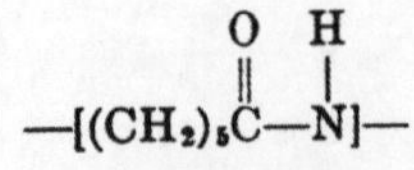
Nylon 66	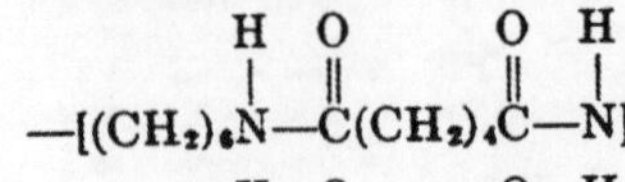
Nylon 610	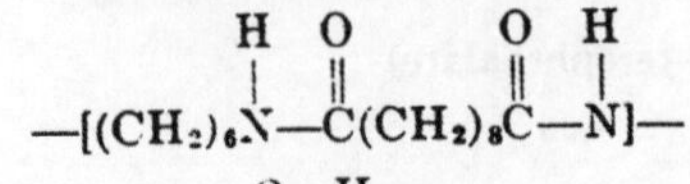
Nylon 11	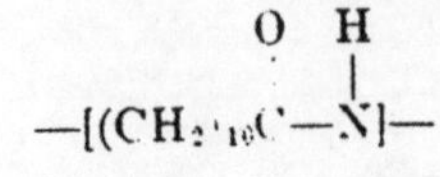
Nylon 12	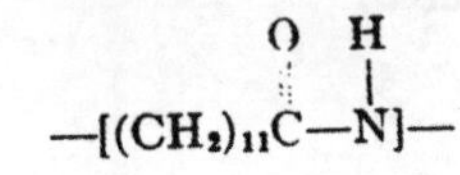
Polyimide	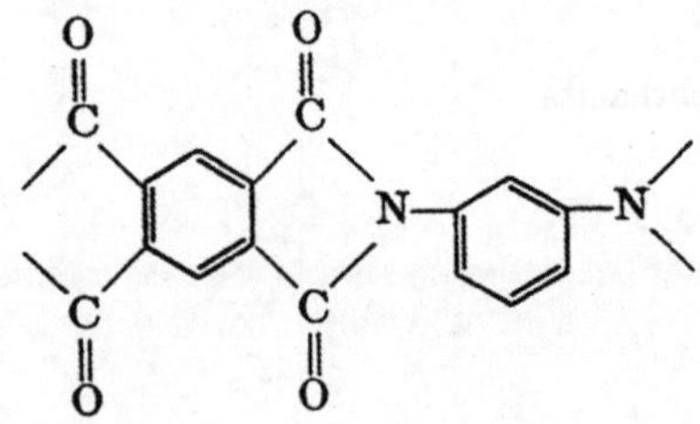
Urea-formaldehyde	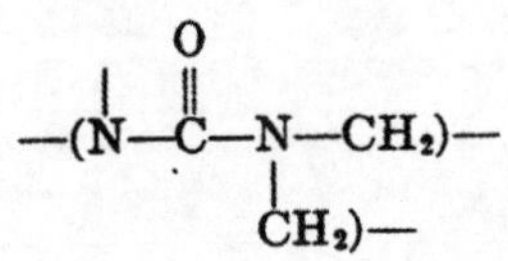
Melamine-formaldehyde	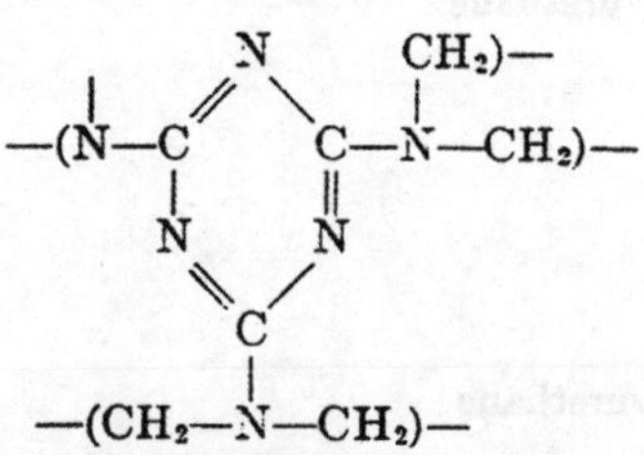
Phenol-formaldehyde	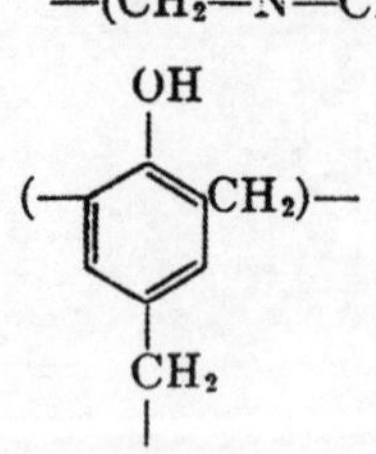
Silicone	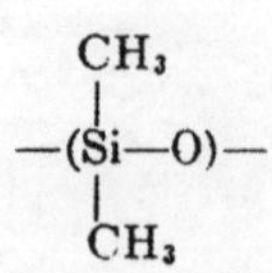

Table 1.11 *(continued)*

General features of chemical composition	Effects on polymer properties
	Chemical properties
Similarity of atoms, groups, polarity, and H-bonding	Solubility and permeability
Bond strength	Reactivity and stabilit
Surface energy	Adhesion
Specific elements and functional groups	Toxicity
	–
	Processability
Reactivity	Pot life, polymerization and cure
Bond strength	Thermal stability
Similarity of atoms, groups, polarity, and H-bonding	Solubility
Volatile impurities	Bubbles and toxicity
	Mechanical properties
Volatile impurities	Low rigidity, strength, and dimensional stability
Surface energy	Friction
Atomic weights	Specific gravity
	Electrical properties
Polarity of bonds	Dielectric constant
Polar and ionic impurities	Conduction
Conjugated unsaturation	Conduction
	Thermal properties
Atomic constants	Specific heat
Bond strength	Thermal stability
Specific elements	Flame resistance
	Optical properties
Polarity of bonds	Refractive index, spectroscopic absorption
Conjugated unsaturation	Color, ultraviolet light sensitivity
	Acoustic properties
Atomic weights	Sound absorption

Table 1.12 Chemical characteristics vs. polymer properties

There are many different catalysts that are usually used for specific chemical reactions. Types include Ziegler-Natta Catalyst (Z-N), metallocene, and others including their combinations. These different systems are available from and used worldwide by different companies.

NANOMETER POLYMER

A team of scientists at the University of Massachusetts Amherst is reconsidering conventional thinking about how polymers harden in hopes of developing finer control over the flexibility of specialty plastics. The theory is based on the fabricating process in which the polymer is heated and then cooled so that it will harden or crystallize. The researchers have been examining the way in which the polymers crystallize and have found that they essentially fold back and forth in tight layers, producing a wide and very thin crystal, perhaps just 10 nm thick (about 10,000 times thinner than a human hair).

The conventional theory suggests that polymers of any length would eventually crystallize entirely if given enough time. Because polymers can be very long, however, the theory could not be tested in a laboratory; it theoretically would have taken an infinite length of time for the longest polymers to crystallize. They report that whether polymers of this size would ever completely crystallize has been a puzzle for 60 years.

To test the theory, the team conducted computer simulations of polyethylene crystallizing. The researchers found that when very lengthy polymers harden, they never actually achieve total crystallinity. The polymers were found to reach a state of equilibrium before all of the necessary folding and assembling of the crystal are completed. They have shown that finite crystallinity is actually the equilibrium state.

MORPHOLOGY/MOLECULAR STRUCTURE/ PROPERTY/PROCESS

Morphology is the study of the physical form or structure of a material (thermoplastic crystallinity or amorphous nature)—the physical molecular structures of a polymer or, in turn, a plastic. As a result of these morphology structures, when processing the plastics into products and completing product designs, great differences are found in a finished part's properties. Table 1.13 provides an example of processing different polymers/plastics based on crystallinity levels.

Three basic molecular properties affect processing performance (flow conditions, etc.), which in turn affect product performance (strength, dimensional stability, etc.). They are (1) mass or density (Table 1.14), (2) molecular weight (MW), and (3) molecular weight distribution (MWD).

In crystalline plastics, such as PE, density has a direct impact on properties such as stiffness and permeability to gases and liquids (Table 1.15). Changes in density may also affect some mechanical properties. For maximum usefulness, density needs to be measured to an accuracy of at least ± 0.001 g/cm^3.

Polymer	Condition	Crystallinity [%]
LDPE	All	40–50
LLDPE	All	60
HDPE	All	60–80
Polypropylene (PP)	Rapidly cooled	45–50
Nylon 6 (PA6)	Slowly cooled	40–50
Nylon 6 (PA6)	Quenched	10
Polyethylene Terephthalate (PET)	Slowly cooled	20–30
Polyethylene Terephthalate (PET)	Quenched	0–10

Table 1.13 Crystallinity levels of different polymers/plastics

Type	Density, g/cm^3 (lb/ft^3)
LDPE	0.910-0.925 (56.8-57.7)
MDPE	0.926-0.940 (57.8-58.7)
HDPE	0.941-0.959 (58.7-59.9)
HMWPE	0.960 & above (59.9 & above)

Table 1.14 Densities of polyethylenes

MOLECULAR WEIGHT

One method of defining plastics melt behavior and property performance is to use information concerning their molecular weight (MW), a reference to the plastic molecules' weight and size. MW is the sum of the atomic weights of all the atoms in a molecule. It represents a measure of the chain length for the molecules that make up the polymer. Atomic weight is the relative mass of an atom of any element based on a scale in which a specific carbon atom (carbon-12) is assigned a mass value of 12. The polymerized material contains molecules having many different chain lengths. For some products, the resulting distribution of molecular weights can be calculated statistically and illustrated by the standard form of frequency distribution.

MW of plastics influences their properties. As an example, with increasing MW, properties increase for abrasion resistance, brittleness, chemical resistance, elongation, hardness, melt viscosity, tensile strength, modulus, toughness, and yield strength. Decreases occur for adhesion, melt index, and solubility.

Adequate MW is a fundamental requirement to meet desired plastics properties. With MW differences in incoming material, the fabricated product performance can be altered: the greater the difference, the more dramatic the change that occurs in the product. Melt flow rate (MFR; chapter 22) tests are used to detect degradation in products. MFR has a reciprocal relationship to melt viscosity. This relationship of MW to MFR is an inverse one; as one drops, the other increases and vice-versa.

Physical properties	Basic molecular properties: If the density (crystallinity) is increased within the range 0.915–0.965	If the melt index is increased (or the av mol wt is lowered)	If the mol wt distribution is narrowed
Melt viscosity	Higher	Lower	Slightly higher
Vicat softening temperature	Much higher	Lower	Slightly higher
Surface hardness (abrasion resistance)	Higher	Slightly lower	
Resistance to blocking and mold sticking	Higher	Slightly lower	
Slip	Higher	No effect	
Minimum heat seal temperature	Higher	Slightly lower	Slightly higher
Heat seal range	Narrower	Narrower	Narrower
Tensile strength			
Yield	Much higher	Slightly lower	
Break	Slightly lower	Lower	Slightly higher
Elongation	Lower	Lower	
Resistance to creep	Higher	Slightly lower	Higher
Flexural stiffness	Much higher	Slightly lower	Higher
Flexibility	Lower		
Film impact strength (toughness)	Lower	Lower	
Film tear strength	Depends on film-making technique and direction of tear		
Resistance to film brittleness	Lower	Slightly lower	Slightly higher
Resistance to brittleness at low temperature	Lower	Lower	Higher
Resistance to environmental stress cracking	Lower	Lower	Slightly higher
Optical properties			
Transparency (transmittance)	Higher [a]	Higher	Higher
Freedom from haze	Higher [a]	Higher	Lower
Gloss	Higher [a]	Higher	Lower
Draw-down rate	Slightly higher	Much higher	Lower
Barrier properties			
MVT rate	Lower		
Gas and liquid transmission	Much lower		
Grease proofness	Much higher	Slightly lower	
Neck-in and beading	Higher	Much higher	Higher
Adhesion to substrates (especially physical adherence to porous substrates)	Slightly lower	Slightly higher	
Shrinkage	Higher	Lower	
Warpage	Slightly higher	Lower	Higher
Electrical properties	Slightly higher	No effect	No effect

[a] Not true in the high-density range. The relations between these three basic molecular properties and the physical and other properties of polyethylene are not always simple and clear. Often, molecular structural features other than the above three basic properties exert some influence on certain physical properties. Processing conditions may radically change the orientation (alignment) of the molecules in the resin and thus, too, affect some of its properties. High density polyethylene resins do not always exhibit the same degree of property change with density change that low density resins do. For these reasons, this table should not be considered final.

Table 1.15 How three basic molecular properties affect essential polyethylene plastic or end product properties

MOLECULAR WEIGHT DISTRIBUTION

MW refers to the average weight of plastics that is always composed of different weight molecules. These differences are important to the processor, who uses the molecular weight distribution (MWD) to evaluate materials. A narrow MWD enhances the performance of plastic products, while a wide MWD permits easier processing (Fig. 1.11). The processing and property characteristics of plastics are partly a function of the MWD, which may vary widely, even among plastics of identical composition, density, average molecular weight, and melt index.

VISCOSITY AND MELT FLOW

Viscosity is a measure of the internal friction resulting when one layer of fluid is caused to move in relationship to another layer. When reviewing the subject of plastic melt flow, the subject of viscosity is involved. Basically, viscosity is the property of the resistance of flow exhibited within a body of material (Figs. 1.12 and 1.13). Ordinary viscosity is the internal friction or resistance of a plastic to flow. It is the constant ratio of shearing stress to the rate of shear. Shearing is the motion of a fluid, layer by layer, like a deck of cards. When plastics flow through straight tubes or channels, they are sheared and the viscosity expresses their resistance.

The melt index (MI) or melt flow index (MFI) is an inverse measure of viscosity (chapter 22). High MI implies low viscosity and low MI means high viscosity. Plastics are shear thinning, which means that their resistance to flow decreases as the shear rate increases. This is due to molecular disentanglements and alignments in the direction of flow.

NEWTONIAN AND NON-NEWTONIAN

Viscosity is usually understood to mean Newtonian viscosity, in which case the ratio of shearing stress to the shearing strain is constant. In non-Newtonian behavior, which is the usual case for plastics, the ratio varies with the shearing stress. Such ratios are often called the apparent viscosities

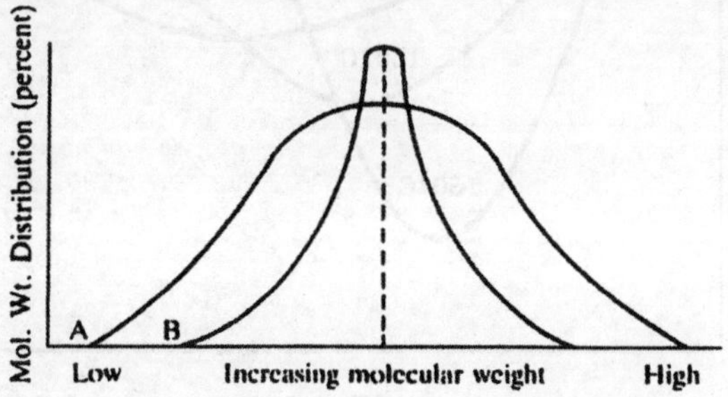

Figure 1.11 Examples of narrow and wide molecular weight distributions

at the corresponding shearing stresses. Viscosity is measured in terms of flow in Pa.s (P), with water as the base standard (value of 1.0). The higher the number, the lower the flow.

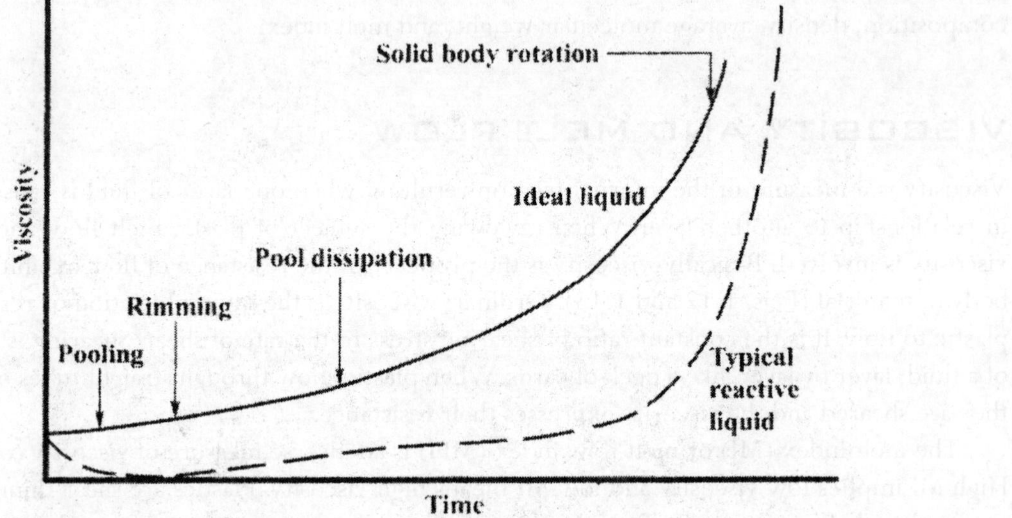

Figure 1.12 Time-dependent viscosities for an ideal fluid applicable to rotationally moldable reactive liquid and typical fluid flow

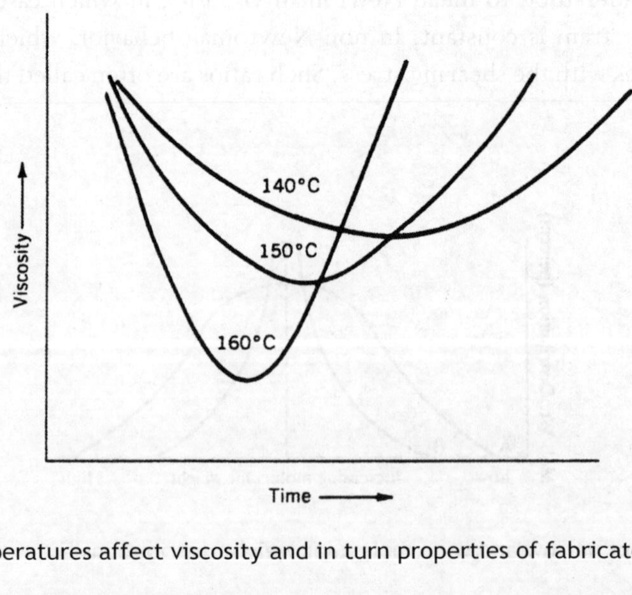

Figure 1.13 Melt temperatures affect viscosity and in turn properties of fabricated products

RHEOLOGY

Rheology is a phenomenon of time-dependent, in addition to elastic, deformation (or recovery) in response to load. This property, possessed by all plastics to some degree, dictates that while plastics have solidlike characteristics such as elasticity, strength, and form stability, they also have liquidlike characteristics such as flow depending on time, temperature, rate, and amount of loading. Thus, plastics are said to be viscoelastic. The mechanical behavior of these viscoelastic plastics is dominated by such phenomena as tensile strength, elongation at break, stiffness, and rupture energy, which are often the controlling factors in a design. The viscous attributes of plastic melt flow are also important considerations in the fabrication of plastic products.

When discussing melt flow, the subject of rheology or flow of matter is involved. It is concerned with the response of plastic melts to mechanical force. An understanding of rheology and the ability to measure rheological properties such as molecular weight and melt flow is necessary before flow behavior can be controlled during processing. Such control is essential for the fabrication of plastic materials to meet product performance requirements.

With plastics, there are two types of deformation or flow: viscous, in which the energy causing the deformation is dissipated, and elastic, in which that energy is stored. The combination produces viscoelastic plastics. Not only are there two classes of deformation, there are also two modes in which deformation can be produced: simple shear and simple tension. The actual action during melting, as in the usual screw plasticator, is extremely complex, with all types of shear-tension combinations. Together with engineering design, deformation determines the pumping efficiency of a screw plasticator and controls the relationship between output rate and pressure drop through a die system or into a mold.

Figure 1.14 shows the different flow behaviors of plastic in comparison with water. The volume of a so-called Newtonian fluid, such as water, when pushed through an opening is directly proportional to the pressure applied (the straight dotted line). The flow rate of non-Newtonian fluids such as plastics when pushed through an opening increases more rapidly than the applied pressure (the solid curved line). Different plastics have their own flow rates so that their non-Newtonian curves are different.

VISCOELASTICITY

This property of viscoelasticity, possessed by all plastics to some degree, dictates that while plastics have solidlike characteristics such as elasticity, strength, and form-stability, they also have liquidlike characteristics such as flow depending on time, temperature, and rate as well as amount of loading (Fig. 1.15). Factually, it is the mechanical behavior in which the relationships between stress and strain are time dependent, as opposed to the classical elastic behavior of steel in which deformation and recovery both occur instantaneously on application and removal of stress, respectively.

On stressing a viscoelastic material, three deformation responses may be observed. They are an initial elastic response, followed by a time-dependent delayed elasticity (also fully recoverable),

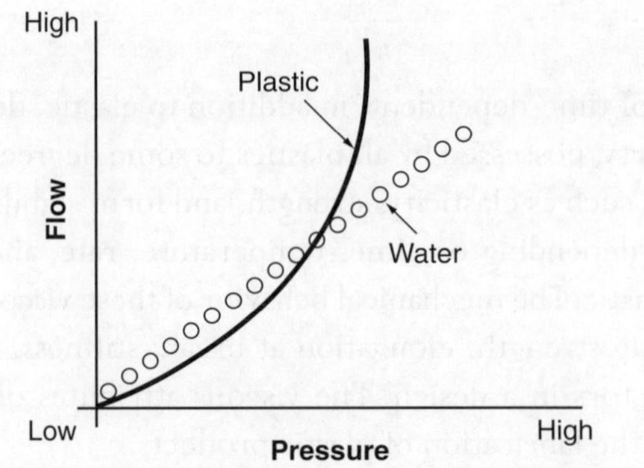

Figure 1.14 Comparing flow of plastic and water subjected to pressure

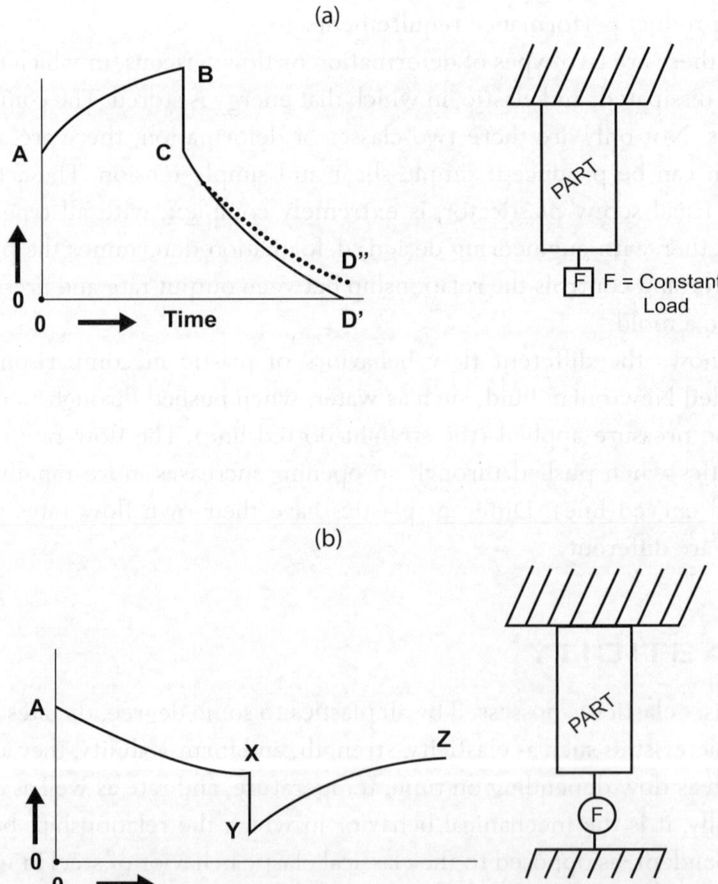

Figure 1.15 Viscoelasticity of plastics behavior of: (a) stress-strain-time in creep and (b) strain-stress-time in stress relaxation

and finally, a viscous, nonrecoverable flow component. Most plastic-containing systems (solids, melts, gels, dilute, and concentrated solutions) exhibit viscoelastic behavior because of the long-chain nature of the constituent molecules. The mechanical behavior of these viscoelastic plastics is dominated by such properties as tensile strength, elongation at break, stiffness, and rupture energy, which are often the controlling factors in a design.

PROCESSING-TO-PERFORMANCE INTERFACE

Different plastic characteristics influence the processing and properties of plastic products. Table 1.16 reviews these different characteristics. Important are glass transition temperature (T_g) and melt temperature (T_m).

GLASS TRANSITION TEMPERATURE

The T_g relates to temperature characteristics of plastics (Table 1.17). It is the reversible change in phase of a plastic from a viscous or rubbery state to a brittle glassy state (Fig. 1.16). T_g is the point below which plastic behaves like glass and is very strong and rigid. Above this temperature it is not as strong or rigid as glass, but neither is it as brittle as glass. At and above T_g the plastic's volume or length increases more rapidly, and rigidity and strength decrease. As shown in Figure 1.17, the amorphous TPs have a more definite T_g when compared to crystalline TPs. Even with variation, T_g is usually reported as a single value.

The thermal properties of plastics, particularly its T_g, influence the plastic's processability performance and cost in different ways. The operating temperature of a TP is usually limited to below its T_g.

A more expensive plastic could cost less to process because of its T_g location, which results in a shorter processing time, requires less energy for a particular weight, and so on (Figs. 1.18 and 1.19).

The T_g generally occurs over a relatively narrow temperature span. Not only do hardness and brittleness undergo rapid changes in this temperature region, but other properties such as the coefficient of thermal expansion and specific heat also change rapidly. This phenomenon has been called *second-order transition*, *rubber transition*, or *rubbery transition*. The word *transformation* has also been used instead of *transition*. When more than one amorphous transition occurs in a plastic, the one associated with segmental motions of the plastic backbone chain, or accompanied by the largest change in properties, is usually considered to be the T_g.

It is important for designers to know that above T_g, many mechanical properties are reduced. Most noticeable is a possible reduction in stiffness by a factor of 1,000.

MELT TEMPERATURE

Crystalline plastics have specific melt temperatures (T_m) or melting points. Amorphous plastics do not. They have softening ranges that are narrow temperature-wise, keyed to when solidification of

Plastic and (morphology)[a]		Density, $g\,cm^{-3}$ ($lb\,ft^{-3}$)	Melt temperature (T_m), °C (°F)	Glass transition temperature (T_g), °C (°F)	Thermal conductivity, $10^{-4}\,cal\,s^{-1}\,cm^{-1}\,°C^{-1}$ ($Btu\,lb^{-1}\,°F^{-1}$)	Heat capacity, $cal\,g^{-1}\,°C^{-1}$ ($Btu\,lb^{-1}\,°F^{-1}$)	Thermal diffusivity, $10^{-4}\,cm^2\,s^{-1}$ ($10^{-3}\,ft^2\,h^{-1}$)	Thermal expansion, $10^{-6}\,°C^{-1}$ ($10^{-6}\,°F^{-1}$)
PP	(C)	0.9 (56)	168 (334)	5 (41)	2.8 (0.068)	0.9 (0.004)	3.5 (1.36)	81 (45)
HDPE	(C)	0.96 (60)	134 (273)	−110 (−166)	12 (0.290)	0.9 (0.004)	13.9 (5.4)	59 (33)
PTFE	(C)	2.2 (137)	330 (626)	−115 (−175)	6 (0.145)	0.3 (0.001)	9.1 (3.53)	70 (39)
PA	(C)	1.13 (71)	260 (500)	50 (122)	5.8 (0.140)	0.075 (0.003)	6.8 (2.64)	80 (44)
PET	(C)	1.35 (84)	250 (490)	70 (158)	3.6 (0.087)	0.45 (0.002)	5.9 (2.29)	65 (36)
ABS	(A)	1.05 (66)	105 (221)	102 (215)	3 (0.073)	0.5 (0.002)	3.8 (1.47)	60 (33)
PS	(A)	1.05 (66)	100 (212)	90 (194)	3 (0.073)	0.5 (0.002)	5.7 (2.2)	50 (28)
PMMA	(A)	1.20 (75)	95 (203)	100 (212)	6 (0.145)	0.56 (0.002)	8.9 (3.45)	50 (28)
PC	(A)	1.20 (75)	266 (510)	150 (300)	4.7 (0.114)	0.5 (0.002)	7.8 (3.0)	68 (38)
PVC	(A)	1.35 (84)	199 (390)	90 (194)	5 (0.121)	0.6 (0.002)	6.2 (2.4)	50 (128)
Aluminum		2.68 (167)	1,000		3,000 (72.5)	0.23	4,900 (1900)	19 (10.6)
Steel		7.9 (493)	2,750		800 (21.3)	0.11	1,000 (338)	11 (6.1)

[a]C = crystalline resin, A = amorphous resin.

Table 1.16 Thermoplastic melt temperatures and other thermal properties

the melt occurs or when the solid softens and becomes a fluid-type melt. They start softening as soon as the heat cycle begins. Regardless, the melting temperature reported usually represents the average softening range.

The T_m of crystalline plastics occurs at a relatively sharp point going from solid to melt. It is the temperature at which the melt softens and begins to have a flow tendency (Table 1.18). They have a

	°F	°C
Polyethylene	−184	−120
Polypropylene	−6	−22
Polybutylene	−13	−25
Polyvinyl fluoride	−4	−20
Polyvinyl chloride	185	85
Polyvinylidene chloride	−4	−20
Polystyrene	203	95
Polyacetal	−112	−80
6-Nylon	158	70
Polyester	230	110
Polycarbonate	302	150
Polytetrafluoroethylene	−175	−115
Silicone	−193	−125

Table 1.17 Range of T_g for different thermoplastics

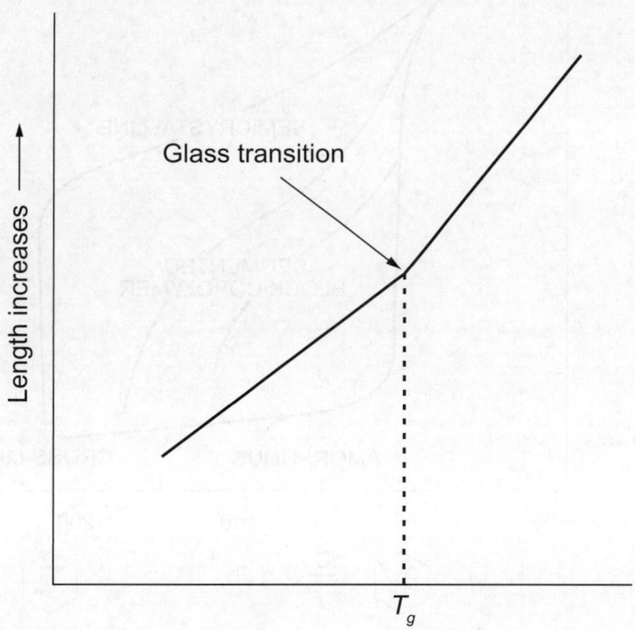

Figure 1.16 Thermoplastic volume or length changes at the glass transition temperature

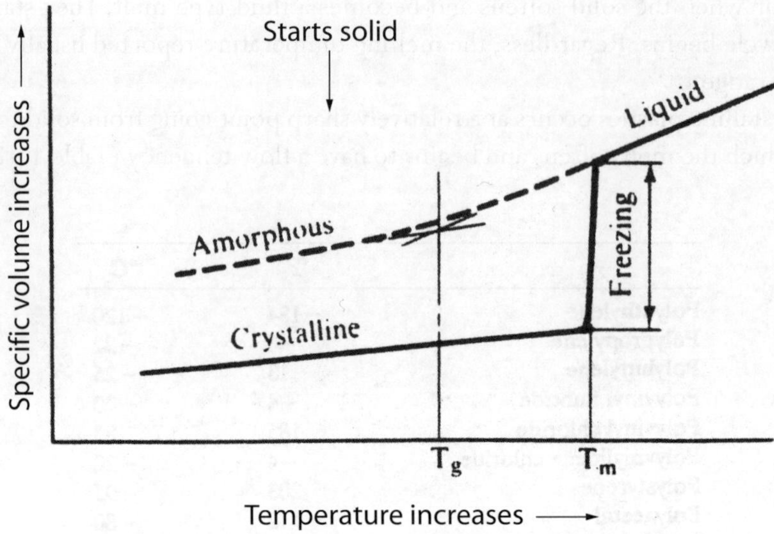

Figure 1.17 Change of amorphous and crystalline thermoplastic's volume at T_g and T_m

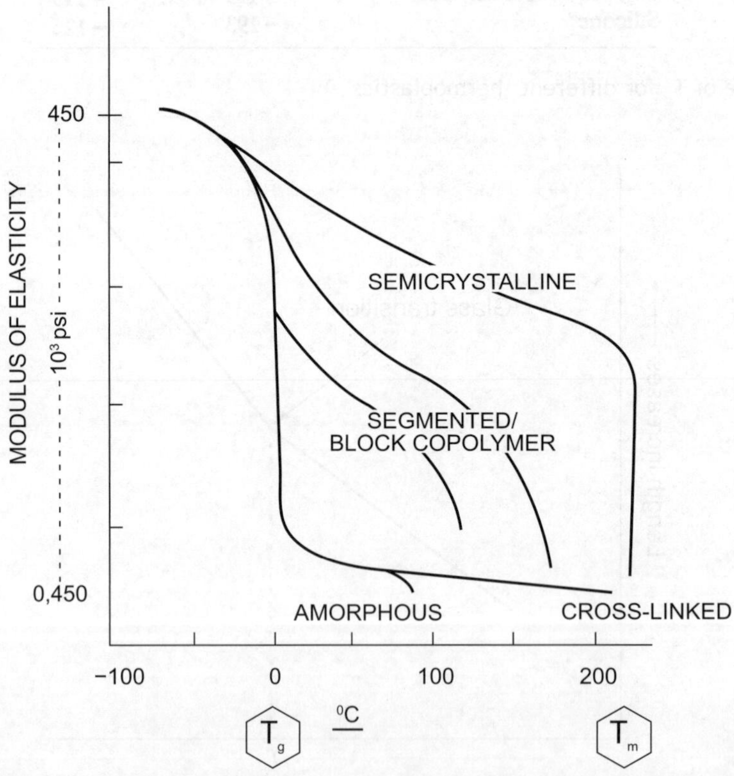

Figure 1.18 Examples of dynamic properties of crystalline and amorphous thermoplastics as well as cross-linked thermoset plastics

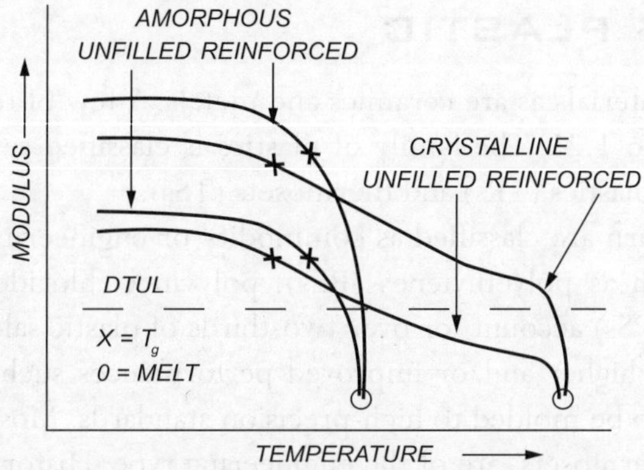

Figure 1.19 Modulus behavior with increase in temperature (DTUL = deflection temperature under load) (courtesy of Bayer)

	°F	°C
Low-density polyethylene	230	110
High-density polyethylene	266	130
Polypropylene (isotactic)	347	175
6-Nylon	419	215
66-Nylon	500	260
Polyester	500	260
Polytetrafluoroethylene	626	330
Polyarylamides	716	380

Table 1.18 Crystalline thermoplastics melt temperatures

true T_m with a latent heat of fusion associated with the melting and freezing process and a relatively large volume change during fabrication. Crystalline plastics have considerable molecular order in the solid state, indicating that many of the atoms are regularly spaced. The melt strength of the plastic occurs while in the molten state. It is an engineering measure of the extensional viscosity and is defined as the maximum tension that can be applied to the melt without breaking.

The T_m is dependent on the processing pressure and the time under heat, particularly during a slow temperature change for relatively thick melts during processing. Also, if the temperature of the melt is too low, the melt's viscosity will be high and costlier power will be required in processing it. Also, if the viscosity is too high, degradation will occur. There is a correct processing window used for the different melting plastics.

CLASSIFYING PLASTIC

Plastics are a family of materials as are ceramics and metals. A few of these different materials are reviewed in Tables 1.19 to 1.21. The family of plastics is classified several ways. The two major classifications are thermoplastics (TPs) and thermosets (TSs).

The TPs and TSs in turn are classified as commodity or engineering plastics (Tables 1.22 and 1.23). Commodities such as polyethylenes (PEs), polyvinyl chlorides (PVCs), polypropylenes (PPs), and polystyrenes (PSs) account for over two-thirds of plastic sales. Engineering plastics are characterized as meeting higher and/or improved performances such as heat resistance, impact strength, and the ability to be molded to high-precision standards. Most of the thermoset plastics, particularly reinforced thermosets, are of the engineering type. Historically, as more competition and/or production occur for certain engineering plastics, their costs go down and they become commodity plastics. Fifty years ago, the dividing line costwise was about \$0.15/lb; now it is just above \$1.00/lb.

Different types of plastics are usually identified by their composition and/or performance. For example, virgin plastics are plastic materials that have not been subjected to any fabricating process. These are called NEAT polymers (*N*othing *E*lse *A*dded *T*o). They are true virgin polymers since they do not contain additives, fillers, and so forth. They are rarely used. Table 1.24 compares a NEAT plastic with the same glass- and talc-reinforced versions (wt %).

Plastic materials to be processed are in the form of pellets, granules, flakes, powders, flocks, liquids, and so on. Of the 35,000 types available worldwide there are about 200 basic types or families that are commercially recognized, with fewer than 20 that are popularly used. Examples of these plastics are shown in Table 1.25.

Within these 20 popular plastics there are five major families of thermoplastics that account for about two-thirds of all thermoplastics consumed. They are the low-density polyethylenes (LDPEs), high-density polyethylenes (HDPEs), polypropylenes (PPs), polystyrenes (PSs), and polyvinyl chlorides (PVCs).

THERMOPLASTIC: CRYSTALLINE OR AMORPHOUS

There are crystalline and amorphous thermoplastics (TPs). TPs during processing soften and upon cooling harden into products that are capable of being repeatedly softened by reheating, with their morphology (molecular structure) being crystalline or amorphous (Fig. 1.20). Their softening temperatures vary. An analogy would be a block of ice that can be softened (turned back to a liquid), poured into any shape mold or die, and then cooled to become a solid again. This cycle repeats. During the heating cycle care must be taken to avoid degrading or decomposition. With some TPs no change or practically no significant property changes occur. However, some may have significant changes.

The crystalline plastics (polymers), such as polyethylene (PE) and polypropylene (PP), tend to have their molecules arranged in a relatively regular repeating structure. This behavior identifies their morphology, that is, the physical form or structure of the materials. They are usually translucent or opaque and generally have higher softening points than the amorphous plastics (Table 1.26).

- **Plastics**
  - Polyethylene
  - Polyvinylchloride
  - Polytetrafluoroethylene
  - Polymethylmethacralate
  - Polyacetal
  - Phenolics
  - Polyimides
  - Polyamides
  - Polycarbonate
  - Elastomers
    - Urethanes
    - Silicones
    - Polychloroprene
- **Ceramics**
  - Aluminum oxide
  - Plasma coatings
    - Chromium oxide
    - Zirconia
    - Titanium carbide
    - Chromium carbide
  - Cemented carbides
  - Vitrified mica
  - Carbon products
- **Metals**
  - Carbon steels
    - 1010
    - 1020
    - 1040
    - 1050
    - 1090
    - B1112
  - Tool Steels
    - 01
    - A2
    - D2
    - S1
    - H13
    - M2
  - Cast irons
    - Class 20
    - Class 35
    - Ductile 60-45-10
  - Copper alloys
    - ETP copper CDA 110
    - DHP copper CDA 122
    - Tin bronze CDA 905
    - Be copper CDA 172
    - Yellow brass CDA 360
    - Phosphor bronze CDA 521
  - Nickel alloys
    - Monel
    - Hastelloys
    - Ni, Cr, B alloys
    - Pure nickel
  - Alloy steels
    - 4140
    - 4340
    - 4620
    - 9310
  - Special steels
    - Nitralloy
    - Marage 200-350
    - Weathering
  - Stainless steels
    - 303
    - 304 (CF-8
    - 316 (CF-8M)
    - 420
    - 440C
    - 17-4 PH (CB-7Cu)
  - Aluminum alloys
    - 3003
    - 5052
    - 6061
    - 7075
    - 355
    - 380
  - Zinc
    - ASTM B86-48
    - No XX5
  - Titanium
    - Pure Ti (Gr 1-3)
    - Ti-6A1-4V
  - Magnesium
    - AZ51
    - AZ63

Table 1.19 Plastic, ceramic, and metal families of materials

Matrix Material	Reinforcement Material	Properties Modified
Metal	Metal, ceramic, carbon, glass fibers	Elevated temperature strength Electrical resistance Thermal stability
Ceramic	Metallic and ceramic particles and fibers	Elevated temperature strength Chemical resistance Thermal resistance
Glass	Ceramic fibers and particles	Mechanical strength Temperature resistance Chemical resistance Thermal stability
Plastic Thermosets, Thermoplastics	Unreinforced and reinforced with carbon, glass, organic fibers, glass beads, flakes, ceramic particles, metal wires	Mechanical strength Elevated temperature strength Chemical resistance Antistatic Electrical resistance EMF shielding Flexibility Wear resistance Energy absorption Thermal stability

Table 1.20 Introduction to properties of metals, ceramics, glasses, and plastics

They can be made transparent with chemical modification. Polymer molecules that can be packed closer together can more easily form crystalline structures in which the molecules align themselves in some orderly pattern. Since commercially perfect crystalline polymers are not produced, they are identified technically as semicrystalline TPs. These semicrystalline TPs normally have up to 80% crystalline structure, and the rest is amorphous.

Amorphous plastic, a term meaning formless plastic, describes a TP having no crystalline plastic structure. They form no pattern whereby their structure tends to form like spaghetti with their molecules going in all different directions. These TPs, such as PS and PMMA (polymethyl methacrylate), have no sharp melting point and are usually glassy and transparent. Amorphous plastics soften gradually as they are heated. If they are rigid they may be brittle, unless modified with certain additives.

Plastics during processing are normally in the amorphous state with no definite order of molecular chains. TPs that normally crystallize (crystalline plastics) may not be properly quenched (when hot melt is cooled to solidify the plastic), resulting in an amorphous or partially amorphous solid state usually having inferior properties. Compared to crystalline types, amorphous polymers undergo only small volumetric changes when melting or solidifying during processing (Fig. 1.21). This action influences the degree of dimensional tolerance that can be met after the heating/cooling process.

As symmetrical molecules approach within a critical distance during melt processing, crystals begin to form in the areas where they are the most densely packed. A crystallized area is stiffer

Material	Density (kg/m³)	Specific heat (cal/g °C)	Thermal conductivity (W/m/K)	Coeff. of therm. exp. (μm/m/K)	Thermal diffusivity (m²/s) × 10⁻⁷	Max. operating temp. (°C)
ABS (high impact)	1040	0.35	0.3	90	1.7	70
Acetal (homopolymer)	1420	0.35	0.2	80	0.7	85
Acetal (copolymer)	1410	0.35	0.2	95	0.72	90
Acrylic	1180	0.35	0.2	70	1.09	50
Cellulose acetate	1280	0.36	0.15	100	1.04	60
CAB	1190	0.35	0.14	100	1.27	60
Epoxy	1200	—	0.23	70	—	130
Modified PPO	1060	—	0.22	60	—	120
Nylon 66	1140	0.4	0.24	90	1.01	90
Nylon 66 (33% glass)	1380	0.3	0.52	30	1.33	100
PEEK	1300	—	—	48	—	204
PEEK (30% carbon)	1400	—	—	14	—	255
PET	1360	—	0.14	90	—	110
PET (36% glass)	1630	—	—	40	—	150
Phenolic (mineral filled)	1690	—	—	22	—	185
Polyamide-imide	1400	—	—	36	—	260
Polycarbonate	1150	0.3	0.2	65	1.47	125
Polyester (TP)	1200	—	0.2	100	—	—
Polyetherimide	1270	—	0.22	56	—	170
Polyethersulphane	1370	—	1.18	55	—	180
Polyimide	1420	—	—	45	—	260
Polyphenylene sulfide (30% carbon)	1460	—	—	16	—	200
Polypropylene	905	0.46	0.24	100	0.65	100
Polysulphane	1240	—	—	56	—	170
Polystyrene	1050	0.32	0.15	80	0.6	50
Polythene (LD)	920	0.55	0.33	200	1.17	50
Polythene (HD)	950	0.55	0.63	120	1.57	55
PTFE	2100	—	0.25	140	0.7	50
PVC (rigid)	1400	0.24	0.16	70	1.16	50
PVC (flexible)	1300	0.4	0.14	140	0.7	50
SAN	1080	0.33	0.17	70	0.81	60
DMC (polyester) (TS)	1800	—	0.2	20	—	130
SMC (polyester) (TS)	1800	—	0.2	20	—	130
Polystyrene foam	32	—	0.032	—	—	—
Stainless steel	7855	—	90	10	—	800
Nickel chrome alloy						900
Molybdenum						1000

Table 1.21 Examples of plastics temperature behavior

and stronger (Fig. 1.22); a noncrystallized (amorphous) area is tougher and more flexible. With increased crystallinity, other effects occur. As an example, with polyethylene (crystalline), there is increased resistance to creep.

In general, crystalline types of plastics are more difficult (though controllable) to process, requiring more precise control during fabrication; have higher melting temperatures; and tend to shrink and warp more than amorphous types. They have a relatively sharp melting point. That is, they do not soften gradually with increasing temperature but remain hard until a given quantity of heat has been absorbed, then change rapidly into a low-viscosity liquid. If the correct amount of heat is not applied properly during processing, product performance can be drastically reduced and/or an increase in processing cost occurs.

	Nylon	Polyacetals	Polycarbonate	Polysulfone	PPO
Heat distortion temp. (264 psi), °F	152	230	275	345	375
Resistance to heat (cont.), °F	270–300	230–255	250	300	250
Creep resistance	Fair	Good	Very good	Exc.	Exc.
Impact strength, ft-lb/in.	1.2–4	1.2–2.3	12–16	1.3	1.5–1.9
Tensile modulus, 10^5 psi	2.6–3.8	3.8–4.8	3.2	3.6	3.6–3.8
Elongation, %	25–300	30–75	60–100	50–100	50–80
Tensile strength, 10^3 psi	7–11.8	8.8–9.3	8–9	10.2	9–10
Hardness, Rockwell	R115	M82	R115	R120	R119
Water absorption, (24 hr), pct	.4–1.6	.22–.25	.15	.22	.06
Strong attack by	Oxidizing agents, mineral acids	Oxidizing agents, mineral acids	Alkalies, polar organic solvents	Polar organic solvents	Polar organic solvents
Additional selection factors	Lubricity; good general purpose material	Best surface; colors; solvent resistance	Color matching; clear transparency	Long-time heat resistance; chemical resistance under stress	Steam resistance; dielectric strength; brittle point −275°F

Table 1.22 Examples of engineering thermoplastic properties

Resin	Properties	Uses	Limitations
Diallyl phthalate polymers	Good electricals, dimensional stability, chemical and heat resistance	Prepregs, ducting, radomes, aircraft, and missiles	
Epoxies	Good electricals, chemical resistance, high strength	Printed circuit board tooling, filament winding	Require heat curing for maximum performance.
Melamines	Good electricals, chemical and heat resistance	Decorative, electrical (arc and track resistance), circuit breakers	
Phenolics	Low cost, chemical resistance, good electricals, heat resistance; nonflammable; can be used to 350–400°F	Work horse—diverse mechanical and electrical applications	Dissolve in caustic unless specially treated
Polyesters	Good all-around properties, ease of fabrication, low cost, versatile	Corrugated sheeting, seating, boats, automotive, tanks and piping, aircraft, tote boxes	Not for strong oxidizers, aromatic solvents, concentrated caustic
Silicones	Heat resistance, good electricals	Electrical, aerospace	

Table 1.23 Examples of engineering reinforced thermoset plastic properties

Unreinforced (neat)	40% Glass Fiber*	40% Talc*
180,000 psi (1,240 MPa)	1,100,000 psi (7,600 MPa)	575,000 psi (3,970 MPa)

*Glass fiber and talc content are by weight.

Table 1.24 Comparison Polypropylene NEAT and filled (flexural modulus of elasticity data)

Acetal (POM)
Acrylics
  Polyacrylonitrile (PAN)
  Polymethylmethacrylate (PMMA)
Acrylonitrile butadiene styrene (ABS)
Alkyd
Allyls
  Diallyl isophthalate (DAIP)
  Diallyl phthalate (DAP)
Aminos
  Melamine formaldehyde (MF)
  Urea formaldehyde (UF)
Cellulosics
  Cellulose acetate (CA)
  Cellulose acetate butyrate (CAB)
  Cellulose acetate propionate (CAP)
  Cellulose nitrate
  Ethyl cellulose (EC)
Chlorinated polyether
Epoxy (EP)
Ethylene vinyl acetate (EVA)
Ethylene vinyl alcohol (EVOH)
Fluorocarbons
  Fluorinated ethylene propylene (FEP)
  Polytetrafluoroethylene (PTFE)
  Polyvinyl fluoride (PVF)
  Polyvinylidene fluoride (PVDF)
Ionomer
Ketone
Liquid crystal polymer (LCP)
  Aromatic copolyester (TP polyester)
Melamine formaldehyde (MF)
Nylon (Polyamide) (PA)
Parylene
Phenolic
  Phenol formaldehyde (PF)
Polyamide (nylon) (PA)
Polyamide-imide (PAI)
Polyarylethers
  Polyaryletherketone (PAEK)
  Polyaryl sulfone (PAS)
Polyarylate (PAR)
Polycarbonate (PC)
Polyesters
  Saturated polyester (TS polyester)
  Thermoplastic polyesters
    Polybutylene terephthalate (PBT)
    Polyethylene terephthalate (PET)
  Unsaturated polyester (TS polyester)
Polyetherketone (PEK)
Polyetheretherketone (PEEK)
Polyetherimide (PEI)
Polyimide (PI)
  Thermoplastic PI
  Thermoset PI
Polymethylmethacrylate (acrylic) (PMMA)
Polyolefins (PO)
  Chlorinated PE (CPE)
  Cross-linked PE (XLPE)
  High-density PE (HDPE)
  Ionomer
  Linear LDPE (LLDPE)
  Low-density PE (LDPE)
  Polyallomer
  Polybutylene (PB)
  Polyethylene (PE)
  Polypropylene (PP)
  Ultra-high-molecular weight PE (UHMWPE)
Polyurethane (PUR)
Silicone (SI)
Styrenes
  Acrylic styrene acrylonitrile (ASA)
  Acrylonitrile butadiene styrene (ABS)
  General-purpose PS (GPPS)
  High-impact PS (HIPS)
  Polystyrene (PS)
  Styrene acrylonitrile (SAN)
  Styrene butadiene (SB)
Sulfones
  Polyether sulfone (PES)
  Polyphenyl sulfone (PPS)
  Polysulfone (PSU)
Urea formaldehyde (UF)
Vinyls
  Chlorinated PVC (CPVC)
  Polyvinyl acetate (PVAc)
  Polyvinyl alcohol (PVA)
  Polyvinyl butyrate (PVB)
  Polyvinyl chloride (PVC)
  Polyvinylidene chloride (PVDC)
  Polyvinylidene fluoride (PVF)

Table 1.25 Examples of the major plastic families with their abbreviations

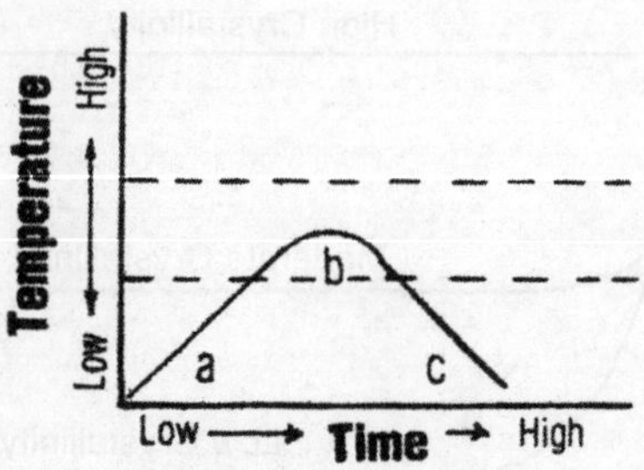

Figure 1.20 Temperature-time melting characteristic and cycle for processing thermoplastics: (a) start of melting process, (b) plastic melts, and (c) plastic hardens

Property	Crystalline	Amorphous
Melting or softening	Fairly sharp melting point	Softens over a range of temperature
Density (for the same material)	Increases as crystallinity increases	Lower than for crystalline material
Heat content	Greater	Lower
Volume change on heating	Greater	Lower
Effect of orientation	Greater	Lower
Compressability	Often greater	Sometimes lower
Strength	High	Low
Stress-craze	No	Yes
Transparent	No	Yes
Chemical resistant	Excellent	Poor
Viscosity	Low	High
Critical temperature-time	Yes	No

Table 1.26 Features of crystalline and amorphous thermoplastics

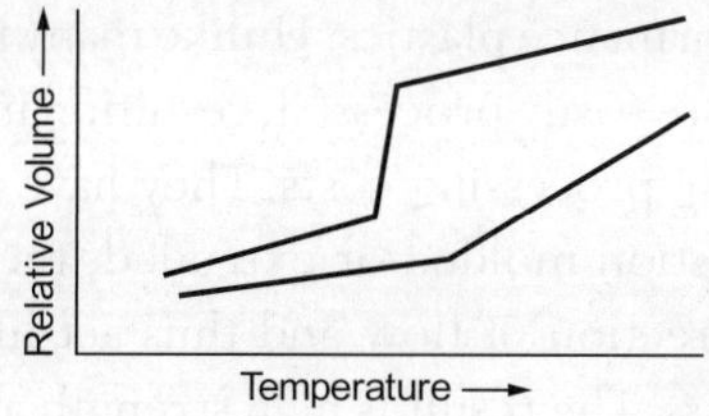

Figure 1.21 During processing, volume changes of crystalline (top) and amorphous TPs differ

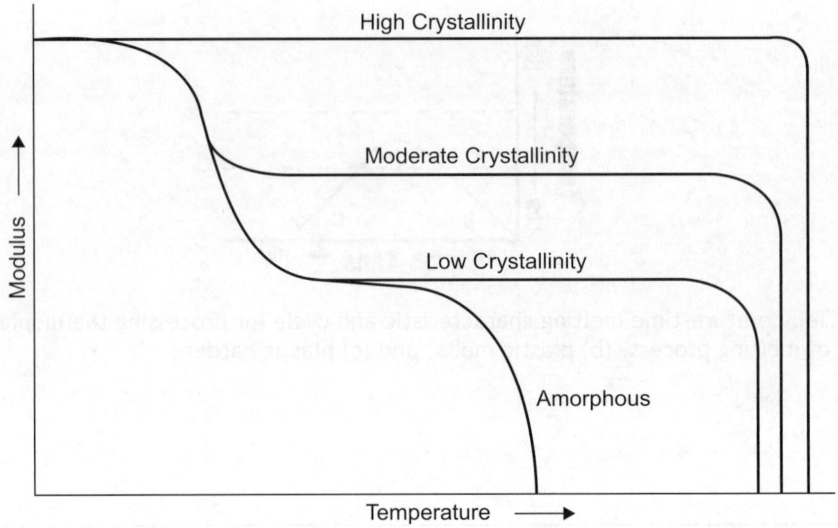

Figure 1.22 Thermoplastic morphologies subjected to different temperatures influence their properties such as tensile modulus of elasticity

Different processing conditions influence the performance of plastics. For example, the effects of time are similar to those of temperature in the sense that any given plastic has a preferred or equilibrium structure in which it would prefer to arrange itself over time. However, it is prevented from doing so instantaneously or at least in a short time. If given enough time, the molecules will rearrange themselves into their preferred pattern. Proper heating time causes this action to occur sooner. Otherwise, with a fast action, severe shrinkage and property changes could occur in all directions in the processed plastic products. This characteristic morphology of plastics can be identified by tests. It provides excellent control as soon as material is received in the plant, during processing, or after part fabrication.

Liquid Crystalline Polymer

Self-reinforcing TP liquid crystal polymers (LCPs) with molecules that are rodlike structures in parallel arrays were commercialized in the early 1980s by Celanese's Rosato, Jaarsma, and McChesney, ushering in the age of ultrahigh-performance polymers. These densely packed fibrous polymer chains result in high-performance plastics. Unlike many high-temperature TPs, LCPs have a low melt viscosity and are thus more easily processed, resulting in faster cycle times than those with a high melt viscosity, thus reducing processing costs. They have the lowest warpage and shrinkage of all the TPs. When they are injection molded or extruded, their molecules align into long, rigid chains that in turn align in the direction of flow and thus act like reinforcing fibers, giving LCPs both very high strength and stiffness. The result is high strength at extreme temperatures, excellent mechanical property retention after exposure to weathering and radiation, good dielectric strength as well as arc resistance and dimensional stability, low coefficient of thermal expansion, excellent flame resistance, and easy processability.

Their UL continuous-use rating for electrical properties is as high as 240°C (464°F). A high heat deflection value permits LCP-molded products to be exposed to intermittent temperatures as high as 315°C (600°F) without affecting their properties. Their resistance to high-temperature flexural creep is excellent, as are their fracture-toughness characteristics. This family of LCPs resists most chemicals and weathers oxidation and flame, making them excellent replacements for metals, ceramics, and other plastics in many product designs.

LCPs with their high strength-to-weight ratios are particularly useful for weight-sensitive products (Table 1.27). They are exceptionally inert and resist stress cracking in the presence of most chemicals at elevated temperatures, including the aromatic and halogenated hydrocarbons as well as strong acids, bases, ketones, and other aggressive industrial products. Their hydrolytic stability in boiling water is excellent, but high-temperature steam, concentrated sulfuric acid, and boiling caustic materials will deteriorate LCPs. With regard to flammability, LCPs have an oxygen

Property	Crystalline	Amorphous	Liquid crystalline
Specific gravity	higher	lower	higher
Tensile strength	higher	lower	highest
Tensile modulus	higher	lower	highest
Ductility, elongation	lower	higher	lowest
Resistance to creep	higher	lower	high
Max. usage temerature	higher	lower	high
Shrinkage and warpage	higher	lower	lowest
Melt flow	higher	lower	highest
Chemical resistance	higher	lower	highest

Table 1.27 Liquid crystal polymer properties compared to other thermoplastics

index ranging from 35% to 50%. When exposed to open flame they form an intumescent char that prevents dripping.

THERMOSET

When processing thermosets (TSs), heat is applied to make them flowable. At a higher temperature they solidify and become infusible and insoluble (Fig. 1.20). Cured TSs cannot be resoftened with heat. Its curing cycle is like boiling an egg that has turned from a liquid to a solid and cannot be converted back to a liquid. They undergo a cross-linking chemical reaction of their molecules by the action of heat and pressure (exothermic reaction), oxidation, radiation, and/or other means often in the presence of curing agents and catalysts. They can be granulated and used as filler in TSs as well as TPs. In general, with their tightly cross-linked structure there are TSs that resist higher temperatures and provide greater dimensional stability and strength than do most TPs.

The cure A-B-C stages identify their cure cycle (Fig. 1.23), where A-stage is uncured, B-stage is partially cured, and C-stage is fully cured. The typical B-stage is TS molding compounds and prepregs, which in turn are processed to produce C-stage fully cured plastic material products.

CROSS-LINKED THERMOPLASTIC

Certain TPs can readily be converted to TSs, providing improved properties by being chemically cross-linked. Cross-linking is an irreversible change that occurs through a chemical reaction. Curing is usually accomplished by the addition of cross-linking (curing) agents with or without heat and pressure. Cross-linking improves resistance to thermal degradation of physical properties and improves resistance to cracking effects by liquids and other harsh environments, as well as resistance to creep and cold flow, among other effects. Prime interest has been in aliphatic polymers, such as the olefins, which include the polyethylenes and polypropylenes; also popular is polyvinyl chloride. The cross-linked PE, identified as XLPE or PEX, is recognized as a standard within the industry. Use includes electrical cable coverings, cellular materials (foams), rotationally molded articles, and piping.

High-intensity radiation from electron beams or UV (ultraviolet) sources has been used to initiate polymerization in TS systems of oligomers capped with reactive methacrylate (acrylic)

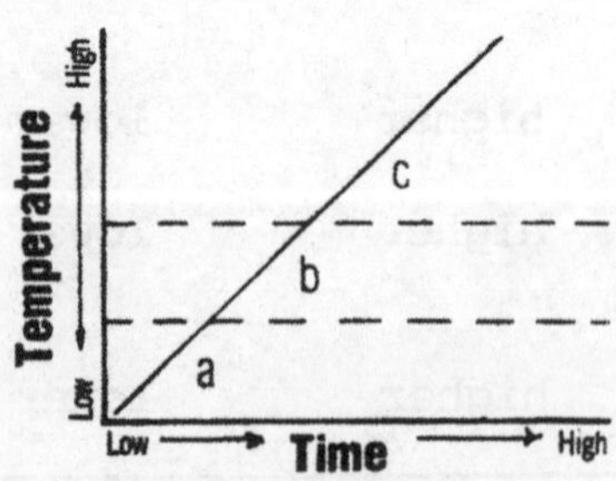

Figure 1.23 Thermoset A-B-C stages from melt to solidification

groups or isocyanates. Using this cross-linking polymerization technique, films with low shrinkage and high adhesion properties have been used in such applications as pressure-sensitive adhesives, glass coatings, and dental enamels.

For crystalline olefin polymers, the structures of the crystallite (regions where the long-chain molecules are closely aligned) and the connected/cross-linked noncrystalline regions contribute to polymer properties, including density (Table 1.28). For example, polyethylenes with densities in the range of 0.94 to 0.96 g/cc may be expected to contain 65% to 90% crystalline materials (crystallite), while polyethylene with a density of about 0.92 g/cc commonly contains 50% to 60% crystalline materials. The proportion of noncrystalline (amorphous) material is increased by high-energy radiation, an effect that is evidently due to the cross-linking of the molecules that occurs. Exposure of low-density polyethylene to radiation sufficient to cause cross-linking of about 10% of the carbon atoms yields a product that is entirely amorphous at room temperature. This loss of crystalline structure is easily accomplished, as it takes only a few of the hundreds of carbon atoms in each chain to be cross-linked to an adjacent chain for the structure to be completely cross-linked.

Polyethylene, like other non-cross-linked crystalline thermoplastic macromolecules, melts over a wide temperature range; as the temperature rises, the proportion of amorphous material increases until all the crystalline regions are melted and disappear. High-density polyethylenes (with a relatively high crystallite proportion) melt at 125° to 131°C, while lower density polyethylenes melt at 110° to 115°C. These results are expected. The presence of structural or stereo-chemical irregularities in macromolecules makes for a lower degree of crystallinity and a corresponding lower melting point. The molecular weight spread in the molecular weight of a specimen does not typically influence the final melting point or the melting range unless there is an appreciable proportion of molecules with a molecular weight below 1,500, which is less likely in commercial processes.

Polymer	Density Range (kg/m^3)	Crystallinity (%)
Polypropylene	920–940	45–55
LDPE	910–925	45–65
LLDPE	918–920	35–45
MDPE	925–940	65–75
HDPE	940–965	75–90
PA-12 (nylon 12)	1020	10–25
PA-6 (nylon 6)	1130	40–50
PA-66 (nylon 66)	1140	50–60*
PET	1130–1450	0–40*

* Upper values achieved by slow cooling, annealing

Table 1.28 Degree of crystallinity of crystalline plastics

COMPOUNDING AND ALLOYING

Converting polymers to almost 35,000 plastics includes mechanical mixing/blending of one or more polymers with additives, fillers, and/or reinforcement (364). They do not normally depend on chemical bonds, but do often require special compatibilizers. Mechanical compounding is extensively used as illustrated in Table 1.29.

Using a postreactor technique, plastics can be compounded by alloying or blending polymers in addition to using additives such as colorants, flame retardants, plasticizers, biocides, heat or light stabilizers, lubricants, fillers, reinforcements, and/or many more. With combinations of two or more polymers, synergistic property improvements beyond those that are purely additive in effect develop. Among the techniques used to combine dissimilar polymers are cross-linking to form what are called interpenetrating networks (IPNs), grafting to improve the compatibility of the plastics, and reactive polymerization in which molecular structure changes occur (Fig. 1.24) (108).

INTRODUCTION TO PROPERTIES

Throughout this book many different properties are reviewed, particularly in chapters 2, 19, and 20. What follows is a primer that provides some degree of familiarity with the variations of properties existing in plastics. For example, polystyrene, styrene-acrylonitrile, polyethylene, acrylic, ABS, polysulfone, EVA, polyphenylene oxide, and many other TPs are satisfactorily odor-free. FDA approvals are available for many of these plastics. There are food packaging and refrigerating conditions that will eliminate the use of certain plastics. Melamine and urea compounds are examples of plastics unsuitable for this service. If the product requires flexibility, examples of the choices available include polyethylene, vinyl, polypropylene, EVA, ionomer, urethane-polyester, fluorocarbon, silicone, polyurethane, plastisols, acetal, nylon, or certain rigid plastics that have limited flexibility in thin sections.

There are TPs that will craze or crack under certain environmental conditions, and products that are highly stressed mechanically must be checked very carefully. Polypropylene, ionomer, chlorinated polyether, phenoxy, EVA, and linear polyethylene offer greater freedom from stress crazing than some other TPs. Solvents may cause products held under stress to crack.

The subject of strength can be complex since so many different test conditions exist: short or long term, static or dynamic, and so forth. (chapter 19). Some strength aspects are interrelated with those of toughness. The crystallinity of TPs is important for their short-term yield strength. Unless the crystallinity is impeded, increased molecular weight generally also increases the yield strength. However, while the cross-linking of TSs increases their yield strength substantially, it has an adverse effect upon toughness (chapter 19).

Increasing the secondary bond strength and crystallinity, while increasing the primary bond strength, aids in optimizing long-term rupture strength in TPs much more readily. Fatigue strength is similarly influenced, and all factors that influence thermal dimensional stability also affect fatigue strength. This is a result of the substantial heating that is often encountered with fatigue, particularly in TPs.

	Thermoplastics																		Thermosets				
Filler/reinforcement	Acetal	Acrylic	ABS	Cellulosics	Fluoroplastics	Nylon	Polycarbonate	Polyester, thermoplastic	Polyethylene	LDPE	HDPE	Polypropylene	PPO, modified	Polystyrene	Polysulfone	PPS	PVC	flexible	rigid	Epoxy	Phenolic	Polyester, unsaturated	Polyurethane (inc. RRIM)	Comment
Alumina trihydrate			●							●							●			●	●	●	●	Extender; flame retardant; smoke suppressant
Aramid fibers						●														●	●	●	●	Reinforcement; strongest and stiffest of any organic fiber; lightweight
Asbestos fibers																	●				●	●		High-strength-to-weight-ratio reinforcement; noncombustible; considered health hazard
Barium sulfate																						●	●	Used as a filler and white pigment; increases specific gravity, frictional resistance, chemical resistance
Boron fibers																				●		●		High tensile strength and compressive load bearing capacity; expensive
Calcium carbonate			●			●			●					●			●			●	●	●	●	Most widely used extender/pigment or filler for plastics
Calcium sulfate						●					●						●					●		Extender; also enhances physical properties, increases impact, tensile, compressive strength
Carbon black			●						●					●			●							Filler; used as pigment, antistat agent, or to aid in crosslinking; conductive
Carbon/graphite fibers						●	●					●								●		●		Reinforcement; high modulus and strength; low density; low coefficient of expansion; low coefficient of friction; conductive
Ceramic fibers																				●		●		Reinforcement; very high temperature resistance; expensive
Feldspar and nepheline syenite		●		●								●					●			●		●		Specialty filler; easily wet and dispersed; enables transparency and translucency; weather and chemical resistance
Glass reinforcement (fiber, spheres, cloth, etc.)	●		●		●	●	●	●	●				●	●	●	●	●			●	●	●	●	Largest volume reinforcement; high strength, dimensional stability, heat resistance, chemical resistance
Kaolin						●			●								●					●	●	Second largest volume extender/pigment, with largest use in wire and cable, SMC and BMC, vinyl flooring
Metal fillers, filaments			●			●	●					●		●			●				●	●		Used to impart conductivity (thermal and electrical) or magnetic properties or to reduce friction; expensive
Mica			●		●	●	●	●	●				●		●					●	●	●	●	Flake reinforcement; improves dielectric, thermal, mechanical properties; low in cost
Organic fillers				●													●			●	●	●	●	Extenders/fillers, like wood flour, nutshell, corncobs, rice, peanut hulls

Table 1.29 Example of mechanically compounding materials used with polymers to develop many different properties of plastics

Silica		●							●					●			●					●	●	Fillers/extenders/reinforcements; functions in thickening liquid systems and making them thixotropic, in helping avoid plateout with PVC, in acting as flatting agents
Sodium aluminum silicates			●			●			●			●					●			●		●		Synthetic color extender/reinforcement: controllable structure level; rheological properties
Talc						●			●					●			●				●		●	Extenders/reinforcements/fillers; higher stiffness, tensile strength and resistance to creep
Wollastonite						●	●	●	●					●						●	●	●	●	High loadings possible; can improve strength, lower moisture absorption, elevate heat and dimensional stability, improve electricals

Table 1.29 *(continued)*

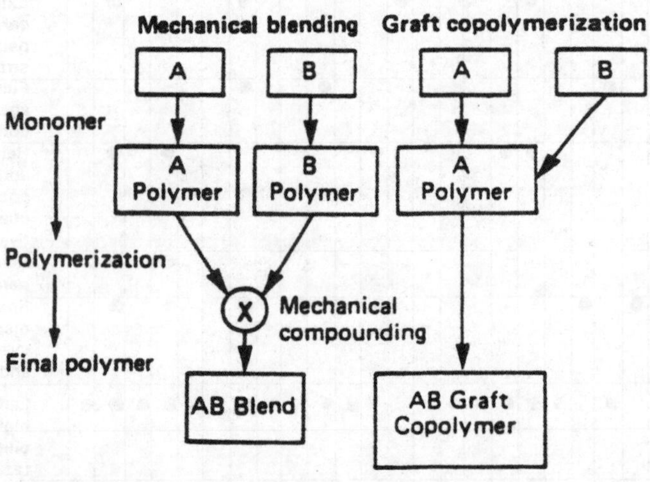

Figure 1.24 Examples of combining polymers

Thermal considerations will eliminate many materials. Examples of products operating above 450°F (232°C) include silicones, fluoroplastics, polyirnides, hydrocarbon resins, and methylpentene; cold mold processing or glass-bonded-mica-filled plastics may also be required. A few of the organic plastic-bonded inorganic fibers such as bonded ceramic wool perform well in this field. Epoxy, diallyl phthalate, and phenolic-bonded glass fibers may be satisfactory in the 450° to 550°F (232° to 288°C) ranges (chapter 15). A limited group of ablation material is made for outer space reentry use (chapter 20).

Between 250° and 450°F (121° and 232°C) glass or mineral-filled phenolics, melamine, alkyd, silicone, nylon, polyphenylene oxide, polysulfone, polycarbonate, methylpentene, fluorocarbon, polypropylene, and diallyl phthalate can be considered. The addition of glass fillers to the thermoplastics can raise the useful temperature range as much as 212°F (100°C) and at the same time shorten the fabricating cycle.

In the 0° to 212°F range, a broad selection of materials is available. Low temperature considerations may eliminate certain thermoplastics. Polyphenylene oxide can be used at temperatures as low as –275°F. Thermosetting materials exhibit minimum embrittlement at low temperatures.

There are plastics to meet different temperatures (Fig. 1.25). Figure 1.26 provides a guide to and comparison of the temperature capabilities for commodity and engineering plastics as well as steel (tensile yield strength versus temperature).

Urea, melamine, polycarbonate, polyphenylene oxide, polysulfone, polypropylene, diallyl phthalate, and the phenolics are needed in the temperature range above 200°F (93°C) for good color stability. Most TPs will be suitable below this range.

The Underwriters Laboratories' ruling on the use of self-extinguishing plastics for contact-carrying members and many other components introduces critical material selection problems. All thermosets are self-extinguishing. Nylon, polyphenylene oxide, polysulfone, polycarbonate, vinyl, chlorinated polyether, chlorotrifluoroethylene, vinylidene fluoride, and fluorocarbon are thermoplastics that may be suitable for applications requiring self-extinguishing properties. Cellulose acetate and ABS are also available with these properties. Glass reinforcement improves these materials considerably.

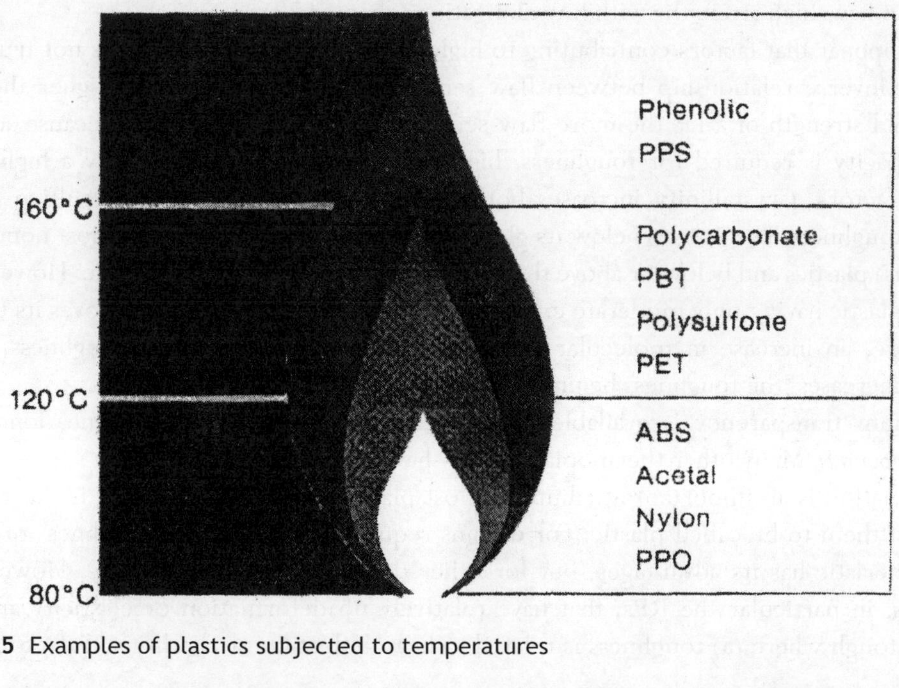

Figure 1.25 Examples of plastics subjected to temperatures

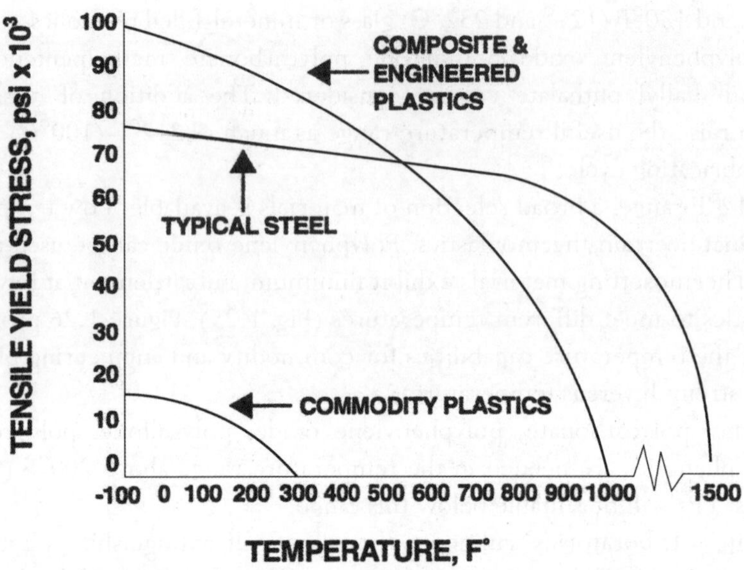

Figure 1.26 Strength vs. temperature of steel and plastics (courtesy of PlastiSource)

Toughness is usually the most complex factor to technically define and understand. Tough TPs are usually described as ones having a high elongation to failure or ones in which a large amount of energy must be expended to produce failure. For high toughness a plastic needs both the ability to withstand load and the ability to elongate substantially without failing, except in the case of RPs. The RPs can have high strengths with low elongation (chapter 22).

It may appear that factors contributing to high stiffness are required. This is not true because there is an inverse relationship between flaw sensitivity and toughness; the higher the stiffness and the yield strength of a TP, the more flaw-sensitive it becomes. However, because some load-bearing capacity is required for toughness, high toughness can be achieved by a high trade-off of certain factors. Crystallinity increases both stiffness and yield strength, resulting usually in decreased toughness. This is true below its glass transition temperature (T_g) in most noncrystalline (amorphous) plastics and below or above the T_g in a substantially crystalline plastic. However, above the T_g in a plastic having only moderate crystallinity, increased crystallinity improves its toughness. Furthermore, an increase in molecular weight from low values increases toughness, but with continued increases, the toughness begins to drop.

Maximum transparency is available in acrylic, polycarbonate, polyethylene, ionomer, and styrene *compounds*. Many other thermoplastics may have adequate transparency.

Deformation is an important attribute in most plastics, so much so that it is the very factor that has led them to be called plastic. For designs requiring such traits as toughness or elasticity, this characteristic has its advantages, but for other designs it is a disadvantage. However, there are plastics, in particular the RPs, that have relatively no deformation or elasticity and yet are extremely tough where (a) toughness is related to heat deflection or rigidity and (b) toughness or impact is related to temperature for polystyrene (PS) and high-impact polystyrene (HIPS). This type of behavior characterizes the many different plastics available. Some that are tough at room temperature are brittle at low temperatures. Others are tough and flexible at temperatures far below freezing but become soft and limp at moderately high temperatures. Still others are hard and rigid at normal temperatures but may be made flexible by copolymerization or adding plasticizers.

Toughness means resistance to fracture. However, there are those materials that are nominally tough but may become embrittled due to processing conditions, chemical attack, prolonged exposure to constant stress, and so on. A high modulus and high strength with ductility is the desired combination of attributes. However, the inherent nature of plastics is such that their having a high modulus tends to associate them with low ductility, and the steps taken to improve the one will cause the other to deteriorate.

Soft, weak materials have a low modulus, low tensile strength, and only moderate elongation to break. According to American Society for Testing and Materials (ASTM) standards, the elastic modulus or the modulus or elasticity is the slope of the initial straight-line portion of the curve. Hard, brittle materials have high moduli and quite high tensile strengths, but they break at small elongations and have no yield point. Hard, strong plastics have high moduli, high tensile strengths, and elongations of about 5% before breaking. Their curves often look as though the material broke about where a yield point might have been expected.

Soft, tough plastics are characterized by low moduli, yield values or plateaus, high elongations of 20% to 1,000%, and moderately high breaking strengths. The hard, tough plastics have high moduli/yield points, tensile strengths, and elongations. Most plastics in this category show cold drawing or necking during the stretching operation. The RPs will have at least one modulus, but some materials can have two or three.

From a practical viewpoint, toughness is readily understood, but technically there tends to be no scientific method of measuring it. One definition of toughness is simply the energy required to break the plastic. This energy is equal to the area under the stress-strain curve. The toughest plastics should be those with very great elongations to break, accompanied by high tensile strengths; these materials nearly always have yield points. One major exception to this rule is RPs that make use of reinforcing fibers such as glass and graphite.

Although impact strength of plastics is widely reported, this property has no particular design value and can be used only to compare relative response of materials (toughness, etc.). Even this comparison is not completely valid, because it does not solely reflect the capacity of the material to withstand shock loading, but it can pick up discriminatory response to notch sensitivity.

A better value is tensile impact, but unfortunately this property is not generally reported. The impact value can broadly separate those that can withstand shock loading versus those that are poor in this response. Therefore, only broad generalizations can be obtained on these values. Comparative tests on sections of similar size that are fabricated in accordance with the proposed product must be performed to determine the impact performance of a plastic material. Laminated plastics, glass-filled epoxy, melamine, and phenolic are outstanding in impact strength. Polycarbonate and ultrahigh molecular weight PE also have excellent impact strength.

Deteriorating/damaging effects of moisture can occur with certain plastics. For high-moisture applications, polyphenylene oxide, polysulfone, acrylic, butyrate, diallyl phthalate, glass-bonded mica, mineral-filled phenolic, chlorotrifluoroethylene, vinylidene, chlorinated polyether chloride, vinylidene fluoride, and the fluorocarbons should be satisfactory. Diallyl phthalate, polysulfone, and polyphenylene oxide have performed well with moisture/steam on one side and air on the other (a troublesome combination), and they also will withstand repeated steam autoclaving. Long-term studies of the effect of water have disclosed that chlorinated polyether gives outstanding performance. Impact styrene plus 25 wt% graphite and high-density polyethylene with 15% graphite give long-term performance in water.

Depending on what is required, the different plastics can provide different rates of permeability properties. As an example, certain polyethylenes will pass wintergreen, hydrocarbons, and many other chemicals. It is used in certain cases for the separation of gases since it will pass one and block another. Chlorotrifluoroethylene and vinylidene fluoride, vinylidene chloride, polypropylene, EVA, and phenoxy merit evaluation.

There are materials with low or no permeability to different products in various environments. Barrier technology is not becoming more complex but more precise. Different factors influence performance such as lack of pinholes, chemical composition, cross-linking, modification, molecular orientation, density, and thickness. Coextrusion and coinjection processes are used to reduce permeability while retaining other desirable properties. Total protection against vapor transmission by a single barrier material increases linearly with increasing thickness, but it usually is not economical. Thus, extensive use is made of multiple-layer constructions. These composites can include low-cost as well as recycled plastics that provide mechanical support, and so on, while the thickness of an expensive barrier material is significantly reduced (chapter 20).

With crystalline plastics, the crystallites can be considered impermeable. Thus, the higher the degree of crystallinity, the lower the permeability is to gases and vapors. Permeability in an amorphous plastic below or not too far above its glass transition temperature (T_g) is dependent on the degree of molecular orientation. It is normally reduced when compared to higher temperatures, although small strains sometimes increase the permeability of certain plastics. The orientation of elastomers well above their T_g has relatively less effect on the overall transport property. Cross-linking thermoplastics will decrease permeability owing to the decrease in their diffusion coefficient. The effect of cross-linking is more pronounced for large molecule size vapors. The addition of a plasticizer usually increases the rates of vapor diffusion and permeation.

The permeation of vapors includes two basic processes: the sorption and diffusion of vapors in the plastic. For example, in the packaging industry, the resistance of moisture is essential for the preservation of many products. The loss of moisture, flavor, and so on through packaging materials may damage foodstuffs. The prevention of the ingress of moisture by a barrier is essential for the storage of dry foods and other products. In other applications, the degree of resistance to water and oxygen is important for the development of corrosion-resistant coatings, electrical and electronic parts, and so forth.

Fluorination is the process of chemically reacting a material with a fluorine-containing compound to produce a desired product. For example, fluorination can improve the gasoline barrier of PE to nonpolar solvents. A barrier is created by the chemical reaction of the fluorine and the PE, which form a thin (~30 mm) fluorocarbon layer on the surface. Two different systems can be used to apply the treatment, depending on the results desired. With the "in-process" system, such as that used during blow molding PE gasoline tanks, fluorine is used as a part of the parison expanding gas in the blowing operation (resulting in no gasoline leakage when tank is used). The barrier layer is created only on the inside. In a posttreatment system, bottles and other products are placed in an enclosed chamber filled with fluorine gas. This method forms barrier layers on both the inside and outside surfaces.

In general, plastics are superior to elastomers in radiation resistance but are inferior to metals and ceramics. The materials that will respond satisfactorily in the range of 1,010 and 1,011 erg per gram are glass and asbestos-filled phenolics, certain epoxies, polyurethane, polystyrene, mineral-filled polyesters, silicone, and furane. The next group of plastics in order of radiation resistance includes polyethylene, melamine, urea-formaldehyde, unfilled phenolic, and silicone plastics. Those materials that have poor radiation resistance include methyl methacrylate, unfilled polyesters, cellulosics, polyamides, and fluorocarbons (chapter 20).

PLASTICS CHARACTERISTICS

To understand plastics, one must first appreciate and accept the polymer chemist's ability to literally rearrange the molecular structure of the polymer (that becomes the plastic) to provide an almost infinite variety of compositions that differ in form, melt behavior (Fig. 1.27 and Table 1.30), thermal

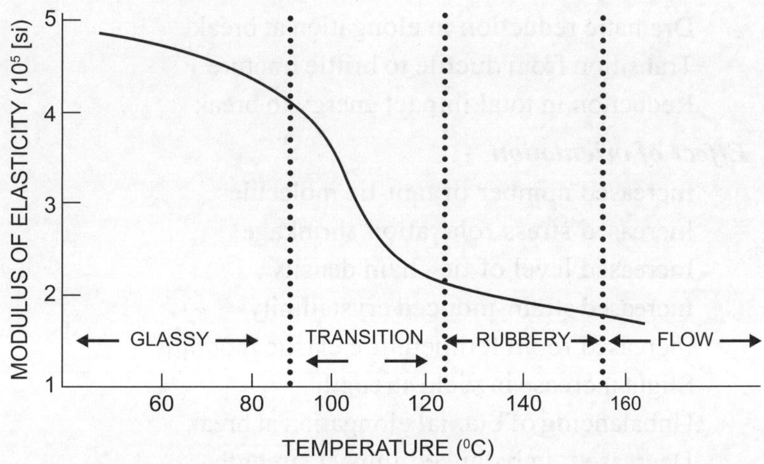

Figure 1.27 Modulus behavior with increase in temperature (DTUL = deflection temperature under load) (courtesy of Bayer)

behavior (Table 1.16), appearance, properties, cost, and other characteristics. One must also approach the subject with a completely open mind that will accept all the contradictions that could make it difficult to pin common labels on the different families of plastics or even on the many various types within a single family that are reviewed in this book. Since each plastic has distinctive characteristics such as performance properties and/or fabricating procedures, they are labeled by their many different characteristics.

Effect of decreased cooling rate
Increased degree of crystallinity
Increased level of crystalline perfection
Increased lamellar thickness
Increased spherulitic size
Increase in b-spherulites (mp 147°C)
Increased elastic modulus
Increased yield strength
Increased molecular diffusion
Increased level of segregation of uncrystallizable impurities at intercrystalline boundaries
Increased weakness of intercrystalline boundaries
Decreased tie chain density
Decreased ductility on deformation
Fewer lamellae interconnections
Higher stress concentrations at surfaces of crystallites
Reduction in room temperature tensile strength
Dramatic reduction in elongation at break
Transition from ductile to brittle fracture
Reduction in total impact energy to break
Effect of orientation
Increased number of taut-tie molecules
Increased stress relaxation shrinkage
Increased level of tie chain density
Increased strain-induced crystallinity
Increased room temperature elastic modulus
Slight increase in yield strength
Unbalancing of biaxial elongation at break
Decreased, unbalanced impact strength

Table 1.30 Example of morphology effects on cooling melts during processing

One must be aware of the many different routes that the starting materials for plastics can take on the way to the consumer or the industry. There are those plastics that are supplied to the processor in the form of granules, powder, pellets, flakes, or liquids and are transformed into plastic products. However, the same starting materials used to make these plastics can take other routes and end up in the textile industry (nylon fibers share common roots with a molded nylon gear; acrylic fibers share common roots with acrylic sheet for glazing; etc.). Similar parallels exist in the paint industry, adhesives industry, and other industries with their special requirements.

THERMAL BEHAVIOR

In order to select materials that will maintain acceptable mechanical characteristics and dimensional stability, one must be aware of both the normal and extreme thermal operating environments to which a product will be subjected. TS plastics have specific thermal conditions when compared to TPs that have various factors to consider that influence the product's performance and processing capabilities. The properties and processes of TPs are influenced by their thermal characteristics such as melt temperature (T_m), glass-transition temperature (T_g), dimensional stability, thermal conductivity, specific heat, thermal diffusivity, heat capacity, coefficient of thermal expansion, and decomposition (T_d). Table 1.31 provides thermal conductivity data in this regard on different plastics.

There is a maximum temperature, or to be more precise a maximum time-to-temperature relationship, for all materials preceding loss of performance or decomposition. Figure 1.28 provides a temperature guide for continuous heating of plastics. The effects of temperature on plastics are discussed throughout this book. There is also the effect of heat during recycling plastics.

Aging at elevated temperatures typically involves exposing test specimens or products at different temperatures for different extended time periods. Tests are performed at room or the respective testing temperatures for whatever mechanical, physical, or electrical property is of interest. These tests of aging can be used as a measure of thermal stability in design as is done with other materials. Data presented for different plastics in Figure 1.29 and Table 1.32 show 50% retention of mechanical and physical properties obtainable at room temperature, with plastics exposure and testing at elevated temperatures.

If a plastic product is free to expand and contract, its thermal expansion property will usually be of little significance. However, if it is attached to another material that has a lower CLTE (coefficient of linear thermal expansion), then the movement of the part will be restricted. A temperature change will then result in developing thermal stresses in the part. The magnitude of these stresses will depend on the temperature change, the method of attachment and relative expansion, and the modulus characteristics of the two materials at the point of the exposed heat.

In its simplest form, thermal stresses can be calculated by using the following equation:

$$\sigma = E_p (\alpha_1 - \alpha_2) (\Delta T),$$

where σ = thermal stress, E_p = elastic modulus of elasticity, α_1 = CLTE of material #1, α_2 = CLTE of material #2, and ΔT = temperature change.

The goal is to eliminate or significantly reduce all sources of thermal stress. This can be achieved by considering the following: (a) when adding material for local reinforcement, select a material with the same or a similar CLTE; (b) where plastic is to be attached to a more-rigid material, use mechanical fasteners with slotted or oversized holes to permit expansion and contraction to occur; (c) do not fasten dissimilar materials tightly; and (d) adhesives that remain ductile, such as urethane and silicone, through the product's expected end-use temperature range can be used without causing stress cracking or other problems.

In addition to dimensional changes from changes in temperature, other types of dimensional instability are possible in plastics, as in other materials. Water-absorbing plastics, such as certain nylons, may expand and shrink as they gain or lose water, or even as the relative humidity changes. The migration or leaching of plasticizers, as in certain PVCs, can result in slight dimensional changes.

RESIN	(THERMAL COND.) Btu/hr ft °F	(SPEC HEAT) Btu/lb °F
ABS	0.108 - 0.192	0.490
ABS/PC	0.142 - 0.217	
CA	0.100 - 0.192	0.410
CAB	0.100 - 0.192	0.390
HIPS	0.024 - 0.073	0.501
IONOMER	0.142 - 0.142	0.645
LDPE	0.192 - 0.192	0.760
MDPE	0.192 - 0.242	0.765
HDPE	0.267 - 0.300	0.870
PA 11	0.167 - 0.167	0.583
PA 6	0.133 - 0.133	0.731
PA 6/6	0.142 - 0.142	0.735
PA 6/10	0.125 - 0.125	
PC	0.108 - 0.108	0.438
PC GF 40%	0.117 - 0.125	0.348
PET, PBT	0.167 - 0.167	0.490
PMMA	0.092 - 0.142	0.454
PMMA IMP	0.100 - 0.125	
POM	0.133 - 0.133	0.715
PP	0.066 - 0.079	0.667
PPO	0.111 - 0.111	
PPO GF	0.092 - 0.092	
PPO/PS	0.125 - 0.125	0.519
GF PPS	0.167 - 0.259	0.497
PS G.P.	0.058 - 0.080	0.480
PSU	0.150 - 0.150	0.520
PVC	0.083 - 0.083	0.383
PVC RIGID	0.092 - 0.092	0.340
SAN	0.070 - 0.070	0.471
TFE	0.142 - 0.142	
TPU	0.041 - 0.041	

Table 1.31 Examples of plastics' thermal conductivity and specific heat

RESIDENCE TIME

The process of heating and cooling TPs can be repeated indefinitely by granulating and recycling scrap, defective products, and so on. During the heating and cooling cycles of injection, extrusion, and other plastics-forming processes, the material develops a heat history. With a limited number of recycles, the properties of certain plastics are not significantly affected by the resulting increased heat history. However, some TPs can undergo a significant loss of certain properties. If incorrect methods have been used in granulating recycled material, even more degradation will occur.

PLASTIC MEMORY

When most materials are bent, stretched, or compressed, they somehow alter their molecular structure or grain orientation to accommodate the deformation permanently. Not so with plastics. Plastics temporarily assume the deformed shape but always maintain internal stresses that want to force the material back to its original shape. This desire to return to an earlier shape is called plastic memory. For example, TPs can be bent, pulled, or squeezed into various useful shapes, but

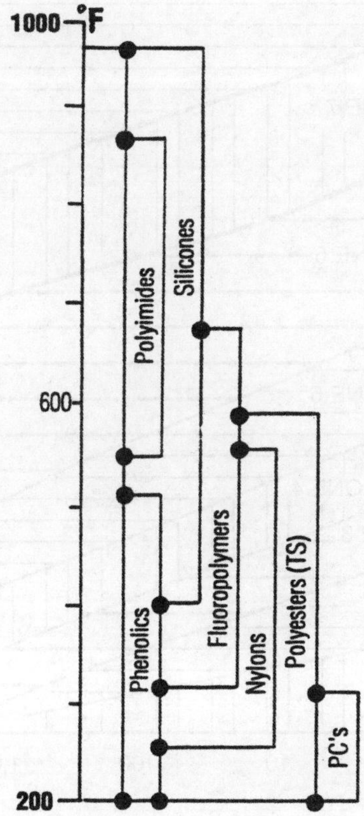

Figure 1.28 Continuous heat data (courtesy of PlastiSource)

eventually, especially when heat is added, they return to their original form. Plastic memory can be annoying, but when properly applied, it offers some interesting design possibilities for plastic products. The time/temperature-dependent change in mechanical properties results from stress relaxation and other viscoelastic phenomena typical of plastics. When the change is an unwanted limitation, it is called creep.

Most plastic products can be produced with a built-in memory. That is, the tendency to move into a desired shape is included as an integral part of the design. So then, after the products are

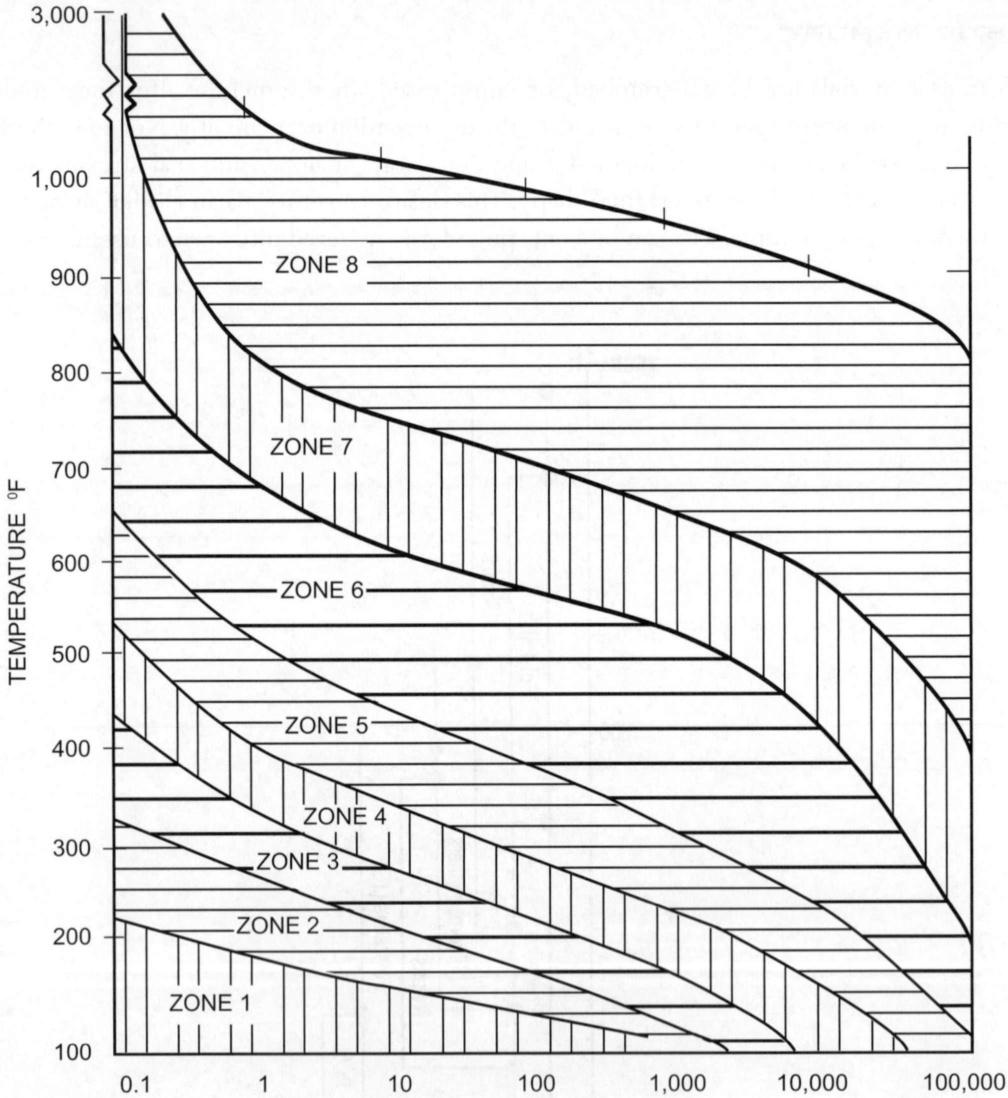

Figure 1.29 Guide to temperature vs. plastic properties; Table 1.32 identifies plastics (courtesy of PlastiSource)

assembled in place, a small amount of heat can coax them to change shape. TP products can be deformed during assembly and then allowed to return to their original shape. In this case, products can be stretched around obstacles or made to conform to unavoidable irregularities without permanent damage.

Potential memory exists in all TPs. Polyolefins, neoprene, silicone, and other cross-linkable polymers can be given a memory either by radiation or by chemically curing. Fluorocarbons, however, need no curing. And when the phenomenon is applied to fluorocarbons such as TFE, FEP, ETFE, ECTFE, CTFE, and PVF, interesting high-temperature or wear-resistant applications are possible.

Thermal Conductivity

Thermal conductivity (TC) is the rate at which a material will conduct heat energy along its length or through its thickness. ASTM tests give an indication of how much heat must be added to a unit

Zone 1 :	acrylic, cellulose esters, crystallizable block copolymers, LDPE, PS, vinyl polymers, SAN, SBR, urea-formaldehyde.
Zone 2 :	acetal, ABS, chlorinated polyether, ethyl cellulose, ethylene-vinyl acetate copolymer, furan, ionomer, phenoxy, polyamides, PC, RDPE, PET, PP, PVC, urethane.
Zone 3 :	polychlorotrifluoroethylene, vinylidene fluoride.
Zone 4 :	alkyd, fluorinated ethylene-propylene, melamine-formaldehyde, phenol-furfural, polysulfone.
Zone 5 :	acrylic, diallyl phthalate, epoxy, phenol-formaldehyde, TP polyester, polytetrafluoroethylene.
Zone 6.:	parylene, polybenzimidazole, polyphenylene, silicone.
Zone 7 :	polyamide-imide, polyimide;
Zone 8 :	plastics now being developed using rigid linear macromolecules rather than crystallization and cross-linking.

Table 1.32 Identification of plastics in Figure 1.29

mass of plastic in order to raise its temperature 1°C (chapter 22). This is an important factor, since plastics are often used as effective heat insulation in heat-generating applications and in structures where heat dissipation is important. The high degree of the molecular order for crystalline TPs makes their values tend to be twice those of the amorphous types.

For certain products, skill is required to estimate a product's performance under steady-state heat-flow conditions, especially those made of RPs (Fig. 1.30). The method and repeatability of the processing technique can have a significant effect. In general, TC is low for plastics, and the plastic's structure does not alter its value significantly. To increase TC the usual approach is to add metallic fillers, glass fibers, or electrically insulating fillers such as alumina. Foaming can be used to decrease thermal conductivity.

The TC of plastics depends on several variables and cannot be reported as a single factor. But it is possible to ascertain the two principal dependencies of temperature and molecular orientation (MO). In fact, MO may vary within a product, producing a variation in thermal conductivity. It is important for the product designer and processor to recognize such a situation. Certain products require personal skill to estimate a part's performance under steady-state heat flow.

Several factors make thermally conductive TPs attractive for different market segments. In the electronics market, the trend is toward smaller, lighter, and faster. As chip fabrication becomes faster, the amount of heat generated by the chip increases; a typical 486 chip generates about 5 watts of power while the newer Pentium 11 chips can generate more than 30 watts. The inability to remove the heat generated by these chips greatly reduces their operating life. The design flexibility afforded by thermally conductive TPs provides solutions to increased demands on chip cooling systems.

In the lighting market, thermally conductive plastics are useful. Here, improving TP thermal capabilities with product integration and lower fabricating costs can improve the operating life span of fluorescent fixtures. Thus, improvements in thermal performance could drive the replacement of traditional metals in these applications.

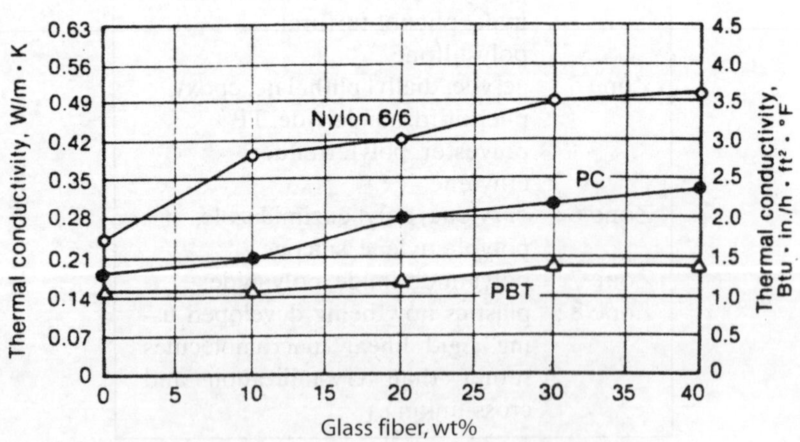

Figure 1.30 Thermal conductivity vs. glass fiber content in reinforced plastics

To replace metals in thermally conductive applications, an understanding of the performance of the current materials and their application requirements is important. TC of various materials in W/m°K are 400, copper; 225, extruded aluminum; 105, brass; 60, cast aluminum; 45, steel; 0.3, PPS; 0.24, nylon (PA); and 0.2, PP. Based on these values, the unmodified TPs do not possess the needed thermal conductivity to provide advanced thermal solutions. However, a significant spread exists in the conductivity performance of the metals currently being used in thermal applications, suggesting that in some applications, metals may have more thermal conductivity than the application requires.

In the past, engineering TPs have replaced metal in numerous products in many industries by providing improvements in thermal properties (140). The ability to prepare and compound material properties through the choice of plastic resins with additives, fillers, and reinforcements maximizes the flexibility inherent in TPs to meet the performance requirements in these different applications.

The key to providing thermally conductive TPs is to understand how much conductivity is needed. To clarify the conductivity requirements for thermal applications, a simple model has been developed by LNP Engineering Plastics to aid in understanding and to quantify the balance between the thermal conductivity as measured in the plane of the plate versus through the plane. Various fillers and reinforcements (nonelectrically conductive powdered ceramics and short glass or carbon fibers) have been studied with different TPs, including polyphenylene sulfides (PPSs), PAs, and PPs. The particulate powdered ceramic filler has no reinforcing properties and serves to lower the tensile and flexural strength. The carbon fiber filler adds additional reinforcement (R), resulting in the increase of tensile and flexural strength. Results show a five-fold increase in the thermal conductivity versus the base plastics, without a significant effect on the physical properties. These compounds provided conductivities ranging from 1 to 10 W/m°K.

Using this heat-transfer model with the compounded PPS, PA (nylon 6), and PP, the thermal performance of materials in air-cooled applications shows that convective heat transfer often governs the overall equilibrium temperature gradient. Because convective heat transfer becomes the limiting factor, TPs with a conductivity of 1 to 10 W/m°K can transfer as much heat as a metal having a higher thermal conductivity.

SPECIFIC HEAT

The specific heat or heat capacity of a unit mass of material is the amount of energy required to raise its temperature 1°C. It can be measured at either constant pressure or constant volume. At constant pressure it can be larger than at constant volume, because additional energy is required to bring about a volume change against external pressure. The specific heat of amorphous plastics increases with temperature in an approximately linear fashion below and above T_g, but a steplike change occurs near the T_g. No such step change occurs with crystalline types. For plastics, specific heat is usually reported during constant pressure heating. Plastics differ from traditional engineering materials because their specific heat is temperature sensitive.

THERMAL DIFFUSIVITY

Whereas specific heat is a measure of energy, thermal diffusivity is a measure of the rate at which energy is transmitted through a given plastic. It relates directly to processability. In contrast, metals have values hundreds of times larger than those of plastics. Thermal diffusivity determines plastics' rate of change with time. Although this function depends on thermal conductivity, specific heat at constant pressure, and density, all of which vary with temperature, thermal diffusivity is relatively constant.

COEFFICIENT OF LINEAR THERMAL EXPANSION

Like metals, plastics generally expand when heated and contract when cooled. Usually, for a given temperature change, many TPs have a greater change than metals. The coefficient of linear thermal expansion (CLTE) is the ratio between the change in a linear dimension relative to the original dimension of the material per unit change in temperature, according to ASTM standards. It is generally given as cm/cm/°C or in./in./°F.

The CLTE is an important consideration if dissimilar materials, such as two different plastics, a plastic and a metal, and so forth, are to be assembled where material expansion or contraction is restricted. The type of plastic and reinforcing material, particularly the glass fiber content and its orientation, influence the CLTE. It is especially important if the temperature range includes a thermal transition such as T_g. The design of products has to take into account the dimensional changes that can occur during fabrication and during its useful service life. With a mismatched CLTE there could be failure of the plastic part through cracking or buckling.

Normally, CLTE data and other information regarding dimensional changes for a given plastic material, which the designer can take into consideration when designing a plastic part or product, are readily available from the material supplier.

Expansion and contraction can be controlled in plastic by its orientation, cross-linking, adding fillers or reinforcements, and so on. With certain additives the CLTE value could be zero or near zero. For example, plastic with a graphite filler contracts rather than expands during a temperature rise. RPs with only glass fiber reinforcement can be used to match those of metal and other materials. In fact, TSs can be specifically compounded to have little or no change.

In TSs, the ease or difficulty of thermal expansion is dictated for the most part by the degree of cross-linking as well as the overall stiffness of the units between the cross-links. The less flexible units are also more resistant to thermal expansion. Such influences as secondary bonds have much less effect on the thermal expansion of TSs.

TEMPERATURE INDEX

The Underwriters Laboratories (UL) tests are recognized by various industries to provide continuous temperature ratings, particularly in electrical applications. These ratings include separate listings for electrical properties, mechanical properties including impact, and mechanical properties without impact. The temperature index is important if the final product has to receive UL recognition or approval.

Corrosion Resistance

Due to complex corrosive environments, at least 30% of the total yearly plastics production is required in buildings, chemical plants, transportation, packaging, and communications. The use of plastics saves some of the billions of dollars lost each year by industry due to the many forms of corrosion.

Corrosion is fundamentally a problem associated with metals. Since plastics are electrically insulating, they are not subject to this type of damage. Plastics are basically noncorrosive. However, there are those that can be affected when exposed to corrosive environments. The material deterioration or destruction of materials and properties are brought about through electrochemical, chemical, and mechanical actions.

Basically, corrosion resistance is the ability of a material to withstand contact with natural ambient conditions or a particular artificially created atmosphere without degradation or change in properties. Since plastics (not containing metallic additives) are not subjected to electrolytic corrosion, they are widely used where this property is required by the product or as coatings and linings for materials subjected to corrosion, such as in chemical and water filtration plants, mold/die, and so on. Plastics find use as protective coatings on products such as steel rods, concrete steel reinforcements, mold cavity coatings, plasticator screw coatings, and other areas.

Chemical Resistance

Part of the wide acceptance of plastics results from their relative compatibility with chemicals, particularly moisture, as compared to other materials. Because plastics, unlike metals, are largely immune to electrochemical corrosion, they can frequently be used profitably to contain water and corrosive chemicals that would normally attack metals.

Plastics are often used in corrosive environments for chemical tanks, water treatment plants, and piping to handle drainage, sewage, and water supply. Structural shapes for use under corrosive conditions often take advantage of the properties of RPs. Today's underground tanks must last 30 or more years without undue maintenance. To meet these criteria, they must be able to maintain their structural integrity and resist the corrosive effects of soil and gasoline, including gasoline that has been contaminated with moisture and soil. Figures 1.31 and 1.32 use glass fiber TS polyester RPs.

An example of an underground tank is a Chicago service station's May 1963 RP underground gasoline storage tank installation. It experienced no leakage and was still leaktight and structurally sound when unearthed in May 1988. The tank was one of 60 developed by RP Amoco Chemical Co. It was fabricated in two semicylindrical sections of fiberglass-woven roving and chopped strand mat impregnated by an unsaturated isophthalic TS polyester plastic selected for its superior resistance to acids, alkalis, aromatics, solvents, and hydrocarbons. The two sections were bonded to each other and to end caps with RP lap joints. Today the cylinder would be a single, unified construction.

Demand for this type of petroleum storage tank has grown rapidly, as environmental regulations have become more stringent. The RP tank just reviewed that was removed in 1988 met these requirements, but two steel tanks unearthed from the same site at that time failed to meet them. There was no record of how long the steel tanks had been in service, but one was dusted with white metal oxide and the other showed signs of corrosion at the weld line. Rust had weakened this joint so much that it could be scraped away with a pocketknife.

Tests and evaluations were conducted on the RP tank that had been 25 years in the ground and also on similarly constructed tanks unearthed at five-and-a-half and seven-and-a-half years that showed the RP tanks could more than meet the service requirements. Table 1.33 provides useful data from these tests.

FIRE PROPERTY

Like other materials, hot enough fires can destroy all plastics. Some burn readily, others slowly, others only with difficulty; still others do not support combustion after the removal of the flame. There are certain plastics used to withstand the reentry temperature of 2,500°F (1,370°C) that

Figure 1.31 Large water filtration tank

Figure 1.32 Underground RP 4,000-gallon gasoline tank (courtesy of Owens Corning Fiberglass)

Property		5.5 years	7.5 years	25.0 years
Flexural strength:	Psi	19,500	24,200	22,400
	MPa	134	167	154
Flexural modulus:	Psi	725×10^3	795×10^3	635×10^3
	MPa	4,992	5,482	4,378
Tensile strength:	Psi	10,700	13,600	10,500
	MPa	74	94	72
Tensile modulus:	Psi	$1{,}160 \times 10^3$	$1{,}053 \times 10^3$	$1{,}107 \times 10^3$
	MPa	7,260	8,000	7,630
Tensile elongation:	%	1.11	1.25	1.13
Notched Izod impact strength:	ft.-lb./in.	9.7	11.0	14.1
	J/m	518	587	753

Table 1.33 Unearthed underground gasoline storage tank data (courtesy of BP-Amoco)

occurs when a spacecraft returns into the earth's atmosphere; the reentry time exposure is parts of a millisecond. Different industry standards and tests can be used to rate plastics at these various degrees of combustibility (chapter 22).

STEEL AND PLASTIC

Plastics' behavior in fire depends upon the nature and scale of the fire as well as the surrounding conditions and how the products are designed. For example, the virtually all-plastic 35-mm slide projectors use a very hot electric bulb. When designed with a metal light and heat reflector with an air-circulating fan, the all-plastic projector operates with no fire hazard.

Steel structural beams cannot take the heat of a fire operating at and above 830°C (1,500°F); they just lose all their strength, modulus of elasticity, and so on. Steel can temporarily be protected from this environment for a short time by being covered with products such as concrete and certain plastics. To significantly extend the life of structural beams, hardwood (thicker, etc.) can be used, allowing people greater time to escape even though the wood slowly burns. In a fire situation, structural beams made of reinforced plastics (RPs) that meet structural performance requirements would extend supporting life beyond that of wood. To date these RPs are not used for protection in a fire environment because their costs are extremely high.

PERMEABILITY

Different plastics can provide different permeability rates. Depending on product requirements, plastic materials are available that exhibit low or no permeability to various products for a range of environments (chapter 20). Various factors influence performance, such as lack of pinholes, chemical composition, cross-linking, modification, molecular orientation, density, and thickness. The coinjection and coextrusion molding processes that combine different plastics, including those with specific permeability capabilities, are examples of methods used to reduce permeability while retaining other desirable properties (chapters 4 to 6).

FLUORINATION

This is an example of providing a barrier by chemically reacting a material with a fluorine-containing compound to produce the required product (Fig. 1.33). For example, it can improve the barrier properties of PE to nonpolar solvents such as gasoline. A chemical reaction of the fluorine with the PE causes a thin (20- to 40-mm) fluorocarbon layer to form on the surface of the PE to create a permeation barrier.

The fluorination can be applied using one of two different treatment methods (in-process or posttreatment) depending on the process and the finished product requirements. With the in-process system typically used during blow molding, the necessary fluorine is introduced as a part of the parison expanding gas during the blowing operation. This process creates a barrier layer on the inside only. Often used to produce PE gasoline tanks, the barrier enhancement achieved with

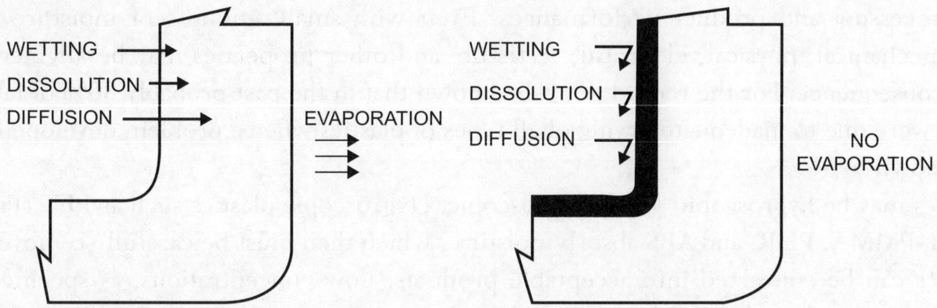

Figure 1.33 Comparing permeation behaviors with solvent (left) and fluorination

fluorination significantly reduces gasoline leakage from gasoline tanks. The posttreatment method forms barrier layers on both the inside and outside surfaces of the plastic part or product. In this system bottles or other products are placed in an enclosed chamber that is then filled with fluorine gas, such that all exposed plastic surfaces become fluorinated.

Radiation

Though the radiation resistance of plastics is generally inferior to that of metals and ceramics, their resistance to radiation is typically superior to elastomers. Plastic materials that respond satisfactorily to radiation, in the order of 10^{10} to 10^{11} erg per gram, include glass and asbestos-filled phenolics, certain epoxies, polystyrene, polyurethane, mineral-filled polyesters, silicone, and furane. Plastics groups with radiation resistance in the intermediate range are formaldehyde, melamine, polyethylene, unfilled phenolic, urea, and silicone resins. Plastics families with the poorest resistance to radiation include cellulosics, fluorocarbons, methyl methacrylate, polyamides, and unfilled polyesters (chapter 21).

Craze/Crack

Many TPs will craze or crack under certain environmental conditions, and products that are highly stressed mechanically must be checked very carefully. Crazing is the development of a multitude of very fine cracks in the matrix material resulting from stresses that exceed the tensile strength of plastic. Such stresses may result from shrinkage or machining, flexing, impact shocks, temperature changes, the action of solvents, or other degrading agents such as ultraviolet radiation. Chlorinated polyether, EVA, ionomer, linear polyethylene, phenoxy, and polypropylene offer greater freedom from stress crazing than some other TPs. TSs are generally preferable for products under continuous loads.

DRYING PLASTIC

Plastics absorb moisture that may be insignificant or damaging. All plastics, to some degree, are influenced by the amount of moisture or water they contain before processing. Moisture may reduce processing and product performances. Even with small amounts of moisture in many plastics, mechanical, physical, electrical, aesthetic, and other properties may be affected or may be of no consequence. For the record, let it be known that in the past probably 80% of fabricating problems were due to inadequate drying of all types of plastics when a problem developed. Now it could be down to 50%.

Plastics may be hygroscopic or nonhygroscopic. Hygroscopic plastics such as PET (Fig. 1.34), PC, nylon, PMMA, PUR, and ABS absorb moisture, which then must be carefully removed before the plastics can be converted into acceptable products. Low concentrations, as specified by the plastic supplier, can be achieved through efficient drying systems and proper handling of the dried plastic prior to and during molding, extrusion, and so forth. When desired, the processor may have these hygroscopic plastics properly dried and shipped in sealed containers.

Tray dryers or mechanical convection hot-air dryers that are adequate for nonhygroscopic plastics are not capable of removing water to the degree necessary for the proper processing of hygroscopic types or their compounds, particularly during periods of high humidity (Table 1.34).

The drying operation for nonhygroscopic plastics is less demanding. These products collect moisture only on the surface. Drying this surface moisture can be accomplished by simply passing warm air over the material. Moisture leaves the plastic in favor of the warm air, resulting in drying of the nonhygroscopic plastics.

Compounding certain plastics with certain additives such as color could result in serious effects. Day-to-night temperature changes can be a source of moisture contamination problems when plastic materials are exposed to the air and must be controlled; otherwise it has a cumulative effect. The critical moisture content (average material moisture content at the end of the constant-rate-drying period) is a function of material properties, the constant rate of drying, and particle size.

Although it is sometimes possible to select a suitable drying method simply by evaluating variables such as humidities and temperatures when removing unbound moisture, many plastic drying processes involve removal of bound moisture retained in capillaries among fine particles or moisture actually dissolved in the plastic (Table 1.35). Internal liquid and vapor mass-transfer mechanisms apply. Measuring drying-rate behavior under controlled conditions best identifies

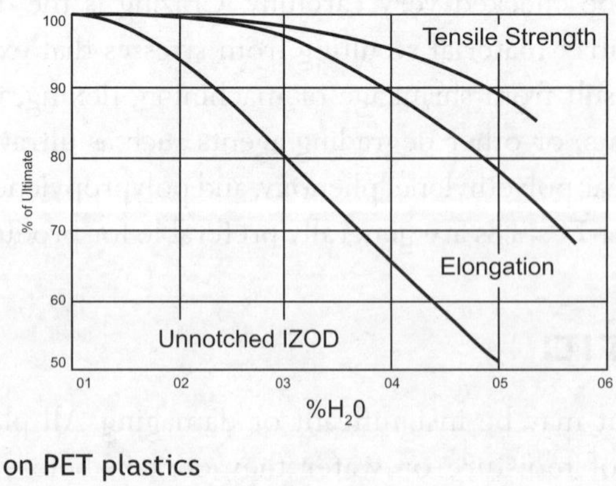

Figure 1.34 Moisture effect on PET plastics

MATERIAL	DRYING TEMP. (°F)	DRYING TIME (HRS.)
ABS	180	3-4
Acetal	210	2
Acrylic	160-180	2
Barex	160	6
Cellulosics	160	6
Ionomer	150	8
Nylon	160	6
PC	250	3-4
PE w/40% black	195	3
PET	325-375	4-6
PBT	250	2-3
PETG	160	3-4
Polyamide	250	2
Polyester Elastomer	225	3
PEM	300	4
PES	300	4
PPS	300	6
PP	195	1
PS (GP)	180	1
HIPS	180	1.5
Polysulfone	250	4
PU	180	3
PPO	255	2
Rynite	250	2
SAN	180	2
Styrene	180	2
Vinyls (PVC)	160	1

Table 1.34 Examples of drying different plastics (courtesy of Spirex Corp.)

these mechanisms. A change in the material handling method or any operating variable, such as heating rate, may affect mass transfer.

At ambient temperature and 50% relative humidity, the vapor pressure of water outside a hygroscopic plastic can be greater than within. Moisture migrates into the plastic, increasing its moisture content until a state of equilibrium exists inside and outside the plastic. But conditions are very different inside a drying hopper that has a controlled environment. At a temperature of 350°F (175°C) and –40°F (–40°C) dew point, the vapor pressure of the water inside the plastic is much greater than the vapor pressure of the water in the surrounding area. As a result, moisture migrates out of the plastic and into the surrounding air stream, where it is carried away to the desiccant bed of the dryer.

Before drying can begin, a wet material must be heated to such a temperature that the vapor pressure of the contained liquid exceeds the partial pressure of the corresponding vapor in the surrounding atmosphere. A psychometric chart can conveniently be used to study the effect of the atmospheric vapor content on the drying rate as well as the effect of the material temperature. It shows plots of moisture content at dry-bulb, wet-bulb or saturation temperature, and enthalpy at saturation.

Drying or maintaining moisture content at designated low levels is important, particularly for hygroscopic types of plastics. They must be carefully dried prior to processing. Usually the moisture

Material	Dryer	Time (hr.)	Temp (°F)	Dew Point (°F)	Moisture Content (%)
POLYESTER:					
- PBT	D	2 - 4	250 - 280	< 0	< 0.02
- PBT	D	8	215 - 225	< 0	< 0.02
- PET	D	2 - 4	250 - 275	< 0	< 0.02
- PET	D	> 4	225	< 0	< 0.02
- ELASTOMER	D	2 - 3	225	0	< 0.10
- ELASTOMER	A	4 - 6	225	0	< 0.10
PVC					
- CUBE	A	2	140 - 150	0	NH
- PELLET	A	2	170 - 180	0	NH
SMA	A	1 - 2	180 - 200	0	0.20
SAN	A	2 -3	180 - 200	0	0.20
TPE:					
- MPR	A	1 - 2	212	0	NH
- ETE	D	2 - 3	225	0	
- TPO	A	1 - 2	120 - 150	0	NH
LCP	D	2 - 3	250 - 300	< 0	< 0.01
PEEK	D	3	300	- 40	< 0.10
PES	D	3 - 4	275 - 300	- 25	0.05
PEI					

Table 1.35 Examples of drying plastics using hot air (A) or desiccant (D) dryer

content is >0.02 wt%. In practice, a drying temperature 30°C below the softening temperature has proved successful in preventing caking of the plastic in a dryer. Drying time varies in the range of 2–4 h, depending on moisture content. As a rule of thumb, the drying air should have a dew point of –30°F (–34°C) and the capability of being heated up to 250°F (121°C). It takes about 1 ft^3 min–1 of plastic processed when using a desiccant dryer. The pressure drop through the bed should be less than 1 mm H_2O per mm of bed height. Simple tray dryers or mechanical convection, hot-air dryers, while adequate for certain plastics, are incapable of removing enough water for the proper processing of hygroscopic plastics, particularly during periods of high humidity.

Hygroscopic plastics are commonly passed through dehumidifying hopper dryers before entering a screw plasticator. However, except where extremely expensive protective measures are taken, the drying may be inadequate, or moisture may be regained too rapidly to avoid product defects unless barrel venting is provided (chapter 3). To ensure proper drying for "moisture sensitive" parts such as lenses and compact disks, the combination of drying the plastics and using vented barrels provides a double check. However, just using vented extruders can be suitable. Plastic usage for a given process should be measured so as to determine how much plastic should be loaded into the hopper. Usually the hopper should hold enough dried plastic for a half to an hour's production. Care should be taken to ensure that hygroscopic plastics are in an unheated hopper for no more than one-half to 1 h, as dictated by experience or as specified by the material supplier.

Hygroscopic plastics predried to a range of 0.15 to 0.3 wt% (or as low as 0.002% for PET) can be delivered to a fabricating plant in sealed containers. The plastic often is used just as it comes out of a sealed container properly dried. However, after a container has been opened and the plastic is exposed to the air, it starts picking up moisture, and some type of drying may be required. In damp weather (relative humidity of 75%), the moisture level of the plastic can rise by 0.35% within 1 h, which is sufficient to cause product performance to be significantly reduced. Proper handling equipment must be used once the seals are broken so that the plastic does not pick up moisture.

VARIABILITY

Continuous progress has been made to reduce the variability of existing plastic material and equipment (as is also the case for steel and other materials). The target is always to improve plastics manufacturing and process control capabilities. However, variabilities still exist. To minimize material and process variations, different tests, specification ranges, and operating limits are important (chapters 22 to 24). Even when operating within established limits, processing the materials could result in inferior products. As an example, the material specification from a supplier will include minimum to maximum values for variables such as molecular weight distribution (MWD). It is determined that when material arrives all on the maximum side, it produces acceptable products. However, when all the material arrives on the minimum side, process control has to be adjusted in order to produce acceptable products (chapter 3).

In order to judge performance capabilities within the controlled variables, there must be a reference to measure performance against. For example, the injection mold cavity pressure profile is easily influenced by material variations. In injection molding there are four groups of variables related to mold cavity pressure profile that together influence the processing profile. These are (a) melt viscosity and fill rate, (b) boost time, (c) pack and hold pressures, and (d) recovery of plasticator. Thus, material variations may directly impact cavity pressure variation.

Even though equipment operations have understandable but controllable variables that influence processing, the usual most uncontrollable variable in the process can be the plastic material. A specific plastic will have a range of performances. However, more significant is the proper compounding or blending by the plastic manufacturer, by the converter, or in-house by the fabricator (Fig. 1.35). Most additives, fillers, and/or reinforcements when not properly compounded will significantly influence processability and molded product performances.

A very important factor that should not be overlooked by a designer, processor, analyst, statistician, and others is that most conventional and commercial tabulated material data and plots for variables such as tensile strength are average or mean values. They would imply a 50% survival rate when the material value below the mean processes unacceptable products. The target is to obtain some level of reliability that will account for material variations and other variations that can occur during the various stages from designing the product to processing the plastics. In addition to material variables, there are a number of factors in equipment hardware and controls that cause processing variabilities. They include factors such as accuracy of machining component equipment parts, method and degree of accuracy during the assembly of component parts, and temperature/pressure control capability, particularly when interrelated with time and heat-transfer uniformity in metal components such as those used in molds and dies.

These variables are controllable within limits to produce useful and cost-efficient products. What is important to appreciate is that over the past decades, improvements in equipment have made exceptional strides in significantly reducing operating variabilities or limitations. These advances will continue into the future since there is a rather endless improvement in performance of steels and other materials and methods of controlling such as fuzzy control. Growth is occurring in the application of fuzzy logic, which, in 1981, was first used to mimic the control actions of the human operator.

Unfortunately, these variables and problems exist in all industries. As an example, a major catastrophe occurred in the aerospace industry. Nothing is perfect in, on, or around the earth

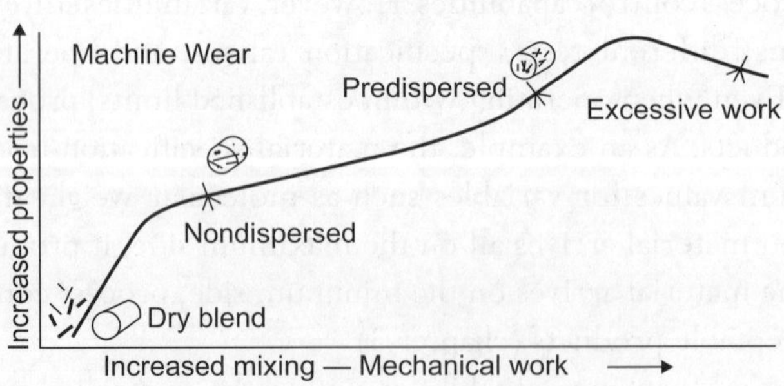

Figure 1.35 Advantages of properly dispersing plastic compounds

Figure 1.36 View when the Challenger shuttle spacecraft exploded January 28, 1986; photo taken by D. V. Rosato from Route 95, Florida

(Fig. 1.36). The Challenger shuttle spacecraft exploded January 28, 1986, above Cape Canaveral, Florida; certain parts did not function properly (see **perfection** in the glossary, volume 4).

ADVANTAGE AND LIMITATION

As a construction material, plastics provide practically unlimited benefits to the design and fabrication of products, but unfortunately, as with other materials, no one specific plastic exhibits all these positive characteristics. The successful application of their strengths and an understanding of their limitations will allow designers to produce useful products. As with any material (plastic, steel, etc.), products often fail not because of their disadvantage(s) but rather because someone did not perform their material selection in the proper manner and/or incorrectly processed the plastic.

There is a wide variation in properties among the over 35,000 commercially available materials classified as plastics. They now represent an important, highly versatile group of commodity and engineering plastics. Like steel, wood, and other materials, specific groups of plastics can be characterized as having certain properties. As with other materials, for every advantage cited for a certain material, a corresponding disadvantage can probably be found in another.

Many plastics that are extensively used worldwide are typically not as strong or as stiff as metals, and they are prone to dimensional changes especially under load or heat. They are used instead of metals (in millions of products) because their performance meets product requirements. However, there are many other plastics that meet tight dimensional requirements, have good dimensional stability, and are stronger or stiffer based on product shape than other materials.

Highly favorable properties such as low density, strength through shape, good thermal insulation, a high degree of mechanical dampening, high resistance to corrosion and chemical attack, and exceptional electrical resistance exist for certain plastics. There are also those that will deteriorate when exposed to sunlight, weather, or ultraviolet light, but then there are those that resist such deterioration.

For room-temperature applications, most metals can be considered to be truly elastic. When stresses beyond the yield point are permitted in the design, permanent deformation is considered to be a function only of applied load and can be determined directly from the usual tensile stress-strain diagram (chapter 22). The behavior of most plastics is much more dependent on the time of application of the load, the past history of loading, the current and past temperature cycles, and the environmental conditions. Ignorance of these conditions has resulted in the appearance on the market of plastic products that were improperly designed (chapters 19 and 20).

FALLO APPROACH

Many factors contribute to plastics' global success. One of these factors is the broad range of fabricating processes. All processes require proper control of the different operations consistent with the FALLO approach (Follow ALL Opportunities; Fig. 1.37).

Individuals with knowledge of the behavior of plastics and their application to the fabrication of the designed product have contributed significantly to the success worldwide of millions of plastic products. Many "tools" are now available that make it easier for us to design and fabricate plastic products, and new or improved materials and processing techniques continue to expand the application of plastic materials. What is still needed, as usual, is to have a design plan properly conceived and subsequently executed by the proper method. Many plastic product designers and fabricators successfully produce products at the lowest cost by making use of a basic concept similar to the FALLO approach. This approach highlights the many steps involved to be successful, all of which must be coordinated and interrelated. The FALLO approach starts with a product design that includes specification of the plastic and manufacturing process (injection, extrusion, blow molding, thermoforming, and so forth) that is an important part of the overall scheme. Basically, the FALLO approach consists of

1. designing a product to meet performance and manufacturing requirements at the lowest cost;
2. specifying the proper plastic materials that meet product performance requirements after processing;
3. specifying the complete equipment line by
   a. designing the tool (die, mold) "around" the product,
   b. putting the "proper performing" fabricating process "around" the tool,

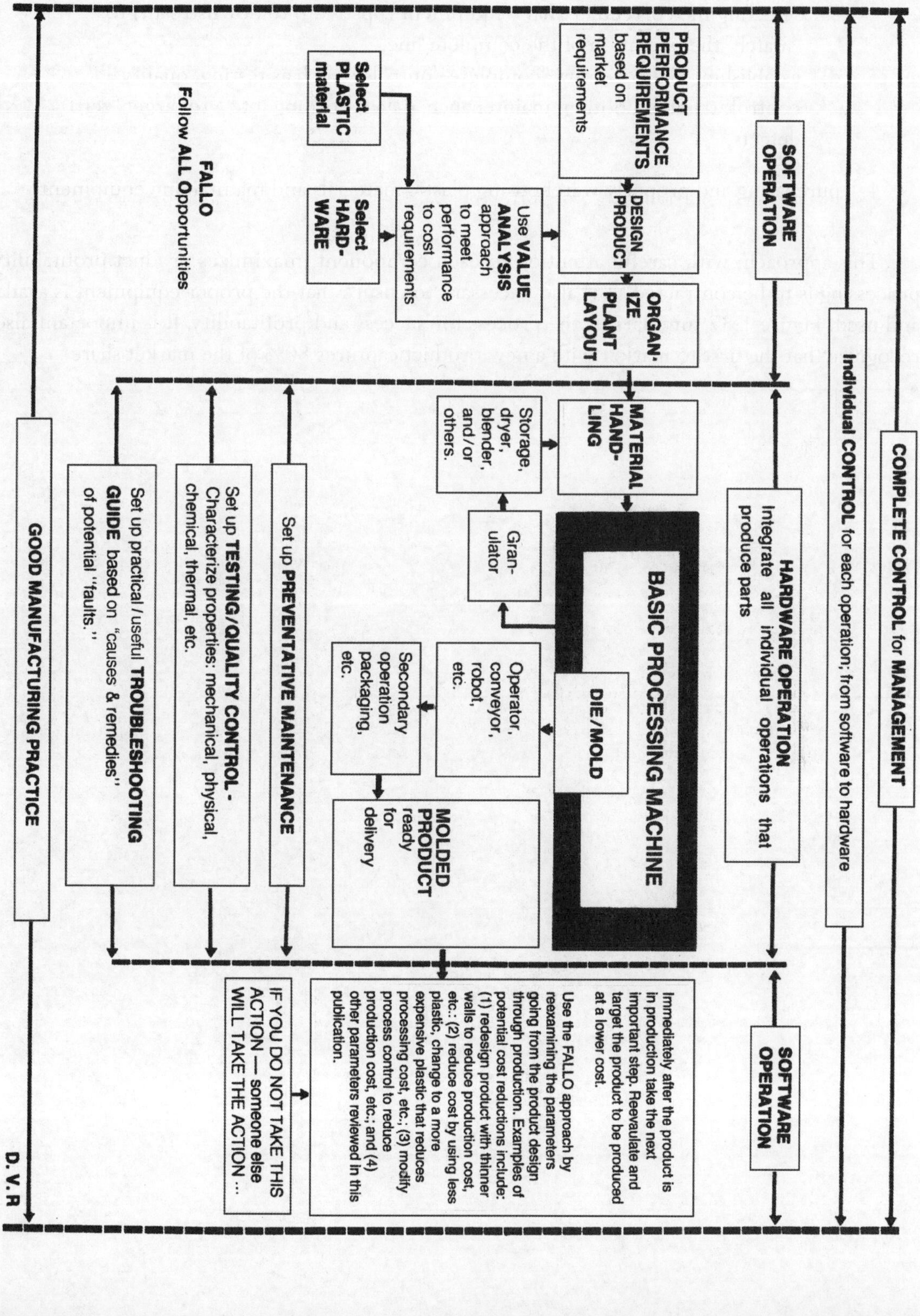

Figure 1.37 The FALLO complete processing approach

c. selecting the correct auxiliary equipment (upstream to downstream) to "match" the operation of the complete line,
d. establishing the necessary "complete controls" (such as testing, quality control, troubleshooting, maintenance, data recording, etc.) to target "zero defects,"

4. purchasing and properly warehousing plastic materials and maintaining equipment.

This approach, with careful attention to each component, maximizes product profitability. If processing is to be contracted, it is also necessary to ensure that the proper equipment is available and used. Figure 1.37 summarizes the process for success and profitability. It is important also to recognize that the first to market with a new product captures 80% of the market share.

CHAPTER 2

PLASTICS PROPERTY

OVERVIEW

The plastics material properties information and data presented in Tables 2.1 to 2.11 provide comparative guides to thermoplastics (TPs) and thermosets (TSs). They show different capabilities of meeting different performance requirements. Different polymerization reactions used to initiate the start of plastics (chapter 1) can be related to their properties. As an example, different polymerizations occurred to produce the polyethylene (PE) materials that range from liquids to stiff plastics (Fig. 2.1). The molecular weight and amount of crystallinity relate to the properties of these materials (chapter 1).

For many materials (plastics, metals, etc.), production can be a highly complex process if not properly approached, particularly when using recycled plastics. For example, its methodology ranges from a high degree of subjective intuition in some areas to a high degree of sophistication in other areas. It runs the gamut from highly systematic value engineering or failure analysis such as in aerospace to a telephone call for advice from a material supplier in the decorative housewares business (chapter 28).

An endless amount of data is available for many plastic materials. Unfortunately, as with other materials, there does not exist only one plastic material that will meet all performance requirements. However, it can be stated that for practically any product requirement(s), particularly when not including cost for very few products, and more so than with other materials, there is a plastic that can be used. Plastics provide more property variations than any other material.

Readers can obtain the latest and more detailed information from suppliers and/or software programs (chapter 25). Since new developments in plastic materials are always on the horizon, it is important to keep up to date. It is important to ensure that the fabricating process to be used to produce a product provides the properties desired. Much of the market success or failure of a plastic product can be attributed to the initial choices of material, process, and cost.

Property	Thermoplastics	Thermosets
High Temperature	TFE, PPS, P1, PAS	Silicones
Gasoline Resistance	Acetal	Phenolic
Impact	UHMW PE	Epoxy-glass fiber
Cold Flow	Polysulfone	Melamine-glass fiberglass
Chemical Resistance	TFE, FEP, PE, PP	Epoxy
Scratch Resistance	Acrylic	Allyl diglycol carbonate (C-39)
Abrasive Wear	Polyurethane	Phenolic-canvas
Colors	Acetate, PS	Urea, melamine
Moisture Absorption	Chlorotrifluorethylene	Alkyd-glass fiber
Steam Resistance	Polysulfone	DAP
Flame Resistance	TFE, P1	Melamine
Water Immersion	Chlorinated polyether	DAP
Stress Craze Resistance	Polypropylene	All
Low Temperature	TFE	DAP
Low Cost	PP, PE, PVC, PS	Phenolic
Low Gravity	Polypropylene methylpentene	Phenolic/nylon
Thermal Expansion	Phenoxy glass	Epoxy-glass fiber
Volume Resistivity	TFE	DAP
Dielectric Strength	PVC	DAP, polyester
Elasticity	EVA, PVC, TPR	Silicone

Table 2.1 Introduction to TP and TS plastics

Each family of plastics has its own special advantages. An example is polyethylene (PE) with its many types that include low-density PE (LDPE), high-density PE (HDPE), and high molecular weight PE (HMWPE). The major consideration for a designer and/or fabricator is to analyze what is required regarding performances and develop a logical selection procedure from what is available.

Recognize that most of the plastic products produced only have to meet the usual requirements we humans have to endure, such as the environment (temperature, etc.). Thus, there is no need for someone to identify that most plastics cannot take heat like steels. Also recognize that most plastics in use do not have a high modulus of elasticity or long creep and fatigue behaviors, characteristics that, in any case, are not required in their respective product designs. However, there are plastics with extremely high heat resistance, high modulus, and very long creep and fatigue behaviors. These types of products have performed in service for long periods of time, with some performing well over a half century. Certain plastic products have definite properties (modulus of elasticity, temperature, chemical resistance, load, etc.) that give them far better performance than steels and other materials (chapter 21).

Use can be made of conventional-type plastics that are available in sheet form, in I-beams, or in other forms as is common with many other materials. Although this approach with plastics has its place, the real advantage with plastics lies in the ability to process them to fit the design shape, particularly when it comes to complex shapes. Examples include two or more products with

Plastic	DIN 7728 Bl. 1	Density g/cm³ DIN 53479	Density lb/in³ D 792	Mechanical: Tensile strength N/mm² DIN 53455	Tensile strength psi ASTM D 638	Elongation at break % DIN 53455 ASTM D 638	Tensile modulus of elasticity N/mm² DIN 53457	Tensile modulus of elasticity kpsi ASTM D 638
Plasticized polyethylene	PE-LD	0.914/0.928	0.0329–0.0330	8/23	1140/3270	300/1000	200/500	28.4/71.1
Rigid polyethylene	PE-HD	0.94/0.96	0.0338–0.0345	18/35	2560/4980	100/1000	700/1400	99.6/199
EVA	EVA	0.92/0.95	0.0331–0.0341	10/20	1420/2840	600/900	7/120	0.99/17.1
Polypropylene	PP	0.90/0.907	0.0324–0.0327	21/37	2990/5260	20/800	11000/1300	156/185
Polybut-1-ene	PB	0.905/0.920	0.0325–0.0331	30/38	4270/5400	250/280	250/350	35.6/49.8
Polyisobutylene	PIB	0.91/0.93	0.0327–0.0334	2/6	284/853	> 1000	–	–
Poly-4-methylpent-1-ene	PMP	0.83	0.0298	25/28	3560/3980	13/22	1100/1500	156/213
Ionomers	–	0.94	0.0338	21/35	2990/4980	250/500	180/210	25.6/29.9
Rigid PVC	PVC-U	1.38/1.55	0.0496–0.0557	50/75	7110/10670	10/50	1000/3500	142/498
Plasticized PVC	PVC-P	1.16/1.35	0.0417–0.0486	10/25	1420/3560	170/400	–	–
Polystyrene, normal	PS	1.05	0.0378	45/65	6400/9240	3/4	3200/3250	455/462
Styrene/acrylonitrile copolymer	SAN	1.08	0.0392	75	10670	5	3600	512
Styrene/polybutadiene graft polymer	SB	1.05	0.0378	26/38	3700/5400	25/60	1800/2500	256/356
Acrylonitrile/polybut./styrene graft polymer	ABS	1.04/1.06	0.0374–0.0381	32/45	4550/6400	15/30	1900/2700	270/384
AN/AN elastomers/styrene graft polymer	ASA	1.04	0.0374	32	4550	40	1800	256
Polymethylmethacrylate	PMMA	1.17/1.20	0.0421–0.0431	50/77	7110/10950	2/10	2700/3200	384/455
Polyvinylcarbazole	PVK	1.19	0.0428	20/30	2840/4270	–	3500	498
Polyacetal	POM	1.41/1.42	0.0507–0.0511	62/70	8820/9960	25/70	2800/3200	398/455
Polytetrafluoroethylene	PTFE	2.15/2.20	0.0774–0.0791	25/36	3560/5120	350/550	410	58.3
Tetrafluoroethylene/hexafluoropropylene copolymer	FEP	2.12/2.17	0.0763–0.0781	22/28	3130/3980	250/330	350	49.8
Polytrifluorochlorethylene	PCTFE	2.10/2.12	0.0755–0.0762	32/40	4550/5690	120/175	1050/2100	149/299
Ethylene/tetrafluoroethylene	E/TFE	1.7	0.0611	35/54	4980/7680	400/500	1100	156
Polyamide 6	PA 6	1.13	0.0406	70/85	9960/12090	200/300	1400	199
Polyamide 66	PA 66	1.14	0.0410	77/84	10950/11950	150/300	2000	284
Polyamide 11	PA 11	1.04	0.0374	56	7960	500	1000	142
Polyamide 12	PA 12	1.02	0.0367	56/65	7960/9240	300	1600	228
Polyamide 6-3-T	PA-6-3-T	1.12	0.0403	70/84	9960/11950	70/150	2000	284
Polycarbonate	PC	1.2	0.0432	56/67	7960/9530	100/130	2100/2400	299/341
Polyethyleneterephthalate	PET	1.37	0.0492	47	6680	50/300	3100	441
Polybutyleneterephthalate	PBT	1.31	0.0471	40	5690	15	2000	284
Polyphenyleneether modified	PPE	1.06	0.0381	55/68	7820/9670	50/60	2500	356
Polysulfone	PSU	1.24	0.0446	50/100	7110/14200	25/30	2600/2750	370/391
Polyphenylenesulfide	PPS	1.34	0.0483	75	10670	3	3400	484
Polyarylsulfone	PAS	1.36	0.0490	90	12800	13	2600	370
Polyethersulfone	PES	1.37	0.0492	85	12090	30/80	2450	348
Polyarylether	PAE	1.14	0.0411	53	7540	25/90	2250	320
Phenol/formaldehyde, grade 31	PF	1.4	0.0504	25	3560	0.4/0.8	5600/12000	796/1710
Urea/formaldehyde, grade 131	UF	1.5	0.0540	30	4270	0.5/1.0	7000/10500	996/1490
Melamine/formaldehyde, grade 152	MF	1.5	0.0540	30	4270	0.6/0.9	4900/9100	697/1294
Unsaturated polyester resin, grade 802	UP	2.0	0.0720	30	4270	0.6/1.2	14000/20000	1990/2840
Polydiallylphthalate (GF) molding compound	PDAP	1.51/1.78	0.0543–0.0640	40/75	5690/10670	–	9800/15500	1394/2200
Silicone resin molding compound	SI	1.8/1.9	0.0648–0.0684	28/46	3980/6540	–	6000/12000	853/1710
Polyimide molding	PI	1.43	0.0515	75/100	0570/14200	4/9	23000/28000	3270/3980
Epoxy resin, grade 891	EP	1.9	0.0683	30/40	4270/5690	4	21500	3060
Polyurethane casting resin	PU	1.05	0.0378	70/80	9960/11380	3/6	4000	569
Thermoplastic PU-elastomers	PU	1.20	0.0432	30/40	4270/5690	400/450	700	99.6
Linear polyurethane (U_{50})	PU	1.21	0.0435	30 (σ_s)	4270 (σ_y)	35 (ε_s)	1000	140

Table 2.2 Thermoplastic and thermoset properties compared

Impact strength	Notched impact strength		Optical		Water absorption	
kJ/m² DIN 53453	kJ/m² DIN 53453	ft lb/in of notch ASTM D 256	Refractive index n_D^{20} DIN 53491	Transparency	mg (4 d) DIN 53492	% (24 h) ASTM D 570
no break	no break	–	1.51	transparent	<0.01	<0.01
no break	no break	–	1.53	opaque	<0.01	<0.01
no break	no break	no break	–	transparent/opaque	–	0.05/0.13
no break	3/17	0.5/20	1.49	transparent/opaque	<0.01	0.01/0.03
no break	4/no break	no break	–	opaque	<0.01	<0.02
no break	no break	no break	–	opaque	<0.01	<0.01
–	–	0.4/0.6	1.46	opaque	–	0.01
–	–	6/15	1.51	transparent	–	0.1/1.4
no break/>20	2/50	0.4/20	1.52/1.55	transparent/opaque	3/18	0.04/0.4
no break	no break	–	–	transparent/opaque	6/30	0.15/0.75
5/20	2/2.5	0.25/0.6	1.59	transparent	–	0.03/0.1
8/20	2/3	0.35/0.5	1.57	transparent	–	0.2/0.3
10/80	5/13	no break	–	opaque	–	0.05/0.6
70/no break	7/20	2.5/12	–	opaque	–	0.2/0.45
no break	18	6/8	–	translucent/opaque	–	
18	2	0.3/0.5	1.49	transparent	35/45	0.1/0.4
5	2	–	–	opaque	0.5	0.1/0.2
100	8	1/2.3	1.48	opaque	20/30	0.22/0.25
no break	13/15	3.0	1.35	opaque	–	0
–		no break	1.34	transparent/translucent	–	<0.1
no break	8/10	2.5/2.8	1.43	translucent/opaque	–	0
–		no break	1.40	transparent/opaque		0.03
no break	no break	3.0	1.53	translucent/opaque	–	1.3/1.9
no break	15/20	2.1	1.53	translucent/opaque	–	1.5
no break	30/40	1.8	1.52	translucent/opaque	–	0.3
no break	10/20	2/5.5	–	translucent/opaque	–	0.25
no break	13	–	1.53	transparent	–	0.4
no break	20/30	12/18	1.58	transparent	10	0.16
no break	4	0.8/1.0	–	transparent/opaque	18/20	0.30
no break	4	0.8/1.0	–	opaque	–	0.08
no break	–	4	–	opaque	–	0.06
–	–	1.3	1.63	transparent/opaque	–	0.02
–	–	0.3	–	opaque	–	0.02
–	–	1/2	1.67	opaque	–	1.8
–	–	1.6	1.65	transparent	–	0.43
–	–	8.0	–	translucent/opaque	–	0.25
>6	>1.5	0.2/0.6	–	opaque	<150	0.3/1.2
>6.5	>2.5	0.5/0.4	–	opaque	<300	0.4/0.8
>7.0	>1.5	0.2/0.3	–	opaque	<250	0.1/0.6
>4.5	>3.0	0.5/16	–	opaque	<45	0.03/0.5
–	–	0.4/15	–	opaque	–	0.12/0.35
–	–	0.3/0.8	–	opaque	–	0.2
–	–	0.5/1.0	–	opaque	–	0.32
>8	>3	2/30	–	opaque	<30	0.05/0.2
–	–	0.4	–	transparent	–	0.1/0.2
no break	no break	no break	–	translucent/opaque	–	0.7/0.9
no break	3	–	–	translucent/opaque	130	–

Table 2.2 *(continued)*

Thermal															
Service temperature						Heat deflection temperature				Coefficient of linear expansion		Thermal conductivity		Specific heat	
						°C		°F							
max./short time		max./continuous		min./continuous		VSP (Vicat 5 kg) DIN 53460	ASTM D 648 1.86/ 0.45 N/ mm²	VSP (Vicat lb) DIN 53460	ASTM D 648 264/ 66 psi	K⁻¹·10⁶	in/in/°F 10⁻⁶	W/mK	BTU in/ ft² h °F	kJ/ kgK	BTU/ lb °F
°C	°F	°C	°F	°C	°F										
80/90	176/194	60/75	140/167	−50	−58	–	35	–	95	250	140	0.32/0.40	2.2/2.8	2.1/2.5	8.8/10.5
90/120	194/248	70/80	158/176	−50	−58	60/70	50	140/158	122	200	110	0.38/0.51	2.6/3.5	2.1/2.7	8.8/11.5
65	149	55	131	−60	−76	–	34/62	–	93/144	160/200	90/110	0.35	2.4	2.3	9.5
140	284	100	212	0/−30	32/−22	85/100	45/120	185/212	113/248	150	83	0.17/0.22	1.2/1.5	2.0	8.3
130	266	90	194	0	32	70	60/110	158	140/230	150	83	0.20	1.4	1.8	7.5
80	176	65	149	−50	−58	–	–	–	–	120	67	0.12/0.20	0.8/1.4	–	–
180	356	120	248	0	32	–	–	–	–	117	65	0.17	1.2	2.18	9.1
120	248	100	212	−50	−58	–	38/45	–	100/113	120	67	0.24	1.7	2.20	9.2
75/100	167/212	65/85	149/185	−5	21	75/110	60/82	167/230	140/180	70/80	39/45	0.14/0.17	1.0/1.2	0.85/0.9	3.55/3.75
55/65	131/149	50/55	122/131	0/−20	32/−4	40	–	104	–	150/210	83/110	0.15	1.05	0.9/1.8	3.75/7.5
60/80	140/176	50/70	122/158	−10	14	78/99	110/80	172/210	230/176	70	39	0.18	1.25	1.3	5.4
95	203	85	185	−20	−4	–	104/90	–	219/194	80	45	0.18	1.25	1.3	5.4
60/80	140/176	50/70	122/158	−20	−4	77/95	104/82	171/203	219/180	70	39	0.18	1.25	1.3	5.4
85/100	188/212	75/85	167/185	−40	−40	95/110	80/120	203/230	176/248	60/110	33/61	0.18	1.25	1.3	5.4
85/90	188/194	70/75	158/167	−40	−40	92	100/110	198	212/230	80/110	44/61	0.18	1.25	1.3	5.4
85/100	188/212	65/90	149/194	−40	−40	70/100	60/100	158/212	140/212	70	39	0.18	1.25	1.47	6.15
170	338	160	320	−100	−148	180	–	356	–	–	–	0.29	2.0	–	–
110/140	230	90/110	194/230	−60	−76	160/173	110/170	320/344	230/338	90/110	50/61	0.25/0.30	1.7/2.1	1.46	6.1
300	572	250	482	−200	−328	–	–/121	–	–/250	100	56	0.25	1.7	1.0	4.20
250	482	205	401	−100	−148	–	–/70	–	–/158	80	45	0.25	1.7	1.12	4.65
180	356	150	302	−40	−40	–	–/126	–	–/259	60	33	0.22	1.5	0.9	3.75
220	428	150	302	−190	−310	–	71/104	–	160/219	40	22	0.23	1.6	0.9	3.75
140/180	284/356	80/100	176/212	−30	−22	180	80/190	356	176/374	80	44	0.29	2.0	1.7	7.1
170/200	338/392	80/120	176/248	−30	−22	200	105/200	392	221/392	80	44	0.23	1.6	1.7	7.1
140/150	284/302	70/80	158/176	−70	−94	175	150/130	347	302/266	130	72	0.23	1.6	1.26	5.25
140/150	284/302	70/80	158/176	−70	−94	165	140/150	329	284/302	150	83	0.23	1.6	1.26	5.25
130/140	266/284	80/100	176/212	−70	−94	145	140/80	293	284/176	80	45	0.23	1.6	1.6	6.70
160	320	135	275	−100	−148	138	130/145	280	266/293	60/70	33/39	0.21	1.45	1.17	4.90
200	392	100	212	−20	−4	188	–	280	–	70	39	0.24	1.65	1.05	4.40
165	329	100	212	−30	−22	178	50/190	352	122/374	60	33	0.21	1.45	1.30	5.40
150	302	80	176	−30	−22	148	100/140	298	212/284	60	33	0.23	1.60	1.40	5.85
200	392	150	302	−100	−148	–	175/180	–	347/356	54	30	0.28	1.95	1.30	5.40
300	572	200	392	–	–	–	137/–	–	277/–	55	31	0.25	1.70	–	–
300	572	260	500	–	–	–	–	–	–	47	26	0.16	1.10	–	–
260	500	200	392	–	–	–	–	–	–	55	31	0.18	1.25	1.10	4.6
160	320	120	248	–	–	–	150/160	–	302/320	65	36	0.26	1.80	1.46	6.1
140	284	110	230	–	–	–	150/190	–	302/374	30/50	17/28	0.35	2.40	1.30	5.40
100	212	70	158	–	–	–	130/–	–	266/–	50/60	28/33	0.40	2.75	1.20	5.0
120	248	80	176	–	–	–	180/–	–	356/–	50/60	28/33	0.50	3.45	1.20	5.0
200	392	150	302	–	–	–	230/–	–	446/–	20/40	11/22	0.70	4.85	1.20	5.0
190/250	374/482	150/180	302/356	−50	−58	–	220/–	–	428/–	10/35	55/19	0.60	4.15	–	5.0
250	482	170/180	338/356	−50	−58	–	480/–	–	896/–	20/50	11/28	0.3/0.4	2.05/2.75	0.8/0.9	3.35/3.75
400	752	260	500	−200	−239	–	240/–	–	464/–	50/63	28/35	0.6/0.65	4.15/4.50	–	–
180	356	130	266	–	–	–	200/–	–	392/–	11/35	6.1/19	0.88	6.1	0.8	3.35
100	212	80	176	–	–	–	90/–	–	194/–	10/20	5.5/11	0.58	4.0	1.76	7.30
110	230	80	176	−40	−40	–	–	–	–	150	83	1.7	1.15	0.5	2.10
80	176	60	140	−15	3	100	–	212	–	210	12	1.8	1.25	0.4	1.65

Table 2.2 *(continued)*

Electrical								
Volume resistivity Ω cm DIN 53482	Surface resistance Ω DIN 53482	Tracking resistance DIN 53480			Dielectric constant DIN 53483		Dissipation (power) factor tan δ DIN 53483	
		KA	KB	KC	50 Hz	10^6 Hz	50 Hz	10^6 Hz
$>10^{17}$	10^{14}	3b	>600	>600	2.29	2.28	$1.5 \cdot 10^{4}$	$0.8 \cdot 10^{-4}$
$>10^{17}$	10^{14}	3c	>600	>600	2.35	2.34	$2.4 \cdot 10^{-4}$	$2.0 \cdot 10^{-4}$
$<10^{15}$	10^{13}	–	–	–	2.5/3.2	2.6/3.2	0.003/0.02	0.03/0.05
$>10^{17}$	10^{13}	3c	>600	>600	2.27	2.25	$<4 \cdot 10^{-4}$	$<5 \cdot 10^{-4}$
$>10^{17}$	10^{13}	3c	>600	>600	2.5	2.2	$7 \cdot 10^{-4}$	$6 \cdot 10^{-4}$
$>10^{15}$	10^{13}	3c	>600	>600	2.3	–	0.0004	–
$>10^{16}$	10^{13}	3c	>600	>600	2.12	2.12	$7 \cdot 10^{-5}$	$3 \cdot 10^{-5}$
$>10^{16}$	10^{13}							
$>10^{15}$	10^{13}	2/3b	600	600	3.5	3.0	0.011	0.015
$>10^{11}$	10^{11}	–	–	–	4/8	4/4.5	0.08	0.12
$>10^{16}$	$>10^{13}$	1/2	140	150/250	2.5	2.5	$1/4 \cdot 10^{-4}$	$0.5/4 \cdot 10^{-4}$
$>10^{16}$	$>10^{13}$	1/2	160	150/260	2.6/3.4	2.6/3.1	$6/8 \cdot 10^{-3}$	$7/10 \cdot 10^{-3}$
$>10^{16}$	$>10^{13}$	2	>600	>600	2.4/4.7	2.4/3.8	$4/20 \cdot 10^{-4}$	$4/20 \cdot 10^{-4}$
$>10^{15}$	$>10^{13}$	3a	>600	>600	2.4/5	2.4/3.8	$3/8 \cdot 10^{-3}$	$2/15 \cdot 10^{-3}$
$>10^{15}$	$>10^{13}$	3a	>600	>600	3/4	3/3.5	0.02/0.05	0.02/0.03
$>10^{15}$	10^{15}	3c	>600	>600	3.3/3.9	2.2/3.2	0.04/0.06	0.004/0.04
$>10^{16}$	10^{14}	3b	>600	>600	–	3	$6/10 \cdot 10^{-4}$	$6/10 \cdot 10^{-4}$
$>10^{15}$	10^{13}	3b	>600	>600	3.7	3.7	0.005	0.005
$>10^{18}$	10^{17}	3c	>600	>600	<2.1	<2.1	$<2 \cdot 10^{-4}$	$<2 \cdot 10^{-4}$
$>10^{18}$	10^{16}	3c	>600	>600	2.1	2.1	$<2 \cdot 10^{-4}$	$<7 \cdot 10^{-4}$
$>10^{18}$	10^{16}	3c	>600	>600	2.3/2.8	2.3/2.5	$1 \cdot 10^{-3}$	$2 \cdot 10^{-2}$
$>10^{16}$	10^{13}	3c	>600	>600	2.6	2.6	$8 \cdot 10^{-4}$	$5 \cdot 10^{-3}$
10^{12}	10^{10}	3b	>600	>600	3.8	3.4	0.01	0.03
10^{12}	10^{10}	3b	>600	>600	8.0	4.0	0.14	0.08
10^{13}	10^{11}	3b	>600	>600	3.7	3.5	0.06	0.04
10^{13}	10^{11}	3b	>600	>600	4.2	3.1	0.04	0.03
10^{11}	10^{10}	3b	>600	>600	4.0	3.0	0.03	0.04
$>10^{17}$	$>10^{15}$	1	120/160	260/300	3.0	2.9	$7 \cdot 10^{-4}$	$1 \cdot 10^{-2}$
10^{16}	10^{16}	2	–	–	4.0	4.0	$2 \cdot 10^{-3}$	$2 \cdot 10^{-2}$
10^{16}	10^{13}	3b	420	380	3.0	3.0	$2 \cdot 10^{-3}$	$2 \cdot 10^{-2}$
10^{16}	10^{14}	1	300	300	2.6	2.6	$4 \cdot 10^{-4}$	$9 \cdot 10^{-4}$
$>10^{16}$	–	1	175	175	3.1	3.0	$8 \cdot 10^{-4}$	$3 \cdot 10^{-3}$
$>10^{16}$	–	–	–	–	3.1	3.2	$4 \cdot 10^{-4}$	$7 \cdot 10^{-4}$
$>10^{16}$	–	–	–	–	3.9	3.7	$3 \cdot 10^{-3}$	$13 \cdot 10^{-3}$
10^{17}	–	–	–	–	3.5	3.5	$1 \cdot 10^{-3}$	$6 \cdot 10^{-3}$
$>10^{10}$	–	–	–	–	3.14	3.10	$6 \cdot 10^{-3}$	$7 \cdot 10^{-3}$
10^{11}	$>10^{8}$	1	140/180	125/175	6	4.5	0.1	0.03
10^{11}	$>10^{10}$	3a	>400	>600	8	7	0.04	0.3
11^{11}	$>10^{8}$	3b	>500	>600	9	8	0.06	0.03
$>10^{12}$	$>10^{10}$	3c	>600	>600	6	5	0.04	0.02
$10^{13}/10^{16}$	10^{13}	3c	>600	>600	5.2	4	0.04	0.03
10^{14}	10^{12}	3c	>600	>600	4	3.5	0.03	0.02
$>10^{16}$	$>10^{15}$	1	>300	>380	3.5	3.4	$2 \cdot 10^{-3}$	$5 \cdot 10^{-3}$
$>10^{14}$	$>10^{12}$	3c	>300	200/600	3.5/5	3.5/5	0.001	0.01
10^{16}	10^{14}	3c	–	–	3.6	3.4	0.05	0.05
10^{12}	10^{11}	3a	>600	>600	6.5	5.6	0.03	0.06
10^{13}	10^{12}	–	–	–	5.8	4.0	0.12	0.07

Table 2.2 *(continued)*

mechanical and electrical connections, living hinges, colors, and snap fits that can be combined into one product (chapter 20).

The ranges of properties in plastics encompass all types of environmental and load conditions, each with its own individual, yet broad, range of properties. These properties can take into consideration wear resistance, integral color, impact resistance, transparency, energy absorption, ductility, thermal and sound insulation, weight, and so forth.

Material		Polyarylsulfone	Polyether sulfone		Polyetherimide	Polyetherketone	
Abbreviation		PAS	PES		PEI	PEEKK	PEKEKK
Commercial grade (Manufacturer)		Radel R 7000 (Amoco)	Ultrason E 3000 (BASF)	Vitrex 3600 G (ICI)	Ultem 1000 (GEP)	Hostatek (Hoechst)	Ultrapek KR 4176 (BASF)
Mechanical properties							
Elastic modulus	N/mm²	2723	2900	2900	3000	4000	4000
Tensile strength	N/mm²	76	85	84	105	108	104
Modulus in flexure	N/mm²	2861	–	2600	3300	3600	3710
Flexural strength	N/mm²	107	–	129	145	120	135
Ultimate strength	N/mm²	–	–	–	90	86	~ 100
Elongation at break	%	23	20 to 40	40 to 80	60	36	53
Izod impact resistance	J/m	–	–	–	1300	–	–
Izod impact resistance, notched	J/m	–	80 to 90	76	50	51	52
Thermal properties							
Melt temperature	°C	360	–	–	–	365	375
Glass transition temperature	°C	–	227	225	215	160	175
Heat resistance Vicat-procedure B	°C	–	222	222	215	> 300	> 250
F_{ISO} method A, 1.8 N/mm²	°C	> 204	215	203	200	160	170
Maximum service temperature	°C	–	180	180	170	260	250
Thermal linear-expansion coefficient lengthwise	10^{-5} K^{-1}	–	5,5	5,5	5,6	4,5	4,2
crosswise	10^{-5} K^{-1}	–	–	–	5,6	–	–
Thermal conductivity	W/(k · m)	–	0,18	–	0,22	0,21	0,22
Electrical properties							
Electric strength	kV/mm	–	≥ 63	16	33	20	–
Specific breakthrough resistance	10^{16} Ω · cm	–	≥ 1	10 to 100	0,67	1	–
CTI surface leakage	V	–	150-0	–	–	175	150.0
Arc resistance	sec	–	–	20 to 120	128	–	–
Other properties							
Density	g/cm²	1,28	1,37	1,37	1,27	1,3	1,311
Water absorption in 24 h	%	–	–	–	0,25	0,5	0,07
equilibrium value	%	–	2,2	–	1,25	–	0,8
Oxygen index, LOI	%	> 42	41	38 to 41	47	40	37
at specimen thickness	mm	–	3,2	1,6	–	4	3,2
Combustibility (UL),							

Table 2.3 High-performance thermoplastic data

Polyetherketone		Liquid crystalline polymers		Polyamide imide	Polyphenylenesulfide	Polycarbonate		Test Procedure
PEEK	PEKK	LCP		PAI	PPS	PC		
Vitrex 450 G (ICI)	Declar (Du Pont)	KUI-9231 (Bayer)	Vectra A 950 Hoechst	Terlon 5030 (Amoco)	Tadur KUI-9511 (Bayer)	Makrolon 6870 (Bayer)	Lexan 950 (GEP)	ASTM/DIN/ISO
3600	4500	1700	9700	14600	19000	2300	2300	D638/53457
92	102	–	–	221	–	> 55	60	D638/53455
3660	–	12000	9000	11700	14000	–	2500	D790/53457
170	–	150	169	338	260	–	100	D790/53452
–	–	160	165	–	200	> 65	70	D638/53455
50	80	1.1	3.0	2.3	1.5	90	–	D638/53455
–	–	560	–	504	–	–	–	D256/ISO 180-IC
83	–	280	520	79	–	> 750	700	D256/ISO 180-1A
334	338	330	280	–	280	–	–	
143	156	–	–	–	–	–	–	
			145	–	–	–	145	D1525/53460
140	–	250	180	182	260	126	135	D648/53461
250	–	240	220	220	240	130	–	UL 746B
4,7	–	–0,4	–0,5	1,62	–	6,5	7	D696/52612
–	–	8,0	7,5	–	–	–	–	
0,25	–	–	–	0,37	0,29	0,21	0,2	C177/52612
19	–	53	39	–	–	> 30	17	D149/53481
4 to 9	–	1	1	0,2	0,1	> 1	10	D257/53482
–	–	150	175	–	125	200 to 225	–	D3636/53480
–	–	101	137	–	–	–	120	D495
1,32	1,28	1,38	1,4	1,61	1,7	1,2	–	D792/53479
0,5	–	–	0,02	0,24	–	–	0,1	D570/53495
0,5	–	–	–	–	–	0,36	0,35	
35	40	48	35	51	48	36	35	D2865
3,2	–	–	–	–	–	–	–	

Table 2.3 *(continued)*

Highly favorable characteristics such as less density, strength through shape, good thermal insulation, a high degree of mechanical dampening, high resistance to corrosion and chemical attack, and exceptional electric resistance exist for certain plastics. There are also those that will deteriorate when exposed to sunlight, weather, or ultraviolet light, but then there are those that resist such deterioration.

Alloy	Properties
PVC/acrylic	Flame, impact, and chemical resistance
PVC/ABS	Flame resistance, impact resistance, processability
Polycarbonate/ABS	Notched impact resistance, hardness, heat-distortion temperature
ABS/polysulfone	Lower cost
Polypropylene/ethylene-propylene-diene	Low-temperature impact resistance and flexibility
Polyphenylene oxide/polystyrene	Processability, lower cost
Styrene acrylonitrile/olefin	Weatherability
Nylon/elastomer	Notched Izod impact resistance
Polyphenylene sulfide/nylon	Lubricity
Acrylic/polybutylene rubber	Clarity, impact resistance

Table 2.4 Examples of plastic alloy properties

	Deformation-under-Load (Tensile)	Static Coefficient of Friction
Test Method **ASTM**	D621 % @23°C (Load)	
Polystyrene	0.5 (13.8 MPa)	-
ABS	0.4 (10 MPa)	-
Polypropylene	2 (6.9 MPa)	-
Polyethylene	3.2 (6.9 MPa)	-
Polyacetal	0.25 (6.9 MPa)	0.38
Polyester	-	0.37
Polyamide, nylon 66	1.5 (10 MPa)	0.50
Polycarbonate	0.45 (20.6 MPa)	0.51
Polysulfone	0.9 (20.6 MPa)	-
Test Method **ASTM**	D621 (obsolete) @23°C (Load)	
PTFE	3.5 (6.9 MPa)	0.1
PFA	2 (0.69 MPa)	0.2
FEP	1.8 (6.9 MPa)	0.2
ETFE	0.3 (6.9 MPa)	0.4
PVDF [@140°C]	1-2 (0.69 MPa)	-

Table 2.5 Mechanical properties of plastics

Different plastics can be combined to produce a product meeting different properties. When compounding or alloying certain plastics, synergistic effects can occur (Fig. 2.2 and Table 2.13). As reviewed in chapter 1, practically all plastics include different additives, fillers, and/or reinforcements providing all kinds of properties, including those with synergistic effects.

	Heat Distortion Temperature, °C@0.45 MPa	Coefficient of Thermal Expansion, cm/cm/°Cx10^5	Continuous-use Temperature,°C	Flammability	Volume Resistivity, Ω·cm	Dielectric Strength, kV/mm
Test Method ASTM	D648	D696	UL-Sub 94	UL 94	D257	D149
Polystyrene	84.4	6.8	45	HB	10^{16}	
ABS	90.5	7.6	65	HB	10^{16}	16.3
SAN	93.3	6.5	50	HB	10^{16}	17.6
Polypropylene	104.4	9.0	105	HB	10^{16}	26.2
Polyethylene	71.1	10.8	85	HB	10^{16}	
Polyacetal	157.8	8.5	90	HB	10^{14}	19.0
Polyester	154.4	9.5	107	HB	10^{16}	18.2
Polyamide, nylon 6	185.0	8.3	95	V-2	10^{14}	15.9
Polyamide, nylon 66	182.2	8.1	105	V-2	10^{14}	19.8
Polycarbonate	137.8	6.7	121	V-0	10^{16}	15.0
Polysulfone	173.9	5.6	140	V-0	10^{15}	16.9
Test Method ASTM	D648	D696,E831 E228		UL 94	D257	
PTFE	122	12.6-18	260	V-0	$>10^{18}$	19.7
PFA	74	13.7-20.7	260	V-0	$>10^{18}$	19.7
FEP	70	8.3-10.4	204	V-0	$>10^{18}$	19.7
ETFE	81	13.1-25.7	150	V-0	$>10^{17}$	14.6
PVDF	140-168	7-15	150	V-0	$>10^{14}$	63-67

Table 2.6 Thermal and electrical properties of plastics

Material	Weak Bases and Salts		Strong Bases		Strong Acids		Strong Oxidants		24-hour Water Absorption
	25°C	93°C	25°C	93°C	25°C	93°C	25°C	93°C	Weight Change, %
Polyacetal	1-3	2-5	1-5	2-5	5	5	5	5	0.23
ABS	1	2-4	1	2-4	1-4	5	1-5	5	0.1-0.4
Nylons	1	2	2	3	5	5	5	5	0.2-1.9
Polyester	1	3-4	2	5	3	4-5	2	3-5	0.06-0.09
Polyethylene	1	1	1	1	1	1	1	1	<0.01
Polystyrene	1	5	1	5	4	5	4	5	0.03-0.60
Polysulfone	1	1	1	1	1	1	1	1	0.2-0.3
PVC[f]	1	5	1	5	1	5	2	5	0.04-1.0
PTFE	1	1	1	1	1	1	1	1	0
PFA	1	1	1	1	1	1	1	1	<0.03
FEP	1	1	1	1	1	1	1	1	<0.01
ETFE	1	1	1	1	1	1	1	1	<0.03
PVDF	1	1	1	2	1	2	1	2	0.04
PCTFE[g]	1	1	1	1	1	1	1	1	0.01-0.10

Legend

* Effect is defined by ASTM D543 and D2299

1 = No effect or inert

2 = Slight effect

3 = Mild effect

4 = Softening or swelling

5 = Severe degradation

[f] PVC = Polyvinylchloride

[g] PCTFE = Polychlorotrifluoroethylene

Table 2.7 Water absorption (ASTM D 543) and the effect of inorganic chemicals (ASTM D 2299) on plastics

	Specific gravity	Flame color (copper wire) As is	Melts/soft	Color	Smoke density	Odor	Solvents
Polypropylene	0.85–0.9	Blue yellow	Yes (trans.)	White	Very little	Heavy	Toluene (slowly slight)[b]
LDPE	0.91–0.93	Blue yellow	Yes (trans.)	White	Very little	Candle wax	Dipropylene glycol[b]
HDPE	0.93–0.96	Blue yellow	Yes (trans.)	White	Very little	Candle wax	Toluene[b]
Epoxy	1–1.25	Orange yellow (green)	No	Black		Phenolic	
Chlorinated PE	1–24	Green	Yes				Toluene[b]
Polystyrene	1.05–1.08	Orange yellow	Yes	Black	Dense	Sweet marigolds	Diethyl benzene
Polyvinyl butyral	1.07–1.08	Blue mantle yellow	Yes (trans.)			Rancid butter	
Nylon	1.09–1.14	Blue mantle yellow	Yes			Burnt hair	
Ethyl cellulose	1.1–1.16	Blue white	Yes			Sweet	See-amyl alcohol
Polyester	1.12–1.46	Yellow	No	Black	Dense	Sweet (resinous)	
Vinyl chloride	1.15–1.65	(Green) yellow orange	Yes, softening	White to green	Little	Acrid chlorine	Toluene[b]
Arcylic	1.18–1.19	Blue mantle yellow orange	Yes (trans.)	Some black		Floral burnt fat	Toluene[b]
Vinyl acetate	1.19	Dark yellow	Yes	Black		Acetic	Sec-hexyl alcohol cyclohexanol acetionitrile
Polycarbonate	1.20	Orange yellow	No	Black		Phenolic sweet	Toluene[b]

Table 2.8 General properties of plastics

Table 2.8 *(continued)*

Cellulose acetate	1.27–1.34	Dark yellow, mauve blue	Yes	Black		Acetic vinegar	Furfuryl alcohol and acetionitrile
Casein	1.35	Yellow	No	Gray		Burnt milk	
Cellulose nitrate	1.35–1.40	Intense white	Yes			No odor	Dipropylene glycol and acetionitrile
Acetal	1.41–1.42	Blue mantle yellow	Yes			Formaldehyde	
Urea formaldehyde	1.47–1.52		No			Urinous	
Melamine formaldehyde	1.50–2.20		No			Fish	
Phenol formaldehyde	1.55–1.90		No			Phenolic	
Saran	1.58–1.75		Yes				
Vinylidene chloride	1.62–1.72	(Green) yellow	Yes			Sweet	
Chlorinated rubber	1.64		Softnens	Black	Dense	Rubber	
Alkyd	1.80–2.24		No				
Tetrafluoroethylene	2.1–23		No			Burnt hair	
Neoprene		(Green) orange	Softens	Black		Rubber	

Different plastics can just be stacked together, but with available processes they can also be fabricated so that each material retains its individuality yet has a bond with the adjoining plastics. These processes include coinjection, coextrusion, laminating, and coating. Each plastic can provide such characteristics as wear resistance, water resistance, electrical conductivity, and additional strength. Low-cost recycled plastics can be "sandwiched" between other expensive, high-performance plastics so they only act as a filler, increase strength, and so forth.

To meet fabricated dimensional tolerances, different approaches are used. They include use of specific fillers and reinforcements and process control (chapter 3). A popular filler used is short glass fibers (chapter 15). In fact, over 50 wt% of all types of glass fibers used with different plastics and by different processes are used in molding compounds. Table 2.14 shows the shrinkage of different unreinforced plastics with glass fiber reinforcements based on American Society for Testing and Materials (ASTM) testing procedures (chapter 22).

Different barrier plastics meet different performance requirements. A very popular barrier plastic is EVOH (ethylene-vinyl alcohol copolymer), which can be tailored to the needs of packages and other product requirements (chapter 20). Generally, the thickness ranges from 0.5% to 3.0% of the wall thickness; it can be thicker if a higher barrier is required. Generally, an EVOH thickness

Polymer	Glass transition temperature (°C)	Crystalline melting temperature (°C)
Acetal	–50	181
Polyethylene	–33	110–140
Polypropylene	–10	165–176
Polyamide 12	37	179
Polyamide 6/10	40	227
Polyamide 11	46	194
Polyamide 6	50	220
PBT	50	240
Polyamide 6/6	50	265
PET	69	267
Polystyrene	100	240
PMMA	90–105	160
PPO	104–120	-
ABS	80–125	190
SAN	115–125	-
PVC-U	87–135	212
Polycarbonate	150	220
PES	230	-

Table 2.9 Glass transition and crystalline melting points of thermoplastics

greater than 8% of the container sidewall can lead to internal structural failures that can fail on drop tests. Also, a very thick layer is difficult to process consistently. The EVOH's crack resistance improves as its ethylene content increases.

Plastics can also be combined with other materials such as aluminum, steel, and wood to provide specific properties. Examples include extruded polyvinyl chloride (PVC)/wood window frames and extruded plastic film/aluminum-foil packaging material. All combinations may require that certain aspects of compatibility such as processing temperature, bondability, and coefficient of expansion or contraction exist.

PROPERTY RANGE

Even though the range of plastics continues to be large and the levels of their properties so varied, in any proposed application only a few of the many plastics will be suitable. A compromise among properties, cost, and manufacturing process generally determines the material of construction.

Polymer	Thermal conductivity at 20°C (W/m.K)
PVC-U	0.14–0.17
ABS	0.18
PES	0.18
PMMA	0.18
Polystyrene, general purpose	0.18
SAN	0.18
PBT	0.21
Polycarbonate	0.21
Polypropylene	0.17–0.22
Polyamide 11	0.23
Polyamide 12	0.23
Polyamide 6/6	0.23
PPO	0.23
PET	0.24
Polyamide 6	0.29
Acetal	0.25–0.30
LD polyethylene	0.32–0.40
HD polyethylene	0.38–0.51

Table 2.10 Thermal conductivity of thermoplastics

Material	Property	Glass Fiber By Weight	Specific Gravity	Density	Tensile Strength	Tensile Modulus
	Units	%		g/cm³	MPa	GPa
	ASTM Test Method		D792		D638	D638
Glass Fiber Reinforced Thermosets	Polyester SMC (compression)	30.0	1.85	1.827	82.20	11.72
	Polyester SMC (compression)	20.0	1.78	1.772	36.31	11.72
	Polyester SMC (compression)	50.0	2.00	2.000	157.55	15.65
	Polyester BMC (compression)	22.0	1.82	1.799	41.10	12.07
	Polyester BMC (injection)	22.0	1.82	1.799	33.29	10.55
	Epoxy (filament wound)	80.0	2.08	1.910	548.00	27.58
	Polyester (pultruded)	55.0	1.69	1.661	205.50	17.24
	Polyurethane, milled fibers (RRIM)	13.0	1.07	1.052	19.18	—
	Polyurethane, flaked glass (RRIM)	23.0	1.17	1.163	30.21	—
	Polyester (spray-up/lay up)	30.0	1.37	1.356	85.61	6.895
	Polyester, woven roving (lay up)	50.0	1.64	1.633	253.45	15.51
Glass Fiber Reinforced Thermoplastics	Acetal	25.0	1.61	1.605	126.73	8.62
	Nylon 6	30.0	1.37	1.356	164.40	7.24
	Nylon 6/6	30.0	1.48	1.467	157.55	8.27
	Polycarbonate	10.0	1.26	1.246	82.20	5.17
	Polypropylene	20.0	1.04	1.024	44.53	3.72
	Polyphenylene sulfide	40.0	1.64	1.633	150.70	14.13
	Acrylonitrile–butadiene–styrene (ABS)	20.0	1.22	1.218	75.35	6.21
	Polyphenylene oxide (PPO)	20.0	1.21	1.190	99.33	6.34
	Polystyrene acrylonitrile (SAN)	20.0	1.22	1.218	99.33	8.62
	Polyester (PBT)	30.0	1.52	1.495	130.15	8.27
	Polyester (PET)	30.0	1.56	1.550	143.85	8.96
Unreinforced Thermoplastics	Acetal	—	1.41	1.412	60.28	2.83
	Nylon 6	—	1.12	1.107	80.83	2.62
	Nylon 6/6	—	1.13	1.135	78.78	2.76
	Polycarbonate	—	1.20	1.190	65.08	2.34
	Polypropylene	—	0.89	0.886	34.25	0.69
	Polyphenylene sulfide	—	1.30	1.246	65.08	3.31
	Acrylonitrile–butadiene–styrene (ABS)	—	1.03	1.024	41.10	2.07
	Polyphenylene oxide (PPO)	—	1.10	1.080	53.43	2.62
	Polystyrene acrylonitrile (SAN)	—	1.05	1.052	65.08	2.76
	Polyester (PBT)	—	1.31	1.301	56.17	1.93
	Polyester (PET)	—	1.34	1.329	58.23	2.76

Table 2.11 Unreinforced and reinforced plastics

Elongation	Flexural Strength	Flexural Modulus	Compressive Strength	Impact Strength IZOD	Hardness	Flammability	Specific Heat
%	MPa	GPa	MPa	kJ/m notched @ 23°C	Rockwell (except where noted)		kJ/kg·K
D638	D790	D790	D695	D256	D785	UL-94	
<1.0	179.27	11.03	165.48	0.85	Barcol 68	5V	12.56
0.4	110.32	9.65	158.59	0.43	Barcol 68	5V	12.56
1.7	310.28	13.79	220.64	1.03	Barcol 68	5V	12.56
0.5	88.26	10.89	137.90	0.23	Barcol 68	5V	12.56
0.5	87.22	9.93	—	0.15	Barcol 68	VO	12.56
1.6	689.50	34.48	310.28	2.39	M98	VO	9.63
—	206.85	11.03	206.85	1.33	Barcol 50	VO	11.73
140.0	—	0.26–0.37	—	—	S.D.* 65–75	VO	—
38.9	—	1.03	—	0.11	—	VO	—
1.3	193.06	5.17	151.69	0.69–0.80	Barcol 50	VO	12.98
1.6	317.17	15.51	186.17	1.75	Barcol 50	VO	—
3.0	193.06	7.58	117.22	0.10	M79	HB	—
3.0	199.96	7.65	165.48	0.12	R121	HB	—
1.9	241.33	5.52	182.72	0.12	M95	HB	12.56
9.0	110.32	4.14	96.53	0.11	M80	V-1	12.15
3.0	57.23	3.59	172.38	0.06	R103	HB	—
3.0	255.12	13.10	144.80	0.08	R123	V-O/5V	10.47
2.0	106.87	6.00	96.53	0.06	R107	HB	—
5.0	127.56	5.17	121.35	0.10	R107	HB	8.38–16.75
1.8	131.01	7.58	120.66	0.06	R122	HB	—
4.0	193.06	8.07	124.11	0.10	R118	HB	4.61
6.6	220.64	8.62	172.38	0.10	R120	HB	—
40.0	89.64	2.62	110.32	0.05	M78–M80	HB	14.66
30.0	108.25	2.69	89.64	0.03	R119	HB	16.75
60.0	117.22	2.90	103.43	0.04	R120, M83	V-2	12.56
110.0	93.08	2.34	86.19	0.85	M70	V-2	12.56
200.0	34.48	0.90–1.38	24.13	0.10–1.01	R50-96	HB	18.85
1.0	96.53	3.79	110.32	<0.03	R123	V-O	—
5.0	75.85	2.41–2.76	68.95	0.16–0.32	R107-115	HB	—
50.0	88.26	2.28–2.76	82.74	0.27	R115	V-1	8.38–16.75
0.5	96.53	3.79	96.53	0.02–0.02	M80-85	HB	13.82
50.0	82.74	2.28–2.76	59.30	0.04	M68-78	HB	—
50.0	96.53	2.41–3.10	75.85	0.01–0.03	M94-101	HB	14.24

Table 2.11 *(continued)*

Material	Thermal Coefficient of Expansion	Heat Deflection (DTUL)	Thermal Conductivity	Dielectric Strength	Volume Resistivity	Relative Permittivity
	mm/m·K × 10^{-6}	°C @ 1820 kPa	W/mK	mV/m	Ohm-cm	60 Hz
	D696	D648	C177	D149	D257	D150
Glass Fiber	—	205 +	—	19.695.	5.7 × 10^{14}	4.40
Reinforced	—	205 +	—	—	—	4.40
Thermosets	16.9	205 +	—	—	—	4.40
	11.9	260	0.70	14,770.	27 × 10^{14}	4.20
	11.9	260	0.70	14,770.	—	4.20
	3.6	205 +	0.28	11,815.	> 10^{12}	—
	9.0	—	0.58	7,880.	10^{13}	4.40
	140.4	30	—	—	—	—
	95.6	—	—	—	—	—
	21.6	205 +	0.22	9,850.	—	—
	7.2	205 +	—	13,785.	10^{14}	4.20
Glass Fiber	8.5	161	—	22,845.	10^{14}	4.12
Reinforced	2.7	211	0.84–1.64	19,695.	10^{13}	3.90
Thermoplastics	3.2	255	0.22	15,755.	10^{15}	3.80
	3.2	141	0.66	19,695.	10^{16}	3.10
	4.3	132	1.21	17,330.	10^{15}	2.70
	2.0	266	0.29	14,970.	4 × 10^{15}	3.00
	3.8	99	0.20	18,315.	10^{15}	3.20
	3.6	143	0.55	19,695.	10^{16}	3.20
	3.8	102	0.40	19,300.	10^{15}	3.50
	2.5	213	1.01	14,770.	3.2 × 10^{16}	3.80
	3.1	216	0.94	20,485.	10^{16}	3.60
Unreinforced	8.5	110	0.22	19,695.	10^{15}	3.70
Thermoplastics	8.6	68–85	0.17	12,015.	4.5 × 10^{13}	4.00
	8.1	75	0.24	15,165.	1 × 10^{15}	4.00
	6.7	132	0.19	14,970.	> 10^{16}	2.96
	6.8	46–60	0.17	23,635.	> 10^{17}	2.20
	—	135	0.24	14,970.	10^{16}	—
	5.8	93–105	0.14	13,785.–19,695.	2.7 × 10^{16}	2.80
	122.4	100	0.13	15,755.	> 10^{16}	2.65
	64.8	105	0.10	20,285.	4.4 × 10^{16}	3.50
	8.1	50–85	0.15–0.24	16,545.–21,665.	4 × 10^{16}	3.10
	—	38–41	0.13	—	> 10^{16}	—

Table 2.11 *(continued)*

Arc Resistance	Water Absorption	Mold Shrinkage
seconds	Percent 24 hours	in/in
D495	D570	D955
188	.25	—
188	.10	.002
188	.50	—
190	.20	.001
190	.20	.004
—	.50	.008
80	.75	—
—	—	—
—	—	—
—	1.30	—
—	.50	—
142	.29	.004
120	1.30	.004
120	.50	.002
125	.14	.005
120	.05	.003
125	.01	.002
80	.30	.002
70	.24	.003
70	.06	.002
135	.06	.003
90	.05	.003
129	.22	.020
—	1.3–1.9	.005
120	1.0–1.3	.008
120	.15	.005–.007
125	.03	.020
—	< .02	.007
—	.20–.45	.004–.009
75	.07	.005–.007
65	.20–.35	—
184	.08–.09	.015–.020
—	.1–.2	.02–.025

Table 2.11 *(continued)*

	Polyethylene		
	Low-density	**Medium-density**	**High-density**
General			
Clarity	Transparent to translucent	Transparent to translucent	Transparent to translucent
Yield (sq. in./lb./ 0.001-inch)	30,000	29,500	29,000
Specific gravity	0.910-0.925	0.926-0.940	0.941 0.965
Mechanical			
Tensile strength (lb/sq.in.) ASTM D-882	1,000-3,500	2,000-5,000	3,000-7,500
Elongation (per cent) ASTM D-882	225-600	225-500	10-500
Impact strength (kg-cm)	7-11	4-6	1-3
Tear strength (gm/0.001-inch Elmendorf) ASTM D-1922	100-400	50-300	15-300
Heat seal range (°F.)	250-350	260-310	275-310
Chemical			
WVTR (gm/24hr/100 sq. in. @ 100° F. 90 per cent RH) ASTM E-96	1.2	0.5-1.0	0.3-0.65
Gas transmission (cc/0.001-inch/100 sq. in./ 24 hr. @ atm 73°F. & 0 per cent RH) ASTM D-1434	O_2-250-840 CO_2-495-5000	O_2-165 335 CO_2-500-840	O_2-33-250 CO_2-250-645
Resistance to grease and oils	Varies	Good	Good
Permanence			
Maximum use temperature (°F.)	150	180-220	250
Minimum use temperature (°F.)	−60	−60	−60
Dimension change at high RH (per cent)	None	None	None
Printability	Good if treated	Good if treated	Good if treated
Heat shrinkable	Some types	Some types	Some types

Table 2.12 Examples of thermoplastic film properties

		Polypropylene			**Polyester (PET)**		
Linear low density	**Low density/ EVA copolymer (12% EVA)**	**Non Oriented**	**Oriented/ (balanced)**	**Oriented/ metallized**	**Uncoated**	**Coated-saran one side**	**Metallized**
Transparent to translucent	Transparent	Transparent	Transparent	Opaque	Transparent	Transparent	Opaque
30,000	29,500	30,800	30,600	30,600	20,000-22,000	20,500	20,000-22,000
.925	0.94	0.88-0.90	0.905	0.905	1.35-1.39	1.4	1.35-139
MD-1540 TD-1620	3000-5000	3000-9000	10,000-30,000	MD-19,000 TD-32,000	25,000-33,000	26,000-31,000	25,000-33,000
MD-640 TD-680	300-500	400-800	50-400	50-400	120-140	90-125	120-140
1.3	11-15	1-3	5-15	5-15	25-30	25-30	25-30
MD-280 TD-400	50-100	40-330	4-6	4-6	13-80	10-20	13-80
250-350	200-300	325-400	Requires coating	Requires coating	—	275-400	—
1.2	3.9	0.5-0.65	0.26-0.65	0.1	1.3	0.9 @0.0005-inch	0.1
O_2-250-840 CO_2-495 5000	O_2-515-645 CO_2-2260-2900	O_2-84-415 CO_2-495-1350	O_2-155 CO_2-540	O_2-3.0	O_2-5	O_2-0.4 @0.0005-inch	O_2-0.08
Good	Varies	Good	Good	Good	Good	Good	Good
170-180	140	250	250	250	400	Coating softens at 180°	400
−60	−60	Not recommended for low temps.	−60	−60	−80	−60	−80
None	None	None	None	None	None	None	None
Good if treated	Good if treated	Good if treated	Good if treated	Good if treated	Good	Good	Good
Not recommended	Some types	No	Some types	No	Some types	Some types	No

Table 2.12 *(continued)*

Nylon			Others		
Uncoated	**Saran coated one side**	**Metallized Oriented**	**Polystyrene**	**Polyvinyl chloride**	**Polyvinyliden chloride**
Transparent to translucent	Transparent	Opaque	Transparent	Transparent to Translucent	Transparent
23,500-24,500	21,200	24,500	26,300	19,000-22,000	16,300
1.13-1.14	1.13-1.14	1.13-1.14	1.05	1.23-1.37	1.59-1.71
7,000-18,000	7,000-18,000	28,000	9,000-12,000	2,000-16,000	8,000-20,000
250-500	250-500	90	10-50	5-500	40-80
4-6	7	7	1-5	12-20	10-15
20-50	20-60	20-60	4-20	Varies Widely	10-20
350-500	350-500	350-500	250-325	200-350	280-300
24-26	0.2	0.2	6.5 and up	>8	0.1-0.3
$O_2$2.6 CO_2-4.7 N_2-0.9	O_2-0.5 CO_2-1.4 N_2-0.1	O_2-0.05	O_2-165-495 CO_2-645 1680	O_2-5-500 CO_2-50-3500	O_2-0.5-1.7 CO_2-3.3-10
Impermeable	Impermeable	Impermeable	Good	Good	Good
350-450	200	200	175	200, depends on plasticizer	Softens at about 290°
−75	−40	−75	<0	Depends on plasticizer	<0
1.3	1.3	1.3	Little to none	None	None
Good	Good	Good	Special inks	Special inks	Special inks
No	No	No	Yes	Some types	Some types

Table 2.12 *(continued)*

Others					
Poly-Butylene	Poly-Carbonate	Nitriles	Cellulose acetate	Flurohalo-carbon	Ionomer
Transparent to translucent	Transparent	Transparent	Transparent	Transparent	Transparent
21,200	23,100	24,000	22,000	13,000	28,600-29,500
1.3	1.2	1.15	1.25-135	2.2	0.94-0.96
4,500-5,000	8,200-9,300	9,500	7,000-15,000	5,000-10,000	3,000-5,000
200-400	95-115	5	15-50	50-400	350-450
20-30	—	High	2-8	2-15	6-11
400-800	20-25	High	2-15	3-40	15-150
325-400	400-430	160-300	350-450+	350-400	190-400
1.2	9.7	5	5	0.4-0.7	1.3-2.1
O_2-385 CO_2-825 N_2-110	O_2-258 CO_2-775	O_2-0.8 CO_2-1.1	O_2-23 CO_2-105 N_2-6.5	O_2-240 CO_2-500	O_2-226-484 CO_2-626-1150
Good	Good	Impermeable	Good	Good	Good
220	265	160-210	200	300	160
−30	−100	−50	About 0	−50	−100
None	None	—	0.2-0.6@ 80 per cent	None	None
Good if treated	Good	Good	Good	Good if treated	Good if treated
Some types	No		No	No	Special Types

Table 2.12 *(continued)*

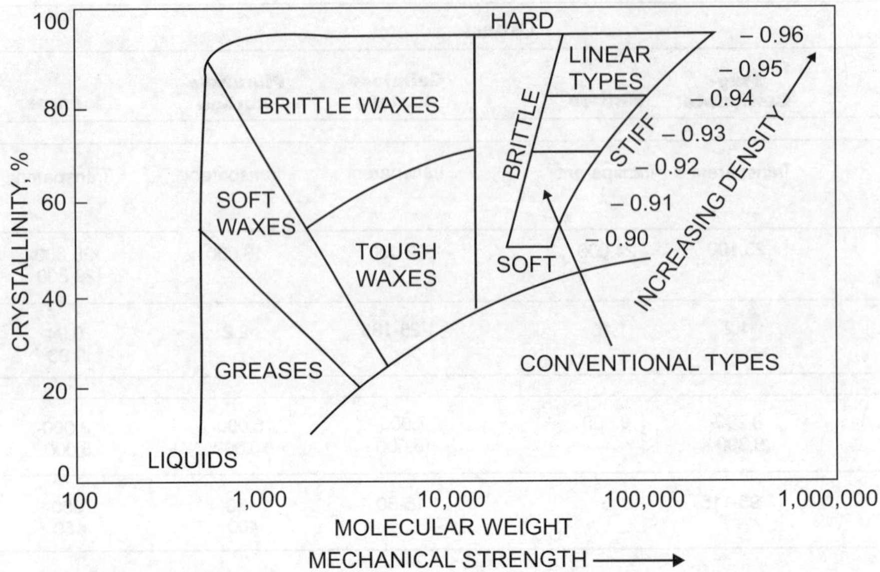

Figure 2.1 Polymerization behavior influences properties of PE

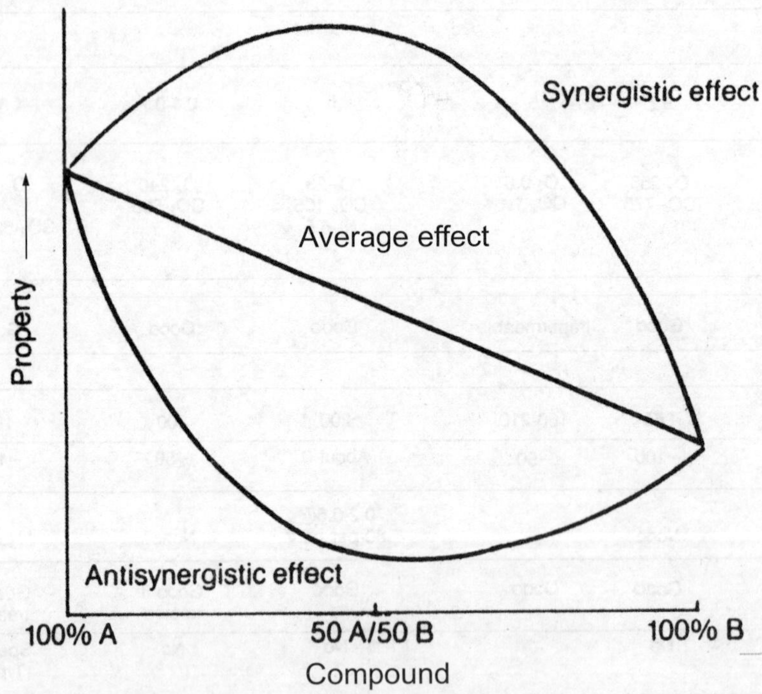

Figure 2.2 Combining certain plastics or a plastic with an additive can result in synergism

Plastics combined	Properties
PVC/acrylic	Flame, impact, and chemical resistance
PVC/ABS	Flame resistance, impact resistance, processability
Polycarbonate/ABS	Hardness, heat-distortion temperature
ABS/polysulfone	Lower cost
Polypropylene/ethylene-propylene-diene	Low-temperature impact resistance and flexibility
Polyphenylene oxide/polystyrene	Processability, lower cost
Styrene acrylonitrile/olefin	Weatherability
Nylon/elastomer	Notched Izod impact resistance
Polyphenylene sulfide/nylon	Lubricity
Acrylic/polybutylene rubber	Clarity, impact resistance

Table 2.13 Example of properties obtained by combining different plastics

As an initial step, the product designer must know and/or anticipate the conditions of use and the performance requirements of the product, considering such factors as life expectancy, size, conditions of use, shape, color, strength, and stiffness. These end-use requirements can be ascertained through market analysis, surveys, examinations of similar products, testing, and/or experience. A clear definition of product requirements will often lead directly to the choice of the material of construction. At times, incomplete or improper product requirement analysis is the cause of a product's failure.

As a general rule, it is considered desirable to examine the properties of three or more materials before making a final choice. Material suppliers should be asked to participate in type and grade selection so that their experience is part of the input. The technology of manufacturing plastic materials, as with other materials (steel, wood, etc.), is not uniform; the same plastic compounds supplied from various sources will generally not deliver the same results in a product. As a matter of record, even each individual supplier furnishes their product under a batch number, so that any variation can be tied down to the exact condition of the raw-material production. Taking into account manufacturing tolerances of the plastics, plus variables of equipment and procedure, it becomes apparent that checking several types of materials from the same and/or from different sources is an important part of material selection (chapter 1).

Experience has proven that the so-called interchangeable grades of materials have to be evaluated carefully as to their effect on the quality of a product. Another important consideration as far as equivalent grade of material is concerned is its processing characteristics. There can be large differences in properties of a product and test data if the processability features vary from grade to grade. It must always be remembered that test data have been obtained from simple and easy-to-process shapes and do not necessarily reflect results in complex product configurations. This situation is similar to that encountered with other materials (steel, wood, glass, etc.).

Most plastics are used to produce products because they have desirable mechanical properties at an economical cost. For this reason, their mechanical properties may be considered the most

Plastic	Avg. rate per ASTM D 955 0.125 in. (3.18 mm)	0.250 in. (6.35 mm)
ABS		
Unreinforced	0.004	0.007
30% glass fiber	0.001	0.0015
Acetal, copolymer		
Unreinforced	0.017	0.021
30% glass fiber	0.003	NA
HDPE, homo		
Unreinforced	0.015	0.030
30% glass fiber	0.003	0.004
Nylon 6		
Unreinforced	0.013	0.016
30% glass fiber	0.0035	0.0045
Nylon 6/6		
Unreinforced	0.016	0.022
30% glass fiber	0.005	0.0055
PBT polyester		
Unreinforced	0.012	0.018
30% glass fiber	0.003	0.0045
Polycarbonate		
Unreinforced	0.005	0.007
30% glass fiber	0.001	0.002
Polyether sulfone		
Unreinforced	0.006	0.007
30% glass fiber	0.002	0.003
Polyether-etherketone		
Unreinforced	0.011	0.013
30% glass fiber	0.002	0.003
Polyetherimide		
Unreinforced	0.005	0.007
30% glass fiber	0.002	0.004
Polyphenylene oxide/PS alloy		
Unreinforced	0.005	0.008
30% glass fiber	0.001	0.002
Polyphenylene sulfide		
Unreinforced	0.011	0.004
30% glass fiber	0.002	NA
Polypropylene, homo		
Unreinforced	0.015	0.025
30% glass fiber	0.0035	0.004
Polystyrene		
Unreinforced	0.004	0.006
30% glass fiber	0.005	0.001

Table 2.14 Example of plastic shrinkage without and with glass fiber

important of all the physical, chemical, electrical, and other considerations for most applications. Thus, everyone designing with such materials needs at least some elementary knowledge of their mechanical behavior and how they can be modified by the numerous structural geometric shape factors that can be in plastics (chapters 19 and 20).

PLASTICS PERFORMANCE

In the following sections, many different types of plastics will be reviewed and their main characteristics highlighted. In addition to the performance data to be presented, extensive data have already been presented in Tables 2.1 to 2.14. Some of these plastics are reviewed in other chapters providing additional information, since they provide special characteristics applicable to that chapter's subject (for example, see chapter 20). Details on their properties can be obtained from plastics producers, different available software packages (chapter 25), and literature (92, 93, 105, 145, 158, 164, 173, 178, 212, 221, 249, 287, 288, 304, 306, 358, 374, 375). What is presented will provide familiarity with the variations of properties existing in plastics. As reviewed in chapter 1, plastics' performances can be changed by compounding, alloying, and so on, such as summarized in Table 2.15.

HEAT-RESISTANT PLASTIC

There have been intensive research efforts to develop more stable and oxidation-resistant polymer systems for continuous service at 300°C (572°F) and above. The impetus for these efforts has been supplied largely by continuous and anticipated military and aerospace requirements. The remarkable rapidity with which some of these polymers have moved from the laboratory to the marketplace provides some insight into the probable pace of future developments.

If any polymer (plastic) is heated to a sufficiently high temperature, both reversible and irreversible changes in its structure occur. These changes can be either useful or highly undesirable. Useful changes include (a) the reversible softening of thermoplastics that permits them to be converted into useful shapes and (b) irreversible chemical changes occurring during the curing of thermoset materials. The undesirable changes include chain scissions, the development of strength-destroying degrees of cross-linking, the formation of low molecular weight volatile products, or various combinations of these modes of degradation.

For some years, modest increases in the maximum service temperatures of thermoplastic materials have been achieved by modifying the polymer chains to render them more rigid at elevated temperatures. This has been accomplished by building crystallinity into the polymers and by the use of polar groups in, or polar groups or bulky substituents as pendent groups on, the chains. However, the thermal decompositions of these, as well as thermoset materials, depend ultimately on the stabilities of the structural units and linkages that make up the polymer chains.

Polymer research has revealed certain structural requirements that the more heat-resistant polymeric materials must possess. These requirements include

1. a high molecular weight to provide strength, flexibility, toughness, and general plastic properties;
2. a minimum of readily oxidizable hydrogen atoms and a maximum of highly stable structural units (aromatic and/or heterocyclic rings) in the chains;

Desired Modification	How Achieved	Sacrifice (from Base Resin) Amorphous	Crystalline	Comments
Increased Tensile Strength	Glass fibers Carbon fibers Fibrous minerals	Ductility, cost Ductility, cost	Ductility, cost Ductility, cost Ductility	Glass fibers are the most cost-effective way of gaining tensile strength. Carbon fibers are more expensive; fibrous minerals are least expensive but only slightly reinforcing. Reinforcement makes brittle resins tougher and embrittles tough resins. Fibrous minerals are not commonly used in amorphous resins.
Increased Flexural Modulus	Glass fibers Carbon fibers Rigid minerals	Ductility, cost Ductility, cost Ductility	Ductility, cost Ductility, cost Ductility	Any additive more rigid than the base resin produces a more rigid composite. Particulate fillers severely degrade impact strength.
Flame Resistance	FR additive	Ductility, tensile strength, cost	Ductility, tensile strength, cost	FR additives interfere with the mechanical integrity of the polymer and often require reinforcement to salvage strength. They also narrow the molding latitude of the base resin. Some can cause mold corrosion.
Increased Heat-Deflection Temperature (HDT)	Glass fibers Carbon fibers Fibrous minerals	Ductility, cost Ductility, cost	Ductility, cost Ductility, cost Ductility	When reinforced, crystalline polymers yield much greater increases in HDT than do amorphous resins. As with tensile strength, fibrous minerals increase HDT only slightly. Fillers do not increase HDT.
Warpage Resistance	5 to 10% glass fibers 5 to 10% carbon fibers Particulate fillers	 Ductility, cost, tensile strength	Cost Cost Ductility, cost, tensile strength	Amorphous polymers are inherently nonwarping molding resins. Only occasionally are fillers such as milled glass or glass beads added to amorphous materials, because they reduce shrinkage anisotropically.
Reduced Mold Shrinkage (Increased mold-to-size capability)	Glass fibers Carbon fibers Fillers	Ductility, cost Ductility, cost Tensile strength, ductility, cost	Ductility, cost Ductility, cost Tensile strength, ductility, cost	Reinforcement reduces shrinkage far more than fillers do. Fillers help balance shrinkage, however, because they replace shrinking polymer. The sharp shrinkage reduction in reinforced crystalline resins can often lead to warpage. The best "mold-to-size" composites are reinforced amorphous composites.
Reduced Coefficient of Friction	PTFE Silicone MoSe Graphite	Cost	Cost	These fillers are soft and do not dramatically affect mechanical properties. PTFE loadings commonly range from 5 to 20%; the others are usually 5% or less. Higher loadings can cause mechanical degradation.
Reduced Wear	Glass fibers Carbon fibers Lubricating	— — —	— — —	The subject of plastic wear is extremely complex and should be discussed with a composite supplier.

Table 2.15 Perspectives on changing properties of plastics

Summation

Perspective	Mechanism	Structural and morphological modification
Dispersing and diffusing	**Fine grinding**	**Enlarging the phase interface**
	High overall shearing	**Enlarging the regions of molecular penetration**
		Reducing diffusion paths; degrading internal stresses
	High overall shearing with quenching	**As above**
		Freezing thermodynamic imbalances
	High local shearing	**Generating local tribochemically or chemically activated conditions**
Active interlayers	**Radical forming substances**	**Corsslinking via the interlayer**
	Substances initiating radical formation	**Crosslinking the components**
Matrix modification	**Elastomer phases**	**Matrix formation; promoting the absorption of fillers and reinforcing agents**
	Filling and reinforcing	**Modifying the matrix properties**
	Foaming	**Structuring the matrix**

Table 2.15 *(continued)*

3. connecting linkages for the stable chains units, which provide flexibility and are also highly stable;
4. no mechanism by which the chains can readily undergo thermal depolymerizations;
5. the polymer must, at some time, exist as a soluble or formable intermediate to permit its fabrication into useful forms;
6. technologically useful properties for a prolonged period of use at the intended service temperature.

Most of these requirements are common to polymers in general. The breakthrough providing a new generation of heat-resistant organic polymers has come from research on criteria 2 and 3. Clues to the identities of the required stable chain units and connecting linkages were provided largely by studies of the stabilities of model organic compounds and polymers. A brief consideration of some of the interesting new polymer systems will serve to illustrate the progress that has been made in the development of more heat-resistant materials.

An infinite variety of polymers containing stable aromatic and heterocyclic rings in their chains can be synthesized, and numerous structural modifications of the polymer types reviewed can and do exist. Consequently, the chemical structures shown (Fig. 2.3) are simply intended to illustrate representative examples of the subject polymer types.

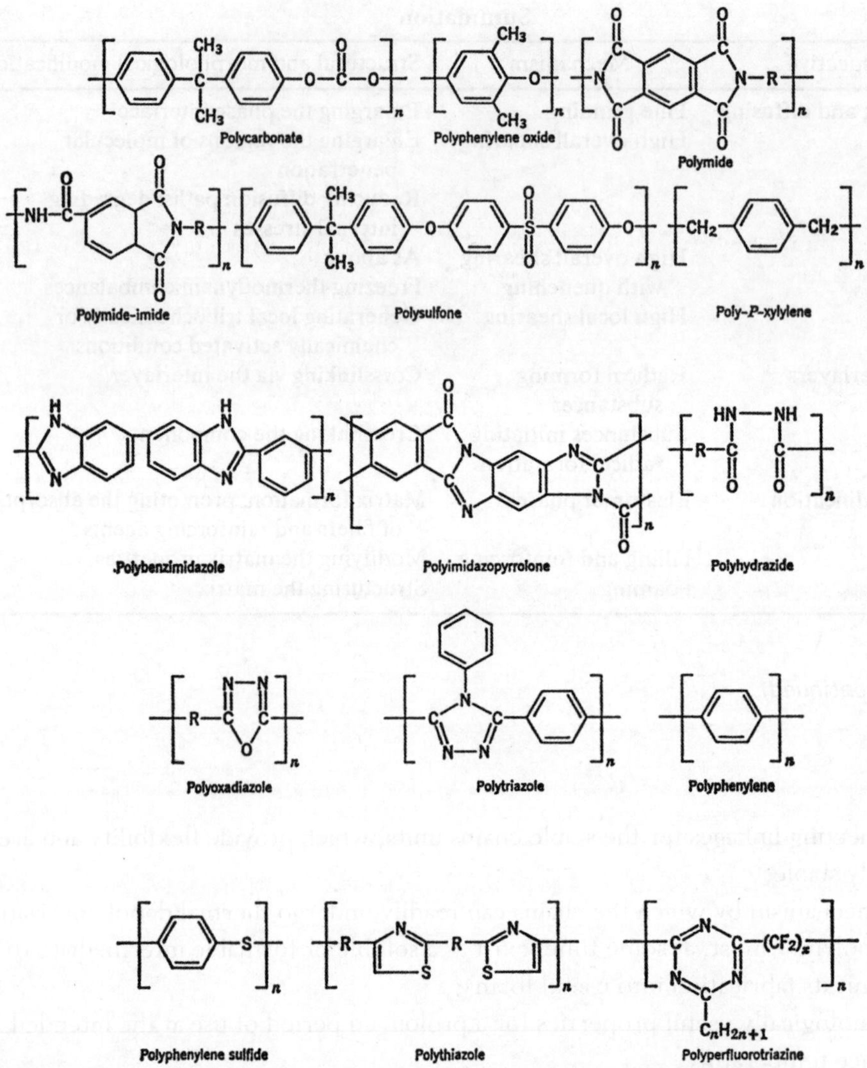

Figure 2.3 Examples of chemical structures of heat-resistant organic polymers

THERMOPLASTICS

These plastics represent approximately 90 wt% of all plastics consumed worldwide. They tend to be more amenable to specialty compounding as copolymers, multipolymers, alloys, and blends, and they are often customized to be cost-effective. Unlike thermoset plastics, they are in many cases reprocessable without any or serious losses of properties.

They can have limitations of heat-distortion temperatures, cold flow, and creep, and they are more likely to be damaged by chemical solvent attack from paints, adhesives, and cleaners. When injection molded, dimensional integrity and ultimate strength are more dependent on the proper process control molding parameters than is generally the case with TSs (in which cross-linking offsets such problems).

Polyolefin

Within the family of polyolefins there are many individual families that include low-density polyethylenes, linear low-density polyethylenes, very low polyethylenes, ultralow polyethylenes, high molecular weight polyethylenes, ultrahigh molecular weight polyethylenes, polyethylene terephthalates, ethylene-vinyl acetate polyethylenes, chlorinated polyethylenes, cross-linked polyethylenes, polypropylenes, polybutylenes, polyisobutylene, ionomers, polymethylpentene, thermoplastic polyolefin elastomers (polyolefin elastomers, TP), and many others. Even though they can have significantly different processing and performance behaviors, some of these plastics often compete for the same applications. Strength, modulus of elasticity, impact strength, and other properties vary greatly with type, degree of crystallinity, and their preparations that result in different densities. Their stress-crack resistance and useful service temperature ranges may also vary with the type of polyolefin, their crystalline structure, and so on.

Polyolefin Elastomer, Thermoplastic

The properties of thermoplastic olefins (TPOs) depend upon the types and amounts of polymers used, the method by which they are combined, and the use of additives such as oils, fillers, antioxidants, and colors. Blends and reactor-made products compete primarily with plastics and metals, whereas vulcanizates have higher elastomeric properties and compete primarily with thermoset elastomers (chapter 1).

Vulcanizates and reactor-made products were expected to account for most of the growth in consumption of TPOs through year 2003. However, blends still accounted for 39% of the total TPO market in 2003 (versus 44% in 1998). Vulcanizates' share of the world TPO market remained at 28% in 2003, whereas reactor-made products increased their share from 28% to 33%. The following factors added elements of uncertainty to the 2003 forecast: (a) the health of the global automotive industry, which accounted for almost 75% of global TPO consumption in 1998; (b) the continuing evolution of new high-impact polypropylene polymers; and (c) market introduction of metallocene-based ethylene copolymers with elastomeric properties, such as Dow's Engage™ polyolefin elastomers and Exxon's Exact™ plastomers.

The thermoplastic characteristics of TPOs allow easy recycling of production waste and final products made of TPOs. Studies have shown that products can be manufactured from recycled TPOs without significant loss of performance. This can be seen as a big advantage compared with traditional rubber, which is relatively difficult to recycle and usually leads to inferior products. The disposal of used rubber products has become an ecological problem in the developed countries. The continued trend toward recycling, particularly in the United States and in Western European countries such as Germany and the Netherlands, is likely to accelerate the substitution of rubber with TPOs in the future.

POLYETHYLENE

Polyethylene is the leading plastic family in total volume sold worldwide. These polyolefin materials are relatively inexpensive, easy to process, and so versatile that they dominate the packaging and disposable fields. PEs are thermoplastic materials produced from polymerized ethylene molecules. Copolymer grades also contain other hydrocarbon molecules for property customization. Polyethylene resins have excellent chemical and impact resistance. There are several types of polyethylenes produced using different manufacturing processes.

Crystalline in structure, these thermoplastics are varied by chain length or molecular weight into very low-density (VLDPE), low-density (LDPE), low-density linear (LDLPE), linear low-density (LLDPE), medium-density (MDPE), high-density (HDPE), ultrahigh-density molecular weight (UHMWPE), and so on. Some are flexible, others rigid, and some have low impact strength, whereas others are nearly unbreakable. Some have good clarity, others are opaque, and so on. The service temperatures for PEs range from –40° to 93°C. In general, toughness, excellent chemical resistance and electrical properties, low coefficients of friction, near-zero moisture absorption, and good ease of processing characterize them. They are basically classified according to their density (Tables 2.16 to 2.21 and Fig. 2.4).

Properties	0.910-0.925 (low)	0.926-0.940 (medium)	0.941-0.965 (high)
Heat softening point		Higher	Highest
Yield strength		Higher	Highest
Tensile strength at rupture	Highest	Higher	
Elongation at rupture	Highest	Higher	
Stiffness		Higher	Highest
Resistance to shrinkage	Highest	Higher	
Resistance to warpage	Highest	Higher	
Film impact strength (toughness)	Highest	Higher	
Resistance to brittleness at low temperatures	Highest	Higher	
Resistance to environmental stress cracking	Highest	Higher	
Impermeability to gases and liquids		Higher	Highest
Resistance to grease and oil absorption		Higher	Highest
Transparency	Higher	Highest	—[a]
Freedom from haze	Higher	Highest	—[a]
Gloss	Higher	Highest	—[a]
Film draw-down		Higher	Highest
Film hot tear resistance		Higher	Highest
Molding cycle time		Shorter	Shortest

[a] Not applicable to high density polyethylene.

Table 2.16 Differences in properties between polyethylene plastics of different densities

PE Property	Density	Melt Index	Molecular Weight
Tensile strength (at yield)	Increases	Decreases	
Stiffness	Increases	Decreases slightly	Decreases slightly
Impact strength	Decreases	Decreases	Decreases
Low-temperature brittleness	Increases	Increases	Decreases
Abrasion resistance	Increases	Decreases	
Hardness	Increases	Decreases slightly	
Softening point	Increases		Increases
Stress-crack resistance	Decreases	Decreases	
Permeability	Decreases	Increases slightly	
Chemical resistance	Increases	Decreases	
Melt strength		Decreases	Increases
Gloss	Increases	Increases	Decreases
Shrinkage	Decreases	Decreases	Increases

Table 2.17 Density, melt index, and molecular weight influence PEs performances

Physical Properties	If Density Increases	If Melt Index Increases	If MWD Broadens
Melt viscosity	Higher	Lower	——
Vicat softening point	Much higher	Lower	Lower
Surface hardness	Higher	Slightly lower	Lower
Tensile strength			
Yield	Much higher	Slightly lower	Lower
Break	Slightly higher	Lower	Lower
Elongation	Lower	Lower	Higher
Creep resistance	Higher	Slightly lower	Higher
Flexural stiffness	Much higher	Slightly lower	——
Toughness	Lower	Lower	Lower
Low-temperature brittleness	Lower	Lower	Higher
Stress crack resistance	Lower	Lower	Higher
Optical properties			
Transparency	Higher*	Higher	——
Freedom from haze	Higher*	Higher	——
Gloss	Higher*	Higher	——
Barrier properties			
MVT rate	Lower	——	——
Gas and liquid			
Transmission	Much lower	——	——
Greaseproofness	Much higher	Slightly lower	——
Electrical properties	Slightly higher	No effect	——

*Not true in the high-density range.

Table 2.18 Polyethylene properties vs. densities

Properties	0.910-0.925 (low)	0.926-0.940 (medium)	0.941-0.965 (high)
Heat softening point		Higher	Highest
Yield strength		Higher	Highest
Tensile strength at rupture	Highest	Higher	
Elongation at rupture	Highest	Higher	
Stiffness		Higher	Highest
Resistance to shrinkage	Highest	Higher	
Resistance to warpage	Highest	Higher	
Film impact strength (toughness)	Highest	Higher	
Resistance to brittleness at low temperatures	Highest	Higher	
Resistance to environmental stress cracking	Highest	Higher	
Impermeability to gases and liquids		Higher	Highest
Resistance to grease and oil absorption		Higher	Highest
Transparency	Higher	Highest	—[a]
Freedom from haze	Higher	Highest	—[a]
Gloss	Higher	Highest	—[a]
Film draw-down		Higher	Highest
Film hot tear resistance		Higher	Highest
Molding cycle time		Shorter	Shortest

[a] Not applicable to high density polyethylene.

Table 2.19 Differences in properties between polyethylenes of different densities

In contrast with linear PE, branched PE has both short- and long-chain branches. This is one reason why properties of LLDPE, for example, differ somewhat from those of its branched analog LDPE. LLDPE is more linear (more crystalline) and thus processes differently and exhibits different product performances. In processing, as in blown-film or blow-molding products, these differences can be significant enough to require equipment modification.

Three basic characteristics of PE determine its processing and end-use properties: density, melt index, and molecular weight distribution (Table 2.18 and Fig. 2.5). Their range in density, from 0.890 to at least 0.96 g/cm^3, is a result of their crystalline structures. This difference accounts for their property variations. As one example, reducing PE's crystallinity increases its impact resistance, cold flow, tackiness, tear strength, environmental stress-crack resistance, and heat-seal range. However, decreases occur in stiffness, shrinkage, brittleness temperature, and chemical resistance.

The crystalline melting transition (T_m) decreases from a maximum of about 135°C (275°F) to a low of about 110°C (230°F) as the degree of crystallinity drops. The very low glass transition temperature (T_g = –110°C [–166°F]) is associated with good retention of mechanical properties, including flexibility and impact resistance at low temperatures. The first method of production of PE involved very high pressures and led to a highly branched polymer with a correspondingly low crystallinity and low density. The terms *high pressure*, *branched*, and *low-density* PE were, therefore, generally associated with this type of flexible PE (density in the range of 0.910 to 0.926 g/cm^3).

Another method was subsequently developed that required much lower pressures with special catalysts and led to much less branched linear PE with higher crystallinity and density. The terms *low pressure*, *linear*, and *high density* (HDPE) became associated with the more rigid PE (density in the

Physical Properties	Basic Molecular Properties: If the *Density (Crystallinity)* is increased within the range 0.915 to 0.965	Basic Molecular Properties: If the *Melt Index* is increased (or the *Average Molecular Weight* is lowered)
Melt viscosity	—	lower
Vicat softening temperature	much higher	lower
Surface hardness (abrasion resistance)	higher	slightly lower
Resistance to blocking and mold sticking	higher	slightly lower
Minimum heat seal temperature	higher	lower
Heat seal range	narrower	narrower
Tensile strength		
Yield	much higher	slightly lower
Break	slightly lower	lower
Elongation	lower	lower
Resistance to creep	higher	slightly lower
Flexural stiffness	much higher	slightly lower
Flexibility	lower	—
Film impact strength (toughness)	lower	lower
Film tear strength	depends on film-making technique and direction of tear	depends on film-making technique and direction of tear
Resistance to film brittleness	lower	slightly lower
Resistance to brittleness at low temperatures	lower	lower
Resistance to environmental stress cracking	lower	lower
Optical properties		
Transparency (transmittance)	higher*	higher
Freedom from haze	higher*	higher
Gloss	higher*	higher
Draw-down rate	slightly higher	much higher
Barrier properties		
MVT rate	lower	—
Gas and liquid transmission	much lower	—
Grease resistance	much higher	slightly lower
Neck-in and beading	higher	much higher
Adhesion to substrates (especially physical adherence to porous substrates)	slightly lower	slightly higher
Shrinkage	higher	lower
Warpage	slightly higher	lower
Electrical properties	slightly higher	no effect

Table 2.20 Examples of molecular properties' effects on essential PE or end products

range of 0.940 to 0.965 g/cm^3). Now, numerous modifications of the methods of preparation, including copolymerization and alloying, and progress in the elucidation of molecular architecture render this simple classification somewhat insufficient. This is illustrated in particular with LLDPE.

Density remains a good indicator of the general level of crystallinity, which affects many properties, such as stiffness, resistance to permeation by gases, or resistance to oils and greases, which increase with density. PE grades are also classified according to their melt viscosity or melt index, which strongly reflect the molecular weight of the polymer. This is important for processing, in which different techniques often call for different melt viscosities. For example, injection molding is generally associated with an easily flowing grade, while thermoforming requires a high melt consistency or viscosity. Molecular weight does not have such a direct effect on solid-state properties, but it is established that high molecular weight is often beneficial, for example, in obtaining adequate environmental stress-cracking resistance.

The properties of PE can be adjusted in a variety of ways besides homopolymerization. They include copolymerization and polyalloys (chapter 1). Numerous additives have also been developed that can change considerably some properties. For example, stabilizers or antioxidants can reduce high-temperature or ultraviolet (UV)-radiation-induced oxidation. Flammability can be controlled through flame retardants. Surface properties can be dramatically modified by the use of slip agents or cling agents or oxidative surface treatment. In some instances, biodegradability or photodegradability has been obtained with suitable modifications. PE is used in practically

Chemical	% Weight increase (30 days immersion)		Remarks	
	Low density	High density	Low density	High density
Acetic acid (glacial)	0·85	0·75	Some attack	Some attack
Acetone	1·25	0·65	Some loss of strength	Some loss of strength
Alcohol (ethyl)	0·20	0·15	Some loss of strength	Some loss of strength
Ammonium hydroxide	Negligible	Negligible	Unaffected	Unaffected
Carbon tetrachloride	36·0	14·5	Swelling and loss of strength	Swelling and loss of strength
Ethyl acetate	3·1	2·3	Slight attack	Slight attack
Motor oil	3·4	0·7	Slight attack	Satisfactory
Nitric acid (conc.)	0·35	0·3	Some attack (yellowing)	Some attack (yellowing)
Petroleum	13	6	Swelling and loss of strength	Slight swelling and loss of strength
Sulfuric acid (conc.)	Negligible	Negligible	Some attack	Some attack

Table 2.21 Effect of various chemicals on polyethylene (at normal temperature)

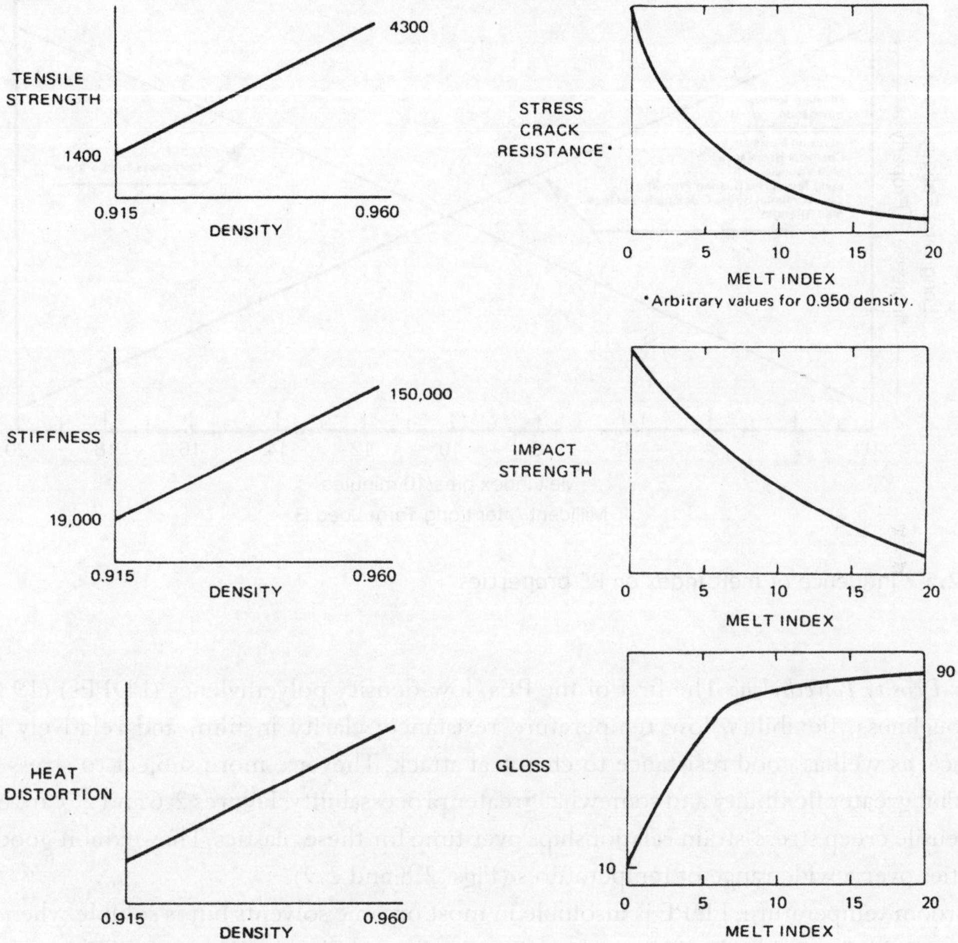

Figure 2.4 Examples of PE properties with variation of density and melt index

all thermoplastic processing techniques. The applications of PE are numerous and involve many properties that are rated through standard tests (chapter 22).

Linear polyethylene (LPE), including ultralow-density PE (ULDPE), linear low-density PE (LLDPE), high-density PE (HDPE), high molecular weight high-density PE (HMWHDPE), and ultra-high molecular weight PE (UHMWPE), is polymerized in reactors maintained at pressures far lower than those for making branched PE. In making branched PE, the crucial plastic parameter of density is varied through changes in reactor pressure and heat. In turn they relate to the closeness and regularity (or crystallinity) of the packing of the long polymer backbones. However, LPE density varies with the quantity of comonomer used with ethylene. The comonomer forms short chain branches along the ethylene backbone; the greater the quantity of comonomer, the lower the density of the plastic.

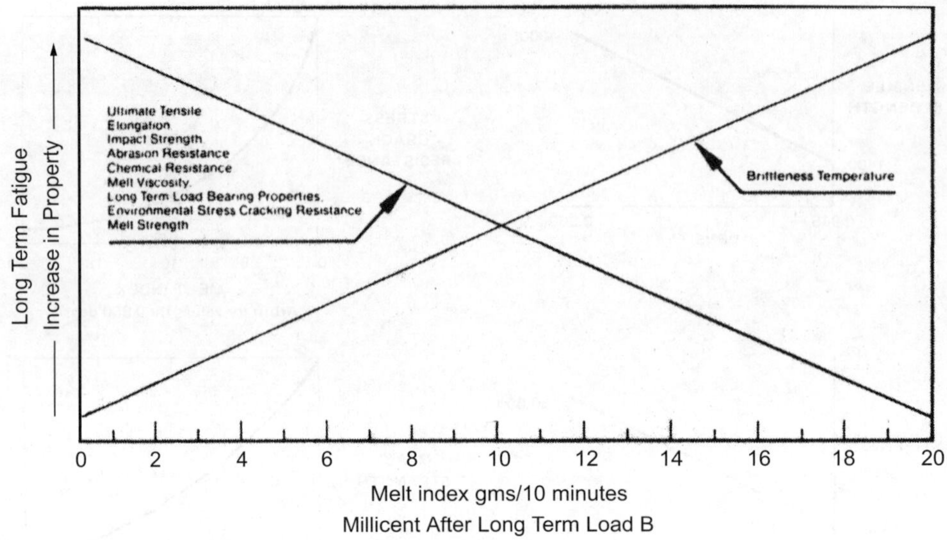

Figure 2.5 Influence of melt index on PE properties

Low-Density Polyethylene The first of the PEs, low-density polyethylenes (LDPEs) (1930s) have good toughness, flexibility, low temperature resistance, clarity in film, and relatively low heat resistance, as well as good resistance to chemical attack. They are more subject to stress cracking but exhibit greater flexibility and somewhat greater processability. Figures 2.6 and 2.7 are examples of the tensile creep stress-strain relationships over time for these plastics. They exhibit good electric properties over a wide range of temperatures (Figs. 2.8 and 2.9).

At room temperature, LDPE is insoluble in most organic solvents but is soluble when attacked by strong oxidizing acids. At high temperatures it becomes increasingly susceptible to attack by aromatic, chlorinated, and aliphatic hydrocarbons. The LDPEs are susceptible to environmental and some chemical stress cracking. For example, wetting agents such as detergents accelerate stress cracking. Some copolymers of LDPE are available with an improved stress-cracking resistance.

The thermal properties of LDPE include a melting range with a peak melting point of 106° to 112°C (223° to 234°F). Its relatively low melting point and broad melting range characterize LDPE as a plastic that permits fast, forgiving heat-seal operations. The glass-transition temperature (T_g) of LDPE is well below room temperature, accounting for the plastic's soft, flexible nature. The combination of crystalline and amorphous phases in LDPE can make determination of T_g difficult. It is reported that the molecular transitions in LDPEs are about –20° and –125°C (–4° and –193°F).

Primarily, molecular weight (MW) and MW distribution (MWD) affect the mechanical properties of LDPE. The average MW is routinely measured by the melt index or gel permeation chromatography (chapter 22). The melt index is a measure of the molten polymer flow rate at a set of conditions specified in ASTM D 1238. The high molecular weight results in a low flow rate and low melt index values, so the MW is inversely proportional to the melt index.

In gel permeation chromatography, a size exclusion technique can be used to measure the range of molecular sizes (lengths) present in a plastic, in order to determine the average MW and MWD. The latter is a graphic representation of the relative amounts of each molecular size present in the

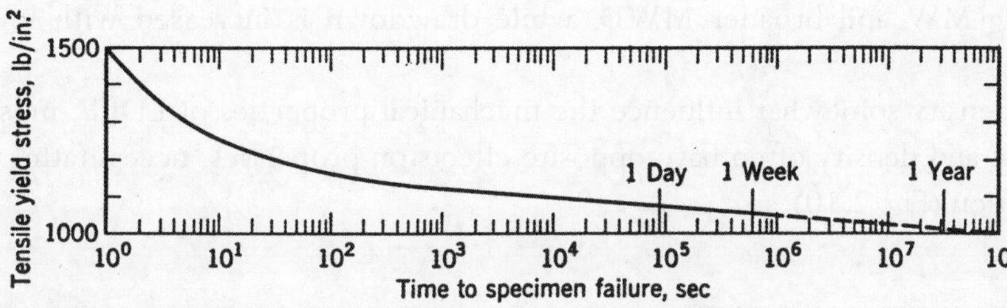

Figure 2.6 LDPE tensile yield stress vs. time to failure

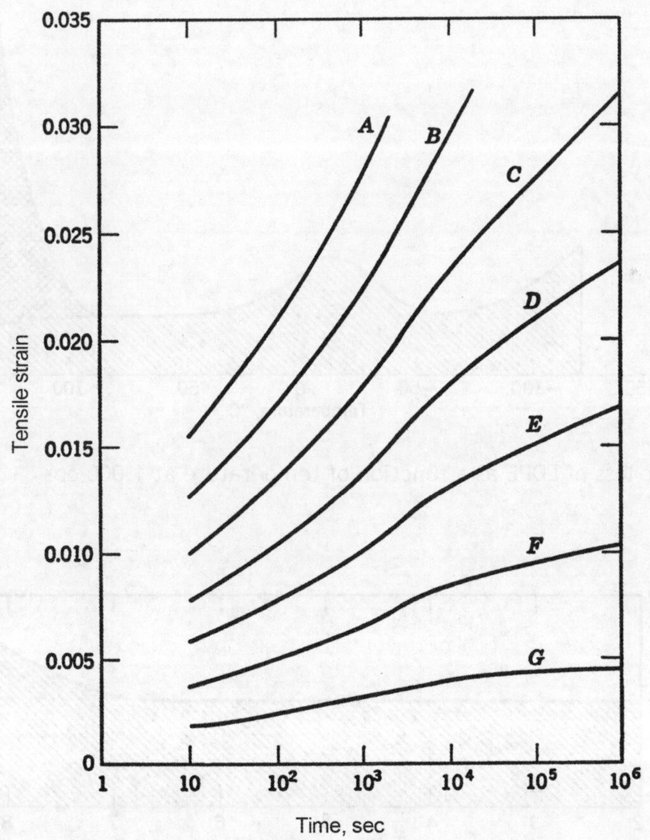

Figure 2.7 LDPE creep in tension at 20°C at various stress levels (density 0.922 g/cc, *A* @ 560 psi, *B* @ 480 psi, *C* @ 400 psi, *D* @ 320 psi, *E* @ 260 psi, *F* @ 180 psi, and *G* @ 100 psi)

plastic. Although LDPEs may have the same average MW, the narrow MWD curve indicates that its distribution of molecular sizes is more uniform than that of the broad MWD curve (chapter 1).

Such molten state properties of LDPE as melt strength and MW and MWD affect drawdown. Melt strength is an indication of how well the molten plastic can support itself, and drawdown is a measure of how thin the molten plastic can be drawn before breaking. Melt strength is increased with increasing MW and broader MWD, while drawdown is increased with lower MW and narrow MWD.

MW and density somewhat influence the mechanical properties of LDPE, mostly by MWD. The melt index and density often have opposite effects on properties, necessitating compromises in plastic selection (Fig. 2.10).

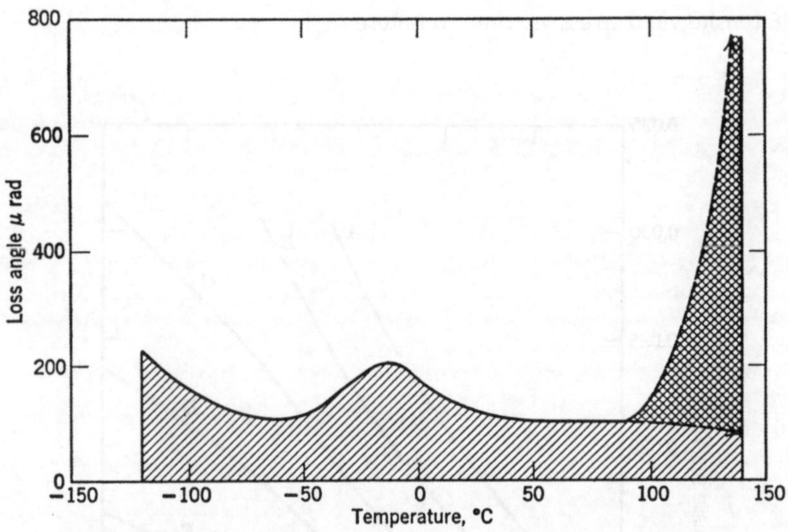

Figure 2.8 Dielectric loss of LDPE as a function of temperature at 1,000 cps

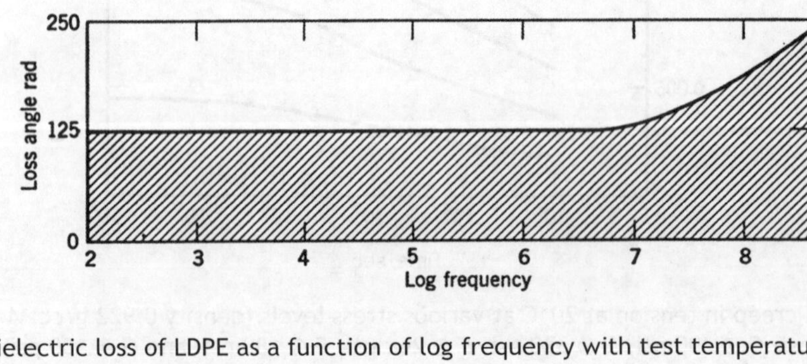

Figure 2.9 Dielectric loss of LDPE as a function of log frequency with test temperature at 20°C

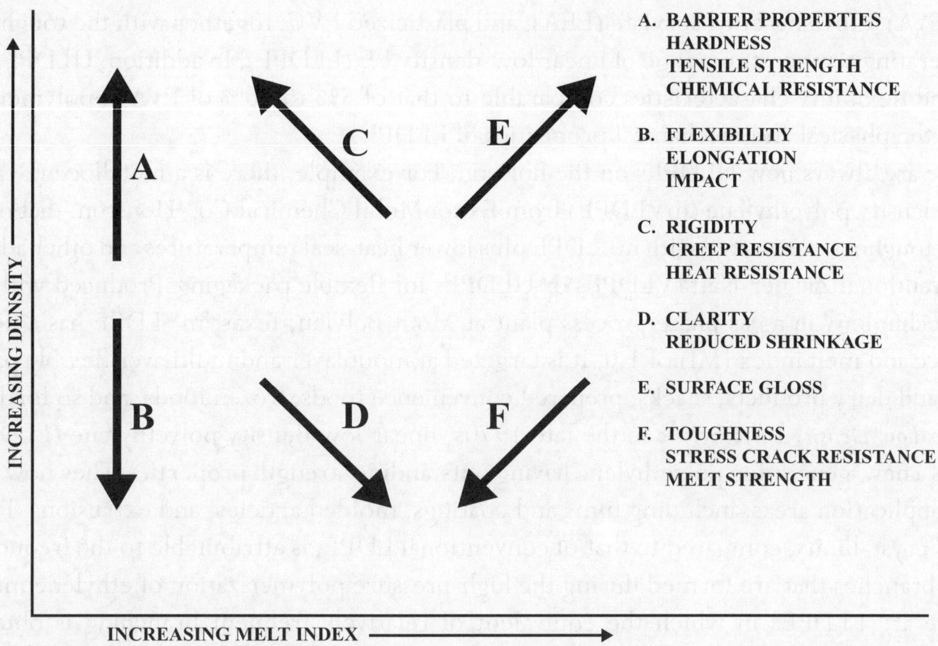

Figure 2.10 Example of how melt index and density influence PE performances; properties increase in the direction of arrows

Other product performance properties are influenced by the MW characteristics of LDPE. MW and density affect the optical properties of LDPE. High MW molecules produce a rough, low-gloss surface; higher-density polyethylenes contain more or larger crystalline areas that scatter light and cause a hazy appearance. Fabrication conditions have a significant effect on optics. Also, the environmental properties of LDPE are subject to thermal and ultraviolet degradation. However, additives are available that can extend outdoor service up to several years.

Different material suppliers prepare different formulations. An example is a fractional-melt LDPE for heavy-duty sacks from Nova Chemicals, Pittsburgh, Pennsylvania. It improves processability of thick-blown films for packaging sand, rock salt, wood chips, mulch, and other heavy bulk materials (204).

LDPE has a good balance of mechanical and optical properties with easy processability and low cost. It can be fabricated by many different methods for a broad range of applications, making it one of the highest-volume plastics in the world. By comparison, other plastics may excel in a specific property but be restricted to specialty applications by cost, processing limitations, or specific property deficiencies. LDPE may not be suitable for applications that require extreme stiffness, good barrier properties, outstanding tensile strength, or high-temperature resistance.

UltraLow-Density Polyethylene Ultralow-density polyethylene (ULDPE) is also called very low density PE (VLDPE) that has densities in the range of 0.85 to 0.92 g/cm^3. ULDPEs offer the flexibility previously available only in generally lower-strength materials such as ethylene-vinyl acetate (EVA), ethylene-ethyl acrylate (EEA), and plasticized PVC, together with the toughness and broad operating temperature range of linear low-density PE (LLDPE). In addition, ULDPE exhibits sealing and flexibility characteristics comparable to that of 5% to 20% of EVA copolymers, while retaining the physical and mechanical properties of LLDPE.

There are always new ULDPEs on the horizon. For example, there is a metallocene-catalyzed, very low density polyethylene (mVLDPE) from ExxonMobil Chemical Co., Houston, that offers the excellent toughness associated with mLLDPE plus lower heat-seal temperatures and other advantages over conventional Ziegler-Natta VLDPEs or ULDPEs for flexible packaging. Produced with Exxpol catalyst technology in a gas-phase process plant at Mont Belvieu, Texas, mVLDPE has a density of 0.912 g/cc and melt index (MI) of 1.0. It is targeted at monolayer and multilayer flexible packaging for meat and dairy products, snacks, prepared convenience foods, frozen foods, and so forth (324).

Linear Low-Density Polyethylene In the late 1970s, linear low-density polyethylene (LLDPE) was offered as a new-generation polyethylene having outstanding strength properties. They now are used in many application areas, including films and coatings, molded articles, and extrusions. Their low degree of crystallinity, compared to that of conventional LDPE, is attributable to the frequent short and long branches that are formed during the high-pressure polymerization of ethylene monomer.

There are LLDPEs in which the equivalent of relatively frequent branching is obtained by copolymerizing ethylene at low pressures and in the presence of catalysts with small amounts of so-called olefin comonomers (butene, hexene, octene), which play the role of uniform short branches along a nearly linear backbone. The plastic's density has a significant effect on the flexibility, permeability, tensile strength, and chemical and heat resistance.

LLDPE is an extremely versatile, low-cost polymer adaptable to many fabrication techniques. When comparing LLDPEs to conventional LDPEs of the same density and melt index in applications, such as films or flexible molded products, they have better impact, tear, toughness, heat-seal strength, or puncture properties and improved environmental stress-cracking resistance (ESCR), and they are chemically inert and resistant to solvents, acids, and alkalies.

Their good dielectric and barrier properties allow them to be used in down-gauging of films. Its major uses are for grocery bags, bread bags, sandwich bags, stretch cling films, industrial trash bags, liners, heavy-duty shopping bags, shrink wrap, and garment bags. LLDPE films perform well in packaging applications because of excellent heat-seal strength and hot-tack properties. They can be pigmented and UV stabilized through conventional means. Formulations are available for specific coefficient of friction and blocking resistance requirements.

LLDPE processing generally requires adjustments when compared to other PEs already in production. LLDPE films are produced using extrusion-blown and cast-film equipment, injection molding, rotational molding, and extrusion coating. The processing characteristics of LLDPE for each process are determined by the rheology of the plastic and by the processing conditions.

HIGH-DENSITY POLYETHYLENE

High-density polyethylene (HDPE) has rigidity and tensile strength considerably higher than LDPE and medium-density PE (MDPE). Figures 2.11 and 2.12 show the mechanical properties of HDPE.

HDPE's impact strength is slightly lower, as is to be expected in a stiffer material, but its overall values are high, especially at low temperatures, compared to the other thermoplastics (TPs). This crystalline TP has a good balance of chemical resistance, low-temperature impact strength, light weight, low cost, and processability.

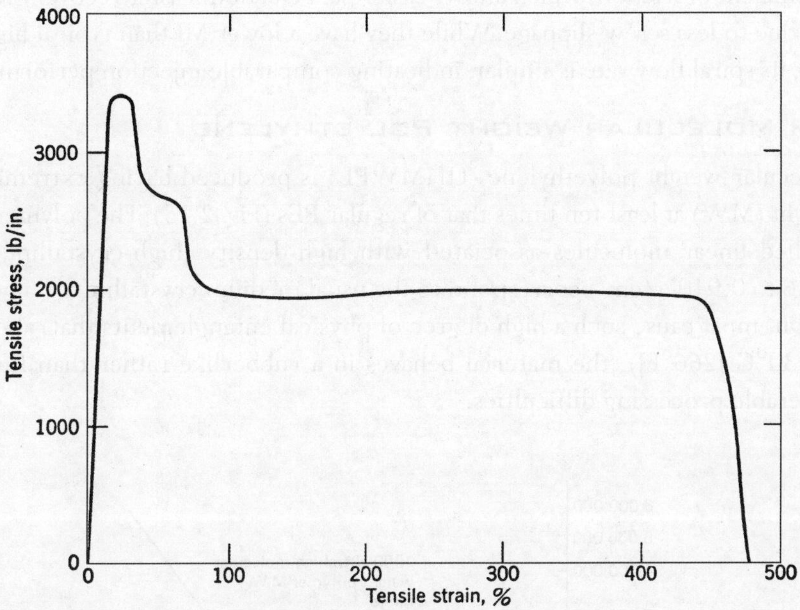

Figure 2.11 Tensile stress-strain for HDPE of density 0.947 g/cc and molecular weight approximately 150,000. ASTM extension rate at 5 in/min

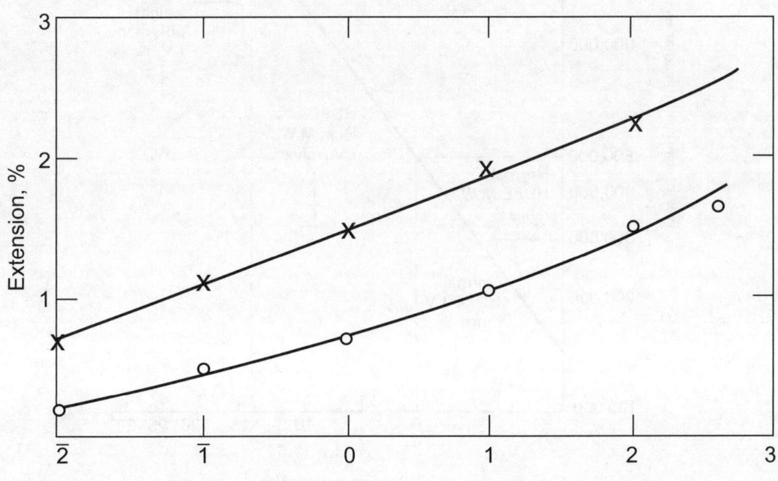

Figure 2.12 Creep curves for HDPE at tensile stress of 600 psi where X is at 60°C and O is at 20°C

An example of a new HDPE on the horizon is a high-flow HDPE that is suited to injection-molded thin-wall products such as food containers, drink cups, and over-caps. Developed by Equistar Chemicals in Houston, these Alathon resins have a 0.956 density, MI of 56, and higher stiffness than most conventional high-flow HDPEs. Its flexural modulus is 1,170 MPa (170,000 psi). The higher-than-usual stiffness and crystallization temperature are said to allow shorter molding cycles. Also, it has a lower coefficient of friction, which allows easier part ejection. Faster recovery rates are reportedly attainable due to less screw slippage. While they have a lower MI than typical high-flow HDPEs of 65 to 80 MI, its spiral flow rate is similar, indicating comparable injection performance (204).

Ultrahigh Molecular Weight Polyethylene

Ultrahigh molecular weight polyethylene (UHMWPE) is produced having extremely (ultra) high molecular weight (MW) at least ten times that of regular PEs (Fig. 2.13). The polymerization process leads to so-called linear molecules associated with high-density (high-crystalline) PE, although densities (0.926 to 0.940 g/cm^3) correspond to the usual medium crystallinity range (MDPE). The molecular weight must cause such a high degree of physical entanglements that, above the melting point (T_m = 130°C [266°F]), the material behaves in a rubberlike rather than fluidlike manner, causing considerable processing difficulties.

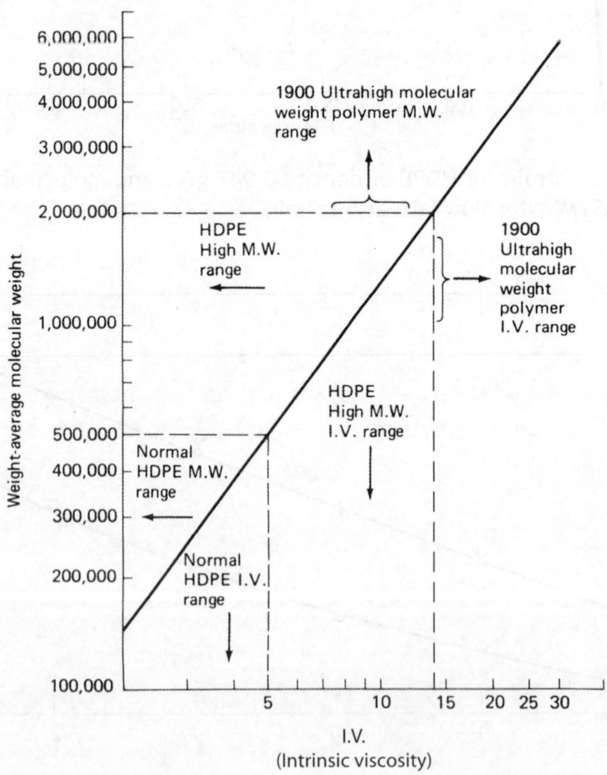

Figure 2.13 UHMWPE compared to other polyethylenes

UHMWPE has certain outstanding properties that qualify it as an engineering or specialty plastic. Its chemical inertness is nearly unmatched and includes environmental stress-cracking (ESC) resistance and resistance to foods and physiological fluids. A very important and outstanding property is wear or abrasion resistance. It is associated with the chemical inertness, a very low coefficient of friction, excellent impact resistance (toughness), and fatigue resistance. These properties and a moderate cost explain the growing use of UHMWPE in large-scale materials-handling equipment (chemical, mining, underwater, etc.), blow-molded drums, as well as in many specialized applications (gears, pulleys, pen tips, prosthetic wear surfaces, etc.) using conventional processing methods.

Because of its high melt viscosity, it does not register on a melt flow index. Conventional screw plasticizing extrusion and injection molding would break the long molecular chains that give this plastic its excellent properties. The processing methods used for UHMWPE are basically compression molding, ram extrusion, ram injection, and warm forming of extruded slugs from powdered plastic. In turn, many components are machined from semifinished products.

Cross-Linked Polyethylene See under "Thermoset Plastic" and "Allyl."

Polyethylene Wax PE with a molecular weight in the range of 2,000 to 4,000 has the properties of high molecular weight hydrocarbon wax (Fig. 2.1). These special LDPEs are generally produced by the high-pressure polymerization of ethylene. They have a specific gravity of 0.91 to 0.96, depending on operating conditions. Melt index is close to 3.5, tensile strength is about 1,500 psi (6.9 MPa), melting point is 99° to 100°C, and needle penetration test at 25°C is 1 to 10. Just over 10 wt% of LDPE produced in the United States find use in typical wax applications, such as paper coatings and floor polishes. A major use is coated paperboard for milk cartons.

Chlorinated Polyethylene Elastomers The moderate random chlorination of polyethylene suppresses crystallinity and yields chlorinated polyethylene elastomer (CPE), a rubberlike material that can be cross-linked with organic peroxides. Also called CM, CPE has a chlorine (Cl) content in the range of 36% to 42%, compared to 56.8% for PVC. Such an elastomer has good heat and oil resistance. It is also used as a plasticizer for PVC.

CPEs provide a very wide range of properties ranging from soft/elastomeric to hard. They have inherent oxygen and ozone resistance, improved resistance (compared to PEs) to chemical extraction, ability to resist plasticizers, volatility, and weathering capabilities. Products do not fog at high temperatures as do PVCs and can be made flame-retardant.

Polymethylpentene The major advantages of polymethylpentene (PMP) over other polyolefins are its transparency in thick sections, its short-time heat resistance up to 200°C (400°F), and its lower specific gravity. Unlike the other polyolefins, however, it is transparent, because its crystalline and amorphous phases have the same index of refraction. Almost clear optically, PMP has a light transmission value of 90%, which is just slightly less than that of the acrylics. It retains most of its physical properties under brief exposure to heat at 200°C (400°F), but it is not stable at temperatures for an extended time over 150°C (300°F) without an antioxidant. In a clear form, it is not recommended where it will have to undergo long-term exposure to UV environments.

The chemical resistance and electrical properties of PMP are similar to those of the other polyolefins, except that it retains these properties at higher temperatures than do either PE or PP. In this respect, PMP tends to compare well with polytetrafluoroethylene (PTFE) up to 150°C (300°F). Molded parts made of this resin are hard and shiny, yet their impact strength is high at temperatures down to –29°C (–20°F). Their specific gravity of 0.83 is the lowest of many commercial solid plastics.

Polyolefin Elastomer Polyolefin elastomers (POE) and polyolefin plastomers (POP) are ethylene alpha olefin copolymers produced using constrained geometry and metallocene catalyst. They differ from traditional polyolefins in that they have a narrow molecular weight distribution and a regular placement of the octene comonomer on the ethylene backbone. This highly uniform distribution allows for some unique resin characteristics.

Polyolefin Thermoplastic Elastomer Polyolefin thermoplastic elastomers (TPEs) essentially are blends of various amorphous rubbers, such as ethylene-propylene, and polyolefin semicrystalline plastics, such as polypropylene. They are part of the family of TPOs that were first produced commercially in 1974. TPOs are mechanical blends consisting of a hard plastic and softer rubber. They are considered different from blends that are dynamically thermoplastic vulcanized (TPV), a process in which the elastomer phase is cured during mixing of the polymers (232).

POLYPROPYLENE

Polypropylene (PP), a member of the polyolefin family, is a major plastic used worldwide providing different performances (Tables 2.22 to 2.29 and Fig. 2.14 and 2.15). PP is produced via polymerization with organometallic stereospecific catalysts. This resin has propylene units that are aligned in a head-to-tail fashion, with methyl groups attached to one side of the polymer backbone. Polypropylene is a thermoplastic polymer with low specific gravity and good resistance to chemicals and fatigue (221).

PP made with metallocene catalysts (mPP) has increased characteristics such as toughness, stiffness, heat resistance, clarity, barrier properties, high melt flow, and high melt strength (135, 326). Their densities are slightly lower than that of polyethylenes, but they are much stiffer and more heat resistant and have the same chemical and electrical resistance. They are semitranslucent and milky white in color, with excellent colorability. A stereo-selective catalyst that produces mPPs orders their molecular configuration so that the basic plastic has a predominantly regular, uniform structure. This action results in the molecules' crystallizing into compact bundles, which makes them stronger than other members of the polyolefin family. Hundreds of formulations are produced.

These high-volume plastics have superior resistance to flexural-fatigue stress cracking, with excellent electrical and chemical properties. This versatile polyolefin overcomes poor low-temperature performance and other shortcomings through copolymer, filler, and fiber additions. It is widely used in packaging (film and rigid), automobile interiors, under-the-hood and underbody applications, dishwashers, pumps, agitators, tubs, filters for laundry appliances and sterilizable medical components, and more.

PPs are an extremely versatile plastic available in many grades as well as copolymers such as ethylene propylene. NEAT (Nothing Else Added To) PP has a low density of 0.90, which—combined with its good balance of moderate cost, strength, and stiffness as well as excellent fatigue, chemical resistance, and thermal and electrical properties—makes PP extremely attractive for many indoor and outdoor applications.

The strength, rigidity, heat-resistance, and dimensional-stability properties of PP can be increased significantly with glass fiber reinforcements. Increased toughness is provided in special

Material family					polypropylene homopolymer	polypropylene homopolymer, flame retardant	polypropylene homopolymer, fibers	polypropylene homopolymer, cast film	polypropylene, 10% talc	polypropylene, 20% talc	polypropylene, 20% talc, easy-flow
Supplier					BASF	Vestolen GMBH	Vestolen GMBH	BASF	Hoechst	Hoechst	Hoechst
Material Trade Name					Novolen 1100L	Vestolen P 7006 S	Vestolen P 4000	Novolen 1127 N	Hostacom M1 001	Hostacom M2 N01	Hostacom M2 U01
Source ID					756, CAMPUS	711, CAMPUS	711, CAMPUS	756, CAMPUS	326, CAMPUS	326, CAMPUS	326, CAMPUS
Test Notes	Test condition	Test Specimen	Test Method	(Unit)							
PROCESSING PROPERTIES											
melt volume rate (230°C, 2.16 kg)			ISO 1133, DIN 53735, CAMPUS	ml/10min	8	3	23	15	20	2.3	18
melt volume rate (190°C, 5 kg)			ISO 1133, CAMPUS	"	12			30	38	4.2	34
melt volume rate (230°C, 5 kg)		test specimen note: granules	DIN 53479	"					85	10	78
melt flow rate (230°C, 2.16 kg)			ISO 1133, DIN 53735	g/10min	5	2.4	17.5	11	16	2	16
melt flow rate (190°C, 5 kg)			"	"	9	4.5	33	22	28	3.8	28
melt flow rate (230°C, 5 kg)			"	"		11	80		67	9	67
molding shrinkage (parallel)		parallel to flow	ISO 2577, CAMPUS	%	1.3						
molding shrinkage (normal)		perpendicular to flow	"	"							
processing shrinkage				"							
molecular weight distribution			GPC								
MECHANICAL PROPERTIES											
tensile modulus (secant, 1 mm/min)	test temperature: 21-25°C; relative humidity: 50%; strain rate: 1 mm/min; elongation: 0.05-0.25%; atmosphere according to ISO 291	ISO 3167 multi-purpose test specimen	ISO 527-1, ISO 527-2, CAMPUS, DIN 53457	MPa	1500	1500	1650		2300	2500	2600
stress at yield (50 mm/min)	test temperature: 21-25°C; relative humidity: 50%; strain rate: 50 mm/min; atmosphere according to ISO 291	"	ISO 527-1, ISO 527-2, CAMPUS, DIN 53455	"	35	34	35		35	32	32
strain at yield (50 mm/min)	"	"	"	%	10	8	8		6	6	5
strain at break (50 mm/min)	"	"	"	"	>50	>50	>50		8	20	3
tensile strength at break (5 mm/min)	test temperature: 21-25°C; relative humidity: 50%; strain rate: 5 mm/min; atmosphere according to ISO 291	"	"	MPa							
strain at break (5mm/min)	"	"	"	%							
shear modulus	test temperature: 21-25°C; relative humidity: 50%		ISO 537, DIN 53445	MPa	750	800	850				
tensile creep modulus (1 hour)	test temperature: 21-25°C; relative humidity: 50%; elongation: <=0.5%; atmosphere according to ISO 291	ISO 3167 multi-purpose test specimen	ISO 899-1, CAMPUS	"	700				1600	1650	
tensile creep modulus (1000 hour)	"	"	"	"	380				800	850	
Charpy impact strength (23°C)	test temperature: 23°C; relative humidity: 50%; atmosphere according to ISO 291	80 x 10 x 4 mm	ISO 179/1eU, CAMPUS	kJ/m2	140	no failure	80		50	50	45
Charpy impact strength (-30°C)	test temperature: -30°C	"	"	"	16	20	15				
Charpy notched impact strength (23°C)	test temperature: 23°C; relative humidity: 50%; atmosphere according to ISO 291	80 x 10 x 4 mm, V notch, r= 0.25 mm	ISO 179/1eA, CAMPUS	"	3	4.5	3		2.5	4	3.2
Charpy notched impact strength (-30°C)	test temperature: -30°C	"	"	"	1.5	2.2	1.7				
Izod impact strength (23°C)	test temperature: 23°C; relative humidity: 50%	80 x 10 x 4 mm	ISO 180/1C	"	65	90	60				
Izod impact strength (-30°C)	test temperature: -30°C	"	"	kJ/m2	13	15	15				
notched Izod impact strength (23°C)	test temperature: 23°C; relative humidity: 50%	80 x 10 x 4 mm	ISO 180/1A	kJ/m2	3.5	4	2.2				
notched Izod impact strength (-30°C)	test temperature: -30°C	"	"	"	1.3	2	1.6				
notched tensile impact strength	test temperature: 23°C; relative humidity: 50%; atmosphere according to ISO 291	80 x 10 x 4 mm, 45° double V notch, r= 1.0 mm recommended	ISO 8256, CAMPUS	"					35	40	35
impact strength	test temperature: 23°C; relative humidity: 50%	small size standard test bar	DIN 53453	"					33	38	25
notched impact strength	"	"	"	"					2.5	5	3
Ball indentation hardness	load: 358 N; note: H 358/30	>=10 x >=10 x 4 mm	ISO 2039T1, DIN 2039T1	MPa	70	75	78		95	81	90

Table 2.22 Polypropylene data

high molecular weight, rubber-modified grades. The electrical properties of PP moldings are affected to varying degrees by their service temperatures. Its dielectric constant is essentially unchanged, but its dielectric strength increases and its volume resistivity decreases as temperature increases.

Basic PPs have limited heat resistance, but heat-stabilized grades are available for applications requiring prolonged use at elevated temperatures. The useful life for products molded from such grades may be at least as long as 5 years at 120°C (250°F), 10 years at 130°C (230°F), and 20 years at 99°C (210°F). Specially stabilized grades are UL rated at 120°C (248°F) for continuous service. Basically, PP is classified as a slow-burning material, but it can also be supplied in flame-retardant grades.

PPs are unstable in the presence of oxidation conditions and UV radiation. Although all its grades are stabilized to some extent, specific stabilization systems are often used to suit a formulation to a particular environment, such as where it must undergo outdoor weathering.

Polymer	Tensile strength (MPa)	Flexural modulus (Gpa)	Elongation at break (%)	Strain at yield (%)	Notched Izod impact strength (kJ/m)
SAN	72	3.6	2.4	3.5	0.02
Polystyrene, general purpose	34	3.4	1.6	1.4	0.02
PET	75	3	70	4	0.02
PVC-U	51	3	60	3.5	0.08
PMMA	70	2.9	2.5		0.02
Polycarbonate	65	2.8	110	7.5	0.7
ABS	47	2.7	8	1.9	0.2
PES	84	2.6	60	6.6	0.084
Acetal	73	2.6	65	8	0.069
PPO	65	2.5	60	4.5	0.16
PBT	52	2.1	250	4	0.06
Polystyrene, high impact	42	2.1	2.5	1.8	0.1
Polyamide 6/6	59	1.6	60	4.5	0.11
Polypropylene homopolymer	33	1.5	150	10	0.07
Polyamide 12	50	1.4	200	6	0.06
HD polyethylene	32	1.3	150	15	0.15
Polyamide 6	40	1.2	60	4.5	0.25
Polypropylene copolymer	25	1.2	300	12	0.1
Polyamide 6/10	58	1.1	300	15	0.09
Polyamide 11	52	0.9	320	20	0.05
LD polyethylene	10	0.3	400	19	no break

Table 2.23 Mechanical properties of PP compared with other thermoplastics

PPs resist chemical attack and staining and are unaffected by aqueous solutions of inorganic salts or mineral acids and bases, even at high temperatures. Most organic chemicals do not attack them, and there is no solvent for this plastic at room temperature. Halogens, fuming nitric acid, and other active oxidizing agents attack the plastics. PPs are also attacked by aromatic and chlorinated hydrocarbons at high temperatures.

PP films and other plastics as well as products when heated can be stretched to increase performances (chapter 5). Table 2.30 provides tensile data based on different rates of stretching. Table 2.31 compares the tensile impact of unoriented and oriented PP with steel.

Although PPs retain their strength and stiffness at elevated temperatures, their performance at low temperatures leaves much to be desired. However, copolymers of PP that are available offer as much as two to three times the impact strength of general-purpose PP, even at temperatures as low as –29°C (–20°F).

Because of its excellent fatigue resistance, PP is widely used in the living hinges that are integral to all types of applications (289).

POLYPROPYLENE BLENDS

There are three large groups available within the class of propylene copolymers: homopolymers, random copolymers, and block copolymers. The macromolecular chains of homopolymers are built from propylene units exclusively. The random copolymer chain contains propylene and comonomer units in a statistical distribution. Essentially, the comonomers are ethylene and butene. In contrast

Polypropylene type	Tensile strenfth (MPa)	Fluxural modulus (Gpa)	Elongation at break (%)	Strain at yield (%)	Notched Izod impact strength (kJ/m)
Homopolymer	33	1.5	150	10	0.07
Homopolymer, UV stabilized	32	1.5	150	8	0.04
Copolymer	25	1.2	300	12	0.1
Copolymer, UV stabilized	24	1.2	300	12	0.08
20% talc filled	32	2.3	15	7	0.03
40% talc filled	30	3.2	8	4	0.03
20% calcium carbonate filled	26	2.0	80		0.05
40% calcium carbonate filled	24	2.8	60		0.04
20% glass fiber reinforced	35	3.8	2		0.08
30% glass fiber coupled	90	6.0	3		0.1
Fire retardant	25	1.8	15	6	0.05
Elastomer modified	23	0.75	350	17	0.5
Elastomer modified, UV stabilized	23	0.75	300	17	0.5
Structural foam	16	0.8	40		0.05

Table 2.24 Mechanical properties of PP with various fillers, reinforcements, and modifiers

to the single-phase homopolymers and random copolymers, the block copolymers consist of a multiple phase system of the PP homopolymer and ethylene propylene rubber (EPM).

For a large variety of special products, these PP reactor blends, as well as others, can be derived from the above three basic groups. PP reactor blends belong to the group of block copolymers that exhibit a high impact resistance at low temperatures. These can be subdivided, depending on the proportion of elastomer, into grades with moderate, high, and superior impact resistance and thermoplastic elastomers (TPEs).

Exceptionally high demands for toughness are set by many applications, particularly in the automotive industry. Materials containing elastomeric components in excess of 25 wt% can meet these. However, such a high proportion of elastomers in blends with PP is barely possible when using most of the existing units. In order to make products with a higher proportion of elastomers, the energy-intensive method of extruder compounding is used. Products prepared by this process are referred to as extruder blends.

Polymer	Maximum operating temperature (°C)	Heat distortion temperature at 0.45 MPa (C°)	Heat distortion temperature at 1.80 MPa (C°)	Linear expansion (mm/°C × 10^{-5})
PES	180	260+	203	5.5
PBT	120	150	60	12
Polycarbonate	115	143	137	7
PET	115	115	80	8
Polypropylene homopolymer	100	105	65	10
Acetal	90	160	110	11
Polypropylene copolymer	90	100	60	10
Polyamide 6	80	200	80	10
Polyamide 6/6	80	200	100	8
PPO	80	137	129	6
Polyamide 6/10	70	157	66	14
Polyamide 11	70	150	55	9
Polyamide 12	70	150	55	11
ABS	70	98	89	8
SAN	55	96	84	7
HD polyethylene	55	75	46	12
PMMA	50	103	95	7
Polystyrene, general purpose	50	90	80	7
Polystyrene, high impact	50	85	75	7
PVC-U	50	70	67	6
LD polyethylene	50	50	35	20

Table 2.25 Thermal properties of PP compared with other thermoplastics

Polypropylene type	Maximum operating temperature (°C)	Heat distortion temperature at 0.45 MPa (°C)	Heat distortion temperature at 1.80 MPa (°C)	Linear expansion (m/m/°C × 10^{-5})
Homopolymer	100	105	65	10
Homopolymer, UV stabilized	100	110	57	10
Copolymer	90	100	60	10
Copolymer, UV stabilized	90	100	60	10
20% talc filled	100	130	78	7
40% talc filled	100	130	80	5
20% calcium carbonate filled	100	105	68	6
40% calcium carbonate filled	100	110	71	4
20% glass fiber reinforced	100	122	93	4
30% glass fiber coupled	100	160	150	4
Fire retardant	100	110	70	9
Elastomer modified	90	82	55	11
Elastomer modified, UV stabilized	90	82	55	11
Structural foam	95	72	46	16

Table 2.26 Thermal properties of polypropylenes with various fillers, reinforcements, and modifiers

Property	With Increasing Molecular Weight
Impact Resistance	Increases
Elongation	Increases
Moduli	Decrease
Strength	Decreases
Die Swell	Increases
Shear Rheology	Increases
Melt Strength	Increases
Heat Seal Strength	Increases
Heat Distortion Temperature	Decreases
Irradiation Tolerance	Decreases
Haze	Decreases
Extractables (solubility)	Decreases
Crystallization Temperature	Decreases

Table 2.27 Effect of increasing molecular weight on properties of polypropylene

Property	Application
Low thermal conductivity	Insulating material
Rot resistance	Civil engineering fabric
Hydrophobicity	Diaper or sanitary napkin cover stock
Chemical resistance	Blood filters
Stain resistance	Upholstery
Resistance to insects	Agricultural bags
Low static build-up	Carpet face fiber

Table 2.28 Useful properties of polypropylene in fiber applications

Property	Metallocene polypropylene	Conventional polypropylene
Molecular weight distribution	2.0	3.5–6.0
Molecular weight capability	0.01–5000	0.01–1000
Extractables	0.7%	3.3%
Melting point	147–158°C	160–165°C

Table 2.29 Comparison of conventional and metallocene PPs

PP reactor blends are formed by gas-phase polymerization. Although products with intermediate and high impact resistance at low temperatures can be called PP reactor blends, only block copolymers with high elastomer content are referred to as PP reactor blends for historical reasons.

POLYBUTYLENE

PBs are crystalline thermoplastics in the polyolefin family. They are similar to polypropylenes and high-density polyethylene but exhibit a more crystalline structure. This crystallinity produces unusual high strength and extreme resistance to deformation over a temperature range of –10° to 190°F. Table 2.32 and Figures 2.16 and 2.17 illustrate the tensile stress-life and stress-strain relationships and physical properties of the materials. Figure 2.16 shows that the stress does not plateau after reaching its yield point (chapter 19), making possible films that look like polyethylene but act more like polyester (TP) films.

Compared to other polyolefins, PB films have superior resistance to creep and stress cracking. They have high tear resistance, toughness, and flexibility. Their chemical and electrical properties are similar to those of the PEs and PPs, but their degree of crystallinity is much lower. This structure results in a rubberlike, elastomeric material with low molded-in stress.

The main applications of PBs are in pipe, packaging, hot-melt adhesives, and sealants. PB piping for cold-water use has a higher burst strength than pipe made from any other polyolefin. Large-diameter pipe has been successfully used in mining and power generation systems to convey abrasive materials. PBs can be alloyed with other polyolefins to provide its inherent advantage. Film made into industrial trash bags gives improved resistance to bursting, puncturing, and tearing.

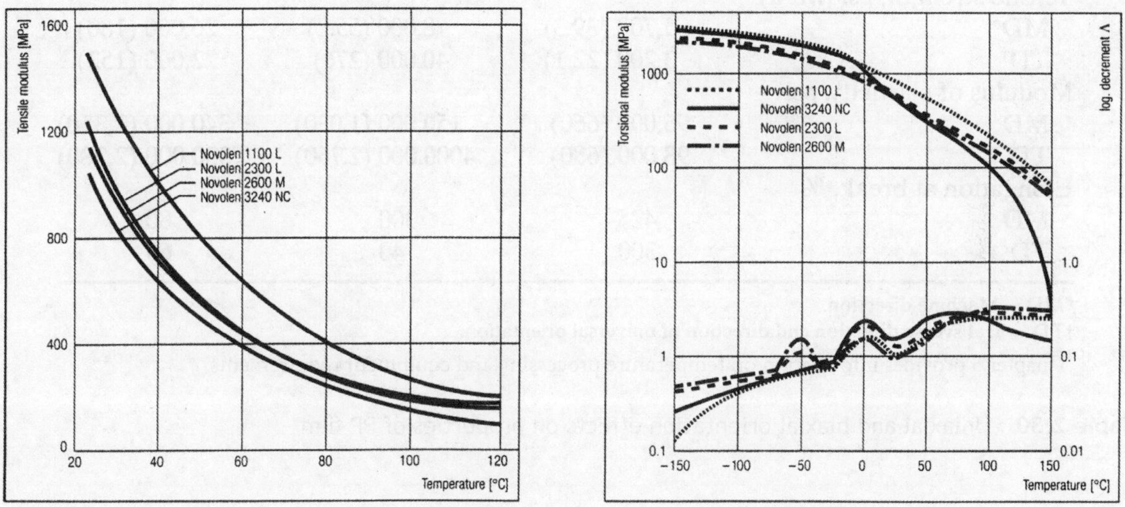

Figure 2.14 Temperature dependence of tensile modulus (left) and torsional shear modulus for BASF PPs

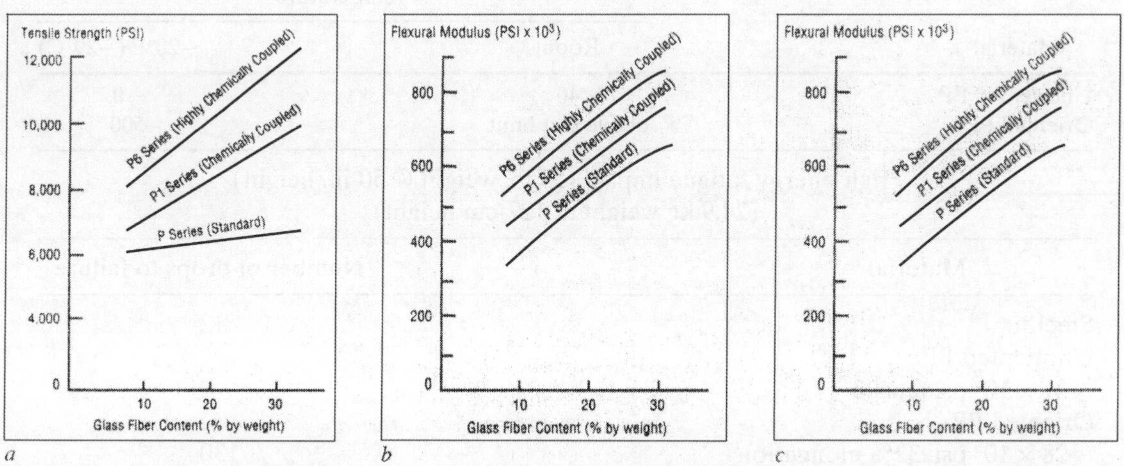

Figure 2.15 Effect of adhesive coupling agents (plastic to glass fiber; chapter 15) on tensile strength, flexural modulus, and heat deflection temperature of glass-fiber-reinforced polypropylene

Properties	Stretch (%) None	200	400	600	900
Tensile strength, psi	5,600	8,400	14,000	22,000	23,000
(MPa)	(38.6)	(58.0)	(96.6)	(152.0)	(159.0)
Elongation at break, %	500	250	115	40	40

Properties	As Cast	Uniaxial Orientation	Biaxial Orientation
Tensile strength, psi (MPa)			
MD*	5,700 (39.3)	8,000 (55.2)	26,000 (180)
TD†	3,200 (22.1)	40,000 (276)	22,000 (152)
Modulus of elasticity. psi			
MD	96,000 (660)	150,000 (1,030)	340,000 (2,350)
TD	98,000 (680)	4000,000 (2,760)	330,000 (2,280)
Elongation at break, %			
MD	425	300	80
TD	300	40	65

*MD = Machine direction.
†TD = Transverse direction and direction of universal orientation.
[1]Chapter 5 provides information on temperature processing and equipment requirements.

Table 2.30 Uniaxial and biaxial orientation effects on properties of PP film

ASTM tensile impact, ft.-lb./in.^{2}		
Material	Temperature: Room	−20°F (−29°C)
Unoriented PP	40	0
Oriented PP	above test limit	500

High-energy fatigue impact (55 lb. weight @ 50-in. height) (24.9 kg weight at 127-cm height)

Material	Number of drops to failure
Steel	12
Unoriented PP	
41 × 10^3 psi tensile	1
Oriented PP	
28 × 10^3 psi, 32% elongation	130

Table 2.31 Tensile impact comparison of oriented PP with steel

VINYL

The vinyls constitute one of the most versatile families of plastics. This family of plastics has basic molecules with two carbon atoms and three pendant hydrogens. They have a fourth pendant group that is a unique group of atoms depending on the particular vinyl. It is this fourth group of atoms that provides the special qualities of the particular vinyl plastic (108, 110). The basic molecule repeats in a chain to form the plastic. In the plastics literature, each member of the vinyl family is properly identified by name in their respective classifications. The term *vinyl* usually identifies the major, very large production of polyvinyl chloride (PVC) plastics.

Tensile strength, psi	3800
Yield strength, psi	2200
Ultimate elongation, %	350
Hardness, Shore D	60
Flexural modulus, psi	26×10^3
Density, g/cc	0.912
Melt index, g/10 min	0.5
Impact resistance, izod	no break
Carbon black, %	0
Melting point, °C	135
Vicat softening point, °F	235
Linear coeff. of thermal expansion, in./in.-°C	15×10^{-5}
Low temp. brittleness, °C	−25
Dielectric constant, 10^3 cps	2.25
Dissipation factor, 10^6 cps	0.005

Table 2.32 Properties of polybutylene

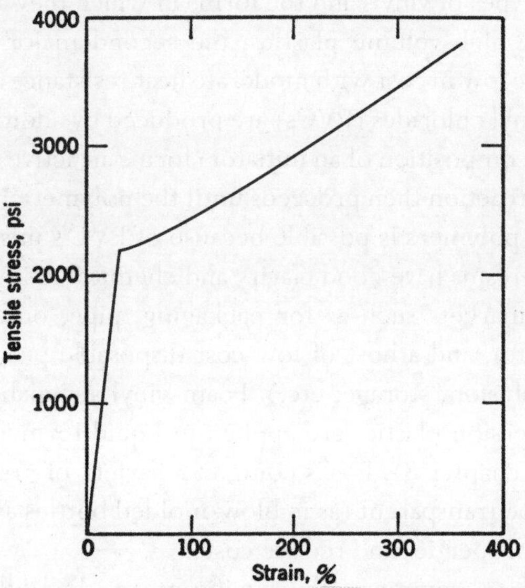

Figure 2.16 Tensile stress-strain curve for polybutylene with strain rate at 20 in/min

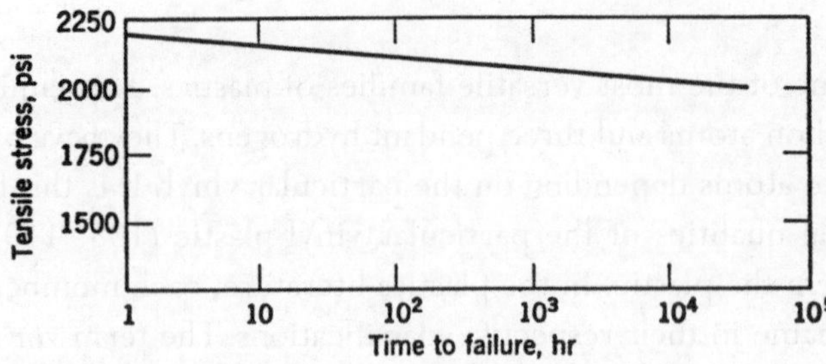

Figure 2.17 Tensile stress-life curve (cold flow) at 73°F for polybutylene

Included in the vinyl family, in addition to polyvinyl chlorides, are polyvinyl acetals, polyvinyl acetates, polyvinyl alcohols, polyvinyl carbazoles, polyvinyl chloride-acetates, and polyvinylidene chlorides.

As a family, they are strong and abrasion resistant. They are unaffected, for the most part, by prolonged exposure to water, common chemicals, and petroleum products. (However, they should be kept away from chlorinated solvents, such as many household cleaning fluids.) Vinyls can withstand continuous exposure to heat up to 130°F and perform satisfactorily at food freezing temperatures. They are slow burning and some types are self-extinguishing, but they should be kept away from direct heat. The vinyls may be given a wide range of colors and may be printed or embossed. They generally have excellent electrical properties, but because of their relatively poor weathering qualities, they are recommended for indoor use only. Vinyls literally can be processed by more techniques than any other plastic, because they contain a relatively polar polymer that allows a large range of formations. A discussion of the various types of vinyls and the forms in which they are available follows.

Polyvinyl Chloride These high-volume plastics, the second major plastic produced based on worldwide consumption, are low in cost with moderate heat resistance and good chemical, weather, and flame resistance. Polyvinyl chlorides (PVCs) are produced by addition polymerization in which a free radical (formed by decomposition of an initiator) forms an active center with the surrounding vinyl chloride monomer. A reaction then proceeds until the polymer chain is terminated. The manufacture of a wide range of polymers is possible because of PVC's miscibility with an amount and variety of plasticizers. PVC resins have good clarity and chemical resistance (Figs. 2.18 and 2.19).

They qualify in many markets such as for packaging, pipe, outdoor construction products (siding, window profiles, etc), and a host of low-cost disposable products (including FDA-grade medical uses in blood transfusion, storage, etc.). Foam-vinyl strippables are used for metal parts packaging. These PVC dispersion plastics are applied in liquid form. Foaming takes place during their cure cycle (Fig. 2.20; chapter 8). PVCs come in a variety of grades, flexible to rigid (Table 2.33). They are tough, can be transparent (as in blow-molded bottles and jugs), and are also a good alloying plastic to improve properties and reduce costs.

The PVCs are the most commercially significant of the different vinyl polymers and copolymers. For example, PVC can be chlorinated (CPVC) and alloyed with other polymers such

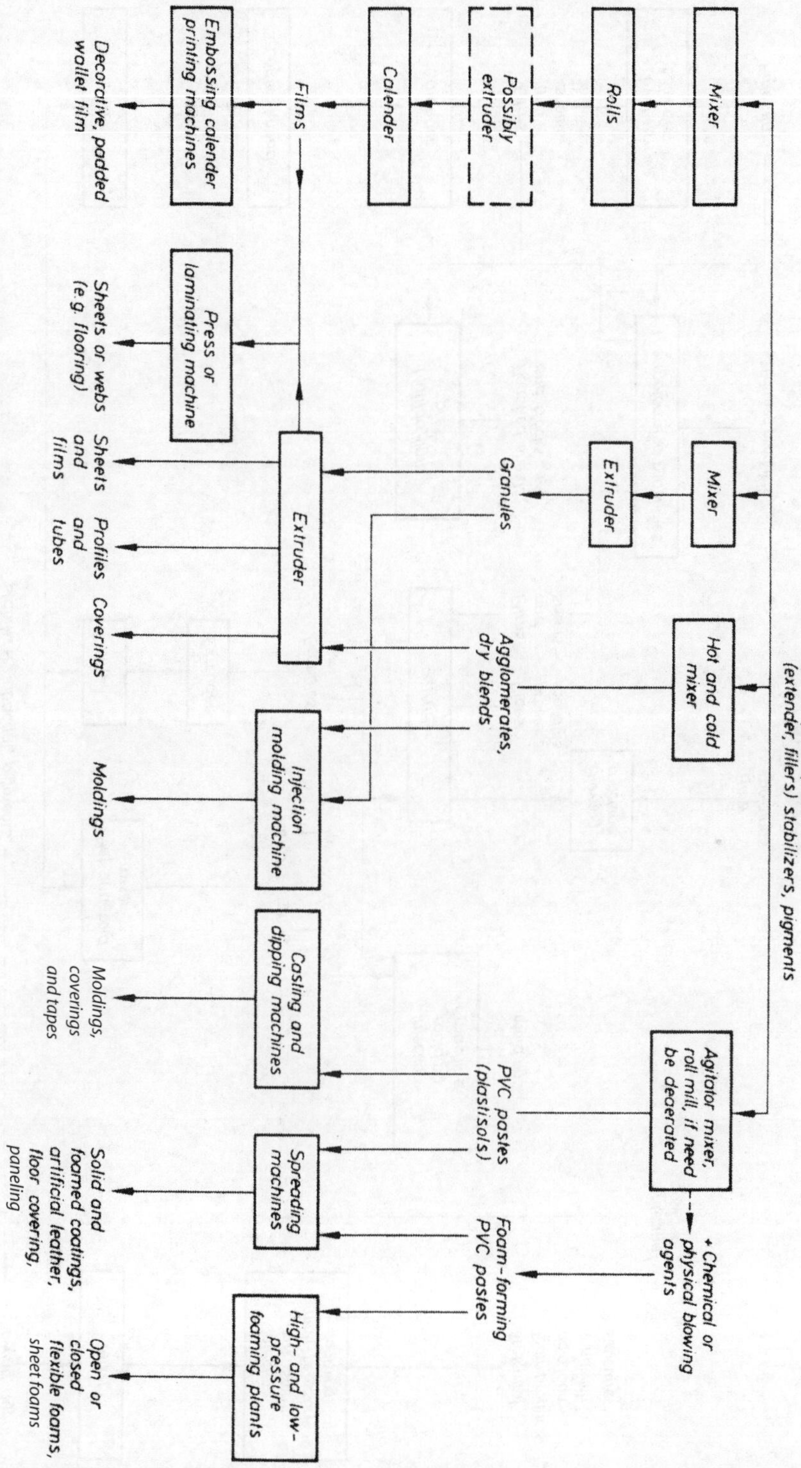

Figure 2.18 Flow chart for plasticized polyvinyl chloride

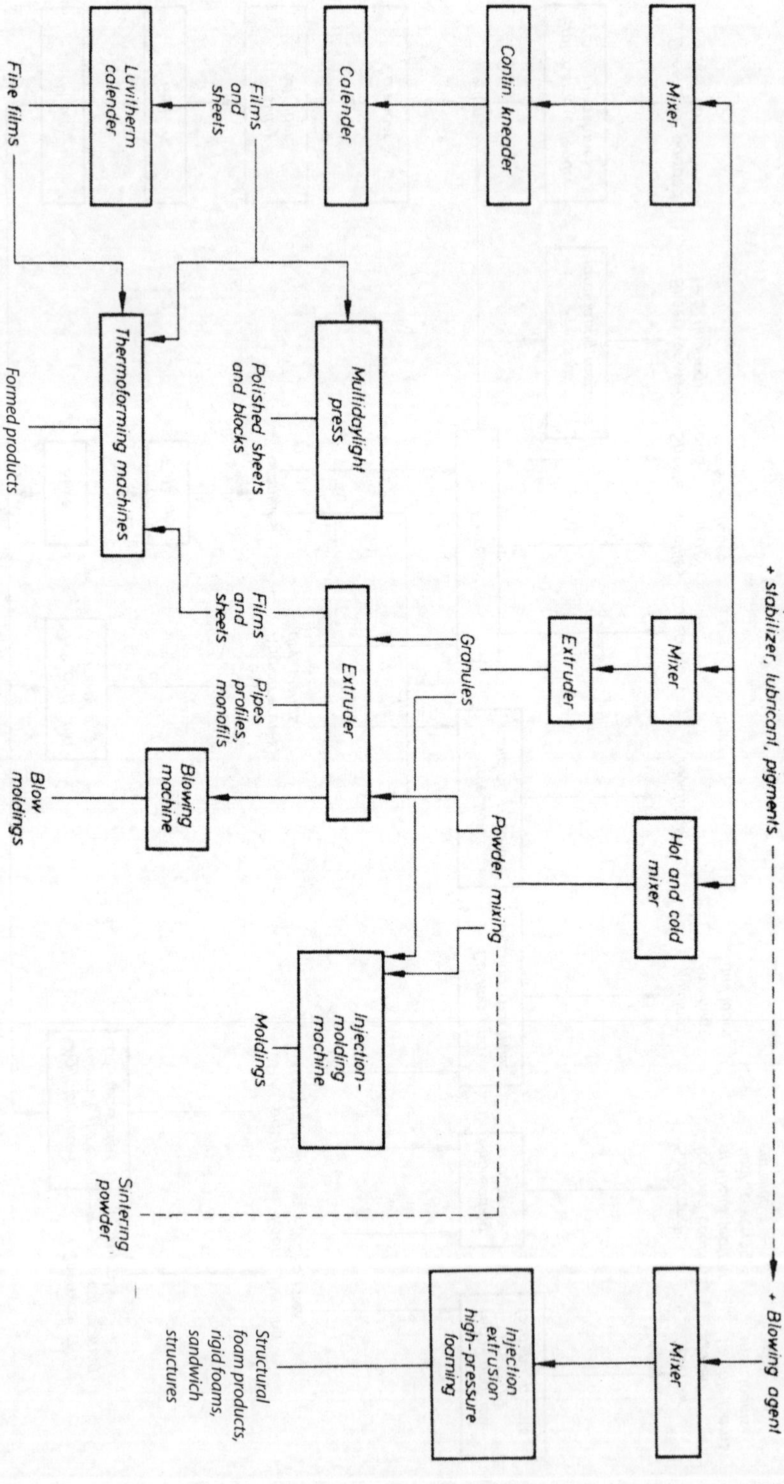

Figure 2.19 Flow chart for rigid polyvinyl chloride

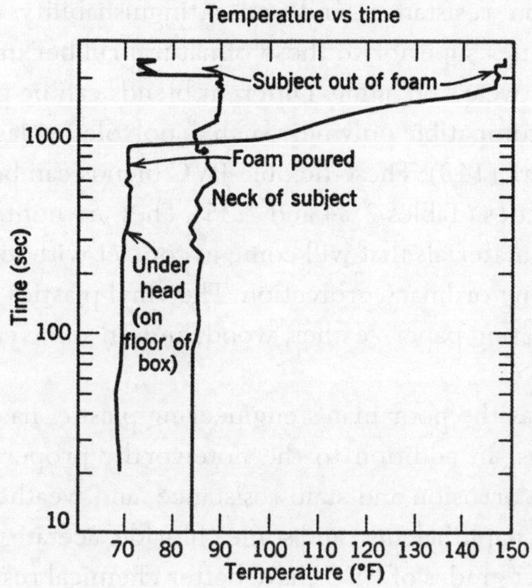

Figure 2.20 Temperature distribution in foam-vinyl strippables

	Flexible	Rigid	
	General purpose	Regular	High impact
Specific gravity	1.4	1.5	1.35
Tensile strength, lb/in.²	2300	6500	9200
Elongation at break, %	3000	25	35
Compression strength, lb/in.²	1250	10,000	7500
Maximum operating temperature, %	80	70	70

Table 2.33 Typical properties of PVC and copolymers

as acrylonitrile butadiene styrene (ABS), acrylics, polyurethanes, and nitrile rubbers to improve its impact resistance, heat deflection, and processability.

Although these vinyls differ because there are literally thousands of varying compositions and properties, there are certain general characteristics that are common to nearly all these plastics. Most materials based on vinyls are inherently thermoplastic and heat sealable. The exceptions are the products that have been purposely compounded with TS plastics or cross-linking agents.

In general, vinyls can be plasticized to give them a wide range of hardness ranging from thin, flexible, free films to rigid molded pieces. Most vinyls are naturally clear, with an unlimited color range for most forms of the materials. They generally have in common excellent water and chemical resistance, strength, abrasion resistance, and self-extinguishability. In their elastomeric form vinyls usually exhibit properties superior to those of natural rubber in their flex life, resistance to acids, alcohols, sunlight, and wear and aging. Different blends can be prepared providing different properties. Blends with noncompatible polymers such as polyolefin elastomers (POEs) are made to blend by using compatibilizers (143). These flexible PVC blends can be made with no plasticizers, resulting in improved properties (Tables 2.34 and 2.35). They are nontoxic, tasteless, odorless, and suitable for use as packaging materials that will come in contact with foods and drugs, as well as for decorative packaging requiring ordinary protection. The vinyl plastics can be used in printing inks and be effectively used in coating paper, leather, wood, and, in some cases, plastics. In most forms vinyl can be printed.

Rigid PVC, referred to as the poor man's engineering plastic, has a wide range of properties for use in different products. In addition to the noteworthy properties mentioned, it has high resistance to ignition, good corrosion and stain resistance, and weatherability. However, aromatic solvents, ketones, aldehydes, naphthalenes, and some chloride, acetate, and acrylate esters attack it. In general, the normal-impact grades of PVCs have better chemical resistance than the high-impact grades. Most PVCs are not recommended for continuous use above 60°C (140°F). Chlorination to form CPVC increases its heat resistance, flame retardancy, and density, depending on the amount

Properties	Flexible PVC Standard Compound PVC + Plasticizer	Alloy 1 PVC/POE + 60 phr Plasticizer	Alloy 2 PVC/POE + 20 phr Plasticizer	Alloy 3 PVC/POE (No Plasticizer)
Hardness, Shore A	70	70	94	94
Specific gravity	1.21	1.21	1.14	1.14
Melt index, g/10 min				
Tested @ 190 °C / 2160 g	49.5	3.5	0.57	—
Tested @ 190 °C / 5000 g	—	—	—	1.55
Tensile strength, psi	1980	1686	2828	2962
Elongation, %	350	411	326	346
Elongation retention, %				
After 7 days @ 121 °C	1	53	80	—
After 7 days @ 136 °C	—	—	52	102
Brittle point, °C				
Original	-29	-40	<-60	<-60
After 7 days @ 121 °C	-3	<-60	-55	—
After 7 days @ 136 °C	—	—	<-60	<-60
Dielectric constant @ 1MHz	3.405	3.067	2.767	2.885
Dissipation factor	0.116	0.065	0.040	0.028

Table 2.34 PVC/POE blend properties improve without plasticizers (Courtesy of Teknor Apex Co.)

Property Improvement	Blending Polymer
Impact resistance	ABS, methacyrylate-butadiene-styrene, acrylics, polycaprolactone, polyimide, polyurethanes, PVC-ethyl acrylate
Tensile strength	ABS, methacyrylate-butadiene-styrene, polyurethanes, ethylene-vinyl acetate
Low-temperature toughness	Styrene-acrylonitrile, polyurethanes, polyethylene, chlorinated polyethylene, copolyester
Dimensional stability	Styrene-acrylonitrile, methacrylate-butadiene-styrene
Heat-distortion temperature	ABS, methacyrylate-butadiene-styrene, polyimide, polydimethyl siloxane
Processability	Styrene-acrylonitrile, methacrylate-butadiene-styrene, chlorinated polyethylene, PVC-ethyl acrylate, ethylene-vinyl acetate, chlorinated polyoxymethylenes (acetals)
Moldability	Acrylics, polycaprolactone
Plasticization	Polycaprolactone, polyurethanes, nitrile rubber, ethylene-vinyl acetate, copolyester, chlorinated polyoxymethylenes (acetals)
Transparency	Acrylics, polymide
Chemical/oil resistance	Acrylics
Toughness	Nitrile rubber, ethylene-vinyl acetate
Adhesion	Ethylene-vinyl acetate

Table 2.35 Examples of PVC mixes/blends to improve properties

of chlorination introduced. Note that in regard to flammability, the vinyls release a limited amount of hydrochloric acid.

PVCs' inherent characteristics generally require special tooling (molds, dies, etc.) and considerations to ensure the best melt-processing conditions so that the tool will not be damaged by the PVC (chapter 17). One such consideration is specifying the correct tool steel in order to meet the products' demanding appearances, long-run production, and so on.

UltraHigh Molecular Weight Polyvinyl Chloride Ultrahigh molecular weight polyvinyl chlorides (UHMWPVCs) are versatile plastics that can provide superior mechanical properties and be formulated to produce a variety of products. Because changes in formulations or equipment conditions may be required to facilitate processing, these plastics are generally used in plasticized applications; it is in highly plasticized uses that they show the greatest advantages in producing compounds with improved properties. They can bring to flexible vinyls improved tensile, modulus, abrasion, and solvent resistance; low and high temperature performance; and retention of properties during aging.

Polyvinyl Acetate The polyvinyl acetate (PVAc) copolymers are odorless, tasteless, nontoxic, slow burning, lightweight, and colorless, with reasonably low water absorption. They are soluble in organic ketones, esters, chlorinated hydrocarbons, aromatic hydrocarbons, and alcohols, but insoluble in water, aliphatic hydrocarbons, fats, and waxes. Water emulsions have extended the use of this plastic. Used perhaps most extensively as adhesives, they are also employed as coatings for paper sizing for textiles, finishes for leathers, as well as bases for inks and lacquers, heat-sealing films, and flashbulb linings.

Polyvinyl acetate includes vinyl acetate homopolymers and all copolymers in which vinyl acetate is the major constituent (50% or greater). The major PVAc copolymers are vinyl acetate-ethylene (VAE) and vinyl acetate-acrylic ester (vinyl acrylic). Vinyl acetate-versatic acid (vinyl versatate) and vinyl acetate-maleate are the major PVAc copolymer emulsions used.

Polyvinyl Chloride Acetate Polyvinyl chloride acetate (PVCA) is a copolymer of vinyl chloride and vinyl acetate. It is a colorless thermoplastic solid with good resistance to water as well as concentrated acids and alkalis. It is obtainable in the form of granules, solutions, and emulsions. Compounded with plasticizers, it yields a flexible material superior to rubber in aging properties. It is widely used for cable and wire coverings, in chemical plants, and in protective garments.

Chlorinated Polyvinyl Chloride Chlorinated polyvinyl chloride (CPVC) is a plastic produced by the postchlorination of PVC. Adding more chlorine raises the glass transition temperature of CPVC from 115° to 135°C (239° to 275°F) and the resultant heat deflection under load from that of PVC at 70°C (158°F) to a level of 82° to 102°C (180° to 219°F), depending on formation. CPVC has improved resistance to combustion and smoke generation with higher tensile strength and modulus while maintaining all the good properties that rigid PVC possesses. Traditional uses are hot and cold water distribution piping and fittings and industrial-chemical liquid-handling pipe, fittings, valves, and other different applications.

Polyvinyl Alcohol

Polyvinyl alcohol (PVOH, or the trade name PVAL) is a crystalline, white powder soluble in water and alcohols. It is the world's largest in volume of a synthetic, water-soluble plastic. In its cast form, it is characterized by water solubility, low gas permeability, high resistance to organic solvents other than alcohol, and crystallinity when stretch oriented. Crystallinity allows the material to polarize light. A series of hydrolysis levels of the plastic are available that range from room-temperature solubility to those not soluble at all. The major applications of the PVOHs are in elastomeric products, adhesives, films, and finishes. Extruded PVOH hoses and tubing are excellent for use subjected to contact with oils and other chemicals. PVOH is used as a sizing in the manufacture of nylon.

Polyvinyl Butyral

This thermoplastic material is derived from a polyvinyl ester in which some or all of the hydroxyl groups are replaced by butyl groups by reaction with butyraldehyde. It is a colorless, flexible, and very tough solid plastic. The polyvinyl butyrals (PVBs) are soluble in esters, ketones, alcohols, and chlorinated hydrocarbons but insoluble in the aliphatic hydrocarbons. They are stable in dilute alkali, but they slowly decompose in dilute acids.

Since the year 1938, PVBs have been widely used as safety-glass interlayers and between sheets of acrylic to protect the enclosures of pressurized cabins in aircraft against shattering. PVB film in interlayers ranges from 10 to 40 mils and continues to be used as an important resource for glass, automotive, architectural, and other industries. PVBs are also used as coatings for textiles and paper and also as adhesives.

Polyvinyl Carbazole Polyvinyl carbazole (PVCB) is a thermoplastic, brown in color, and obtained by reacting acetylene with carbazole. The plastic has excellent electrical properties and good heat and chemical resistance. Use includes high-frequency dielectrics, impregnant for paper capacitors and photoconductive plastics.

Polyvinyl Pyridine Polyvinyl pyradine (PVP) is primarily used as a constituent in copolymers as adhesives.

Polyvinyl Pyrrolidone Polyvinyl pyrrolidone (PVPO) is a highly polar and water-soluble plastic. It finds applications in adhesives and as a water thickener. Water solutions can be used as blood plasma substitute or artificial blood.

Polyvinyl Fluoride PVF products are strong and tough, with good abrasion and staining resistance up to fairly high temperatures of 100° to 150°C (212° to 302°F), and they are classified as slow burning. They are generally less chemical resistant than fully fluorinated plastics but show excellent UV resistance and good color retention. They are not affected by water. Their excellent weatherability has made them a choice material for exterior applications such as coatings for metals (slides, gutters, etc.), plywood finishes, architectural sheets, lighting panels, and glazing for solar energy collection. PVFs are also used for electrical wrapping tape and parting layers for laminates.

Polyvinyl Formal Polyvinyl formal (PVFO), which corresponds to formaldehyde in the presence of PVOH, finds applications as temperature-resistant coatings for containers and electric wires. It is also resistant to greases and oils.

Polyvinylidene Chloride Polyvinylidene chloride (PVDC) may be made into both flexible and rigid forms. Molded parts in PVDC have high strength, abrasion resistance, strong welds, dimensional stability, toughness, and durability. This material is especially suited for high-speed injection molding for products with heavy, thick cross-sections. Molded fittings and parts are particularly valuable in industries involving the use of chemicals. For example, pipes of this material are superior to iron pipes to dispose of waste acids. As an extruded monofilament, it is woven into upholstery fabric and screening. Films produced from PVDC exhibit an extremely low water-vapor transmission rate, as well as flexibility over a wide range of temperatures and heat sealability. They are particularly suitable for various types of packaging, including medical products, metal parts, and food. Food packaging for the home refrigerator includes the highly popular Saran wrap from Dow.

Polyvinylidene Fluoride Polyvinyl fluoride (PVDF) is a fluorine-containing TP. It is unlike other plastics, being a crystalline. PVDF is a high molecular weight polymer of vinylidene fluoride *(CH2* = CF_2). Compounds are available that contain 59% fluorine. This nonflammable plastic is mechanically strong and tough, thermally stable, resistant to almost all chemicals and solvents, and is stable under UV and extreme weather conditions. It has a higher strength and abrasion resistance than PTFE; however, it does not match the high chemical and temperature resistance of PTFE.

Where unfavorable combinations of chemical, mechanical, and physical environments may preclude the use of other materials, PVDF has been successfully used. Examples include valve and pump parts, heavy wall pipe fittings, gears, cams, bearings, coatings, and electrical insulation. Its limitations include lower service temperatures than the highly fluorinated fluoropolymers, no antistick qualities, and the fact that it produces toxic products upon thermal decomposition.

POLYSTYRENE

Polystyrenes are one of the five high-volume plastics. Polystyrene (PS) is noted for its sparkling clarity, hardness, low water absorption, extreme ease of processing general-purpose PS (GPPS), excellent colorability, dimensional stability, and relatively low cost (Fig. 2.21). This amorphous TP often competes favorably with higher-priced plastics. It is available in a wide range of grades for all types of processes.

The basic form (crystal PS) is brittle, with low heat and chemical resistance and poor weather resistance. High-impact polystyrene is made with butadiene modifiers, which provide significant improvements in impact strength and elongation over crystal polystyrene, accompanied by a loss of transparency and little other property improvement. PS is used in many different formulations.

Modifications available to basic GPPS include grades for high heat and for various degrees of impact resistance (Table 2.36). Clarity and gloss are reduced, however, in the impact grades. Some examples of members in the PS family are compounds of ABS, styrene-acrylonitrile copolymer

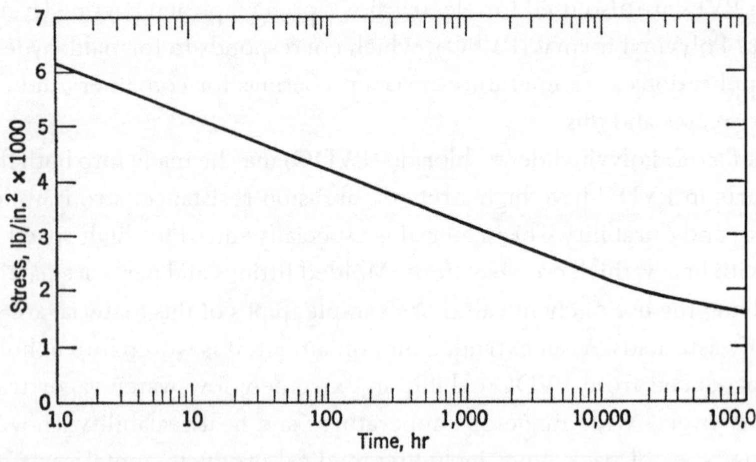

Figure 2.21 Tensile stress at failure vs. time for a general-purpose polystyrene

Property	Units	Medium impact	High impact	High heat	Super high impact
Ultimate tensile strength	Psi	3500	2400	4000	1900
Ultimate elongation	%	20	40	22	70
Izod impact strength	F-16/in. of notch				
at 73°F		0.6	1.2	1.0	2.8
at 0°F		0.5	0.65	0.7	1.9
Rockwell hardness	M	60	22	33	51
Softening point	°F	207	195	215	200

Table 2.36 Average properties of impact- and-heat resistant polystyrene

(SAN), and SMA (styrene maleic anhydride). The structural characteristics of these copolymers are similar, but SMA has the highest heat resistance.

Expandable polystyrene (EPS) is a specialized form of polystyrene plastics. Solid PS processed products have low heat resistance, as compared to most TPs. Their maximum recommended continuous service temperature is below 93°C (200°F). Their electrical properties, which are good at room temperature, are affected only slightly by higher temperatures and varying humidity.

PS is soluble in most aromatic and chlorinated solvents but insoluble in such alcohols as methanol, ethanol, normal heptane, and acetone. Most fluids in households, as well as drinks and foods, have no effect, but the oils in citrus-fruit rinds, gasoline, turpentine, and lacquer attack PS. PSs are available in FDA-approved grades.

Waste that occurs during the manufacturing and processing of PS has practically always been fed back into the production or processing cycle. The reuse of municipal waste is feasible without any problems with uncontaminated and contaminated materials. Each is used in their respective new market products.

Polystyrene Copolymer Copolymers of styrene include a large group of random, graft, and block copolymers; those with a high proportion of acrylonitrile used in barrier films as well as others such as methacrylic-butadiene-styrene copolymer (MBS) plastic are used as modifiers in PVC, SAN, ABS, ASA, and others. SAN is the most important when considering volume and number of applications.

Expandable Polystyrene Expandable polystyrene (EPS) is a modified PS prepared as small beads containing pentane gas, which, when steamed, expand to form lightweight, cohesive masses for forms used to mold cups and trays, package fragile products for shipment, and so on. Similar dimensionally stable forms molded from EPS are used as cores for such products as automobile sun visors with surface overlays.

Polystyrene Maleic Anhydride Polystyrene maleic anhydride (SMA) is a copolymer made with or without rubber modifiers. They are sometimes alloyed with ABS and offer good heat resistance, high impact strength and gloss, but with little appreciable improvement in weatherability or chemical resistance over other styrene-based plastics.

Crystal Clear Polystyrene The styrene-butadiene styrene block copolymers with a polybutadiene content of up to 30 wt%, which are referred to as impact-resistant crystal clear, are completely different from standard impact-resistant PS. These are frequently mixed with standard PS on price grounds and also in order to achieve greater rigidity. Only certain defined grades of standard PS will give a mixture that is crystal clear.

Crystal-clear PS is a clear, amorphous polymer that is normally manufactured by continuous mass polymerization of styrene monomer. These resins have excellent clarity, gloss, and stiffness, but are very brittle. Crystal polystyrene is sometimes blended with other resins to optimize physical properties or coextruded onto impact polystyrene to provide a high-gloss surface.

Flame-Retardant Polystyrene The market for flame-retardant polystyrene plastics has been disconcerted by announcements that thermoplastics containing polybrominated diphenyl ethers can release brominated dioxins or furanes under certain conditions. While investigators began looking into the formation of these compounds and their toxicity, a precautionary substitution of the products in question commenced. The advantages of these halogen-containing materials, namely, an improved flowability and light stability, do not make up for their drawback of a lower temperature stability in a large number of cases. For example, product grades containing polybrominated diphenyl ether are still being used to meet the UL 94 V-O rating. The manufacturers of these products are not bound by any obligation to find substitutes.

There is a basis of the findings regarding the formation and toxicity of brominated dioxins and furanes that have been established to date. The nature, quantity, and toxicity of the substances that develop during the manufacture and processing of PS with additives of brominated diphenyl do not present any dangers, providing that the standard industrial safety regulations are respected. As an example, under no circumstances does a running television set pose a danger to health. However, the debate regarding the potential danger of brominated diphenyl ethers (simply referred to under the general designation of halogen compounds in many cases) led to a call for halogen-free flame retardants for PS. A V-O product was presented, marking the first of others in this direction. It is particularly difficult to meet the market requirements on processing and flame-retardant performance, physical and mechanical properties, and price.

Polystyrene Film, Heat-Sealable

From the viewpoint of recycling, or suitability for recycling, it was particularly important that a substitute be found for the standard heat-sealable film in aluminum or PVC that had been used on thermoformed PS packs. Recycling was not worthwhile unless there was complete compatibility between the material used for the heat-sealable film and that used for the packaging. Material combinations were found for two-layer film that additionally provided good sealing-seam strength, tear resistance, and tear propagation strength.

High-Gloss Polystyrene High-gloss products with a medium or high impact strength have established themselves on the market. They incorporate particles of rubber that are several times smaller than those used for standard impact-resistant PS, referred to as capsule or small cell particles, as a function of their structure. With their combination of high gloss, high rigidity, high impact strength, and high heat resistance, which was directed at ABS high-gloss products, the products have replaced traditional impact-resistant PS in different markets.

High-Impact Polystyrene This is PS modified by the incorporation of rubber such as polybutadiene particles to improve impact strength. A large volume of high-impact polystyrene (HIPS) is produced in this form, since unmodified PS is rather brittle. It is usually produced by the mass polymerization of styrene containing dissolved rubber. Products obtained merely by milling together rubber and PS or by coagulation of mixed lattices of these two materials have inferior impact properties. As the rubber content increases so does the impact strength, but correspondingly, the softening point and stiffness decrease and the products are no longer transparent. These modified resins exhibit enhanced impact properties and stress-crack resistance. They are translucent to opaque. End uses for these resins include foamed food trays, packaging, disposable cups, and printed displays.

The rubber should have a low T_g value, polybutadiene often being preferred, and it is usually somewhat cross-linked. Some grafting of the rubber to the PS occurs during the polymerization, and this action increases interfacial adhesion and impact strength. The rubber is dispersed as particles about 1 to 10 μm in diameter but containing inclusions of PS particles. This effectively increases the rubber volume fraction. The study of particle morphology by osmium staining and transmission electron microscopy is a classical application of this technique to synthetic plastics.

Syndiotactic Polystyrene

Syndiotactic PS (SPS) is a crystalline plastic with far higher heat resistance than standard amorphous PS, lower moisture pickup, improved warp resistance, and outstanding dimensional stability (eliminating the need for mineral fillers commonly used to counter warpage in other plastics such as nylon and PBT). It is made with metallocene catalyst technology. This plastic has the highest melting point (518°F [270°C]) of any styrenic homopolymer. It also has high chemical, water, and steam resistance, exceptional electrical properties, and well-balanced impact resistance and stiffness (204).

Polystyrene Acrylonitrile Related to ABS, SAN is hard, rigid, and transparent. It has no butadiene. Excellent chemical and heat resistance, good dimensional stability, and ease of processing characterize it. Special grades are available that have improved UV stability, vapor-barrier characteristics, and weatherability. SAN is used for tinted drinking glasses, low-cost blender jars and water pitchers, and other consumer goods with longer life expectancies than ordinary PS.

Polystyrene-Polyethylene Blend

Mixing these two is nothing new. The target of combining the lower water-vapor permeability and good stress cracking of PE (or PP) with the problem-free processing and high rigidity of PS has proven to be unattainable because of their incompatibility. This situation can be reduced through the use of mixing agents made up of styrene/olefin copolymers and others. PS-PE blends are primarily used as a substitute for PVC and ABS in the form of monofilm or multilayer film to produce thermoformed packaging for foods such as those that contain fat.

Polystyrene-Polyphenylene Ether Blend

The good compatibility of PS and polyphenylene ether (PPE) has been used for a long time to make blends that even with a PS content in excess of 50% still count as modified PPE. The addition of PPE results in the increase of PS's heat resistance, which can be raised to the same range as that for ABS. The result is a lower-cost plastic.

Advanced Styrenic Resins Advanced styrenic resins (ASRs) are produced either chemically in a reactor or by blending GPPS (general-purpose polystyrene) and rubber in downstream operations. This family of resins has good toughness and gloss, and very good processability. ASRs can be processed on conventional sheet extrusion and thermoforming equipment. They are recommended for applications where intermediate properties between HIPS and ABS are desired. It has been used in coextruded refrigerator liners where ASR is the cap layer.

Styrene Butadiene Styrene-butadiene copolymers (SBs) are produced via anionic polymerization by sequential addition of styrene and 1,3-butadiene monomers. These clear, tough plastics are easily processed with conditions and equipment similar to those used for high-impact polystyrene. Various amounts of general-purpose polystyrene are usually mixed with the SB copolymers during sheet extrusion to maximize rigidity and economics in the thermoformed parts. Typical applications include drinkware, medical packaging, lids, containers, and blister packages.

Polyalphamethylstyrene Polyalphamethylstyrenes (PAMS) has not achieved much commercial importance as a homopolymer despite a relatively high glass transition temperature that gives it much better temperature resistance than PS. It finds significant use in copolymers.

ACETAL

This crystalline plastic is strong and stiff and has exceptional resistance to abrasion, heat, chemicals, creep, and fatigue. With a low coefficient of surface friction, it is especially useful for mechanical products such as gears, pawls, latches, cams, cranks, and plumbing parts. It is chrome platable.

ACRYLIC

This family of commercial thermoplastics involves repeat units whose names feature the term *acryl*, such as polymethyl methacrylate (PMMA), polyacrylic acid, polymethacrytic acid, poly-R acrylate, poly-R methacrylate, polymethylacrylate, polyethylmethacrylate, and cyanoacrylate plastics. They are produced by stepwise polymerizing of methyl, ethyl, and butyl acrylates with methacrylates. PMMA is the major and most important homopolymer in the series of acrylics with a sufficient high glass transition temperature to form useful products. Repeat units of the other types are used. Ethylacrylate repeat units form the major component in acrylate rubbers.

These polymethyl methacrylate (PMMA) plastics have high optical clarity, excellent weatherability, very broad color range, and the hardest surface of any untreated thermoplastic. Chemical, thermal, and impact properties are good to fair. Normally, PMMA is used as an exterior material on optical lenses, automotive taillights, decorative nameplates, aircraft glazing, illuminated signs, medical devices, and so on. It is also used as an opaque-colored sheeting thermoformed to produce an outer coating, behind which glass-fiber-reinforced TS polyester plastics are sprayed, to produce camper tops, swimming-pool steps, and plumbing fixtures with weatherability and repairability reported superior to that of polyester gel coats. Like plywood, there are outdoor weather-resistant grades and indoors non-weather-resistant grades.

Acrylics come in a wide variety of forms: rigid plastics made by the polymerization of the esters of acrylic or methacrylic acid, fibers, acrylic rubbers, and coating resins. Acrylic molding powders are available for injection and extrusion processes. Their mold shrinkage is quite low, and a semiviscous liquid casting syrup may be poured into a mold and cured at temperatures of 150° to 250°F to convert it into a hard, rigid solid. Acrylic sheets of excellent clarity are made this way. Among the other forms of acrylics, coatings for protecting metal and acrylic enamels for cars and appliances are available in great variety. Water emulsion acrylic paints give excellent service, both indoors and out, and acrylic adhesives are used to bond many carpeting fibers to their backing and provide nonskid properties and dimensional stability.

Acrylic materials are manufactured into clear or opaque films, which are subsequently laminated onto a variety of resins such as acrylic, ABS, HIPS, PVC, and PC. Acrylic films can be decorative or UV resistant for outdoor applications. Typical products include siding and shutters, automotive and recreational vehicle (RV) exteriors, marine and leisure craft parts, sanitary wares, interior decorative panels, tractor canopies, spas, glazing, and outdoor signs and profiles.

Acrylic resins will fail in a brittle manner, independent of the temperature. They will suffer crazing when loaded at stress about halfway to the failure level. This effect is enhanced by the presence of solvents.

Acrylic Elastomer Under the heading acrylic elastomer (or acrylate or polyacrylate rubber or elastomer), the plastic literature has included a broad spectrum of carboxy-modified rubbers that have as a minor portion of the comonomers acrylic acid, its derivatives, or both. However, in more recent usage, the term *acrylic elastomer* is used to designate those rubbery products that contain a predominant amount of an acrylic ester, such as ethyl acrylate or butyl acrylate, in the polymer chain. Fluoroacrylate elastomers are based on plastics prepared from the acrylic acid ester-dihydroperfluoro alcohols.

Polymethacrylic Acid Polymethylacrylic acid (PMAA) is water soluble and essential in the formation of ionomers.

Polymethylacrylate Polymethylacrylate (PMA) is used in adhesives, paints, and other products.

Polyethylmethacrylate This is a special plastic in the acrylic family; polyethylmethacrylate (PEMA) provides the usual properties with flexibility.

Polyglutarimide Acrylic Copolymer This family of plastics can be used in hot-fill and retort packaging applications that provide clarity and heat resistance.

ACRYLONITRILE

Acrylonitrile resins contain a majority of nitrile polymers. Acrylonitrile provides good gas barrier, chemical resistance, and low taste and odor transfer. Acrylonitrile has good impact properties when modified with rubber. Sheet extruded from these resins is used extensively in food packaging and rigid packaging applications.

This crystalline thermoplastic is most useful in copolymers. Its copolymer with butadiene is nitrile rubber. Acrylonitrile-butadiene copolymers with styrene (e.g., SAN) exist that are tougher than PS. It is also used as a synthetic fiber and a chemical intermediate.

Acrylonitrile-Butadiene Rubber Also called nitrile rubber (NBR), it is a copolymer with 20% to 50% of acrylonitrile that was developed as a general-purpose, oil-resistant rubber. Its resistance to oils, fuels, and solvents is superior to that of polychloroprene (CR). Heat aging is rather good, and abrasion resistance is high. Cost is moderate. Ozone and weathering resistance is not very good, and low-temperature flexibility is quite limited in some grades. Applications include fuel and oil tubing and hose, gaskets and seals, conveyor belting, and printing rolls and pads.

Acrylonitrile-Butadiene-Styrene Acrylonitrile-butadiene-styrene (ABS) is an amorphous terpolymer composed of acrylonitrile, butadiene, and styrene. ABS is produced from suspension, emulsion, and continuous mass polymerization. These materials are extruded and thermoformed on typical processing equipment and exhibit a wide processing window.

ABS provides a tough, hard, not brittle, rigid plastic with adequate chemical, electrical, and weathering characteristics; low water absorption; and resistance to hot-and-cold water cycles (Fig. 2.22). They offer good dimensional stability, good electrical properties, and high abrasion resistance. Some grades are easily electroplated. It is a terpolymer of the three monomers, often considered a modified polystyrene, since its properties resemble PS, except that its impact strength is much higher. A very wide variety of commercial products are available, varying not only in composition but also in morphology, owing to the different procedures used in copolymer preparation.

In the past, mechanical mixtures of styrene-acrylonitrile copolymer and nitrile rubber were made, or they were obtained from co-coagulating lattices of the polymers. Present products are almost entirely produced by the copolymerization of styrene and acrylonitrile in the presence of polybutadiene latex—the so-called latex-grafting method. This results in considerable grafting, and the products have higher impact strengths. The morphology is similar to that of high-impact PS, but the dispersed rubber particles are much smaller (1 to 10 mm) and contain styrene-acrylonitrile copolymer inclusions. Other rubbers that are sometimes used are ethylene-propylene-diene monomer rubber and polybutadiene solution.

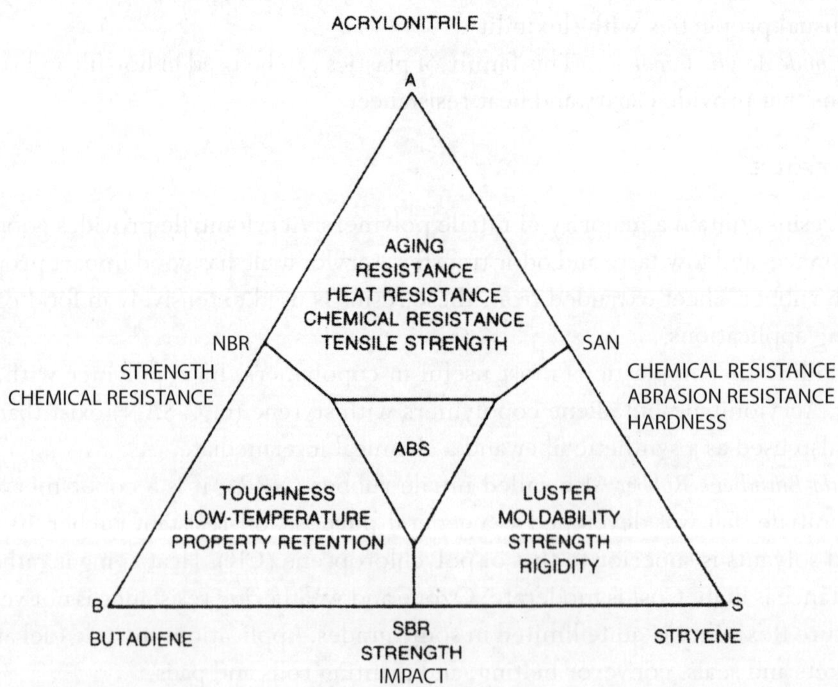

Figure 2.22 Components of ABS provide different properties

A typical ABS contains about 20 wt% rubber, about 25% acrylonitrile, and about 55% styrene, having a T_g value of about 105°C, a tensile modulus of 2.5 GPa, and an impact strength of about 4 J $(12.7\ mm)^{-1}$ on an Izod test.

However, owing to the many variations possible, the properties can vary considerably (e.g., Super ABS can have a notched Izod impact strength of up to 8 J $[25\ mm]^{-1}$). Also, some or all of one of the comonomers may be replaced. Methylmethacrylate may replace acrylonitrile, as in MBS and MABS, with improvement in transparency, or a saturated rubber may replace the polybutadiene, as in ASA and ACS, with an improvement in oxidation resistance. ABS is used where good appearance and strength, together with reasonable stiffness and softening point, are required. It is used for electronic instrument housings, telephones, sports gear, automotive grilles, furniture, and so on. It is electroplatable, serves well as a structural foam, and is available as a tinted transparent. Other applications include luggage, truck caps, spas, RV and automotive interior and exterior panels and trim, appliances, refrigerator liners, tabletops, and leisure crafts.

Transparent Acrylonitrile-Butadiene-Styrene By matching the refractive indices of the elastomer and the SAN phases, usually by incorporating methyl methacrylate, transparent products are possible. Progress in product development is achieved by further matching the properties of those of the standard ABS and also by increasing the light transmission up to 88%. Another gain is better melt flowability of the products. An example of an application is in transparent trays for freezers. Other applications include paper feeds for copy machines, watch crystal, transparent building blocks for toy systems, products for medical technology, and packaging for cosmetics.

Acrylonitrile-Chlorinated Polyethylene-Styrene Copolymer This ACS is a terpolymer obtained by the copolymerization of acrylonitrile and styrene in the presence of chlorinated polyethylene. Properties are similar to ABS, except that it is more resistant to embrittlement owing to oxidative degradation, and it has better fire resistance. It has very high flame retardance; ACS is classified as UL 94V-0 (1/16-in-thick specimen). This styrenic plastic inherently resists the electrostatic deposition of dust; thus there is no need for the addition of antistatic agents to the formulation. The material's deflection temperature under load ranges from 78° to 90°C (172° to 194°F). Products made of ACS can be adhered to each other and hot stamped and painted, and they find their greatest use in cabinets and housings.

Acrylonitrile-Ethylene/Propylene-Styrene Copolymer Acrylonitrile-ethylene/propylene-styrene copolymer (AES) is a terpolymer obtained by grafting styrene-acrylonitrile copolymer to ethylene-propylene or ethylene-propylene-diene monomer rubber. It is similar to ABS except with better weathering resistance.

Acrylonitrile-Ethylene-Styrene Acrylonitrile-ethylene-styrenes or AESs are sometimes called OAS—olefin-modified acrylonitrile-styrene or SAN. They are amorphous, opaque terpolymers produced by suspension, emulsion, or continuous mass polymerization. Properties are similar to ABS, with the addition of weatherability or UV protection for outdoor use. These materials are usually coextruded over ABS. Typically, applications include exterior automotive and RV parts, truck caps, pool steps, outdoor signs, camper shells, and sidings.

Acrylonitrile-Methylmethacrylate The use of copolymers involving about 70 wt% of acrylonitrile and 30% of methylmethacrylate repeat units is used for thick transparent products, such as glazing, which require high-impact properties and good chemical and weathering resistance.

Acrylonitrile-Styrene This is a high-nitrile plastic of styrene-acrylonitrile or other monomer acrylonitrile copolymer with a high acrylonitrile content of about 70 wt% to 80 wt%. Such copolymers have a very low permeability to gases (including carbon dioxide), and liquids; thus they are called barrier plastics. Acrylonitrile styrenes (ANSs) also provide good chemical resistance. They are used in packaging, especially for bottles containing carbonated drinks. To provide higher softening points, the ANS may be partially replaced by methacrylonitrile.

This was the first plastic used to produce stretched two-liter Coca-Cola carbonated beverage bottles during 1958; production was by Monsanto Co. using AN plastic "Barex" (Sohia of BP Chemical International) and DuPont's stretched-injection blow-molding process. Unfortunately, AN, after a successful development project, was not permitted by the FDA to be used because of possible food contamination, thus stopping production in about eight plants on the East Coast, including recycling operations. However, it was the forerunner of an avalanche of containers to appear commercially in the 1970s and thereafter. Following decades of taxpayer and industry money being expended, AN was cleared to be used. In the meantime, Polyethylene terephthalate (PET) plastic became very successful as its replacement.

Acrylonitrile-Styrene-Acrylate Acrylonitrile-styrene-acrylate (ASA) terpolymer is similar to ABS but with a saturated acrylate rubber replacing the unsaturated polybutadiene. This blend improves its resistance to oxidative degradation. This UV-resistant material offers an excellent combination of color, in both light and dark colors, and property retention after outdoor exposure. There are three major markets for ASA: building and construction, leisure and recreation, and automotive.

Polymethacrylonitrile Polymethacrylonitrile (PMAN) is transparent, is an excellent barrier to carbon dioxide (CO_2), and possesses good creep resistance, making it a suitable material for carbonated beverage containers.

Polyvinylcyanide Polyvinylcyanide (PAN) homopolymer is a polar crystallizing polymer that undergoes chemical decomposition before crystalline melting at temperatures above 300°C (572°F). It is therefore not a true thermoplastic material. The almost pure homopolymer, however, can be processed into fiber (acrylic fibers) through its spinning from solutions. PAN is also used as a base material or precursor in the manufacture of certain carbon and graphite fibers.

CELLULOSIC POLYMERS

These are tough, transparent, hard or flexible natural polymers made from plant cellulose feedstock. With exposure to light, heat, weather, and aging, they tend to dry out, deform, embrittle, and lose gloss. Molding applications include tool handles, control knobs, and eyeglass frames. Extrusion uses include blister packaging, toys, and holiday decorations. Cellulosic types, each with their specialty properties, include cellulose acetates (CAs), cellulose acetate butyrates (CABs), cellulose nitrates (CNs), cellulose propionate (CAPs), and ethyl celluloses (ECs).

CHLORINATED POLYETHER

Chlorinated polyethers (CPs) are corrosion- and chemical-resistant thermoplastics whose prime use has been to manufacture products and equipment for the chemical and processing industries. Uses also include molding components for pumps and water meters, pump gears, bearing surfaces, and the like. This plastic resists both organic and inorganic agents, except fuming nitric acid and fuming sulfuric acid, at temperatures up to 121°C (250°F) or higher. Its heat-insulating characteristics, dimensional stability, and outdoor exposure resistance are also excellent.

Ethylene-Vinyl Acetate

Ethylene-vinyl acetate (EVAs; one of many in the polyolefin copolymer family) have exceptional barrier properties, good clarity and gloss, stress-crack resistance, low-temperature toughness, flexibility retention, adhesion, resistance to UV radiation, and so forth. They have low resistance to heat and solvents as well as exceptional weathering resistance. Table 2.37 lists some of the properties of EVA, EEA (or ethylene ethyl acrylate, another polyolefin copolymer), and low-density polyethylene.

Ethylene-Vinyl Alcohol

Ethylene-vinyl alcohol copolymer (EVOH) plastics are specialty products that have superior gas-barrier properties. They are often used as the internal layer in multilayer food-packaging films and blow-molded rigid containers for a variety of purposes, including gasoline tanks for automobiles. EVOH can be fabricated by the usual melt-processing methods. The barrier properties of films decrease in the presence of moisture, so coextrusion with protective polypropylene (especially biaxially oriented material), low-density polyethylene, nylons, or other moisture-barrier films provides films or articles that are useful even with liquids. The ethylene contents of commercial ethylene-vinyl alcohol polymer ranges from about 30% to 50% mole (20 wt% to 40 wt%) ethylene.

Fluoroelastomer

These types of elastomers, also called fluorine rubbers, contain fluorine and have the highest resistance to heat deformation and chemical attack. By far the most important are the FPM rubbers that are mainly made from vinylidene fluoride, TFE, and hydropentetrafluoropropylene or perfluoromethyl vinyl ether. Alternative copolymers made from propylene TFE (TFE/P), polyfluoroalkoxyphosphazene, and fluorosilicone elastomer (FMQ or FSIE) have also been used.

Property	Units	EVA	EEA	Low density polyethylene
Melt index	g/min	2.5	6	2.0
Density	g/cc	0.943	0.931	0.920
Stiffness, modulus of elasticity at 23°C	psi	5700	5500	20,000
Tensile strength, ultimate	psi	2800	2000	1850
Elongation	%	650	650	600

Table 2.37 Comparative properties of EVA, EEA, and LDPE

These fluorine-containing plastics are noted for their exceptional resistance to heat, weathering, and a wide variety of fluids and chemicals as well as for their excellent sealing and mechanical properties. They are used in highly demanding applications in which extreme temperature ranges and chemical attacks are encountered. The chief limitations on the use of fluoroelastomers are cost and the relatively different technology of compounding and curing, which may involve long postcures.

The cured products generally have fairly low temperature flexibility and low resilience. Solvent resistance is generally high; however, certain organic liquids can cause considerable swelling. In addition, fluoroelastomers are very sensitive to moisture and require a strictly controlled environment to process, thereby making the processing of fluoroelastomers difficult to handle when compared to conventional commodity plastics.

Fluoroelastomers are used in a variety of demanding applications in such industries as transportation, Chemical and food processing, electronics, and aerospace. Environmental concerns in automotive exhausts are driving the growth of these compounds. Components made with these materials are becoming increasingly popular in sealant applications because they can withstand temperatures of car engines (which have been increasing in recent years) and the corrosive nature of fuel and engine fluids. As automotive manufacturers require long-lasting components, demand for fluoroelastomers will increase.

Fluoroplastic

Fluoroplastics (FPs) are also called fluorocarbon plastics. They are based on polymers made of monomers composed of fluorine and carbon; they may also include chlorine atoms in their structure. This family of plastics includes polytetrafluoroethylene (PTFE), polytetrafluoroethylene-cohexafluoropropylene or fluorinated ethylene propylene (FEP), polyfluoroethylene-copperfluoropropylvinyl ether (PFA-CFVE), ethylenetetrafluoroethylene (ETFE), polychlorotrifluoroethylene (PCTFE), ethylene-chlorotrifluoroethylene (ECTFE), polyvinylidene fluoride (PVDF), polyvinyl fluoride (PVF), perfluoromethylvinylether (PFMV), and perfluoroalkoxy (PFA).

Their common characteristics are outstanding chemical inertness, superior resistance to temperatures from –220°C (–364°F) to as high as 260°C (500°F), low coefficient of friction, good electric properties, low permeability, practically zero moisture absorption, and good resistance to weathering and ozone. They have only moderate strength. There are amorphous and crystalline fluoroplastics: perfluoroplastics and fluoroplastics with stabilized end groups that enhance surface properties and advance processing. Figure 2.23 provides examples of properties influenced by fluorine content in fluoroplastics. Their mechanical properties normally are low, but they change dramatically when the fluorocarbons are reinforced with glass fibers, molybdenum disulfide fillers, and so forth.

The higher-performing fluoroplastics cannot be processed by the usual procedures, since they have very low melt-flow behavior; these are called non-melt-processable fluoroplastics. When modified, they can use conventional fabricating processes. As an example, PTFE is extremely difficult to process via melt extrusion and molding. The material usually is supplied in powder

Properties (LOW FLUORINE CONTENT)	Designations (HIGH FLUORINE CONTENT)	
←Coefficient of Friction—	PTFE or TFE	Poly tetrafluoroethylene
←Adhesive Character—	FEP	Copolymer of hexafluoropropylene and tetrafluoroethylene or fluorinated ethylene propylene
—Thermal Stability→	CTFE or PTFCE	Polychlorotrifluoroethylene
—Mechanical Strength at High Temp.→	PVF	Polyvinylfluoride
—Softening Temperature→	PVF_2 or PVDF	Polyvinylidenefluoride
—Antistick→	ETFE	Copolymer of ethylene and tetrafluoroethylene
←Cohesive Forces—	ECTFE	Copolymer of ethylene and chlorotrifluoroethylene
—Creep→	PFA	Polyperfluoroalkoxyethylene
←Dielectric Constant—		
—Chemical Resistance→		
—Solvent Resistance→		
←Mechanical Strength at Ambient Temp.—		
—Permeability→		
←Processing Ease—		
—Oxidative Stability→		

Figure 2.23 Different properties of fluoroplastics

form for compression and sintering or in water-based dispersions for coating and impregnating. Information on properties of fluoroplastics compared to other plastics as well as special conditions is reviewed in Tables 2.38 to 50 and Figure 2.24. (146).

Polychlorotrifluoroethylene In terms of processability, polychlorotrifluoroethylene (PCTFE) can be molded and extruded by conventional thermoplastic processing techniques. For example, it can be made into transparent film and sheet with extremely low water vapor transmission.

The repeat unit of PCTFE differs from that of PTFE by the replacement of a fluorine atom with a chlorine atom, which reduces the flexibility of the chains (higher glass-transition temperature) and their tendency to crystallize (maximum 40% to 65%), as well as the melting point. PCTFE can be made optically clear, an amorphouslike characteristic, with a thickness as high as 3 mm (1/8 in). The reduced melting point and a generally lower molecule weight (100,000 to 200,000) make it a true thermoplastic. It has better mechanical properties than PTFE (stiffer, stronger, etc.), but its frictional properties do not come near those of PTFE. Chemical resistance is generally good, although not comparable to PTFE and other fluoroplastics. Chlorinated solvents, in particular, can affect PCTFE, and it can be subjected to stress cracking at elevated temperatures.

The presence of chlorine atoms, noted to enhance flame retardancy, keeps PCTFE in the nonburning category with PTFE. It causes, however, a serious increase in dielectric losses, particularly at high frequencies. Processing PCTFE is by the conventional methods (extrusion, injection, etc.); processing temperatures are rather high, and any degradation of PCTFE can cause severe corrosion and environmental problems. It crystallizes on slow cooling, while quenching, for films in particular, and can yield transparent products. Applications include wire and cable insulation and electronic flexible printed circuits. In the chemical industry, PCTFE is used as gaskets, O-rings, valve seats, chemical tank liners, and so on. Its extremely low water-vapor transmission, even with very thin transparent film, and its thermosealability explain its use as a packaging material, in the pharmaceutical industry in particular (strip and blister packs for tablets and capsules).

Polytetrafluoroethylene PTFE (or TFE) has a unique position in the plastics industry (as well as others) because of its chemical inertness (Tables 2.51 and 2.52), heat resistance, excellent electrical insulation properties, and low coefficient of friction in a wide temperature range. Polymerization of TFE monomer produces this perfluorinated straight-chain high polymer. It is a highly crystalline

PROPERTY	POLYTETRA-FLUORO-ETHYLENE	POLYETHY-LENE
Densitv	2.2-2.3	0.92–1
Melting Temperature,°C	342 (first) 327 (second)	105–140
Dielectric Constant (1 kHz)	2.0	2.3
Dynamic Coefficient of Friction	0.04	0.33
Surface Energy. dynes/g	18	33
Resistance to Solvents & Chemicals	Excellent No known solvent	Susceptible to hot hydrocarbons
Thermal Stability[1] $T_{1/2}$,°C k_{350}, %/min E_{act}. kJ/mol	 505 0.000002 339	 404 0.008 264
Melt Viscosity[2]. Poise	10^{10}-10^{12}	-
Refractive Index	1.35	1.51
Chain Branching Propensity	No	Yes

[1] $T_{1/2}$ is the temperature a which 50% of the polymer is lost after 30 minutes heating in vacuum; k_{350} is the rate of volatilization, i.e., weight loss, at 350°C; E_{act} is the activation energy of thermal degradation.

[2] Melt Creep Viscosity for PTFE at 380°C.

Table 2.38 Comparing properties of PTFE and PE

	Specific Gravity	Mold Shrinkage, %	Tensile Strength, MPa	Break Elongation, %	Tensile Modulus, MPa	Flexural Strength, MPa	Flexural Modulus, MPa 23°C	Impact Notched Izod J/m (23°C)	Compressive Strength, MPa
Test Method ASTM	D792	D955	D638	D638	D638	D790	D790	D256	D695
Polystyrene	1.06	0.5	46	2.2	3,172	96	3,103	10.8	96.6
ABS	1.05	0.6	48	8.0	2,069	72	2,621	242	69
SAN	1.08	0.4	72	3.0	3,862	103	3,793	27	103
Polypropylene	0.90	1.5	32	15.0	1,310	41	2,069	27	34.5
Polyethylene	0.96	2.0	30	9.0	1,034	38	1,517	70	27.6
Polyacetal	1.41	1.8	61	60.0	2,827	90	2,552	70	36
Polyester	1.30	2.0	55	200.0	2,758	88	2,345	10.8	90
Polyamide, nylon 6	1.13	1.3	81	200.0	2,758	103	2,759	53.8	90
Polyamide, nylon 6,6	1.14	1.8	79	300.0	1,310	103	1,310	53.8	34
Polycarbonate	1.20	0.6	62	110.0	2,379	93	2,345	161	86
Polysulfone	1.24	0.7	70	75.0	2,482	106	2,690	32	96.6
Test Method ASTM	D792	Measured on Parts	D4894-5 D1708 D638	D4894-5 D1708 D638	D638	D790	D790	D256	
PTFE[a]	2.14-2.22	2-10	20-35	300-550	550	No Break	340-620	188	34.5
PFA[b]	2.15	3.5-6.0	20-26	300	276		551	No Break	
FEP[c]	2.15	3.5-6.0	20-28	300	345	No Break	655	No Break	15.2
ETFE[d]	1.71	1-7	45	150-300	827	38	1,034-1,171	No Break	17.2
PVDF[e]	1.78	0.2-3	31-52	50-250	1,040-2,070	45-74	1,140-2,240	107-427	55-110

Abbreviations:

[a] PTFE = Polytetrafluoroethylene

[b] PFA = Perfluoroalkoxy resin (copolymer of tetrafluoroethylene and perfluorovinylethers)

[c] FEP = Fluorinated ethylene-propylene copolymer

[d] ETFE = Ethylene-tetrafluoroethylene copolymer

[e] PVDF = Polyvinylidene fluoride

Table 2.39 Comparing physical and mechanical properties of fluoroplastics with other plastics

plastic. PTFE is extremely resistant to attack by corrosive reagents or solvents. Only alkali metals (either molten or dissolved in liquid ammonia) have been found to attack the plastic, presumably by removing fluorine atoms from the chain. Its high thermal stability is due to the strong carbon-fluorine bond that characterizes the plastic (Tables 2.53 to 2.58 and Fig. 2.25). Figure 2.26 provides its high limiting oxygen index (146).

The rheological properties are so different from that of the usual thermoplastics that the common melt-processing methods (extrusion, injection, etc.) are not feasible; thus, PTFE is treated like a

FLUOROPOLYMERS	FORMULA	COEFICIENT OF FRICTION (DYNAMIC)	CRITICAL SURFACE TENSION, [12] dyne/cm	SURFACE TENSION, [21] (Harmonic-Mean Method) dyne/cm
Polyethylene	$-CH_2-CH_2-$	0.33	31	36.1
Polyvinylfluoride	$-CHF-CH_2-$	0.3	28	38.4
Polyvinylidenefluoride	$-CF_2-CH_2-$	0.3	25	33.2
Polytrifluoroethylene	$-CF_2-CHF-$	0.3	22	-
Polytetrafluoroethylene	$-CF_2-CF_2-$	0.04	18	22.5
Polyvinylchloride	$-CHCl-CH_2-$	0.5	39	41.9
Polyvinylidenechloride	$-CCl_2-CH_2-$	0.9	40	45.4

Table 2.40 Coefficient of friction and surface energy of unfilled fluoropolymers

Filler	Material Description	Particle Size (μm)	Particle Shape	Sensity, g/cm^3
Glass	E glass	Diameter 13 mm Length 0.8 mm Aspect ratio>10	Milled fibers	2.5
Carbon	Amorphous petroleum coke	Diameter <75 μm	Roundish	1.8
Carbon Fiber	Pitch or PAN* based		Short fibers	
Graphite	>99% C, Synthetic or natural	<75 μm	Irregular shape	2.26
Bronze	9/1 Copper to tin ratio	<60 μm	Spherical or irregular shape	
Molybdenum Disulfide	Mineral (98% pure)	<65 μm		4.9

*PAN is an abbreviation for polyacrylonitrile. PAN fibers are thermally carbonized to obtain carbon fiber.

Table 2.41 Properties of common fillers used with fluoroplastics

POLYMER	MONOMER UNIT	MELT POINT °C[1]	MELT VISCOSITY Pa.Sec.[2]	MELT FLOW RATE, dg/min	FABRICATION TECHNIQUE
Granular PTFE	$-CF_2-CF_2-$	327	$10^{10}-10^{12}$ (3)		Compression Molding
Fine Powder PTFE	$-CF_2-CF_2-$	327	$10^{10}-10^{12}$ (3)		Paste Extrusion
Dispersion PTFE	$-CF_2-CF_2-$	317-337	$10^{10}-10^{12}$ (3)		Coating Methods
FEP	$-CF_2-CF_2-C(CF_3)F-CF_2-$	260-282	10^4-10^5 (4)	0.8-18 (5)	Melt Processing
PFA	$-CF_2-CF_2\ C(O-R_f)F-CF_2-$	302-310	$4\times10^3-3\times10^4$ (4)	1-18 (6)	Melt Processing
ETFE	$-CF_2-CF_2-CH_2-CH_2$	254-279	$0.7-10\times10^3$ (7)	3.7-16 (8)	Melt Processing
PCTFE	$-CClF-CF_2-$	210-215	1-10 (9)		Compression Molding/ Melt Processing
ECTFE	$-CClF-CF_2-CH_2-CH_2-$	240	-	1-50	Melt Processing
PVDF	$-CF_2-CH_2-$	155-192	$0.2-17\times10^3$	5-180 (10)	Melt Processing
PVF	$-CHF-CH_2-$	190	-		Melt Extrusion of Dispersion in Latent Solvent
Soluble Fluoropolymers	$-[CF_2-CF_2]_{n-m}-[C_f]_m$ (11)	305	-	-	Latex Application Methods

(1) Second melt point for PTFE.

(2) From Ref. 5.

(3) Melt creep viscosity by Dynamic Mechanical Analysis at 380°C.

(4) By capillary rheometer ASTM D 2116.

(5) Melt flow rate by ASTM D 2116.

(6) By capillary rheometer ASTM D 3307.

(7) By capillary rheometer ASTM D 3159.

(8) Melt flow rate by ASTM D 3159.

(9) By capillary rheometer at 230°C.

(10) Melt flow rate at 265°C.

(11) C_f is the comonomer $m > 0.2$ weight fraction.

Table 2.42 Summary of structural-rheology-fabrication process for commercial fluoropolymers

ceramic. Basically, nearly all PTFEs are processed by forming the plastic to an approximate final shape at or near room temperature and then completing the operation by heating (sintering) the material at temperatures above the melting point and cooling to adjust the crystalline content. Each type of operation has its own specific method, such as billet molding and skiving, sheet molding, automatic preforming and sintering, and ram extrusion.

PTFE is a very useful high-temperature plastic. Even though it is a waxy plastic of the general appearance of polyethylene, it has heat resistance up to 288°C (550°F) and remarkable lubricity, and it is attacked only by sodium and other hot alkali metals. For all practical purposes, the plastic is unaffected by water and its thermal stability is such that its electrical and mechanical properties do not change for long intervals (months) at temperatures as high as 250°C (Tables 2.59 to 2.62).

Molded PTFE articles have high impact strength but are easily strained beyond the point of elastic recovery. The plastic is not hard but is slippery to the touch and has a very low coefficient of friction. Its density is unusually high (2.1 to 2.3). PTFEs' electrical properties are excellent with a dielectric constant of 2.0 and a loss factor that is one of the lowest known for any solid substances.

The uses of PTFE are largely those applications requiring a combination of excellent toughness, electrical properties, heat resistance, a low coefficient of friction, wear resistance, and corrosion resistance. Among the electrical applications of the plastic are wire and cable insulation, insulation for motors and generators, and high-frequency electronic uses. Chemical equipment such as gaskets, pump and valve packing, and valve parts are also made from the polymer. Low-friction and antistick applications include nonlubricated bearings and linings for bakery trays and home and institutional food-processing equipment.

Polyhexafluoropropylone Polyhexafluoropropylone (PHF) is a fluoropolymer that has a repeat unit corresponding to a fully fluorinated polypropylene repeat unit and is significantly more rigid than the PTFE repeat unit with a glass transition temperature about 11°C (52°F).

Polyvinyl Fluoride Polyvinyl fluoride (PVF) is a highly crystalline plastic that is commercially available in the form of a tough but flexible film. This film has outstanding chemical resistance and

FABRICATION PROCESS	SHAPE	PART DIMENSION	PART WEIGHT
Billet/Block Molding	Rectangular, cylindrical	One centimeter in diameter or height to 5 meters in diameter or 1.5 meters in height	Ten grams to several hundred kilograms
Sheet Molding	Flat sheet	One centimeter to 1 meter in width and thickness of 3–75 millimeters	A few hundred grams to a few tens of kilograms
Automatic Molding	Small round	A few millimeters to a few centimeters in diameter	A few grams to a few hundred grams
Isostatic Molding	Complex geometry	A few centimeters to 0.5 meters in the major dimension	A few tens of grams to a few tens of kilograms
Ram Extrusion	Rod or tube	2–400 millimeter in diameter	Continuous process

Table 2.43 Selection of granular fabrication process based on part geometry

Property [a]	Units	Test method ASTM	TFE-fluoro-carbon
Mechanical			
Tensile strength, 73°F	psi	D-638	1,500–5,000①
Elongation, 73°F	%	D-638	75–350
Modulus of elasticity, 73°F	psi	D-638	50,000–90,000
Flexural strength, 73°F	psi	D-790	No break
Hardness, Rockwell, 73°F		D-785	R10–20
durometer, 73°F		D-676	D55–70
Tensile impact	ft-lb/in.^{2}		30–200
Thermal			
Coef. of linear thermal exp.	in./in.-°F	D-696	5.5×10^{-5}②
Deformation under load 122°F., 2,000 psi	%	D-621	3–7
Heat distortion temperature, 264 psi	°F	D-648	120
Heat distortion temperature, 66 psi	°F		270
Melting point	°F	D-789	621 ± 9
Electrical			
Dielectric strength short time (.080 in. thickness)	V/mil	D-149	400–500⑥
Volume resistivity	OHM-CM	D-257	10^{15}
Dielectric constant, 60 cy		D-150	2.0–2.1
10^3 cy			2.0–2.1
10^6 cy			2.0–2.1
10^8 cy			2.0–2.1
Dielectric loss tangent, 60 cy		D-150	<0.005
10^3 cy			<0.005
10^6 cy			<0.005
10^8 cy			<0.005
Chemical			
Resistant to:			All chemicals and solvents, except molten alkali metals and fluorine
Not resistant to:			Halogenated solvents absorbed to minor degree at elevated temperatures and pressures
Miscellaneous			
Water absorption			
Immersion (24 hr)	%	D-570⑦	0.00–0.05
Saturation (Boiled to constant wt)	%		
equilibrium (50% RH 73°F)	%		
Specific gravity		D-792	2.1–2.3

[a] Property data shown are typical average values and will vary on specific production lots and by size and configuration of product. They, therefore, should be used only as a guide to primary selection for application of a given material, and never for purchase specifications. Further technical information is available for specific application requirements. All values shown are based on bone dry specimens. Where no value is listed, sufficient details are not available to present a usable figure.

① ASTM-D-1457 used for thin specimens. ② Value applies from 73–140° F. ③ Value applies from 73–300° F. ④ 200°F, 1200 psi. ⑤ 500° F, 1200 psi. ⑥ Up to 500V/mil on 1 mil tape. ⑦ Specimen ⅛ in. thick, 2 in. diameter.

Table 2.44 TFE film properties

excellent weatherability and maintains its strength from –180° to 150°C. It has low permeability to most gases and vapors and resists abrasion and staining. Moldings and fibers can be made from PVF, but the major application of the material is in the building industry as a protective coating, in

Dose,[a] Mrads	Yield stress, psi	Tensile strength, psi	Ultimate elongation, %
	Tested at 23°C		
0	1920	3330	354
2.6	2030	3010	383
6.5	2160	2450	274
13	2170	2770	263
130	3430	3420	42
	Tested at 225°C		
0	333	333	160
2.6	290	316	498
6.5	269	409	480
13	—	286	220

[a] Irradiated at 250°C under nitrogen with 2-MeV electrons.

Table 2.45 Tensile properties of irradiated FEP

Dose,[a] Mrads	Yield stress, psi	Tensile strength, psi	Ultimate elongation, %
	Initial Properties		
0	1920	3330	334
2.6	2030	2010	383
6.5	2160	2450	274
13	2170	2770	263
	After 1000 Hr in Atlas Model XW Weather-Ometer (without optical filters)		
0	1890	2890	289
2.6	2160	2740	350
6.5	—	—	—
13	2080	2480	220
	After 960 Hr in Circulating Air Oven at 240°C		
0	2060	3170	400
2.6	2120	2860	430
6.5	2230	2450	290
13	2320	2400	260

[a] Irradiated at 250°C under nitrogen with 2-MeV electrons.

Table 2.46 Tensile effect of aging on FEP

Exposure [b]	Temperature	0.010 in. Air	0.010 in. Oronite 8515 hydraulic fluid	0.021 in. Air	0.021 in. Helium	0.041 in. Air	0.115 in. Air	0.115 in. Helium	0.125 in. Air	0.125 in. Oronite 8515 hydraulic fluid
		Tensile strength (psi)								
Control	Amount	2495		3503		3040	3165		2545	
N 1.6×10^{14}	−65°F	1481								
G $1.3–1.4 \times 10^{6}$	Ambient	1265	1452						1217	1229
	−275°F	1107	1490							
N 2.8×10^{14} G 4.8×10^{6}	Ambient	1371		1293		1197	1035			
N 4.5×10^{14} G 6.8×10^{6}	Ambient	1128		1171		1226	1356			
N $5.9–6.6 \times 10^{14}$ G $0.8–1.0 \times 10^{7}$	Ambient	88	1396						1183	1333
N 7×10^{14} G 1.2×10^{7}	Ambient				1571	797	867	1545		
N 1.6×10^{15} G 2.3×10^{7}	Ambient	1286	1664							
N 5.7×10^{15} G 1.1×10^{8}	Ambient			569		591	638			
N $5.9–6.5 \times 10^{15}$	Ambient	Negligible measurable strength	1641						832	1302
N 1×10^{16} G 1.7×10^{8}	Ambient				1749					

[a] The performance of Teflon (FEP) in vacuum is comparable to that of Teflon (TFE); in air, the effects are similar but slightly less severe for Teflon (FEP).
[b] N, integrated neutron flux $E > 0.33$ MeV (n/cm^2); G, gamma dose (rad).

Table 2.47 TFE tensile properties vs. irradiation in mixed environments

which it is bonded to wood, metal, or asphalt-based materials in 0.001- to 0.002-in thickness. PVF should outlast most paints, enamels, and other surface coatings.

Polyvinylidene Fluoride Related to PVF is polyvinylidene fluoride (PVDF), a crystalline plastic with a melting point of about 170°C. It has good strength properties and resists distortion and creep at both high and low temperatures. PVDF has very good weather, chemical, and solvent resistance. Extrusion, compression molding, and injection molding can process the material. Uses include as a coating, gasketing, and wire- and cable-jacketing material.

Material	Wear rate (mg/hr)	Test conditions
Unfilled Teflon	200	Bearings: 1-in. long by 0.75 in. I.D.
Teflon and 25% graphite	0.7	Shaft: 410 stainless steel Load: 42 lb
Teflon and 22% glass	0.2	Speed: 60 ft/min
Teflon and 25% copper	1.2	
Teflon and 25% graphite	0.7	Bearings: 1-in. long by 0.75 in. I.D.
Teflon and 22% glass	0.4	Shaft: 410 stainless steel Speed: 215 ft/min
Unfilled Teflon	0.74	Bearings: 0.50-in. long by 0.25-in. I.D.
Teflon and 15 to 25% graphite	0.0015	Shaft: 303 stainless steel
Teflon and 15 to 25% glass	0.0015	Load: 250 lb Speed: 9.5 ft/min
Teflon and 4 to 8% copper	0.032	

Table 2.48 Wear rates for sleeve bearings of molded TFE with various fillers

Composition (% by volume)			Coefficient of friction		Wear rate (mil/hr)
Teflon [a]	Resin [b]	Filler	Initial	After 10 hr	
100	0	0			69
40	22	38 [c]	0.25	0.35	3
30	26	44 [c]	0.25	0.27	5
20	30	50 [c]	0.25	0.38	4
37	31	32 [d]	0.15	0.15	

[a] Coarse, general purpose PTFE particles.
[b] Phenolic type.
[c] Asbestos.
[d] Powdered lead.

Table 2.49 Friction and wear characteristics of molded plastics including TFE (Teflon) as an additive

Fluorinated Ethylene Propylene The discovery of fluorinated ethylene propylene (FEP) was first announced in 1960. It is closely related to PTFE but has a lower melt viscosity and may therefore be processed by conventional thermoplastic processing techniques and possesses most of the PTFE. FEP is a tough, resilient material with an Izod value of 2.9 ft-lb/in at –70°F, no break at 73°F, and 95,000 ft-lb/in at 170°F. FEP is nonflammable and melts at 545° to 563°F. It has excellent chemical and solvent resistance and is largely used in such electrical applications as terminal blocks and valve and tube holders. FEP is also used for a variety of nonstick applications in food-processing equipment.

FEP-fluorocarbon resin, a copolymer of tetrafluoroethylene and hexafluoropylene, cross-links when exposed at elevated temperatures to radiation, resulting in interesting and important effects; other fluoropolymers also have interesting and important effects. The properties, including chemical inertness, excellent weatherability, and outstanding electrical properties, are unique and

			After 1000 hr exposure in Weather-Ometer	
Dose,[a] Mrads	Dielectric constant	Dissipation factor	Dielectric constant	Dissipation factor
0	2.04	<0.0001	2.04	<0.0001
2.6	2.06	<0.0001	2.06	<0.0001
6.5	2.06	<0.0001	2.08	<0.0001
13	2.06	<0.0001	2.05	<0.0001

[a] Irradiated at 250°C under nitrogen with 2-MeV electrons, 60-mil films.

Table 2.50 Electrical properties of irradiated FEP

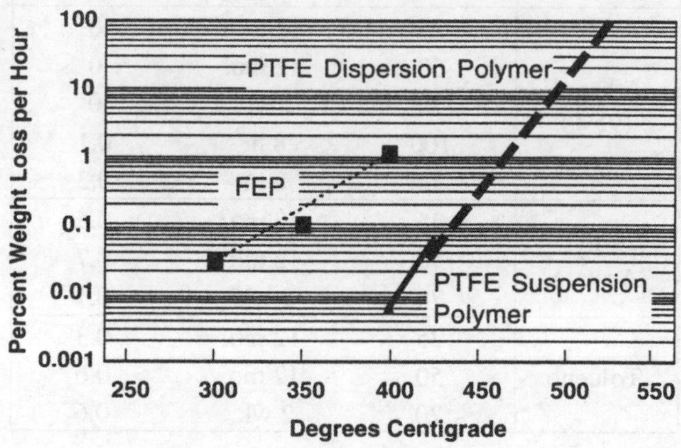

Figure 2.24 Comparison of thermal degradation of PTFE and FEP

are a result of their perfluorocarbon structure, the fluorine atoms being bonded strongly to the carbon atoms.

FEP resin, like PTFE resins, degrades when exposed to high-energy radiation, with a resultant adverse effect upon the physical properties. At elevated temperatures it can be cross-linked by use of ionizing and ultraviolet radiation. With the introduction of cross-linking reactions, two types of resin became available. One, in which a small amount of cross-linking is introduced, is a resin of altered melt-processing characteristics due to a changed distribution of molecular weights. The other type is cross-linked to the extent that it is incapable of melt processing and, in general, has the high-temperature properties associated with a three-dimensional network.

When FEP resin is irradiated at room temperature, net degradation results, as shown by the decrease in the melt viscosity of the resin after irradiation (Fig. 2.27). If its temperature is

Solvent	Exposure Temp., °C	Exposure Time	Weight Gain, %
Acetone	20	12 mo.	0.3
	50	12 mo.	0.4
	70	2 wk.	0
Benzene	78	96 hr	0.5
	100	8 hr	0.6
	200	8 hr	1.0
Carbon Tetra-chloride	25	12 mo.	0.6
	50	12 mo.	1.6
	70	2 wk.	1.9
	100	8 hr	2.5
	200	8 hr	3.7
Ethanol (95%)	25	12 mo.	0
	50	12 mo.	0
	70	2 wk.	0
	100	8 hr	0.1
	200	8 hr	0.3
Ethyl Acetate	25	12 mo.	0.5
	50	12 mo.	0.7
	70	2 wk.	0.7
Toluene	25	12 mo.	0.3
	50	12 mo.	0.6
	70	2 wk.	0.6

Table 2.51 Chemical resistance of PTFE to common solvents

Chemical	Effect on PTFE Sample
Abietic Acid	Insoluble at boiling point
Acetic Acid	Wets
Acetophenone	Insoluble – 0.2% weight gain after 24 hr. at 150ºC
Acrylic Anhydride	No effect at room temperature
Allyl Acetate	No effect at room temperature
Allyl Methacrylate	No effect at room temperature
Aluminum Chloride	Insoluble in solution with NaCl; 1-5% anhydrous $AlCl_3$ affects mechanical properties
Ammonium Chloride	Insoluble at boiling point
Aniline	Insoluble – 0.3% weight gain after 24 hr. at 150ºC
Borax	No wetting or effect by 5% solution
Boric Acid	Insoluble at boiling point
Butyl Acetate	Insoluble at boiling point
Butyl Methacrylate	No effect at room temperature
Calcium Chloride	No effect by saturated solution in methanol
Carbon Disulfide	Insoluble at boiling point
Cetane	Wets, insoluble at boiling point
Chromic Acid	Insoluble at boiling point
Cyclohexanone	No effect observed
Dibutyl Phthalate	Wets, no effect at 250ºC
Diethyl Carbonate	No effect at the room temperature
Dimethyl Ether	No effect observed
Dimethyl Formamide	No effect observed
Ethyl Ether	Wets
Ethylene Glycol	Insoluble at boiling point
Ferric Chloride	1-5% $FeCl_3.6H_2O$ reduces mechanical properties
Ferric Phosphate	No effect by 5% solution
Formaldehyde	Insoluble at boiling point after 2 hr.
Formic Acid	Insoluble at boiling point
Hexane	Wets
Hydrogen Fluoride	Wets, no effect by 100% HF at the room temperature
Lead	No effect
Magnesium Chloride	Insoluble at boiling point
Mercury	Insoluble at boiling point
Methacrylic Acid	No effect at the room temperature
Methanol	Wets
Methyl Methacrylate	Wets above melting point
Naphthalene	No effect
Nitrobenzene	No effect
2-Nitro-Butanol	No effect
Nitromethane	No effect
2-Nitro-2-Methyl Propanol	No effect
n-Octadecyl Alcohol	Wets
Phenol	Insoluble at boiling point
Phthalic Acid	Wets

Table 2.52 Chemical compatibility of PTFE with various chemicals

Property	Test method	Units	Temperature °C	Unfilled[1]	Compound 65% Bronze	Compound 15% Carbon	Compound 25% Glass Fibre
Tensile yield strength	ASTM D 1708	N/mm²	−253	123	–	–	30
			−196	92	47	44	24
			−183	(84)	(44)	(41)	(23)
			−129	53	(35)	(32)	(18)
			− 79	36	(26)	(24)	(14)
			+ 23	18	12	8	10
Ultimate tensile strength	ASTM D 1708	N/mm²	−253	124	–	–	31
			−196	103	47	47	25
			−183	(95)	(45)	(45)	(24)
			−129	63	(35)	(35)	21
			− 79	42	(26)	(27)	20
			+ 23	36	15	16	19
Tensile modulus	ASTM D 1708	N/mm²	−253	4300	–	–	3200
			−196	3200	–	–	2600
			−183	3100	–	–	(2500)
			−129	2100	–	–	(2000)
			− 79	1400	–	–	(1600)
			+ 23	750	–	–	800
Elongation	ASTM D 1708	%	−253	3	0	0	1
			−196	10	4	2	4
			−183	(12)	(5)	(4)	(6)
			−129	70	(34)	(23)	(35)
			− 79	130	(70)	(50)	(80)
			+ 23	400	200	140	240
Flexural strength	ASTM D 790	N/mm²	−253				
			−196	Did not break	Did not break	Did not break	Did not break
			−183				
			−129				
			− 79				
			+ 23				

Note: Parenthesis indicate values estimated by interpolation

(1) Crystallinity 41–71%

Table 2.53 Mechanical properties of PTFE compounds

Property	Test method	Units	Temperature °C	Unfilled[1]	Compound 65% Bronze	Compound 15% Carbon	Compound 25% Glass Fibre
Flexural modulus	ASTM D 790	N/mm²	−253	5100	–	–	2800
			−196	4700	–	–	2500
			−183	4600	–	–	2300
			−129	3200	–	–	1900
			− 79	1600	–	–	1700
			+ 23	700	–	–	1000
Compressive strength	ASTM D 695	N/mm²	−253	220	–	–	188
			−196	171	–	–	152
			−183	(168)	–	–	(140)
			−129	158	–	–	(109)
			− 79	143	–	–	(79)
			+ 23	118	–	–	24
Compressive modulus	ASTM D 695	N/mm²	−253	6200	6200	6100	6800
			−196	5500	5700	6000	5900
			−183	(5400)	(5400)	(5500)	(5600)
			−129	4100	(4100)	(4300)	(4400)
			− 79	1900	(3000)	(3200)	(3200)
			+ 23	700	550	900	860
Torsional modulus of rigidity	ASTM D 1043	N/mm²	−253	2200	–	–	1700
			−196	1500	–	–	720
			−183	(1400)	–	–	(690)
			−129	1000	–	–	(520)
			− 79	500	–	–	(390)
			+ 23	160	–	–	110
Izod impact strength (notched)	ASTM D 256	J/m	−253	75	67	51	53
			−196	80	70	60	59
			−183	(85)	(73)	(64)	(63)
			−129	(97)	(86)	(103)	(98)
			− 79	133	(121)	(139)	(143)
			+ 23	160	161	169	173

Table 2.53 *(continued)*

raised before irradiation, cross-linking predominates, as shown by an increase in melt viscosity. The temperature at which the effects of cross-linking equal the effects of degradation is the glass transition temperature (approximately 80°C [176°F]), as determined by internal friction measurements. Cross-linking overcomes degradation above 80°C. Above 80°C at a constant dose of radiation, the melt viscosity increases with temperature up through the crystalline melting point. At higher temperatures (>300°C [572°F]), thermal degradation becomes a factor and net increases in viscosity are smaller.

Given high doses of radiation (100 Mrads) at elevated temperatures in the presence of air, FEP becomes embrittled, while specimens irradiated under nitrogen do not. At low doses, 1 Mrad or less, differences in melt viscosity and toughness are not observed. However, all irradiations carried out under a nitrogen atmosphere are maintained at the temperature of the plastic.

After irradiation, FEP resin still exhibits the excellent electrical properties associated with fluorocarbons. In addition, the electrical properties of the irradiated resin are not changed after accelerated aging (Table 2.63).

Radiation doses up to 13 Mrads cause minor reductions in the tensile properties of flat sheets of resin, as shown in Table 2.64. At 130 Mrads, the yield stress increases from 1920 to 3,430 psi, and the ultimate elongation decreases from 354% to 42%. This 80% increase in yield stress will be of value in certain high-stress/low-strain design considerations.

Temperature: 23°C	**Unfilled PTFE**	**15% Glass**	**25% Glass**	**60% Bronze**	**23C/2Gr**	**29C/2Gr**
Tensile strength, N/mm^2, MD	35,5	25,5	19,5	17,0	11,0	7,0
CD	36,5	24,5	18,5	17,0	15,0	10,0
Ultimate elongation, %, MD	400	285	235	250	65	35
CD	450	290	250	235	105	50
Temperature: 150°C	**701-N**	**1103-N**	**1105-N**	**1146-N**	**1191-N**	**1192-N**
Tensile strength, N/mm^2, MD	13,5	8,6	6,3	5,4	3,8	3,0
CD	14,0	8,4	5,8	6.4	5,6	3,8
Ultimate elongation, %, MD	440	350	305	305	120	60
CD	480	340	300	330	185	85

C = Carbon, Gr = Graphite

Table 2.54 Tensile properties of filled PTFE compounds (ASTM D 1708)

Property	Test	Unfilled TFE	15% Glass Fiber	25% Glass Fiber	15% Graph-ite	60% Bronze	20% Glass 5% Graph-ite	15% Glass 5% MoS_2
Filler loading vol.%			13.3	22.2	14.6	27.1	17.6, 5.0	13.3, 2.4
Bulk density, lb./cu. in.		0.0129	0.0103	0.0107	0.0087	0.0103	0.0104	0.0107
Spec. gravity								
Theoretical			2.22	2.26	2.19	3.97	2.25	2.28
Measured	ASTM D1457-66T	2.18	2.20	2.22	2.12	3.85	2.18	2.27
Coeff. of linear therm. exp. (x 10^{-5}), m./in./°F.	ASTM D696-44							
78°-200°F	MD		8.02	6.97	6.97	5.40	7.73	8.33
	CD	6.81	2.94	4.19	4.38	4.38	2.61	3.51
78°-300°F	MD		8.40	7.33	7.46	5.72	7.73	8.77
	CD	7.04	2.96	4.19	4.67	4.37	3.01	3.55
78°-400°F	MD		9.05	8.00	8.14	6.35	8.12	9.61
	CD	7.60	3.42	4.72	5.12	4.99	3.22	3.85
78°-500°F	MD		10.31	9.35	9.50	7.82	9.24	11.11
	CD	9.12	4.15	5.55	5.97	5.75	3.88	4.45

MD = Parallel to molding direction
CD = Cross section perpendicular to molding direction

Table 2.55 Effect of fillers on the linear thermal expansion of PTFE

When the tensile measurements are made at 225°C (435°F), a slight decrease in yield stress is observed, and the tensile strength increases up to a dose of 6.5 Mrads, then decreases. At elevated temperatures, the ultimate elongation is increased by as much as a factor of 3 over the unirradiated resin. The increased elongation at elevated temperatures can be advantageous in the shrink-fitted type of encapsulations where higher shrinkage ratios are desired. For example, a piece of irradiated tubing can be expanded at elevated temperature and chilled while stretched, and then the material to be encased is positioned in the tubing. Upon reheating to a slightly higher temperature than the expansion temperature, the tubing will shrink and tightly encase the article.

Property	Definition	Reference ASTM Method
Bulk Density[1]	Mass of one liter of resin measured under the test conditions	D 1895
Particle Size[1]	Average particle size and distribution by sieving	E 11
Melting Characteristics[1]	Heat of Fusion and Melting Peak Temperature of resin as determined by Differential Scanning Calorimetry	D 4591
Water Content[1]	Water present in the PTFE resin	
Standard Specific Gravity (SSG)[2]	Specific gravity of a sample of molded and sintered PTFE according to this method	D 792, D 1505
Thermal Instability Index (TII) [2]	A measure of decrease in molecular weight of PTFE material determined by the difference between ESG and SSG: TII = (ESG-SSG) × 1000	
Tensile Properties[2]	Elongation and strength at break of a sample made according to the specified method	D 638
Shrinkage and Growth	The change in the diameter of SSG preform due to sintering	
Extended Specific Gravity (ESG)	The specific gravity of a PTFE specimen molded for SSG after sintering for an extended period of time, compared to the sintering time of SSG	D 792, D 1505
Electrical Properties	Dielectric constant, dissipation factor, dielectric breakdown voltage, dielectric strength	D 149

[1] Properties required for resin specification.
[2] Properties required ıor on molded specimen.

Table 2.56 Definition of basic properties of granular PTFE (ASTM D 4894)

FEP resin has outstanding resistance to aging. After irradiation, this property is substantially unchanged, as shown by comparison of results of tensile measurements before and after aging in the Weather-Ometer and in a high-temperature air oven (Table 2.65).

The flexural moduli of FEP resin measured over the range of 23° to 200°C (73° to 392°F) exhibit only minor changes (Table 2.66) when given radiation doses up to 13 Mrads.

Property	Definition	Reference ASTM Method
Bulk Density[1]	Mass of one liter of resin measured under the test conditions	D 1895
Particle Size[1]	Average particle size and distribution by sieving	E 11
Melting Characteristics[1]	Heat of fusion and melting peak temperature of resin as determined by differential scanning calorimetry	D 4591
Water Content[1]	Water present in the PTFE resin	
Standard Specific Gravity (SSG)[2]	Specific gravity of a sample of molded and sintered PTFE according to this method	D 792, D 1505
Thermal Instability Index (TII)[1,2]	A measure of decrease in molecular weight of PTFE material determined by the difference between ESG and SSG: TII = (ESG-SSG) × 1000	
Tensile Properties[1,2]	Elongation and strength at break of a sample made according to the specified method	D 638
Extrusion Pressure[1]	The pressure measured while extruding a paste of fine powder PTFE made with an iso-paraffin under specified conditions	
Stretch Void Index[1,2] (SVI)	A measure of change in specific gravity of a PTFE specimen as a result of being subjected to tensile strain	
Strained Specific Gravity	specific gravity of a PTFE specimen after being subjected to tensile strain	
Untrained Specific Gravity	specific gravity of a PTFE specimen before being subjected to tensile strain	
Shrinkage and Growth	The change in the diameter of SSG preform due to sintering	
Extended Specific Gravity (ESG)	The specific gravity of a PTFE specimen molded for SSG after sintering for an extended period of time, compared to the sintering time of SSG	D 792, D 1505

[1] Properties required for resin specification.
[2] Properties required for on molded specimen.

Table 2.57 Definition of basic properties of fine-powder PTFE (ASTM D 4895)

CHEMICAL	FILLER CARBON/ GRAPHITE	GLASS	BRONZE
Acetaldehyde	A	A	A
Acetone	A	A	A
Aluminium sulphate	A	A	B
Ammonium chloride	A	A	C
Ammonium hydroxide	A	B	C
Aniline	A	A	C
Benzene	A	A	A
Boric acid	A	A	A
Brine	A	A	A
Bromine (anhydrous)	C	B	C
Carbon disulphide	A	A	A
Chloroacetic acid	A	A	B
Chlorobenzene	A	A	A
Chloroform	A	A	A
Chromic acid	B	B	C
Citric acid	A	A	A
Diethyl ether	A	A	A
Ethylene glycol	A	A	A
Fatty acids	A	A	A
Ferric chloride	A	A	C
Ferric sulphate	A	A	C
Fluorosilicic acid	B	C	C
Formic acid	A	A	A
Hydroboric acid	A	B	C
Hydrochloric acid	A	B	C
Hydrocyanic acid	A	B	C
Hydrofluoric acid	A	C	C
Hydrogen sulphide (solution)	A	C	C
Lead acetate	A	A	C
Maleic acid	A	A	B
Mercury salts	A	A	C
Molasses	A	A	B

Table 2.58 Chemical resistance of filled PTFE compounds

CHEMICAL	FILLER		
	CARBON/ GRAPHITE	GLASS	BRONZE
Naphtha	A	A	B
Naphthalene	A	A	B
Nickel salts	A	A	A
Nitric acid (0-50%)	C	B	C
Nitro benzene	A	A	A
Phenol	A	B	A
Phosphoric acid	A	A	C
Phthalic acid	A	A	A
Picric acid	A	A	A
Pyridine	A	A	C
Salicylic acid	A	A	B
Silver nitrate	A	A	C
Sodium carbonate	A	A	A
Sodium hydroxide	A	B	A
Sodium nitrite	A	A	A
Sodium peroxide	B	A	C
Sodium silicate	A	C	A
Sodium sulphide	A	A	C
Starch	A	A	A
Sulphuric acid	B	A	C
Tallow	A	A	A
Tannic acid	A	A	A
Tartaric acid	A	A	A
Trichloroethylene	A	A	B
Zinc chloride	A	A	C

A = excellent
B – Fair
C – Unsatisfactory

Table 2.58 *(continued)*

Radiation at elevated temperatures does cause some reduction in toughness, as shown in Table 2.67. Tensile impact resistance, flexural life, and tear resistance decrease with increased dose.

Small amounts of cross-linking produce a branched polymer structure in addition to an increased molecular weight. The increase in molecular weight and broadening of the molecular weight distribution cause a change in the melt-flow characteristics of the plastic.

Chlorofluorohydrocarbon Chlorofluorohydrocarbons are plastics made from monomers composed of chlorine, fluorine, hydrogen, and carbon only.

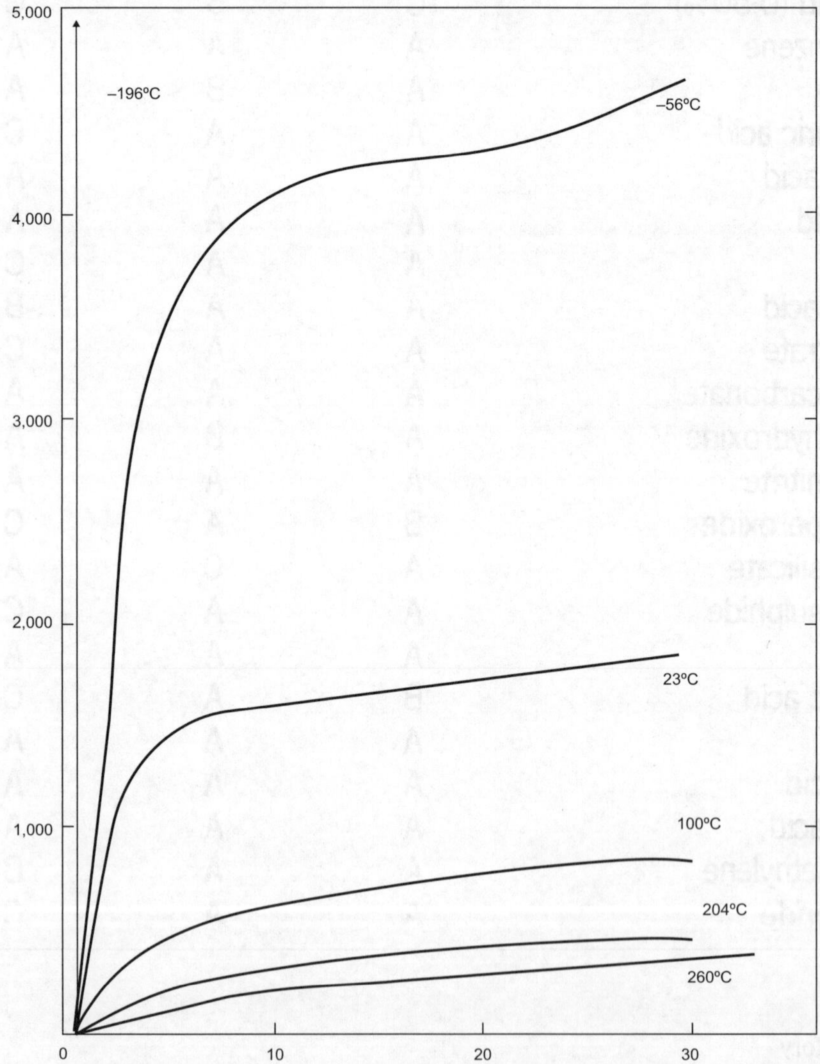

Figure 2.25 Tensile stress-strain curves at different temperatures for PTFE

Polyfluoroalkoxyphosphazene Polyfluoroalkoxyphosphazene (PNF) is a very expensive product. It achieved its first mass-production applications in the automobile industry. Although its heat resistance is less than that of FPM, this is offset by its considerably better low temperature flexibility. It has good resistance to fuels.

Other Other fluorocarbon plastics include perfluoroacrylates and fluorinated silicones in which the addition of the fluorine atom improves the chemical and solvent resistance of the original material. The addition of fluorine also generally increases the temperature stability of the physical properties of the original plastics.

IONOMER

The ionomers were introduced in 1964. They are in the polyolefin family. Their interchain ionic bonding distinguishes them from the other plastics. Performances include being extremely tough, very high tensile strength, excellent abrasion resistance, and outstanding transparency. Clarity, strength, and good adhesion of ionomer films to metal surfaces are the important properties that have led them into widespread use in food packaging. They are often used as a heat-seal layer in thermoplastic structures (coextruded films, etc.).

These characteristics have led to their use as coatings, film, and sheet. The ionomer plastic can be processed basically by the conventional thermoplastic processing techniques. Spirex Corp. has a patented vented melt plasticizer system that ensures a constant moisture level, reducing variability in product dimensions and mechanical properties. The ionomers have a relatively low softening point (160°F) but offer high resistance to stress cracking and a constant dielectric loss over their useful temperature range (–160° to 160°F). Table 2.68 summarizes the properties of ionomers.

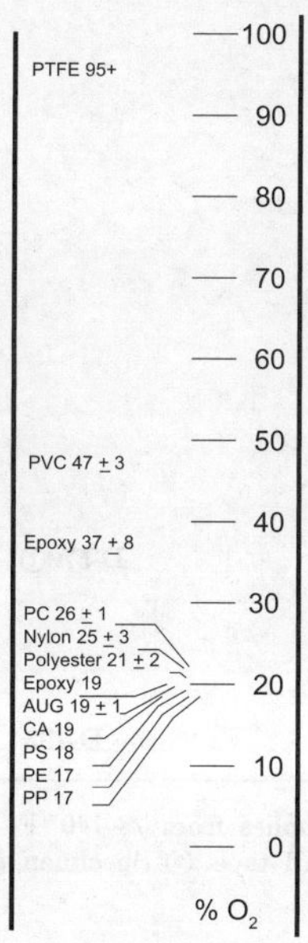

Figure 2.26 Examples of plastics limiting oxygen index.

Property	Units	Test method ASTM	TFE-fluoro-carbon
Mechanical			
Tensile strength, 73°F	psi	D-638	1,500–5,000①
Elongation, 73°F	%	D-638	75–350
Modulus of elasticity, 73°F	psi	D-638	50,000–90,000
Flexural strength, 73°F	psi	D-790	No break
Hardness, Rockwell, 73°F		D-785	R10–20
durometer, 73°F		D-676	D55–70
Tensile impact	ft-lb/in.^{2}		30–200
Thermal			
Coef. of linear thermal exp.	in./in.-°F	D-696	5.5×10^{-5}②
Deformation under load 122°F., 2,000 psi	%	D-621	3–7
Heat distortion temperature 264 psi	°F	D-648	120
Heat distortion temperature 66 psi	°F		270
Melting point	°F	D-789	621 ± 9
Electrical			
Dielectric strength short time (.080 in. thickness)	V/mil	D-149	400–500⑥
Volume resistivity	OHM-CM	D-257	10^{15}
Dielectric constant, 60 cy		D-150	2.0–2.1
10^3 cy			2.0–2.1
10^6 cy			2.0–2.1
10^8 cy			2.0–2.1
Dielectric loss tangent, 60 cy		D-150	<0.005
10^3 cy			<0.005
10^6 cy			<0.005
10^8 cy			<0.005
Chemical			
Resistant to:			All chemicals and solvents, except molten alkali metals and fluorine
Not resistant to:			Halogenated solvents absorbed to minor degree at elevated temperatures and pressures
Miscellaneous			
Water absorption			
Immersion (24 hr)	%	D-570⑦	0.00–0.05
Saturation (Boiled to constant wt)	%		
equilibrium (50% RH 73°F)	%		
Specific gravity		D-792	2.1–2.3

① ASTM-D-1457 used for thin specimens. ② Value applies from 73–140° F. ③ Value applies from 73–300° F. ④ 200°F, 1200 psi. ⑤ 500° F, 1200 psi. ⑥ Up to 500V/mil on 1 mil tape. ⑦ Specimen ⅛ in. thick, 2 in. diameter.

Table 2.59 TFE properties

NYLON (POLYAMIDE)

Nylons or polyamides (PAs) are crystalline plastic. They were the first of the thermoplastic engineering plastics (first developed in the 1930s). They were originally developed as high-strength textile fibers for stockings. There are different types available. The usual numerical classification is based upon the number of carbon atoms in the polybasic acid and the polyfunctional amines. Nylon 6/6, for example, is polyhexamethyleneadipamide.

Nylons contain an amide group as a recurring part of the chain. They are produced by condensation polymerization. Nylons exhibit good chemical resistance and toughness and retain their physical properties over a wide range of temperatures. They also have good surface lubricity.

PROPERTY	VALUE
Molecular weight	100.02
Boiling point at 101.3kPa, °C	-76.3
Freezing point, °C	-142.5
Liquid density vs. temperature (°C), g/mL	
$-100 < t < -40$	$1.202 - 0.0041t$
$-40 < t < 8$	$1.1507 - 0.0069t - 0.000037t^2$
$8 < t < 30$	$1.1325 - 0.0029t - 0.00025t^2$
Vapor pressure at T K, kPa	
$196.85 < T < 273.15$	$\log_{10}P_{kPa} = 6.4593-875.14/T$
$273.15 < T < 306.45$	$\log_{10}P_{kPa} = 6.4289-866.84/T$
Critical temperature, °C	33.3
Critical pressure, MPa	39.2
Critical density, g/mL	0.58
Dielectric constant at 28°C	
at 101.3 kPa	1.0017
at 858 kPa	1.015
Thermal conductivity at 30°C, mW/(m·K)	15.5
Heat of formation for ideal gas at 25°C, ΔH, kJ/mol	-635.5
Heat of polymerization to solid polymer at 25°C, ΔH, kJ/mol	-172.0
Flammability limits in air at 101.3 kPa, vol.%	14-43

Table 2.60 Properties of PTFE

Most sheet applications are in automotive and industrial markets where parts are stamped or machined into final shape.

Tough and slippery, these families of TPs have good electrical properties but are hygroscopic and have dimensional stability lower than that of most other engineering types. They are also offered in reinforced and filled grades as a moderately priced metal replacement.

These crystalline plastics, including the amorphous types developed more recently, are available for processing by different methods. There are different nylons, but as a family their characteristic

	Test condition [a]	Static coefficient of friction: Ground cast iron (20 μin. rms)	Scraped cast iron (20 contact points/in.²)	Ground steel (20 μin. rms)
Teflon TFE resin	Dry	0.079	0.070	0.087
	Lubricated	0.046	0.098	0.079
Glass-filled Teflon TFE resin	Dry	0.148	0.091	
	Lubricated	0.090	0.094	0.095
Graphite-filled Teflon TFE resin	Dry	0.076	0.130	0.110
	Lubricated	0.075	0.084	0.087
Bearing bronze	Dry	0.260	0.250	— [b]
	Lubricated	0.150	0.211	0.098
Cast iron	Dry	0.350	0.202	— [b]
	Lubricated	0.142	0.30	0.123

[a] Period at rest, 1 min; load, 50 lb.
[b] Denotes increasing coefficient of friction in successive tests.

Table 2.61 Static coefficients of friction for PTFE and other materials

Composition (% by volume): Teflon [a]	Resin [b]	Filler	Coefficient of friction: Initial	After 10 hr	Wear rate (mil/hr)
100	0	0			69
40	22	38 [c]	0.25	0.35	3
30	26	44 [c]	0.25	0.27	5
20	30	50 [c]	0.25	0.38	4
37	31	32 [d]	0.15	0.15	

[a] Coarse, general purpose PTFE particles.
[b] Phenolic type.
[c] Asbestos.
[d] Powdered lead.

Table 2.62 Friction and wear characteristics of moldings using PTFE as a filler

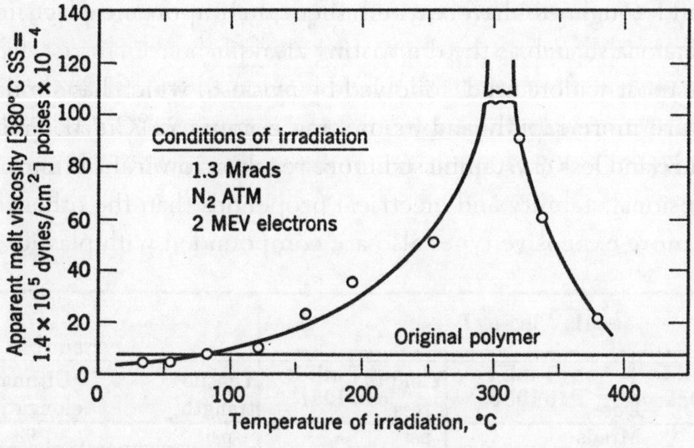

Figure 2.27 Effect of temperature of irradiation on apparent melt density of FEP

Dose,[a] Mrads	Dielectric constant	Dissipation factor	After 1000 hr exposure in Weather-Ometer: Dielectric constant	Dissipation factor
0	2.04	<0.0001	2.04	<0.0001
2.6	2.06	<0.0001	2.06	<0.0001
6.5	2.06	<0.0001	2.08	<0.0001
13	2.06	<0.0001	2.05	<0.0001

[a] Irradiated at 250°C under nitrogen with 2-MeV electrons, 60-mil films.

Table 2.63 Electrical properties of irradiated FEP

Dose,[a] Mrads	Yield stress, psi	Tensile strength, psi	Ultimate elongation, %
	Tested at 23°C		
0	1920	3330	354
2.6	2030	3010	383
6.5	2160	2450	274
13	2170	2770	263
130	3430	3420	42
	Tested at 225°C		
0	333	333	160
2.6	290	316	498
6.5	269	409	480
13	—	286	220

[a] Irradiated at 250°C under nitrogen with 2-MeV electrons.

Table 2.64 Tensile properties of irradiated FEP

strength, stiffness, and toughness have earned them an important place in the plastics when compared to other materials, such as the die-casting alloys.

Nylon 6/6 is the most widely used, followed by nylon 6, which has similar properties except that it absorbs moisture more rapidly and its melting point is 21°C (70°F) lower. Also, its lower processing temperature and less crystalline structure result in lower mold shrinkage. Nylons 11 and 12 have better dimensional stability and electrical properties than the others because they absorb less moisture. These more expensive types also are compounded with plasticizers to increase their

Dose,[a] Mrads	Yield stress, psi	Tensile strength, psi	Ultimate elongation, %
Initial Properties			
0	1920	3330	334
2.6	2030	2010	383
6.5	2160	2450	274
13	2170	2770	263
After 1000 Hr in Atlas Model XW Weather-Ometer (without optical filters)			
0	1890	2890	289
2.6	2160	2740	350
6.5	—	—	—
13	2080	2480	220
After 960 Hr in Circulating Air Oven at 240°C			
0	2060	3170	400
2.6	2120	2860	430
6.5	2230	2450	290
13	2320	2400	260

[a] Irradiated at 250°C under nitrogen with 2-MeV electrons.

Table 2.65 Effect of aging on FEP tensile properties

Dose,[a] Mrads	Temperature of test Flexural modulus, psi 23°C	100°C	200°C
0	102,000	15,900	6310
2.6	110,000	20,900	5290
6.5	79,000	18,300	5550
13	77,400	21,500	4900

[a] Irradiated at 250°C under nitrogen with 2-MeV electrons.

Table 2.66 Effect of radiation on FEP flexural modulus

flexibility and ductility. With nylon toughening and technology advancements, supertough nylons became available. Their notched Izod impact values are over 10 J/m (20 ft-lb/in), and they fail in a ductile manner.

Other important types of nylons include the castable types, liquid monomers (not the usual solid) that polymerize and become solid at atmospheric pressure. From these nylons complex products several inches thick and weighing hundreds of pounds can be cast. Another castable liquid monomer is a moldable transparent material. This amorphous type offers better chemical resistance than other transparent thermoplastics.

Property comparisons among the commercial grades of different nylons vary widely because so many formulations are available. In general, they all have excellent fatigue resistance, low coefficients of friction, and good toughness (depending on their degree of crystallinity), and they resist a wide spectrum of fuels, oils, and chemicals. Mechanical and other properties of nylons are increased at room and elevated temperatures by incorporating glass fibers. An example of elevated mechanical properties when reinforced is shown in Table 2.69.

Dose,[a] Mrads	Elemendorf tear strength, g/mil	MIT flex life test, flexes to failure	Tensile impact, ft-lb/cu in.
0	116[b]	1885[b]	908[b]
2.6	70	1770	814
6.5	50	1970	813
13	26	1410	348

[a] Irradiated at 250°C under nitrogen with 2-MeV electrons.
[b] Annealed samples. Values for tear strength and flex life are lowered for non-radiated samples because of increased crystallinity. Annealing has little effect on tensile impact resistance.

Table 2.67 Effect of radiation on FEP toughness

Property	Unit	Valve
Specific gravity		0.93–0.95
Izod impact	ft-lb/in. of notch	5.7–14.6
Brittleness temperature	°F	—160
Softening point	°F	160
Tensile strength	psi	3500–5500
Elongation	%	300–500

Table 2.68 General Properties of Ionomer plastics

Property	Short Glass Fiber 30%	Short Glass Fiber 50%	Long Glass Fiber 30%	Long Glass Fiber 50%
	At 300°F			
Tensile strength (10^3 psi)	12.8	13.8	14.3	19.2
Elongation (%)	9.3	7.8	5.3	5.6
Flexural strength (10^3 psi)	13.8	14.3	17.4	23.7
Flexural modulus (10^5 psi)	4.64	5.27	5.50	8.90
	At 400°F			
Tensile strength (10^3 psi)	6.3	7.3	7.8	8.3
Elongation (%)	8.6	9.5	6.2	6.8
Flexural strength (10^3 psi)	6.9	7.4	8.9	10.0
Flexural modulus (10^5 psi)	3.96	4.80	5.19	7.51

[1]Chapter 15 reviews fiber lengths.

Table 2.69 Nylon 6/6-glass fiber reinforcement properties at different temperatures

Nylon 6/6 has the lowest permeability by gasoline and mineral oil of all the nylons. The 6/10 and 6/12 types are used where lower moisture absorption and better dimensional stability are needed. All nylons are inert to biological attack and have electrical properties adequate for most voltages and frequencies. The crystalline structure of nylons that can be controlled to some degree by processing affects their stiffness, strength, and heat resistance. Low crystallinity imparts greater toughness, elongation, and impact resistance but at the sacrifice of tensile strength and stiffness.

All nylons absorb moisture that may be present in the application's environment. An increase in moisture content decreases a material's strength and stiffness and increases its elongation and impact resistance. Type 6/6 nylon usually reaches equilibrium at about 2.5 wt% moisture when the relative humidity (RH) reaches 50%. The equilibrium moisture at 50 RH in nylon 6 is slightly higher. In general, nylon's dimensions increase by about 0.2% to 0.3% for each 1% of moisture absorbed. However, performing moisture conditioning prior to putting products into service can compensate for dimensional changes caused by moisture absorption. Such formulations as 6/12, 11, and 12 are considerably less sensitive to moisture than others.

When UV stabilizers are compounded in the nylon, they become insensitive to UV light. Carbon black is the most effective stabilizer. UV stabilizers also increase tensile strength and hardness and slightly decrease ductility and toughness.

Nylons have good resistance to creep and cold flow as compared to many of the less-rigid thermoplastics. Usually, creep can be accurately calculated using apparent modulus values. They also have outstanding resistance to repeated impact. Nylons can withstand a major portion of a breaking load almost indefinitely.

Nylons are used in many different markets, the largest being the automotive market. Their performance capabilities make them suitable for different mechanical and electrical hardware, particularly for such under-the-hood products as timing sprockets, speedometer gears, cooling fans, wire connectors, windshield-wiper parts, door latches, fender extensions, steering-column-lock housings, brake-fluid reservoirs, and other uses. Their low friction, good abrasion resistance, and ability to operate without lubricants qualify nylons for use in many bearing applications, business machines, and appliances. For extra protection, occasional lubrication can be applied (Table 2.70).

Extruded nylon tubing and hoses are used in hydraulic and other fluid systems because of their resistance to different fluids. The applications for castings are mostly in industrial equipment, including large rollers, bearings, gears, cams, sheaves, guide blocks, wear plates, and the like. Nylon powder, which can be applied either electrostatically or by a fluid bed, provides tough, wear-resistant coatings.

PARYLENE

The melting point of these film and coating plastics ranges from 290° to 400°C (554° to 752°F), and T_g from 60° to 100°C (140° to 212°F). Their cryogenic performances are excellent. Physical properties are unaffected by thermal cycles from 2°K to room temperature. They have good thermal endurance in air, the absence of air, and inert atmosphere. They are generally insoluble up to 150°C (302°F). Weather resistance is poor.

These poly-*p*-xylylenes combine the stability of aromatic phenylene units with the flexibility of aliphatic connecting linkages. Pyrolyzing *p*-xylene at 950°C in the presence of steam produces them. The cyclic dimer di-*p*-xylylene forms initially and is further pyrolyzed to *p*-xylylene vapor. This vapor, upon being cooled below 50°C, spontaneously polymerized to form linear polymers that are usually deposited as thin films on a suitable substrate. Solid chemicals can be encapsulated by this process, and free films can be prepared by coating surfaces from which the films can be conveniently stripped.

The thermal stabilities of the poly-*p*-xylylene are good but not exceptional. In an inert atmosphere, the materials can provide long-term service up to 222°C and short-term service to 350°C. However, service temperatures below about 80°C are more realistic in air. Under cryogenic conditions, the polymers show excellent flexibility at –201°C and good electrical insulation properties near absolute zero. The electrical properties also remain good at elevated temperatures, even after the mechanical properties begin to degrade.

The poly-*p*-xylylenes are insoluble in all organic solvents below 150°C, and they resist permeation by all except aromatic hydrocarbons. Thus, the polymer films exhibit excellent barrier properties against gases and moisture. Tough, homogeneous moldings of poly-*p*-xylylenes have been prepared at 400°C/2,000 psi.

PHENOXY

An outgrowth of epoxy plastic technology, phenoxy is a high molecular weight thermoplastic polyester plastic based on bisphenol-A and epichlorohydrin. Phenoxies have moderately good impact resistance, relatively high strength, and good elongation and creep resistance. They are thermoplastics, but they can also be thermosets, through cross-linking chemical reactions.

Material	Type of lubrication	Type of material	Load (psi)	Surface speed (ft/min)	Coefficient of friction	Total wear[a] (mil)	Pressure-velocity product (PV)[b]	Power consumption (FV)[c]	Running time (hr)
Nylasint 66	Sintered in Gulf Crest 55 Petroleum base oil (no other lubrication)		135	240	0.12 (smooth, no change)	1.5	32,400	3,890	48
Nylasint 66	Sintered in Gulf Crest 55 (no other lubrication)		195	240	0.10 (smooth, no change)	1.5	46,800	4,680	34
Nylasint 66	Sintered in Gulf Crest 55 (no other lubrication)		195	370	0.09 (smooth, no change)	1.5	72,200	6,500	24
Nylasint 64M	Sintered in Gulf Crest 55 (no other lubrication)		195	240	0.10 (smooth, no change)	1.5	46,800	4,680	24
Nylon 101 injection-molded	2 drops Gulf Crest 55		195	240	0.11 (smooth, no change)	2.7	46,800	5,150	21
Nylon 101 injection-molded	2 drops Gulf Crest 55		195	370	0.11–0.32 (very rough and variable)	0.6	72,200	8,000 to 23,000	Failed-0.5
Nylon 101 injection-molded	None		29	240	0.26–1.18 (steady increase, rough during latter part of test)	1.6	6,900	1,800 to 8,200	Failed-0.5
Nylon 101 injection-molded	None		52	240	0.19–0.67 (steady increase, very rough)	20.0	12,500	8,370	Failed-3 min

[a] Wear based upon 24-hr test unless otherwise noted; values including deformation of samples.
[b] PV = Pressure in psi × velocity in ft/min.
[c] FV = Frictional force in psi × velocity in ft/min.

Table 2.70 Accelerated wear test results of different types of nylon

As with most of the aromatic polymers, phenoxy resins' color retention, UV resistance, and weatherability are generally poor. They have limited thermal exposure with a recommended operating temperature range from –60° to 80°C (–76° to 176°F). Phenoxy resins are resistant to acids and alkalis and have poor solvent resistance (especially in ketones), but they have good resistance to aliphatic hydrocarbons and resist staining from common household agents rather well. Their permeability, particularly to oxygen and other gases, is the lowest of any melt-processable plastic. Their oxygen permeability is one-twentieth that of HDPE, and their moisture-vapor transmission is about the same as that of rigid PVC, higher than PE, but much lower than the styrenes.

The major use of phenoxies is as a vehicle for coating formulations. They are exceptionally useful in primer applications in which drying speed, compatibility with various kinds of topcoats, and high adhesive strength are required. Phenoxies are used in automotive and marine primers as well as in heavy-duty maintenance primers.

Their molding compounds are limited to applications requiring service temperatures not in excess of about 80°C (176°F). Their combination of good impact strength, clarity, and impermeability makes these resins attractive for molding, including blow molding of cosmetic, foodstuff, and household-chemical bottles.

POLYALLOMER

The polyallomers are a family of crystalline propylene-ethylene copolymers that exhibit rigidity and high impact strength combined with light weight (density of 0.902). These plastics are copolymers, and the size of the blocks allows crystallization to take place, although seemingly slower than in either HDPE or PP, thus affording easier control of the crystalline morphology resulting from processing. Properties are quite similar to those of HDPE and PP.

The material provides the greatest yield per pound of any noncellular commercial plastic. The useful temperature range of the polyallomers is –40° to 210°F. Both frozen food and heat-sterilizable containers can be made from the material. The surface hardness of polyallomer is slightly less than that of polypropylene, but its abrasion resistance is greater. Polyallomer is superior to linear polyethylene in processability and stress-crack resistance.

The polyallomers have a tensile strength of 2,900 to 4,200 psi, an Izod impact strength of 1.5 to 12.0 ft-lb/in of notch, and an elongation greater than 400% at fracture. The material has excellent flexural strength and exhibits a better "living hinge" characteristic than polypropylene. This property is being used in a number of applications including fishing tackle boxes and typewriter cases.

POLYAMIDE

See Nylon.

POLYAMIDE-IMIDE

Polyamide-imides (PAIs) are engineering thermoplastics characterized by excellent dimensional stability, high strength at high temperatures, and good impact resistance. Molded parts in this material can maintain their structural integrity in continuous use at temperatures of 260°C (500°F). Different grades are available from BP-Amoco (Torlon) such as general-purpose, injection-moldable, PTFE/graphite wear-resistant compounds, 30% graphite-fiber-reinforced compounds, 30% glass-fiber-reinforced components, and so on. Properties of BP-Amoco's Torlon polyamide-imides are reviewed in Tables 2.71 to 2.74 (164).

Property	Testing Temp, °F	4203 and 4203L	4301	4275	4347	5030	5430	7130	9040
Tensile Strength, ksi									
ASTM D1708	-321	31.5	—	1818	—	29.5	—	22.8	—
	73	27.8	23.7	22.0	29.7	29.7	27.7	29.4	25.5
	275	16.9	16.3	16.3	15.1	23.1	19.0	22.8	24.0
	450	9.5	10.6	8.1	7.8	16.3	12.9	15.7	19.8
Tensile Elongation, %									
ASTM D1708	-321	6	—	3	—	4	—	3	—
	73	15	7	7	9	7	12	6	4
	275	21	20	15	21	15	10	14	8.3
	450	22	17	17	15	12	8	11	7.7
Tensile Modulus, msi									
ASTM D1708	73	7.0	9.5	11.3	8.7	15.6	15.8	3.2	—
Flexural Strength, ksi									
ASTM D790	-321	41.0	—	29.0	—	54.4	—	45.0	—
	73	34.9	31.2	30.2	27.0	48.3	39.7	50.7	49.0
	275	24.8	23.5	22.4	20.5	35.9	29.7	37.6	39.5
	450	17.1	16.2	15.8	14.3	26.2	20.7	25.2	22.8
Flexural Modulus, msi									
ASTM D790	-321	11.4	—	13.9	—	20.4	—	35.7	—
	73	7.3	10.0	10.6	9.1	17.0	16.7	22.8	21.1
	275	5.6	7.9	8.1	6.4	15.5	14.2	27.2	20.9
	450	5.2	7.2	7.4	6.2	14.3	13.2	22.8	18.3
Compressive Strength msi, ASTM D695	73	32.1	21.7	17.8	18.3	34.8	34.1	32.7	—
Shear Strength, ksi ASTM D732	73	18.5	16.1	11.1	11.5	20.1	19.1	17.3	—
Izod Impact, ft lbs/in ASTM D256									
notched	73	2.7	1.2	1.6	1.3	1.5	1.3	0.9	
unnotched	—	20.0	7.6	4.7	—	9.5	—	6.4	1.8

Table 2.71 Mechanical properties of polyamide-imide compositions

Property	4203L	4301	4275	4347	5053	5430
Dielectric constant						
at 10^3 Hz	4.2	6.0	7.3	6.8	4.4	4.2
at 10^6 Hz	3.9	5.4	6.6	6.0	6.5	4.0
Dissipation factor						
at 10^3 Hz	0.026	0.037	0.059	0.037	0.022	0.020
at 10^6 Hz	0.031	0.042	0.063	0.071	0.023	0.036
Volume resistivity, ohm-in, D257	$8x10^{16}$	$8x10^{16}$	$3x10^{15}$	$3x10^{15}$	$6x10^{16}$	$2x10^{16}$
Surface resistivity, ohm, D257	$5x10^{18}$	$8x10^{17}$	$4x10^{17}$	$1x10^{18}$	$1x10^{18}$	$3x10^{17}$
Dielectric strength, V/mil	580	—	—	—	835	650

Table 2.72 Electrical properties of polyamide-imide compositions

Property	4203 and 4203L	4301	4275	4347	5030	5430	7130	9040
Deflection temperature °F at 264 psi, D646	532	534	536	532	539	534	540	536
Coefficient of linear thermal expansion 10^{-6} in/in, °F, D696	17	14	14	15	9	10	6	7
Thermal conductivity Btu in/hr, ft², °F, C117	1.8	3.7	—	—	2.5	—	—	—
Flammability Underwriters Lab 94	94VO	94VO	94VO	94VO	94VO	94VO	94VO	94VO
Limiting oxygen index %, D2863	45	44	45	46	51	48	48	50
Density, g/cc, D792	1.38	1.43	1.44	1.44	1.56	1.62	1.42	1.66
Hardness Rockwell E, D792	86	72	70	66	94	93	94	107
Water absorption, % D57	0.33	0.28	0.33	0.17	0.24	0.31	0.26	0.21

Table 2.73 Thermal and general properties of polyamide-imide compositions

Grade Code Number	Description
TORLON 4000T	This powdered form of the resin is used for compression molding. It is the base resin used to produce most other grades of TORLON.
TORLON 4203L	This grade contains TORLON 4000T plus 3 % TiO_2 and 0.5 % of pTFE compounded and pelletized. It is used primarily for injection molding. pFTE is added to enhance moldability and mold release. 4203L is used where elongation and impact resistance are important considerations.
TORLON 4301	This grade contains TORLON plus 12 % graphite powder and 3 % pTEF compounded and pelletized. It was designed to reduce the coefficient of friction and to increase its wear resistance in lubricated and non-lubricated bearing applications.
TORLON 4275	This grade contains TORLON 4000T plus 20 % graphite powder and 3 % pTFE compounded and pelletized. It was specifically designed to provide maximum wear resistance and a lower coefficient of friction.
TORLON 4347	This grade contains 12 % graphite powder and 8 % pTFE compounded and pelletized. This new material is designed for the lowest coefficient of friction and should be useful for bearings under high load or surfaces subject to reciprocating motion.
TORLON 5030	This grade contains TORLON 4000T plus 30 % glass fiber and one % pTFE compounded and pelletized. It is used primarily in applications requiring high modulus, low shrinkage, and high creep resistance to deformation at high temperatures, or good electrical properties.
TORLON 7130	This grade contains TORLON 4000T plus 30 % graphite fibers and 1 % pTFE compounded and pelletized. The most metal-like of all TORLON resins, this grade was specifically developed as a structural metal replacement in the aerospace industry. Incorporation of graphite fibers resulted in a material with a high modulus that is highly resistant to creep and has the highest fatigue resistance available in a TORLON resin.
TORLON 9040	This grade is the first in a new series of TORLON resins made possible by developments in the basic polymer chemistry. This version contains 40 % glass and carries a volume tag of $5.40 per pound. Even higher heat resistance has been achieved with a minor loss of impact strength. Ground forms of 4203, 5030, and 4301 are also available for compression molders.
TORLON 5430	This grade is compounded with 30 % glass fibers. It offers superior resistance to creep and excellent stiffness at elevated temperatures, as well as improved processibility.

Table 2.74 Grades of commercially available polyamide-imide

The room-temperature tensile strength of an unfilled PAI is about 192 MPa (28,000 psi), its compressive strength about 220 MPa (32,000 psi). At 232°C (450°F) its tensile strength is about 65 MPa (9,500 psi), or as strong as many engineered plastics at room temperature. Continued exposure at 260°C (500°F) for up to 8,000 hours produces no significant decline in its tensile properties.

PAIs' flexural modulus of 5,000 Ma (730,000 psi) in an unfilled grade can be increased with graphite fiber reinforcement to 20×10^3 MPa (2.9×10^6 psi). The degree of retention of its modulus at temperatures to 260°C (500°F) is on the order of 80%. Its creep resistance, even at high temperatures and under load, is among the best of the thermoplastics, and its dimensional stability is extremely good.

The unfilled grade of PAI is rated UL 94 V-0 at thicknesses as low as 0.008 in and has an oxygen index of 45%. PAIs are extremely resistant to flame and have quite low smoke generation. Some reinforced grades have surpassed the U.S. Federal Aviation Administration requirements for flammability, smoke density, and toxic gas emission.

PAIs' radiation resistance is good with a tensile strength that drops only about 5% after exposure, to 110^9 rads of gamma radiation. Its chemical resistance is very good, virtually unaffected by the aliphatic and aromatic hydrocarbons as well as halogenated solvents and most acid and base solutions. PAI is attacked, however, by some acids at high temperatures, by steam at high pressure and temperatures, and by strong bases. PAI moldings absorb moisture in humid environments or when immersed in water, but the rate is low and the process reversible. For example, at 50% relative humidity and 23°C (73°F), PAIs absorb about 1% weight in 1,000 hours. Parts can be restored to their original dimensions by drying.

One important area for the use of PAIs is in structural parts requiring high strength at high temperatures: aerospace products, business equipment, industrial chemical plants, heavy-duty trucks, underground environments, and other such markets. Some of the specific parts for which they are used include electrical connectors, switches, relays, gears, ball bearings, marine winches, highroad thrust bearings, and so on. Automotive parts using PAIs include power and valve trains, piston skirts, tappets, piston rings, valve stems, and timing gears.

POLYANILINE

Polyanilines are processable, electrically conductive plastics.

POLYARYLATE

Polyarylates (PARs) are a form of aromatic polyester (amorphous) exhibiting an excellent balance of properties such as stiffness, UV resistance, combustion resistance, high heat-distortion temperature, low notch sensitivity, and good electrical insulating values. Uses include solar glazing, safety equipment, electrical hardware, transportation components, and in the construction industry.

They are often defined as copolyesters involving bisphenol-A (BA) and a mixture of terephthalic acid (TA) and isopthalic acid (IA), resulting in two distinct repeat units. This source of irregularity along the chains prevents crystallization, yielding amorphous plastics that are thus naturally transparent. The aryl groups are associated with stiff chains resulting in a glass transition temperature of 185°C (365°F). Performance includes good resistance to heat, steam, and radiation and good weatherability and fire resistance without additives.

They are used primarily in electronic, electrical, and automotive applications. Examples are light housings and reflectors. They also find use in microwave cookware, exterior lighting, and safety and medical equipment.

POLYARYLESTER

A number of specialty plastics have repeat units that feature only aromatic-type groups (phenyl, etc.) between ester linkages. Polyarylester is a general name for this class of plastics that are often referred to as wholly aromatic polyesters.

POLYARYLETHERKETONE

Polyaryletherketone (PAEK), when compared with the temperature-stable thermoset plastics and fluoropolymers, have clear economic and processing advantages, particularly for new materials that can be highly stressed thermally and mechanically. PAEK is a leading material among the high-temperature stable thermoplastics. This family of plastics allows continuous operating temperature of 250°C (480°F) and, depending on the type of short-term load, up to 350°C (660°F). The glass transition and melting temperatures are thermodynamic quantities that depend on the ratio of the ketone to the ether groups. Various complicated configurations can be obtained, such as polyetherketoneetherketoneketone (PEKEKK).

The properties of this family of plastics include a tensile strength at break of 85 MPa (12,300 psi), an elongation at break of 56%, a tensile modulus of elasticity of 0.6×10^6 psi, a tensile stress at yield at 23°C (74°F) of 104 MPa (15,100 psi), and at 160°C (320°F) of 37 MPa (5,400 psi), an elongation at yield of 6% at 23°C (74°F) and of 2% at 160°C (320°F), and no break using an unnotched Izod impact test.

The processing flow behavior of PAEK does not differ fundamentally from that of other partially crystalline TPs. The shear rate is similar to that of nylon 6 and PBT at 25°C (77°F) above the melting point. Besides having good mechanical and rheological properties, PAEK is characterized by its favorable behavior in fire. Without additives it has a UL 94 V-0 rating down to a test specimen thickness of 0.030 in. The density of PAEK fumes in a fire is the lowest of the TPs, and it has exceptionally low corrosive and toxic fumes. The quantity of heat released upon the outbreak of a fire is quite low, and it meets aviation regulations for interior use. PAEK has high hydrolysis resistance and good resistance to many different chemicals.

PAEKs are plastics in which phenylene rings are linked together via oxygen bridges (ether and carbonyl groups [ketone]) and may be viewed as the family name of this class of plastics. It includes PEEK, PEK, PEEKK, PEKEKK, and PEKK. The various representatives of this class of plastics differ in their chemical composition by the sequence and proportion of ether and ketone units.

Since their ratio influences the glass transition temperature (T_g) and the melt temperature of the polyether ketones, they also differ in features that are of such importance to the processor as heat resistance and processing temperature. As an example, a high ketone content leads to a higher T_g and a higher melting point.

POLYARYLSULFONE

PAS consists mainly of phenyl and biphenyl groups linked by thermally stable ether and sulfone groups. Its most outstanding property is resistance to low and high temperatures from –240° to 260°C (–400° to 500°F). It also has good impact resistance; resistance to chemicals, oils, and most solvents; and good electrical insulating properties. It can be processed by conventional fabricating methods (injection, extrusion, ultrasonic welding, etc.).

POLYBUTYLENE TEREPHTHALATE

Polybutylene terephthalate (PBTs), in the family of polyester plastics, are crystalline polymers with excellent engineering properties. They have marginal chemical resistance but resist moisture, creep, fire, fats, and oils. Molded items are hard and brightly colored, and they retain their impact strength at temperatures as low as –40°F (–40°C). PBT can crystallize much faster than PET. The properties of the highly crystalline PBT (as much as 60%) are fairly similar to unoriented crystalline PET; PBT is not as conveniently oriented as PET.

Excellent dimensional stability, principally resulting from low moisture-absorption rates, is a distinguishing feature of PBT plastics. Notched Izod impact strength ranges from 1.0 ft-lb/in (53 J/m) for unreinforced grades to 3.5 and 16.0 ft-lb/in (187 and 854 J/m) for reinforced and impact-modified, unreinforced grades. Glass-reinforced PBT provides good resistance to creep at both ambient and elevated temperatures.

PBT molding materials are available in formulations such as unmodified, glass-reinforced, and flame-retarded grades. They offer high strength and rigidity, low moisture absorption, excellent electrical properties and chemical resistance, rapid molding cycles, and excellent reproducible mold shrinkage. Unreinforced PBT has a tensile strength of 8,000 psi (55 MPa). With 40 wt% glass reinforcement, tensile strength increases to 21,300 psi (147 MPa). Corresponding flexural moduli are 330,000 psi (2,280 MPa) and 1,500,000 psi (10,340 MPa), respectfully. Mineral-filled and mineral/glass-filled grades provide intermediate strength and stiffness.

A fast crystallization rate enhances moldability, producing smooth surfaces that can be painted, printed, and ultrasonically welded. UL temperature indexes are relatively high because of a 227°C (440°F) melting point and a 40°C (104°F) glass-transition temperature. Unreinforced PBT has a heat-deflection temperature of 154°C (310°F) at 66 psi (0.5 MPa). Aside from improving strength and stiffness, glass reinforcement also enhances thermal properties, making possible heat-deflection temperatures to 217°C (422°F).

Water absorption is less than 0.1 wt% after 24 hours. However, PBT is not recommended for extended use in water or aqueous solutions above 52°C (125°F). PBT plastics are also intrinsically resistant to detergents, weak acids and bases, aliphatic hydrocarbons, fluorinated hydrocarbons, alcohols, ketones, ethylene glycol, carbon tetrachloride, oils, and fats at room temperature. They also exhibit good resistance to motor oil, gasoline, transmission fluids, and brake fluids at temperatures to 60°C (140°F).

The electrical properties of PBT are stable over a wide range of temperatures and humidities. They are good insulators; volume resistivity is greater than 10^{16} ohm-cm. Arc resistance and dielectric strength are high and the dissipation factor low. The different grades of PBT are rated UL 94 HB. Grades are available to meet UL 94 V-0 requirements, and many of them also meet the more restrictive flammability requirements of UL 94 5V. Strength, stiffness, and better thermal properties are gained by reinforcing PBT with glass fibers. If very low warpage is desired, reinforcement with mineral or mineral and glass fillers is used.

PBT resins are used in many types of products: appliances, automobiles (louvers, under-the-hood electricals, mechanical parts, etc.), electrical and electronic parts, industrial components, consumer items, and others. Their resistance to heat and chemicals, plus their wide range of color possibilities, dictate their use in appliances, where they have been found easier to injection mold (IM) and are less expensive than thermosetting plastics. Prime use is in IM where fast cycles can be achieved with minimal precautions other than a thorough drying of the hygroscopic plastic; less than 0.005 wt% of water is usually required.

POLYCARBONATE

Polycarbonate (PC) is a polyester of carbonic acid. It is produced using an interfacial reaction between dihydric or polyhydric phenols and a suitable carbonate precursor such as dichlorocarbonate. Polycarbonate is an amorphous engineering resin that exhibits excellent clarity and impact strength, plus high heat-deflection temperature. It is extruded into sheet for vandal-resistant glazing applications, display signs, and RV windshields. It is also thermoformed into a variety of medical parts and business machine housings.

PCs are highly ductile (more than 100% in standard tests) and are self-extinguishing, have heat deflection temperatures (264 psi) in the 130° to 138°C temperature range, and can withstand continuous service at temperatures up to 122°C. The electrical properties of the PCs remain relatively constant over a wide range of temperatures and humidities. The polymers resist water, acids, and oxidizing and reducing agents, but they can be dissolved in aromatic and chlorinated solvents.

They represent over 50 wt% of all engineering plastics consumed worldwide. Among the engineering thermoplastics, these amorphous PCs are one of the most exceptional materials, as distinguished by their highly versatile combination of properties. Most PCs can exist individually in various other materials but generally not in their entirety (Fig. 2.28). Different grades of PCs provide specific properties and processing characteristics. These include the flame-retardant and reinforced grades, resistance to weather and UV radiation, EMI and RFI shielding, FDA-approved grades for use in food-contact and medical applications, and grades for different processes such as injection, structural foam and blow molding, the extrusion of film and sheet, and others. As shown in Figure 2.29, melt flow has an effect on properties.

Typical mechanical properties for the PCs include tensile strength, 55 to 65 MPa (8,000 to 9,500 psi); tensile modulus, 24×10^5 kPa (3.5×10^5 psi); flexural strength, 90 MPa (13,000 psi); and impact strength, 640 to 853 HJ/m (12 to 16 ft-lb/in) of notch. They are moldable from 249° to 300°C into parts having very close tolerances that are exceptionally dimensionally stable and machinable.

PC is a tough, heat- and flame-resistant, and dimensionally stable, transparent plastic. It withstands boiling water but is less resistant to weather and scratching than acrylics. It is notch sensitive and has poor solvent resistance in stress-molded products.

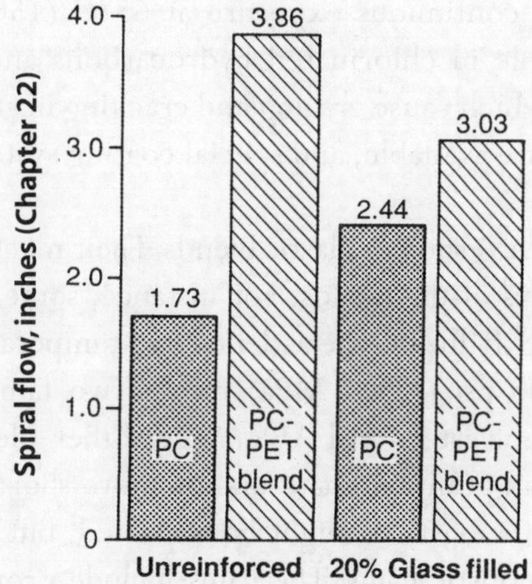

Figure 2.28 Example to improve processing of PC/PET blend

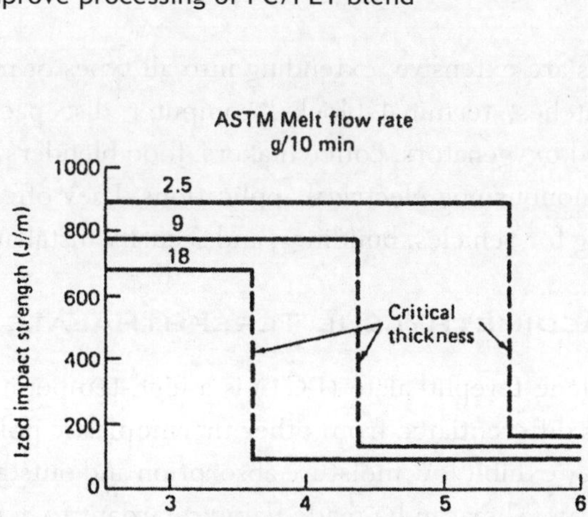

Figure 2.29 Polycarbonate properties vs. melt index (courtesy of Bayer)

Thick, unreinforced PC resists breakage at temperatures down to –54°C (–65°F). Grades are available to provide high impact strength, based on different thicknesses at room temperature and a notched Izod impact strength of 6.4 to 8.5 J/cm (12 to 16 ft-lb/in). Even in thick sections, a properly designed PC product has more impact strength at –54°C (–65°F) than most plastics generally do at RT. Many plastics are not tough at 18°C (65°F), but there are plastics that are tough even at much lower temperatures. Creep resistance, which is already excellent throughout a broad temperature range, can be further improved by a factor of 2 to 3 when PC is reinforced with glass fibers.

PCs' insulating and other electrical properties are excellent and remain almost unchanged by temperature and humidity conditions. One exception is arc resistance, which in PCs is lower than in many other plastics. They are generally unaffected by greases, oils, and acids. Water at RT has no effect on PCs, but continuous exposure at 65°C (150°F) to water causes gradual embrittlement. They are soluble in chlorinated hydrocarbons and attacked by most aromatic solvents, esters, and ketones, which cause crazing and cracking in stressed products. Grades with improved chemical resistance are available, and special coating systems can be applied to provide additional chemical protection.

PCs are major additions to the group of plastic blends. Their mechanical and thermal properties make them the natural mainstays of the blends. For instance, some PC grades have notched Izod impacts of up to 960 J/m (18 ft-lb/in), heat-deflection temperatures of 138°C (280°F), and flexural moduli in the 300,000 psi (2,067 MPa) range. Two families of blends dominate: the thermoplastic PBT and PET polyesters and ABS. Many other blends exist to provide specific desirable characteristics. Blending can eliminate one or more shortcomings by the type of blend and the mixture ratio. Trade-offs exist, as might be expected, but with certain blends there are overall net gains and even synergistic gains. The gains include providing for an easier melt flow during processing. In the blending process, the natural water clearness and transparency of PCs can be reduced or lost.

The applications of PCs are extensive, extending into all types of markets. Examples include electronic connectors, switches, terminal blocks, computer disc packs, storage modules and housings, power tools, blood oxygenators, coffee makers, food blenders, automobile lenses, safety helmets, lenses, and many nonburning electrical applications. They offer resistance to bullets and thrown projectiles in glazing for vehicles, buildings, and security installations.

Polycyclohexylenedimethylene Terephthalate

Polycyclohexylenedimethylene terephthalate (PCT) is a high-temperature, semicrystalline thermoplastic polyester. PCT is differentiated from other thermoplastic polyesters by its higher heat-deflection temperature. They exhibit low moisture absorption and outstanding chemical resistance to a wide variety of chemicals. They can be made flame retardant to achieve UL 94 V-O recognition down to a thickness of 1/16 in (1.6 mm). Different copolyesters of PCT exhibit a wide range of properties depending on the particular copolyester incorporated.

Polyelectrolyte

Polyelectrolyte denotes a class of macromolecular compounds that when dissolved in a suitable polar solvent (generally water) spontaneously acquire or can be made to acquire a large number of elementary charges distributed along the macromolecular chain. In the former case and when the charge that appears spontaneously has its maximum value, these macromolecules are termed *low molar-mass (LMM) electrolytes* and in other cases *weak polyelectrolytes*.

Thermoplastic Polyester

The family of polyesters has a widely varying and significant range of properties. In the family of polyesters, there are two major groups: (a) thermoplastics, which are highly crystalline with comparatively high melting points, and (b) thermoset plastics, which are usually typified by a cross-linked structure. Thermoplastic polyesters are often called saturated polyesters to distinguish them from unsaturated polyesters that are the thermoset plastics. TS polyesters will be reviewed later.

This family of saturated plastics includes thermoplastic general polyesters, polyalkylene terephthalates (PATs), polycyclohexanedimethanol terephthalates (PCTs), acid-modified PCT (PCTAs), glycol-modified PET (PETGs), liquid crystal polymers, and the production materials (as reviewed) PBTs and (to be reviewed) polyethylene terephthalates (PETs).

Polyester Thermoplastics and the Environment

Environmental compatibility has played a major role in key developments in the field of polyesters. Flame retardancy and disposal problems were the two principal topics on which attention was focused. The conceivable risks involved in the manufacture and use of flame-retardant plastics incorporating brominated diphenylether formed the subject of a report (1990) by the Federal German Office of Environment. Thanks to the development work in which many producers had already been engaged for a long time, positive progress has been made. As a result, the full-scale substitution of this material in polyester molding compounds, which had commenced as soon as the problem became known, has already been completed.

Despite the fact that it only accounts for a small percentage of plastic refuse in weight terms, the PET bottle, because it is a conspicuously large-volume polyester application, has attracted much discussion as a form of packaging that involves the familiar problems of domestic refuse disposal. Worldwide, large-scale systems are being used to recover and reuse the materials.

Polyester-Reinforced Urethane

Polyester-reinforced urethanes are a poromeric, microporous material that may have a urethane impregnation or a silicone coating for shoe uppers and industrial leathers.

Water-Soluble Polyester

Water-soluble polyesters (WSPs) have increased significantly in the surface-coatings industry. They provide ease of application, ease of cleanup, freedom from toxicity, and freedom from flammability when compared to conventional solvent-based paint. WSPs are obtained by incorporating hydrophilic groups in the polymer backbone during polymerization. They are subsequently diluted in water for application, optionally in the presence of neutralizing agents and water-miscible cosolvents.

They can be utilized as hydrophilic plastics in paper coating and textile coating. In most surface coatings, clear or pigmented solutions are converted to water-insoluble coatings by condensation or oxidative polymerization. Their largest use is in surface coating as pigmented electrodeposition and conventional dipping primers. Other applications include sprayed-primer surfaces; semigloss trade-sales paints; coil-coating vehicles; and enamels applied by dip, flow-coat, electrodeposition, and spray methods.

Polyetherketone

Polyetherketone (PEK) is a partially crystalline, high-performance, aromatic, ketone-based thermoplastic that is heat stable. As a member of the ketone family, it shares with PEEK good chemical resistance; exceptional toughness, strength, rigidity, and load-bearing capabilities; good radiation resistance; the best fire-safety characteristics of any thermoplastic; and the ability to be easily melt processed.

Super PEKs designed for advance composites have a continuous-service temperature rating of 260°C (500°F), a glass transition temperature (T_g) of 200°C (400°F), and a slow crystallization rate that suits them for processes with slow rates of cooling from the melt.

Polyetheretherketone

Polyetheretherketones (PEEKs) are high-temperature, crystalline engineering TPs used for high-performance applications such as wire and cable for aerospace applications, military hardware, oil wells, and nuclear plants. They hold up well under continuous 323°C (450°F) temperatures with up to 316°C (600°F) limited use. Their fire-resistance rating is UL 94 V-0; they resist abrasion and long-term mechanical loads.

An example of their use was in the design of a low-speed air motor for dental attachments. Star Dental of Lancaster, Pennsylvania, sought a material for the sliding vanes and bushings that would not require lubrication. Thus, the time and expense of lubricating the motor between dental patients would be eliminated. The PEEK vanes are located in the motor, which operates at 20,000 rpm; the bushings are in the adapter. To ensure the optimum power of the equipment, both components are molded to exacting standards with dimensional tolerances of less than 0.0005 in, including flash.

PEEK from Victrex USA, Inc. (Greenville, SC 29615) was selected not only for its inherent lubricity but also for its ability to withstand repeated sterilizations, noting that the components must withstand being autoclaved at 121°C (250°F) to 135°C (275°F), or chemiclaved at 132°C (270°F). The PEEK durability was tested up to 1,000 cycles, and the absence of residual lubricating-oil coating was found to facilitate the sterilization process.

CHLORINATED POLYETHER

Chlorinated polyether is a corrosion-resistant thermoplastic that is primarily used to manufacture products and equipment for the chemical and processing industries. It has also found use in molding components for pumps and water meters, pump gears, bearing surfaces, and the like. This plastic resists both organic and inorganic agents, except fuming nitric acid and fuming sulfuric acid, at temperatures up to 120°C (250°F) or higher. Its heat-insulating characteristics, dimensional stability, and outdoor exposure resistance are also excellent.

The material has excellent weathering resistance. Creep resistance at elevated temperatures is also excellent. The material exhibits an elongation under 1,000 psi of 4.5% after 5 hours at 280°F and of 1.5% after 12 hours at 75°F. In addition, the plastic has good impact strength, low moisture absorption, and high dimensional stability. It retains its properties at elevated temperatures and can be used at 100° to 125°C.

POLYETHERIMIDE

Polyetherimides (PEIs) are amorphous engineering thermoplastics characterized by superior strength, high modulus, heat resistance, flame resistance, UV resistance, and transparency, despite their amber-brown color. They are characterized by excellent electrical properties that remain stable over a wide range of temperatures and frequencies, and very easy processability. NEAT (unmodified) PEI is transparent, with inherent flame resistance and low smoke evolution. Its UL continuous-use listing is 170°C (338°F) and its heat-deflection temperature is 200°C (392°F) at 1,820 kPa (264 psi). It has a T_g of 215°C (419°F), a UL rating of 94 V-O at a thickness of 0.016 in and of 5v at 0.075 in without additives.

PEI has a limited oxygen index of 47% (one of the highest among the engineering TPs). Its smoke evolution as measured in an NBS chamber test per ASTM E 662 is low, and its dielectric constant remains virtually unchanged between frequencies of 60 to 10^9 Hz and temperatures of 28° to 82°C (73° to 180°F). It has a high-volume resistivity of 6.7×10^{17} ohm/cm and a dielectric strength from 830 V/mil at 1/16 in in air to greater than 6,500 V/m in submil film thicknesses. Its arc resistance exceeds 120 s, meeting one of the UL electrical requirements for the sole support of "live" parts.

A key feature of PEI is its ability to maintain properties at elevated temperatures. For example, at 180°C (356°F), its tensile strength and flexural modulus are 41 and 2.067 MPa (6,000 and 300,000 psi), respectively. The strength and modulus of the glass-reinforced grades are higher. The modulus is 9×10^3 MPa (1.3×10^6 psi) with 3 wt% of glass to over 80% retained at 180°C (356°F). PEI also has good creep resistance, as indicated by its apparent modulus of 2,410 MPa (350,000 psi) after 1,000 hours at 82°C (180°F) under an initial load of 35 MPa (5,000 psi).

This plastic resists a broad range of chemicals under varied conditions of stress and temperature. PEI's resistance to mineral acids, for example, is outstanding. It is, however, attacked by such partially halogenated solvents as methylene chloride, trichloroethane, and strong acids. Its resistance to UV radiation is good with a negligible change in tensile strength after 1,000 hours of xenon-arc exposure. PEI's resistance to gamma-ray radiation is also good, there being a strength loss of less than 6% after 500 megarads of exposure to cobalt 60 at the rate of one Mrad/hour. Hydrolytic stability tests show that more than 85% of PEI's tensile strength is retained after 10,000 hours of immersion in boiling water. This material is suitable for short-term or repeated steam exposure.

POLYETHYLENE NAPHTHALATE

Polyethylene naphthalates (PENs) are polyester plastics that can penetrate different markets such as the market for stretch-blow-molded bottles. When compared to the very popular PETs, processing with the more expensive PENs, based on the monomer naphthalate, does not require the use of energy-consuming air cooling. None of the processing variations alter the PEN bottle's dimensional stability after filling at 98°C (208°F), a temperature that indicates the plastic's thermal stability. As an example, bottles compete in markets such as glass beer, hot-fill juice, pharmaceutical, and returnable soft drink due to their hot-fillable, high barrier, and other properties.

Very importantly, PENs are competing in markets previously off limits to plastics. With an oxygen barrier 5.6 times better than PET, it will give the necessary protection to beer and will extend the shelf life of food and fruit juices. For hot-filled and fruit juices, PEN allows hot-fill temperature of 98°C (208°F). This greater temperature resistance makes PEN more acceptable for packaging beer, which is pasteurized in the container. For products in which flavor is crucial, from beer to mineral water, acetaldehyde extractables tend to run only 20% as high as PET. For pharmaceuticals, a benefit is that it almost totally blocks UV light. Returnable and recycled bottles better resist caustic washing.

Resultant crystallinity in clear PEN bottles is high. It results in a fast-rate, strain-induced crystallization that, with its high T_g, allows a significant reduction in the filling and cooling periods in the preform injection blow-molding process.

When first introduced in 1995, the cost to use PEN was about three times that of PET, but PEN has better processing characteristics. For example, preforms are injection molded in 23 s on a machine that requires a 39 s for PET. Stretch blow molding is performed on the same equipment as is currently used with PET. PEN can be alloyed with Lower-cost PET to optimize cost and performance.

POLYETHYLENE TEREPHTHALATE

Polyethylene terephthalate (PET) is produced from dimethyl ester (DMT) or terephthalic acid (PTA) and ethylene glycol by a continuous melt-phase step, followed by a solid-state polymerization process. PET can be extruded, blow molded, and thermoformed on conventional equipment if dried to a moisture content of less than 50 ppm. PET exhibits good clarity, toughness, and gas barrier properties.

PETs, in the family of polyester plastics, are thermoplastic polymers that are available in engineering grades providing high-performance properties. PET's broad range of mechanical and electrical properties make them ideal replacements in certain applications for such metals as die-cast aluminum or zinc, as well as in thermoset switches, relays, sensors, and so on.

They can be specifically formulated to meet specific product requirements such as increased toughness, decreased warpage, or flame retardance. PET plastics are hygroscopic and so require proper drying before processing. Different PET grades are used in different processes; they are used in applications such as packaging (particularly stretched-injection blow-molded beverage bottles and film), appliances, and electrical products and electronics.

There are amorphous (APET), crystalline (CPET), and copolyester plastics. The name *copolyester* is applied to those thermoplastic polyesters that use more than one glycol, more than one dibasic acid, or both in their synthesis. Its chain is less regular than the monopolyester chain and therefore has reduced tendency to crystallize and provide different properties. An example of a copolyester is PETG, a glycol-modified polyester (108). After processing, it normally remains amorphous, clear, and virtually colorless even in very thick sections. Recycling all types of PET materials can be easily accomplished.

The rates of crystallization of PET are low compared to plastics such as polypropylene and polyethylene. Its ultimate extent of crystallization is about 0.0012 g/cm^3 or about 0.09% change with each 1% change in crystallinity. As a result, when PET is injection molded into a crystalline state, the molded product is under very high residual stress and is thus very brittle.

Generally, PET is injection molded with a high mold temperature for dimensionally stable products. High mold temperature reduces the cooling rate, thereby allowing the plastic to spend a longer time at 165° to 175°C (329° to 352°F), which is the temperature range over which the rate of PET crystallization is known to be highest. The primary consideration in specifying the mold temperature is that it correlates with crystallization behavior.

PET homopolymer and some of its lower-degree copolymer modifications are known as crystallizable polymers. When held for a period of time in a given temperature range above their T_g of 80°C (176°F), they will crystallize spontaneously (Fig. 2.30). The maximum rate of crystallization occurs in the temperature range of 140° to 190°C (285° to 375°F); the melting point is about 254°C (489°F). The crystallization rate can be decreased by chemical modification. Sufficient modification (copolymerization) of PET with an additional glycol (or acid) can result in a plastic that will not crystallize. Such plastics are commonly referred to as having an "amorphous" molecular structure.

PET is known for its clarity and toughness when it is used for the manufacture of oriented film or stretch-blown bottles. It is also a good barrier to gases, such as oxygen and carbon dioxide. PET can be reinforced with glass fibers to produce an injection-molded product with high heat resistance and high stiffness. The resistance of PET to chemicals and solvents is greatly enhanced as the crystallinity of the polymer increases.

PET finds applications in such diverse end uses as fibers for clothing, films, bottles, food containers, and engineering plastics for precision-molded parts. A wide range of applications is possible because of the excellent balance of properties PET possesses and because the degree of crystallinity and the level of orientation in the finished product can be controlled.

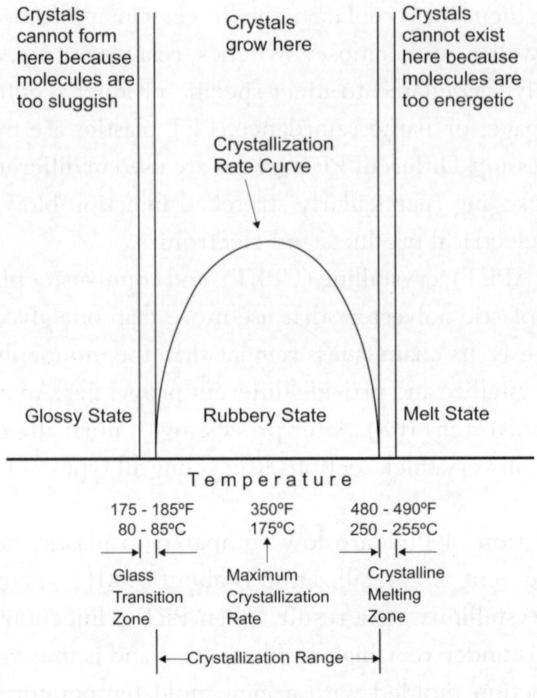

Figure 2.30 Effect of temperature on the crystallization of PET that influences processing requirements

As noted, PET is a versatile plastic because of its excellent physical properties and because it can be converted into either amorphous or semicrystalline products. When properly processed, it can be made into oriented and crystallized articles that still possess excellent clarity. Outstanding dimensional stability can be obtained in PET film by controlling orientation and by heat setting during processing (chapter 3). Very few other materials offer such a range of processing and property variables. For packaging applications, PET is used because it combines optimum processing, mechanical, and barrier properties.

The barrier properties of PET have been found to be adequate for the retention of carbon dioxide for packaging carbonated beverages. The good oxygen barrier properties of PET allow for its use in many types of food containers. However, PET barrier properties are deficient for packaging beer and certain foods that are highly sensitive to oxygen unless coinjected with suitable materials.

PET homopolymer is injection molded primarily as a crystallized, reinforced material. Applications for glass-reinforced PET include electrical connectors, auto body products, and other applications in which very-high-temperature environments are not encountered. PET homopolymer is not generally acceptable for injection-molded, amorphous (clear) products because of its tendency to crystallize when used in thick sections and when exposed to temperatures above its T_g. For clear moldings, PET copolymers are more useful. Typical injection-molded uses for copolymers include medical devices, toys, and cosmetic jars.

Polyethylene Terephthalate Glycols Polyethylene terephthalate glycol (PETG) resin is a clear, amorphous, glycol-modified polyester that is polymerized from dimethyl ester (DMT) or terephthalic acid (PTA) and ethylene glycol. The addition of glycol prevents polymer crystallization, providing a broader process window than PET. PETG parts such as blow-molded bottles have good clarity, toughness, and ESCR performance, as well as good gas-barrier properties. Major sheet markets include medical trays, rigid packaging, and point-of-purchase displays.

POLYHYDROXYBUTYRATE

In 1992, researchers passed what was called a crucial milestone in efforts to create farm crops that would make it possible to grow high-quality biodegradable plastics. Agricultural scientists at Michigan State University in East Lansing, Michigan, and James Madison University in Harrisonburg, Virginia, used genetics to create an experimental plant that makes a plastic called polyhydroxybutyrate (PHB). It is much like PP except that it is biodegradable.

It is reported to be the first plant that has been genetically engineered to make something other than a protein. Britain's ICI previously made PHB using the soil bacteria *Acaligenes eutrophus*, which is being used in blow and injection molding. Researchers at Michigan State University used three genes identified and cloned from the bacteria ICI used in 1987.

POLYIMIDAZOLE

A variety of polyimidazoles can be prepared by aromatic nucleophilic displacement, from the reactions of bisphenol imidazoles with activated difluoro compounds. These plastics have good mechanical properties that make them suitable for use as films, moldings, and adhesives.

POLYIMIDE

Polyimides (PIs) were among the first so-called high-heat-resistant thermoplastics. They in fact retain a significant portion of their room temperature mechanical properties from –240° to 315°C (–400° to 600°F) in air. Kapton is DuPont's trade name for their polyimide (Tables 2.75 to 2.83 and Fig. 2.31; 164). PIs, which are available in both TPs and TSs, are a family of some of the most heat- and fire-resistant polymers known.

Moldings and laminates are generally based on TSs, though some are made from TPs. PIs are available as laminates and in various shapes, as molded parts, stock shapes, and resins in powders and solutions. PI parts are fabricated by techniques ranging from powder-metallurgy methods to conventional injection, transfer, and compression molding as well as extrusion methods. Porous PI parts are also available. Generally, the compounds that are the most difficult to fabricate are also the ones that have the highest heat resistance. The service temperature for the intermittent exposure of PIs can range from cryogenic to as high as 480°C (900°F). Glass-fiber-reinforced PIs retain 70% of their flexural strength and modulus at 250°C (480°F). Creep is almost nonexistent, even at high temperatures. Their deformation under a 28-MPa (4,000-psi) load is less than 0.05% at room temperature for 24 hours.

These materials have good wear resistance and a low coefficient of friction, both of which are factors that can be further improved by including additives like PTFE. Self-lubricating parts containing graphite powders have flexural strengths above 69 MPa (10,000 psi). Their electrical

Physical Properties	Typical Properties			Test Method
	-195°C	23°C	200°C	
Ultimate tensile strength, psi	35,000	25,000	17,000	ASTM D882-64T
Yield point at 3%, psi	—	10,000	6,000	ASTM D882-64T
Stress to produce 5% elongation, psi	—	13,000	8,500	ASTM D882-64T
Ultimate elongation, %	2	70	90	ASTM D882-64T
Tensile modulus, psi	510,000	430,000	260,000	ASTM D882-64T
Impact strength, j/mm	—	23	—	Du Pont pneumatic impact test
Folding endurance MIT	—	10,000 cycles	—	ASTM D2176-63T
Tear strength-propagating (Elmendorf), g	—	. 8	—	ASTM D1922-61T
Tear strength-initial (Graver), g	—	510	—	ASTM D1922-61T
Density, g/cm³	—	1.42	—	ASTM D1505-63T
Coefficient of friction, Kinetic (film to film)	—	0.42	—	ASTM D1894 63
Refractive index (Becke line)	—	1.78	—	Encyl. Dict. of Phys., Vol; Ave. 3 samples, elongated at 5%, 7%, 10%
Poisson's ratio	—	0.34	—	

Table 2.75 Physical properties of 1 mil DuPont type H Kapton (polyimide) film

properties are also outstanding over wide temperature and humidity ranges. They are unaffected by exposure to dilute acids, aromatic and aliphatic hydrocarbons, esters, alcohols, hydraulic fluids, JP-4 fuel, and kerosene. They are, however, attacked by dilute alkalis and concentrated inorganic acids.

PI film has useful mechanical properties, even at cryogenic temperatures. At –269°C (–453°F), this film can be bent around a ¼-in mandrel without breaking. At 500°C (932°F) its tensile strength is 30 MPa (4,500 psi).

Property	Typical Values		
	Film Type 120F616*	Film Type 150F019*	Film Type 250F029*
Ultimate tensile strength, psi			
23°C	24,000	17,000	25,000
200°C	16,000	11,000	16,000
Yield point at 32 psi			
23°C	9,000	7,300	10,000
200°C	5,500	4,000	8,000
Stress at 5% elongation, psi			
23°C	12,500	9,000	—
200°C	7,500	5,500	—
Ultimate elongation, %			
23°C	65	75	80
200°C	95	85	—
Tensile modulus, psi			
23°C	415,000	320,000	—
200°C	215,000	173,000	—
Impact strength at 23°C, Kg cm/mil	6.0	4.6	—
Tear strength-propagating Elmendorf, g/mil	10	13.5	12
Tear strength-initial (graves), g/mil	750	435	—
Weight, % polyimide	80	57	73
Weight, % FEP	20	43	27
Density gm/cm³	1.53	1.67	1.57
*Film engineering:			

120F616 1.2-mil nominal thickness; 0.1-mil Teflon® FEP; 1-mil Kapton® Type H, 0.1-mil Teflon® FEP

150F019 1.5-mil nominal thickness, 0.0-mil Teflon® FEP; 1-mil Kapton® Type H, 0.5-mil Teflon® FEP

250F029 25-mil nominal thickness, 0.0-mil Teflon® FEP; 2-mil Kapton® Type H; 0.5-mil Teflon® FEP

Table 2.76 Mechanical properties of DuPont type F Kapton (polyimide) film

The polyimides show good radiation resistance and retain useful properties after absorbing radiation doses as high as 6,000 Mrads. Polyimide laminates' curing conditions are 200 psi at 370°C for 1 hour. Laminates have shown 25% strength retention after 1,000 hours at 315°C. The resins do lose considerable strength in air at service temperatures above 315°C. Typical polyimide film properties include tensile strength, 24,000 psi; modulus, 415,000 psi; and elongation, 65%.

They have a density of 1.41 to 1.43, tensile strength of 12,000 psi at 73°F, and an elongation of 6.8% at that same temperature. They have a low coefficient of expansion and are difficult to process by conventional means. Uses include critical engineering parts in aerospace, automotive and electronics components subject to high heat, and parts used in corrosive environments.

Gas	(cc/100in^2)(24hr)(atm/mil)	Test Method
Carbon dioxide	45	ASTM D14340
Hydrogen	250	at 23°C
Nitrogen	6	
Oxygen	25	
Helium	415	
Water vapor	5.4 g/(100in^2)(24 hr)/min	ASTM E96-63

Table 2.77 Gas permeability of DuPont type H Kapton (polyimide) film

Property	Typical Values	Test Conditions	Test Method
Dielectric strength			
2 mil	5,400 V/mil	60 Hz	ASTM D149-64
3 mil	4,600 V/mil		
5 mil	3,600 V/mil	1/4-in. electrode	
Dielectric constant			
2 mil	3.6	1 kHz	ASTM D149-64
3 mil	3.7		
5 mil	3.7		
Dissipation factor			
2 mil	0.0025	1 kHz	ASTM D149-64
3 mil	0.0025		
5 mil	0.0025		
Volume resistivity			
2 mil	8×10^{15} ohm-cm	125 V	ASTM D257
3 mil	5×10^{15} ohm-cm		
5 mil	1×10^{15} ohm-cm		

Table 2.78 Electric properties of DuPont type V Kapton (polyimide) film

A few more of their many uses include polyimide-based wire enamel, insulating varnish, and coated glass fabrics. The insulating varnish possesses good electrical properties in the –190° to 340°C temperature range. In 1965, DuPont began producing commercial polyimide film (Kapton), which

Property	Typical Values	Test Conditions	Test Method
Dielectric strength			
1 mil	7,000 V/mil	60 Hz	ASTM D149-61
2 mil	5,400 V/mil	1/4-in. electrode	
3 mil	4,600 V/mil		
5 mil	3,600 V/mil		
Dielectric constant			
1 mil	3.5	1 kHz	ASTM D150-59T
2 mil	3.6		
3 mil	3.7		
5 mil	3.7		
Dissipation factor			
1 mil	0.0025	1 kHz	ASTM D150-59T
2 mil	0.0025		
3 mil	0.0025		
5 mil	0.0027		
Volume Resistivity			
1 mil	$1x10^{15}$ ohm-cm	125 Volt	ASTM D257-61
2 mil	$8x10^{15}$ ohm-cm		
3 mil	$5x10^{15}$ ohm-cm		
5 mil	$1x10^{15}$ ohm-cm		

Table 2.79 Electric properties of DuPont type H Kapton (polyimide) film

Dielectric strength	
Short time (80 mils)	570 V/mil
(3 mils)	4100 V/mil
Volume resistivity	1.6×10^{15} ohm-cm
Arc resistance	185 sec (tracks)
Dielectric constant, dry, 10^6 cps at	
73°F	3.4
347°F	3.4
437°F	3.4
527°F	3.4
572°F	3.4
Dissipation factor, dry, 10^6 cps at	
73°F	.003
347°F	.003
437°F	.003
527°F	.003
572°F	.003

Table 2.80 Electrical properties of polymide at elevated temperature

has useful mechanical and electrical properties from –269° to 400°C. This film can be Teflon-coated to render it heat sealable. Since polyimides are essentially nonfusible and difficult to fabricate by conventional polymer-shaping processes, DuPont has employed special processes, including a high temperature-pressure procedure similar to that used in powder metallurgy, to fabricate its polyimide into finished parts (called Vespel). This process is useful for producing parts for low-friction, high-temperature applications.

Temperature, °F	Time, hr	Tensile shear strength,[a] psi: Unformulated[b] polyimide	Formulated[b] polyimide
77	—	2500	2200
550	1	1800	1400
550	100	2200	1900
550	200	1800	1400
700	1	1400	600
700	10	1300	1100
700	24	1300	1200
700	60	0	1100

[a] Substrate tested was 17–7 PH stainless steel, phosphate etched. Bonds were formed at 575°F, under 200 psi pressure for 2 hr.
[b] Formulated material contains an antioxidant and various fillers; unformulated material is the straight polyimide.

Table 2.81 Strength of polyimide adhesives

Apparatus: High temperature performance test apparatus
Specimen: 20 mm radial ball bearing
Load: 5 lb thrust, 3 lb radial
Speed: 10,000 rpm
Lubricant: Self lubricating aromatic polyimide retainer

Bearing	Retainer	Temp., °F	Performance life (hr)
AISI–M10	Polyimide	450	151, 153, 159
AISI–M10	Polyimide-graphite	450	473
AISI–M10	Polyimide-graphite	600	183, 205, 341
AISI–440C	Polyimide-graphite	600	200, 238
AISI–M10	Polyimide-WSe_2	600	79
AISI–M10	Polyimide-MoS_2/graphite	600	90
AISI–M10	Polyimide-MoS_2	600	122
AISI–M10	Polyimide-graphite	650	180, 155
AISI–M10	Polyimide-graphite	700	95, 81

Table 2.82 Comparison of polyimide lubricant bearing performance life

POLYIMIDE POWDER

Powders with desirable melt-flow behavior and properties are formed in reaction vessels, without grinding (NASA Langley Research Center 1994). Semicrystalline polyimides that have controlled molecular weights can be synthesized in a process that yields the polyimides in powder form. These PIs exhibit a transparent crystallinity that gives rise to high melt flow at processing temperatures of about 300° to 350°C (572° to 662°F), which are slightly above their melt temperatures. They

Property		ASTM No.		Polyimide		Graphite filled polyimide
Sp. grav.				1.41–1.43		1.49
Ten. str., psi at	73°F	D 638		13,400		9,120
	302°F			9,410		6,530
	482°F			—		5,200
	617°F			5,370		3,390
Elong., % at		D 638				
	73°F			10		4.9
	302°F			10		4.4
	482°F			—		4.1
	617°F			6		2.7
Flex. mod., psi at		D 790				
	−310			515,000		—
	73°F			450,000		627,000
	392°F			225,000(302°F)		448,000
	482°F			210,000		376,000
	707°F			100,000(608°F)		325,000
Flex. str., psi at		D 790				
	73°F			14,700		12,500
	509°F			8,000(482°F)		7,520
Shear str., psi at		D 732		11,400		6,265
			Notched	Unnotched	Notched	Unnotched
Impact str., izod ft-lb/in. at		D 256–56				
	−112°F		0.50	—	—	—
	73°F		0.70	9.6	0.23	1.8
	482°F		0.90	11.8	0.47	2.8
Deformation under load, % at 2000 psi and 122°F		D 621		0.13		0.14
Comp. str., psi		D 695		24,400		19,000
Heat dist. temp. °F at 264 psi		D 648		>473°F		>464°F
Sp. heat, Btu lb/°F		D 648		0.27		—
Thermal conductivity, Btu/hr-ft²-°F-in.				2.20		—
Rockwell hardness				H 83–89		H 73–75
Coefficient of linear thermal exp. per °F (73–752°F)				30×10^{-6}		22.5×10^{-6}
Taber abrasion (1000 g load, CS–17 wheel) mg/1000 cycles		D1044–56		6.3		7.8

Table 2.83 Summary of polyimide properties

are commercially attractive for processes such as compression molding of complex-shaped parts, adhesive bonds, and deposition onto reinforcing fibers for subsequent hot pressing into PI-matrix and fiber-reinforced plastics.

A PI powder of this type is formed directly by precipitation from the reaction vessel in which polymerization occurs; no grinding is used. In the process, the monomeric constituents are allowed to react in a solvent such as N-methylpyrrolidene. Next, the polymer precursor is dehydrated with glacial acetic acid, which is also a nonsolvent for the resulting plastic.

The molecular weight of the PI polymerized in this process depend primarily on the stoichiometric offset between the monomeric constituents and the amount of end-capping agent. Moreover, it is necessary to add an excess of end-capping agent about 1 hour after the beginning of the reaction to afford desirable melt stability in the powder end product.

POLYESTERIMIDE

Polyesterimides show good physical and electrical properties at high temperatures and 20,000 hours of service life at 230°C. The thermal-oxidative stabilities of the polyesterimides are described as being less than the best polyimides but better than the aromatic polyesters.

POLYKETONE

Polyketones (PKs) are semicrystalline engineering thermoplastics that have strengths in their thermal, mechanical, chemical, and electrical properties. They are used in a variety of products for

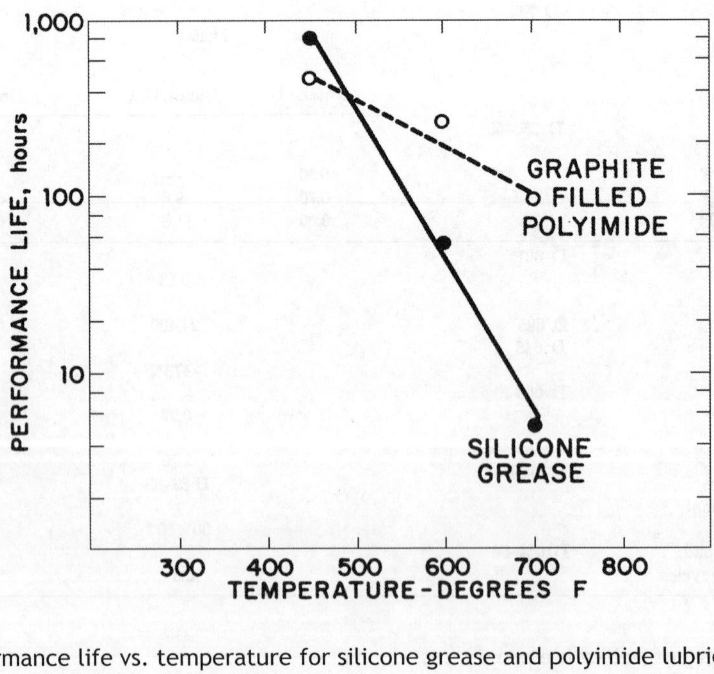

Figure 2.31 Performance life vs. temperature for silicone grease and polyimide lubricating ball bearings

the electrical, automotive, aerospace, chemical, and oil industries. They compete for applications with ceramics, glass, metals, thermoset plastics, and heat-tolerant and chemical-resistant engineering thermoplastics such as polysulfone, polyimide, polycarbonate, fluoropolymer, and some nylons.

The family of polyketones, also called polyaryletherketones (PAEKs), consists of polyether-etherketone (PEEK), polyetherketone (PEK), and polyetherketoneetherketoneketone (PEKEKK).

They share similar molecular structures based upon repeating ether and ketone groups in various ratios.

Polyetheretherketone With its flexibility, PEEK behaves like a TP and has the ability to crystallize (25% to 50%). Its high-glass transition temperature (T_g) and the high melting point (T_m), combined with high-temperature chemical stability, rate this plastic among the most temperature-resistant TPs. As with other crystallizing TPs, crystallinity can develop only at temperatures between T_m and T_g, a fact that must be taken into account for processing (extrusion, injection, etc.). PEEK retains good mechanical properties at high temperatures, such as 200°C (392°F) for periods of time. It is practically insoluble in any solvents and particularly resistant to hydrolysis by steam or high-temperature pressurized water, absorbs little moisture, and has excellent resistance to nuclear radiation. As with other crystallizing materials, it is resistant to environmental-stress cracking. They have a very low flammability and very low smoke and toxic gas emission.

Polyetheretherketoneketone The development of PEEKK was the result of selective industrial research for high-performance plastics that would meet the growing requirement for thermal stability and mechanical strength in the electronics, automotive, and mechanical-engineering industries. Their chemical bonds rank among the most stable ones in organic chemistry. The molecules are closely packed over wide areas, forming crystalline regions. This crystallinity, along with the chemical nature of PEEKK, provides exceptional performance. Their most important properties are resistance to dimensional changes (softening) when exposed to high temperatures and resistance to oxidation as they age.

POLYLACTIDE

Polylactide (PLA) is a biodegradable plastic. The first worldwide production facility for PLA was opened by the Cargill Dow LLC joint venture at the end of 2001 in Blair, Nebraska. The plastic is made from lactic acid, which is produced from cornstarch and other plant carbohydrates by bacterial fermentation. PLA shows good clarity, gloss, deadfold, twist retention, heat sealability, and is a good flavor and aroma barrier. After use, the material is fully compostable. Using material from a semiworks plant, Cargill Dow has already developed applications in fibers, thermoformed containers, cups, and packaging films. Their commercial uses include film for packaging, golf balls, and mini-CDs. Thermoformed trays and film for packaging fresh pasta and salads are already in place.

Polyorganophosphazene Polyorganophosphazene (PPZ) is a plastic involving inorganic chains. In this case, atoms of nitrogen (N) and phosphorus (P) form the chain of a variety of organic side groups that can be attached to the phosphorus atom. The resulting high molecular weight chains are often flexible (low glass transition temperature, T_g) and thus are suitable as elastomers. Types with higher T_g and crystallizing types have good applications as coatings, fibers, and biomedical materials. PPZs are inherently fire resistant and offer good weatherability and water and oil repellency.

Polyoxymethylene Polyoxymethylene (POM) is usually called acetal or polyacetal. See the section titled "Acetal."

Polyparamethylstyrene Polyparamethylstyrene (PPMS) homopolymer is fairly similar to PS with a slightly higher glass-transition temperature. It is a low-cost alternative for styrene in homopolymers and copolymers.

Polyperfluoroalkoxy A so-called fully fluorinated plastic, polyperfluoroalkoxy (PPFA) has properties quite close to those of PTFE while being a true thermoplastic processwise.

Polyphenylene Ether Polyphenylene ether (PPE) is frequently referred to as polyphenylene oxide. Modified PPE products are used in many applications. Their unique compatibility with PS, particularly HIPS, results in a wide range of high-temperature, tough, and dimensionally stable products. They can be processed by conventional equipment that produces either solid or foam products. PPE compounds are characterized by outstanding dimensional stability, low water absorption in the engineering TPs, broad temperature ranges, excellent mechanical and thermal properties from –46° to 121°C (–50° to 250°F), and excellent dielectric characteristics over a wide range of frequencies and temperatures. Several injection-molded and extrusion grades are rated UL 94 V-1 or V-O, including the glass-reinforced compounds. Foamable grades have service temperature ratings of up to 96°C (205°F) in ¼-in sections.

Because of their hydrolytic stability, both at room and elevated temperatures, blended parts in PPE can be repeatedly steam sterilized with no significant change in their properties. When exposed to aqueous environments, their dimensional changes are low and predictable. PPE's resistance to acids, bases, and detergents is excellent. However, it is attacked by many halogenated or aromatic hydrocarbons.

POLYPHENYLENE OXIDE

PPO is a polyether through the –O- group in the repeat unit and thus the term *polyethylene ether (PPE)* is sometimes used. PPO has a high glass-transition temperature (T_g). There are commercially available aromatic polyethers prepared from 2,6-xylenol by a rapid oxidation process known as oxidative coupling. Both transparent and opaque grades are available.

They are capable of crystallizing between T_g and the melting point (T_m) of about 260°C (500°F). The cost and certain processing difficulties associated with a high melt viscosity originally led to the use of blends (polyalloys) with PS or HIPS resulting in a single T_g of about 150°C (302°F) to blends from 100° to 135°C (212° to 275°F). These lower T_g blends are often referred to as modified PPOs (MPPOs). The mechanical properties of MPPO are generally good with high stiffness and low creep over a wide temperature range. Good toughness extends to low temperatures. Excellent dimensional stability is associated with the noncrystalline structure, a low coefficient of thermal expansion, and very low moisture absorption. Electrical properties are generally good and unaffected by moisture. Dielectric properties, in particular, are good and stable. They are classified as self-extinguishing and nondripping. Hydrolytic stability is exceptionally high. MPPO is also highly resistant to water, including hot water and steam. It can be repeatedly sterilized in steam autoclaves.

The polyphenylene oxides have good hydrolytic resistance, are soluble in chlorinated and aromatic hydrocarbons, and have good mechanical and electrical properties over a wide temperature range (–170° to 190°C), but they are not as thermally stable as polyimides or polybenzimidazoles. The material has a brittle point of –170°C. Representative properties of a PPO include a heat-deflection temperature, 192° to 194°C at 264 psi; tensile strength at yield, 75 MPa (11,000 psi); tensile modulus, 0.03 MPa (3.8×10^5 psi); tensile elongation at break, 5% to 6%; and flexural strength at yield, 100 MPa (14,500 psi). The PPOs can be injection molded (343°C/8,000 to 12,000 psi) or extruded (288°C) on standard equipment, as well as machined like brass.

Polyphenylene Sulfide

This crystalline, high-performance engineering TP is characterized by outstanding high-temperature stability, flame resistance, and a broad range of chemical resistance. PPS plastics and compounds provide various combinations of high mechanical strength, impact resistance, and electrical insulation, with their high arc resistance and low arc tracking. The pigmented PPS compounds include several grades that are suitable to support current-carrying parts in electrical components. They are essentially transparent to microwave radiation.

Unreinforced PPS is available for use in slurry coating and electrostatic spraying. Coatings are suitable for food-contact applications as well as chemical-processing equipment. PPS is also available with long-fiber glass, carbon, or other reinforced forms. The stampable sheet type contains fiber-mat reinforcement and can be processed by compression molding. Other forms can contain predesigned reinforcement patterns for different processes, such as laminating and thermoforming. These cross-linked types of resins are more crystalline than any of the sulfones, which are generally classified as being more amorphous. They are quite stiff, with a flexural modulus ranging from 11.7×10^3 to 17.2 MPa (1.7×10^6 to 2.5×10^6 psi). Their tensile strengths range from 69 to 172 MPa (10,000 to 25,000 psi). Their heat-deflection temperatures (HDTs) are up to 275°C (525°F) at 1,820 kPa (264 psi).

PPS has excellent resistance to a broad range of chemicals, even at high temperatures. In fact, below 200°C (400°F) the resin has no known solvent. PPAs (polyphthalamides) are flame retardant without additives, being rated at UL 94 V-O. The oxygen index of the neat resin is 44, with the indexes of the compounds ranging from 47 to 53. Because flame retardance is inherent in it, a regrind will be as flame resistant as a production in the virgin material.

More recently developed linear PPSs have far lower proportions of inorganic impurities than the conventional cross-linked material. They are also characterized by higher ultimate strength and an elongation at break of 4% as compared to 1%, as well as higher flexural strength and notched impact strength. However, the cross-linked product is somewhat more rigid. Linear PPS is partially crystalline, with pronounced thermal-transition ranges similar to those of PET that run from 85° to 100°C (185° to 212°F) for the T_g. Its melting point (T_m) is 280° to 285°C (535° to 545°F).

Polyphosphazene

Polyphosphazene icephobics as coatings offer increased anti-icing effectiveness and durability relative to fluorocarbon plastics and silicone elastomers. These icephobic coats can reduce the accumulation of ice on products such as rooftops, aircraft, radomes, antennas, ships, and power-transmission lines. The weight of accumulated ice has resulted in aircraft crashes, fallen power lines, and so on.

The icephobic coats reduce the adhesive force between ice and a surface. As a consequence, the increasing weight of the ice, wind loading, or vibration of the surface causes the ice to be shed. Studies of icephobic response have shown that materials with low surface tension alone (for example, polytetrafluoroethylene) are insufficiently icephobic and that a rubbery material (for example, silicone and neoprene) is more effective. An effective icephobic material should combine low surface tension with a low modulus of elasticity (a rubbery character). Polyphosphazene elastomers possess these desired properties. In addition, they have low glass-transition temperatures (T_g), good environmental stability, and curability. They also are moderately priced.

These plastics/elastomers have surface tensions of about 15 dynes/cm and tensile moduli of elasticity of 1,000 psi (7 MPa); they can be formulated as icephobic coating materials by the addition of plastics, reactive compounds, and initiators. Coating additives include vinyl-terminated siloxanes, acrylic monomers, and unsaturated urethanes, or TS polyesters. Other compounds that react with polyphosphazene might also be used.

The monomer/additive mixtures can be cured by heating or by using chemical initiators such as Michler's ketone in the presence of UV. In addition, an interpenetrating network (IPN) plastic can be made from a blend of the phosphazene elastomer with isocyanate (polyurethane) prepolymers. These compounds then cure upon exposure to moisture, trapping and bonding the phosphazene to the coated surface. Physical properties can be varied over a wide range by the use of either coreaction or IPN-type cures; silane coupling agents bond the coat to the coated fabric.

POLYPHTHALAMIDE

These heat-resistant, high-performance plastics have been used successfully in different applications such as the auto engine block. BP Amoco Polymer reports that, after 7 years and 85,000 miles of actual vehicle use, a Dodge Neon 2.0-liter thermostat housing molded from BP's PPA (polyphthalamide) showed no measurable dimensional variations, no surface degradation, and no decline in the performance of the plastic. In 1994, BP Amodel A-113 3 HS (a 33 wt% glass-filled, heat-stabilized grade) replaced die-cast aluminum and competing plastics in this application because of the PPA's superior corrosion resistance, superior chemical resistance to long-term exposure of engine coolants at 135°C (275°F), lower moisture absorption, improved hydrolytic stability, higher thermal resistance, approximate 20% weight reduction, and overall cost savings. The thermostat housing was designed and developed by the Powertrain Division of LDM Technologies (formerly HPG), headquartered in Auburn Hills, Michigan. The component is currently manufactured in Port Huron, Michigan (281).

POLYSULFIDE

SRs contain sulfur and carbon linkages, produced from organic dihalides and sodium polysulfides. They are thermoset elastomers (TSEs). These polymers have outstanding resistance to oils, greases, and solvents but have unpleasant odors. Their resiliency is poor and their heat resistance is only fair. The abrasion resistance of polysulfides is half that of natural rubber (NR), and their tensile strength runs only from 8.3 to 9.7 MPa (1,200 to 1,400 psi). However, these values are still retained even after extended immersion in oil. Their increased sulfur content improves their resistance to solvents and oils, but reduces their permeability to gases.

POLYSULFONE

The polysulfones (PSUs) are a family of aromatic, moderately heat-resistant thermoplastics. These polymers, prepared from the sodium salt of bisphenol-A and p,p'-dichlorodiphenylsulfone, have good electrical properties and are resistant to acids, bases, detergents, oils, and so on. They have useful properties in the –100° to 150°C temperature range. They can be injected, extruded, or blow molded at 344° to 400°C using standard equipment.

PSUs are amorphous engineering thermoplastics noted for their high heat-deflection temperatures, outstanding dimensional stability and electrical properties, and excellent chemical resistance, as well as for being biologically inert, rigid, strong, and easily processed by different methods. They are stable and self-extinguishing in their completely natural, unmodified NEAT forms (chapter 1). In most plastics, these qualities must be obtained by using chemical modifiers.

They are also heat resistant and maintain their properties in a range from –100° to over 150°C (–150° to over 300°F). These strong, rigid plastics are the only type that will remain transparent at service temperatures as high as 200°C (400°F). The name *polysulfone* has been assigned to polymers with SO_2 groups in their backbone. The basic types are the standard PSUs: polyaryl, polyether, and polyphenyl. Other similar types in this group include polyethersulfone, polyphenylsulfone, and polyarylsulfone.

Products are basically transparent and a slightly clouded amber color. PSUs are available in opaque colors and in mineral-filled and glass fiber (and other reinforced compounds) to provide higher strength, stiffness, and thermal stability. For example, reinforced carbon fiber PSU is used in hip replacements.

The tensile strengths of PSUs go up to 110 MPa (16,000 psi), their flexural moduli to more than 1.0×10^6 psi, and their HDTs up to 200°C (400°F). A high percentage of their physical, mechanical, and electrical properties are maintained at elevated temperatures. For example, their flexural moduli remain above 0.3×10^6 psi at service temperatures as high as 160°C (320°F). Even after prolonged exposure to such temperatures, the plastics do not discolor or degrade. Their thermal stability and oxidation resistance are also excellent at service temperatures well above 150°C (300°F). Heat aging a PSU increases its tensile strength, HDT, and modulus appreciably. However, prolonged heat aging for about a year or so decreases its toughness, tensile strength, and elongation.

A PSU's creep, compared with other TPs, is exceptionally low at elevated temperatures and under certain continuous loads. For example, its creep at 99°C (210°F) is less than that of acetal or heat-resistant acrylonitrile-butadiene-styrene (ABS) at room temperature.

The hydrolytic stability of these materials makes them resistant to water absorption in aqueous acidic and alkaline environments. Their combination of hydrolytic stability and heat resistance results in exceptional resistance to boiling water and steam, even under autoclave pressures and cyclic hot-to-cold and wet-to-dry exposure.

Materials exhibit notch sensitivity and are attacked by ketones, esters, and aromatic hydrocarbons. The PSUs also have the drawback of absorbing UV rays, giving them poor weather resistance. Thus, they are not recommended for outdoor use unless they contain a UV stabilizer, are protected with paint, or are plated.

PSUs are used in medical equipment, solar-heating applications, and other performance applications in which flame retardance, autoclavability, and transparency are needed.

POLYETHERSULFONE

Polyethersulfone (PES) is a high-temperature engineering TP in the polysulfone family. PES is recommended for load-bearing applications up to 182°C (360°F). Even without flame retardants, it offers low flammability, and it has little change in its electrical properties in the temperature range from 0° to 200°C (32° to 390°F). At room temperature, it behaves like a traditional engineering TP, being rough, rigid, and strong with outstanding long-term load-bearing properties. Of far greater importance are its high-temperature properties.

PES can be used for tens of thousands of hours at 200°C (390°F) without a significant loss of strength. The UL temperature index is 200°C (390°F) for a PES compound that is 30 wt% glass fiber. Its dimensional changes at 200°C (390°F) are negligible, although small variations occur in its electrical performance from 0° to 200°C (32° to 390°F). Compared with most TPs, PES has low flammability without flame-retardant additives and carries a UL 94 V-0 rating at a thickness of 0.017 in with glass reinforcement. Its limiting oxygen index is in the range of 34 to 41. Its smoke and gas emissions are very low.

PES possesses good resistance to X-rays, beta rays, and gamma rays. Even though it does not have good resistance to outdoor weathering, with a carbon-black filler it is acceptable. Chemical resistance in PES is dependent on the external and molded-in stress levels and the temperature. Its chemical resistance can be improved by annealing it at 200°C (390°F). It has good resistance to aqueous acids, bases, and most inorganic solutions. Most of the common sterilizing solutions and anesthetics can safely be used with PES. (Many cleaning and degreasing solvents are based on chlorinated and fluorinated hydrocarbons.) Unless PES is heavily stressed, it can be cleaned by most of these solvents.

Polyphenyl Sulfone Polyphenyl sulfones (PPSUs) balance the following properties: excellent toughness, excellent hydrolytic stability, excellent creep resistance, a high heat-deflection temperature (204°C [399°F] at 1.82 MPa [264 psi]), good electrical properties, good environmental stress-crack resistance relative to other amorphous plastics, and low flammability based on standard laboratory tests.

Polyphenylethersulfone Generally, properties of polyphenylethersulfone (PPESU) are similar to the common polysulfone. Different formulations include those with a glass-transition temperature of 220°C (428°F). Temperature resistance is higher, and it is less sensitive to stress cracking and oxidative attack.

POLYPHTHALAMIDE

Polyphthalamide is basically a crystalline aromatic nylon, combining the high strength and stiffness of nylon with the thermal stability of polyphenylene sulfide. Its molding characteristics are similar to nylon 6/6, with similar or better chemical resistance, but its 24 hours' water absorption is only 0.2% compared with 0.7% for nylon 6/6. A key behavior is high heat resistance.

POLYSACCHARIDE

This is a naturally occurring polymer consisting of simple sugars. An example is starch arid cellulose.

POLYTERPENE

Polyterpene is a thermoplastic obtained by the polymerization of turpentine in the presence of catalysts. These plastics are used to manufacture adhesives, coatings, and varnishes, and in food packaging. They are compatible with waxes, natural and synthetic rubbers, and polyethylene.

POLYTHIOPHENE

This is a melt-processable plastic that is electrically conductive.

POLYURETHANE, THERMOPLASTIC

Polyurethane plastics (PURs) produced by the reaction of polyisocyanates with polyester or polyester-based resins can be either TPs or TSs. There are a wide variety of PURs in form and physical or mechanical properties, and they exhibit an extraordinary range of toughness (Fig. 2.32), flexibility, and abrasion resistance. Their grades can range in density from 16 kg/m^3 (½ lb/ft^3) in cellular forms to 1,120 kg/m^3 (70 lb/ft^3) in solid forms. PUR's hardness runs from rigid, solid forms at 85 Shore D to soft elastomers (209).

PUR materials are available in three forms: rigid foam, flexible foam, and as an elastomer. They are characterized by high strength, good chemical and abrasion resistance, and superior resistance to ozone, oil, gasoline, and many solvents. The rigid foam type is widely used as an insulation material in buildings, appliances, and other applications. The flexible foam is an excellent cushioning material for furniture, and the elastomeric type is used in solid tires, shock absorbers, and so on. It has outstanding flexural life, cut resistance, and abrasion resistance. Some formulations are as much as twenty times more resistant to abrasion than metals are.

The T_g of the flexible foams is well below room temperature; for rigid foams, the T_g is actually higher than room temperature. The cells of a rigid foam are about the same size and uniformity as those of a flexible foam, but rigid foams usually are 90% closed cells. For this reason, their water absorption is low. Compressing the foam beyond its elastic limit will damage its cellular structure. Rigid foams are blown with either carbon dioxide or fluorocarbons, although a nondamaging ozone-blowing agent is now replacing the fluoros. The blowing agent used and the density grade of the foam impact its thermal conductivity.

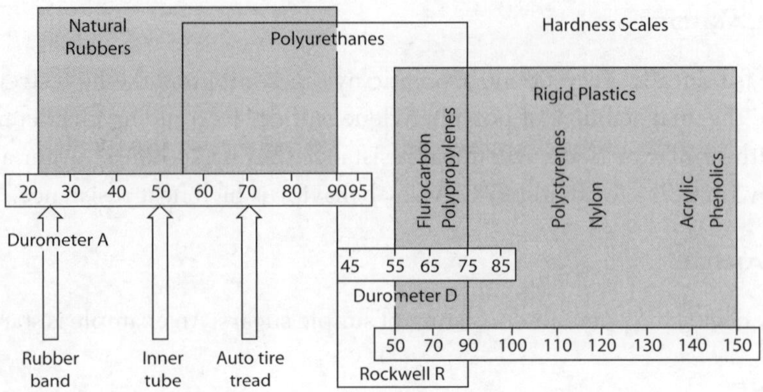

Figure 2.32 Extensive range of toughness with PURs

PURs are manufactured by reacting a polyfunctional alcohol with an isocyanate in a simple blend-reaction operation. With the proper choice of alcohol and isocyanate, polyester or polyether urethanes can be made. Similarly, either thermoplastic or thermoset urethanes can be made with the proper choice of isocyanate. As an example, PURs based on toluene diisocyanate (TDI) continue to hold a large proportion of the soft-foam market. However, TDI was generally outstripped by methane diisocyanate (MDI), since the more flexible chemistry of these isocyanates, as compared to TDI's, made a wider spectrum of applications possible. MDI offers a highly developed building-block system with elements that allow widely differing materials (soft, semirigid, rigid) to be tailored to specific requirements.

Polyurethane Elastomer

This is a PUR with elastomeric properties. The term covers a very wide range of PURs, classified according to the method of manufacture. The main types are cast PUR, millable PUR elastomer, and thermoplastic PUR elastomer.

Polyurethane Isoplast

Isoplast thermoplastic polyurethane plastics have a unique "live" nature that combines the ease of processing low-molecular-weight plastics and the engineering properties of higher-molecular-weight (or lower-melt-index) plastics. Dow's isoplast plastics achieve this combination through depolymerization during melt processing and repolymerization when cooling the final part. This unique nature significantly affects processing.

Isoplast resins are amorphous; as a result they feature clarity, toughness, and dimensional stability. In addition, due to their extremely high molecular weight, they resist chemical attack, creep, and fatigue at a level at least equal to workhorse crystalline resins such as nylon and acetal.

Impact-modified grades offer toughness without notch sensitivity, and this toughness does not decrease with increasing thickness. With a properly designed mold, parts up to 1-in thick can be molded without embrittlement or voids. Reinforcement with long glass fibers at 40 wt% and 60 wt% loadings results in easy flowing composites with toughness, chemical resistance, and high modulus.

Isoplast resins have very low viscosity melts and can be molded with low injection pressures 3.5 to 14 MPa (500 to 2,000 psi) and low clamping pressures (approximately 2 tons/sq in), even in large, difficult-to-fill parts or with high loadings of glass fiber. During cooling, the molecular weight will increase approximately tenfold. Compared to most other thermoplastics, isoplast resins require rigorous drying, moderately low shear conditions, and good moisture control.

Polyurethane Rubber This is an alternative name for polyurethane elastomer.

Polyurethane Virtually Cross-linked This thermoplastic PUR (TPU) is in a unique physical state. It has the properties of a thermoset elastomer without being cross-linked. Strong intramolecular forces—such as hydrogen bonds, van der Waals, London forces, and intramolecular entanglement of chains—all contribute to the virtually cross-linked (VC) state. This state, however, depends on temperature. On heating, the action of these forces disappears, permitting the plastic to be processed by standard methods used for a thermoplastic system. On cooling, these forces reappear. The intramolecular forces of TPUs can be temporarily destroyed by salvation, which enables them to be employed in adhesives and coatings. When the solvents are evaporated, the original properties of the TPUs are restored.

THERMOSET PLASTIC

These plastics, after final processing into products, are substantially infusible and insoluble. Examples of their properties are reviewed in Tables 2.84 to 2.90.

ALKYD

Alkyds are polyesters formed by repeated esterification reactions (polycondensation) between polyhydric alcohols and di- or poly-basic carboxylic acids (or their anhydrides). The name *alkyd* was coined to express that these resins are products of an alcohol and aCID; altering "cid" to "kyd" adjusts the pronunciation and avoids similarity to the word *acid*.

Alkyd resins are condensation polymers used primarily in organic paints and some molding applications. The molding powders find use in the electrical and electronics industries because of their high strength, excellent electrical properties, dimensional stability, and heat resistance. They may be light-colored. Mineral and glass fiber materials are often used to further strengthen the alkyd resins (Table 2.91).

Alkyd compounds are available in "putty" forms and are used in the encapsulation of small electronic components. These include resistors, capacitors, coils, transformers, and other small devices where delicate parts are to be molded inside a protective shell. The glass-fiber-filled alkyd formulations are used for molding large, heavy-duty parts when combinations of the alkyd properties are required. Initial uses were in defense equipment because their high impact strength permitted molded parts to operate effectively under gunfire.

Property	Polyesters	Epoxies	Phenolics	Melamines	Silicones	Polyimides
Specific gravity color	1.10–1.45	1.10–1.4	1.30–1.86	1.40–1.48	1.30–1.34	1.42–1.90
Possibilities	Very good	Good	Limited	Very good	Good	——
By-products of cure	None	None	H_2O	H_2O	H_2O, RCOOH	H_2O
Molding pressures	0–high	0–high	Low–high	Medium–high	Low–high	Low–high
Molding temp., °F	<70–320	<70–330	270–320	270–360	280–360	330–660
Shrinkage, %	2–8	0.1–0.4	1–1.2	1–1.5	1–1.5	0.1–0.3
Tensile strength, 10^3 PSI	6–10	4–13	6–9	7–9	4–5	5–27
Elongation, %	5	2.6	1.5–2.0	——	——	<1–10
Modulus of elasticity tension, 10^3 PSI	3–6.5	4–6	4–5	7.5–10	——	1.9–28.5
Compressive strength, 10^{-3} PSI	13–36.5	15–21.5	12–15	40–45	9–15	4.7–40.0
Flexural strength 10^3 PSI	8.5–18.5	13–21	11–17	11–14	9–14	7.1–49.5
Impact strength (Izod)	0.2–4	0.2–1.0	0.25–0.40	0.24–0.40	0.25–6.0	0.25–0.8
Heat resistance °F (continuous) (°C)	250–500 (121–260)	250–500 (121–260)	160.250 (71–121)	210 (99)	>600 (315)	>600 (315)
Heat distortion, °F (°C)	140–400 (60–204)	115–550 (46–288)	165–260 (74–127)	298 (148)	>900 (482)	270–680 (132–360)
Water absorption, %, (24 hrs.)	0.15–0.60	0.08–0.15	0.03–0.04	0.3–0.5	0.1–0.5	0.11–0.60
Dielectric strength V per mil	200–420	400–500	360–400	300–400	200–500	250–725
Dielectric constant ($60–10^6$ CPS)	2.8–5.2	3.3–5.0	4.0–7.5	4.3–7.6	3.2–5.2	3.2–131.5
Dissipation (power factor $60–10^6$ CPS)	0.003–.028	0.002–.050	0.01–.15	0.015–0.080	0.0008–0.01	0.0018–0.013
Arc resistance, sec.	125	45–120	Tracks	100–145	250–360	230
Burning rate	Slow to self-extinguishing	Slow to self-extinguishing	Very low	Self-extinguishing	None to slow	——

Table 2.84 General properties of thermoset plastics

Table 2.84 *(continued)*

Effect of sunlight	Yellows	None	Darkens	Slight color change	None to slight	——
Effect of weak acids	None	None	None	None	None to slight	Resistant
Effect of strong acids	None to considerable	Attacked by some	Decomposed by oxid. acids, none to slight with regular org. acids	Decomposes	Slight	Slowly attacked–resistant
Effect of weak alkalis	None to slight	None	Slight to marked function of pH	None	None to slight	Slowly attacked
Effect of strong alkalis	Attacked	Slight	Decomposes	Attacked	Slight	Attacked
Effect of organic solvents	Attacked (ketones and chlorinated solvents)	Generally resistant	Attacked by some	None	Attacked by some	Very resistant
Machining qualities	Good	Good	Excellent	Fair	Fair to good	——
Major limitations	Cure shrinkage	Dermatitis, difficult to mold release	Colors limited	High cost	Very high cost	High cost
Major advantages	Ease of fabrication, clarity with flame retardancy, moderate dissipation (power) factor	Low shrinkage, excellent adhesion	Good general properties, low cost	Arc resistance, Scratch resistance, colorability	Heat resistance, low dissipation (power) factor	Heat resistance

Reinforcing material	Specific gravity	Tensile Strength 10³ psi	Tensile modulus 10⁶ psi	Compressive strength, 10³ psi	Flexural strength, 10³ psi	Izod impact strength, ft-lb/in. notch	Heat resistance continuous, °F	Arc resistance, sec.
Polyester								
Glass cloth	1.5–2.1	30–70	1–3	25–50	40–90	5–30	300–350	60–120
Glass mat	1.3–2.3	20–25	$\frac{1}{2}$–2	15–50	25–40	2–10	300–350	120–180
Asbestos	1.6–1.9	30–60	1–3	30–50	50–70	2–8	300–450	100–140
Paper	1.2–1.5	6–14	$\frac{1}{2}$–1$\frac{1}{2}$	20–25	13–28	1–2	220–250	28–75
Cotton cloth	1.2–1.4	7–9	$\frac{1}{2}$–1$\frac{1}{2}$	23–24	13–18	1–4	230–250	70–85
Epoxy								
Glass cloth	1.9–2.0	20–60	2–4	50–70	70–100	11–26	330–500	100–110
Glass mat	1.8–2.0	14–30	1–3	30–38	20–26	8–15	330–500	110–125
Paper	1.4–1.5	10–19	$\frac{1}{2}$–1	20–28	19–24	$\frac{1}{2}$–1	260–300	30–100
Phenolic								
Glass cloth	1.8–2.0	40–60	1–3	35–40	65–95	10–35	350–500	20–130
Glass mat	1.7–1.9	5–20		17–26	10–60	8–16	350–500	400–150
Asbestos	1.7–1.9	40–65	2–5	45–55	50–90	1–6	350–600	120–200
Paper	1.3–1.4	8–20	1–2	20–40	10–30	$\frac{1}{3}$–1	225–250	Tracks
Cotton cloth	1.3–1.4	7–16	$\frac{1}{2}$–1$\frac{1}{2}$	30–44	14–30	$\frac{1}{2}$–3	225–250	Tracks
Nylon cloth	1.1–1.2	5–10	$\frac{1}{4}$–$\frac{1}{2}$	28–36	9–22	2–4	150–165	Tracks
Silicone								
Glass cloth	1.6–1.9	10–35	1–2	25–46	10–38	5–13	400–700	150–250
Asbestos cloth	1.7–1.8	10–25	1–2	40–50	12–20	6–9	450–730	150–300

Table 2.85 Properties of reinforced thermoset plastics

	Room temperature tensile strength (10^3 psi)	Room temperature compressive strength (10^3 psi)	Ultimate flexural strength (10^3 psi)				Flexural modulus (10^3 psi)		
			At room temperature	After 30-day water immersion	At 300°F	At 500°F	At room temperature	At 300°F	At 500°F
Fabric Laminates									
Epoxy-glass									
Bisphenol A	50	55	85–89	65–80	40–60	8–11	3600–3800	2900–3300	3200
Novalac	42	57	85–90		58–63	25	3200–3400	2700	
Peracetic acid	57–59	49–53	85–90		60–65	47	4000–4200	3700–3900	
Bisphenol A and high-modulus glass	64	61	90	76	44		5000	3700	
High-temperature epoxy and high-modulus glass	55	59	86			23	5300		3000
Phenolic-glass									
Standard grade	50	35	50			40	3000		3000
Heat-resistant	65	48	86	52	78	47–52	4800	4000	3300–3600
Silane-modified	46	51	80	73	65	65	4300	3500	3500
Phenolic-leached glass	14	24	25		20		2600	2600	2600
Phenolic-quartz	35		68		37	50	3700	3400	3200
Polyesters									
Styrene-modified	50–55	40–45	50–75	64	30	24	3300	2500	2000
TAC-modified	59	51	65	59	43	30	3000	2000	1800
Styrene-modified and high-modulus glass		52	59	49			4200		
Silicone-glass	41	29	50	36	23	21	2900	2400	2200
Filament-Wound Structures									
Epoxy-glass	270	70	271		245				
Epoxy and high-tensile glass	285								
Mil P-27327 (proposed)	200	50	220	210	200	180	7000	6800	6800
Molding Compounds									
Phenolic-glass	10	35	20		16.4		3300		
Phenolic-leached-glass	8.5	36	14				3000		
Phenolic-asbestos	9	24	24			15	2000		1400
Silicone-glass	4.5	16	14		6	6	2100	1100	500
Polyester-glass	9.5	29.5	19		12.7	6.7	1400	910	910

Table 2.86 Mechanical properties of thermoset-reinforced plastics at ambient and elevated temperature

Resin material	Ultimate tensile strength (10^3 psi) 70°F	−110°F	−320°F	−423°F	Ultimate flexure strength (10^3 psi) 70°F	−110°F	−320°F	−423°F	Ultimate compression strength (10^3 psi) 70°F	−110°F	−320°F	−423°F
Epoxy												
Epon 828	40.9	62.3	91.6	97.9	77.9	99.3	150.4	145.0	43.5	85.4	109.9	108.8
Epon 1001	49.0	71.4	94.9	100.7	83.2	105.2	154.0	122.0	58.6	86.2	89.4	78.9
Phenolic												
Narmco 506	38.1	58.3	74.4	72.2	63.4	76.2	84.5	83.0	46.5	55.1	81.8	76.7
CTL 91-LD	51.7	65.7	65.1	64.9	77.4	100.1	109.0	110.0	70.4	88.1	99.1	91.6
Polyester												
Hetron 92	39.2	63.2	71.7	84.9	71.4	82.1	87.2	81.9	22.9	35.8	43.6	41.4
Paraplex P43	47.9	58.5	81.3	78.2	68.8	79.4	84.2	84.9	31.5	43.2	51.3	47.9
Triallylcyanurate												
Laminac 4232	41.0	60.7	56.9	55.5	63.3	74.7	70.0	65.4	28.4	42.0	47.6	44.1
Vibrin 135	32.7	57.6	52.0	54.6	52.0	76.2	78.1	75.1	26.3	33.5	38.1	36.1
Silicone												
Narmco 513	29.6	47.8	76.3	80.5	34.1	40.7	52.1	52.8	25.3	28.7	39.1	43.3
Trevarno F130	25.7	46.7	67.9	74.6	36.1	43.5	63.7	72.8	17.7	23.6	34.9	30.8
Epoxy												
Epon 828	3.51	3.84	4.15	4.54	3.54	3.96	5.04	4.66	3.35	3.33	4.07	4.26
Epon 1001	3.34	3.65	3.93	4.26	4.22	3.78		3.76	3.75	3.30	3.90	3.96
Phenolic												
Narmco 506	3.65	3.88	4.74	3.83	3.56	3.61		4.06	4.39	7.46	4.5	5.12
CTL 91–LD	3.85	4.31	4.25	4.48	4.32	4.45		5.18	4.69	3.61	8.43	4.48
Polyester												
Hetron 92	2.46	3.50			3.48	3.52	3.04	2.96	1.57	1.65	2.48	2.28
Paraplex P43	2.45	3.30		3.18	3.17	3.76	3.39	2.96	2.75	3.18	3.78	3.17
Triallylcyanurate												
Laminac 4232	3.08	3.38	3.67	3.85	3.29	3.82	4.37	4.30	3.14	2.76	3.09	3.25
Vibrin 135	3.36	3.50	3.72	3.76	3.58	4.01	4.06	4.00	1.79	1.99	2.45	2.54
Silicone												
Narmco 513	2.73	2.73	2.75	3.09	2.85	3.18	3.19	3.08	2.72	2.50	4.61	3.28
Trevarno F130	2.52	2.66	2.58	3.34	2.69	2.72	2.85	3.32	2.06	2.15	2.41	2.44

Table 2.87 Examples of glass-fiber-reinforced plastics at low temperatures

ALLYL

Allyls have high heat and moisture resistance, good electrical performance, good chemical resistance, dimensional stability, and low creep. Allyl resins retain their physical and electrical characteristics under prolonged exposure to severe environmental conditions. These plastics have been used in a number of civilian, military, and aerospace applications where their environmental resistances are important.

There are two major allyl resins: diallyl phthalate (DAP) and diallyl isophthalate (see "Diallyl Phthalate"). Both of these resins are widely used in fiber-reinforced forms. The allyl resins can be

Material composition	Phenolic (33%), carbon fabric (67%)	Phenolic(31%), graphite fabric (69%)	Epoxy-novolac (37%), carbon fabric (63%)	Epoxy novolac (35%) graphite fabric (65%)
Density, lb/ft³	84	89	85	85
Tensile strength, psi				
Parallel				
75°F	21,600	12,900	18,500	10,100
350°F	19,250	12,325	2620	3820
Perpendicular				
75°F	2450	680	4000	1000
350°F	230	574	160	100
Tensile elastic modulus, million psi,				
Parallel				
75°F	3.50	1.58	2.63	2.16
350°F	2.80	1.70	0.64	0.58
Perpendicular				
75°F	2.39	0.81	1.46	1.13
350°F	—	0.48	0.09	0.14
Tensile elongation at failure, %				
Parallel				
75°F	1.08	0.97	1.15	0.77
350°F	1.14	1.10	0.67	0.47
Perpendicular				
75°F	0.10	0.11	0.29	0.10
350°F	0.32	0.16	0.19	0.10
Shear strength, psi				
Parallel				
75°F	4000	1820	—	2600
350°F	3230	1590	—	1500
Compressive strength, psi	35,000	10,800	16,700	—
Compressive elastic modulus, million psi	1.57	1.29	1.19	—
Flexural strength, psi	30,000	23,000	16,000	—
Flexural elastic modulus, million psi	2.4	1.9	0.93	—
Hardness, Barcol	73	52	—	—
Specific heat, Btu/lb–°F	0.24	0.26	—	—
Thermal conductivity, Btu/hr/ft²–°F/ft				
Parallel	0.82	1.30	0.59	1.50
Perpendicular	0.27	0.80	0.35	0.66
Thermal expansion coefficient, 10^{-6} in./in./°F				
Parallel	3.8	2.4	7.5	7.0
Perpendicular	8.5	9.0	60.0	53.0
Emittance	0.8	0.8	0.8	0.8

All properties are parallel to the fabric warp unless otherwise noted.

Table 2.88 Properties of carbon/graphite-reinforced plastics

Test conditions	Flexural modulus (10^6 psi) Polyester	Epoxy	Phenolic
Atmospheric pressure and room temperature	4.67	3.54	4.17
Low pressure (10^{-6} torr) at room temperature	4.67	3.54	4.17
Low pressure (10^{-6} torr), ultraviolet (0.036 W/cm²) and moderate temperature (250°F)	3.84	2.14	3.94
Atmospheric pressure, ultraviolet (0.036 W/cm²) and and moderate temperature (250°F)	3.38	2.00	3.86
Atmospheric pressure, ultraviolet (0.036 W/cm²) and and room temperature	3.75	3.54	3.68
Low pressure (10^{-6} torr) and moderate temperature (250°F)	3.61	2.38	3.96
Atmospheric pressure and moderate temperature (250°F)	3.75	2.25	3.81

Table 2.89 Flexural modulus of glass-fiber-reinforced plastics when exposed to various elements

Resin type	Test	Exposure (rad)	Temperature (°F)	Exposure time (hr)	Ultimate strength (psi)	Flexural modulus (10^6 psi)
Silicone [a]	Flexure	None	Room	None	31,760	3.06
		8.3×10^7	Room	200	31,460	2.94
		None	500	50	12,390	1.90
		2.1×10^7	500	50	13,625	2.0
		None	500	100	13,410	2.0
		4.15×10^7	500	100	11,720	2.0
		None	500	200	14,060	2.0
		8.3×10^7	500	200	9,860	1.9
Heat-resistant epoxy [b]	Compression	None	Room	None	46,680	
		8.3×10^7	Room	200	46,660	
		None	500	50	3,705	
		2.1×10^7	500	50	3,780	
		None	500	100	4,090	
		4.15×10^7	500	100	5,490	
		None	500	200	4,720	
		8.3×10^7	500	200	6,360	
Phenolic [b]	Flexure	None	Room	None	84,525	4.22
		8.3×10^7	Room	200	84,040	4.35
		None	500	50	27,300	3.14
		2.1×10^7	500	50	55,020	3.46
		None	500	100	17,660	2.62
		4.15×10^7	500	100	47,015	3.61
		None	500	200	12,330	2.13
		8.3×10^7	500	200	15,645	2.41

[a] 181 glass fabric (112 finish).
[b] 181 glass fabric (Volan-A finish).

Table 2.90 Mechanical properties of glass-fabric-reinforced plastics after irradiation at elevated temperature

Property	Unit	Regular molding powder	"Putty" formulation	Glass-fiber reinforced
Molding temperature	°F	270–330	270–330	270–330
Specific gravity		2.17–2.20	2.05–2.15	2.02–2.10
Water absorption, 24 hr at 23°C	%	0.05–0.08	0.10–0.15	0.07–0.10
Long term heat resistance	°F	275	250	300
Flammability		Self-extinguishing or nonburning		
Impact strength, izod	Ft-lb/in. of notch	0.30–0.35	0.25–0.35	8–12
Compressive strength	psi	18,000	23,000	30,000
Tensile strength	psi	4000	5000	9000

Table 2.91 Properties of alkyd molding compounds

compressed or transfer molded and perform well in automated equipment. The reinforced allyls can be fabricated by a variety of methods and have been used for large sections such as radomes and for aircraft and missile components that require heat and chemical resistance combined with excellent electrical characteristics.

Amino In the family of aminos there are melamine and urea-formaldehydes (to be reviewed later in this section). Examples of their performances are provided in Table 2.92. Melamine formaldehydes (MFs) have excellent electrical properties, heat and moisture resistance, and abrasion resistance (good for dinnerware and buttons); in high-pressure laminates, they are resistant to alkalies and detergents. They are used as the plastic for countertops. Urea-formaldehydes (UFs) have properties similar to melamines and are used for wall switch plates, light-colored appliance hardware, buttons, toilet seats, and cosmetics containers. Unlike MFs, they are translucent, giving them a brightness and depth of color somewhat similar to opal glass.

Chlorosulfonated Polyethylene Elastomer CSPE (or CSM) refers to the moderate random chlorination chlorine (Cl) incorporated with infrequent chlorosulfonic groups (SO_2Cl) as preferred cross-linking sites. This thermoset elastomer has excellent combinations of properties that include total resistance to ozone; excellent resistance to abrasion, heat, flame, solvents, oxidizing chemicals, and crack growth; excellent dielectric properties, low moisture absorption, and resistance to oil similar to neoprene; and fair low temperature flexibility at –40°F. It also has excellent weather resistance even in light colors, low gas permeability for an elastomer, and good adhesion to substrates. CSPE can be made into a wide range of colors, because carbon black is not required for reinforcement (as usually required in high-performance elastomers). Uses include hoses, roll covers, tank liners, wire and cable covers, footwear, and building products.

Cross-linked Polyethylene XLPE is PE (also called PEX) that by chemical or irradiation treatment becomes a thermoset plastic with significant improvements in properties such as strength, chemical, and temperature resistance (outstanding heat resistance; Table 2.85). XLPE can be produced by the

Property	Urea, general purpose Wood-filled	Urea, general purpose α-Cellulose-filled	Melamine, general purpose Wood-filled	Melamine, general purpose α-Cellulose-filled	Melamine, electrical grades Mineral-filled	Melamine, electrical grades Glass-filled
Density	1·4–1·5	1·5–1·55	1·5–1·55	1·5–1·55	1·8–2·1	1·9–2·0
State	Granular powder	Gran. or fine powder	Granular powder	Gran. or fine powder	Granular powder	Granular powder
Appearance	Opaque colors	Translucent or opaque colors	Opaque colors	Translucent or opaque colors	Opaque grey or brown	Opaque grey
Processing temperature range, °F	~275°–330°F	~275°–350°F	~280°–340°F	~280°–340°F	~270°–340°F	~290°–325°F
Processing temperature range, °C	~135°–165°C	~135°–175°C	~140°–170°C	~140°–170°C	~130°–170°C	~145°–160°C
Mold shrinkage (in./in.)	0·006–0·010	0·005–0·010	0·007–0·008	0·004–0·009	0·0015–0·007	0·002–0·005
Mechanical properties						
Impact strength [a] (ft lb/in.)	0·22–0·28	0·24–0·30	0·26–0·38	0·24–0·32	0·20–0·45	0·35–0·50 [c]
Yield stress [b]	—	—	—	—	—	—
Distortion [b]	—	—	—	—	—	—
Maximum working temperature, °F	170	170	250	210	300	300
°C	75	75	120	100	150	150
Tensile strength lb/in.^{2}	5500–8000	7000–11,000	5500–8500	7000–12,000	4000–6500	6000–10,000
Tensile modulus lb/in.^{2}	$1·0–1·4×10^6$	$1·0–1·4×10^6$	$\sim1·1×10^4$	$1·2–1·6×10^6$	$1·5–3·5×10^6$	$2·0–4·0×10^6$
Electrical properties						
Power factor at 60 cps	0·025–0·045	0·025–0·035	0·020–0·040	0·040–0·080	0·100–0·300	0·040–0·100
10^6 cps	0·020–0·040	0·020–0·030	0·010–0·030	0·020–0·050	0·040–0·070	0·020–0·030
10^6 cps	0·025–0·040	0·020–0·030	0·025–0·040	0·025–0·045	0·010–0·050	0·020–0·030
Dielectric constant at 60 cps	7·0–9·0	7·5–8·0	6·0–7·0	6·0–9·0	7·0–10·0	7·0–14·0
10^3 cps	6·5–8·5	6·5–7·5	6·0–7·0	8·0–10·0	8·0–9·0	7·0–9·0
10^6 cps	6·0–8·5	6·5–7·0	5·5–6·0	7·0–9·0	5·5–6·5	5·5–7·5
Dielectric strength [d]						
Short time at 25°C	300–450	300–400	350–400	300–400	400–450	150–250
at 100°C	200–300	200–300	300–350	150–250	250–300	50–150
Step by step at 25°C	200–300	200–250	200–300	200–300	250–300	150–200
at 100°C	100–200	100–150	150–250	100–150	150–250	50–100
Volume resistivity (ohm cm)	$10^{11}–10^{14}$	$10^{11}–10^{14}$	$10^{11}–10^{13}$	$10^{11}–10^{13}$	$10^{10}–10^{13}$	$10^{11}–10^{14}$
Track resistant	Yes	Yes	Yes	Yes	Yes	Yes
Arc resistant	Yes	Yes	Yes	Yes	Yes	Yes
Thermal properties						
Coefficient of linear thermal expansion [e] (°C)	$3–6×10^{-5}$	$2–5×10^{-5}$	$3–5×10^{-5}$	$2–5×10^{-5}$	$2–5×10^{-5}$	$2–5×10^{-5}$
Thermal conductivity (cal/cm^2/°C/cm)	$6–9×10^{-4}$	$7–10×10^{-4}$	$7–10×10^{-4}$	$7–10×10^{-4}$	$10–15×10^{-4}$	$11–10×10^{-4}$
Flammability	S.E.[g]	S.E.[g]	S.E.[g]	S.E.[g]	S.E.[g]	S.E.[g]
Chemical properties						
Water absorption, 24 h at 23°C [b] (%)	00·5–1·2	0·4–1·3	0·3–0·4	0·1–0·4	0·1–0·3	0·03–0·1
Gas permeabilities	—	—	—	—	—	—
Resistance to:						
Mineral acids (dil)	Fair	Fair	Good	Good	Fair	Very good
Mineral acids (conc)	Poor	Poor	Poor	Poor	Poor	Poor
Alkalis	Fair	Fair	Good	Good	Fair	Good
Solvents:						
Alcohols	Good	Good	Good	Good	Fair	Good
Ketones	Good	Good	Good	Good	Fair	—
Aromatic hydrocarbons	Good	Good	Good	Good	Fair	—
Chlorinated hydrocarbons	Good	Good	Good	Good	Fair	—
Detergents	Very good	Very good	Very good	Very good	Good	—
Greases and oils	Good	Good	Good	Good	Fair–good	—

[a] These materials are brittle and notch-sensitive at room temperature and below.
[b] These materials have a linear stress–strain relationship until fracture occurs at about 1% elongation. Maximum temperatures for continuous use are indicated.
[c] Special "high impact" material containing long glass fibers has an "impact strength" of 4.0–6.0 ft-lb/in.
[d] Preheating the material before molding usually improves the dielectric strength.
[e] These materials shrink on continuous heating.
[f] Equilibrium water absorption is not attained for a considerable time—more than a year for melamine materials.
[g] Self-extinguishing.

Examples of satisfactory applications for urea formaldehyde: domestic electrical accessories (switch covers, plugs, etc.); toilet seats; radio cabinets and closures (bottle caps, etc.)

Examples of satisfactory use for melamine formaldehyde: tableware—cups, saucers, large and small plates, etc.; kitchen utensils. Mineral filled types are used primarily for industrial electrical applications requiring arc or track resistance.

Table 2.92 Properties of amino molding compounds (urea- and melamine formaldehydes)

addition of small amounts of organic peroxides (dicumyl, peroxide, etc.), which does not cause significant cross-linking before the polymer has acquired its final shape in processing.

A process such as rotational molding is suited to this cross-linking method. Another method involves the irradiation of finished products in high-energy fields. It is used particularly for extruded products, such as films (shrink-wrap films in particular), pipes, electrical wire, and cable insulation. It is also used to fabricate XLPE foams (chapters 8 and 13).

Diallyl Phthalate

Diallyl phthalate (DAP) and diallyl isophthalate (DAIP) are the principal thermosets in the allyl family; DAP is predominantly used. DAP's major use is in electrical connectors since it performs well in electrical circuits. DAP is also used in reinforced plastics (RP) laminates and molding compounds (Tables 2.94 and 2.95 and Figs. 2.33 to 2.35).

DAPs are used for preimpregnated glass cloth and paper, which in turn go through a heat, time, pressure cycle to produce products. Molding compounds are reinforced with fibers to improve mechanical and physical properties. Glass fibers impart high mechanical performance, acrylic fibers provide improved electrical performance, polyester fibers improve impact resistance, and other fibers as well as fillers can impart different performances.

In some applications, DAPs are competitive with TS polyester compounds (bulk molding compounds [BMCs], etc.). They can offer longer shelf life (in the B-stage), less shrinkage during curing, somewhat better chemical or electrical properties, and higher heat resistance. In general, these allyls are more expensive and, therefore, find few uses in consumer products. They can be molded at lower pressures and in faster molding cycles. With the triallyl cyanurate formulation, temperatures as high as 260°C (500°F) are obtained; uses include high-speed radomes on aircraft, among others. Major advantages of all allyls over TS polyesters are freedom from styrene odor,

Property	Crosslinked high density unfilled	Crosslinked high density, 55% carbon black	Crosslinked low density, 27½% carbon black	Crosslinked low density, 10% carbon black
Density	0.956	1.27	1.13	1.42
Tensile strength, psi	3250	4700	3050	3200
Elongation, %	460	46	290	70
Tensile strength, psi @ 100° C	—	2000	800	1100
Hardness, Shore D	59	78	58	67
Brittle point, °C	<—65°	—	<—70°	—15°
Impact strength, izod, ft lb/in. of notch	22.8	5.5	12.6	5.0
Stress crack resistance	Not tested	No cracks in >1000 hr	No cracks in >1000 hr	No cracks in >1000 hr

Table 2.93 Properties of cross-linked polyethylene plastics

low toxicity, low evaporation losses during fabricating evacuation cycles, no subsequent oozing or bleed-out, and long-term retention of electrical-insulating performances.

DAP's major use is in electrical connectors used in communications, computer, and aerospace systems. Its high thermal resistance permits its use in vapor-phase soldering. Other uses include arc-track-resistant switchgear, TV components, and circuit boards.

EPOXY

Epoxy plastics have perhaps the highest performance of all the thermosets. Properties include very high strength in tension, compression, flexural loadings, very low shrinkages, and hard, superior

Property	Orlon	Dacron	Long Glass	Asbestos	Short Glass	Short Glass*
Tensile strength, psi.....	6,000	5,000	10,000	5,500	7,000	7,000
Compressive strength, psi	25,000	25,000	25,000	25,000	25,000	28,000
Flexural strength, psi....	10,000	11,500	16,000	9,600	12,000	12,000
Flexural modulus psi X 10^{-6}.....	0.71	0.64	1.3	1.2	1.2	1.3
Impact strength, Izod, ft-lb/in of notch.....	1.2	4.5	6.0	0.4	0.6	0.5
Hardness, Rockwell M.....	108	108	100	100	105	110
Specific gravity at 25°C..	1.31-1.45	1.39-1.62	1.55-1.70	1.55-1.65	1.6-1.8	1.65-1.75
Dielectric constant:						
At 1 kHz.....	3.7-4.0	3.79	4.2	....	4.4	4.1
At 1 MHz	3.3-3.6	3.4	4.2	4.5-6.0	4.4	3.4
Dissipation factor:						
At 1 kHz....	0.020-.025	0.008	0.004-.006	0.04-0.08	0.006	0.004
At 1 MHz.....	0.015-0.02	0.012	0.008	0.04-0.08	0.008	0.008
Mold shrinkage, in./in.....	0.009	0.010	0.002	0.006	0.003	0.003
Postmold shrinkage, in./in....	0.001	0.0006	0.0007	0.001	0.0007	0.0002
Heat-deflection temperature, °F......	265	290	392	325	400	500+
Heat resistance, continuous, °F.....	300-500	350-400	350-400	350-400	350-400	450

*Based on diallyl isophthalate

Table 2.94 Properties of several DAP compounds with various fillers (7)

GRADE	1-503	1-501N	1-501AN	1-510N	2-501N	1-520	1-540	1310	2-520
Reinforcement	Orlon	Mineral	Mineral Nylon	Mineral	Mineral	Short Glass	Short Glass Mineral	Short Glass	Short Glass
Resin (DAP)	Ortho	Ortho	Ortho	Ortho	Iso	Ortho	Ortho	Ortho	Iso
Form	Granular	Granular	Granular	Granular	Granular	Granular	Granular	Granular	Granular
Bulk Factor	3.0	2.3	2.6	2.4	2.1	2.4	2.3	2.4	2.5
Specific Gravity (g/cm^3)	1.40	1.80	1.73	1.66	1.90	1.80	1.82	1.80	1.75
Shrinkage (Comp) in/in	.010-.012	.003-.006	.007-.010	.003-.006	.003-.006	.001-.003	.0015-.003	.001-.003	.001-.003
Izod Impact (ft. lb./in.) (C/T)	0.65	0.45	0.45	0.45	0.45	1.0	0.70	1.0	1.0
Flexural Strength (psi)	9000/ 12,000	11,000/ 14,000	10,000/ 13,000	12,000/ 15,000	10,000/ 13,000	15,000/ 18,000	10,000/ 13,000	16,000/ 20,000	15,000/ 18,000
Flexural Modulus (psi)	0.7×10^6	1.6×10^6	1.1×10^6	1.3×10^6	2.0×10^6	1.8×10^6	1.8×10^6	1.9×10^6	1.7×10^6
Tensile Strength (psi)	5000	6500	5500	7500	5500	8000	6500	9000	7500
Compressive Strength (psi)	29,000	24,000	25,000	21,000	27,000	28,000	23,000	28,000	30,000
Water Absorption, %	0.40	0.30	0.40	0.50	0.25	0.30	0.35	0.25	0.25
Deflection Temperature °F	290	330	300	325	525	500	425	500	525
Continuous Use Temperature °F	225	230	200	225	425	375	325	375	425
Flammability (IGN/Burn), sec.	87/440	90/373	–	110/279	115/222	70/340	107/297	100/350	98/262
UL Rating 1/8"	–	–	94 HB	–	94 HB	–	94 HB	–	94 HB
Dielectric Strength 60Hz, ST/SS, wet (vpm)	375/350	375/350	400/380	400/390	400/400	375/350	390/370	400/380	380/360
Dielectric Constant 1 khz/1 mhz, wet	3.4/3.1	4.2/3.6	4.3/3.7	4.2/3.7	4.3/3.8	4.1/3.9	4.4/4.2	4.1/3.9	3.9/3.7
Dissipation Factor 1 khz/1 mhz, wet	.016/.021	.013/.014	.015/.019	.015/.017	.015/.019	.010/.017	.013/.018	.009/.016	.011/.015
Arc Resistance (sec.)	125	140	135	135	150	135	180	140	150
MIL-M-14G Type	SDI-5	MDG	MDG	MDG	MDG	SDG	–	SDG	SDG
Reinforcement	Mineral	Short Glass	Short Glass Mineral	Short Glass	Short Glass	Long Glass	Long Glass	Long Glass	Long Glass
Resin (DAP)	Ortho	Iso	Ortho	Ortho	Ortho	Ortho	Iso	Ortho	Iso
Form	Granular	Granular	Granular	Granular	Granular	Coarse Granular	Coarse Granular	Coarse Granular	Coarse Granular
Bulk Factor	2.0	2.4	2.2	2.3	2.4	3.5	3.5	3.5	3.5
Specific Gravity (g/cm^3)	1.83	1.90	2.08	1.87	1.86	1.74	1.73	1.76	1.76
Shrinkage (Comp) in/in	.003-.006	.001-.003	.0015-.003	.0015-.003	.0015-.003	.0015-.004	.0015-.004	.0015-.004	.0015-.004
Izod Impact (ft. lb./in.) (C/T)	0.60	1.0	0.70	1.0	1.1	3.0-7.0	3.0-7.0	3.0-7.0	3.0-7.0
Flexural Strength (psi)	11,000/ 14,000	14,000/ 17,000	13,000/ 16,000	16,000/ 19,000	16,000/ 20,000	17,000/ 20,000	17,000/ 20,000	18,000/ 21,000	18,000/ 21,000
Flexural Modulus (psi)	1.7×10^6	1.4×10^6	2.4×10^6	1.8×10^6	1.7×10^6	1.8×10^6	1.4×10^6	1.6×10^6	1.7×10^6
Tensile Strength (psi)	6500	7500	7500	8000	9000	8500	7500	8500	8500
Compressive Strength (psi)	25,000	26,000	28,000	28,000	26,000	27,000	25,000	27,000	28,000
Water Absorption, %	0.30	0.25	0.25	0.25	0.25	0.35	0.35	0.35	0.35
Deflection Temperature °F	325	525	400	450	400	500	550	500	550
Continuous Use Temperature °F	225	425	300	350	350	400	450	400	450
Flammability (IGN/Burn), sec.	106/23	119/40	159/32	110/40	120/20	95/275	95/339	110/40	110/40
UL Rating 1/8"	–	94V-0	94V-0	94V-0	94V-0	–	–	94V-0	–
Dielectric Strength 60Hz, ST/SS, wet (vpm)	400/375	375/350	370/350	400/375	400/375	400/350	400/350	400/350	400/350
Dielectric Constant 1 khz/1 mhz, wet	4.3/3.7	4.1/3.9	4.4/4.3	4.3/4.1	4.4/3.9	4.1/4.0	3.8/3.6	4.2/3.9	4.1/3.9
Dissipation Factor 1 khz/1 mhz, wet	.013/.016	.010/.013	.011/.015	.010/.016	.009/.017	.011/.019	.010/.019	.010/.018	.010/.019
Arc Resistance (sec.)	135	150	180	130	140	135	135	130	135
MIL-M-14G Type	MDG	SDG-F	–	SDG-F	SDG-F	GDI-30	GDI-30	GDI-30F	GDI-30F

Test methods used are according to A.S.T.M., Fed. Std. 406, or Military Specifications.

Table 2.95 DAP molding material properties (6)

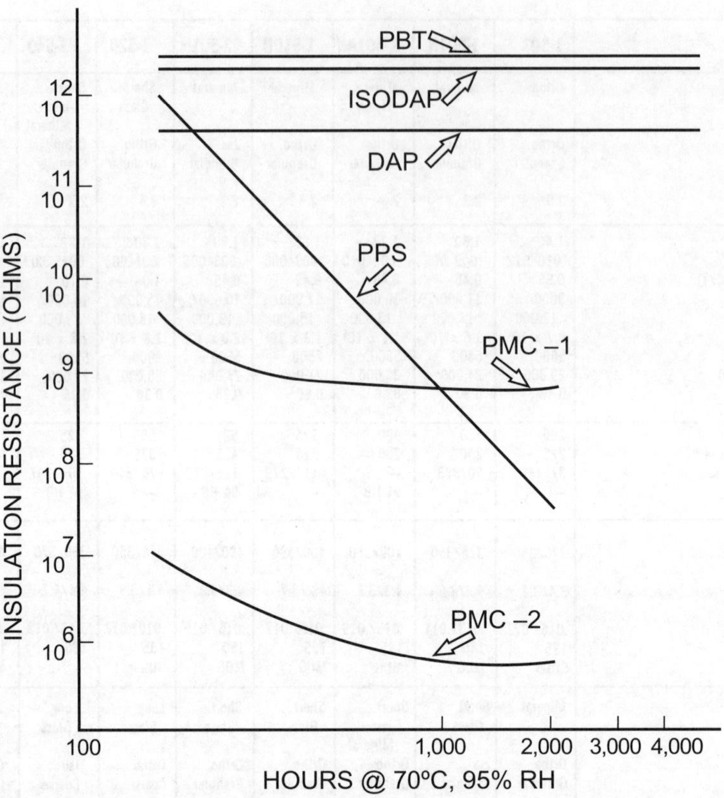

Figure 2.33 Insulation resistance vs. exposure to high humidity

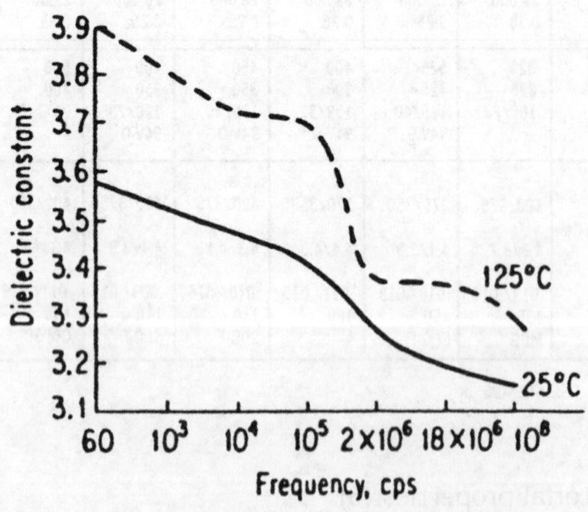

Figure 2.34 Effect of frequency and temperature on the dielectric constant of unfilled DAP

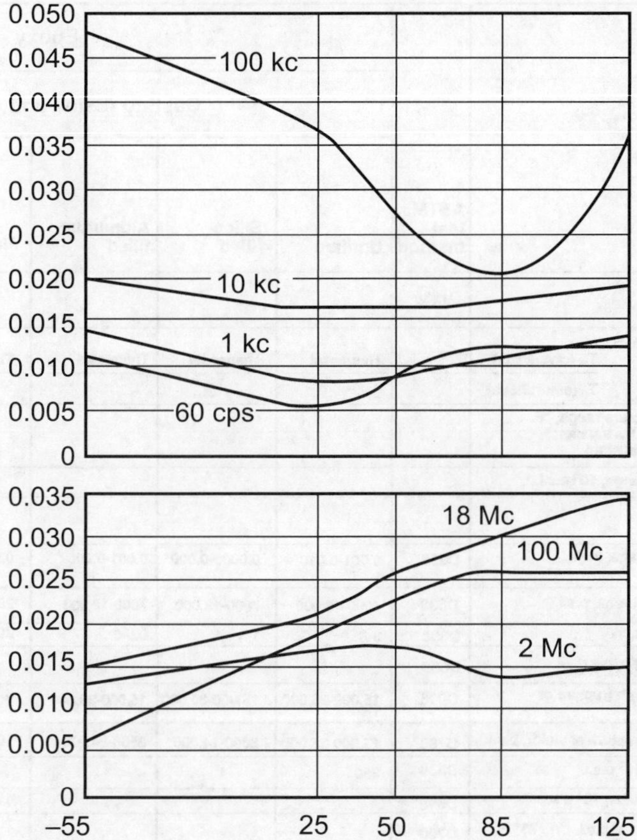

Figure 2.35 Effect of frequency and temperature on the dissipation factor of unfilled DAP

adhesion to other materials (Tables 2.96 and 2.97 and Fig. 2.36). They are used with glass cloth to make RP circuit boards, tooling surfaces for metals, and RP castings. They can be cured chemically with or without heat.

The major uses of the epoxy plastics are as surface-coating materials where they combine toughness, flexibility, adhesion, and chemical resistance to a degree unmatched by almost any other polymeric material.

Tables 2.98 and 2.99 provide condensed information on epoxy plastics. These types of condensed descriptions are available on more than 2,800 epoxy plastics, curing agents, epoxy compounds, and miscellaneous modifiers. They are of value to technical and managerial personnel involved in the manufacture and use of the final products made from the various epoxy plastics and curing agents (144).

Epoxy resins have experienced tremendous growth since their introduction in the 1950s. Rather than become a traditional commodity, epoxy resin has become more of a specialty. New molecules are regularly being developed and existing ones constantly modified to improve performance in

Materials			Epoxy				
			Casting resins and compounds				
	Properties	ASTM test method	Unfilled	Silica-filled	Aluminum-filled	Flexibilized	Cyclo-aliphatic
Processing	1a. Melt flow (gm./10 min.)	D1238					
	1. Melting temperature, °C. T_m (crystalline)		Thermoset	Thermoset	Thermoset	Thermoset	Thermoset
	T_g (amorphous)						
	2. Processing temperature range, °F. (C = compression; T = transfer; I = injection; E = extrusion)						
	3. Molding pressure range, 10^3 p.s.i.						
	4. Compression ratio						
	5. Mold (linear) shrinkage, in./in.	D955	0.001-0.010	0.0005-0.008	0.001-0.005	0.001-0.010	
Mechanical	6. Tensile strength at break, p.s.i.	D638	4000-13,000	7000-13,000	7000-12,000	2000-10,000	8000-12,000
	7. Elongation at break, %	D638	3-6	1-3	0.5-3	20-85	2-10
	8. Tensile yield strength, p.s.i.	D638					
	9. Compressive strength (rupture or yield), p.s.i.	D695	15,000-25,000	15,000-35,000	15,000-33,000	1000-14,000	15,000-20,000
	10. Flexural strength (rupture or yield), p.s.i.	D790	13,000-21,000	8000-14,000	8500-24,000	1000-13,000	10,000-13,000
	11. Tensile modulus, 10^3 p.s.i.	D638	350				
	12. Compressive modulus, 10^3 p.s.i.	D695				1-350	495
	13. Flexural modulus, 10^3 p.s.i. 73° F.	D790					
	200° F.	D790					
	250° F.	D790					
	300° F.	D790					
	14. Izod impact, ft.-lb./in. of notch (1/8-in. thick specimen)	D256A	0.2-1.0	0.3-0.45	0.4-1.6	2.3-5.0	
	15. Hardness Rockwell	D785	M80-110	M85-120	M55-85		
	Shore/Barcol	D2240/ D2583				Shore D65-89	
Thermal	16. Coef. of linear thermal expansion, 10^{-6} in./in./°C.	D696	45-65	20-40	5.5	20-100	
	17. Deflection temperature under flexural load, °F. 264 p.s.i.	D648	115-550	160-550	190-600	73-250	200-450
	66 p.s.i.	D648					
	18. Thermal conductivity, 10^{-4} cal.-cm./sec.-cm.^{2}-°C.	C177	4.5	10-20	15-25		
Physical	19. Specific gravity	D792	1.11-1.40	1.6-2.0	1.4-1.8	0.96-1.35	1.16-1.21
	20. Water absorption (1/8-in. thick specimen), % 24 hr.	D570	0.08-0.15	0.04-0.1	0.1-4.0	0.27-0.5	
	Saturation	D570					
	21. Dielectric strength (1/8-in. thick specimen), short time, v./mil	D149	300-500	300-550		235-400	

Table 2.96 General properties of epoxies unfilled and with different fillers

Materials			Epoxy								
			Bisphenol molding compounds			Sheet molding compound (SMC)		Novolak molding compounds			
	Properties	ASTM test method	Glass fiber-reinforced	Mineral-filled	Low density glass sphere-filled	Glass fiber-reinforced	Carbon fiber-reinforced	Mineral- and glass-filled, encapsulation	Mineral- and glass-filled, high temperatue	60% graphite fiber-reinforced	Glass-filled, high strength
Processing	1a. Melt flow (gm./10 min.)	D1238									
	1. Melting temperature, °C. T_m (crystalline)		Thermoset	Thermoset	Thermoset	Thermoset	Thermoset	Thermoset	Thermoset	Thermoset	Thermoset
	T_g (amorphous)							145-155	155-195		
	2. Processing temperature range, °F. (C = compression; T = transfer; I = injection; E = extrusion)		C: 300-330 T: 280-380	C: 250-330 T: 250-380	C: 250-300 I: 250-300	C: 250-330 T: 270-330	C: 250-330 T: 270-330	C: 280-360 I: 290-350 T: 250-380	T: 340-400	C: 290-350	C: 290-330 T: 290-330
	3. Molding pressure range, 10^3 p.s.i.		1-5	0.1-3	0.1-2	0.5-2.0	0.5-2.0	0.25-3.0	0.5-2.5	3-6	2.5-5.0
	4. Compression ratio		3.0-7.0	2.0-3.0	3.0-7.0	2.0	2.0		1.5-2.5		6-7
	5. Mold (linear) shrinkage, in./in.	D955	0.001-0.008	0.002-0.010	0.006-0.010	0.001	.001	0.004-0.008	0.004-0.007	to max. 0.0001	0.0002
Mechanical	6. Tensile strength at break, p.s.i.	D638	5000-20,000	4000-10,800	2500-4000	20,000-35,000	40,000-50,00	5000-12,500	6000-15,500	20,000	18,000-27,000
	7. Elongation at break, %	D638	4			0.5-2.0	0.5-2.0			0.4	
	8. Tensile yield strength, p.s.i.	D638									
	9. Compressive strength (rupture or yield), p.s.i.	D695	18,000-40,000	18,000-40,000	10,000-15,000	20,000-30,000	30,000-40,00	24,000-48,000	30,000-48,000	28,000	30,000-38,000
	10. Flexural strength (rupture or yield), p.s.i.	D790	8000-30,000	6000-18,000	5000-7000	50,000-70,000	75,000-95,00	10,000-21,800	10,000-21,800	40,000	50,000-70,000
	11. Tensile modulus, 10^3 p.s.i.	D638	3000	350		2000-4000	10,000	2100	2300-2400	6000	
	12. Compressive modulus, 10^3 p.s.i.	D695		650					660		
	13. Flexural modulus, 10^3 p.s.i. 73° F.	D790	2000-4500	1400-2000	500-750	2000-3000	5000	1400-2400	2300-2400	5500	2.8-4.2
	200° F.	D790				1500-2500					
	250° F.	D790									
	300° F.	D790				1200-1800					
	14. Izod impact, ft.-lb./in. of notch (1/8-in. thick specimen)	D256A	0.3-10.0	0.3-0.5	0.15-0.25	30-40	15-20	0.3-0.5	0.4-0.45	10	
	15. Hardness Rockwell	D785	M100-112	M100-M112				M115		R110	
	Shore/Barcol	D2240/ D2583				B: 55-65	B: 55-65	Barcol 70-75	Barcol 78		Barcol 60-74
Thermal	16. Coef. of linear thermal expansion, 10^{-6} in./in./°C.	D696	11-50	20-60		12	3	18-43	35	1.0	
	17. Deflection temperature under flexural load, °F. 264 p.s.i.	D648	225-500	225-500	200-250	550	550	300-500	500	500	
	66 p.s.i.	D648									
	18. Thermal conductivity, 10^{-4} cal.-cm./sec.-cm.^{2}-°C.	C177	4.0-10.0	4-35	4.0-6.0	1.7-1.9	1.4-1.5	10-31	17-24	16	
Physical	19. Specific gravity	D792	1.6-2.0	1.6-2.1	0.75-1.0	0.10	0.10	1.6-2.05	1.85-1.94	1.46-1.50	1.84
	20. Water absorption (1/8-in. thick specimen), % 24 hr.	D570	0.0-4-0.20	0.03-0.20	0.2-1.0	1.4	1.6	0.04-0.29	0.15-0.17	0.4	
	Saturation	D570						0.15-0.3		0.8	
	21. Dielectric strength (1/8-in. thick specimen), short time, v./mil	D149	250-400	250-420	380-420			325-450	440-450		380-400

Table 2.96 *(continued)*

traditional applications, as well as for new applications in high-technology areas. Future growth appears to lie in the evolution of new markets in the specialty performance areas.

EPOXY VINYL ESTER

These plastics extensively used in RPs can withstand many of the world's most aggressive chemical environments. Different formulated epoxies permit meeting exceptional corrosion resistance and different performance such as the ability to withstand exothermic stresses that are built up during

DRH-504 or 506/DRH-550

Feed temperature = 60 C
Mould temperature = 120–135 C
Closed mould time = 1.5–5 min.

[Properties determined longitudinally with major fibre axis, except those marked superscript t which are transverse.]

Reinforcement			%w	Tensile, Kpsi		Elongation	Flexural, Kpsi		Izod impact	CLTE in/in/°F
Type	Layers	Direction[e]	Glass	Strength	Modulus	%	Strength	Modulus	ft-lb/in	× 10^6
None	–	–	0	6.5	450	6	17	450	~0.5	(~30)
Random,[a] 1.5 oz.[d]	2	–	40	22.7	1730	1.2	35	1330	–	10
Unid,[b] 1.44 oz.	2		53	64.2	3480	2.1	114	4200	–	
Random, 1 oz.	1	–								
Unid, 1.44 oz.	2									
Unid, 1.44 oz.	1		44	27.9	2180	1.5	72	2550	–	8.7
	1	....								
Random, 1 oz.	1	–								
Unid, 1.44 oz.	1									
	1	....								
Triax,[c] 3.78 oz.	1		52	45.1	2610	2.1	97	3650	38	
Random, 1 oz.	1	–								
Triax 3.78 oz.			1							
Same	Same	Same	52	42.4	2300	2.3	95	3600	–	9.2
Same, plus 2 veils			54	15.8[t]	1530[t]	2.4[t]	27[t]	1330[t]	38/26[t]	11.7[t]

a OCF continuous fibre mat.
b Proform unidirectional mat.
c Proform triaxial mat.
d Oz/ft².
e Orientation of main fibres relative to main flow in mould; i.e., –, parallel, or . . . ., perpendicular.
t Transverse

Table 2.97 Properties of epoxy with glass-fiber fillers

curing (Tables 2.100 to 2.102). Fabricators have a longer shelf life, particularly in warmer climates. The epoxy resins have been the workhorses of the industry for many years. Different structural types have been synthesized for use at various temperature ranges.

ETHYLENE-PROPYLENE ELASTOMER

EP elastomers are the third-largest synthetic rubber consumed worldwide, after styrene-butadiene rubber and polybutadiene rubber. World consumption of EP elastomers in 1998 was about 800,000 metric tons.

EP elastomers are random, amorphous polymers with outstanding resistance to ozone, aging, and weathering, mainly because of the saturated structure in their hydrocarbon backbone. They also possess good low-temperature flexibility and heat resistance and have excellent electrical properties. Their resistance to hydrocarbons and solvents is poor. The low density of these elastomers plus their ability to accept very high levels of extender oils and fillers often give them a cost advantage over other elastomers in many applications. Principal applications are in automotive products, single-ply roofing, thermoplastic olefins, and viscosity index improvers for lubricating oils.

Automotive applications are the major use for EP elastomers (including the EP content of thermoplastic olefins). EP elastomers are used for various body and chassis products, radiator and heater hoses, and door and window weather stripping. Passenger-car tires use EP elastomers in white sidewall compounds.

Single-ply roofing membranes and oil viscosity index improvers have been major growth areas for EP elastomers, particularly in the United States. Both these markets suffered declines in the

Figure 2.36 Complete helicopter canopy consists of high-performance epoxy-glass fiber engineering reinforced plastics

early 1990s because of a combination of economic conditions and competitive products. Long-term growth in both areas, however, is expected to resume.

Fluorosilicone Elastomer

These are silicone elastomers based on polytrifluoropropylmethyl-siloxane with a small amount of vinylmethyl units for ease of peroxide vulcanization. The vulcanizates have a much-improved solvent resistance, especially to fuels and other hydrocarbons, compared with other silicone rubbers.

```
ECN 1235:
    Melting Point C: 34-42
    W.P.E. (EEW): 200-227
    Functionality: 1.7
    Epoxy cresol novolac for high temperature adhesives, coatings,
electrical and laminating product areas.

LT 8052:
    Melting Point C: 37-47
    W.P.E. (EEW): 284-352
    Functionality: 2.0
    36-40% brominated flame retardant epoxy for impregnating,
casting where non-burning properties are desired.

LT 8049:
    Melting Point C: 45-60
    W.P.E. (EEW): 322-417
    Functionality: 2.0
    47-50% brominated flame retardant epoxy resin.

ECN 1273:
    Melting Point C: 68-78
    W.P.E. (EEW): 217-233
    Functionality: 3.8
    Higher functionality, higher melting ECN 1235 for higher
heat resistance.

MT 0163:
    Melting Point C: 55-95
    W.P.E. (EEW): 179-220
    Functionality: 4.0
    Tetra-functional-phenol-based-epoxy resin for exceptional
strength at elevated temperatures, as well as, improved
thermal aging characteristics for molding, laminating and
adhesives.

ECN 1280:
    Melting Point C: 75-85
    W.P.E. (EEW): 213-233
    Functionality: 4.1
    Slightly higher melting, higher functionality ECN 1273 type.

ECN 1282:
    Melting Point C: 75-85
    W.P.E. (EEW): 213-233
    Functionality: 4.1
    Higher purity ECN 1280
```

Table 2.98 Information on specialty solid Ciba-Geigy Corp. epoxies

They also retain their extreme-low-temperature flexibility and the excellent high-temperature resistance of silicones and fluoroplastics. Uses include seals and O-rings for fuel pumps, aerospace applications, and underground use.

Products	EEW	Viscosity -cps at 25°C	Color*	Type
D.E.R. 736	175-205	30-60	1	Polyglycol
D.E.R. 732	305-335	55-100	1	Polyglycol
Epon 871	390-470	400-900	12 max	
Epon 872	650-750	1,500-3,800	10 max	
Epon 872-x-75	625-700	2,000-2,800	6	75%NV in xylene
Araldite GY508	400-455	2,000-5,000	5	Polyglycol
Epotuf 37-151	450-550	30,000-70,000	4	
Heloxy 505	550-650	300-500	8	

*Gardner

Table 2.99 Flexible epoxy resins (courtesy of Dow)

	Standard Vinyl Ester	Acid Modified Vinyl Ester
Stage I:*		
Epoxy Resin (DGEBA)**	795	795
Bisphenol A	237	237
Trimethyl ammonium chloride	0.825	0.825
Stage II:*		
Glacial methacrylic acid	195.2	171.3
Hydroquinone	0.45	—
Trimethyl ammonium chloride	6.6	6.6
Maleic acid	-	32.25
Styrene	970	970

*Stage I was processed for 1 hour at 340°F (171°C). Stage II was processed for 3 hours at 240°F (116°C) with a nitrogen-air sparge.

**DGEBA: Av. MW 350, equiv. wt. about 170-190.

Table 2.100 Maleic acid modified vinyl ester SMC resin

Melamine Formaldehyde

The melamine formaldehydes (MFs) are in the amino family of plastics created by the reaction between melamine and formaldehyde (Table 2.103; 164). They are available in a wide variety of forms, including molding powders, adhesives, laminating resins, and surface coatings. MFs may be strengthened by the addition of reinforcing materials. Table 2.104 summarizes the properties of some of the melamine and urea-formaldehyde plastics.

The major uses for amino plastics molding powders are in plastic dishware and utensils and in electrical fixtures, appliances, and instruments. Their properties may be summarized as follows: (a) heat resistance of 250°F, (b) good electrical resistance, (c) high moisture resistance, (d) good dimensional stability, (e) excellent mechanical strength and wear resistance, and (f) easy to color.

Almost one-half of the production of these plastics goes into adhesive applications. Melamine resins are excellent plywood bonding agents, resulting in "boil-resistant" boards that meet extreme specifications for exterior use. Both melamine and urea adhesive resins have found many uses in the furniture, clipboard, aircraft, boat-building, and paperboard industries. Melamines are also used as the bonding agent in the manufacture of a wide variety of laminates.

Styrene content, %	50-45
NVM, %	50-55
Viscosity, cps, 77°F (25°C)	80-600
Color, Gardner	2-3
Wt. per gallon, lbs.	8.6 - 8.7
Flash point,, tag open cup, °F (°C)	95 (35)
Reactivity	
Gel time at 77°F (25°C) min.*	20-25
SPI gel time, min.	10-19

*Promoted with 0.5 PHR 12% cobalt octoate and catalyzed with 1.0 PHR MEK peroxide catalyst.

Table 2.101 Styrenated vinyl ester resin liquid properties

Tensile strength, psi, 77°F (25°C)	10,000-11,000
Tensile modulus, psi, 77°F (26°C)	$0.36\text{-}0.44 \times 10^6$
Tensile elongation, %	5.3-5.8
Flexural strength, psi, 77°F 25°C)	18,000-19,000
Flexural modulus, psi, 77°F((25°C)	$0.45\text{-}0.46 \times 10^6$
Heat deflection temperature, °F (°C)	190-210 (88-99)
Notched Izod impact - ft. lbs./in. of notch	0.5-0.55
Barcol hardness	40-50

Table 2.102 Physical properties of cast vinyl ester resin

Materials	Properties	ASTM test method	Urea	Melamine formaldehyde		Melamine phenolic	Furan
			Alpha cellulose-filled	Cellulose-filled	Glass fiber-reinforced	Woodflour-and cellulose-filled	Asbestos-filled
Processing	1. Melting temperature, °C. T_m (crystalline)		Thermoset	Thermoset	Thermoset	Thermoset	Thermoset
	T_g (amorphous)						
	2. Processing temperature range, °F. (C = compression; T = transfer; I = injection; E = extrusion)		C: 275-350 I: 290-320 T: 270-300	C: 280-370 I: 200-340 T. 300	C: 280-350	C: 300-350 I: 350-400	C: 275-300
	3. Molding pressure range, 10^3 p.s.i.		2-20	8-20	2-8	5-20	0.1-0.5
	4. Compression ratio		2.2-3.0	2.1-3.1	5-10	2.1-4.4	
	5. Mold (linear) shrinkage, in./in.	D955	0.006-0.014	0.005-0.015	0.001-0.006	0.009-0.010	
Mechanical	6. Tensile strength at break, p.s.i.	D638	5500-13,000	5000-13,000	5000-10,500	6000-8000	3000-4500
	7. Elongation at break, %	D638	<1	0.6-1	0.6	0.4-0.8	
	8. Tensile yield strength, p.s.i.	D638					
	9. Compressive strength (rupture or yield), p.s.i.	D695	25,000-45,000	33,000-45,000	20,000-35,000	26,000-30,000	10,000-13,000
	10. Flexural strength (rupture or yield), p.s.i.	D790	10,000-18,000	9000-16,000	14,000-23,000	8000-10,000	600-9000
	11. Tensile modulus, 10^3 p.s.i.	D638	1000-1500	1100-1400	1600-2400	800-1700	1580
	12. Compressive modulus, 10^3 p.s.i.	D695					
	13. Flexural modulus, 10^3 p.s.i. 73° F.	D790	1300-1600	1100		1000-1200	
	200° F	D790					
	250° F.	D790					
	300° F	D790					
	14. Izod impact, ft.-lb./in. of notch (1/8-in. thick specimen)	D256A	0.25-0.40	0.2-0.4	0.6-18	0.2-0.4	
	15. Hardness Rockwell	D785	M110-120	M115-125	M115	E95-100	R110
	Shore/Barcol	D2240/ D2583					
Thermal	16. Coef. of linear thermal expansion, 10^{-6} in./in./°C.	D696	22-36	40-45	15-28	10-40	
	17. Deflection temperature under flexural load, °F 264 p.s.i.	D648	260-290	350-390	375-400	285-310	
	66 p.s.i.	D648					
	18. Thermal conductivity, 10^{-4} cal.-cm / sec-cm.^{2}-°C.	C177	7-10	6.5-10	10-11.5	4-7	
Physical	19. Specific gravity	D792	1.47-1.52	1.47-1.52	1.5-2.0	1.5-1.7	1.75
	20. Water absorption (1/8-in. thick specimen), % 24 hr	D570	0.4-0.8	0.1-0.8	0.09-1.3	0.3-0.65	0.01-0.2
	Saturation	D570					
	21. Dielectric strength (1/8-in. thick specimen), short time, v./mil	D149	300-400	270-400 175-215 @ 100° C	130-370	220-325	

Table 2.103 Properties of amino (urea, melamine, furan) molding compounds

Property	Urea, general purpose Wood-filled	Urea, general purpose α-Cellulose-filled	Melamine, general purpose Wood-filled	Melamine, general purpose α-Cellulose-filled	Melamine, electrical grades Mineral-filled	Melamine, electrical grades Glass-filled
Density	1·4–1·5	1·5–1·55	1·5–1·55	1·5–1·55	1·8–2·1	1·9–2·0
State	Granular powder	Gran. or fine powder	Granular powder	Gran. or fine powder	Granular powder	Granular powder
Appearance	Opaque colors	Translucent or opaque colors	Opaque colors	Translucent or opaque colors	Opaque grey or brown	Opaque grey
Processing temperature range, °F	~275°–330°F	~275°–350°F	~280°–340°F	~280°–340°F	~270°–340°F	~290°–325°F
Processing temperature range, °C	~135°–165°C	~135°–175°C	~140°–170°C	~140°–170°C	~130°–170°C	~145°–160°C
Mold shrinkage (in./in.)	0·006–0·010	0·005–0·010	0·007–0·008	0·004–0·009	0·0015–0·007	0·002–0·005
Mechanical properties						
Impact strength [a] (ft lb/in.)	0·22–0·28	0·24–0·30	0·26–0·38	0·24–0·32	0·20–0·45	0·35–0·50 [c]
Yield stress [b]	—	—	—	—	—	—
Distortion [b]	—	—	—	—	—	—
Maximum working temperature, °F	170	170	250	210	300	300
°C	75	75	120	100	150	150
Tensile strength lb/in.^{2}	5500–8000	7000–11,000	5500–8500	7000–12,000	4000–6500	6000–10,000
Tensile modulus lb/in.^{2}	1·0–1·4×10^6	1·0–1·4×10^6	~1·1×10^4	1·2–1·6×10^6	1·5–3·5×10^6	2·0–4·0×10^6
Electrical properties						
Power factor at 60 cps	0·025–0·045	0·025–0·035	0·020–0·040	0·040–0·080	0·100–0·300	0·040–0·100
10^6 cps	0·020–0·040	0·020–0·030	0·010–0·030	0·020–0·050	0·040–0·070	0·020–0·030
10^6 cps	0·025–0·040	0·020–0·030	0·025–0·040	0·025–0·045	0·010–0·050	0·020–0·030
Dielectric constant at 60 cps	7·0–9·0	7·5–8·0	6·0–7·0	6·0–9·0	7·0–10·0	7·0–14·0
10^3 cps	6·5–8·5	6·5–7·5	6·0–7·0	8·0–10·0	8·0–9·0	7·0–9·0
10^6 cps	6·0–8·5	6·5–7·0	5·5–6·0	7·0–9·0	5·5–6·5	5·5–7·5
Dielectric strength [d]						
Short time at 25°C	300–450	300–400	350–400	300–400	400–450	150–250
at 100°C	200–300	200–300	300–350	150–250	250–300	50–150
Step by step at 25°C	200–300	200–250	200–300	200–300	250–300	150–200
at 100°C	100–200	100–150	150–250	100–150	150–250	50–100
Volume resistivity (ohm cm)	10^{11}–10^{14}	10^{11}–10^{14}	10^{11}–10^{13}	10^{11}–10^{13}	10^{10}–10^{13}	10^{11}–10^{14}
Track resistant	Yes	Yes	Yes	Yes	Yes	Yes
Arc resistant	Yes	Yes	Yes	Yes	Yes	Yes
Thermal properties						
Coefficient of linear thermal expansion [e] (°C)	3–6×10^{-5}	2–5×10^{-5}	3–5×10^{-5}	2–5×10^{-5}	2–5×10^{-5}	2–5×10^{-5}
Thermal conductivity (cal/cm^2/°C/cm)	6–9×10^{-4}	7–10×10^{-4}	7–10×10^{-4}	7–10×10^{-4}	10–15×10^{-4}	11–10×10^{-4}
Flammability	S.E.[g]	S.E.[g]	S.E.[g]	S.E.[g]	S.E.[g]	S.E.[g]
Chemical properties						
Water absorption, 24 h at 23°C [b] (%)	00·5–1·2	0·4–1·3	0·3–0·4	0·1–0·4	0·1–0·3	0·03–0·1
Gas permeabilities	—	—	—	—	—	—
Resistance to:						
Mineral acids (dil)	Fair	Fair	Good	Good	Fair	Very good
Mineral acids (conc)	Poor	Poor	Poor	Poor	Poor	Poor
Alkalis	Fair	Fair	Good	Good	Fair	Good
Solvents:						
Alcohols	Good	Good	Good	Good	Fair	Good
Ketones	Good	Good	Good	Good	Fair	—
Aromatic hydrocarbons	Good	Good	Good	Good	Fair	—
Chlorinated hydrocarbons	Good	Good	Good	Good	Fair	—
Detergents	Very good	Very good	Very good	Very good	Good	—
Greases and oils	Good	Good	Good	Good	Fair–good	—

[a] These materials are brittle and notch-sensitive at room temperature and below.
[b] These materials have a linear stress–strain relationship until fracture occurs at about 1% elongation. Maximum temperatures for continuous use are indicated.
[c] Special "high impact" material containing long glass fibers has an "impact strength" of 4.0–6.0 ft-lb/in.
[d] Preheating the material before molding usually improves the dielectric strength.
[e] These materials shrink on continuous heating.
[f] Equilibrium water absorption is not attained for a considerable time—more than a year for melamine materials.
[g] Self-extinguishing.

Table 2.104 Properties of melamine and urea-formaldehyde plastics

NEOPRENE

Neoprene, or polychloroprene rubber (CR), was one of the very first synthetic rubbers produced that had desirable properties in many applications. It was a material of choice for exterior applications such as profiles used in vehicles and buildings, and for cables. Many more marketable products have benefitted from this plastic.

Except for styrene-butadiene (SBR) elastomers and polyisoprenes (IRs), neoprene (CR) elastomers are perhaps the most rubberlike of all materials, particularly with regard to their dynamic responses. CRs are a family of elastomers with a property profile that approaches that of NRs but has better resistance to oils, ozone, oxidation, and flame. CRs age better and do not soften upon exposure to heat, although their high-temperature tensile strength may be lower than that of NRs. Neoprenes come in grades suitable for service at 250°C (480°F) and have maximal resistance to oils and greases.

These materials, like NRs, can be used to make soft, relatively high-strength compounds. One important difference is that, in addition to neoprenes being more costly by the pound than NRs, their density is about 25% greater than NRs. CRs also do not have the low-temperature flexibility of NRs, which detracts from their use in low-temperature shock or impact applications.

PHENOL-FORMALDEHYDE (PHENOLIC)

These types of thermoset (TS) plastics are obtained by the condensation of phenol or substituted phenols with aldehydes such as formaldehyde and furfural. Phenol-formaldehydes (PFs) are typical and constitute the chief class of phenolics. PFs have been the low-cost workhorses of the electrical industry (particularly in the past). They have low creep, excellent dimensional stability, good chemical resistance, and good weatherability (Tables 2.105 and 2.106). Molded black or brown opaque handles for cookware are familiar applications. They are also used as caramel-colored impregnating plastics for wood or cloth laminates, and (with reinforcement) for brake linings and many under-the-hood automotive electricals.

Since 1909, phenol-formaldehyde has continued to be a generally low-cost, general-purpose thermoset compound meeting a multitude of applications. A wide range of fillers is used, with each intended to fulfill the needs of particular fabricated product service environments. Typical among them is wood flour, for general use in electrical wall plates, industrial switch gears, circuit breakers, handles and knobs for small appliances, and so on. For slightly better impact strength, cotton flock is used. Mica, glass, and other minerals provide better electrical properties, heat resistance, and dimensional stability for use in automotive power brakes, industrial electrical terminal strips, and so on.

Phenolics are formulated with a one- or two-stage curing system. In general, one-stage plastics are slightly more critical to process. They have a time-to-temperature-to-viscosity behavior (Fig. 2.37). Although phenolics have properties that are somewhat inferior to those of the more expensive TSs, they are usually more easily molded. They can be processed by compression, transfer, and injection

Property	Units	Test method	Materials				
Reinforcing material			Wood flour	Cotton	Asbestos	Mica	Glass fiber
Density	g/ccc	D792	1.39	1.40	1.70	1.80	1.45
Mechanical Properties							
Impact			Brittle	Tough	Brittle	Brittle	Strong
Tensile strength	lb/in.^{2}	D651	8500	6000	6500	6700	8500
Elongation	%	D638	0.60	0.50	0.34	0.30	0.20
Modulus of elasticity	lb/in.^{2}	D638	10^6	10^6	—	—	—
Compressive strength	lb/in.^{2}	D695	29,000	23,000	27,500	28,000	28,500
Flexural strength	lb/in.^{2}	D790	10,000	12,000	11,000	10,000	35,000
Rockwell hardness		D785	M116	M110	M105	M105	M100
Electrical Properties							
Volume resistivity, 50% RH, 23°C	ohm/cm	D257	10^{11}	10^{12}	10^{12}	10^{13}	10^{12}
Track resistance	(relative IEC. 112) method		No	No	No	No	No
Arc resistance		D495	No	No	Yes	No	No
Thermal Properties							
Coefficient of linear expansion	/°C×10^{-5}	D696	3.6	2.5	2.4	2.2	1.2
Conductivity	10^{-4} cal/s/cm^2/°C/cm	C177	5.5	6	15	12	10
Flammability		D635	Slight	Slight	None	None	None
Chemical Properties							
Water absorption	% wt/24 hr immersion	D570	0.60	0.70	0.03	0.03	0.15

Table 2.105 Phenolic molding materials

Material composition	Phenolic (35%), asbestos mat (65%)	Phenolic (35%), glass fabric (65%)	Phenolic (35%), silica fabric (65%)	Phenolic (50%), nylon fabric (50%)
Density, lb/ft^3	108	112	102	72
Tensile strength, psi				
75°F	59,000	45,800	9225	6100
350°F	48,000	43,200	7700	—
Tensile elastic modulus, million psi				
75°F	5.06	4.02	2.71	1.00
350°F	4.50	3.70	2.06	
Tensile elongation at failure, %				
75°F	—	—	0.52	2.00
350°F	—	—	0.50	—
Shear strength, psi				
75°F	2210	3300	4410	—
350°F	1360	1300	3370	—
Compressive strength, psi	26,900	65,800	31,000	15,800
Compressive elastic modulus, million psi	5.33	3.64	2.4	0.4
Flexural strength, psi	52,700	67,500	23,000	13,100
Flexural elastic modulus, million psi	5.43	3.78	2.9	0.44
Hardness, Barcol	62	75	70	30
Specific heat, Btu/lb-°F	0.19	0.21	0.25	0.47
Thermal conductivity, Btu/hr/ft^2-°F/ft	0.10	0.16	0.36	0.10
Thermal expansion coefficient, 10^{-6}in./in./°F				
Parallel	2.0	2.1	7.0	36.
Perpendicular	17.3	12.0	10.0	89.

All properties are parallel to the fabric/mat warp unless otherwise noted.

Table 2.106 Phenolic fiber/fabric-reinforced plastics

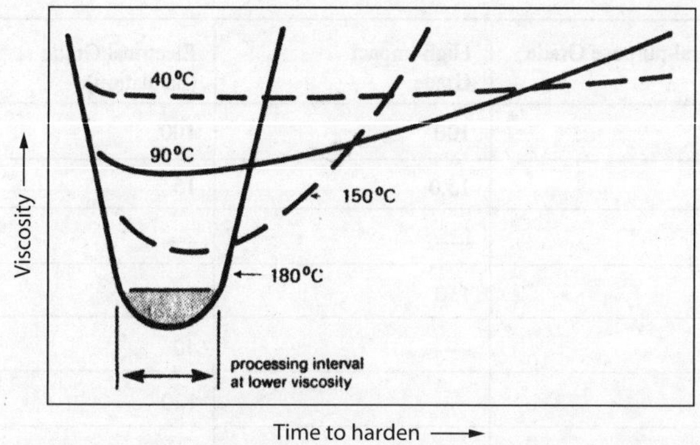

Figure 2.37 Examples of phenolics' relationship of time-to-temperature-to-viscosity behavior

molding and to a limited degree by extrusion. The colors of these compounds are limited to black or brown. As is typical of many TSs, they are postcured to obtain maximum performance.

Phenolic molding compounds are generally characterized as being low in cost and having superior heat resistance, a high heat-deflection temperature, good electrical properties and flame resistance, excellent moldability and dimensional stability, and good water and chemical resistance. Specialty (more expensive) compounds can provide higher-performance heat resistance, impact strength, electrical properties, and creep resistance. Data on phenolics are reviewed in Tables 2.107 and 2.108.

POLYBENZIMIDAZOLE

Polybenzimidazoles (PBIs) show (via thermal gravimetric analysis [TGA]) that their thermal breakdown temperatures are as high as 600°C in nitrogen, but their thermal resistance is reduced to about 300°C in air. They are generally infusible after curing and must be converted as prepolymers into useful forms prior to the final cures. Phenols further complicate fabrication of the polybenzimidazoles, with water side products emitted during cures. Most are insoluble in all but exotic solvents and show excellent resistance to oils, acids, and alkalies. They are reported to show useful stress and moduli properties in a –253° to 650°C temperature range (for at least short periods of time). Some polybenzimidazole resins have also been made that are more soluble and fusible.

PBIs have no known melting point. They have a glass-transition temperature (T_g) of 427°C (800°F). They have an ultrahigh heat-distortion temperature of 435°C (815°F), retard flame, and will not burn in air. The material can withstand steady temperatures up to 427°C (800°F) and short bursts up to 760°C (1,400°F). The material is reported to resist steam at 343°C (650°F) and 15 MPa (2,200 psi) pressure. When exposed to saturated steam, PBIs absorb only 0.4% moisture. They resist a wide range of chemicals, including harsh ones.

These plastics have high mechanical and physical strength properties, including high compression strength. PBIs can bear loads for short periods at temperatures up to 650°C (1,200°F). When

Composition (phr)*	General-purpose Grade	High-impact Grade	Electrical Grade (Insulation)	(Pipe) Coating Grade
Novolac/resin	100	100	100	100
HEXA	15.0	15.0	15	----
Wood flour	120	-----	----	----
Fabric shreds	-----	150	----	----
Fiber glass	-----	-----	40	----
Mica	-----	-----	120	----
Lime/magnesium oxide	2.0	2.0	2.0	5.0
Clay	-----	-----	-----	5.0
Magnesium stearate (lubricant)	2.0	3.0	2.0	----
Plasticizer	-----	-----	-----	10.0
Colorants (dyes or pigments)	4.0	3.0	3.0	73.0 (iron oxide)
Alcohol	-----	-----	-----	55.00

*phr = parts per hundred resin

Table 2.107 Typical formulations (phr) of phenolic molding compounds

Composition (phr)*	Plywood	Particle Board	Fiber Board
Phenolic resin	100	100	100
Water	0 - 20	50 - 110	0 - 120
Filler	5 - 10		
Extender	0 - 10		
Hardener (CH_2O)	0 - 5.0	0.2 - 0.5	0 - 0.5
Solubilizer (NaOH)	3 - 10	8 - 12	5 - 10
Fungicides/insecticides		2 - 4	2 - 5
Process temp (°C)	100 - 150	160 - 220	
Process pressure (psi)	730 -5880	290 - 590	

* phr = parts per hundred resin

Table 2.108 Typical formulations for adhesives used in composite wood products

reinforced with silicon carbide fibers, they can thwart an attack even from laser weapons. They have a low coefficient of both friction and thermal expansion (0.000013 in/in/F).

These wholly aromatic heterocyclic polymers are fabricated by sintering under high pressures. Low molecular weight PBIs flow better than their high molecular weight (HMW) counterparts. However, HMW PBIs outgas less during processing, making them more suitable to mold large products.

PBIs have been targeted to replace metals, ceramics, carbon, and other materials where an industry needs materials that are highly resistant to heat and corrosion, as in chemicals and oil processing, aerospace, and transportation industries. Their main commercial applications include use as valve seats, seals, electrical connections, thrust washers, bearings, and other mechanical components. PBIs may become the material of choice for products that need to resist temperatures up to 400°C (750°F) or higher, such as clothing for astronauts and fire fighters.

POLYBENZOBISOXAZOLE

PBZs are a family of high-performance, liquid crystalline, rigid rod plastic compositions (including polybenzobisoxazole) that are characterized by a unique combination of high molecular weight and concentration resulting in high tensile strength and stiffness, as well as resistance to moisture, heat, and UV exposure. Uses include structural composites for lightweight aircraft, military transportation, industrial products, and sporting goods.

POLYBUTADIENE

These general-purpose, crude-oil-based elastomers are more resilient than NRs. Polybutadienes (BRs) are the elastomer materials that made the modern solid golf balls possible. They are superior to NRs in their low-temperature flexibility and in having less dynamic heat buildup. However, they lack NRs' toughness, durability, and cut-growth resistance. BRs can be used as blends in NRs, SBRs, and other materials to improve their low-temperature flexing. However, this is achieved at a much higher price, and it sacrifices key properties such as tensile strength, tear resistance, and general durability.

Advantages of polybutadiene or modified copolymer versions of the butadiene resins are good heat resistance, good electricals, and good physicals. Properties and applications are shown in Tables 2.109 and 2.110. Disadvantages include slow cures, high cure temperatures, poor filler and glass wet-out, the need for specific high-temperature monomers, and high-pressure molding conditions.

POLYCHLOROPRENE

See Neoprene.

Polydicyclopentadiene PDCPD is a highly cross-linked thermoset polyolefin formed by the reaction of two liquid streams. It is chemically based on the low-cost petroleum by-product dicyclopentadiene that constitutes about 90% of the plastic; the balance is made up of tougheners, antioxidants, processing aids, and a catalyst/activator. The catalyst is in one liquid and the activator

1. Excellent thermal resistance
2. Outstanding electrical properties
3. Good chemical resistance
4. High surface gloss
5. Craze and crack resistance
6. Trouble-free mold release
7. Fast, efficient cures
8. High monomer and filler acceptance
9. Low weight loss
10. High heat distortion values
11. Mechanical strength at high temperatures
12. Retention of strength at intermediate and high temperatures
13. Room temperature curability
14. Excellent response to being prethickened
15. Available with excellent fire resistance
16. Excellent track resistance
17. UV Stability and weather resistance

Table 2.109 Properties of polybutadiene

1. Laminating
2. Casting
3. Premixing
4. Prepreging
5. Prethickening
6. Filament winding
7. Impregnating
8. Potting and encapsulating
9. Varnishes
10. Track resistant applications
11. Hand layup uses
12. Coatings
13. Pultrusion applications

Table 2.110 Examples of polybutadiene applications

in the other. The plastic is processed by reaction injection molding at low temperatures and pressures. Its low viscosity and potential to extend gel times to a matter of minutes allow the production of very large parts.

It produces parts with high gloss (automotive Class A) and relatively high mechanical properties. This 1.03 specific gravity plastic (unreinforced) is very rigid, having a flexural modulus in the order of 1,895 MPa (275,000 psi); reinforcing with glass fibers doubles the value. Notched Izod impact strength is in the 5 to 6 ft-lb/in (267 to 320 HJ/m) and remains above 3 ft-lb/in (160 HJ/m) to at least –40°F. The high-gloss surface is quite hard (Rockwell R113 to 114) and, unlike thermoplastic polyolefins, exhibits excellent paint adhesion. Heat-distortion temperature at 264 psi is 104° to 116°C (220° to 240°F). Like other polyolefins, it is resistant to a wide range of chemicals, including alcohols, acids, and bases. It is attacked by some organic solvents, particularly chlorinated and aromatic hydrocarbons. A corrosion-resistant formulation designed for long-term use with strong acids or bases is available.

POLYESTER, THERMOSET

The term *unsaturated polyester resins* is restricted to products consisting of straight-chain polyesters having reactive double bonds at intervals along the chain. They are supplied as solutions in a vinyl monomer, generally styrene. Before application, an initiator and a promoter are added, usually together with extra vinyl monomer. A copolymerization is thereby initiated. The polyester chains become linked together in a cross-link network by means of bridges formed by polymerized vinyl monomers. The major uses for these resins are in reinforced laminates, moldings, and coatings. The polyesters can be cured under low pressure and at low temperatures.

In the family of polyesters there are two major groups: (a) the thermoset plastics, which are usually typified by a cross-linked structure and (b) the thermoplastics, which are highly crystalline with comparatively high melting points (previously reviewed). Thermoplastic polyesters are often called saturated polyesters to distinguish them from unsaturated polyesters that are the thermoset plastics (Tables 2.111 to 2.114).

There have been several instances in which polyesters have been separated into three groups and called polyesters, diallyl phthalates, and alkyds. This review only concerns thermoset polyesters; the others were discussed earlier in the chapter. In the family of polyester plastics, there are thermosets and thermoplastics. Thermoset polyesters have an excellent balance of properties and a room-temperature cure. These major plastics are used to make glass-fiber-reinforced parts for automobiles, boats, and aircraft; they are also used with other types of reinforcing agents. Commodity types have moderate weatherability and high molding shrinkage with wavy surfaces and warpage. Low-profile (polystyrene, etc.) additives reduce shrinkage and surface waviness to almost nil. This has led to major growth in such applications as automobile exterior body panels, instrument housings, and microwave dinnerware in the form of bulk molding compounds (BMCs) and sheet molding compounds (SMCs).

Unsaturated-polyester (UP) condensation-type plastics are principally used with glass fibers and other fibers (Table 2.115) as the reinforcing matrix material (chapter 15). Glass-reinforced plastics were developed in the early 1940s in the United States for military applications (50). Styrene, which by far has been the most important monomer, has contributed as a reactive solvent to UP's versatile combination of properties. Consequently, a very large number of different applications has been realized since the 1940s.

Processing with the styrene monomer requires precautions to ensure the proper removal and handling of this toxic material. Legal limits in the workplace have been set up by regulations. Other monomers used include diallyl phthlate (DAP), para-methylstyrene (PMS), and vinyl toluene (VT).

Suppliers and fabricators continue to target the reduction of styrene monomer quickly, effectively, and economically. Information regarding the most desirable styrene concentration is available from the suppliers. The different methods available include the following:

a. *Low-Styrene-Content Plastics* Although many TS polyester plastics are commercially available, this method does not lend itself to every molding process. The physical properties of the final product can be affected. In some instances, the reduction of styrene is not significant enough to make a difference, and in other cases, styrene emissions may even increase. Fabricators have found that by changing their processing methods they have been able to use low-styrene-content plastics and reduce styrene emissions to some extent.

b. *Controlled Spray-on Technique* This method is very effective and works by controlling the amount of surface area of the wet plastic that is exposed to the air, whether

Materials		Cast		Glass Fiber-reinforced						Alkyd Molding Compounds	
Properties	ASTM Test Method	Rigid	Flexible	Performed, Chopped Roving	Premix, Chopped Glass	Woven Cloth	SMC	SMC Low-shrink	BMC, TMC	Granular and Putty, Mineral-filled	Glass Fiber-reinforced
Processing											
1a. Melt flow (gm/10 min)	D1238										
1. Melting temperature, °C T_m (crystalline) T_g (amorphous)		Thermoset	Thermoset	Thermoset	Thermoset	Thermoset	Thermoset	Thermoset	Thermoset	Thermoset	Thermoset
2. Processing temperature range, °F (C = compression; T = transfer; I = injection; E = extrusion)				C: 170–320	C: 280–350	C: 73–250	C: 270–380 I: 280–310 T: 280–310	C: 270–330 I: 270–380	C: 280–380 I: 280–370 T: 280–320	C: 270–350 I: 280–390 T: 320–360	C: 290–350 I: 280–380
3. Molding pressure range, 10^3 psi				0.25–2	0.5–2	0.3	0.3–2	0.5–2	0.4–1.0	2–20	2–25
4. Compression ratio				1.0	1.0		1.0	1.0	1.0	1.8–2.5	1–11
5. Mold (linear) shrinkage, in./in.	D955			0.0002–0.002	0.001–0.012	0.0002–0.002	0.0005–0.004	0.0002–0.001	0.0003–0.004	0.003–0.010	0.001–0.010
Mechanical											
6. Tensile strength at break, psi	D638[b]	600–13,000	500–3,000	15,000–30,000	3,000–10,000	30,000–50,000	7,000–25,000	4,500–20,000	3,000–10,000	3,000–9,000	4,000–9,500
7. Elongation at break, %	D638[b]	<2.6	40–310	1–5	<1	1–2	3	3–5			
8. Tensile yield strength, psi	D638[b]										
9. Compressive strength (rupture or yield), psi	D695	13,000–30,000		15,000–30,000	20,000–30,000	25,000–50,000	15,000–30,000	15,000–30,000	14,000–30,000	12,000–38,000	15,000–36,000
10. Flexural strength (rupture or yield), psi	D790	8,500–23,000		10,000–40,000	7,000–20,000	40,000–80,000	10,000–36,000	9,000–35,000	11,000–24,000	6,000–17,000	8,500–26,000
11. Tensile modulus, 10^3 psi	D638[b]	300–640		800–2,000	1,000–2,500	1,500–4,500	1,400–2,500	1,000–2,500	1,500–2,500	500–3,000	2,000–2,800
12. Compressive modulus, 10^3 psi	D695									2,000–3,000	
13. Flexural modulus, 10^3 psi 73°F 200°F 250°F 300°F	D790 D790 D790 D790	490–610		1,000–3,000	1,000–2,000	1,000–3,000 660 430 270	1,000–2,200	1,000–2,500		2,000	2,000
14. Izod impact, ft-lb/in. of notch (1.8-in.-thick specimen)	D256A	0.2–0.4	>7	2–20	1.5–16	5–30	7–22	2.5–15	2–13	0.3–0.5	0.5–16
15. Hardness Rockwell Shore/Barcol	D785 D2240/ D2583	Barcol 35–75	Shore D84-94	Barcol 50–80	Barcol 50–80	Barcol 60–80	Barcol 50–70	Barcol 40–70	Barcol 50–65	E98	E95
Thermal											
16. Coefficient of linear thermal expansion, 10^{-6} in./in./°C	D696	55–100		20–50	20–23	15–30	13.5–20	6–30		20–50	15–33
17. Deflection temperature under flexural load, °F 264 psi 66 psi	D648 D648	140–400		>400	>400	>400	375–500	375–500	320–400	350–500	400–500
18. Thermal conductivity, 10^{-4} cal-cm/sec-cm^2-°C	C177								18–22	12–25	15–25
Physical											
19. Specific gravity	D792	1.04–1.46	1.01–1.20	1.35–2.30	1.65–2.30	1.50–2.10	1.65–2.6	1.6–2.4	1.72–2.10	1.6–2.3	2.0–2.3
20. Water absorption (1.8-in.-thick specimen), % 24-hour Saturation	D570 D570	0.15–0.6	0.5–2.5	0.01–1.0	0.06–0.28	0.05–0.5	0.1–0.25	0.01–0.25	0.1–0.45	0.05–0.5	0.03–0.5
21. Dielectric strength (1/8-in.-thick specimen), short time, V/mil	D149	380–500	250–400	350–500	345–420	350–500	380–500	380–450	380–390	350–450	259–530

b – Tensile test method varies with material: D638 is standard for thermoplastics; D651 for rigid thermoset plastics; D412 for elastomeric plastics; D882 for thin plastics sheeting.
c – Dry, as molded (approximately 0.2% moisture content).
d – As conditioned to equilibrium with 50% relative humidity.
e – Test method is ASTM D4092.
f – *Pseudo* indicates that the thermoset and thermoplastic components were mixed in the form of pellets or powder prior to fabrication.
– Reprinted with permission from *Modern Plastics Encyclopedia*, 1994-5, Copyright 1994, McGraw-Hill. All rights reserved.

Table 2.111 Physical properties of unsaturated polyesters

spraying gel-coat or plain plastic. This method can be used with the spray-on fabrication techniques.

c. *Paraffin Wax* A popular method that has been used for many years is using paraffin wax as a surface film-forming additive to aid in the curing process of the RPs. It also suppresses styrene emissions through the film it provides, but in doing so, can create the problem of poor secondary bonding between the layers of laminates, which can cause delamination, resulting in a structural failure. The newer suppressant formulations, such as Styrid from Scott Bader Co., create a film similar to that provided by paraffin wax, but with one major difference. The film, after migrating

	Contributes
Glycols:	
Propylene Glycol (PG)	Low cost, styrene compatibility
Ethylene Glycol (EG)	Low cost, rigidity
Dipropylene Glycol (DPG)	Flexibility, toughness
Diethylene Glycol (DEG)	Flexibility, toughness
Neopentyl Glycol (NPG)	UV, water and chemical resistance
Trimethylpentanediol (TMPD)	Water and chemical resistance
Cyclohexane Dimethanol (CHDM)	Electrical properties
Propoxylated Bisphenol A (PBPA)	Water and chemical resistance
Hydrogenated Bisphenol A (HBPA)	Water and chemical resistance
Dibromoneopentyl Glycol (DBNPG)	Flame retardance
Acids:	
Phthalic Anhydride (PA)	Low cost, styrene compatibility
Maleic Anhydride (MA)	Lowest cost unsaturation
Adipic Acid (AA)	Flexibility, toughness
Isophthalic Acid (IPA)	Toughness, water and chemical resistance
Terephthalic Acid (TPA)	Higher heat deflection point
Fumaric Acid (FA)	Maximum reactive unsaturation
Glutaric Acid	Flexibility, toughness
Dimer Acids	Flexibility, toughness
Azelaic Acid	Flexibility, toughness
Chlorendic Acid	Flame retardance, chemical resistance
Tetrabromophthalic Anhydride	Flame retardance, chemical resistance
Tetrachlorophthalic Anhydride	Flame retardance, chemical resistance
Endomethylenetetrahydro-phthalic Anhydride	Air drying properties

Table 2.112 Common raw materials for TS polyesters

to the surface of the RP, has very minute spaces containing a special component that permits the next superimposed RP layer to penetrate the film and adhere properly to the previous layer.

d. *Other Monomers* Monomers other than styrene are used (Table 2.116). This includes the use of methyl methacrylate, vinyl toluene, butyl styrene, or possibly oligomers, which consist of two or three molecules of styrene that have been combined. The styrene combination approach is effective but can be very expensive; in addition, some can even be more toxic than styrene, being considered hazardous air pollutants (HAPs) or volatile organic compounds (VOCs). Furthermore, if used in

Polyester types	Performances	Applications
General purpose	Rigid moldings, craze resistant	Trays, boats, tanks, boxes, luggage, seating
Flexible resins and semirigid resins	Tough, good impact resistance, high flexural strength, low flexural modulus	vibration damping: machine covers and guards, safety helmets, electronic parts encapsulation, gel coats, patching compounds, auto bodies, boats
Light stable and weather resistant	Resistant to weather and ultra-violet degradation	Structural panels, skylighting, glazing
Chemical resistant	Highest chemical resistance of polyester group, excellent acid resistance, fair in alkalies	Corrosion-resistant, applications such as pipes, tanks, ducts, fume stacks
Flame resistant	Self-extinguishing, rigid	Building panels (interior), electrical components, fuel tanks
High-heat distortion	Service up to 500 °F, rigid	Aircraft parts
Electrical	Fast cure, good hot strength, good electrical properties	Where good electrical properties are required
Hot strength	Fast rate of cure, "hot" moldings easily removed from die	Containers, trays, housings
Low exotherm	Void-free thick laminates, low heat generated during cure	Encapsulating electronic components, electrical premix parts – switchgear
Extended pot life	Void-free and uniform, long flow time in mold before gel	Large complex moldings
Air dry	Cures tack free at room temperature	Pools, boats, tanks
Thixotropic	Resists flow or drainage when applied to vertical surfaces	Boats, pools, tank linings

Table 2.113 Performance of different polyester types

Property	Glass fiber, wt %						
	0	10	20	30	40	50	60
specific gravity	1.14	1.21	1.28	1.37	1.46	1.57	1.70
specific volume, cm^3/g	0.86	0.82	0.77	0.72	0.68	0.63	0.58
tensile strength, MPa[a]	83	90	131	172	214	221	228
tensile elongation, %	60	3.5	3.5	3.0	2.5	2.5	1.5
flexural strength, MPa[a]	103	138	200	234	290	317	345
flexural modulus, GPa[b]	2.8	4.1	6.2	9.0	11.0	15.1	19.3
compressive strength, MPa[a]	34	90	159	186	193	200	207
heat-deflection temp at 1.8 MPa,[a] °C	66	243	246	252	260	260	260
thermal expansion	4.5	1.6	1.4	1.3	1.2	1.0	0.9
water absorption, 24 h, %	1.6	1.1	0.9	0.9	0.6	0.5	0.4
mold shrinkage, %	15	6.5	5	4.0	3.5	3.0	2.0

Table 2.114 Examples of properties due to different concentrations of glass fibers in reinforced TS polyester plastic

Fiber	Axial Modulus GPa (Msi)	Transverse Modulus GPa (Msi)	Inplane Shear Modulus GPa (Msi)	Poisson's Ratio	Axial Tensile Strength MPa (Ksi)	Transverse Tensile Strength MPa (Ksi)	Axial Compressive Strength MPa (Ksi)	Transverse Compressive Strength MPa (Ksi)
E-glass	23 (3.4)	23 (3.4)	9.0 (1.3)	0.28	550 (80)	550 (80)	330 (48)	330 (48)
Aramid	29 (4.2)	29 (4.2)	11 (1.6)	0.32	460 (67)	460 (67)	190 (28)	190 (28)
Boron	80 (11.6)	80 (11.6)	30 (4.3)	0.33	480 (69)	480 (69)	1100 (160)	1100 (160)
SM carbon a	54 (7.8)	54 (7.8)	21 (3.0)	0.31	580 (84)	580 (84)	580 (84)	580 (84)
UHS carbon a	63 (9.1)	63 (9.1)	21 (3.0)	0.31	1350 (200)	1350 (200)	580 (84)	580 (84)
UHM carbon a	110 (16)	110 (16)	41 (6.0)	0.32	490 (71)	490 (71)	270 (39)	70 (39)
UHM carbon b	165 (24)	165 (24)	63 (9.2)	0.32	310 (45)	310 (45)	96 (14)	96 (14)
UHK carbon b	165 (24)	165 (24)	63 (9.2)	0.32	310 (45)	310 (45)	96 (14)	96 (14)

a. PAN type
b. Pitch type

Table 2.115 Examples of reinforced polyester plastic properties with different fibers

Monomer type	Shrinkage, %
Styrene (most popular)	17
Vinyltoluene	12.6
Diallyl phthalate	11.8
Methyl methacrylate	21
Vinyl acetate	27
t-butylstyrene	7

Table 2.116 Examples of monomers that can be used with polyester plastic

large quantities, there is the likelihood that, in the future, their use may be regulated as well.

e. *Closed Molding Process* This method can be extremely effective in lowering styrene emissions, but equipment cost is a potential disadvantage. Changing fabrication methods from open to the closed molding process will reduce styrene emissions considerably.
f. *Styrene Suppressant Additives* These additives are now available. They are most effective when using the open molding process during the curing stage. When properly used, they can reduce styrene emissions during the curing stage of the RP. They will reduce the amount of styrene vapors escaping from the RPs. Styrene-suppressant formulations, in addition to the wax, can contain components that allow bonding between laminates while effectively reducing the amount of styrene vapor emitted as the RP cures. Styrene-suppressant additives are supplied in either semiliquid or solid form. They can be used as supplied or diluted in organic solvents. Some require heating to make them flow before mixing into the plastic, while others do not. They are incorporated directly into the plastic before it is catalyzed. The treated plastic can be used immediately or kept in storage for further use.

POLYESTER, WATER-EXTENDED

Special casting polyesters allow the use of water as an extender (filler). Water as an extender offers advantages over conventional fillers, such as lower density, high temperature control (fast cure), easy mixing, and lower material cost. These plastics form an emulsion with water, providing improved physical properties. They can be thought of as microcellular water-filled foams; they are plastic cellular composite structures.

Water acts as a heat sink, facilitating the casting and curing of thick parts without high temperature and cracking problems. A water-in-oil emulsion is prepared in which the plastic is the continuous phase, and small-diameter (2-mm) water droplets are the discontinuous phase. Before curing, the emulsion is a low-viscosity liquid that flows easily into the mold, filling it completely. The emulsion gels and cures quickly.

A wide variation in properties can be readily achieved by changes in polyester formulation, water level, catalysts, and additives. Compressive strengths approaching those of concrete are obtainable at 50% to 60% water. Tensile and flexural strengths correspond to at least those of wood and related materials of construction. Impact strengths, even at 60% water, are significantly improved over plaster. When substituted for plaster, these polyesters provide significant reduced breakage of decorative art objects during shipping and improved durability for the consumer.

Polyester, Water-Soluble These polyesters (WSPs) have increased use significantly in the surface coatings industry. They provide ease of application, ease of cleanup, freedom from toxicity, and freedom from flammability when compared with conventional solvent-based paint. They are obtained by incorporating hydrophilic groups in the polymer backbone during polymerization. WSPs are subsequently diluted in water for application, optionally in the presence of neutralizing agents and water-miscible cosolvents. They can be utilized as hydrophilic plastics in paper coatings and textile coatings. In most surface coatings, clear or pigmented solutions are converted to water-insoluble coatings by condensation or oxidative polymerization. Their largest use is in surface coatings as pigmented electrodeposition and conventional dipping primers. Other applications include sprayed primer surfaces, semigloss trade-sales paints, coil-coating vehicles, and enamels applied by dip, flow-coat, electrodeposition, and spray methods.

Polyester Reinforced Urethane This poromeric, microporous material may have a urethane impregnation or a silicone coating for shoe uppers and industrial leathers.

POLYIMIDAZOPYRROLONE

The polyimidazopyrrolones, also known as pyrrones, are experimental materials prepared from aromatic dianhydrides and aromatic tetraamines. The polymer syntheses provide soluble prepolymers that are converted to the pyrrone structures by thermal or chemical dehydration. The precursors can be used to cast films or coatings, or can be molded under very high pressures into filled or unfilled forms. The pyrrones combine some of the best properties of the polybenzimidazoles and polyimides. The pyrrone films are exceptionally radiation resistant and retain their strength properties after 10,000 megarads of 1-MeV electrons.

POLYISOBUTYLENE

This is the polymerization product of isobutylene. This thermoplastic varies in consistency from a viscous liquid to a rubberlike solid with corresponding variation in molecular weight from 1,000 to 4,000.

POLYISOBUTYLENE BUTYL

PIB is remarkably different from most common rubbers in two respects. It has very low gas permeability and very high damping properties (energy absorption). Since the molecules do not contain unsaturation (C=C), they cannot be cross-linked by the most conventional methods.

However, networks have good weathering, ozone, and heat aging resistance. PIB is capable of strain crystallization and can have good mechanical resistance. Resistance to oils and fuels is not good.

If isobutylene is copolymerized with a small fraction of isoprene (1 wt% to 2 wt%), the isoprene unsaturations can be used for conventional vulcanization. Such an isobutylene-isoprene copolymer is referred to as *butyl rubber (IIR)*. A major limitation of butyl rubber is its poor compatibility with other common rubbers/elastomers and generally low adhesion properties. Partial chlorination or bromination brings a marked improvement and allows faster cure. Such modified butyl rubbers are referred to as *chlorobutyl rubber*, *bromobutyl rubber*, or *halobutyl rubber*.

POLYISOPRENE

These elastomers (IRs) are chemistry's nearest approach to synthesizing rubber equal to natural rubber (NR). They are made from a colorless, volatile liquid derived from propylene or from coal gases or tars. IRs are chemically similar to the natural rubber molecules.

The synthetic rubber that comes closest to duplicating the chemical composition of natural rubber is synthetic polyisoprene (sometimes abbreviated PI). It shares with NR (a thermoset material) the properties of good uncured tack, high unreinforced strength, good abrasion resistance, and characteristics that provide good performance in dynamic applications. However, because polyisoprene has some inherent impurities, NR is somewhat better overall. Its low hysteresis and high tensile strength distinguish NR. However, solvents, gasoline, and ozone readily attack it. Its tensile strength is in the range of 24 to 31 MPa (3,500 to 4,500 psi) and its elongation is 550% to 650%.

One significant disadvantage of IRs is lack of green strength, which is during processing and prior to curing. An IR can be used interchangeably with an NR in all but the most demanding products. The applications are about the same as for NRs.

NATURAL RUBBER AND OTHER ELASTOMERS

Extensive data on properties and compounding (Fig. 2.38) are available regarding natural rubber (NR) and other elastomers (chapter 1). Examples are reviewed in Figures 2.39 to 2.42 (93).

POLYNORBORNENE

PNB has a glass-transition temperature of 35°C (95°F). This plastic can easily be plasticized with large amounts of oil, subsequently vulcanized into an elastomer with a low service temperature of –65°C (–85°F). The damping properties of the elastomer can be conveniently adjusted to meet different performance requirements.

POLYURETHANE, THERMOSET

The durometers of these polyurethanes range from soft cushion to glass hard with superior wear resistance. Uses include skateboard wheels, solid tires, floor coatings, and marine finishes. A major use for soft foam is automotive bumpers; another is upholstery. Basically, property improvements

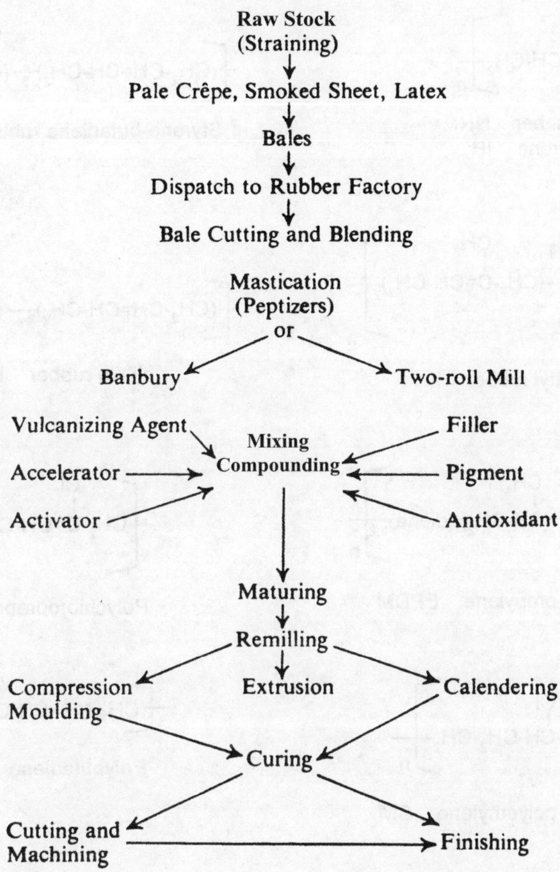

Figure 2.38 Compounding natural rubber

are made with added fibers and fillers in reaction-molded products to improve cut strength resistance, provide stiffer moduli, reduce waviness caused by heat and weathering, and so on. See "*Polyurethane, Thermoplastic*" for more details.

RUBBER, NATURAL

In addition to the information already presented on *polyisoprene*, the following is offered. The worldwide importance of natural (not synthetic) rubber (NR) lies in providing materials with inherently useful and required properties using fabricating techniques that the synthetic plastics industry inherited. In fact, the NR processes, principally compression molding and vulcanization, were expanded in order to process certain plastics. Even though the major-marketed products are tires, NR is used in many other products (seals, gaskets, etc.).

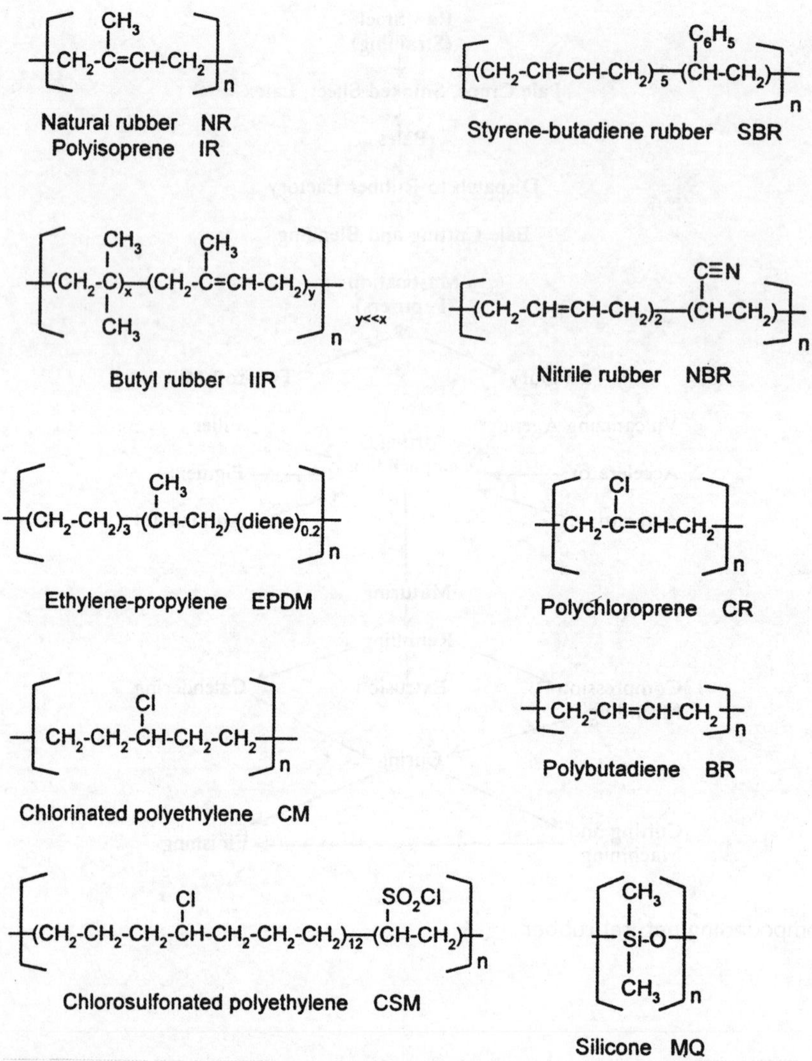

Figure 2.39 Examples of common elastomers

The worldwide shift to radial tires, which use a higher percentage of NR than bias-ply tires, has resulted in a significant increase in NR consumption over the past fifteen years. Increased rubber consumption in the NR-producing countries has also been a factor.

The productivity of NR trees is slowly increasing, primarily through the maturing of higher-yielding clones, the use of chemical stimulants, and the improved cultivation of seedlings. Trees are now being tapped six years after planting versus seven years just a decade ago. Increased replanting may be necessary to avoid future NR shortages.

NR's share of total rubber consumption is not expected to rise much above the current 40%, and some sources are projecting a decline, based primarily upon supply constraints. Worldwide consumption of NR is estimated to have been over six million metric tons in 1998. A 2% annual growth resulted in consumption of about 7 million metric tons in 2003.

Asia continues to dominate the world supply of NR, consistently averaging 94% of total production. (Over a century ago, when the rubber industry started, the only source worldwide was Brazil. For an interesting history on how their trees were exported, see chapter 29.) However, there has been a significant shift among the major producing countries. Malaysia, which accounted for 32 wt% of world production ten years ago, now produces only 13% and continues to shift emphasis to other crops and nonagricultural investments. Thailand has become the world's largest producer of NR, with a growth rate averaging 8% yearly since 1988. Indonesia also has potential for long-term growth because of the availability of land and labor. India and China have also steadily increased their share of world production. Malaysia now imports some of its requirements for tires and latex from other developing countries.

RUBBER LATEX, NATURAL

The NR latex obtained by tapping the bark of the *hevea brasiliensis* tree contains about 35 wt% NR hydrocarbon as particles that are about 1 μm in diameter and about 5% nonrubber components

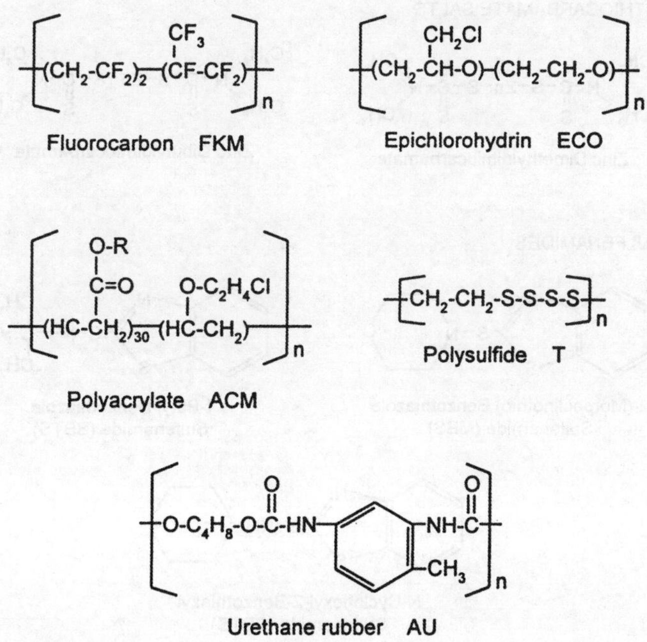

Figure 2.40 Examples of common specialty elastomers

THIURAMS

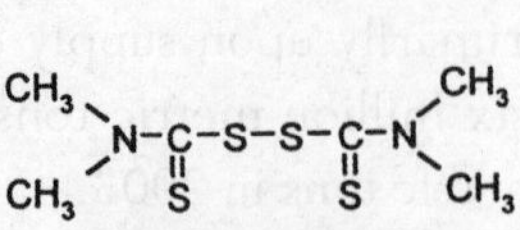

Tetramethyl Thiuram Disulfide (TMTD)

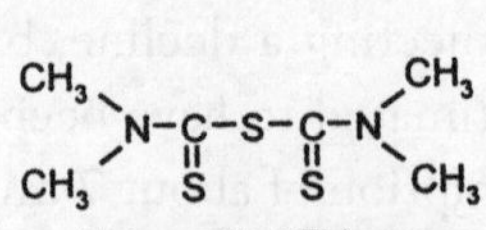

Tetramethyl Thiuram Monosulfide (TMTM)

GUANIDINES

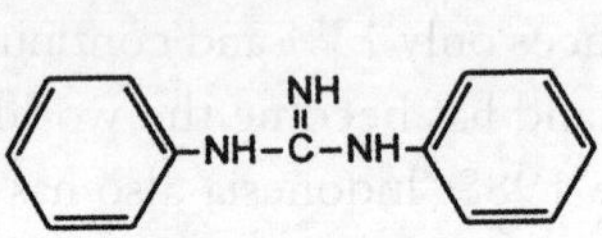

Diphenyl Guanidine (DPG)

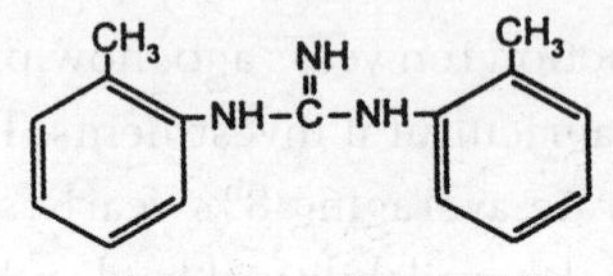

Di-o-tolylguanidine (DOTG)

THIAZOLES

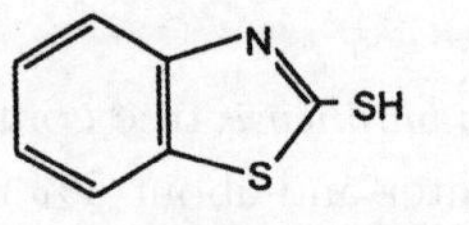

2-Mercaptobenzothiazole (MBT)

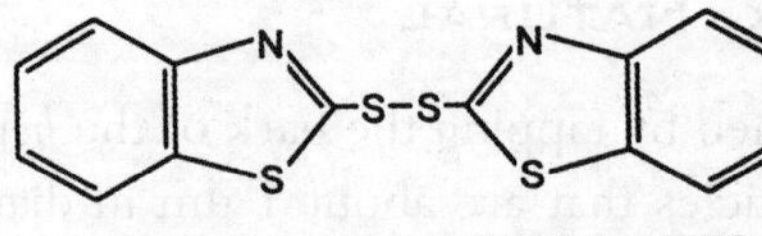

Mercaptobenzothiazole Disulfide (MBTS)

DITHIOCARBAMATE SALTS

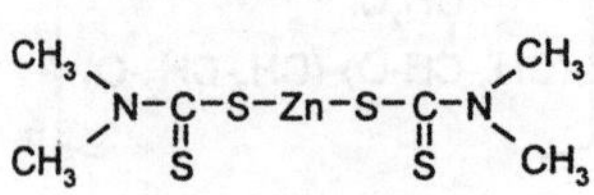

Zinc Dimethyldithiocarbamate

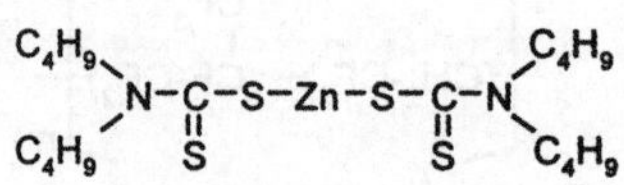

Zinc Dibutyldithiocarbamate

SULFENAMIDES

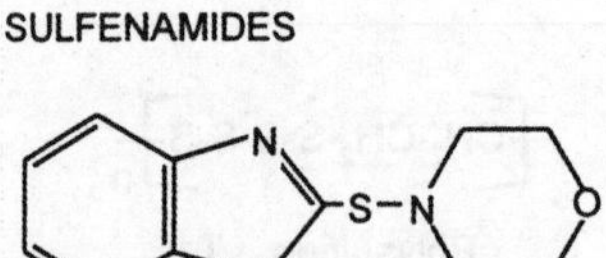

2-(Morpholinothio) Benzothiazole Sulfenamide (MBS)

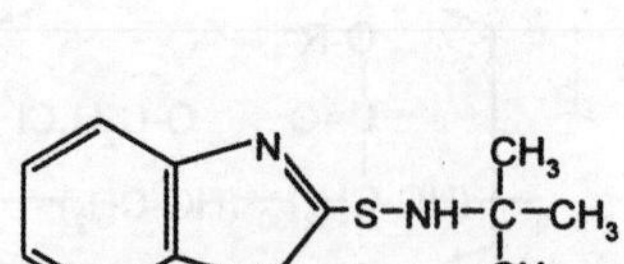

t-Butyl Benzothiazole Sulfenamide (BBTS)

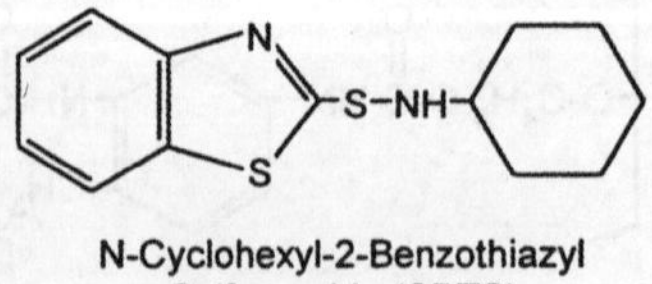

N-Cyclohexyl-2-Benzothiazyl Sulfenamide (CBTS)

Figure 2.41 Common vulcanization accelerators

consisting of protein, lipid, sugar, and salt. Most latex is coagulated by the addition of acetic or formic acid to produce solid natural rubber. Some latex is used by itself after being centrifuged or creamed to a concentration of about 60% rubber content. The stability of concentrated latex is preserved by the addition of about 1.5% ammonia. This releases the fatty acids from the lipids that stabilize the latex after the protein has broken down by natural microbiological attack.

SILICONE

Silicones are a class of engineering materials developed during World War II. They are semiorganic, high molecular weight polymers ranging in physical form from rigid plastics to inert fluids. They have excellent heat resistance up to 260°C (500°F), excellent chemical resistance, and good electricals; are compatible with human body tissues, and so forth; and have a high cost (Tables 2.117 to 2.119). Different formulations are available such as the room-temperature-vulcanizing (RTV) types that cure and cross-link at ambient temperatures and are catalyzed by moisture in the air. Silicone is a good sealant and excellent for making flexible molds for casting. It is widely used for human implants.

These plastics are based on the main polymer chains consisting of alternating silicone and oxygen atoms. Basically, their behavior is similar to silicone elastomers, except they provide increased mechanical properties. Silicone alloys are used to provide special properties (Table 2.120). As with glass-fiber/silicone-reinforced plastics, retention of mechanical and electrical properties up to 316°C (600°F) for relatively long time periods is obtained. RPs can be made by vacuum bag

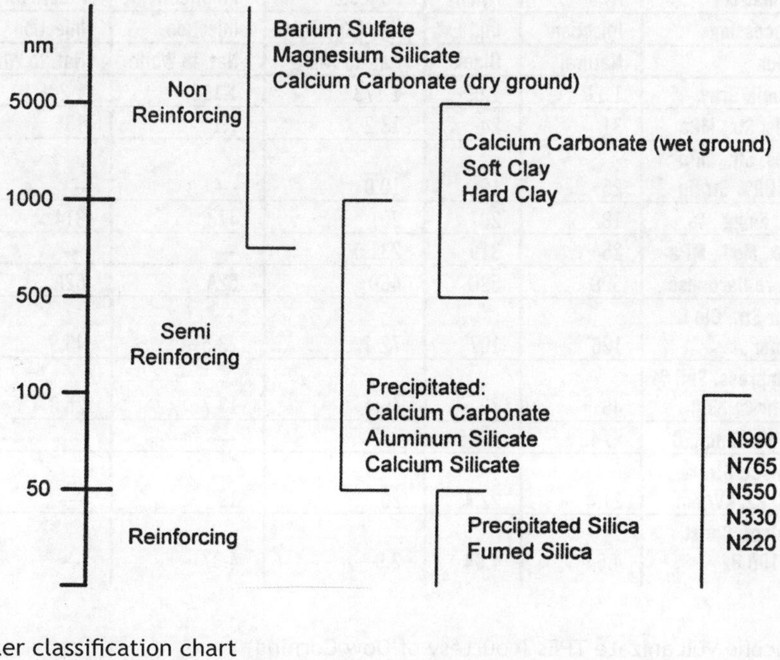

Figure 2.42 Filler classification chart

Structure	Name	Application
CH_3–	methyl	Basic substitution found in virtually all silicones; a cross-linking point with peroxide cures
C_6H_5–	phenyl	Increases modulus, thermal, and UV stability; raises refractive index; decreases reactivity with peroxides
$CF_3CH_2CH_2$–	trifluoropropyl	Increases stability, increases solvent resistance
H–	hydride	Introduces metal-catalyzed and vinyl-addition cross-link sites
HO–	silanol	Cross-linking point for condensation and metal-catalyzed cross-linking
CH_2=CH–	vinyl	Increases peroxide reactivity; introduces cross-link points for vinyl addition

Table 2.117 Silicone substitutions

	TPSiV X1120	TPSiV X1180	TPSiV X2100	TPSiV X3000-50A	TPSiV X3000-60A
TP Matrix	Nylon	Nylon	Polyester	Proprietary	Proprietary
Processing	Injection	Inj., Ext.	Inj., Ext., Blow	Injection	Injection
Color	Natural	Black	Nat. to White	Nat. to White	Nat. to White
Specif. Grav.	1.11	1.09	1.173	1.118	1.206
Tens. Str., MPa	31	24	13.2	7.1	13
Tens. Str., MPa @100% Strain	25	19	10.9	—	—
Ult. Elong., %	182	207	201	473	814
Flex. Mod., MPa	254	319	231.5	—	—
Shore Hardness	59D	52D	46D	52A	62A
Tear Str., Die C N/mm	126	107	72.4	23.5	43.9
Compress. Set, % 22 hr @ 23 C	46	70	46.4	13.7	25.3
Brittle Temp., C	<-45	<-45	—	—	—
Dielect. Str. @ 100 Hz, kV/mm	21.4	22.4	20	19	—
Dielect. Const. @ 100 Hz	4.64	4.04	2.9	4.37	—

Table 2.118 Silicone vulcanizate TPEs (courtesy of Dow Corning)

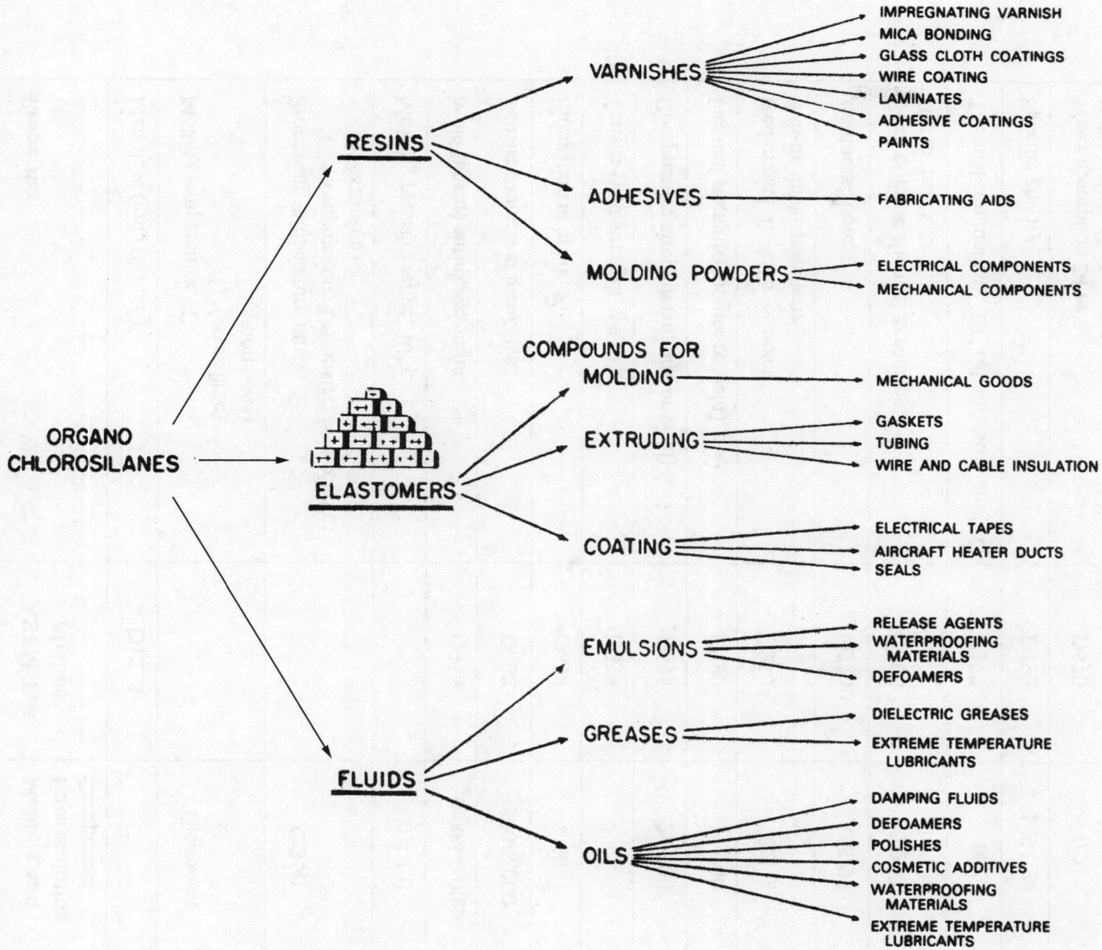

Table 2.119 Examples of silicone's diverse applications

Properties	ASTM Test Method	Molding and Encapsulating Compound
Melt flow (gm./10 min.)	D1238	
Melting temperature, °C T_m (crystalline) T_g (amorphous)		Thermoset
Processing temperature range, °F. (C=compression; T=transfer; I=injection; E=extrusion)		C:350
Molding pressure range, 10^3 p.s.i.		0.4-1.0
Mold (linear) shrinkage, in./in.	D955	0.005-0.006
Tensile strength at break, p.s.i.	D638	500-8,000
Elongation at break, %	D638	60
Tensile yield strength, p.s.i.	D638	
Compressive strength (rupture or yield) p.s.i.	D695	28,000
Flexural strength (rupture or yield), p.s.i.	D790	17,000
Izod impact, ft.-lb./in. of notch (1/8-in. thick specimen)	D256A	0.3
Hardness Shore	D2240	A68-95
Coef. of linear thermal expansion, 10^{-6} in./in./ °C.	D696	30-200
Thermal conductivity, 10^{-4} cal.-cm./sec.- cm.^{2} °C	C177	16
Specific gravity	D792	1.2-1.84
Water absorption 24 hr. (1/8-in. thick specimen), %	D570	0.2
Dielectric strength (1/8-in. thick specimen), short time, v./mil	D149	246-500

Table 2.120 Silicone-epoxy performances

molding to Higher-pressure molding techniques. Cure is generally at temperatures in excess of 150°C (300°F), depending on the particular thermoset silicone plastic used.

Silicones are any of various polymeric organic silicon compounds obtained as oils, greases, or plastics. Silicone is an alternate name for polyorganosiloxane, being more frequently used when a plastic of this type is used as the basis of useful commercial products. Such products span a diverse range from low molecular weight or branched plastics useful as oils and greases to high molecular weight cross-linkable plastics useful as elastomers (rubbers) to rigid-type plastics. The name arises from the assumed but erroneous analogy between the structure of silicones (R_2SiO) and ketones ($R_2C = 0$).

There are silicone laminating resins especially designed for fabricating RP parts by low pressure forming methods. Included in such methods are filament winding, vacuum bag molding, matched metal molds, and mandrel wrapping combined with cellophane lagging techniques. The resin is usually used in combination with an inorganic reinforcing material such as glass roving or fabric, or quartz fibers. Laminated parts made from these resin-reinforcement combinations have outstanding thermal stability. They remain serviceable almost indefinitely at temperatures up to 260°C (500°F); they also withstand many hours of exposure at temperatures up to 370°C (700°F) and shorter periods at temperatures up to 555°C (1,000°F). They have low dielectric losses, good electrical properties under wet conditions, and very stable electrical properties over a wide range of frequencies and temperatures.

Silicone Elastomer Also called silicone rubber, silicone elastomer was first produced in 1940 and has found substantial use ever since in areas where retention of properties at both high and low temperatures is required. A temperature range of –51° to 316°C (–70° to 600°F) is generally covered with little loss of properties (Table 2.121). Silicone elastomer (Q) is a high molecular weight polyorganosiloxane rubber. The basic plastic is polydimethylsiloxane (MQ) and has a molecular weight in the range of 300,000 to 1 million. Such plastics are produced by ring opening polymerization of octamethylcyclotetrasiloxane.

Several copolymers are also of interest, notably those containing about 0.5% of methylvinylsiloxane groups (vinylsilicone elastomer, VMQ), which are more readily vulcanized and give vulcanizates with better compression set resistance. Copolymers with 5% to 15% of the methyl groups replaced with phenyl groups (phenylsilicone elastomer, PMQ) give vulcanizates with better low-temperature properties, while related fluorosilicone elastomers (FVMQs) have improved solvent resistance (Fig. 2.43). Nitrilesilicone elastomers have also been of interest.

The unvulcanized polymer (dimethylsilicone gum) is usually cross-linked to a useful elastomer by heating it with an organic peroxide, such as benzoyl peroxide or 2,4-dichlorobenzoyl peroxide. Special low-molecular-weight plastics with reactive end groups are room-temperature-vulcanizing (RTV) silicone elastomers. RTVs are two-component liquids that are important to the plastics and other industries. This system combines the advantages of simple and rapid processing, analogous to that for thermoplastic elastomers, with those of an elastomer that can be chemically cross-linked and can be used over a wide (low-to-high) temperature range. The high reactivity allows injection molding with contact times as short as a few seconds, enabling small elastomer components to be produced in large numbers. Molds can quickly and easily be produced, and many other applications exist for RTVs. Despite their relatively high cost, they cost little to process, making them economically viable (Table 2.122).

Silicone elastomer vulcanizates have relatively poor mechanical properties compared with other elastomers and plastics (for example, tensile strength of 600 to 1,350 psi [4 to 9 MPa]). However, these properties are retained over a very wide temperature range. Typically, they have a useful

Service Temperature	Useful Life*
250°F	10 to 20 years
300°F	5 to 10 years
400°F	2 to 5 years
500°F	3 months to 2 years
500°-600°F	1 week to 2 months
600° -700°F	6 hours to 1 week
700°-800°F	10 minutes to 1 week
800°-900°F	2 to 10 minutes

* Retention of 50% elongation.

Table 2.121 Estimated useful life of silicone rubber at elevated temperatures

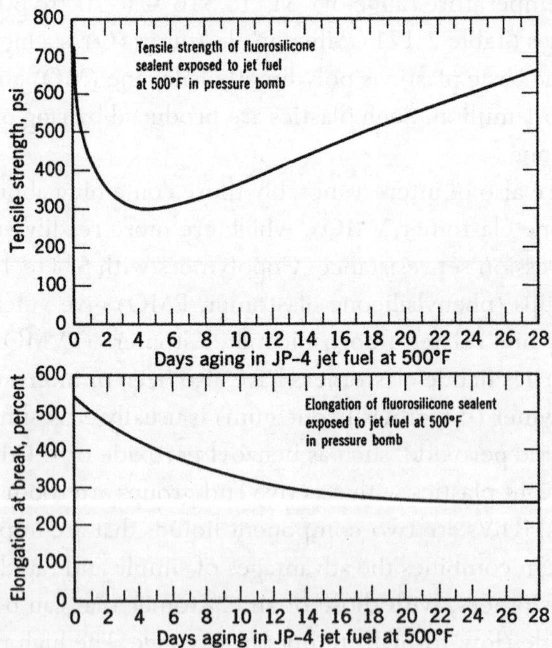

Figure 2.43 Retention of room-temperature mechanical properties of a fluorosilicone elastomer sealant after aging in JP-4 jet fuel vapor at 260°C (500°F) for periods up to 28 days

life of about two years at 150°C (300°F) and retain their flexibility to about –50°C (–60°F) or lower in the phenyl-substituted plastics. They retain excellent electrical properties under extreme temperatures and moisture, but they have poor abrasion resistance. They swell moderately in oils, fuels, and many solvents, although the fluorosilicones are much better in this respect.

STYRENE-BUTADIENE ELASTOMER

These elastomers (SBRs) emerged as high-volume substitutes for natural rubber (NR) during World War II because of their suitability for use in tires, belts, hoses, rubber floor tiles, and the like.

As Supplied	
Color	Black. White
Flow, sag or slump, in.	0.1
Specific gravity	1.51
Extrusion rate (1/8 in. Orifice, 90 psi), grams/min.	300
Cure Characteristics – exposed to air, 77° F (25°C) and 50% RH	
Skin-over time, min.	20 to 30
Working time, min.	30
Task-free time, hr	2 to 3
Cure time (1/8 in. Thickness), hr.	24
Physical Properties – cured 3 days at 77°F (25°C) and 50% RH	
Durometer hardness, shore A, points	28
Tensile strength, psi	150
Elongation, percent	550
Tear strength, die B, ppi	27
Adhesion, lap shear (to glass and aluminum), psi	100
Adhesion, peel strength, * lb/in width	20
Electrical Properties – cured 7 days at 77 °F (25 °C) and 50% RH	
Volume resistivity, ohm-cm	4.7×10^{14}
Dielectric strength, V/mil	500
Dielectric constant at 100 Hz	3.6
Dielectric constant at 100 KHz	3.6
Dissipation factor at 100 Hz	0.0021
Dissipation factor at 100 KHz	0.0010

* Laboratory tests and market tests have demonstrated superior adhesion to many substrates, including :

- Plastic surfaces–acrylic, polycarbonate, polyvinylidene fluoride, polyvinyl chloride, polystyrene, and acrylonitrile-butadiene-styrene
- Metals– milled aluminum, anodized aluminum, steel, galvanized steel, and stainless steel
- Other–glass, wood, cement, and painted surfaces

Table 2.122 Typical properties of general-purpose RTV silicone rubber

Tensile strength after compounding it with carbon black and vulcanizing it is 17 to 24 MPa (2,500 to 3,500 psi), which is less than NR's, but it has an elongation of 500% to 600%. In abrasion and skid resistance it is superior to NR and has better resistance to solvents and weathering.

This general-purpose rubber-type material continues to be used in many applications where it has replaced NR, even though it does not have NR's overall versatility, because it meets performance requirements and has a cost advantage over NR. For most uses, SBR must be reinforced to have acceptable strength, tear resistance, and general durability. It is significantly less resilient than NR, so it has higher heat buildup upon flexing. It also lacks NR's green strength and tack.

Urea-Formaldehyde

UF thermoset plastic compounds are in the amino family of plastics. They are available in a wide range of colors, from translucent colorless and white to a lustrous black. Unlike the colored phenolic compounds, the molded UFs can be made with a considerable degree of translucency, giving them a brightness and depth of color somewhat similar to, although better than, opal glass. Although the UFs' appearance can be duplicated in some TPs, the fact that they are not affected by heat within their range of operating temperatures makes their use necessary in many products.

These noninflammable (self-extinguishing), odorless, and tasteless materials char at about 200°C (395°F). Temperatures from –21° to 80°C (–70° to 175°F) have no effect on them, but higher temperatures over prolonged periods will cause fading and eventual blistering. When used within their temperature limitations, UFs have good electrical properties. They have high dielectric strength, high arc resistance, no tendency to track after arcing, and a low-order power factor. Their electrical properties are not greatly influenced by high humidity, and they resist static electricity buildup.

Under dry conditions, UF moldings are remarkably resistant to corrosive fumes and have no effect on organic solvents in terms of absorption, swelling, or changes in appearance. However, their water absorption is relatively high. For this reason, they are not recommended for applications involving continuous or intermittent exposure to water, although occasional exposure will have little effect if the material has been well cured. The UF intermediate water-soluble products are the starting materials for the production of adhesives, surface coatings, paper conditioners, and other such special items that require heat and catalysts for their final curing. They are used to bond laminated sheets, sometimes only on the surface, for color effects.

The applications of UFs include sanitary wares such as toilet seats, knobs, closures, buttons, electrical accessories such as housings and switches, laminates, and so on. Compounds of UFs use different additives, fillers, and reinforcements to provide different characteristics and permit processing in different equipment, principally by compression, transfer, and injection molding. Like many other TSs, such as the phenolics and melamines, they are easily preformed and preheated, either by RF preheaters or with screw plasticators. The higher bulk factor grades of the melamine compounds require using special equipment for these operations, because of their lack of easy pourability.

Commercial production of UF dates from the early 1930s, when industrial applications included modifying compounds, laminating plastics, and applying coatings. At one time, urea adhesives were popular for use with plywood, but more highly water-resistant phenolics largely replaced them. Owing to their light color, particularly when reinforced with alpha cellulose, the urea plastics were once very popular for appliance housings. Highly translucent pearl-like qualities in a variety of attractive colors are quite possible with ureas. However, as the tougher thermoplastics with rapid molding cycles were developed in the 1930s, urea plastics were largely displaced from their early markets. Buttons continue to remain a major market.

Even though UF plastics have been manufactured over a half century, their chemistry is not completely understood, and many variations of the basic molecule can result by slight changes in the manufacturing process. Basically, urea is reacted with formaldehyde in the presence of a catalyst. The resultant plastic is then filtered out and spray-dried, after which it is blended with appropriate fillers and packaged. The molecule of urea has a backbone of carbon and nitrogen atoms, while pendant from the backbone are carbon, oxygen, nitrogen, and hydrogen.

ELASTOMER

Elastomer materials are a separate, important segment in the plastics industry. New elastomers are always being developed to meet new industry requirements. Information on thermoplastic and thermoset plastic elastomers has been included in the sections "Thermoplastic" and "Thermoset Plastic." This section summarizes their technical information.

An elastomer is a rubberlike material (natural or synthetic) that is generally identified as a material which at room temperature stretches under low stress to at least twice its length and snaps back to approximately its original length on release of the stress (pull) within a specified time period. This term is often used interchangeably with the term *rubber*. Although rubber originally meant a thermoset elastomeric material obtained from a rubber tree, it identifies any TS elastomer (TSE) or thermoplastic elastomer (TPE) material.

Differences in elastomers can be based on the materials' rates of recovery, the time required for deformed materials to return to their approximate original size after the deforming forces are removed. Their elastomeric shape factor identifies an elastomer loaded in compression; it is the ratio of the loaded area to the force-free area. Different properties also identify the elastomers, such as tear resistance, strength and stiffness, abrasion resistance, solvent resistance, shock and vibration control, electrical and thermal insulation, waterproofing, and cost. For an elastomeric slab loaded in compression, the elastomeric shape factor is the ratio of the loaded area to the force-free area.

Many different mixtures with two or more polymers are used, with many containing proprietary treatments. As an example, there are thermoplastic vulcanizates (TPVs) and melt-processable rubbers (MPRs). A TPV is essentially a fine dispersion of highly vulcanized rubber in a continuous phase of a polyolefin. Its cross-linking gives it high tensile strength, high elongation, resistance to compression and tension set, oil resistance, resistance to flex fatigue, and a maximum service temperature of 135°C (275°F).

As explained in the previous sections of this chapter, elastomers are used to fabricate many different products to meet many different requirements. Practically all fabricating processes (such as extrusion, injection molding, blow molding, compression molding, casting, and encapsulation) are used to process elastomers.

Special and different applications also exist. For example, there are elastomeric molds. These are tooling systems that use the thermal expansion and particularly the flexibility of elastomers to form molds. The product being molded could have a complicated cavity whereby the mold does not require any special action to remove a complex product because the mold flexes (chapter 17).

Additional general and technical information is provided in Tables 2.123 to 2.139.

REINFORCED PLASTIC

Reinforced plastics (RPs) are a separate, important segment in the plastics industry. Information on thermoplastic and thermoset RPs has been included in "Thermoplastic" and "Thermoset Plastic." A major review is in chapter 15. This section extends and summarizes their technical information.

RPs are also called *plastic composites* and *composites*. The term RP refers to combinations of plastic (matrix) and reinforcing materials that predominantly come in fiber forms, such as chopped, continuous, and woven and nonwoven fabrics, and in other forms such as powders and flakes. They provide significant property and/or cost improvements relative to the individual components;

ASTM designation	Common or trade name	Chemical designation
IR	Synthetic natural rubber	Synthetic polyisoprene
IIR	Butyl, Chloro-butyl	Isobutylene-isoprene
ABR	Acrylic	Polyacrylate
AU	Urethane (UR)	Polyurethane (polyester)
EU	Urethane (UR)	Polyurethane (polyether)
BR	CBR, PBd	Polybutadiene
CO	Hydrin (CO, ECO)	Polyepichlorohydrin
CR	Neoprene	Chloroprene
CSM	Hypalon (HYP)	Chlorosulfonyl polyethylene
EPM	EP elastomer	Ethylene propylene copolymer
EPDM	EP elastomer	Ethylene propylene terpolymer
ET	Thiokol A	Ethylene polysulfide
EOT	Thiokol B	Ethylene ether polysulfide
EPM	Viton, Fluorel, Kel-F	Fluorinated hydrocarbon
NBR	Buna N, Nitrile	Butadiene-acrylonitrile
SBR	GR-S, Buna S	Styrene-butadiene
Si	Silicone	Organopolysiloxane
FVSi	Silastic LS	Fluorosilicone
—	Plaskon CPE	Chlorinated polyethylene
NR	Natural rubber	Natural polyisoprene

Table 2.123 Generic classification of elastomers

Type	Temperature, °C (°F)	Class	Volume Swell, Max. %
A	70 (158)	A	Not required
B	100 (212)	B	140
C	125 (257)	C	120
D	150 (302)	D	100
E	175 (347)	E	80
F	200 (392)	F	60
G	225 (437)	G	40
H	250 (482)	H	30
J	275 (527)	J	20
		K	10

*Per ASTM D 2000 and SAE J 200; based on type and class designations.

Table 2.124 ASTM elastomer type requirements

primary benefits include high strength, oriented strength, light weight, high strength-to-weight ratio, high dielectric strength and corrosion resistance, and long-term durability.

The term *composite* denotes the thousands of different combinations of two or more materials considered RPs. If referring to composites that incorporate plastics, consider calling them *plastic composites*. However, the more descriptive and popularly used worldwide term is *reinforced plastic*. In the United States, annual consumption of all forms of RPs is over 3½ billion lb (1.6 billion kg).

Both thermoset (TS) and thermoplastic (TP) are used. At least 90 wt% use glass fiber and about 45 wt% use TS polyester plastic. The RP market started in 1940, using contact or low-pressure TS polyester plastics–glass-fiber fabricating systems that were practically all hand lay-ups with bag molding. Since then, many different plastics with different reinforcements have been used with many different RP processes. All these combinations meet different requirements. RP products have gone worldwide into the deep ocean waters, on land, and into the air, including landing on the moon and in spacecraft.

It is important to recognize that a major cost in the production of RPs, from the design concept to the finished product, is from the materials of construction. They can range from 40% to 90% of the total cost. Thus, it is important to understand how best to use the materials based on their design and processing requirements. An important criteria is the understanding and proper application of the interrelations of design requirements with materials of construction and fabricating methods. RPs have some mechanical, formability, and other characteristics that differ from other materials (unreinforced plastics, steel, aluminum, wood, etc.). RPs have not come near to realizing their great potential in a multitude of applications, usually because of cost limitations that particularly involve the use of expensive fiber reinforcements (graphite, silica, etc.).

Additional general and technical information in different formats is provided in Tables 2.140 to 2.146.

Elastomer	Typical structure	Comments
cis-Polyisoprene	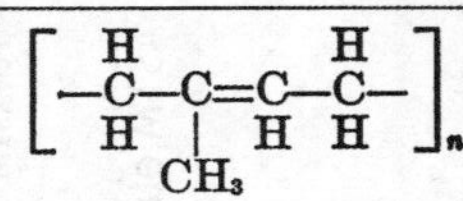	Historically natural rubber (Hevea) now also prepared synthetically; configurations other than *cis* of minor importance
Styrene-butadiene copolymers	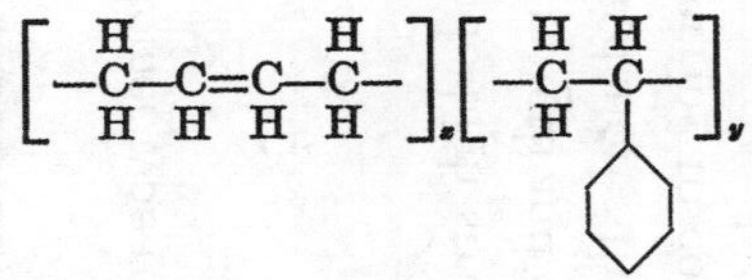	Largest volume elastomer in the U.S.; $x:y$ usually 3:1, but varied from about 10:1 to 1:1
Polysulfides	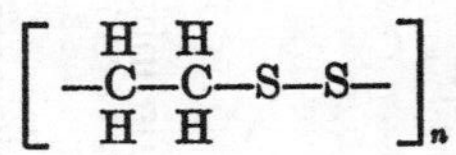	Numerous structural variations obtainable available as solids, liquids, and water dispersions; first commercial oil-resistant elastomer
Polychloroprene	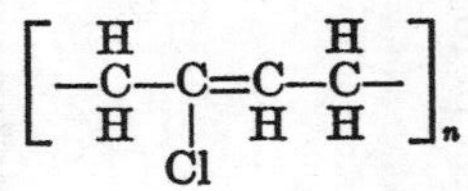	First general purpose elastomers, combining high mechanical properties with good resistance to weathering, heat, and oils.
Acrylonitrile-butadiene copolymers	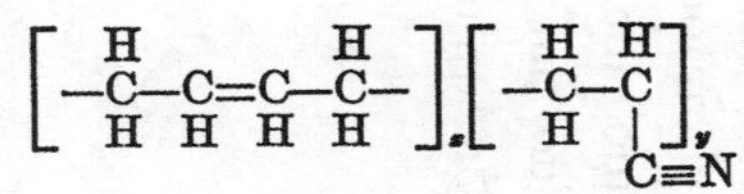	Largest volume highly oil-resistant elastomer; $x:y$ varied from about 8.5:1.5 to 5.5:4.5
Isobutylene-isoprene copolymers	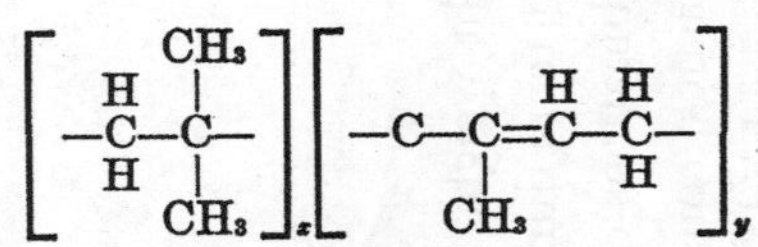	Highly resistant to oxygen and ozone; low air permeability; $x:y$ high, typically 98:2
Ethyl acrylate copolymers	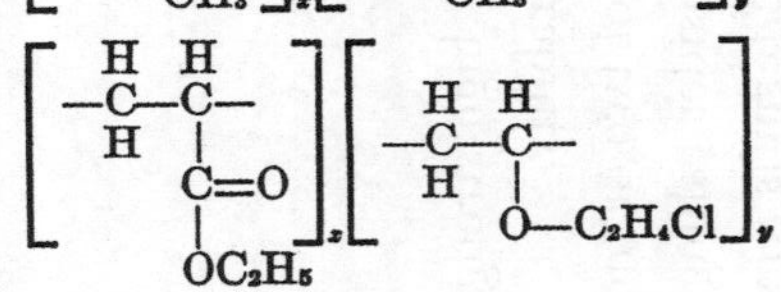	Highly resistant to oils, oxygen, and ozone; $x:y$ typically 95:5; several polyacrylates and copolymers have been available

Table 2.125 Elastomers by type

Chlorosulfonated polyethylene	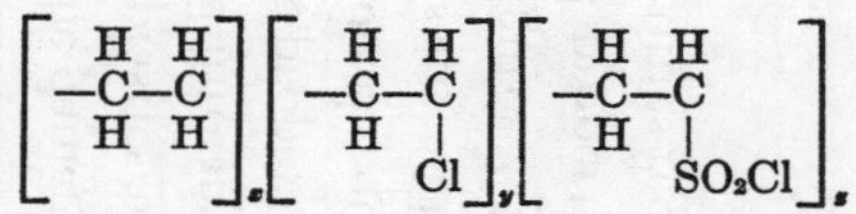	Very high oxygen and ozone resistance; widely used as coatings; $y:y:z$ about 72:26:2
Polysiloxanes	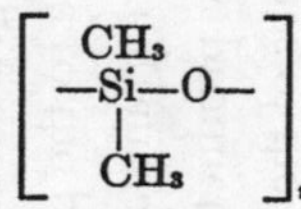	Excellent thermal stability and low temperature flexibility; available as solids and liquids
Polyfluorosiloxanes	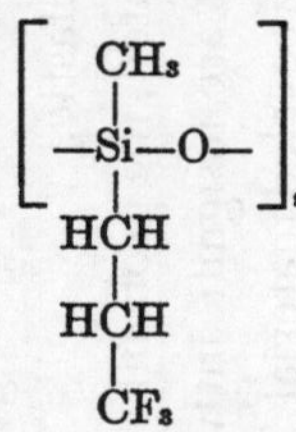	Similar to the polysiloxanes, but with substantially better oil resistance
Ethylene-propylene copolymers or terpolymers	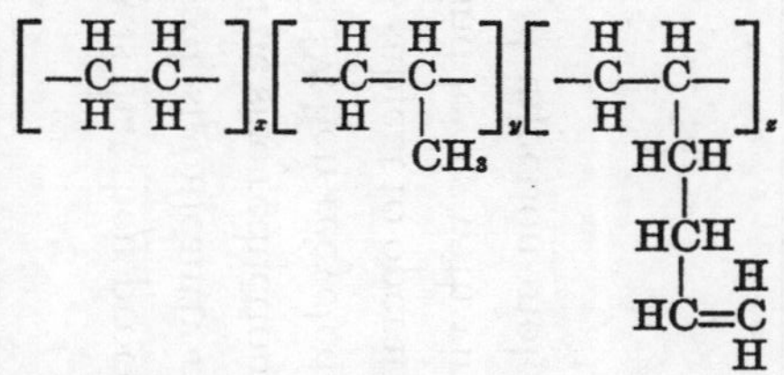	$x:y$ typically 45:55, z minor component for vulcanization of terpolymers; excellent ozone and oxygen resistance
Fluoroelastomers	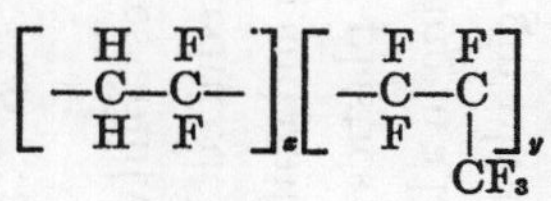	$x:y$ about 82:18; variations include chlorinated segments and perfluorotermonomers; excellent oil and thermal stability
cis-Polybutadiene	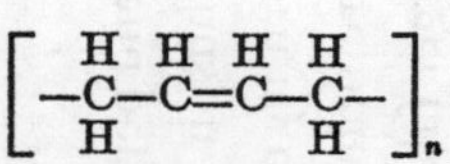	Primarily used in blends with other elastomers for tires

Table 2.125 *(continued)*

RECYCLED PLASTIC

These include fabricating scrap and pre- and postconsumer plastics. In North America, about 5¼ billion pounds are recycled, with total recycling capacity at 7 1/8 billion pounds (270). Examples of recycling scrap from fabricating processes are shown in Figures 2.44 and 2.45.

Different systems are used for different materials and products based on thickness, degree of hardness, and so forth. They include mechanical granulators, energy recovery systems (energy thermal reclamation), chemical recycling systems, and others (Fig. 2.46). Granulators are predominantly used (chapter 18). There are different sizes and types of granulators used that process thin to thick plastics; flexible, soft, or hard plastics; unreinforced to reinforced; and so on. If large and thick plastic is to be granulated, two or more granulators are used to gradually reduce their size to minimize the damage to the plastic.

When plastics are granulated, the probability is that their processability and performance when reprocessed into any product are insignificantly or significantly reduced. Thus, it is important to evaluate the properties of the recycled material. The size reduction and uniformity exert substantial influences on the quality of the recycled plastics. When recycled plastics are nonuniform in size, their processing with or without virgin plastics is subject to operating in a larger fabricating process window. Figures 2.47 and 2.48 show how regrind levels with virgin plastic mix (wt%) affect the mechanical properties of certain formulations of injection-molded plastics after the first and a number of times recycled, respectively.

Type ↓ ↓ class	Typical
AA	Natural rubber, styrene butadiene, butyl, ethylene propylene, polybutadiene, polyisoprene
AK	Polysulfide
BA	Ethylene propylene, styrene butadiene (high temperature), butyl
BC	Chloroprene, chlorinated polyethylene
BE	Chloroprene, chlorinated polyethylene
BF	Nitrile
BG	Nitrile, urethane
BK	Polysulfide, nitrile
CA	Ethylene propylene
CE	Chlorosulfonated polyethylene, chlorinated polyethylene
CH	Nitrile, epichlorohydrin
DA	Ethylene propylene
DE	Chlorinated polyethylene chlorosulfonated polyethylene
DF	Polyacrylate (butyl-acrylate type)
DH	Polyacrylate
FC	Silicone (high strength)
FE	Silicone
FK	Fluorinated silicone
GE	Silicone
HK	Fluorinated rubber

Per ASTM D 2000 and SAE J 280

Table 2.126 Elastomers by class

Generic Name				Natural Rubber	Styrene-Butadiene (GRS) (SBR)	Butyl	EP Rubber
				Non-Oil-Resistant Rubbers			
Weight of Base Elastomer			Lb./in.³	0.033	0.034	0.033	0.031
			(g/cm³)	(0.91)	(0.94)	(0.91)	(0.86)
			Specific Gravity	0.93	0.94	0.92	0.86
Physical Properties for Elastomer Compounds		Durometer range		30–100	40–100	30–100	30–90
		Resilience		O	G	F	G
		Tensile strength, psi (reinforced) (MPa)		4,000 (27.5)	2,000 (13.8)	2,000 + (13.8)	2,000–3,000 (13.8–20.7)
		Elongation % (reinforced)		500	450	300–800	500
		Creep, room temperature		E	E	F	F
		Compression set		G	G	F	F
		Electrical resistivity		E	E	E	E
		Impermeability, gas		G	F	O	G
Resistance Properties	Mechanical	Resistance to	Impact	E	E	G	G
			Abrasion	E	E	G	G
			Tear	E	F	G	P
			Cut growth	E	G	E	G
	Temperature	Tensile strength; Psi (M Pa), at	250°F (121°C)	1,800 (12.4)	1,200 (8.27)	1,000 (6.90)	2,000 (13.8)
			400°F (204°C)	125 (0.862)	170 (1.17)	350 (2.41)	400 (2.76)
		elongation, %, at	250°F (121°C)	500	250	250	300–500
			400°F (204°C)	80	60	80	0–120
		Creep at 212°F (100°C)		G	G	F (−23 to −40)	F
		Heat aging at 212°F (100°C)		G	G (−18 to −45)	E	E (−29 to −46)
		Flame resistance		P	P	P	P
		Low temperature	Stiffening, °F (°C)	−20 to −50 (−29 to −45)	0–50 (−18 to −45)	−10 to −40 (−23 to −40)	−20 to −50 (−29 to −46)
			Brittle pt., °F (°C)	−80 (−62)	−80 (−62)	−80 (−62)	−90 (−68)
	Environmental	Weather		F	F	E	E
		Oxidation		G	G	E	G
		Ozone		P	P	E	E
		Radiation		F–G	G	P	P
		Water		E	E	E	G–E
		Acid		F–G	F–G	E	G–E
		Alkali		F–G	F–G	E	G–E
		Gasoline, kerosene, etc. (aliphatic HCs)		P	P	P	P
		Benzol, toluol, etc. (aromatic HCs)		P	P	F–G	F
		Degreaser solvents (halogenated HCs)		P	P	P	P
		Alcohol		G	F	E	P
		Synthetic lubricants (diester)		P–F	P	F	P–F
		Hydraulic fluids	Silicates	P	P–F	F	F–G
			Phosphates	P–F	P	G	G–E
Subjective Properties		Taste		F–G	F–G	F–G	G
		Odor		F–G	G	G	G
		Nonstaining		P–G	P–G	G	G
Bonding to rigid materials				E	E	F–E	P
Typical applications				Elastic cord, tires	Tires	Air and gas containers	Weather-resistant parts

*Ratings are in this decreasing order: Outstanding, Excellent, Good, Fair, Poor.
Note: O = outstanding; E = excellent; G = good; F = fair; P = poor.

Table 2.127 Physical and mechanical properties of elastomers in different environments

Oil-Resistant Rubbers				Heat-Resistant Rubbers			
Neoprene	Acrylonitrile (BUNA-N)	Polysulfide	Poly-urethane	Silicone	Fluoro-silicone	Fluoro-elastomers	Acrylics
0.044	0.036	0.048	0.039	0.036	0.049–0.06	0.05–0.07	0.040
(1.22)	(1.00)	(1.33)	(1.08)	(1.00)	(1.36–1.66)	(1.38–1.94)	(1.11)
1.23	1.00	1.34	1.05	0.95	1.35–1.65	1.4–1.95	1.10
40–95	20–90	20–80	55–100	25–80	40–80	65–90	40–90
E	G	F	G–E	P–E	P–G	F	G
3000+	1,000–3,500	500–1,500	4,000–8,000	600–1,500	800–1,350	1,500–3,000	1,700
(20.7)	(6.90–24.1)	(3.44–10.3)	(27.6–55.2)	(4.14–10.3)	(5.52–9.31)	(10.3–20.7)	(11.7)
650–85	400–600	200–550	250–800	90–800	100–480	1 00–450	450
F–G	G	P	G–E	F–E	F–G	G	F
F–G	G	P–F	E	G–E	G–E	G–E	F–G
F	P	F	G	E	E	G	F
G	G	E	G	P	P	E	G
G	F	P	E	P–F	P–F	P–G	P
G–E	E	P–F	E	P–G	P–G	G	F–G
G	G	P–F	E	P–G	P–G	P–G	F
G	G	P	F–E	P–G	P–G	P–G	G
1,500	700	700	1,800	850	850	300–800	1,300
(10.3)	(4.83)	(4.83)	(12.4)	(5.86)	(5.86)	(2.07–5.52)	(8.96)
180	130	Under 25	200	400	350	150–300	225
(1.24)	(0.90)	(0.17)	(1.38)	(2.76)	(2.41)	(1.03–2.07)	
350	120	140	300	350	160	100–350	400
0–100	20	Under 25	140	200	150	50–160	150
F–G	E	P	E	E	E	G–E	F
G	G	G	F–G	E	E	O	E
G	P–F	P	P–F	F–G	F–G	E	P–F
+10–20	+30– –20	–10– –45	–10– –30	–60– –120	–70	+20– +30	+35– +10
(–12– –29)	(–1.1– –29)	(–23– –43)	(–23– –34)	(–51– –84)	(–57)	(–6.6– –1.1)	(1.7– –12)
–45	–65	–60	–60	–90– –140	–90– –120	+10– +20	–10
(–43)	(–54)	(–51)	(–51)	(–68– –96)	(–68– –84)	(–12– –6.7)	(–23)
E	G	E	E	E	E	E	E
G	F–G	E	E	E	E	O	E
E	P	E	E	E	E	O	E
F–G	F–G	F–G	G	F–E	F	F–G	P–G
G	E	G	G	G	G	G	F
G	G	F	P–F	P–G	G	G–E	F
G	F–G	G	P–F	P–F	G	P–G	P
G	E	E	E	P–F	G	E	E
P	G	E	P–F	P	F	E	P
P	P	F–G	F–G	P–F	F–G	G	P
F	E	G	G	G	G	E	P
P	F–G	G	P	P–F	G	F–G	G
G	F	P–F	——	P	G	G	G
P	P	P–F	P	G	G	P	P
F–G	F–G	P–F	G	G	G	F–G	F–G
F–G	G	P	G	G	G	G	F–G
G–E	P–G	P–F	G	O	O	P–G	G
G–E	G–E	F–G	F–G	F–E	G	P–G	G
Weather- and oil-resistant parts	Gasoline- and oil-resistant parts	Solvent-resistant parts	Abrasion-resistant parts	High- and low-temperature, non-oil-resistant parts	High- and low-temperature, oil-resistant parts	High-temperature solvent-resistant parts	High-temperature, oil-resistant parts

Table 2.127 *(continued)*

			Weather-Resistant Rubbers	Specialty Rubbers
Generic Name			Chlorosulfonated Polyethylene (Hypalon)	Various Types
Weight of Base Elastomer		Lb./in.^{3} (g/cm^3)	0.040	Electrically conductive,
		Specific Gravity	1.10	
Physical Properties for Elastomer Compounds	Durometer range		50–95	Thermoplastic,
	Resilience		G	Thermally conductive,
	Tensile strength, psi (reinforced) (MPa)		1500–2500 (10.3–17.2)	Damping,
	Elongation % (reinforced)		200–500	High and low permeability
	Creep, room temperature		F	For specific details consult manufacturer.
	Compression set		F–G	
	Electrical resistivity		G	
	Impermeability, gas		E	
Resistance Properties — Mechanical	Resistance to	Impact	F–G	
		Abrasion	G	
		Tear	F–G	
		Cut growth	F–G	
Resistance Properties — Temperature	Tensile strength; Psi, (MPa) at	250°F (121°C)	500 (3.45)	
		400°F (204°C)	200 (1.38)	
	elongation, %, at	250°F (121°C)	60	
		400°F (204°C)	20	
	Creep at 212°F (100°C)		F	
	Heat aging at 212°F (100°C)		E	
	Flame resistance		G	
	Low temperature	Stiffening, °F (°C)	−30 to −50 (−34 to −46)	
		Brittle pt., °F (°C)	−60 (−51)	
Resistance Properties — Environmental	Weather		E	
	Oxidation		E	
	Ozone		O	
	Radiation		F–G	
	Water		G	
	Acid		E	
	Alkali		E	
	Gasoline, kerosene, etc. (aliphatic HCs)		F	
	Benzol, toluol, etc. (aromatic HCs)		P–F	
	Degreaser solvents (halogenated HCs)		P–F	
	Alcohol		G	
	Synthetic lubricants (diester)		P	
	Hydraulic fluids	Silicates	G	
		Phosphates	P–F	
Subjective Properties	Taste		F–G	
	Odor		G	
	Nonstaining		E	
Bonding to rigid materials			F–G	
Typical applications			Weather-resistant coatings	

Table 2.127 *(continued)*

Thermoplastic Elastomers				
Styrene-butadiene	Olefinic	Urethanes	Polyester	EVA and EEA
0.033–0.038	0.031–0.037	0.042–0.045	0.043	0.033–0.34
0.93–1.113	0.88–1.02	1.15–1.25	1.20	0.91–0.95
35–95A	65–90A, 50D	75A–65A	92A–63D	80A–60D
High	Low	High-Damp	Low	Medium
500–3,000 (3.45–20.7)	650–2,000 (4.48–13.8)	3,000–7,000 (2.07–48.3)	5,800–6,400 (40.0–44.1)	800–4,000 (5.52–27.6)
400–1,350	180–250	200–500	500–800	50–1,200
—	—	—	—	—
P	P–G	F–G	G	P
G	High	P	G	F
—	—	—	—	—
—	—	—	—	—
F–G	F–G	O	G	F–G
G	G	E	E	F–G
—	—	—	—	—
—	—	—	—	—
—	—	—	—	—
—	—	—	—	—
—	—	—	—	—
	—	—		—
Maximum use temp, °F (°C)	250 (121)	200 (93.3)	250–300 (121–149)	140 (60)
150 P	P	P	P	P
−40 (−40)	−60 (−51)	−40 (−40)	−70 (−57)	−20 (−29)
—	−76 (−60)	—	−94 (−70)	—
F–P	E	E–G	G	G–F
F	E	G–F	G	G–F
F	E	G–F	G	G–F
—	—	—	—	—
G	E	P	G	E
F	G–F	F	G	G
F	—	—	G	—
Very poor	G–F	E	G	P
Very poor	P	G	G–F	P
Very poor	P	F	P	P
Very poor	G–F	F–P	G	F
—	—	—	—	—
—	—	—	—	—
—	—	—	—	—
—	—	—	—	—
—	—	—	—	—
G	P	G	G	F
Gaskets, seals, footware	Gaskets, seals, toys	Household items, toys	Gears, industrial parts	Household items, toy

Table 2.127 *(continued)*

Since scrap can be a mixture ranging from fine dust to large irregular chunks of different shapes, thicknesses, and so forth, it is important to use a granulator that provides the most uniformity and the least heat and mechanical damage to the scrap. Material having this range of size when granulating will influence product performance where critical requirements exist. The material influences the processing characteristics that in turn affect the product's performance and cost.

	Natural rubber	Isoprene rubber	Styrene-butadiene rubber	Butadiene rubber	Butyl rubber
Abbreviation(s)	NR	IR	SBR	BR	IIR
Density, Mg/m³	0.92	0.92	0.94	0.93	0.92
Glass transition temp., °C	−70	−70	−60	−105	−65
Temperature serviceability, °C					
lower limit	−55	−55	−45	−70	−50
upper limit, continuous	+70	+70	+70	+70	+100
intermittent	+100	+100	+100	+100	+125
Physical properties					
Hardness rang, IRHD	30–100	35–100	40–100	45–90	35–85
Tensile strength, MPa					
gum	24	21	3	3	10
reinforced	28	24	24	17	17
Resilience	E	E	G	E	P
Resistance to:					
tear	E	G-E	G	F	G
abrasion	E	E	E	E	F
compression set	G	G	G	G	F
creep/stress relaxation	E	E	G	G	F
gas permeation	F	F	F	F	E
Electrical resistivity	E	E	E	E	E
Environmental resistance to:					
heat	F	F	F	F	G–E
oxidation	F	F	F	F	G
ozone	P	P	P	P	G
flame	P	P	P	P	P
water	G	G	G	G	E
dilute acids	G	G	G	G	G
concentrated acids	F–G	F–G	F–G	F–G	F–G

Per ASTM D 2000 and SAE J 200

Table 2.128 Examples of elastomer performances (E = Excellent, G = Good, F = Fair, and P = Poor)

Overheating is the major cause of damage to the plastic during the cutting action of the granulator. To eliminate heat damage, cryogenic granulating is used on heat-sensitive plastics. A granulator that handles soft plastics will not work well when granulating hard plastics. One that handles thin plastics is not the proper type to handle the size and shape of thick plastics (bottles, solid handles, etc.). Two or more different granulators may be required to process thick material to

Ethylene-propylene	Chloroprene rubber	Nitrile rubber	Polyurethane	Silicone	Fluorocarbon
EPM, EPDM	CR	NBR	AU, EU	Q	FKM
0.86	1.23	1.00	1.05	0.98-1.6	1.85
−58	−49	−24	−50	−120	−22
−40	−35	−20	−50	−60	−20
+125	+100	+100	+70	+200	+200
+150	+125	+125	+100	+250	+250
30–90	35–95	40–100	50–100	40–90	50–95
3	17	4	35	7	17
21	21	21	—	10	17
G	G	P–F	F	F	F
F	F	P–F	E	P	F
F	E	F	E	F	G
G	G	G	G	G	G
F	F	F	F	F–G	G
G	G	G–F	G	P–F	E
E	F	F	G	E	G
E	G	G	F	E	E
E	G	G	G	E	E
E	G	P	E	E	E
P	F	P	P	F	F
E	F	F–G	F	G	E
E	G	G	F	F	E
G	G	F–G	P	F	

Table 2.128 *(continued)*

Property	Ethylene-propylenes	Polysulfides	Nitriles	Polyurethanes	Neoprenes	Colorosulfonated polyethylenes	Propylene oxides	Silicones	Acrylates	Fluoroelastomers
Hardness (Shore A)	40–80	40–85	40–95	50–100	20–95	50–90	40–80	40–85	40–90	60–90
Tensile strength, 1000 psi										
Pure gum	<1000	<1000	<1000	>4000	>3000	>1000	>1000	<1000	<1000	>2000
Reinforced	>2000	>1000	>2000	>4000	>3000	>2000	>2000	>1000	>1000	>2000
Compression set resistance	G	P	VG	G	G	G	F	G	G	VG
Modulus of elasticity										
Dynamic	G	—	G	VG	G	G	VG	G	G	—
Static	G	—	G	VG	G	G	VG	G	G	—
Resiliency	G	P–F	F–G	P–G	VG	F	VG	G	P	VG
Hysteresis resistance	G	—	G	G	G	—	VG	—	G	—
Flex cracking resistance										
Slow rate	G	F	G	VG	E	G	VG	F	G	G
Fast rate	G	P	P	VG	E	G	VG	P	G	G
Tear strength	G	F	F	E	G	F–G	E	P–G	F	F
Abrasion resistance	G	P	VG	O	VG	G	G	P	F	G
Dielectric strength	E	—	—	E	G	E	—	—	—	G
Electrical insulation	VG	F	P	F	F–G	G	—	E	F	F–G
Service temperature, F										
Min. continuous use	−65	−65	−65	−65	−50	−40	−90	−180	−20	−10
Max. continuous use	<350	<200	<250	<250	<200	<300	<250	<500	<350	<500
Aging										
Sunlight	E	VG	P	VG	VG	E	VG	G	E	E
Oxidation	E	VG	G	VG	G	E	VG	E	E	E
Ozone	E	VG	F	E	VG	E	VG	VG	E	VG
Water resistance	VG	E	E	G(RT),P(212F)	G	G	E	E	P	VG
Acid resistance										
Dilute (10%)	E	G	G	P	VG	E	G	G	P	E
Concentrated	VG	F	F	P	G	VG	G	E	P	VG
Alkali resistance										
Dilute (10%)	VG	G	G	F–G	G	VG	VG	F	P	VG
Concentrated	VG	G	F	P	G	VG	E	P	P	F–G
Solvent resistance										
Aliphatic hydrocarbons	P	E	E	E	F	G	F	P	G	E
Aromatic hydrocarbons	P	E	G	G	P	P	P	E	P	E
Oxygenated (ketones, etc)	VG	VG	P	P	F	P	P	F	P	P
Halogenated	P	F	VP	—	VP	F–G	P	VP	VP	F–G
Oil resistance										
Low aniline	P	O	F–E	F–G	F	G	F–G	P	G	VG
High aniline	P	O	F–E	F–G	G	G	F–G	G	G	VG
Gasoline resistance										
Aromatic	P	E	F–G	F–G	F	G	P	P–G	P	VG
Nonaromatic	P	E	G–E	F–G	G	G	F–P	G	G	VG

[a] O = outstanding; E = excellent; VG = very good; G = good; F = fair; P = poor.

Table 2.129 Comparative properties of elastomeric vulcanizates[a]

ensure that overheating is minimized; a cascading action occurs by granulating the thick scrap, and the next granulator further reduces the output from the first granulator.

There are different approaches to eliminate overheating, particularly when granulating heat-sensitive plastics. One method is to expose the scrap to cryogenic temperatures. The low-temperature freezing action on plastics can keep them from overheating during granulating. It can also aid to separate plastics or different materials, and so forth. For example, a cryogenic recycling line can be used to clean PET beverage bottles and separate HDPE-based cup flakes. Cryogenic cleaning uses liquid nitrogen at –93°C (–200°F) to freeze PET and HDPE flakes, which are hurled by high-speed vanes against a steel impact plate. This action pulverizes any frozen glue and releases labels. At such a low temperature, PET and HDPE are not affected, but PVC (from other bottles) crumbles to the consistency of sugar, allowing it to be mechanically screened out of the PET and HDPE. When required, metal detectors, aspirators, hot-water float/sink tanks, and other equipment are used.

If possible, the goal is to significantly reduce or eliminate any trim, scrap, rejected products, and so on, in an industrial plant, because it has already cost money and time to go through a fabricating process. Recycled scrap handling and granulating just adds more money and time. Also, it usually requires resetting the process to handle it alone (or in most cases, when blending with virgin plastics and/or additives); this is because recycled scrap usually does not have uniform particle sizes, shapes, and melt-flow characteristics, owing to its overheating during the cutting action of a granulator, and so on. Keeping the scrap clean before/after granulating is an important requirement.

Methods used for handling mixed plastics parts include (a) selective collection of parts (selective picking by IR spectroscopic identification, light optical identification, geometrical identification by hand), (b) separation of granulates and ground stock (floating-settling trough or separating column, liquid cyclone, wind sifting, flotation, electrostatic separation, extraction, melt separation), (c) degradation (pyrolysis, reduced-oxygen incineration, hydrolysis, methanalysis, glycolysis, hydrogenation), (d) material recycling inconsistent with plastics characteristics (building and construction aggregates, soil-improving additives), (e) processing in a mixed phase (physical

Elastomer	Vulcanization system
Neoprene	Magnesium oxide, zinc oxide, ethylene thiourea
Nitrile, SBR, natural rubber, EPT (ethylene propylene terpolymer)	Sulfur, zinc oxide, stearic acid thiazole
Butyl	(*1*) Sulfur, thiazole, zinc oxide; (*2*) quinoid (quinone dioxime and lead oxide); (*3*) methylol phenol resin
Silicone, ethylene propylene	Peroxides
Fluoroelastomer	Blocked diamines, metal oxide

Table 2.130 Examples of vulcanization systems for elastomers

Secondary requirement	Primary requirement				
	Material cost	Oil resistance	Heat resistance	Weather resistance	Wear resistance
Minimum material cost		1. NBR 2. PS 3. ACR	1. EPR 2. IIR 3. HYP	1. EPR 2. CR, IIR 3. HYP	1. IR, SBR 2. BR, NR
Oil resistance	1. CR 2. NBR 3. PS		1. ACR, FLU 2. HYP 3. NBR, PS, PU	1. PS, PU, ACR, FLU 2. CR, HYP	1. NBR, PU 2. CR 3. ACR
Heat resistance	1. EPR 2. IR, SBR 3. NR, BR	1. FLU, ACR 2. NBR 3. PU, PS		1. EPR, SIL, HYP, ACR, FLU, IIR	1. PU 2. CR, NBR, NR, IR, SBR, CR
Weather resistance	1. EPR 2. IR, SBR 3. NR, BR	1. PS, PU, ACR, FLU 2. NBR, POR	1. EPR, SIL, HYP, ACR, FLU, IIR		1. PU, CR 2. NBR, NR, IR, SBR, BR, POR
Wear resistance	1. IR, SBR 2. EPR 3. NR, BR	1. NBR, PU 2. CR 3. ACR	1. EPR, HYP, FLU, IIR 2. ACR	1. CR, PU 2. EPR, HYP, ACR, FLU, IIR	
Tear resistance	1. IR, SBR 2. NR 3. BR	1. PU 2. NBR, CR 3. ACR, HYP, POR	1. HYP, ACR, IIR	1. PU 2. HYP, ACR, CR, IIR	1. PU, NR 2. IR, SBR, BR, CR, NBR
Cold flow resistance	1. IR, SBR, EPR 2. NR, BR	1. PU, FLU 2. NBR, ACR 3. CR, HYP	1. SIL, FLU 2. EPR, HYP, ACR, IIR	1. PU, SIL, HYP 2. CR, ACR, EPR, IIR	1. PU 2. NR, IR, SBR
Resilience	1. IR 2. EPR, SBR 3. NR, BR	1. PU 2. NBR, PS, ACR, FLU	1. EPR, SIL, HYP, ACR, FLU, IIR	1. CR, PU 2. PS, SIL, IIR, HYP, ACR, FLU	1. CR, PU, NR, IR, BR, POR
High tensile strength	1. EPR, IR, SBR 2. NR 3. BR	1. PU 2. NBR 3. FLU	1. EPR, FLU 2. HYP, IIR	1. PU 2. CR 3. EPR, FLU, IIR	1. PU 2. NR, NBR 3. CR
High hardness	1. SBR 2. EPR 3. IR	1. PU 2. NBR, ACR, FLU	1. IIR 2. HYP 3. EPR, ACR, FLU	1. PU, IIR 2. CR, HYP	1. PU 2. SBR, NR, NBR
Electrical resistance	1. IR, SBR, EPR 2. NR, BR	1. PS, PU, ACR, FLU 2. HYP, CR	1. EPR, SIL, IIR 2. HYP, ACR	1. EPR, SIL, IIR 2. CR, PS, HYP, PU, ACR, FLU	1. NR, IR, BR, SPR 2. CR, PU

Table 2.131 Selection of elastomeric vulcanizates for combined environmental effects

	Primary requirement					
	Tear resistance	Cold flow resistance	Resilience	High tensile strength	High hardness	Electrical resistance
Minimum material cost	1. NR 2. PU 3. EPR, IR, SBR	1. PU 2. SIL 3. FLU	1. IR 2. NR, BR 3. CR	1. NR 2. CR 3. PU	1. SBR 2. NR 3. IIR	1. EPR, IR, SBR 2. NR, BR
Oil resistance	1. PU 2. NBR, ACR, POR 3. CR, HYP	1. PU, FLU 2. ACR, NBR 3. CR, HYP	1. PU 2. CR 3. NR, PS, ACR FLU, POR	1. PU 2. NBR, FLU 3. CR	1. PU 2. HYP 3. NBR, ACR, FLU	1. PS, PU, ACR, FLU 2. HYP, CR
Heat resistance	1. PU 2. HYP, ACR, IIR, POR	1. SIL, FLU 2. PU	1. CR, PU, NR IR, BR, POR	1. PU 2. NR 3. EPR, FLU	1. IIR 2. PU, NR, SBR	1. EPR, SIL, IIR 2. NR, IR, SBR
Weather resistance	1. PU 2. NR 3. CR, HYP, ACR	1. PU, SIL, FLU 2. EPR, CR, HYP, ACR, IIR	1. CR, PU, POR 2. NR, IR, BR	1. PU 2. CR 3. NR	1. IIR, PU 2. NR, SBR	1. EPR, SIL, IIR 2. NR, IR, SBR
Wear resistance	1. PU, NR 2. NBR, CR, IR, SPR, BR, POR	1. PU 2. FLU 3. CR, NBR, NR, IR	1. CR, PU, NR BR, IR, POR	1. PU 2. NR, NBR 3. CR	1. PU, NR, SBR 2. IIR, NBR	1. NR, IR, SBR, BR 2. EPR, IIR
Tear resistance		1. PU 2. NR 3. NBR	1. PU, NR OR, IR, BR, PO	1. PU 2. NR 3. CR	1. PU, NR 2. SBR, IIR	1. NR 2. IR, SBR, IIR, BR
Cold flow resistance	1. PU 2. NR, IR, SBR		1. PU 2. CR, NR, IR, BR	1. PU 2. NR 3. CR	1. PU 2. NR, SBR, IR	1. SIL 2. EPR, NR, IR, SBR, IIR, BR
Resilience	1. PU, NR 2. CR, IR, BR, POR	1. PU 2. SIL, FLU 3. NR, IR, BR, CR		1. PU 2. NR 3. CR	1. PU, NR 2. CR	1. NR, IR, BR 2. EPR, SIL, SBR
High tensile strength	1. PU 2. NR 3. CR	1. PU 2. FLU 3. SIL	1. PU 2. NR 3. CR		1. PU 2. NR 3. SBR, IIR	1. NR 2. IR, SBR, IIR, BR, EPR
High hardness	1. PU, NR 2. BR, IIR	1. PU 2. FLU 3. SIL	1. PU, NR 2. CR 3. IR, BR	1. PU 2. NR 3. CR		1. NR, SBR, IIR 2. EPR
Electrical resistance	1. NR 2. PU 3. IIR, SBR, IR, BR	1. SIL 2. PU, FLU 3. EPR, NR, IR, SBR, IR, IIR	1. NR, IR, BR 2. CR, PU	1. NR 2. PU 3. CR	1. NR, IIR, SBR 2. PU	

Numbers indicate first, second, and third choices.

Key: NR—natural rubber, IR—synthetic rubber, SBR—styrene butadiene rubber, IIR—butyl rubber, BR—polybutadiene rubber, EPR—ethylene propylene rubber, CR—neoprene, NBR—nitrile rubber, PS—polysulfide, PU—polyurethane, SIL—silicone, HYP—chlorosulfonated polyethylene, ACR—acrylate, FLU—fluoroelastomers, POR—propylene oxides.

Table 2.131 *(continued)*

	Nitrile	EPR [b]	Butyl	EPT [c]
Fluid				
70 hr at room temp.				
Acetone	+125	+2	+2	+1
Aniline	+225	+1	+1	+10
Benzene	+155	+82	+92	+80
Benzaldehyde	+195	+5	+3	+5
Butyl acetate	+115	+18	+27	+12
Butylamine	+102	+30	+58	+23
Butyl carbitol	+40	+1	NC	−1
Butyl Cellosolve	+38	+2	+1	NC
Carbon disulfide	+56	+135	+139	+136
Carbon tetrachloride	+81	+152	+173	+167
Chlorobromomethane (CBM)	+257	+33	+54	+31
Chloroform	+279	+113	+139	+119
Cyclohexane	+11	+148	+195	+122
Dichloroethylene	+159	+108	+153	+112
Diethyl sulfate	+110	−1	NC	+2
Dioxane	+155	+9	+10	+7
Diphenyl (Dowtherm A)	+145	+16	+10	+16
Ethyl chloroformate	+140	+14	+16	+8
Ethylene dichloride (dichloroethane)	+230	+16	+19	+11
Ethyl ether	+44	+58	+51	+64
Formaldehyde	+2	+1	NC	+1
Isooctane	+2	+103	+112	+71
30% Aromatics	+31	+128	+145	+93
Methylene dichloride	+243	+44	+74	+47
Nitrobenzene	+246	+4	+4	+2
Nitromethane	+75	NC	NC	NC
Orthodichlorobenzene	+212	+116	+118	+125
Phenyl Cellosolve	+83	NC	NC	−1
Pyridine	+208	+10	+16	+14
Sodium hydroxide	+1	NC	NC	−1
Styrene	+180	+87	+78	+88
Tetrachloroethane	+305	+61	+88	+59
Tetrachloroethylene	+56	+153	+197	+160
Trichloroethane	+256	+49	+49	+23
Trichloroethylene	+178	+163	+178	+191
Turpentine	+9	+163	+182	+146

Table 2.132 Volume change of elastomers in various fluids

Neoprene	SBR	Thiokol	Poly-acrylate	Urethane	Silicone	Fluoro-silicone	Viton[d]
+31	+18	+7	+201	+87	+18	+205 S	+165
+51	+24	+325	+272	+261	+2	—4	+1
+168	+180	+113	+214	+109	+70	+23	+10
+131	+94	+203	+210	+100	+6	+17	+67
+107	+99	+34	+175	+100	+78	+175	+294
+85	+81	S	+183	D	+145	D	D
+42	+18	+15	+84	+81	+4	+5	+67
+31	+23	+7	+89	+72	+100	+7	+8
+163	+191	+51	+51	+42	+48	+20	+1
+142	+207	+53	+214	+76	+103	+12	+1
+175	+185	+298	+260	+216	+44	+24	+6
+191	+217	+311	+296	+263	+99	+36	+9
+73	+143	+3	+24	+21	+102	+15	NC
+137	+174	+78	+234	+137	+104	+58	+27
+5	+4	D	+208	+127 S	+2	+31	+297
+120	+107	+175	+207	+176	+23	+32	+100
+185	+174	+150	+375	+110	NC	+4	NC
+74	+65	D	+256	+172	D	+185	+164
+126	+113	+224	+250	+171	+25	+39	+11
+64	+82	+12	+99	+36	+101	+60	+97
+2	+2	+3	+13	+9	+2	+2	+1
+14	+52	NC	—2	NC	+93	+3	NC
+65	+112	+10	+39	+12	+109	+24	+1
+169	+165	+249	+255	+257	+72	+68	+22
+143	+66	+195	+197	+162	+3	+68	+22
+6	+3	+39	+173	+63	+2	+25	+127
+190	+192	+155	+226	+143	+40	+15	+8
+5	+84	+80	+192	+85	+2	+2	NC
+144	+125	+220	+233	+304	+14	+28	+118
+1	—2	NC	NC	+2	NC	NC	NC
+178	+181	+132	+212	+131	+52	+16	+4
+193	+205	+335	+302	+301	+42	+4	+1
+150	+187	+29	+108	+151	+82	+11	+1
+156	+165	+293	+277	+244	+36	+23	+4
+187	+212	+150	+248	+152	+98	+26	+4
+60	+153	—4	+22	+21	+98	+13	+1

Table 2.132 *(continued)*

	Nitrile	EPR[b]	Butyl	EPT[c]
Acids				
Acetic acid, conc.	+55	+16	+1	+6
Benzoic acid	+2	+1	NC	NC
Carbolic acid (phenol)	+250	NC	NC	NC
Chromic acid (10%)	D	−3	NC	NC
Cresylic acid	+200	+1	NC	+1
Hydrochloric acid, conc.	+7	+2	NC	+10
Nitric acid, conc.	D	+2	+4 S	+4
Phosphoric acid, conc.	+3	NC	NC	NC
Sulfuric acid, conc.	D	+7	S	+8
Sulfuric acid, 50%	+1	−1	+1	−6
70 hr at 158°F				
Diacetone alcohol	+119	−1	+1	−6
Furfural alcohol	+163	NC	+1	−4
Hexyl alcohol	+31	+12	+10	+6
Methyl alcohol	+18	+1	−2	−5
Soal sol. (conc. Castile)	+4	+2	NC	+4
70 hr at 212°F				
ASTM Oil #1	−2	+83	+104	−89
ASTM Oil #3	+15	+120	+217	+140
Benzyl alcohol	+180	+1	−2	−5
Brake fluid (Wagner)	+47	NC	+3	−5
Castor oil	+6	−2	NC	−8
Chlorobenzene	+192	+137	+239	+158
Detergent sol (conc surf)	+3	+22	NC	+5
Diethylene glycol	+3	+2	−2	−1
Ethylene glycol	+3	+2	−1	−1
Glycerol	+1	−1	NC	−1
Linseed oil	+5	+22	+11	+17
Cyclohexanone	+164	+54	+62	+52
Neatsfoot oil	+11	+126	+216	+132
Salt sol (conc NaCl)	−1	+1	NC	−1
Sperm oil	+2	+69	+106	+66
Tri normal butyl phosphate	+134	+15	+18	+8
Water	+11	+6	+1	+1
70 hr at 300°F				
Diphenyl (Dowtherm A)	D	+77	D	+133
EP Lube (Parapoid 10C)	+9 H	+124	D	+211
MIL-L-7808	+27	+81	S	+112
Silicate ester	+3	+36	D	+27
Silicone oil (DC-200)	−4	−5	−5	−10

Table 2.132 *(continued)*

Neoprene	SBR	Thiokol	Poly-acrylate	Urethane	Silicone	Fluoro-silicone	Viton [d]
+16	+19	+23	+154	+172	+3	+12	+112
+3	+2	+5	+15	+6	+2	+1	NC
+13	+15	+466 S	+323	+336	+2	+2	NC
+3	D	D	+6	S	NC	−1	NC
+11	+32	+260	+266	+404	+4	+2	NC
+2	+3	D	+4	+50 S	+12 S	+5	+4
D	D	D	+52	D	+25 S	+2	+5
−1	+1	D	+30	+50	NC	−1	NC
D	D	D	D	D	D	D	NC
NC	+3	D	+3	+36	NC	NC	NC
+27	+11	+35	+190	+131	+6	+40	+218
+37	+13	D	+215	+153	+3	+14	+80
+14	+17	+14	+104	+77	+27	+7	+5
+8	+2	+15	+140	+28	NC	+8	+68
+6	+5	+5	+33	−1	+4	+1	+1
+8	+44	−2	+2	−8	+7	NC	NC
+68	+105	+5	+16	+5	+23	+3	+1
+52	+30	D	+255	+428	+2	+4	+8
+13	+10	+23	+116	D	+8	D	+50S
+10	+4	−3	+11	+19	+2	NC	NC
+202	+219	+526	+276	+443	+77	+29	+27
+7	+4	+11	+31	+15	+2	+3	+2
+6	−1	+3	+27	+12	+2	+1	NC
+7	+4	+10	+37	+8	+3	+1	+1
+4	+1	+4	+16	+2	+1	NC	NC
+40	+56	+1	+4	NC	+3	NC	NC
+99	+107	D	D	+386	+28	+109	+211
+65	+94	+8	+9	+59	+26	+1	NC
NC	NC	+1	+2	+1	NC	+1	+16
+40	+69	−1	+6	−4	+5	+1	NC
+138	+72	+78	+158	+234	+36	D	+430S
+7	+10	+45	+23	+15	+3	NC	NC
D	D	D	D	D	+15	+9	+13
S	+154	S	+30	D	+16 S	D	D
+87	D	−8	+36	S	+16	+8	+10
+34	D	S	+8	−6	+35	+5	+4
−2	−3	S	−3	−9	+17	NC	NC

[a] D—Indicates that material disintegrated.
H—Indicates that material hardened excessively.
S—Indicates that material has excessive softening and loss of physical properties.
NC—Indicates no change.

[b] EPR—Ethylene-propylene rubber.

[c] EPT—Ethylene-propylene terpolymer.

[d] Viton—Copolymer of vinylidene fluoride and perfluoropropylene.

Table 2.132 *(continued)*

compatibility improved, cross-linking, fine dispersion), and (f) energy recycling (refuse and waste incineration, energy recycling of the pyrolysis products).

When required, different approaches are used to improve performances or properties of mixed plastics, such as (a) additives, fillers, and/or reinforcements (use specific types, such as processing agents, talc, short glass fibers), (b) active interlayers (cross-linking, molecular wetting), and (c) dispersing and diffusing (fine grinding, enlarging molecular penetration via melt shearing).

Different methods are used to recycle materials to provide plastics with a continuing life. The method used is influenced by factors such as quantity involved, weight involved, size and shape, costs,

Thermoplastic elastomers

	Specific gravity	Hardness Shore A	Tensile strength, 1000 psi	Weather resistance	High temperature service, °F	Low temperature service, °F	Electrical properties	Typical trade names and suppliers
Polyurethane	1.24	65A-80D	3-10	excellent	180	−80	good	Roylar (Uniroyal), Esthane (B.F. Goodrich), Castethane (Upjohn)
Copolyester	1.17-1.22	90A-72D	5.9-6.4	very good	250+	−94	fair-good	Hytrel (DuPont)
Styrene-butadiene block copolymer	0.94-1.15	40-95	0.6-2	fair	150	−80	good	Kraton (Shell), Solprene LR (Phillips)
Polyolefin	0.88-1.02	65-92A	0.65-3.8	good	250-300	−95	very good	TPR (Uniroyal), Somel (DuPont), Renflex (Ciba-Geigy

Thermoset elastomers

	Specific gravity	Hardness Shore A	Tensile strength, 1000 psi	Weather resistance	High temperature service, °F	Low temperature service, °F	Electrical properties	Typical trade names and suppliers
Natural rubber	0.93	30-90	1-4	poor	180	−65	good	Many suppliers
Styrene-butadiene	0.94	30-90	0.2-3.5	fair	180	−60	good	Diene (Firestone), Ameripol CB, (B.F. Goodrich)
Polyisoprene (synthetic rubber)	0.91	30-90	2.5-4	fair	180	−65	good	Ameripol SN, (B.F. Goodrich), Isoprene, (Shell)
Chloroprene (neoprene)	1.23	30-90	3.0+	very good	240	−40	good	Neoprene (DuPont)
Nitrile	1.0	40-90	3.5-4	fair	300	−60	poor	Hycar (B.F. Goodrich), Paracril, (Uniroyal)
Butyl	0.92	40-90	2.5-3	good	300	−50	good	Butyl, (Exxon)
Ethylene propylene (EPM, EPDM)	0.86	30-90	3.5	very good	350	−60	good	Nordel (DuPont), Epcar (B.F. Goodrich), Vistalon (Exxon), Royalene (Uniroyal)
Chlorosulfonated polyethylene	1.12	50-90	1.5-2.5	very good	325+	−40	good	Hypalon (DuPont)
Polyacrylate	1.09	40-95	2.0-2.5	fair	350	−20	fair	Hycar (B.F. Goodrich)
Silicone	1.1-1.6	30-90	0.6-1.3	very good	600	−180	very good	Silastic (Dow Corning), SE Rubber, (General Electric), (Union Carbide)
Urethane	1.05-1.3	40-100	10.0	good	240	−65	very good	Adriprene (DuPont), Vibrathane (Uniroyal), Castethane (Upjohn), Cyanaprene (American Cyanamid)
Fluoroelastomer	1.4-1.95	60-90	1.5-3	good	500	−10	fair-good	Viton (DuPont)
Epichlorohydrin	1.32-1.49	30-95	2-3	very good	300	−80	fair	Hydrin (B.F. Goodrich), Herclor, (Hercules)
Fluorosilicone	1.4	40-70	1.2	very good	400	−90	good	Silastic (Dow-Corning)
Propylene oxide	1.0-1.02	40-80	1.0	very good	300	−80	good	Parel (Hercules)
Polysulfide	1.35	40-85	0.25-1	good	250	−65	fair	LP (Thiokol Chemical)

Table 2.133 Examples of selected elastomers

Properties of selected thermoset elastomers							
	Specific gravity	Hardness shore A	Tensile strength, 1000 psi	Weather resistance	High temperature service, °F	Low temperature service, °F	Electrical properties
Natural rubber	0.93	30–90	1–4	poor	180	−65	good
Styrene-butadiene	0.94	30–90	0.2–3.5	fair	180	−60	good
						−70	
Polyisoprene (synthetic rubber)	0.91	30–90	2.5–4	fair	180	−65	good
Chloroprene (neoprene)	1.23	30–90	3.0+	very good	240	−40	good
Nitrile	1.0	40–90	3.5–4	fair	300	−60	poor
Butyl	0.92	40–90	2.5–3	good	300	−50	good
Ethylene propylene (EPM, EPDM)	0.86	30–90	3.5	very good	350	−60	good
Chlorosulfonated polyethylene	1.12	50–90	1.5–2.5	very good	325+	−40	good
Polyacrylate	1.09	40–95	2.0–2.5	fair	350	−20	fair
Silicone	1.1–1.6	30–90	0.6–1.3	very good	600	−180	very good

Table 2.134 Thermoset elastomer performances

Urethane	1.05–1.3	40–100	10.0	good	240	–65	very good
Fluoroelastomer	1.4–1.95	60–90	1.5–3	good	500	–10	fair-good
Epichlorohydrin	1.32–1.49	30–95	2–3	very good	300	–80	fair
Fluorosilicone	1.4	40–70	1.2	very good	400	–90	good
Propylene oxide	1.0–1.02	40–80	1.0	very good	300	–80	good
Polysulfide	1.35	40–85	0.25–1	good	250	–65	fair
Polyurethane	1.24	65A–80D	3–10	excellent	180	–80	good
Copolyester	1.17–1.22	90A–72D	5.9–6.4	very good	250+	–94	fair-good
Styrene-butadiene block copolymer	0.94–1.15	40–95	0.6–2	fair	150	–80	good
Polyolefin	0.88–1.02	65–92A	0.65–3.8	good	250–300	–95	very good

Table 2.134 *(continued)*

		Aged at 400°F test temperature (°F)				Aged at 500°F test temperature (°F)			
		73		400		73		400	
Elastomer	Aging time (hr)	Tensile strength (psi)	Elongation (%)	Tensile strength (psi)	Elongation (%)	Tensile strength (psi)	Elongation (%)	Tensile strength (psi)	Elongation (%)
Viton A									
	0	1600	230			1600	230		
	100	1625	150	275	60	1600	130	230	50
	200	1700	170	250	60	1550	125	230	40
	300	1600	160	280	60	1000	80	190	20
	400	1725	140	275	60	1325	50	190	30
	500	1725	130	270	50	1400	30		
Viton A–HV	0	2200	380			2200	380		
	100	2175	360	280	140	1425	450	190	160
	200	2125	410	300	150	1075	580	200	160
	300	2000	350	290	150	1050	560	180	130
	400	2250	420	270	140	1150	420	170	120
	500	2025	380	280	150	1025	420	130	80
Acrylon EA–5	0	2000	400						
	100	1300	20	240	15				
	200	1600	10						
	300	300	10						
	400	625	10						
	500								

Table 2.135 Effect of aging at elevated temperatures on the tensile strength and elongation of high-temperature elastomers

Poly(ethyl acrylate)	0	2075	330						
	100	1350	20						
	200	850	25						
	300	175	10						
	400	575	10						
	500								
Silicone SE 555U	0	1250	290			1250	290		
	100	1050	320	375	80	750	230	310	80
	200	1050	250	350	80	725	220	330	75
	300	1025	240	380	80	625	180	350	70
	400	1125	250	330	80	675	150	330	60
	500	1000	225	380	80	650	120	360	40
Silicone (Silastic 916U)	0	1650	600			1650	600		
	100	1175	460	350	180	875	310	280	70
	200	1225	475	340	150	700	200	260	45
	300	1275	440	370	150	700	160	320	35
	400	1275	480	380	160	575	110	210	35
	500	1175	400	320	130	500	60	190	35
Kel–F 3700	0	4375	430			4375	430		
	100	3025	570	250	180	1750	750	290	160
	200	2175	620	240	190	750	690	130	110
	300	3450	620	640	150	725	400	100	60
	400	3475	650	490	110	400	260	60	30
	500	3350	630	530	170	500	50	20	10
Kel–F 550	0	2500	590			2500	590		
	100	1375	810	140	230	500	1500	50	500
	200	1356	820	140	200	250	1600	50	580
	300	1400	880	140	250	350	850	50	350
	400	1250	810	130	450	325	550	40	140
	500	1250	880	130	270	400	300	10	30

Table 2.135 *(continued)*

	Specific gravity of base	Durometer hardness range	Tensile strength (psi at room temp.)	Elongation (per cent at room temp.)	High temp (max service, °F.)	Low temp (min service, °F.)	Oil resistance	Fuel resistance	Water swell resistance	Tear resistance	Abrasion resistance	Ozone resistance at room temp., 50 ppm stressed;	Weather resistance	Electrical properties	Adhesion to metal	Corrosion of metal	Principal advantages
POLYACRYLATE	1.10	40-100	1000-2200	100-400	300	0	Very good	Fair	Very poor	Fair	Good	Excellent	Very good	Poor	Good	Poor	Oil and heat resistance to 350°F.
POLYISOBUTYLENE	0.92	30-100	1000-3000	100-700	212	—65	Very poor	Very poor	Very good	Very good	Good	Excellent	Very good	Very good	Good	Good	Ozone resistance; good heat resistance
URETHANE	0.85	62-95	1000-8000	100-700	212	—65	Excellent	Very Good	Fair	Excellent	Superior	Excellent	Excellent	Fair	Excellent	Good	Oil resistance; excellent abrasion resistance
EPDM (ethylene propylene)	0.85	30-100	1000-3000	100-300	300	—40	Very poor	Very Poor	Very good	Good	Excellent	Excellent	Very good	Very good	Fair	Excellent	Lightest weight polymer; ozone resistance
FLUORO ELASTOMERS	1.4-1.95	60-90	1000-2400	100-350	450	—40	Excellent	Excellent	Excellent	Fair	Good	Excellent	Excellent	Excellent	Very good	Fair	Maximum resistance to dry heat and oils
CHLOROSULFONATED POLYETHYLENE	1.10	50-95	1000-2800	100-500	250	—65	Good	Fair	Excellent	Good	Excellent	Excellent	Excellent	Fair	Very good	Fair	Ozone resistance; chemical resistance
NATURAL RUBBER	0.93	20-100	1000-4000	100-700	180	—65	Very poor	Very poor	Very good	Very good	Excellent	Poor	Fair	Excellent	Excellent	Excellent	High resilience; good abrasion resistance
POLYCHLOROPRENE	1.23	20-90	1000-4000	100-700	212	—65	Good	Fair	Good	Good	Excellent	Good	Very good	Good	Excellent	Fair	Ozone and oil resistance; heat resistance to 212°F.
NITRILE	1.00	30-100	1000-4000	100-600	250	—65	Very good	Good	Very good	Good	Excellent	Fair	Poor	Fair	Excellent	Fair	Oil resistance to 250°F.
POLYBUTADIENE	1.93	30-100	1000-3000	100-700	212	—80	Very poor	Very poor	Very poor	Very good	Excellent	Poor	Fair	Good	Excellent	Good	Low temperature and good abrasion resistance
POLYISOPRENE	0.94	20-100	1000-4000	100-750	180	—65	Very poor	Very poor	Very good	Very good	Excellent	Poor	Fair	Good	Excellent	Good	Synthetic natural rubber
POLYSULFIDE	1.34	20-80	500-1250	100-400	180	—65	Excellent	Excellent	Very good	Good	Poor	Good	Excellent	Very good	Excellent	Fair	Excellent weather and oil resistance
SBR (styrene butadiene)	0.94	40-100	1000-3500	100-700	225	—65	Very poor	Very poor	Good	Good	Excellent	Fair	Fair	Fair	Excellent	Fair	Good heat resistance to 212°F.
SILICONE	0.98	20-95	500-1500	50-800	450	—120	Fair	Poor	Excellent	Poor	Poor	Excellent	Excellent	Excellent	Excellent	Good	Can be made white with no loss in tensile strength

Table 2.136 Overview guide to selecting elastomers

Type	Polymer and Source	Typical Applications	Useful Temperature Range, °F	Major Characteristic
Non-oil Resistant (swelling or decomposition in presence of oil, fuel, etc.)	Natural rubber	Elastic cord, tires	−65 to 220	High resilience, abrasion resistance, tear strength
	Styrene-Butadiene GRS (thermoset)—various	Tires	−65 to 220	Good abrasion resistance
	Butyl—Exxon	Inner tubes	−65 to 220	Very low gas permeability
	Ethylene-Propylene (EPR and EPDM)—various	Automobile weather stripping, seals	−60 to 300	Excellent weather resistance
	TPEs			
	Styrene-Butadienes: Kraton–Shell Solprene–Phillips	Footwear, toys, household items	−80 to 150	Tough, rubbery materials; 35 to 95 Shore A; inexpensive; can be blended with polyethylene or polypropylene to increase toughness
	Ethylene-Ethyl Acetate (EEA), Ethylene-Vinyl Acetate (EVA)—U.S. Ind. Chem.	Toys, household items, wire insulation, flexible hose, toys, household items, medical devices	−60 to 150	High flexibility; can be clear; can be filled; moderate price
Oil Resistant (slight swelling in oils, fuels, etc.)	Perfluoroelastomer: Kalrez–Du Pont	Seals, O-rings for special applications	−20 to 550	Excellent high-temperature, oil, and chemical resistance; very expensive; 65–80 Shore A
	Fluorocarbon: Viton–Du Pont	Seals, O-rings for high-temperature applications	−10 to 400	Very good resistance to chemical deterioration; poor low-temperature properties; expensive; 60–80 Shore A

Table 2.137 Examples of elastomers' property-to-application

Table 2.137 *(continued)*

Acrylonitrile (Buna N)—various	Hydraulic O-rings, aromatic fuel seals	− 65 to 250	Low swell and good strength in oil, fuel, etc.; good abrasion resistance; inexpensive; 40–80 Shore A
Neoprene (Chloroprene)–Du Pont	Gasoline hose, cable covers	− 20 to 300	High strength, good abrasion resistance, excellent weathering, best general-purpose rubber, 40–95 Shore A, moderate price
Polysulfide–Thiokol Corp.	Sealing material	− 65 to 225	Good resistance to many solvents, low mechanical and thermal properties, 40–85 Shore A, inexpensive
Fluorosilicone–Dow Corning	Seals, O-rings	− 120 to 350	Very good flexibility at low temperatures, good resistance to high temperatures, oils and gasolines, expensive, 50–80 Shore A
Chlorinated Polyethylene (CPE)–Dow Chemical		− 10 to 300	High temperature and combustion resistant, good weathering, 50–95 Shore A, moderate price
Acrylic, Ethylene	Automotive transmission seals	0 to 400	Good chemical resistance, poor low-temperature properties, 55 Shore A, moderate price
Acrylic: Cyanacryl–American Cyanamid Vamac–Du Pont	Seals	− 30 to 250	50 Shore A, moderate price

Polymer and Source	Typical Applications	Useful Temperature Range, °F	Major Characteristic
TPEs			
Polyurethane: Estane–Goodrich, Pellethane–Upjohn, Plastothane–Thiokol Corp., Roylar–Uniroyal, Texin–Mobay, Cyanaprene–American Cyanamid	Seals, cable jacketing, abrasion-resistant parts	−65 to 200	High strength and excellent abrasion resistance, 75 shore A to 60 Shore D, moderate price
Polyester: Hytrel–Du Pont	Couplings, gears, mechanical parts	−65 to 300	Very tough, high-performance material; moderate to high-priced; 40–50 Shore D
Olefinic Rubbers: Somel–Du Pont, Telca–Goodrich, TPR–Uniroyal, Vistaflex–Exxon	Household items, mechanical parts	−60 to 250	Medium hardness and performance materials, 50 Shore A to 50 Shore D, inexpensive
Perfluoroelastomer: Kalrez–Du Pont	Seals, O-rings	−20 to 550	See above
Silicones: Dow Corning, General Electric Co.	Seals, O-rings, etc., not requiring oil resistance	−100 to 450	Available as curable gum stock, RTV materials, and one-part air-curing; excellent weathering and electrical properties; 30–90 Shore A; moderately expensive
Fluorocarbon: Viton–Du Pont, Fluorel–3M Co.	Seals, O-rings, etc., not requiring low-temperature flexibility below −20°F	−20 to 400	Poor low-temperature flexibility, high heat resistance, good weathering and electrical properties, good oil resistance

Table 2.137 *(continued)*

Fluorosilicone: Dow Corning	Seals, O-rings, etc., in contact with oil	−120 to 350	See above, except good low-temperature properties, fair mechanical properties
Acrylic: Cyanacryle–American Cyanamid	Auto transmission seals	0 to 450	See above
TPEs			
Polyester: Hytrel–Du Pont	Mechanical parts	−65 to 300	See above
Ethylene–Propylene: Nordel–Du Pont, Epcar–Goodrich, Vistalon–Exxon	Dust seals, fairings exposed to weather	−65 to 300	Excellent weather resistance, poor oil resistance, good thermal and meachenical properties, resistant to phosphate ester fluids, inexpensive
Chlorosulfonated Polyethylene: Hypalon–Du Pont	Flexible coatings, tarps, etc.	−50 to 300	Excellent weather resistance, need not be black, relatively inexpensive
Silicones and Fluorosilicons–Dow Cornings and G.E. Co.	Fairings, seals, etc. requiring flexibility at low temperatures	−100 to 450	See above, expensive
Fluorocarbons: Viton–Du Pont	Fairings, seals, O-rings, etc., requiring good weather resistance and chemical resistance	−20 to 400	See above, expensive
Neoprene–Du Pont	Cable covers, etc., where good all-around properties are required	−20 to 300	See above, moderately expensive
TPEs			
Polyester: Hytrel–Du Pont	Mechanical parts, seals	−65 to 300	See above
Olefinics: Somel–Du Pont, Telcar–Goodrich, TPR–Uniroyal	Seals, noncritical items	−60 to 250	See above

Table 2.137 *(continued)*

Type	Polymer and Source	Typical Applications	Useful Temperature Range, °F	Major Characteristic
Specialty Rubbers Low gas permeability	Butyl	Seals, O-rings, not in contact with oil	−65 to 250	Best for gas retention, medium physical properties, medium cost
	Fluorocarbons: Viton–Du Pont	Seals, O-rings, in contact with oil and solvents	−20 to 400	See above
High gas permeability	Silicones and Fluorosilicones	"Breathable" diaphragms and seals	−100 to 450	See above
Damping and energy absorbing	Silicone, Fluorosilicone, Chlorobutyl Polyurethane	Shock absorbers, mounts, and vibration dampers	−65 to 300	Very low resilience, "dead" materials, moderate physicals. Silicones and fluorosilicones can be used over wider temperature range than chlorbutyl or polyurethane
Electrically conductive	Silicone filled with electrically conductive particles	RFI shielding	−100 to 400	Conductive materials with fair physical properties
Thermally conductive	Silicone and other polymers filled with thermally conductive material	Conforming gaskets between hot parts and heat sink	−100 to 450	High thermal conduction and conformance to rough surfaces
Room-temperature vulcanizable (RTV)	Silicones	Form-in-place seals, coatings, prototype parts	−120 to 450	Many types differing mainly in hardness and mechanical properties
Thermoplastic elastomers TPEs	Styrene-Butadiene, Ethylene Polymers, Polyesters Urethanes, Olefinic Elastomers	Various parts not to be operated at temperatures over 120–140°F (except polyester, 300°F)	——	Thermoplastics, injection moldable at production rates. Used for high production of light-colored or transparent parts. Polyesters for high-strength harder parts, urethanes for abrasion resistance
Extremely low Shore	Polynorborene: Norsorex–American Cyanamid	Very soft dampers, etc.	−65 to 160	Very soft to hard, good mechanical properties

Table 2.137 *(continued)*

Rubber	Advantages	Disadvantages	Typical Applications
Natural rubber (NR)	Building tack, resilience, and flex resistance	Reversion at high molding temperatures	Tires, engine mounts
Styrene butadiene rubber (SBR)	Abrasion resistance	Poor ozone resistance	Tires, general molded goods
Ethylene propylene diene monomer (EPDM)	Good ozone resistance	Poor hot-tear resistance	Door and window seals, wire insulation
Nitrile butadiene rubber (NBR) or nitrile	Good solvent resistance	Poor building tack	O-rings and hose
Thermoplastic rubber	Short injection molding cycle (thin parts)	Poor creep characteristics	Shoe soles, wire insulation
Polyurethane	Short molding cycle (RIM) and low molding pressure	Adhesion to mold	Cushioning, rolls, exterior automotive parts
Isobutylene isoprene rubber (IIR) or butyl	Low air permeation in finished parts	Voids caused by trapped air during molding	Inner tubes, body mounts for automobiles
Chloroprene rubber (CR) or neoprene	Moderate solvent resistance	Sticking during processing and premature crosslinking (scorch) with some types	Hose tubes and covers, V belts

Table 2.138 Examples of general performances and applications for elastomers

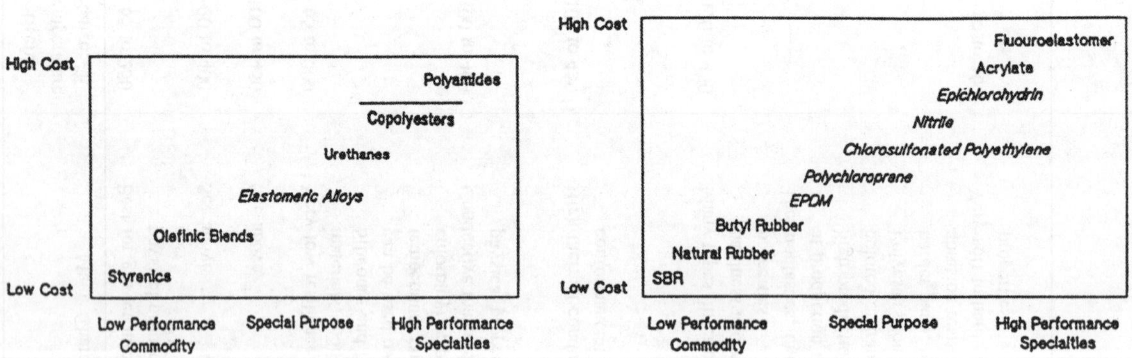

Table 2.139 Comparison of properties and costs of TP and TS elastomers

and continued availability of material. Solid waste volume reduction is an important consideration. This is the decreasing of the volume of solid waste through compaction or incineration. A 50% to 80% reduction is possible through compaction, 90% to 98% through incineration.

In the industry, *fines* are the pieces that are substantially smaller than the bulk of the regrind that fall through the granulator screen. Fines are bigger than dust but smaller than regrind. Too many fines can cause feeding and processing problems. There are also *longs*, which are oversized raw materials that can result from loose or broken granulating screens.

Desired Modification	Type Fiber	Gains or Losses		Performance
		Amorphous	Crystalline	
Increased Tensile Strength	Glass fibers Carbon fibers	Ductility, cost Ductility, cost	Ductility, cost Ductility, cost Ductility	Glass fibers provides cost-effective way of gaining tensile strength. Carbon fibers are more expensive; fibrous minerals are least expensive but only slightly reinforcing. Reinforcement makes brittle resins tougher and embrittles tough resins.
Increased Heat-Deflection Temperature (HDT)	Glass fibers Carbon fibers Fibrous minerals	Ductility, cost Ductility, cost	Ductility, cost Ductility, cost Ductility	When reinforced, crystalline polymers yield much greater increases in HDT than do amorphous resins. As with tensile strength, fibrous minerals increase HDT only slightly. Fillers do not increase HDT.
Warpage Resistance	5 to 10% glass fibers 5 to 10% carbon fibers Particulate fillers	 Ductility, cost, tensile strength	Cost Cost Ductility, cost, tensile strength	Amorphous polymers are inherently nonwarping molding resins. Only occasionally are fillers such as milled glass or glass beads added to amorphous materials, because they reduce shrinkage anisotropically.

(Continues)

Table 2.140 Properties of reinforced amorphous and crystalline thermoplastics

When fiber RPs are granulated, the lengths of the fibers are reduced. Upon reprocessing, with virgin materials or alone, their processability and performance definitely change. Thus, it is important to determine if the change will affect final product performances.

Profitable business opportunities are gradually developing to produce many different products from recycled plastics, such as boat docks, park benches, detergent bottles, office equipment, highway barriers, wastebaskets, pallet strapping, toolboxes, fast-food trays, wetland walkways, signs, hampers, carpeting, irrigation pipes, and many others. Figure 2.49 is an example of a product

Desired Modification	Type Fiber	Gains or Losses: Amorphous	Gains or Losses: Crystalline	Performance
Reduced Mold Shrinkage (Increased mold-to-size capability)	Glass fibers Carbon fibers	Ductility, cost Ductility, cost	Ductility, cost Ductility, cost	Reinforcement reduces shrinkage far more than fillers do. Fillers help balance shrinkage, however, because they replace shrinking polymer. The sharp shrinkage reduction in reinforced crystalline resins can often lead to warpage. The best "mold-to-size". composites are reinforced amorphous composites.
Reduced Coefficient of Friction	PTFE Silicone MoSe Graphite	Cost	Cost	These fillers are soft and do not dramatically affect mechanical properties. PTFE loadings commonly range from 5 to 20 wt% the others are usually 5% or less. Higher loadings can cause mechanical degradation.
Electrical Conductivity	Carbon fibers Carbon powders	Ductility, cost Tensile strength, ductility, cost	Ductility, cost Tensile strength, ductility, cost	Resistivities of 1 to 100,000 ohm-cm can be achieved and are proportional to cost. Various carbon fibers and powders are available with wide variations in conductivity yields in composites.

Table 2.140 *(continued)*

Reinforcing material	Specific gravity	Tensile Strength 10³ psi	Tensile modulus 10⁶ psi	Compressive strength, 10³ psi	Flexural strength, 10³ psi	Izod impact strength, ft-lb/in. notch	Heat resistance continuous, °F	Arc resistance, sec.
Polyester								
Glass cloth	1.5–2.1	30–70	1–3	25–50	40–90	5–30	300–350	60–120
Glass mat	1.3–2.3	20–25	$\frac{1}{2}$–2	15–50	25–40	2–10	300–350	120–180
Asbestos	1.6–1.9	30–60	1–3	30–50	50–70	2–8	300–450	100–140
Paper	1.2–1.5	6–14	$\frac{1}{2}$–1$\frac{1}{2}$	20–25	13–28	1–2	220–250	28–75
Cotton cloth	1.2–1.4	7–9	$\frac{1}{2}$–1$\frac{1}{2}$	23–24	13–18	1–4	230–250	70–85
Epoxy								
Glass cloth	1.9–2.0	20–60	2–4	50–70	70–100	11–26	330–500	100–110
Glass mat	1.8–2.0	14–30	1–3	30–38	20–26	8–15	330–500	110–125
Paper	1.4–1.5	10–19	$\frac{1}{2}$–1	20–28	19–24	$\frac{1}{2}$–1	260–300	30–100
Phenolic								
Glass cloth	1.8–2.0	40–60	1–3	35–40	65–95	10–35	350–500	20–130
Glass mat	1.7–1.9	5–20		17–26	10–60	8–16	350–500	400–150
Asbestos	1.7–1.9	40–65	2–5	45–55	50–90	1–6	350–600	120–200
Paper	1.3–1.4	8–20	1–2	20–40	10–30	$\frac{1}{3}$–1	225–250	Tracks
Cotton cloth	1.3–1.4	7–16	$\frac{1}{2}$–1$\frac{1}{2}$	30–44	14–30	$\frac{1}{2}$–3	225–250	Tracks
Nylon cloth	1.1–1.2	5–10	$\frac{1}{4}$–$\frac{1}{2}$	28–36	9–22	2–4	150–165	Tracks
Silicone								
Glass cloth	1.6–1.9	10–35	1–2	25–46	10–38	5–13	400–700	150–250
Asbestos cloth	1.7–1.8	10–25	1–2	40–50	12–20	6–9	450–730	150–300

Table 2.141 Properties of thermoset-reinforced plastics per ASTM tests

Material composition	Phenolic (25%), phenolic microspheres (25%), nylon powder (50%)	Epoxy–Novolac (38%), phenolic microspheres (44%), silica fibers (9%), glass fibers (9%)	Silicone (75%), microspheres phenolic (10%), silica (15%)	Silicone (66%), microspheres phenolic (16%), glass (10%), quartz fibers (7%), silica powder (1%)	Plasticized phenolic (12%), granular cork (78%)
Density, lb/ft^3	37	24	37	42	30
Tensile strength, psi					
—100°F	1180	1550	275	544	—
75°F	1000	640	72	167	300
300°F	925	300	109	131	—
Strain at failure, psi					
—100°F	0.8	1.1	6.5	8.6	—
75°F	0.9	2.9	10.0	5.0	16.0
300°F	1.7	0.8	15.0	4.5	—
Elastic modulus, psi					
—100°F	150,000	165,000	23,000	32,000	—
75°F	125,000	40,000	2100	5500	6200
300°F	49,000	36,000	3400	4000	—
Specific heat, Btu/lb–°F	0.38	0.46	0.35	0.37	0.47
Thermal conductivity, Btu(hr)(ft^2)(°F/ft)	0.07	0.04	0.06	0.07	0.05
Thermal expansion coefficient, 10^{-5} in./(in.)(°F)	3.05	1.71	5.91	4.40	5.10

Table 2.142 Properties of thermoset-reinforced plastics with different reinforcements

made from recycled plastic. It is Goodyear's two-piece suit and matching tie made in 1978 from recycled 2-L polyethylene terephthalate (PET) beverage bottles. In 1990, it was donated to the then new Ripley's Believe It or Not Museum in Wisconsin Dells, Wisconsin.

Different organizations in the United States and worldwide have developed identification systems for recycled plastics. Examples are shown in Tables 2.147 and 2.148.

RECYCLE DEFINITION

Different definitions exist to meet different industry and commercial requirements. ASTM defines *recycled plastics* as those plastics composed of postconsumer material and/or recovered material that may or may not have been subjected to additional processing steps used to make products such as recycled regrind or processed or reconstituted plastics. *Industry scrap* includes what is commonly referred to as *trim* or *regrind* in plastic production; this is not considered recycled material.

Recycling commingled plastics are plastics not sorted by type in a waste system. The unsorted plastics are combined or blended into one harmonious material.

Postconsumer material identifies those products generated by businesses or consumers that have served their intended end uses. They have been separated or diverted from solid waste for the purposes of collection, recycling, and disposition.

Postconsumer plastic color-measurement technology control is used in recycling of postconsumer plastics (PCPs). As the demand for consistent colors of recycled PET and HDPE increases, the need for standardization and codification of color measurement based on analytical data continues to be critical. In the past, the human eye was the only means of determining color quality.

Test conditions	Flexural modulus (10^6 psi)		
	Polyester	Epoxy	Phenolic
Atmospheric pressure and room temperature	4.67	3.54	4.17
Low pressure (10^{-6} torr) at room temperature	4.67	3.54	4.17
Low pressure (10^{-6} torr), ultraviolet (0.036 W/cm^2) and moderate temperature (250°F)	3.84	2.14	3.94
Atmospheric pressure, ultraviolet (0.036 W/cm^2) and and moderate temperature (250°F)	3.38	2.00	3.86
Atmospheric pressure, ultraviolet (0.036 W/cm^2) and and room temperature	3.75	3.54	3.68
Low pressure (10^{-6} torr) and moderate temperature (250°F)	3.61	2.38	3.96
Atmospheric pressure and moderate temperature (250°F)	3.75	2.25	3.81

Table 2.143 Flexural modulus of glass-fiber-thermoset-reinforced plastics exposed to various environments

	Ultimate tensile strength (10^3 psi)				Ultimate flexure strength (10^3 psi)				Ultimate compression strength (10^3 psi)			
Resin material	70°F	−110°F	−320°F	−423°F	70°F	−110°F	−320°F	−423°F	70°F	−110°F	−320°F	−423°F
Epoxy												
Epon 828	40.9	62.3	91.6	97.9	77.9	99.3	150.4	145.0	43.5	85.4	109.9	108.8
Epon 1001	49.0	71.4	94.9	100.7	83.2	105.2	154.0	122.0	58.6	86.2	89.4	78.9
Phenolic												
Narmco 506	38.1	58.3	74.4	72.2	63.4	76.2	84.5	83.0	46.5	55.1	81.8	76.7
CTL 91-LD	51.7	65.7	65.1	64.9	77.4	100.1	109.0	110.0	70.4	88.1	99.1	91.6
Polyester												
Hetron 92	39.2	63.2	71.7	84.9	71.4	82.1	87.2	81.9	22.9	35.8	43.6	41.4
Paraplex P43	47.9	58.5	81.3	78.2	68.8	79.4	84.2	84.9	31.5	43.2	51.3	47.9
Triallylcyanurate												
Laminac 4232	41.0	60.7	56.9	55.5	63.3	74.7	70.0	65.4	28.4	42.0	47.6	44.1
Vibrin 135	32.7	57.6	52.0	54.6	52.0	76.2	78.1	75.1	26.3	33.5	38.1	36.1
Silicone												
Narmco 513	29.6	47.8	76.3	80.5	34.1	40.7	52.1	52.8	25.3	28.7	39.1	43.3
Trevarno F130	25.7	46.7	67.9	74.6	36.1	43.5	63.7	72.8	17.7	23.6	34.9	30.8
Epoxy												
Epon 828	3.51	3.84	4.15	4.54	3.54	3.96	5.04	4.66	3.35	3.33	4.07	4.26
Epon 1001	3.34	3.65	3.93	4.26	4.22	3.78		3.76	3.75	3.30	3.90	3.96
Phenolic												
Narmco 506	3.65	3.88	4.74	3.83	3.56	3.61		4.06	4.39	7.46	4.5	5.12
CTL 91–LD	3.85	4.31	4.25	4.48	4.32	4.45		5.18	4.69	3.61	8.43	4.48
Polyester												
Hetron 92	2.46	3.50			3.48	3.52	3.04	2.96	1.57	1.65	2.48	2.28
Paraplex P43	2.45	3.30		3.18	3.17	3.76	3.39	2.96	2.75	3.18	3.78	3.17
Triallylcyanurate												
Laminac 4232	3.08	3.38	3.67	3.85	3.29	3.82	4.37	4.30	3.14	2.76	3.09	3.25
Vibrin 135	3.36	3.50	3.72	3.76	3.58	4.01	4.06	4.00	1.79	1.99	2.45	2.54
Silicone												
Narmco 513	2.73	2.73	2.75	3.09	2.85	3.18	3.19	3.08	2.72	2.50	4.61	3.28
Trevarno F130	2.52	2.66	2.58	3.34	2.69	2.72	2.85	3.32	2.06	2.15	2.41	2.44

Table 2.144 Strength and moduli for some glass-fiber laminates at low temperatures

Samples were graded A, B, or C, and judgment calls were the standard of the day. But today's recycled-plastic customers and fabricators have very specific color needs that must be met. Visual standards are not satisfactory; standardization and codifying provide credibility and professionalism to an increasingly sophisticated industry.

Colorimetric instruments provide benchmarks of quality in recycled clear and colored PCPs. The colorimeter is also used to meet other requirements, such as determining the most effective cleaning agents for PCPs. Because it is primarily dirt that gives recycled plastic an unattractive appearance, the portable colorimeter was a significant advance in determining the dirt levels of raw materials during the recycling process. These measurements provide early data on PCP appearance prior to pelletizing.

PLASTIC SELECTION

Any attempt to compare plastic with other conventional materials (metal, wood, glass, etc.) on a straight property-for-property or a straight cost-for-cost basis is doomed to failure from the very start. There are just too many different types of grades and formulations grouped under the overall heading of plastics, just as there are a wide variety of grades grouped under the category of woods or metals, to make any such comparisons valid.

About all one can be certain of is the fact that the technology of plastics has become so sophisticated that plastics are virtually the most versatile group of materials available to industries

Resin material	Property	Temperature	Gamma ray dose (rad) 0	1×10^8	1.2×10^8	5×10^8
Phenolic Conolon 506 [a]	Ultimate tensile strength (10^3 psi)	Ambient	43	42		42
		—320°F	77		78	79
		—423°F	67		93	
	% Total elongation	Ambient	1.7	1.8		1.7
		—320°F	3.5		3.5	3.8
		—423°F	3.8		3.7	
Polyester (Paraplex P–43) [b]	Ultimate tensile strength (10^3 psi)	Ambient	28	32		31
		—320°F	48		46	53
		—423°F	41		42	
	% Total elongation	Ambient	1.5	1.8		1.7
		—320°F	2.8		2.3	2.5
		—423°F	2.2		2.2	

[a] Manufactured by Narmco, Inc.
[b] Manufactured by Rohm and Haas Co.

Table 2.145 Mechanical properties of glass-fiber-reinforced plastics after irradiation at elevated temperatures

	Room temperature tensile strength (10^3 psi)	Room temperature compressive strength (10^3 psi)	Ultimate flexural strength (10^3 psi)				Flexural modulus (10^3 psi)		
			At room temperture	After 30-day water immersion	At 300°F	At 500°F	At room temperature	At 300°F	At 500°F
Fabric Laminates									
Epoxy-glass									
Bisphenol A	50	55	85–89	65–80	40–60	8–11	3600–3800	2900–3300	3200
Novalac	42	57	85–90		58–63	25	3200–3400	2700	
Peracetic acid	57–59	49–53	85–90		60–65	47	4000–4200	3700–3900	
Bisphenol A and high-modulus glass	64	61	90	76	44		5000	3700	
High-temperature epoxy and high-modulus glass	55	59	86			23	5300		3000
Phenolic-glass									
Standard grade	50	35	50			40	3000		3000
Heat-resistant	65	48	86	52	78	47–52	4800	4000	3300–3600
Silane-modified	46	51	80	73	65	65	4300	3500	3500
Phenolic-leached glass	14	24	25		20		2600	2600	2600
Phenolic-quartz	35		68		37	50	3700	3400	3200
Polyesters									
Styrene-modified	50–55	40–45	50–75	64	30	24	3300	2500	2000
TAC-modified	59	51	65	59	43	30	3000	2000	1800
Styrene-modified and high-modulus glass		52	59	49			4200		
Silicone-glass	41	29	50	36	23	21	2900	2400	2200
Filament-Wound Structures									
Epoxy-glass	270	70	271		245				
Epoxy and high-tensile glass	285								
Mil P–27327 (proposed)	200	50	220	210	200	180	7000	6800	6800
Molding Compounds									
Phenolic-glass	10	35	20		16.4		3300		
Phenolic-leached-glass	8.5	36	14				3000		
Phenolic-asbestos	9	24	24			15	2000		1400
Silicone-glass	4.5	16	14		6	6	2100	1100	500
Polyester-glass	9.5	29.5	19		12.7	6.7	1400	910	910

Table 2.146 Properties of reinforced plastics at ambient and elevated temperatures

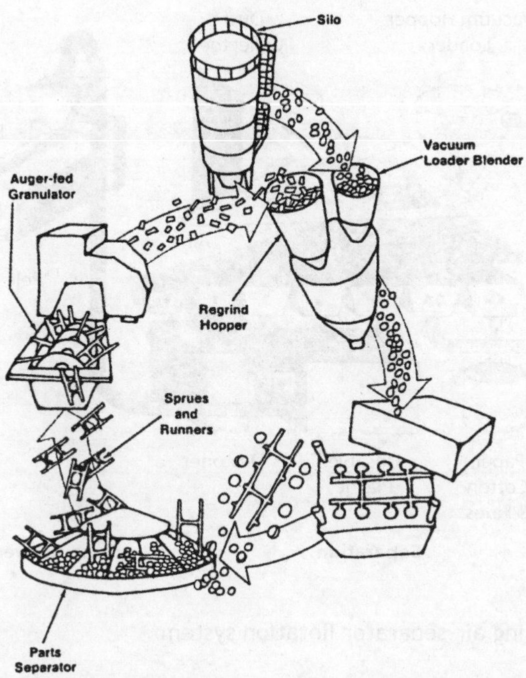

Figure 2.44 Recycling plastic scrap

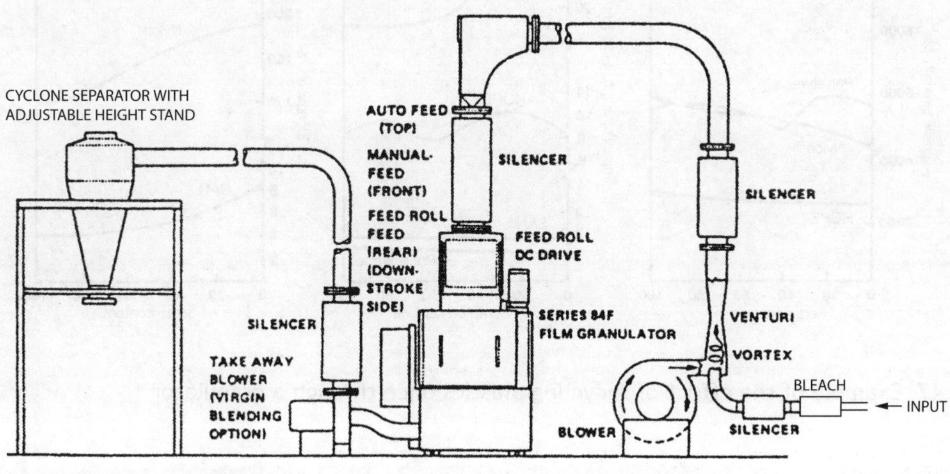

Figure 2.45 Recycling plastic film

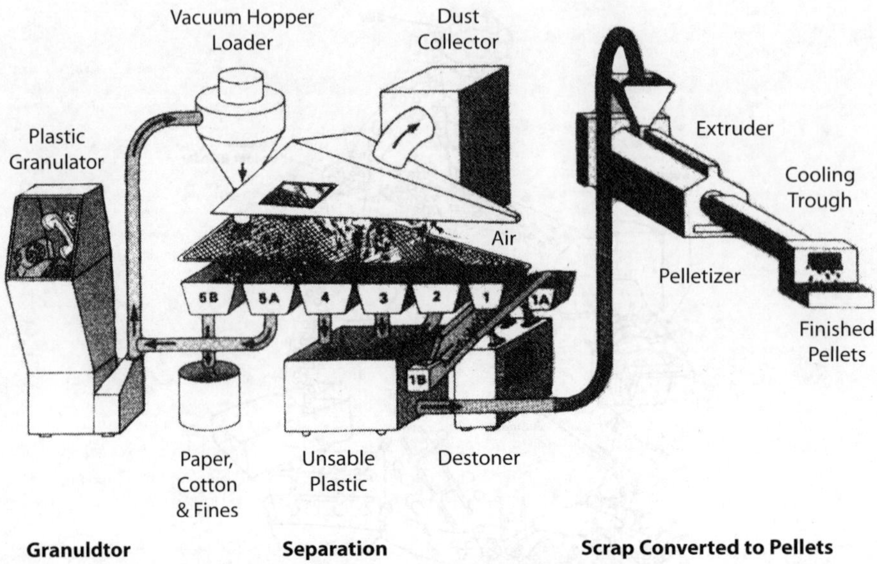

Figure 2.46 ABS recycled using air-separator flotation system

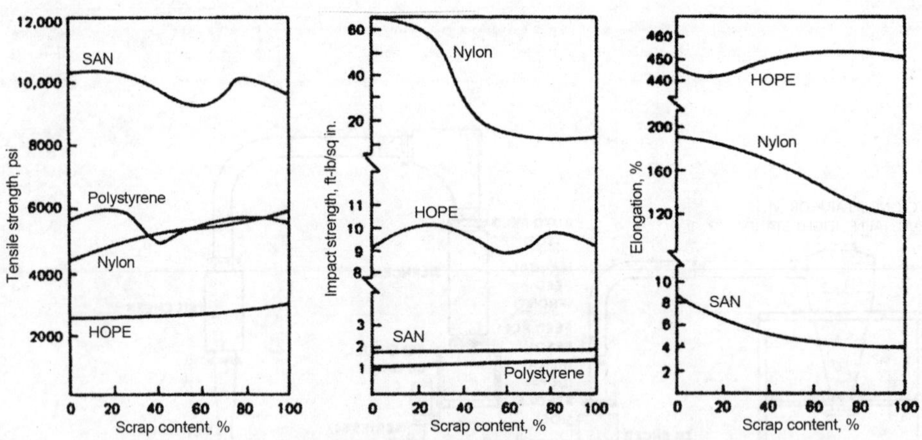

Figure 2.47 Example of the effect of recycling plastics once through a granulator

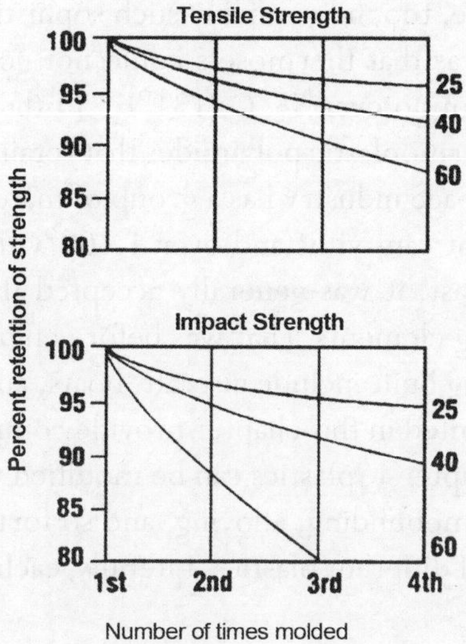

Figure 2.48 Examples of the effect of recycling plastics more than once through a granulator where the mix of virgin plastic is with wt% of regrind

Figure 2.49 Suit and matching tie made from recycled PET bottles (courtesy of Goodyear)

today. It has been interesting to note, too, how quickly such sophistication has taken place. Almost a century ago, the rule of thumb was that thermosets could not go much over 200°C (392°F) in service and thermoplastics could not go over 55°C (131°F). In the meantime, many new plastics have been developed, such as the thermoplastic polyimides that retain properties at temperatures up to 400°C (750°F) (266). The aerospace industry has a group of materials known as ablative plastics (based on reinforced phenolics) that can withstand over 1,100°C (2,000°F) for short periods of time (less than a second). In the past, it was generally accepted that plastics could not compete with other materials as load-bearing elements. That was before engineering designs were properly applied; plastic structures now being built include aircraft, boats, and bridge products.

The information and data presented in this chapter provide comparative guides about the properties of plastics. As reviewed in chapter 1, plastics can be modified to meet all kinds of properties, performances, and processes by compounding, alloying, and so forth. Figure 2.50 provides a simplified summary of the properties of different plastics. Literally, each of the plastics can be modified to exist in nearly in the diagram.

Industry plastic classification systems, such as ASTM D 4,000, are shown in Tables 2.149 to 2.153. There are also ISO-1043 and others to meet different industry requirements. They provide a way to identify plastics for purchasing, quality control, and other needs. Plastics are classified according to their origin and method of synthesis as well as their place in systematic group categories, such as material type, modulus, and rigidity.

The first step in selecting a plastic for a product to be fabricated is to determine the product's complete requirements. This is important; there could be a tendency to overlook certain properties because they may appear to be insignificant, such as ensuring that the product will perform during packaging, shipping, and/or service. Selecting an optimal material for a given product must be based on analyzing the requirements to be met. A simplified approach involves comparing

Code	Abbreviation	Material
1	PETE	Polyethylene terephthalate (PET)
2	HDPE	High-density polyethylene
3	V	Vinyl/polyvinyl chloride (PVC)
4	LDPE	Low-density polyethylene
5	PP	Polypropylene
6	PS	Polystyrene
7	Other	All other resins and layered multimaterial

Table 2.147 Bottle and container code plastic identification system

the specific service requirements to the potential properties of a plastic. What follows is a simplified but practical material-selection approach. This "longhand" system has been used for almost a half century, during which time it has become the basis of fast computerized software material-selection databases.

In a simplified approach, one starts by selecting the design criteria as well as potential plastics of interest, incorporating them into a table format (Table 2.154) and checking off only the major

G	Granulate
M	Mechanically processed molded material, without after-treatment (bill base)
T	Thermally processed molded material, not thermally degraded or disintegrated
GM	Granulate obtained by mechanical processing (i.e., a regenerate or regranulated)
GT	Granulate made of molten, extruded, and then regranulated molded material
I	Of one grade and quasi identical with the original molded compound; quasi identical because the molecular and reinforcing structures may differ from the new material as a result of at least one processing operation
N	Not of the same grade as the original molded compound
X	Molded compound of an unknown category
20	20% admixed recyclate in the new material
100	No admixture, but exclusively mechanically or thermally processed and regranulated molded compound
1	First processing
X	Unknown number of processing operations

Table 2.148 Coding system for recycled plastics

criteria across the worksheet. Follow by setting up a comparison of the performance requirements for the plastics being considered (Table 2.155) and transferring the bold-faced numerical rating in each selected criteria column to the worktable. Add these numbers across the worktable to determine the plastic group with the lowest point subtotal; that will be the best type of plastic for a given application on a performance basis. Next, add in the cost factor and total it to find the plastic group with the lowest number; that will be the best choice based on a cost-performance evaluation.

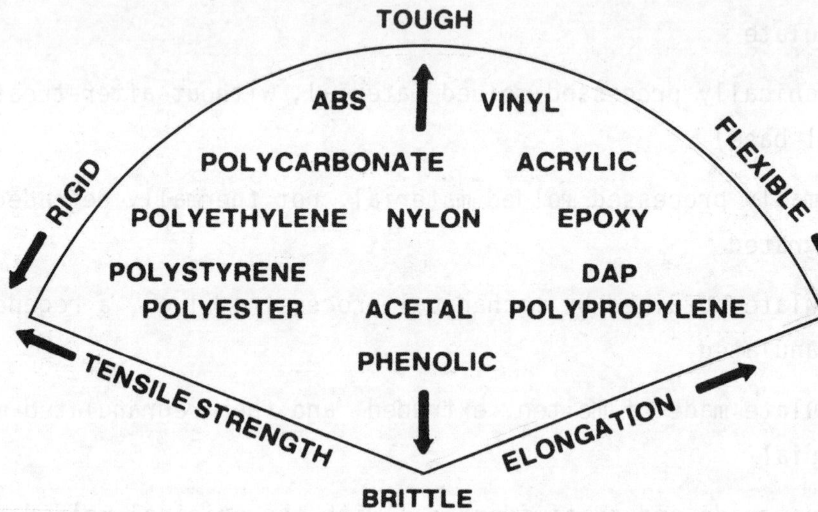

Figure 2.50 With modifications, each of these plastics can meet different requirements and thus be moved into literally any position in the diagram

ASTM D 4000 PA122 G33 A65480 Hxx Fxx

Material (plastic)	Basic Material						Special Material Requirements							Special Material for Application			
	Broad generic type	Specific type	Class	Grade	Filter type	% Filler	Cell table	Tensile strength	Flexural modulus	Izod Impact	Heat deflection temperature	TBD*	Suffix	Special test requirements	Suffix	Special test requirement
ASTM D4000	PA	1	2	2	G	33	A	6	5	4	8	0	H	xx	F	xx

Table 2.149 Classification of plastics (ASTM D 4000)

Follow by determining the specific plastic within the plastic group selected. The plastic with the lowest final total will be the best for the application on a cost-performance basis. Tables 2.156 and 2.157 are examples of the selection of plastics for a chain saw and pump.

ASTM D 1600 Standard Plastics Abbreviations

Term	Abbreviation	Term	Abbreviation
Cellulose propionate	CP	Polyethylene terephtalate	PETP
Chorinated poly(vinyl chloride)	CPVC	Poly(methyl-α-chloroacrylate)	PMCA
Cresoi-formaldehyde	CF	Poly(methyl methacrylate)	PMMA
		Polyoxymethylene, polyacetal	POM
Diallyl phthalate	PDAP	Polypropylene	PP
		Polystyrene	PS
Epoxy, epoxide	EP	Polytetrafluoroethylene	PTFE
Ethyl cellulose	EC	Polyurethane	PUR
		Poly(vinyl acetate)	PVAC
Melamine-formaldehyde	MF	Poly(vinyl alcohol)	PVAL
		Poly(vinyl butyral)	PVB
Perfluoro(ethylene-propylene) copolymer	FEP	Poly(vinyl chloride)	PVC
		Poly(vinyl chloride-acetate)	PVCA
Phenol-formaldehyde	PF	Poly(vinyl fluoride)	PVF
Poly(acrylic acid)	PAA	Poly(vinyl formal)	PVFM
Polyacrylontirle	PAN		
Polyamide (nylon)	PA	Silicone plastics	SI
Polybutadiene-acrylonitirle	PBAN	Styrene-acrylonitrile	SAN
Polybutadiend-styrene	PBS	Styrene-butadiene	SB
Polycarbonate	PC	Styrene-rubber plastics	SRP
Poly(dially phthalate)	PDAP		
Polyethylene	PE	Urea-formaldehyde	UF

Table 2.150 Examples of symbols for the families of plastic

Symbol	Material	Tolerance
C	Carbon and graphite fiber-reinforced	± 2 Percentage points
G	Glass reinforced	± 2 Pertange points
L	Lubricants (i.e., TFE, graphite, silicone, and molybdenum disulfide)	By agreement between the supplier and the user
M	Mineral reinforced	± 2 Pertange points
R	Reinforced-combinations/mixtures or reinforcements or other fillers/reinforcements	± 3 Percentage points (based on the total reinforcement)

Table 2.151 Additive, filler, and reinforcement symbols with tolerances

SELECTION APPROACH

The physical, mechanical, electrical, and/or other properties of plastics enable them to perform in a wide variety of end uses and environments, often at a lower cost than other design materials, such as metal or wood (Table 2.158). However, in order to take the fullest advantage of plastics when producing products, the fabricators and buyers of plastic products must insist upon careful design and material selection (Table 2.159) (197).

The final design of the article will often hinge on what is the best fabricating process for the product in question (Tables 2.160 and 2.161). Sometimes the necessity for certain elements in the design, such as thin sections, long delicate inserts, exact concentricity, or extremely high accuracy of dimensions, makes it desirable to use one technique of fabricating rather than another. Therefore, it is frequently necessary to reverse the usual design procedure and determine the material to be used and the specific process and technique to be employed before designing the product in question.

As reviewed in chapter 1, plastic compounding by alloying produces plastics with different properties (Tables 2.162 to 2.166). Special compounds or alloys can be prepared to meet specific properties when a plastic is not available. Cost would be reasonable if the quantity was sufficient, or it would be high if a limited amount was required. In some cases, high cost is warranted, as in injection molding (and other processes) in which the cost of a mold to produce just a few parts is high.

Much of the market success or failure of a plastic article can be attributed to the initial choice of material. Although there are literally thousands of plastics available, usually no single one will exhibit all the desired properties in their proper relationships. Therefore, a compromise among properties, cost, and manufacturing process generally determines the material of construction. Selecting a plastic is very similar to selecting a metal. Even within one class, plastics differ because of varying formulations, just as steel compositions vary (tool steel or stainless steel, for example). Many developments in industries, such as electronics, transportation, packaging, and domestic goods, have been expanded because of the availability of suitable plastics.

CELL A	0	1	2	3	4	5	6	7	8	9
Tensile strength D-638 MPa (psi)	UNSPECIFIED	96.3 (14000)	110 (16000)	123.8 (180000)	137.5 (20000)	151.3 (22000)	165 (24000)	192.5 (28000)	206.3 (30000)	220 (32000)
Flexural modulus D-790 GPa (10^6 psi)		3.44 (0.5)	4.82 (0.7)	5.51 (0.8)	6.88 (1.0)	7.57 (1.1)	8.26 (1.2)	8.95 (1.3)	9.64 (1.4)	11.0 (1.6)
Izod impact (notched) D-256 J/m (ft-lb/in.)		42.7 (0.8)	53.4 (1.0)	85.4 (1.6)	96.1 (1.8)	106.8 (2.0)	117.4 (2.2)	128.1 (2.4)	138.8 (2.6)	149.5 (2.8)
Heat deflection temperature (264 psi) D-648°C (°F)		138 (280)	143 (290)	149 (300)	160 (320)	193 (380)	204 (400)	227 (440)	238 (460)	249 (480)

Table 2.152 Example of an ASTM D 4000 cell table

As an initial step, the product designer must anticipate the conditions of use and the performance requirements of the product. Factors to consider include chemical resistance, permeability, life expectancy, size, conditions of use, shape, strength, stiffness, environment, color, and/or other conditions. These end-use requirements can be ascertained through market analysis, surveys, examinations of similar products, testing, and general experience. A clear definition of product requirements will lead directly to the choice of material of construction (and process to be used).

There is a logical, workable elimination approach to the selection of the correct plastic. Examples of specific properties have been reviewed in this chapter and will be in other chapters.

Type	Class	Grade									
Desc-ription	*Desc-ription	**Desc-ription	Relative Viscosity D 789 min	Melting point D 789°C	Specific gravity D 792	Flexural modulus D 790 MPa min.	Tensile strength D 638 MPa min	Elong. (ult.) D 638 /min.	Izod impact (notched) D 256 J/m min.	Heat defl. temp. D 648 455 kPa °C min.	1820 kPa °C min.
6:6 Nylon	1 GP	1 LV	35	250-260	1.13-1.15	2600	76	10	40	180	65
		2 MV	45	250-260	1.13-1.15	2600	76	50	50	180	65
		3 HV	60	250-260	1.13-1.15	2600	76	50	50	180	65
		4 E	120	250-260	1.13-1.15	2600	76	50	50	180	65
		5 BM	300	250-260	1.13-1.15	2600	76	50	50	180	65
		0 Other									
	2 HS	1 LV	35	250-260	1.13-1.15	2600	76	10	40	180	65
		2 MV	45	250-260	1.13-1.15	2600	76	40	50	180	65
		3 HV	60	250-260	1.13-1.15	2600	76	40	50	180	65
		4 E	120	250-260	1.13-1.15	2600	76	40	50	180	65
		5 BM	300	250-260	1.13-1.15	2600	76	40	50	180	65
		0 Other									
	3 N	1 LV	35	250-260	1.13-1.15	2800	90	5	40	185	70
		2 MV	45	250-260	1.13-1,15	2800	90	5	40	185	70
		3 HV	60	250-260	1.13-1.15	2800	90	5	40	185	70
		4 E	120	250-260	1.13-1.15	2800	90	5	40	185	70
		5 BM	300	250-260	1.13-1.15	2800	90	5	40	185	70
		0 Other									
	4 IM	1 MI	–	250-260	1.09-1.11	1700	58	55	150	145	50
		2 HI	–	250-260	1.09-1.11	1500	48	50	800	145	55
		0 Other									
	5 IMHS	1 MI	–	250-260	1.09-1.11	1700	58	55	130	145	50
		0 Other									
	6 WS	1 MV	45	250-260	1.14-1.16	2700	80	20	40	180	65
		0 Other									
	0 O(s)	0 Other									
2 6 Nylon	1 GP	1 LV	30	210-225	1.12-1.14	2600	75	40	50	–	65
		2 MV	40	210-225	1.12-1.14	2500	75	40	50	–	65
		3 HV	50	210-225	1.12-1.14	2500	75	100	50	–	65

*GP = General Purpose; HS = Heat Stabilized; N= Nucleated; IM = Impact Modified; IMHS = Impact Modified Heat Stablized; WS = Weather Stabilized; O(s) = Other (stabilized)

**LV = Low Viscosity; MV = Medium Viscosity; HV = High Viscosity; E = Extrusion; BM = Blow Molding; MI = Medium Impact; HI = High Impact

Table 2.153 Example of the data developed based on using ASTM D 4000

What follows is a summary of certain properties. They are listed not in any order of priority, but rather on the requirements of the product to be fabricated.

CHEMICAL RESISTANCE

Most plastic producers have developed long-term data for commonly used chemicals. Some plastics, such as HDPE, are immune to almost all commonly found solvents. PTFE (polytetrafluoroethylene) in particular is noted principally for its resistance to practically all chemical substances. It has been generally identified as the most inert material known worldwide.

Others exhibiting good chemical resistance are all of the fluorocarbon plastics, ethylpentene, polyolefins, some phenolics, and diallyl phthalate compounds. Chlorinated polyether is formulated particularly for products requiring good chemical resistance. Additives, such as fillers, plasticizers, stabilizers, colorants, and catalysts, can increase the chemical resistance of unfilled plastics (Tables 2.167 to 2.171 and Fig. 2.51). Careful tests must be made under actual use conditions in final selection studies, and the use of the required amount and proper blending of fillers, and so forth, must be ensured.

Material Characteristics	Strength and Stiffness	Toughness	Short-Term Heat Resistance	Long-Term Heat Resistance	Environmental Resistance	Dimensional Accuracy in Molding	Dimensional Stability	Wear and Frictional Properties	Point Subtotal	Cost	Point Total
G/R \ Resing Groups											
Styrenics ABS SAN Polystyrene											
Olefins Polyethylene Polypropylene											
Other Crystallines Resins Nylons 6 6/6 6/10, 6/12 Polyacetal											
Acylates Modified Polycarbonate Polusulfone Polythersulfone											
High Temp. Resins PPS Polyamide-imids											
Fluorocarbons FEP ETFE											

Ratings: 1—most desirable; 6—least desirable; Large numbers indicate group classification; small numbers the specific resins within that group.

Table 2.154 Worktable format related to requirements

Extensive and important data is available concerning the chemical resistance (of the properties and characteristics) of plastics and elastomers (Tables 2.172b and 2.173b). A detailed review prepared by the Plastics Design Library (PDL) is provided in their two books (totaling 2,070 pages) *Chemical Resistance: Volume I—Thermoplastics* and *Volume II—Thermoplastic Elastomers, Thermosets, and Rubbers*. They identify materials by specific trade names and manufacturers (91).

These books put a wealth of information at your fingertips. They represent an exhaustive compilation of data on how chemicals and other exposure environments affect the properties and characteristics of thermoplastics, thermoplastic elastomers, thermosets, and rubber materials. How a plastic will perform in its end-use environment is a critical consideration, and the information presented here gives useful guidelines. However, this or any other informational resource should not serve as a substitute for actual testing in determining the applicability of a given part or material in a given end-use environment.

Volume I—Thermoplastics of the Handbook Series is an update to the original *Chemical Compatibility and Environmental Stress Crack Resistance* published by Plastics Design Library in 1990. This updated version contains information on more than twice as many material and exposure condition combinations. In addition, more information is presented for each combination. Other

G/R Resin Groups \ Design Criteria	Strength and Stiffness	Toughness	Short-Term Heat Resistance	Long-Term Heat Resistance	Environmental Resistance	Dimensional Accuracy in Molding	Dimensional Stability	Wear and Frictional Properties	Cost
Styrenics	**3**	**6**	**6**	**6**	**6**	**1**	**5**	**6**	**2**
ABS	2	1	1	1	1	3	2	3	3
SAN	1	2	2	2	2	1	1	1	2
Polystyrene	3	3	3	3	3	2	3	2	1
Olefins	**5**	**4**	**4**	**5**	**3**	**5**	**5**	**3**	**1**
Polyethylene	2	2	2	2	2	1	1	2	1
Polypropylene	1	1	1	1	1	1	1	1	2
Other Crystalline Resins	**1**	**1**	**2**	**4**	**4**	**4**	**4**	**2**	**3**
Nylons									
6	2	2	2	2	5	1	4	3	1
6/6	1	3	1	1	4	2	3	2	2
6/10, 6/12	3	1	3	3	3	2	2	3	4
Polyester	4	4	2	1	2	2	1	4	1
Polyacetal	5	5	5	2	1	3	2	1	1
Arylates	**3**	**2**	**3**	**3**	**5**	**1**	**2**	**4**	**4**
Modified PPO	4	3	4	4	3	4	4	4	1
Polycarbonate	2	1	3	3	4	1	3	3	2
Polysulfone	2	2	2	2	2	2	2	1	3
Polyethersulfone	1	3	1	1	1	3	1	2	4
High Temp. Resins	**2**	**4**	**1**	**1**	**2**	**4**	**1**	**4**	**5**
PPS	1	2	2	2	1	1	2	2	1
Polyamide-imide	2	1	1	1	2	2	1	1	2
Fluorocarbons	**6**	**2**	**2**	**1**	**1**	**6**	**6**	**1**	**6**
FEP	2	1	2	1	1	2	2	1	2
ETFE	1	2	1	2	2	1	1	2	1

Ratings: 1—most desirable; 6—least desirable. Large numbers indicate group classification, small numbers the specific resins within that group.

Table 2.155 Selection approach is targeted to obtain the best choice plasticwise

significant improvements include a greatly expanded section 2 in which data are sorted by exposure environment, CAS Registry Numbers are added, and the PDL Resistance Rating throughout the product is used.

If you require detailed information on how a specific exposure environment influences the properties and characteristics of a specific thermoplastic, then section 1 is the place to review. On the other hand, for an indication of how various thermoplastics are influenced by given exposure environments, section 2 is the place to look.

Volume II—Thermoplastic Elastomers, Thermosets, and Rubbers of the Handbook Series is an update to the original *Chemical Compatibility and Environmental Stress Crack Resistance* published by Plastics Design Library in 1990. This volume contains information not previously published by Plastics Design Library. It provides new data on thermoplastic elastomers and introduces information on thermosetting elastomers (i.e., rubbers) and thermoset materials. Permeation data, including permeation rate, breakthrough time (BTT), and lower detection limit (LDL), are included for the first time. Additional results of exposure fields include hardness change and maximum recommended continuous service temperature (Max. Temp.). As in the first volume, section 2 of this book also

Material Characteristics	Strength and Stiffness	Toughness	Short-Term Heat Resistance	Long-Term Heat Resistance	Environmental Resistance	Dimensional Accuracy in Molding	Dimensional Stability	Wear and Frictional Properties	Point Subtotal	Cost	Point Total
Design Criteria / G/R Resin Groups	X	X	X		X						
Styrenics: ABS, SAN, Polystyrene	3	6	6		6				21	2	23
Olefins: Polyethylene, Polypropylene	5	4	4		3				16	1	17
Other Crystalline Resins: Nylon	1	1	2		4				8	3	11
6	2	2	2		5				11	1	12
6/6	1	3	1		4				9	2	11
6/10, 6/12	3	1	3		3				10	4	14
Polyester	4	4	2		2				12	1	13
Polyacetal	5	5	5		1				16	1	17
Arylates: Modified PPO, Polycarbonate, Polysulfone, Polyethersulfone	3	2	3		5				13	4	17
High Temp. Resins: PPS, Polyamide-imide	2	4	1		2				9	5	14
Fluorocarbons: FEP, ETFE	6	2	2		1				11	6	17

Table 2.156 Nylon 6 or 6/6 provides the best choice for a gasoline-powered chain saw

includes data sorted by exposure environment, CAS Registry Numbers, and use of the PDL Resistance Rating throughout the product.

If you require detailed information on how a specific exposure environment influences the properties and characteristics of a specific thermoplastic elastomer, thermoset, or rubber, then section 1 is the appropriate reference. On the other hand, section 2 is recommended for an indication of how various thermoplastic elastomers, thermosets, and rubbers are influenced by given exposure environments.

The philosophy of the Plastics Design Library is to provide as much information as is available. This means that complete information for each test is provided. At the same time, an effort is made to provide information for as many exposure environment and material combinations as possible. Therefore, even if detailed test results are not available (i.e., the only information available is that a material is resistant or has limited resistance), information is still provided. The belief is that some limited information serves as a reference point and is better than no information. Flexibility and ease of use were also carefully considered in designing the layout of the books. Tables 2.172a and 2.173a provide listings of what these two books review.

Material Characteristics	Strength and Stiffness	Toughness	Short-Term Heat Resistance	Long-Term Heat Resistance	Environmental Resistance	Dimensional Accuracy in Molding	Dimensional Stability	Wear and Frictional Properties	Point Subtotal	Cost	Point Total
G/R Resin Groups \ Design Criteria	X		X	X	X						
Styrenics ABS SAN Polystyrene	3		6	6	6				21	2	23
Olefins Polyethylene Polypropylene	5		4	5	3				17	1	18
Other Crystalline Resins Nylons 6 6/6 6/10, 6/12 Polyester Polyacetal	1		2	4	4				11	3	14
Arylates Modified PPO Polycarbonate Polysulfone Polyethersulfone	3		3	3	5				14	4	18
High Temp. Resins PPS Polyamide-imide	1/2 2		2/1 1	2/1 1	1/2 2				6/6 6	1/2 5	7/8 11
Fluorocarbons FEP ETFE	6		2	1	1				10	6	16

Table 2.157 PPS provides the best choice for the impeller used in a chemical-handling pump

COLOR

Because coloring plastics is an important subject, it is reviewed throughout this book. Color has always been an important component of the aesthetic value of plastics. In the beginning, putting color into plastics justified the cost by eliminating, in many cases, the need to paint a fabricated part. This is no longer sufficient. Today, color stylists seek to exploit the entire appearance experience to help create a desired product image. Package design has been an interactive process; stylists demand more from the technology, and at the same time, advances in plastics coloring technology stir creative and innovative approaches (173).

The plastics industry has created more value by making plastic look more like the traditional materials they replaced: glass, aluminum, wood, stone, metal, and so on. The industry has responded with special-effects colorants to give plastic the luster and iridescence of pearl, the rich sheen of gold and silver, or the natural appearance of wood, stone, or leather.

There are pigment and dye colorants. Pigments are defined as colorants that do not dissolve in the plastic matrix of interest (Tables 2.174 and 2.175). Dyes are colorants that go into the solutions (Table 2.176). Different colorants are used to meet different requirements, ranging from surface to transmission conditions. A few examples of the extensive data on colorants are presented in Tables 2.177 to 2.186 and Figures 2.52 to 2.56. The testing was conducted with dry pigment dispersed at a concentration of 3.0% in a nitrocellulose lacquer having a solids content of 9.5%. Films that were approximately 0.003 in were formed of the dispersed particles on glass slides. They were analyzed using a Leres Trilac Goni–spectrophotometer that measured reflection and transmission (173).

Urea, melamine, polycarbonate, polyphenylene oxide, polysulfone, polypropylene, diallyl phthalate, and the phenolics are examples of plastics needed in the temperature range above 93°C (200°F) for good color stability. Most thermoplastics will be suitable below this range.

Special colors have been produced for specific products. As an example, a line of green and amber color concentrates are used for PET beer bottles that can be used with the many different methods of processing the bottles (96).

CRAZING/CRACKING

Many thermoplastics will craze or crack under certain environmental conditions; therefore, products that are highly stressed mechanically must be checked very carefully. Polypropylene, ionomer, chlorinated polyether, phenoxy, EVA, and linear polyethylene offer greater freedom from stress crazing than some other thermoplastics. Solvents may crack parts held under stress. Thermosets are generally preferable for parts under continuous loads.

ELASTICITY

If the product requires flexibility, the choice is limited to plastics such as EVA, ionomer, polyethylene, vinyl, polypropylene, fluorocarbon, silicone, polyurethane, plastisols, acetal, nylon, or some of the rigid plastics that have limited flexibility in thin sections; for example, thin laminations are quite flexible.

	Nylon 6	Nylon 66	Nylon Copolymer	Polypropylene	Syndio-tactic poly styrene (SPS)	PBT poly-ester	Liquid - crystal polymer (LCP)	Polyph-enylene sulfide (PPS)	Polysul-fone (PSU)	PPOm	ABS	BMC	Phenolic	Epoxies
Flammability ratings	E	E	E	E	E	E	E	E	E	E	E	E+	E	E
Comparative tracking index	G/E	G/E	G	E	G/E	F/G	P	P	P	F/G	G	E+	G	G
Ultrasonic weldability	G	G	G	F	G	G	F	G	E	E	E	P	P	P
Temperature resistance	G	G	G	F	G	G	E	E	E	F/G	F	E+	E+	E+
Chemical resistance	G/E	G/E	G/E	E	G	G/E	E	E	F/G	F/G	F/G	E+	E	E
Processability (mold shrinkage, warpage, flow and fill)	G/E	G/E	G/E	E	E	G/E	E	E	F	G	E	E+	E	G
Snap-fit characteristics	F/G	F/G	F/G	G	F	F/G	P/F	P/F	E	E	E	P	P	P
Laser markability	G/E	G/E	G/E	G/E	G	G/E	G/E	P/F	E	G/E	G/E	N/A	N/A	F
Cost index	3	3	3	2	3	3	5	4	5	2	2	2/1	3/1	5/3
Mechanical properties High temperature mechanical properties	G	G	G	F	E	G	E	E	E	F	F	E+	G	E
Elongation	F/G	F/G	F/G	G	P	F/G	P	P	E	E	G	P	P	P
Impact strength	G	G	G	G	P	G		P	E	E	E	E	P	P
Typical applications	Connectors, circuit breaker components, switches, relays	Connectors, circuit breaker components, switches relays	Connectors, fuse blocks, switches, fluorescent lamp housings	Connectors, appliance light housings, junction boxes, appliance components, lamp holders, switch supports, insulators	Connectors, fuse holders	Connectors, circuit breaker components, control housings, fuse holders	Connectors	Switch components bobbins, lamp holders, connector	Fuse holders, light reflectors, clips	Circuit breaker components, control housings	Computer housings	Circuit breaker brush plates, holders, bobbins, arc chutes, solenoids, insulators	Commutators, switches, bobbins, motor & plates	Semi-conductors, capacitors, transformer bushings, encapsulants
Amorphous, semicrystalline	SC	SC	SC	SC	SC	SC	SC	SC	A	A	A	Thermoset	Thermoset	Thermoset.
Warpage	F/G	F/G	F/G	F/G	E	p		E	E	E	E	E+	G	E++

Key: E=Excellent G=Good F=Fair P=Poor

Table 2.158 Example of a plastic material chart

ELECTRIC/ELECTRONIC

Plastic materials and their properties are extensively used and extremely important in the electric and electronics markets. Applicable data have been presented throughout this book. Additional information is provided in Figures 2.57 to 2.62 and Tables 2.187 to 2.202.

FLAME RESISTANCE

The Underwriters' ruling on the use of self-extinguishing plastics for contact-carrying members and many other components introduces critical material-selection problems. All thermosets are self-extinguishing. Nylon, polyphenylene oxide, polysulfone, polycarbonate, vinyl, chlorinated polyether, chlorotrifluoroethylene, vinylidene fluoride, and fluorocarbon are thermoplastic examples that may be suitable for applications requiring self-extinguishing properties. Cellulose acetate and ABS are also available with these properties. Glass reinforcement improves these materials considerably.

IMPACT

Although impact strength of plastics is widely reported, the properties have no particular design values and can be used only to compare the relative response of materials. Even this comparison is not completely valid because it does not solely reflect the capacity of the material to withstand

Points for Comparison	Die-casting Alloys	Nylon
Cost of raw material/ton	Low	High
Cost of mold	High	Can be lower—no higher
Speed of component production	Slower than injection molding of nylon	Lower component production costs
Accuracy of component	Good	Good
Postmolding operations	Finishing—painting Paint chips off easily	Finishing—not required
Surface hardness	Low—scratches easily	Much higher. Scratch resistant.
Rigidity	Good to brittleness	Glass-reinforced grades as good or better
Elongation	Low	GR grades comparable; unfilled grades excellent
Toughness (flexibility)	Low	GR grades comparable; unfilled grades excellent
Impact	Low	All grades good
Notch sensitivity	Low	Low
Young's modulus (E)	Consistent	Varies with load
General mechanical properties	Similar to GR grades of 6/6 nylon	Higher compressive strength
Heat conductivity	High	Low
Electrical insulation	Low	High
Weight	High	Low
Component assembly	Snap fits difficult	Very good

Table 2.159 Comparing cost and performance of nylon and die-cast alloys

Material Family	Injection	Compression	Transfer	Casting	Cold Molding	Coating	Structural Foam	Extrusion	Laminating	Sheet Forming	RP Molding FRP	Filament	Dip and Slush	Blow	Rotational
ABS	X			!			X	X		X				X	
Acetal	X						X	X		X				X	X
Acrylic	X			X		X		X	X	X				X	
Allyl		X	X	X	X	X			X			X			
ASA	X						X	X		X					
Cellulosic	X					X	X	X	X	X				X	X
Epoxy		X	X	X	X	X			X			X			
Fluoroplastic	X	X	X		X	X		X		X					X
Melamine-formaldehyde	X	X	X	X		X			X						
Nylon	X			X		X	X	X						X	X
Phenol-formaldehyde	X	X	X	X	X	X			X						
Poly (amide-imide)	X	X	X					X							
Polyarylether	X							X							
Polybutadiene	X		X					X						X	
Polycarbonate	X	X					X	X		X				X	X
Polyester (TP)	X					X		X						X	X
Polyester-fiberglass (TS)		X	X					X	X		X	X			X
Polyethylene	X	X				X	X	X	X	X				X	X
Polyimide	X	X				X		X(TP)							
Polyphenylene oxide	X						X	X							X
Polyphenylene sulfide	X	X				X									
Polypropylene	X						X	X		X				X	X
Polystyrene	X						X	X		X				X	X
Polysulfone	X	X						X		X					
Polyurethane (TS) (TP)	X	X	X	X		X	X	X(TP)	X	X				X	X
SAN	X						X	X	X	X					X
Silicone		X		X				X							
Styrene butadiene	X					X		X	X						
Urea formaldehyde		X							X						
Vinyl	X	X				X	X	X	X	X			X	X	X

Table 2.160 Examples of processes for plastic materials

Plastic	Properties	Processes
Thermoplastics		
Polystyrene	Low cost, moderate heat distortion, good dimensional stability, good stiffness, impact strength	Injection molding Continuous laminating
Nylon	High heat distortion, low water absorption, low elongation, good impact strength, good tensile and flexural strength	Injection molding Blow molding, Rotational molding
Polycarbonate	Self-extinguishing, high dielectric strength, high mechanical properties	Injection molding
Styrene-acrylo-nitrile	Good solvent resistance, good long-term strength, good appearance	Injection molding
Acrylics	Good gloss, weather resistance, optical clarity, and color; excellent electrical properties	Injection molding Vacuum forming Compression molding Continuous laminating
Vinyls	Excellent weatherability, superior electrical properties, excellent moisture and chemical resistance, self-extinguishing	Injection molding Continuous laminating Rotational molding
Acetals	Very high tensile strength and stiffness, exceptional dimensional stability, high chemical and abrasion resistance, no known room temperature solvent	Injection molding
Polyethylene	Good toughness, light weight, low cost, good flexibility, good chemical resistance; can be "welded"	Injection molding Rotational molding Blow molding
Fluorocarbons	Very high heat and chemical resistance, nonburning, lowest coefficient of friction, high dimensional stability	Injection molding Encapsulation Continuous pultrusion
Polypropylene	Excellent resistance to stress or flex cracking, very light weight, hard, scratch-resistant surface, can be electroplated; good chemical and heat resistance; exceptional impact strength; good optical qualities	Injection molding Continuous laminating Rotational molding
Polysulfone	Good transparency, high mechanical properties, heat resistance, electrical properties at high temperatures; can be electroplated	Injection molding
Thermoset		
Polyesters Properties shown also apply to some polyesters formulated for thermoplastic processing by injection molding	Simplest, most versatile, economical and most widely used family of resins, having good electrical properties, good chemical resistance, especially to acids	Compression molding Filament winding Hand lay-up Mat molding Pressure bag molding Continuous pultrusion Injection molding Spray-up Centrifugal casting Cold molding Comoform[1] Encapsulation
Epoxies	Excellent mechanical properties, dimensional stability, chemical resistance (especially alkalis), low water absorption, self-extinguishing (when halogenated), low shrinkage, good abrasion resistance, very good adhesion properties	Compression molding Filament winding Hand lay-up Continuous pultrusion Encapsulation Centrifugal casting
Phenolics	Good acid resistance, good electrical properties (except arc resistance), high heat resistance	Compression molding Continuous laminating
Silicones	Highest heat resistance, low water absorption, excellent dielectric properties, high arc resistance	Compression molding Injection molding Encapsulation
Melamines	Good heat resistance, high impact strength	Compression molding
Diallyl phthalate	Good electrical insulation, low water absorption	Compression molding

Table 2.161 Examples of properties and processes for plastic materials

shock loading, but it can pick up a discriminatory response to notch sensitivity. A better value is tensile impact, but unfortunately this property is not generally reported. The impact value, with its limitation, can broadly distinguish plastics that can withstand shock loading from those that are poor in this response; however, only broad generalizations can be obtained from these values.

Comparative tests on sections of similar size that are molded in accordance with the proposed product must be run to determine the impact performance of a plastic material. The laminated plastics, glass-filled epoxies, melamines, and phenolics are outstanding in impact strength. Polycarbonate and ultrahigh molecular weight PEs are examples of plastics with outstanding impact strength.

ODOR/TASTE

Many different plastics are used in food packaging and refrigeration conditions. Thermoplastics that are odor- and taste-free include polyethylene, polystyrene, styrene-acrylonitrile, acrylic, ABS,

Concept	Mechanism	Structural and morphological modification
Dispersing and diffusing	Fine grinding	Enlarging the phase interface
	High overall shearing	Enlarging the regions of molecular penetration Reducing diffusion paths; degrading internal stresses
	High overall shearing with quenching	As above Freezing thermodynamic imbalances
	High local shearing	Generating local tribochemically or chemically activated conditions
Active interlayers	Radical forming substances	Crosslinking via the interlayer
	Substances initiating radical formation	Crosslinking the components
	Amphiphilic low -molecular -weight substances	Improved intermolecular interaction; promotion of wetting; particle refining
Matrix modification	Elastomer phases	Matrix formation; promoting the absorption of fillers and reinforcing agents
	Filling and reinforcing	Modifying the matrix properties
	Foaming	Structuring the matrix

Table 2.162 Examples of modifying plastics

polysulfone, EVA, polyphenylene oxide, and many others (Table 2.203). Thermoset melamine and urea compounds are suitable for these services. FDA approvals are available for many plastics.

PERMEABILITY

Certain plastics are impervious to different materials. However, there are those that have certain poor permeability properties. As an example, polyethylene will pass wintergreen, hydrocarbons, and many other chemicals. It is used in certain cases for the separation of gases since it will pass one and block another. Chlorotrifluoroethylene, vinylidene fluoride, vinylidene chloride, polypropylene, EVA, and phenoxy merit special study (Tables 2.204 to 2.208) (157).

Extensive data is available concerning the permeability (of the properties and characteristics) of plastics and elastomers. A detailed review prepared by the Plastics Design Library is provided in their 706-page book *Permeability and Other Film Properties of Plastics and Elastomers*. It is a unique reference and data bank on the barrier and film properties of plastics and elastomers. The basic physical characteristics of these materials are generally well defined by manufacturers. However, data on the more capricious phenomenological issues, such as permeability, are difficult to find, especially in a comprehensive compilation. This PDL publication serves to turn disparate information from

Plastic	Reinforcements	Fillers
Amorphous **ABS** **SAN** **Amorphous Nylon** **Polycarbonate** **Modified PPO** **Polystyrene** **Polysulfones**	**Can more than double tensile strength** **Can increase flexural modulus fourfold** **Raise HDT slightly** **Toughen brittle resins, embrittle tough resins** **Can provide 1,000 Ω-cm resistivity** **Reduce shrinkage** **Raise cost**	**Lower tensile strength** **Can more than double flexural modulus** **Raise HDT slightly** **Embrittle resins** **Can impact special properties such as lubricity conductivity, flame retardance** **Reduce and balance shrinkage** **Reduce melt flow** **Can lower cost**
Crystalline **Aceals** **Nylon 6, 6/6, 6/10, 6/12, 11, 12** **Polypropylene** **Polyphenylene sulfide** **Thermoplastic Polyesters** **Polyethylene**	**Can more than triple tensile strength** **Can raise flexural modulus sevenfold** **Can nearly triple HDT** **Toughen brittle resins, embrittle tough resins** **Can provide 1 Ω-cm resistivity** **Reduce shrinkage** **Raise cost**	**Lower tensile strength** **Can more than triple flexural modulus** **Raise HDT slightly** **Embrittle resins** **Can impart special properties such as lubricity, conductivity, magnetic properties, flame retardance** **Reduce shrinkage** **Can lower cost**

Table 2.163 Examples of adding reinforcements and fillers to thermoplastics

Plastic	Glass Fiber Content, (wt %)	Specific Gravity, D 792	Tensile Strength (MPa), D 638	Tensile Elongation (%), D 638	Tensile Modulus (GPa), D 638	Flexural Strength (MPa), D 790	Flexural Modulus (GPa), D 790	Compressive Strength (MPa), D 695	Impact Strength, Izod Notched, (J/m), D 256
ABS	10	1.10	65	3.0	4.6	102	4.5	83	64
	30	1.28	90	1.4	6.3	116	6.4	104	53
Acetal	10	1.54	72	2.4	6.6	107	6.1	69	53
	30	1.63	83	2.0	7.7	114	7.2	81	43
Nylon 6	15	1.25	104	4.0	5.9	159	5.4	97	80
	30	1.37	166	3.0	7.2	200	6.9	166	117
Nylon 6/6	13	1.23	97	4.0	6.2	173	4.5	93	53
	30	1.37	173	3.0	9.0	235	9.0	186	107
Nylon 6/12	30	1.30	135	4.0	8.3	193	7.6	138	117
Polycarbonate	10	1.26	83	9.0	5.2	110	4.1	97	107
	30	1.43	121	2.0	8.6	141	6.9	117	128
Polyester, TP	30	1.52	131	4.0	8.3	193	7.9	124	96
Polyethylene	10	1.04	36	4.0	2.5	46	2.5	35	75
	30	1.18	59	3.0	5.0	89	4.9	41	91
Polyphenylene sulfide	40	1.64	152	3.0	14.1	255	13.0	145	80
Polypropylene	10	0.98	43	4.0	2.5	54	2.4	41	43
Polystyrene	30	1.12	47	2.0	4.4	63	4.3	47	69
High heat co-polymer	20	1.22	90	1.2	8.3	131	7.9	110	59
High heat ter-polymer	30	1.35	83	1.8	6.5	123	5.7	76	80
Polysulfone	20	1.38	97	2.5	6.0	138	5.9	124	64
	40	1.55	124	1.5	11.6	173	10.7	138	80
Polyurethane	10	1.22	33	48.0	0.7	43	0.6	35	747
PVC	20	1.58	97	3.0	0.8	145	6.9	83	80
SAN	20	1.22	100	1.8	8.6	131	7.6	121	64
	35	1.35	110	1.4	10.4	155	9.3	45	53

Table 2.164 Mechanical properties of glass-fiber-reinforced thermoplastics per ASTM procedures

Filler or reinforcement	Properties improved												
	Chemical resistance	Heat resistance	Electrical insulation	Impact strength	Tensile strength	Dimensional stability	Stiffness	Hardness	Lubricity	Electrical conductivity	Thermal conductivity	Moisture resistance	Process-ability
Alumina, tabular	•	•	•	•		•	•					•	•
Aluminum powder										•	•		
Aramid	•	•	•	•	•	•	•	•	•				•
Bronze							•	•		•	•		•
Calcium carbonate	•	•	•	•		•	•	•					
Carbon black		•				•	•			•	•		•
Carbon fiber										•	•		
Cellulose				•		•	•	•					
Alpha cellulose			•			•						•	
Coal, powdered	•											•	
Cotton			•	•	•	•	•	•					
Fibrous glass	•	•	•	•	•	•	•	•				•	
Graphite	•				•	•	•	•	•	•	•		
Jute				•			•						
Kaolin	•	•	•			•	•	•	•			•	•
Mica	•	•	•			•	•	•	•			•	
Molybdenum disulfide							•	•	•			•	•
Nylon	•	•	•	•	•	•	•	•	•				•
Orlon	•	•	•	•	•	•	•	•				•	•
Rayon			•	•	•	•	•	•					
Silica, amorphous			•									•	•
Sisal fibers	•			•		•	•	•				•	
Fluorocarbon						•	•	•	•				
Talc	•	•	•			•	•	•	•			•	•
Wood flour			•			•							

Table 2.165 Effects of filler or reinforcement on plastic properties

wide-ranging sources (i.e., conference proceedings, test laboratories, materials suppliers, monographs, trade and technical journals) into useful engineering knowledge (277). It provides detailed test results on the plastics and elastomers listed in Table 2.209a. An example of the data is provided in Table 2.209b (277).

The information provided ranges from a general overview of the barrier properties of plastics and elastomers to detailed discussions and test results. For users to whom the study and use of permeability and film property data are relatively new, the primer on barrier properties and the detailed glossary of terms, including descriptions of test methods, will prove useful. For those who wish to delve beyond the data presented, source documentation is presented in detail.

Data presented detail differences in permeation between generic families of plastic and elastomer materials. Also covered are differences within the same generic family due to environmental factors, such as temperature and humidity, as well as material characteristics, such as sample preparation and material composition. The data indicate how a material is likely to behave relative to another material or relative to the same material exposed under different conditions.

The permeation of gases and vapors through thin films is dependent on the molecular size, shape, wettability, and soundness of the fabricated membranes. Since permeation in well-made items is a molecular transport phenomenon, it is affected by orientation, degree of crystallinity, and temperature. Attempts have been made to relate permeation rates through thin films to absorption of thicker films, sheets, pipes, and so on. This has been generally unsuccessful. Thicker films and sheets represent average sets of properties obtainable from many thin films produced under a variety of conditions. Producing a thin film representative of this average is not practical.

The philosophy of the Plastics Design Library is to provide as much information as is available. This means that complete information corresponding to each test result is provided. At the same time, an effort is made to provide information for as many tests, conditions, penetrants, and material combinations as possible. Therefore, even if detailed test results are not available, information is still provided. The belief is that some limited information serves as a reference point and is better than no information. This PDL publication provides information that is as complete and detailed as it was in the source documentation.

Material	Temperature					After cooling to 86°F
	59°F	212°F	392°F	482°F	572°F	
Sintered copper, impregnated to a depth of about 1 mil	0.05	0.05	0.05	0.05	~0.12	0.05
Sintered copper, untreated	0.24	0.24	~0.7	~0.8	~1.2	~1
Solid copper	1.3	~1.3	~1	~1	~1	~1

Table 2.166 Coefficient of friction of impregnated fluoroplastic materials for unlubricated sliding against steel

PLASTIC MATERIAL	Aromatic solvents		Aliphatic solvents		Chlorinated solvents		Weak bases and salts		Strong bases		Strong acids		Strong oxidants		Esters and ketones		24-hr. Water absorption
	temperature F (Best = 1)																
	77	200	77	200	77	200	77	200	77	200	77	200	77	200	77	200	% change by weight
Acetals	1-4	2-4	1	2	1-2	4	1-3	2-5	1-5	2-5	5	5	5	5	1	2-3	0.22 - 0.25
Acrylics	5	5	2	3	5	5	1	3	2	5	4	4-5	5	5	5	5	0.2 - 0.4
Acrylonitrile-Butadiene-Styrenes (ABS)	4	5	2	3-5	3-5	5	1	2-4	1	2-4	1-4	5	1-5	5	3-5	5	0.1 - 0.4
Cellulose Acetates (CA)	2	3	2	3	3	4	2	3	3	5	3	5	3	5	5	5	2 - 7
Cellulose Acetate Propionates (CAP)	4	5	1	3	3	4	1	2	3	5	3	5	3	5	5	5	1.3 - 2.8
Diallyl Phthalates (DAP, filled)	1-2	2-4	2	3	2	4	2	3	2	4	1-2	2-3	2	4	3-4	4-5	0.2 - 0.7
Epoxies	1	2	1	2	1-2	3-4	1	1-2	1	2	2-3	3-4	4	4-5	2	3-4	0.01 - 0.10
Ethylene Copolymers (EVA) (Ethylene-Vinyl Acetates)	5	5	5	5	5	5	1	2	1	5	1	5	1	5	2	5	0.05 - 0.13
Fluorocarbons																	
Ethylene/Tetrafluoroethylene Copolymers (ETFE)	1	1	1	1	1	1	1	1	1	1	1	1	1	1	1	1	<0.03
Fluorinated Ethylene Propylenes (FEP)	1	1	1	1	1	1	1	1	1	1	1	1	1	1	1	1	<0.01
Perfluoroalkoxies (PFA)	1	1	1	1	1	1	1	1	1	1	1	1	1	1	1	1	<0.03
Polychlorotrifluoroethylenes (CTFE)	1	1	1	1	3	4	1	1	1	1	1	1	1	1	1	1	0.01 - 0.10
Polytetrafluoroethylenes (TFE)	1	1	1	1	1	1	1	1	1	1	1	1	1	1	1	1	0
Ionomers	2	4	1	4	4	4	1	4	1	4	2	4	1	5	1	4	0.1 - 1.4
Melamines (filled)	1	1	1	1	1	1	2	3	2	3	2	3	2	3	1	2	0.01 - 1.30
Nitriles (high barrier alloys of ABS or SAN)	1	4	1	2-4	1-4	2-5	1	2-4	1	2-4	2-5	5	3-5	5	1-5	5	0.2 - 0.5
Nylons	1	1	1	1	1	2	1	2	2	3	5	5	5	5	1	1	0.2 - 1.9
Phenolics (filled)	1	1	1	1	1	1	2	3	3	5	1	1	4	5	2	2	0.1 - 2.0
Polyallomers	2	4	2	4	4	5	1	1	1	1	1	3	1	4	1	3	<0.01

Table 2.167 Chemical resistance of plastics (courtesy of Plastics FALLO)

PLASTIC MATERIAL	Aromatic solvents		Aliphatic solvents		Chlorinated solvents		Weak bases and salts		Strong bases		Strong acids		Strong oxidants		Esters and ketones		24-hr. Water absorption
	temperature F (Best =1)																
	77	200	77	200	77	200	77	200	77	200	77	200	77	200	77	200	% change by weight
Polyamide-imides	1	1	1	1	2	3	1	1	3	4	2	3	2	3	1	1	0.22 - 0.28
Polybutylenes (PB)	3	5	1	5	4	5	1	2	1	3	1	3	1	4	1	3	<0.01 - 0.3
Polycarbonates (PC)	5	5	1	1	5	5	1	5	5	5	1	1	1	1	5	5	0.15 - 0.35
Polyesters (thermoplastic)	2	5	1	3-5	3	5	1	3-4	2	5	3	4-5	2	3-5	2	3-4	0.06 - 0.09
Polyesters (thermoset-glass fiber filled)	1-3	3-5	2	3	2	4	2	3	3	5	2	3	2	4	3-4	4-5	0.01 - 2.50
Polyethylenes (LDPE-HDPE — low-density to high-density)	4	5	4	5	4	5	1	1	1	1	1-2	1-2	1-3	3-5	2	3	0.00 - 0.01
Polyethylenes (UHMWPE — ultra high molecular weight)	3	4	3	4	3	4	1	1	1	1	1	1	1	1	3	4	<0.01
Polyimides	1	1	1	1	1	1	2	3	4	5	3	4	2	5	1	1	0.3 - 0.4
Polyphenylene Oxides (PPO) (modified)	4	5	2	3	4	5	1	1	1	1	1	2	1	2	2	3	0.06 - 0.07
Polyphenylene Sulfides (PPS)	1	2	1	1	1	2	1	1	1	1	1	1	1	2	1	1	<0.05
Polyphenylsulfones	4	4	1	1	5	5	1	1	1	1	1	1	1	1	3	4	0.5
Polypropylenes (PP)	2	4	2	4	-3	4-5	1	1	1	1	1	2-3	2-3	1-5	2	4	0.01 - 0.03
Polystyrenes (PS)	4	5	4	5	5	5	1	5	1	5	4	5	4	5	4	5	0.03 - 0.60
Polysulfones	4	4	1	1	5	5	1	1	1	1	1	1	1	1	3	4	0.2 - 0.3
Polyurethanes (PUR)	3	4	2	3	4	5	2-3	3-4	2-3	3-4	2-3	3-4	4	4	4	5	0.02 - 1.50
Polyvinyl Chlorides (PVC)	4	5	1	5	5	5	1	5	1	5	1	5	2	5	4	5	0.04 - 1.00
Polyvinyl Chlorides-Chlorinated (CPVC)	4	4	1	2	5	5	1	2	1	2	1	2	2	3	4	5	0.04 - 0.45
Polyvinylidene Fluorides (PVDF)	1	1	1	1	1	1	1	1	1	2	1	2	1	2	3	5	0.04
Silicones	4	4	2	3	4	5	1	2	4	5	3	4	4	5	2	4	0.1 - 0.2
Styrene Acrylonitriles (SAN)	4	5	3	4	3	5	1	3	1	3	1	3	3	4	4	5	0.20 - 0.35
Ureas (filled)	1	3	1	3	1	3	2	3	2	3	4	5	2	3	1	2	0.4 - 0.8

Table 2.167 *(continued)*

RADIATION

In general, rigid plastics are superior to elastomers in radiation resistance but are inferior to metals and ceramics. Examples of materials that will respond satisfactorily in the range of 10^{10} and 10^{11} erg per gram are fluoroplastics (Fig. 2.63), glass-filled phenolics, certain epoxies, polyurethane, polystyrene, mineral-filled polyesters, silicone, and furane resins. The next group of resins in order of radiation resistance includes polyethylene, melamine, urea-formaldehyde resins, unfilled phenolics, and silicone resins. Materials having poor radiation resistance include methyl methacrylate, unfilled polyesters, cellulosics, polyamides, and fluorocarbons (Table 2.210; chapter 21).

TEMPERATURE RESISTANCE

Thermal considerations will quickly eliminate certain materials. For products operating above 232°C (450°F), the silicones, polyimides, fluoroplastics, hydrocarbons, methylpentene cold molds, and glass-bonded mica are examples of plastics that may be used. A few of the organic resin-bonded inorganic fibers such as bonded ceramic wool perform well in this field. Epoxy, diallyl phthalate, and phenolic-bonded glass fiber may be satisfactory in the 232° to 288°C (450° to 550°F) range (224).

Material	Aromatic Solvents		Aliphatic Solvents		Chlorinated Solvents		Esters and Ketones	
	25ºC	93ºC	25ºC	93ºC	93ºC	25ºC	93ºC	
Polyacetal	1-4	2-4	1	2	1-2	4	1	2-3
ABS	4	5	2	3-5	3-5	5	3-5	5
Nylons	1	1	1	1	1	2	1	1
Polyester	2	5	1	3-5	3	5	2	3-4
Polyethylene	3	4	3	4	3	4	3	4
Polystyrene	4	5	4	5	5	5	4	5
Polysulfone	4	4	1	1	5	5	3	4
PVC	4	5	1	5	5	5	4	5
PTFE	1	1	1	1	1	1	1	1
PFA	1	1	1	1	1	1	1	1
FEP	1	1	1	1	1	1	1	1
ETFE	1	1	1	1	1	1	1	1
PVDF	1	1	1	1	1	1	3	5
PCTFE	1	1	1	1	3	4	1	1

Table 2.168 Effects of organic chemicals on plastics

Between 121° and 232°C (250° and 450°F), glass- or mineral-filled phenolic, melamine, alkyd, silicone, nylon, polyphenylene oxide, polysulfone, polycarbonate, methylpentene, fluorocarbon, polypropylene, and diallyl phthalate can be evaluated. The addition of glass fillers to the thermoplastics can raise the useful temperature range to as much as 100°C (212°F) and at the same time shorten the molding cycle. In the –18° to 100°C (0° to 212°F) range, a broad selection of

Material	Commercial designation	Fuel[a]						Oxidizer[a]									Recommended application
		Unsymmetrical demethyl-hydrazine			Hydrazine			Nitrogen tetroxide			Inhibited red fuming nitric acid			Propyl nitrate			
		Splash[b]	Short term[c]	Long term[d]		Short term	Long term		Short term	Long term		Short term	Long term		Short term	Long term	
Plastics																	
Polyvinyl Fluoride	Teslar[e]	S	S	U	S	S	U	S	U	U	U	U	U				Not for seal service (film)
Polycarbonate	Lexan[f]	S	U	U	U	U	U	U	U	U	U	U	U	S	U	U	Not for seal service (sheet or molding rigid)
Polytetrafluorethylene	Teflon FEP[e]	S	S	S	S	S	S	S	S	S	S	S	S	S	S	S	Dynamic or static seals. Long-term service with fuel and oxidizer
Polytetrafluorethylene	Teflon, TFE filled and unfilled)[e]	S	S	S	S	S	S	S	S	S	S	S	S	S	S	S	Dynamic or static seals. Long-term service with fuel and oxidizer
Fluorosilicone rubber	LS–53[g]	S	M	U	S	M	U	M	U	U							Static seals. Short-term fuel service only
Polyethylene (branched)	Alathon (low density)[e]	S	S	U	S	S	U	S	U	U	S	S	U	S	S		Dynamic or static seals, retainers. Short-term fuel service only
Polyethylene (linear)	Marlex or Hi-Fax (high density)[h]	S	S	U	S	S		S	M	U				S	S		Dynamic or static seals, retainers. Short-term fuel service only
Polyethylene (irradiated)	Irrathene[f]	S	S	U	S	S	U	S	U	U	S	U	U	S	S		Dynamic or static seals, retainers. Short-term fuel service only
Polytrifluorochlorethylene (modified)	Genetron modified (Kel-F)[i]	S	U	U	S	U	U	S	M	U	S	S		S	S		Dynamic or static seals. Short-term oxidizer service only
Polytrifluorochlorethylene	Trithene (Kel-F)[j]	S	S	U	S	U	U	S	M	U	S	S		S	S		Dynamic or static seals. Short-term oxidizer service only
Polyvinyl chloride (plasticized)	Tygon[k]	S	M	U	S	M	U	S	U	U	S	U	U	U	U	U	Static seals. Short-term service with fuels only
Polyvinylidene chloride	Saran[g]	M	U	U	U	U	U	M	U	U	U	U	U	U	U	U	Not for seal service (film)
Polyethylene terephthalate	Mylar[e]	M	U	U	U	U	U	M	U	U	U	U	U	S	S		Not for seal service (film)
Polystyrene	Styrene[l]	S	U	U	S	U	U	M	U	U	U	U	U				Not for seal service (sheet or molding, rigid)
Hexamethylenedianuine adipimide	Nylon (Zytel)[e]	S	S	S	S	U	U	U	U	U	U	U	U	S	U	U	Dynamic or static seals. Long-term with UDMH fuel only
Elastomers																	
Fluororubber	Viton-A[e]	S	S	U	S	U	U	S	U	U	S	U	U				Static seals. Short-term fuel service only
Isobutylene-isoprene	Butyl Rubber[m]	S	S	U	S	S	U	S	U	U	S	U	U	S	S	U	
Polytrifluorochorethylene	Kel-F 5500 Elastomer[n]	S	U	U	S	U	U	S	M	U	S	M					
Styrene-butadiene rubber	SBR (GRS)[o]				S	S											
Isoprene	Natural Rubber	S	M	U	S	M	U	U	U	U							
Acrylonitrile-butadiene rubber	Buna-N (Hycar)[o]	S	U	U	S	U	U	U	U	U							
Chloroprene	Neoprene[e]	S	U	U	S	U	U	U	U	U							

[a] S, Satisfactory; M, Marginal; U, unsatisfactory.
[b] I min at 77°F.
[c] 1–60 days at 77–160°F.
[d] <1 year.
[e] E. I. du Pont Co.
[f] General Electric Co.
[g] Dow-Corning Corp.
[h] Hercules Powder Co., Inc., Phillips Petroleum Co.
[i] Allied Chemical and Dye Corp.
[j] Visking Corp.
[k] U. S. Stoneware Co.
[l] Monsanto Chemical Co.
[m] Enjay Co.
[n] Minnesota Mining and Mfg. Co.
[o] B. F. Goodrich Chemical Co.

Table 2.169 Compatibility of plastics and elastomers with liquid propellant fuels and oxidizers

Chemical Class	Acetal Copolymer	Acetal Homopolymer	Nylon 6/6	Thermoplastic Polyester (PBT)	Thermoplastic Polyester (PET)	Polyester Elastomer	Liquid Crystal Polymer*	Polyphenylene Sulfide*	Polyarylate	Polycarbonate	Polysulfone*	Modified Polyphenylene Oxide,	Polypropylene	ABS	316 Stainless Steel	Carbon Steel	Aluminum
Acids and bases																	
Acids, weak	A	B	C	A	A	A	A	A	A	A	B	A	A	A	A	A	C
Acids, strong	C	C	C	B	—	C	B	A	—	C	C	—	A	A	B	C	C
Bases, weak	A	C	A	B	B	A	B	A	—	C	A	A	A	A	A	B	C
Bases, strong	A	C	C	—	—	B	C	A	—	C	A	—	A	A	B	B	C
Acids, organic, weak	A	B	C	A	A	A	A	A	A	A	B	A	A	A	A	C	C
Acids, organic, strong	C	C	C	B	—	C	B	A	—	C	C	A	A	A	B	C	C
Automotive																	
Automotive, fuel	A	A	A	A	A	A	A	A	C	C	A	C	C	A	A	A	A
Automotive, lubricants	A	A	A	A	A	A	A	A	C	C	A	A	A	A	A	B	A
Automotive, hydraulic	A	A	—	A	A	—	A	A	C	C	C	A	A	A	—	—	—
Solvents																	
Aliphatic hydrocarbons	A	A	A	A	A	A	A	A	A	A	A	B	C	A	A	A	A
Aliphatic hydrocarbons, halogenated	A	B	C	B	B	A	A	A	C	C	C	—	—	—	B	B	B
Alcohols	A	A	B	A	A	A	A	A	A	A	A	A	A	A	A	A	B
Aldehydes	A	A	A	A	B	B	A	A	—	C	B	—	A	—	A	B	A
Aromatic hydrocarbons, halogenated	—	—	—	—	—	C	—	A	C	C	C	—	—	—	A	A	A
Aromatic, hydroxy	C	C	C	C	—	C	A	A	—	C	C	—	A	—	B	C	A
Esters	B	B	A	B	B	B	A	A	C	C	C	—	C		B	B	B
Ethers	B	—	A	A	—	—	—	A	—	A	B	—	C	—	A	A	A
Ketones	B	B	A	B	B	B	A	A	C	C	C	—	B	C	A	A	A
Miscellaneous																	
Detergents	A	—	A	—	B	—	—	A	A	A	—	B	A	—	A	A	B
Inorganic salts	B	B	B	—	A	—	—	A	—	A	—	—	A	A	B	B	B
Oxidizing agents, strong	C	C	C	—	C	—	B	B	—	C	—	—	A	—	C	C	C
Oxidizing agents, weak	C	C	C	A	—	A	A	A	—	A	—	A	A	A	B	C	A
Water, ambient	A	A	B	A	A	A	A	A	A	A	A	A	A	—	A	C	B
Water, hot	B	C	B	C	C	B	A	A	—	C	—	A	C	—	A	C	B
Water, steam	C	C	C	C	C	C	B	A	—	C	—	—	C	—	A	C	—

A—minimal effect.

B—some effect.

C—generally not recommended.

Room temperature except hot water, steam and "*" Generally, extended exposure (more than a week) data were used

* – 200F

Table 2.170 Comparing resistance of plastics with other materials

Reagent	Concentration	Temp. 70° F	Temp. 140° F
Acetaldehyde	100%	O	U
Acetic Acid*	1-10%	S	S
Acetic Acid*	10-60%	S	O
Acetic Acid*	80-100%	O	U
Acetic Anhydride		U	U
Acetone		U	U
Acrylic Emulsions*		S	S
Allyl Alcohol		U	U
Allyl Chloride		U	U
Aluminum Chloride	Dilute	S	S
Aluminum Chloride	Conc.	S	S
Aluminum Fluoride	Conc.	S	S
Aluminum Sulfate	Conc.	S	S
Alums (all types)	Conc.	S	S
Ammonium Carbonate		S	S
Ammonium Chloride	Sat'd	S	S
Ammonium Fluoride	20%	S	S
Ammonium Hydroxide	35%	S	S
Ammonium Metaphosphate	Sat'd	S	S
Ammonium Nitrate	Sat'd	S	S
Ammonium Persulfate	Sat'd	S	S
Ammonium Sulfate	Sat'd	S	S
Ammonium Sulfide	Sat'd	S	S
Ammonium Thiocyanate	Sat'd	S	S
Amyl Acetate	100%	U	U
Amyl Alcohol*	100%	S	S
Amyl Chloride	100%	U	U
Aniline*	100%	S	U
Aniline Hydrochloride	Sat'd	U	
Antimony Chloride		S	S
Aqua Regia		U	U
Arsenic Acid	100%	S	S
Barium Carbonate	Sat'd	S	S
Barium Chloride	Sat'd	S	S
Barium Hydroxide		S	S
Barium Sulfate	Sat'd	S	S
Barium Sulfide	Sat'd	S	S
Beer		S	S
Benzene		U	U
Benzene Sulfonic Acid*		S	S
Benzoic Acid	All Conc.	S	S
Bismuth Carbonate	Sat'd	S	S
Black Liquor		S	S
Bleach Lye	10%	S	S
Borax	Cold Sat'd	S	S
Boric Acid	Dilute	S	S
Boric Acid	Conc.	S	S
Bromic Acid	10%	S	S
Bromine Liquid	100%	U	U
Bromine Water		U	U
Butanediol*	10%	S	S
Butanediol*	60%	S	S
Butanediol*	100%	S	S
Butyl Alcohol	100%	S	S
Butyric Acid	Conc.	U	U
Calcium Bisulfide		S	S
Calcium Carbonate	Sat'd	S	S
Calcium Chlorate	Sat'd	S	S
Calcium Chloride	Sat'd	S	S
Calcium Hydroxide		S	S
Calcium Hypochlorite	Bleach Sol'n	S	S
Calcium Nitrate	50%	S	S
Calcium Sulfate		S	S
Camphor Oil		U	U
Carbon Dioxide	100% dry	S	S
Carbon Dioxide	100% wet	S	S
Carbon Dioxide	Cold Sat'd	S	S
Carbon Disulfide		U	U
Carbon Monoxide		S	S
Carbon Tetrachloride		U	U
Carbonic Acid		S	S
Castor Oil*	Conc.	S	S
Chloracetic Acid	100%	U	U
Chlorine Dry Gas	100%		
Chlorine Moist Gas		O	U
Chlorine Liquid		U	U
Chlorine Water	2% Sat'd Sol.	U	U
Chlorobenzene		U	U
Chloroform		U	U
Chlorosulfonic Acid	100%	U	U
Chrome Alum	Sat'd	S	S
Chromic Acid	20%	S	S
Chromic Acid	up to 50%	S	S
Chromic Acid & Sulfuric Acid*		S	O

Key:
S-satisfactory
O-some attack
U-unsatisfactory

Table 2.171 Chemical resistance of low- and medium-density polyethylene to various reagents

Reagent	Concentration	Temp. 70° F	Temp. 140° F	Reagent	Concentration	Temp. 70° F	Temp. 140° F
Cider*		S	S	Fluoboric Acid		S	S
Citric Acid*	Sat'd	S	S	Fluorine		S	U
Coconut Oil Alcohols*		S	S	Fluosilicic Acid	32%	S	S
Cola Concentrates*		S	S	Fluosilicic Acid	Conc.	S	O
Copper Chloride	Sat'd	S	S	Formaldehyde*	40%	S	S
Copper Cyanide	Sat'd	S	S	Formic Acid*	0-20%	S	S
Copper Fluoride	2%	S	S	Formic Acid*	20-50%	S	S
Copper Nitrate	Sat'd	S	S	Formic Acid*	100%	S	S
Copper Sulfate	Dilute	S	S	Fructose	Sat'd	S	S
Cottonseed Oil*		S	S	Fruit Pulp		S	S
Cresol	100%	S	S	Fuel Oil		O	U
Cresylic Acid	50%	S	S	Fufural	100%	U	U
Cuprous Chloride	Sat'd	S	S	Furfuryl Alcohol		U	U
Cyclohexanol*		S	S	Gallic Acid*	Sat'd	S	S
Cyclohexanone		U	U	Gasoline		U	U
Detergents, Synthetic*		S	S	Gin		U	U
Developers, Photographic		S	S	Glucose		S	S
				Glycerine*		S	S
Dextrin	Sat'd	S	S	Glycol*		S	S
Dextrose	Sat'd	S	S	Glycolic Acid*	30%	S	S
Dibutylphthalate		O	O	Grape Sugar	Sat'd Aq.	S	S
Disodium Phosphate		S	S	Heptane		U	U
Diazo Salts		S	S	Hexanol, Tert.*		S	S
Diethylene Glycol*		S	S	Hydrobromic Acid	50%	S	S
Diglycolic Acid*		S	S	Hydrocyanic Acid	Sat'd	S	S
Dimethylamine		U	U	Hydrochloric Acid	10%	S	S
Dioctyl Phthalate		O	U	Hydrochloric Acid	30%	S	S
Emulsions, Photographic*		S	S	Hydrochloric Acid	35%	S	S
Ethyl Acetate	100%	O	U	Hydrochloric Acid	Conc.	S	S
Ethyl Alcohol*	100%	S	S	Hydrofluoric Acid	40%	S	S
Ethyl Alcohol*	35%	S	S	Hydrofluoric Acid	60%	S	S
Ethyl Butyrate		O	U	Hydrofluoric Acid	75%	S	O
Ethyl Chloride		U	U	Hydrogen	100%	S	S
Ethyl Ether		U	U	Hydrogen Bromide	10%	S	S
Ethylene Chloride		U	U	Hydrogen Chloride Gas	Dry	S	S
Ethylene Chlorohydrin		U	U	Hydrogen Peroxide	30%	S	O
Ethylene Dichloride		U	U	Hydrogen Peroxide	90%	S	U
Ethylene Glycol*		S	S	Hydrogen Phosphide	100%	S	S
Ferric Chloride	Sat'd	S	S	Hydrogen Sulfide		S	S
Ferric Nitrate	Sat'd	S	S	Hydroquinone		S	S
Ferrous Chloride	Sat'd	S	S	Hypochlorous Acid	Conc.	S	S
Ferrous Sulfate		S	S	Inks*		S	S
Fish Solubles*		S	S	Iodine (in KI Sol'n.)		O	U

Table 2.171 *(continued)*

Reagent	Concen-tration	Temp. 70° F	Temp. 140° F
Lactic Acid*	10%	S	S
Lactic Acid*	90%	S	S
Latex*		S	S
Lead Acetate	Sat'd	S	S
Lead Tetra-Ethyl	100%	S	
Linseed Oil		O	U
Lube Oil		U	U
Magnesium Carbonate	Sat'd	S	S
Magnesium Chloride	Sat'd	S	S
Magnesium Hydroxide	Sat'd	S	S
Magnesium Nitrate	Sat'd	S	S
Magnesium Sulfate	Sat'd	S	S
Maleic Acid	Sat'd	S	S
Mercuric Chloride	Sat'd	S	S
Mercuric Cyanide	Sat'd	S	S
Mercurous Nitrate	Sat'd	S	S
Mercury		S	S
Methyl Alcohol*	100%	S	S
Methyl Bromide		O	U
Methyl Chloride		O	U
Methyl Ethyl Ketone	100%	U	U
Methylene Chloride	100%	U	U
Methylsulfuric Acid*		S	S
Milk		S	S
Mineral Oils		O	U
Molasses	Comm.	S	S
Naptha		S	U
Naphthalene		U	U
Nickel Chloride	Sat'd	S	S
Nickel Nitrate	Conc.	S	S
Nickel Sulfate	Sat'd	S	S
Nicotine*	Dilute	S	S
Nicotinic Acid*		S	S
Nitric Acid	0-30%	S	S
Nitric Acid	30-50%	S	O
Nitric Acid	70%	S	O
Nitric Acid	95-98%	U	U
Nitrobenzene	100%	U	U
Octyl Cresol		O	U
Oils and Fats		O	U
Oleic Acid	Conc.	O	U
Oleum	Conc.	U	U
Orange Extract		S	S
Oxalic Acid*	Dilute	S	S
Oxalic Acid	Sat'd	S	S
Oxygen	100%	S	
Ozone	100%	O	U
Perchloric Acid	10%	S	S
Petroleum Ether		U	U
Phenol	90%	U	U
Phosphoric Acid	up to 30%	S	S
Phosphoric Acid	30-90%	S	S
Phosphoric Acid	90%	S	U
Phosphorus (Yellow)	100%	S	
Phosphorus Pentoxide	100%	S	S
Phosphorus Trichloride		S	
Photographic Solutions		S	S
Pickling Baths			
Hydrochloric Acid*		S	S
Sulfuric Acid*		S	S
Sulfuric-Nitaric*		S	
Picric Acid*	1%	S	
Plating Solutions			
Brass*		S	S
Cadmium*		S	S
Chromium*		U	U
Copper*		S	S
Gold*		S	S
Indium		S	S
Lead*		S	S
Nickel*		S	S
Rhodium*		S	S
Silver*		S	S
Tin*		S	S
Zinc*		S	S
Potassium Bicarbonate	Sat'd	S	S
Potassium Borate	1%	S	S
Potassium Bromate	10%	S	S
Potassium Bromide	Sat'd	S	S
Potassium Carbonate		S	S
Potassium Chlorate	Sat'd	S	S
Potassium Chloride	Sat'd	S	S
Potassium Chromate	40%	S	S
Potassium Cyanide	Sat'd	S	S
Potassium Dichromate	40%	S	S

Table 2.171 *(continued)*

Reagent	Concentration	Temp. 70° F	Temp. 140° F
Potassium Ferri/ Ferro Cyanide	Sat'd	S	S
Potassium Fluoride		S	S
Potassium Hydroxide	20%	S	S
Potassium Hydroxide	Conc.	S	S
Potassium Nitrate	Sat'd	S	S
Potassium Perborate	Sat'd	S	S
Potassium Perchlorate	10%	S	S
Potassium Permanganate	20%	S	S
Potassium Persulfate	Sat'd	S	S
Potassium Sulfate	Conc.	S	S
Potassium Sulfide	Conc.	S	S
Potassium Sulfite	Conc.	S	S
Propargyl Alcohol*		S	S
Propyl Alcohol*		S	S
Propylene Dichloride	100%	U	U
Propylene Glycol*		S	S
Rayon Coagulating Bath*		S	S
Sea Water		S	S
Selenic Acid		S	S
Shortening*		S	S
Silicic Acid		S	S
Silver Nitrate Sol'n.		S	S
Soap Solution*	Any Conc'n	S	S
Sodium Acetate	Sat'd	S	S
Sodium Benzoate	35%	S	S
Sodium Bicarbonate	Sat'd	S	S
Sodium Bisulfate	Sat'd	S	S
Sodium Bisulfite	Sat'd	S	S
Sodium Borate		S	S
Sodium Bromide	Dil. Sol'n	S	S
Sodium Carbonate	Conc.	S	S
Sodium Chlorate	Sat'd	S	S
Sodium Chloride		S	S
Sodium Cyanide		S	S
Sodium Dichromate	Sat'd	S	S
Sodium Ferricyanide	Sat'd	S	S
Sodium Ferrocyanide	Sat'd	S	S
Sodium Fluoride	Sat'd	S	S
Sodium Hydroxide	Conc.	S	S
Sodium Hypochlorite		S	S
Sodium Nitrate		S	S
Sodium Sulfate		S	S
Sodium Sulfide	25%	S	S
Sodium Sulfide	Sat'd Sol'n	S	S
Sodium Sulfite	Sat'd	S	S
Stannic Chloride	Sat'd	S	S
Stannous Chloride	Sat'd Sol'n	S	S
Starch Solution*	Sat'd	S	S
Stearic Acid*	100%	S	S
Sulfur	Colloidal	S	
Sulfur Dioxide	Dry, 100%	S	S
Sulfur Dioxide	Wet, 100%	S	
Sulfur Trioxide		S	S
Sulfuric Acid	0-50%	S	S
Sulfuric Acid	70%	S	O
Sulfuric Acid	80%	S	U
Sulfuric Acid	96%	O	U
Sulfuric Acid	98% (Conc.)	O	U
Sulfuric Acid, Fuming		U	U
Sulfurous Acid		S	S
Tallow		S	O
Tannic Acid*	10%	S	S
Tanning Extracts*	Comm.	S	S
Tartaric Acid	10%	S	S
Tartaric Acid	Sat'd	U	U
Tetrahydrofuran		U	U
Titanium Tetrachloride	Sat'd	U	U
Toluene		U	U
Transformer Oil		O	U
Trichloroethylene		U	U
Triethanolamine	100%	S	U
Trisodium Phosphate	Sat'd	S	S
Turpentine		S	U
Urea*	up to 30%	S	S
Urine		S	S
Vinegar	Comm.	S	S
Vanilla Extract*		S	S
Wetting Agents*		S	S
Whiskey*		S	
Wines		S	S
Xylene		U	U
Yeast		S	S
Zinc Chloride	Sat'd	S	S
Zinc Sulfate	Sat'd	S	S

Table 2.171 *(continued)*

Figure 2.51 This large, corrosion-resistant, filament-wound, glass-fiber-reinforced TS polyester plastic stack and breach is used in a chemical plant. It uses bell and spigot joints for ease of installation.

materials is available. Low-temperature considerations may eliminate many of the thermoplastics. Polyphenylene oxide can be used at temperatures as low as –170°C (–273°F). Thermosetting materials exhibit minimum embrittlement at low temperatures (Fig. 2.64).

Different plastics behave differently when exposed to temperatures; most plastics can take greater heat than humans or buildings can take. Some cannot take boiling water, others operate at 150°C (300°F), and a few can withstand up to 540°C (1,000°F). The flexible (elastomer) plastics at room temperature become less flexible as they are cooled, finally becoming brittle at a certain low temperature. Then there are plastics that reach 1,370°C (2,500°F) with exposures in fractions of a second. An excellent test to determine whether a plastic can take heat is to put the plastic in an automobile trunk or a railroad boxcar where temperatures can reach 55°C (130°F).

It is important to understand that there is a temperature transition in plastics, also called ductile-to-brittle transition temperature. It is the temperature at which the properties of a material change. Depending on the material, the transition change may or may not be reversible. A few other characteristics are presented. The plastic softening range temperature is the temperature at which a plastic is sufficiently soft to be distorted easily. A number of tests exist (ASTM), and the temperatures arrived at may vary according to the particular test method. Softening range is sometimes erroneously referred to as the *softening point*. Temperature stability is the percent change, usually in tensile strength or in percent elongation, as measured at a specified temperature and compared with values obtained under the standard conditions of testing.

Temperature can influence short- and long-time static and dynamic mechanical properties (chapter 19), aesthetics, dimensions, electronic properties, and other characteristics. The so-called high-temperature types can take various degrees of continuous use above 149°C, and there are

Table 2.172a Table of contents in the PDL book *Chemical Resistance: Volume I—Thermoplastics*, 2/e. Example is provided in Table 2.172b.

Table 2.172a *(continued)*

Table 2.172a *(continued)*

Vinyl Resin

Table 2.172a *(continued)*

Reagent	Reagent Note	Conc. (%)	Temp. (°C)	Time (days)	Load	PDL Rating	% Change Weight	% Retained Tensile Strength	% Retained Modulus	Resistance Note	Material Note
Acetaldehyde			23			2				not resistant	Miles Makrolon
Acetic Acid		5	23	7		8	0.56			no effect	GE Lexan
		10	23	1		1		37			GE Lexan 141
		20	93			8				resistance	Miles Makrolon
		25	23	365		6	0.58			no effect, embrittles	GE Lexan
	concentrated		23			2				not resistant	Miles Makrolon
			23	1		2				attacked, cracking, distort	GE Lexan
			23	7	0,25%	8		90.1			
			23	7	0,25%	8		90.6			LNP; 30% gl. fib.
			23	7		8		92.2			"
			23	7		8		92.7			
	glacial		93	1		3		67			GE Lexan 141
Acetic Anhydride			23	6		2				attacked, crystallizing	GE Lexan
			93	1		0		0			GE Lexan 141
Acetone			23			2				no resistant	Miles Makrolon
			23	0	1%	0				instant rupture	Miles Makronlon 2608
	5 min. immersion		23	0.0035	0.32%	2		52			Miles Makrolon; 3.2 mm thick bars
	stressed										(Bisphanol A PC)
	"		23	0.0035	0.63%	1		39			"
	"		23	0.0035	0.95%	0				specimen broken	"
	5 min. immersion		23	0.0035		4		70			"
			23	3		2				attacked	GE Lexan
Acetonitrile			93	1		1					GE Lexan
Acetophenone			93	1		0					"
Acetyl Chloride			93	1		0					"
Acrvlonitrile			23			2				not resistant	Miles Makrolon
Adhesives	acrylic cement		23	3	1%	2				attacked	GE Lexan
	silicone rubber		23	3	1%	8				not effect	"
	acrylic cement		23	3		2				attacked	"
	silicone rubber		23	3		8				no effect	"
	"		85	3	1%	4				crazing	"
	"		85	3		8				no effect	"
Adiprene L100			23	3	1%	8				"	"
			23	3		8				"	"
			85	3	1%	8				"	"
			85	3		8				"	"
After Shave Lotion	Old Spice		23	2	1%	4				crazing	"
	"		23	2		8				no effect	"
	"		85	2	1%	4				crazing	"
	"		85	2		8				not effect	"
Aircraft Turbine Oils			23	30		8				"	"
			85	30		8				"	"
Ally Alcohol			23	182	0.5%	7	0.9			"	"
			23	182		7	0.9			"	"
Alluminum Ammonium Sulfate			23	365	0.5%	8	0.29			"	"
			23	365		8	0.29			"	"
Aluminum Chloride			23			8				resistant	Miles Makronlon
			23	365	0.5%	8	0.3			no effect	GE Lexan
			23	365		8	0.3			"	"

Table 2.172b Chemical resistance of polycarbonates (Vol. I, first page of twenty-three pages on PC)

Table 2.173a Table of contents in the PDL book *Chemical Resistance: Volume II—Thermoplastic Elastomers, Thermosets, and Rubbers*, 2/e. Example is provided in Table 2.173b.

Table 2.173a *(continued)*

Table 2.173a *(continued)*

Table 2.173a *(continued)*

plastics that reach at least 538°C (1,000°F). Examples of temperature influence on plastics are provided in Tables 2.211 to 2.215 (143).

With a temperature change, the short-term static strength, elastic modulus, and elongation behavior of a material will be similar for its tensile, compressive, flexural, and shear properties. A material's strength and modulus will decrease and its elongation increase with an increasing temperature at constant strain.

Reagent	Reagent Note	Conc. (%)	Temp. (ºC)	Time (days)	PDL Rating	% Change Weight	% Change Length	% Change Volume	% Retained Tensile Strength	% Retained Elongation	Hardness Change	Resistance Note	Material Note
Acetic Acid	3n		23	7	9			0 - 3				excellent resistance	BASF Elasiollan 1164D (TPEU); annealed 20hrs @ 100C
	"		23	7	7			3 - 15				good resistance	BASF Elasiollan 1185A (TPEU); annealed 20hrs @ 100C
	"		23	7	9			0 - 3				excellent resistance	BASF Elasiollan C64D (TPEU); annealed 20hrs @ 100C
	"		23	7	7			3 - 15				good resistance	BASF Elasiollan C85A (TPEU); annealed 20hrs @ 100C
		5	22	270	4		<0.5		15.6	64	A - 6	white retains 21,7% of tear str	Miles Texin 591AR (TPAU) 3.2 mm slabs, cured 16hrs. @ 93 deg C
		5	22	365	3		<0.5		28	13.3	D - 18	retains 8% of tear str.	Miles Texin 355DR (TPAU) 3.2 mm slabs, cured 16hrs. @ 93 deg C
		5	22	365	3		<0.5		6.4	25.5	A - 13	minor effect	Miles Texin 480AR (TPAU) 3.2 mm slabs, cured 30days. @ 22 deg C
		20			6	5 - 15		5 - 15	80 - 90	80 - 90		little or no effect	Dow Pellethane 2102 - 55D (TPAU)
		20			8	<5		<5	>90	80 - 90		minor effect	Dow Pellethane 2102 - 65A (TPAU)
		20			6	5 - 15		5 - 15	80 - 90	70 - 80		"	Dow Pellethane 2102 -80A (TPAU)
		20			6	5 - 15		5 - 15	80 - 90	70 - 80		moderate effect	Dow Pellethane 2102 -90A (TPAU)
		20			4	15 - 30		15 - 30	70 - 80	70 - 80		"	Dow Pellethane 2103-55D (TPAU)
		20			4	15 - 30		15 - 30	70 - 80	70 - 80		"	Dow Pellethane 2103-80AE (TPEU)
		20			4	15 - 30		15 - 30	70 - 80	70 - 80		"	Dow Pellethane 2103-90A (TPEU)
		20			4	15 - 30		15 - 30	70 - 80	70 - 80		"	Dow Pellethane 2103-90AE (TPEU)
Acetone			20		6							little resist, OK only under certain condit.	BASF Elastollan C64D (TPAU)
			20		5			40				"	BASF Elastollan C85A (TPAU)
			22	365	3		12.5		46	115.6	D - 24	white, retains 16.7% of tear str.	Miles Texin 355DR (TPAU) 3.2 mm slabs, cured 16 hrs @ 93 deg-C
			22	365	1		37.7		21.6	127.3	A - 48	retains 11.4% of tear str.	Miles Texin 480AR (TPAU) 3.2 mm slabs, cured 30 days @ 22 deg-C
			23	7	2			>30				poor resistance	BASF Elastollan 1164D (TPEU): annealed 20hrs @ 100C
			23	7	2			>30				"	BASF Elastollan 1185A (TPEU): annealed 20hrs @ 100C
			23	7	2			>30				"	BASF Elastollan C64D (TPAU): annealed 20hrs @ 100C
Aluminum Chloride		10	23	7	7			3 - 15				good resistance	BASF Elastollan 1164D (TPEU): annealed 20hrs @ 100C
		10	23	7	7			3 - 15				"	BASF Elastollan 1185A (TPEU): annealed 20hrs @ 100C
		10	23	7	9			0 - 3				excellent resistance	BASF Elastollan C64D (TPAU): annealed 20hrs @ 100C
		10	23	7	9			0 - 3				"	BASF Elastollan C85A (TPAU): annealed 20hrs @ 100C
Ammonia	3n		23	7	9			0 - 3				"	BASF Elastollan 1164D (TPEU): annealed 20hrs @ 100C
	"		23	7	9			0 - 3				"	BASF Elastollan 1185A (TPEU): annealed 20hrs @ 100C
	"		23	7	9			0 - 3				"	BASF Elastollan C64D (TPAU): annealed 20hrs @ 100C
	"		23	7	9			0 - 3				"	BASF Elastollan C85A (TPAU): annealed 20hrs @ 100C
		10	22	270	4		3		27.3	112	A - 6	snow-white: retains 62.3% of tear str.	Miles Texin 591AR (TPAU) 3.2 mm slabs, cured 16 hrs @ 93 deg-C
		10	22	365	3		1.3		28	13.3	D- 10	blistered, whitened, retains 10.7% of tear sstr.	Miles Texin 355DR (TPAU) 3.2 mm slabs, cured 16 hrs @ 93 deg-C
		10	22	365	4		<0.5		5.9	14.5	A - 6	snow-white: reatains 20.5% of tear str.	Miles Texin 480AR (TPAU) 3.2 mm slabs, cured 30 days @ 22 deg-C
Ammonium		1	20		9			1				resist., little wgt. or vol cng	BASF Elastollan C85A (TPAU)
		10	20		3							not resist., short term contact demages parts	BASF Elastollan C64D (TPAU)
		10	20		9			1				resist., little/no wgt or vol. chg.	BASF Elastollan C85A (TPAU)
Ammonium Chloride		10	23	7	7			3 - 15				good resistance	BASF Elastollan 1164D (TPEU): annealed 20hrs @ 100C
		10	23	7	7			3 - 15				"	BASF Elastollan 1185A (TPEU): annealed 20hrs @ 100C
		10	23	7	9			0 - 3				excellent resistance	BASF Elastollan C64D (TPAU): annealed 20hrs @ 100C
		10	23	7	9			0 - 3				"	BASF Elastollan C85A (TPAU): annealed 20hrs @ 100C
Aniline			20		3							not resist., short term contact demages parts	BASF Elastollan C64D (TPAU)
			23	7	2			>30				poor resistance	BASF Elastollan 1164D (TPEU): annealed 20hrs @ 100C
			23	7	2			>30				"	BASF Elastollan 1185A (TPEU): annealed 20hrs @ 100C
			23	7	2			>30				"	BASF Elastollan C64D (TPAU): annealed 20hrs @ 100C
			23	7	2			>30				"	BASF Elastollan C85A (TPAU): annealed 20hrs @ 100C

Table 2.173b Chemical resistance of urethane thermoplastic elastomer (Vol. II, first page of twenty pages)

Curves for creep isochronous stress and isometric stress are usually produced from measurements at a fixed temperature. Complete sets of these curves are sometimes available at temperatures other than the ambient. It is common, for instance, to find creep rupture or apparent modulus curves plotted against log time, with temperature as a parameter. Figure 2.65 shows time-temperature shifting of apparent modulus curves to estimate the extended time values at lower temperatures (chapter 19).

These curves suggest that it would be reasonable to estimate moduli at somewhat longer times than the data available from the lower temperatures. However, a set of creep-rupture curves from various temperatures, as in Figure 2.66, would suggest that projecting the lowest-temperature curves to longer times could produce a dangerously high prediction of rupture strength, so this approach is not recommended (chapter 19). As previously reviewed, one advantage of conducting complete creep-rupture testing at elevated temperatures is that although such testing for endurance requires

Pigment	Strength	Heat stability	Weather resistance	FDA status	Cost $/lbs
Simple metal oxides					
Titanium dioxide (Pigment White 6)	G	G-E	G-E	yes	1.0
Zinc oxide (Pigment White 4)	P	G-E	G-E	yes	1.0
Iron oxides (Pigment Red 101)	F	G-E	G-E	yes	1.0
Chrome (III) oxide (Pigment Green 17)	F	E	E	yes	2.0
Mixed metal oxides					
Nickel titanate (Pigment Yellow 53)	P	E	E	no	4.0
Nickel chrome titanate (Pigment Brown 24)	P	E	E	no	4.0
Copper chromate (Pigment Black 28)	P	E	E	no	6.0
Cobalt chromite (Pigment Green 26)	F	E	E	no	8.0
Cobalt aluminate (Pigment Blue 28)	F	E	E	yes	15.0
Metal sulfides					
Zinc sulfide (Pigment White 7)	F	G-E	G	yes	1.5
Cadmium sulfide (Pigment Yellow 35,37)	G	E	F	no	12.5
Cadmium sulfoselenide (Pigment Orange 20)	G	E	G	no	17.0
Cadmium sulfoselenide (Pigment Red 108)	G	E	E	no	22.0
Ultramarine pigments					
Ultramarine blue (Pigment Blue 29)	F	E	P-F	yes	2.0
Ultramarine violet (Pigment Violet 15)	P	E	P-F	yes	4.5

Table 2.174 Inorganic pigments

long times, the strength levels of the plastic at different temperatures can be developed in relatively short times, usually just 1,000 to 2,000 hours. Underwriters Laboratories and other such organizations have employed such a system for many years.

Testing different impact properties at various temperatures produces a plot that looks very much like an elongation versus temperature curve. As temperatures drop significantly below the ambient temperatures, most TPs lose much of their room-temperature impact strength. A few, however, are on the lower, almost horizontal portion of the curve at room temperature and thus show only a gradual decrease in impact properties with decreases in temperature. One major exception is provided by the glass-fiber RPs, which have relatively high Izod impact values, down to at least –40°F. The S-N (fatigue) curves for TPs at various temperatures show a decrease in strength values with increases in temperature. In comparison, the TSs, specifically the TS RPs, can have very low losses in strength.

Pigment	Strength	Heat stability	Weather resistance	FDA status	Cost $/lbs
Carbon black					
Furnace black (Pigment Black 7)	E	E	G-E	no	0.7
Channel black (Pigment Black 7)	E	E	G-E	yes	4.0
Phthalocyanines					
Red shade phthalo blue (Pigment blue 15:1)	E	G	G-E	yes	12.0
Green shade phthalo blue (Pigment Blue 15:3)	G-E	G	G-E	yes	10.0
Blue shade phthalo green (Pigment Green 7)	G	E	G-E	yes	15.0
Yellow shade phthalo green (Pigment Green 36)	G	E	G-E	no	18.0
Quinacridones					
Quinacridone red (Pigment Violet 19)	G	F	G-E	yes	30.0
Quinacridone violet (Pigment Violet 19)	G	F	G-E	yes	30.0
Quinacridone magenta (Pigment Red 122)	G	G	G-E	no	35.0
Quinacridone (Pigment Red 202)	G	G-E	G-E	no	40.0
Other organic pigments					
Disazo yellow (Pigment Yellow 93,95)	G	G	G-E	PY 95 yes	32.0
Quinophthalone yellow (Pigment Yellow 138)	G	G	F	no	27.0
Disazo red (Pigment Red 220)	G	G	G	yes	45.0
Azoic red (Pigment Red 177)	G	G	G	yes	60.0

Table 2.175 Organic pigments

Dye	Strength	Heat stability	Weather resistance	FDA status	Cost $/lbs
Antraquinone dyes					
Solvent Red 111	G	E	E in mass F in tint	no	25.0
Solvent Red 52	G	E	G	no	100.0
Solvent Blue 59	G	F-G	F	no	35.0
Solvent Blue 97	G	E	F	no	55.0
Solvent Yellow 163	G	E	F	no	30.0
Solvent Green 28	G	E	F	no	65.0
Solvent Green 3	G	G	F	no	40.0
Solvent Violet 13	G	F	F	no	45.0
Perinone dyes					
Solvent Red 135	G	G	E in mass F in tint	yes, PET	35.0
Solvent Red 179	G-E	E	F	no	43.0
Solvent Orange 60	G	G	E in mass F in tint	no	40.0
Fluorescent dyes					
Solvent Yellow 160	E	G-E	F	no	100.0
Vat Red 41	E	G	F	no	85.0
Solvent Orange 63	E	G	F	no	115.0

Table 2.176 Dyes

Product	Pigment content, %	Carrier
Powder	100%	
Encapsulated pigments	100%	Silane (R-types)
Pastes	86%	Plasticizer, mineral oil, solvents
Pellets	95%	Pending

Table 2.177 Gold bronze pigments

Product	Pigment content, %	Carrier
Powder	100%	
Encapsulated pigment	100%	Silica (PCR-types), polymer (PCA-types)
Pastes	65-80%	Plasticizer, mineral oil, polyol, solvents
Pellets	80-95%	Polyolefin, proprietary blends

Table 2.178 Aluminum pigments

Type of treatment	Applications	Performance characteristics
Aluminum pigments, silica encapsulated, (PCR-type)	PVC, powder coating, thermoplastic resins, thermosetting resins	Improved chemical resistance and outdoor durability
Aluminum pigments, polymer coated (PCA-type)	PVC, powder coating, thermoplastic resins, thermosetting resins	Improved chemical resistance and outdoor durability; better resin compatibility
Bronze pigments, silane coated (Resist type)	PVC, powder coating, thermoplastic resins, thermosetting resins	Improved heat and storage stability; better resin compatibility

Table 2.179 Encapsulated metallic pigments

Color	Resin	Relative concentration	L*	C*	Relative strength, %
Blue	POM	X	37.05	55.01	
	PPS	3X	47.78	34.34	42
	LCP	3X	61.60	27.93	16
	LCP	6X	56.14	35.09	26
Red	PA	X	35.91	41.61	
	PPS	3X	44.43	33.67	48
	LCP	3X	56.87	34.13	23
Yellow	POM	X	77.52	77.94	
	PPS	2X	69.35	52.45	50
	LCP	15X	78.13	51.03	31

Table 2.180 Relative color strength in various plastics

TRANSPARENCY

Maximum transparency is available in acrylic, polycarbonate, polyethylene, ionomer, polysulfone, and styrene compounds (Fig. 2.67). They permit transferring light in different forms, such as shown in Figure 2.68. Many other thermoplastics may have adequate transparency. Polycarbonate and polysulfone provide high-temperature transparency (Table 2.216).

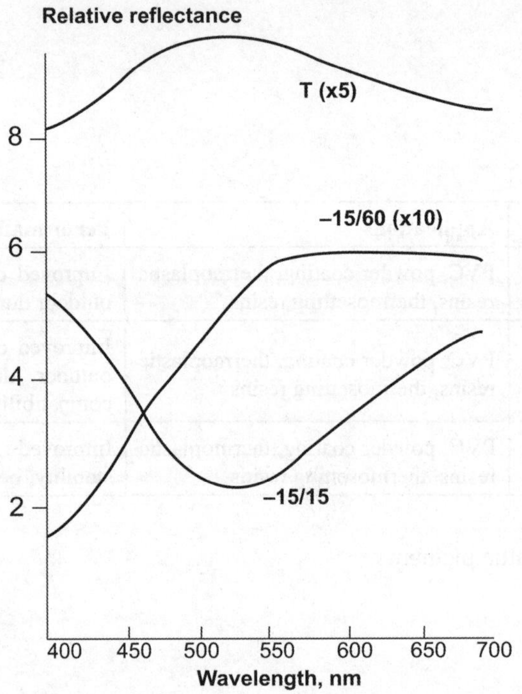

Colorant added	Reflection color	Transmission color	Third color
Yellow Yellow	Red Green	Green Red	Yellow-Green Orange
Red Red	Yellow Blue	Blue Yellow	Violet Orange
Blue Blue	Red Green	Green Red	Blue-Green Violet
Green Green	Yellow Blue	Blue Yellow	Blue-Green Yellow-Green

Table 2.181 Colorants and transmission colors differ

WEATHERING

Many plastics have short lives when exposed to outdoor conditions. The better materials for weatherability include acrylic, chlorotrifluorethylene, vinylidene fluoride, chlorinated polyether, polyester, alkyd, and black linear polyethylene (Fig. 2.69). Some of these plastics have successfully

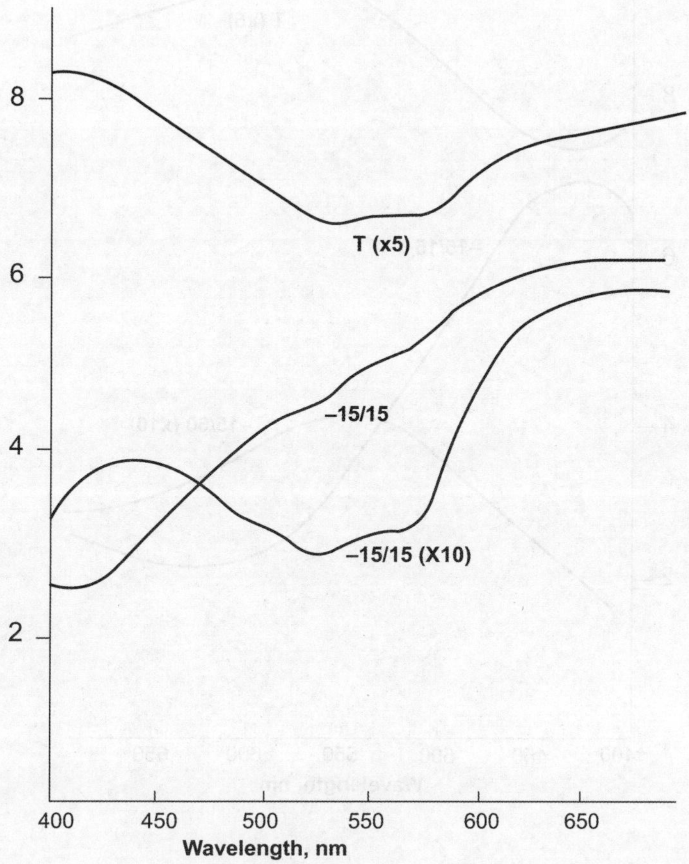

Colorant added	Reflection color	Transmission color	Third color
Yellow	Blue	Yellow	Reinforced yellow
Red	Green	Red	Reinforced Red
Blue	Yellow	Blue	Reinforced blue
Green	Red	Green	Reinforced green

Table 2.182 Colorants and transmission colors are the same

performed outdoors for over a half century. Black materials are best for outdoor service. Some of the styrene copolymers are suitable for certain outdoor uses (chapter 21).

Moisture For high-moisture applications, polyphenylene oxide, polysulfone, acrylic, butyrate, diallyl phthalate, glass-bonded mica, mineral-filled phenolic, chlorotrifluoroethylene, vinylidene,

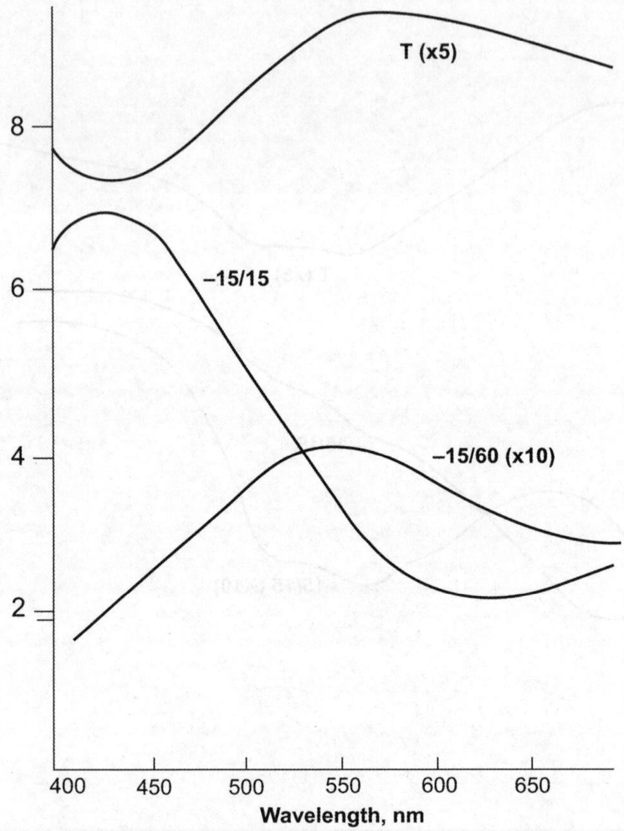

Color added	Reflectance color	Transmission color	Third color
Yellow	Yellow	Blue	Neutral
Red	Red	Green	Neutral
Blue	Blue	Yellow	Blue-Green
Green	Green	Red	Orange

Table 2.183 Colorants and transmission colors are complementary

chlorinated polyether chloride, vinylidene fluoride, and the fluorocarbons should be satisfactory. Diallyl phthalate, polysulfone, and polyphenylene oxide have performed well with moisture/ steam on one side and air on the other (a troublesome combination). They also will withstand repeated steam autoclaving. Long-term studies of the effect of water have disclosed that chlorinated

Color	Meaning	Color	Meaning
Red	Fire, heat, excitement, strength	Light Green	Freshness, crispness, coolness
Rose	Daintiness, softness, fragrance, freshness	Purple	Royalty, stateliness, opulence
Orange	Warmth, action, power, tastiness	Lavender	Fragrance, richness, refinement
Maroon	Richness, solidity, luxury, quietness	Brown	Wholesomeness, mellowness, utility
Yellow	Brightness, airiness, refreshment	Gray	Mildness, softness, reserve primness
Dark Blue	Coldness, formality, haughtiness	White	Purity, professionalism, cleanliness, chasity
Light Blue	Coolness, fragility, daintiness, youthfulness	Black	Strength, mystery, heaviness, coldness
Dark Green	Unhealthfulness, cheapness, coldness		

Table 2.184 Color meanings

Color	*Miles*
Red	3.0
Green	2.5
White	2.0
Yellow	1.0
Blue	.5
Violet	.5

Table 2.185 Comparative visibility at a distance

Test Results per Heat Performance Test	Stabilizer Formulation, Minutes		
	Standard Medical (Calcium-Zinc)	Best Non-Medical (Barium-Cadmium)	New Medical (Proprietary)
Dynamic stability[a] *(at 205 °C and 100 rpm)*	21	48	>60
Static stability[b] *(at 210 °C)*			
Initial color	10	20	20
Decomposition	40	60	50

a: Relates to behavior of PVC during extrusion when it is subject to shear.
b: Relates to behavior of PVC when it encounters "dead spots" and material "hang-ups" during processing or when it is stored under high-heat conditions.

Table 2.186 Time before onset of discoloration or degradation in three 80 Shore vinyl compounds (courtesy of Teknor Apex)

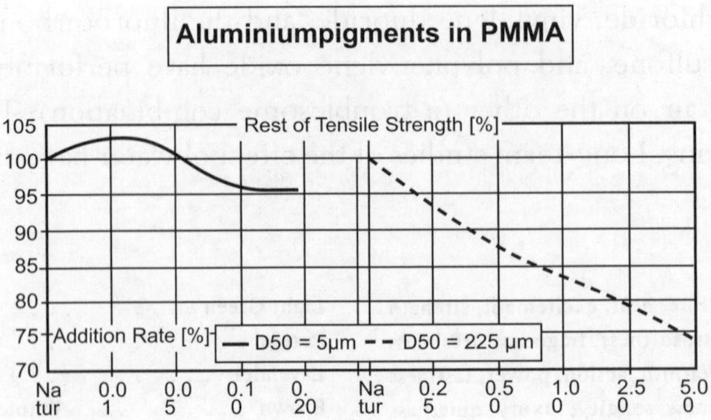

Figure 2.52 Tensile strength vs. pigment concentration

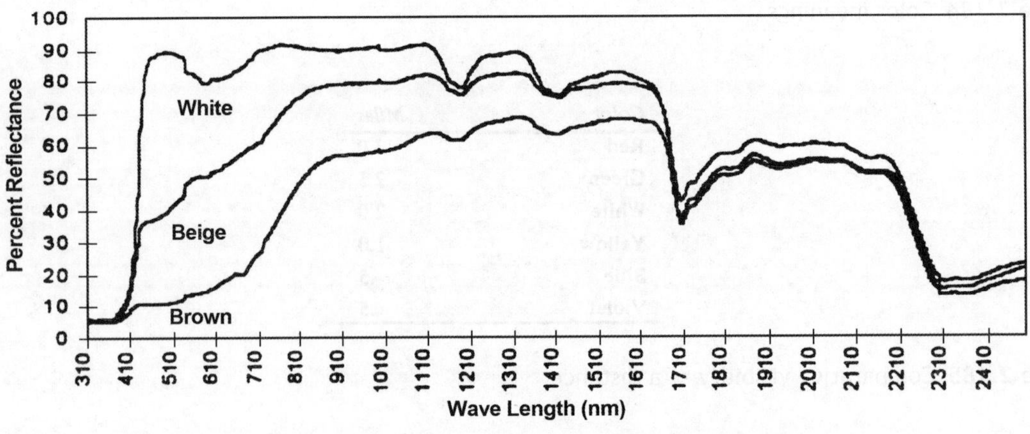

Figure 2.53 Spectral reflectance curves for three colors of rigid vinyl

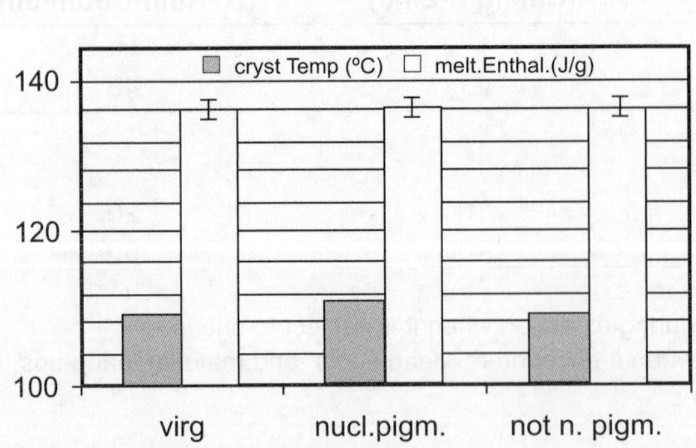

Figure 2.54 Effect of pigmentation on the thermal properties of turbo-blended PE

polyether gives outstanding performance. Impact styrene plus 25 wt% graphite and high-density polyethylene with 15% graphite give long-term performance in water.

The effect of having excess moisture manifests itself in various ways, depending on the process being employed. The common result is a loss in both mechanical and physical properties for hygroscopic and nonhygroscopic plastics. During injection molding, splays, nozzle drool between shot-size control, sinks, and other losses may occur. The effects during extrusion can include gels, trails of gas bubbles in the extrudate, arrowheads, wave forms, surging, lack of size control, and poor appearance.

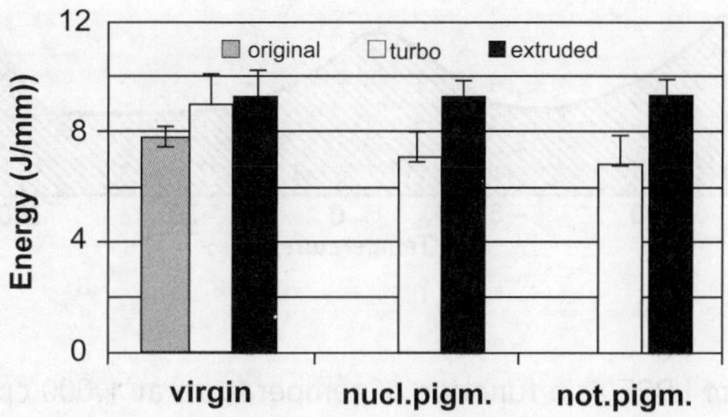

Figure 2.55 Effect of pigmentation and mixing on the impact strength of PE

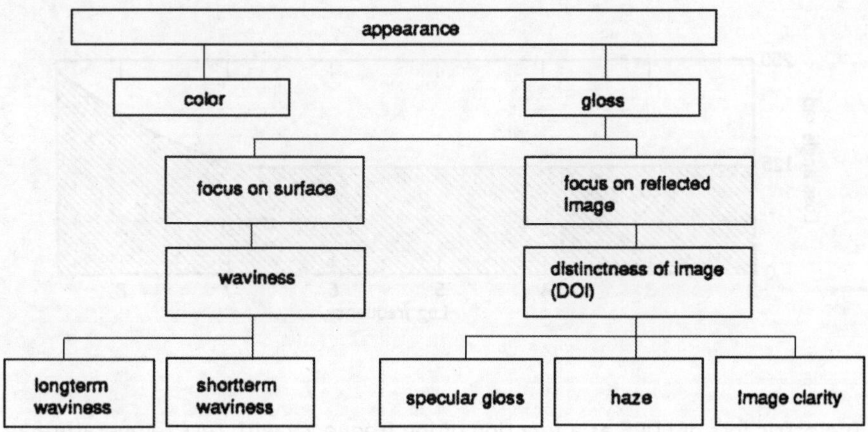

Figure 2.56 Different types of surface appearance

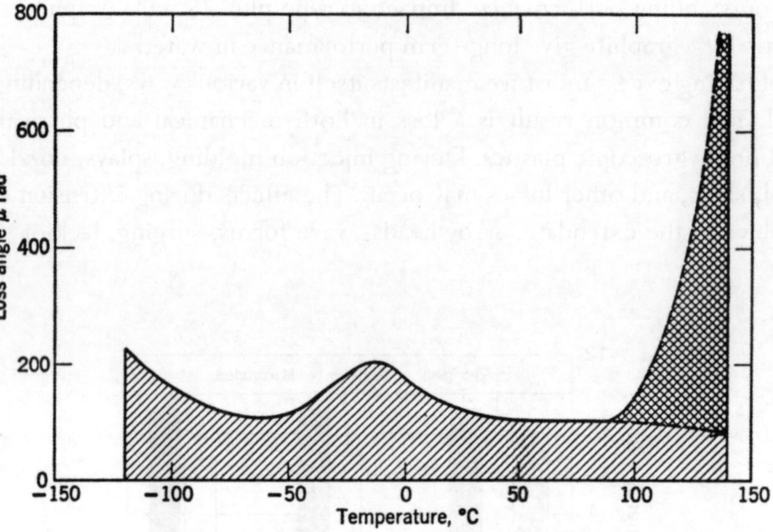

Figure 2.57 Dielectric loss of LDPE as a function of temperature at 1,000 cps

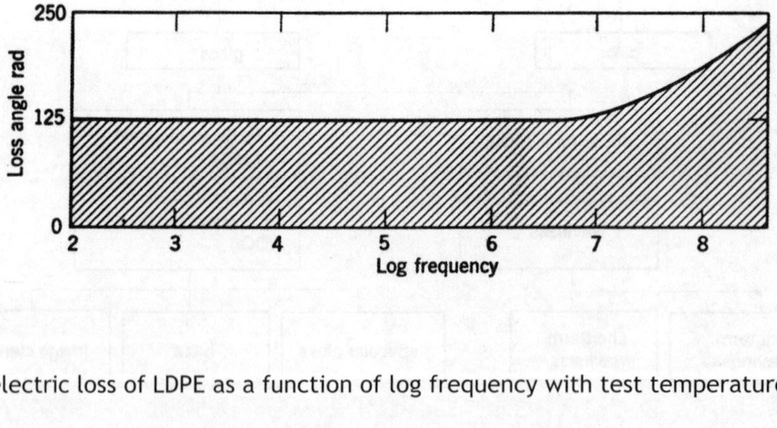

Figure 2.58 Dielectric loss of LDPE as a function of log frequency with test temperature at 20°C

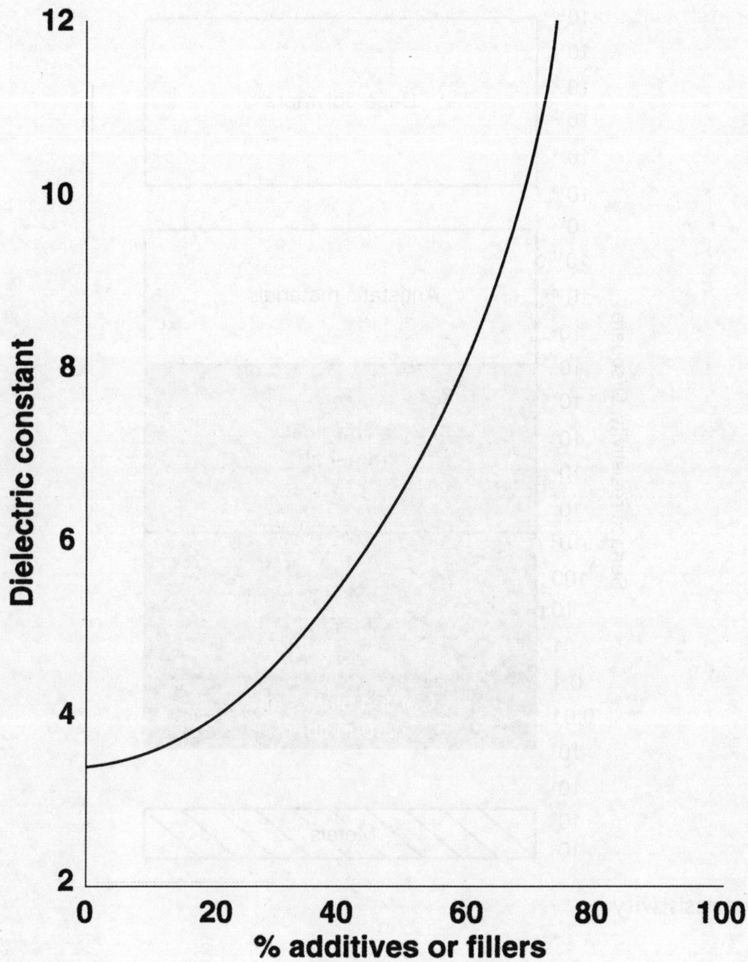

Figure 2.59 Dielectric constant

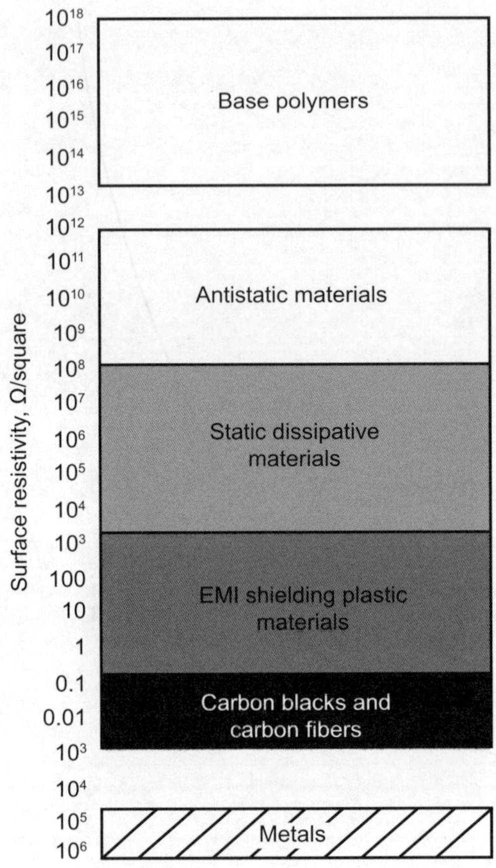

Figure 2.60 Surface resistivity

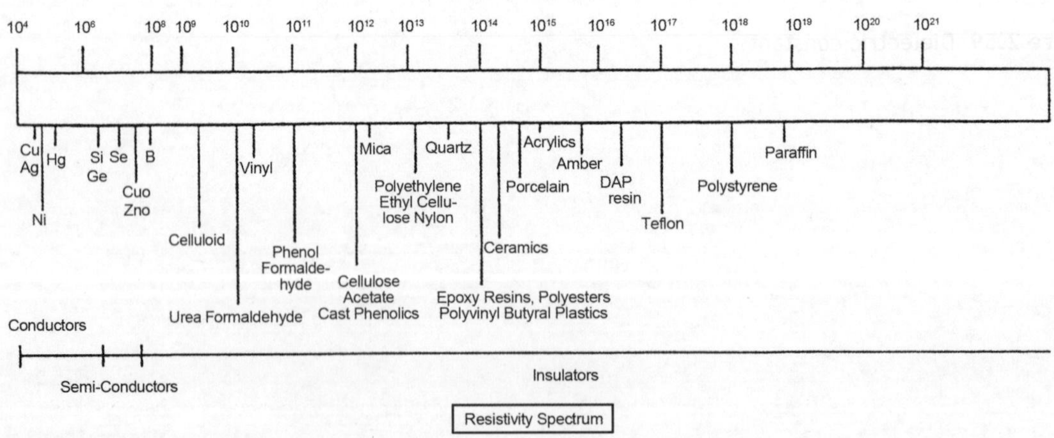

Figure 2.61 Volume resistivity

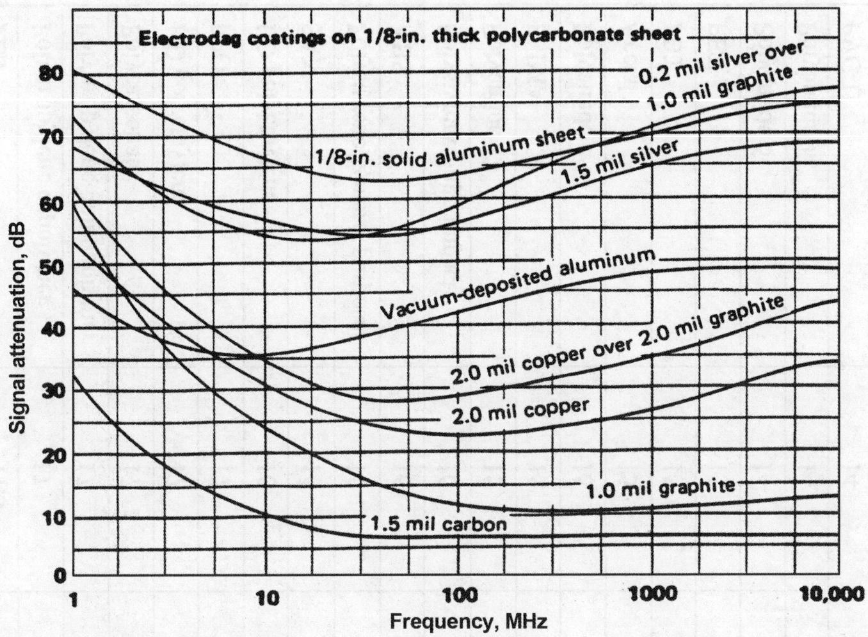

Figure 2.62 Conductive coating shielding

Polymer	Volume reistivity (log ohm cm)	Dielectric strength (MV/m)	Dielectric constant (1KHz)	Dissipation fector (1kHz)
PES	17.5	16	3.5	0.0021
Polypropylene copolymer	17	28	2.3	0.0005
Polypropylene homopolymer	17	28	2.28	0.0001
Polycarbonate	17	23	3	0.001
HD polyethylene	17	22	2.3	0.0005
PPO	17	21	2.6	0.0004
LD polyethylene	16	27	2.3	0.0003
SAN	16	25	3	0.01
Polystyrene, general purpose	16	20	2.6	0.0002
ABS	16	20	2.8	0.007
Polystyrene, high impact	16	15	2.8	0.0006
Polyamide 12	15	60	3.6	0.05
PMMA	15	25	3.3	0.03
Polyamide 6/6	15	25	8	0.2
Acetal	15	20	3.7	0.0015
PBT	15	20	3.2	0.002
PET	15	17	3.3	0.002
Polyamide 6	14	25	8	0.2
Polyamide 11	14	20	4	0.05
PVC-U	14	14	3.1	0.025
Polyamide 6/10	13	20	4.7	0.09

Table 2.187 Electrical properties of thermoplastics

	ASTM	PEI	PES	PPS	PSF	PC
Electrical						
Dielectric strength, v/mil	D 149	710	400	380	425	425
Arc resistance, sec	D 495	128	120	34	39	115
Volume resistivity, ohm/cm	D 257	10^{18}	10^{16}	10^{15}	10^{17}	10^{17}
Physical						
Specific gravity, g/c³		1.27	1.37	1.30	1.55	1.20
Water absorption, % by weight	D 570	0.25	0.43	0.02	0.3	0.26
Thermal						
HDT at 264 psi. °F	D 648	392	397	275	345	265
Long-term service temp. UL index, °F		338	356	—*	302	255
Oxygen index, %		40	38	47	38	25
Mechanical						
Flexural modulus, psi (MPa)	D780	480,000 (3,310)	375,000 (2,586)	550,000 (3,792)	390,000 (2,689)	340,000 (2,344)
Impact strength, notched Izod, ft./lb./in.	D 256	1.0	1.6	0.4	1.3	2.2
Tensile strength, psi (MPa)	D 638	15,200 (104.8)	12,200 (84.12)	9,500 (65.5)	10,200 (70.33)	9,500 (65.5)

Table 2.188 Electrical and other properties of electrical-grade plastics

Material	Dielectric Strength (MV/m)	Dielectric Constant
ABS (high impact)	25	2.8
Acetal (homopolymer)	20	3.8
Acetal (copolymer)	20	3.8
Acrylic	25	3.5
Epoxy	16	3.5
Modified PPO	22	2.6
Nylon 66 (33% glass)	15	8.0
PEEK	19	3.2
PET (crystalline)	17	3.3
PET (36% glass)	50	4.0
Phenolic (mineral filled)	12	5.0
Polyamideimide	23	3.5
Polycarbonate	23	3.0
Polyetherimide	33	3.2
Polyethersulphone	16	3.5
Polyimide	22	3.5
Polypropylene	28	2.3
Polysulphane	16	3.0
Polystyrene	20	2.5
Polythene (LD)	27	2.3
Polythene (HD)	22	2.3
PTFE	45	2.1
PVC (rigid)	14	3.1
PVC (flexible)	30	6.0
SAN	25	3.0
BMC (polyester)	15	5.0
SMC (polyester)	12	5.0

Table 2.189 Plastics' dielectric strength and constant

	Resistivity		Dielectric Constant/Dissipation Factor				
Material	Volume	Surface	100 Hz	1 kHz	1 mHz	100 mHz	1,000 mHz
ABS	2×10^{16}	10^{14}	.005/2.9	.006/2.8	.008/2.8	.005/2.7	.001/2.7
Acrylic	10^{18}	10^{14}	.062/3.6	.058/3.2	.045/3.1		
Cellulose Ester	3×10^{15}	10^{14}	.006/3.8	.011/3.6	.024/3.3	.020/3.0	.014/2.1
FEP	10^{18}	10^{16}	.0005/2.1	.0005/2.1	.0005/2.1	.0008/2.09	.0007/2.05
Nylon 6	10^{15}	10^{13}	.031/4.2	.024/3.8	.031/3.8		
Polycarbonate	10^{16}	10^{15}	.001/3.1	.0013/3.1	.007/3.1	.015/3.1	
Polyethylene	10^{19}	10^{16}	.0001/2.34	.0001/2.34	.0001/2.34	.0001/2.34	.0001/2.34
Alkyd	10^{13}	10^{14}	.02/6.0	.02/5.8	.015/5.4		
DAP	10^{16}	10^{13}	.026/3.8	.020/3.7	.016/3.6		
Phenolic	10^{14}	10^{9}	.013/5.4	.013/5.3	.033/4.9		
Epoxy	10^{16}	10^{14}	.004/3.22	.004/3.25	.004/3.25		

Table 2.190 Plastics' resistivity and dielectric constant at different frequencies

Material	Arc Resistance, s	Critical Tracking Index, V
ABS	89	400+
ABS-PC	91	250+
DAP	150	600+
POM	220	600+
PAR	78	200
LCP	192	175+
MF	180	600+
Nylon 6	60	600+
Nylon 6/6	60	600+
Nylon 12	120	600+
PBT	184	600+
PC	10	100+
PBT-PC	99	260
PEI	126	100+
PESV	20	150
PET	125	250+
PF	190	100+
PPO	34	400+
PPS	60	130
PSU	120	100+
SMA	—	600+
UP	—	600+

Table 2.191 Plastics' arc resistance and tracking index

Material	Dissipation factor	
Methylpentene Polymer (TPX)	.000025	–
Polystyrenes (General Purpose)	.0001	– .0004
Polypropylene (Copolymer filler)	>.0001	– .0018
Polypropylene Impact Rubber modified	>.0003	–
Polystyrene (Medium and High Impact)	.0004	– .0020
Polyallomer	>5	x 10^{-4}
Polyethylenes (Low Density)	>.0005	–
Polyethylenes (Medium Density)	>.0005	–
Polyethylenes (High Density)	>.0005	–
Polypropylene (Unmodified)	>.0005	– .0018
Polystyrenes (Heat and Chemical)	.0005	– .0050
Polyphenylene Sulphides (Unfilled)	.0007	–
Polyphenylene Oxides (Unfilled) PPO	.0009	–
Polyphenylene Oxides (Modified Unfilled)	.0009	–
Polyethylenes Crosslinkable Grades	.001	– .002
Polystyrenes (20-30% Glass filler)	.001	– .003
Styrene Butadiene Thermoplastic Elastomers	.001	– .003
Polyphenylene Oxides (Modified 20-30% Glass filler)	.0015	–
Ionomers	.0019	–
Silicones (Mineral filler)	.002	– .005
Silicones (Glass filler)	.002	– .020
Polypropylene (Inert filler)	.0021	–
Polysulfone	.0034	–
Polypropylene (Glass filler)	.0035	–
Acrylics (Impact)	.004	– .02
Polyphenylene (40% Glass filler)	.0041	–
Acetal Homopolymer	.0048	–
Polyamides Aromatic	.005	–
Polybutylene	.005	–
Epoxy (Low Density)	.005	– .012
Phenolic (Mica filler)	.005	– .013
Acetal Copolymer	.006	–
ABS Polycarbonate Alloy	.006	– .013
PVC Rigid	.006	– .019
Polyester Alkyd Granular Putty Type (Mineral filler)	.006	– .04
Acetal (20% Glass filler)	.0062	–
Styrene Acrylonitrile (Glass filler)	.00668	–
Polycarbonate (10-40% Glass filler)	.0067	– .0075
Polyaryl-ether	.007	–
Styrene Acrylonitrile Copolymer (SAN)	.007	– .010
ABS (High Impact)	.007	– .015
ABS (Heat Resistant)	.007	– .015
ABS (Medium Impact)	.007	– .015
ABS (Self Extinguishing)	.007	– .015
ABS (20-40% Glass filler)	.007	– .015
Polycarbonate (10% Glass filler)	.0075	–
Polyester (Premixed Chopped Glass filler)	.008	– .022
Polyester Alkyd Granular Putty Type (Glass filler)	.008	– .023
Diallyl Phthalate (Glass filler)	.009	– .014
Chlorinated Polyether	.01	–
Epoxy (Glass filler)	.01	–
Epoxy (Mineral filler)	.01	–
Epoxy Encapsulating (Glass filler)	.01	–
Epoxy Encapsulating (Mineral filler)	.01	–
Phenolic Acrylonitrile Copolymer (Asbestos filler)	.010	–
Polycarbonate (Unfilled)	.010	–
Polyester Alkyd Granular Putty type (Synthetic filler)	.010	– .016
Ethylene – Ethyl Acrylate Copolymer (EEA)	.01	– .02
Polypropylene Modified Vinyl Chloride (Rigid)	.01	– .02
(Rigid)	.01	– .02
Phenolic (Glass filler)	.010	– .026
Polyester Woven Cloth	.01	– .03
Polyester Preformed chopped rovings	.01	– .03
Cellulose Acetate Butyrate	.01	– .04
Cellulose Propionate	.01	– .05
Ethyl Cellulose	.010	– .060
Cellulose Acetate	.01	– .10
Diallyl Phthalate (Synthetic Fiber filler)	.012	– .020
Polyaryl sulfone	.013	–
Nylons (Polyamide Type 6/10 (20-40% Glass filler)	.016	– .022
Polyester Sheet Molding Compound	.016	– .024
Nylons (Polyamide) Type 6/6 (20-40% Glass filler)	.017	– .018
Polyester (Low Shrink)	.019	–
Nylons (Polyamide) Type 6 (20-40% Glass filler)	.019	– .021
Acrylics Multipolymer	.02	–
Nylons (Polyamide) Type 6/12	.02	–
Polyesters (18% Glass filler)	.0120	–
Acrylics	.02	– .03
Diallyl Phthalate (Mineral filler)	.02	– .04
Polyesters (Thermoplastic)	.0208	–
Melamine Alpha Cellulose	.027	– .045
Acrylic PVS Alloy (Kydene)	.03	–
Nylons (Polyamide) Type 12	.03	–
Ethylene Vinyl Acetate Copolymer EVA	.03	– .05
Phenolic Formaldehyde (Cotton filler)	.03	– .07
Phenolic Formaldehyde (Woodflour filler)	.03	– .07
Phenolic (Fabric and Cord filler)	.03	– .09
Melamine (Cellulose filler)	.032	– .060
Melamine (Fabric filler)	.036	– .041
Melamine (Phenolic filler)	.04	– .06
Polyester Alkyd Granular Putty type (Asbestos filler)	.04	– .06
Melamine (Asbestos filler)	.041	– .050
PVC (Flexible unfilled)	.04	– .14
Melamine (Phenolic Modified Fabric filler)	.050	– .065
Urethanes Elastomers	.050	– .075
Cellulose Nitrate	.06	– .09
Cold Mold Nonrefractory (Organic)	.07	–
Phenolic Acrylonitrile Copolymer (Woodflour and Flock filler)	.08	– .10
Phenolic Acrylonitrile Copolymer (Rag filler)	.08	– .15
PVC (Flexible filler)	.09	– .10
Nylon (Polyamide) Type 6/12 (Glass filler)	.16	– .17
Urea Formaldehyde	.25	– .35
Phenolic Formaldehyde (Asbestos filler)	.35	– .80

Table 2.192 Plastics' dissipation (power) factor at 10^6 cycles

Material	Continuous material temperature range (°F)	Radiation damage threshold (rads)	Stability-space UV and vacuum	Relative rating	Remarks
Electrical Insulation					
Polyethylene expanded tubing, heat shrinkable Types 1, 2, and 3	to 215	2×10^7 to 10^8	Fair	A	Type 1: Heat shrinkable jacket insulation for equipment Type 2: Material for repair of jacket defects Type 3: Dual-wall shrinkable material for embedding and parts (~75%) Excellent electrical properties, for wire connector insulation and membrane insulation
Modified neoprene rubber tubing, heat shrinkable, flexible	—65 to 250	3×10^6 to 3×10^8	Fair	C	For repair of jacket where oil resistance is needed; softens in benzene group solvents and in ketones
Polytetrafluorethylene coated tape, glass braid lacing	to 500	10^6	Good	B	Teflon coating limits radiation resistance, excellent electrical properties at elevated temperature
Vinyl coated tape, glass braid lacing and tying	to 215	2×10^7 to 5×10^8			Excellent electrical properties

Table 2.193 Electrical insulation and dielectric plastic materials

Irradiated modified polyolefin, insulation for vehicle wiring	—65 to 275	5×10^8	Fair	A	Used for vehicle wiring and all flight items, thermally stabilized, and irradiated modified polyolefin insulated wire
Polyamide (Nylon)	—65 to 300	5×10^6 to 10^8	Good	C	Electrical insulation
Polyester (Mylar), glass filled	—70 to 300	2×10^9 to 5×10^9	Good	B	Electrical insulation
Polytetrafluorethylene (Teflon)	to 400	10^5 (air) to 10^6 (vacuum)	Good	B	Special applications requiring unique electrical properties or heat resistance, where radiation total dose is low
Natural rubber (antirad type)	—40 to 225	2×10^7 to 2×10^8	Poor	B	Elastomer for vibration damping; wire insulation and sheathing
Neoprene rubber (polychloroprene)	—65 to 250	3×10^6 to 2×10^8	Fair	C	Elastomeric cable sheathing, molded parts, bumpers, and shock absorbers, jacketing and sheathing only, not insulation
Synthetic rubber (Buna-N)	—40 to 250	10^7	Fair	C	Cable jacket and tubing material; wire insulation
Neoprene sheath, Buna-N insulated	—40 to 250	6×10^6 to 3×10^7	Fair	C	Cable sheath; wire insulation
Butyl rubber	—65 to 250	10^6 to 10^7	Good	C	Elastomer for vibration damping; wire insulation or sheathing

Table 2.193 *(continued)*

Material	Continuous material temperature range (°F)	Radiation damage threshold (rads)	Stability-space UV and vacuum	Relative rating	Remarks
Electrical Insulation (continued)					
Polyvinyl formal (Formvar)	—75 to 325	2×10^7 to 3×10^8	Good	B	Magnet wire insulation; solar array application
Polyethylene	—70 to 250	2×10^7 to 2×10^8	Fair	A	Embrittled in UV radiation
Polyester (Mylar)	—100 to 300	5×10^7 to 5×10^8	Fair in UV unless metallized Good in vacuum	B	Electrical insulation in manufacture of cables. Thermosetting
Polyvinyl chloride	—500 to 200	2×10^7 to 2×10^8	Poor	C	Electrical insulation and vibration protection. Plasticized types outgas and embrittle above 10^8 rad. Liberates HCl above 10^7 rad
Kel-F (monochlorotrifluorethylene) AMS 3650	—80 to 300	3×10^6 to 2×10^7	Fair	C	Electrical insulation, thermoplastic elastomer; embrittled in UV radiation; may evolve corrosive gases
Tetrafluorethylene (Teflon)	—150 to 500	10^5 air to 10^6 (vacuum)	Good	B	Electrical insulation, thermoplastic; low radiation resist-fluoride gases at higher ance, releasing corrosive doses
Polymethyl methacrylates	—80 to 190	1×10^7	Good in vacuum; fair in UV	C	Magnet wire insulation; thin coating

Table 2.193 *(continued)*

Circuit Boards and Other Electrical Laminates					
Copper-clad epoxy, glass laminate	—65 to 300	5×10^9	Good	A	Printed circuit board
Epoxy-glass laminate	—65 to 300	5×10^9	Good	A	Material in thickness of 7 mils up to 0.25 in.
Epoxy-glass laminate	—65 to 250	5×10^9	Good	A	Satisfactory where used within temperature limits; antenna window and related uses
Phenolic glass laminate	—65 to 450	8×10^9	Excellent	A	Elevated temperature resistance to 700°F for short time less than 10 min; brackets, insulating blocks, and strips
Silicon glass laminated sheet	—65 to 500	3×10^8	Fair	A	Use only where continuous temperature is over 300°F
Silicone glass laminate	—65 to 500	3×10^8	Fair	A	Good electrical properties and temperature resistance; antenna application
Diallyl phthalate	—100 to 400	10^8		B	Used for molded potting containers; specify type and size
Copper-clad Teflon glass laminate	—65 to 400	10^5 (air) 10^6 (vacuum)	Good	C	Restricted Use: Do not use in general applications; use only for VHF or UHF where other material is not satisfactory or a very precise or low dielectric constant is required. Specify laminate and copper thickness

Table 2.193 *(continued)*

Material	Resistivity Volume	Resistivity Surface	Dielectric Constant/Dissipation Factor 100 Hz	1 kHz	1 mHz	10 mHz	100 mHz	1.000 mHz
ABS	2×10^{16}	10^{14}	.005/2.9	.006/2.8	.008/2.8	.007/2.8	.005/2.7	.001/2.7
Acrylic	10^{18}	10^{14}	.062/3.6	.058/3.2	.045/3.1	.033/2.9		
FEP	10^{18}	10^{16}	0005/2.1	.0005/2.1	.0005/2.1	.0005/2.1	.0008/2.09	.0007/2.05
Nylon 6	10^{15}	10^{13}	.031/4.2	.024/3.8	.031/3.8	.020/4.0		
Polycarbonate	10^{16}	10^{15}	.001/3.1	.0013/3.1	.007/3.1	.011/3.1	.015/3.1	
Polyethylene	10^{19}	10^{16}	.0001/2.34	.0001/2.34	.0001/2.34	.0001/2.34	.0001/2.34	.0001/2.34
Alkyd	10^{13}	10^{14}	.02/6.0	.02/5.8	.015/5.4			
DAP (SD15)	10^{16}	10^{13}	.026/3.8	.020/3.7	.016/3.6			
Epoxy	10^{16}	10^{14}	.004/3.22	.004/3.25	.004/3.25			

Table 2.194 Plastic resistivity and dielectric constant/dissipation factor data

Material	Conductivity σ S/cm	Density o g/cm^3
Copper	5.9×10^5	8.9
Silver	6.3×10^5	10.4
Aluminum	3.6×10^5	2.7
Polyacetylene with iodine	2.0×10^4	0.8
Polypyrrole with phenylsulfonate	1.5×10^2	1.3
Polystyrene	10^{-16}	1.05

Table 2.195 Plastics' and other materials' electrical conductivity

Material	Application	Continuous material temperature range (°F)	Radiation damage threshold (rad)	Remarks
Polyurethane Encapsulants				
Type I				
Eccofoam FP (Class A, B, and C) Emerson and Cuming, Inc.; Hitco R (Class A, B, and C) H. I. Thompson Co.	Foam-in-place	—65 to 165	10^8	Rigid foam encapsulant for mechanical shock and vibration damping; type is determined by maximum temperature of use, class is determined by density: Class A: 2 lb/ft^3, Class B: 10 lb/ft^3, Class C: 20 lb/ft^3
Type II				
Eccofoam FP (Class A, B, and C) Emerson and Cuming, Inc.; Hitco MT (Class A, B, and C) H. I. Thompson Co.	Foam-in-place	—65 to 250	10^9	
Type III				
Eccofoam FPH (Class A, B, and C) Emerson and Cuming, Inc.; Hitco HT (Class B and C) H. I. Thompson	Foam-in-place	—65 to 350	10^9	
PE–102 American Latex Co.				
Type I	Foamed slab	—65 to 165		Flexible foam packing material; specify density and softness; not to be used for structural appli-
Type II	Foam-in-place			

Table 2.196 Electrical encapsulating materials

Table 2.196 *(continued)*

Silicone Rubber				
Dow-Corning Corporation Class 1 Q–9–0031–1/2 Class 2 Q–9–3003–1/2 Class 3 Q–9–0006–1/2 Class 4 RTV–503 Products Research Company Class 1 PR–1930–1/2 Class 2 PR–1920–1/2 Class 3 PR–1910–1/2	Potting	—85 to 500	5×10^6	Requires rigid mold for mechanical shock and vibration; class determinded by viscosity; should not be used at 400–500°F unless given a progressive cure in 100° steps to 500°F; use for elevated temperature where limited to room temperature to cure
Silicone Rubber Primer				
General Electric Co. SS–4004 Products Research Company 1902	Primer to RTV rubber	—85 to 500	5×10^6	Primer used silicone rubber; 1-hr cure at room temperature
Polysulfide				
3C–737, Class A 3C–747T, Class B Churchill Chemical Co.	Potting	—70 to 300	10^6	Use for cable connectors; not for general purpose potting or embedding of electric modules
Epoxy Encapsulants				
Insulating Lacquer 1162 Dennis Chemical Company	Moisture resistant coating (thin)	—65 to 200	10^7	Spray or dip coat (<0.005 in.); printed wiring board coating and preencapsulation check for nonwelds in welded modules
Coating, IRFNA and UDMH Resistant Furane Plastics, Inc., H2C–101	Coating, missile fuel and acid	—65 to 200	10^7	Packaged in aerospray bomb; acid and missile-fuel-resistant overspray coating

Epocast H2E-011 Furane Plastics, Inc. Scotchcast No. 3 3M Co.	Impregnating	—65 to 300	5×10^8	Low-viscosity coil impregnant for inductors; not recommended for welded modules; rigid, unfilled epoxy.
Epocast H2E-012 Furane Plastics, Inc.	Potting	—65 to 300	5×10^8	Rigid, filled epoxy.
Hysol 4238/3475 Class A Hysol 4239/3487 Class B Eccobond 70C Class C	Conformal coating	—65 to 250		Electrically conductive; r-f interference applications
Stycast 2762/Cat 17 Emerson and Cuming, Inc.	Potting and embedding	—65 to 500		Require elevated temperature cure for 500°F use
Epoxy Encapsulants Epocast 202/9615, Class A and B Epocast 202/9647, Class C and D Epocast 202-11/9615A, Class E Furane Plastics, Inc.	Potting, embedding, conformal coating	—65 to 200	10^7	Preferred material in its type for radiation protected applications in electronic enclosures in vehicles; clear amber-liquid resin; class is determined by hardness; hardener/resin ratio can be varied to produce Shore D hardness from 3 to 95; low Shore D will have decreased resistanct to radiation and vacuum; shore D range of 50–85 is recommended for encapsulation of soldered and welded modules
Epocast H2E-037 Furane Plastics, Inc.	Embedding	—65 to 250	10^8	Filled opaque compound for electrical insulation; semirigid, good thermal shock resistance

Table 2.196 *(continued)*

Fillers	Conductivity σ S/cm
Carbon black	0.01 to 0.1
Aluminum platelets	1 to 50
Steel fibers	1 to 50
Carbon fibers	0.1 to 10
Mica coated with nickel	1 to 10
Glass flakes with nickel	1 to 10

Table 2.197 Conductivity of fillers

Material	Form	Characteristics	Applications
Diallyl Phthalate	Powder	Excellent humidity resistance. Withstands temperature over 400 F. Easily processed. Dimensional stability excellent. May be compression, transfer and injection molded. Low cost.	Precision potentiometers, RF connectors, waveguide auxili aries, attenuators, heating panels, heated battery cases.
Epoxies	Powder, one and two-part liquids and paste	Many types of resins available, providing wide spectrum of properties. Easy to compound. Low shrinkage and excellent dimensional stability. Good to excellent adhesion. May be cast or molded.	Coatings, sealants, adhesives, solderless PC boards.
Fluorocarbons (PTFE, FEP, PVF_2)	Powder, emulsions	Excellent high temperature properties. TFE 500 F. FEP is easier to mold, but maximum use temperature is 400 F. Nearly inert chemically. Nonflammable. Loading with conductive filler improves creep resistance. Low coefficient of friction.	High-temperature cable shielding, gaskets, heat-shrinkable tubing.
	Powder, solutions	Excellent thermal stability to over 300 F generally, and over 400 F in special formulations. Broad choice of resins. May be cast or compression, transfer, or injection molded.	Precision potentiometers, RF connectors, heating panels.
Polyimides	Powder, solutions	Excellent high temperature properties, 400 to about 700 F. Difficult to process.	High-temperature cable shielding, conductive film.
Polyolefins (Polyethylene, Polypropylene)	Powder, pellets	Tough and chemical resistant. Weak in creep and thermal resistance. Polyethylene maximum use temperature 210 F, polypropylene 260 F. May be injection and extrusion molded, vacuum formed. Low cost.	Antistatic sheet and tiles, heat-shrinkable tubing, deicer boots.
Natural rubber	Solid	Good physical properties and resistance to cutting and abrasion. Low heat and ozone resistance.	Gaskets.
Silicone	Paste, liquid for foaming	Excellent chemical resistance. High temperature capability to 500 F. New formulations have higher tear strength and lower compression set.	Gaskets, antistatic rollers, RF shielding, heat shrinkable tubing, setting and nonsetting calking, deicer boots, flexible heater tape.
Urethane	Liquid	Exceptional abrasion, cut, and tear resistance. Poor moisture and heat resistance. Variety of formulations leading to different properties including range of durometers without plasticizers.	Antistatic rollers and tires, hose for transfer of flammables, strain gages, pressure transducers.
Vinyls (PVC, PVA, PVAC, Copolymers)	Powder, pellets, organosols	General-purpose material. Many forms available including hard and flexible types. Properties are highly dependent on plasticizer used. May be injection, extrusion, compression molded, vacuum formed. Low cost.	Cable shielding, antistatic sheet and hose, RF gaskets, heat-shrinkable tubing.

Table 2.198 Examples of magnetic field shielding coatings at different frequencies

Shielding System	Advantages	Disadvantages
Vacuum Metalization/Ion Plating	Highly conductive (less than 0.1 ohm per square foot); controllable film thickness; not limited to simple housing designs.	Requires primer coat; entire part must be done, forcing exterior painting; not field repairable; specialized application equipment; vacuum chamber size a limiting factor; requires specialized knowledge.
Electrolysis Deposition	Highly conductive (both nickel and copper less than 0.1 ohm per square foot).	Requires specialized equipment/knowledge; entire part must be coated, forcing exterior painting; if copper is used it must be protected by a nickel coating or some other coating.
Conductive Plastics	Good thermal transfer; elimination of secondary operation for shielding.	Requires a secondary operation for grounding.
Conductive Coatings		
Silver	Highly conductive (0.1 ohm per square foot or less); applied by conventional spray equipment; easy application; electrically stable (minimal change in resistance with environmental cycling).	High cost
Nickel	Low cost (15–30 cents per square foot); good conductivity (less than 1.0 ohm per square foot); applied by conventional spray equipment; easy application; relatively stable (differs with manufacturer).	Lasser quality formulations available; some are stable, some are not.
Copper	Highly conductive (less than 0.5 ohm per square foot); easy application; low cost (15–30 cents per square foot).	Oxidation can reduce conductivity (resistance can change to effectively make copper an insulator); some may be alloys—if layered with silver, cost will rise.
Graphite	Low cost (5–15 cents per square foot); easy application; excellent ESD (electrostatic discharge) performance.	Less conductivity (ranging from 2 ohms to the thousands per square foot, depending upon the amount of graphite); modest shielding capability (up to 30–40 dB).
Arc/Flame Spray	Highly conductive (less than 0.1 ohm per square foot; hard, dense coating.	Requires grit blasting to promote mechanical bonding to plastic; special applications equipment required; requires special applicator safety procedures for dust and fumes; warps thermoplastics; not suitable for thin-walled designs; not field repairable.

Table 2.199 Electromagnetic radiation shielding plastic techniques

Shielding System	Advantages	Disadvantages
Conductive Coatings Silver	Highly conductive (0.1 ohm per square foot or less); applied by conventional spray equipment; easy application; electrically stable (minimal change in resistance with environmental cycling); easily applied to selected area; field repairable.	High cost
Nickel	Low cost; good conductivity (less than 1.0 ohm per square foot); applied by conventional spray equipment; easy application; relatively stable (differs with manufacturer); easily applied to selected area; field repairable.	Lesser quality formulations available; some are stable, some are not.
Copper	Highly conductive (less than 0.5 ohm per square foot); easy application; low cost.	Oxidation can reduce conductivity (resistance can change to effectively make copper an insulator); some may be alloys.
Graphite	Low cost; easy application; excellent ESD (electrostatic discharge) performance.	Less conductivity (ranging from 2 ohms to the thousands per square foot, depending upon the amount of graphite); modest shielding capability (up to 30–40 dB).
Arc/Flame Spray	Highly conductive (less than 0.1 ohm per square foot; hard, dense coating.	Requires grit blasting or chemical etching to promote mechanical bonding to plastic; special applications equipment required; requires special applicator safety procedures for dust and fumes; warps thermoplastics; not suitable for thin-walled designs; not field repairable.
Vacuum Metalization/ Ion Plating	Highly conductive (less than 0.1 ohm per square foot); controllable film thickness; not limited to simple housing designs.	Requires primer coat; entire part must be done, forcing exterior painting; not field repairable; specialized application equipment; vacuum chamber size a limiting factor; requires specialized knowledge; subject to corrosion in humid atmosphere unless protected.
Electrolysis Deposition	Highly conductive (both nickel and copper less than 0.1 ohm per square foot).	Generally requires specialized equipment/knowledge; entire part must be coated, forcing exterior painting; if copper is used it must be protected by a nickel coating or some other coating.
Conductive Plastics	Good thermal transfer; elimination of secondary operation for shielding.	Requires a secondary operation for grounding.

Table 2.200 Examples of conductive coating systems

Conductive coatings silver	Highly conductive (0.1 ohm per square or less); applied by conventional sprayequipment; easy application; electrically stable (minimal change in resistance with environmental cycling); easily applied to selected area; field repairable.	High cost.
Nickel	Low cost (15–30 cents per square foot); good conductivity (less than 1.0 ohm per square); applied by conventional spray equipment; easy application; relatively stable (differs with manufacturer); easily applied to selected area; field repairable.	Lesser quality formulations available; some are stable, some are not.
Copper	Highly conductive (less than 0.5 ohm per square); easy application; low cost.	Oxidation can reduce conductivity (resistance can change to effectively make copper an insulator); some may be alloys—if layered with silver, cost will rise.
Graphite	Low cost (5–15 cents per square foot); easy application; excellent ESD (electrostatic discharge) performance.	Less conductivity (ranging from 2 ohms to the thousands per square depending upon the amount of graphite); modest. shielding capability (up to 30–40 dB).
Arc/flame spray	Highly conductive (less than 0.1 ohm per square); hard dense coating.	Requires grit blasting to promote mechanical bonding to plastic; special applications equipment required; requires special applicator safety procedures for dust and fumes; warps thermoplastic; not suitable for thin-walled designs; not field repairable.
Vacuum metalization/ ion plating	Highly conductive (less than 0.1 ohm per square); controllable film thickness; not limited to simple housing designs.	Requires primer cost; entire part must be done, forcing exterior painting; not field repairable; specialized application equipment; vacuum chamber size a limiting factor; requires specialized knowledge; subject to corrosion in humid atmosphere unless protected.
Electrolysis deposition	Highly conductive (both nickel and copper less than 0.1 ohm per square).	Requires specialized equipment/ knowledge; entire part must be coated, forcing exterior painting; if copper is used it must be protected by a nickel coating or some other coating.
Conductive plastic	Good thermal transfer; elimination of secondary operation for shielding	Requires a secondary operation for grounding.

Table 2.200 *(continued)*

Material	Conductivity σ S/cm	Density o g/cm^3	σo S/(cm^2 g)
Copper	6.1×10^5	8.9	6.6×10^4
Silver	6.3×10^5	9.5	6.0×10^4
Aluminum	3.6×10^5	2.7	1.3×10^5
Polyacetylene with iodine	2.0×10^4	0.8	2.5×10^4
Polypyrrole with phenylsulfonate	1.5×10^2	1.3	1.2×10^2
Polystyrene	10^{-16}	1.05	9.5×10^{-17}

Fillers	Conductivity σ S/cm
Carbon black	0.01 to 0.1
Aluminum platelets	1 to 50
Steel fibers	1 to 50
Carbon fibers	0.1 to 10
Mica coated with nickel	1 to 10

Table 2.201 Examples of material and filler conductivities

Emission Frequency (mHz)	Shielding Effectiveness (dB)		
	Nickel	Copper	Solid Steel
0.01	0	0	50
0.10	2	2	
0.50	5	7	
1.0	8	13	
10	33	37	
30	39	47	

Table 2.202 Examples of conductive coatings subjected to magnetic field shielding

Film	Methyl Ethyl Ketone	Methyl Salicylate	Toluene	Styrene
HDPE/EVOH/EVA (1.25 mil)	0.65	0.0011	0.027	0.0610
PP/EVOH/PP (1.0 mil)	<0.01	0.009	0.0003	<0.0002
PP/PET/PP (1.0 mil)	2.40	2.16	1.310	0.0018
PVDC-coated oriented PP (1.8 mil)	0.44	0.871	0.470	0.0046

Table 2.203 EVOH odor permeability

Type of Polymer	Specific Gravity (ASTM D 792)	Water Vapor Barrier	Gas Barrier	Resistance to Grease and Oils
ABS (acrylonitrile butadiene styrene)	101–1.10	Fair	Good	Fair to good
Acetal—homopolymer and copolymer	1.41	Fair	Good	Good
Acrylic and modified acrylic	1.1–1.2	Fair	——	Good
Cellulosics acetate	1.26–1.31	Fair	Fair	Good
Butyrate	1.15–1.22	Fair	Fair	Good
Propionate	1.16–1.23	Fair	Fair	Good
Ethylene vinyl alcohol copolymer	1.14–1.21	Fair	Very good	Very good
Ionomers	0.93–0.96	Good	Fair	Good
Nitrile polymers	1.12–1.17	Good	Very good	Good
Nylon	1.13–1.16	Varies	Varies	Good
Polybutylene	0.91–0.93	Good	Fair	Good
Polycarbonate	1.2	Fair	Fair	Good
Polyester (PET)	1.38–1.41	Good	Good	Good
Polyethylene				
Low density	0.910–0.925	Good	Fair	Good
Linear low density	0.900–0.940	Good	Fair	Good
Medium density	0.926–0.940	Good	Fair	Good
High density	0.941–0.965	Good	Fair	Good
Polypropylene	0.900–0.915	Very good	Fair	Good
Polystyrene				
General purpose	1.04–1.08	Fair	Fair	Fair to good
Impact	1.03–1.10	Fair	Fair	Fair to good
SAN (styrene acrylonitrile)	1.07–1.08	Fair	Good	Fair to good
Polyvinyl chloride				
Plasticized	1.16–1.35	Varies	Good	Good
Unplasticized	1.35–1.45	Varies	Good	Good
Polyvinylidene chloride	1.60–1.70	Very good	Very good	Good

Table 2.204 Permeability of plastics

Plastic film	Temperature		Film thickness	Water vapor	N_2	Air	O_2	CO_2	H_2	Ar	He
	SI	US									
	°C	°F	µm [1]	g/cm² day	cm³/m² day bar						
PE-LD	23	73	100	1	700	1100	2000	10000	8000	–	–
PE-HD (ρ = 0.95 g/cm³, unstretched)	25	77	40	0.9	525	754	1890	7150	6000	–	–
	30	86	40	1.7	720	960	2270	8600	7600	–	–
	40	104	40	3.5	1220	1660	3560	13100	11400	–	–
	50	122	40	8.1	2140	2650	5650	19500	16800	–	–
PE-HD (ρ = 0.95 g/cm³, stretched)	25	77	40	1.0	430	680	1210	5900	5000	–	–
	30	86	40	1.6	560	830	1530	7200	6000	–	–
	40	104	40	4.3	1050	1490	2650	11200	9400	–	–
	50	122	40	10.5	1870	2670	4650	18100	14800	–	–
E/VA copolymer, VAC 20%	23	73	100	455	1400	–	4000	17000	–	–	–
Polypropylene (unstretched)	25	77	40	2.1	430	700	1900	6100	17700	1480	19200
	30	86	40	3.2	600	960	2500	8400	18200	2100	21700
	40	104	40	7.4	1280	1820	5100	14800	28100	4100	29800
	50	122	40	19.0	2800	3600	9200	27300	46600	8000	43500
Polypropylene (stretched)	25	77	40	0.81	200	350	1000	3300	6700	–	7300
	30	86	40	1.2	260	480	1200	3900	8200	–	8500
	40	104	40	3.3	560	940	2300	7050	12300	–	12100
	50	122	40	8.4	1200	1850	4150	13200	19800	–	17800
PVC-U (unstretched	20	68	40	7.6	12	28	87	200	–	–	–
PVC-U (stretched)	20	68	40	4.4	13	13	43	110	–	–	–
PVC-P	20	68	40	20	350	550	1500	8500	–	–	–
Polyvinylidenechloride	25	77	25	0.1/0.2	1.8/2.3	5/10	1.7/11	60/700	630/1400	–	–
Polystyrene (stretched)	25	77	50	14.0	27	80	235	800	1260	–	–
Polyacetal	20	68	40	2.5	10	16	50	96	420	–	–
PFEP copolymers	40	104	25	2.0	375	–	3000	6500	2000	–	–
CTFE	40	104	25	0.38/0.85	39	–	110/230	250/620	3400/5200	–	–
E/TFE copolymers	23	73	25	0.6	470	–	1560	3800	–	–	–
E/CTFE copolymers	23	73	25	9.0	150	–	39.0	1700	–	–	–
PVF	23	73	25	50.0	3.8	–	4.7	170	900	–	–
Polyamide 6	25	77	25	80/110	14	–	40	200	1500	–	–
Polyamide 66	25	77	25	15/30	11	–	80	140	–	–	–
Polyamide 11	25	77	25	1.5/4.0	50	–	540	2400	5000	–	–
Polyamide 12	25	77	25	0.35	200/280	–	800/1400	2600/5300	–	–	–
Polycarbonate	23	73	25	4	680	–	4000	14500	22000	–	–
Polyethyleneterephthalate (stretched)	23	73	25	0.6	9/15	–	80/110	200/340	1500	–	–
Polysulfone	23	73	25	6	630	–	3600	15000	28000	–	–
PU elastomers	23	73	25	13/25	550/1600	–	1000/4500	6000/22000	–	–	–
Polyimide	23	73	25	25	94	–	390	700	3800	–	–

[1]) 1 µm = 0.0394 mil.

Table 2.205 Plastic film permeability based on DIN 53380 for gases and DIN 53122 for water

Elastomer	Permeability (10^{-7} cm^3/sec-cm^2/cm) (STP) 75°F	176°F	250°F	350°F	400°F
Natural rubber	0.49	4.4	7.1	20.7	26.2
Buna–S (SBR)	0.25	2.9	4.7	15.4	
Neoprene	0.09–0.10	0.98–1.7	2.6–3.0	7.3	
Butyl	0.02	0.32–0.46	1.3–1.8	5.6–6.1	
Nitrile	0.13	0.8	2.2	6.6	
Polysulfide	0.02	0.37	1.6	Melted	
Polyurethane (polyester type)	0.05	0.97	3.1	7.1	Melted
Silicone	11.33	35–47		69–113	74
Acrylic	0.16	1.5	3.7	10.2	
Polyester	0.19	1.8	4.8	9.4	
Methacrylate					16.3
Polyurethane		2.3	3.8	16.6	
Carboxy		2.4–2.6	2.3–6.2	7.1–14	

Table 2.206 Air permeabilities of elastomers at various temperatures

	Water Vapor Transmission gm-mil. t00 in¹/24 hrs. @ 100ºF @ 90% RH (gm-mm/m²/24 hrs. @ 37.8ºC @ 90% RH)	Gas Transmission cc (STP) mil/100 in² 24 hr-ATM @ 77ºF (cc [STP]-mm/m⁴/ 24 hr-ATM @ 25ºC)		
		O^2	N^2	CO^2
ACLAR UltRx, SupRx, Rx	0.016 (0.006)	7 (2.8)	1 (0.4)	14 (5.5)
ACLAR 33C	0.020 (0.008)	7 (2.8)	1 (0.4)	16 (6.3)
ACLAR 22C	0.026 (0.010)	15 (5.9)	2.5 (1.0)	40 (15.7)
ACLAR 22A	0.027 (0.011)	12 (4.7)	2.5 (1.0)	30 (12.0)
PVC. PVdC Copolymer	0.20–0.6 (0.08–0.24)	0.8–6.9 (0.3–2.7)	0.12–1.5 (0.05–0.6)	38–44 (15–17)
Polyethylene				
Low Density	1.0–1.5 (0.39–0.59)	500 (195)	180 (71)	2700 (1060)
Medium Density	0.7 (0.28)	250–535 (100-210)	85–315 (35–125)	100–2500 (40–985)
High Density	0.3 (0.12)	185 (73)	42 (17)	580 (230)
Nylon 6	19–20 (7.5–7.9)	2.6 (1.0)	0.9 (0.35)	4.7 (1.9)
Fluorinated Ethylene Propylene	0.22 (0.08)	715 (281)	320 (125)	1670 (660)
Polyvinyl Fluoride	2.1 (0.81)	7.5 (3)	0.25 (0.10)	11 (4.3)
Polyvinylidene Fluoride	2.5 (1.0)	3.4 (1.34)	9 (3.5)	5.5 (2.2)
Polyester–PET Oriented	1.0–1.3 (0.39–0.51)	3.0–6.0 (1.2–2.4)	0.7–1.0 (0.28–0.39)	15–25 (5.9–9.8)

Table 2.207 Water and gas permeability through plastic films

	Metalized Coextruded Polyolefin	Foil Laminate*
Tensile strength MD	18–19	18–19
CD	12–13	11
Mullen strength	19–20	17
WVTR (gm/csl/24 hrs.)	Approx. .05	.0006
Oxygen transmission (cc/csl/24 hrs.)	Approx. 10	Less than .004
Light transmission	Slightly less than 1%	Approx. 0%
Seal type	Fin only†	Fin or Lap
Seal range (40 psi, 15 sec.)	350–500°F	160–350°F
Deadfold (subjective) (1–10 Scale)	4	7
Flex crack resistance (subjective) (1–10 Scale)	8	5

*.0003" gauge foil, wax laminated to $12^1/_2$# paper, wax laminated to $8^1/_2$# tissue.

†While a lap seal is technically possible, the bond is too weak to be considered commerical.

Table 2.208 Permeability of metalized coextruded LDPE and aluminum-foil laminate

Thermoplastics

Table 2.209a Table of contents in the PDL book *Permeability and Other Film Properties of Plastics and Elastomers*

Table 2.209a *(continued)*

Table 2.209a *(continued)*

Table 2.209a *(continued)*

Table 2.209a *(continued)*

Table 2.209a *(continued)*

Table 2.209a *(continued)*

Table 2.209a *(continued)*

Appendices

Indicies

Table 2.209a *(continued)*

Material Family	ETHYLENE-VINYL ALCOHOL COPOLYMER
Material Supplier/ Grade	EVAL COMPANY EVOH 44
Product Form	FILM
Features	barrier properties
Reference Number	255

MATERIAL CHARACTERISTICS

Sample Thickness	0.04 mm

TEST CONDITIONS

Penetrant	d-limonene
Temperature (°C)	25
Relative Humidity (%)	dry

PERMEABILITY (source document units)

Vapor Transmission (10^{-20} kg · m/m² · sec · Pa)	4700

PERMEABILITY (normalized units)

Vapor Transmission Rate (g · mm/m² · day)	0.41

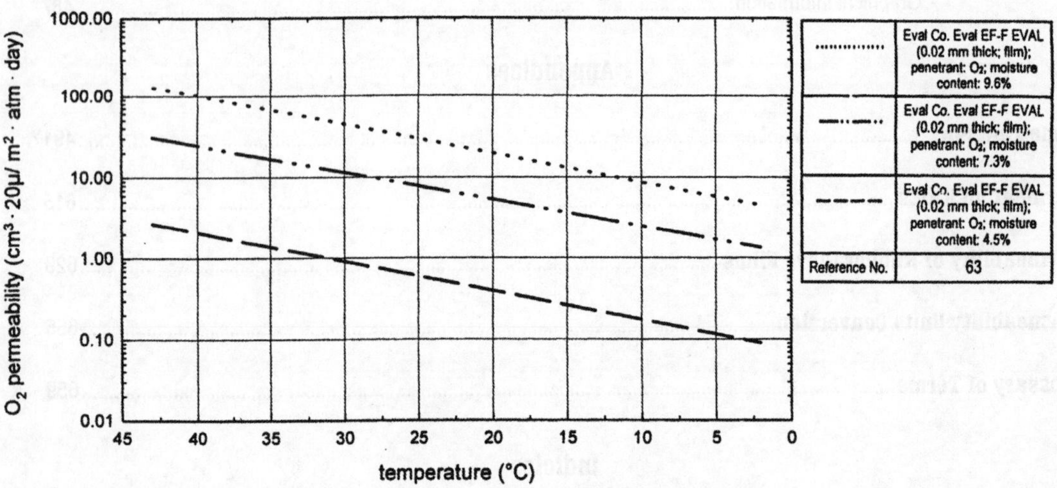

Table 2.209b Ethylene-vinyl alcohol copolymer (one page from thirty-four pages in EVAL section)

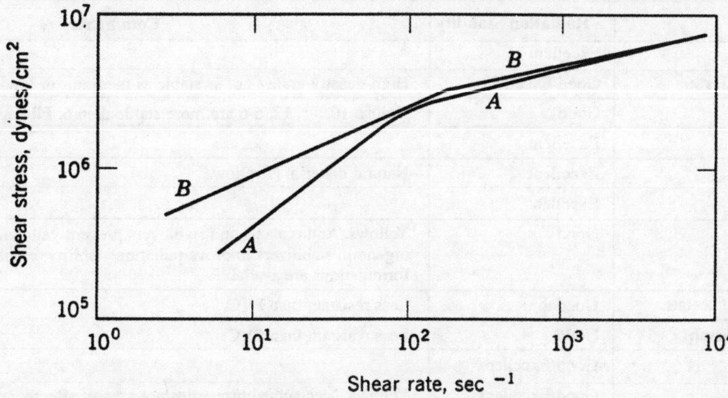

Figure 2.63 Effect of irradiation on FEP before (A) and after (B) exposure to 0.7 Mrad at 250°C under nitrogen

Material	Radiation Stability	Comments
Polystyrene	Excellent	
Polyethylene, various densities	Good/Excellent	High-density grades not as stable as medium- or low-density grades.
Polyamides (nylon)	Good	Nylons 10, 11, 12, 6-6 are more stable than 6. Film and fiber are less resistant.
Polyimides	Excellent	
Polysulfone	Excellent	Natural material is yellow.
Polyphenylene sulfide	Excellent	
Polyvinyl chloride (PVC)	Good	Yellows. Antioxidants and stabilizers prevent yellowing. High-molecular-weight organotin stabilizers improve radiation stability; color-corrected radiation formulations are available.
Polyvinyl chloride/Polyvinyl acetate	Good	Less resistant than PVC.
Polyvinylidene dichloride (Saran)	Good	Less resistant than PVC.
Styrene/acylonitrile (SAN)	Good/Excellent	
Polycarbonate	Good/Excellent	Yellows. Mechanical properties not greatly affected; color-corrected radiation formulations are available.
Polypropylene, natural Polypropylene, stabilized	Poor/Fair	Physical properties greatly reduced when irradiated. Radiation-stabilized grades, utilizing high molecular weights and copolymerized and alloyed with polyethylene, should be used in most radiation applications. High-dose-rate E-beam processing may reduce oxidative degradation.
Fluoropolymers:		When irradiated, PTFE and PFA are significantly damaged. The others show better stability. Some are excellent.
Polytetrafluoroethylene (PTFE)	Poor	
Perfluoro alkoxy (PFA)	Poor	
Polychlorotrifluoroethylene (PCTFE)	Good/Excellent	
Polyinyl fluoride (PVF)	Good/Excellent	
Polyvinylidene fluoride (PVDF)	Good/Excellent	
Ethylene-tetrafluoroethylene (ETFE)	Good	
Fluorinated ethylene propylene (FEP)	Fair	
Polyacetals	Poor	Irradiation causes embrittlement. Color changes have been noted (yellow to green).
ABS	Good	High-impact grades are not as radiation resistant as standard-impact grades.
Acrylics (PMMA)	Fair/Good	
Polyurethane	Good/Excellent	Aromatic discolors; polyesters more stable than esters. Retains physical properties.
Liquid crystal polymer (LCP)	Excellent	Commercial LCPs excellent; natural LCPs not stable.
Polyesters	Good/Excellent	PBT not as radiation stable as PET.
Thermosets:		
Phenolics	Excellent	Includes the addition of mineral fillers.
Epoxies	Excellent	All curing systems.
Polyesters	Excellent	Includes the addition of mineral or glass fibers.
Allyl diglycol carbonate (polyester)	Excellent	Maintains excellent optical properties after irradiation.
Polyurethanes:		
Aliphatic	Excellent	
Aromatic	Good/Excellent	Darkening can occur. Possible breakdown products could be derived.
Elastomers:		
Urethane	Excellent	
EPDM	Excellent	
Natural rubber	Good/Excellent	
Nitrile	Good/Excellent	Discolors.
Polychloroprene (neoprene)	Good	Discolors. The addition of aromatic plasticizers renders the material more stable to irradiation.
Silicone	Good	Phenyl-methyl silicones are more stable than are methyl silicones. Platinum cure is superior to peroxide cure; full cure during manufacture can eliminate most postirradiation effects.
Styrene-butadiene	Good	
Polyacrylic	Poor	
Chlorosulfonated polyethylene	Poor	
Butyl	Poor	Friable, sheds particulates.

Table 2.210 Examples of radiation's effect on plastics

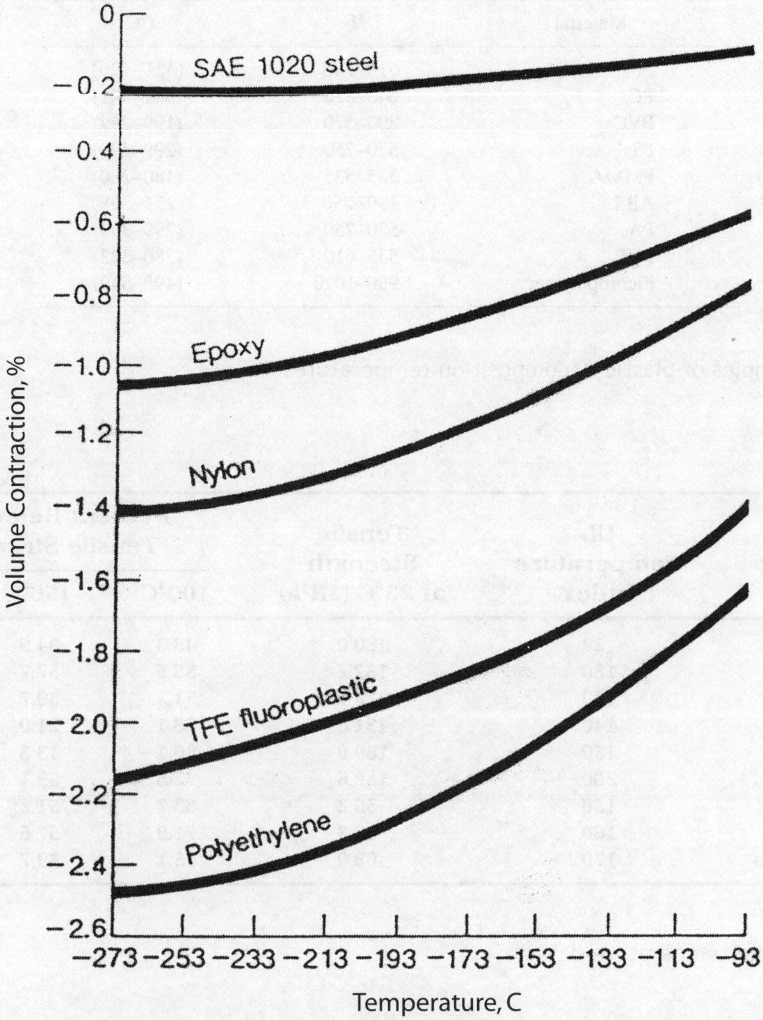

Figure 2.64 Examples of plastic contraction at low temperatures

Material	°F	(°C)
PP	610–788	(321–420)
PC	645–825	(341–441)
PVC	390–570	(199–299)
PS	570–750	(299–399)
PMMA	355–535	(180–280)
ABS	490–750	(254–399)
PA	570–750	(299–399)
PET	535–610	(280–322)
Fluoropolymer	930–1020	(499–549)

Table 2.211 Examples of plastic decomposition temperatures

Engineering Polymer	UL Temperature Index	Tensile Strength at 23°C (MPa)	Percent Retention of Tensile Strength at		
			100°C	150°C	200°C
PPS	22	160.0	48.3	34.9	4.7
PES	180	157.2	85.5	57.7	13.7
Nylon 6/6	130	213.7	51.3	39.7	7.2
PET	140	131.0	38.1	21.0	
P-Sulfane	150	120.0	86.1	13.3	
PAI	200	189.6	72.8	59.1	29.9
Phenolic—GP	150	55.2	83.7	51.2	50.1
Phenolic—HR	160	58.6	71.9	57.6	52.0
Phenolic—glass	170	69.0	75.1	59.7	69.9

Table 2.212 Tensile-temperature data

Engineering Polymer	UL Temperature Index (°C)	Flexural Strength at 23°C (MPa)	Percent Retention of Flexural Strength at		
			100°C	150°C	200°C
Nylon 6/6	115	120.0	36	29	23
Nylon 6/6	120	246.8	51	41	36
PET	150	201.3	46	30	24
PET	155	259.3	51	33	26
Phenolic—GP	150	60.7	67	62	42
Phenolic—HR	160	70.3	77	64	48
Phenolic—glass	170	109.6	82	70	54

Table 2.213 Flexural-temperature data

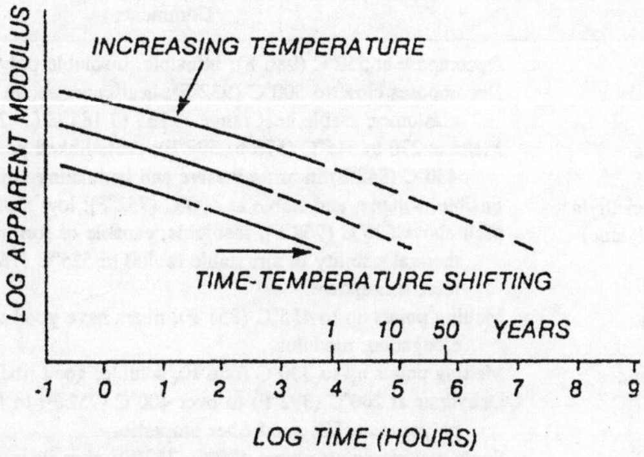

Figure 2.65 Influence of temperature on apparent modulus

Material	ASTM D 648 Deflection Temperature at 1820 kPa (°C)	Test Results at Deflection Temperature: Maximum Stress Load (MPa)	Test Results at Deflection Temperature: Degrees Deflection in 4 hours	Test Results at Deflection Temperature Plus 10°: Maximum Stress Load (MPa)	Test Results at Deflection Temperature Plus 10°: Degree Deflected/ When Occurred
P-Sulfane	175°C	7.24	5	1.90	Failed/< 1 hour
PBT	200°C	3.62	9	1.90	Failed/< 2 hours
PET	210°C	7.24	8	3.62	Failed/< 4 hours
PEI	210°C	7.24	8	1.90	Failed/< 1 hour
PAS	215°C	1.90	Failed		
PPS	260°C	1.90	4	1.90	Failed/< 2 hours
Phenolic HR	210°C	14.48	4	14.48	7°/4 hours
Phenolic glass	270°C	14.48	1	14.48	2°/4 hours

Table 2.214 Deflection-temperature data

Polymer	Comments
Polyphenyls	Decompose at 530°C (986°F); infusible, insoluble polymers.
Polyphenylene oxide	Decomposes close to 500°C (932°F); heat cures above 150°C (302°F) to elastomer; usable heat range −135 to 185°C (−211 to 365°F)
Polyphenylene sulfide	Melts at 270 to 315°C (578 to 599°F); cross-linked polymer stable to 450°C (842°F) in air; adhesive and laminating applications.
Polybenzyls; polyphenethyls	Fusible, soluble, and stable at 400°C (752°F); low molecular weight.
Parylenes (poly-*p*-xylylene)	Melt above 520°C (968°F); insoluble; capable of forming films; poor thermal stability in air; stable to 400 to 525°C (752 to 977°F) in inert atmosphere.
Polyterephthalamides	Melting points up to 455°C (851°F); fibers have good tenacity, elongation, modulus.
Polysulfanyldibenzamides	Melting points up to 330°C (626°F); soluble; good fiber properties.
Polyhydrazides	Dehydrate at 200°C (392°F) to over 400°C (752°F) to form polyoxadiazoles; good fiber properties.
Polyoxamides	Some melting points above 400°C (752°F); give clear, flexible films.
Phenolphthalein polymers	Melting points of 300°C (572°F) to over 400°C (752°F); formable into fiber and film.
Hydroquinone polyesters	Soluble polymers with melting points of 335°C (635°F) to over 400°C (752°F).
Polyhydroxybenzoic acids	Films melt at 380 to 450°C (716 to 842°F); stable to oxidation but not to hydrolysis; tough, flexible films; good thermal stability.
Polyimides	Commercial film, coating, and resin stable up to 600°C (1,112°F); continuous use up to 300°C (572°F).
Polyarylsiloxanes	Good thermal stability 400 to 500°C (752 to 932°F); coatings, adhesives.
Carboranes	Stable in air and nitrogen at 400 to 450°C (752 to 842°F); elastomeric properties for silane derivatives up to 538°C (1,000°F); adhesives.
Polybenzimidazoles	Developmental laminating resin, fiber, film; stable 24 hours at 300°C (572°F) in air.
Polybenzothiazoles	Stable in air at 600°C (1,112°F); cured polymer soluble in concentrated sulfuric acid.
Polyquinoxalines	Stable in air at 500°C (932°F); tough, somewhat flexible resins; make film, adhesive.
Polyphenylenetriazoles	Thermally stable to 400 to 500°C (752 to 932°F); make film, fiber, coatings.
Polydithiazoles	Decompose at 525°C (977°F); soluble in concentrated sulfuric acid.
Polyoxadiazoles	Decompose at 450 to 500°C (842 to 932°F); can be made into fiber or film.
Polyamidines	Stable to oxidation up to 500°C (932°F); can make flexible elastomer.
Pyrolyzed polyacrylonitrile	Stable above 900°C (1,625°F); fiber resists abrasion with low tenacity.
Polyvinyl isocyanate ladder polymer	Soluble polymer that decomposes at 385°C (725°F); prepolymer melts above 405°C (761°F).
Polyamide-imide	Service temperatures up to 288°C (550°F); amenable to fabrication.
Polysulfone	Thermoplastic; use temperature −102°C (−152°F) to greater than 150°C (302°F); acid and base resistant.
Polybenzaylene benzimidazoles (pyrrones)	Thermally stable to 600°C (1,112°F); insoluble in common solvents; good mechanical properties.
Polybenzoxazoles	Stable in air to 500°C (932°F); insoluble in common solvents except sulfuric acid; nonflammable; chemical resistant; film.
Ionomer	High melt and tensile strength; tough; resilient; oil and solvent resistant; adhesives, coatings.
Diazadiphosphetidine	Thermoplastic up to 350°C (662°F); thermosetting at 357°C (707°F); cured material has good thermal stability to 500°C (932°F); amenable to fabrication.
Phosphorous amide epoxy	Soluble B-staged material; amenable to fabrication; good thermal stability.
Phosphonitrilic	Retention of properties in air up to 399°C (750°F).
Metal polyphosphinates	Polymers stable to better than 400°C (752°F).
Phenylsilesesquioxanes (phenyl-T ladder polymers)	Soluble; high molecular weight; infusible; improved tensile strength; high thermal stability to 525°C (977°F) in air; film forming.

Table 2.215 Examples of plastics operating in extreme temperatures

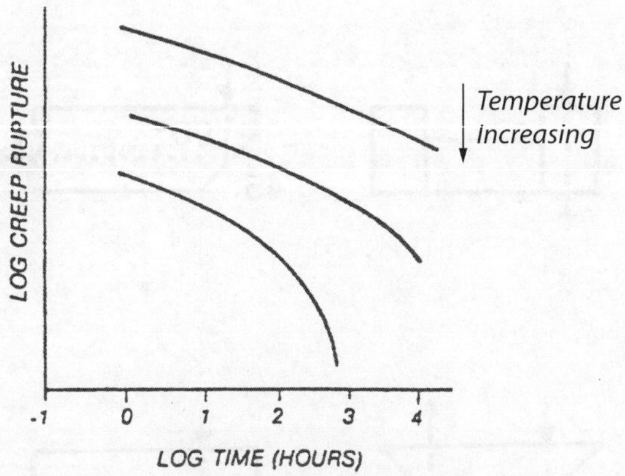

Figure 2.66 Influence of temperature on creep-rupture curves

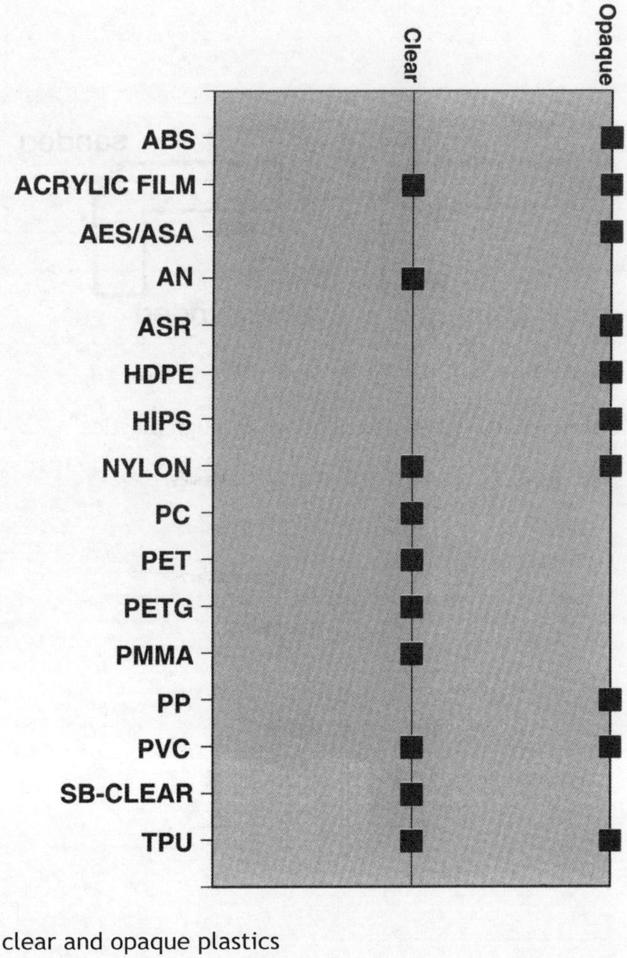

Figure 2.67 Guide to clear and opaque plastics

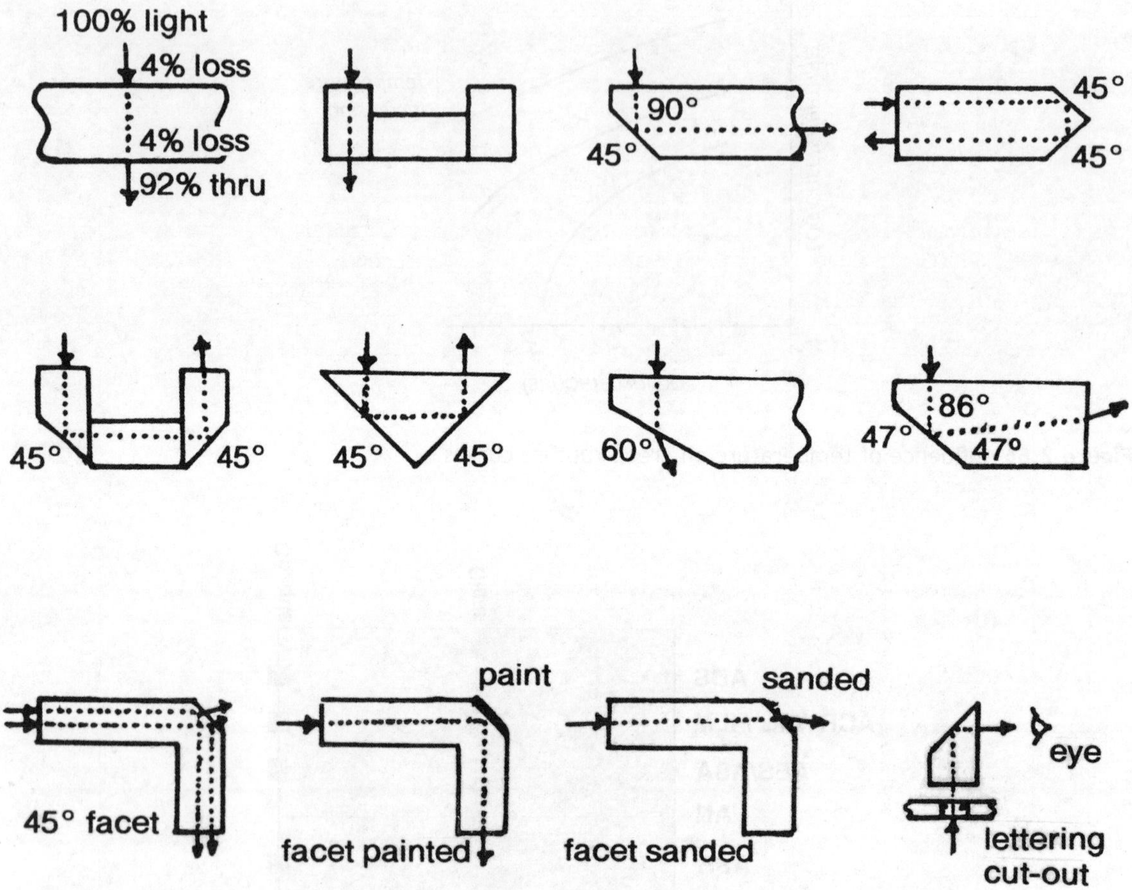

Figure 2.68 Example of transfer light rays (edge lighting) through plastics

Generic family	Notable characteristics
Transparent ABS	Good impact properties, good processibility
Acrylic (PMMA)	Excellent resistance to outdoor exposure, crystal clarity
Allyl diglycol carbonate	Good abrasion/chemical resistance, thermoset
Cellulosics	Heat sensitive, limited chemical resistance, good toughness
Nylon, amorphous	Excellent abrasion resistance, moisture sensitive
PET, PETG	Good barrier properties, not weatherable, clarity dependent on processing, orientation greatly increases physical properties
Polyarylate	Excellent UV resistance, high heat distortion
Polycarbonate	Excellent toughness, good thermal/flammability characteristics
Polyetherimide	Good chemical/solvent resistance, good thermal/flammability properties, inherent high color
Polyphthalate carbonate	Good thermal properties, autoclavable
Polyethersulfone	Excellent thermal stability, resists creep
Poly-4–methylpentene–1	UV/moisture sensitive, high crystalline melting point, lowest density o all thermoplastics
Polyphenylsulfone	Excellent thermal stability, resists creep, inherent high color
Polystyrene	Excellent processibility, poor UV resistance, brittle
Polysulfone	Excellent thermal/hydrolytic stability, poor weatherability/impact strength
PVC, rigid	Excellent chemical resistance/electrical properties, weatherable, decomposition evolves HCl gas
Styrene acrylonitrile	Good stress-crack and craze resistance, brittle
Styrene butadiene	Good processibility, no stress whitening
Styrene maleic anhydride	Higher-heat styrenic, brittle
Styrene methyl methacrylate	Good processibility, slightly improved weatherability
Thermoplastic urethane, rigid	Excellent chemical/solvent resistance, good toughness

Table 2.216 Examples of transparent plastics

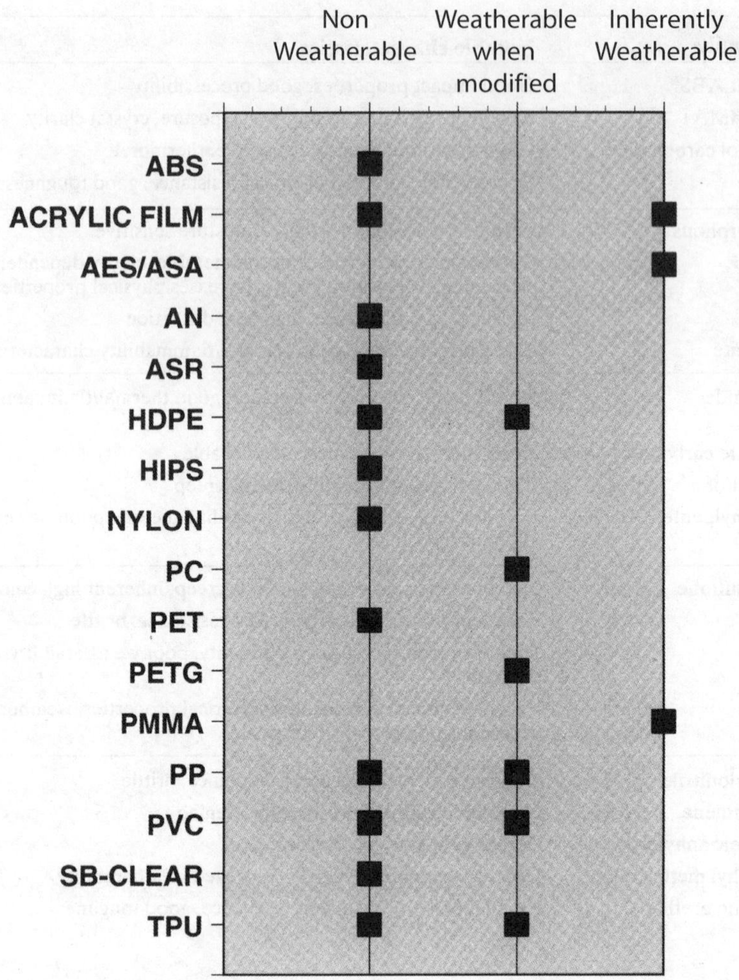

Figure 2.69 Examples of the weatherability of plastics

CHAPTER 3

Fabricating Product

OVERVIEW

Plastics have a profound impact on people worldwide and in all industries. The intelligent application of processing plastics includes utilizing the versatility and vast array of properties inherent in these materials as well as high-speed/relative low-energy processing techniques. The result has been the development of millions of cost-effective products that provide exceptional benefits for people and industries worldwide.

In a market economy, the world is ruled by competition and processed plastics will be employed only in applications where they can be expected to bring an overall economic advantage compared with other competing products. It is worth noting that the biggest competitor to a given plastic may be another plastic and its respective processing techniques. On the basis of an overall benefits assessment that considers the full service of a processed plastic product, it has been shown in millions of cases worldwide that the use of processed plastics not only makes economic sense but also makes a contribution toward conserving resources.

Many important factors have contributed to the worldwide success of plastics. One of these factors involves the use of the availability of different fabricating processes. All processes fit into an overall scheme that requires interaction and proper control of different operations.

Fabricating is an important part of the overall project to produce acceptable plastic products. It highlights the flow pattern for the fabricator (manufacturer) to be successful and profitable. Recognize that the first to market with a new product captures 80% of the market share. Factors such as good engineering, process control, and so on are very important but only represent pieces of the "pie." Many different ingredients blend together to produce profitable products, but fabricating is one of the main ingredients.

In order to understand potential problems and solutions of fabrication, it is helpful to consider the relationships of machine capabilities, plastics processing variables, and product performance (chapters 27 and 28). A distinction has to be made here between machine conditions and processing variables. Machine conditions include the operating temperature and pressure, mold and die temperature, machine output rate, and so on. Processing variables are more specific, such as the melt condition in the mold or die, flow rate versus temperature, and so on (chapter 1).

Processing of plastic is the art of detail; the more attention paid to details, the fewer problems in the process. A process that is up and running will continue running well unless a change occurs. Correct the problem and do not compensate. It may not be an easy task, but understanding what you have in the way of equipment, material, environment, and/or people can help.

Fabricating products involves conversion processes that may be described as art. Like some art, they have a basis in science; a relatively quick way to improve processing is to study the relevant sciences. The plastic-processing target is to take the plastic in the form of pellets, powders, granules, liquids, and so on, and convert them into useful products, usually through a screw plasticator. Common features of these different processes are as follows:

1. *Mixing and melting*: This stage takes the plastic and produces a homogeneous melt (chapter 1). This is often carried out in a screw plasticator or compounder, where melting takes place as a result of heat conducted through the barrel wall and heat generated in the plastic by the action of shear via the screw. Homogeneity is called for at the end of this stage, not only in terms of material but also in respect to temperature.
2. *Tooling*: When processing plastics, some type of tooling is required. These tools include molds and dies for shaping and fabricating products. They have some type of female and/or male cavity into or through which a molten or rigid plastic moves, usually under heat and pressure. Molds and dies are used in processing many different materials to form desired shapes and sizes. They can comprise many moving parts requiring high-quality metals and precision machining. Some molds and dies cost more than the primary processing machinery, but most approach half the cost of the primary machine (chapter 17).
3. *Melt transport and shaping*: In a screw plasticator, the next step would be to build up an adequate pressure in the plasticator so that it will produce the desired shape to be fabricated. In an injection-molding process, pressure is applied to force the melt into a mold that defines the product shape in three dimensions (chapter 4). In an extruder, the die (that initiates the shape) can vary from a simple cylindrical shape to a complex cross-head profile shape (chapters 5 and 9).
4. *Drawing, blowing, and forming*: There are processes that use a screw plasticator melt to stretch the melt to produce orientation and desired shape, as in blow molding, vacuum forming, rotational molding, and foaming (chapters 6, 7, 8, and 13).
5. *Coating and casting*: With screw plasticators or other systems, the melt provides coatings and castings, as reviewed in chapters 10, 11, and 16.
6. *Nonscrew plasticating*: Reactive mixing provides the melt in reaction injection molding (chapter 12). In compression molding, the usual material is precompounded or preimpregnated prior to being placed in or around a mold (chapters 14 and 15).

7. *Finishing*: The final stage after a process fabricates a solid product usually does not require secondary operations. However, there are materials that may require annealing, sintering, coating, assembling, decorating, and so forth (chapter 18).

These common features of the different processes fit into an overall fabricating pattern, as shown in Table 3.1 (chapter 18).

Processing techniques range from the unsophisticated (high labor costs with low capital costs) to sophisticated (zero or almost zero labor costs with very high capital costs). Production quantity, the material being processed, the available equipment, and the total cost govern decisions on the appropriate technique. Small quantities are usually produced with an unsophisticated approach.

Many fabricating processes are employed. Which process to use depends on the nature and requirements of the plastic to be processed, the properties required in the finished product, the cost of the process, speed, and the volume to be produced. Some processes can be used with many kinds of plastic, while some plastics require specialized processes.

The final actual properties of a processed plastic for an application are directly related to how the Plastic is processed. If process controls are not properly set up, followed, and continually rechecked to ensure meeting part performance requirements, products could be improperly processed.

This quality control requirement (chapters 13 and 24) on processing plastics applies to all products.

Practically all plastics can be processed by some method. About a century ago, it was initially found to be sufficient to adapt existing equipment to process plastics. (At the time, that was primarily for processing rubber.) The new plastics had a wider range of working temperatures and viscosities compared with rubber, so it was soon recognized that they required individual treatments in process design because of their special processing characteristics and their wide application potential.

It eventually became clear that very little was really known about what was happening inside the familiar processing machines, whether in a mold cavity, extruder die, gap of a roll mill or calendar, or the flight of a ram or screw. Only with the beginning of a deeper understanding of process mechanisms and their underlying physical laws, made possible through close cooperation between theorists and practical people, has processing technology and machinery design made any real progress. This progress will always continue as new plastics and processing techniques are being developed.

Despite what has been achieved so far, the industry continually faces new hurdles of systematic advanced development. The industry must not regard the present state of processing as the last word in progress. On the contrary, the great possibilities in development, many of which are dormant, must be recognized and opened by close cooperation between theorists and technologists.

There are over 600 different processes (Table 3.2). These can be grouped in subbasic and basic families of processes, such as those in Tables 3.3 and 3.4. Tables 3.5 to 3.7 relate processes to specialty situations (178). The term *macroscopic* refers to dimensions that are on the order of thin-part dimensions such as 2 mm; *microscopic* refers to dimensions two orders of magnitude or smaller, such as a 300-layer laminate as the part thickness.

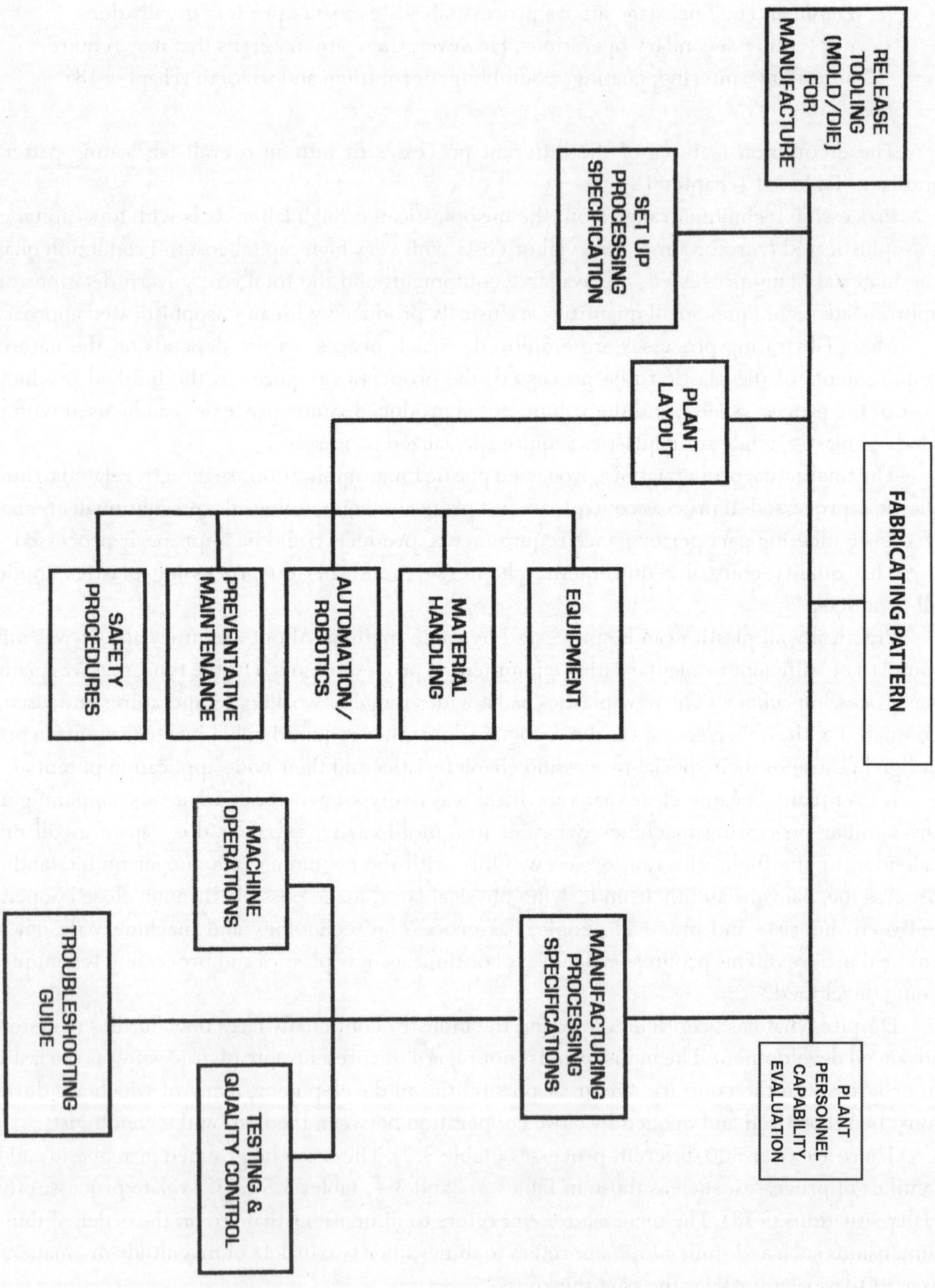

Table 3.1 Fabricating product flow pattern in a manufacturing operation

Names include the following: adiabatic extrusion, adiabatic injection molding, adiabatic processing, advanced composite molding, air floatation, airmold gas-assist injection molding, autoclave adhesive bonding, autoclave molding, autogeneous extrusion, automatic extrusion, automatic molding, automatic processing, auxiliary equipment; backmolding (Hinterspritzen), bag molding, biaxially-oriented extrusion, biaxially-oriented molding, bladder molding, blister process, blow molding (different types), blown film, BMC injection molding, bridge reinforced plastic, bulk molding compound; cable extrusion, calendering (different types), carded package, carousel molding, casting (different types), C-clamp injection molding, cellular plastic molding, cellular chemical blow molding, centrifugal casting, centrifugal molding, ceramic-plastic molding, chemical vapor deposition, cladding, closed molding, coating (different types), coextruded foamed blow molding, coextrusion, coextrusion capping, coining, coinjection foam molding, coinjection molding, cold flow molding, cold forming, cold heading, cold molding, cold press molding, cold stamping, cold working, combiform, comoforming cold molding, compounding, compound molding, composite molding, Compreg™ molding, compression-injection molding, compression molding (different types), computer-aided extrusion, computer-aided molding, computer aided processing, contact molding, contact pressure molding, continuous coating, continuous fiber spinning, continuous injection molding, continuous laminating, continuous molding, continuous strip molding, controlled density molding, copolymer molding, corrugated pipe extrusion, corrugated multilayer pipe extrusion, counter pressure intrusion, counter pressure molding, crossflow molding, cross laminating; decompression molding, devolatilizing extrusion, devolatilizing molding, die casting, die-slide molding, dip casting, dip forming, dip blow molding, dip molding, dip coating, doctor blade coating, dose molding, dosing extrusion, dosing molding, double-daylight molding, double shot molding, draw working, dry blend molding; elastomer molding, electric operating injection molding, electroforming, electron beam polymerization, electroplating, electrostatic spray, embedding, embossing, encapsulation, expandable polystyrene (and other plastics), extruder (different types), extrusion blow molding, extrusion compounding, extrusion molding; female forming, fiber forming, fiber placement molding, fiber reinforced molding, fiber spinning (wet, dry, jet, etc.), fibrillation, FIFO injection molding, filament placement, filament winding (different types), film casting, film extrusion, flame

Table 3.2 Examples of names of plastic fabricating processes spraying, flat film, flexible plunger molding, flocculation, flocking or floc spraying, flow molding, fluidized bed, foamed casting (different types), foamed extrusion, foamed-in-place, foamed-in-place gasketing, foamed molding (many different types such as injection, extrusion, calendering, casting, blow molding, etc.), foamed reservoir molding, forging, forming (different types),forming plastic-metal, forming scrapless, forming solid phase pressure, foundry molding, Fourdrinier, four-station molding, free extrusion, free molding, fusible core molding; gas assist molding, gas assist molding without gas channels, gas blow molding, gas counter-pressure injection molding, gas counter pressure molding, gas injection foam molding, gas injection molding, gear pump extrusion, gear pump injection molding, geometric forming, geometric molding, glass fiber spinning, glass mat reinforced molding, granular paint injection, graphitized fiber spinning, grease-free injection molding, group transfer polymerization, grow molding; hand layup molding, heat-cured rubber molding, heat sealing, high density molding, high frequency molding, high pressure foam molding, high pressure injection molding, high pressure molding, horizontal extrusion, horizontal injection molding, horizontal wheel blow molding, horizontal wheel extrusion, horizontal wheel forming, horizontal wheel molding, hot melt molding, hot stamping, hot working, hybrid-electric operating injection molding, hydroclave molding, hydromechanical clamp injection molding; impregnation molding, impulse sealing, infusion molding, injection blow molding, injection compounding, injection-compression molding, injection-die pultrusion, injection molding (different types), injection molding-prepressurized cavity, injection molding stamping, injection transfer molding, in-line slot extrusion/thermoforming, in-mold coat molding, in-mold decorating, intermediate pressure molding, interpenetrating blend molding, intrinsic molding, inplace molding, insert injection molding, insert molding, intrusion-flow molding, inverse lamination, investment casting, isotactic molding/pressure; jet molding, jet spinning; lagging molding, laminated molding, layup molding, leatherlike molding, Lego molding, LIFO injection molding, liquid crystal extrusion, liquid crystal molding, liquid curing extrusion, liquid injection molding, liquid silicone rubber injection molding, liquid transfer molding, lost wax molding, low pressure foam molding, low-pressure injection molding, low-pressure inverted-force injection molding, low pressure molding, low-profile resin molding; machining, male forming, manifold molding, manual extrusion, manual molding, manual processing, marbleize molding, Marco pressure molding, Marco vacuum molding, Marco vacuum-pressure molding, matched die molding, mechanical clamping injection molding, melt lamination, melt roll, metal injection molding, metallizing,

Table 3.2 *(continued)*

metal powder injection molding, metal powder molding, metal spraying, microencapsulation, molding with rotation, melt processable rubber process, melt processable wood process, metal-plastic molding, molding (compression, injection, bag, and so on molding), molecular density molding, multi-color injection molding, multi-component injection molding, multi-compound molding, multi-injection molding, multilayer blow molding, multilayer foam extrusion, multilayer foam injection molding, multilayer solid-foam extrusion, multilayer solid-foam molding, multilayer solid extrusion, multilayer solid molding, multilive feed molding, multi-material molding, multi-station forming, multi-station molding, multiwall molding; netting, netting extrusion, non-porous metal-plastic molding, notched die molding; off-center injection molding, offset extrusion, offset molding, one-shot molding, open molding, orientation process (different types), oriented extrusion, open frame forming, oriented molding, oscillating die extrusion, overcoat extrusion, overcoat lamination, overcoat molding; packaging (different types), parallel laminating, pelletizing extrusion, perforating, photopolymerization, physical blow molding, pinhole-free coating, pipe blow molding, pipe extrusion, plastic-concrete process, plastic-metal molding, plunger molding, polyurethane foam molding, poromeric molding, post-consumer extrusion, post-consumer molding, post forming, powder molding, potting, powder injection molding, preform molding, premolding, prepolymer molding, prepreg molding (different types), press lamination, pressure bag molding, pressure fabrication, pressure forming, pressure lamination, processing-artistic, processing-basics, profile extrusion, pullforming, pulp molding, pulse molding, pultrusion molding, pyrolysis carbon fiber spinning; ram extrusion, ram injection molding, ram molding, rapid prototype molding, radio frequency molding, reaction injection molding, reactive polymer processing, recycled compound molding, reinforced foam molding, reinforced plastics (different types), reinforced reaction injection molding, reinforced reaction molding, reinforced rotational molding, resin transfer molding, rock-and-roll molding, roll covering, rolling, roll milling, room temperature molding, rotary core molding, rotary molding, rotary table molding, rotating die extrusion, rotating mold turret injection molding, rotational casting, rotational molding, rotomold, rotomold ovenless, rotovinyl sheet, rubber insert molding; salt bath process (different types), sandwich molding, scrapless forming, scrapeless molding, screw molding, screw plunger transfer molding, Scorim molding, scrimp, semiautomatic extrusion, semiautomatic molding, semiautomatic processing, sheet extrusion, sheet molding compound, shell molding, shrink wrap, shrink wrap bag processing, shuttle forming, shuttle molding, sintering, skin molding, skiving,

Table 3.2 *(continued)*

sliding insert molding, slip forming, slot extrusion, slush molding, smart-card/closed-loop controlled injection molding, SMC continuous fiber molding, SMC directionally oriented molding, SMC randomly oriented molding, soluble core injection molding, soluble core molding, solution casting, solvent bonding, solvent casting, solvent molding, spin casting, spinneret fiber forming, spinning, spline process, spraying (different types), spray-up molding, spread coating, spreader molding, squeeze molding, stack blow molding, stack injection molding, stamping, staple fiber spinning, steam chamber-filament spinning, stretch blow molding, strip molding, structural casting, structural foam molding, structural reaction injection molding, stuffer injection molding, supperplastic forming, syntactic foaming; tape placement wrapped molding, tenter frame forming, thermal expansion molding, thermoforming (different types), thermoplastic extrusion, thermoplastic injection molding, thermoplastic molding, thermoplastic structural foam molding, thermoset extrusion, thermoset injection molding, thermoset molding, thermoset structural foam molding, thick compound molding, thin-wall injection molding, thixomolding, three-platen injection molding, three-station molding, toggle clamp injection molding, tooling, torpedo molding, transfer molding, trickle impregnation, tube extrusion, tubing-heat shrinkable, turnkey injection molding, twin-sheet forming, twin-sheet thermoforming (different types), two-color injection molding, two-color molding, two-platen clamp injection molding, two-stage injection molding, two-station molding; ultrasonic fabrication, ultrasonic vacuum bag molding, ultraviolet molding; vacuum bag molding, vacuum casting, vacuum coating, vacuum forming, vacuum hot forming, vacuum press molding, vacuum pressure bag molding, variable pressure foaming, vented extrusion, vented injection molding, vertical extrusion, vertical injection molding, vertical wheel extrusion, vertical wheel forming, vertical wheel injection molding, vibration gas injection molding, vibration molding, vinyl dispersion, vinyl plastisol forming, viscous molding, void-plastic impregnation, vulcanization; waste molding, welding, wet layup molding, wire coating, wire coating extrusion, wheel blow molding, wood-plastic impregnation molding, wood pulp-plastic extrusion; and others

Table 3.2 *(continued)*

Autoclave molding
Auxiliary equipment
Bag molding
Blister package
Blow molding (different types)
Bridge, reinforced plastic
Bulk molding compound
Calendering (different types)
Carded package
Casting
Centrifugal casting
Centrifugal molding
Cladding
Coating (different types)
Coextrusion
Coining
Coinjection
Cold forming
Cold heading
Cold molding
Cold-press molding
Cold stamping
Cold working
Compounding
Compreg molding
Compression–injection molding
Compression molding (different types)
Contact molding
Contact pressure molding
Continuous coating
Continuous injection molding
Continuous laminating
Die casting
Dip casting
Dip coating
Dip forming
Dip molding
Doctor blade
Double shot molding
Draw working
Electroforming
Electroplating
Electrostatic coating
Embedding
Embossing
Encapsulation
Expandable polystyrene
Extruder (different types)
Extrusion blow molding
Fiber spinning
Filament placement
Filament winding (different types)
Film casting
Flame spraying
Flocculation
Flocking or floc spraying
Flow molding
Fluidized-bed coating
Foamed casting (different types)
Foamed-in-place fabrication
Foamed molding (different types)
Foamed reservoir molding
Forging
Forming, plastic–metal
Forming, scrapless
Forming, solid-phase pressure
Forming, various
Foundry molding
Fourdiner
Fusible-core molding
Gas counterpressure molding
Gas-assisted injection molding
Hand layup molding
Heat sealing
High-pressure molding
Hot-melt molding
Hot stamping
Hot working
Impregnation molding
Impulse sealing
Injection blow molding
Injection–compression molding
Injection molding, powder
Injection molding, various
Insert molding
Inverse lamination
Investment casting
Isotactic molding/pressure
Jet molding
Jet spinning
Joining, various
Lagging molding
Laminated molding
Leatherlike molding

Table 3.3 Subbasic families of plastic fabricating processes

Liquid injection molding	Screw-plunger transfer molding
Lost-wax molding	Scorim
Low-pressure molding	Sheet molding compound
Machining	Shell molding
Marco vacuum molding	Shrink-wrapping
Melt roll	Sintering
Metallizing	Skiving
Metal powder molding	Slip forming
Metal spraying	Slot extrusion
Microencapsulation	Slush molding
Multilive feed molding	Spline process
Netting extrusion	Soluble-core molding
Notched-die molding	Solvent bonding
One-shot molding	Solvent casting
Open molding	Solvent molding
Orientation process, various	Spinning
Packaging, various	Spraying, various
Perforating	Sprayup molding
Pinhole-free coating	Spread coating
Plastic–concrete	Squeeze molding
Plunger molding	Stamping
Poromeric molding	Stretch blow molding
Potting	Structural foam molding
Powder molding	Superplastic forming
Preform molding	Syntactic molding
Premolding	Tape-placement wrapped molding
Prepolymer molding	Thermal expansion molding
Prepreg molding	Thermoforming, various
Pressure bag molding	Thixomolding
Pressure fabrication	Transfer molding
Processing, art of	Trickle impregnation
Processing, fundamental	Tubing, heat shrinkable
Pulp molding	Two-color molding
Pultrusion	Ultrasonic process
Ram molding	Vacuum bag molding
Reaction injection molding	Vacuum coating
Reactive processing	Vacuum casting
Reinforced plastic, various	Vacuum hot pressing
Resin transfer molding	Vacuum press
Rock-and-roll processing	Vinyl dispersion
Rolling	Viscous process
Rotary molding	Void, plastic impregnated
Rotational casting	Vulcanization
Rotational molding	Welding
Rotomolding	Wet layup molding
Rotovinyl sheet	Wood–plastic impregnation molding
Salt bath process	

Table 3.3 *(continued)*

Process	Description	Limitations
Blow Molding	An extruded parison tube of heated elastomer is positioned between two halves of an open split mold and expanded against the sides of the closed mold via air pressure. The mold is opened and the part ejected. Low tool and die costs, rapid production rates, and ability to mold fairly complex hollow shapes in one piece.	Limited to hollow or tubular parts; wall thickness and tolerances can be difficult to control.
Calendering	Dough-consistent elastomer mass is formed into a sheet of uniform thickness by passing it through and over a series of heated or cooled rolls. Calenders are also utilized to apply elastomer covering to the backs of other materials. Low cost, and sheet materials are virtually free of molded-in stresses.	Limited to sheet materials, and very thin films are not possible.
Casting	Stock is poured into a mold without pressure, cured, and taken from the mold. Relatively low mold cost, capability to form large parts with thick cross sections, good surface finish, and convenient to low volume production.	Limited to relatively simple shapes. Most elastomers are not suitable for this method. Except for cast tires, method becomes uneconomical at high volume production rates.
Coating	Process methods vary. Various elastomers widely used in coating of numerous materials. Roller coating similar to calendering process. Spread coating employs blade in front of roller to position resin on material. Coatings also applied via brushing, spraying, and dipping.	Economics dependent upon close tolerance control generally.
Compression Molding	Principally, polymerized thermoset compound, usually preformed, is positioned in a heated mold cavity; the mold is closed (heat and pressure are applied) and the material flows and fills the mold cavity. Heat completes polymerization and the part is ejected. Large bulky parts can be molded; and process is adaptable to rapid automation (racetrack techniques, etc.)	Extremely intricate parts containing undercuts, side draws, small holes, delicate inserts, etc., very close tolerances difficult to produce.
Encapsulation	Mixed compound is poured into open molds to surround and envelope components; cure may be at room temperature with heated postcure. Generally, encapsulation includes several processes such as potting, embedding, and conformal coating.	Low volume process subject to inherent limitations on materials which can lead to product defect caused by exotherm, curing, or molding conditions, low thermal conductivity, high thermal expansion and internal stresses.
Extrusion Molding	Widely used for continuous production of sheet, tube, and other profiles; also can be used in conjunction with blow molding. Elastomer compound is fed from a hopper to a screw and barrel where it is heated to plasticity and then forwarded, usually via a rotating screw, through a nozzle possessing the desired cross section configuration. Production lines require input and takeoff equipment that can be complex. Low tool cost, numerous complex profile shapes possible, very rapid production rates, can apply coatings or jacketing to core materials (such as wire).	Limited to sections of uniform cross section.

Table 3.4 Families of plastic fabricating processes

Process	Description	Limitations
Filament Winding	Excellent strength-to-weight here. Continuous, reinforced filaments are saturated with resin and machine wound onto mandrels having shape of desired finished part. Once winding is completed part and mandrel are cured; mandrel can then be removed. High strength reinforcements can be oriented precisely in direction where strength is required. Good uniformity of resin distribution in finished part; mainly circular objects such as pipes.	Limited to shapes of positive curvature; openings and holes can reduce strength if not properly designed into molding operations.
Injection Molding	Very widely utilized method. High automation of manufacturing is standard practice. Stock is fed from the hopper and conveyed forward towards the nozzle by a rotating screw. Elastomer is thoroughly and uniformly heated to plasticity in the cylinder at controlled temperature, then forced under pressure through the nozzle into sprues, runners, gates, and cavities of mold. The material is restrained from extruding out the nozzle orifice during plasticizing. As the melted material accumulates at the front of the screw, it forces the rotating screw rearwards (away from the nozzle). Screw rotation and rearward motion stops when a preset position-signal is reached and immediately the screw acts as a ram forcing material to the mold cavity. Final cure occurs in the mold rapidly; the mold is opened; and the part ejected. There is no waiting for the mold to heat the compound. This factor is the basic reason for its high production rates. Efficient preheating of the rubber, just below cure, occurs in the screw. High production runs, low labor costs high reproducibility of complex details, and excellent surface finish exists.	High initial tool and die costs; not economically practical for small runs.
Laminating	Material, usually in form of reinforcing cloth, paper, foil, metal, wood, glass fiber, plastic, etc., preimpregnated or coated with elastomer is molded under pressure into sheet, rod, tube, or other simple shape. Excellent dimensional stability of finished product; very economical in large production of parts.	High tool and die costs. Limited to simple shapes and cross section profiles.
Pultrusion	This process is similar to profile extrusion, however, it does not provide flexibility and uniformity of product control, and automation. Used for continuous production of simple shapes (rods, tubes, and angles) principally incorporating cotton, wire, and fiberglass (or other reinforcement). High output possible.	Close tolerance control requires diligence. Unidirectional strength usually the rule.
Rotational Molding	A predetermined amount of powered or liquid elastomer is poured into mold; mold is closed, heated, and rotated in the exis of two planes until contents have fused to inner walls of mold; mold is then opened and part is removed. Low mold cost, large hollow parts in one piece can be produced, and molded parts are essentially isotropic in nature.	Limited to hollow parts; production rates are usually slow.
Slush Molding	Powdered or liquid elastomer is poured into a mold to capacity; mold is closed and heated for a predetermined time in order to achieve	Limited to hollow parts production rates are very slow; and limited choice of materials that can be processed.

Table 3.4 *(continued)*

Process	Description	Limitations
	a specified buildup of partially cured material on mold walls; mold is opened and unpolymerized material is poured out; and semifused part is removed from mold and fully polymerized in oven. Low mold costs and economical for small production runs.	Limited to hollow parts production rates are very slow; and limited choice of materials that can be processed.
Thermo-forming	Heat softened elastomer sheet is positioned over male or female mold; air is evacuated from between sheet and mold, forcing sheet to conform to contour of mold. Variations include vacuum snap-back, plug assist, drape forming, etc. Tooling costs are generally low, large part production with thin sections possible and often comes out economical for limited part production.	Limited to parts of simple configuration, high scrap, and limited number of materials from which to choose.
Transfer Molding	Related to compression and injection molding processes. Elastomer molding compound is fed from hopper into a transfer chamber where it is then heated to plasticity; it is then fed by means of a plunger through sprues, runners, and gates into a closed mold where it cures; mold is opened and part is ejected. Good dimensional accuracy, rapid production rate, and very intricate parts can be produced.	High mold cost; high material loss in sprues and runners; and size of parts is somewhat limited.

Table 3.4 *(continued)*

Scale	Composition		
	Plastics combined with		
	Other plastics	Gas	Solids
Macroscopic	Blow Molding Coinjection Molding (MMP)* Multimaterial Molding (MMP) In-Mold Coating (MMP) Dual Molding (LPM)	Gas-Assisted Molding Liquid-Gas Molding (LPM) Blow Molding	Laminate Molding (LPM)
Microscopic	Lamellar Molding	Microcellular Plastics Controlled Density (LPM)	Sheet Composites Reactive Liquid Molding

Table 3.5 Processes vs. material compositions

Materials		Geometry: Macroscopic structure — Laminate	Geometry: Macroscopic structure — Segmented	Geometry: Size & shape — Large	Geometry: Size & shape — Hollow
Plastics		In-Mold Coating Coinjection Blow Molding Dual Molding (LPM)	Multimaterial Molding (MMP), Blow Molding	Sheet Composites In-Mold Coating (MMP) Injection-Compression (MMP)	Dual Molding (LPM) Blow Molding
Plastics and non-plastics	Solid	Laminate Molding (LPM) Reactive Liquid Molding		Laminate Molding (LPM) Sheet Composites Reactive Liquid Molding	
	Gas	Blow Molding Gas-Assisted Molding Liquid Gas Molding (LPM) Fusible Core		Blow Molding Controlled Density Molding (LPM)	Blow Molding Gas-Assisted Molding Liquid Gas Molding (LPM) Fusible Core

Table 3.6 Processes vs. material compositions and geometries

COMPLEXITY	FUNCTION: ENCLOSURE (Panel, Container) ←→ LOAD BEARING
SYMMETRIC ↕	In-Mold Coating (LPM) Laminate Molding (LPM) Controlled Density Molding (LPM) Sheet Composites Multimaterial Molding (MMP) Injection-Compression Molding (MMP) Coinjection (MMP) Blow Molding Reactive Liquid Molding Gas-Assist Molding Liquid Gas Molding (LPM) Dual Molding (LPM) Fusible Core Molding
THREE DIMENSIONAL	

Table 3.7 Processes vs. product functions and complexity

The long inventory of methods used to process plastics in Table 3.2 includes all types of basic and specialty processes that have been developed over the past century. The various names for the same process are also included. The following are among the diverse reasons different names are used:

1. Different industries have their own methods of identifying a process based on their market requirements.
2. An old process that may be basically the same or slightly modified requires a more modern name.
3. The promotion of new ideas requires a name to symbolize a new generation.

There is also an overlapping of terms such as *molds*, *dies*, and *tools* as well as *molding*, *embedding*, *casting*, and *potting*. There are continuous and noncontinuous extrusion processing methods. Injection molding includes gas injection, insert molding, and micromolding. The overlap in names for terms and processes does not cause a problem and should not affect anyone's thinking when examining processes throughout the world, particularly in the industrialized nations. One might say that there are words or situations that could have more than one meaning. The main message here is that it is important for you to be very specific when describing processes, materials, designs, and so on.

The major families of processing are based on the amount of plastic processed in the United States and worldwide. Extrusion (EX) consumes approximately 36 wt% of all plastics, injection molding (IM) follows by consuming 32%, blow molding at 10%, calendering at 8%, coating at 5%, compression molding at 3%, and others at 3%. Thermoforming can be considered the fourth major process used; it consumes about 30% of the extruded sheet and film that principally goes into packaging.

At least 65 wt% of all plastics requires some type of specialized compounding. They principally go through compounding extruders, usually twin-screw extruders (chapter 5), before going through equipment such as injection-molding machines (IMMs), extruders, and blow-molding machines to produce products.

It is estimated that in the United States there are about 18,000 extruders, 80,000 injection-molding machines, and 6,000 blow-molding machines, together producing about one-third of the world's plastic products. For the 80,000 IMMs in the United States, approximately 30% are less than five years old, at least 35% are five to ten years old, and the rest are more than ten years old.

In the United States, the yearly demand of machinery sales normally is about $1.5 billion (not taking into account the depressed years that occur every 10 to 20 years). IMM is the largest category and accounts for at least 50% of all the machinery sales. Blow-molding (extrusion- and injection-type) machines are at about $505 million, extrusion reaches $440 million, and thermoforming reaches $455 million. There are over 350 U.S. machinery builders, with about five having over 50% of sales.

The number of processes can be related to the number of plastics processed. As reviewed in chapter 1, there are over 35,000 plastics available to meet different product performance requirements (Fig. 1.8), processing requirements, and/or cost requirements. Only a few of these plastics are used in large quantities. About 90 wt% are thermoplastics and 10% thermosets. These plastics are made up of different families of plastics, such as polyethylenes, polyvinyl chlorides, nylons, fluoroplastics, epoxies, and neoprenes (chapter 2).

These families of plastics are compounded (additives, fillers, and/or reinforcements) into hundreds of thousands of materials to meet different product requirements. Of the 35,000 types available worldwide there are about 200 basic types or families that are commercially recognized, with fewer than twenty that are popularly used. Examples of these plastics are shown in Table 1.25. Within these twenty popular plastics, there are five major families of thermoplastics that consume about two-thirds of all thermoplastics. They are the low-density polyethylenes (LDPEs), high-density polyethylenes (HDPEs), polypropylenes (PPs), polystyrenes (PSs), and polyvinyl chlorides (PVCs).

PROCESS

The plastics industry is characterized by mature practical and theoretical technology. Improved understanding and control of materials and fabricating processes (Table 3.8) have significantly increased product performances and reduced their variability resulting in good to excellent return on investments (ROIs) (141).

Plastic processes permit the fabrication of products whose manufacturing would be very costly or difficult, if not impossible, with other materials. Processors must routinely keep up to date on developments with the more useful plastics and acquire additional information on how to process them. The emphasis throughout this book has been that it is not difficult to design and fabricate with plastics and to produce many different sizes and shapes of thermoplastic (TP) and thermoset (TS) commodities and engineering plastics, whether unreinforced or reinforced. The basics of material and process selection should be product performance requirements, shape, dimensional tolerances, processing characteristics, production volume, and cost.

Most plastic products can be produced using different processes; however, others require specific processes. Process selection can take place before material selection, when a range of materials may be available, or it can be based on certain performance requirements. Usually in the latter situation, only one special process can be used to provide the best performance-to-cost advantages. A particular design group may have its own in-house processing capabilities. Unfortunately, some operations use whatever equipment is available. This situation can be very unprofitable, limit profitability, or restrict product performance. It is important to recognize that the fabrication process can markedly influence all aspects of product performance, including cost.

Compared with other material-based industries, plastics have enjoyed an impressive growth rate since their inception over a century ago, but particularly since about 1940. The product-design and product-material communities were quick to recognize the design freedom and great versatility that plastics' materials and processing techniques afforded (Tables 3.9 to 3.14). Recognizing a

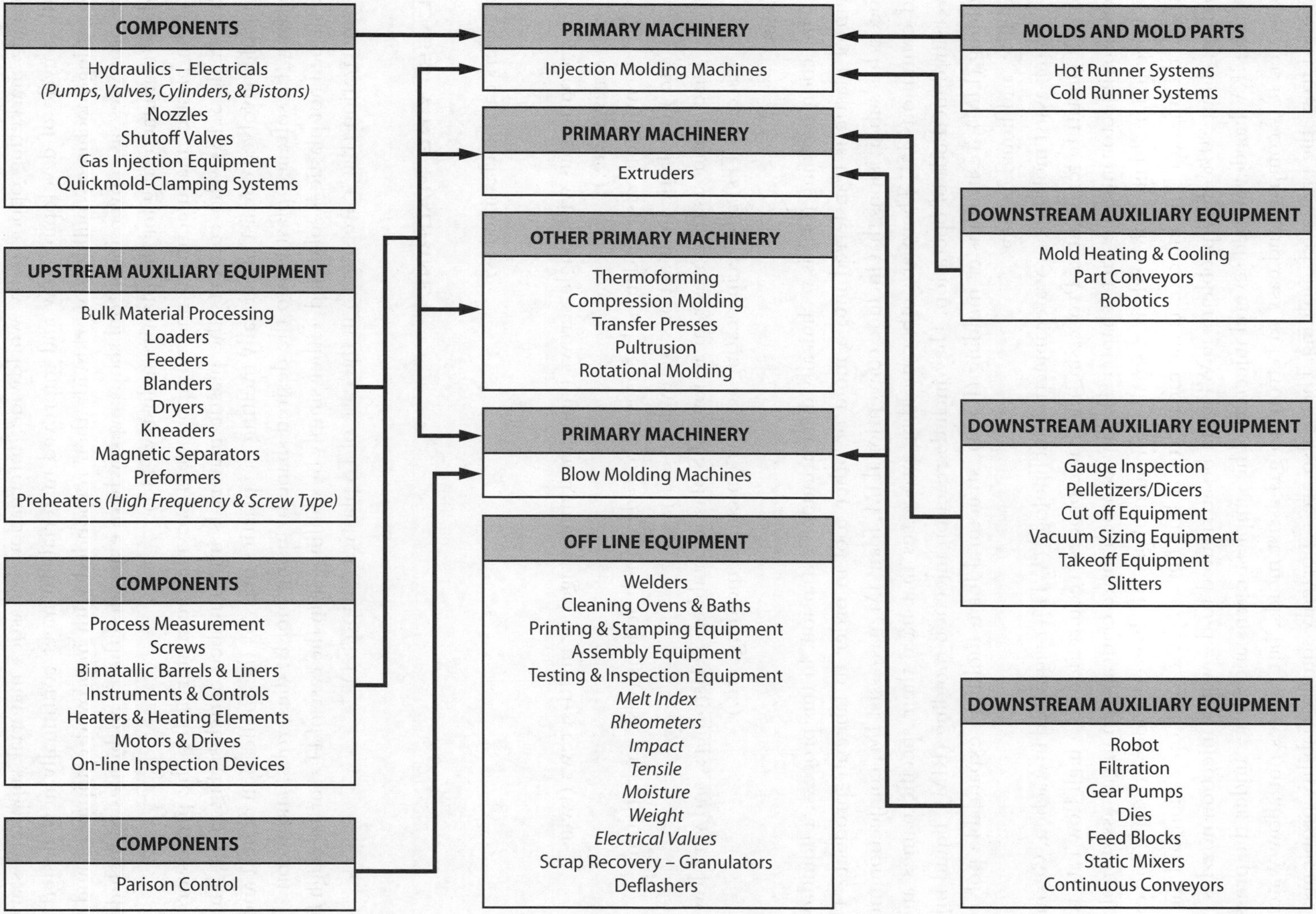

Table 3.8 Flow chart in fabricating plastic products (courtesy of Adaptive Instruments Corp.)

growing marketing opportunity worldwide, international plastics material suppliers started an endless cycle of developing new and improved materials to meet continually new design needs. Processing machinery builders worldwide then responded with improved equipment and totally new processes, as conventional tool shops everywhere expanded their capabilities to include mold and die manufacturing for the plastics industry (415).

The plastics manufacturers and compounders provide the material for the processor (fabricator; Table 3.15) who converts the plastic materials into finished articles (Fig. 3.1). The basic principles involved consist of partially melting the plastics mass by the application of heat and compressing the softened plastic into the desired shape by using molds, dies, rolls, and so on. Finally, the shaped hot plastic is solidified, either simply by chilling, as in the case of TPs, or through further polymerization (the "cure stage") in the case of TS plastics (chapter 1).

Classifying Machine

Extruders can be classified as:

1. Continuous with single-screws (single and multistage) or multi-screws (twin-screws, etc.);
2. Continuous disk or drum that uses viscous drag melt actions (disk pack, drum, etc.) or elastic melt actions (screwless, etc.);
3. Discontinuous that uses ram actions (TS plastics, rubbers/elastomers, and very low viscosity TPs) and reciprocating actions (injection molding, etc.).

Injection molding (IM) is technically discontinuous extrusion. In this process, a liquid or solid form of plastic is transferred into a mold or other tool in order to fabricate products. This IM process has subdivisions that include conventional IM, foam IM, gas-assist IM, coinjection molding, and continuous IM. There are other molding processes that have their specific names and very diversified methods of operation. They include reaction injection molding (RIM), liquid injection molding (LIM), resin transfer molding (RTM), structural foam molding, expandable polystyrene molding, and liquid casting.

Water-assist molding, or water injection technology (WIT), originated in Aachen, Germany, at the IKV (Institute for Plastics Processing). A basic process requirement is a high flow rate through the water injector into the part (chapter 4). This is much more critical than in gas-assist molding or gas-injection molding (GIM) where a smaller amount of gas is delivered under pressure and expands outside the injector. The injector is the most crucial link between the WIT operation and the mold. To date, about 90% of all problems in WIT can be attributed to the water injection nozzles (416).

The IKV has developed several injectors. The simplest opens and closes under the pressure of the water. It is compact and easy to retrofit to gas-assist molds, but it does not allow water to be expelled from the part after molding. There also are reportedly some reliability problems. Other

Part design	Blow molding	Casting	Compression	Extrusion	Filament winding	Injection	Matched die molding	Rotational	Thermo-forming	Transfer compression	Wet lay-up (contact molding)
Molded-in designs and nos.	Yes	Yes	Yes	No	No	Yes	Yes	Yes	Yes	Yes	Yes
Surface finish	1–2	2	1–2	1–2	5	1	4–5	2–3	1–3	1–2	4–5
Threads	Yes	Yes	Yes	No	No	Yes	No	Yes	No	Yes	No
Undercuts	Yes	Yes[1]	NR[2]	Yes	NR	Yes[1]	NR	Yes	Yes	NR	Yes
Inserts	Yes	Yes	Yes	Yes	Yes	Yes	Yes	Yes	NR	Yes	Yes
Built-in cores	Yes	Yes	No	Yes	Yes	Yes	Yes	Yes	Yes	Yes	Yes
Molded-in holes	Yes	Yes	Yes	Yes[4]	Yes	Yes	Yes	Yes	No	Yes	Yes
Bosses	Yes	Yes	Yes	Yes	No	Yes	Yes	Yes	Yes	Yes	Yes
Fins or ribs	Yes	Yes	Yes	Yes	No	Yes	No	Yes	Yes	Yes	Yes
Major shape characteristics	Hollow bodies	Simple configurations	Moldable in one plane	Constant cross section profile	Structure with surfaces of revolution	Few limitations	Moldable in one plane	Hollow bodies	Moldable in one plane	Simple configurations	Moldable in one plane
Limiting size factor	Material	Material	Equipment	Material	Equipment	Equipment	Equipment	Material	Material	Equipment	Mold size
Maximum thickness, in. (mm)	> 0.25 (6.4)	None	0.5 (12.7)	6 (150)	3 (76)	6 (150)	2 (51)	0.5 (12.7)	3 (76)	6 (150)	0.5 (12.7)
Minimum inside radius, in. (mm)	0.125 (3.18)	0.01–0.125 (0.25–3.18)	0.125 (3.18)	0.01–0.125 (0.25–3.18)	0.125 (3.18)	0.01–0.125 (0.25–3.18)	0.06 (1.5)	0.01–0.125 (0.25–3.18)	0.125 (3.18)	0.01–0.125 (0.25–3.18)	0.25 (6.4)
Minimum draft (deg.)	0	0–1	> 1	NR[2]	2–3	< 1	1	1	1	1	0
Minimum thickness, in. (mm)	0.01 (0.25)	0.01–0.125 (0.25–3.18)	0.01–0.125 (0.25–3.18)	0.001 (0.02)	0.015 (0.38)	0.005 (0.1)	0.03 (0.8)	0.02 (0.5)	0.002 (0.05)	0.01–0.125 (0.25–3.18)	0.06 (1.5)

Rating: very smooth (1) to rough (5)

Table 3.9 Interrelating processes and designs

Material	Extrusion	Blow molding	Fiber	Film	Foam
ABS	X	X	—	—	X
SBR	X	X	—	—	X
EVA	X	X	—	X	X
Melamine	—	—	—	—	—
Nylon	X	X	X	—	X
Phenolic	—	—	—	—	—
PC	X	X	—	—	X
LDPE/ HDPE	X	X	—	X	X
PET	—	—	X	X	—
Acetal	X	X	—	—	X
Methyl pentene	X	X	—	—	—
PP	X	X	—	X	X
GPS	X	X	—	—	X
TFE	—	—	—	—	—
Urethanes	—	—	X	X	—
PVC	X	X	—	X	X
Ureas	—	—	—	—	—
PVC/PVA	X	X	—	X	—
Acrylic	X	—	X	—	—

Table 3.10 Interrelating processes and plastics

injectors, for series production, are operated externally. Water can pass both ways so that parts can be drained in the mold. These injectors are said to be more reliable, but they take up more space and, thus, are only appropriate for molds designed for WIT. The IKV is working on optimizing the scaling effect around the water outlet and improving water retrieval during the molding cycle.

There are differences in casting, encapsulation, and potting terms; however, they are often interchangeable and interrelate very closely to describe processes and performances. Both TPs and TSs are used. As an example, there are reactive TS liquids that are often used to form solid shapes. Such plastic systems harden or cure at room temperature or at elevated temperatures because of the irreversible cross-linking of rather complex molecular structures. This is different from the hardening of plastics in solution, which harden when the solvent is evaporated. The hardening of the reactive plastics produces no by-products, such as gases, water, and/or solvents. When reactive plastics are used as impregnates, they are sometimes called solventless systems. However, there are plastics and certain additives that release gases and may require degassing during processing.

To help in quickly evaluating what machinery is available worldwide that will meet your requirements, Plastics Technology has set up a Web site (www.ptonline.com). This is in addition to its annual *Processing Handbook & Buyers' Guide*, which has been published for many decades.

Material family	Injection	Compression	Transfer	Casting	Cold molding	Coating	Structural foam	Extrusion	Laminating	Thermoforming	RP molding FRP	Filament Winding	Dip and slush	Blow molding	Rotational
ABS	X			X			X	X	X	X				X	
Acetal	X						X			X	X			X	X
Acrylic	X			X		X		X	X	X				X	
Allyl		X	X	X	X	X			X			X			
ASA	X						X	X		X		X			
Cellulosic	X						X	X	X	X	X				X
Epoxy		X	X	X	X	X			X			X			
Fluoroplastic	X	X	X		X	X		X		X					X
Melamine-formaldehyde	X	X	X	X		X			X						
Nylon	X			X		X	X	X						X	X
Phenol-formaldehyde	X	X	X	X	X	X			X						
Polyamide-imide	X	X	X					X							
Polyarylether	X							X							
Polybutadiene	X		X					X						X	
Polycarbonate	X	X					X	X		X			X	X	
Polyester (TP)	X					X		X						X	X
Polyester (TS) - fiberglass	X	X	X		X		X	X	X		X	X			X
Polyethylene	X	X				X	X	X	X	X				X	X
Polyimide	X	X				X		X(TP)							
Polyphenylene oxide	X						X	X							X
Polyphenylene sulfide	X	X					X	X							
Polypropylene	X						X	X		X				X	X
Polystyrene	X						X	X		X				X	X
Polysulfone	X	X						X		X					
Polyurethane (TS) (TP)	X	X	X	X		X	X	X(TP)	X					X	X
SAN	X						X	X	X	X					X
Silicone		X		X				X							
Styrene butadiene	X					X		X	X						
Urea formaldehyde		X							X						
Vinyl	X	X				X	X	X	X	X			X	X	X

* Compounding permits using other processes

Table 3.11 Interrelating molding processes and thermoplastics and thermoset plastics

Processes	Plastics	Properties
	Thermoplastics	
Injection molding Continuous laminating	Polystyrene	Low cost, moderate heat distortion, good dimensional stability, good stiffness, impact strength
Injection molding Blow molding Rotational molding	Nylon	High heat distortion, low water absorption, low elongation, good impact strength, good tensile and flexural strength
Injection molding	Polycarbonate	Self-extinguishing, high dielectric strength, high mechanical properties
Injection molding	Styrene-acrylo-nitrile	Good solvent resistance, good long-term strength, good appearance
Injection molding Vacuum forming Compression molding Continuous laminating	Acrylics	Good gloss, weather resistance, optical clarity, and color; excellent electrical properties
Injection molding Continuous laminating Rotational molding	Vinyls	Excelllent weatherability, superior electrical properties, excellent moisture and chemical resistance, self-extinguishing
Injection molding	Acetals	Very high tensile strength and stiffness, exceptional dimensional stability, high chemical and abrasion resistance, no known room temperature solvent
Injection molding Rotational molding Blow molding	Polyethylene	Good toughness, light weight, low cost, good flexibility, good chemical resistance; can be "welded"
Injection molding Encapsulation Continuous pultrusion	Fluorocarbons	Very high heat and chemical resistance, nonburning, lowest coefficient of friction, high dimensional stability
Injection molding	Polyphenylene oxide modified	Very tough engineering plastic, superior dimensional stability, low moisture absorption, excellent chemical resistance
Injection molding Continuous laminating Rotational molding	Polypropylene	Excellent resistance to stress or flex cracking, very light weight, hard, scratch-resistant surface, can be electroplated; good chemical and heat resistance; exceptional impact strength; good optical qualities
Injection molding	Polysulfone	Good transparency, high mechanical properties, heat resistance, electrical properties at high temperatures; can be electroplated

Table 3.12 Interrelating processes and plastic properties

Process	*Thermosets*	Properties
Compression molding Filament winding Hand lay-up Mat molding Pressure bag molding Continuous pultrusion Injection molding Spray-up Centrifugal casting Cold molding Comoform Encapsulation	Polyesters Properties shown also apply to some polyesters formulated for thermoplastic processing by injection molding	Simplest, most versatile, economical and most widely used family of resins, having good electrical properties, good chemical resistance, especially to acids
Compression molding Filament winding Hand lay-up Continuous pultrusion Encapsulation Centrifugal casting	Epoxies	Excellent mechanical properties, dimensional stability, chemical resistance (especially alkalis), low water absorption, self-extinguishing (when halogenated), low shrinkage, good abrasion resistance, very good adhesion properties
Compression molding Continuous laminating	Phenolics	Good acid resistance, good electrical properties (except arc resistance), high heat resistance
Compression molding Injection molding Encapsulation	Silicones	Highest heat resistance, low water absorption, excellent dielectric properties, high arc resistance
Compression molding	Melamines	Good heat resistance, high impact strength
Compression molding	Diallyl phthalate	Good electrical insulation, low water absorption

Table 3.12 *(continued)*

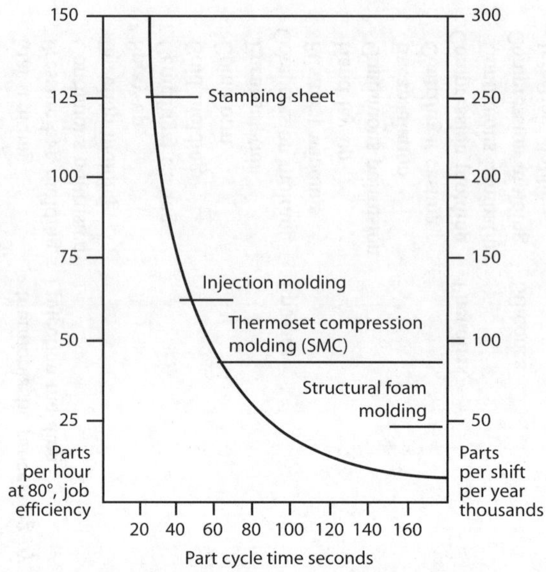

Table 3.13 Interrelating processes and times to fabricate products

COMPLETE OPERATION

Even though modern fabricating machines, with all their ingenious microprocessor control technology, are in principle suited to perform flexible tasks, it nevertheless takes a whole series of peripheral auxiliary equipment to guarantee the necessary degree of flexibility (chapter 18). Examples include (a) raw material supply systems, (b) mold/die transport facilities, (c) mold/die preheating banks, (d) mold/die changing devices that include rapid clamping and coupling equipment, (e) plasticizer cylinder changing devices, (f) fabricated product handling equipment, particularly robots with interchangeable arms allowing adaptation to various types of production, and (g) transport systems for finished products and handling equipment to pass products on to subsequent production stages.

Equipment selection should address the economics of the product being fabricated, the degree of precision necessary, and the production rate required. The mix can be manually proportioned using balancing and weighing scales. Proportioning pumps proportion according to a desired ratio, usually by weight but also by the less accurate method of volume. Diaphragm pumps may be used with adjustable displacement to draw the plastic components from reservoirs and to meter the correct proportions into a mixing chamber. There is the use of gear pumps with drive ratios to deliver the correct portion. They are used only when the mix is relatively free of abrasive fillers or additives (chapter 18).

PROCESSING AND PATIENCE

The start-up of fabricating lines usually requires changing equipment settings. When making processing changes, allow enough time to achieve a steady state in the complete line before

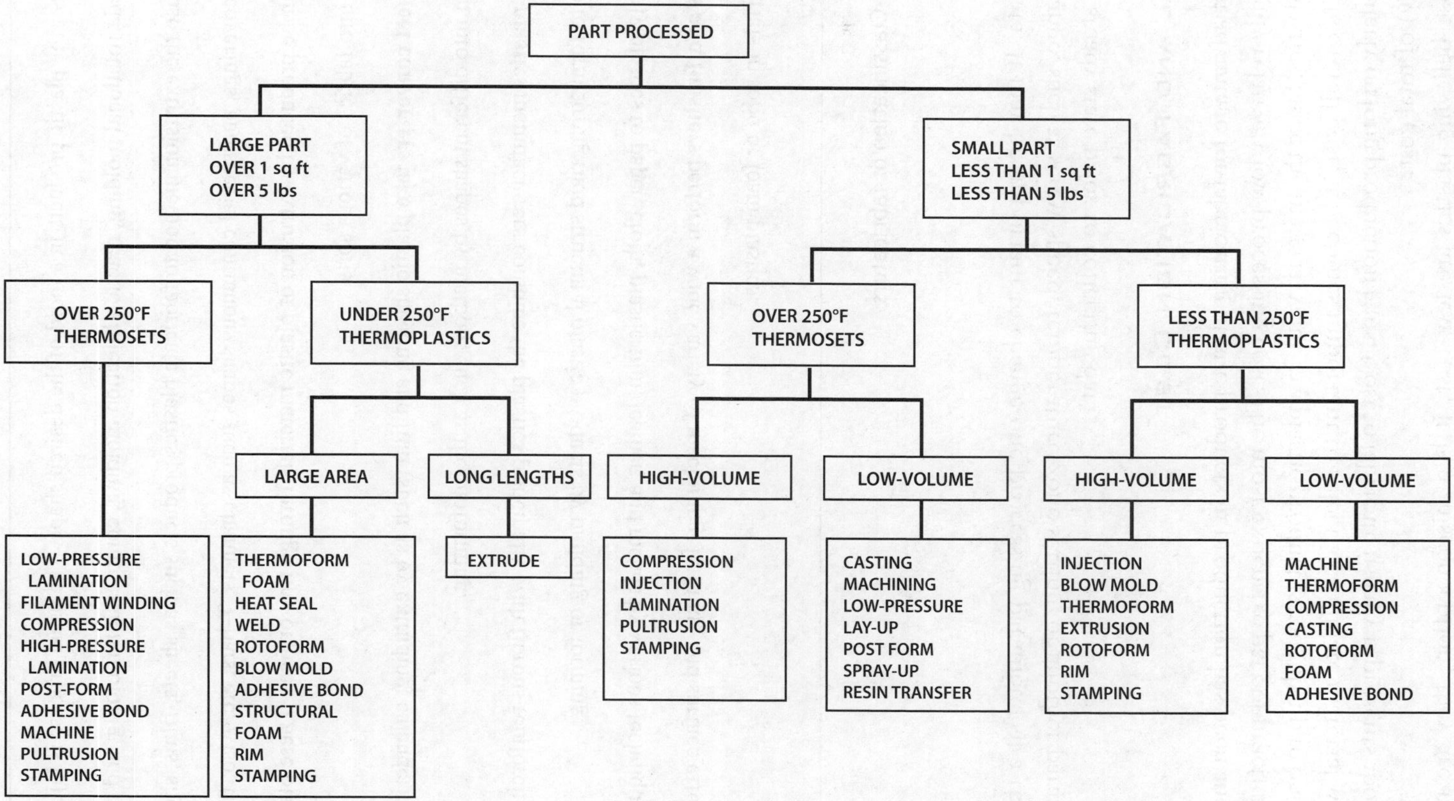

Table 3.14 Large and small part processing guide

Molders form the plastic products in a mold of the desired shape. The usual process is injection molding. Others include rotational molding, reaction injection molding, compression molding, and transfer molding.

Extruders use the extrusion method to fabricate plastics products such as sheets, films, sheets, rods, tubings, pipes, wire coverings, and special continuous shapes such as channels, T-bars, etc. Also, fine threadlike filaments can be produced by extrusion of plastic materials through tiny orifices. These plastic filaments are then woven into fabrics of various sorts.

Film and sheet converters take plastic sheeting and film (such as by extruding, calendering, or casting) convert them into different shapes by methods such as thermoforming.

Reinforced plastics manufacturers combine the plastics principally with fibrous reinforcing agents (usually glass fibers) to form strong rigid structural plastic products by molding or forming.

Coaters apply plastics to paper, cloth, plastic film, leather, and metals by various techniques.

Fabricators and finishers perform a wide variety of machining, forming, and surface effect operations on plastic materials in solid or foam plastics.

Table 3.15 Classification of fabricators

collecting data. It may be important to change one processing parameter at a time. For example, with one change, such as screw speed, temperature-zone setting, or another parameter, allow time to achieve a steady state prior to collecting data.

MATERIAL AND FABRICATION COST

A major cost advantage for fabricating plastic products in production has been and will continue to be their usually relatively low processing cost. The most expensive part of practically all products is the cost of plastics. Since the material value in a plastic product is roughly up to one-half (possibly up to 90%) of its overall cost, it becomes important to select a candidate material with extraordinary care, particularly on long production runs. Cost to fabricate usually represents about 5% (maximum 10%, usually) of total cost.

For those believing plastics are low cost, it is a misconception; they are not. The so-called low-cost types (commodity types) are more economical when compared with the more expensive

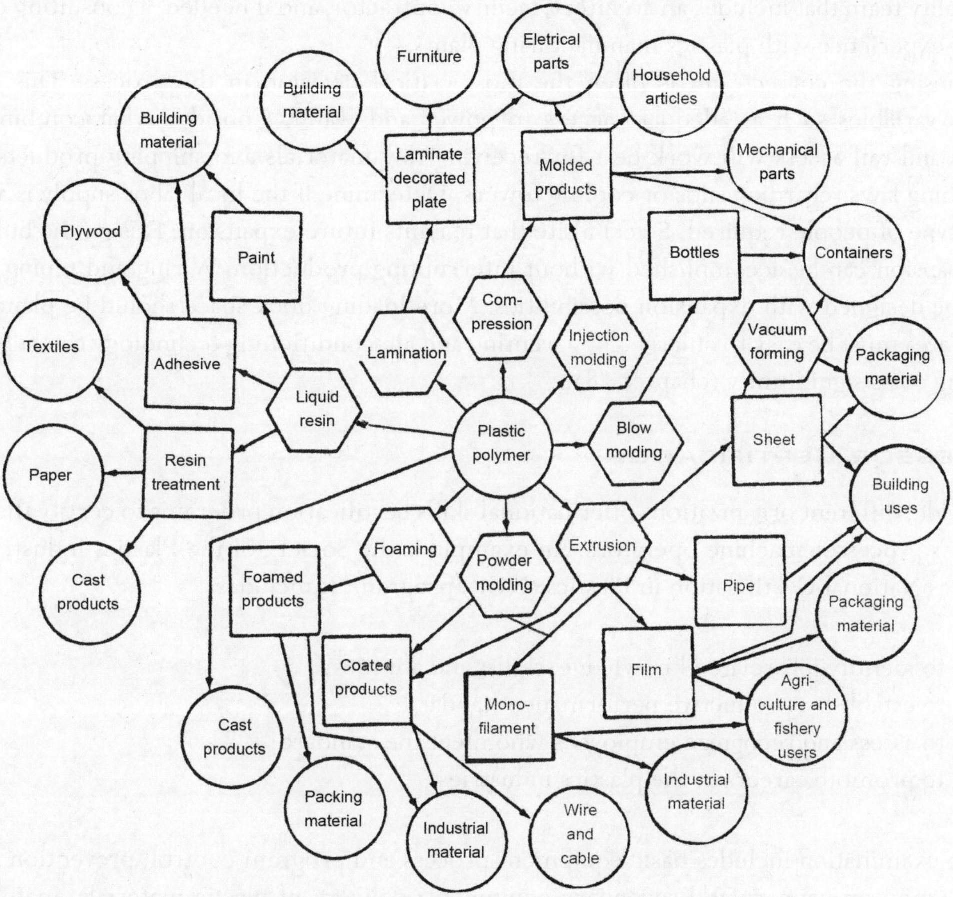

Figure 3.1 Flow chart from plastic materials through processes to products

engineering types (chapter 1). It is important to recognize that it is economically possible to process a more expensive plastic because it provides for a lower processing cost. By far, the real advantages to using plastics to produce many low-cost products are their low weights with their low processing costs.

Upgrading Plant

When a plastic fabricator considers updating a fabricating facility with a state-of-the-art operation, the usual operating factors already in use require reviews and updates, such as materials handling and services to machine safety operations (electric power, water cooling, etc.). Incorrectly estimating cost and incorrectly choosing a site location are two initial pitfalls that must be avoided. One can overestimate difficulties or underestimate challenges, with results ranging from expensive to disastrous financial situations. However, these problems can be avoided by assembling a qualified high-quality team that includes an architect, facility contractor, and if needed, a consulting engineer who has experience with plastics manufacturing plants.

Choosing the correct site is often the most critical decision in the process. This decision includes variables such as adequate access to power and water. Consider what combination of highway and rail access will work best for receiving raw materials and shipping products. Check local zoning laws regarding silos or cooling towers. Determine if the local labor supply is adequate for the type of people required. Select a site that permits future expansion. Design the building so that expansion can be accomplished without interrupting production. Wiring and piping systems should be designed with expansion possibilities. More loading dock space should be planned. The parking area must be easy to enlarge. New venting and air-conditioning technology can help reduce operating costs significantly (chapter 18).

PROCESSOR CERTIFICATION

Worldwide, different organizations offer national skills certification programs to certify the plastics industry's processor machine operators. An example is the Society of the Plastics Industry (SPI's) Industries National Certification in Plastics (NCP) program. It includes:

1. to identify job-related knowledge, skills, and abilities;
2. to establish a productive performance standard;
3. to assess and recognize employees who meet the standard;
4. to promote careers in the plastics industries.

The examination includes basic equipment process and program control; prevention and corrective action on primary and secondary equipment; delivery of plastic materials, material handling, storage, and quality assurance; machinery and plant safety; handling tools and equipment; packaging fabricated products; and general knowledge of plastics. SPE's Plastics Technology Certification is for plastics professionals who have the knowledge and ability to apply mathematics, the physical sciences, and engineering principles and methods to technological problem solving.

Plastics Learning Network SPI's televised training program is important for plastics production workers. It prepares people for SPI's NCP examination. Though mechanisms for distribution vary, state funding for SPI training and certification started to provide training for plastics workers in New York, North Carolina, South Carolina, Pennsylvania, Florida, and Kentucky. Contact SPI (telephone: 202-974-5246; gmoore@socplas.org) for details.

PROCESSING FUNDAMENTALS

While the processes differ, there are elements common to many of them. In the majority of cases, TPs are melted by heat so that they can flow. Pressure is often involved in forcing the molten plastic into a mold cavity or through a die, and cooling must be provided to allow the molten plastic to harden. With TSs, heat and pressure also are most often used, only in this case, higher heat rather than cooling serves to cure or set the TS plastic, usually under pressure. The descriptions of processes that follow this chapter cover the basics of the major fabricating systems. It should be recognized, however, that there are variations in virtually every process in order to service a particular market or a particular plastic that represents some degree of deviation from the basics.

An important factor for the processor is obtaining the best processing temperature for the plastic used. A guide is obtained from past experience and/or the material producer (Table 3.16). The temperature setting is influenced by whether or not the plastics have extensive additives, fillers, and/or reinforcements. Crystalline and amorphous thermoplastics have different melt temperature requirements that influence properties, for example, their mechanical properties. Figure 3.2 shows the difference between the melting point for crystalline TPs, the glass-transition temperature for amorphous TPs, and the temperature when the material is liquid/completely melted (chapter 1). This type of information on initial start-up of the fabricating equipment is important but only provides a guide. The setup person determines the best conditions (usually requires certain temperature, pressure, and time profiles) for the plastic being processed. Recognize that if the same plastic is used with a different machine (with identical operating specifications), then the probability is that new control settings will be required. The reason is that, like the material, machines have variables (chapter 1).

It is important to understand and measure melt flow, or heat behavior of plastics, during processing (see below). Subjects such as rheology, molecular weight distribution (MWD), and viscosity are involved when discussing melt flow (chapter 1).

MELT FLOW ANALYSIS

Measuring melt flow is important for two reasons. First, it provides a means for determining whether a plastic can be formed into a useful product, such as if it can completely fill a mold cavity, be a usable extruded extrudate, provide mixing action in a screw, meet product thickness requirements, and so forth. Second, the flow is an indication of whether the plastic's final properties will be consistent with those required by the product. The target is to provide the necessary homogeneous melt during processing to have the melt operating completely stable and working in equilibrium.

In practice, despite the continual improvements, this perfect, stable situation is never achieved, and there are variables that affect the output. If the process is analyzed, one can decide that two types of variables affect the quality and output rate. They can be identified as (a) the variables of the machine's design and manufacture and (b) the operating or dynamic variables that control how the machine is run.

Software provides simulation of the desired process and comparison with reality (chapter 25). The purpose of flow analysis is to gain a comprehensive understanding of the melt flow filling process based on process controls. The most sophisticated computer models provide detailed information concerning the influence of filling conditions on the distribution of flow patterns as well as flow vectors, shear stresses, frozen skin, temperatures and pressures, and other variables. The less sophisticated programs that model fewer variables are also available. From this data, conclusions

	EXTRUSION TEMPERATURE °F	INJECTION TEMPERATURE °F
ABS - Extrusion	435	
ABS - Injection		500
Acetal - Injection		390
Acrylic - Extrusion	375	
Acrylic - Injection		450
CAB	380	440
Cellulose Acetate - Extrusion	380	
Cellulose Acetate - Injection		450
Cellulose Proprionate - Extrusion	380	
Cellulose Proprionate - Injection		425
CTFE		550
FEP	600	600
Ionomer - Extrusion	500	
Ionomer - Injection		420
Nylon 6	520	550
Nylon 6/6	510	510
Nylon 6/10		450
Nylon 6/12	475	500
Nylon 11	460	450
Nylon 12	450	445
Phenylene Oxide Based	480	525
Polyallomer	405	405
Polyarylene Ether	460	535
Polycarbonate	550	575
Polyester PBT		460
Polyester PET	480	490
HD Polyethylene - Extrusion	410	
HD Polyethylene - Injection		480
HD Polyethylene - Blow Molding	410	
LD Polyethylene - Film	350	
LD Polyethylene - Injection		400
LD Polyethylene - Wire	400	
LD Polyethylene - Ext. Coating	600	
LLD Polyethylene - Extrusion	500	
LLD Polyethylene - Injection		425
Polypropylene - Extrusion	450	
Polypropylene - Injection		490
Polystyrene - Impact Sheet	450	
Polystyrene - G.P. Crystal	410	425
Polystyrene - Injection Impact		440
Polysulfone	650	680
Polyurethane (non-elastomer)	400	400
PVC - Rigid Profiles	365	
PVC - Pipe	380	
PVC - Rigid Injection		380
PVC - Flexible Wire	365	
PVC - Flexible Extruded Shapes	350	
PVC - Flexible Injection		300
PTFE		650
SAN	420	470
TFE		610
Urethane Elastomers (TPE)	390	400

Table 3.16 Examples of thermoplastic processing temperatures for extrusion and injection molding (courtesy of Spirex Corp.)

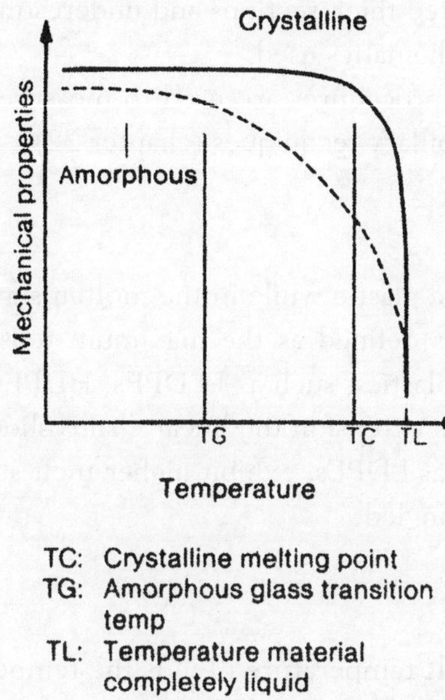

Figure 3.2 Example of the different processing temperatures for crystalline and amorphous thermoplastics

regarding tolerances, as well as part quality in terms of factors such as strength and appearance, can be drawn. Location of weld lines and weld line integrity can be predicted. The likelihood of the occurrence of warping surfaces, blemishes, and strength reductions due to high-shear stress can be anticipated. On this basis, the best filling conditions can be selected. An example of this software is The Molder's Technician from Spirex Corp.

Computer flow analysis programs used throughout the plastics industry worldwide utilize 2-D and 3-D models of a part in conjunction with rheology equations. Models range from a simple Poiseuille's equation for fluid flow to much more complex mathematical models involving differential calculus. (Poiseuille flow is an alternate name for pressure flow in general but often refers to axial streamline flow in a circular cross-section pipe for which Poiseuille law holds for Newtonian fluids.) It is important for the user of this technology to recognize that, from the simplest to the most complex of these models, all are only approximations. The user must also understand that their relational techniques, coupled with the user's assumptions, determine whether the findings of the flow analysis have any real validity. What actually happens is determined after processing the plastic.

When analyzing the results, it is important to determine the type of error. As an example, determine if the pressure loss error is in the thin, thick, or virtually all sections. If the error appears in all, it may mean that there is simply an offset caused by a difference between viscosity data used in the flow analysis and the actual viscosity during processing. If this is true, changing the processing speed should allow the flow analysis data to be duplicated with a different fill or exit time. If, however, the flow analysis overstated thick sections and underestimated thin sections, there could be a serious problem with the mathematics used.

A variety of instruments and procedures are used to measure melt flow and viscosity; these basically include rotational and capillary techniques (chapter 22).

Melt Strength

Melt strength is the strength of the plastic while in the molten state. It is an engineering measure of the extensional viscosity and is defined as the maximum tension that can be applied to the melt without breaking it. Linear plastics, such as LLDPEs, HDPEs, and PPs, generally have poor melt strength. The reason is that in extension the linear chains slide by without getting entangled. However, branched plastics, such as LDPEs, exhibit higher melt strength and elongation viscosity because the long branches get entangled.

Melt Temperature

Also called melting point, the melt temperature (T_m) is the temperature at which a plastic liquefies on heating or solidifies on cooling. T_m depends on the processing pressure and time at heat, particularly during a slow temperature change for relatively thick melts. Also, if T_m is too low, the melt's viscosity is high and more power is required to process the plastic. Degradation can occur if the viscosity is too high. Some plastics have a melting range rather than a single point. Amorphous plastics do not have melting points but rather softening ranges, and they undergo only small volume changes when solidified from a melt or when the solid softens and becomes a fluid. They start melting as soon as the heat cycle begins. T_m is often taken at the peak of a thermal analysis test using a differential scanning calorimeter (DSC) (chapter 22). Crystalline plastics have considerable molecular order in solid states, indicating that many of the atoms are regularly spaced. They have a true melting point with a latent heat of fusion associated with the melting and freezing processes, and a relatively large volume change during fabrication, or the transition from melt to solid.

Newtonian Melt Flow Behavior

This is a flow characteristic in which a material (liquid, etc.) flows immediately on application of force and for which the rate of flow is directly proportional to the force applied. It is a flow characteristic evidenced by viscosity that is independent of shear stress to strain rate (chapter 22). Water and thin mineral oils are examples of materials that exhibit Newtonian flow (Table 3.17).

Non-Newtonian Melt Flow Behavior

This is a flow characteristic in which a material (plastic) has basically abnormal flow response when force is applied. That is, its viscosity is dependent on the rate of shear. It does not have a straight proportional behavior with application of force and rate of flow (Fig. 3.3). When proportional, the behavior has a Newtonian flow.

Water temperature		
F	C	Viscosity
32	0	1.79
50	10	1.30
68	20	1.00
100	38	0.68
150	66	0.43
212	100	0.28

Table 3.17 Newtonian viscosity or coefficient of viscosity in centistokes of water

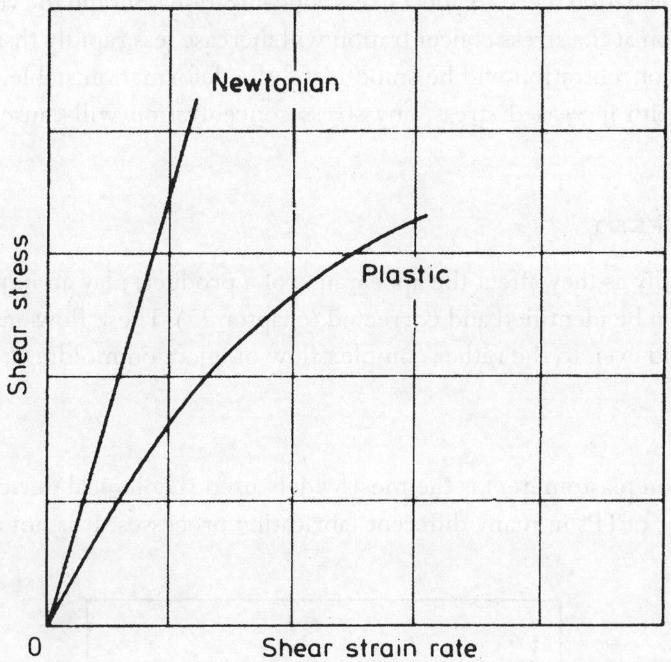

Figure 3.3 Nonplastic (Newtonian) and plastic (non-Newtonian) melt flow behavior (courtesy of Plastics FALLO)

MELT FLOW DEVIATION

The characteristic of the deviation from the ideal behavior may be of several different types. One type, called apparent viscosity, may not be independent of the rate of shear; it may increase with shear rate (shear thickening or shear dilatancy) or decrease with rate of shear (shear thinning or pseudoplasticity). The latter behavior is usually found with plastic melts and solutions. In general, such a dependency of shear stress on shear rate can be expressed as a power law. Another type is where the viscosity may be time dependent, as for material exhibiting thixotropic behavior.

MELT FLOW RATE

Degradation in fabricated products is detected using MFR tests, in which comparisons, for example, are made between the MFR of pellets and the MFR of the products (chapter 22). MFR has a reciprocal relationship to melt viscosity. This relationship of molecular weight (MW) to MFR is an inverse one; as the MFR increases, the MW drops. MW and melt viscosity are also related; as one increases the other increases.

MELT FLOW PERFORMANCE

In practically any deformation there are local stress concentrations. Should the viscosity increase with stress, the deformation at the stress concentration will increase less rapidly than in the surrounding material. The stress concentration will be smooth and the deformation stable. However, when the viscosity decreases with increased stress, any stress concentration will cause catastrophic failure (Fig. 3.4).

MELT FLOW DEFECT

Flow defects, especially as they affect the appearance of a product, play an important role in many processes. Defects can be identified and corrected (chapter 27). These flow analyses can be related to other processes and even to the rather complex flow of injection molding.

MELT INDEX

The MI test (extrusion plastometer) is the most widely used rheological device for examining and studying the behavior of TPs in many different fabricating processes. It is not a true viscometer in

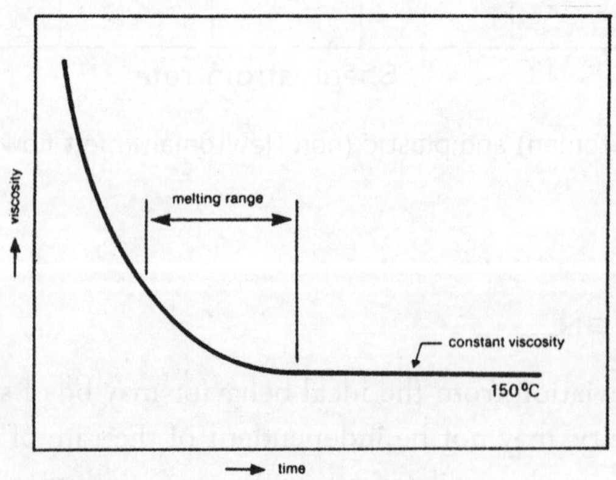

Figure 3.4 Relationship of viscosity to time at constant temperature

that a reliable value of viscosity cannot be calculated from the flow index that is normally measured. However, it does measure isothermal resistance to flow using an apparatus and test method that are standard throughout the world. Details on the MI instrument are provided in chapter 22.

The MI is an indicator of the average molecular weight (MW) of a plastic and is a rough indicator of processability due to molecular weight distribution (MWD) (Fig. 3.5). Low MW materials have high MIs and are easy to process. High MW materials have low MIs and are more difficult to process, as they have more resistance to flow; but, they are processable. End-use physical properties improve as the MI decreases. MI selection for a given application is a compromise between properties and processability. Table 22.16 lists typical MI ranges for the more common plastic processes. Materials with other MIs are still processable, but they usually require more sophisticated start-up procedures and operating process controls. Figure 22.16 shows how MI and density interrelate.

In-line Melt Analysis

There are systems that provide real-time online evaluation of mixing and melt quality. An example is the Spirex Technical Center's (located in Youngstown, Ohio) system. This system is exclusively used in the Technical Center with the Johnson Extruder. The University of Paderborn in Paderborn, Germany, initially developed this system. It uses a custom die with a quartz window, a DC light source, and a special video camera to measure light intensity passing through the melt flow. This light intensity is affected by the efficiency of a screw to disperse a standardized plastic/color concentrate mix. The computer software collects data during a test run and calculates a standard deviation. The lower the standard deviation, the better the mixing; the higher the standard deviation, the poorer the mixing (94).

With the use of this sophisticated system, actual levels of mixing quality can be measured. Spirex can evaluate all of its patented mixing elements along with many of the other mixing elements that are available to the plastics industry today.

Thermodynamics

Basically, thermodynamics is the scientific principle that deals with the interconversion of heat and other forms of energy. Thermodynamics (*thermo* = heat + *dynamic* = changes) is the study of these energy heat transfers (Tables 3.18 and 3.19). The law of conservation of energy is called the

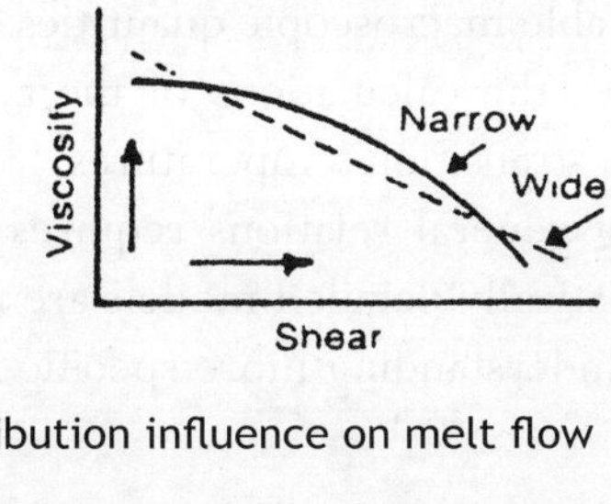

Figure 3.5 Molecular weight distribution influence on melt flow

first law of thermodynamics. It states that energy can be converted from one form to another but cannot be created or destroyed. The second law states that the entropy of the universe increases in a spontaneous process and remains unchanged in a reversible process; it can never decrease. Alternately, entropy is a measure of the unavailable energy in a thermodynamic system, commonly expressed in terms of its exchanges on an arbitrary scale, with the entropy of water at 0°C (32°F) being zero. The increase in entropy of a body is equal to the amount of heat absorbed divided by the absolute temperature of the body.

With the heat exchange that occurs during processing, thermodynamics becomes important. It is the high heat content of melts (about 100 cal/g) combined with the low rate of thermal diffusion (10^{-3} cm^2/s) that limits the cycle time of many processes. Density changes are also important, which for crystalline plastics may exceed 25% as melts cool. Melts are highly compressible; a 10% volume change for a force of 700 kg/cm^2 (10,000 psi) is typical. A surface tension of about 20 g/cm may be typical for film and fiber processing when there is a large surface-to-volume ratio.

Thermodynamic properties provide a means of working out the flow of energy from one system to another. Any substance of specified chemical composition perpetually found in electrical, magnetic, and gravitational fields has six fundamental thermodynamic properties: pressure, temperature, volume, internal energy, entropy, and enthalpy (Table 3.18). All changes in these properties must fulfill the requirements of the first and second laws of thermodynamics. The third law provides a reference point, the absolute zero temperature, for all these properties, although such a reference state is unattainable. The proper modes of applying these laws to the above six fundamental properties of an isolated system constitute the well-established subject of thermodynamics.

Thermodynamic Phase Transformation In thermodynamic equilibrium, a system may be composed of one or several physically distinct macroscopic homogeneous parts called phases, which are separated from one another by well-defined interfaces. These phases are determined by several parameters, such as temperature, pressure, and electric and magnetic fields. By continuously varying the parameters, it is possible to induce the transformation of the system from one phase to another.

Thermodynamics and Statistics This discipline tries to compute macroscopic properties of materials from more basic structures of matter (chapter 24). These properties are not necessarily static properties as in conventional mechanics. The problems in statistical thermodynamics fall into two categories. The first one involves the study of the structure of phenomenological frameworks and the interrelations among observable macroscopic quantities.

The secondary category involves the calculations of the actual values of phenomenological parameters, such as viscosity or phase-transition temperatures, from more microscopic parameters. With this technique, understanding general relations requires only a model specified by fairly broad and abstract conditions. Realistically detailed models are not needed to understand general properties of a class of materials. Understanding more specific relations requires microscopically detailed models.

MACHINES NOT ALIKE

Not all machines are created equal. Identical machine models, including auxiliary equipment, built and delivered with consecutive serial numbers to the same site can perform so differently as to make some completely unacceptable to the customer, assuming they were installed properly.

MACHINERY PERFORMANCE

The plastics industry is characterized by a wide variety of many distinct processing methods and techniques that fabricate many different plastic materials into many different products. The advent

Process	Temperature of Material (°F)	Guideline/ton of Chilling (lb/h)	Enthalpy Btu/lb (Final T_1, 125°F)
Injection molding			
Polyethylene	425	30	300
Polypropylene	400	35	275
PVC	375	45	200
ABS	425	50	175
Polystyrene	425	50	175
Sheet extrusion, calendering			
ABS	435	60	
Polystyrene	410	60	
Pipe, profile extrusion			
Polyethylene	400	50–65	
PVC	380	60–75	
ABS	435	60–75	

lb/h × 0.454 = kg/h.

Table 3.18 Examples of heat-transfer energy for different processes

Process or System	Heat-Transfer Coefficient (Btu/°F/h/sq ft)	Limiting Factors
Injection molding	30	Thickness and area available
Blow molding	25	One-side cooling and pinch-off cooling
Foam injection molding	20	Thickness and cellular structure
Extrusion bath or trough	100	One-side cooling, thickness, and area available

Table 3.19 Process heat-transfer coefficient (cooling characteristic)

of communications compatibility between varying brands and types of electrical and electronic process control (PC) components has furthered the ability of technology to advance in the industry. For example, from the simplest of programmable logic controllers to the most complex of PC-based control systems, all have found a place in the extrusion, injection molding, blow molding, thermoforming, and other manufacturing processes.

There are many different processing factors that could influence the repeatability of meeting performance requirements. Some products may require only the compliance of one or two processing factors, but others may require many. Computer programs have been developed to provide the capability of integrating all the applicable factors, thus replacing traditional trial-and-error methods. These programs improve with time, and providing improved controls is an endless effort.

The successful design and fabrication of good plastic products require a combination of sound judgment and experience. Designing good products requires a knowledge of plastics that includes their advantages and disadvantages (limitations) and some familiarity with processing machines (Table 3.8). Until the designer becomes familiar with processing machines, a qualified fabricator must be taken into the designer's confidence early in development and consulted frequently during those early days. The fabricator and mold or die designer should advise the product designer on materials' behavior and how to simplify processing. Understanding only one process and in particular just a certain narrow aspect of it should not restrict the designer.

PLASTICS PROCESSING PERFORMANCE

This review provides an introduction to processing performances; many more examples are provided throughout this book. The flow patterns resulting from the conditions of a particular fabricating process are very important in affecting product performances. The melting of plastics follows different phases that affect performances. An example is its modulus of elasticity. As the temperature increases, the plastic goes through the phases of glassy, transition, rubbery, and melt flow.

Reinforcing fibers, specifically the glass fibers, are brittle. Thus, when they are used in conjunction with a brittle matrix, as are certain TSs, it might be expected that the plastic composite would have low fracture energy. This is not true, and the impact strength of most glass- reinforced TS plastics is many times greater than the impact strengths of either the fibers or the matrix. Impact strength is higher if the bond between the glass fibers and the matrix is relatively weak, because if it is so strong that it cannot be broken, cracks will propagate across the matrix and fibers, and very little energy will be absorbed. Thus, there is a conflict between the requirements for maximum tensile or flexural modulus or strength (long glass fibers and strong interface bonds) and maximum impact strength (chapters 15 and 22).

To predict product dimensions and the fluctuation of dimensions (tolerances) during a production run, many variables have to be considered. These include the plastic materials with their variabilities (chapter 1); geometry of the product, which includes its thicknesses; toolmaking quality applied in producing the mold/die; and processing conditions and fluctuations, which is very important. Computer programs have made it possible to provide model guides that tend to understand the complex (controllable) interactions of these many factors. This allows fabricators to more accurately predict product dimensions and to model the relationship between the control of the molding process and the product tolerances.

This interplay of the many variables is extremely complex and involves a matrix of the many variables. As an example, in the molding simulation TMconcept system programmed by Molding and Cost Optimization (MCO) of Plastics and Computer Inc., Dallas, Texas, there are well over 300 variables (72). It is not reasonable to expect a person using manual methods to calculate these complex interactions without omissions or errors, even if molding only a simply shaped product. Computerized process simulation is a practical tool to monitor the influence of design alternatives on the processability of the product and to select molding conditions that ensure the required product quality.

The process used provides different control capabilities. For example, closed molding (injection, compression, etc.) provides fine detail on all surfaces. Open molding (blow molding, thermoforming, spray-up, etc.) provides detail only on the one side in contact with the mold, leaving the second side free-formed. Continuous production (extrusion and pultrusion) yields products of continuous length. Hollow molding (rotational or blow) produces hollow products. These processes can be used creatively to make different types of products. For example, two molded or thermoformed components can be bonded together to form a hollow product, or they can be blow molded.

PLASTIC MEMORY

Fabricated TPs can be bent, pulled, or squeezed into various useful formed shapes. But eventually when heat is added, they return to their original forms. This behavior can be annoying or provide interesting design possibilities for all types of fabricated products. When the change is unwanted, it is called *creep*; when the change is skillfully adapted to use in the overall design, it is referred to as *plastic memory*.

Practically all materials (metals, etc.) when bent, stretched, or compressed alter their molecular structure or grain orientation to accommodate the deformation permanently. TPs temporarily assume the deformed shapes and always maintain the internal stresses that want to force the material back to the original shape. After forming the shapes, heat can be applied so that they return to their original shapes. An example of applying this plastic memory behavior is to use stretched film around the mating section of two tubes. With the application of heat, the film shrinks and produces a tight grip where the tubes meet. Examples of applications are flat communication cable wraps, heat-shrinkable furniture webbings, pipe fittings, medical devices, and many other products.

This behavior results from stress relaxation and other viscoelastic phenomena that are typical of TPs. In addition to using heat, radiating or chemically curing TPs, such as polyolefins, neoprenes, silicones, and other cross-linkable TPs, will give them memory. Fluorocarbons need no such curing. When this phenomenon of memory is applied to a fluorocarbon, such as TFE,

FEP, ETFE, ECTFE, CTFE, and PVF, interesting and useful high-temperature or wear-resistant applications become possible.

ORIENTATION

The process of plastic orientation consists of a controlled system for stretching TP molecules in an unoriented (unidirectional [UD]) or bioriented (biaxial direction [BD]) fashion. Orientation in effect provides a means of tailoring and improving the optical, electrical, and other properties of plastics by increasing strength, stiffness, and toughness. In the direction or plane of the orientation, resistance to liquid and gas permeation, crazing, microcracks, and so on also is gained. Overall, the usual result is an improvement in a product's performance-to-cost ratio.

A TP's molecular orientation can be accidental or deliberate. Accidental orientation can occur during the processing in which excessive frozen-in stresses develop; however, with the usual proper process controls, accidental orientation will not occur. The frozen-in stresses with certain TPs can be extremely damaging, with products being subjected to environmental stress cracking or crazing in the presence of heat, chemicals, and so forth.

During processing (extrusion, injection, blow molding, etc.), the molecules tend to be more oriented than relaxed, particularly when the melt is subjected to excessive shearing action. After temperature-time-pressure is applied and the melt goes through restrictions (mold, die, etc.), the molecules tend to be stretched and aligned in a parallel form. The result can be undesirable changes in the directional properties and dimensions immediately when the plastic is processed and/or afterward when it is in use, if stress relaxation occurs.

By deliberate stretching, the molecular chains of a plastic are drawn in the direction of the stretching, and inherent strengths of the chains are more nearly realized than they are in their naturally relaxed configurations. Stretching can take place with heat during or after processing by blow molding, extruding film, thermoforming, and so on. Products can be drawn in one direction or in two opposite directions, in which case many properties significantly increase uniaxially or biaxially.

The amount of change depends on the type of TP, the amount of stretching, and most importantly, its rate of cooling. The faster the rate, the more retention there is of the frozen orientation. After processing, products could be subjected to stress relaxation, resulting in changes in performance and dimensions. With certain plastics and processes, there is an insignificant change. If changes are significant and undesirable, one must take action to change the processing conditions during and/or after processing; for example, increase the cooling rate during processing. Annealing of the product is a postprocessing condition approach.

Practically all TPs can undergo orientation, although certain types find it particularly advantageous, such as PETs, PPs, PVCs, PEs, PSs, PVDCs, PVAs, and PCs. The largest market for plastics worldwide, consuming a total of about 20 wt%, is oriented plastic film. Stretching can take place in-line or off-line, with or without tenter frames (stretching frames), using the appropriate temperature-pull rates. During extrusion of flat individual or multiple layers of different plastic films or sheets, stretching can be performed in-line. Off-line tenter frames are predominantly used. Details on orientation are provided in chapter 5.

DIRECTIONAL PROPERTY

An important type of orienting fabricated products concerns applying directional properties to plastics by using fiber reinforcements. Orientation is the alignment of fiber reinforcement within the product that affects mechanical properties. The reinforced plastic (RP) properties increase in the direction of alignment (Fig. 3.6).

In an RP construction, both plastics and fibers influence orientation properties. For example, with certain TPs, the plastics' molecular orientation can be used to aid in increasing stiffness, strength, and toughness, as well as craze and microcrack resistance in the direction of the plane or the plane of orientation. By far, fibers are the main source of orientation, resulting in positive and significant increased performance (Table 3.20).

Such effects are obvious in continuous filament winding (chapter 15). Not so obvious are the anisotropic materials' properties resulting from many TP processes. For example, viscous melt flow of hot plastic into injection molds produces an oriented structure (chapter 4), usually having greater strength with crystalline TPs in the direction of flow than perpendicular to this direction. Shrinkage also is usually greatest with crystalline TPs perpendicular to the direction of flow. With amorphous TPs, the greatest shrinkage can be in the direction of flow; however, as in IM, melt flow control can readily have even shrinkage in all directions. With fibers, shrinkage is less than that in unreinforced RPs (URPs) in the flow and perpendicular directions.

PLASTIC DEFORMATION

Plastics have certain degrees of elasticity associated with them. As an example, if the plastic stretches within its elastic limit, it will eventually return to its natural shape (chapter 22). When

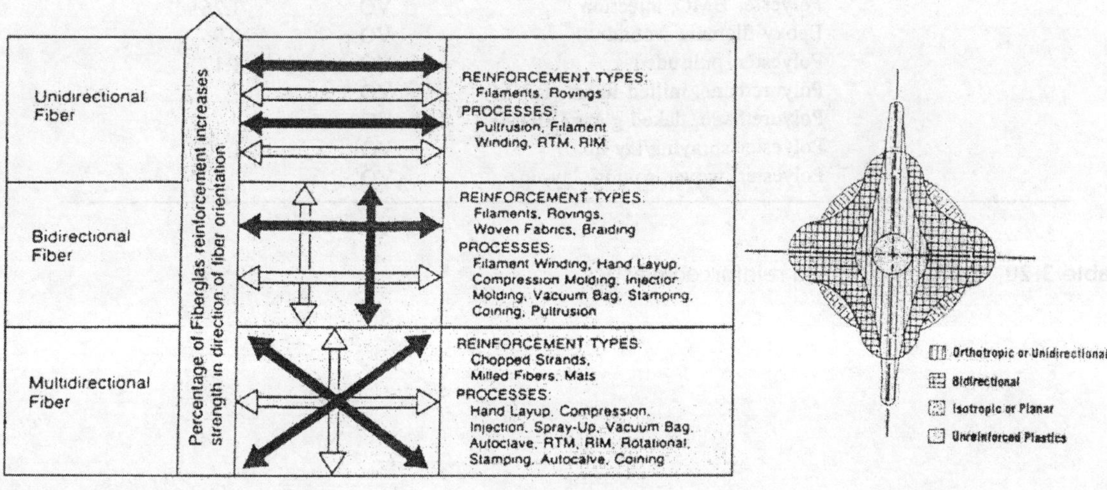

Figure 3.6 Examples of reinforced plastic directional properties

Material		Flammability UL94	Specific heat, J/(kg · K)
Unreinforced thermoplastics (TP)	Acetal resin	HB	1.46
	Nylon-6	HB	1.67
	Nylon-6,6	V-2	1.26
	Polycarbonate	V-2	1.26
	Polypropylene	HB	1.88
	Poly(phenylene sulfide)	V-O	
	Acrylonitrile-butadiene-styrene terpolymer (ABS)	HB	
	Poly(phenylene oxide) (PPO)	V-1	0.84–1.67
	Styrene-acrylonitrile (SAN)	HB	
	Poly(butylene terephthalate)	HB	
	Poly(ethylene terephthalate)	HB	1.42
Glass fiber-reinforced thermoplastics (RTP)	Acetal resin	HB	
	Nylon-6	HB	
	Nylon-6,6	HB	1.26
	Polycarbonate	V-1	1.21
	Polypropylene	HB	
	Poly(phenylene sulfide)	V-O/5V	1.05
	Acrylonitrile-butadiene-styrene terpolymer (ABS)	HB	
	Poly(phenylene oxide) (PPO)	HB	0.84–1.67
	Styrene-acrylonitrile copolymer (SAN)	HB	
	Poly(butylene terephthalate)	HB	0.46
	Poly(ethylene terephthalate)	HB	
Glass fiber-reinforced thermosets (RTs)	Polyester SMC, compression	5V	1.26
	Polyester SMC, commpression	5V	1.26
	Polyester SMC, compression	5V	1.26
	Polyester BMC, compression	5V	1.26
	Polyester BMC, injection	VO	1.26
	Epoxy filament wound	VO	0.96
	Polyester, pultruded	VO	1.17
	Polyurethane, milled fibers (RRIM)	VO	
	Polyurethane, flaked glass (RRIM)	VO	
	Polyester spraying/lay-up	VO	1.30
	Polyester, woven roving, lay-up	VO	

Table 3.20 Unreinforced and reinforced plastics

Thermal coefficient of expansion	Heat deflection at 1.8 MPa, C°	Thermal conductivity, W/(m · K)	Dielectric strength, kV/25 mm	Volume resistivity, ohm-cm	Relative permittivity, 60 Hz	Arc resistance, s
1.4	213	12.1	375	3.2×10^{16}	3.80	135
1.7	216	11.2	520	10^{16}	3.60	90
4.7	110	2.80	500	10^{15}	3.70	129
4.8	68–85	2.08	305	4.5×10^{13}	4.00	
4.5	75	2.94	385	1×10^{15}	&4.00	120
3.7	132	2.34	380	$> 10^{16}$	2.96	120
3.8	46–60	2.10	600	$> 10^{17}$	2.20	125
	135	2.89	380	10^{16}		
3.2	93–104	1.61	350–500	2.7×10^{16}	2.80	
68.0	100	1.59	400	$> 10^{16}$	2.65	75
36.0	104	1.21	515	4.4×10^{16}	3.50	65
4.0	200 +		350	10^{14}	4.20	
4.7	161		580	10^{14}	4.12	142
1.5	211	10.0–19.7	500	10^{13}	3.90	120
1.8	254	2.60	400	10^{15}	3.80	120
1.8	141	7.97	500	10^{16}	3.10	125
2.4	132	14.5	440	10^{15}	2.70	120
1.1	266	3.47	380	4×10^{15}	3.00	125
2.1	99	2.42	465	10^{15}	3.20	80
2.0	143	6.57	500	10^{16}	3.20	70
2.1	102	4.84	490	10^{15}	3.50	70
D696	D648	C177	D149	D257	D150	D495
	200 +		500	5.7×10^{14}	4.40	188
	200 +				4.40	188
9.4	200 +				4.40	188
6.6	260	8.37	375	27×10^{14}	4.20	190
6.6	260	8.37	375		4.20	190
2.0	200 +	1.77	300	$> 10^{12}$		
5.0		6.92	200	10^{13}	4.40	80
78.0	29					
53.1						
12.0	200 +	2.60	250			

Table 3.20 *(continued)*

overstressed, it reaches what is known as plastic deformation, meaning the plastic will not return to its original shape.

This behavior can affect processability. It is important to keep the drag force from melt feed mechanism friction as low as possible and to keep acceleration of the feed material low enough so the product does not permanently deform or tear.

Sinusoidal acceleration profiles reduce the load on the plastic by providing gentle forces near zero speed and top speed while making up for lost ground in the middle, resulting in diminished stretching. Stretching also contributes to reduced dimensional accuracy. This situation can be overcome by either compensating for the stretch in the process controller when the amount of stretch is predictable or by installing a position verification device between the feed mechanism and the postfeed position.

COEXTRUSION/COINJECTION: FABRICATING MULTILAYER PLASTICS

Multilayer fabrication is the simultaneous processing of two to seven or more plastics melt streams meeting in a die/mold to produce laminated or multilayer plastic products. The product performances gained are a combination of what each individual plastic provides, with a potential of obtaining synergistic gains. Each ply of the laminated structure imparts a desired property, such as impermeability or barrier/resistance to some environments, heat stability, toughness, tear resistance, and lower cost.

The layup can include recycled material and/or low-cost material that will "reinforce" more expensive material and provide additional strength, tear resistance, and so on. Production of these products (such as extruded film, sheet, cable coverings, pipe, coinjection bottles, containers, thermoformed packages, etc.) involves two or more different or similar plastics going through a single die/mold. Two or more orifices from each extruder or each injection-molding machine processing the different plastics are arranged so that the extrudates merge and "weld" together into a single structure prior to cooling (chapters 4, 5, 6, 17).

When bonding layers are not desired in a plastic laminated composite structure, a plastic tie layer is used that provides bonding. Choosing the proper "adhesive" layer is by no means a simple task, since evaluation includes processability, bonding capabilities, and performance in the final product. There are many different types with different capabilities, with EVALs being one of the important ones.

When thicker wall thickness can be used, low-density foam core products are produced, such as pipe and paneling. This multilayer sandwich structure provides reduced material costs without sacrificing performance; in fact, properties can be improved by using the sandwich design advantages (chapter 19).

Certain melt processing factors have to be considered in order to eliminate problems. Some of these factors can be compensated by the available plasticator and die/mold adjustments. An unsteady balance of shear forces causes the interfacial instability. Examples of the factors include (a) different melt temperatures of adjacent layers, (b) plastic viscosity differentials that should not be greater than 2 to 4/1, and (c) minimum thickness of a cap (top) layer, which because it is subjected to a high shear stress, it is usually limited to 5% to 10% of the total thickness. There is a tendency for the less viscous plastic to migrate to the region of high shear stress in the flow channel, causing an interface deformation. With a great difference in viscosities existing between adjoining layers, the less viscous tends to surround or encapsulate the other plastic. The result could be fuzzy interfaces, orange peel, and so on.

When running a line, make provisions to adequately alarm the operator if one of the plasticators trips out. This will prevent melt from backflowing into the stopped machine and creating a blockage in the feed section that could become hazardous.

PLASTICATOR MELTING OPERATION

The major processing methods that process well over 80 wt% of all plastics are extrusion, injection molding, and blow molding. These processes, as well as a few others, use a plasticator to melt plastics. With its usual barrel and screw, the plasticator is a very important component in a melting process. If factors such as the proper screw design and/or barrel heat profile are not used correctly, fabricated products may not meet or maximize their performance and, very importantly, not provide a low-cost process.

Plasticators have wide operating ranges to meet different performance requirements of all the different plastics processed. Their rotating drive systems can be via hydraulic and/or electrical motors (Table 3.21). Electrical motors tend to increase melt processing efficiency that in turn increases production rate. They have wide operating ranges to meet different performance requirements of all the different plastics processed. It is important to obtain maximum throughput with as close to a perfect melt quality. This is an endlessly moving target, owing to the limits and/or variabilities of the plastics, machines, and controls (chapter 1). Since the start of using screw plasticators, definite improvements have been occurring in the melt quality. This action continues because advancements tend to be endless in applying advanced screw designs and the changing melting characteristics of plastic materials.

SCREW

Figures 3.7 to 3.10 provide an introduction to the performance of a plasticator where the screw usually has a 17.6 degree flight helical angle.

Figure 3.10 identifies the following three main parts (zones) of a screw that are included in a hypothetical screw design (Table 3.22):

1. Feed Zone: This is the part of the screw that picks up the plastic at the feed opening (throat) plus an additional portion downstream. Many screws, particularly for extruders, have an initial constant lead and depth section, all of which compose the

	With accumulator-assisted hydraulics	With accumulator and electric drive
Part weight, oz	91	91
Cycle time, seconds	70	70
Throughput, lb/hr	293	293
Installed power, hp		
Hydraulic motor	122	61
Servo motor		68
Power consumption, kW-hr/lb		
Hydraulic motor	.24	.086
Servo motor		.077

Table 3.21 Servo-electric screw drive

feed section. This section can be welded onto the barrel or a separate part bolted onto the upstream end of the barrel. The feed section is usually jacked for fluid heating and/or cooling.

2. Transition Zone: It is the section, also called the compression zone, of a screw between the feed zone and metering zone in which the flight depth decreases in the direction of discharge. In this zone, the plastics start in both solid and molten states, with the target to have all molten upon leaving this zone.
3. Metering Zone: This section is a relatively shallow portion of the screw at the discharge end with a constant depth and lead usually having the melt move three or four runs of the flight length.

Many different screw designs are available to meet the desired performance for the different plastics being processed. The features common to all screw plasticators are screw(s) with matching barrel(s) that have at least one hopper/feeder intake entrance for plastics and one discharge port/exiting of the melt. The essential factor in their "pumping" process is the interaction between the rotating flights of the screw and the stationary barrel wall. If the plastic is to be mixed and conveyed at all, its friction must be low at the screw surface but high at the barrel wall. If this basic criterion is not met, the material may rotate with the screw without moving at all in the axial direction and out through the die/mold. The clearance between the screw and barrel is usually extremely small. The difference in the diameters of the screw and barrel bore (diametrical clearance), or more commonly one-half the diametrical clearance, is referred to as the *radial clearance*.

In the output zone, both screw and barrel surfaces are usually covered with the melt, and external forces between the melt and the screw-channel walls have no influence except when processing

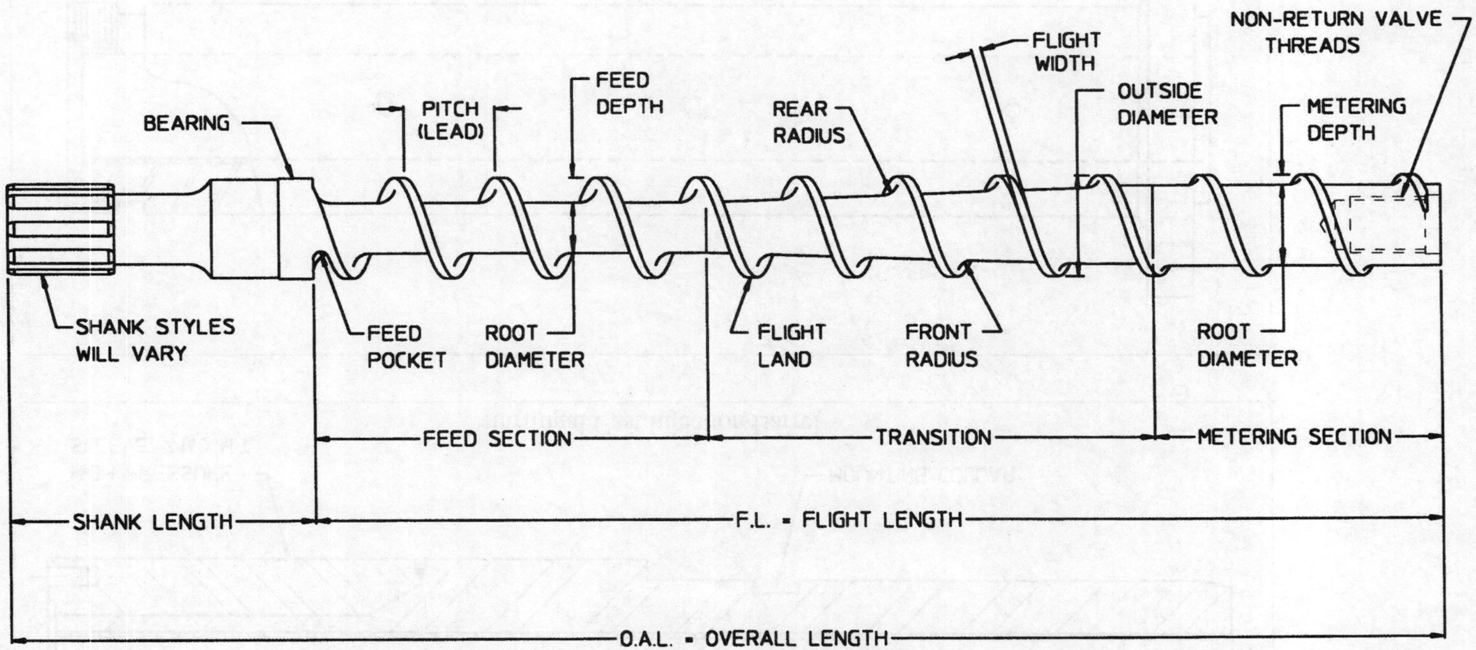

Figure 3.7 Nomenclature of an injection screw (top) and extrusion screw (courtesy of Spirex Corp.)

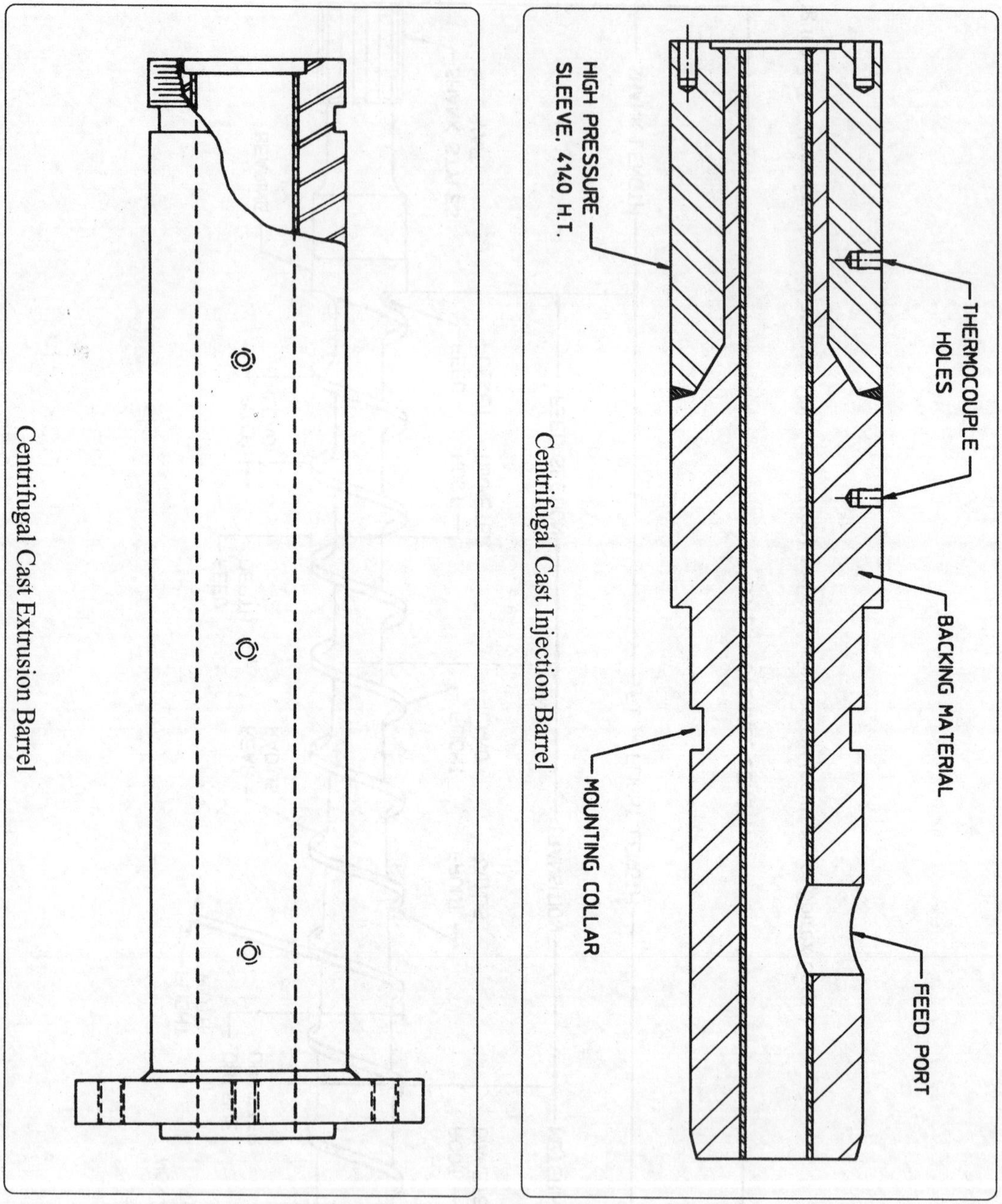

Figure 3.8 Nomenclature of an injection barrel (top) and extrusion barrel (courtesy of Spirex Corp.)

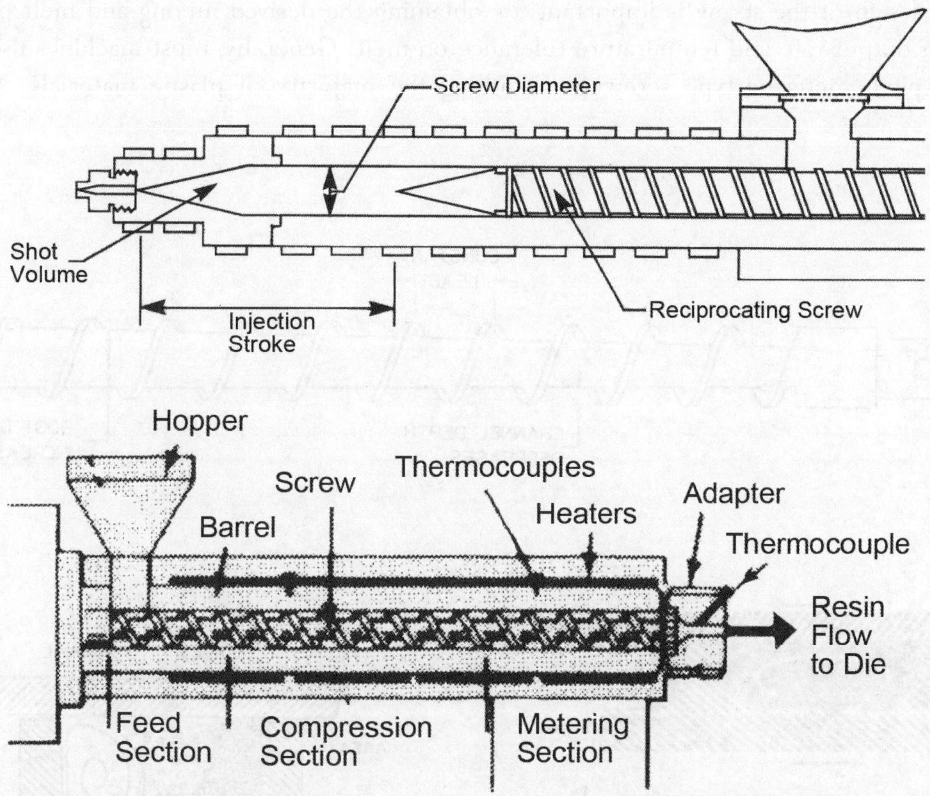

Figure 3.9 Assembled screw-barrel plasticator for injection molding (top) and extruding (courtesy of Plastics FALLO)

extremely high viscosity materials, such as rigid PVCs (polyvinyl chlorides) and UHMWPE (ultra-high molecular weight polyethylenes). The flow of the melt in the output section is affected by the coefficient of internal friction (viscosity), particularly when the die/mold offers a high resistance to the flow of the melt. The constantly turning screw augers the plastic through the heated barrel where it is heated to a proper temperature profile and blended into a homogeneous melt. The rotation causes forward transport. It is the major contributor to heating the plastic via the plastic's shearing action once the initial barrel heat start-up occurs. The melting action through the screw is shown in Figure 3.10.

DESIGN

The primary purpose for using a screw is to take advantage of its mixing action. Theoretically speaking, the motion of the screw should keep any difference in melt temperature to a minimum. It should also permit materials and colors to be blended better, with the result that a more uniform melt is delivered to the mold.

The design of the screw is important for obtaining the desired mixing and melt properties as well as output rate and temperature tolerance on melt. Generally, most machines use a single constant-pitch metering-type screw for handling the majority of plastic materials. A straight

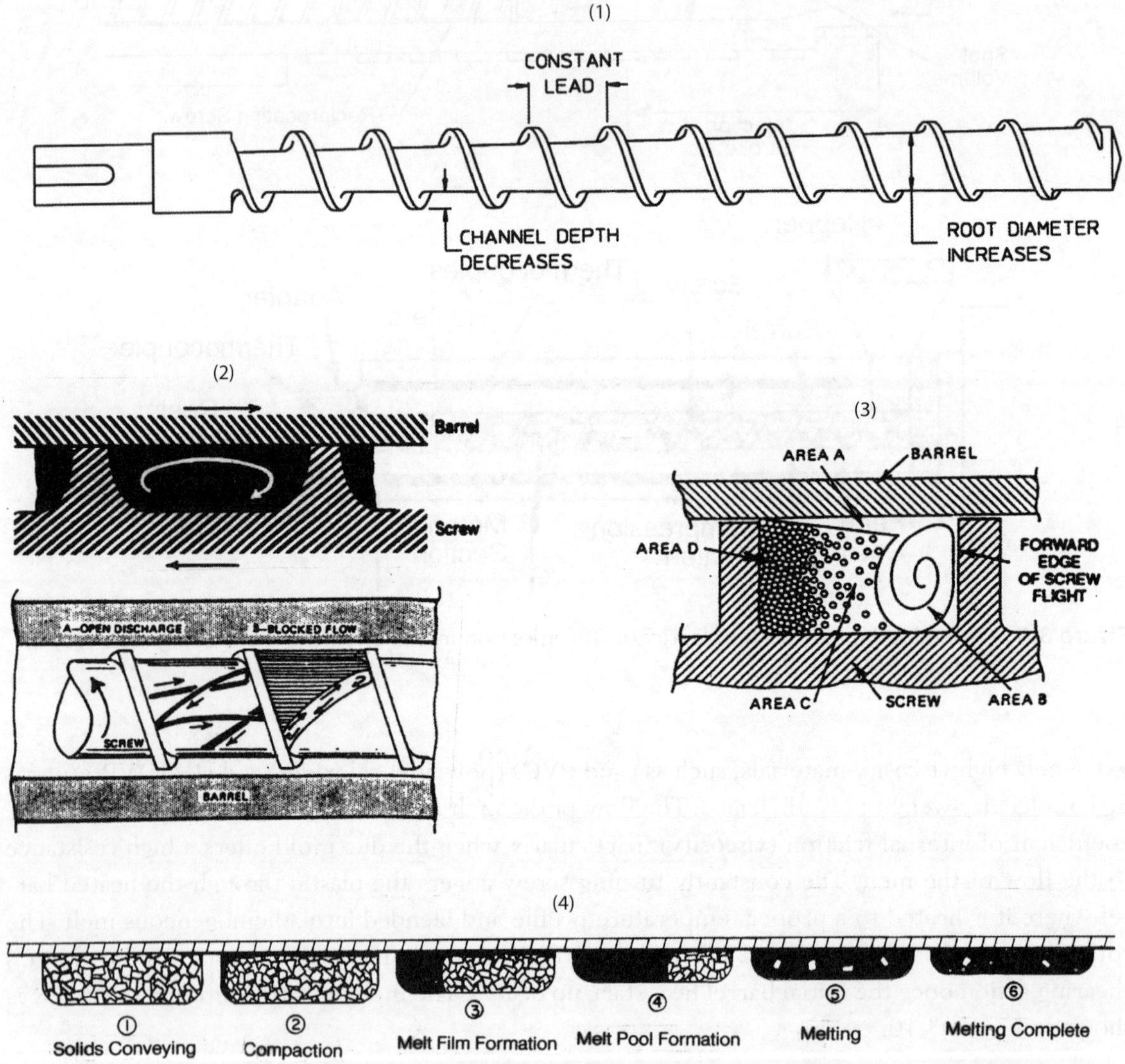

Figure 3.10 Action of plastic in a screw channel during its rotation in a fixed barrel: (1) highlights the channel where the plastic travels; (2) basic plastic drag actions; (3) example of melting action as the plastic travels through the barrel where areas A and B have the melt occurring from the barrel surface to the forward screw surface, area C has the melt developing from the solid plastic, and area D is solid plastic; and (4) melt model of a single screw (courtesy of Spirex Corp.)

Dimensions (in.)	Rigid PVC	Impact polystyrene	Low-density polyethylene	High-density polyethylene	Nylon
Diameter	4½	4½	4½	4½	4½
Total length	90	90	90	90	90
Feed zone (F)	13½	27	22½	36	67½
Compression zone	76½	18	45	18	4½
Metering zone (M)	0	45	22½	36	18
Depth (M)	0.200	0.140	0.125	0.155	0.125
Depth (F)	0.600	0.600	0.600	0.650	0.650

Table 3.22 Hypothetical screw design (courtesy of Plastics FALLO)

compression-type screw or metering screw with special tips (heads) is used to process heat-sensitive thermoplastics, and so on.

The helix angle affects the conveying and the amount of mixing in the channel. Experience has shown that a helix that advances one turn per nominal screw diameter gives excellent results. This corresponds to an angle of 17.8 degrees, a standard that has been universally adopted. The land width is usually 10% of the diameter. The radial flight clearance is the clearance between the screw flight and the barrel. It is specified considering the following effects:

1. Amount of leakage flow over the flights.
2. Temperature rise in the clearance. The heat is generated in shearing the plastic. The amount of heat generated is related to the screw speed, the design of screw, and the material.
3. The scraping ability of the flights in cleaning the barrel.
4. The eccentricity of the screw and the barrel.
5. Manufacturing costs.

The length of the screw is the axial length of the flighted section. An important criterion of a screw design is the ratio of the length over the diameter of the barrel (L/D). As reviewed, a screw has three sections: feed, melting (transition), and metering. The feed section that is at the back end of the screw can occupy from 0% to 75% of the screw length. Its length essentially depends upon how much heat has to be added to the plastic in order to melt it. The pellet or powder is generally fed by gravity into this section and is conveyed some distance down the barrel, during which time it becomes soft. Both conduction and mechanical friction accomplish heating.

The melting (transition) section is the area where the softened plastic is transformed into a continuous melt. It can occupy anywhere from 5% to 50% of the screw length. This compression zone has to be sufficiently long to make sure that all of the plastic is melted. The straight compression-type screw is one having no feed or metering sections.

In the metering section, the plastic is smeared and sheared to give a melt having a uniform composition and temperature for delivery to the mold/die. As high shear action tends to increase the melt's temperature, the length of the metering zone is dependent upon the plastic's heat sensitivity and the amount of mixing required. For heat-sensitive materials, practically no metering zone can be tolerated. For other plastics, it averages out at about 20% to 25% of the total screw length. Both the feed and the metering sections have a constant cross-section. However, the depth of the flight for the feed zone is greater than that in the metering zone. Table 3.23 provides examples of the transition sections based on the type of plastic being processed.

Fixed screw speed, pitch, diameter, and depth of the channels relate to output. A deep-channel screw is much more sensitive to pressure changes than a shallow channel. In the lower pressure range, a deep channel will mean more output; however, the reverse is true at high pressures. Shallower channels tend to give better mixing and flow patterns.

The flow pattern in the screw flights changes with the back pressure. In the flow of a particle in the flights with open discharge and in the blocked flow there is a similar circulatory motion between the flights and no forward motion because the open end is closed. The greatest mixing occurs when the flow is blocked. The importance of this flow concept is that it shows that the more blocked the flow, the better the mixing in the screw. With higher pressure, there is greater pressure flow and lower output. In injection molding, this pressure corresponds to the back-pressure setting of the machine. This is the reason that color dispersion is improved and homogeneity increased by raising the back pressure. Often, warpage and shrinkage problems can be overcome in this manner.

In our highly technological world, the art of screw design is still dominated by experienced trial and error approaches providing the exact capabilities of the screws for a particular plastic operating under specific conditions. Screw design technology is considered to be basically empirical and/or secretive; however, scientific approaches to screw designs based on an analytical melting model can be used. Computer models that are based on proper data input and are impacted by how much experience one has with a setup similar to the one being studied are available and play very important roles. When new materials are developed or improvements in old materials are required, one must go to the laboratory to obtain rheological and thermal properties before computer modeling can be performed effectively.

Production rate of an acceptable melt from a screw is its most important function. It is often limited by its melting capacity. The melting capacity of the screw depends on the plastics' properties, the processing conditions, and the particular geometry of the screw. Once the melting capacity is predicted, the screw can be designed to match the melting capacity.

The computer models based on proper data input and, very important, experience of the person with a set up similar to the one being studied play a very important role. When new materials are developed or improvements in old materials are required, one must go to the laboratory to obtain rheological and thermal properties before computer modeling can be performed effectively. New screws improve one or more of the basic screw functions of melt quality, mixing efficiency, melting performance along the screw, melt heat level, output rate, output stability, and power usage or energy efficiency.

The rotating geometric-helical flighted screw mechanically plasticizes, with the help of heat and pressure at a controlled flow rate, and advances a melt to be pumped through the barrel. Plastic

Material	Transition Length (a)	Compression Ration (b)	Meter Depth (c)
ABS	Long	Low	Deep
Acetal	Short	Medium	Shallow
Acrylic	Long	Low	Deep
Cellulosics	Medium	Low	Deep
FEP	Short	Medium	Medium
HDPE	Medium	Medium	Shallow
LDPE	Medium	Medium	Medium
Nylon	Short	High	Shallow
PAI	Long	Low	Deep
PBT	Medium	Low	Shallow
PC	Medium	Low	Deep
PEI	Medium	Medium	Medium
PES	Medium	Medium	Medium
PET	Medium	Medium	Medium
PP	Medium	Medium	Shallow
PPO	Medium	Low	Medium
PS	Medium	Low	Medium
PVC (F)	Medium	Low	Medium
PVC (R)	Long	Low	Deep

(a) Short – 4 turns or less; Medium – 5 to 7 turns; Long – 8 or more turns
(b) Low – less than 2.5:1; Medium – 2.5 to 3.5:1; High – 3.5:1 or more
(c) Degree of depth: Example – Medium for 2″ diameter 0.100″ to 0.125″

Dimensions, in.	Rigid PVC	Impact polystyrene	Low-density polyethylene	High-density polyethylene	Nylon	Cellulose acet/ butyrate
Diameter	$4\frac{1}{2}$	$4\frac{1}{2}$	$4\frac{1}{2}$	$4\frac{1}{2}$	$4\frac{1}{2}$	$4\frac{1}{2}$
Total length	90	90	90	90	90	90
Feed zone (F)	$13\frac{1}{2}$	27	$22\frac{1}{2}$	36	$67\frac{1}{2}$	0
Compression zone	$76\frac{1}{2}$	18	45	18	$4\frac{1}{2}$	90
Metering zone (M)	0	45	$22\frac{1}{2}$	36	18	0
Depth (M)	0.200	0.140	0.125	0.155	0.125	0.125
Depth (F)	0.600	0.600	0.600	0.650	0.650	0.600

Table 3.23 Examples of screw transition sections based on type of plastic being processed

in the screw channel is subject to changes during operation. Each operation of the screw subjects the plastic to different thermal and shear situations. Consequently, the plasticizing process becomes rather complex. However, it is controllable and repeatable within the limits of the equipment and material capabilities.

Mixing

The action of mixing plastic can be distributive and/or dispersive. They are not physically separated. In dispersive mixing, there will always be distributive mixing. However, the reverse is not always true. In distributive mixing, there can be dispersive mixing only if there is a component exhibiting a yield stress and if the stress acting on this component exceeds the yield stress.

In order for a dispersive mixing device to be efficient, the mixing section should have a region where the plastic is subjected to high stress. Its high-stress region should be designed so that exposure to high stress occurs only for a short time, and all fluid elements should experience the same high stress level to accomplish uniform mixing. In addition, the mixing device should follow the general rules for mixing: minimum pressure drop in the mixing section, streamline flow, and complete barrel surface wiping action. The mixing section should be easy to manufacture.

As to be reviewed, these dynamic mixers are used to improve screw performance. Static mixers are sometimes also inserted at the end of the plasticator. Proof of their success is shown by their extensive use worldwide. Each type of mixer offers its own advantages and limitations. Where practical, they should not be located in a region where the melt viscosity is not too low. With some of these installations, because they may have to operate at a lower speed to avoid problems such as surging, independently driven mixers can be used so that machines can operate at optimum speed. Other benefits of independently driven mixers involve feeding capability and performance. For example, metering pumps can inject with precision liquid additives directly into the mixer.

Static mixers or so-called motionless mixers have no moving parts, and only a small increase in the screw energy is needed to overcome the resistance of their mechanical baffles. The installation of a static mixer increases the effective length to diameter (L/D) of the extruder and usually results in some increase of the melt temperature and head pressure. The overall result is a more homogeneous melt and a more stable extrusion process with less output surge. When using static mixers, sufficient heat-up time should be provided with start-up of the extruders. To prevent excessive pressure and damage to the plasticator during start-up, the screw's rpm should be kept low until plastic flows uniformly from the die.

Shear Rate

Most of the energy that a screw imparts to the plastic material is by means of shear. The velocity of the plastic relates to the shearing action between two surfaces moving in relation to each other (Fig. 3.11). These surfaces are the barrel's inside diameter (ID) and the root diameter of the screw.

The rate of energy imparted to the plastic increases as the shear rate increases. The shear rate increases as the relative speed of the two surfaces increases and as the distance between the surfaces

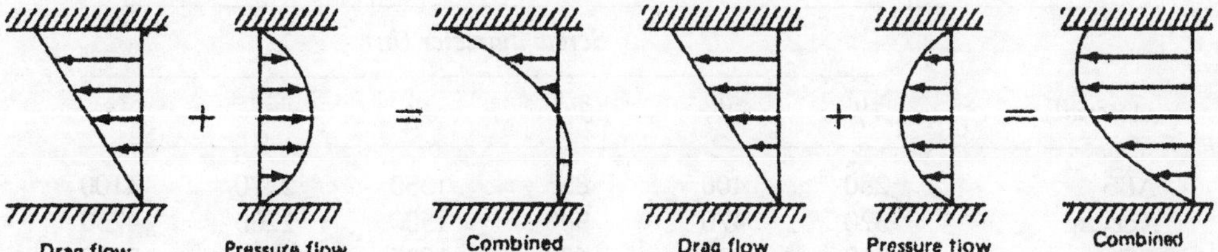

Figure 3.11 Examples of melt flow velocity in a plasticator that relates to positive flow pressure, negative drag flow, and their combined distribution

becomes less. Knowledge of shear rate can be useful when there are problems with excessive shear causing high melt temperatures and burning heat-sensitive plastics. Low shear rate can cause poor mixing, low melt temperatures, and unmelted plastic. The actual shear rate at any single point along a screw can be calculated using the following formula (94):

$$S = DN/19.1\ h,$$

where S = shear rate in seconds (reciprocal seconds), D = screw diameter in inches, N = screw rpm, and h = screw channel depth in inches.

RATE OF OUTPUT

The rate of output (throughput), or the speed at which plastics are moved through the plasticator, continues to be pushed higher as a result of design advances in equipment and materials. Output rates generally range from a few kilograms to over 5 tons per hour on single-screw machines (Table 3.24). Twin-screw and other multiscrew extruder diameters have output rates ranging from a few kilograms to at least 30 tons per hour.

The approximate output of a screw is predictable, provided that the melt is under control and reasonably repeatable. A very rough and preliminary estimate for output rate (OR) in lb/h can be calculated by using the barrel's inside diameter (ID) in inches and using the following equation:

$$OR = 16\ ID^2\ (lb/h \times 0.4536 = kg/h).$$

There is no easy way to calculate the output of extrusion, injection, and blow-molding screws. A multitude of factors affect the output. Many capable people are spending the majority of their time working on mathematical models and analyses of screw performance. They have developed sophisticated computer simulation programs that can be of considerable assistance in predicting performance of a certain screw. Even with all of this knowledge and computer assistance, exact answers to a specific problem can be difficult to obtain. Spirex Corp. has computer systems and screw simulation programs that are helpful. These simulation programs are most helpful in making

Material	Screw diameter (in) 1½	2½	3½	4½	6	8
ABS	280	400	825	1350	2270	4100
Acrylic	320	470	900	1500	2700	4750
PC	210	320	680	1025	1850	3200
PP	280	400	825	1350	2270	4100
HIPS	340	560	1100	1800	3250	5750
PVC, flexible	300	450	900	1500	2700	4750
PVC, rigid	180	250	500	800	1450	2300
LDPE	310	525	1050	1750	3000	5500
LLDPE	200	300	600	1000	1800	3260
HDPE	215	325	725	1175	2150	3750

Table 3.24 Examples of extruder output in lb/h for different plastics

modifications from existing designs and results. There are also some simplified formulas that can lend some assistance in certain cases (94).

A simplified formula that gives the drag flow output of the metering section of a conventional square pitch (conventional screw where the distance from flight to flight is equal to the diameter) screw is

$$R = 2.3\ D^2\ hgN,$$

where R = screw rpm or output in pounds per hour, D = screw diameter in inches, h = depth of the metering section in inches (for two-stage screws use the depth of the first metering section), g = melt density, and N = Newtonian flow.

This formula does not take into consideration backflow or leakage flow over the flights. This is not usually a large factor unless the plastic has a very low viscosity at process temperatures and shear rates. Of course, leakage flow becomes a significant factor when screws or barrels become severely worn. The formula also assumes pumping against zero pressure at the head or die. It also assumes no over- or underpumping from the feed section and gives no consideration to melt quality (94).

With all these and other limitations, the formula can be useful for the following reasons:

1. It provides a general guide to the output of screws.
2. If the actual output of the screw is significantly greater than calculated, it is caused by high compression ratios that overpump the metering section. Sometimes this is desirable, but it can lead to surging and rapid screw wear if it is excessive.
3. If the output is a lot less, it usually indicates a feed problem, a worn screw or barrel, or the inability of the screw to melt at higher screw speeds. The screw or barrel wear

problem can be determined by measurement. A feeding problem can, on occasion, be corrected by changes in barrel temperature settings. More often, the problem is caused by screw design, shape and bulk density of the feedstock, surface condition of the screw root and barrel ID in the feed area, feed-throat design, screw temperature, and other items.

SHOT SIZE

The proper selection of screw diameter and L/D for IMMs is critical to the manufacture of high-quality parts at economical cycle times. Generally, 25% to 60% of the full stroke of the particular injection machine is considered a good operating range. This figure is probably valid if the full stroke of the machine is approximately four diameters long for a 20:1 L/D screw. Many machines have either shorter or longer total strokes. A better determination of a good range for the stroke is 25% to 60% of a four-diameter full stroke on a 20:1 L/D screw. This gives a two-diameter stroke length as midrange. This is appropriate when recovery time is approximately 50% of the overall cycle and given a screw/barrel combination with proper design to melt/mix the succeeding shot (94).

SCREW WEAR

There are basically two methods of evaluating output loss caused by screw wear; one is rather accurate and the other is a rough approximation. The accurate method is to compare the current worn screw output with a production benchmark reference output established when the screw was new. The approximate method involves determining screw wear by measuring the screw, calculating the resultant added screw-to-barrel clearance, and estimating the output loss from the added clearance. There are three major problems with the approximate method. First, about 24 hours machine downtime occurs. Second, it may understate the extent of output loss by as much as two-and-a-half times. Third, it does not account for the commonly encountered problem where the screw's rpm is simply increased yet the output loss due to increased melt temperature is not well established.

SINGLE-STAGE SCREW

When manufacturers began using TPs during the 1930s, they started with plasticators developed earlier by the rubber industry. The rubber extruder or tuber used a screw, and with modification, these machines would melt and pump the TPs. Injection-molding machines used the plunger until the early 1960s when the screw was adopted. The early TP screws were called plastic screws to differentiate them from rubber screws. Plastic screws usually had a constant square pitch (lead = diameter; Fig. 3.10[1]) and no feed or metering section. The root depth of the screw constantly changed from the feed to the discharge end. This was different from the typical rubber screw, which had a constant channel depth but obtained its compression by means of a variable or changing lead. (This review on the plasticator includes abstracts from *Plasticating Components 2000*, a booklet by Spirex Corp., a world leader in the design and manufacturer of plasticators; 94.)

The design of screws for TPs gradually changed, and by 1960 most screws were of the metering type. The metering screw has a constant depth at the discharge end (Fig. 3.12). This metering section was designed to smooth out pressure, temperature, and discharge rate irregularities. It is targeted to meter the plastic at a uniform rate much like a metering gear pump (chapter 5). The metering screw, with many modifications, is the basis for most present-day extrusion and injection screw designs.

Until the 1960s, TSs were molded using compression or transfer presses (chapter 14). At that time, screw injection machines with modifications were developed to process TSs (Fig. 3.13). These modifications included low compression, deeper channel depths, short length to diameter screws (L/Ds), tool steel construction, barrel cooling with heat-transfer fluids, and spiral-down discharge ends in place of nonreturn valves.

The screw gave better mixing than the plunger. Therefore, the TS or TP was superior in color dispersion, physical properties, and so on. It was also possible to mold at lower temperatures, resulting in faster fabricating cycles and greater profits. The stationary screw extruding into a pot or accumulator was also popular in the early stages, but the reciprocating screw is the most common design used to date. Figure 3.14 shows the basic three-stage action in a reciprocating injection-molding machine where (a) melt from the plasticator moves to the front of the screw causing the screw to retract, (b) screw acts as a plunger forcing the melt through a nozzle into a mold cavity, and (c) plastication repeats moving the melt in front of the screw (chapter 4). Figure 3.15 shows examples of plasticators used in two-stage injection-molding machines. The continuously rotating screw directs melt into an accumulator pot where in turn it is rammed into the mold cavity (chapter 4).

Feeding Problem

Generally, the plastic being fed flows by gravity (usually controlled weightwise) from the feed hopper down into the throat of the plasticator barrel. Special measures are taken and devices used for plastics (chapter 18) that do not flow easily or can cause hang-ups (bridging or solidification resulting in plastic not flowing through the hopper). This initial action occurs when the plastic is in a solid state with its temperature below its melting point. As the screw turns in the heated barrel, plastic falls down into its channel. Frictional forces develop in the plastic during plasticizing so that the melt moves forward toward the mold/die.

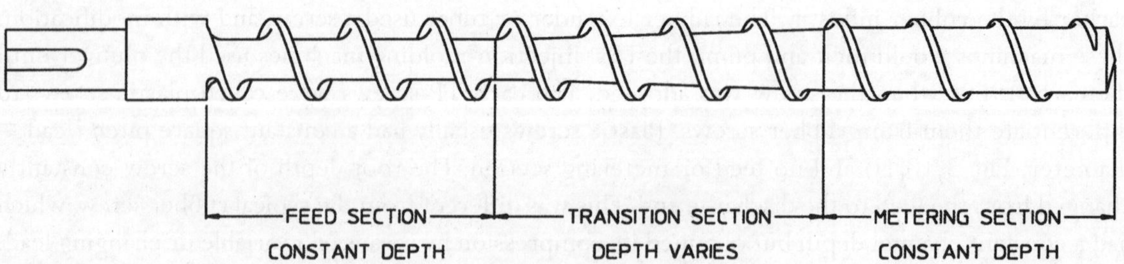

Figure 3.12 Thermoplastic metering screw (courtesy of Spirex Corp.)

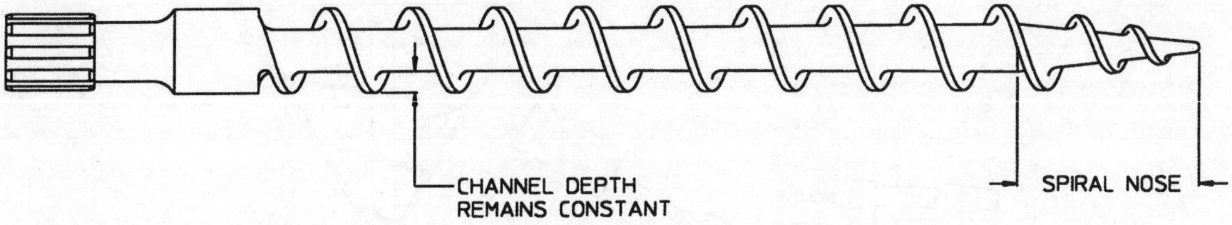

Figure 3.13 Thermoset plastic screw (courtesy of Spirex Corp.)

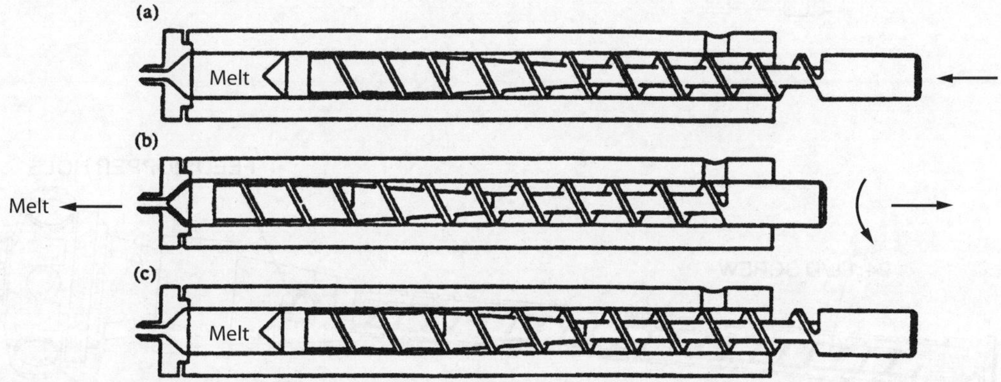

Figure 3.14 Example of a reciprocating plasticator screw injection molding machine

The action that pushes solid particles forward in the feed section of a single-screw extruder, blow, or injection machine has always had the potential for being one of the weakest features of these machines. This forward feeding force near the feed hopper is often weak and erratic and is classified as "nonpositive." It can be so tenuous that a specific screw/barrel combination will feed virgin but little or no additions of regrind, one feedstock shape but not another, and often one family of plastics but not another. This action results in nonuniform feed that will in turn result in poor production rates, nonuniform output (surging), and poor product quality (94).

The feeding mechanism of solid plastics is dependent on the surface friction of the screw surfaces and the inner surface of the barrel. The easier the solid particles of plastic slide on the screw, the better the screw will feed. Also, the greater the friction or resistance to sliding on the barrel wall, the better it will feed. This is perhaps best visualized by considering the worst condition in which it slides on the barrel and literally sticks to the screw. In this case, the plastic will merely go round and round with the screw and never move forward along the barrel. Processors have several immediate time and cost solutions that may help. It is possible that a redesign of the screw by machining or replacement will cure the problem for a specific situation. It is also possible that

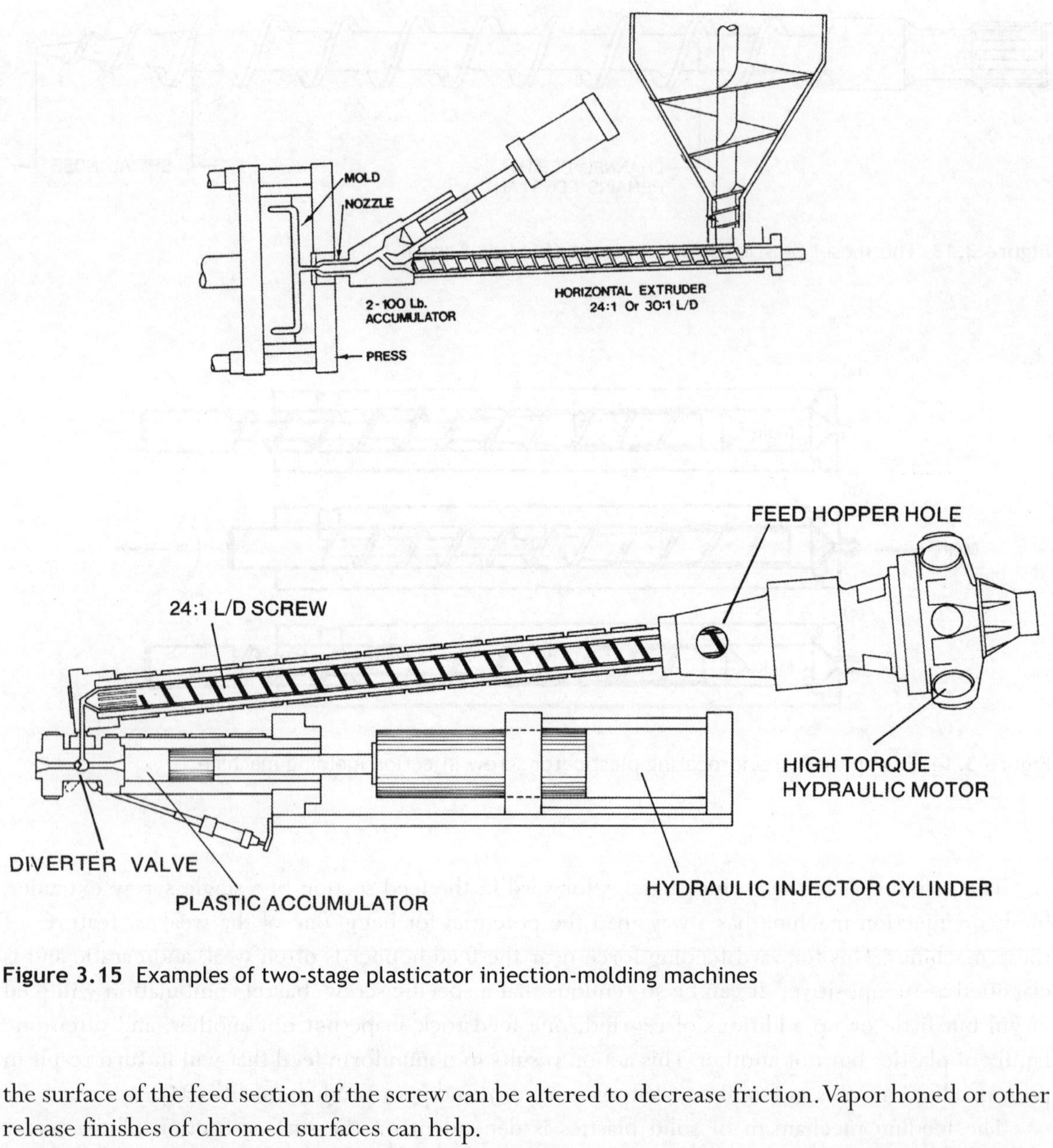

Figure 3.15 Examples of two-stage plasticator injection-molding machines

the surface of the feed section of the screw can be altered to decrease friction. Vapor honed or other release finishes of chromed surfaces can help.

The most immediate tool available to the processor is finding the optimum barrel temperature settings. This optimum setting will give the best feeding temperature at the inside of the barrel for that rpm and plastic combination. These feed critical settings are the rear ones and will vary depending on many things, including rpm, barrel wall thickness, depth of thermocouple recording melt temperature, plastic composition (filler, etc.; chapter 1), and other factors. The intent is to obtain an inside barrel wall temperature that will be hot enough to provide a viscous "sticky" melt film early without overheating to make the plastic so fluid that it flows too easily. A stylized graph shown in Figure 3.16 illustrates this and shows why it is not certain whether raising or lowering the temperature will help.

Sometimes these temperature settings can solve a problem; however, based on experience from different sources, looking for the right settings will usually result in a low probability of success. An important consideration in all of these feed problems is that many are improperly diagnosed and are actually melting problems. Every screw design and plastic combination has a practical limit for the rate at which it can melt the material. If the screw is run at an rpm that exceeds the ability of the screw to melt material at that rate, solids blocks will form with surging and the appearance of poor feeding. This is particularly true of plastics with high specific heats, such as the polyolefins. If you obtain low and erratic output in conjunction with temperature override in the transition, the problem is usually melting and not feed.

TWO-STAGE SCREW

Overview A variation of the metering screw is the two-stage or double-metering screw. It basically is two single-stage screws attached to each other (Figs. 3.17 and 3.18). The two-stage screw was first designed to run with a vented extruder. In an extruder, the plastic is melted and pumped by the first stage into the vent or second feed section. In the deep vent section, the plastic melt is decompressed and the entrapped volatiles (moisture, etc.) escape. The plastic is then compressed again and pumped through the die by the second-stage screw. In the early 1970s, vented two-stage screws were adapted to injection-molding machines.

The two-stage screw has other advantages aside from its venting capabilities. It provides for additional mixing because of the tumbling that the plastic receives in the vent section and because

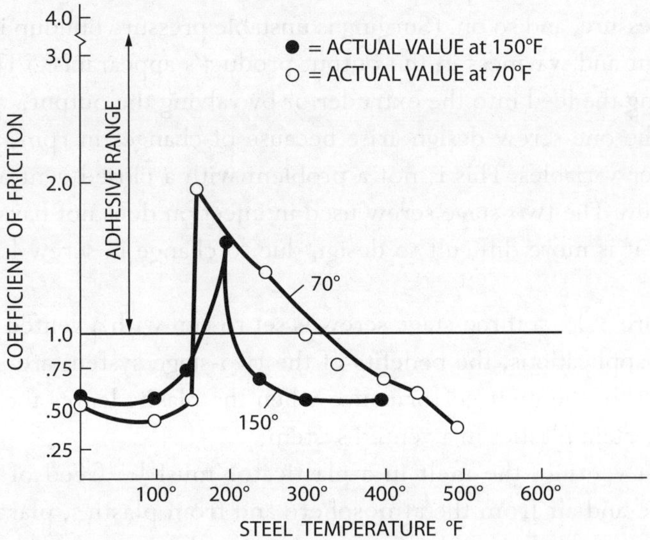

Figure 3.16 Coefficient of friction of LDPE vs. steel at different temperatures (courtesy of Spirex Corp.)

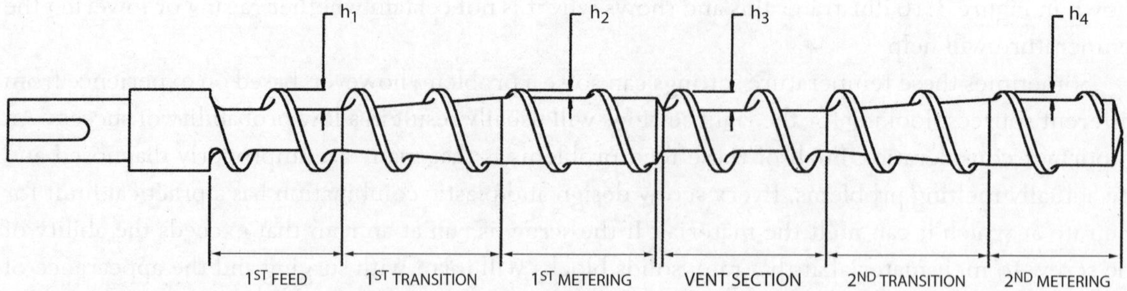

Figure 3.17 Two-stage screw (courtesy of Spirex Corp.)

the material is compressed, decompressed, and compressed again. All of this tends to give some mixing without shear. Because the screw runs partially filled in the vent section and part of the second transition, the torque and horsepower requirements are somewhat reduced for the same output and same screw speed when compared with a single-stage screw of the same diameter and flighted length. Other advantages include elimination of predrying plastic, greater use of regrind, reduced mold venting, elimination of dryer variability, less space required compared with hopper dryers, rapid start-up, and rapid color or plastic changes.

A potential problem with a two-stage screw in vented extrusion is the difficulty in balancing the first-stage output (which can be calculated and is determined by the hour two) with the second-stage output (from hour four) as shown in Figure 3.17. If the first-stage screw delivers more than the amount the second-stage screw pumps through the die, the result is vent flooding. If the second-stage screw tends to take away or pump more than the first-stage screw delivers, the result is surging of output, pressure, and so on. (Surging is unstable pressure buildup in an extruder leading to variable throughput and waviness in the output product's appearance.) This can sometimes be adjusted by controlling the feed into the extruder or by valving the output.

Problems with the one-screw design arise because of changes in rpm, plastic variations, die restrictions, and other variables. This is not a problem with a closed vent and a low pump ratio using a two-stage screw. The two-stage screw used in injection does not have the surging problem described above, but it is more difficult to design due to change in screw location relative to the feed and vent ports.

As shown in Figure 3.19, a three-stage screw is set to run with a vented extruder. For certain plastics and product applications, the benefits of the two-stage system are at least obtained with improved repeatability of the melt performance when the plastic leaves the extruder die, or it is required to process certain plastics in a vented system.

While drying via venting, the melt in a plasticator must be freed of gaseous components that include moisture and air from the atmosphere and from plastics, plasticizers, and/or other additives, as well as entrapped air and other gases released by certain plastics. Retention of gas components such as moisture has always been a potential problem for all processors. All kinds of problems develop in products (splay, poor mechanical properties, dimensions, etc.). This situation is particularly important when processing hygroscopic plastics (chapter 1). One major approach to this plastic degrading situation is to use plasticators that have vents in their barrels to release these contaminants.

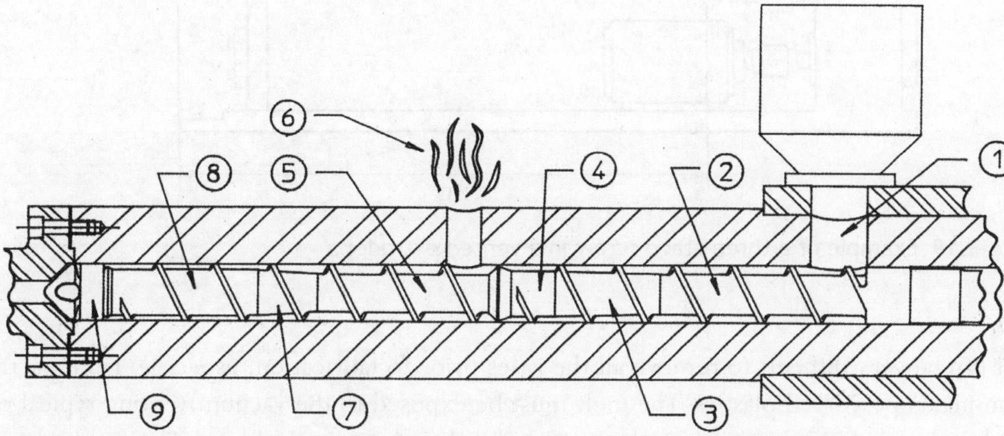

(1) Wet material enters from a conventional hopper or through a controlled feeding device.

(2) The pellets are conveyed forward by the screw feed section and are heated by the barrel and by some frictional heating. Some surface moisture is removed here and escapes rearward through the feed opening.

(3) The compression or transition section does most of the melting.

(4) The first metering section accomplishes final melting and evens flow to the vent section.

(5) Resin is pumped from the first stage metering section to a deep vent or devolatizing section. This vent section is capable of moving quantities well in excess of the material delivered to it by the first metering section. For this reason, the flights in the vent section run partially filled and at zero pressure. It is here that volatile materials such as water vapor and other non-desirable materials escape from the melted plastic. The vapor pressure of water at 500° F is 666 psi. (See Table 1 on page 84). These steam pockets escape the melt and travel spirally around the partially filled channel until they escape out the vent hole in the barrel.

(6) Water vapor and other volatiles escape from the vent.

(7) The resin is again compressed and pressure is built in the second transition section.

(8) The second metering section evens the flow and maintains pressure so that the screw will be retracted by the pressure in the front of the non-return valve.

(9) A low resistance, sliding ring non-return valve works in the same manner as it does with a non-vented screw.

Figure 3.18 Simplified version of the mechanics of a vented injection-molding machine (courtesy of Spirex Corp.)

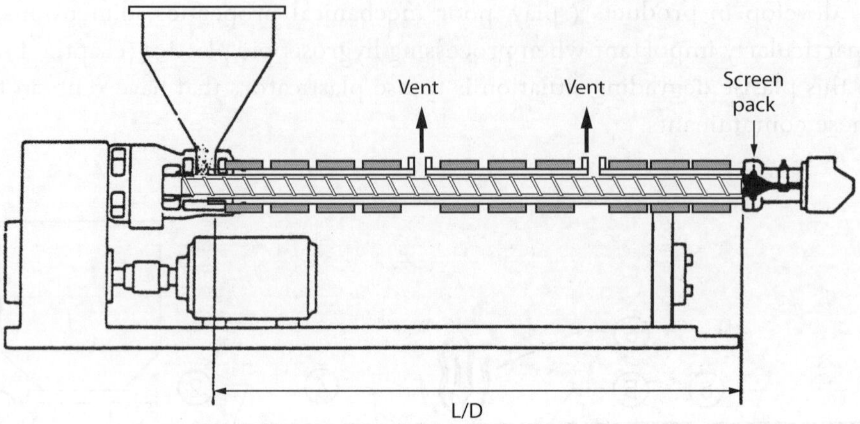

Figure 3.19 Example of a three-stage screw in a vented extruder

It can be very difficult to remove all the gases prior to fabrication, especially from particularly contaminated powdered plastics. The melt must be exposed to the vacuum venting typical of most vented screws. A vacuum is connected to the vent's exhaust port in the barrel. The standard machines operate on the principle of melt degassing. The degassing is assisted by a rise in the vapor pressure of volatile constituents, which results from the high melt heat. Only the free surface layer is degassed; the rest of the plastic can release its volatile content only through diffusion. Diffusion in the nonvented screw is always time dependent, and long residence times are not possible for melt moving through a plasticator. Thus, a vented barrel with a two- or three-stage melting screw is used.

The first-stage screws of the transition and metering zones are often shorter than the sections of a single-stage conventional screw. The melt discharges at zero pressure into the second-stage screw under vacuum instead of pressure. The first-stage extrudate must not be hot enough to become overheated in the second-stage screw; the first-stage screw must not deliver more output per screw rotation at discharge pressure than the second-stage screw can pump through the barrel under the maximum normal operating pressure. This usually means that the second-stage metering section must be at least 50% deeper than the first-stage section.

Based on practical experience, the best metering-section depth ratio (pump ratio) is about 1.8:1. The ratio to be used depends on factors such as screw design, downstream equipment, feedstock performance, operating conditions, and plastic being processed (Table 3.25). There is likely to be melt flow through the vent (avoid this situation) if the compression ratio is high or the metering depth ratio is slightly too low. When the metering depth ratio is moderately high, there is a gradual degradation of the output. With the screw channel in the vent area not filling properly, the self-cleaning action is diminished and the risk of plate-out increases. In any case, sticking or smearing of the melt must be avoided or degradation will accelerate.

It is important to consider that a controlled material feeding device (also called a starve feeder) may be necessary to maximize the operation of a vented system (chapter 18). The controlled feeding

PLASTIC	DEPTH (inches)
ABS	0.0010 - 0.0015
ACETAL	0.0005 - 0.0010
ACRYLIC	0.0015 - 0.0020
ETHYLENE VINYL ACET.	0.0010 - 0.0015
IONOMER	0.0005 - 0.0010
LCP	0.0005 - 0.0007
NYLON	0.0003 - 0.0005
POLYCARBONATE	0.0015 - 0.0025
PET, PBT, POLYESTERS	0.0005 - 0.0007
POLYSULFONE	0.0010 - 0.0020
POLYETHYLENE	0.0005 - 0.0012
POLYPROPYLENE	0.0005 - 0.0012
POLYSTYRENE	0.0007 - 0.0010
POLYSTYRENE (IMPACT)	0.0008 - 0.0012
PVC (RIGID)	0.0006 - 0.0010
PVC (FLEXIBLE)	0.0005 - 0.0007
POLYURETHANE	0.0004 - 0.0008
SAN	0.0010 - 0.0015
T/P ELASTOMER	0.0005 - 0.0007

Table 3.25 Guide for the depth of vent openings for different plastics

device determines the amount of plastic being fed into the screw, thereby controlling the output of the first stage of a two-stage screw. This action should eliminate all cases of possible vent bleeding and plugging of the vent hole. In addition to this necessary function, the device has other uses. When the feeder partially fills the screw flight channels, the surface moisture that is being driven off the plastic is allowed a place to evaporate into the atmosphere. Other uses include its ability to govern the amount of shear and energy that is delivered to the plastic via the screw geometry, ability to provide different shear history, and so forth.

Melt Behavior The first-stage screws of the transition and metering zones are often shorter than the sections of a single-stage conventional screw. The melt discharges at zero pressure into the second-stage screw under vacuum instead of pressure. The first-stage melt must not be hot enough to become overheated in the second-stage screw; the first-stage screw must not deliver more output per screw rotation at discharge pressure than the second-stage screw can pump through the barrel under the maximum normal operating pressure.

This usually means that the second-stage metering section must be at least 50% deeper than the first-stage screw. The pumping ratio (PR) as applied to two-stage vented screws gives a measure of the ability of the second-stage screw to pump more than the first-stage screw delivers to it. Too high a PR will tend to cause a surge and a low PR will tend to cause vent melt flow.

In practice the best metering-section depth ratio (pump ratio) is about 1.8:1. The ratio to be used depends on factors such as screw design, feedstock performance, and operating conditions. There is likely to be melt flow through the vent (avoid this situation) if the compression ratio is high or the metering depth ratio is slightly too low. When the metering depth ratio is moderately high, there is a gradual degradation of the output. With the screw channel in the vent area not filling properly, the self-cleaning action is diminished and the risk of plate-out increases. In any case, sticking or smearing of the melt must be avoided or degradation will accelerate.

Melt Degassing

The standard machines operate on the principle of melt degassing. The degassing is assisted by a rise in the vapor pressure of volatile constituents, which results from the high melt heat. Only the free surface layer is degassed; the rest of the plastic can release its volatile content only through diffusion. Diffusion in the nonvented screw is always time dependent, requiring long residence time. Thus, a vented barrel with a two- or three-stage melting screw is used.

Vent Bleeding

Vent bleeding is the unplanned escape of melt bleeding through the vent during operation of the vented barrel processing. Vent flow problems are usually blamed on the screw design, but more often they are due to bad vent diverter design. The function of this decompression volume or vent section of the screw is simply to generate a partially filled channel with no pressure. The vent diverter function is to accept the moving melt and move it into the next section of the screw.

To determine if the problem is melt flow, two tests can be performed. First, remove the diverter, rotate the screw slowly, and observe the degree of fill. If it is one-third or less filled, the problem is almost certainly the diverter. The other method is to run it for a few minutes at open discharge at normal screw speed. If vent flow begins, the diverter is causing the problem, as the screw is working against no discharge pressure.

Other factors that can cause vent flow are the diverter or screw design. They can cause melt foaming reaction, screw/barrel wear, improper vent location, and excessive pressure. A hopper-feeder venting system can be used to provide starve feeding. It is a controlled material-feeding device that may be necessary to maximize the operation of a vented system. Regardless of its necessity, it is certainly a useful device for many reasons. The controlled feeding device determines the amount of plastic that is being fed into the screw, thereby controlling the output of the first-stage screw of a two-stage screw. This action should eliminate all causes of possible vent bleeding and plugging of the vent hole.

Another advantage of starve feeding is that the screw flight channels are partially filled, allowing the surface moisture that is being driven off the plastic a place to evaporate into the atmosphere. Other advantages include its ability to govern the amount of shear and energy that is delivered to the plastic via the screw geometry, the ability to provide different shear history, and so forth.

Blister Section Blister sections are typically used only with two-stage screws that are being vented. The blister element is usually short and only amounts to about 0.25 L/D (length to diameter of screw; Fig. 3.20). The two basic types are underside outer diameter and full diameter with slots (Fig. 3.21). In both types, one of the functions of a blister is to impose a known back pressure on the first stage of a screw. This can be on either an extrusion or injection screw.

In the case of the undersize diameter blister, the land length and diameter of the blister are determined by calculating the pressure drop and shear that is desired to be placed on the plastic prior to being exposed to the decompressing atmospheric region in the vent section of the screw. Since all of the plastic must flow between the inside diameter of the barrel and the outside diameter of the blister, the plastic is sheared as it passes over the blister and develops a back pressure on the first stage. This back pressure can help stabilize the first-stage screw and therefore stabilize the output of the screw.

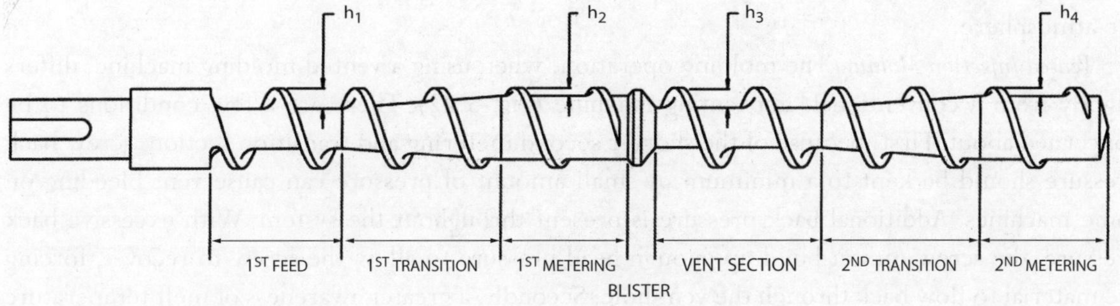

Figure 3.20 Blister-type variation of a two-stage screw (courtesy of Spirex Corp.)

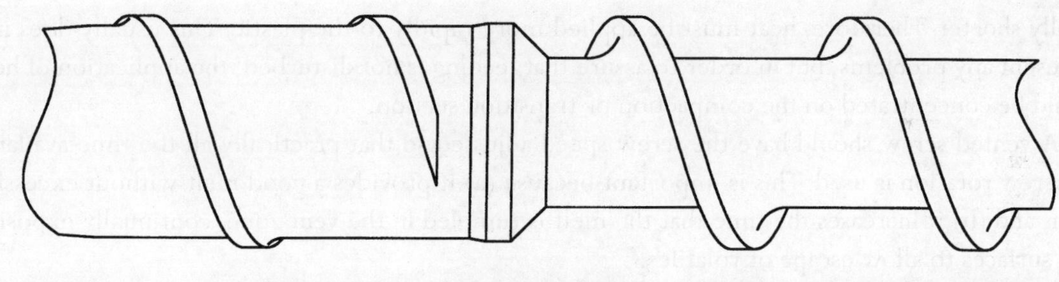

Undersized Blister

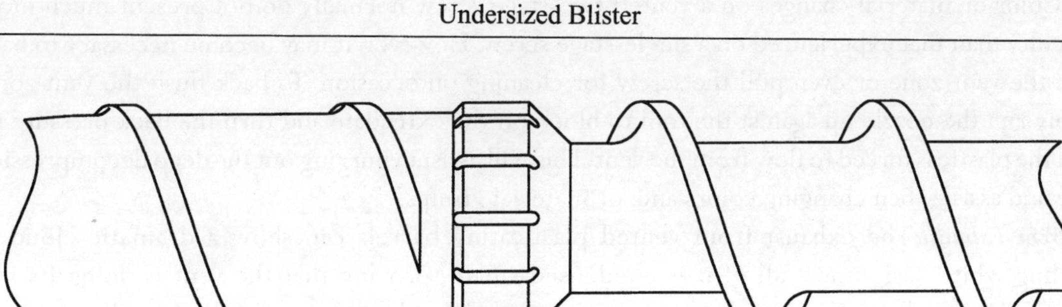

Slotted Blister

Figure 3.21 Examples of the two types of the two-stage blister sections (courtesy of Spirex Corp.)

With the full-diameter and slotted blister, this is done primarily to improve devolatilization of the plastic. Since this type of blister has an outside diameter the same as the screw flights, all of the plastic must be pumped through the slots. The slots or notches are designed in the same manner as the underside blister, so as to develop a predetermined back pressure on the first-stage screw. The primary purpose of the slots is to divide the flow of the plastic into many strands, thereby giving more surface area to the plastic and allowing any volatiles an easier escape into the atmosphere.

Vented Injection Molding The molding operation, when using a vented molding machine, differs slightly from a conventional, nonventing machine (Fig. 3.22). There are a few conditions to be concerned about. First, because of the shorter second metering and transition section, screw back pressure should be kept to a minimum. A small amount of pressure can cause vent bleeding on some machines. Additional back pressure is present throughout the system. With excessive back pressure, the screw cannot build up enough head pressure to allow the screw to recover, forcing the material to flow back through the vent hole. Secondly, a greater awareness of melt temperature is necessary. Excessive melt temperature or barrel heat imbalance can also cause vent bleeding. It must be pointed out that although a small amount of bleeding may occur if the vent hole is still open, proper and efficient devolatilization will take place (94).

Temperature profiles on a vented system are expected to be different from those of a conventional system, especially on a short conversion. The length of the screw available for melting is usually shorter. Therefore, heat must be applied more rapidly to the plastic. This usually does not represent any problems, but in order to assure that feeding is not disturbed, the application of heat should be concentrated on the compaction or transition section.

A vented screw should have the screw speed adjusted so that practically all the time available for screw rotation is used. This is important because (a) it provides a good melt without excessive shear, and (b) it increases the time that the melt is tumbled in the vent zone, continually exposing new surfaces to allow escape of volatiles.

Color or material changes on a vented two-stage screw normally do not present much more difficulty than that experienced on a single-stage screw. However, it may become necessary to back flush the vent zone or even pull the screw for cleaning on occasion. To back flush the vent zone, simply run the nozzle up against the tool to block off the extrudate and turn the back pressure up until the plastic is forced to flow from the vent. This will assist in purging out the deep decompression zone and assist when changing colors and/or material groups.

Vent Exhaust The exhaust from vented plasticating barrels can show a dramatic cloud of swirling white gas; almost all of it is condensed steam, proving that the vent is doing its job. However, a small portion of the vent exhaust can be other materials, such as by-products released by certain plastics and/or additives, and could be of concern to plant personnel safety and/or plant equipment. Purifiers can be attached (with or without vacuum hoods located over the vent opening) to remove and collect the steam and other products. The purifiers include electronic precipitators.

LENGTH-DIAMETER RATIO

L/D is the ratio of the plasticizing screw length to its diameter. The diameter of the ratio is reduced to 1 so that the ratio is defined evenly, such as a 24/1 screw that has a screw length 24 times its diameter. Based on the melt characteristics, there are general reasons for having short or long L/Ds. (There is also a barrel length to diameter.)

To determine L/D, use the following formula as applied to the layout in Figure 3.17: flight length/outside diameter of the screw or FL/D. The flight length (FL) does not include the length of

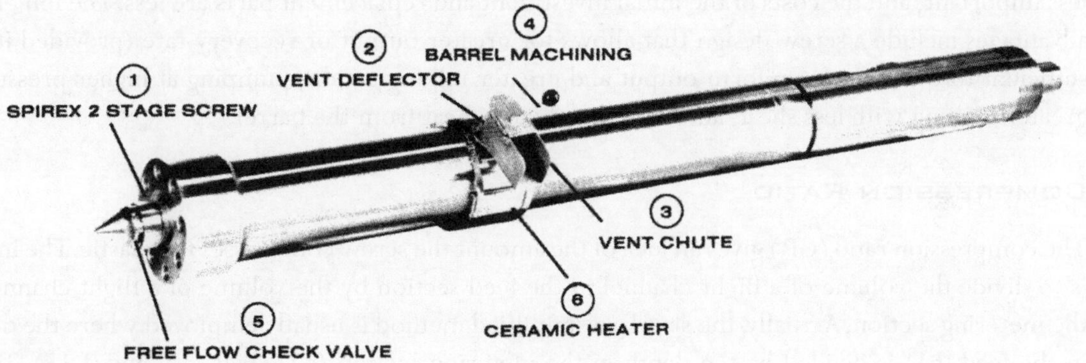

(1) **Spirex Two-Stage Screw**

This unique, length conserving design is successful on a great variety of engineering materials, both filled and unfilled. Spirex offers many materials for screw construction and coating, including: 4140, 17-4 PH stainless, Stellite, Colmonoy, nitrided, chrome or nickel plated, and tungsten carbide coated for resistance to glass abrasion.

(2) **Vent Deflector**

The Spirex vent deflector is designed to prevent vent weepage and to drag material back into the barrel. Our deflectors are made of special materials for durability and maximum heat conductance which minimizes vent hole seal off.

(3) **Vent Chute**

Materials that escape from the vent during startup must be kept from the barrel, heaters and heater wiring. The Spirex chute is constructed of heavy gage stainless steel and is easy to clean.

(4) **Barrel Machining**

The barrel hole is accurately machined and honed to prevent side wall leakage. The deflector is pin-located to prevent incorrect orientation. The deflector is honed in position to match with the barrel I.D. and eliminate dead spots.

(5) **Free Flow Check Valve**

A special, low resistance sliding ring valve is supplied to give optimum venting characteristics. The tip, check ring, and rear seat are made of H-13 for long life and abrasion resistance. All flow passages are streamlined for engineering polymers.

(6) **Ceramic Heater**

A thermally efficient ceramic heater is provided to go around the barrel behind the vent. This high quality heater is designed to operate easily at the high temperatures required by the newer materials. A thermocouple hole for individual vent zone control is provided for use if desired. A safety terminal box with both 220V and 440V connections is standard.

Figure 3.22 Example of an injection-molding two-stage vented plasticator (courtesy of Spirex Corp.)

the check valve in the case of an injection screw. The SPI and SPE have alternate methods for determining the flight length for the L/D ratio. In addition to the one described, they also consider the enclosed and flighted portion of the screw and eliminate that portion exposed in the feed port. This means deducting the axial length of the extruder or injection feed port from the flighted length of the screw.

Reasons exist for using either long or short L/D for screw and barrel length. The short L/D advantages include less residence time in the barrel, keeping heat-sensitive plastics at melt temperature for a shorter time, and lessening the chance of plastic degradation. The machine occupies less space and requires less torque, making the strength of the screw and amount of horsepower less important, and the costs of the initial investment and replacement parts are less. The long L/D advantages include a screw design that allows for greater output or recovery rate (provided it has sufficient torque), more uniform output and greater mixing action, pumping at higher pressures, greater melting with less shear, and more conductive heat from the barrel.

Compression Ratio

The compression ratio (CR) gives an idea of the amount the screw compresses the plastic. The intent is to divide the volume of a flight channel in the feed section by the volume of a flight channel in the metering section. Actually, the standard simplified method is usually employed where the depth in the feed (h1) is divided by the depth in the metering section (h2) (Figs. 3.7 and 3.15). These calculations are incorrect if there are any lead changes between the feed and metering sections.

This approach to obtaining CR via channel depth is the most commonly used, but it is the least accurate to give an idea of the actual compression. It has, however, been used to such a great extent that knowledge of ratios and depths of common usage have evolved over the years. A more accurate method is the volumetric method. This gives a relatively close ratio of volumes, but it is not entirely accurate because it neglects the effect of the radii of the front and trailing edges. The volumetric calculation of compression (VCR) ratio is $[D^2 - (D - 2\,(h1))]^2/[D^2 - (D - 2\,(h2)]^2$, where D = the screw's outer diameter (OD), hl = feed depth, and h2 = metering depth (Figs. 3.7 and 3.15).

The VCR is always a smaller number than the depth compression ratio approach. The effect is greater with higher CRs and with smaller screws. The CR should be high enough to compress the low bulk density unmelted plastic into the solid plastic without air pockets. A compression that is too low will tend to result in bubbles in the melt. High percentages of regrind, powders, and other low bulk density materials will be helped by a high CR. A high CR can overpump the metering section.

A common misconception is that engineering and heat-sensitive plastics should use a low CR. This is true only if it is decreased by deepening the metering section and not having a more shallow feed section. The problem of overheating is more related to channel depths and shear rates than to CRs. As an example, a high CR in polyolefins can cause melt blocks in the transition section, leading to rapid wear of the screw and/or barrel. Tables 3.26 to 3.28 provide guides for CRs of TPs where they may differ based on melt index and heat (shear) sensitivity of the plastics. For TSs, the

Low-compression screw (1.2–1.8 compression ratio)	Acrylics Acrylic multipolymer ABS and SAN Polyvinyl chloride, rigid
Medium-compression screw (2.0–2.8 compression ratio)	Acetal Cellulosics (acetate, propionate) Nylon (low melt index) Polycarbonates Polyethylene (medium to low melt index) Polypropylene (medium to low melt index) Polystyrene (crystal and impact) Polyvinyl chloride (flexible)
High-compression screw (3–4.5 compression ratio)	Acetal Fluoroplastics (Teflon 110) Nylon (high melt index) Polyethylene (high density) Polyethylene (high melt index) Polypropylene (medium to high melt index)

Note: Depending on the melt index and heat (shear) sensitivity of material, compression ratios may differ from those indicated.

Table 3.26 Guide to compression ratios for thermoplastics

CR is usually 1 so that accidental overheating does not occur and cause the plastic to solidify in the barrel. Their barrels are usually heated using a liquid medium so that very accurate control of the melt occurs with no overriding of the maximum melt heat. With overheating, TS melt solidifies. If it solidifies, the CR of 1 also permits ease of removal by just "unscrewing" it from the screw. A CR ratio of 1 is also used for TPs when the rheology so requires.

Pump Ratio

Pump ratio (PR) applies to two-stage vented screws and gives a measure of the ability of the second-stage screw to pump more than the first-stage screw delivers to it. Figure 3.17 relates to the equation PR = h4/h2. In extrusion, a high pump ratio will tend to cause a surge, and a low compression ratio will tend to cause vent flow.

Transition

Two basic types of transitions are shown in Figure 3.23. The conical transition has a root that is cone shaped and is not parallel to the axis of the screw. The involute transition has a root that is always parallel to the screw axis and the channel depth varies uniformly. Actually, the use of the word *involute* is not correct in geometric terms, but most people that work with screws are familiar with it.

Much has been said about the advantage of conical versus involute transitions. The knock on involute transitions is the theory that one side is deeper than the side opposite, and therefore, the screw is unbalanced. It is said that blocks of unmelted plastics form on one side and force the screw against the barrel at very high pressures, causing rapid wear. It is also theorized that surges occur as these solid blocks are formed and released. The disadvantage of conical transitions is that they are more difficult to machine and usually are a bit more expensive (94).

SCREW TORQUE

Screws always run inside a stronger and more rigid barrel. For this reason, they are not subjected to high bending forces. The critical strength requirement is resistance to torque. This is particularly true of the smaller screws with diameters of 6 cm (2.5 in) or less. Unfortunately, the weakest area of a screw is the portion subjected to the highest torque. This is the feed section that has the smallest root diameter. A rule of thumb is that a screw's ability to resist twisting failure is proportional to the cube of the root diameter in the feed section. Finite element analysis (FEA) software has been used to obtain a more accurate determination of the stress levels (chapter 19). The maximum torque that can be applied to a screw can be calculated by the following formula (some variable speed drives can deliver full HP at less than full speed) (94):

$$T = 5{,}250\ (HP)/rpm,$$

Material	Transition length[a]	Compression ration[b]	Meter depth[c]
ABS	Long	Low	Deep
Acetal	Short	Medium	Shallow
Acrylic	Long	Low	Deep
FEP	Short	Medium	Medium
HDPE	Medium	Medium	Shallow
LDPE	Medium	Medium	Medium
Nylon	Short	High	Shallow
PBT	Medium	Low	Shallow
PC	Medium	Low	Deep
PEI	Medium	Medium	Medium
PES	Medium	Medium	Medium
PET	Medium	Medium	Medium
PMP	Medium	Medium	Medium
PP	Medium	Medium	Shallow
PPO	Medium	Low	Medium
PS	Medium	Low	Medium
PVC (F)	Medium	Low	Medium
PVC (R)	Long	Low	Deep
SAN	Medium	Low	Medium

[a]Short—4 turns of less; Medium—5 to 7 turns; Long—8 or more turns.
[b]Low—less than 2.5:1; Medium—2.5 to 3.5 1; High—3.5:1 or more.
[c]Degree of depth: Example—Medium for 2″ diameter 0.100″ to 0.125″.

Table 3.27 Relative rating of compression ratio to other features of a screw for different plastics

where T = torque in foot pounds, HP = horsepower of the drive motor, and rpm = lowest screw speed in revolutions per minute at which the drive can deliver full horsepower.

Once the maximum torque is determined, one can determine whether the screw can take the torque or use the information to determine the feed depth. Of course, other feeding characteristics are also important. An approximate formula used to determine the minimum root diameter of the screw is provided. Spirex Corp. (Youngstown, Ohio; 330-726-4000) has an FEA program available for more accurate determination of stress levels, and so on.

$$D = (T \times 0.6S)\ (1/1 - d)$$

where D = root diameter of the screw in the deepest section (usually the feed), and d = diameter of the cooling hole if there is one. A very minor factor in extrusion screws. Use d = O when there is no hole. T = torque in foot pounds, and S = tensile yield strength in psi (obtain from the usual screw materials chart such as Table 3.29; most of these do not provide torsional strength, so make the assumption that the torsional strength is 60% of the tensile yield strength).

This formula is intended strictly as a guide and is not exact for a number of reasons: (1) Assumption that torsional shear strength is 60% of the tensile yield. Torsional shear strength is inexact at best, and there are very few places where you can find these data. (2) Not taken into consideration is the added strength provided by the flight. The approach is merely to calculate the

		Depth (mm)			Length (diameters)				
Purpose	Plastic	Feed	Transition[a]	Meter	Feed	Transition	Meter	Total	Compression ratio[b]
High-output compounding	LDPE	15	15–6.5T	6.5	6	12	6	24	2.1:1
General purpose	LDPE, HDPE, PS	15	15–5.5T	5.5	4	10	6	20	2.4:1
Film	LDPE	15	15–3.0S	3.0	10	11	9	20	4.3:1
High-output compounding	PP powder	–	15–7.9T	7.9	–	18	2	20	1.8:1
General purpose	PP	–	15–6.0T	6.0	–	16	–	16	2.2:1
General purpose	PP	–	12.7–5.6T	5.6	–	10.9	9.1	20	2.1:1
General purpose	POM	15	15–5.0T	5.0	4	12	4	20	2.7:1
General purpose	PMMA	15	15–4.0S	4.0	14	1	5	20	3.3:1
Compounding	Nylon 66	–	11.1–9.6T/S	2.8	–	12 + ½	7.5	20	3.6:1
Film	PET	15	15–3.0S	3.0	14	1	9	24	4.3:1
Film	UPVC	15	15–7.0T	7.0	4	12	4	20	1.9:1
Low-output compounding	UPVC	15	15–9.0T	9.0	4	7	4	15	1.5:1

[a]T = tapered; S = stepped.
[b]Cross-section basis.

Table 3.28 Measurements of compression ratios and other features of a screw for different plastics

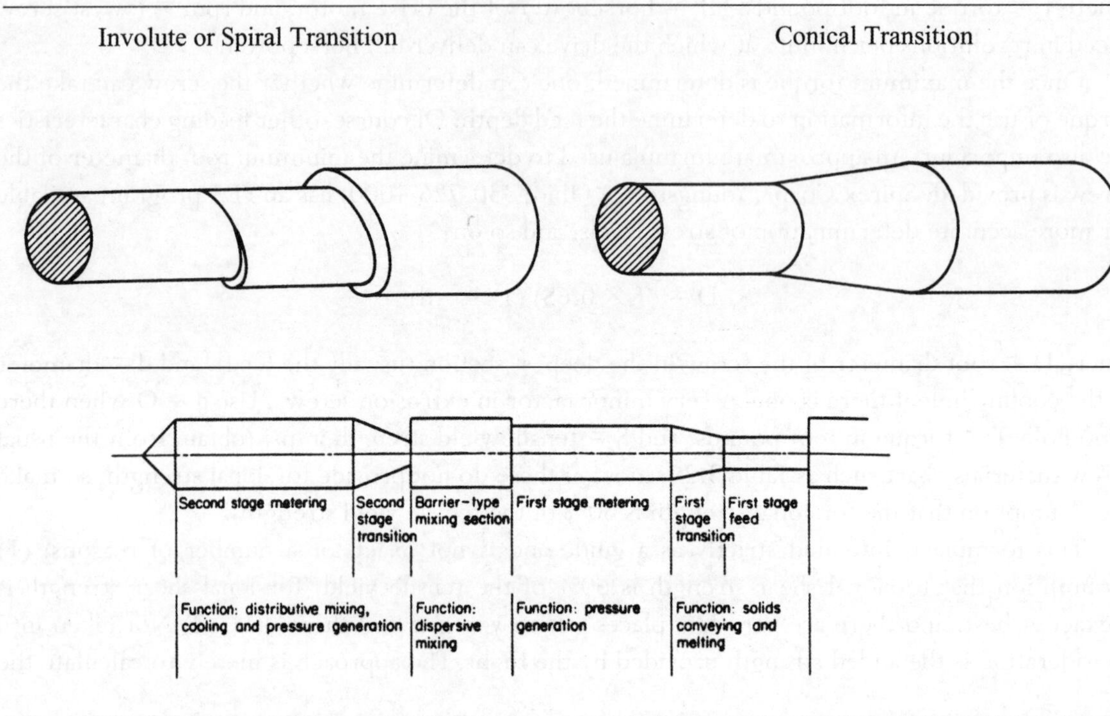

Figure 3.23 Screw transitions with flights omitted

section as a cylinder, which gives a built-in safety factor. (3) No consideration is given to rapid shock loads, which can be a significant factor. (4) Stress concentration factors can vary widely depending on the radius at the bottom of the flight. (5) All strengths given in tables and in most other places are at room temperature. They are bound to be less at operating temperatures. Again, this is only a guide. Also, the user must determine the amount of added safety (94).

In summary, different torque requirements are used to meet the requirements of the different plastics. As an example, much higher torque is required to plasticize PC (polycarbonate) than GPPS (general-purpose polystyrene). The strength used limits the input power. Using little torque to turn the screw can indicate that the heater bands are providing too much of the energy required to melt the plastic, usually as a result of poor or no temperature control. Plastication efficiency suffers under these conditions and mixing problems and/or long, inconsistent recovery times are symptomatic of such.

STANDARD SCREW

The basic single flighted screw is a good conveyor, melter, and pumper of most plastic materials, but it does have some limitations. These limitations become particularly important when (a) the

	Yield Strength psi	Hardness as Machined Rc	Availability of Case Hardness Rc	O.D. Wear Resistance	Root Wear Resistance	Corrosion Resistance	Material Availability	Ease of Machining	Approx. Cost per Pound ($)
ALLOY STEELS									
AISI 4140	100,000	28-32 Rc[1]	48-55 Rc[1]	Fair[1]	Poor	Poor	Excellent	Fair	.90
AISI 4340	110,000	28-32 Rc[1]	48-55 Rc[1]	Fair[1]	Poor	Poor	Fair	Fair	1.30
Nitralloy 135-M	85,000	33 Rc	60-65 Rc[2]	Good[3]	Good	Poor	Fair	Fair	1.65
STAINLESS STEELS									
304	35,000	80 Rb	90 Rb	Poor[3]	Poor	Good	Good	Fair-Poor	3.40
316	35,000	90 Rb	90 Rb	Poor[3]	Poor	Good	Good	Fair-Poor	4.15
17-4 PH	175,000	38 Rc	42 Rc[3]	Poor[3]	Fair-Poor	Good	Fair	Fair-Poor	2.90
15-5 PH	175,000	38 Rc	42 Rc[3]	Poor[3]	Fair-Poor	Good	Poor	Fair-Poor	4.40
TOOL STEELS									
CPM-10V	300,000	98 Rb	58-60 Rc	Excellent	Excellent	Fair	Fair	Fair	15.00
CPM-T420V	300,000	100 Rb	55-57 Rc	Excellent	Excellent	Good	Poor	Fair	15.00
CPM-9V	285,000	100 Rb	52-55 Rc	Excellent	Excellent	Fair	Fair	Fair	15.00
D-2	>300,000	96 Rb	58-60 Rc	Good	Good	Fair-Good	Good	Fair	4.50
H-13	>300,000	96 Rb	50-60 Rc	Good	Good	Poor	Good	Fair	3.95
Elmax	335,000	96 Rb	58 Rc	Good	Good	Excellent	Fair	Fair	18.00
Vanadis 10	>300,000	100 Rb	58-60 Rc	Excellent	Excellent	Fair	Fair	Fair	15.00
SPECIALTY MATERIALS									
Duranickel 301	125,000	30-38 Rc	32 Rc[3]	Poor[3]	Poor	Excellent	Very Poor	Poor	25.00
Hastelloy C-276	80,000	86 Rb	86 Rb[3]	Poor[3]	Poor	Excellent	Very Poor	Poor	25.00

(1) Flame or induction hardened
(2) Nitrided
(3) Usually improved by hardsurfacing

Table 3.29 Common screw materials (courtesy of Spirex Corp.)

plastic is difficult to melt and the screw controls production rates, (b) improved mixing of colors or better product uniformity is needed, and (c) greater output is needed, but higher screw speeds will cause excessive shear and overheating.

An almost universally accepted model of melting in a single-screw extruder exists. This model is the basis for most computer simulations. It has been demonstrated to be correct by many "freeze tests" (such as examining contents in screw channels after melt solidification occurs and the screw is removed from the barrel). An explanation follows relating to the melt model in Figure 3.10, view 4 (94):

1. The feed section initiates solids conveying. The feed sliding on the screw (low friction) and high friction on the barrel enhance this action. When the plastic sticks to the screw and slides on the inside surface of the barrel, it just goes around with the screw and never moves forward. In the feed section, there is also some compaction and a little heating of the plastic.
2. At the beginning of the transition, the plastic is further heated and more compression occurs. The solid plastic is forced against the barrel, causing a sliding friction. This frictional and conductive heat creates a film of melted plastic on the inner barrel surface.
3. As the plastic proceeds down the transition, there is more melting and more compression. Usually most of the melting takes place in the transition. Here, the plastic is divided into three parts: a compacted solids bed, a melt film along the barrel surface, and a melt pool (Fig. 3.10[3]). The melt pool is formed as the melt film is collected by the advancing flight. Most of the melting continues to be the result of sliding friction of the solids bed against the heated barrel. This is rapid, efficient melting something like melting an ice cube by pushing it against a hot grinding wheel.
4. The channel depth continues to decrease as the material progresses down the transition. Melting continues and the width of the solids bed decreases while the width of the melt pool increases. Unfortunately, as the channel gets shallower, shear rate increases. The melted plastic continues to develop heat. This is normally undesirable.
5. Further down, the solids bed breaks up and the unmelted plastics are distributed throughout the channel like ice cubes in water. The efficient melting by friction of the solids bed against the barrel stops; now only less-efficient melting remains. This is something like heating the water to melt the ice cubes. It will finally get the job done, but it is slow and much less efficient. Overheating of the melt continues in the shallow metering section.
6. The plastic continues down the shallow metering section to the discharge. It is possible that there remain unmelted pellets or portions within the melt having higher or lower temperatures and viscosities. With a situation like this, the melt is nonuniform, giving poor properties and color mixing. Greater mixing can be

achieved by reducing channel depth, but this must be done at the expense of more overheating and less output per revolution. The constant-depth metering section is not a good mixer. This is because smooth laminar flow patterns are established, causing the different portions of melt to continue to move in a fairly constant circular pattern. This does not mix the dissimilar portions of the melt.

Marbleizing Screw

Marbleizing or mottling screws are used in instances in which little or no mixing is desired. A typical application would be in making women's cosmetic cases where a swirling effect is wanted in the plastic coloration. In this unique situation, the screw is designed not to mix the plastic after melting. This is in contrast to the typical screw with superior mixing characteristics that give a homogeneous melt and proper color dispersion. Quite often, this screw is designed to simulate the effects obtained with a plunger machine.

Typically, the screw will have a very low compression ratio, with a good portion of the screw consisting of the feed section, followed by a short taper (low compression) and oftentimes a nonflighted metering section. This design allows melting, but once this is achieved, the screw simply conveys the melt forward with little or no additional mixing. Other designs also provide multiflighted screws that convey plastic and colorant in different screw channels and introduce them at the discharge to provide very little mixing action together.

Mixing Device

Different mixing devices and/or mixing screws are designed to overcome the problems described above. Some of the more common mixing devices are illustrated and described here.

Dulmage A Dulmage mixing section is part of the screw usually located several flights back from the discharge end. The Dulmage screw was one of the first mixing screws developed. It was developed by Fred Dulmage of Dow. It has a series of semicircular grooves cut on a long helix in the same direction as the screw flights (Fig. 3.24). There are usually three or more sections interrupted by short cylindrical sections. This interrupts the laminar flow and it divides and recombines the melt many times. In this way, it works something like a static mixer, giving distributive mixing. It is still used for foaming plastics as well as other applications.

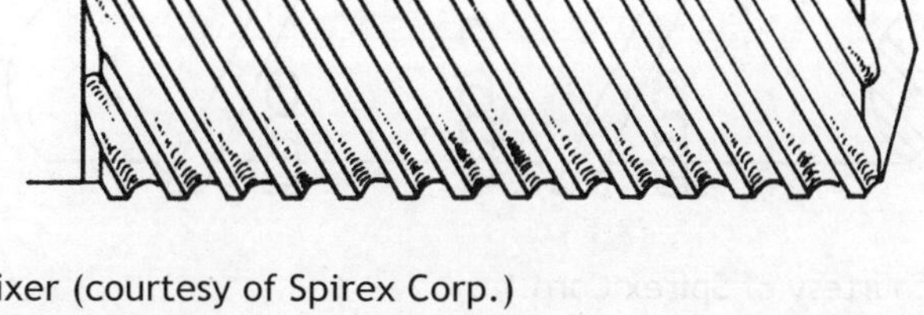

Figure 3.24 Dulmage mixer (courtesy of Spirex Corp.)

MIXING PIN

During the 1960s, several companies started to place radial pins in the screw root. These pins tend to interrupt the laminar flow and do a little better job of mixing. Because these pins improve mixing, it is also possible to design the screw a little deeper to get some more output with the same degree of mixing.

Many patterns and shapes of pins have been used, but in general, they are usually placed in rows around the screw root. They are located in the metering section after most of the melting has taken place. A typical arrangement would have three rows, with one row at the beginning of the meter, another one flight back from the end, and the other halfway between. Figure 3.25 shows the pins located so that the screw flight remains intact. The flight may be milled out so that the pins can go around the root uninterrupted. This layout causes a slight loss in drag flow because of the flight interruption and increased back melt flow.

A pin should be hardened and have an interference fit to prevent dislodgment. Pins, unlike other mixing devices, are easy to install as afterthoughts. This is usually done after the screw has been running and found to need more mixing ability.

PULSAR MIXING SCREW

The Pulsar mixing screw (Patent #4,752,136) was developed by Spirex Corp. to combat the most common problems encountered in mixing plastic melt: (a) high shear and the associated excessive melt temperature (from the high compression of shallow depth screws), which result in low production, poor quality, and degraded plastic and (b) low shear (from the low compression of deeper channel screws) that results in insufficient mixing, trapped air, and poor quality.

The Pulsar is particularly effective with shear-sensitive or temperature-sensitive plastics, such as PVCs. It can provide excellent mixing with little or no temperature rise. In many cases, a lower melt temperature can be achieved. Spirex Pulsars have been successfully designed to run a wide range of plastics. Applications include ABSs, acrylics, HDPEs, LDPEs, LLDPEs, nylons, rigid and flexible PVCs, polypropylenes, polystyrenes, urethanes, and polycarbonates.

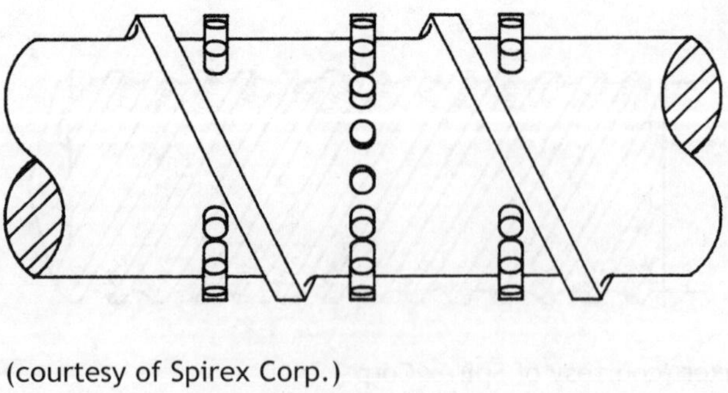

Figure 3.25 Mixing pins (courtesy of Spirex Corp.)

The metering section of a Pulsar mixing screw is divided into alternating sections that are either deeper or shallower than the average metering depth (Fig. 3.26). The material is forced from one section to another, causing a gentle tumbling and massaging action. This results in excellent mixing, distribution, and melt uniformity without high shear. It has no "dead" spots like some mixers, which causes the Pulsar to be self-cleaning and allows for easy plastic changes. Other advantages include low melt temperature, no marbling, strong weld lines, quick color and plastic changes, no splay generation, and less back pressure required.

UNION CARBIDE MIXER

This mixing device is also referred to as the Maddock mixer. It was patented by Union Carbide Corp. (UCC) and developed for practical use by Bruce Maddock of UCC. The patent has been given to the public, so there are no royalties charged. The mixer consists of a series of opposed semicircular grooves parallel to the screw axis (Fig. 3.27). Alternate grooves are open to the upstream entry. The other grooves are open to the downstream discharge. The ribs or lands that divide the alternating entry and discharge grooves also alternate. These lands are called mixing lands and wiping or cleaning lands. The plastic is forced over the mixing land, which is ½ in (2.70 mm) across and undercut about 0.019 in (0.47 mm) below the screw OD. The cleaning land is narrower (approximately 3/16 in) and has a full diameter. This mixer does an effective job of mixing and screening unmelted material.

The plastic is pumped into the inlet groove, and as the screw rotates, the undercut mixing land passes under it. The melted material ends up in the outlet or discharge groove. As it goes over the undercut mixing land, it is subject to high shear but for a very short interval. The material is then pumped out of the discharge groove as new material enters over the mixing land; it cannot escape over the full-diameter cleaning land.

The UCC mixer screens out unmelted materials. This action permits a deeper design that gives greater output. In most cases, a screw can be designed to give improved output over a conventional single-stage screw but still give equivalent or better mixing. This mixing device was developed for low-density polyethylene film and this is still its largest use. It is also used for many other extrusion applications. Injection screws for polypropylene and HDPE also use this mixer.

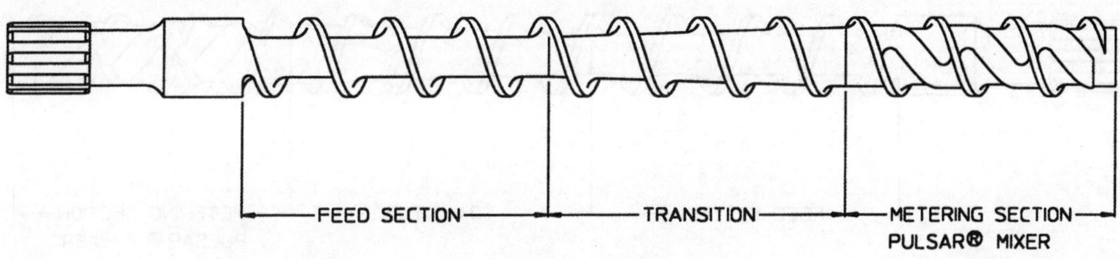

Figure 3.26 Pulsar mixing screw (courtesy of Spirex Corp.)

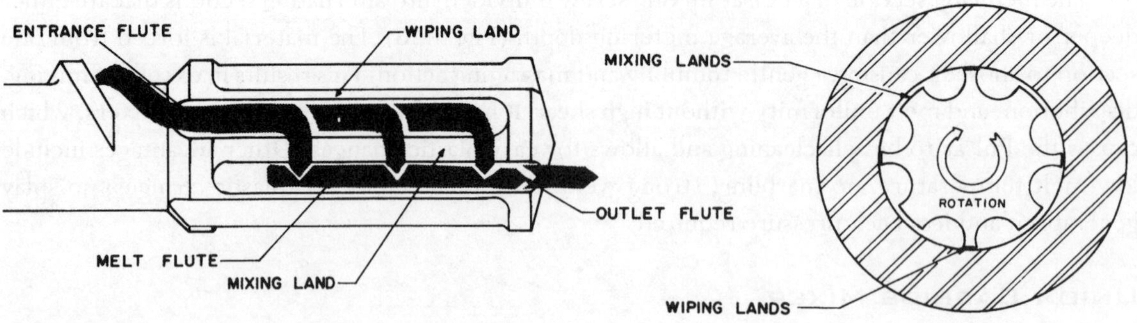

Figure 3.27 Union Carbide mixer (courtesy of Spirex Corp.)

Many screws have been retrofitted with a UCC mixer on the discharge end. The drawback for this mixing device is that it cannot be used for shear-sensitive materials unless the barrier clearance is deepened, in which case performance is sacrificed. UCC mixers can be thought of as a type of barrier screw with multiple barriers parallel to the screw axis.

Pulsar 11 Mixing Screw

The Pulsar 11 (Patent #5,816,698) is a recent development of Spirex Corp. It is a modification of the previously patented Pulsar mixer. The Pulsar 11 was developed after a need for more intense distributive mixing was required. In studying the geometry of the original Pulsar mixer, it became apparent that by cutting a reverse interrupting channel through the pockets of each "pulse" section, more chaotic mixing could be developed (Fig. 3.28). This change also eliminated any dead spots that may occur at the beginning and end of each "pulse" section. This new advanced design also allows for more streamline channeling, further eliminating any dead-spot areas.

The Pulsar 11 has been found to work very well with both crystalline and amorphous TPs. Just like its predecessor, it processes shear sensitive materials, such as PVCs, ABSs, and PCs, very

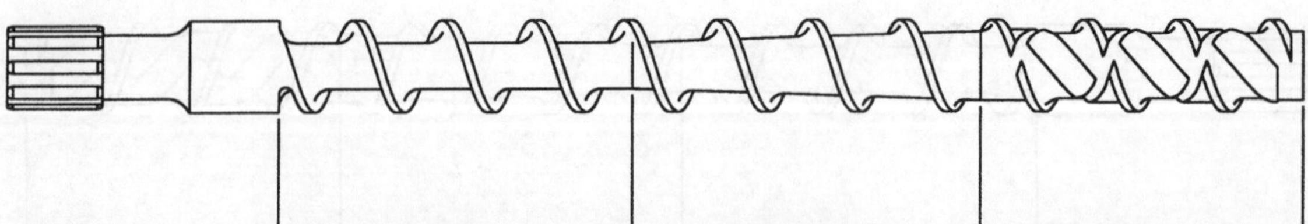

Figure 3.28 Pulsar 11 mixing screw (courtesy of Spirex Corp.)

well and gives excellent distributive mixing. This distributive mixing also allows for superior melt homogeneity. Other advantages include low melt temperature, low fiber breakage, versatility with all amorphous plastics, no marbling, less back pressure applied, and strong weld lines.

Saxton Mixer The Saxton mixer has proven to offer good distributive mixing for a wide range of resins. The mixer was developed by Ronald Saxton and was patented in 1961. The mixer is designed with a plurality of minor flights and channels on a helix angle that is normally greater than that of the primary flight (Fig. 3.29). The minor flights and channels are interrupted by major channels that are cut with an opposite hand lead. These channels and interruptions help in distributive mixing by separating and recombining the plastic, thus breaking up the laminar flow.

The one drawback of this type of mixer, or any type that has nonflighted interruptions, is that the plastic no longer has the positive forward conveyance other than the pressure flow. However, the interruptions on a helix are much better than tangential grooves, as some barrel wiping action still takes place. The Saxton can be designed for various plastics and mixing actions by varying the total length, number of minor flights and channels, number of interruptions, helix angles, depths, and so forth.

Double Wave Screw The double wave screw, patented by HPM Corp., has two equal width channels separated by an undercut barrier flight. The roots of each channel go up and down like a wave (Fig. 3.30). The channel depth on one is shallow while the channel across the barrier is deep. This continually reverses, forcing melted polymer back and forth across the barrier. The material in the channel is alternately subjected to high then low shear.

Usually these double wave mixing sections are located in the metering section where the plastic has already been melted. The channels are open at both ends and run parallel.

Dispersion Disc Dispersion discs are devices of various shapes and sizes that are installed between the endcap/nozzle adapter and the nozzle tip. They are usually round (Fig. 3.31) with holes through which the material can pass. The purpose of the disc is to disrupt the laminar flow of the plastic during injection, thereby creating distributive and dispersive mixing. As a result of the small orifices, there is a tremendous increase in velocity, and hence, shear during injection. There is also a need for using high pressures to assure adequate fill time. In addition, due to the increase in shear, only low viscosity plastics can be used with these discs. Also, there is usually a need to change the discs when a color or material change is required to ensure that there will be no contamination.

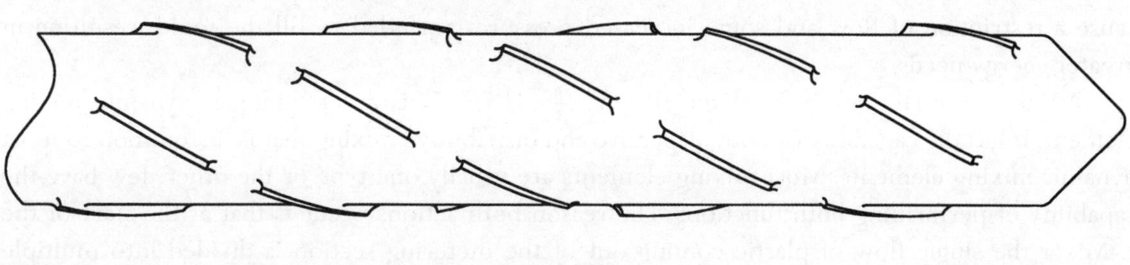

Figure 3.29 Saxton mixer (courtesy of Spirex Corp.)

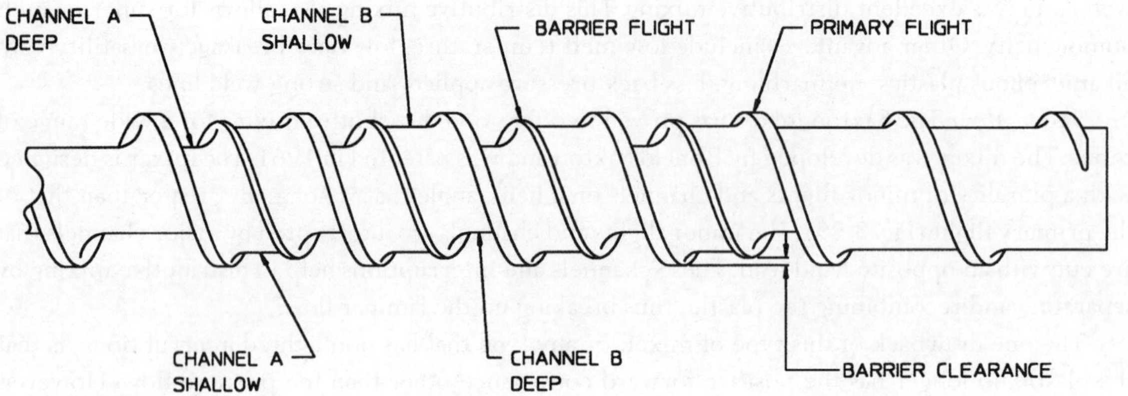

Figure 3.30 Double Wave screw (courtesy of Spirex Corp.)

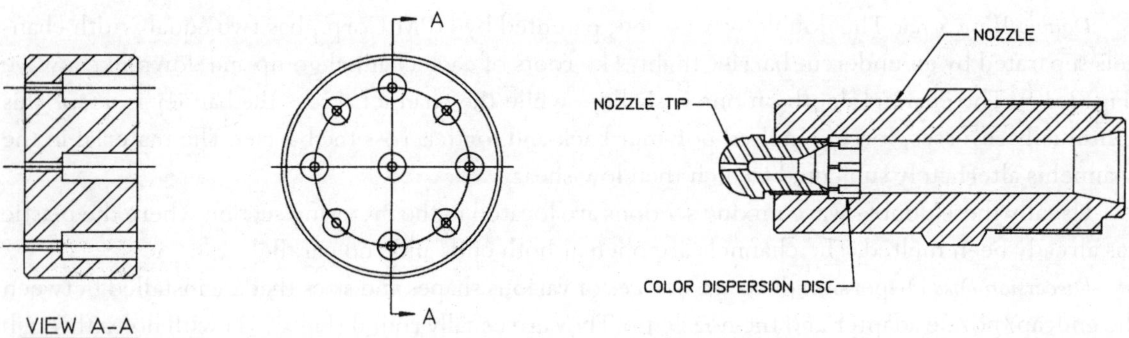

Figure 3.31 Dispersion discs (courtesy of Spirex Corp.)

Static Mixer There are a number of different brands of static mixers (Fig. 3.32). They are attached downstream of the barrel before the injection nozzle or extrusion die. Mixing is almost entirely distributive rather than by shear. The flow is constantly divided or separated and then recombined. They have been more readily accepted by the extrusion industry than by injection molders. They cause a restriction of flow and some increased pressure is needed to fill the mold, resulting in greater energy needs.

Z-Mixer Screw The Spirex Z-Mixer (Patent #5,318,357) is a very unique, dynamic mixing element. It has the capability of both dispersive and distributive mixing that is uncommon to most dynamic mixing elements. Most mixing elements are usually one type or the other; few have the capability of performing both functions. The reason both actions occur is that at the start of the Z-Mixer the single flow of plastic coming out of the metering section is divided into multiple channels. The size of the screw determines the number of channels that the plastic is distributed

Figure 3.32 Static mixers (courtesy of Spirex Corp.)

into. The plastic will both be divided into two, three, or four distinct channels and separated by a full-diameter primary flight, as seen in Figure 3.33. Then, each division of plastic passes over a barrier in each of its distributive channels.

With the passing of the plastic over the barrier, shear or work is applied to the plastic. The shear that is applied is required to give good dispersive mixing and the breaking up of conglomerates. After the dispersive mixing takes place, the plastic is distributed again by means of cross-flow from one distributive channel to the next through the flight interruption in the primary flight of the mixer. Then, the cycle repeats itself. The amount of mixing can be controlled by the height of the barrier gap, the size of the flight interruption, or the number of Zs that are placed in each distributive channel. Empirical studies have shown that typically three cycles are used to give satisfactory mixing.

This mixing device has been found to work very well on polyolefins and styrenics and other nonshear sensitive plastics. It has also been found that the Z-Mixer performs very well when color concentrates are added to the virgin plastic, either in a master batch or at-the-throat form. Other advantages include no marbling, strong weld lines, no unmelts, best color dispersion in the industry, and highest throughputs in the industry.

V-Mixer Screw The Spirex V-Mixer (Patent #5,798,077) screw design is unique because it is a dispersive and distributive mixing device that is installed near the discharge end of a plasticating screw. The feeding and pumping mechanism are the same as a single flighted screw; however, the melting and mixing effects are superior to a single flighted nonmixing design as well as many other

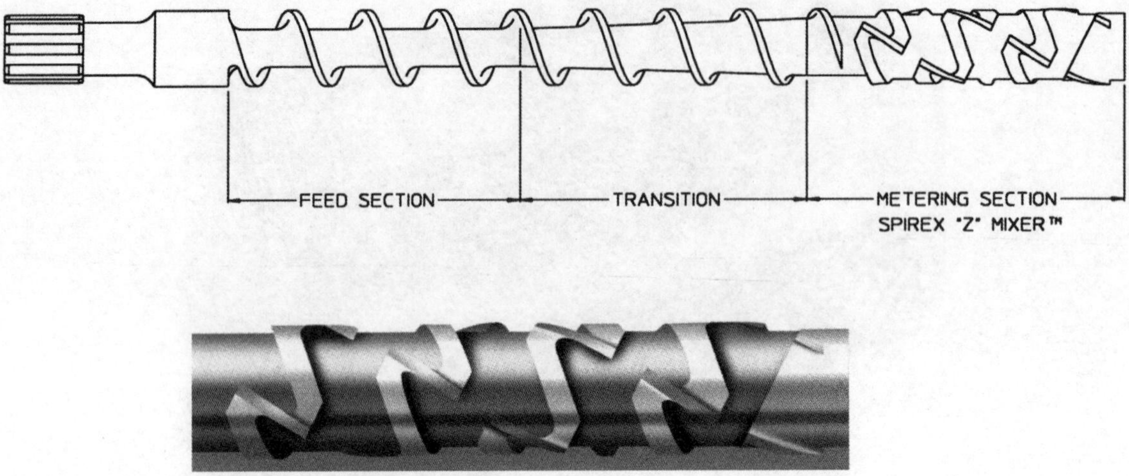

Figure 3.33 Spirex Z-Mixer (courtesy of Spirex Corp.)

mixers. The V-Mixer has many different faces. Nine different versions were tested, and all assist in the melting phase by having a shallower channel depth, thus delivering a degree of increased shear and melting capability.

The homogenizing effect is accomplished with a chaotic mixing action by creating a high shear rate with a side-to-side (wavy wall) movement of the flow as well as an up-and-down (wavy root) movement (Fig. 3.34). By controlling these depths of the channels and creating this flow action, the V-Mixer yields a superior dispersionary melt quality.

This distribution delivers a more uniform, homogenized melt with minimal temperature differential. The more uniform the plastic, the more consistent the viscosity, and hence, the potential for consistent repeatable flow through the die or into the mold. The more repeatable the flow, the greater the molecular orientation that determines the physical properties of the component being manufactured. Other advantages include the ability to process a variety of plastics, low melt temperature, less back pressure required, quick color changeability, and no marbling.

Flex Flight Mixing Screw The Spirex Flex Flight mixing screw (Patent #5,071,256) is an alternate mixing screw that can be used for both high and low shear applications. It consists of a secondary flight (starting in the metering section) that is nonparallel to the primary flight and has a pitch that is alternately greater than and less than that of the primary flight. This barrier flight creates two helical valleys that are constantly changing in width, while the total volume of the two combined remains constant (Fig. 3.35). The melt in the decreasing width channel is squeezed over the barrier and into the adjacent widening path, creating a tumbling and mixing action in the plastic.

The screw can be designed for various plastics by changing the clearance over the barrier flight, the width of the barrier flight, channel depth, and the number of cycles back and forth. This mixing screw varies from a standard barrier screw in the following ways: (a) A standard barrier usually extends in the transition section to aid in melting. The Flex Flight is introduced after most of the melting has occurred and is used primarily for mixing. (b) A standard barrier normally intersects with the primary flight where the Flex Flight stops before intersection so as not to create an unwanted high-shear area.

The Flex Flight can be manufactured on a new screw or can be welded onto an existing screw to increase mixing. This is a big advantage because most mixers that are added as afterthoughts

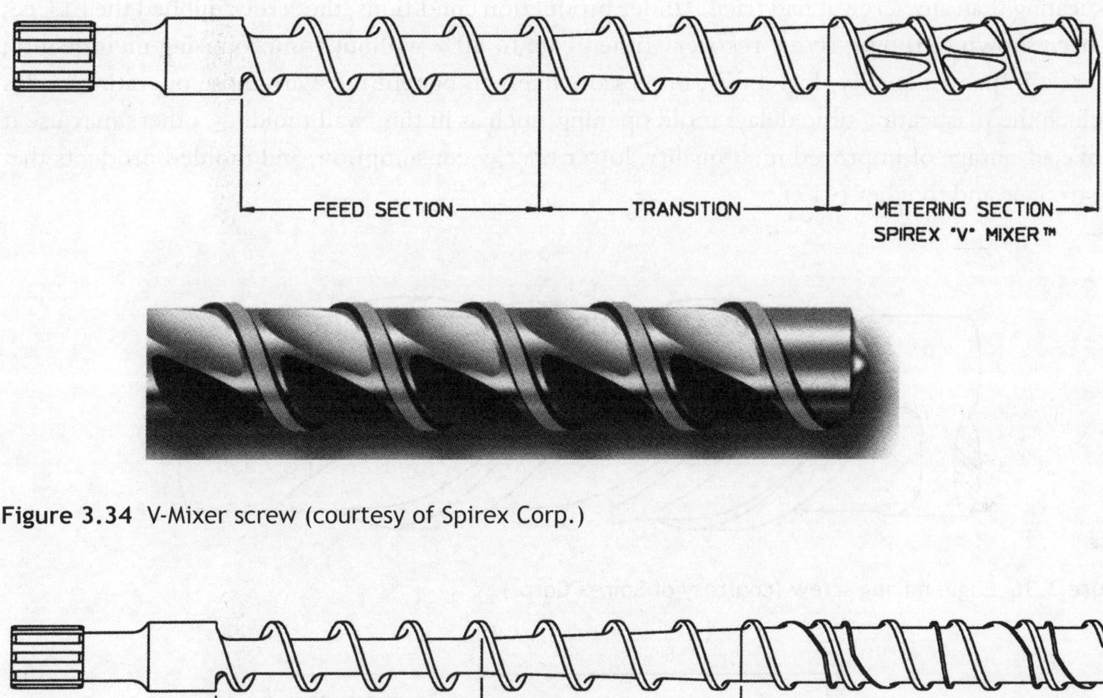

Figure 3.34 V-Mixer screw (courtesy of Spirex Corp.)

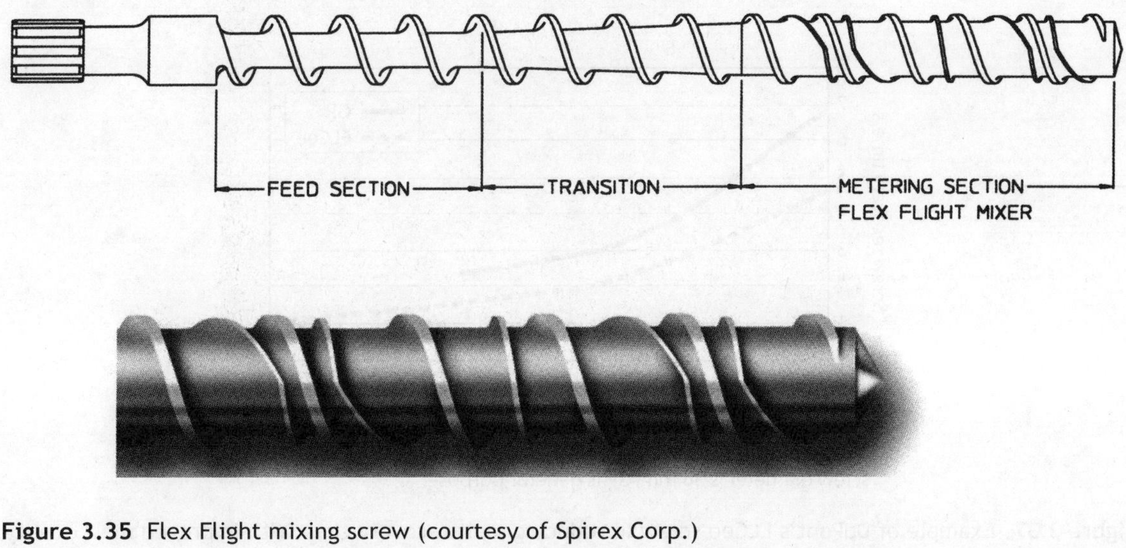

Figure 3.35 Flex Flight mixing screw (courtesy of Spirex Corp.)

require cutting off the existing screw and then adding a new section. The cost is high and the concern is always the strength of the new section where it is fitted to the existing screw. The Flex Flight is relatively easy to add with no reduction in the integrity of the existing screw.

Eagle Mixing Screw The Eagle mixing screw (Patent #5,215,764) is supplied by Westland Corp. of Wichita, Kansas. This mixer combines dispersive and distributive mixing. It is reported to have application in a wide range of plastics and it gives good color mixing with high outputs and low stock temperatures. The design includes wide-entering, deep channels that narrow and converge, plus barrier flights with mixing notches (Fig. 3.36).

ELCee Screw DuPont has an injection molding screw that it claims is significantly better at plasticating than any screw it had tried. Under production conditions, the screw, dubbed the ELCee, has been shown to reduce screw recovery time by up to 50% without compromising melt quality; in fact, it improves quality (Fig. 3.37). Processors likely to benefit most are those operating cycles in which the plasticating time delays mold opening, such as in thin-wall molding; others may use it to take advantage of improved melt quality, lower energy consumption, and molded products that are stronger and tougher (417).

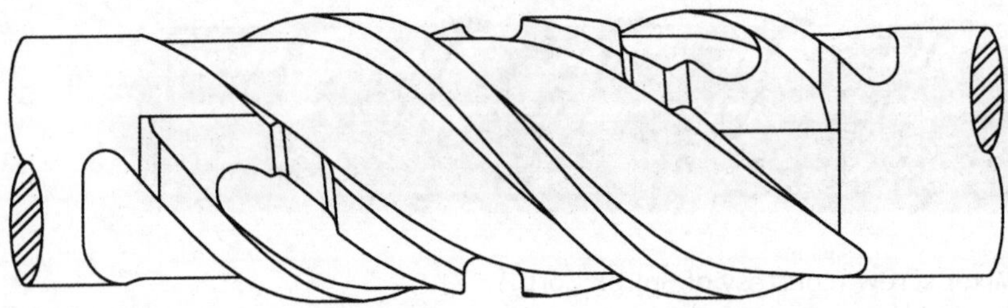

Figure 3.36 Eagle mixing screw (courtesy of Spirex Corp.)

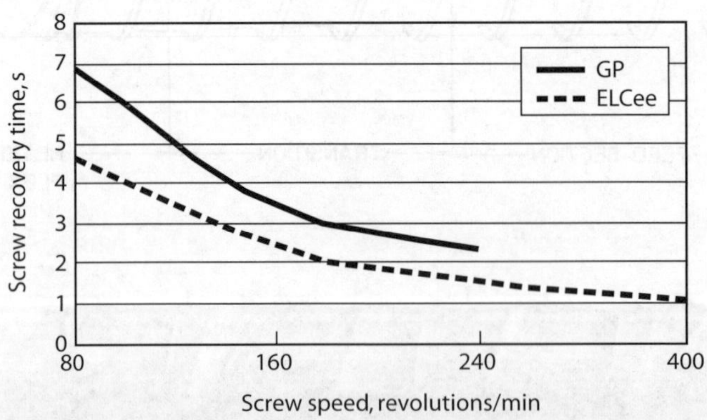

a: resin processed is DuPont Delrin 100 acetal homopolymer; screw diameter is 36 mm; GP is general-purpose.

Figure 3.37 Example of DuPont's ELCee screw in reducing melt recovery time with improved melt quality

After recently receiving a patent, DuPont licensed two screw makers, Spirex Corp. (Youngstown, Ohio) and PMT S.A. (Herstal, Belgium), to build it with a potential third licensee in Asia. The screw is available to all molders, irrespective of whether they are DuPont customers, but they must sign confidentiality agreements since the screw incorporates proprietary know-how. The screw can be retrofitted to most injection machines. Around 300 have been sold worldwide since late 1999. They are used for processing crystalline and amorphous TPs, mostly in Europe, where DuPont had concentrated marketing. The screw is priced similarly to an OEM screw, and return on investment reportedly can be realized in less than one month.

DuPont literature reports that the development derived from a fundamental understanding of polymer characteristics and melt behavior, resulting in a flight geometry that is radically different from that of general-purpose screws. There is no special design since it has no barrier design or twin flights, for example, although a mixing head may be advisable for master-batch processors.

The name of the screw is derived from the term *Low Compression*. DuPont reports that screws with high compression ratios regularly used to boost output often produce unwanted effects, such as plastic degradation, while the ELCee screw delivers high output with a compression ratio that barely meets the minimum requirement for a thermoplastic plasticating screw (around 1.3).

The screw breaks one of the basic rules of thumb in screw design, which is that flight pitch should remain constant along the screw length. In the ELCee screw, which is designed according to the plastic to be processed, the pitch in the metering zone is higher than in the feeding zone and the pitch increases through the compression zone as the flight depth decreases. The result is that the screw provides balanced mass flow, a constant increase in pressure along its length. Melt homogeneity increases and screw recovery time decreases, reducing degradation.

Because the screw design yields little or no deposits, wear resistance is improved. For example, in a production operation, no wear was seen in a nitrided ELCee screw that had been running a glass-reinforced nylon for two years. (Based on past DVR experience, screws are usually replaced in 6 months.)

Barrier Screw

An important development in screw design was the barrier screw. The primary reason for barrier screws is to eliminate the problem of solids bed breakup for more efficient melting. They have been around for over a quarter century. Original developments were for extrusion, but later they were used to solve problems in injection and blow molding. There are many different patented barrier screw designs that fall under the broad claims of the Geyer or Uniroyal U.S. Patent #3,375,549, which expired in 1985. Many of the other screws in use, and those shown here, have patents of their own. Some barrier screws work better than others, but all require knowledge and experience to design properly.

All barrier screws have two channels in the barrier section, which is mostly located in the transition section. A secondary flight is started (usually at the beginning of the transition), creating the two distinct channels: a solids channel and a melt channel. The barrier flight is undercut below the primary flight, allowing melted plastic to pass over it. The theory of most barrier screws is best understood by referring to Figure 3.38, which can be compared with Figure 3.10, view 4.

In Figure 3.38, as the plastic moves from left to right, the following action occurs:

1. The feed section establishes solids by conveying in the same way as a conventional screw.
2. At the beginning of the transition (compression), a second flight is started. This flight is called the barrier or intermediate flight and it is undercut below the primary flight OD. This barrier flight separates the solids channel from the melt channel.
3. As material progresses down the transition, melting continues as the solids are pressed and sheared against the barrel, forming a melt film. The barrier flight moves under the melt film and the melt is collected in the melt channel. In this manner, the solid pellets and melted plastic are separated and different functions are performed on each.
4. The melt channel is deep, giving low shear and reducing the possibility of overheating the melted plastic. The solids channel becomes narrower and/or shallower, forcing the unmelted pellets against the barrel for efficient frictional melting. Breakup of the solids bed does not occur to stop this frictional melting.
5. The solids bed continues to get smaller and finally disappears into the back side of the primary flight.
6. At this position, all of the plastic has melted and gone over the barrier flight. Melt refinement can continue in the metering section. In many cases, mixing sections are also included downstream of the barrier section. In general, the melted plastic is already fairly uniform upon exit from the barrier section.

An introduction to the different barrier screws being used follows in no order of importance or priority. The Uniroyal screw is the original barrier screw. This screw is also sometimes referred to as the Maillefer screw. The barrier flight starts on the front side of the primary flight at a greater lead and it disappears into the back side of the primary flight (Fig. 3.39). The channels are basically close ended and the depths on either side of the barrier are usually the same. There are many ways the channel widths and depths can vary.

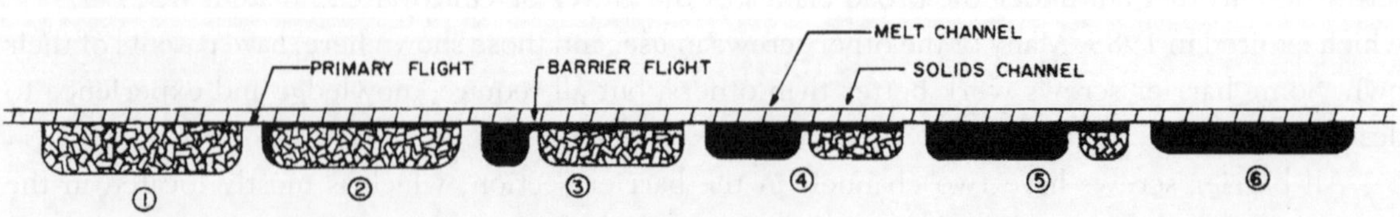

Figure 3.38 Melt model of a barrier screw (courtesy of Spirex Corp.)

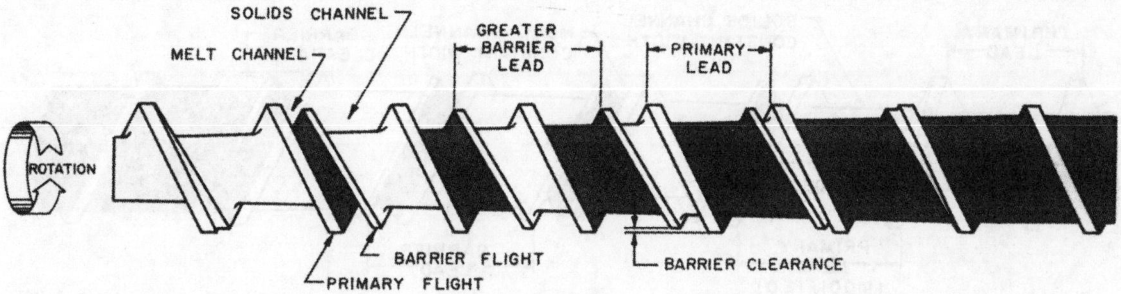

Figure 3.39 Uniroyal screw (courtesy of Spirex Corp.)

The MC-3 screw starts the barrier flight from the front side of the primary flight just like the Uniroyal screw. The greater lead of the barrier makes it move away from the primary flight, creating the melt channel. After it has gone a certain distance, the lead changes back to the same lead as the primary flight, and the two flights run parallel for most of the barrier section (Fig. 3.40). The melt channel becomes deeper and the solids channel becomes progressively shallower. At the end, the barrier flight is terminated and all depths equal the metering depth. The solids channel is open at the discharge end.

The Efficient screw has a conventional feed section, usually with square pitch. At the beginning of the transition, the primary lead increases substantially, providing space for a new barrier flight and melt channel. After the width of the new melt channel has been established, the flights and channels remain parallel through the transition section. The solids channel remains approximately the same width as in the feed section. The barrier flight ends and the open-ended melt channel merges with the solids channel at the end of the transition (Fig. 3.41).

The Barr II screw begins the barrier flight from the root of the screw at the beginning of the transition. The open-ended melt channel is created and the flights run parallel to the end of the

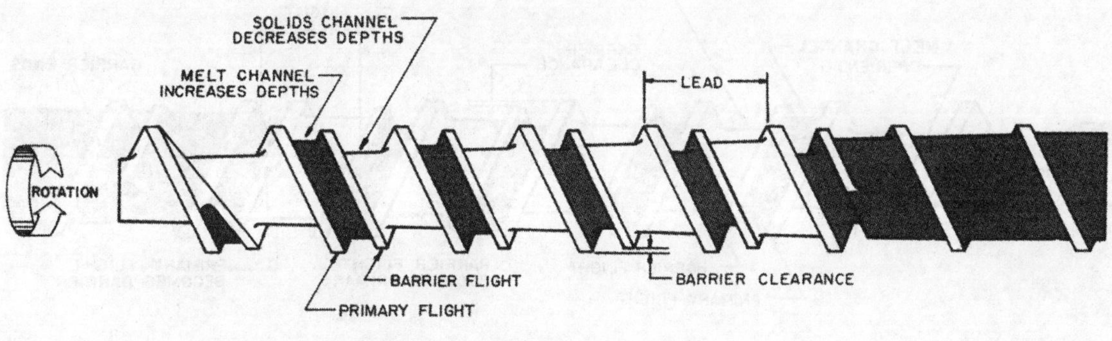

Figure 3.40 MC-3 screw (courtesy of Spirex Corp.)

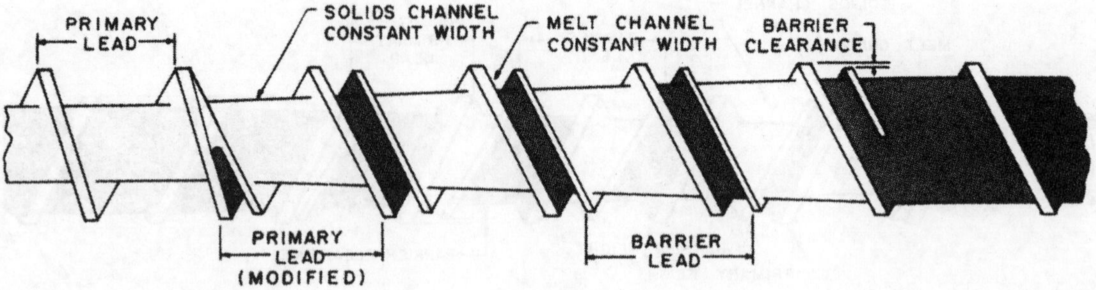

Figure 3.41 Efficient screw (courtesy of Spirex Corp.)

mixing section (Fig. 3.42). The depth of the solids channel decreases and the depth of the melt channel increases. Near the end, there is a flight interchange where the primary flight becomes the barrier flight and vice versa. This promotes mixing. The barrier flight disappears into the channel root and the melt channel is open ended.

The ET (energy transfer) screw is covered by U.S. Patent #4,000,884 and # 4,405,239; it was invented by Robert Barr and sold by Barr Inc. The ET section is incorporated at the downstream end of the Barr II screw, a conventional single-stage screw or a two-stage screw for vented applications. The ET section is double flighted with parallel flights that are alternately deep then shallow (Fig. 3.43). The flights are neither primary nor secondary. Undercuts in the flights are provided at numerous places, allowing both melted and unmelted plastic to pass into the adjacent channel and mix with and be melted by the already melted plastic. This causes the melt not to overheat and assists in melting the pellets. There is also distributive mixing by the frequent changes in channels and direction.

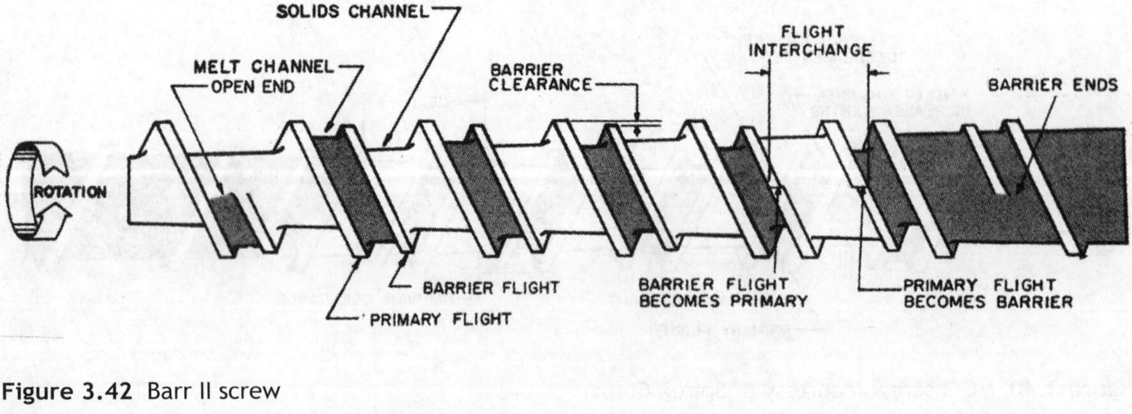

Figure 3.42 Barr II screw

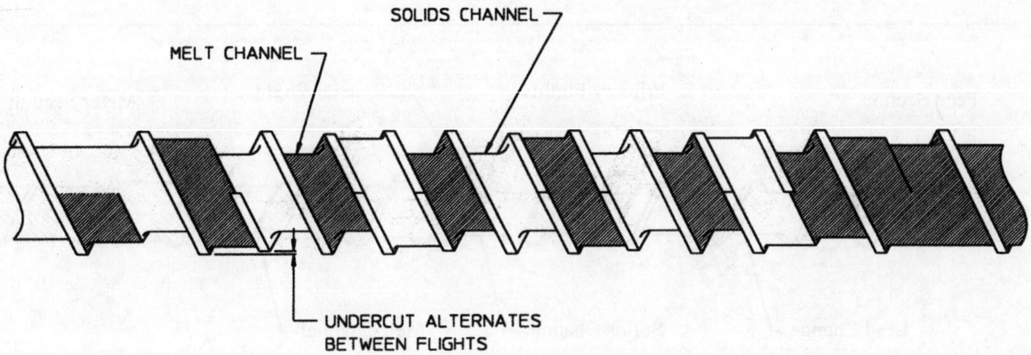

Figure 3.43 Barr ET screw

The MeltPro™ (barrier) screw uses a greater than square pitch lead in the feed section, increases the primary flight pitch even greater when the barrier flight starts, and continues this primary pitch the entire screw (Fig. 3.44). The barrier flight comes off the front of the primary flight pitch, continues away until a set width is established, and thereafter runs parallel with the primary flight. The melt channel depth continually increases as the solids channel depth decreases, until at the end of the barrier where both channels merge at the metering depth.

A description of its operations follows:

1. The feed section establishes solids conveying in the same way as a conventional screw.
2. At the beginning of the transition (compression), a second flight is started. This flight is called the barrier or intermediate flight and it is undercut below the primary flight OD. The barrier flight separates the solids channel from the melt channel.
3. As material progresses down the transition, melting continues as the solids are pressed and sheared against the barrel forming a melt film. The barrier flight moves under the melt film and the melt is collected in the melt channel. In this manner, the solids pellets and melted plastics are separated and different functions are performed on each.
4. The melt channel is deep, giving low shear and reducing the possibility of overheating the melted plastic. The solids channel becomes shallower, forcing the unmelted pellets against the barrel for efficient frictional melting. Breakup of the solids bed does not occur to stop this frictional melting.
5. The solids bed continues to get shallower and finally disappears into the back side of the primary flight.
6. Now, all of the plastic has melted and gone over the barrier flight. Melt refinement can continue in the metering section.

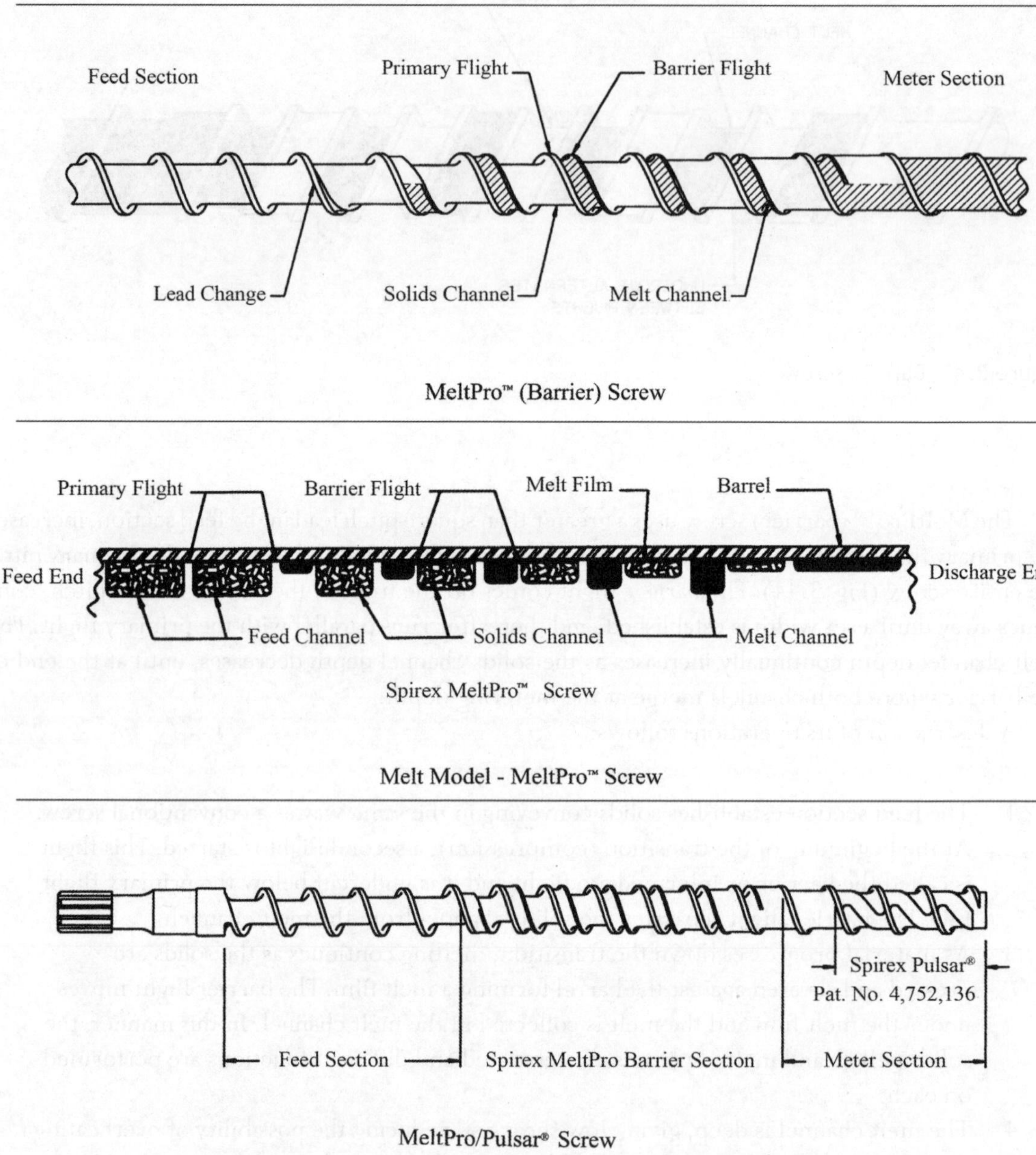

Figure 3.44 Different views of the MeltPro™ (barrier) screw (courtesy of Spirex Corp.)

In some cases, a mixing section, such as the patented Pulsar® section, is also included downstream of the barrier section. In general, the melted plastic is already fairly uniform upon exit from the barrier section.

SCREW/BARREL BRIDGING

When an empty hopper is not the cause of machine output failure, plastic might have stopped flowing through the feed throat because of screw bridging. An overheated feed throat, or start-up followed with a long delay, could build up sticky plastics and stop flow in the hopper throat. Plastics can also stick to the screw at the feed throat or just forward from it. When this happens, plastic just turns around with the screw, effectively sealing off the screw channel from moving plastic forward. As a result, the screw is said to be "bridged" and stops feeding the screw. The common solution is to use a proper rod, such as brass rod, to break up the sticky plastic or to push it down through the hopper.

SCREW TIP

Screw tip valves, popularly called nonreturn valves, ball check valves, or sliding ring valves, are used in reciprocating injection and injection blow-molding machines (IMM and IBMM) to control the melt flow in one direction. (There is also the smearhead for IMM, IBMM, and extrusion.) Backflow will not occur when the screw is used as a ram to push melt through the IMM nozzle and into a mold cavity. Also, melt drooling from the nozzle is prevented. When valves are used, they must be inspected regularly as they can easily become worn or damaged. Shut-off nozzle valves are not widely used nowadays owing to material leakage and degradation taking place within the nozzle assembly. Popular types used are the ball check and sliding ring valves (Table 3.30). Many different valves exist, with each having advantages and disadvantages based on the plastic being processed and the type of IMM and IBMM to be used. Table 3.31 provides a guide to valve materials of construction.

Ball Check Valve The two basic types of ball check valves are the front discharge and side discharge, with modifications to the ball and seat sizes that determine flow rates. Their bodies rotate with the screw. They are designed to handle the melt compressive load. The front discharge ball check is a positive and accurate valve design. It is difficult to clean and the front angle does not match that of the endcap/nozzle adapter. Most have a removable insert or pins so that the ball and wear parts can be replaced. The Side Discharge Ball Check is also a positive and accurate design that has better self-cleaning characteristics over the traditional ball check valve. The nose cone angle matches that of the endcap/nozzle adapter, lending itself to quicker color changes (419). The nose cone is normally removable to replace internal components (94).

Even though the ball check valve provides excellent repeatability performance, it can affect the plastic properties. The traditional ball check forces plastic to take sharp turns through highly restrictive flow passages. The traditional ball check has little or no front nose, which results in stagnation of plastic in the endcap radius. Different modifications have been made, including replacing the ball with a pin to improve melt flow for certain plastics (Fig. 3.45).

BALL CHECK VALVE

Advantages

- More positive shut off
- Better shot control
- Less expensive than sliding ring valve

Disadvantages

- Most designs are less streamlined, resulting in degradation of heat-sensitive materials
- More barrel wear, and a potential for galling
- Side discharge type is more expensive than sliding ring type
- Greater pressure drop creates more heat
- Poor for vented operation
- Harder to clean

SLIDING RING VALVE

Advantages

- Greater streamlining for less degradation of materials
- Best for heat-sensitive materials
- Less barrel wear
- Less pressure drop across valve
- Best for vented operation
- Easier to clean
- Less expensive than side discharge ball check

Disadvantages

- Less positive shut off, especially in 4 1/2" (115mm) diameter and larger sizes
- Less shot control
- More expensive than front discharge ball check

Table 3.30 Popular screw tip valves (courtesy of Spirex Corp.)

	Yield Strength psi	Hardness as Machined RC	Availability of Case Hardness Rc	Uses	Wear Resistance	Corrosion Resistance	Material Availability	Ease of Machining	Approx. Cost per Pound ($)
ENDCAP MATERIALS:									
AISI 4140	100,000	28-32 Rc[1]	48-55 Rc[1]	General	Poor	Poor	Excellent	Fair	.90
AISI 4150				General/Large					
Nitralloy 135-M	85,000	33 Rc	60-65 Rc[2]	General	Fair	Poor	Fair	Fair	1.65
17-4 PH	175,000	38 Rc	42 Rc[3]	Corrosive	Fair-Poor	Good	Fair	Fair-Poor	2.90
Hastelloy C-276	80,000	86 Rc	86 Rc[3]	Corrosive	Poor	Excellent	Very Poor	Poor	25.00
NON-RETURN VALVE MATERIALS AND HARD FACING:									
17-4 PH	175,000	38 Rc	42 Rc[3]	F,D	Fair-Poor	Good	Fair	Fair-Poor	2.90
Aer Met 100	260,000	38 Rc	50 Rc	H[4]	Good	Poor	Fair	Fair-Poor	9.00
CPM-10V	300,000	98 Rb	58-60 Rc	A	Excellent	Fair	Fair	Fair	15.00
CPM-T420V	300,000	100 Rb	55-57 Rc	A	Excellent	Good	Poor	Fair	15.00
CPM-9V	285,000	100 Rb	52-55 Rc	A,B,C	Excellent	Fair	Fair	Fair	15.00
D-2	300,000	96 Rb	58-60 Rc	B,C	Good	Fair-Good	Good	Fair	4.50
H-13	300,000	96 Rb	50-60 Rc	A,C,E	Fair	Poor	Good	Fair	3.95
A9	300,000	96 Rb	50-60 Rc	A,C,E	Good	Poor	Good	Fair	3.50
Stellite 12	N/A	45 Rc	41-47 Rc	D,F	Excellent	Good	Good	Poor	Low
Colmonoy 56	N/A	N/A	50-55 Rc	A	Excellent	Fair-Good	Good	Poor	Medium
Titanium Nitride	N/A	N/A	67 Rc	E	Good	Good	Good	N/A	Medium
Tungsten Carbide	N/A	N/A	>70 Rc	E	Excellent	Good	Good	N/A	Medium
Electroless Nickel	N/A	N/A	45 Rc	G	Fair	Good	Good	N/A	Low
Chrome Plating	N/A	N/A	50 Rc	G	Good	Good	Poor	Poor	High
Hardlube	N/A	N/A	90 Rc	E,A	Good	Good	Good	N/A	Medium

[1] Flame or induction hardened; [2] Nitrided; [3] Usually improved by hardsurfacing; [4] Extreme high temperature torsional strength

A) Retainer; B) Ring; C) Seat; D) A.S. Poppet; E) A.S. Poppet Shaft; F) A.S. Body; G) Endcap

Table 3.31 Guide to valve materials of construction

The poppet valve moves forward during flow from the screw and retracts when melt pressure develops during melt injection into the cavity. It is used for low-viscosity plastics. Spirex's Poly-Check valve provides improved performance over the traditional ball check valve without the usual disadvantages. This starts with the design. In order to eliminate unnecessary shear stress in the plastic, it is designed with large inlet holes angled as much as possible. Included is a nose on the valve that extends into the endcap and cleans the endcap radius out during processing.

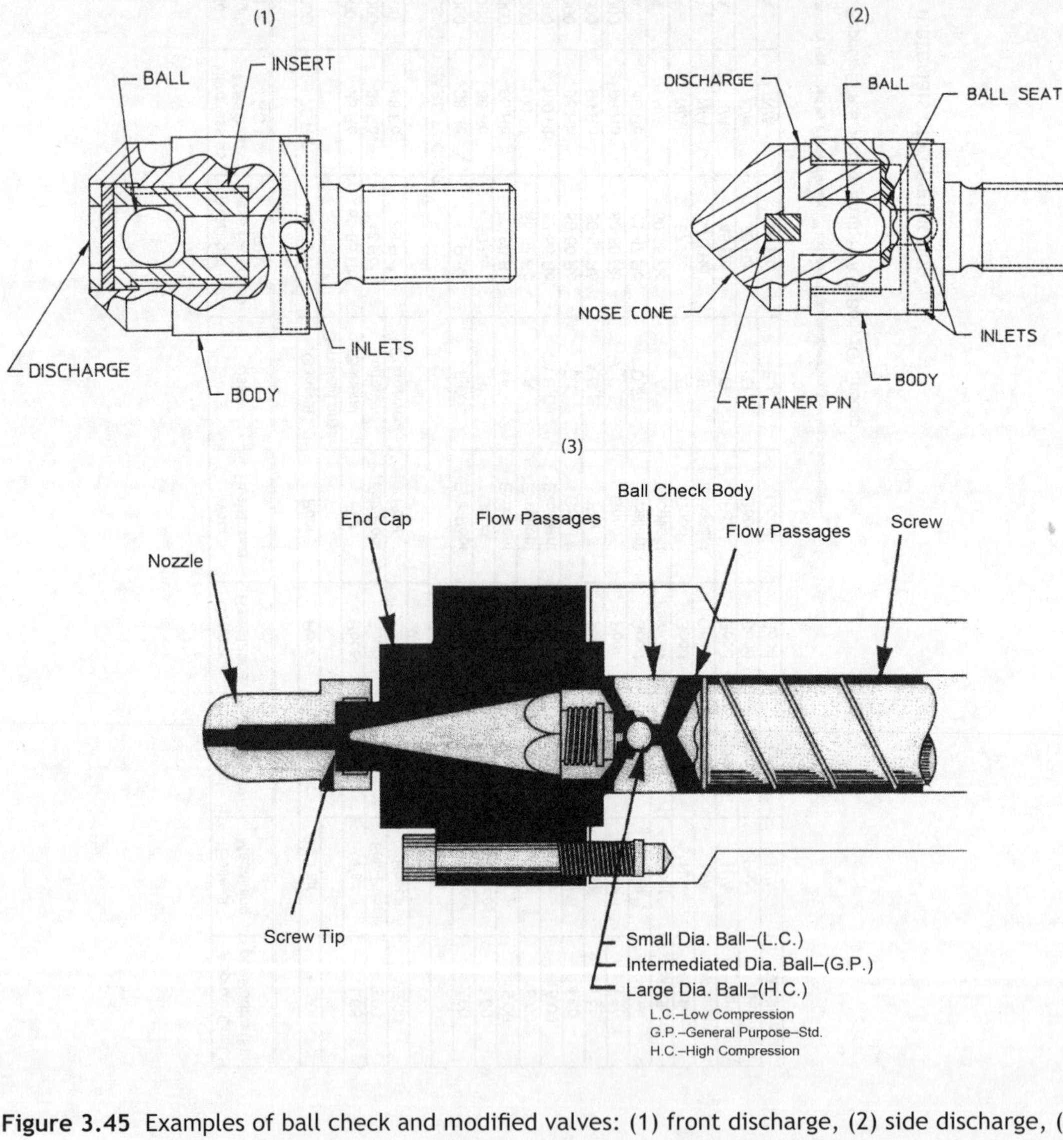

Figure 3.45 Examples of ball check and modified valves: (1) front discharge, (2) side discharge, (3) ball check with nozzle, (4) poppet, (5) Spirex Poly-Check, (6) pin forward/back, (7) Dray DNRV pin, (8) retracting nozzle/sliding pin-ball, and (9) spring operated

Nonreturn Valve Sliding ring nonreturn check valves are the workhorses of the industry. There are many variations of the design supplied by numerous manufacturers. As the name implies, a check ring is used to provide the shutoff. Basically, it is a three-part assembly. The check ring and seat are slipped on the main body that contains the tip. The assembly is then screwed into the reciprocating screw. The sliding ring fits snugly in the barrel. When the screw rotates, the force of the plastic pushes the ring toward the nozzle end and permits the plasticized material to flow under it through flutes or grooves on the main assembly. The screw slides back until the amount

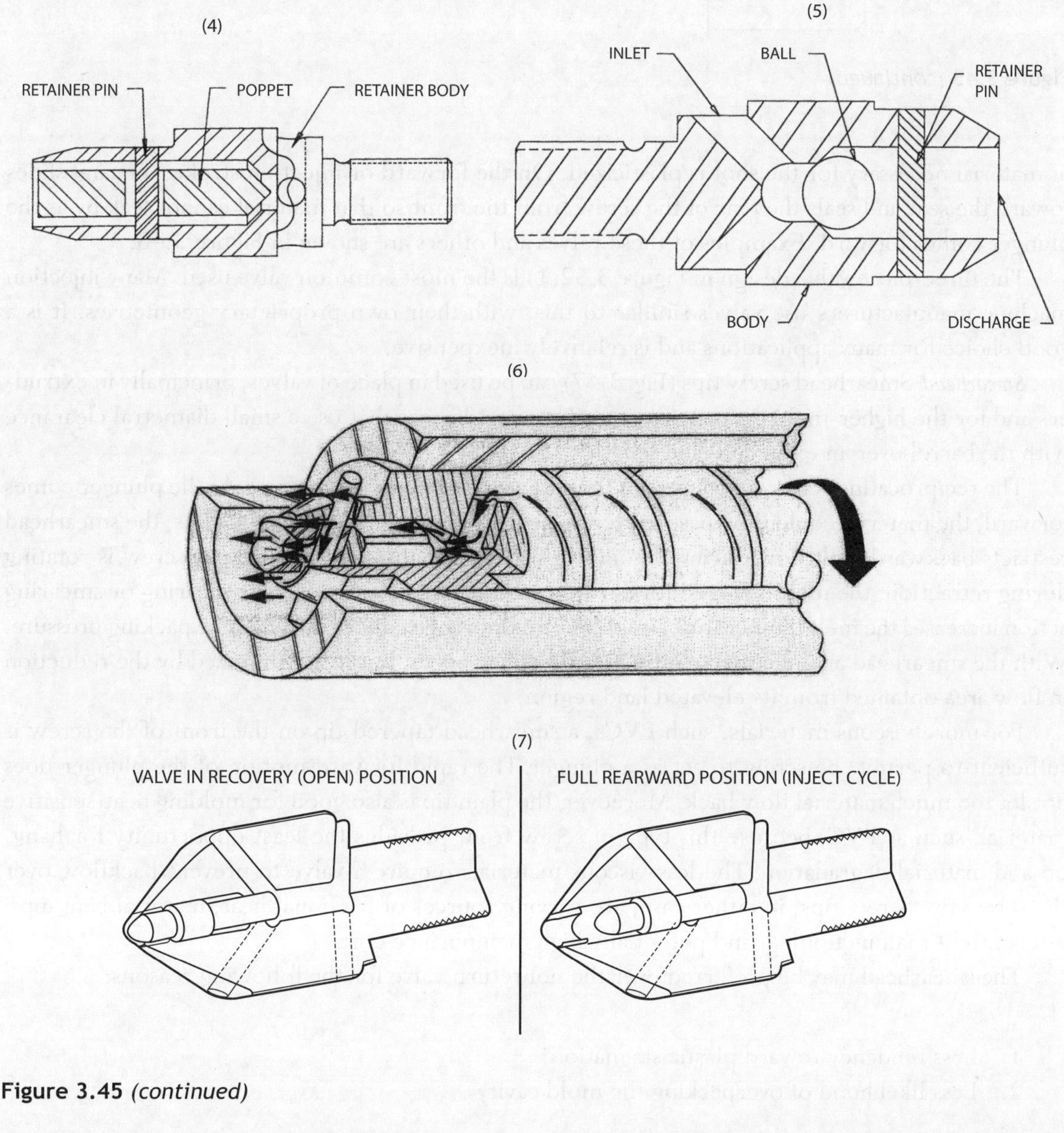

Figure 3.45 *(continued)*

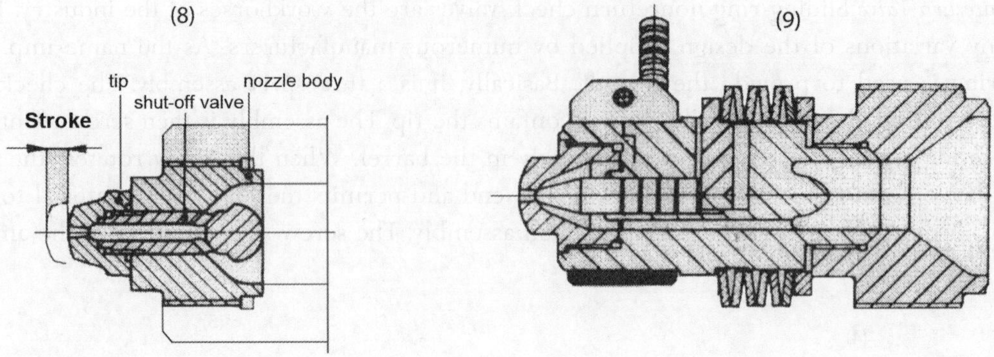

Figure 3.45 *(continued)*

of material necessary for the shot is plasticized. On the forward or injection stroke, the ring slides toward the seat and seals the rear of the screw from the front so that material cannot leak by as the plunger comes forward. Examples of these valves and others are shown in Figure 3.46.

The three-piece valve design in Figure 3.52(1) is the most common valve used. Many injection machine manufacturers use valves similar to this, with their own proprietary geometries. It is a good choice for many applications and is relatively inexpensive.

Smearhead Smearhead screw tips (Fig. 3.47) can be used in place of valves, principally in extruders and for the higher-melt-viscosity plastics. They are devices that use a small diametral clearance with the barrel over an extended land length.

The reciprocating screw machine uses the screw as a plunger (chapter 4). As the plunger comes forward, the material could flow past the screw head and back into the flights. Thus, the smearhead restricts backward melt flow during the injection stroke of the screw. When the screw is rotating during retraction, the melt is forced forward through a narrow annulus. This shearing or smearing action increases the melt temperature, improves mixing, and reduces the effective packing pressure. With the smearhead arrangement, some backflow may occur, but it is minimized by the reduction in flow area obtained from its elevated land region.

For more viscous materials, such PVCs, a smearhead tapered tip on the front of the screw is sufficient to permit the screw to act as a plunger. The rapid forward motion of the plunger does not let too much material flow back. Moreover, the plain tip is also good for molding heat-sensitive material, such as PVC, because this type of screw front provides the least opportunity for hang-up and material degradation. The less viscous materials require a valve to prevent backflow over the screw tip. Screw tips, in either case, are varying sources of frictional heat, material hang-ups, intermittent malfunctioning, and potentially high maintenance costs.

The smearhead may be preferred over the nonreturn valve for the following reasons:

1. Less tendency toward plastic stagnation
2. Less likelihood of overpacking the mold cavity

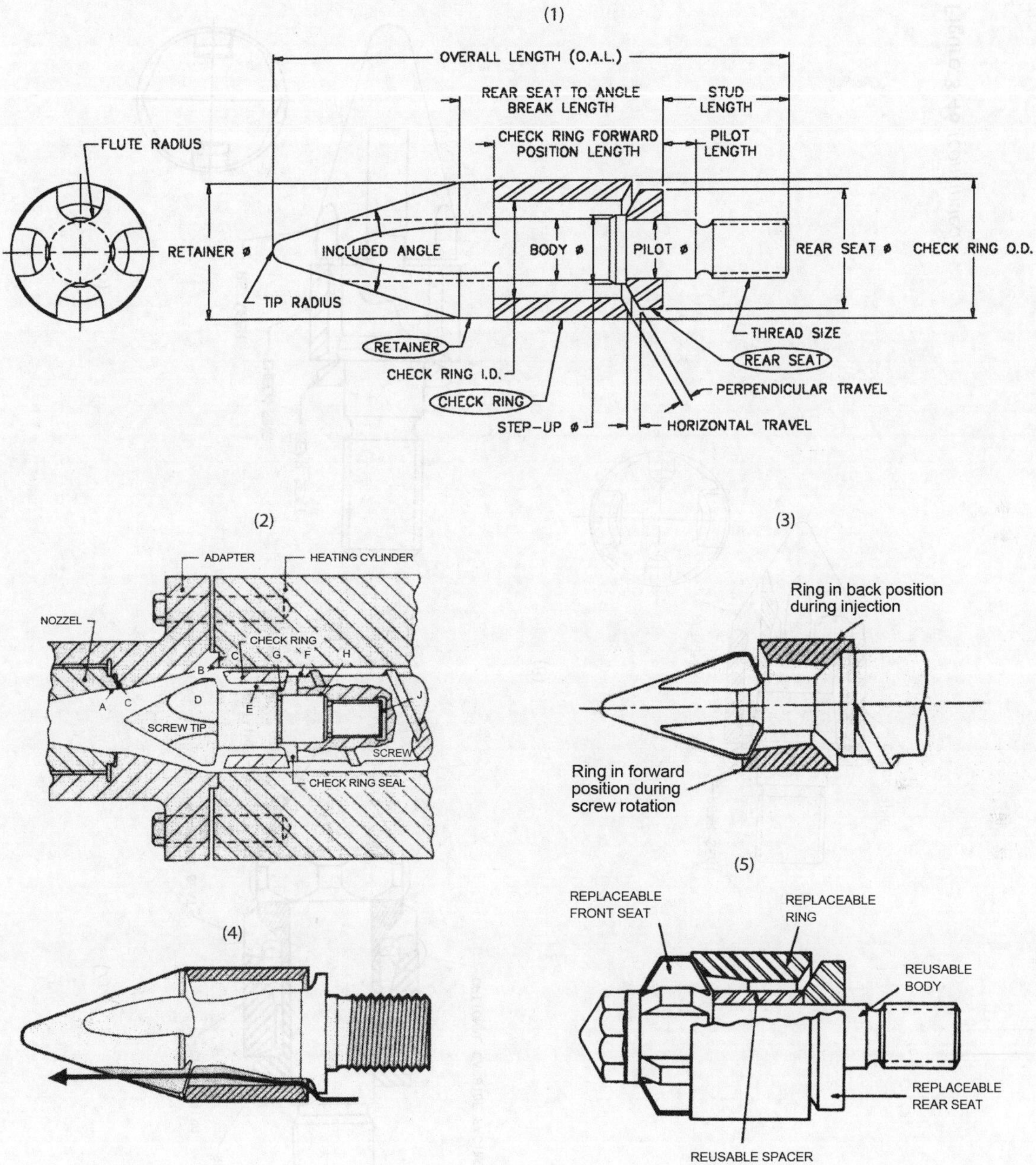

Figure 3.46 Examples of sliding ring and modified valves: (1) nomenclature of three-piece free flow valve (retainer, check ring, and rear seat), (2) valve with adapter, (3) split view showing action of ring, (4) melt flow when ring is in the back position, (5) patented CDM Corp. valve, (6) Zeiger Industries' four-piece Mallard valve, (7) Castle series of fingers design interlocks with slots of the retainer, (8) Spirex's patented F-LOC design with large flow paths prevents shearing problems, and (9) Spirex's patented Auto-Shut valve with positive/quick shutoff mechanism independent of screw travel

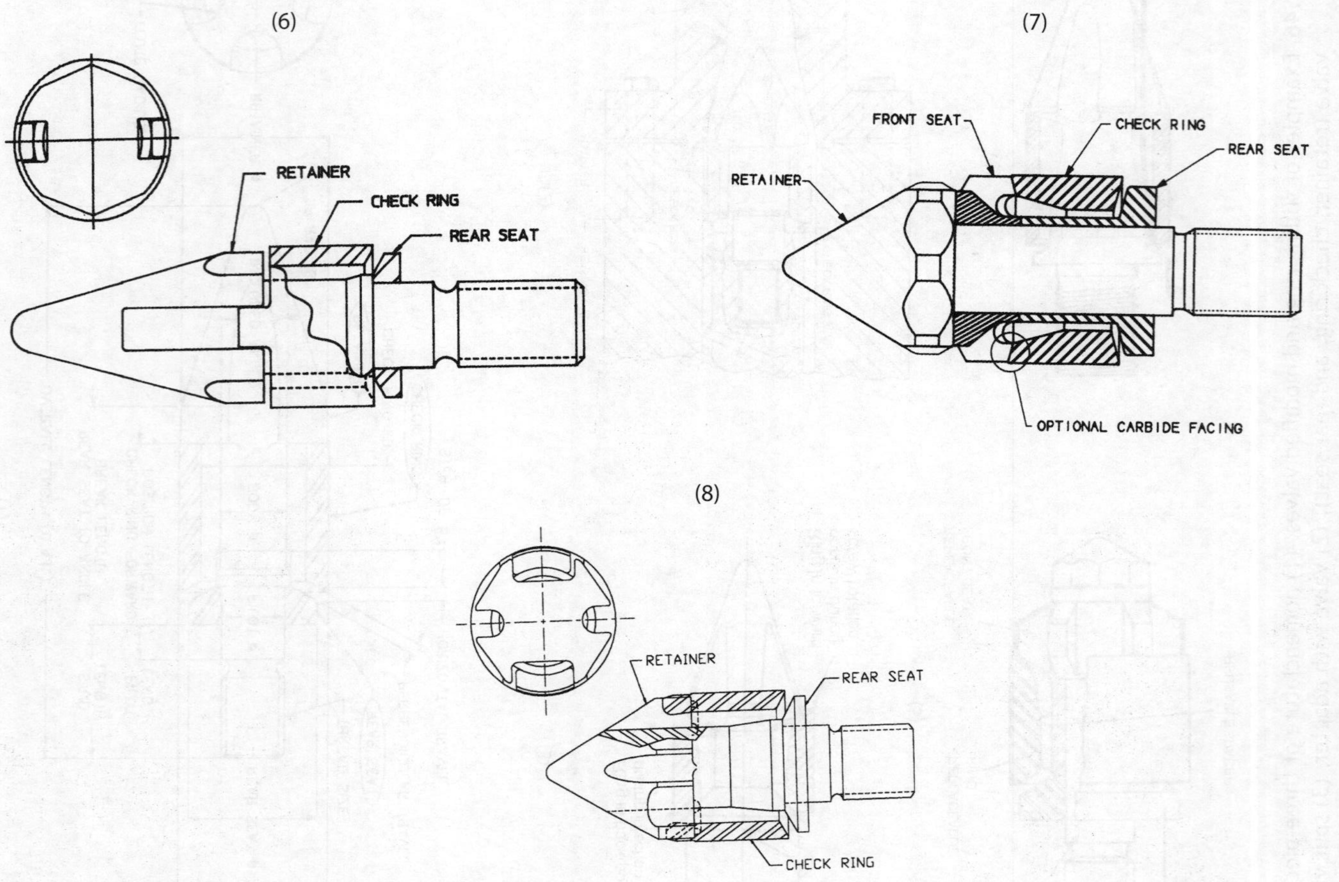

Figure 3.46 *(continued)*

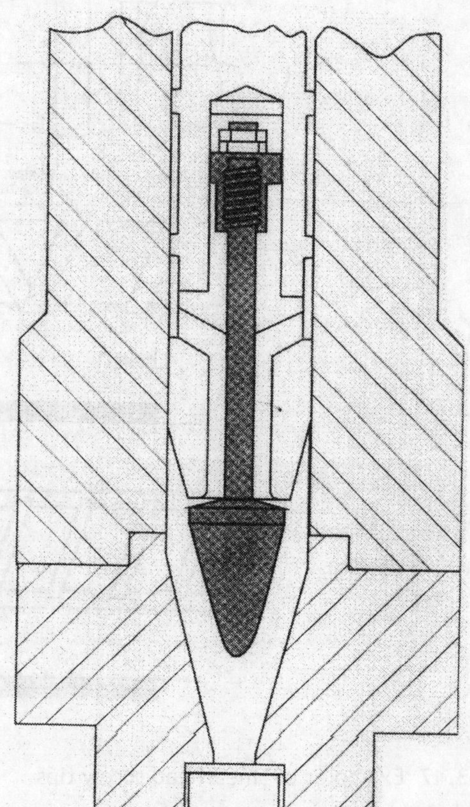

(b)

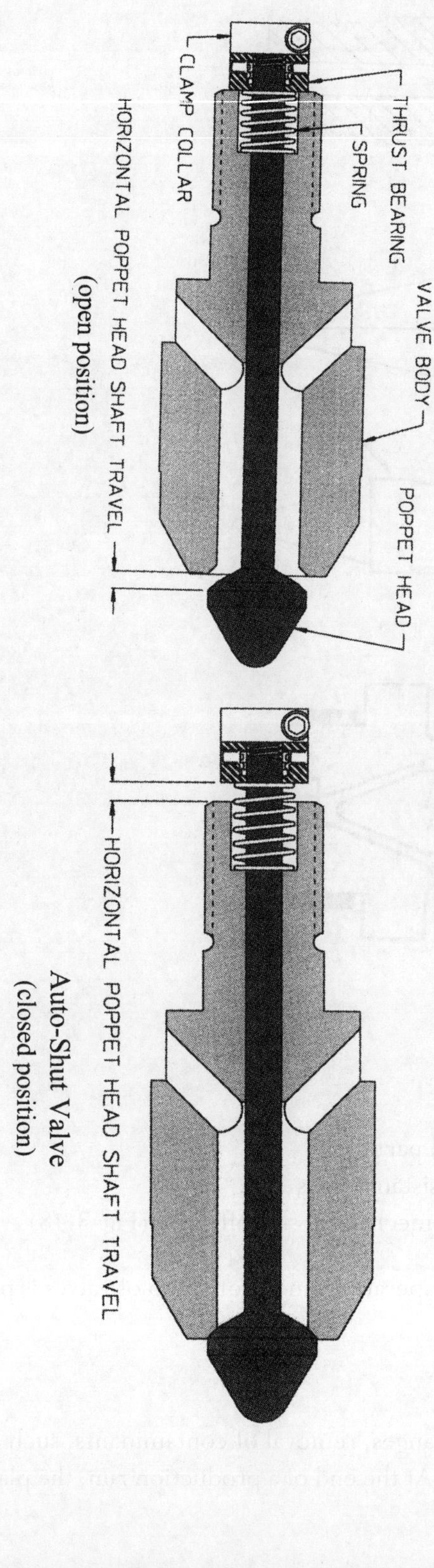

Figure 3.46 *(continued)*

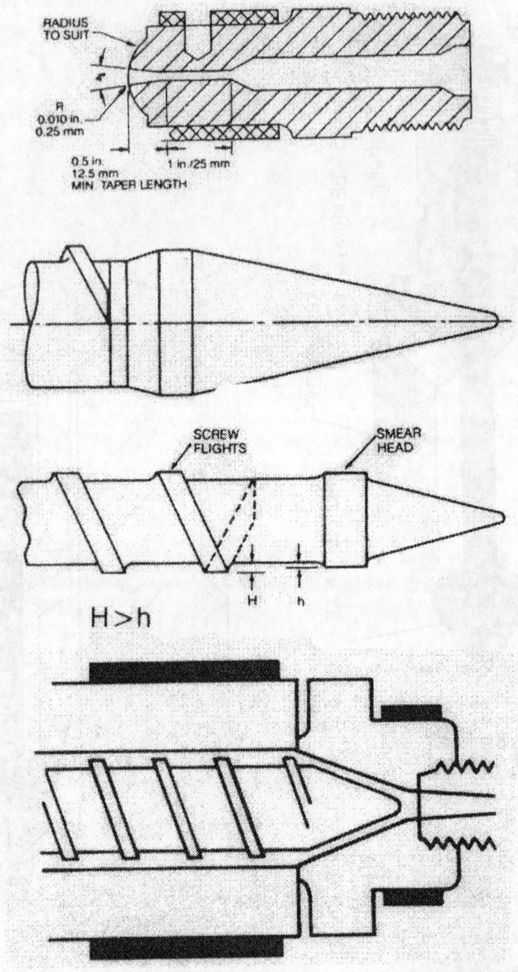

Figure 3.47 Examples of smearhead screw tips

3. Less tendency to form streaks in the molded part
4. Less abrasion on relatively soft corrosion-resistant alloys
5. Elimination of potential melt hang-up using mechanical shutoff valve (Fig. 3.48)

Guidelines and information for safe installation, operation, and protection of valves is provided in Tables 3.32 and 3.33.

PURGING

Purging is an important tool that allows for color changes; removal of contaminants, such as black specks; and adhering of plastic to screws and barrels. At the end of a production run, the plasticator

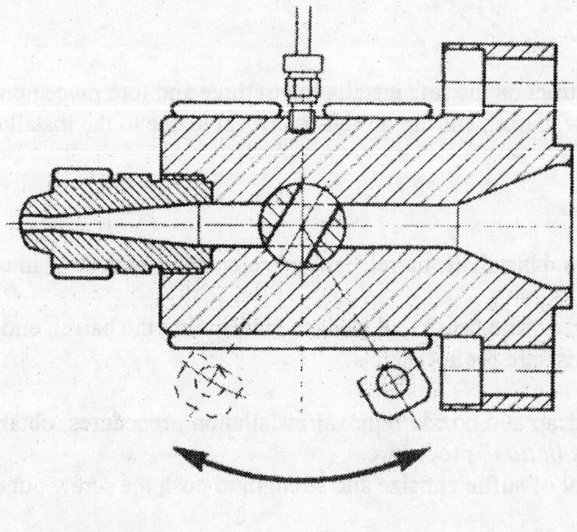

Figure 3.48 Example of a mechanical shutoff valve

may have to be cleared of all its plastics in the barrel/screw to eliminate barrel/screw corrosion (Table 3.34). This action consumes substantial nonproductive amounts of plastics, labor, and machine time. It is sometimes necessary to run hundreds of pounds of plastic to clean out the last traces of a dark color before changing to a lighter one; when a choice exists, process the light color first. Sometimes there is no choice but to pull the screw for a thorough cleaning.

Purging material ranges from certain plastics to chemical purging compounds. The use of ground/cracked cast acrylic and PE-based (typically bottle-grade HDPE) plastics is popular. Others are used for certain plastics and machines. Cast acrylic, which does not melt completely, is suitable for virtually any plastic. PE-based compounds containing abrasive and release agents have been used to purge the softer plastics, such as other polyolefins, polystyrenes, and certain PVCs. These types of purging agents function by mechanically pushing and scouring residue out of the plasticator (Tables 3.35 and 3.36).

The chemical purging compounds are generally used when major processing problems develop. However, to eliminate the major problems with their associated machinery downtimes, regularly scheduled purgings prevent quality problems and can yield operational benefits (418). With proper use, these purging agents help to reduce reject rates significantly. The schedule depends on factors such as plastics being processed, barrel size, and plasticator operational settings with the time schedule that is in use. Repeated equipment shutdowns and start-ups are the most common cause of degraded plastic buildup. Purging compound producers can recommend the time schedule to be used in order to minimize downtime and increase profits.

SAFETY ALARM

When running a line, make provisions to adequately alarm the operator if a plasticator trips out, particularly in a coextrusion or coinjection line. This will prevent melt from backflowing into

Objective

The intent of this procedure is to instruct on the safe installation of three and four piece non-return valves. A safe installation is when no damage occurs to the screw, barrel, endcap or non-return valve due to the installation.

Tools & Equipment

- As necessary, to safely remove and install the barrel, endcap (refer to the owner's manuals or contact your machine supplier) and non-return valve.
- A brass putty knife, copper gauze, wire brush and cleaner for cleaning the barrel, endcap, non-return valve and fasteners. Check to be certain the cleaner is safe for hot metals.
- Anti-seize.
- If you are not familiar with endcap and nozzle removal/installation procedures, obtain a copy of our *"Injection Molding Machine Endcap & Nozzle Installation"* procedure.
- A block of wood or soft material of sufficient size and strength to push the screw out of the barrel.

Removal of Old Non-Return Valve

- Let the barrel reach operating temperature, adequately soak, and start the machine motor(s). Do not rotate the injection unit, but prepare it for purging as your machine operator's manual instructs.
- The injection unit components and the heat profile are identical to a production run.
- The barrel, endcap and non-return valve must be hot for cleaning.
- Shut off the supply of plastic that is running to the machine. Close the supply of plastic running to the feed throat of the injection unit – do this by moving the hopper slide to the closed position. Note: Your system may be different. Refer to your operator's manual for your machine or after-market material handling equipment supplier. Purge the screw of plastic and/or purging compound.
- Bring the screw all the way forward, observing the safety precautions in the owner's and operating manuals. In manual mode it is best to start with a velocity and pressure of twenty percent (20%). In set-up mode the machine automatically reduces pressure and speed. Never select injection-forward with someone in front of the screw.
- Position the injection unit for endcap removal, turn the motor off and lock it out. Safely remove the endcap (*Tools & Equipment*).
- Position the injection unit for screw removal. Disconnect the screw from the drive shaft or quill. Each machine is different and you must refer to your owner's or service manual for the safe and proper procedure for your specific machine. Retract the drive shaft, place the block of wood between the drive shaft and screw and run the screw forward until the non-return valve is out of the barrel. Push the screw forward until approximately six inches of the screw is protruding from the barrel.
- Safely remove the old non-return valve from the screw. Remember that most threads are left handed, so turn the retainer clockwise to loosen. Use a wrench on the wrench flats, or a screw tip removal wrench. Placing a torch on a non-return valve generally voids the warranty offered by most manufacturers. If you must apply heat, it is the screw that should be heated. If a hammer is necessary, a dead blow hammer is less likely to damage the non-return valve.

Inspecting & Cleaning

Inspect and clean the screw threads, pilot, face and bottom of dill cavity prior to assembly. Tools that repair threads can also clean threads. Remove plastic, dirt and old anti-seize from the threads. In the absence of proper tools, a wire brush, cleaner and rag will work once or twice. Obtain the correct tools for future use.

Check that the screw face is flat and straight to 0.001" (0.025mm) TIR (total indicator reading). If not, have a professional experienced in plastic screws machine the face. If machining is required, have the vendor check the entire non-return valve cavity. From time to time, the geometry is out of specification. This may result in early failure of the non-return valve, or damage to the endcap or barrel. The Society of the Plastics Industry maintains standards for this geometry.

Table 3.32 Nonreturn valve installation (courtesy of Spirex Corp.)

Procedure for Checking Clearances

Follow this procedure if you have a new screw, barrel, endcap or non-return valve:

Extra steps are necessary when installing new components. It is always best to check for interference between the endcap and .the non-return valve. Interference infrequently occurs due to design or manufactoring erros. In addition, there are machines where the screw-forward position is adjustable.

1. Let the barrel reach operating temperature, adequately soak, and start the machine motor(s).
2. Bring the screw all the way forward, observing the safety precautions in the owner's and operating manuals. In manual mode, it is best to start with velocity and presure of twenty percent (20%), In set-up mode, the machine automatically reduces pressure and speed. Never select injection forward with someone in front of the screw.
3. Remove the old no-return valve as described above, and install the new valve as described below.
4. Put the endcap in place, but first place a soft material (such as modeling clayor soldering wire) in the endcap so that it is between the head and non-return valve. Evenly seat the head and snug it down–it is not necessary to fully tighten it.
5. Remove the endcap and observe the gap between the endcap and non-return valve by measuring the thickness of the clay. The clearance varies depending on the size of the injection unit and the machine OEM. It is safe to say that less than 0.060 inches (1.5 mm) is too little clearance, adjust the screw back. If there is no adjustment possible, contact the supplier(s) of the new component about the problem.

Procedure for Installing the Non-Return Valve

1. Inspect and clean the screw as stated above
2. Put a light coat of anti-seize on the male threads of the new valve.
3. Screw the non-return valvė into the screw. Do not use a wrench to tighten the valve (hand-tigh is sufficient). Inspect the assembly to be certain the seat is flush against the screw face. Try to rotate the seat by hand—the retainer is sufficiently tight uf the seat does not rotate.
4. Use the wooden block to push the screw inside the barrel. Be prepared to center the slider ring as you push, otherwise the ring will hang up between the retainer and the barrel.
5. **Caution:** If you are installung a new barrel, screw, endcap or non-return valve, perform the above procedure "*Procedure for Checking Clearences.*" It is a good way to check the clearances of new assembly without causing damage to the components.
6. Reattach the screw to the drive shaft, and the endcap to the barrel. **Caution:** Improper installation of the endcap can result in damage to the endcap, barrel and/or non-return valve. Read your owner's operator's manual, or obtain a copy of our endcap installation procedure.

The assembly is now ready for production.

Table 3.32 *(continued)*

the stopped machine and creating a blockage in the feed section, which could result in a hazardous explosion.

MATERIAL OF CONSTRUCTION

The usual alloy steels and special materials used to manufacture screws are listed in Table 3.29. The most common screw material used is the alloy steel AISI 4140. It probably accounts for in excess of 80% of all extrusion and injection screws. Readily available, it is purchased in the heat-treated condition (28 to 32 Rc) and is stress relieved. It has good strength and can be flame hardened or hard surface welded. It is important to make sure that special, easy-to-machine grades are not

Objective

The intent of this procedure is to instruct on the safe installation of a barrel endcap and nozzle. A safe installation is when no damage occurs to the screw, barrel, endcap or non-return valve due to the installation, including tightening of the endcap.

Machine Setup

- The injection molding machine should have the motor off and locked out.
- The injection unit must be positioned so the endcap can be safely removed.
- The injection unit components and the heat profile are identical to a production run.
- If the components are to be cleaned, the barrel and endcap must be hot for cleaning.
- If the components are not to be cleaned, the endcap can be positioned and lightly tightened cold, but must be hot when fully tightened.

Tools & Equipment

- As necessary, to safely install the endcap to the barrel (refer to the owner's manual or contact your machine supplier).
- As necessary, to clean the female threads in the barrel and endcap.
- A brass putty knife, copper gauze, and wire brush for cleaning the barrel, endcap, and male threads.
- Anti-seize.
- A torque wrench if socket head cap screws are used to fasten the endcap.
- Torque specifications from the owner's manual or machine supplier.

Inspecting & Cleaning

Inspect the mating surfaces prior to assembly. Check for the following:

- The face and pilot areas are clean and contain no nicks. It is too dangerous to stone damaged surfaces; machining is necessary to maintain safety, the proper fit, and finish.
- The non-return valve and screw are not in the way of the head. Never hit the screw or non-return valve, always push them with a soft material such as wood.

Inspect and clean the threads prior to assembly. This includes the socket head cap screws, barrel threads, and nozzle threads. Inspect the threads and repair as necessary. The tools used to repair threads can also be used to clean threads. Remove plastic, dirt and old anti-seize from the threads. In the absence of the proper tools, a wire brush, cleaner and rag can be used once or twice. Check to be certain the cleaner is safe for hot metals, and obtain the correct tools for future use.

Clean all mating and interior surfaces using the above tools. Mating parts must be clean in order to properly seal. Interior surfaces must be clean and smooth, or specs and degradation may occur.

Procedure for Checking Clearances

Follow this procedure if you have a new barrel, screw, endcap or non-return valve:

Extra steps are required when new components are involved. It is always best to check for interference between the endcap and the non-return valve. Interference infrequently occurs due to design or manufacturing errors. In addition, there are machines where the screw-forward position is adjustable.

1. Let the barrel reach operating temperature, adequately soak, and start the machine motor(s).

Table 3.33 Valve protection: Injection-molding machine endcap and nozzle installation (courtesy of Spirex Corp.)

2. Bring the screw all the way forward; observing the safety precautions in the owner's and operating manuals. In manual mode it is best to start with a velocity and pressure of twenty percent (20%). In set-up mode the machine automatically reduces pressure and speed. Never select injection-forward with someone in front of the screw.
3. Turn off and lock out the motor(s).
4. Next put the endcap in place, but first place a soft material (such as modeling clay or soldering wire) in the endcap so that it is between the head and non-return valve. Evenly seat the head and snug it down – it is not necessary to fully tighten it.
5. Remove the endcap and observe the gap between the endcap and non-return valve by measuring the thickness of the clay. The clearance varies depending on the size of the injection unit and the machine OEM. It is safe to say that less than 0.060 inches (1.5 mm) is too little clearance, and the screw must be adjusted back. If there is no adjustment possible, contact the supplier(s) of the new component about the problem.
6. Continue to the procedure below if there is no interference.

Procedure for Installing the Endcap and Nozzle

1. Let the barrel reach operating temperature.
2. Put a light coat of anti-seize on the male threads.
3. Put the endcap in place and tighten only until snug. Install the nozzle and tighten only until it is snug. Install the heater bands and thermal couples on the endcap and nozzle, and turn on the heater bands.
4. Wait until the endcap and nozzle are at the operating temperature set point. The barrel, endcap and nozzle should be at the set-point temperature before you proceed.
5. It is important to understand that damage will occur to either the barrel or endcap if the endcap is tightened incorrectly. Incorrect assembly occurs when the endcap is tightened unevenly or too tightly. Both cases result in excessive force applied to the mating surfaces. These surfaces are then permanently damaged, or the uneven force on the endcap results in early failure.

The damage to the mating surfaces is termed *coining*, and often results in a change in the barrel internal diameter. If this does occur, it is best to repair the barrel and/or endcap before the machine is put back into production. Otherwise, the non-return valve will tend to seize to the barrel when it passes through the damaged area. The seizure can result in further damage to the barrel, screw, non-return valve or endcap.

Uneven force on the endcap usually does not result in immediate failure. Instead, the failure is a slower process known as fatigue failure. Since the load is uneven, there is some portion of the endcap that is subjected to significantly higher stress than other portions of the endcap. This higher stress will gradually break down the metal until a crack eventually forms. The crack lengthens until the remaining material fails.

Even if the components do not fail, an incorrect installation may result in contamination of the plastic, or leakage between the endcap and barrel. When the endcap does not seat properly, a gap may occur between the endcap and barrel mating surfaces that result in contamination, specs or leakage. As is often the case, the processor will tighten the endcap if leakage occurs, and this leads to further damage. If leakage occurs on your installation, remove the endcap, inspect the mating surfaces, and clean and/or repair as required. In the long run, this is the most time and cost effective way to deal with such a problem.

Table 3.33 *(continued)*

selected. These grades usually contain either lead or are resulfurized. This makes them, for all practical purposes, impossible to rebuild by hard surface welding because of extreme porosity and stress cracks. 4140L is a grade of 4140 modified by the addition of lead, which makes it easier to machine but impossible to rebuild (94).

The alloy AISI 4340 is similar to 4140 but includes nickel as an alloying element plus a greater percentage of molybdenum. This alloy gives a slightly higher strength, but the major difference is the greater penetration of heat treatment. This gives superior mechanical properties at the core

6. Installation procedure for a threaded endcap:
   - First, snug the endcap.
   - Use a wrench to tighten the endcap. Generally, the length of the wrench is designed to provide adequate tightening torque. If the wrench contains dead blow flats, then a ***small*** (60 ounce) dead blow hammer is permissible. Tighten the endcap.
   - If you undertighten the endcap and it leaks, you will have to remove the endcap and clean the mating surfaces. Then reinstall the assembly, and tighten the endcap a little tighter than you did with the leaking assembly. If you overtighten the assembly, you may have expensive replacement or repair costs.
7. Installation procedure for a flanged endcap:
   - First, snug the fasteners. Use the crisscross method where, for example, you tighten the first screw at twelve o'clock, the second at six o'clock, the third at two o'clock, and the fourth at eight o'clock. Continue until all of the screws are snug.
   - Set the torque wrench for 50% of the recommended tightening torque, and tighten using the above crisscross method. Set the torque wrench for 75% of the recommended tightening torque and tighten. Set the torque wrench for 95% of the recommended tightening torque and tighten. Finally, set the torque wrench to the recommended tightening torque and tighten. Always use the crisscross tightening method or a similar method that is recommended by the machinery manufacturer.
   - If you do not have a torque wrench and cannot rent one, then use a ratchet wrench and hex bit socket. The length of the wrench will determine the maximum torque on the screws. Just as above, tighten the screws using the crisscross method, and tighten to a lower torque and gradually increase the torque until tight. Do not overtighten – just a strong steady pull is sufficient (do not strain) to tighten. If you do not use a torque wrench, there is no guarantee that you have properly tightened the endcap, so seek assistance from a more experienced technician. If you must use a hex wrench and cheater bar, cut the cheater bar to approximately the same length as a wrench for that size screw.

   If you undertighten the endcap and it leaks, you will have to remove the endcap and clean the mating surfaces. Then reinstall the assembly and tighten the endcap a little tighter than you did with the leaking assembly. If you overtighten the assembly, you may have expensive replacement or repair costs.

8. Use a wrench to tighten the nozzle. The wrench length will determine the maximum torque on the nozzle. Do not overtighten – just a strong, steady pull is sufficient (do not strain) to tighten. If you undertighten the nozzle and it leaks, you will have to remove it, clean the mating surfaces, reinstall it, and tighten the nozzle a little more than you did with the leaking assembly. If you overtighten the assembly, you may have expensive replacement or repair costs.

   The assembly is now ready for production.

Table 3.33 *(continued)*

of the bar. It is most beneficial for screws that have very deep flights and smaller diameter screws where strength is more critical.

A number of steel manufacturers produce a product similar or identical to Nitralloy 135-M, but this is the name best known. Nitralloy is similar to 4140 but is slightly lower in physical properties. The major difference is the inclusion of a small portion of aluminum, which forms aluminum nitrides in the nitriding process (to be reviewed).

A number of 300 series stainless steels, including 304 and 316, are occasionally used to manufacture screws for which corrosion resistance is of primary importance. These materials have relatively poor strength and abrasion resistance. They cannot be flame hardened and must have the flights hard surface welded. A better selection is a precipitation hardening stainless steel such as 17-4 PH that gives similar corrosion resistance but mechanical properties superior to 4140 (94).

Objective

The intent of this procedure is to offer guidance in properly pre-heating and soaking the injection front-end components prior to processing.

General Thoughts

One of the best ways to avoid damage to machinery is to use purging compound during shut-down and start-up. Familiarity of this compound will be a guide for the proper soak time. Often it is acceptable to purge with a safer material, such as linear low density polyethylene (LLPE).

Do not overlook obvious sources of information. Draw on the experience of others, or records of previous jobs. The material supplier can provide recommendations from the producer. Industry contacts are a good resource of experience and other contacts. Try to compare the material to other similar materials. Exercise caution as many like-materials have different additives that result in very different properties. The soak time may change, there may be a critical temperature at which damaging gas releases, or heat sensitivity may increase.

Excessive soak time can cause a problem if the material is heat sensitive. If extra caution is necessary, a heat probe can be inserted into the nozzle, or the endcap can be removed for checking the melt directly. Once the endcap is removed, insert a temperature probe inside the non-return valve. The material within a non-return valve is generally the last to reach operating temperature. The Auto-Shut valve can be checked by using pliers to gently pull open the poppet.

Using Soak Timers

New OEM machines come with soak timers. The function of the soak timer is to lock out the injection unit until the timer times out. The timer starts once the heaters reach the operator-set temperature. Our experiences with OEM soak timers are that they are satisfactory for many materials and front-end components. However, exercise caution if you are unfamiliar with the material, or your front-end components. Our belief is that many non-return valves and materials require 40% more soak time than the OEM timer provides.

Older Machines Without Soak Timers

There are still many machines in service without soak timers. Try to compare the machine with other similar machines that do have a timer. Remember the soak time starts after the injection unit reaches the set point temperature. You know the injection unit is at set point by watching the cycling of the heater bands. Add your soak time at this point, and maintain records for your future use. Remember, you can apply the other methods stated above in *General Thoughts*.

Table 3.34 Purging: Preheat/soak time (courtesy of Spirex Corp.)

If you do not have an adequate means to determine the appropriate soak time, there is an old industry rule-of-thumb that can help. This rule is "turn on the heaters and heat for one hour for each inch of barrel wall thickness." This rule seems to work and is actually a little on the safe side, so burning of heat-sensitive materials is a risk.

Comments on the Auto-Shut Valve

The Auto-Shut Valve requires more soak time than conventional non-return valves. This valve has a spring loaded poppet that rides in the body of the valve. There is a pool of plastic contained inside the valve by the poppet. The pool of plastic in the center of the valve, and the poppet shaft, are the last front end components to reach operating temperature. The plastic around the screw will reach operating temperature before the internals of the Auto-Shut Valve. If screw rotation occurs too soon, the plastic from the screw will flow into the valve, and either the poppet will resist opening, or the cold slug of plastic in the valve will block the flow. If the screw continues to rotate, the Auto-Shut valve may become damaged.

Table 3.34 *(continued)*

Others are used that provide different performances, machinability, and cost. Armco Steel's 17-4 PH stainless is much stronger than 4140 and has corrosion resistance similar to the 300 series of stainless steels. This steel eliminates the need for frequent stripping and chrome plating. The D-2 and H-13 tool steels have been primarily used for high wear applications caused by abrasion. The uses include components for nonreturn valves, nozzles, endcaps/nozzle adapters, barrel sleeves, and plasticating screws. Both grades require a heat-treating process to obtain maximum wear properties. The cost to manufacture screws out of these steels is considerably more than 4140 base steel; however, one could expect to receive a longer service life.

CPM is a designation of materials made by Crucible Specialty Metals Division and stands for crucible particle metallurgy. These CPM materials contain very high portions of alloying elements usually not possible in conventional processes starting with ingot casting. The HIP process is used to make these special CPM alloys. HIP stands for hot isostatic pressure and is a process in which powdered metals are combined at extremely high temperatures and pressures to make very dense bars or other shapes with superior properties not possible by the other older methods.

Spirex Corp. pioneered the use of CPM materials for the manufacture of extremely abrasion-resistant screws. Now a variety of CPM materials are used in all types of plastic-related machinery components. They include screws, barrels, barrel sleeves, nonreturn valves, cutter and pelletizer knives, and molds. Materials such as CPM-10V have as much as ten times the abrasion resistance as a good D-2 tool steel, which is far superior to conventional AISI 4140 steel with typical hard surfaced flights. CPM materials are particularly useful as resistance to wear caused by abrasive filters, such as glass fibers. CPM is not a coating but exhibits its abrasion resistance throughout the entire part. This protects screws in the root surfaces where glass erosion is the most severe. Tables 3.37 and 3.38 compare wear and toughness of CPM materials with D-2 tool steel. Table 3.39 provides examples of CPM products used in plastic machinery components.

Plastic in machine	Plastic changing to	Mix with rapid purge and soak	Follow with
ABS	PP	ABS	PP
ABS	PC	ABS	PC
ABS	PBT	ABS	PBT
Acetal	PC	Acetal	PC
Acetal	Any material	PE	New material
Acrylic	PP	Acrylic	PP
Acrylic	Nylon	Acrylic	Nylon
TPE	Any material	PE	New material
Nylon	PC	PC	PC
Nylon	PVC	Nylon	PVC
PBT	ABS	PBT	ABS
PC	Acrylic	PC	Acrylic
PC	ABS	PC	ABS
PC	PVC	PC	PVC
PE	PP	PP	PP
PE	PE	PE	PE
PE	PS	PS	PS
PETG	Polysulfone	PETG	Polysulfone
Polysulfone	ABS	Polysulfone	ABS
Polysulfone	ABS	Cracked acrylic	ABS
PP	ABS	ABS	ABS
PP	Acrylic	Acrylic	Acrylic
PP	PE	PE	PE
PP	PP	PP	PP
PS	PP	PP	PP
PVC	Any material	LLDPE or HDPE	New material
PVC	PVC	LLDPE or HDPE	PVC
PPS	PE	PPS	PE
SAN	Acrylic	Acrylic	Acrylic
SAN	PP	SAN	SAN

Table 3.35 Examples of purging when changing plastic in a plasticator

Heat Treatment To improve the performance of screw materials, various different processes are used, for example, gas/oxygen flame hardening followed with rapid quenching to increase wear resistance of the tops of flights. Nitriding over the entire flighted area and root provides a very hard outside case. It is obtained by subjecting screws and barrels to a high hydrogen atmosphere that is usually ammonia gas at high temperature. Another method is to use precipitation hardening at low temperature (94).

Coating By applying coatings, different advantages are gained, such as wear resistance. Chrome plating permits easier cleaning of the screw after it is removed from the barrel. It also permits the feeding characteristics of the screw to remain constant over a longer period. Nickel plating

Material to be purged	Recommended purging agent
ABS	Cast acrylic, polystyrene
Nylon	Polystyrene, low-melt-index HDPE, cast acrylic
PBT polyester	Next material to be run
PET polyester	Polystyrene, low-melt-index HDPE, cast acrylic
Polycarbonate	Cast acrylic or polycarbonate regrind; follow with polycarbonate regrind; do not purge with ABS or nylon
Acetal	Polystyrene; avoid any contact with PVC
Engineering resins	Polystyrene, low-melt-index, HDPE, cast acrylic
Fluoropolymers	Cast acrylic, followed by polyethylene
Polyphenylene sulfide	Cast acrylic, followed by polyethylene
Polysulfone	Reground polycarbonate, extrusion-grade PP
Polysulfone/ABS	Reground polycarbonate, extrusion-grade PP
PPO	General-purpose polystyrene, cast acrylic
Thermoset polyester	Material of similar composition without catalyst
Filled and reinforced materials	Cast acrylic
Flame-retardant compounds	Immediate purging with natural, non-flame-retardant resin, mixed with 1% sodium stearate
Polyolefins	HDPE
Polystyrene	Cast acrylic
PVC	Polystyrene, general-purpose, ABS, cast acrylic

Table 3.36 Recommended purging agents

behaves somewhat similarly to chrome plating, with potentially better corrosion resistance depending on the chemicals involved. It is softer and has less wear resistance. Other hard surfacing coatings include welding special wear-resistant alloys over the screw surface, such as Stellite and Colmonoy (Fig. 3.49 and Table 3.40).

MULTIPLE SCREW

Most plasticating machines use the single screws that have been reviewed; they dominate the industry. In use worldwide are also plasticators for extruders called compounders, with two (twin-screw) or more screws (Fig 3.50). Although there are fewer twin-screw extruders (TS extruders) than single-screw extruders, they are widely employed for difficult compounding applications, devolatilization, and other special operations. For certain plastics they provide required higher pressures and more uniform or higher temperatures than available with single screws.

They can process some plastics of peculiar consistency that interfere with the feeding and pumping processes. The problem is magnified with bulky materials, certain types of emulsion PVC and HDPE, as well as loosely chopped PE film or sticky pastes such as PVC plastisols. The intensive mixing action of multiscrew extruders is highlighted by the fact that adjacent layers of plastic have different speeds, namely, velocity gradients or shear rates. In addition to the actual mixing effect, owing to displacement of layers with respect to each other, pigment particles or ungelled particles of the plastic (fish-eyes) are broken down by its limited friction action, especially in highly viscous plastics.

The conveyance and flow processes of multiscrew extruders are very different from single-screw extruders. The main characteristics of multiscrew extruders include:

1. high conveying capacities at low speeds;
2. positive and controlled pumping rates over a wide range of temperatures and coefficients of frictions;
3. low frictional (if any) heat generation, which permits low heat operation;
4. relatively low motor-power requirements for self-cleaning action with high degrees of mixing;

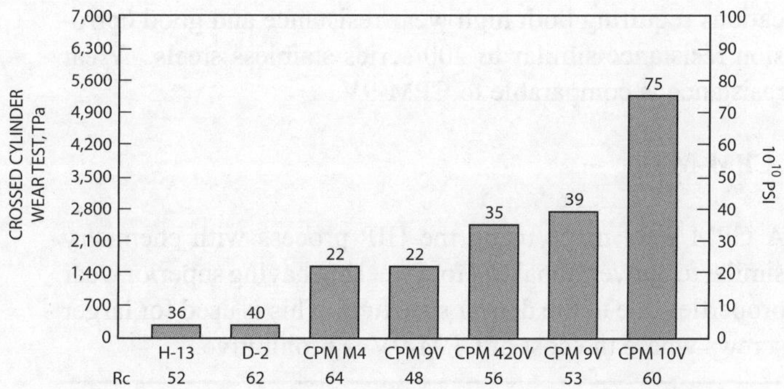

Table 3.37 Examples of wear resistance for different materials (courtesy of Spirex Corp.)

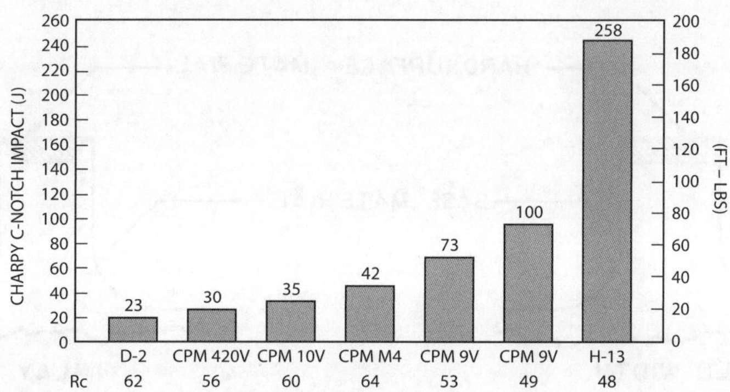

Table 3.38 Examples of toughness for different materials (courtesy of Spirex Corp.)

CPM-10V

A high-vanadium, high-carbon alloy with extreme abrasion resistance. Used for barrel liners and some screws. This alloy exhibits the highest wear resistance available in the particle metal series.

CPM-9V

Slightly less vanadium and carbon giving a little less abrasion resistance but improved toughness over CPM-10V. This is now the standard CPM material for screws by virtue of properties at elevated temperatures.

CPM-420V

This is a high-vanadium, high-chromium tool steel for applications requiring both high wear resistance and good corrosion resistance similar to 400 series stainless steels. Wear resistance is comparable to CPM-9V.

CPM-M4

A CPM steel made using the HIP process with chemistry similar to conventional M4 tool steel but having superior wear properties due to the denser structure. This is used for larger screws where the cost of CPM-9V is prohibitive.

Table 3.39 Examples of CPM products used in plastic machinery components (courtesy of Spirex Corp.)

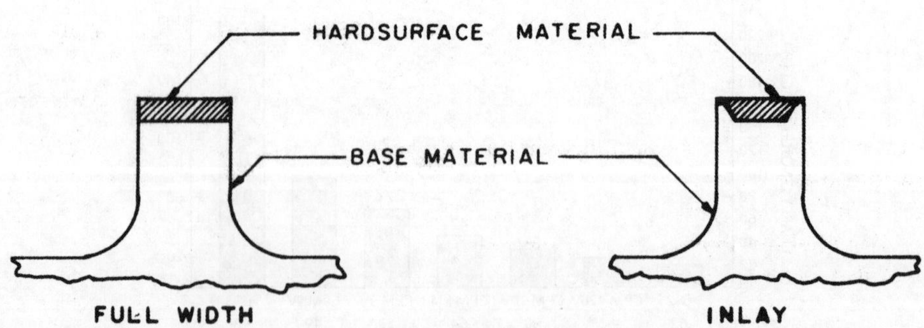

Figure 3.49 Two screw hard surface geometries (courtesy of Spirex Corp.)

Product	Base Material	Typical Hardness Rc	Cracking Tendency	% Carbon	% Chromium	% Tungsten	% Boron	Approx. Cost per Pound ($)
Stellite No. 1	Cobalt	48-54	4	2.5	30.0	12	-	25-40
Stellite No. 6	Cobalt	37-42	3	1.1	28.0	4	-	25-40
Stellite No. 12	Cobalt	41-47	3	1.4	29.0	8	-	25-40
Colmonoy No. 5	Nickel	45-50	3	.65	11.5	-	2.5	15-25
Colmonoy No. 56	Nickel	50-55	4	.70	12.5	-	2.7	15-25
Colmonoy No. 6	Nickel	56-61	5	.75	13.5	-	3.0	15-25
Colmonoy No. 83	Nickel	50-55	4	2.0	20.0	34	1.0	40-50
N-45	Nickel	30-40	3	0.3	11.0	-	2.2	15-25
N-50	Nickel	40-45	3	0.4	12.0	-	2.4	15-25
N-55	Nickel	45-50	4	0.6	13.5	-	2.8	15-25

Table 3.40 Common hard surface materials (courtesy of Spirex Corp.)

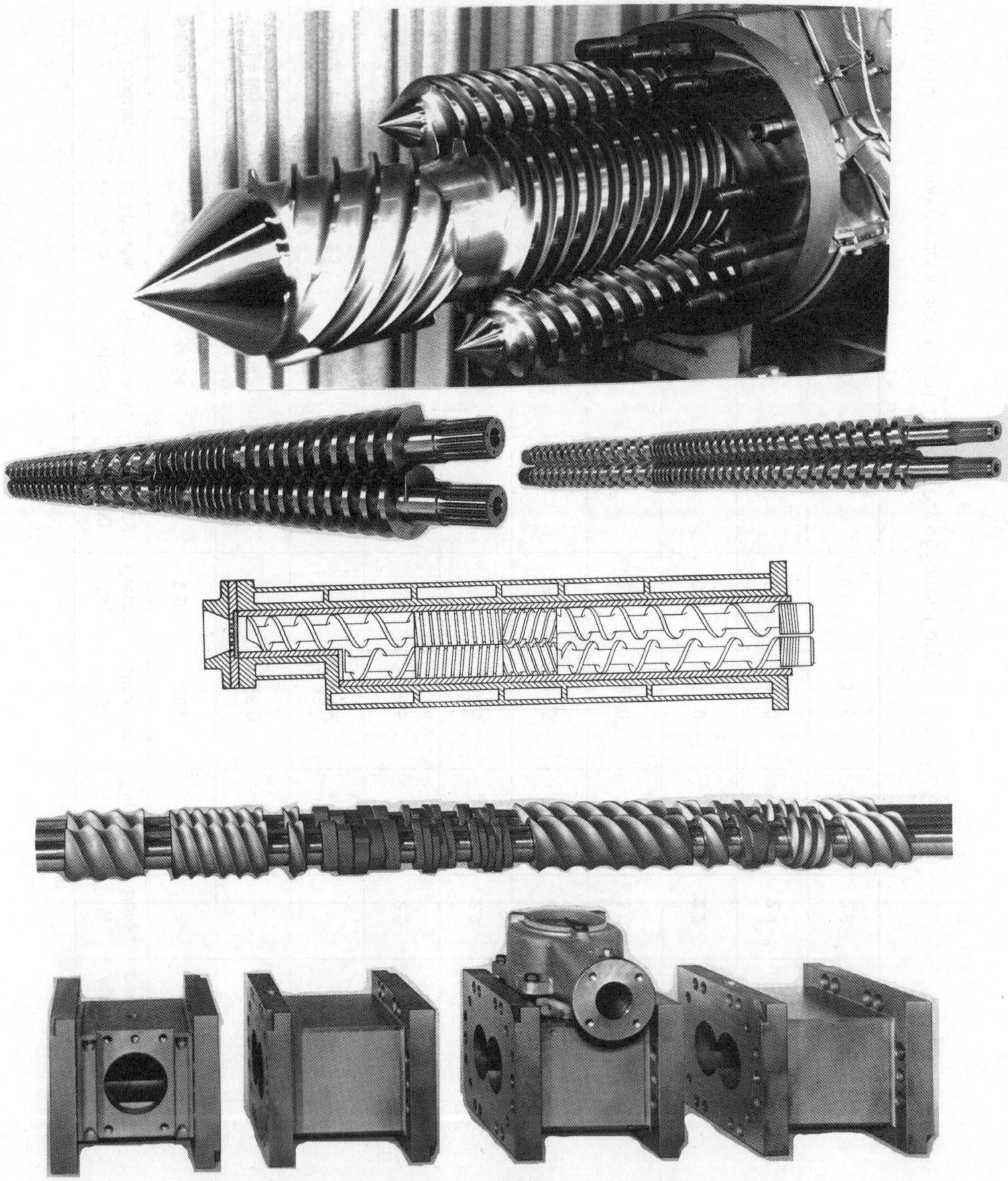

Figure 3.50 Examples of intermeshing multiple screws

5. low contact times in the extruders;
6. positive pumping abilities that are independent of the friction of the plastic against the screws and barrels, which is not reduced by backflow;
7. increased initial cost due to their more complicated construction as well as their more expensive maintenance.

Even though the backflow does not exist, their flow phenomena are more complicated and therefore far more difficult to treat theoretically than single-screw flow. The result has been that the machine designer has to rely mainly on experience.

Figure 3.51 shows the different designs used with the twin-screw (TS) extruders. The market for counter-rotating TS extruders includes the cylindrical-screws (parallel TS) and the extruder fitted with conical screws (Fig. 3.52). The type of design to be used will depend on performance requirements for a specific material to produce a specific product. In the past, major differences existed between corotating or counter-rotating; today, they work equally well in about 70% of compounding applications, leaving about 30% where one machine may perform dramatically better than the other.

Some plastics require very gradual heating using external heaters and sometimes require specific mixing effects not available with single-screw extruders. With these machines, the terms *external* and *internal* are not preferred. They are replaced by the terms *extensive* and *intensive* to avoid confusion with internal mixers. The multiscrew extensive mixing action is characterized by the fact that the plastic constantly changes in flow direction. Each particle moves sometimes on the surface and other times in the body of the mass of plastic. The surface of the mass is constantly renewed, and after a certain time, a statistical distribution of all components is achieved. Typical of these extensive mixers are agitators and most of the kneading machines (Fig. 3.53).

Multiscrew extruders require very exact metered feeding for certain materials; otherwise, output performance will vary. With overfeeding, there is a possibility of overloading the drive or bearings of the machine, particularly with counter-rotating screw designs. For mixing and homogenizing plastics, the absence of pressure flow can be a disadvantage. However, there are many applications where these disadvantages are outweighed by the performance required, such as in compounding (chapter 5). Twin-screw extruders with intermeshing, counter-rotating screws are principally used for compounding, including situations in which volatiles must be removed during plastication. Twin-screw extruders have found a substantial market in difficult compounding and devolatilization processes. To provide specialized compounding and mixing, particularly in the laboratory, different mixing techniques are required, such as using interchangeable screw sections on a splined shaft (Fig. 3.54).

The market for counter-rotating TS extruders is basically dominated by two designs. The parallel TS extruder has cylindrical screws, and the other TS extruder is fitted with conical screws. Regarding performance, the superiority of the conical principle over the parallel does not only appear in the theoretical comparison but in practice as confirmed by users. Flexibility of conical turns results in an extrudate of consistent quality at both low and high output rates and that is not

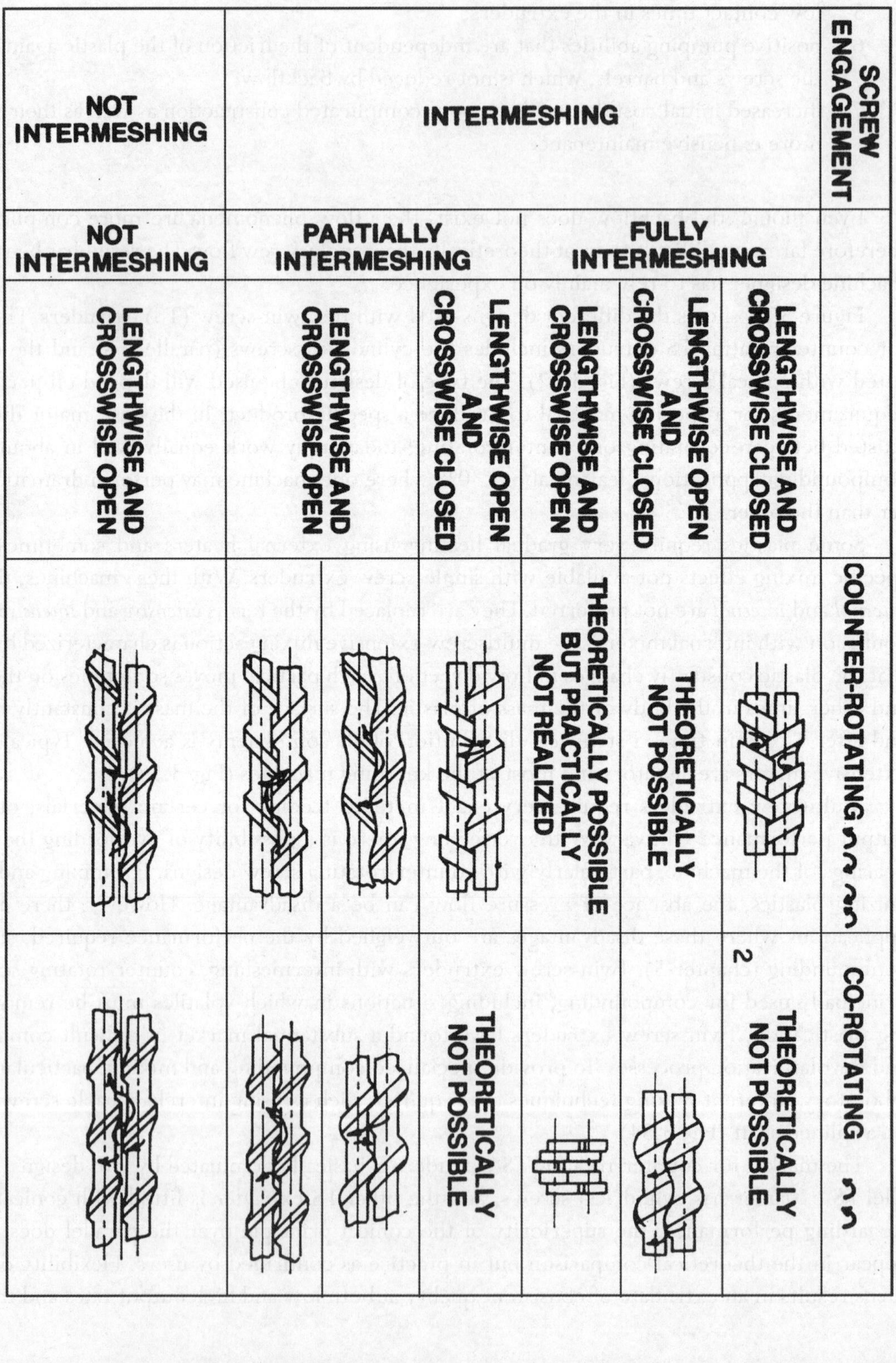

Figure 3.51 Twin-screw operational designs to process different plastic compounds (courtesy of Coperion/Werner & Pfleiderer)

Figure 3.52 Conical twin-screw extruders

sensitive to raw material fluctuations. It appears that the parallel extruder has reached its efficiency limit, unless a means of drastically increasing the screw torsional strength is developed. The conical extruder still offers more benefits through further development.

RECOMMENDED SCREW DIMENSIONAL GUIDELINE

In Tables 3.41 to 3.44, recommendations for single screws for extrusion/injection/blow machinery have been prepared as a guide for manufacturers and processors. These standards provide working tolerances that produce effective performance with economy of manufacture. Information was prepared by the Machinery Component Manufacturers Division of the Society of the Plastics Industry (SPI).

DEFINING/IDENTIFYING SCREW

Tables 3.45 to 3.49 from Spirex Corp. are used by fabricators to order screws to meet their machine requirements and plastic melt processing requirements. These tables provide another means of defining and identifying a screw. When receiving a new machine and/or new screw (or barrel), take measurements such as those in these tables so that you have accurate readings. Later, when inspecting your screw, you can make comparisons to determine if any wear, etc., occurred.

The special screws such as those in Figure 3.55 also have setup sheets.

BARREL

The barrel, also called cylinder or pipe, is used to enclose the screw. This combination provides the control mechanism that targets to produce a uniform plasticized plastic melt of constant composition at the required controllable rate. To achieve this, the barrel must be made very accurately; the total out-of-alignment error, after all machining, must be less than one-half of the screw/barrel clearance. The screw in the barrel provides the bearing surface where shear is imparted to the

(a)

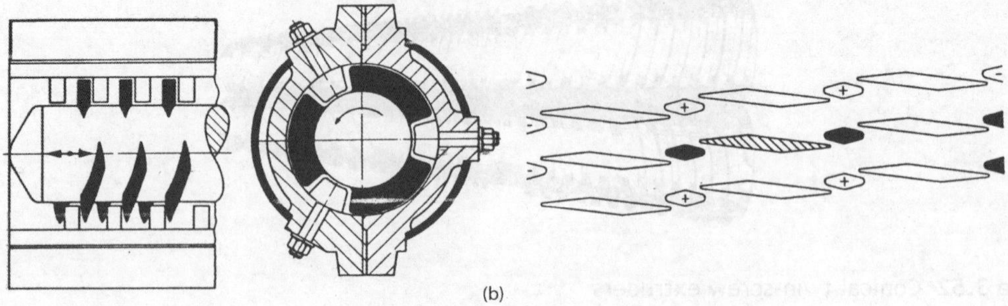

(b)

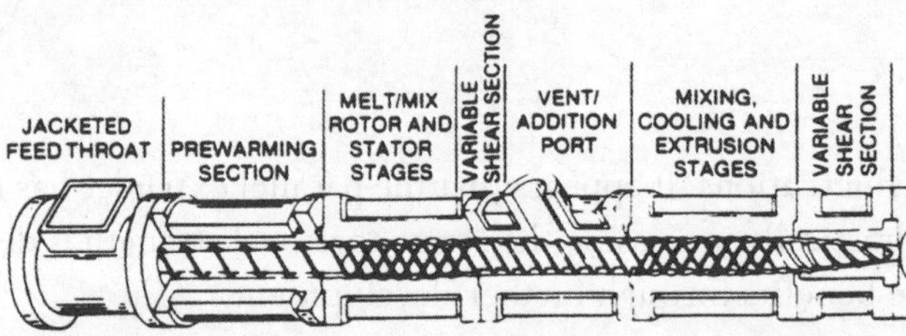

(c)

Figure 3.53 Examples of (a) mixer with screw flights and stationary teeth, (b) concentric screw mixer, and (c) kneader with open split barrel

plastic material. Barrel heating occurs with certain types of cooling media housed around it to keep the melt at the desired temperature profile.

There are many options for barrel material construction, with most extruder barrels designed to withstand up to at least 10,000 psi (69 MPa) internal pressure; higher-pressure units (30,000

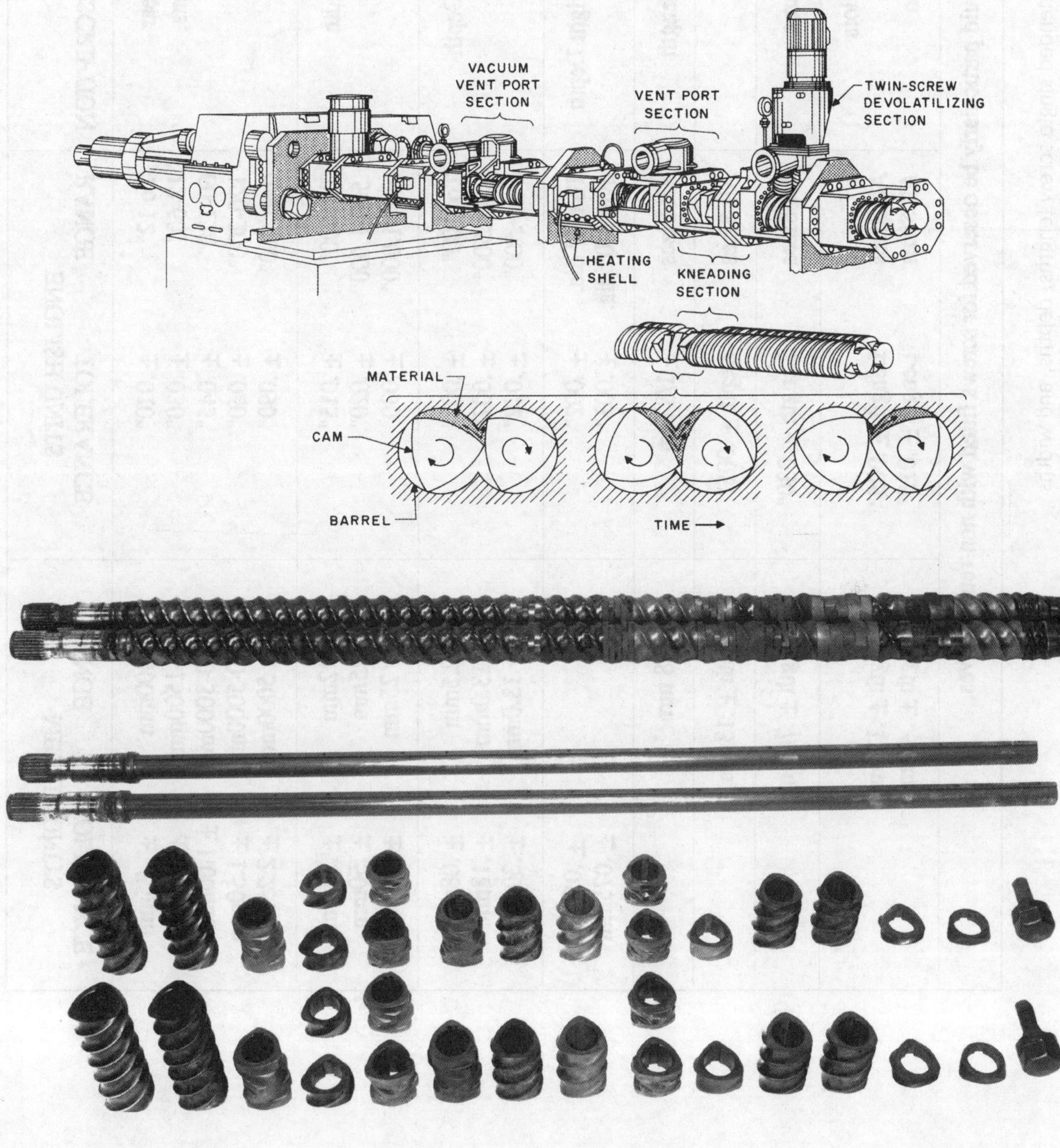

Figure 3.54 Example of using interchangeable screw sections to provide different mixing actions (courtesy of Coperion/Werner & Pfleiderer)

DIMENSION DESCRIPTION		ENGLISH UNITS RANGE	ENGLISH UNITS TOLERANCE	METRIC UNITS RANGE	METRIC UNITS TOLERANCE
I.	Most Linear Dimensions	To 12" 12-60" 60-120" 120-200" over 200"	± .010" ± .030" ± .045" ± .060" ± .090"	To 300mm 300-1500mm 1500-3000mm 3000-5000mm over 5000mm	± .25mm ± .75mm ± 1.00mm ± 1.50mm ± 2.25mm
II.	Flight Width	To .500" .500-1.000" over 1.000"	± .015" ± .020" ± .030	To 12mm 12-25mm over 25mm	± .38mm ± .50mm ± .75mm
III.	Channel Depth	To .100" .100-.500" over .500"	± .003" ± .007" ± .012"	To 2.5mm 2.5-13.0mm over 13.0mm	± .08mm ± .18mm ± .30mm
IV.	Barrier Flight Depth	To 6.0" dia. over 6.0" dia.	± .002" ± .003"		± .05mm ± .075mm
V.	Section Length	All Sizes	± 1/8 turn	± 1/8 turn	
VI.	Keyways	All Sizes	Depth ± .005"	Depth ± .13mm	
VII.	Hollowbores	All Sizes	Length ± .030"	Length ± .76mm	
VIII.	Thread Pilots *Injection Extrusion	 All Sizes All Sizes	 Length ± .005" Length ± .015"	 Length ± .13mm Length ± . 4mm	

*This should particularly be observed for screws fitted with non-return valves.

Table 3.41 Recommended single screw lengths, depths, and widths

DIAMETERS

DIMENSION DESCRIPTION		ENGLISH UNITS RANGE	ENGLISH UNITS TOLERANCE	METRIC UNITS RANGE	METRIC UNITS TOLERANCE
I.	Flight & Bearing (Hub) Diameters	To 6.0" over 6.0"	+ .000-.002" + .000-.004"	To 150mm over 150mm	+ .00-.05mm + .00-.10mm
II.	Shank Diameters (Drive Fit)	All Sizes	+ .0-.002"	All Sizes	+ 0-.50mm
III.	Hollowbores	To 3" 3 to 6" over 6"	± .020" ± .040" ± .060"	To 75mm 75-150mm over 150mm	± .5mm ± 1.0mm ± 1.5mm
IV.	Thread Pilots *Injection Extrusion	 All Sizes All Sizes	 dia. + .001-.000" dia. + .002-.000"		 dia. + .025-.000mm dia. + .050-.000mm

*This should particularly be observed for screws fitted with back-flow valves.

CONCENTRICITY -- OUTSIDE DIAMETERS

Concentricity of outside diameters to include all except root diameter of flighted sections.

ENGLISH UNITS LENGTH	ENGLISH UNITS T.I.R.	METRIC UNITS LENGTH	METRIC UNITS T.I.R.
To 100"	.004"	To 2500mm	.100mm TIR
100-200"	.006"	2500-5000mm	.150mm TIR
200-300"	.010"	5000-7600mm	.250mm TIR
over 300"	.015"	over 7600mm	.400mm TIR

Concentricity of diameters in flighted sections cannot be accurately measured due to the interrupting effect of the flight. Flight depth variations taken from a true O.D. are used as a measure of concentricity in this area.

CONCENTRICITY -- INSIDE DIAMETERS

		ENGLISH UNITS		METRIC UNITS	
I.	Thread Pilots (non-return registers)	All Sizes	.0005" TIR Measured to the flight O.D.	All Sizes	.013mm TIR
II.	Thread Pilots, Threads, and Hollowbores	All Sizes	.001" TIR	All Sizes	.025mm TIR

Inside diameters are measured concentric to all inside diameters and/or the nearest outside flight diameter.

Table 3.42 Recommended single screw diameters and concentricity

Concentricity of Diameters

Concentricity between all inside and outside diameters should be held within .002" TIR or .051mm TIR within one bore diameter of each end. Concentric diameter dimensions further inside of one bore diameter from the ends should be avoided and/or specially tolerated when required.

Straightness

Straightness is generally specified for the bore of the barrel and is measured by TIR.

1 - 3"	bore diameter	(≤ 24 L/D)	.008" TIR
1 - 3"	bore diameter	(≥ 24 L/D)	.010" TIR
3 - 5 1/2"	bore diameter	(≤ 24 L/D)	.010" TIR
3 - 5 1/2"	bore diameter	(≥ 24 L/D)	.012" TIR
5 1/2 - 8"	bore diameter	(≤ 24 L/D)	.012" TIR
5 1/2 - 8"	bore diameter	(≥ 24 L/D)	.015" TIR
8 - 12 1/2"	bore diameter	(≤ 24 L/D)	.014" TIR
8 - 12 1/2"	bore diameter	(≥ 24 L/D)	.016" TIR
25 - 75mm	bore diameter	(≤ 24 L/D)	.203mm TIR
25 - 75mm	bore diameter	(> 24 L/D)	.254mm TIR
75 - 140mm	bore diameter	(≤ 24 L/D)	.254mm TIR
75 - 140mm	bore diameter	(> 24 L/D)	.305mm TIR
140 - 200mm	bore diameter	(≤ 24 L/D)	.305mm TIR
140 - 200mm	bore diameter	(> 24 L/D)	.380mm TIR
200 - 315mm	bore diameter	(≤ 24 L/D)	.355mm TIR
200 - 315mm	bore diameter	(> 24 L/D)	.406mm TIR

To prevent short "kinks" to which the screw could not conform, any two successive measurements taken less than twice the bore diameter apart should have no more than 1/2 the total allowable TIR.

Bar Test

A second way to check both straightness and bore size is by using a precision ground test bar. This method is detailed in the Appendix 1.

Finish

The following surface finishes apply unless specified otherwise.

Outer diameters	125	RMS
Counterbores	32	RMS
Pilots	32	RMS
Cylinder Bores	8-32	RMS
Feed Ports	125	RMS
*Vent Ports	32	RMS

***Required within 1/4" of cylinder bore.**

The bore shall be free from visual surface defects in the bore over its entire length.

Table 3.43 Recommended single screw diameters and concentricity details

psi [210 MPa]) are manufactured for the injection-molding processes since they operate at higher pressures. They have a minimum safety burst pressure of at least 50,000 psi (350 MPa). The need for corrosion and/or wear protection, cost, repair, or their combinations may determine the choice of materials. They can be made from a solid piece of metal. The most common material is carbon steel.

Flight Radii

Unless otherwise specified, the root radius will not be less than 1/2 of the flight depth up to a 1" or 25mm radius.

When specified, the following radius tolerances should be observed: ± .030" ± .75mm

Parallelism & Perpendicularity

All flights, unless otherwise specified, shall be perpendicular to the screw axis from the root radius to the O.D. on both sides.

Other surfaces perpendicular to the screw axis can be tested by use of a surface plate and an adjustable height table indicator or a precision square. The following tolerances should be observed:

All Sizes	**± .001 in/in**	**± .0025 mm/mm**
OR		
All Sizes	**.006 degrees**	**.006 degrees**

Parallel surfaces can be determined by TIR or by using a surface plate and an adjustable height gage. All dimensions meeting the concentricity and/or flight depth standards are acceptable.

Surface Finish

I.	**Unplated Surfaces**	**All Surfaces**	**32 microinch**	**.80 micrometers**
II.	**Chrome Plated Surfaces**	**Channel**	**16 microinch**	**.40 micrometers**
		O.D.	**32 microinch**	**.80 micrometers**
		Bearing/Shank	**32 microinch**	**.80 micrometers**

Screw Threads

The variation in threads used is too broad to be addressed by these recommendations. It is suggested that whenever thread selection is possible that either the ANSI or ISO standards are adhered to for ease of measurement and compatibility.

Hard Surfacing

All hard surfacing materials should be specified as to alloy, width of weld, and depth of weld.

Measurement Temperature

All measurements are to be taken at room temperature of 72 degrees F ± 20 degrees or 22 degrees C ± 11 degrees.

Table 3.44 Recommended single screw details

Name ________ **Title** ________
Company ________ **Address** ________
City ________ **State/Province** ________ **Postal Code** ________
Country ________ **E-mail** ________
Phone ________ **Fax** ________ **Date** ________

SCREW RECOMMENDATION FOR INJECTION PROCESS (Please Supply The Following Information So That SPIREX May Reply To Your Request)

MACHINE SPECIFICATIONS

O.E.M. ________

Model # ________ **Serial #** ________ **Year Mfrd.** ________

Rated Shot Size Styrene ________ **Clamp Tonnage** ________ **Stroke of Injection Unit** ________
Max. RPM ________ **Max. Injection Pressure** ________

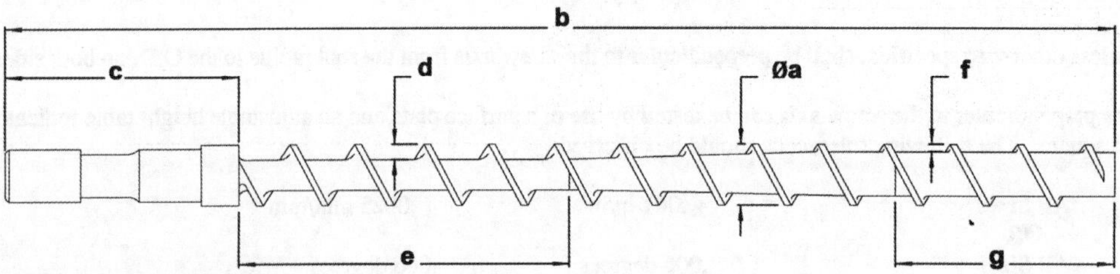

DIMENSIONS

a) ________ **d)** ________ **g)** ________
b) ________ **e)** ________
c) ________ **f)** ________

EXISTING SCREW DATA ☐ **Single Stage** ☐ **Two Stage** ☐ **Other** ________

Screw Construction Materials ________

If there is a particular problem that you are having with a resin or resins, please explain the process conditions. We may have a product(s) to solve your problem.

PROCESS CONDITIONS

Resin(s) Processed ________

Overall Cycle Time ________ **Shot Size (oz., grams)** (including runner) ________ **Screw Recovery Time** ________

Barrel Temperature Settings:

(Set Point): Rear ________ **Center** ________ **Front** ________
(Actual): Rear ________ **Center** ________ **Front** ________

Back Pressure ________ **Screw Speed (RPM)** ________ **Melt Temperature** ________

Melt Decompression? ☐ **Yes** ☐ **No** **If yes, distance** ________ **Scrap Level** ________

Problems with job? List: ________

Table 3.45 Spirex injection screw questionnaire

Name ____ Title ____
Company ____ Address ____
City ____ State/Province ____ Postal Code ____
Country ____ E-mail ____
Phone ____ Fax ____ Date ____

SCREW RECOMMENDATION FOR EXTRUSION PROCESS (Please Supply The Following Information So That SPIREX May Reply To Your Request)

MACHINE SPECIFICATIONS

O.E.M. ____ **Dia.** ____ **L/D** ____ **Model** ____ **Year** ____

Drive HP ____ ☐ **AC** ☐ **DC** **Maximum Screw RPM 0-**____

Motor Wired Voltage and Maximum Motor Amperage ____ ☐ **Water cooled barrel or** ☐ **Air cooled barrel**

Vented? ☐ **Yes** ☐ **No** **Location of Vent from Back of Feed Throat** ____

Screen Changer? ☐ **Yes** ☐ **No**

PROCESS DATA/OUTPUT DATA

☐**Sheet** ☐**Blown Film** ☐**Cast Film** ☐**Pipe** ☐**Profile** ☐**Blowmolding** ☐**Pelletizing** ☐**Wire & Cable** ☐**Compounding** ☐**Medical Tubing** ☐**Other** ____

Resin Type(s) ____ ☐ **Pellets** ☐ **Powder** **Bulk Density** ____
____ ☐ **Pellets** ☐ **Powder** **Bulk Density** ____

Is resin drying used? ☐ **Yes** ☐ **No** **If yes,** ____ **°F for** ____ **Hours**

Is a melt pump being used? ☐ **Yes** ☐ **No** **Grooved Feed Throat?** ☐ **Yes** ☐ **No** **Groove L/D?**____

Groove Geometry? ____ **Groove Depth?** ____ **Groove Width?** ____ **# of Grooves?** ____

Melt Flow Characteristic (MI, MFR, IV, Durometer) ____

% Regrind/Blend ____ **Color Concentrate or Fillers** ____

Present Output (lbs/hr) ____ **@ the following rate specific results:** ____

Melt Temp. (°F) @ Die Exit. ____ **PSI @ Exit** ____ **Screw RPM** ____ **Motor Amps** ____

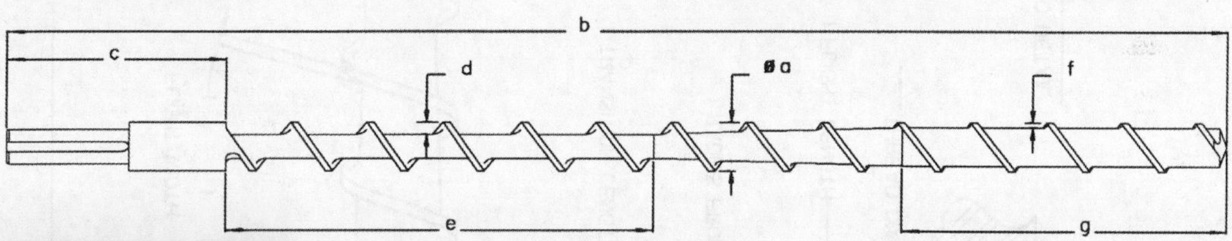

EXISTING SCREW DATA

☐ **Single Stage** ☐ **Two Stage** ☐ **Other** ____

DIMENSIONS

a) ____ **d)** ____ **g)** ____
b) ____ **e)** ____
c) ____ **f)** ____

Screw Cooling: ☐ **Water** ☐ **Oil** **Depth:** ____ ☐ **None**

Design: ☐ **Use Existing** ☐ **Modify Existing** ☐ **Supply New** **Ref. Job/Design** ____

Describe Current Screw Design and Processing Problems That Require Modified or New Screw Design: ____

Table 3.46 Spirex extrusion screw questionnaire

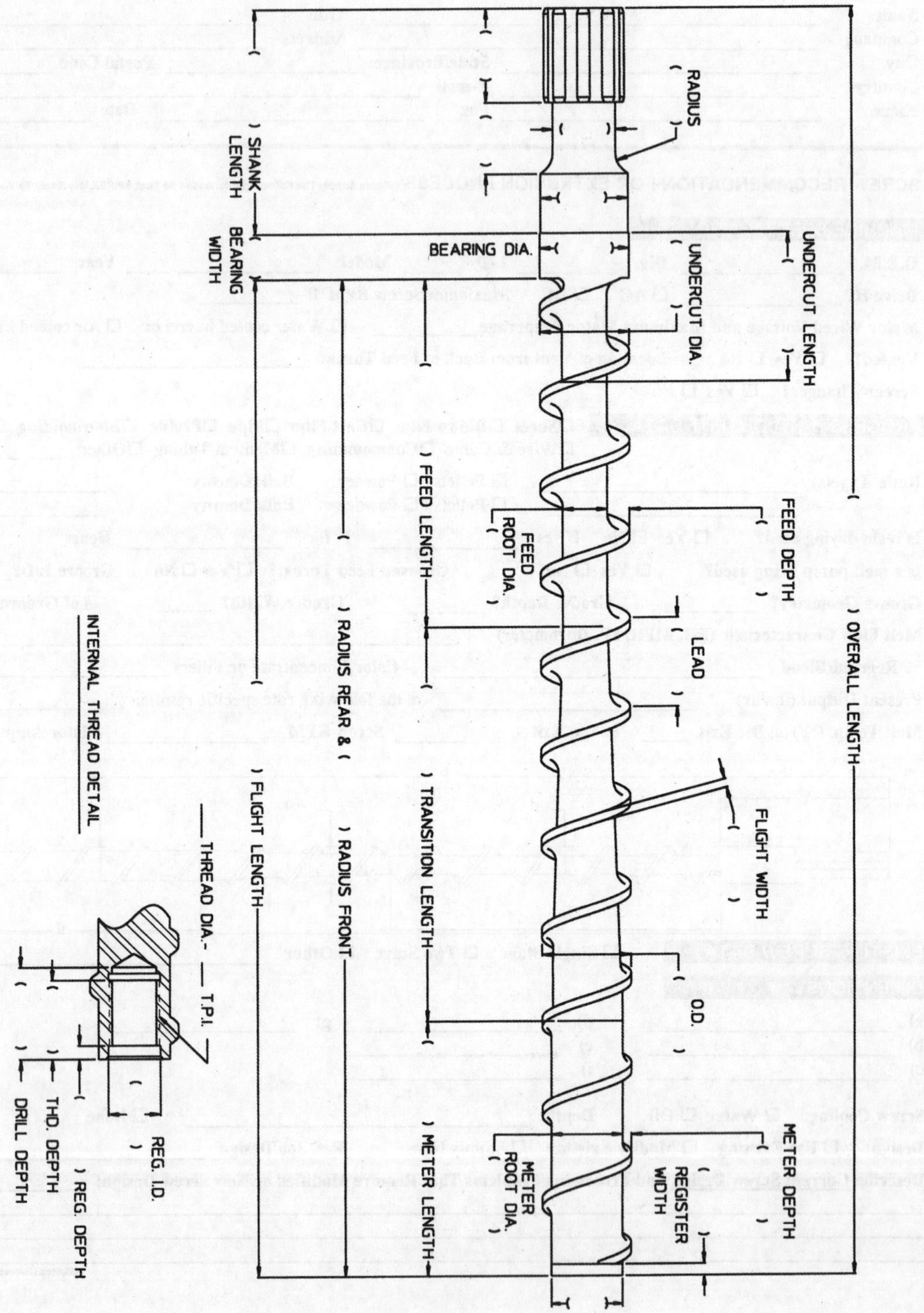

Table 3.47 Spirex injection screw sketch

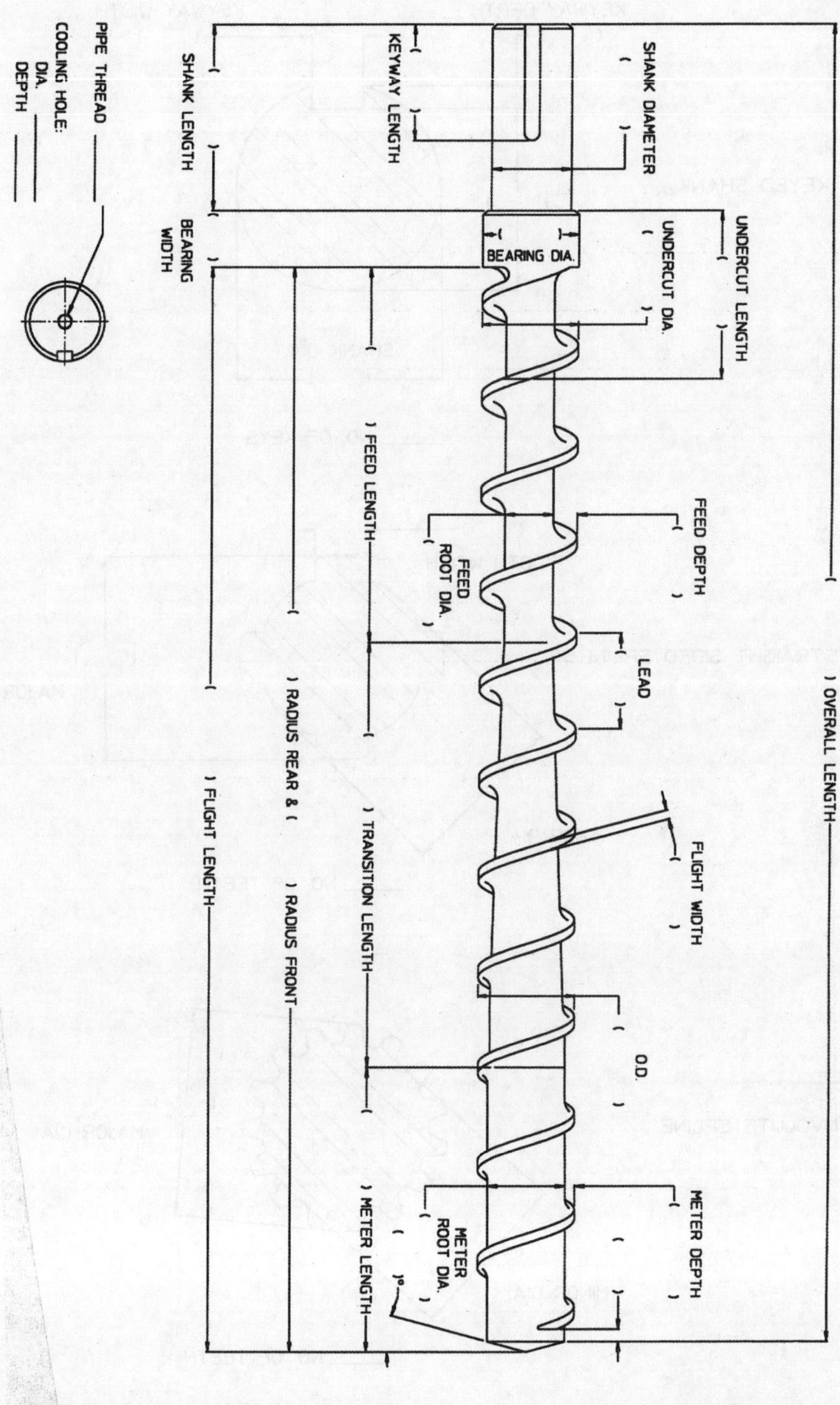

Table 3.48 Spirex extrusion screw sketch

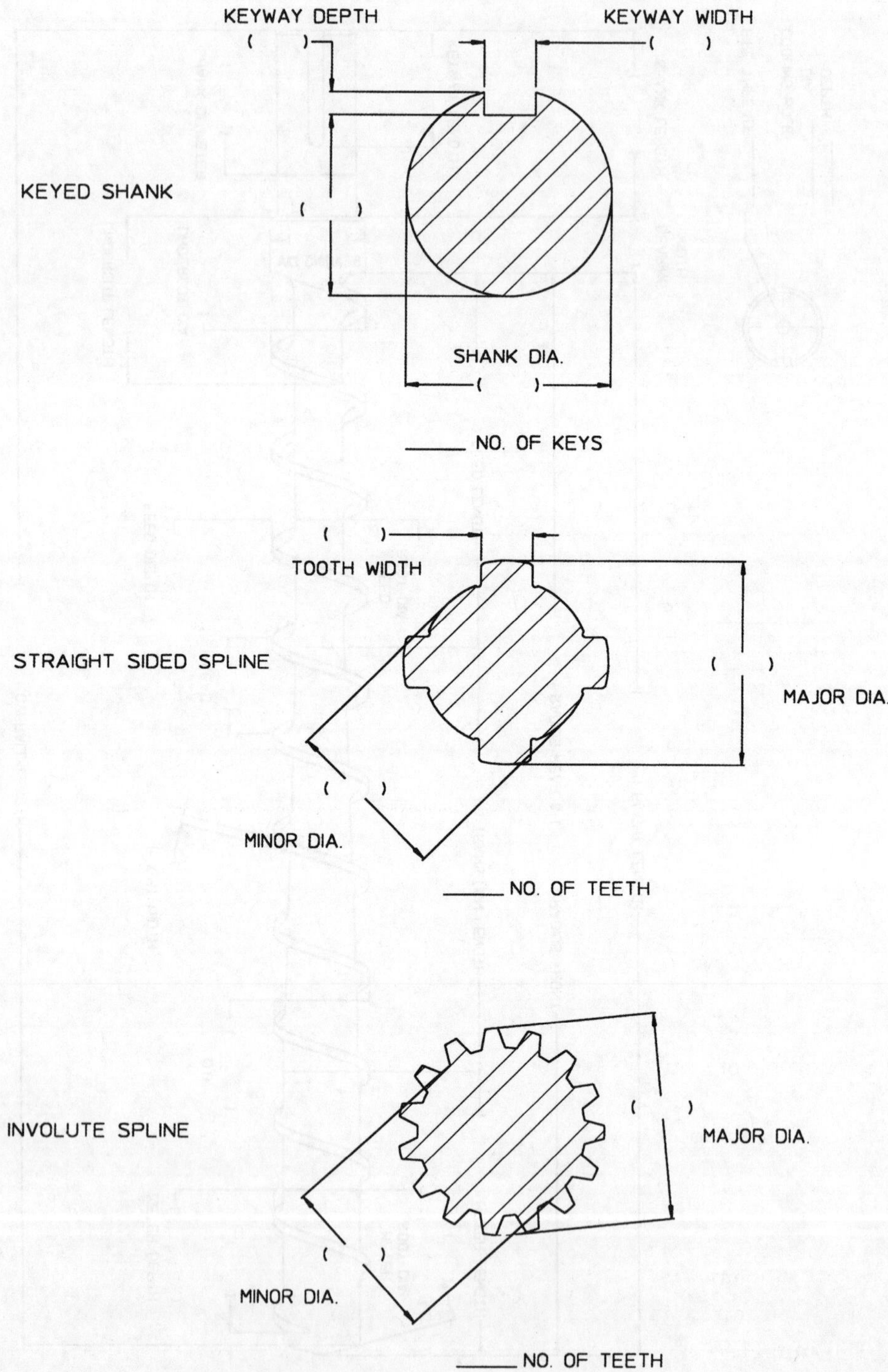

Table 3.49 Spirex screw drive ends

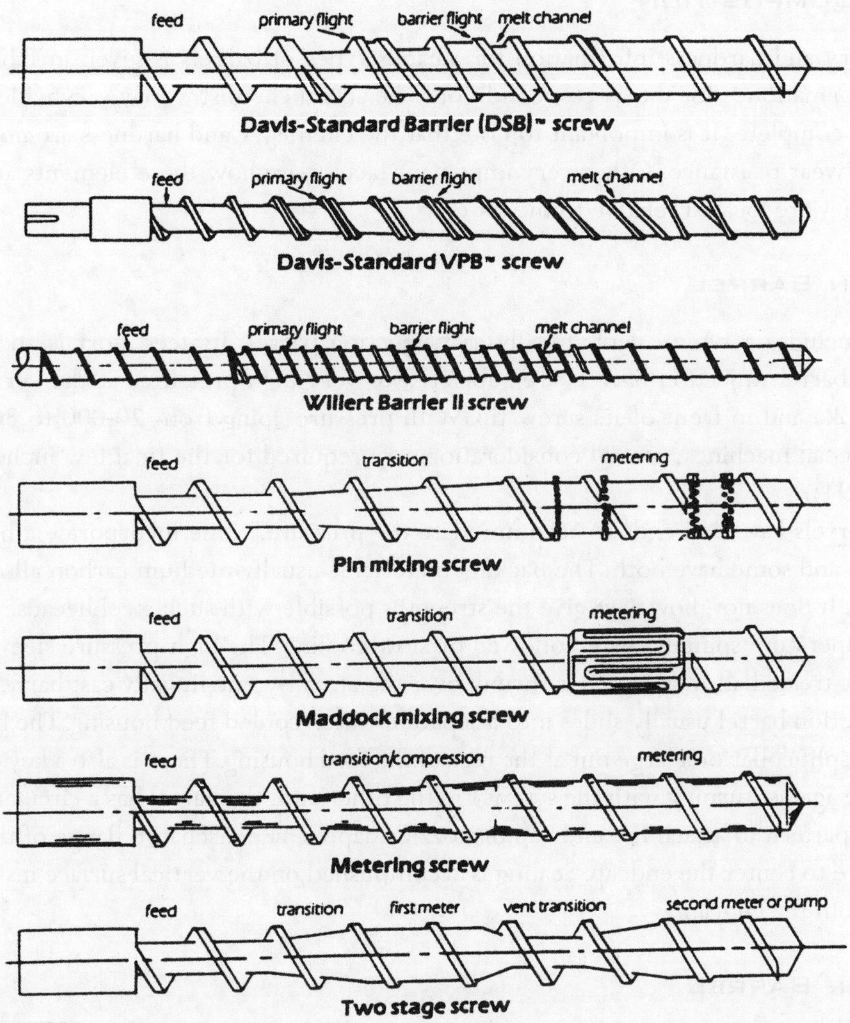

Figure 3.55 Example of special screws

The barrel's ID (inside diameter) with its length classifies sizes. It is common practice to refer to the L/D ratio, that is, barrel length (L) to hole diameter (D) ratio. (As reviewed, there is also a screw length-to-diameter [L/D] ratio.) For low output, such as filament or profile extrusion, 40- to 60-mm (1.6- to 2.3-in) diameter extruders are normally used, whereas for sheet extrusion, 120- and 150-mm (4.7- and 5.8-in) diameter screws are more common. Injection-molding barrel diameters are approximately the same, with the smaller diameters providing the smaller melt shot size and the larger diameters providing the larger melt shot size.

BARREL COMPOSITION

The chemistry and hardness information for various types of barrels is given in Table 3.50. The chemical information is for the as-cast condition. The actual chemistry may vary widely after final machining is complete. It is important to note that the chemistry and hardness are not necessarily indicative of wear resistance. Other very important factors are how these elements are combined and where they are located relative to the bore.

INJECTION BARREL

Figure 3.8 includes a typical centrifugally cast injection barrel. Its feed port is an opening cut through the barrel. Injection barrels are subjected to very high pressures at the very end of the injection stroke and in front of its screw tip. With pressure going from 20,000 to 30,000 psi or higher on special machines, special considerations are required for the final few inches at the discharge end (94).

Some barrels have bell ends to accommodate the pressure, some incorporate a high-pressure outer sleeve, and some have both. The backing material is usually medium carbon alloy steel, such as AISI 4140. It does not, however, give the strengths possible with such steel because it anneals in the high-temperature spinning oven followed by slow cooling. The high-pressure sleeve is made of stronger heat-treated alloy steel and is shrunk over the already centrifugally cast barrel.

The injection barrel usually slides into a separate water-cooled feed housing. The barrel is held in place by a split collar or a large nut at the rear of the feed housing. There is also a key or machined flat to ensure against turning with the screw. On the other end, the barrel has a circular flange with a bolt circle pattern to attach the end cap or nozzle adapter. The discharge flange of the barrel has a counterbore to center the endcap. Sealing is accomplished on the vertical surface inside the pilot, extending from the endcap.

EXTRUDER BARREL

When compared with injection barrels, the trend for extruders has been toward the use of longer barrel lengths. In the past, L/D ratios of 24/1 were the rule. Now 30/1 and longer are frequently specified. Injection-molding L/D ratios are usually up to 24/1. These longer barrels permit more melt inventory time; consequently, a higher output rate is possible without sacrificing the plasticization rate or reducing the homogenization of the melt. To take full advantage of the increased output rates of longer barrels, a more powerful drive unit should be used; actually, the power requirement is based on the plastic-melt torque requirement. Typically for extruders, 150-hp motors are used on 120-mm (3.2-in) 24/1 L/D extruders. For a 30/1 L/D, 200 hp are frequently specified. Smaller 60-mm (1.6-in) machines have drive motors of up to 50 hp.

In addition to the extrusion barrels being different from injection barrels in L/D, they are designed to withstand lower pressures of 10,000 psi (maximum). This means a thinner wall and eliminates the high-pressure sleeve or bell end. Extrusion barrels normally do not have an integral

	ABRASION RESISTANT						
Chemical Composition % by weight	NEW CASTLE AR1000	WEXCO 666	XALOY 101	HITACHI H-10	REILOY 112	WBC 60	IDM 200
Carbon	3.2-3.6	3.2-3.5	2.0-4.0	1.0-3.0	3.0-4.0	3.0-4.0	<1
Manganese	1.4-1.9	1.4-1.7	2.0 max	2.0 max	-	2.0 max	-
Silicon	.75-1.2	0.8-1.1	1.5 max	2.0 max	1.0 max	2.0 max	5
Boron	.90-1.3	.95-1.2	0.5-1.2	0.5-2.0	2.0-2.5	1.0-1.5	3
Chromium	.50-.85	-	0.7-1.4	-	1.0-2.0	.75-1.5	10-13
Iron	Balance	Balance	Balance	Balance	Balance	Balance	4-6
Nickel	4.5-6.0	4.6-5.2	2.5-5.0	2.0-7.0	3.0-5.0	4.0-5.5	72-77
Cobalt	-	-	-	-	-	-	-
Tungsten	-	-	-	-	-	-	-
	CORROSION RESISTANT						
Chemical Composition % by weight	NEW CASTLE CR3000	WEXCO 555	XALOY 306	HITACHI H-503	REILOY 115	WBC 50	IDM 260
Carbon	0.2-0.5	.25-.40	.35 max	0.3 max	0.1-0.3	.38 max	<1
Manganese	.45-1.1	0.5-1.5	1.0 max	2.0 max	-	1.5 max	-
Silicon	1.0-2.5	1.0-2.5	1.5 max	2.0-10	1.0-2.0	3.0 max	5
Boron	3.0-3.5	2.5-3.2	2.0-.0	2.0-10	3.0-4.0	2.5-3.5	3
Chromium	7-11	7-10	5-10	5-10	7-8	7-11	14
Iron	-	1 max	5-25	5-25	<1	0.5 max	<1
Nickel	38-44	39-43	28-40	Balance	Balance	35-43	76
Cobalt	Balance	Balance	Balance	5-35	34-37	Balance	-
Tungsten	-	-	-	-	-	-	-
	CARBIDE PARTICLE						
Chemical Composition % by weight	NEW CASTLE QC9100	WEXCO 777	XALOY 800	HITACHI H-70M	REILOY 130	WBC Titan	IDM Ultramax®
Carbon	.16-.35 max	5.0	1.3-3.5	5.0 max	3.0-4.0	4.0	.02-.03
Manganese	.34-.75	-	1.0 max	2.0 max	-	-	-
Silicon	.75-1.9	1.5-2.9	0.5-3.3	4.0 max	1.0-3.0	1.0-4.0	.02-.03
Boron	2.3-2.9	1.0-2.3	1.0-3.0	4.0 max	2.0-4.0	1.0-3.0	-
Chromium	4.5-7.5	5.1-9.5	5 max	1-7	8-13	5-10	.03-.05
Iron	-	1.5-2.9	5-25	5-25	Balance	4.0 max	.01
Nickel	29-35	28-40	Balance	Balance	4-6	Balance	Balance
Cobalt	29-42	-	3-7	2-15	-	2-5	.05-.02
Tungsten	*	46-57	25-50	40-60	**	45-60	45-55

*New Castle QC9100 contains a mixture of the following metal carbides: vanadium carbide, tungsten carbide, titanium carbide, and tantalum carbide. The total weight of metal carbides is up to 27%.

**Reiloy 130 contains a mixture of vanadium carbide and iron-boride as well as iron-chromium-carbon-boride.

Table 3.50 Barrel material of construction (courtesy of Spirex Corp.)

feed port but are fastened to a separate water-cooled feed throat casting. At the rear, there is a large-diameter flange with a bolt circle to attach the barrel to the feed throat.

The extrusion barrel connects to the die adapter at the discharge end, but the seal is made in a manner slightly different than the injection barrel and endcap (chapter 5). The barrel flange has a female counterbore just like the injection barrel, but the die adapter has a counterbore instead of a pilot. The counterbore is called a breaker plate recess and it allows space for a breaker plate or scaling ring to make the seal. Attachments for the die adapter are designed for more rapid removal than in injection. This allows for ease of screen changes, screw changes, and die changes. The most common closures are split C clamp, swing gate, or bolt circle.

Wear-Resistant Barrel

As it is easier to replace a screw than a barrel, the barrel is made to be harder than the screw. The barrel-and-screw assembly operates in an aggressive environment that can cause minimal to severe wear problems. To improve the wear resistance of the barrel, it may be modified or lined. Modification may be by nitriding or ion implantation, but these treatments are not as good as lining. Lining is done with a wear-resistant alloy. This wear-resistant layer may be cast during barrel manufacture, or the liner may be inserted subsequently or for rebuilding a worn barrel. These bimetallic barrels are used when abrasive compounds are being processed.

Nitrided barrels are almost never used on U.S. extruders, and few are used on domestic injection machines. Nitriding gives a very thin hardened layer of only about 0.3 mm (0.01 in). Nitrided barrels do not wear as well as bimetallic barrels, and because the case hardening is thin, wear becomes progressive after it starts. In general, corrosion resistance is not as good as any of the bimetallic grades. Nitrided barrels are a very poor selection for machines that will run large amounts of glass fiber or other abrasive filled materials. Honing to a uniform oversize is not usually economical because of the high cost of renitriding.

Wexco 666, New Castle AR-1000, WBC 60, IDM 200, and Xaloy 101 are the standard grades of general-purpose abrasion-resistant bimetallic cylinders. They are grades used for most standard extrusion or injection applications. They are very similar or identical in chemical composition. Unless otherwise specified, these are the grades implied when someone refers to bimetallic barrels. A typical composition for this type of liner is carbon 4 wt%, nickel 4%, boron 1%, molybdenum 1%, and the balance iron (94).

The increasing use of plastics with abrasive fillers created a demand for an even more abrasion-resistant barrel than the standard iron/boron type. The use of glass-reinforced compounds for injection molding has been the single most important factor for more resistant barrels, since a fabricator would be lucky if they could reach four to six months of operation. This need has been successfully answered by the development of liner materials containing metallic carbides, such as tungsten carbide and titanium carbide.

These finely divided carbides are suspended in a matrix of nickel, cobalt, chromium, boron, or some combination of these corrosion-resistant materials. Some domestic designations for carbide-type bimetallic barrels are Wexco 777, New Castle QC-9100, WBC Titan, IDM Ultramax, and Xaloy 800. It has been determined that the wear resistance of Spirex's CPM-10V is comparable to any of the centrifugally cast bimetallic grades. They cost more as an initial investment, but they will outlast the general-purpose type many times over. They are also more corrosion resistant than the standard bimetallic barrels. If a lot of glass-filled compounds are expected to be processed, the carbide type is to be used. Selecting a proper screw material also is critical with these carbide-type barrels. Some of the screw hard surfacing materials, such as Stellite 6, will pick up or gall when used in combination with some of these barrel materials. This will cause catastrophic wear. It is best to consult with a plasticator manufacturer before using an unproven combination (94).

Corrosion-Resistant Barrel

The corrosion-resistant grades are usually nickel-/cobalt-based steels. These liners are recommended for use with corrosive materials, such as rigid PVCs, Sarans, and some fluorocarbons. They give improved chemical resistance but sacrifice some wear resistance. Chemically resistant grades offered by domestic suppliers are Wexco 555, New Castle CR-3000, WBC 50, IDM 260, and Xaloy 306. A typical composition for a corrosion-resistant grade is nickel 40 wt%, chromium 8%, boron 4%, and the balance cobalt.

Solid barrels can also be fabricated from stainless steel with a hardened interior. However, the hardening process often lessens the corrosion protection aspects of stainless steel. Other negative aspects to solid stainless steel are poor heat transfer and higher cost. Therefore, this is usually an option for smaller machines of less than 60 mm (2.3 in). With castings applied to the interior in place of hardening, a superior wear and/or corrosion protection occurs, but at a higher cost.

Barrel Feed Throat

When selecting a plasticator barrel, the size and shape of the feed throat is very important. Feed-throat size and shape can have significant effects on output and the barrel's stability. In general, the smaller the hole, the more adverse the effect. Sometimes, smaller feed holes can be compensated by screw designs, but more often, the feed hole geometry must be modified.

The feed throat is the section in a barrel where plastic is directed into the screw channel. It is fitted around the first few flights of the screw. Some do not have a separate feed throat. In these machines, the feed throat is an integral part of the barrel, even though it may not be the best design approach. The feed-throat casting is generally water cooled to prevent an early temperature rise of the plastic.

The feed throat is cut through the barrel wall. The size of the feed opening is approximately the same size as the barrel diameter. At the other end of the barrel, provisions are made to attach the extruder hardware or injection-molding machine nozzle by means of a clamping system or endcap.

Output rates have been observed to vary as much as 25 wt%, with the only variable being the feed-throat geometry. Round feed throats are sufficient for 100% pellet feed, but when 20% or more regrind is added with the virgin feed, rate is reduced; a rectangular or oblong opening will improve the feeding characteristics. An elongated opening also helps in eliminating potential bridging problems in the throat.

Some materials, such as flakes, regrind, and others, have poor flow, so they present their own problems that can be resolved. The feed-throat and feed-hopper units are important in ensuring that plastics are properly plasticized.

Feed zone cooling and/or temperature control is very important. Until recently, the feed zone was simply kept cool by circulating water through channels in the feed throat; if this was not done, then the hopper could become bridged (blocked with plastics). Now, for the processing of plastics, particularly engineered plastics (chapter 1), the temperature of the feed zone is controlled at a relatively high level so as to improve plasticization and/or output rates.

Barrel Grooving

The most positive action that can be done to improve feeding is the installation of axial grooves in the feed area. Grooves in the internal barrel surface in the feed section, particularly for certain materials, permit considerably more friction between the solid plastic particles and the barrel surface. The result is that the process output increases and/or the process stability improves. On the inside surface of the barrel, grooves will make a very dramatic change to the sliding characteristics.

Barrel grooves have been around for a long time but have recently become more important. Many European extruders and blow molders use these grooves as standard. Some grooving systems require special metal carbide in the feed sections, special high-capacity cooling in the grooved areas, and specialized screw designs.

Barrel grooves can be specified when ordering a new machine when it is anticipated that difficult materials will be processed. Unfortunately, this is usually learned after the fact. Grooves can usually be installed on an existing machine to solve a feed problem. However, it is important to determine if the problem is a feed problem or a melting problem. If the problem is melting, grooves will only make it worse.

The usual method for installing these grooves has been to remove the barrel from the machine and ship it to a company that can do the job. This involves the costs for disassembly, crating, two-way shipping, reinstallation, plus the actual grooving cost. It can also involve the loss of the machine for periods up to a month. Spirex has invented and patented a machine (Patent #5,018,917) that will install these grooves in the processor's plant while the barrel is still installed on the machine. This can usually be done in one shift.

Barrel Heating and Cooling Method

Heat to soften the plastic is supplied in two ways: by external barrel heating and internal frictional forces brought about on the plastic via the action of the metal screw in the metal barrel. The amount of such frictional heat supplied in the plasticator is appreciable. In many extrusion operations, it represents most of the total heat supplied to the plastic.

To provide temperature control, the barrel is divided into zones. Each of these zones is fitted with its own heating/cooling system for extruders; injection-molding machines are usually fitted with just heating zones (Fig. 3.56). The smallest machine will have three zones and larger machines may have twelve. Examples for different plastics' temperature profiles are shown in Figure 3.57. A temperature sensor and associated equipment, usually a microprocessor-based (three-term [PID]) controller, are fitted for each of the zones. The purpose of these controllers is to measure the melt temperatures, which are more important than the cylinder temperatures (which interrelate with melt temperatures).

There are three principal methods of barrel heating: electrical, fluid, and steam. Electrical heating can cover a much larger temperature range. It is clean, relatively inexpensive, and efficient. The heaters are generally placed around and along the barrel, grouped in zones. Each zone is usually controlled independently so that the desired temperature profile can be maintained along the barrel. Electrical heating is generally preferred because it is the most convenient, responds rapidly, is easiest to adjust, is easy to clean, requires a minimum of maintenance, covers a much larger temperature range, and is generally the least expensive in terms of initial investments.

Fluid heating, such as the use of heated oil, allows an even temperature over the entire heat-transfer area, avoiding local overheating. If the same heat-transfer fluid is used for cooling, an even reduction in temperature can be achieved. The maximum operating temperature of most fluids is relatively low for processing TPs, generally below 250°C (482°F). With its even temperature, the required fluid heating is desirable with TS plastics so that no accidental overheating occurs to chemically react and solidify in the barrel.

Steam heating was used in the past, particularly when processing natural rubber; now, it is rarely used. Steam is a good heat-transfer fluid because of its high specific heat capacity, but it is difficult to obtain steam to the temperatures required for TP processing of 200°C (392°F) or greater.

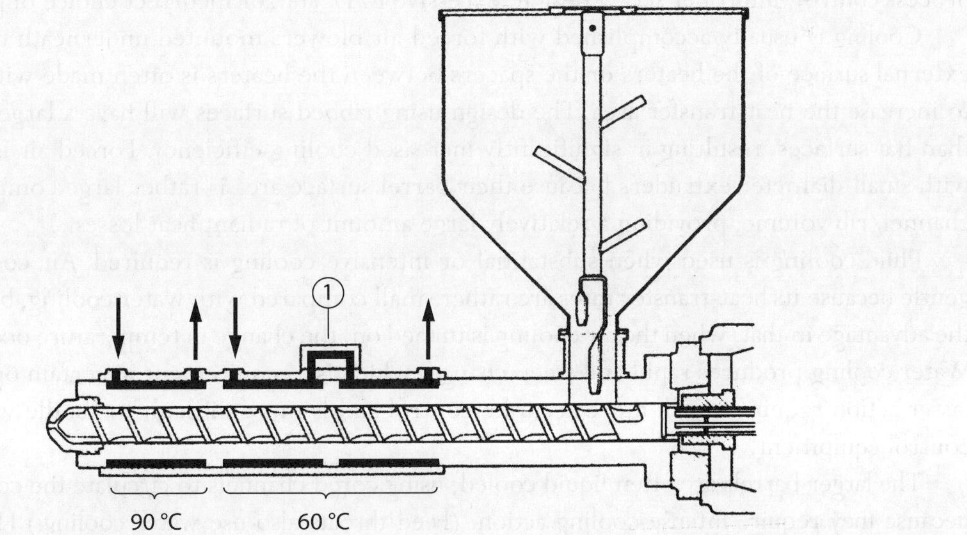

Figure 3.56 Injection-molding machine using hot water zones for heating thermoset plastics (courtesy of Negri Bossi)

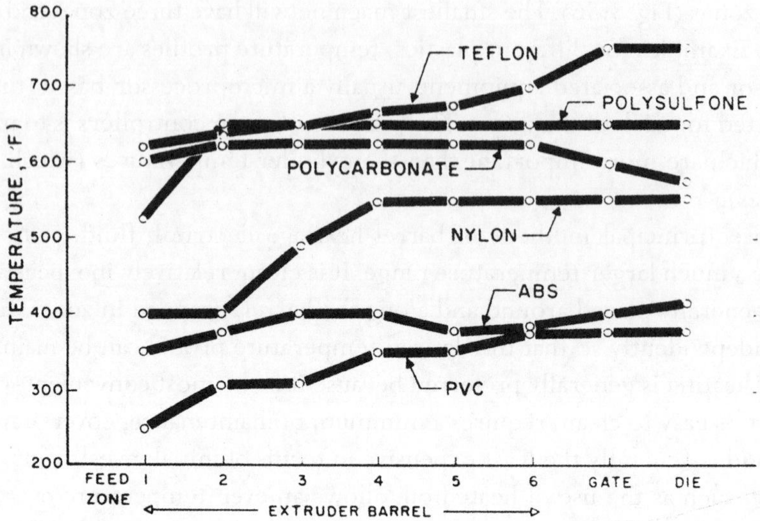

Figure 3.57 Examples of different plastics' temperature profiles (courtesy of Plastics FALLO)

The cooling of barrels is an important aspect. The goal is to minimize any cooling and, where practical, to eliminate it. In a sense, cooling is a waste of money. Any amount of cooling reduces the energy efficiency of the process because cooling directly translates into lost energy; it contributes to a machine's power requirement. If a machine requires a substantial amount of cooling when compared with other machines, it is usually a strong indication of overheating the plastic, improper process control, improper screw design, excessive L/D, and/or incorrect choice of plasticator.

Cooling is usually accomplished with forced-air blowers mounted underneath the barrel. The external surface of the heaters or the spacers between the heaters is often made with cooling ribs to increase the heat-transfer area. The design using ribbed surfaces will have a larger cooling area than flat surfaces, resulting in significantly increased cooling efficiency. Forced air is not required with small-diameter extruders because their barrel surface area is rather large compared with the channel/rib volume, providing a relatively large amount of radiant heat losses.

Fluid cooling is used when substantial or intensive cooling is required. Air cooling is rather gentle because its heat-transfer rates are rather small compared with water cooling, but it does have the advantage in that, when the air cooling is turned on, the change in temperature occurs gradually. Water cooling produces rapid and steep change, which is a requirement in certain operations. This faster action requires much more accurate control and is more difficult to handle without proper control equipment.

The larger barrels are often liquid cooled, using cored channels to circulate the cooling medium because they require intense cooling action. (Feed throats also use water cooling.) However, if not properly controlled, problems could develop. If the water temperature exceeds its boiling point, evaporation can occur, resulting in poor cooling control. The water system is an effective way to extract heat, but it can cause a sudden increase in cooling rate, resulting in a nonlinear control problem and thus more difficulty in regulating temperature. Nevertheless, the water-cooling approach is used very successfully with adequate installation in the extruder and adequate control and start-up procedures.

IMMs for processing thermoset plastics and thermoset rubbers control the barrel temperature most of the time indirectly with an external heat exchanger. It is operated by a liquid heat-transfer medium, such as oil or brine. Curing of the plastic or rubber occurs in the mold cavity(ies) by the application of higher heat than what exists in the barrel melt. A chemical cross-linking action occurs, with the additional heat resulting in the solidification of the thermoset materials (chapter 1).

Depending on the IMM operation capability as well as the type of plastic being processed, the melt passing from the plasticator through the nozzle may require heat control. This action is usually required when processing certain heat-sensitive plastics and/or if a long nozzle is used. Examples of barrel and melt temperatures for plastics are shown in Figures 3.58 to 3.61.

Barrel Temperature Override

The screw/barrel combination tends to be a complex heat-transfer system. To understand something as simple as a zone override can require a complete analysis of the system. Just a few of the factors that can cause a zone override are screw design, barrel mass, thermocouple placement, heating and/or cooling jacket fit, barrel/screw wear, head pressure, overall melt temperature profile, defective temperature controller, and inadequate cooling.

Before assuming that zone override is strictly a screw design problem, analyze the system as a complete heat-transfer mechanism. Although the screw is responsible for most of the heat input, it cannot control the heat distribution in the equipment.

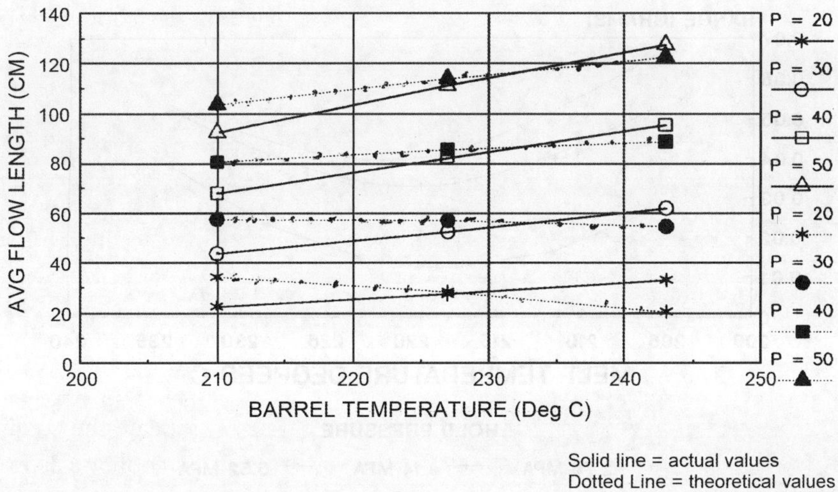

Figure 3.58 Average melt flow length vs. barrel temperature for general polystyrene

BARREL MACHINING OF HOLE

The various barrel holes required for measurement and control are mostly standard. They are shown in Figure 3.62. Holes have been designated as Type A and Type B (94). Type A holes are the blind holes used for measuring temperature and controlling thermocouples. Type B holes extend through the inner wall of the barrel or adapter. Type B holes are used for flush melt thermocouples, immersion melt thermocouples (not in an area wiped by the screw or valve), pressure transducers,

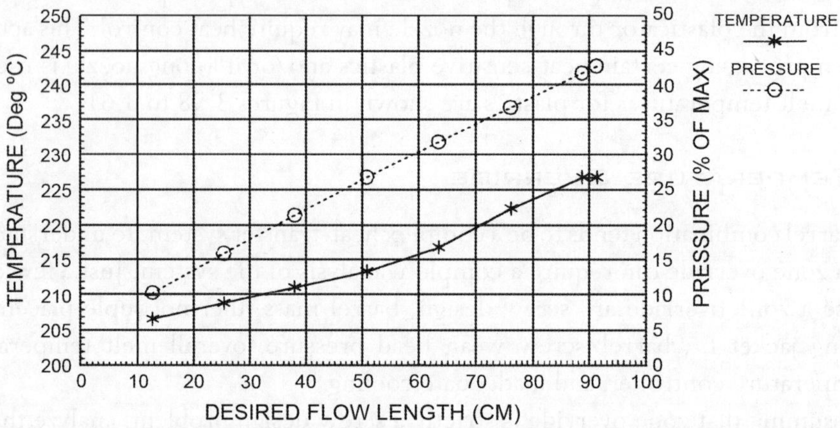

Figure 3.59 Optimum barrel temperature and injection pressure to minimize variation in length

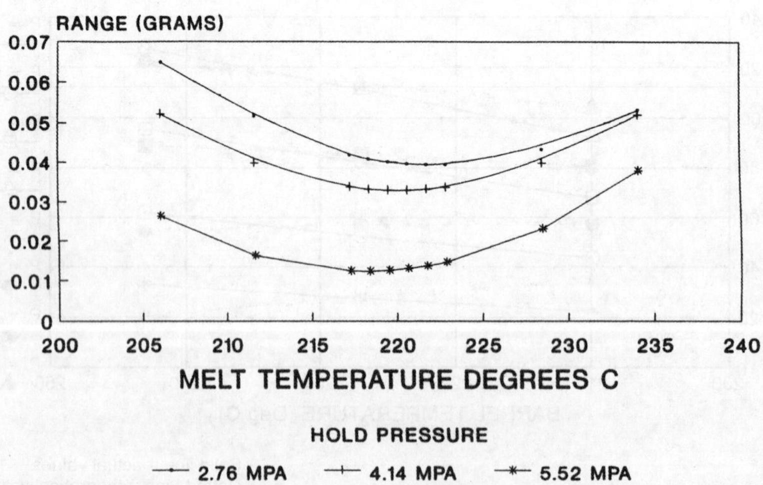

Figure 3.60 Part weight vs. melt temperature at varying hold pressure

and even some safety rupture disks. It is difficult to drill any of the through holes without chipping the brittle liner material and causing a dead spot. It is best to EDM these holes (94).

For control and measurement, the depth of thermocouples is an important factor. A deep hole closer to the inside diameter measures the metal temperature closer to the plastic, therefore, measuring a temperature closer to the plastic's actual temperature. The problem with this deep location is the temperature lag causing control problems. The thermocouple senses a low temperature and tells the heater to pour more heat into the outside of the barrel. By the time this heat penetrates a

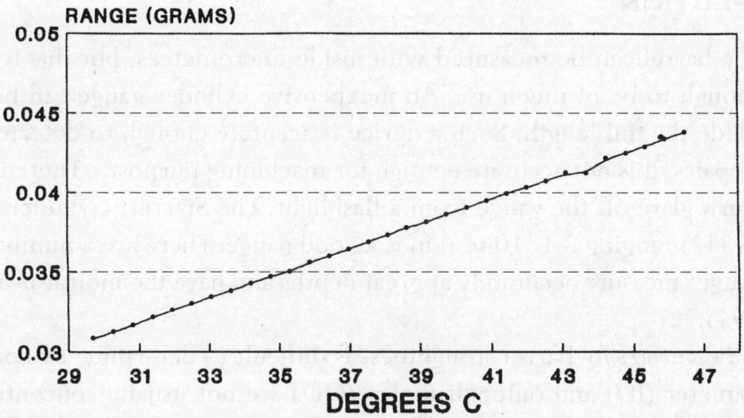

Figure 3.61 Part weight range vs. IMM hydraulic oil temperature

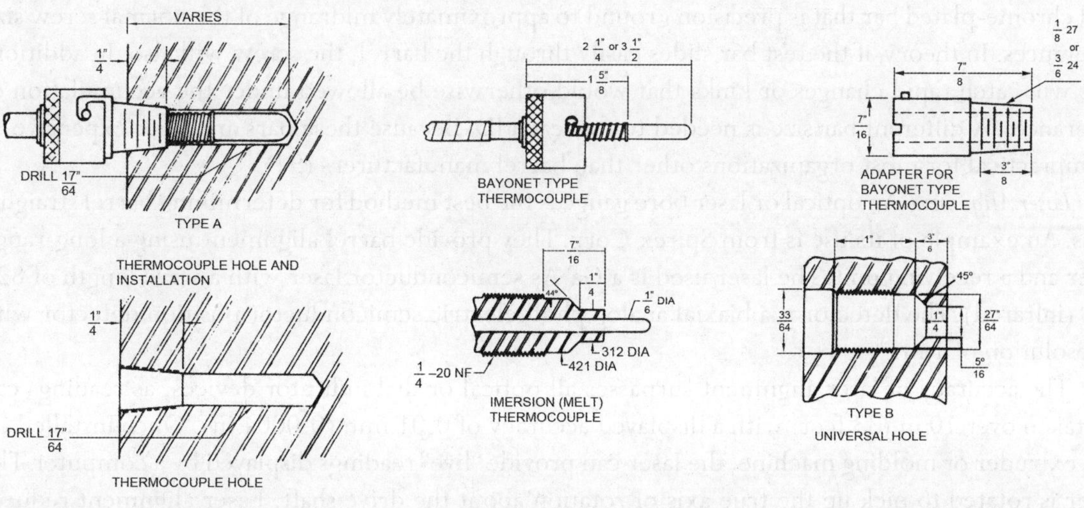

Figure 3.62 Example of machined barrel holes used for measurement and control devices (courtesy of Spirex Corp.)

thick barrel and heats the deep thermocouple, the outside is very hot. This causes an overshoot and wide swings in barrel temperature. This is more pronounced in thicker barrels. Conversely, the shallow thermocouple gives the best control, but set temperatures can be very different from the inner barrel temperature. Actually, the set temperature that produces the best processing results is more important than the one that is closest to the actual plastic temperature. Some compromise in thermocouple depth is desirable, but shallow is usually better than deep. A good alternative solution is the paralleling and averaging of the output of a shallow and a deep thermocouple. Computer controls are used to provide a guide in relating the thermocouple reading directly to the melt temperature.

Barrel Inspection

The very front of a barrel can be measured with inside micrometers, but this is limited and will not show deep enough to be of much use. An inexpensive cylinder gauge can be rigged up with a long handle to slide the full length. Such a device is accurate enough to determine the need for replacements or repairs. It is not accurate enough for machining purposes. There is a problem with light and then return glare off the gauge from a flashlight. The Starrett Cylinder Gauge (Cat. No. 452E EDP, No. 52342) ranging 2 1/10 to 6 in is a good gauge. There are a number of bore gauges available. These gauges measure accurately at great depths and have the indicator outside the barrel for easy reading (94).

Straightness and Concentricity Barrel straightness is difficult to determine by conventional methods. The inner diameter (ID) and outer diameter (OD) are not usually concentric. One method is to set up the barrel on a lathe, indicate both ends, and run in an indicator on a long rod. This is limited in length. Tolerances allow deviation from straight barrels to accumulate total allowable indicated runouts.

Another method is the use of a test bar, usually about 70% of the barrel length. This is a slotted and chrome-plated bar that is precision ground to approximately midrange of the normal screw size tolerances. In theory, if the test bar slides easily through the barrel, the screw will also. In addition, this will catch rapid changes or kinks that would otherwise be allowed under the accumulation of tolerances. A different bar size is needed to test each ID. Because these bars are quite expensive, it is impractical for most organizations other than barrel manufacturers (94).

Laser Alignment An optical or laser bore gauge is the best method for determining barrel straightness. An example of its use is from Spirex Corp. They provide barrel alignment using a long-range laser and a receiving unit. The laser used is a Ga-As semiconductor laser with a wave length of 820 nm (infrared). The detector is a biaxial analog photoelectric semiconductor position detector with a resolution of 1 m.

The accuracy of laser alignment surpasses all optical or dial indicator devices, as readings can be taken over 10 m (33 feet) with a displayed accuracy of 0.01 mm (0.0004 in). Once installed on the extruder or molding machine, the laser can provide "live" readings displayed by a computer. The laser is rotated to pick up the true axis of rotation about the drive shaft. Laser alignment reduces typical operator error involved with an optical system, and accurate diagnoses and corrections can be performed more rapidly (94).

Hardness The standard hardness testers are not able to get inside barrels to measure hardness. The instrument most commonly used for this purpose is the Internal Mobile Hardness Tester. When testing and comparing the hardness of a bimetallic liner, it should be remembered that absolute hardness, as measured, is not necessarily in direct proportion to wear resistance. Sometimes you may be measuring the softer matrix rather than the wear-resistant carbides. The degree of lubricity or how well the screw and barrel slide against each other is very important. This is not always related to hardness, but it is very critical to barrel and screw wear (94).

BARREL BORESCOPING

This technique concerns the alignment of a barrel with the screw. Alignment is important to ensure maximizing the repeatability of melt consistency. Borescoping is not a guarantee that it will correct all potential problems, but it comes close, so it is important to schedule it into maintenance service. Their clearance can range from 0.05 to 0.20 mm (small to large diameter screw) on any side of the screw. With an alignment scope, one can tell what the internal shape of a barrel is at any point, whether it is straight, curved, or even S shaped, which may be due to machining or normal wear. Other areas are also to be examined. Aligning with a scope provides a much better opportunity of producing better products with less downtime, less scrap, and extending the life of the barrel/ screw; otherwise, it will be costly. Most machines can be adjusted in a day at very little cost. The machine will have at least a 25% life extension.

If the barrel of a plasticator tends to wobble when the screw is rotating, it is a symptom of a misalignment. It could be an alignment deficiency between the transmission, feed block, and/ or barrel. This should be investigated immediately, as it can result in a high wear situation, screw breakage, heater creep, and/or shutdown of the machine.

RECOMMENDED BARREL DIMENSIONAL GUIDELINE

Tables 3.51 to 3.58 show recommendations for barrel dimension and tolerance as a guide for manufacturers and processors. These standards provide working tolerances that produce effective performance with economy of manufacture. Information was prepared by the Machinery Component Manufacturers Division of the Society of the Plastics Industry. Upon receiving or replacing machines with barrels, it is best to have them measured so that you can determine if any wear or damage occurs after they are put into operation.

DOWNSIZING MACHINE

Very few of the installed IMMs run shot sizes anywhere near the full shot-size capacity of the injection unit. Typical usage is from 25% to 60%. Most suppliers of injection machines offer several sizes of injection plasticating units for any given press tonnage. Having too much shot capacity can render some IMMs unusable for certain plastics and applications. One problem is excessive residence time

LENGTHS, DEPTHS & WIDTHS

	DIMENSION DESCRIPTION	ENGLISH UNITS RANGE	ENGLISH UNITS TOLERANCE	METRIC UNITS RANGE	METRIC UNITS TOLERANCE
I.	Most Linear Dimensions	To 12"	± .010"	To 300mm	± .25mm
		12-60"	± .030"	300-1500mm	± .75mm
		60-120"	± .045"	1500-3000mm	± 1.00mm
		120-200"	± .060"	3000-5000mm	± 1.50mm
		over 200"	± .090"	over 5000mm	± 2.25mm
II.	Counterbore Depths or Pilot Lengths	All Sizes	± .005"	All Sizes	± .125mm

DIAMETERS

	DIMENSION DESCRIPTION	RANGE		TOLERANCE
		ENGLISH UNITS		
I.	Bores (Including Counterbores)	To 1 1/2" dia.	(<60" long)	+ .001" - .000"
		1 1/2 - 3" dia.	(>60" long)	+ .002" - .000"
		3 - 5 1/2" dia.	(<120" long)	+ .002" - .000"
		3 - 5 1/2" dia.	(>120" long)	+ .0025" - .000"
		5 1/2 - 8" dia.	(<180" long)	+ .0025" - .000"
		5 1/2 - 8" dia.	(>180" long)	+ .003" - .000"
		8 - 12 1/2" dia.	(<180" long)	+ .003" - .000"
		8 - 12 1/2" dia.	(>180" long)	+ .004" - .000"
		METRIC UNITS		
		0-38mm dia.	(<1500mm long)	+ .025mm - .000mm
		3-75mm dia.	(<1500mm long)	+ .050mm - .000mm
		25-75mm dia.	(>1500mm long)	+ .038mm - .000mm
		75-140mm dia.	(<3000mm long)	+ .051mm - .000mm
		75-140mm dia.	(>3000mm long)	+ .063mm - .000mm
		140-200mm dia.	(<4500mm long)	+ .063mm - .000mm
		140-200mm dia.	(>4500mm long)	+ .076mm - .000mm
		200-315mm dia.	(<4500mm long)	+ .076mm - .000mm
		200-315mm dia.	(>4500mm long)	+ .102mm - .000mm
II.	Outside Diameters Except Pilots	All Sizes and Lengths	± .005"	± .102mm
III.	Pilot Diameters	All Sizes	± .001"	± .038mm

Table 3.51 Recommended single-barrel lengths, depths, and widths

Concentricity of Diameters

Concentricity between all inside and outside diameters should be held within .002" TIR or .051 mm TIR within one bore diameter of each end. Concentric diameter dimensions further inside of one bore diameter from the ends should be avoided and/or specially tolerated when required.

Straightness

Straightness is generally specified for the bore of the barrel and is measured by TIR.

1 - 3"	bore diameter	(≤ 24 L/D)	.008" TIR
1 - 3"	bore diameter	(≥ 24 L/D)	.010" TIR
3 - 5 1/2"	bore diameter	(≤ 24 L/D)	.010" TIR
3 - 5 1/2"	bore diameter	(≥ 24 L/D)	.012" TIR
5 1/2 - 8"	bore diameter	(≤ 24 L/D)	.012" TIR
5 1/2 - 8"	bore diameter	(≥ 24 L/D)	.015" TIR
8 - 12 1/2"	bore diameter	(≤ 24 L/D)	.014" TIR
8 - 12 1/2"	bore diameter	(≥ 24 L/D)	.016" TIR

25 - 75mm	bore diameter	(≤ 24 L/D)	.203mm TIR
25 - 75mm	bore diameter	(> 24 L/D)	.254mm TIR
75 - 140mm	bore diameter	(≤ 24 L/D)	.254mm TIR
75 - 140mm	bore diameter	(> 24 L/D)	.305mm TIR
140 - 200mm	bore diameter	(≤ 24 L/D)	.305mm TIR
140 - 200mm	bore diameter	(> 24 L/D)	.380mm TIR
200 - 315mm	bore diameter	(≤ 24 L/D)	.355mm TIR
200 - 315mm	bore diameter	(> 24 L/D)	.406mm TIR

To prevent short "kinks" to which the screw could not conform, any two successive measurements taken less than twice the bore diameter apart should have no more than 1/2 the total allowable TIR.

Bar Test

A second way to check both straightness and bore size is by using a precision ground test bar. This method is detailed in Fig. 3.53

Finish

The following surface finishes apply unless specified otherwise.

Outer diameters	125	RMS
Counterbores	32	RMS
Pilots	32	RMS
Cylinder Bores	8-32	RMS
Feed Ports	125	RMS
*Vent Ports	32	RMS

***Required within 1/4" of cylinder bore.**

The bore shall be free from visual surface defects in the bore over its entire length.

Table 3.52 Recommended single-barrel diameters and concentricity

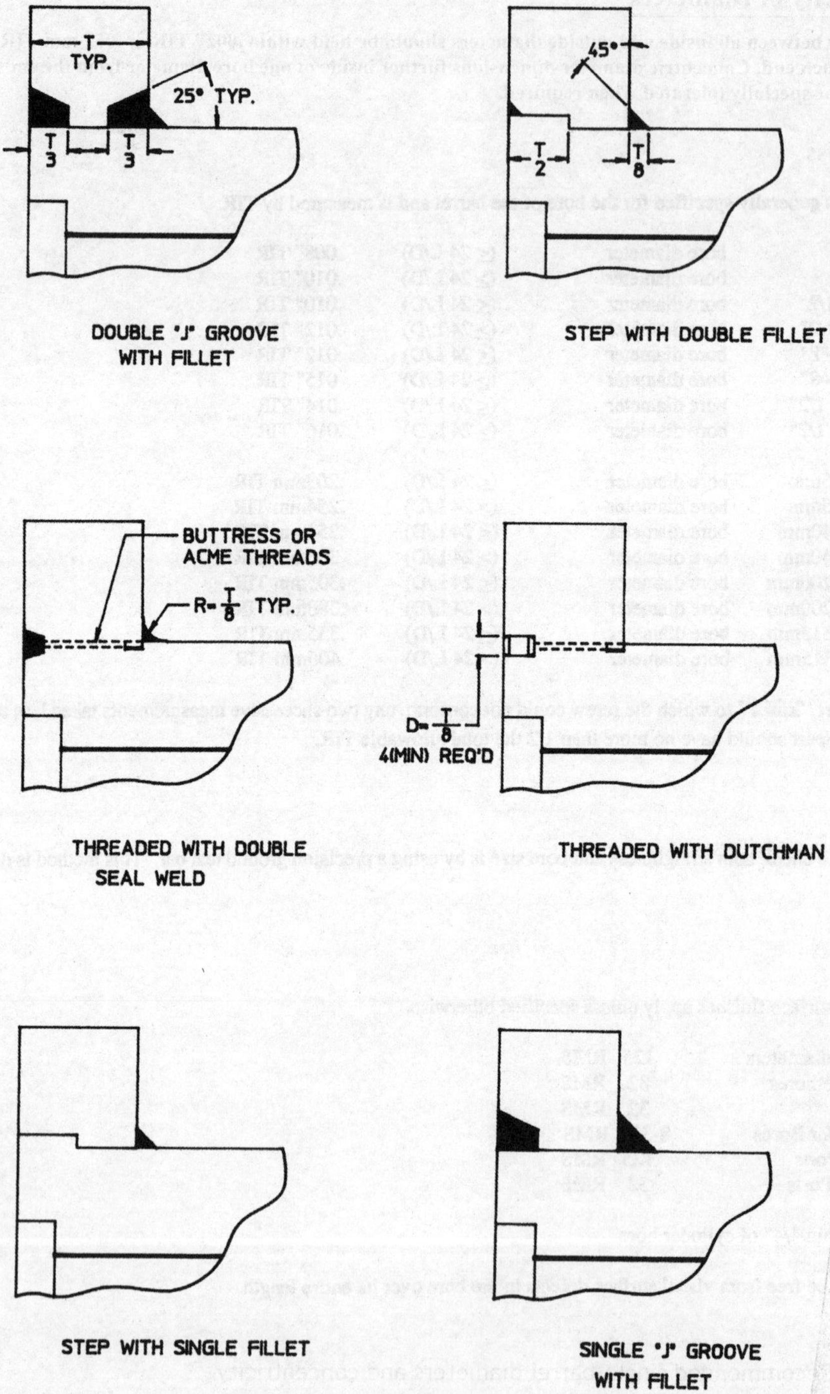

Table 3.53 Precision-ground test bars applicable to Figure 3.52

Parallelism & Perpendicularity

Parallel surfaces can be determined by TIR and all dimensions meeting the concentricity recommendations are acceptable.

Perpendicular surfaces to the cylinder bore can be tested by placing the barrel on a surface plate and indicating with an adjustable height table indicator or with a precision square. All mating surfaces should be perpendicular as follows:

Pilots and Counterbores	.0015" TIR
Flanges	.001" TIR per 10" of dia.
Pilots and Counterbores	.038mm TIR
Flanges	.038mm TIR per 250mm of dia.

Threads

The variation in threads used is too broad to be addressed in these recommendations. It is suggested that whenever thread selection is possible that either ANSI or ISO standards are adhered to for ease of measurement and compatibility.

Flange Attachments

Barrel flanges may be affixed in accordance with the procedures in Figures 1-6 shown in the Appendix depending on the resultant thrust from internal pressure and other design factors. These procedures apply for flanges at either end of the barrel.

Venting

The following warning label should be affixed to a vented extruder or injection barrel supplied without a venting plug.

Warning

This vented injection/extrusion barrel was designed for operation with the vent OPEN ONLY.

Operation of this machine with the vent plugged or otherwise closed off may result in serious injury to persons in the vicinity. Check with your supervisor if this plug is intentionally or unintentionally plugged.

Downsizing Injection Barrels

When downsizing (reducing the barrel/screw bore size and resultant shot capacity) an injection molding machine must have the pressure capability reduced or be redesigned for the new resultant pressure. This is necessary to reduce the screw thrust pressure in proportion to the reduction in area of the bore to prevent the barrel from being over-pressurized. All units that are downsized should be equipped with a warning label on the barrel where it can be easily seen.

Table 3.54 Recommended single-barrel parallelism check and other details

for the plastic, particularly the engineering materials. Any plastic that will degrade when held at injection temperatures for long periods will have problems with small shots, long cycles, and large injection units. These types of plastics include PCs, ABSs, nylons, acetals, cellulosics, PESs, and most fire-retardant grades.

Another problem associated with very large injection units and small shot sizes is relative to the plasticating screw design. In order to properly plasticize, the screw should impart approximately 40% of the energy needed to melt the plastic via the drive motor. If the screw rpm is too low and the meter zone flight depth is too deep relative to the throughput needed, very little energy will come from the screw drive. This situation will result in very poor homogenization of the melt pool,

Barrel Construction

I. Bimetallic Centrifugally Cast and Hot Isostatically Pressed Barrels

The inlay thickness is to be 1/16" (1.6mm) nominal, with a 1/8" (3.2mm) maximum and 1/32 (0.8mm) minimum.

Hardness Standards are those of the manufacturer and are based on the arithmetic means of five hardness measure ments taken at random points along the bore. The hardness values vary greatly with the type of inlay. The inlay should have a 100% metallurgical bond at all points determined by ultrasonic techniques.

II. Nitrided Barrels or Other Single Metal Barrels

Depending on the material used in the outer tube and the bore hardening method, both the hardness and depth of hardness will vary. As a result the hardness standards are those of the manufacturer.

III. Sleeved Barrels (New)

All sleeves should be inserted by either shrink fit or press fit with no less than the interference required to maintain the stress levels applying to all barrels detailed in the Appendix. In all cases a minimum interference of .001" (.025mm) diametral shrunk per inch of diameter should be used.

Hardness standards are those of the manufacturer and vary with the material and hardening method selected for the sleeve.

IV. Sleeved Barrels (Relined)

The liner material should have a co-efficient of thermal expansion within ± 30% of the base (outer tube) material from room temperature through to maximum expected operating temperature.

All internal sleeves should blend to any remaining bore with no visible lip and with no more than the following mismatch in TIR.

To 2"	**bore dia..001"**	**To 50mm**	**bore dia..025mm**
2 - 3 1/2"	**bore dia..002"**	**50-90mm**	**bore dia..051mm**
> 3 1/2"	**bore dia..003"**	**> 90mm**	**bore dia..076mm**

In the case of injection barrels, relining of the discharge end should extend a minimum of 2" (50mm) past the maximum backward travel of the rear edge of the check ring on the non-return valve.

Table 3.55 Recommended single-barrel construction

CRITERIA

Clearance specified in the following two tables are based upon a minimum of 0.001" to a maximum of 0.0015" per inch if diameter between the barrel inside diameter and the screw flight outside diameter. The test-bar outside diameters are sized to the midway point between the screw flight outside diameter and the barrel inside diameter. The lengths of the test-bars specified are based on approximately 15 to 20 times the nominal screw/barrel diameter. All table dimensions are based on the parameters established in the first sentence of this criteria.

NOMINAL DIAMETER	SPECIFIED BARREL I.D. SIZE	SPECIFIED DIAMETER TEST BAR	SPECIFIED SCREW O.D. SIZE	TEST BAR LENGTH	BARREL/TEST BAR DIAMETRAL CLEARANCE	BARREL/SCREW DIAMETRAL CLEARANCE
1-1/8"	1.125 +.001/-.000	1.1235/1.1230	1.1225/1.1220	22"	.0015/.003	.0025/.004
1-1/4"	1.250 +.001/-.000	1.2485/1.2480	1.2475/1.2465	24"	.0015/.003	.0025/.0045
1-1/2"	1.500 +.001/-.000	1.4985/1.4980	1.497/1.496	26"	0015/.003	.003/.005
1-3/4"	1.750 +.001/-.000	1.7480/1.7475	1.7465/1.7455	30"	.002/.0035	.0035/.0055
2"	2.000 +.001/-.000	1.9980/1.9975	1.996/1.995	36"	.002/.0035	.004/.005
2-1/4"	2.250 +.001/-.000	2.2480/2.2475	2.246/2.245	38"	.002/.0035	.004/.005
2-1/2"	2.500 +.001/-.000	2.4975/2.4970	2.495/2.494	42"	.0025/.004	.005/.007
2-3/4"	2.750 +.001/-.000	2.7475/2.7470	2.745/2.744	46"	.0025/.004	.005/.007
3"	3.000 +.001/-.000	2.9970/2.9965	2.994/2.993	50"	.003/.0045	.006/.008
3-1/4"	3.250 +.002/-.000	3.2465/3.2460	3.2435/3.2425	56"	.0035/.006	.0065/.0095
3-1/2"	3.500 +.002/-.000	3.4965/3.4960	3.493/3.492	60"	.0035/.006	.007/.010
3-3/4"	3.750 +.002/-.000	3.7465/3.7460	3.743/3.742	62"	.0035/.006	.007/.010
4"	4.000 +.002/-.000	3.9960/3.9955	3.992/3.991	68"	.004/.0065	.008/.011
4-1/4"	4.250 +.002/-.000	4.2460/4.2455	4.242/4.241	72"	.004/.0065	.008/.011
4-1/2"	4.500 +.002/-.000	4.4960/4.4955	4.491/4.490	76"	.004/.0065	.009/.012
4-3/4"	4.750 +.002/-.000	4.7455/4.7450	4.741/4.740	82"	.0045/.007	.009/.012
5-1/4"	5.250 +.002/-.000	5.2445/5.2440	5.240/5.238	90"	.0055/.008	.010/.014
6"	6.000 +.002/-.000	5.994/5.9935	5.988/5.986	96"	.006/.0085	.012/.016
8"	8.000 +.003/-.000	7.993/7.992	7.984/7.982	108"	.007/.011	.016/.021

NOMINAL DIAMETER	SPECIFIED BARREL I.D. SIZE	SPECIFIED DIAMETER TEST BAR	SPECIFIED SCREW O.D. SIZE	TEST BAR LENGTH	BARREL/TEST BAR DIAMETRAL CLEARANCE	BARREL/SCREW DIAMETRAL CLEARANCE
30mm	1.181 +.001/-.000	1.1795/1.1790	1.1785/1.1780	22"	.0015/.003	.0025/.004
35mm	1.378 +.001/-.000	1.3765/1.3760	1.3750/1/3745	24"	.0015/.003	.003/.0045
38mm	1.496 +.001/-.000	1.4945/1.4940	1.493/1.492	26"	.0015/.003	.003/.005
40mm	1.575 +.001/-.000	1.5735/1.5730	1.572/1.571	28"	.0015/.003	.003/.005
50mm	1.969 +.001/-.000	1.9670/1.9665	1.965/1.964	36"	.002/.0035	.004/.005
60mm	2.362 +.001/-.000	2.3595/2.3590	2.357/2.356	38"	.0025/.004	.005/.007
65mm	2.559 +.001/-.000	2.5565/2.5560	2.554/2.553	42"	.0025/.004	.005/.007
70mm	2.756 +.001/-.000	2.7535/2.7530	2.751/2.750	46"	.0025/.004	.005/.007
75mm	2.953 +.001/-.000	2.9500/2.9495	2.947/2.946	50"	.003/.0045	.006/.008
80mm	3.150 +.001/-.000	3.1465/3.1460	3.1435/3.1425	56"	.0035/.005	.0065/.008
90mm	3.543 +.002/-.000	3.5395/3.5390	3.536/3.535	60"	.0035/.006	.007/.010
100mm	3.937 +.002/-.000	3.9330/3.9325	3.929/3.928	68"	.004/.0065	.008/.011
105mm	4.134 +.002/-.000	4.1300/4.1295	4.126/4.125	72"	.004/.0065	.008/.011
115mm	4.528 +.002/-.000	4.5240/4.5235	4.519/4.518	76"	.004/.0065	.009/.012
120mm	4.724 +.002/-.000	4.7195/4.7190	4.715/4.714	82"	.0045/.007	.009/.012
135mm	5.315 +.002/-.000	5.3095/5.3090	5.305/5.303	90"	.0055/.008	.010/.014
150mm	5.906 +.002/-.000	5.900/5.8995	5.894/5.892	96"	.006/.0085	.012/.016

Table 3.56 Barrel/test-bar/screw clearance criteria

Brinell Indentation Dia. mm	Brinell Hardness No. 10-mm Tungsten Carbide Ball, 3,000-Kg Load	Rockwell Hardness No. C Scale 150-Kg Load Brale Penetrator	B Scale 100-Kg Load 1/15-in Diam. Ball	A Scale 60-Kg Load Brale Penetrator	Diamond Pyramid Hardness No., Vickers	Shore Sclero-scope Hardness No.	Tensile Strength (Appx) in 1,000 psi
2.25	745	65.3	–	84.1	840	91	–
2.30	712						
2.35	682	61.7	–	82.2	737	84	–
2.40	653	60.0	–	81.2	697	81	–
2.45	627	58.7	–	80.5	667	79	–
2.50	601	57.3	–	79.8	640	77	–
2.55	578	56.0	–	79.1	615	75	–
2.60	555	54.7	–	78.4	591	73	298
2.65	534	53.5	–	77.8	569	71	288
2.70	514	52.1	–	76.9	547	70	274
2.75	495	51.0	–	76.3	528	68	264
2.80	477	49.6	–	75.6	508	66	252
2.85	461	48.5	–	74.9	491	65	242
2.90	444	47.1	–	74.2	472	63	230
2.95	429	45.7	–	73.4	455	61	219
3.00	415	44.5	–	72.8	440	59	212
3.05	401	43.1	–	72.0	425	58	202
3.10	388	41.8	–	71.4	410	56	193
3.15	375	40.4	–	70.6	396	54	184
3.20	363	39.1	–	70.0	383	52	177
3.25	352	37.9	(110.0)	69.3	372	51	171
3.30	341	36.6	(109.0)	68.7	360	50	164
3.35	331	35.5	(108.5)	68.1	350	48	159
3.40	321	34.3	(108.0)	67.5	339	47	154
3.45	311	33.1	(107.5)	66.9	328	46	149
3.55	293	30.9	(106.0)	65.7	309	43	141
3.60	285	29.9	(105.5)	65.3	301	–	138
3.65	277	28.8	(104.5)	64.6	292	41	134
3.70	269	27.6	(104.0)	64.1	284	40	130
3.75	262	26.6	(103.0)	63.6	276	39	127
3.80	255	25.4	(102.0)	63.0	269	38	123
3.85	248	24.2	(101.0)	62.5	261	37	120
3.90	241	22.8	100.0	61.8	253	36	116
3.95	235	21.7	99.0	61.4	247	35	114
4.00	229	20.5	98.2	60.8	241	34	111
4.05	223	(18.8)	97.3	–	234	–	–
4.10	217	(17.5)	96.4	–	228	33	105
4.15	212	(16.0)	95.5	–	222	–	102
4.20	207	(15.2)	94.6	–	218	32	100
4.25	201	(13.8)	93.8	–	212	31	98
4.30	197	(12.7)	92.8	–	207	30	95
4.35	192	(11.5)	91.9	–	202	29	93
4.40	187	(10.0)	90.7	–	196	–	90
4.45	183	(9.0)	90.0	–	192	28	89
4.50	179	(8.0)	89.0	–	188	27	87

Table 3.57 Hardness conversion table applicable to barrel and screw

which will lead to poor part quality. When the injection unit is too large, the travel of the screw needed to fill the mold is also very short, sometimes not allowing the machine drive system and electronics to be utilized effectively. One solution is to purchase a completely new smaller injection unit from the original machine supplier. This is very expensive, both in investment and labor.

NOMINAL PIPE SIZE	THREADS PER INCH	INSIDE DIA. (INCHES)	OUTSIDE DIA. (INCHES)
1/8	27	.269	.405
1/4	18	.364	.540
3/8	18	.493	.675
1/2	14	.622	.840
3/4	14	.824	1.050
1	11 1/2	1.049	1.315
1 1/4	11 1/2	1.380	1.660
1 1/2	11 1/2	1.610	1.900
2	11 1/2	2.067	2.375
2 1/2	8	2.469	2.875
3	8	3.068	3.500
3 1/2	8	3.548	4.000
4	8	4.026	4.500
5	8	5.047	5.563
6	8	6.065	6.625
8	8	7.981	8.625
10	8	10.020	10.750
12	8	12.00	12.750
14	8	13.250	14.000
16	8	15.250	16.000
18	8	17.250	18.000

Table 3.58 Standard pipe data applicable to barrels

A few organizations, such as Spirex, have developed a regular business of downsizing existing injection units. They supply a new barrel, screw, valve, endcap, and either a screw adapter or a new drive. In some cases, the existing heater bands can be used after considering barrel wall thickness. Often it is possible to utilize greater injection pressures acquired through the downsize process. If not, the injection pressure relief valve must be adjusted to make sure excessive pressure is not developed at the discharge end as well as the screw drive motor. All of this action is a lot less expensive than a new injection unit. Consideration to limiting torque to the screw drive motor should also be given to reduce breakage of the smaller screws.

UPSIZING MACHINE

To increase shot size, upsizing the plasticator can be done. A number of items have to be considered for the upsizing process, such as barrel wall thickness, resultant screw L/D, injection speed reduction, screw drive torque, and injection pressure drop. Before considering the upsizing process, one has to determine whether the molds can be filled properly using the decreased pressure and injection speeds. The injection pressure and speeds will decrease directly proportional to the difference in the barrel ID projected areas. If this poses no problems, the L/D and structural integrity of the barrel have to be considered before proceeding (94).

REBUILDING VERSUS BUYING

This review is subject to pros and cons. With a logical approach, the outcome depends on various factors, such as extent of damage, professional feasibility to rebuild, time to be back in production, and availability of money. Even though the initial capital expenditure is much lower than for new screws and barrels (plus other equipment), the long-term economical value can be questionable. As an example, machine retrofits can be tailored to meet the customer's performance requirements at 40% to 70% of a new tool. In order to provide a good basis for a decision, a technical evaluation matrix system using weighted criteria and a time-related method for judging the economical value of an investment are required.

REPAIR

Major rebuilding and repairs involve screws and barrels. Screws and barrels are expensive and can cause downtime when damaged or worn. It may be practical (cost-effective) to repair rather than replace. It is common practice to rebuild a worn screw with hard surfacing materials (cobalt, etc.). Quite often the rebuilt screw will outlast the original screw time in service. The larger the screw, the more economical screw repairing becomes. Usually it does not pay to rebuild 50-mm (2-in) diameter or smaller screws. To be practical as to repairing depends on the location and degree of damage.

SCREW REPAIR

After a certain period of service, most screws become scratched, carburized, and/or discolored owing to the plastic's temperature and pressure melting action. They are difficult to clean and tend to lose their original operating characteristics. If they have been plated (usually chrome), the plating may be gone in some places or peeling in others. It is best to refurbish in this condition by stripping the old chrome, polishing, buffing, plating, and buffing again. The part will look much better and will also perform better for little cost and a short delivery time. Most parts that are rebuilt are also stripped and so on.

Tables 3.59 and 3.60 and Figure 3.63 provide information applicable to screw repair.

1) Wear determination: abrasive, abrasive or corrosive
2) Visual inspection on rollers
3) Measurement of the root diameter or outside flighted diameters using specialized equipment
4) Channel depth measurement using a "screw depth indicator"
5) Check straightness and concentricity
6) Measure the pilot I.D. (for injection screws) using a digital bore gauge
7) Test the hardness using an electronic hardness tester
8) Measure the surface finish using a profilometer
9) Issue an inspection report as well as a quotation

Table 3.59 Screw inspection process (courtesy of Spirex Corp.)

- Screw, barrel and drive alignment
- Straightness of screw and barrel
- Screw design
- Uniformity of barrel heating
- Material being processed
- Abrasive fillers, reinforcing agents and pigments
- Screw surface materials
- Barrel liner materials
- Combination of screw surface and barrel liner
- Improper support of the barrel
- Excessive loads on barrel discharge end
- Corrosion caused by polymer degradation
- Corrosion caused by additives such as flame retardants
- Excessive back pressure on injection recovery
- High screw speeds (rpm)

Table 3.60 Typical factors affecting screw, barrel, and other components (courtesy of Spirex Corp.)

Barrel Repair

Table 3.61 goes through steps for rebuilding a barrel.

Injection barrels can become worn in the front 4 to 5 diameters where the ball check or sliding ring valve reciprocates. Thus, a 3-in ID barrel would normally become worn in the front 12 to 15 in. With wear in the front only, an economical and reliable (if done properly) repair is

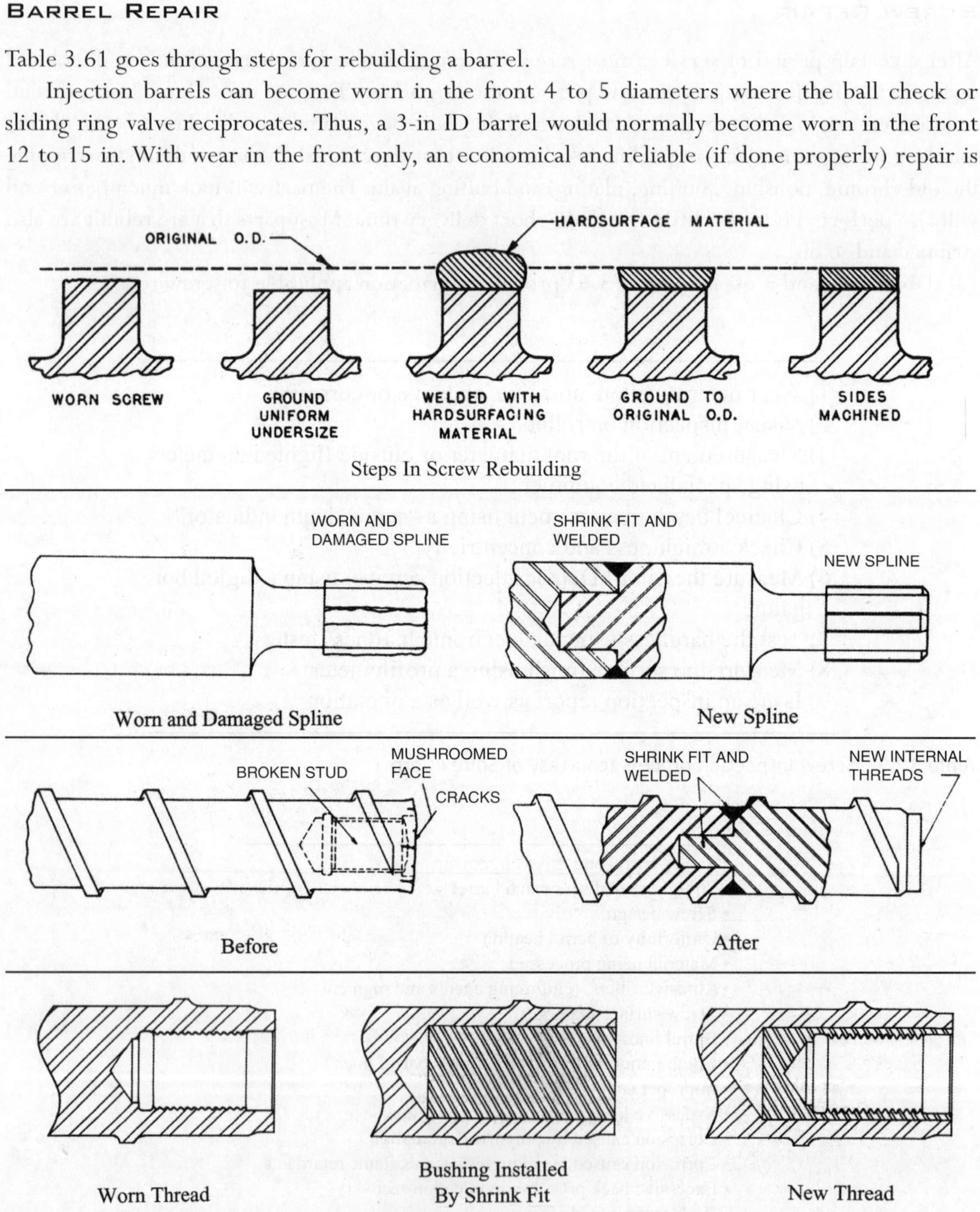

Figure 3.63 Examples of repairing screws (courtesy of Spirex Corp.)

- The screw is set up in a lathe and a center is found. At this time it is also checked for straightness and concentricity;
- It is then polished and prepped for stripping off the existing chrome;
- Entire screw is submerged in an acid bath to remove the chrome plating;
- Screw then moves to the grinder where it is ground undersize;
- Screw flights are welded with a hardsurface material, i.e. Colmonoy 56 or Stellite 12;
- Back to the grinder for rough grind after weld; at this time, screw is also checked for straightness;
- Sent to the flight grinder to trim the sides of the flight;
- Sent to the polishing booth for a rough polish;
- Inspected and buffed for chrome plating if needed;
- Chrome plated coating of .0015" is applied to the entire root and bearing surface;
- Buffed after chrome;
- Ground to final O.D. specification;
- Final polish and buff as needed;
- Grind front face, size register and board the O.D.;
- Final inspection.

Table 3.61 Steps for rebuilding a barrel (courtesy of Spirex Corp.)

accomplished by sleeving. Wear from glass fibers or other abrasive fillers will cause barrel wear in the middle where the screw transition reciprocates. In order not to waste money, it is important to check the remainder of the barrel before sleeving the front. This can be done accurately with a bore gauge or close enough with a cylinder gauge. It is also important that the bore of the new bimetallic sleeve be aligned with the remainder of the existing ID. This requires special procedures because the new sleeve must be installed accurately with the proper interference fit in order to eliminate any possibility of movement in operation. Full-length sleeves are available for some barrels smaller than 3 inches in diameter. Usually the cost is close to, or more than, a new bimetallic barrel (94).

Lined barrels are used where repair or high-performance alloys are considerations. The liners are thin walled and inserted into an opening in the larger load-bearing barrel. Liners can be made from various hardened carbon steel, stainless steel, nickel-based alloys, and so on. They can be used with a bimetallic coating. With liners, heat transfer can become a problem if air gaps between liner and barrel materials exist.

Bimetallic coatings are used when wear and corrosion are of concern. They can be applied by traditional welding techniques for large barrels. Centrifugal casting is the most common technique for both single- and twin-bore barrels. A wide variety of bimetallic coatings are used. The choice of coating depends on the plastic to be processed. For noncorrosive processes with up to about 30 wt% to 40 wt% of filler, an iron-based alloy is usually considered. Highly corrosive systems using materials such as PVCs and fluoropolymers can benefit from the Ni/Co family of alloys. Processes compounding high levels of abrasive fillers should use the family of alloys with tungsten.

STORAGE

Usually a plant has more than one screw to process different plastics. During both short- and long-term storage, ranging from hours to months or longer, screws must be protected from water and humidity. Unprotected steel can almost immediately begin to corrode, resulting in damage that will require repolishing, regrinding, and/or repair of at least the surface. The net result is a cost in both labor and machine downtime. It is most cost-effective to protect the tools. There are excellent rust protectants on the market that last for different lengths of time. However, most of these anticorrosion treatments must be completely removed before using the screw. Some may require special cleaners, including toxic solvents.

TOOLING

When processing plastics, some type of tooling is required. These tools include dies, molds, mandrels, jigs, fixtures, punch dies, and perforated forms for shaping and fabricating products (chapter 17).

PROCESS CONTROL

OVERVIEW

In the past, because melts have different properties and there are many ways to control processes, detailed factual predictions of final output have been difficult to arrive at prior to prototyping or fabricating, or without prior experience. Research and hands-on operations have been directed mainly at explaining the behavior of melts or plastics with other materials (steel, glass, and so on). Modern equipment and process controls (PCs) are overcoming some of this unpredictability. Ideally, processes and equipment should be designed to take advantage of the novel properties of plastics rather than to overcome them.

This important area has to be thoroughly analyzed and studied to obtain the desired performance of the complete line and/or its parts, such as the injection-molding machine, extruder puller, and so on. The first task is to determine what is required and how to approach any potential problem. Adequate PC and its associated instrumentation are essential. Sometimes the goal is precise adherence to a control point; other times it is sufficient to maintain a control within a comparatively narrow range. For effortless controller tuning and lowest initial and operating cost, the processor should select the simplest controller (temperature, time, pressure, flow rate, etc.) that will produce the desired results. For the complete line, the controllers can range from unsophisticated to extremely sophisticated devices that interrelate information.

Machine control operation and the control behavior of the plastic are involved. The interaction between the machine's operation and the plastic's behavior is most important. Example of controls used with injection molding (a rather complex process when compared with others), extrusion, and other processes are reviewed in chapters 4 to 16. Basically, the processing pressure and temperature versus time determine the quality of the fabricated product. The design of the control system has to take into consideration the logical sequence of all these basic functions and their ramifications. Developing a PC flow diagram requires a combination of experience (at least familiarity) with the process and a logical approach to meet the objective that has specific target performance requirements. PCs range from very simple/standard types to advance complex types. It should be noted that none of the PC solutions address the problem of the lack of skilled setup people.

There is a continuous stream of improvements in PC. Control of machines continually enters new eras that dramatically improve ease of machine setup, ensure ease of meeting fabricated product requirements, allow more uninterrupted operation, simplify remote handling, reduce fabricating times, cut energy costs, boost part quality, and so on. As an example, the National Research Council of Canada's Industrial Materials Institute's (IMI's) system uses computerized ultrasonic technology for accurate, nonintrusive, and nondestructive measurement of the surface and interior of molding materials during the filling, packing/holding, and cooling phases during injection molding. The system uses pulse-echo ultrasonic techniques, similar in principle to those used for an expectant mother's sonogram, to listen through tool steel and see parts as they are being molded. For example, when the ultrasonic waves meet the acoustic-resistance boundary between the two different media of the mold and plastic, the air gap formed from shrinkage of the cooling part causes some of the energy to be transmitted through the boundary. The rest of the energy is reflected back to an ultrasonic transmitter. No mold modification is needed.

The process of making a product has many dynamic fragments that must come together properly for successful results. Lack of sufficient PC over each of these fragments will result in a less-than-desirable product and/or reduction/loss of profit. For success, there are three key ingredients: sufficient dynamic performance, sufficient repeatability, and very importantly, selection of proper control parameters. A lack of these ingredients can result in unacceptable products, higher scrap rate, lower output, higher part cost, and so on. Figure 3.64 shows a control unit that is composed of input, signal processing, and power stages.

Properly set PCs can monitor the process variables (chapter 1), compare them with values known to be acceptable, and make appropriate corrections without operator intervention. The acceptable range of values can be determined by using melt flow analysis software (chapter 25) and/or trial and error when the machine is first starting its production. Using the applicable software approach, the acceptable process values are known before the mold or die is ever built. Many of the PC systems available today are rather complex and require skilled people to use them efficiently.

For example, like others in the plastic industry, Phillips Plastics Corp., Phillips, Wisconsin, provides full automated monitoring of the molding process (381). The task is performed by an off-the-shelf unit that monitors process variables, such as injection speed, fill speed, cavity pressure, and mold temperature. Processors and even customers need this data to prove that their production processes are in control. The data can have other uses as well. For one thing, they can help molders optimize their processes. In addition, the data can be fed into an automated control system that rejects bad parts or shuts down an inefficient process.

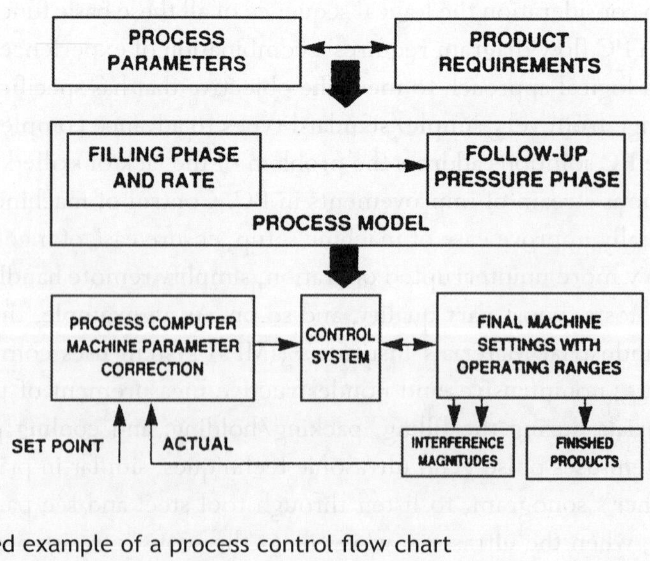

Figure 3.64 Simplified example of a process control flow chart

Controls are not something to be toyed with, as they demand a high level of expertise from the processor. There are those that (a) provide closed-loop control of temperature, pressure, thickness, and so on; (b) maintain preset parameters; (c) monitor and/or correct equipment operations; (d) constantly fine-tune equipment; (e) provide consistency and repeatability in the operations; and (f) provide self-optimization of the process. Most processes operate more efficiently when functions must occur in a desired time sequence or at prescribed intervals of time. In the past, mechanical timers and logic relays were used. Now electronic logic and timing devices are used based on computer-software-programmable logic controllers. They lend themselves to easy setup and reprogramming, and they provide more accurate control.

Adaptive PCs are control systems that change the settings in response to changes in machine performance to bring the product back into its preset requirements or specification. The shift is maintained so that the control has adapted to changing conditions. It is a technique typically used to modify a closed-loop control system. The process control comparator is the portion of the control elements that determines the feedback error on which a controller acts.

Computer PCs usually are a requirement for the production lines. They control a range of simple to very complex equipment systems, meeting different requirements for the processors. Deciding whether to computerize a line and, more importantly, what type to use, requires clearly defining the control needs, which include many conditions, such as improving profitability. Thus, based on requirements that have to be met, appropriate action is taken. Note that changes occurring in control programmers and data acquisition systems continually simplify the controlling processes and provide more accurate operations.

Purchasing a sophisticated PC system is not a foolproof solution that will guarantee perfect products. Solving problems requires a full understanding of their causes, and these may not be as obvious as they first appear (chapters 27 and 28). Failure to identify contributing factors when problems arise can easily result in the microprocessor not doing its job. The conventional place to start troubleshooting a problem is with the basics of temperature, time, and pressure requirement limits. Often a problem may be very subtle, such as a faulty control device or an operator making random control adjustments. PCs cannot usually compensate for such extraneous conditions; however, they may be included in a program that provides the capability to add functions as needed.

There are two basic approaches to problem solving: (a) Find and correct the problem applying only the control needed and (b) overcome the problem with an appropriate PC strategy. The approach one takes depends on the nature of the processing problem and whether enough time and money are available to correct it. PCs may in most cases provide the most economical solutions. Before investing in a more expensive system, the processor should methodically determine the exact nature of the problem to decide whether a better control system is available and will solve the problem. For example, the temperature differential across a mold (or die) can cause uneven thermal mold growth. The mold growth can also be influenced by uneven heat on equipment that has tie bars for platens. With injection molding, the uppers can be hotter, causing platens to bend where the change could be reflected on the mold operation. Perhaps all that is needed to correct the mold heat variation is to close a nearby large garage door to eliminate the flow of air upon the mold. With air-conditioning, all that may be required is to change its direction of airflow.

SENSOR

The PC is dependent on the ability to measure parameters such as the variability of temperature, pressure, and output rate. Sensors have traditionally played an important role in measuring and monitoring these broad ranges of parameters (427). All sensors perform the same basic function of the conversion of one type of measurable quantity, such as temperature, into a different but equally quantifiable value, usually an electrical signal. Although the basic function remains the same, the technologies used to perform that function vary widely (Tables 3.62 to 3.65).

Sensors can be categorized generally as being either physical or chemical in nature. Physical sensors are used to measure a range of physical responses, such as temperature and pressure. In addition to the most common types of physical sensors, optical and electrical, the category includes geometric, mechanical, thermal, and hydraulic types. Sensors that detect electrical activity include electrodes. Optical sensors are being used in a number of applications in which light is used to collect physical data. They are a key element of certain new technologies.

To select the correct sensor, you should know something about how the different sensors work (accuracy, repeatability, environmental effect, etc.) and which is used for what application. This is important since not all sensors measure the same way. The three most common sensors used downstream are nuclear, infrared, and caliper. There are also specialized types, such as microwave, laser, X-ray, and ultrasonic. They sense different conditions for operating equipment (temperature, time, pressure, dimensions, output rate, etc.) and sense color, smoothness, haze, gloss, moisture, dimensions, and much more (Fig. 3.65).

Process variables	Instrumentation
Temperature	Thermoelements Resistance elements, PT 100 Radiation recorder
Pressure	Strain gauges Resistance elements
Rotational speed	A-c/d-c-generators Incremental counters
Weight	Strain gauges Piezo elements
Velocity	A-c/d-c-generators Incremental counters Correlators
Torque	Strain gauges Scales
Time	Impulse counters
Logic state (on/off)	Switches

Table 3.62 Examples of process variables and sensors

Sensitivity and complexity increased as advances were made in electronics technology, including innovative circuit designs and more efficient power sources that followed the invention of vacuum tubes and transistors. Sensors today are evolving more rapidly than ever, resulting in microcircuits and nanotechnology, improved materials, and new design capabilities. Microfabrication and nanotechnology in recent years have had a significant impact on sensors. Microfabrication technology can be used to produce geometrically well-defined, highly reproducible structures and surface areas. Consequently, this may simplify or minimize the need for individual calibration (424–427).

Chemical sensors are designed to detect or measure the presence of specific chemical compounds. This category includes gas and electrochemical devices. Photometric sensors, which are optical sensors used to measure chemical presence, are also included in this category.

Type of sensor	Accuracy	Dimensional range	Transverse measurement	Reflex measurement	Freedom from interference	Material dependency	Sensor-size limitations	Complexity of equipment	Ease of calibration	Sensitivity
Rolling-contact	Good	Wide	Yes	No	Good	No	Low	Low	Easy	Med.
Air	Good	Wide	Yes	No	Good	No	Some	Med.	Easy	High
Magnetic-reluctance	Fair	To $\frac{1}{4}$″	Yes	Poss.	Fair	Some	Some	High	Easy	Med.
Sonic	Good	To 1″	Yes	Yes	Fair	Yes	Some	High	Fair	Med.
Optical	Fair	Wide	Yes	Yes	Good	Some	High	Med.	Fair	Med.
Laser-intercept	Good	Wide	Yes	Yes	Good	Some	High	Med.	Fair	High
Laser-interferometry	Exc.	Ltd.	No	Yes	Good	Yes	Some	High	Easy	High
Capacitance	Good	Med.	Yes	No	Fair	Yes	Low	High	Easy	Med.
Proximity	Good	Wide	Yes	No	Fair	Some	Low	High	Easy	High
Beta-ray	Good	Ltd.	No	Yes	Good	Some	Low	High	Easy	High

Table 3.63 Guide to performance of different sensors

Temperature	Time	Distance	Speed	Pressure or force	Profile
Melt	Hold changeover	Hold changeover	Screw rotation	Hold changeover	Injection pressure
Mold	Open	Fast open	Injection stroke	Injection	Injection speed
Nozzle	Carriage forward delay	Slow open	Carriage	Peak	Holding pressure
Barrel zones	Hold	Eject	Slow break	Holding	
Feed throat	Carriage back delay	Sprue break	Fast open	Back	
Pellets	Screw start delay	Suck back	Slow open	Nozzle hold-on	
	Cooling	Shot	Eject	Clamp close	
	Eject	Cushion		Clamp open	
	Cycle delay				

Table 3.64 Examples of injection-molding control factors

Parameter	Sensor type	Physical principle
Temperature	• Bimaterial thermometer • Resistance temperature detectors	• Thermal expansion of metal • Electric resistance of metals changes with temperature
Pressure	• Diaphragm • Bourdon tube • Piezoelectric	• Mechanical displacement • Mechanical/Electric resistance • Deformation/Electric charge
Proximity/Position	• Electro-optical	• Light waves • Sound waves • Magnetic fields • Electric fields
Photoelectric/Light sensing	• Photodetector	• Reflected, transmitted, scattered light
Flow	• Orifice plate, venturi tube, nozzle • Rotameter • Positive displacement • Turbine	• Differential pressure • Variable area • Volume • Revolutions per minute
Load cells/Strain gauges	• Piezoresistive foil gauges	• Electric resistance change
Electrical properties	• Voltmeters • Current (amps) devices • Wattmeters	• Mechanical torque • Frequency response, Hall effect • Power
Displacement	• Potentiometer	• Restive displacement
Vibration	• Accelerometer	• Seismic mass displacement
Acceleration	• Accelerometer	• Displacement, magnetic force
Velocity	• Proximity transducers, timing devices • Doppler shift radar	• Distance divided by time • Sound wave frequency
Moisture	• Capacitive humidity	• Permittivity
Mass	• Balance • Strain gauge load cell	• Standard comparison • Electric resistance
Level	• Float • Ultrasonic • Electrode	• Density • Time of flight • Capacitance
Gas/Vapor	• Cross duct monitors • Optic/Ceramic filter	• Beer-Lambert law • Ultraviolet and infrared wave
Vision	• Vacuum tubes • Charge-coupled devices • Photodiodes	• Light/Photoconductivity • Silicon/Light-sensitive photosite • Silicon
Chemical	• Electrodes • Thermobalance • Gas chromatograph	• Chemical potential (Nernst equation) • Physical parameter as function of temperature • Adsorption and partition
Tilt	• Tiltmeters • Accelerometer	• Liquid, vertical and horizontal pendulum • Lateral acceleration
Acoustic emissions	• Pressure microphone • Sound-level meter	• Sound waves, pressure • Sound frequency
Tactile	• Resistive transducer • Parallel-plate capacitor	• Force and electric resistance • Force and capacitance
Radioactivity	• Ionization chamber • Scintillators • Silicon wafer	• Ions, electrons, detector gas • Gamma rays • Alpha, proton, ion

Compiled by Nicholas Sheble, nsheble@isa.org

Table 3.65 Examples of sensor operations

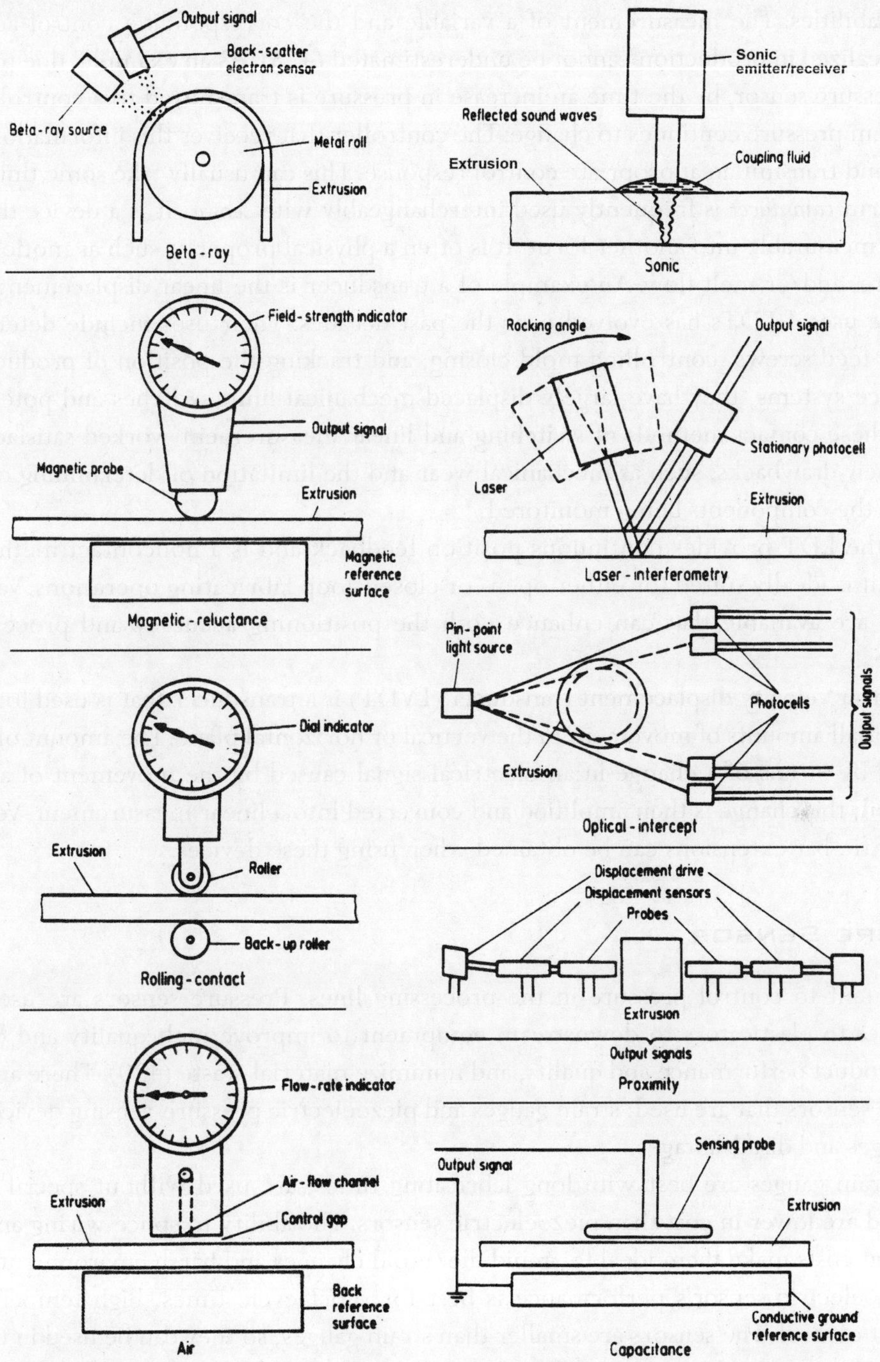

Figure 3.65 Different types of sensors

There is a very broad selection of sensors with extremely different sensitivities, capabilities, and repeatabilities. The measurement of a variable and the corresponding control action to the precision realized in production cannot be underestimated (423). As an example, due to a response lag in a pressure sensor, by the time an increase in pressure is transmitted to a control device, the actual system pressure continues to change. The controller that receives this information then must process it and transmit an appropriate control response. This can usually take some time.

The term *transducer* is frequently used interchangeably with *sensor*. It is a device that converts something measurable into another form. It is often a physical property, such as motion, pressure, temperature, and/or melt flow. An example of a transducer is the linear displacement transducer (LDT). The use of LDTs has evolved over the past decades. Their uses include determining the location of feed screws, controlling mold closing, and tracking the position of products. In high-performance systems, they have largely displaced mechanical limit switches and potentiometers. Although these contact methods of switching and linear measurement worked satisfactorily, they did have their drawbacks, such as mechanical wear and the limitation of determining only the end position of the components being monitored.

Since the LDT provides continuous position feedback and is a noncontact method of measurement, it is ideally suited for either open- or closed-loop fabricating operations. Various types of outputs are available that can enhance both the positioning accuracy and processing speed of machines.

The linear velocity displacement transducer (LVDT) is a transducer that is used for measuring relatively small amounts of movement in the vertical or horizontal plane. The amount of movement is detected by means of a change in an electrical signal caused by the movement of an iron core within a coil; this change is then amplified and converted into a linear measurement. Very accurate readings of tie-bar extensions can be obtained when using these devices.

PRESSURE SENSOR

It is important to control pressure in the processing lines. Pressure sensors are used from the feeding lines to plasticators to downstream equipment to improve melt quality and output rate, enhance product performance and quality, and minimize material waste (420). There are two types of pressure sensors that are used: strain gauges and piezoelectric pressure-sensing devices. Each has its advantages and disadvantages.

The strain gauges are best with long fabricating times, are used without special wiring, are rugged, and are lower in cost than piezoelectric sensors. The ability to splice wiring and their low replacement costs make them ideal for rapid die/mold changes and harsh environments.

A piezoelectric sensor's performance is best for short cycle times, high temperatures, and critical part control. The sensors are smaller than strain gauges, so they can be used in tight spaces, and they are immune to high electrical discharge and radio frequency noise. They require balanced wiring, charge amplifiers, and careful attention to ground loops. Piezoelectric sensors have quicker response times, typically greater than 20 kHz compared with 0.2 kHz for strain gauges.

These sensors play a critical role at every stage of the fabricating process. For example, because pressure affects the physical properties of a plastic material, controlling pressure is critical. Pressure translates into heat and shear, which can significantly change physical properties, and even chemical properties, in the finished product. Pressure sensors can help solve these control problems. In many processes, measuring the critical pressure points depends on one's technical knowledge and ability to justify the expense of time, personnel, and equipment. An important aspect is to optimize the number and location of pressure sensors. This ensures that plastics and machine time are put to optimum use and that production of quality parts is maintained.

Because pressure transducers from different manufacturers can vary significantly, it is important to understand their performances, such as accuracy. An ideal device would have a direct linear relationship between pressure and output voltage. In reality, there will always be some deviations; this is referred to as *nonlinearity*. The best straight line is fitted to the nonlinear curve. The deviation is quoted in the specifications and expressed as a percent of full scale. The nonlinear calibration curve is determined in an ascending direction from zero to full rating. This pressure will be slightly different from the pressure measured in descending mode. This difference is termed *hysteresis*; it can be reduced via electrical circuits.

TEMPERATURE SENSOR

Fabricating plastic products is a thermal process, with the major task to ultimately control temperature. Too much or too little heat at the wrong place can cause many problems. Understanding these temperature characteristic behaviors is important to successful fabrication. Pinpointing temperature accuracy is essential. In order to achieve it, microprocessor-based temperature controllers can use a proportional-integrated-derivative (PID) control algorithm acknowledged to be accurate. The unit will instantly identify varying thermal behavior and adjust its PID values accordingly (6, 425).

It is generally recognized that increasing the temperature of plastics increases their atomic vibration and molecular mobility, resulting in reduced melt viscosity. Thus, for example, during plastication when a plastic melt is too viscous, the first reaction could be to increase the temperature of the melt. The effect of molecular weight distribution (MWD) on this relationship becomes complex. With PEs broadening, the MWD decreases the sensitivity of melt viscosity to temperatures, whereas with PSs broadening, the MWD increases temperature sensitivity. Methods of expressing molecular averages and distributions, and the combined effects of branching, may be responsible for the discrepancy (chapter 1).

One cannot see thermal energy, only its effects. Thermal energy radiates in the IR spectrum, outside the spectrum of visible light. Use has been made of IR video cameras to detect energy color patterns in all locations around the machine and auxiliary equipment. With IR thermography, every plastic has its own wavelength and temperature readings are related to the IR color patterns. It also provides IR signatures for each plastic using Fourier transform infrared (FTIR) spectroscopy.

Temperatures can be measured with thermocouples (TCs) or resistance temperature detectors (RTDs). The RTD provides for stability; its variation in temperature is both repeatable and predictable. TCs tend to have shorter response times, while RTDs have less drift and are easier to calibrate. The RTD contains a temperature sensor made from a material such as high-purity platinum wire; resistance of the wire changes rapidly with temperatures. These sensors are about 60 times more sensitive than thermocouples.

The thermistor (TM) is a semiconductor device with a high resistance dependence on temperature. They may be calibrated as thermometers. The semiconductor sensor exhibits a large change in resistance that is proportional to a small change in temperature. Normally, TMs have negative thermal coefficients. Like RTDs, they operate on the principle that the electrical resistance of a conductive metal is driven by changes in temperatures. Variations in the conductor's electrical resistance are interpreted and quantified as changes in temperature occur.

A TC is a thermoelectric heat-sensing instrument used for measuring temperature in or on equipment, such as the plasticator, mold, die, preheater, and melt. The TC depends on the fact that every type of metallic electrical conductor has a characteristic barrier potential. Whenever two different metals are joined together, there will be a net electrical potential at the junction. This potential changes with temperature.

Trying to measure the melt temperature can be deceiving. For example, an extrudate with a room temperature TC pyrometer probe will often give a false reading, because when the cold probe is inserted, it becomes sheathed with the plastic that has been cooled by the probe. A more effective method is using what some call the 30/30 method. One simply raises the temperature of the probe about 30°F above the melt temperature and then keeps the probe surrounded with hot melt for thirty seconds. The easiest way to preheat the probe is to place the probe on, near, or in a hole in the die. By preheating above the anticipated temperature just prior to inserting it into the melt, the probe is actually cooled by the melt. The lowest temperature reached will be the stock temperature. It also helps to move the probe around in the melt to have the probe more quickly reach a state of equilibrium. To be more accurate, repeat the procedure.

Traditionally, PID controls have been used for heating and on-off controls for cooling. To control temperature, the more recent use is the fuzzy logic control (FLC). One of FLC's major advantages is the lack of overshoot on start-up, resulting in achieving the set point more rapidly. Another advantage is in its multivariable control where more than one measured input variable can affect the desired output result. This is an important and unique feature. With PID, one measured variable affects a single output variable. Two or more PIDs may be used in a cascade fashion, but with more variables they are not practical to use.

Fuzzy Logic Different logic control systems are used. An example is fuzzy logic control (FLC), which provides a way of expressing nonprobabilistic uncertainties. Fuzzy theory has developed and found application in database management, operations analysis, decision support systems, signal processing, data classifications, computer vision, and so forth. However, the application that has attracted most attention is control. FLC is being applied industrially in an increasing number of processing plants. The early work in FLC was motivated by a desire to directly express the control actions of an experienced operator in the controller and to obtain smooth interpolation between discrete controller outputs.

The FLC system approach can be used to solve problems. Many applications of FLC are related to simple control algorithms, such as the PID controller. In a natural way, nonlinearities and exceptions are included; these are difficult to realize when using conventional controllers. In conventional control, many additional measures have to be included for the proper functioning of the controller: anti-resist windup, proportional action, retarded integral action, and so on. These enhancements of the simple PID controller are based on long-lasting experience and the interface of continuous control and discrete control. The fuzzy PID-like controller provides a natural way to applied controls. The fuzzy controller is described as a nonlinear mapping.

Temperature Controller

For temperature controllers to obtain quality fabricated products, accuracy in their dynamic behavior, such as response time and transient performances, is required. Most controllers provide a derivative term that is called the *rate term*. This is an anticipatory characteristic that shortens the response time to changing conditions. There is also a circuit that limits overshooting. Controllers can differ widely, depending on requirements such as short response time, minimal overshooting, high circuit stability even with system (equipment and plastic) variations, transient suppression, and sluggishness to keep temperature variations small.

The art of trade-off has to be accepted in the adjustment of controllers and in the selection of suitable control elements. Not all demands can be met in an optimal manner at the same time. Controllers operate in a digital mode and employ microprocessors. The input signal is converted into a numerical value and mathematically manipulated. The results are summarized to obtain an output signal. This signal regulates the power output in such a way that the temperature is maintained at the set value.

Most processes operate more efficiently when functions must occur in a desired time sequence or at prescribed intervals of time (422). In the past, mechanical timers and logic relays were used. Now electronic logic and timing devices predominate. These programmable logic control devices provide sophisticated operations with controllable functions.

A calibration check of controllers should be made on a regular basis. The ISO 9000 standard reviews the development of the frequency of calibration checks. A visual examination should be made before proceeding with the check to determine that no damage exists. Some of the more common problems caused by a plant's hostile environment that can affect equipment, such as sensors/transducers, are noise interference, mounting holes (must be concentric and clean), installation, diaphragm considerations, and transducer calibration. Zero balance, full-scale sensitivity and R-cal at 80% parameter reference points for calibration can be used. The sensor/transducer manufacturer provides these parameters (426).

Processing Window

Regardless of the type of controls available, the processor setting up a machine uses a systematic approach based on experience or one that should be outlined in the machine and/or control manuals. It is a defined area or volume in a processing system's PC pattern. The processing window for a specific plastic part can vary significantly if changes are made in its design and the fabricating equipment used. Note that a major cause for problems with any process is not poor product design but the operation of processes outside their required operating window.

Once the machine is operating, the processor methodically makes one change at a time to determine the result for each change. It provides a range of processing conditions, such as melt temperature, pressure, and shear rate, within which a specific plastic can be fabricated with acceptable and optimum properties. As reviewed in chapter 4, windows such as molding area diagrams (MADs) and molding volume diagrams (MVDs) can be used during injection molding. The same approach can be applied to the other processes (extrusion, blow molding, etc.).

By plotting at least injection pressure (ram pressure) with mold temperature, an MAD will provide the best combination of pressure and mold temperature necessary to produce quality parts (Fig. 4.54). Developing this 2-D MAD approach results in a dramatic and easily comprehensive visual aid in analyzing these variables. Within the diagram area, all parts meet performance requirements. However, rejects could occur at the edges since materials and machine capability are not perfect; variability exists (chapter 1). Other controllable parameters can be added to improve quality, such as melt temperatures (in the plasticator, nozzle, and cavity), rate of injection, and so on.

After a 3-D MVD is constructed, it can be analyzed to find the best process settings of three combinations evaluated during start-up, such as melt temperature, mold temperature, and injection or ram pressure (Fig. 4.55).

This type of procedure can be used in setting up, as an example, a complicated IM product. As shown in Figure 3.66, IMM control settings involve melt temperature, mold cavity filling speed, shot size, melt pressure packing, and so on.

The term *PC* is often used when machine control is actually performed. As the knowledge base of the fundamentals of the fabricating process continues to grow, the control approach is moving away from press control and closer to real process control where material response is monitored and then moderated or even managed. The fabricator should note that changes in process parameters, such as injection rate, could have dramatic effects on moldings, especially mechanical properties, meeting tolerances, and surface properties.

Process Control and Patience

When making processing changes, have patience by allowing enough time to achieve a steady state in the complete fabricating line before collecting data. It may be important to change one processing parameter at a time. As an example, with one change such as extruder screw speed, temperature zone setting, cooling roll speed, blown film internal air pressure, or another parameter, allow four time constants to achieve a steady state prior to collecting data.

Process Control Trade-Off

There are units that provide independent control loops and usually only control one major variable (Fig. 3.67 to 3.72). There are also those that interact with different variables so that the operation

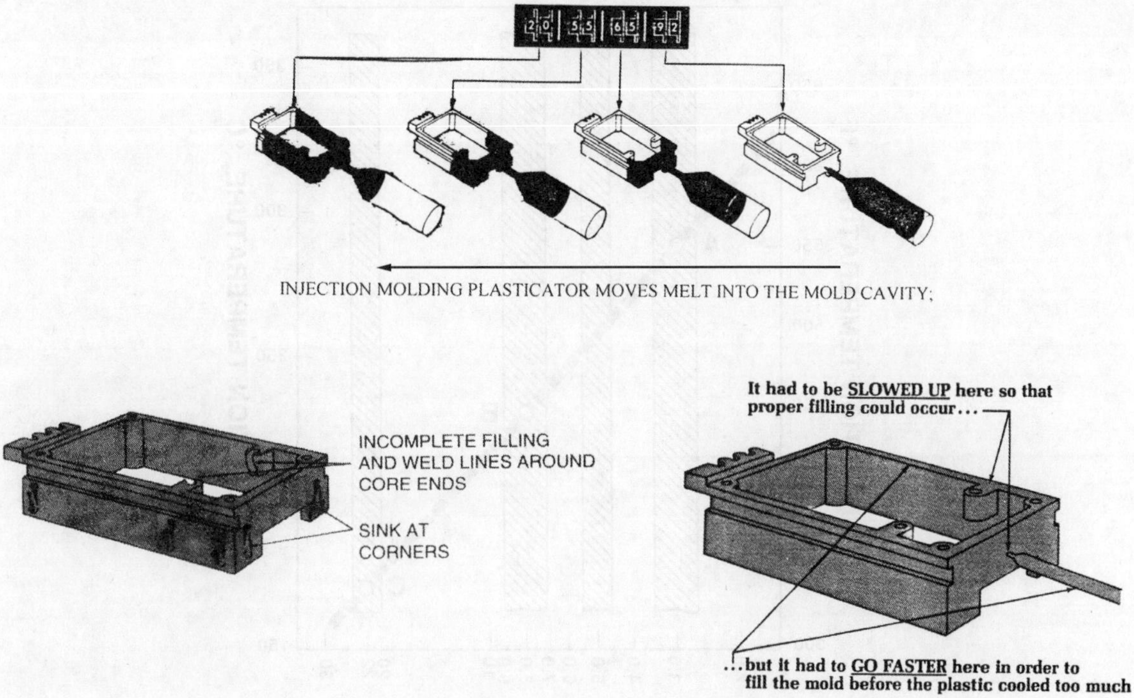

Figure 3.66 Example of setting process controls for a melt going from an IM plasticator into the mold cavity

of a complete line requires the skill of an operator(s). The available controls properly installed and used have been extremely useful to operator(s) by simplifying the setting up, operating, and shutdown of the line. Trade-offs are inevitable in these complex operations where controls provide major input on the action to be taken.

Lines can operate with different degrees of automation via computer-integrated PCs that provide improvements in operating procedures and quality assurance; rejects are reduced (if not eliminated) and fabricating costs are usually reduced. These closed-loop systems maintain long-term repeatability of factors such as melt velocity and pressure. All this action occurs independently of what could be occurring with equipment component wear, unbalance of equipment in the line, and/or plastic material variations.

Usually, elaborate control systems cannot correct for problems such as those caused by (a) a worn screw and barrel, (b) inadequate drive torque, and/or (c) poor screw design. For example, such systems will not yield good temperature control unless all features essential to good control are well maintained. Obviously, burnt-out heating elements cannot be tolerated. Other factors of these types also exist.

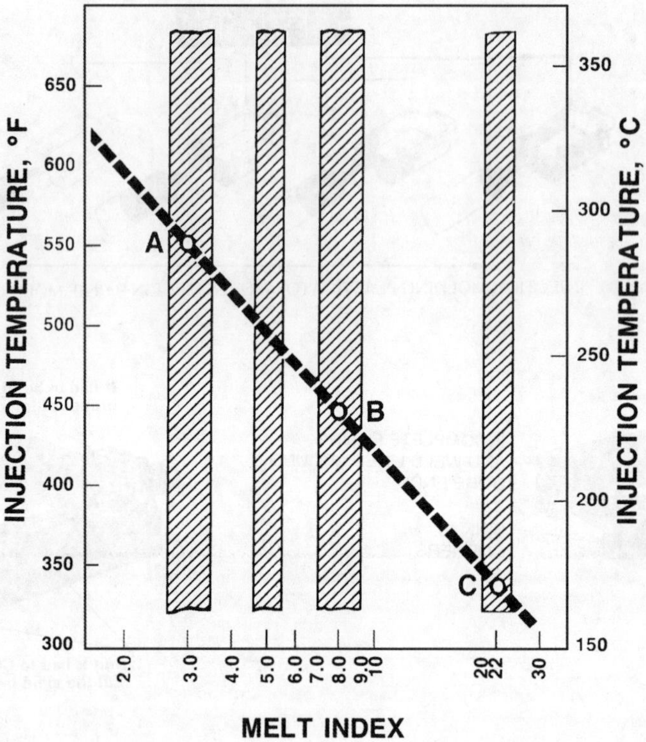

Figure 3.67 Effect of melt index (chapter 22) for a polyethylene on injection temperature

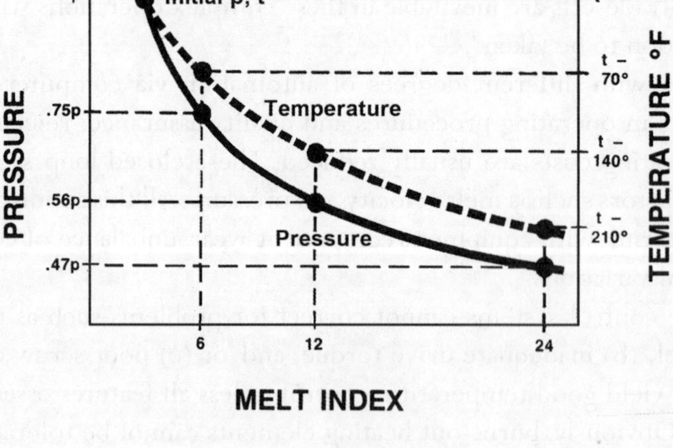

Figure 3.68 Effect of melt index (chapter 22) for a polyethylene on injection pressure and temperature

CONTROL AND MONITORING

The different fabricating processes have their own process controls whose common purpose is to produce products that meet performance requirements at the lowest cost. The following review of injection-molding (IM) controllers considers many factors that can also be related to other processes (304). It describes Moldflow Corp.'s Manufacturing Solutions for the automation, control, and monitoring of the IM process; the problems they solve; the methods used to solve those problems; and the ways in which they add value to the IM process (421).

The Manufacturing Solutions suite consists of three distinct products:

1. The Moldflow Plastics Xpert® (MPX™) process automation and control system decreases mold setup time, cycle time, and scrap, and improves molded part

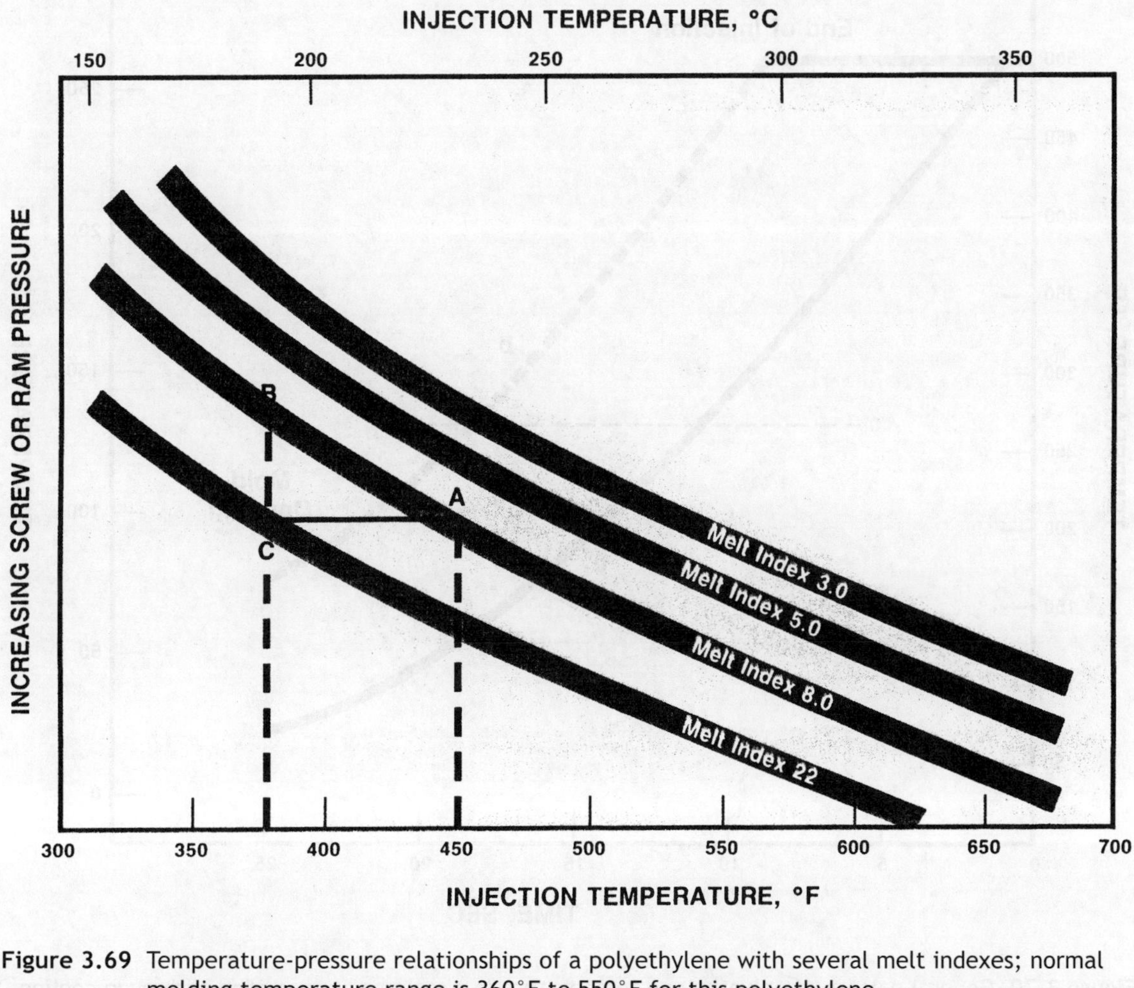

Figure 3.69 Temperature-pressure relationships of a polyethylene with several melt indexes; normal molding temperature range is 360°F to 550°F for this polyethylene

quality and labor productivity. Unlike traditional trial-and-error methods, MPX tools provide a consistent and systematic method for improving and optimizing the molding process.

2. The Moldflow Shotscope® process monitoring and analysis system collects critical data in real time from injection-molding machines on the factory floor, then records, analyzes, reports, and allows access to the information for use in critical decision making.
3. The Moldflow EZTrack™ production monitoring and reporting system is a product for real-time, plant-wide production monitoring and reporting. The EZTrack system can be attached to virtually any cyclic manufacturing equipment and machinery.

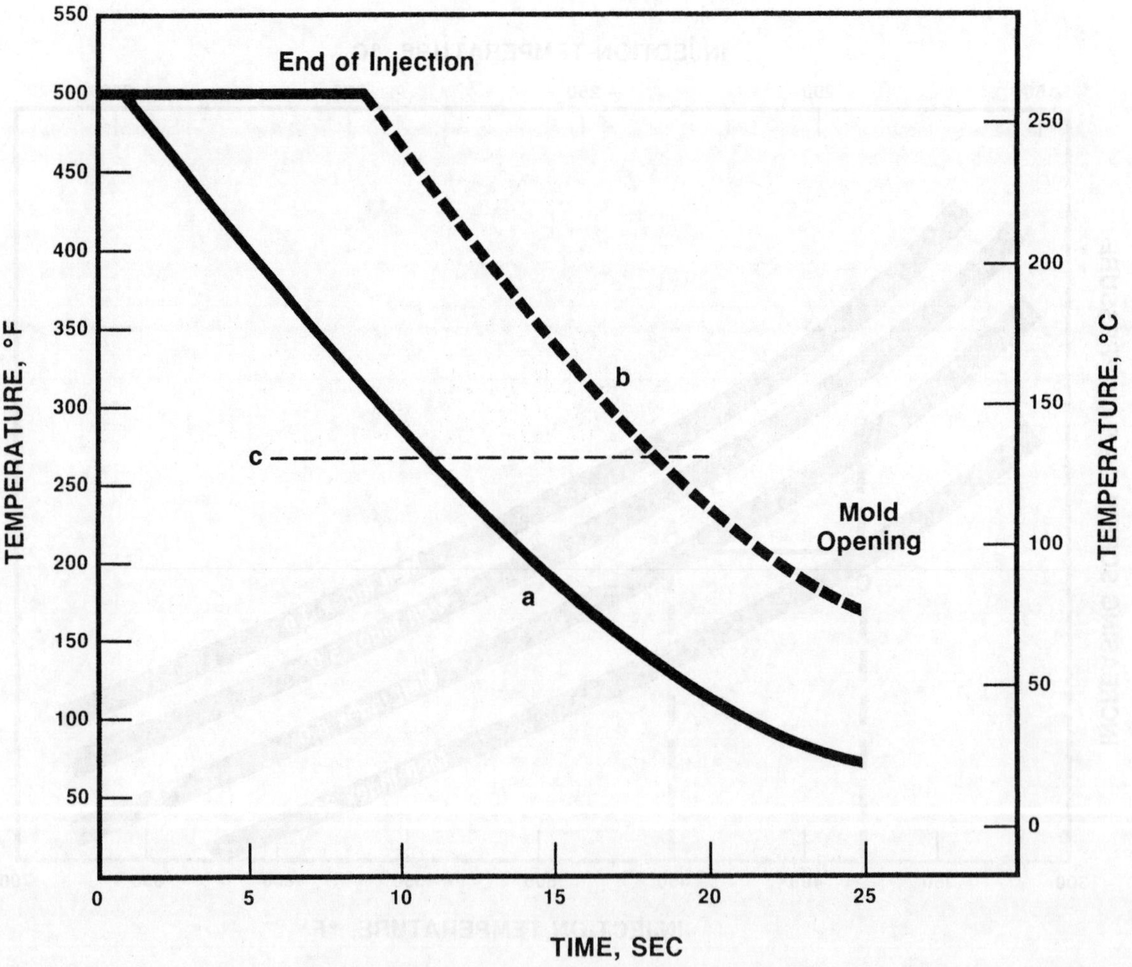

Figure 3.70 General pattern of polyethylene temperature in a mold cavity provided with even cooling

Moldflow Development In response to market feedback regarding existing plastic manufacturing practices, Moldflow developed a complete suite of manufacturing solutions for the automation, control, and monitoring of the injection-molding process. Custom and captive injection molders wanted a suite of products from one global supplier that would provide IM manufacturing personnel with all the tools necessary for the scheduling, setup, optimization, control, and reporting of the IM process. Specifically, customers pointed out that existing molding practices often resulted in:

1. inefficient scheduling of molds, machines, and labor resources;
2. long process setup times and associated scrap;
3. nonoptimized cycle times;

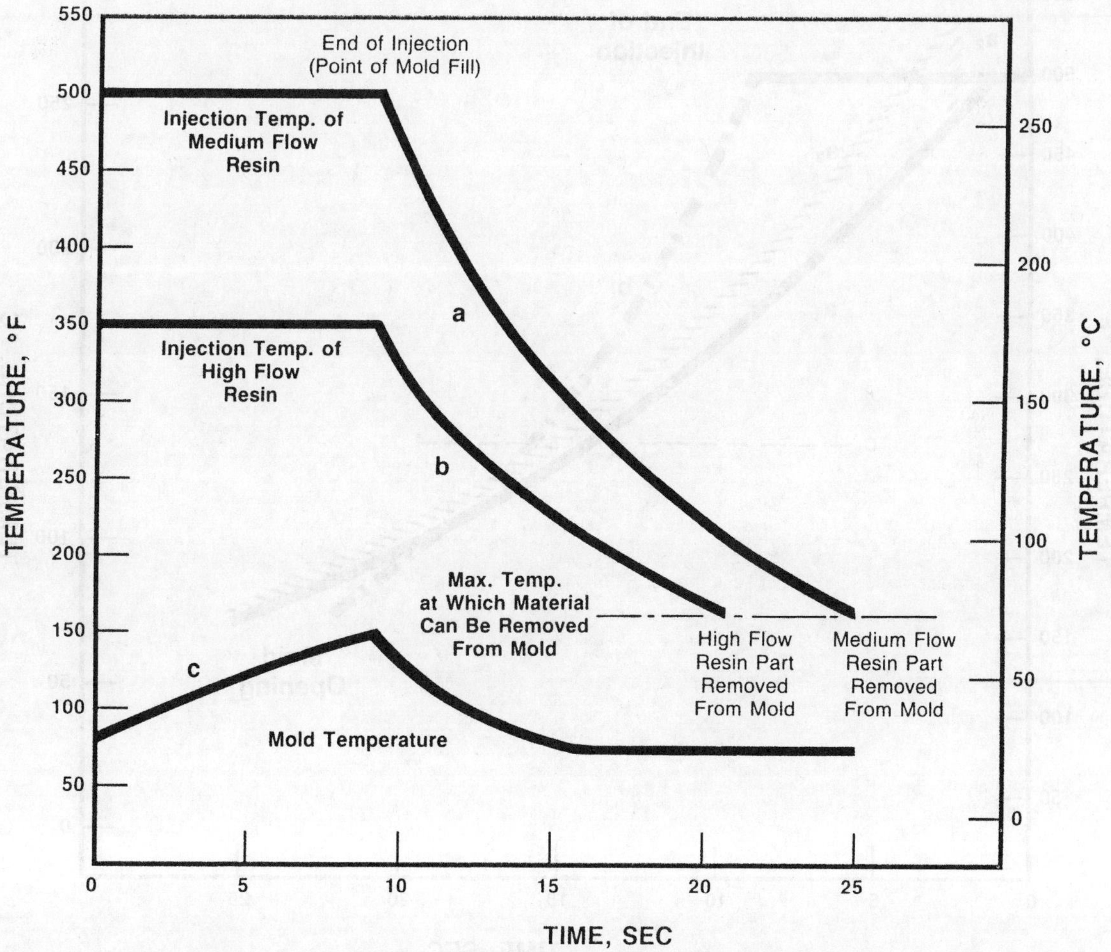

Figure 3.71 Curves a and b between the end of the injection and ejection of the molded product related to the cooling pattern (c) of the melt in the cavity

4. unacceptable molded parts quality;
5. unacceptable production scrap rates;
6. poor or inconsistent control of the molding process;
7. lack of part traceability;
8. lack of manufacturing management information.

These Manufacturing Solutions products address all of these problems with scalable solutions that will work for small custom injection molders and large multinational corporations alike. The following sections provide an overview of the IM industry and descriptions of how the Manufacturing Solutions products address these problems.

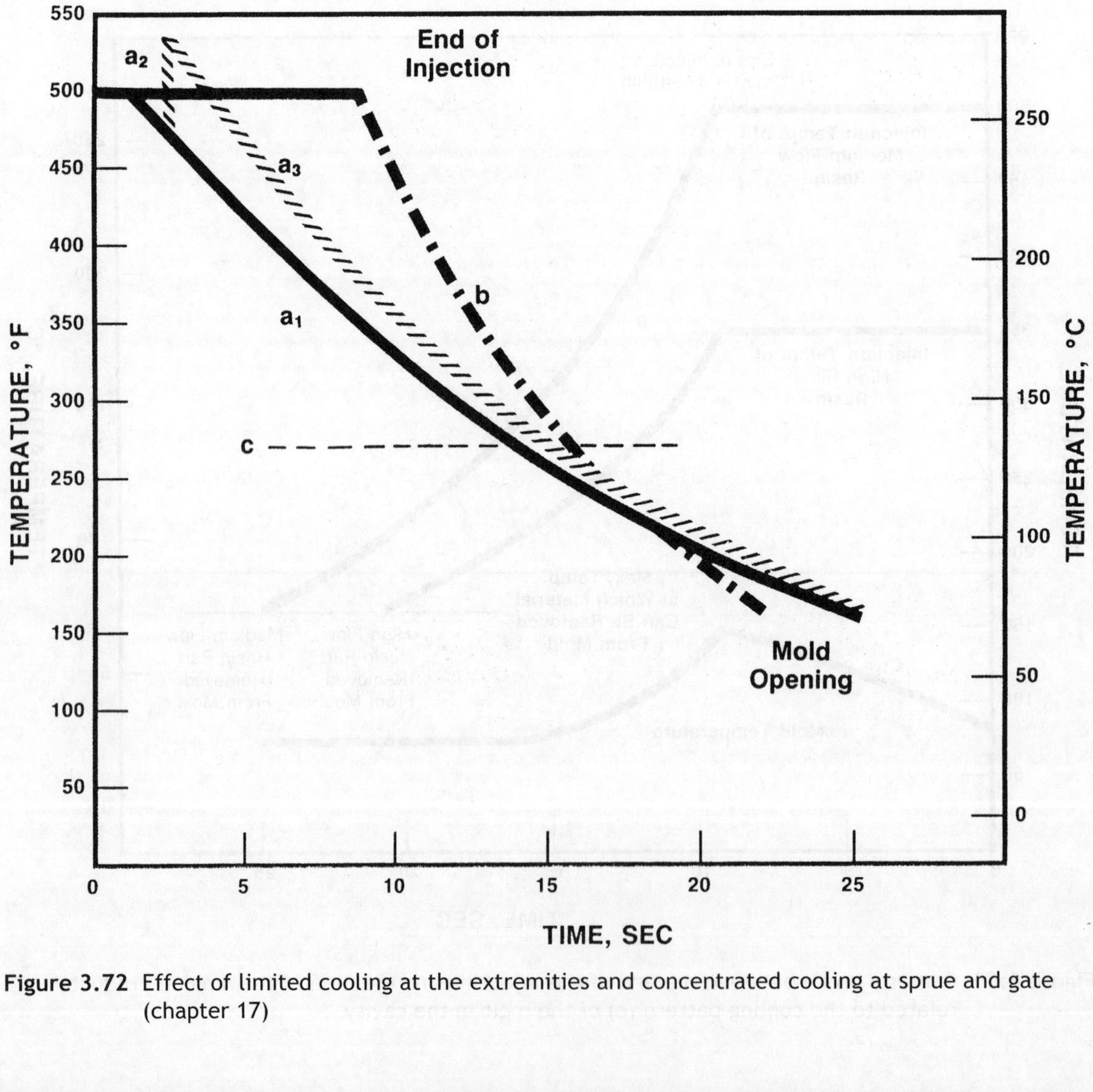

Figure 3.72 Effect of limited cooling at the extremities and concentrated cooling at sprue and gate (chapter 17)

Industry Overview The plastic IM process is integral to many of today's mainstream manufacturing processes. While demand for IM plastic parts is increasing, the problems associated with the process often can cause significant time delays and cost increases. This is because the IM process is a complex mix of machine variables, mold complexity, operator skills, and plastic material properties; in addition, there are constant pressures to reduce mold setup times and scrap, improve part quality, and maximize the productivity of every IMM. Because of this, it is becoming increasingly important to have systems in place to allow the molding process to be scheduled, set up, optimized, controlled, and monitored with an intuitive, systematic, documentable, and globally supported method.

Such a system must be intuitive so that machine operators can maximize productivity by not having to be experts on every machine/mold combination they are responsible for running. It must be systematic so that the process of setting up and optimizing the molding process can be done with a scientific method that does not rely solely on the skills of the machine operator. It must be documentable in order to meet the strict quality-control reporting requirements that are commonplace today. Finally, the system must be globally supported so that large multinational corporations can source these solutions from one supplier and implement companywide standards across their enterprises.

The following sections present a brief overview of each of the Manufacturing Solutions products, the problems they solve, the methods used to solve those problems, and the way in which they add value to the IM process.

Moldflow Plastics Xpert In an all-too-common occurrence throughout the IM industry today, the number of molds that must be set up and optimized for high-volume part production is far outpacing the number of process engineers or trained technicians qualified to do so. It is not uncommon for a molding operation to have a small number of individuals with the education or experience to set up the injection-molding process. Even those who can set up the process often do not have time to optimize it due to production pressures. This results in problems, such as long process setup times and associated scrap, nonoptimized cycle times, unacceptable molded parts quality, unacceptable production scrap rates, and poor or inconsistent control of the molding process.

People Moldflow Plastics Xpert (MPX) process automation and control technology provides machine operators with an easy-to-learn and easy-to-use tool for the setup, optimization, and control of the IM process. Finally, a tool exists that allows a less-experienced operator to set up molds, optimize the process, and control production.

Setup Xpert MPX functionality is arranged into three modules, the first of which is the Setup Xpert, a module that allows users to perform a variety of injection-, velocity-, and pressure-phase-related setup routines to fix certain defects, such as short shots, flash, burn marks, and sink marks. The objective of Setup Xpert is to achieve one good molded part with no defects. The basic process is that a user molds a part and then provides feedback to the MPX system regarding molded part quality. The MPX system then processes this feedback along with data being collected from the machine, and if necessary, determines a process change that will improve the result.

After completing Setup Xpert and determining a combination of processing parameters that results in a single good molded part, the user still does not know if these parameters are within a robust processing window. For example, any process parameter drift or variation could easily result in parts of unacceptable quality. In the IM process, variation is inherent. Whether the material, the machine, the process, the operator, or the environment causes it, there will always be some variation. This variation may or may not result in the production of bad parts. Since variation is normal, the processing window must be robust enough to compensate for it without producing bad parts.

Optimize Xpert It is common knowledge that design of experiments (DOE) is a useful tool in the fight to find a robust processing window. The process window basically is defined as the maximum amount of allowable process variation that still will not result in the production of bad parts. However, the historical perception of DOE is that it can be complicated, resulting in extensive training requirements and costs for those responsible for running it, and time consuming, thus increasing the time required to put a given mold into production.

The second module in the MPX system is the Optimize Xpert. Simply put, the Optimize Xpert is an automated DOE that can be run quickly and easily, and it does not require any special training in statistical process control. The goal of Optimize Xpert is to obtain a robust processing window that will compensate for normal process variation and ensure that acceptable quality parts are produced consistently.

While the Optimize Xpert DOE is automated, easy to use, and relatively fast to complete, it is far from simple. There are five process parameters that can be used as DOE factors: packing pressure, mean (or average) injection velocity, velocity stroke, packing time, and cooling time. In addition, there are a number of molding defects that can be used to measure part quality criteria, including short shots, flash, sink marks, burn marks, poor weld line appearance, weight, dimension, and warpage. Assuming a robust processing window is determined using the Optimize Xpert, control mechanisms are still required to make sure that the process stays within its specified limits.

Production Xpert The third module in the MPX system is the Production Xpert. The Production Xpert is a comprehensive process control system that maintains the optimized processing conditions determined with Optimize Xpert. Production Xpert allows the user to maintain the production process consistently, resulting in reduced reject rates, higher part quality, and more efficient use of machine time. If desired, the Production Xpert will correct the process automatically should it drift or go out of control.

Statistical Process and Quality Control One goes through MPX to set up, optimize, and control the IM process. There are still many functions required in a manufacturing operation, including production scheduling, process monitoring, statistical process control (SPC), statistical quality control (SQC), scrap tracking, production monitoring and reporting, and preventive maintenance scheduling. Moldflow meets these requirements with two additional Manufacturing Solutions products, which are described in the following sections.

Moldflow Shotscope The Shotscope process monitoring and analysis system collects critical data in real time from IMMs on the factory floor, then records, analyzes, reports, and allows access to the information for use in critical decision making.

Preventative Maintenance The Shotscope system allows injection molders to maximize their productivity by providing the tools necessary to schedule mold and machine resources efficiently and to monitor the status and efficiency of any mold/machine combination. By monitoring the efficiency of a given mold/machine combination, molders can schedule jobs based on a number of criteria, including minimum cycle times, highest production yields, and so on. Users also can define periodic maintenance schedules for molds and machines, and after a predetermined number of cycles or operating hours, Shotscope will signal that preventive maintenance is required.

Statistical Process Control The Shotscope system also maintains and displays SPC data in a variety of formats, including trend charts, X-bar and R charts, histograms, and scatter diagrams. This information provides molders with the knowledge that their processes are in control, and should they go out of control, Shotscope can alert to an out-of-control condition and divert suspect-quality parts. Furthermore, because the Shotscope system can measure and archive up to fifty process parameters (pressures, temperatures, times, etc.) for every shot monitored, the processing "fingerprint" for any part can be stored and retrieved at any time in the future. This functionality is extremely important to any manufacturer concerned with the potential failure of a molded part in its end-use application (for example, medical devices).

Finally, the Shotscope system contains a reporting mechanism that allows all the data collected and entered into the system to be communicated across a manufacturing enterprise so that informed decisions can be made. Users can generate production, scrap, downtime, efficiency, and job summary reports, any of which also can be used as documentation that accompanies part shipments.

Moldflow EZTrack The third product in the Manufacturing Solutions suite is the Moldflow EZTrack system, a product for real-time, plantwide production monitoring and reporting. It is truly plantwide because the EZTrack system can be attached to virtually any cyclic manufacturing equipment and machinery, such as ultrasonic welders, assembly machines, and packaging equipment, in addition to injection-molding machines. The EZTrack system provides a scalable solution for production monitoring, which can be used by small custom molders with fewer than ten machines or by large multinational corporations with distributed IM and manufacturing operations around the world. There are extensive setup capabilities that allow for the complete definition of resources and flexible customization of most displays and reports.

The EZTrack system collects data on cycle times, cycle/part counts, and number of rejects; it uses this data as the foundation to perform powerful scheduling tasks. The EZTrack scheduler can check for mold conflicts and machine feasibility, highlight any problems, and continuously update estimates of job completion times based on actual cycle time, downtime, rejects, and cavitation. In addition, the scheduler also supports family molds.

Monitoring The EZTrack system monitors machine status, downtime, scrap, raw material usage, and labor activity. It can also be used to track machine efficiencies and compute yield efficiencies. Labor, time, and attendance can be tracked by employee and associated with machines, jobs, and activities. In this way, manufacturing managers can determine what jobs, machines, or activities require more labor resources than others require. This then allows managers to investigate areas where more efficiency, possibly in the form of process automation, could be introduced into their manufacturing operations.

The EZTrack system also can be used to count good parts, diverted parts, packed cases, and so forth. Downtime is measured automatically and can be classified into an unlimited number of causes. Once production data are collected, an extensive set of Web-based reports can be generated that incorporate trend charts, tabular reports, pie charts, and Pareto charts. Finally, it is possible to interface the EZTrack system to ERP/MRP systems via an advanced SQL database that is open, fully documented, and ODBC-compliant. Not only is the EZTrack system scalable from small to large numbers of molding machines and other types of cyclical manufacturing equipment, it also can play an important role in sending real-time production data to companywide ERP/MRP systems.

Conclusion There are many companies today across a broad range of industries for which plastic injection molding and related upstream and downstream manufacturing processes are on the critical path to achieving successful and profitable product launches. These companies face a variety of issues that make it difficult to remain competitive:

1. Product life cycles are decreasing while short-term volume requirements are increasing exponentially.
2. Customers continue to demand increased quality at lower costs.
3. There is a shortage of skilled labor to run ever more sophisticated IM equipment.
4. Inefficiencies in the scheduling, monitoring, and reporting of production do not allow for efficient manufacturing management.
5. Molded part process documentation and traceability increasingly are becoming standard requirements.

These companies require tools that are intuitive, systematic, documentable, and globally supported to remain competitive on a global scale. Moldflow's Manufacturing Solutions products directly address these needs.

The Plastics Xpert system applies and automates the process of scientific molding to decrease mold setup time, cycle time, and scrap, and to improve molded part quality and labor productivity.

The Shotscope process monitoring and analysis system collects critical data in real time from injection-molding machines on the factory floor and then records, analyzes, reports, and allows access to the information for use in critical decision making.

The EZTrack production monitoring and reporting system enables real-time, plantwide production monitoring and reporting.

For the latest information on all of the Moldflow Manufacturing Solutions and Design Solutions products, visit www.moldflow.com.

PROCESS CONTROLLER

In summarizing the technology of PCs, it can be said that the solid-state controller consists of a set of base functions. Each requires a programming method, a logic engine, an operator interface, communication interfaces to I/O with a programming system, an operating system, and a hardware platform/package. To make an informed decision on which controller is best for an application, it is important to understand that for each controller there is a choice regarding its different levels of risk or responsibility. With low user risk, there is higher vendor responsibility; with higher risk, there is lower vendor responsibility.

This range of vendor-to-user responsibility is an essential consideration for making an informed hardware choice. There are choices available for both controllers and the functions within them. The fundamental requirements involve speed, scale, packaging, reliability, and peripherals. These requirements must be satisfied before successfully applying any control system technology.

Control Choice Control choices continue to provide many novel approaches. The user has to recognize what to specify, such as soft, programmable, hybrid, or entirely new architectures. One should define control not by the PC box performing it, but rather by virtue of the problem it solves. To do this, users must consider the choices available at each level within a controller. The choice requires personal knowledge or help from a reliable source to determine which type of controller is appropriate for a specific application.

Today's programmable microprocessor controller operating systems are the result of about a half century of evolution to provide the available repeatability and reliability required on the plant floor. In the past, achieving these objectives meant choosing a vendor-specific operating system and choosing that vendor's entire control system. This can be a benefit if risk is to be avoided, but it can be detrimental if a high degree of in-house customization and integration is desired.

Intel Corp. basically started logical circuits of a microprocessor using a single silicon chip in 1969 in controlling IMMs. Up to that time, in conventional logic control all sequential functions were programmed by means of an electrical wiring matrix. In a microprocessor control system, this matrix is replaced by software that contains the instructions.

The processor is told how to process the signals from the operator panel. These signals are linked with other process parameters that are stored in electronic memory modules. This sequence logic can be divided into individual independent blocks with separate microprocessors running their programs. By this multiprocessor technique, which is a modular design, the microprocessor takes over the entire logic of the IMM whereby it has complete control in operating the machine, producing products that meet performance-to-cost requirements (194).

PCs are fairly simple devices to operate. If they do not function properly, all kinds of problems develop. A checklist for eliminating problems includes nonuniform processed plastic; heater element burnout; location and depth of sensors as related to response times; type of on/off control action, such as proportional controller; set point control; and basic electrical component proper selection. It is important to have the proper depth of the sensor in a barrel in order to obtain the best reading for the melt; the deeper, the better.

Where water is involved in a PC (as in mold/die cooling controllers), water leaks are the most common problem as a result of improper construction (due to expansion/contraction not being properly incorporated). With their external pressure relief valves, they ensure discharge outside the cabinet. With inside discharge, severe damage can occur to mechanical and electrical components.

Different computer designs are used, such as computer coordinator controllers. These groups of controllers are connected together so that they may all be changed at the same time from a single point. Multizone microprocessors that monitor temperature, pressure, output rate, and so on, as well as signals from several sensors to achieve more reliable and efficient performance (either independently or coordinated) also are used. The microprocessor has to carry out functions, including:

1. standard sequence control, such as timer, temperature, malfunction indication, and so on;
2. monitoring functions, such as self-diagnosis of malfunctions, control of setup procedures, calculation of operating data, and troubleshooting;
3. control functions, such as different temperatures or pressures in a mold/die, output speeds, melt consistencies, etc.

What is required of this equipment is to ensure that all its actions include what has to be carried out to produce the required products. Determining operating points by trial and error is inevitably eliminated.

Intelligent Processing

The aim during fabricating (design, etc.) is to cut inefficiency, such as the variables, and in turn cut the costs associated with them. The intelligent processing (IP) approach is a technology that utilizes sensors, expert systems, and process models that control processing conditions as materials are produced and processed without the need for human control or monitoring. Sensors and expert systems are not new in themselves. What is new is the manner in which they are put together (temperature, pressure, and other variables). In IP, new nondestructive evaluation sensors are used to monitor the development of a material's microstructure as it evolves during production in real time. These sensors can indicate whether the microstructure is developing properly. Poor microstructure will lead to defects in materials. In essence, the sensors are inspecting the material online before the product is produced.

Information from these sensors and data from conventional monitoring sensors are gathered and sent to a computerized decision-making system. This decision maker includes an expert system and a mathematical model of the process. The system then makes any changes necessary in the production process to ensure the material's structure is forming properly. These might include changing temperature or pressure profiles, or altering other variables that will lead to a defect-free fabricated product.

With IP, different benefits occur, such as a marked improvement in overall product quality and a reduction in the number of rejected parts, thus reducing cost. It is important to note that IP involves building in quality rather than attempting to obtain it by inspecting a product after it is manufactured (chapter 23). The result is reducing postmanufacturing inspection costs and time.

Being able to change manufacturing processes or the types of material being produced is another potential benefit of IP.

Intelligent Settings and Controls IP is applicable to different processes. The following review concerning IM can be related to other processes. When compared with other processes, IM requires rather sophisticated controls to produce products meeting performance requirements at the lowest cost. The need to change the way IMMs are initially set for a new job was recognized by the industry in the mid-1980s. As reviewed by Anne Bernhardt of Plastics & Computer Inc. (P&C) (72), originally CAE analysis with machine settings was used. To further advance PC, an expert system was developed to assist the settings of IMMs by an operator interface that requires a description of the mold and the part features.

This action occurred because the IM industry has always been pushed by demands for higher levels of product quality and productivity, both obtained and maintained without the costly errors and corrections experimented with by many users and processors. Unfortunately, its implementation is very challenging due to the complexity of the IM process. The governing input variables (material and equipment) are not generally measured directly, and the output (properties in terms of dimensions, appearances, strengths, etc.) is difficult to measure directly, especially in real time.

The solution to the problem require the following steps: (a) for each specific application determine all parameters of the process that influence part quality, (b) compare relationships of these parameters with the ones used by the IMM controls, (c) inform the machine controls of all possible quality problems in the shortest possible time, and (d) program the IMM controls to react to each changing situation.

The evolution in this development has taken place by understanding two conditions. For machine capability, there are requirements of more repeatability, more precise (multipoint) settings, self-tuning (closed-loop) controls, statistical process controls (SPC), and the possibility to fully automate a molding cell with all the necessary auxiliaries. The other condition concerns the determination of the molding conditions that involve more precise modeling of the relationship between part/mold design, processing parameters, and quality/productivity values. It became clear that the self-regulation of machines could be effective only when the part/mold designs were optimized and when at least the major parameters required for the correct processing conditions were predetermined. It opened the way to computerized process simulation to replace the traditional trial-and-error method for molding optimization.

Program Analysis This approach resulted in the development by C&P of a complete TMconcept analysis of filling, packing, and cooling in a microprocessor-controlled IMM. It is important to understand that such activity is not confined to a simple reorganization of data but requires an additional set of calculations.

For example, whereas the CAE calculations for the filling phase are based on the melt temperature entering the nozzle (in the best case where the finite element analysis [FEA] algorithm allows it; chapter 19), the machine settings to reach this temperature are affected by barrel temperature profile, screw speed, and back pressure. It calls for the use of empirical models to set parameters not considered by the typical CAE programs. A statistical method based on experimental data, previously developed for the *TMconcept-MCO* package for molding optimization, was taken for these calculations. The machine characteristics database, already part of the package to estimate the moldability, was enlarged to define more precisely the process capability essential to the settings, for example, plasticating rate at different rpm.

The requirements were (a) the need for an interface between the FEA workstation and the molding machine, (b) the need to replace the IMM with new generation machines equipped with advanced microprocessor controls, (c) working around the lack of the complete set of CAE analyses for many commercial parts, and (d) a solution for the recurrent problem of part design modifications during mold construction not being included in the CAE analysis. The first two points became more moot since the advent of PCs made the power of a workstation affordable to any molder. Many IM shops use PCs for production management systems, thereby simplifying the connection problem to the microprocessor of new machines that are gradually replacing the old ones. More CAE analyses are made today, so the third point is less critical and the fourth one is improving.

Stand-alone systems were studied for the computation of the initial IMM settings. There was an inherent advantage in limiting errors when using a computerized system. Errors in machine settings are difficult to identify, and they may cause not only production inefficiencies but may even damage the mold and the machine. Two types of errors can be recognized. They are mistakes in data entry and mistakes due to insufficient knowledge.

The basic concept behind the development of the program was the realization that the initial setting of IMM parameters should be based on a description of the mold and part specifications instead of the process. This has the potential to eliminate two steps from the current traditional three-step process that requires the operator to understand the features of the mold, material, and part requirements: (a) relying on one's experience to set the parameters for the above and (b) needing to enter the settings in the IMM controller.

The resulting program created a link between IMM capabilities, machine settings, and mold/part characteristics. The need to detect if a particular machine was certainly not or may not be suitable for a specific job was clear from the beginning. This situation is not typical, since the causes of machine inadequacy are not always directly evident. While the size of the tool is an obvious factor in the machine suitability, the clamping force requirements can remain questionable even with a complete set of the most advanced FEA analysis (chapter 19). The latter must be integrated with part quality, cost considerations, and skilled judgment of the complete project.

Warning messages were an important part of the program from the beginning of its development. These messages indicate, with various degrees of probability, the potential for the occurrence of a particular problem, helping the operator make a decision on how to handle a specific situation. Machine settings were integrated with problem analysis and strategy for continuous improvements. Along with the program development, a quick setup was added to the machine microprocessor so that the operator did not have to adjust all the parameters not strictly relevant to the molding process, but only those key to machine operation. In this way, to avoid the risks due to unnecessary machine overcharging, reasonable values are always set. Whenever necessary, the operator can fine-tune these parameters.

This procedure highlights that no system can claim any initial settings are the best IMM settings, regardless of the quality of previous calculations and operator experience; the complexity of the molding process makes this claim inappropriate for any experienced engineer. Fine-tuning is always possible. For critical production, one needs to consider the possibilities for continuous productivity improvement and molding process robustness.

Fuzzy Logic To accomplish the above, two major developments were made in the CAE programs. One development was the application of fuzzy logic in expert software to examine mold-filling simulations for potential processing problems and advise on ways to rectify them. The fuzzy logic was required because the correct judgment of a specific part quality parameter (e.g., shear stress) cannot be made without weighting other parameters, such as local melt temperature and its variability.

Troubleshooting The other development related to troubleshooting of defects to provide guidance to test and debug start-up and also solve problems in operations that had been running well. These defects are considered qualitative in nature because the machine cannot detect them and they need operator judgement. The system forces the operator to define the problem in terms of occurrence (such as in one cavity), nonoccurrence (not in other symmetrical locations), where did it occur (near the gate, etc.), and when it occurred (from time to time).

Experience has shown that once the problem is well described, the ability to find the right solution instead of providing a set of possible remedies is strictly connected to the knowledge of existing molding conditions. Therefore, the goal is to have a way to keep track of the evolution of these conditions. The program starts the settings like its predecessor, provides a direct comparison of evolving parameters whenever required by the operator, and asks to enter the quality conditions.

At current development levels, the solution must be implemented by the operator even in cases where it could be made directly by the machine controls. For example, too low a shrinkage can be corrected by a decrease in holding pressure.

Because of the risks of automatic implementation, there will be, at least for a long time to come, problems that require manual operations. An example is material clogging the throat of the hopper, a problem that can be minimized by proper equipment design but never eliminated.

Regardless, one thinks that an intelligent machine control should always explain the reasoning behind its suggestion. This is the essence of all intelligent actions, which by definition require judgment to solve situations that do not have simple, exact answers. The strategy for initial settings and fine-tuning includes consideration of energy requirements and optimizes this important aspect.

PROTOTYPING MODEL

When a plastic product is to be fabricated, usually the first step is to create a model. This approach is similar to design in other materials. A design is prepared and revised until it is an acceptable product with regard to function and cost. Many times, the products being designed incorporate features that exist in other products, such as bosses, ribs, and snap fits, and standard design practices, such as using standard mold/die components and allowing for shrinkages (chapter 19). Following the product model is creating the mold/die geometry (chapter 17).

The basic point of the preceding illustration is that there is a great amount of repetitive information flow in the design process. As a result, what is perceived to be an extremely creative process is actually very repetitive in nature. The types of analytical problems that are encountered in the mold/die design process generally fall into the sciences of fluid mechanics and heat-transfer theory.

These fields encompass many complicated mathematical functions and relationships that were too time-consuming to evaluate in manual or conventional designs. The ability of the computer to remember and execute these computations quickly has added a new dimension to the mold/die design process, allowing prospective design alternatives to be evaluated and simulated by computer rather than in the molding press. Computer aids (CAD/CAM/CAE/CIM; chapter 25) are enabling the creative energies of plastic product and mold/die designers to be spent in producing better designs in shorter time periods rather than performing repetitive design tasks. CADs/CAMs/CAEs/CIMs have bridged the gap between designer and tool (mold/die) maker, benefiting the plastics industry as a whole (chapter 17).

ENERGY

When examining energy consumed or lost, the equipment used in the complete production line as well as the plastic is involved (chapter 31). Regarding electric energy, 1998 was the year that the U.S. government subsidy to manufacturers stopped. Since then, the cost of energy has doubled and tripled, setting market forces in motion that make energy-efficient all-electric equipment desirable. As an example, an injection molder facing a mandatory energy cutback realizes that two or three all-electric IMMs could run on the same power as it takes to run one hydraulic machine of the same size. As the electric injection machines become larger, their energy advantage over hydraulics increases (208).

SAFETY

Throughout this book, different aspects of safety have been presented. Safety in designing products, handling plastic materials, operating processing equipment, and fabricating products exists from the past into the future as an issue of ongoing concern. Figure 3.73 highlights the trend of accidents in fabricating plants based on the National Safety Council's data. Different software and Web sites are available about safety concerns when working with plastics (chapter 25).

MACHINE SAFETY

Operator safety is given high priority in the design of equipment. For example, devices are equipped with safety gates that are operated electrically, hydraulically, mechanically, or a combination thereof.

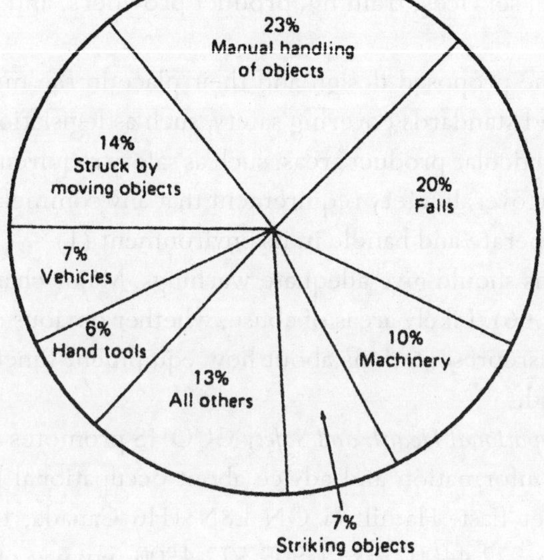

Figure 3.73 Examples of accidents in fabricating plants

Many manufacturers design their machinery so that all machinery movements halt if a safety gate is opened during machine operation. On some equipment, the mold/die area rear guards and other sliding guards that usually are not opened during normal use are designed in such a way that opening them during the operation will cause all the drive systems to stop. Electrically interlocked purge shields prevent purging of the barrel unless the safety gates, sliding guards, and the purge shield are all closed. Programmable controllers are designed with the safety interlock features hardwired to prevent inadvertent bypassing of the safeties when a program is modified or updated.

Manufacturers take great care to design machines that offer high levels of protection for the operators. Employers must also take steps to provide information and instruction to operating personnel covering safe operating and maintenance procedures, understanding of safety signs on the machines, and safety recommendations contained in the manuals supplied with the machines that pertain to ANSI, National Safety Council, and other safety bulletins/standards. ISA Society has a CD on process safety standards. Employers should take full advantage of the information and formal training on machine and operator safety available from manufacturers, the Society of the Plastics Industry Inc., ISO, and other sources.

Material Safety Data Sheet A national repository is dedicated to providing the most comprehensive source of information related to the document known as a material safety data sheet. MSDSSEARCH, Inc., supplies MSDSs directly from manufacturers on its Web site: www.msdssearch.com. As the largest centralized reference source on the Internet, this site offers free access to over 1 million MSDSs, including 350,000 MSDSs from over 1,600 manufacturers and

700,000 MSDSs from public access databases. The site also provides links to government information and MSDS software, services, training, product providers, and a discussion forum where one can ask questions.

The safety aspects of the proposed design and their place in the market must be considered. There may be legislation and standards covering safety, such as legislation on product liability, and specifications relevant to particular product areas, such as safety requirements on medical electrical equipment. Important is the overall safety requirement that any commercial and industrial product designed has to be safe to operate and handle in its environment (1).

Labeling and safety signs should give adequate warnings. Major changes are occurring in U.S. standards via ANSI (Table 3.66). Likely areas of abuse, whether obvious or not, should also be considered, as well as likely misrepresentations about how equipment functions. Definitive operating instructions may be required.

Canadian Center for Occupational Health and Safety CCOHS promotes a safe and healthy working environment by providing information and advice about occupational health and safety. Contact CCOHS at 250 Main Street East, Hamilton ON L8N 1H6 Canada, tel: 1-800-263-8466 (toll-free in Canada only)/1-905-572-4400, Fax: 1-905-572-4500, www.ccohs.ca/products/databases/msds.html.

Injection Molding Safety Issue

Since injection molding is a high-pressure, high-speed process, it is reasonable to assume that the molding machine generates a great deal of force and heat. Thus, machine safety is a must to ensure operator safety. A machine without adequate safety guards is significantly dangerous to the operator and other personnel working in the area. Processing equipment has standard procedures to operate and meet safety requirements (428).

Safety information and standards are available from various sources, including equipment suppliers, National Safety Council, Society of the Plastics Industry (SPI), American National Standards Institute (ANSI), Occupational Safety & Health Administration (OSHA), International Organization for Standardization (ISO), and European Machinery Safety Directive (EMD).

For a long time, there has been increasing activity on the part of equipment manufacturers to upgrade safety and fabricating plants. Examples of safety features vary for different equipment. Safety interlocks ensure that equipment will not operate until certain precautions have been taken. Safety machine-lockout procedures are set up for action to be taken to properly lock out the machine's electrical and mechanical circuits.

The operating environment is continuously upgraded with reduced sound and noise in the operating areas. The National Safety Council annually provides general statistics on where accidents occur in all types of U.S. manufacturing plants, including plastic plants.

All processing equipment, including the different auxiliary equipment, should have safety devices and procedures to operate and meet safety requirements. They should include a checklist that reviews preparation (moving material, etc.), start-up and shutdown procedures, tooling

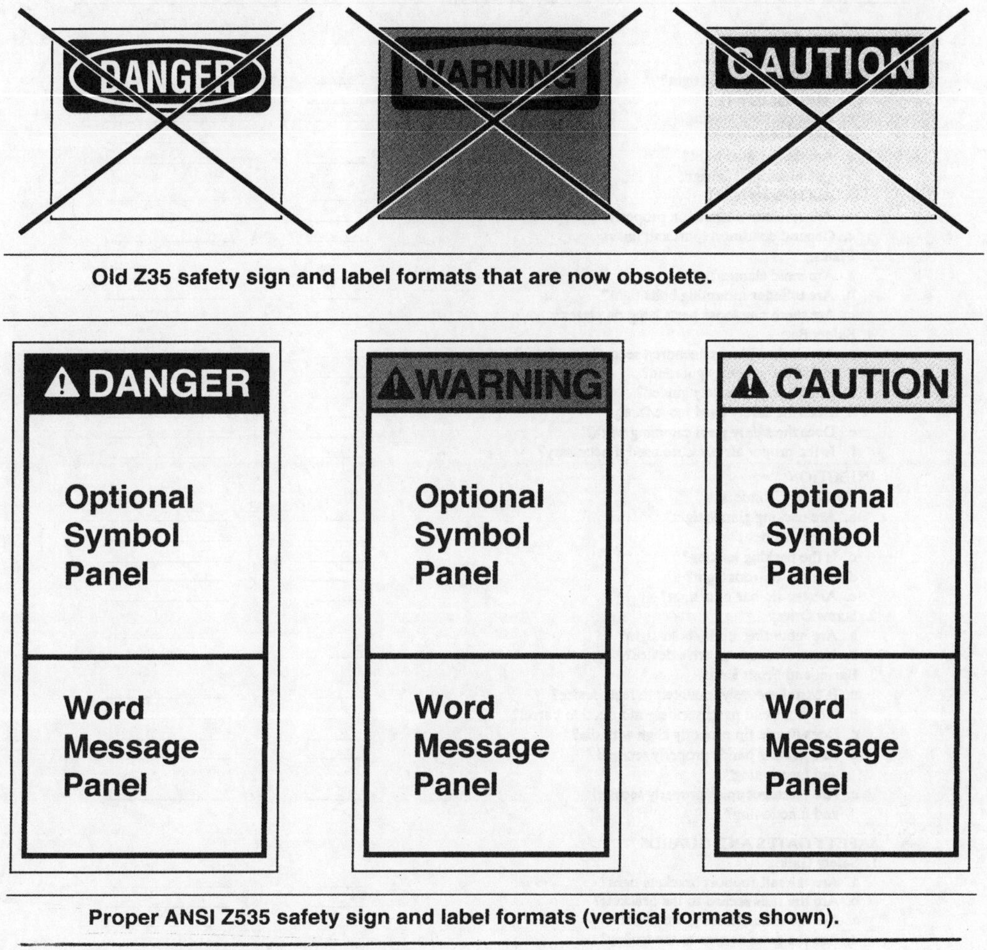

Table 3.66 Examples of safety signs for machines per ANSI Z535

changes, and cleanup (Table 3.67). Most equipment generates high heat and pressure. The units are built to run safely, but they must be treated with respect. With plastics that decompose, there may be hazards such as burns, wounds, air contamination, and major equipment problems. Faulty controllers and/or freeze-off can cause an overheating situation from a burned-out heater. In addition, safety devices that alert the plant when problems develop should be installed. Workers have to be aware of these possible situations.

Barrel Cover Safety To avoid electrical shock from the heater barrel, keep the barrel guard in place. Consider using integral armored leads or ceramic terminal covers on all adapter-zone and nozzle-heater bands. In addition to being important to operator safety, barrel cover can have important bonuses in melt quality and energy savings of the melt (Fig. 3.74).

	O.K.	NEEDS REPAIR
CLAMP:		
1. Hydraulic Cylinders:		
a. Are packing glands tight?		
b. Are bolts tight?		
c. Is the packing leaking?		
d. Are tie-rods tight?		
e. Are tie-bar nuts tight?		
2. Toggle Machine Linkage:		
a. Are all bolts tight?		
b. Are retainer washers on properly?		
c. General condition (pins and links).		
3. Plates:		
a. Are mold clamps tight?		
b. Are cylinder mounting bolts tight?		
c. Are there any loose parts lying on plates?		
4. Safety Bar:		
a. Are anchor blocks anchored securely to plates?		
b. Is the bar properly guarded?		
c. Is the bar adequately guided?		
d. Does the safety pawl move freely?		
e. Does the safety pawl camming work?		
f. Is the proper air pressure used if necessary?		
INJECTION:		
1. Hydraulic Cylinders:		
a. Are packing glands tight?		
b. Are bolts tight?		
c. Is the packing leaking?		
d. Are the tie-rods tight?		
e. Are the tie-bar nuts tight?		
2. Screw Drive:		
a. Are mounting studs/bolts tight?		
b. Is screw secure to drive device?		
3. Barrel and Front End:		
a. Is barrel securely mounted to feed device?		
b. Are front end parts securely mounted to barrel?		
c. Does nozzle tip properly align with die?		
d. Are heating bands properly secured and functioning?		
e. Are thermocouples properly secured and functioning?		
SAFETY GATES AND GUARDS		
1. Safety Gate:		
a. Are the rail support brackets tight?		
b. Are the rails secure to the brackets?		
c. Are the roller/trolley, etc., tight?		
d. Are the gates secure to the trolley?		
e. Are the gates/windows in good condition?		
f. Is the door edge safety working properly?		
1. g. Does the hydraulic/pneumatic interlock work properly?		
h. Does the electrical interlock work properly?		
i. Does the gate prevent access to pinch points?		
2. Rear Guard:		
a. Are the mounting brackets secure and tight?		
b. Are guards secure to the brackets?		
c. Are the guards in good condition?		
d. Does the electrical interlock work properly?		
e. Does the rear guard prevent access to pinch points?		
3. Fixed Guards:		
a. Are the guards securely mounted to the machine?		
b. Do the guards prevent access to pinch points?		
c. If guards are removed for reasons other than maintenance, are they interlocked to prevent machine operation?		

Table 3.67 Example of checklist for safety requirements

4. Top Guards:
   a. Is the top of the machine adequately protected by either a guard or height to prevent someone standing on the floor from reaching over the top of the safety gate?
   b. Is the top guard, if needed, properly interlocked?
5. Purge Guard:
   a. Is purging prevented by machine circuitry when the safety gate is open?
   b. Is the purge guard securely mounted to the machine?
   c. Is the purge guard in good condition?
   d. Does the purge guard contain a safety-glazed window in good condition?
   e. Does the purge guard protect the front, rear, and top of the purging area?
6. Pump Coupling Guards:
   a. Are guards in place?
   b. Do guards adequately cover rotating shaft?
7. Feed Openings:
   a. Are feed openings guarded against accidental insertion of hands?

SAFETY TAGS:

1. Are tags properly located?
2. Are tags legible and understandable?

ELECTRICAL:

1. Controls and Operator's Panel:
   a. Is the inside clean and neat?
   b. Is the disconnect working properly?
   c. Is the panel door kept closed?
   d. Are there any uncovered openings?
   e. Are all tags legible?
   f. Are all buttons and switches working properly?
   g. Do all components work freely?
2. Wire Ways and Junction Boxes:
   a. Are all covers on boxes and connectors?
   b. Is any sealtite broken, or are connectors loose?
3. Switches:
   a. Are all covers in place?
   b. Are switches free of oil and water?
   c. Are all switches working freely?
4. Electrical Circuit:
   a. Are circuit drawings legible?
   b. Are the circuit drawings up-to-date for the machine?
   c. Have any circuit changes been made, and have they been approved by the machine builder?
   d. Does the circuit conform to the latest state of the art?
5. Machine and Auxiliary Equipment:
   a. Is electrical interface wiring done safely?
   b. Is there duplication or confusion of terms on various pieces of equipment?
   c. Is the overall electrical circuit safe?
   d. Has the interface created any electrical, hydraulic, or mechanical safety hazards?

OPERATOR SAFETY:

1. Has the operator been trained?
2. Can the operator read all tags?
3. Can the operator understand the tags?
4. Has the operator had time to become familiar with the machine?
5. Is the operator's manual easily accessible to the operator?

Table 3.67 *(continued)*

Figure 3.74 A safety aspect is the plasticator cover over a hot barrel (courtesy of Plastics FALLO)

Barrel Venting Safety It is a common but dangerous practice to plug a vented barrel and use it the same way as a solid barrel machine. It is rare, but the internal pressure can exceed the strength limit of the bolts retaining the plug, and in turn, the plug could be released violently from the barrel.

To prevent this potential hazard, a number of safety precautions are taken, such as ensuring retaining bolts with more than enough strength are used. Also, the barrel should be rotated downward or away from the operator (even with no plug, in case the vent opening becomes overloaded with melt that is forced out).

A pressure gauge at the head of the barrel can provide a preliminary warning at a minimum safety pressure value, followed with shutoff of the machine at higher pressures (if practical; otherwise, alert all in the plant). Install shear pins and/or a rupture disk (if not already installed) and ensure the machine is heated adequately at the forward barrel end.

Machine Lockout Operators of machines should know steps to be taken to ensure personal safety. An example is the proper lockout of the machine's electrical circuit before starting repairs; this protects the maintenance worker from accidental start-ups that would result in severe injury. The National Safety Council offers the following steps for proper lockout:

1. Shut down all possible switches at the point of operation, then open/disconnect the wires in the main switch box.
2. Snap your own lock on the main switch box so that only you can open the box.
3. Check the lockout device to make sure the switch cannot be operated.
4. Place a name tag on the shank of the locked box to indicate that the machine has been locked out by you.
5. Notify the supervisor when the repair work has been completed; wait for the supervisor to give the go-ahead to remove your lock.
6. Take off the name tag, remove the lock, and reset all disconnected electrical circuits.

SAFETY AGENCY

Safety agencies are responsible for establishing product safety testing standards. Manufacturers must comply with these standards. UL, TOV, CSA, IEC, VDE, and BSI are some of the better known safety agencies in the world. As an example, all electrical consumer products need to be tested to one or more product safety standards, ensuring that the product is safe for consumer use. Once a product has passed all the specified tests in a standard, it becomes listed by the agency. The safety agency's listing mark must then be displayed on the product. Products carrying these marks are recognized throughout the world as safe for consumer use (430).

There is a difference between electrical safety testers that test to safety agency specifications and those that are actually listed by a safety agency. In order to obtain safety agency listing, an electrical safety tester must be submitted to the safety agency itself for testing. It then needs to pass all the necessary tests required by the safety agency. Although unlisted electrical safety testers can be used for testing to agency specifications, there is no assurance for the manufacturer that it is safe for operator use.

The Occupational Safety & Health Administration (OSHA) of the U.S. Department of Labor is responsible for ensuring safe working environments. Safety agencies that carry the Nationally Recognized Testing Laboratory (NRTL) designation have been recognized by OSHA as being certified to perform safety testing and certification. UL is one safety agency that has received the OSHA recognition of NRTL. This means that an electrical safety tester with the UL mark is compliant to OSHA standards for safe work environments.

CHAPTER 4

INJECTION MOLDING

INTRODUCTION

The process of injection molding (IM) is used principally for processing unreinforced or glass-fiber-reinforced thermoplastics (TPs); however, it also processes thermosets (TSs; Fig. 4.1). Up to at least 90 wt% of all plastics processed are TPs. There are many different types or designs of IM machines (IMMs) that permit molding many different products based on factors such as quantity, size (ranging from auto bumpers to medical microproducts), shape (ranging from simple to complex), product performance, and/or economics (1, 4, 8, 10, 16, 77, 134, 186, 304, 330, 378, 381).

Temperature melting/solidifying profiles are different for TPs and TSs (Fig. 4.2). TPs simply melt in the plasticator and solidify in the cooled mold. TSs melt in the plasticator and cure to a harden state in a mold that operates at a higher temperature than the plasticator (chapter 1). For the best controlled machines used for molding TSs, the heated plasticator usually uses an L/D ratio of one (chapter 3) and includes a water jacket to ensure that the melt temperature does not become too hot (Fig. 4.3).

The term *injection molding* is an oversimplified description of a quite complicated process that is controllable within specified limits. Melted or plasticized plastic material is injected or forced into a mold (Fig. 4.4) The mold may consist of a single cavity or a number of similar or dissimilar cavities, each connected to flow channels or runners that direct the flow of the melted plastic to the individual cavities. The process is one of the most economical methods for mass producing of products.

Three basic operations exist:

1. Raising the temperature of the plastic to a point where it will flow under pressure. This is usually done by simultaneously heating and masticating the granular solid until it forms a melt at an elevated and uniform temperature and viscosity (Table 4.1).

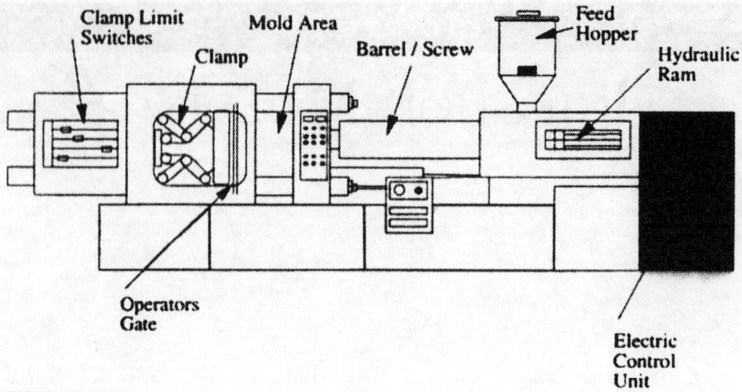

Figure 4.1 IM machine schematic

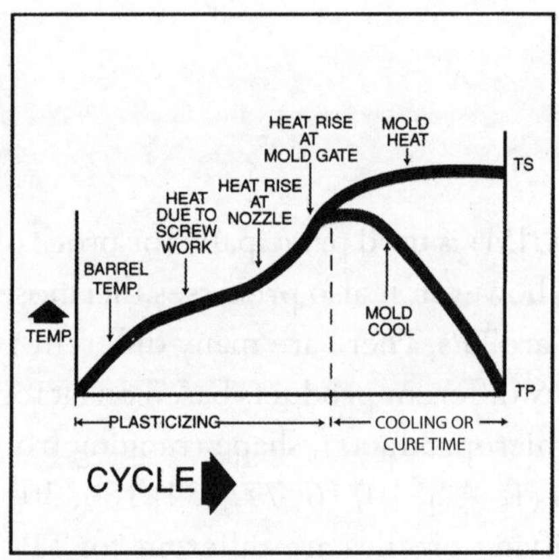

Figure 4.2 Melt to solidification of thermoplastics and thermosets during the injection-molding process (courtesy of Plastics FALLO)

This is accomplished in the cylinder of an IMM equipped with a screw. The latter provides the mechanical working required in conjunction with the heating of the material. This overall process is called plasticizing (or plastication) of the material (chapter 3).

2. Allowing the plastic to solidify in the mold, which the clamping system of the machine keeps closed. The liquid molten plastic from the injection cylinder of the IM is transferred through various flow channels into the cavities of a mold where it is finally shaped into the desired object by the confines of the mold cavity. What makes this apparently simple operation complex are the limitations of the hydraulic or electrical circuitry used in the actuation of the injection plunger and the complicated flow paths involved in the filling of the mold.

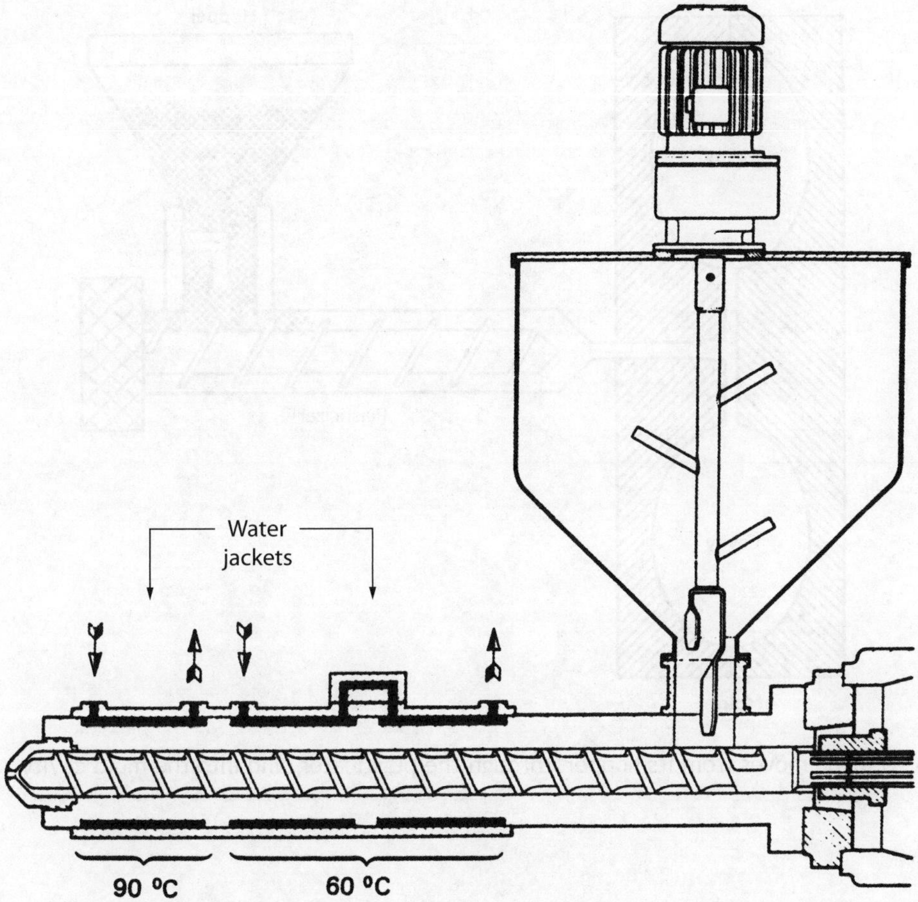

Figure 4.3 Example of a plasticator barrel (in an IMM used for thermoset plastics) that has electric heaters and water-cooling control jackets (courtesy of Negri Bossi)

3. Opening the mold to eject the plastic after keeping the material confined under pressure; it is removed to solidify the plastic and freeze it permanently into the shape desired.

These three steps are the only operations in which the mechanical and thermal inputs of the injection equipment must be coordinated with the fundamental properties of the plastic being processed. These three operations also are the prime determinants of the productivity of the process, since manufacturing speed will depend on how fast the plastic can be heated to molding temperature, how fast it can be injected, and how long it takes to cool the product in the mold.

The other operations in the injection process, such as clamping the mold, ejecting the part, and feeding the machine, are also important in the injection process. Their speed and effect on

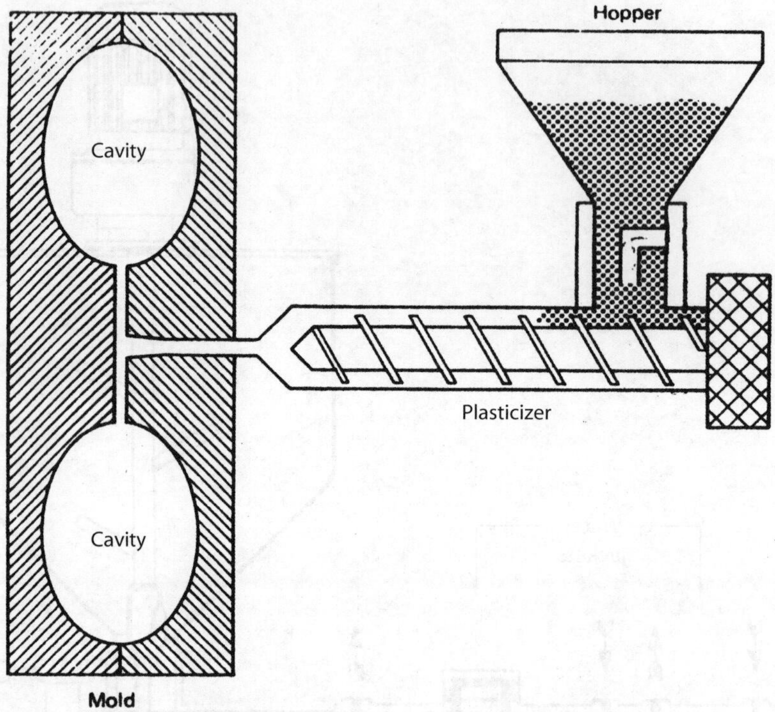

Figure 4.4 Plastic moves from its hopper, through the plasticator, and into the mold cavity

Material	Processing temperature rate °C	°F
ABS	180–240	356–464
Acetal	185–225	365–437
Acrylic	180–250	356–482
Nylon	260–290	500–554
Polycarbonate	280–310	536–590
LDPE	160–240	320–464
HDPE	200–280	392–536
Polypropylene	200–300	392–572
Polystyrene	180–260	356–500
PVC	160–180	320–365

Table 4.1 Examples of IM thermoplastic processing temperatures

productivity will normally depend on factors other than the material being processed. They depend on factors such as hydraulic and/or electric drives of the clamp system and design of the ejection system.

The basic machine is made up of the clamping unit on one end, and on the other end, the injection unit of an in-line reciprocating screw. The mold area is located in the center section (Fig. 4.5). The empty space in the mold is filled with melted plastic under high pressure. The moving platen rides on steel bars called tie rods (or tie bars) or on rails (without tie bars) The clamping force is generated by the hydraulic and/or electric mechanism pushing against the moving platen.

The part taken from the mold is, in most cases, a finished product ready to be packed and shipped or ready to be used as a part of an assembled unit. In contrast to metal forming, there is very little if any wasted material in injection molding. Most runners and sprues are reground and reused. By using hot runner molds, the sprue and runner system remains in a melted state in the mold and becomes part of the next finished part (chapter 17). The hot runners can be thought of as extensions of the plasticizing chamber. Usually, the part and mold can be designed so that there is

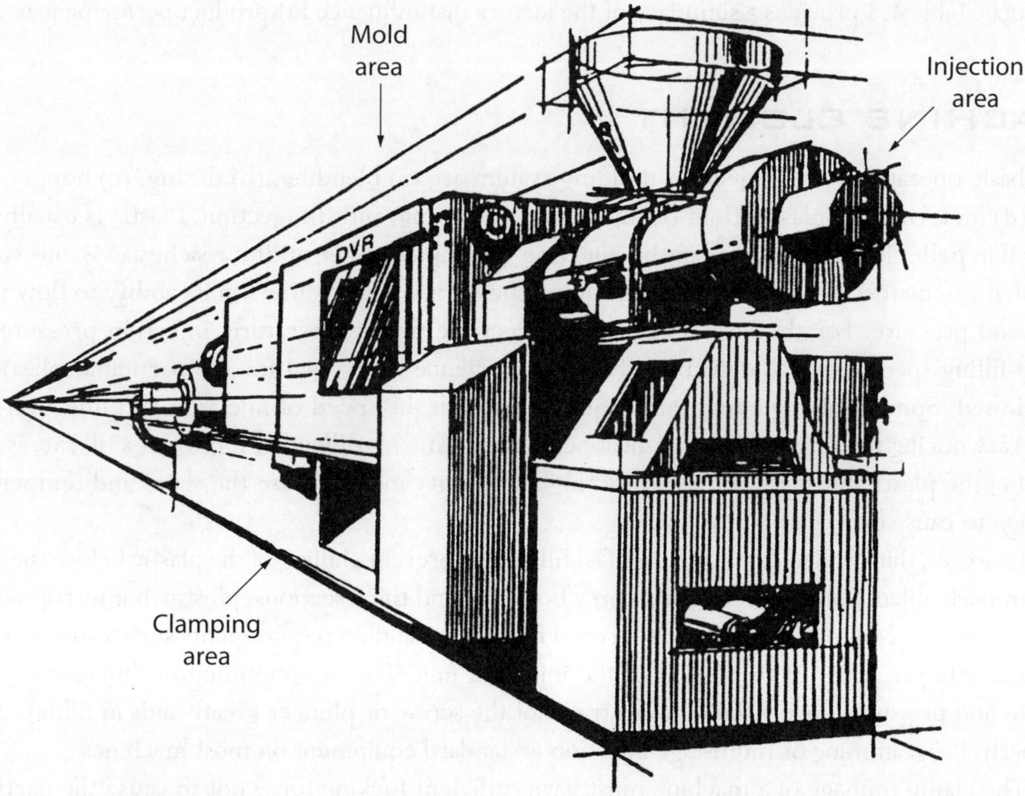

Figure 4.5 Three basic parts of an injection-molding machine

no subsequent machining required. If this is not possible for a certain design, conventional cutting tools can be used (chapter 18).

Materials having all shades of coloring can be purchased or colorants can be preblended with clear pellets to obtain desired colors. Postformed coloring is eliminated and the color exists throughout the entire molded part.

IM lends itself readily to automation in varying degrees, depending upon the ingenuity of the machine, mold designer, and molder (Table 4.2). The machine manufacturer can usually add components to the basic machine to implement any desired automatic arrangement. Molding cycles are relatively fast, and with new mold design developments constantly in progress and centering on faster heat transfer within the mold, the molding cycle is continually being reduced.

Modern methods of material handling, both of the raw plastic and the finished product, are becoming more generally employed, further reducing costs of the finished part. Efficient use of leftover runners and sprues plays an important part in the overall economy of the molding plant, and increased attention is continually being given to this area. The reduction of raw material inventory is seriously considered, and studies of production control methods are no longer uncommon, even by the small molder (chapter 18). All these economies are important in making plastic parts less expensive to manufacture and show plastics to be even more attractive to manufacturers dealing in all kinds of products. Table 4.3 provides a summary of the factors that influence IM product performances.

MACHINE ELEMENT

The basic operations of an injection-molding system are (a) blending, (b) drying, (c) hopper feeding, (d) metering, (e) plastication, (f) injection, (g) cooling, and (h) ejection. Plastic is usually purchased in pellet form and heated in the injection-heating chamber until it reaches a viscous state in which it can be forced to flow into the mold cavities. Each plastic differs in its ability to flow under heat and pressure. For the best result, the correct melting temperature, injection pressure, and mold-filling speed must be determined from experience or by trial for the particular plastic and mold used. Some molding conditions require that both the speed of injection and injection pressure vary during the filling process. A heat-sensitive plastic may degrade if too fast a fill rate is used. Forcing the plastic through orifices at too high a velocity may increase the shear and temperature enough to cause overheating and burning.

However, thin-walled parts require a fast fill rate to prevent chilling of the plastic before the cavity has properly filled. Some molded parts carry both thin and thick sections, plus such interrupted flow patterns as are required to move around cored holes. Demanding requirements such as these require considerable versatility in the design of the injection unit. The programming of different injection speeds and pressures during the forward travel of the screw or plunger greatly aids in filling cavities properly. Programming or multistage injection is standard equipment on most machines.

The clamp tonnage of a machine must have sufficient locking force not to cause the parting of mold halves; it resists the force of melted plastic moving at high pressures into the mold halves. If

Flexible manufacturing cell (FMC)	Single CNC injection molding machine without interlinkage Automatic mold changing system Automatic mold preheating Automatic raw material feed Automatic demolding and/or removal
Flexible manufacturing system (FMS)	Several interlinked FMCs with freely selectable material flow System computer for information interlinkage and monitoring of the components Automatic mold tramsport Tie-up with mold store Automatic transport of molded products Single and/or multi-stage production Modular method of construction
Flexible manufacturing line (FML)	One or several CNC injection molding machines Automatic raw material feed system One or more downstream operations Rigid interlinkage of the stations with rectified material flow Automatic demolding and/or removal Automatic transport of molded products Setting up of individual components only when the system is stopped

Table 4.2 Flexible automated manufacturing concepts with IM

PART DESIGN:	Part configuration (size/shape). Relate shape to flow of melt in mold to meet performance requirements that should at least include tolerances.
MATERIAL:	Chemical structure, molecular weight, amount and type of fillers/additives, heat history, storage, handling.
MOLD DESIGN:	Number of cavities, layout and size of cavities/runners/gates/cooling lines/side actions/knockout pins/etc. Relate layout to maximize proper performance of melt and cooling flow patterns to meet part performance requirements; preengineer design to minimize wear and deformation of mold (use proper steels); lay out cooling lines to meet temperature to time cooling rate of plastics (particularly crystalline types).
MOLDING CYCLE:	Set up the complete molding cycle to repeatedly meet performance requirements at the lowest cost by interrelating material/machine/mold controls.
MACHINE CAPABILITY:	Accuracy and repeatability of temperature/time/velocity/pressure controls of injection unit, accuracy and repeatability of clamping force, flatness and parallelism of platents, and even distribution of clamping on all tie rods. In hydraulic system check repeatability of controlling pressure and temperature of oil, oil temperature variation minimized, no oil contamination (by the time you see oil contamination damage to the hydraulic system could have already occurred), machine properly leveled.

Table 4.3 Simplified approach to injection-molding plastic products

the mating surfaces of the mold are forced apart, even a few thousandths of an inch (depending on type of plastic), fluid plastic will flow out and produce flash. The amount of clamp force depends upon the projected area of the molded part; concern should also be given to a plastic's viscosity at the final fill stage (chapter 17).

MOLDING SYSTEM

The IMM process can be identified by the three most basic, popular methods of operation: hydraulic, electrical, and hybrid. Fifty percent to 70% of machines in the 30- to 300-clamp tonnage ranges now represent about 60% of sales. The two basic plasticizing systems used are the single-stage and the two-stage molding systems (Figs. 4.6 to 4.8); there are also three-stage molding units and others. The single-stage system is more popularly known as the reciprocating screw IMM. The two-stage systems include other types, such as the piggyback IMM, which can partially be related more to a continuous extruder.

Different IMMs meet different qualitative requirements to mold specific parts (Tables 4.4 to 4.9). These features relate to factors such as the type of plastic to be processed, shot size, mold clamping force, type of process control, operational convenience, operating cost, maintenance cost, and cost of machine. The feature of shot size or capacity represents the maximum usable volume of melt that is injected into the mold. It is usually about 30% to 70% of the actual available volume in the plasticator. The difference basically relates to the plastic materials' melt behavior, and it provides a backup safety factor to meet different mold packing conditions. Shot size capacity may be given in

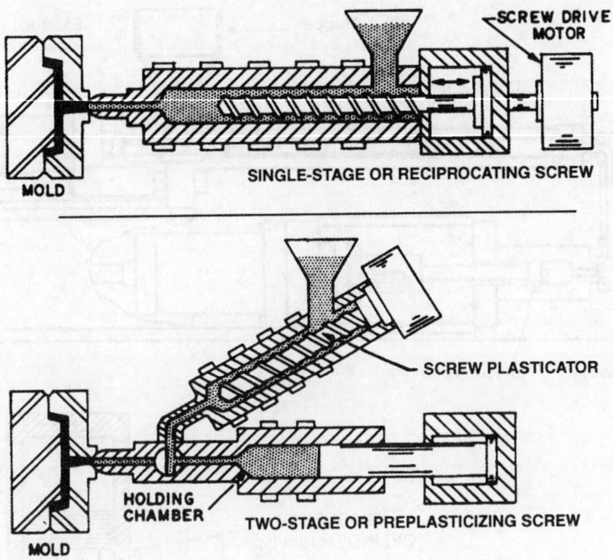

Figure 4.6 Schematics of single- and two-stage plasticators

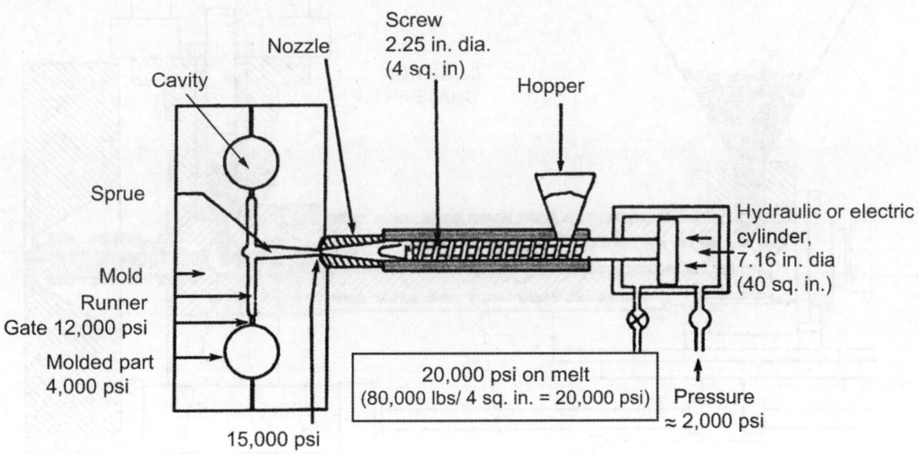

Figure 4.7 Simplified plastic flow through a single-stage IMM

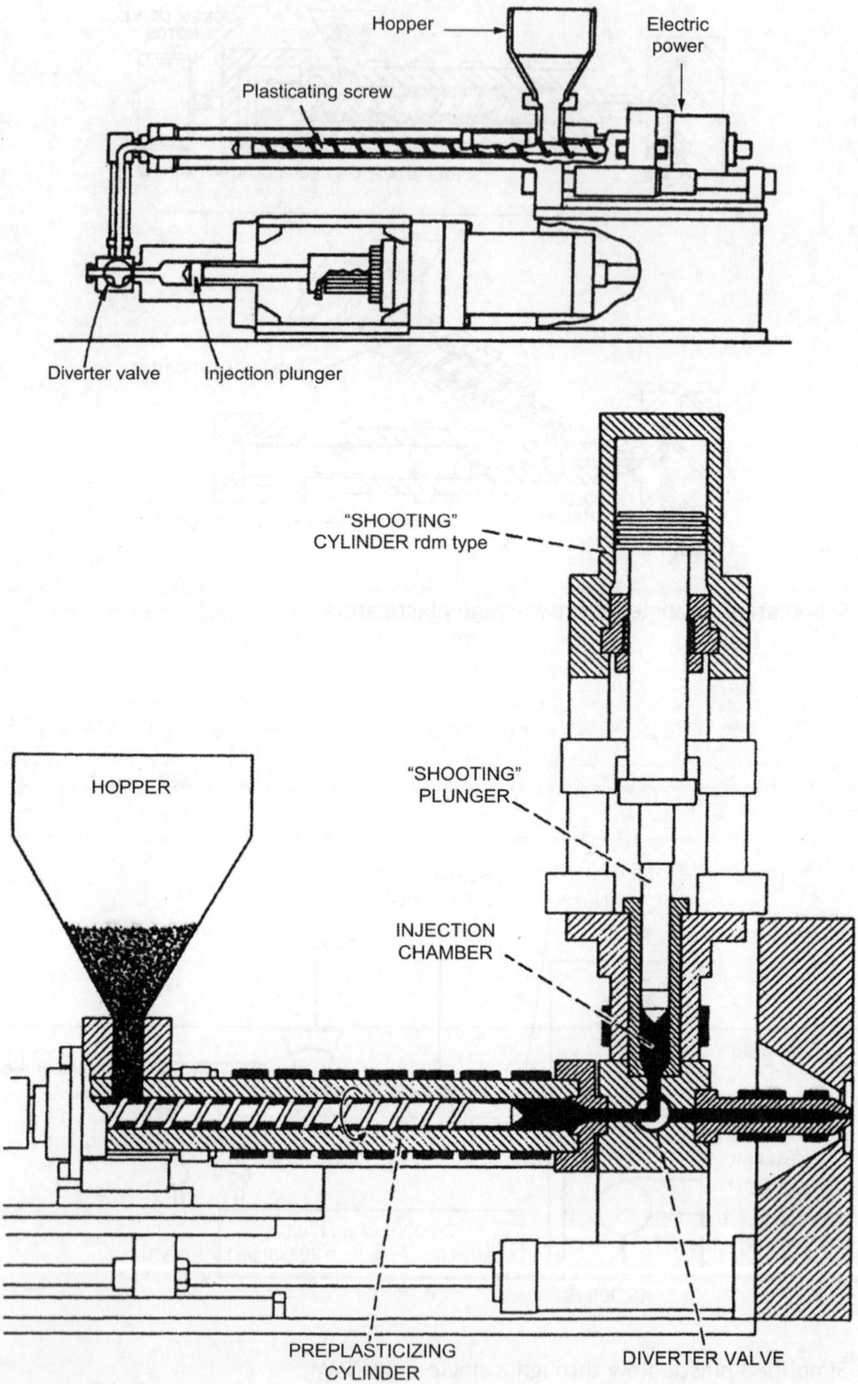

Figure 4.8 Simplified plastic flow through parallel- and vertical-designed two-stage IMMs

terms of the maximum weight that can be injected into a mold cavity(ies), usually quoted in ounces or grams of general purpose polystyrene (GPPS). Since plastics have different densities, the better way to express shot size is in terms of the volume (in^3 or cm^3) of melt that can be injected into a mold at a specific pressure. The rate of injecting the shot relates to the IMM's speed and the process control capability of cycling the melt to move fast-slow-fast, slow-fast, and so forth, into the mold cavity(ies).

Injection pressure in the barrel can range at least from 2,000 to some plastics up to 45,000 psi (14 to 310 MPa). The characteristic of the plastic being processed defines what pressure is required in the mold to obtain acceptable products. Based on what cavity pressure is required, the barrel pressure has to be high enough to meet pressure flow restrictions going from the plasticator into the mold cavity(ies).

The clamp force on the mold halves required in the IMM also depends on the plastic being processed. A specified pressure is required to retain the pressure in the mold cavity(ies). The clamp force is based on the cross-sectional area of any melt located on the parting line of the mold, including the cavities and mold runner(s) (if they solidify) that are located on the mold parting line. The cross-sectional area of a TP hot melt runner located within the mold half is not included in the parting line area. By multiplying the pressure required on the melt by the melt cross-sectional area, the clamping force required can be determined (chapter 17). To provide a safety factor in case more clamp force is needed, consider a 10% to 20% increase.

The molding cycle is the complete sequence of operations in the process (Fig. 4.9). One cycle represents the time period or elapsed time between a certain point in one cycle and the same point in the next. It is the time for fabricating a part from start to finish. Prior to determining the actually cycle time, consider the use of a formula, regardless of the plastic processed, where the average wall thickness (inches in decimal form) is multiplied by 250. Actual cycle time will depend on factors such as part thickness (Fig. 4.10) and the time required to melt, inject, and solidify the plastic and to eject the molded part.

Feature	Unit
Dry cycle	s
Recovery rate	g/s
Injection rate	cm^3/s
Available injection power	kW
Injection or shot capacity	cm^3 or g
Torque of screw	Nm
Injection pressure	MPa
Clamping force	kN
Opening force	kN
Platen/size (height × width)	mm × mm
Distance between tie bars (horizontal × vertical)	mm × mm
Max./min. daylight	mm
Max. stroke	mm
Drop chute	m^2
Floor space	m^2
Weight	kg

Table 4.4 Injection-molding features

Nominal shot volume (cm^3)	Equiv. Volume (in^3)	Equiv. shot weight in GPPS (gm)	Equiv. shot weight in GPPS (oz)	Nominal shot volume (in^3)	Equiv. Volume (cm^3)	Equiv. shot weight in GPPS (gm)	Equiv. shot weight in GPPS (oz)
25	1.5	22.0	0.8	2	32.8	28.8	1.0
50	3.1	44.0	1.6	3	49.2	43.3	1.5
75	4.6	66.0	2.3	4	65.5	57.7	2.0
100	6.1	88.0	3.1	5	81.9	72.1	2.5
125	7.6	110.0	3.9	6	98.3	86.5	3.1
150	9.2	132.0	4.7	7	114.7	100.9	3.6
175	10.7	154.0	5.4	8	131.1	115.4	4.1
200	12.2	176.0	6.2	9	147.5	129.8	4.6
250	15.3	220.0	7.8	10	163.9	144.2	5.1
300	18.3	264.0	9.3	20	327.7	288.4	10.2
350	21.4	308.0	10.9	30	491.6	432.6	15.3
400	24.4	352.0	12.4	40	655.5	576.8	20.3
450	27.5	396.0	14.0	50	819.4	721.0	25.4
500	30.5	440.0	15.5	60	983.2	865.2	30.5
750	45.8	660.0	23.3	70	1147.1	1009.4	35.6
1000	61.0	880.0	31.0	80	1311.0	1153.6	40.7
1500	91.5	1320.0	46.6	90	1474.8	1297.9	45.8
2000	122.0	1760.0	62.1	100	1638.7	1442.1	50.9
2500	152.6	2200.0	77.6	150	2458.1	2163.1	76.3
3000	183.1	2640.0	93.1	200	3277.4	2884.1	101.7
4000	244.1	3520.0	124.2	250	4096.8	3605.2	127.2
5000	305.1	4400.0	155.2	500	8193.5	7210.3	254.3
10000	610.2	8800.0	310.4	750	12290.3	10815.5	381.5
15000	915.4	13200.0	465.6	1000	16387.1	14420.6	508.7

Table 4.5 Shot volume conversion

Polymer	Multiply volume (cm^3) by this factor to obtain shot weight (gm)	Multiply volume (cm^3) by this factor to obtain shot weight (oz)	Multiply volume (in^3) by this factor to obtain shot weight (gm)	Multiply volume (in^3) by this factor to obtain shot weight (oz)
LD polyethylene	0.79	0.028	12.95	0.46
HD polyethylene	0.81	0.029	13.27	0.47
Polypropylene	0.85	0.030	13.93	0.49
Polystyrene	0.88	0.031	14.42	0.51
ABS	0.89	0.031	14.58	0.51
PPO	0.92	0.032	15.08	0.53
SAN	0.92	0.032	15.08	0.53
Polyamide 6/6	0.95	0.034	15.57	0.55
Polyamide 6/6	0.97	0.034	15.90	0.56
PMMA	1.01	0.036	16.55	0.58
Polycarbonate	1.01	0.036	16.55	0.58
PBT	1.12	0.040	18.35	0.65
PET	1.15	0.041	18.85	0.66
PVC-U	1.15	0.041	18.85	0.66
Acetal	1.22	0.043	19.99	0.71
PES	1.48	0.052	24.25	0.86

Table 4.6 Shot weight conversion

Nominal value (tonne)	Equiv. value (ton)	Equiv. value (kN)
25	27.6	245.2
50	55.1	490.3
75	82.7	735.5
100	110.2	980.7
150	165.3	1471.0
200	220.5	1961.3
250	275.6	2451.7
300	330.7	2942.0
350	385.8	3432.3
400	440.9	3922.7
450	496.0	4413.0
500	551.2	4903.3
600	661.4	5884.0
700	771.6	6864.7
800	881.8	7845.3
1000	1102.3	9806.7
1500	1653.5	14710.0
2000	2204.6	19613.3
2500	2755.8	24516.6
3000	3306.9	29420.0

Nominal value (ton)	Equiv. value (tonne)	Equiv. value (kN)
25	22.7	222.4
50	45.4	444.8
75	68.0	667.2
100	90.7	889.6
150	136.1	1334.5
200	181.4	1779.3
250	226.8	2224.1
300	272.2	2668.9
350	317.5	3113.8
400	362.9	3558.6
450	408.2	4003.4
500	453.6	4448.2
600	544.3	5337.9
700	635.0	6227.5
800	725.7	7117.2
1000	907.2	8896.4
1500	1360.8	13344.7
2000	1814.4	17792.9
2500	2268.0	22241.1
3000	2721.6	26689.3

Nominal value (kN)	Equiv. value (tonne)	Equiv. value (ton)
250	25.5	28.1
500	51.0	56.2
750	76.5	84.3
1000	102.0	112.4
1500	153.0	168.6
2000	203.9	224.8
2500	254.9	281.0
3000	305.9	337.2
3500	356.9	393.4
4000	407.9	449.6
4500	458.9	505.8
5000	509.9	562.0
6000	611.8	674.4
7000	713.8	786.8
8000	815.8	899.2
10000	1019.7	1124.0
15000	1529.6	1686.1
20000	2039.4	2248.1
25000	2549.3	2810.1
30000	3059.2	3372.1

Table 4.7 Clamp force conversion

Polymer	Melt temperature (°C)	Melt temperature (°F)	Mold temperature (°C)	Mold temperature (°F)
ABS	200–260	390–500	50–80	120–175
Acetal	180–320	355–445	60–120	140–250
HD polyethylene	200–300	390–570	10–60	50–140
LD polyethylene	160–270	320–520	20–60	70–140
PBT	230–280	445–535	40–80	105–175
PES	320–380	610–715	90–160	195–320
PET	260–300	500–570	130–150	265–300
PMMA	190–290	375–555	40–90	105–195
Polyamide 11	200–270	390–520	40–80	105–175
Polyamide 12	190–270	375–520	20–100	70–210
Polyamide 6	240–290	465–555	40–120	105–250
Polyamide 6/10	230–290	445–555	40–120	105–250
Polyamide 6/6	260–300	500–570	40–120	105–250
Polycarbonate	270–380	520–715	80–120	175–250
Polypropylene	200–300	390–570	20–90	70–195
Polystyrene	170–280	340–535	10–60	50–140
PPO	250–300	480–570	30–110	85–300
PVC-U	170–210	340–410	20–60	70–140
SAN	200–260	390–500	50–80	120–175

Table 4.8 Melt and mold temperature ranges

Nominal value (psi)	Equiv. value (bar)	Equiv. value (MPa)	Nominal value (bar)	Equiv. value (psi)	Equiv. value (MPa)	Nominal value (MPa)	Equiv. value (psi)	Equiv. value (MPa)
10000	689.5	68.9	600	8702.3	60.0	60	8702.3	600.0
11000	758.4	75.8	700	10152.7	70.0	70	10152.7	700.0
12000	827.4	82.7	800	11603.0	80.0	80	11603.0	800.0
13000	896.3	89.6	900	13053.4	90.0	90	13053.4	900.0
14000	965.3	96.5	950	13778.6	95.0	95	13778.6	950.0
15000	1034.2	103.4	1000	14503.8	100.0	100	14503.8	1000.0
16000	1103.2	110.3	1050	15229.0	105.0	105	15229.0	1050.0
17000	1172.1	117.2	1100	15954.2	110.0	110	15954.2	1100.0
18000	1241.1	124.1	1150	16679.4	115.0	115	16679.4	1150.0
19000	1310.0	131.0	1200	17404.6	120.0	120	17404.6	1200.0
20000	1379.0	137.9	1250	18129.8	125.0	125	18129.8	1250.0
21000	1447.9	144.8	1300	18854.9	130.0	130	18854.9	1300.0
22000	1516.8	151.7	1350	19580.1	135.0	135	19580.1	1350.0
23000	1585.8	158.6	1400	20305.3	140.0	140	20305.3	1400.0
24000	1654.7	165.5	1450	21030.5	145.0	145	21030.5	1450.0
25000	1723.7	172.4	1500	21755.7	150.0	150	21755.7	1500.0
26000	1792.6	179.3	1600	23206.1	160.0	160	23206.1	1600.0
27000	1861.6	186.2	1700	24656.5	170.0	170	24656.5	1700.0
28000	1930.5	193.1	1800	26106.8	180.0	180	26106.8	1800.0
29000	1999.5	199.9	1900	27557.2	190.0	190	27557.2	1900.0
30000	2068.4	206.8	2000	29007.6	200.0	200	29007.6	2000.0

Table 4.9 Injection pressure conversion

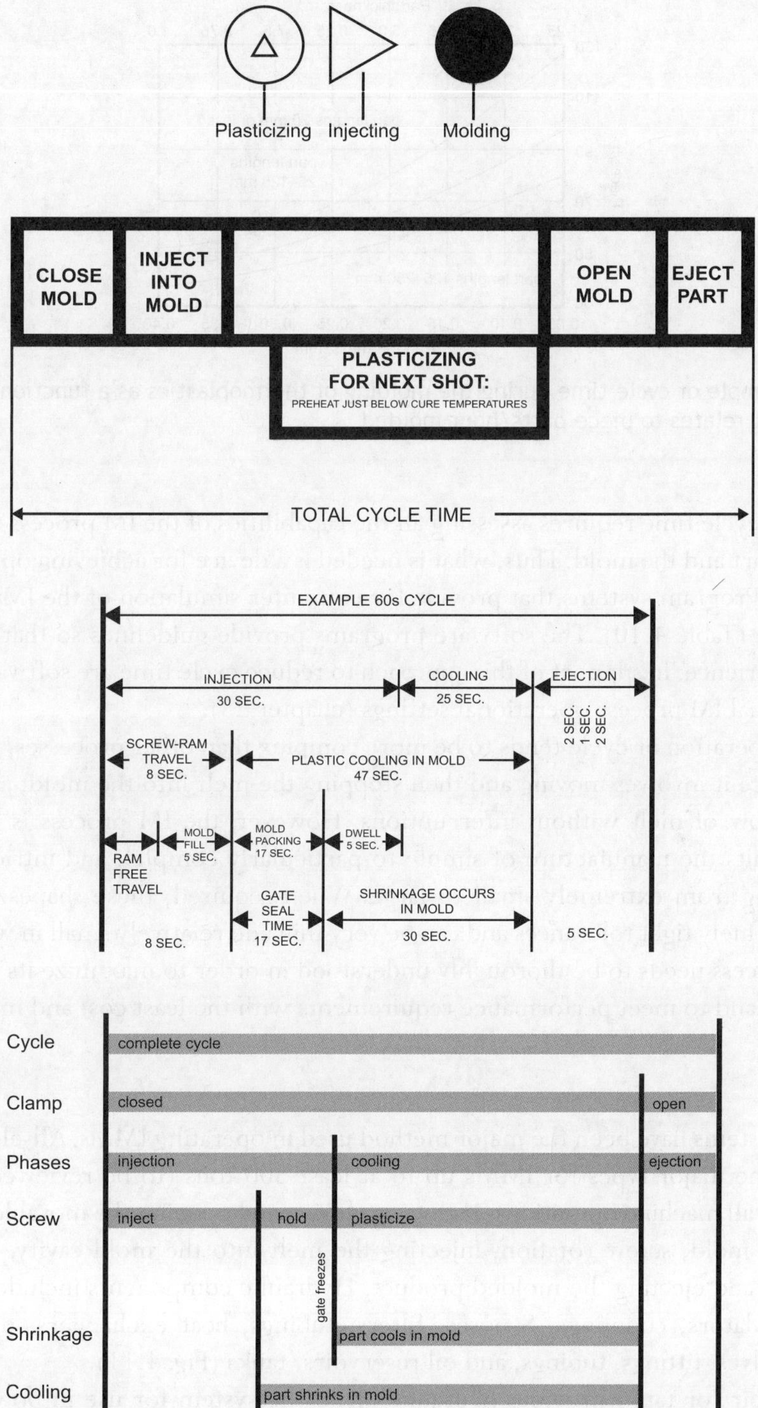

Figure 4.9 Overview of IM with cycle time that could include about 60% cooling time

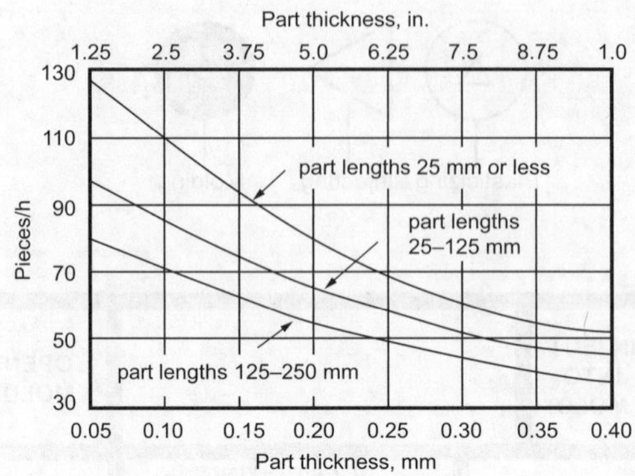

Figure 4.10 Example of cycle time during the molding of thermoplastics as a function of part thickness as it relates to piece parts/hour molded

Shortening cycle time requires assessing all the capabilities of the IM process in addition to the designs of the part and the mold. Thus, what is needed is a device for achieving optimum designs of part and mold. Program systems that provide for computer simulation of the IM process are used for this purpose (Table 4.10). The software programs provide guidelines so that one can develop continuing experience. In support of this approach to reduce cycle time are software programs that evaluate the actual IM process operational settings (chapter 25).

The IMM operation or cycle tends to be more complex than other processes, such as extrusion (chapter 5), since it involves moving and then stopping the melt into the mold, rather than having a continuous flow of melt without interruptions. However, the IM process is extremely useful because it permits the manufacture of simple to particularly complex and intricate 3-D shapes/products ranging from extremely small to large. When required, these shapes/products can be molded to extremely tight tolerances and can be very thin and relatively small in weights to parts of a gram. The process needs to be thoroughly understood in order to maximize its performance and mold products, and to meet performance requirements with the least cost and most ease.

HYDRAULIC

Oil hydraulic systems have been the major method used in operating IMMs. All-electrical machines are becoming the major types for IMMs up to at least 500 tons (to be reviewed). The hydraulic system controls all machine operations, including closing and opening the movable platen, applying pressure to the mold, screw rotation, injecting the melt into the mold cavity, maintaining hold clamp tonnage, and ejecting the molded product. Hydraulic components include motors, control pumps, accumulators, couplings, S hoses, filters, tubings, heat exchangers, directional valves, proportional valves, fittings, tubings, and oil reservoirs/tanks (Fig. 4.11).

The reservoir (or tank) provides hydraulic oil to the system for use in powering the IMM's different actions. The reservoir must be sized to ensure that an adequate supply of oil is available to the system and allow sufficient capacity for the system to return oil to the reservoir. In the past,

Name of program system	Cadmould	Femould	I-Deas Plastics	Moldflfow	PC-Mold	Simuflow	Strim 100 – Injection moulding simulation	TMConcept
Supplier	Simcon	Hahn & Kolb	SDRC	Mòldflow	KIMW	C- Tech	Cisigraph	Plastics & Computer
A) Simulation of rheological behavio								
A1: Filling phase	C2D C3D-Mefisto C3D-Mestro	C2D C3D-Mefisto C3D-Mestro	Moldfilling[1] (A1 and A2)	MF-2D MFLA-3D	PC-Flow	Simuflow- 2D	2D-rheology[1] 3D-rheology[1]	TMC-FA
A2: Pressure holding phase	C2D C3D-Mehold	C2D C3D-Mehold		MFLP-3D		Simuflow-3D		TMC-FA Best
B) Simulation of thermal behaviour	C2D C3D-Mecool	C2D	Moldcooling[2]	MF-Cool	PC-Therm	Simucool	Cooling Analysis	TMC-MTA
C) Simulation of mechanical behavioı								
C1: of moulding		MEF	FE modelling[3] model solution optimization	Fenas			polymer structural analysis	
C2: of mould	C2D C3D-Meclamp (clamp force calculation)	C2D C3D-Meclamp (clamp force calculation)		Moldmech	PC-Mech			
D) Other program modules			warp (shrinkage, distortion)	MF-warp (shrinkage, distortion)		Optimold (design macros from DME and Hasco standards)	3D modelling (for rheological simulation)	TMC-MCO (mould cost optimization) TMC-CSE (shrinkage calculation) PP (postprocessor) SAS (mould fault analysis)
E) Comments	C = Cadmould	C = Cadmould	[1] antecedent: Polyfill [2] antecedent: Polycarb [3] antecedent: Superlab				[1] antecedent: Procop	TMC = Total Molding Concept

Table 4.10 Examples of injection-molding software

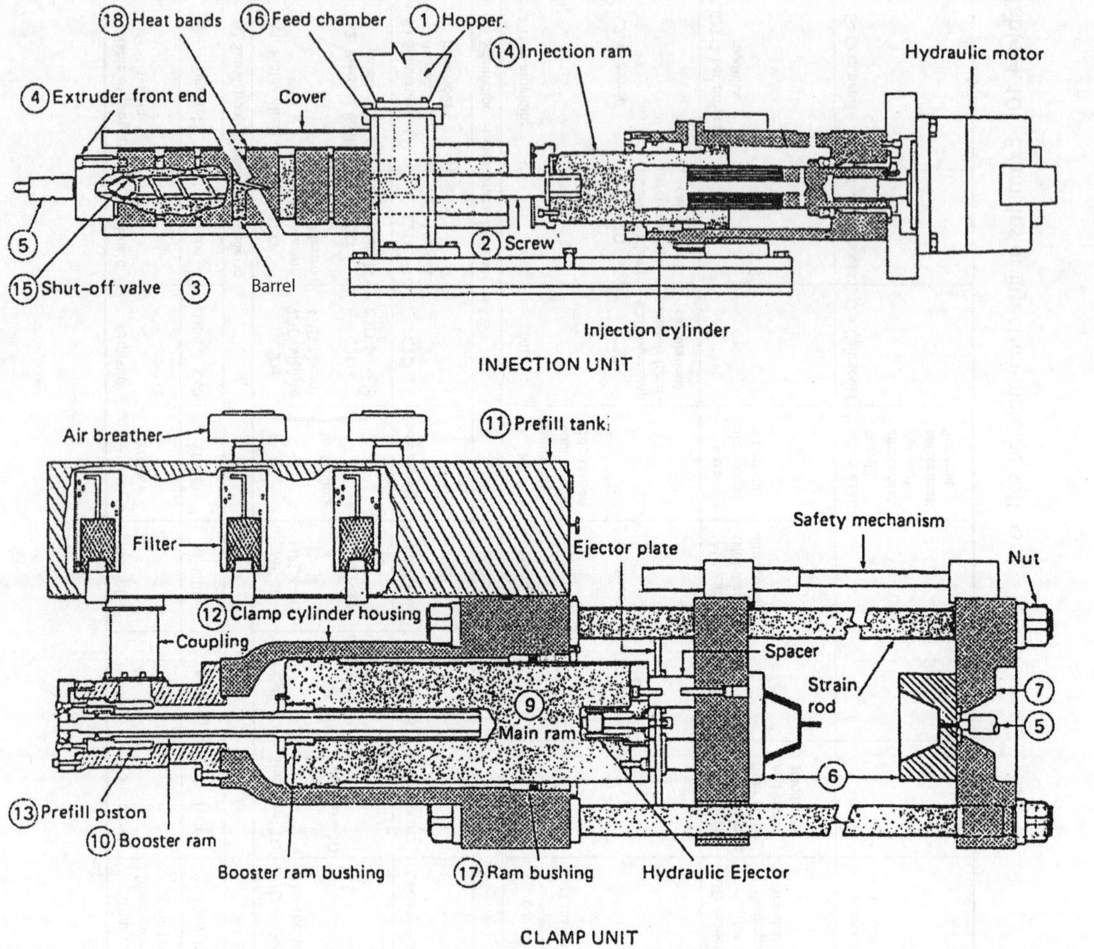

Figure 4.11 Examples of hydraulic IMM components

most oil systems were too complex to meet the different operations, such as moving platens. More recently, owing to simplifying the actions of the IMMs, very little volume of oil is used, resulting in eliminating (or practically eliminating) oil leaks with much less maintenance work.

The hydraulic pump receives oil from the reservoir at low pressure and increases the pressure to that required by the system. The pump provides hydraulic flow and pressure to the system. Several different types are used. The most common are fixed- and variable-displacement pumps. Different designs are available, with the most common being vane, piston, and gear types.

Variable-volume and variable-pressure compensating pumps are used more frequently in an attempt to conserve energy. These pumps are capable of varying output to meet a particular flow requirement or will dispense only enough flow to develop a particular pressure requirement. There is no single type pump that is perfect for every class and size of IMM.

As an alternative to closed-loop control, servo valves or hydraulic logic elements can be driven by digital electrical signals. The hydraulics perform an analog conversion. Separate control manifolds are required for pressure and volume flow control. The number of logic elements (bit number) determines the resolution. With a seven-bit pressure manifold, 128 steps can be achieved, including the value zero. For a system pressure of 160 bar, the resolution is 1.25 bar. This resolution is sufficient for all requirements in injection molding.

Hydraulic oil fluid is a mixture of liquids designed to transfer pressure from one point to another in a system on the basis of Pascal's law wherein pressure on a confined liquid is transmitted equally in all directions throughout the liquid. The pressure-flow work produces excessive heat in the operation of the machine hydraulic system, such as an IMM, and can have a broad-based negative effect on the machine operation of the entire system. Heat can and does affect the following five major areas of the machine hydraulics, which in turn affects the cost and/or performance of the molded plastic product(s): (a) hydraulic fluid life, (b) energy loss, (c) erratic operation of components, (d) formation and removal of sludge and varnish, and (e) operating conditions that cause overheating. Cooling systems are used to keep the oil fluid temperature under control. The overheating causes leakage of check valves, relief valves, and so on.

FLUID POWER BASICS

Fluid power refers to energy transmitted via a fluid under pressure. In hydraulic systems, that fluid is a liquid (oil or water). Pneumatic fluids are typically compressed air or inert gases. Fluid power's motive force results from the principle that pressure applied to a confined fluid is transferred uniformly and undiminished to every portion of that fluid and to the walls of its container. A surface (cylinder piston) will move if the difference in force across the surface is larger than the total load plus frictional forces. The resulting net force can then accelerate the load proportionately to the ratio of the force divided by the mass (262).

Fluid power is used in diverse applications ranging from mobile construction and aerospace equipment to powering industrial machinery (IMMs, etc.) and offers several advantages over other types of motive force. As an example, a single source of fluid pressure (compressor or pump) can power many motion axes or fluid-power devices. The power source can be located where space is not critical. Because much of the system's size and weight is off-loaded onto the power unit, the individual actuators can be small compared with the power they produce. In addition, they are often quieter and generate less heat than electric actuators. Fluid-power actuators can also be used in hazardous environments where electric sparks must be avoided.

By using accumulators to store energy, a hydraulic power unit needs to provide only slightly more than the average demand, increasing efficiencies for machines with varying load cycles. In applications requiring a constant holding pressure or torque (clamp press), hydraulic actuators have big advantages because they use no energy while they are stationary. In contrast, electric motors draw considerable current to maintain torque, even while stopped. Most motors will overheat and fail under these conditions.

Hydraulic cylinders are very smooth and efficient for linear movement. There are no poles that cause cogging and there is no need for backlash compensation.

Designing the power circuit and selecting and physically placing system components are critical to maximizing the system's performance. Figure 4.12 shows a typical hydraulic system with an electronic motion controller. Following is a discussion of key system elements.

The motion controller gets its position and pressure inputs either from transducers mounted in the cylinders (linearly actuated systems) or via encoders mounted on axes (rotational systems). Pressure transducers in linear activation systems are typically mounted on either side of the piston (Fig. 4.12) to measure differential pressure. To measure linear position, systems can use magnetostrictive displacement transducers (MDTs), which do not require homing steps.

Hydraulic and pneumatic power offer certain advantages over electric motors, especially for systems that require high-speed linear travel, moving or holding heavy loads, or very smooth position or pressure control. In comparison, hydraulic and pneumatic actuators are smaller and quieter. They also produce less heat and electromagnetic interference at the machine than electric actuators do. Selecting appropriate system components and programming the appropriate motion control algorithms can result in systems that offer high performance and precise motion at a reasonable cost.

ELECTRICAL

Electric molding technology (EMT) eliminates many variables from the hydraulically operating IMMs. The EMT machine will produce more good parts per day for a lower cost. There are also significant operating advantages related to energy and environmental issues. During the past few decades, complete operating electrical IMMs have become more popular worldwide. About 60%

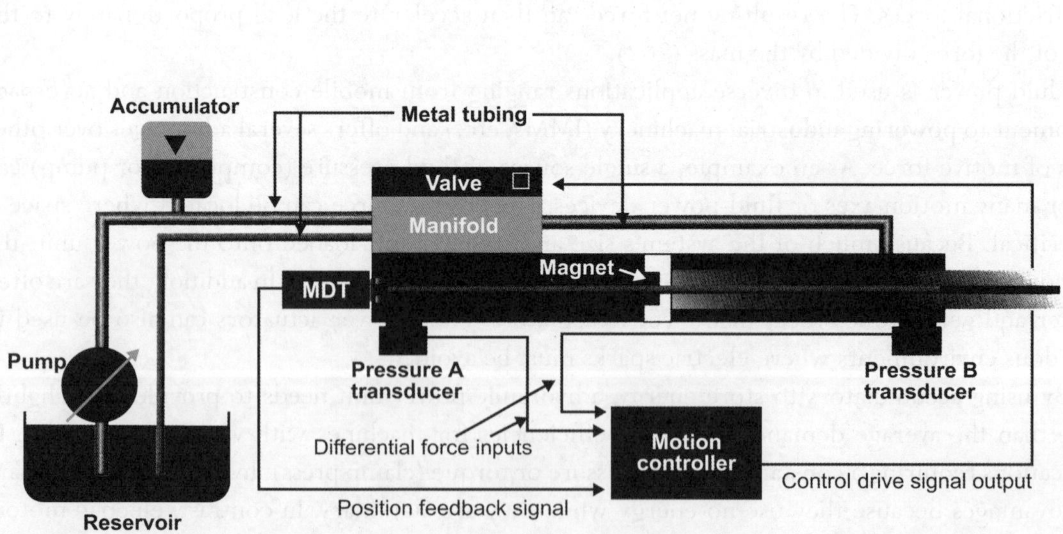

Figure 4.12 Example of fluid power–control hydraulic system

of IMMs up to 300-clamp tonnage now sold are electrically operated. The transition to electric IM will proceed rapidly, and most machines below 1,500 tons in developed countries will probably be all electric in the near future, though the basic configuration for clamp and injection functions will change relatively little. Hydraulic machines will, at least for the foreseeable future, probably have a secure market in high-tonnage applications, primarily over 1,500 tons, because of the present cost premium for high-horsepower servomotors.

When compared with hydraulics, the different electrical designs provide advantages such as improved molding accuracy resulting in improved quality control (repeatability accuracy exists at ±0.001 in), shot-to-shot repeatability, decreased energy consumption of 50% to 90% (Fig. 4.13), cleanliness by elimination of oil fluid problems, higher power, more variable controlled power, dynamic braking resistors, quick start-up and setup, increased productivity, repeatable operation without operator attention, elimination of closed-loop liquid cooling, no need for extensive air-conditioning, environmental benefits, lower number of components/easy maintenance, and low noise level (below 70 dB).

The Japanese were first with an electric machine, because environmental issues are a high priority on that densely populated island nation. Compact size, low noise, and elimination of oil as an environmental and fire hazard led the Japanese to create the first commercially viable EMT. In the European market, speed and precision are high priorities, along with environmental benefits. The first electric machine developed in Germany reflected those priorities, with dry cycle times less than one second. This machine configuration is ideal for quick precision molding of thin-walled parts, such as CDs, jewelry boxes, and medical disposables.

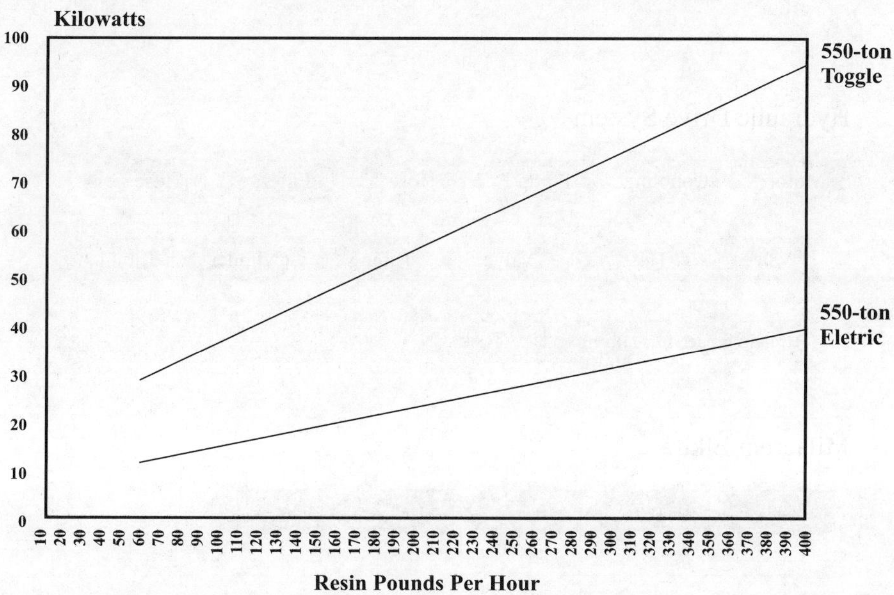

Figure 4.13 Energy usage vs. throughput (courtesy of Milacron)

The United States wanted EMT benefits within a market of custom-molding requirements with mid- to high-tonnage machines and larger shot sizes. The first U.S.-built machine (Milacron) was designed to appeal to the toggle machine users, with features, capabilities, and controls analogous to popular hydraulically powered toggle machines. They required high process repeatability and removal from the process of variables that required operators to monitor and adjust the machines (Fig. 4.14), hydraulic oil, and all the hardware needed to manage the machines. Another goal was to reduce labor costs, or even more desirable, provide unattended production.

Repeatability improvement is why electric drives completely supplanted hydraulics on machine tools in the 1970s (and on robots in the 1980s). It is repeatability improvement that is pulling EMT into the mainstream of injection molding.

The earliest applications of electric motors were on extruders, providing continuous processes compared with the intermittent nature of injection molding (chapter 5). This experience, combined with developments in drives and controls, served as a foundation. Today, the demand for servomotors of all sizes is attracting development funding and driving down the cost of manufacturing.

EMT's simpler solid-on-solid power train eliminates the major causes of molding variations in hydraulic oil IMMs with all their hardware. Electric systems include servomotors, pulleys, belts, and ball screws. Basically, they provide decentralized power generation with individual electrical drives for the main and secondary functions. They use servo drives and main-spindle drives that are comparable with the drive technology already used for many years in machine tools (330, 331).

Using adjustable speed drives (ASDs) reduces energy. At the same time, they can improve process efficiency and minimize machine wear and tear. Energy savings and efficiency stem from ASDs' precise electronic control of motor speed. They provide soft starts, thus extending the life of the components they drive, such as pumps, fans, and seals on rotating shafts.

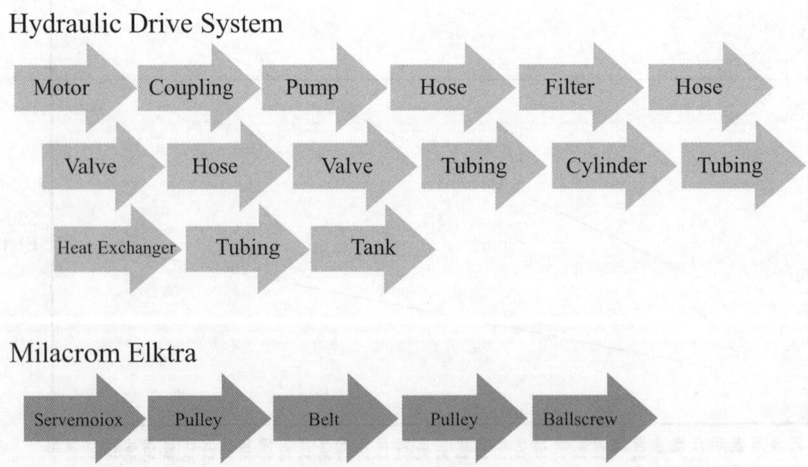

Figure 4.14 Electric-machine power train eliminates the major cause of variation in conventional IMMs (courtesy of Milacron)

Electric drive DC motors are controlled through solid-state circuitry that rectifies the AC supply. Apart from being one of the most efficient motors in the speed range of 20% to 100% of maximum, DC motors give a wide range of controllable speeds of better than 30:1.

Electric designs provide various engineering combinations. There are high-speed, directly connected rack-and-pinion clamp drive systems. Also in use are 64-bit microprocessor and digital communication between the AC servomotor and controller to create a closed-loop feedback circuit for each of the four axes of motion—namely, clamping, injection, screw recovery, and injection. The four electrical servomotors work independently so that control can overlap the motion of each axis to shorten cycle time. As an example, the IMM need not wait for full screw recovery prior to opening the mold.

Servo-electric drive systems provide screw rotation independent of other machine functions, replacing the more conventional hydraulic drive at significant operating energy savings. They allow screw recovery simultaneously with other machine functions. Their high positioning accuracy and high repeatability meet the users' increased demands, in this respect on the clamping unit and ejector mechanism. Most electrics use toggles.

MACHINE CAPABILITY

The Hunkar Class standard is a formalized ASME-defined test regimen for machine tools (186). It establishes machine capability before making parts rather than inferring it from statistical process control (SPC) studies on parts already made. One graphic result of a process capability study is a bell-shaped curve, which many processors with SPC backgrounds use (chapter 24). Figure 4.15 shows electric process capability that provides savings in material, scrap, energy, and so on.

The higher process capability is inherent to electric machine design. Even general-application electric machines have process capability that is significantly better than the Hunkar Class 1 category. Table 4.11 provides typical Hunkar test results from a series of 60 shots on general-purpose, 300-ton, U.S.-designed Milacron Elektra IMM molding pipe elbows.

SUMMARY

EMT machines at the present time are more costly, but this situation is changing. When comparing prices with hydraulic machines, specifications and capabilities should be balanced. By the time the hydraulic circuits and controls on a hydraulic machine are enhanced to approach the performance of a general-application electric machine, the cost difference narrows significantly and hydraulic technology remains, by its nature, less precise. For example, in some molding applications it is advantageous to have simultaneous operations occur in the clamp and injection ends of the machine. With its independent axis drives, this capability is inherent to the electric machine, but it significantly affects the cost of a hydraulic machine. However, electrical systems to date do not work for all applications. Hydraulic machines will still be in favor in undeveloped countries with low-cost, low-skilled labor, and low-quality requirements.

Within the near future, expect to see 70% to 75% of all IMMs up to 800 tons in developed countries to be electric. EMTs will enable developed countries to remain competitive and retain jobs through higher quality and productivity, even while labor costs and environmental regulations add to overall costs. Hydraulic oil, because it impacts molding costs in so many ways, will be seen as a business hazard as well as an environmental hazard. Any increase in electricity prices will also drive demand for EMT.

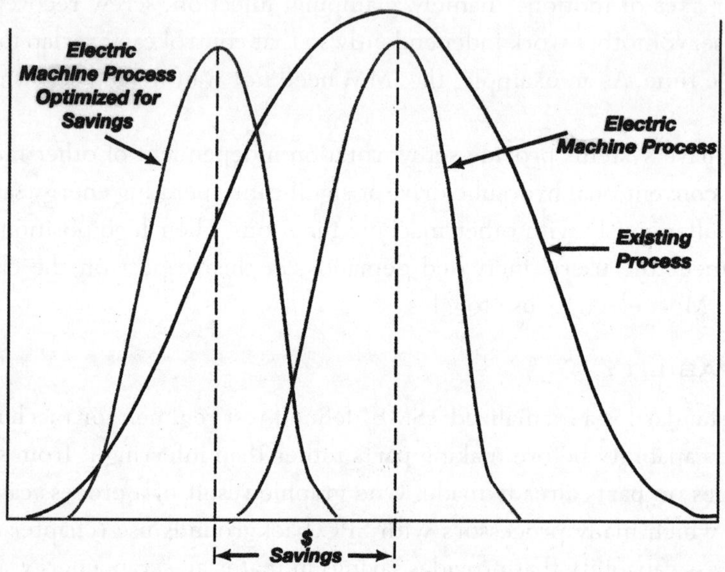

Figure 4.15 Guide in comparing economics of good parts for electric vs. hydraulic IMMs (courtesy of Milacron)

Parameter	Max.	Min.	Range
Cycle Time	36.95 sec	36.70 sec	0.25 sec
Hold Time	7.02 sec	7.01 sec	0.01 sec
Fill Time	2.83 sec	2.81 sec	0.02 sec
Plasticate	11.14 sec	10.92 sec	0.22 sec
Peak Pressure	11289 psi	11154 psi	135 psi
Hold Pressure	8549 psi	8542 psi	7 psi
Back Pressure	1686 psi	1665 psi	21 psi

Table 4.11 Molded product Hunkar test results (courtesy of Milacron)

HYBRID

Many different combinations of hydraulic and electrical machine operating systems provide advantages, such as fast-moving platens, reduced-size hydraulic cylinders, and/or reduced operating costs. There are many examples of these hybrid operating systems meeting the molders' different molding requirements. Popular examples that have been used for many decades are the electric screw drive system designs in hydraulic-operating IMMs.

All-hydraulic drive components to date offer good price-performance ratios and numerous technical advantages. It is therefore beneficial to develop combinations of the hydraulic and electrical systems so as to have the advantages of both. Based on one's requirements, a hybrid machine may be the ideal IMM.

OPERATING CHANGE

HYDRAULIC TO ELECTRICAL

Changes in IMM operation occur when using hydraulic, electrical, or hybrid (77, 78). The usual variable requirements for each system have to be determined, such as clamp tonnage, shot size and control, tie bar spacing, daylight opening, platen size (mold use at least 70% of the surface area between tie bars), plasticizer performance (temperature and screw rpm), and others.

The three primary differences are the need to watch out for hydraulic clamp versus toggle clamp issues, the requirement of working in plastic pressure versus hydraulic pressure, and the advantages offered by electrics for more precise control of screw position for filling and shot size. However, with either hydraulic or electric systems, the plastic variables need to be identical to make identical parts (chapter 1).

Changes that exist in some machines today involve different factors. For example, the plasticizer's nozzle contact force is generally not an issue with most hydraulics. With electrics, make sure there is enough nozzle contact force. In an electric system, there is no hydraulic pull-in cylinder to pull the device forward. When the nozzle is in contact with the mold sprue bushing, a two- to fifteen-ton force is required to keep the seal tight so that the high pressure melted plastic does not leak. Proper mechanical design can ensure the proper nozzle contact pressure. When an electric motor is trying to turn while holding the nozzle tip up against the sprue bushing, there is the possibility that the motor will overheat, limiting torque.

Ejection force in hydraulics is not a problem. With electrics, there can be sufficient ejection force, but there may be a problem with how the ejection device stops. Hydraulics use high ejection force and the ejection stroke is longer than the travel of the ejector plate; the ejection rod may bend or damage the mold. With a high ejection force on an electric machine and the stroke longer than the travel allows, the drive device might fall out of synchronism as it tries to push the ejector plate farther than the mold will allow. The belt-drive ejection device is similar to a hydraulic timing chain. Damage can occur, requiring extensive downtime during its expensive repair.

Setting mold protection in an electric machine is easy and often more sensitive than on most hydraulics. Setting mold protection can be made for the last several millimeters of travel. Some electrics can monitor the pressure to close the clamp for the entire travel distance. At fast closing speeds, machines are available with controls that allow a foam polystyrene cup to be dropped between the faces of the mold without crushing the cup by more than about 5 mm.

Screw positioning and suck back may be more accurate because control of inertia on electrics is better than on hydraulics. Back-pressure settings (screw pushing plastic forward generates back pressure) from 50 to 200 psi of hydraulic pressure are usually used during screw rotation. The difference between hydraulic and plastic pressure is known as the intensification ratio (I_g).

Electric back pressures usually are from 1,000 to 1,500 psi. All electric machines work in plastic pressure, not hydraulic pressure. Most use a pressure transducer behind the screw drive mechanism to measure force directly.

In all hydraulic injection units, hydraulic power is converted or amplified into plastic pressure. Its force (F) is equal to pressure (P) times area (A) or $F = P \times A$. The large hydraulic piston acts on the screw, or essentially the nonreturn valve. The hydraulic piston has a large surface area such as 100 cm^2. The nonreturn valve during forward injection acts like a smaller-area piston, such as a 10-cm^2 piston. This causes the hydraulic pressure to be converted to plastic pressure in the barrel. In this example, hydraulic pressure is intensified or multiplied by a factor of 10. That means 100 psi of hydraulic pressure provides about 1,000 psi of melt pressure inside the nozzle of the machine. Machines are available with intensification ratios that range from 6:1 to 43:1. Different IMMs in a fabricating facility most likely have different intensification ratios. It is because of these different ratios that one cannot use the same back pressure. Electrics have 1:1 ratio and have to be calibrated in plastic pressure.

There are machines that do not provide processing values with set points but instead provide percent values. With those using percents, the extra calculation is required of multiplying it by whatever 100% is found to be its equal. Unfortunately, finding what value correlates to 100% can be time consuming. As John Bozzelli explains, it is like retrofitting a car's speedometer with 0% to 100%. With today's computers, there is no acceptable reason for a machine controller to be using percents. The computer can easily multiply the percent times the maximum and provide that number in bar or psi, mm/sec, or in/sec.

When comparing most hydraulics with electrics, pressure-position transfer differs. With hydraulics, the screw does not stop. It continues to coast forward and then sometimes bounces back. There is screw overtravel from momentum action. There is no braking action on most hydraulic machines. The faster the injection speed, the greater the inertia; at slower speeds, the less inertia or kinetic energy. With electric machines, the drive mechanism for forward injection is different. The drive is an electric motor coupled to the screw via a mechanical device (belt, ball screw, etc.). When the motor stops, there is a braking action that is not found on most hydraulics. The inertia is controlled because of the mechanical link. Each revolution of these drive motors can be broken into 2,000 to 8,000 segments. This allows precise control of transfer position.

With this pressure-position transfer in electrics, more plastic is forced into the cavity during the first stage because the inertia will not overtravel the screw. More or better control of first stage occurs, especially at fast injection speeds. On some electrics, the rate of acceleration can even be adjusted to set point velocity in milliseconds. Be careful not to set this acceleration too fast, as it is a common cause of overtorquing on servomotors. This control over inertia or braking action is one of the major advantages of electric machines.

When moving from the first- to second-pressure stage in either machine, the part being molded in the cavity begins to pack out. The forward injection pressure is held for some accepted amount of time, which depends on whether the mold gate needs to be sealed or unsealed. For both first and second stages on electric machines, the pressure values are plastic pressure, just like the back pressure reviewed. To switch from the hydraulics to an electric, the hydraulic pressure values should be converted to plastic pressures via the intensification ratio. There should be little difference in pressures when viewed in plastic pressure values.

When the high pressure required with the second stage calls for long hold times, there is a concern for electrics. For hydraulics, where the forward injection pressure is maintained by the hydraulic system often through a relief valve, there is no excessive strain on the pump(s). However, for electrics, which are driven by electric motors, a situation is created in which the motor is trying to turn during a phase when ram travel is minimal. The result is a heated motor; fans are used to provide cooling action. Action can be taken (pack and hold) to adjust maximum second-stage pressures to lower than what was available during the first stage. It limits the torque on the electric motor to prevent overload and overheating.

The final molding stages with both drive systems result in the cavity being filled, packed, and cooled, followed by the start of screw recovery to develop the next shot, suck back if necessary, and wait a second or two before opening the mold. Molded-part ejection takes place and the next cycle starts.

CLAMPING DESIGN

Controllable action of IMM clamps exists (Fig. 4.16). Their operating mechanisms are identified as mechanical or toggle, hydraulic, electrical, and hybrid (hydromechanical) (Table 4.12).

TOGGLE

Toggle clamps are more popular in smaller-tonnage machines, because the mechanisms are inexpensive to manufacture and require less-complicated hydraulic or electrical circuitry than hydraulic or hydromechanical designs. Most electric IMMs use toggles. A typical compound double-toggle clamp is shown in Figures 4.17 and 4.18. This balanced design is intricate and sophisticated; it features a very high mechanical advantage at lockup, built-in and geometrically determined clamp slowdown, slow mold breakaway, and rapid traverse at the appropriate points along the clamp stroke. In addition, there are no side thrusts to cause tic rod or platen bushing wear. By comparison, a simple mono-toggle is shown in Figure 4.19. Offsetting the obvious simplicity of the design is the side thrust generated by the actuating piston that loads the moving platen at a right angle to the tie

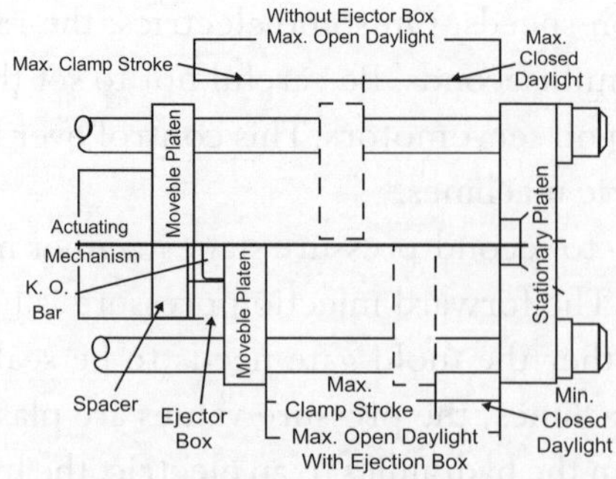

Figure 4.16 Example of basic clamp action in this split schematic showing maximum and minimum daylight openings to meet mold open and close requirements

	Mechanical	Hydraulic	Electrical	Hybrid
Fewest moving parts		X	X	
Direct tonnage readout		X	X	X
Sensitive mold protection		X	X	X
Economical long stroke		X	X	X
Ease of setup		X	X	X
Locked clamp force	X		X	
Build-in speed profile	X		X	
Daylight independent of stroke	X		X	X
Daylight independent of ejector	X		X	X
Low power	X		X	X
No overhead reservoir	X		X	X
Evenly loaded tie-rods			X	X
Least mold flashing			X	X
Least overall length			X	X

Table 4.12 Examples of clamp design performances

rods while moving it forward along them. In all but the smallest tonnage presses, this force may cause added wear, requiring frequent attention to platen bushings and possible occasional replacement of the tie rods. The price advantage of a machine with this design should be weighed against other designs.

With the double-toggle design, the actuating device moves the cross-head forward, which in turn extends the toggle links and moves the platen forward. In the initial clamp closing motions, the toggle link geometry dictates a very low mechanical advantage that translates as rapid platen movement with little force behind it (desired situation of accomplishing rapid clamp closing simultaneously with mold protection in the form of low closing force). As the moving platen

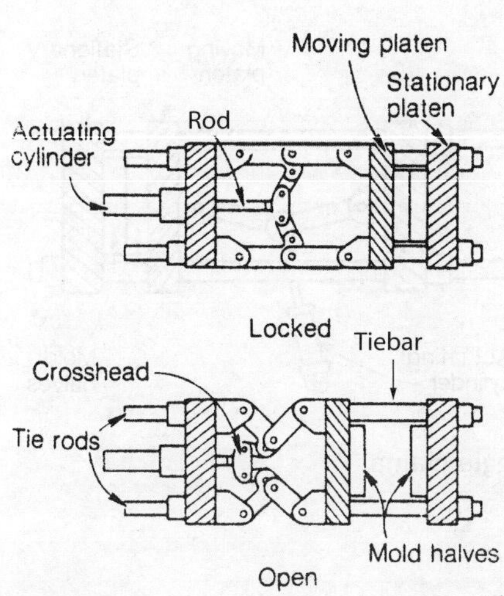

Figure 4.17 Example of double-toggle clamp

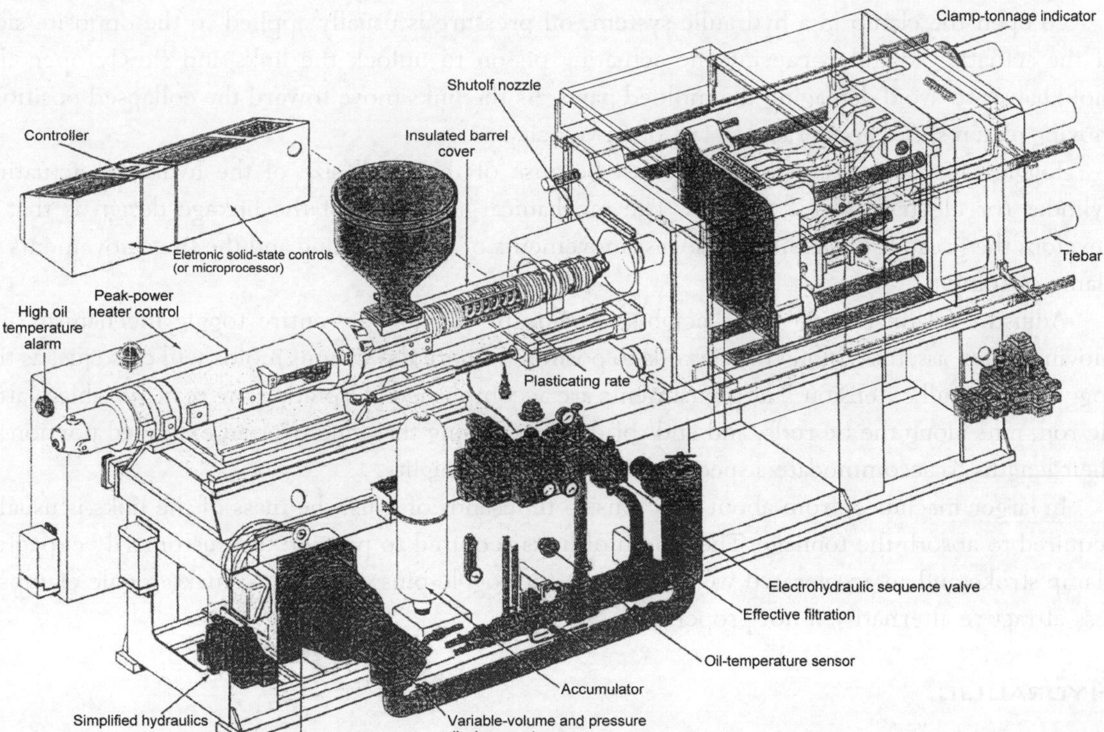

Figure 4.18 Machine schematic with a double-toggle clamping system

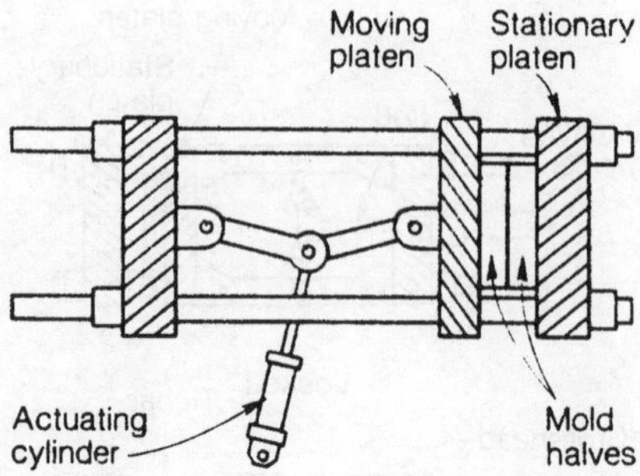

Figure 4.19 Example of mono-toggle clamp

approaches the mold-closed position, the toggle links reach their fully extended configuration, and the toggle geometry completes a smooth transition from high speed/low advantage to a high mechanical advantage (for tonnage buildup) combined with low clamp speed to avoid sudden force or slamming the mold shut. At full extension, the links are almost straight out, thus creating a lockup situation in which there is minimum tendency for the toggle mechanism to open.

To open the clamp in a hydraulic system, oil pressure is usually applied to the opposite side of the actuating cylinder, causing the actuating piston to unlock the links and slowly open the mold halves to avoid damaging the molded parts. As the links move toward the collapsed position, moving platen speed accelerates to fast open velocity.

Toggle mechanisms are inherently fast because of the small size of the hydraulic actuating cylinder (or electrical drive system). The mechanical advantage of the linkage design is that it develops the greatest force during the last movements of clamp closing and the first movements of clamp opening.

Adjustment for varying mold heights is made by moving the entire toggle mechanism and moving platen assembly along the tie rods to positions where the various molds will close just as the toggles reach full extension. The adjustments are accomplished by rotating the rear movable platen tie rod, nuts along the tie rods, and ends of the latter being threaded over an extended portion of their lengths to accommodate a specified range of mold heights.

In larger machines, from about 400 tons to thousands of tons, the mass of the links is usually required to absorb the tonnage. The length of links required to provide the customarily expected clamp stroke and an accelerated wear factor on the toggle pins combine to make toggle clamps a less-attractive alternative if not properly designed.

HYDRAULIC

Hydraulic clamps are used extensively on machines in the medium-capacity range of about 150 through 1,000 tons, with highest predominance in the 250- to 700-ton range (Fig. 4.20). They offer flexibility in machine setup and operation. Since tonnage can be developed at any point along the clamp stroke, just setting limit switches at the desired points along a calibrated scale during mold setup is required. Clamp slowdown, mold close, slow-mold breakaway, and fast open functions are all adjustable. This versatility is particularly useful in complicated molding applications: molds with core pulls or unscrewing dies, multiple plate molds, or delicate parts requiring careful mold handling. In addition to what is being reviewed, new hydraulic design systems are developing to compete with the advanced electrical design systems.

Rapid clamp close is obtained through the use of a small jack ram that usually is built into the center of the hydraulic ram. A large volume of oil acting on the small area of this cylinder achieves rapid advance of the clamp. A prefill valve is an example that allows oil from the elevated reservoir to flow by gravity into the chamber behind the main hydraulic ram. Before the mold reaches the closed position, oil from the high-volume pumps is diverted out of the clamp-close circuit, leaving only a low-volume pump to complete mold closing at slow speed and low pressure. Low-pressure closing protects the mold from damage by a foreign object, part of a flash, and/or a part that failed to eject after the previous cycle. A preset timer gauges the time taken by the mold-close function; if full close is not achieved within the prescribed time, the timer signals the machine to open the mold (the presumption being that something has improperly blocked the mold from closing). When the mold faces touch, the prefill valve shifts to close the port to the reservoir and a high-pressure pump builds tonnage over the full area of the clamp ram.

After plastic injection and time for the molded part to cool to a solid, oil from a low-volume, high-pressure pump acts against the pullback area of the ram to gently separate the mold halves. The prefill valve opens to direct oil from the main cylinder back to the hydraulic reservoir.

Within a short distance, another switch initiates fast clamp open by activating a high-volume pump. Additional sensors slow the ram and finally bring it to a full stop. Mechanical or hydraulic ejector rods are usually employed to complete the sequence by ejecting the finished parts; although, some molded parts may use air ejection on parts like pails and buckets or for manual removal in the case of large precision parts requiring extra care in their removal from the mold.

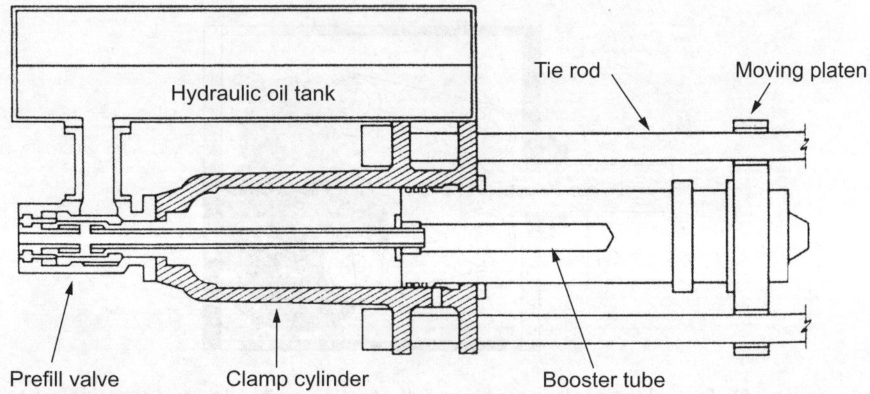

Figure 4.20 Example of a hydraulic clamp

Electrical

Electrical operating clamps (as well as the complete IMMs) are available from many sources worldwide. In the past few decades, the all-electrical IMMs have been producing all types, shapes, and sizes of molded products. Different electrical designs are used. For example, there are the clamp (most electrics use toggles), ejection, injection, and sled pull-in motion systems that are driven through computer-optimized ball screws developed for the machine tool industry. Servomotors are connected to the ball screws through a heavy-duty timing belt and pulleys. Die height is set by a servo-driven, chain-and-sprocket arrangement. The plasticator is directly driven through a timing belt. Its design objective is high speed, which it meets with dry cycle times of less than one second. It is ideal for thin-walled parts and packaging applications.

Key to this design is the use of mechanical transmission devices to amplify torque or speed and generate specific force/velocity profiles for injection and ejection. The clamp is driven through a high-speed double rack-and-pinion with an upstream two-stage spur-gear set. As shown in Figure 4.21, the crank arm is used for injection. The crank arm is able to deliver high force or high speed, depending on its position. The injection crank arm, powered through a multistage gear drive, is able to double the velocity on the screw at the beginning of the move and double the pack pressure at the end of injection. The crank arm performs only about one-third of a revolution. Die height on this machine is adjusted automatically through a servo-driven ring gear. The plasticizing screw is driven through a two-speed gearbox.

Different clamp designs are used in the all-electric machines. An example is Nissei of Japan that uses a three-stage clamping mechanism (Fig. 4.22). It uses both a toggle and a central locking ram to eliminate any possible deflection in the center of a clamping platen (121).

Hybrid

The hybrid is a combination of hydraulic and electrical. In turn, these basic systems provide many different IMM designs to meet different product requirements. A major use is electrical-hydraulic

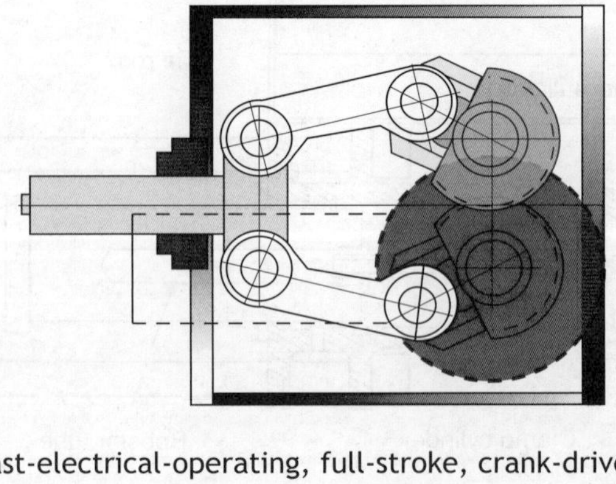

Figure 4.21 Example of a fast-electrical-operating, full-stroke, crank-driven injection system (courtesy of Milacron)

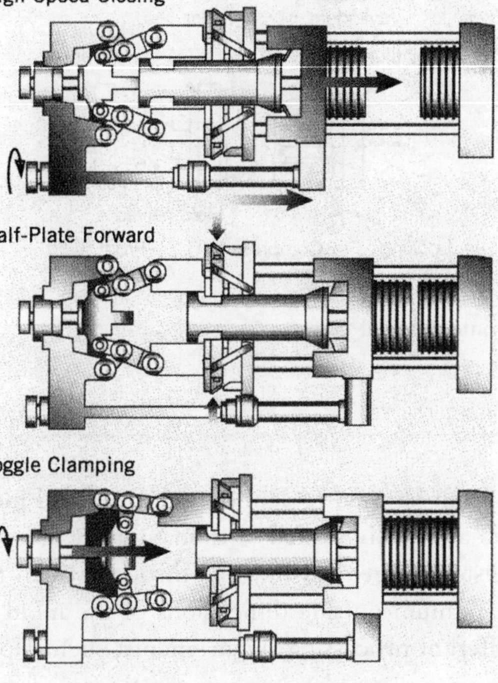

Figure 4.22 Triple-clamp all-electric design (courtesy of Nissei)

hybrid IMMs. Each system provides advantages, such as fast-moving platens, reduced-size hydraulic cylinders, and/or reduced operating costs. Examples of these hybrid operating systems meet the molders' different molding requirements. A popular example that has been used for many decades is the electric screw drive system design in hydraulic-operating IMMs. It is beneficial to develop combinations of the hydraulic and electrical systems so as to have the advantages of both. Based on the requirements, a hybrid may be the ideal IMM.

Hydromechanical or electromechanical clamps utilize a combination of mechanical and hydraulic and/or electrical functions to achieve clamp movement and tonnage buildup (Fig. 4.23). Hydromechanical designs evolved in clamp sizes from about 1,000 to 1,500 tons and up because of problems with moving the large volumes of oil needed by fully hydraulic designs in those size ranges. Clamp speeds were limited by the flow capacity of the prefill valve. Examples of designs include (a) hydraulic traversing cylinders to bring the moving platen to within very close proximity of the mold-closed position, (b) a mechanical latching or blocking device to prevent any subsequent rearward movement, and (c) a short-stroke hydraulic ram that moves the mold the final fractional inch to complete closing and obtain the required tonnage buildup. After injection and part cooling, oil working against a pull-back area of the short-stroke ram opens the mold, then the blocking device is removed and traversing cylinders pull the ram/moving platen and mold assembly open for part(s) removal.

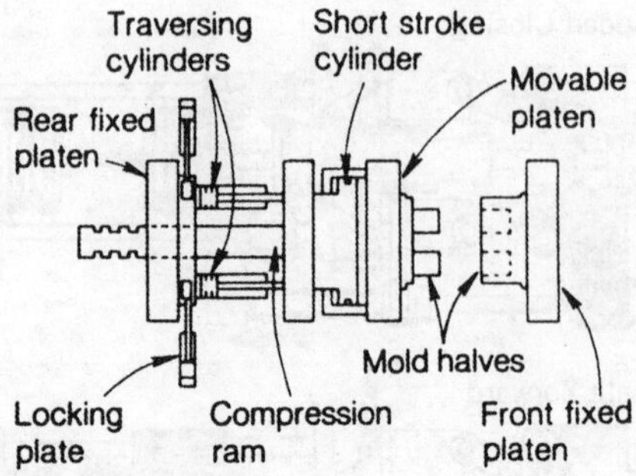

Figure 4.23 Example of hydromechanical clamp

TIE BAR

The clamping tie bars (rods) can be used to support the fixed and movable platens on which the mold is attached. They serve as equally loaded tension support members of the clamp when the mold is closed. The open distance between tie rods through which the mold must fit and eject molded parts sets up the maximum outside dimensions of the mold that can be used. Different designs are used to meet different processing requirements, such as permitting the installation of molds that would occupy the complete platen minus the tie rod circular areas. There are designs used to unlock one to all tie bars, those with one to four retractable tie bars, three tie bars, and where no tie bars are used. A design without tie bars is of a C frame (also called U frame, open frame, etc.) construction targeted to provide clamping pressure and proper parallelism as well as operating platens. The fundamental purpose of these different actions is to provide faster automated mold changes (in and out of the IMMs). Each system provides its own advantages (and limitations) for specific operations required in the different operating IM plants.

During clamping and when applying pressure on the molds, the tie rods stretch. If everything is in balance, the platens and mold stretch evenly. The distance the rods stretch is directly proportional to the applied load. Sensors, such as electrical strain gauges, can be used to detect the stretch or load applied, and if an unbalanced situation occurs, an indicator can alert the operator or the process control system. Bar sensing can also be used as a means of signaling the switch from pack to hold pressures, a potential alternate or support to pressure transducer use.

THERMAL MOLD INSULATION

A problem that controls may not consider involves the effect of heat on tie bars. Unbalanced heat of tie bars can directly influence mold performance, particularly at start-up. If the heat differs from top to bottom bars, different degrees of expansion can occur. Insulating between mold and platen can help or eliminate the problem. The insulator pad used can also confine heat more to the mold, producing savings in heat and/or better heat control. There are different insulation materials available that provide different performances (Table 4.13).

Property	Units	Glastherm Grade S	Sheet Grade HT	Asbestos Concrete	Calcium Silicate
Compressive strength					
@ 75F	psi	45,000	49,000	14,000	2,350
@ 302F	psi	16,000	27,000	-	-
@ 392F	psi	11,000	18,000	-	-
@ 425F	psi	10,000	15,500	-	-
Maximum services temp.	F	425	550		
Thermal conductivity	Btu/hr/ ft^2/in/F	1.8	1.9	4.5	.88
Water absorption	%	.4	.2	22	89
Thickness tolerance	inch	±.002	±.002		
Resistance to lubricants & hydraulic fluids		no effect	no effect		

Table 4.13 Mold heat-insulation properties (courtesy of Dielectric Corp.)

Factors to consider when selecting the insulation material start with compression strength, that is, the maximum force required to deform a material prior to reaching its yield point. The importance of this property is for maintaining mold and press alignments. Another requirement is service temperature that usually ranges from 275° to 450°F. It is recommended to select insulation with a service temperature 25% higher than the operating temperature of the mold.

PLASTICIZING

The function of the IMM heating cylinder is to thoroughly and uniformly convert (plasticize) the plastic feed material into a homogeneous heated plastic melt (Tables 4.14 to 4.17). With controlled viscosity, the melt is forced into the clamped mold where the end product is formed. Basically, the heating cylinder is a simple heat exchanger (chapter 3).

It is important to note that only the cylinder temperature is directly controlled. The actual temperature of the plastic melt within the screw and as it is ejected from the nozzle can vary considerably, depending on the efficiency of the screw design and the method by which it is operated. Factors that affect the melt temperature include the time the material remains in the cylinder; the internal surface heating area of the cylinder and screw per volume of material being heated; the thermal conductivity of the cylinder, screw wall, and plastic material (chapter 1); the differential

These are strictly typical, average values for a resin class. Consult your resin supplier for values and more accurate information. We have not included many of the newer engineering materials because they vary considerably within a family of materials and mostly they have fillers such as glass fillers which affect all of the data considerably.	MELT DENSITY SPECIFIC GRAVITY GRAMS/CM³ AT PROCESSING TEMPERATURE	DENSITY SPECIFIC GRAVITY GRAMS/CM³ AT 70° F (SOLID)	DENSITY POUNDS/FT³	SPECIFIC VOLUME INCHES³/POUND	SPECIFIC VOLUME CM³/GRAM	INJECTION TEMPERATURE °F	LINEAR MOLD SHRINKAGE INCHES/INCH	SPECIFIC HEAT BTUs/POUND/°F	WATER ABSORPTION % IN 24 HRS.	MAXIMUM WATER CONTENT ALLOWABLE FOR MOLDING
ABS - Injection	.97	1.05	65.0	26.6	.952	500	.005	.40	.40	.20
Acetal - Injection	1.17	1.41	88.0	19.7	.709	390	.020	.35	.25	
Acrylic - Injection	1.04	1.16	72.0	24.1	.868	450	.005	.35	.20	.08
CAB	1.07	1.20	74.6	23.1	.833	440	.004	.35	1.50	.15
Cellulose Acetate - Injection	1.13	1.26	79.0	21.9	.794	450	.005	.36	2.40	.20
Cellulose Proprionate - Injection	1.10	1.22	75.5	22.9	.828	425	.004	.40	2.00	.25
CTFE	1.49	2.11	134.0	13.1	.473	550	.008	.22	.01	
FEP	1.49	2.11	134.0	12.9	.465	600	.010	.28	<.01	
Ionomer - Injection	.73	.95	59.1	29.2	1.060	420	.007	.54	.20	
Nylon 6	.97	1.13	70.5	24.5	.886	550	.013	.40	1.60	.15
Nylon 6/6	.97	1.14	71.2	24.3	.878	510	.015	.40	1.50	.15
Nylon 6/10	.97	1.08	67.4	25.6	.927	450	.011	.40	.40	.15
Nylon 6/12	.97	1.07	66.8	25.9	.935	500	.011	.40	.40	.20
Nylon 11	.97	1.04	64.9	26.6	.962	450	.005	.47	.30	.10
Nylon 12	.97	1.02	63.7	27.1	.980	445	.003		.25	.10
Phenylene Oxide Based	.90	1.08	67.5	25.6	.926	525	.006	.32	.07	
Polyallomer	.86	.90	56.2	30.7	1.110	405	.015	.50	.01	
Polyarylene Ether	1.04	1.06	66.2	30.7	.940	535	.006		.10	
Polycarbonate	1.02	1.20	74.9	23.1	.832	575	.006	.30	.20	.02
Polyester PBT	1.11	1.34	83.6	20.7	.746	460	.020		.08	.04
Polyester PET	1.10	1.31	81.8	21.1	.746	490	.002	.40	.10	.005
HD Polyethylene - Injection	.72	.95	59.3	29.1	1.050	480	.025	.55	<.01	
HD Polyethylene - Blow Molding	.73	.95	56.9	28.8	1.040		.025	.55	<.01	
LD Polyethylene - Injection	.76	.92	57.4	30.1	1.090	400	.032	.55	<.01	
LLD Polyethylene - Injection	.70	.93	58.0	29.8	1.075	425	.022		<.01	
Polypropylene - Injection	.75	.90	56.2	30.7	1.110	490	.018	.50	<.01	
Polystyrene - Impact Sheet	.96	1.04	64.9	26.6	.963		.005	.34	.10	
Polystyrene - G.P. Crystal	.97	1.05	65.5	26.2	.943	425	.004	.32	.03	
Polystyrene - Injection Impact	.96	1.04	64.9	26.6	.968	440	.006	.34	.10	
Polysulfone	1.16	1.25	77.4	22.3	.807	680	.007	.28	.30	.05
Polyurethane (non-elastomer)	1.13	1.20	74.9	23.1	.834	400	.020	.40	.10	.03
PVC - Rigid Injection	1.20	1.29	83.6	21.0	.756	380	.025	.25	.10	.07
PVC - Flexible Injection	1.20	1.29	80.5	21.4	.776	300	.025	.25	.10	
PTFE	1.50	2.16	134.8	12.9	.464	650	.006	.25	<.01	
SAN	1.00	1.08	67.4	25.6	.927	470	.005	.31	.03	.02
TFE	1.50	1.70	106.1	16.3	.589	610	.040	.46	.01	
Urethane Elastomers (TPE)	.82	.83	51.6	33.5	1.210	400	.001	.46	.07	.03

Table 4.14 Injection temperature processing guide (courtesy of Spirex Corp.)

Polymer	Type	T_g deg. F (°C)	Processing Temperature, deg. F (°C)
Polyetheretherketone (PEEK)	Semicrystalline	290 (143)	650 (343)
Polyphenylene sulfide (PPS)	Semicrystalline	185 (85)	630 (332)
Polyaryleneketone	Semicrystalline	400 (204)	700–780 (371–416)
Polyarylene sulfide	Amorphous	410 (210)	625–650 (329–343)
Polyetherimide (PEI)	Amorphous	Varies	Varies
		450 (232)	575–650 (302–343)
		545 (285)	650–700 (343–371)
Polyarylether	Amorphous	476 (247)	650 (343)
Polyethersulfone (PES)	Amorphous	510 (266)	575 (302)
Polyamide-imide (PAI)	Amorphous	470 (243)	650 (343)
Polyimide	Pseudothermoplastic	480 (249)	680 (360)
		482 (250)	660 (349)
		536 (280)	660 (349)
		536 (280)	660 (349)

Commodity thermoplastics processing temperatures average 204° to 288°C (400° to 550°F).

Table 4.15 Heat-resistant engineering thermoplastics processing temperatures

in temperature between the cylinder and the plastic; the wall thickness of the cylinder and the stationary film (on the inner cylinder wall) of the plastic being heated; and the amount of turbulence in the cylinder.

Since plastics, due to their molecular structures, have low thermal conductivities, it is difficult to transmit heat through them rapidly (chapter 1). In addition, plastic melts are very viscous and it is difficult to create any turbulence or mixing action in them without the positive application of some form of mechanical agitation in the screw. The problem is further complicated by limitations on the length of time the plastic may be allowed to remain in the cylinder. In designing the screw, a balance must be maintained between the need to provide adequate time for proper heat exposure of plastic in the cylinder and the need to process maximum quantities of materials for the most economical operation.

In general, the heat-transfer problems have led injection screw designers to concentrate on making the screw a more efficient heat-transfer device. The internal design and performance of these units varies considerably based on material to be processed.

IM is a thermal process with its major task ultimately being the control of temperature. Too much or too little heat at the wrong place can cause many problems (short shot, galling, splay, brittleness, plastic degradation, etc.). One cannot see this thermal energy, only its effects. Thermal energy radiates in the infrared (IR) spectrum, outside the spectrum of visible light. Use has been made of IR video cameras to detect energy color patterns in all locations around the IMM and auxiliary equipment. With this IR thermography, each plastic has its own wavelength; temperature readings are related to the IR color patterns. It also provides IR signatures for each plastic using Fourier transform infrared (FTIR) spectography.

MACHINE CONTROL

Machine process controls coordinate individual functions of the clamp, injection unit, ejector mechanism, and mold systems and accessories, such as core pulls and unscrewing dies for threaded parts (Fig. 4.24). The more advanced controls employ a feedback system (closed loop) to provide much tighter control over actual parameters versus set points. High-level controls are capable of communicating with auxiliary equipment, such as chillers, hopper loaders, mold temperature controllers, and robots, and displaying all machine parameters and conditions on a CRT (chapter 3).

MATERIAL	PROCESS MELT TEMP	MOLD TEMP
ABS....med, high impact	440-510	90-180
ABS....high heat grade	510-540	140-180
ACETAL	380-420	140-220
ACRYLIC	420-485	120-180
ETHYLENE VINYL ACETATE	350-400	50-100
IONOMER	420-460	50-100
NYLON....6/6	500-560	120-180
NYLON....6	470-530	100-180
NYLON....6/10	460-510	100-180
NYLON....11	420-480	80-140
NYLON....12	400-450	80-140
POLYCARBONATE	590-640	180-240
POLYESTER - PBT	460-490	100-140
POLYESTER - PET	520-560	60-120
POLYESTER, TP	560-590	300-320
POLYETHERIMIDE	680-720	220-300
POLYETHYLENE..low density	340-440	50-100
POLYETHYLENE..med density	390-490	50-120
POLYETHYLENE.high density	420-540	50-150
POLYPHENYLENE SULFIDE	590-670	190-230
POLYPROPYLENE	420-520	60-150
POLYSTYRENE....G.P.	380-460	50-120
POLYSTYRENE..impact modified	400-480	50-120
POLYSULFONE	650-750	200-320
POLYURETHANE	390-440	70-120
PVC.	320-420	50-120
SAN	400-500	120-180
TPE	340-440	70-120

Table 4.16 Examples of melt and mold temperatures for various plastics

These never-ending advanced controls allow interfacing many machines to a common host computer, allowing plantwide monitoring of the overall production status. The many software developments are rapidly changing the character of the molding machine. The following are some examples:

Plastics flowability Bulk density Particle size distribution	Feeding ease Feeding accuracy
Material density Material height	Pre-heating
Melt flow behavior and characteristic, Water content	Process: Compression molding transfer molding injection molding pre-heating (temperature, time) mold filling (time, pressure) curing (temperature, time)
Shrinkage characteristic Water content Stickiness (adhesion)	Dimensional stability Demolding behavior
Wear	Mold life Machine wear

Table 4.17 Processing flow chart for IM

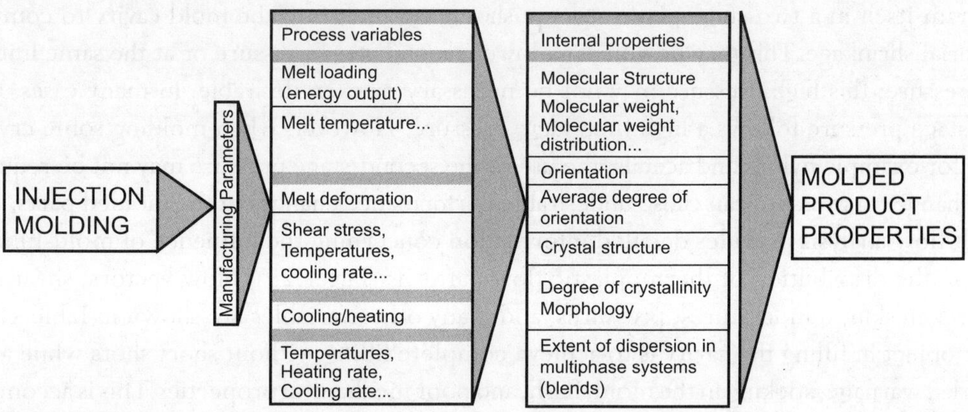

Figure 4.24 Examples of functions that are controllable

1. SPC-Light Software from Lighthouse Systems Inc. (located in Fairport, New York; telephone: 585-223-0600, www.lighthousesystems.com) eliminates use of manual data collection and chart presentation, and makes it easier to spot problematic trends and correct them while they are small. It provides for an instantaneous understanding of what is going on in an injection-molding process. It is versatile enough to monitor and analyze what is going on in a specific mold cavity. Information gathered through digital gauges and measuring devices is directly fed through a multiplexer into the database. With a couple of clicks, anything that is out of spec or out of control can quickly be spotted (379).
2. SPC-OnLine has a suite of predesigned reports that can be edited to suit the needs of an individual operation. It is a Web-friendly reporting system. The database structure makes it easy to trace details and enables an injection molder's customers to keep track of all the process improvements being made.

DEVELOPING MELT AND FLOW CONTROL

The mechanism for melting starts after the plastics move from the hopper to the plasticator screw (Fig. 4.6). Plastic touches the barrel to form a thin film of melted plastic on the barrel surface. The relative motion of the barrel and screw drag this melt, which is picked up by the leading edge of the advancing flight of the screw. This flushes the plastic down in front of it, forming a pool that circulates. Heat is first conducted from the barrel through the film of plastic attached to it. Heat then enters the plastic by the shearing action, with energy being derived from the turning of the screw. The width of the melted plastic increases as the width of the solid bed decreases. Melting is complete at the point where the width of the solid bed is zero.

The melt is forced into the mold cavity. During the initial mold-filling phase of the molding cycle, high injection pressure may be needed in order to maintain the desired mold-filling speed. Once the mold is full, the cycle enters its "holding phase" and the reciprocating screw acting as a ram (or the ram itself in a two-stage plasticator) pushes extra melt into the mold cavity to compensate for material shrinkage. This may be done at a lower second-stage pressure or at the same initial high filling pressure; this high pressure may not be necessary or even desirable. In many cases, a lower second-stage pressure follows a high first-stage pressure. However, when molding some crystalline plastics, for example nylons and acetals, the use of the second-stage pressure may not be required, as abrupt changes in pressure can cause undesirable performance changes in the molded parts.

Melt flow analysis provides detailed information concerning the influence of mold-filling conditions on the distribution of thermoplastic flow patterns (Fig. 4.25). Flow vectors, shear stresses (214), frozen skin, temperatures, pressures, and many other variables are shown in Table 4.18.

The object in filling the cavity is to achieve complete filling without short shots while avoiding sink marks, warpage, sticking in the mold, flash, and poor mechanical properties. This is accomplished by delivering the correct amount of melt to the cavity while avoiding overpressurization, high thermal stresses, and high residual orientation. Some of the factors that favor complete filling, however, also promote overpressurization and residual stresses so care must be taken in selecting operating conditions for a given mold and plastic.

As melt flows into the cavity, the situation cannot be described in terms of pressure flow between parallel plates with a gap equal to the mold clearance because a frozen layer forms immediately at the cavity wall. Moreover, the melt in the center has a lower viscosity owing to its higher temperature, and as a result, the maximum shear rate occurs not at the surface of the frozen layer but closer to the center. The shear rate in the cavity generally falls in the range of 8,000 to 15,000 sec^{-1}.

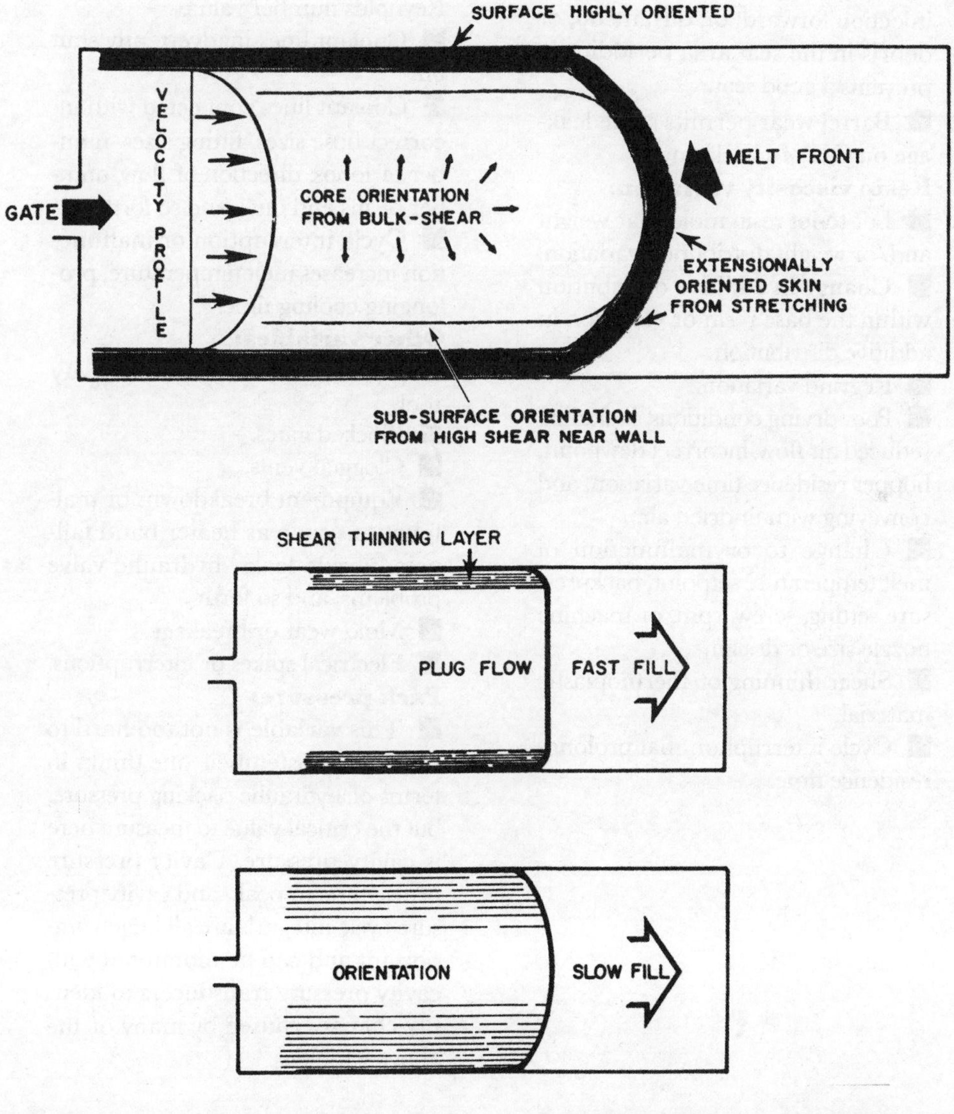

Figure 4.25 Melt flow fountain (or balloon) pattern across the thickness in a mold cavity

Shot variation:

■ Insufficient backpressure and/or incorrect screw rpm results in poorly compacted plasticated shot volume.

■ Decompression after screw run permits a small amount of resin to pass through to the front side of the check ring.

■ Check ring does not seat during injection forward; or, dirt, metal, or debris in the seat area periodically prevents a good seat.

■ Barrel wear permits some leakage on OD of check ring.

Resin viscosity variation:

■ Lot-to-lot resin molecular weight and/or weight distribution variation.

■ Change in additive distribution within the base resin or variation in additive distribution.

■ Regrind variation.

■ Poor drying conditions, including reduced air flow, incorrect dewpoint, hopper residence time variation, and conveying with undried air.

■ Change to or malfunction of melt temperature setpoint, backpressure setting, screw, rpm, or machine nozzle size or design.

■ Shear thinning of thermoplastic material.

■ Cycle interruption that prolongs residence time.

Cooling rate variation:

■ Poor coolant flow rates, which lead to increased backpressure in return lines.

■ Blocked or clogged coolant lines and bubblers.

■ Coolant line IDs fouled with rust and scale, affecting heat transfer.

■ Pumps become worn, affecting flow rates, heat transfer, and the Reynolds number value.

■ Coolant lines inadvertently shut off.

■ Coolant lines connected with incorrect hose sizes, fitting sizes, number of loops, direction of flow, number of ins and outs, and so forth.

■ Cycle interruption or malfunction increases melt temperature, prolonging cooling time.

Other variables:

■ Imbalanced fill in a multicavity tool.

■ Blocked gates.

■ Clogged vents.

■ Equipment breakdowns or malfunctions such as heater band failures, nozzle leaks, hydraulic valve problems, and so forth.

■ Mold wear or breakage.

■ Electrical spikes or interruptions.

Pack pressure:

■ This variable is not too hard to control consistently if one thinks in terms of hydraulic packing pressure, but the critical value to measure here is cavity pressure. Cavity pressure peaks, time to peak, and cavity pressure cycle integrals are all hugely important and can be monitored with cavity pressure transducers to identify changes caused by many of the above variables.

Table 4.18 Processing variables (courtesy of The Tech Group, Scottsdale, Arizona)

Another important phenomenon that causes the flow to deviate from two-dimensional flow between parallel plates is termed the *fountain effect* (Fig. 4.25). Here the melt does not reach the wall or surface of the frozen wall layer by simple forward advance, but rather it tends to flow down the center of the cavity to the melt front and then flow out toward the wall. This can have an important effect on the direction of the flow-induced orientation of the polymer molecules.

If the melt must flow around an obstacle of any kind in the cavity, a weld line will result. A phenomenon that can lead to a complex pattern of weld lines is "jetting." This term refers to the tendency of the melt to spurt into the cavity without wetting the walls near the gate; the result is that the cavity fills by a piling up of the jet at the end of the cavity rather than the smooth advancement of a melt front starting at the gate.

Within different-shaped cavities, melt flow directions can vary and influence shrinkage and strength properties. Figures 4.26 and 4.27 provide examples of the relationship between melt flow and shrinkage as particularly occurs with crystalline TPs (chapter 1). This relationship also influences directional properties. Figure 4.28 highlights the influence and effect of a mold gate location on the direction of melt flow, affecting directional properties. Figure 4.29 relates strength properties to melt flow directions. The control of shrinkage and strength properties can be made to meet design requirements by factors such as the design of the mold with its singular or multiple gate

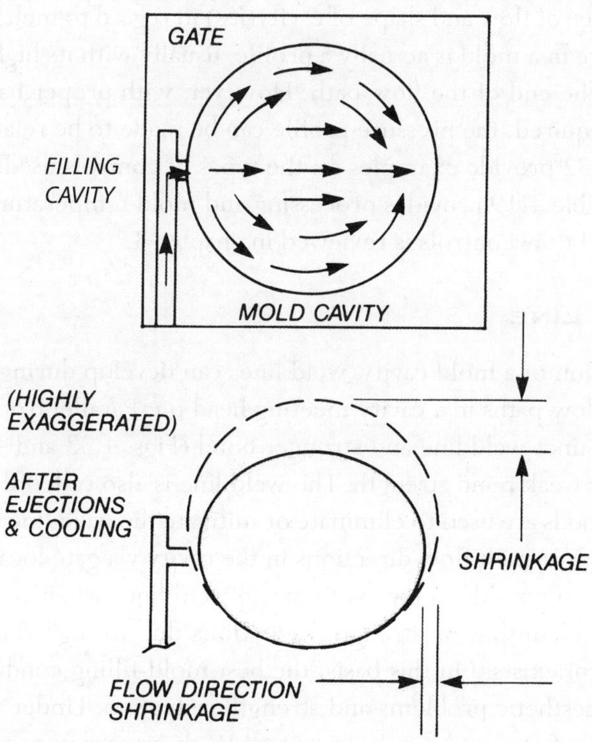

Figure 4.26 Relation of melt flow to shrinkage

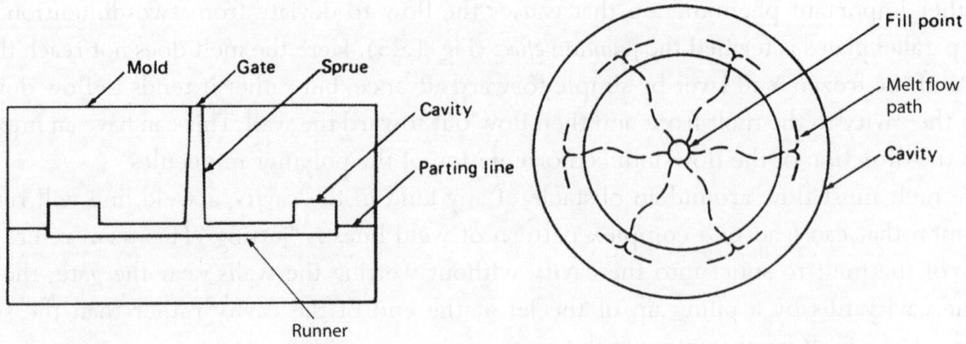

Figure 4.27 Melt flow pattern in a center gated disc

locations, the processing machine controls, the change of product shape, and the type of plastic. When properly coordinated, these factors will result in balanced shrinkages and strength performances, or they may be unbalanced to meet the product requirements. Different software programs are available to meet different product requirements.

In monitoring and controlling melt pressure, one of the problems is that it is not a constant for all positions in the cavity. Pressure in the mold cavity varies significantly from the applied pressure. It also varies with distance of flow and shape of cavity(ies) in regard to melt flow pattern. The result is that the cavity pressure in a mold is actually a profile, usually with its highest pressure at the gate and lowest pressure at the end of the flow path. However, with proper balance of pressure, temperature, and time, if required, the pressure profile can be made to be relatively even (Fig. 4.30).

Figures 4.31 and 4.32 provide examples on the types of controls used and how they influence properties of plastics. Table 4.19 provides processing and mold temperatures for different plastics. Information on melt and flow controls is reviewed in chapter 3.

Weld and Meld Line

Based on the configuration of a mold cavity, weld lines can develop during molding. A weld line is the result of two melt flow paths in a cavity meeting head on. Parallel fronts meet to form a meld line producing a less distinct weld line and stronger bond (Figs. 4.33 and 4.34). The weld line can be very visible and have weak bond strength. The weld line is also called the weld mark, flow line, or striae. Different methods are used to eliminate or mitigate the weld line by changing the product design, mold design, and/or melt flow directions in the cavity via gate locations.

The melt flow pattern provides a means to locate weld lines when two melt fronts meet (Fig. 4.35). Warping surface blemishes or strength reductions due to high shear stresses can occur if improper process control exists. On this basis, the best mold-filling conditions can be selected to eliminate or minimize aesthetic problems and strength reduction. Under the best molding conditions, up to about 90% of strength can be retained. With improper conditions, strength can be reduced to 15%.

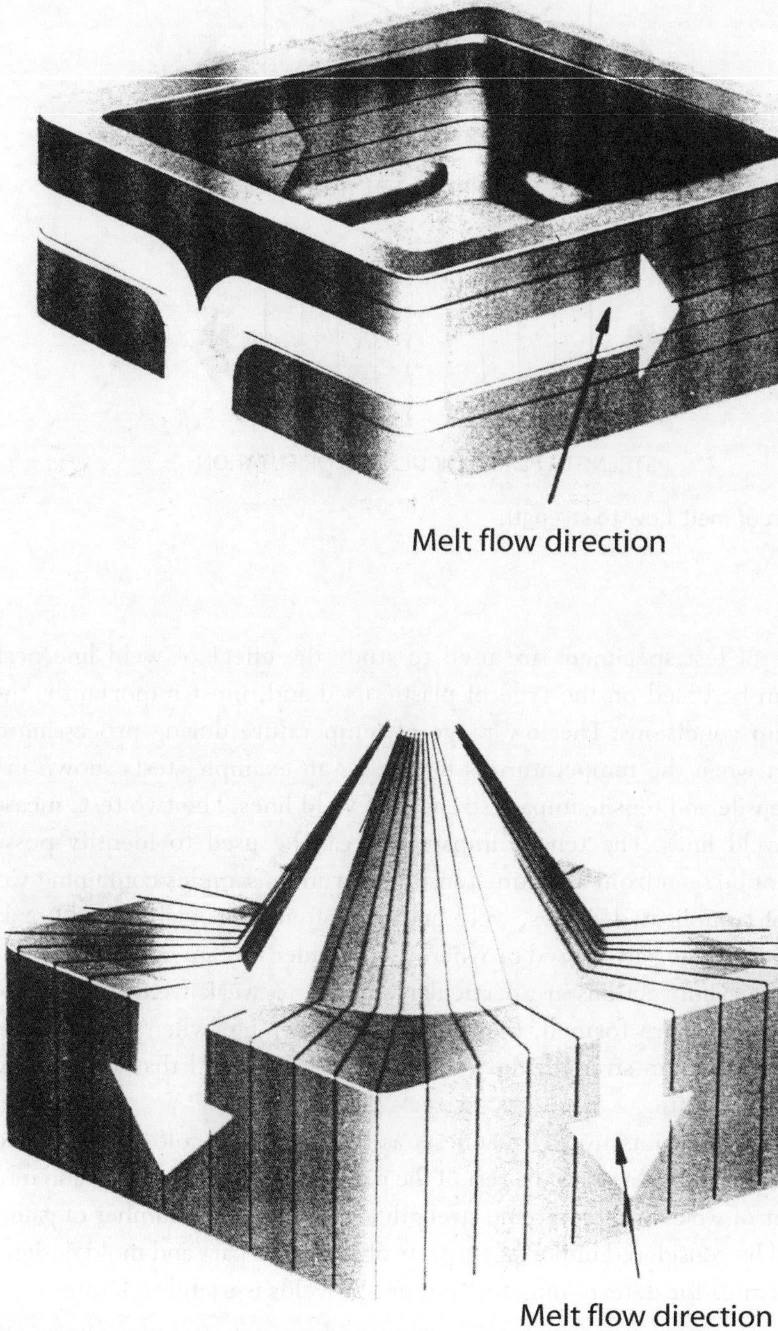

Figure 4.28 Examples of side and center gate locations influencing melt flow and property direction

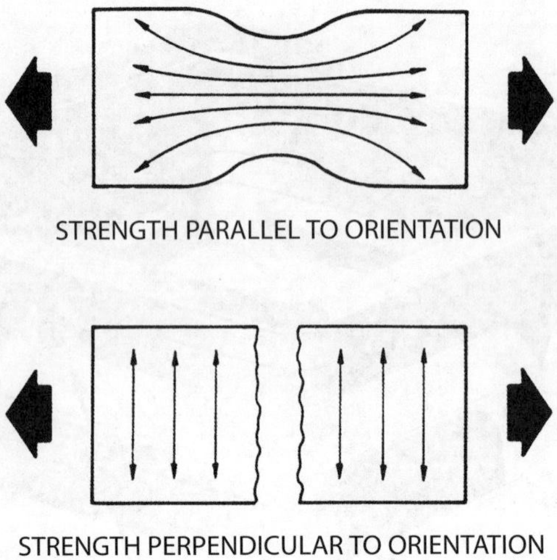

Figure 4.29 Relation of melt flow to strength

Different types of test specimens are used to study the effect of weld line performance in a product; these can be based on the type of plastic used and, most importantly, the processing minimum/maximum conditions. The low range of temperature during processing can result in lower strength than when the temperature is higher. As an example, tests shown in Figure 4.35 are used to study tensile and tensile impact strength of weld lines. The two tests measure different characteristics of weld lines. The tensile impact test can be used to identify possible product weakness that cannot be seen from a routine tensile test run on samples containing weld lines.

In the design of complicated shapes, weld lines are often unavoidable. Weld or knit lines are formed when more than one gate is used or wherever a divided stream of plastic joins after flowing around a pin or core. Thin sections are particularly prone to weak welds because of rapid melt solidification. When welds are formed, they should be sweeping; when unavoidable, butt welds must be vented for maximum strength. In these cases, it is essential that air at the weld escapes before the melt streams unite.

Poor venting at weld points usually manifests as burning or discoloration. At such spots, the strength of the welds will be inferior to the rest of the part. Obviously, good part and mold design call for the least number of welds when extreme strength is necessary. The number of gates and internal shutoff cores should be considered important aspects of the initial part and mold designs.

The general formula for determining the number of welds is useful to know:

$$N = G - \mathrm{I} + P,$$

where N = number of weld points, G = number of gates, and P = number of shutoff cores. Examples of using this formula are shown in Figure 4.36.

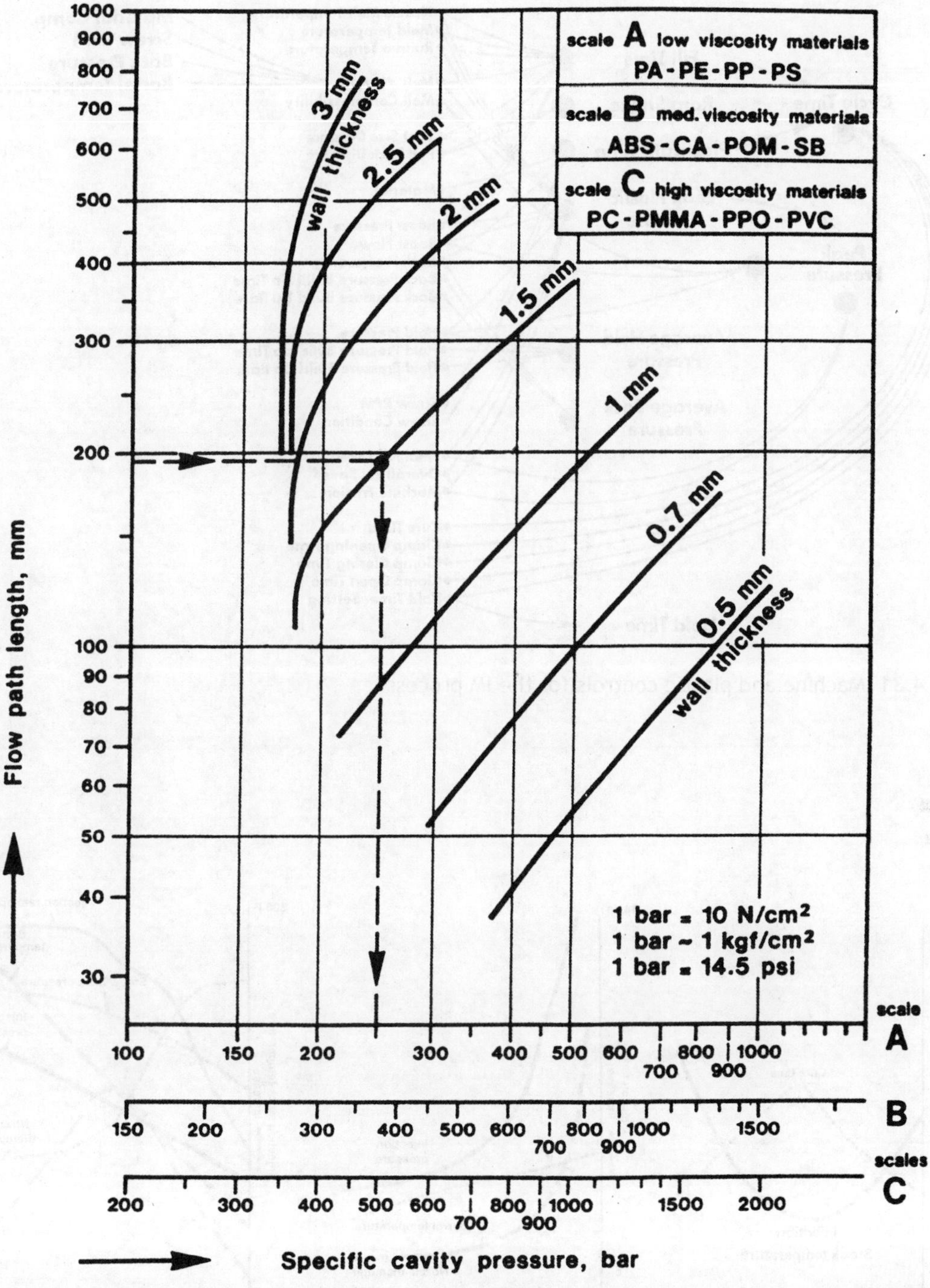

Figure 4.30 Relation of melt flow (viscosity), cavity pressure, and product thickness (courtesy of Negri Bossi)

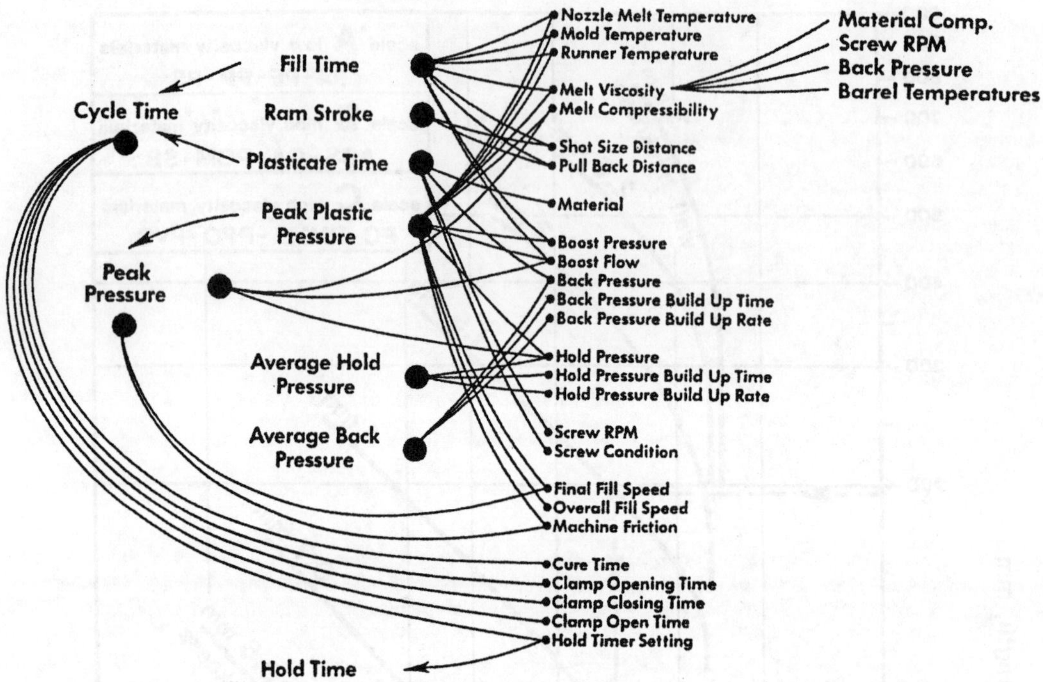

Figure 4.31 Machine and plastic controls for the IM process

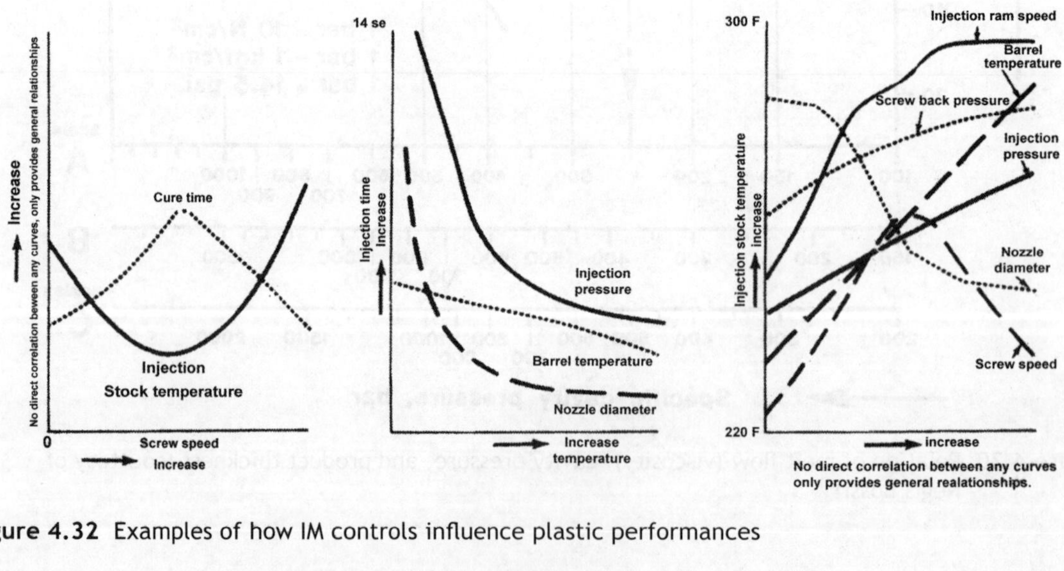

Figure 4.32 Examples of how IM controls influence plastic performances

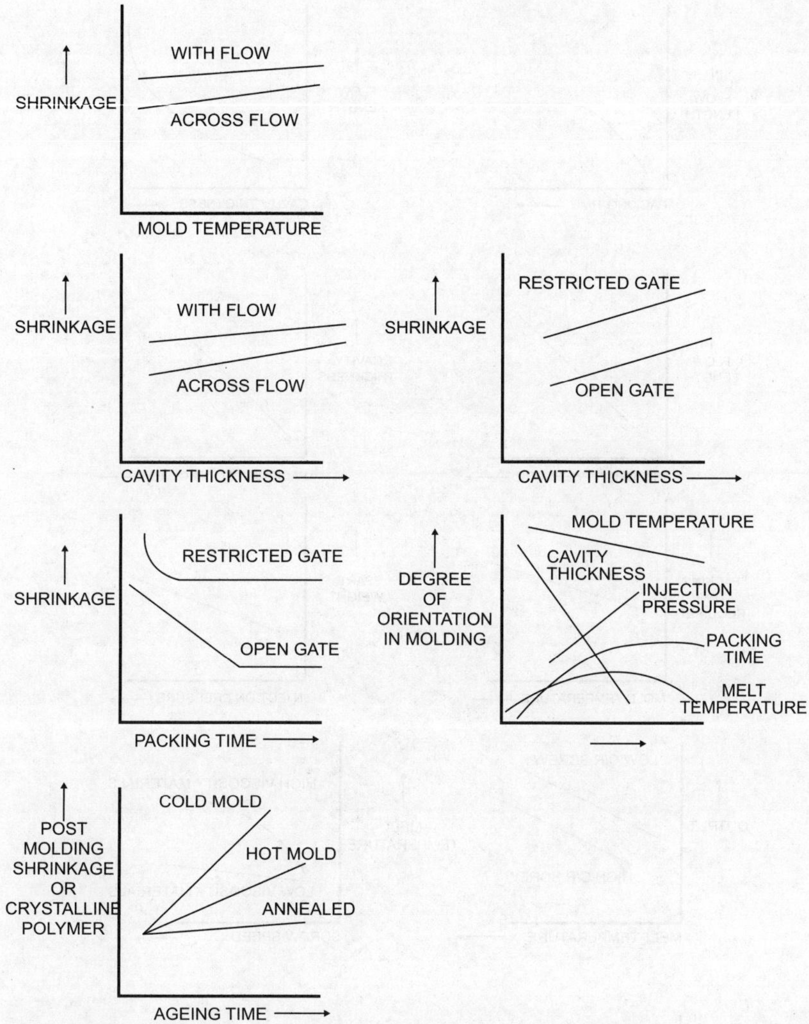

Figure 4.32 *(continued)*

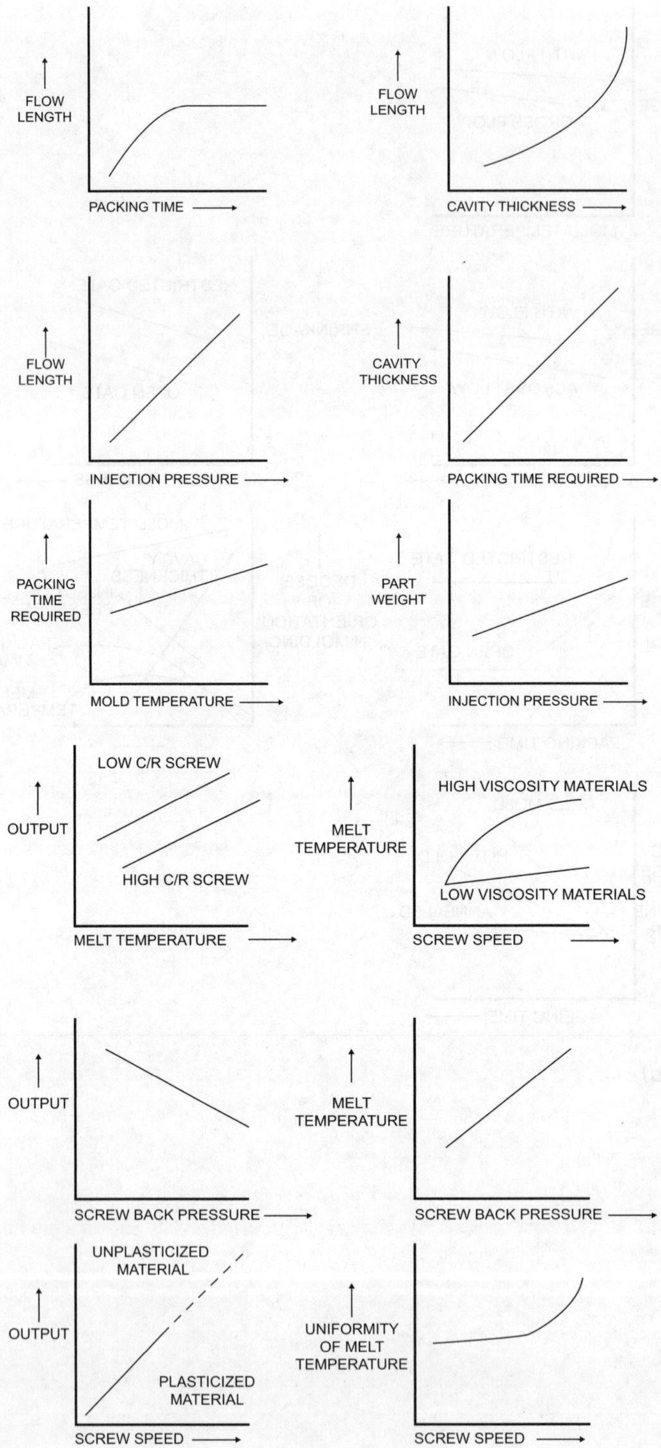

Figure 4.32 *(continued)*

Material	Symbol	Density [g/cm³]	Glass fiber content [%]	Average specific heat [kJ/(kg × K)]	Processing temperature [°C]	Mold temperature [°C]	Shrinkage [%]
Polycarbonate	PC	1.2		1.3	280 - 320	80 - 100	0.8
Polycarbonate-GR	PC-GR	1.42	10 - 30	1.1	300 - 330	100 - 120	0.15 - 0.55
Poly(ethylene terephthalate)	PET	1.37			260 - 290	140	1.2 - 2.0
Poly(ethylene terephthalate)-GR	PET-GR	1.5 - 1.53	20 - 30		260 - 290	140	1.2 - 2.0
Poly(butylene terephthalate)	PBT	1.3			240 - 260	60 - 80	1.5 - 2.5
Poly(butylene terephthalate)-GR	PBT-GR	1.52 - 1.57	30 - 50		250 - 270	60 - 80	0.3 - 1.2
Polyetheretherketone	PEEK	1.32			350 - 390	120 - 150	1.1
Polyetheretherketone-GR	PEEK-GR	1.49	30		350 - 400	120 - 150	0.2 - 1
Polyamide 6 (nylon-6)	PA 6	1.14		1.8	240 - 260	70 - 120	0.5 - 2.2
Polyamide 6-GR	PA 6-GR	1.36 - 1.65	30 - 50	1.26 - 1.7	270 - 290	70 - 120	0.3 - 1
Polyamide 66 (nylon-66)	PA 66	1.15		1.7	260 - 290	70 - 120	0.5 - 2.5
Polyamide 66-GR	PA 66-GR	1.20 - 1.65	30 - 35	1.4	280 - 310	70 - 120	0.5 - 1.5
Polyamide 11	PA 11	1.03 - 1.05		2.4	210 - 250	40 - 80	0.5 - 1.5
Polyamide 12	PA 12	1.01 - 1.04		1.2	210 - 250	40 - 80	0.5 - 1.5
Polyamide-imide	PAI	1.4		-	330 - 380	230	-
Poly(phenylene sulfide)	PPS	1.64	40		370	> 150	0.2
Poly(etherimide)	PEI	1.27		-	340 - 425	65 - 175	0.4 - 0.7
Poly(ether sulfone)	PES	1.6	30	-	360 - 390	140 - 190	0.2 - 0.5
Polyether ketone	PEK	1.3		-	360 - 420	120 - 160	1.1
Polysulfone	PSU	1.24		-	330 - 380	110 - 180	0.7
Polyurethane (thermopl.)	PUR	1.2		1.85	195 - 230	20 - 40	0.9
Phenol-formaldehyde resin	PF	1.4		1.3	60 - 80	170 - 190	1.2
Melamine-formaldehyde resin	MF	1.5		1.3	70 - 80	150 - 165	1.2 - 2
Melamine/phenol-formaldehyde resin	MPF	1.6		1.1	60 - 80	160 - 180	0.8 - 1.8
Unsaturated polyester	UP	2.0 - 2.1		0.9	40 - 60	150 - 170	0.5 - 0.8
Epoxy, epoxide	EP	1.9	30 - 80	1.7 - 1.9	ca. 70	160 - 170	0.2
Polystrene	PS	1.05		1.3	180 - 280	10 - 40	0.3 - 0.6
Styrene-butadiene	SB	1.05		1.21	170 - 260	5 - 75	0.5 - 0.6
Styrene-acrylonitrile	SAN	1.08		1.3	180 - 270	50 - 80	0.5 - 0.7
Acrylonitrile-butadiene-styrene	ABS	1.06		1.4	210 - 275	50 - 90	0.4 - 0.7
Acrylonitrile-styrene-acrylate	ASA	1.07		1.3	230 - 260	40 - 90	0.4 - 0.6
Low-density polyethylene	LDPE	0.954		2.0 - 2.1	160 - 260	50 - 70	1.5 - 5.0
High-density polyethylene	HDPE	0.92		2.3 - 2.5	260 - 300	30 - 70	1.5 - 3.0
Polypropylene	PP	0.917		0.84 - 2.5	250 - 270	50 - 75	1.0 - 2.5
Polypropylene-GR	PP-GR	1.15	30	1.1 - 1.35	260 - 280	50 - 80	0.5 - 1.2
Polyisobutylene	PIB	0.93		-	150 - 200	50 - 80	-
Poly(4-methylpentene-1)	PMP	0.83		-	280 - 310	70	1.5 - 3.0
Poly(vinyl chloride)	PVC soft	1.38		0.85	170 - 200	15 - 20	< 0.5
Poly(vinyl chloride)	PVC rigid	1.38		0.83 - 0.92	180 - 210	30 - 50	~ 0.5
Poly(vinylidene flouride)	PVDF	1.2			250 - 270	90 - 100	3 - 6
Polytetraflouroethylene	PTFE	2.12 - 2.17		0.12	320 - 360	200 - 230	3.5 - 6.0
Poly(methyl methacrylate)	PMMA	1.18		1.46	210 - 240	50 - 70	0.1 - 0.8
Polyoxymethylene, polyacetal	POM	1.42		1.47 - 1.5	200 - 210	> 90	1.9 - 2.3
Poly(phenylene oxide)	PPO	1.06		1.45	250 - 300	80 - 100	0.5 - 0.7
Poly(phenylene oxide)-GR	PPO-GR	1.27	30	1.3	280 - 300	80 - 100	> 0.7
Cellulose acetate	CA	1.27 - 1.3		1.3 - 1.7	180 - 230	50 - 80	0.5
Cellulose acetate butyrate	CAB	1.17 - 1.22		1.3 - 1.7	180 - 230	50 - 80	0.5
Cellulose propionate	CP	1.19 - 1.23		1.7	180 - 230	50 - 80	0.5

Table 4.19 Plastics guide: plasticizing and mold temperatures, specific heat, and shrinkage data provided

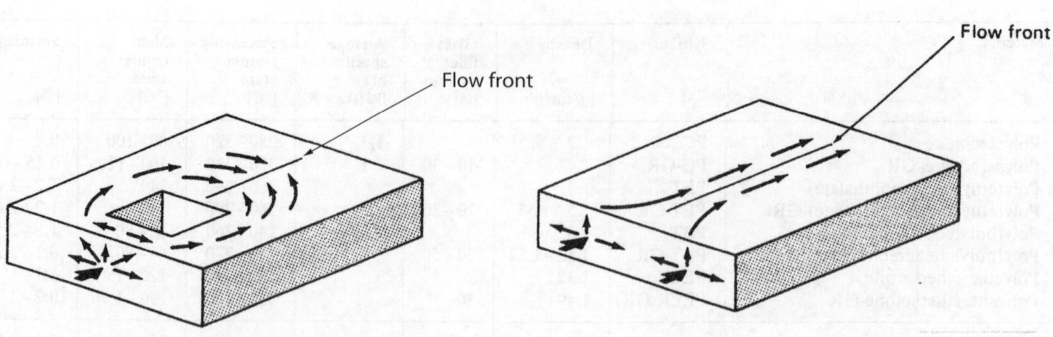

Figure 4.33 Examples of weld line (left) and meld line

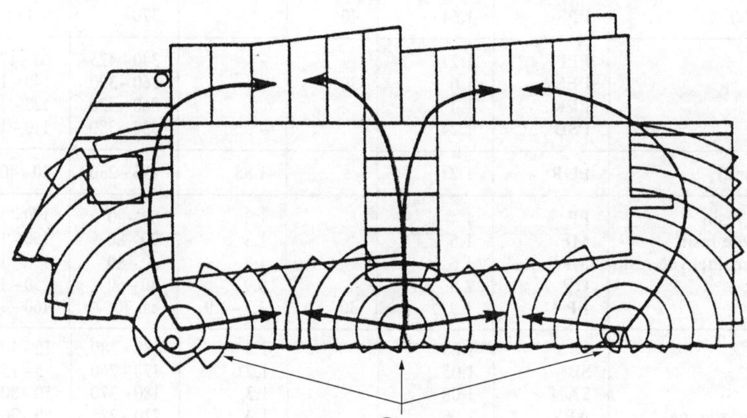

Figure 4.34 Examples of the melt flow weld lines in a mold with three gates

When adequate venting exists, conventional techniques to improve weld strength include (a) increase melt temperature, (b) increase mold temperature, (c) increase injection pressure, and (d) avoid use of external release mold lubricant. Often, mold release sprays are pushed into the weld area by the advancing front of the melt and prevent good fusion.

The goal is to increase the melt flow at the junction, that is, to push the melt into the mold faster so that the two (or more) fronts can fuse or unite properly before melt solidification occurs. Typical weld strengths, in tension, range from 50% to 95% of base material strength, although poor weld strength usually shows up as failure in repeated flexure or shear. To ensure maximum part strength when welds are involved in the design, examine Table 4.20. It assumes that cavity venting is not limiting and that melt at 277°C (530°F) flows into a cold mold at 16°C (60°F). Higher melt or mold temperatures improve welding. The usual basis for optimum welding is two-thirds of the part freeze time. Using DuPont nylon 6/6, Figures 4.37 to 4.40 provide guides to factors such as cavity fill rates and pressure.

MOLDING VARIABLES

COOLING

As reviewed by Husky Injection Molding Systems Inc. (that provides, worldwide, parts for IM production lines or complete lines [IMMs, molds, chiller systems, robots, etc.] to mold specific

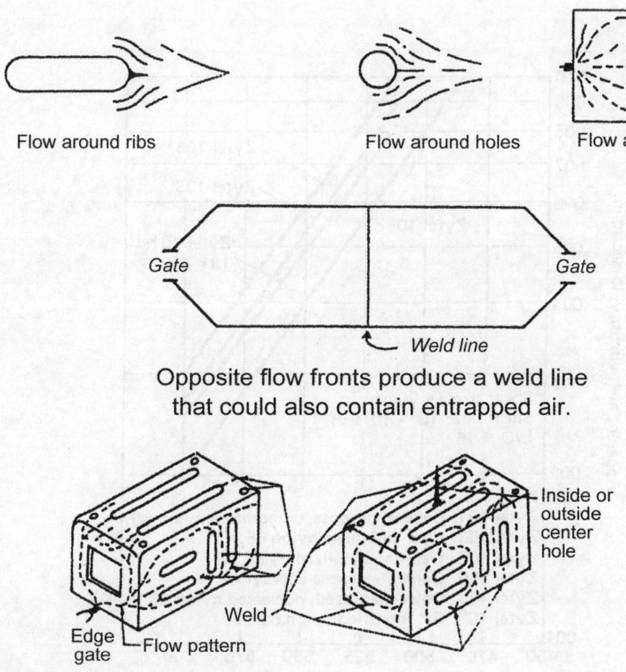

Figure 4.35 Examples of weld line formations

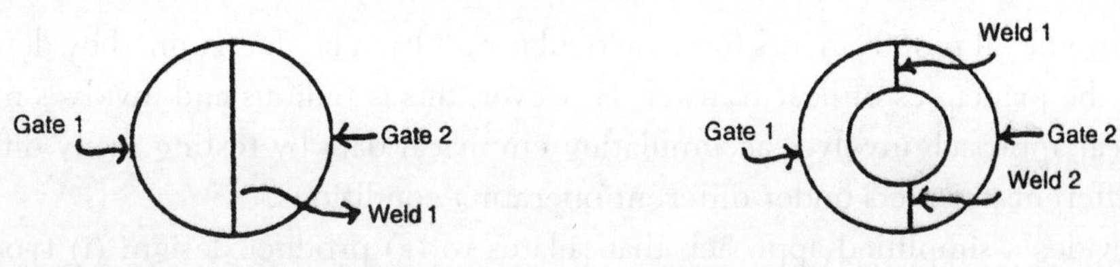

Figure 4.36 Determining weld lines

Part Thickness (in.)	Maximum Fill Time (s)	Maximum Part Weight/Gate (oz)
0.020	0.14	0.05
0.030	0.37	0.05
0.040	0.64	0.50
0.050	0.95	1.00
0.060	1.30	2.25
0.070	1.90	13.00
0.080	2.80	19.00

Table 4.20 Maximum weld strength in thin nylon 6/6 sections

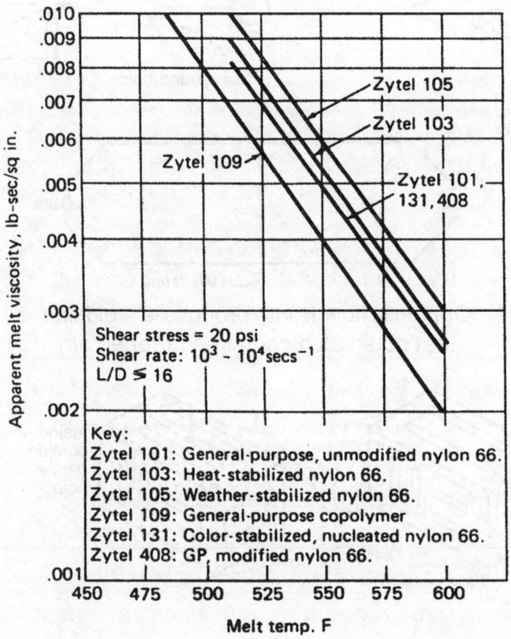

Figure 4.37 Nylon 6/6 melt viscosity vs. temperature

products), heat removal requirements for a particular mold can be determined by detailed calculations based on the principles of heat transfer. However, this is tedious and involves many assumptions. A practical approach involves accumulating empirical data by testing many different molds and running different products under different operating conditions.

Husky provides a simplified approach that relates to (a) product design: (i) type of material and (ii) maximum part thickness; (b) mold design: (i) size and location of cooling channels and

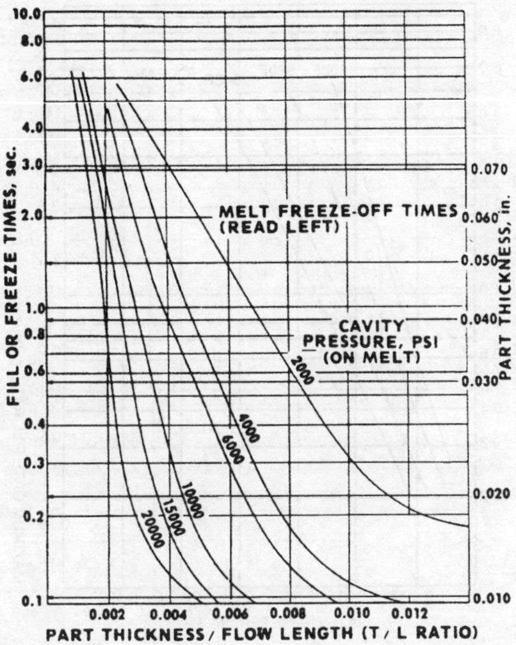

Figure 4.38 Nylon 6/6 relation of fill time, cavity dimensions, and pressure in estimating fill at a melt temperature of 550° ±10°F and mold temperature of 120° ±20°F

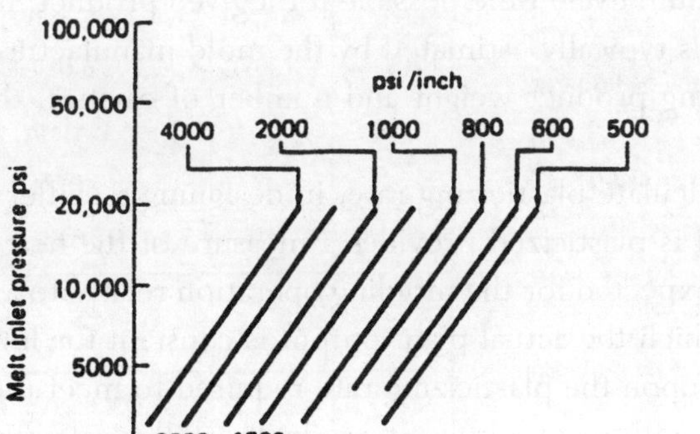

Figure 4.39 Nylon sprues, round runners, and gate pressure drops (psi/in of length)

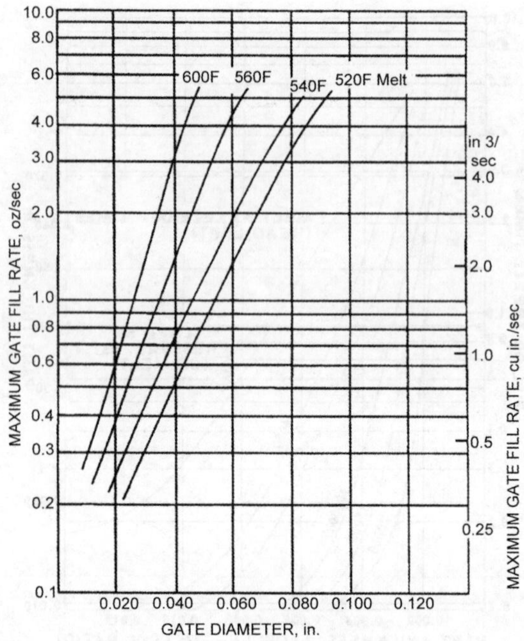

Figure 4.40 Nylon 6/6 maximum fill rates through round gates

(ii) size and location of mold heaters; (c) IMM design: (i) injection pressure, (ii) injection speed, (iii) maximum plasticizing, (iv) clamp speed, and (v) others; (d) chiller: (i) flow rate and (ii) water temperature; and (e) process: (i) injection temperature, (ii) ejection temperature, and (iii) others (Figs. 4.41 to 4.43).

In this approach, the first step is to estimate cycle time. From experience in design and manufacturing molds, the optimum cycle time possible for a given product, mold design, and IMM can be estimated. Cycle time is typically estimated by the mold manufacturer and included in a mold quotation. Then by knowing product weight and number of cavities, the plasticizing rate can be calculated.

The next step is to calculate plasticizing rate. In designing a chiller system for mold cooling, the rate at which material is plasticized provides a measure of the heat input into the mold. The length of production run expected for the molding operation relates to a chiller system design. For long-run operations, in which the actual plant output is constant for long periods of time, chiller selection should be based upon the plasticizing rate required to meet the manufacturing volumes. For short-run operations, with many different products, chiller selection should be based on the maximum plasticizing rates of the IMMs selected.

The third step is to calculate rate of heat removal. Once the plasticizing rate has been established, the total heat load for the chiller can be calculated. The calculation must take into account three heat sources: (a) plastic in the mold, (b) mold heaters (hot runner mold applications), and (c) circulating pump, chilled water piping, and reservoir.

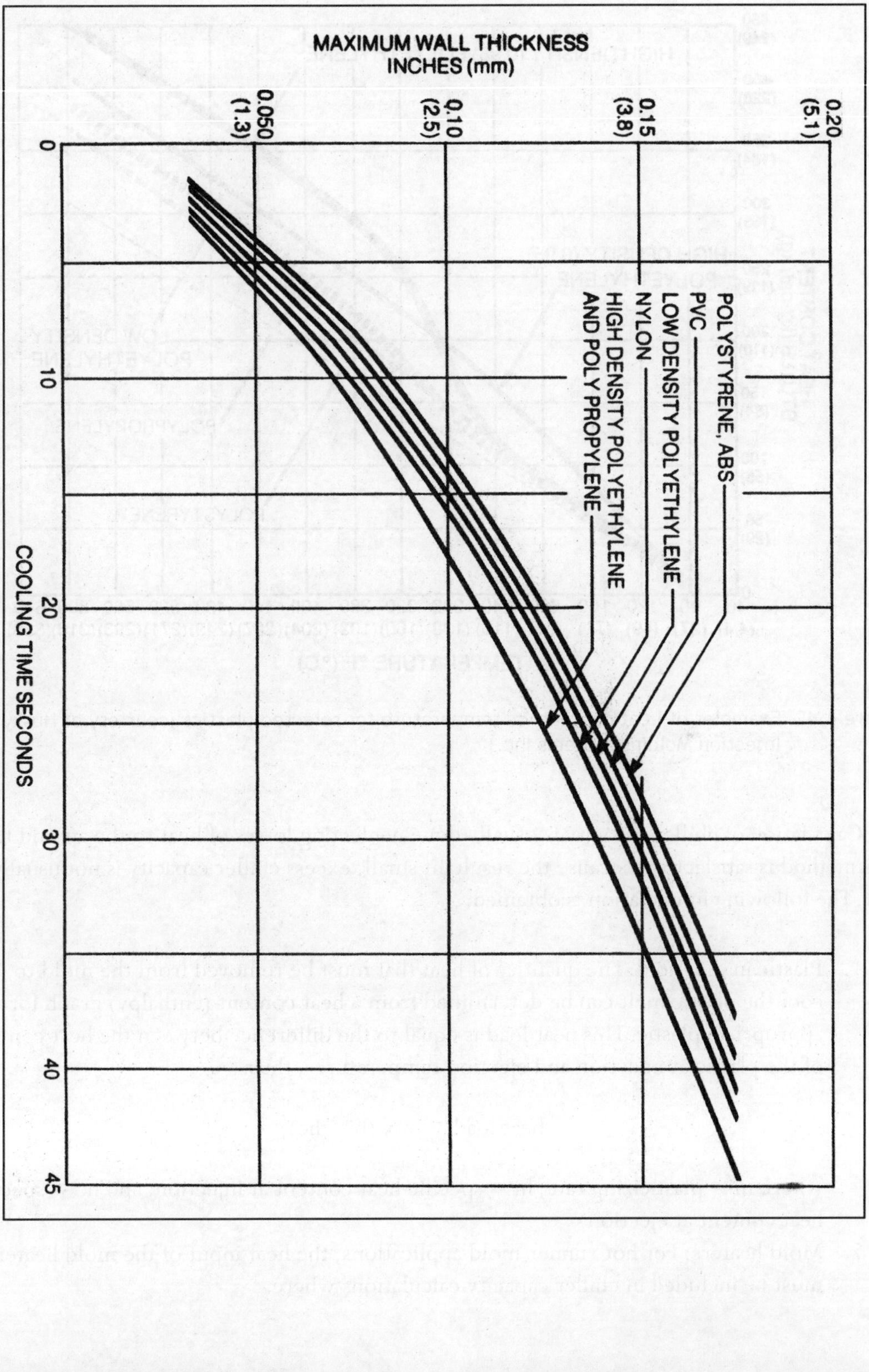

Figure 4.41 Examples of minimum cooling time for selected plastics (courtesy of Husky Injection Molding Systems Inc.)

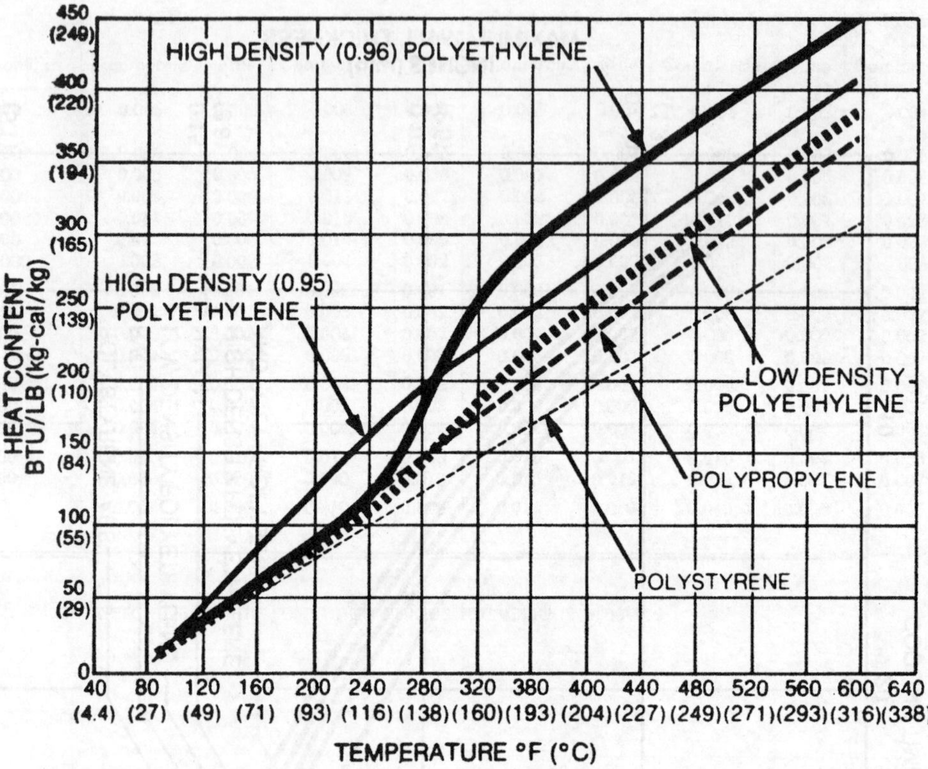

Figure 4.42 Examples of heat content vs. temperature for selected plastics (courtesy of Husky Injection Molding Systems Inc.)

Calculation of chiller capacity is usually done neglecting losses of heat to the air and the IMM. This method is satisfactory because the resultant small, excess chiller capacity is not usually significant. The following information is obtained:

1. Plastic in the mold: The quantity of heat that must be removed from the mold to cool the plastic melt can be determined from a heat content (enthalpy) graph for the appropriate plastic. This heat load is equal to the difference between the heat content of the plastic at injection and ejection temperatures, thus:

$$\text{heat load} = m \times (hi - he),$$

where m = plasticizing rate, hi = specific heat content at injection, and he = specific heat content at ejection.
2. Mold heaters: For hot runner mold applications, the heat input of the mold heaters must be included in chiller capacity calculations where

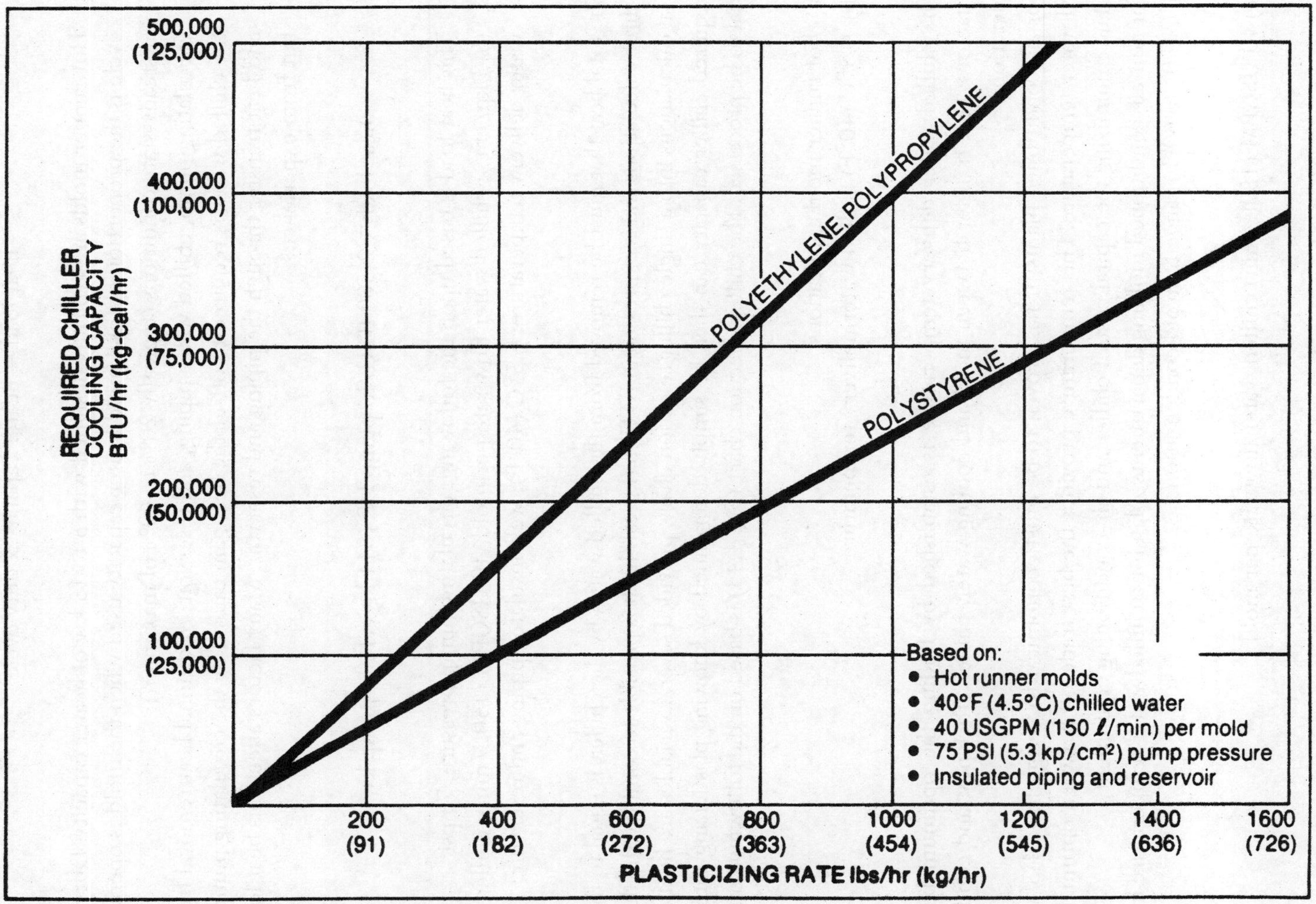

Figure 4.43 Chiller selection guide (courtesy of Husky Injection Molding Systems Inc.)

heat load = heater capacity x duty cycle.

Hot runner molds are typically designed with 5 to 15 kw of heater capacity. Duty cycle is the proportion of total mold heater capacity used when the mold is operating at steady-state conditions (typically 20% to 60% of maximum).

3. Circulating pump, chilled water piping, and reservoir: Additional heat removal from the chilled water is required to offset the heat gain caused by the circulating pump and heat transfer through the piping and reservoir. Many factors affect this heat load, but from experience:

heat load = 2,500 BTU/h/pump HP (844 kg-cal /h/pump kw).

This heat load is based upon the following: circulating pump pressure, 75 psi (5.3 kp/cm^2); chilled water flow rate per mold, 40 USGPM (150L/min); chilled water delivery temperature: 4.5°C (40°F); well-insulated piping and reservoir.

The total heat that must be removed from the chilled water, by the chiller, is equal to the sum of these three calculations and is expressed in BTU/h (kg-cal/h). It is possible to combine the three heat-removal calculations into a single chiller selection chart, provided several variables are held constant. The required chiller capacity can then be simply determined by knowing plasticizing rate and the material to be processed. The chiller selection chart (Fig. 4.43) is based on the following conditions:

(a) hot runner mold applications
(b) 4.5°C (40°F) chilled water delivery temperature

The chilled water delivery temperature is established by finding the optimum balance of three factors: heat transfer between mold and cooling water, chiller efficiency, and condensation considerations.

For air-conditioned plants, it has proven to be most economical to design for a 4.5°C (40°F) chilled water temperature. This temperature provides the best trade-off between optimizing cycle times and minimizing the capital cost of chiller and air-conditioning equipment.

For non-air-conditioned molding departments, plan for a chilled water temperature in the 10° to 16°C (50° to 60°F) range to avoid condensation.

(c) 40 USGPM (150L/min) chilled water flow rate per mold

Optimum heat transfer occurs when fully developed turbulent flow is established within the mold cooling channels. To ensure turbulent flow conditions, this chilled water flow rate is typically required.

(d) 75 psi (5.3 kp/cm^3) circulating pump pressure

To overcome the pressure losses in the molds and piping, at the required flow rate, the circulating pump pressure must typically be in this range.

(e) well-insulated chilled water piping and reservoir

Proper insulation will minimize the heat gain from the ambient air within the plant.

When selecting and specifying chiller systems, four operating parameters must be determined: (a) heat load to chiller, (b) required chilled water delivery temperature, (c) required chilled water flow rate, and (d) pump pressure necessary to achieve required flow rate. These four parameters define the requirements of the chiller, pump, and reservoir.

SHRINKAGE/TOLERANCE

Many plastics require a time period of twenty-four or forty-eight hours to become dimensionally stabilized. With proper machine control and control of the plastic to be used, repeatable close shrinkages and tolerances are achieved. Shrinkage varies based on factors such as specific type of plastic (with specific amounts and types of fillers, additives, and/or reinforcements) and processing conditions (temperature, pressure, etc.). Therefore, shrinkage data usually presented, such as what follows, are guides. Certain thermoplastics (TPs) and thermoset (TS) plastics can produce products with exceptionally tight tolerances (Tables 4.21 and 4.22). Tables 4.23 to 4.33 include guides to shrinkage of different TPs. Figure 4.44 shows the effect of glass fiber content (wt %) on reinforced TPs. Figure 4.45 is a simplified nomogram to determine mold shrinkage based on part thickness. The Drawing Codes A, B, and C listed in Tables 4.22 and 4.28 to 4.32 describe the following:

- Class A tolerances apply to parts molded in precision molds.
- Class B use less precise and costly molds.
- Class C refers to less precise and costly molds using a relatively nonhigh-quality molding compound since the product requirements do not require precision molding.

A simplified initial approach to determining shrinkage of molded plastics is reviewed in chapter 17.

COOLING/CURE TIME

The actual injection of melt (ram-in-motion time) typically takes only a fraction of a second. Figure 4.46 shows that most of the cycle involves cooling (TP) or curing (TS) the plastic, which, until gate freeze-off, can be considered the packing of the cavity (chapter 1). After a gate seals, while the part is cooling until it is stiff enough to be ejected, the screw normally rotates and retracts to produce

	Minimum Thickness in. (mm)	Average Thickness in. (mm)	Maximum Thickness in. (mm)
Alkyd—glass filled	.040 (1.0)	.125 (3.2)	.500 (13)
Alkyd—mineral filled	.040 (1.0)	.187 (4.7)	.375 (9.5)
Diallyl phthalate	.040 (1.0)	.187 (4.7)	.375 (9.5)
Epoxy-glass filled	.030 (0.76)	.125 (3.2)	1.000 (25.4)
Melamine—cellulose filled	.035 (0.89)	.100 (2.5)	.187 (4.7)
Urea—cellulose filled	.035 (0.89)	.100 (2.5)	.187 (4.7)
Phenolic—general purpose	.050 (1.3)	.125 (3.2)	1.000 (25.4)
Phenolic—flock filled	.050 (1.3)	.125 (3.2)	1.000 (25.4)
Phenolic—glass filled	.030 (0.76)	.093 (2.4)	.750 (19)
Phenolic—mineral filled	.125 (3.2)	.187 (4.7)	1.000 (25.4)
Silicone glass	.050 (1.3)	.125 (3.2)	.250 (6.4)
Polyester premix—glass filled	.040 (1.0)	.070 (1.8)	1.000 (25.4)

Table 4.21 Thickness guides for thermoset plastics

the next shot. On very fast cycles, the screw sometimes must continue to rotate while the mold is open in order to produce the required melt.

Figures 4.47 and 4.48 give minimum cure times for nylon 6/6 at various melt and mold temperatures. Data are plotted for two thicknesses: 0.1 and 0.2 in. The times include both time to freeze and time to cool to a temperature where the modulus (stiffness) of the part is suitable for ejection. Cooling time varies with part thickness approximately as the square of the thickness difference. Thus, if part thickness is reduced by half, cooling is four times faster. Cooling times for a nucleated nylon 6/6 (Zytel 131) are shown in both figures as dotted lines. About 12% to 15% cycle reductions are possible at any thickness using this type of resin.

Figures 4.47 and 4.48 can also be used to predict the changes in cooling time as melt or mold temperatures are varied. Obviously, the coldest melt and mold temperatures that can be used successfully result in the fastest cooling times and shortest cycles. These figures also indicate the necessity of careful planning for maximum mold cooling when fast cycles are involved.

Tolerance/Fast Cycle

Like moldings of any thermoplastic, parts injection molded from nylon 6/6 resins are subject to some variation in dimensions from shot to shot. The allowable variations in the dimensions of an injection-molded part are called the tolerances for the part.

Molding tolerances include the total variations in a part dimension that are caused by deviations in the overall molding operations. These deviations may be found in the mold or molding conditions and may be short or long term. Good quality-control records are invaluable in determining the source of trouble.

The part or mold designer should be aware of a few general conditions in this regard.

Drawing Code	Dimensions (Inches)	Plus or Minus in Thousands of an Inch	
	0.000		
	0.500		
A = Diameter	1.000		
	2.000		
B = Depth	3.000		
C = Height	4.000		
	5.000		
	6.000		
		Comm. ±	**Fine ±**
	6.000 to 12.000 for each additional inch add (inches)	0.003	0.002
D = Bottom Wall	(See note #3)	0.006	0.004
E = Side Wall	(See note #4)	0.004	0.003
F = Hole Size Diameter (See note #1)	0.000 to 0.125	0.002	0.001
	0.126 to 0.250	0.003	0.002
	0.251 to 0.500	0.004	0.003
	0.501 & over	0.005	0.003
G = Hole Size Depth (See note #5)	0.000 to 0.250	0.004	0.002
	0.251 to 0.500	0.005	0.003
	0.501 to 1.000	0.007	0.004
H = Corners, Ribs, Fillets	(See note #6)	0.030	0.015
Flatness (See note #4)	0.000 to 3.000	0.014	0.008
	3.001 to 6.000	0.021	0.014
Thread Size (Class)	Internal	1	2
	External	1	2
Concentricity	(See note #4) (F.I.M.)	0.007	0.004
Draft Allowance Per Side	(See note #5)	1.0°	0.5°
Surface Finish	(See note #7)		
Color Stability	(See note #7)		

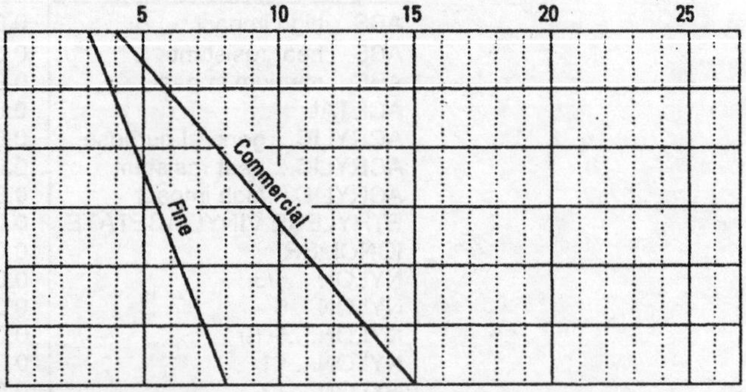

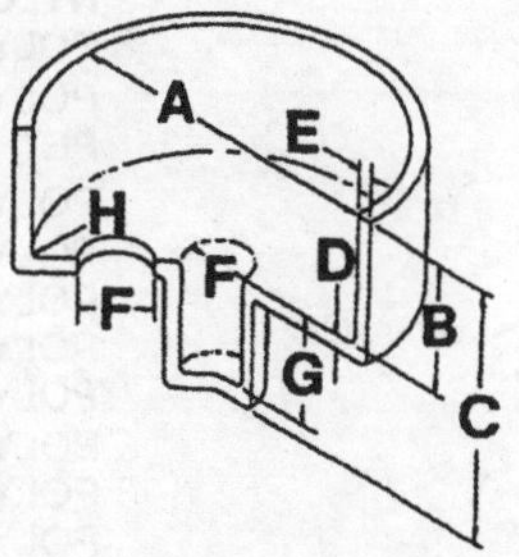

REFERENCE NOTES

1. These tolerances do not include allowance for aging characteristics of material.
2. Tolerances are based on 0.125 inch wall section.
3. Parting line must be taken into consideration.
4. Part design should maintain a wall thickness as nearly constant as possible. Complete uniformity in this dimension is sometimes impossible to achieve. Walls of non-uniform thickness should be gradually blended from thick to thin.
5. Care must be taken that the ratio of the depth of a cored hole to its diameter does not reach a point that will result in excessive pin damage.
6. These values should be increased whenever compatible with desired design and good molding techniques.

Table 4.22 Commercial and fine tolerances for phenol-formaldehyde thermoset plastic (courtesy of Society of the Plastics Industry)

MATERIAL	SHRINK (in/in)
ABS....high impact	0.005-0.007
ABS....heat resistant	0.004-0.005
ABS....medium impact	0.005
ACETAL	0.020-0.035
ACRYLIC....general purpose	0.002-0.009
ACRYLIC....heat resistant	0.003-0.010
ACRYLIC....high impact	0.004-0.008
ETHYLENE VINYL ACETATE	0.010-0.030
IONOMER	0.003-0.020
NYLON....6/6	0.010-0.025
NYLON....6	0.007-0.015
NYLON....6/10	0.010-0.025
NYLON....11	0.010-0.025
NYLON....12	0.008-0.020
NYLON....glass filled	0.005-0.010
POLYBUTYLENE	0.020-0.040
POLYCARBONATE	0.005-0.007
PET	0.003-0.005
POLYESTER PBT	0.010-0.020
POLYESTER PBT 30% GF	0.003-0.005
POLYETHERIMIDE	0.005-0.007
POLYETHYLENE....low dens	0.015-0.035
POLYETHYLENE....high dens	0.015-0.030
POLYPROPYLENE	0.010-0.030
POLYSTYRENE....G.P.	0.002-0.008
POLYSTYRENE..heat resistant	0.002-0.008
POLYSTYRENE.impact mod	0.003-0.006
POLYSULFONE	0.008
POLYURETHANE	0.010-0.020
PVC....rigid	0.002-0.004
PVC....semi-rigid	0.005-0.025
PVC....flexible	0.015-0.030
SAN	0.002-0.006

Table 4.23 Examples of thermoplastics shrinkages

1. Tolerances set on any dimension by the designer usually represent a compromise between part function and its cost of fabrication.
2. An important factor (often overlooked) is that plastic parts can usually operate satisfactorily with wider tolerances than metal parts. It does not pay to specify closer tolerances than necessary.
3. A part that has many critical dimensions will be more difficult to mold to tolerance than a part with fewer such dimensions. Tight tolerances should not be put on every dimension, particularly those across a parting line or sections formed by movable cores or sliding cams.
4. Minimum tolerances are easiest to achieve in a single-cavity mold. Several sources of variability are introduced when multicavity molds are used (such as cavity-to-cavity differences and nonuniformity of runners and gates leading to the individual

Materials	Shrinkage 0.125 in. wall.	Specific gravity	Drying time	Drying temp °F	Mold temp °F	Melt temp °F
ABS	0.005–0.009	1.05	2–4 hrs.	180–200	100–180	390–540
ABS-PC	0.006–0.009	1.13	3–4 hrs.	175–230	140–210	486–562
Acetal	0.018–0.025	1.42	1–4 hrs.	160–185	180–220	380–430
Acrylic	0.002–0.008	1.19	2–3 hrs.	165–200	150–200	420–510
ASA	0.004–0.008	1.05	2 hrs.	180–190	100–180	486–552
EVA*	0.002–0.005	0.937–0.964	2–4 hrs.	140	60–110	350–450
IONOMER	0.002–0.006	0.957–0.967	8 hrs.	140	60–100	400–500
LCP	0.000–0.008	1.5–1.81	2–4 hrs.	275–300	85–280	660–710
Nylon 4/6	0.004–0.030	1.17–1.70	2–4 hrs.	250	180–300	580–590
Nylon 6	0.015–0.025	1.13	2–4 hrs.	180	160–220	460–550
Nylon 6/6	0.015–0.022	1.14	2–4 hrs.	175	160–240	520–580
Nylon 6/10	0.003–0.016	1.07–1.63	2–4 hrs.	180	165–220	480–530
Nylon 6/12	0.001–0.020	1.03–1.46	2–4 hrs.	180–200	165–220	450–580
PBT	0.001–0.023	1.31–1.72	2–4 hrs.	200–250	150–250	450–490
PC	0.005–0.008	1.2	4 hrs.	250	150–225	520–610
PC+PET	0.006–0.008	1.22	4–8 hrs.	210–250	65–200	530–560
PE	0.015–0.050	0.88–0.965	1–2 hrs.	120–150	60–110	360–525
PEEK	0.0005–0.007	1.37–1.70	2–16 hrs.	300–350	250–400	700–750
PEI	0.0005–0.002	1.27–1.61	2–15 hrs.	300	275–325	670–710
PES	0.001–0.005	1.37–1.70	2–16 hrs.	300	280–325	650–750
PET	0.001–0.023	1.28–1.84	3–4 hrs.	250–275	190–250	520–590
PP-CP	0.012–0.024	0.9	1–2 hrs.	120–150	70–120	375–525
PP-HP	0.012–0.024	0.9	1–2 hrs.	120–150	70–120	400–500
PPO	0.005–0.008	1.04–1.06	2–4 hrs.	170–230	150–220	520–600
PPS	0.001–0.012	1.4–2.0	2–4 hrs.	300–350	275–325	600–650
PS-GP	0.003–0.005	1.05	2 hrs.	150	70–120	400–480
PS-HI	0.005–0.008	1.05	1–2 hrs.	150–170	60–180	400–475
PSU	0.0005–0.008	1.38–1.57	4–16 hrs.	275–325	250–325	650–750
PVC (RIGID)	0.002–0.006	1.23–1.54	2 hrs.	140–180	70–100	350–410
PVC (FLEX)	0.010–0.025	1.13–1.72	2 hrs.	120–150	60–120	320–390
SAN	0.001–0.007	1.07	2–3 hrs.	180–200	100–200	425–500
TPE	0.008–0.050	0.875–1.34	2–4 hrs.	170–230	60–175	350–500
TPO	0.009–0.018	0.83–1.15	1–2 hrs.	120–180	60–150	350–500
TPU	0.006–0.020	1.04–1.30	2–4 hrs,	180–230	60–140	380–440
TPUR	0.001–0.006	1.19–1.70	4–12 hrs.	180–280	150–250	420–530
SMA	0.002–0.007	1.07	2 hrs.	180–200	120–180	470–550

Table 4.24 Shrinkage of different plastics related to processing conditions

	ABS		Acetal		Acrylic		Nylon		Polycarbonate	
Dimensions, in	Commercia	Fine	Commercial	Fine	Commercial	Fine	Commercial	Fine	Commercial	Fine
Up to 1.000	0.005	0.003	0.006	0.004	0.005	0.003	0.004	0.002	0.004	0.0025
1.000–2.000	0.006	0.004	0.008	0.005	0.006	0.004	0.006	0.003	0.005	0.003
2.000–3.000	0.008	0.005	0.009	0.006	0.007	0.005	0.007	0.005	0.006	0.004
3.000–4.000	0.009	0.006	0.011	0.007	0.008	0.006	0.009	0.006	0.007	0.005
4.000–5.000	0.011	0.007	0.013	0.008	0.009	0.007	0.010	0.007	0.008	0.005
5.000–6.000	0.012	0.008	0.014	0.009	0.011	0.008	0.012	0.008	0.009	0.006
6.000–12.000	0.003	0.002	0.004	0.002	0.003	0.002	0.003	0.002	0.003	0.015
0.000–0.125	0.002	0.001	0.002	0.001	0.003	0.001	0.002	0.001	0.002	0.001
0.125–0.250	0.002	0.001	0.003	0.002	0.003	0.002	0.003	0.002	0.002	0.015
0.250–0.500	0.003	0.002	0.004	0.002	0.004	0.002	0.003	0.002	0.003	0.002
0.500 and over	0.004	0.002	0.006	0.003	0.006	0.003	0.006	0.003	0.003	0.002
0.000–0.250	0.003	0.002	0.004	0.002	0.004	0.002	0.004	0.002	0.002	0.002
0.250–0.500	0.004	0.002	0.005	0.003	0.004	0.002	0.004	0.003	0.003	0.002
0.500–1.000	0.005	0.003	0.006	0.004	0.006	0.003	0.005	0.004	0.004	0.003
0.000–3.000	0.015	0.010	0.011	0.006	0.010	0.007	0.010	0.004	0.005	0.003
3.000–6.000	0.030	0.020	0.020	0.010	0.015	0.010	0.015	0.007	0.007	0.004
TIR	0.009	0.005	0.010	0.006	0.010	0.006	0.010	0.006	0.005	0.003

Table 4.25 Commercial and fine tolerance guides for various plastics

	Minimum in. (mm)	Average in. (mm)	Maximum in. (mm)
Acetal	.015 (0.38)	.062 (1.6)	.125 (3.2)
ABS	.030 (0.76)	.090 (2.3)	.125 (3.2)
Acrylic	.025 (0.63)	.093 (2.4)	.250 (6.4)
Cellulosics	.025 (0.63)	.075 (1.9)	.187 (4.7)
FEP fluoroplastic	.010 (0.25)	.035 (0.89)	.500
Nylon	.015 (0.38)	.062 (1.6)	.125 (3.2)
Polycarbonate	.040 (1.0)	.093 (2.4)	.375 (9.5)
Polyester T.P.	.025 (0.63)	.062 (1.6)	.500 (12.7)
Polyethylene (L.D.)	.020 (5.1)	.062 (1.6)	.250 (6.4)
Polyethylene (H.D.)	.035 (0.89)	.062 (1.6)	.250 (6.4)
Ethylene vinyl acetate	.020 (0.51)	.062 (1.6)	.125 (3.2)
Polypropylene	.025 (0.63)	.080 (2.0)	.300
Polysulfone	.040 (1.0)	.100 (2.5)	.375 (9.5)
Polystyrene	.030 (0.76)	.062 (1.6)	.250 (7.6)
SAN	.030 (0.76)	.062 (1.6)	.250 (7.6)
PVC—Rigid	.040 (1.0)	.093 (2.4)	.375 (9.5)
Polyurethane	.025 (0.63)	.500 (12.7)	1.500 (38.1)
Ionomer	.025 (0.63)	.062 (1.6)	.750 (19.1)

Table 4.26 Minimum/maximum thickness guides for thermoplastics

Parameter	Increase	Maximum shrinkage variation (%)
Mold temperature	20°C to 90°C	+ 0.6
Wall thickness	1mm to 6mm	+ 0.5
Duration of holding pressure	up to 20 sec	- 0.3
Magnitude of injection/holding pressure	600 bar to 1400 bar	- 0.3
Melt temperature	220°C to 280°C	+ 0.3
Melt flow rate	1 g/10 min to 50 g/10 min	- 0.3

Table 4.27 Some factors influencing polypropylene shrinkage

cavities). Fine tolerances usually cannot be achieved in molds that have more than one type of cavity.

In general, a large variation in part dimensions is introduced by the molding operation itself. Molding variables must be controlled closely if fine tolerances are specified, because slight variations in molding conditions can affect part shrinkage. To attain high-dimensional reproducibility, it is essential to mold on a fully repetitive cycle (chapter 3).

The ability to maintain close tolerances is dependent on part design, mold design, injection-molding equipment used, and the ability of the molder. All areas must be optimized to maintain tight tolerances. Without doubt, finer tolerances can be achieved in many cases by resorting to improved control of these problem areas.

Mold Release

It can also be seen from Figure 4.46 that part ejection affects the overall cycle. The production cycle can be seriously lengthened if parts hang up and do not fall free from the mold or if the operator must frequently remove stuck parts. Mold release agents often help minimize these problems, but be aware that excessive mold release can cause mold vents to plug (leading to part burning) or contribute to surface defects on the molded part. Also, the best mold releases cannot obviate serious mold-design limitations, such as excessive undercuts, too little draft or taper, improperly cooled cores, hot spots, overpacking of sprues, ejector pin penetration, and insufficient knockout.

Mold release is particularly affected by injection pressure and mold temperature. Higher injection pressures and mold temperatures usually necessitate higher ejection forces.

The nucleation of nylon 6/6 also improves part release. For example, note in Table 4.34 that the nucleated nylon requires 40% less ejection pressure than the unnucleated nylon. As a result, it experiences less deformation on ejection than the unnucleated nylon when they are molded under

Drawing Code	Dimensions (Inches)	Plus or Minus in Thousands of an Inch	
	0.000		
	0.500		
A = Diameter	1.000		
	2.000		
B = Depth			
	3.000		
C = Height			
	4.000		
	5.000		
	6.000		

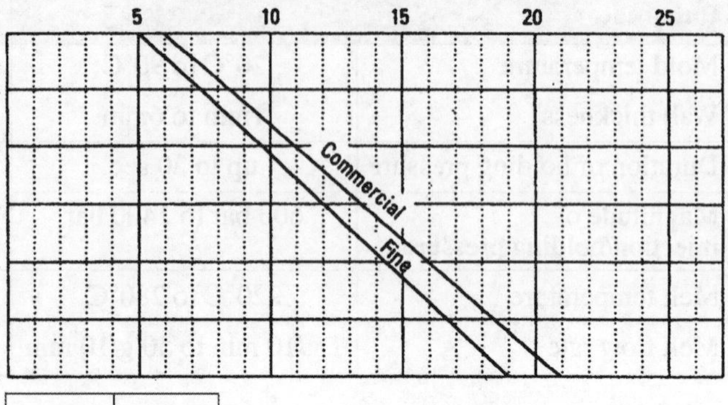

Drawing Code	Dimensions (Inches)	Comm. ±	Fine ±
	6.000 to 12.000 for each additional inch add (inches)	0.006	0.003
D = Bottom Wall	(See note #3)	0.006	0.004
E = Side Wall	(See note #4)	0.006	0.004
F = Hole Size Diameter (See note #1)	0.000 to 0.125	0.003	0.002
	0.126 to 0.250	0.004	0.002
	0.251 to 0.500	0.006	0.004
	0.501 & over	0.008	0.005
G = Hole Size Depth (See note #5)	0.000 to 0.250	0.005	0.003
	0.251 to 0.500	0.007	0.004
	0.501 to 1.000	0.009	0.006
H = Corners, Ribs, Fillets	(See note #6)	0.025	0.010
Flatness (See note #4)	0.000 to 3.000	0.023	0.015
	3.001 to 6.000	0.037	0.022
Thread Size (Class)	Internal	1	2
	External	1	2
Concentricity	(See note #4) (F.I.M.)	0.027	0.010
Draft Allowance Per Side	(See note #5)	2.0°	0.75°
Surface Finish	(See note #7)		
Color Stability	(See note #7)		

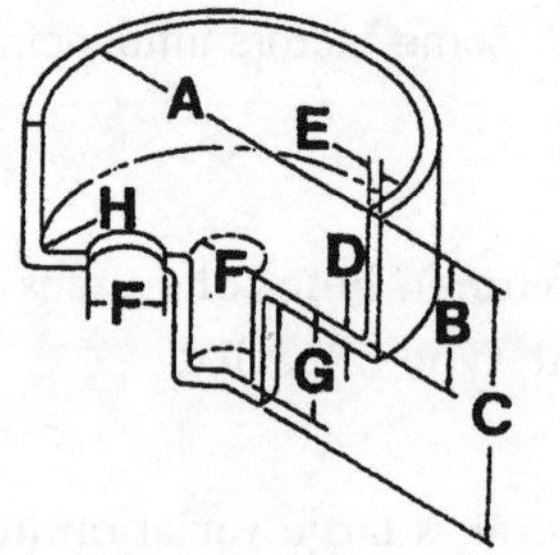

REFERENCE NOTES

1. These tolerances do not include allowance for aging characteristics of material.
2. Tolerances are based on 0.125 inch wall section.
3. Parting line must be taken into consideration.
4. Part design should maintain a wall thickness as nearly constant as possible. Complete uniformity in this dimension is sometimes impossible to achieve. Walls of non-uniform thickness should be gradually blended from thick to thin.
5. Care must be taken that the ratio of the depth of a cored hole to its diameter does not reach a point that will result in excessive pin damage.
6. These values should be increased whenever compatible with desired design and good molding techniques.

Table 4.28 Commercial and fine tolerances for high-density polyethylene plastic (courtesy of Society of the Plastics Industry)

Drawing Code	Dimensions (Inches)	Plus or Minus in Thousands of an Inch	
	0.000		
	0.500		
A = Diameter	1.000		
	2.000		
B = Depth	3.000		
C = Height	4.000		
	5.000		
	6.000		
	6.000 to 12.000 for each additional inch add (inches)	**Comm. ±**	**Fine ±**
		0.005	0.003
D = Bottom Wall	(See note #3)	0.006	0.003
E = Side Wall	(See note #4)	0.006	0.003
F = Hole Size Diameter (See note #1)	0.000 to 0.125	0.003	0.002
	0.126 to 0.250	0.004	0.003
	0.251 to 0.500	0.005	0.004
	0.501 & over	0.008	0.006
G = Hole Size Depth (See note #5)	0.000 to 0.250	0.005	0.003
	0.251 to 0.500	0.006	0.004
	0.501 to 1.000	0.009	0.006
H = Corners, Ribs, Fillets	(See note #6)	0.029	0.016
Flatness (See note #4)	0.000 to 3.000	0.022	0.014
	3.001 to 6.000	0.036	0.021
Thread Size (Class)	Internal	1	2
	External	1	2
Concentricity	(See note #4) (F.I.M.)	0.015	0.012
Draft Allowance Per Side	(See note #5)	1.5°	0.5°
Surface Finish	(See note #7)		
Color Stability	(See note #7)		

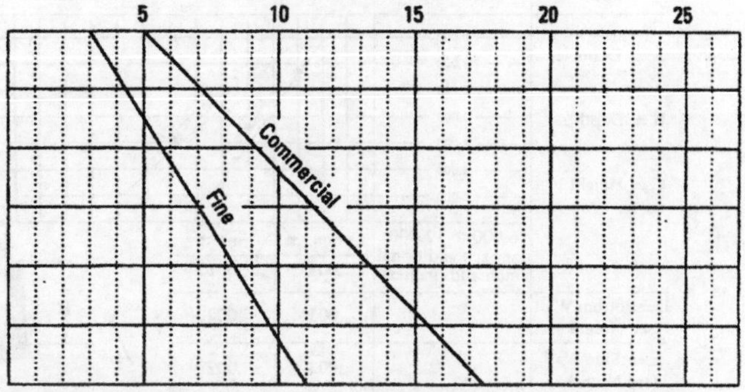

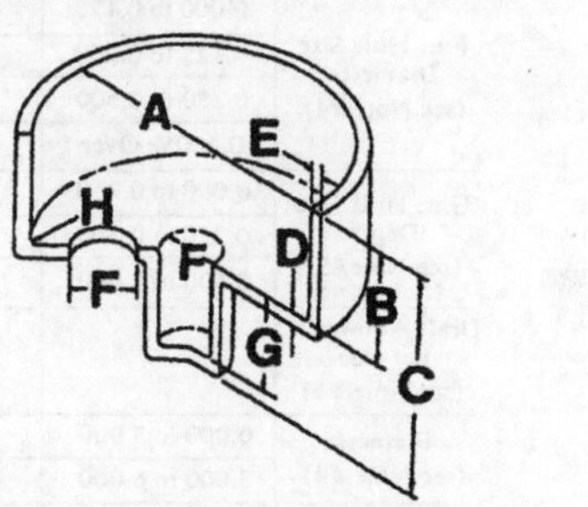

REFERENCE NOTES

1. These tolerances do not include allowance for aging characteristics of material.
2. Tolerances are based on 0.125 inch wall section.
3. Parting line must be taken into consideration.
4. Part design should maintain a wall thickness as nearly constant as possible. Complete uniformity in this dimension is sometimes impossible to achieve. Walls of non-uniform thickness should be gradually blended from thick to thin.
5. Care must be taken that the ratio of the depth of a cored hole to its diameter does not reach a point that will result in excessive pin damage.
6. These values should be increased whenever compatible with desired design and good molding techniques.

Table 4.29 Commercial and fine tolerances for polypropylene plastic (courtesy of Society of the Plastics Industry)

Drawing Code	Dimensions (Inches)	Plus or Minus in Thousands of an Inch 1 2 3 4 5 6 7 8 9 10 11 12 13 14 15 16 17 18 19 20 21 22 23 24 25 26 27 28	
A = Diameter	0.000 / 0.500 / 1.000		
B = Depth	2.000 / 3.000 / 4.000		
C = Height	5.000 / 6.000		
	6.000 to 12.000 for each additional inch add (inches)	Comm. ±	Fine ±
		.003	.0015
D = Bottom Wall (see Note #3)		.003	.002
E = Side Wall (see Note #4)		.003	.002
F = Hole Size Diameter (see Note #1)	0.000 to 0.125	.002	.001
	0.125 to 0.250	.002	.0015
	0.250 to 0.500	.003	.002
	0.500 & Over	.003	.002
G = Hole Size Depth (see Note #5)	0.000 to 0.250	.002	.002
	0.250 to 0.500	.003	.002
	0.500 to 1.000	.004	.003
Draft Allowance per side (see Note #5)		1°	½°
Flatness (see Note #4)	0.000 to 3.000	.005	.003
	3.000 to 6.000	.007	.004
Thread Size (class)	Internal	1B	2B
	External	1A	2A
Concentricity (see Note #4)	(T.I.R.)	.005	.003
Fillets, Ribs, Corners (see Note #6)		.015	.015
Surface Finish	(see Note #7)		
Color Stability	(see Note #7)		

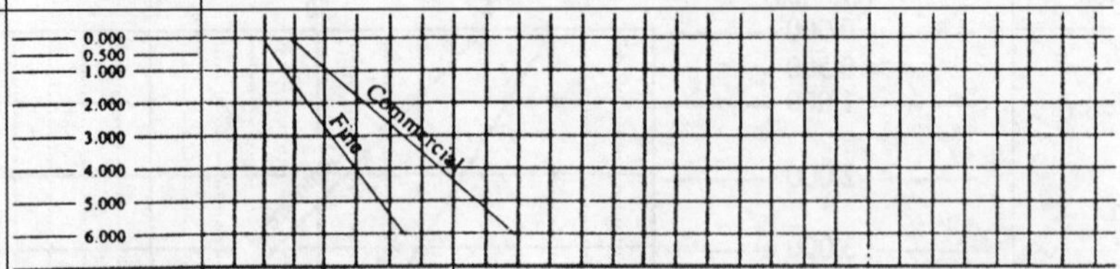

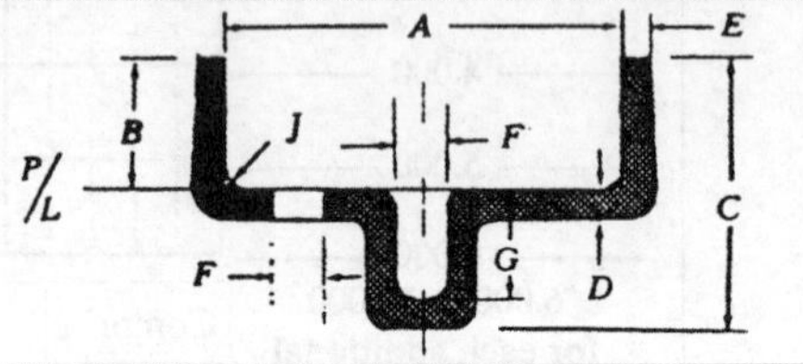

REFERENCE NOTES

1 – These tolerances do not include allowance for aging characteristics of material.

2 – Tolerances based on ⅛" wall section.

3 – Parting line must be taken into consideration.

4 – Part design should maintain a wall thickness as nearly constant as possible. Complete uniformity in this dimension is impossible to achieve.

5 – Care must be taken that the ratio of the depth of a cored hole to its diameter does not reach a point that will result in excessive pin damage.

6 – These values should be increased whenever compatible with desired design and good molding technique.

Table 4.30 Commercial and fine tolerances for polycarbonate plastic (courtesy of Society of the Plastics Industry)

identical conditions. The lower ejection pressure stems from the fact that the nucleated nylon, such as Zytel 131, sets up faster and offers less drag resistance during part ejection.

Being stiffer, nucleated compositions also resist pin penetrations. In the practical mold-design case, nucleated nylons can be ejected using mall-diameter pins, whereas unnucleated nylons frequently require sleeves, blades, or other types of more costly stripper knockouts, especially on fast cycles. On balance, in molds that are prone to cause part sticking, nucleated nylon 6/6 will run on faster cycles.

Surface coating nylon 6/6 resins with about 0.1 wt% aluminum stearate will reduce the ejection pressure significantly. Surface coating a standard nylon 6/6 molding resin with aluminum stearate reduces ejection pressure about 35%. Although effective, surface coating unnucleated nylons does not give a product equivalent to a nucleated nylon in regard to mold-release characteristics. However, surface coating nucleated nylon with about 0.1% aluminum stearate can further reduce ejection pressures by 10% to 15%.

Plus or Minus in Thousands of an Inch

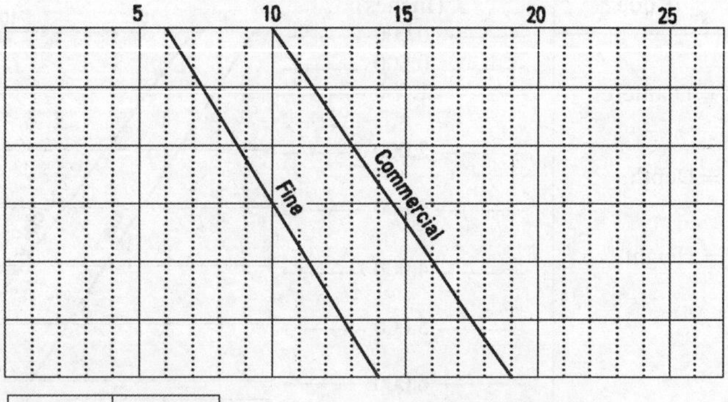

Drawing Code	Dimensions (Inches)	Comm. ±	Fine ±
	0.000		
	0.500		
A = Diameter	1.000		
	2.000		
B = Depth	3.000		
C = Height	4.000		
	5.000		
	6.000		
	6.000 to 12.000 for each additional inch add (inches)	0.005	0.003
D = Bottom Wall	(See note #3)	0.007	0.003
E = Side Wall	(See note #4)	0.007	0.003
F = Hole Size Diameter (See note #1)	0.000 to 0.125	0.004	0.003
	0.126 to 0.250	0.005	0.004
	0.251 to 0.500	0.006	0.005
	0.501 & over	0.008	0.006
G = Hole Size Depth (See note #5)	0.000 to 0.250	0.004	0.003
	0.251 to 0.500	0.005	0.004
	0.501 to 1.000	0.006	0.005
H = Corners, Ribs, Fillets	(See note #6)	0.030	0.010
Flatness (See note #4)	0.000 to 3.000	0.010	0.007
	3.001 to 6.000	0.020	0.015
Thread Size (Class)	Internal		
	External		
Concentricity	(See note #4) (F.I.M.)	0.015	0.010
Draft Allowance Per Side	(See note #5)	1.5°	1.0°
Surface Finish	(See note #7)		
Color Stability	(See note #7)		

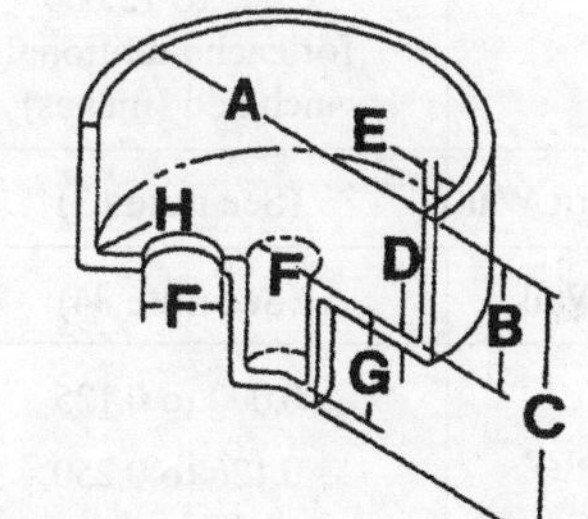

REFERENCE NOTES

1. These tolerances do not include allowance for aging characteristics of material.
2. Tolerances are based on 0.125 inch wall section.
3. Parting line must be taken into consideration.
4. Part design should maintain a wall thickness as nearly constant as possible. Complete uniformity in this dimension is sometimes impossible to achieve. Walls of non-uniform thickness should be gradually blended from thick to thin.
5. Care must be taken that the ratio of the depth of a cored hole to its diameter does not reach a point that will result in excessive pin damage.
6. These values should be increased whenever compatible with desired design and good molding techniques.

Table 4.31 Commercial and fine tolerances for polyvinyl chloride plastic (courtesy of Society of the Plastics Industry)

Drawing Code	Dimensions (Inches)	Plus or Minus in Thousands of an Inch	
	0.000	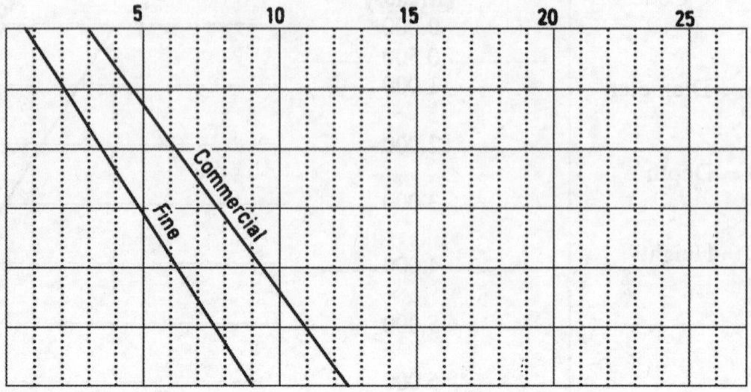	
	0.500		
A = Diameter	1.000		
	2.000		
B = Depth	3.000		
C = Height	4.000		
	5.000		
	6.000		
		Comm. ±	Fine ±
	6.000 to 12.000 for each additional inch add (inches)	0.003	0.002
D = Bottom Wall	(See note #3)	0.004	0.003
E = Side Wall	(See note #4)	0.005	0.003
F = Hole Size Diameter (See note #1)	0.000 to 0.125	0.002	0.001
	0.126 to 0.250	0.003	0.002
	0.251 to 0.500	0.003	0.002
	0.501 & over	0.005	0.003
G = Hole Size Depth (See note #5)	0.000 to 0.250	0.004	0.002
	0.251 to 0.500	0.004	0.003
	0.501 to 1.000	0.005	0.004
H = Corners, Ribs, Fillets	(See note #6)	0.021	0.013
Flatness (See note #4)	0.000 to 3.000	0.010	0.004
	3.001 to 6.000	0.015	0.007
Thread Size (Class)	Internal	1	2
	External	1	2
Concentricity	(See note #4) (F.I.M.)	0.005	0.003
Draft Allowance Per Side		1.5°	0.5°
Surface Finish	(See note #7)		
Color Stability	(See note #7)		

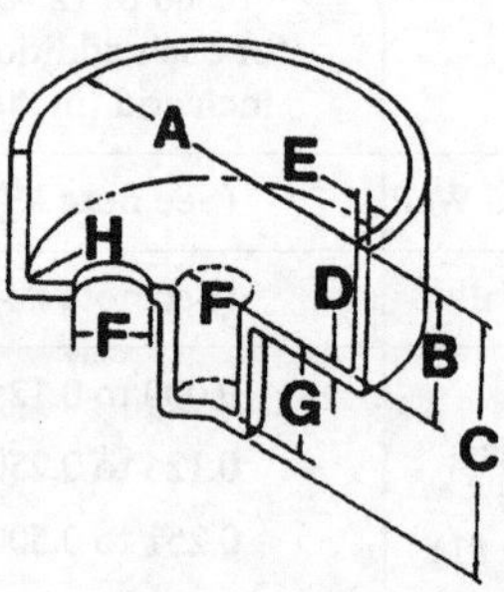

REFERENCE NOTES

1. These tolerances do not include allowance for aging characteristics of material.
2. Tolerances are based on 0.125 inch wall section.
3. Parting line must be taken into consideration.
4. Part design should maintain a wall thickness as nearly constant as possible. Complete uniformity in this dimension is sometimes impossible to achieve. Walls of non-uniform thickness should be gradually blended from thick to thin.
5. Care must be taken that the ratio of the depth of a cored hole to its diameter does not reach a point that will result in excessive pin damage.
6. These values should be increased whenever compatible with desired design and good molding techniques.

Table 4.32 Commercial and fine tolerances for nylon (polyamide) plastic (courtesy of Society of the Plastics Industry)

RECYCLING PLASTIC

It is common practice for injection molders to recycle reject parts, along with sprues and runners (cold runner systems), through their molding machines (chapter 2). To the molder, this reuse of material frequently means the difference between profit and loss on a job; to the designer, it is often the economic incentive to injection mold a part. In a typical mold design, it is an unusual occurrence when the sprues and runners amount to less than 25 wt% of the shot weight. This percentage can occasionally run as high as 75%.

It is possible to reuse previously molded plastics without undue sacrifice in physical properties or quality, provided that proper precautions are taken in initial and subsequent moldings and, most

Dimension, mm (in.)	Class A (fine tolerance), mm (in.)	Class B (normal tolerance), mm (in.)	Class C** (coarse tolerance), mm (in.)
0–25 (0–1)	±0.12 (±0.005)	±0.25 (±0.010)	±0.4 (±0.016)
25–100 (1–4)	±0.2 (±0.008)	±0.4 (±0.016)	±0.5 (±0.020)
100–200 (4–8)	±0.25 (±0.010)	±0.5 (±0.020)	±0.8 (±0.030)
200–400 (8–16)	±0.4 (±0.016)	±0.8 (±0.030)	±1.3 (±0.050)
400–800 (16–32)	±0.8 (±0.030)	±1.3 (±0.050)	±2.0 (±0.080)
800–1,600 (32–64)	±1.3 (±0.050)	±2.5 (±0.100)	±3.8 (±0.150)

Table 4.33 Guide for reinforced plastic tolerances

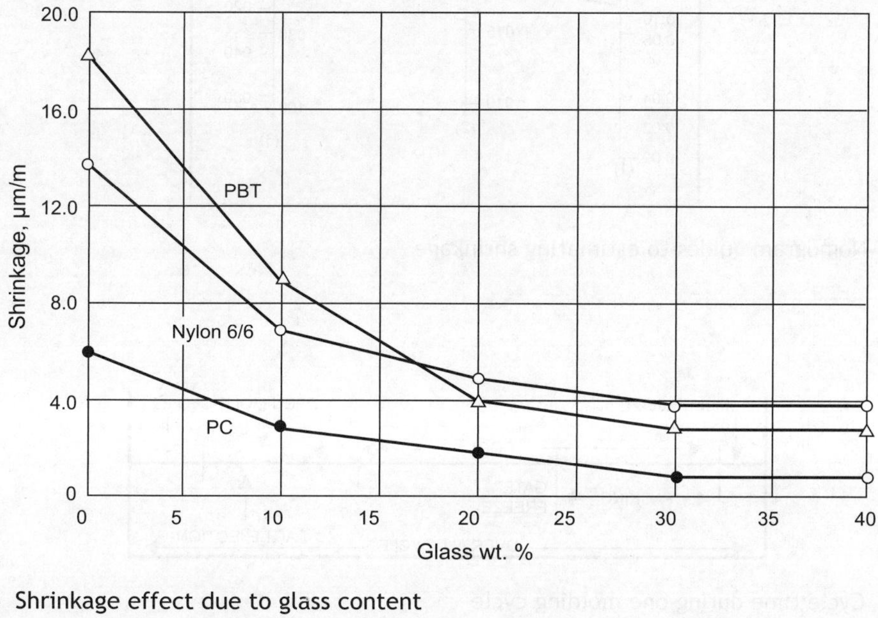

Figure 4.44 Shrinkage effect due to glass content

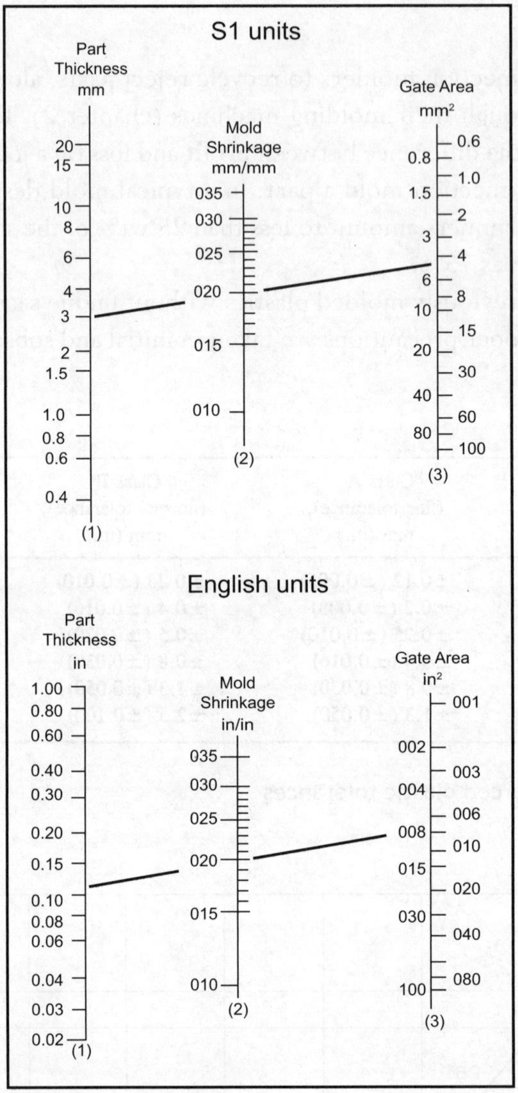

Figure 4.45 Nomogram guides to estimating shrinkage

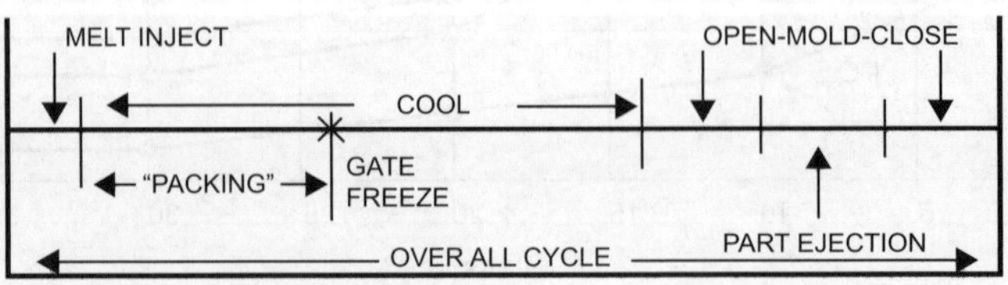

Figure 4.46 Cycle time during one molding cycle

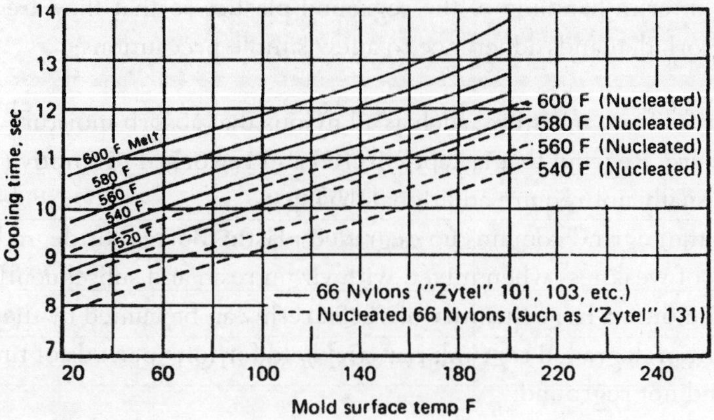

Figure 4.47 In-mold cooling times for 0.1-in-thick parts

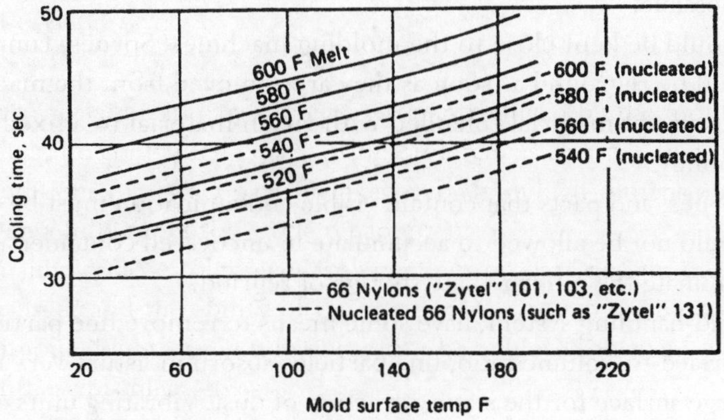

Figure 4.48 In-mold cooling times for 0.2-in-thick parts

Typical Ejection Pressures (psi)	Resin
1,900–2,000	Standard nylon 66 molding resins
1,300–1,400	Surface-coated with mold release
1,100–1,200	Nucleated
900–1,000	Nucleated and surface-coated with mold release

Table 4.34 Mold release behavior

importantly, during interim handling of the reground plastics so that they are not contaminated. Profitable use of rework demands adherence to a few simple precautions:

1. Protect regrind from moisture, such as all nylons that absorb moisture rapidly from the atmosphere. Regrind that is kept covered and reworked promptly (within a half hour) will usually not require additional drying.
2. Ensure that the regrind contains no degraded plastic. Burned or degraded plastic can form points of weakness when mixed with virgin resin and subsequently molded into new parts. Because a large quantity of virgin resin can be ruined by the inclusion of a small amount of degraded regrind, material held for long periods of time should be discarded and not reground.
3. Prevent contamination of rework from other sources. Good housekeeping procedures and limited exposure of rework to dirty surroundings are keys to prevention of contamination.
4. The equipment and area in which the regrind is produced and handled should be kept as clean as possible.
5. Grinders should be kept close to the molding machines. Sprues, runners, and rejects should be reground as soon as they are removed from the machines; continuous reuse of material (blended with virgin material to a fixed proportion) is the best policy.
6. Runners, sprues, and parts that contain visible contamination must be discarded.
7. Regrind should not be allowed to accumulate in uncovered containers. Whenever possible, eliminate the intermediate storage of regrind.
8. In any regrind-handling system, have some means to remove fine particles. Because of a large surface-to-volume ratio, fine particles absorb moisture very rapidly and present a large surface for the static attraction of dust. Vibrating units equipped with sixteen to twenty mesh screens have been found useful for separating fine particles. Keeping grinder blades sharp, with proper clearance and screen sizes, will also minimize fine particles.

The ratio of reground nylon 6/6 that may be blended with virgin nylon 6/6 will depend on the quality of the regrind and specifications of the part. If careful regrind-handling procedures are followed, high percentages of regrind can be used without difficulty.

The amount used should be established by the ratio of the weight of the sprues and runners to the weight of the parts. It is important that the regrind and virgin materials be mixed before molding and that a constant proportion of regrind be maintained. Figure 4.49 shows the number of successive passes (in a screw injection machine operating at normal conditions) that various blends of regrind and virgin materials can withstand before the as-molded elongation of nylon 6/6 is reduced in half; this corresponds to a drop from 60% to 30%. It is easily seen that as the percentage of regrind in the blend increases, the number of passes decreases. Another practical way of looking at rework stability occurs when regrind is immediately fed back to the machine and proportioned to an exact sprue-runner/shot weight ratio. In this system, an equilibrium feed composition will be quickly attained.

If a shot that is 50% sprue and runner is recycled with virgin resin, a condition rapidly exists in which 25% of any shot will have been through the machine one time; 12.5%, two times; 6.25%, three times; 3.12%, four times; and so on. Based on experimental data, this composition will have an as-molded elongation of 50% (vs. 60% if only virgin resin was used). An 80% regrind/20% virgin blend will yield on immediate recycle an elongation of 35% (vs. 60%), whereas a 20% regrind/80% virgin blend yields 58% (vs. 60%). Thus, using less than 50% regrind and recycling immediately results in minimal reduction in elongation. Runner layout should be designed with the weight ratio (to total shot weight) in mind whenever possible.

MACHINE START-UP/SHUTDOWN

For IMM start-up, experience provides a guide to setting plasticizer heat profiles as well as other settings; otherwise, start with the plastic manufacturer's recommendations. There are different starting points for the various types of plastics that have to be interfaced with the different capabilities of IMMs to be used. The time and effort on start-up make it possible to achieve maximum efficiency of performance versus cost for the processed plastics. Information on process control settings developed can be stored and applied to future setups. Recognize that two identical IMMs usually require slightly different settings to maximize their performance. Figures 4.50 to 4.52 and Tables 4.35 to 4.41 (also Table 4.14) provide an overview on this subject.

The term *process control* has often been used when machine control is actually performed. As the knowledge base of the fundamentals of the molding process continues to grow, the control approach is moving away from principally press control and closer to real process control in which material response is monitored and then moderated or even managed.

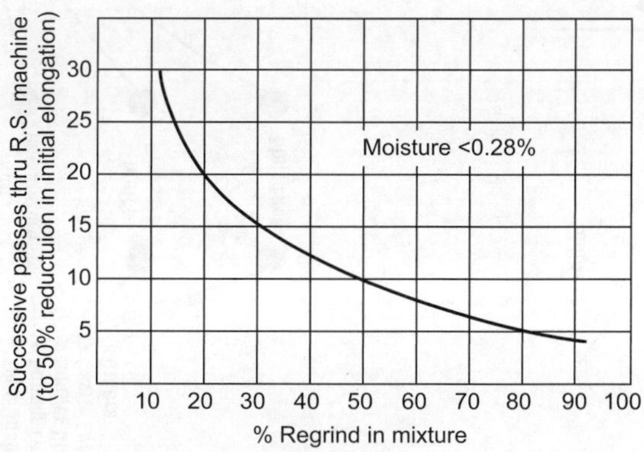

Figure 4.49 Example of virgin and recycled plastic stability

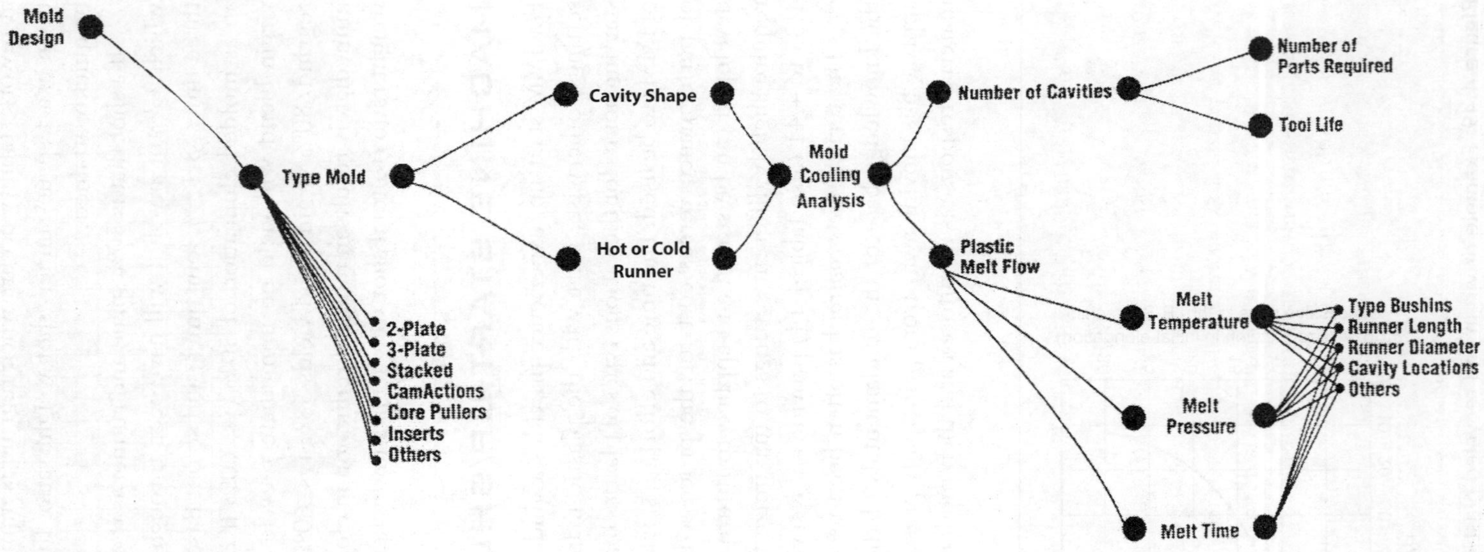

Figure 4.50 Basic mold process controls

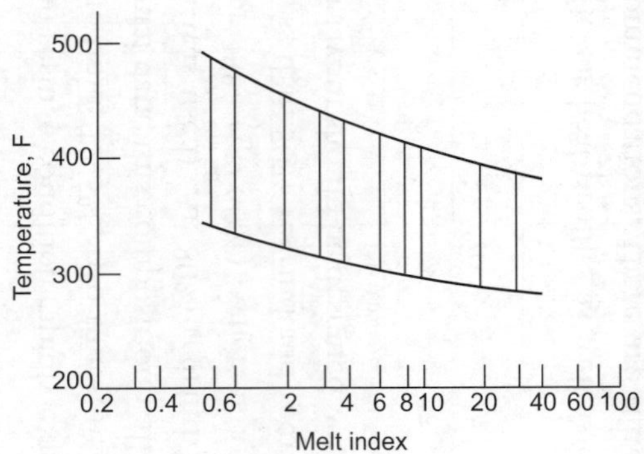

Figure 4.51 Example of melt temperature range for an LDPE

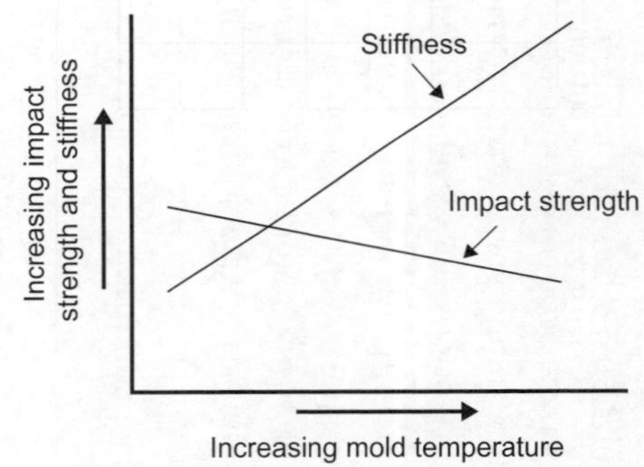

Figure 4.52 Effect of mold temperature on a PP

Melt Index	1,000 psi	1,400 psi	1,800 psi
7	Would not flow	450°F	400°F
20	475°F	375°F	350°F
40	425°F	350°F	350°F

Table 4.35 LDPE minimum melt temperatures at different injection pressures

Melt Index	1,000 psi	1,400 psi	1,800 psi
7	Would not flow	450°F	425°F
20	500°F	425°F	400°F
50	425–450°F	350°F	350°F

Table 4.36 LDPE melt temperature at which optimum shot weight is obtained based on injection pressure

Flow Rate	Melt Temperature for Injection Molding (F)
4.5	380–450
9.0	370–440
18–30	360–430
50–70	340–410
Talc-filled	380–450
Calcium carbonate	380–450

Table 4.37 Examples of melt temperature range for a PP

For start-up, mold setup is important. It includes (a) determining plastic requirements based on type of mold to be used (cold runner [includes nozzle into cavity] or hot runner [only cavity]) (386); (b) locating proper KO bars with all having equal lengths; (c) selecting eyebolt hole that yields a level hang/lift; (d) leveling mold and clamp to fixed platen; (e) lining up locating ring; (f) slowly closing mold; (g) opening moving platen and installing KO bars if KOs are acting as pullbacks; (h) tightening bars making certain they bottom out against the ejector plate; (i) closing platen; (j) clamping mold to moving platen, removing safety straps, and unhooking hoist; (k) opening mold to desired daylight and set slowdown switches so that no high impact on the mold will occur; (l) fine-tuning the final switch positions by repetitive small adjustments; (m) connecting all required power (electric, hydraulic, and/or pneumatic); (n) checking powered functions (such as electrical

Molding Conditions	Thickness of Section 0.063 in. (1.6 mm)	0.25 in. (6.4 mm)
Temperatures		
Rear cylinder, F	380–420	380–400
C	193–216	193–204
Middle cylinder, F	400–450	380–420
C	204–232	193–216
Forward cylinder, F	420–480	400–420
C	216–249	204–216
Nozzle, F	380–420	380–420
C	193–216	193–216
Melt, F	400–480	380–420
C	204–249	193–216
Mold coolant,	50–80	50–80
	10–27	10–27
Hydraulic injection pressure		
psi	600–1,500	600–1,500
MPa	4–10	4–10
Typical cycle time, s		
Plunger forward	5–10	15–20
Total cycle	15–25	35–60
Shrinkage, %	1–2	1–2

Table 4.38 Examples of melt temperature range for a PP based on part thickness

Cylinder temperatures, F (C)	
Rear	420 (216)
Center	450 (323)
Front	470 (243)
Nozzle	470 (243)
Melt temperature, F (C)	470 (243)
Mold temperature, F (C)	80 (27)
Injection pressure, psi (MPa)	12,000 (82.7)
Cycle time, s	
Inject	20
Cooling	20
Recycle	2
Overall	42
Screw speed, rpm	60
Injection speed	Slow

Table 4.39 Molding conditions for a ¼-in PETG

Temperature profile (F)	Screw rpm 20–50
Zone 1 300–330	Injection pressure (psi) 1,200–1,500
Zone 2 320–340	Hold pressure (psi) 800–1,000
Nozzle 320–350	Back pressure (psi) 0–100
Mold temperature 50–100 F	Injection rate
Stock temperature 395–410 F	Large machine (750 tons) 3–5 s/in.
	Small machine (375 tons) 1.5–2.0 s/in.

Table 4.40 Example of PVC molding conditions

Nominal Cavity Thickness (in.)	Mold Temperature		
	60 F	120 F	180 F
0.020	3.2	3.4	3.8
0.030	6.8	7.3	8.2
0.040	11.2	12.0	13.4
0.050	17.0	18.0	20.0

Table 4.41 Melt flow distances for uniform physical properties of a nylon 6/6 molding compound

heaters) to ensure they are operating correctly and just long enough to provide functionality yet avoid excessive heat buildup before water is connected; (o) connecting water lines; and (p) turning water on (electric heaters off) and examining for leaks.

Start-up process control involves the machine operation and behavior of the plastic; the interaction between the machine's operation and the plastic's behavior from the plasticator into the mold cavity(ies) is most important. Principally, the processing pressure and temperature versus time determine the quality of the molded product. The design of the control system has to take into consideration the logical sequence of all the basic functions, including injection speed (pressure dependent), clamping and opening the mold, opening and closing actuating devices, barrel temperature profile (Table 4.42), melt temperature, mold temperature, cavity pressure (387), holding pressure, and mold cooling rate. These controls are essential to produce molded quality products and minimize cycle time. Quality features include mechanical properties, dimensional accuracy, absence of distortion, and surface quality (chapter 23).

Molding a product (part) involves the three stages of fill, pack, and hold. The following guide provides a simplified example for IM plastics.

1. Start with the plastic melt temperature at the midpoint of your supplier's recommended range. Know the actual melt temperature (not the barrel temperature set points).
2. As with the melt, the mold temperature should be centered to the recommended range.

Zone	Temperature setting (°C)	Temperature setting (°F)
1	150–210	300–410
2	210–250	410–480
3	220–250	430–480
4	220–250	430–480
Nozzle	240–260	465–500

Table 4.42 Example of barrel zone temperature settings

3. Fill the mold as fast as you can and as far as you can. Separate speed from pressure. (Peak pressure during fill should never reach the injection pressure set point.)
4. Packing should be as slow as possible via separating speed from pressure. The priority for termination of pack is the same as fill. The ability to pack on velocity is dependent on the hydraulic and/or electric architecture of the machine. Few presses are able to do this, which requires the operator to pack on machine pressure. Unfortunately, a constant pressure applied to a variable plastic leads to a product that varies.
5. Hold with enough pressure and time to prevent plastic melt discharge from the mold until the gate seals.

   Ideally, hold should be a zero velocity setting with whatever pressure needed being available.
6. The criteria for determining cooling time now occurs when the product can accept the force of ejection and does not distort (hold time is cooling time).
7. For the majority of plastic materials, you should always run with a cushion.
8. Back pressure and screw rpm should be minimized. The goal should be to plasticate to the shot limit just before the cooling timer times out.
9. Once the process has been optimized, plastic conditions should be recorded, such as fill time, peak pressure at fill, cavity pressure (387), melt temperature, mold temperature, melt flow rate, and gate seal time. Record all basic machine set points on the setup sheet, such as the transfer time (fill time) and weight, overall cycle time, and total shot weight, part weight, and % runner (Table 4.43).

Start molding short shots and gradually increase the shot size as the injection speed increases while watching for flash or burning. Short shots that exhibit flash and/or burns indicate problems with tooling. Processing around a tool problem is a temporary resolution at best. The goal is to fill the mold as fast as possible. An ideal approach would provide a product 95% filled using 90% of the maximum injection rate of the press to operate with maximum efficiency. With the cavity

JOB
OPERATORS
ENGINEERS

MOLD DESCRIPTION
SCREW USED
PRESS NO.

MACHINE SETUP INSTRUCTIONS
NOZZLE #

SPECIAL INSTRUMENTATION
SAFETY CHECK

DATE TIME	Run Number	Resin	Lot Number	TEMPERATURES (°F) (°C)								PRESSURES (PSI) (kp/cm²)				CYCLE TIMES (SEC)									WEIGHTS (GMS)		REMARKS
									MOLD																		
				Rear	Center		Front	Nozzle	Fixed	Movable	Melt	Injection 1st Stage	Injection 2nd Stage	Clamp (Tons)	Back	Injection	Hold	Open	Overall	Booster	Ram in Motor	Screw Retraction	Pad (inches)	RPM	Full Shot	Part Only	

COMMENTS ON MOLDING OPERATION STARTUP, ETC.

Table 4.43 Molding data record

approximately 95% filled, leave the shot limit alone and start to lower the cutoff position. This will allow completing the fill portion of the cycle and using the inertia of the ram to pack out the product. With certain hydraulic IM machines that use servo-valve technology for injection speed and pressure, it is not possible to completely separate fill from pack. This is best accomplished on machines that use a dual-valve system. Lower the cutoff position until the product cavity is packed out, making sure that a melt cushion exists. It may be necessary to increase the speed setting on the last step, but packing should be done as slowly as possible.

After completing the packing start, add hold pressure and time. Pressure should be high enough to keep plastic from discharging and time adequate to allow the gate to freeze (melt to solidify). Gate seal time can be determined by looking at cavity pressure at the gate or by weighing the product without the runner. After hold time is complete, delay the start of screw rotation (or decompress before starting the screw) to allow the pressure ahead of the screw to decay. Once the optimum cooling time is determined, screw rpm should be adjusted to minimize residence time (Fig. 4.53). Profiled back pressures are not recommended. The slower the screw is allowed to turn, the better the mixing action. Look at the peak pressure reached during fill and set the system pressure about 10% higher than this peak pressure. Minimize the time used to open the mold, eject the product, and close the mold.

Additional details are provided in the following summary. If required, purge barrel free of degraded resin. Set machine for semiauto and start cycle; observe screw. Set barrel temperature profile based on experience or start with resin supplier's recommendation. Determine or estimate the shot size (386) and set machine for approximately two-thirds of the mold's full shot. Set decompression stroke. Set a position transfer point (if machine is so equipped) approximately an inch from the bottom. Initially, set first-stage pressure at 50%; ultimately, set at 100%. Estimate and set second-stage time with pressure at zero. Set melt injection velocity to maximum. Adjust velocity and/or pressure as needed; if the fill was fast and short, the pressure can be increased. The fill pressure should be set high enough so that the fill speed is not pressure limited but controlled by velocity set points. Estimate and set cooling time. Set back pressure at 50 psi and gradually increase if necessary.

After observing each cycle, the shot size and transfer point will be adjusted frequently to set the process so that the first stage accomplishes 95% to 98% of the fill as measured by shot weight. Once the first-stage shot size, transfer, velocity, and pressure are set, second-stage packing pressure can be set. Adjust pack pressure as needed, but do not overpack. Recheck cushion. Some cushion should be maintained. Set screw speed so that recovery is completed just prior to the next cycle but not limiting cycle time. If flash occurs, slow velocity.

Additional guides to IMM start-ups and shutdowns as well as related procedures are presented in Tables 4.44 to 4.47.

MAXIMIZING PROCESSING WINDOW CONTROL

For start-ups, a processing window is determined that sets controls to fabricate acceptable products. It sets up the range of processing conditions, such as melt temperature, pressure, and shear rate

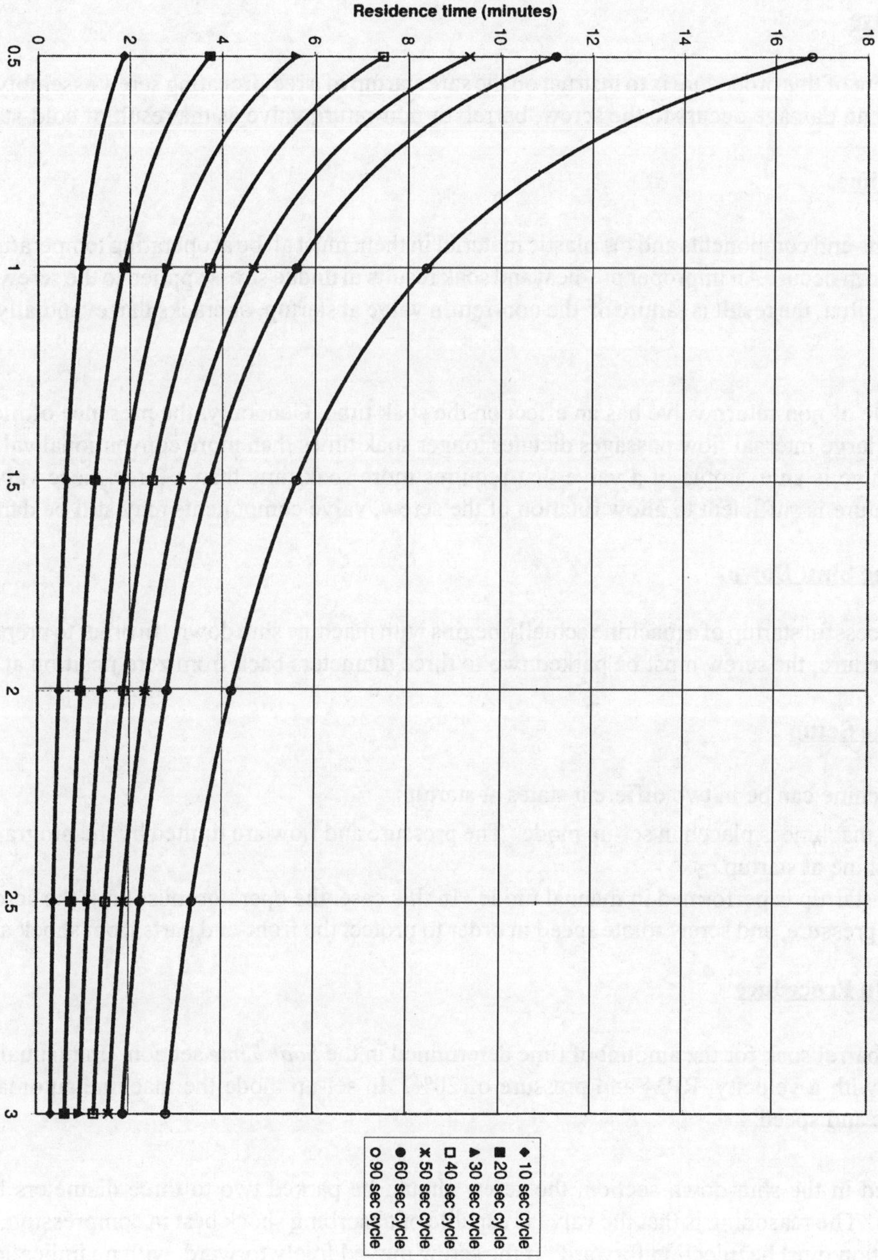

Figure 4.53 Plastic residence time

Objective

The intent of this procedure is to instruct on the safe startup of a reciprocating screw assembly. A safe startup is when no damage occurs to the screw, barrel, or non-return valve as the result of cold-start conditions.

Soak Time

The front-end components and the plastic material in them must all be at operating temperature before screw motion can occur. An improper pre-heat and soak results in undue stress applied to the screw and non-return valve. Often, the result is failure of the non-return valve at startup, or cracks that eventually lead to failure.

The style of non-return valve has an effect on the soak time. Generally, the presence of moving precision fits and large internal flow passages dictates longer soak times than more conventional valves. The Auto-Shut Valve is an example of a valve that requires more soak time than a three-piece valve. Even if the temperature is sufficient to allow rotation of the screw, valve components may still be damaged.

Machine Shut Down

The successful startup of a machine actually begins with machine shut down. In order to prepare for the start-up procedure, the screw must be parked two to three diameters back from zero position at shut down.

Machine Setup

The machine can be in two different states at startup:

1. The machine is placed in set-up mode. The pressure and flow are limited by the program to protect the machine at startup.
2. The startup is performed in manual mode. In this case, the operator must limit the screw travel speed and pressure, and screw rotate speed in order to protect the front-end parts from shock at initial startup.

Start-Up Procedure

Let the barrel soak for the amount of time determined in the *Soak Time* section. In manual mode it is best to start with a velocity, RPM and pressure of 20%. In set-up mode the machine automatically reduces pressure and speed.

As stated in the shut-down section, the screw should be parked two to three diameters back from zero position. The reasoning is that the valve is capable of absorbing shock best in compression. Therefore, the first motion must be injection forward. If the screw moved freely forward, with no indication of an unmelt situation, then it is safe to run screw rotate.

Table 4.44 IMM start-up procedure (courtesy of Spirex Corp.)

If the screw output is too low or motion is erratic, stop plasticating immediately. A low output is an indication that the Auto-Shut Valve is not opening completely. If the Auto-Shut Valve is not opened completely, the tip assembly (the portion of the Auto-Shut that moves to open the valve) will be subjected to extremely high forces that may damage the tip assembly. Let the valve soak for another 15 minutes and repeat the start-up procedure.

If screw operation is satisfactory, then it is safe to operate the screw at full speed and pressure. Operate the unit at production capacity and look for reduced output or erratic operation. Soak the machine longer if there is any doubt as to the quality of the melt. Otherwise, the valve and screw are at the correct temperature for production startup.

Table 4.44 *(continued)*

within which a specific plastic can be fabricated with acceptable and optimum properties by a particular fabricating process. It is a defined area in a system's process control pattern. This window for a specific plastic part can vary significantly if changes are made in its design and the fabricating equipment used.

Once the machine is operating, the processor uses a systematic approach by making one change at a time, allowing the change to occur and then determining the result for each change. As an example, a 2-D molding diagram can be used to relate to the two controls of injection ram pressure and mold temperature (Fig.4.54). The molding area diagram (MAD) identifies molded products that meet requirements. By operating in the center of this diagram, you are guaranteed to continuously mold acceptable products. To produce products at the lowest cost, set the machine where maximum ram pressure and minimum mold temperature exist. However, to compensate for potential variables that exist in machine and plastic performances (chapter 1), carefully analyze molded products and if necessary reduce the maximum settings to ensure acceptable products.

A similar approach can be applied using three controls. Figure 4.55 presents a 3-D molding volume diagram (MVD) using injection ram pressure, mold temperature, and melt temperature. This approach simulates a processor's approach in start-up of a machine when making control changes. Because of machine and plastic variations, rejects can develop at the volume (or area) edges. As the IMM is very complex with all the controls required to set it up, as well as the variables that exist with plastic materials, these examples refer to the overall injection molding process. Note that a major cause for problems with any process is not poor product design but that the process operated outside its required operating window.

Many different processing window studies are conducted. For example, an injection-molded radar application requires critical controls for its parabolic form that have to be maintained through stringent application and environment conditions while satisfying a number of other functional and quality requirements. A design of experiment (DOE) analysis was run to identify an optimal process window in the four-parameter design space, within which ten very critical and tight-tolerance

Objective

The intent of this procedure is to offer guidance in properly pre-heating and soaking the injection front-end components prior to processing.

General Thoughts

One of the best ways to avoid damage to machinery is to use purging compound during shut-down and start-up. Familiarity of this compound will be a guide for the proper soak time. Often it is acceptable to purge with a safer material, such as linear low density polyethylene (LLPE).

Do not overlook obvious sources of information. Draw on the experience of others, or records of previous jobs. The material supplier can provide recommendations from the producer. Industry contacts are a good resource of experience and other contacts. Try to compare the material to other similar materials. Exercise caution as many like-materials have different additives that result in very different properties. The soak time may change, there may be a critical temperature at which damaging gas releases, or heat sensitivity may increase.

Excessive soak time can cause a problem if the material is heat sensitive. If extra caution is necessary, a heat probe can be inserted into the nozzle, or the endcap can be removed for checking the melt directly. Once the endcap is removed, insert a temperature probe inside the non-return valve. The material within a non-return valve is generally the last to reach operating temperature. The Auto-Shut valve can be checked by using pliers to gently pull open the poppet.

Using Soak Timers

New OEM machines come with soak timers. The function of the soak timer is to lock out the injection unit until the timer times out. The timer starts once the heaters reach the operator-set temperature. Our experiences with OEM soak timers are that they are satisfactory for many materials and front-end components. However, exercise caution if you are unfamiliar with the material, or your front-end components. Our belief is that many non-return valves and materials require 40% more soak time than the OEM timer provides.

Older Machines Without Soak Timers

There are still many machines in service without soak timers. Try to compare the machine with other similar machines that do have a timer. Remember the soak time starts after the injection unit reaches the set point temperature. You know the injection unit is at set point by watching the cycling of the heater bands. Add your soak time at this point, and maintain records for your future use. Remember, you can apply the other methods stated above in *General Thoughts*.

Table 4.45 Preheat/soak time (courtesy of Spirex Corp.)

If you do not have an adequate means to determine the appropriate soak time, there is an old industry rule-of-thumb that can help. This rule is "turn on the heaters and heat for one hour for each inch of barrel wall thickness." This rule seems to work and is actually a little on the safe side, so burning of heat-sensitive materials is a risk.

Comments on the Auto-Shut Valve

The Auto-Shut Valve requires more soak time than conventional non-return valves. This valve has a spring loaded poppet that rides in the body of the valve. There is a pool of plastic contained inside the valve by the poppet. The pool of plastic in the center of the valve, and the poppet shaft, are the last front end components to reach operating temperature. The plastic around the screw will reach operating temperature before the internals of the Auto-Shut Valve. If screw rotation occurs too soon, the plastic from the screw will flow into the valve, and either the poppet will resist opening, or the cold slug of plastic in the valve will block the flow. If the screw continues to rotate, the Auto-Shut valve may become damaged.

Table 4.45 *(continued)*

performance criteria were satisfied simultaneously. Prediction models generated based on the DOE analysis were shown to accurately represent the actual molding process. These models were then coded in a program to be utilized by molding engineers in Process-sensitivity analyses. Table 4.48 summarizes the results of the analysis for each of the performance criteria considered (112).

It lists the process parameters that affect each criterion and describes the nature of the effect. As can be seen in this table, each of the performance criterion considered is affected by one or more of the four investigated process parameters. Furthermore, some of these effects are in terms of interactions of multiple process parameters (effect of one process parameter is dependent upon the value of another process parameter). This table points out that for those performance criteria that are affected by multiple process parameters the level of effect from each process parameter is different. For example, it is pointed out that foiling quality at edges of the part is affected by melt temperature, injection speed, and hold time. Then, the table further points out that melt temperature has the strongest effect on this performance criterion, injection speed has somewhat lesser effect, and hold pressure has the least effect.

The processing window has been used to optimize the required uniform quality of optical components (315, 316). Part weight, dimensions, shrinkage, and birefringence are a few important measurable parameters that are used to define the quality of plastic optical components. The quality of a plastic part can be assured by determining the proper and optimized set of injection-molding process variables. Online cavity pressure data as functions of time for dual-cavity optical molds were analyzed for establishing the PVT (or PTV [pressure, temperature, volume]) relationship. The PVT data were then used in an empirical model to determine the optimized set of process variables for the expected quality of a part (Fig. 4.56).

Objective

The intent of this procedure is to instruct on the safe installation of a barrel endcap and nozzle. A safe installation is when no damage occurs to the screw, barrel, endcap or non-return valve due to the installation, including tightening of the endcap.

Machine Setup

- The injection molding machine should have the motor off and locked out.
- The injection unit must be positioned so the endcap can be safely removed.
- The injection unit components and the heat profile are identical to a production run.
- If the components are to be cleaned, the barrel and endcap must be hot for cleaning.
- If the components are not to be cleaned, the endcap can be positioned and lightly tightened cold, but must be hot when fully tightened.

Tools & Equipment

- As necessary, to safely install the endcap to the barrel (refer to the owner's manual or contact your machine supplier).
- As necessary, to clean the female threads in the barrel and endcap.
- A brass putty knife, copper gauze, and wire brush for cleaning the barrel, endcap, and male threads.
- Anti-seize.
- A torque wrench if socket head cap screws are used to fasten the endcap.
- Torque specifications from the owner's manual or machine supplier.

Inspecting & Cleaning

Inspect the mating surfaces prior to assembly. Check for the following:

- The face and pilot areas are clean and contain no nicks. It is too dangerous to stone damaged surfaces; machining is necessary to maintain safety, the proper fit, and finish.
- The non-return valve and screw are not in the way of the head. Never hit the screw or non-return valve, always push them with a soft material such as wood.

Inspect and clean the threads prior to assembly. This includes the socket head cap screws, barrel threads, and nozzle threads. Inspect the threads and repair as necessary. The tools used to repair threads can also be used to clean threads. Remove plastic, dirt and old anti-seize from the threads. In the absence of the proper tools, a wire brush, cleaner and rag can be used once or twice. Check to be certain the cleaner is safe for hot metals, and obtain the correct tools for future use.

Clean all mating and interior surfaces using the above tools. Mating parts must be clean in order to properly seal. Interior surfaces must be clean and smooth, or specs and degradation may occur.

Procedure for Checking Clearances

Follow this procedure if you have a new barrel, screw, endcap or non-return valve:

Extra steps are required when new components are involved. It is always best to check for interference between the endcap and the non-return valve. Interference infrequently occurs due to design or manufacturing errors. In addition, there are machines where the screw-forward position is adjustable.

1. Let the barrel reach operating temperature, adequately soak, and start the machine motor(s).

Table 4.46 IMM endcap and nozzle installation (courtesy of Spirex Corp.)

2. Bring the screw all the way forward; observing the safety precautions in the owner's and operating manuals. In manual mode it is best to start with a velocity and pressure of twenty percent (20%). In set-up mode the machine automatically reduces pressure and speed. Never select injection-forward with someone in front of the screw.
3. Turn off and lock out the motor(s).
4. Next put the endcap in place, but first place a soft material (such as modeling clay or soldering wire) in the endcap so that it is between the head and non-return valve. Evenly seat the head and snug it down – it is not necessary to fully tighten it.
5. Remove the endcap and observe the gap between the endcap and non-return valve by measuring the thickness of the clay. The clearance varies depending on the size of the injection unit and the machine OEM. It is safe to say that less than 0.060 inches (1.5 mm) is too little clearance, and the screw must be adjusted back. If there is no adjustment possible, contact the supplier(s) of the new component about the problem.
6. Continue to the procedure below if there is no interference.

Procedure for Installing the Endcap and Nozzle

1. Let the barrel reach operating temperature.
2. Put a light coat of anti-seize on the male threads.
3. Put the endcap in place and tighten only until snug. Install the nozzle and tighten only until it is snug. Install the heater bands and thermal couples on the endcap and nozzle, and turn on the heater bands.
4. Wait until the endcap and nozzle are at the operating temperature set point. The barrel, endcap and nozzle should be at the set-point temperature before you proceed.
5. It is important to understand that damage will occur to either the barrel or endcap if the endcap is tightened incorrectly. Incorrect assembly occurs when the endcap is tightened unevenly or too tightly. Both cases result in excessive force applied to the mating surfaces. These surfaces are then permanently damaged, or the uneven force on the endcap results in early failure.

The damage to the mating surfaces is termed *coining*, and often results in a change in the barrel internal diameter. If this does occur, it is best to repair the barrel and/or endcap before the machine is put back into production. Otherwise, the non-return valve will tend to seize to the barrel when it passes through the damaged area. The seizure can result in further damage to the barrel, screw, non-return valve or endcap.

Uneven force on the endcap usually does not result in immediate failure. Instead, the failure is a slower process known as fatigue failure. Since the load is uneven, there is some portion of the endcap that is subjected to significantly higher stress than other portions of the endcap. This higher stress will gradually break down the metal until a crack eventually forms. The crack lengthens until the remaining material fails.

Even if the components do not fail, an incorrect installation may result in contamination of the plastic, or leakage between the endcap and barrel. When the endcap does not seat properly, a gap may occur between the endcap and barrel mating surfaces that result in contamination, specs or leakage. As is often the case, the processor will tighten the endcap if leakage occurs, and this leads to further damage. If leakage occurs on your installation, remove the endcap, inspect the mating surfaces, and clean and/or repair as required. In the long run, this is the most time and cost effective way to deal with such a problem.

Table 4.46 *(continued)*

6. Installation procedure for a threaded endcap:
   - First, snug the endcap.
   - Use a wrench to tighten the endcap. Generally, the length of the wrench is designed to provide adequate tightening torque. If the wrench contains dead blow flats, then a ***small*** (60 ounce) dead blow hammer is permissible. Tighten the endcap.
   - If you undertighten the endcap and it leaks, you will have to remove the endcap and clean the mating surfaces. Then reinstall the assembly, and tighten the endcap a little tighter than you did with the leaking assembly. If you overtighten the assembly, you may have expensive replacement or repair costs.
7. Installation procedure for a flanged endcap:
   - First, snug the fasteners. Use the crisscross method where, for example, you tighten the first screw at twelve o'clock, the second at six o'clock, the third at two o'clock, and the fourth at eight o'clock. Continue until all of the screws are snug.
   - Set the torque wrench for 50% of the recommended tightening torque, and tighten using the above crisscross method. Set the torque wrench for 75% of the recommended tightening torque and tighten. Set the torque wrench for 95% of the recommended tightening torque and tighten. Finally, set the torque wrench to the recommended tightening torque and tighten. Always use the crisscross tightening method or a similar method that is recommended by the machinery manufacturer.
   - If you do not have a torque wrench and cannot rent one, then use a ratchet wrench and hex bit socket. The length of the wrench will determine the maximum torque on the screws. Just as above, tighten the screws using the crisscross method, and tighten to a lower torque and gradually increase the torque until tight. Do not overtighten – just a strong steady pull is sufficient (do not strain) to tighten. If you do not use a torque wrench, there is no guarantee that you have properly tightened the endcap, so seek assistance from a more experienced technician. If you must use a hex wrench and cheater bar, cut the cheater bar to approximately the same length as a wrench for that size screw.

   If you undertighten the endcap and it leaks, you will have to remove the endcap and clean the mating surfaces. Then reinstall the assembly and tighten the endcap a little tighter than you did with the leaking assembly. If you overtighten the assembly, you may have expensive replacement or repair costs.
8. Use a wrench to tighten the nozzle. The wrench length will determine the maximum torque on the nozzle. Do not overtighten – just a strong, steady pull is sufficient (do not strain) to tighten. If you undertighten the nozzle and it leaks, you will have to remove it, clean the mating surfaces, reinstall it, and tighten the nozzle a little more than you did with the leaking assembly. If you overtighten the assembly, you may have expensive replacement or repair costs.

   The assembly is now ready for production.

Table 4.46 *(continued)*

Another example of a processing window study concerns thin-wall molding. In this study, the effects of IM conditions and critical design parameters on the filling, dimensional stability, and crystallization of syndiotactic polystyrene (SPS) parts were studied (103). Part wall thickness was the primary factor affecting filling, shrinkage, and crystallization. While injection velocity was a secondary influence during filling, mold temperature was the minor factor for crystallization and shrinkage. Melt temperature and gate dimensions had little or no effect on filling or part properties.

In creating a process window for the combination of material, machine, and heater used in this study, the first consideration was moldability, which was dictated by the maximum injection pressure and the maximum clamp force. No changes in melt or mold temperature, injection velocity, or gate size could remedy this situation. Thus, an IMM with higher injection speeds, pressures, and clamp force was required to mold thin-walled SPS.

Objective

The intent of this procedure is to instruct on the safe installation of three and four piece non-return valves. A safe installation is when no damage occurs to the screw, barrel, endcap or non-return valve due to the installation.

Tools & Equipment

- As necessary, to safely remove and install the barrel, endcap (refer to the owner's manuals or contact your machine supplier) and non-return valve.
- A brass putty knife, copper gauze, wire brush and cleaner for cleaning the barrel, endcap, non-return valve and fasteners. Check to be certain the cleaner is safe for hot metals.
- Anti-seize.
- If you are not familiar with endcap and nozzle removal/installation procedures, obtain a copy of our *"Injection Molding Machine Endcap & Nozzle Installation"* procedure.
- A block of wood or soft material of sufficient size and strength to push the screw out of the barrel.

Removal of Old Non-Return Valve

- Let the barrel reach operating temperature, adequately soak, and start the machine motor(s). Do not rotate the injection unit, but prepare it for purging as your machine operator's manual instructs.
- The injection unit components and the heat profile are identical to a production run.
- The barrel, endcap and non-return valve must be hot for cleaning.
- Shut off the supply of plastic that is running to the machine. Close the supply of plastic running to the feed throat of the injection unit – do this by moving the hopper slide to the closed position. Note: Your system may be different. Refer to your operator's manual for your machine or after-market material handling equipment supplier. Purge the screw of plastic and/ or purging compound.
- Bring the screw all the way forward, observing the safety precautions in the owner's and operating manuals. In manual mode it is best to start with a velocity and pressure of twenty percent (20%). In set-up mode the machine automatically reduces pressure and speed. Never select injection-forward with someone in front of the screw.
- Position the injection unit for endcap removal, turn the motor off and lock it out. Safely remove the endcap (*Tools & Equipment*).
- Position the injection unit for screw removal. Disconnect the screw from the drive shaft or quill. Each machine is different and you must refer to your owner's or service manual for the safe and proper procedure for your specific machine. Retract the drive shaft, place the block of wood between the drive shaft and screw and run the screw forward until the non-return valve is out of the barrel. Push the screw forward until approximately six inches of the screw is protruding from the barrel.
- Safely remove the old non-return valve from the screw. Remember that most threads are left handed, so turn the retainer clockwise to loosen. Use a wrench on the wrench flats, or a screw tip removal wrench. Placing a torch on a non-return valve generally voids the warranty offered by most manufacturers. If you must apply heat, it is the screw that should be heated. If a hammer is necessary, a dead blow hammer is less likely to damage the non-return valve.

Inspecting & Cleaning

Inspect and clean the screw threads, pilot, face and bottom of dill cavity prior to assembly. Tools that repair threads can also clean threads. Remove plastic, dirt and old anti-seize from the threads. In the absence of proper tools, a wire brush, cleaner and rag will work once or twice. Obtain the correct tools for future use.

Check that the screw face is flat and straight to 0.001" (0.025mm) TIR (total indicator reading). If not, have a professional experienced in plastic screws machine the face. If machining is required, have the vendor check the entire non-return valve cavity. From time to time, the geometry is out of specification. This may result in early failure of the non-return valve, or damage to the endcap or barrel. The Society of the Plastics Industry maintains standards for this geometry.

Table 4.47 Three- and four-piece nonreturn valve installation (courtesy of Spirex Corp.)

Follow this procedure if you have a new screw, barrel, endcap or non-return valve:

Extra steps are necessary when installing new components. It is always best to check for interference between the endcap and the non-return valve. Interference infrequently occurs due to design or manufacturing errors. In addition, there are machines where the screw-forward position is adjustable.

1. Let the barrel reach operating temperature, adequately soak, and start the machine motor(s).
2. Bring the screw all the way forward, observing the safety precautions in the owner's and operating manuals. In manual mode, it is best to start with a velocity and pressure of twenty percent (20%). In set-up mode, the machine automatically reduces pressure and speed. Never select injection forward with someone in front of the screw.
3. Remove the old non-return valve as described above, and install the new valve as described below.
4. Put the endcap in place, but first place a soft material (such as modeling clay or soldering wire) in the endcap so that it is between the head and non-return valve. Evenly seat the head and snug it down – it is not necessary to fully tighten it.
5. Remove the endcap and observe the gap between the endcap and non-return valve by measuring the thickness of the clay. The clearance varies depending on the size of the injection unit and the machine OEM. It is safe to say that less than 0.060 inches (1.5 mm) is too little clearance, adjust the screw back. If there is no adjustment possible, contact the supplier(s) of the new component about the problem.

Procedure for Installing the Non-Return Valve

1. Inspect and clean the screw as stated above.
2. Put a light coat of anti-seize on the male threads of the new valve.
3. Screw the non-return valve into the screw. Do not use a wrench to tighten the valve (hand-tight is sufficient). Inspect the assembly to be certain the seat is flush against the screw face. Try to rotate the seat by hand — the retainer is sufficiently tight if the seat does not rotate.
4. Use the wooden block to push the screw inside the barrel. Be prepared to center the slider ring as you push, otherwise the ring will hang up between the retainer and the barrel.
5. **Caution:** If you are installing a new barrel, screw, endcap or non-return valve, perform the above procedure "*Procedure for Checking Clearances*." It is a good way to check the clearances of a new assembly without causing damage to the components.
6. Reattach the screw to the drive shaft, and the endcap to the barrel. **Caution:** Improper installation of the endcap can result in damage to the endcap, barrel and/or non-return valve. Read your owner's or operator's manual, or obtain a copy of our endcap installation procedure.

The assembly is now ready for production.

Table 4.47 *(continued)*

Crystallization and shrinkage were influenced by cooling rate (part thickness and mold temperature). While the oil-heated mold maximized crystallization, cooler (water-heatable) molds produced crystallinity levels of 25% or better (chapter 1). Parts molded with the high mold temperature did exhibit better surface finishes. Shrinkage was relatively low for all processing conditions and design variables.

PLASTICS BEHAVIOR

As shown in Figure 4.2, the heat profiles for TPs and TSs differ. A major difference is that after plasticizing (melting in the plasticator), solidification for TPs occurs by cooling the melt in the mold cavity, whereas TSs solidify by applying higher heat (chapter 1). Another difference is that the plasticizing temperatures for TPs are usually higher than those for TSs.

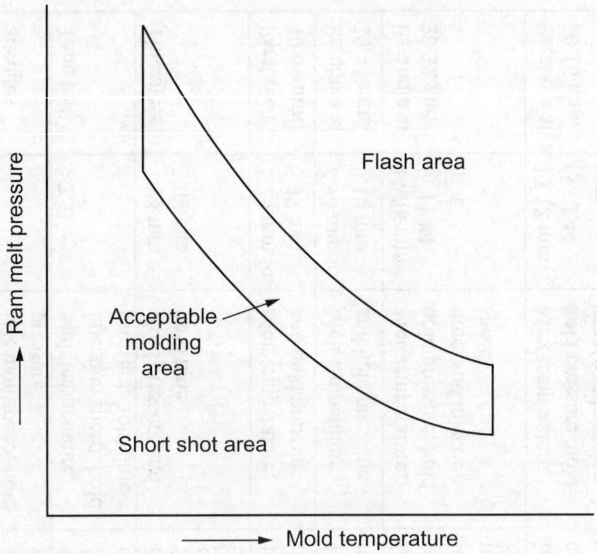

Figure 4.54 Molding area diagram processing window concept

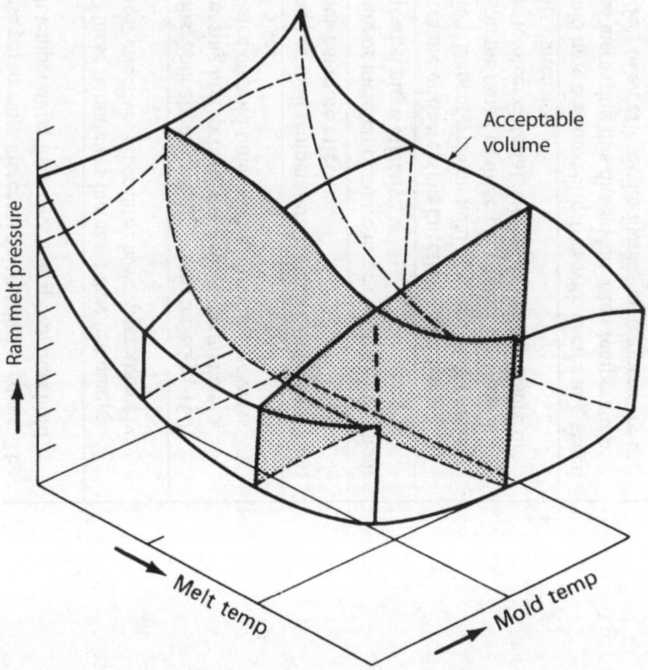

Figure 4.55 Molding volume diagram processing window concept

Quality Criteria	Spec/ Target	Critical Process Parameters	How the process parameters affect the quality criteria & other comments
Heat Deflection	<0.178 mm	Injection speed & hold time	Increasing hold time decreases Heat Deflection (HDT). Decreasing speed decreases HDT.
Foil quality at center	>2.5	Injection speed	Increasing speed decreases foil quality at center. Worst condition (=1) results in washing away of the foil in a 3.175 mm. to 3.429 mm. diameter at center.
Foil quality at edges	>2.5	Melt temperature, injection speed, hold pressure	Increasing any of these parameters increases foil quality at edges. The most effective is melt temperature, then injection speed, then hold pressure.
Tape test	>2.5	Melt temperature, injection speed, hold pressure, hold time	Increasing any one of these parameters increases tape test quality. Speed is by far the most effective, other three have approximately equal effect.
Thickness	1.98 mm to 2.02 mm.	Hold pressure and hold time	Measured at 2 locations, both about 25.4 mm. from the edge, 180° from each other. Increasing hold pressure increases the thickness. Second order dependence on hold time, with maximum thickness occurring at around 7.5 sec.
Parabola at 40.64 mm.	5.23 mm to 5.49 mm.	Melt temperature, and hold pressure	Increasing hold pressure or melt temperature increases this dimension; melt temperature somewhat more effective.
Parabola at 30.48 mm.	8.94 mm to 9.19 mm.	Melt temperature, and hold pressure.	Increasing hold pressure or melt temperature increases this dimension; melt temperature somewhat more effective.
Parabola at 20.32 mm.	11.58 mm to 11.84 mm.	Melt temperature, hold pressure, hold time, and injection speed	Increasing melt temperature or decreasing injection speed linearly increases this dimension. Dependence on hold time and hold pressure is quadratic; however, in the range of these parameters the team is most interested in (due to other criteria), increasing hold pressure or hold time increases this dimension.
Parabola at 10.16 mm.	13.18 mm to 13.44 mm.	Melt temperature, hold pressure, hold time, and injection speed	Injection speed and hold time are most effective with decreasing speed or increasing hold time increasing this dimension. Melt temperature and hold pressure has less of an effect with increasing temperature or decreasing hold pressure increasing parabola height at 10.16 mm.
Parabola at center	13.87 mm to 13.97 mm.	Melt temp., hold pressure, hold time, and injection speed	Same trends as above (for "parabola at 10.16 mm.) with, in this case, hold time having the biggest effect, followed by injection speed and hold pressure; melt temperature has the least effect.
Parabola Constant	193.92 mm to 195.92 mm.	Hold time, hold pressure, & injection speed	Increasing injection speed or hold pressure increases parabola constant; increasing hold time decreases it.

Table 4.48 Processing window analysis

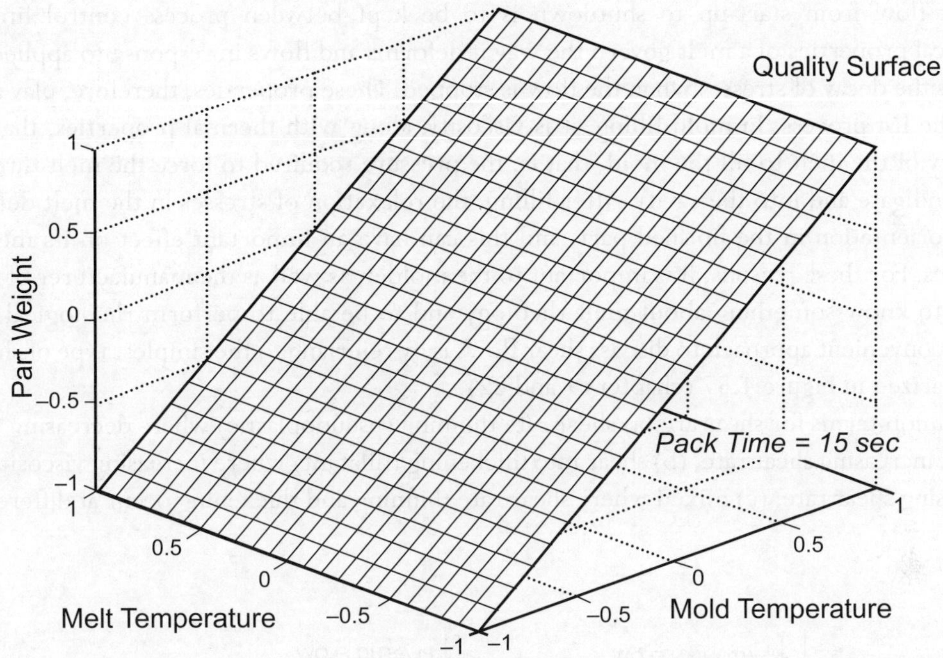

Figure 4.56 Quality surface as a function of process variables

With start-ups, it is important that the quality of the raw materials always conforms to specification. Certain properties must be checked when the goods are received. In view of the wide variety of applications for plastic articles, a testing schedule of general validity cannot be submitted here. Each case must be treated individually. Over the years, many hours have been devoted to designing methods for testing materials to develop values for their properties. These tests, conducted under procedures established by organizations such as the American Society for Testing and Materials (ASTM), are a means of extracting basic knowledge about materials (chapter 22).

In checking goods received, make sure that they conform to the type ordered. In other words, they must be checked to ensure that they agree with samples of former deliveries. This check includes examination for contamination and is followed by specific tests, such as simply determining bulk density. Often, samples are sent in advance of the materials dispatcher in tank cars or large containers. In this case, statistical rules must be observed in selecting the random samples (chapter 24).

A preliminary check must proceed without loss of time, so rapid tests with specific aims are frequently used. Since injection molding continues to be caught up in the automation trend, it is feasible for the checking of goods received to become part and parcel of the actual production start-up process. However, this entails that any deviations from standard must remain within narrow limits. For technical and economic reasons, this adaptive process control, as it is called, is the processor's target.

Melt flow from start-up to shutdown is to be kept between process control limits. The rheological properties of a melt govern the way it deforms and flows in response to applied forces, as well as the decay of stresses when the flow is stopped. These properties, therefore, play a central role in the IM process. In mold filling, it is viscosity, along with thermal properties, that govern the ability of the melt to fill the mold, that is, the pressure required to force the melt through the runner and gate and into the cavity. After filling, the relaxation of stresses in the melt determines residual orientation in the finished part, and this can have an important effect on its mechanical properties. For these reasons, it is important to the molder, as well as the manufacturer of molding plastics, to know something about melt rheology and to be able to perform rheological tests on melts. A convenient approach to discuss shear flows is by referring to the simplest type of shear flow as summarized in Figure 4.57 (chapters 1 and 22).

Common terms for shear are (a) shear rate thinning (pseudoplastic) where decreasing viscosity occurs at increasing shear rate, (b) shear rate thickening (dilatant) where increasing viscosity occurs at increasing shear rate, (c) mixed where shear rate thinning and thickening occur at different shear

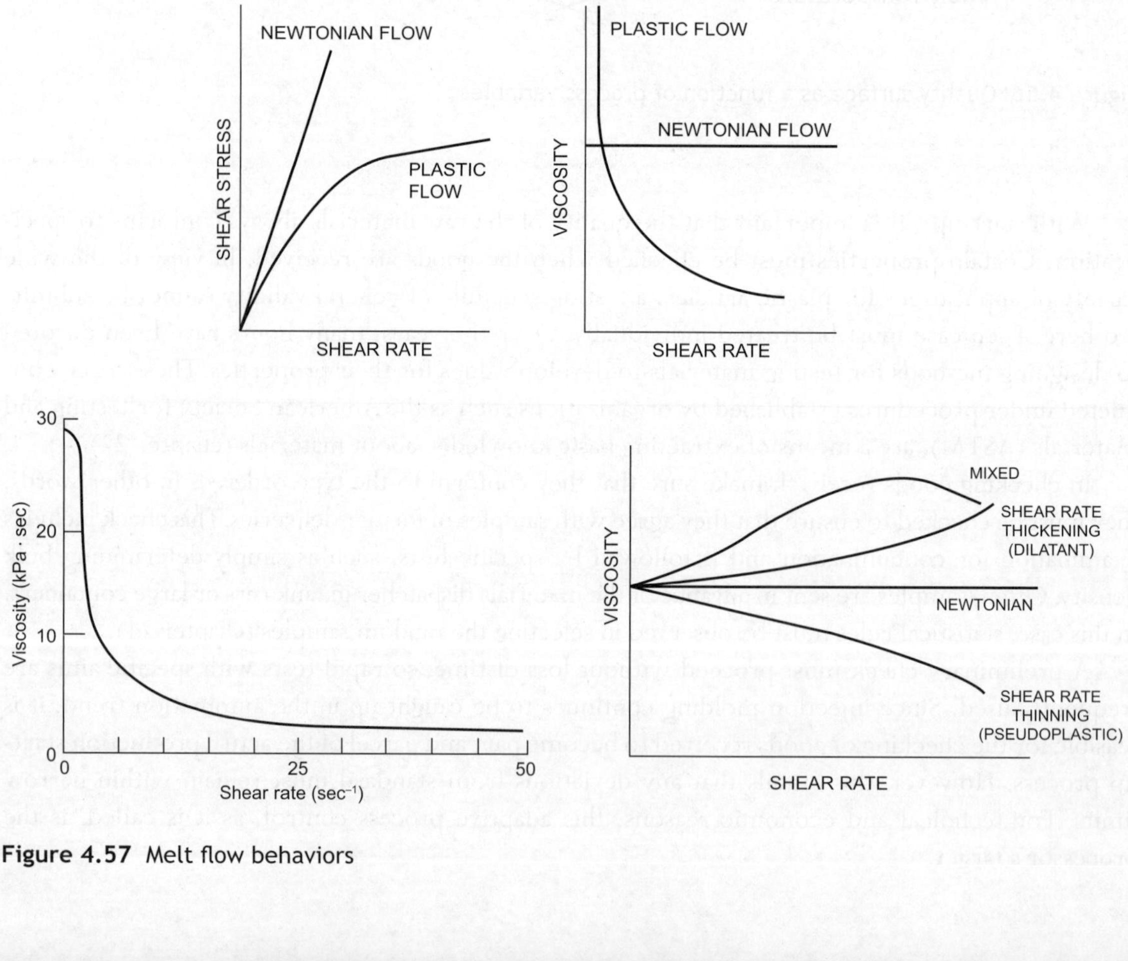

Figure 4.57 Melt flow behaviors

rate ranges, (d) rheopectic where increasing viscosity with time of agitation occurs at constant shear rate, and (e) thixotropic where decreasing viscosity occurs with time of agitation at constant shear rate.

Type and amount used can influence start-ups (chapter 1). Plastics are usually obtained in the form of granules, powders, flakes, or pellets, or sometimes in liquid form or tapes. Each has certain advantages, such as providing specific molded part performances, ease in compounding, convenient processing, and/or reduced cost. They can be delivered in small to large amounts from all kinds of materials suppliers worldwide. Usually, delivery is in 55-lb (25-kg) sacks, gaylords of 1,000 lb (455 kg), semibulk containers (half to a full ton), or in truck or railcar tanks of up to 25 tons. Since plastics have plus to minus ranges of performance per their individual specifications, start-up settings when all the materials arrive on the plus side can be significantly different when another lot arrives meeting the minus range.

MACHINE DEVELOPMENT

Much of the innovation potential continues to rely on the IM process. Special IM processes, such as coinjection, gas-assisted, thin-wall, micro, macro, hybrid, multicomponent, two-color, counterflow, multilive, oscillatory, reaction, liquid, foam, fusible and soluble core, tandem, injection blow, IM with rotation, continuous IM (Velcro® strips, etc.), metal-plastic, vacuum IM, and injection blow molding (chapter 6), are playing increasingly large roles in the processing of plastics. At the same time, the requirements placed on product quality, production quality, and part precision are rising while costs are expected to be reduced wherever possible. This combination of objectives can only be achieved by an appropriate increase in productivity from process-engineering innovations and a reduction in the number of process steps(178).

Each machine design provides different advantages, such as reducing product weight (reducing plastic consumption), eliminating or minimizing molded-in stresses, molding extremely small to very large products, and/or improving performances.

From the microeconomic approach, the challenge facing a plastic-processing company is one of retaining its market position through readiness and ability to innovate. This applies particularly to the opening of new application potential and the application technology that this requires, as well as to the crucial aspect of employee skills. Specialist know-how is essential in order to survive in the marketplace. IM companies are called upon to acquire know-how on new plastics as well as process variants and special IM processes.

COINJECTION MOLDING

The review of coextrusion (chapter 5) advantages also applies to coinjection. Two or more injection-molding barrels are basically joined together by a common manifold and nozzle through which melts flow before entering the mold cavity by a controlled device such as an open closed valve system. The plastics can include the same material but with different colors. In some systems, one material with two shots is made from one plasticator, thus imparting certain advantages over the usual single-shot IM, such as reducing pin holes and/or strengthening the product. The nozzle is usually designed with a shutoff feature that allows only one melt to flow through at a controlled time. Other designs are used (6, 384).

In the usual coinjection method, two or more different plastics are bonded or laminated together. Figure 4.58 shows the action in which two or three plasticators can be used. With a two-layer system, one delivers melt to both sides of the part.

Proper melt flow and compatibility of the plastics are required in order to provide the proper adhesion. Table 4.49 provides a guide on compatibility of plastics ranging for good adhesion, no adhesion, poor adhesion, and limited adhesion. These plastics can be modified so that their adhesion capabilities change (chapter 1). Some of the melt flow variable factors can be compensated by the type of the available plasticator and mold process control adjustments.

LOW-PRESSURE COINJECTION FOAM MOLDING

Using the coinjection procedure, a solid melt is injected to form the solid, smooth skin against the cavity surface. Simultaneously, a second short shot melt with blowing agent is injected to form the foamed core. With a full second shot, the mold can incorporate pins or a mold that opens similarly to high-pressure foam molding.

GAS-ASSISTED MOLDING

There are different processes for gas-assisted injection molding (GAIM). Other terms include injection molding gas-assisted (IMGA), gas injection molding (GIM), gas injection-molding machine (GIMM), or injection gas pressure (IGP). Most of the gas-assisted molding systems are patented.

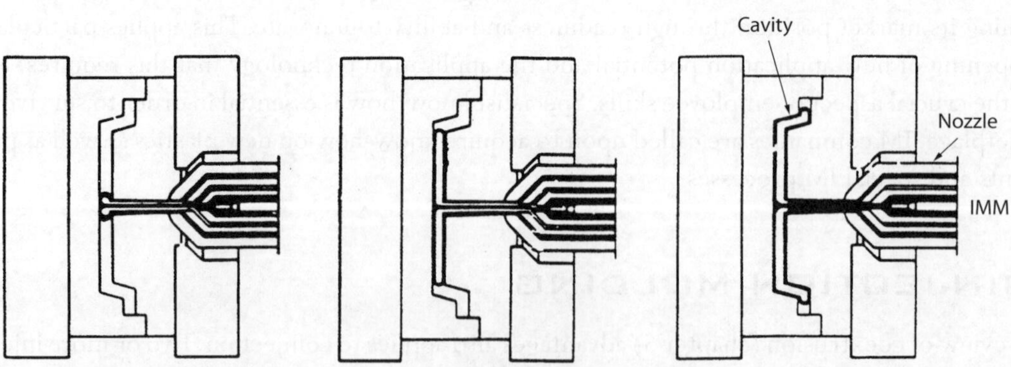

Figure 4.58 Example of a three-layer coinjection system (courtesy of Battenfeld of America)

	ABS	ASA	CA	EVA	PA 6	PA 66	PC	PE-HD	PE-LD	PMMA	PP	PS-GP	PS-HI	PBT	TPU	PVC	SAN	TPR	PET	PVAC	PPS
ABS	+	+	+				+	◆	◆	+	◆	◆	◆	+	+	+	+		+	○	
ASA	+	+	+	+			+	◆	◆	+	◆	◆	◆	+	+	+	+			○	
CA	+	+	+	○				◆	◆		◆	◆	◆	+	+	+	+				
EVA		+	○	+				+	+		+	+			◆	◆					
PA 6					+	+		○	○		○	◆	◆		+						
PA 66					+	+	○	○	○		○	◆	◆		+						
PC	+	+				○	+	◆	◆	(+)	◆	◆	◆	+	+	+	+		+		+
PE-HD	◆	◆	◆	+	○	○	◆	+	+	○	◆	◆	◆	◆	◆	○	◆				
PE-LD	◆	◆	◆	+	○	○	◆	+	+	○	+	◆	◆	◆	◆		◆				
PMMA	+	+					+)	○	○	+	○	◆	◆			+	+				
PP	◆	◆	◆	+	○	○	◆	◆	+	○	+	◆	◆	◆	◆	○	◆	+			
PS-GP	◆	◆	◆	+	◆	◆	◆	◆	◆	◆	◆	+	+	◆	◆	○	◆				
PS-HI	◆	◆	◆		◆	◆	◆	◆	◆	◆	◆	+	+	◆	◆	○	◆				
PBT	+	+	+				+	◆	◆		◆	◆	◆	+	+	+	+		+		
TPU	+	+	+	◆	+	+	+	◆	◆		◆	◆	◆	+	+	+	+		+		
PVC flex.	+	+	+	◆			+	○		+	○	○	○	+	+	+	+				
SAN	+	+	+				+	◆	◆	+	◆	◆	◆	+	+	+	+		+	○	
TPR											+							+			
PET	+						+							+	+		+		+		
PVAC	○	○															○			+	
PPS							+														+

+ good adhesion ◆ no adhesion ○ poor adhesion (+) limited adhesion

Table 4.49 Examples for evaluating adhesion between coinjection plastics

This review concerns the use of gas (Table 4.50); however, there are others, such as water-assisted injection molding (178).

The processes use inert gases (usually nitrogen) with pressures up to 20 to 30 MPa (2,900 to 4,400 psi). Within the mold cavity, the gas in the melt forms channels. Gas pressure is maintained through the cooling cycle. In effect, the gas packs the plastic against the cavity wall. Gas can be injected through the center of the IMM nozzle as the melt travels to the cavity or it can be injected separately into the mold cavity. In a properly designed tool run under the proper process conditions, the gas with its much lower viscosity than the melt remains isolated in the gas channels of the part without bleeding out into any thin-walled areas in the mold. The gas produces a balloonlike pressure on the melt.

The gas channels are those areas that have been thickened to achieve functional utility in the part or to promote better melt flow during cavity filling. This action provides a high degree of packing the melt against the cavity walls. Gas pressure is held until the melt solidifies. This coring action results in reducing cycle time and quantity of plastic used while developing a more structurally sound part

PROCESS: Nitrogen gas under high pressure is injected through the nozzle or mold wall into plastic partially filling a mold. The gas flows preferentially through local thick sections with hot interiors and pushes the plastic ahead to fill the mold.
SIMULATION: Commercial software now available to predict gas flow paths, polymer thickness, clamp force, and contraction during cooling for various geometries and process variables including gas pressure, injection time, and prefilled polymer volume.
Simulations and experiment generally show

- increasing gas pressure decreases fill time, gas penetration distance, and (by conservation of mass) polymer wall thickness,
- melt temperature has a variable effect on gas penetration length,
- increasing delay time before the start of gas injection increases wall thickness and gas penetration length,
- increasing gas injection time increases gas penetration distance,
- decreasing the prefilled polymer volume fraction increases the penetration length until a critical level when gas blows through.
- increasing gas pressure level and time decreases shrinkage.

Simulations are generally able to predict undesirable air traps and gas penetration into thin sections. Simulation of a freezer bottom part converted to gas-assisted molding showed a 70% reduction in packing pressure, feasibility of using a less expensive material, and reduced warpage due to lower, more uniform pressure and higher part stiffness. Eight design guidelines are given based on both experiment and computer simulation [2.1].

Advantages/Disadvantages	Applications	Materials
PROCESS: Part weight and cooling time can be reduced up to 50%. Sink marks are eliminated. Warpage is reduced. Clamp force and injection pressure are lower. Part stiffness is increased because of the higher moment of inertia. Licensing is necessary. SIMULATION: Simulation helps identify optimal process conditions including runner layout and size, and location and timing of gas introduction. Software is available but new.	Handles, Panels with Ribs, Appliance/machine Housings (TV benzels) Automotive Parts (Clutch Pedals, Mirror Housings)	ABS, PA, PE, PP, PS, PPO, PC, PBTP, PC/PBTP, SAN, TPE, TPU

Table 4.50 Gas-assisted injection-molding process

(increases section stiffness), improving surface flatness, reducing warps and sinks over thick sections, and so on. Thick parts can easily be made without voids, sink marks, and so forth.

The gas-assisted approach is a solution to many problems associated with conventional IM and structural foam molding (chapter 8). It significantly reduces volume shrinkage that causes the sink marks in injection molding. Products are stiffer in bending and torsion than equivalent conventional IM products of the same weight. The process is very effective in products of different sizes and shapes, especially the larger, longer, thick molded products. It offers a way to mold products with only 10% to 15% of the clamp tonnage that would be necessary in conventional injection molding.

The mold is designed for optimum material flow and gating. It is also designed for gas (or water) injection and venting. The mold must also have shutoffs for the gas (and water) and another shutoff valve for the overflow.

GAS-ASSISTED WITHOUT GAS CHANNEL MOLDING

The Battenfeld Airmould Contour process provides a gas-assisted alternative when it is not possible to inject gas directly into the molten plastic or where moldings with gas channel openings are not desirable. It can be used where parts require only one smooth or high-gloss surface with different wall thickness or complex geometry on the other side. The process has gas enter between the melt and mold cavity surface where external pressure is applied to the melt. The gas can be applied within specific part sections.

GAS COUNTERFLOW MOLDING

A conventional IM system is used with a separate entrance to the mold cavity providing pressurized gas (usually nitrogen) prior to injecting the melt shot. This back-pressure action can provide an even distribution of melt packing during its cooling cycle. When producing foamed plastic parts, this gas back pressure prevents the blowing agent from expanding until its part skins form followed by releasing (venting) the gas pressure. Controlled foam expansion is provided.

WATER-ASSISTED MOLDING

A variation of gas-assisted molding is water-assisted molding. To a degree, water injection molding will compete with gas-assisted molding. Instead of gas, water is used to mold plastic into a hollow, thick-walled part, whereby the interior plastic of the part is cored out by water instead of gas. This action forces the surrounding plastic into contact with the mold walls as well as assists in the cooling phase of the molding cycle. Its first stage is the complete filling and packing of the part. This typically means that the gate must be located on one end of the part, providing a means of positively closing after plastic injection (199).

Materials used in the mold must resist corrosion. This makes stainless steel and copper alloys the materials of choice for water injection molds. The mold is designed with an overflow to provide positive control material flow at the opposite end of the cavity. When the main cavity has been filled, the valve to the overflow is opened and the injection of the secondary material takes place. At this point in the cycle, a gas (typically nitrogen) might be injected to core out the hollow feature and force the displaced plastic into the overflow. If a gas is used, then water injection follows. Following water injection, another gas or even plain air is used to discharge the water from the part.

LOW-PRESSURE MOLDING

Table 4.51 provides a review on this subject (178). See also chapter 15.

INJECTION-COMPRESSION MOLDING

Also called coining or injection stamping, injection-compression molding produces molded products with virtually no internal stresses and the ease of meeting very tight tolerances. Instead of using

Low pressure molding, as developed by Siebolt Hettinga, enables a number of other molding technologies.In Low Pressure Molding (LPM) the mold cavity is filled at low speed through large gates with a controlled pressure profile in the shape of a broad inverted U. LPM has no packing stage and no cushion. The melt temperature profile is controlled by adjusting screw speed and flow resistance during plastication. LPM works better with low viscosity semicrystalline materials and is not suitable for thin-wall parts. Slow injection, lower melt and higher mold temperatures reduce residual stress to allow demolding at a higher temperature to maintain or reduce cycle times. For larger parts, low clamp force can be achieved using multiple valve gates with programmed opening [4.2]. Lower clamp force allows use of self clamping molds and multistation injectors. Laminate Molding involves molding plastic at low pressure directly behind textile, film, or metal. In Liquid Gas Injection Molding, a volatile liquid is injected at low pressure into the melt and then vaporizes to form hollow channels in the part. The liquid condenses and is absorbed in the part. Dual Molding, similar to Bayer's Multishell Molding, forms an integrated hollow part by overmolding at low pressure an assembly formed from separately molded parts. In Controlled Density Molding the mold is partially opened once a skin has formed to give a low density interior.		
Advantages/Disadvantages	**Applications**	**Materials**
Substantial capital costs savings result from the use of presses with a lower camp force or self clamping molds. Laminate Molding saves on assembly and adhesive costs in fabric/plastic laminates.	*Low Pressure Molding*: Interior Vehicle Panels, Bumper Fascia. *Laminate Molding*: Fabric/Plastic Seats, Vehicle Trim Panels. *Liquid-Gas Assist Molding*: Large Chairs, Chair Bases. *Dual and Shell Molding*: Manifolds, Pump Bodies, Valves, and Fittings. *Low Density Molding*: Fittings, Electronic Enclosures, Table Tops.	Thermoplastics, especially polyolefins, thermosets

Table 4.51 Low-pressure molding

the conventional IM mold with its two halves meeting face to face, it uses a compression-designed mold (chapter 14) with a male plug fitting into a female cavity. The essential difference in these molds lies in the manner in which the thermal contraction in the mold cavity occurs during cooling (shrinkage). With conventional injection molding, the reduction in material volume in the cavity due to thermal contraction is compensated basically by forcing in more melt during the pressure-holding phase. By contrast, with a compression mold, a full shot size enters the preopened closed cavity mold but does not completely fill the cavity (Fig.4.59). The mold closes, subjecting the melt to a relatively even melt flow. The required pressure is applied, providing a molded part without stresses. This type of molding has many advantages in favor of molded part performances.

TWO-SHOT MOLDING

This is also called overmolding, in-mold assembly, two-color rotary, or two-color shuttle. Two materials are molded so that the first molded shot is overmolded by the second molded shot; the

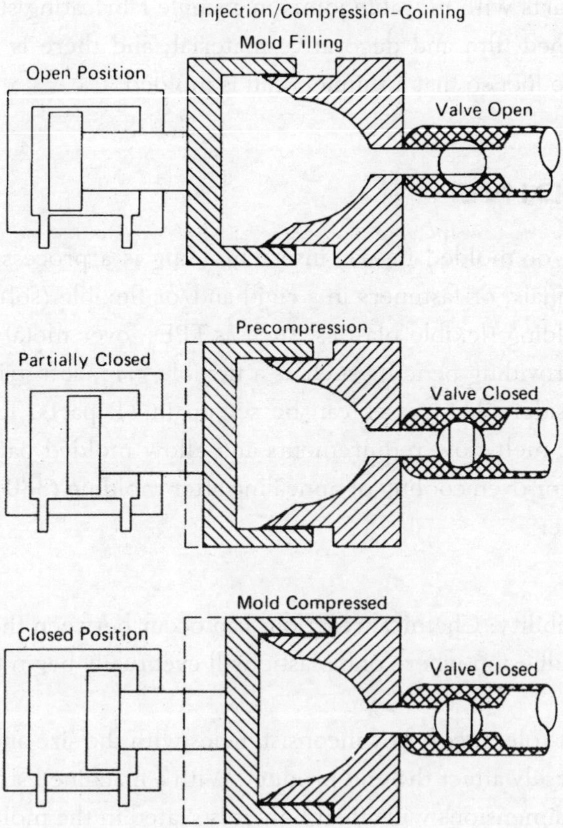

Figure 4.59 Example of action during injection-compression molding (courtesy of Plastic FALLO)

first molded part is positioned so that the second material can be molded around, over, or through the first molded part. The two materials can be the same or different and they can be molded to bond together or not bond together. If materials are not compatible, the materials will not bond, so that a product such as a universal or ball-and-socket joint can be molded in one operation. If they are compatible, controlling the processing temperature can eliminate bonding. A temperature drop at the contact surfaces can occur in relation to the second hot melt shot to prevent the bond.

IN-MOLD MOLDING

This involves the decorating of the plastic part while it is in the process of being molded (also called back molding). Performing operations such as decorating material, assembling, painting, labeling, printing film or foil, and/or laminating in the mold usually can result in cost savings compared with postmold operations. Some part designs require materials that do not share any adhesive properties. In these cases, in-mold assembly not only allows use of such incompatible materials but also facilitates molding parts with movable joints in a single fabricating step. With plastic labeling, this includes thermoformed film and decorative material, and there is usually the possibility of adding strength to the product so that a thinner wall is molded.

INSERT MOLDING

Also called overmolding or molded insert, insert molding is a process that molds components such as pins, studs, terminals, or fasteners in a rigid and/or flexible (soft) plastic to eliminate the cost of postmolding. Molding flexible plastics, such as TPEs, over metal and/or rigid plastics has been extensively used, providing benefits such as a flexible grip, soft grip, and visually appealing products. However, considerable stresses can be set up in TP parts. To relieve stresses, design the mold to meet plastic melt flow requirements and allow molded parts to cool slowly during molding and/or provide for oven cooling or annealing after molding (380). The following are other considerations to examine:

1. Material compatibility: Chemical reactions can occur between the different materials and affect their adhesion where one plastic will eventually begin to peel away from the other.
2. Consistent insert tolerances: Any inconsistencies with the size and shape of the inserts will adversely affect the end product. With a machined steel or metal insert, any variation in dimensions will affect how it is seated in the mold prior to plastic injection. Prior to overmolding plastics cover the plastic until it has met its final dimensions after sufficient cooling (this may take up to several days).

THIN-WALL MOLDING

As reviewed in this chapter, smaller, lighter thin-walled products have been in demand. Thickness of a thin-wall molding varies depending on the product, such as electronics with walls of 1 mm and large automotive walls at 2 mm. Conventional melt delivery systems are usually not well suited for thin-wall molding. This is particularly true using a hot runner system with engineering plastics. Extremely high injection pressures, properly controlled cavity fill speeds at the start of fills, and stringent gate quality requirements associated with these applications require specialized melt deliveries. As an example, without high pressures and uniform high heat at the gate(s), melt tends to freeze at the gate(s). Pressure changes occur with fill rates; faster fill tends to produce parts with lower pressure gradients from the gate(s) to the last area to fill in the cavity (Table 4.52).

Thermoplastics have non-Newtonian melt flow characteristics; this means that their viscosity will change depending on their velocity or the amount of shear that occurs in the melt. This

KEY FACTORS	CONVENTIONAL	THIN-WALL	
Typical Wall, in	0.080–0.120	0.050–0.080	<0.050
Machinery	Standard	High-end	Custom
Inject, Pressure, psi	9,000–14,000	16,000–20,000	20,000–35,000
Hydraulic System	Standard	Standard	Accumulators on injection and clamp units. Servo valves.
Control System	Standard	Closed-loop on Injection speed, hold pressure, decompression speed, screw rpm, backpressure, and all temperatures	Same as at left, with resolution of 0.40 in. on speed, 14.5 psi on pressure, 0.004 in. on position, 0.01 sec on time, 1 rpm on rotation, 0.10 ton on clamp force, 2º F on temperature.
Processing			
Fill Time, sec	>2	1–2	0.1–1
Cycle Time, sec	40–60	20–40	6–20
Tooling	Standard	Better venting, heavier construction, more ejector pins, better polish	Extreme venting, very heavy construction, mold interlocks, precise surface preparation, extensive ejection features, mold costs 30–40% higher than standard.

Table 4.52 Comparing conventional and thin-wall processing (courtesy of GE Plastics)

non-Newtonian characteristic is a key in thin-wall molding. As in any molding setup, one cannot just simply ram the melt into the cavity. Its flow characteristic, gate size as well as position, and venting have to be balanced in order to obtain the desired structural part and meet tight tolerance requirements.

Computer programs have been successfully used to provide 3-D simulation of thin-walled IM products. An example is CAE using 3-D Timon (Japan) software. The model used in this simulation was a speaker grille that has thin walls and many tiny openings (nets) for sound to pass through them. These openings cause unfavorable weld lines. The effects on weld lines and effects of the number and location of gates were reviewed (Table 4.53). This study resulted in using shorter flow length to thickness ratio, which resulted in more stable filling and more uniform temperature distribution along the product. The multigating system makes weld lines more favorable than the single-gating system (87).

	Single gate	Double gate	Center gate
Node/elements	14090/5040		
Injection time	0.600 sec		
Injection pressure, MPa	14.72	6.01	7.33
Injection temperature	280.0 (deg. C)		
Mold temperature	60.0 (deg. C)		
Maximum shear rate (1/sec)	1.51E+04	7.53E+03	6.78E+03
Maximum shear stress (MPa)	2.38E+05	7.53E+05	1.65E+05

Table 4.53 Processing conditions and simulation data for speaker grille model

SOLUBLE CORE MOLDING

Soluble core technology (SCT) is also called soluble fusible metal core technology (FMCT), fusible core technology, lost-core technology, or lost-wax techniques (Table 4.54). In this process, a core (usually molded of a low-melting alloy [eutectic mixture] but can also use water-soluble TPs, wax formulations, etc.) is inserted into a mold, such as an injection-molding mold (6, 12, 178). This core can be of thin wall or solid construction. If the part design permits, it can be supported by the mold halves or spider-type pin supports that are used to having it located within the cavity; during

Complex hollow parts are formed by injection molding plastic around a fusible alloy core which is subsequently removed by melting. The fusible (or lost) core typically is cast form a bismuth-tin alloy with a eutectic melting point of 138°C. The molten metal fills a split steel mold from the bottom and then cools for 2 min to produce a heavy core with a mirrorlike surface. The still-hot core is positioned by a robot in a steel mold and plastic is injected.
Flow channels are designed to balance forces around the core during filling to prevent core movement. For thermoplastics the injection temperature, e.g. 290°C for polyamide, can be well above the melting point of the core since the relatively high thermal diffusivity of the metal maintains a low interface temperature. After demolding, cores are melted out in a large bath or by induction heating or by injecting heat transfer fluid inside hollow cores.

Advantages/Disadvantages	Applications	Materials
Plastic parts made by fusible core technology have a weight and cost advantage over metal parts. Fusible core molding eliminates the need for mechanically complex molds or joining separately molded parts. Interior surfaces of fusible core parts are smooth which increases gas flow. Disadvantages are loss or oxidation of expensive core metal and need for robots to handle heavy cores.	Air intake manifolds, tennis racquets, pump parts	PA Poly(etherarylketone)

Table 4.54 Fusible core injection-molding process

plastic molding, the pins will melt. After the plastic solidifies, the core is removed by applying a temperature below the melting point of the plastic. Core material is poured through an existing opening or requires drilling a hole in the plastic. This technique is a variation of and is similar to the lost-wax molding process used by the ancient Egyptians to fabricate jewelry. Also, the 1944 all-plastic airplane used the lost-wax process to bag mold its RP sandwich construction (chapter 20).

CONTINUOUS MOLDING

Machines to mold parts using continuously operating extruders (chapter 5) have been designed, built, and used in different major production lines. The extruders are conventional types. They melt the plastic and have different techniques to deliver the melt into mold cavities. These continuous screw-rotating machines use many molds. The molds are usually located on a rotating circular platform that can operate as Ferris wheels or carousals. Feeding a melt (through special nozzle adapters to the contour of the molds) onto a rotating mold is not new. Products made include shoe soles, shoe sandals, boats, Velcro® strips, electrical male and female mini-connectors, and photographic film containers.

TANDEM MACHINE MOLDING

When a large enough IMM is not available and/or limited production exists, two IMMs side by side can operate in tandem. A large mold is located across both sets of platens.

MICROMOLDING

Overview

Micromolding is precision molding of extremely small parts and components, usually in unfilled, filled, and/or reinforced engineering plastics. Parts usually weigh less than 20 milligrams (0.020 g), with some even less than 0.1 mg and others as small as 1 mm^3. Products are measured in microns and have tolerances of $\pm$10 microns or less. A micron (μm) is one-millionth of a meter; 25.4 μm make up one-thousandth of an inch. In comparison, a human hair is 50 to 100 μm in diameter. A mil, which is about 25 times smaller than a micron, is one-thousandth of an inch.

To date, no standard micromolding definition has been adopted. Some refer to a micromolded part as anything smaller than a single plastic pellet or weighing less than a tenth of a gram, even though micromolding technology can do better by turning out parts just a few ten-thousandths of an inch thick and weighing less than a thousandth of a gram.

Micromolders predict that demand and technology will continue to drive part sizes down, probably to the point where molds cannot be used for molding. New processing technology is being used to ensure that the right amount of plastic enters the mold.

Molding small parts requires material moving in and out of cavities fast so that degradation and so forth does not occur. IMMs operate at very high injection pressures. Cycle times typically are about 1 to 8 seconds. Parts are produced with zero flash. They usually mold in multistage vertical or horizontal specialty IMMs (plasticating slanted screw, vertical dosing unit, horizontal injection plunger, etc.).

IMMs have been used in a clean room, molding 0.8-mg acetal watch gears, 2-mg PC housing for hearing-aid implants, 16-mg-glass-reinforced LCP automotive microswitch actuator pins, and so on (chapter 1). Care has to be taken in handling parts. Some parts are so small and lightweight that static electricity can make them float in the air.

To obtain proper filling of microstructured parts, significant modifications to the standard process are made. Because of the extremely low surface roughness of the mold cavity walls, demolding without a standard draft of 3° becomes feasible. However, during demolding any lateral offset has to be avoided. Otherwise, structures are ripped or sheared off the ground plate. To support the filling of small cavities, especially with a high aspect ratio (height against smallest lateral dimension) so-called variothermal heating is used (205).

In contrast to standard cooling, which keeps the mold temperature at a certain temperature below transition temperature, the surface of the mold is heated up with the help of inductive heating almost to the melt temperature in order to gain a lower melt viscosity during filling. Compressed air causes problems, so the air in the mold must be evacuated by a vacuum pump. This is necessary to provide complete part filling as well as to prevent the burning of plastic by the compressed hot air at the bottom of the cavity.

Molding machines and tooling for small parts are not just smaller versions of their regular larger molding counterparts. IM in micron sizes can be performed in specially designed IMMs and molds (265, 381). Some micromolders use screw-and-barrel molding machines similar to conventional machines except that the components are much smaller. There are plasticators (screw-and-barrel injection systems) with diameters that range from 14 to 20 mm and nozzle orifices of 1.5 mm.

To mold micron-scale precision parts with shot weights of only 0.0022 g, electrical operating IMMs are being used. Those who desire precise shot size (dosing) can buy electric injection machines with servomotors capable of very accurate screw positioning. With a hydraulic IMM, the molder can install a valve gate that shuts when the correct amount of plastic has been injected into the mold.

Precision molding involves proper process control, particularly high injection speed and extremely low residence time. Proper venting usually has to include precision venting in the cavity as well as possibly removing air prior to entering the cavity. Product handling has to be considered. Approaches used include floating air systems, parts placed on film carriers, molded-in carrier strips made of the same material followed with automated separation, and so on.

The ideal types of plastics to process are those with high melt flow indexes, low viscosity at processing temperatures, and consistent melt flow from one lot to the next from a material supplier. Reinforced plastics can set up problems, such as meeting dimensions and/or tolerances and damage to molds.

Though material costs are usually negligible, to date the cost of specialized machines, molds, and secondary equipment makes micromolding an expensive process when compared with larger fabricated products.

SUMMARY

To manufacture microparts as small as 0.0008 g, the molding industry depends on new designs of IMMs, molds, and part-handling systems. The key to molding on this scale is getting the right amount of plastic into the mold. When molders do not accurately dose their shots, they risk overpacking their parts, which can result in the sticking of parts in the mold, flash, and/or destruction of the parts themselves.

A new option for micromolders is an injection machine that replaces the traditional screw with a motor-driven plunger. Named for the part sizes it produces, the Sesame is a small electropneumatic micromolding machine made by Hull Corp. (Warminster, Pennsylvania). The machine can mold twenty thermoplastic "sesame seeds" from a single pellet. Instead of a screwing action, the Sesame uses a tiny plunger to push material into the mold. The plunger, which can be as small as 1.5 mm in diameter, is driven by an electric servomotor. To ensure accurate shot sizes, the servomotor can control plunger position to within 5 μm. Total injection time can be as quick as 0.020 second (Medical Murray Inc., Buffalo Grove, Illinois, developed the Sesame technology and has licensed it to Hull).

To keep plastic flowing through its small passages, the Sesame relies on a combination of high injection pressures (up to 50,000 psi) and high melt temperatures. High temperatures degrade materials during the time they spend in a conventional machine, but the Sesame is designed to minimize residence time. Material is in the Sesame for only a couple of minutes compared with a couple of hours in a regular IMM. What accounts for the difference in residence times is that the Sesame holds an extremely small amount of material, far less than the smallest screw-and-barrel machine. A short screw would hold twenty times more material than the Sesame. The less material there is to heat, the shorter the required residence time.

A conventional method for removing material from a machine before it degrades is using a large runner and sprue. Sometimes the runner and sprue are so large that the part material makes up less than 1% of the shot. That means that more than 99% of the shot material is wasted. Worse, the manufacturer loses control of the part-molding process. The sprue and runner are molded, with the part becoming a kind of by-product.

The Sesame is designed so that molders can use a smaller runner and sprue, which gives them more control over the amount of plastic and pressure used to form the part itself. A smaller runner and sprue also means less material waste. While screw-and-barrel systems waste as much as 99.7% of the shot material, the Sesame wastes less than 80%. This is particularly important when molding expensive materials like biodegradable plastics, which cost as much as $10 per gram. The Sesame can handle any type of moldable plastic, as well as silicone rubber. Super-small medical parts that have been molded by the machine include (a) polyethylene catheter tips with a volume of 0.16 mm^3, (b) 0.0001 g thermoplastic elastomer tubes for a microsurgery device, and (c) silicone rubber tear-duct plugs with an outside diameter of 0.61 mm.

Technology advances are working into ideas about where the basic polymer raw materials would be inserted into molds. Different monomers could be separately injected into molds, where they would be subjected to form polymers (chapter 1). There is also the concept of using several lasers in a box of gas. At the point where the laser beams intersect, the energy would produce a reaction that causes a part to form without a mold (381).

MONOSANDWICH MOLDING

Two-component injection machines, using separate injection units for each of the two components, perform the standard sandwich molding process. This process is characterized by a sequential injection of the two components using the same gating system. After filling the mold partially with the skin material with one injection unit, the core component will be injected by the second injection unit. In order to avoid flow marks on the part surface, a simultaneous phase is inserted between the injection of skin and core components. The final packing phase may be performed either with the core or the skin component.

In 1992, a new sandwich injection molding concept, called monosandwich, was developed by Jaroschek and Thoma (Ferromatic Milacron). In contrast to the standard sandwich molding process, this technology uses only one injection unit for both skin and core components with an extruder (Table 4.55).

The cycle starts with the plastification of the core component in the injection unit. The extruder moves to the bottom position, then the injection unit moves forward to the extruder nozzle to link the nozzles of the extruder and the injection unit. The extruder starts plastification of the skin component and extrudes the melted skin component into the screw antechamber of the injection unit. Thus, the skin and core components are located one after the other in the screw antechamber. After the extruder moves back to the top position, the injection unit moves forward to the mold followed by a conventional filling phase. Because of the fountain flow effect, the first injected material forms the skin layer followed by the second component forming the core. Compared with the standard sandwich process, the injection phase of the monosandwich process is less complicated as it is identical to the conventional injection-molding process (306).

DOUBLE-DAYLIGHT MOLDING

This type of molding combines hydroelastic metal forming on the moving platen and a hot runner IM system on the fixed platen (Arburg GmbH). A center mold plate separates the two processes and is supported by guide arms on the tie bars. During processing a robot loads a metal blank into the hydroform section and moves the shaped blank from the previous cycle to the IM side. The mold's closing action drives the hydroforming process with the already shaped blank overmolded with plastics.

FOAMED GAS COUNTER PRESSURE MOLDING

An airtight sealed mold can be pressurized to 2.8 to 3.5 MPa (400 to 500 psi) with an inert gas, usually nitrogen, or enough pressure to suppress foaming as the plastic mix enters the cavity. After

The use of multiple materials and processes is the overarching technology in achieving one-hot manufacturing for large and/or complex parts. Material thermal expansion differences can be dealt with by flexible joints, (adhesives), process sequence to minimize distortion, sliding at interfaces (incompatible materials), and design for minimum distortion. Common multimaterial multiprocess technologies include Injection Compression Molding in which resin is injected into a partially open mold which closes, requiring less clamp force and producing less residual stress in the part. Multimaterial Molding in which a material is shaped, the mold is altered, and a second (or subsequent) material is shaped. Shaping processes are combinations of injection and compression molding and stamping. In-Mold Coating in which a thin thermoset coating is injected onto an injection or compression molded part in a closed mold. 'Mono-Sandwich' Coinjection Injection Molding in which a small extruder, operated intermittently, pumps a skin layer into the front of the main injection unit for subsequent coinjection. The Alpha 1 machine at GE Plastics, with two injection units, a long stroke vertical press, and shuttle table, allows combinations of compression molding, (gas-assisted) injection molding, and stamping.

Advantages/Disadvantages	Applications	Materials
The advantages of more than one material and/or process include design flexibility, tailored performance, effective material use, lower labor costs, improved quality through automation, reduced secondary operations, less auxiliary equipment, and more recycle use. Multiple materials allow advantageous combinations such as multiple colors (automotive lens), flexible/rigid (conduits with connectors), and consolidated/strong (plastic/metal composite) and cost /barrier or strength (laminate structure).	Telephone booth molded from 132 lbs of structural foam on a 2500-ton press with three injection units Air vent with molded movable louvers made from incompatible materials. Automotive bumper with injection molded fascia over a stamped beam. Multi-color automotive taillights.	Combinations of thermoplastic, thermoset, and reinforcements subject to constraints of product performance, limitations on distortion, and interface requirements (adherent or incompatible)

Table 4.55 Multimaterial multipurpose technology

the measured shot is injected, pressure is released, allowing the instantaneous foaming to form a core between the already formed solid skins. A conventional IM system is used with a separate entrance to the mold cavity providing gas pressurization prior to injecting the melt shot. This back-pressure action can provide an even distribution of melt packing during its cooling cycle. When producing foamed plastic parts, this gas back pressure prevents the blowing agent from expanding until its part skins form followed with releasing (venting) the gas pressure. Controlled foam expansion is provided.

HIGH-PRESSURE FOAM MOLDING

The plastic mix is injected into the mold, creating a cavity pressure higher than the blowing agent gas pressure; usually, it is much higher, ranging from 34.5 to 138 MPa (5,000 to 20,000 psi). As soon as the part skin surface hardens to the desired thickness, cavity melt pressure is reduced to produce a foamed core. Different techniques are used to release pressure, such as withdrawing core pins in the cavity or by special press motions that partially open the mold halves using 2-D or 3-D motions (chapter 8).

LOW-PRESSURE FOAM MOLDING

Low-pressure or short-shot conventional IM are the most commonly used because they are easy and simple to operate. They are suited for economical production, particularly of large, complex 3-D parts. Usual cavity pressure is 1.4 to 3.5 MPa (200 to 500 psi). An accumulator can be used between the plasticator and mold, providing a means of injecting a very large shot. Since low pressures are used, the parts' surface appearance does not approach those in processes using higher pressures. Attractive surface patterns can be molded, such as swirls.

LIQUID MOLDING

Also called reaction injection molding (RIM), the liquid injection molding (LIM) process involves proportioning, mixing, and dispensing two liquid plastic formulations. This compound is directed into a closed mold. It can be used for encapsulating electrical and electronic devices, decorative ornaments, medical devices, auto parts, and so on. It uses a mechanical mixer rather than a RIM high-pressure impingement mixer. Flushing the mix at the end of a run is easily handled automatically. Plastics used include silicones, acrylics, and so on. To avoid liquid injection hardware from becoming plugged with plastics, consider using a spring-loaded pin-type nozzle. The spring loading allows you to set the pressure so that it is higher than the pressure inside the extruder barrel, thus keeping the port clean and open.

COUNTERFLOW MOLDING

This is used to eliminate or reduce weld lines. Two separate injectors or one injector with a splitting device are used to move melt in and out of the cavity from opposite sizes. This type of action repeats and is programmed to maximize the best melt flow patterns.

MELT FLOW OSCILLATION MOLDING

Scorim Process or SP (Cinpres-Scorim), Rheomolding Process or RP (Thermold's), and the Press Alpha Process or PAP (Sumitomo Heavy Industries and Sankyo Chemical Engineering of Japan) processes are examples of this method. The SP is a multilive feed molding process in which two packing pistons oscillate 180° out of phase and eliminate weld lines and so forth. The RP system provides 3-D orientation based on the concept of melt rheology as a function of vibration frequency and amplitude as well as temperature and pressure. The equipment utilizes piston-type melt accumulators set up adjacent to the melt stream of the plasticator; the piston oscillates back and forth. The PAP system uses compression pins that are actuated when the cavity fills. These pins protrude into the cavity and begin oscillating to create localized compressions. The action eliminates weld lines, sinks, and warpage; reduces filling pressures; minimizes localized thin-wall molding; and allows for gate positioning flexibility.

SCREWLESS MOLDING

Practically all IMMs worldwide use screws to melt plastics. However, the originally designed ram hydraulic IMM (screwless) of over a century ago has limited use (Fig. 4.60). This IMM ram can process certain plastics with very little or no melt flow capability that a screw system cannot. Ultrahigh molecular weight polyethylene (UHMWPE) with its rather exceptional superior properties, such as exceptional wear resistance, is ram injection molded (also compression press sintered or ram extruded). For long runs of fairly small products (2 to 300 g) with complicated shapes, conventional screw injection molding can be used as long as one recognizes that this highly viscous plastic can cause high pressure loss along the melt flow path through the plasticator. The molecular weight and end properties can be severely degraded by the screw.

NONPLASTIC MOLDING

These processing methods go by different names, such as metal injection molding (MIM), ceramic injection molding (CIM), and powder injection molding (PIM). They include aluminum, ceramics, concrete, copper, dynamite, food, magnesium, stainless steel, wood, zinc, tungsten carbide, and other alloys of metals. Most of the materials have been in very fine powdered form. These green materials have been used alone but usually with plastic binders (Fig. 4.61).

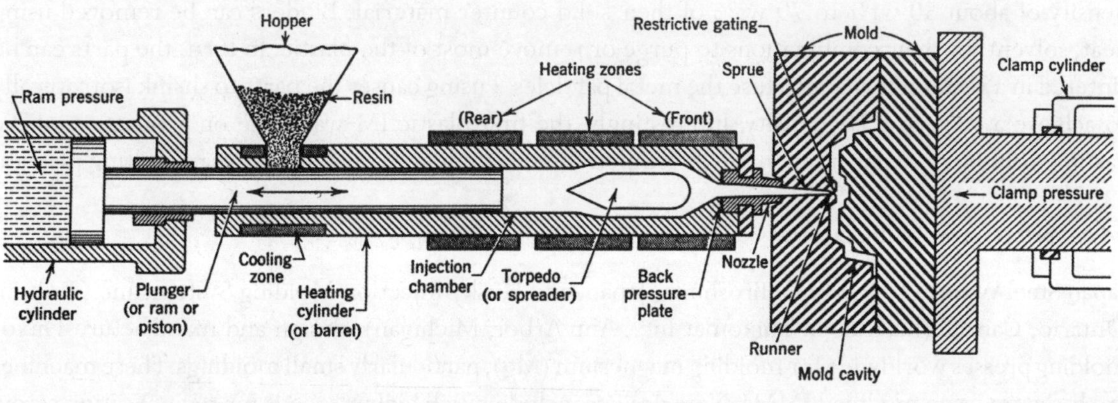

Figure 4.60 Schematic of a ram (plunger) injection molding machine

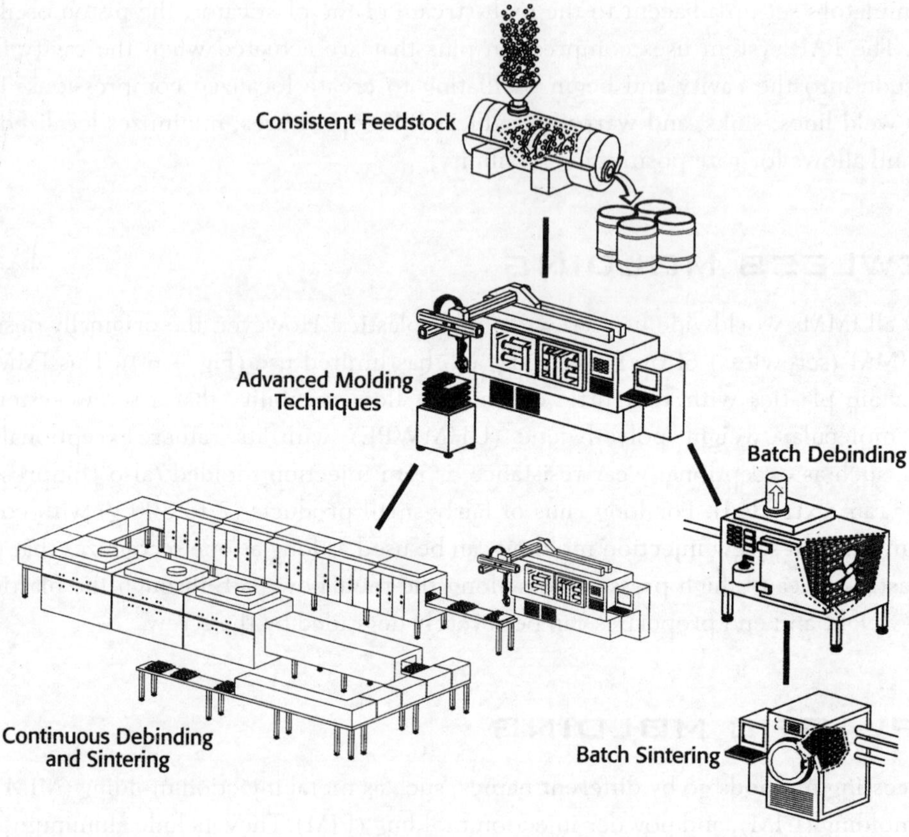

Figure 4.61 Metal injection-molding cycle (courtesy of Phillips Plastics)

Precise and complex parts have been molded that are usually small in size, such as micro size. Micromolding requires more accuracy than macromolding. The green material can have a high bulk density of about 50 wt% to 70 wt% of their solid counter material. Binders can be removed using heat, solvent, or their combinations to purge or remove most of the plastic. In turn, the parts can be sintered in a vacuum furnace to fuse the metal particles. Fusing causes the parts to shrink isotropically to achieve a 95% to 99% density. Interestingly, the first plastic IM was done on a variation of die casting machinery (using a much lower temperature melting pot that is higher than melting plastic).

MAGNESIUM MOLDING

Japan Steel Works Ltd. (JSW), Hiroshima, Japan, and Husky Injection Molding Systems Inc., Bolton, Ontario, Canada (licensee of Thixomat Inc., Ann Arbor, Michigan), design and manufacture Thixomolding presses worldwide for molding magnesium (Mg), particularly small moldings. These machines are based on using modified IMMs. New designs include much higher speeds for reciprocating screw action and higher heated barrel zones. For injection, Mg has to blow out a magnesium plug that is in a semi-solidified state. Owing to its very high injection speed (compares to firing a cannon), the machine is subjected to very high shock and vibration, requiring a much stronger machine than a plastic IMM.

Thixotropic Molding

The patented thixotropic technology is called Thixomolding. Traditional die casting machines use a large pot of molten Mg that can be hazardous to workers and creates sludge that is foundry waste. By contrast, thixotropic IMMs use chips of Mg similar to plastic pellets. All the heating (up to 1,200°F) and melting is done in the barrel, totally enclosed and airtight to keep oxygen away from the Mg, which can catch on fire in its molten stage. The screw does not melt the Mg; it just conveys it forward and breaks up the jagged dendrites into round shapes that can be pushed into the mold with ease. Injection speeds can reach 3,000 mm/s. After each molding shot, the mold cavity has to be sprayed with a mold release agent to eliminate the Mg fusing to the cavity.

SUMMARY

As IM technology expands, requirements for precision injection-precise IMMs have followed suit. Profitable molding in today's highly technical and competitive world depends on precise clamping action coupled with both high plasticizing performance and reliability. This review has examined basic physical processes affecting the speed for productivity of an IMM. These are the processes that must occur in any IMM regardless of its specific design or of the refinements it contains in control systems or hydraulic circuitry.

Today, the actual IMs that may be found on the market come in a myriad of sizes and designs. Because it is possible to combine a variety of different sizes of injection or plasticating systems with almost as many different sizes of clamp systems, there is almost no limit to the number of combinations available considering both the domestic and foreign sources of equipment. However, in every case, each machine must provide the necessary facilities to melt the plastic, inject it, and cool it.

A glimpse of future IMMs has been introduced that will use laser and microwave plasticators, fiber-optics monitoring, quiet electromagnetic drives, and voice-activated controls; permit quick plastic changes without purging; eliminate hoppers by storing plastics in modular tanks on the machine's bed and feeding by vacuum pumps behind the plasticators; and introduce more innovations. Benefits to be gained include more energy savings, increased process efficiencies, simplified controls so that IMMs will be easier to operate, and improved repeatability of melts. A program called Mother Project was started in 1999 and is targeted to be completed by 2017. Studies are being conducted by MIR, SpA, Italy, in cooperation with the University of Turin's Plasturgy Department. The U.S. agent is MIR USA, Leominster, Massachusetts.

CHAPTER 5

EXTRUSION

INTRODUCTION

Extrusion is the process of heating practically only thermoplastics (TPs) that may be in the shape of powders, beads, flakes, pellets, or combinations of these forms. This plastic enters the extruder's hopper. The extruder utilizes a plasticator (spiral screw that rotates within a heated barrel [cylinder]) to melt the plastic. The melted plastic is then forced through a die to produce the desired continuous product shape. Figure 5.1 shows a very simplified schematic of the extrusion process. Figures 5.2 and 5.3 highlight the major parts of a single-screw extruder.

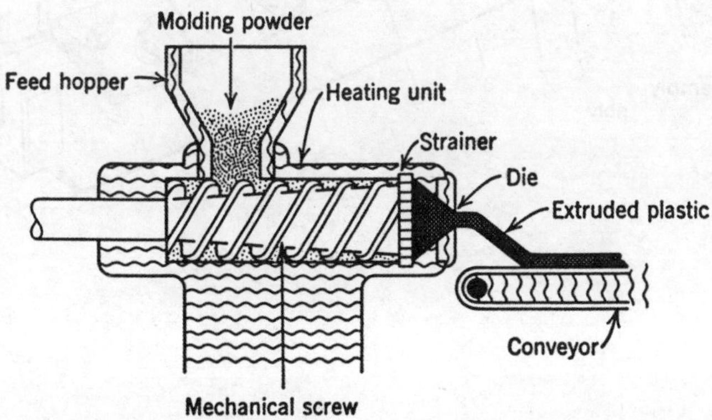

Figure 5.1 Basic concept of extrusion process

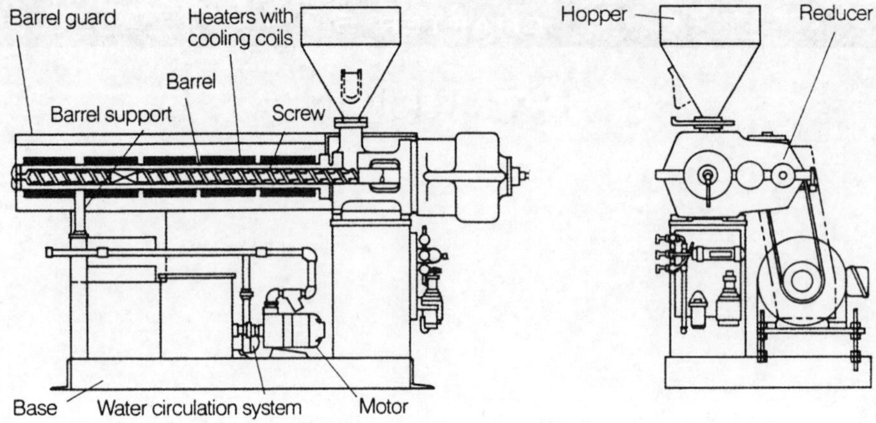

Figure 5.2 Simplified example of a single-screw extruder

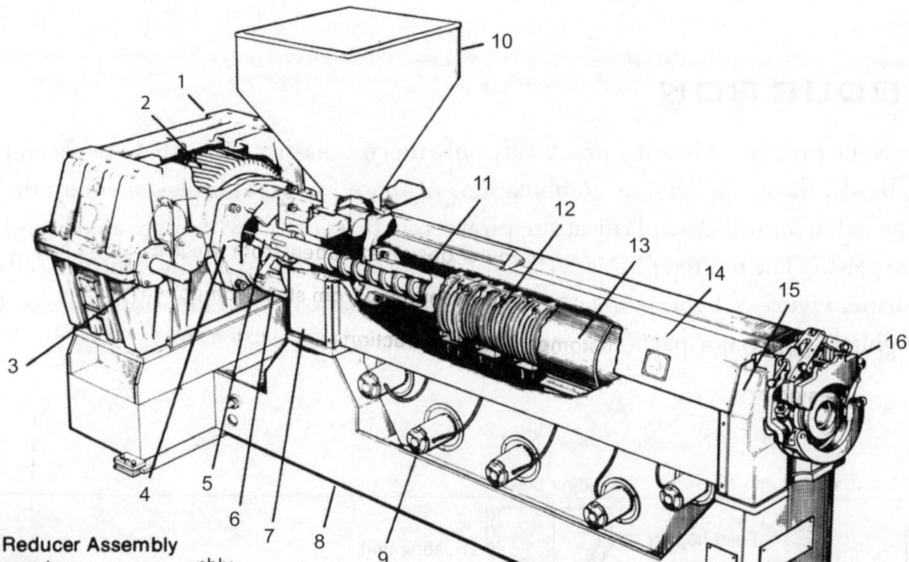

1. Gear Reducer Assembly
2. Low speed gear
3. Input shaft
4. Thrust housing
5. Stock screw
6. Spacer
7. Feed section
8. Extruder base
9. Blower motor
10. Hopper assembly
11. Barrel
12. Heater
13. Inner hood
14. Girt
15. Barrel support
16. Clamp assembly

Figure 5.3 Detailed summary of an extruder (courtesy of Davis Standard)

Extrusion is the single most popular process for forming thermoplastics (TPs). It processes over 36% of all plastics consumed worldwide. Most of plastics used are the polyethylenes to produce film and other products (chapter 2). Other plastics are used. About 45% of all polypropylenes go through extruders. Unlike injection molding, the second major process consuming plastics (32%), extrusion usually results in semifinished or intermediate products. They require further processing to produce industry-usable products (221).

Extruder designs principally include rotating single- and twin-screw designs. Extruders can be classified as (a) continuous with single-screws (single and multistage) or multiscrews (twin-screw, etc.), (b) continuous disks or drums that use viscous drag melt actions (disk packs, drums, etc.) or elastic melt actions (screwless, etc.), and (c) discontinuous that use ram actions (very low viscosity thermoplastics [TPs], thermoset [TS] plastics, and rubbers/elastomers) and reciprocating actions (injection molding, etc.).

Extrusion is a continuous process that involves forming a product in two dimensions. These X and Y dimensions determine the cross-sectional form of the extrudate (exiting plastic from the extruder die), which can be almost unlimited in scope, ranging from rather simple films to very complex forms used in window profiles and so on. The third dimension (Z) is the length of the extrudate. In principle, it can be infinite. In fact, it is limited by practical considerations governing winding, reeling, storage, or transport.

The product cross-section is formed in a die; the extrusion process consists of raising a TP to its melt temperature and forcing it through this die at a constant rate. The heating and pressurizing device involves one or more screws operating in a heated barrel and is known as a plasticator (chapter 3). Downstream of the die, the extrudate is calibrated, cooled, and packaged by an array of ancillary auxiliary devices, including vacuum calibrators, water tanks, cooling rolls, haul-offs, cutters, and winders (chapter 18). Upstream of the die, a melt pump may be interposed between the extruder and the die. The exact selection and arrangement of these component parts in an extrusion system will depend on the end product and tolerance requirements that have to be met. The principle of the extruder is common to all extrusion process lines, such as fabricating films, sheets, pipes, tubes, profiles, and fibers (9, 221, 252, 263, 332, 450, 451, 453, 456, 486, 500).

Figure 5.4 is an example of a single-screw plasticator extruder that produces coextruded sheet. Figure 5.5 is a twin-screw extruder that produces custom profiles, including a vacuum calibration table.

Different sizes and types of extruders are manufactured by different companies worldwide. Table 5.1 lists some of these extruder manufacturers (501).

The basic processing concept has plastic material passing from a hopper into a plasticating cylinder (chapter 2) in which it is melted and moved forward by the rotating screw. When the melt reaches the end of the cylinder, it is forced through a die that gives the desired shape with no break in continuity. Quality plastic extrusions require maximizing homogeneity in terms of the melts' heat profiles and mixes, accurate and sustained flow rates, a good die designs, and accurately controlled downstream equipment for cooling and handling the products.

Figure 5.4 Coextruder sheet line showing two single-screw plasticators feeding melts to its flat sheet die (courtesy of Welex Inc.)

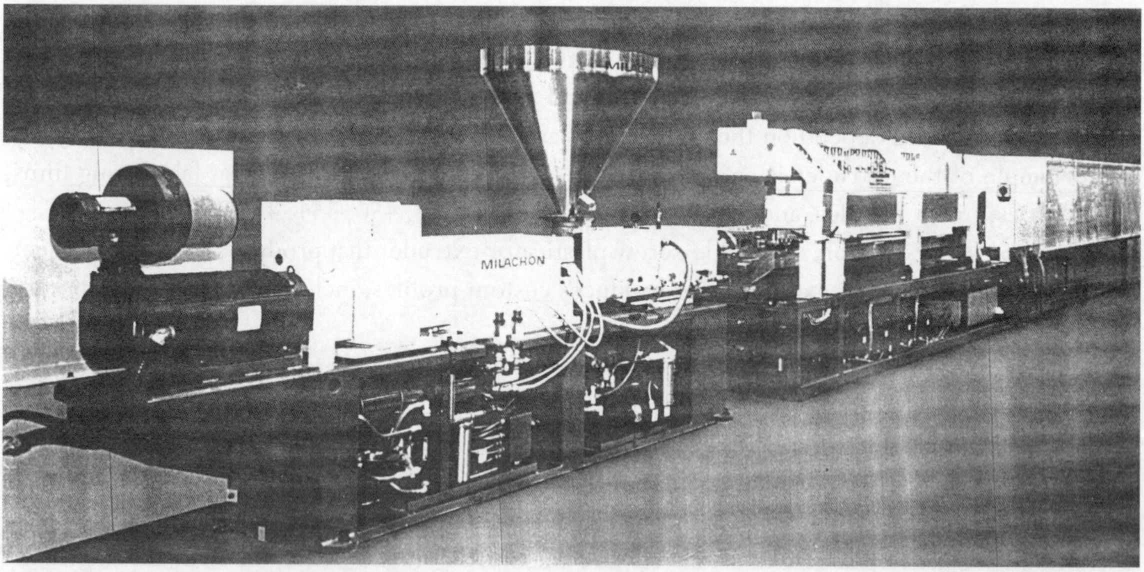

Figure 5.5 Twin-screw profile extruder line that includes a vacuum calibration table (courtesy of Milacron)

COMPANY	See Ad on Page	Reader Service Number	0-25 mm	26-75 mm	76-150 mm	151 + mm	Single screw	Vented	Twin-screw, co-rotating	Twin-screw, counter-rotating	Multiple-screw, planetary	Recycling	Laboratory	Compounding/pelletizing	Production	Pipe	Profile	Tubing	Blown film	Cast film	Sheet	Coating	Wire & cable	Parison (blowmolding)	Foam	WEBSITE
			SIZE				CONFIGURATION						FUNCTION			END PRODUCT										
			1	2	3	4	5	6	7	8	9	10	11	12	13	14	15	16	17	18	19	20	21	22	23	
Aaron Equipment Co., IL, (630) 350-2200			1	2	3	4	5	6	7	8	9	10	11	12	13	14	15	16	17	18	19	20	21	22	23	www.aaronequipment.com
Adept Corp., PA, (800) 451-2254, (717) 266-3606					3					8				12			15									www.adeptcorp.com
A+G Extrusion Technology GmbH, Austria +43 7586 6010			1	2			5								13	14	15					20			23	www.gruberextrusion.com
Akron Milacron, OH, (513) 536-3303			1	2	3	4	5	6				10	11	12	13	14	15	16	17	18	19	20	21		23	www.milacron.com
American Kuhne Inc., CT, (860) 886-7745			1	2	3	4	5	6				10	11	12	13	14	15	16	17	18	19	20	21	22	23	www.americankuhne.com
Axon Plastics Machinery AB, Sweden, +46 4257080			1	2	3		5	6				10	11	12	13	14	15	16	17	18	19	20	21			www.axonmachinery.se
B&P Process Equipment & Systems, MI, (989) 757-1300			1	2	3	4	5		7			10	11	12	13											www.bpprocess.com
Battenfeld Gloucester Engineering Co., MA (978) 282-1800	35	39	1	2	3	4	5	6							13				17	18	19				23	www.bge.battenfeld.com
Berlyn Extruders Inc., MA, (800) EXTRUDE			1	2	3	4	5	6				10	11	12	13	14	15	16	17	18	19	20	21		23	www.berlyn.com
Boston Matthews Inc., NJ, (201) 767-7111			1	2	3		5	6						12	13	14	15	16					21			www.bostonmatthews.com
Brampton Engineering Inc., ON, (905) 793-3000				2	3		5								13	14		16	17	18		20	21	22		www.be-ca.com
CarolMac Corp., SC, (800) 772-5355, (864) 297-5355				2	3	4	5	6				10														www.carolmac.com
CPM Milacron Inc., OH, (513) 536-3303				2	3	4				8		10	11		13	14	15	16			19				23	www.milacron.com
Davis-Standard, CT, (860) 599-1010			1	2	3	4	5		7	8		10	11	12	13	14	15	16	17	18	19	20	21	22	23	www.davis-standard.com
Diamond America Corp., OH, (330) 633-9959			1	2	3		5	6				10	11	12	13	14	15	16	17	18	19	20	21	22	23	
Dual Spiral Systems Inc., ON, (905) 524-1679			1	2	3	4	5	6				10	11		13	14	15	16	17	18		20	21	22		www.dualspiralsystems.com
Future Design Inc., ON, (905) 361-9978			1	2			5						11		13				17						23	www.saturn2.com
Harrel Inc., CT, (203) 866-2573			1	2	3		5	6					11	12	13	14	15	16			19		21		23	www.harrel.com
Hosokawa Alpine American, MA, (508) 655-1123				2	3		5								13				17							www.halpine.com
HPM Division, OH, (419) 949-2206			1	2	3	4	5	6				10	11	12	13	14	15	16	17	18	19	20	21	22	23	www.taylorsind.com
Leistritz, NJ, (908) 685-2333			1	2	3			6	7	8		10	11	12	13	14	15	16	17	18	19	20	21		23	www.leistritz-extrusion.de
Processing Technologies Inc., IL, (630) 585-5800				2	3	4	5	6				10	11	12	13	14	15	16	17	18	19	20	21	22	23	www.ptiextruders.com
Randcastle Extrusion Systems Inc., NJ, (973) 239-1150			1	2			5	6				10	11	12	13		15	16	17	18	19	20	21		23	www.randcastle.com
Tex America Inc., NC, (704) 552-5404			1	2	3	4	5	6	7			10		12		14	15	16		18	19				23	www.texamerica.com
Thermoplastics Engineering Corp., MA, (800) 241-4311			1	2	3		5						11	12	13		15	16		18			21		23	www.thermoplaticseng.com
Theysohn America Inc., NC, (704) 455-8829			1	2	3	4			7				11	12	13											www.theysohn.com
Wayne Machine & Die Co., NJ, (973) 256-7374			1	2	3	4	5	6	7			10	11	12	13	14	15	16	17	18	19	20	21		23	www.waynemachine.com
Welex Inc., PA, (215) 542-8000				2	3	4	5	6					11	12	13		15	16	17		19		21			www.welex.com

Table 5.1 Examples of extruder manufacturers

On leaving the extruder, the melt (extrudate) is drawn by a pulling action or device, at which stage it is subject to cooling, usually by water and/or air blown from devices. This is an important aspect of downstream control if tight dimensional requirements exist or conservation of plastics is desired. The processor's target is to determine the tolerance required for the pull rate and to see that the downstream equipment meet the total line requirements. Even if tight dimensional requirements are not required, the probability is that better control of the pull speed will permit tighter tolerances so that a reduction in the material's output will occur, resulting in lower product cost.

By far, practically all extruders use screws in their plasticators to melt the plastics. However, the ram, a popular melting system for over a century, is used to process plastics that cannot be melted by a screw. The ram devices are essentially batch devices, and although it is possible to achieve a constant output by sequentially operating two or more rams, the method is of virtually no importance for practically all TPs. The ram is used for processing plastics that come close to not being meltable. Another possibility is the rotary extruder, a device in which rotating discs or rotors are used to generate shear. However, TP extrusion depends almost entirely on the rotating screw as a melt delivery device.

TPs are characterized by low thermal conductivity, high specific heat, and high melt viscosity. Preparation of a uniform homogeneous melt and its delivery at adequate pressure and a constant rate could pose considerable problems if not properly processed (chapter 3). The principal extruder variants are the single-screw and the twin-screw types. Of these, the single-screw extruder is by far the most versatile and popular in use.

The single-screw extruder consists essentially of a screw that rotates in an axially fixed position within the close-fitting bore of a barrel. Extruder sizes are identified by the inside diameter of their barrel. Sizes range from ¼- to 24-in diameters with the usual from 1 to 6 in (European sizes range from 20 to 600 mm with the usual from 25 to 159 mm).

The screw is motor driven through different devices, such as a gear-reduction train or belt to meet different performance and cost requirements. Figures 5.6 to 5.9 and Table 5.2 show examples of these mechanical devices. These gear reducers are rated in mechanical horsepower and thermal horsepower as defined by the American Gear Manufacturers (AGMA). The AGMA rating system is based on the understanding that not all gear reducers are used the same way. A service factor is assigned as an aid in sizing machines properly.

Support for the screw is by a thrust bearing that opposes the force exerted on the plastic melt (Fig. 5.10). A helical flight on the screw provides the drag-induced conveying motion that develops this force. The barrel is equipped with both heating and cooling means, and its downstream end is provided with an attachment device for a shaping die that determines the cross-section of the extruded product, known as an extrudate or extrusion.

The upstream or inlet end of the barrel is equipped with a feed throat or aperture in the barrel wall where plastic is input, generally in the form of granules, pellets, and so on (Figs. 5.11 to 5.14). During its passage along the helical screw flight, this material is heated by a combination of conducted heat received from the barrel and mechanical shear heat derived from the mixing and kneading action of the screw (chapter 3). The output rate of the extruder is a function of screw

Figure 5.6 Example of a motor-driven belt drive system (courtesy of Welex Inc.)

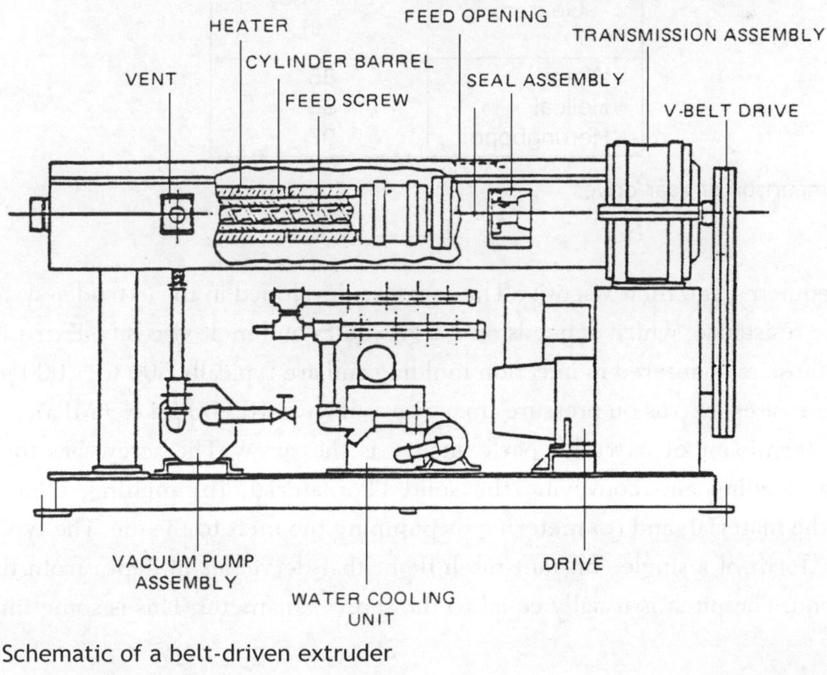

Figure 5.7 Schematic of a belt-driven extruder

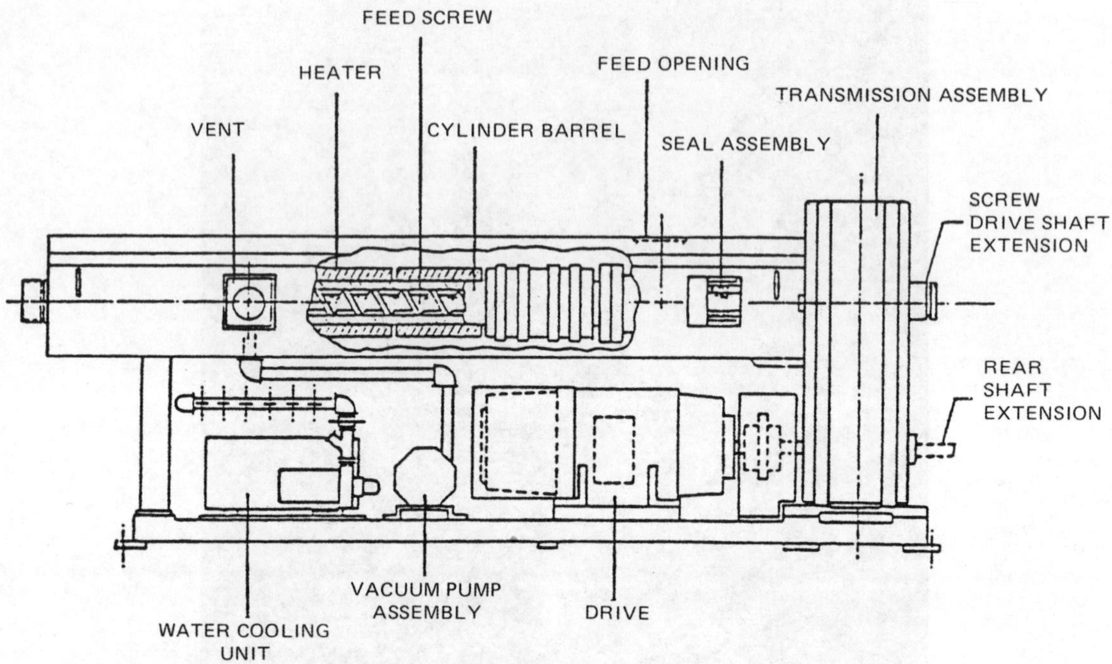

Figure 5.8 Schematic of a direct-driven extruder

Gearbox	Efficiency (%)
Worm	85
Helical	95
Herringbone	97

Table 5.2 Comparison of gear drives

speed, screw geometry, and melt viscosity. The pressure developed in the extruder system is largely a function of die resistance, which depends on die geometry and melt viscosity. Extrusion pressures are lower than those encountered in injection molding and are typically 500 to 5,000 psi (3.5 to 35 MPa). In extreme cases, extrusion pressures may rise as high as 10,000 psi (69 MPa).

The key determinant of extruder performance is the screw. The screw has three functions to perform: (a) feeding and conveying the solid TP material; (b) melting, compressing, and homogenizing the material; and (c) metering or pumping the melt to the die. The typical extruder screw takes the form of a single constant-pitch flight that decreases in depth from the input end to the output end. The pitch is usually equal to the screw's diameter. This is sometimes known as

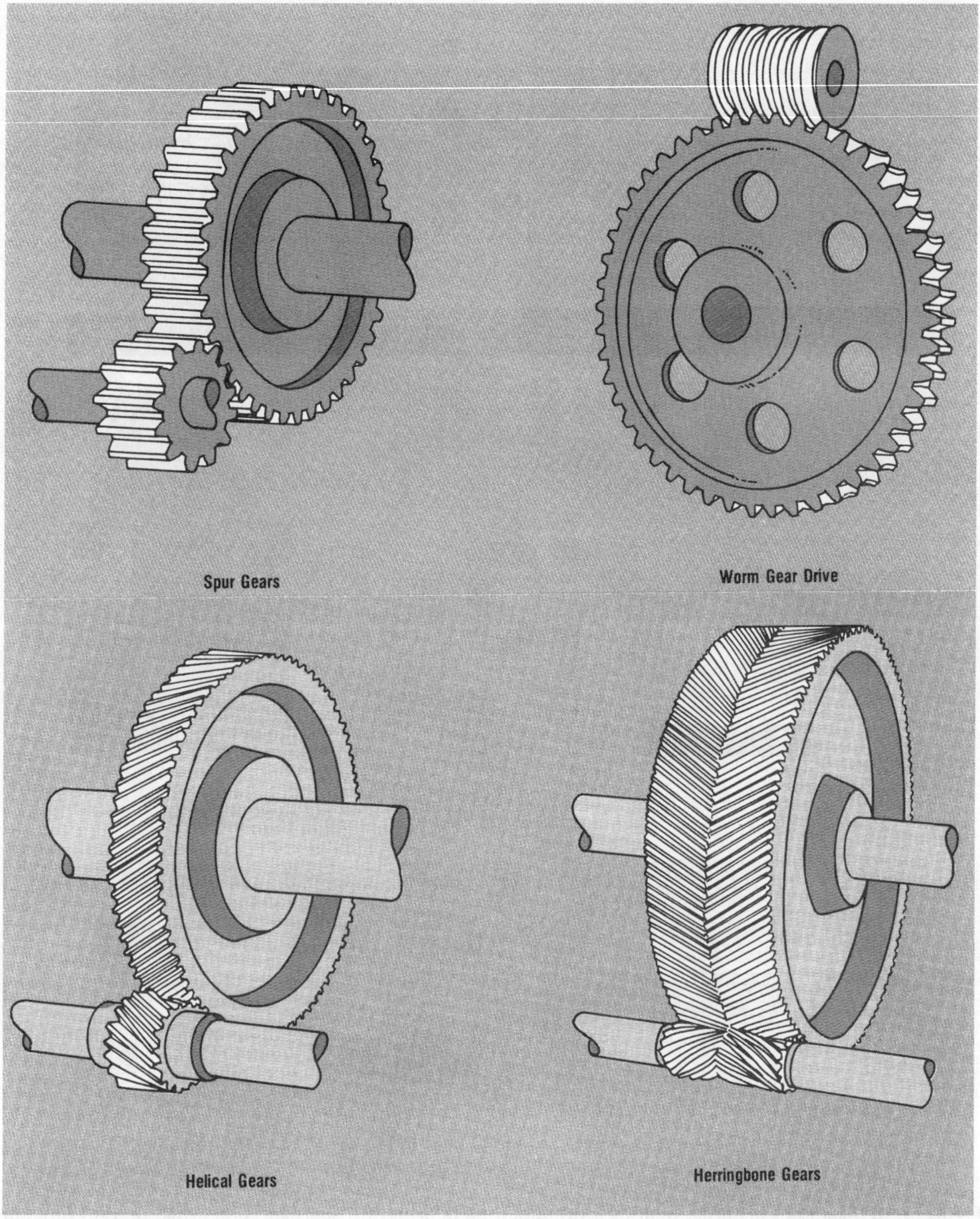

Figure 5.9 Various gear reducers

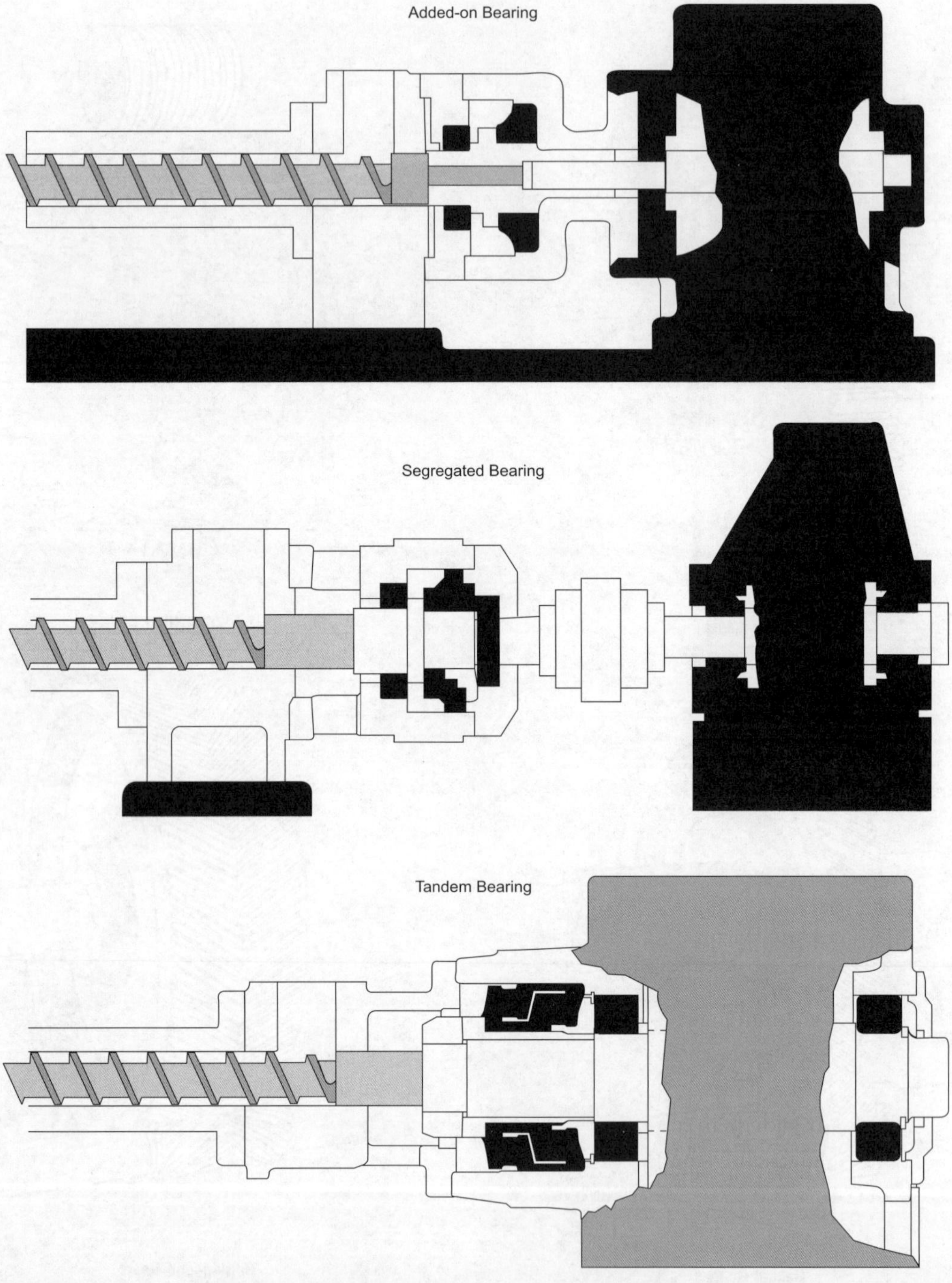

Figure 5.10 Examples of thrust bearings: (a) added-on bearing, (b) segregated bearing, and (c) tandem bearing

Figure 5.11 Example of an extruder with a crammer feeder to handle low-bulk plastics (courtesy of Welex Inc.)

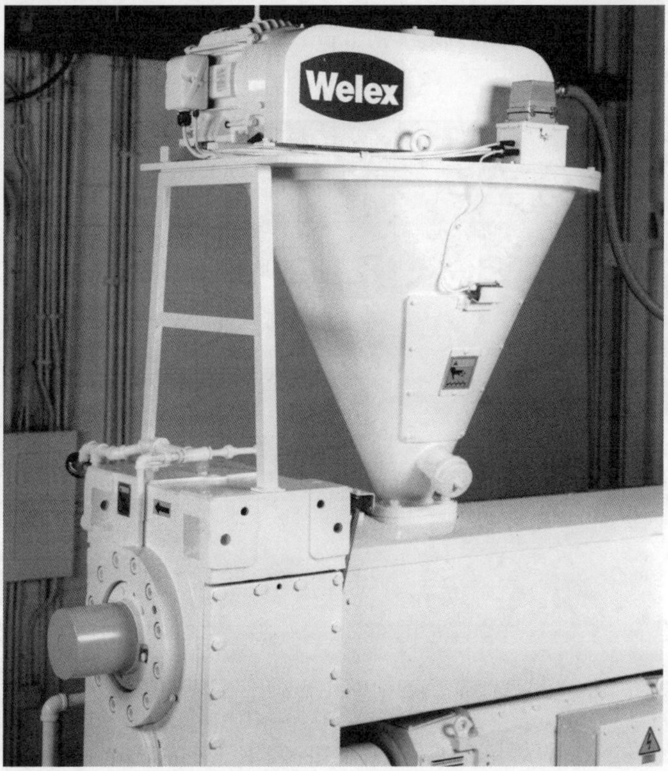

Figure 5.12 Close-up of extruder crammer feeder (courtesy of Welex Inc.)

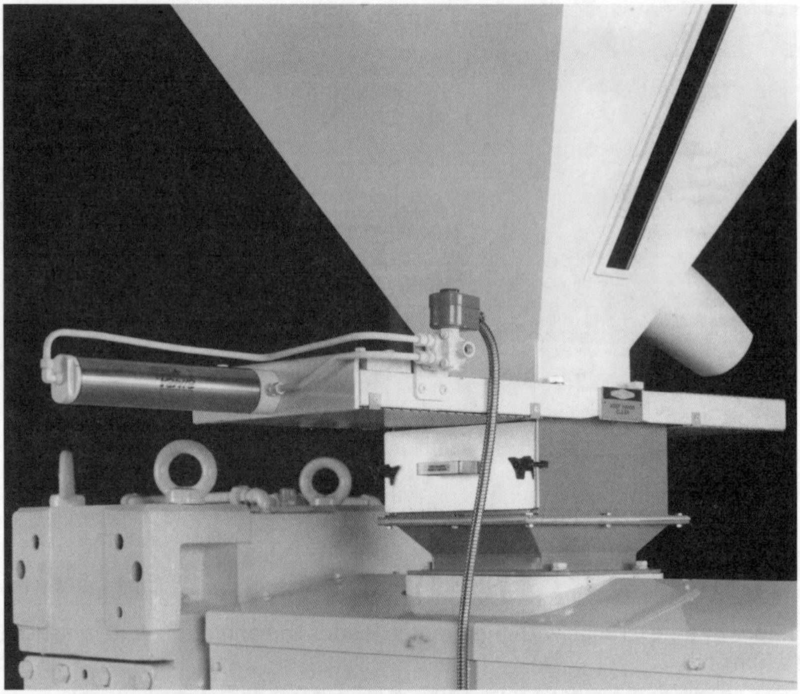

Figure 5.13 Example of extruder feed hopper with pneumatic sliding shutoff and magnetic drawer (courtesy of Welex Inc.)

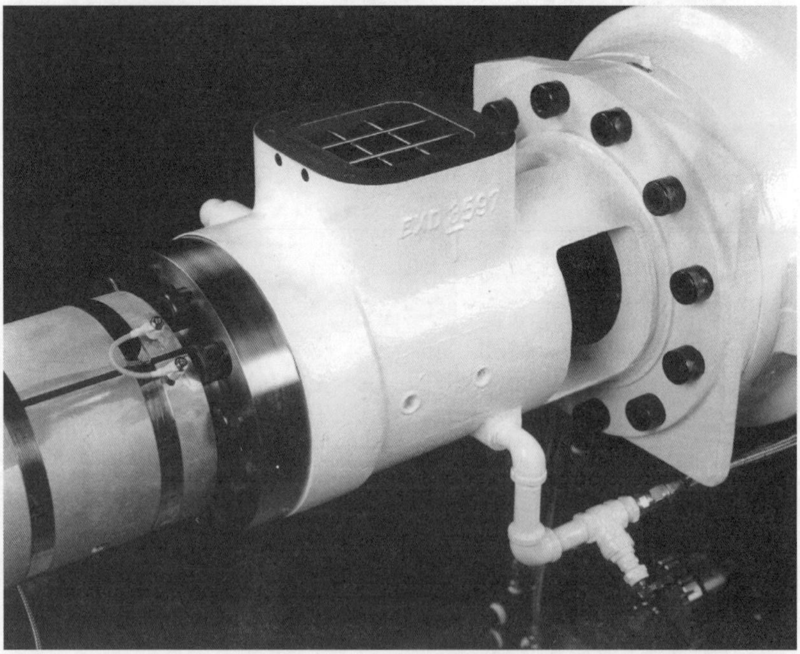

Figure 5.14 View of an extruder feed section with guards removed (courtesy of Welex Inc.)

a square pitch screw; the resulting helix angle is 17.8°. The screw features three sequential zones corresponding to the three functions of feeding, compression, and metering. Figure 5.15 features an extrusion screw where L = screw length, D = diameter, h = initial flight depth, h1 = final flight depth, L/D = length/diameter ratio, and h/hl = compression ratio.

The rotating screws in the barrels, which identify plasticators, are required to provide the best melts for processing the different melt behaviors of the different plastics. As reviewed in chapter 3, the important details of screw designs include flight depths, feeding/compression/metering zone lengths, compression ratios (ratio of the flight depths in the feeding and metering zones), L/D ratios (ratio of the screw length to its diameter), mixing elements, static elements, axial grooves and other devices to control/improve feeding in the hopper and barrel (Figs. 5.16 to 5.20), and so forth. Attention to the plasticator can improve the material's conveying performance.

Variants on the single screw include the barrier or melt-extraction screw and the vented screw. The barrier screw has separate input and output flights that overlap in the midsection of the screw where the plastic is only partially melted. The barrier flights are slightly smaller in diameter than

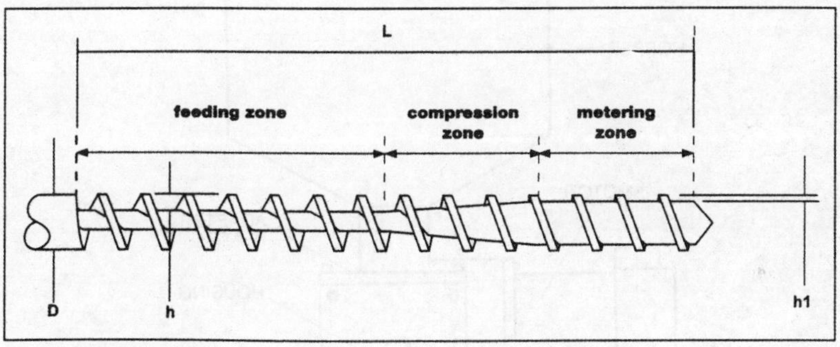

Figure 5.15 Example of an extrusion screw

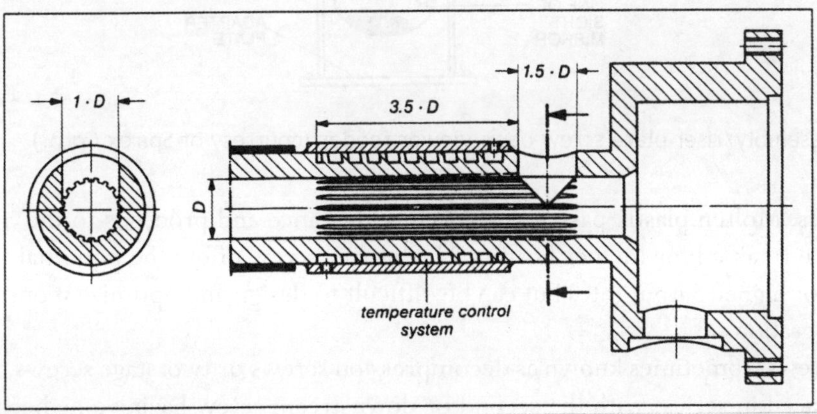

Figure 5.16 Example of a grooved feed section in a barrel

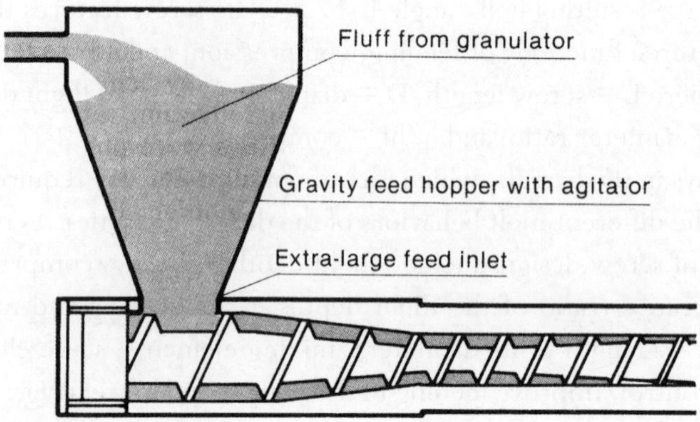

Figure 5.17 Dual-diameter barrel feed

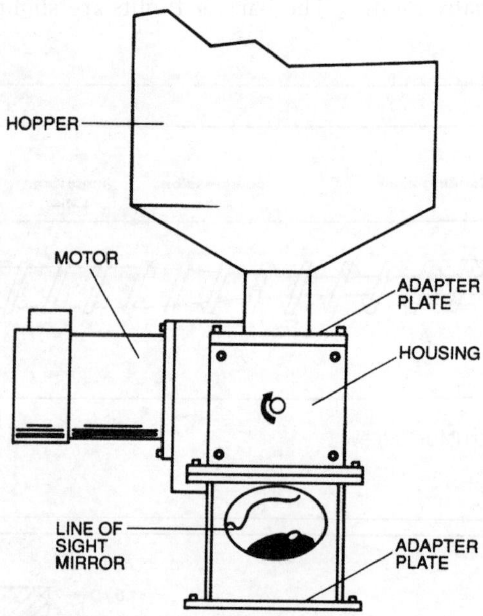

Figure 5.18 Assembly/riser plate screw open-viewer feeder (courtesy of Spirex Corp.)

the main flights. Molten plastic passes through this clearance and proceeds to the metering zone, while solids are retained for further heating. The barrier screw offers the potential for lower melt temperatures or higher output rates but can be difficult to design and optimize if one is not familiar with screw operations.

Vented screws, sometimes known as decompression screws or two-stage screws, are effectively two screws placed in series, with the second or downstream screw having a higher transport rate than the first (chapter 3). The upstream end of the screw is provided with conventional feeding,

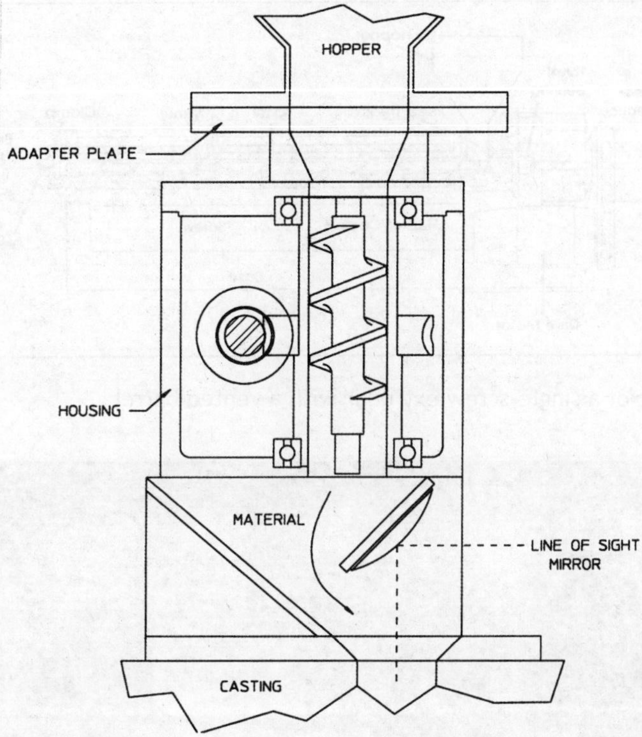

Figure 5.19 Controlled-feeding open-viewer feeder (courtesy of Spirex Corp.)

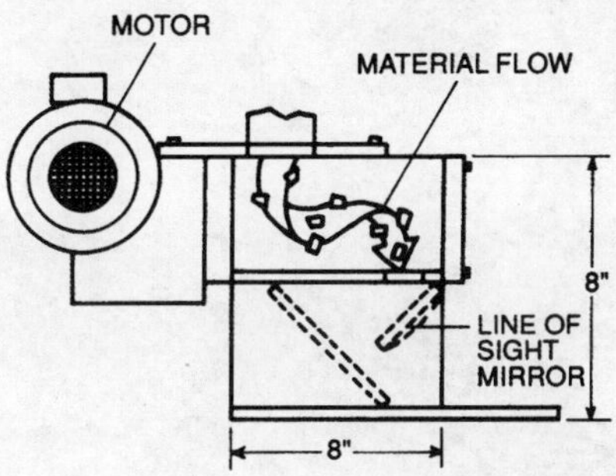

Figure 5.20 Material motor-speed-controlled open-viewer feeder (courtesy of Spirex Corp.)

compression, and metering zones; these are followed by a sudden increase in flight depth, then by further compression and metering zones. The midscrew increase in flight depth causes a sudden pressure drop in the plastic melt; this allows any dissolved volatiles to boil out. The lack of compression at this point, coupled with the differential transport rates of the two screw sections, makes it possible to extract the volatiles through a port in the extruder barrel without the plastic's melt emerging at the same time (Figs. 5.21 to 5.23).

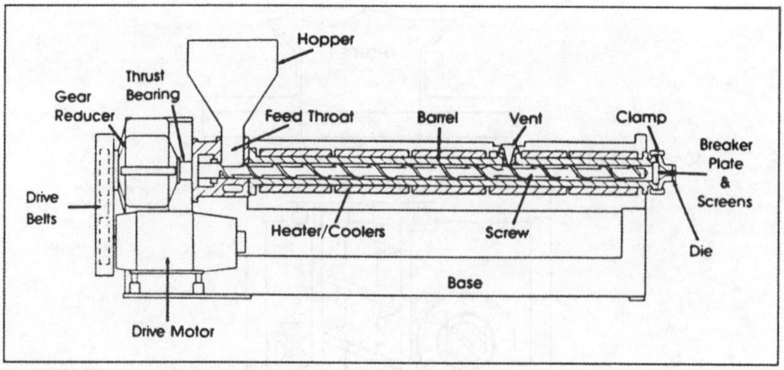

Figure 5.21 Schematic of a single-screw extruder with a vented barrel

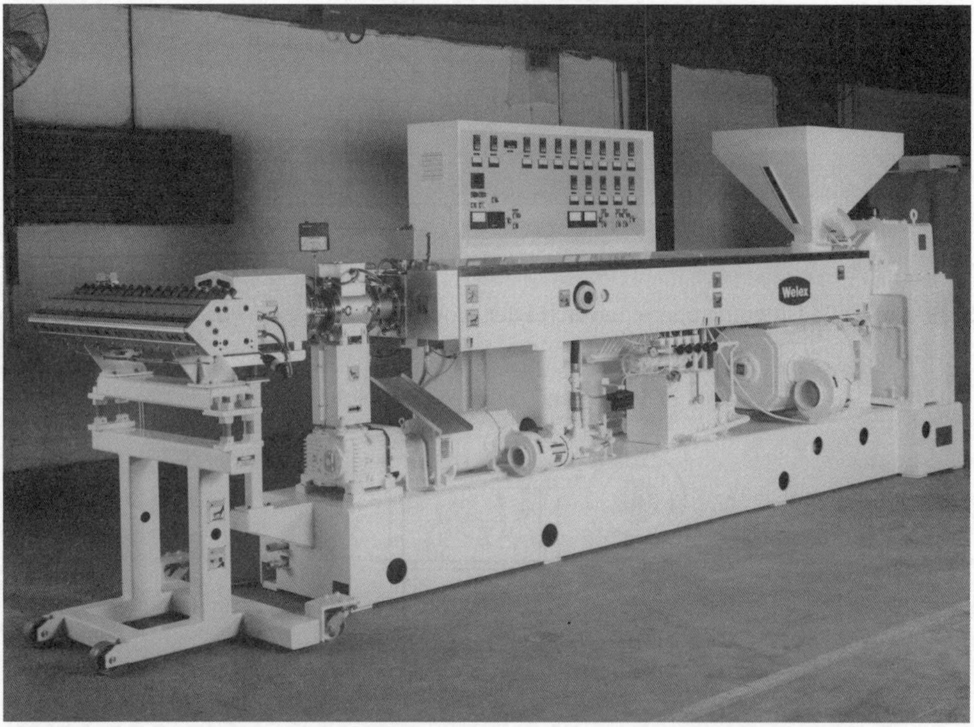

Figure 5.22 The extruder's barrel cover guard is closed over the exhaust vent port; the screen changer, gear pump, static mixer, and sheet die are located toward the end (exit) of the extruder (courtesy of Welex Inc.)

Vented screws typically have an L/D ratio of greater than 30:1 and can suffer from some instability in pumping output. Operating conditions are also constrained by the need to ensure that the vent does not plug with plastic melt. The use of vented screws is usually confined to materials likely to contain moisture, volatiles, or entrained air.

Figure 5.23 Barrel cover guard over the exhaust vent port (courtesy of Welex Inc.)

The twin-screw extruder may have parallel or conical screws, and these screws may rotate in the same direction (corotating) or in opposite directions (contrarotating). Extruders with more than two screws are known as multiple-screw extruders. These extruders are normally used when mixing and homogenization of the melt is very important, and in particular where additives, fillers, and/or reinforcements are to be included in the plastic (chapter 3).

They are used for plastic compounding of heat-sensitive materials (such as polyvinyl chlorides [PVCs]), processing of difficult-to-feed materials (such as certain powders), reactive processing, and for plastic devolatilization. Twin-screw extruders particularly offer a wide processing variability. They can be starve-fed so that residence time, amount of shear, and control of melt temperature can be controlled by means of their segmental modular designs.

Twin screws can have various design configurations to achieve varying degrees of shear using bilobe or trilobe kneading blocks that melt, shear, and mix. Pigments, additives, and fillers can be dispersed and distributed using corotating twin-screw extruders. Generally, multiscrew machines for compounding applications have a higher screw rpm (up to 400) than single-screw machines. Proper design of their thrust bearings is important because of the limited center-to-center distance between screws. Other differences are higher initial cost and higher repair costs, but superior performance is a very important balancing factor.

As in all other processes, an extensive theoretical analysis has been applied to facilitate understanding and maximize the extrusion manufacturing operation. However, the real world must be understood and appreciated as well. The operator has to work within the limitations and variabilities of the plastic materials and equipment (the basic extruder and all auxiliary upstream and downstream equipment). The interplay and interchange of process controls can help to eliminate problems and aid the operation in controlling the variables that exist (chapter 1). The greatest degree of instability is initiated by nonuniform delivery of plastic to the extruder, having an improper screw design or using the wrong screw for the plastic being processed, and restrictions occurring via the die. For uniform and stable extrusion, it is important to periodically check the drive system, the take-up device, and other equipment and compare them to their original performances. If variations are excessive, all kinds of problems will develop in the extruded product. An elaborate process-control system can help, but it is best to improve stability in all facets of the extrusion line. As shown in Figure 5.24, the extrusion FALLO approach definitely applies; details are provided with Figure 1.37.

Extruder Basics

An overview of what happens in an extruder can be summarized as follows:

1. Screws for single-screw extruders are usually right-hand thread and turn counterclockwise when looking from the rear or power drive mechanism. Their rotation appears as if to unscrew itself backward out of the material. The machine's thrust bearing prevents this backward action. Twin-screws may turn in either or both directions, but the principle is the same in moving the plastic.

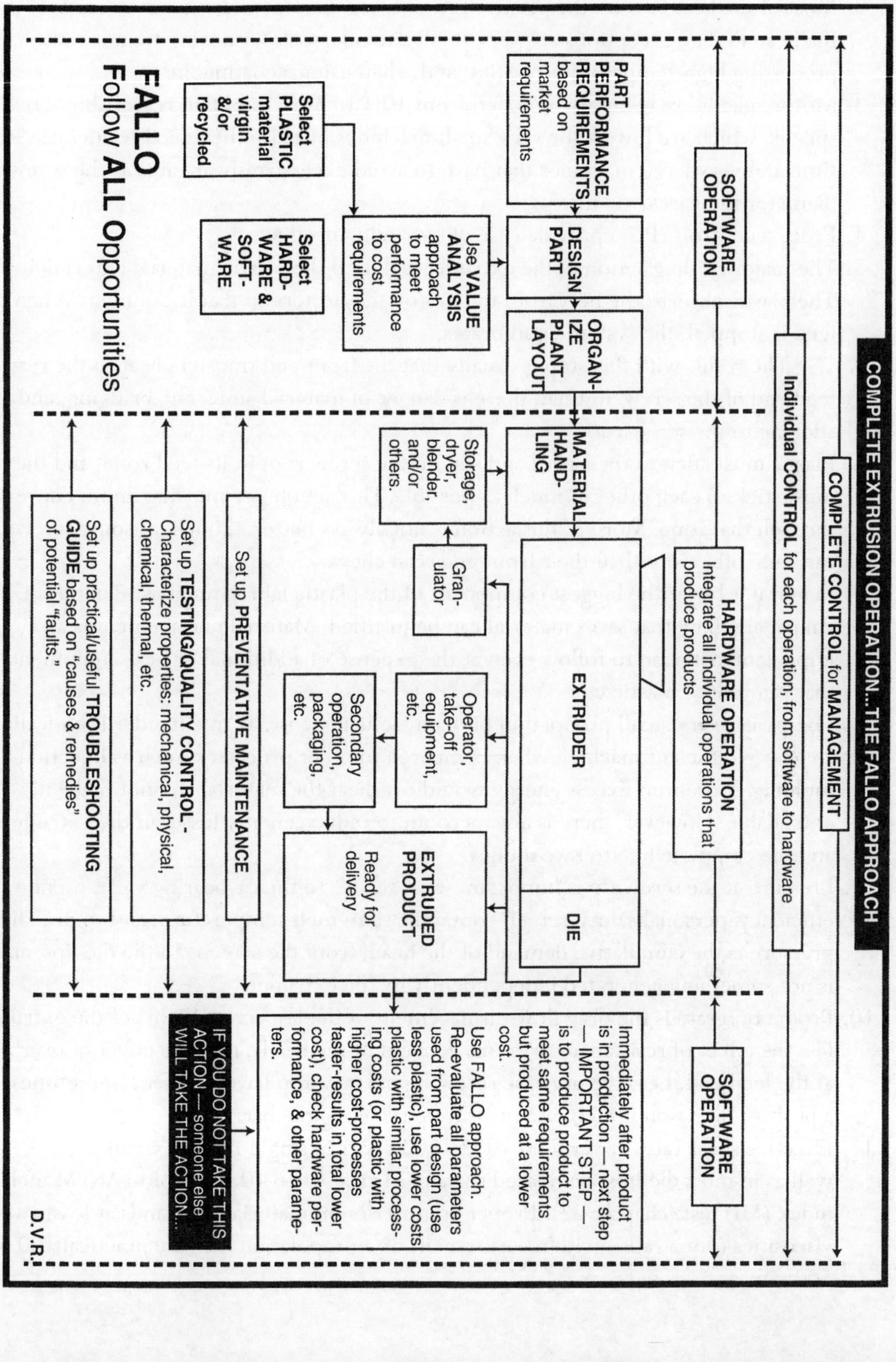

Figure 5.24 This FALLO approach is a guide to meeting product performance and cost requirements

2. Most of the heat to melt the plastics comes from the motor drive system, not the heaters. The exceptions are very small extruders, slow-moving twin-screws, processing high-temperature plastics, and when using a coating line.
3. Motor speeds are generally reduced from 10:1 to 30:1 to obtain reasonable screw speeds, which are lowest for very small machines because they need residence time and very large machines that have to avoid excessive shear rates as the screw diameter increases.
4. Process controls (PCs) provide ease of extruding products.
5. The major cooling action in the extruder is due to the plastic material entering it. The plastic absorbs the heat from the motor and heaters so that it can melt. When the feed is stopped, the system equilibrates.

   The result with this stop is usually that the front end transfers heat to the rear feed end of the screw. In turn, there is danger of material sintering, bridging, and/or sticking to the screw root.
6. Plastic must stick to the barrel and slip on the screw root in its feed zone, and they must stick to each other as much as possible. This action permits maximum conveying through that zone. More of this action is not always better, though, as some screws can "bite off" more than their front ends can chew.
7. Material is by far the biggest component of the plastic fabricating cost. Therefore, almost anything that saves material can be justified. Material management is an important criterion to follow even at the expense of additional or new equipment that can reduce plastic use.
8. Energy is a very small proportion of manufacturing cost, as an extruder is basically an energy-efficient machine when compared to other processes, such as injection molding. Moreover, excess energy would overheat the material and not make it extrudable. However, there is always room to reduce energy losses in the extruder and the complete line to save money.
9. Pressure at the screw tip is important, as it relates to thrust-bearing wear, mixing efficiency, personal safety, screen contamination, melt temperature, and so on. This pressure is the cumulative demand of the head, from the screens to the die lips, and it is not something generated independently by the extruder.
10. Production rate is the drag flow displacement of the last screw flights of the extruder, less the effect of resistance (pressure demand of the head), plus the effect of overbite at the feed end. Leakage over the screw flights may also have an effect, sometimes in a positive direction if there is a pressure peak along the barrel.
11. Effective shear rates are around 100 to 500 rev/s in single-screw extruders, as well as in most die lips, compared to values from 0.1 to 10 in a typical ASTM melt index (MI) test (chapter 22). Proper comparison of materials demands at least two viscosities (not a ratio) with measurement or extrapolation into the practical 100 to 500 range.

12. In the numerical simulation of an extrusion process, it is essential to use physical data that is as accurate as possible before one can evaluate the usefulness of the program.

COMPONENTS

Different components make up the extruder, each with its specific important function. All components have to operate efficiently; otherwise, the extruder's operation is inefficient. A very important parameter in the extruder is the plasticator's pumping process. This is the interaction between the rotating flights of the screw and the stationary barrel wall. For the plastic material to be conveyed, its friction must be low at the screw surface but high at the barrel wall. If this basic criterion is not met, the plastic will usually rotate with the screw and not move in the axial output direction.

In the plasticator's output zone, both screw and barrel surfaces are usually covered with the melt, and external forces between the melt and the screw channel walls have no influence except when processing extremely high viscosity plastics, such as rigid PVCs and UHMWPEs. The flow of the melt in the output section is affected by the coefficient of internal friction (viscosity), particularly when the die offers a high resistance to the flow of the melt. Details on the plasticator are in chapter 3.

Some of the extruder components are reviewed in this section. Each plays an important part in ensuring the extrusion process meets its performance requirements. Figure 5.11 shows some of these components on an extruder. Figure 5.25 shows the extruder's components where the following identifications are listed:

1. Drive motor from 20 to 2,000 hp with infinitely variable speed drives directly coupled to the reducer for maximum efficiency designed to save floor space.
2. Gears to provide high-efficiency capability to process plastics.
3. Performance-efficient heat-treated helical or herringbones gears equipped with shaft-driven oil pumps and oil cooler.
4. Thrust bearing with long life expectancy of well in excess of 30 years' continuous operation.
5. Large rectangular standard feed opening (round with lining, optional, for use with crammer feeders).
6. Long-lasting barrel heater/cooler elements that heat quickly.
7. Cooling tubes that run parallel with heating elements. The cast-in stainless steel tubes closed-loop system provides nonferrous distilled water that is automatically adjusted via microprocessor-based temperature controllers providing uniform, efficient cooling.

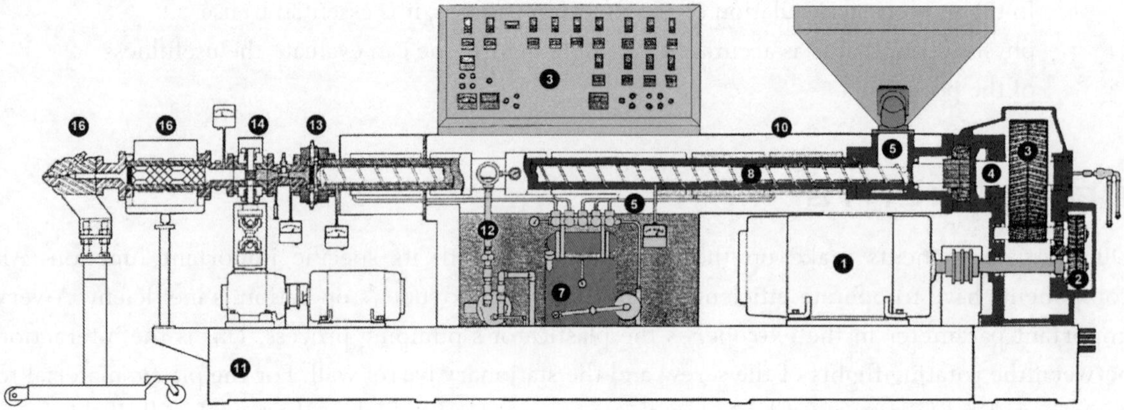

Figure 5.25 Schematic identifies the different components in an extruder (courtesy of Welex Inc.)

8. High-performance screw with bimetallic-lined cylinder designed for processing a specific plastic; can be cored for cooling.
9. Prepiped and prewired power installation.
10. Safety heat conservation and heat protection guard; this piece is hinged and doesn't have insulated loose parts.
11. Heavy single-unit steel-based machine foundation preassembled so all parts are in place ready to be used.
12. When required, patented two-stage vented plasticator that can be plugged in for use in minutes.
13. Screen changer that uses standard hinged-swing bolt gate, allowing continuous operation without shutdown.
14. Gear pump to ensure absolute volumetric output stability.
15. Static mixer to provide thermal and viscosity homogeneity.
16. Die designed to produce single or multilayer sheet without modifications, strand dies, and so on.

Extruder and Injection Barrel Compared

An extruder barrel differs from an IM barrel in several ways. It is usually longer, with a minimum L/D of 24 in and can go to at least 36 in. In an IM, the L/D is usually 18 or 20 in, with a possible few at 24 in. Some vented barrels are 32 in, but the trend is toward shorter lengths. The extrusion barrel is designed to withstand lower melt pressures of usually 500 to 5,000 psi (3.5 to 35 MPa) to possibly 10,000 psi (69 MPa); an IMM is up to 20,000 psi (138 MPa) as the usual standard and can go up to 30,000 (207 MPa). This means a thinner wall and eliminates the high-pressure sleeve or bell end. The extruder barrel connects to the die adapter, but the seal is made slightly different. It has a female counterbore, just like IM, but the die adapter has a recess for a rapid-removal breaker plate (Fig. 5.26).

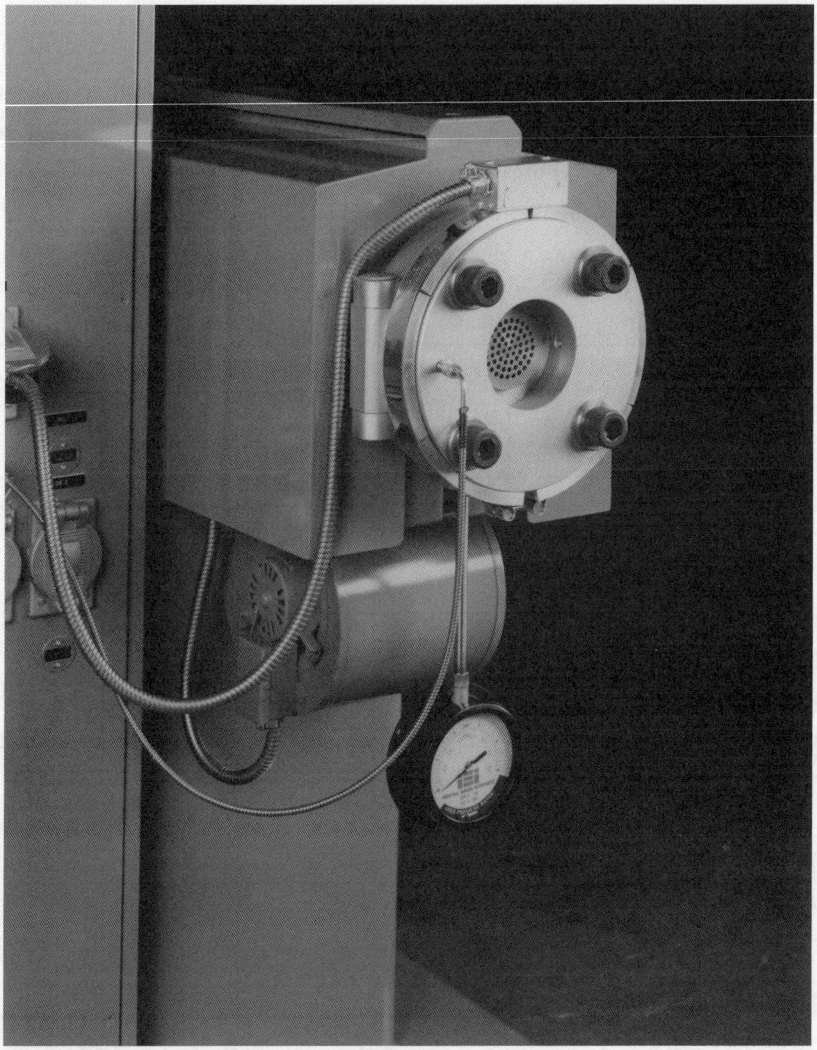

Figure 5.26 Four-bolt swing-gate die-clamping system (courtesy of Welex Inc.)

DRIVE SYSTEM

It is important that the drive system perform accurately within limits. The drive turns the screw at the required rotating speed, with a capability to provide a constant speed. The drive also provides the required amount of torque to the shank of the screw (502). Various drive systems are used to meet performance requirements at the lowest cost. Plastics with their low-to-high force and output requirements dictate the torque needed in the machine. Too high or too low a torque will be damaging to the plastic. A meaningful way to express this torque is in hp per 100 rpm of screw speed.

The power to turn the screw comes from the drive system with an often very large motor. Power requirements are usually 5 to 10 lb/h/motor hp. Thus, an extruder processing 100 lb/h (45 kg/h) output requires a 10- to 20-hp motor. The speed of electric motors is usually much too fast, such as 1,700 rpm, for direct connection to the screw. Common screw speeds range from 20 to 200 rpm. For this reason, a gear reducer is used between the motor and screw. Reduction can also be achieved with pulleys and belts connecting the motor and gearbox (Tables 5.3 to 5.6).

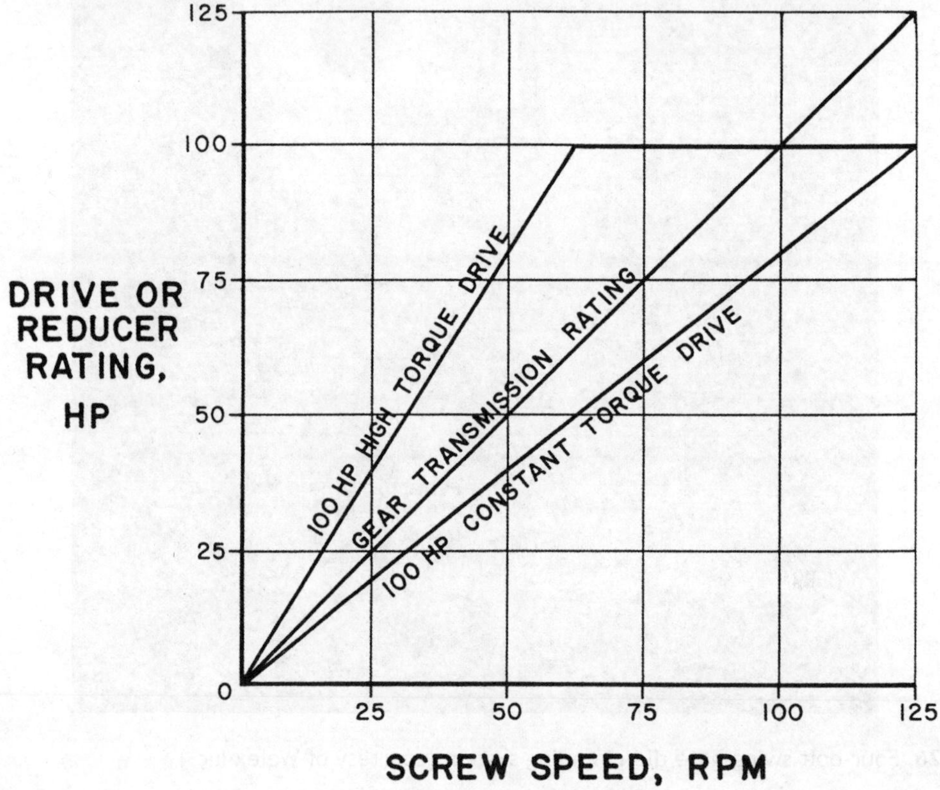

Table 5.3 Comparison of power speed for speed reducers and drive

Screw Diameter, Inches	1¾	2	2½	3½	4½	6	8	10
Drive Horsepower* per 100 rpm Screw Speed	5	10	20	60	120	210	550	800

*Based on practical flight depths in the feed sections and reasonable service factors.

Table 5.4 Torque as expressed in hp per 100 rpm of screw speed

	Conventional DC with tach feedback	Brushless DC	Variable frequency AC with tach feedback	Flux vector AC
Power range	1/4–2,000 hp*	1/4–600 hp	1/4–1,000 hp	1/4–300 hp
Speed regulation	1.0–0.01%*	0.5–0.01%	0.5–0.03%	0.5–0.01%
Constant torque	20:1	100:1	40:1	1,000:1
Breakaway torque	1.5	1.25	0.6	1.5
Power factor	20–85%	0.98	0.98	0.98

Table 5.5 Performance of different drive motors

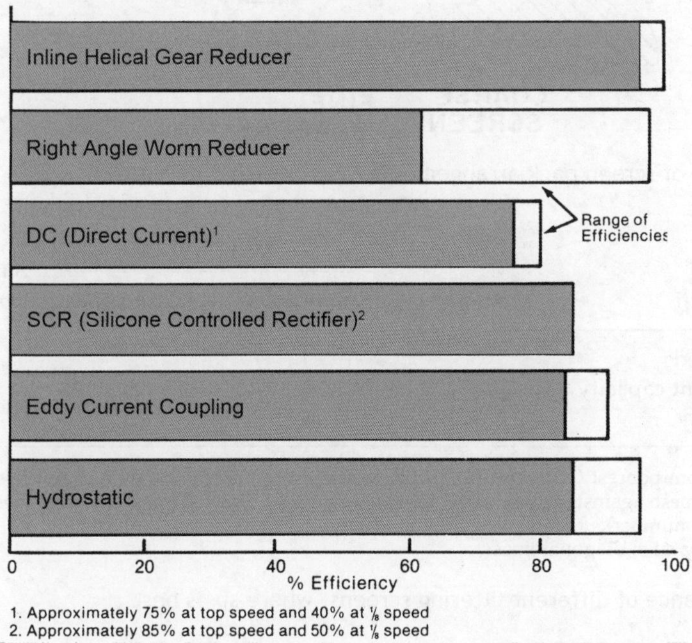

Table 5.6 Performance of different drive systems

SCREEN PACK

The purpose of the screens is primarily twofold: (a) to change the melt's spiraling motion (caused by the screw rotation) and (b) to filter contaminants out of the melt. Most plastics contain contaminants, and these particles can be conveniently removed by means of a screen placed after the extruder barrel and before the melt flow reaches the extrusion die. The simplest means for filtering plastic melts are woven wire mesh disks of about the same diameter as that of the extruder barrel. Several layers of different screens are usually made up into one screen pack. The innermost layer is the finest mesh screen that determines the particle size that will be caught by the screen pack. The outer and inner coarser layers only lend support to the flimsy innermost screen (Fig. 5.27 and Tables 5.7 and 5.8).

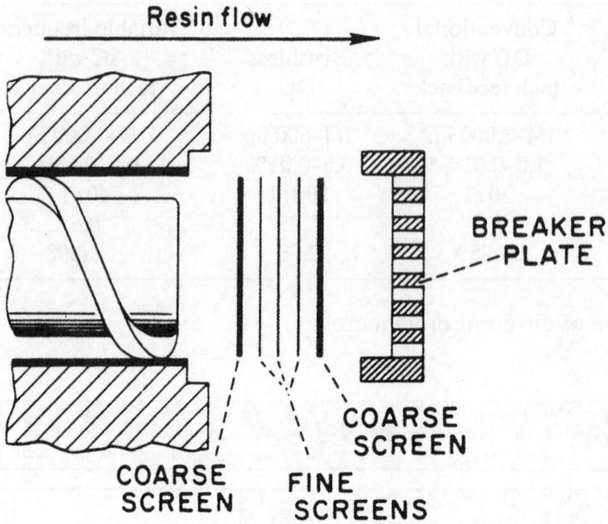

Figure 5.27 Example of screen pack arrangements

Contaminant	Metal fibers	Wire mesh: Square weave	Wire mesh: Dutch twill[b]	Sintered powder
Gel captured	5	1	2	3
Contaminant capacity	6	2	3	3
Permeability	4	4	1	2

[a] Range is from poorest (1) to best (6). Multiple screens are normally used; example screen pack has 20 mesh against breaker plate, followed by 40, 60, and 100 mesh (coarsest mesh has lowest mesh number).
[b] Woven in parallel diagonal lines.

Table 5.7 Performance of different filtering screens, where six is best

Against the forces exerted by the melt flow, the screen packs are backed by a thick, densely perforated steel disk called a breaker plate. The outer rims of the breaker plate and of the screen pack fit into a round recess in the end of the extruder barrel and are clamped in place by the adapter flange of the adjoining piece of equipment, usually that of the extrusion die. To change a clogged screen pack, the die adapter flange has to be removed, the old pack taken out and replaced with a new one, and the equipment reassembled.

Screen changers are mechanical devices that permit changing screens in faster, more convenient ways. Screen changers fall into three main categories: (a) manual, (b) intermittent (reciprocating), and (c) continuous (Figs. 5.28 and 5.29).

Reciprocating screen changers hold two identical sets of screen packs. Only one of these is in the path of the melt flow and, when needed, the screen changer switches the other (clean) one into its place. The most commonly used types have the two breaker plate and screen pack assemblies

Mesh	Diameter of wire, decimal of an inch	For closer sizing sieves from 0.0015 in to 3.000 ratio $\sqrt[4]{2}$ or 1.189
2½	0.099	0.312
3	0.070	0.263
3½	0.065	0.221
4	0.065	0.185
5	0.044	0.156
6	0.036	0.131
7	0.0328	0.110
8	0.032	0.093
9	0.033	0.078
10	0.035	0.065
12	0.028	0.055
14	0.025	0.046
16	0.0235	0.0390
20	0.0172	0.0328
24	0.0141	0.0276
28	0.0125	0.0232
32	0.0118	0.0195
35	0.0122	0.0134
42	0.0100	0.0168
48	0.0092	0.0116
60	0.0070	0.0097
65	0.0072	0.0082
80	0.0056	0.0069
100	0.0042	0.0058
115	0.0038	0.0049
150	0.0026	0.0041
170	0.0024	0.0035
200	0.0021	0.0029
250	0.0016	0.0024

Mesh	*10*	*16*	*20*	*35*	*50*	*120*
mil	78.7	46.9	33.1	19.7	11.7	4.9
micron	2000	1190	840	500	297	125

Table 5.8 Classification of screens with conversion of mesh to particle size

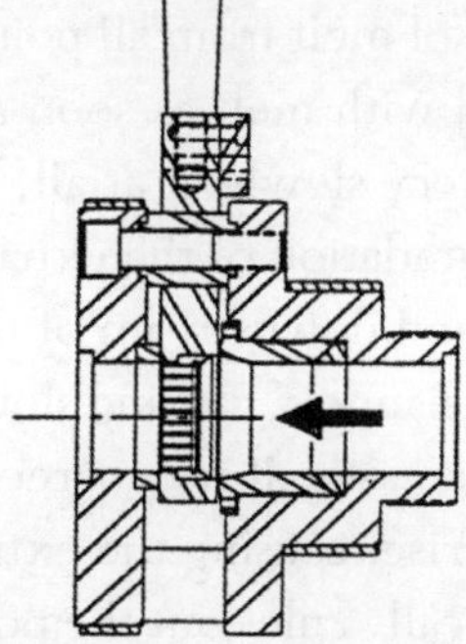

Figure 5.28 Example of a manual screen changer (courtesy of Spirex Corp.)

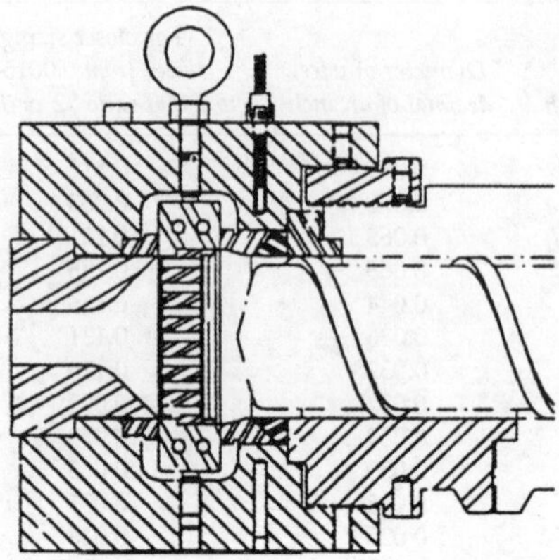

Figure 5.29 Example of an intermittent screen changer (courtesy of Spirex Corp.)

mounted into rectangular plates that move along slide guides. They use hydraulic mechanisms to swiftly shuttle the slide plates in and out of the melt flow. Other types of reciprocating screen changers employ valves by means of which the melt flow may be diverted from one screen pack to the other and back again.

The ever-changing pressure conditions that are inherent in all intermittently operating machines can be eliminated by the use of continuous screen changers. Continuous screen changers employ screens in the form of long ribbons that move slowly and continuously. The speed of the filter ribbons is infinitely variable. It can be chosen either to remain at a steady rate or it can be made to automatically adjust so as to maintain steady extruder head pressure conditions despite widely varying levels of contamination in the plastic being processed.

With improper screw design, contaminated feedstock, poor control of regrind, overheating melt, and so on, the screens may be helpful. Sometimes screen packs are used to control the operating pressure of extruders. However, there are advantages in processing with matched and controlled back pressure and operating within the required melt pressure and temperature, as this can facilitate mixing, effectively balancing out melt heat problems. Extra heat develops when melt goes through the screens, so some heat-sensitive plastics should not use a screen pack.

If it is at all possible to do without screen packs, they should not be used. Various reasons exist. Complete and continuing displacement of melt from all points in the screen pack is rather difficult. Hundreds of small dead spots are filled with melt as soon as the pack is put into service, and the material in these spots is only moved very slowly, if at all, by the drag of neighboring melts. This action can cause contamination and degradation of the extrudate.

The spaces immediately upstream and downstream of the screen pack are usually much larger in cross-section than the screw and die channels, causing slower melt flow. As the screen pack clogs, its resistance to the passage of melt increases. If no corrective actions were taken, output would go down and melt temperature would rise, causing the extruder to develop processing problems. The presence of screens, therefore, basically rules out the possibility of reaching a truly steady-state operation and necessitates closer supervision of the process.

However, the screens are required because a large amount of plastic to be processed needs cleaning due to being contaminated. What can help is to have excellent in-plant housekeeping rather than using screen packs.

Where screens are proved necessary, the automatic screen changer is the best way of handling the problem of screen changing and cleaning. This device soon repays its cost through material savings and increased production.

Gear Pump

The gear pump is a component that has been standard equipment since the 1930s in textile fiber production. During the 1980s, they established themselves in all kinds of extrusion lines. The gear pump is also called the melt or metering pump. It consists of a pump, drive for the pump, and pump controls located between the screen pack (or screw) and die. Two counter-rotating gears transport a melt from the pump inlet (extruder output) to the pump discharge outlet. Gear rotation creates a suction that draws the melt into a gap between two teeth. This continuation action from tooth to tooth develops a surface drag that resists flow so some inlet pressure is required to fill the cavity. This pump device is basically a closely intermeshing counter-rotating twin-screw extruder (Fig. 5.30). However, since gear pumps are solely used to generate pressure, they are seldom called extruders, even though they are extruders.

Static Mixer

A static mixer is also called a motionless mixer. It is designed to achieve a homogeneous mix by flowing one or more plastic streams through geometric patterns formed by mechanical elements in a tubular tube or barrel; the mixers contain a series of passive elements placed in a flow channel. These elements cause the plastic compound to subdivide and recombine in order to increase the homogeneity and temperature uniformity of the melt. There are no moving parts and only a small increase in the energy is needed to overcome the resistance of the mechanical baffles. These mixers are located at the end of the screw, before the screen changer, or between the screw and gear

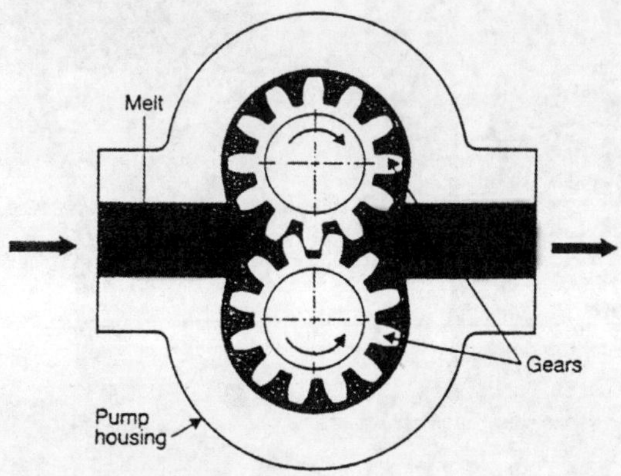

Figure 5.30 Example of melt flow through a gear pump

pump (Figs. 5.31 and 5.32). They can be used to mix different plastics as well as plastics with their component ingredients, such as colors and additives.

HEATING AND COOLING

Most of the drive motor energy supplied to the screw is used to heat and melt the plastic to be extruded. In order to bring the melt back to the solid state, a large fraction of heat must be removed from the extrudate within a short distance from the die. Shape and appearance of the hot molten extrudate have

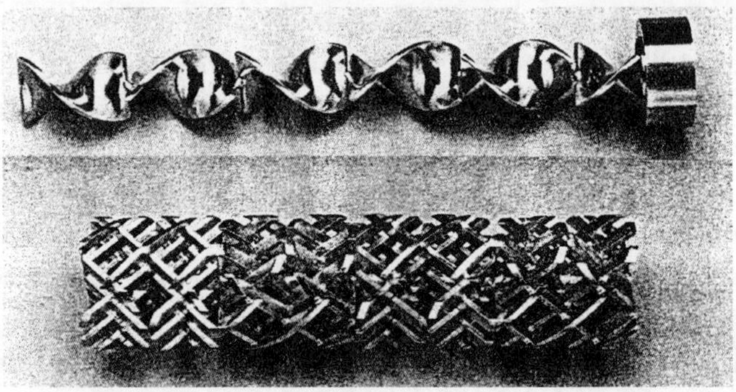

Figure 5.31 Two examples of available static mixers

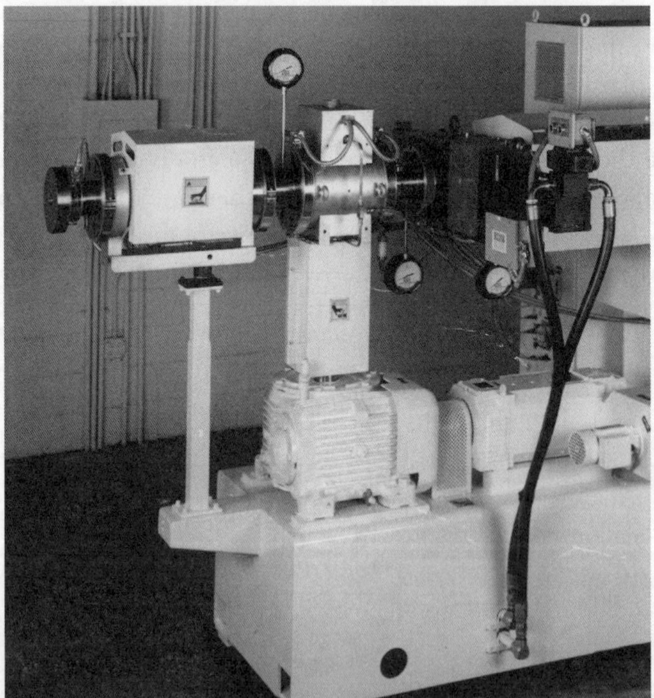

Figure 5.32 View of an extruder with a static mixer located after the screen changer and gear pump prior to the die adapter (Courtesy of Welex Inc.)

to be precisely maintained until sufficient cooling and setting have taken place. Output can increase for a given shape, resulting in higher lineal takeoff speed. This action can reduce the available cooling time near the die that is essential for proper shape molding. Based on this combination of facts, highly developed cooling methods and takeoff procedures are indispensable for high-speed extrusion production.

The temperature profile required along a barrel, adapter, and die largely depends on the plastic used and the specific extrusion process line with its screw design and available process controls. The thermal condition of the plastic is essentially determined for a given material by screw geometry with its rotational speed and the total restriction or pressure existing in the die. The electrical heaters are normally placed along the barrel grouped in separate and adjoining zones; each zone is controlled independently. Small machines usually have two to four zones. Larger machines have five to ten zones. Tables 5.9 and 5.10 provide information on the different types of heater bands.

Electric heat is used on all types of plasticating equipment to soften and melt plastic material. The type of heater and the power output (wattage) used depends on the material being processed, the cycle time expected, and the amount of material being molded. The following paragraphs will outline the different types of heaters currently available and provide a guideline for selecting the correct heater.

All electric heaters are constructed from resistance wire, an electric insulating material, and steel sheath. Insulating materials include ceramic, mica, and magnesium oxide (MGO). The element is supported by a metal sheath and power is connected through various electrical terminations. Heaters are rated according to total wattage output over the surface area of the heater or Watts per Square Inch. For example, a mica band heater 3" ID x 2" W at 300 watts is rated at 300/, (3 x 3.1416 x 2) or 16 watts per square inch.

There are three methods of heating: Radiant, Convection, and Conduction. Radiant heat is used primarily in thermoforming applications, convection for drying material, and conduction for melting. The types of heaters mentioned in this article are all conduction heaters since conduction heat is the most efficient way to get electric heat into the plasticating process.

Energy efficiency is defined as the energy absorbed into the process versus the energy (wattage) put into the heater. Energy efficiency impacts the overall cost of running the heaters and the response time of the process to the heaters. Obviously, the more heat that conducts from the Nichrome element into the process, the lower the cost and the shorter the response time. Energy efficiency is determined by the conductive characteristics of the heater, the fit between the heater and the surface, and the loss of heat to the air. Conductive heaters include Mica Insulated, Ceramic Insulated, Cast Aluminum, Tubular, Mineral Insulated, and Heat/Cool Assemblies.

•MICA INSULATED HEATERS

Mica insulated heaters are constructed from 20 - 24 gauge rust resistant steel, Nichrome resistance wire, and plate mica. The resistance wire is wound onto the plate mica and supported by the metal sheath. These heaters are capable of a maximum operating temperature of 900 degrees Fahrenheit and are typically rated from 25 to 50 watts per square inch. Since these heaters transfer energy through conduction, tight fit between the heater and the barrel or nozzle is critical. The higher the watt density the more critical the fit becomes. Any voids, holes, or grooves in the surface being heated can cause hot spots and burn out the element.

The advantages to this type of heater is its lower cost, durability, quick response times, and design flexibility. Mica heaters can be made in a wide variety of shapes including round, rectangular, cone, and plate all with different electrical connections and lock-ups. The disadvantages are the lower operating temperature, which makes it unsuitable for many of the newer high temperature resins, and its lower energy efficiency. Without an insulating shroud or blanket, as much as 40% of the wattage input into the heater is lost to the surrounding atmosphere.

Mica bands are flexible in design but stiff in construction. They should not be opened up to wrap around the barrel but should be slipped on over the end. Flexing can damage the mica sheet and cause premature failure. Mica bands can be made in 2, 3, and 4 piece segments. All mica bands which are larger than 20" diameter should include compression springs to help take up the expansion of the element under heat. Mica heaters should be slightly undersized to insure a tight fit on the surface being heated. They should be retightened after the first heat up. Lastly, Mica bands can be made up to 24" wide but they tend to work better and last longer when the width is held to 3" or less. Mica bands are best suited for applications where the melt/process temperature is less than 550 degrees Fahrenheit.

Table 5.9 Types of barrel heater bands (courtesy of Spirex Corp.)

•CERAMIC HEATER BANDS

Ceramic heaters are constructed from 24-26 gauge Stainless or Aluminized steel. The element is a helically wound Nichrome resistance wire which is supported by interlocking ceramic blocks. Depending on the application the ceramic blocks are made from either a cordierite, steatite, or silicon carbide material. A ceramic fiber insulating material is sandwiched between the ceramic pieces and the outer steel shell to insulate against heat loss. The insulating blanket is typically 1/4" thick but can be manufactured up to 1" thick to provide additional energy efficiency and conservation, as much as 30- 40% over standard mica bands.

Besides being more energy efficient, ceramic band heaters have a maximum operating temperature of 1,400 degrees Fahrenheit and are typically rated at 35 to 50 watts per square inch. Ceramic heaters heat through both conduction and radiation and therefore do not need as integral a fit as mica bands. This characteristic tends to give ceramic heaters a longer service life than mica. They are flexible in construction and can be opened up to wrap around a barrel or die. The disadvantages are the higher cost (50% more than mica) and that ceramic heaters are prone to contamination. Ceramic heaters are best suited for applications requiring high process temperatures 500 degrees and higher, faster cycle times, or in locations with high electric rates.

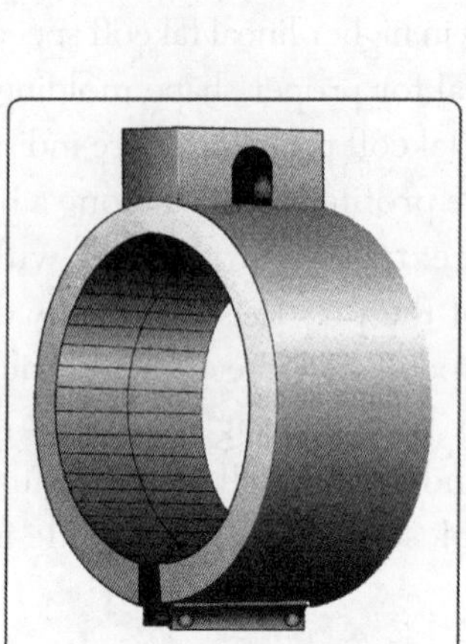

Ceramic Heater Band

Ceramic can also be made in 2, 3, and 4 piece segments but given the flexibility of the band this is usually unnecessary. Larger ceramic bands (>20" ID) should include compression springs in the lock up. Ceramic heaters are commonly manufactured in widths greater than 3" to take advantage of the insulating blanket. They can be made with either lead wire or screw terminals, however, given the high operating temperatures, screw contacts help extend service life.

•MINERAL INSULATED HEATERS

Mineral insulated heaters are manufactured using a stainless steel sheath with a sinuated Nichrome element sandwiched between thin layers of mineral insulation. The mineral insulation is a magnesium oxide material pressed into sheet form. Once the heater is assembled it is pressed to compact the sinuated element into the sheet, then shaped into a cylinder, and finally baked to remove any organic binders from the mineral insulation.

This process creates excellent heat transfer properties allowing for high operating temperatures and high watt densities. Mineral insulated heaters can operate up to 1,400 degrees Fahrenheit and are rated as high as 230 watts per square inch. With low mass and high watt densities, mineral insulated heaters heat up and cool down very quickly. The disadvantages are that magnesium oxide is a hydroscopic material which can reduce dielectric strength. Further mineral insulated heaters are stiff and limited in design variations. Lastly, mineral insulated heaters have energy efficiencies similar to mica heaters.

Table 5.9 (*continued*)

•TUBULAR AND CAST-IN HEATERS

Tubular band heaters are constructed using either stainless steel or Inconel tubing. A helically wound element is stretched the length of the tube. The heater is filled with magnesium oxide for insulation and compressed. The heaters are then shaped into a cylinder. Lastly, a stainless steel strap is used to tighten the heater onto the die or barrel. Similar to the mineral insulated heaters, tubular heaters have good heat transfer characteristics and because of the tubing they are highly resistant to contamination. Tubular heaters can reach operating temperatures of 1,200 degrees Fahrenheit and are rated up to 100 watts per square inch. Watt density is limited by application and heat transfer.

The disadvantage to tubular heaters is that while the heat will radiate from the element efficiently, it radiates 360 degrees from the surface of the element. The heater only makes tangential contact with the surface being heated. This creates a very inefficient heat transfer from the element to the surface and most of the energy is lost to the open air. To overcome this problem the tubular heaters can either be set into an aluminum extrusion which is formed into a cylinder or it can be cast directly into molten aluminum. Using either of these variations creates a heater with an excellent heat profile. The heat will be transferred uniformly across the width of the heater. Also, these heaters are very rugged and have quick heat up and cool down times. However they are the most expensive heaters on the market and have a maximum operating temperature of 650 degrees Fahrenheit. At 650 degrees, the aluminum starts to get soft and will melt at 750-800 degrees.

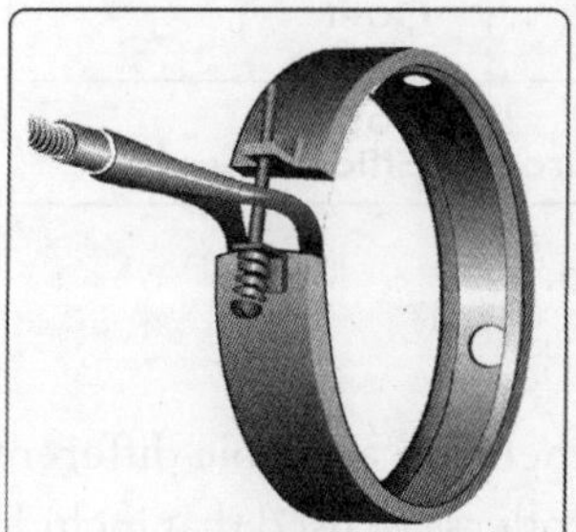

Tubular Heater

Tubular heaters without the aluminum extrusion or casting are best suited for applications with high levels of contamination and/or fumes that would quickly attach the Nichrome wire and cause failure in standard mica or ceramic heaters. Cast-in heaters are best suited for situations that require good heat profiles, like blown film dies and extrusion barrels where cooling is required.

with cooling tubes cast into the heaters. Water lines are attached to the cooling tubes. Water cooled units cool more efficiently than air cooled. However the piping can be messy and leak water, and the cooling tubes can get clogged with mineral deposits. Ceramic air cooled units are capable of higher operating temperatures than cast aluminum and are lower cost.

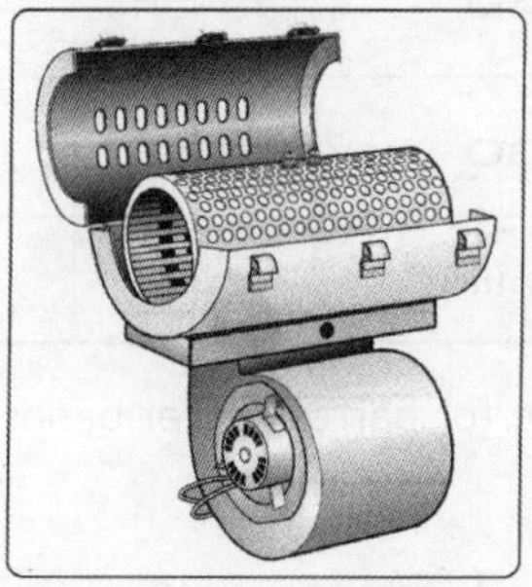

•HEAT/COOL APPLICATIONS

The extrusion process generates heat through friction as the screw shears the resin against the inside of the barrel. Heaters are used at start-up to get the process going and then in many applications cooling is required to offset the heat generated by the shear affect. Cooling is accomplished with either water or forced air.

Air cooled units use either cast aluminum or ceramic heaters. The heaters are enclosed in a sheet metal shroud, and a blower is attached. Water cooled heaters are cast aluminum heaters

Table 5.9 (*continued*)

STYLE	INSULATION	MAX. TEMP.	MAX. WSI	ADVANTAGES	DRAWBACKS
Mica	Plate Mica	900 F	50	Low cost, Versatile	Low temperature
Ceramic	Cordierite Steatite Silicon Carbide	1400 F	50	High temperature, Flexible, Energy efficient	Prone to contaminants
Mineral Insulated	MGO	1400 F	230	High temperature Response time	Cost, Versatility, Energy efficiency
Tubular	MGO	1200 F	100	Durability	Energy efficiency
Cast Aluminum	MGO	650 F	35	Uniform heat	Cost, Low temperature
Cast Water Cooled	MGO	650 F	35	Efficient cooling	Cost, Water leaks, Scaling
Cast Air Cooled	MGO	650 F	35	Durability, Cost	Cost
Ceramic Air Cooled	Steatite	1200 F	50	Cost, High temperature	Cooling, Efficiency

Table 5.10 Selection guide for barrel heater bands (courtesy of Spirex Corp.)

Where it is desirable to have repeatedly close temperature control, there are available different barrel designs, such as those with different shaped fins. Temperature controllers are used that include the cooling water- and air-pumping systems. Different manufacturers meet different requirements based on the extruder's requirements.

Adapter

If the exit opening of the extruder barrel does not match up with the entry opening of the die, an adapter is used between the barrel and die (Fig. 5.32). The usual arrangement is in the horizontal position; a 90° adapter is used in a vertical blown-film die (Figs. 5.33 and 5.34). Any adapter used can help reduce complex melt flow patterns entering the die. The adapter is also called an offset adapter, cross-head adapter, or die feedblock. A fishtail die adapter describes the transition of a round block of melt as it exists from the barrel and is then flattened while being spread prior to entering the die.

Barrel-Die Coupling

The coupling between an extruder barrel and die can be carried out in various ways using bolts or locking devices. They include a flange fitting with a clamp ring on the barrel and a fixed flange on the die, flanges on the barrel and die with tapered links and two bolted half-clamps, or a ring clamp

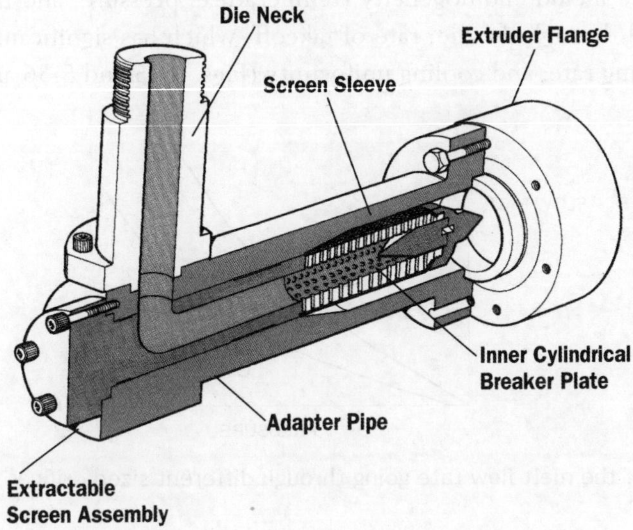

Figure 5.33 Example of a 90° adapter

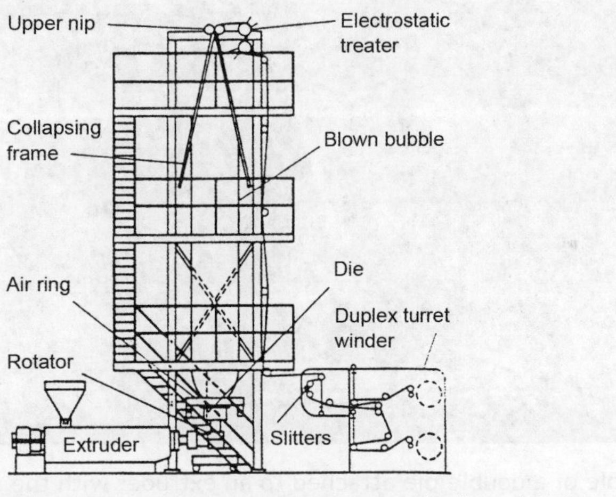

Figure 5.34 Example of a blown-film line that uses an adapter attached to the die

hinged at one side and bolted to the other side with a swing-bolt flange connection between the barrel and die flange. In most extruders, a breaker plate with screens is used between the barrel and die assembly.

DIE

Information on die designs and manufacturers is in chapter 17. This review will provide a summary. Basically, the size of the die orifice initially controls the thickness, width, and shape of any extruded product dimension. In general, the orifice is developed oversized to allow for the drawing and shrinkage that occur during conveyor-pulling and cooling operations. Factors that influence extrudate performance include homogeneity (temperature, pressure, and time/rate of travel) of the melt being pushed through the die; rate of takeoff, which has significant influence on dimensions and shapes; cooling rate; and cooling uniformity (Figs. 5.35 and 5.36 and Table 5.11).

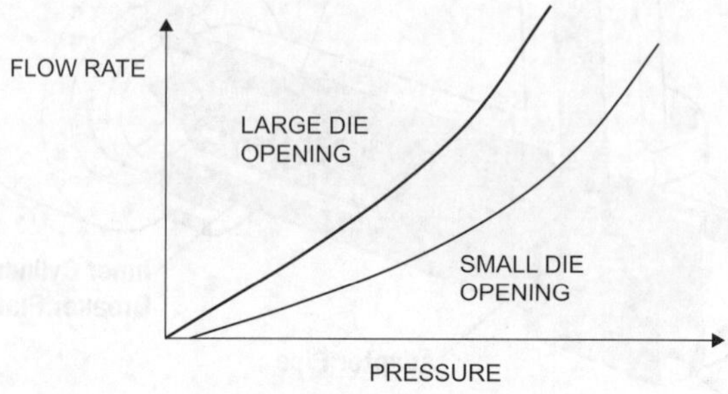

Figure 5.35 Example of the melt flow rate going through different sized orifices

Figure 5.36 Example of a double die attached to an extruder with the required output capacity

Extruded shape	Melt pressure at the die	
	psi	MPa
Film, blown	1,000–5,000	6.9–34.5
Film, cast	200–1,500	1.4–10.4
Sheet	200–1,500	1.5–10.4
Pipe	400–1,500	2.8–10.4
Wire coating	1,000–5,000	6.9–34.5
Filament	1,000–3,000	6.9–20.7

Table 5.11 Range of melt pressures required in different designed dies

PROCESS CONTROL

Information on process control is included in chapter 3. Different control systems are used to process the different extruded products. Examples of different controls for different and same products are provided in Figures 5.37 to 5.55.

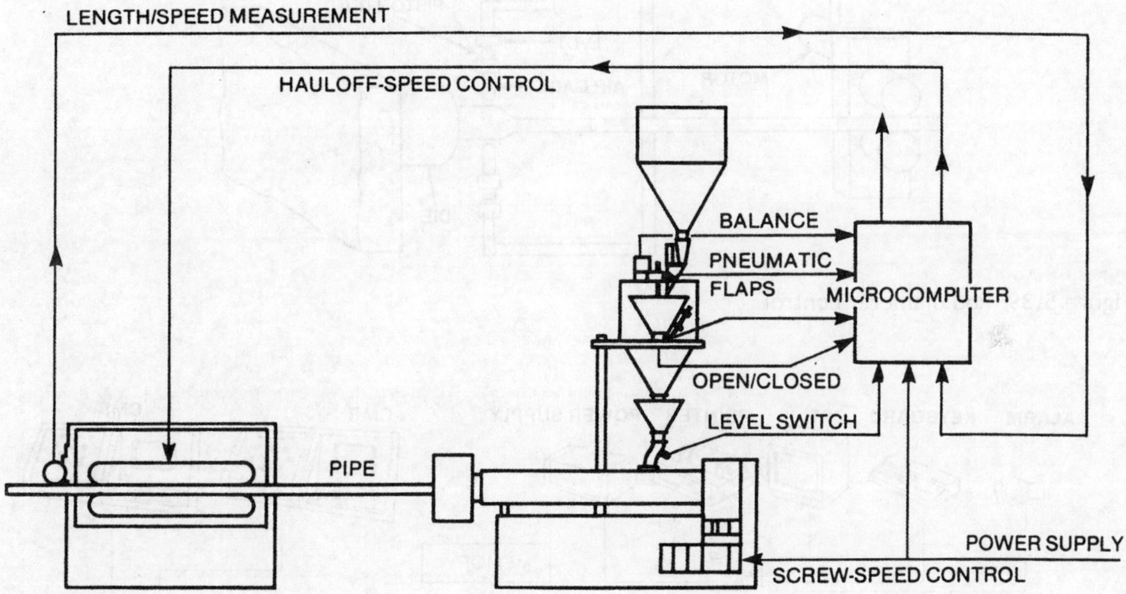

Figure 5.37 Pipeline control

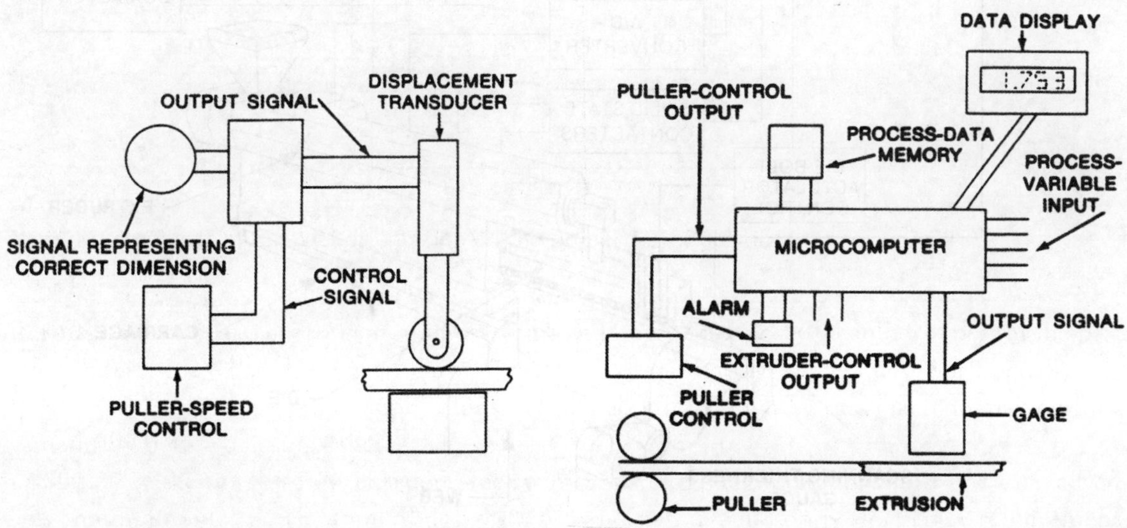

Figure 5.38 Sheet line speed control

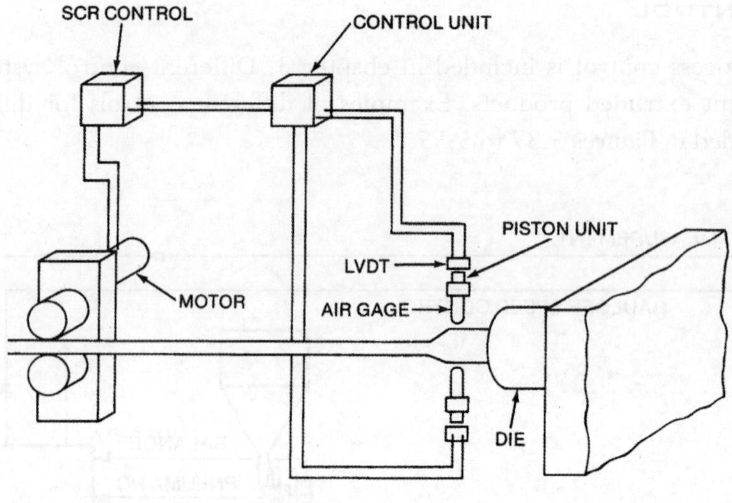

Figure 5.39 Rod diameter control

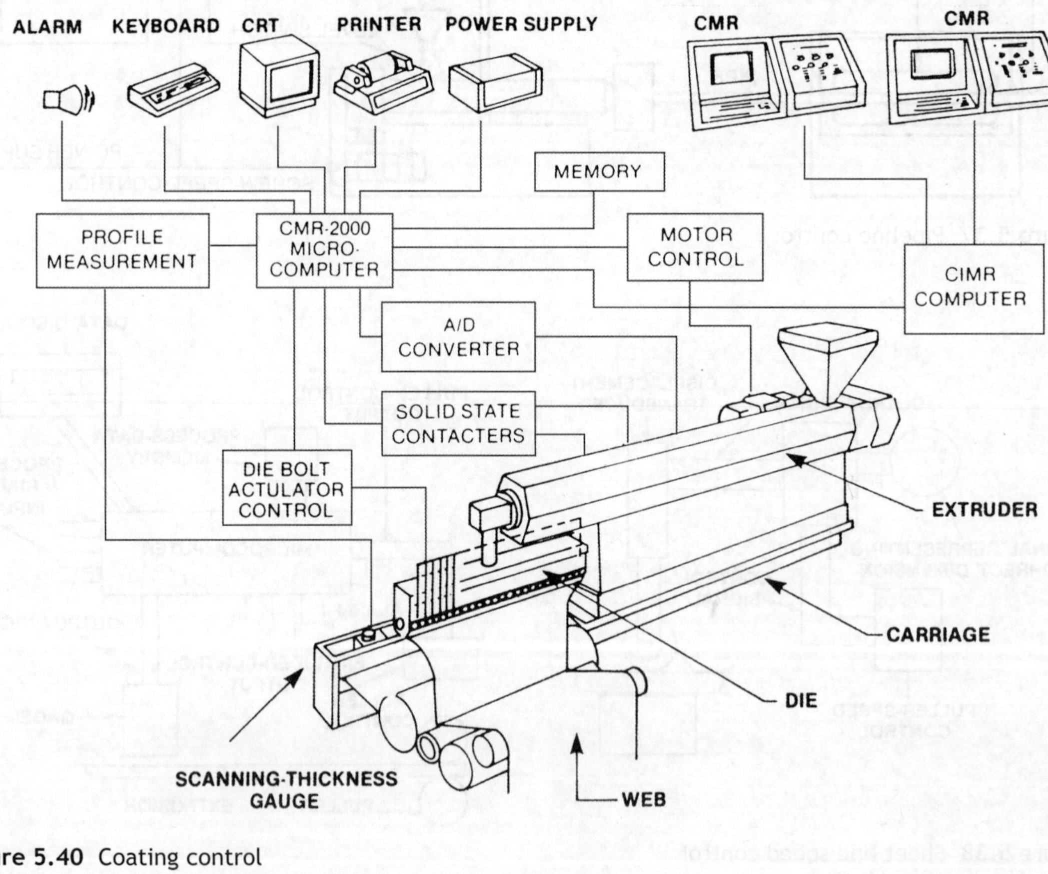

Figure 5.40 Coating control

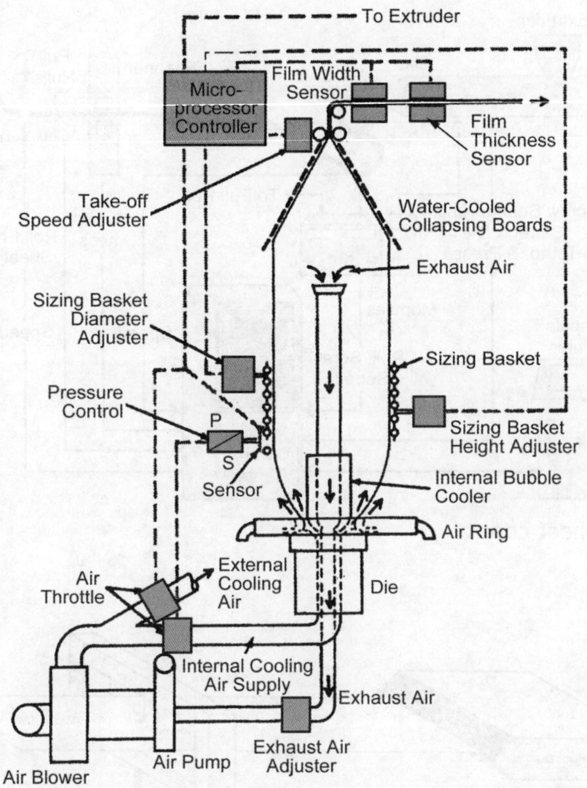

Figure 5.41 Blown-film control

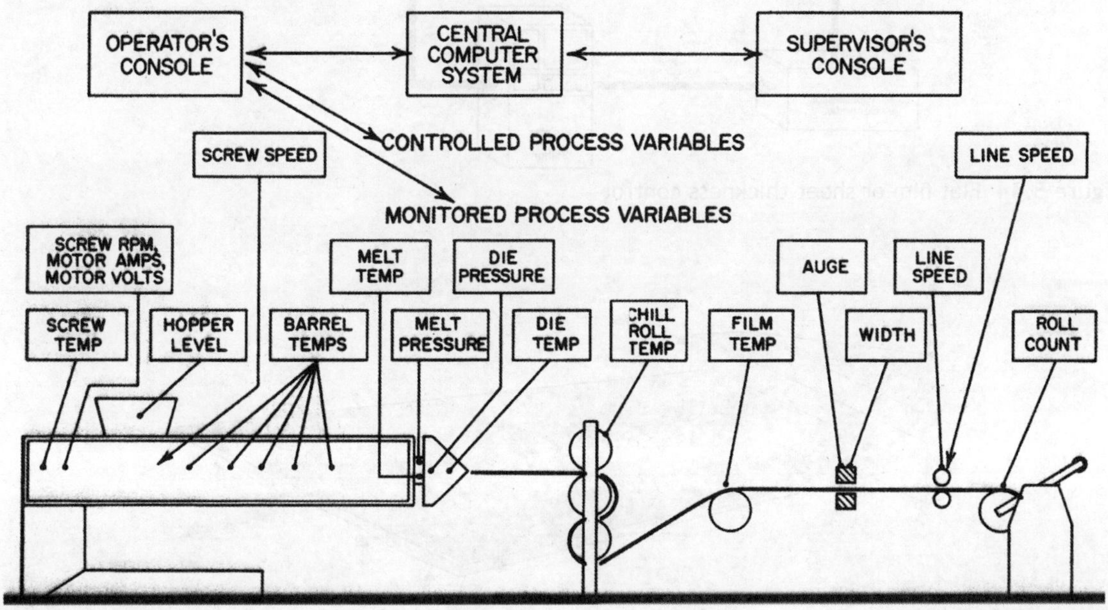

Figure 5.42 Overall sheet control

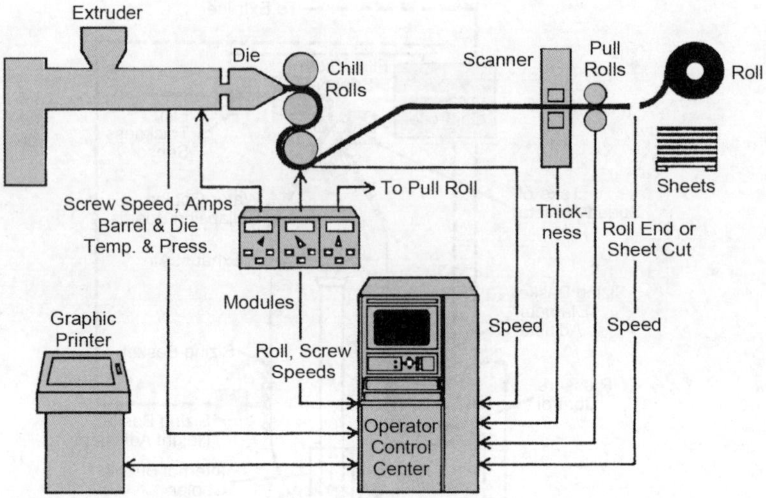

Figure 5.43 Simplified sheet control

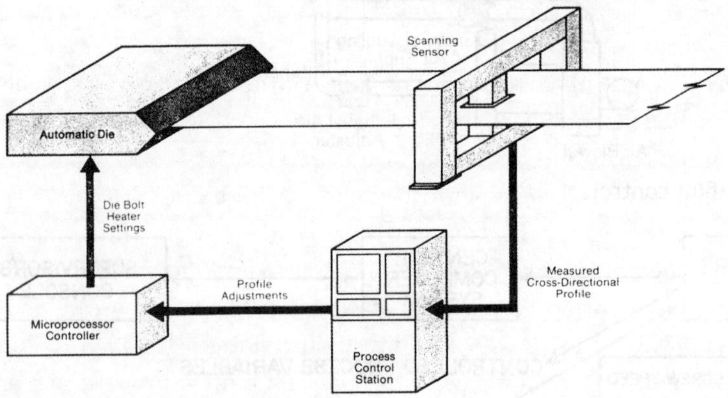

Figure 5.44 Flat film or sheet thickness control

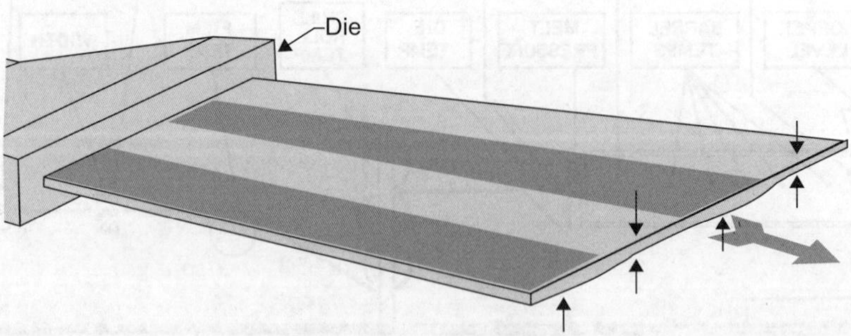

Figure 5.45 Flat film or sheet profile control

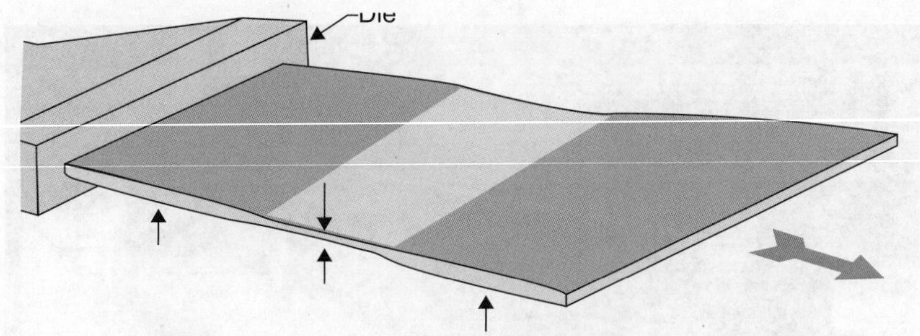

Figure 5.46 Flat film or sheet long-term machine direction control

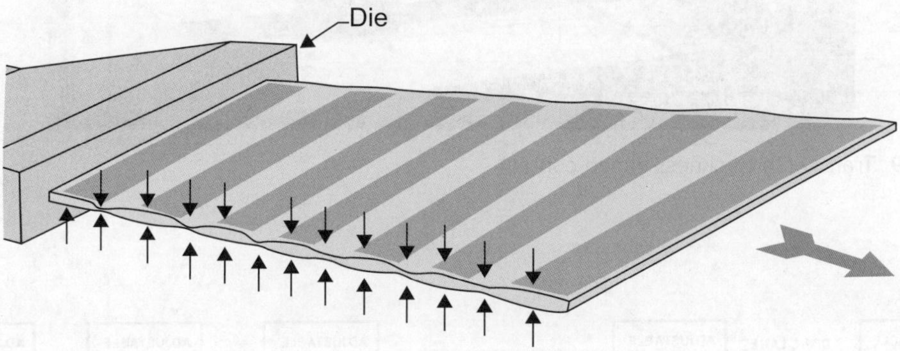

Figure 5.47 Flat film or sheet short-term machine direction control

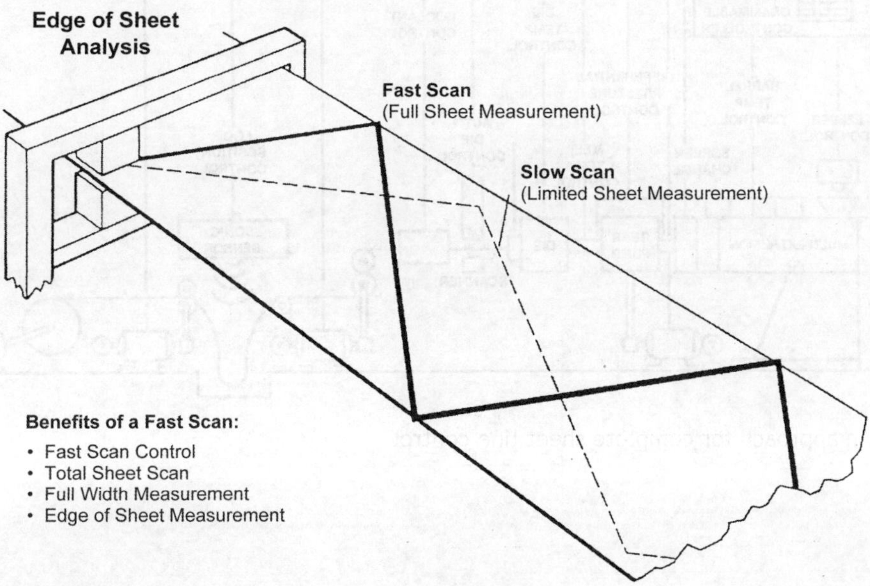

Figure 5.48 Flat film or sheet more accurate control at higher production rate

Figure 5.49 Transverse thickness gauge control

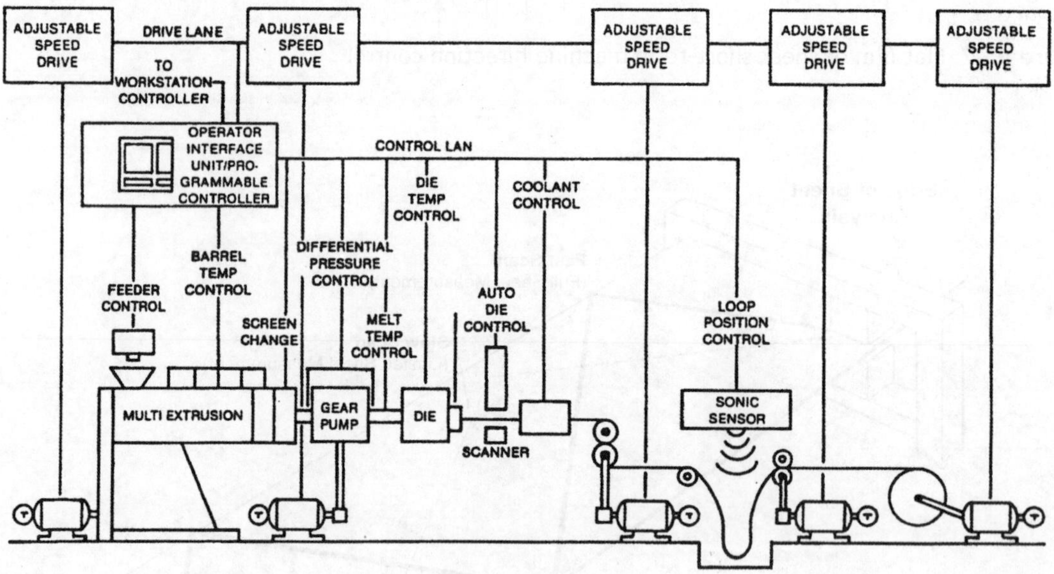

Figure 5.50 An approach for complete sheet line control

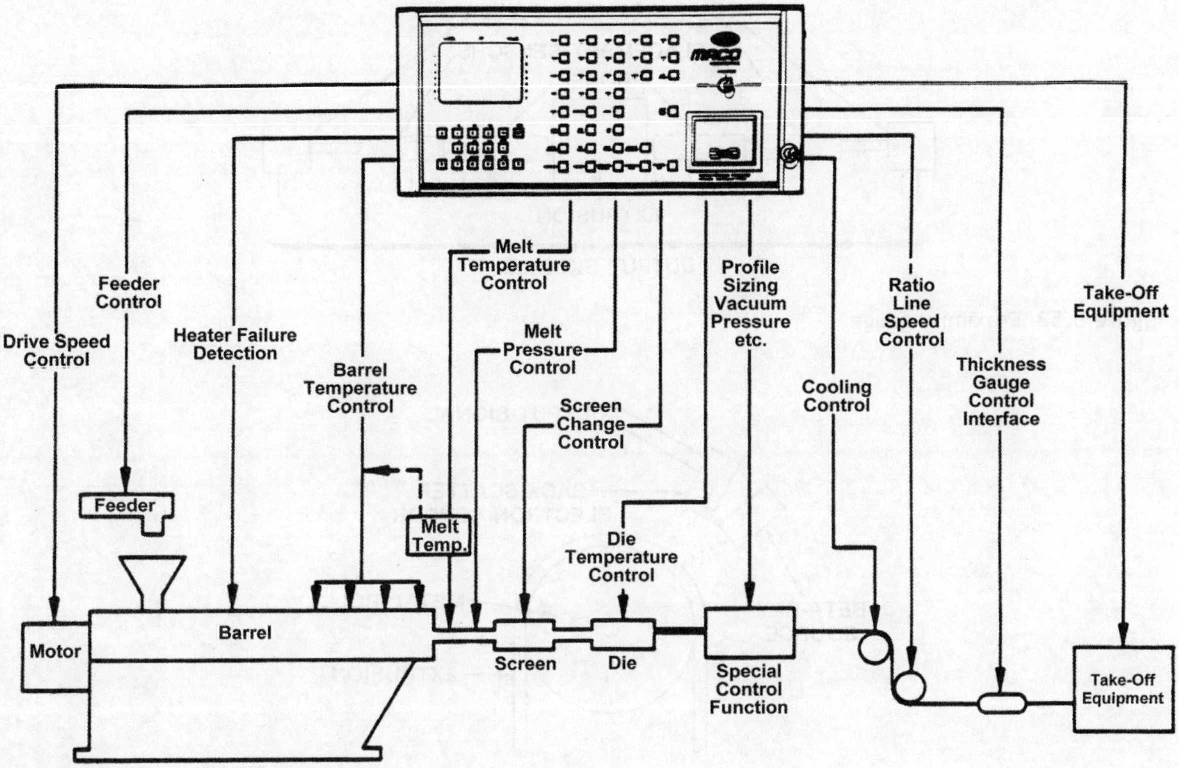

Figure 5.51 Another approach for complete sheet line

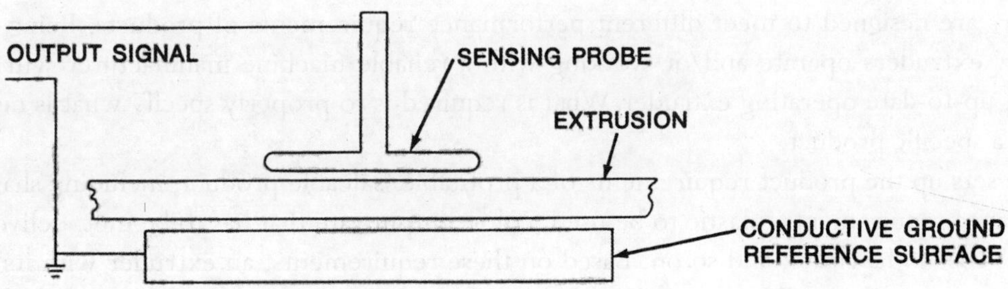

Figure 5.52 Capacitance thickness gauge

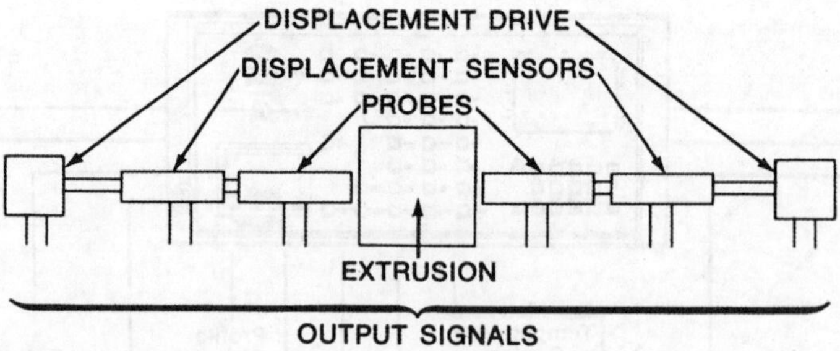

Figure 5.53 Proximity gauge

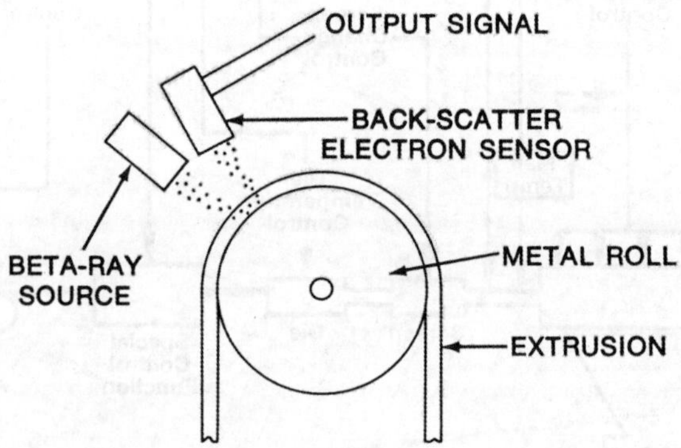

Figure 5.54 Beta ray gauge control

MACHINE DESIGN/PERFORMANCE

Extruders are designed to meet different performance requirements of products. Being familiar with how extruders operate and/or working with a reliable machine manufacturer will provide the most up-to-date operating extruder. What is required is to properly specify what is needed to produce a specific product.

One sets up the product requirements of a profitable/saleable product, including shape, size, performance requirements, plastic to be used, color, output rate that has to be met, delivery time schedule, cost to fabricate, and so on. Based on these requirements, an extruder with its die and other components that will be required (screen pack, motionless mixer, and/or others) is literally designed around the product. You should have familiarity with what may appear as minor requirements in the equipment; the machine manufacturer will be familiar with these requirements as well. These include the hopper shape, extrudate output direction, ease of equipment alignment (barrel [Fig. 5.56], auxiliary equipment, etc.), ease of screw removal, and location of control systems.

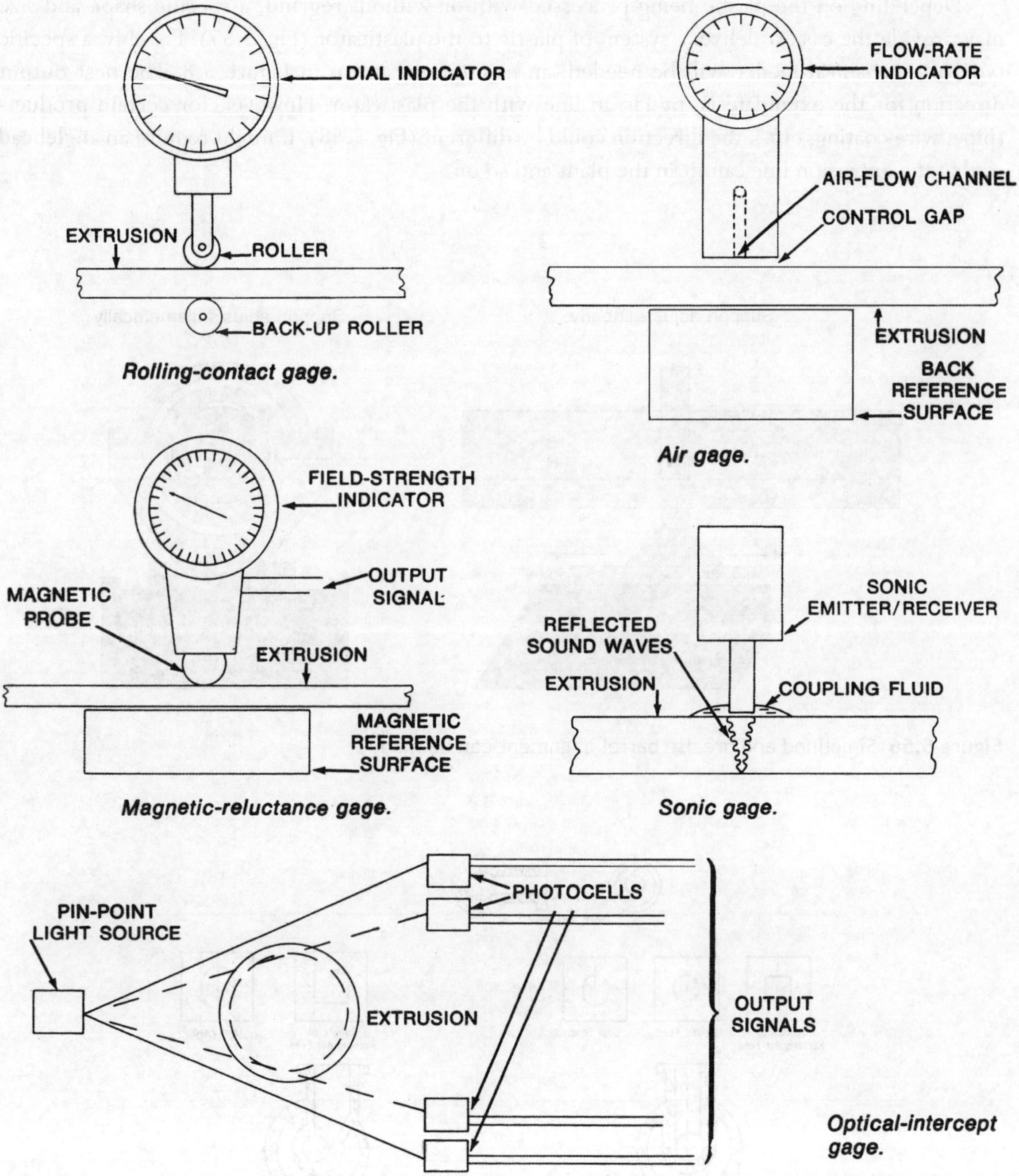

Figure 5.55 Different type dimensional controls

Depending on the plastic being processed with or without regrind, a specific shape and size may provide the easiest delivery system of plastic to the plasticator (Fig. 5.57). Possibly, a specific type of mechanical feeder will be needed; an example is shown in Figure 5.8. The best output direction for the extrudate is, by far, in line with the plasticator. However, for certain products (pipe, wire coating, etc.), the direction could be different (Fig. 5.58). It might require an anglehead so that the extrusion line can fit in the plant and so on.

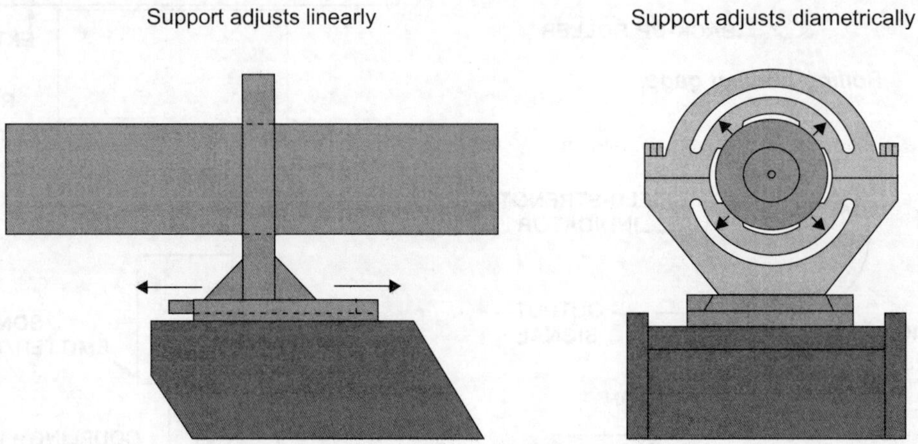

Figure 5.56 Simplified and precise barrel alignment can be made

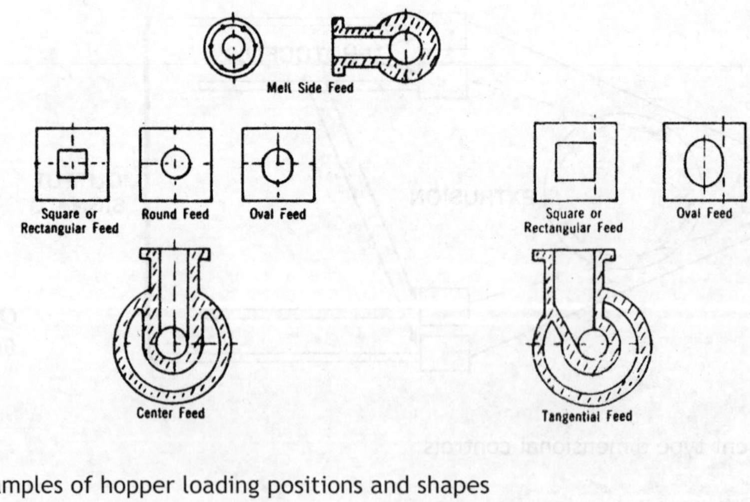

Figure 5.57 Examples of hopper loading positions and shapes

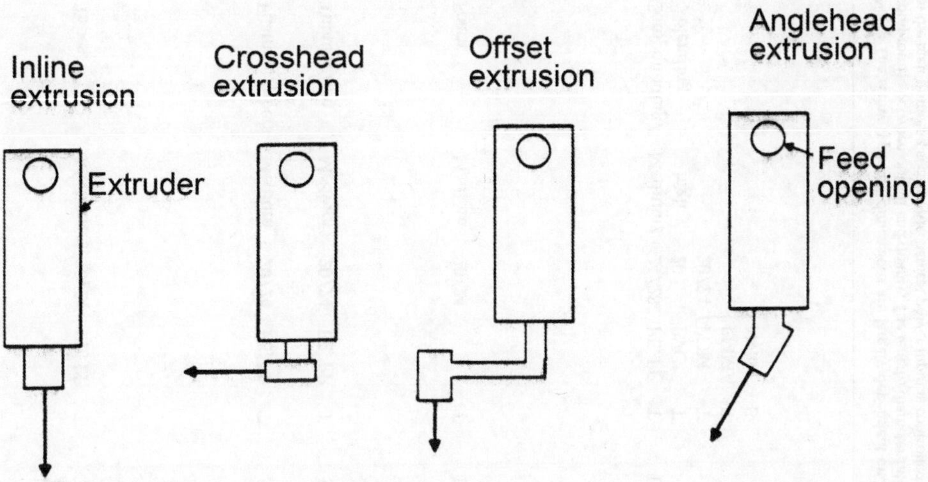

Figure 5.58 Examples of the extrudate exiting an extruder in different positions

PLASTIC

In maximizing performance, or at least meeting performance requirements, of extruded plastics as well as minimizing cost to extrude products, it is important to understand the processing behavior of the different plastics. In producing the different extruded products (film, sheet, profile, etc.) certain plastics can be used. An understanding of factors such as the plastics' rheological and decomposition behaviors, as well as problems that can develop, provides information that will make it easier to extrude products. This type of information will make it easier to understand what is happening when the plastic goes from a solid state to a melt stage to the final shaped solid state. This information is in chapters 1, 2, 3, 5, 17 to 19, 22 to 25, 27, and 28. In this section, Tables 5.12 to 5.26 and Figures 5.59 to 5.70 provide an overview on processing plastics.

EXTRUDER TYPE/PERFORMANCE

The popularly used single-screw and multiscrew types have their differences. Each has its benefits, depending on the plastic being processed and the products to be fabricated. At times their benefits can overlap so that either type could be used. When that is the case, the type to be used would depend on cost factors, such as cost to produce a quality product, cost of equipment, and cost of maintenance.

In the past, with the development of single-screw extrusion techniques for newer TP materials, it was found that some plastics with or without additives required higher pressures (torque) and

Paper coating	Medium	24/28	PE, PP	T	Shallow	Medium	High	×××	×××	×	Output relative to drum/haul-off speed for thickness. Medium output for rapid cooling. Very high temperature for flow and surface finish.
Tapes	Medium	20/24	PE, PP	T	Medium	Medium	Medium	×××	×××	××	Medium temperature for crystallization and drawdown. Uniform output for dimensions and crystallization. Uniform temperature for dimensions.
Filament/fibers	Medium	20/24	PE, PP	T	Shallow	Medium	High	×××	×××	×	High temperature for drawing.
Tubular film	Medium	20/24	PE, PP	T	Shallow	Medium	Medium	×××	×××	×××	High pressure for die. Medium temperature for drawing and blowing. Medium speed for uniform elastic memory.
Foam	Medium	20/24		T	Medium	Low	Low	×	××	××	Temperature depends on blowing agent. Low shear heating for decomposition. Medium pressure and mixing for uniform distribution and pore size.
Compounding/ blending	Medium/ high	24/28 20 20/24	PE, PP PVC {POM PMMA Nylon}	T T T S S	Deep	Medium	Low	×	××	××	Deep screw at high pressure for mixing. Large diameter and medium speed for high output. Low temperature for mixing and heat history.

Notes: The screw type and dimensions are largely dependent on the plastic and temperatures. Screw diameter and length are a compromise with capital cost. Screw speed is a compromise between output and quality. The length/diameter ratio and channel depth depend largely on the plastic used. 'Temperature' under 'Control' refers to set temperatures in the melt pumping section. Mechanical power input is controlled mainly by temperatures in the melting zone. Screw cooling is excluded since it is always very sensitive, requiring precise control.

Suggested *running* limits are:

Speed:		Temperature:		Mechanical power:	
×	±1%	×	±2°C	×	±3%
××	±0.3%	××	±0.5°C	××	±1%
×××	±0.1%	×××	±0.1°C	×××	±0.4%

Table 5.12 Relating product to extruder to control

Process	Screw diameter (relative)	Screw length L/D ratio		Screw type	Metering depth	Screw speed (relative)	Melt temperature	Degree of control: Speed	Degree of control: Temperature	Degree of control: Mechanical power	Comments
Rod/section and profiles	High	20/24 15/20 20/24	PE, PP PVC Nylon	T T S	Shallow	Low	Medium	××	××	×××	Output relative to haul-off speed, for dimensions. Minimum elastic memory. Medium temperature for rigidity while cooling. Add screens to promote mixing.
Tube	High	20/24 15/20 20/24	PE, PP PVC Nylon	T T S	Medium	Medium	Medium except die	××	××	××	Minimum elastic memory. High die temperature for gloss. Medium temperature for drawdown of small tubes.
Wire covering/ sheathing	Low	20/24 15/20 20/24	PE, PP PVC Nylon	T T S	Shallow	High	High	××	×	×	High temperature for good drawing and avoiding melt fracture. High pressure for thin insulation. Output relative to haul-off speed for thickness
Sheet/ laminating (including continuous vacuum forming)	High	20/24 15/20 20/24	PE, PP PVC Nylon	T T S	Medium	Medium	Medium	××	××	××	Medium speed is compromise for high output and uniformity of *Q* and *T*. Steady output to avoid ripples. High output for thick sheet. Uniform temperature for die distribution. Medium temperature for drawdown.
Flat film	Medium	24/28 20	PE, PP PVC	T T	Shallow	Medium	High	×××	×××	×××	High temperature for drawdown. High pressure for die. Medium speed for uniform elastic memory. Uniform output for thickness.

Product	HP	Barrel temperature, C Rear	Front	Die temperature	Pressure (psi)	Material temperature
Pipe	40	150	160	165	1500	165
Tubular film	40	150	160	165	1500	165
Flat film	40	200	240	250	1000	250
Coating	100	250	315	325	1000	320
Wire coating	50	220	240	240	3000	240
Contours	20	175	200	205	1500	200

Table 5.13 Guide to extruder settings to produce different LDPE products

Machine size (Screw diam. = inches)	Output (No./h)	Horsepower (hp)
2.5	100–250	25–100
3.5	300–800	50–150
4.5	700–1500	100–250
6	1200–2000	150–350

Table 5.14 Extruders' output rates and power requirements for ABS

Plastic	Temperature (C)
Cellulose acetate	160–205
Cellulose acetate butyrate	160–205
Cellulose propionate	160–205
Copolyester	200–250
Chlorotrifluoroethylene	260–320
Ethylcellulose	190–230
FEP fluorocarbon	380–400
Ionomer	230–290
Nylon-6	250–270
Polycarbonate	290–300
Polyethylene, low-density	230–260
Poly(ethylene terephthalate)	285
Poly(methyl methacrylate)	180–210
Polypropylene	270–300
Polystyrene	240
Polyurethane elastomer	165–220
Poly(vinyl chloride)	
Rigid	160–165
Nonrigid	190–205
Vinylidene chloride-vinyl chloride copolymers	175–190

Table 5.15 Melt temperatures for thermoplastics

Type of material	T_d (F)
PP	610–750
PC	645–825
PVC	390–570
PS	570–750
PMMA	355–535
ABS	480–750
PA	570–750
PET	535–610
Fluoropolymer	930–1020

Table 5.16 Decomposition temperatures for thermoplastics

Resin data[b]	*Specific gravity* ($g\,cm^{-3}$)	*Density* ($lb\,ft^{-3}$)	*Specific volume* ($in^3\,lb^{-1}$)	*Specific volume* ($cm^3\,g^{-1}$)	*Extrusion temperature* (°F)	*Specific heat* ($BTU\,lb^{-1}\,°F^{-1}$)	*Water absorption in 24h* (%)	*Maximum water content allowable*
ABS, extrusion	1.02	64.0	27.0	0.980	435	0.34	0.25	
ABS, injection	1.05	65.0	26.0	0.952		0.40	0.40	0.20
Acetal, injection	1.41	88.0	19.7	0.709		0.35	0.25	
Acrylic, extrusion	1.19	74.3	23.3	0.839	375	0.35	0.30	
Acrylic, injection	1.16	72.0	24.1	0.868		0.35	0.20	0.08
CAB	1.20	74.6	23.1	0.833	380	0.35	1.50	0.15
Cellulose acetate, extrusion	1.28	80.2	21.6	0.781	380	0.40	2.50	
Cellulose acetate, injection	1.26	79.0	21.9	0.794		0.36	2.40	0.20
Cellulose proprionate, extrusion	1.22	76.1	22.7	0.821	380	0.40	1.70	
Cellulose proprionate, injection	1.22	75.5	22.9	0.828		0.40	2.00	0.25
CTFE	2.11	134.0	13.1	0.473		0.22	0.01	
FEP	2.11	134.0	12.9	0.465	600	0.28	<0.01	
Ionomer, extrusion	0.95	59.6	29.0	1.050	500	0.54	0.07	
Ionomer, injection	0.95	59.1	29.2	1.060		0.54	0.20	
Nylon-6	1.13	70.5	24.5	0.886	520	0.40	1.60	0.15
Nylon-6,6	1.14	71.2	24.3	0.878	510	0.40	0.15	0.15
Nylon-6,10	1.08	67.4	25.6	0.927		0.40	0.40	0.15
Nylon-6,12	1.07	66.8	25.9	0.935	475	0.40	0.40	0.20
Nylon-11	1.04	64.9	26.6	0.962	460	0.47	0.30	0.10
Nylon-12	1.02	63.7	27.1	0.980	450		0.25	0.10
Phenylene oxide based	1.08	67.5	25.6	0.926	480	0.32	0.07	
Polyallomer	0.90	56.2	30.7	1.110	405	0.50	0.01	
Polyarylene ether	1.06	66.2	30.7	0.940	460		0.10	
Polycarbonate	1.20	74.9	23.1	0.832	550	0.30	0.20	0.02
Polyester PBT	1.34	83.6	20.7	0.746			0.08	0.04
Polyester PET	1.31	8.18	21.1	0.746	480	0.40	0.10	0.005

Table 5.17 Guide to extruder control for different thermoplastics (courtesy of Spirex Corp.)

HD polyethylene, extrusion	0.96	59.9	28.8	1.040	410		<0.01	
HD polyethylene, injection	0.95	59.3	29.1	1.050		480	<0.01	
HD polyethylene, blow molding	0.95	56.9	28.8	1.040	410		<0.01	
LD polyethylene, film	0.92	57.44	30.1	1.090	350		<0.01	
LD polyethylene, injection	0.92	57.4	30.1	1.090		400	<0.01	
LD polyethylene, wire	0.92	57.4	30.1	1.090	400		<0.01	
LD polyethylene, ext. coating	0.92	57.1	30.0	1.090	600		<0.01	
LLD polyethylene, extrusion	0.92	57.4	30.1	1.087	500			
LLD polyethylene, injection	0.93	58.0	29.8	1.075		425		
Polypropylene, extrusion	0.91	56.8	30.4	1.100	450		0.03	
Polypropylene, injection	0.90	56.2	30.7	1.110		490	<0.01	
Polystyrene, impact sheet	1.04	64.9	26.6	0.963	450		0.10	
Polystyrene, gp crystal	1.05	65.5	26.2	0.943	410	425	0.03	
Polystyrene, injection impact	1.04	64.9	26.6	0.968		440	0.10	
Polysulfone	1.25	77.4	22.3	0.807	650	680	0.30	0.05
Polyurethane	1.20	74.9	23.1	0.834	400	400	0.10	0.03
PVC, rigid profiles	1.39	86.6	19.9	0.720	365		0.02	
PVC, pipe	1.44	87.5	19.7	0.714	380		0.10	
PVC, rigid injection	1.29	83.6	21.0	0.756		380	0.10	0.07
PVC, flexible wire	1.37	85.5	20.2	0.731	365			
PVC, flexible extruded shapes	1.23	76.8	22.5	0.814	350			
PVC, flexible injection	1.29	80.5	21.4	0.776		300		
PTFE	2.16	134.8	12.9	0.464			<0.01	
SAN	1.08	67.4	25.6	0.927	420	470	0.03	0.02
TFE	1.70	106.1	16.3	0.589		610	0.01	
Urethane elastomers	0.83	51.6	33.5	1.210	390	400	0.07	0.03

[a]Specific information on all machine settings and plastic properties is initially acquired by using the resin supplier's data sheet on the particular compound or resin to be used.
[b]These are strictly typical average values for a resin class; consult your resin supplier for values and more accurate information.

Table 5.17 *(continued)*

needed higher temperatures. There was also the tendency for the plastic to rotate with the screw. The result was degraded plastics. The peculiar consistency of some plastics interfered with the plasticator's feeding and pumping processes. The problems were magnified with bulky materials, certain types of emulsion PVC and HDPE (high-density polyethylene), chopped PE (polyethylene) film, and sticky pastes (such as PVC plastisols).

Twin and other multiscrew extruders were originally developed to correct the problems that existed with single-screw extruders. Later, single-screw designs and material developments practically eliminated all the original problems.

The conveyance and flow processes of multiscrew extruders are very different from those in the single-screw extruders. The main characteristics of multiscrew extruders include: (a) high conveying capacities at low speeds, (b) positive and controlled pumping rates over wide ranges

Plastic	*Density*	*Application*
mPE plastomers	0.900–0.915 g/cc.	Coextrusion, heat-seal layers
mPE elastomers	0.854–0.900 g/cc.	Stretch-cling films
mLLDPE	0.916–0.925 g/cc.	Stretch films, industrial sacks, shopping bags, form/fill/seal applications, frozen food packaging
mMDPE	0.932 g/cc.	Good-optics food packaging films, personal care, agriculture
High-performance, high-clarity LLDPE	0.909–0.912 g/cc.	Liner, stretch film, lamination, highly transparent food packaging
Ethylene-propylene resins, random heterophasic PP copolymer	0.890 g/cc.	Coextruded form/fill/seal film, hot fill, heavy duty, medica land personal care films, blends with other PP or PE grades

Table 5.18 Extruded plastic product applications

	Rigidity	Heat resistance	Dimensional stability	Tensile strength	Impact resistance	Thermal conductivity	Acoustic properties	Price reduction by direct extrusion of the individual components
Chalk ($CaCO_3$)	+	+	+	–	+	+	+	+
Talcum	+	+	+	–	–	+	+	+
Barium sulfate ($BaSO_4$)	+	+	+	–		+	+	+
Wood dust	+		+	–	–		+	+
Glass fibers	+	+	+	+	–			

+Property improvement.
–Property loss.

Table 5.19 Effect of additives on properties and cost

of temperatures and coefficients of frictions, (c) low frictional (if any) heat generation, which permits low heat operations, (d) low contact times in the extruders, (e) relatively low motor-power requirements and self-cleaning actions with high degrees of mixing, and (f) very importantly, positive pumping abilities that are independent of the friction of the plastics against the screws and barrels, which is not reduced by backflow. Even though the backflow theoretically does not exist, multiscrew extruders' flow phenomena are more complicated and, therefore, far more difficult to treat (theoretically) than flow in single-screw extruders. The result has been that the machine designer has to rely mainly on experience.

Central blending	*At-the-throat coloring*
If the color blend is not correct (parts too light or more color than necessary in the mix), the mix must be emptied from the hopper and reblended, a costly and time-consuming process. Even then, parts may not have the exact depth of color you desire.	Adjustments to color are made with immediate results. If parts are light in color, you can make an adjustment and see the results in minutes.
Although some parts, as a cost-saving measure, could be with less color than others, the mixing of several batches with different let-down ratios is generally considered impractical.	Every part can be custom blended while in production for optimum color usage and cost savings. Settings are recorded and used for exact repeatability on future runs.
Conveying of blended material often produces separation of the color pellets from the natural material due to the much heavier bulk density of color pellets. The result is inconsistent coloring of parts.	Material is conveyed without color; separation is not a factor.
Normal machine vibration and flow patterns in the hopper cause some separation of the color pellets during residence time in the hopper. Again, the result is inconsistent coloring of parts.	Color is metered only as parts are fabricated. There is no residence time for separation to occur.
When a production run for one color is over, your hopper must be emptied in preparation for the next color. Time is lost. Sometimes material is lost. If hopper and conveying system are not adequately cleaned, color contamination and rejected parts will result.	Color changes require no emptying of the main material hopper. Production does not stop. Transition time to the new color is several minutes.
If your material must be dried before use, the drying of color blends presents many logistical problems: more driers, careful planning for transition to the next color, cleanout of dryers, etc.	You dry only natural material without having to concern yourself about color changes.

Table 5.20 Different methods of color blending

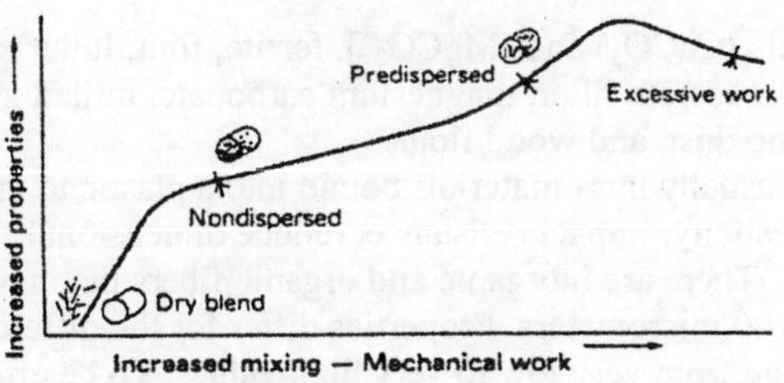

The use of a virtually endless array of materials permits compounding from the raw material suppliers to the fabricators imparting specific qualities to the basic raw materials (polymers) and expanding opportunities for plastics. Compounding relies on the polymerization chemistry to mechanical mixing to combine a base polymer with modifiers, additives, and other plastics to develop new plastics. Clearly these many combinations are endless so that new materials are always on the horizon to meet new industry requirements.

Compounding to change and improve the physical and mechanical properties of plastics makes use of a wide variety of materials. The major and large market for these materials, such as additives, fillers, and reinforcements, continues to expand as the demand for reducing the cost of plastics, plastics to function in wider or more extreme markets, and under stricter regulatory regimes continue to expand.

Additive They are substances compounded into a plastic to modify its characteristics. They are basically physically dispersed in a plastic matrix without affecting significantly the molecular structure of the TPs. In TS plastics additives such as crosslinking; catalyst; and other agents do purposely affect their structure. They are classified according to their specific functions rather than a chemical basis, While some additives have broad applications and are adaptable to many TPs and TS plastics, others are used exclusively with specific plastics. Examples of classifications are:

1. assist processing (processing stabilizers, processing aids and flow promoters, internal and/or external lubricants, thixotropic agents, etc.);
2. modify the bulk mechanical properties (plasticizers or flexibilizers, reinforcing agents, toughening agents, etc.);
3. reduce formulation costs (diluents and extenders, particle fillers, etc.);
4. surface properties modifiers (antistatic agents, slip additives, anti-wear additives, anti-block additives, adhesion promoters, etc.);
5. optical properties modifiers (colorants, pigments, dyes, nucleating agents, etc.);
6. anti-aging additives (anti-oxidants, UV stabilizers, fungicides, etc.);
7. others (blowing agents, flame retardants, etc.).

Examples of only a few additives are carbon black, carnauba wax, coconut shell, coke dust, macerated filler, shell flour, vermiculite, and wax. Many additives, especially those that are conductive may affect electrical properties. Most plastics, which are poor conductors of current, build up a charge of static electricity. Antistatic agents can be used to attract moisture, reducing the likelihood of a spark or discharge.

Filler These low cost fillers, also called extenders, with their many different inert substances (organic and inorganic with low to high molecular weights) are added to plastics principally to reduce costs. They may also improve processing and physical and mechanical properties, particularly hardness, thermal insulation and stiffness. The particles are usually small, in contrast to those of reinforcements.

The mineral fillers are a large subclass of inorganic fillers comprised of ground rocks as well as natural, refined, or synthetic minerals. Commodity minerals are relatively inexpensive and are used mostly as additive extenders. Other fillers, so-called specialty minerals, are usually the reinforcing types. There are also inherently small particle size fillers such as talc and surface chemically modified fillers. The inert filler are those added to plastics to alter the properties of a product through physical rather than chemical means.

Examples of a few are alpha cellulose, ash, calcium carbide, calcium carbonate, carborundum, channel black, china clay, coral, coke dust, diatomaceous earth, dolomite [double carbonate of lime and

Table 5.21 Better mixing of compounds results in improved processing

magnesia filler having the formula ($CaCO_3$) and ($MgCO_3$)], ferrite, flint, fuller's earth, glass spheres, hemp, keratin, lampblack, leather-dust, macerate filler, magnesium carbonate, milled glass, mica, pumice, quartz, sawdust, talc, vermiculite, volcano dust, and wood flour.

Reinforcement They are strong, usually inert materials bound into a plastic to improve its properties such as strength, stiffness/modulus of elasticity, impact resistance, reduce dimensional shrinkage, etc. They include fiber and other forms of material. There are inorganic and organic fibers that have the usual diameters ranging from about one to over 100 micrometers. Properties differ for the different types, diameters, shapes and lengths. Their properties range from very low to very high values. To be effective, the reinforcement must form a strong adhesive bond with the plastic; for certain reinforcements special cleaning, sizing, finishing, etc. treatments are used to improve the bond.

Types of reinforcements include fibers of glass, carbon, graphite, boron, nylon, polyethylene, polypropylene, cotton, sisal, asbestos, metals, whiskers, etc. Other types and forms of reinforcements include whiskers, bamboo fibers, burlap fibers, carbon blacks, platelet forms (includes mica, glass, and aluminum), and hemp fibers.

Mixing General theory of mixing usually considers a non-random or segregated mass of at least two components and their deformation by a laminar or shearing deformation process. The object of the shearing is to mix the mass in such a way that samples taken from the mass exhibit minimal variations, ultimately tending to be zero. The three basic principles are:

1. interfacial area between different components must be greatly increased to decrease striation thickness,
2. elements of the interface must be distributed uniformly,
3. ratio of mix within any unit or the whole is the same.

A mixture can be described in terms of the statistical deviation of a suitable number of samples from a mean, the sample sizes being dependent upon some length, volume, or area characteristics of the mixture or its properties (Chapter 24). Types of mixtures and methods of evaluation range from simple to complex systems. There is the usual dispersive and/or distributive mixing.

Dispersive Mixing It is the mixing of a fluid/plastic melts with a solid/unmelted plastic that exhibits a yield point. It involves the final melting of a plastic or breaking down an additive such as a pigment in the manufacture of a color concentrate.

Distributive Mixing Distributive mixing is the commingling of two fluid/plastic melts so that the scale of fluid separation reduces to where another process (diffusion or a chemical reaction) can occur. The mixing is in a laminar flow regime which is characteristic of NEAT plastics (Chapter 1). It is distinguished by the deformation of the fluid interfaces as a result of the applied shear strain. Distributive mixing relates the amount of interfacial area growth to the fluid strain rate, as distinguished from dispersive mixing, that is a function of the magnitude of the stress. The latter accomplishes droplet and agglomerate breakup; the former is the distribution of those components.

Table 5.21 (*continued*)

Machine conditions	*To increase output*	*To improve end product quality*
Barrel pressure through valving (back pressure)	Decrease	Increase
Screen pack (Screen mesh or number of screens)	Decrease	Increase
Screw temperature	Increase	Decrease
Screw speed	Increase	

Table 5.22 Guides for increasing extruder output and product performance

Class	Material types	properties (rad)	range (°F)	radiation stability [a]	Vacuum stability	Remarks
I	*Films*					
	Polyimide (H-film)	5×10^8	—423 to 600	Good	Good	Preferred materials for internal and external structural and related applications (e.g., inflatable space structures) particularly where long-term resistance to UV radiation (films) is required and moderate temperatures and radiation doses are encountered; good vacuum stability and mechanical strength properties
	Polyvinylfluoride (Teslar)	10^8	—100 to 250	Good	Good	
	Mylar	10^8	—423 to 300	Fair [b]	Good	
	Foams (closed cell, rigid thermoset types)					
	Polyurethane (alkyd-isocyanate)	10^9	—423 to 300	Not applicable if protected by sandwich skin	Good (Poor for open cell and some flexible types)	
	Polyether	10^9	—423 to 300			
	Phenolic	10^9	—423 to 450			
	Epoxy	10^9	—423 to 250			
	Silicone	10^9	—423 to 500			
	Fibers					
	Dacron (polyester)	10^8	—423 to 300	Fair [b]	Good	
	Nylon	5×10^7	—320 to 300	Fair	Good	
II	*Films*					
	Polyethylene (high density)	10^8	—320 to 225	Poor	Good	Not preferred for long-term exposures to UV radiation unless surface is metalized but may be used for shorter missions internally or externally at low to moderate exposures and moderate temperatures; good vacuum stability
	Polypropylene	10^8	—423 to 275	Fair	Good	
	Kel-F	10^6	—423 to 350		Good	
	Teflon (TFE)	5×10^4	—423 to 500	Good	Good	
	Teflon (FEP)	5×10^5	—423 to 450	Good	Good	
III	*Films*					
	Polyethylene (low density)	10^8	—320 to 180	Poor [b]	Good	Not recommended for external applications because of UV radiation susceptibility; may be used internally in applications where low-to-moderate temperatures are encountered
	Polyvinyl chloride (PVC)	10^8	—100 to 165	Poor [b]	Poor	

[a] This is probably the most severe environment affecting the properties of thin films in external usage in space. It is beneficial to metallize (vacuum deposit reflective coatings such as Al) the surface of such films whenever possible to prevent UV degradation. This practice also improves vacuum stability.
[b] Discolors and embrittles.

Table 5.23 Properties of extruded films, foams, and fibers

Material in machine	*Material changing to*	*Mix with rapid purge and soak*	*Temperature-bridging material*	*Follow with*
ABS	PP	ABS	–	PP
ABS	SAN	SAN	–	SAN
ABS	Polysulfone	ABS	PE	Polysulfone
ABS	PC	ABS	PE	PC
Acrylic	PP	Acrylic	–	PP
Acrylic	Nylon	Acrylic	–	Nylon
TPE	Any material	PE	–	New material
Nylon	PC	PC	–	PC
Nylon	PVC	Nylon	PE	PVC
PBT	ABS	PBT	PE	ABS
PC	Acrylic	PC	–	Acrylic
PC	ABS	PC	PE	ABS
PC	PVC	PC	PE	PVC
PE	PP	PP	–	PP
PE	PE	PE	–	PE
PE	PS	PS	–	PS
PETG	Polysulfone	PETG	–	Polysulfone
Polysulfone	ABS	Polysulfone	PE	ABS
Polysulfone	ABS	Cracked acrylic	–	ABS
PP	ABS	ABS	–	ABS
PP	Acrylic	Acrylic	–	Acrylic
PP	PE	PE	–	PE
PP	PP	PP	–	PP
PS	PP	PP	–	PP
PVC	Any material	LLDPE or HDPE	–	New material
PVC	PVC	LLDPE or HDPE	–	PVC
PPS	PE	PPS	PE	PE
SAN	Acrylic	Acrylic	–	Acrylic
SAN	PP	SAN	–	SAN

Table 5.24 Approaches to changing plastic being extruded to eliminate or reduce processing problems

Although there are very few twin-screw (TS) extruders in comparison to the many more single-screw extruders, they are used to produce products such as window and custom profile systems. Their major use is in compounding applications (to be reviewed). The popular, common twin-screw extruders (in the family of multiscrew extruders) include tapered screws or parallel cylindrical screws with at least one feed port through a hopper; a discharge port to which a die is attached; and process controls for temperature, pressure, screw rotation (rpm), and melt output rate (2).

Twin-screws with intermeshing counter-rotating screws are principally used for compounding. Different types have been designed that include corotating and counter-rotating intermeshing twin screws. The nonintermeshing twin screws are offered only with counter-rotation. There are fully

Material to be purged	*Recommended purging agent*
Acetal	Polystyrene; avoid any contact with PVC
Engineering resins	Polystyrene, low melt index, HDPE, cast acrylic
Fluoropolymers	Cast acrylic, followed by polyethylene
Polyphenylene sulfide	Cast acrylic, followed by polyethylene
Polysulfone	Reground polycarbonate, extrusion-grade PP
Polysulfone/ABS	Reground polycarbonate, extrusion-grade PP
PPO	General-purpose polystyrene, cast acrylic
Thermoset polyester	Material of similar composition without catalyst
Polystyrene	Cast acrylic
PVC	Polystyrene, general-purpose, ABS, cast acrylic
ABS	Cast acrylic, polystyrene
Nylon	Polystyrene, low melt index HDPE, cast acrylic
PBT polyester	Next material to be run
PET polyester	Polystyrene, low melt index HDPE, cast acrylic
Polycarbonate	Cast acrylic or polycarbonate regrind; follow with polycarbonate regrind; do not purge with ABS or nylon

Table 5.25 Examples for purging different plastics

There are four essential instruments that enable you to know and control the final state of the extruded melt. These are

1) a pyrometer indicating the temperature of the melt entering the die,
2) a wattmeter or ammeter in the drive-motor circuit,
3) a pressure gage just upstream from the breaker plate or valve,
4) a screw-speed indicator.

The readings of these instruments, through proper interpretation, tell even the inexperienced operator a great deal about what is happening inside the extruder. In two-stage machines, whether vented or not, an additional pressure gage located at the start of the first metering zone should be used.

If the temperature and pressure of the melt at the die are closely controlled, then the properties and dimensions of the extrudate can be controlled. The temperature is measured by means of a stock thermocouple whose tip is immersed in the melt stream. This temperature may or may not be close to the set cylinder temperature, which has a relatively minor influence on the stock temperature. In the near-adiabatic operation of a modern extruder the stock temperature is almost always higher than that of the cylinder.

It is important to keep the die temperature closely equal to the stock temperature. Since there is virtually no mixing in the die passages, differences in temperature between metal and melt can cause troublesome variations in extrudate dimensions.

Table 5.26 Simplified procedure for examining melt performance

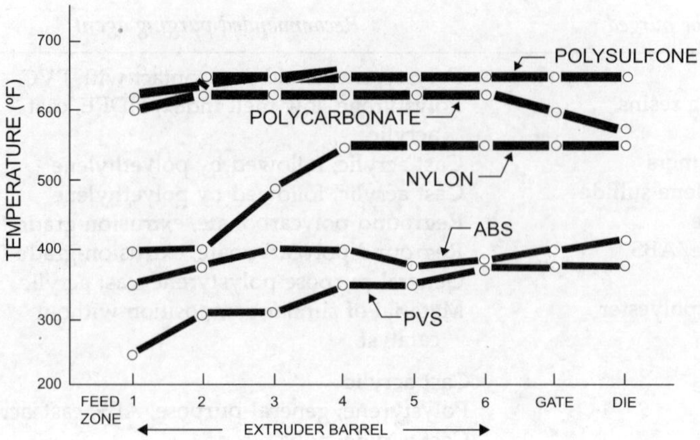

Figure 5.59 Temperatures for different plastics in different zones of extruder barrels

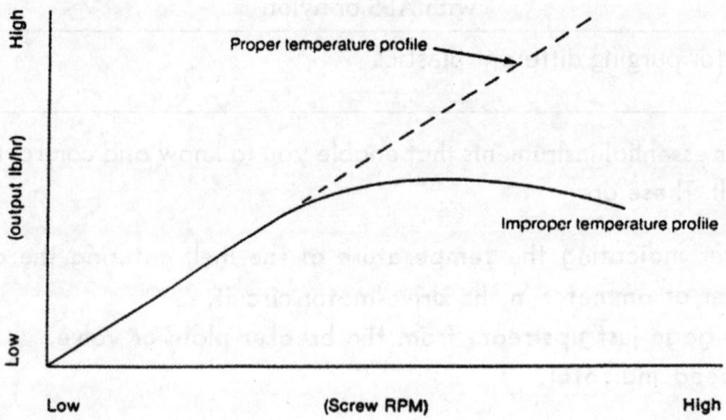

Figure 5.60 Example of barrel throat temperature influencing plastic output

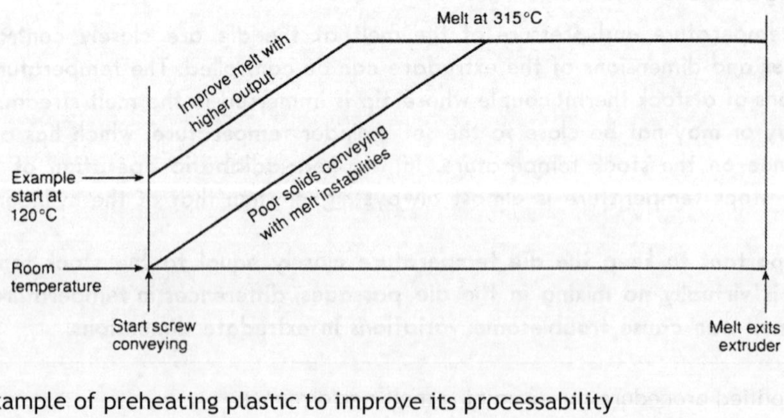

Figure 5.61 Example of preheating plastic to improve its processability

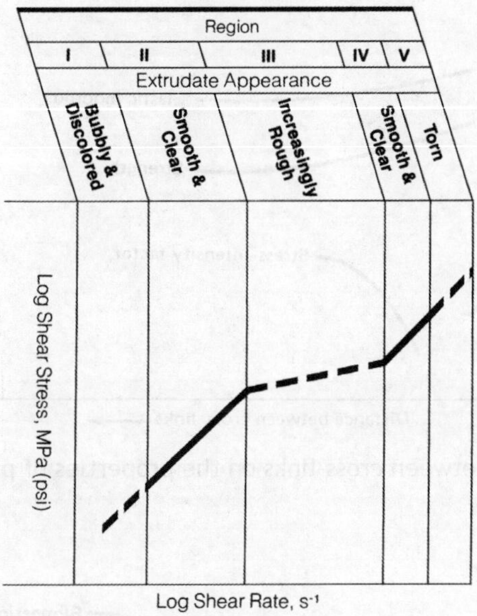

Figure 5.62 Example of melt's shear stress vs. shear rate

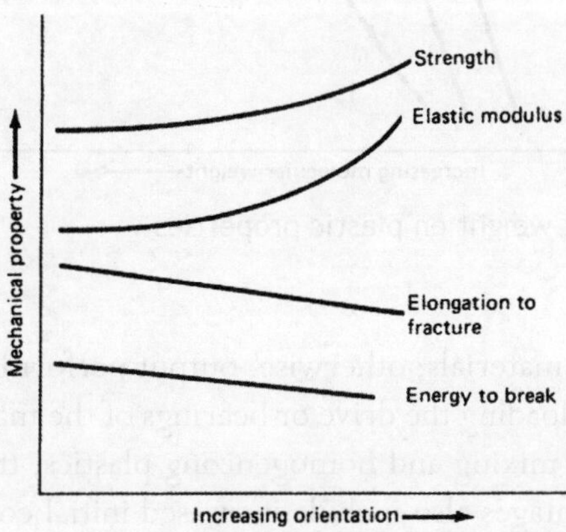

Figure 5.63 Effects of uniaxial orientation on the properties of plastics

intermeshing and partially intermeshing systems and open- and closed-chamber types. In the past, major differences existed between corotating or counter-rotating; today, they work equally well in about 70% of compounding applications, leaving about 30% where one machine may perform dramatically better than the other.

Similar to the single-screw extruder, the twin-screw extruder, including multiscrew, has advantages and disadvantages. The type of design to be used will depend on performance requirements for a specific material to produce a specific product. With the multiscrews, very exact metered

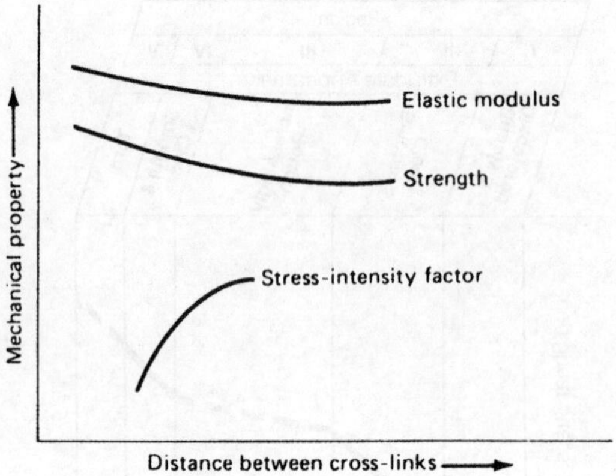

Figure 5.64 Effects of distance between cross-links on the properties of plastics

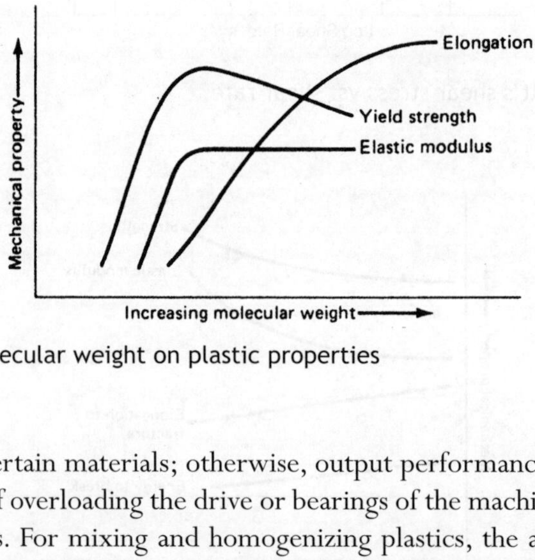

Figure 5.65 Effects of molecular weight on plastic properties

feeding is necessary for certain materials; otherwise, output performance will vary. With overfeeding, there is a possibility of overloading the drive or bearings of the machine, particularly with counter-rotating screw designs. For mixing and homogenizing plastics, the absence of pressure flow is usually a disadvantage. Disadvantages also include increased initial cost due to the extruder's more complicated construction, higher maintenance costs, and the potential for difficulty with heating.

The market for counter-rotating twin-screw (TS) extruders is basically dominated by two designs. The one called parallel TS extruder has cylindrical screws and the other TS extruder is fitted with conical screws. Regarding performance, the superiority of the conical principle over the parallel does not only appear in the theoretical comparisons but in practice as confirmed by users. The flexibility of the conical screw results in an extrudate of consistent quality at both low and high output rates that is not sensitive to raw material fluctuations. It appears that the parallel screws have reached their efficiency limits unless a means of drastically increasing the screws' torsional strength is developed. Conical screws continue to offer what appears to be endless improved benefits through further development.

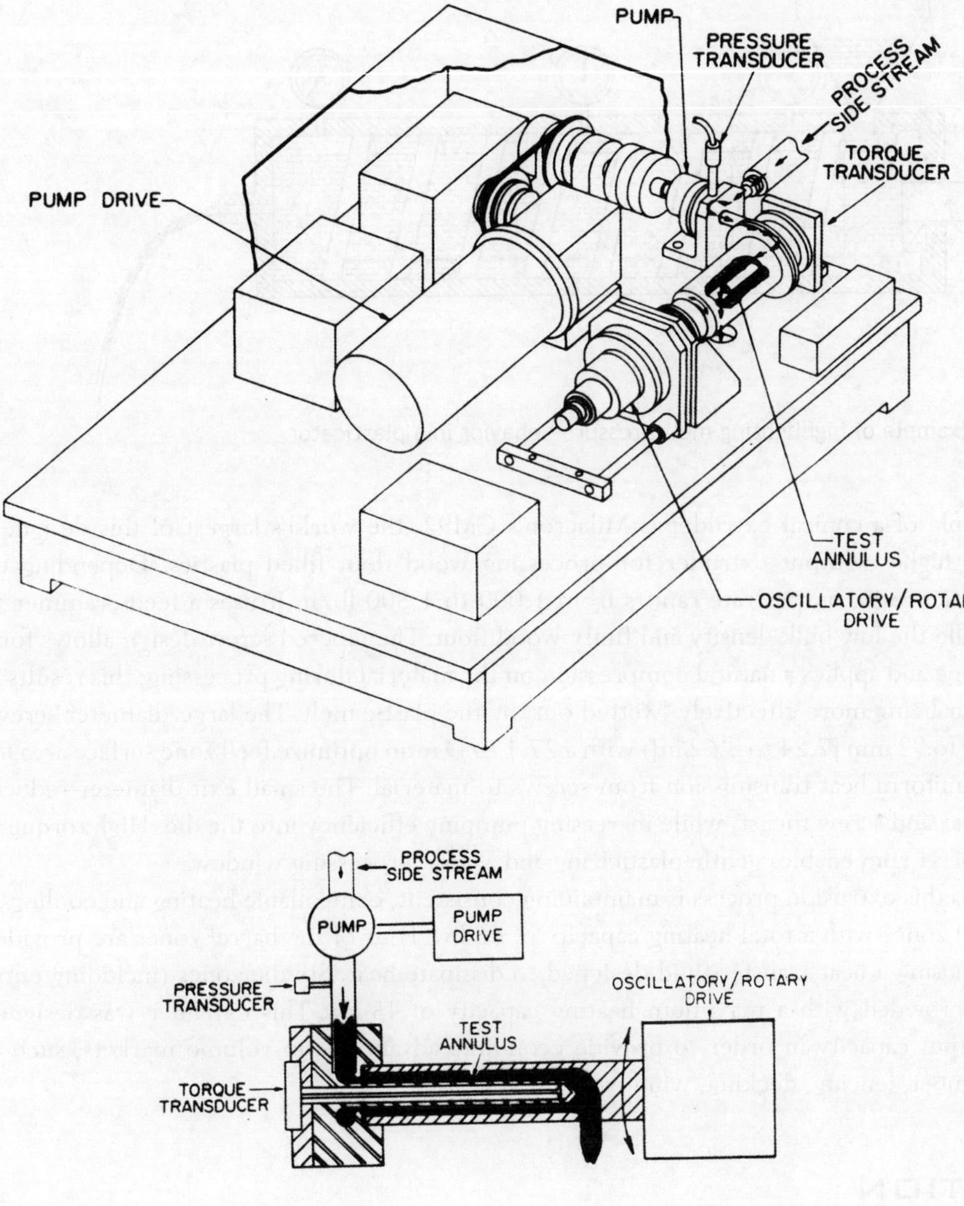

Figure 5.66 Example of in-line rheometer to obtain instant melt behavior during extrusion

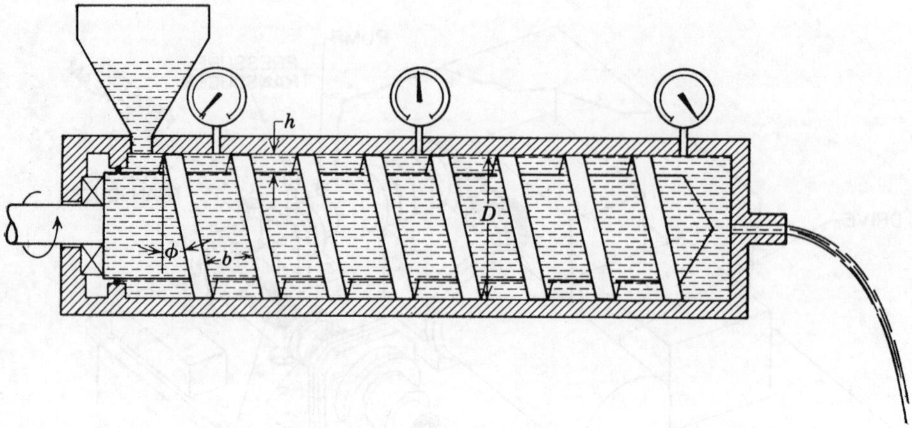

Figure 5.67 Example of highlighting melt pressure behavior in a plasticator

An example of a conical extruder is Milacron's CM92, the world's largest of this design. It produces the highest output extruder for processing wood flour-filled plastics. Depending on the flour-plastics ratio, output rate ranges from 1,000 to 1,800 lb/h. It uses a feed crammer to properly handle the low bulk density and fluffy wood flour. The tapered screw design allows for a larger feed zone and applies a natural compression on the material during processing; this results in the wood flour being more effectively "wetted out" by the plastic melt. The large-diameter screws (184 tapering to 92 mm [7.24 to 3.62 in]) with a 27:1 L/D ratio optimize feed zone surface area for faster, more uniform heat transmission from screws to material. The small exit diameter reduces rotational shear and screw thrust, while increasing pumping efficiency into the die. High torque at a low speed of 34 rpm enables gentle plasticizing and a wide processing window.

Critical to this extrusion process is maintaining consistent, controllable heating and cooling. It has five barrel zones with a total heating capacity of 86 kw. Four of the barrel zones are provided with cooling, using a heat-transfer fluid designed to dissipate heat. Six die zones (including entry adapter) are provided with a maximum heating capacity of 45 kw. This extruder was designed with high output capacity in order to provide economic advantages in volume markets, such as composite lumber, fencing, decking, windows, and doors.

OPERATION

START-UP

Machine operation can take place in three stages, going from start-up to shutdown. The first stage requires operating the extruder for warm-up with operational settings of upstream and downstream equipment. The next stage involves setting the required processing conditions to meet product

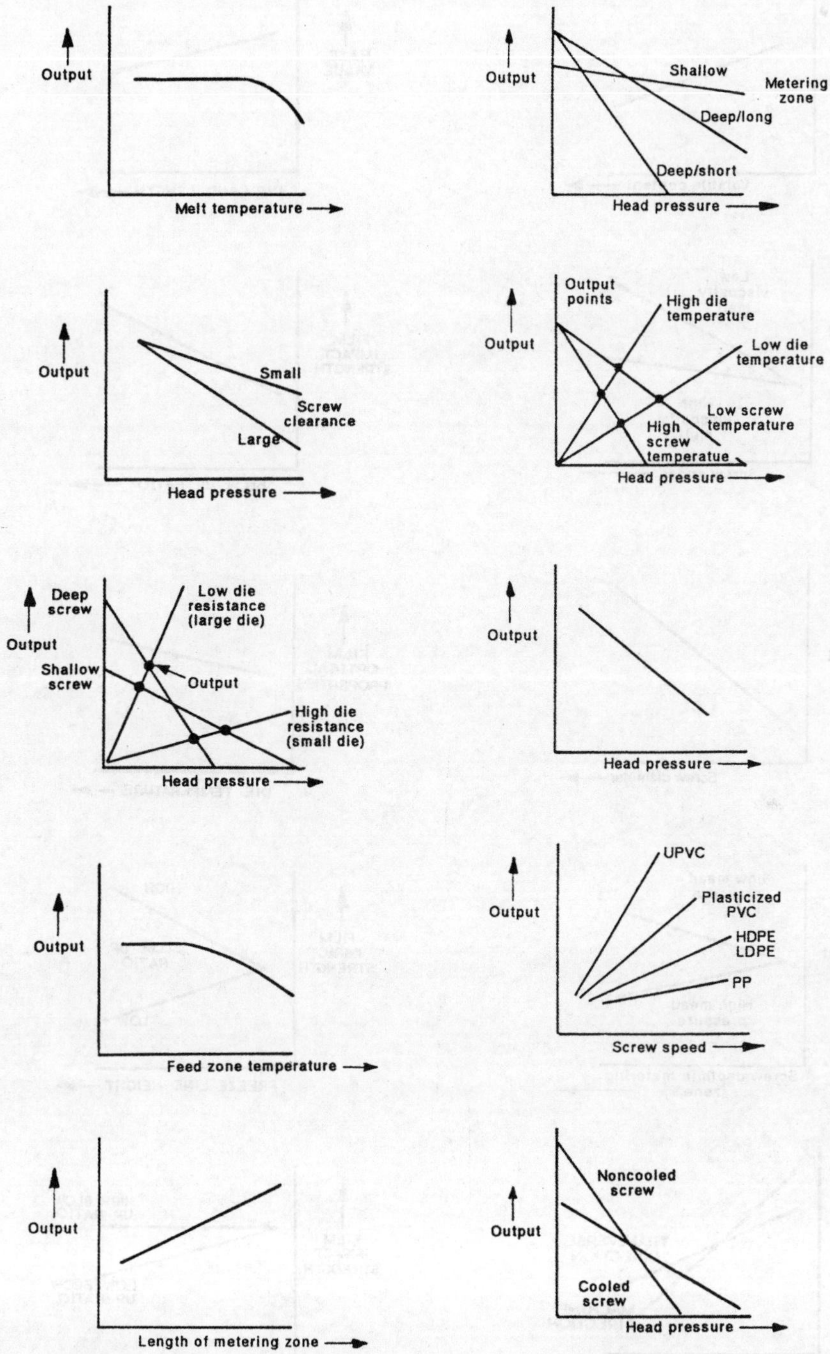

Figure 5.68 Examples of properties vs. changes in process performances

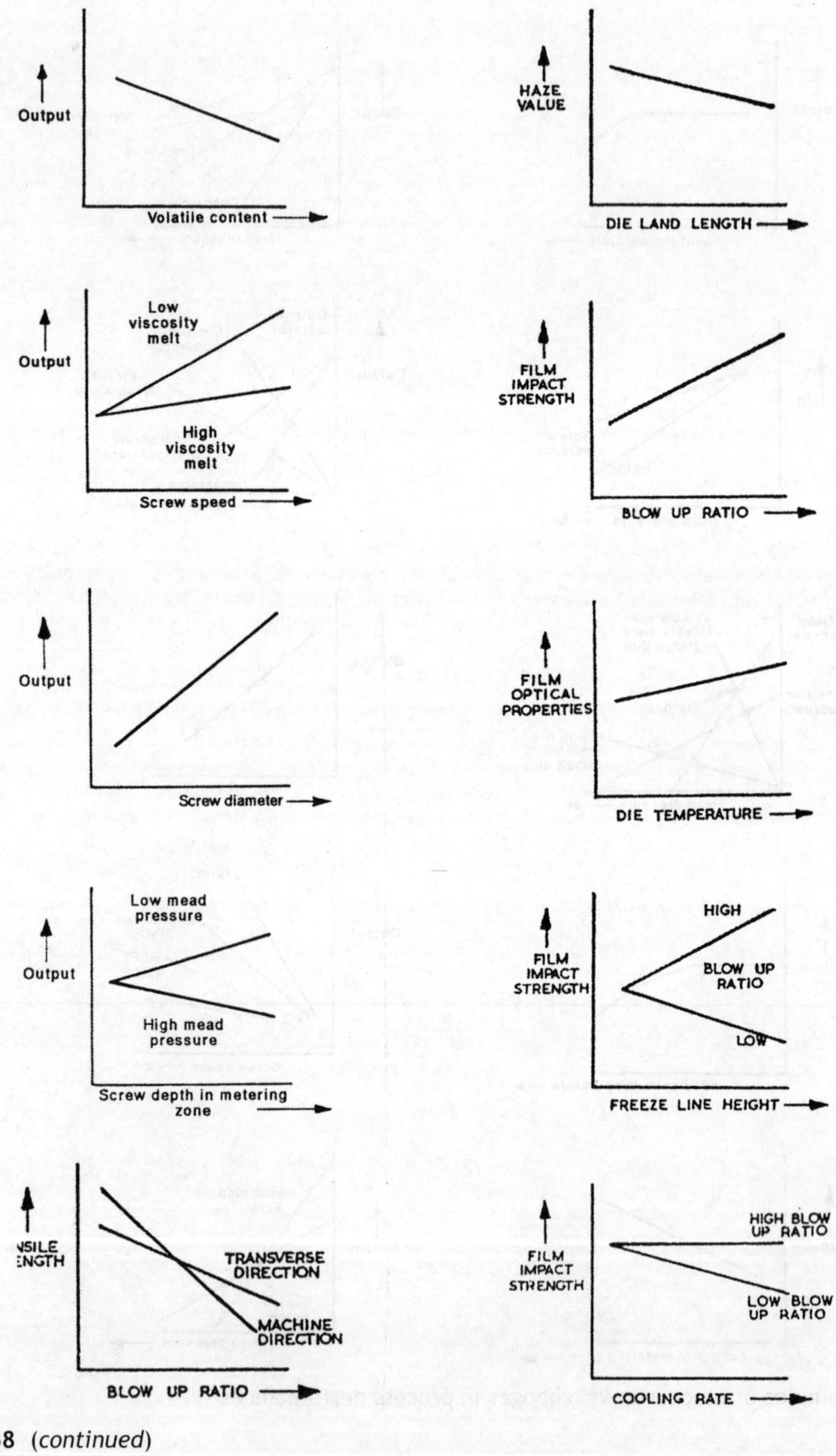

Figure 5.68 *(continued)*

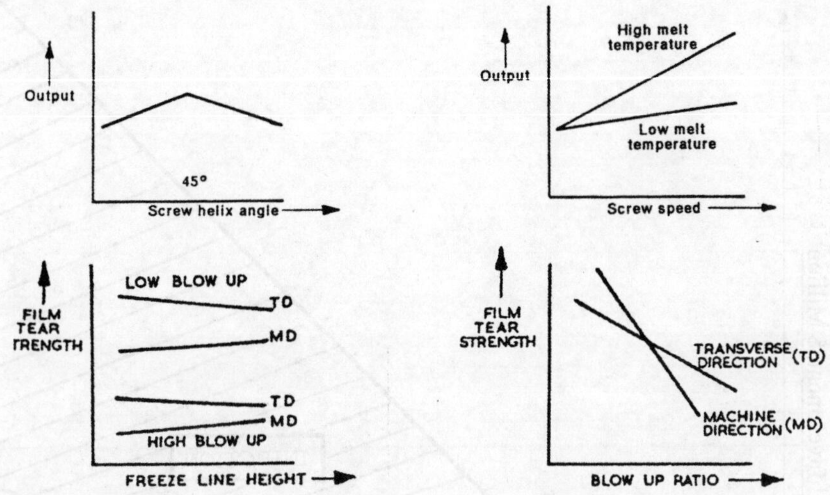

Figure 5.68 (*continued*)

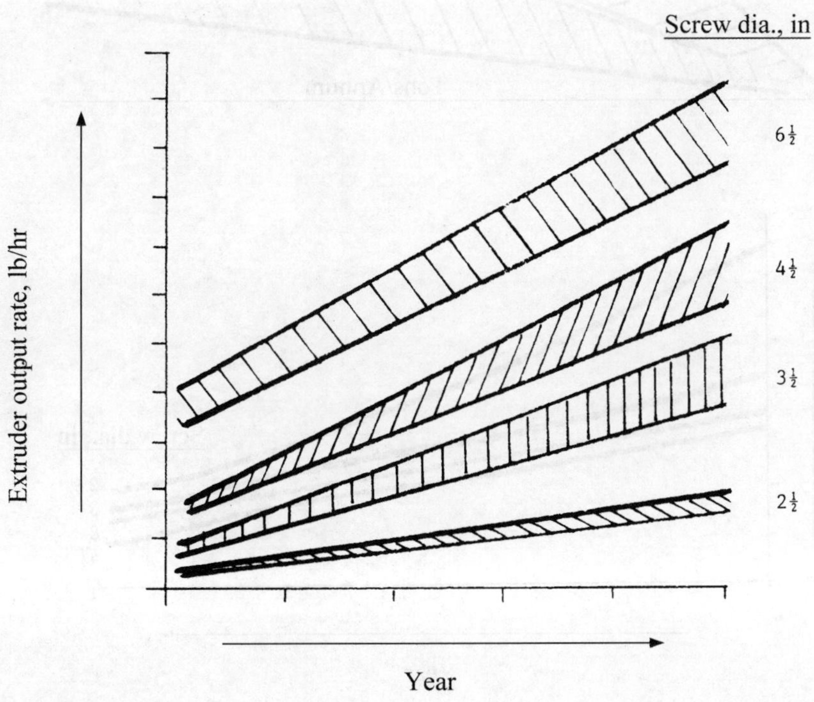

Figure 5.69 Example of extruder output increases vs. time

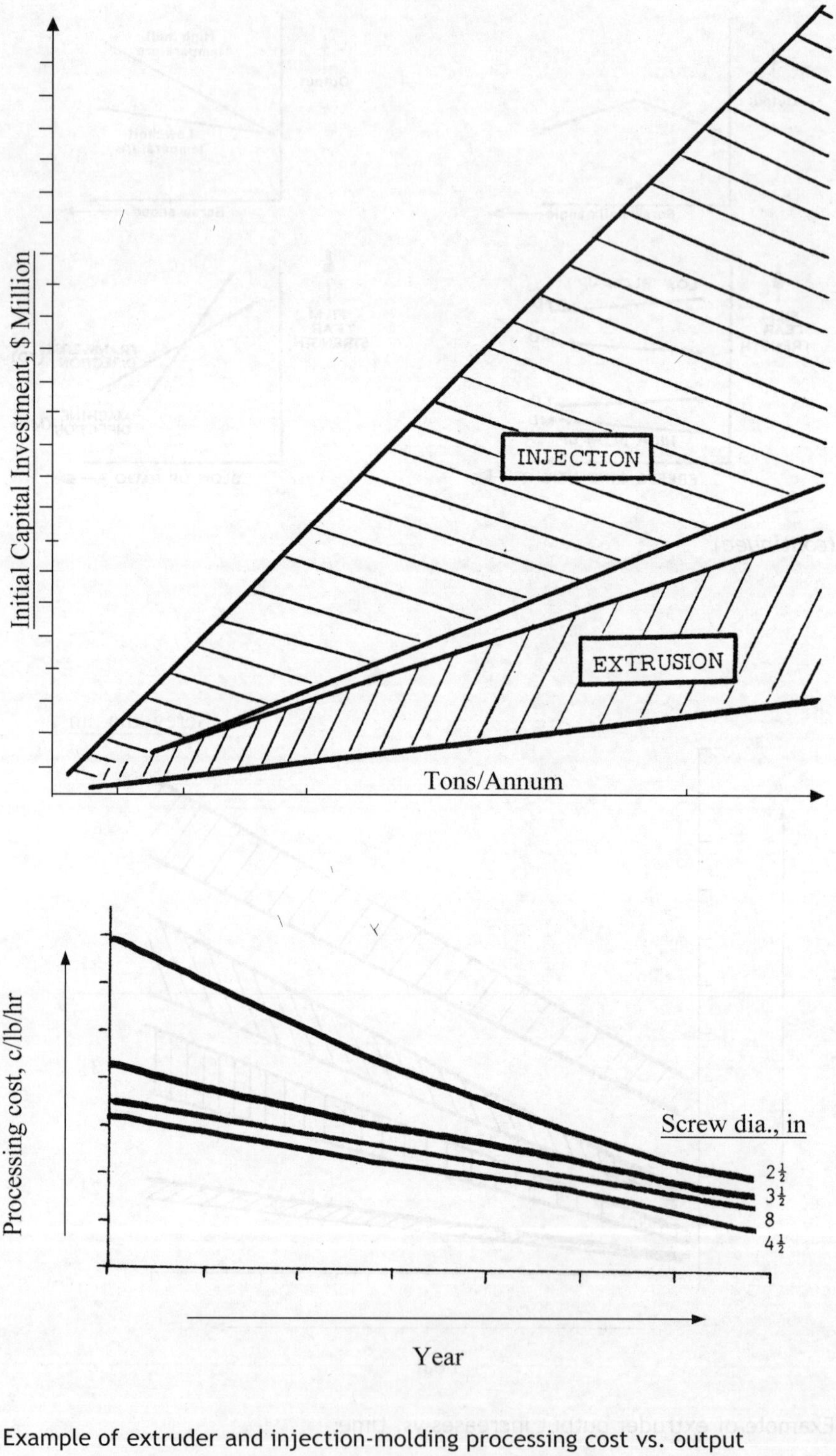

Figure 5.70 Example of extruder and injection-molding processing cost vs. output

requirements at the lowest cost. The final stage is devoted to fine-tuning and problem solving the complete line. A successful operation requires close attention to many details, such as the melt quality, a temperature profile adequate to the melt but that does not degrade the plastic, production of a minimum of scrap, and procedures for start-up and shutdown that will not (or will minimally) degrade the plastic. Processors must also become familiar with troubleshooting guides (chapter 27).

Extrusion operation differs based on the type of product to be produced and the plastic to be processed. However, on start-ups there are some aspects that all processes have in common. The process differs somewhat if one has to clean and empty the machine compared with a machine that contains plastic and is reheated. A main source of difficulty in starting an extrusion run is impatience of people (as previously reviewed throughout this book). It is necessary to wait until the barrel and die are at the correct operating temperatures before starting; otherwise, problems develop, such as having hot or cool melt spots, overstressed melt sections, overloading the screw with plastics, plastic bridging at the hopper, and degrading plastic. Starve feeding plastic on start-up at a low screw speed and until melt is pumped from the die helps prevent bridging of the screw.

Consider purging the extruder plasticator when it contains plastic that can be detrimental to start-up and/or produces unacceptable products (chapter 3). If a plastic was left in the barrel for awhile with the heat off, the processor must determine if the material is subject to shrink. This could have caused moisture entrapment from the surrounding area, producing contamination that would require cleanup (this situation could also be a source of corrosion in/on the barrel/screw. Even with the same plastic in the machine from a previous run, the entire machine should be cleaned and/or purged, including the hopper, barrel, breaker plate, die, and downstream equipment.

When starting up a new extrusion setup, start the screw rotation at about 5 rpm. In the meantime, look into the air gap between the feed throat and throat housing and make sure the screw is turning. Screws have been installed without having their keys in place or keys have fallen out during installation. Also make sure that antiseize material is applied to the drive hub to help installation and removal. If the key is left out and the drive quill is turning and the screw is not, the screw will not gall to the drive quill.

Prior to start-up, one must check certain machine conditions that should be listed on some kind of worksheet from the machine manufacturer, plastic supplier, and/or, more importantly, the plant setup person with experience. The previous section on process control and chapter 3 can provide guides. A careful checkup includes:

1. inspecting heater bands and electrical connections;
2. inspecting thermocouples, pressure transducers, and their connections;
3. inspecting all machine heating, cooling, and ventilation systems to ensure adequate flow;
4. ensuring flow path through the extruder is not blocked;
5. having a bucket or drum, half filled with water, to catch extrudate whenever purging or initial processing of plastics is contaminated with gaseous by-products;

6. reviewing the machine's operating manual for other start-up checks and requirements that have to be met, such as motor load (amperage) readings.

Details for operating the machine prior to operating the extrusion downstream line are based on what the plastic being processed requires, such as temperature settings and screw rpm, which is available from the material supplier and/or experience. Start-up procedures involve certain precautions.

Within various types of the family of plastics (polyethylenes [PEs], polyvinyl chlorides [PVCs], polypropylenes [PPs], etc.), each type usually has different heat profiles and other settings (Figs. 5.59 to 5.61 and Tables 5.12, 5.13, and 5.17). Experience shows how to set the profile and/or obtain preliminary information from the material supplier. Degrading or oxidizing certain plastics is a potential hazard that occurs particularly when the extruder is subject to frequent shutdowns. In this respect, the shutdown period is even more critical than the start-up period.

In setting up the barrel temperature profile, start with the front-to-rear zones (die end to feed section). The heat controllers are set slightly above the plastic melting point prior to turning on the heaters. Heat up should be gradual from the ends to the center of the barrel to prevent pressure buildup from possible melt degradation. With this start-up, take the following steps:

1. Gradually increasing heaters, check for deviations that might indicate burned-out or runaway heaters by slightly raising and lowering the controller set point, checking if power goes on and off.
2. Following with all heaters slightly above the melt point, adjust to the desired operating heats; time required to reach temperature equilibrium may be ½ to 2 hours, depending on the size of the extruder.
3. If overshooting occurs, it is usually observed with the on/off controllers.
4. After set heats have been reached, put the plastic in the hopper and start the screw at a low speed such as 2 to 5 rpm; some plastics, such as nylons, may require 10 to 20 rpm.
5. Observe the amperage required to turn the screw, stop the screw if the amperage is too high, and wait a few minutes before restart.
6. Observe and maintain the required melt pressure; the extruder barrel pressure should not exceed 1,000 psi (7 MPa) during the start-up period
7. Run the machine for a few minutes and purge the initial run until a good quality extrudate is obtained visually. Experience shows what it should look like; for example, a certain size and amount of bubbles or fumes may be optimum for a particular melt based on one's experience (or the trainer's experience) after setting up all controls.
8. Turn up the screw to the required rpm if not already properly set, checking to see that maximum pressure and amperage are not exceeded.

9. When time permits, after running for awhile, consider stopping the machine, let it start cooling, and remove the screw to evaluate how the plastic performed from the start of feeding to the end of metering. Thus, one can see if the melt is progressive and can relate it to screw and product performances.
10. Adjust the die with the controls it contains, if required, at the desired running speed.

Once the extruder is running at maximum performance, the setup controls for takeoff/downstream equipment may require more precision settings and/or changes in the extruder to meet downstream equipment requirements.

The extrudate can start its tract from the die by threading (also called stringing up) through the cooling and takeoff downstream equipment to its haul-off initially at a slower speed than production operation. When possible, rather than taking the extrudate from the die and being directed through the equipment, the hot melt is made to weld to the thread-up end already in the equipment that is usually a leftover from the previous run. In turn, the thread-up is pulled through the line carefully and safely. If a welding action does not occur, a metal hook may be pushed into the melt. Cooling of this joint is required to give it strength. Care is needed to avoid making a lump too large to go through the line.

This operation requires the personal skill of the start-up person. That person is required to integrate/interrelate the extruder and downstream equipment. Extruder screw speeds and haul-off rates may then be increased. Downstream equipment is adjusted to meet its maximum operating performance, such as having the vacuum water tank operate with its proper level and vacuum applied.

The extruder can be fine-tuned to obtain the final required setting for meeting the desired output rate and product size.

Start-up operations are made at rates people can handle. The process is very slow compared to standard operating speeds. The puller starts its movement at just about the same speed the person has been pulling, or there may be a pileup or tear-off of melt at the die. That usually will mean threading up again.

The operator's skill will involve pulling continuously at a steady rate. The person acts like a machine or robot for a few minutes. Skill on the part of a good operator is very evident at start-up.

Cooling the extrudate during hand pulling is important. It gives strength and form stability to the extrudate. Without cooling, the melt will string out and pull apart. The steadiness of pulling and the evenness of cooling determine what the hand-pulled product will look like and how easily it can be threaded and fed into the takeoff.

After this operation, follow up on the product's dimensions or the factors determining that the product is meeting requirements. Minor changes in speed may be needed. Adjustments to centering the die or die opening may be necessary if there are thick or thin spots. Product measurement and die adjustment are continued until a satisfactory product is made. Frequently, this process may take an hour or more. During this time, scrap is produced; when practical, it should be used as a regrind and reused. To reduce this time schedule significantly, program controllers provide quick means to balance out all the control settings to produce the desired product.

SHUTDOWN

It is common to run the extruder to an empty condition when one is shutting down. This action ensures that there is no start-up with cold plastic; an improper start-up condition, such as this, could overload the extruder. Some extruders, such as those processing PE film, are shut down with the screw full of plastic. This prevents air from entering and oxidizing the plastic. Because PVC decomposes with heat, to ensure that this material is completely removed at shutdown, purge processed material, such as low-melt PE, that can remain in the barrel (chapter 3). On start-up, it is preferable to raise barrel heat slightly above its normal operating temperatures. The higher temperature ensures that unmelted plastic will not produce excessive torque in the screw. In regard to the downstream equipment, such as with film or sheet lines, consider leaving some "threading" for an easy start-up, as reviewed.

The shutdown is usually very simple. Procedures for shutdown without cleanout starts by stopping the feeding of plastic into the plasticator and reducing all heat settings to the melt heat. Reduce the screw speed to 2 to 5 rpm, purging the plastic if required into a water bucket or drum prior to reducing the melt heat. The screw rotation continues until no more plastic exits the die. Rotations of the screw as it stops will result in so-called pumping the screw dry of plastics.

With the screw stopped, shut off the heaters and disconnect the cross-head (or die) heaters. Reduce other heaters to about 170° to 330°C (400° to 625°F), depending on the plastic's temperature at the melt point.

If a screen pack with breaker plate is used, disconnect the cross-head (or die) from the extruder and remove the breaker plate and screen. If necessary, appropriate action is taken to clean the screen pack (chapter 17).

For cleanout of the extruder, continue by disassembling the cross-head and clean it while it is still hot. Remove the die, and gear pump if used, and remove as much plastic as possible by scraping with a copper spatula or brushing with a copper wire brush. Remove all heaters, thermocouples, pressure transducers, and so on. Consider using an exhaust duct system above the disassembly and cleaning area, even if the plastic is not a contaminating type. This procedure keeps the area clean and safe.

Follow by pushing the screw out gradually while cleaning with a copper wire brush and copper wool (94). Care should be exercised if a torch is used to burn and remove plastic; tempered steel may be altered and the screw distorted or weakened as well as subjected to excessive wear, corrosion, or even failure (broken).

After screw removal, continue the cleaning, if necessary. Follow by turning off the main electric power switch.

Final cleaning of products, particularly disassembled parts, is best done manually, or much better, in ventilated burnout ovens, if available, operating at about 540°C (1,000°F) for about ninety minutes. For certain parts with certain plastics, useful life could be shortened by corrosion; check with the part manufacture. After burnout, remove any grit that is present with a soft, clean cloth. If water is used, air-blasting to dry is required. With precision-machined parts, water cleaning could be damaging because of the potential of corrosion when certain metals are used.

EXTRUDER LINE

The extruder and its components (die, gear pump, etc.) are costly. However, the other equipment in the line (also called train) can include more expensive equipment; potentially, even one piece of equipment is more costly. Major costs may be incurred at the end of the extrusion line, such as at finishing, trimming, or postforming, as well as with the windup or stacking equipment, which usually is very expensive. As an example, up to one-third of the capital equipment cost in a high-speed film line could go for the winder. Sophisticated materials-handling equipment is costly. Unfortunately, these costs are high, but normally there is a good return on this type of investment provided you know (or someone you trust knows) what is required to be purchased.

All in-line equipment has to be properly interfaced so it operates efficiently. Much of the equipment used in the past did not properly interface, so operations were rather less efficient than they could have been. Rules and communication protocols (chapter 25) have been developed that govern the communication and transfer of data along the complete line. The processor must determine what is needed based on what has to be accomplished, what controls are required, ease of operation and maintenance, safety devices, energy requirements, compatibility with existing equipment (speed, etc.), and so on.

Each of the processes (blown film, sheet, tube, etc.) to be reviewed will contain some equipment information applicable to its specific line; other equipment is reviewed in chapter 18. New generations of equipment provide meaningful contributions to plant productivity via greater reliability of operation and pinpoint equipment control, making these products more reliable and profitable. So it is important to keep up to date. As an example, fluid chillers and temperature control systems are major beneficiaries of the computer revolution. They became more energy efficient, reliable, and cost-effective. The application of microprocessor and computer-compatible controls that can communicate with the extruder results in pinpoint control of the line.

A major part of many extruder lines involves rolls for windup. There are many different types of rolls to handle films, sheets, coatings, pipes, and so on. They include winders, dancer rolls, lip rolls, spreader rolls, textured rolls, engraved rolls, and cooling rolls. All are required to be extremely precise in all their measurements, surface conditions that include commercial-grade mirror finishes, center lines with tolerance operations, bearings and all ancillaries mounted on journals, and rotating speeds (chapter 18).

FILM AND SHEET

Practically all thermoplastics can be formed into sheets and films, and the amount consumed in these forms represents the largest proportion of plastic production worldwide. Films and sheets are produced in several ways, including extrusion, calendering, and casting. The method used involves the properties required of the basic plastics and finished products as well as cost usually based on quantity. Films may be divided, generally, into three markets: (a) packaging, (b) industrial (tape, photo film, etc.), and (c) soft goods (raincoats, shower curtains, upholstering material, etc.).

The following classifications can be helpful as guides to film and sheet thicknesses: (a) film is generally less than 0.010 in (0.003 mm) and sheet at 0.010 in or more. (b) In turn, sheet can be classified as (c) intermediate sheet in the range of 0.040 to 0.250 in (0.01 to 0.06 mm). (d) Thin-gauge sheet is up to 0.060 in (0.015 mm). (e) Heavy-gauge sheet is at 0.080 to 0.500 in (0.02 to 0.13 mm). Most commercial plastic films are produced with a thickness of less than 0.005 in and most packaging films are less than 0.003 in. Different groups within the different industries (plastic, packaging, aluminum, clothing, etc.) may have their own thickness definitions; they use the terms used by their buyers/customers. Some of them use 0.004 in (0.10 mm) as the dividing line between film and sheet.

The different plastics extruded into film and sheets provide different markets with different performing products. An introduction to their properties and other performances is provided in Tables 5.27 to 5.32.

FILM

Film can be produced either by the basic extrusion tubular-blowing or flat processes. Each has its advantages and disadvantages. These basic processes result in film with a molecular orientation predominantly in the machine direction (MD). To be reviewed later, orienting the film can be in two orthogonal directions to develop superior optical, mechanical, and physical properties. The process is known as biaxial orientation and it can be applied to both tubular and flat film.

Regardless of process, film production lines include common downstream equipment, such as haul-off, tensioning, and reeling stations. Other common features include static control units (chapter 22) and corona discharge treaters to prepare the film surface for subsequent printing processes (Fig. 5.71). A high-purity melt free of inclusions is essential for film production. This is achieved by filtering the melt through a screen pack upstream of the die. The screen pack should be a fine mesh with a rating up to 3,000 mesh/cm^2.

BLOWN FILM

Overview The blown-film process involves extruding a relatively thick tube that is then expanded or blown by the usual internal air pressure or the water quench process to produce a relatively thin film (Figs. 5.72 and 5.73). The tube can be collapsed to form double-layer film that lies flat or can be slit to make one or two single-layer film webs. The air-cooled blown-film process is in very widespread use for polyethylene films, but it presents, in comparison, some difficulty when used with certain other plastics, such as polypropylenes. The water quench process is the generally preferred method of producing blown polypropylene film.

Plastic	Specific Gravity	Sq In Per Lb One Mil Thickness	Thickness Range	Manufacturing Method (extrusion, biaxial, orientation, casting, etc.)	Availability (in sheets, rolls, tapes)	Colorability
Acrylonitrile butadiene styrene	1.01-1.30	25,600-27,400	0.0075-2.000	calendering, extrusion, press laminated	sheets, rolls	good
Acrylonitrile styrene copolymer	1.08-1.09	25,400	0.001-0.020	biaxially oriented	rolls	excellent
Cellulose acetate	1.28-1.43	21,000-22,000	0.0005-0.250	casting, extrusion	sheets, rolls, tapes	excellent
Cellulose acetate butyrate	1.19-1.22	23,800	0.0011-0.150	casting, extrusion	sheets, rolls	good to excellent
Cellulose propionate	1.20	23,000	0.003-0.250	extrusion	sheets, rolls	good
Cellulose triacetate	1.28-1.31	21,200-22,000	0.0005-0.200	casting	sheets, rolls	....
Ethyl cellulose	1.14-1.16	23,800	0.003-0.015	casting, extrusion	sheets, rolls, tapes	excellent
Fluorinated ethylene	2.15	12,900	0.0005-0.060	extrusion	rolls, tapes	excellent
Polyamide	1.08-1.14	24,350-25,700	0.0005-0.030	extrusion	rolls	good
Polycarbonate	1.20	23,000-23,100	0.001-0.250	extrusion	sheets, rolls	excellent
Polyester	1.380-1.399	20,000-21,000	0.00015-0.014	biaxially oriented, extrusion	rolls, tapes, sheets	....
Polyethylene (low density)	0.910-0.925	30,000	0.0004-0.250	calendering, casting, extrusion, biaxially oriented	sheets, rolls, tapes, tubes	excellent
Polyethylene (medium density)	0.926-0.940	29,500	0.001-0.250	calendering, casting, extrusion, biaxially oriented	sheets, rolls, tapes, tubes	excellent
Polyethylene (high density)	0.941-0.965	29,000	0.0004-1.000	calendering, extrusion, stress relieving	sheets, rolls, tapes, tubes	excellent
Polyimide	1.42	19,400	0.0005-0.005	casting	rolls, tapes	....
Polymethyl methacrylate	1.18-1.19	23,400	0.005-5.000	extrusion, biaxially oriented, casting	sheets, rolls	excellent
Polypropylene, Cast	0.9	30,700	0.0003-1.000	casting, extrusion	sheets, rolls, tapes	excellent
Polypropylene, balanced, oriented	0.91	30,500-30,600	0.0005-0.0015	orientation	rolls	....
Polypropylene, balanced, oriented, coated	0.94	29,500	0.0009-0.0011	orientation	rolls	....
Polystyrene (casting, extrusion)	1.05-1.06	26,100	0.0004 and up	casting, extrusion	sheets, rolls	excellent
Polystyrene (oriented)	1.05-1.06	26,100-26,300	0.0003-0.020	oriented, biaxially oriented	sheets, rolls	excellent
Polytetrafluoroethylene	2.14-2.17	12,700-13,000	up to 0.250	skiving, casting, extrusion	sheets, rolls, tapes	good
Polytrifluorochloroethylene	2.1-2.2	13,000	0.0005-0.020	casting, extrusion, molding	sheets, rolls, tapes, tubes	good
Polyvinyl alcohol	1.21-1.31	21,600	0.0005-0.012	casting, extrusion	sheets, rolls, tapes	excellent
Polyvinyl chloride (rigid)	1.35-1.48	20,000-23,000	0.001-2.000	calendering, press, casting, extrusion	sheets, rolls, tapes	excellent
Polyvinyl chloride (flexible)	1.24-1.45	20,000-24,000	0.0005-0.375	calendering, press, casting, extrusion	sheets, rolls, tapes	good to excellent
Polyvinyl fluoride	1.38	20,000	0.0005-0.002	extrusion	rolls	good
Regenerated cellulose (cellophane)	1.40-1.55	19,500	0.0008-0.0018	extrusion into bath	sheets, rolls, ribbons	excellent
Rubber hydrochloride	1.11	25,000	0.0004-0.0025	casting	sheets, rolls, tapes	....
Vinyl chloride-acetate copolymers (rigid)	1.30-1.39	21,000	0.001-0.030	calendering, extrusion, casting	sheets, rolls	excellent
Vinyl chloride-acetate copolymers (flexible)	1.20-1.35	21,000	0.001-0.030	calendering extrusion, casting	sheets, rolls, tapes, tubes	excellent
Vinylidene chloride-vinyl chloride copolymer	1.57-1.68	16,300-23,000	0.004-0.010	calendering, biaxially extrusion, casting, oriented	rolls	excellent
Vinylnitrile rubber	1.18-1.21	22,800-23,700	0.001-0.020	casting, extrusion, calendering	sheets, rolls, tapes, tubes	....

Table 5.27 Examples of properties and manufacturing methods for films and sheets

Clarity	Shelf Life	Flammability	Maximum-minimum Use Temperature Range °F	U.V. Resistance	Heat Sealing Range °F	Dimensional Change At High R-H%
poor	good	slow burning to self-extinguishing	−65 to 215	poor to good	....	....
excellent	good	slow burning	−80 to 185	fair	....	....
excellent	indefinite	slow burning	−15 to 200	good	350-450	1.5
good to excellent	indefinite	slow burning	−30 to 180	good	....	....
good	indefinite	self-extinguishing	−30 to 200	good	350-450	0.5
excellent	indefinite	self-extinguishing	0 to 400	fair	non-sealing	0.5
excellent	....	slow burning	−75 to 250	good	non-sealing	....
good	indefinite	non-flammable	−415 to 400	good	575-700	none
fair to good	indefinite	self-extinguishing	−100 to 300	poor to fair	400-450	3-4
excellent	excellent	self-extinguishing	−150 to 275	fair to good	350-440	none
excellent	excellent	slow burning to self-extinguishing	−75 to 300	fair to good	325-450	11×10^{-6}
good to excellent	indefinite	slow burning	−70 to 180	fair	230-400	none
excellent	indefinite	slow burning	−70 to 200	fair	250-400	none
good	indefinite	slow burning	−70 to 250	fair	260-425	none
good	excellent	self-extinguishing	−450 to 600	good	....	22×10^{-6}
excellent	excellent	slow burning	...... to 190	good to excellent	375-450	0.35
excellent	indefinite	slow burning	−0 to 275	fair	260-375	none
excellent	indefinite	slow burning	−60 to 240	fair	narrow	none
excellent	indefinite	slow burning	−60 to 225	fair	200-275	none
excellent	....	slow burning	−55 to 200	....	....	....
excellent	indefinite	slow burning	−80 to 175	fair	280-330	none
poor	indefinite	non-flammable	−450 to 500	good	....	....
good	indefinite	non-flammable	−423 to 300	good	325-375	....
excellent	....	slow burning	−40 to 420	good	240-380	none
good to excellent	excellent	slow burning to self-extinguishing	−50 to 200	fair to good	340-400	negligible
good to excellent	indefinite	slow burning to self-extinguishing	−50 to 200	poor to good	275-350	very slight
good	indefinite	slow burning to self-extinguishing	−100 to 225	excellent	400-425	none
excellent	good	slow burning	0 to 300	good	200-350	2.5
good	poor	self-extinguishing	−20 to 205	poor	225-250	....
excellent	....	slow burning to self-extinguishing	−50 to 200	poor to good	300-350	none
excellent	....	slow burning to self-extinguishing	−60 to 200	poor to good	280-350	none
excellent	indefinite	self-extinguishing	0 to 200 (dry) 300 (wet)	good	230-315	none
....	....	slow burning	32 to 200	....	....	....

Table 5.27 *(continued)*

Gas Permeability cc/100 sq in 24 hour/mil (O_2,N_2,CO_2,SO_2)	Water Vapor Permeability Gm/24hr/100 sq in/mil @ 95°F.90%R.H.	Water Absorption 24 hr Immersion %	Resistance to Acids	Resistance to Alkalies	Resistance to Greases and Oils	Resistance to Organic Solvents
....	....	0.2-0.3	fair to good	good	excellent	fair to excellent
CO,210,$N_2$9,$O_2$36	12.0-14.0	0.5	good	excellent	good	excellent to poor
$O_2$117,$N_2$40,$CO_2$1,000	100-200	4-9	poor to good	poor	good	poor
$CO_2$6000,$N_2$250,$O_2$950	4.5-6.5 / 730[x]	0.1-3.4	poor to good	poor	good	poor
High	100-200	1.6	poor	poor	good	poor
CO,880,$N_2$15$O_2$30,0	50-150	2.0-4.5	good to poor	poor	good	fair
$O_2$3,500,$CO_2$7,800,$N_2$300	4.8-14.2~50	2.5-7.5	poor to fair	excellent	good	poor
$O_2$3,250,$N_2$1,500,$CO_2$3,150	....	0.01	excellent	excellent	excellent	excellent
$CO_2$160,$O_2$25,(50%R.H.)	19.0	9.5	poor	excellent	excellent	excellent
$CO_2$1,075,O_2-300,$N_2$50	11.0	0.15	excellent	fair	good	poor
$CO_2$16,$O_2$6,$N_2$1	1.7	<0.8	fair to excellent	fair to good	excellent	good to excellent
$CO_2$2,900,$O_2$550,$N_2$180	1.2-1.5	0-0/8	excellent	excellent	poor	good (60°C)
$CO_2$990,$O_2$280	0.7-0.9	nil	excellent	excellent	good	good
$CO_2$6,200-3,300,$O_2$800-450, $N_2$100-50	0.3-0.4	nil	excellent	excellent	good	good (80°)
$CO_2$450,$O_2$25,$N_2$6	5.4	2.9	good to excellent	poor	excellent	excellent
....	0.5	negligible	good	good	good	fair
$CO_2$800,$O_2$240,$N_2$60	0.7	0.005 or less	excellent	excellent	good	good
$CO_2$540,$O_2$160,$N_2$15	0.25-0.45	nil	excellent	excellent	good to excellent	good
$CO_2$520,$O_2$130	0.40	nil	excellent	coating attacked	excellent	good
....	0.5-0.7,6.0	....	good	excellent	good	poor to excellent
AIR 105,$CO_2$1,080,$H_2$2,420, $N_2$49.5,$O_2$310	7.0-10.0	0.04-0.06	good	excellent	good	poor to excellent
....	....	....	excellent	excellent	excellent	excellent
He600-1,150,$O_2$1.3-58, $N_2$9.3-19,$CO_2$11.7-225	.025	nil	excellent	excellent	excellent	excellent
He.02,H_2nil,CO_2nil,O_2nil, (0 o/o R.H.)	10.0 est.	55	poor	poor	excellent	excellent
$CO_2$970,$O_2$150,(50%R.H.)	4.0	negligible	excellent	excellent	good	poor to excellent
$CO_2$5000,$O_2$500(50%R.H.)	0.8-20.0	negligible	good to excellent	good to excellent	good to excellent	poor to good
$CO_2$15,$O_2$3,$N_2$0.25	5.2	0.5	excellent	excellent	excellent	excellent
$CO_2$3,O_2 <0.5,N <0.5	1.4-2.7(NMP) 0.008-0.03(MP)	45,115	poor	poor	excellent	insoluble
$CO_2$6-13,000,$O_2$3.2-3.250	....	5	good	good	excellent	good
$CO_2$970,$O_2$150,(50%R.H.)	4.4	negligible	very good to excellent	excellent	excellent	poor to excellent
$CO_2$970,$O_2$150,(50%R.H.)	0.28	negligible	excellent	excellent	fair to excellent	poor to excellent
$CO_2$3.8-150,O_2.8-24,NE.12-10	0.02-0.6	negligible	excellent	good to excellent	good to excellent	good to excellent
low	....	negligible	good	good	excellent	good

Table 5.27 *(continued)*

Resistance to Detergents	Resistance to Oxidizing Agents	Tensile Strength lbs sq in	Elongation %	Tearing Strength (Elmendorf GM/mil)	Folding Endurance
....	fair	1,700-7,800	....	....	....
....	good	10,000-15,000	20-60	....	....
good	poor	8,000-16,400	25-35	2-25	250-2,000
excellent	poor	4,100-9,700	40-100	3-10	1,200-1,500
excellent	poor	5,000	80	25	1,000-2,000
....	poor	9,000-16,000	10-40	4-10	3,500-4,000
....	poor	6,800-10,600	25-40	7-36	>2,700
excellent	good	3,000	300	125	4,000
good	poor to fair	9,000-12,000	300-400	50-90	>250,000
fair	good	8,000-10,500	85-105	20-25	250-400
excellent	good	25,000-40,000	50-120	12-40	230,000
....	good	up to 12,000	200-800	50-300	very high
fair	fair	up to 17,000	50-650	50-300	very high
....	poor	2,400-6,100	10-650	15-300	very high
....	good	25,000	70	8	10,000
....	poor	8,000-12,000	4-12	....	....
good	fair	3,200-10,000	300-500	5-1,800	very high
good	fair	25,000-30,000	45-85	5-7	very high
good	fair	30,000	70	8	very high
....	good	7,100-12,100	3-40	25	....
good	good	7,100-12,100	3-40	....	....
excellent	good	1,500-4,000	100-350	10-100	very high
excellent	excellent	6,300-10,000	50-400	8-26	....
excellent	good	6,500-12,000	200-600	250-800	very high
excellent	very good	7,000-14,000	10-175	10-700	very high
good to very good	good to very good	1,400-17,000	100-510	60-1,400	very high
excellent	excellent	7,000-19,000	110-260	12-48	excellent
good	fair	7,000-18,000	15-65	2-30	500 cycles
....	....	3,500-5,000	stretches	60-1,600	250,000
excellent depends on plasticizer	good to very good	3,000-8,000	depends on plasticizer	10-30	very high
very good	very good	2,000-4,500	250-300	30-1,400	....
excellent	excellent	8,000-20,000	35-110	10-7,100	>500,000
....	....	2,500-4,000	....	200-960	very high

Table 5.27 *(continued)*

Plastic	Specific Gravity	Sq. In. Per Lb. One Mil Thickness	Thickness Range	Manufacturing Method
ASTM No.	D-1505	—	—	—
1. Acrylic	1.14–1.26	22,000–24,000	0.002–0.006	extrusion
2. Acrylonitrile butadiene styrene	1.01–1.30	25,600–27,400	0.0075–2.000	calendering, extrusion, press laminated
3. Acrylonitrile–styrene copolymer	1.08–1.09	25,400	0.001–0.020	biaxially oriented
4. Cellophane	1.40–1.50	11,600–25,000	0.0008–0.0017	extrusion
5. Cellulose acetate	1.28–1.43	21,000–22,000	0.0005–0.250	casting, extrusion
6. Cellulose acetate butyrate	1.19–1.22	23,800	0.0011–0.150	casting, extrusion
7. Cellulose propionate	1.20	23,000	0.003–0.250	extrusion
8. Cellulose triacetate	1.28–1.31	21,200–22,000	0.0005–0.200	casting
9. Ethyl cellulose	1.15	23,800	0.002–0.015	extrusion
10. Ethylene chlorotrifluoroethylene copolymer (E-CTFE)	1.66–1.68	16,600	0.005–0.090	extrusion
11. Ethylene methacrylate	0.942	29,280	0.04–0.250	extrusion, blown, casting
12. Ethylene–tetrafluoroethylene copolymer (ETFE)	2.15	16,300	—	extrusion
13. Ethylene–vinylacetate copolymer	0.924–0.920	29,000	0.00075 and up	extrusion
14. Fluorinated ethylene propylene copolymer	2.15	12,900	0.0005–0.060	extrusion
15. Ionomer	0.94	29,400	0.001–0.01	extrusion
16. Polyamide (nylon 6/6)	1.14	24,350–25,700	0.0005–0.030	extrusion
17. Polyamide (nylon 6) (biaxially oriented)	1.16	23,870	0.0005–0.001	extrusion, orientation
18. Polyamide (nylon 11 and 12)	1.03	26,630–27,200	0.0005–0.030	extrusion
19. Polycarbonate	1.20	23,000–23,100	0.001–0.250	extrusion
20. Polyester	1.380–1.399	20,000–21,000	0.00015–0.014	biaxially oriented, extrusion
21. Polyether sulfone	1.37	20,000	0.001–0.250	casting, extrusion
22. Polyethylene (low density)	0.910–0.925	30,000	0.0004–0.250	calendering, casting, extrusion, biaxially oriented
23. Polyethylene (medium density)	0.926–0.940	29,500	0.001–0.250	calendering, casting, extrusion, biaxially oriented
24. Polyethylene (high density)	0.941–0.965	29,000	0.0004–1.000	calendering, extrusion, stress relieving

Table 5.28 Examples of mechanical, physical, and electrical properties for films and sheets

Clarity	Shelf Life	Flammability	Maximum-minimum Use Temperature Range °F	U.V. Resistance	Heat Sealing Range °F
——	——	CS-192	D-759	D-1435	——
excellent	indefinite	slow burning	− 65 to 215	excellent	yes
poor	good	slow burning to self-extinguishing	− 65 to 215	poor to good	——
excellent	good	slow burning	− 80 to 185	fair	——
good	indefinite	slow burning	0 to 350	good	250–300
excellent	indefinite	slow burning	− 15 to 200	good	350–450
good to excellent	indefinite	slow burning	− 30 to 180	good	——
good	indefinite	self-extinguishing	− 30 to 200	good	350–450
excellent	indefinite	self-extinguishing	0 to 400	fair	nonsealing
excellent	indefinite	slow burning	− 75 to 250	good	nonsealing
good	indefinite	self-extinguishing	− 80 to 400	good	475–500
fair	indefinite	slow burning	− 80 to 150	fair	175–350
good	indefinite	self-extinguishing	——	good	525
——	——	burns	120 to 180	fair	150–300
good	indefinite	nonflammable	− 415 to 400	good	575–700
good	indefinite	——	——	——	——
fair to good	indefinite	self-extinguishing	− 100 to 300	poor to fair	400–450
good	indefinite	slow burning	− 100 to 300	good	400–450
good	indefinite	slow burning	——	——	320–390
excellent	excellent	self-extinguishing	− 150 to 275	fair to good	350–440
excellent	excellent	slow burning to self-extinguishing	− 75 to 300	fair to good	325–450
fair	indefinite	slow burning to self-extinguishing	− 100 to 400–450	poor	500–550
good to excellent	indefinite	slow burning	− 70 to 180	fair	230–400
excellent	indefinite	slow burning	− 70 to 200	fair	250–400
good	indefinite	slow burning	− 70 to 250	fair	260–425

Table 5.28 *(continued)*

Dimensional Change at High RH%	Gas Permeability cc/100 sq. in. 24 hr./mil (O_2, N_2, CO_2, SO_2)	Water Vapor Permeability Gm/24 hr./100 sq. in./mil @ 95° F. 90% RH	Water Absorption 24 hr. Immersion %	Resistance to Acids	Resistance to Alkalies
——	D-1434	E-96 (B)	D-570	D-543	D-543
——	——	8 (3 mil)	1.4–16	good	good
——	——	——	0.2–0.3	fair to good	good
——	CO, 210, N_2 9, O_2 36	12.0–14.0	0.5	good	excellent
2.5	CO_2 0.4–0.6, H_2 1.2–2.2 N_2 0.5–0.6, O_2 0.5–0.8	0.4–1.34	45–115	poor	poor
.5	O_2 117, N_2 40, CO_2 1000	100–200	4–9	poor to good	poor
——	CO_2 6,000, N_2 250, O_2 950	4.5 – 6.5 730[x]	0.1–3.4	poor to good	poor
0.5	High	100–200	1.6	poor	poor
0.5	CO_2 880, N_2 15, O_2 30, 0	50–150	2.0–4.5	good to poor	poor
——	CO_2 7,800, N_2 100 O_2 3,500	4.8–14.2	2.5–7.5	fair to poor	excellent
2.5	CO_2 110, N_2 10, O_2 25	0.6	<0.02	good	good
nil	O_2 335, CO_2 4,780	4.1	nil	fair to good	good
——	CO_2 250, N_2 30, O_2 10	0.04	<0.02	good	good
——	CO_2 6,000, N_2 400, O_2 840	2–3	<0.01	fair	fair
none	O_2 3,250, N_2 1,500, CO_2 3,150	——	0.01	excellent	excellent
——	O_2 600	——	0.4	good	good
3.4	CO_2 160, O_2 25 (50% RH)	19.0	9.5	poor	excellent
3.4	O_2 1.2–1.4	10–11	——	poor	fair
——	CO_2 153–336, N_2 3.4–18, O_2 34–92, H_2 323	4.0	0.27–0.25	poor	good
none	CO_2 1,075, O_2 300, N_2 50	11.0	0.15	excellent	fair
11×10^{-6}	CO_2 16, O_2 6, N_2 1	1.7	<0.8	fair to excellent	fair to good
——	CO_2 950, N_2 40, O_2 230	——	2.1	good	good
none	CO_2 2,900, O_2 550, N_2 180	1.2–1.5	0–0/8	excellent	excellent
none	CO_2 990, O_2 280	0.7–0.9	nil	excellent	excellent
none	CO_2 6,200–3,300, O_2 800–450, N_2 100–50	0.3–0.4	nil	excellent	excellent

Table 5.28 *(continued)*

Resistance to Greases and Oils	Resistance to Organic Solvents	Resistance to Water and/or Detergents	Tensile Strength lbs. sq. in @ RT	Elongation %	Tearing Strength (Elmendorf GM/mil)	Folding Endurance
D-722	D-546	——	D-882	D-882	D-1004	D-2172
good	paramatrics, (aliphatics)	good	4,400	130	25	6,000 +
excellent	fair to excellent	——	1,700–7,800	——	——	——
good	excellent to good	——	10,000–15,000	20–60	——	——
excellent	poor	good	7,000–18,000	10–50	2–20	500
good	poor	good	8,000–16,400	25–35	2–25	250–2,000
good	poor	excellent	4,100–9,700	40–100	3–10	1,200–1,500
good	poor	excellent	5,000	80	25	1,000–2,000
good	fair	——	9,000–16,000	10–40	4–10	3,500–4,000
good	poor	——	8,000–10,000	20–30	7–36	250,000
good	good	good	8,000–10,000	150–250	900–1,300	250,000
poor	poor	excellent	1,500–1,800	600–1,000	60 MD 150 TD	6,000 +
good	good	good	7,000–8,000	300	600–900	——
poor	fair	good	1,000–3,500	400–800	——	very high
excellent	excellent	excellent	3,000	300	125	4,000
good	good	good	5,000	250–450	30	very high
excellent	excellent	good	9,000–12,000	300–400	50–90	>250,000
excellent	good	good to poor	32,600–35,000	100–120	——	——
good	good	good to poor	7,000–12,000	250–400	——	good
good	poor	fair	8,000–10,500	85–105	20–25	250–400
excellent	good to excellent	excellent	25,000–40,000	50–120	12–40	230,000
good	good to poor	good	10,000–12,000	20–150	7–16	——
poor	good (60°C)	——	up to 12,000	200–800	50–300	very high
good	good	fair	up to 17,000	50–650	50–300	very high
good	good (80°C)	——	2,400–6,100	10–650	15–300	very high

Table 5.28 *(continued)*

Generic Name	Dielectric Constant 10³ cps	Dielectric Constant 10⁶ cps	Dielectric Constant 10⁹ cps	Dissipation Factor 10³ cps	Dissipation Factor 10⁶ cps	Dissipation Factor 10⁹ cps	Dielectric Strength, V/mil	Volume Resistivity ohm-cm
ASTM No.	D-150	D-150	D-150	D-150	D-150	D-150	D-149	D-257
1.	4.4	3.5	—	0.046	0.040	—	2,400	10^{16}
2.	2.5–3.0	2.5–3.0	—	—	—	—	350–400	10^{16}
3.	—	—	—	—	—	—	—	—
4.	3.2 (uncoated)	—	—	0.015 (uncoated)	—	—	2,000–2,000 (uncoated)	10^{11}
5.	3.6	3.2	3.2	0.016	0.33	—	3,200–5,000	10^{10}–10^{15}
6.	2.9	2.5	2.8	0.013	0.03–0.04	—	3,400	10^{11}–10^{15}
7.	—	3.2	—	—	0.015	—	—	—
8.	3.2–4.5	3.3–3.8	3.2	0.016	0.033	—	3,700	10^{13}–10^{15}
9.	3.1	3.0	—	0.002–0.20	0.010–0.06	0.046	1,500 (10 mil)	10^{15}
10.	2.6	2.6	—	0.002	0.013	—	5,000–6,000	10^{16}
11.	—	3.1	—	—	0.15	—	300–500	10^{15}
12.	2.6	2.6	—	0.008	0.015	—	3,500	10^{16}
13.	2.7–2.9	2.7–2.9	—	0.01–0.02	0.01–0.02	—	5,000	10^{9}
14.	2.0–2.50	2.0–2.05	2.05	<0.002	0.0004	0.0015	7,000	10^{19}
15.	—	—	—	—	—	—	—	—
16.	4.0	3.4	—	—	—	—	850	4.5×10^{13}
17.	—	—	—	—	—	—	—	—
18.	3.8	3.1	—	0.05	0.03	—	1,800	10^{13}
19.	2.99	2.93	2.89	0.0015	0.010	0.012	1,500	10^{16}
20.	3.2	3.0	2.8	0.005	0.016	0.003–0.008	7,500	10^{18}
21.	3.5	3.5	—	0.0035	0.006	—	2,400 (4 mil)	10^{17}–10^{18}
22.	2.2	2.2	2.2	0.0003	0.0003	0.0003	470	10^{16}
23.	2.2	2.2	2.2	0.0003	0.0003	0.0003	500	10^{16}
24.	2.3	2.3	2.3	0.0005	0.0005	0.0005	500	10^{16}

Table 5.28 *(continued)*

Blown film is usually extruded vertically, upward through a circular die. This forms a tube that is then blown into a bubble that thins or draws down to the required final gauge. Orientation takes place in two directions: horizontally (transverse direction/TV) as the bubble is formed and in the machine's direction (MD) as controlled by adjustable-speed haul-off nip rolls.

An air ring, either with single lips or two or more lips, directs air to cool the bubble at the die exit. Internal bubble cooling is used to cool the inner surface of the extruded bubble to gain high production rates (typically 50% higher than with external air only).

In a typical blown-film system, the bubble enters a collapsing frame and, after passing through upper nip rolls, becomes a tube that can then be processed into bags, flat film by slitting open the tube, and so on. Because convection cooling is relatively slow, blown film tends to be hazier than cast film.

Plastic	Specific Gravity	Sq. In. Per Lb. One Mil Thickness	Thickness Range	Manufacturing Method
25. Polymimide	1.42	19,400	0.0005–0.005	casting
26. Polymethyl methacrylate	1.18–1.19	23,400	0.005–5.000	extrusion, biaxially oriented
27. Polymethyl pentene	0.83	33,360	0.003–0.030	extrusion
28. Polyparabanic acid	1.3	20,800	0.001–0.005	casting
29. Polypropylene, cast	0.9	30,700	0.0003–1.000	casting, extrusion
30. Polypropylene, balanced, oriented	0.91	30,800–30,600	0.0005–0.0015	orientation
31. Polypropylene, balanced, oriented, coated	0.94	29,500	0.0009–0.0011	orientation
32. Polystyrene (casting, extrusion)	1.05–1.06	26,100	0.0004 and up	casting, extrusion
33. Polystyrene (oriented)	1.05–1.06	26,100–26,300	0.0003–0.020	oriented, biaxially oriented
34. Polysulfone	1.24–1.25	22,355	0.0001–0.250	extrusion
35. Polytetrafluoroethylene (Teflon TFE)	2.14–2.17	12,700–13,000	up to 0.250	skiving, casting, extrusion
36. Poly(perfluoroalkoxytetra-fluorethylene) (PFA)	2.13–2.16	—	—	casting, extrusion, molding
37. Polytrifluorochloroethylene (Kel-F)	2.1–2.2	13,000	0.0005–0.020	casting, extrusion
38. Polyurethane	1.11–1.24	23,000	0.0005 and up	calendering, casting extrusion
39. Polyvinyl alcohol	1.21–1.31	21,600	0.0005–0.012	casting, extrusion
40. Polyvinyl chloride (PVC) (rigid)	1.35–1.48	20,000–23,000	0.0001–2.000	calendering, press, casting, extrusion
41. Polyvinyl chloride (PVC) (flexible)	1.24–1.45	20,000–24,000	0.0005–0.375	calendering, press, casting, extrusion
42. Polyvinyl fluoride (Tedlar)	1.38–1.57	20,000	0.0005–0.002	extrusion
43. Polyvinylidene fluoride (Kynar)	1.75–1.78	16,000	0.002–0.020	extrusion
44. Rubber hydrochloride	1.11	25,000	0.0004–0.0025	casting
45. Vinyl chloride–acetate copolymers (rigid)	1.30–1.39	21,000	0.001–0.030	calendering, extrusion, casting
46. Vinyl chloride–acetate copolymers (flexible)	1.20–1.35	21,000	0.001–0.030	calendering, extrusion, casting
47. Vinylidene chloride–vinyl chloride copolymer	1.57–1.68	16,300–23,000	0.004–0.010	calendering, biaxially oriented, extrusion, casting
48. Vinylnitrile rubber	1.18–1.21	22,800–23,700	0.001–0.020	casting, extrusion, calendering

Table 5.28 *(continued)*

Plastic	Clarity	Shelf Life	Flammability	Maximum-minimum Use Temperature Range °F	U.V. Resistance
25.	good	excellent	self-extinguishing	−450 to 600	good
26.	excellent	excellent	slow burning	− to 190	good to excellent
27.	fair	indefinite	burns	——	poor
28.	good	indefinite	slow burning	−320 to 500	——
29.	excellent	indefinite	slow burning	0 to 275	fair
30.	excellent	indefinite	slow burning	−60 to 240	fair
31.	excellent	indefinite	slow burning	−60 to 225	fair
32.	excellent	——	slow burning	−55 to 200	——
33.	excellent	indefinite	slow burning	−80 to 175	fair
34.	good	indefinite	——	−100 to 350	poor
35.	poor	indefinite	nonflammable	−450 to 500	excellent
36.	good	indefinite	self-extinguishing	−450 to 500	excellent
37.	good	indefinite	nonflammable	−423 to 300	excellent
38.	good	indefinite	slow burning	−100 to 190	poor to good
39.	excellent	——	slow burning	−40 to 420	good
40.	good to excellent	excellent	slow burning to self-extinguishing	−50 to 200	fair to good
41.	good to excellent	indefinite	slow burning to self-extinguishing	−50 to 200	poor to good
42.	good	indefinite	slow burning to self-extinguishing	−100 to 225	excellent
43.	fair	indefinite	self-extinguishing	−80 to 275	good
44.	good	poor	self-extinguishing	−20 to 205	poor
45.	excellent	——	slow burning to self-extinguishing	−50 to 200	poor to good
46.	excellent	——	slow burning to self-extinguishing	−60 to 200	poor to good
47.	excellent	indefinite	self-extinguishing	0 to 200 (dry) 300 (wet)	good
48.	——	——	slow burning	32 to 200	——

Table 5.28 *(continued)*

Heat Sealing Range °F	Dimensional Change at High RH%	Gas Permeability cc/100 sq. in. 24 hr./mil (O_2, N_2, CO_2, SO_2)	Water Vapor Permeability Gm/24 hr./100 sq. in./mil @ 95° F. 90% RH	Water Absorption 24 hr. Immersion %	Resistance to Acids
does not heat seal	22 × 10^{-6}	CO_2 450, O_2 25, N_2 6	5.4	2.9	good to excellent
375–450	0.35	——	0.5	negligible	good
285–400	——	——	——	0.005	good
250–350	——	——	——	2.8	good
260–375	none	CO_2 800, O_2 240, N_2 60	0.7	0.005 or less	excellent
narrow	none	CO_2 540, O_2 160, N_2 15	0.25–0.45	nil	excellent
200–275	none	CO_2 520, O_2 130	0.40	nil	excellent
——	——	——	0.5–0.7, 6.0	——	good
280–330	none	AIR 105, CO_2 1,080, H_2 2,420, N_2 49.5, O_2 310	7.0–10.0	0.04–0.06	good
——	——	CO_2 405, N_2 40, O_2 90	18	0.3	good
does not heat seal	——	——	——	<0.01	excellent
600–700	——	——	——	0.03	excellent
325–375	——	He 600–1,150, O_2 1.3–5.8, N_2 9.3–19, CO_2 11.7–225	0.025	nil	excellent
275–375	——	CO_2 450–1650, N_2 41–119, O_2 75–327	40–75	0.55–0.77	poor to good
240–380	none	He 0.2, H_2 nil, CO_2 nil, O_2, nil (0% RH)	10.0 est	55	poor
340–400	negligible	CO_2 970, O_2 150 (50% RH)	4.0	negligible	excellent
275–350	very slight	CO_2 5,000, O_2 500 (50% RH)	0.8–20.0	negligible	good to excellent
400–425	none	CO_2 15, O_2 3, N_2 0.25, H_2 58.1	5.2	0.5	excellent
375	——	CO_2 5.5, N_2 9, O_2 14	2.6	0.04	excellent
225–250	——	CO_2 6–13,000, O_2 3.2–3.250	0.008–0.03 (MP)	5	good
300–350	none	CO_2 970, O_2 150 (50% RH)	4.4	negligible	very good to excellent
280–350	none	CO_2 970, O_2 150 (50% RH)	0.28	negligible	excellent
230–315	none	CO_2 3.8–150, O_2 8–24, NE, 12–10	0.02–0.6	negligible	excellent
——	——	low	——	negligible	good

Table 5.28 *(continued)*

Plastic	Resistance to Alkalies	Resistance to Greases and Oils	Resistance to Organic Solvents	Resistance to Water and/or Detergents	Tensile Strength lbs. sq. in. @ RT	Elongation %	Tearing Strength (Elmendorf GM/mil)
25.	poor	excellent	excellent	—	25,000	70	8
26.	good	good	fair	—	8,000–4,000	4–12	—
27.	good	fair	fair	good	2,700–4,000	10	—
28.	poor	good	good	—	16,000	10	500
29.	excellent	good	good	good	3,200–10,000	300–500	5–1,800
30.	excellent	good to excellent	good	good	25,000–30,000	45–85	5–7
31.	coating attacked	excellent	good	good	30,000	70	8
32.	excellent	good	poor to excellent	—	7,100–12,100	3–40	25
33.	excellent	good	poor to excellent	good	7,100–12,100	3–40	—
34.	good	good	poor	good	8,400–10,600	64–110	9–12
35.	excellent	excellent	excellent	excellent	1,500–4,000	100–350	10–100
36.	excellent	excellent	excellent	excellent	4,000–4,500	300–480	—
37.	excellent	excellent	excellent	excellent	6,300–10,000	50–400	8–26
38.	poor to good	good	good	fair to good	5,000–12,000	200–700	220–710
39.	poor	excellent	excellent	excellent	6,500–12,000	200–600	250–800
40.	excellent	good	poor to excellent	excellent	7,000–14,000	10–175	10–700
41.	good to excellent	good to excellent	poor to good	good to very good	1,400–17,000	100–510	60–1,400
42.	excellent	excellent	excellent	excellent	7,000–19,000	110–260	12–48
43.	fair to poor	excellent	fair to good	good	5,200–7,400	25–500	—
44.	good	excellent	good	—	3,500–5,000	stretches	60–1,600
45.	excellent	excellent	poor to excellent	excellent depends on plasticizer	3,000–8,000	depends on plasticizer	10–30
46.	excellent	fair to excellent	poor to excellent	very good	2,000–4,500	250–300	30–1,400
47.	good to excellent	good to excellent	good to excellent	excellent	8,000–20,000	35–110	10–7,100
48.	good	excellent	good	—	2,500–4,000	—	200–960

Table 5.28 *(continued)*

Air Cooled In this process, the die is usually side fed from an extruder (Fig. 5.74) (chapter 17). The melt exiting from a circular or ring orifice is air inflated via the center of the die to the required diameter as it moves vertically (usually). Figures 5.75 and 5.76 show the expanding shapes of LDPE and HDPE film, respectively. The inflated film is then usually cooled through an air-cooling ring. Different designs are used to deliver the forced air at different angles so that the film starts its cooling cycle. The rings can have adjustable open lips or adjustable air holes (503–505). Figure 5.77 shows airflow from single and dual-lip air-cooling ring designs. Blown-film throughput as a function of the diameter of the die's orifice is shown in Figure 5.78.

Folding Endurance	Dielectric Constant 10³ cps	10⁶ cps	10⁹ cps	Dissipation Factor 10³ cps	10⁶ cps	10⁹ cps	Dielectric Strength, V/mil	Volume Resistivity ohm-cm
10,000	3.5	3.4	——	0.003	0.010	——	7,000	10^{18}
——	3.5–4.0	3.0–3.5	2.58	0.040	0.030	0.009	400	10^{15}
——	2.1	2.1	——	0.003	——	——	——	10^{16}
25,000	3.4	3.4	——	0.004	——	——	5,700	5×10^{16}
very high	2.2	2.2	2.2	0.0003	0.0003	0.0003	3,000–4,500	3×10^{15}
very high	2.2	2.2	2.2	<0.0002	<0.0002	<0.0002	7,000–10,000	3×10^{18}
very high	2.2	2.2	2.2	<0.0002	<0.0002	<0.0002	7,000–10,000	3×10^{18}
——	2.4–2.7	2.4–2.7	2.4–2.7	0.0005	0.0005	0.0005	5,000	$10^{7} \times 10^{16}$
——	2.4–2.7	2.4–2.7	2.4–2.7	0.0005	0.0005	0.0005	5,000	$10^{7} \times 10^{16}$
——	3.05	3.05	3.00	0.0008	0.0034	0.0041	7,500	5×10^{16}
very high	2.0–2.1	2.0–2.1	2.0–2.1	0.0002	0.0002	0.0002	≈2,500 (1 mil)	10^{18}
——	2.06	2.06	2.06	0.000035	0.000080	0.00115	2,000	10^{18}
——	2.5–2.7	2.3–2.4	2.3	0.022–0.24	0.009–0.017	0.004	1,000–3,700	10^{18}
——	5.2–7.5	5.5–7.1	——	0.040–0.060	——	——	600–1,300	$1\text{–}3.0 \times 10^{11}$
very high	——	——	——	——	——	——	——	——
very high	3.0–3.3	2.8–3.1	2.8	0.009–0.017	0.006–0.017	0.019	425–1,300	10^{16}
very high	4.0–8.0	3.3–4.5	——	0.07–0.016	0.04–0.14	——	250–1,000	$10^{11}\text{–}10^{14}$
excellent	——	——	——	——	——	——	——	——
——	7.46	6.1	2.98	0.019	0.059	0.110	260–1,280	2×10^{14}
250,000	3.0	3.0	——	——	0.006	——	——	10^{13}
very high	3.0–3.3	2.8–3.1	2.8	0.0009–0.017	0.006–0.019	0.019	425–1,300	10^{16}
——	4.0–8.0	3.3–4.5	——	0.070–0.160	0.04–0.140	——	250–1,000	$10^{11}\text{–}10^{14}$
>500,000	3.9–4.5	3.0–4.0	2.7	0.052–0.063	0.050–0.080	0.016	3,000–5,000	$10^{12}\text{–}10^{15}$
very high	——	——	——	——	——	——	——	——

Table 5.28 *(continued)*

Film size and thickness are controlled by the combinations of die and cooling-ring sizes, internal air pressure, and takeoff speed. The blown film is usually moving vertically through several guide roll or sizing supports (Figs. 5.79 to 5.82) to keep it aligned with the machine. After a few yards (meters) of free suspension, it is flattened via some type of collapsing device that directs the flattened film through pressure-controlled pull nip rolls (Figs. 5.83 and 5.84).

The rotating speed of the nip rolls is a major source for controlling the rate that the bubble is drawn. At the end of the line, winder technology allows the selection of surface winding, center winding, and a combination of surface/center winding to suit the behavior of the film being run. The film may be wound directly as a layflat tube, slit at both sides and wound into two flat reels, slit

	Cellophane		Cellulose acetate	Flurohalo-carbon	Ionomer
	Nitrocellulose coated	Saran coated two sides			
General					
Clarity	Transparent	Transparent to translucent	Transparent	Transparent	Transparent
Yield (sq. In./lb./ 0.001-inch)	19,500	19,500	22,000	13.000	28,600-29,500
Specific gravity	1.44	1.44	1.25-135	2.2	0.94-0.96
Mechanical					
Tensile strength (lb/sq.in.) ASTM D-882	MD-18,000 TD-9,000	MD-18,000 TD-9,000	7,000-15,000	5,000-10,000	3,000-5,000
Elongation (per cent) ASTM D-882	15-23	15-25	15-50	50-400	350-450
Impact strength (kg-cm)	8-15	8-15	2-8	2-15	6-11
Tear strength (gm/0.001-inch Elmendorf) ASTM D-1922	2-10	7-15	2-15	3-40	15-150
Heat seal range (°F.)	200-350	200-350	350-450+	350-400	190-400
Chemical					
WVTR (gm/24hr/100 sq. in. @ 100° F. 90 per cent RH) ASTM E-96	0.5	0.45	5	0.4-0.7	1.3-2.1
Gas transmission (cc/0.001-inch/100 sq. in./ 24 hr. @ atm 73°F. & 0 per cent RH) ASTM D-1434	O_2-2.0	O_2-0.6	O_2-23 CO_2-105 N_2-6.5	O_2-240 CO_2-500	O_2-226-484 CO_2-626-1150
Resistance to grease and oils	Impermeable	Impermeable	Good	Good	Good
Permanence					
Maximum use temperature (°F.)	Depends on RH	Depends on RH	200	300	160
Minimum use temperature (°F.)	Depends on RH	Depends on RH	About 0	−50	−100
Dimension change at high RH (per cent)	2-3	2-3	0.2-0.6@ 80 per cent	None	None
Converting characteristics					
Machine performance	Excellent	Excellent	Good	Good	Good
Printability	Good	Good	Good	Good if treated	Good if treated
Heat shrinkable	No	No	No	No	Special Types

Table 5.29 Examples of general properties for films

Nitriles	Nylon			Poly-Butylene	Poly-Carbonate
	Uncoated	Saran coated one side	Metallized Oriented		
Transparent	Transparent to translucent	Transparent	Opaque	Transparent to translucent	Transparent
24,000	23,500-24,500	21,200	24,500	21,200	23,100
1.15	1.13-1.14	1.13-1.14	1.13-1.14	1.3	1.2
9,500	7,000-18,000	7,000-18,000	28,000	4,500-5,000	8,200-9,300
5	250-500	250-500	90	200-400	95-115
High	4-6	7	7	20-30	
High	20-50	20-60	20-60	400-800	20-25
160-300	350-500	350-500	350-500	325-400	400-430
5	24-26	0.2	0.2	1.2	9.7
O_2-0.8 CO_2-1.1	$O_2$2.6 CO_2-4.7 N_2-0.9	O_2-0.5 CO_2-1.4 N_2-0.1	O_2-0.05	O_2-385 CO_2-825 N_2-110	O_2-258 CO_2-775
Impermeable	Impermeable	Impermeable	Impermeable	Good	Good
160-210	350-450	200	200	220	265
−50	−75	−40	−75	−30	−100
—	1.3	1.3	1.3	None	None
Good	Good	Good	Good	Fair	Good
Good	Good	Good	Good	Good if treated	Good
	No	No	No	Some types	No

Table 5.29 *(continued)*

Polyester (PET)			Polyethylene		
Uncoated	Coated-saran one side	Metallized	Low-density	Medium-density	High-density
Transparent	Transparent	Opaque	Transparent to translucent	Transparent to translucent	Transparent to translucent
20,000-22,000	20,500	20,000-22,000	30,000	29,500	29,000
1.35-1.39	1.4	1.35-139	0.910-0.925	0.926-0.940	0.941 0.965
25,000-33,000	26,000-31,000	25,000-33,000	1,000-3,500	2,000-5,000	3,000-7,500
120-140	90-125	120-140	225-600	225-500	10-500
25-30	25-30	25-30	7-11	4-6	1-3
13-80	10-20	13-80	100-400	50-300	15-300
—	275-400	—	250-350	260-310	275-310
1.3	0.9 @0.0005-inch	0.1	1.2	0.5-1.0	0.3-0.65
O_2-5	O_2-0.4 @0.0005-inch	O_2-0.08	O_2-250-840 CO_2-495-5000	O_2-165 335 CO_2-500-840	O_2-33-250 CO_2-250-645
Good	Good	Good	Varies	Good	Good
400	Coating softens at 180°	400	150	180-220	250
−80	−60	−80	−60	−60	−60
None	None	None	None	None	None
Good	Good	Good	Fair	Fair	Good
Good	Good	Good	Good if treated	Good if treated	Good if treated
Some types	Some types	No	Some types	Some types	Some types

Table 5.29 *(continued)*

		Polypropylene			Polystyrene/ oriented	Polyvinyl chloride	Polyvinylidene chloride
Linear low density	Low density/ EVA copolymer (12% EVA)	Non Oriented	Oriented/ (balanced)	Oriented/ metallized			
Transparent to translucent	Transparent	Transparent	Transparent	Opaque	Transparent	Transparent to Translucent	Transparent
30,000	29,500	30,800	30,600	30,600	26,300	19,000-22,000	16,300
.925	0.94	0.88-0.90	0.905	0.905	1.05	1.23-1.37	1.59-1.71
MD-1540 TD-1620	3000-5000	3000-9000	10,000-30,000	MD-19,000 TD-32,000	9,000-12,000	2,000-16,000	8,000-20,000
MD-640 TD-680	300-500	400-800	50-400	50-400	10-50	5-500	40-80
1.3	11-15	1-3	5-15	5-15	1-5	12-20	10-15
MD-280 TD-400	50-100	40-330	4-6	4-6	4-20	Varies Widely	10-20
250-350	200-300	325-400	Requires coating	Requires coating	250-325	200-350	280-300
1.2	3.9	0.5-0.65	0.26-0.65	0.1	6.5 and up	>8	0.1-0.3
O_2-250-840 CO_2-495 5000	O_2-515-645 CO_2-2260-2900	O_2-84-415 CO_2-495-1350	O_2-155 CO_2-540	O_2-3.0	O_2-165-495 CO_2-645 1680	O_2-5-500 CO_2-50-3500	O_2-0.5-1.7 CO_2-3.3-10
Good	Varies	Good	Good	Good	Good	Good	Good
170-180	140	250	250	250	175	200, depends on plasticizer	Softens at about 290°
−60	−60	Not recommended for low temps.	−60	−60	<0	Depends on plasticizer	<0
None	None	None	None	None	Little to none	None	None
Fair to good	Fair	Good	Good	Good	Good	Fair	Fair
Good if treated	Good if treated	Good if treated	Good if treated	Good if treated	Special inks	Special inks	Special inks
Not recommended	Some types	No	Some types	No	Yes	Some types	Some types

Table 5.29 *(continued)*

Plastic film	Temperature SI	Temperature US	Film thickness	Water vapor	N_2	Air	O_2	CO_2	H_2	Ar	He
	°C	°F	µm [1]	g/cm² day	cm³/m² day bar						
PE-LD	23	73	100	1	700	1100	2000	10000	8000	–	–
PE-HD ($\rho = 0.95$ g/cm³, unstretched)	25	77	40	0.9	525	754	1890	7150	6000	–	–
	30	86	40	1.7	720	960	2270	8600	7600	–	–
	40	104	40	3.5	1220	1660	3560	13100	11400	–	–
	50	122	40	8.1	2140	2650	5650	19500	16800	–	–
PE-HD ($\rho = 0.95$ g/cm³, stretched)	25	77	40	1.0	430	680	1210	5900	5000	–	–
	30	86	40	1.6	560	830	1530	7200	6000	–	–
	40	104	40	4.3	1050	1490	2650	11200	9400	–	–
	50	122	40	10.5	1870	2670	4650	18100	14800	–	–
Polypropylene (unstretched)	25	77	40	2.1	430	700	1900	6100	17700	1480	19200
	30	86	40	3.2	600	960	2500	8400	18200	2100	21700
	40	104	40	7.4	1280	1820	5100	14800	28100	4100	29800
	50	122	40	19.0	2800	3600	9200	27300	46600	8000	43500
Polypropylene (stretched)	25	77	40	0.81	200	350	1000	3300	6700	–	7300
	30	86	40	1.2	260	480	1200	3900	8200	–	8500
	40	104	40	3.3	560	940	2300	7050	12300	–	12100
	50	122	40	8.4	1200	1850	4150	13200	19800	–	17800
PVC-U (unstretched	20	68	40	7.6	12	28	87	200	–	–	–
PVC-U (stretched)	20	68	40	4.4	13	13	43	110	–	–	–
PVC-P	20	68	40	20	350	550	1500	8500	–	–	–
Polyvinylidenechloride	25	77	25	0.1/0.2	1.8/2.3	5/10	1.7/11	60/700	630/1400	–	–
Polystyrene (stretched)	25	77	50	14.0	27	80	235	800	1260	–	–
Polyacetal	20	68	40	2.5	0	16	50	96	420	–	–
PFEP copolymers	40	104	25	2.0	375	–	3000	6500	2000	–	–
CTFE	40	104	25	0.38/0.85	39	–	110/230	250/620	3400/5200	–	–
E/TFE copolymers	23	73	25	0.6	470	–	1560	3800	–	–	–
E/CTFE copolymers	23	73	25	9.0	150	–	39.0	1700	–	–	–
PVF	23	73	25	50.0	3.8	–	4.7	170	900	–	–
Polyamide 6	25	77	25	80/110	14	–	40	200	1500	–	–
Polyamide 66	25	77	25	15/30	11	–	80	140	–	–	–
Polyamide 11	25	77	25	1.5/4.0	50	–	540	2400	5000	–	–
Polyamide 12	25	77	25	0.35	200/280	–	800/1400	2600/5300	–	–	–
Polycarbonate	23	73	25	4	680	–	4000	14500	22000	–	–
Polyethyleneterephthalat (stretched)	23	73	25	0.6	9/15	–	80/110	200/340	1500	–	–
Polysulfone	23	73	25	6	630	–	3600	15000	28000	–	–
PU elastomers	23	73	25	13/25	550/1600	–	1000/4500	6000/22000	–	–	–
Polyimide	23	73	25	25	94	–	390	700	3800	–	–
Celluloseacetate	25	77	25	150/600	470/630	–	1800/2300	13000/15000	14000	–	–
Celluloseacetobutyrate	25	77	25	460/600	3800	–	15000	94000	–	–	–

[1]) 1 µm = 0.0394 mil.

Table 5.30 Examples of gas permeabilities

very wide on one side (so that it can be opened with a visible line due to the fold), or used for other constructions, such as in-line grocery bag lines.

Trapped air that forms the continuous tube is directed through a mandrel via the die. Once the bubble has been formed, the controlled air pressure required to keep the bubble stable is kept constant. Usual pressure is about 40 ft^3/min (1.1 m^3/min). As the hot tube/melt leaves the die, a cooling system is used to uniformly cool the melt. This cooling action has a major influence on the

Application	Demands made	Rate of stretching	Dimensions (width b in mm thickness d in mm and denier Tt in dtex)	Thermoplastic
Carpet basic weave	low shrinkage high strength temperature stability specific splicing tendency matt surface	1 : 7 1 : 5	b = 1.2 and 2.4 d = 0.05 Tt = (550 and 1100)	PP PETP
Tarpaulins	high strength	1 : 7	b = 2.4 d = 0.04 Tt = 850	PP PE
Sacks	high strength high friction value specific elongation weather resistant	1 : 7	b = 3.0 d = 0.03 Tt = 800	PP PE
Ropes	high tensile strength specific elongation good tendency to splicing	1 : 9 to 1 : 11 (15)	b = 20 to 60 d = 0.04 to 0.1 Tt = 15000 to 50000	PP
Twine	high tensile strength high knotting strength	1 : 9 to 1 : 11	b = 30 to 80 d = 0.03 to 0.06 Tt = 14000 to 30000	PP PP/PE-mixture
Separating weave	high strength	1 : 7	b = 2.1 d = 0.04 Tt = 750	PP
Filter weave	low shrinkage abrasion resistant	1 : 7 1 : 5	b = 1.0 to 2.0 d = 0.04 Tt = 350 to 700	PP PETP
Reinforcing weave	low shrinkage specific elongation temperature resistance	1 : 7 1 : 5	b = 2.0 d = 0.03 Tt = 550	PP PETP
Tapestry and home textiles	UV-resistance low static charge uniform colouration textile type handle	1 : 7	b = 1.2 to 3.0 d = 0.035 Tt = 350 to 900	PE
Outdoor carpets	low shrinkage wear resistance weather resistance elastic recovery	1 : 7	b = 1.0 to 12 d = 0.03 to 0.06 Td = 300 to 3500	PP
	uniform colouration defined splicing	1 : 5	b = 1.0 to 3.0 d = 0.02 Tt = 300 to 1000	PETP
Decorative tapes	effective surface low specific gravity	1 : 6	b = 4 to 12 d = 0.03 to 0.05 Tt = 800 to 3000	PP with blowing agent
Knitted tapes, sacks and other packagings seed and harvest protective nets	high knotting strength low splicing tendency suppleness UV – resistance	1 : 6.5	b = 2.4 d = 0.025 Tt = 550	PP PE
Packaging tapes	high strength low splicing tendency	1 : 9 1 : 7	b = 5 to 16 d = 0.3 to 0.6	PP PETP
Fleeces	fibre properties	1 : 7	b = 20 to 300 d = 0.025 Tt = 10 to 70	PP and polymer mixtures

Table 5.31 Examples of film tapes

bubble's thickness and uniformity. The usual dual-lip or three-chamber air-cooling ring, located outside the bubble as it exits the die, has airstreams gently cooling the bubble. To speed up a line and improve output performance, an internal bubble cooling (IBC) air system is extensively used. It directs cool air at a low velocity to enter and exit the inside of the bubble (Figs. 5.85 and 5.86).

The bubble's diameter is normally always much greater than the die's diameter. The bubble's diameter divided by the diameter of the die's orifice is called the blowup ratio (BUR). The BUR is usually 1.5 to 4.0, depending on the plastic being processed and the thickness required. The bubble's diameter must not be confused with the width of the flattened double layer of film between the nip rolls. The width of this double layer is 1.57 times the bubble's diameter and is called the blown-film width (BFW).

Film type	*Advantage*	*Possible problem*
Polyethylene (low density)	Strong heat seals Low-temperature shrink Medium shrink force for broad application Lowest cost	Narrow shrink temperature range Low stiffness Poorer opticals Sealing wire contamination
Polypropylene	Good optical appearance High stiffness High shrink force No heat-sealing fumes Good durability	High shrink temperature High shrink force, not suitable for delicate or fragile products Brittle seals High sealing temperature
Copolymers	Strong heat seals Good optical appearance High shrink force No heat-sealing vapors	High shrink force, not suitable for fragile products Higher shrink temperature Higher heat-seal temperature Lower film slip may give machine problems
Polyvinyl chloride	Lowest shrink temperature range Wide shrink temperature range Excellent optical appearance Controlled stiffness by plasticizer-content control Lowest shrink force for wrapping fragile products	Weakest heat seals Least durable after plasticizer loss Toxic and corrosive gas emission from heat sealing, good ventilation required Durability problem at low temperature Low shrink force inhibits use as a multiple-unit bundling film Low film slip causes machine wrapping difficulties

Table 5.32 Examples of shrink films

Figure 5.71 Example of antistatic bath (cover guard removed) at the end of a sheet extruder line following the line's takeoff unit (courtesy of Welex Inc.)

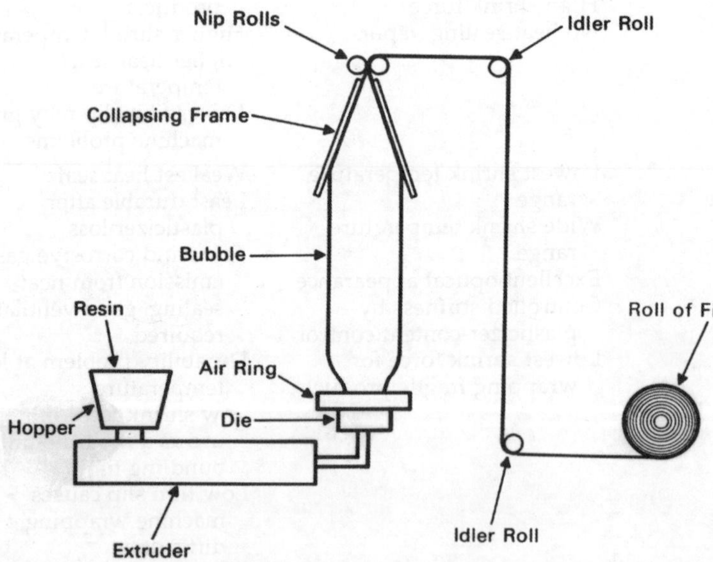

Figure 5.72 Simplified schematic of a blown-film line

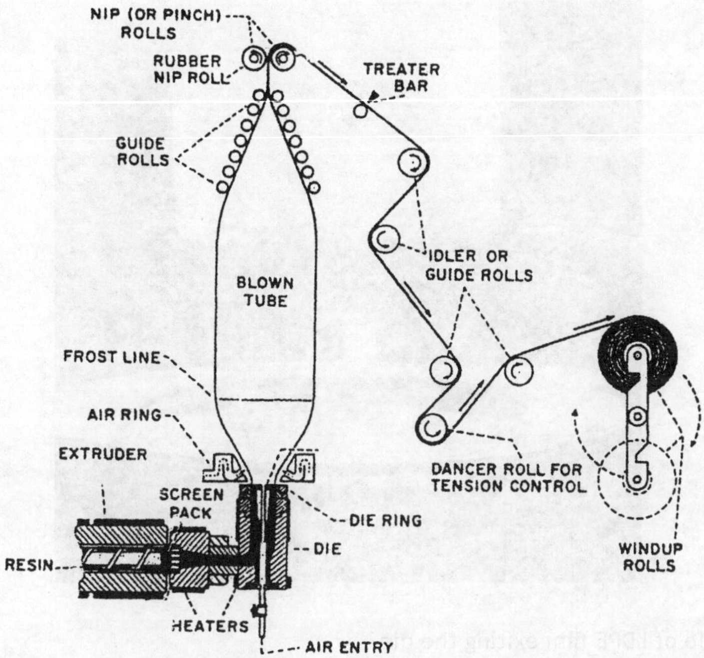

Figure 5.73 More detailed schematic of a blown-film line

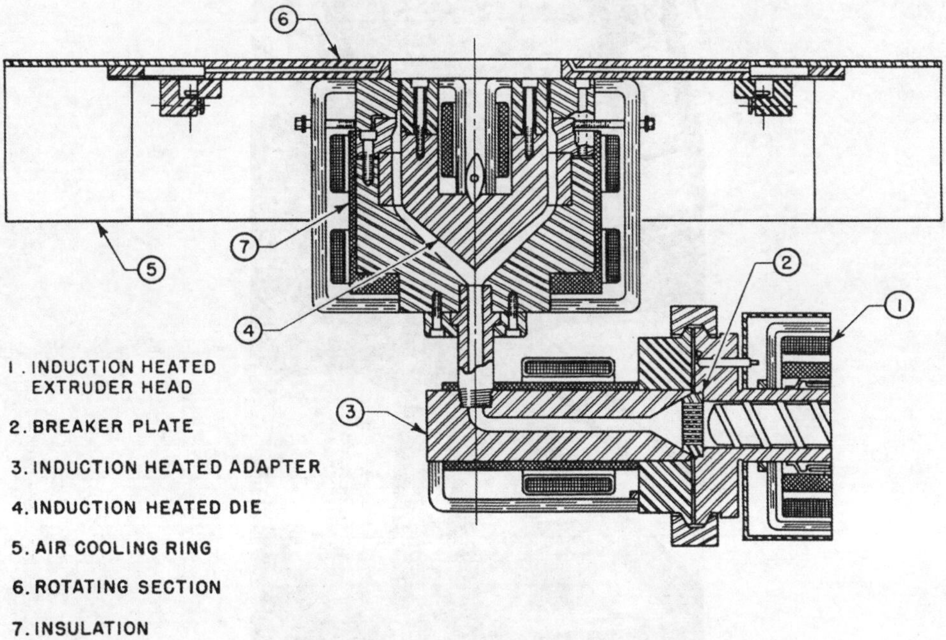

Figure 5.74 Example of a blown-film die

Figure 5.75 Example of LDPE film exiting the die

Figure 5.76 Example of HDPE film exiting the die

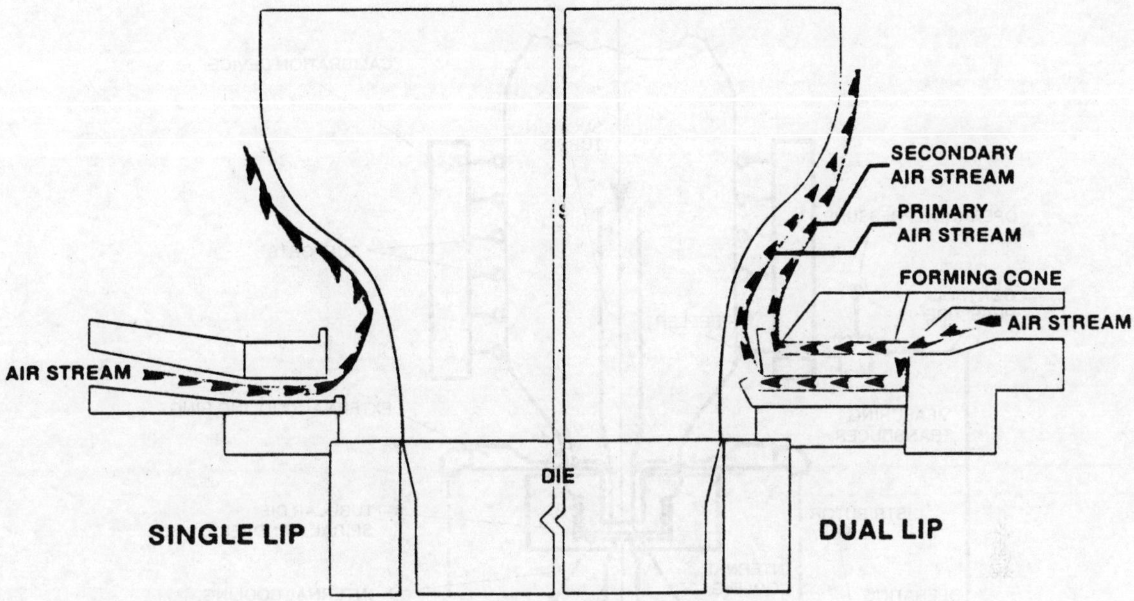

Figure 5.77 Examples of air-cooling ring designs

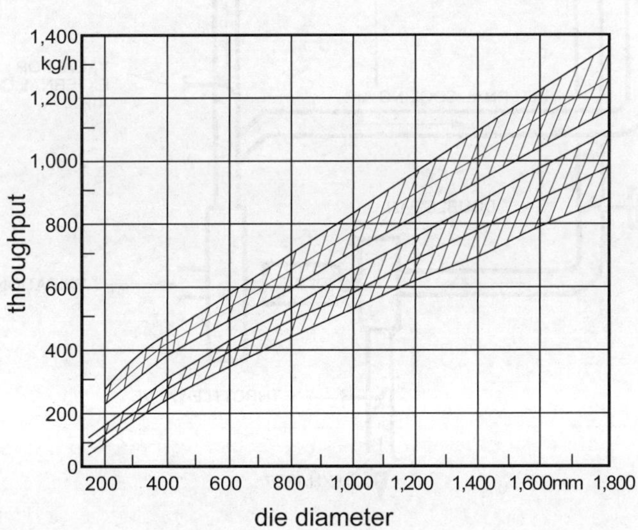

Figure 5.78 Blown-film throughput as a function of the diameter of the die's orifice

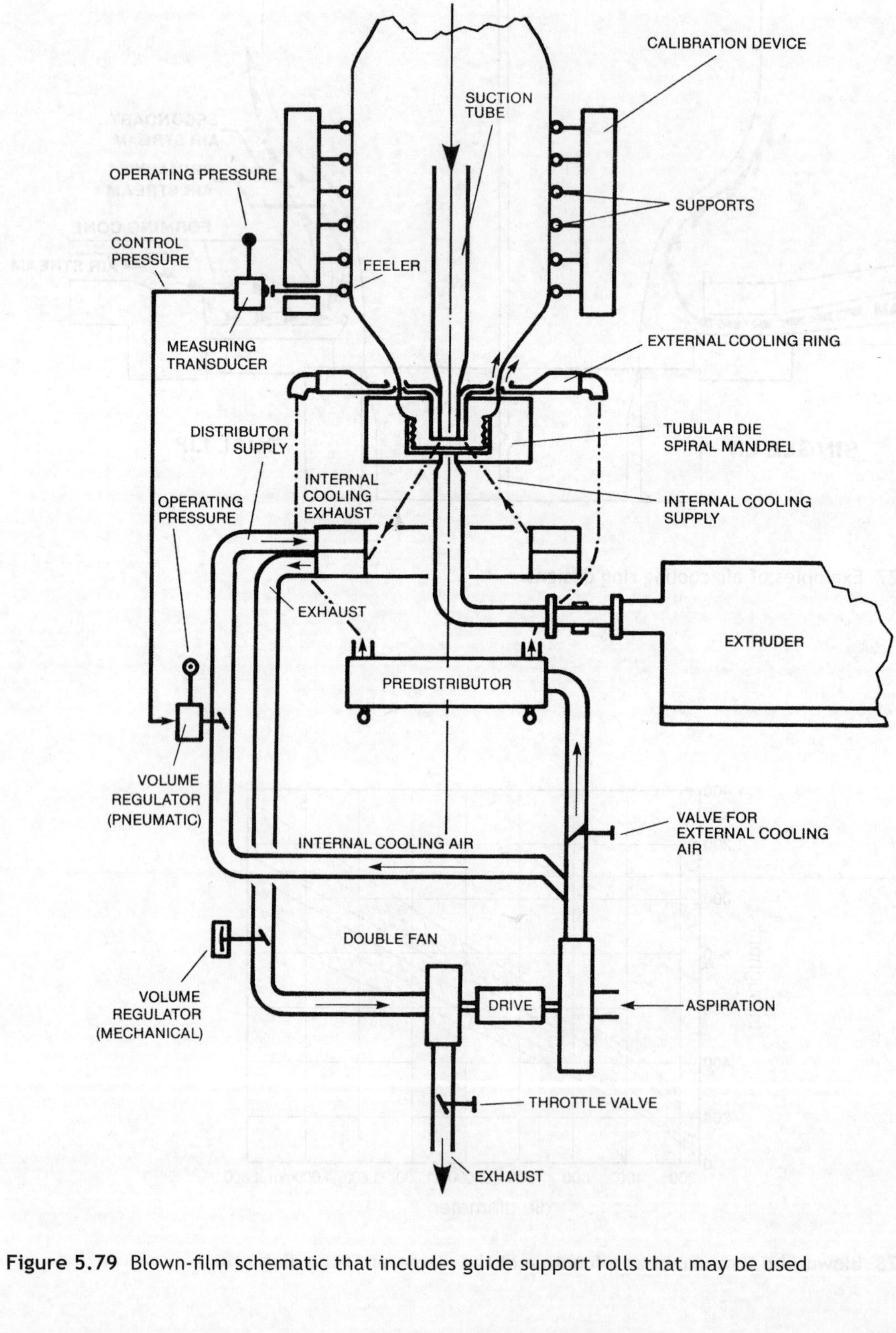

Figure 5.79 Blown-film schematic that includes guide support rolls that may be used

With a crystalline type of plastic (not amorphous [chapter 1]), a melt leaving the die (and moving to a ring-shaped zone where the film approaches its diameter) changes from a hazy to a transparent (amorphous) condition. The level at which this transition occurs is the frost line. This

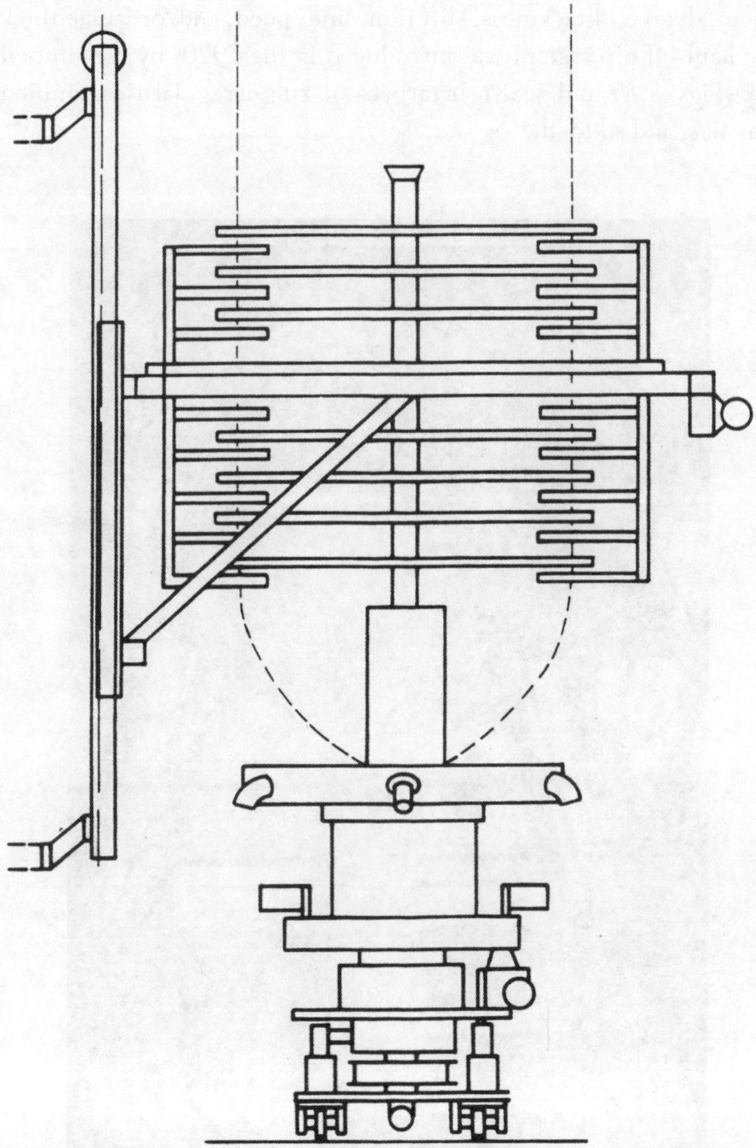

Figure 5.80 Schematic of basket-type height- and width-motorized adjustable sizing support

zone is characterized by a "frosty" appearance to the film caused by the film temperature falling below the softening range of the plastic.

Gauge thickness can be extremely nonuniform, owing to melt's behavior on exiting the die and/or distortions of the collapsing frame. To provide uniformity, controlling the melt is required. Different techniques are used to handle the film, such as oscillating or rotating dies and oscillating film haul-offs. The different systems available meet different requirements, such as web width, cooling system effect, degree of tackiness, stiff film, line speed, and/or gauge thickness.

The oscillating haul-off nip system was introduced in the 1970s by Windmoeller & Hoelscher Corp. of Germany (Figs. 5.87 and 5.88). It targets air-ring irregularities, ambient drafts, bubble alignment irregularities, and so forth.

Figure 5.81 View of basket-type blown-film sizing support

Other systems have been used, such as processing equipment for HMWHDPE blown film that included a rotating extruder to further randomize gauge variations and an adjustable-height collapsing frame to eliminate wrinkles. Optimum HMWHDPE equipment has also used grooved intensively cooled extruders (20:1 L/D) to maximize rate and minimize the plastic's molecular degradation. These extruders are also used to process other polyolefins. The typical melt temperature of a polyolefin in the blown-film process is 350° to 450°F, with typical line speed up to 600 fpm.

The blown tubular film process, by its nature, gives orientation to the plastic. Systems have been designed to increase the degrees of orientation in order to obtain film of improved clarity, strength, heat resistance, and so on. Except for special applications, where greater strength in one

Figure 5.82 View of basket-type blown-film sizing support with internal bubble cooler

Figure 5.83 Collapsing frame with two opposite sets of flat bars in a V form

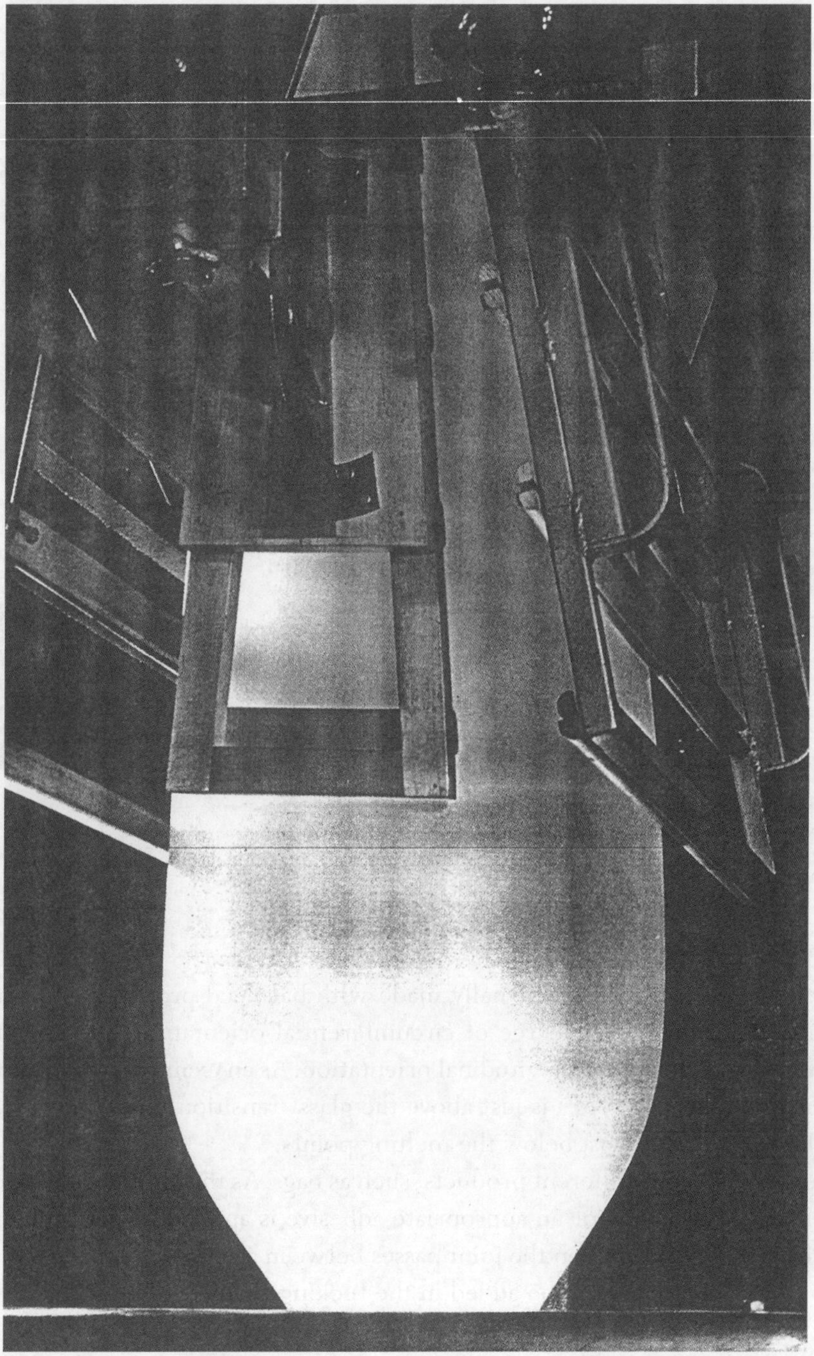

Figure 5.84 Collapsing frame with four opposite sets of flat bars in V forms

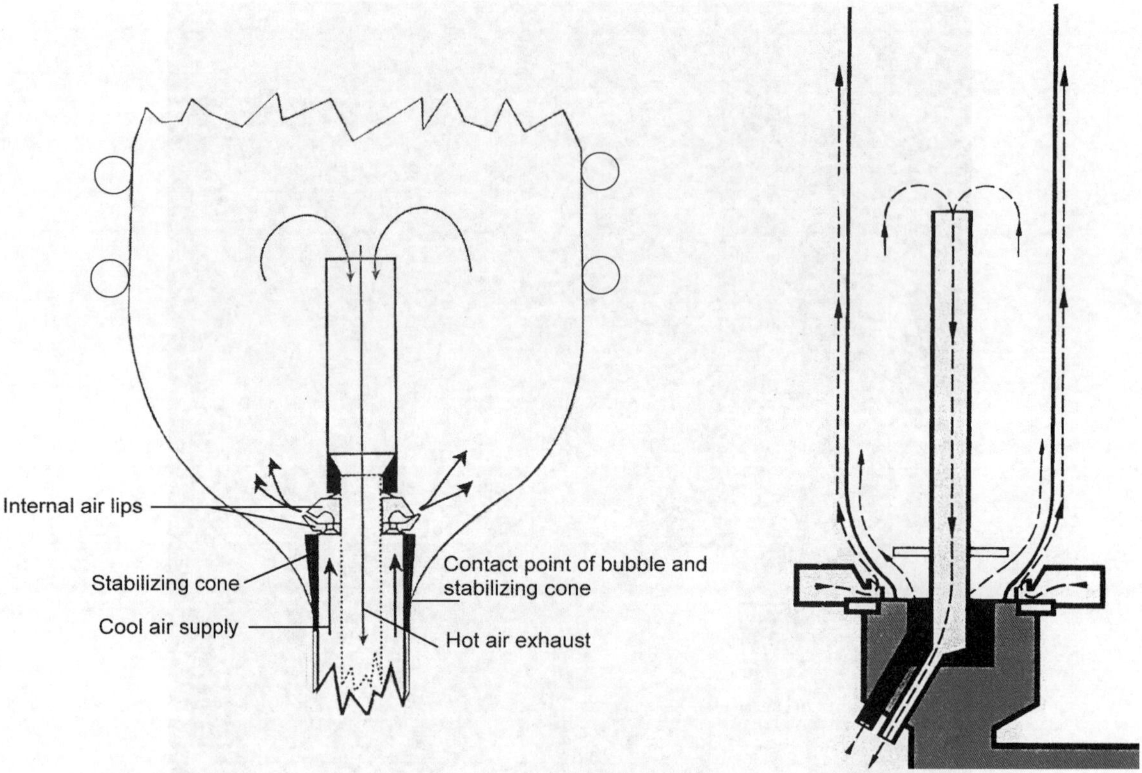

Figure 5.85 Schematic of an air-operated internal bubble cooler

direction may be needed, films are normally made with balanced properties. During processing, the blowup ratio determines the degree of circumferential orientation, and the pull rate of the bubble by the nip rolls determines longitudinal orientation. As an example, the optimum stretching for amorphous plastics (PVCs, etc.) is just above the glass-transition temperatures; for crystalline plastics (PEs, PETs, etc.), it is just below the melting points.

This process can produce different products, such as bags. As the flat tube passes over a forming mandrel or former, heat sealing or an appropriate adhesive is applied to the edge of the web. The seam is completed by pressure when the joint passes between a set of rollers. As the web advances, gussets with or without zippers can be added in the tucking-former section. A cutoff converts the web into bags. Bottom heat sealing or adhesive is automatically applied to the trailing edges. Bottom folds and seams are made, completing the in-line operation. Most common bags are self-opening styles (SOS).

Water Quench The water quench blown-film process, also known as the tubular water quench process, uses a tube die that is arranged to extrude vertically downward (Fig. 5.89). The process

Figure 5.86 Example of a combination of an external film cooling-air ring and an internal bubble cooler

needs melt strength greater than that required by other film processes, so lower melt temperatures are used, typically in the range of 190°C to 230°C for polypropylene.

The die should be designed for constant output rates and thickness at every point around the annular die gap. This requires streamlined internal melt flow paths and precise multipoint means of centering the mandrel within the die ring. The die gap is typically about 0.4 mm, with a short land length and a die entry angle of about 10°. Close thickness tolerances are difficult to achieve with the blown-film process; variations around the bubble can be limited to perhaps ±10% of the targeted film thickness. To distribute these variations evenly, blown-film lines may include (as in air-cooled systems) rotating or oscillating motions applied to the die, extruder, or haul-off.

As with an air-cooling system, blown film emerges from the die and the tube passes through an air ring that lays a large volume of low-velocity cooling air over the external surface. At the same time, the tube is pressurized internally by air supplied through the die mandrel. Downstream nip rollers confine the air so it inflates the still-soft tube to form an enlarged bubble. This distends and thins the tube walls to the final film thickness. The bubble size is limited and calibrated by an annular water-cooling weir that covers the outer bubble surface with a stream of cooling water.

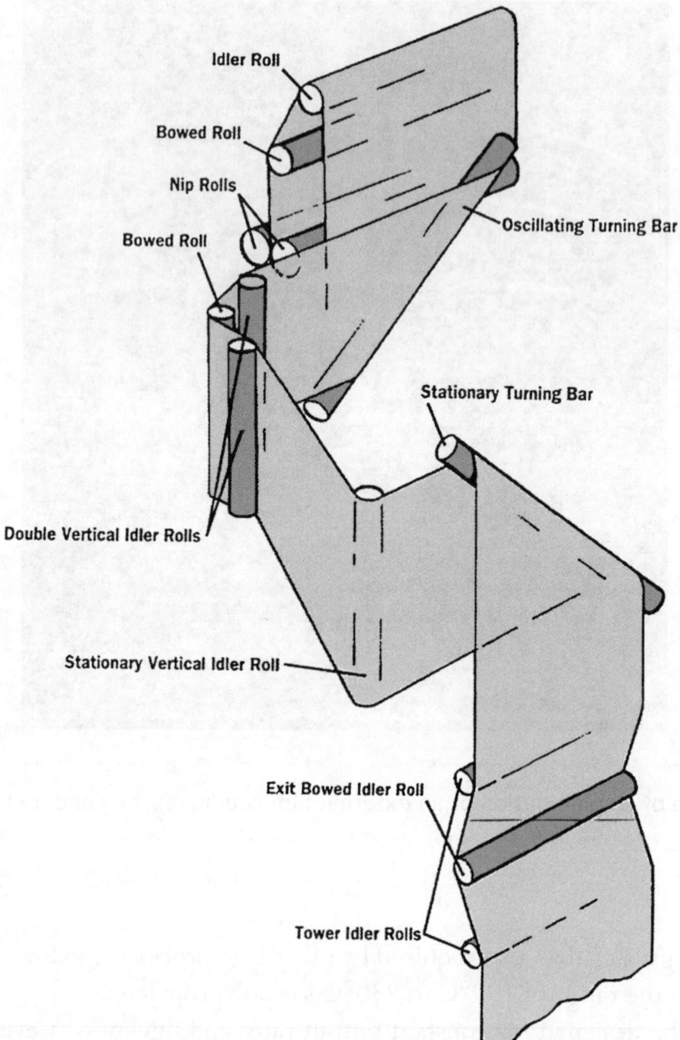

Figure 5.87 Schematic of the oscillating 360-degree haul-off system (courtesy of Windmoeller & Hoelscher)

The water contact is directed and controlled by an annular flexible skirt. Depending on the plastic being processed, the water temperature could be at the low range of 5°C to 10°C to 10°C to 30°C. The rapid bubble cooling induced by the water stream limits crystallinity and produces a clearer film than is possible by the air-cooled blown process (chapter 1). After cooling, the film passes through a collapsing frame and proceeds through de-watering, haul-off, and winding stations. The water quench blown-film process is normally used with smaller extruders having screw diameters up to 65 mm.

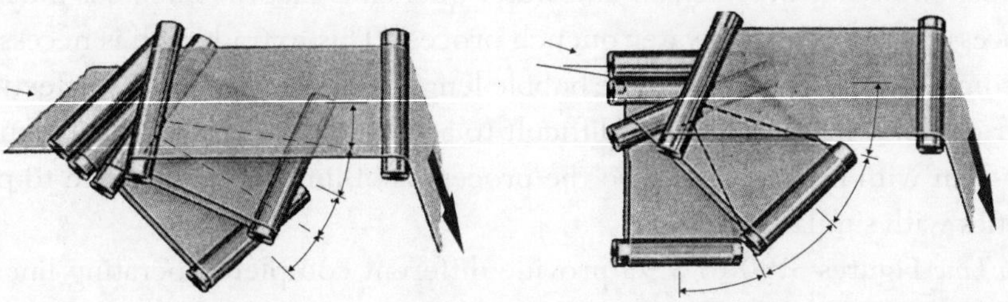

Figure 5.88 Simplified schematic using turning bars in the oscillating haul-off system (courtesy of Windmoeller & Hoelscher)

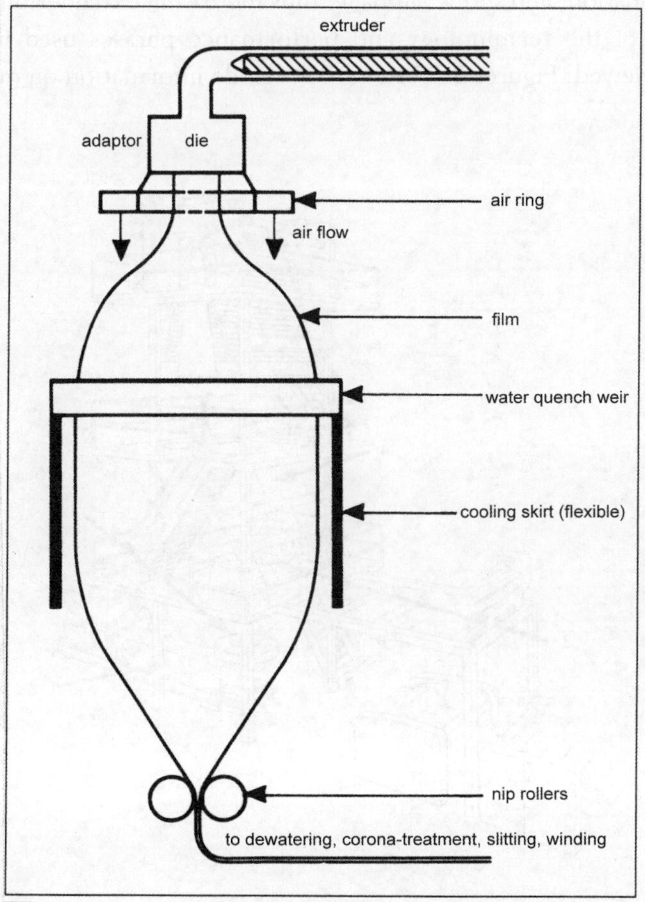

Figure 5.89 Example of water quench process for blown film

The principal difference between air and water quench is that the bubble is much longer in the air-cooled process than it is in the water quench process. This extra length is necessary for bubble cooling to be completed by air contact. The bubble length available for cooling effectively limits line speed. With air cooling, stability is more difficult to achieve with certain plastics, such as polypropylenes rather than with polyethylenes, so the process is of limited significance to polypropylenes and other plastics with similar behaviors.

Blown-Film Line Figures 5.90 to 5.95 provide different complete operating lines that produce film. Tables 5.33 to 5.37 provide an introduction to properties and behaviors of films.

Process Control As reviewed earlier in this chapter and elsewhere (chapter 3), setting and controlling the extruder output influence blown-film properties. This is summarized in Figure 5.96a. As reviewed in chapter 1, different additives are used to simplify and provide more efficient melt flow, resulting in improvements of the film's performance. As an example, there are processing aids to improve the extrusion process and reduce operating cost. Figure 5.96b provides data on the DuPont Dow Elastomers processing aid. This family of fluoroelastomers significantly reduces melt fractures, risk of degradation, and screw slippage, thus increasing extruder output.

Terminology Some of the terminology and performance phrases used in the blow-molding industry have been reviewed. Figures 5.97 to 5.103 include information on more terms.

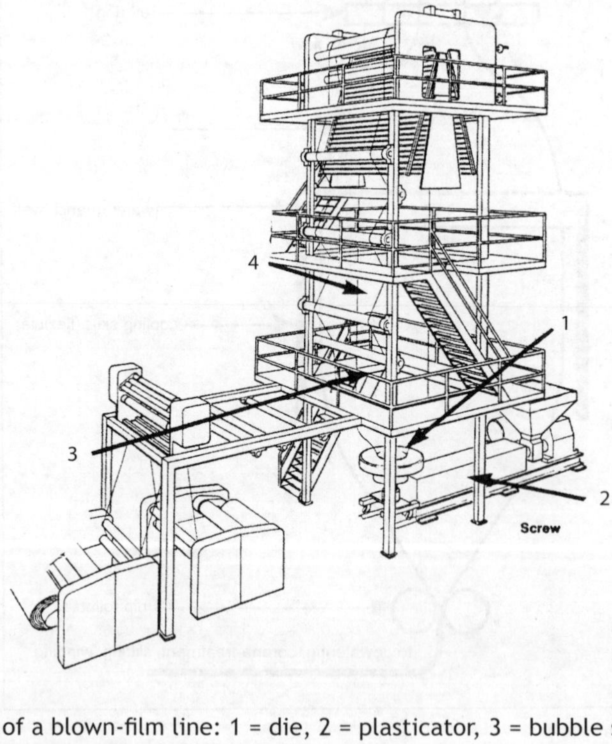

Figure 5.90 Schematic of a blown-film line: 1 = die, 2 = plasticator, 3 = bubble stabilizer, and 4 = tension control roll

Figure 5.91 Three-platform, 40 ft high, with 10-ft-wide nip rolls

FLAT FILM

Flat film identifies cast film. Other names used include chill-roll film, roll cast film, slot cast film, water quench, and water-chill film. These cast-film lines require dies that yield a wide range of diverse products that include food packaging, display packaging, building materials, automotive parts, capacitor film for computer chips, X-ray films, and blood bags. These tooling systems are central components in the processing of cast film, oriented film, and sheet, as well as in extrusion and proximity (slot-die) coating. Extruded structures may be monolayers or coextrusions of up to seven layers or more. Widths may range from less than 6 in (15 cm) to more than 33 ft (10 m) for geomembranes.

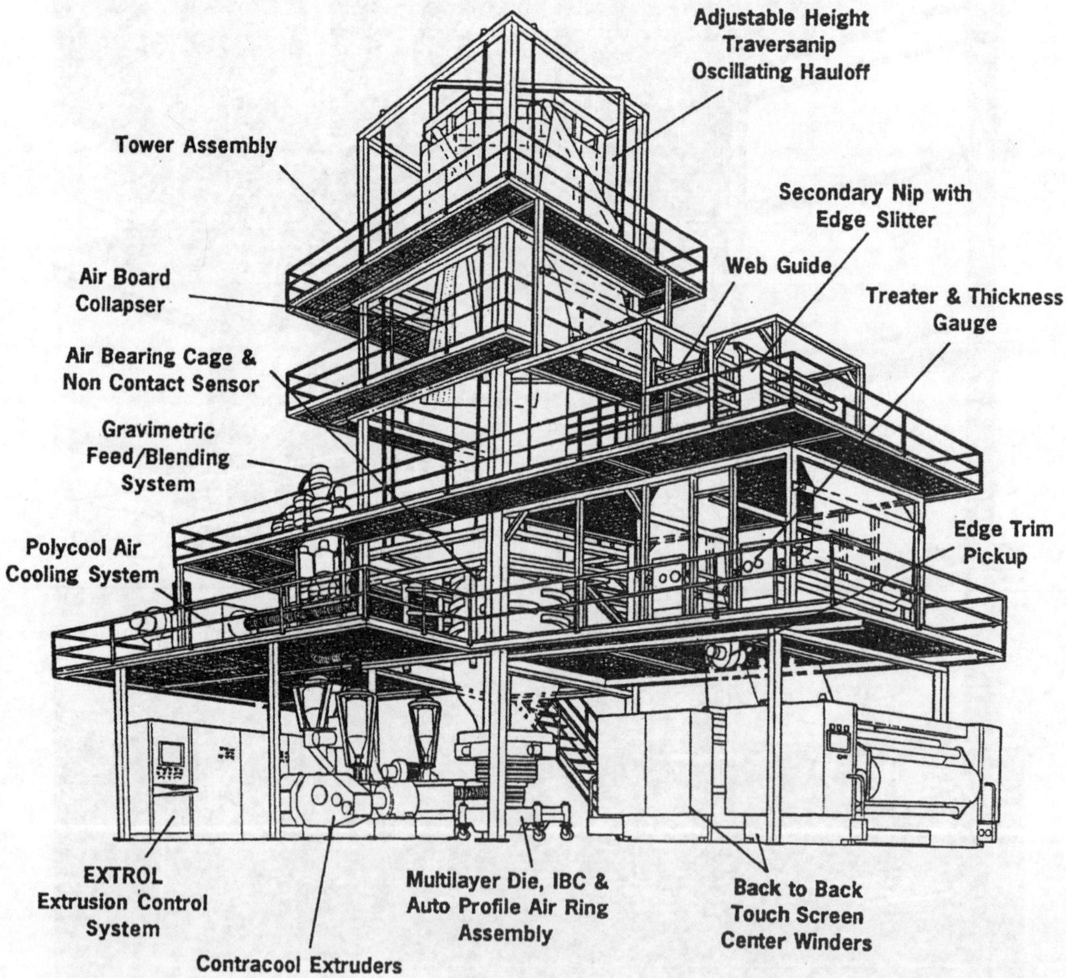

Figure 5.92 Assembled blown-film line (courtesy of Battenfeld Gloucester)

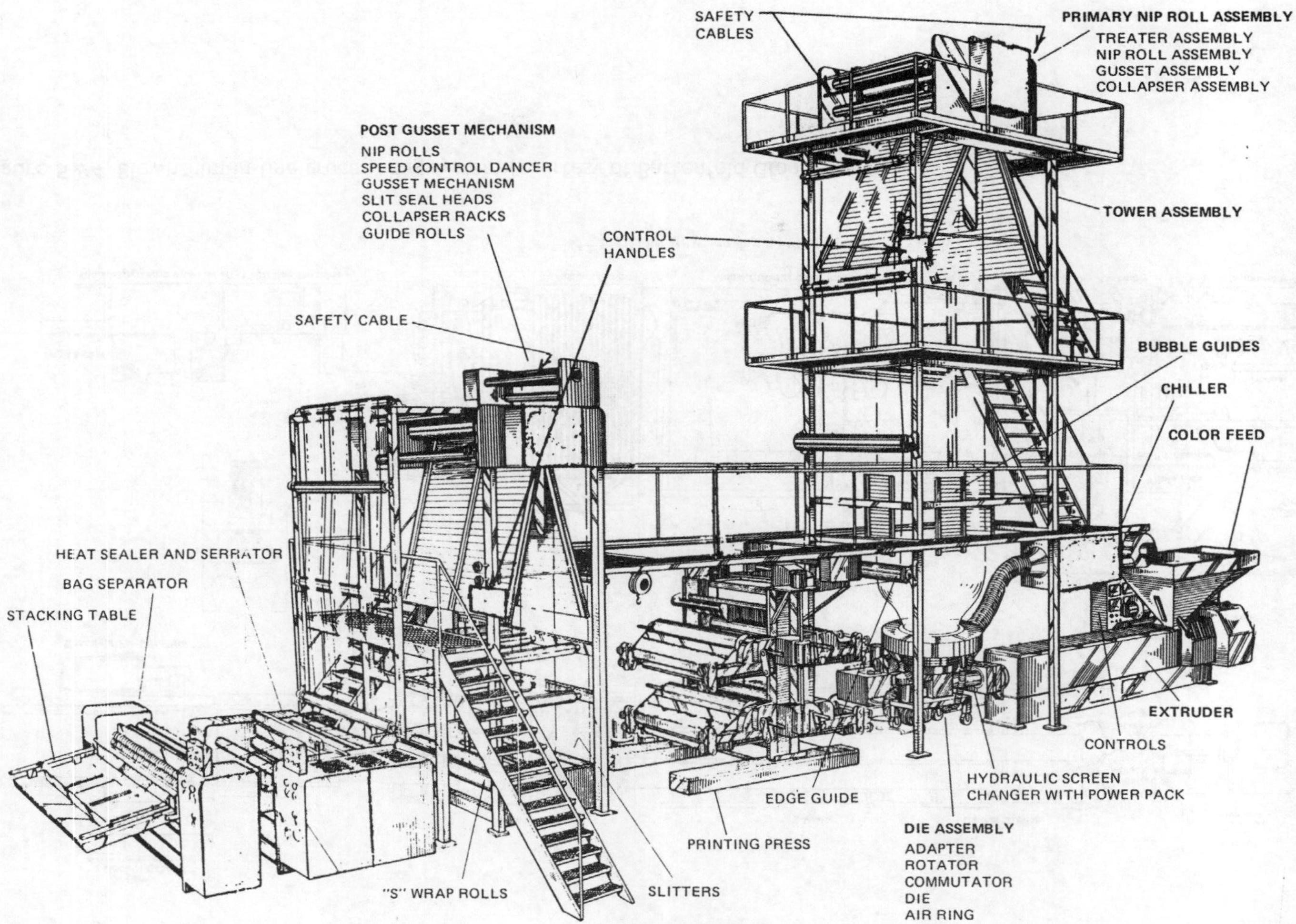

Illustration 6
Typical Blown Film Tower
and Takeoff Equipment

Figure 5.93 Example of blown-film tower and takeoff equipment

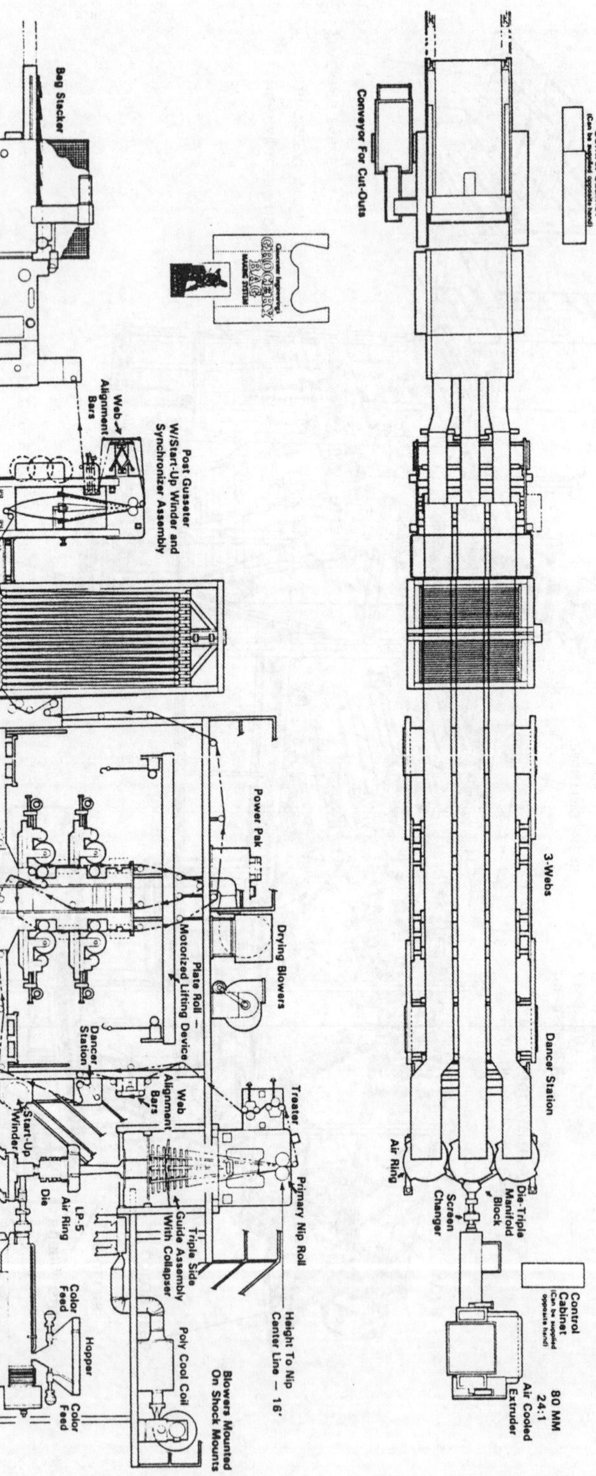

Figure 5.94 Blown-film in-line grocery bag system (courtesy of Battenfeld Gloucester)

Figure 5.95 Blown-film line using oscillating haul-off

Melt index of plastic	Extrusion temperature (°C)	Minimum film thickness (in)	Minimum film thickness (mm)
0.2	150	0.0020	0.051
	170	0.0015	0.038
1.0	150	0.0012	0.030
	170	0.0009	0.023
2.0	150	0.0010	0.025
	170	0.0007	0.018
4.0	150	0.0007	0.018
	170	0.0004	0.010

Density 0.922

Table 5.33 Guide to LDPE film thickness

Die gap		*Film thickness*	
(mm)	*(in)*	*(μm)*	*(mil)*
0.5	0.02	25	1
1.0	0.04	250	10

Table 5.34 Example of relating die gap with film thickness

		Haze value (%)		
		Die land length		
	(in)	*0*	*0.375*	*0.75*
Die entry angle (°)	*(mm)*	*0*	*9.5*	*19*
6		11.3	9.5	10.9
13		10.6	9.0	8.9
33		10.8	8.4	7.8

Table 5.35 Effect of die melt entry angle on film haze

The function of a flat die is to distribute the molten plastic pumped into the die's center by the extruder to the desired end-product width, develop a uniform flow pattern, and establish the desired product thickness. Successful operation begins with a good flow-channel design. The flow channel of the die is the most intricate part of the die to machine because the geometry is complex, the tolerances are critical, and the highly polished surface finish requirements are stringent (chapter 17).

Cast film is produced by extruding the melt from a slit die and cooling it either by contact with a chill roll or by quenching in a water bath. The most popular process used to produce the flat film is with the chill rolls. Chill-roll lines can be arranged in different layouts to meet different requirements. Examples are shown in Figures 5.104 and 5.105. Water chill tank or quench film is also a popular process.

In flat film extrusion, particularly at high takeoff rates, there is high orientation of the film in the machine direction (MD) of the extrudate flow and low orientation in the transverse direction (TD). However, the film can be made to be bidirectional (BD). The effects of biaxial orientation are important in film manufacture of PP, PE, PET, and PVDC. The result is stronger film of high clarity. The orientation-induced crystallization of the plastic produces molecular structures that do not interfere with light waves. For example, when biaxially stretching PS using a 4:1 ratio in both

	FP120-A	FP120-B	FP120-C	FP120-D	FP120-F
Tear Str., g					
MD	400	390	460	420	363
TD	710	740	620	605	631
Dart Impact,					
g	337	303	270	270	271
Puncture,					
in.-lb/mil	12	6	15	8	7
Tens. Break, psi					
MD	6815	7250	6525	6960	6525
TD	6380	5945	5510	4640	5075
Yield Str., psi					
MD	1450	1450	1450	1450	1450
TD	1450	1450	1450	1450	1450
Elongation, %					
MD	430	440	470	480	460
TD	805	740	700	670	688
1% Secant Modulus, psi					
MD	26,825	26,100	26,100	25,375	23,200
TD	29,725	28,275	29,725	28,275	26,800
Haze, %	14	15	7	10	10
Gloss @45°	45	47	65	59	54

Table 5.36 Blown-film properties of 1-mil-thick octene LLDPE film (courtesy of Nova Chemicals)

directions, the higher tensile strength of 10,000 to 12,000 psi (70 to 80 MPa) is achieved compared to 6,000 to 8,000 psi (40 to 55 MPa) for unstretched PS film.

Chill Roll This cast-film system extrudes melt from a slit die and cools it by contact with chill rolls (Fig. 5.106). Relatively high melt temperatures characterize the process and its rapid rate of film cooling. This results in film with low haze, good clarity, and high gloss. Cast-film grades typically have melt flow indexes in the range 5.0 to 12.0 grams/10 min and contain slip and antiblock agents to improve handling performance on packaging machinery.

The melt is extruded downward through a flat die onto a rotating polished chill roll. The molten plastic, formed in the shape of a flat web, is usually conveyed over several cooling rolls in a

Material	Yield			
	(yd^2/lb) *0.001 in thickness*	*(m^2/kg)* *0.025 mm thickness*	*($lb/1000\,yd^2$)* *0.001 in thickness*	*(g/m^2)* *0.025 mm thickness*
Cellulose acetate	17	31	59	32
Cellulose tri-acetate	16	30	62	33
Nylon	18.5	34	54	29
Polyethylene				
low density	23	43	43	23
high density	22	41	45	24
Polyethylene terephthalate	15.5	28	65	36
Polypropylene	24	44	42	23
Polystyrene	20	37	50	27
Polyvinylidene chloride	17	31	59	32
PTCFE	10	18	99	55
PTFE	10	18	99	55
PVC				
flexible	14.5–17	27–31	59–68	32–37
rigid	15.5	28	65	36

Table 5.37 Examples of film yields

series. The cooling rolls are thin-shelled, spirally baffled, double-walled designs that extract heat uniformly and most effectively across the width of the film.

Since this contact-type cooling is faster and more uniform than air-cooled systems used for blown film, higher production rates are achieved with cast film. Cast film tends to be clearer and with less thickness variation than blown film. Film produced on the cast process is used in high-speed converting operations, such as laminating and multicolor printing, as well as in packaging applications where high clarity and gloss are desirable. Figure 5.106 is a typical chill-roll line.

A method of pinning/locating the film on the chill roll is normally required to realize high line speeds. This is achieved by the use of an air knife to force the molten web in contact with the casting drum. A vacuum box that removes the layer of air from the surface of the chill roll and draws the molten web to the casting roll by a vacuum can also be used. The edges of the cast film also have to be pinned to the casting roll to achieve high production rates. Air-jet edge pinners or electrostatic pinners do this. Figure 5.107 highlights the following: 1 = die, 2 = air knife, 3 = cleaning roll, 4 = vacuum box, 5 = first chill roll, and 6 = second chill roll.

Stretching the film in the cast process direction will result in uniaxial orientation in the machine direction. Orientation methods will be reviewed latter. As an example, typical melt temperatures used in cast polyolefins are 475° to 575°F, and line speeds are at least 600 to 1,500 ft/min.

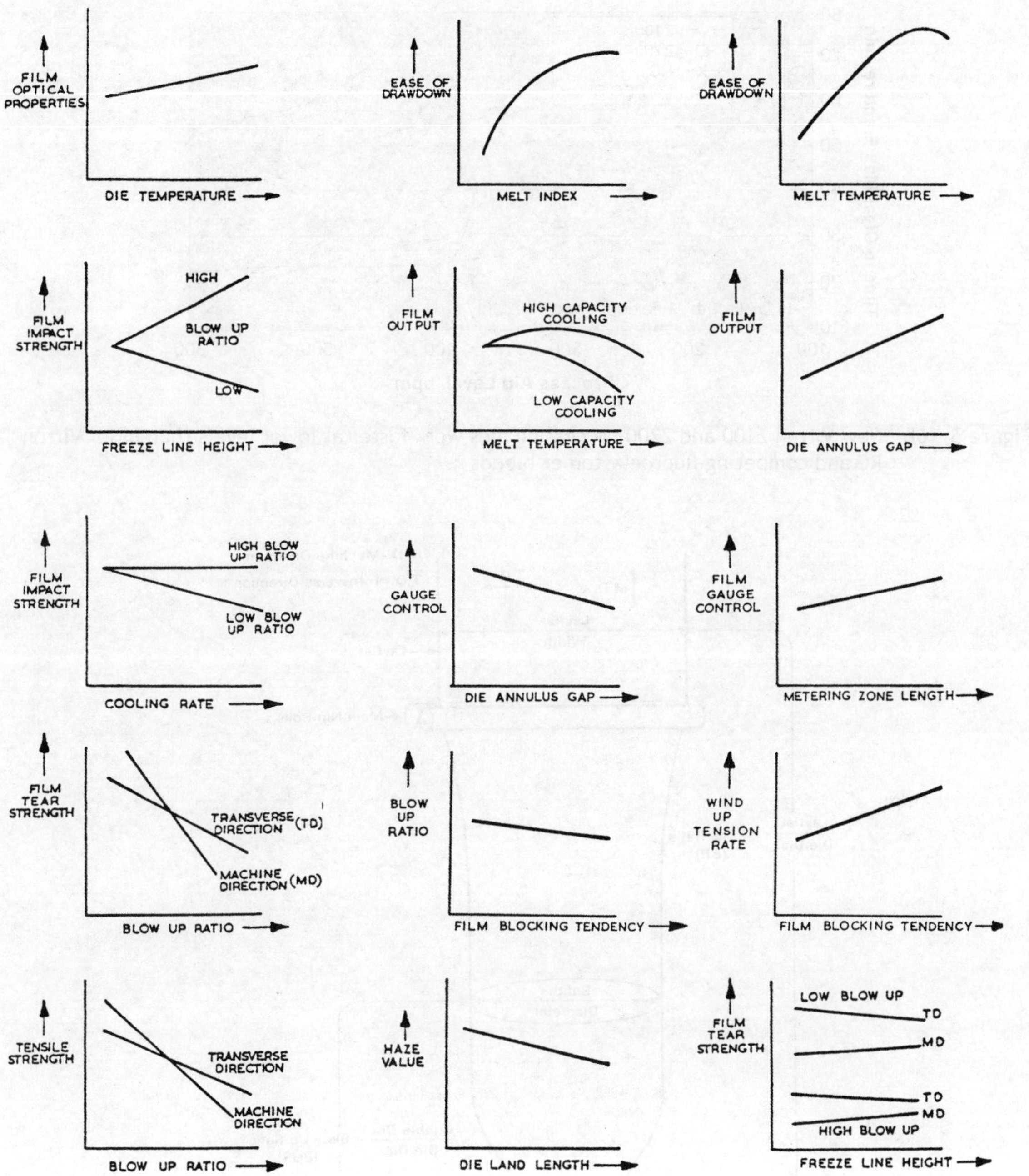

Figure 5.96a Examples of blown-film properties based on the extruder's operations

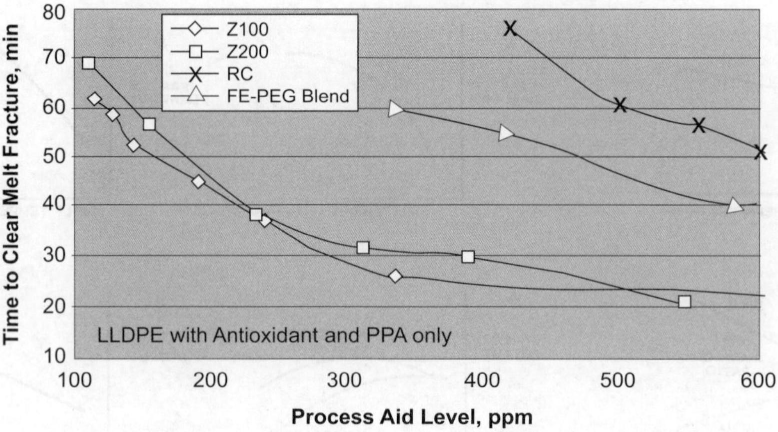

Figure 5.96b New Vitron Z100 and Z200 processing aids work faster at lower levels than older Vitron RC and competing fluoroelastomer blends

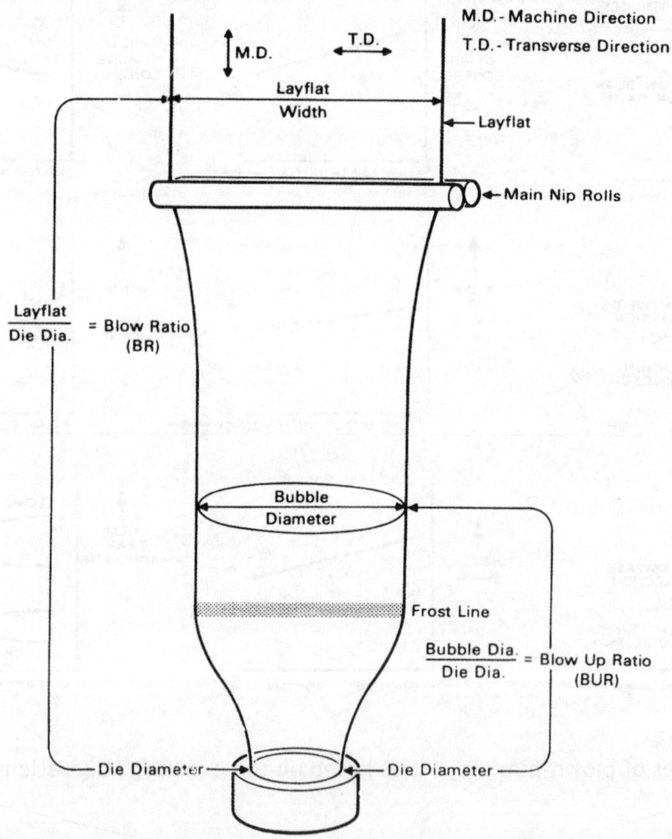

Figure 5.97 Schematic highlighting blown-film terms

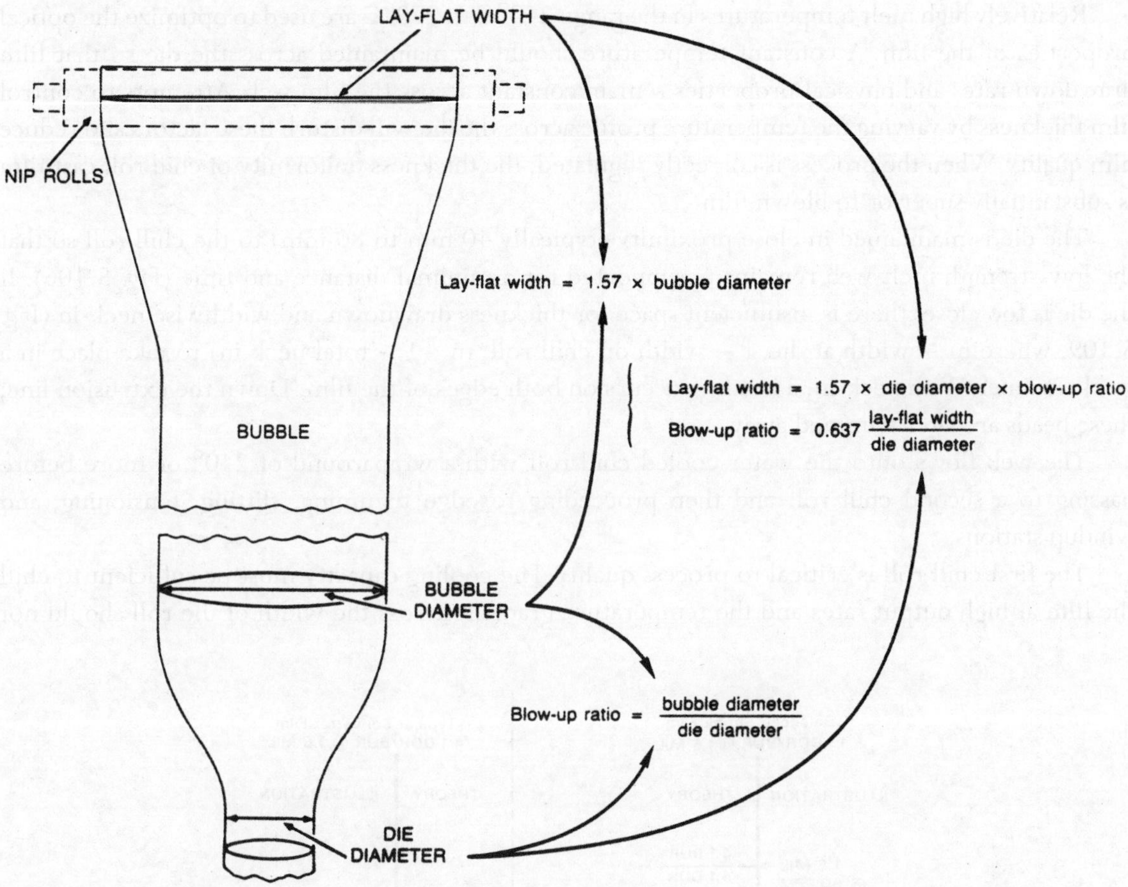

Figure 5.98 Schematic highlighting blown layflat film terms

In this process, a plastic web is extruded from a slit die (Fig. 5.108) against the surface of a water-cooled chill roll. The die is arranged to extrude vertically or obliquely downward so that the film web is delivered approximately tangentially to the roll surface.

The die is similar in principle to a sheet die but will usually not include a restrictor bar (chapter 17). Film thickness is partially regulated by the gap between the die lips but also by the rotational speed of the chill roll, which is arranged so as to draw down and thin the melt web. Consequently, the die gap is set in excess of the desired film thickness. Examples of die gap settings are 0.4 mm for films up to 0.25 mm thick and 0.75 mm for film gauges in the range 0.25 mm to 0.6 mm. Die lip adjusters should allow the die gap to be varied at each adjustment point across the die width in order to allow control of the transverse film thickness. Film reel quality will suffer if the transverse thickness tolerance exceeds ±5% of target thickness.

Relatively high melt temperatures in the range 240°C to 270°C are used to optimize the optical properties of the film. A constant temperature should be maintained across the die so that film drawdown rates and physical properties remain constant across the film web. Attempts to control film thickness by varying the temperature profile across the die will disturb these factors and reduce film quality. When the process is correctly regulated, the thickness uniformity of chill roll cast film is substantially superior to blown film.

The die is maintained in close proximity (typically 40 mm to 80 mm) to the chill roll so that the low-strength melt web remains unsupported for a minimal distance and time (Fig. 5.106). If the die is too close, there is insufficient space for thickness drawdown and widthwise neck-in (Fig. 5.109, where m = width at die, f = width on chill roll, m – f = total neck-in) to take place in a stable manner. With neck-in, a beading occurs on both edges of the film. Down the extrusion line, these beads are later trimmed away.

The web flows onto the water-cooled chill roll with a wraparound of 240° or more before passing to a second chill roll and then proceeding to edge trimming, slitting, tensioning, and windup stations.

The first chill roll is critical to process quality. The cooling capacity must be sufficient to chill the film at high output rates and the temperature gradient across the width of the roll should not

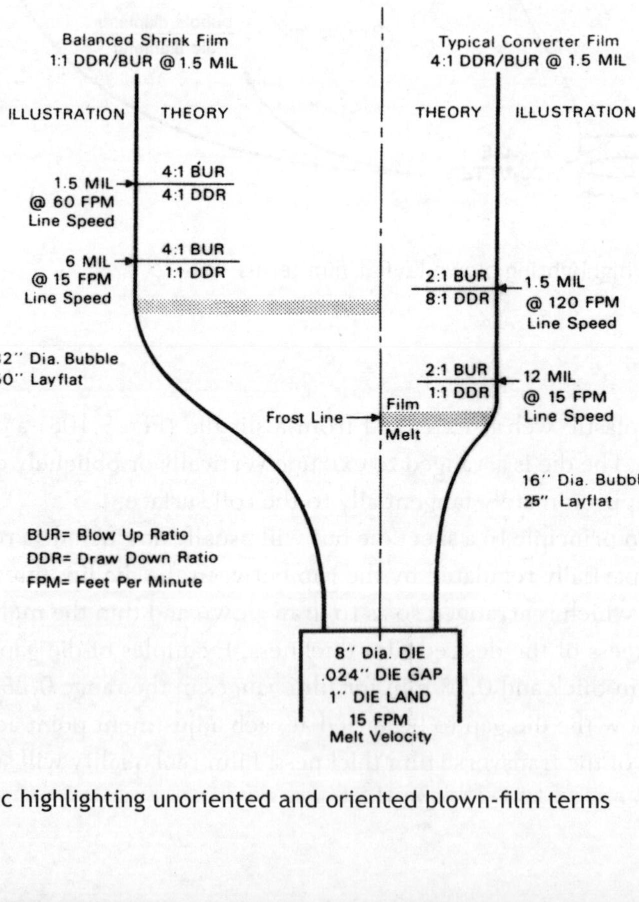

Figure 5.99 Schematic highlighting unoriented and oriented blown-film terms

exceed 1°C. The actual roll temperature depends on film gauge, line speed, and roll diameter; a temperature of about 20°C is common. The chill roll drive speeds must be very precisely regulated to control film drawdown and finished thickness. The chill rolls normally have mirror finishes, but matte finish rolls are often used for producing stationery films.

At line speeds greater than 30 m/min, there is a tendency for a thin cushion of air to become trapped between the film and the chill roll. This results in slow and uneven cooling, affecting the appearance and properties of the film. Two measures are adopted to counteract the problem (Fig. 5.107). An air knife delivers a streamlined jet of filtered air that impinges on the film just beyond the point of first contact with the chill roll and presses it against the roll. The air is supplied through a narrow slit, about a 1.5-mm gap, and controlled at a low differential above atmospheric pressure. An optimum air knife setting improves film clarity and gloss, but excessive pressure induces melt vibration and mars the film surface.

The second measure is the provision of a vacuum box, associated with the die, and operating close to the chill roll surface just ahead of the extruded melt web. The vacuum box removes any condensates that have been deposited on the roll by the film and tends to draw air out of the interface between the film and roll.

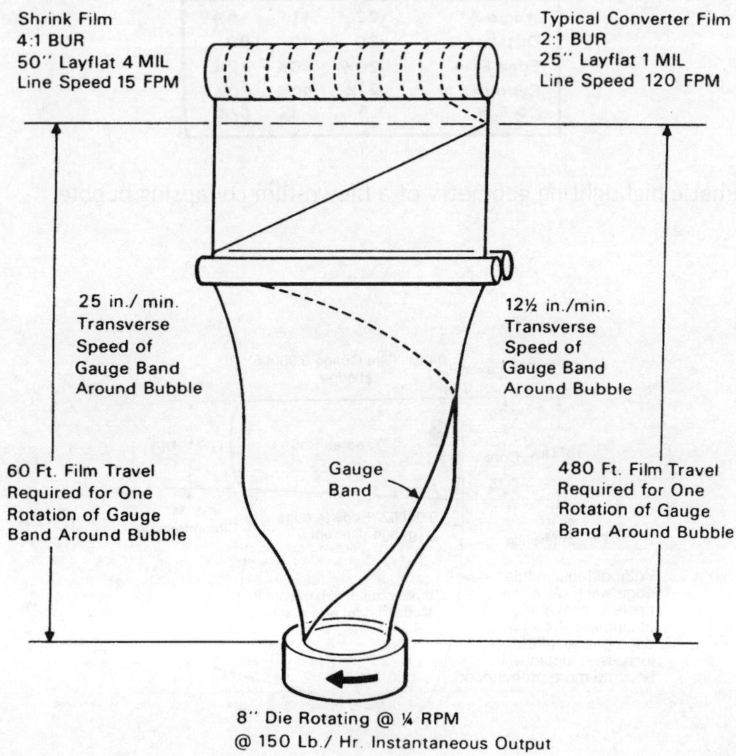

Figure 5.100 Schematic highlighting blown-film die rotation terms to average out thickness

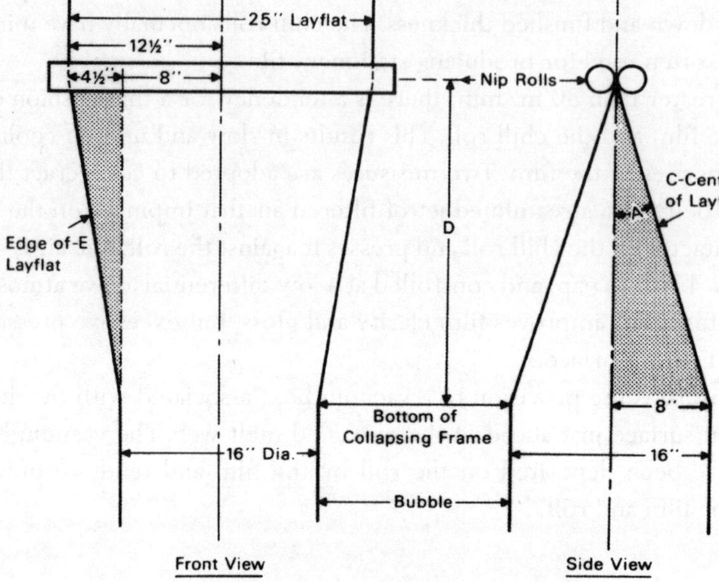

Angle A°	22	11	5½
Distance D, in.	20	40	80
Edge E, in.	20½	40¼	80¼
Center C, in.	21½	40⅜	80½
C > E, %	5	1¼	5/8

Figure 5.101 Schematic highlighting geometry of a blown-film collapsing bubble

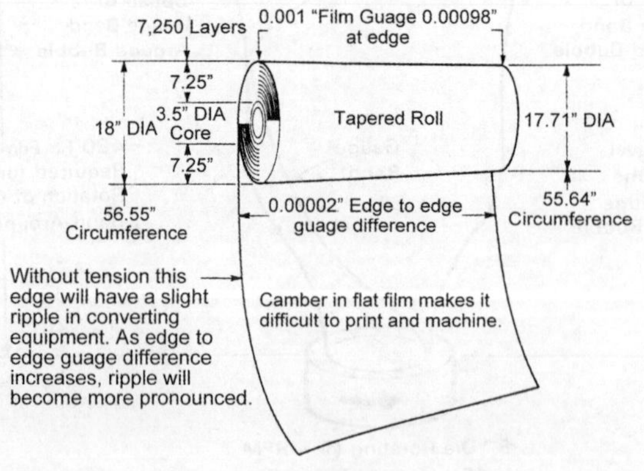

Figure 5.102 Schematic showing slight influences that affect performance of film during windup

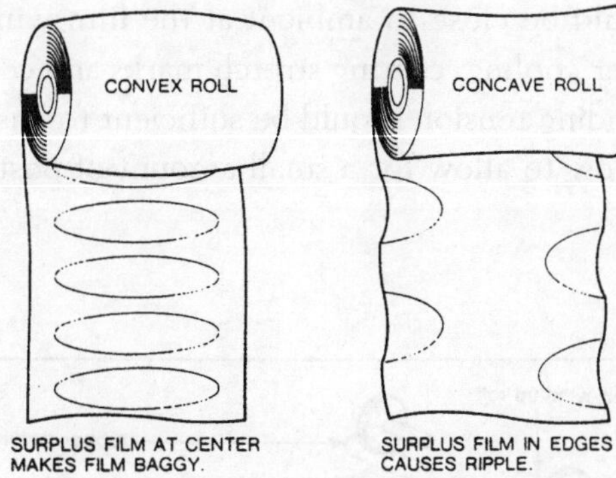

Figure 5.103 Schematic showing major influences that affect performance of film during windup

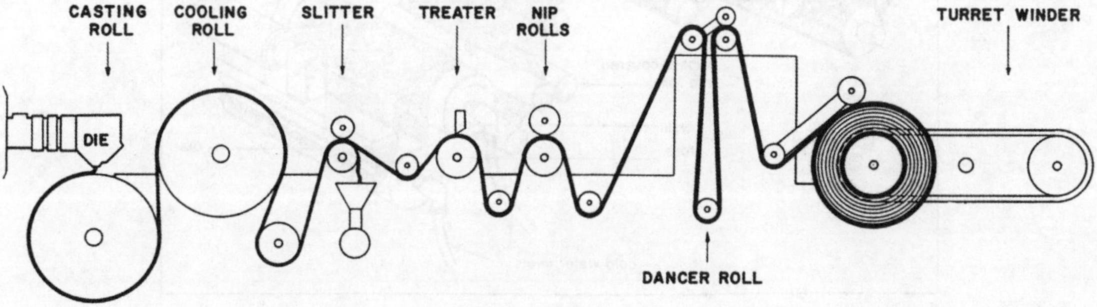

Figure 5.104 Chill-roll film relatively flat processing line

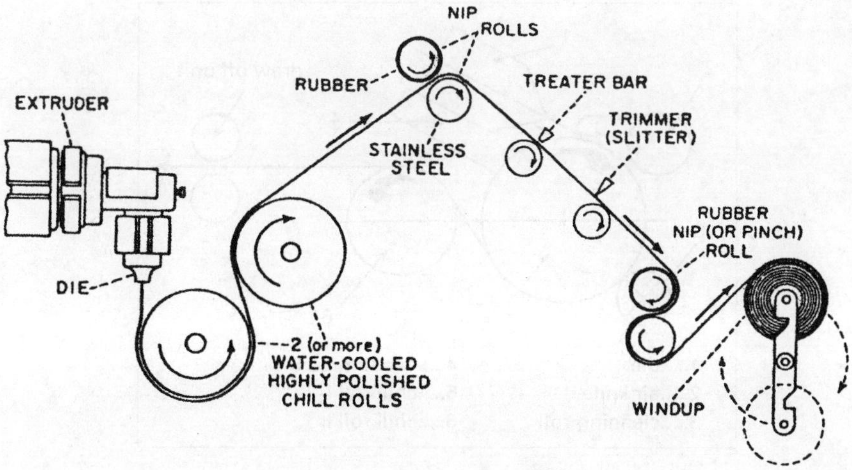

Figure 5.105 Chill-roll film relatively vertical peak processing line

The film temperature should be close to ambient at the film windup station; otherwise, the reel will continue to shrink after cooling, causing stretch marks and corrugations and accentuating any variations in thickness. Winding tension should be sufficient to ensure the integrity of the reel but otherwise should be kept low to allow for a small amount of postwinding shrinkage that will tighten the reel.

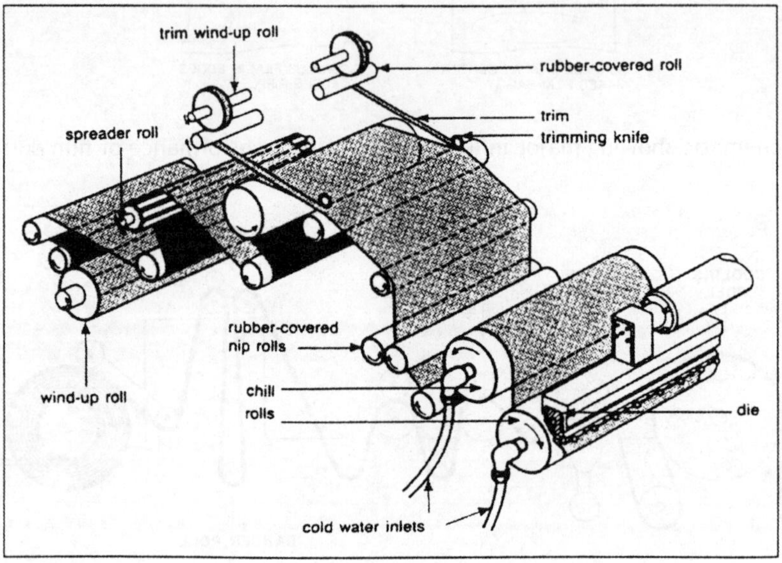

Figure 5.106 A 3-D view of a typical chill-roll line

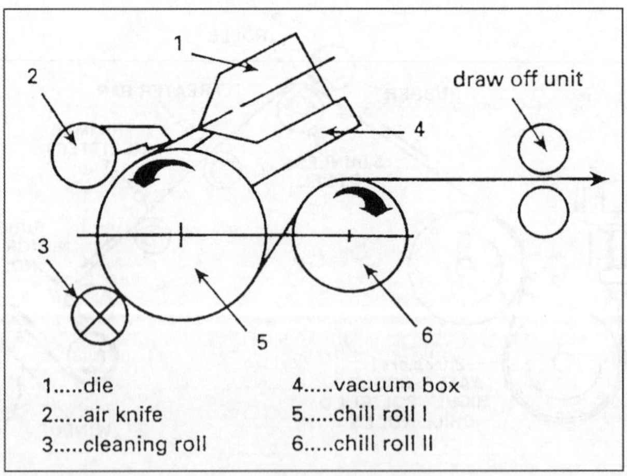

Figure 5.107 Important details of the chill-roll film process

As in other plastics processes, the cast-film process depends on many interactive variables, so that any defect may have one of several causes. A remedy for one defect may introduce another, so process problem solving is not straightforward. With these provisos, the troubleshooting chart (Table 5.38) provides a guide for problem solving (chapter 27).

Water Quench The water quench cast-film process (Figs. 5.110 and 5.111) is similar in concept to the chill-roll process and uses similar downstream equipment. A water bath takes the place of the chill rolls for film cooling, and by cooling both sides of the film equally, it produces a film with slightly different properties compared to chill-roll cast film. Its slit die is arranged vertically and extrudes a melt web directly into the water bath at close range. The bath is typically maintained at

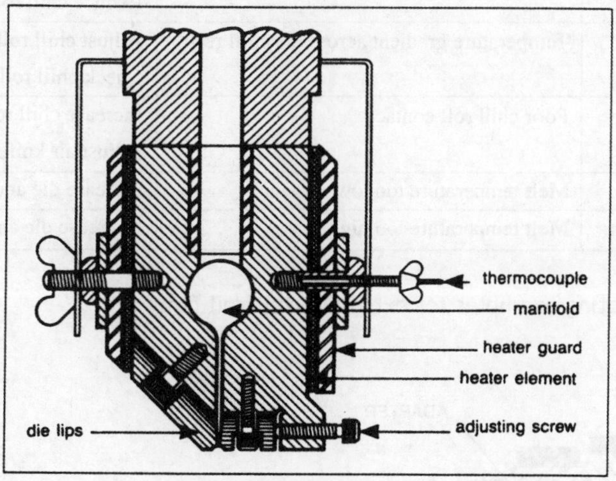

Figure 5.108 Example of a slit die for cast film

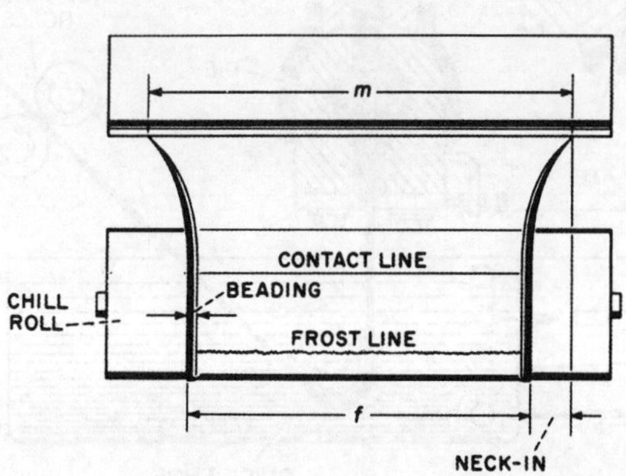

Figure 5.109 Example of neck-in and beading that occur between the die's orifice and the chill roll

20°C. The film passes under a pair of idler rollers in the bath and, for any given rate of extrusion, it is the rate of downstream haul-off that regulates film drawdown and finished thickness.

The speed of the process is limited by the tendency of the film to carry over water from the quench bath. The presence in the film of slip and antistatic additives increases this tendency. Surface defects can also arise from rippling in the water bath. The very rapid quenching induced by the

Problem	Possible cause	Suggested remedy
Poor clarity	Cooling rate too low	Increase melt temperature Reposition air knife Increase air knife pressure Reduce chill roll temperature
Variable optical properties	Temperature gradient across the chill roll	Adjust chill roll coolant supply Check chill roll cooling circuits
Plateout	Poor chill roll contact	Increase chill roll wrap-around Adjust air knife
Flow defects	Melt temperature too low	Increase die and adapter temperatures
Brittle or fibrous heat seals	Melt temperature too high	Decrease die and adapter temperatures

Table 5.38 Troubleshooting examples for extruded chill-roll film

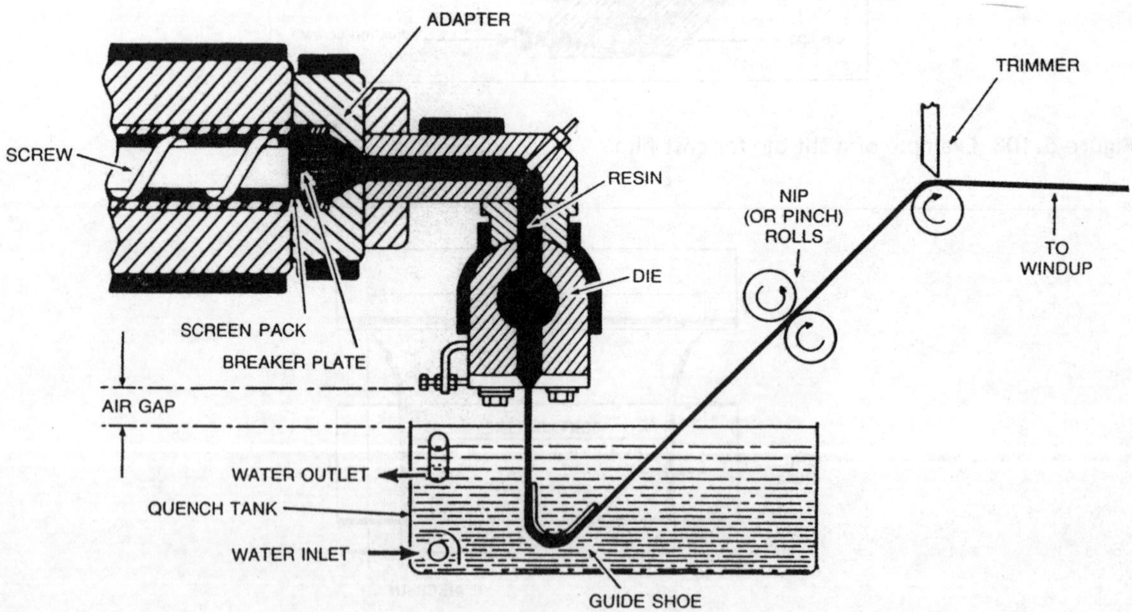

Figure 5.110 Simplified water quench film line

water bath reduces crystallinity and produces a tough film, so the process is often preferred to make films for strapping and packaging tapes, generally in the thickness range 0.07 mm to 0.25 mm (Fig. 5.112 and Table 5.39).

FILM WINDING

Center winding and surface winding are two fundamental winding processes. In center winding, the wrapping force or tension is transmitted through previously wound layers through the core to the winding shaft or spindle. In surface winding, the tension is generated at the nip formed by the driven drum and the wound roll.

Surface winding is generally used for nonextensible film where tension control is not critical. High-quality rolls can be produced on these inexpensive winders. Surface winders are generally limited to 500 fpm. Center winding is usually used for extensible film, where control of tension is

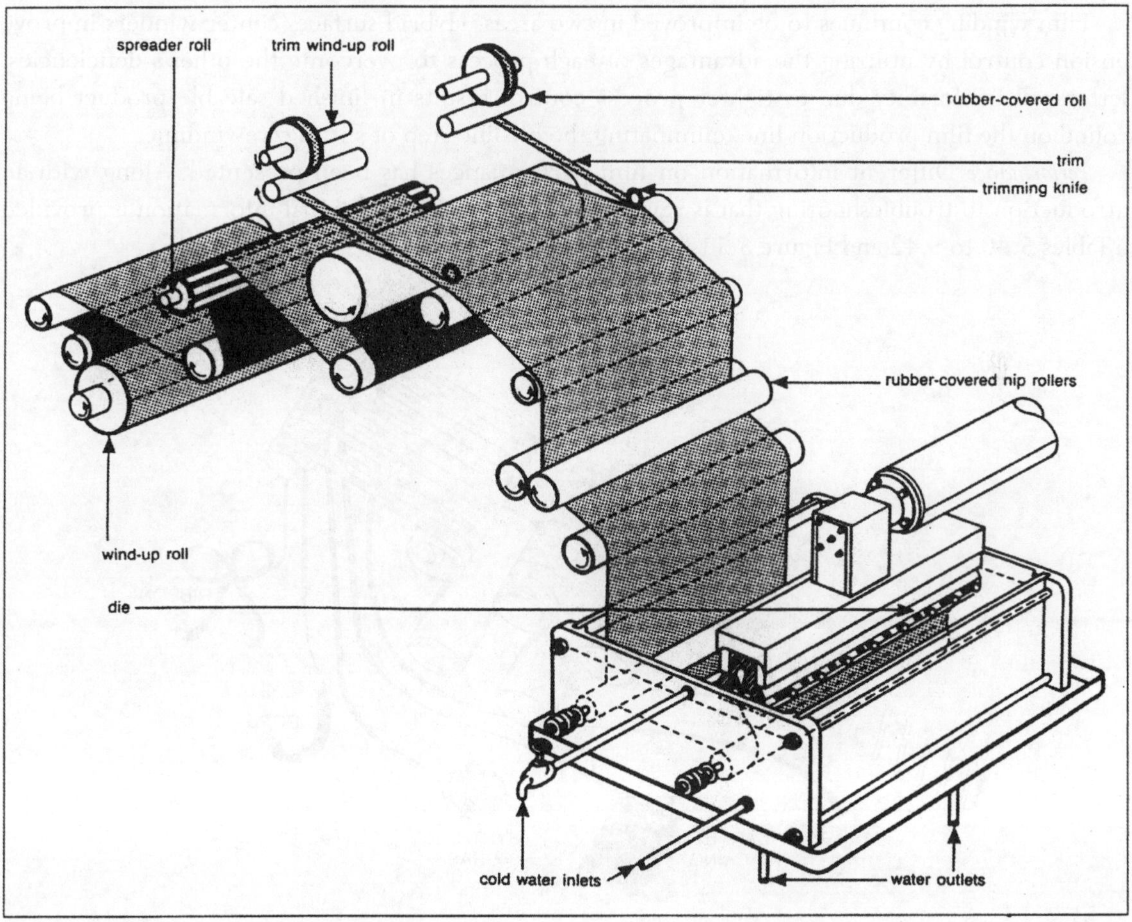

Figure 5.111 Detailed water quench film line

required. Three methods of regulating winder motor speeds to control tension are torque, dancer, and force transducer.

Torque-regulated drives assume that the film tension is directly related to the motor torque. Inexpensive, they utilize pneumatic slip clutches, AC torque motors, or current-regulated DC motors. These are normally applicable only to high tension levels used with thick or nonextensible films.

A dancer is a roll in contact with the film and is free to move depending on tension levels. The film tension can be established by air loading the dancer roll. A feedback loop from the position potentiometer to the winder drive is employed to keep the dancer in its neutral position. Any displacement of the dancer is compensated for by increasing or decreasing the drive speed, and this restores the dancer to its neutral position. Dancers are used in most film applications but not in extremely high tension winding.

The force transducer is a more sensitive device for tension control. Consisting of a roll mounted on a strain gauge, it is a fast reacting, accurate method of tension control because there is virtually no mechanical inertia or friction as compared with the other methods.

Film winding continues to be improved in two areas. Hybrid surface/center winders improve tension control by utilizing the advantages of each process to overcome the other's deficiencies. Better roll conformity due to tighter process control results in finished saleable product being wound on the film production line, eliminating the off-line step of slitting/rewinding.

Performance Different information on film performances has been presented. Along with an introduction to troubleshooting that is reviewed in chapter 27, additional information is provided in Tables 5.40 to 5.42 and Figure 5.113.

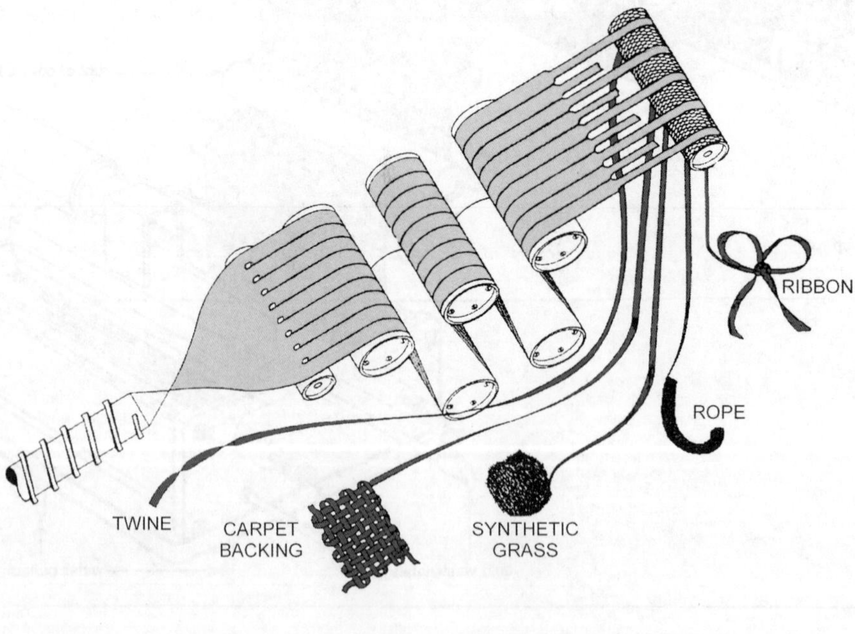

Figure 5.112 Example of tapes being slit from film that are used in different markets

Application	Rate of stretching	Thermoplastic	Requirements
Carpet basic weave	1:7	PP	Low shrinkage
	1:5	PETP	High strength
			Temperature stability
			Specific splicing tendency
			Matt surface
Tarpaulins	1:7	PP	High strength
		PE	
Sacks	1:7	PP	High strength
		PE	High friction value
			Specific elongation
			Weather resistant
Ropes	1:9 to 1:11 (15)	PP	High tensile strength
			Specific elongation
			Good tendency to splicing
Twine	1:9 to 1:11	PP	High tensile strength
		PP/PE	High knotting strength
Separating weave	1:7	PP	High strength
Filter weave	1:7	PP	Low shrinkage
	1:5	PETP	Abrasion resistant
Reinforcing weave	1:7	PP	Low shrinkage
	1:5	PETP	Specific elongation
			Temperature resistance
Tapestry and home textiles	1:7	PE	UV resistance
			Low static charge
			Uniform coloration
			Textile-type handle
Outdoor carpets	1:7	PP	Low shrinkage
	1:5	PETP	Wear resistance
			Weather resistance
			Elastic recovery
			Uniform coloration
			Defined splicing
Decorative tapes	1:6	PP with blowing agent	Effective surface
			Low specific gravity
Knitted tapes, sacks, and other packagings, seed and harvest protective nets	1:6.5	PP	High knotting strength
		PE	Low splicing tendency
			Suppleness
			UV resistance
Packaging tapes	1:9	PP	High strength
	1:7	PETP	Low splicing tendency
Fleeces	1:7	PP and blends	Fiber properties

Table 5.39 Tapes identified by type of plastic, amount of stretching/orienting during processing film for each application, and examples of performance requirements

Property		Unit	Test Method	Block copolymer	Random copolymer		Homopolymer		
Grade				2309KX	3200 MCX	3520 LX	1125N	1127MX	1325L
Additives				None	None	antiblock agents	slip, antibloc k agents	slip, antibloc k agents	slip, antibloc k agents
Melt flow rate		g/10 min	ISO 1133	4	8	5	11	8	5
Gloss	20°	%	DIN 67530	5	130	130	110	95	115
Haze		%	ASTM 1003	28	0.5	0.3	2.3	3.0	1.7
Tensile strength at break	machine direction	MPa	DIN 53455	55	37	32	42	44	35
	transverse direction	MPa	DIN 53455	36	35	30	38	40	33
Elongation at break	machine direction	%	DIN 53455	750	730	750	680	700	750
	transverse direction	%	DIN 53455	720	750	760	720	730	780
Dart drop impact resistance F50		g	ASTM 1709	500	450	>800	280	300	800
Modulus of elasticity in flexure	machine direction	MPa	DIN 53121	650	480	300	700	680	370
	transverse direction	MPa	DIN 53121	640	470	310	670	650	350
Coefficient of friction			DIN 53375	0.90	Not measurabl e	Not measurabl e	0.25	0.17	0.13

[1] Film properties depend considerably on processing conditions. This must be taken into account when comparing these data with data obtained under different processing conditions.

Table 5.40 Properties of cast polypropylene film with µm gauge

Film Property		Test Method	Increasing Process Variables: Melt Temp. 480°-520°F (250°-270°C)	Chill Roll Temp. 100°-150°F (40°-65°C)	Take-Off Speed 50-200 ft/min. (15-60 m/min.)	Increasing Resin Variables: Density 0.916-0.933 g/cm³	Melt Index 1.0-5.0 g/10 min.
Gloss (%)		ASTM D2457	Increase	Increase	NSE	Increase	Increase
Haze (%)		ASTM D1003	Decrease	Increase	NSE	Decrease	Decrease
Film Tensile Strength		ASTM D882					
Yield (lb/sq. in.)	MD*		Decrease	Increase	NSE	Increase	NSE
	TD*		NSE**	Increase	NSE	Increase	NSE
Break (lb/sq. in.)	MD		Decrease	NSE	Increase	NSE	Decrease
	TD		NSE	NSE	Increase	NSE	Decrease
Elongation (%)	MD		NSE	NSE	NSE	See (1)[1]	NSE
	TD		NSE	NSE	NSE	See (2)[2]	NSE
Elmendorf Tear Strength (g/mil)	MD	ASTM D1922	Slight Decrease	Increase	Increase	See (3)[3]	Decrease
	TD		NSE	Increase	NSE	Increase	Decrease
Impact Strength (Dart drop) (g)		ASTM D1709	NSE	Decrease	Not Determined	Decrease	Decrease
Film Density		ASTM D1505	Slight Increase	Increase	Not Determined	Increase	NSE

*MD = Machine direction
*TD = Transverse direction
**NSE = No significant effect

[1](1) Decrease 0.915 to 0.924
Increase 0.925 to 0.933

[2](2) NSE 0.916 to 0.924
Increase 0.925 to 0.933

[3](3) Increase 0.916 to 0.924
Decrease 0.925 to 0.933

Table 5.41 Effects of processing and variables on polypropylene cast-film properties

SHEET

Sheet extrusion is usually defined as thicker than 0.25 mm (0.01 in); anything thinner is classed as film. The thickness range of extruded sheet is normally between 0.010 and 3 in. Generally, 0.500 in is the upper limit in the conventional range, with higher thicknesses associated with specialty products and process techniques. Widths can be up to at least 10 ft (3 m). Sheet extrusion technology can produce single (mono) layer products and coextruded or multilayer engineered products. Also popular is orientation to improve properties and so forth.

The sheet material can be thermoformed (estimated that 60% of sheets are thermoformed) or fabricated by blanking, punching, machining, and welding. A hinge or crease can be formed in certain plastics, such as polypropylene sheet, with a cold coining technique. Key characteristics of sheet include a good ratio of strength and rigidity to thickness, toughness, resistance to moisture, resistance to sterilization procedures, good moisture barrier properties, chemical resistance, and/or nontoxicity (Table 5.43).

The thinner gauges are used principally for thermoformed food packaging, packaging blisters, tote trays, panels, and other rigid objects that hold shape. Other applications include stationery and office products, tanks, ducting, and wall cladding for the food industry. For example, sheet grades of plastics may have a low melt flow index (MFI), typically in the range of 2.0 to 6.0 grams/10 min for thin-gauge grades. Grades for products of intermediate thickness have an MFI of 1.8 to 2.2 grams/10 min, while thicker sheets are usually produced from grades having an MFI of 0.8 to 1.5 grams/10 min.

PRODUCTION

The most popular method to produce sheets is using an extruder with polished stacked rolls that could include the use of an air knife. There is the extrusion process in which an annular pipelike

Problem	Possible cause	Suggested remedy
Poor clarity	Cooling rate too low	Increase melt temperature Reposition air knife Increase air knife pressure Reduce chill roll temperature
Variable optical properties	Temperature gradient across the chill roll	Adjust chill roll coolant supply Check chill roll cooling circuits
Plateout	Poor chill roll contact	Increase chill roll wrap-around Adjust air knife
Flow defects	Melt temperature too low	Increase die and adapter temperatures
Brittle or fibrous heat seals	Melt temperature too high	Decrease die and adapter temperatures

Table 5.42 Guide to troubleshooting cast film

cross-section die is used. The extrudate is slit in one or more places and then flattened via rolls into a sheet. In addition to extruders, this system is one of the methods used in foam sheet production. With large production orders, use is made of calenders (chapter 9). Figures 5.114 to 5.118 provide an introduction to lines that include those going from the extruder to a winder or stacker.

Extruded sheet is produced with a die having a wide slitlike outlet (Fig. 5.119). The internal flow geometry is designed to transform the cylindrical melt flow received from the extruder to the slit form, while at the same time ensuring that the melt flow velocity at the slit exit is constant

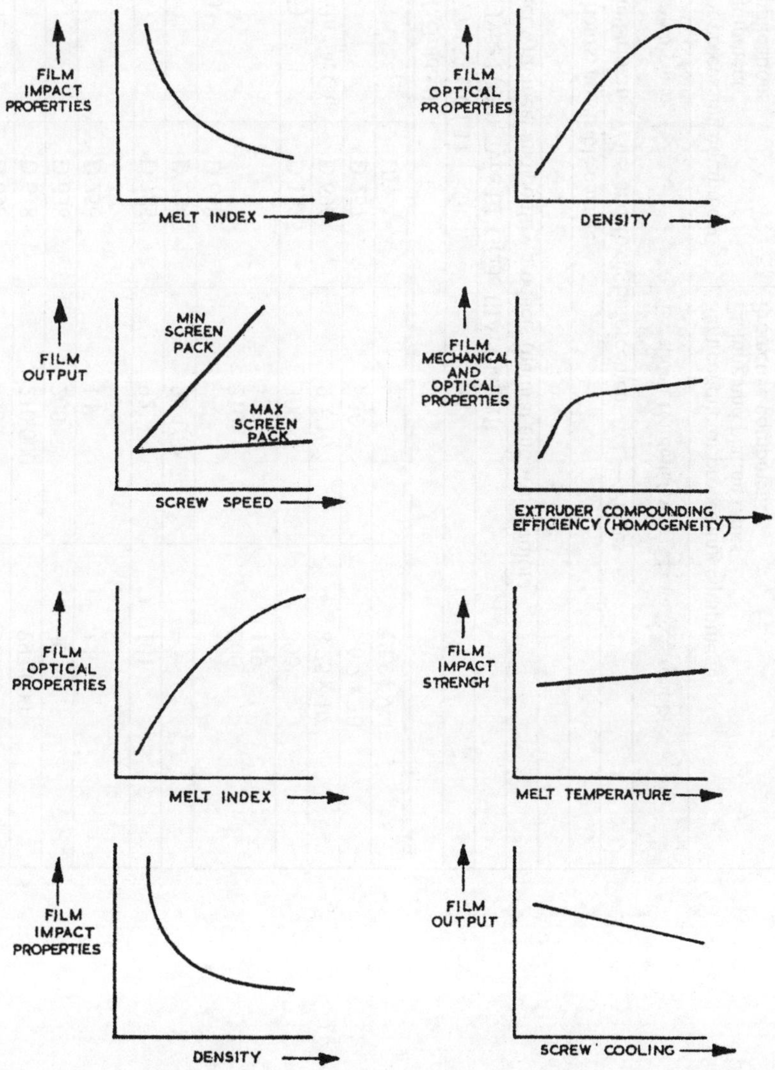

Figure 5.113 Examples of properties vs. changes in flat-film process performances

Property	ASTM Test Method	Stress relieved	Flame Retardant, Stress-Relieved
Density (g/cm^3)	D 792	0.91	0.98–1.02 at 23°C
Water Absorption (%)	D570	0.02	<0.01
Tensile Strength at Yield (psi)	D 638	4900	3770
Modulus of Elasticity (psi)	D 638	210,000	297,000
Ultimate Elongation (%)	D 638	200	30
Izod Notch Impact Strength (@ 73°F; ft. $lbs/in.^2$)	D 256	1.9	1.8
Rockwell Hardness (R)	D 785	92	101
Durometer (Shore A/Shore D)	D 2240	98.1/85.2	–
Heat Distortion Temperature (°F)	D 648		
at 66 psi		210	266
at 264 psi		135	149
Vicat Softening of Expansion (°F)	D 1525	300	306
Linear Coefficient of Expansion (in./in. C)	D 696	9×10^{-5}	8.25×10^{-5}
Volume Resistivity (ohm-cm)	D 257	$>10^{16}$	$>6.2 \times 10^{16}$
Flammability (1/8" +)	UL	–	UL94 V–0
ASTM D–4101, Group I, Class I, Grade II			
FDA Regulation Title 21 CFR 177.1520 (C1.1)			
Federal Specification LP–394B, Type I (GP), Type III, Grade IIIA, Class III			
USDA Approval for direct contact with meat and poultry products (natural and white colors			
Military Standard D / N / A			
UV Radiation Exposure — 500 hours, no visible change			
California Proposition 65 — Safe Drinking Water and Toxic Enforcement Act — Passes			

Applications	Extruded Homopolymer	Extruded Homopolymer, Flame Retardant
	Semiconductor processing equipment Laboratory equipment Food grade applications Prosthetics Etching and plating lines Protective equipment Acid tanks	Semiconductor processing equipment Etching and plating tanks Protective equipment

*Versadur polypropylene sheet from HPG International, Inc.

Table 5.43 Properties of polypropylene sheet

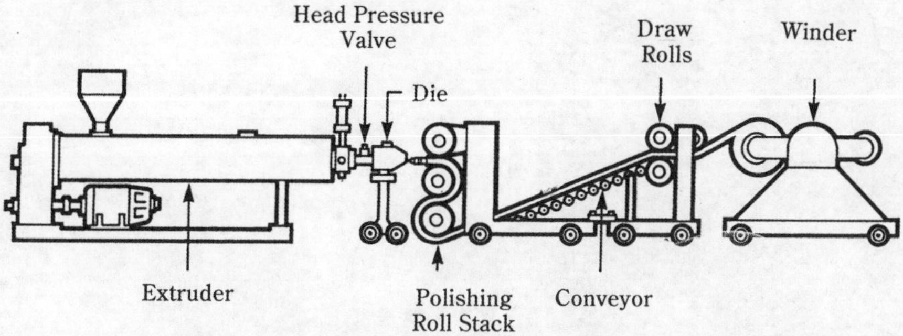

Figure 5.114 Schematic of sheet line processing plastic

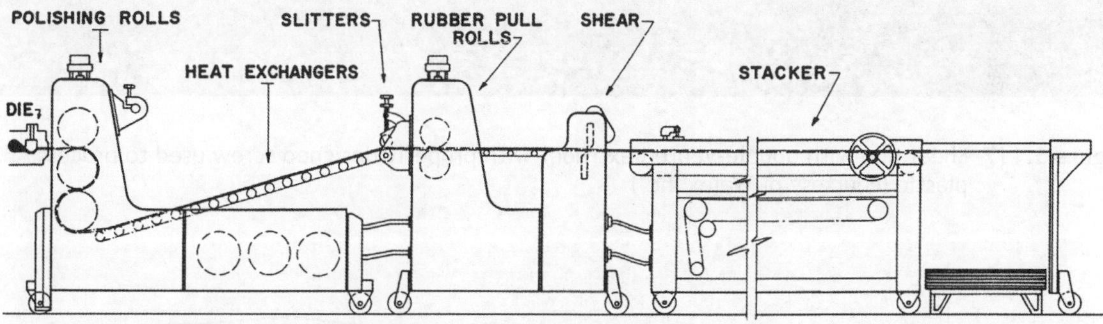

Figure 5.115 Schematic of sheet line processing elastomer

Figure 5.116 Sheet line with double-vented extruder with properly designed screw used to process PET plastic (courtesy of Welex Inc.)

Figure 5.117 Sheet line with double-vented extruder with properly designed screw used to process ABS plastic (courtesy of Welex Inc.)

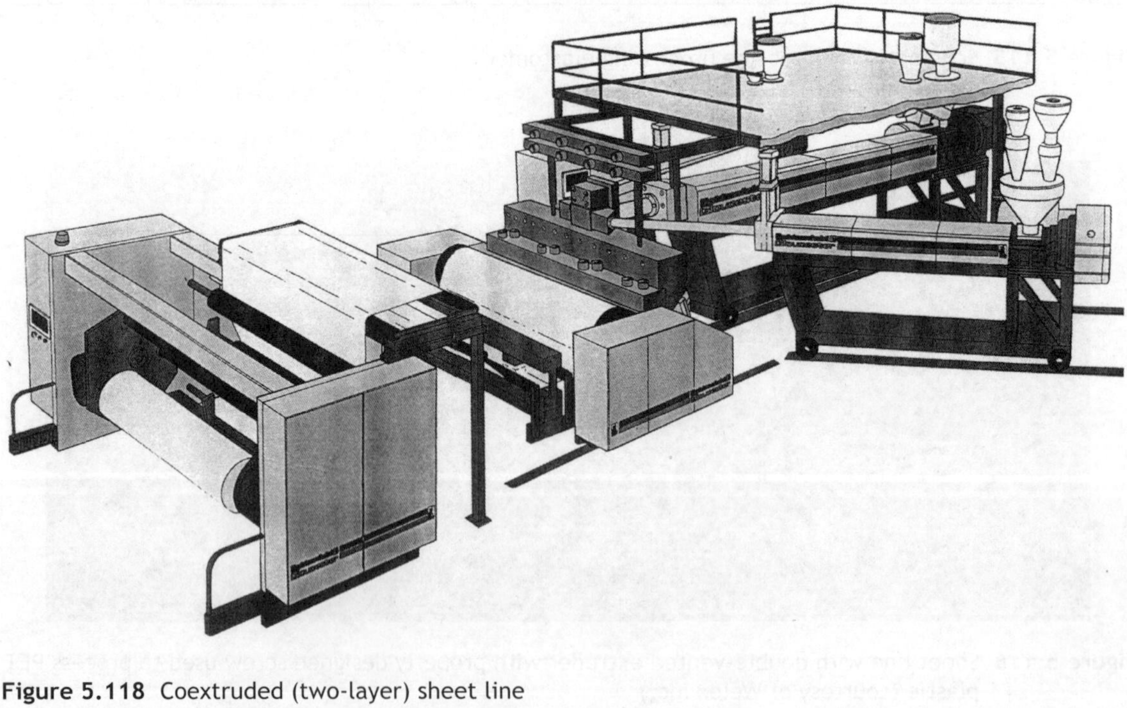

Figure 5.118 Coextruded (two-layer) sheet line

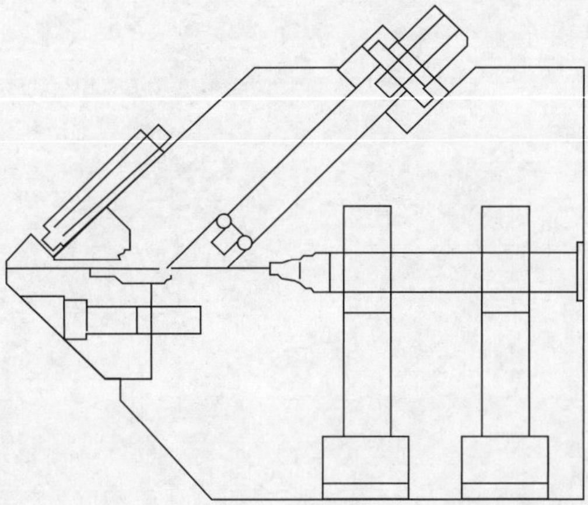

Figure 5.119 Example of a sheet die

at all points across the slit width. This is difficult to achieve, particularly in wide sheet dies. A number of solutions have been devised to deal with the problem to maximize the operation of the dies (chapter 17).

The usual approach is known as the coat hanger die, so called because the internal flow passage tapers from the die center to the edges. By offering a greater resistance to flow in the die center, this tends to equalize flow rates across the die. Most sheet dies include two other regulating devices. An internal adjustable dam known as a restrictor bar or choker bar is positioned in the flow path to the die exit and acts as a coarse regulator. Fine regulation is achieved by adjustment of the die lips at the die exit. The upper die lip includes a thin section that can be flexed elastically by the action of manual or automatic adjuster bolts so that the die exit gap can be varied across the die width. The die gap should be set in the range of 0% to 10% greater than the desired sheet thickness (chapter 17).

An air knife can be used to ensure contact between the extrudate and roll (Fig. 5.120). It provides a means to produce plastics, such as thin sheets and PP sheets, on a conventional three-roll sheet line that can have difficulty in obtaining a proper contact. (They are also used in other extrusion lines that include cast film, coating, and lamination.) The air knife ensures positive and firm positioning of the sheet against the roll. Automatic die lip adjustment is used. High-efficiency internal design provides precise pressure uniformity and optimum air impingement at suitable angles.

From the die, the sheet passes immediately to a cooling (Table 5.44) and finishing device in the form of a roll cooling stack. The usual configuration is a three-roll vertical stack with the sheet entering at the nip between the upper two rolls. Variants to meet certain processing requirements include an up-stack working where the sheet enters between the lower two rolls, a horizontal roll stack used with a vertical die for low-viscosity melts, a two-roll stack for thinner sheet gauges, and others to meet different material requirements (Figs. 5.121 to 5.130 and Tables 5.45 and 5.46).

Figure 5.120 Air knife located next to the heated roll (courtesy of Welex Inc.)

	ABS, F	*Polystyrene, F*
Top roll	175–225	150–200
Middle roll	150–175	110–180
Bottom roll	175–225	110–210

Table 5.44 Example of three-roll down-stack temperatures

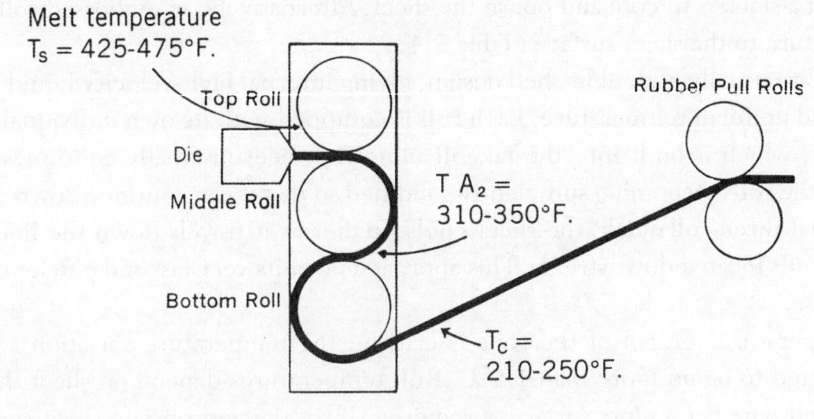

Figure 5.121 Schematic of a three-roll sheet cooling stack

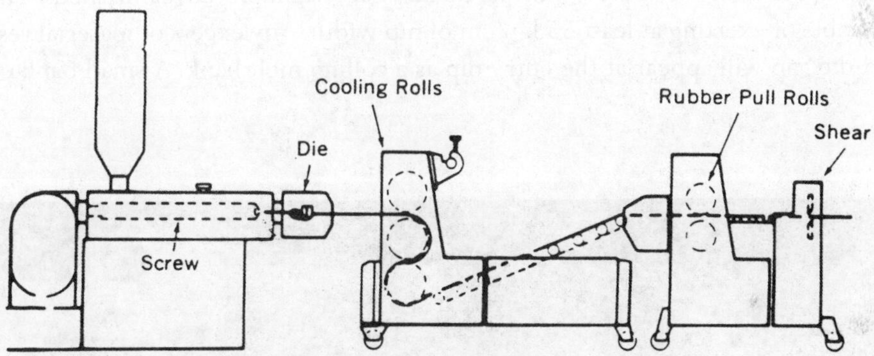

Figure 5.122 Schematic of a three-roll sheet cooling stack in line with other equipment

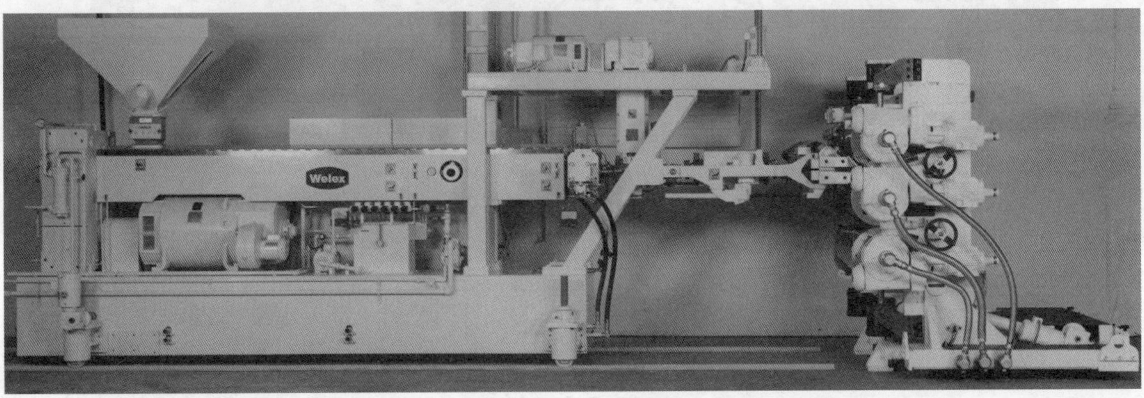

Figure 5.123 Example of a three-roll down-stack in a sheet line (courtesy of Welex Inc.)

The function of a stack is to cool and polish the sheet. Alternatively, an embossed roll may be used to impart a texture to the sheet surface (Table 5.45).

Stack rolls are usually of double-shell design, giving internal high-velocity liquid circulation at a controlled and uniform temperature. Each roll is equipped with its own individual temperature control system, which is built into the takeoff unit. The sheet gradually continues to cool as it travels around the rolls, becoming sufficiently solidified so that it can continue down the line. After leaving the usual three-roll stack, the sheet cools further as it travels down the line while being pulled by pull rolls located downstream. This approach permits very fast and efficient temperature control response.

Close temperature control of the rolls is critical; the temperature variation across the roll should be targeted to be no more than 1.5°C. Roll temperatures depend on sheet thickness, production rate, and whether a gloss surface is required. If so, the appropriate sheet surface must be maintained at a required plastic behavior. For polypropylene the sheet is at 110°C until it contacts the polishing roll. For the polishing roll it is typically 100°C to 120°C. Polypropylene requires a high roll-closing force in the nip region, particularly for thinner gauges of sheet. The roll stack should be capable of exerting at least 55 kg/cm of nip width. Any excess of material resulting from an oversized die gap will appear at the entry nip as a rolling melt bank. A small bank compensates

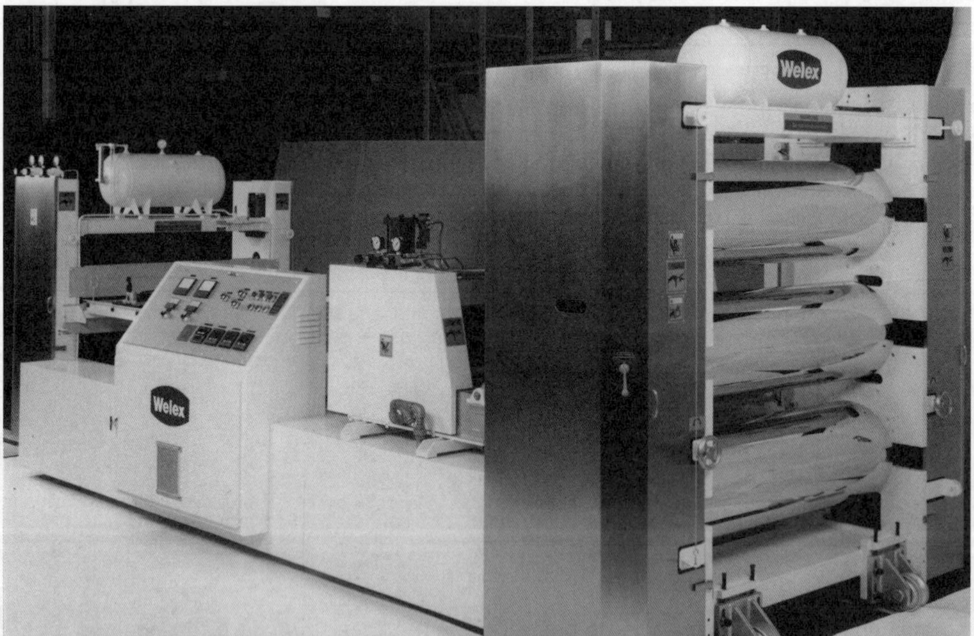

Figure 5.124 Example of opened three-roll stack in a sheet line (courtesy of Welex Inc.)

for output variations from the extruder; too large a bank leads to poor thickness distribution. The sheet extrusion line is completed by a continuous edge trimmer, a pair of traction rolls for maintaining sheet tension, and a reeling device for thin sheet or a cutter and stacker for thick sheet. Table 5.47 reviews some of the effects of die and roll variables on sheet characteristics.

Figure 5.125 Example of silent chain-driven three-roll sheet stack (courtesy of Welex Inc.)

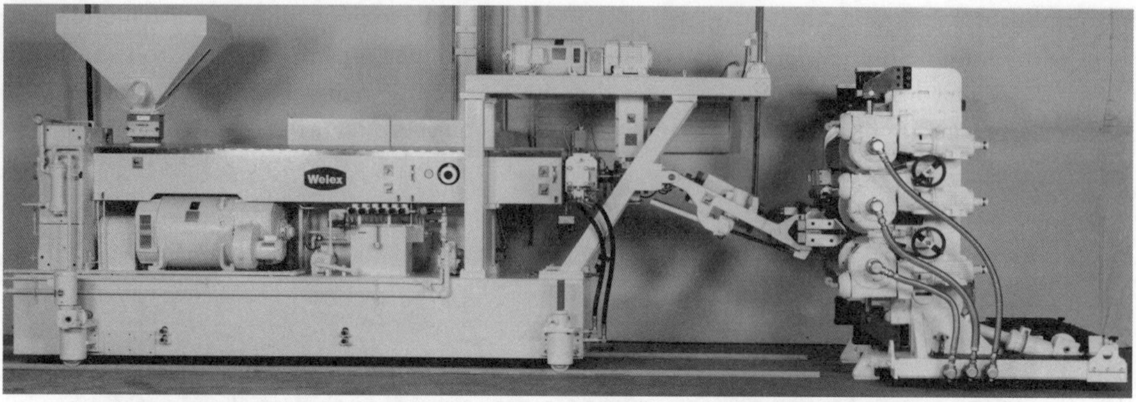

Figure 5.126 Example of a three-roll up-stack in a sheet line (courtesy of Welex Inc.)

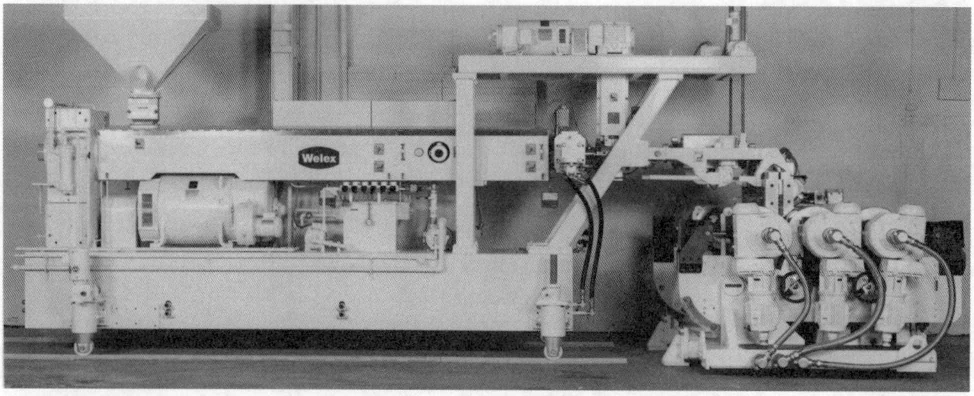

Figure 5.127 Example of a three-roll horizontal stack in a sheet line (courtesy of Welex Inc.)

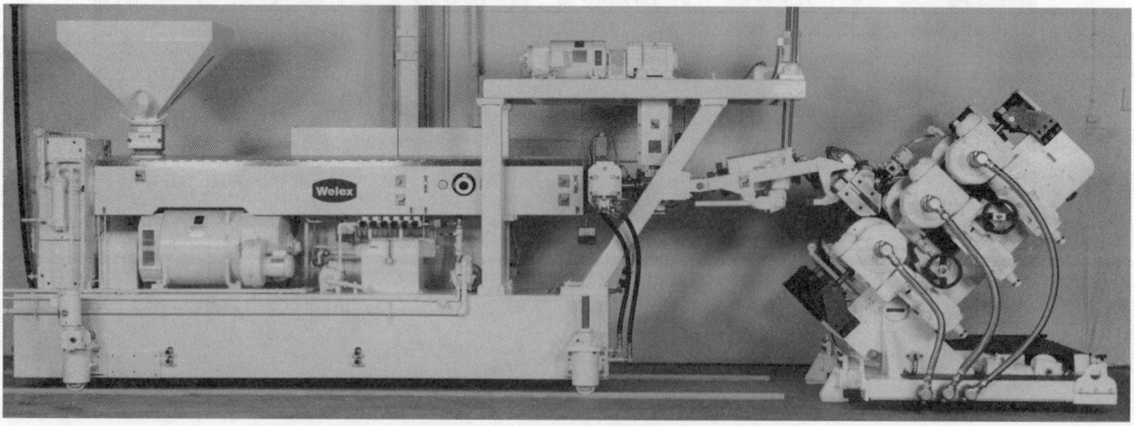

Figure 5.128 Example of a three-roll inclined stack in a sheet line (courtesy of Welex Inc.)

Figure 5.129 Example of a two-roll down-stack in a sheet line (courtesy of Welex Inc.)

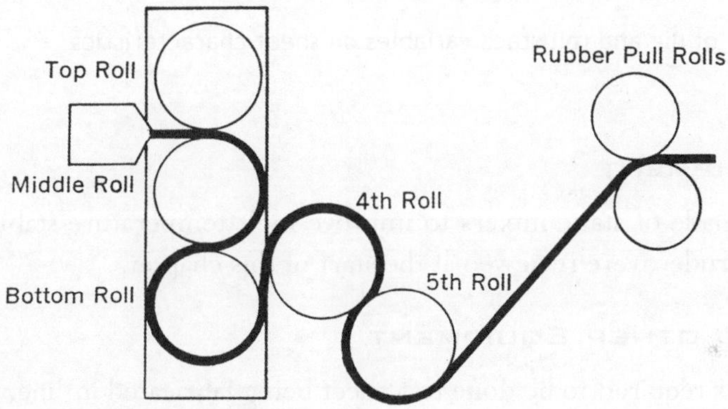

Figure 5.130 Schematic of a five-roll stack

Problem	*Cause*
Continuous lines in direction of extrusion	Die contamination or scored rollers or moisture in the polymer
Continuous lines across the sheet	Jerky operation; maybe too low a temperature
Curved lines across a colored sheet	Color not mixed well enough
Discoloration	Too high a mass temperature or too high a percentage of reground material added
Bad gloss	Roll temperature too low or defective rolls
Dull strip	Die set too narrow at this point

Table 5.45 Examples of troubleshooting sheet problems (chapter 27)

	F
Bottom roll	150–170
Middle roll (embossing roll)	200–250
Top roll	180–220

Table 5.46 Example of embossed three-roll up-stack temperatures

Variable	Orientation	Transparency	Gloss	Stiffness	Impact resistance
Die gap	>	–	–	–	–
Melt temperature	<	–	>	–	–
Roll temperatures	–	<	>	>	<
Roll speeds	>	<	>	>	<

Table 5.47 Influence of die and roll stack variables on sheet characteristics

Auxiliary Equipment

Extruder Part Use is made of static mixers to improve melt temperature stability. This device and others used in an extruder were reviewed at the start of this chapter.

Trim, Cut, and Other Equipment

Depending on what is required to be done to a sheet being fabricated in-line, different equipment is available. Examples include a trimmer (Fig. 5.131) and slitter (Fig. 5.132).

The sheet leaving the roll stack can move through an antistatic bath or a water dip tank and through a gauging system that monitors overall thickness. Razor blades trim the edges of the sheet, and then the sheet enters a pull-roll station. Both the cooling/polishing rolls and the pull rolls are driven by individual DC motors with separate controls to maintain tension between each point. Variable-speed transmissions are used with each motor to provide multiple speed ranges to produce a wide range of sheet thickness.

Automatic shears or cutoff saws cut thicker-gauge sheet to required lengths. Thinner-gauge sheet is wound on two-shaft or three-shaft winders (either fixed shaft or turret winders depending on application). Large rolls up to at least 7 ft in diameter permit long runs without roll transfers, which is advantageous in in-line thermoforming operations. Throughput capacities for these lines can be up to at least 900 to 2,500 lb/h (400 to 1,200 kg/h). Winders can run up to at least 250 ft/min (75 m/min).

Glazier To improve the sheet gloss, heat glazers are used (Fig. 5.133).

Multilayer/Coextrusion Multilayer or coextrusion of sheet is the simultaneous extrusion of two or more layers of plastics through a common die. Examples of conventional two-layer coextrusions

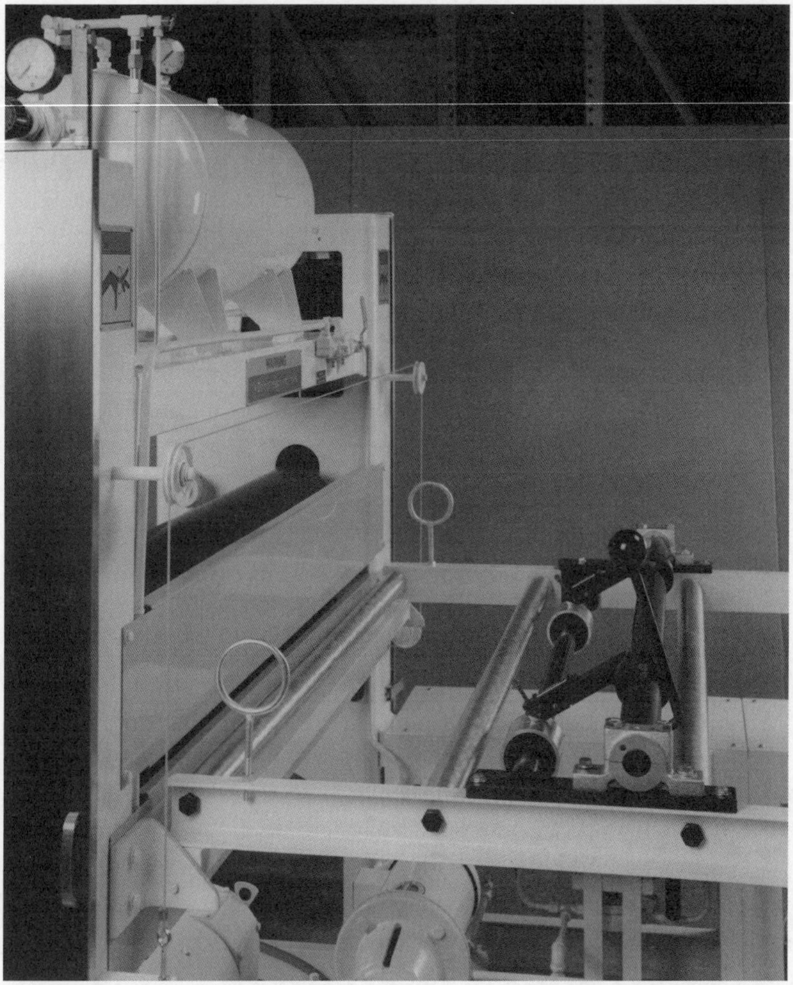

Figure 5.131 Example of a razor edge-trim unit in a film line (courtesy of Welex Inc.)

are adding a crystal layer of gloss over a high-impact base and extruding a sheet with two surface colors. Coextrusion to bury an off-color or off-grade scrap plastic between two layers of virgin material is also common.

Continual food packaging concepts and demand for freezer-to-microwave containers focus attention on more high-barrier sheet coextrusion with as many as nine layers. A typical barrier structure consists of a virgin plastic layer, a scrap layer, an adhesive layer, a barrier layer, an adhesive layer, a scrap layer, and a virgin plastic layer. Multiple extruders are arranged in a cluster and often include a smaller adhesive-layer extruder mounted overhead to save space. Additional extruders are needed when outer layers have special food-contact requirements, such as PP over PS for an odor barrier.

Figure 5.132 Example of a rotary slitting unit in a film line (courtesy of Welex Inc.)

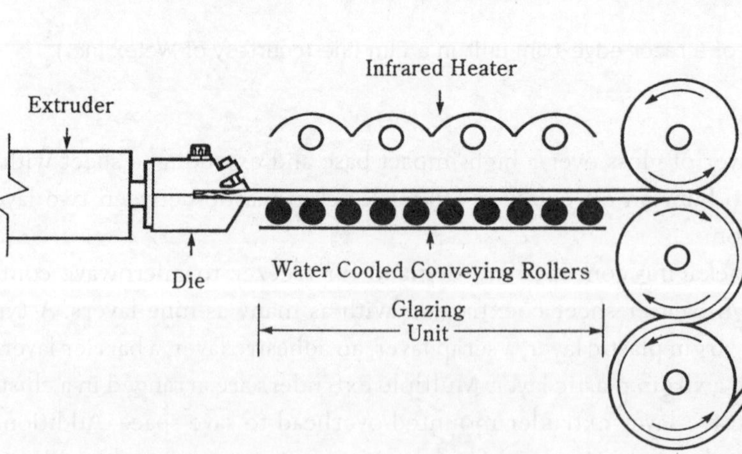

Figure 5.133 Example of heat being applied to the surface of a sheet (film, etc.) to provide surface gloss

Polyvinylidene chloride (PVDC) and ethylene vinyl alcohol (EVOH) are used as barrier materials. PVDC is highly heat sensitive and requires corrosion-resistant materials in the extruder's subsequent flow passages. Shorter extruder barrels and special screw designs are also required for PVDC. The coextruded outputs are most often fed to a combining coextrusion feedblock. The system assembles outputs into one structure that then flows into a single manifold die.

The use of multimanifold dies is another method of coextruding flat sheet. The various extruded layers are each fed to a separate manifold in the die. They combine into one structure just before they leave the die. It offers the advantage of using plastics of vastly differing viscosities and melt temperatures. A combination of a feedblock and a multimanifold die can be used to produce complex multilayer structures. Coextrusion is reviewed later in this chapter.

LAMINATING AND CAPPING

Different combinations of extruded sheets (films, etc.) are laminated, usually by pressure bonding or adhesive bonding. These layups provide different commercial products. When possible and practical, such as having sufficient quantity, it is usually more economical to have these coextruded (to be reviewed). Laminated sheet products also can be made using the extruder three-roll stacks when extruding plastic (Fig. 5.118). Laminating or capping a roll of sheet, film, tape, or any web material can be accomplished by unwinding a roll supported above the stack (Figs. 5.134 and 5.135).

The film is fed into the nip together with the molten web that exited the extruder die. The film bonds to the sheet and transfers its special surface properties. The payoff stations maintain the laminating film under adjustable tension and can provide lateral adjustments. To eliminate problems, it is necessary to maintain uniform sheet thickness, keep orientation low, avoid wrinkles, and prevent air bubbles between laminations.

Extrusion and Reinforced Plastic To date, this process continues to be insignificant insofar as RPs (plastic composites) are concerned; less than 1 wt% of all RPs are processed by laminating. These products can provide high-performance RTP products, such as sheets, films, pipes, rods, and tubes (chapter 15). Gradually, markets should expand using these types of materials such as stampable Azdel (chapter 19; Table 5.48).

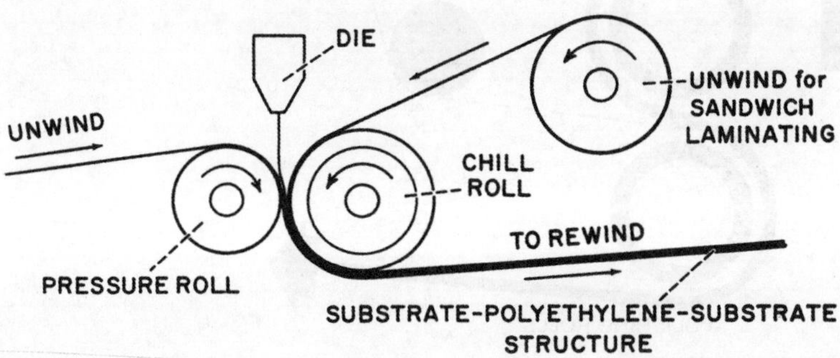

Figure 5.134 Example of laminating a substrate in an extrusion line

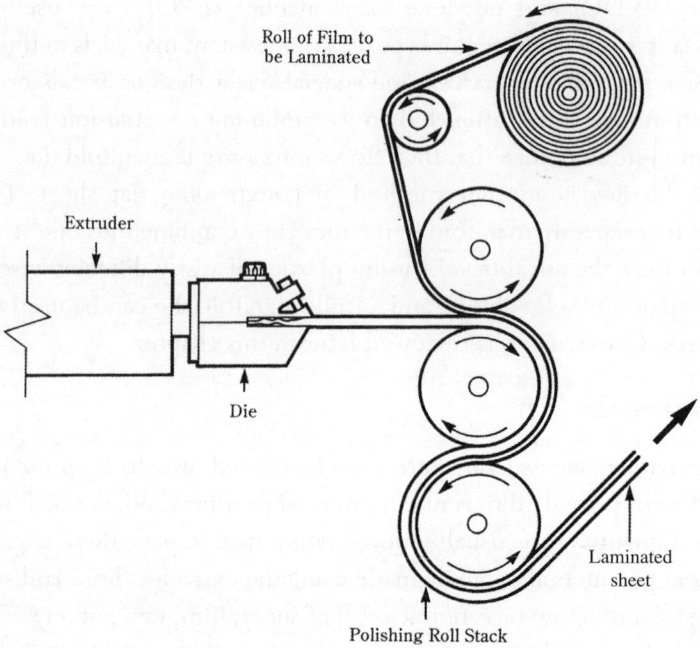

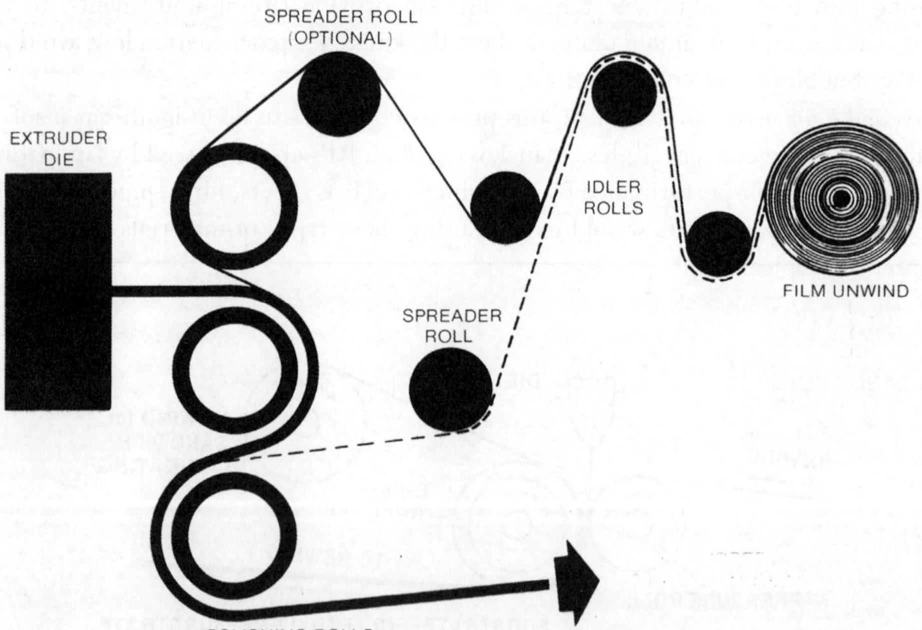

Figure 5.135 Example of capping a substrate with extra tension-control rolls in an extrusion line

Production of thermoplastic sheet composites involves two steps: (1) FORMATION of fiber reinforced sheets (prepreg) by polymer impregnation and sheet consolidation and (2) SHAPING of the sheets. Large volume competitive sheet FORMATION processes are continuous Melt Impregnation (e.g. Azdel sheet by extruding polypropylene onto a continuous "swirled' fiber glass mat), and Slurry Deposition in which long fibers and polymer powder with dispersing agents are deposited on a moving screen similar to paper making. Other processes are Powder Impregnation (powder and fiber consolidated by pultrusion, double belt press, or compression molding), Reactive Pultrusion, and Commingling (intertwining different fibers) and Coweaving polymer fibers, and reinforcing fibers. SHAPING techniques for consolidated prepreg include Melt-Phase Stamping (prepreg covering the mold cavity is heated with infrared and shaped in a fast closing press), Fast Compression Molding (thick charge flows during mold closing), and Solid State Stamping (semicrystalline plastics below their melting point are stamped into parts with simple geometries in 15 sec. Other shaping methods include Pultrusion of prepreg tapes, Diaphragm Molding (preform between plastically deformable diaphragms shaped by hard tooling), Rubber Pad Molding, Hydroforming (rubber bladder inflated hydraulically, Vacuum Forming in an autoclave, and Flexible Resin Transfer Molding (sheets of resin and fiber between elastomeric diaphragms are consolidated, then shaped).		
Advantages/Disadvantages	**Applications**	**Materials**
Extrusion melt impregnation allows high fiber contents and longer fibers which give improved mechanical properties. Slurry deposition employs shorter fibers which allow greater flow and more complex parts. Cycle times are short.	Automotive Body Panels, Other Components, Aircraft Components	PP, PE, PA, PBT, PET, PVC, PC, PEEK, PSU, PPS

Table 5.48 Reinforced thermoplastic sheet

FOAM SHEET

Extruded polystyrene (major plastic used) sheet can be produced on systems utilizing specially designed single- or tandem-screw extruders (Figs. 5.136 to 5.138 and Tables 5.49 and 5.50). Most production is done with tandem (two) single-screw equipment consisting of a primary extruder with a blowing agent injection system delivering melt to the second cooling extruder then the annular die and sizing mandrel followed by a pull-roll unit and one or more winders.

A typical primary extruder is an electrically heated water-cooled machine designed to melt the plastic and uniformly mix in a nucleator and blowing agent. The nucleator, usually a filler such as talc calcium silicate, is a cell-size control agent. The nucleator is mixed with the plastic in the first stage of the screw. The second stage begins with a decompression area where the blowing agent is introduced by a pressurized injection system (chapter 8).

The primary extruder mixes the blowing agent uniformly with the plastic and pumps the mixture into the cooling extruder. The cooling extruder cools the mixture to a lower temperature necessary to stabilize the foaming process outside the die lips. To achieve high rates, these extruders use multiflighted screws with deep channels operating at low speeds and cooled barrels.

The foam tube exits the die and is cooled and sized by a mandrel. Single or multiple slitting units (depending on rate and width) cut the foam tube into single or multiple webs that are transferred to

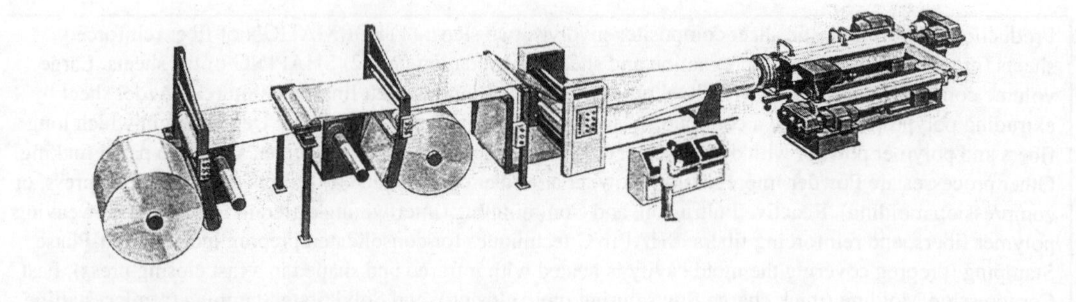

Figure 5.136 Example of single extruder foam sheet line

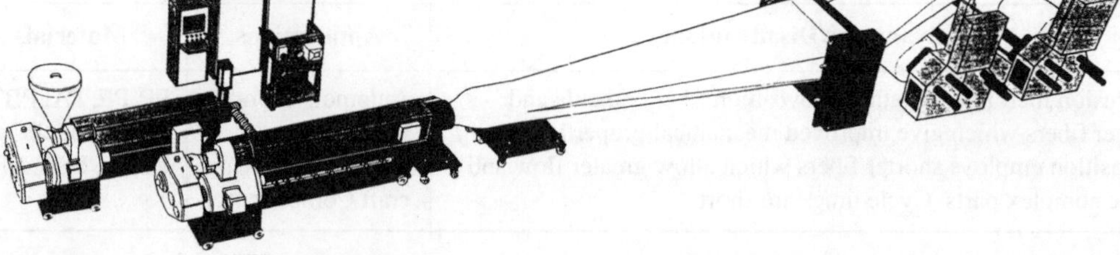

Figure 5.137 Example of tandem extruder foam sheet line (courtesy of Battenfeld Gloucester)

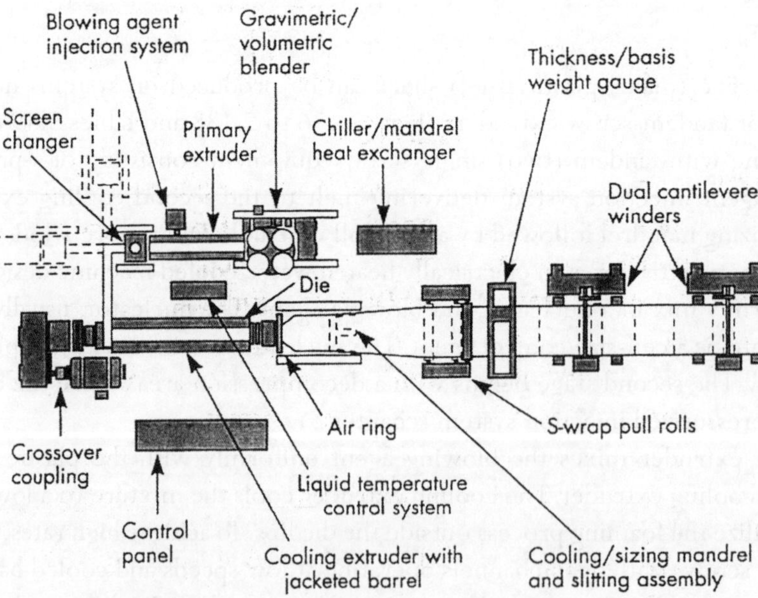

Figure 5.138 Terminology used in a tandem extruder foam sheet line

Extruder Sizes Primary / Secondary in/mm.	Polystyrene		Polyethylene	
	HP/Kw	Lb/Hr Kg/Hr	HP/Kw	Lb/Hr Kg/Hr
2.5/65 3.5/90	50/37 50/37	200–250 90–115	40/30 40/30	175–200 80–90
3.5/90 4.5/115	100/75 75/56	400–500 185–225	75/56 60/44	350–400 160–185
4.5/115 6.0/150	150/112 100/75	900–1000 410–455	150/112 100/75	800–850 365–385
6.0/150 8.0/200	300/224 200/150	1500–1600 680–725	300/224 200/150	1200 545

Blowing agent: butane, pentane, typically 4-6% gas
Nucleating agent: talc, hydrocerol typically 0.4-0.8% by weight
Polyethylene outputs based on 4.0 lb./cu. ft. density

Table 5.49 Example of plastic output for a tandem extruder foam sheet line

Die Size in/mm	Mandrel Diameter in/mm	Sheet Width	
		PS in/mm	PE in/mm
5–6" 127–152 mm	10–24" 254–610 mm	31–75" 787–1905 mm	28–68" 711–1727 mm
6–9" 152–229 mm	25–51" 635–1295 mm	78–160" 1981–4064 mm	71–146" 1803–3708 mm

Table 5.50 Example of die, mandrel, and foam sheet web relations

turret or cantilevered winders through a pull-roll station. Multimotor DC drives are used in drawing the foam from the die and winding a roll under proper tension as the roll enlarges with foamed sheet.

Microprocessor control of the process provides closer tolerances, more consistent final products, and substantial cost savings. Many sophisticated engineering plastics have narrow temperature processing windows. Microprocessor control can maintain the melt temperature of a plastic within a narrow heat range.

Microprocessor controllers can be separated into three general levels of performance: (a) control of extruder barrel temperatures only, (b) control of an entire extrusion line, and (c) control of an entire plant including all lines plus inventory, goods shipped and received, and payrolls.

Automatic die control of web-extruded material typically utilizes a closed-loop system. A scanning gauge measures the moving film's thickness and signals the computer, which converts the readings to degrees of temperature in thermally actuated die bolts. Adjustment is virtually instantaneous and continuous. The microprocessor can be used with a wide variety of gauging systems, such as infrared, beta, X-ray, caliper, gamma backscatter, X-ray fluorescence, IR backscatter, capacitance, and IR transmission, depending on the product. Systems with at least ten gauges on one line are used. Other important advantages of the automatic die are much faster start-up and less scrap generation. Coextrusion lines that utilize multiple extruders require a centralized process control system as there are far too many conditions to be monitored and adjusted on individual control panels.

PIPE AND PROFILE

As an introduction, pipes and profiles are produced from both single-screw and twin-screw extruders. The extruder may be single stage or vented, depending on devolatilization requirements. Products range from a simple pipe or profile to highly complex shapes utilizing coextrusion, in-line welding of two shapes, foamed core products, corrugated pipes, thin-walled medical tubings, and high-strength close-tolerance tubes for the aerospace industry.

A typical pipe or profile line requires an extruder, die, sizing device, cooling tank, puller, and either a cutoff or coiling device. The dies must be dimensionally stable and thus must have accurate temperature controls. A pipe die may have an internal mandrel that is cooled to give good diameter control and surface finish. Another arrangement uses a vacuum sizing sleeve downstream of the die to control the outer diameter of the pipe.

Pipe and profile dies must be streamlined to prevent material hang-up, and they are designed for the particular polymers' viscoelastic characteristics (chapter 17). Construction is of carbon steel, tool steel in some applications, and stainless steel where corrosive plastics are used. The internal flow surfaces are normally chrome plated to aid in cleaning. The manner with which tubing is sized determines the type of die used. Some products are run directly into a water tank, and the puller speed determines the size and wall thickness. Other products are sized by cooling the inner or outer walls.

When extremely precise dimensions are required on an extruded shape, the output of the extruder may not be sufficiently stable to get the desired results. A gear pump is then used at the extruder discharge for stable throughput. Gear pumps are calibrated to forward precisely metered amounts of plastics. This system is used in the production of small thin-walled medical tubing, optical fibers, and similar applications.

Takeoff equipment for rigid pipes or profiles usually consists of a cat track puller and a knife cutter or saw to cut sections to proper lengths. Flexible shapes can be coiled on winders.

PIPE AND TUBE

Pipes and tubes are extruded in a wide range of sizes, from drinking straws up to pipes of many feet in diameter. Plastic pipes and tubings have different definitions that are usually associated with the different industries (plumbing, gas transmission, beverage, medical, mining, and so on). A popular definition for pipes is that they are rigid, hollow, long, and larger in diameter than tubes. Tubings are basically the same except flexible and smaller in diameter, such as up to 0.5 in (0.13 mm). Practically all pipes are extruded using thermoplastics. Single-screw extruders are usually used, but with PVC twin-screw extruders are also used. Some of the lines use the same basic types of dies and plastic melt temperature ranges used in wire coating.

The extruder and die as well as downstream devices for the outside and inside calibration of the pipe's cross-sectional area, if required, use air pressure and/or a vacuum to contain the pipe's shape. A wall-thickness measuring device, mandrel design (inside a thin spiral gap between the fixed mandrel attached to the die to provide cooling air, such as when water cools outside), cooling tank, and automatic cutting with pallet equipment for rigid pipe or a windup unit for flexible pipe are downstream. The line could include a marking device, testing device, and so on. An important requirement is to cool the extrudate rather fast near the die while keeping control of its dimensions and properties.

Included in the processes are various techniques to control the dimensions/sizes that are either free-drawn melts (usually for small-diameter tubes) or sizing fixtures. Dimensional- and/or thickness-calibrating disks of different designs are used. There are small-diameter tube lines using drawdown control (free extrusion) sizing technology where the extruded tubular melt has no calibrating device after leaving the die. It could have internal air pressure so that the tube does not collapse upon leaving the die. Devices are also used with differently designed calibrating/sizing plates or tubes, with or without pressure or a vacuum assist, in and/or outside the tubes.

Regarding the extruder, what has been reviewed in this chapter on size and capabilities applies to pipes and tubes. Figure 5.139 presents information on the variability of extruders. Different takeoff equipment downstream of the extruder is available depending on factors such as plastic to be used, size, tolerance, and quantity.

DIE/MANDREL

Dies for pipe production consist essentially of a female die ring that shapes the pipe's outside diameter and a male mandrel that shapes the inside diameter. The difficulty is to support the

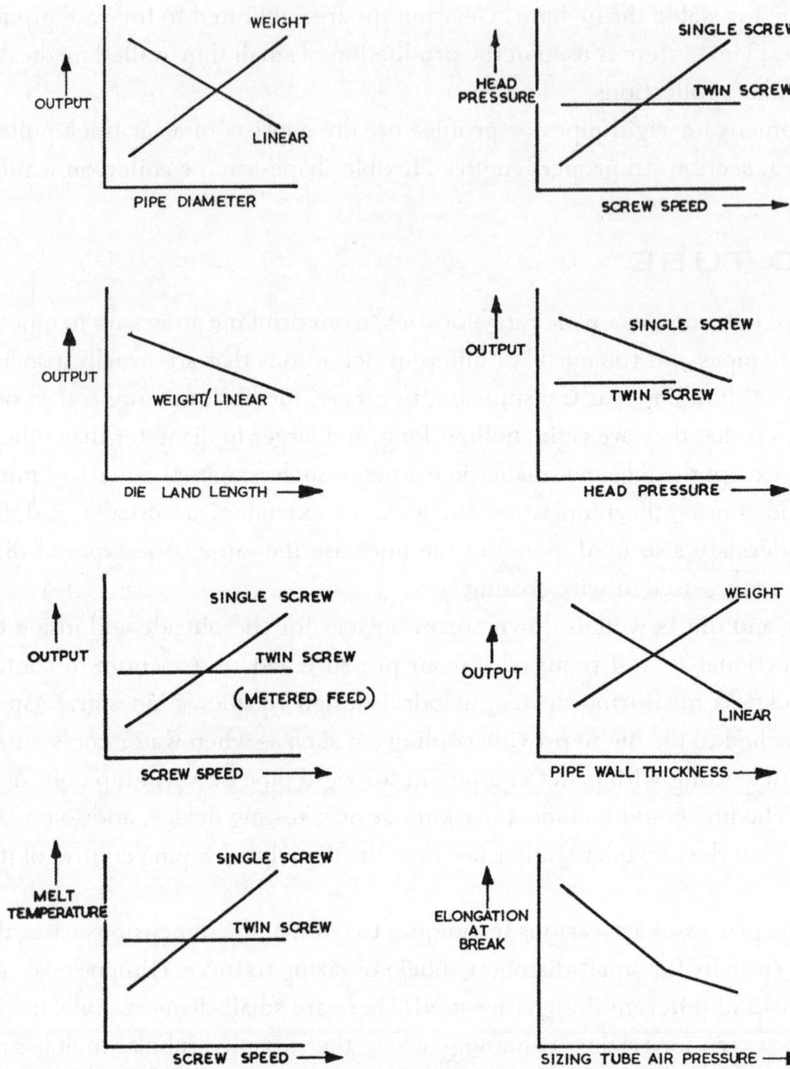

Figure 5.139 Examples of operational changes in an extrusion line that influence pipe performances

mandrel in rigid and accurate alignment with the die ring without compromising the product. The spider type (Fig. 5.140) uses three or four spider legs to support the mandrel, but these legs cause axial weld lines as the melt flows around them. Descriptors in this figure are A = die body, B = mandrel, pin, male die part, C = die, die bushing, female die part, D = die retaining ring, E = die retaining bolt, F = die centering bolt, G = spider leg, H = air hole, I = seat for breaker plate, J = ring for attachment to extruder, and K = die land.

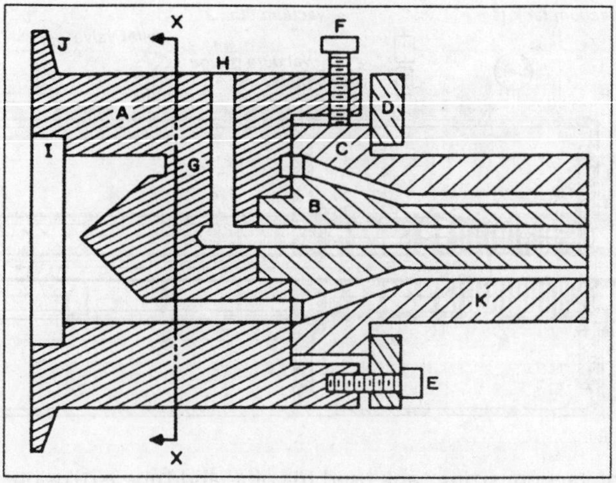

Figure 5.140 Example of a spider-type die for pipe and tube extrusion

An alternative is the breaker-plate design in which the mandrel is supported by a disk pierced by a multitude of small holes. A better solution, particularly for larger pipe dies, is the spiral mandrel (chapter 17). With that, the mandrel is rigidly supported and the melt flow is directed around it in an initially helical pattern that is converted to axial flow by the geometry of the flow channels and lands. Weld lines are eliminated, but the difficulty can be to ensure an even flow rate at all points of the die outlet. As an example, PP requires a long die land length to increase die pressure, minimize flow rate variations, and improve the weld line strength of spider-type dies. The length of the die land should be fourteen to twenty-five times the wall thickness of the pipe.

The outside diameter of the pipe may be calibrated either by an internal air pressure of 1.3 to 2.0 bar or by external calibration in a vacuum sizing tank (Fig. 5.141) operating at a pressure of 0.3 to 0.5 bar. For calibration by internal pressure, the diameter of the die ring is made approximately equal to the internal diameter of the finished pipe. For vacuum tank calibration, the die ring is made up to 25% larger and the haul-off is set for a wall thickness drawdown of up to 30%.

The match between die size and extruder size is inexact, but certain limits should be observed in the interests of economical and consistent production. Figure 5.142 gives the recommended relationships for conventional single-screw extruders.

Plastic

The consumption of plastic is always targeted to be minimized. Costs to produce the pipe could include material costs of up to 80%. The goal is always tighter wall thickness tolerance control to reduce material consumption. Virtually all TPs can be used, including cross-linked TPs that become TSs (XLPEs, etc.) (chapter 1) and some reinforced thermoset plastics (chapter 15). The major

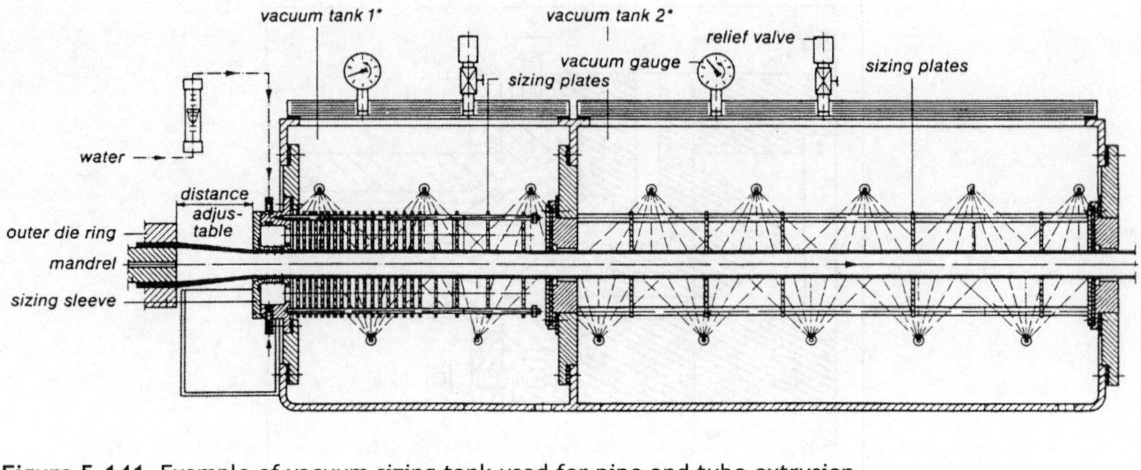

Figure 5.141 Example of vacuum sizing tank used for pipe and tube extrusion

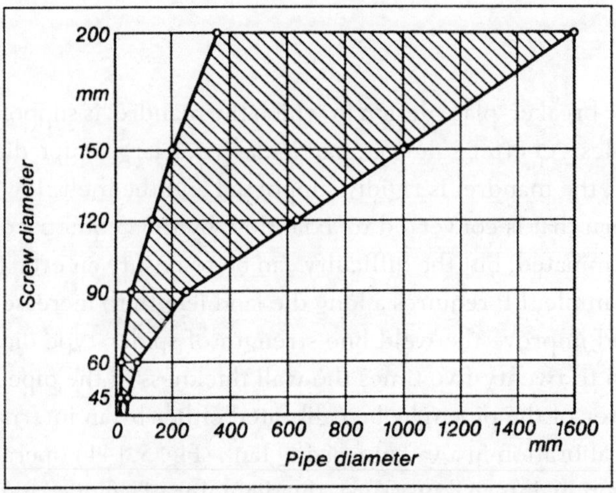

Figure 5.142 Recommended relationships between pipe diameter and screw diameter

plastics used are PVCs, HDPEs, PPs, and ABSs. Since different plastics have different performance capabilities, use is made of the plastic that meets the pipe requirement (chapter 2).

For example, PP's principal characteristics are taking internal and/or external pressure loads, very good resistance to heat aging, high heat-distortion temperature, low friction losses, being lightweight, and high resistance to attack by chemicals and soils. Heat-stabilized grades are available, particularly for use with hot aqueous solutions. Flame-retardant grades cater to construction use. These are not acceptable for use with potable water, but PP is otherwise very suitable for this application and many grades conform to the requirements of the various water authorities.

PP pipes are subject to degradation by ultraviolet light, so UV-stabilized grades should be used if the pipes are to be stored in the open for any length of time. This is particularly true for thin-walled pipes. The melt flow index for pipe grades is relatively low, typically ranging from 0.8 to 2.0 grams/10 min. Applications for PP pipes include potable water supplies, domestic waste systems, under-floor heating circuits, chemical-effluent disposal systems, and thermal spring conduits.

Like other TPs, PP pipes creep when exposed to long-term stress arising from internal pressure. The effect is more marked at higher temperatures. Pipe creep is determined by long-term pressure tests at a range of temperatures (chapter 19). The resulting curves in Figure 5.143 can be used to establish a safe working stress for design purposes.

The wall thickness necessary to withstand a given internal pressure can be calculated from the expression:

$$s = \frac{p}{10}\left(\frac{D}{2\sigma + \frac{p}{10}}\right)$$

where s = wall thickness (mm), p = internal pressure (bar), D = outside diameter of pipe, and σ = safe working stress at the service temperature (N/mm^2).

A safety factor of 2 should be used when designing PP pressure pipes, so a representative safe working stress at 20°C would be 5 N/mm^2 (Table 5.51). The table suggests representative design

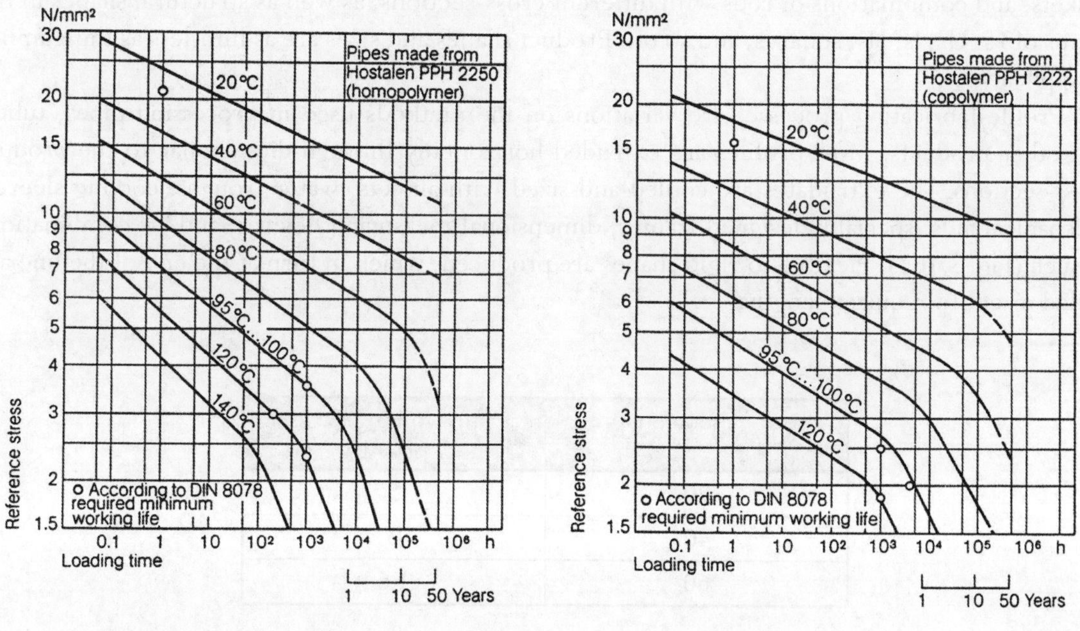

Figure 5.143 Creep rupture strength of PP pipes (Hoeschst Hostalen homopolymer PPH 2250 and copolymer PPH 222) with the pressure medium being water

figures, but the exact values are grade specific and should always be confirmed before designs are finalized. The quoted figures are based on a design life of ten years or more. At higher temperatures, the design life must be shorter (Fig. 5.143).

PP under-floor heating pipes typically operate at about 40°C and temperatures should not exceed 65°C. Similarly, effluent waste pipes may operate intermittently at up to 80°C, but sustained service should be limited to 60°C. The smallest recommended bending radius for a PP pipe, such as an under-floor heating duct, is eight to ten times the pipe's diameter.

EXTRUSION LINE

Figures 5.144 to 5.160 provide pictorial views of downstream equipment pertaining to rigid and flexible pipes and tubes.

PROFILE

Profiles are in a variety of shapes (Figs. 5. 161 and 5.162). Pipes and tubes could be included but the industry has had them classified as a separate category because they represent major large markets on their own. Profiles normally identify shapes that are noncircular or are not symmetrical. However, there are exceptions where extruded products, such as capillary tubing and rod, are usually called profiles. They can be solid, hollow, or a combination of solid and hollow. Popular shapes include many hollow sections, such as window frame profiles, tapes, edgings, gaskets, and combinations of rods with different cross-sections, as well as structural shapes in the forms of Ts, Us, Is, Hs, squares, and so on. Product shapes and sizes are as limitless as the number of applications.

Profile fabricating processes are variations on the methods used in processing pipes, tubes, and other products. Most profiles are extruded horizontally through dies similar to the product cross-sections. The extrudates are cooled and sized with air jets, water troughs, cooling sleeves, mechanical aids, specially designed frames, dimensional measuring devices, and/or combinations of such (Fig. 5.163). Flexible to rigid shapes are produced, which in turn at the end of the line are coiled or cut to required lengths.

Service temperature (°C)	Safe working stress (N/mm^2)
20	5
40	3
60	1.6

Table 5.51 Suggested safe working stresses for PP pipes. The quoted figures are based on a design life of ten years or more.

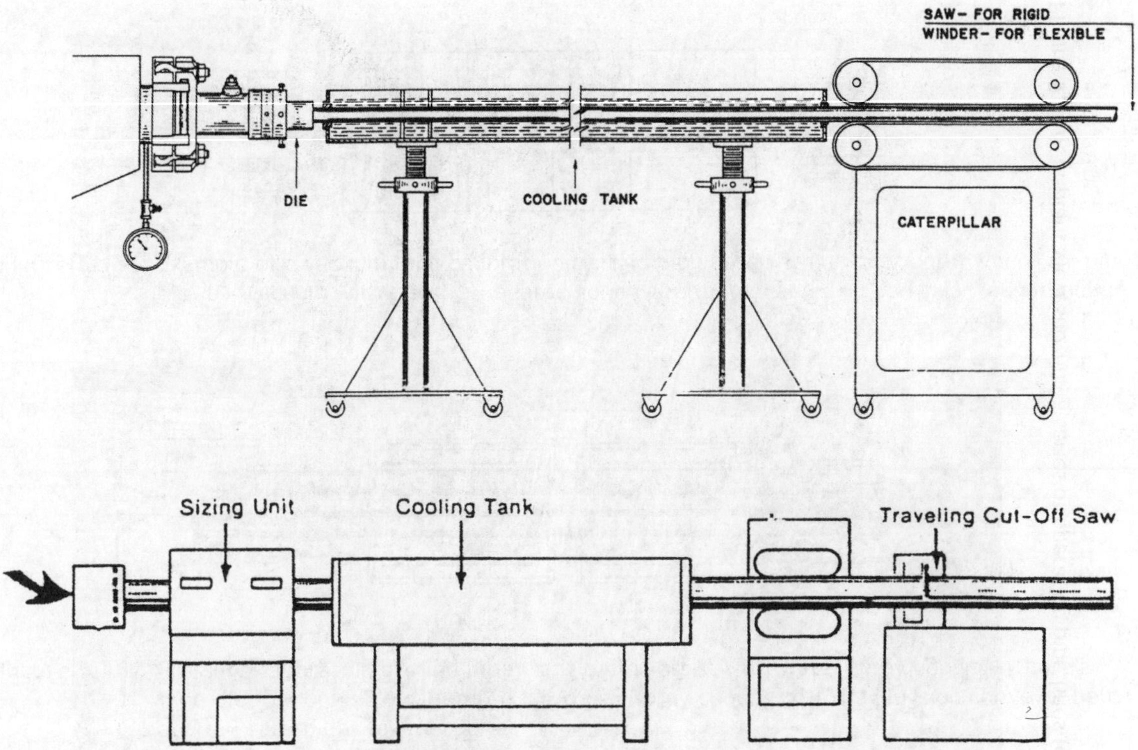

Figure 5.144 Introduction to downstream pipe/tube line equipment

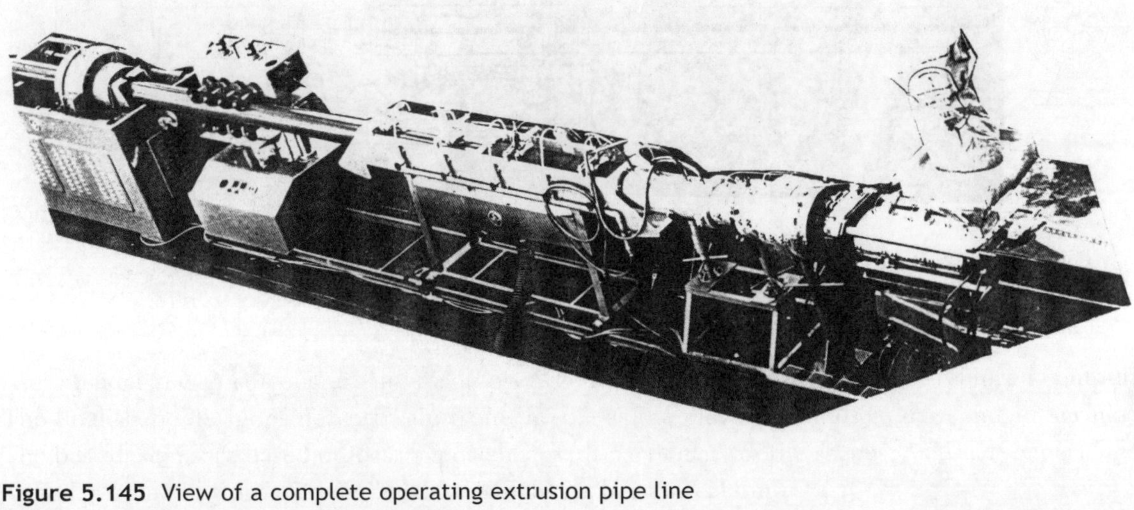

Figure 5.145 View of a complete operating extrusion pipe line

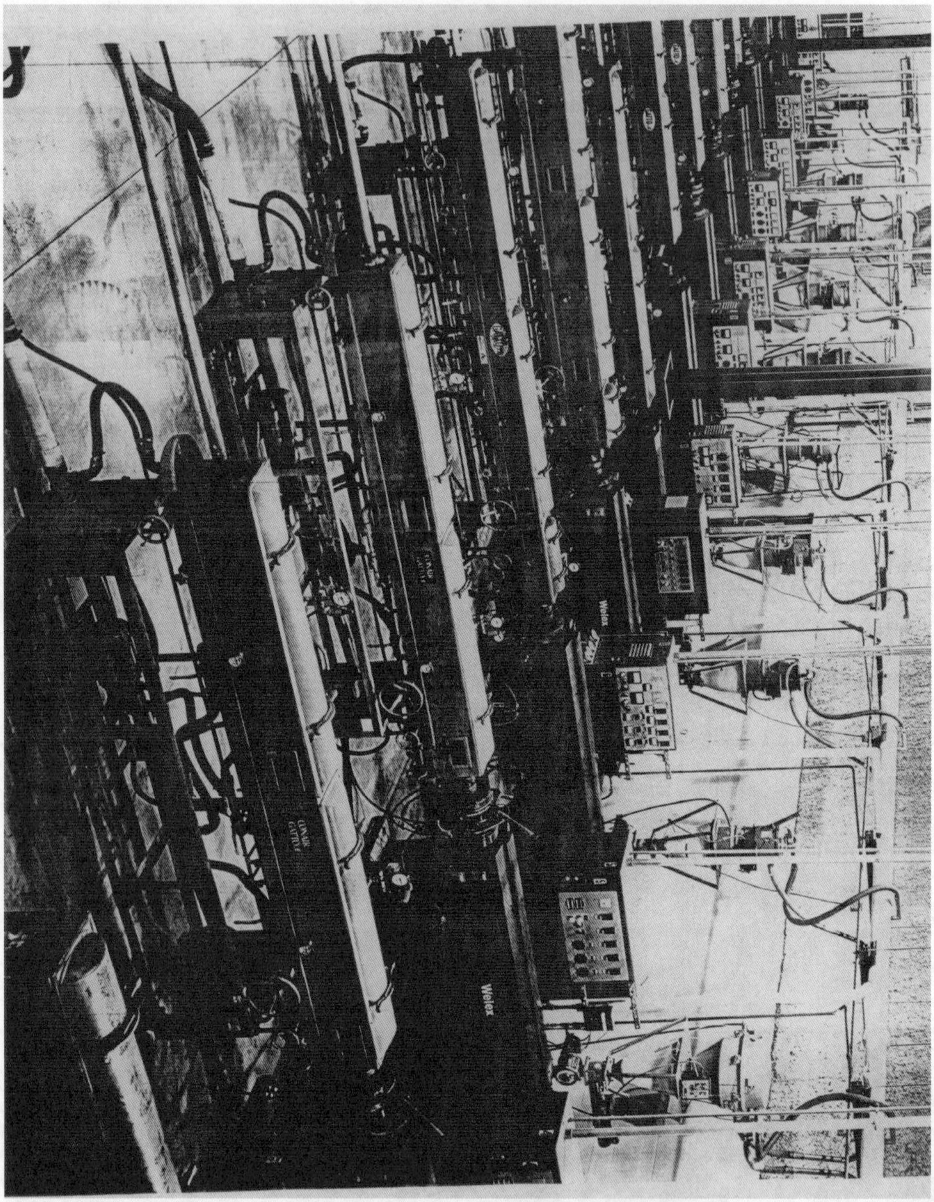

Figure 5.146 In-plant view showing a series of operating pipe lines (courtesy of Welex Inc.)

Figure 5.147 Example of a 2½-in (60-mm), 24/1 L/D extruder used to produce tubes and profiles (courtesy of Welex Inc.)

Figure 5.148 The Figure 5.147 extruder with plasticator safety guard removed (courtesy of Welex Inc.)

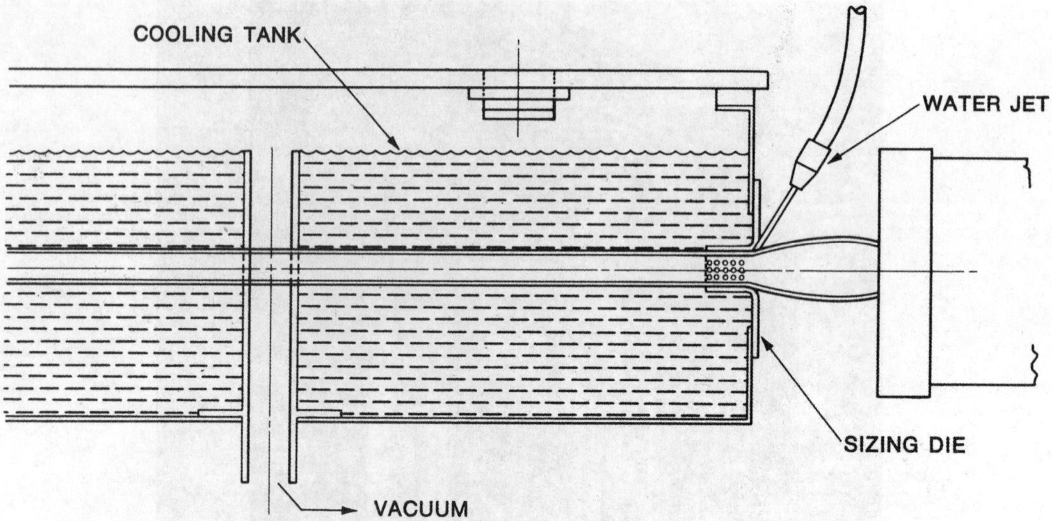

Figure 5.149 Example of water lubrication when pipe is entering a water tank (may be required)

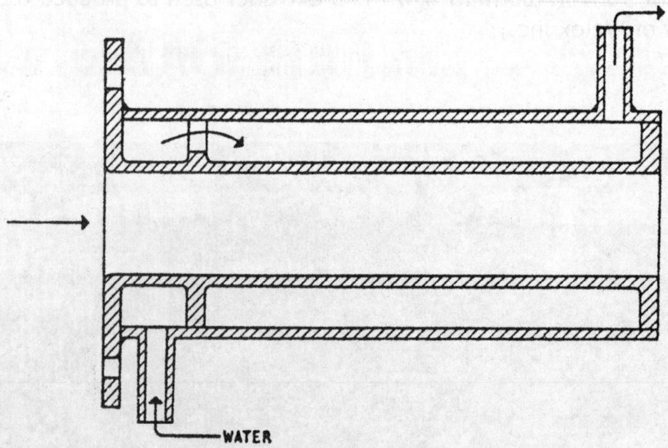

Figure 5.150 Example of a basic water-cooled sizing calibrator

Industry requirements for profiles range from very loose to very tight tolerances. Extruder specifications for a tight process capability include three critical areas of consideration. First, the temperature control on the barrel is paramount for a stable process (Table 5.52). Many PLCs (programmable logic controllers) exceed requirements; however, PID (proportional-integral-differential) instrumentation will also hold a ±1°F on barrel control (chapter 3). Second, a tight drive-control screw coupled with a motor capable of holding 0.01% speed control with 1% drift is

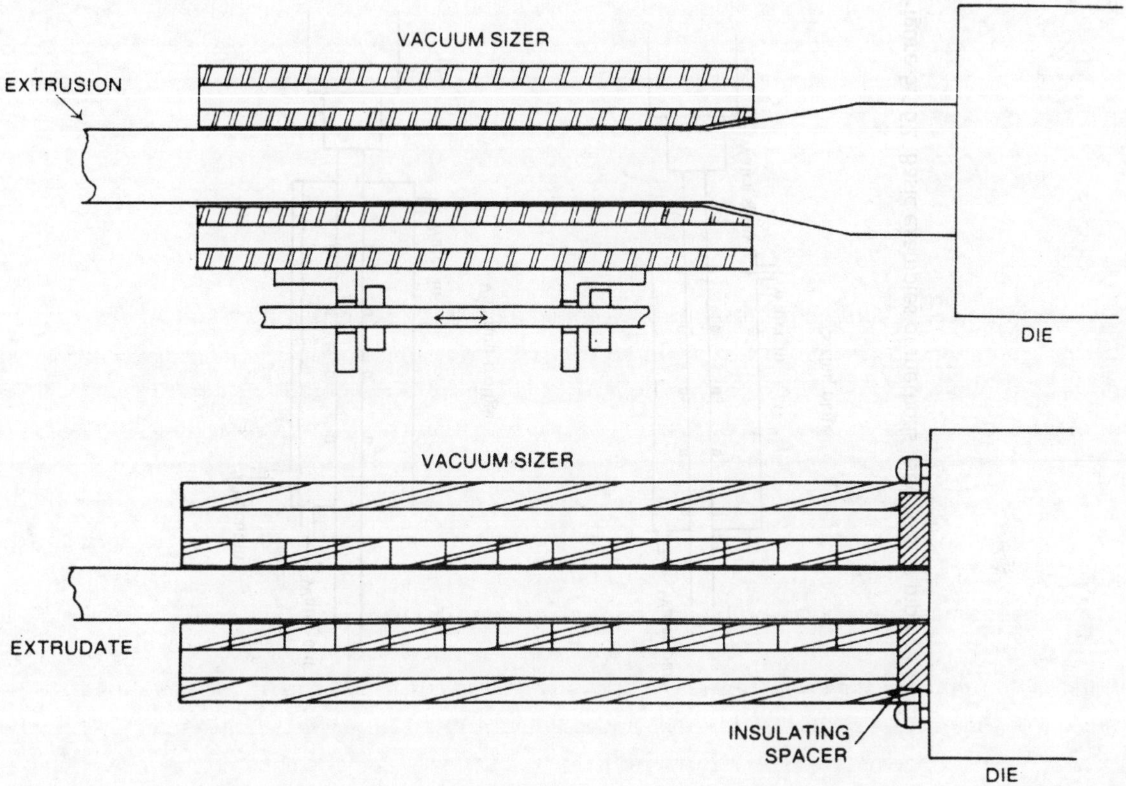

Figure 5.151 General views of vacuum sizer with or without extrudate drawdown

used. Third, the screw design should be specific to the thermoplastic being extruded (chapter 3). Table 5.53 provides a guide to tolerance.

The big profile market is in many small specialized plants. This end of the business represents a large market that requires short runs. Usually, the profiles are small dimensionally and only require small extruders. Most of these plants design their dies and have their own machine shops to make rather inexpensive plate dies. They may also do their own plastic compounding, since many runs are limited regarding material, but exchanges in profile shapes may be extensive. Setting up these lines for small operations generally requires personnel with experience and skill.

There are large production runs that are made by rather large profile fabricating plants. These plants operate like the major extrusion plants extruding film, sheet, pipe, and so forth. They can produce long runs providing low costs for products, such as building sidings, building tongue-and-groove panels, and strappings. Building and other profiles are used with and without fillers (cement, plastic foams, etc.) to provide different aesthetics and/or structural benefits.

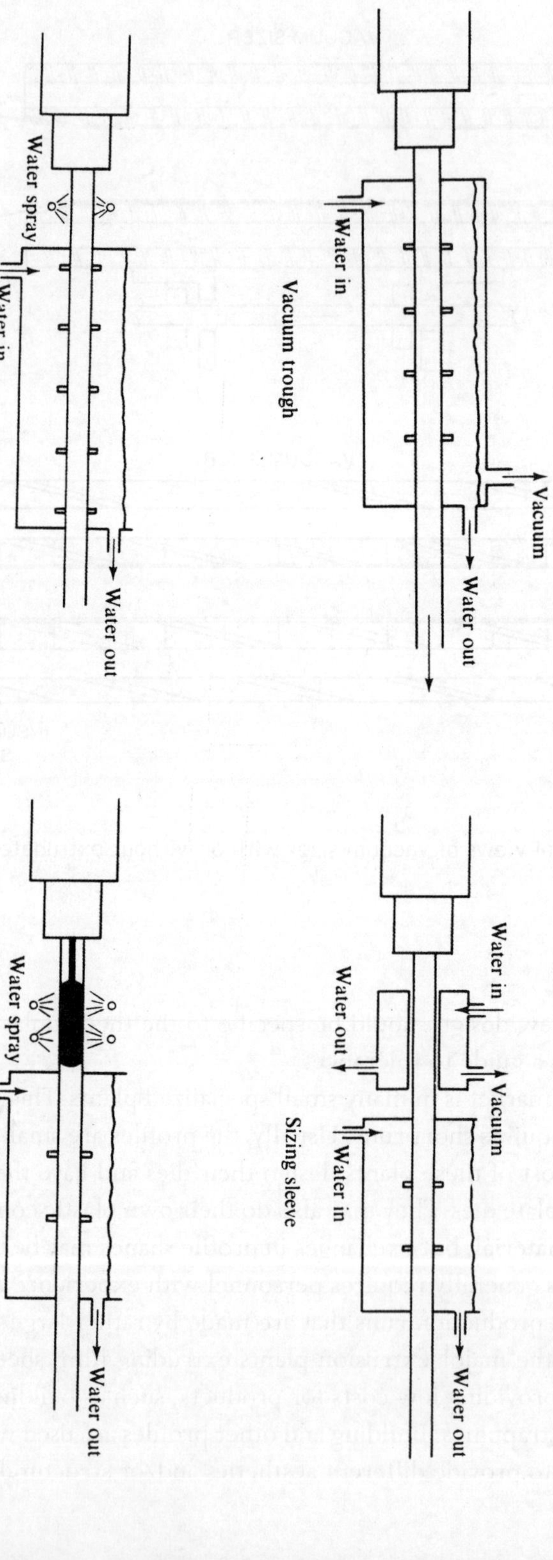

Figure 5.152 Basic examples of methods used to size pipe

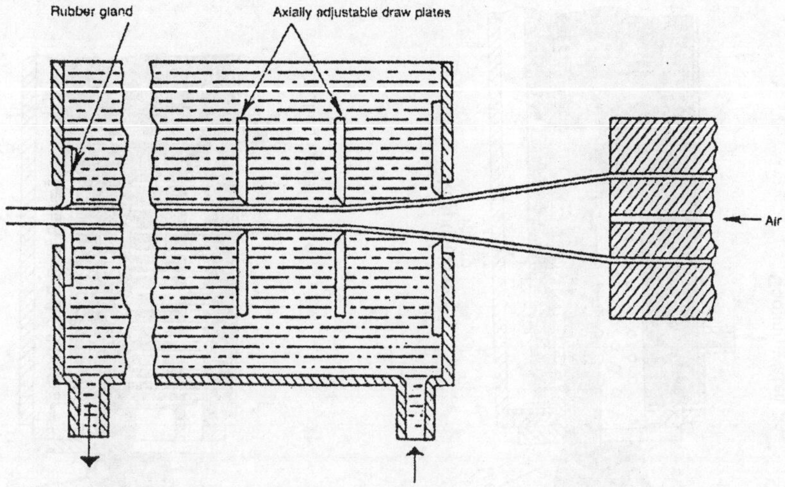

Figure 5.153 Approach to making tubes or small pipes using sizing draw plates

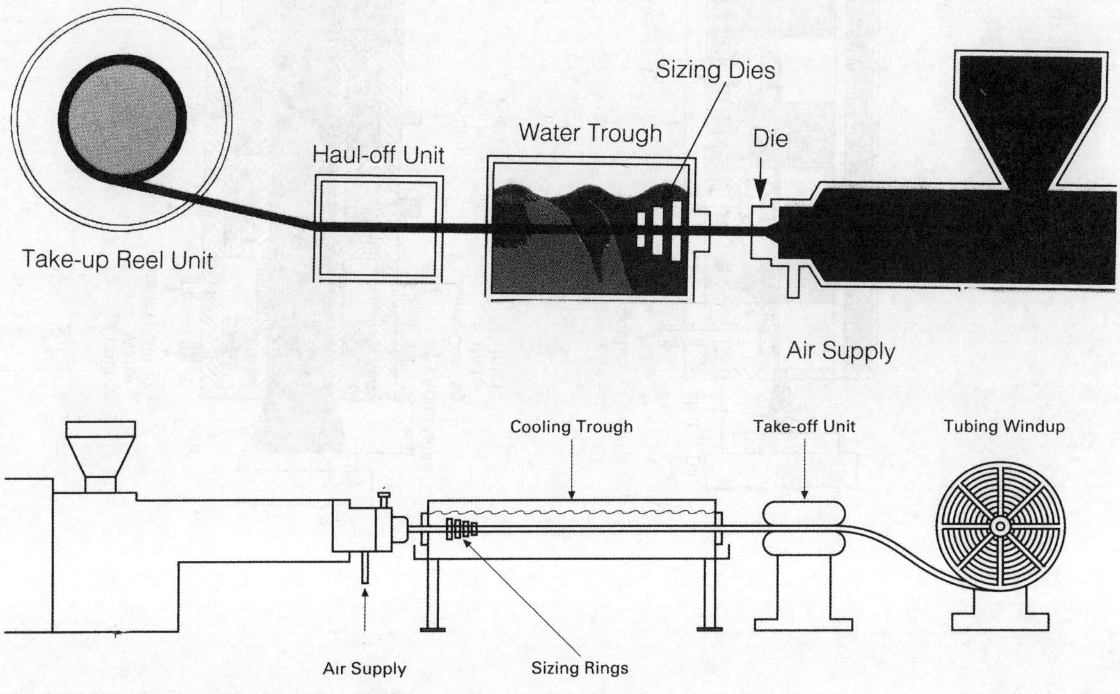

Figure 5.154 In-line tube/pipe using sizing draw plates

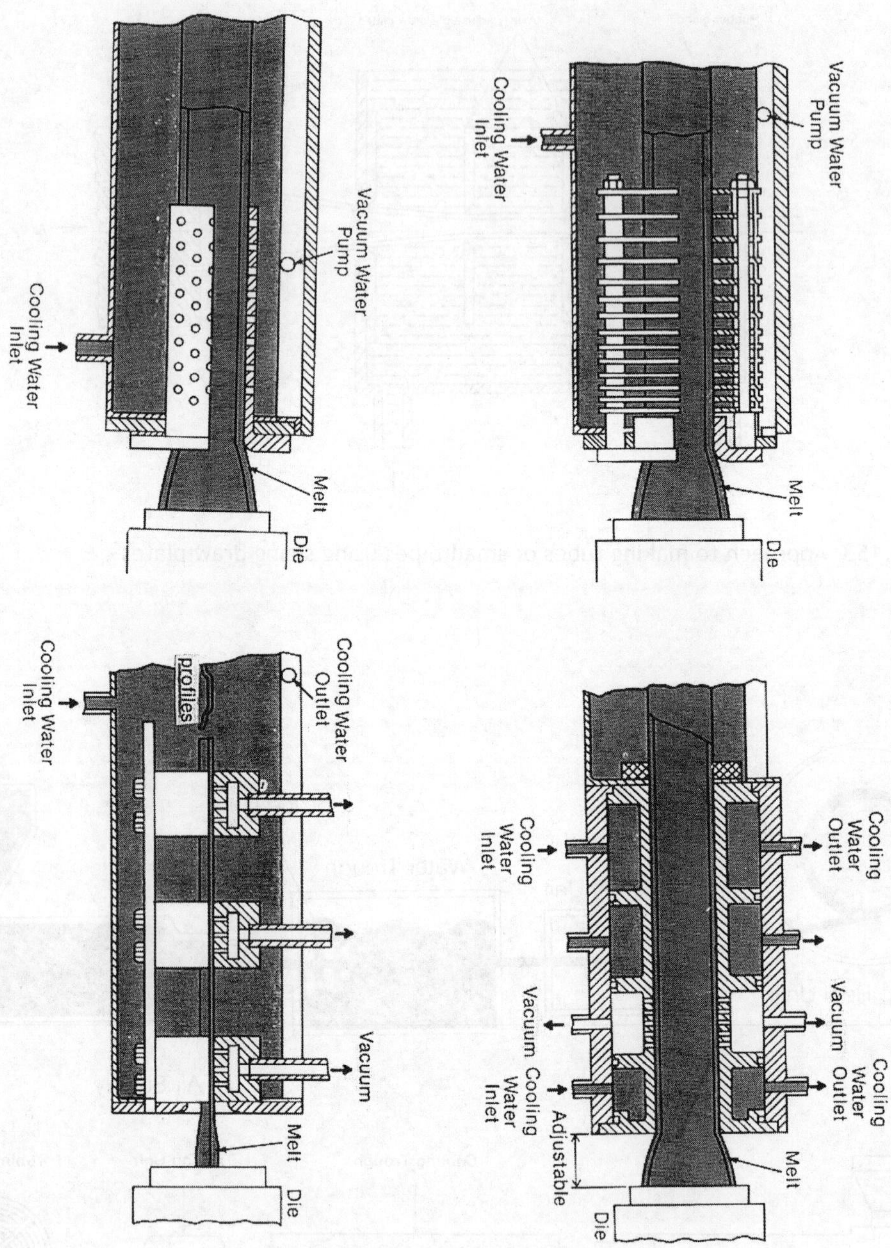

Figure 5.155 Details provided on vacuum use with spacers or holes to size pipe

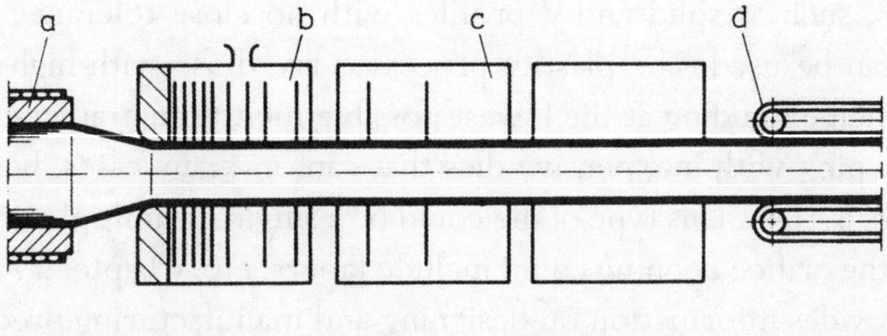

Figure 5.156 Example of a vacuum tank calibration of rigid pipe used with a water bath, where a = pipe die, b = vacuum with discs, c = heated zone water baths, and d = caterpillar takeoff puller

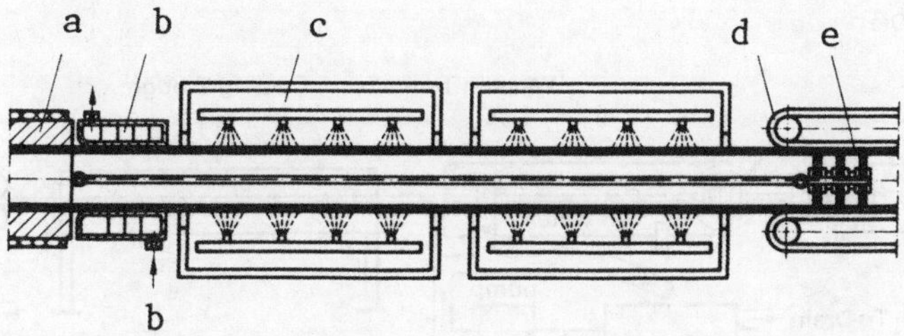

Figure 5.157 Pressure calibration of rigid pipe using a plug assist with water spray cooling, where a = pipe die, b = pressure calibrator, c = water spray cooling, d = drag lugs on conveyor belt, and e = caterpillar takeoff puller

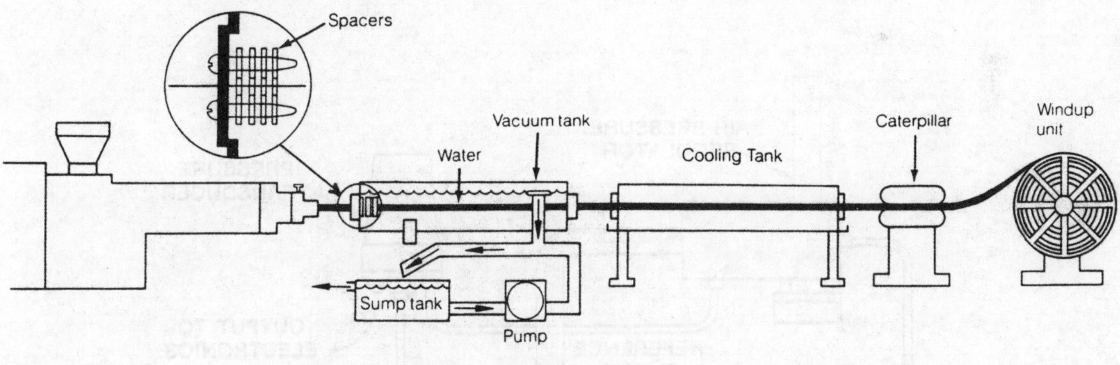

Figure 5.158 Extruder line using spacers to size pipe

Die

An extruder die's orifice is usually made oversized to the shape and size of the required contour. The final shape develops downstream from the die as the plastic expands, warps, and/or shrinks. The best designs are based on past experience with particular profile shapes. As usual, experience or being trained by one with experience has always been a real plus.

For simple shapes, such as solid and V profiles with no close tolerance requirements, a free extrusion technique can be used. The plastics processed are those with high viscosities and softer grades, particularly when extruding at the lowest possible melt temperatures. They can be used for short runs and prototyping with inexpensive dies that can, in many cases, be made from single or multiple flat plates (Fig. 5.164). This type of die could be a single or multiple array of thin plates with the orifice machined; the orifice opening could include tapers, etc. Chapter 17 relates to this streamlined approach and provides information on designing and manufacturing precision dies to produce tight-tolerance-controlled profiles. Examples of these types of dies are shown in Figure 5.165. Even

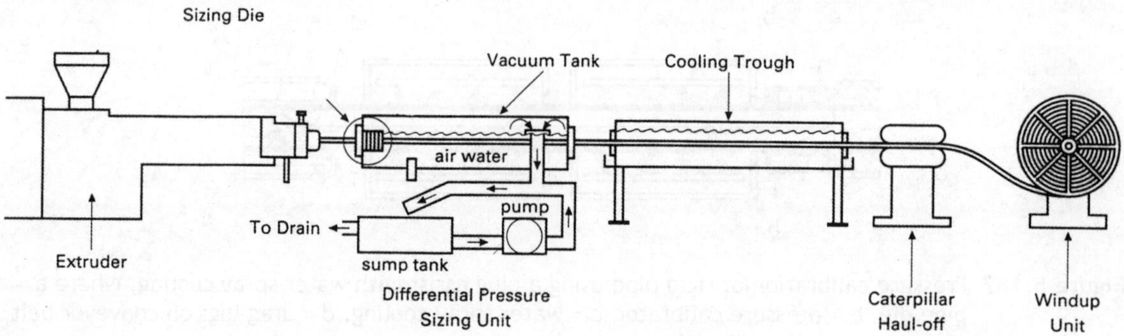

Figure 5.159 Extruder line using differential pressure to size tube

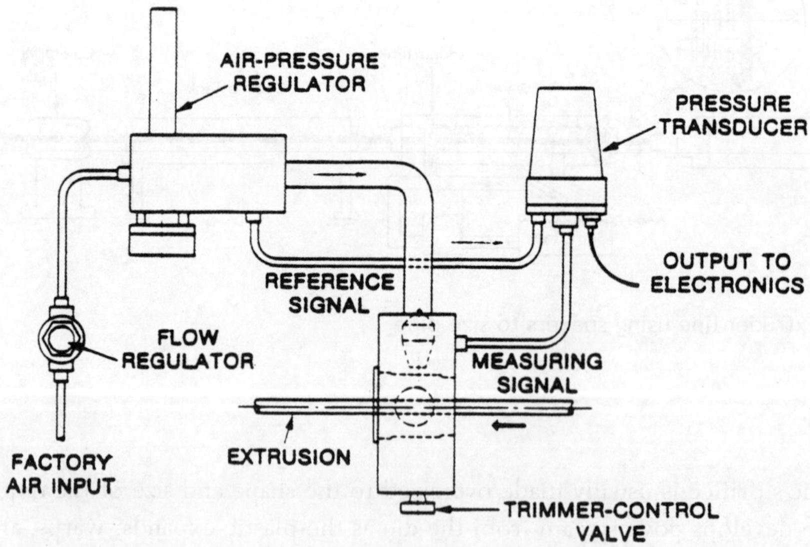

Figure 5.160 Schematic of a controlled air pressure system used in the pipe/tube line

for this very simple design and manufacturing requirement, it may be desirable to approaching the die plate with a streamlined entrance from the plasticator, particularly if you are using a heat-sensitive plastic.

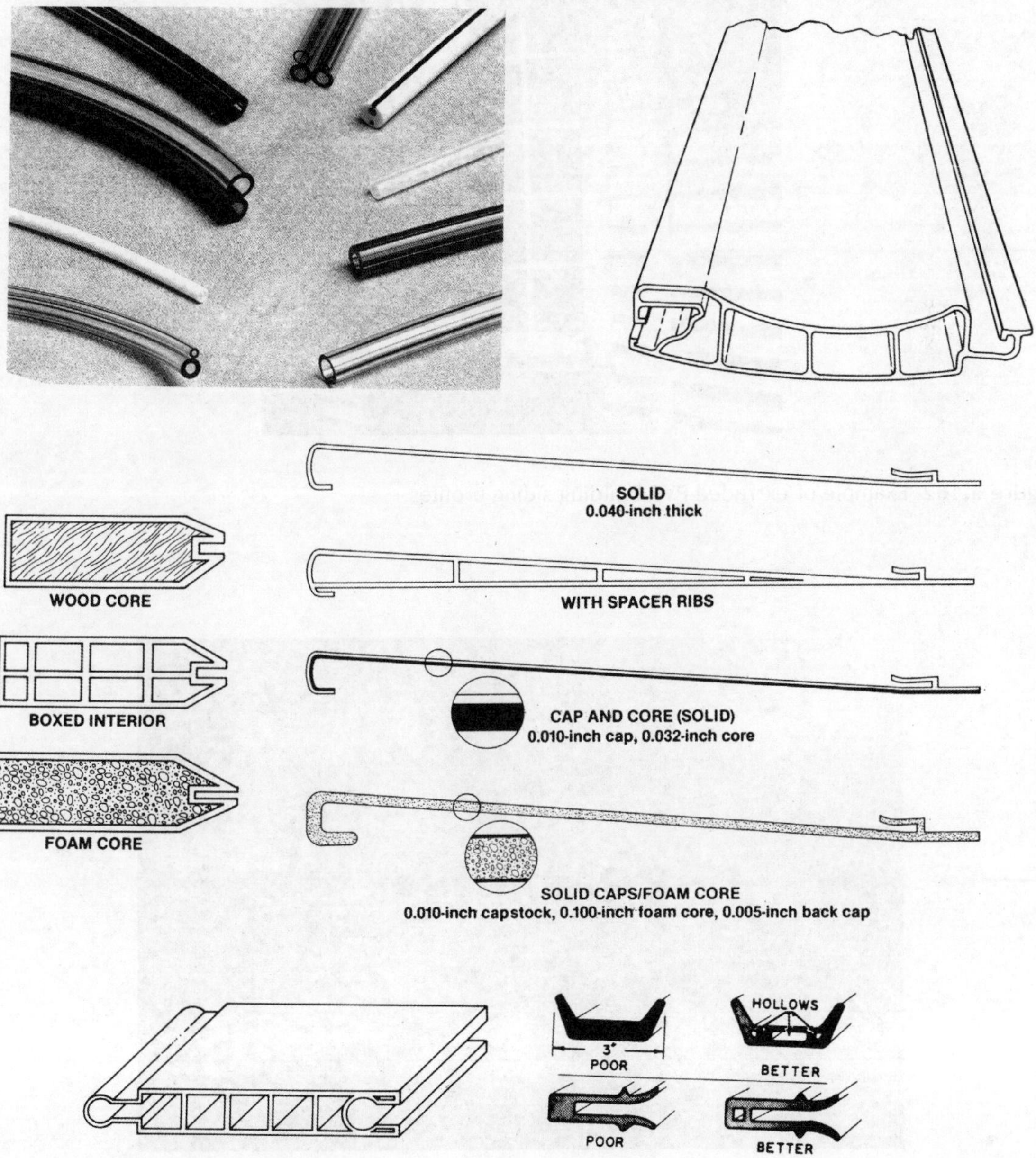

Figure 5.161 Examples of extruded profiles

Figure 5.162 Example of extruded PVC building siding profiles

Figure 5.163 Window extrusion profile line (courtesy of Battenfeld Gloucester)

Starting points	Zone 1 Temp °F	Zone 2 Temp °F	Zone 3 Temp °F	Zone 4 Temp °F	Die Temp °F	Melt Temp °F
ABS	375–400	390–425	410–435	435–460	400–475	400–475
ACRYLIC SHEET	320–450	350–470	350–480	350–480	350–450	370–480
ACRYLIC PROFILE	375–390	390–410	400–410	400–410	350–410	370–430
ACETAL	350–370	360–380	370–390	380–400	390–400	400–430
ASA & ASA/AES (CENTREX[1])	375–400	390–425	410–435	435–460	410–465	420–465
CPE (TYRIN[2])	220–225	225–290	270–300	300–325	320–330	340–360
HDPE FILM (0.17 MI)	370–430	380–450	390–470	400–480	410–480	410–480
HDPE FILM (0.53 MI)	370–390	380–410	390–430	400–440	400–450	400–450
HDPE PIPE (0.43 MI)	370–390	380–400	380–430	390–450	390–450	390–450
HDPE SHEET (0.60 MI)	350–390	360–400	370–410	380–420	370–420	370–420
IONOMER (SURLYN[3])	300–320	300–400	300–450	300–450	300–450	300–450
LDPE METALLOCENE	250–290	300–370	350–400	390–450	390–510	380–500
LDPE BLOWN FILM	320–340	320–340	330–350	340–360	350–370	350–370
LDPE CABLE COATING	330–350	340–380	350–390	360–390	380–410	380–410
LDPE EXTRUSION COATING	460–690	480–540	530–590	550–625	600–640	600–640
NYLON 6	450–460	450–460	430–450	430–450	430–440	445–460
NYLON 6 FILM	450–480	460–490	470–500	480–510	500–510	500–510
NYLON 6/12	450–520	440–510	430–500	430–500	425–495	430–500
NYLON 6/6	510–550	525–545	510–540	500–530	520–550	520–550
NYLON/ABS (TRIAX[1] 1000)	430–450	440–460	450–470	460–480	460–480	485–510
PAX-PURGE[4]	350–380	360–390	370–410	390–450	375–450	390–450
PC	530–550	540–560	550–570	560–580	550–580	510–575

Table 5.52 Guide to setting the temperature zones for different plastics to fabricate profiles

Dimension	*Rigid vinyl (PVC)*	*Polystyrene*	*ABS*	*Polypropylene*	*Flexible vinyl (PVC)*	*Polyethylene*
Wall thickness	±8%	±8%	±8%	±8%	±10%	±10%
Angles	±2°	±2°	±3°	±3°	±5°	±5°
Profile dimensions. ± mm (in.)						
0–3 (0–1/8)	0.18 mm (0.007 in)	0.18 mm (0.007 in)	0.25 mm (0.010 in)	0.25 mm (0.010 in)	0.25 mm (0.010 in)	0.30 mm (0.012 in)
3–13 (1/8–1/2)	0.25 mm (0.010 in)	0.30 mm (0.012 in)	0.50 mm (0.020 in)	0.38 mm (0.015 in)	0.38 mm (0.015 in)	0.63 mm (0.025 in)
13–25 (1/2–1)	0.38 mm (0.015 in)	0.43 mm (0.017 in)	0.63 mm (0.025 in)	0.50 mm (0.020 in)	0.50 mm (0.020 in)	0.75 mm (0.030 in)
25–38 (1–1 1/2)	0.50 mm (0.020 in)	0.63 mm (0.025 in)	0.68 mm (0.027 in)	0.68 mm (0.027 in)	0.75 mm (0.030 in)	0.90 mm (0.035 in)
38–50 (1 1/2–2)	0.63 mm (0.025 in)	0.75 mm (0.030 in)	0.90 mm (0.035 in)	0.90 mm (0.035 in)	0.90 mm (0.035 in)	1.0 mm (0.040 in)
50–75 (2–3)	0.75 mm (0.030 in)	0.90 mm (0.035 in)	0.94 mm (0.037 in)	0.94 mm (0.037 in)	1.0 mm (0.040 in)	1.1 mm (0.045 in)
75–100 (3–4)	1.1 mm (0.045 in)	1.3 mm (0.050 in)	1.3 mm (0.050 in)	1.3 mm (0.050 in)	1.7 mm (0.065 in)	1.7 mm (0.065 in)
100–125 (4–5)	1.5 mm (0.060 in)	1.7 mm (0.065 in)	1.7 mm (0.065 in)	1.7 mm (0.065 in)	2.4 mm (0.093 in)	2.4 mm (0.093 in)
125–180 (5–7)	1.9 mm (0.075 in)	2.4 mm (0.093 in)	2.4 mm (0.093 in)	2.4 mm (0.093 in)	3.0 mm (0.125 in)	3.0 mm (0.125 in)
180–250 (7–10)	2.4 mm (0.093 in)	3.0 mm (0.125 in)	3.0 mm (0.125 in)	3.0 mm (0.125 in)	3.8 mm (0.150 in)	3.8 mm (0.150 in)

Table 5.53 Guide to dimensional tolerances of different plastics for extruded profiles

For long runs, these plate dies have the disadvantage of encouraging buildup of stagnant plastic on the rear of the plates that eventually degrade the plastic, producing unacceptable profiles. Usually, no melt-degrading problem develops before the short run is completed; if the problem starts developing, the plate die is dismantled and cleaned.

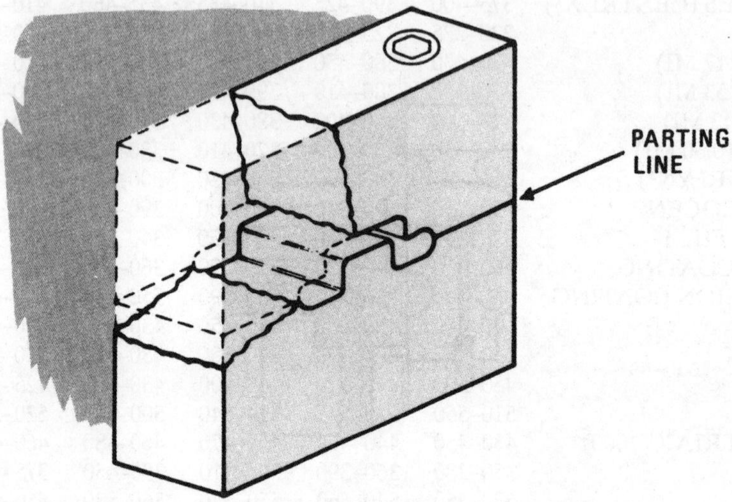

Figure 5.164 Example of an inexpensive plate die

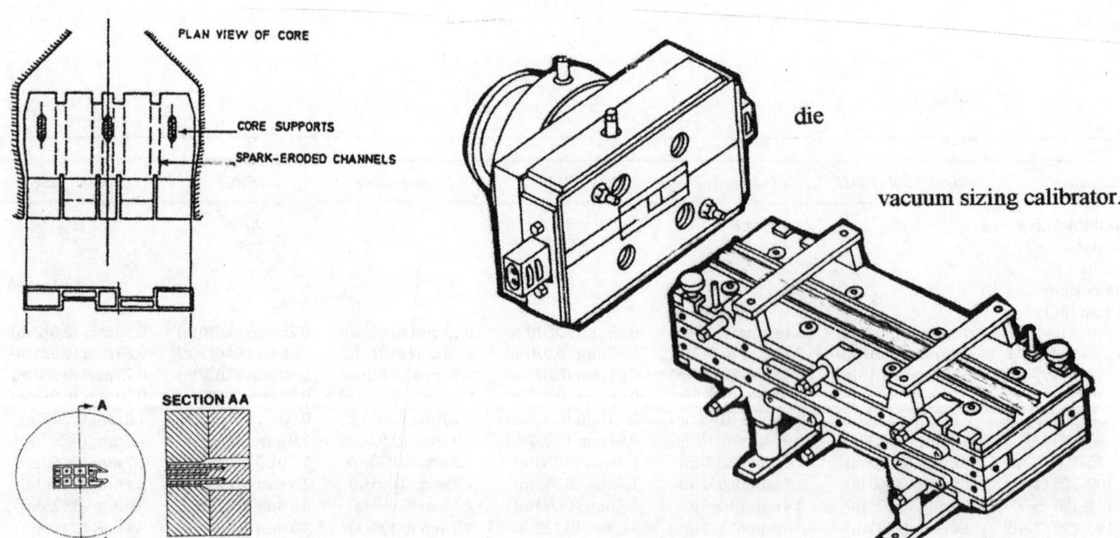

Figure 5.165 Examples of precision dies to produce close tolerance profiles

As the fluid plastic melt leaves the die, it usually has to be supported by a shaping fixture or sizing fixture. It retains a desired shape through the period when the plastic cools. A metal cooling device (steel or brass sleeve, flat sizing plates, etc.) can be used. Cooling is provided by a water trough/tank and/or a water cascade, except for those rigid plastics that can be cooled in air. With certain plastics, water cooling sets up internal stresses and gives surfaces poor appearances. Other techniques are used to speed up the line, such as cryogenic or nitrogen liquid cooling.

Extruding thick/large rods can set up problems. They require special care in order to eliminate voids. (Some plastic rods, particularly those with small diameters, have no problems.) Different techniques are used. One well-established approach is to ensure that the plastic melt is packed and moves at a sufficiently low output rate so the relationship of pressure, time, and output rate is closely controlled. A die with a sizing tube is attached to the die face. This sizing tube is also called a cooling die or forming die. Surrounding the sizing tube is a jacket that continually has water flowing through and around it. It could include baffles and restrictions to direct the water's entrances and exits in different sections so that a temperature profile cascading flow exists.

Plastic melt is pumped into and through this sizing tube. The melt starts to "freeze"/solidify as soon as it touches the inner cool wall of the sizing tube. A hardened skin forms and slides along the metal wall surface. While it moves, the skin becomes thicker. In the meantime, the plastic shrinks; however, the cooling action has not reached the rod's core. With the proper rate of flow, the extruder pumps more plastic melt to pack the interior with more plastic.

The packing action eliminates the voids. The plastic continues to go downstream while it starts with surface cooling and internal packing. To ensure cooling action from the sizing tube's internal wall to the plastic's wall surface (avoiding shrinking away from the jacket wall), the jacket inside wall diameter is reduced downstream to permit contact with the plastic. This action is usually not required if the packing action causes the plastic to continually press and slide along the straight-bored jacket hole. Before the rod leaves the sizing tube, it is sufficiently cooled along the complete cross-section. This packing action is similar to what happens when filling a mold during injection molding.

Rod extrusion is a slow process because the plastic is a poor conductor of heat. Output rate may be about 9 to 23 kg/h (20 to 50 lb/h). After extrusion, these rods are usually annealed for stress relief and stabilization of dimensions. If required by design, they can be centerless ground to final size.

Different extrusion profile processes can be used, such as robotic profiles. A robot delivers the hot melt over or around a substrate. This robotic extrusion process uses a flexible, heated high-pressure hose that is connected to the extruder's die exit. The hot melt travels through the hose. At the end of the hose is a nozzle; the tip of this nozzle is, in effect, the actual profile die. A computer-regulated multiaxis robot controls the positioning of the profile nozzle die. The nozzle/die is guided by the robot to deposit the profile's hot melt on, as an example, a substrate that is on a multistation rotating table supporting other substrates to be covered.

COATING

INTRODUCTION

Many different processes are used to coat or laminate; there are literally over hundreds of different processes used just to coat materials (Table 5.54; chapter 10). A coating method that involves the extruder is particularly important for long runs. The extruded dedicated fast lines are used since they provide quality-controlled products and are very cost effective.

Extrusion coating is the process of applying a thin-bonded film of a plastic to a continuous substrate (web), such as paper, board, foil, fabric, or plastic. The process is similar to chill-roll film extrusion. A high melt temperature is employed with a downwardly directed slit die to produce a low-viscosity melt web that adheres to the substrate (Figs. 5.166 and 5.167). The plastic must adhere well to the substrate and exhibit good sealing properties. If necessary, the substrate may be given a pretreatment to improve adhesion.

The extruder is on a movable track or platform. On start-up, the extruder is in the retracted position, that is, the die is not over the web and its operating mechanism (Fig. 5.168). The reason for this pullback is to permit the extruder's start-up away from the web and its mechanism. Thus, the web and its mechanism are not contaminated with the initial poor-grade melt that has to exit the extruder. After the melt develops the desired melt viscosity, the extruder is moved forward so the die is in the proper position to drop its curtain exactly where required.

In this process, the plastic is directed against the surface of the substrate immediately before its entry into the nip between a pressure roll and a chill roll (Fig. 5.169). The coating thickness is

Thickness (mm)	Mass (g) required by 1 m² of substrate	Length (m) covered by 0.45 kg PE
0.001	5.8	175
0.002	11.6	85
0.004	23.2	42.5
0.008	46.4	21

Coating thickness			Weight required		Surface coverage	
(in)	(mil)	(mm)	for 1,000 ft² (lb)	for 1 m² (g)	per lb of resin (ft²)	per kg of resin (m²)
0.00025	0.25	0.006	1.2	5.8	840	10.8
0.0005	0.5	0.013	2.4	11.5	420	21.5
0.001	1	0.025	4.8	23	210	43
0.002	2	0.051	9.6	46	105	86

Table 5.54 Information pertaining to different coating methods

determined by the extrusion rate and the line speed of the substrate. The chill roll controls the surface finish of the coating. Extrusion coatings are typically applied in thin layers, ranging down to 0.005 mm. A closeup view of the extruder die over the moving substrate is shown in Figure 5.170. Figure 5.171 has views from opposite ends of a very small coater (and laminator). It consumes a paper substrate at 30 miles per hour, applying half a ton of PE coating. Table 5.55 provides a guide of the amount of average PE coating that has a density of 0.92 g/cm^3.

A coating can be applied to a substrate and in turn another substrate applied to the coating, producing a laminated product. Laminating involves combining two or more substrates together by literally bonding them together with a material, such as a plastic coating.

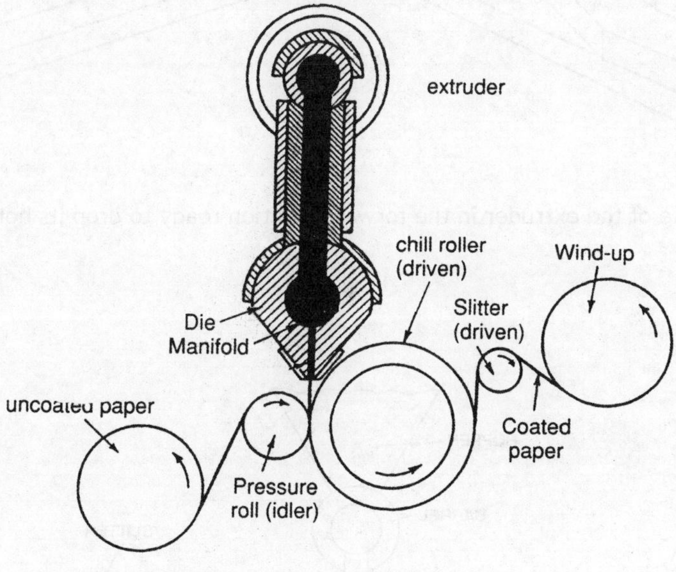

Figure 5.166 Closeup of the coating web contacting the substrate

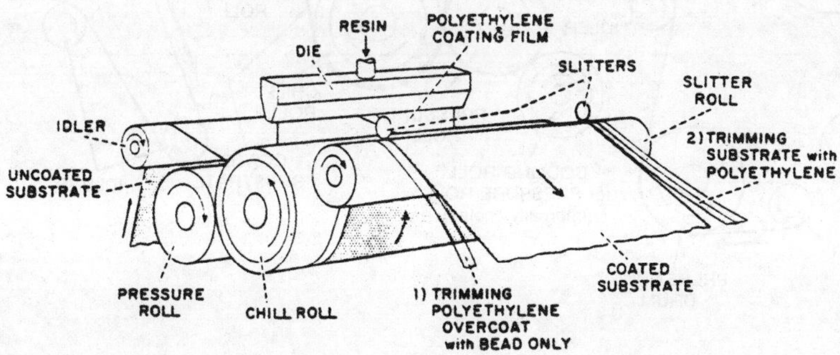

Figure 5.167 A 3-D view of the coating process

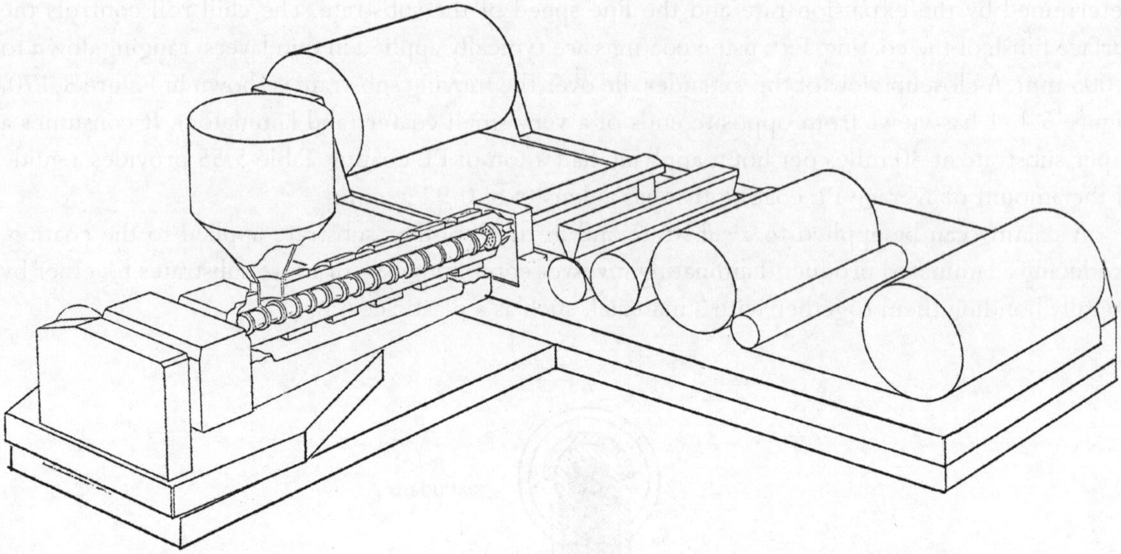

Figure 5.168 Example of the extruder in the forward position ready to drop its hot melt

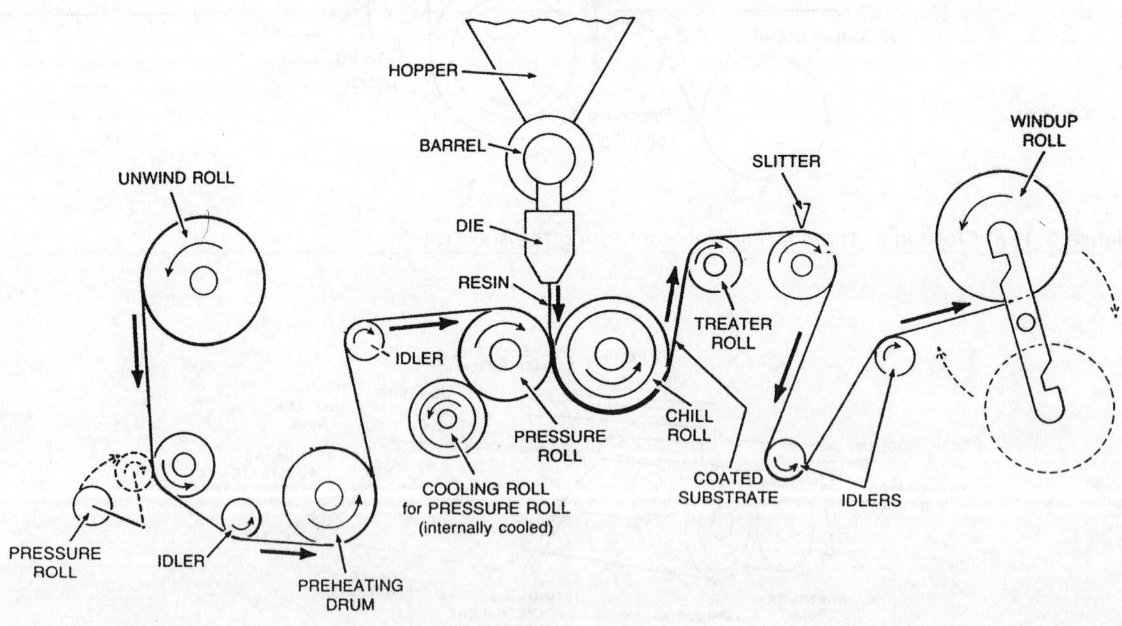

Figure 5.169 Coating extruder line that highlights the hot melt contacting the substrate just prior to entry into the nip of the pressure chill rolls

Figure 5.170 View of the extruder die over the moving substrate

Depending on the coating performance requirements, the appropriate plastic is used (chapter 2). As an example, extrusion coatings are available that impart water and grease resistance, heat scalability, and/or pinhole resistance to the substrate. Since LLDPE has a narrow MWD of the LLDPE controlling the process, it is important to obtain a good bond.

PRODUCTION

The coating process itself tends to be complex since it contains many controllable variables where each, as well as their interrelation, affects the quality of the coated or laminated product (chapter 1). In general, the processing temperature used in extrusion coating is higher than those in the other extrusion processes. Melt temperatures for the conventional plastics are normally 288° to 300°C (550° to 575°F). The higher temperatures are required to affect adhesion to the substrate. These higher temperatures put the plastics closer to being degraded. Thus, the extruder has to operate at its top efficiency controlwise.

The usual objective is to adhere relatively thin plastic of uniform thickness to a coated substrate or between substrates. This is accomplished by extruding a thin molten plastic from a slit die very similar to the extruded cast-film die. This very high-temperature-melt film curtain is drawn down to its final thickness in the air gap between the die lips and the point of contact on the substrate(s) in the nip between the rolls. Pressing and quenching between the pressure and chill rolls aid adhesion between the film and substrate(s). This process follows a three- or four-roll coating line that is similar in principle to the sheeting extrusion line.

Figure 5.171 Views of an extrusion coating line

Coating method	*Base material*[a]	*Coating composition*[b]	*Usual coating speed* ($m\,min^{-1}$)	*Viscosity range,* (*mPa s*)	*Wet-coating thickness range* (μm)
Air knife	B, D	P, T, X	15–600	1–500	25–60
Brush	B, C, E, F, G	R, S, X, Z	30–120	100–2000	50–200
Calender	A, B, D, E	U, V, W	5–90		100–500
Cast-coating	A, B, D	Q, R, S, T, V, Y	3–60	1000–5000	50–500
Curtain	A, B, C, D, E, F	R, S, V, X, Z	20–400	100–20 000	25–250
Dip	A, B, D, E, F, G	R, S, V, X, Y, Z	15–200	100–1000	25–250
Extrusion	A, B, D, E	T, U, V, W	20–900	30 000–50 000	12–50
Blade	A, B	R, S, T, V, X, Y, Z	300–600	5000–10 000	12–25
Floating knife	A, B, D	R, S, T, V, X, Y, Z	3–30	500–5000	50–250
Gravure	A, B, D, E	R, S, T, U, V, Y, Z	2–450	100–1000	12–50
Kiss roll	A, B, C, D, E, F	R, S, V, X, Z	30–300	100–2000	25–125
Knife-over-blanket	A, B, D	R, S, T, V, X, Y, Z	3–30	500–5000	50–250
Knife-over-roll	A, B, C, D, E	R, S, T, U, V, X, Y, Z	3–60	1000–10 000	50–500
Offset gravure	B, D	R, S, T, Z	30–600	50–500	12–25
Reverse roll	A, B, C, D, E, F	R, S, T, U, V, X, Y, Z	30–300	50–20 000	50–500
Reverse-smoothing roll	A, B	R, T, X	15–300	1000–5000	25–75
Rod	B, D	R, S, T, V, X, Y, Z	3–150	50–500	25–125
Sprays					
Airless spray	A, B, C, D, E, F, G	S, T, V, X, Y, Z	3–90	c	2–250
Air spray	A, B, C, D, E, F, G	S, T, V, X, Y, Z	3–90	c	2–250
Electrostatic	A, B, C, D, E, F, G	S, T, V, X, Y, Z	3–90	c	2–250
Squeeze roll	A, B, C, D, E, F	R, S, T, U, V, X, Y	30–700	100–5000	25–125
In situ polymerization	A, B, C, D, E, F, G	Y, Z	undetermined	liquid or vapor	6–2.5
Powdered resin	A, B, C, E, F, G	Q	3–60		25–250[c]
Electrostatic spray		Q			20–75[c]
Fluidized bed	E, G	Q			200–2000[c]

[a] Key: A = woven and nonwoven textiles; B = paper and paperboard; C = plywood and pressed fiberboards; D = plastic films and cellophane; E = metal sheet, strip, or foil; F = irregular flat items; G = irregularly shaped

[b] Key: Q = powdered resin compositions; R = aqueous latexes, emulsions, dispersions; S = organic lacquer solutions and dispersions; T = plastisol and organosol formulations; U = natural and synthetic rubber compositions; V = hot-melt compositions; W = thermoplastic masses; X = oleoresinous compositions; Y = reacting formulations, e.g., epoxy and polyester; Z = plastic monomers

[c] Dry thickness.

Table 5.55 Guide to surface PE coating coverage

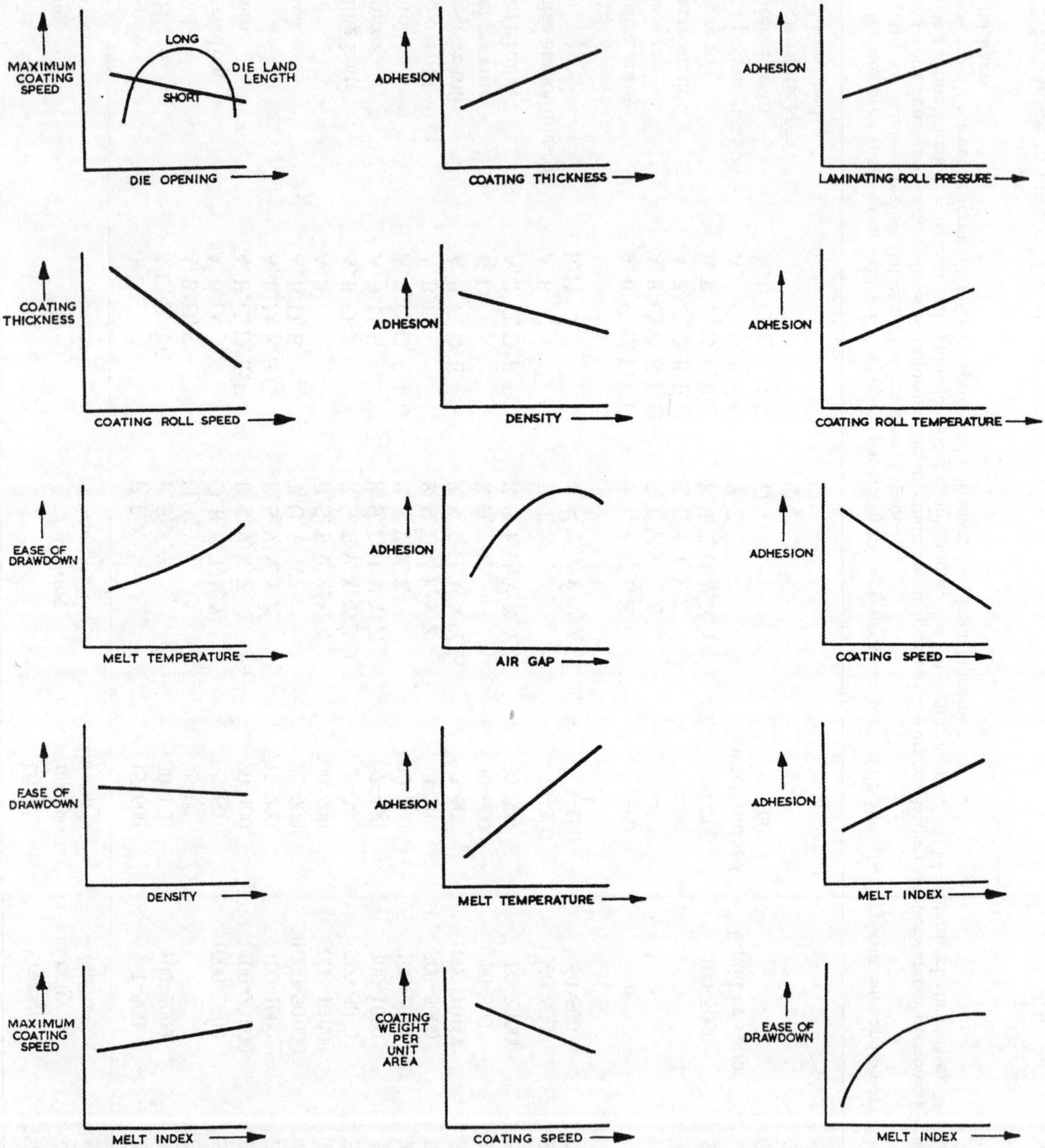

Figure 5.172 Examples of the influence of temperature and other controls on extrusion coating performances

Owing to the presence of the relatively stable substrate and the rapid cooling, very high speed can be obtained during processing, such as at least 300 m/min (1,000 ft/min). In order to obtain these speeds and to avoid fluctuations, the equipment is designed to be extra sturdy. As in the case of other extruded products, but more so, the types of plastic will have direct effects on the products' properties. The important processing requirements involve adhesion between plies, gauge variation, neck-in, surface finish, dimensional stability, and heat-seal range.

Different TP materials are used, such as PEs, PPs, nylons, ionomers, and EVAs. These types of plastics are more popular based on processing, product performances, and cost. Each plastic, these and others, has its own processing requirements and capabilities. As an example, an LDPE (a very popular material) is more stable than PP, since PP tends to have more fluctuations in extruder output. Although PP has good heat-sealing strengths, it tends to be more sensitive to processing conditions. It bonds well with porous substrates but generally has problems with nonporous substrates. However, with plastic modifications, these types of problems have practically disappeared.

One of the problems encountered in extrusion coating is that of beading at the edges of the hot film where neck-in occurs after the extrudate leaves the die opening. This local thickening of the film has to be removed during the trimming operation. Otherwise, when the roll is wound up, the beads at either end only support the weight of the roll, leaving a loosely wound and sagging roll in the middle. The extra trimming required means a reduction in the usable web width, thus increasing the costs (this situation relates to the problems reviewed for film).

Neck-in occurs when the melt leaves the die; the width of the melt film shrinks, causing neck-in. In the case of PE, the incidence of neck-in can be reduced by decreasing the melt index or the density, but the risk of voids in the hot film at low coating speeds increases because of imperfect melt flow. An optimum combination of these factors has to be found. One should determine the minimum neck-in and bead size based on processing conditions and use those observations as control parameters. Sensors for neck-in and beads can be used in automatic process controls.

Neck-in with nylon is greater than with PE. Typically, the neck-in is about 15 cm (6 in) or 7.5 cm (3 in) per edge. Unlike PE, the neck-in with nylon increases with increasing line speed. The amount of neck-in will lead to excessive trimming and waste of nylon. The use of air jets located just below the melt exiting the die will minimize this problem. They are effective in reducing neck-in to less than 5 cm (2 in) per edge and reduce the amount of trim to about 2 cm (0.75 in) per edge. A low flat of air directed down and out has been beneficial. Preheating of the air is effective in reducing the changes in the freeze-off when the web reaches the chill roll.

The uncoated substrate is fed from an unwinding roll or reel into the coater. After being trimmed, with or without being slit, the coated stock is rewound on a core roll. There is a variety of handling equipment for the unwinding and rewinding stations; turret or center winders are usually used. Which type to use depends on the kind of substrate being coated, the volume of production, and other factors. Large production will require fully automatic equipment (with control of tension in the windup, etc.) that must not be stopped to change rolls.

From the unwind step to the rewind-rolls stage, there are a number of rolls over which uncoated and coated substrates travel. They are necessary to guide the web at an accurate position through the line. The rolls are running in ball or other types of bearings that are precision ground. They must be kept clean and accurately fitted into the line on sturdy frames. They are kept in perfect alignment in the line. When the substrate is a very light or smooth material, it is frequently necessary for several of the rolls to be driven. For paper, however, most auxiliary rolls are idlers.

Because the extrusion coating process requires such a high operating temperature, shutdown requires special considerations and care. One cannot just flip a switch for an immediate shutdown. This approach would probably result in degraded plastic in the extruder through to the die. Probably the usual approach to purging would not be adequate. At the time it is planned to shut down the complete line, the extruder is pulled away from over the coating line. At that time, the coating line is shut down by releasing or opening rolls, such as the pressure and chill rolls, with probably coated material in the system so it can be used for start-up.

Keep pumping plastic through the extruder while reducing the temperature from the usual running at 370°C (700°F) to 260°C (500°F). When the temperature has reached 260°C (500°F), it is time to turn off the extruder. At that temperature, the plastic is not usually subjected to rapid, degrading oxidation with its usual formation of hard, discolored plastics that adhere, with exceptional strength, to any adjoining metals.

WIRE AND CABLE

From the very beginning of the electrical industry, plastics have been used for their insulation, semiconductivity, and jacketing capabilities. Different plastics are used to meet different performance requirements that include electrical, mechanical, chemical, thermal, and/or environmental. Different conditions exist for each of these categories, such as operating at different frequencies, underground, and in ocean depths (Table 5.56). Extensive amounts of different grades of PVC and PE plastics are used. The PEs include low- to high-density compounded grades, cross-linked grades, and dry swellable grades that prevent water penetration.

Electrical insulation is of major importance to the plastics industry. Large quantities of plastics are consumed that have literally covered many billions of miles worldwide. As shown in Figure 5.173, a wire extrusion line requires different equipment to operate. The line starts with a wire or cable unwinding roll followed with a tension-controlled input capstan, possibly a wire straightener, and wire preheater. The wire proceeds through the usual extruder's 90° cross-head die where the plastic coating is applied (Table 5.57). It continues through a water and/or air cooling system for thermoplastics (TPs) or a heating system for thermosets (TSs). Other equipment includes an electrical tester, gauge controller, tension output capstan, tension controller, and windup roll. There is also special equipment included for use with certain plastics, such as nylons, to maximize their performance (toughness, stress relaxation, etc.) by using in-line annealing and moisture-conditioning

Elastomers	*Abbreviation*
Natural rubber	NR
Butyl rubber	IIR
Styrene–butadiene rubber	SBR
Nitrile–butadiene rubber	NBR
Ethylene–propylene rubber	EPM
Ethylene–propylene terpolymer rubber	EPDM
Polychloroprene	CR
Chlorosulfonated polyethylene	CSM
Chlorinated polyethylene	CM
Silicone rubber	SR
Epichlorohydrin rubber	ECO
Ethylene–vinyl acetate copolymers	EVA
Thermoplastics	*Abbreviation*
Poly(vinyl chloride)	PVC
Polyethylene	PE
Ethylene–vinyl acetate copolymers	EVA
Ethylene–alkyl acrylate copolymers	
ethylene–ethyl acrylate copolymers	EEA
ethylene–butyl acrylate copolymers	EBA
Polypropylene	PP
Polyamide	PA
Ethylene–tetrafluoro-ethylene copolymers	ETFE
Tetrafluoroethylene–hexafluoropropylene copolymers	FEP

Table 5.56 Examples of thermoplastics and elastomers used for wire and cable insulations

equipment. As shown in Figure 5.174, variations in extruder and plastic performances, such as melt index (MI) (chapter 22), influence output and properties of the insulation.

This coating process produces many different types and sizes of product. Different TP and TS plastics are used to meet different performance requirements, in processing operations and service environments. Because of the potential dangers of the misuse of wires and cables in electrical products, there are many governmental and industrial codes and specifications, including tests for in-line extruding processes and so forth.

Insulation is essential, but there are applications in which adhesion of plastic to wire is also important to reduce breakdown due to corona discharge in high-voltage AC applications. As an example, in submarine telephone cables, where a 15-mm (0.6-in) conductor may be covered with a

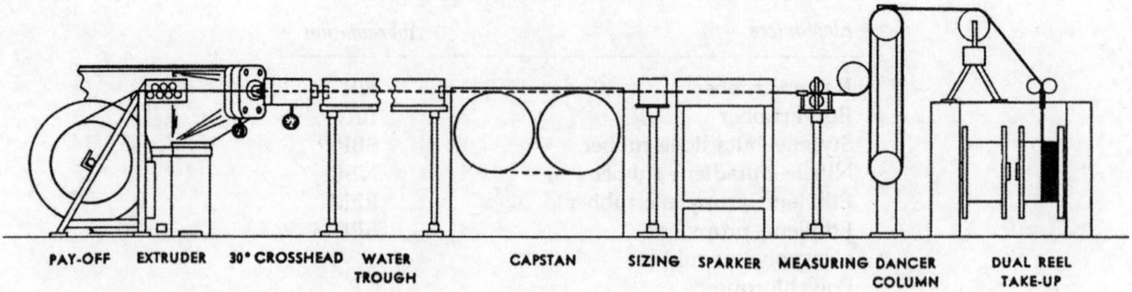

Figure 5.173 Example of a wire coating extrusion line

Extruder size		*Output*	
(in)	*(mm)*	*(lb/h)*	*(kg/h)*
$1^1/_2$	38	30	14
2	51	35–40	16–18
3	76	70–80	32–36
$4^1/_2$	114	150–160	68–73
10	254	1000–1500	450–680

Table 5.57 Examples of LDPE output in wire and cable lines

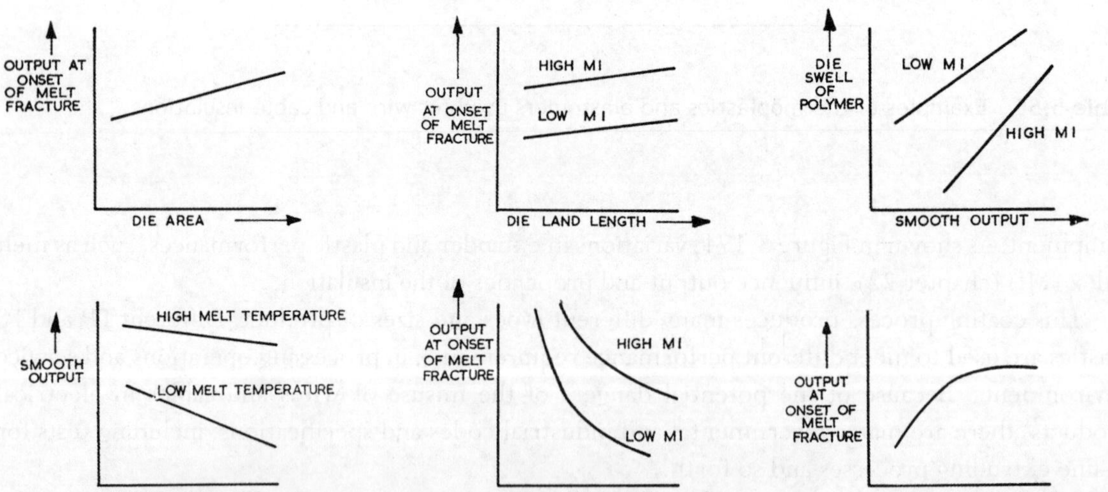

Figure 5.174 Examples of the influence of extruder and plastic on wire insulation

25-mm (1-in) thickness of LDPE, slow extrusion rates are used. The line downstream will include water cooling in stages. Sometimes cooling will include programmed temperatures to minimize temperature and shrinkage differentials within the insulation and maintain adhesion to the core. With such thick insulation or sheathing, haul-off tension will not be critical. However, haul-off speed will determine thickness and a constant speed is essential to eliminate surface lumps or ripples.

Multilayer insulation constructions are used. With the high-voltage cable, three separate layers may cover the center conductor. The innermost layer will function as an electric screen, the intermediate layer is usually a PE insulation material, and the outer layer serves as an electric screen. The processing technique can be used for primary and secondary coating where the covering is on metallic conductors or coating previously insulated wires.

PRODUCTION

The melt shear conditions when coating wire or cable differ from processes that use no substrate in that the inner surface is moving with the conductor causing additional drag flow. Cooling is usually in air, sometimes in a festoon, but water baths have been used for the higher speeds.

The usual die used is a cross-head with a 90° angle between the wire line and the extruder body axis (Fig. 5.175). With this setup, the entire length of the extruder projects sideways from the coating lines. To help melt flow from developing dead spots in the melt channels, 30° or 45° cross-heads can be used. They provide more streamlined interiors and the extruder location is better adapted to some plant layouts. They are sometimes preferred when processing PVC because of

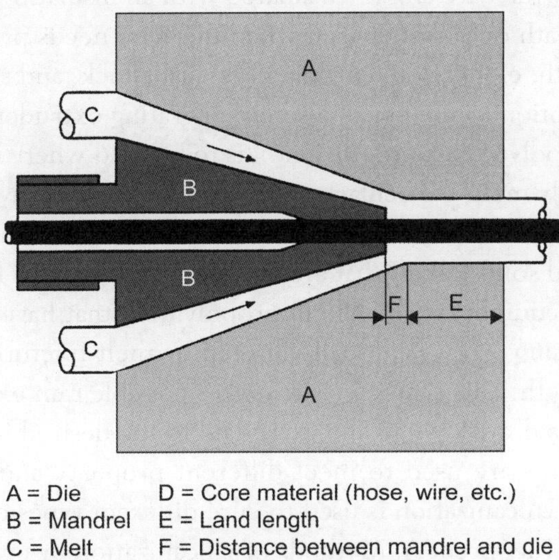

Figure 5.175 Schematic of a wire and cable die

the streamlining and better control of the melt's heat profile. Most dies are subjected to very high internal pressures, since the uncommon pressure of over 5,000 psi (35 MPa) is required.

As explained in chapter 17, die configurations meet certain melt flow requirements, such as a drawdown ratio and draw ratio balance. The drawdown ratio (DDR) in a circular die, such as a wire die, is the ratio of the cross-sectional area of the die orifice/opening to the final extruded shape. Another guide for setting uniformity and best repeatable references is the draw ratio balance (DRB), which aids in determining the minimum and maximum values that can be used for different plastics.

The specific equipment used in-line depends upon the kind of wire that is being handled and the product to be made. When plastic covers bare electric wire, the coating is called primary insulation. A secondary layer of plastic extruded over a wire or group of wires is called a sheath or jacket. The sheath may provide only mechanical protection, and its electrical insulation quality is of less importance. The cores that refer to either the wires or cables are handled in the extrusion process. The core is the part in the center that is to be coated.

TP, TS, and elastomer TP or PS plastics are used to meet the different requirements. Important TPs based on quantity used include PEs, PPs, PAs, and PVCs. The epoxy (EP) and polyurethane (PU) plastics are popular TS plastics. Also extensively used are cross-linked polyolefins, such as cross-linked PEs (XLPEs) and ethylene copolymers that behave similarly to TS elastomers above the crystalline melting point.

PP is used for its good temperature resistance, other performance, and economics. It must contain a stabilizer/inhibitor additive to prevent any reaction with copper, which would otherwise hasten its degradation. Stabilizers are also needed in PVC to prevent degradation under heat during processing and in service. Generally, nylon (PA) is utilized as the outer cover in wire coating applications. It is extruded over wire already insulated with an insulating plastic, such as PVC, PE, PP, and PBT. This nylon sheath or jacket is used when the wire needs protection from mechanical abrasion, toughness, strength, elongation, cutting, chemical attack, and so forth.

An example of an exception to the use of a screw-operating extruder occurs when a very low-melt plastic, such as PTFE (polytetrafluoroethylene), is to be used where its specialty properties are needed. PTFE does not easily melt, particularly in the production of heavy-walled PTFE wire coatings, so ram extrusion is used. Ram extrusion is used for these wire and cable jobs and for producing rods, tubes, and large, round solid "cakes" that are later skived to make PTFE film (chapter 4).

With developments in equipment and the fluoropolymers that have high melt viscosities, the maximum extrusion rate using screws is usually limited by melt fracture. However, these plastics have exceptional melt strength. This characteristic makes possible ram extruding using a die with a large opening and applying a drawdown of the extrudate to the desired insulation thickness.

Other processing systems are used to meet different property and/or cost requirements of TS plastics. As an example, vulcanization is used to cure different types of TS elastomers/rubbers, such as SBRs, IRs, and CRs (neoprenes). Methods of vulcanization or curing include steam, liquid (eutectic mixtures), microwave, and hot air. There are systems used to cure peroxide-based cross-linkable polyethylene (XLPE) compounds. They include steam cure, nitrogen cure, and pressurized liquid continuous vulcanization (CV). All these methods involve higher heat than the melt heat to cause the peroxide in the plastic to decompose into a reactive radical and initiate the curing cycle. Examples of these different wire coating systems are reviewed in Figures 5.176 to 5.180.

FIBER

OVERVIEW

In the United States alone during 2009, the man-made plastic fiber industry had over ninety plants with sales of $13 billion and about 45,000 employees. Production processes are diverse both in technology and equipment design. They have common steps, which include preparation of reactants, polymerization, polymer recovery, polymer extrusion (if in pelletized form), and supporting operations (chapter 1).

In some preparation operations, solvents are used to dissolve or dilute monomers and reactants. Solvents are also used to facilitate the transportation of the reaction mixture throughout the plant, to improve heat dissipation during the reaction, and to promote uniform mixing. Solvent selection is optimized to increase monomer ratios and to reduce polymerization costs and emissions. The final polymer may or may not be soluble in the solvent (374).

The following combinations of polymers (with additives) and solvents are commonly used: (a) HDPE (isobutane and hexane), (b) LDPE (hydrocarbons), (c) LLDPE (octene, butene, or hexane), (d) PP (hexane, heptane, or liquid propylene), (e) PS (styrene or ethylbenzene), and (f) PMMA (dimethylacetamide or aqueous inorganic salt solutions). These examples show that there are options available. Excess monomer may replace solvent or water can be used as the solvent. During recovery of polymers, unreacted monomers and solvents are separated from polymers (monomers and solvents are flashed off by lowering the pressure and sometimes degassing under vacuum), liquids and solids are separated (the polymers may be washed to remove solvents), and residual water and solvents are purged when the polymers are drying. Further drying and extrusion remove residual solvents. Figure 5.181 shows a schematic diagram of potential emissions during polymer manufacture.

Manufacture of man-made fibers involves polymerization (usually the core part of the process), preparation of the solution, spinning, washing and coagulation, drying, and other operations. Forcing the viscous liquid through small-bore orifices produces and forms fibers. A suitable viscosity can be achieved either by heating or dissolution. The rheological properties of the solution are governed to a large degree by the solvents selected. Wastes generated during the spinning operation include evaporated solvent and wastewater contaminated by solvent. The typical solvents used in the production of fibers are dimethylacetamide (acrylic), acetone or chlorinated hydrocarbon (cellulose acetate), and carbon disulfide (rayon).

In the dry spinning process, a solution of polymer (with additives) is first prepared. The solution is then heated above the boiling temperature of the solvent and the solution is extruded through spinneret. The solvent evaporates into the gas stream. With wet spinning, the fiber is directly extruded into a coagulation bath where solvent diffuses into the bath liquid and the coagulant

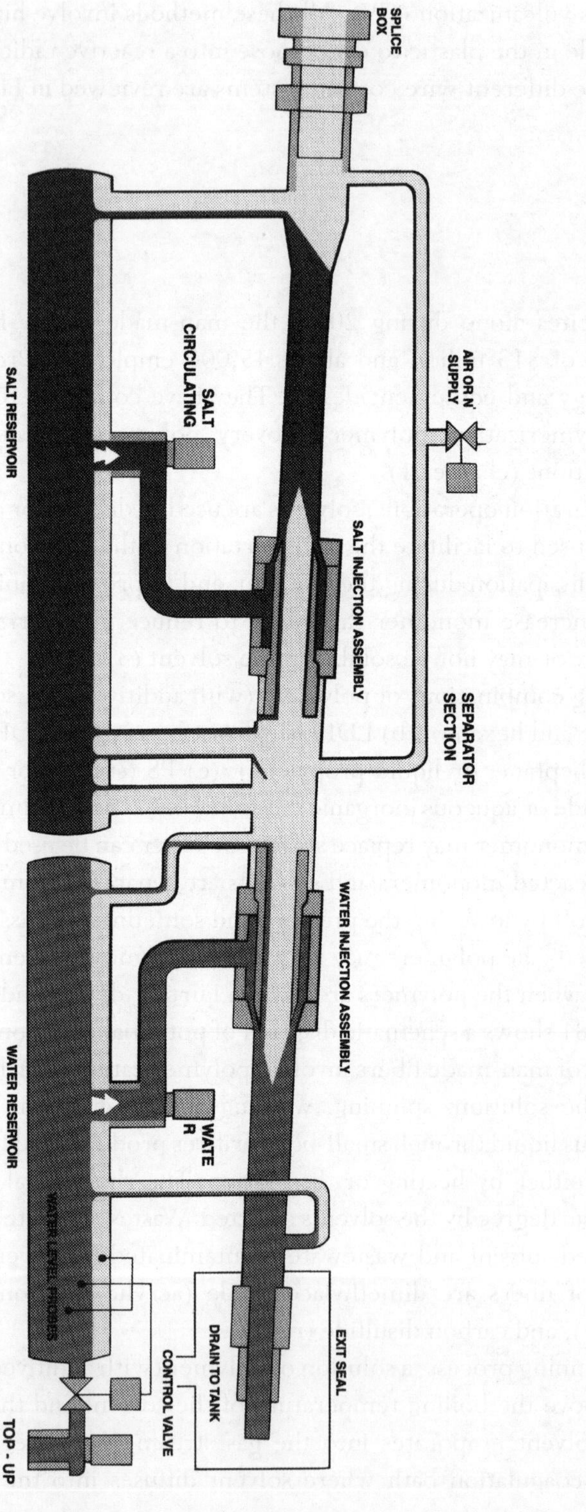

Figure 5.176 Example of continuous vulcanization pressurized liquid salt wire coating system

diffuses into the fiber. The fiber is washed free of solvent by passing it through an additional bath. Each process step generates emissions or wastewater. Solvents used in production are recovered by distillation. Figure 5.182 is a schematic of fiber production showing that almost all stages of production generate emissions.

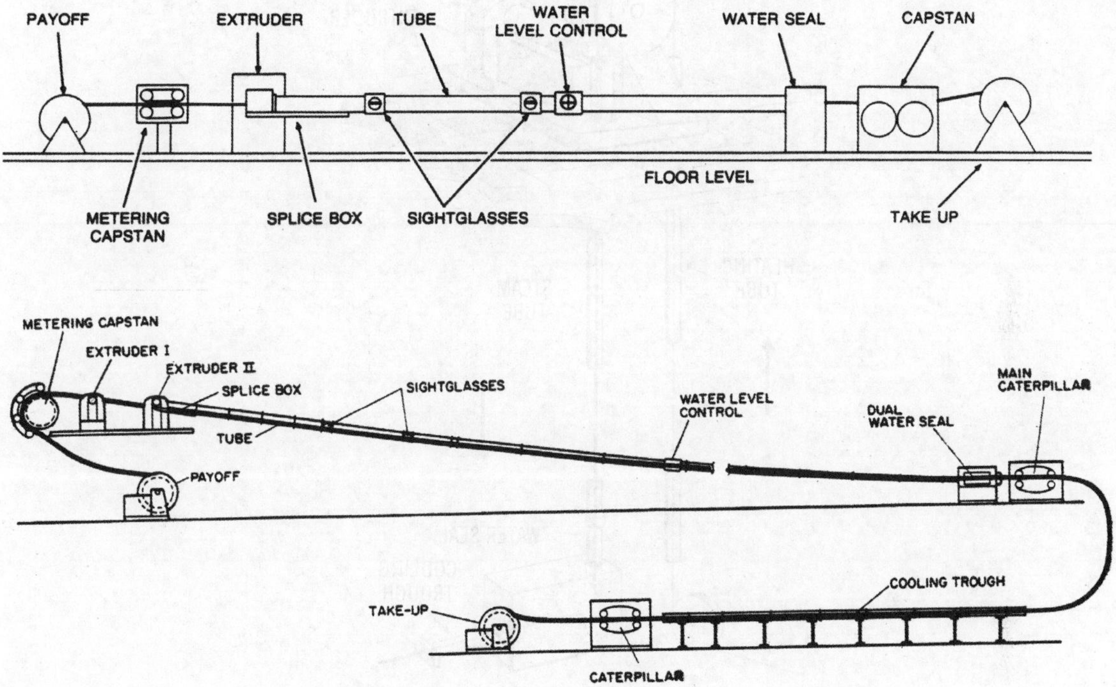

Figure 5.177 Examples of horizontal continuous vulcanization wire coating systems

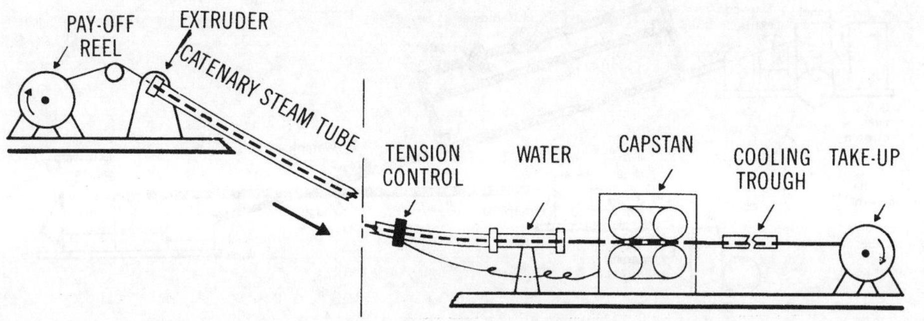

Figure 5.178 Examples of catenary continuous vulcanization wire coating system

New technology is emerging to reduce solvent use. Recent inventions disclose that, in addition to reducing solvents, the stability of ethylene polymers can be improved with the newly developed process. A proper selection of solvent improved a stripping operation and contributed to the better quality of cyclic esters used as monomers. Solvent was used for the recovery of fine particles of

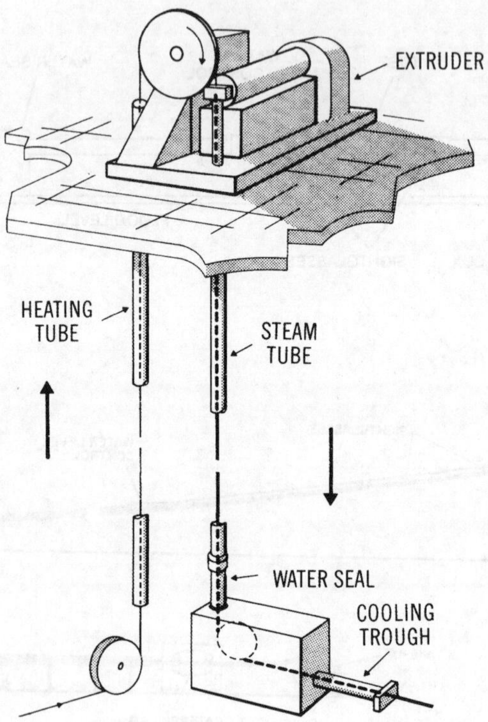

Figure 5.179 Example of vertical continuous vulcanization wire coating system

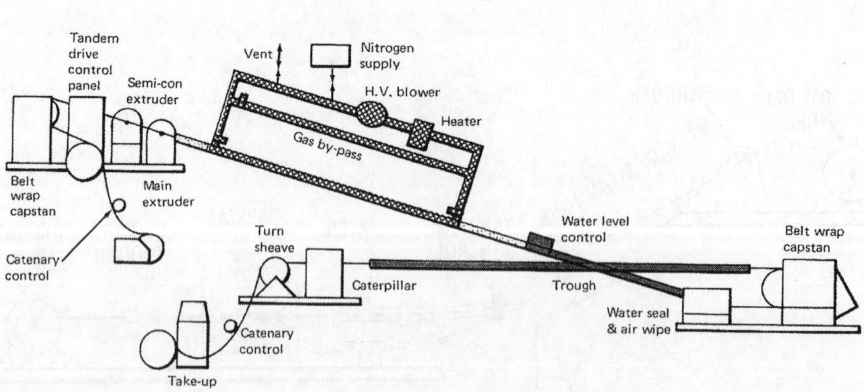

Figure 5.180 Examples of thermoset gas-curing wire coating system

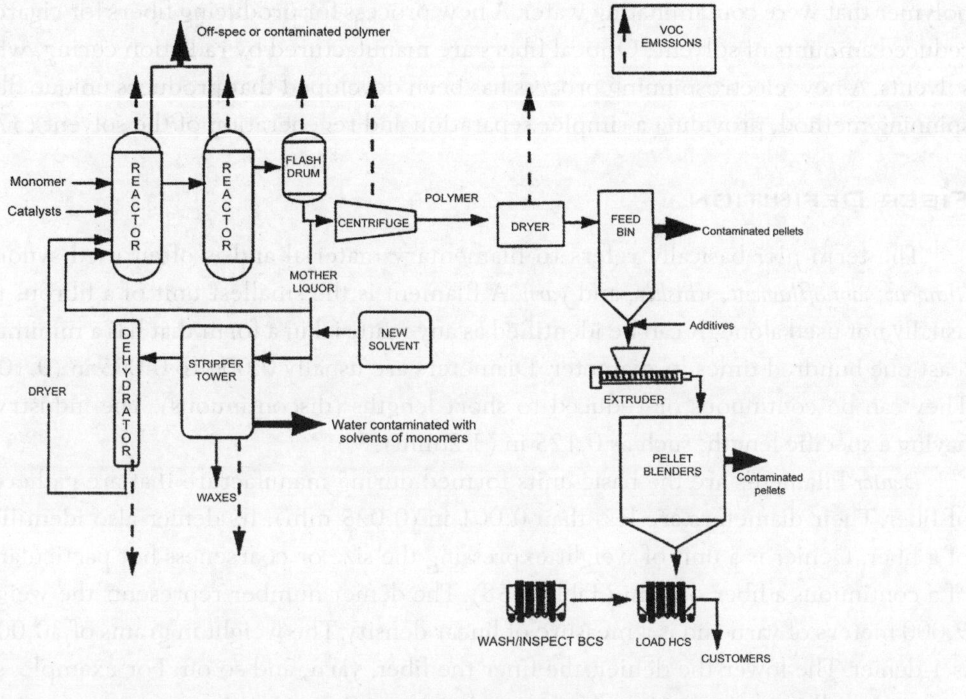

Figure 5.181 Schematic diagram of emissions from the polymer fiber manufacturing industry

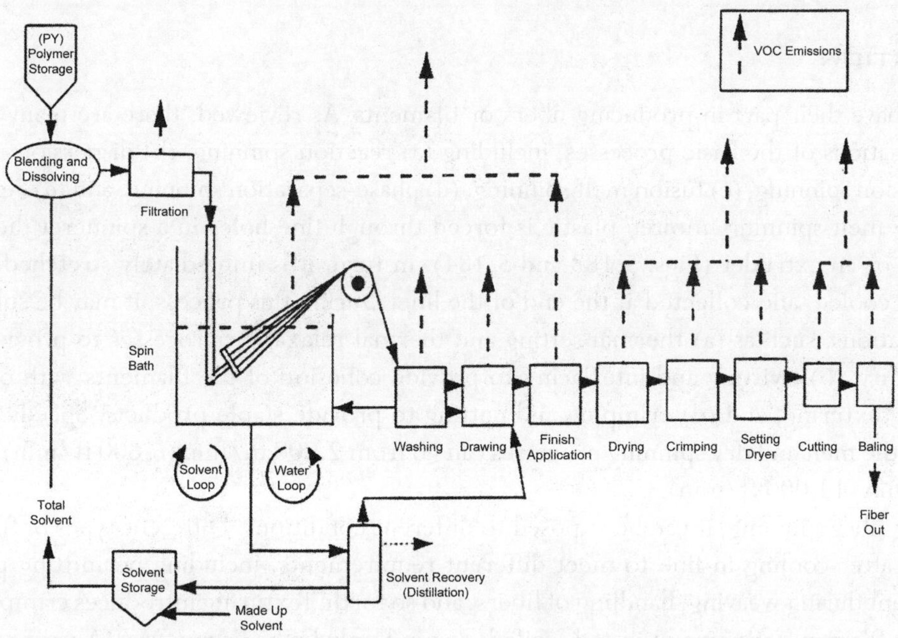

Figure 5.182 Schematic of emissions from the man-made fiber manufacturing industry

polymer that were contaminating water. A new process for producing fibers for cigarette filters uses reduced amounts of solvent. Optical fibers are manufactured by radiation curing, which eliminates solvents. A new electrospinning process has been developed that produces unique fibers by the dry spinning method, providing a simpler separation and regeneration of the solvent (374).

FIBER DEFINITION

The term *fiber* basically refers to filamentary material and is often used synonymously with *filament*, *monofilament*, *whisker*, and *yarn*. A filament is the smallest unit of a fibrous material and is usually not used alone. It can be identified as any material in a form that has a minimum length of at least one hundred times its diameter. Diameters are usually 0.004 to 0.005 in (0.10 to 0.13 mm). They can be continuous or reduced to short lengths (discontinuous). The industry lists fibers as having a specific length, such as 0.125 in (3.2 mm).

Denier Filaments are the basic units formed during manufacture that are gathered into strands of fiber. Their diameters are less than 0.001 in (0.025 mm). Its denier also identifies the fineness of a fiber. Denier is a unit of weight expressing the size or coarseness but particularly the fineness of a continuous a fiber or yam (Table 5.58). The denier number represents the weight in grams of 9,000 meters of yarn and is a measure of linear density. The weight in grams of 30,000 ft (9,000 m) is 1 denier. The lower the denier, the finer the fiber, yarn, and so on. For example, sheer women's hosiery usually runs from 10 to 15 denier. An alternative unit is the tex, representing the weight in grams of 1,000 meters. Yam linear density is sometimes expressed in decitex where 1 dtex = 0.1 tex. Different fiber products, such as staple fibers, will be reviewed.

PRODUCTION

Extruders have their part in producing fibers or filaments. As reviewed, there are many variations and combinations of the basic processes, including (a) reaction spinning, (b) dispersion, emulsion, and suspension spinning, (c) fusion melt spinning, (d) phase-separation spinning, and (e) gel spinning.

During melt spinning, molten plastic is forced through fine holes in a spinneret die by just a gear pump or an extruder (Figs. 5.183 and 5.184). In turn, it is immediately stretched or drawn (oriented), cooled, and collected at the end of the line. During this process, it may be subjected to other operations, such as (a) thermal setting and thermal relaxation processes to provide dimensional stability, (b) twisting and interlacing to provide cohesion of the filaments with or without sizings, (c) texturing, and (d) crimping and cutting to provide staple products. Speeds of certain lines using the melt and dry spinning processes can go from 2,000 m/min (6,600 ft/min) to at least 4,000 m/min (13,000 ft/min).

During fiber spinning, it can be exposed to different conditions. Different types of finishes can be applied after cooling in-line to meet different requirements, including permitting processing improvement during weaving, handling of fibers, and so forth. Texturing introduces crimp, whereby the straight filaments are given twisted, coiled, or randomly kinked structures. A yarn made up of these filaments is softer and more open in structure; it is more pleasing to the touch.

Denier	Number of filaments	Yield (yd/lb)	Yield (m/kg)
55	25	81175	163636
195	90	22895	46155
195	134	22895	46155
380	180	11749	23684
380	267	11749	23648
720	490	6200	12500
750	490	5952	12000
840	534	5314	10714
1000	666	4464	9000
1140	768	3916	7895
1420	1000	3144	6338
1500	1000	2976	6000
2160	1000	2097	4228
2250	1000	1984	4000
2840	1333	1572	3169
3000	1333	1488	3000
4320	2000	1048	2110
4560	3072	979	1974
7100	5000	630	1268
8640	4000	524	1057
10800	5000	413	833
11400		391	789
15000	10000	298	600

Table 5.58 Example of the relationship of denier to filaments and their weights

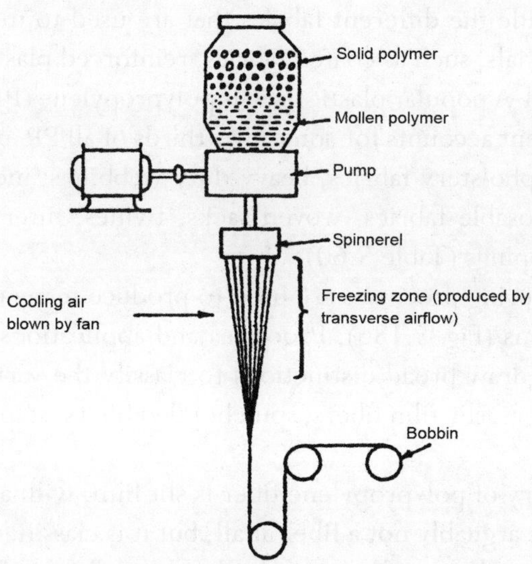

Figure 5.183 Example of using a gear pump to produce fibers

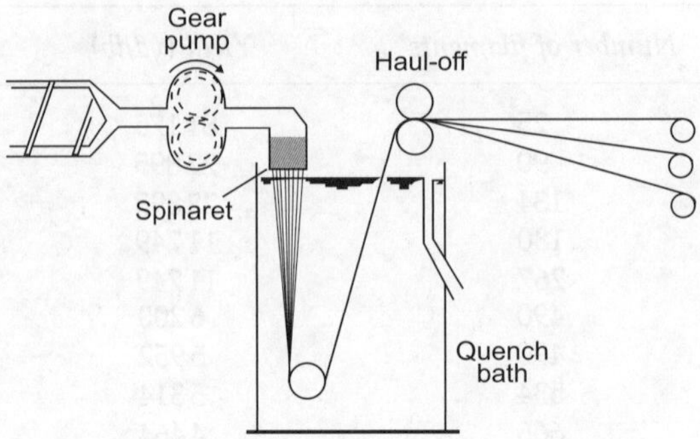

Figure 5.184 Example of using an extruder and gear pump to produce fibers

Fibers are twisted during fabrication. Twist is the number of spiral turns about its axis per unit of length. Twist may be expressed as turns per inch (tpi). The letters S and Z indicate the direction of the twist, in reference to whether the direction conforms to the middle-section slope of the particular letter (Fig. 5.185). A fiber, yarn, or strand has what is known as an S twist if when held in a vertical position, the spirals conform in slope to the central position of the letter S. It has a Z twist if the spirals conform in slope to the central portion of the letter Z. Fibers that are simply twisted (greater than 1 turn/in or 40 turns/m) will kink, corkscrew, and/or unravel because their twists are only in one direction. The plying operation normally eliminates this problem. For example, single yarns having Z twists are plied with S twists, thus resulting in balanced yarn. Depending on the twisting and plying operations, different yarn strengths, diameters, and flexibilities can be obtained. These actions provide the different fabrics that are used to meet different performance requirements of plastic materials, such as coated fabrics, reinforced plastics, and pultrusions.

Different plastics are used. A popular plastic used is polypropylene (PP) because of its properties (Table 5.59). Fiber and filament accounts for some two-thirds of all PP extruded. The principal end products are carpet yarns, upholstery fabrics, heavy-duty webbings, industrial fabrics, nonwoven upholstery, geotextiles, disposable fabrics, woven sacks, twines, filter fabrics, wearing apparel, ropes, tapes, cords, and strappings (Table 5.60).

Wide arrays of manufacturing processes are used to produce many different fiber types aimed at specific end-use applications (Fig. 5.186). Processes and applications overlap to an extent, but it is nevertheless possible to draw broad distinctions to classify the various options. They include filaments, spinneret-spun fibers, slit-film fibers, spun bonded fibers, staples, nonwoven staples, and melt-blown fibers.

The largest single category of polypropylene fiber is slit film, with a 30% share of the total. In its nonfibrillated form, this is arguably not a fiber at all, but it is classified as such because it is used entirely in textile applications. Filament fibers (multifilament and monofilament) account for about 22% of polypropylene fiber production. Air-attenuated fibers constitute about 20% of production.

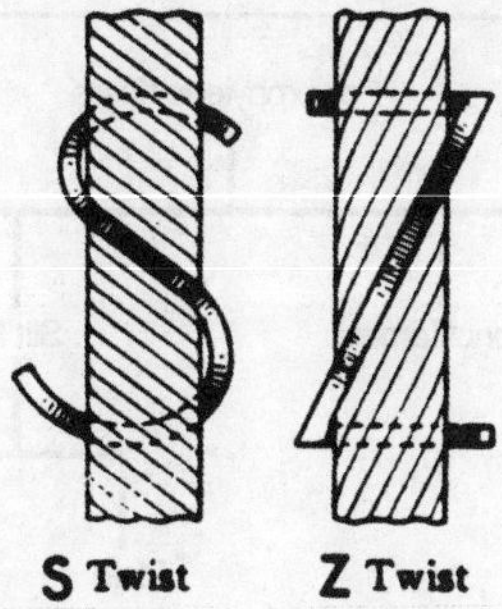

Figure 5.185 Views of the S and Z strand twists for fibers, yarns, and other textiles

Property	Application
Low thermal conductivity	Insulating material
Rot resistance	Civil engineering fabric
Hydrophobicity	Diaper or sanitary napkin cover stock
Chemical resistance	Blood filters
Stain resistance	Upholstery
Resistance to insects	Agricultural bags
Low static build-up	Carpet face fiber

Table 5.59 Useful properties of polypropylene in fiber applications

Properties	Applications
Good skin tolerance Excellent color fastness Light weight Good insulation capacity Soft textile handle Low electrostatic discharge Hydrophobic	Floorcoverings Sports and leisure wear Belts Ropes and cords Furniture upholstery Woven screening

Table 5.60 Properties and applications for multifilament polypropylene yarn

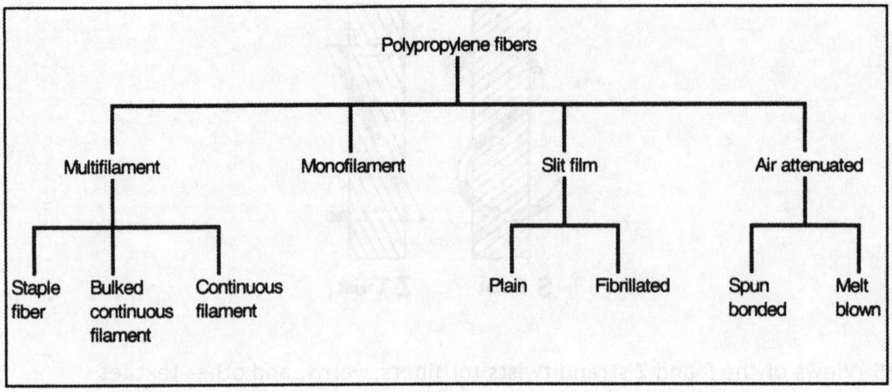

Figure 5.186 Relationship between polypropylene fiber processes

MULTIFILAMENT

The multifilament yarns are composed of bundled individual filaments of 75 denier or less. All multifilament fiber extrusion processes involve the basic steps of spinning, quenching, and stretching, with the subsequent treatments that vary. Chopping the multifilaments into short lengths produces staple fibers. Bulked continuous filament (BCF) and continuous filament yams are not chopped. The BCF is given a treatment to impart a 3-D bulk. Multifilament fiber is usually produced from polypropylene homopolymer with a high melt flow index from 12.0 to 38.0 grams/10 min.

CONTINUOUS FILAMENT

The multifilament extrusion process uses an extruder to prepare a plastic melt stream for extrusion through a special type of die known as a spinneret. This involves passing the melt through a very large number of very small holes. This action gives the process a number of distinctive features. The process demands a low melt viscosity. High melt flow index grades are used at relatively high melt temperatures in the range 230°C to 260°C. Because the holes in the spinneret are very small, an exceptionally high degree of melt filtration is necessary to screen out foreign particles that would otherwise clog the spinneret or cause the filament to fracture in the stretching operation.

Unlike other forms of extrusion in which the die is mounted on an adapter at the extruder outlet, the fiber-forming spinneret is mounted at some distance from the extruder and is connected to it by a heated-melt manifold system. This arrangement makes it possible to build down-working (Fig. 5.187) or up-working melt spinning systems. Pressure losses in the melt distribution system are the reason for another distinctive feature that uses melt pumps to develop the necessary pressure flow rate for spinning.

From the extruder, the melt stream passes through a filter system operating typically at the 40 μm level. The filtered stream then proceeds to the spinning beam, consisting of a heated- melt

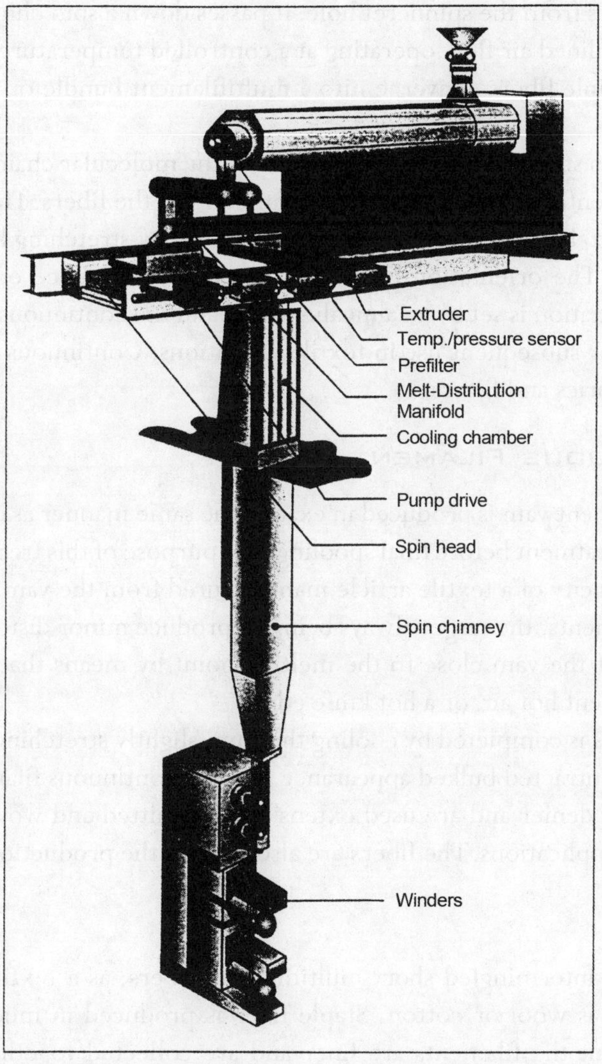

Figure 5.187 Example of a multifilament melt spinning system

distribution manifold feeding a number of spinning packs via melt-metering gear pumps. Each spinning pack is effectively a separate extrusion head, consisting of a final filter, ahead of a streamline flow chamber housing the spinneret. The spinneret is a rectangular plate die, typically provided with 200 or more holes arranged in a grid formation.

The spinneret holes may be simple cylindrical bores but may also be profiled to create fibers with improved wicking properties. Common profiles include trilobal, quadrilobal, delta, and dogbone shapes. The land length of the spinneret holes is typically five to fifteen times the holes' diameters.

As the melt emerges from the spinneret hole, it passes down a spin chimney where it is cooled or quenched in a streamlined air flow operating at a controlled temperature. Toward the exit of the spin chimney, the multiple fibers converge into a multifilament bundle or yam and are hauled off by godet rolls.

The final step is yarn stretching or drawing to orient the molecular chains to high degrees in the machine direction. Orientation also reduces the diameters of the fibers. The process is performed at a temperature close to but less than the melt temperature by stretching between rolls operating at a speed differential. The orientation process has a critical influence on the properties of the finished fiber. The orientation is set by an annealing step, and the continuous filament yam is wound on spools or bobbins for subsequent use in textile operations. Continuous filament yams are used for woven industrial fabrics and filters.

Bulked Continuous Filament

Bulked continuous filament yam is produced in exactly the same manner as continuous filament but is given an additional treatment before final spooling. The purpose of this treatment is to improve the handling, feel, and elasticity of a textile article manufactured from the yam. There are a number of different bulking treatments, the target always being to produce minor distortions to the filaments. The principle is to heat the yam close to the melting point by means that perturb the filaments, using steam jets, turbulent hot air, or a hot knife edge.

The bulking process is completed by cooling the yam, slightly stretching it, and then relaxing it to produce a slightly contracted bulked appearance. Bulked continuous filament yams are typically produced from 3 to 30 denier and are used extensively in knitted and woven fabrics for clothing, upholstery, and other applications. The fibers are also used in the production of tufted carpets.

Staple Fiber

Staple fiber consists of intermingled short multifilament fibers; as a textile concept, it is similar to natural fibers, such as wool or cotton. Staple fiber is produced in much the same manner as continuous filament, but the filaments are finer and are collected together in large numbers in loose ropes or tows. The spinneret contains a very large number of holes, such as 10,000 or more and sometimes ranging as high as 50,000. The tow is passed through a crimper that imparts a small zigzag configuration to the filaments and is then chopped into short lengths from 5 to 60 mm. The chopped fibers are baled in random orientations for use in any textile process conducted with staple fiber. The denier range is typically 1 to 10. Uses include clothing and other fine fabrics.

Monofilament

As the name implies, a monofilament yam consists of a single filament. The filament size is much larger than those found in multifilament yarn. Consequently, monofilament is relatively stiff and is used mainly for the production of rope and twine. Table 5.61 provides the breaking strengths of ropes made with different plastics; the safe load ratio is the ratio of breaking strength to safe loading condition that takes into account a safety factor (chapter 19). The fiber size range is typically 75 to 5,000 denier. Monofilament fiber is usually produced from polypropylene homopolymer with a relatively low melt flow indexes in the range of of 3.5 to 5.0 grams/10 min.

The monofilament process (Fig. 5.188) includes aspects of both the melt spinning and slit-film processes. The extruder is equipped with a melt pump immediately upstream of a spinneret that extrudes relatively thick filaments ranging from 0.08 mm to 3.0 mm in diameter. The filaments are quenched in a water bath and are hauled off by godet rolls. Orientation is performed by reheating and stretching in an oven under tension applied by a second haul-off. An alternative orientation process uses heated godet rolls in place of the oven.

SLIT FILM

Slit-film fiber is produced by slitting an extruded cast film into narrow tapes that are then uniaxially oriented in the machine direction. The film may be cast by the chill-roll or water quench processes

Diameter Inches	Polypropylene Break Lbs	Lbs/100 feet	Nylon Break Lbs	Lbs/100 feet	Manila Break Lbs	Lbs/100 feet	Safe Load Ratio
3/16	800	0.7	1000	1.0	406	1.5	10:1
1/4	1250	1.2	1650	1.5	540	2.0	10:1
5/16	1900	1.8	2550	2.5	900	2.9	10:1
3/8	2700	2.8	3700	3.5	1220	4.1	10:1
7/16	3500	3.8	5000	5.0	1580	5.3	10:1
1/2	4200	4.7	6400	6.5	2380	7.5	9:1
9/16	5100	6.1	8000	8.3	3100	10.4	8:1
5/8	6200	7.5	10400	10.5	3960	13.3	8:1
3/4	8500	10.7	14200	14.5	4860	16.7	7:1
13/16	9900	12.7	17000	17.0	5850	19.5	7:1
7/8	11500	15.0	20000	20.0	6950	22.4	7:1
1	14000	18.0	25000	26.4	8100	27.0	7:1
1-1/16	16000	20.4	28800	29.0	9450	31.2	7:1
1-1/8	18300	23.8	33000	34.0	10800	36.0	7:1
1-1/4	21000	27.0	37500	40.0	12200	41.6	7:1
1-5/16	23500	30.4	43000	45.0	13500	47.8	7:1
1-1/2	29700	38.4	53000	55.0	16700	60.0	7:1
1-5/8	36000	47.6	65000	66.5	20200	74.5	7:1
1-3/4	43000	59.0	78000	83.0	23800	89.5	7:1
2	52000	69.0	92000	95.0	28000	108	7:1
2-1/8	61000	80.0	106000	109	...	...	7:1
2-1/4	69000	92.0	125000	129	...	...	6:1
2-1/2	80000	107	140000	149	...	...	6:1
2-5/8	90000	120	162000	168	...	...	6:1
2-7/8	101000	137	180000	189	...	...	6:1
3	114000	153	200000	210	...	...	6:1

Table 5.61 Different plastics used to produce rope

(Fig. 5.189). The slit-film fiber (plain tape) may be used as it is in weaving processes or it may be given a subsequent fibrillation treatment to impart a fibrous character.

Plain Tape

The cooled cast-film web is slit into narrow, multiple parallel tapes at a tension point immediately upstream of the first set of godet rolls that anchor the tapes ahead of the orientation oven. A second set of godet rolls applies a stretching tension to the tapes that are heated to a temperature just below the melting point. Draw ratios typically range from 1:5 to 1:8. Heated rolls may be used as an alternative to the draw oven. In the final step in this operation, the tapes are annealed to set the orientation and are wound on spools for subsequent use in textile operations.

Slit-film fiber is used in woven form mainly for heavy-duty applications, including carpet backing, coating substrates, geotextiles, and industrial sacks and bags.

Fibrillated Tape

Fibrillated tape fiber is produced in the same manner as slit-film fiber, except that a great draw ratio of up to 1:10 is used and a fibrillation step is added to the process. The slit-film tapes are fibrillated by passing them over a rapidly rotating roll fitted with staggered rows of pins. As the pins are traveling faster than the tapes, they make a series of short discrete cuts in the longitudinal direction of the tapes. Fibrillated tape fiber is used for producing packaging and baling twines and in the manufacture of artificial grass.

Air-Attenuated

An air-attenuated fiber process uses an airstream to draw and orient very fine fibers that are then deposited as a randomly oriented mat to form a nonwoven fabric. Uses include diapers and sanitary items, disposable clothing, industrial and domestic wipes, tablecloths, battery separators, carpet backing, and geotextiles. The two principal variants of the air-attenuation method are the spun-bonded and melt-blown processes. These processes require polypropylene grades with high melt flow indexes of 24.0 to 38.0 and narrow molecular weight distributions.

Spun-Bonded

In the spun-bonded fiber process, an extruder and spinneret are used to produce filaments in the same manner as the continuous multifilament process. The filaments are initially quenched in air and then passed to an air venturi aspirator where they are exposed to a high-velocity stream of heated air that stretches and orients the filaments. Drawdown takes place at very high speeds up to 5,000 m/min (Fig. 5.190).

At the outlet of the aspirator, the fibers are randomly blown onto an endless mesh belt where, with suction assistance, they form a mat or web. The web is subsequently stabilized by a variety of means, including hot-air bonding, adhesive bonding, stitching, or needle punching. The web

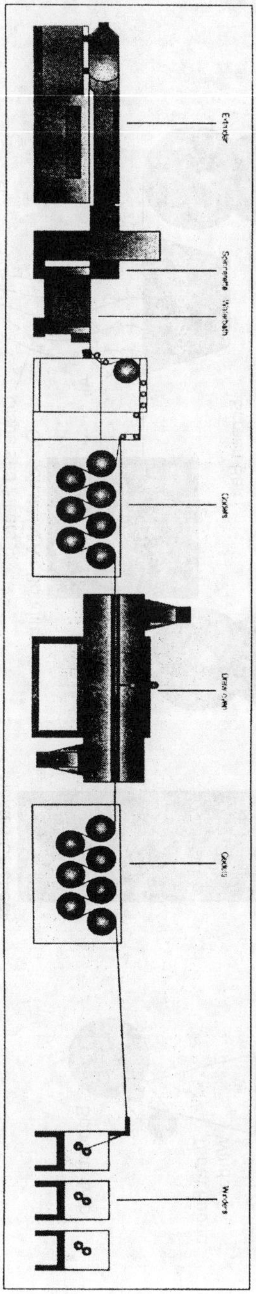

Figure 5.188 Example of a monofilament extrusion yarn line

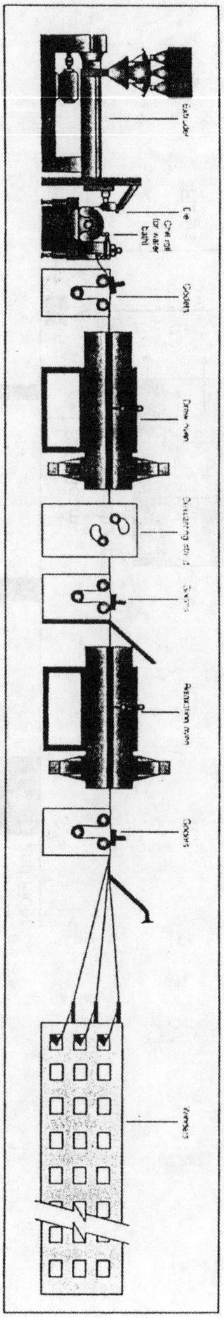

Figure 5.189 Example of a slit-film tape line

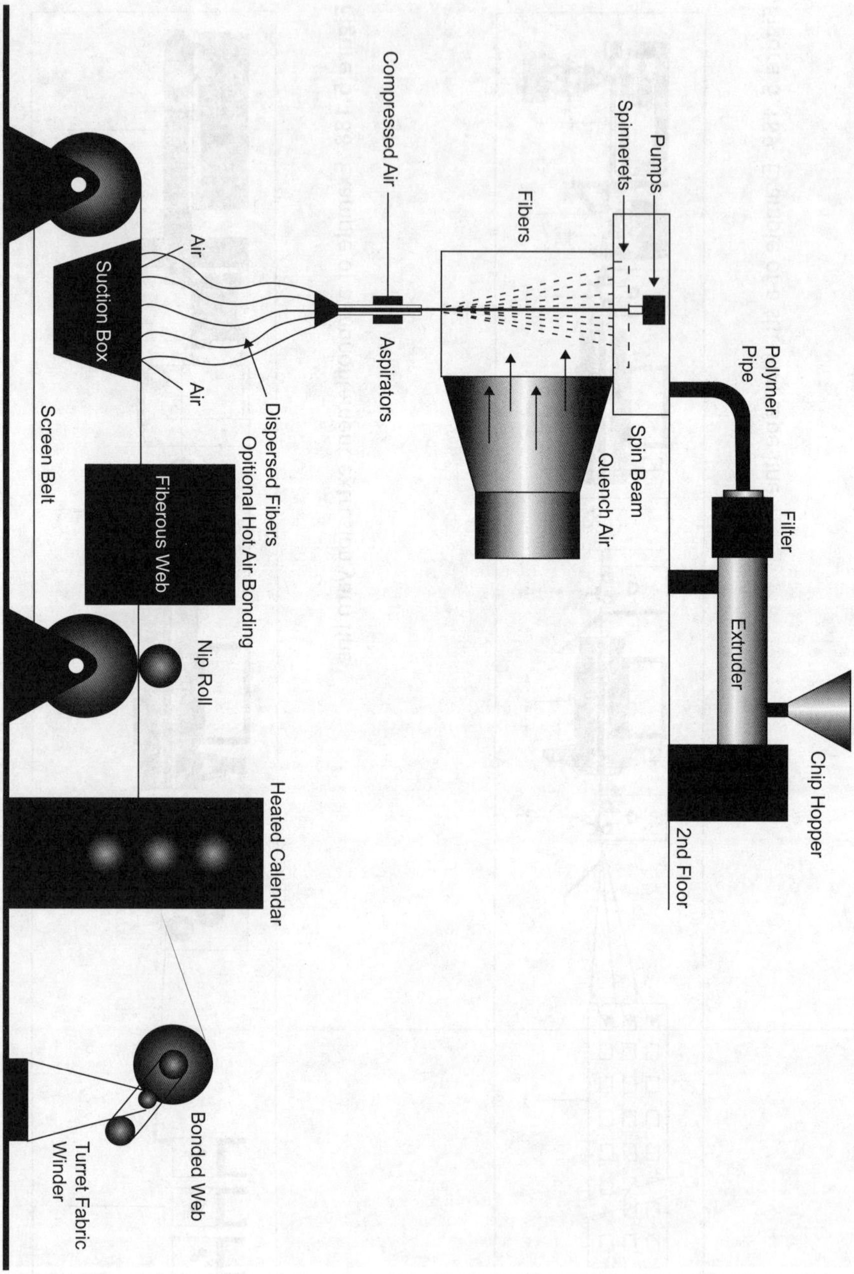

Figure 5.190 Example of spun-bonded fiber extrusion line

may be calendered to provide a smooth or textured surface and may be oriented by tentering. The constituent fibers of a spun-bonded web may be as fine as 3 dtex.

MELT-BLOWN

The melt-blown process produces a nonwoven web in a single operation by omitting the initial air quench step. As they leave the spinneret, the melt streams are entrained by high-speed, obliquely aimed hot-air jets that produce very fine filaments of as little as 0.1 dtex. Web formation and stabilization are performed by the same means used for spun-bonded webs. The process is sometimes combined with a spun-bonded line to produce a nonwoven fabric with different characteristics on either face.

COEXTRUSION

Throughout this chapter, coextruded, also called multilayer, products have been reviewed. This section provides additional information. This process forms an extrudate composed of more than one thermoplastic melt stream. The process came about over a half century ago because some service demands, particularly from the packaging industry, could not be satisfied by a single plastic although they could be met by a combination of plastics. Coextrusion was first practiced in the production of cast film followed with blown-film, sheet, and other extruded shapes. The intention is normally to produce a laminar structure in which each layer contributes a key property to the overall product performance. The different products reviewed in this chapter can be extruded using two or more plasticators (Figs. 5.191 to 5.196).

Coextruded films may appear to be very complex structures because they are composed of many different functional layers, which may include tie layers with the purpose of bonding neighboring layers of limited compatibilities. Seven or more layers are not uncommon. For those involved with this type of expensive equipment, it can be said, "It's a piece of cake."

Coextrusion can provide multiple molten layers using only one plastic (one plasticator) that provides two separate layers through its die. This type of structure can provide improved performance when compared to a single layer, particularly in thick parts.

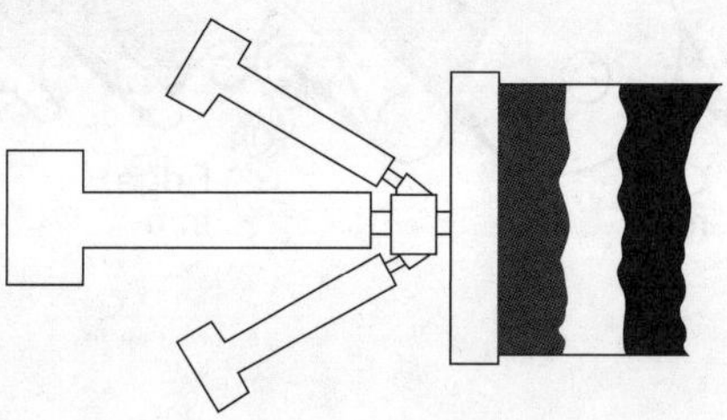

Figure 5.191 Schematic of a basic three-layered coextrusion system

These different multilayer products meet diversified applications at relatively low cost. When compared to other products, they provide increased versatility, productivity, and quality. Separate extruders are required for each distinct plastic in the coextrusion. Practically all coextruded products have the layers in orderly forms on top of each other (parallel). However, the layout can be different, such as shown in Figure 5.197. By sectionalizing, advantages can be gained such as in thermoforming (chapter 7). Single recycled plastic on the outside can be used just for clamping.

DIE

The process has two variants, depending on the point at which the separate melt streams are brought together. In feedblock coextrusion, the streams are merged into a single laminar melt flow in a feedblock that is positioned immediately upstream of the extrusion die. Figure 5.198 shows the melt flow patterns for three- and five-layer feedblocks. The process depends on the high viscosity of plastic melts to prevent intermingling of the layers as they pass through the extrusion die. The flow rate of each component layer can be controlled by valves in the feedblock, and the capital cost is relatively low.

The alternative to the feedblock die is a more complex die construction in which separate melt path manifolds are arranged to merge at a point close to the die exit. Figure 5.199 shows examples of a three-layered multimanifold die. The thickness and flow rate of individual layers can be independently controlled, and it is possible to handle plastics with substantially differing viscosities and melt temperatures. The capital and maintenance costs of such multimanifold coextrusion dies are high (Table 5.62).

Figure 5.200 provides a few examples of different types of dies to meet different requirements (such as encasing or enveloping plastic) and having equal amounts of plastics. Factors to consider

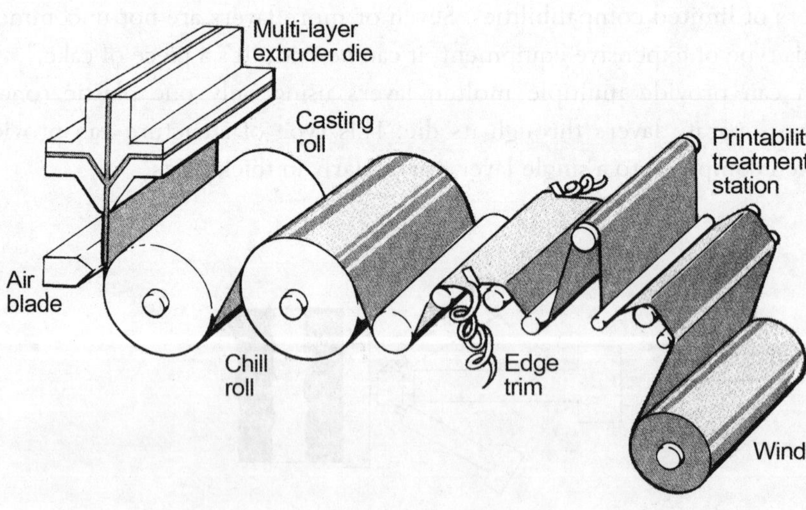

Figure 5.192 Schematic of a three-layered cast film coextrusion system

Figure 5.193 View of two of seven plasticators feeding a coextruded film line (courtesy of Davis Standard)

Figure 5.194 View of three-layer coextrusion sheet line (courtesy of Welex Inc.)

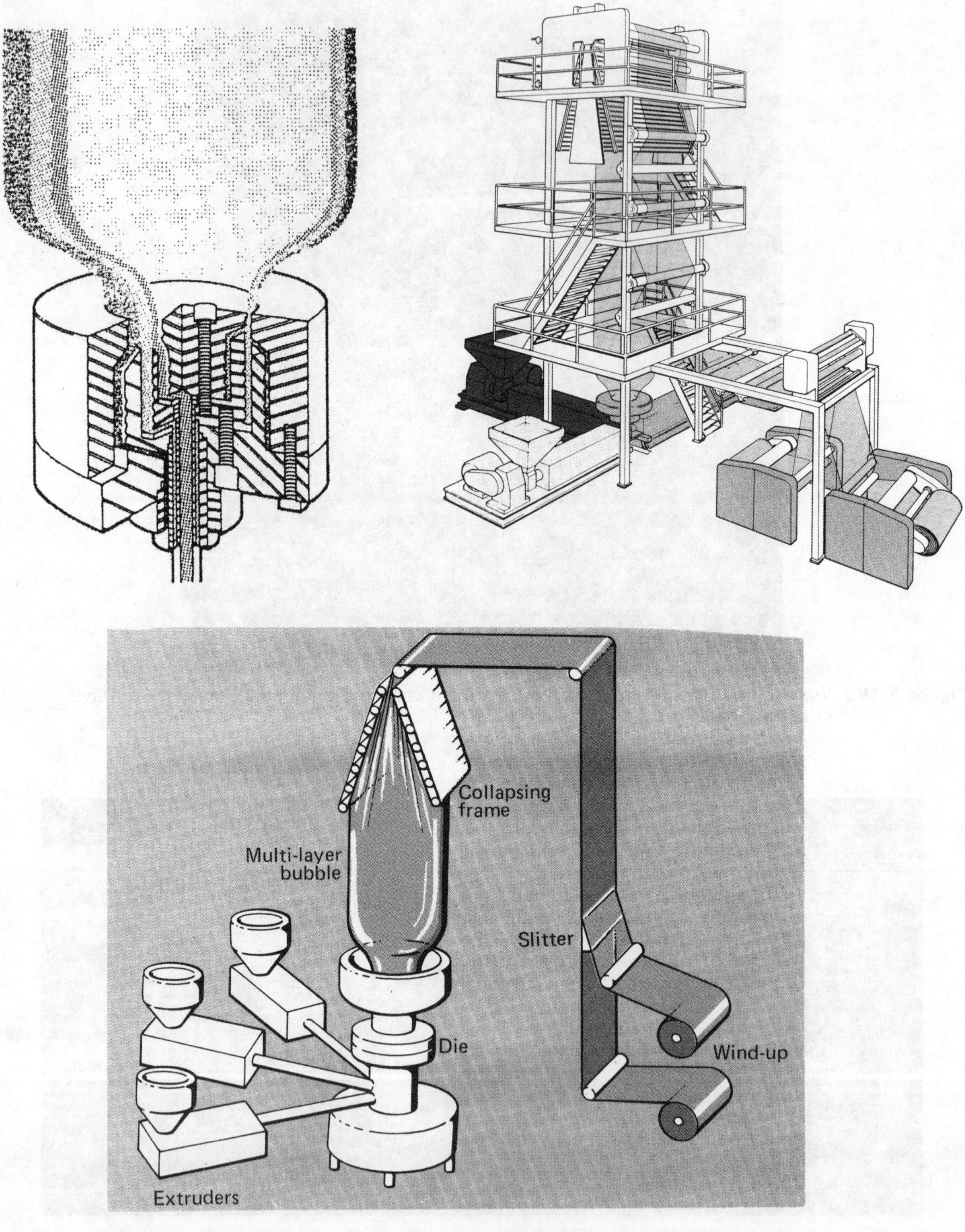

Figure 5.195 Example of coextrusion three-layered blown-film die and lines

when designing dies include orifice shape, land length, and restricter bar. Information on designing and manufacturing dies is provided in chapter 17. It includes an over 115-layer coextrusion of two plastics (Fig. 17.127), etc. This multilayer construction with its refractive indexes creates multilike reflection similar to the interference phenomenon in pearlescent pigments (Fig. 5.201).

PLASTIC

Coextrusion is an economical competitor to conventional laminating processes by virtue of its reduced materials-handling costs, raw materials costs, and machine-time cost. Pinholing is also reduced with coextrusion, even when it uses one extruder and divides the melt into at least a two-

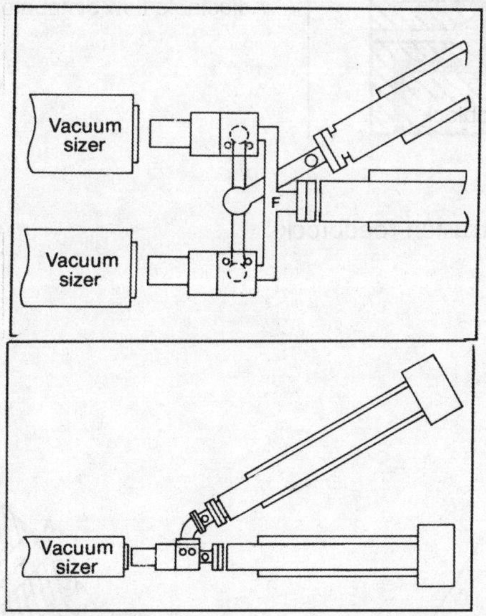

Figure 5.196 Examples of two-layered single- and dual-pipe coextrusion systems

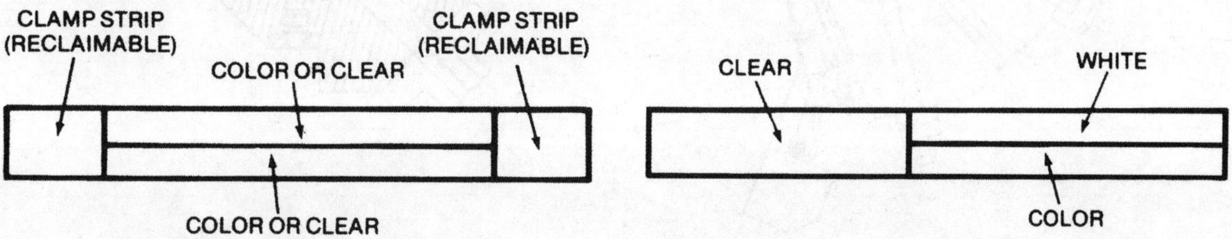

Figure 5.197 Nonconventional coextruded construction (courtesy of Welex Inc.)

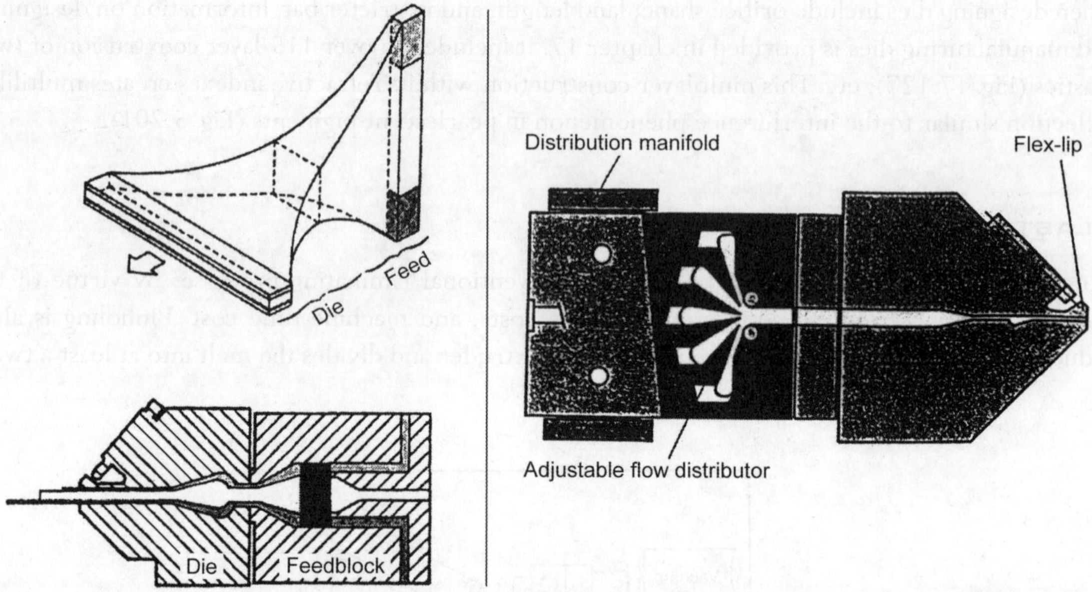

Figure 5.198 Examples of coextrusion feedblocks

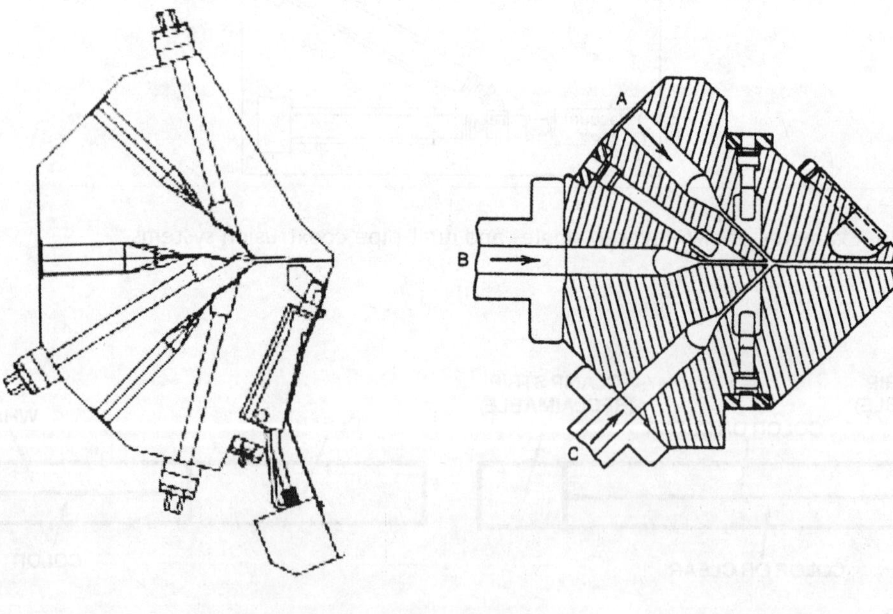

Figure 5.199 Examples of multimanifold coextrusion dies

layer structure. Other gains include elimination or reduction of delamination and air entrapment. It provides an excellent way to integrate/entrap recycled contaminated plastic with one side or both sides using virgin plastic. It combines the different properties of the different plastics, resulting in performance and cost advantages.

The different materials used in the coextruded structure meet different performance requirements, such as strength, stiffness, toughness, noise absorption, resistance of permeation (air, oxygen, moisture, gas, odor, etc.), adhesive qualities, and economics (Table 5.63). A single expensive plastic could be used to meet performance requirements; however, it is possible to provide combinations of plastics to reduce the cost. As an example of its value in the marketplace, coextrusion has been adapted to the production of products like building profiles, pipes, and packaging films that incorporate in a single extrusion structure several layers of different plastics, each offering varying degrees of different performance requirements.

Compatible plastics can flow through a single manifold, reducing any potential problems downstream in the multimanifold. Combinations can be made to provide different laminated designs. However, the final exiting layer thickness distributions can be affected by the amount of die body defection if the die is not properly designed to take the required pressure loads. Any deflection causing distortion influences the melt flow channels.

There are many types of coextruded structures that range from two to at least seven layers. Common structures include PP (polypropylene)/filled PP/PP, which is used for many food trays and containers, oriented PP (OPP) products generally produced in copolymer/homopolymer/

Characteristic	Feedblock	Multimanifold
Basic difference	Melt streams brought together outside die body (between extruder and die) and flow through the die as a composite	Each melt stream has a separate manifold; each polymer spreads independent of others; they meet at die preland to die exit
Cost	Lower	Higher
Operation	Simplest	
Number of layers	Not restricted; seven- and eight-layer systems are commercial	Generally restricted to three or four layers
Complexity	Simpler construction; no adjustments	More complex
Control flow	Contains adjustable matching inserts, no restrictor bar	Has restrictor bar or flow dividers in each polymer channel; but with blown-film dies, control is by individual extruder speed or gearboxes
Layer uniformity	Individual layer thickness correction of ±10%	Restrictors and manifold can meet ± 5%
Thin skins	Better on dies >1 m (40 in)	Better on dies <1 m (40 in)
Viscosity range	Usually limited to 2/1 or 3/1 viscosity range of materials	Range usually much greater than 3/1

Table 5.62 Performances of coextrusion feedblocks and multimanifold dies

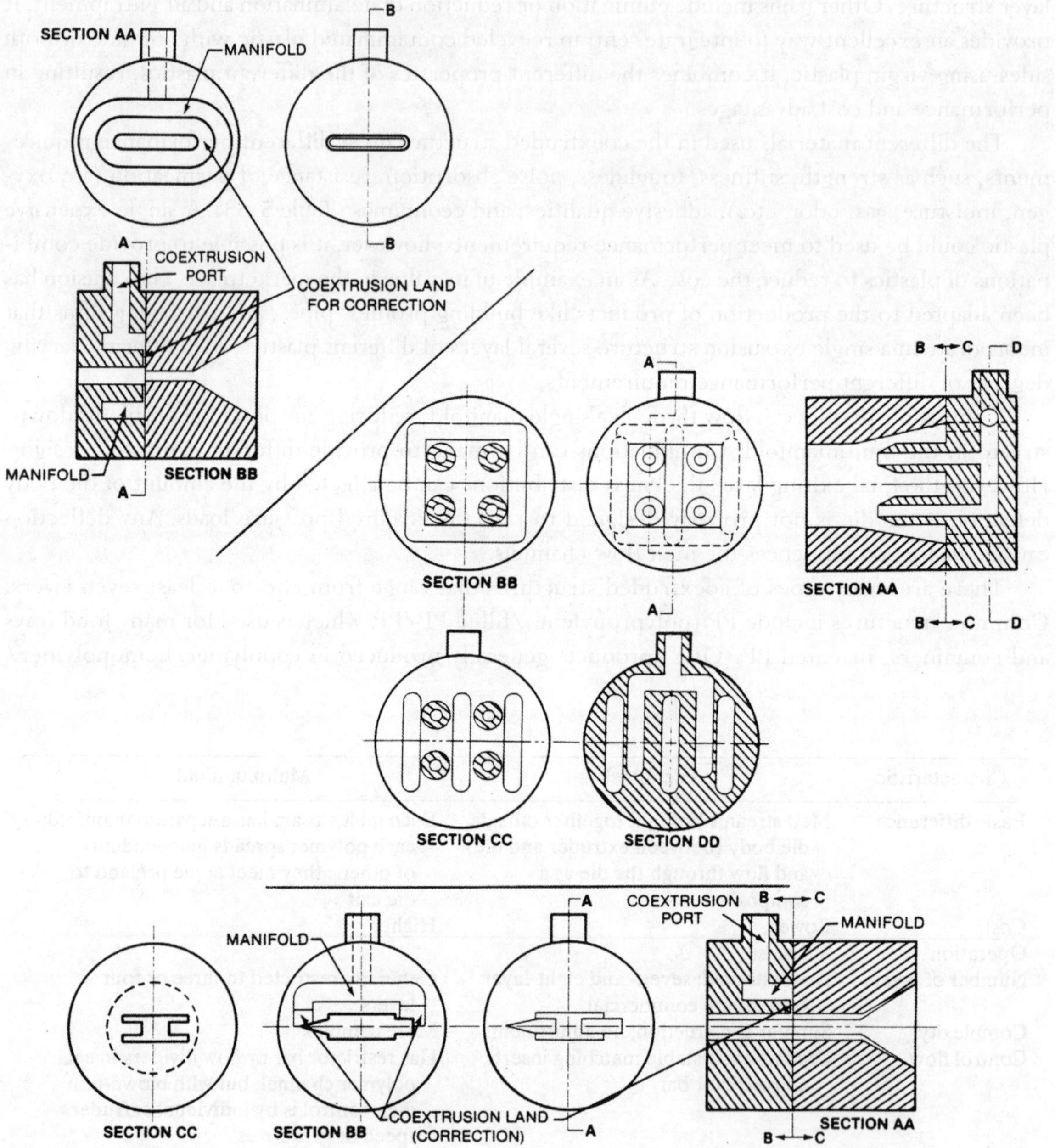

Figure 5.200 Examples of coextruded dies

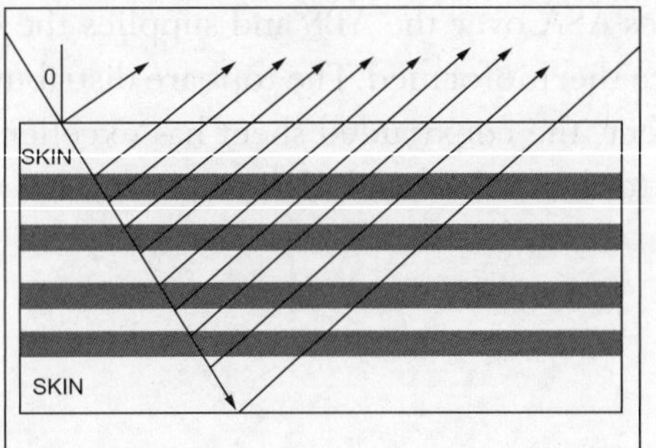

Figure 5.201 Coextrusion of at least 115 plastic layers produces light reflection similar to pearlescent pigments

copolymer configurations, PP/adhesive/HIPS often used for fruit-juice containers, and PP-virgin/PP reclaim/adhesive/EVOH/adhesive/PP-reclaim/PP-virgin for rigid food containers. PP normally bonds to other olefins without a tie layer. There are certain plastics that do not bond. A tie layer is generally required with these plastics.

Choosing an adhesive layer is by no means a simple operation. There are many different types, each with specific capabilities (Tables 5.64 and 5.65). EVAs form the bulk being used. Proper choice can improve performance, such as increasing melt strength and bubble stability in blown film. High melt strength can also help in cast film used in thermoforming or coating. Good melt draw is required to run higher take-up speeds and/or thinner structures without causing flow distribution or edge weave problems.

Melt flow instabilities, such as interfacial instability, melt fracture, surging, and/or layer nonuniformity, can cause problems that can lead to quality issues with the coextruded product. There are several options to correct these problems. The key to success is to select the best option or combination of options to eliminate the problem and minimize other possible adverse effects. Flow instabilities can cause limited production rates. Appropriate operating conditions of the coextrusion process, combined with the proper plastic selections, can successfully be used to solve the problems.

APPLICATION

Many applications exist and are presented (Table 5.66). Here is one interesting application. For the tops of their aftermarket golf carts, Primex Plastics Corp. (Oakwood, Georgia) and Piper Plastics (Orlando, Florida) sought materials that would provide sturdy construction coupled with attractive appearance. Bayer Plastics recommended coextruded sheets of its Centrex ASA (acrylonitrile-styrene-acrylate) and Lustran ABS (acrylonitrile-butadiene-styrene) plastics, based on the materials' outdoor performances. The ABS offered high rigidity and better aesthetics than other resins that were considered. Coextruding over an ASA capping layer provided high weatherability and a high-gloss finish that resisted fading, cracking, and yellowing. The coextruded plastics provided UV protection, which was crucial to their suitability for this application.

Primex Plastics coextrudes ASA over the ABS and supplies the 0.188-in-thick sheets to Piper Plastics, where the cart tops are thermoformed. The tops are distributed in 48-, 56-, and 80-in sizes. From the thermoforming aspect, the coextruded sheet has excellent processing characteristics. It heats well and conforms fully to the mold detail (281).

ORIENTATION

INTRODUCTION

Orientation consists of a controlled system of stretching thermoplastic molecules in unioriented (unidirectional [UD]) or bioriented (biaxial direction [BD]) orientation to improve their strength,

Plastic	Product characteristics and applications
PMMA / PMMA	Ultra-high-transparency sheets with various tints, e.g., for use as glazing.
PMMA/ABS	Sheets with scratch-resistant surface for sanitary applications, e.g., sanitary blocks and baths.
PMMA/ABS with MB/ABS-scrap	Colored three-layer sheets containing a high proportion of recycled material (70–80%), for sanitary applications.
ABS/ABS-scrap/ABS	Three-layer sheets containing a high proportion of recycled material (70–80%), e.g., for the production of cases. Characteristics of the ABS layer are its resistance to Freon and its gloss finish. These sheets are particularly suitable for refrigerators/freezers.
PS/HIPS with MB/HIPS-scrap	Three-layer sheets incorporating regrind from production scrap (65–80%); particularly suitable for refrigerators/freezers.
PC(UV)/PC/PC(UV)	Transparent or colored sheets for building applications, suitable for use as safety glazing and billboards. Characteristics of these polycarbonate sheets are their high impact resistance and an integral UV-resistant layer.

Property	Metalized coextruded polyolefin	Foil laminate*
Tensile strength MD	18–19	18–19
CD	12–13	11
Mullen strength	19–20	17
Gurley stiffness MD	70–75	117–112
CD	42–47	72–77
WVTR (gm/csl/24 hrs.)	Approx. 0.05	0.0006
Oxygen transmission (cc/csl/24 hrs.)	Approx. 10	Less than 0.004
Light transmission	Slightly less than 1%	Approx. 0%
Seal type	Fin only	Fin or Lap
Seal range (40 psi, 15 sec.)	350–500°F	160–350°F

Table 5.63 Examples of the performances of coextruded materials

stiffness, optics, electrical, and/or other properties with the usual result that improved product performance-to-cost occurs (Tables 5.67 to 5.71). This technique is used during the processing of many different products, such as films, sheets, pipes, fibers, and tapes.

Depending on the properties of a specific plastic, the stretch ratio may vary from 2½:1 to as high as 10:1. Some specialty films may have an even higher stretch ratio. Used for almost a century, orientation became prominent during the 1930s for stretching fibers up to ten times. Later,

Plastic materials	*Rigid vinyls*	*Flexible vinyl*	*ABS*	*Polyethylene*	*Polypropylene*	*EVA*
Rigid vinyls	Yes	Yes	Yes	No	No	No
Flexible vinyls	Yes	Yes	No	No	No	No
ABS	Yes	No	Yes	No	No	No
Polyethylene	No	No	No	Yes	Yes	Yes
Polypropylene	No	No	No	Yes	Yes	No
EVA	No	No	No	Yes	No	Yes

Table 5.64 Information on plastics' compatibilities

	Material	LDPE	HDPE	PP	Low-Impact PS	Medium-Impact PS	High-Impact PS	Crystal PS	ABS[b]	ABS	Rigid PVC	Flexible PVC	Polycarbonate	Polyurethane	Acrylic	Nitriles	Nylon 6	EVA	Ionomers	SBS[c]
		OLEFINS			STYRENES						PVC's		MISC.					ADHESIVES		
OLEFINS	LDPE	G	G	G	P	P	P	P	P	P	P	P	P	P	P	P	P	G	G	G
	HDPE	G	G	G	P	P	P	P	P	P	P	P	P	P	P	P	P	G	G	G
	PP	G	G	G	P	P	P	P	P	P	P	P	P	P	P	P	P	G	P	G
STYRENES	Low-Impact PS	P	P	P	G	G	G	G	G	P	P	P	P	F	P	P	P	G	?	G
	Medium-Impact PS	P	P	P	G	G	G	G	G	P	P	P	?	F	P	P	P	G	?	G
	High-Impact PS	P	P	P	G	G	G	G	G	P	P	P	?	F	P	P	P	G	?	G
	Crystal PS	P	P	P	G	G	G	G	G	P	P	P	?	F	P	P	P	G	?	G
	ABS[b]	P	P	P	G	G	G	G	G	G	G	G	?	F	G	P	P	G	?	G
	ABS	P	P	P	P	P	P	P	G	G	G	G	?	F	G	P	P	G	?	G
PVC's	Rigid PVC	P	P	P	P	P	P	P	G	G	G	G	U	G	G	P	U	G	?	?
	Flexible PVC	P	P	P	P	P	P	P	G	G	G	G	U	G	G	P	U	G	?	?
MISC.	Polycarbonate	P	P	P	P	?	?	?	?	?	U	U	G	?	G	P	?	?	?	?
	Polyurethane	P	P	P	F	F	F	F	F	F	G	G	?	G	?	P	?	?	?	F
	Acrylic	P	P	P	P	P	P	P	G	G	G	G	G	?	G	P	?	?	?	?
	Nitriles	P	P	P	P	P	P	P	P	P	P	P	P	P	P	G	P	F	?	F
	Nylon 6	P	P	P	P	P	P	P	P	P	U	U	?	?	?	P	G	P	G	P
ADHESIVES	EVA	G	G	G	G	G	G	G	G	G	G	G	?	?	?	F	P	G	?	?
	Ionomers	G	G	P	?	?	?	?	?	?	?	?	?	?	?	?	G	?	G	?
	SBS[c]	G	G	G	G	G	G	G	G	G	?	?	?	F	?	F	P	?	?	G

[a]G = good, F = fair, P = poor, ? = unknown, U = undesirable. [b]Less than 20% acrylonitrile content. [c]Styrene butadiene styrene.

Table 5.65 More information on plastics' compatibilities

it was adapted principally to films and other products, such as stretched blow-molded bottles. Practically all TPs can undergo orientation, although certain types find it particularly advantageous (PETs, PPs, PVCs, PEs, PSs, PVDCs, PVAs, PCs, etc.). Many different markets use oriented plastic

Applications	Properties	Combination of Layers					
		1 Inside	2	3 Middle	4	5	6 Outside
Detergents Cosmetics	Surface gloss	HDPE	—	—	—	—	LDPE
Motor oil	Clear-stripe	HDPE LDPE Stripe	—	—	—	—	—
Insecticides Pesticides Chemicals	Resistant to aggressive substances	PA	—	Bonding agent	—	—	PE Filler Regrind Color
Milk	UV-protection	HDPE White	—	HDPE Black	—	—	HDPE White
Cosmetics Hand lotions Hair cosmetics Pharmaceuticals Toothpaste Skin oils Medicine Infusions	Squeezable Printable Scratch-resistant Aroma-barrier O_2 barrier	HDPE LDPE PP Regrind Color	—	Bonding agent	—	—	PA EVAL
Food Soy sauce Ketchup	Short shelf life; hot fill	PP	—	Bonding agent	—	—	PA EVAL
Beverages Milk Liquor Handleware	Transparent Hot fill Gas barrier Rigid	PC	—	EVAL PET	—	—	PC
Food Ketchup Baby food Mayonnaise Edible oil Salad dressing Spaghetti sauce Fruit juice Soup concentrate	Squeezable Transparent (5) Translucent (6) Long shelf life Hot fill Gas barrier Moisture barrier	PP	Bonding agent	EVAL PET	Bonding agent	(Regrind)	PP
Sterilized milk Carbonated drinks Liquor Tea Coffee Peanut butter Handleware	Rigid Returnable Gas barrier (N_2, O_2, CO_2) Moisture barrier Transparent	PC PP	Bonding agent	EVAL PET PETG PAN PVDC	Bonding agent	(Regrind)	PC

Table 5.66 Examples of common commercial coextruded applications

products. Table 5.72 provides examples of film tapes. The largest market for plastics worldwide, consuming about 20 wt% of the total, is oriented plastic film.

HEAT-SHRINKABLE

Examples of many oriented products include the heat-shrinkable material found in flat or tubular films or sheets and fibers. The orientation in these cases is terminated downstream of an extrusion stretching operation when a cold-enough temperature is achieved. Reversing the operation, shrinkage occurs when a sufficiently high temperature is introduced. The reheating and subsequent shrinking of these oriented plastics can result in useful properties. It is used, for example, in heat-shrinkable flame-retardant PP tubular or flat communication cable wraps, heat-shrinkable furniture webbings, pipe fittings, medical devices, and many other products.

PLASTIC BEHAVIOR

Mechanical properties depend directly upon the relationship between the axis of orientation of the plastic molecules and the axis of mechanical stress upon the molecules. Modulus, strength, and other properties increase in the direction of stretch and decrease in the perpendicular direction. This is the mechanism of pseudoplastic and thixotropic rheology typical of a non-Newtonian plastic's flow behavior. After processing, some loss in properties may occur (insignificant) when subjected to heat during further processing, such as thermoforming, heat sealing, and solvent sealing.

Property	ASTN Test Method	Cast, Uniaxial Orientation	Biaxial Orientation
Area Factor in 2/lb./1 mil film	–	30,400–31,300	30,600
Specific Gravity, g/cc	D1505	0.885–0.905	0.902–0.907
Tensile Strength, psi	D882	4500–7000	7500–40,000
Elongation, %	D882	550–1000	35–475
Tear Strength, g/mil Propagation	D1922	25 MD* 600 TD*	3–10
Fold Endurance	D2176	Very High	Excellent
24 hr. % Water Absorption	D570	0.005	0.005
Water Vapor Transmission Rate, g/mil/100 sq. in./24 hr. @ 100°F	E96	0.7	0.25
Oxygen Permeability, cc/100 sq. in./mil/24 hr./atm. @ 77°F	D1434	150–240	160
Heat Sealing Temperature Range, °F	–	285–400	–
Gloss, %	–	90	95
Haze, %	–	1–2	1–2

*MD = Machine Direction *TD = Transverse Direction

Table 5.67 Properties of oriented polypropylene

Property		Unit	Test Method	Block copolymer	Random copolymer		Homopolymer		
Grade				2309KX	3200 MCX	3520 LX	1125N	1127MX	1325L
Additives				None	None	antiblock agents	slip, antibloc k agents	slip, antibloc k agents	slip, antibloc k agents
Melt flow rate		g/10 min	ISO 1133	4	8	5	11	8	5
Gloss	20°	%	DIN 67530	5	130	130	110	95	115
Haze		%	ASTM 1003	28	0.5	0.3	2.3	3.0	1.7
Tensile strength at break	machine direction	MPa	DIN 53455	55	37	32	42	44	35
	transverse direction	MPa	DIN 53455	36	35	30	38	40	33
Elongation at break	machine direction	%	DIN 53455	750	730	750	680	700	750
	transverse direction	%	DIN 53455	720	750	760	720	730	780
Dart drop impact resistance F50		g	ASTM 1709	500	450	>800	280	300	800
Modulus of elasticity in flexure	machine direction	MPa	DIN 53121	650	480	300	700	680	370
	transverse direction	MPa	DIN 53121	640	470	310	670	650	350
Coefficient of friction			DIN 53375	0.90	Not measurabl e	Not measurabl e	0.25	0.17	0.13

[1] Data was obtained using film specimens prepared by internal standards. Film properties depend considerably on processing conditions. This must be taken into account when comparing these data with data obtained under different processing conditions.

Table 5.68 Properties of Novolen (BASF) 50-μm-gauge cast polypropylene film

ASTM Tensile Impact, ft.-lb./in.^{2}		
	Temperature	
Material	Room	−20°F (−29°C)
Unoriented PP	40	0
Oriented PP	above test limit	500

High-Energy Fatigue Impact (55 lb. weight @ 50-in. height) (24.9 kg weight at 127-cm height)	
Material	Number of drops to failure
Unoriented PP 41 × 10^3 psi tensile	1
Oriented PP 28 × 10^3 psi, 32% elongation	130

Table 5.69 Examples of drop impact tests on unoriented and oriented polypropylene film

			Modulus of elasticity (GPa)					
			Biaxially oriented film			Oriented fiber		
Plastic	Melting temperature, T_m (°C)	Glass transition temperature, T_g (°C)	Longitudinal	Transverse	Unoriented film	Melt spun	Gel spun	Density (g cm^3)
PVC		70 to 90	2.5–2.7	3–3.5	2.2	5.5	—	1.35
HDPE	138	−70 to −110	3–4	3–4	1.2	5	170	0.96
PP	134	−5 to −20	3–4	2–3	0.9	5	18	0.90
PA	260	50 to 75	2–2.5	2–2.5	0.5	4.5	19	1.13
PET	250	70 to 110	4–5.7	4.5–8.5	1.5	15	28	1.35

Table 5.70 Examples of tensile modulus of elasticity on polypropylene unoriented and oriented film as well as fibers (always oriented)

Biaxial orientation of crystalline plastics generally improves clarity of films. This occurs because stretching breaks large crystalline structures smaller than the wavelength of visible light. With uniaxial orientation, the result is an anisotropic refractive index and thus birefringence, especially in crystalline plastics.

Orientation decreases electrical dissipation factors in the direction of the orientation, and it increases these factors perpendicular to orientation. Since modulus changes in the opposite way, this indicates that polar vibrations along a stretched plastic molecule are decreased, while transverse vibrations between the stretched molecules are increased.

Property	Stretch (%) 0	200	400	600	900
Tensile strength, psi	5600	8400	14000	22000	23000
(MPa)	(38.6)	(58.0)	(96.6)	(152.0)	(159.0)
Elongation at break (%)	500	250	115	40	40

Property[a]	As cast	Uniaxial orientation	Balanced orientation
Tensile strength, psi (MPa)			
MD	5700 (39.3)	8000 (55.2)	26000 (180)
TD	3200 (22.1)	40000 (276)	22000 (152)
Modulus of elasticity, psi (MPa)			
MD	96000 (660)	150000 (1030)	340000 (2350)
TD	98000 (680)	400000 (2760)	330000 (2280)
Elongation at break (%)			
MD	425	300	80
TD	300	40	65

[a]MD = machine direction, TD = transverse direction and direction of uniaxial orientation.

Table 5.71 General mechanical properties of polypropylene film from zero to a 9:1 stretch

Orienting by stretching is influenced by factors such as (a) the lowest temperature will give the greatest orientation (tensile strength, modulus, etc.), (b) the highest rate of stretching will give the greatest orientation at a given temperature and percent stretch, (c) the highest percent stretch will give the greatest orientation at a given temperature and rate of stretching, and (d) the greatest quench rate will preserve the most orientation under any stretching condition.

With orientation, the thickness is reduced and the surface area enlarged. If film is longitudinally stretched, its thickness and width are reduced in the same ratio. If lateral/transverse contraction is prevented, stretching reduces the thickness only. Orientation temperature is normally 60% to 75% of the range between the plastic's glass-transition temperature (T_g) and melting point (T_m). Equipment can be used in-line or off-line to gain output yield and properties.

An on-line ordinary (common) blown- or cast-film line uses a machine direction orienter (MDO) on the front end of a biaxially oriented film heated-chamber extrusion line. If only the machine direction is to be stretched, a series of precision-controlled heated rolls can be used. Film is fed through a series of rolls where it is sequentially heated, drawn around rolls that increase in rotational speed providing the stretching action, annealed around a larger-diameter roll or rolls, and cooled on a final roll or rolls (chapter 5).

Biaxially Oriented Polypropylene Film Biaxially oriented polypropylene film (BOPP) is a highly transparent stiff film produced in thin gauges ranging typically from 0.012 mm to 0.040 mm.

Application	Requirement	Rate of Stretching	Thermoplastic
Ropes	High tensile strength	1:9 to 1:11 (15)	PP
	Specific elongation		
	Good tendency to splicing		
Twine	High tensile strength	1:9 to 1:11	PP
	High knotting strength		PP/PE
Separating weave	High strength	1:7	PP
Filter weave	Low shrinkage	1:7	PP
	Abrasion resistant	1:5	PETP
Reinforcing weave	Low shrinkage	1:7	PP
	Specific elongation	1:5	PETP
	Temperature resistance		
Tapestry and home textiles	UV-resistance	1:7	PE
	Low static charge		
	Uniform coloration		
	Textile-type handle		
Outdoor carpets	Low shrinkage	1:7	PP
	Wear resistance		
	Weather resistance		
	Elastic recovery		
	Uniform coloration	1:5	PETP
	Defined splicing		
Decorative tapes	Effective surface	1:6	PP with
Knitted tapes, sacks, and other packagings, seed and harvest protective nets	High knotting strength	1:6.5	PP
	Low splicing tendency		PE
	Suppleness		
	UV-resistance		
Packaging tapes	High strength	1:9	PP
	Low splicing tendency	1:7	PETP
Carpet basic weave	Low shrinkage	1:7	PP
	High strength	1:5	PETP
	Temperature stability		
	Specific splicing tendency		
	Matt surface		
Tarpaulins	High strength	1:7	PP
			PE
Sacks	High strength	1:7	PP
	High friction value		PE
	Specific elongation		
	Weather resistant		

Table 5.72 A few of the uses for oriented flat-film tapes

It is widely used for the packaging of foods, cigarettes, and other items. The film is produced by stretching an extruded sheet or tube in two orthogonal directions, namely, the machine direction and the transverse direction. Stretching is carried out at a temperature below the melting point of the plastic in the direction of stretch.

In principle, biaxially oriented film is isotropic. Its properties are the same in both the machine and transverse directions. In practice, film produced by certain tenter processes tends to be more highly oriented in the machine direction, whereas the blown process produces a film that is more nearly isotropic. Capital costs for a blown BOPP film line are substantially less than those of a tenter line. Biaxially oriented film is typically produced from plastics with melt flow index in the range 2.5 to 4.5 grams/10 min (chapter 22).

Biaxially oriented polypropylene film is not readily heat sealable. This is a prime requirement for packaging films, so BOPP is normally given a surface coating of a heat-sealable plastic, such as a coextruded polypropylene random copolymer.

ACCIDENTAL OR DELIBERATE ORIENTATION

A TP's molecular orientation can be accidental or deliberate. Accidental can occur during the processing ofTPs where excessive frozen-in stresses develop; however, with the usual proper process control, there is no accidental orientation (chapter 3). The frozen-in stresses with certain TPs can be extremely damaging if products are subjected to environmental stress cracking or crazing in the presence of heat, chemicals, and so forth. Initially, the molecules are relaxed. Molecules in the amorphous regions are in random coils; those in crystalline regions are relatively straight and folded.

During processing (extrusion, injection, blow molding, etc.), the molecules tend to be more oriented than relaxed, particularly when the melt is subjected to excessive shearing action (chapter 1). After temperature-time-pressure is applied and the melt goes through restrictions (mold, die, etc.), the molecules tend to be stretched and aligned in a parallel form. The result can be undesirable changes in the directional properties and dimensions immediately when processed and/or thereafter when in use if stress relaxation occurs.

By deliberate stretching, the molecular chains of a plastic are drawn in the direction of the stretching, and inherent strengths of the chains are more nearly realized than they are in their naturally relaxed configurations. Stretching can take place with heat during or after processing by extruding film, blow molding, injection molding, thermoforming, and so on. Products can be drawn in one direction or in two opposite directions, in which case many properties significantly increase uniaxially or biaxially.

The amount of change depends on the type ofTP, the amount of restriction, and, most important, its rate of cooling. The faster the rate, the more retention there is of the frozen orientation. After processing, products can be subjected to stress relaxation, with changes in performance and dimensions. With certain plastics and processes, there are insignificant changes. If changes are significant and undesirable, action must be taken to change the conditions during and/or after processing. For example, during processing increase the cooling rate. Annealing of the product is an after-processing approach.

The processing hardware of orientation is fairly expensive. It increases the cost per unit weight of the product; however, the yield increases considerably, its quality improves greatly, and product cost reduction occurs. Many products are made much stronger, flexible, tougher, and so on, resulting in significant cost advantages.

PRODUCTION

BLOWN FILM

The blown products, such as upward blown film, are basically naturals for providing orientation (Fig. 5.202). The blowup ratio determines the degree of circumferential orientation, and the pull rate of the bubble determines the longitudinal orientation.

The optimum stretching heat for amorphous plastics (PVCs, etc.) is usually just above their glass-transition temperatures (T_g) (chapter 1). Generally, the orientation temperature is 60% to 75% between the T_g and T_m (melt temperature). For crystalline plastics (PEs, PETs, etc.), generally,

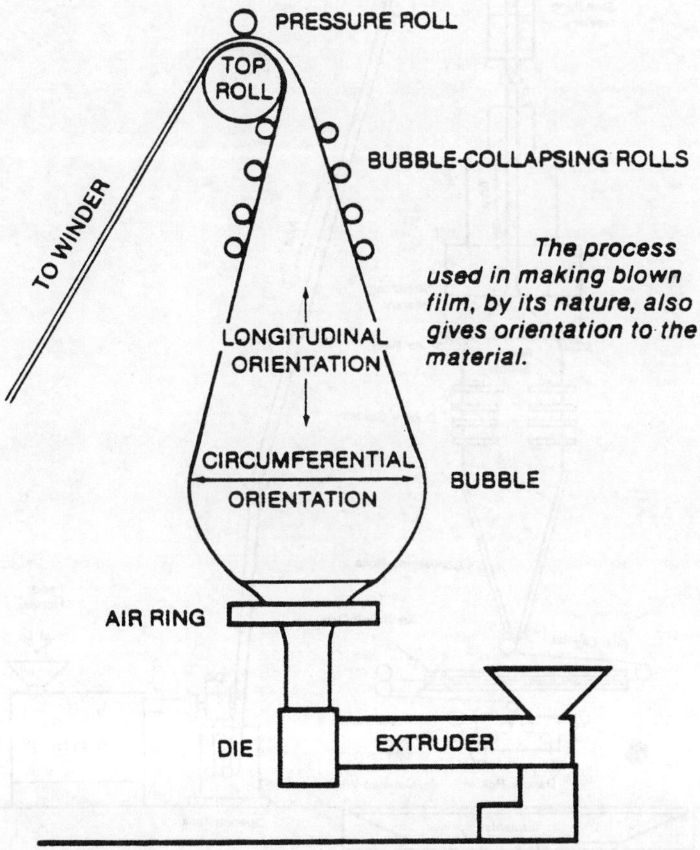

Figure 5.202 Example of upward extruded blown-film process for biaxially orienting film

it is below the T_g. Stretching can take place in-line or off-line, with or without tenter frames, using the appropriate temperature pull rates as the plastic travels first through a series of heated rolls. For unidirectional orientation, just the rolls are used.

In addition to the upward blown-film extrusion system, there is also the downward extrusion system (Fig. 5.203). It allows the tube to be quenched rapidly in a water bath, after which it is collapsed as a layflat for passage over nip and idler rollers.

In this example, the film passes through a reheating tunnel where it is raised to a temperature above the softening point but below the melting point. The heated tube is then inflated by internal

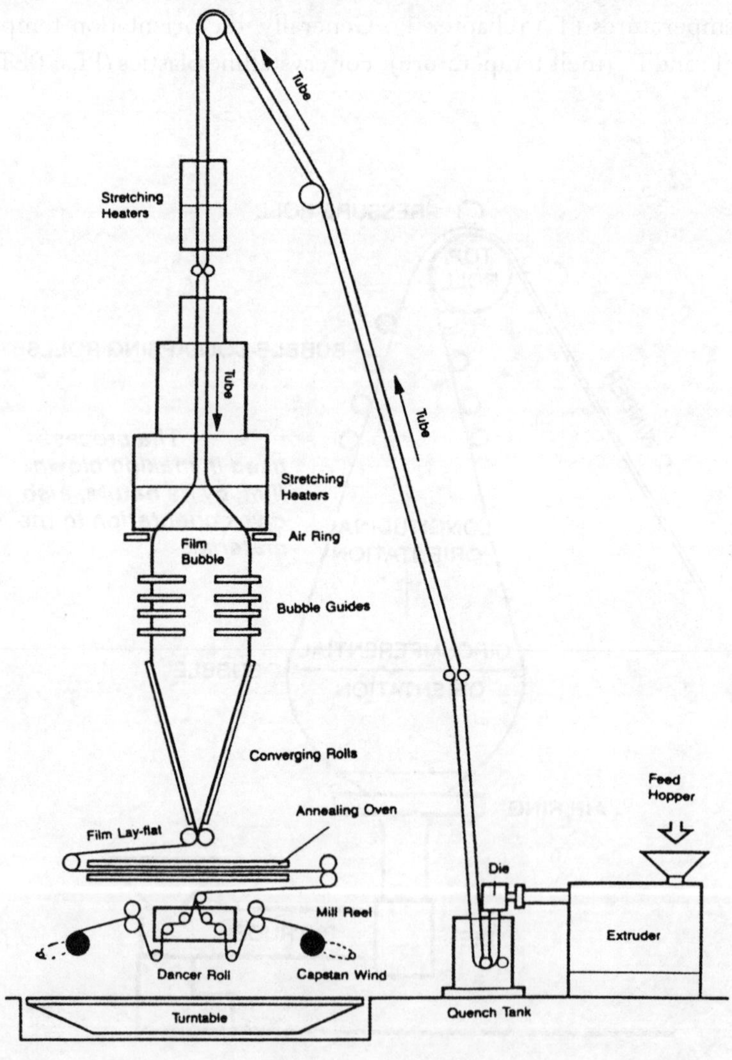

Figure 5.203 Example of downward extruded blown-film process for biaxially orienting film

air pressure that forms a bubble in which the film is stretched in all directions. Some machine-direction stretch may take place in the ovens upstream of the bubble; the haul-off rate can be adjusted if necessary to secure an orientation balance. Bubble cooling is provided by an air ring similar to that used in other blown-film processes. Subsequent calibration and bubble-collapsing operations are also similar.

Flat Film Flat products can be biaxially oriented by using some type of tenter frame (Fig. 5.204). Orienting film is accomplished by mechanically stretching the film in the tenter machine. This takes its name from the tenter frame originally used for stretching cloth between grips known as tenterhooks.

Simultaneous tentering is possible. It involves complex movements of the film edge grips so that the film is stretched in the machine and transverse directions at the same time. However, the process can be mechanically complicated, and it can be difficult to adjust the balance between the stretch directions for certain plastics. A steel stretching tenter frame is in the process of being assembled; it will eventually have heavy-duty clamps with gentle high-tensile grip moving along with a rotating chain (Fig. 5.205). The tenter frames' bulkiness is necessary to manipulate the transverse gradual, uniform pull on heated film.

To eliminate potential problems, the two-step tenter process is used (Figs. 5.204 and 5.206). The process starts with the production of a relatively thick cast film. This is then stretched in the machine direction by passing it around heated rollers rotating at controlled and increasing speeds in excess of the extrusion speed. Varying the roll speeds controls the degree of stretch. Typically, stretch in the machine direction is about 4.5:1.

When machine direction stretching is completed, the tenter machine applies transverse stretching. This consists essentially of a temperature-regulated tunnel in which the film edges are gripped by chain-driven tension clips running on divergent paths. As the film passes through the tunnel, it is progressively stretched in the transverse direction as the clips diverge. The edge grip mechanism must withstand high cross loads of up to 3.5 kN per clip and be capable of operating at line speeds of up to 450 m/min. Transverse stretch is controlled by varying the divergence of the edge clip paths. A ratio of 8:1 is typical.

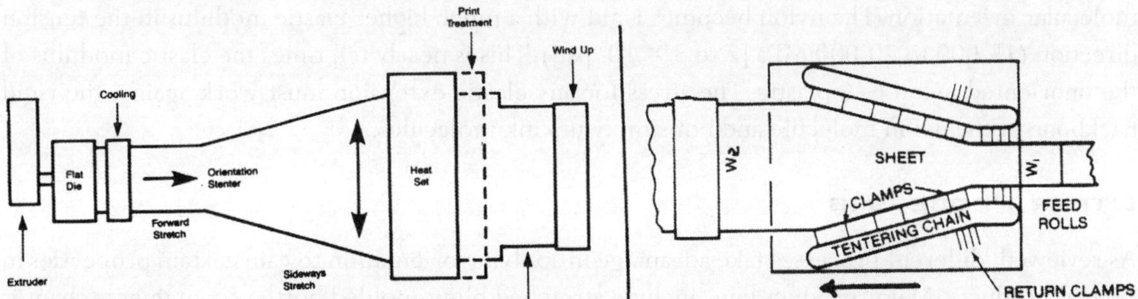

Figure 5.204 Example of a tenter process for biaxially orienting flat film

Figure 5.205 Transverse tenter frames being assembled

FIBER

Fiber orientation has been popular since the 1930s. It is accomplished usually by just stretching when the fibers are produced. They can also be stretched by a set of heated rolls, where each of their rotating speeds (rpm) is increased (Fig. 5.207). A fiber or thread of nylon 6/6, that is, an unoriented glassy polymer, has a modulus of elasticity of about 2,000 MPa (300,000 psi). Above the T_g (glass-transition temperature), its elastic modulus drops even lower because small stresses will readily straighten the kinked molecular chains (chapter 1). However, once it is extended and has its molecules oriented in the direction of the stress, larger stresses are required to produce added strain. The elastic modulus increases.

The next stop is to cool the nylon below its T_g without removing the stress, retaining its molecular orientation. The nylon becomes rigid with a much higher elastic modulus in the tension direction (15,000 to 20,000 MPa [2 to 3 × 10^6 psi]). This is nearly ten times the elastic modulus of the unoriented nylon 6/6 plastic. The stress for any elastic extension must work against the rigid backbone of the nylon molecule and not simply unkink molecules.

OTHER PROCESSES

As reviewed, different processes take advantage in applying orientation to gain certain properties in certain products. Major product lines include stretched blow-molded bottles/containers (chapter 6), thermoformed-oriented containers (used for such products as fruits, vegetables, and baked goods) (chapter 7), and tapes (Fig. 5.208).

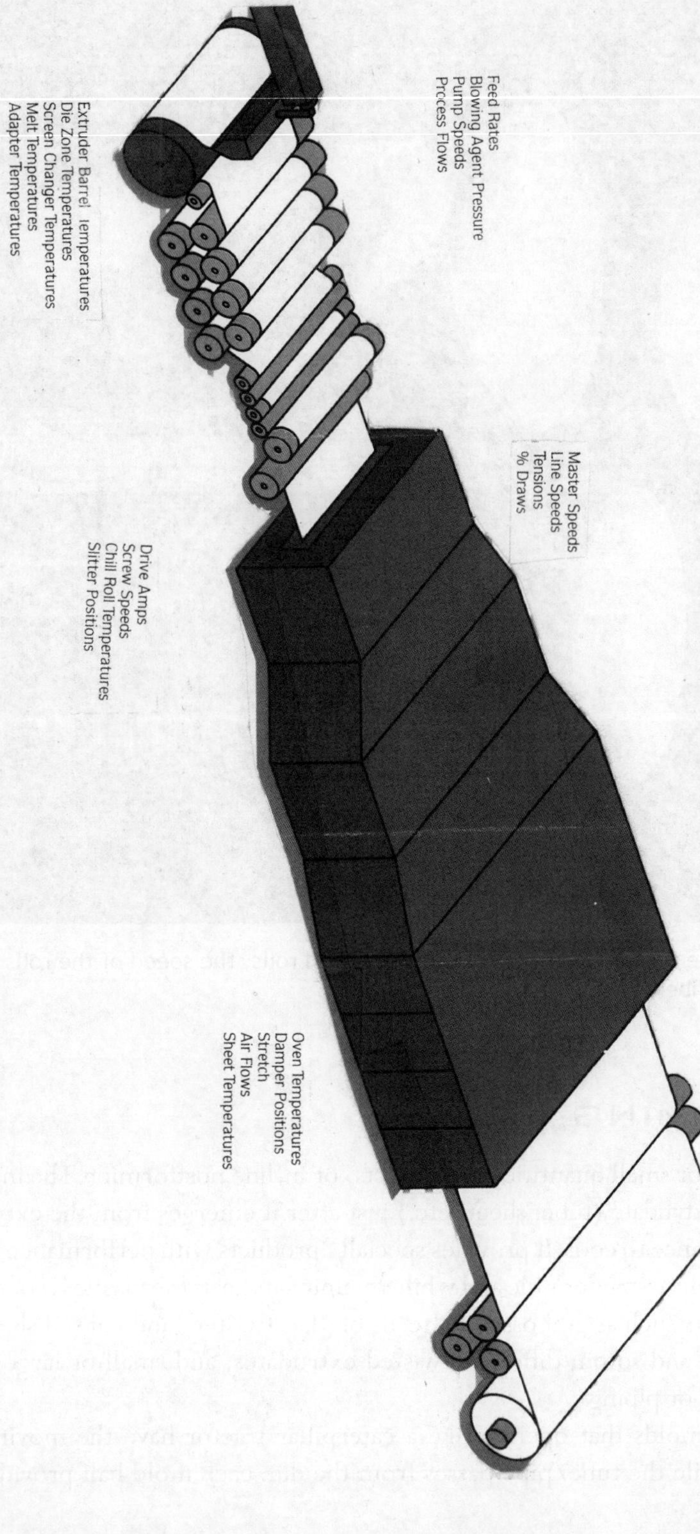

Figure 5.206 Example of two-step tenter process

Figure 5.207 As the fibers roll over the heat-controlled rolls, the speed of the rolls increases, stretching the fibers

POSTFORMING

There is off-line (for small quantities of products) or in-line postforming. The in-line refers to forming/shaping the extrudate (tube, sheet, etc.) just after it emerges from the extruder die but before the plastic has a chance to cool. It provides specialty products with performance and cost advantages.

Upon leaving the extruder's die and while retaining its heat, the plastic is continually postformed, producing products such as embossed patterns on flat, twisted, and curved shapes; coiled forms of tube, rod, profile, and so on; different twisted extrudates; and small or large (up to at least 6 ft) corrugated tubing or piping.

Moving split molds that operate like a caterpillar tractor have the moving corrugated mold cavities closed while the tube/pipe passes from the die; each mold half provides vacuum pressure

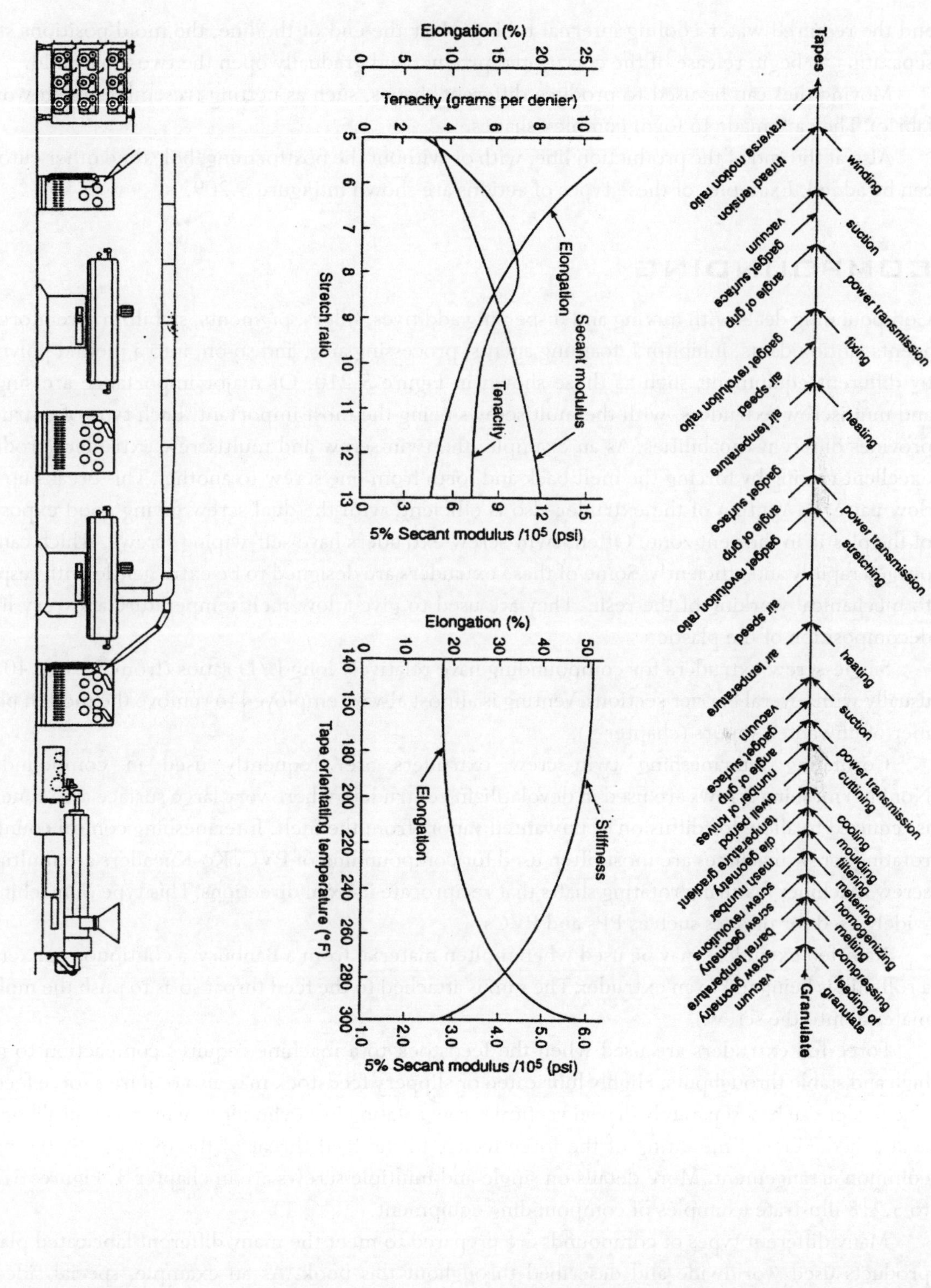

Figure 5.208 Example of orienting film tape with property-temperature profiles and stretch ratios

and the required water-cooling internal tubing. Near the end of the line, the mold positions start separating to begin release of the corrugated product and gradually open the two-part mold.

Moving dies can be used to produce different shapes, such as netting (resembles open woven fabric). They are made to form complex shapes.

Also at the end of the production line, with or without the postforming, holes and other cutouts can be added. Examples of these types of actions are shown in Figure 5.209.

COMPOUNDING

Compounding deals with mixing and dispersing additives, fillers, pigments, stabilizers, reinforcing agents, antioxidants, inhibitors, foaming agents, processing aids, and so on, into a melted polymer by different equipment, such as those shown in Figure 5.210. Of major importance are single- and multiscrew extruders, with the multiscrews being the most important. Each type of extruder provides different capabilities. As an example, the twin-screw and multiscrew extruders produce excellent mixing by forcing the melt back and forth from one screw to another. This breaks up the flow patterns. Venting of the extruder also is efficient, with the dual screw giving good exposure of the plastic in the vent zone. Often, twin-screw extruders have self-wiping screws, which can be purged rapidly and efficiently. Some of these extruders are designed to be extra gentle with respect to mechanical working of the resin. They are used to give a low melt temperature and very little decomposition of the plastic.

Single-screw extruders for compounding have relatively long L/D ratios (from 32:1 to 40:1), usually with several barrier sections. Venting is almost always employed to remove the molten polymer of unwanted vapors (chapter 3).

Corotating intermeshing twin-screw extruders are frequently used in compounding. Nonintermeshing screws are used in devolatilizing extruders where very large surface area renewal is required to allow for diffusion of unwanted vapors from the melt. Intermeshing conical counter-rotating screw machines are most often used for compounding of PVC. Ko-Kneaders are multiple-screw machines featuring rotating shafts that reciprocate in axial directions. This type of machine is widely used for plastics such as PPs and PVCs.

Ram-fed extruders may be used when molten material from a Banbury, a continuous mixer, or a roll mill is being fed to an extruder. The ram is attached to the feed throat so as to push the molten material into the screw.

Force-fed extruders are used when the feedstock to a machine requires compaction to give high and stable throughputs. Highly lubricated or slippery feedstock may also require a force feeder. The feeder can be a separately driven vertical screw rotating in a cylinder to which a conical hopper is attached. Vertical mounting of the force feeder to the feed throat of the extruder is the most common arrangement. More details on single and multiple screws are in chapter 3. Figures 5.211 to 5.218 illustrate examples of compounding equipment.

Many different types of compounds are prepared to meet the many different fabricated plastic products used worldwide and described throughout this book. As an example, special side-fed

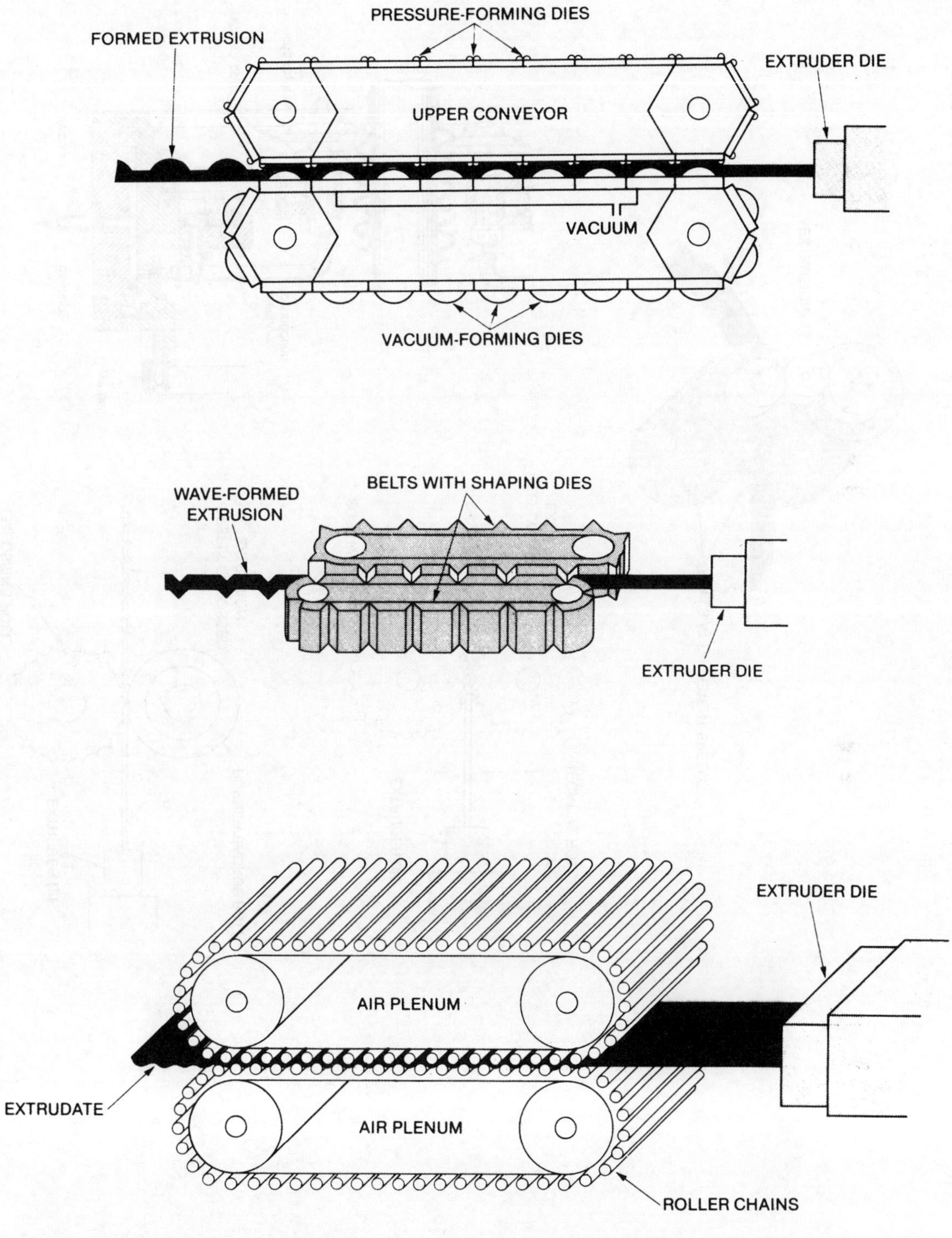

Figure 5.209 Examples (some showing dies) of different postformed shapes and cuts

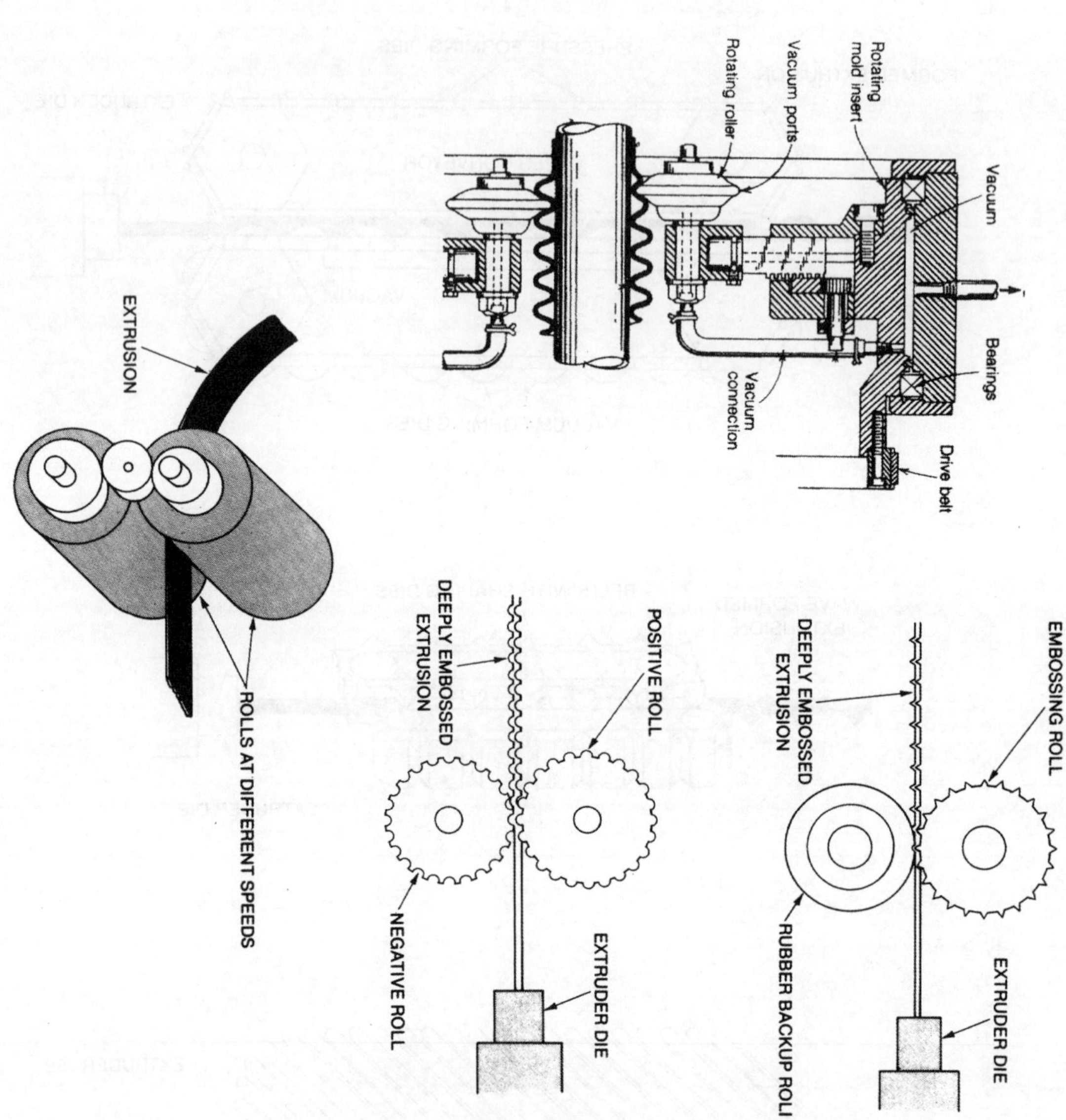

Figure 5.209 *(continued)*

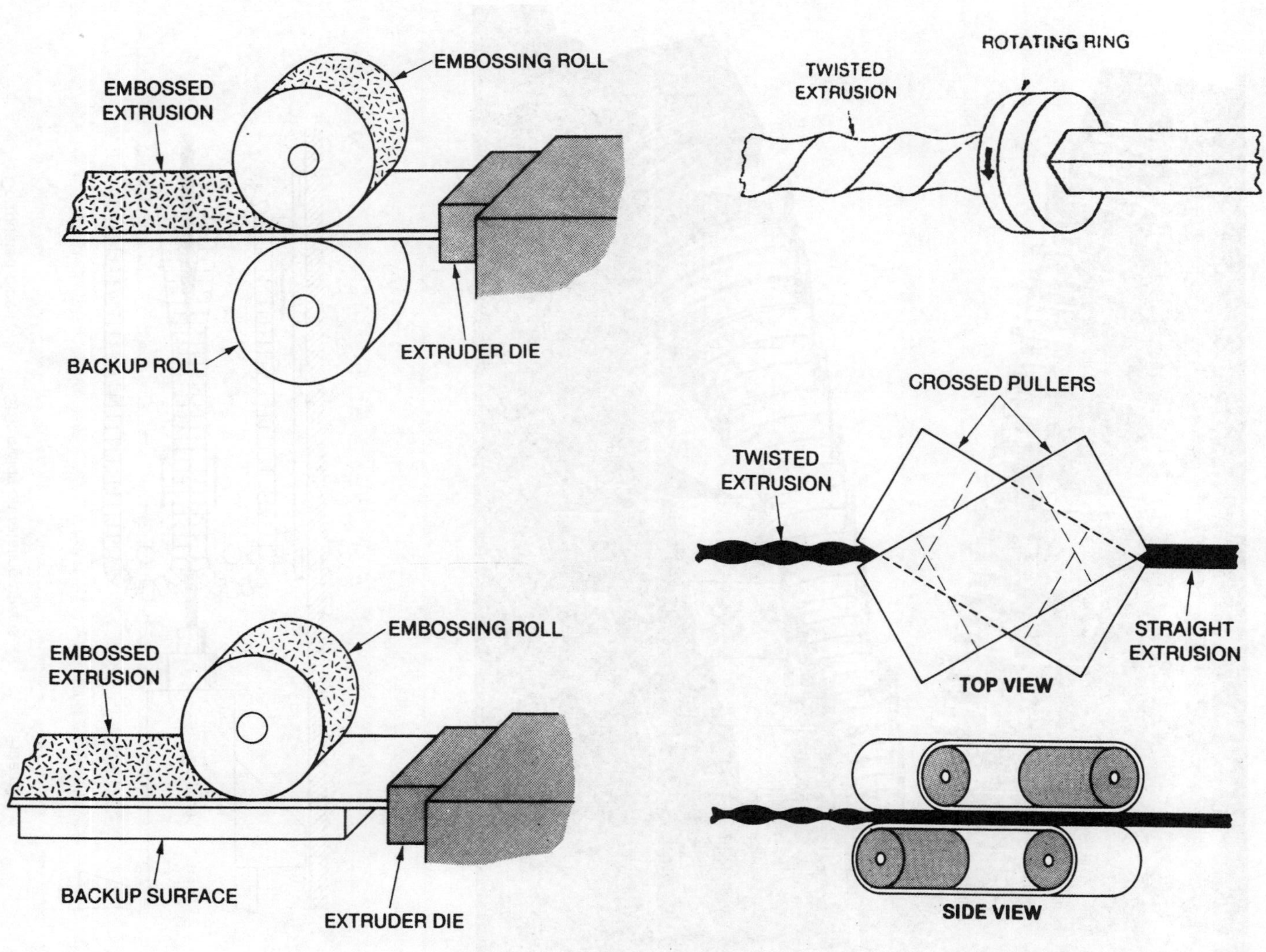

Figure 5.209 *(continued)*

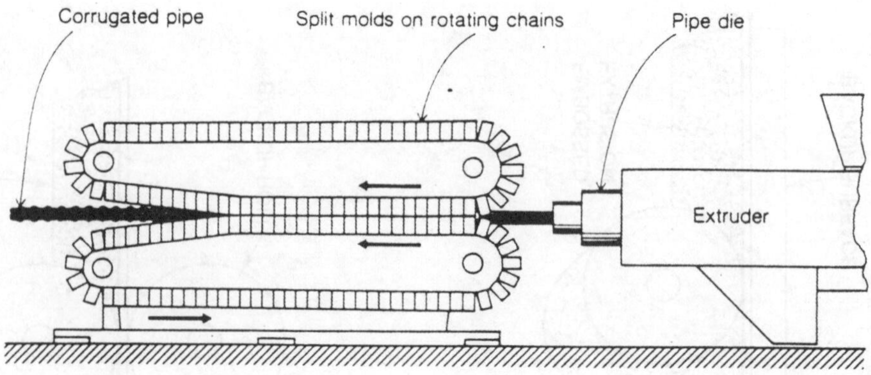

Figure 5.209 *(continued)*

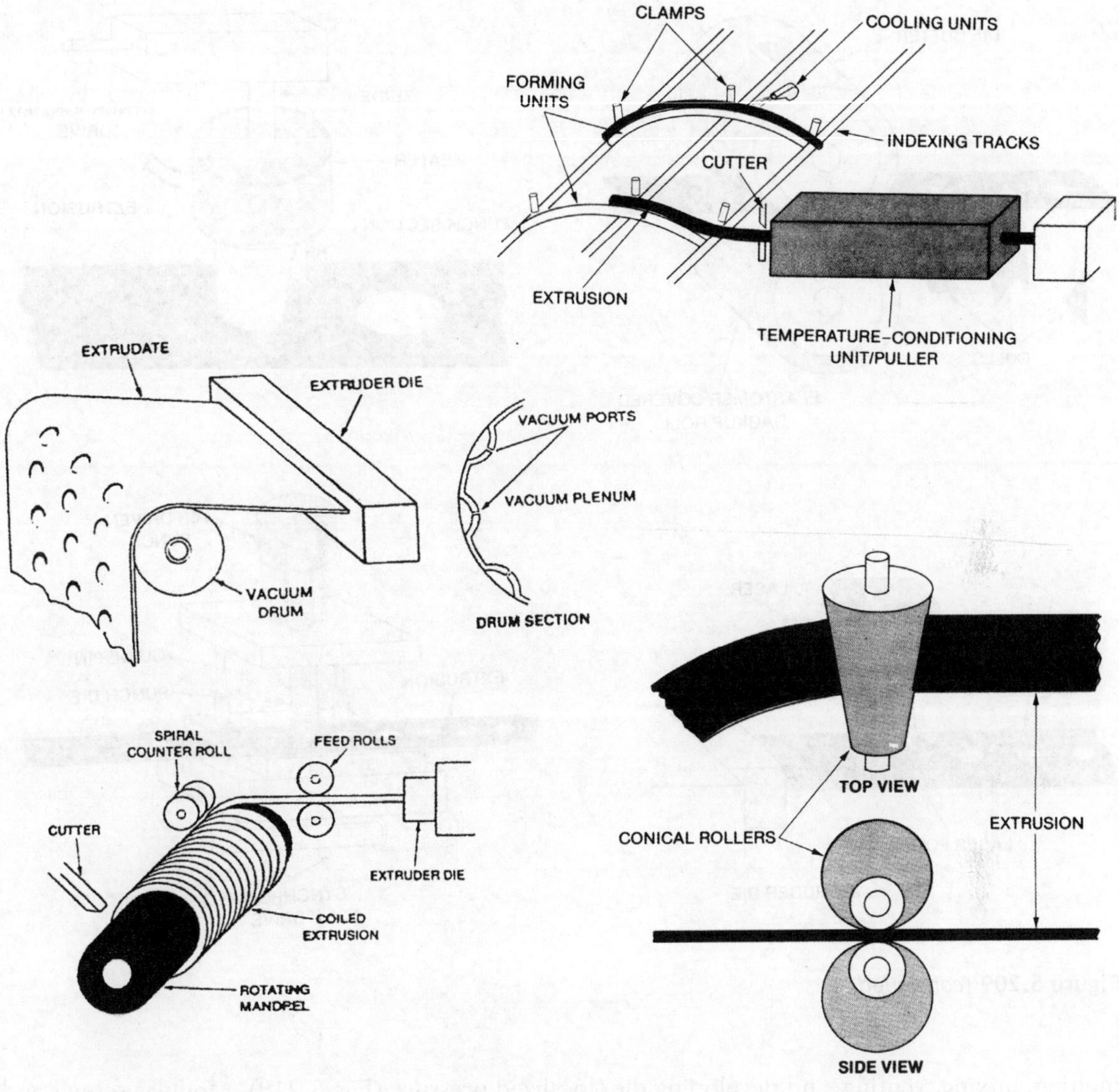

Figure 5.209 *(continued)*

extruders are used to add short and long fibers as well as fillers to TPs to improve their structural or physical properties. Equipment used includes the three-stage screw. The plastic is melted in the first stage. Chopped fibers or fillers are conveyed into the side of the barrel and mixed in the second stage. Venting and pressure generation are accomplished in the third stage (383).

Twin-screw extruders mix, devolatilize, and perform reactions in a continuous manner with the end product being either a pellet or extruded part (506). Whether the screw rotation is corotating or counter-rotating, the twin-screw extruder performs the same basic process tasks: feeding,

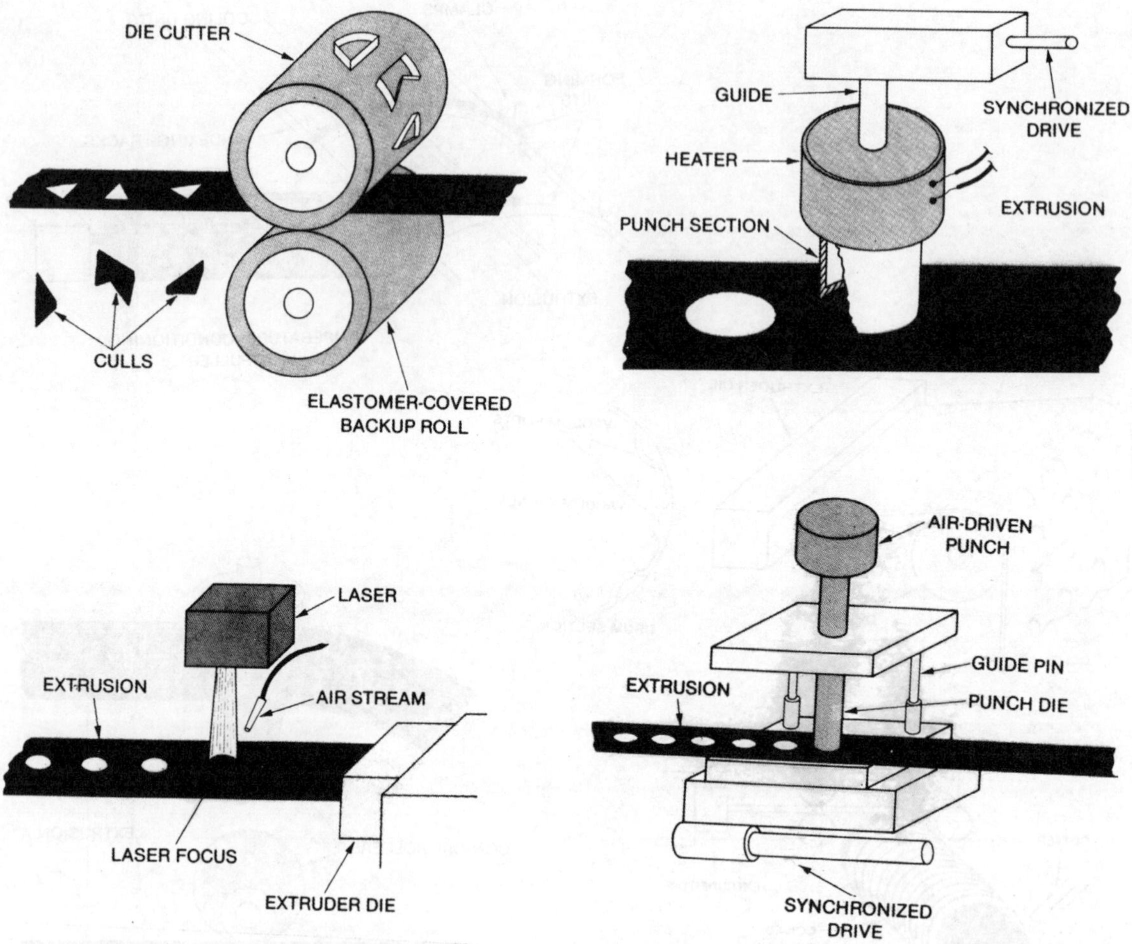

Figure 5.209 *(continued)*

melting, mixing, venting, and developing die/localized pressure (Fig. 5.219). Modular screws and barrels make an infinite number of process configurations possible.

There are, however, only three families of screw elements: (a) flighted, (b) mixing, and (c) zoning. Flighted elements feed and forward material past barrel ports, through mixers, and out of the die. Mixers can operate under dispersive and/or distributive conditions (chapter 2), depending on the amount of elongational acceleration or shear versus the number of melt divisions in the design. Zoning elements isolate two operations within the extruder, such as for providing a vent seal. Screws can be shear intensive or passive, based on the elements used and machine operating.

Twin-screw extruders are usually run starve fed. Both the feed rate (Q) and screw speed (N) are independently controlled. Q is a measure of how much is added into the extruder and N is proportional to how fast it is taken away (507). A wide range of Q and N values can be run, but

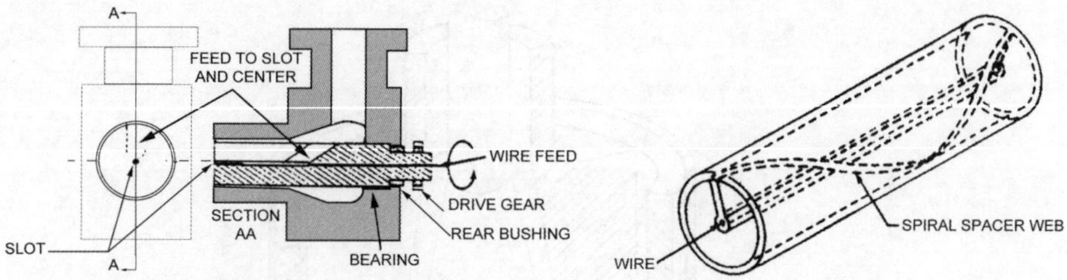

Coaxial cable with a spiral spacer web that keeps coated wire in the center

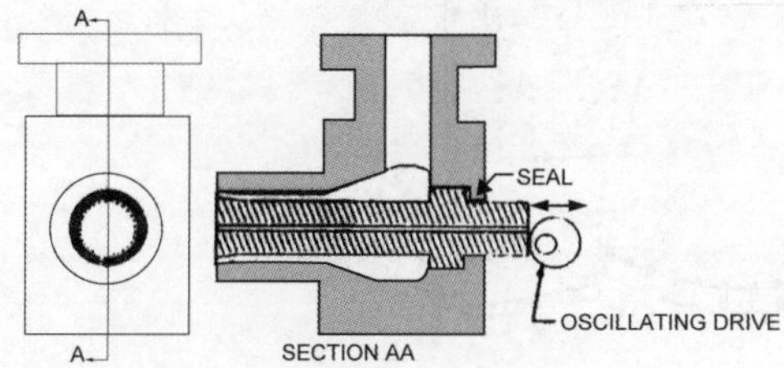

Varying tube wall thickness using an oscillating mandrel in cross head tube die

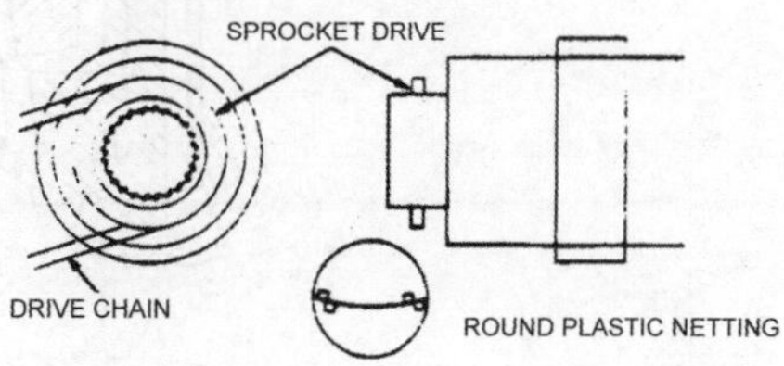

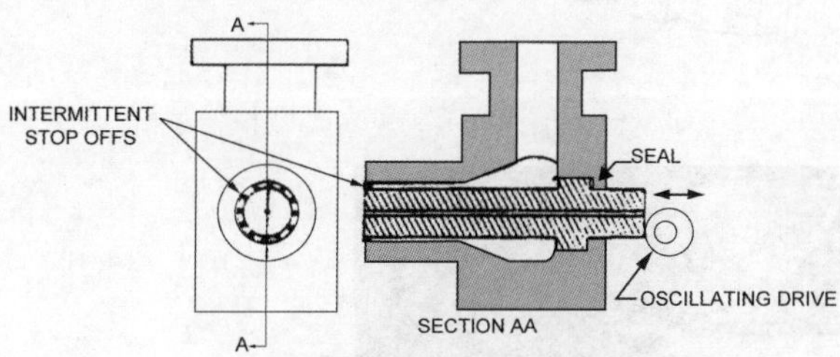

Different perforated tubing pattern; using oscillating mandrel die

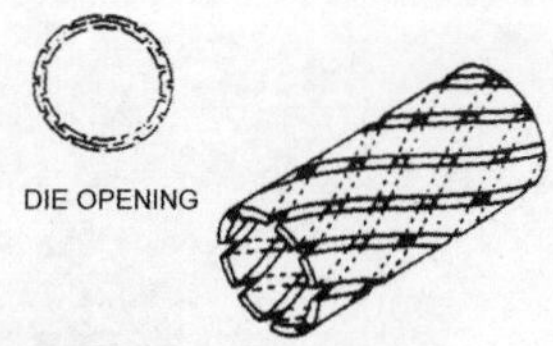

Rotating mandrel die makes perforated tubing

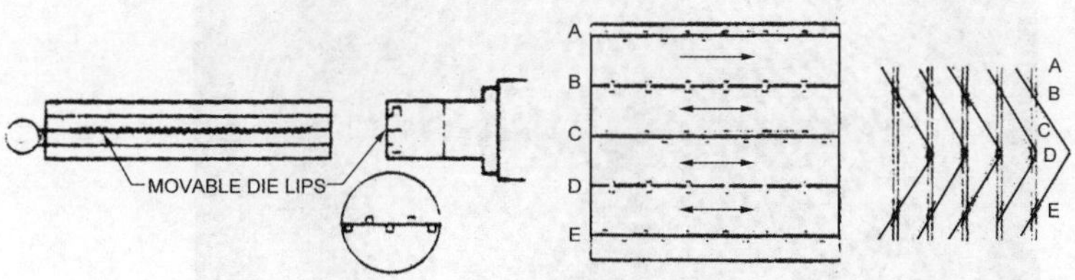

Flat netting ....with different positions of die lips

Figure 5.209 *(continued)*

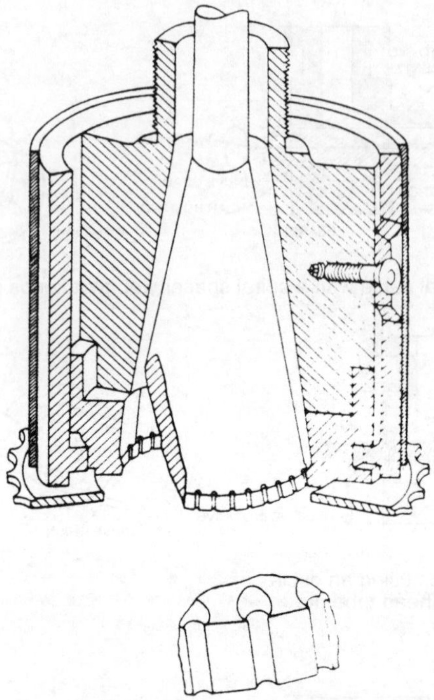

Figure 5.209 *(continued)*

these values must conform to a certain range of the Q/N ratio. An example for a 2.8-in (71-mm) twin-screw extruder might be a feed rate Q of 900 pph and a screw speed N of 300 rpm. The Q/N would be 3.0 pph/rpm. Higher screw speeds (even up to 1,000 rpm) would give proportionally higher rates with the same 3.0 pph/rpm.

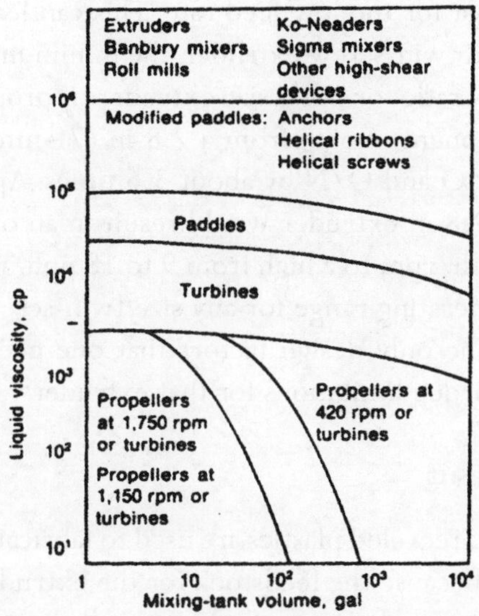

Figure 5.210 Examples and performances of compounding equipment

Figure 5.211 Two-stage vented single-screw compounding extruder (courtesy of Welex Inc.)

A maximum limit on Q/N is reached when the screw becomes filled and the screw speed is not high enough to take away the feed. The maximum limit can also be reached by the mechanical torque limitation of the motor, gearbox, or screws. For the 2.8-in example, the maximum Q/N ranges from about 2.5 to 5.0 pph/rpm, depending on many other extruder factors, such as the torque limit of the screws and gearbox, screw design, plastic viscosity, head pressure, backflow, and so on. A minimum for the Q/N ratio is also seen. With this example, the screw fill will be very low because the speed is too fast for the low feed rate. This can lead to surging, overheating, and unsteady operation. For a 2.8-in twin-screw extruder, the minimum Q/N is about 0.75 pph/rpm.

The feed rate Q or the Q/N ratio for other size extruders is proportional to the screw diameter ratio raised to about a 2.7 exponent. Going from a 2.8-in (71-mm) to a 4.5-in (114-mm) twin-screw extruder would increase Q and Q/N by about 3.6 times. Applying this factor to the above operating Q/N range for the 2.8-in extruder would result in an operating range of Q/N for the 4.5-in extruder, a low of 2.7 pph/rpm to a high from 9 to 18 pph/rpm, the high depending on the torque limitations. Thus, the operating range for any size twin-screw extruder can be determined from the above information. The only design factors that one needs to know are the maximum screw speed possible and the torque limitations for that extruder.

Reclamation/Recycling

As reviewed in chapters 1 and 2, recycled plastics are used to fabricate products. Reclaiming plastics can require special equipment because the feedstock for the extruder comes in many forms. Since the bulk density of most of the scrap plastic used is low, a dual-diameter extruder is often used. The

Figure 5.212 Twin-screw compounding extruder (courtesy of Coperion/Werner & Pfleider)

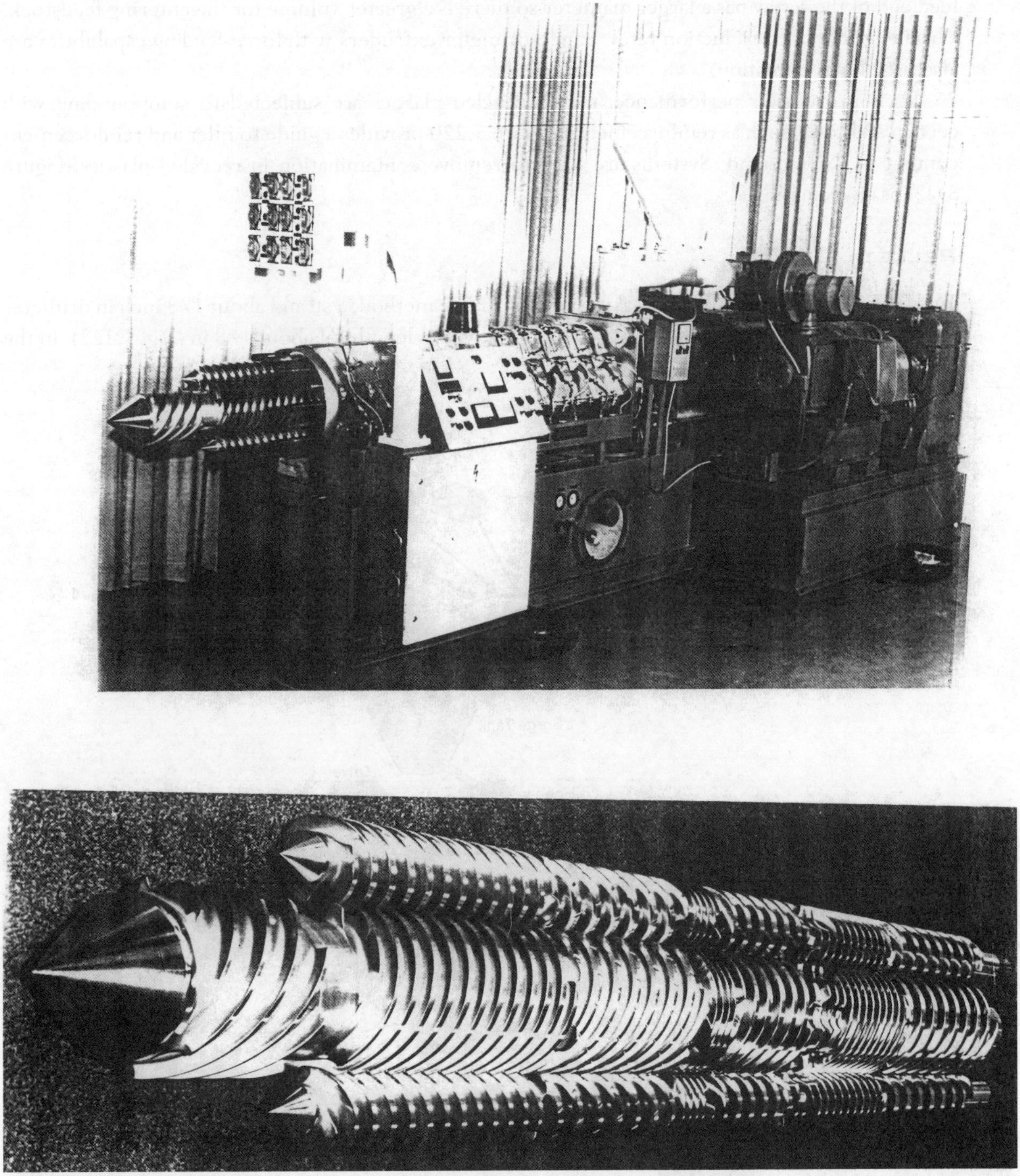

Figure 5.213 Multiscrew compounding extruder (courtesy of Milacron)

feed end of the screw has a larger diameter so there is a greater volume for the entering feedstock. This helps increase production rates. Single-diameter extruders with force-feeding capabilities are also used for reclamation.

To improve their performance, many recycled plastics are subjected to compounding with certain additives, such as reinforcements. Figure 5.220 provides a guide to filler and reinforcement content in a compound. Systems are used to remove contamination in recycled plastics. Figure 5.221 is an example.

PELLET

Two basic methods are used for making pellets. In one method, a strand about 1/8 inch in diameter is extruded, and after cooling in a water trough, is cut in lengths of about 1/8 in (Fig. 5.222). In the

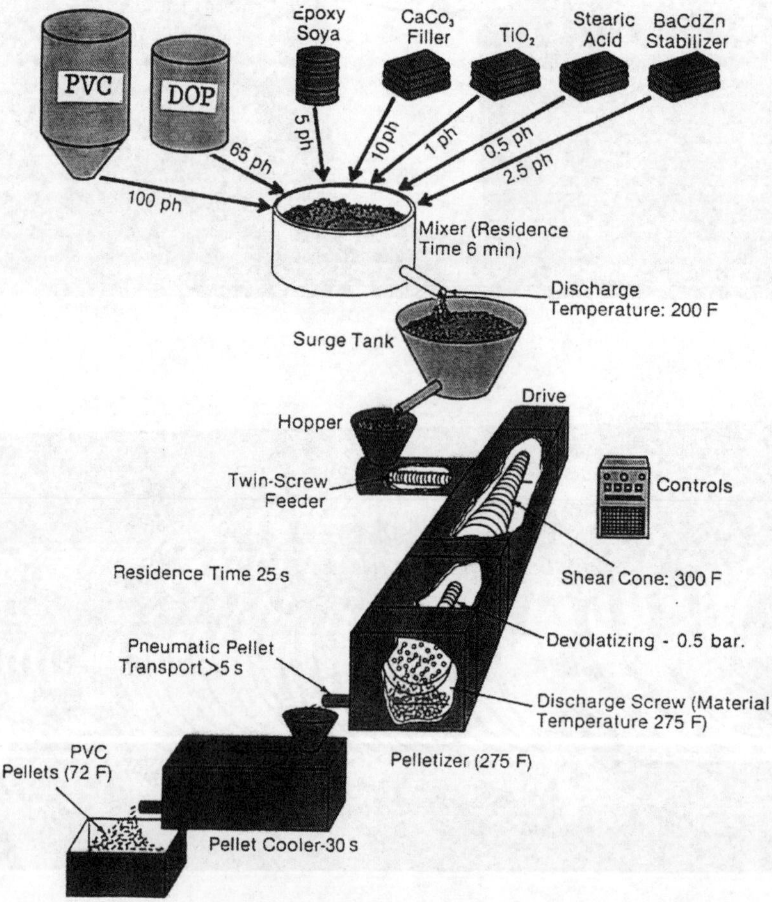

Figure 5.214 Schematic of compounding PVC

other, more popular method, called the underwater pelletizing method, the die opening is actually under water and a fast rotating knife almost touching the die cuts the melt into pellets as it exits the die. This method is used at high rates with large extruders and with dies having many openings. Pellets cut are carried away by water circulated past the die at high rates. They are collected on a conveying screen and are blown free of water by air. Different shaped pellets can be produced (Table 5.73). Different versions of these pelletizing methods are used. They include dry and wet (water), hot and cold die face cutting, and so on (Table 5.74).

The underwater pelletizing process has been in use for about a half century and was developed to eliminate longs and fines and to eliminate handling a large number of strands. As it has been reported by many, technically, it is considered an engineering marvel that the process actually works (500). The molten plastic is on one side of a very hot die and flowing water is on the other side where the cutting occurs. With improper operation, the die is frozen solid by the water. Start-up of this process means strict attention to established procedures.

EXTRUDER CLASSIFICATION

As reviewed at the start of this chapter, there are different types of extruders that can be classified as:

1. continuous with single screws (single and multistage) or multiscrews (twin-screw, etc.),
2. continuous disk or drum that uses viscous drag melt actions (disk pack, drum, etc.) or elastic melt actions (screwless, etc.),

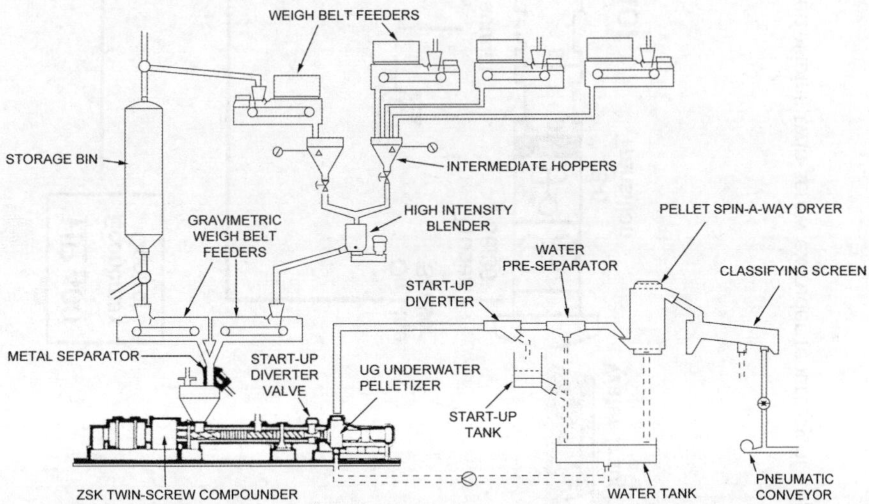

Figure 5.215 Schematic for compounding polyolefins using twin-screw extruder (courtesy of Coperion/Werner & Pfleiderer)

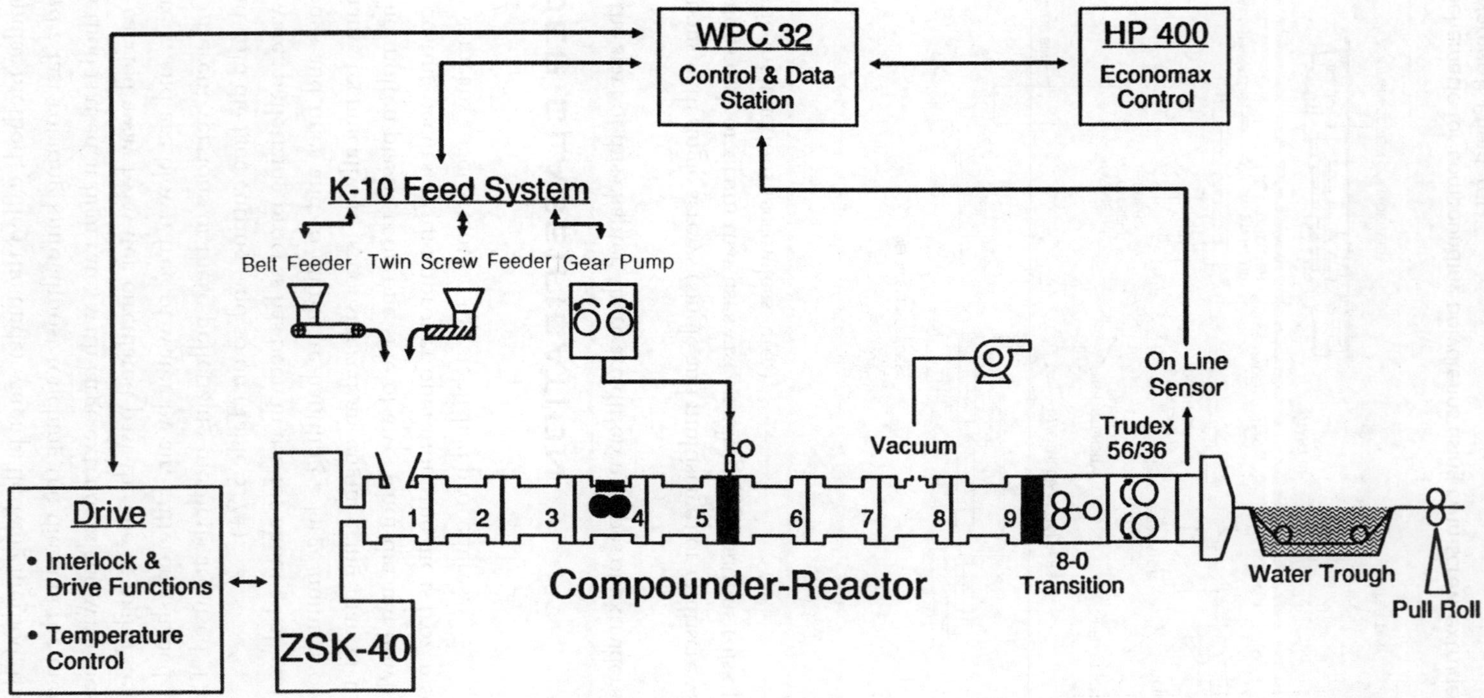

Figure 5.216 Schematic for reactive compounding (chapter 1) using corotating, self-wiping twin-screw extruder (courtesy of Coperion/Werner & Pfleiderer)

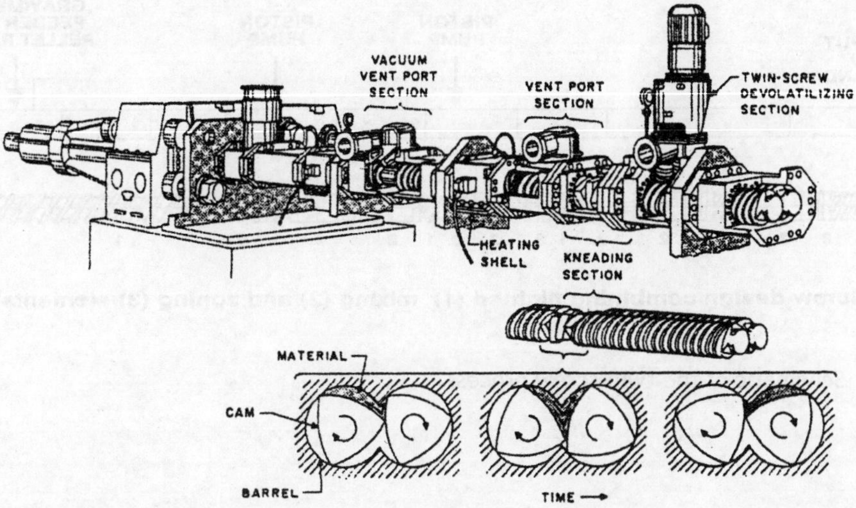

Figure 5.217 Schematic of twin-screw extruder that operates in different modes by changing screw and vent sections (courtesy of Coperion/Werner & Pfleiderer)

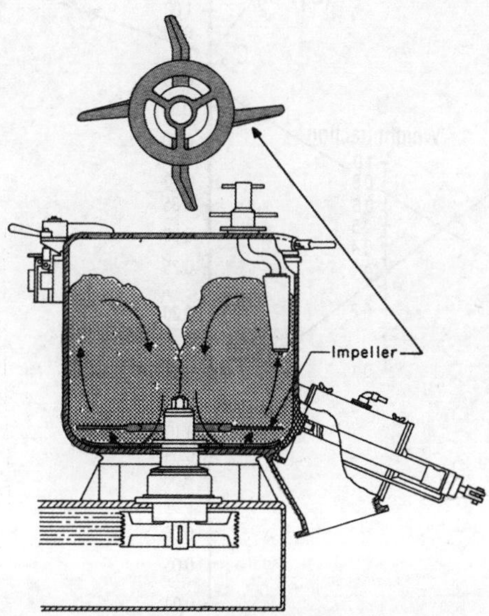

Figure 5.218 Example of removing heat and volatiles from a compound using an internal mixer with high-speed impeller

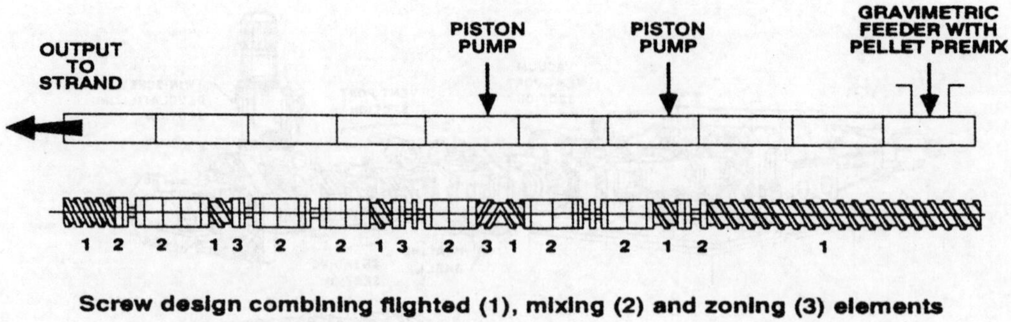

Figure 5.219 Schematic of the twin-screw process

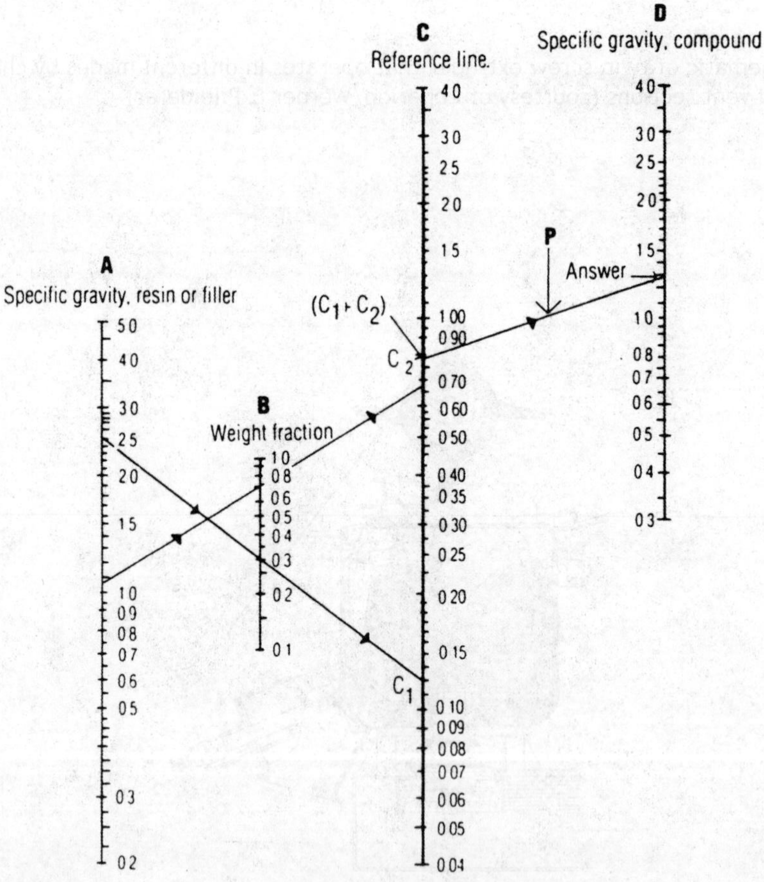

Figure 5.220 Nomograph for determining the specific gravity of compounds filled with fillers and reinforcements

3. discontinuous that uses ram actions (very low viscosity thermoplastics [TPs], thermoset [TS] plastics, and rubbers/elastomers) and reciprocating actions (injection molding, etc.).

HORIZONTAL/VERTICAL EXTRUDER

The continuous extruders with single screws (single and multistage) or multiscrews (twin-screw, etc.) reviewed in this chapter are used to fabricate practically all extruded products. These extruders are of the horizontal type; that is, the plasticizing screws are in horizontal positions. Vertical extruders (Fig. 5.223) have had very limited use in special applications, for example, extruding PVC that is cut into disks and compression molding musical playing records. This extruder has interesting engineering aspects, including providing high torque on the screw where it is subjected to high mixing forces.

INJECTION MOLDING/NONCONTINUOUS EXTRUDER

The injection-molding process, which is rarely identified as a discontinuous extruder, is a major method used to fabricate plastic products. This subject is reviewed in chapter 4.

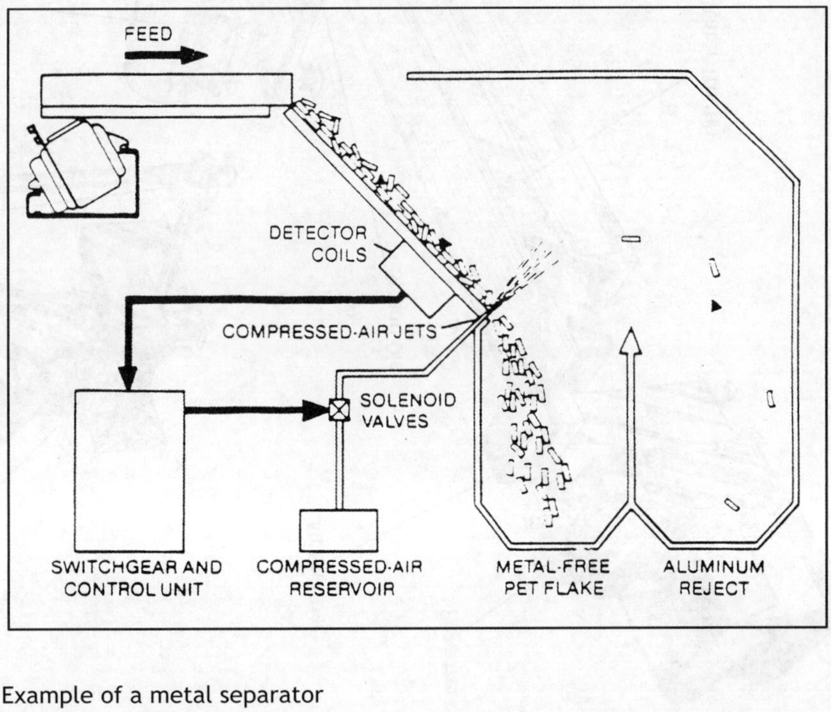

Figure 5.221 Example of a metal separator

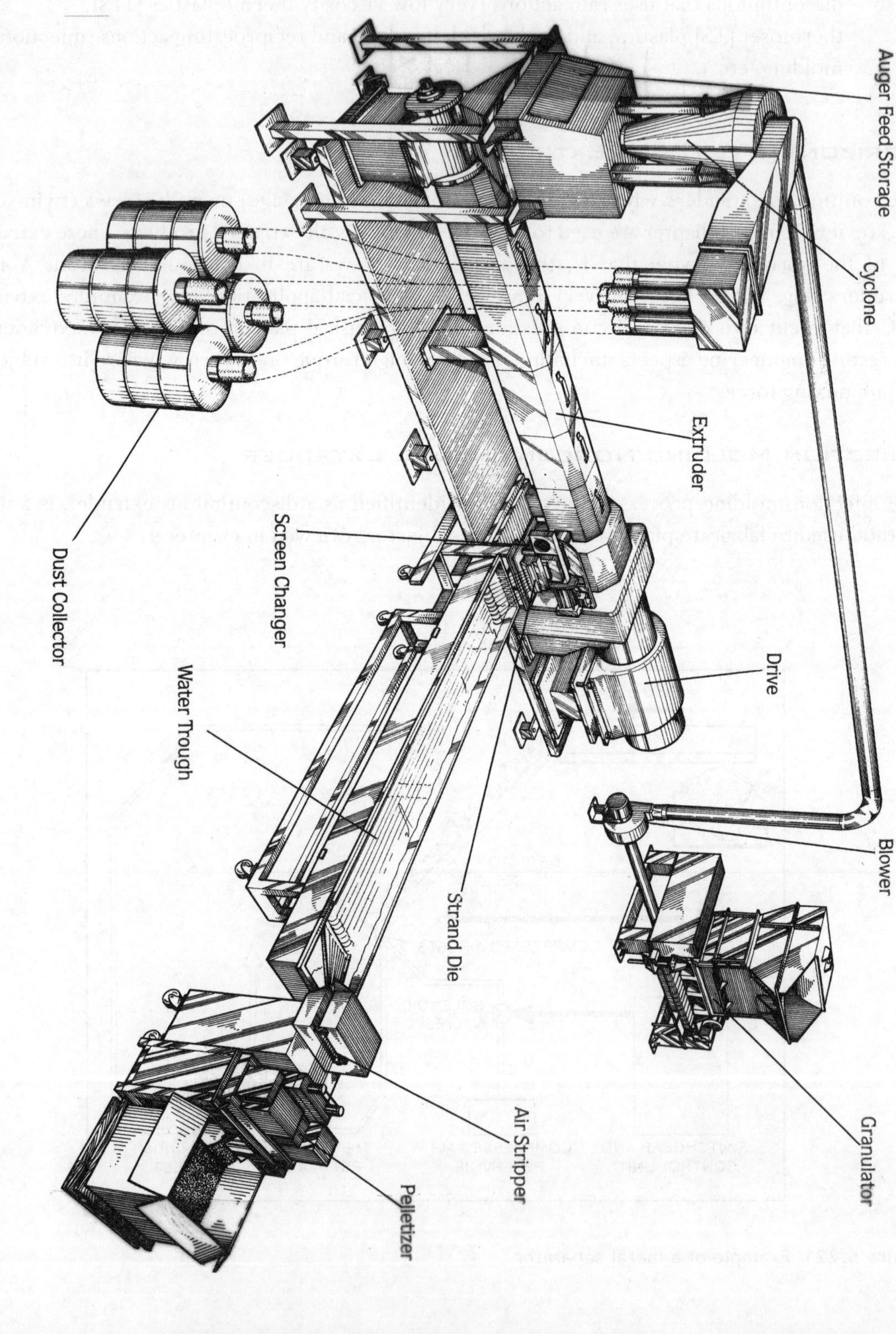

Figure 5.222 Example of pelletizing plastic extruded strands

GEOMETRIC DESCRIPTION	CUBE	CYLINDER	CYLINDER WITH ROUNDED EDGES	CYLINDER FLAT/ CONVEX	CYLINDER BICONVEX	CYLINDER CONCAVE/ CONVEX	CYLINDER BICONCAVE	SPHERES	SHAPE D WITH FLANGE	SHAPE E WITH FLANGE	SHAPE F WITH FLANGE
PROCESSING INFLUENCES	NO PARTICULAR INFLUENCES	NO PARTICULAR INFLUENCES	VISCOUS FLOWING	VISCOUS FLOWING, SHAPE IS DEPENDENT ON FLOW INDEX,m, AND SURFACE STRESS	VISCOUS FLOWING AND ELASTIC STRESSES - THE COLDER THE ORIFICE THE MORE CONVEX THE SHAPE	DENSITY CHANGE AND CONTRACTIONS WITH COLD FACE	VISCOUS FLOWING AND ELASTIC STRESSES - THROUGH HOT ORIFICE	SPECIAL SHAPE OF E — SURFACE STRESSES	SHAPE AS D, BUT WITH FLANGED EDGE CAUSED BY KNIFE STROKE	SHAPE AS E, BUT WITH FLANGED EDGE CAUSED BY KNIFE STROKE	SHAPE AS F, BUT WITH FLANGED EDGE CAUSED BY KNIFE STROKE
PELLETIZING SYSTEM	STRIP DICING	STRAND PELLETIZERS OR AIR PELLETIZER OR FORCED CONVEYANCE PELLETIZER	AIR PELLETIZER	AIR PELLETIZER OR AIR/WATER PELLETIZER OR UNDERWATER PELLETIZER OR KNIFE ROTOR PELLETIZER	AIR/WATER PELLETIZER OR UNDERWATER OR KNIFE ROTOR PELLETIZER	AIR/WATER PELLETIZER OR UNDERWATER OR KNIFE ROTOR PELLETIZER	AIR/WATER PELLETIZER	AIR/WATER PELLETIZER OR UNDERWATER	AIR/WATER PELLETIZER OR UNDERWATER	AIR/WATER PELLETIZER OR UNDERWATER	AIR/WATER PELLETIZER OR UNDERWATER

Table 5.73 Examples of different pellets

RAM EXTRUDER

As explained in this chapter, the ram extruder is used to process plastics that basically are difficult to melt because they have very narrow melting ranges, such as certain fluoroplastics and UHMWPEs. Examples of three different types of ram extruders are shown in Figures 5.224 and 5.225. Figure 5.226 provides an example of a ram speed process control system.

Two types of common commercial ram extruders are vertical and horizontal extruders, where the direction of the ram motion and the extrudate are, respectively, vertical and horizontal. The fundamental working principles of the two pieces of equipment are the same. The key difference between them is the method of extrudate support. In horizontal ram extrusion, a tray, moving belt, or other similar means can support the extrudate. In the vertical machine, a mechanical brake may be needed to protect the extrudate and provide back pressure for coalescence. This brake is usually a chuck, collet, or a gland that grips the extrudate and applies a controlled amount of pressure (146).

In both vertical and horizontal machines, a metered quantity of granular powder is charged to the feed section of the die. This end of the die is cooled to allow easy flow of the plastic into the die. For example, polytetrafluoroethylene (PTFE) particles and agglomerates become sticky and powder flow suffers when heated above approximately 25°C. The next step is compaction of the powder and pushing it into the heated segment of the die by the forward action of the ram. Repetition of these steps advances the compacted plastic through the heated length of the die where sintering of PTFE takes place. The adjacent charges of the plastic are welded to each other

Type of pelletizing system	Typical materials processed	Pellet shapes	Throughput rate, pounds per hour
Cold cutting			
Strand granulators			
With water baths	Polystyrene, polyamide, polycarbonate, and polyethylene terephthalate	Cylinders	8,000
With automatic strand pulling	polystyrene, polyamide, polycarbonate, and polyethylene terephthalate	Cylinders	6,500
Underwater	Polyethylene terephthalate, polyamides, polypropylene, and polycarbonate	Cylinders	26,000
Strip dicing			
With longitudinal and cross cut	Polyvinyl chloride, ABS, and polyethylene terephthalate	Cubes	11,000
With stairstep cut	Polyvinyl chloride, ABS, and polyethylene terephthalate	Cubes	11,000
Hot die-face cutting			
Air			
Centrifugal	Rigid and flexible polyvinyl chloride	Lenses, spheres, ellipsoids and cylinders with rounded faces	4,500
Eccentric	Rigid and flexible polyvinyl chloride, and crosslinkable polyethylene	Lenses, spheres, ellipsoids and cylinders with rounded faces	9,000
Air/water			
Eccentric with flat water stream	Low- and high-density polyethylene, polystyrene, ABS, and polymethyl methacrylate.	Lenses, spheres, ellipsoids and cylinders with rounded faces	12,000
Centrifugal with water ring	Low- and high-density polyethylene, poly styrene, ABS, and polypropylene	Lenses, spheres, ellipsoids and cylinders with rounded faces	15,000
Underwater	Low- and high-density polyethylene, polypropylene, and polystyrene	Lenses, spheres, ellipsoids and cylinders with rounded faces	50.000

Table 5.74 Descriptions of various pelletizing methods

under pressure while sintering is taking place. The essential elements of ram extrusion processes are plastic feed and compaction, sintering, and cooling.

Feed The main function of the feed section is to provide repeated individual charges of plastic with uniform weight during every cycle of the process. The PTFE plastic must be distributed evenly in the feed port at the cold end (start) of the die (Fig. 5.227). It is important to water cool the feed section and the beginning section of the die where the ram penetrates, preferably to 21°C. Temperature must be kept above the dew point, particularly if the extruder is operated in a humid area. Failure to maintain the temperature above dew point will lead to condensation of moisture in the feed cup and result in defects in the part. Increasing the temperature of the plastic accelerates the decrease in plastic flow.

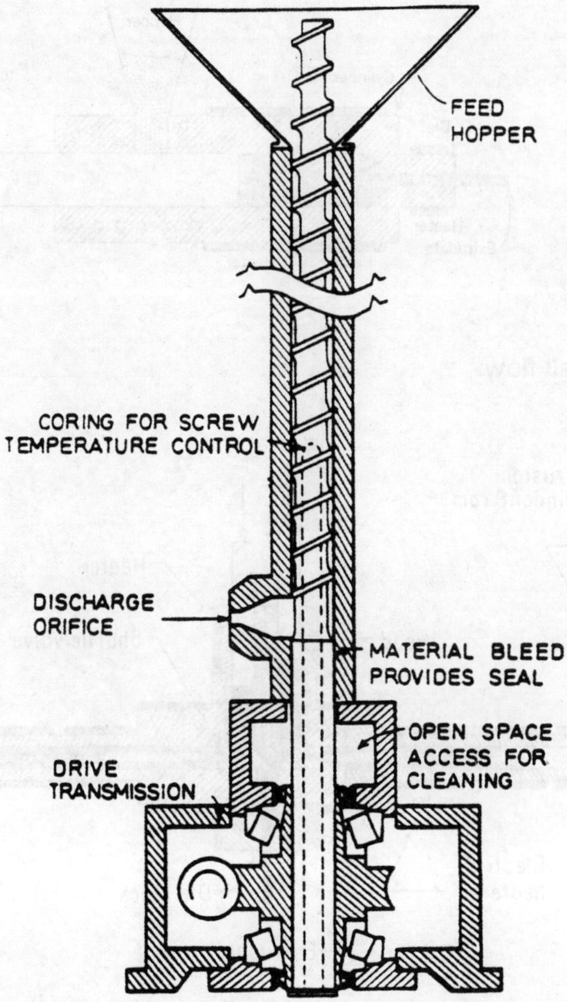

Figure 5.223 Schematic of a vertical extruder

Both uniform distribution and uniform charge weight of the feed plastic are important. The uniformity of the distribution is especially critical for products such as the extrusion of thin-walled (< 2-mm-thick) tubing. The presence of the mandrel in the middle of the die forms an annulus, which is where plastic must enter. The mandrel blocks the free movement of the plastic. An increase in the size of the mandrel to produce larger-diameter tubing intensifies the importance of uniform plastic distribution. The other factor is the poor flow of plastic under pressure, which means that little flow can be expected inside the die during compaction by the ram. To extrude a high-quality

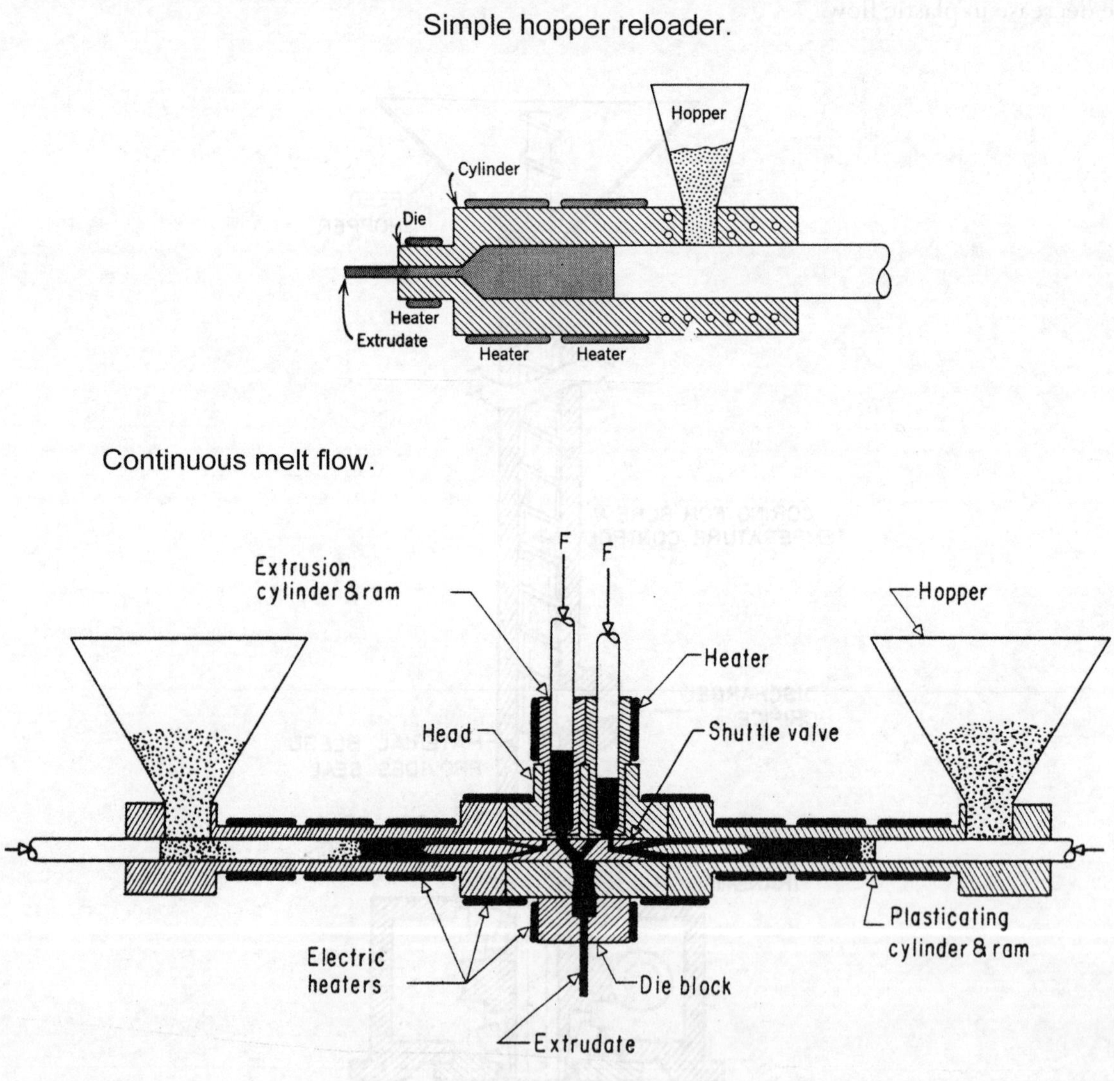

Figure 5.224 Examples of continuous ram extruders using a single hopper reloader and a two-hopper loader

tubing, the plastic must be uniformly fed to the die; that requires excellent plastic flow in addition to uniform charge weight.

A solid rod die, where no mandrel is present to restrain flow, is less sensitive to the uniformity of plastic distribution than a tubing die. The rod properties are sensitive to the uniformity of the weight of each charge. Variations in plastic distribution across the cross-section can also lead to the curving of the rod, which should be normally straight. Variations in the charge weight cause variabilities in the extrusion rate and loss of productivity. Variation of apparent density will give nonuniform properties.

Compaction The ram has two modes of motion: advancing and retracting. After the plastic has been fed into the feed section, the advancing motion of the ram compresses it. The single charge of plastic is squeezed between the bottom surface of the ram and the top surface of the previous charge (Fig. 5.227). This is the mechanism of preform development from the plastic inside the die.

The force of the ram is not the key factor in generating the compaction pressure as long as the ram has sufficient tonnage or force, which is determined by the capability of the hydraulic piston/cylinder. The main determinant of the preforming force is back pressure that is generated by what

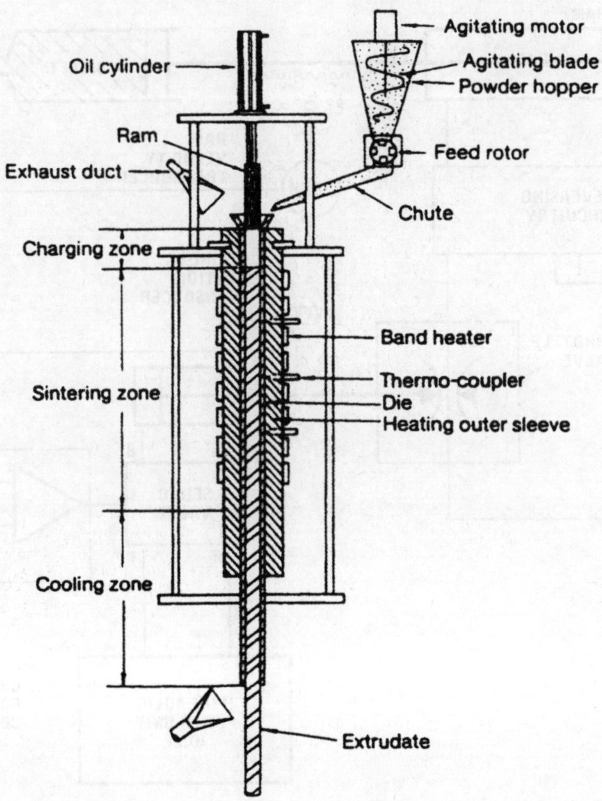

Figure 5.225 Vertical ram extruder

occurs farther down inside as well as outside the die. Back pressure in ram extrusion can be defined as the resistance of the part inside and outside the die to advancing movement. This resistance comprises two forces: the friction force developed between the molten PTFE and the inside wall of the die, and the mechanical break, which can be installed near the exit point of the part from the die (Fig. 5.228). At some point, the plastic does not compact and the ram force overcomes the back pressure because of the continued advancing movement to complete its stroke. At this point, the length of one plastic charge advances the product, giving the ram extrusion process its name.

The compaction pressure of the ram should be in the right range. Estimation and experimentation best define this range. The pressure required to compact the powder into a dense preform with minimum void content defines the lower end of the range. The pressure at which poker chipping (fracturing at the interface of adjacent charges) in the part begins determines the upper end of the range. This means that increasing the pressure does not help the welding of the charges to each other beyond a certain limit. The pressure limits vary from one plastic to another. Acceptable operating compression pressures of presintered PTFE powders may be as low as 5 MPa to over 100 MPa. General-purpose plastics can exhibit poker chipping at pressures as low as 10 MPa.

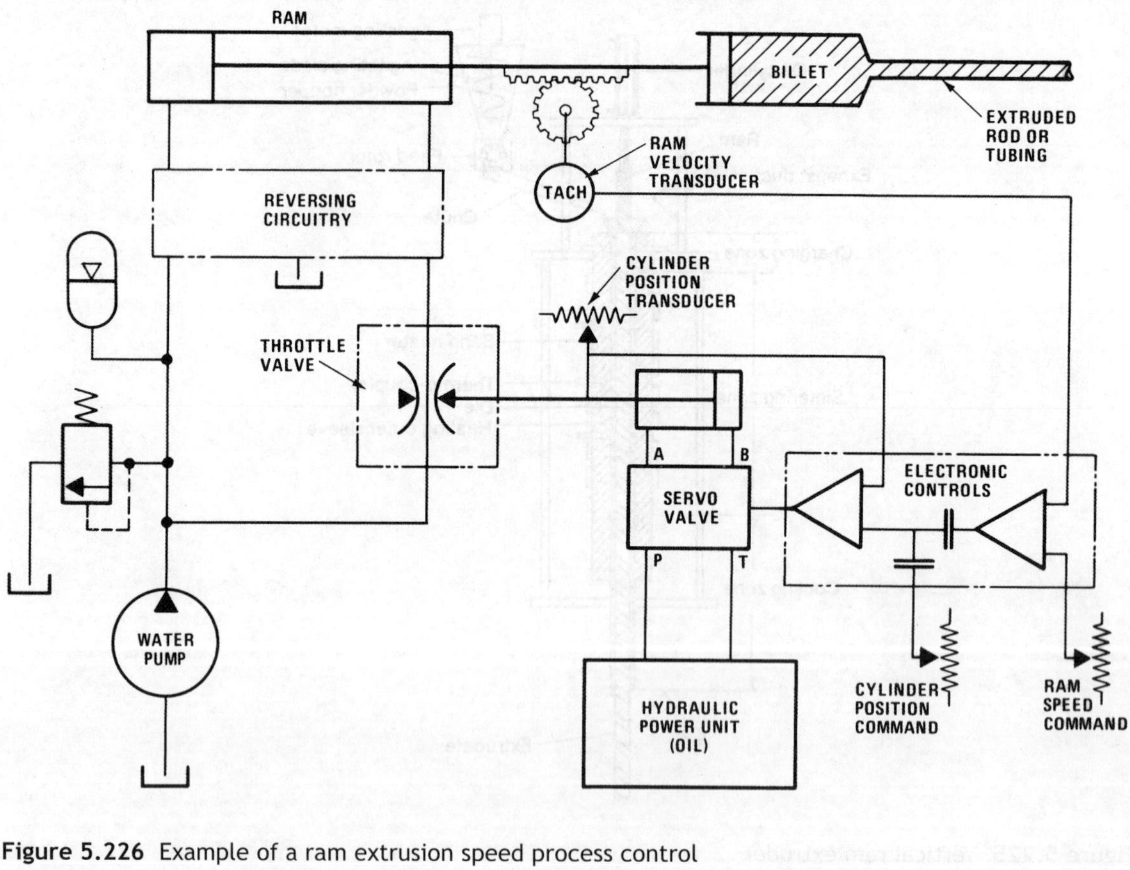

Figure 5.226 Example of a ram extrusion speed process control

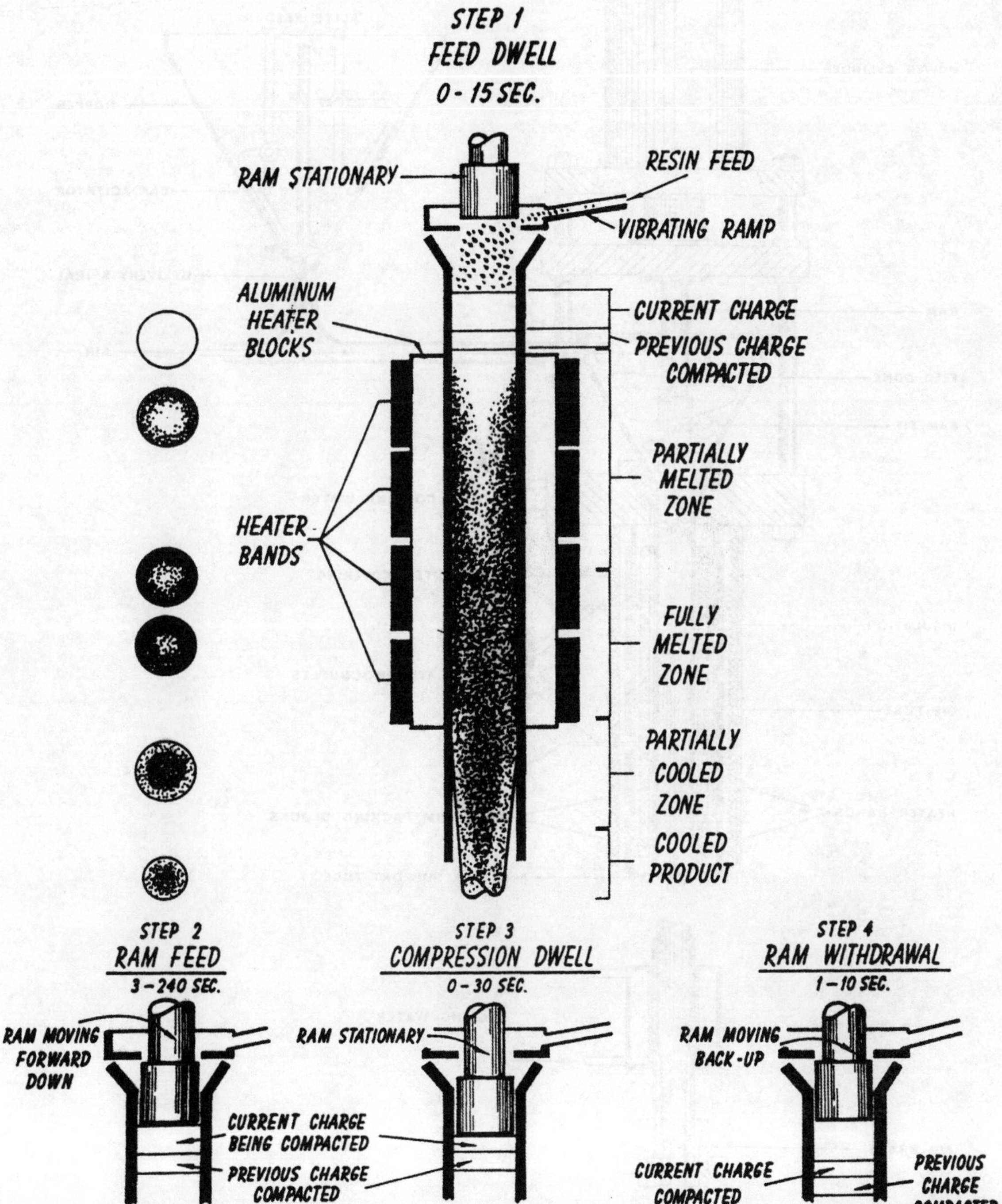

Figure 5.227 Ram extrusion cycles

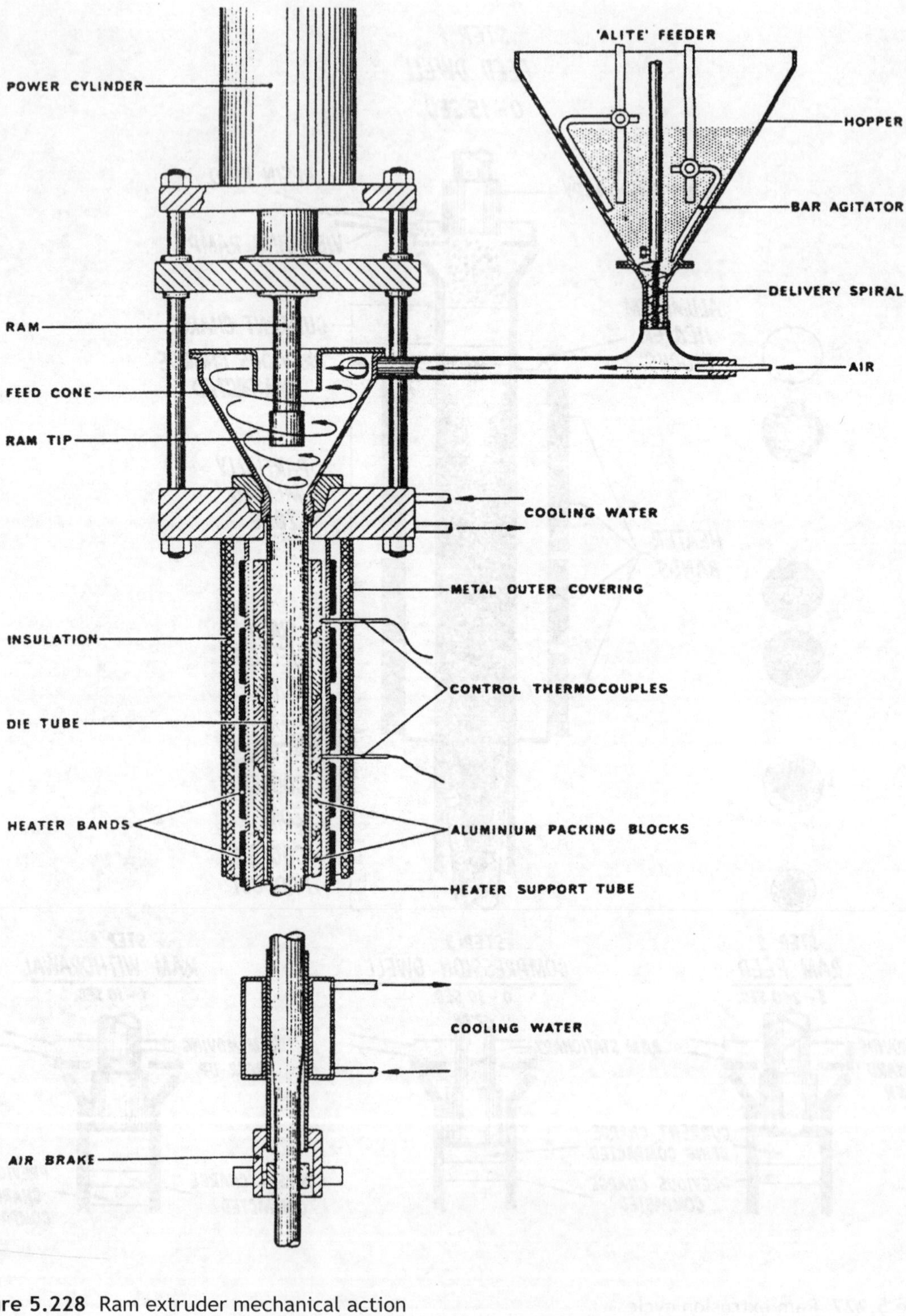

Figure 5.228 Ram extruder mechanical action

General-purpose free-flow plastics tend to build up too much back pressure (compression pressure), which causes poker chipping. Presintered materials do not build high compression pressure under most conditions. Even at high pressures, parts made from these plastics resist poker chipping. Parts look grainy or chalky when compression pressure is insufficient because of the presence of voids. The same problem will be encountered if the plastic is not given sufficient time to sinter, that is, the extrusion rate is too high. Presintered plastic cannot overcome an extrusion rate that is too rapid.

Another way to enhance the weld strength between the charges is to keep a small length of the die below the end of the ram stroke cool. This will keep cool the top surface of the previous charge, against which the plastic is being compacted. Other benefits include the prevention of certain types of plastic skin formation, that is, plastic buildup on the die wall, and the elimination of "springback" of the plastic after the removal of the compaction force at the onset of ram retraction. Springback is simply the expansion of the compressed plastic opposite the pressure direction, that is, opposite the extrusion direction. Presintered plastics have a greater springback than the general-purpose powders. It also increases with an increase in the ratio of cross-section to the perimeter of the part. This means that springback is more important with large-diameter rods and tubes.

Sintering and Cooling After compaction, the preform is advanced into the first heated zone of the die for sintering (Fig. 5.25). Sufficient heat must be supplied to the preform to raise its temperature to above the melting point of the plastic. There are two considerations in the selection of the die temperature. First, melting temperature of PTFE increases as a function of the plastic pressure. Second, the temperature used should be high enough for PTFE to melt and sinter completely during the residence time in the heated length of the die. Adequate pressure should also be applied to the part to eliminate the voids and obtain a strong welded bond to the adjacent charges. Finally, the part must be cooled at a rate that will lead to the desired crystallinity content (chapter 1).

Sintering Heat Requirement Heating up in ram extrusion takes place by a conduction mechanism. Thermal conductivity and heat capacity of PTFE, die temperature, and preform size determine the time required to reach the sintering temperature. The preform has to be held at the sintering temperature for a period of time until densification and void elimination processes have been completed. The extrusion rate can be calculated by dividing the heated length of the die by the minimum time required for achieving acceptable sintering. For example, assume the minimum required sintering time for a solid rod is two minutes at 398°C and the heated length of the die is 600 mm. The maximum extrusion rate is 300 mm/minute (600 mm divided by two minutes) or 18 m/h. Typically, extrusion rates in the industry are 50% to 60% of the maximum rate (Table 7.10). The excess residence time helps reduce the temperature gradient in the part, thus minimizing residual stress. It also allows void reduction and elimination of charge lines.

Table 5.75 gives a summary of minimum sintering times required for solid PTFE rods from 3 to 25 mm in diameter. The minimum required time increases with the diameter of the rod, as does the shrinkage. The rise in the minimum time is not linear with diameter but proportional to the square of the diameter. The selected temperature of the die is 395°C (743°F) in this table. The recommended

Rod Diameter, d mm (in)	Average Shrinkage, S %	Minimum Required Heating Time, min, 395°C (743°F)	Representative Die Length, mm (in)*	For Representative Die Length: Maximum Extrusion Rate, m/h (ft/h), 395°C (743°F)	For Representative Die Length: Typical Extrusion Rate, m/h (ft/h), 395°C (743°F)
3 (0.12)	6	1	250 (10)	15 (50)	10 (33)
6 (0.25)	8	1.25	500 (20)	24 (80)	12 (40)
13 (0.50)	11	5	1,000 (40)	12 (40)	8 (25)
19 (0.75)	12.5	11	1,500 (60)	8 (25)	4.5 (15)
25 (1.0)	13.5	20	2,000 (80)	6 (20)	3 (10)

* *Die lengths considerably greater or smaller than the values given can be used with ram extrusion resins. Generally, shorter dies must be used when ram extruding general-purpose resins.*

Table 5.75 Average shrinkage, required heating times, and representative die lengths for ram extruders used with PTFE fluoropolymer plastics for ram extrusion

range is 370° to 425°C, but the temperature should be kept below 400°C to minimize the degradation of the plastic. Overly long residence time in the die can also lead to degradation.

Sintering temperature and residence time have profound effects on the quality of parts made by ram extrusion. Figure 5.229 shows tensile strength of a rod, which is a measure of quality, as a function of extrusion rate. Residence time has an inverse relationship with extrusion rate and clearly has a strong positive effect on tensile strength. Residence time is also a function of the shape of the part being extruded in addition to size and extrusion conditions. For example, extrusion of a rectangular part with the same sized side as the diameter of a rod requires more heat or a higher residence time (20% to 30%) than the rod.

Pressure Requirement Pressure is required while the plastic is being heated for the sintering to take place properly. The magnitude of pressure needed is much smaller than the preforming pressure in compression or isostatic molding techniques. Heat and pressure act interchangeably, that is, lower pressure is required when the plastic is being compacted at an elevated temperature. A simple way of visualizing the interaction of pressure and temperature is to note that raising the temperature softens the plastic, which allows the application of a lower pressure.

Another aspect of ram extrusion is the much shorter sintering time. Here is how the sintering takes place. The plastic expands axially and builds up internal pressure as it is heated without the ability to expand radially because of the die. Expansion of the plastic is driven by a 15% increase in the specific volume of the plastic from 0.434 cm^3/g for a 100% crystalline phase to 0.500 cm^3/g for a 100% amorphous phase (presintered plastic has a crystallinity of approximately 50%). The expansion occupies the voids and reduces the void content of the preform. This internal pressure becomes the driving force for rapid coalescence of the molten particles and consolidation of the preform into a mass of gel. The internal pressure of the resin helps in the welding of the charge-to-charge interfaces. Insufficient pressure leads to voids being left in the part. For example, the center of a solid rod will look chalky, which is evidence of voids; away from the axis of the rod there will be fewer voids. A void-free rod appears somewhat translucent.

It is important to compact the plastic sufficiently to reduce its void content in the feed end of the die. It is virtually impossible to sinter a preform that has a large void content. The molten plastic restrained from expansion wets the surface of the die and exerts a significant friction force on the surface resisting motion. Adequate ram force (or pressure) to overcome the frictional force developed between the molten PTFE and the die wall will allow the extrusion to take place and not stall. This force is the source of back pressure, which can be calculated by dividing the force by the area of exertion.

The resistance force of a mechanical brake is added to the back pressure if one is present in the system. Ram force must equal the total resistance presented by back pressure and the brake. Back pressure depends on the friction force, which itself depends on the following factors:

1. Temperature profile of the hot length of the die, surface smoothness of the die, and the type of metal from which the die is constructed all impact the coefficient of

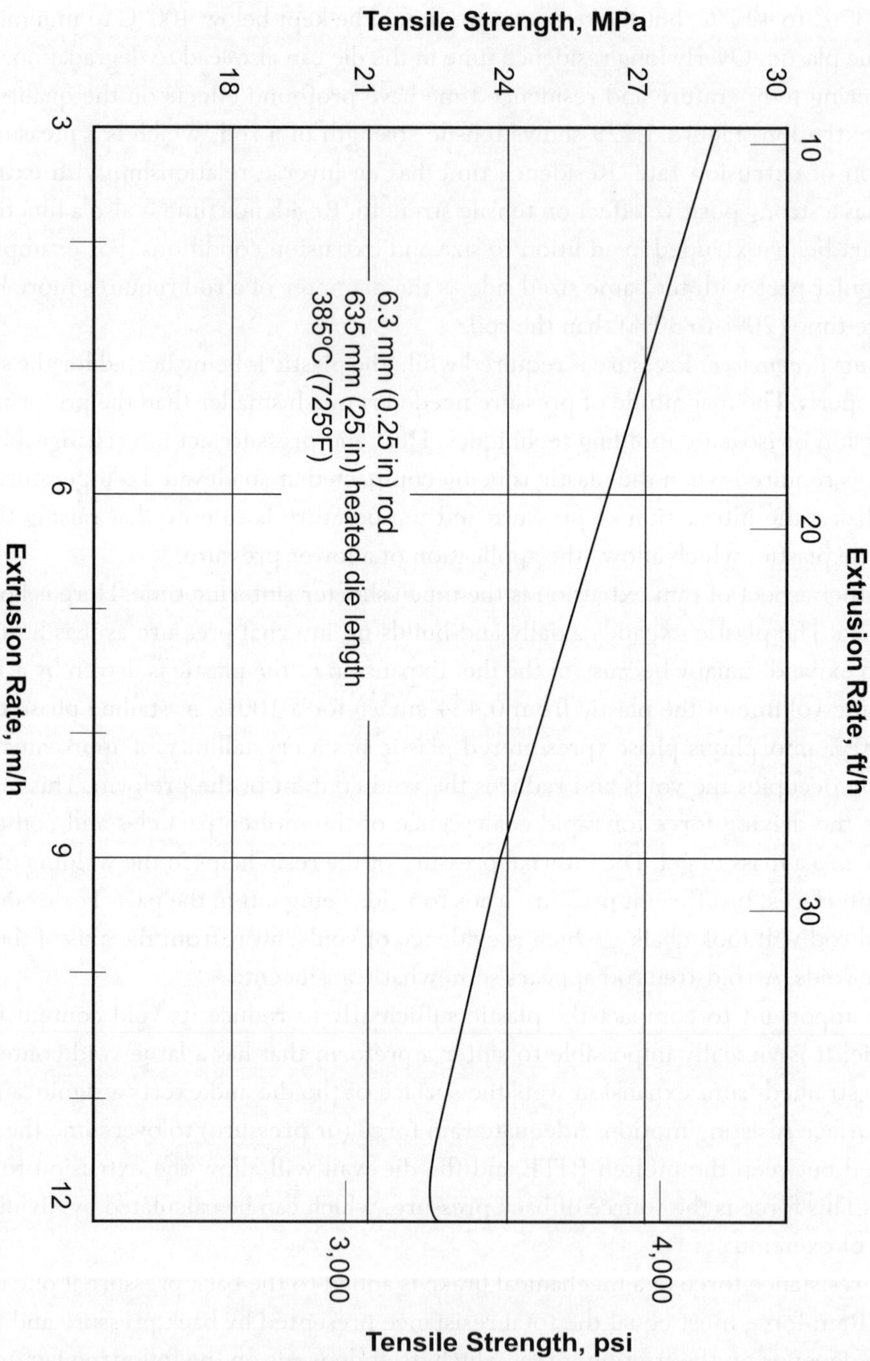

Figure 5.229 Ultimate tensile strengths vs. ram extrusion rates

friction, which along with the internal pressure developed by the plastic, determines the quantity of the friction force.

2. The pressure buildup of the plastic upon melting, which depends on temperature and void content of the plastic.
3. An increase in the area of the contact between the PTFE part and the die, excluding the cooling zone, elevates back pressure. In the cooling zone, the part shrinks and separates from the die wall, therefore developing no back pressure. Higher back pressure is obtained by an increase in the length of the inside die diameters (L/D) of any of the three zones of ram penetration, cooled length below the penetration zone, and the heated length of the die.
4. Increasing the extrusion rate results in a higher back pressure.
5. The longer the charge length, the higher the back pressure, particularly if the ram penetration zone is water cooled.

Temperature of the die has a fairly complex effect on back pressure development. As temperature increases, back pressure decreases, as discussed previously about the entry end of the die. The force necessary to extrude the powder at room temperature is very high because of the large back pressure. It will begin to rise at some point above the melting temperature of the plastic, on the order of 40°C. This value varies depending on the type of the plastic and other extrusion parameters.

In vertical ram extrusion of large cross-section solid parts, supporting the weight of the extrudate is important because it is suspended and its weight counteracts the back pressure. For example, a 50-mm-diameter extrudate is heavy and must be either cut in small sections or it must be supported by a higher back pressure. External means, such as a mechanical brake, would be required if longer lengths are desired. In the absence of weight support, the weight of the extrudate strains the molten plastic, which can lead to melt fracture or rupture of the melt.

Extrusion of hollow parts, such as pipe liners, requires the use of a mandrel, as seen in Figure 5.230. There are two kinds of mandrels: the ram and mandrel are connected and reciprocate together (driven) or the ram reciprocates and the mandrel is fixed (Fig. 5.231). The latter is generally used to extrude presintered PTFE. Back pressure is a strong function of the ratio of the area of contact between the plastic and the die-to-ram cross-section area in the extrusion of tubing. Most of the same factors affecting the back pressure of the rod also impact the tubing back pressure.

There are unique mandrel effects absent from the extrusion of solid rods and other parts. The length of the mandrel can affect the back pressure profoundly. It often has a different temperature profile from the die. For instance, consider the case in which the length of the mandrel exceeds the length of the die. Near the end of the die, the tube begins to cool and shrink away from the die and toward the mandrel. At some point, back pressure begins to build up because of the friction force between the mandrel surface and the tube. Choosing the appropriate length of the mandrel can control this type of back pressure. No back pressure from this source will be experienced if the mandrel ends at the beginning of the die cooling zone.

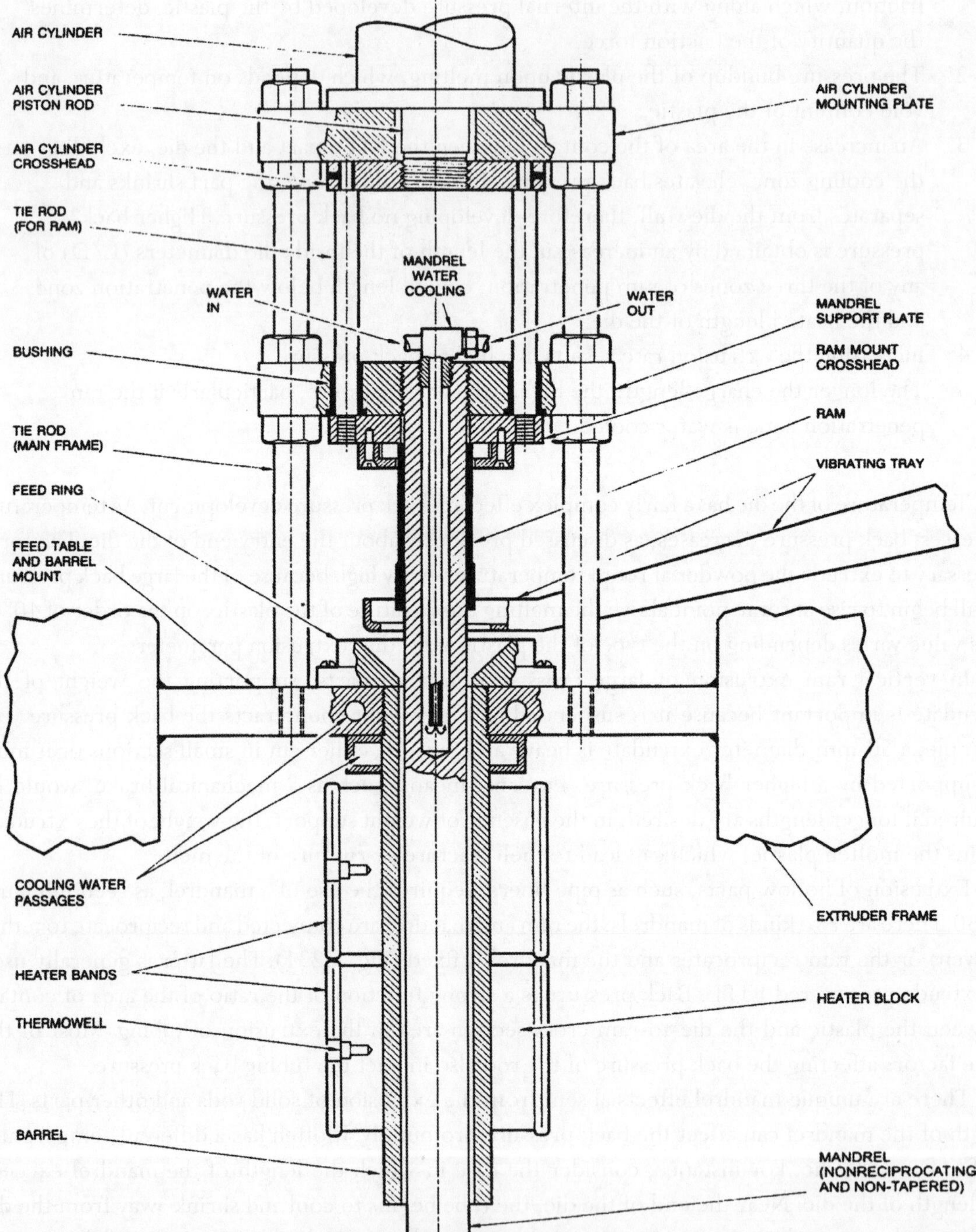

Figure 5.230 Vertical ram extruder for fabricating PTFE tubing

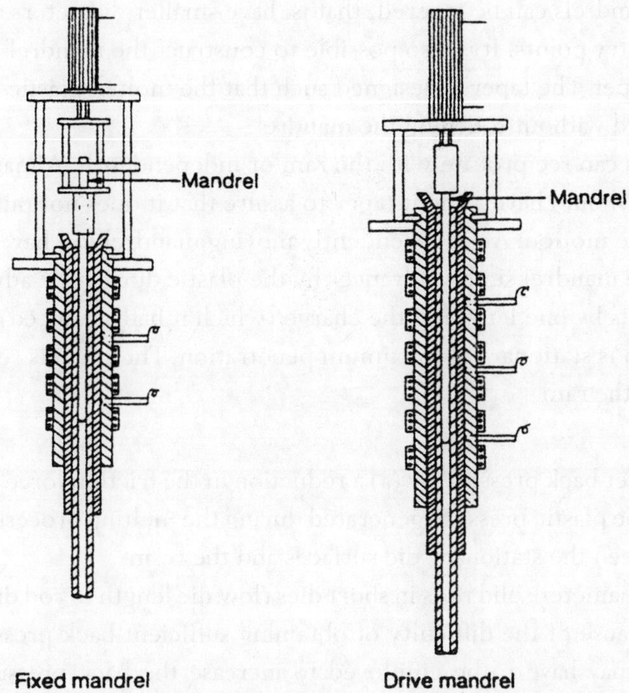

Figure 5.231 Mandrels for ram extruding pipe

Another method for controlling the back pressure of thin-walled tubing is to retain the cooling zone temperature above 200°C, which limits the extent of shrinkage. The tube is sufficiently cool to shrink away from the die but not cool enough to shrink tightly onto the mandrel. The rest of the cooling can be accomplished outside the die, assuming the mandrel does not extend more than 5 to 10 cm beyond the die exit. The best setup for extruding a thin-walled tube is to employ a nontapered, nonreciprocating mandrel that extends a few centimeters outside the die. The shrinkage of the tube onto the mandrel over a short length does not significantly increase the back pressure but improves the appearance of the inside surface of the tube.

If the length of the mandrel is so short that the extrudate passes over the end of the mandrel while the plastic is molten, poor dimensional uniformity is obtained. The cross-section may not be round, and the diameter of the tube may alternate between larger and smaller than the desired tube diameter, known as "hour glassing." A similar phenomenon is observed when the tube exits the die, supported by the mandrel, in a gel state. The surface defects of hour glassing may resemble poker chipping. The severity of the defects grows with a decrease in the tube's wall thickness. Tube stock shapes should be ram extruded from presintered plastic. If tubing is to be made from general-purpose powder, the following machine designs will help reduce back pressure and prevent poker chipping:

1. Stationary mandrels can be tapered, that is, have smaller diameters at the exit ends than at the entry points. It is also possible to construct the mandrel for two cylinders joined by a taper. The taper is designed such that the molten plastic would be able to expand inward without touching the mandrel.
2. The mandrels can reciprocate with the ram or independently. A mandrel moving with the ram should have a slight taper to assure that it does not pull back the tube during reverse motion. An independently moving mandrel can have a constant diameter. The mandrel simply advances by the plastic during the advancement of the ram. It retracts by one length of the charge (which it had advanced) during the dwell, while the ram is stationary at maximum penetration. The cycle is repeated after the retraction of the ram.

These designs lower back pressure by (a) a reduction in the friction force exerted on the die and mandrel surfaces by the plastic pressure generated during the melting process and (b) a reduction in the contact area between the stationary die surfaces and the resin.

Producing large-diameter solid rods in short dies (low die length to rod diameter ratio) presents a special problem because of the difficulty of obtaining sufficient back pressure. All the strategies previously discussed may have to be employed to increase the back pressure adequately for the compaction of the powder. They include cooling a length of the die below the entry point, tapering the die to have a smaller diameter at the exit than at the die entry, and application of a mechanical brake to the extrudate outside the die.

Cooling Cooling is the last step in ram extrusion, as in other fabrication techniques. It determines the crystallinity and shrinkage of the part. Crystallinity determines the properties of the rod, tube, and other cross-sections. Shrinkage determines the final dimensions of the parts. Often, air quenching is the means of cooling, in which the gel leaves the die and is exposed to forced air. This method yields minimum crystallinity and shrinkage. Typically, crystallinity is in the range of 55% to 60%. Shrinkage of the solid rods ranges from 10% to 14% of the die cavity diameter. It can be held constant by adjusting the die temperature profile, extrusion pressure, cooling rate, and sintering time. Presintered plastics shrink less than the general-purpose free-flow powders. Hollow parts, such as tubing, shrink less (4% to 6%) than solid parts, such as rods.

Internal stresses build up in ram-extruded parts depending on the cooling rate, part geometry, die temperature profile, and other process variables. Length and temperature profiles of the cooling section of the die can be designed to achieve controlled cooling rates. Large rods (> 50 mm) are susceptible to shock and even internal fracture if cooled too rapidly. Control can be achieved by extending the cooling section of the die by adding an insulated tube. This tube may have to be heated to assure a controlled cooling rate.

It is possible to design the cooling step to obtain precise sizes required for parts that cannot be machined or ground to size. In some applications, particularly when precise machining is to take place, the products of ram extrusion can be annealed in a second process step. The residual stresses in the part are reduced or eliminated by annealing, which may also cause some size alteration of the part.

Die Skin Formation An inevitable event during the ram extrusion of PTFE is the buildup of a thin plastic skin on contact surfaces of the die and the mandrel. The plastic, which is fairly soft even in the solid phase, passes through the die under high pressure to overcome the back pressure generated by friction. This force gives rise to a shear stress at the interface of the die wall and the resin surface. PTFE has fairly low shear strength and cannot stand up to the wall's shear stress, particularly at the minute pores arising from the imperfect smoothness of the die wall. Consequently, plastic is sheared off from the gel body and is deposited on the die wall. The same argument can be applied to the mandrel surface, where shear stress exceeds the shear strength of the molten PTFE.

The plastic coats the wall of the die, mandrel, and other contact surfaces. Once the surface pores have been coated, no more coating will take place until some of the plastic is degraded and is removed. At this point, rough pores are exposed that are refilled by shearing off new plastic. This skin formation is normal and affects the die because it resides in the surface pores. PTFE degradation generates corrosive products, such as hydrofluoric acid, which can corrode the die to different extents, depending on the material of construction and the temperature of the die (chapter 17). The skin immediately next to the metal wall turns black over time because of carbonization of PTFE. The layers closer to the extrudate are renewed continuously and thus stay degradation free.

Die Maintenance The formation of normal skin during the normal operation of the ram extruder does not interfere with process performance or product quality. After shutdown and cooling, the skin shrinks and pulls away from the die wall, taking away parts of the immediate black buildup. If the extrusion is resumed without cleaning the die, the skin will be carried off by the extrudate and appears as black spots on the PTFE part. After a new skin is established, the black spots will disappear and the extrudate appearance will return to a white color.

In commercial operations, the small dies (less than 2 to 3 cm in diameter) are cleaned at the beginning of the extrusion by manually inserting a small cylinder of soft metal wool in the die ahead of the plastic. The extruder is then started and the metal wool is allowed to be driven by the plastic through the die. One or two doses of metal wool usually abrade most of the die wall buildup, and after a short transition, the extruded product will be back to normal. Larger dies have to be cleaned off-line by mechanical techniques or by placing them in a hot oven and burning the skin off at a temperature in excess of 450°C.

Normal operations take place when the die and ram clearing is roughly in the range of 100 to 200 μm. If the plastic contains particles that are smaller than the ram clearing, they may enter the clearing space and get smeared, that is, highly sheared by the motion of the ram. The repetition of this process will lead to the buildup of these smeared particles at the entry area of the die. The sheared particles will enter the plastic being compacted and form defect sites because of the difficulty of coalescing these damaged particles. The flaws form voids and weaken the extrudate mechanically. The best way to prevent this problem and avoid excessive ram clearing is to cool the die entry section with water and control the particle size distribution of the plastic to avoid particles smaller than the clearing.

Another type of skin formation is observed in the form of periodic bands of loose particles around the extrudate. The particles are highly sheared and accumulate somewhere in the sintering section of the die. Typical symptoms observed sometimes before any bands are seen include a reduction in the diameter of the extrudate and a roughening of its surface. The cause of the formation of these bands is primarily cooling the gel too rapidly. Other factors include using certain die metals (such as unplated steel), too much ram clearing, and the presence of fines in the powder.

Ram Extrusion Equipment Design A key decision in the design of a ram extruder is to select the type(s) of plastics that may be extruded using the equipment. Recognize that general-purpose free-flow plastics require short dies because of the large back pressure associated with these plastics. High back pressure leads to poker chipping in the extrudate, which renders it commercially useless. Short dies can only be operated at low extrusion rates because of sintering time limitations. For example, a 13-mm-diameter solid rod requires a minimum sintering time of five minutes and cannot be extruded in a die longer than 500 mm. This means that a maximum rate of 100 mm/min can be reached, which is about double (Table 5.75) the practical rate that allows production of rod with acceptable quality. In contrast, the maximum die length for a presintered plastic to make the same rod is 1,000 mm. This indicates that for a minimum sintering time of five minutes, the maximum rate is 200 mm/min, which is double that attainable with the general-purpose powder. More interestingly, presintered plastics can withstand much higher back pressure; 100-MPa back pressure has been reported in a die with a die-cavity-surface-to-cross-section-surface ratio of 400:1, without encountering poker chipping.

Table 5.75 can be used as a reference for selecting the die lengths for various solid rod sizes for extruding presintered PTFE powders. Ordinary free-flow plastics must be extruded in shorter dies at slower rates. Conversely, presintered plastics can be extruded through short dies as long as sufficient back pressure can be generated to compact the material. Cooling the die below the entry section can be used to heighten the back pressure for short dies. The recommendations of Table 5.75 can be altered by varying the process conditions (cooling techniques) to produce parts with good properties from presintered powder.

Two types of equipment are commercially used: vertical extruders (Fig. 5.230) and horizontal extruders (Fig. 5.232). In principle, both types are capable of producing rods. Each has certain advantages and disadvantages. For example, it is easier to center the mandrel in a vertical machine, by gravity, than in a horizontal piece of equipment. Horizontal extruders allow production of very long parts more easily than a vertical extruder. Table 5.76 summarizes a comparison of the capabilities of the two types of extruders.

The choice of the extruder depends on all these factors and one cannot be automatically selected over the other. Disregarding type, a ram extruder consists of four basic components:

1. Drive system (hydraulic or pneumatic)
2. Ram
3. Die and mandrel (if one is needed) and heaters
4. Plastic feed system

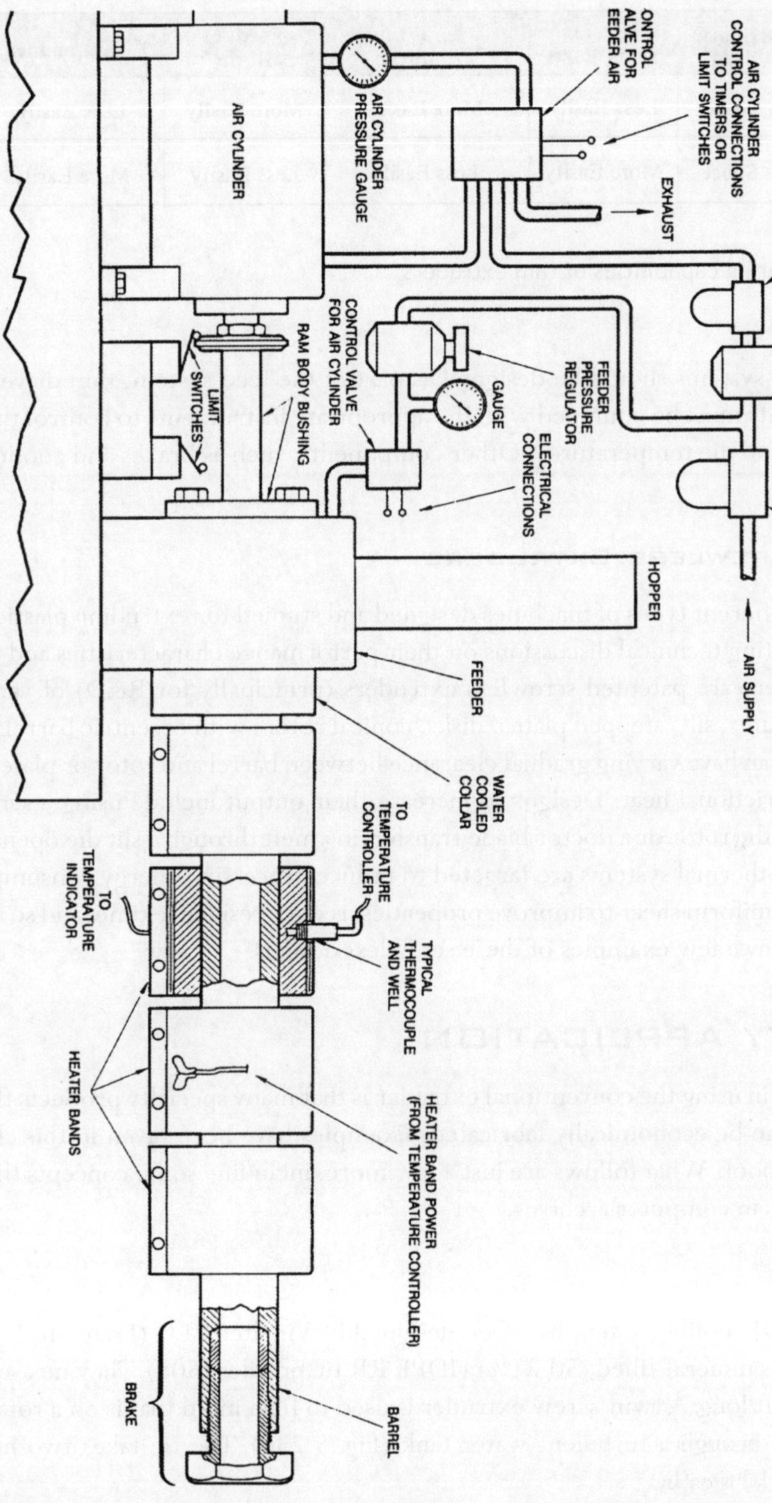

Figure 5.232 Example of horizontal ram extruder for processing PTFE plastic

Extruder	Floor Space Required	Producing Long Parts	Centering Mandrel	Feeding Resin	Maintenance	Cooling the Die
Vertical	Less Space	Less Easily	More Easily	More Easily	Less Easily	Less Easily
Horizontal	More Space	More Easily	Less Easily	Less Easily	More Easily	More Easily

Table 5.76 Comparing capabilities of ram extruders

Water cooling systems should be designed as part of the feed system, ram drive, and the die. All four components must be equipped with the appropriate instruments to control the plastic feed rate, ram timing, and die temperatures. Other components, such as brakes and cooling tubes, may be required.

Disk and Screwless Extruders

There have been different types of machines designed and studied for extruding plastics that usually provide for interesting technical discussions on their performance characteristics and potentials for being realized. There are patented screwless extruders (principally for R&D) of various designs, such as rotating drums, slit/stepped plates/disks, conical rotors with eccentric barrels, and/or ram plasticators. They can have varying gradual clearance between barrel and rotor or plates to plasticize (melt) the TP by frictional heat. Designs to increase their output include using a screw extended from the center of the rotor or a doctor blade transferring melt through a slit die opening. Different designs of these isothermal systems are targeted to reduce processing energy consumption, reduce and provide more uniform shear to improve properties, reduce residence time, and so forth. Figures 5.233 to 5.235 show a few examples of these screwless designs.

SPECIALTY APPLICATION

A major advantage in using the conventional extruder is that many specialty products that have large production runs can be economically fabricated. Examples have been given in this chapter and in other parts of this book. What follows are just a few more, including some concepts that are still on the drawing boards in computer archives.

Railroad Tie

A fabricating special cooling system has been designed by Vinyltec LLC (Franklin, Tennessee) that extrudes and molds mineral-filled (50 wt%) HDPE RR tie profiles (508). They measure 7 in thick, 9 in wide, and 8.5 ft long. A twin-screw extruder is used to fill a mold that is on a rotating carousel that passes slowly through a turbulent water tank (Fig. 5.236). The tie takes two hours to cool. Machine output is 10 ties/h.

VELCRO STRIP

In this process, an extruder feeds plastic melt into a continuous molding system producing Velcro® strips (Velcro USA, Manchester, New Hampshire, a subsidiary of the European Velcro Group). The

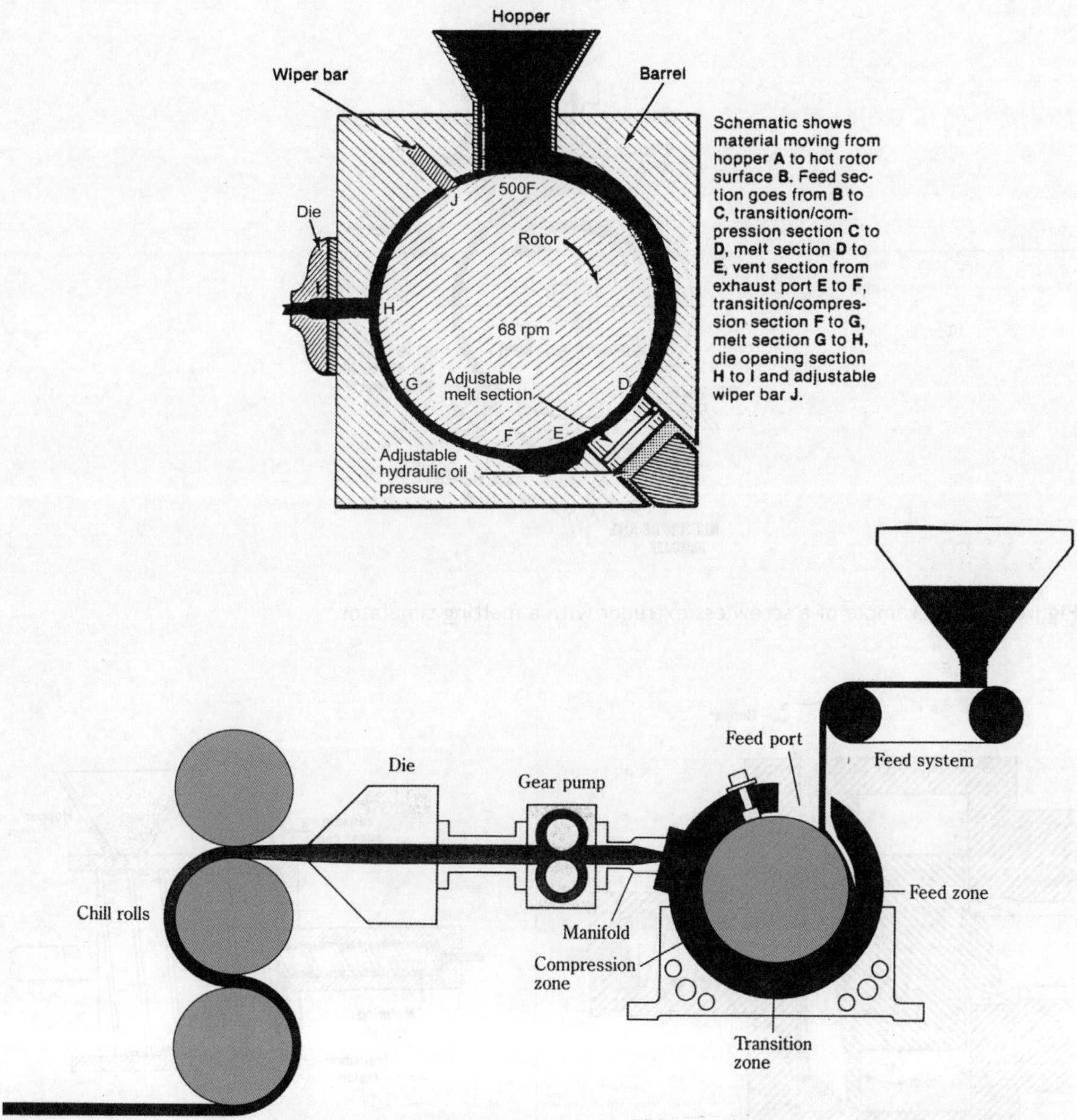

Figure 5.233 Example of a screwless extruder; top view shows cross-section of its rotor shape and bottom view shows a sheet line

equipment molds the strip from the plastic, trims it, conditions it for flatness, applies an adhesive backing, and winds it on a reel (Fig. 5.237). The key elements are a conventional extruder that combines extrusion with injection and a rotating ferris wheel mold. An extruder runs continuously, feeding the melt into the continuously rotating mold through a special adapter mounted on the extruder barrel melt outlet. As an example, a 2-ft (61-cm) diameter mold turns at about 10 rpm

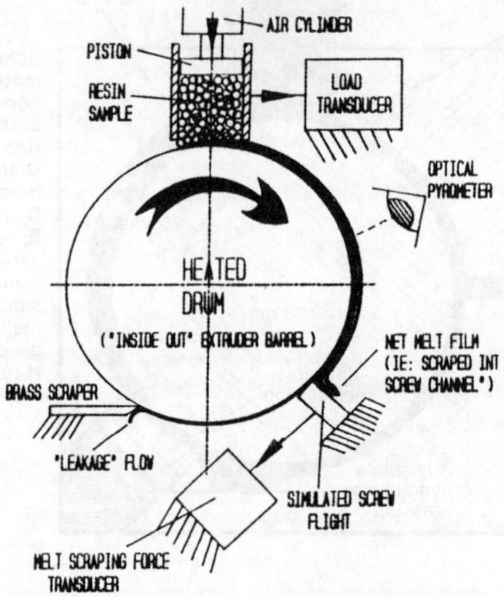

Figure 5.234 Example of a screwless extruder with a melting simulator

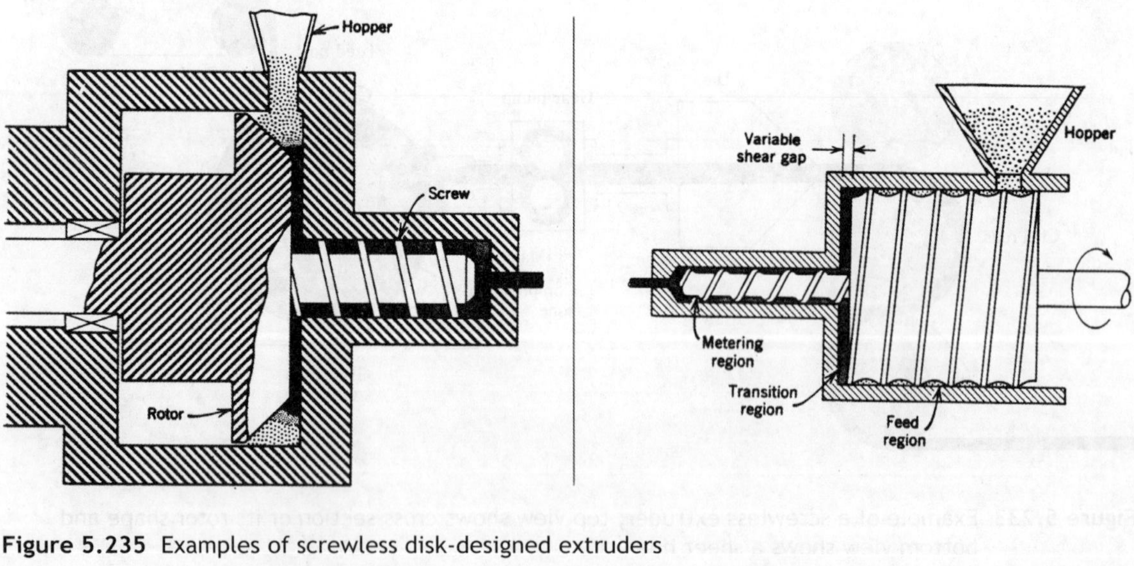

Figure 5.235 Examples of screwless disk-designed extruders

(Fig. 5.238), delivering Velcro® at 60 to 70 ft/min (18.3 to 21.3 m/min). The mold contains more than 15,000 cavities meeting very tight tolerances (Fig. 5.239). Plates (forming cavities) in the mold stack are surface ground to be flat within 0.002 in (0.050 mm) across a 2-ft (0.6-m) diameter. Also, plate thickness, which determines the spacing between adjacent rows of splines as well as the quality of the seals between adjacent plates, is controlled to within 0.0001 in (0.0025 mm).

As the mold rotates past the extrusion injection die head, the melt flows onto the circumference of the mold and is forced by the melt pressure into the cavities. Enough additional melt is supplied to create the base strip at the same time. By the time the mold has completed about half a revolution, the plastic has been cooled and solidified. Mold temperature is controlled by an air plenum located about one-quarter of a revolution before the injection point; the molded Velcro itself is cooled by another plenum ahead of the strip-off point. The airflow is adjusted to keep the nylon (usual plastic used) sufficiently warm and flexible so that the projections can pull free from the undercuts in the cavities without damage during strip-off.

NONCONVENTIONAL EXTRUDING

Figures 5.240 to 5.243 show examples of a continuous extruded tube that travels to a rotary drum thermoformer producing thermoformed products.

Figure 5.243 is an example of exploring new fields of fabricating film directly on farmland. Here, two extruders are producing film that can in turn incorporate seeds, fertilizers, and so on so that they are applied uniformly or at a controlled rate. This type of system, as well as those for extruding other products (pipe, rails, etc.), has been in development for decades, but to date no known commercial applications exist.

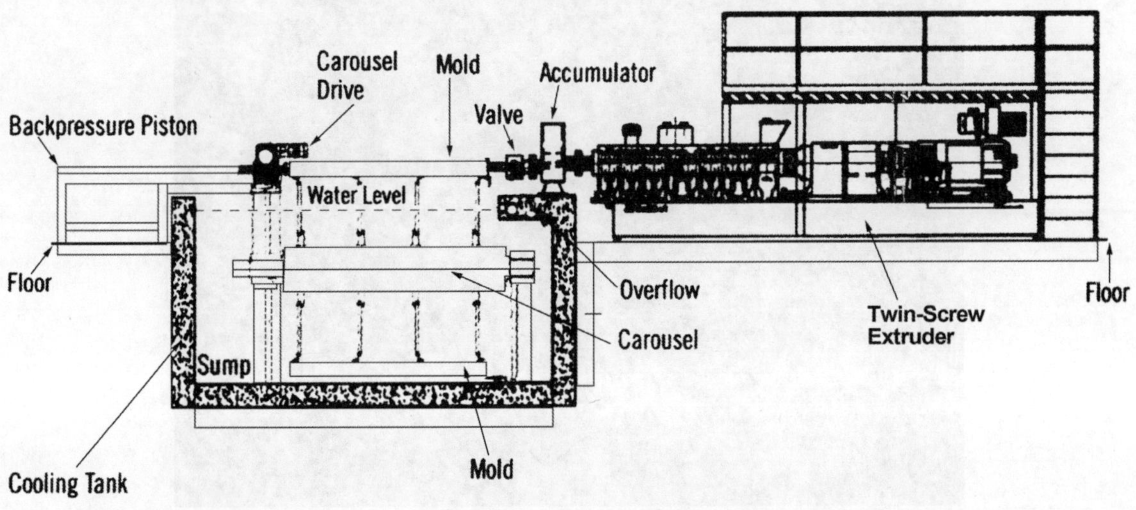

Figure 5.236 Example of combining extrusion and molding PVC railroad ties

TROUBLESHOOTING

Troubleshooting extrusion guides have been reviewed in this chapter and in chapter 27. The following is a summation on running an extruder:

1. Safety gates, guards, and displays should be in place (Fig. 5.244).
2. Wear safety glasses at all times.

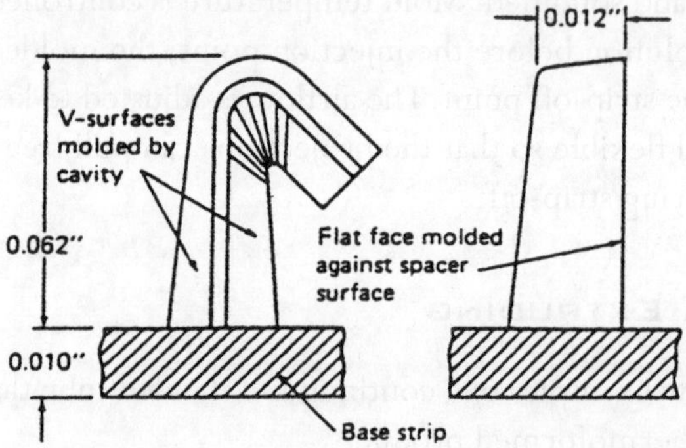

Figure 5.237 Example of a Velcro® spline

Figure 5.238 View of a rotating mold being fed by an extruder

3. Wear hearing protection, if required.
4. Keep the floor free of pellets as they act as ball bearings and can cause bad falls.
5. Avoid standing opposite the die, melt pipe, and head flanges as they can unexpectedly squirt hot plastic, especially during the heat-up and start-up stages.
6. Avoid standing downstream from head or flange bolts and rupture disks, especially at start-up. An overpressure condition can cause a bolt to break and shoot out like a bullet. Contamination can cause rapid screen pack plugging, which the overpressure and shutdown circuits may not safely detect.
7. Keep hands and metal items out of the feed throat. To break a pellet bridge or plastic jam, use a plastic or wooden rod.
8. When checking an open vent, wear a full-faced clear polycarbonate shield.

An excellent source for information on troubleshooting guides is the SPE Extrusion Newsletter. Look into these newsletters, which go back many decades, as well as the updates (called "Hints") in the issues. Excerpts from recent issues are shown in Table 5.77.

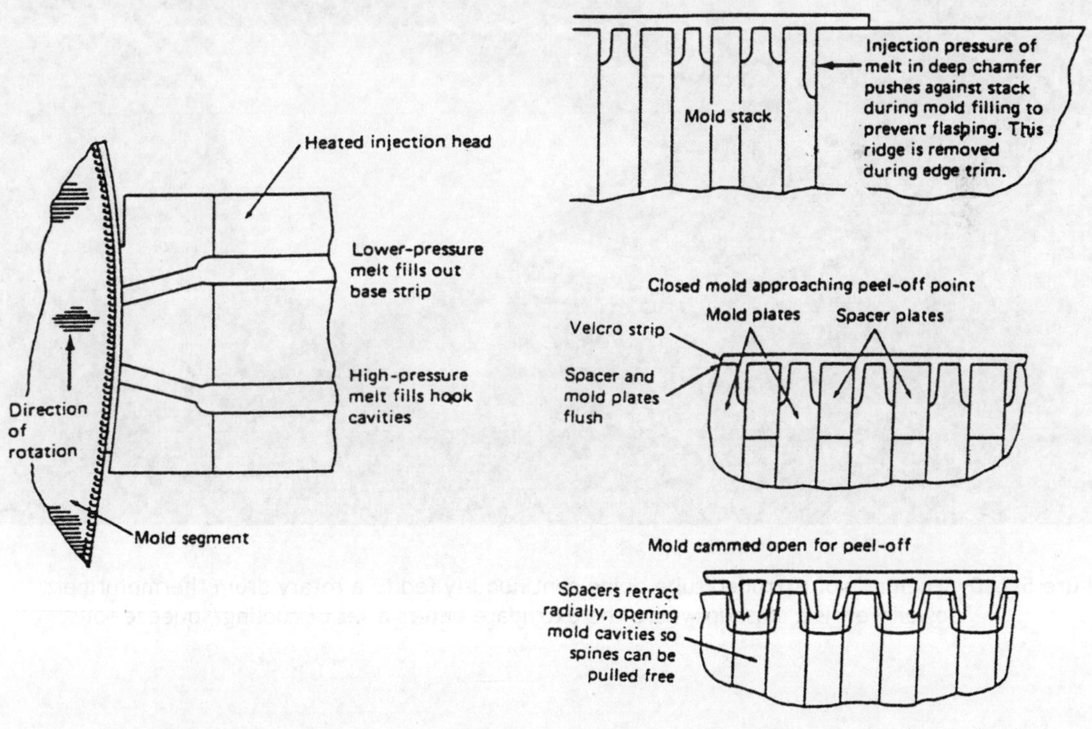

Figure 5.239 Examples of mold cavity filling actions and product release from the cavities

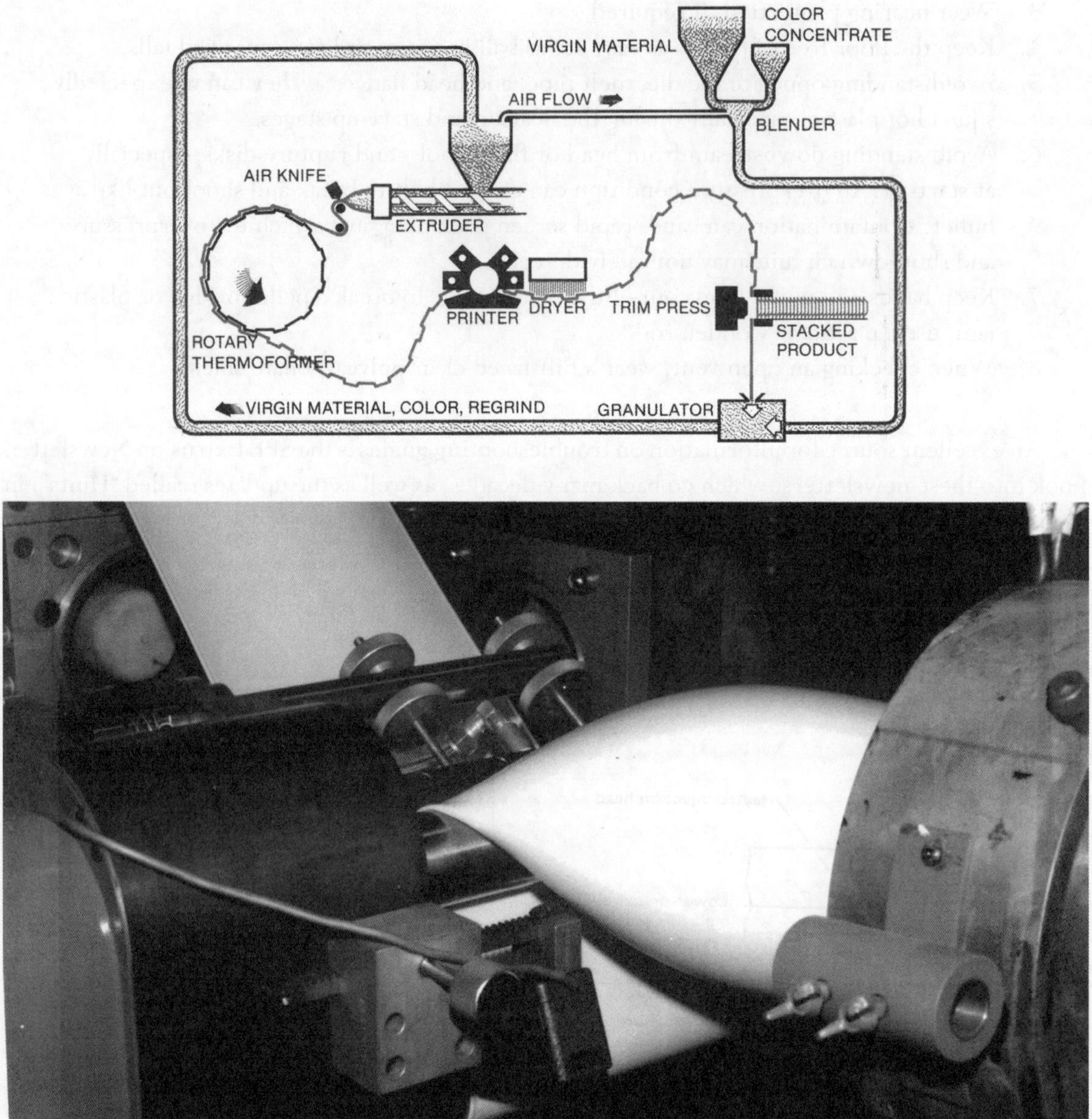

Figure 5.240 Schematic of extruded tube being continuously fed to a rotary drum thermoformer; lower view is a closeup where the extrudate enters a set of cooling/squeeze rolls

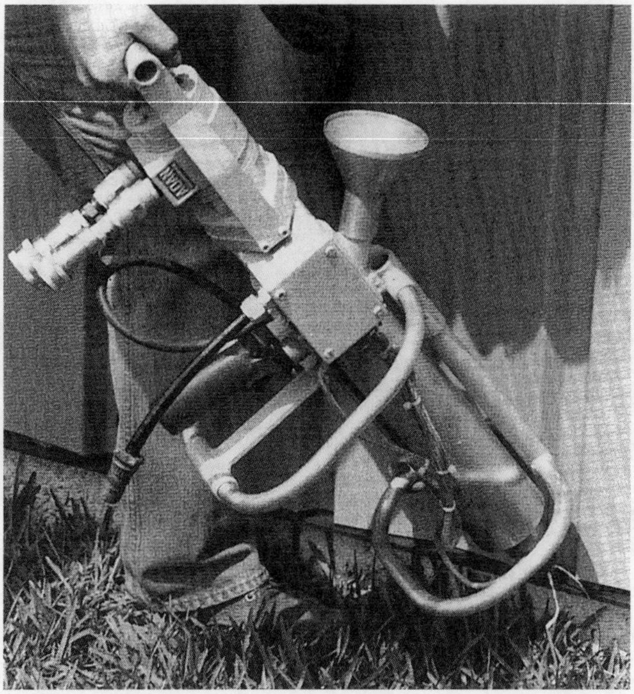

Figure 5.241 Example of an extruder caulking gun

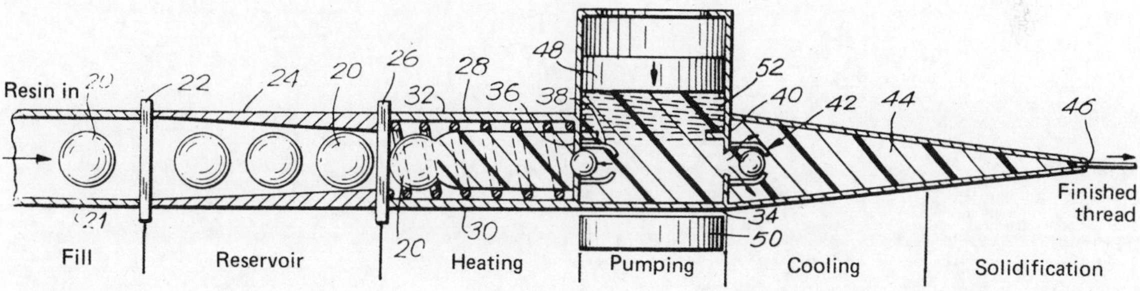

Figure 5.242 Example of sewing machine threading head

Figure 5.243 Example of extrusion film being produced and laid on the ground

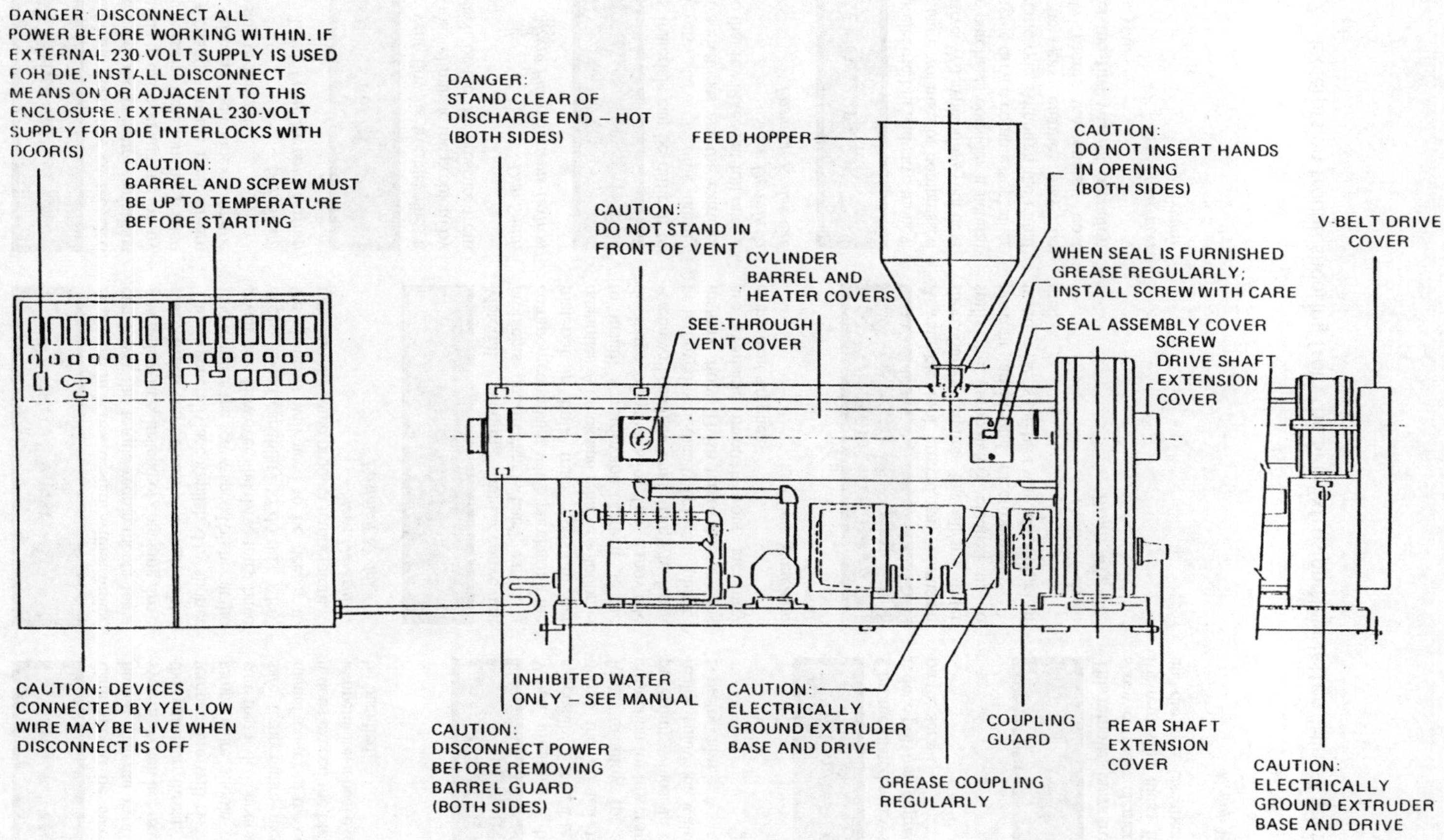

Figure 5.244 Examples of safety warning signs and guards for an extruder

DIE PRESSURE VARIATION

When significant quantities of recycled plastics are used, die pressure variations may not be due to extruder surging. Many times, these pressure variations are related to viscosity changes in the melt stream due to the different melt indexes of the materials being processed.

- Dan Smith
Maag Pump Systems Textron

VENT FLOW

If extruder vent flow is troublesome, a melt pump may eliminate the problem by isolating much of the die pressure from the extruder.

- Dan Smith
Maag Pump Systems Textron

EXTRUDING AMORPHOUS MATERIALS

Amorphous materials can be difficult to extrude. Severe starve feeding during startup can sometimes get the extruder "primed" so that it will feed at full rate.

- Steve Derezinski
Eastman Kodak Co.

GAS COMPRESSOR

Lower your electric bill and save money. Generating large volumes of compressed air is expensive. When planning your next compressor purchase, consider a natural gas fired engine driven compressor rather than an electric one. Although the initial cost of the gas engine may be substantially more, the energy cost savings over the first year of operation will surprise you.

- Pete Stoughton
Conair

DRIVE RATIO

A change in screw design, within limits, can lower the torque requirements for a given flow rate, instead of changing drive ratio, which can be costly or impossible. This is generally accomplished by shorter lead to make the screw speed higher at the given rate. However, the drive must have excess speed capability to do this. Also, the channels will need to be deeper to maintain the same product temperature.

- Steve Derezinski
Eastman Kodak Co.

ANGEL HAIR

Material handling and conveying problems, due to "angle hair", can be a common complaint when purchasing the material in bulk truck and rail car quantities. A quick and reliable solution is to install a removable screen at the discharge of your silo. The removable screen will catch the angle hair, which can be removed periodically before it inhibits material flow. These removable screens are available through your auxiliary equipment supplier.

- Pete Stoughton
Conair

PORTABLE WET/DRY VACUUM CLEANER

A valuable tool around any extrusion shop. Safety requires that all/any water spills are immediately cleaned up to prevent accidental electrocutions! A wet/dry vac is the best way to accomplish this.

- Daniel Cykana
Bemis Mfg. Co.

HEAT EXCHANGER

Care should be taken when conveying heated material in a vacuum conveying system. When hot material is conveyed to the extruder throat, the conveying air temperature will rise. The conveying air temperature should be monitored to be sure that it does not exceed the maximum safe temperature recommended by the manufacturer of the conveying blower. If it does exceed the blowers safe operating temperature, a heat exchanger will need to be installed.

- Pete Stoughton
Conair

ALIGNING A BLOWN FILM TOWER

When aligning a blown film tower, especially if it is a tall one that pops up above the normal factory roof, don't forget to allow for a small amount of correction for tower movement, based on prevailing winds. In other words, if it is really windy out, wait until a day where it is pretty calm.

- Bill Davis
Poly-America

POLYMER DEGRADATION

Chrome plated screws are a good way to reduce polymer degradation and plate out on the screw.

- John Wagner
JVW Enterprises

LAP STONE

This removes burrs from the faces of twin screw extruder elements. It is also useful for removing nicks from breaker plates and die parts. Use them to sharpen knives.

- Paul Andersen
Krupp Werner & Pfleiderer

Table 5.77 Excerpts on troubleshooting from the SPE Extrusion Newsletter "Hints" section

COEXTRUSION

In coextrusion in a single manifold flat die, the ability to maintain good layer uniformity depends greatly on the design of the die. However, it is also affected by the ratio of the viscosities of the poly mers and the ratio of their flow rates. For an A/B or A/B/A structure, the ten dency of the cap layer, A to encapsulate the base or core layer, B will be higher when the viscosity of A is low and/or the flow rate of A is high relative to those of polymer B. When either of these ratios differs by more than a factor of ten, a multi-manifold die is usually recom mended.

- Wayne A. Gifford
Dieflow

REMOVING POLYMER MELT

A useful trick for removing polymer melt from deep channels and intricate die or extruder parts is the use of com pressed air. Use a compressed air noz zle to direct cooling at the melt while pulling the melt away from the part. The polymer solidifies as it is being removed. In this way, polymer can often be completely removed from hard to reach areas of the part.

- Mike Becraft
Sealed Air Corp.

DEHUMIDIFYING DRYERS

The dehumidifying dryers used to pre-dry hygroscopic polymers operate with a closed loop dry air circuit. Any air leaks in the drying air circuit will allow ambi ent moisture to enter the drying system and prematurely saturate the dryer's desiccant, resulting in fluctuating dew point readings.

- Pete Stoughton
Conair

INTERACTIONS

Interaction effects on the extrusion process are tied to many things, with one example being material feed pro perties. Such things as moisture content and particle size can and do impact the feed accuracy of ingredients which in turn affect product properties. There fore, care must always be given to the quality and condition of feed ingredi ents.

- Bill Corcoran
Dupont Co.

PURGE - CLOSE VENT

Many extrusion purges are required to "soak" in the barrel and downstream equipment. If the extruder is vented, remember to turn off the vacuum and keep the vent sealed closed.

- Mark Sewak
LCI Corp

DRYING TEMPERATURE

When drying moisture sensitive materi als, material temperature has the greatest affect on drying performance. There fore, the drying hopper should always be insulated.

- Pete Stoughton
Conair

DOCUMENT/DOCUMENT/ DOCUMENT

If it is not in writing, it never happened. Also, when developing checklists, always make the checks objective, not subjective. A piece of equipment is not "cool, warm, hot" - it is ________°C.

- Bill Davis
Poly-America

MELT PRESSURE INDICATOR

Always include a melt pressure indicator to assist in the resolution of extrusion problems. Absolute extrusion pressure indication and variation in extrusion pressure are key in the diagnosis of prob lems, as well as improvement of the overall extruded product.

- Dan Smith
Maag Pump Systems

TWIN SCREW EXTRUDER

When assembling twin screw extruder elements on a shaft, lop them to remove burrs. This assures proper fit to mini mize the possibility of polymer getting onto the shaft. The element should be lopped in a "figure 8" pattern, to keep the element faces "square".

- Paul Andersen
Krupp Werner & Pfleiderer Corp.

IMPROVE MELT QUALITY

When blending polymers with different melting points and unmixed material is evident in the extrudate from the die, add 300-1000 ppm of a PPA to improve melt quality.

- Tom Butler
Dow Chemical Co.

EXTRUDER SCREW WON'T PUSH OUT

When an extruder screw won't push out due to an interference (foreign object) in the barrel, reverse the direction of the extruder motor/screw rotation. This will normally break the screw loose and allow the pusher to push it out.

- Bill Davis
Poly-America

Table 5.77 *(continued)*

DEVIATION ALARMS

Most extruders are equipped with instrumentation with alarm capability that is often under utilized. A simple alarm setup can provide a powerful process status indicator. For example, setting a 10° deviation alarm on an extruder's barrel zone temperature controller can often alert an operator to a feed problem before the output rate is affected enough to go off size.

- Bill Kramer
American Kuhne Corp.

CROSS CONTAMINATION

Screws should be removed from the extruder for cleaning when changing polymers, to avoid cross contamination, especially for R&D tests. Elevate the barrel temperatures to make sure all residual polymer left in the screw is melted to make screw removal easier. Also clean the barrels with a long rod brush before installing the screw, whether or not the same or a new screw is used.

-Bill Corcoran
Dupont Co.

PRE-DRYING MOISTURE SENSITIVE POLYMER

When pre-drying a moisture sensitive polymer, it is important to remember that any component that is volatile at less than the specified drying temperature may also be removed. This would include additives such as plasticizers, lubricants, and fire retardants. The removal of these additives may affect the properties of the product being extruded and they may also contaminate the dryer's desiccant beds.

- Pete Stoughton
Conair

CONDENSATION

Condensation can cause surface imperfections, "chicken tracks", craters, and if severe, can cause voids, blowouts and product/part breakdown. Condensation is normally a problem where the pellets are stored outside or in colder areas before using. Condensation on the pellets can also occur during periods of high humidity. Be sure material is properly dried to prevent this occurrence.

- Mark Sewak
LCI Corp.

EXTRUDING AMORPHOUS MATERIALS

A hollow screw can make extrusion of amorphous materials more robust. If needed, the bore in the screw can be used for some screw cooling in the feed section to assist feeding. However, just the reduced heat flow from the melting section by virtue of the reduced screw cross section is often adequate to make feeding stable for some difficult to extrude amorphous polymers.

- Steve Derezinski
Eastman Kodak Co.

CHECKING EXTRUSION OUTPUT RATES WITHOUT THE USE OF A CALCULATOR

Take an output rate for a 36 second interval. Weigh it on a scale in lbs or kg. Multiply this figure by 100. This will be the output rate of the extruder in lbs/hr or kg/hr. Report two or three times, to check output consistency. Average these output checks to calculate the average output rate.

- John Christiano
Davis-Standard

PROFILE DIE DESIGN

Good profile die design starts with an understanding of the rheology (study of melt flow) of the thermoplastic that will ultimately be used to extrude product from the profile die. Two key factors that one must study are die swell (actually the swell of the thermoplastic as it exits the die lip) and land length ratio (the length of steel that runs parallel to the melt path). Material suppliers will share the shear/stress curves for predictability of a thermoplastic at a given rate.

- Dan Cykana
Bemis Mfg. Co.

CORONA TREATMENT

Corona treatment is a method of oxidizing a film surface by generating free radicals on a film surface by election bombardment and ozone generation in the treater gap. Because of this chemistry, the surface can be degraded giving low molecular weight and weak surfaces. This leads to lower ink or lamination bond strengths even though the treatment level has increased, i.e., you can overtreat.

- Eldridge Mount
Mobil Chemical Co.

EXTRUDER OUTPUT

When considering the theoretical and actual output of an extruder, the density change of any specific polymer from solid to melt phase must be considered. For instance, the solid density of polyethylene is .92-.96, however, the melt density ranges from .72-.76.

- Dan Smith
Maag Pump Systems

Table 5.77 *(continued)*

CHAPTER 6

Blow Molding

INTRODUCTION

Blow molding (BM) is a process for converting thermoplastics into hollow objects. Like injection molding, the process is discontinuous or batchwise in nature, involving a sequence of operations that culminates in the production of a molding. This sequence or cycle is repeated automatically or semiautomatically to produce a stream of molded parts (3, 9, 23, 109, 221, 456). The three basic processes are shown in Figure 6.1. Blow molding is a very highly developed process with variant forms as summarized in Figure 6.2.

An advantage in blow molding is that of manufacturing molded products economically, in unlimited quantities, with virtually no finishing or secondary equipment required for most products. The basic process is to inflate a softened thermoplastic hollow form against the cooled surface of a closed mold where the plastic solidifies into a hollow product. The surfaces of the moldings are smooth and bright or as grained and engraved as the surfaces of the mold cavities in which they are processed.

With plastics in blow-molded products, to a greater extent than other materials, an opportunity will always exist to optimize their use since new and useful developments in materials and processing continually are on the horizon. Blow molding permits potential of consolidation of parts.

Selecting the right plastic requires applying certain factors such as, setting up performance target requirements, choosing and adapting the process to be used, and intelligently preparing a specification purchase document and work order (Fig. 1.37). Recognize that with the many varying properties of the different plastics, there are those that meet high performance requirements, such as long-time creep resistance, fatigue endurance, toughness, and so on (chapter 19). Conversely, there are those plastics that are volume- and cost-driven in their uses (chapter 30).

Applications for blow moldings are used to contain many different products, including foodstuffs, beverages, household products (appliances, air-conditioners, furniture at homes/offices/hospitals/sports arenas, etc.), personal care products, medicines/pharmaceutical products, automotive parts (bumpers, spoilers, seat backs, etc.), construction panels, tote-boxes, trays, leisure items (toys, flotation devices, marine buoys, canoes, sailboards, sports goods, etc.), industrial parts (business machines, toolboxes, trash containers, hot-water tanks, etc.), and so on. Examples of products are shown in Figures 6.3 to 6.16. Methods of processing these and other products will be reviewed.

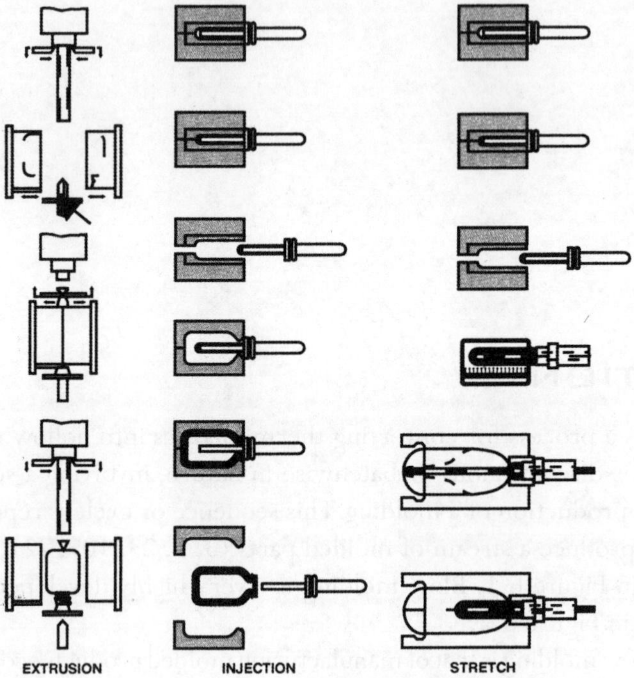

Figure 6.1 Examples of extrusion, injection, and stretch blow-molding techniques

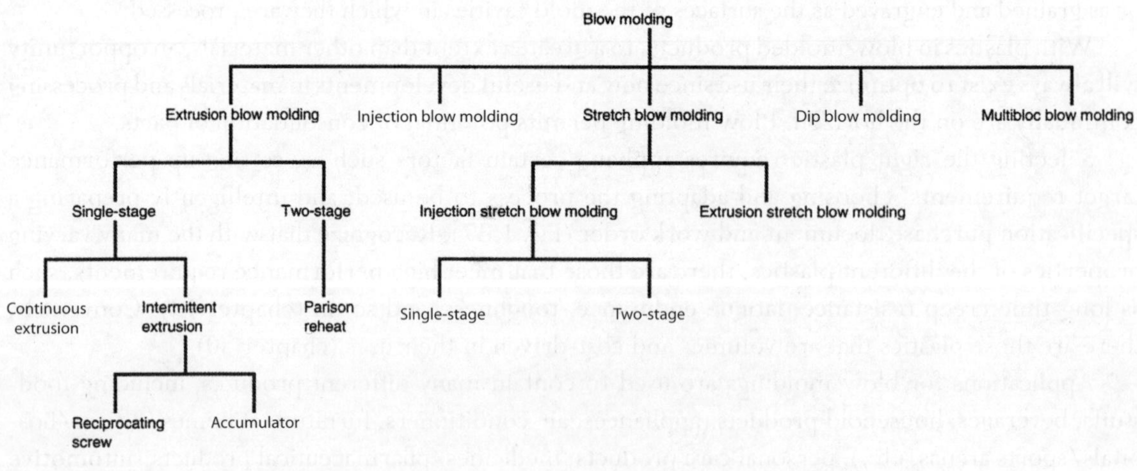

Figure 6.2 Examples of the different forms of blow molding

Although there is more market activity and sales in food and beverage containers, the blow-molded packaging market continues to be active. It consumed 2.5 billion lb of blow-molding resins (25% of the total) in 2001. By 2006, this sector consumed approximately 2.7 billion lb. The largest segments are household and industrial chemical (HIC) containers, drums, intermediate bulk containers (IBCs), tight-head pails (THPs), and motor-oil containers. The HIC market represents 50% of blow-molded industrial packaging. Drums, IBCs, and THPs together amount to 18%. Motor-oil packaging accounts for another 14% of the market. (This study was prepared by Mastio & Co., St. Joseph, Missouri, a well-known consulting firm specializing in industrial-consumer opinion research and market trends in the plastics industry. For more information, call 816-364-6200 or visit www.mastio.com/pt/outlook.html.)

Mastio's report shows that consumption of HDPE far exceeds that of any other resin used for blow-molded industrial packaging. HDPE accounts for nearly 90% of the poundage in this product segment. PET resin, in addition to being used in beverage bottles, has started to appear in bottles for motor oil and other auto and marine fluids. PET is also being utilized more frequently in HIC containers. PET's high clarity imparts greater shelf appeal. Growth for PET in food and beverage containers will average 9% to 10% per year over the next five years, but growth is expected to be only 2% annually for PET industrial blow-molded containers owing to HDPE's lower price and greater chemical resistance. If PET can be price competitive with HDPE, then it is likely that PET will capture market share from HDPE bottles. Polyvinyl chloride (PVC) is losing popularity but is

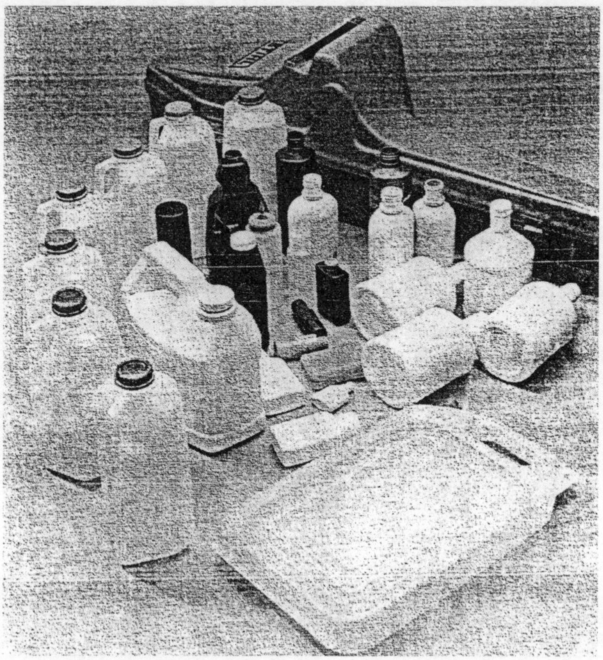

Figure 6.3 Montage of commercial and industrial blow-molded products

Figure 6.4 Examples of blow-molded foodstuff containers

Figure 6.5 Example of longneck blow-molded products

still used in some industrial packaging. A small volume of automotive and marine fluid bottles and HIC bottles are molded of PVC because it provides hydrocarbon resistance and barrier properties that are not possible with untreated HDPE.

CONTAINER

Large containers ranging from at least 5 gallons (18.9 L) to 55 gallons (208.2 L) are fabricated, continually replacing metal (109). The U.S. Department of Transportation (DOT) continues to

Figure 6.6 Blow-molded containers for potato chips

Figure 6.7 Examples of two sizes of blow-molded containers

Figure 6.8 Blow-molded ribbed-panel automotive floor

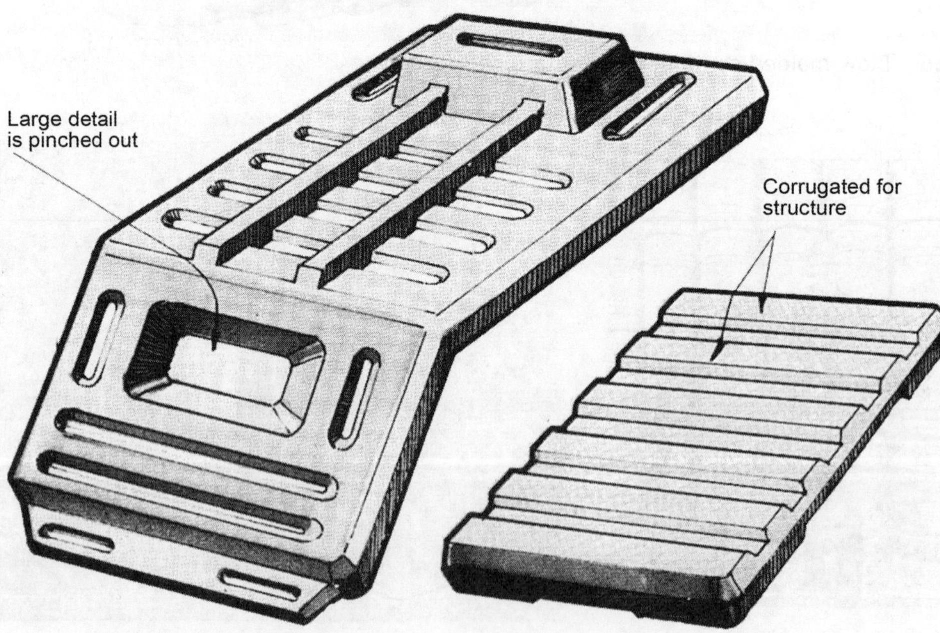

Figure 6.9 Complex 3-D blow-molded products

approve more of these plastic containers for hazardous materials. The light weight of the plastic as well as the number of trips that it can make continues to provide an attractive 55-gallon drum for the industry.

There is no reconditioning necessary, as in steel drums. Plastic containers up to the 55-gallon size can be made on-site as opposed to metal containers that are made and shipped to the consumer. This type of BM is done with accumulator-type BM machines, since the large weight of the parison precludes the use of continuous or intermittent-type machines. The accumulator machines extrude the parison at a high speed so that the parison does not sag.

Blow-molded components have significantly eroded the market for traditional materials, particularly in liquid packaging applications. Since the era of the introduction of polyethylene (PE) squeeze bottles for washing liquids, polyvinyl chloride (PVC) for cooking oils and fruit squash bottles, polyethylene terephthalate (PET) for carbonated beverage bottles, and others, there have been rapid advances in the last few decades in the process machinery as well as the characteristics and range of materials available (chapter 2). Although blow molding of plastics dates from about 1880, rapid developments in machinery and technology only began in the late 1930s, virtually contemporary with the introduction of PE. Blow molding is actually a technique adopted from the

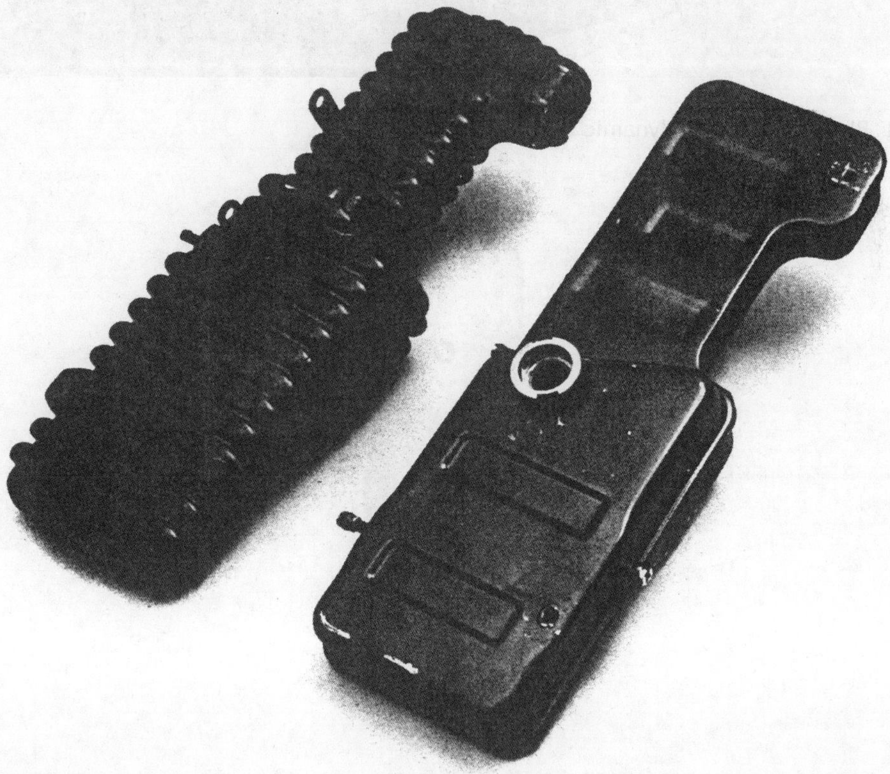

Figure 6.10 Plastic blow-molded fuel tank (left) compared to a metal fuel tank

Figure 6.11 Blow-molded aerodynamic truck wind spoiler

Figure 6.12 Blow-molded 52-gallon hot-water heater that is jacketed by filament winding (chapter 15) to meet UL burst strength requirements

glass industry for the molding of plastic bottles and other containers from thermoplastic materials (chapter 29).

Gasoline Fuel Tank Extrusion blow-molded plastic tanks have been in use by the U.S. military since the 1940s. Gradually, the European and Asian automobiles started using these types of tanks, and later the United States started to adopt them. Different techniques are used to eliminate or reduce gasoline permeability to meet federal standard requirements. They include the use of coextruded structures with barrier plastics, chemically modifying the plastic's surface during or after fabrication (sulfonation or fluorination), and so forth (chapter 20).

Figure 6.13 Blow-molded water flotation wheels

Figure 6.14 Blow-molded swimming pool (courtesy of Vogue Pool Products, La Salle, Quebec, Canada)

Figure 6.15 Blow-molded bellow boots for automotive and other markets

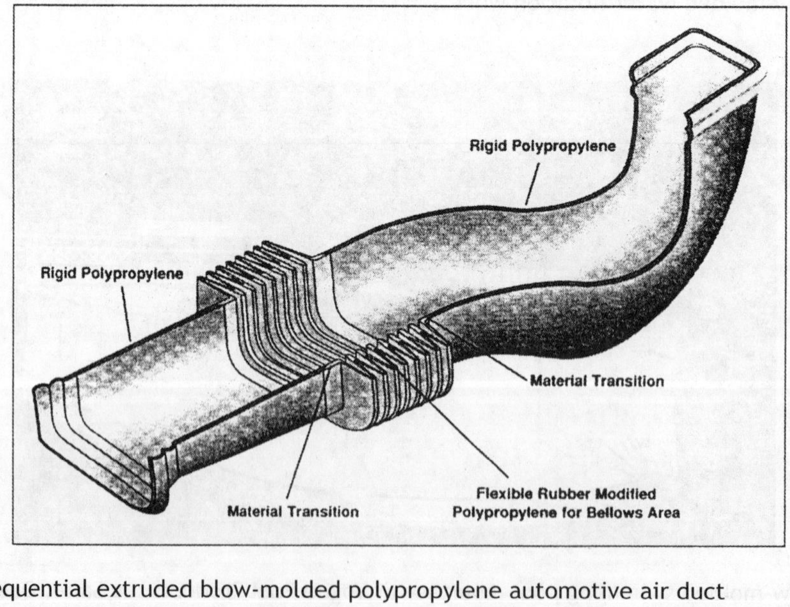

Figure 6.16 Sequential extruded blow-molded polypropylene automotive air duct

In addition to blow molding, there are thermoformed fuel-tank developments. For example, over the past few years, components and systems suppliers, among them Delphi, TI Automotive, Visteon, and Kautex Textron, had interest in thermoforming automotive fuel tanks rather than blow molding them (chapter 7). As reported, recent indications are that thermoforming may have lost some of its luster, while some steel-tank producers are shifting to blow-molded tanks (538). Steel tanks are getting another look, since stricter emissions laws are developing. However, plastic tanks have increasingly gained market share.

Some advantages of thermoforming include its lower capital costs and better wall-thickness control, as well as attachment of secondary components to meet emissions requirements. Advances in blow molding have also occurred. The largest tank manufacturers, Inergy Automotive Systems (Paris) and Kautex Textron (Bonn, Germany), which together control over 65% of the noncaptive tank market, remain firmly in favor of blow molding. Kautex Textron did acquire a twin-sheet thermoforming machine for testing purposes, but the machine has been idle because they are no longer considering thermoforming.

Thermoforming has the limitation of requiring a weld around the complete tank to seal the two halves, resulting in greater emission potentials. They are not as flexible for complex-shaped tanks designed to save space and increase vehicle safety. As TI Automotive reports, with blow molding different designs are being used (vent lines, valves, etc.) so that tanks emissions can be under 20 mg/day compared to the standard blow-molded tanks that emit up to 240 mg/day. TI is also lowering emissions by welding a cover over most of the exterior connections.

An indication that blow-molded tanks will prevail is the reported Fuel Total Systems (FTS) joint venture between Toyota, Horie Metal, and Toyoda Gosei Co. Toyota's vehicles have been using steel tanks almost exclusively. Fewer than 15% of cars made in Japan have plastic fuel tanks. FTS began production during mid-2009 at Horie Metal's plant in Tahara, Japan, and new plant construction started in Lathrop, California.

Industry Size

Blow molding is the third largest plastic processing technique worldwide used for producing many different products. It consumes about 10 wt% of all plastics. Extrusion consumes 36%, injection molding 32 wt%, calendering 8%, coating 5%, compression molding 3%, and others 3%. In the United States alone, there are about 6,000 BM machines, 18,000 extruders, and 80,000 injection-molding machines (IMMs) operating to process the many different types of plastics. U.S. annual sales in BM machines are about $350 million (not accounting for periods of economic downturns).

The BM industry is one of the fastest-growing industries worldwide. The demand for commercial and industrial molding of small to large containers, irregular hollow-shaped industrial products, and bottles is predicted to continue growing. The market for these thermoplastic (TP) products comprises about 22 wt% food, 20% beverages, 15% household chemicals, 12% toiletries and cosmetics, 8% health, 7% industrial chemicals, 5% auto, and 11% others.

BLOW-MOLDING PROCESS

Blow molding can be divided into three major processing categories (Fig. 6.1):

1. Extrusion blow molding (EBM) with continuous or intermittent melt producing a parison from an extruder (chapter 5)
2. Injection blow molding (IBM) with noncontinuous melt from an injection-molding machine (chapter 4) that fabricates a preform supported by a metal core pin
3. Stretched/oriented EBM and IBM to obtain biaxially oriented products providing significantly improved performance-to-cost advantages

Almost 75% of processed plastics are EBM, almost 25% are IBM, and about 1% use other techniques, such as dip blow molding. About 75% of all IBM products are stretched biaxially; there are also stretched EBM products. With stretched blow molding, orientation takes place simultaneously in the hoop and longitudinal directions. These blow-molding processes offer different advantages in producing different types of products based on the plastics to be used, performance requirements, production quantity, and costs (Table 6.1).

BLOWING REQUIREMENTS

Overview The nature of these processes requires the supply of clean compressed air to blow the hot melt located within the BM female cavity. Other gases can be used, such as carbon dioxide (CO2), to speed up cooling of the blown melt in the mold. Production can increase usually by at least 20%

Extrusion blow molding	*Injection blow molding*
Used for larger parts	Use for smaller parts
Best process for HDPE, PVC and others can be used provided adequate melt strength is available	Best process for GPPS and PP; most resins can be and are used
Much fewer limitations on part proportions, permitting extreme dimensional ratios: long and narrow, flat and wide, double-walled, offset necks, molded-in handles, odd shapes	Scrap-free: no flash to recycle, no pinchoff scars, no postmold trimming
Low-cost tooling often made of aluminum; ideal for short-run or long-run production	Injection-molded neck provides more accurate neck-finish dimensions and permits special shapes for complicated safety and tamper-evident closures
	Accurate and repeatable part weight and thickness control
	Excellent surface finish or texture

Table 6.1 Examples of extrusion vs. injection blow-molding performances

to 40% by using turbulent chilled air at about –30°F that is allowed to escape. This action provides several changes of air through the blow pin during a single blowing cycle.

The gas for blowing usually requires at least a pressure of 30 to 90 psi (0.20 to 0.62 MPa) for EBM and 80 to 145 psi (0.55 to 1 MPa) for IBM. Some of the melts may be exposed as high as 300 psi (2 MPa). Stretch EBM or IBM often requires a pressure up to 580 psi (4 MPa). For IBM of the preform, the pressure is usually 2,000 to 20,000 psi (14 to 138 MPa) and in some cases up to 30,000 psi (207 MPa).

The lower blowing pressures generally create lower internal stresses in the solidified plastics and more proportional stress distribution. The result is improved resistance to all types of stresses (tension, impact, bending, environment, etc.). The higher pressures provide faster molding cycles and ensure excellent conformance to complex shapes.

It is important that dry air be used. Moisture in the blowing air can cause pockmarks on the inside product surface. This defective appearance is particularly objectionable in thin-walled items, such as milk bottles. A system of separators and traps to dry the air taken into the air compressor can prevent this problem.

AIRFLOW CONTROL

Blowing bottles and other products requires air pressures that depend on the plastic being processed (Table 6.2). The air performs three functions: It expands the parison or preform against the closed mold cavity, exerts pressure on the expanding plastic to produce surface details on the cavity, and aids in cooling the parison. Various techniques are used to introduce air into the parison, with the usual going through the extrusion die mandrel (over which the top of the parison has dropped) or the injection mold core pin that supports the preform. In extrusion blow molding, it can also enter from the pinch-off end of the mold or through the side of the mold piercing the parison, usually using a needle that resembles a hypodermic needle (Fig. 6.17). This side action can be located at the parting line of the mold halves.

The type of molding machine available has a strong influence upon the blow opening used. Some machines, for instance, use needle blowing exclusively and excess material is machined off

Plastic	*Pressure (psi)*
Acetal	100–150
PMMA	50–80
PC	70–150
LDPE	20–60
HDPE	60–100
PP	75–100
PS	40–100
PVC (rigid)	75–100
ABS	50–150

Table 6.2 Examples of air blowing pressure required for certain plastics

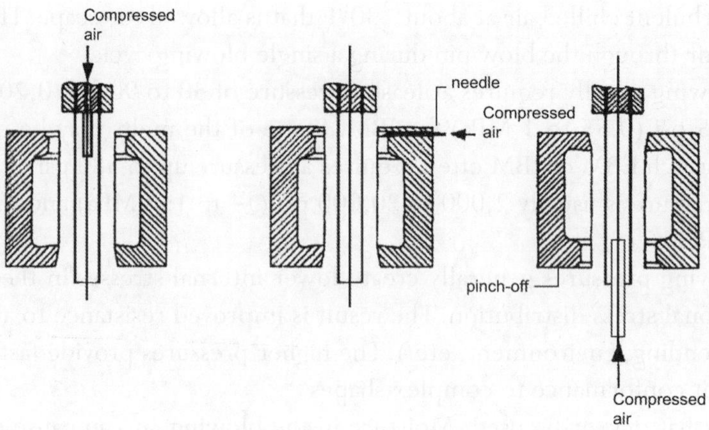

Figure 6.17 Three locations for air to enter extrusion blow molds

to provide larger openings when required. Hollow needle insertion at the mold parting line is considered when a very small opening is required that may even need closing or if the hollow object must be blown up on the side, for example, on carousels with a series of molds. The needle is led right up to the mold cavity by the retainer and moved by a cylinder. This device should be fastened to one mold half. This arrangement should be provided close to a firmly held section of the parison.

During the parison drop between mold halves, the air path is used, allowing for air to be introduced in the parison as it is being extruded to prevent collapse of the parison. This allows for certain functions of blowing products where the parison can be closed after it is cut and inflated for certain processes. The air used for blowing serves to expand the parison tube against the mold walls, forcing the material to assume the shape of the mold cavity and forcing it into the surface details, such as raised lettering, surface designs, and ribs. During the expansion phase of the blowing process, it is desirable to use as high a volume of air as is available so that the expansion of the parison against the mold walls is accomplished in the minimum amount of time.

The aim is to develop a high volumetric airflow at a low linear velocity. A high volumetric flow gives the parison a minimum time to cool before coming in contact with the mold and provides a more uniform rate of expansion. A low linear velocity is desirable to prevent a venturi effect from collapsing a portion of the parison while the remainder is expanding. Volumetric flow is controlled by line pressure and orifice diameter. The flow control valves, which are located as close as possible to the orifice, control linear velocity.

The maximum volumetric flow rate into the cavity at a low linear velocity can be achieved by making the air inlet orifice as large as possible. In the case of blowing inside the neck, this is sometimes difficult. Small air orifices may create a venturi effect, producing a partial vacuum in the tube and causing it to collapse. If the linear velocity of the incoming blow air is too high, the force of this air can actually draw the parison away from the extrusion head end of the mold. This results in an unblown parison. Control valves placed as close as possible to the blow tube must carefully regulate air velocity.

Often, too high a blow pressure will blow out the parison. Too little pressure, on the other hand, will yield end products lacking adequate surface detail. As high a blowing air pressure as possible is desirable to give both minimum blow time (resulting in higher production cycle rates) and finished products that faithfully reproduce the mold surfaces. The optimum blowing pressure is generally found by experimentation on the machinery with the product being produced.

When air contacts the hot plastic, it can result in freeze-off and stresses in the plastic at that point. Air is a material just as is the parison, and as such, it is limited in its ability to blow through an orifice. If the air entrance channel is too small, the required blow time will be excessively long or the pressure exerted on the parison will not be adequate to reproduce the surface details of the mold. A guide to air entrance orifice size when blowing is summarized in Table 6.3.

As reviewed, pressure to be used depends on the plastic being processed. As an example, some PE products with heavy walls can be blown and pressed against the mold walls by air pressures as low as 30 to 40 psi. Low pressure can be used since items with heavy walls cool slowly, giving the plastic more time at a lower viscosity to flow into the indentations of the mold surface. Thin-walled products cool rapidly; therefore, the plastic reaching the mold surface will have a high melt viscosity and higher pressures (in the range of 50 to 100 psi) will be required. Larger products, such as 1-gallon bottles, require increased air pressure (100 to 150 psi). The plastic has to expand farther and takes longer to get to the mold surface. During this time, the melt temperature will drop somewhat, producing a more viscous plastic mass, which in turn requires more air pressure to reproduce the details of the mold.

There is a difference between blowing time and cooling time. Blowing time is much shorter than the time required to cool the thickest section to prevent distortion on ejection. Blow time for an item may be computed from the formula in Table 6.4. Knowing the line pressure and orifice diameter from Table 6.3 will allow you to convert blow time to cubic feet per second. The final mold pressure is assumed to be the line pressure for purposes of this calculation.

During the blowing action, air is heated by the mold, raising its pressure. Calculations ignoring this temperature effect will be satisfactory when blow times are under one second (for small to medium products), but if blow times are longer, the air will have time to pick up heat, resulting in a more rapid pressure buildup and shorter calculated blow time. As shown in Figure 6.18, the blow pin can provide a release channel for the blown air.

Orifice diameter (in)	(mm)	Product capacity (vol.)	
1/16	1.6	Up to one quart	(0.95 liter)
1/4	6.4	1 quart – 1 gallon	(0.95–3.8 liter)
1/2	12.7	1 gallon – 54 gallons	(3.8–205 liter)

Table 6.3 Guide to air entrance orifice size

Gauge (psi)	Orifice Diameter (inch/mm)			
	1/16; 1.6	1/8; 3.2	1/4; 6.4	1/2; 12.7
5	0.993	3.97	15.9	73.5
15	1.68	6.72	26.9	107
30	2.53	10.1	40.4	162
40	3.10	12.4	49.6	198
50	3.66	14.7	58.8	235
80	5.36	21.4	85.6	342
100	6.49	26.8	107.4	429

This is the free air per minute, but since there will be a pressure build up as the parison is inflated, the blow rate of air has to be adjusted.

$$\text{Blow time} = \frac{\text{Mold Vol. Cu. Ft.}}{\text{Ft.}^3\text{/sec.}} \times \frac{\text{Final Mold psi} - 14.7\text{ psi}}{14.7\text{ psi}}$$

Table 6.4 Discharge cu ft/s @ 14.7 psi and 70°F with extrusion blow time formula

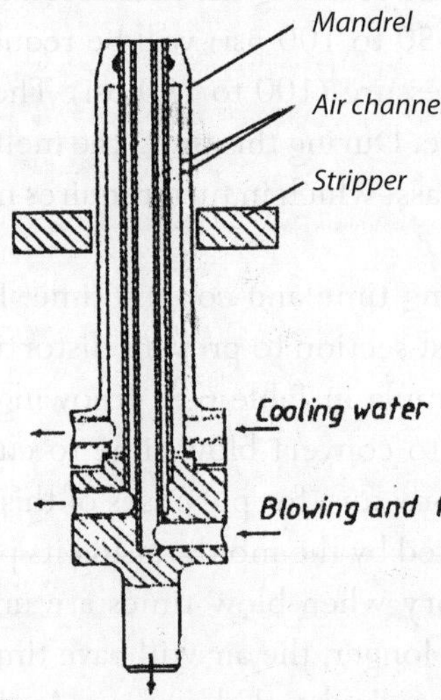

Figure 6.18 Blow-molding pin with escape channel for the blown air

With multicavity molds it is usual to employ a cross-head on which the blowing mandrels are arranged in series. Mandrel chains can also be used for product ejection. The channel for the blowing medium should be as large as possible. Angled radial orifices at the tapered end of the mandrel achieve a fast distribution of the cooling media and turbulence.

The blowing chamber above the neck of the hollow article is known as the lost head. This always becomes a necessity when the width of the neck no longer allows for calibration. In these instances one is concerned with wide-neck containers. A height should be chosen that ensures uniform wall thickness of the neck. A diameter-to-height ratio of 2:1 should be used. When blowing with escaping air, and for fast venting, it is recommended that a tear-off line in the blowing chamber be provided.

Extrusion versus Injection Blow Molding

Compared to IBM, the advantages of EBM include lower tooling costs, incorporation of blown handle-ware, and so on. Disadvantages or limitations can include the need to control parison swell, the production of scrap, limited wall thickness control, and limited plastic distribution control. If desired, blown solid handles can be molded during the BM process. Trimming can be accomplished in the mold for certain designed molds or secondary trimming operations are included in the production lines. With 3-D molding, scrap is significantly reduced.

With IBM, a main advantage is that no flash or scrap occurs during processing. It gives the best of all wall thicknesses and plastic distribution control, and critical bottleneck finishes are easily molded to higher accuracies. The initial IBM preforming cavities are designed to have the exact dimensions required after blowing the plastic melt as well as accounting for any shrinkage, and so on, that may occur. Neck finishes, internally and externally, can be molded with an accuracy of at least ±4 mil (±0.10 mm). It also offers precise weight control in the finished product accurate to at least ±0.1 g.

Disadvantages can include its high tooling costs and solid-only handle-ware. It was reported in the past that EBMs were restricted or usually limited to very small products; however, large and complex-shaped products were fabricated once the market developed. Similar comparisons exist with biaxial orienting EBM or IBM. With respect to coextrusion, the two methods also have similar advantages and disadvantages, but mainly more advantages for both.

When compared to EBM, the IBM procedure permits the use of plastics that are suitable for EBM, but more importantly, those unsuitable for EBM (other than when certain types are modified). Specifically, it is those with no controllable melt strength, such as the conventional polyethylene terephthalate (PET), that predominantly use the stretch IBM method for large quantities of carbonated beverage bottles (liters and other sizes).

Basics in Processing

The blow-molded parts are formed in molds that define the external shapes only. As the name implies, the inner shape is defined by fluid pressure, normally compressed air. In this respect, blow molding differs radically from many molding processes where mold members (male and female cavities are used) determine both inner and outer forms. A major advantage of blow molding is that the inner form is virtually free of constraints because there is no core to extract. The main drawback is that the inner form is only indirectly defined by the mold so high precision and independent internal features, such as in injection molding, are impossible. This has a bearing on wall thickness that can never attain the consistency and accuracy of a full-mold process, such as injection molding (chapter 4).

EXTRUSION BLOW MOLDING

At its most basic, the process involves melt processing a thermoplastic into a tube that is generally referred to as a parison. While still in a heated ductile and firm plastic melt state, the parison is clamped between the halves of a cooled mold so that the open top and bottom ends of the parison are trapped, compressed, and sealed by the mold faces. A blowing tube is also trapped in one parison end, creating a channel through which air pressure is introduced within the sealed parison. Air pressure causes the parison to expand like a balloon so that it takes up the form of the mold cavities (Figs. 6.19 and 6.20). Contact with the cooled mold chills the thermoplastic to its solid state, so the form is retained after the mold is opened and the part removed.

EXTRUDER

The extrusion blow molding process starts by preparing a parison (tube) that is the plastic melt exiting an extruder. Figure 6.21 is a schematic of the plastic melting action in an extruder that has two exiting parisons. This plastic melt has to be a controlled material based on having its basic temperature, pressure, and melting time properly set during its mixing action. Mixing occurs in the extruder's plasticator (screw within a barrel). An example of just setting temperatures in the different zones of a plasticator for different plastics is shown in Table 6.5 As shown in Table 6.6, the plastic output rates of extruders are based on the diameters of screws.

The extruder is usually of the single-screw type. The screw should have a length/diameter (L/D) ratio of 24:1 to 30:1 or more. Screws at the high end are preferred. The compression ratio should be about 3.5:1. This type of information is described with other important information on operating extruders in chapter 5. It is important that one is familiar with the extruder operation in order to ensure that the plastic heated preform is properly prepared to provide blow-molded products that meet performance and cost requirements. Improper preform preparation is damaging to the operation of the blow-molding operation.

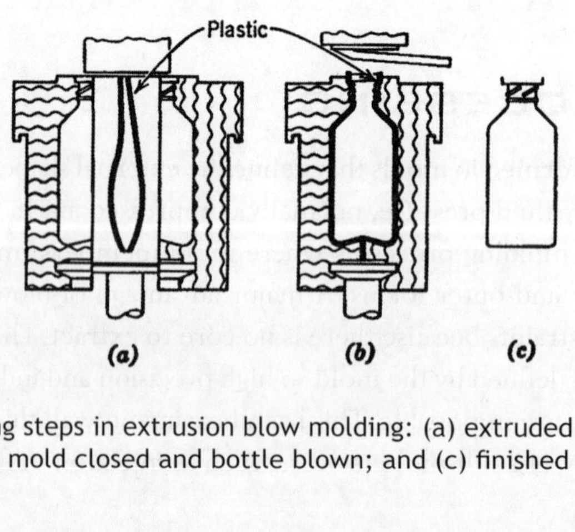

Figure 6.19 Basic processing steps in extrusion blow molding: (a) extruded heated plastic parison, mold open; (b) mold closed and bottle blown; and (c) finished bottle removed from mold

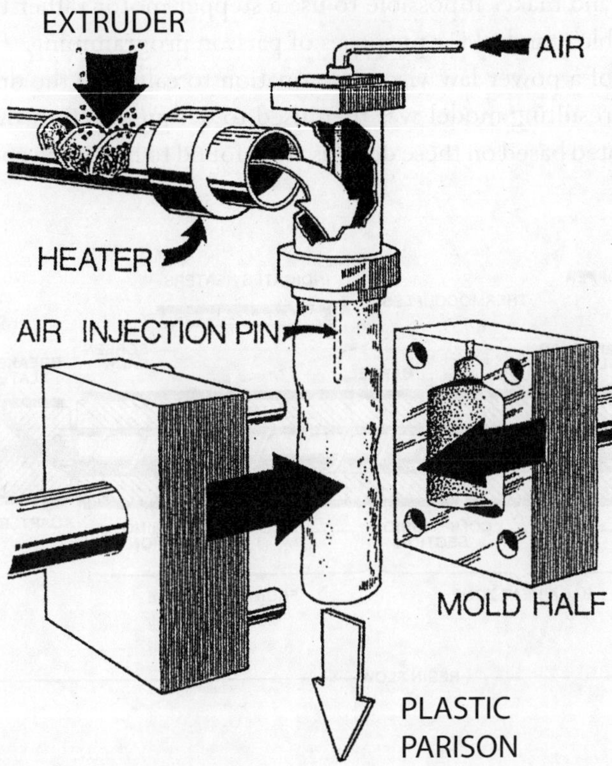

Figure 6.20 Schematic of extrusion blow molding a single parison

MELT FLOW

Understanding and controlling plastic melt flow in the die are important to be successful in blow molding. Factors directly related to die flow include pressure drop, forces acting on the mandrel, and the distribution of melt velocity around the annular flow gap. Die flow also influences the swelling behavior of the melt after it leaves the die. Approximate methods have been proposed to calculate the important aspects of flow in a blow-molding die.

It has been found that the calculation of the major forces and velocity distribution requires only knowledge of the viscous properties of the melt, that is, the viscosity as a function of shear rate and temperature. There have been suggestions that the neglect of the temperature change in the die leads to a large error in the calculated pressure drop in the case of rigid PVC; models proposed for polyolefin flow generally assume the flow to be isothermal (chapter 17).

Such models have been used, for example, to calculate the total force exerted by the melt on the mandrel. For a converging die, the normal pressure acts in a direction opposite to the shear force. This makes it possible to design a die so that the net force on the mandrel is quite small. This simplifies the die design and makes it possible to use a stepper motor rather than a servo hydraulic actuator to drive a movable mandrel for purposes of parison programming.

Use has been made of a power law viscosity equation to calculate the distribution of velocity around the die gap. The resulting model was then used to design dies to provide uniform velocity distributions; dies fabricated based on these designs were found to have very good flow distributions.

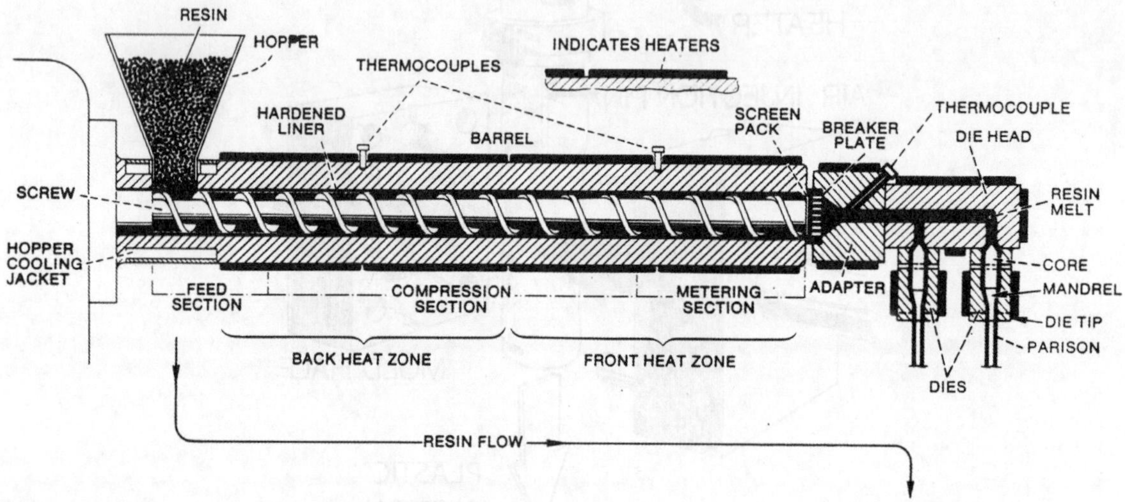

Figure 6.21 Schematic of the plastic melting action in an extruder that has two exiting parisons

Suggested starting points only Material	Zone 1 temp °F	Zone 2 temp °F	Zone 3 temp °F	Zone 4 temp °F	Melt temp °F	Mold temp °F
ABS	380–440	390–440	400–440	400–440	400–440	110–170
ACRYLIC	370–460	380–460	390–460	400–460	400–460	130–200
LDPE	320–340	320–350	330–360	340–370	340–380	40–100
HDPE	350–390	360–400	370–410	370–420	370–420	40–100
NYLON 6	440–480	450–480	460–480	470–490	450–510	140–200
NYLON 6/6	470–550	480–540	480–540	490–530	510–540	140–200
PC	440–480	450–480	460–480	470–490	460–510	140–220
PC/ABS	420–445	430–455	440–465	450–475	450–490	120–220
PP	340–430	350–440	360–450	370–450	390–450	40–120
PS	340–360	360–390	390–420	400–440	360–440	40–120
PS HIGH IMPACT	370–390	390–420	420–450	430–470	420–460	40–120
PVC FLEXIBLE	250–290	260–300	270–310	280–320	310–375	50–100
PVC RIGID	280–300	300–320	320–350	340–370	340–390	50–100

Table 6.5 Example of temperature conditions in an extruder plasticator based on processing different plastics

Diameter		2½ inch (64 mm)	3½ inch (89 mm)	4½ inch (114 mm)	6 inch (152 mm)	8 inch (203 mm)
LLDPE	Pounds per hour	275–340	550–650	900–1100	1600–1950	2840–3475
	Kg per hour	125–155	250–300	400–500	725–890	1290–1580
LDPE	Pounds per hour	475–575	925–1125	1530–1870	2700–3300	4800–5900
	Kg per hour	210–260	420–510	695–850	1225–1500	2200–2675
HDPE	Pounds per hour	330–425	650–800	1080–1320	1920–2350	3400–4170
	Kg per hour	150–195	300–360	490–600	870–1065	1550–1895
PP	Pounds per hour	360–440	700–865	1170–1430	2080–2540	3700–4500
	Kg per hour	160–200	320–390	530–650	945–1150	1680–2050
FPVC	Pounds per hour	415–510	815–1000	1350–1650	2400–2930	4260–5200
	Kg per hour	190–230	370–450	610–750	1090–1330	1940–2370
RPVC	Pounds per hour	220–270	435–530	720–880	1280–1560	—
	Kg per hour	100–120	195–240	325–400	580–710	
ABS	Pounds per hour	360–440	700–865	1170–1430	2080–2540	3700–4500
	Kg per hour	160–200	320–390	530–650	945–1150	1680–2050
HIPS	Pounds per hour	490–615	965–1200	1600–2000	2840–3550	5050–6320
	Kg per hour	225–280	440–550	725–900	1290–1615	2300–2870
Acrylic	Pounds per hour	415–510	815–1000	1350–1650	2400–2930	4260–5200
	Kg per hour	190–230	370–450	610–750	1090–1330	1940–2370
Poly-carbonate	Pounds per hour	285–350	560–680	925–1125	1650–2000	2900–3500
	Kg per hour	130–160	250–310	420–510	750–900	1325–1600

Table 6.6 Examples of extruder output rates based on processing HDPE

Thus, the forces acting in the die and the velocity distribution are governed by the viscosity function, and variations in temperature need to be taken into account, at least for polyolefins with melt indexes in the range of 0.5 to 1.5 (chapter 3).

Parison Formation Die design influences the behavior of the parison that is formed at the die lips. First, if there is a spider to center the mandrel, weld lines will be formed as the melt flows around the legs (chapter 17). The number of molecules bridging the weld lines to return the melt to a homogeneous state increases rather slowly with time, and for this reason, there will be some lateral weakness in the parison and in the molded product in the neighborhood of a weld line.

Another way in which die flow influences the parison properties is through the process of shear modification or shear refining. The effects of this process are most pronounced in highly branched and high molecular weight linear polyolefins. The effect can be particularly pronounced in the case of blends. The molecular mechanisms underlying the process are not fully understood, but shearing at high rates alters the structure of the melt, decreasing the strength of the interactions between molecules, a process sometimes referred to as disentanglements. The effect is a reversible one, and if a shear modified plastic is annealed for a period of time at a temperature above its melting point, it will recover its preshear structure. However, the time required for this recovery is often much longer than the time a parison is formed and the time it is inflated. Thus, the melt that forms the parison and is inflated into the mold may be a material highly altered by shear in the die. Shear modification reduces elasticity and can result in lower viscosities, especially in the case of branched polymers.

Parison Swell The die flow influences parison behavior through molecular orientation, which manifests itself at the die exit as extrudate swell. Flow at the inlet of a die, where the streamlines are converging rapidly, involves a high rate of stretching in the flow direction. This produces a high degree of molecular orientation, and if the melt is permitted to exit immediately, as for flow through an orifice plate, there would be a very high degree of swell. However, if a long straight section, for example a capillary or straight annular die, follows the entrance, then molecular relaxation processes will lead to a disappearance of the orientation generated at the entrance, and as the die is made longer, the degree of swell is reduced. At the same time, however, the shearing in the die produces some axial orientation, and for a very long die, there is still significant swell resulting from this shear flow. If the die has an expansion or contraction section, this will introduce stretching, and this will introduce orientation whose direction depends on the details of the die design. For these reasons, parison swell is very sensitive to die design.

The melt leaving the die may exhibit sharkskin or melt fracture, which are irregularities in the surface of the extrudate that can affect the surface finish of an extrusion blow-molded container. This effect occurs above a critical shear stress in the die and is often the factor that limits the speed of an extrusion process. Extrudate distortion is most severe in the case of narrow molecular weight distribution (MWD) (chapter 1) high-viscosity plastics. Increasing the temperature or reducing the extrusion rate can sometimes eliminate it, but either of these actions will increase the cycle time. The use of an internal heater in the mandrel is a method to reduce the tendency toward melt fracture on the inside of the parison. The detailed origins of this phenomenon are not fully understood, but the shape and material of construction of the die and the formulation of the plastic are known to be contributing factors.

The cross-sectional shape and surface appearance of an extrudate are governed by many factors in addition to the dimensions of the die lips, and there is currently no reliable method for predicting these characteristics for a given die, resin, or operating condition. A further complication is that as soon as melt leaves the die to form part of a parison, it becomes subject to the force of gravity, which leads to sag or drawdown. This tends to make the parison smaller at the top than at the bottom. The shape of the parison at the moment of inflation is thus the result of the simultaneous processes of sag and swell.

Because of the complexity of this situation, together with the complexity of the rheological properties of the melt (chapter 1), it is not possible to design a blow-molding machine to produce prescribed parison shapes and dimensions at the moment of inflation. However, these dimensions are of crucial importance as they govern the thickness distribution in the finished products. For this reason, machines being used to produce large or irregularly shaped objects are usually equipped with parison programming devices to alter the geometries of the dies (annuli) during parison extrusion to produce parisons that have the prescribed shapes and dimensions. However, parison programming cannot compensate for plastic properties that are basically unsuited for the process, and for this reason, it will be useful to examine in more detail the processes of parison swell and sag.

The wall thickness of the BM product can be related to the swelling ratio of the parison. Referring to Figure 6.22, the swelling thickness of the parison is given by Bt = hp/hd and the

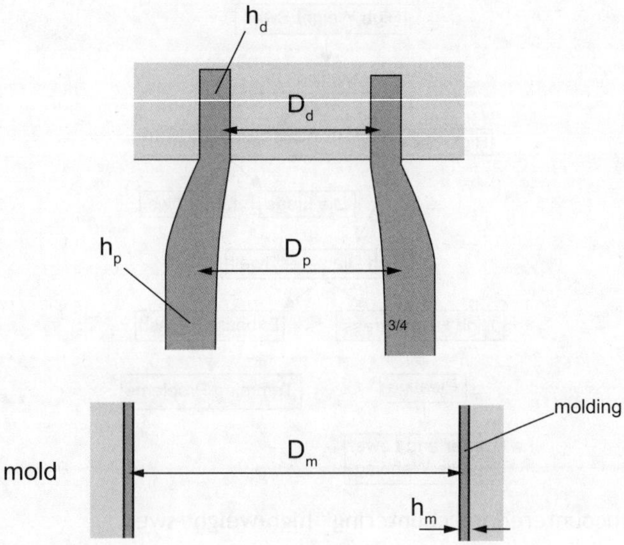

Figure 6.22 Relating thicknesses of swell ratio of parison and BM product

swelling of the parison diameter as Bp = Dp/Dd. Using the relationship Bt = B2p, it follows that hp = (hd) (B2p). Thus the swell ratio Bp depends on recoverable strain that can be measured (69).

Examples of average parison swell for some plastics are given in Table 6.7.

Both the quality and cost of a blow-molded container are strongly dependent on the parison swell ratios. If the diameter swell is too small, incomplete handles, tabs, or other asymmetrical features may result. On the other hand, if the diameter swell is too large, polymer may be trapped in the mold relief or pleating may occur. Pleating, in turn, can produce webbing in a handle. Swell governs a product's weight and thus the material cost of the molded product. What is desired is the minimum weight that provides the necessary strength and rigidity (Fig. 6.23).

Because swell is a manifestation of the viscoelasticity of the melt, it is time dependent. For example, for high-density polyethylene at 170°C (338°F), 70% to 80% of the swell occurs during the first few seconds after the melt leaves the die, but the remainder can occur over a period of two

Plastics	*Swell, percent*
HDPE	15–40
LDPE	30–65
PVC (rigid)	30–35
PS	10–20
PC	5–10

Table 6.7 Examples of plastic melt parison swell

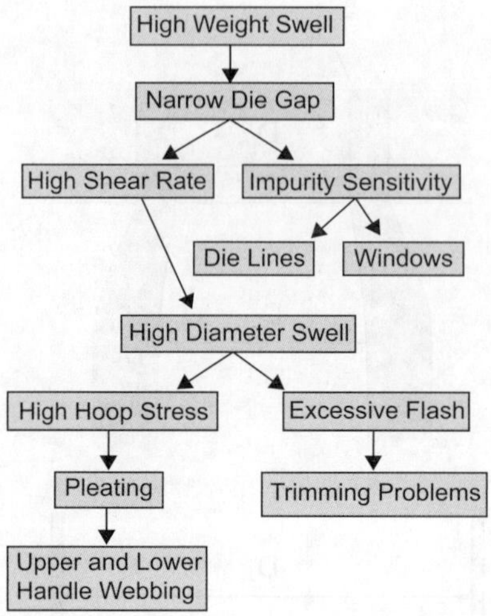

Figure 6.23 Problems encountered in "countering" high-weight swell

to three minutes. For polypropylene at 190°C (374°F), only 50% of the swell occurs in the first few seconds, while more than ten minutes are required to reach an ultimate value. Another source of time dependency in the case of intermittent extrusion is the so-called cuff effect.

The first melt to be extruded after a no-flow period has had an opportunity to relax in the die and experiences relatively little shear just prior to being extruded. Thus, there is less molecular orientation, and thus less swell, than for melt that has just flowed through the entire length of the die. Due to the combined effects of the cuff and the natural viscoelasticity of the melt, the swell is least at the bottom of the parison (the cuff), increasing to a maximum and then decreasing. Of course, sag will accentuate the decrease in swell toward the top of the parison.

Swell increases as the temperature decreases, although it takes place somewhat more slowly. It has been suggested that this effect might be used to control swell by the manipulation of the power for the die heater. Since the time during which the parison is exposed to the air before the mold closes is rather short, and the air surrounding the parison is warm and relatively still, the decrease in temperature that occurs during parison formation is generally rather small, usually less than 5°C (9°F). For this reason, parison formation is generally assumed to be an isothermal process.

Swell increases as the flow rate increases, owing to the enhanced orientation resulting from molecular orientation in the die. Swell varies greatly from one plastic to another and is strongly affected by the extent of branching and the molecular weight distribution. For linear plastics, a broader MWD generally results in a higher swell. However, plastics with very similar measured molecular weight distributions can have significantly different swell behavior, and this probably reflects the fact that swell is highly sensitive to amounts of high molecular weight material that are too small to be detected using standard analytical methods. Highly branched plastics tend to swell more, but it is not possible to generalize when both branching and MWD are changing.

Because swell is an elastic recoil process that results from molecular orientation in the die, the shape of the die channel has a strong effect on both diameter and thickness swells. The simplest case is a long straight annular die. Here, we have only shear flow with no stretching and we would expect to see some orientation in the axial direction. This suggests that there would be no preferential direction of swell and that the diameter swell should be equal to the thickness swell.

For the more complex BM dies, the swell ratios are strongly influenced by two die geometry features, the angle of divergence or convergence of the outer die wall and the variation of gap spacing along the flow path. A diverging die stretches the melt in the hoop direction, and this should reduce diameter swell by counteracting the axial orientation generated by the shear flow.

Basics of Melt Flow The non-Newtonian behavior of a plastic (chapters 1 and 17) makes its flow through a die somewhat complicated. When a plastic melt is extruded from the die, there is the usual condition of swelling (Fig. 6.24 and Table 6.8). After exiting the die, it is usually stretched down to a size equal to or smaller than the die opening. The dimensions are reduced proportionally so that, in an ideal plastic, the drawn-down section is the same as the original section but smaller proportionally in each dimension. The effects of melt elasticity mean that the material does not draw down in a simple proportional manner; thus the drawdown process is a source of errors in the profile. The errors are significantly reduced in a circular extrudate, such as in a BM parison. These errors must be corrected by modifying the die and take-off equipment (5).

There are substantial influences on the material because of the flow orientation of the molecules, so there are different properties parallel and perpendicular to the flow direction. These differences have significant effects on the performance of the products.

Another important characteristic is that melts are affected by the orifice shapes. The effect of the orifice is related to the melt condition and the die design (land length, etc.). A slow cooling rate can have a significant influence, especially with thick parts. Cooling is more rapid at the corners; in fact, a hot center section can cause a part to "blow" outward and/or include visible or invisible vacuum bubbles.

PARISON SAG

Sag (drawdown) can cause a large variation in thickness and diameter along the parison, and in an extreme case, can cause the parison to break off. For a Newtonian fluid, sag can be kept under

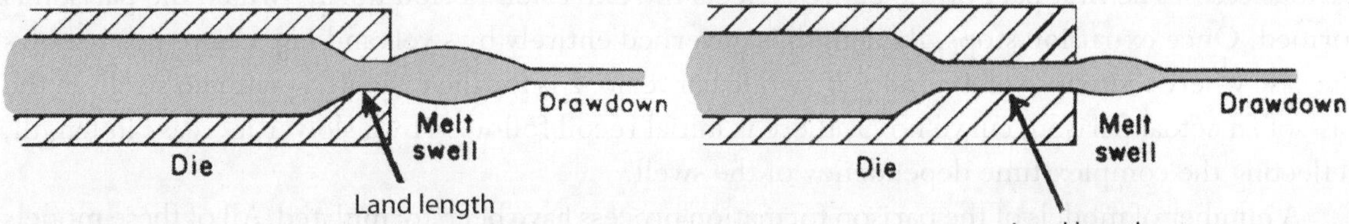

Figure 6.24 Effect of land length on swell

SHEAR RATE, S^{-1} / PLASTIC	DIE SWELL RATIO AT 392 F (200 C)			
	10	100	400	700
PMMA-HI	1.17	1.27	1.35	—
LDPE	1.45	1.58	1.71	1.90
HDPE	1.49	1.92	2.15	—
PP—copolymer	1.52	1.84	2.1	—
PP—homopolymer	1.61	1.9	2.05	—
HIPS	1.22	1.4	—	—
HIPVC	1.35	1.5	1.52	1.53

Table 6.8 General effect of shear rate on die swell of various thermoplastics

control simply by using a material with a sufficiently high viscosity. Plastics sag because they are non-Newtonian and generally have melt indexes of less than 1 (chapter 1). Sag becomes more severe as the temperature is increased. Since plastic melts are viscoelastic, resistance to sag cannot be quantitatively correlated with viscosity.

A number of proposals have been made as to which viscoelastic property (chapter 1) of a melt governs sag. There is the linear creep compliance, as calculated from a tensile relaxation modulus, to predict sag. Another suggests that the extensional stress growth function can be correlated with sag behavior. J. M. Dealy (McGill University) points out that the sag process is neither a constant stress nor a constant strain rate process, so neither of these functions is directly relevant (540). Noting that the strain rate is small, it is proposed that the use of a relatively simple general model of a viscoelastic liquid be used.

Reliable methods for predicting sag behavior on the basis of well-defined rheological properties are still lacking. In their absence, empirical techniques for evaluating sagging tendencies are employed. An example is to make video records of parison length versus time. A simpler procedure that might be useful is to determine the extensibility of film plastic by simply clamping a weight to a specified length of extrudate from a melt indexer and monitoring the resulting behavior.

When we consider the combined effects of swell and sag, the situation becomes quite complex from a rheological point of view. Figure 6.25 is a sketch of parison length versus time curves for various cases. The first part of the curve reflects the extrusion period during which the parison is formed. Once extrusion stops, the length is governed entirely by swell and sag. Curve no. 1 shows the case where there is swell but not sag, while curve no. 2 is for the case of sag with no swell. In the case of an actual parison, curve no. 3, there is initial recoil followed by a slower increase in length, reflecting the complex time dependency of the swell.

A number of models of the parison formation process have been formulated. All of these models assume that swell and sag are in some way additive. Unfortunately, none of them can be used to predict with confidence the behavior of the parison for a given die, resin, or operating condition.

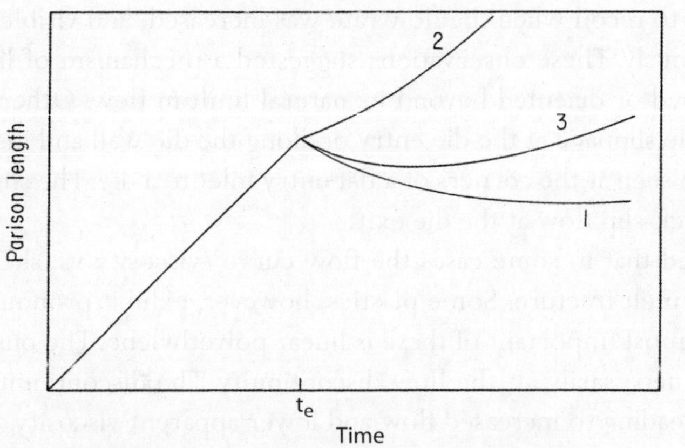

Figure 6.25 Parison length vs. time curves for three different situations

Pleating The action of pleating (also called draping or curtaining) is a buckling of the parison that occurs when the melt at the top of the parison is unable to withstand the hoop stresses due to the weight of the parison suspended from it. Pleating is undesirable because it can cause webbing in the blown product.

Obviously, large diameter swell and a small die gap are factors that will increase the probability of pleating. It is not clear exactly which rheological property governs the ability of a melt to resist pleating, but the viscosity is a rough guide to the melt stiffness that is desired. Thus, high-viscosity plastics are less likely to pleat, and increasing the temperature increases the likelihood of pleating.

Parison Inflation The behavior of the closed parison during the inflation process is a manifestation of its extensional-flow rheological properties. It has been observed that the parison does not inflate uniformly and tends to bulge out in the center. A blowout can occur if the ratio of the mold diameter to the parison diameter (the blowup ratio) is too high.

The deformation is not the same as that which occurs in the uniaxial extension experiment, but the results of such an experiment are still thought to be relevant to inflation performance (18). In particular, it is thought that extension thickening behavior implies that a plastic will be easy to inflate and unlikely to exhibit blowout, even when the blowup ratio and inflation pressure are high. Extension thinning, on the other hand, is thought to imply unstable inflation and an increased likelihood of blowouts. This hypothesis is consistent with the observation that low-density (branched) polyethylene is easier to inflate than HDPE.

Melt Fracture Melt fracture is a loosely defined term that has been applied to various forms of extrudate roughness or distortion encountered at high extrusion rates for all plastic melts. In some cases, there may be small-scale roughness, rippling, or sharkskin, in others a very regular helical screw-thread extrudate. Generally, at very high rates the extrudate becomes completely irregular.

The term *melt fracture* originated from early observations of marker threads or of flow birefringence at the entrance of a die. A line that was continuous at low extrusion rates could be seen to snap suddenly and to recoil when the flow rate was increased, and visible evidence of roughness was noted simultaneously. These observations suggested a mechanism of literal breakage of melt that had been stretched or oriented beyond its natural limit in flow. Other mechanisms that have been proposed include slippage at the die entry or along the die wall and instability of the circulating vortex flow, often seen at the corners of a flat entry inlet to a die. The small-scale roughness has been attributed to stick-slip flow at the die exit.

It should be noted that in some cases the flow curve (viscosity vs. shear rate) is not affected at all by the onset of melt fracture. Some plastics, however, exhibit pronounced discontinuities of the flow curves. The most important of these is linear polyethylene. The onset of visible roughness occurs near, but not necessarily at, the flow discontinuity. The discontinuity has been ascribed to slippage at the wall, leading to increased flow and lower apparent viscosity.

The effect of slippage on the extrudate depends on the nature of the extrusion apparatus. A very stiff apparatus, such as a piston driven by a motor, will give a constant extrusion rate but severe pressure fluctuations. A soft apparatus, for example a gas-driven rheometer, will produce large variations in output. The most serious effect occurs when the extrusion is done at a rate very close to the critical one for the onset of slippage. The output can then oscillate between the two branches of the flow curve as shown in Figure 6.26, causing large variations in flow rate and gross distortions of the extrudate. The severity of distortion is also markedly dependent on die geometry.

These effects may be rationalized by considering that in the absence of slippage the melt is highly oriented by the flow. When the melt slips suddenly, it creates room for the relatively unoriented stagnant melt that has been circulating in the entry vortexes to enter the die and to be extruded without the usual stretching at the die entry. The alternating of highly oriented extrudate (with large elastic recovery) with the unoriented melt accounts for the extrudate distortion.

Also, there is a pronounced effect of inlet taper angle on the extent of the distortion. A correctly tapered inlet suppresses the formation of a stagnant vortex. Therefore, even if slippage occurs, there is no stagnant material to cause nonuniformity and the external appearance of the extrudate is relatively smooth. It has also been reported that the use of a slightly tapered conical die, rather than a cylindrical one, delays the onset and minimizes the severity of melt fracture.

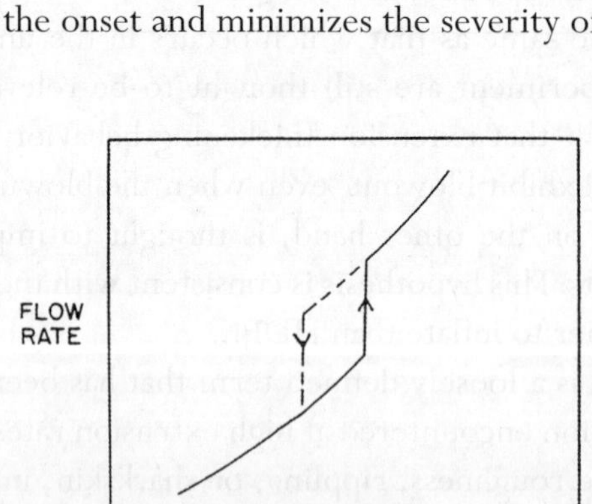

Figure 6.26 Oscillating melt flow rate near slip discontinuity of flow curve

It is generally agreed that shear stress is a better criterion for the onset of melt fracture than shear rate. For example, when the extrusion temperature of an LDPE is varied from 130° to 230°C (266° to 446°F), the shear stress at the onset of melt fracture varies by only about 30%, whereas the critical shear rate varies by a factor of 100. The critical stress generally ranges from about 10^6 dynes/cm^2 (14.5 psi) for a nonpolar polymer (such as polystyrene, polypropylene, and polyethylene) to about 10^7 dynes/cm^2 (145 psi) for a polar polymer, such as nylon, polyacetal, and PET.

The effect of molecular weight and MWD is not as clear-cut. The most comprehensive study indicates that the critical stress increases linearly with the reciprocal of the average molecular weight; that is, the critical stress is lowest for high molecular weight polymers.

The effect of MWD on the critical stress is less well defined, with many conflicting reports in the literature. Because of the difficulty of measuring the extremes of the MWD, it is probably not practical to use such a correlation for quantitative predictions in any case. As with die swell, it is easier and more reliable to measure the critical shear stress for the polymer of interest.

Extensional Flow Historically, shear flow is important because it is the simplest way to deform a fluid plastic, particularly a low-viscosity fluid plastic, in a controlled manner that can be analyzed and used to measure the viscosity. Of course, it is a flow that occurs in nearly all plastic melt processing operations. However, shear flow is not the only deformation that can be applied to a fluid. In general, the flow that occurs in melt processing machines may be quite complex.

In addition to simple shear flow, extensional flow usually occurs. The most elementary type of extensional flow is viscosity, and end correction is needed in any case to characterize the processing parameters, thus specifying the die swell for any extrusion condition. It should be noted, of course, that a die swell curve depends upon the molecular weight and distribution of the plastic, as do rheological properties in general.

Next is a review on the dependence of die swell on molecular parameters where the situation is not completely clear. With one exception, the literature indicates that die swell increases with increasing breadth of MWD. The breadth of the MWD is usually specified in terms of a ratio of two types of average molecular weight. However, these averages are strongly affected by the shape of the MWD, especially at the extremes of high and low molecular weight. It has been found that a blend containing a very low molecular weight fraction shows unexpectedly high swell. Conversely, it has been found that the addition of as little as 0.1% of an ultrahigh molecular weight polyethylene increases the die swell of an HDPE.

Perhaps it is best to say that die swell varies with the MWD in a manner not predictable by a simple parameter but depends in some complicated way on the entire shape of the MWD. Another possible complication is that an HDPE may contain small amounts of long chain branching (LCB), and the amount can vary with the details of the polymerization process. It is known that LCB has a strong influence on melt rheology, especially elasticity.

Fillers, particulate or fibrous, have the effect of reducing or suppressing die swell. Small amounts of crystallinity have similar effects, and this may account for the surprisingly small die swell of rigid polyvinyl chloride.

PARISON HEAD

The parison head, sometimes called the die head or simply the die, is a specialized form of tube extrusion die (chapter 17). Its function is to deliver a straight parison in the correct diameter, length, and wall thickness and at the correct temperature for blow molding. Prior to being clamped in the mold, the parison is suspended unsupported in free air. To avoid undue deformation, it is necessary to extrude the parison vertically, downward.

The blow-molder extruder is almost always arranged in a horizontal attitude, so the first task for the parison head is to turn the melt flow stream through a right angle. This is basically difficult and undesirable but necessary to achieve in a way that meets the essential requirement for a constant flow rate at every point in the annular die gap. A second and related requirement is that the parison should carry as little evidence as possible of the weld line (chapter 17) formed when the melt stream from the extruder flows around the torpedo. Many parison die-head designs have been evolved to deal with these problems. An example, known as the heart-shaped cross-head, is shown in Figures 6.27 and 6.28. Figure 6.27 provides a simplified view. More details are given in Figure 6.28; this figure's call-outs are as follows: 1 = housing, 2 = torpedo, 3 = seating, 4 = melt flow channel, 5 = die, 6 = mandrel, 7 = die gap, 8 = retainer ring, 9 and 10 = die gap adjustment means, AG = plane of annular flow channel, and E = flow divider edge. Head design has benefited

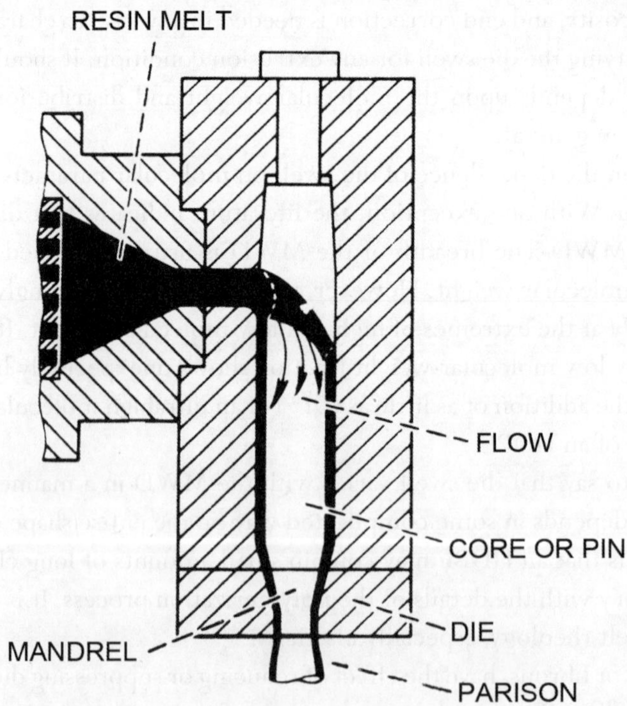

Figure 6.27 Simplified view of a heart-shaped parison die head

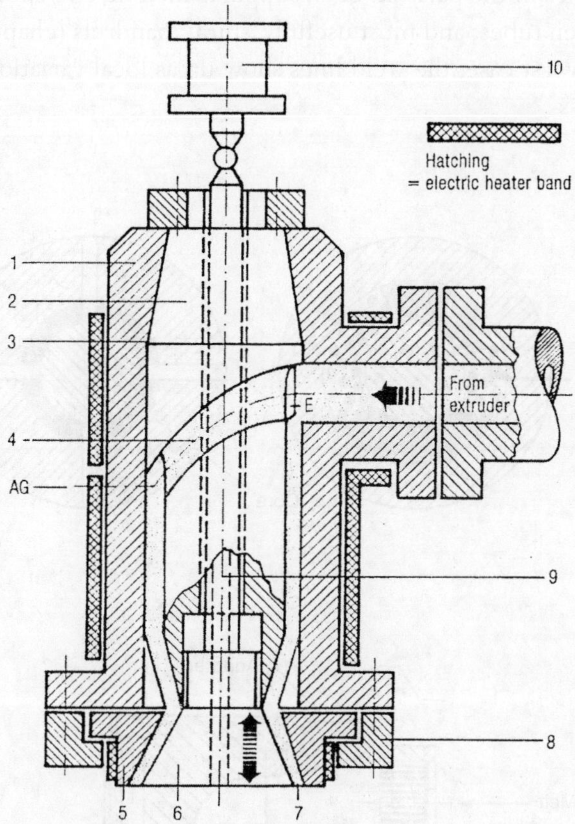

Figure 6.28 Details of a heart-shaped parison die head

from computer analysis of the internal flow channels, which has removed much of the trial and error and given a greater assurance of success.

In addition to Figure 6.27 and Figure 6.28, Figures 6.29 to 6.32 show different continuous parison die heads that provide different capabilities to minimize or eliminate weld lines.

PARISON WALL THICKNESS

A potential blow-molding problem is one shared with pipe and tube extrusion dies (chapter 17). This is the difficulty of ensuring that the mandrel (also called core or torpedo) remains coaxial with the die, without dividing the melt stream with mechanical supports or too much excessive wall thickness deviation. The parison die heads are examples of different design approaches. Those without supports (no spiders) in the melt stream rely instead on accurate seating between the mandrel and die. The parts need to be relatively massive for this design to be sufficiently rigid. Smaller torpedoes will need supports in the melt stream and here the challenge is to minimize the formation of weld lines in the parison. Such supports include two spider legs, staggered spider legs, breaker plates, screen tubes, and most usefully, spiral mandrels (chapter 17). The problem is a difficult one and in the worst case, the weld lines show up as local variations and/or weaknesses in the parison wall thickness.

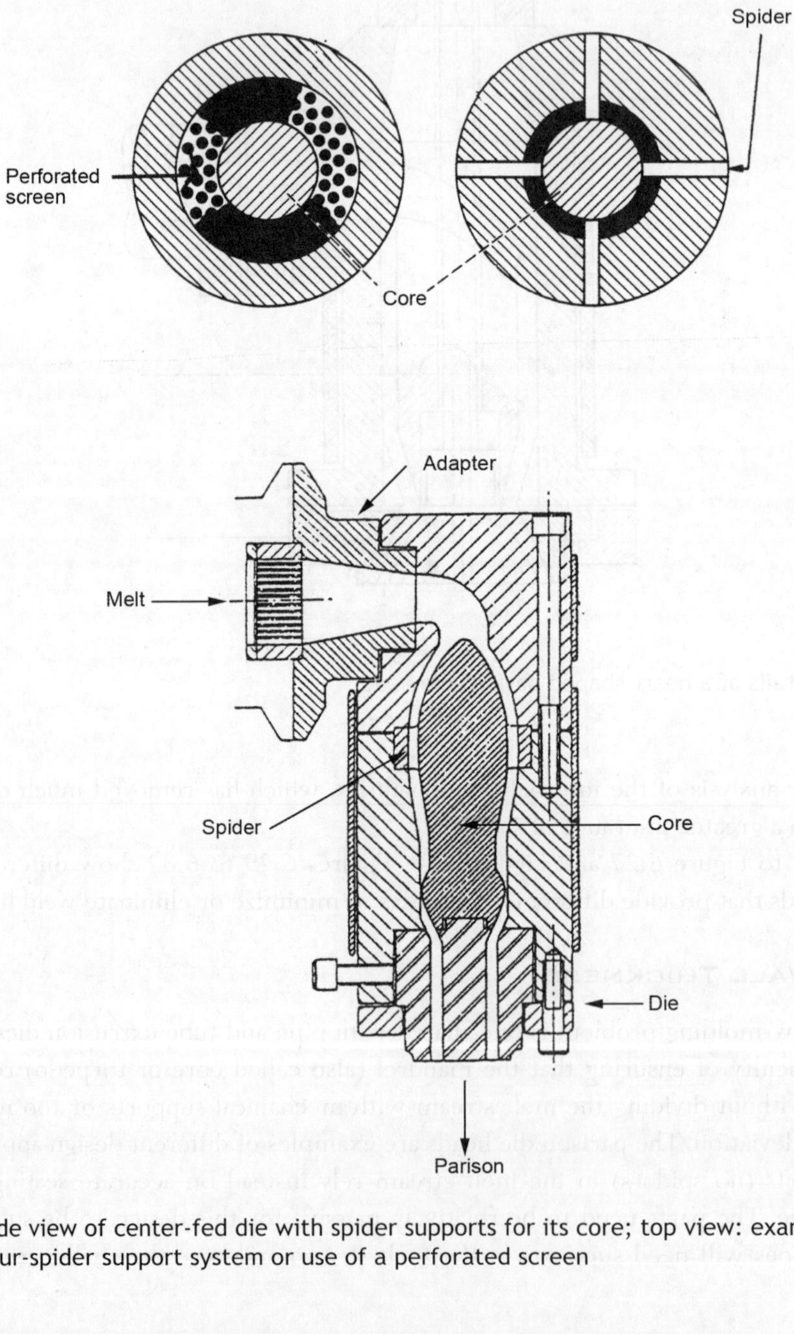

Figure 6.29 Side view of center-fed die with spider supports for its core; top view: examples of four-spider support system or use of a perforated screen

The basic parison is a tube that in principle at least has a relatively constant wall thickness at all points around its circumference and along its length. In practice, circumferential thickness may vary as a result of weld lines or if the mandrel is not truly coaxial with the die. Thickness along the parison length may vary because of tensile thinning caused by the weight of the dependent parison and/or uneven melt flow from the extruder as well as restrictions in the parison die head. However, the ideal parison with a constant wall thickness is not necessarily the optimum for blow molding. Even simple molded shapes encompass considerable variations in profile.

For example, the body of a bottle is much greater in diameter than the neck. For some containers, there may be several significant variations along the axis, and the effect is usually pronounced in technical blow moldings. If such articles are blow molded from a parison of constant wall thickness, then the finished molding will be thinner where the parison has expanded the most and thicker where it has expanded the least. In most cases, the ideal is a finished product with a constant wall thickness at all points. To approach this ideal, what is used is a parison in which the wall thickness varies along the length so that it is thickest where it must expand the most. This is achieved by parison programming or profiling.

As shown in Figure 6.33, the usual way to do this is by moving the mandrel in an axial direction relative to the die. If both die and mandrel are provided with conical outlet features, this movement will increase or decrease the annular die gap between the two. A servo system acting on a preprogrammed thickness profile controls the mandrel movements. The system may include a feedback loop to adjust parison profiling in response to screw speed variations in the extruder.

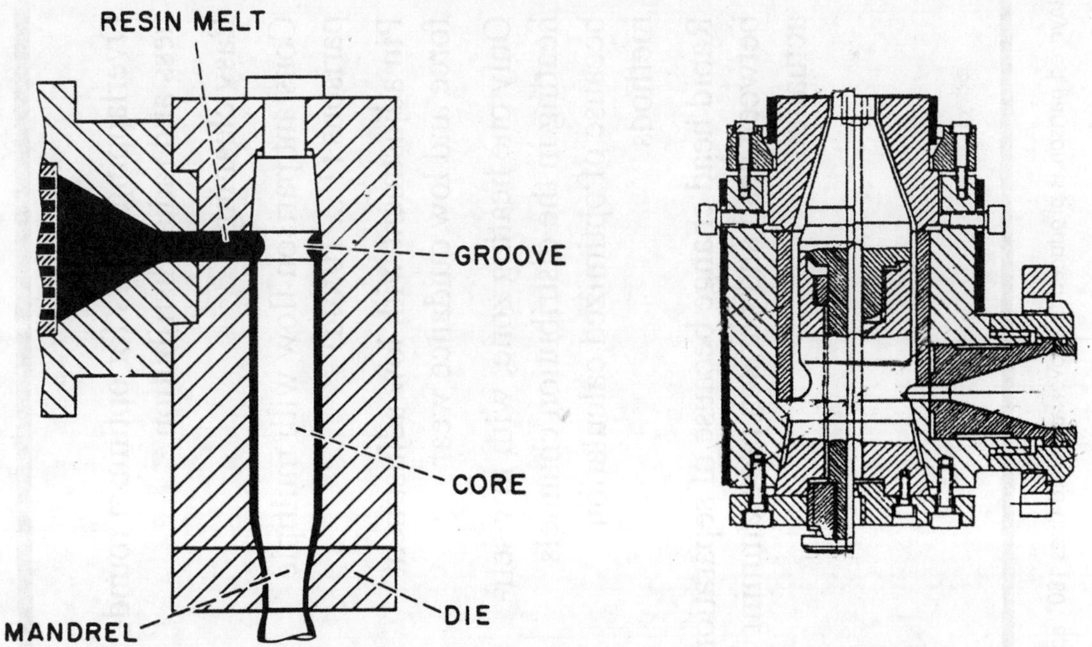

Figure 6.30 Examples of a grooved-core parison die head

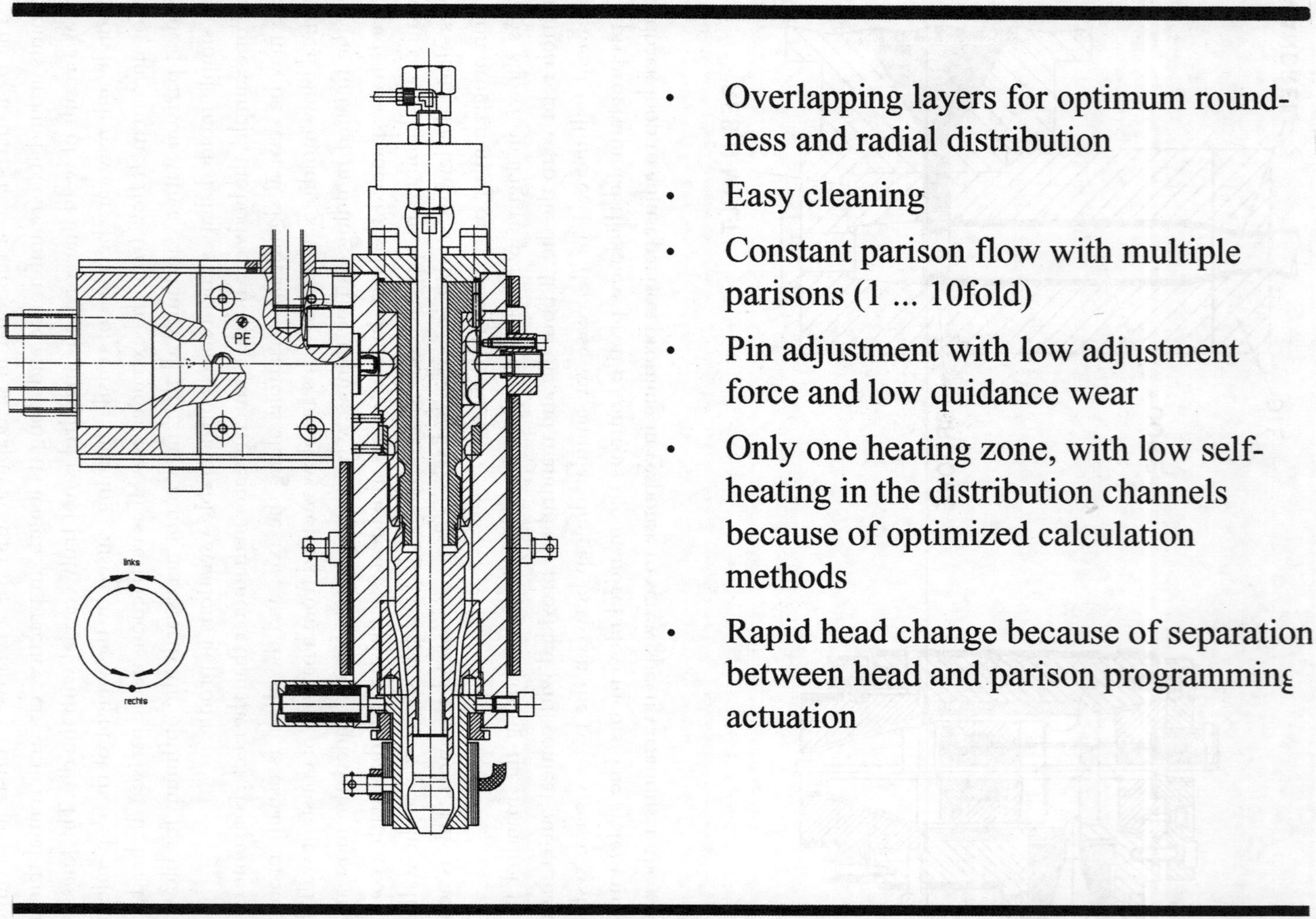

Figure 6.31 Example of double-sided parison feedhead so that a double-layered parison is produced that overlaps weld lines 180° apart (courtesy of Graham Machinery Group)

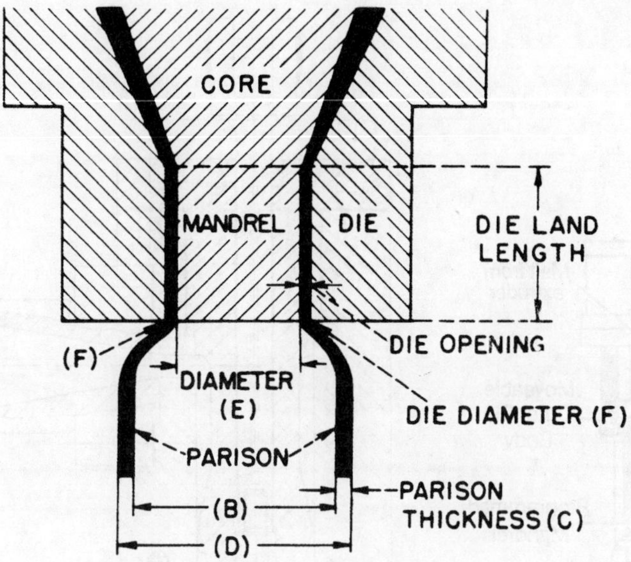

$$\text{Tube wall swell, \%} = \frac{C-A}{A} \times 100$$

$$\text{Die swell, \%} = \frac{D-F}{F} \times 100$$

$$\text{Mandrel swell, \%} = \frac{B-E}{E} \times 100$$

Figure 6.32 Explanations of a parison die head

There are two types of mandrel and die bushings. One is the convergent type and the other is the divergent type (Fig. 6.34). The selection of which mandrel and die bushing (commonly known in blow molding as head tooling) is made by the size of the parison necessary. Convergent tooling is the easiest to control and is used whenever possible. Divergent tooling is used when the parison needs to be larger so the tooling causes the parison to flare out as it is extruded. Both convergent and divergent mandrels are mounted directly to the programming mandrel and are programmed up and down to open or close the gap between the pin and bushing (Fig. 6.35).

For product shapes that are not circular, variations in the preform thickness perpendicular to the preform at the die opening are designed to provide the final dimensions. Figure 6.36 is an example of a rectangular shape.

Machine Design

In extrusion blow molding, an extruder feeding a parison head is used to produce a parison as a precursor or preform for the molding process. The configuration of individual machines may vary greatly, but some essential elements can be distinguished, such as those shown in Figure 6.37. The

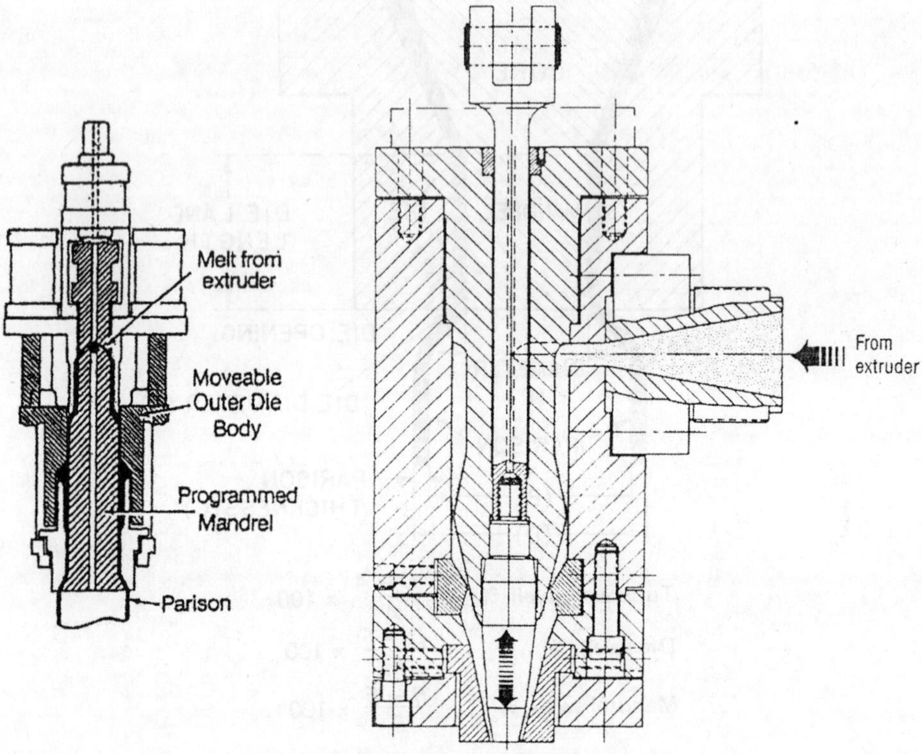

Figure 6.33 Examples of parison wall thickness control by axial movement of the mandrel

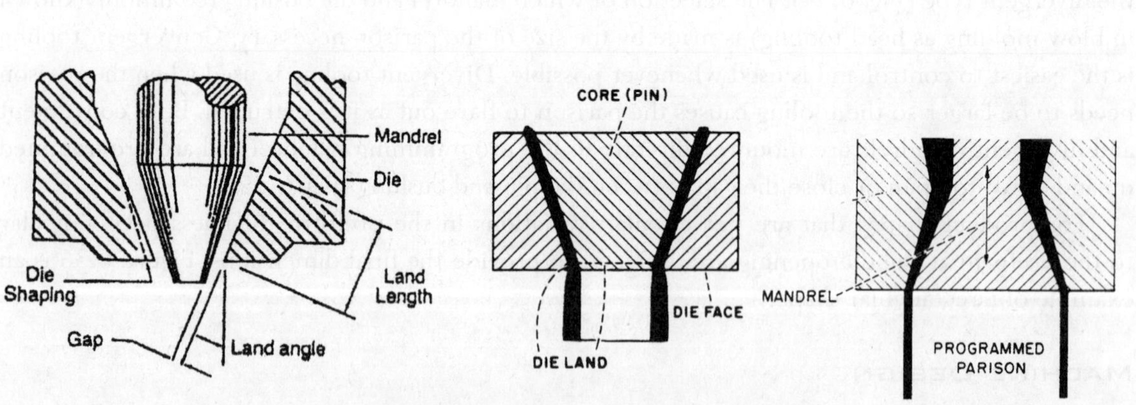

Figure 6.34 Examples of convergent and divergent die-head tooling

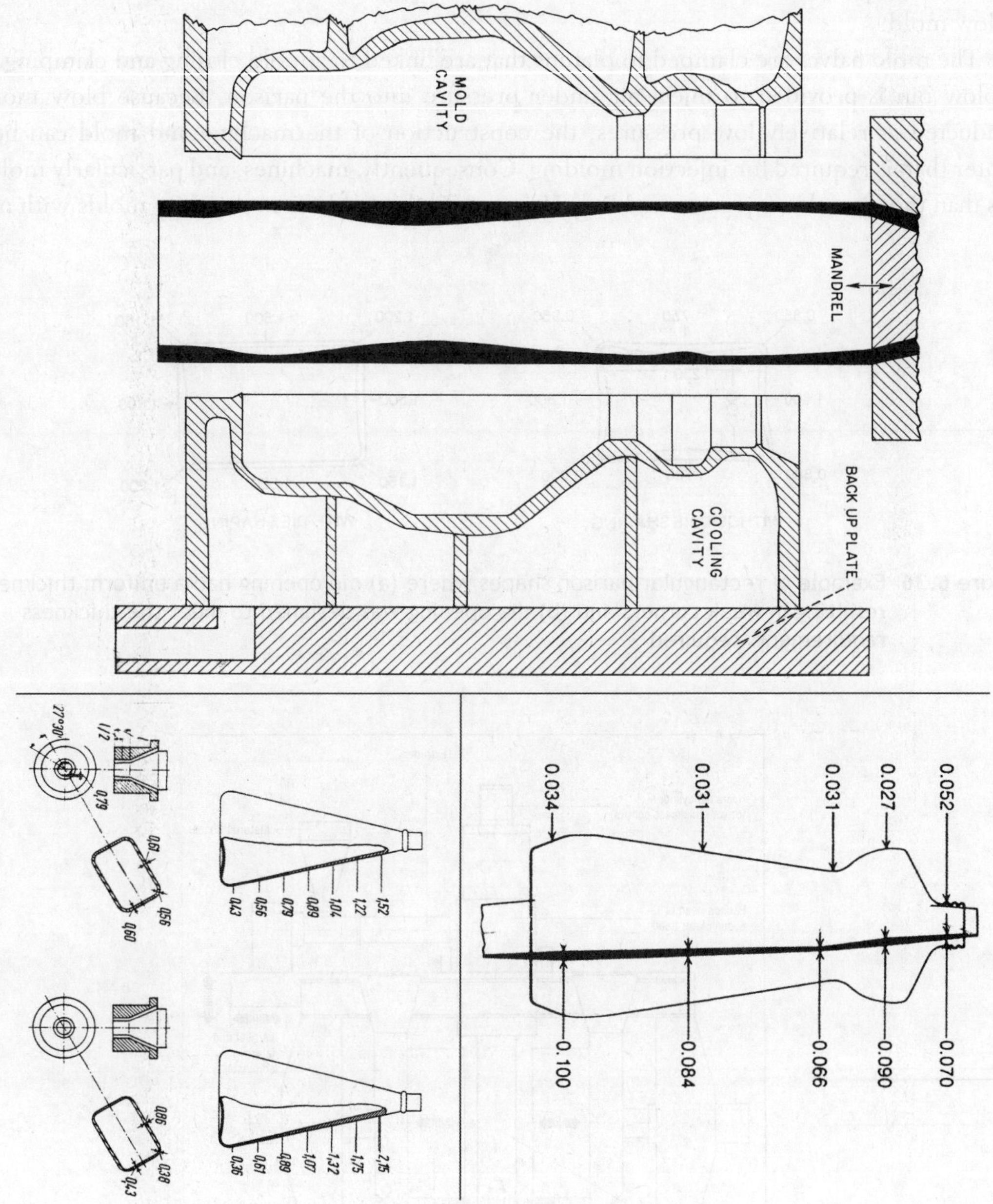

Figure 6.35 Examples of programmed parisons

extruder and parison head are arranged to extrude a parison vertically between the two halves of a blow mold.

The mold halves are clamped to platens that are linked to a mold closing and clamping device. A blow pin is provided to inject air under pressure into the parison. Because blow molding is conducted at relatively low pressures, the construction of the machine and mold can be much lighter than is required for injection molding. Consequently, machines, and particularly molds, cost less than those used in injection molding. Using multiple molds or multicavity molds with multiple

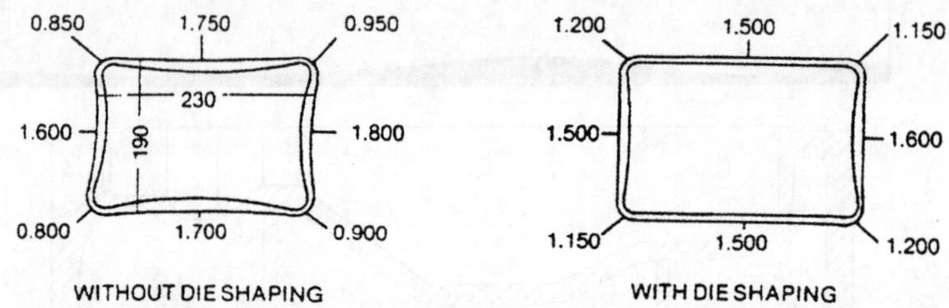

Figure 6.36 Example of rectangular parison shapes where (a) die opening had a uniform thickness resulting in weak corners and (b) die opening was designed to meet the thickness requirements required

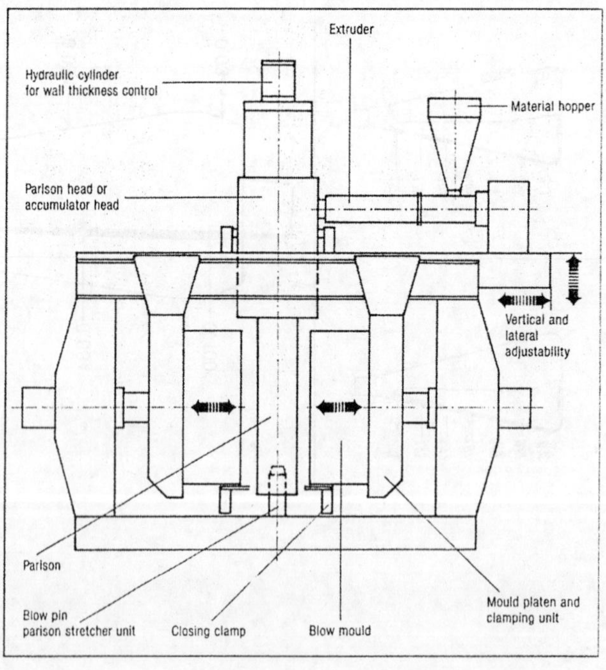

Figure 6.37 Simplified schematic showing parts of a blow-molding machine

parisons can increase machine production rates. These general principles apply to a number of distinct extrusion blow-molding machine designs.

SINGLE-STAGE DESIGN

Single-stage extrusion blow molding produces a blown product in a single integrated process cycle. Blow molding immediately follows parison extrusion and relies on the melt condition of the parison for the deformation and flow necessary to take up the shape of the mold. There is no reheating of the parison before molding. The principal variants are the continuous extrusion and the intermittent extrusion processes. Extrusion in this distinguishing sense applies to extrusion of the parison rather than the operation of the extruder.

TWO-STAGE DESIGN

Two-stage extrusion blow molding treats the parison as a true preform by separating the functions of parison preparation and blow molding. Parisons are produced by conventional tube-extrusion methods (chapter 5). The tube is cooled and cut into parison lengths that are stored before being reheated and blow molded in the normal way (Fig. 6.38).

The reheating process can be selective, leading to a degree of control over wall thickness in the blown product. The reheating process also tends to relax stress in the parison, resulting in a stronger product. However, the process has a number of disadvantages. The plastic heat history is increased by the use of two heating cycles, energy use is increased, and costs arise from the storage and handling of parisons. The process is not suited to high production speeds and is little used.

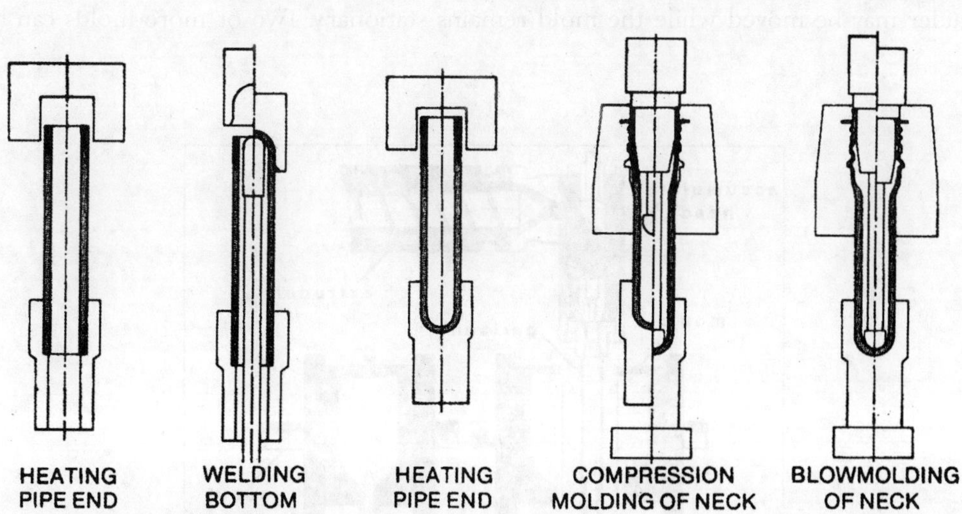

Figure 6.38 Examples of preparing cut-to-size parisons for a two-stage extrusion blow-molding process (courtesy of SIG Plastics International)

CONTINUOUS EXTRUSION DESIGN

In this process, the parison is continuously extruded between the open mold halves from an accumulator head. When the required length of parison has been produced, the mold is closed, trapping the parison that is severed by a hot knife from the die. Figure 6.39 provides a simplified schematic of a continuous blow-molding process. More details are given in Figure 6.40.

The continuous EBM (Fig. 6.40) shows two different techniques for blow molding a container with a calibrated neck. The top schematics (a and b) illustrate a top mold closing system where a = mold, b = calibrating blow mandrel, c = parison, d = parison head, e = extruder, f = neck insert, g = pinch-off blade, h = pinch-off, and i = clearance for scrap waste. The bottom schematic (a, b, and c) illustrates a movement by the mandrel where a = blow mold, b = calibrating blow mandrel, c = parison, d = parison head, e = extruder, f = cutting insert in mold, g = scrap waste, and h = pinch-off.

Land or pinch-off areas on the mold compress and seal the upper and lower ends of the parison to make an elastic airtight part. Compressed air is introduced through the blow pin into the interior of the sealed parison that expands to take up the shape of the mold cavities. The cooled mold chills the blown object that can then be ejected when the mold opens.

The blowing air can be introduced in a variety of ways. In the simplest case, the parison is extruded downward so that the open end slips over a blow pin. Alternatively, the blow pin can be introduced from above the mold after the parison is severed, or blow needles can be built into the mold, where they pierce the parison as the mold closes.

Shuttle Relative movement between the parison head and mold is necessary so that parison extrusion can proceed continuously while the mold is closed. This is achieved in many different ways. The mold may be lowered, moved aside laterally, or swung aside on an accurate path. Alternatively, the extruder may be moved while the mold remains stationary. Two or more molds can be used

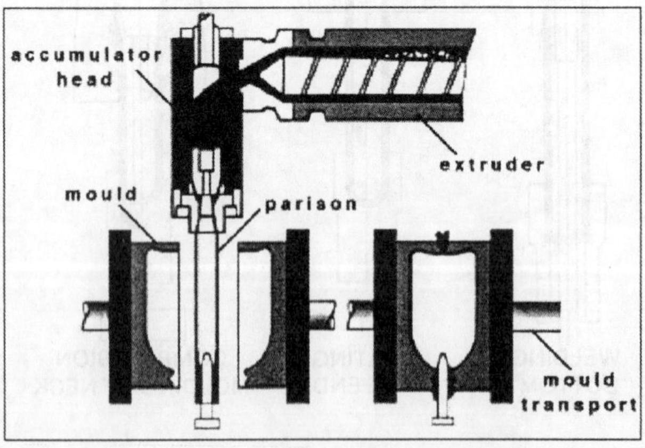

Figure 6.39 Introduction to a continuous extruded blow-molding system with its accumulator die head

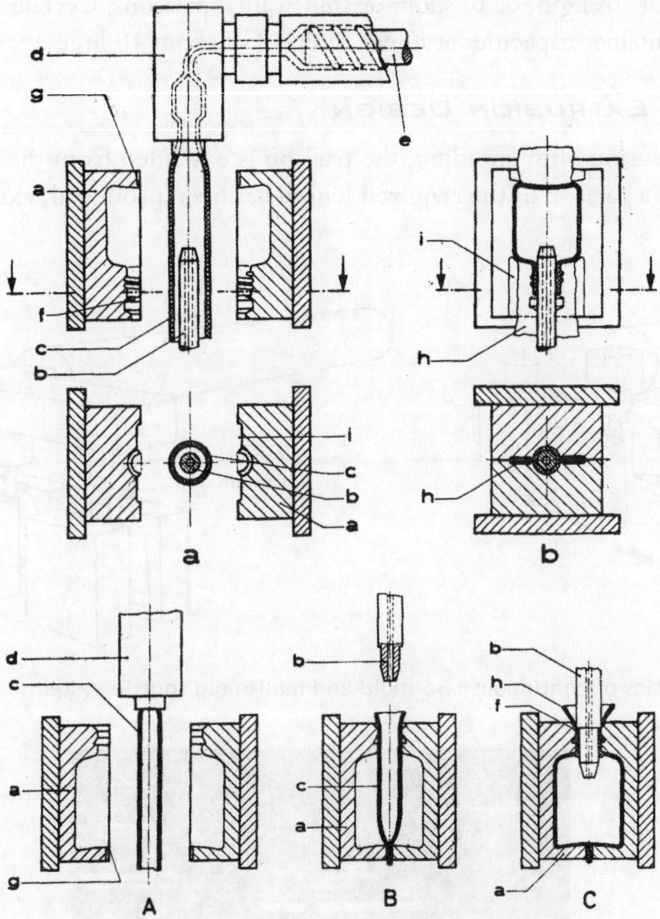

Figure 6.40 Examples of continuous extruded blow-molding systems with calibrated necks

in a shuttle arrangement, so that one or one set is open for parison extrusion while the others are performing the blowing and cooling cycle (Fig. 6.41 to Fig. 6.47; 539).

Rotary There is a method of mold rotary movement that results in very high production rates. In such machines, known as horizontal or vertical wheel machines (also called carousel or ferris wheel machines), a number of molds are mounted on the machine's rotary table. Movement of the table carries the closed mold away and presents a new open mold to the die head, allowing extrusion to continue (Fig. 6.48 to Fig. 6.52).

Except in special cases such as the shuttle and wheel machines, the extrusion rate of the parison must be synchronized with the blow cycle for a single mold. This may involve a relatively slow extrusion rate, running the risk of thinning as the parison stretches under its own weight. For this reason, continuous extrusion blow molding is best suited to thermoplastics with a high melt viscosity or high melt strength, or to short or thin-walled parisons. Certain plastics, such as polypropylenes, have container capacities at upper limits of of about 10 liters.

INTERMITTENT EXTRUSION DESIGN

In intermittent extrusion blow molding, the parison is extruded from the parison die head discontinuously. When a parison of the required length has been produced, extrusion is interrupted

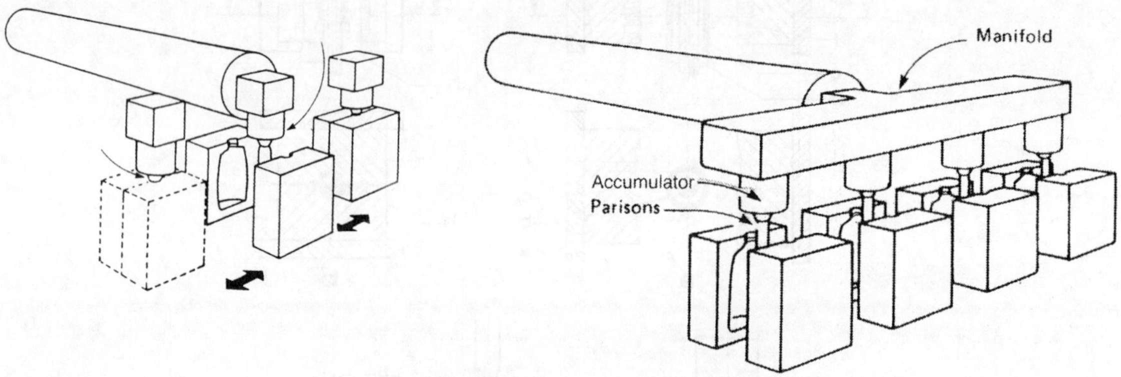

Figure 6.41 Schematics of continuous two-mold and multimold shuttle systems

Figure 6.42 View of a three-milk bottle mold shuttle system

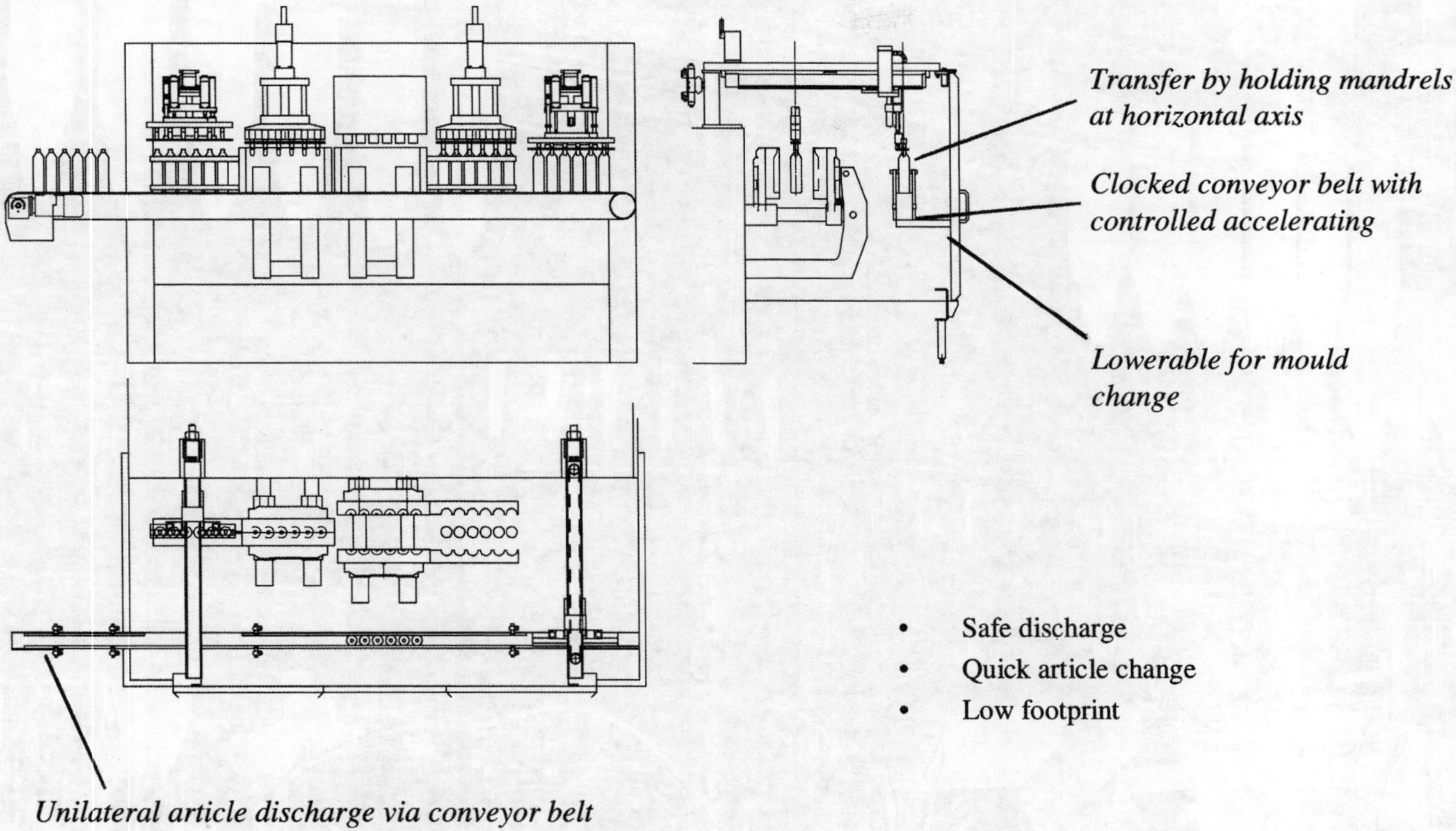

Figure 6.43 Schematic of dual-sided shuttle with six parisons (courtesy of Graham Machinery Group)

Figure 6.44 Closeup of dual-sided shuttle with six parisons (courtesy of Graham Machinery Group)

Figure 6.45 Dual-sided shuttle with six parisons with safety doors opened (courtesy of Graham Machinery Group)

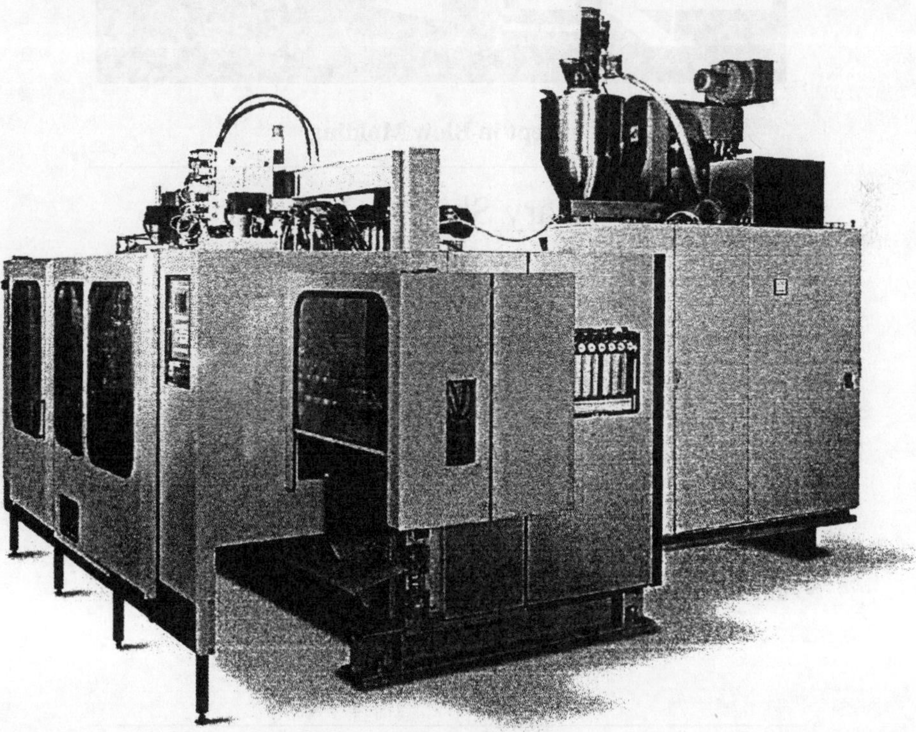

Figure 6.46 Dual-sided shuttle with six parisons with safety doors closed (courtesy of Graham Machinery Group)

Overcoming Shuttle Machine Limitations

- Another possible approach is to utilize not two, but four molds, somehow shuttling the molds along multiple axes.
- This is also not reasonable, due to the impracticality of adding such mechanical complexity to the machine.

The Rotary Shuttle Machine

A Concept in Blow Molding

The Rotary Shuttle Concept

- Incorporate four molds by utilizing a rotary axis of movement
- Use a single flowhead for four molds
- Shuttle the molds perpendicular to the axis of rotation to avoid hitting the flowhead
- Index in 90° increments

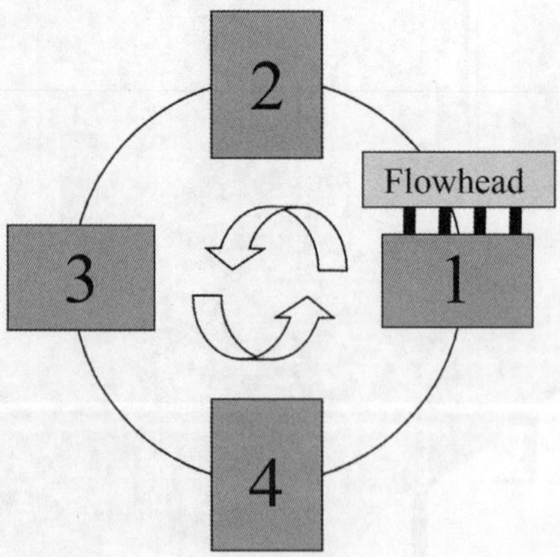

Figure 6.47 Overcoming shuttle machine limitations (courtesy of Graham Machinery Group)

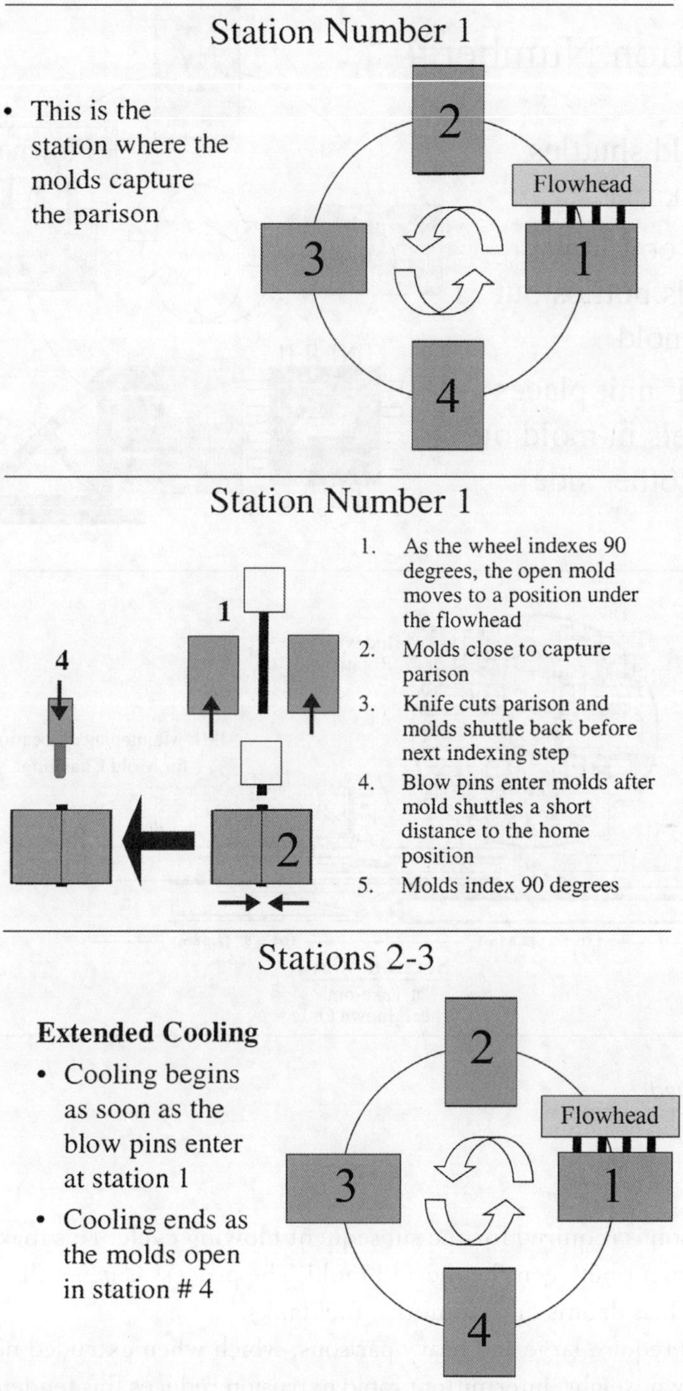

Figure 6.47 *(continued)*

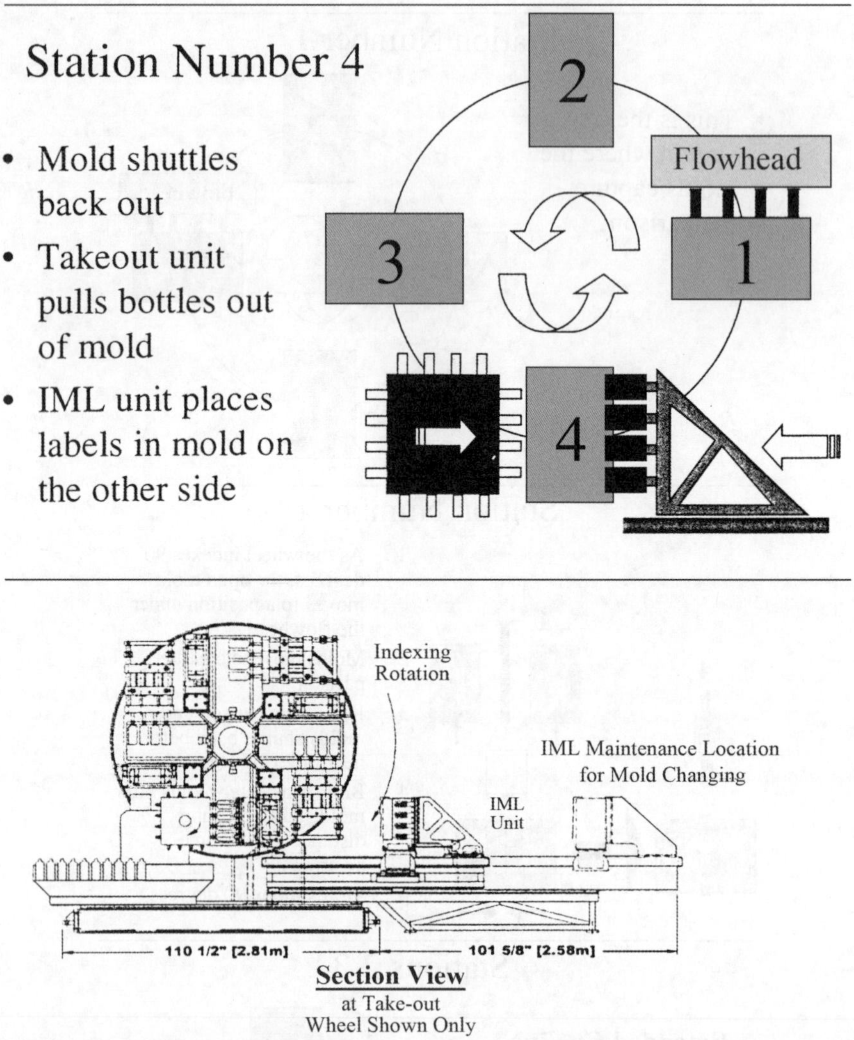

Figure 6.47 *(continued)*

until the next parison is required for the subsequent blowing cycle. This makes it unnecessary to have relative movement between the die and mold. The process is generally used to make larger blow moldings, such as drums or automotive fuel tanks.

These products require large and heavy parisons, which when extruded normally, sag and thin down under their own weight. Intermittent rapid extrusion reduces this tendency, and it is achieved either by reciprocating screw or accumulator means. Intermittent extrusion blow molding is best suited to long or heavy parisons and to thermoplastics with a low melt viscosity or low melt strength.

Reciprocating Screw Reciprocating screw intermittent extrusion blow-molding machines use extruder units in which the screws are capable of axial as well as rotational movement. The function is identical to that of the injection unit on an injection-molding machine (Fig. 6.53). The injection-molding machines described in chapter 4 are used. This intermittent extrusion blow-molding machine operates as an injection-molding machine except at a lower melt pressure and flow rate.

The parison is prepared in two steps. The first step is melt preparation. As explained in chapter 4, the screw rotates, thus heating and melting the material. It is conveyed along the screw flights to

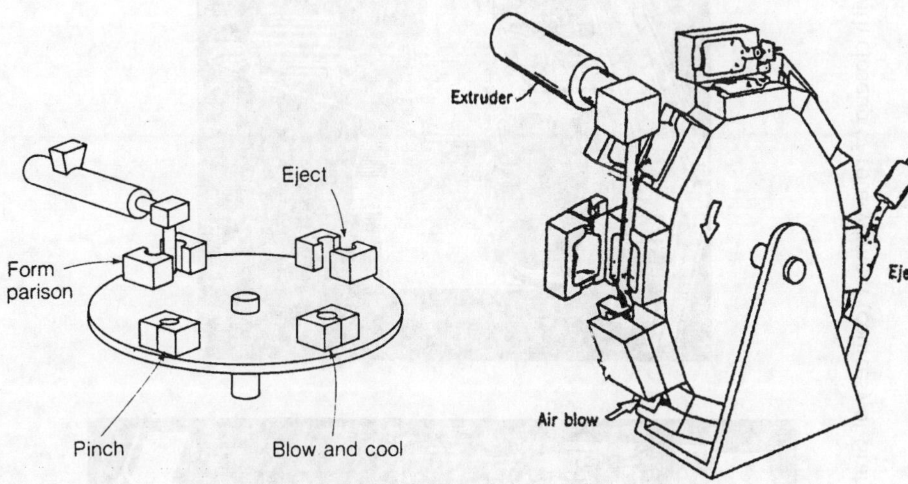

Figure 6.48 Schematics of continuous horizontal or vertical wheel machines

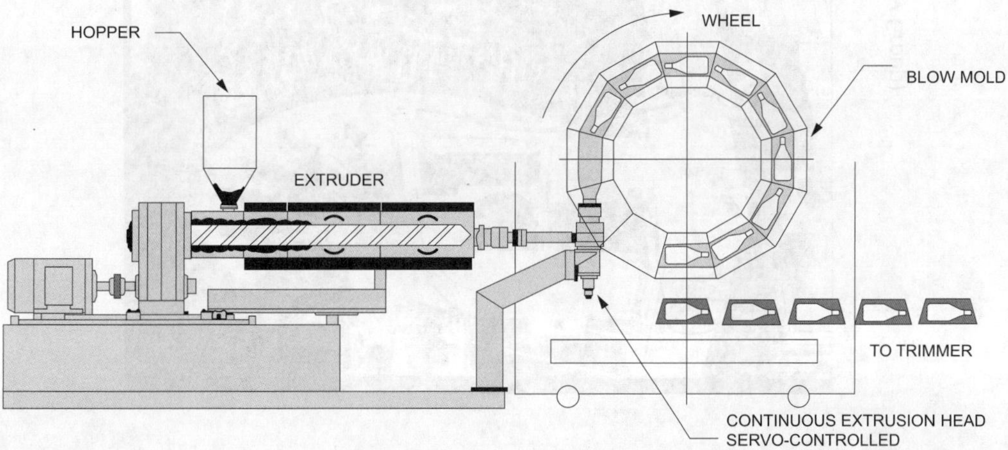

Figure 6.49 Schematics of vertical wheel machine in a production line (courtesy of Graham Machinery Group)

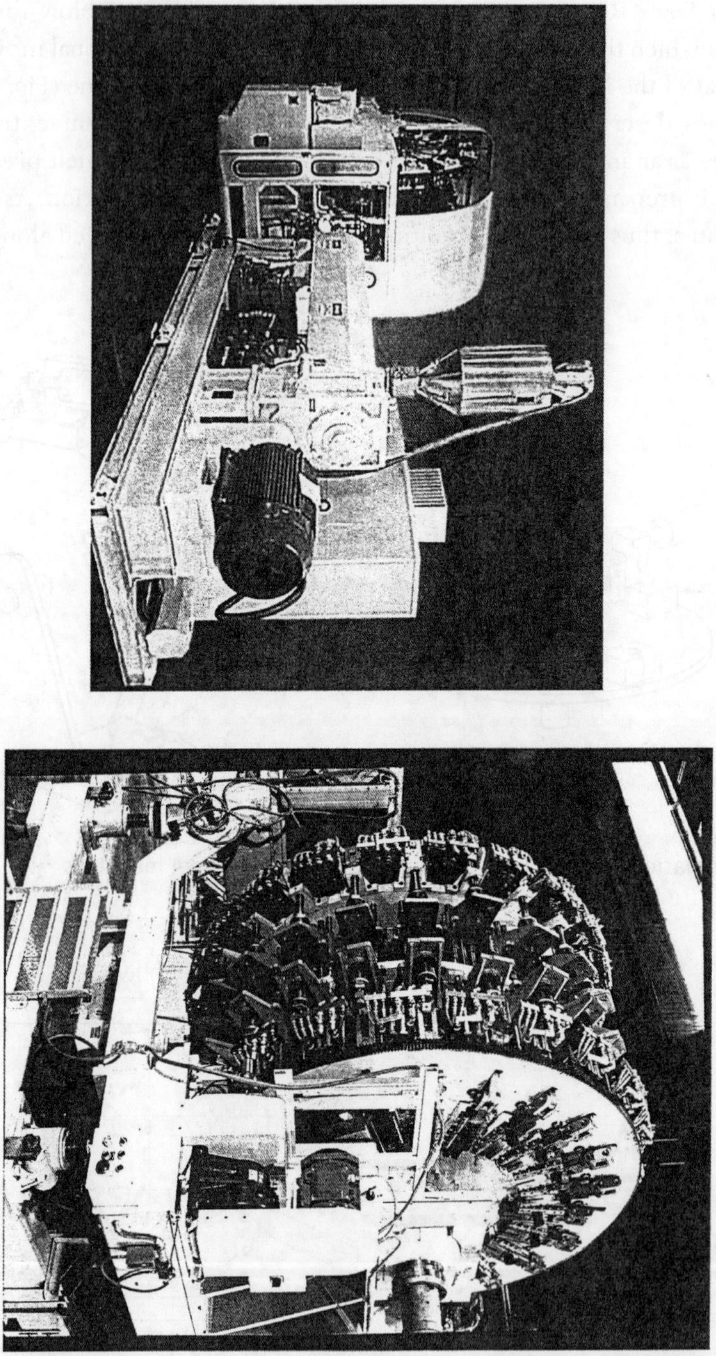

Figure 6.50 Rotary machine with closeup of rotary wheel (courtesy of Graham Machinery Group)

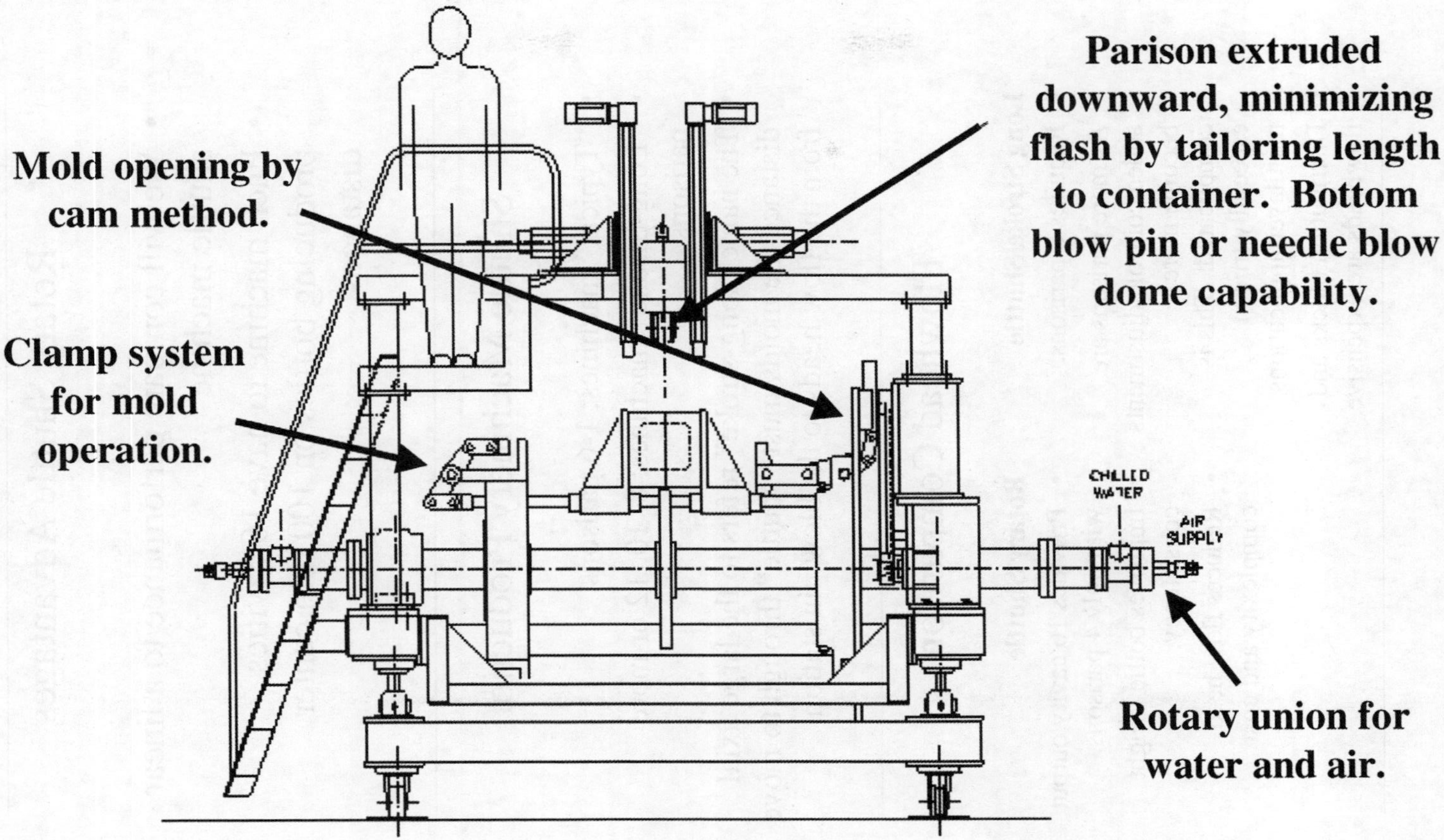

Figure 6.51 Schematic side view of five-station rotary wheel (courtesy of Graham Machinery Group)

Rotary Shuttle Advantages

- We will compare performance to a linear shuttle machine
- Each machine to have 16 cavities, producing bottles on 100 mm center distances

Shuttle Machinery Production

- "Typical" machines: 1-6 parisons
- "Long-Stroke" machines: 8, 10, 12 or more parisons.
- The name "long-stroke" refers to the large axial distance the molds must "shuttle" through to move from the flow heads to the blow pin station.

Flowhead Comparison

Long Stroke Shuttle

- Requires 8 parisons.
- As more parisons are added control adjustments become more complicated. This is especially true in multilayer applications.
- Distribution blocks and flowheads are expensive.

Rotary Shuttle

- Provides 16 cavity output with only 4 parisons.
- Improves bottle weight consistency.
- Reduces flow head complexity and cost.

Figure 6.52 Rotary shuttle advantages (courtesy of Graham Machinery Group)

Long Stroke Shuttle

Rotary Shuttle

Mold Comparison

Long Stroke Shuttle
- Two large eight-cavity molds

Rotary Shuttle
- Four smaller four-cavity molds

The rotary shuttle is versatile- it can accept existing shuttle molds. This allows converters to run small runs using two molds on a small shuttle machine, then ramping up to four molds on a single rotary shuttle machine if orders increase.

Figure 6.52 *(continued)*

Cooling Efficiency

- The shorter shuttling distance of the rotary shuttle allow a larger percentage of the cycle to be used for cooling.
- For example, if a cooling time of 11 seconds is required, the overall cycle might be about 16.25 seconds on a linear shuttle, and 15 seconds on the rotary shuttle.

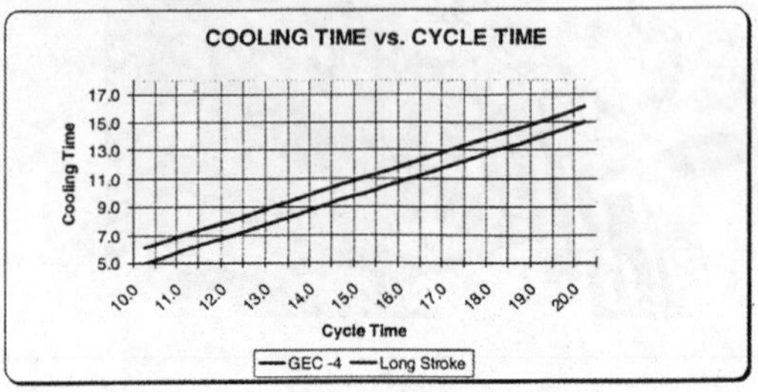

In Mold Labeling – Long Stroke Shuttle

- Long horizontal motion- the IML heads must travel the entire length of the mold horizontally, slowing cycle time significantly. Total cycle time delay of 3-5 seconds typical.
- Label placement accuracy may be compromised due to vibrations in the cantilevered IML unit.

In Mold Labeling – Rotary Shuttle

- Short horizontal motion- the IML heads move into position as the molds index, and travel a short distance. Cycle time delay of less than 0.5 seconds per index, or 2 seconds for complete cycle time.
- The IML heads can actually move into the mold before the bottles are fully extracted.
- High label placement accuracy.

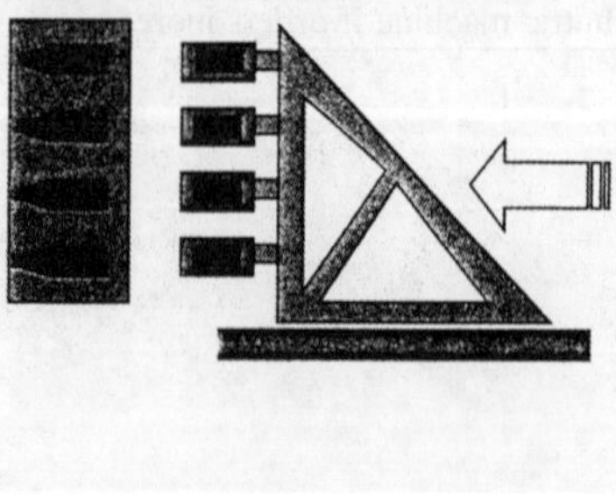

Figure 6.52 *(continued)*

In-Mold Labeling Unit

Custom Designed for GMG by:

AUTOTEC
Sylvania, Ohio

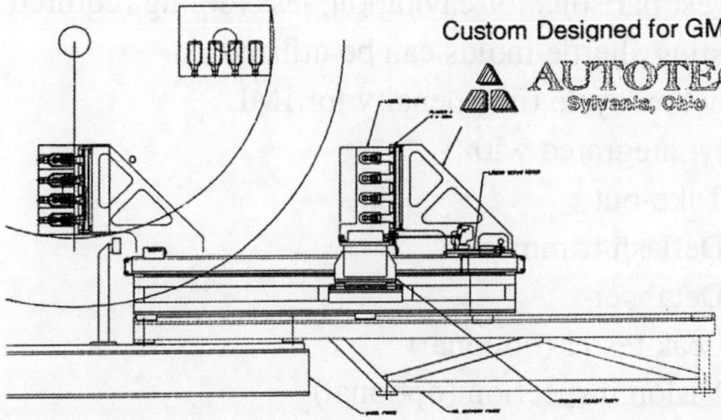

Footprint: Rotary Shuttle vs. Long-Stroke

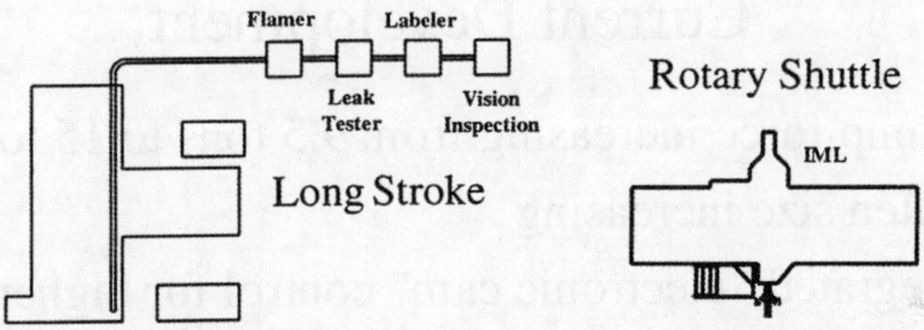

Comparison of equipment and space required to produce labeled, leak tested bottles including vision inspection

Comparison of Tooling Components Required for 16 Cavity Moldset

Tooling Component	"Long stroke"	GEC-4
Die pin and bushing Sets	8	4
Cut-off knives	8	4
Blow pins	16	16
Transfer holders / assemblies	2	1
Trim dies and punches	16	4
Leak tester heads	If available - 16	4
IML Machine Units	2	1
IML label baskets	32	8
IML deployment heads	32	8

Figure 6.52 *(continued)*

Summary: Rotary Shuttle vs. Long-Stroke Shuttle

- Fewest parisons for cavitation, less tooling required
- Existing shuttle molds can be utilized
- Very low cycle time penalty for IML
- Fully integrated with:
  - Take-out
  - Deflash trimmer
  - Detabber
  - Leak tester (optional)
  - Vision inspection (optional)
- Compact design requires minimum floor space
- Lowest energy consumption

Current Development

- Clamp force increasing from 9.5 tons to 15 tons
- Platen size increasing
- Integrated “electronic cam” control for higher speed and reduced cycle time capability from 12 seconds to 10 seconds
- Mold shuttle motion accomplished by linear motors instead of force tubes

Figure 6.52 *(continued)*

the forward end of the screw. The extruder die block is closed by a valve, so the accumulating melt forces the rotating screw back against a controlled resistance until sufficient melt has accumulated to make the next parison shot (this action identifies the reciprocating machine action). At this point, usually screw rotation stops.

The second step is the parison extrusion phase. The extruder die block valve is opened and the screw performs the action of a ram by moving forward in the axial direction without rotating. This forces the accumulated melt at the forward end of the screw through the parison head, where it is

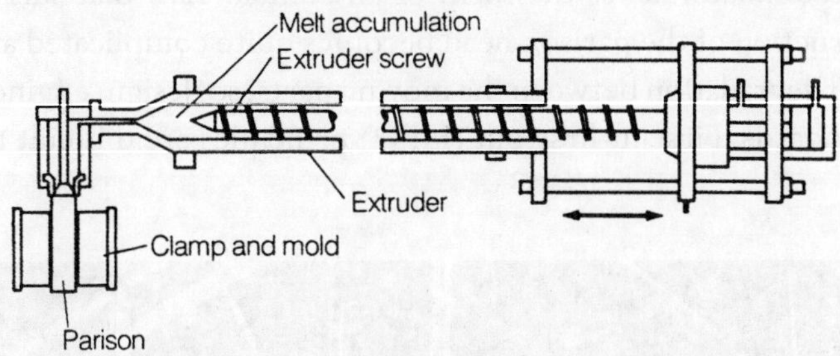

Figure 6.53 Example of a reciprocating screw intermittent extrusion blow-molding machine

extruded at a relatively high flow rate. In practice, the flow rates are limited by the onset of shear phenomena, such as sharkskin and melt fracture.

Thousands of conventional reciprocating injection-molding machines have been built and used worldwide to fabricate blow-molded containers of all sizes. These machines used conventional horizontal-injection molding machines with appropriate continuous operating dies delivering the parisons between all kinds of molds (Fig. 6.54). Mold designs range from simple to those that have all kinds of actions for blow molding complicated products. Decades ago, the IMPCO Machinery Division of Ingersoll-Rand was probably the world leader in manufacturing these types of machines.

Accumulator An intermittent accumulator extrusion blow-molding machine uses a conventional, axially fixed continuously operating extruder to prepare the melt. The accumulator is a heated reservoir where the melt is temporarily stored in the intervals between parison extrusions (similar action of a two-stage injection-molding machine as reviewed in chapter 4) (Figs. 6.55 to 6.57).

The parison is extruded by a ram action on the stored melt. This means that the melt volume in the accumulator fluctuates in a cyclic manner. It is substantially empty immediately after the parison is ram extruded and then it gradually is refilled with melt as the extruded parison enters its blowing phase. The accumulator reaches a maximum fill, based on the parison requirement, immediately after the blown parison leaves the mold.

Accumulators introduce a number of necessary undesirable conditions. They increase the heat history of the melt, and they involve valves and complex flow paths that offer flow resistance and may also lead to difficulties with cleaning and even degradation of the plastic. Accumulators are designed on the FIFO principle (the first material in must be the first out).

Blow-molding accumulators use one of two systems. The separate accumulator is a heated chamber that is an integral part of the blow-molding machine. The capacity of such a chamber can be very large and can, if necessary, be served by several extruders. Figures 6.58 and 6.59 show the other type of accumulator that is built into the parison head. Figure 6.59 call-outs are as follow: 1 = housing, 2 = torpedo, 3 = cardioid groove, 4 = guide legs, 5 = guide ring, 6 = merging zone, 7 = annular ram, 8 = accumulator melt chamber, 9 = die retainer ring, 10 = die, 11 = axially adjustable mandrel, and 12 = axial bore. Other accumulator designs are shown in Figures 6.60 to 6.63.

This type of accumulator takes the form of an annular ram that surrounds the mandrel and torpedo. The construction of the parison head becomes quite complicated and there is the potential for melt leakage and degradation between the moving parts, so design, engineering, and manufacture must be of high standards. First-in-first-out (FIFO) principles are difficult to apply.

Figure 6.54 Series of conventional horizontal injection-molding machines with appropriate blow-molding dies

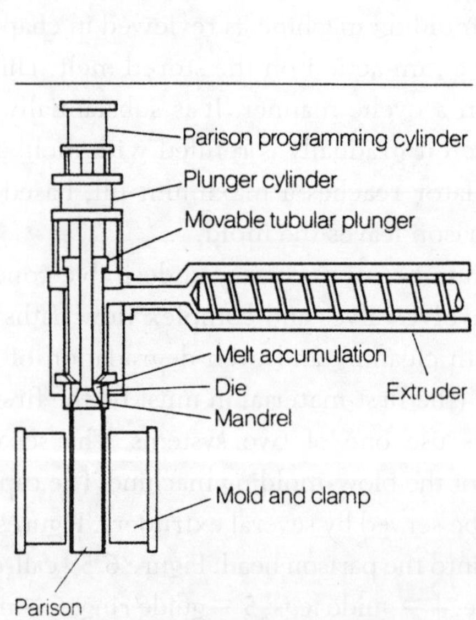

Figure 6.55 Example of an intermittent accumulator head extrusion blow-molding machine

INJECTION BLOW MOLDING

The injection blow-molding (IBM) process uses an injection-molding machine rather than extruder to produce the precursor (chapter 4). This precursor is called a preform rather than a parison as in extrusion blow molding. A major advantage of IBM compared to EBM is that the preform shape can be designed to obtain a uniform (or whatever is desired) wall thickness when blow molded (Fig. 6.64).

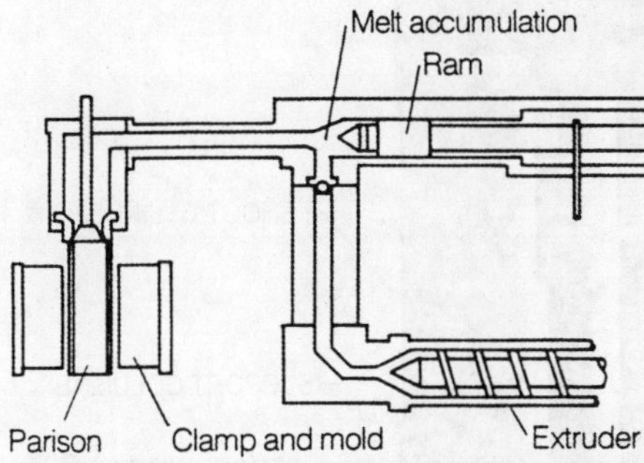

Figure 6.56 Example of an intermittent ram-accumulator extrusion blow-molding machine

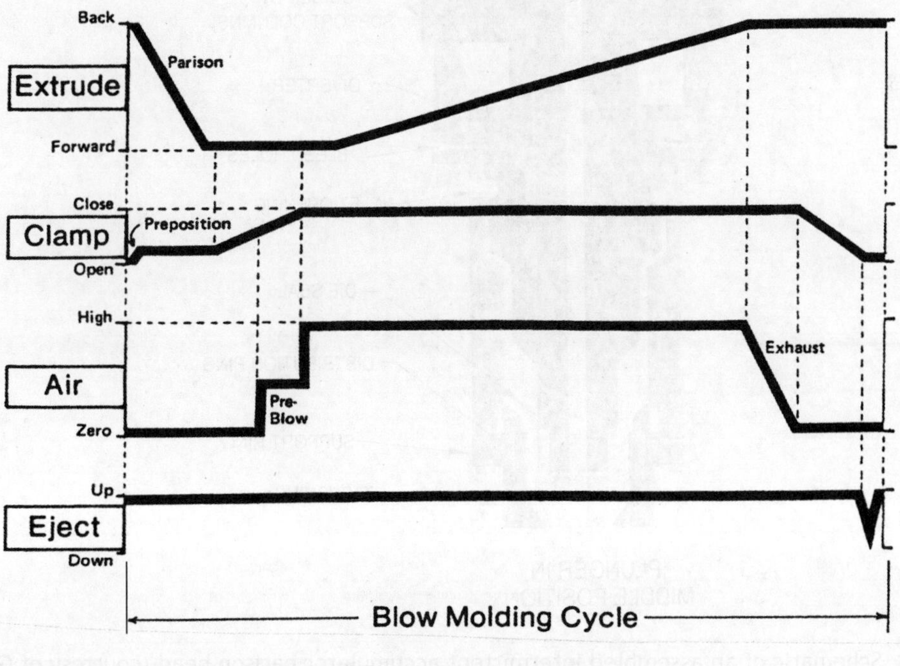

Figure 6.57 Example of the extrusion blow-molding cycle with an accumulator

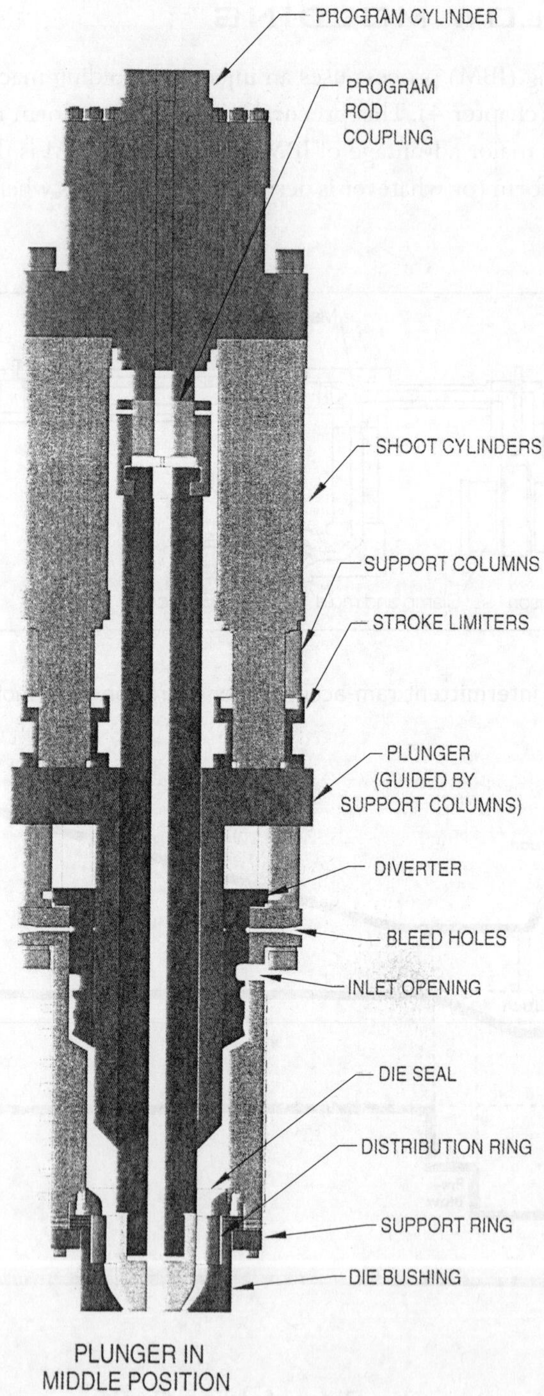

Figure 6.58 Schematic of an assembled intermittent accumulator parison head (courtesy of Graham Machinery Group)

Figure 6.58 *(continued)*

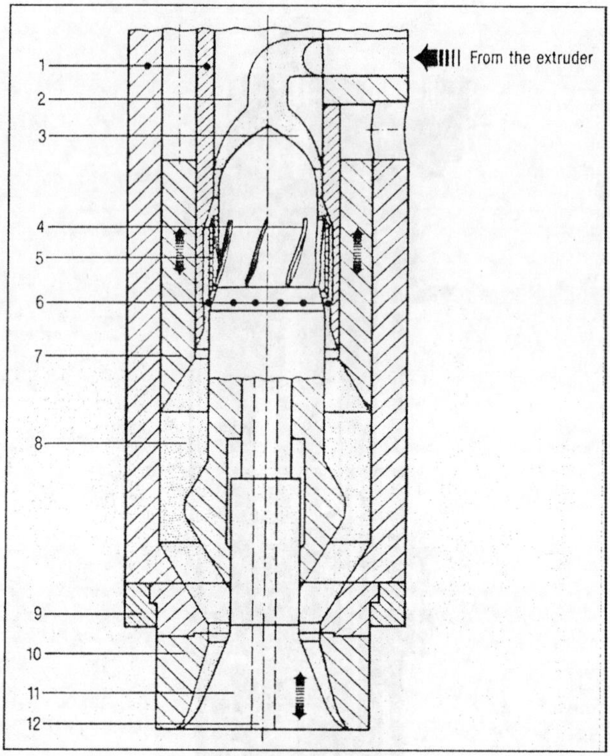

Figure 6.59 Example of intermittent accumulator parison head (courtesy of Bekum)

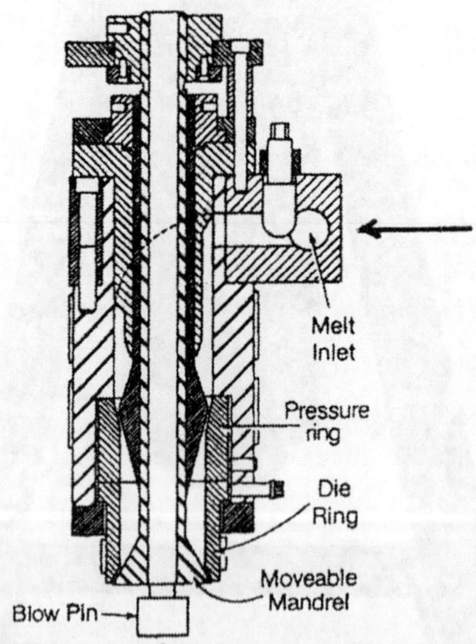

Figure 6.60 Example of intermittent accumulator parison head with a calibrated neck finish

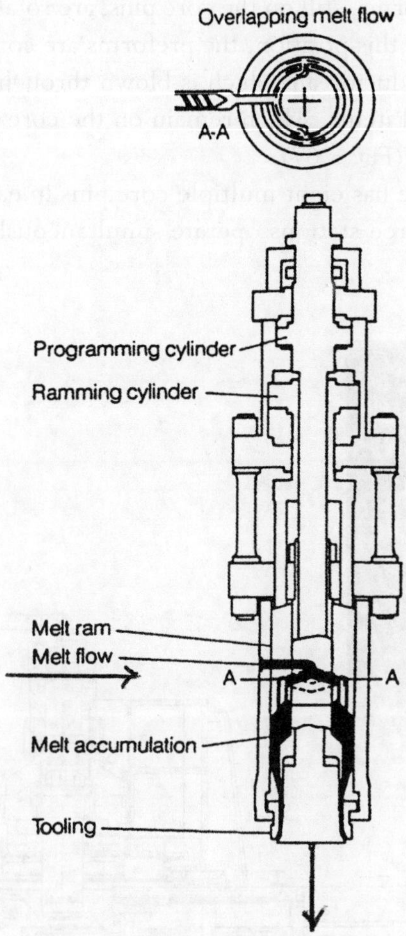

Figure 6.61 Example of intermittent accumulator parison head with overflow melts in the parison to eliminate weld lines

The process consists of blowing a molten thermoplastic against the inside walls of a female mold cavity and chilling it to a rigid solid product. The molding cycle for IBM is shown in Figure 6.65.

The injection blow-molding equipment includes an integral injection component and a multi-impression mold assembly in which mold cores are typically installed on a rotary table. The cores also serve as blowing pins that index in 120° increments between injection, blowing, and ejection stations (Fig. 6.66).

The first station usually has multiple preform injection molds where preforms are formed over core pins. The preforms have hemispherical closed ends (resembling laboratory test tubes). The other ends have open bores, formed by the core pins.

External features, for example the neck flange and threads of a screw-top container, can be produced by injection molding. With the preform still hot and plastic, the mold of the injection machine is opened and the preforms, still on the core pins, are rotated to the second station, which serves as the blowing station. In this position, the preforms are contained by the blow molds with blow molding initiated by introducing air which is blown through the core pins. Next, the blow mold is opened and the finished items, which remain on the core pins, are rotated to an ejection station where they are removed (Fig. 6.67).

In Figure 6.68, the machine has eight multiple core pins in each of the three stations so that after the initial start-up, the three stations operate simultaneously. Most machines in operation

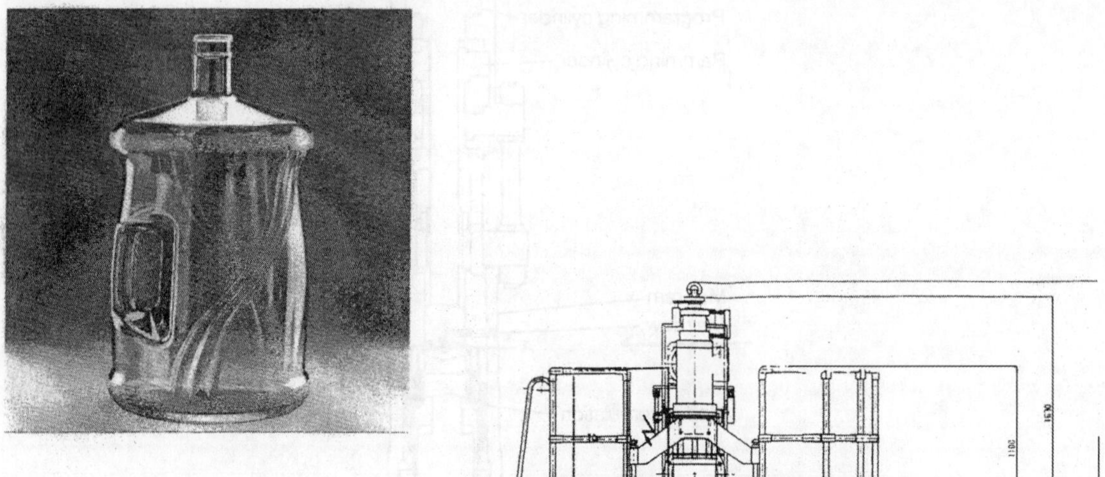

Figure 6.62 Schematic of an EBM with an intermittent accumulator that is fully automatic; insert is an example of a 20-liter (5-gallon) PC plastic bottle fabricated in this machine (courtesy of SIG Blowtec 2-20/30 of SIG Plastics)

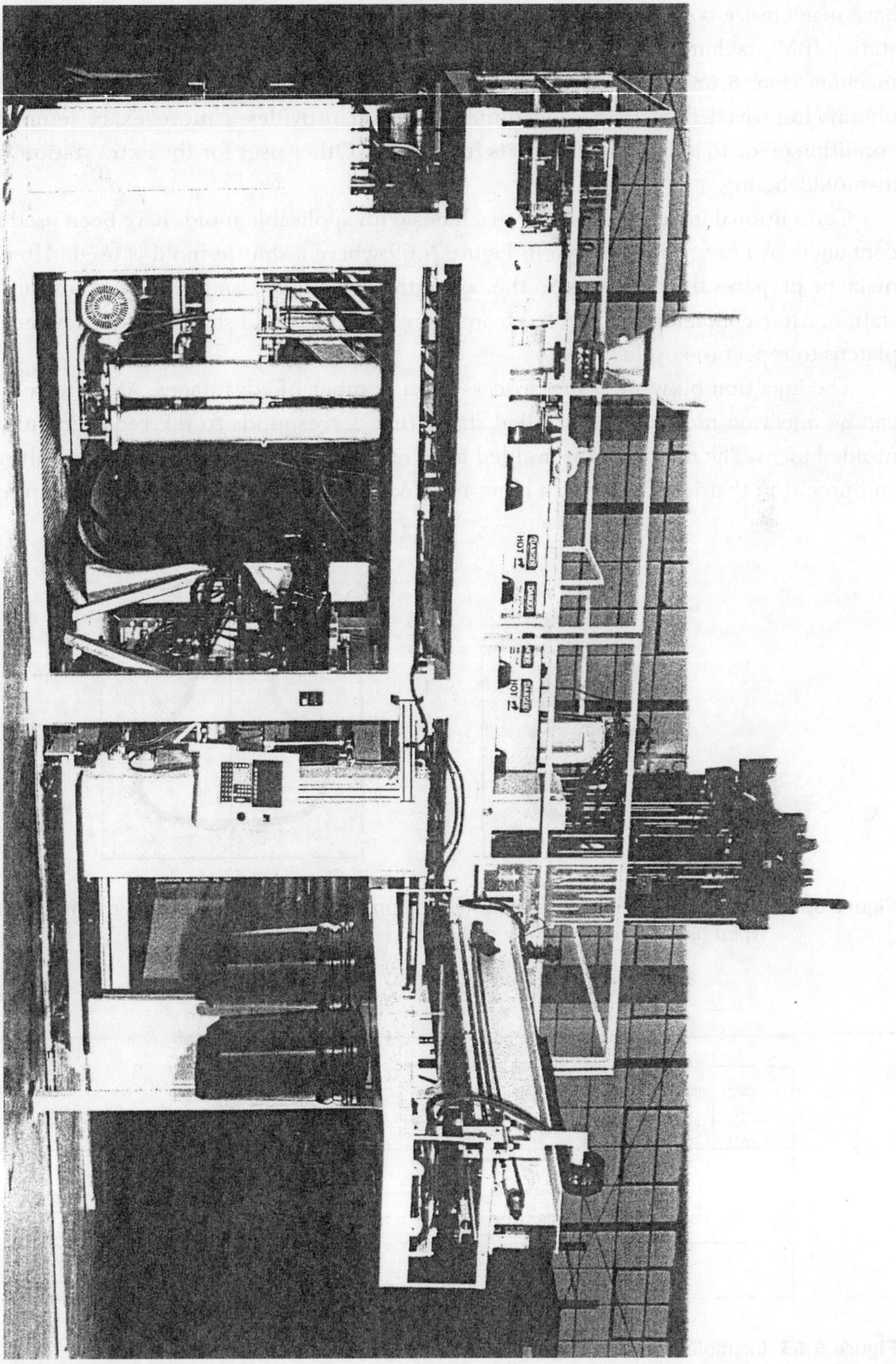

Figure 6.63 Intermittent extrusion blow-molding machine with accumulator molding large tanks (courtesy of Graham Machinery Group)

have many more core pins so that output can be significantly increased. In addition to the three-station IBM machines, there are those that have more stations, with the usual being a four-station machine (Fig. 6.68). The extra station (shown in this figure) is where the preblown preform obtains improved final shape and dimensions and provides a more exact temperature profile condition prior to being blown into its final shape. Other uses for the extra station would include in-mold labeling.

Conventional injection-molding machines with applicable molds have been used to blow mold containers. An example is shown in Figure 6.69 where a shuttle mold is used. After the injection machine prepares the preforms on the cores, the mold opens and is shuttled upward to a blowing station. After containers are ejected from their pins, the mold shuttles back between the machine platens to repeat the cycle.

The injection blow-molding process has a number of advantages. As reviewed, the preform can be injection molded in a profiled shape that corresponds to the requirements of the blow-molded form. The neck form is molded in its entirety at the injection stage, resulting in a quality and precision that is superior to a blow-molded neck. There is no pinch-off scrap to be removed

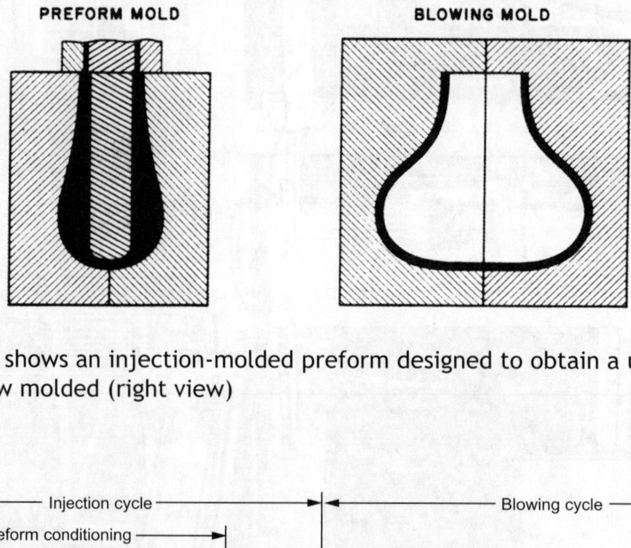

Figure 6.64 Left view shows an injection-molded preform designed to obtain a uniform wall thickness when blow molded (right view)

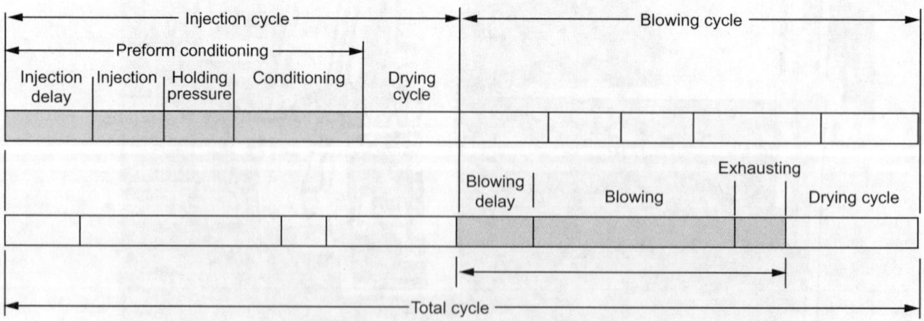

Figure 6.65 Example of the injection blow-molding cycle

and recycled and no bottom seam. There are some drawbacks. The process is difficult to use with high melt viscosity materials. View strips and multiwall constructions are impractical. Integral hollow handles to date are not practical; however, solid integral handles can be included during fabrication. Examples of solid handles are shown in Figures 6.70 and 6.71.

A patent was issued during 1913 (Fig. 6.70) that produced an integral solid handle during unstretched (or stretched) injection blow molding of a bottle. French patent #1,192,475 was issued to the Italian company Manifattura Ceramica Pozzi SpA. The figure shows (a) precision-molded neck that includes the plastic solid handle, (b) preform core and blow pin, (b) basic water-cooled bottle female mold, and (d) injection nozzle of the injection-molding machine.

STRETCH BLOW MOLDING

Stretch blow molding may be performed by the injection or extrusion processes (541). Stretching during blow molding produces biaxial orientation in the blown article (chapters 3 and 5). Conventional blow molding imparts a degree of circumferential orientation, caused by the expansion of the parison into the mold cavity, but there is little or no axial expansion and, correspondingly, no axial orientation. Stretch blow molding provides for axial orientation by stretching the preform or parison axially before or during blowing. This is normally accomplished by means of a stretch rod that is advanced axially inside the preform or parison at a controlled rate. Another axial-stretching system literally clamps the ends and stretches them apart.

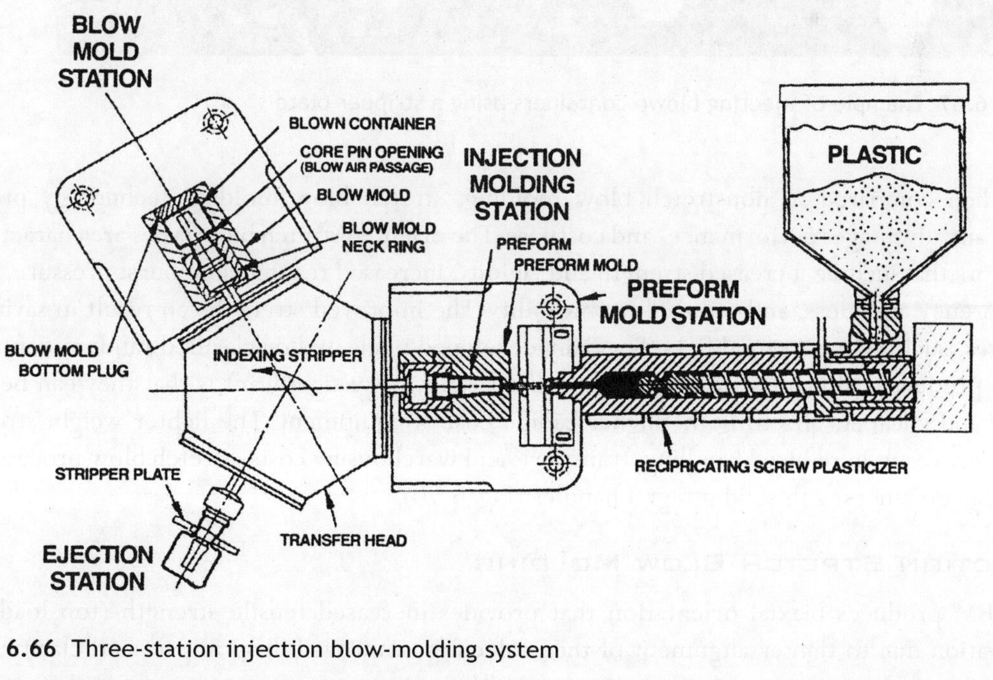

Figure 6.66 Three-station injection blow-molding system

Figure 6.67 Example of ejecting blown containers using a stripper plate

When compared to nonstretch blow molding, stretch blow molding technology provides significant advantages performance- and costwise. The oriented stretch blow types are characterized by factors that include increased strength and rigidity, increased resistance to burst pressure, better transparency and gloss, and reduced permeability. The improved strength can result in savings on additives, such as impact modifiers. Specially developed nucleated or clarified random copolymer grades produce a blown bottle of glasslike transparency. Very significant is that they can be made lighter and cheaper, offsetting the higher capital cost of equipment. The lighter weight stretched containers result in reduced handling, transport, and warehousing costs. Stretch blow processes can produce containers with solid integral handles (Fig. 6.70).

Injection Stretch Blow Molding

ISBM produces biaxial orientation that provides increased tensile strength (top load); less permeation due to tighter alignment of the molecules; improved drop impact and clarity; and a lightweight container. The injection-molded preform is usually short and thick-walled relative to the finished blown article. The neck profile complete with screw thread is entirely formed by injection molding and is generally not modified by the blowing process. The other end of the preform is closed and typically dome shaped. The design and precision of the preform have critical influences on the degrees of orientation and quality of the blown articles. The stretch blow-molding processes continue to be in widespread use worldwide for producing injection blow-molded polyethylene terephthalate (PET) carbonated drink bottles and other plastics, such as PVCs. The injection-molded preform may be converted to a blow molding either by the single-stage or the two-stage process.

Not all thermoplastics can be oriented. The major thermoplastics used are polyethylene terephthalate (PET), polyacrylonitrile (PAN), polyvinyl chloride (PVC), and polypropylene (PP). PET is by far the largest volume material, followed by PVC, PP, and PAN. The amorphous materials with a wide range of thermoplasticity, such as PETs, are easier to stretch blow than the partially crystalline types, such as PPs (chapter 1). Approximate melt and stretch temperatures to yield maximum container properties are shown in Table 6.9.

Single-Stage In the single-stage (or one-step) injection stretch blow-molding process, the perform injection-molding step is integrated with the stretch blow machinery. The machines are generally arranged for rotary operations so that the preforms pass directly from the injection-molding station to a thermal-conditioning station and then to a stretch blow-molding station (Figs. 6.72 to 6.76). The Figure 6.72 call-outs are as follows: 1 = preform injection-molding station, 2 = thermal-conditioning station, 3 = stretch blow-molding station, and 4 = ejection station.

Single-stage equipment is capable of processing PVC, PET, and PP. Once the parison is formed (either injection or extruded), it passes through conditioning stations that bring it to the proper

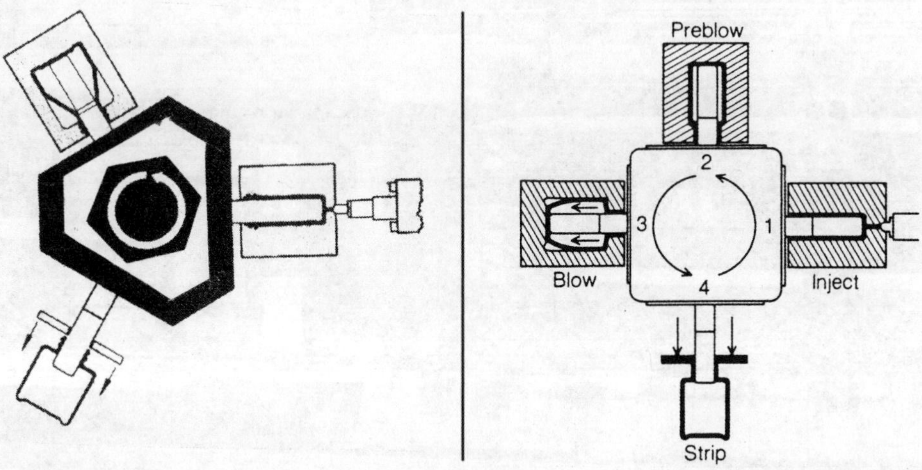

Figure 6.68 Examples of three-station and four-station injection blow-molding machines

orientation temperature. The single-stage system allows the process to proceed from raw material to finished product in one machine, but since tooling cannot be easily changed, the process is best suited for dedicated applications and low volumes.

Many oriented PET containers are produced on single-stage machines. The preform is first injection molded, then transferred to a temperature-conditioning station, then to the blow-molding operation where it is stretch blown into a bottle, and finally to an eject station.

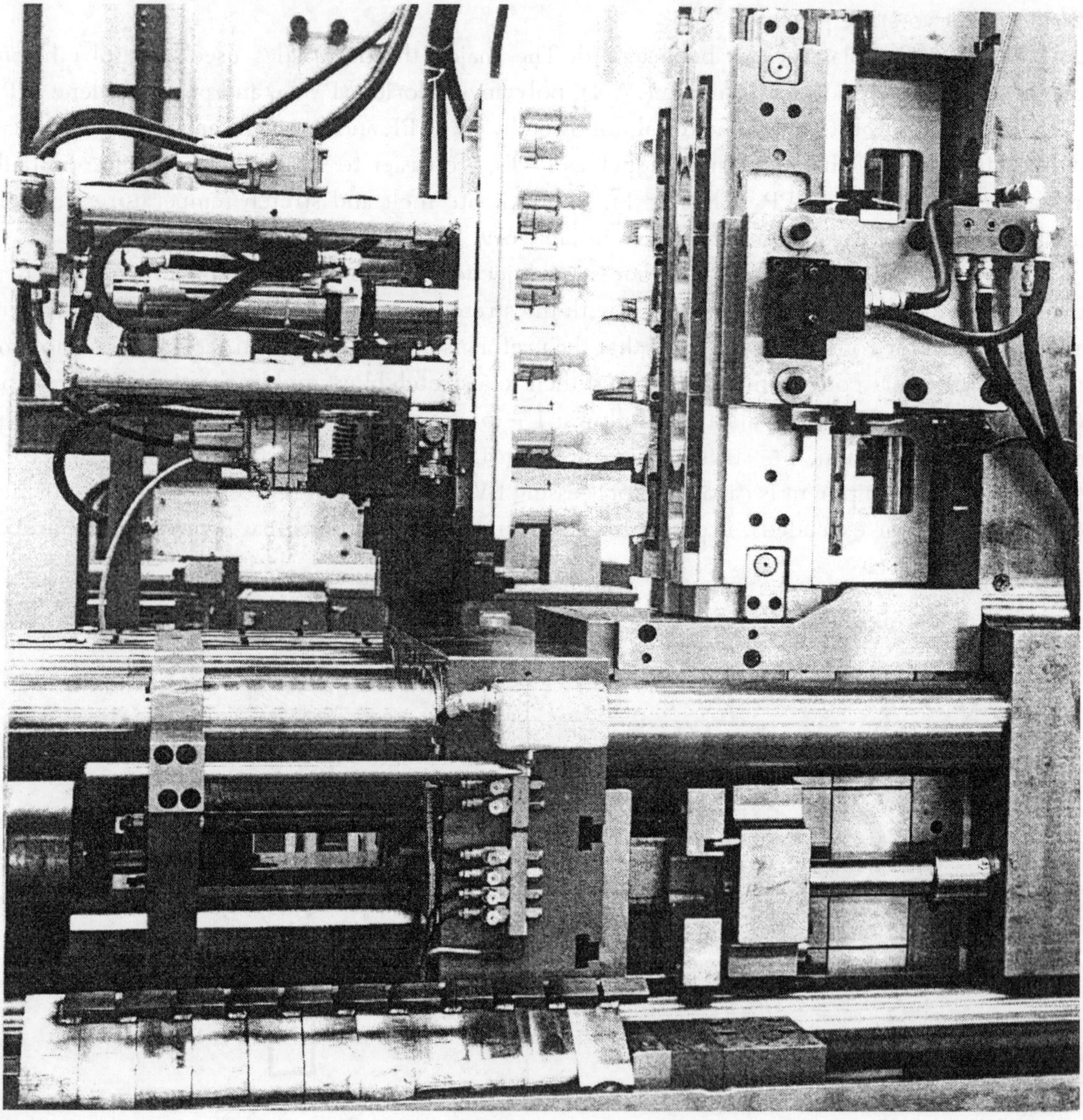

Figure 6.69 View of a shuttle mold to fabricate injection-molded containers

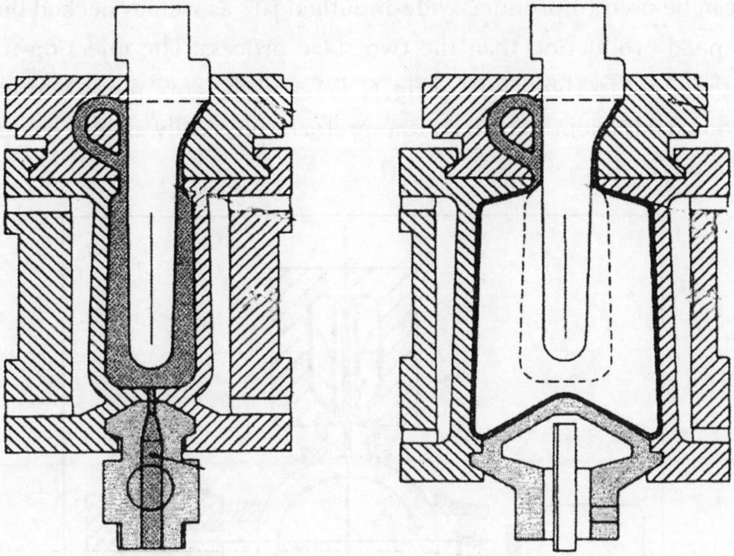

Figure 6.70 Schematic of injection blow mold with a solid handle

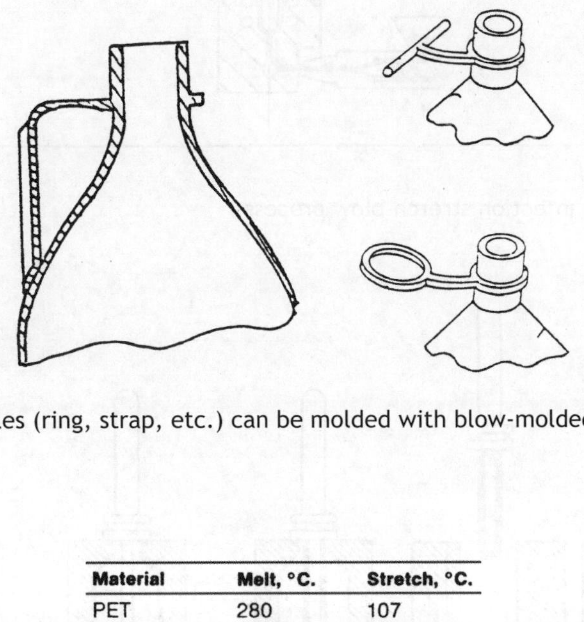

Figure 6.71 Simple handles (ring, strap, etc.) can be molded with blow-molded bottles and other products

Material	Melt, °C.	Stretch, °C.
PET	280	107
PVC	180	120
PAN	210	120
PP	240	160

Table 6.9 Examples of plastic melt and stretch temperatures

This process can be used to produce wide-mouthed jars as well as necked bottles, but it is less suitable for high-speed production than the two-stage process. The injection-molded preform is cooled rapidly in the mold to about 90°C (Fig. 6.77), and the core pin is extracted as soon as the surface skin has cooled sufficiently to make the preform physically stable. The pin has a generous draft angle to make this easier.

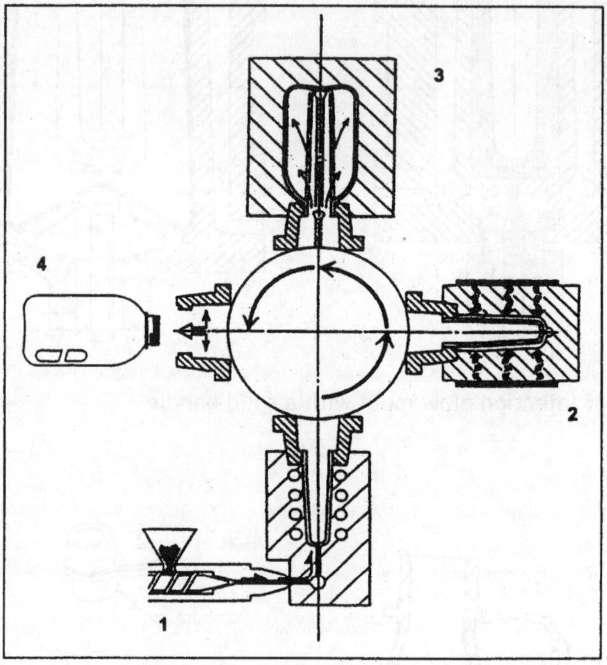

Figure 6.72 Single-stage injection stretch-blow process

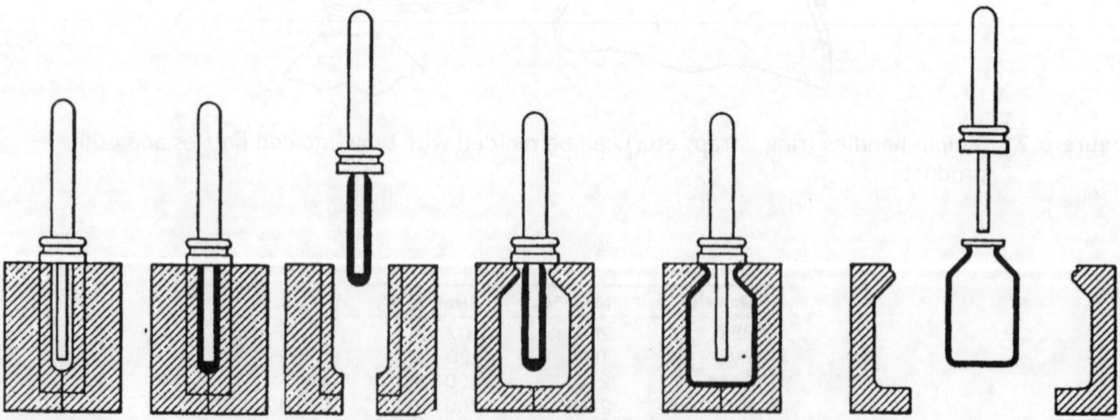

Figure 6.73 Schematic of the steps taken for injection stretch blow molding

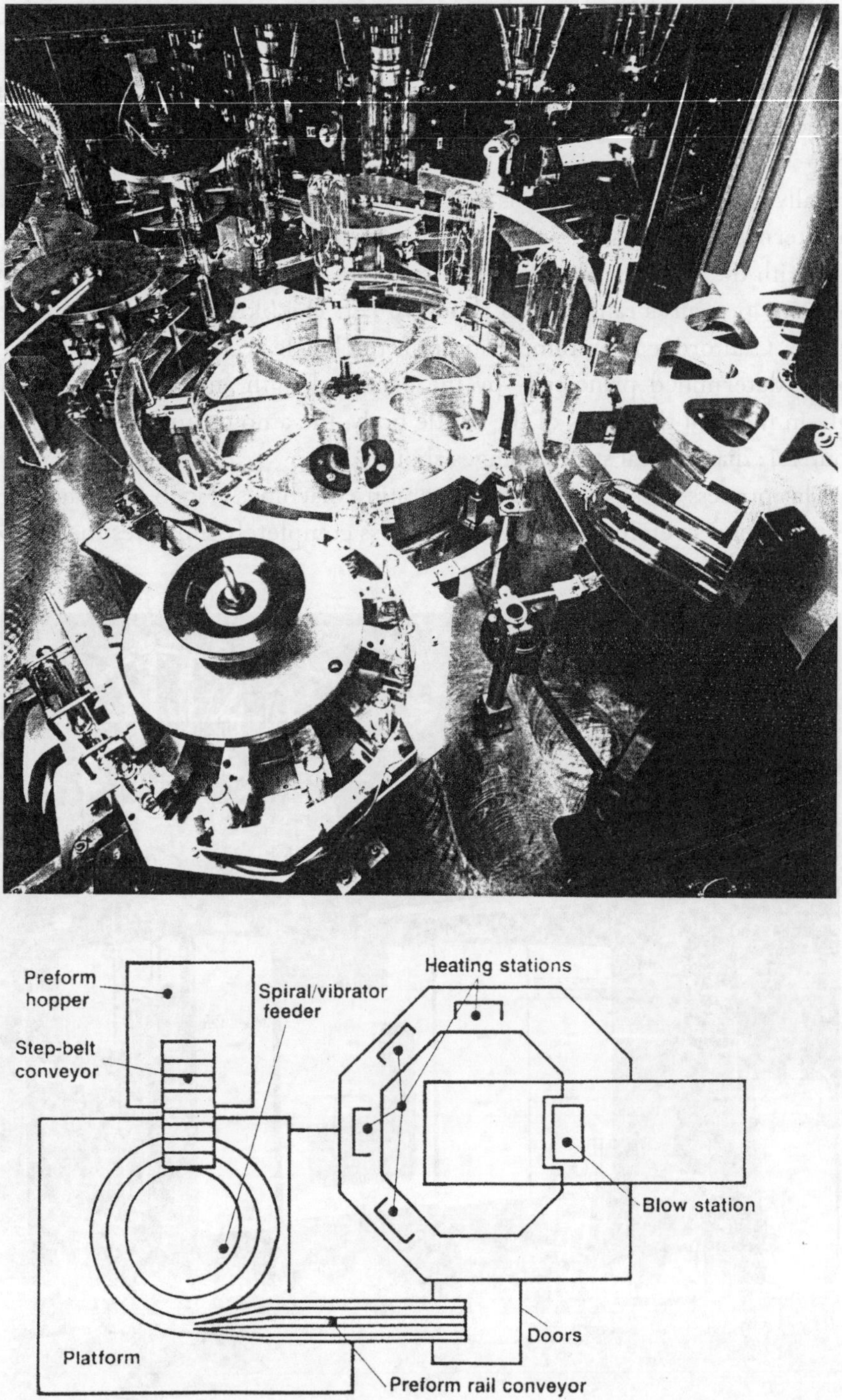

Figure 6.74 Schematic and internal view of a fast-operating reheat preform for stretched IBM (courtesy of SIG Plastics International)

Thermal conditioning is the most critical part of this process. The optimum stretch-blow temperature span can be very narrow, ranging from within about 10°C. Uniform heating of the order of ±1°C is important. Thermal-conditioning stations include up to thirty-six independently controlled heat zones, typically using a combination of infrared radiant external heaters and hot air for internal heating.

The thermally conditioned preform is transferred to a blow mold and stretched and oriented axially by an internal stretch rod or mechanical grip stretching, either immediately prior to or simultaneously with the blowing operation that provides radial stretch and orientation (Figs. 6.78 and 6.79). Blowing pressures range up to about 40 bar. The blow mold temperature is relatively high at 35° to 65°C in order to minimize strain in the bottle. For a given bottle size, the degree of orientation is determined principally by the parison length and diameter. Stretch ratios are relatively high. In the wall thickness of the bottle body, the amount may be as high as 15:1. Axial stretch is about 4:1; diametrical stretch ranges about 3.5:1.

Two-Stage This process uses two different operating machines (Fig. 6.80). The two-stage (two-step) or reheat injection stretch blow-molding process completely separates the preform injection

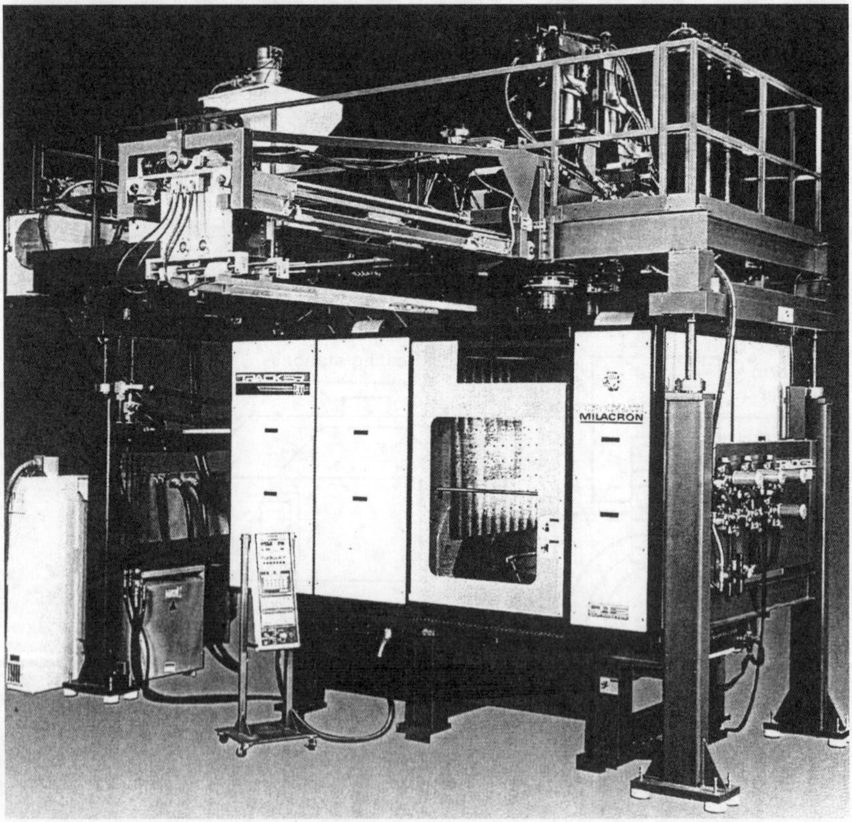

Figure 6.75 Easy-to-operate and control in-line stretch IBM (courtesy of Milacron)

molding operation from the blow-molding operation. The two operations may be carried out in different locations by different manufacturers, with a substantial time interval between them. The process lends itself to high rates of production and has the advantage that preform production may be entrusted to a specialist processor (Figs. 6.81 and 6.82).

This process requires two heating operations, resulting in a greater energy use and increased heat history in the plastic. It also requires the need to store and handle preforms. Before being

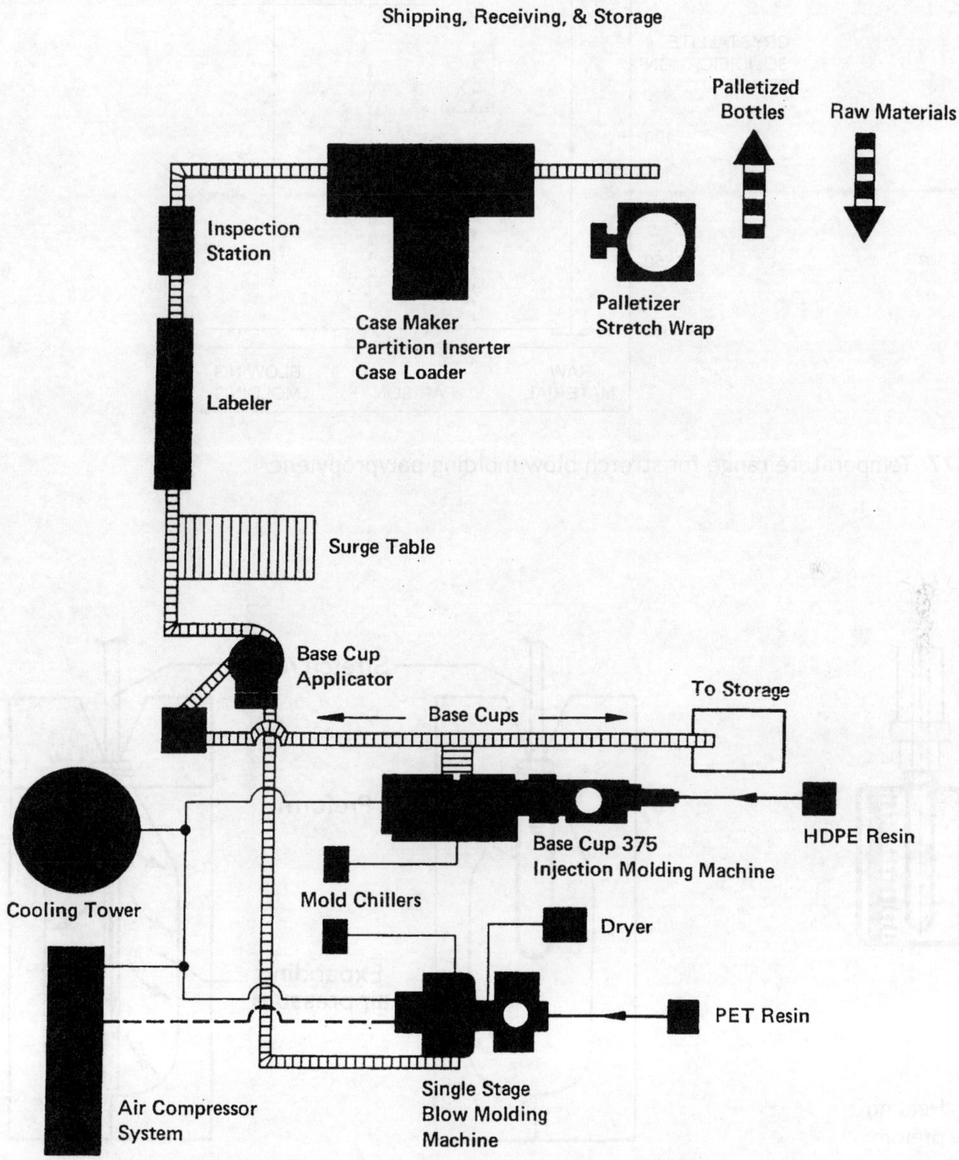

Figure 6.76 Example of a single-stage injection stretch blow-molding production line

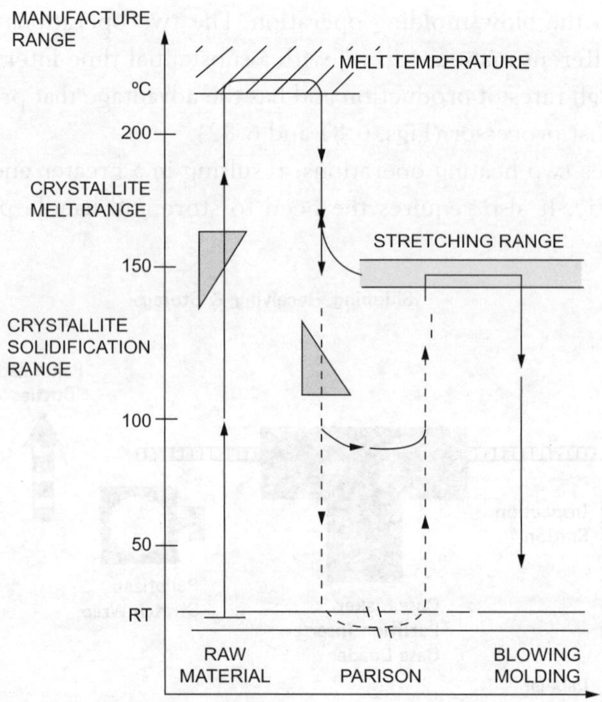

Figure 6.77 Temperature range for stretch blow molding polypropylene

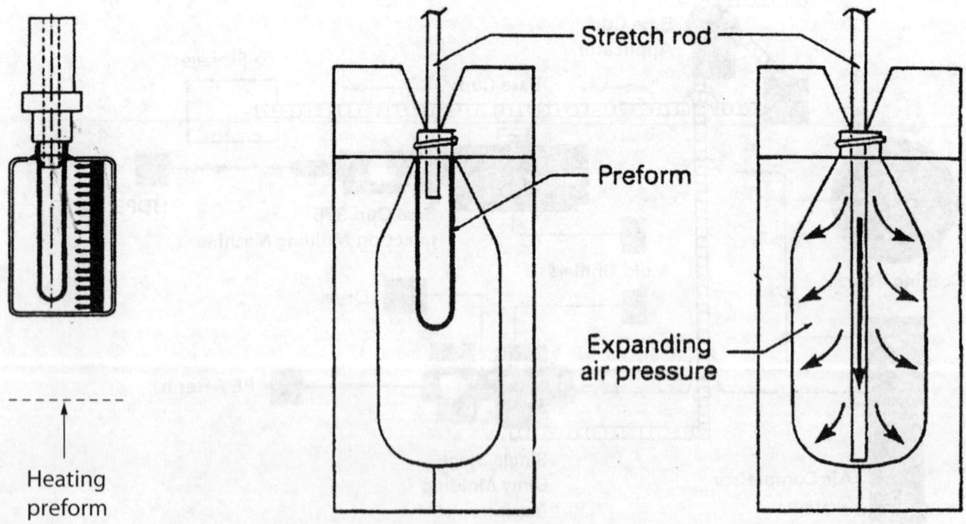

Figure 6.78 Example of stretched injection blow molding using a rod

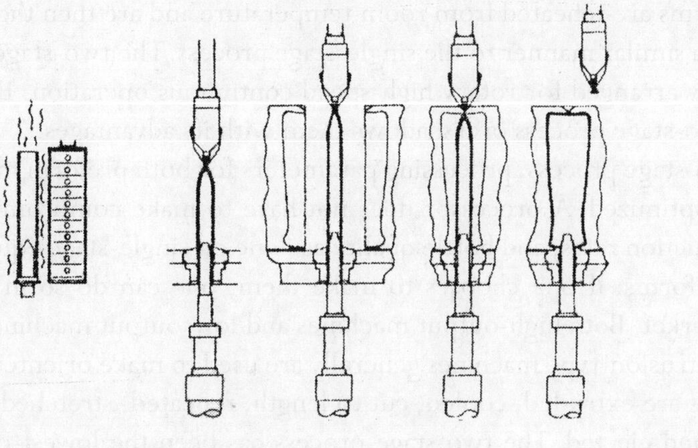

Figure 6.79 Example of stretched injection blow molding by gripping and stretching the preform

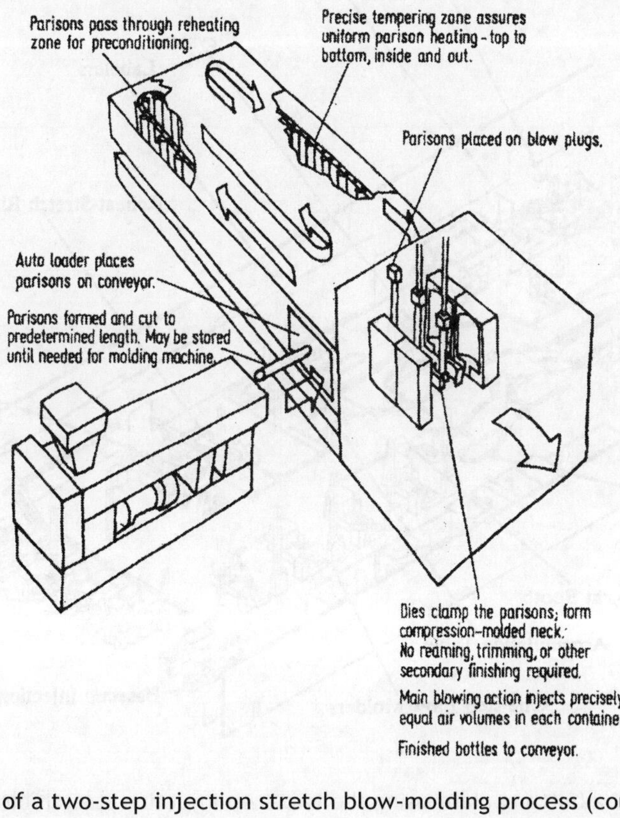

Figure 6.80 Schematic of a two-step injection stretch blow-molding process (courtesy of Milacron)

blown, the preforms are reheated from room temperature and are then thermally conditioned and stretch blown in a similar manner to the single-stage process. The two-stage or reheat stretch-blow machine is usually arranged for rotary high-speed continuous operation. Even with these types of situations, the two-stage process overshadows them with its advantages.

With the two-stage process, processing parameters for both preform manufacturing and bottle blowing can be optimized. A processor does not have to make compromises for preform design and weight, production rates, and bottle quality as done on single-stage equipment. One can either make or buy preforms. If one chooses to make them, one can do so in one or more locations suitable to the market. Both high-output machines and low-output machines are available.

Two-stage extrusion-type machines generally are used to make oriented PP bottles. In a typical process, preforms are extruded, cooled, cut to length, reheated, stretched while the neck finish is being trimmed, and ejected. The two-stage process has been the lowest-cost method to produce

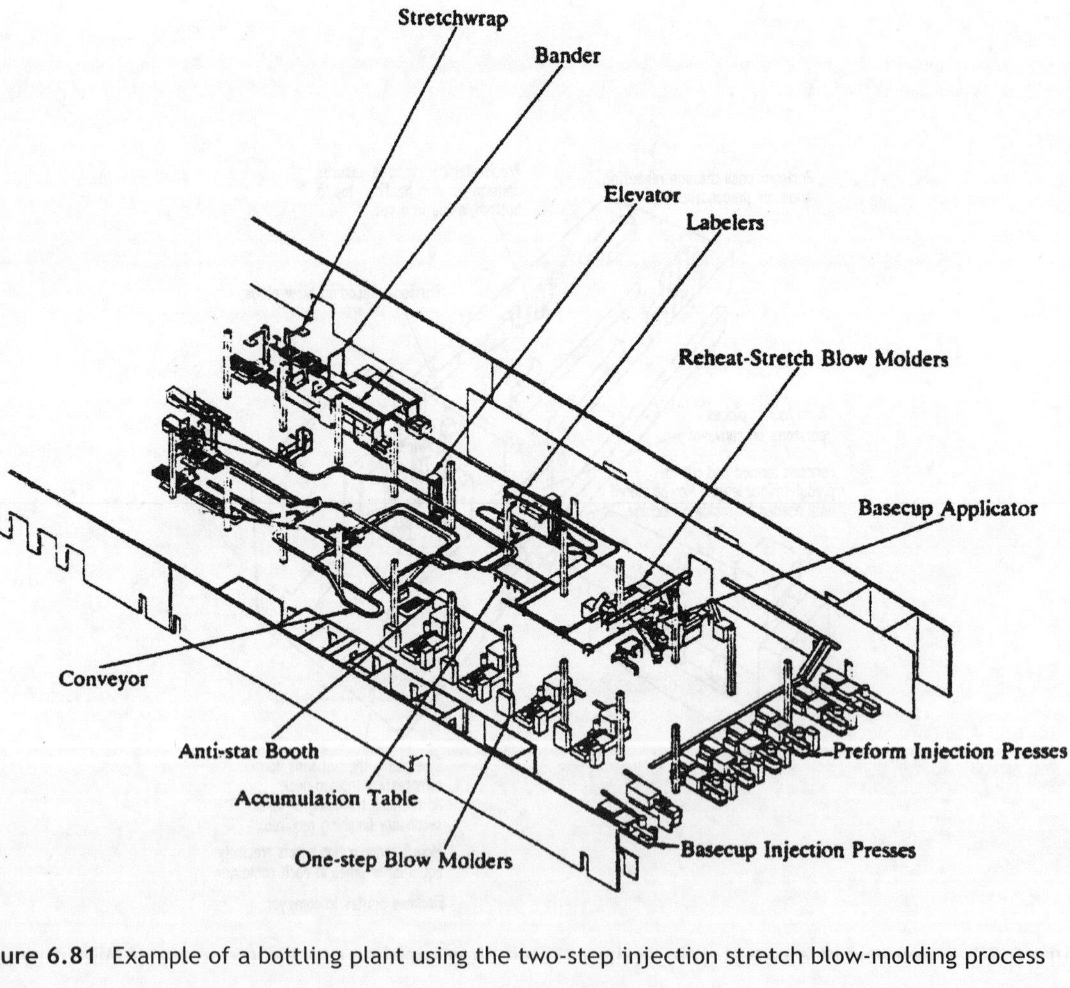

Figure 6.81 Example of a bottling plant using the two-step injection stretch blow-molding process

oriented PET containers. Two-stage stretch blow molding also is being used for production of oriented PVC containers.

The two-stage process, which permits injection molding of the preform and then shipping to blow-molding locations, has allowed companies to become preform producers and to sell to blow-molding producers. Thus companies that wish to enter the market with oriented containers can minimize their capital requirements.

Preform design and its relationship to the final container remain very critical factors. The proper stretch ratios in the axial and hoop directions must be met if the container is to properly package its intended product (Table 6.10).

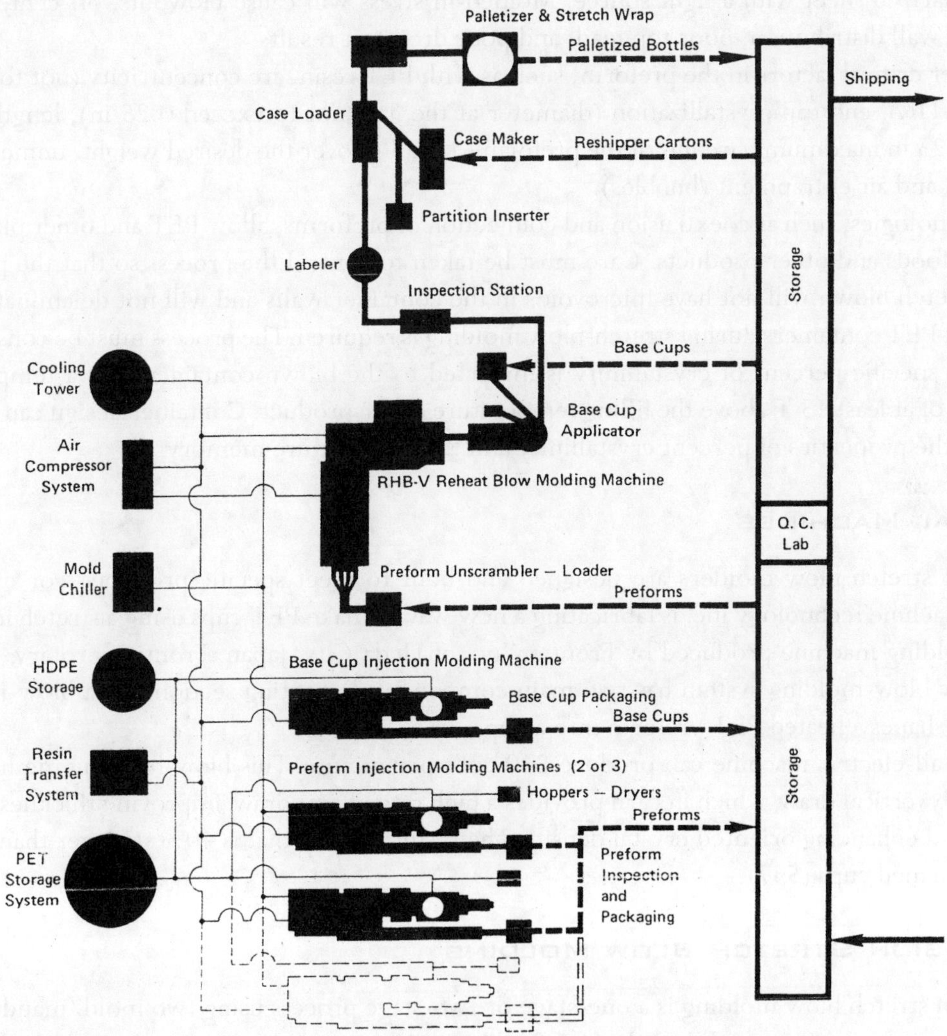

Figure 6.82 Example of a two-stage injection stretch blow-molding production line

Material	Stretch ratios	Orientation temp., ° F.
PET	16/1	195-240
PVC	7/1	210-240
PAN	9/1	220-260
PP	6/1	260-280

Table 6.10 Examples of stretch ratios for different plastics

Specifications should be established on the preform, which should be produced with as little molded-in stress as possible. The molded-in stress can be observed by looking at the preform using cross-polarized sheet with a light source. Molded-in stress will cause blowouts, off-center gates, irregular wall distribution, poor top load, and poor drop test results.

Other critical factors in the preform, such as with PET resin, are concentricity (not to exceed 0.005 in TIR), end cap crystallization (diameter at the gate not to exceed 0.25 in), length of the gate (0.125 in maximum), weight of the preform (±0.25 g) over the desired weight, unmelt in the preform, and air entrapment (bubbles).

Technologies, such as coextrusion and coinjection of preforms, allow PET and other plastics to package foods and other products. Care must be taken to control the process so that the preform when stretch blown will not have microvoids in the container walls and will not delaminate. Heat setting of PET containers during stretch-blow molding is required. The process must be constructed so that a specific percent of crystallinity is imparted to the blown container plus a temperature memory of at least 15°F above the filling temperature of the product. Container design can be used to assist the properties of percent crystallinity and the temperature memory.

SPECIAL MACHINES

Different stretch blow molders are designed and built to meet specific products. For example, Mitsui Machine Technology Inc. is fabricating a new way to make PET cups using a stretch injection blow-molding machine produced by Frontier Inc. of Ueda City, Japan. Frontier's rotary, double-axis draw blow-molding system has two main components: a heating section and a blow-molding unit that shapes a heated PET preform into a cup.

This all-electric machine can produce 240 cups per minute. This blow-forming method provides high vertical draw, which in turn provides a high rate of axis draw, improving thickness distribution and enhancing oriented crystal density. The result is a cup that is 40% stronger than similar thermoformed cups (537).

EXTRUSION STRETCH BLOW MOLDING

Extrusion stretch blow molding is a one-stage or two-stage process using two mold/mandrel sets, where one is for preblow and the other for final blow. An extruded parison is first pinched off and blown conventionally in a relatively small preblow mold to produce a closed-end preform. The preform is then transferred to the final blow mold, where usually an extending stretch rod within the blowing mandrel bears on the closed preform end to stretch it axially. The stretched preform is then blown to impart circumferential stretch. Standard blow-molding machines can be converted for extrusion stretch blow molding. The process is most often used for PVC bottles.

Oriented PVC containers most commonly are made on single-stage, extrusion-type machines. The parison is extruded on either single- or double-head units. Temperature conditioning, stretching, and thread forming are done in a variety of ways depending on the design of the machine. Many of the processes in use are proprietary.

DIP BLOW MOLDING

The dip blow molding process bears some resemblance to injection blow molding in that it is a single-stage process performed with a preform on a core/blow pin (Fig. 6.83). This figure's call-outs are as follows: 1 = accumulator filled, 2 = blow pin inserted, 3 = neck mold filled, 4 = blow pin partially withdrawn, 5 = blow pin fully withdrawn at variable speed, 6 = preform trimmed, 7 = preform clamped in mold, and 8 = preform blow molded.

The difference is in the way the preform is made. The process uses an accumulator cylinder that is fed by an extruder. The cylinder has an injection ram at one end while the other is a free fit over the blow pin. The blow pin is dipped into the melt so that a neck mold on the pin seals the end of the accumulator cylinder. The injection ram is advanced to fill the neck mold; then, the blow pin is withdrawn at a controlled rate so that it is coated with a melt layer extruded through the annular gap between the pin and the accumulator cylinder. The thickness of the coating can be varied or profiled to an extent by varying the speed of the blow pin and the pressure on the injection ram. After trimming, the preform is blow molded in the same manner used for injection blow molding.

The process results in a seam- and flash-free container with a high-quality molded neck. The preform is produced at a lower pressure than that used for injection molding, so the machine can be

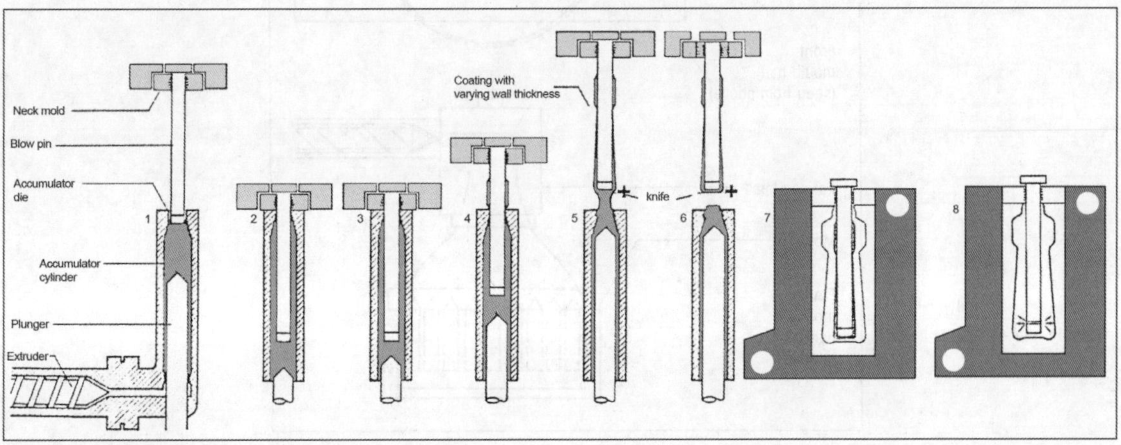

Figure 6.83 Stages in the dip blow-molding process

more lightly constructed and at lower cost. The preform is formed under relatively low stress. Like injection blow molding, the process is best suited to the production of smaller containers.

Multibloc Blow Molding

The multibloc process is used for high-volume blow molding of very small containers, such as pharmaceutical vials and whiskey bottles. A multicavity mold is used with an extruded parison whose circumference approaches twice the total width of the closely spaced cavities. Before the mold closes, the parison is stretched and semi-flattened laterally so that it extends across the full width of the cavities (Fig. 6.84). The process is usually combined with blow/fill/seal techniques.

Other Blow-Molding Processes

Coextrusion and Coinjection Coextrusion and coinjection (or multilayer processes) are essential techniques in the production of high-performance blow-molded products (chapters 4 and 5). The parison or preform is coextruded with a number of different layers, each of which contributes an important property to the finished product. Increasingly, a midlayer may consist of recycled material that is encapsulated between inner and outer layers of virgin plastics.

Blow moldings commonly include from two to seven layers, although more are also used. The construction usually includes one or more barrier layers. These are plastics that have particular resistance to the transmission of water vapor or gases, such as oxygen or carbon dioxide. Examples

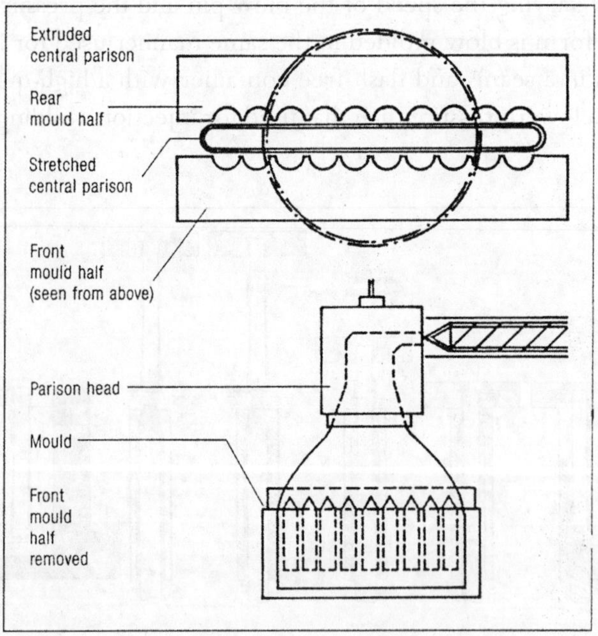

Figure 6.84 Multibloc blow-molding process

are ethylene vinyl alcohols (EVOHs), nylons, and polyvinylidene chlorides (PVDCs). Their presence greatly enhances the performance of the blow molding as a package for foodstuffs, beverages, and other critical products. Each barrier material is deficient in some respect, such as price, mechanical strength, and moisture resistance, and thus are not used as a material of sole construction for blow molding. Their use is in thin layers shielded by other more robust and economical body plastics.

Two considerations lead to high-layer counts in coextruded or coinjection blow-molded products. The barrier materials may be incompatible with body materials, such as polyethylenes and polypropylenes, so intermediate or tie layers of a mutually compatible adhesive must be included to bring about a bond between the layers. Some barrier layers are so deficient in other respects that they need protective layers before they can be used effectively.

As an example, ethylene vinyl alcohol has outstanding resistance to oxygen transmission but has high rates of moisture absorption and water vapor transmission, both of which damage its oxygen barrier performance. The plastic needs both protective layers and adhesive layers, which contribute five layers whenever used. An example using a tie layer with ethylene vinyl alcohol is an ethylene-vinyl acetate-maleic anhydride terpolymer. As shown in Figure 6.85, a typical coextruded polypropylene bottle will include body, regrind, barrier, and tie layers. This figure's call-outs are as follows: 1 = inner body layer of polypropylene, 2 = reground material, 3 = tie layer, 4 = ethylene vinyl alcohol barrier layer, 5 = tie layer, and 6 = outer body layer of polypropylene. Figures 6.86 and 6.87 show examples of coinjection blow-molded bottles.

Such containers are fabricated by performing conventional extrusion or injection blow-molding operations on a multilayer parison or preform. A coextrusion parison head, served by a separate extruder for each distinct component material, is used to produce the parison. The barrier and tie layers are usually very thin, so the flow engineering in the head is critical in order to preserve the integrity of the layers. For this reason, the various melt streams are merged as close to the die exit as possible, even though this complicates die-head construction (Figs. 6.88 and 6.89). Other disturbing influences, such as parison profiling and intermittent extrusion, are often avoided. If

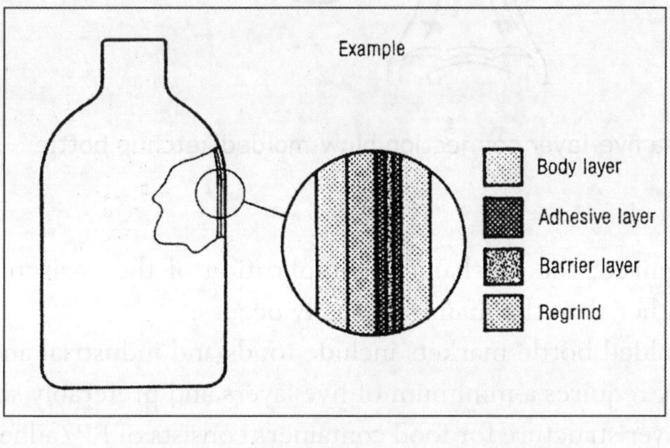

Figure 6.85 Example of a six-layer coextruded blow-molded bottle

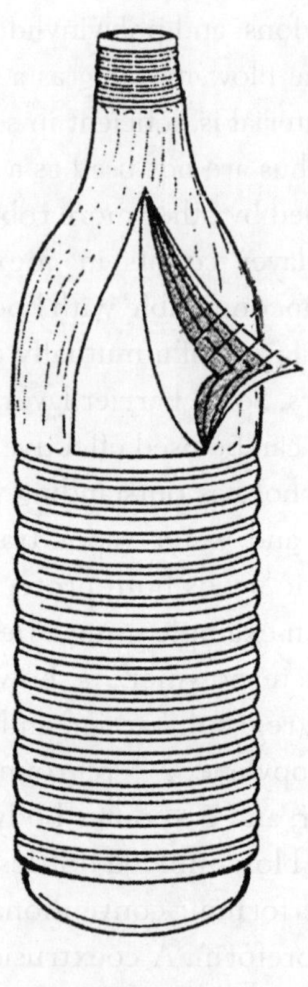

Figure 6.86 Example of a five-layer coinjection blow-molded bottle

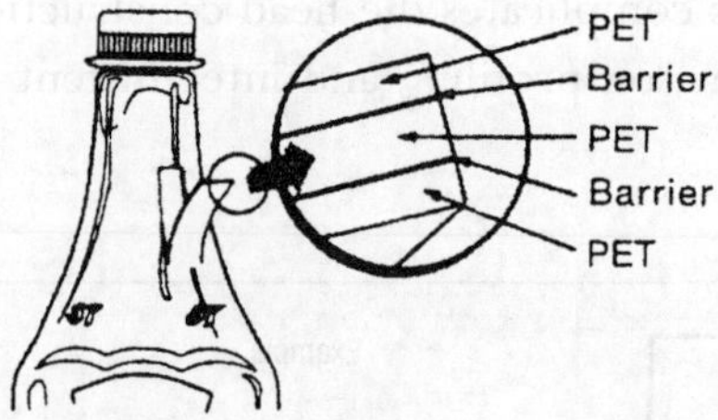

Figure 6.87 Example of a five-layer coinjection blow-molded ketchup bottle

parison profiling is required, the mechanical complication of the parison head is such that axial movement of the die rather than the mandrel usually occurs.

Multilayer blow-molded bottle markets include foods and industrial and household chemicals. The food market usually requires a minimum of five layers and preferably six where regrind can be used. An example of a layer structure for food containers consists of PP/adhesive/EVOH/adhesive/PP. This produces a container that can be hot filled (Fig. 6.90), is relatively inexpensive, has decent

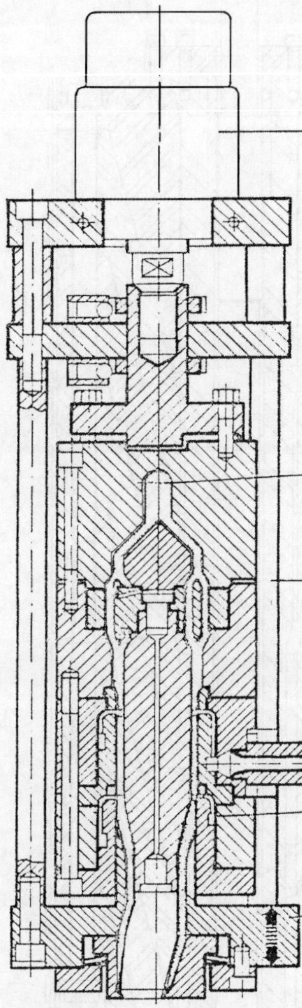

Figure 6.88 Example of a three-layer coextrusion parison blow-molded head with die profiling

contact clarity, and will have a shelf life of approximately one year without refrigeration. The industrial and household chemical markets can be met with three-layer combinations where barriers, such as nylons or acrylonitriles, are placed on the inside and the inexpensive HDPE resin on the outside.

The machinery used primarily for blow molding multilayer containers is extrusion blow-molding equipment of a proprietary type. These machines are the rotary wheels, where multiple extruders feed a single parison head. From four to eighteen cavities may be used on these rotary wheels. These machines have the advantage of using only a single parison, since multilayer structures are difficult to split for multiple parisons.

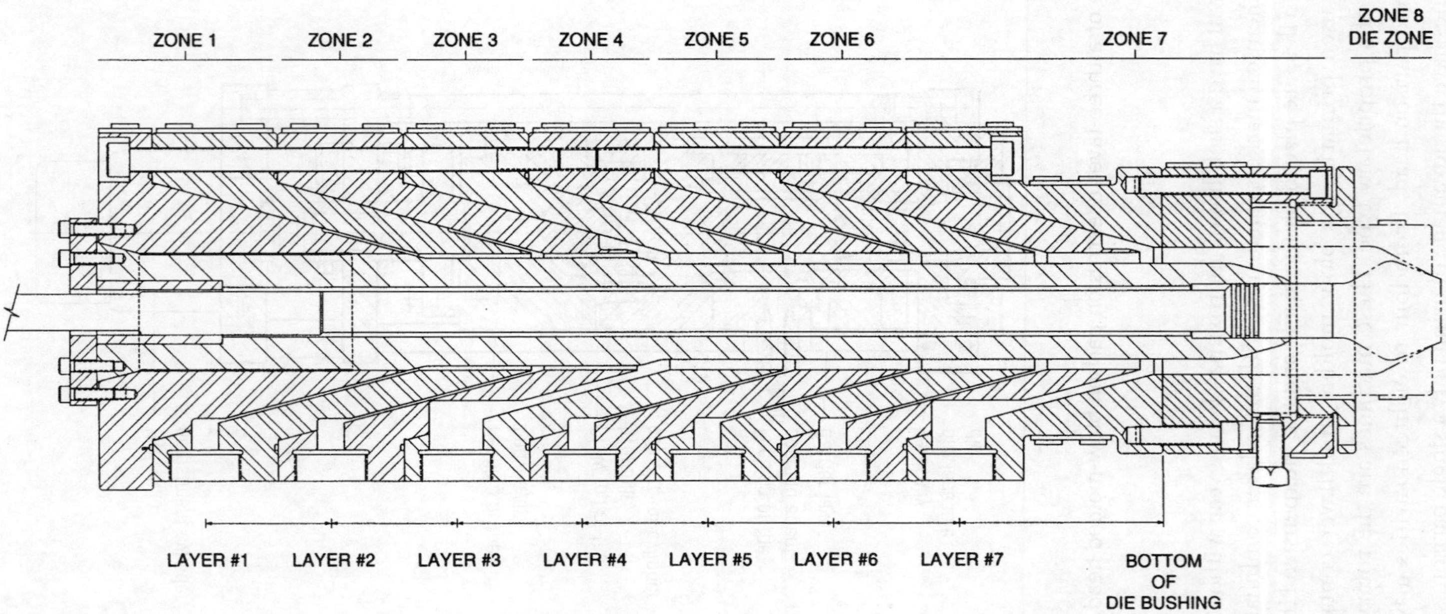

Figure 6.89 Example of a five-layer coextrusion parison blow-molded head with die profiling (courtesy of Graham Machinery Group)

The output that can be achieved on rotary wheels is at least up to 5,000 per hour. The disadvantages are that neck finishing must be done in a secondary operation and there is generally more scrap generated since most wheels do not use cut parisons. There is flash between the molds that must be reused in the process. The shuttle-type platen machines can be equipped quite easily with extruders and a multilayer head to produce bottles with calibrated neck finishes and a minimum amount of scrap. The disadvantage of a platen shuttle-type machine is the output requirements since the multilayers are difficult to split in flow to produce more than two parisons. The output that can be achieved on platen-type machines, with two parisons, is up to 1,200 per hour. The tie layers used are extrudable adhesives and work well in small extruders. The key factor in the head design is to keep the adhesive layers and the barrier layer at a minimum since these materials are expensive.

As reviewed, the other method for producing multilayer bottles is by coinjection blow molding where a system of two or more layers produces bottles. With the use of a barrier layer, it is encapsulated between other structured layers up to the preform neck ring, which ensures barrier integrity and prevents air entrapment. The injection-blow multilayer system is primarily used to fabricate PET soft drink bottles where the barrier layer achieves a longer carbonation retention.

Sequential Extrusion Sequential extrusion blow molding is a special multimaterial technique used for the production of specially designed products. The different plastics are chosen typically to

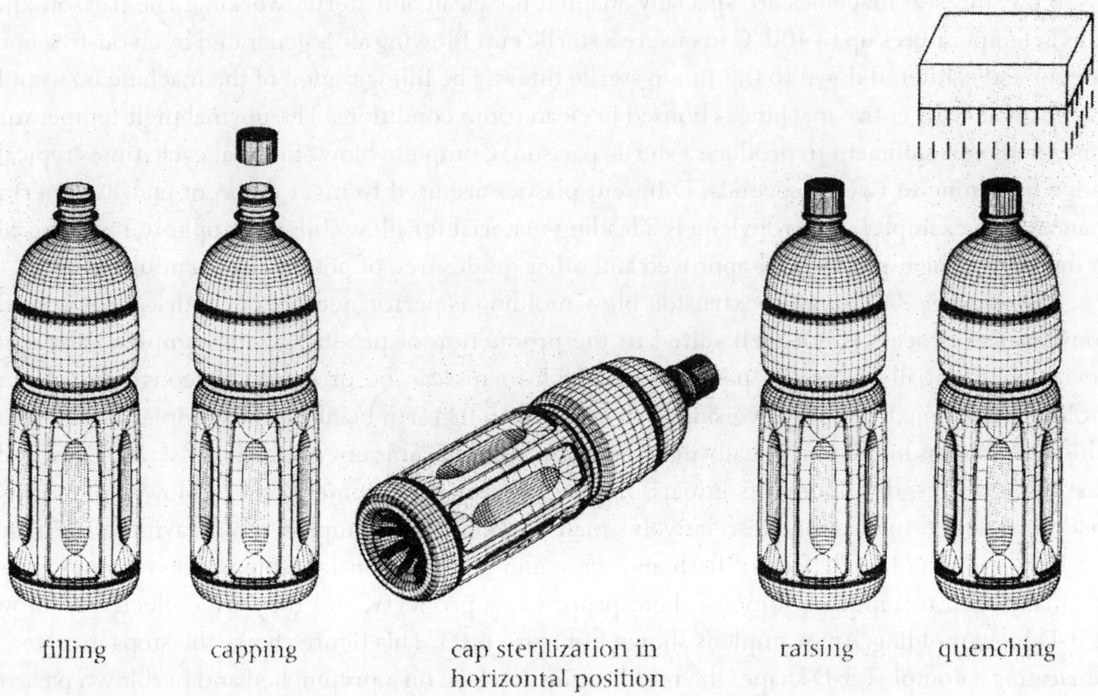

Figure 6.90 Example of hot-filling PET bottle at 80° to 95°C (courtesy of SIG Plastics International)

contribute complementary mechanical properties and are present in distinct sequential zones in the finished part. Normally, two materials are used but three or more are also used. Separate external ram accumulators for each material serve the die head. These are operated sequentially, typically in A-B-A sequence, to produce a parison with three distinct material zones in axial succession. The parison is subsequently blow molded by normal techniques.

An example for sequentially blow molding polypropylene is an automotive air duct in which a central flexible zone (Figs. 6.16 and 6.91) joins rigid end sections. The flexible zone allows for installation mismatches, accommodates thermal expansion, and damps vibration noise. The rigid portions allow for direct connections to other mechanical elements in the assembly.

Blow/Fill/Seal The blow/fill/seal process is a complete packaging technique that integrates the extrusion or injection blow-molding and container-filling steps. This provides for aseptic filling of the hot as-blown container and is used for pharmaceutical, food, and cosmetic products. The process employs a two-part mold in which the container body mold cavity blocks are separate from the neck-forming members (Fig. 6.92).

The body mold closes on the parison that is blown normally by a neck-calibrating blow pin. Immediately, with the mold still closed, the liquid contents are injected through the pin. The pin is then withdrawn and the neck is formed and sealed under vacuum by the neck-forming members. Both mold parts then open to eject a filled and sealed container. Small containers may be formed entirely by vacuum rather than blowing.

Blow/fill/seal machines are specially adapted for clean and sterile working. The parison knife runs at temperatures up to 400°C to ensure a sterile cut. Blowing air is generated by an oil-free compressor and is filtered down to 0.2 μm in sterile filters. The filling region of the machine is swept by sterile air and the entire machine is housed in clean room conditions. The normal melt temperature for extrusion is sufficient to produce a sterile parison. Complete blow/fill/seal cycle times typically range from nine to twenty seconds. Different plastics are used to meet different end-item performances. For example, polypropylene is a leading material for blow/fill/seal applications, generally in the form of high-purity food-approved and other grades free of potentially harmful additives.

Blow Molding 3-D Because extrusion blow molding is performed on a cylindrical parison, the conventional process is not well suited to the production of products with complex forms that deviate substantially from the parison axis. Such forms can be produced by conventional blow molding equipment but only by using a parison that in its form blankets the complex mold cavity. This 3-D process in the past usually developed an excessive amount of pinch-off scrap. During the past few decades, developments in parison-handling robot equipment and in blow-mold designs made it possible to manipulate relatively small parisons into complex mold cavities. The result is a blow molding largely free of flash and scrap and offering considerable process savings. There are many such techniques, some of them proprietary property, and they are collectively known as 3-D blow molding. An example is shown in Figure 6.93. This figure shows the steps involved in fabricating a complex 3-D shape: the mold begins to close on a prepinched and preblown parison, next is to mold fully closed with the core forward and the side walls hinged, and the final step is to open the mold and retract the core.

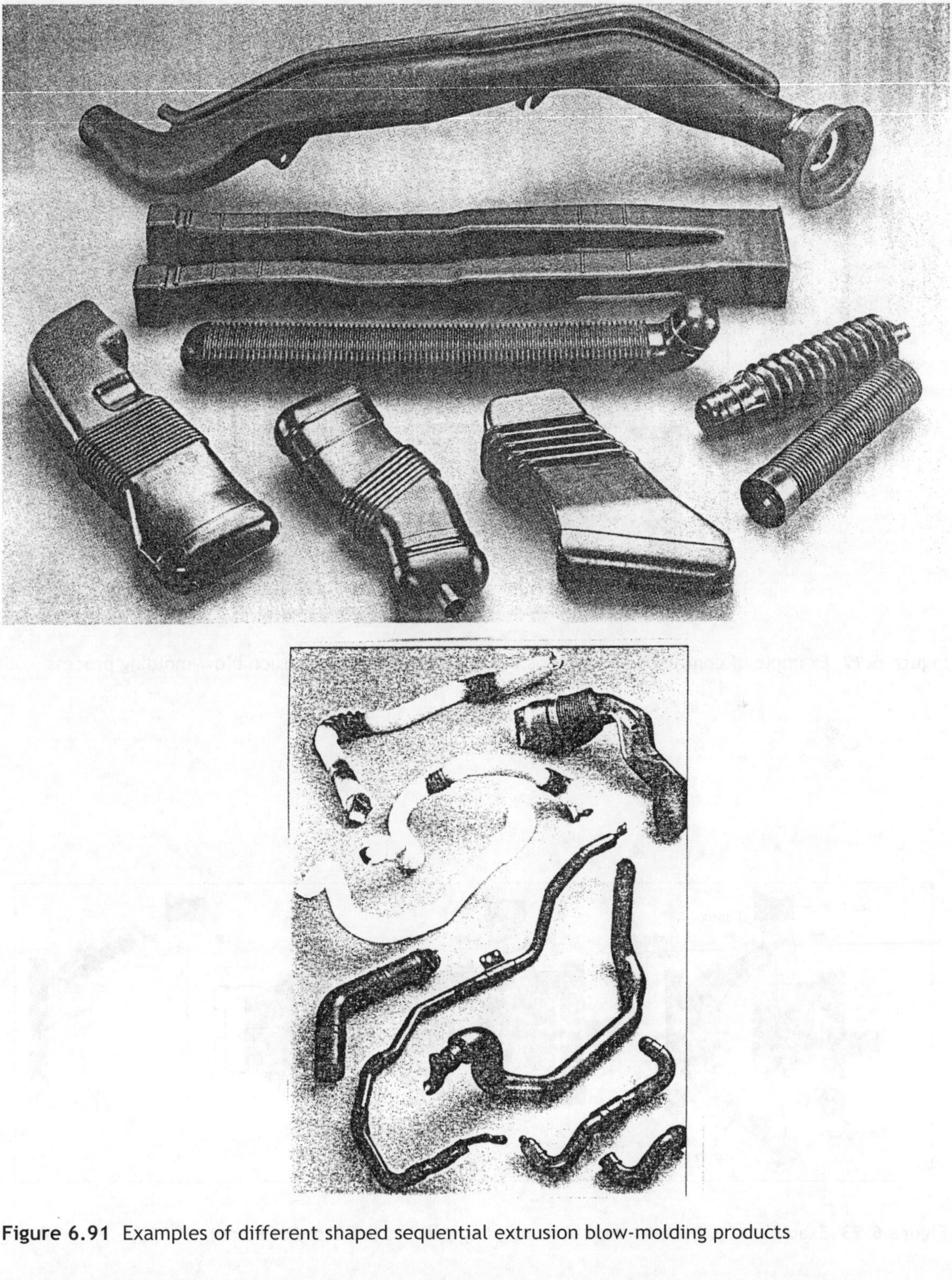

Figure 6.91 Examples of different shaped sequential extrusion blow-molding products

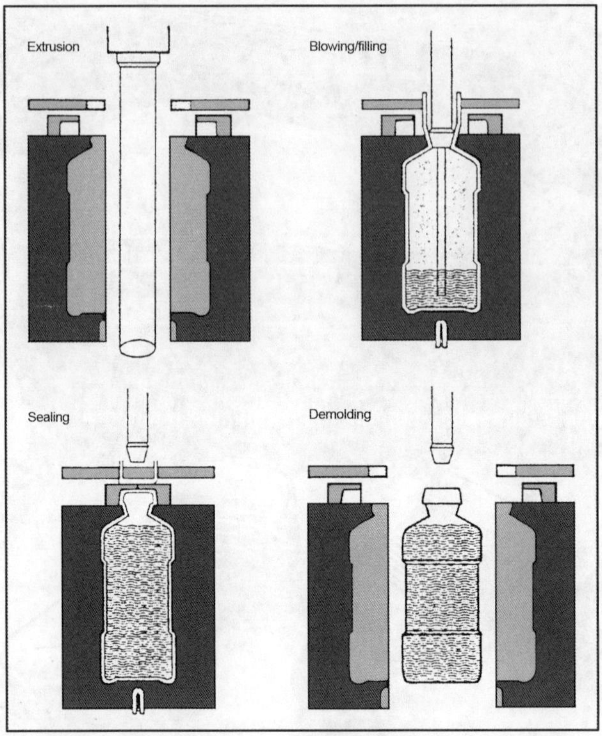

Figure 6.92 Example of container-filling steps in the blow/fill/seal extrusion blow-molding process

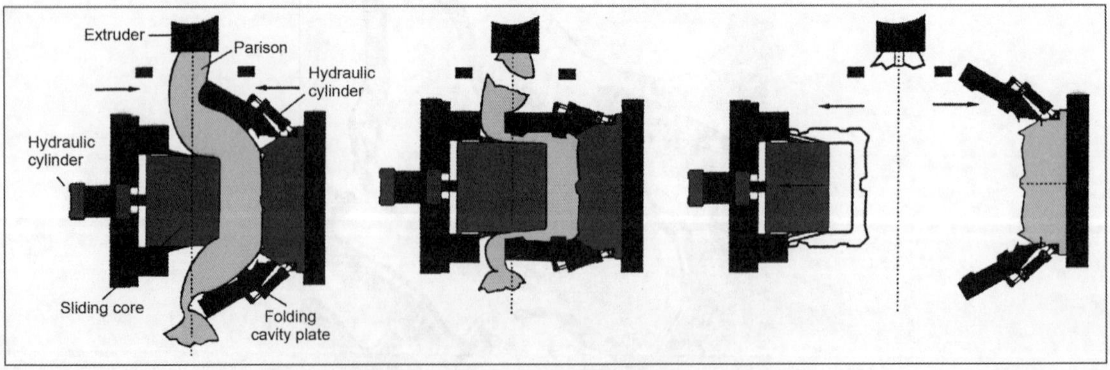

Figure 6.93 Example of a 3-D extrusion blow molding process (courtesy of Placo)

Other 3-D blow-molding methods are shown in Figures 6.94 to 6.98.

BLOW MOLDING WITH ROTATION

Injection molding with rotation (MWR) is an example of processing at lower temperatures, pressures, and so on. It is also called injection spin molding or injection stretched molding. This blow-molding process combines injection molding and injection blow molding (and performs like injection blow molding, as reviewed) but has an additional step of melt orientation (Dow Chemical patent). The equipment used is what is commercially available for IM except the mold is modified so that either the core pin or outside cavity rotates. The rotated melt on its preform pin is transferred to a blow mold (Fig. 6.99). The end product can come directly from the IMM mold or be a result of two-stage fabrication: making a parison and blow molding the parison.

This technology is most effective when employed with an article that has a polar axis of symmetry and reasonably uniform wall thickness as well as dimensional specifications and part-to-part trueness that meets market acceptance. The MWR process requires no sacrifice of either cycle time or surface finish. Both laboratory and past commercial runs identify good potentials for reducing cycle times, for reducing the amounts of plastic required or improving properties with the same amount of plastic (or both), and for substituting less expensive plastics while achieving adequate properties in the fabricated products.

During fabrication using the MWR process, two forces act on the plastic: injection (longitudinal) and rotation (hoop). The targeted balanced orientation is a result of those forces. As the product's walls cool, additional high-magnitude cross-laminated orientation is frozen in and develops throughout the walls. Orientation on molecular planes occurs as each layer cools after injection. This orientation can change direction and magnitude as a function of wall thickness. The result is analogous to plywood or reinforced plastics (chapter 15) and the strength improvements are as dramatic. In the MWR process, there are an infinite number of microscopic layers, each of which has its own controlled direction of orientation. Using appropriate processing conditions, the magnitude and direction of the orientation and strength properties can be varied and controlled throughout the wall thickness.

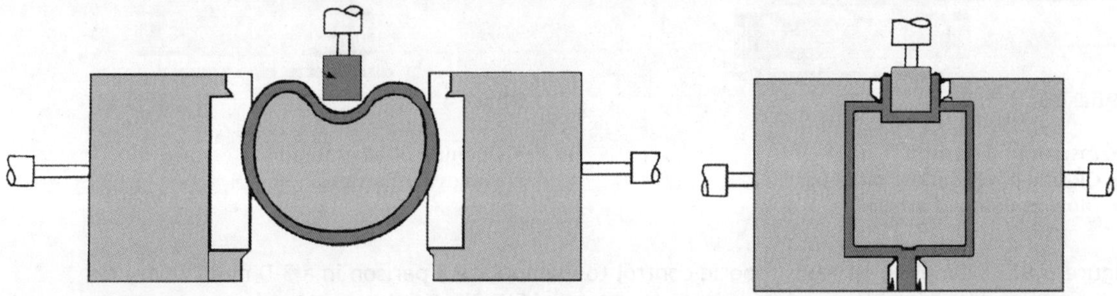

Figure 6.94 Examples of multiple side action 3-D extrusion blow-molding molds

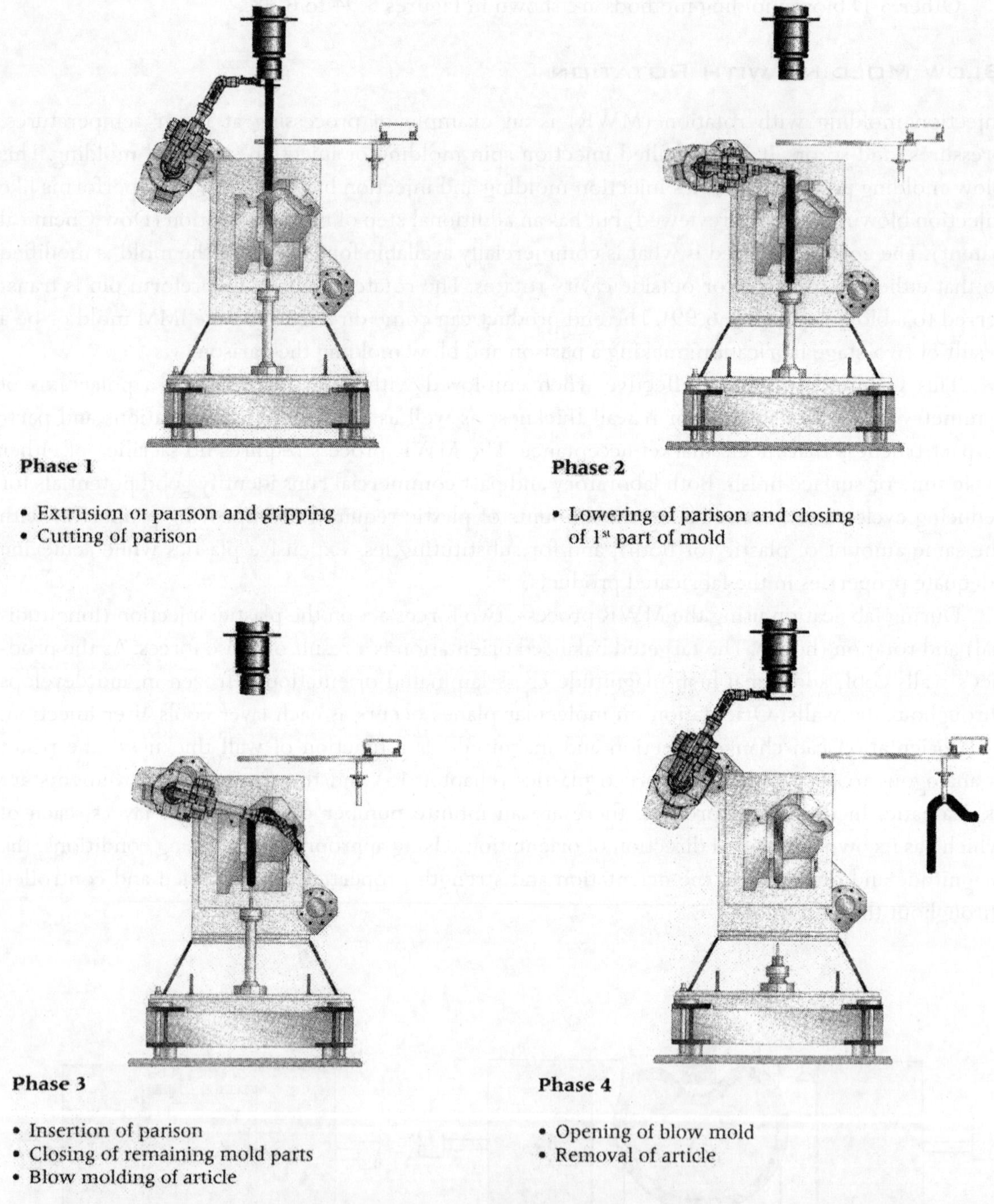

Figure 6.95 Example of six-axis robotic control to manipulate a parison in a 3-D mold cavity to extrusion blow mold products (courtesy of SIG Plastics International)

MOLD

Essentially, the blow mold consists of two halves, each containing cavities that, when the mold is closed, define the exterior shape of the blow molding. Because the process produces a hollow article, there are no cores to define the inner shape. The blow-molding process is carried out at relatively low pressures, so blow molds can be more lightly engineered than injection molds and

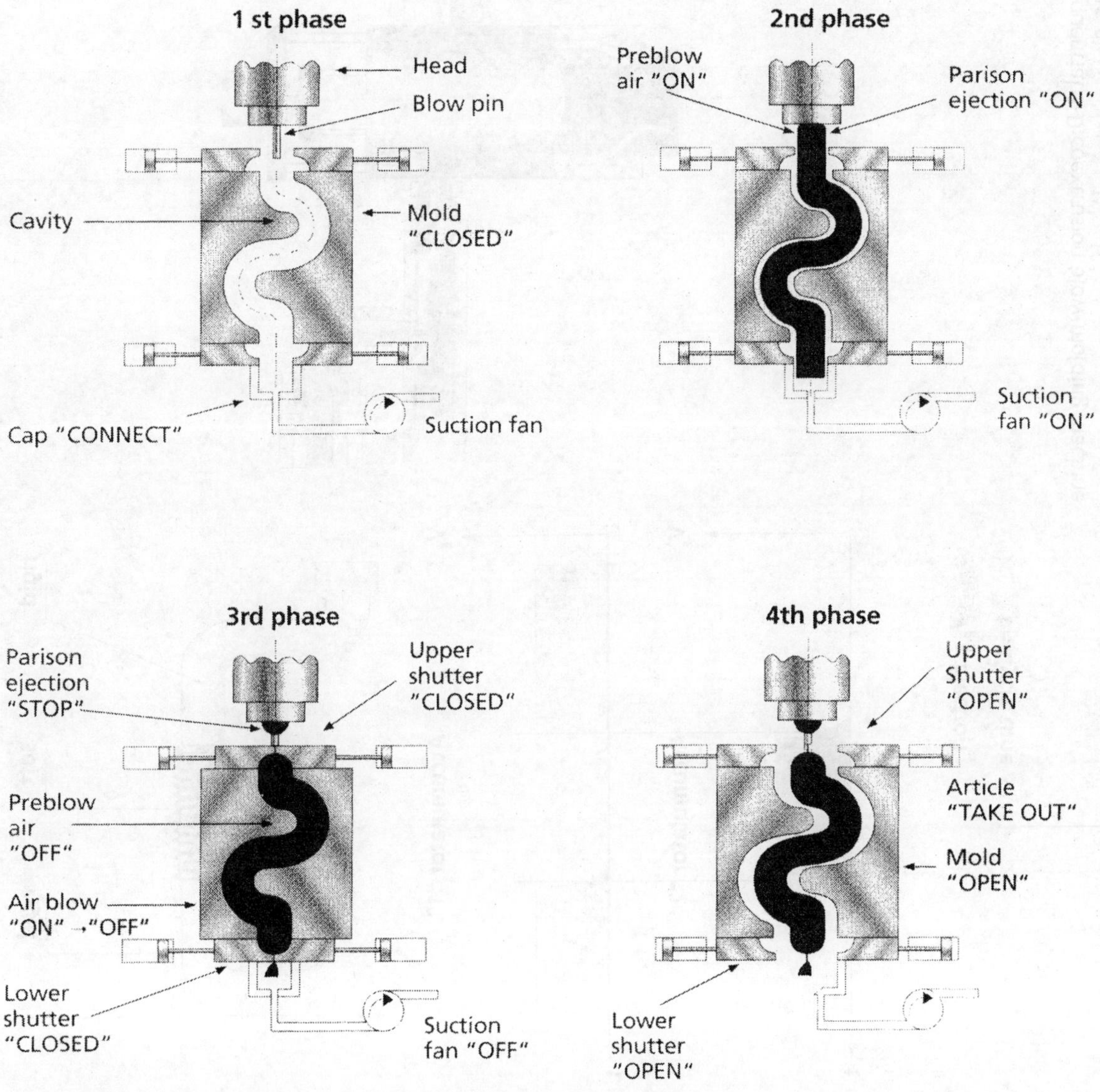

Figure 6.96 Example of a suction 3-D extrusion blow-molding process (courtesy of SIG Plastics International)

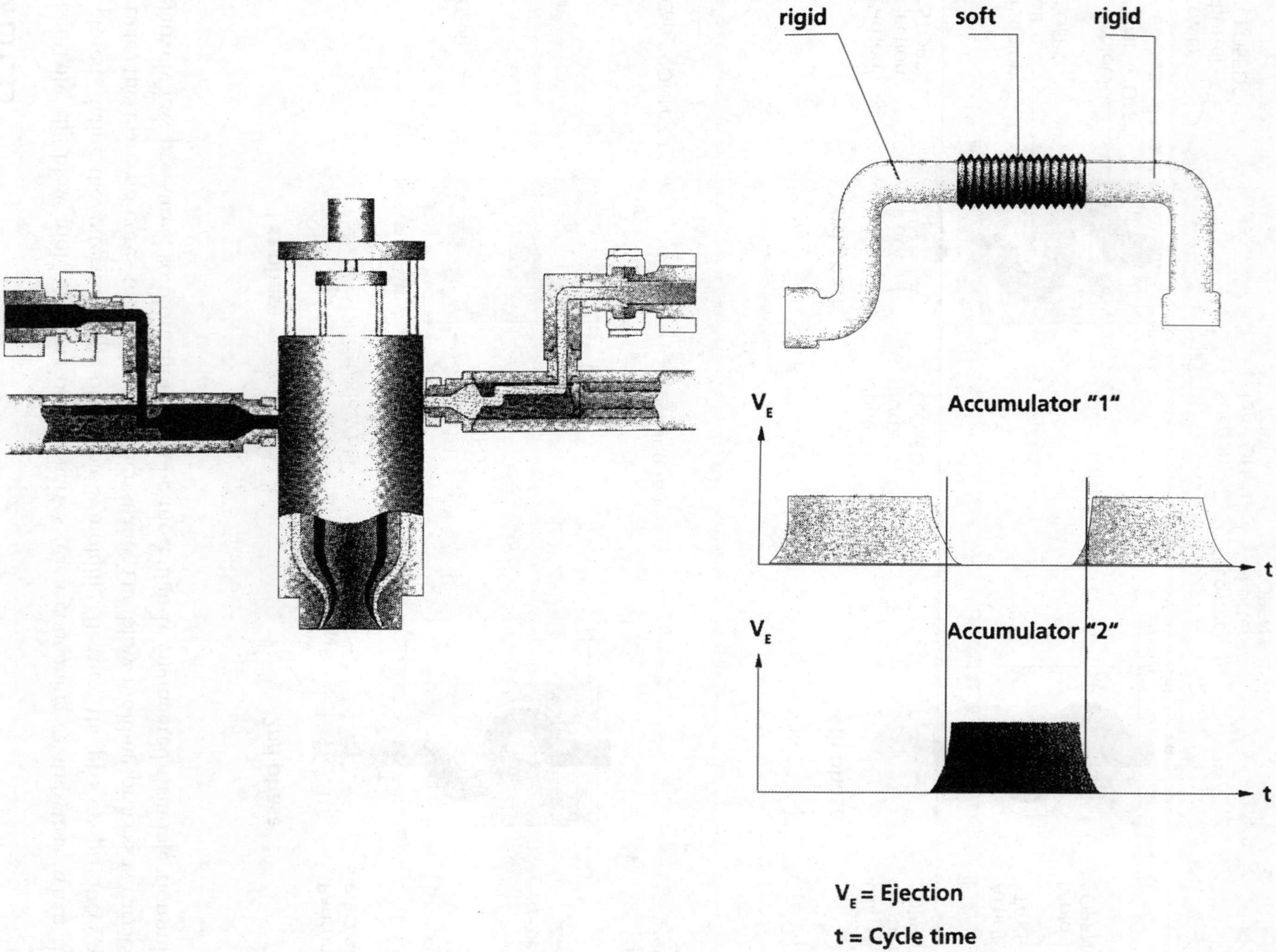

Figure 6.97 Example of sequential 3-D coextrusion blow-molding machine (courtesy of SIG Plastics International)

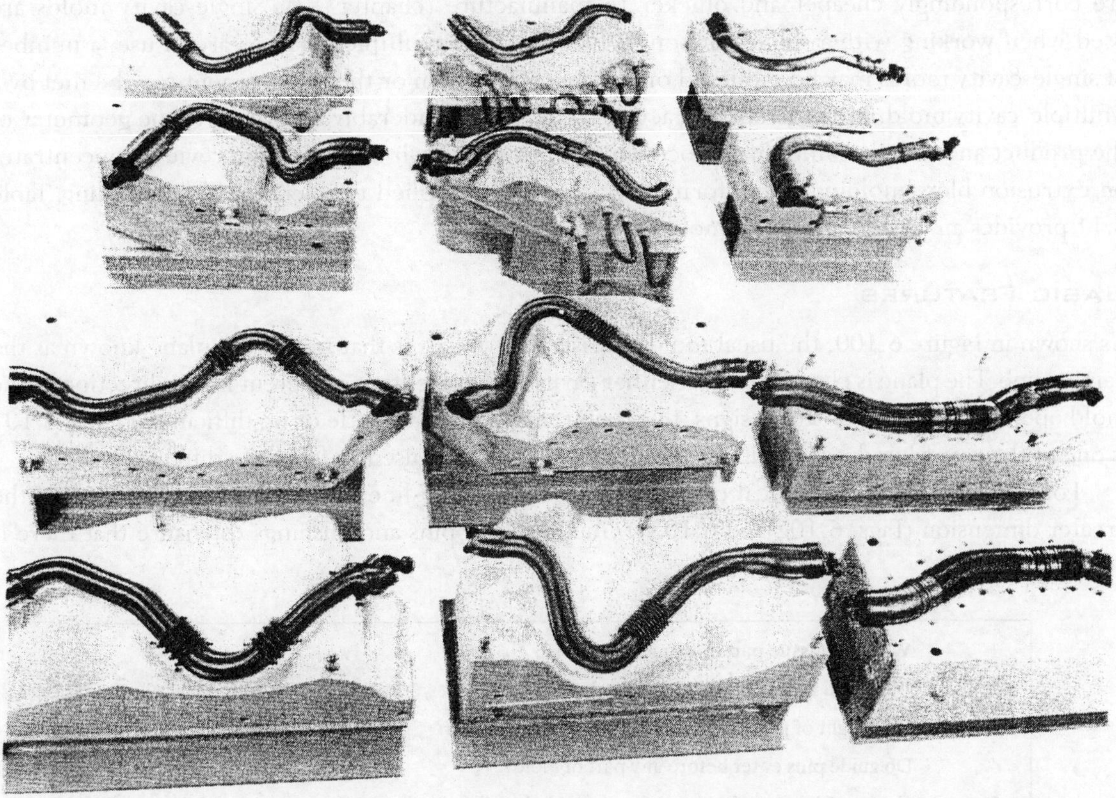

Figure 6.98 Examples of 3-D extrusion blow-molded products in their mold cavities (courtesy of SIG Plastics International)

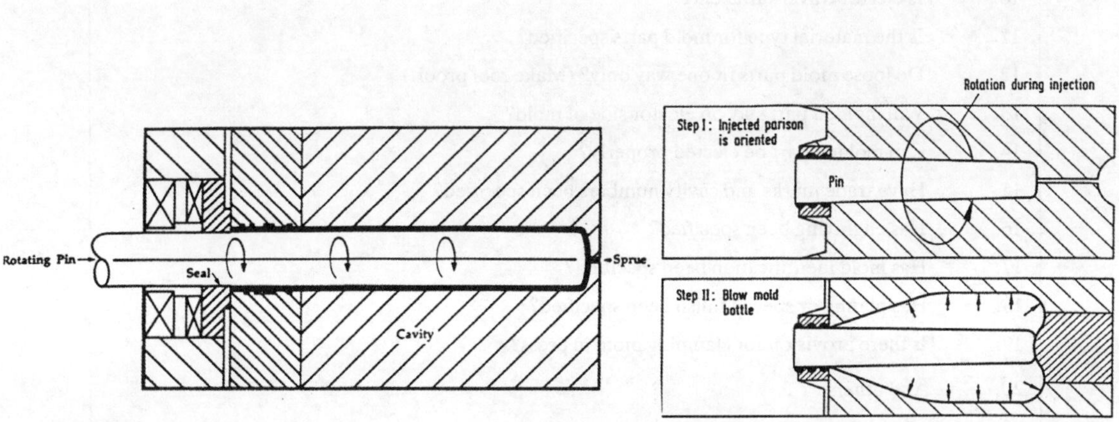

Figure 6.99 Schematic for molding with rotation using a two-stage blow-molding procedure

are correspondingly cheaper and quicker to manufacture (chapter 17). Single-cavity molds are used when working with a single parison. When the usual multiple parisons are in use, a number of single-cavity molds may be mounted on the machine platen or the requirement may be met by a Multiple-cavity mold. Mold details and actions will vary considerably according to the geometry of the product and the blow-molding process in use. Even though the following review concentrates on extrusion blow molding, the information can also be applied to injection blow molding. Table 6.11 provides an introduction to a mold design checklist.

Basic Features

As shown in Figure 6.100, the usual mold consists of two halves that meet on a plane known as the parting line. The plane is chosen so that neither cavity half presents an undercut in the direction of the mold opening. For most bottle designs, this requirement presents little or no difficulty. Figure 6.101 is one of the many hundreds of different blow-molding molds used to fabricate different products.

For products of asymmetrical cross-sections, the parting line is placed in the direction of the greater dimension (Figs. 6.102 and 6.103). Guide pillars/pins and bushings to ensure that there is

1.	Was latest issue part drawing used?
2.	Will mold fit press for which intended?
3.	Is daylight of press sufficient for travel and ejection?
4.	Do guide pins enter before any part of mold?
5.	Can mold be assembled and dis-assembled easily?
6.	Has allowable draft been indicated?
7.	Is plastic material and shrinkage factor specified?
8.	Are mold plates heavy enough?
9.	Are waterlines, air lines, thermo-couple holes or cartridge holes shown and specified?
10.	Is ejector travel sufficient?
11.	Is the material type for mold parts specified?
12.	Do loose mold parts fit one way only? (Make fool proof.)
13.	Will molded part stay on ejector side of mold?
14.	Can molded part be ejected properly?
15.	Have trade marks and cavity numbers been specified?
16.	Has engraving been specified?
17.	Has mold identification been specified?
18.	Has plating or special finish been specified?
19.	Is there provision for clamping mold in press?
20.	And others . . .

Table 6.11 Mold design checklist

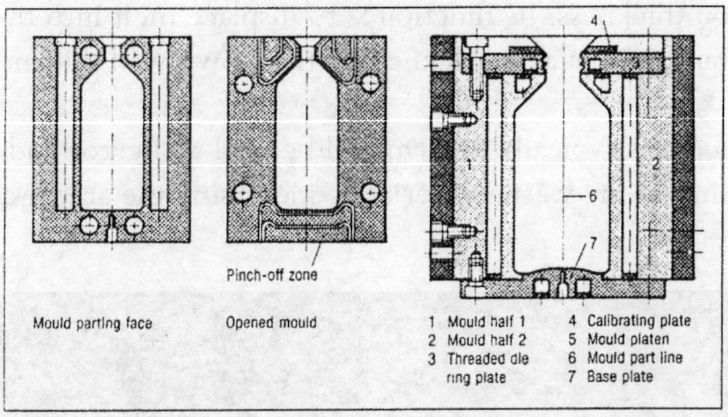

Figure 6.100 Example of an extrusion blow mold

no mismatch between the cavities align the two mold halves. With extrusion blow molding, the parison passes across the mold in the axis of the cavity and is pinched and compressed between the faces of the closing mold at the neck and base regions of the cavity. These are known as the pinch-off zones. Separate inserted mold blocks typically form the base and neck regions of the mold. The mold includes channels for the circulation of cooling water. With injection blow molding the preform only has a pinch-off at the neck.

MATERIALS OF CONSTRUCTION

The key factors influencing the choice of materials for a blow mold, as with molds for other processes, are durability, thermal conductivity, finished cost, and surface finish. The main cavity body for blow molds is usually produced in machined aluminum alloy, cast zinc, or another nonferrous alloy. For long production runs, steel or cast beryllium copper may be preferred (Table 6.12; chapter 17). The pinch-off zones are subject to compressive stress and wear, and these are usually furnished as inserts of alloy steel or hard beryllium copper.

PINCH-OFF ZONE

In extrusion blow molding, the pinch-off zone performs two functions. It must weld the parison to make a closed vessel that will contain blowing air and it must leave pinched-off waste material in a condition to be removed easily from the blown product. To accomplish this, the pinch-off zone has a three-stage profile: pinch-off edge, pressing area, and flash chamber (Figs. 6.104 and 6.105).

The pinch-off edge is flush with the mold parting line and forms part of the cavity periphery. The edge should be as narrow as possible depending on the material of construction. For example, steel edges can be from 0.3 to 1.5 mm wide for rigid plastics; for softer materials, the dimensions can be 0.8 to 2.5 mm. The pressing area is recessed into the mold parting face by an amount depending on the parison thickness. Its function is to displace melt into the weld area to ensure a strong weld that is substantially equal in thickness to the blown wall thickness at the same point on the parison axis.

The pressing area depth is typically determined by trial and error and is often given a ribbed surface to improve cooling of the waste material by increasing the area available for heat transfer.

Figure 6.101 Blow-molded corrugated bellow part between its mold halves

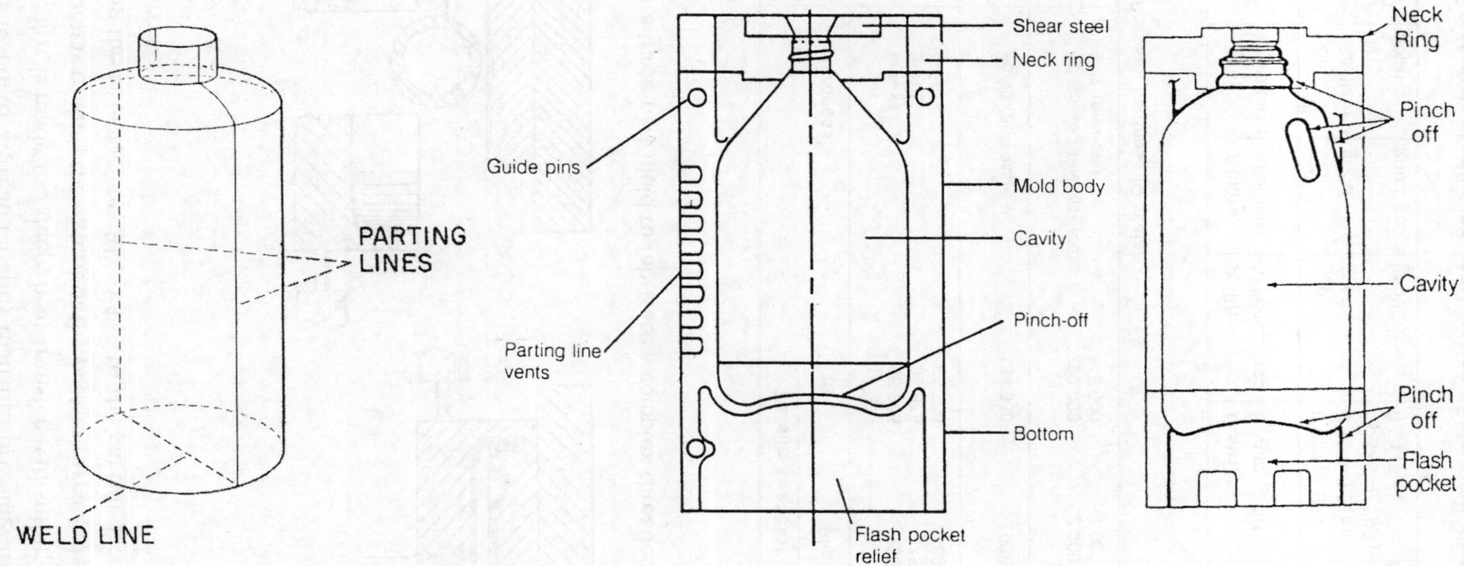

Figure 6.102 Examples of parting line locations and other parts of a mold

The flash chamber is recessed to a greater depth than the pressing area, typically to about the thickness of the parison. If it is made too deep, heat transfer will suffer and the parison scrap will cool too slowly. Its function is to limit the extent of the pressing area and so limit the force opposing the mold closing. Angling the riser face by 30° to 45° to the parting face will strengthen the steps between the different levels.

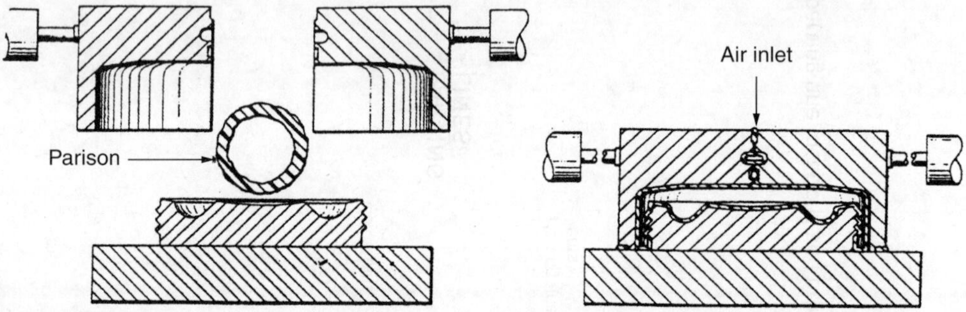

Figure 6.103 Example of a three-part mold to fabricate a complex threaded lid

Material	Hardness[1]	Tensile strength psi	Tensile strength MPa	Thermal conductivity Btu/in./ft² h F
Aluminum				
A346	BHN-80	36.975	255	1.047
6061	BHN-95	39.875	275	1.165
7075	BHN-150	66.700	460	905
Beryllium copper				
23 and 165	RC-30 (BHN-285)	134.850	930	728
Steel				
0-1 and A-2	RC 52-60 (BHN-530-650)	290.000	2.000	243
P-20	R-32 (BHN-298)	145.000	1.000	257

BHN = Brinell and RC = Rockwell hardness (C scale).

Material	Wear resistance	Ability to be cast	Ability to be repaired	Density lb/cum (g/cm³)	Ability to be machined and polished
Aluminum	poor	fair	good	0.097 (2.699)	excellent
Beryllium-copper	excellent	good	good	0.129–0.316 (3.589–8.793)	fair
Cast iron	good	good	good	0.24 (7.6)	good

Table 6.12 Examples of materials used in the construction of blow-molding molds

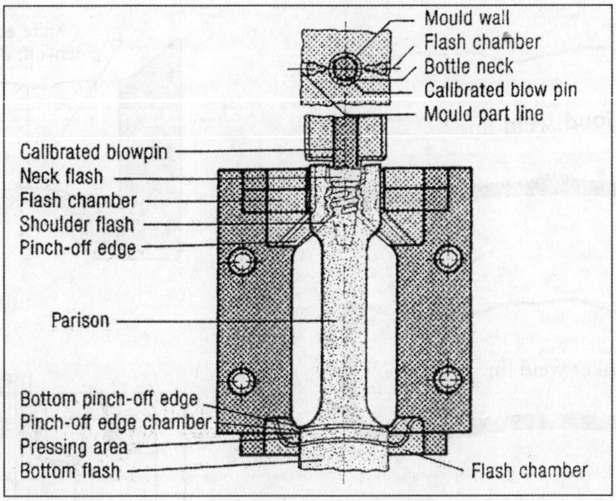

Figure 6.104 Examples of pinch-off zones in an extrusion blow mold

FLASH CONTROL

Flash caused by the pinch-off is an unavoidable evil in extrusion blow molding. The ability to control the adverse effects of the flash is critical to success of the process (543). Pinch-off generates excess material in the form flash that is usually twice the thickness of the part's wall. This thicker plastic cools slower than the blown product. It is subject to fold-over and adheres to the blown product.

Flash imposes costly limits on the efficiency of blow molding. It has potential for significantly extending the molding cycle, primarily by increasing the time needed to cool the thick flash. This cycle increase could approach twice what would normally be required. Removal calls for a postmolding trim step that requires secondary equipment and poses a risk of damaging good parts.

To reduce cycle time, a fabricator has some damaging options, such as ejecting the part before the flash is sufficiently cooled. Because it is still soft and pliable when ejected, it can create other problems, such as a fold over on itself and adhering to adjoining surfaces of the part after ejection of the molding. Flash is also considerably more difficult to handle and trim while hot. In either case, the resultant penalty may be a significant increase in the part reject rate.

By locating cooling lines as close as possible to the flash, heat transfer to the cooling water will reduce cycle time. It is critical to appreciate maximizing the heat transfer as much as possible to the flash area. Keep the water turbulent by operating the water in the proper Reynold's number (chapter 17).

In most bottle molds where flash occurs, it always has a separate insert. That is because the pinch-off area traditionally requires a metal insert (such as QC-7 aluminum) that provides resistance to wear and increased heat-transfer efficiency. When the insert starts to become inefficient, it is replaced with a new insert.

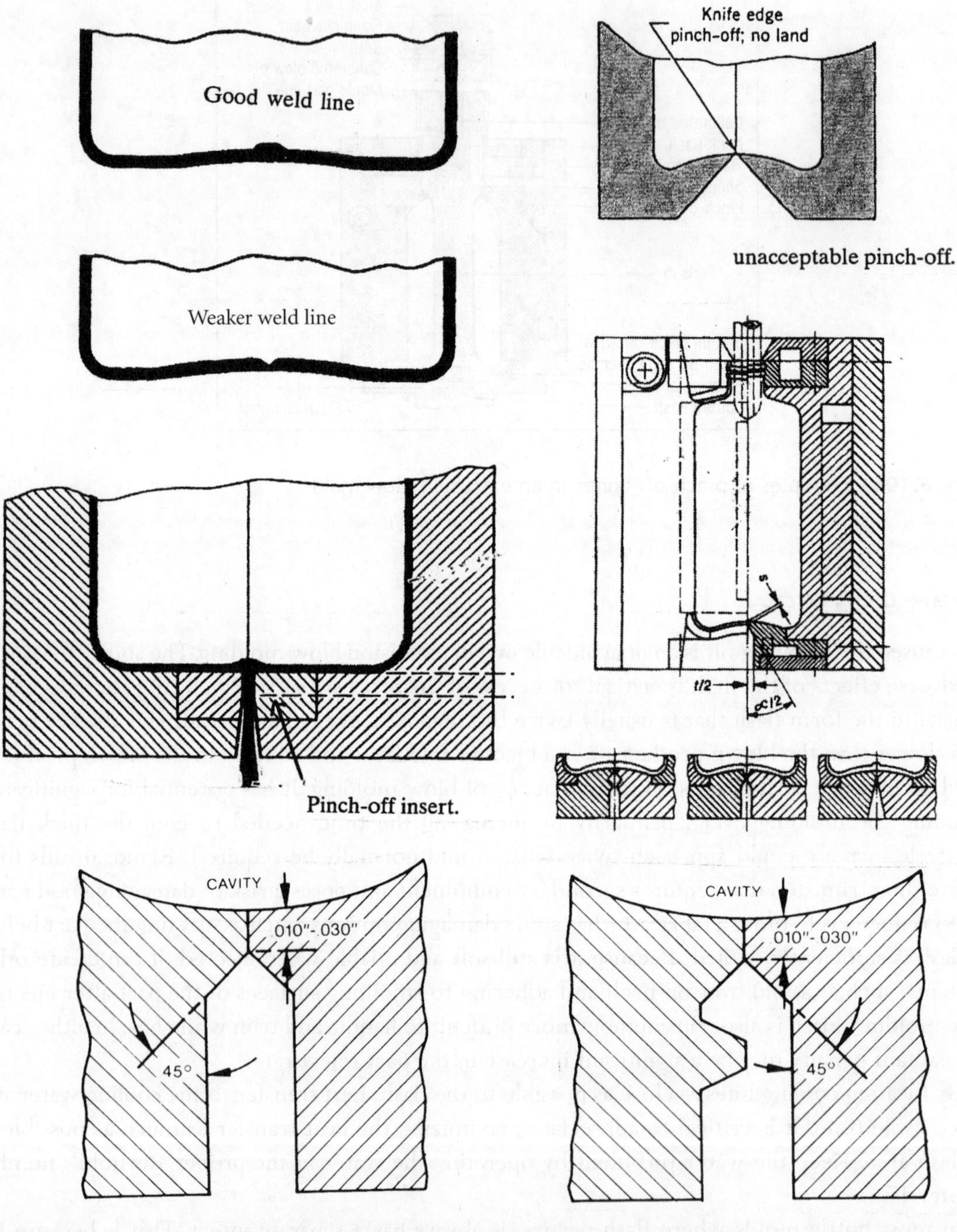

Figure 6.105 Examples of pinch-off designs to meet requirements for different plastics and contours

The cross-section of the insert's parting line in a conventional blow mold traditionally takes the form of a horizontal line, the main reason being that it is easier and less costly for a tool builder to machine in straight lines. However, a trapezoidal cross-section would expand the surface area of the mold in contact with the flash, thereby removing heat more efficiently (Fig. 6.106). A trapezoidal insert cross-section can also increase the flash's stiffness, making the hot flash more resistant to folding over.

BLOWING AND CALIBRATING DEVICE

The blow pin is the means by which blowing air is inserted into the parison through what will become a hole in the finished blow molding. In the case of a bottle (Fig. 6.107), the blow pin is inserted through the neck before or after mold closing. The blow pin body has the secondary function of calibrating the bore of the bottle neck. If the blow pin is plunged into the neck after mold closing, it is possible to produce a flash-free bottle mouth.

After the mold opens, the finished bottle is stripped off by retracting the blow pin through a stripper plate. The blowing pin contains channels for the circulation of blowing air and cooling water. There are blow-molded products, such as toys, decorations, and medical components, that do not have internal apertures that can accept blow pins. In such cases, blowing air can be supplied through a blow needle, such as a hypodermic medical needle, that punctures the parison wall after the mold closes (Fig. 6.108).

The needle can be advanced and retracted by means of an air or hydraulic cylinder and is located at a point close to a pinch-off edge. The parison is gripped by the mold at this point and so is unable to deflect away from the advancing needle point. The blow needle leaves an extremely small witness mark that may not be noticeable, or if visible, is often acceptable within the function of the finished part. If not, it can be sealed by a secondary operation.

VENTING AND SURFACE FINISH

When a parison is blown, a large volume of air must be displaced from the mold cavity in a short time. Because blowing is carried out at relatively low pressure, it is essential to provide venting

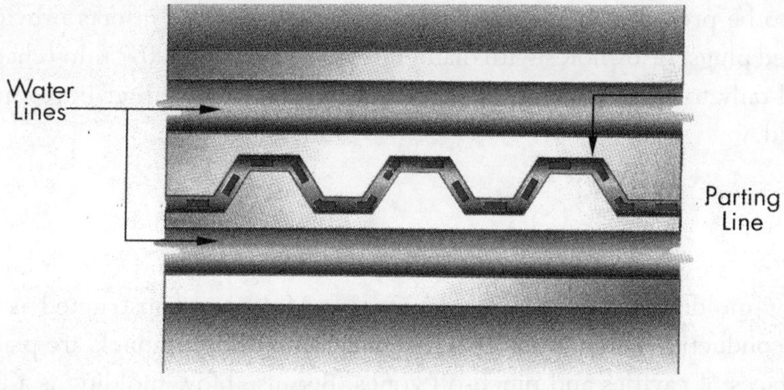

Figure 6.106 Example of a trapezoidal cross-section insert at the parting line

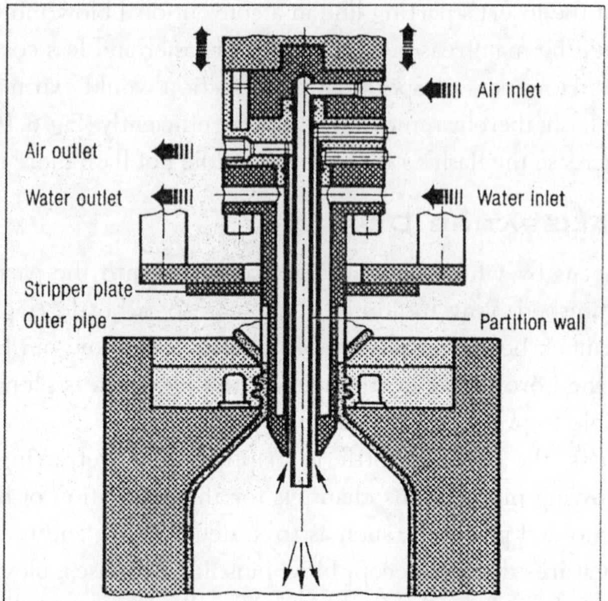

Figure 6.107 Example of a calibrating blow pin

to allow this air to escape without resistance. Unless a gloss finish is required on the molding, it is common practice to sandblast the cavity to a fine matte finish. This helps air to escape as the expanding parison touches the cavity face, but it is not sufficient in itself. Vent slots may be cut at appropriate points into the mold parting face to a depth of 0.05 to 0.15 mm. The appropriate point is where there is a possibility for air to collect as the hot plastic expands in the cavity. Figure 6.109 shows a schematic of an injection molding of a preform mold including venting slots. This type of mold is used after preforms are molded, usually in a multicavity mold as shown in Figure 6.110. Multicavity preform molds for PET can have up to 144 cavities.

Venting can also be provided within the mold cavity by means of inserts provided with vent slots, porous sintered plugs, or by holes with diameters not greater than 0.2 mm (chapter 17). Such holes are machined only to shallow depths and are relieved by much larger bores machined from the back of the mold.

Cooling

Efficient mold cooling is essential for economical blow molding. As in injection molding, typically up to 80% of a blow-molding cycle is devoted to cooling. Molds are constructed as far as possible from high thermal conductivity aluminum alloys, and water-cooling channels are placed as close as possible to the surfaces of cavities and pinch-off zones. Because blow molding is a relatively low-pressure process, the channels can be quite close to the surface and quite closely spaced before mold strength is compromised. The actual dimensions will depend on the heat-transfer rates and cooling temperature requirements for the materials of construction and the plastics being processed (Tables 6.13 to 6.16). As a guide, channels may approach within 10 mm of the cavity and center spacing should not be less than twice the channel diameter.

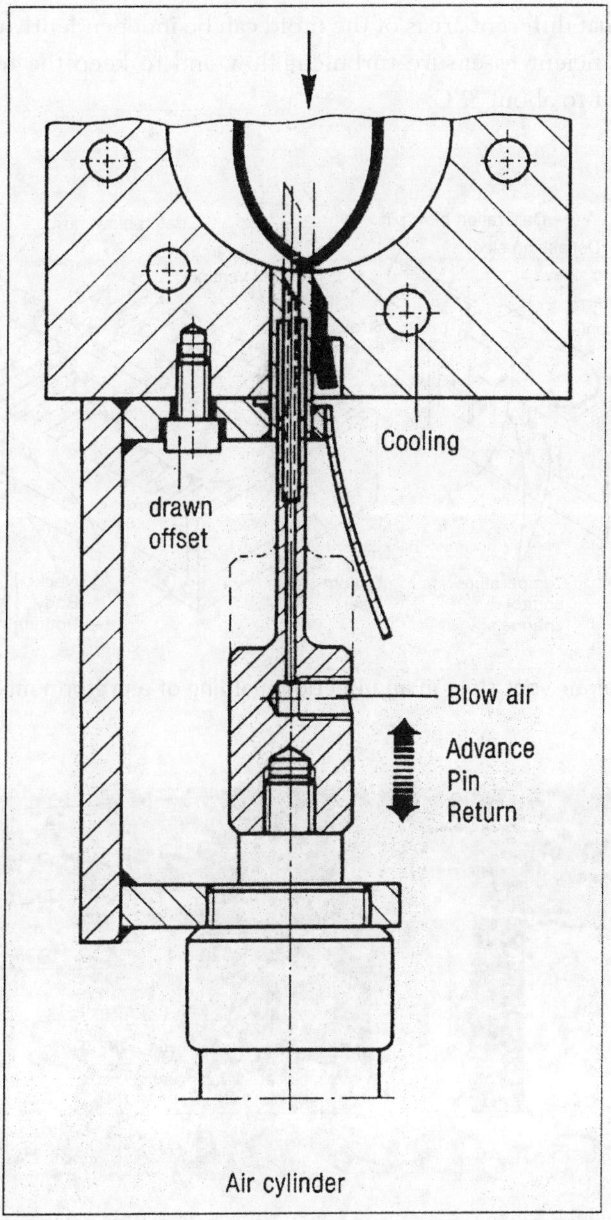

Figure 6.108 Example of blow needle

If the mold body is cast, the cooling channels can be fabricated in copper pipe to closely follow the cavity contours before being cast in place. If the mold is machined, drilling and milling will produce channels and it is not usually possible to follow the cavity contours so closely.

An alternative in cast molds is a large flood chamber (Fig. 6.111). However, efficient water cooling requires turbulent flow, and this may not be attained in a flood chamber or in large coolant channels (chapter 17). Many small channels are better than a few large ones. The cooling circuits will normally be zoned so that different areas of the mold can be independently controlled. The coolant flow rate should be sufficient to ensure turbulent flow and to keep the temperature differential between inlet and outlet to about 3°C.

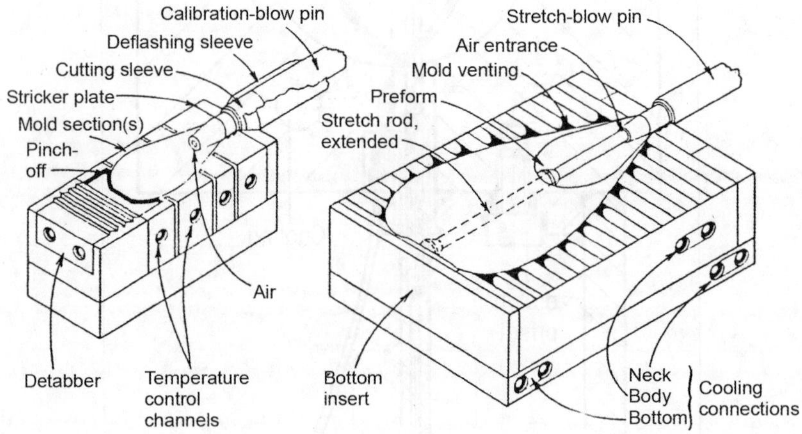

Figure 6.109 Example of air vent slots in an injection molding of a preform mold

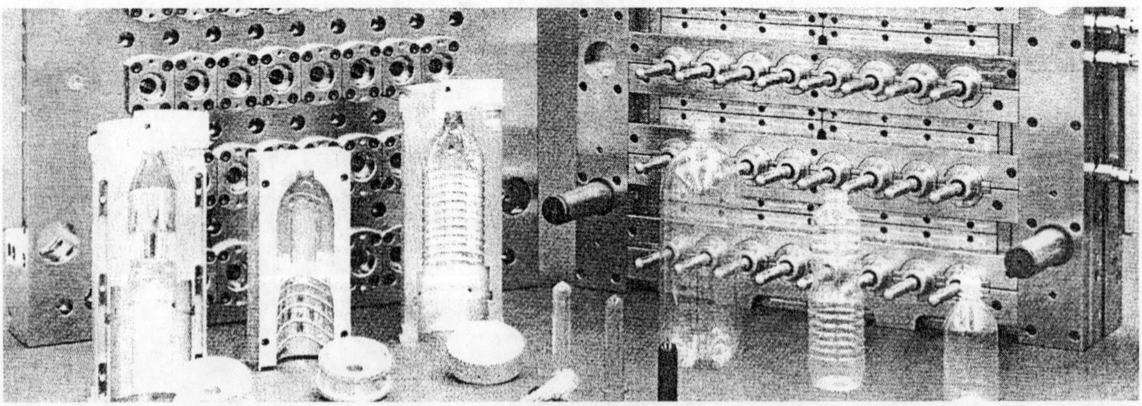

Figure 6.110 View of a multicavity preform mold in the background with blow molds and molded bottles in front (courtesy of SIG Plastics International)

Process or System	Heat-Transfer Coefficient (Btu/°F/h/sq ft)	Limiting Factors
Injection molding	30	Thickness and area available
Blow molding	25	One-side cooling and pinch-off cooling
Foam injection molding	20	Thickness and cellular structure
Jacketed vessels	50	Agitation and area available
Bath or trough	100	One-side cooling, thickness, and area available
Shell and tube exchanger		
Oil to water	100	Area and temperature difference
Water to water	250	Area and temperature difference

Table 6.13 Cooling characteristics

Molds on blow molding machines	20–55°F
Molds on bottle molding machines	50–55°F
Eddy current drive	80–85°F
Extruder barrels	80–85°F
Hydraulic oil coolers	80–85°F
Air compressors	80–85°F
Vacuum pumps	80–85°F
Refrigeration condensers	80–85°F
Temperature-control units	80–85°F
Molds on injection molding machines	20–55°F
Extruder troughs and cooling baths	50–55°F

20°F (−7°C); 50°F (10°C); 55°F (13°C); 80°F (27°C); 85°F (29°C); 250°F (121°C).
Note: High-temperature cooling requirements for special compounds may require temperatures in excess of 250°F.

Table 6.14 Cooling temperature requirements

Plastic	Recommended temperature	
	(°C)	(°F)
Polyacetates	80–100	176–212
Polyamides	20–40	68–104
Polyethylenes and PVC's	15–30	59–86
Polycarbonates	50–70	122–158
Polymethyl methacrylates	40–60	104–140
Polypropylenes	30–60	86–140
Polystyrenes	40–65	104–149

Table 6.15 Examples of blow-molding mold cavity temperatures based on plastic being processed

Cooling Analysis	
Mold surface	Part quality: even cooling prevents distortion
Temperature	
Freeze time	Minimizes cycle time.
Coolant temperature and flow rate	Optimizes coolant: can eliminate need to chill coolant.
Metal temperature	Cooling efficiency: ensures optimum circuit design.
Warpage Analysis	
Warped shape	Tendency to warp.
Single variant	Indicates fundamental causes of warpage.
Warpage shape	
Flow Analysis	
Fill pattern	Weld line position. Air trap position. Position of vents. Overpacking: excessive costly part weight. Overpacking: warpage due to differential shrinkage. Underflow: structural weaknesses.
Pressure distribution	Pressure required to fill. Clamp force required. Overpacking: ribs, etc. sticking in mold.
Temperature	Poor surface finish. Weak weld lines. Distortion due to differential cooling.
Shear stress distribution	Quality of part: tendency to distort. Quality of part: tendency to crack. Cycle time: low stresses permit hotter demold temperature.
Shear rate	Avoids degradation of material.
cooling time	Shows tendency to distort due to uneven cooling.
Flow-angle	Quality of part: molecular orientation.
packing pressure	Under/overpack to poor packing.
volumetric shrinkage	Dimensional variations due to poor packing.

Table 6.16 Examples of computer software information generated and typical problems it can solve (chapter 25)

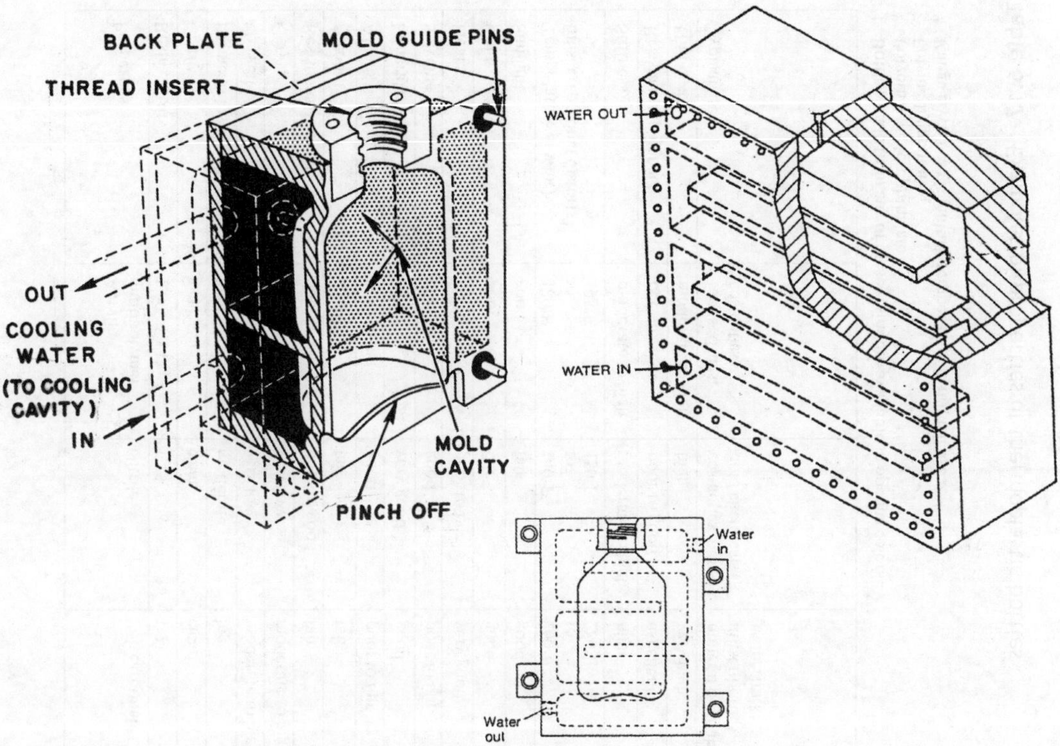

Figure 6.111 Examples of water flood cooling blow-molding molds

PLASTIC MATERIAL

Many thermoplastics can be blow molded. Different properties can be obtained with the different available and useful plastics that are now used to blow mold different sized and shaped products, providing the machines are designed to process the plastics (Tables 6.17 and 6.18). About 50 wt% is HDPE, 25% PET, 20% PVC, 2% polypropylene, and 6% other plastics are blow molded.

Properties can include formability, weight, strength, toughness, clarity, gas barriers, chemical resistance, heat resistance, color, and many others. Examples of these properties are in Tables 6.19 to 6.26 and Figures 6.112 and 6.113. Different combinations of properties can be achieved by combining two or more plastics. Parisons and preforms can be coextruded, coated, dipped, coinjected, and so on in order to provide the combinations of different properties available in different plastics. Stretch blow molding also provides an important processing technique to improve properties and reduce the quantity of plastics to improve cost/performance characteristics.

Examples of properties that can be obtained with the different plastics are as follows: clarity (crystal PS, AN, styrene AN, PVC, stretched PET, PSF), moisture barrier (PP, HDPE, transparent nylon), oxygen barrier (PET, PVOH, PVDC, PVC, AN, PSF, transparent nylon), and so on.

Properties	Acrylic multipolymer	Nitrile	Polycarbonate	Polyester (oriented) (PET)[b]	PETG copolyester	Polyethylene	
						Low density	High density
Resin density	1.09–1.14	1.15	1.2	1.35–1.40	127	0.91 to 0.925	0.94 to 0.965
Clarity[a]	clear	clear	clear	clear	clear	hazy transparent	hazy translucent
Permeability to water vapor	high	moderate	high	moderate	moderate	low	very low
Oxygen	low	very low	moderate to high	low	low	very high	high
CO_2	moderate	very low	moderate to high	low	low	very high	high
Resistance to acids	poor to good	poor to good	fair	fair to good	fair	fair to very good	fair to very good
Alcohol	good	fair	fair	good	good	fair to very good	good
Alkalis	poor to fair	good	poor to fair	poor to fair	poor to fair	good to very good	good to very good
Mineral oil	good	very good	good	good	good	poor	fair
Solvents	poor	good	poor to fair	good	poor to good	poor to fair	poor to good
Heat	fair	poor to fair	very good	poor to fair	poor to fair	fair	fair to good
Cold	poor	fair	good	good	good	very good	very good
Sunlight	good	fair	good	good	fair	fair	fair
Temperature at which finished product distorts[d]	180°F to 195°F	140°F to 150°F	260°F to 280°F	100°F to 160°F	160°F to 220°F	160°F to 220°F	160°F to 250°F
Stiffness	moderate to high	moderate to high	high	moderate to high	moderate to high	low	moderate
Resistance to impact	poor to good	poor to good	excellent	good to excellent	poor to fair	excellent	good to very good
Unit cost	moderate to high	high	very high	moderate	moderate to high	low	low
Typical uses	foods, drugs, cosmetics	cosmetics, household chemicals	mineral oil, baby nurser bottles, water, milk	carbonated beverages, mouthwash, liquor, edible oil, drugs, cosmetics	cosmetics, foods, mineral oil, personal care	cosmetics, personal products, mustard, drugs	detergents, bleaches, milk, chocolate, syrup, industrial cleansing powders, drugs, cosmetics, lube oil, edible oil

[a] Bottles made from any of these resins are available in opaque colors
[b] Polyethylene terephthalate
[c] Polymethylpentane
[d] Range given here is limiting temperature in normal use

Table 6.17 Examples of properties of thermoplastic bottles

Properties	Polypropylene Regular	Polypropylene Oriented	Polystyrene	Polysulfone	SAN (styrene acrylonitrile)	TPX[c]	PVC
Resin density	0.89 to 0.91	0.90	1.0 to 1.1	1.24	1.07 to 1.08	0.83	1.35
Clarity[a]	moderate haze	clear	clear	clear	clear	moderate haze	clear
Permeability to water vapor	very low	very low	high	high	high	very low	moderate
Oxygen	high	high	high	low to moderate	high	very high	low
CO_2	moderate to high	moderate to high	high	low to moderate	high	high	low
Resistance to acids	fair to very good	fair to very good	fair to good	fair to very good	fair to good	fair to good	good to very good
Alcohol	good	good	fair	fair to good	poor	good	good to very good
Alkalis	very good	very good	good	very good	good	very good	good to very good
Mineral oil	fair	fair	fair	very good	fair	fair to good	good
Solvents	poor to good	poor to good	poor	fair to good	poor	fair to good	poor to good[e]
Heat	good	good	fair	very good	fair	very good	poor to fair
Cold	poor to fair	very good	poor	very good	poor	poor to fair	fair
Sunlight	fair to good	fair to good	fair to poor	fair	fair to poor	fair	poor to good
Temperature at which finished product distorts[d]	250°F to 260°F	250°F to 260°F	200°F to 220°F	330°F to 365°F	170°F to 190°F	230°F to 260°F	140°F to 150°F
Stiffness	moderate to high	moderate to high	moderate to high	high	moderate to high	moderate to high	moderate to high
Resistance to impact	poor to good	very good	poor to good	good to very good	poor to good	poor to good	fair to good
Unit cost	moderate	moderate to high	moderate	very high	moderate to high	high	moderate to high
Typical uses	drugs, cosmetics, syrups, juices, detergents, mouthwash, shampoo, IV solutions	detergents, drugs, mouthwash, shampoo, household chemicals, IV solutions, liquid soaps	dry drugs, petroleum, jellies, vitamins, spices	laboratory ware, microwave cooking	dry drugs	laboratory ware, microwave cooking	cosmetics, personal care, household chemicals, edible oils, vinegar

[a] Bottles made from any of these resins are available in opaque colors
[b] Polyethylene terephthalate
[c] Polymethylpentane
[d] Range given here is limiting temperature in normal use
[e] Ketones, esters, chlorinated and aromatic solvents: poor; aliphatic solvents: good

Table 6.17 *(continued)*

Criterion	Resin				
	PET	PET copolymer	Lowmonomer PVC	Nitriles	Poly-carbonate
Taste					
Brown goods	A	A	UA	M	M
White goods	A	A	UA	–	–
Odor	A	A	UA	M	M
Appearance: clarity, color, transparency	E	A	A	M	A
Strength	A	A	A	–	A
Ability to produce handleware	No	No	Yes	Yes	Yes
Product retention					
Alcohol loss	A	Note[b]	A	A	Note[b]
Proof rise (water loss)	A	M	Note[c]	UA	UA
Approval by					
FDA	Yes	Note[b]	Yes	–	Yes
BATF	Yes	No	No	No	No
Resin cost, cents per pound (December 1982)	60	–	70	92 to 108[d]	169[d]

[a] E = excellent; A = acceptable; UA = unacceptable; M = marginal
[b] Unknown, [c] Slightly higher, [d] Excessive

Table 6.18 Examples of various plastics suitable for plastic liquor bottles

Properties	Desired goals
Stiffness	Varies with product requirements
Environmental stress crack resistance	Generally maximum, though sometimes not important
Gloss	Usually high, but varies with product requirements
Overall appearance	High gloss, smooth surface, no flow lines
Wall thickness uniformity	Minimum gauge variation in walls
Weld-line thickness	Equal to or greater than the side wall thickness
Parting-line difference	Minimum gauge variation from side wall
Mold cycle time	Minimum

Table 6.19 Important properties of extrusion blow-molded products and the desired goal(s) for each

Extrusion blow molding represents the biggest outlet for blown containers, with HDPEs being the major plastics used. Recognize that there are different grades of HDPE that include those with changes in density, stiffness, strength, hardness, toughness, stress crack resistance, permeability, creep resistance, softening temperature, and gloss.

	Higher Density	Higher Melt Index
Bottle[1] Properties		
Stiffness	Large Increase	NSE*
Environmental Stress Crack Resistance	Decrease	Large Decrease
Gloss	Large Increase	Increase @ 0.915 g/cm³; NSE @ 0.932 g/cm³
Overall Appearance[2]	Improvement	NSE
Wall Thickness Uniformity	NSE	NSE
Weld Line Thickness	Large Decrease	NSE
Parting Line Difference	NSE	Large Increase at high blow pressure
Mold Cycle Time	Large Decrease	NSE
Test limited to range specified:	Densities of 0.915 to 0.932 g/cm³	Melt Indexes of 1.3 to 4.7 g/10 min.

**NSE = no significant effect*
[1]tests made on 4 oz. (115 ml) Boston round bottles
[2]quantitative rating of gloss, surface smoothness and flow lines in bottles

Table 6.20 Changes in extrusion blow-molded bottle properties resulting from resin properties

Bottle[1] Properties	Extrusion Conditions: Higher Melt Temperature	Extrusion Conditions: Higher Annular Die Clearance	Extrusion Conditions: Higher Extrusion Rate	Molding Conditions: Higher Mold Temperature	Molding Conditions: Higher Molding Time	Molding Conditions: Higher Blow Pressure
Stiffness	NSE*		Slight Decrease			
Gloss	Increase	NSE		NSE	Increase	NSE
Overall Appearance[2]	Slight Improvement	NSE	NSE	Improvement	Improvement	NSE
Wall Thickness Uniformity	Slight Decrease	Decrease	NSE	Slight Decrease	NSE	
Weld Line Thickness	NSE	Large Increase	NSE	NSE	NSE	Large Increase from 10 to 50 lb/in.^{2}
Parting Line Difference	Slight Increase	NSE	NSE	Increase	NSE	Large Increase
Mold Cycle Time	Increase	Increase		Large Increase	Large Increase	NSE
Tests limited to ranges specified	Temperatures of 150-170^{0}C (300-340^{0}F)	54.5 to 79.5 mils (1.4 to 2.0 mm)	0.2 to 0.4 in./sec. (5 to 10 mm/sec)	15 to 70^{0}C (60 to 150^{0}F)	12 to 24 sec	10 to 80 lb/in.^{2} (0.7 to 6.5 kg/cm^2)

*NSE = no significant effect

[1]tests made on 4 oz. (115g) Boston round bottles

[2]quantitive rating of gloss, surface smoothness and flow lines in bottles

Table 6.21 Changes in extrusion bold-molded blow properties resulting from changes in extrusion and molding conditions

Type of bottle	Rate (m^2 day^{-1}) Oxygen (cm^3)	Water vapor (g)
PET (oriented)	10.2	1.10
Extrusion blow molded PVC	16.4	2.01
Stretch blow molded PVC (impact-modified)	11.9	1.8
Stretch blow molded PVC (nonimpact-modified)	8.8	1.3

At 38 °C (100 °F).

Table 6.22 Gas barrier transmission comparisons for a 24 fl oz (689 cm^3) container weighing 40 g

Type of bottle	Percent
Extrusion blow molded PVC	–
Impact-modified PVC (high orientation)	4.2
Impact-modified PVC (medium orientation)	2.4
Impact-modified PVC (low orientation)	1.6
Nonimpact-modified PVC (high orientation)	1.9
Nonimpact-modified PVC (medium orientation)	1.2
Nonimpact-modified PVC (low orientation)	0.9
PET	1.2

Seven days at 80 °F (27 °C).

Table 6.23 Volume shrinkage of stretch blow-molded bottles

Position	Direction	Yield point (kg/mm^2)	Ave.	Break point (kg/mm^2)	Ave.	Elongation at break (%)	Ave.
Upper panel	Axial	9.30 9.47	9.30	8.82 9.72	9.27	55.0 57.5	56.3
	Hoop	– –	–	18.29 19.37	18.83	22.5 25.0	23.8
Lower panel	Axial	9.33 9.28	9.31	11.86 8.62	10.24	92.5 32.5	62.5
	Hoop	– –	–	18.58 19.66	19.12	20.0 25.0	22.5

Table 6.24 Tensile test data of PET plastic

Beverage bottles provide the major reason for PET being the second major plastic consumed in stretched injection blow molding. PET properties meet performance requirements for pressurized beverage bottles.

PVC is a major plastic used in great quantity in the United States for extrusion blow molding due to its low oxygen transmission, water-clear clarity, relative ease of processing, low cost, and so on. It is being challenged by stretch-blown PP and stretch-blown PET. PP and PP copolymers are used in packaging containers for cosmetics, pharmaceuticals, and some foods (syrup, etc.), especially for medical-surgical fluids, where hot filling is required, and so on.

This brief review on the different plastics is used to emphasize the importance of using the correct plastic to meet specific requirements. The real test on any material is when it is blown, either unoriented or stretched oriented. Details on properties and melt processabilities are reviewed in chapters 1 to 3.

BLOW MOLDING AND PLASTIC

Different plastics can require different blow molding equipment, such as feed mechanisms, screw designs, and mold designs (chapters 3 and 17). Here is an example concerning polypropylene (PP). Blow molding processes only about 2 wt% of PP. The combination of economy and good properties offered by PP leads to its widespread use in films, fibers, and injection moldings. So, why are the packaging advantages of PP films not exploited in the containers and bottles that the blow-molding process produces? Why has the outstanding environmental stress crack performance of PP, along with its chemical resistance, not made it the material of choice for blow molding?

The answer lies in the development of blow-molding machinery and the differences in melt rheology between PP and the popularly used PE. Blow molding's rapid developments in machinery

Plastic	*Temperature (°C)*
LDPE	130–180
MDPE	150–200
HDPE	160–220
HMWPE	180–230
PVC	190–205
PP	200–220
PS	280–300
PA	240–270
POM	150–280
ABS	180–230
ABS/PC	230–250
PPE	240–250
PBT	245–260
PBT/PC	240–260
PUR	180–190

Table 6.25 Guide to plastics processing temperatures for blow molding

and technology began in the late 1930s, virtually contemporary with the introduction of PE. PP was not commercialized until the mid-1950s, so BM technology was largely optimized on the requirements for PE (chapter 29).

Blow molding is a relatively low-pressure process, using only 1% or less of the pressure levels employed in injection molding. This requires a low melt viscosity, produced by high shear rates and temperatures. The difficulty with PP is that its melt viscosity is far more sensitive to temperature and shear rate than is the case with PE, so there are problems in working PP on PE equipment (221).

To change this condition, machinery and technology are now being developed specifically for PP. At the same time, developments in PP are making it a more suitable material for the process. The increased transparency of nucleated, clarified, and random copolymer grades is making them more attractive for bottle and container applications. Advances in polymerization technology have

Increase in these factors → affects these properties ↓	Resin Properties		Extrusion Conditions		
	Density	Melt Index	Melt temperature	Annular die clearance	Extrusion rate
Stiffness	Large increase	—	—	—	Slight decrease
Environmental stress crack resistance	Decrease	Large decrease	—	—	—
Gloss	Large increase	Increase at 0.915 density No effect at 0.932 density	Increase	—	—
Overall appearance	Improvement	—	Slight improvement	—	—
Wall thickness uniformity	—	—	Slight decrease	Decrease	—
Weld-line thickness	Large decrease	—	—	Large increase	—
Parting-line difference	—	No effect at low blow pressure Large increase at high blow pressure	Slight increase	—	—

Table 6.26 Examples of fabricating conditions on blow-molded PE bottles

made it possible to produce grades with melt strengths and shear dependencies that are more tolerant of the blow-molding process. Consequently, blow molding PP is a growth area. Its market share is currently small, but its growth rate is second only to PET. PP blow molding is targeted to take market share from PVC and high-density polyethylene. The continuous extrusion blow-molding process is the most widely used for working with PP.

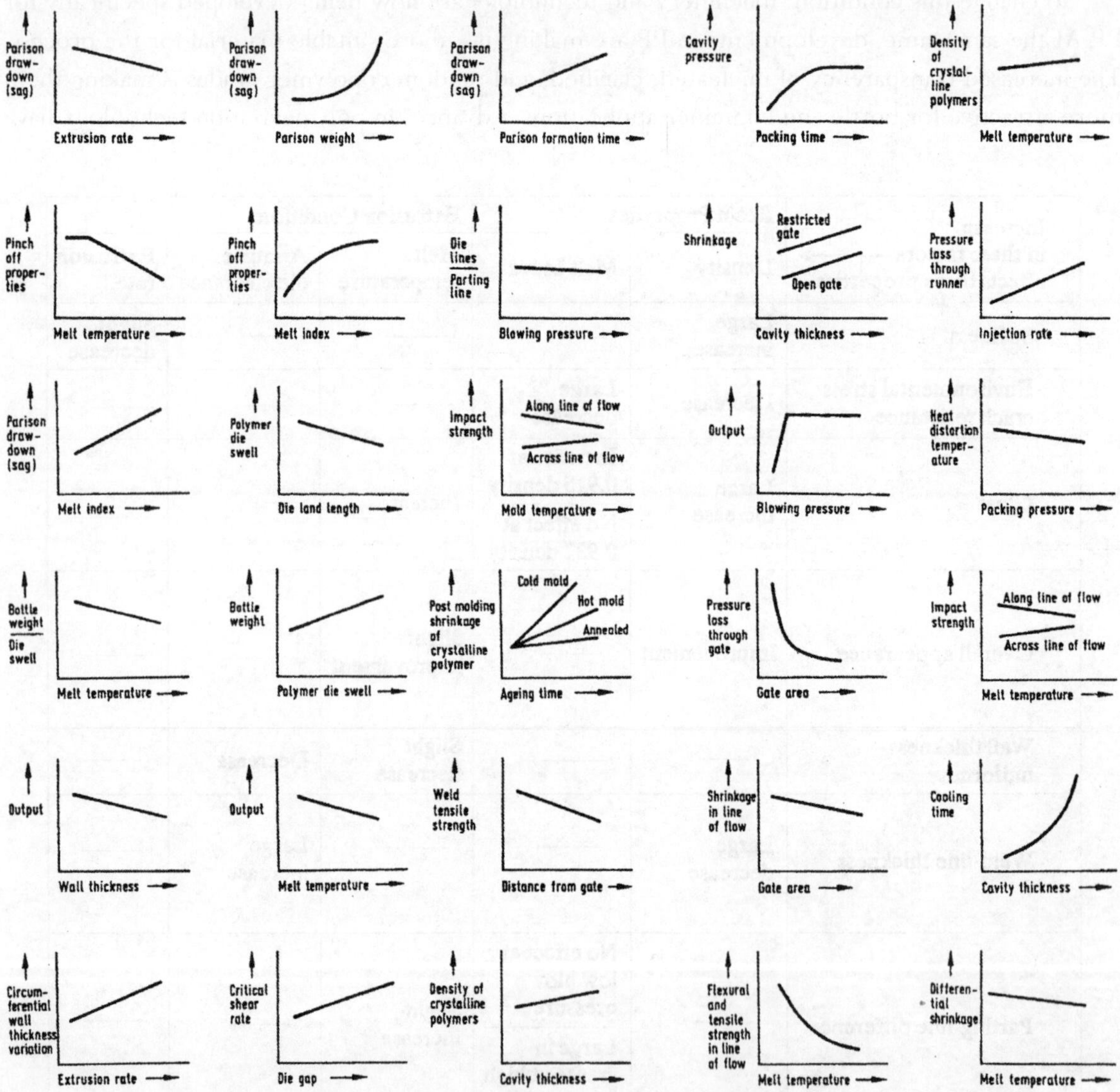

Figure 6.112 Examples of effects of the blow-molding extruder and plastic variables on product performances

Stretch blow-molded PP containers have some important advantages. They can be sterilized at higher temperatures than competitive containers and are not prone to environmental stress cracking. Stretched PP containers can be hot filled at about 90°C.

BEHAVIOR OF PLASTICS

A representative cross-section of the myriad of plastics available to designers has been described in this chapter and throughout this book. This section is to acquaint the designer with the structural behaviors of plastics. It provides concepts as background for estimating and anticipating behaviors in design situations. Obviously, there are no universal methods to describe the behavior of all plastics, just as there is no single reference that completely covers the behavior of all metals and their alloys. Nonetheless, all plastics show many behavioral similarities. The differences are frequently in terms of magnitude, not of kinds.

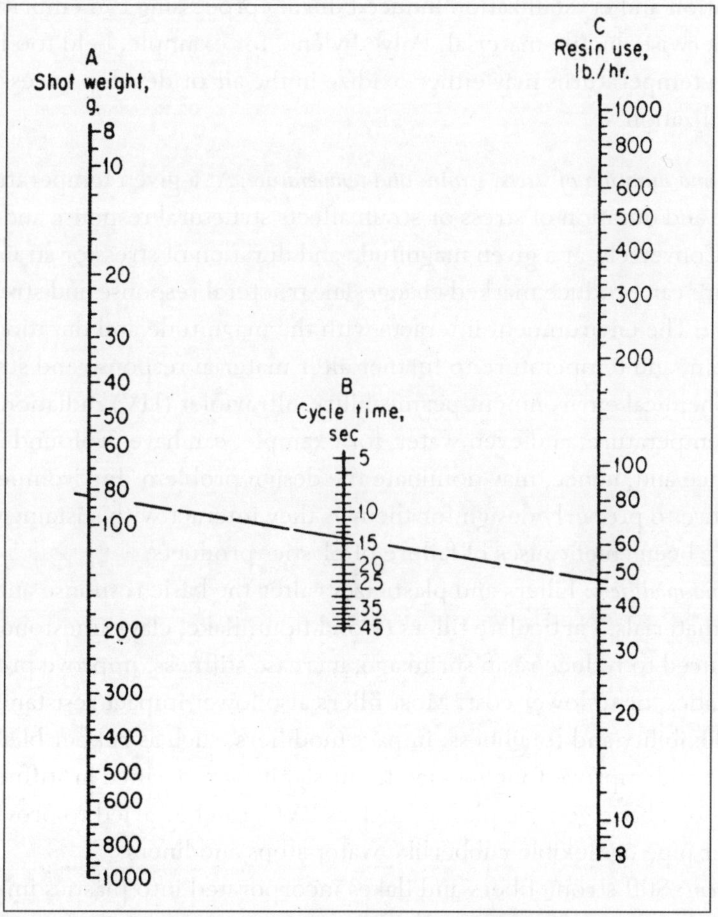

Figure 6.113 Nomogram for injection blow-molded preform shot weight, cycle time, and resin use

The stress strain and strength behavior of plastics vary widely, depending on the generic type or family of plastic and on the specific composition of compounds within that family. Many factors interact to alter behavior of a given compound over a very wide range. The following are key parameters that must be considered in structural design of plastics:

1. *Process*: The blow-mold process used to convert plastic materials into structures may dictate the structural performance of the finished product, based on the following:
   a. Orientation of the molecular structure of a thermoplastic may strengthen the product in the direction of orientation. Such preferential orientation can be used to advantage; however, the designer must be on guard to ensure that orientation does not prove to be the cause of failure. For example, flow patterns and knit and weld lines developed during molding can cause orientation that creates zones of weakness.
   b. Oxidation and crystallization induced during processing can embrittle an otherwise ductile material. Polyethylene, for example, held too long at high temperatures may either oxidize in the air or develop excessive crystallization.
2. *Magnitude and duration of stress, strain, and temperature*: At a given temperature, both the magnitude and duration of stress or strain affects structural response and strength behavior. Conversely, at a given magnitude and duration of stress or strain, a shift in temperature can produce marked changes in structural response and strength behavior.
3. *Environment*: The environment interacts with the magnitude and duration of stress, strain, and temperature to further alter material response and strength of plastics. Chemical environment permeability, ultraviolet (UV) radiation, sustained elevated temperature, and even water, for example, can have profound influence on performance and, hence, may dominate the design problem. Environmental effects, or the failure to properly design for them as they interact with sustained stress or strain, have been chief causes of failure of plastics products.
4. *Additives and modifiers*: Fillers and plasticizers alter the basic response and strength of plastic materials. Particulate fillers (wood flour, flake, clay, limestone, etc.) are introduced to reduce resin shrinkage, increase stiffness, improve processing characteristics, or to lower cost. Most fillers also lower impact resistance. Plasticizers increase flexibility and toughness; impact modifiers, such as rubber blends, are more permanent and improve toughness without significant sacrifice in stiffness. Thus, the properties of a basic generic plastic, such as PVC, can be varied to provide either rigid sewer pipe or flexible rubberlike water stops and liners.
5. *Reinforcement*: Stiff strong fibers and flakes incorporated into plastics improve stiffness, strength, and dimensional stability. An increase in their proportion relative to the resin matrix results in a corresponding improvement in these properties.

6. *Similarities with conventional materials*: The plastics behavior introduced here is not necessarily any more complicated than the behavior of conventional structural materials. Overall, the engineer must design both plastics and conventional materials to meet the criteria of time, temperature, and environment. The similarities are as follows:
   a. Time: In designing with conventional materials (steel, aluminum, glass, wood, and concrete), the effects of load duration are recognized in terms of creep and ultimate strength. Time effects are frequently more prominent in plastics, particularly in terms of strength.
   b. Temperature: In designing with steel and glass, brittleness at low temperatures and loss of yield strength at high temperatures must be considered. Similarly, in designing with plastics, both stress strain and strength behavior can vary with temperature, and the low and high temperature limits of certain plastics are frequently in the range of the normal outdoor environmental limits.
   c. Environment: The strength, stiffness, dimensional stability, and useful life of wood are strongly impacted by moisture content. Steel rusts, glass cracks, aluminum corrodes, and so on. The problem is similar for plastics, but because composition can vary widely, each plastic compound must be considered for its strengths and weaknesses on exposure to various environments.

While there are certain similarities, there are significant differences between designing with plastics and with conventional materials (steel, wood, etc.). Stringent specifications define the characteristics, and formal detailed rules define design procedures, for a limited number of aluminums, steels, and glass as well as a broad range of timber species. Concrete and steel quality is checked by fairly simple tests; wood quality is assessed visually in accordance with long established "grading" rules. Working rules of thumb, which may be highly empirical, have evolved from both experience in the field and research using these materials of construction (chapter 19).

Design criteria for plastics are established in different standards and specifications (chapter 22). A myriad of different materials and variations are available (chapter 2).

BARRIER PLASTIC

As briefly reviewed ("*Coextrusion and Coinjection*"), there are barrier requirements for blown products that range from packaging foodstuff to gasoline. The barrier requirements to meet package needs vary depending on the product being packaged. Properties that may need to be retained or excluded can include gases, water vapors, aromas, tastes, and solvents. The result has been the development and use of a wide diversity of barrier plastics in blow molding. Information on coextruding or coinjecting these barrier materials is reviewed in chapters 3 to 5.

Many oxygen-sensitive food products can be either hot filled or retorted. In these cases, the container must have requisite thermal as well as barrier properties. Further, if the container is to be subjected to elevated temperatures during filling or processing, it must be designed to either resist or anticipate paneling. Both of these factors can considerably increase the complexity of developing satisfactory rigid barrier containers.

An example is the success of packaging carbonated beverages using PET plastic. The plastic bottle industry continues its research into expanding plastic bottles in other large-volume applications to replace glass and metal with blow-molded plastic containers. There are more high-volume markets that could be expanded and/or switched into plastic containers, such as beer (United States).

To date, it appears that Americans prefer their beer in glass bottles, but that could change if various development programs take off, such as PINTA. As reported, when Rob Peabody had a few friends over to watch a football game on TV, he served them beer in plastic bottles. As head of the newly formed Polyester Intermediates Americas (PINTA), Peabody figures on plastic beverage bottles to be a strong growth market in North America (275). The BP PINTA's business unit is to produce purified terephthalic acid (PTA), a raw material used to make polyethylene terephthalate (PET), a primary ingredient in plastic bottle manufacturing.

Although plastic bottles do not yet dominate U.S. beer sales, they reign supreme as holders of everything from juice to water. In fact, 60% of any plastic beverage bottled in North America is made with BP PTA. The driving force for this application is to keep bottle contents fresh longer. That includes BP's dimethyl 2,6-naphthalenedicarboxylate (NDC), which provides heat resistance for pasteurizable plastic beer bottles, and Amosorb, which keeps oxygen out of sensitive beverages, such as juice. Both are produced by PINTA in Naperville, Illinois.

There are at least four basic ways in which to increase the barrier properties of a rigid plastic container. The first is to improve the barrier properties of the plastic being used in a monolayer container. The second is to place a barrier coating on either the inside or the outside of the monoplastic layer. The third is to employ several distinct plastics in a multilayer coextruded structure with each providing a different specific barrier or performance quality.

An example of coating is from Tetra Pak USA in Vernon Hills, Illinois, which has FDA acceptance for its Glaskin silicon dioxide barrier coating technology for PET containers (542). Special plasma technology deposits a clear, glasslike coating, thinner than a human hair, on the bottle interior. Studies show that the coating increases the oxygen barrier of a PET bottle seventeenfold and the $C0_2$ barrier twenty-five-fold. The coating can extend the shelf lives and maintain the taste and nutritional values of beverages, such as beer, carbonated soft drinks, and fruit juices. The Glaskin coating has a shelf life of more than a year. The very thin coating offers high elasticity and perfect clarity with no haziness. Since it is inside the bottle, it cannot be damaged by conveyors and packaging lines. Initial use started in Europe, where Spendrups and Bitburger beer are bottled in PET with the Glaskin coating.

The following plastics are being stretched coinjected blow molded with PET plastic for developing beer applications: EVOH (ANC), nylon MXD6 (Schmalbach Lubeca), nylon MXD6 (Claropak), oxygen scavenger (Continental), and Amosorb 02 Scavenger (Kortec).

Monolayer Chemical reactive systems are used to change the monolayer barrier behavior of certain plastics. As an example, fluorination, a secondary process used on monolayer HDPE containers, involves the additional step of treating the container with fluorine gas. This treatment adds side chains of fluorine to the molecular structure of the HDPE, resulting in improved barrier properties.

Examples of fluorination processes used include Air Products & Chemicals' process that injects fluorine gas during the blowing cycle. The Union Carbide/Linde process is a postmolding treatment in which molded bottles are placed in a chamber and exposed to fluorine gas at elevated temperatures. Both processes are accepted by the industry, and fluorinated HDPE bottles are being used for packaging pesticides, paint thinners, charcoal lighter fluid, kerosene, and so on.

There is also Linde's Surface Modified Plastics (SMP) process. It was a spin-off from their aerosol container research. The process is unique among current fluorination methods for polyethylene because it allows precise control over the degree of treatment. In effect, the user is able to custom build a barrier that is specifically suited for the application at hand.

The development of barrier plastics that can be used in monolayer containers to pack oxygen-sensitive food products has not, to date, been commercially successful. As an example, plastic producers have developed barrier PET copolymers, but in all cases, they have had to trade off some other critical performance characteristics, such as impact strength or creep resistance. Further, barrier plastics, even when they do achieve acceptable balances of characteristics, are too expensive, running at least between $1 and $3 a pound. Very little development work is being done in this area and the likelihood of an inexpensive barrier PET is several years off at best.

Coating The concept of coating a plastic to improve its barrier performance is not new. Some of the first and still most widely used flexible barrier packages are PVDC-coated polyester and oriented polypropylene. Coating is perceived by many as the most reasonable low-cost method to improve the barrier properties of rigid monolayer containers.

Coating has several drawbacks. The primary functional problem is adhesion. Another is the cost of high-speed continuous coating lines. Also, there is the problem of achieving an adequate barrier with a single pass. Assuming one is coating the outside of a container, there are three basic types of coating application systems: roll coating, dip coating (with or without a wipe step), and spray coating.

Roll and dip coating are used in Europe to apply a layer of PVDC directly on oriented PET (OPET) bottles for beer. In the United States, these application methods have been used to coat both preforms and finished PET bottles. Coating the preform appeared originally to have some advantages because the subsequent stretch-blowing step oriented the PVDC and increased its barrier properties. During 1980, wine bottles used coated preform. There were, however, resultant adhesion problems and an overthinning of the PVDC layer during blowing, which effectively reduced the net barrier effectiveness despite the biaxial orientation achieved. In addition, the lines were too slow to be effective commercial processes.

There is also the approach to achieve an adequate barrier with a single pass. Efforts have been made to develop dual-pass systems or coating systems in which a superior oxygen barrier (EVOH or EVA) is covered with a water barrier, but the cost has outweighed the benefits. The critical point to make is that although coating does not produce a totally successful barrier container, it is a commercially available technology that offers a modest degree of barrier improvement.

Some of the other emerging coating technologies that may generate interest in the future are other sprayable barrier materials, including EVOH, and sinter coating techniques that would permit the use of certain plastics that can be powdered but not sprayed.

Multilayer Most of the development efforts to produce high-quality rigid barrier containers involve the use of multilayer structures. There are two basic approaches to the production of multilayer structures. These are coextrusion and coinjection. The basic issues to consider when developing a multilayer barrier container are which barrier material to use and which manufacturing technology to employ. Blow molding multilayer hollow products continues to experience a phase of worldwide interest, including a rapid upsurge. It utilizes the material-specific properties of various plastics for the production of high-grade multilayer containers.

The introduction of the five-layer to seven-layer technique in the United States brought the breakthrough in 1984 for commercial production of these containers, particularly in the foods industry (ketchup, mayonnaise, soup concentrates, baby food, and the like). A crucial factor was the development of a completely new multilayer head extrusion technology permitting wall thicknesses down to less than 0.03 mm (0.001 in) with excellent, constant coaxial wall thickness distribution. This was the key to using expensive barrier and bonding agent materials at commercially viable production costs.

The advantages of these bottles are many. Here are some of them, related to different material combinations: (a) excellent barrier properties against gases, such as nitrogen, carbon dioxide, and air, as well as against water vapor (foodstuffs, pharmaceuticals), (b) resistance to aroma transmissions (cosmetics), (c) protection against UV radiation (mild), (d) higher strength (greater product stackability), (e) lightweight (lower transport costs), (f) scratch-resistant glossy surfaces (cosmetics), (g) resistance to aggressive media (insecticides, herbicides), (h) barrier to solvents (paint industry), (i) easy printability and clear printed impressions (cosmetics), and (j) prevention of static charging (explosion protection).

Figure 6.114 is an example that shows the relation between comonomer concentration, the moisture transmission, and gas transmission properties of these copolymers. These exceptional barrier properties allow the use of extremely thin layers of these materials in combination with other inexpensive plastics to BM economical high-performance products. Table 6.27 indicates the range of commercially available EVOH plastics. Within this range, conventional melt BM processes readily process EVOH plastics.

The method that has the greatest application potential is to arrange the barrier layer in the middle between the inner and outer body layers. Whether reclaimed material is used in the outer layer is a quality question that the container filler has to answer based on product performance requirements. The seven-layer combination with reclaim in the penultimate layer from the outside, without any further bonding agent to the body material, is the most satisfactory solution (economically) to date.

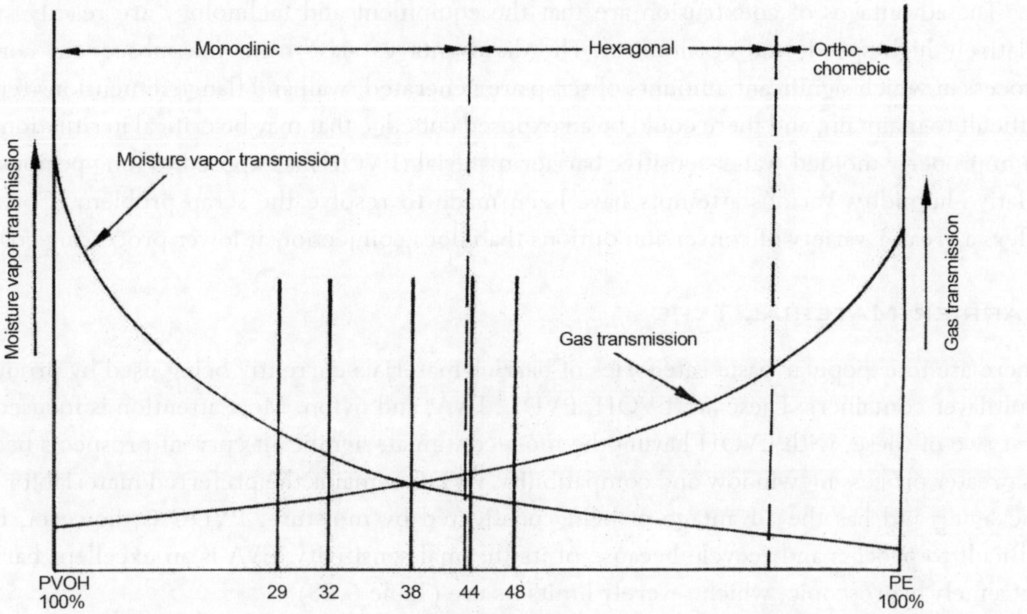

Figure 6.114 Comonomer concentrations vs. barrier properties of crystalline structures

Melt index (g/10 min)	Density (g/cc)	Ethylene content (mol %)	Melting point (°C)
0.7–20	1.13–1.21	29–48	158–189

Table 6.27 EVOH plastic range of properties

With the coextruded technology well established and the hardware in position, the variety of achievable properties is readily extendable through the correct combination of different materials. The potential for BM packaging products to be much more than simple bottles has been known for some years. Now the know-how and economics are such that many previously futuristic ideas are much closer to reality.

The potential in coextruded bottles, coupled with the machinery platform being established, means that glass and metal will come under increasing pressure. Often, the higher initial costs of the plastic materials are counterbalanced by the total processing and packaging economics that can be obtained. The impetus behind this rate of growth is the combination of the inherent versatility of plastics with blown shapes (cylindrical to noncylindrical, irregular hollow-shaped industrial products, etc.) that provides cost advantages over other processes.

The advantages of coextrusion are that the equipment and technology are readily available, relatively high speed, and economical. The disadvantages arise from the subsequent converting process in which significant amounts of scrap are generated, wall and flange dimensions tend to be difficult to maintain, and there could be an exposed cut edge that may be critical in situations where an improperly molded water-sensitive barrier material (EVOH) is exposed to long periods of high relative humidity. Various attempts have been made to resolve the scrap problem. Coextrusion offers a greater variety of conversion options than does coinjection at lower processing cost.

Barrier Material Type

There are four popular basic categories of barrier materials currently being used by producers of multilayer containers. These are EVOH, PVDC, EVA, and nylon. Most attention is focused on the first two of these, with EVOH having the most continual, significant current prospects because of its greater processing window and compatibility. PVDC remains the preferred material for flexible packaging and has the advantage of being unaffected by moisture. PVDC is, however, basically difficult to process and recycle because of its thermal sensitivity. EVA is an excellent barrier but extremely hygroscopic, which severely limits its use (Table 6.28).

To date, there is no dominant technology, product, market, or set of performance specifications directing the development of rigid plastic barrier containers. They deal with a number of variables that include choice of materials, choice of conversion processes, choice of filling and processing technology, and redefinition of product performance requirements. There is no single optimum container. Rather, there is an increasing diversity of rigid barrier plastic containers, each designed for a specific end-use application. This situation does not exist with the OPET (oriented PET) soft drink container business.

Blow Molding Reinforced Plastic

Practically all BM uses unreinforced TPs. In the past, different systems were used to produce RP BM products. One of the latest is from Krupp Kautex, which uses two extruder blow-molding

Polymer	O_2 transmission rate (25°C, 65% RH cc.mil/100 in^2/24 hrs)	Moisture vapor trans. rate (40°C, 90% RH gm.mil/100 in^2/24 hrs)
Ethylene vinyl alcohol	0.05–0.18	1.4–5.4
Nitrile barrier resin	0.80	5.0
High barrier PVDC	0.15	0.1
Oriented PET	2.60	1.2
Oriented nylon	2.10	10.2
Low density polyethylene	420	1.0–1.5
High density polyethylene	150	0.69
Rigid PVC	5–20	0.9–5.1

Table 6.28 Examples of barrier properties of commercially available plastics

(EBM) methods to provide increased structural properties through the use of relatively long glass-fiber reinforcement, such as with PP plastic (chapter 15). As an example, the 15 wt% content uses 7-mm-long (2-in-long) fibers.

One approach extrudes unreinforced melt into a low-shear Z-blade or sigma-blade batch kneader. Dry, chopped glass fibers are blended into the melt under vacuum, and a discharge screw in the mixer's bottom feeds the compound to the blow molder's accumulator head. Another approach uses commercially available glass-fiber/PP-compounded pellets made by the pultrusion process. Pellets are plasticized using a special screw that melts by heat conduction rather than by shear in order to minimize fiber damage. The parison is blown initially at very low air pressure; after complete expansion, full pressure is used.

DESIGN

Designers should become aware of the potential that blow molding offers, since shapes ranging from simple to intricate can be fabricated. The process is especially amenable to the designer's goal of consolidating as much function as possible into a single product. Some of the features that can be incorporated include threads, inserts, fasteners, snap fits, hinges, and others somewhat similar to those covered under injection molding.

Design is the process of devising a product that fulfills as completely as possible the total requirements of the user, and at the same time, satisfies the needs of the fabricator in terms of cost effectiveness (return on investment). The efficient use of the best available material and production process should be the goal of every design effort, including tool design. Product design is as much an art as a science. Design guidelines for blow-molding plastics have existed for over a century, pro-ducing many thousands of parts meeting service requirements, including those requiring long lives.

Designing acceptable products requires knowledge of plastics that includes their advantages and disadvantages (limitations) and some familiarity with processing methods. Until the designer becomes familiar with processing, a qualified fabricator must be taken into the designer's confidence early in development and consulted frequently during those early days. The fabricator and mold or die designer should advise the product designer on the materials' behaviors and how to simplify the product in order to simplify and/or reduce cost of processing. Understanding only one process, and in particular just a certain narrow aspect of it, should not restrict the designer.

The effective exploitation of product design opportunities is the key start to success. In turn, success hinges on other factors, such as the proper selection of materials and using the best available processing equipment. Detailed analysis to the design approach is given in chapter 19.

Because new materials and equipment continue to be more productive and produce better quality products, one should stay abreast of new material and equipment developments and evaluate them logically. With designing, there is an extremely vast area for improving profitability by ensuring that the best available material and best processing method are used to meet specific design performance requirements. Fabricating plastic products is analogous to modern living: change continually occurs.

The aim of perfection requires constant change; however, recognize that perfection is never reached (chapter 30).

As reviewed at the start of this chapter, blow-molded plastic products have certain inherent characteristics, which in most cases are the principal reasons for their application. An important aspect concerns shape configurations, ranging from very simple to extremely complex, resulting in manufacturing economies that are now being used more frequently for cost reduction. To date, most of the innovation has taken place in commercial products, since those markets took immediate advantage of the performance/cost gains. The shapes that are produced to meet packaging, medical, auto, electronic, toy, leisure, and other markets include mostly complex shapes. They include integral handle lids, snap fits (chapters 19 and 20), and others. Hollow and structural industrial shapes are blow molded into such forms as automotive sun visors, chair seats, and fuel tanks.

Bottle Design

Beverage suppliers with requests for attractive, lightweight, and more hot-fill, single-serve bottles continue to focus on new blow-molded designs. Heat-set PET bottles typically weigh 5% to 10% more than other bottles, owing to extra material used for vacuum panels that are located on their sides. These panels prevent bottles from warping or crumpling from the vacuum created by hot-fill contents; but, these panels have limited PET bottle designs for familiar shapes. Hot-filling remains the prevalent bottling method in North America, so processors are forced to be creative in developing containers that meet performance requirements and offer appealing designs.

An example of this development is from Snapple Beverage Corp. They will replace the glass packaging it uses for its flagship products (ice teas and juice drinks) with 20-oz hot-fillable stretch blow-molded PET bottles from Graham Packaging, York, Pennsylvania. They are equal in appearances, with novel bell-shaped upper profiles, and offer greater appeal to consumers with plastics' lighter weights and resistance to breakage. The bottles feature Graham's Active Cage vacuum panel technology. An Active Cage panel can be smaller in size, thinner, and allows the frame of the bottle walls to move or flex while absorbing the hot-fill vacuum, as opposed to the traditional rigid frame that encases the panel. The Active Cage panel also tightens as the vacuum is absorbed, increasing bottle strength.

A newcomer to the plastic packaging industry is multinational water bottler Ty Nant, Bethania, Wales. It enlisted industrial designer Ross Lovegrove, renowned for Apple's Mac computer. He created what are believed to be the first asymmetric PET bottles having no visible ribs. Ribs for strengthening the bottle walls are concealed within the contours of the package. The bottles, which have ripple effects and are organic shaped, come in different sizes. They are blow molded at a new plant using six-cavity stretch injection blow machinery from ADS, Cergy-Pontoise, France (536).

Industrial Products

Industrial blow molding continues to take advantage of blowing complex parts in one piece, resulting in a number of benefits. Blow molding continues to challenge injection molding, thermoforming, and rotomolding. Industrial blow molding by the automotive industry has led everyone who works in or is involved in plastic development to exploit and expand its market opportunities. There are three basic design approaches in industrial blow molding: hollow, double-wall, and structural shapes. Today, they are used to market a variety of industrial as well as consumer parts. All these design methods offer protection, versatility, and economy for applications with which other processes have difficulty to compete.

Whether one discusses uneconomical or impractical approaches to tool up large structural parts in injection molding, whether we are designing products with limited volumes that need energy-absorbing or insulating properties, or whether we simply want to reduce part weight, the blow-molding process is an excellent choice to consider. Further, in comparison with metal forming or some other plastic processes, such as thermoforming, industrial blow molding offers a definite advantage through the elimination of secondary operations (consolidation of parts, modular approaches).

The development of double-wall structure concepts is directly attributable to the work of Peter T. Schurman. In December 1963, Schurman developed double-wall cases that required a substantial male mold member or "core" instead of just two flat cavities with incidental contours therein. Schurman explained that Illinois Tool Works molded, on a developmental basis, flat step treads for use on Hamilton Cosco kitchen stools. They actually were double-wall parts, about 13 × 8 × 1 in, with a level textured top surface and a ribbed underside. Joining two treads with an integral hinge created the first double-wall product, with the smooth side on the outside cavity and the ribbed side on the inside core. From this exciting beginning, the concept of double-wall cases has been used wherever there was a need for special products, such as those to store and protect high-value instruments and major tools.

In addition, the sales appeal of a permanent storage case developed the market opportunities that corrugated packaging could not offer. Double-wall cases are being used for chain saws, small power tools, monitors for heart pacers, computers, security chests, musical instruments, carrying cases (Fig. 6.115), and many other applications. Possibilities are almost unlimited. HDPE and propylene copolymers, with or without fillers, are the resins of choice, concurrent with new resin developments, such as advanced engineered thermoplastic alloys.

COMPLEX IRREGULAR SHAPE

With regard to development and use, the products now on the increase are those that have extremely complex 3-D shapes. They are targeted for all markets, particularly irregular hollow-shaped industrial parts. Often a single blow-molded part can replace an assembled part made of other materials, substantially reducing costs. The process adapts to both low- and high-volume production runs. The process is flexible as to part size and shape. Part sizes range from extremely small to very large, with the only restriction being the size of the machine available.

Other important characteristics that relate to the plastic used include low weight, thermal and electrical insulation, corrosion resistance, chemical resistance, permeability resistance, transparency, low coefficient of friction, color, aesthetics, and capability of monolayer and multilayer. In regard

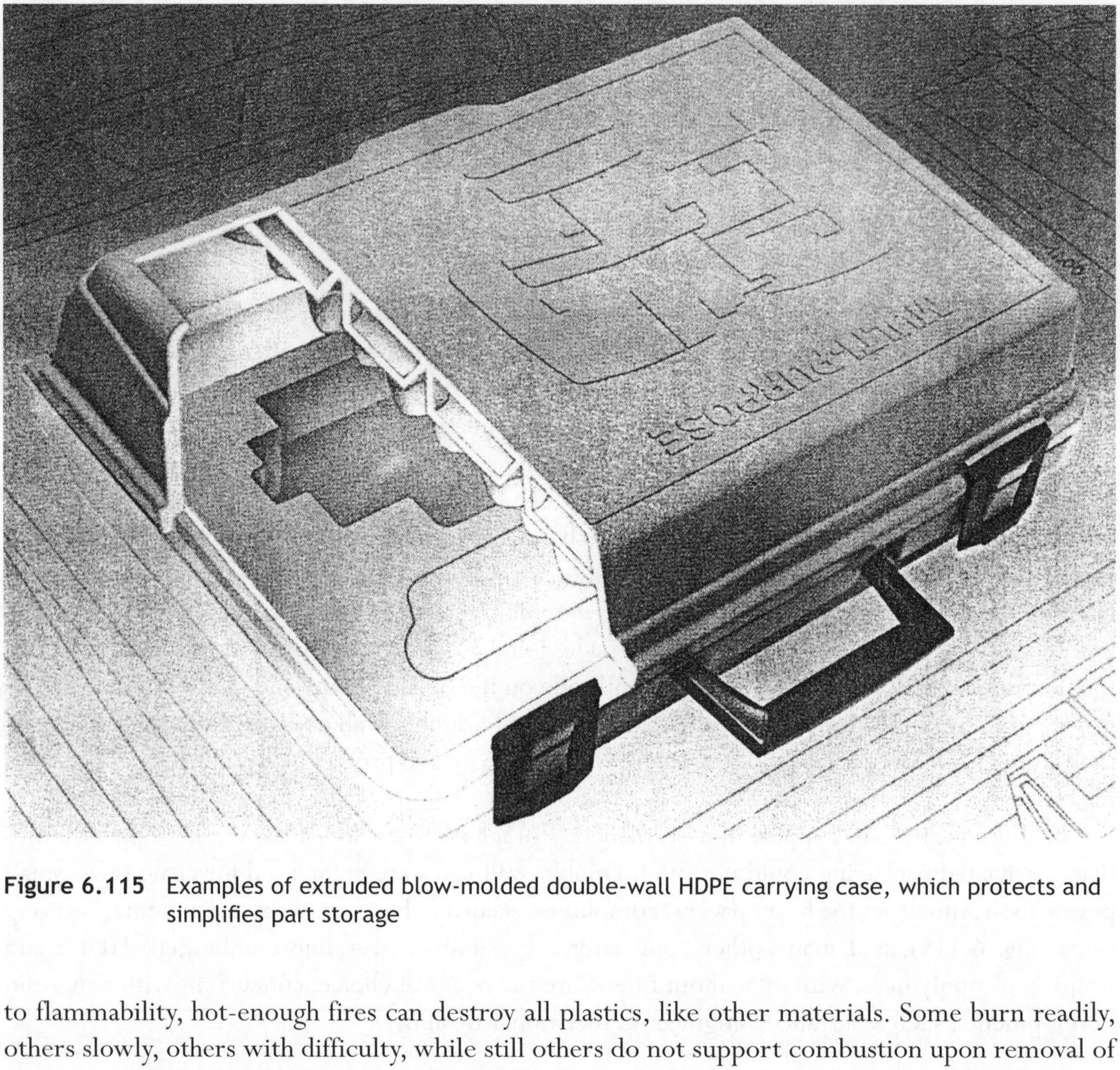

Figure 6.115 Examples of extruded blow-molded double-wall HDPE carrying case, which protects and simplifies part storage

to flammability, hot-enough fires can destroy all plastics, like other materials. Some burn readily, others slowly, others with difficulty, while still others do not support combustion upon removal of the flame (chapters 21 and 22).

Very complex irregularly shaped products have been extending state-of-the-art blow-molding technology, because conventional mold designs fail to produce acceptable wall thicknesses. Using mold halves with moving sections, some complex products are routinely and successfully molded. When the product geometry involves reverse folds or inwardly protruding channels and narrow projections, conventional molds cannot guarantee acceptable wall thicknesses. In plastic-starved areas, particularly corners, product walls are thinner than in other sections and are thus subject to rupture. The solution is in using an alternative to the conventional moving section.

Complex Mold Action Equipment has been developed and applied to moving mold sections to blow mold highly irregular product shapes (similar action is used in other molds, such as injection-molding molds). The significance of the technology is that it provides a means to BM complex products in one piece. Although it is fairly common to move mold sections to release undercuts, which permits product removal, molding techniques that affect or control wall thicknesses are used by relatively few blow molders. The ability of these new molders to blow mold uniquely shaped products has paid dividends by eliminating or reducing costly secondary operations.

ORIENTED 3-D PARISON

The 3-D extrusion blow-molding processes have been reviewed. In conventional extrusion blow molding, the parison enters the mold vertically rather in a straight tube. In 3-D BM, the parison is oriented in an open or closed mold. It is manipulated in the tool cavity, providing complex geometric products that can have uniform or nonuniform wall thicknesses, corrugated and noncorrugated sections, and so on. It provides a means to significantly reduce scrap/flash, waste, and quality (3).

Dr. Michael Thielen and Frank Schuller introduced this process some years ago (528 to 535). In the meantime, a certain number of different systems became established in the market using single and multilayer plastics with hard-soft-hard plastics. Systems permit different materials (single or coextrusion) to be sequentially extruded, alternating one after the other so that a parison with different sections of material in the extrusion direction can be produced. The processes include suction blow molding, 3-D blow molding with parison manipulation, and a split-mold horizontal machine with a vertically opening mold and a six-axis robot laying the parison into the cavity or a machine without a closing unit.

With 3-D, substantial savings are possible, which may be achieved by using blow-molding machines specifically designed for the production without (or at least with significantly reduced) pinch lines. In these processes, the extruded parisons (with diameters smaller than the articles' diameters) are deformed and manipulated then moved directly into the mold cavities so that the remaining lengths of the pinch lines are reduced to minimums. The 3-D manipulation of the parison with programmable manipulators or up to six-axis robots and special devices allows the pinchless production of complex articles.

The process has advantages that include lower clamping force required, less effort for deflashing, very little if any flash, no refinishing work on outer article diameter, improved quality of the article due to wall thickness distribution, and no reduction of strength due to pinch lines.

Molds are comparably expensive; however, the suction blow-molding process requires simple and inexpensive molds. Here, the parison is ejected into the closed mold and "sucked" through the mold via an airstream provided by a blower system. After exiting at the lower end of the mold, the parison is squeezed off by closing devices and the inflation and cooling process can follow.

A simple and inexpensive mold can be installed on a horizontal 3-D machine. This mold is installed in a vertically opening clamp. The lower mold half slides out under the head, where a manipulating robot places the parison into the cavity. After laydown, the mold half slides back under the top mold half. The clamp closes and the inflation and cooling processes follow.

Major problems involved in high-strength/strong-curved conventional blow moldings can produce irregular wall thicknesses at inner and outer radii, a result of the different stretch ratios.

This was solved with a radial wall-thickness distribution system (RWDS). It permits a balancing out of the wall thicknesses between the inner and outer radii of the curved sections, even with parisons that have small diameters. To guarantee a regular wall-thickness profile of the final product, the wall thickness is adjustable not only concerning the length of the parison but also as to its circumference.

OTHER DESIGN APPROACHES

In addition to products, processes, and design approaches reviewed in this book, there are many more. Figures 6.116 and 6.117 provide examples of BM-stacked (two or more) products or parts during a single BM operation followed with manual or automatic cutting action to separate the parts. They can include snap-fit, in-mold labeling, and others.

During extrusion blow molding, molds can be moved and flash removal at the bottleneck using the blow pin is quick and efficient (Fig. 6.118).

To reinforce an injection or extrusion blow-molded container, opening plastic inserts that are injection molded are used (Fig. 6.119).

The shape of an extruded parison can be used to provide for a living hinge (chapter 20) (Fig. 6.120).

There are bellows-collapsible bottles produced in conventional extrusion blow-molding equipment (Patent #4,492,313). These foldable bottles provide advantages and conveniences such as (a) reducing storage, transportation, and disposal space, (b) prolonging product freshness by reducing oxidation and loss of carbon dioxide when contents are removed, and (c) providing continuous surface access to foods, such as mayonnaise and jams (Fig. 6.121) (3).

SUMMARY

The plastics industry is one of the most important business sectors providing significant contributions to the economy and standard of living across all sectors worldwide. In this industry, blow molding is the third major process following extrusion and injection molding. Plastic blow-molded products are used in many different markets, including foodstuff, beverage, packaging, medical, toy and game, recreation, transportation, appliance, marine, agriculture, water filter, missile and rocket, U.S. postal service, toilet and water conservation, and others.

HISTORY

Extensive information is available concerning the history of BM worldwide. Only a few are presented here. The first patent for extruded polymer parison molding was U.S. Patent #237168, filed May 28, 1880, and issued February 1, 1881. The patent was issued to Celluloid Novelty Co. and Celluloid Manufacturing Co., New York. For over the past century, there have been extensive technological and new product developments, with many more developments and innovations occurring now.

The major commercial development in BM started with the commercial availability of low-density polyethylene (LDPE) during World War II. In the 1940s, Monsanto blow molded an LDPE squeeze bottle. The industry started its growth with the liquid bleach and detergent market. This market was rapidly converted and replaced glass, which is bulky and dangerous to health and

Shuttle Machine Limitations

- The key limitation of shuttle machines is that only one or two molds are available to deliver the desired output.
- To raise the output of the machine, the number of parisons and mold cavities must be raised.
- The axial mold shuttle distance increases greatly.
- Parison adjustment and control become more difficult, especially in multi-layer applications.
- The number of blow pins, knives, trimming stations and IML baskets also increases proportionally.

Overcoming Shuttle Machine Limitations

- One possible approach is to "stack" mold cavities vertically, forming bottles in a tail-to-tail approach.
- This would require both top and bottom blow pin calibration and more complex trimming and handling mechanisms.
- Extruding and capturing the longer parison accurately is more difficult.

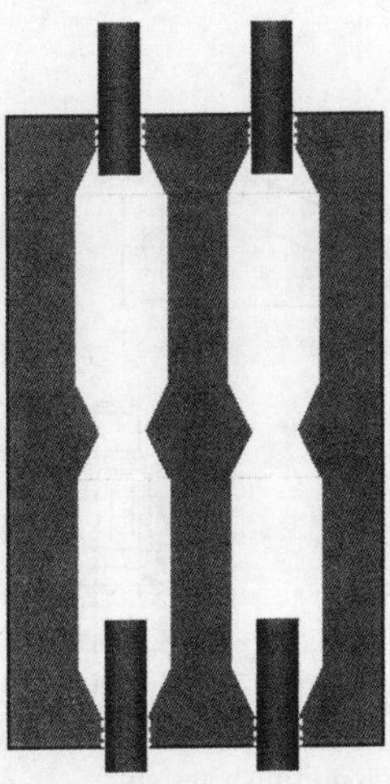

Figure 6.116 A shuttle EBM machine limitation and solution (courtesy of Graham Plastics Group)

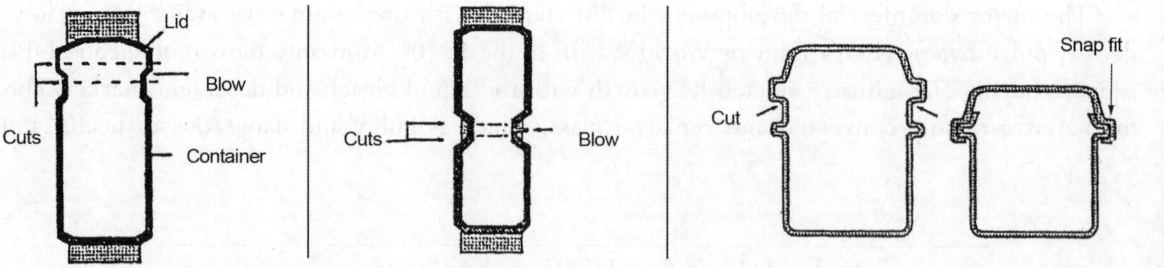

Figure 6.117 Views of multiple action extrusion blow-molding containers

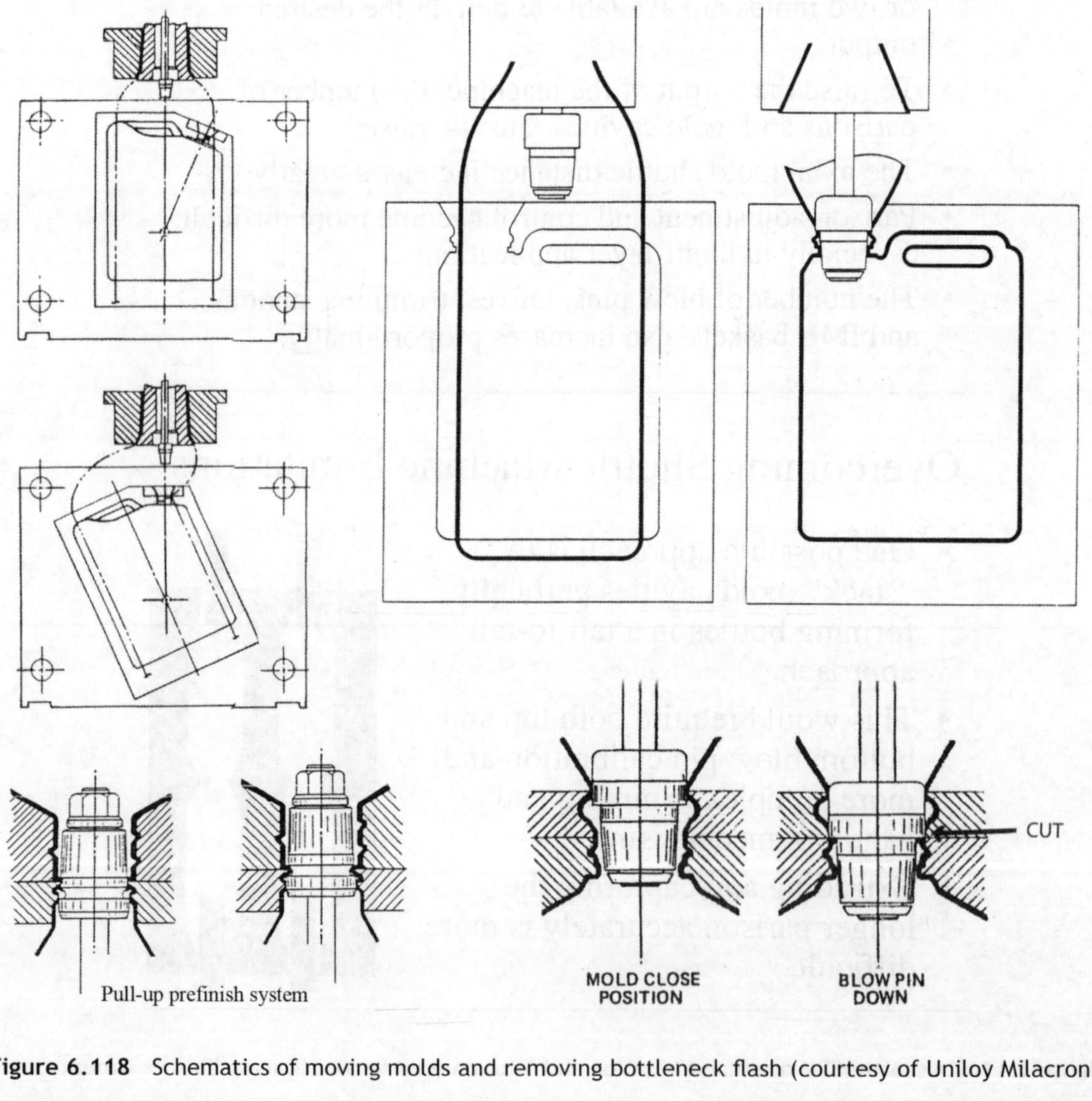

Figure 6.118 Schematics of moving molds and removing bottleneck flash (courtesy of Uniloy Milacron)

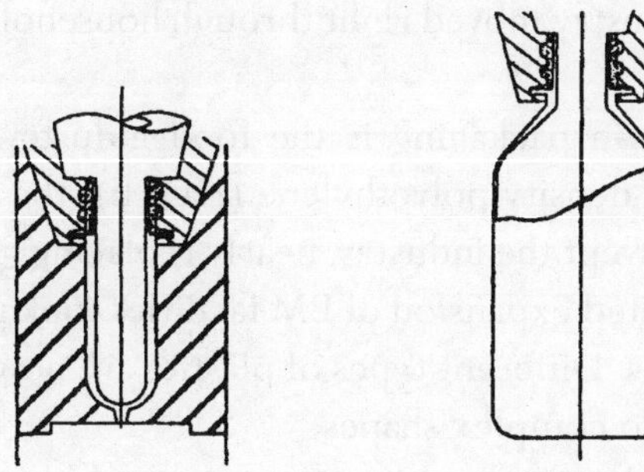

Figure 6.119 Example of inserting a plastic injection-molded reinforcement into a blow mold

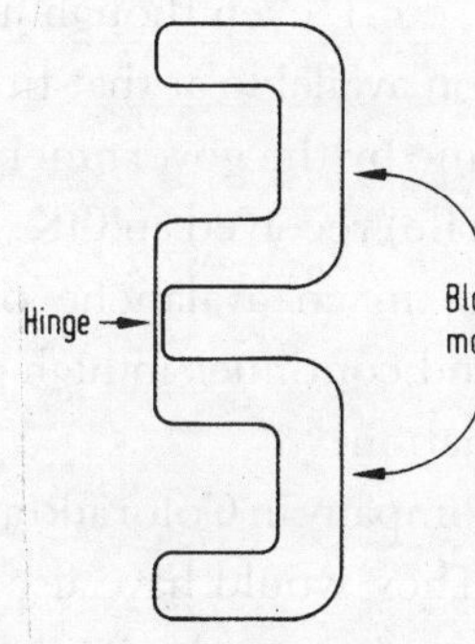

Figure 6.120 Living hinge is part of the extruded blow-molding parison

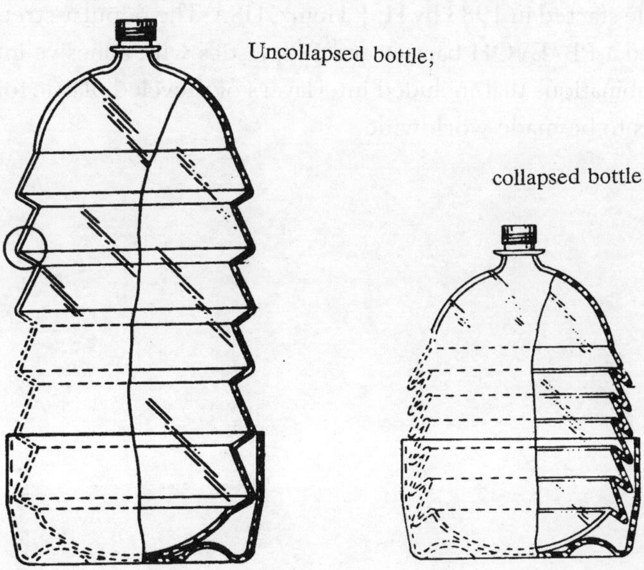

Figure 6.121 Collapsible bottle capable of 85% size reduction or 75% volume reduction

property due to breakage. The industry moved right through household products, such as shampoos and liquid soaps.

The largest potential market in packaging is the food industry, and in the late 1950s with commercial development of high-density polyethylene (HDPE), the milk market emerged with a plastic gallon (3.8-liter) jug that swept the industry, nearly replacing glass and plastic-treated paper. Early in the 1960s, an unprecedented expansion of BM facilities took place accompanied by general public acceptance of plastic bottles. Different types of plastics are now used to produce various BM products that range from simple to complex shapes.

During 1958, the Coca-Cola bottle was injection stretched blow molded using acrylonitrile-styrene (AN) plastic. Unfortunately, after production started in about eight plants with efficient recycling facilities on the East Coast, AN was not permitted to be used by the FDA because of possible food contamination (carcinogenic, etc.), even though its permeability requirements really could not be determined by instrumentation available at that time.

After decades of wasting money and time by the government (taxpayers) and the cost and time used by the plastics industry, AN (acrylonitrile) received an OK. In the meantime, that development was the forerunner in using PET plastics in an avalanche of bottles to appear commercially worldwide that started during the 1970s and continues in high gear. Originally, rumors had it that "competitors" fed FDA the "wrong" information.

During the mid-1950s, Coor's (beer) Company in Colorado almost went into using commercially stretched injection blow-molded bottles. They would have used the injection blow molding with rotation process. Unfortunately, they were using acrylonitrile-styrene plastic (from Barex plastic of Sohio of BP Chemical International; D.V.R. IMM equipment project). As just reviewed with the Coca-Cola bottle, AN could not be used per the FDA.

The ketchup bottle started in 1983 by H. J. Heinz, USA. The popular stretched squeezable coextruded BM bottle used a PP/EVOH barrier and PP plastics with adhesive interlayers. Later, PET/EVOH and other combinations that included interlayers of recycled plastic followed.

History continues to be made worldwide.

CHAPTER 7

THERMOFORMING

INTRODUCTION

Thermoforming is a process for converting thermoplastics into shell forms, using sheet (or film) material as a preform. Many different methods of thermoforming are used. Figure 7.1 provides an introduction to the thermoforming methods: (a) vacuum forming into male mold, (b) drape forming into male mold, (c) blow-back molding, (d) plug-assisted vacuum forming, (e) plug and ring vacuum forming, (f) snap-back vacuum forming, and (g) pressure snap-back forming.

With the exception of a few, such as matched mold and twin-sheet thermoforming, the forming process uses an open mold that defines only one surface of the thermoformed part. The second surface is only indirectly defined by the mold. This second surface will lack precision definition of features to an extent, dependent partly on the sheet thickness and thickness tolerance as well as the uniformity in heat subjected to the sheet prior to forming (1, 13, 86, 153, 221, 333, 350, 356, 538, 544–547).

There is no direct control over wall thickness of the formed part; this will vary from feature to feature according to the degree of stretch and thinning experienced at that point. Normally, a goal in thermoforming is to obtain as even a wall thickness as possible in the finished part. Because the process uses a sheet preform and a single-surface mold, it is not possible to create independent features on the second surface.

These processing considerations confine most thermoformed parts to relatively simple shapes; however, there are complex 3-D parts formed. As an introduction to thermoforming, clamping a heated thermoplastic sheet over a mold cavity and drawing a vacuum in the cavity performs it. This causes atmospheric air pressure to press and deform the sheet into the mold cavity, where it is cooled to retain the formed shape. This process, one of many, is known as vacuum forming.

In practice, thermoforming processes all have two or three optional forms (Table 7.1). As an example, the forming force can be supplied by vacuum, by positive air pressure, or by a power

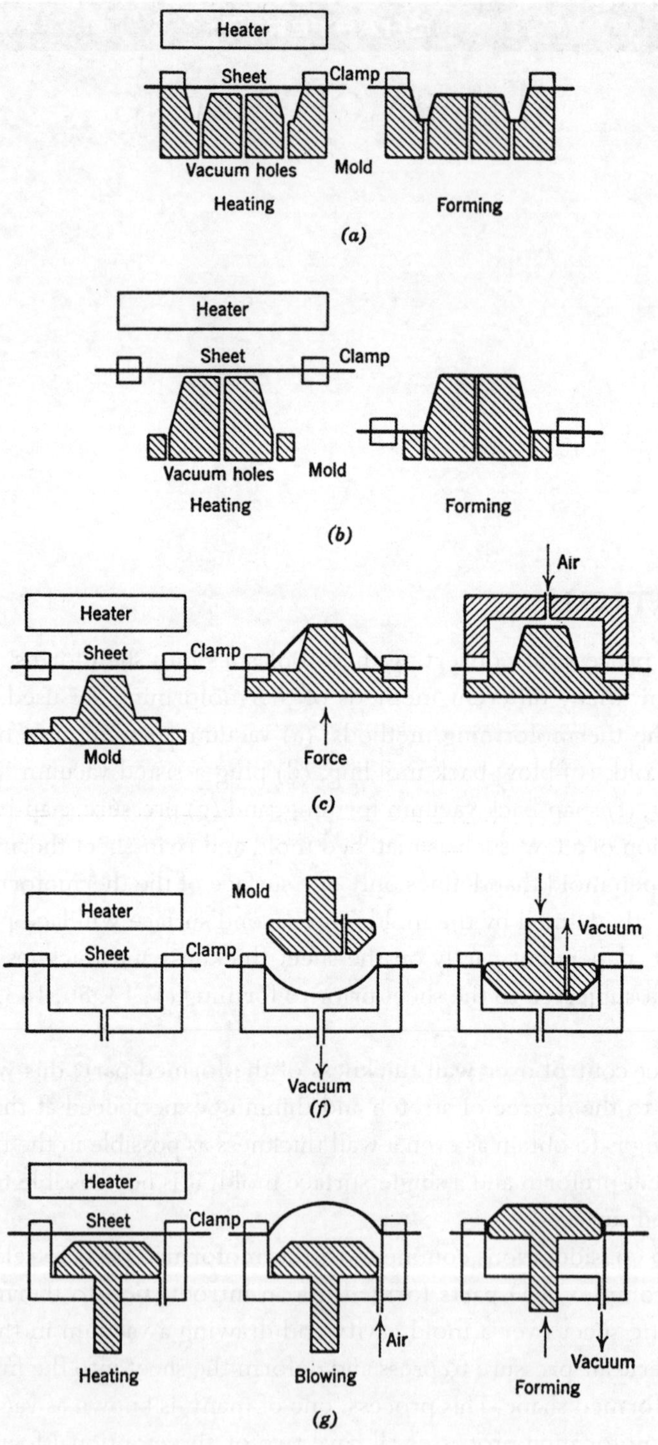

Figure 7.1 Examples of thermoforming methods

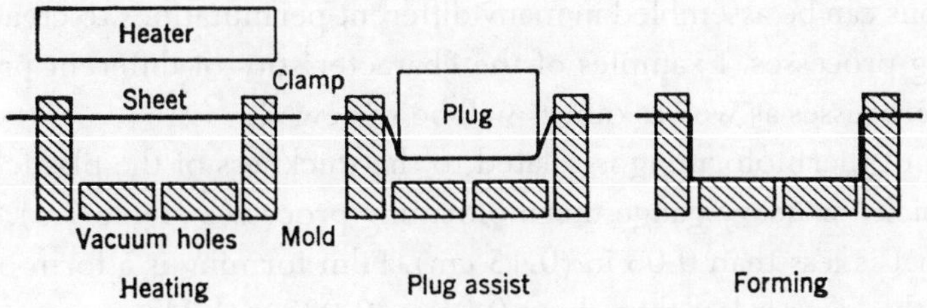

(d)

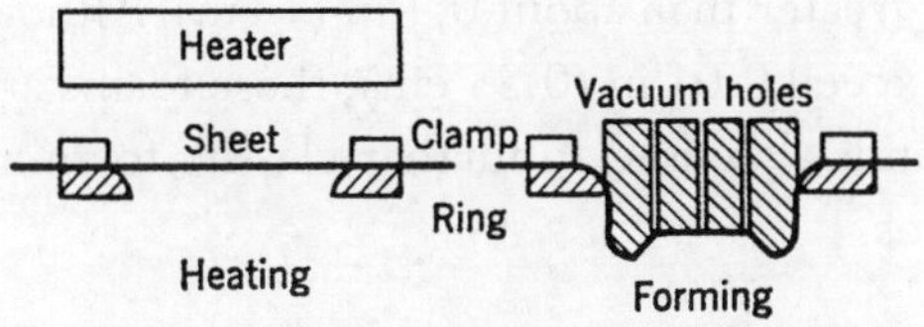

(e)

Figure 7.1 *(continued)*

Process Factor	Options
Forming force	Vacuum Positive air pressure Power press
Mold type	Female Male Matched female/male
Sheet pre-stretch	Vacuum Positive air pressure (billow) Mechanical plug
Material input	Extruder (in-line, hot forming) Reel (reheat, cold forming) Cut sheet (reheat, cold forming)
Process phase	Solid phase Melt phase
Heating mode	One side of sheet Both sides of sheet (sandwich)
Heating means	Radiation: rod, ceramic, quartz or infra-red heaters Convection: hot air oven Conduction: hot roller, contact panel or hot oil bath

Table 7.1 Options available in thermoforming processes

press. These options can be assembled in many different permutations to create a very wide variety of thermoforming processes. Examples of the characteristics of different processes are shown in Table 7.2. These processes as well as others will be reviewed.

Identification of thermoforming is related to the thickness of the plastic processed. There are thin-gauge and thick- or heavy-gauge thermoforming processes (Table 7.3). Thin-gauge identifies sheet thickness that is less than 0.06 in (0.15 cm). Film forming is a form of thin-gauge forming where the plastic thickness is less than about 0.01 in (0.025 cm). Heavy-gauge means that the sheet thickness is greater than about 0.1 in (0.25 cm). There is also plate forming or heavy-gauge forming where the sheet thickness is greater than about 0.4 in (1 cm). Although polystyrene and polyolefin foam sheet thicknesses can exceed 0.01 in (0.25 cm), these foams are usually treated as thin-gauge sheet stock. This classification is also used to further classify forming by machinery type, product type, and processing problems.

PROCESS

Basically, to form thermoplastic sheets or films, they must be heated to the drawing temperature just prior to and during the drawing cycle. The material can be heated on the mold plate or preheated on a hot plate, oven, or heated tunnel. Combinations of preheating with mold heating have their advantages, particularly in production runs. The material when formed in the mold or die is held in position by some mechanical device, such as clamps or pressurized hold-down plates. During forming, the heated flexible/rubbery sheet is stretched against a rigid surface mold. After being drawn, the sheet material must be cooled to harden. Frequently, chill boxes, cold plates, or cool-air systems are included in the forming equipment.

Typical steps are clamping, heating, shaping, cooling, and trimming. It is classified as a low-pressure, relatively low-temperature, nearly solid-phase forming process, with the sheet being

Process	Forming force	Mold type	Sheet pre-stretch
Basic vacuum	Vacuum	Female	None
Basic pressure	Positive air pressure	Female	None
Drape	Vacuum	Male	None
Snap back	Vacuum	Male	Vacuum
Billow	Vacuum	Male	Positive air pressure
Plug assist	Vacuum	Female	Mechanical plug
Billow plug assist	Vacuum	Female	Positive air pressure and mechanical plug
Air slip	Vacuum	Male	Positive air pressure
Air slip plug assist	Vacuum	Male	Positive air pressure and mechanical plug
Matched mold	Power press	Matched male/female	None
Twin sheet	Positive air pressure	Matched female/female	None

Table 7.2 Introduction to some of the thermoforming processes

Characteristic	Thin-Gauge	Heavy-Gauge
Initial sheet thickness	<0.060 in or 1.5 mm	> 0.120 in or 3 mm
Dominant products	Packaging, disposables	Cabinetry, industrial
Sheet handling characteristics	Roll-fed	Palletized cut sheet
Machine type	In-line extruder-former	Shuttle or rotary press
Machine control aspects	Automated	Semi-automated to manual
Controlling aspect about heating	Heater output, W/in²	Conduction into sheet
Pattern heating	Difficult, not usually done	Common
Part size tendency	Small	Medium to very large
Number of mold cavities	Many	One or two, usually
Mechanical assist	Plug	Plug, billow, vacuum or draw box
Mold type	Female, usually	Male, female, androgynous
Mold materials	Aluminum, machined	Wood, plaster, syntactic foam, white metal, cast aluminum
Mold cooling	Active, controlled	Active to none for prototype
Free surface cooling	Ambient, usually	Forced air, water mist or fog
Trimming aspects	Punch and die, matched die, rim rolling	Multi-axis routering
Non-product trim level	Nom. 50%	Nom. 25%–30%
Wall thickness tolerance, normal	20%	20%
Wall thickness tolerance, high control	10%	10%
Pressure forming application	Deep draw, foam	Textured surfaces, deep draw

Table 7.3 Thin-gauge and thick-gauge thermoforming materials

heated only a few degrees above its glass-transition temperature (T_g) or melt temperature (T_m) (chapter 1). This process is basically a unique process for producing parts that have high surface area-to-thickness ratios. It is limited by high residual stress in parts, high levels of trim, high energy costs, and extensive trim requiring regrinding and resulting in nonuniform wall thickness.

The temperature used to form sheets varies with material type, thickness, size, and depth of draw. Other important factors include technique to be used and speed of operation. The most efficient temperature for a specific package is generally determined by a combination of drawing temperature previously experienced and/or experimenting. Too high a temperature may cause sags, heat marks, or tearing. With too low a temperature, wrinkles and fractures can occur.

Vacuum forming methods use atmospheric pressure of air to force a heated sheet into a form. In the free blowing method, the shaped plastic does not touch any solid surface. In the snap-back method, after the plastic is expanded, a mold formed to the desired contour is inserted. In turn, the vacuum is released so that the plastic contracts and snaps back on the mold. This method permits forming reverse curves and difficult contours that can not be formed by other standard methods. In the vacuum-mold method, the plastic expands against a mold surface. The mold surface contains small holes through which air can pass.

One of the oldest methods for forming sheet plastics consists of dropping a hot pliable sheet over a form. This dropping and hot bending method permits forming 2-D shapes over inexpensively constructed molds. In stretch forming, the hot sheet can be stretched and simultaneously put over a mold. This method can produce both simple and compound curvatures.

Cold bending of rigid thermoplastic sheets continues to be important since it is one of the simplest and fastest forming methods. The sheets can be easily fastened to other materials, such as metals or cardboard, by means of plastic fasteners or adhesives. Cold bending has limitations based on the type of material to be used and degree of bending radii. Minimum cold bending radii for different plastics versus sheet thickness are developed.

There appears to be more room for developing new methods of fabricating plastic sheets. The possibility for continual development of this technique exists because new plastic materials and equipment continue to be produced and, most importantly, new consumer products are being packaged.

Growth

Leading the growth was the expansion of twin-sheet and pressure-formed plastic products. Fueled mostly by advances in mold technology, material developments, and thermoforming machinery capabilities, technology improvements in the form of machine controls have led to machine designs that are faster and more consistent than was previously possible. As an example, advanced materials and machinery enable manufacturers to significantly expand their culinary market share by producing high-performance containers for fresh food.

Control improvements have provided more consistent heating capabilities. Some heater manufacturers have developed or refined their manufacturing processes, while others have developed new heating systems, such as gas catalytic panel heating systems, to provide the industry with more effective heaters and more heater-to-heater temperature consistency. This type of technology permits innovative mechanical designs, such as adding a third forming station to a double-ended thermoformer and a six-station rotary thermoformer to meet customer cost/performance requirements.

Other technology improvements, such as position control for electric platens, as well as the speed capabilities of electric drive systems for platen-drive systems, sheet-wheel rotate systems, and sheet-car transfer systems, have provided faster and more consistent machinery for fabricators to operate.

Use is made of the latest in computer design technology to ensure both the electrical and mechanical design integrity of equipment and to upgrade the systems provided for customers. This includes a full array of services for cut-sheet and roll-fed customers, including new equipment, complete rebuilding, and upgrade kits that use a variety of proven modern technologies. The result is providing all fabricators with effective and efficient thermoforming solutions.

Product

Durable products are made in many sizes and various shapes. The various forming techniques permit manufacture of products individually or on mass production lines via continuous belt–type machines that are used in many different markets. Products include machinery and tool housings, industrial pallets, boat hulls, computer housings, transportation components (auto [550], bus, aircraft, etc.), and refrigerator door liners. Typical products are high-production items, such as plates, cups, lids, trays, and thin-walled containers. Figures 7.2 to 7.8 illustrate just a few thermoformed products.

OPERATING BASICS

The principal factors in the thermoforming process are the forming force, heating the plastic, mold type, method of prestretching the sheet, the material input form, and the process phase

Figure 7.2 Thermoformed TPO front bumper fascia for a Colombian-built Renault car (551)

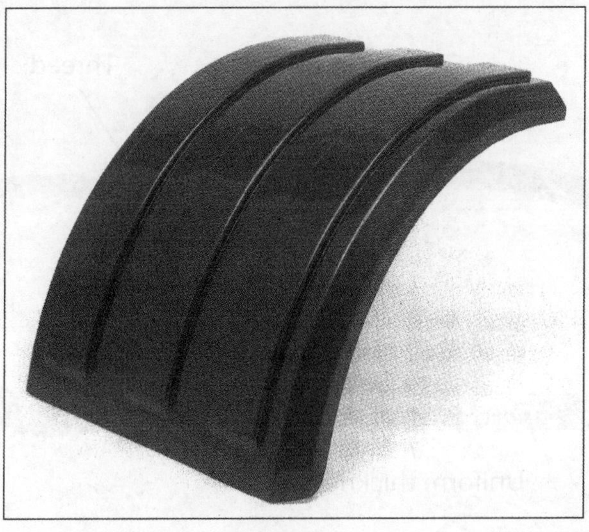

Figure 7.3 Thermoformed TPO truck fender (551)

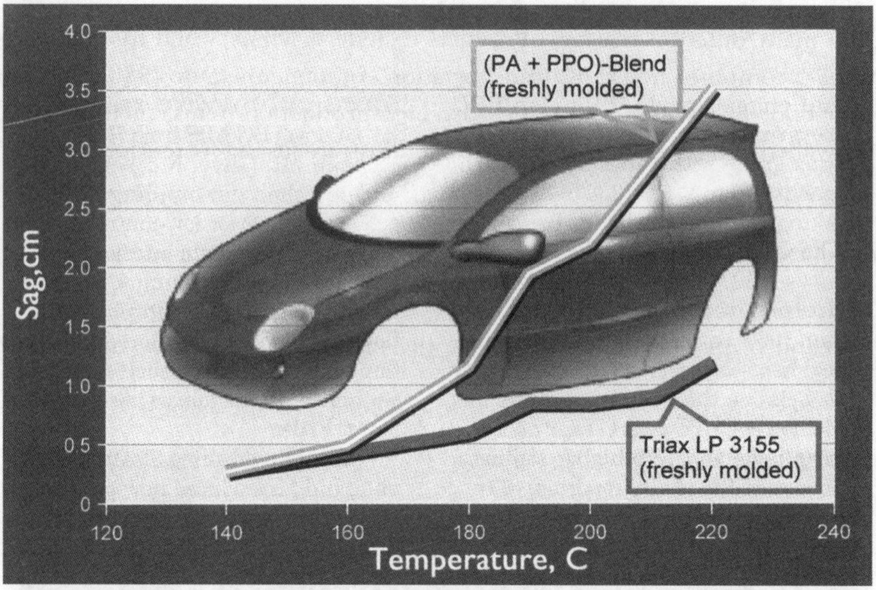

Figure 7.4 Thermoformed Bayer's Triax nylon/ABS auto panel heat sag test results (552)

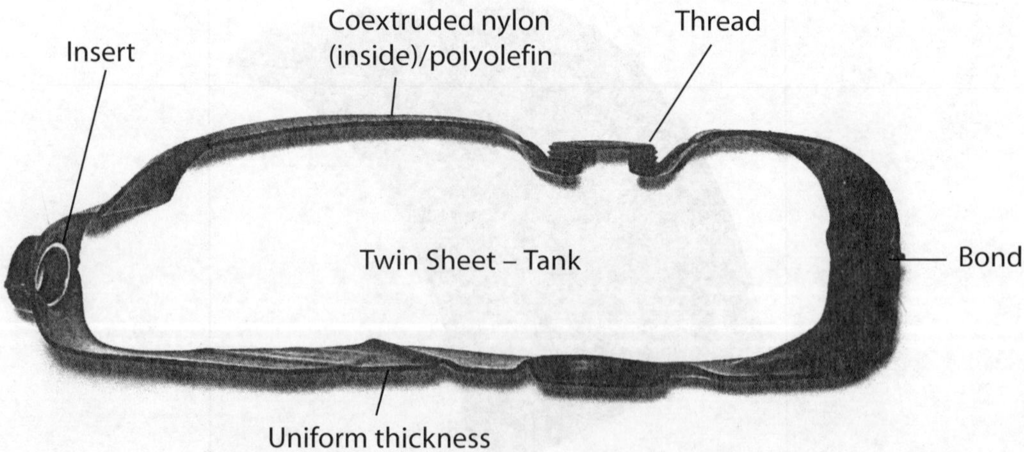

Figure 7.5 Thermoformed automotive gasoline tank

Figure 7.6 Thermoformed electronic printer housings

Figure 7.7 Thermoformed polystyrene foam food container

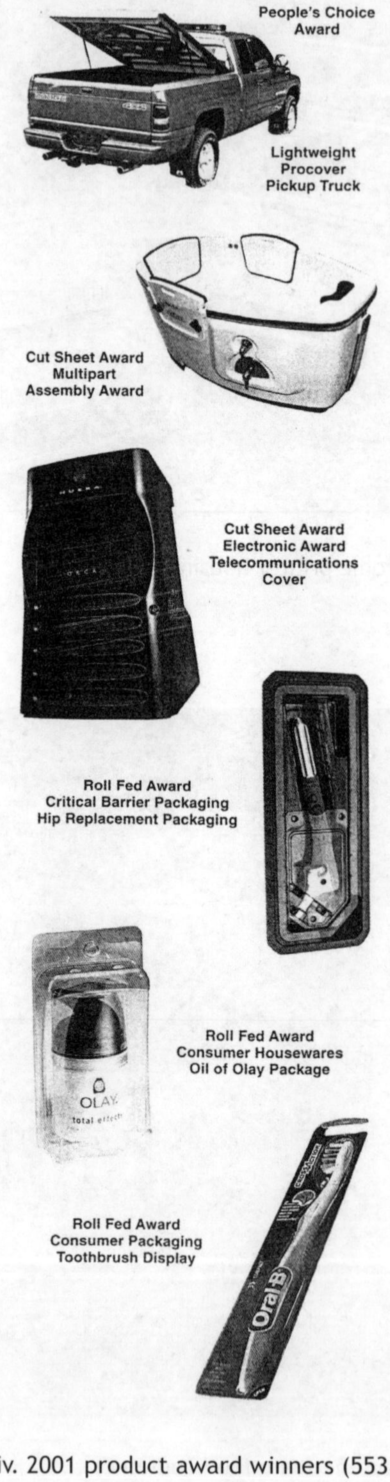

Figure 7.8 SPE Thermoformed Div. 2001 product award winners (553)

condition (554). These factors are critical to the appearance, quality, and performance of the thermoformed part.

Forming Pressure

Thermoforming is essentially a low-pressure process in which the forming force is generated either by vacuum, positive air pressure, or a combination of both. However, in the special case of matched mold thermoforming, the force is supplied by a power press and is usually higher. The usual methods are identified as follows:

1. Pneumatic differentials in air pressure are created by pulling a vacuum between the heated plastic and the mold.
2. Pneumatic pressure created by compressed air is used to force the hot plastic against a mold surface.
3. Mechanical means are used, such as a plug, matched (male and female) mold cavities, and so on.
4. A combination of vacuum and/or pressure (pneumatic) and mechanical means is used.

The process involves pressure with heat and can be summarized as:

1. Heat the plastic in a separate oven, then transfer the heated sheet to a forming press.
2. Use automatic equipment to combine heating and forming in a single unit, or use a continuous operation feeding the plastic from a roll or directly from an extruder line where the sheet retains some of its heat.
3. Evaluate the use of a mold where some heat is applied to the mold; this approach can produce parts with no or relatively very little frozen internal stress after cooling the formed product.

In vacuum forming, the heated sheet forms part of a closed cell together with the mold. When the air in this cell is evacuated, atmospheric pressure forces the sheet into contact with the mold. The force so generated clearly cannot exceed atmospheric pressure and in practice is rather less because of the difficulty of creating a perfect vacuum in the mold/sheet cell. A vacuum of 85% to 90% of theoretical maximum is more likely, and this will generate a forming force not greater than about 90 kPa (13 psi).

Pressures, particularly in the low pressure range used in thermoforming, may be expressed not only in a wide variety of units but also by two systems of measurement, namely gauge pressure and absolute pressure. Gauge pressure takes atmospheric pressure as the datum or zero point, while zero pressure is the datum for absolute pressure measurements. In other words, absolute pressure values exceed gauge pressure values by a quantity equal to atmospheric pressure. The distinction is sometimes made clear by adding a suffix to the unit of pressure; for example, psia and psig refer respectively to absolute and gauge pressures expressed in pound force per square inch. Table 7.4 compares absolute and gauge values for some common units of pressure.

Pressure forming overcomes the force limitations of vacuum forming by applying air under pressure on the nonmold side of the sheet in place of a vacuum on the mold side. The sheet is forced into contact with the mold at pressures greater than atmospheric, and this necessitates venting any air that is trapped between the sheet and the mold. Venting may be provided by simple passages connecting to the atmosphere but is often improved by using a vacuum on the mold side of the sheet. In this case, the venting vacuum increases the forming force by an increment approaching 1 atmosphere.

Industrial compressed-air supply systems normally operate at about 550 kPa to 710 kPa (80 psi to 100 psi), and this pressure may be sufficient for many forming applications. However, pressure thermoforming equipment usually operates at pressures up to about 2,500 kPa (360 psi), with some processes operating up to at least 4 MPa (580 psi). These pressures are very low when compared to those commonly used for injection molding or extrusion (chapters 4 and 5).

The forming force in matched mold forming is supplied by a power press and so, in principle, is almost unlimited as performed in compression molding (chapter 14). In practice, the force is limited by the design of mold construction as well as the needs of the process and is usually in the range of 1.5 MPa to 4 MPa (218 psi to 580 psi).

CONTROLLING PRESSURE

When using vacuum pressure, the target is not to operate below 20 in of mercury (Hg), which at sea level is 68 kPa (9.82 psi) of atmospheric pressure on the formed plastic. What is desired is at least 25 in Hg (85 kPa). In Table 7.5, the gauge pressure is the amount of pressure exceeding atmospheric pressure. In a vacuum system, absolute pressure (psia) is equal to the negative gauge pressure (psig) subtracted from the atmospheric pressure.

When prestretching the hot sheet before forming on a mold, usually 3 to 5 psi compressed air is used, resulting in a greater amount of air being at atmospheric pressure than in the processing of nonstretched parts. Two important requirements in the forming cycle are to sustain the pressure and to maintain uniform heat in the plastic. Faster air evacuation produces higher-quality products.

It is important to have the correct mold heating so the fast vacuum will produce a part with very little or no internal stress. As a thermoplastic cools, this pressure may not provide sufficient force to form the part and will not hold the plastic tight against the mold. This difference has to set up an acceptable tolerance that is incorporated in the mold cavity shape. Matched mold forming can eliminate or significantly reduce any existing tolerance problem.

For proper pressure regulation, a vacuum storage or surge tank is necessary to retain a minimally even vacuum. To determine the vacuum surge tank size in cubic feet, refer to Table 7.6 (544–547).

The objective is to have the vacuum surge tank as close as possible to the forming station and the vacuum control valve. A flexible vacuum hose with connections eliminates elbows, tees, and

Atmospheres gauge	Bar absolute	Bar gauge	kPa absolute	kPa gauge	kgf/cm² absolute	kgf/cm² gauge	Psi absolute	Psi gauge
-1.00	0.00	-1.01	0.00	-101.33	0.00	-1.03	0.00	-14.70
-0.90	0.10	-0.91	10.13	-91.19	0.10	-0.93	1.47	-13.23
-0.85	0.15	-0.86	15.20	-86.13	0.15	-0.88	2.20	-12.49
-0.50	0.51	-0.51	50.66	-50.66	0.52	-0.52	7.35	-7.35
0.00	1.01	0.00	101.33	0.00	1.03	0.00	14.70	0.00
0.50	1.52	0.51	151.99	50.66	1.55	0.52	22.04	7.35
1.00	2.03	1.01	202.65	101.33	2.07	1.03	29.39	14.70
1.50	2.53	1.52	253.31	151.99	2.58	1.55	36.74	22.04
2.00	3.04	2.03	303.98	202.65	3.10	2.07	44.09	29.39
3.00	4.05	3.04	405.30	303.98	4.13	3.10	58.78	44.09
4.00	5.07	4.05	506.63	405.30	5.17	4.13	73.48	58.78
5.00	6.08	5.07	607.95	506.63	6.20	5.17	88.18	73.48
6.00	7.09	6.08	709.28	607.95	7.23	6.20	102.87	88.18
7.00	8.11	7.09	810.60	709.28	8.27	7.23	117.57	102.87
8.00	9.12	8.11	911.93	810.60	9.30	8.27	132.26	117.57
9.00	10.13	9.12	1013.25	911.93	10.33	9.30	146.96	132.26
10.00	11.15	10.13	1114.58	1013.25	11.37	10.33	161.66	146.96
15.00	16.21	15.20	1621.21	1519.88	16.53	15.50	235.14	220.44
20.00	21.28	20.27	2127.83	2026.51	21.70	20.66	308.62	293.92
25.00	26.34	25.33	2634.46	2533.13	26.86	25.83	382.10	367.40
30.00	31.41	30.40	3141.09	3039.76	32.03	31.00	455.58	440.88
35.00	36.48	35.46	3647.71	3546.39	37.20	36.16	529.06	514.36
40.00	41.54	40.53	4154.34	4053.02	42.36	41.33	602.54	587.84

Table 7.4 Comparison of pressure scales for thermoforming

Pressure (psi)		
Gauge	Absolute	Pressure (in. Hg)
0.0	14.7	0.0
−1.0	13.7	2.04
−2.0	12.7	4.07
−4.0	10.7	8.14
−6.0	8.7	12.20
−8.0	6.7	16.30
−9.9	4.9	20.00
−10.0	4.7	20.36
−11.0	3.7	22.40
−12.0	2.7	24.43
−12.3	2.4	25.00
−13.0	1.7	26.47
−13.7	1.0	27.89
−14.0	0.7	28.50
−14.2	0.5	28.91
−14.6	0.1	29.73
−14.7	0.0	29.92

Table 7.5 Pressure measurements comparing gauge, absolute, and inches of mercury

tubing reducers. The valves must be capable of operation at their fully open position. To capitalize on the rapid vacuum capability of the surge tank, the mold should be able to exploit all the available vacuum pressure. Vacuum holes should be drilled as large as possible, and a maximum number should be used in the molds.

Among the techniques used to aid the vacuum action is to back drill large holes (to 0.125 in [3.18 mm]) on the underside when smaller holes are required on the part side. Male molds can be mounted on a vacuum plate with thin washers or shims, and large vacuum holes can be drilled under the mold. Narrow slots can also be used, and they offer much less resistance than holes when air is evacuated through the mold. Flat areas, segmented sections, or male portions of a mold can be joined with shims that will provide long slots (chapter 17).

The compressed air requirements for controls and restrictions are similar to those of vacuum, such as an adequate accumulator tank positioned as close to the mold as possible. Air is required to be very dry, such as having a dew point of –40°F. The air is also required to be oil free to eliminate contamination of the forming processes. Excellent filter systems are available from equipment suppliers.

MOLD CONSTRUCTION

Thermoforming is a shell process performed in an open mold that defines only one surface of the product. Definition of the second surface is indirect and not precise as in injection molding. This has

$V_o \times P_o + V_m \times P_m = V_1 \times P_1$

where: V_o = surge tank volume, including piping to vacuum control valve

V_m = volume of the molding area

$V_1 = V_o + V_m$

P_o = absolute pressure in surge tank [0.5 psi (3.4 kPa)]

P_m = initial pressure in the mold at sea level 14.7psi (101 kPa); with prestretched forming use 17.7 psi (122 kPa)

P_1 = desired atmospheric working pressure

In an example where the volume of the mold and piping is 4 ft^3 (0.11 m^3), the vacuum pump can pull about 29 in. Hg (98 kPa), so the surge tank pressure is 0.5 psi (3 kPa).

The desired working pressure is 2.42 psi (16.7 kPa) in the tank, and the initial mold pressure is 14.7psi (101 kPa).

$$V_o \times 0.5 + 4 \times 14.7 = (V_o + 4) \times 2.42$$

$$0.5\ V_o + 58.8 = 2.42\ V_o + 9.68$$

$$V_o = 25.58\ ft^3\ (191\ gal\ or\ 723\ dm^3)$$

Suppose a lower pressure of 20 in. Hg (68 kPa) is used, which is 4.88 psi (33.6 kPa) in the tank.

$$V_o \times 0.5 + 4 \times 14.7 = (V_o + 4) \times 4.88$$

$$0.5\ V_o + 58.8 = 4.88\ V_o + 19.52$$

$$V_o = 8.97\ ft^3\ (671\ gal\ or\ 2540\ dm^3)$$

Table 7.6 Formula to determine the vacuum surge tank size in cubic feet

a direct bearing on the mold that must be designed so that it directly defines the more important of the product's two surfaces.

Product geometries are limited to forms attainable by deformed sheets and consist mainly of recesses and projections orthogonal to and arranged on either side of the plane of the sheet preforms. This makes it convenient and reasonably accurate to classify thermoforming molds as having female or male cavities. The female mold consists of a recessed cavity or cavities formed in a mold block. This is the most common mold form, because it is easy to clamp the heated sheet across the unobstructed mold face. The resulting product has a clearly defined male form and not a precise female form.

Often the simplicity of the female mold outweighs the cosmetic requirements of the product; thus, it is common to see a packaging product, such as a nest-type box liner, in which the recessed presentation face is the imprecise second surface.

The male mold consists of a form or punch projecting from a mold block. The punch interferes with the plane of the sheet preform so that sheet clamping takes place outside the mold. Subsequently, the sheet is moved into contact with the mold. This action of a male mold requires using an increase in the mechanical action of the process. The resulting product has a clearly defined female form and a not too precise male form.

The third mold type used in thermoforming is the matched mold (chapter 17). This is a closed mold consisting of two halves, each defining one surface of the finished article. The matched mold is similar in concept to a compression mold (chapter 14), but the cavity geometries remain limited to shapes attainable by sheet preforms.

SHEET PRESTRETCH

It is often advantageous in thermoforming to prestretch the heated sheet immediately prior to forming it in the mold. There are two principal reasons for this. The first is that the available force may be insufficient to stretch and form the sheet efficiently, particularly when simple vacuum forming is combined with solid-phase forming. The second is that some shapes, for example cup or box forms of relatively high aspect ratio (deep draw in relation to their diameter), give rise to excessive variations in wall thickness when formed without prestretch. This arises because the sheet touches down first on the edges and side walls of the mold cavity and then in the center of the base.

The combined mechanisms of chilling and friction tend to anchor the sheet in these positions so that all the remaining deformation needed to complete the forming is contributed by a minor proportion of the sheet area. In this case, the remedy is selective prestretching in a manner related to the geometry of the mold form and providing mechanical means to control contacting the mold wall to develop a more uniform contact.

Selective prestretching is achieved by means of specially shaped plugs that are mechanically advanced into the heated sheet to produce local stretching in a way calculated to counteract the thinning tendency imposed by the mold geometry. The technique is generally referred to as plug assist. The plug shape exerts a considerable influence on wall thickness variation (Fig. 7.9), and plug design is as much an art as a science. A blunt-nosed plug tends to produce products with a thick base and thinner side walls. Tapered canlike plugs produce a forming with thicker and stronger corners.

Plug design is optimized for each individual application, but the general rule is for smooth surfaces and blended shapes without sharp transitions in form. As an example, the coefficient of friction for heated polypropylene sheet is relatively high, so plugs should be made from low-friction materials. The usual choices are nylons, PTFEs, or filled epoxy resins.

The sequencing of the plug prestretch also influences the wall thickness distribution of the thermoformed part. The plug should contact the sheet virtually at the moment it is clamped against the mold. If contact is made before clamping, the stretching will not be sufficiently selective and localized. There may also be a tendency to crease or puncture the sheet. However, the sheet will cool quite rapidly after clamping, so if plug contact is delayed, material thickness distribution will be adversely affected.

Plug speed is also significant. A high-speed plug prestretch will result in highly oriented side walls, but the wall thickness distribution will suffer because too much sheet material will be drawn to the bottom of the mold cavity. The imbalance can be observed by measuring the crush-resistance performance of thermoformed containers (Fig. 7.10).

When the sheet prestretch is to be generalized over the whole sheet rather than localized in the area of individual mold features, the effect is achieved either by partial vacuum or positive air

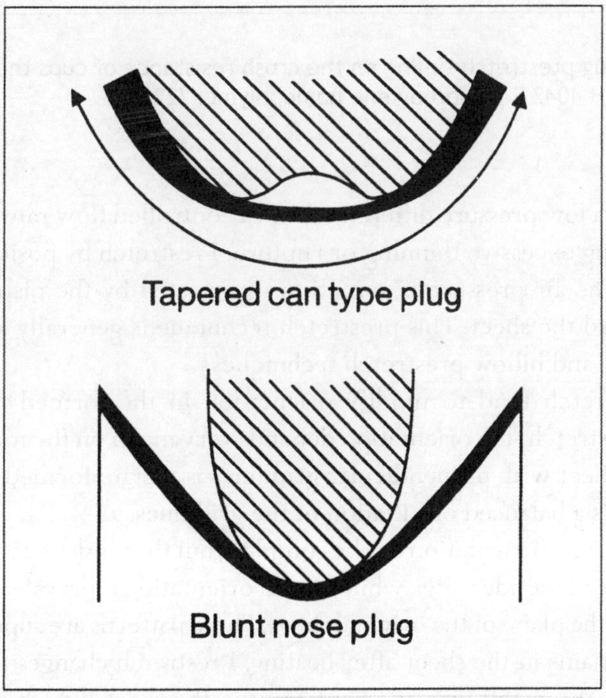

Figure 7.9 Influence of plug profile on sheet thinning

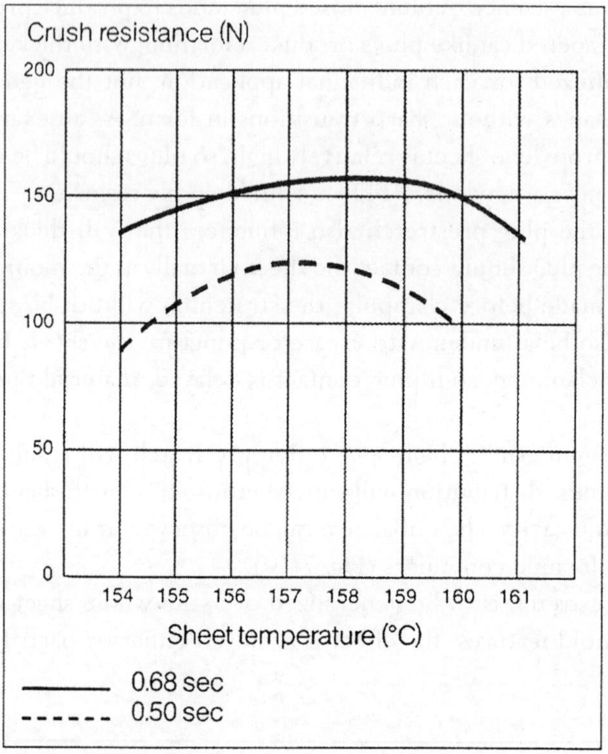

Figure 7.10 Effect of plug prestretch timing on the crush resistance of cups thermoformed from Fina-pro PPH 4042 S polypropylene homopolymer (221)

pressure. In both cases, a low pressure differential and a controlled flow rate are employed to inflate the sheet without causing excessive thinning or rupture. Prestretch by positive air pressure is often referred to as billow. The air pressure is sometimes generated by the piston effect, produced by advancing a mold toward the sheet. This prestretch technique is generally known as air slip. Some processes combine plug and billow prestretch techniques.

All forms of prestretch tend to modify orientation in the formed part. When forming is performed without prestretch, the orientation depends very much on the mold form geometry. For example, where a flat sheet with perpendicular grid lines is thermoformed in a relatively balanced shape (cup), the result is a balanced orientation of the grid lines.

The situation concerning orientation can be complex, but the tendency is for uniaxial orientation to predominate in the product side walls, while biaxial orientation is prevalent in regions that remain substantially parallel to the plane of the original sheet. These patterns are superimposed on whatever residual orientation remains in the sheet after heating. Prestretch changes the forming action. The billow process imposes an overall biaxial orientation on the sheet, which is later modified by the forming process. Plug prestretch, on the other hand, tends to emphasize the orientation patterns arising from forming. Information on uniaxial and biaxial orientation of plastics when using other processes is reviewed in chapters 5 and 6.

PLASTIC

OVERVIEW

Literally any thermoplastic can be used. However, certain types make it easier to meet certain forming requirements, such as deep draws without tearing or excessive thinning in areas such as corners. Ease of thermoforming basically depends on the type of plastic used and minimizing plastic's thickness tolerance. Certain plastics provide ease of forming.

More than 80 wt% of the thermoformed plastics is amorphous. The styrenic family of plastics represents approximately 80 wt% of the thermoformed amorphous plastics. Disposable thin-gauge products represent approximately two-thirds of their consumption, with the rest being permanent heavy-gauge products.

Factors that influence thermoforming include the available minimum tolerance (minimum to maximum thickness range) of the sheet. The aim is to minimize the variation of the sheet thickness so that a more uniform heat occurs in the sheet. To accomplish that, eliminate or minimize pinholes in the purchased sheet, control the rate and depth of draw, understand that the product ratio depth geometry is related to the capability of the type of plastic being processed, and determine the temperature profile required for the specific plastic being formed (Table 7.7).

Plastics' forming capabilities relate to their pressure stretch-and-draw ratios, which identify their depth-to-draw ratios. The stretch-to-draw ratio is the ratio of the surface of the formed part to the net starting area of the original sheet. As an introduction to this subject, with an appropriate plastic, an average stretch ratio is 3:1 for pneumatic forming. The draw ratio is the maximum depth of the forming mold to the minimum distance across the open face at any given location on the mold; the usual draw ratio is 1:1.

The linear draw ratio is the ratio of the length of a scribed line on the formed product to a scribed line on the unformed sheet used to form the product. It is a measure of the overall uniaxial elongation the plastic must have at the forming opening. It can be defined only for simple symmetrical shapes in respect to an axis. This temperature-dependent draw ratio is used primarily to screen candidate plastics and to help define potential forming processing windows.

Plastic sheet	*Typical starting temperature (°F)*	*Typical finish temperature (°F)*
HDPE/LDPE	310	120
PP	330	120
PS	300	110
PVC	295	110

Table 7.7 Forming temperature profiles for various plastics

The thermoplastics used include many different types, such as polyethylenes, polyvinyl chlorides, polystyrenes, acetates, acrylics, butadiene-acrylonitrile-styrenes, and cellulose derivatives. The thermoplastic material for thermoforming is in sheet (or film) form. Two types are used wherein one type processes easier than the other type. An example of the other type is polypropylene; this plastic will be reviewed so that one can understand the different behaviors of plastics being thermoformed.

The sheet material that is used is produced usually by extrusion (chapter 5). A small amount is calendered (chapter 9) or cast (chapter 16). The sheet can either pass directly from the extruder to the thermoformer (Figs. 7.11 and 7.12) or can pass through an intermediate storage phase. During storage, the sheet is held at room temperature and is reheated before forming; this two-stage process is known as reheat forming or cold forming. The alternative single-stage process is known as in-line forming or hot forming.

Reheat forming is the most widely used process. Thermoforming is entirely separated from sheet production, which can thus be left in the hands of specialists. The capital cost of equipment is lower and the process is easily switched between different materials and sheet sizes. Material input for reheat forming is either in the form of bulk reeled sheets or cut sheets; the choice depends mainly on the sheet thickness.

In-line forming integrates thermoforming with sheet production and the two must proceed at synchronized rates. The alternative name of hot forming is something of a misnomer because the extruded sheet reaches the thermoformer with only residual heat remaining after passing through chilling and polishing rolls. Reheating is necessary before forming can take place, but there is some savings in heat input and heat history, and the sheet line can be controlled to produce a degree of crystallinity (chapter 1) favorable to the product to be thermoformed.

In-line production has been successfully operated where the hot sheet from an extruder (that may go through a higher temperature two-roll mill) goes directly into a thermoformer with the uniform heat it requires.

Heating through the sheet thickness is more uniform than that achieved by the conventional reheat process, so the stress levels in formed containers are lower and the resistance to distortion under hot-filling or microwave heating is better. The in-line process eliminates sheet windup and stacking stations, sheet inventories, and the associated handling. It also removes the risk of postextrusion crystallization on the reel. Process scrap can be returned directly to the extruder, but the process is relatively inflexible and is best suited to very long production runs and dedicated product lines.

When extrusion and thermoforming are separate operations, the high-heat energy supplied for extrusion is completely lost by chilling the sheet. Reheating for thermoforming requires additional heat energy. With the in-line process, a high percentage of energy/heat is already contained in the sheet, which conditions it to the forming heat. Savings of about 30% to 40% can actually be obtained.

The in-line process also provides a more even heat distribution followed with weight distributions that can be reduced without changing physical properties. At equal output rates, an in-line process needs only at least half the floor space when compared to the separate operations.

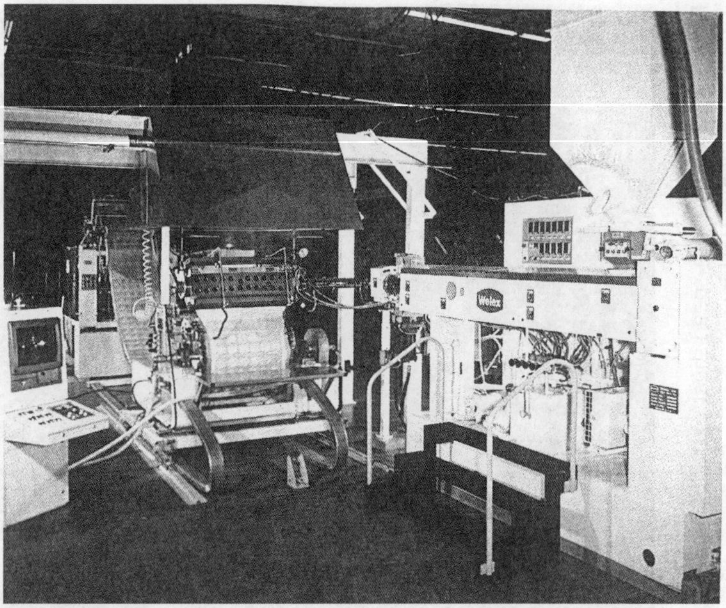

(1)

(2)

Figure 7.11 (1) In-line high-speed sheet extruder feeding a rotary thermoformer and (2) view of the thermoforming drum (courtesy of Welex/Irwin)

Figure 7.12 In-line high-speed sheet extruder feeding a stamping/trimming thermoformer (courtesy of Brown Machinery)

PROPERTY/PERFORMANCE

The thermoplastic material used is an important factor in the selection of the process. The properties and characteristics of a particular material will determine its thermoforming qualities. As the thermoforming process continues to be increasingly important, the emphasis of extruders (calenders, etc.) has turned to producing sheet specifically to meet the demands of the thermoforming markets.

Certain types make it easier to achieve deep draws without tearing or excessive thinning in areas such as corners. Ease of forming depends on the material's characteristics. It is influenced by minimum and maximum thickness, pinholes, the ability of the material to retain heat gradients across the surface and the thickness, the controllability of applied stress, the rate and depth of draw, the mold geometry, the stabilizing of uniaxial and/or biaxial deformation, and most important, the minimizing of thickness variation of the sheet.

The following TPs are the main thermoforming materials processed: high-impact and high-heat PSs, HDPEs, PPs, PVCs (polyvinyl chlorides), ABSs, CPETs, and PMMAs. Other plastics of lesser usage are transparent styrene-butadiene block copolymers, acrylics, polycarbonates, cellulosics, thermoplastic elastomers (TPEs), and ethylene-propylene thermoplastic vulcanizates. Coextruded structures of up to seven layers include barriers of EVAL, Saran, or nylon (Fig. 7.5), with polyolefins and/or styreneics for functional properties and decorative aesthetics at reasonable costs (330).

Some plastic sheets stretch as much as 600%, others as little as 15%. This behavior directly influences what shapes can be formed and their quality. Those with a puttylike appearance respond to very small pressures; others, which tend to be stiff, require heavier operating equipment. The pressure response is somewhat related to the ability to be stretched while hot, but the correspondence is not exact.

The most useful formable plastics do not have sharp melting points. Their softening with increasing heat is gradual. Each material has its own range of heat, wide or narrow, within which it can be effectively formed. This single property is one of the most important of all the factors involved in forming.

Thus, one plastic may have a forming heat of 135° to 204°C (275° to 400°F), whereas another may become soft enough for forming at 177°C (350°F) but melt at 204°C (400°F). A plastic may stretch well at a given heat but tear easily if heated a few degrees or cooled a few degrees.

Films (< 10 mil, < 250 grn) of formable plastics exhibit different behaviors, depending on the plastic. Examples include where PS is unstable with heat and requires extra cooling, PVC and PVDC are excellent (with no restrictions), nylon is difficult, PCTFE is sensitive to heat and pressure fluctuations, HDPE is difficult without a support film, and PP has a very narrow heat range. In fact, PP is extremely unstable within the conventional range of forming heat so it is processed by other techniques.

Conventional PP has the major deficiency of lacking a rubbery plateau at the forming heat; it just sags and falls apart. A process was developed to form PP just below its softening point, avoiding any sag. Known as solid-phase pressure forming (SPPF), it forces PP into the desired shape by mechanical plugs and pressure. In turn, researchers changed PP to overcome its deficiencies. They developed a proprietary catalyst and reactor technology to extrude thermoformable sheet and film. Their material has a rubbery plateau region and a high dynamic modulus so it is processible in conventional thermoforming machines.

Similar changes have also been made to plastics that were difficult or impossible to form but that had properties desirable in a formed product. PET is an example involving large production quantities. To make it formable, researchers produced crystallized PET (CPET).

Other important materials are coextruded sheets. These multilayer extruded materials provide synergism between physical properties and chemical resistance. They include barrier layers of ethylene-vinyl alcohol (EVOH) copolymers and others, including those required for aseptically packaged food products with a long shelf life at room temperature. Crystallized polyethylene terephthalate (CPET) has been used in dual-ovenable thermoformed trays for packaging frozen foods. These trays can be heated in microwave and conventional ovens; the formed parts are not affected.

Maximum dimensions for thick stock sheet are limited by available manufacturing facilities, such as sheet up to 10 ft wide × 16 ft long. Some thermoformers will purchase a thicker sheet than required and "stretch" it with a thermoforming machine to obtain a larger mold area, such as swimming spas. Mold sizes may be almost unlimited in these types of product applications. Thinner sheeting produced by extrusion, calendering, or continuous casting is available in rolls of several hundred feet in length.

Various modifications to the body or surface of the plastic sheet can be incorporated during the extrusion process to produce specialized effects. The modified plastic material may be coextruded, flocked, textured, tinted, pigmented, metallized, filled, and foamed to meet different performance and/or aesthetic requirements.

Amorphous thermoplastics, such as ABSs, acrylics, polystyrenes (PSs), polycarbonates (PCs), and vinyls, do not have melting points. Their softening with increase of temperature is gradual; each material has its own range of specific processing temperatures. Selection of the forming temperature comes by knowing the degradation temperature and then determining the highest temperature under which the sheet has enough hot strength to handle and still form properly without degrading.

Crystalline thermoplastics, such as polyethylenes (PEs), polypropylenes (PPs), and nylons (PAs), have sharp melting points. Unfortunately, most forming temperatures are the same as the melting temperatures. Forming these plastics requires more precise control of temperature and the forming operation. These plastics can be modified to relax their forming requirements.

Plastics Thermal Expansion

Thermal expansion of plastics is required to provide a guide in designing a mold so that the formed product meets its dimensions. The linear thermal expansion coefficient of plastics refers to the plastic's dimensional change when exposed to heat (chapter 22). Values are listed in in/in (mm/mm) of length per degree change in temperature (555). It provides a method to determine the change in length of a part by multiplying the part length by the change in temperature and by the thermal expansion coefficient of the plastic used. Thus, using ABS plastic as an example for thermal analysis and given a 48-in-long part subjected to a temperature reduction from 70° to –40°F, the expected change in length for an unrestrained part would be determined as follows:

High-impact ABS has a thermal expansion coefficient of 6.0×10^{-5} in/in °F. Its change in length is given as:

length change = part length × temperature change × thermal xxpansion coefficient

or

$$0.32 \text{ in} = 48 \text{ in} \times 110°\text{F} \times (6 \times 10^{-5} \text{ in/in °F}).$$

For HDPE with a coefficient of 7×10^{-5} in/in °F, the length change is

$$0.37 \text{ in} = 48 \text{ in} \times 110°\text{F} \times (7 \times 10^{-5} \text{ in/in °F}).$$

If the part is free to expand or contract, then its thermal expansion property is usually of little significance in the formed part. If being attached to another part having a lower or different thermal expansion coefficient restricts the part, then part movement will be restricted. A change in temperature will then result in the development of thermal stresses in the part. The magnitude of the stresses will depend on the temperature change, the method of attachment, and the relative expansion and modulus characteristics of the two materials at the temperature of operation (chapter 19).

Examples of the thermal expansion coefficients and moduli for a few materials commonly used with thermoplastics are given in Table 7.8. In comparing these values, it shows that ABS has a significantly higher thermal expansion coefficient and a much lower modulus than the others. This combination of properties highlights the development of substantial thermal stress in an ABS part that is rigidly fastened to one of the materials.

For example, assume that an ABS part is securely fastened to a heavy steel member and the total assembly is subjected to a 100°F (70° to –40°F) temperature reduction. Using the equation above, the contraction of the steel member is determined as follows:

Material	Thermal Expansion	Modulus @ 73F
Wood	0.3×10^{-5} in/in F	12.4×10^5 psi
Steel	0.6×10^{-5} in/in F	300×10^5 psi
Aluminum	1.3×10^{-5} in/in F	100×10^5 psi
High Impact ABS	6.0×10^{-5} in/in F	2.5×10^5 psi

Table 7.8 Examples of coefficients of thermal expansion for different materials

$$0.032 \text{ in} = 48 \text{ in} \times 110°\text{F} \times (0.6 \times 10^{-5} \text{ in/in °F}).$$

Comparing the contraction expected for the two materials shows that the ABS part will contract approximately ten times as much as that of the steel member. Because of the modulus advantage of the heavy steel member, the contraction of the plastic will be restrained and thermal stresses will develop. The level of stress developed in the ABS will be determined by its coefficient of expansion and modulus of elasticity where:

thermal stress = thermal expansion coefficient differential × temperature change × modulus at –40°F of thermoplastic (modulus of elasticity at the temperature extreme is used, not that at 73°F)

or

$$1{,}663 \text{ psi} = [(6.0 - 0.6) \times 10^{-5} \text{ in/in °F}] \times 110°\text{F} \times 280{,}000 \text{ psi}.$$

While the level of stress produced is below the yield stress of the thermoplastic ABS, the presence of stress risers can lead to ultimate product failure. Stress risers can magnify the effect of the thermally induced stress to the point that the material tensile stress will be exceeded and the part will fail. Stress risers may be in the form of sharply reduced section thickness in the parts, poor forming practices, notches from poor trimming techniques, and drilled fastener holes.

When thermoformed (or other processed plastic) parts are attached to dissimilar materials, the above information should be carefully considered. The designer of the part or assembly needs to take expansion and contraction into consideration. When selecting the right material, using good design, and forming the material with minimum stress, parts are not liable to fail when the environment is at a constant temperature. When parts are subjected to rapid or extensive thermal cycling, failure can be much greater. The broader the temperature range becomes, the higher the thermal stresses become.

THERMOFORMING POLYPROPYLENE

Polypropylenes for thermoforming typically have a melt flow index in the range 1.8 to 8.0. Homopolymers are widely used; nucleated grades are used for improved clarity and stiffness and for melt-phase processing. Block copolymers are employed for improved impact strength at low temperatures, while random copolymers are selected for high transparency and improved heat-sealing qualities (chapter 1). Stiff high tensile modulus grades can be used to replace standard grades when wall thickness downsizing is important. Thick-walled parts are typically produced from low MFI homopolymer grades that have high molecular weights.

Polypropylene sheet for thermoforming should be produced with a highly uniform degree of crystallinity across the web. Heating, too, must be very uniform, both across and through the sheet thickness. Preheating is performed at about 120°C, and the final forming temperature depends on whether solid-phase or melt-phase processing is being undertaken. For unfilled grades of polypropylene, the solid-phase forming range is typically 155° to 165°C, with 160°C frequently quoted as the optimum. However, the true optimum depends on grade and on the required properties of the forming (Table 7.9).

Nucleated grades, for example, will normally require a slightly higher temperature than basic grades. The sheet-forming temperature substantially influences the wall thickness, orientation, and mechanical properties of thin-walled polypropylene packaging products, so precise temperature control is required. Temperature variations across the sheet should not be greater than 3°C.

The temperature range for melt-phase forming is 170°C to 180°C; in this case the variation in sheet elastic response across the temperature range is relatively small and the concept of an optimum temperature has less significance.

Thermoforming molds for use with polypropylene sheet should be constructed from materials of high thermal conductivity and should be efficiently cooled. Plug assist is recommended, using low-friction plugs of low thermal conductivity to avoid premature chilling of the sheet.

The shrinkage of polypropylene thermoformings is complex and is a function of sheet crystallinity and orientation as well as thermoforming conditions and orientation arising from stretching during forming. Allied to the high basic shrinkage characteristic of polypropylene, this makes it impossible to quote a single reliable thermoforming shrinkage figure.

THERMOFORMING REINFORCED PLASTIC

Practically all thermoforming materials used are thermoplastics. However, B-staged TS plastics with reinforcements can be thermoformed (chapter 14). Specialty products, such as TS polyester glass-fiber-reinforced plastics, have been made into boat hulls and so forth.

HEATING

The heating process is of critical importance in thermoforming, both performancewise and economically. Heating accounts for some 80% of the total energy demand in thermoforming, so

Polypropylene type	Forming temperature range (°C)	Optimum forming temperature (°C)	Minumum forming strain (% yield)	Maximum forming strain (% break)	Forming cycle time for 0.05in thickness (sec)
Homopolymer	160-170	160	24	3360	28
Homopolymer, 10% calcium carbonate	155-165	160	23	3280	22
Homopolymer, 30% calcium carbonate	150-165	160	21	3190	14
Copolymer	155-165	160	22	3360	26
Copolymer, 10% calcium carbonate	150-165	160	23	3180	22
Copolymer, 30% calcium carbonate	130-155	150	80	1500	30

Table 7.9 Typical solid-phase forming conditions for selected types of polypropylene

heating efficiency exerts an important influence on process economics. Technically, the heating system must satisfy exacting demands of the plastics being formed. Heat can be transferred to the sheet only through its outside surfaces, but forming quality and consistency require uniform heating throughout the sheet thickness and at its core. Heating that results in a nonuniform temperature profile across the thickness of the sheet does not maximize development of the most efficient formed part with regard to performance or economics.

The target in heating plastic material to be thermoformed is to heat rapidly with a minimum temperature gradient from its edge to center and throughout the sheet thickness (Fig. 7.13). When areas do not require heating, baffle shields can be used (Fig. 7.14). The heating phase in the forming cycle is important to the product performance and particularly cost (heating energy cost followed by cooling cost). There are different factors that influence the heating cycle. They include the type of plastic used (including those with additives, fillers, etc.), the temperature required for forming, the plastic's thermal conductivity, the material's absorption characteristics, the plastic's specific heat, the type and intensity of heat source, and the plastic's thermal degradation characteristics. There are exceptions to completely uniform heating when using special thermoforming techniques, such as profile or area heating, where forming does not take place or is limited.

Any variation in the temperature of the hot stock dramatically affects the hot tensile strength or elasticity of the plastic. It is essential that the sheet material be heated very uniformly. With uniform

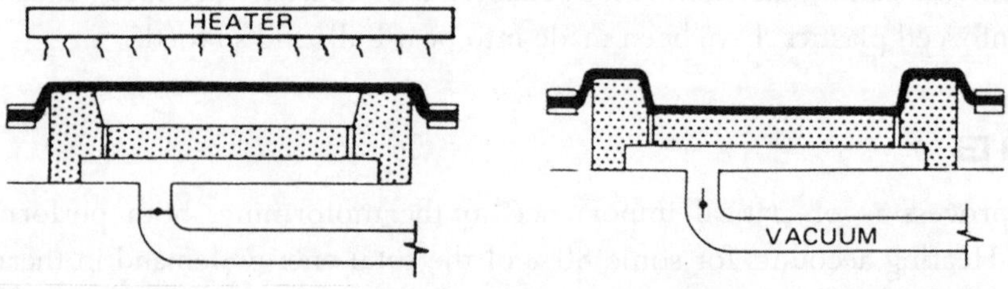

Figure 7.13 Example of applying uniform heat to a sheet that will be vacuum formed

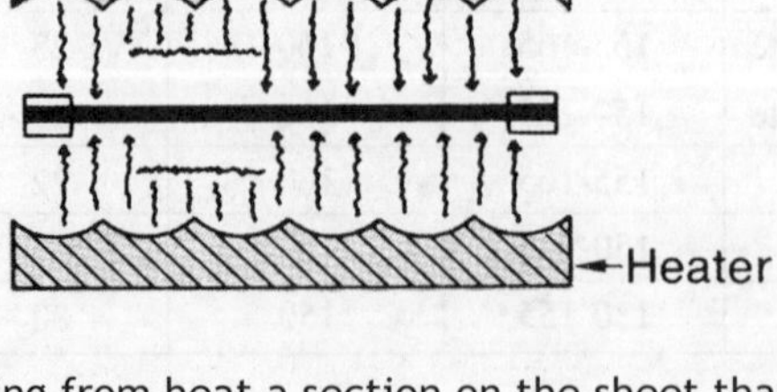

Figure 7.14 Example of shielding from heat a section on the sheet that will remain flat after thermoforming

heating, the faster the forming pressure, the better the material distribution because the material does not have a chance to cool off as it is being formed. The resulting product exhibits a minimum of internal stress and the best possible physical properties. Some materials, such as cast acrylic in very deep draws, do not process well with fast vacuum forming. This material has a great deal of hot tensile strength, which permits the processor to use a slower vacuum. However, a very hot mold must be employed to maximize its performance, particularly transparency.

When pressure forming is used, the material is moved even faster than it is with vacuum forming. Under optimum conditions, the material distribution will improve and the products will be even more stress free. All thermoplastic materials have specific processing temperatures that are available from material suppliers, experience, and/or computer software data. Examples of processing temperatures generally used for thermoforming different plastics are given in Table 7.10.

The cycle time is controlled by the heating and cooling rates, which in turn depend on factors such as the temperature of the heater(s) and the cooling medium(s), the initial temperature of the sheet, the effective heat-transfer coefficient, the sheet thickness, and thermal properties of the sheet. The degree of control necessary for proper heating will vary with the type of material and the amount of orientation induced during the extrusion of the sheet. Most plastic materials

MATERIAL	MOLD & SET TEMP.	LOWER PROCESSING TEMP.	ORIENTING TEMP.	NORMAL FORMING TEMP.	UPPER TEMP. LIMIT
ABS	185	260	280	300	360
Acetate	160	260	280	310	360
Acrylic	185	300	325	360	380
Acrylic/PVC	175	290	310	340	360
Butyrate	175	260	275	295	360
FEP	300	450	490	550	620
Polycarbonate	280	335	350	375	400
Polyester, Thermoplastic	170	250	275	300	330
Polyethersulfone	400	525	560	600	700
Polyethylene, high density	180	260	270	295	360
Polypropylene	257	270	280	310-330	340
Polypropylene, glass filled	195	265	280	400+	450
Polysulfone	325	375	415	475	575
Polystyrene	185	260	275	300	360
Vinyl, rigid	150	220	245	280-285	310
Vinyl, rigid foam	160	240	260	300	350

Table 7.10 Thermoformed mold and plastic temperature processing guide

tend to relax their internal strain or orientation during the forming heating cycle. The result is movement of the sheet that can cause its thinning before the forming operation takes place. A method to determine degree of uniform draw on a sheet is to put perpendicular lines on the sheet prior to heating and forming. After the forming, examine the lines (Fig. 7.15). If the curved lines are relatively uniform, the sheet has produced a uniform draw.

Heating is particularly difficult to achieve with certain plastics, such as polypropylene, which is characterized by high enthalpy and low thermal conductivity (Table 7.11). Additionally, and particularly for polypropylene, the process temperature window may be very narrow, requiring precise temperature control linked to and proportional with sheet throughput rates.

To counteract sagging, such as in polypropylene sheet, the final heating to the forming temperature should be as rapid as possible without compromising temperature uniformity throughout the sheet thickness. Some thermoformers use a sheet conveyor system similar to a tenter chain to orient extruded sheet or film (chapter 5). The conveyor system is used to counteract sagging by tensioning the sheet. Such systems are equipped with independent edge heating zones to compensate for the heat sink effect of the conveyor. Alternatively, hot air ovens may also be used to heat the sheet by convection, but it is more difficult to arrange for distinct and progressive zones of temperature control.

Different arrangements are necessary for cut-sheet machines. With these machines, the usual procedure is to preheat sheets by convection in static ovens followed by final heating of both surfaces in a radiant sandwich heater operating close to the mold. Conduction may also be used for sheet heating and particularly for preheating, either in the form of heated rolls or as a flat contact plate that the sheet passes over. Hot immersion baths filled with a heat-transfer oil or other suitable fluid can also be used for sheet heating, but these introduce practical operational difficulties. They are not widely employed.

The thermoforming temperature for polypropylene sheet depends on whether solid-phase or melt-phase forming is being used. In either case, the recommended process window covers a

Figure 7.15 Relatively uniform curved lines indicate a uniform thermoformed wall thickness

Plastic and (morphology)[a]		Density, g cm^{-3} (lb ft^{-3})	Melt temperature (T_m), °C(°F)	Glass transition temperature (T_g), °C (°F)	Thermal conductivity, 10^{-4} cal s^{-1} cm^{-1} °C^{-1} (Btu lb^{-1} °F^{-1})	Heat capacity, cal g^{-1} °C^{-1} (Btu lb^{-1} °F^{-1})	Thermal diffusivity, 10^{-4} cm^2 s^{-1} (10^{-3} ft^3 h^{-1})	Thermal expansion, 10^{-4} °C^{-1} (10^{-4} °F^{-1})
PP	(C)	0.9 (56)	168 (334)	5 (41)	2.8 (0.068)	0.9 (0.004)	3.5 (1.36)	81 (45)
HDPE	(C)	0.96 (60)	134 (273)	−110 (−166)	12 (0.290)	0.9 (0.004)	13.9 (5.4)	59 (33)
PTFE	(C)	2.2 (137)	330 (626)	−115 (−175)	6 (0.145)	0.3 (0.001)	9.1 (3.53)	70 (39)
PA	(C)	1.13 (71)	260 (500)	50 (122)	5.8 (0.140)	0.075 (0.003)	6.8 (2.64)	80 (44)
PET	(C)	1.35 (84)	250 (490)	70 (158)	3.6 (0.087)	0.45 (0.002)	5.9 (2.29)	65 (36)
ABS	(A)	1.05 (66)	105 (221)	102 (215)	3 (0.073)	0.5 (0.002)	3.8 (1.47)	60 (33)
PS	(A)	1.05 (66)	100 (212)	90 (194)	3 (0.073)	0.5 (0.002)	5.7 (2.2)	50 (28)
PMMA	(A)	1.20 (75)	95 (203)	100 (212)	6 (0.145)	0.56 (0.002)	8.9 (3.45)	50 (28)
PC	(A)	1.20 (75)	266 (510)	150 (300)	4.7 (0.114)	0.5 (0.002)	7.8 (3.0)	68 (38)
PVC	(A)	1.35 (84)	199 (390)	90 (194)	5 (0.121)	0.6 (0.002)	6.2 (2.4)	50 (128)

Table 7.11 Thermal conductivity and other thermal properties of a few plastics

temperature range of only 10°C. In solid-phase forming, the force required to form the sheet varies greatly within the recommended temperature range (Fig. 7.16). The variation is much smaller in melt-phase thermoforming.

When forming is performed at temperatures toward the lower end of the solid-phase range, the product will have a higher degree of orientation and improved transparency. It will have a less uniform distribution of wall thickness when compared to one made at the higher end of the range. These variations combine to affect the product crush resistance in a complex manner. For example, wall thickness variations decrease crush resistance, while orientation increases it. This means that optimum settings depend to an extent on individual product geometries and any generalizations must be approached with care.

As shown in Figure 7.17, a very small variation in sheet temperature can have a marked effect on the properties of the finished part and confirms the importance of precise temperature control when thermoforming polypropylene. The plastics identified in this figure are (a) Fina-pro polypropylene PPH 5042 S, which is a nucleated homopolymer with narrow molecular weight distribution and antistatic protection (MFI 6.0), (b) PPH 4042 S, a nucleated homopolymer with antistatic protection (MFI 3.0), and (c) PPH 4060 S, a general-purpose homopolymer (MFI 3.0) (221). Information on the melt flow index (MFI) is provided in chapter 22.

Most plastics are good insulators, so there is a danger of overheating the surface, causing degradation of material and possible color change prior to the entire sheet being brought to the forming temperature. To avoid this, lower heating temperatures or use sandwich heating with sheets of 0.060 in (0.15 cm) thickness or greater. Sandwich heating also substantially reduces the total

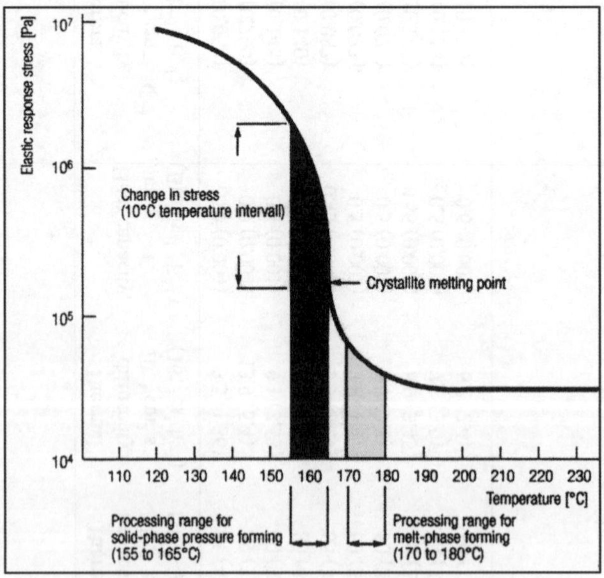

Figure 7.16 Process phases for thermoforming polypropylene

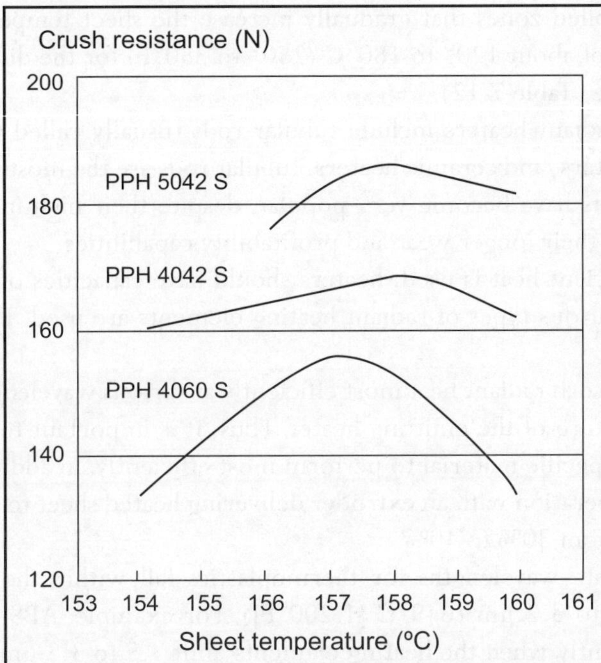

Figure 7.17 Effect of sheet-forming temperature on the crush resistance of cups thermoformed from Fina-pro polypropylenes

heating cycle time. The distance between heater and sheet is usually kept at a minimum to reduce edge losses. Six to eight inches is considered a practical minimum for rod or spot heaters if localized overheating is to be avoided.

There are occasions when a portion of the sheet may be kept cooler to prevent excessive thinning in that area during forming. This localized underheating, called shading or preferential heating, is accomplished by suspending metal screens or sheets between the heaters and the plastic. Screens are usually placed over those portions of the sheet that will be drawn into deep cavities or corners.

Heating Method

Methods used to heat the plastic sheets for thermoforming are radiation, convection, or conduction. When radiation or conduction is used, there is the option to apply heat either to one or both surfaces of the sheet. In the case of convection, both sides of the sheet will be heated; for thin films, the usual approach is to heat only one side. Given the difficulty of heating the sheet core uniformly, it is preferable to heat both surfaces of the sheet. The upper limit of sheet thickness for effective single-sided heating is about 1 mm.

Infrared radiation (IR) heating is the most widely used means of sheet heating for reel-fed or in-line thermoforming machines. The heaters are normally arranged in a heating tunnel in a series of independently controlled zones that gradually increase the sheet temperature first to a stress relaxation temperature of about 120° to 180°C (250° to 350°F) for the different plastics, then to the forming temperature (Table 7.12).

Elements used in radiant heaters include tubular rods (usually called Calrod® heaters), flat strip heaters, quartz heaters, and ceramic heaters. Tubular rods are the most widely used. However, ceramic heating elements have become very popular, despite their high initial cost compared to tubular rods, because of their longer wear and profitability capabilities.

To ensure that sufficient heat is used, heaters should have capacities of at least 4 to 6 kw/sq ft (43 to 65 kWm^{-2}). Various types of radiant heating elements are used. Their performances are shown in Table 7.13.

Different plastics absorb radiant heat most efficiently at various wavelengths, which in turn are affected by the temperature of the emitting heater. Thus, it is important to use the proper wavelength in order for the specific material to perform most efficiently. In addition to having reduced floor space, an in-line operation with an extruder delivering heated sheet to the thermoformer can achieve energy savings from 30% to 40%.

The most appropriate wavelengths for thermoplastics fall within the infrared spectrum of 6 μm (204°C [400°F]) to 3.2 μm (649°C [1,200°F]). For example, ABS, HIPS, and PE absorb radiant heat most efficiently when the heating elements emit 3.5 to 3.3 μm, whereas PC requires 3.4 μm. Materials that have wide softening ranges do not require the same degree of temperature control as materials that have narrow melting ranges. Some plasticized materials require careful temperature control, because rapid or excessive heating can cause internal vaporization of volatiles or plasticizers. The result would be blistering or excessive exudation. Since different types of additives and fillers are used in plastic sheet manufacture, the actual wavelengths may vary for generic plastics, such as PEs, PCs, and ABSs.

Evaluating heaters involves factors that interrelate and overlap, such as initial cost, operating cost, energy efficiency, durability, profile control capability, plant operating conditions, and how the cycle time is affected. Most of the heaters used are considered throwaways. There are heaters that can be repaired, but their designs typically shorten the heating element's life.

Material	Typical starting temperature (°F)	Typical finish temperature (°F)	Specific heat (Btu/lb × °F)
HDPE/LDPE	310	120	.90
PP	330	120	.72
PS	300	110	.41
PVC	295	110	.24

Table 7.12 Examples of the range of temperatures and specific heats required for thermoforming

Element	Efficiency (%) When new	Efficiency (%) After 6 months	Average life (h)	Performance
Ceramic panel	65	55	12000–15000	Best buy; heats uniformly and is efficient and capable for profiling heat
Quartz panel trademark	58	50	8000–10000	Same as ceramic heaters
Coiled Nichrome wire	18–20	8–10	1500	Initially lowest cost; is very inefficient, and heats nonuniformly with use
Tubular rods	45	20	3000	Inexpensive; heats nonuniformly with use and is difficult to screen or mask for profiling heat
Gas-fired infra-red	40–45	25	5000–6000	Lowest cost to operate; has many disadvantages, including wavelength variations and frequent maintenance

Table 7.13 Examples of types of radiant heating elements

A heater's average life span increases with proper maintenance, including the replacement of parts as follows: 15,000 hours for quartz tubes, 10,000 to 15,000 hours for ceramic elements, 15,000 to 25,000 hours for Calrods®, 25,000 hours for panels, and 10,000 to 15,000 hours for gas catalysts.

Radiant Heating As reviewed, the radiant heating method is the most popular and useful. Its infrared rays radiate from a heat source at temperatures ranging from 260° to 9,450°C (500° to 17,000°F). Radiant heaters are operated electrically, with the electrical energy converted to heat. The amount of energy emitted from the heater is determined by the electrical input measured in watts or watt density.

Since the heater's total intensity of radiation increases with the fourth power of the absolute temperature, a slight increase in the heater's temperature requires a substantial increase in wattage input and results in greatly increased heating intensity. Much of the radiant energy may pass through thin transparent plastic materials without being absorbed. To divert the energy back through the sheet, a reflecting shield can be placed on the opposite side. Opaque materials 10 mils or greater in thickness will absorb almost all of the radiant energy. Different pigments will have less effect on heating time than might be expected, because the particles of pigment are completely surrounded by plastic and tend to absorb the energy. Sheets having excessive orientation should be avoided when using conventional radiant heating equipment.

Convection Heating Even though it is the slowest heat-transfer system, convection heating has its advantages. The big advantages are its superb uniformity of heating and its ability to keep the sheet surfaces from becoming hotter than the oven temperature. It has important uses for certain applications, such as for heating heavy-gauge foam sheet, very thick solid sheet, sheet stock where the gauge is difficult to control accurately (off-gauge), sheet where surfaces might degrade easily if they are overheated, and so forth.

Convection heating is accomplished utilizing a recirculating hot-air oven. The most accurate hot-air-recirculating ovens are powered by electricity. The oven temperature is carefully maintained at the thermoforming temperature of the particular material to be formed. Air is a great insulator and plastic materials absorb heat slowly, which is why this method is relatively slow. The specific heat of the particular material governs the heating cycle. For example, a 0.125-in-thick acrylic sheet has a specific heat of 0.35. In a well-baffled hot-air-recirculating oven running at the forming temperature of 182°C (360°F), acrylic requires approximately 12.5 minutes of soak time to uniformly bring this mass up to the forming temperature. This is about 1 minute for every 10 mils (0.01 in) of thickness of the sheet. Using radiant heat and the proper wavelength, 0.125-in acrylic sheet can be brought to a core temperature of approximately 177° to 182°C (350° to 360°F) in 2.1 minutes.

Conduction Heating The speed of this mode of heat transfer is between radiant and convection heating. Heat transfer by conduction takes place when a temperature gradient exists within a material. Most conduction heater plates used in thermoforming are electrically heated PTFE-coated aluminum plates. Sandwich heater panels are used, except when thin-gauge material is being processed. A very common use of hot plates is in the packaging field using roll-fed or in-line thermoformers.

HEAT CONTROL

To develop precise controlled heating, two factors are considered: the temperature output rate of the heater elements and the time exposure of the sheet to the heat. Lower heat levels slow cycle rates but can improve forming condition and final product outcome. All heating elements must have a means of control, whether it be a simple on/off switch or some type of thermostat. A device offering more control is the percentage timer, which relays predetermined on/off settings to the heating elements.

For more precise temperature controls, thermocouples are used, which actually sense the temperature and activate the on/off control or a more complicated but more precise electronic relay system. Proportional controls provide variable power inputs that increase up to set-point temperatures, then decrease power input proportionately (chapter 3). In turn, solid-state relays offer even more control.

The ideal controller is the microprocessor when it is justified to meet performance and cost requirements. Precise and accurate temperature readings are made automatically by the system, calculated in accordance with the acceleration and deceleration deviations, and adjustments are made accordingly to bring the sheet to set-point temperatures. With the microprocessor, the ultimate in repeatability from cycle to cycle can be achieved as well as on-screen readouts of oven functions.

The controllers are fairly simple devices, but if they do not function properly, all kinds of problems develop. A checklist for eliminating problems includes improper insertion of heater element in its well, heater element burnout, location and depth of sensor as related to response time, type of on/off control action (such as proportional controller or set-point control), and proper selection of basic electrical components.

Forming requires thorough and controlled uniform radiant heating of the surface to the plastic core to the other surface. As a general guide to achieve these conditions, sheet plastics over 0.040 in (1.02 mm) should use sandwich-type heater banks; heaters are located above and below the sheet being subjected to heat. The sandwich heating system can typically also use conventional electric heaters. The sheet is heated from both sides simultaneously (under and over the sheet) prior to forming to ensure fast and uniform heating throughout the sheet.

As required when heated sheets bellow downward, make sure they do not touch and are not close to the heaters; this will help eliminate problems in the material processing and/or product performance as well as fires. Some type of protective interface, such as a wired screen, may be required between the heater and sheet.

Heater Type

Examples of different types of heaters are listed in Table 7.14 and Table 7.15 provides a comparison of different heaters.

Annealing

Annealing takes place after the formed part is produced. This heat treatment is directed at improving performance by removal of those parts that contain stresses or strains set up in the material during its fabrication. Depending on the plastic used, it is brought up to a required temperature for a definite time period; then, it is liquid- (usually with water, but oils and waxes also are used) and/or air-cooled (quenched) to room temperature at a controlled rate. The temperature is near the melting point. At the specified temperature, the molecules have enough mobility to allow them to orient to a configuration removing or reducing residual stress.

Annealing is generally restricted to thermoplastics, either amorphous or crystalline. The result is increasing density, thereby improving the plastic's heat resistance and dimensional stability when exposed to elevated temperatures. It frequently improves the impact strength and prevents crazing and cracking of excessively stressed products. The magnitude of these changes depends on the nature of the plastic, the annealing conditions, and the geometries of the parts.

The most desirable annealing temperatures for amorphous plastics, certain blends, and block copolymers are above their glass-transition temperatures (T_g), where the relaxation of stress and orientation is the most rapid. However, the required temperatures may cause excessive distortion and warping. The plastic is heated to the highest temperature at which dimensional changes owing to strain are released. This temperature can be determined by placing the plastic part in an air oven or water liquid bath and gradually raising the temperature by intervals of 3° to 5°C until the

Catalytic Gas These systems have lower operating cost (50 to 80% energy saving) but with a higher cost. It is a stainless steel panel containing a perforated steel plate to distribute the gas over ceramic-fiber mat impregnated with a platinum catalyst. The catalyst converts propane or natural gas into carbon dioxide and water without generating a flame. A metal-sheathed tubular electric heater is used to preheat the panel to 120C (250F), the thresh-hold at which the catalytic conversion begins. Since they operate at a lower watt density and temperature than electric heaters, they cannot be used with materials that require forming temperatures above 260C (500F).

Ceramic These flat or concave (with reflectors) element focuses energy rather than emitting parallel rays like a flat heater. Flat panels do not require a reflector and can be placed closer to each other and to the sheet. About 98% of the emitted energy is IR. Ceramic heaters are fairly rugged and cost slightly more than metal-sheathed tubular elements. Replacing a tubular with a ceramic of the same watt density will save energy because they hold heat longer.

Electric Panel They give off 80 to 90% of their energy as useful heat. The panel radiates heat across its entire surface, providing uniform heat over a broader area than is provided by separate smaller heaters. Both striping and no-heat zones between elements are less of a concern with panel heaters.

Halogen These heaters have the fastest heat-up of all heaters (as little as 0.2 s) and emit most of their energy at a wavelength of 1 micron at their maximum temperature of 1100C (2000F). Compared to quartz-tube heaters they provide energy savings of 18 to 20%. They heat only the plastic sheet, not the ambient air. The complete machine that includes the molds remains cool enough to touch. Risk of fire is lower because of no ambient heat. Finer temperature control exists than most quartz heaters.

Metal-Sheathed/Calrod These tubular heating elements are the inexpensive workhorse of the industry commonly just known as calrod. They have a relatively fast heat-up time (1.5 min), quick response to setpoint changes, take a lot of abuse, etc. However they radiate in every direction resulting in low efficiency.

Tubular Quartz Energy source is a helical resistance wire within a quartz tube. A reflector is used to concentrate the heat on the plastic sheet. Their greatest assist is quick heat-up and cool-down ability permitting complete shut-off between the heating cycles saving energy. They are about 60% efficient when new with a slight drop-off after use. They are fragile and are not appropriate for bottom heating.

Table 7.14 Examples of different types of heaters

Heater Type	Advantages	Disadvantages
Tubular metal rod (Calrod)	Low cost, durable, rapid heat-up, easy to clean	Rapid loss in efficiency, difficult to control, needs reflectors, reflectors must be cleaned, rust, difficult to zone
Tubular quartz	Fastest heat-up, excellent zoning, wide wattage range	Fragile, can be pitted
Ceramic	Durable, good zoning, lower cost, good temperature control	High installation cost, hard to find burned-out elements, below average temperature -response time
Electric panel	Stable heat, available with quartz cloth, metal, ceramic faces; installation easy	Large size, high replacement cost
Catalytic gas	Uniform heat, low operating cost, gas company may subsidize	Large size, difficult to zone, very slow temperature response, very high installation cost
Non-catalytic gas	Inexpensive energy source, very durable	High installation cost, intense energy source, can cause fires, restricted to heavy-gauge
Halogen	Pulsed heat, fastest heat-up, excellent zoning, very small elements	Fragile, very expensive, high installation cost, unknown reliability

Table 7.15 Comparison of thermoformer heaters

maximum allowable change in shape or dimension occurs. This distortion temperature is dictated by the thermomechanical processing history, geometry, thickness, and size. Usually, the annealing temperature is set about 5°C lower using careful quality-control procedures.

Rigid amorphous plastics, such as polystyrenes (PSs) and acrylics (PMMAs), are frequently annealed for stress relief. Annealing a crystalline plastic, in addition to the usual stress relief, may also bring about significant changes in the nature of its crystalline state. The nature of the crystal structure, degree of crystallinity, size and number of spherulites, and orientation control it. In cases when proper temperature and pressure are maintained during processing, the induced internal stresses might be insignificant and annealing might not be required.

Plastic blends and block copolymers typically contain other low and intermediate molecular weight additives, such as plasticizers, flame retardants, and UV or thermal stabilizers. During annealing, phase and microphase separation may be enhanced and bleeding of the additives may be observed. The morphologies of blends and block copolymers can be affected by processing and quenching conditions. If their melt viscosities are not matched, compositional layering perpendicular to the direction of flow may occur. As in the case of crystalline plastics, the skin may be different both in morphology and composition. Annealing may cause more significant changes in the skin than in the interior.

COOLING

A time period is required for cooling the heat-softened plastic sheet below its heat-deflection temperature while it is in contact with the mold in order to properly remove the product from the mold. Conductive heat loss to the mold and convective heat loss to the surrounding air accomplish cooling. The cooling rate depends upon the tooling because, in all methods except matched molds, the plastic is in contact with the mold on one side only. The opposite side is cooled convectively by forced air and ambient air. Water sprays are sometimes used but often pose as many problems (such as water spotting, which magnifies uneven plastic surfaces). Pressure forming helps to minimize cooling time, because the higher air pressure keeps the sheet in more intimate contact with the mold surface.

The cooling action after completing the forming operation is critical to the shape and/or performance of the formed product as well as reducing the cycle time. This part of the forming cycle also requires careful consideration, because the time required to cool formed products properly is usually longer than other phases in the cycle other than heating. When too short, product performances usually suffer. Products are cooled on the side opposite the mold by convection and on the side facing the mold by conduction through the mold.

When practical, molds should be made of materials having high thermal conductivity to obtain maximum cooling. Most of the heat can be removed through the mold by coring the mold to circulate temperature-controlled water. As the part cools, the rate of heat transfer to the mold decreases because of the decreasing temperature differential between the product and mold. The cooling rate of cavity-formed products is further retarded, because the sheet tends to shrink away from the mold and lose surface contact for conduction.

There is a tendency to hold temperatures in temperature-controlled molds to less than 38°C (100°F) and to shorten the cooling phase. However, this rapid cooling and the induced thermal stresses it may cause can result in a reduction of physical and other properties in the product. Mold temperatures at least over 55°C (130°F) should be used to avoid this situation; actual temperature depends on type of plastic used and product shape. To obtain optimum products, it may be necessary to operate molds at 77°C (170°F) or higher even though cooling time is increased.

Cooling from the top side can be obtained with fans or blowers. This method is needed particularly when molds of nonconductive material, such as wood, plastic, or plaster, are used. Water sprays usually cause spotting on the part surface and are not generally recommended.

When the products are cool enough to prevent warpage, they are removed from the molds. Different methods are used to release the products from the molds. From a female cavity, owing to shrinkage, it literally just falls out. When required, the formed product is usually released from the mold by building up air pressure under the product.

Once the product is released, either the clamping frame is raised or the mold is lowered to separate the product from the mold. It may be necessary to strip the product from the mold, either when the product is draped over a mold of insufficient wall taper or when there are undercuts. Stripping may be accomplished by ejection pins or by rings mechanically actuated from within the mold (chapter 17). Other methods are used, such as building a light grid or framework onto the clamping frame.

With the formed products removed, they may be placed in jigs or racks for further cooling (postcooling); they hold the contour of the products during the final cooling. With in-line continuous forming that does not have trimming accomplished in the forming equipment, they are fed directly to a trimming press.

Heat-Transfer Requirement

Heat transfer is a fundamental element of plastics processing. In a variety of plastics processes, heat is transferred from plastic material (typically by a recirculating water system, such as extrusion, injection molding, or blow molding) to induce its conversion from a liquid to a solid state. The thermoforming process has unique characteristics, specifically less extreme material temperatures than extrusion, injection molding, or blow molding, that one must understand to successfully apply or design a water cooling system.

The heat-transfer requirements for solidification of a plastic material are most easily defined by start temperature for the material, finish temperature, rate at which the material is to be processed (lb/h), and inherent physical characteristic of the material, namely, its heat capacity (BTU/lb × °F; Table 7.11). The equation used to express the relationship of these parameters in the definition of a cooling load is:

$$\text{(start temperature} - \text{finish temperature)}^{\circ}\text{F} \times \text{rate} \times \text{specific heat} \times \text{safety factor} = \text{load (BTU/h)}.$$

A typical safety factor used is 1.2 (chapter 19). Other typical parameter values utilized for the process are shown in Table 7.12.

When considering the thermoforming of 300 lb/h of polypropylene, one could calculate the cooling load as follows:

$$(330 - 120)^{\circ}\text{F} \times 300\ \text{lb/h} \times 0.72\ \text{BTU/lb} \times 1.2 = 54{,}432\ \text{BTU/h}.$$

A chilled water ton is expressed as 12,000 BTU/h; thus, the application of a 5-ton chiller should be considered. It is important that vendors of cooling equipment be supplied with data on the highest processing rates anticipated from thermoforming equipment. Although cooling equipment design typically allows for low load conditions, physical limitations restrict the effective operation of cooling equipment upon overloading.

EQUIPMENT

Thermoforming machines range from small to very large that can handle precut sheets to continuous sheet feed from a roll (Fig. 7.18) or directly from an extruder into a continuous operating thermoforming machine (Figs. 7.19 and 7.20). Classification of machines is usually by the number of operations they perform, such as single-stage (Fig. 7.1), double-stage, three-stage (Figs. 7.21 to 7.23), five-stage (Fig. 7.24), rotary (Fig. 7.25), and continuous forming machines (Fig. 7.26). There is a specially designed thermoforming machine (Fig. 7.27) that starts with a plastic extruded tube, flattens it with rolls, then forms the molds on a rotary wheel.

A single-stage precut sheet-fed machine can perform only one operation at a time; its total cycle time will be the sum of the times required for loading, heating, forming, cooling, and unloading. In a typical operation, the sheet is clamped in a frame and the frame is moved between the heaters (or under a single heater) and back to the forming station for thermoforming.

A two-stage precut sheet-fed machine can perform two operations simultaneously. It usually consists of two forming stations and a bank of heaters that move from one station to the other. A machine with three or more stages is usually built on a horizontal circular frame and is called a rotary. The rotary thermoformer operates like a merry-go-round, indexing through the various stations. A three-stage rotary machine has a loading and unloading station, a heating station, and a forming station (where cooling also takes place) and indexes 120° after each operation. Because there is always a sheet in each of the stations, it provides higher output than the other machines.

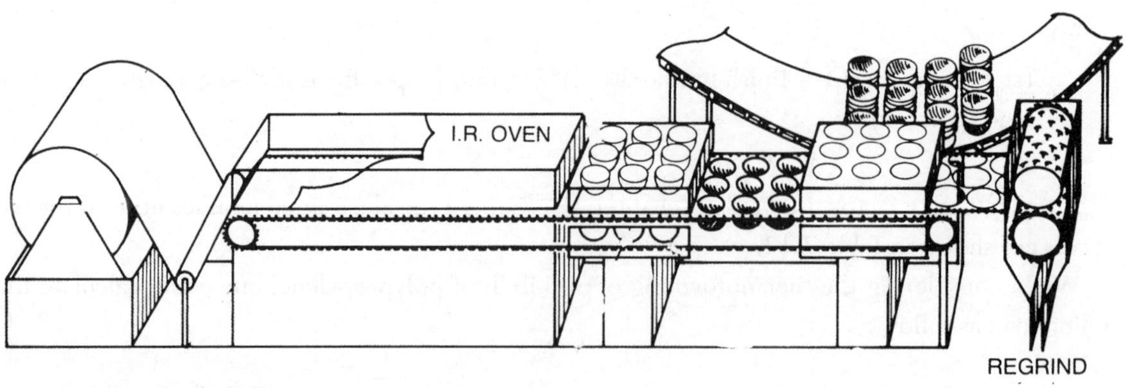

Figure 7.18 Schematic of roll-fed thermoforming line

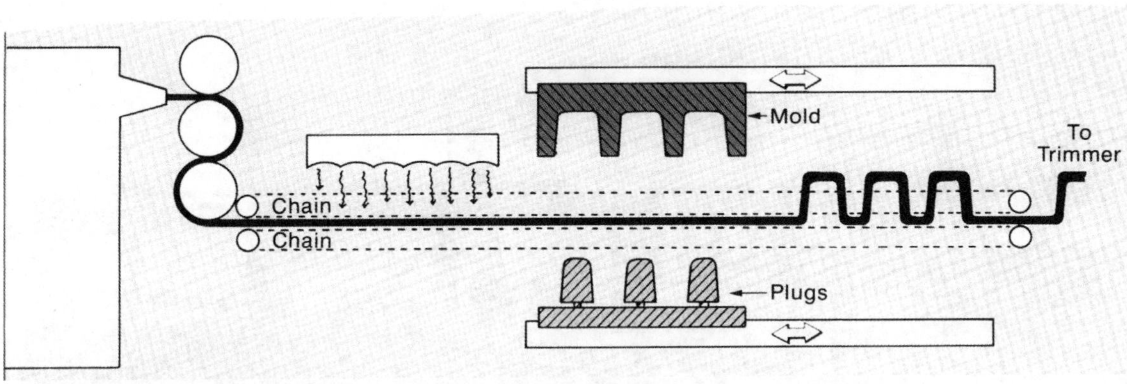

Figure 7.19 Schematic of simplified in-line thermoforming line

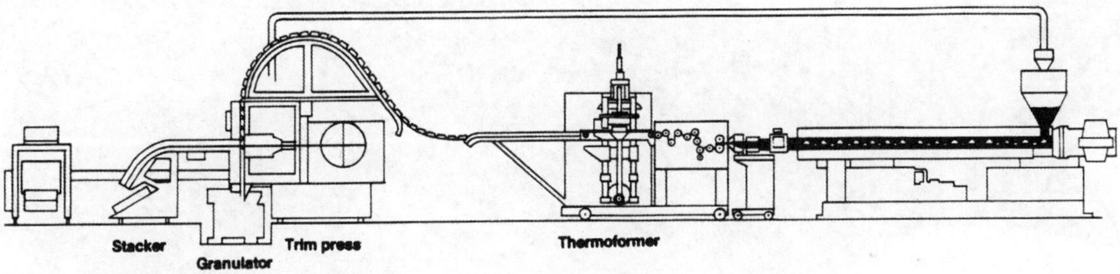

Figure 7.20 Schematic of in-line thermoforming line including auxiliary equipment

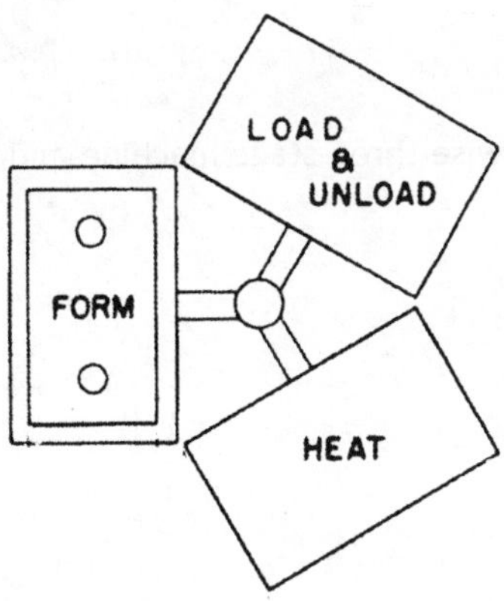

Figure 7.21 Schematic of rotating clockwise three-stage machine

Figure 7.22 View of a rotating clockwise three-stage machine midway in being manufactured

Figure 7.23 View of a rotating clockwise three-stage machine

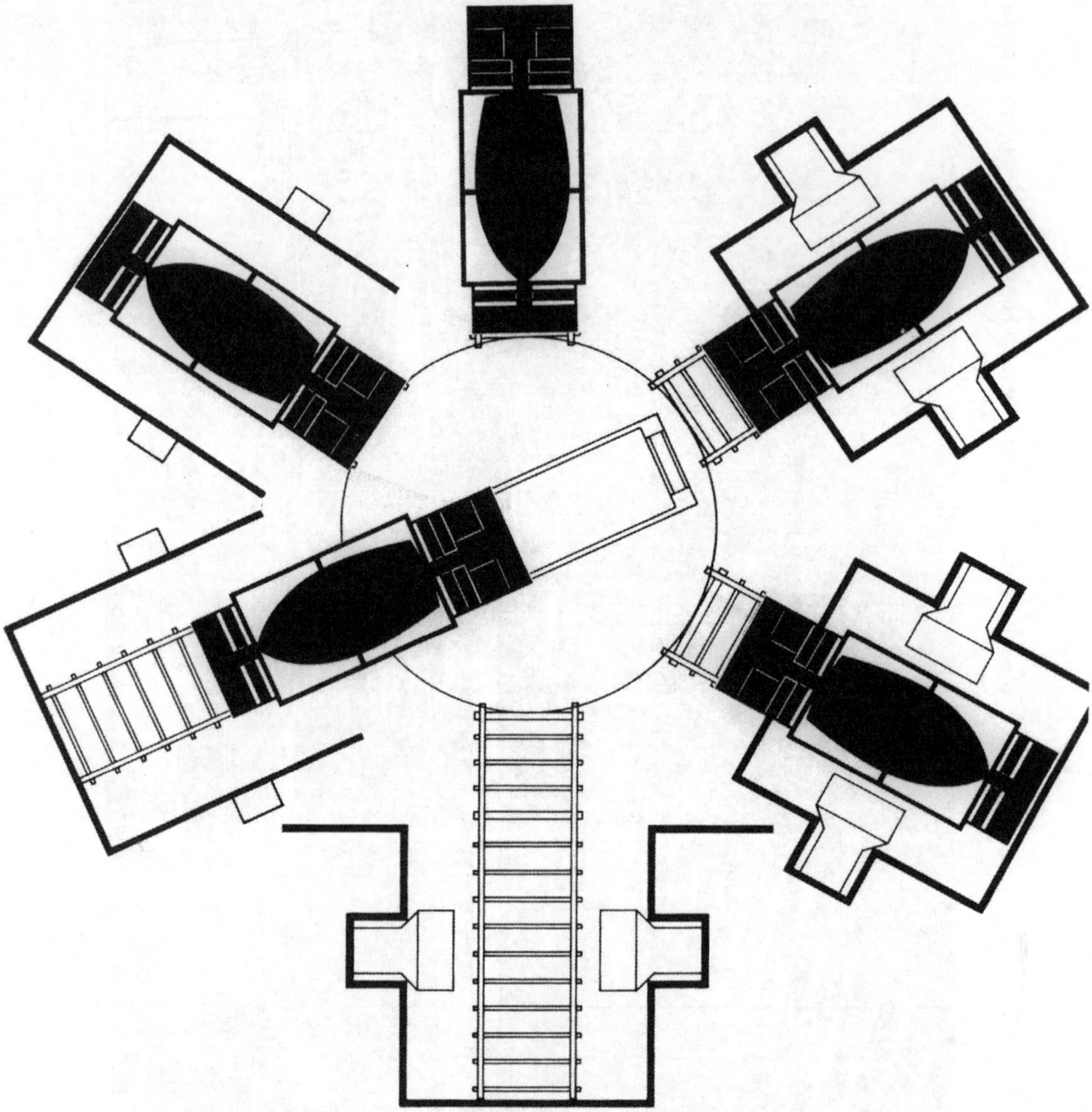

Figure 7.24 View of a rotating clockwise five-stage machine (courtesy of Wilmington Machinery)

Figure 7.25 Rotary thermoformer (courtesy of Welex Inc.)

Figure 7.26 Compact in-line sheet extrusion thermoforming machine provides more heat retention for the thermoformer (courtesy of Welex Inc.)

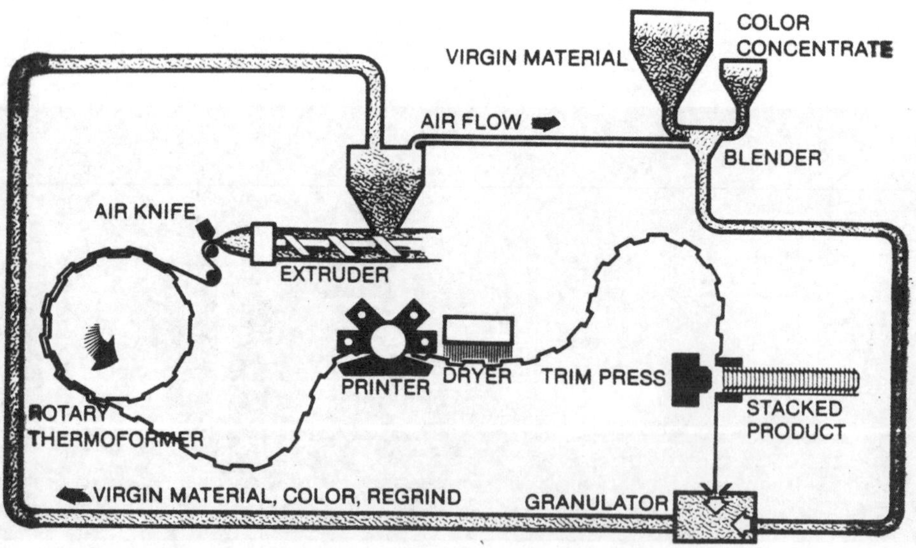

Figure 7.27 Thermoforming machine starts with a plastic extruded tube, flattens it with rolls, then forms the molds on a rotary wheel (courtesy of Brown Machinery)

Four-stage (or more) rotaries can incorporate various station configurations, such as load, preheat, heat, form, decorate, and unload.

An in-line sheet-fed machines has the sides of a sheet clamped in a frame that travels into the heating station, indexes through the forming station, then moves to the unloading station, where the product is removed by the operator or usually automatically. This type of machine can have many clamping frames in use at all times.

A continuous roll-fed machine feeds off a roll of plastic or directly from an extruder using either a rotating cylinder as the mold (drum former) or a conventional mold that travels horizontally at the same speed as the sheet. The upper and lower platens move with the sheet with a reciprocating motion so that they can index to the next unformed section of sheet once the forming cycle is complete.

Roll-fed thermoforming machinery can produce beverage cups at rates of at least 75,000 to 100,000 cups per hour while consuming plastic sheet in excess of 1 ton per hour. Sheet consumption for larger and heavier containers can exceed 2 tons per hour (1,800 kg per hour), but unit production rates may be lower because fewer mold cavities can be mounted in the machines and more time is required to cool the thicker walls. Cost of equipment (extruder, thermoformer, and trimmer) can be related to their output capacities (Fig. 7.28).

FUNCTION

Machines are required to provide certain functions, such as clamping to hold the plastic sheet for heating and positioning for forming. Use is to be made of adequate source and control of heaters.

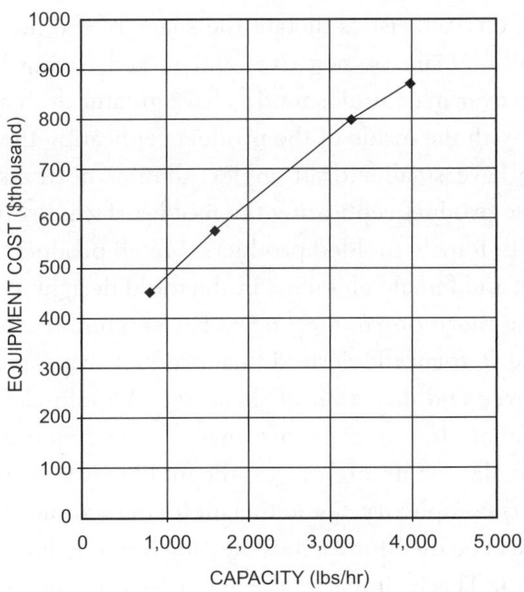

Figure 7.28 Example of the cost of equipment compared to the forming line output

For certain automatic processes, an operation is required to raise or lower the mold into the plastic or vice versa where a clamped sheet is moved over the mold. The vacuum and/or pressure control system needs to be carefully monitored with sensor controls. It is important to have safety devices to ensure operators are not burned or otherwise injured.

Usually, either air pressure or electric motors are used to provide the motion of mold, plug, and clamping frame. They are designed to provide ample force to raise and lower platens rapidly when operating with the largest possible molds and sheet materials having high hot strength. With plug-assisted forming of large area parts, where the plug is driven by air pressure, there may be some plug hesitation unless ample driving force is provided. It may be necessary to lock the mold platen to prevent its movement during plug travel. Plug speeds up to 30 in/s may be required for proper forming.

Frames must be capable of exerting sufficient clamping force to prevent materials having high hot strength from pulling out of the frames during rapid draping action. The frame should also be able to provide some stripping action as the sheet is removed from the mold. Mechanical stripping by the frame, however, should not be substituted for air ejection and properly designed easy-release molds.

A clamping frame should be designed to allow a maximum temperature buildup in the sheet under the clamp. Its construction sometimes will cause a substantial difference in sheet temperature from the relatively cool edges to the heated center of the sheet. After forming and cooling, shrinkage in the center is greater than at the edges, which can induce stresses likely to cause warpage. That is particularly true in shallow drawn parts formed from heavy sheet.

MOLD

OVERVIEW

In many cases, the product characteristics dictate the shape of the mold. The three general mold shapes are male or positive, female or negative, and mixed having both positive and negative characteristics. Parts drawn over male molds tend to have greater draft angles, heavier bottoms and corners, and thinner rims, with the inside of the product replicating the mold surface. Parts drawn into female molds tend to have smaller draft angles, thinner bottoms and corners, and heavier rims, with the outside of the product replicating the mold surface. Prestretching, particularly plug assisting, is easier to use with female molded products. Mixed products must be designed with the characteristics of both male and female elements in the mold designs.

Many molds used in the thermoforming process have common assembly and operating parts. The tool's cavity is designed to form the desired final product shape and size based on the plastic's characteristics, such as degree and direction of shrinkage. A mold can be a highly sophisticated/expensive piece of equipment. It can comprise many parts requiring high-quality metals and precision machining. To capitalize on its advantages, the mold may incorporate many male or female cavities, further adding to its complexity. Some thermoforming molds have been preengineered as standardized products that can be used for cavities, heating/cooling lines, mechanisms for trimming and/or unscrewing, and so on. This is similar to the extensive preengineered standardized molds for injection molding (chapter 17).

DETAIL

The thermoforming mold performs two equally important functions. It defines one surface of the product and it acts as heat exchanger to cool the product rapidly from the forming temperature to ejection temperature. The cooling function has a direct relation on process economics. It continues to be more efficiently performed even though difficulties exist. One basic difficulty is that all heat must be extracted through one surface of the product. Because contact between the mold and the product is not intimate, an insulating space develops. This space significantly reduces the rate of cooling and can cause problems, such as nonuniform cooling with uneven stresses remaining in the plastic. Slowing up the cooling rate will eliminate problems, but extending the cooling cycle time results in increased costs.

The heat exchange function of the mold leads to a conflict of interest. Thermoforming is carried out at relatively low pressures, very low pressures in the case of vacuum forming, so molds can be constructed from light, inexpensive easily shaped materials, such as wood, plaster, or epoxy resins. However, these materials have poor thermal conductivity. Thus, such molds do not function well as heat exchangers.

Consequently, their use is best confined to short runs or prototypes. In normal production, the improved heat-transfer capability of a metal mold will more than repay the greater cost. Aluminum is most commonly used for thermoforming molds; other options include cast or sprayed low-melting-point alloys, porous sintered metals, and copper alloys.

The molds can include channels for the circulation of a cooling water. Because the forming process is performed at relatively low pressure, the channels can, depending on the material of construction and the actual working pressure, approach within about 10 mm of the forming surface for greater cooling efficiency. In the case of a cast or sprayed mold, the cooling channels can be prefabricated in copper pipe to closely follow the cavity contours.

In machined molds, the channels will be drilled or milled and it is usually impossible to follow the contours so closely. The greatest cooling efficiency is achieved when the flow of water is turbulent. This is a function of the Reynolds number and in practice it means that channels should not be too large in relation to the coolant flow rate. This runs counter to the instinctive feeling that larger cooling channels must always be better (chapter 17).

The principal decision to be made when designing a thermoforming mold is to determine which of the two product faces is to be defined by the mold. As so many thermoformings are containers and are substantially cup or box shaped, this decision determines whether the mold is to be of the male or female type. The shape of a thermoforming mold is sharply defined only on the surface in contact with the mold, so whichever is the primary presentation face of the product may determine the mold strategy.

When the forming may be produced from a sheet with a grained or textured surface, the mold may be designed to define the second surface of the product in order to preserve the sheet surface. Another consideration is to simplify the production process and machinery, and here the best choice is usually a female mold, because sheet clamping is easier to arrange and machine. Female molds also have the advantage when multiple cavities are to be placed close together. When male cores are closely spaced, there is a risk of the sheet bridging between cores and either failing to conform fully to the mold or becoming excessively thinned or even ruptured. Shrinkage also makes it easier to release the forming from a female mold than from a male mold.

The cavity surfaces should be finely sandblasted to prevent formings from sticking in the mold. The mold requires a multiplicity of vacuum or vent ports, and these should be distributed across the forming surface and must be carefully sited in ribs, slots, and other features that are likely to become isolated as the forming sheet progressively seals off the other ports. The ports must be small in diameter so that little or no trace of their presence is transferred to the product as witness marks. One rule is to make the vent diameter smaller than the thickness of the forming at that point, subject to a maximum diameter of 0.3 mm. The vents are relieved from the back of the mold by a much larger bore or a series of diminishing bores drilled to within about 2 mm of the mold face.

The thermoforming mold may include two other important features. Prestretch plugs are designed specifically for particular mold contours and should be regarded as integral parts of the mold designs. There is an increasing tendency, too, for product trimming to be built into the thermoform tooling, either by means of a peripheral knife edge around the mold form or by providing an integral punch that separates the product by shearing it from the sheet. The advantage is that there is no possibility of sheet shrinkage producing a misalignment between the product and the trimming action. The disadvantage is that in-mold trimming increases the mechanical complication and cost of tooling, so the technique tends to be confined to high-volume packaging applications.

DESIGN

Certain basic facts must be observed, or at least targeted, in the manufacture of molds. Flat surfaces should be avoided if possible, because slight domes or dish effects will allow the sheet to stretch over the entire surface. The curved surface prevents the slight bumps that usually may appear in flat sections. Maximum allowable vent hole diameters will vary with materials and sheet thickness. Air evacuation holes should be as small as possible to minimize restriction of plastic flow through vent holes, and the openings should be back drilled on the mold surface, as reviewed above.

The total number of vent holes depends on the desired rate of drawing. Since it is usually desirable to form as rapidly as possible, a number of holes should be provided based on trial and error and/or experience. A combination mold with several cavity sections (such as with large formed products) requires an increase in the ratio of holes to a cavity chamber's volume in order to permit evacuation of a deep chamber more rapidly than an adjacent shallow section.

Rapid plug-assisted forming also requires an increase in the number of vent holes. Vent holes should be at least be located in those areas into which the sheet will be drawn last. In vacuum forming, they project downward into a common chamber at the bottom of the mold. In parts where fine details or textured patterns must be accurately reproduced, vent holes less than 0.5 in (0.013 cm) apart are usually necessary.

Undercuts used in a mold require them to provide a means to easily remove the formed product. Split-section molds can be designed to be disassembled, permitting removal of the product. Other approaches include hinged sections that unfold as the part is removed using straight line undercuts in the mold. Cam-actuated protruding sections may be withdrawn into the mold before the product is removed. Mechanical strippers may be used in the mold to provide positive mold release for products having slight undercuts, particularly in high-speed production operations.

Where applicable, an undercut insert in the mold can remain as an insert in the finished part. Such inserts can be held in place by the undercut. If possible, they should be made from the same material used in forming the part so there will be no difference in thermal expansion and contraction.

Careful design and manufacturing planning are necessary to assure proper channels for cooling water are cored in the mold. These molds may be built either with alternate channels for air and water or with an upper and a lower section. The upper section would contain the water channels, and the lower section would have a vacuum chamber. Vent holes are carefully drilled into the vacuum chamber through areas in the upper section that are not channeled for water.

In the design of molds, consider using the largest cavity draft angle; the larger, the better for the formed product removal. A guide for draft angle is at least 2° to 3° per side for female molds and 5° to 7° for the male molds. A straight-sided angle in the direction of draw makes release of products more difficult. This is especially true for male molds where the natural mold shrinkage is toward the mold. When possible, to remove any release difficulty, use split molds or collapsible molds, which can even permit the use of negative drafts.

Male molds provide for tighter tolerance controls. What causes female molds to have more difficulty in controlling tolerances is the plastic shrinking away from the mold during cooling. Prestretching or extended localized heating can be used with either type of mold to provide a more uniform wall thickness.

The female molded product has the greatest wall thickness with the thinnest bottoms. The reverse occurs when using a male mold. During the forming, the part of the hot sheet that touches any part of the mold will start to cool, resulting in a thicker wall with possible frozen stresses. With multiple cavities, the female mold permits the cavities to be spaced closer together. The male mold costs less than the female mold.

Molds for forming are of all sizes and shapes. For long production runs, they are generally made from cast or machined aluminum that include many vent holes to allow for air movement from between the heated sheet and the mold cavity surface. Back drilling of larger holes on the underside to approximately 0.1 in (0.25 cm) can be used to aid in the air or vacuum action. Narrow slots also can be used, since they offer much less resistance than holes when air is evacuated through the mold. Also used is porous or sintered metal providing exceptional air removal capability. These vents are small enough that no surface imperfections occur.

Size openings should be kept as small as possible so they do not interfere with surface requirements. Different types of plastics tend to have different requirements. Careful placement of the holes will be helpful in providing fast, efficient airflow during forming. Logic and experience basically provide guidelines for the placement and size of openings. For impact PS up to 0.030 in (0.08 cm), openings may be used for material thicknesses greater than 0.060 in (0.15 cm). For thinner sheet, especially where both sides of the part will be viewed, it may be necessary to reduce hole diameters to 0.010 in (0.03 cm) to avoid hole marks.

Polyethylene-formed products exhibit more severe hole markings, so it is usually necessary to hold the diameter under 0.020 in (0.05 cm) regardless of sheet thickness; however, consider 0.010 in (0.03 cm) for the thinner gauges. Often, it is possible to split molds, leaving a slot with a width of 0.010 to 0.01 in (0.03 to 0.04 cm) extending around the entire part. This allows greater passage of air and will present no more of a vent mark than a hole of comparable diameter.

Material of Construction

Different materials of construction are used to meet different requirements, such as the product quantity, type of plastic used, mold requirement (pressure load, heat-transfer rate, surface finish, wear resistance, etc.), forming technique, product performance requirements, and cost. Many materials, including woods, cements/plasters, epoxy plastics, polyester thermoset plastics, gypsums, aluminums, and metals, are used in the manufacture of molds for forming (chapter 17).

For long production runs, nearly all molds are made from cast aluminum, brass, magnesium, or steel. Aluminum, cast or machined, is the most widely used metal. These materials are cast by conventional sand- or shell-casting techniques. For smoother molds that require less finishing, plaster patterns may be used. The molds are then buffed or polished to the degree of finish desired. Channels for temperature-controlled water lines can either be cored or cast as integral units, or copper tubing can be cast into the mold and/or against the bottom side. The molds with cored channels have greater capacities for removing heat.

Aluminum-machined metal molds are recommended where greater thermal conductivity and dimensional accuracy are required. The air holes and slight slag entrapments found in cast molds are eliminated, increasing the thermal conductivity significantly. Machined aluminum inserts are often used in multiple molds for higher-speed container production requiring minimum cooling cycles.

The plastic molds principally use phenolic epoxies or TS polyester compounds. They offer tough machinable molds that give excellent reproductions of the patterns. However, like woods and plasters, these materials are excellent insulators and build up heat with extended production. The problem can be alleviated by filling the materials with aluminum filings or shavings and by casting in cooling lines to increase conductivity. Copper or aluminum tubing is often cast within the plastic to serve as channels for the temperature-controlled water.

Where maximum strength is required, glass-fiber matting, wire screen, or some other reinforcing material is incorporated in the mold material. Vent holes may be machined. It is often easier, however, to cast vents into the mold by placing lubricated wires into the pattern. Once the plaster has set up around the pattern, the wires may be removed, leaving vacuum channels. With these materials, there is a problem of heat buildup during continuous operation, resulting in slow cooling rates and gradual embrittlement of the forming molds.

Wood and plaster are usually limited to short runs or sample preparation, since they tend to chip and break with extended production. Plastic molds, because they are easily molded, also are used for short runs and sample/prototype preparation. Also, they provide sufficient strength for most extended production runs. The major consideration between plastic and metal, as well as between metal and metal, for production molds is the difference in thermal conductivity.

Different types of wood can be used, but those having high strength and close grains are usually preferred. These include most of the hardwoods, such as maple, mahogany, birch, and oak. Mahogany is especially suitable because it has a close grain, is fairly hard, is low in pitch content, and it machines readily. Wood molds should be sanded to smooth finishes and then waxed to fill the pores. The wax will facilitate mold release and will help to provide a better surface on the formed part. A more durable surface can be obtained by coating the wood with epoxy plastic. Wood is used for either male or female cavity molds, and for plugs in plug-assisted forming. When used in continuous in-line production, it can build up excessive heat that may cause the mold to warp or the plastic to adhere and chip away portions of the mold.

Plaster of Paris or gypsum cements provide excellent short-run molds. These cements are available in several formulations, having different relationships of hardness and toughness and varying degrees of shrinkage during the hardening process. Gypsum molds will accurately reproduce the details from the patterns with good surface finishes.

A popular construction is with metal molds that are built up by spraying molten metal against suitable patterns. A variety of metals, ranging from low-melting lead alloys to bronze or steel, may be used. Pattern requirements vary with the metal used. The metal can be built up to any desired thickness, while thin shells can be reinforced with filled or laminated thermosetting resins. Spray-metal molds are often porous, making them useful to evacuate air without drilling additional vents in the molds.

There are electroplated molds that are made by depositing a thin layer of metal on a conductive surface in an electrolytic bath. The metal is then backed up with spray metal or thermosetting plastic; usually, a reinforced TS polyester-glass-fiber compound is used. The electroplated mold surface gives excellent reproduction of the pattern and provides a durable mold finish.

PROCESSING

The options available for the principal process factors can be combined in different ways to produce an almost infinite variety of thermoforming processes (Table 7.16). There are thermoforming machines that are designed to adapt to a number of different processes. Others, particularly those designed for high production rates and long product runs, are dedicated to specific processes (221). Table 7.17 provides general information to determine size of cut sheet and draw ratio.

Many different techniques are used to thermoform plastic products. Choosing the optimum process encompasses a broad spectrum of possibilities (Fig. 7.29). Sometimes, only one process can be used, but generally there are options. Influencing the selection are quantity, size, thickness,

Single Step Forming	Prestretching Before Forming	Matched Mold Forming	Twin-Sheet Forming
Vacuum forming	Billow-assist	Foam	Hollow shape
Drape forming	Plug-assist	Filled polymer	Foam-filled shape
Contact forming	Pressure forming	Reinforced polymer	
Pressure forming			

Table 7.16 Examples of different thermoforming processes

tolerances, type of plastic used, performance requirements, design optimized, cost limitations, and so on. For example, plastics with fillers and/or reinforcements are generally far more stable in meeting tighter tolerances. Processing may involve equipment that is simple to operate or it may require extensive specialized equipment along with all types of auxiliary equipment.

The equipment available ranges from manual to fully automatic. With manual operation, all operations, such as pressure application, timing, trimming, and packaging, are controlled manually by an operator. Although complex automated equipment is used for production forming, the need for economical prototypes and short-run production of plastic parts has always created markets for manual-control thermoforming machines. They are adaptable for initial educational purposes. A major use is to augment production equipment by enabling research and development of molds and techniques for high-speed production equipment. With these machines, techniques for proper forming can be developed to test tooling and various types of plastics. They provide a means to develop the next equipment and/or set up the next product line without interruption of normal production on the assembly line.

Semiautomatic operations perform a complete cycle of programmed forming functions automatically and then stop. An operator is required to manually start another cycle. This approach permits conducting detailed preliminary studies prior to going with a completely automatic system.

With automatic operation, a machine automatically performs all operations. The types of operations and their synchronizations will be provided when reviewing specific thermoforming processes. The machine will continually repeat the fabricating cycle. It stops only in the event of a malfunction or if it is manually interrupted. The machinery and mechanisms are so self-controlled and automatic that manual input is not necessary during operation once preliminary operating setup procedures are installed into the computerized process controls.

The most elementary thermoforming techniques include basic vacuum forming, drape forming, pressure forming, and matched-mold forming. These terms describe the single operation of clamping the heated sheet against the male or female mold and removing air between the sheet and mold surface by vacuum, external air pressure, or both. All thermoforming molds contain some type of system for air evacuation. The usual is using fine holes and/or slots.

Special processes were developed to accommodate the handling of certain plastics that are normally unavailable with conventional thermoforming machines. Most of these methods tend to

The blown bubble prestretches the plastic to uniform thickness, or the thickness of the material of the height of the bubble is equal to the thickness of the bottom of the finished part. Clearance between mold and clamp frame is usually gaged as 12% of the width of the part.

Width of the part is 23″ x 12% = 2-¾″

Usually ¾″ is required for clamping and mold clearance.

Formed part 48″ x 23″ x 20″ deep

Blank size is 55″ x 30″

Blank thickness is figured "area equivalent" in ratio to the area to be covered. Expect to get 100% use of center area, and 50% of the surplus area.

Center area 48 x 23 = 1104 sq. in.	1104
Surplus area (53.5 x 28.5) − 1104 ÷ 2	421
"Area-equivalent" total	1525 sq. in.

The finished part represents 3944 sq. in.

The ratio of part to "area equivalent" of stock is
3944:1525 or 2.5 ratio

The finished part requires a thickness of .090
.090 x 2.5 = .225

The above part requires a blank 55″ x 30″ x .225 thickness.

Table 7.17 Guide to determine size of cut sheet and draw ratio

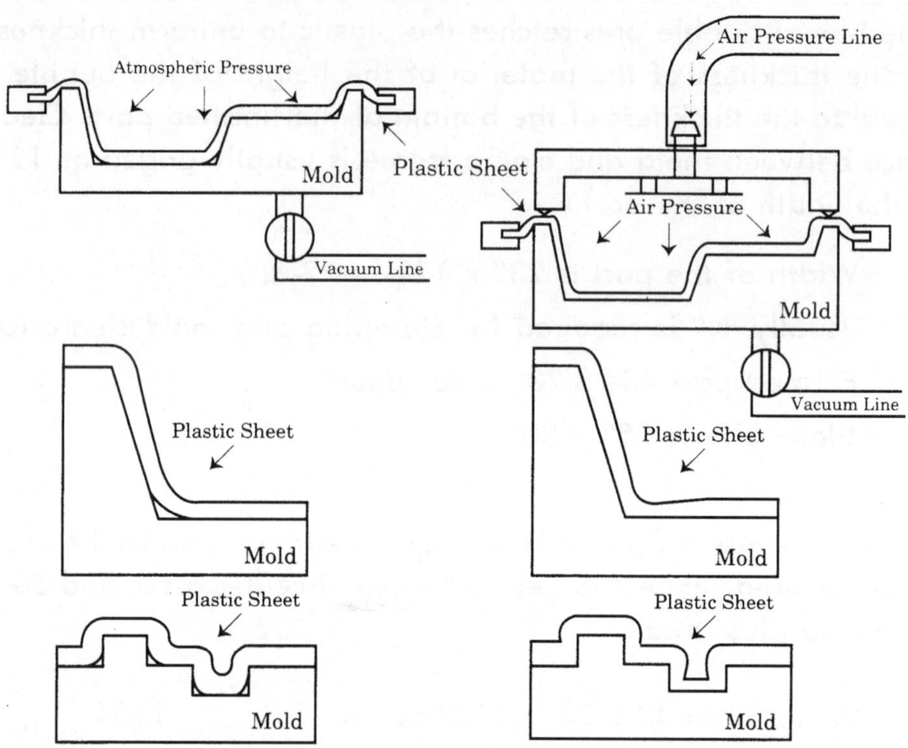

Figure 7.29 Comparison of vacuum and pressure-forming processes

reduce the required heat or even eliminate it entirely. One popular technique is high-pressure forming, which is like conventional compression forming or compression molding. Many of the techniques that are used have been modified in the past for conventional metal-working forming tools.

These methods are cold forming (performed at room temperature with unheated tools), solid-phase forming (plastic is heated below the melting point and formed), and compression molding of reinforced/composite sheets (using heat). Other methods are classified as forging (including closed-die forming, open-die forming, and cold pressing), stamping, rubber pad or diaphragm forming, fluid forming, coining, spinning, explosive forming, scrapless forming, and so on.

Cold forming and solid-phase forming can be used to process plastics, such as ABSs, PCs, conventional PPs, and ultrahigh molecular weight high-density PEs (HMWHDPEs). By using solid-phase forming, processors can make more efficient use of HMWHDPE plastics that are difficult or impossible to process by other methods. Thermoplastic-reinforced plastic composites can be formed by stamping to produce high-performance products. Different fiber reinforcements are used (chapter 15).

Solid-state (compression-molding approach) forming uses a male metal plug mold that matches a female metal cavity mold and can be used only with crystalline plastics. Below their glass- transition temperatures (T_g), amorphous resins are generally too stiff to be rapidly formed into stable products. Crystalline types can be permanently deformed at temperatures between their T_g and their melting-point temperatures (T_m). Molecular orientation, the mechanism that allows this to occur, relates to the draw ratio. Examples of draw ratios can vary from 5:1 for PET and nylon to 10:1 for low molecular weight PP. The major advantage of solid-state forming is that parts can be produced in very fast cycle times, usually ten to twenty seconds. The surface finishes of these products are rather smooth, as the fibers do not surface.

Either crystalline or amorphous types (chapter 1) can be used with flow molding, because the plastic is melted prior to forming. The forming temperature is usually lower than for injection molding or extrusion. Plastics need not be trimmed, as the composite is compression molded to completely fill the mold cavity (chapter 14). It is very important to note that the flow-molding process permits more complex parts to be formed than solid-state forming. The process cycle time is usually about one minute.

Processing Phase

There are two distinct process strategies for thermoforming thermoplastics. As an example, polypropylene is a crystalline material that can be processed either below or above its crystalline melting point (chapter 1). When forming is carried out below the crystalline melting point, it is known as solid-phase forming; melt-phase forming is performed at temperatures above the crystalline melting point.

Solid-phase forming of polypropylene is carried out at temperatures ranging from 155° to 165°C; the range midpoint is usually quoted as the optimum setting. Under these conditions, the sheet is relatively strong and resistant to sagging. The disadvantage is that a simple vacuum may be insufficient to form the sheet fully, so plug assistance is advisable to prestretch the sheet. The magnitude of force needed to form the sheet in the solid phase is markedly dependent on temperature, and even within the narrow recommended working range, may vary by an order of magnitude. Solid-phase forming is effective with all thermoforming grades of polypropylene.

Melt-phase thermoforming is performed at temperatures ranging from 170° to 180°C and requires polypropylene grades with high melt strength. Conventional grades have insufficient sag resistances to be formed successfully by this method. Sag resistance is related to molecular weight and molecular weight distribution and is expressed by the extensional viscosity of the material (chapter 1). High melt strength thermoforming grades are formulated to optimize these characteristics, generally using long-chain branching to impart the necessary behavior.

The two techniques result in products with rather different characteristics (Table 7.18). Melt-phase thermoforming requires less force so it generates lower residual stress levels in the finished part. The lower tensile strength of the sheet during forming also tends to produce a more even thinning, therefore reducing wall thickness variations.

However, heating above the crystalline melting point produces an amorphous sheet structure that recrystallizes during cooling to produce a forming with decreased transparency but better heat

Property	Solid phase forming	Melt phase forming
Wall thickness distribution		More even
Residual stress		Lower
Optical properties	Clearer	
Mechanical properties	Stronger	
Heat resistance		Greater

Table 7.18 Comparison of product behavior in solid-phase and melt-phase thermoforming

resistance. Performance under heating also benefits from the lower stress level, resulting in a reduced tendency for distortion and reversion to the original flat sheet form. For these reasons, melt-phase thermoforming is often preferred for retortable and microwavable polypropylene containers. However, modern solid-phase thermoforming can also produce containers capable of service under these conditions, particularly when carried out with multilayer (coextruded) sheet or film (chapter 5).

Process Control

Various systems are available to provide automated control. Microprocessors are a desired means of control. Repeatability of operations from cycle to cycle and accurate control of all functions are essential. Preprogrammed setups can decrease cycle times, scrap, labor, and cost. Processing conditions are continually readjusted during operation to keep unacceptable products from being formed. The productivity savings can defray the initial investment over a short period of time.

Use is made of the programmable logic controller (PLC) system as a means of computer control. Providing the same capabilities of repeatability, programmed operation, and increased productivity as a microprocessor system, the PLC has several advantages. Because a PLC uses ladder logic, less technically trained people are required and troubleshooting is simplified. A microprocessor system uses computer language (chapter 3).

Electromechanical control is used. It is the oldest method of automatic control in thermoforming equipment. Multiple relays, individual timers, and limit switches initiate successive steps in the sequence. This method is used by a small percentage of thermoformers performing simple, nondemanding operations.

Vacuum Forming

The oldest method of thermoforming, vacuum forming relies on the sheet's self-sealing ability and evacuation of trapped air by applying a vacuum. Natural atmospheric pressure fills the mold cavity, forcing the heated sheet into the evacuated space. A vacuum pump is necessary in this process.

Vacuum forming is the simplest of the thermoforming processes. A female mold is used with an unencumbered planar upper surface against which the sheet is clamped (Fig. 7.30). The sheet is first gripped around its periphery in a frame clamp and is then heated to the final forming temperature, preferably in a sandwich heater that heats both surfaces of the sheet. The frame is then lowered to contact and seal against the mold face, where together with the mold cavity, it forms a closed cell.

The mold is provided with a series of small ports through which the air in this cell can be evacuated by a controllable connection to a vacuum pump, usually augmented by a vacuum surge tank. The vacuum in the mold cavity causes atmospheric pressure to force the heated sheet into contact with the cooled mold cavity, where it cools in the shape determined by the cavity. It is a characteristic of thermoforming that the product is formed within a region of waste sheet material arising from the peripheral clamped areas and from points between the cavities of multicavity molds. Trimming away this waste material completes the product. The basic vacuum-forming process is the foundation on which more flexible and elaborate thermoforming processes are built by adding elements, as to be reviewed, such as billow and plug prestretching of the sheet.

Pressure Forming

This process is faster than vacuum forming and provides more clearly defined details on finished parts than straight vacuum forming. Certain plastics require this faster forming method, such as oriented polystyrene (OPS). High air pressure is used, thus requiring molds capable of withstanding the pressures applied. Also, a sealable pressure chamber is needed, as well as equipment capable of withstanding the pressure forces. An ample supply of pressurized air with uniform pressure levels is required. The compressed air system supplies dry air to prevent machinery and mold damage.

The pressure-forming process is a variation on vacuum forming. It uses positive air pressure instead of a vacuum to form the sheet, therefore overcoming the force ceiling of less than 1 atmosphere, which is a limitation of vacuum forming. The sheet is frame clamped and heated in the same way as in vacuum forming but is then trapped and sealed between the mold and a pressure chamber (Fig. 7.31).

When air under pressure is admitted to this cell chamber, the sheet is forced into contact with the mold form. During this operation, air must be expelled from the space between the sheet and the mold form. This is done either by venting the space to atmosphere through a series of small ports in the mold, by evacuating the space by vacuum means, or by a combination of both. The vacuum increases the forming force, but its primary purpose is as a venting device. It is convenient to use a vacuum in dual-purpose thermoforming machines designed to operate both the vacuum and pressure-forming processes. Most of the process variations available to vacuum forming can also be incorporated in the pressure-forming process, adding to the scope available to the thermoformer.

When exerting forming forces in excess of atmospheric pressure, pressure-forming processes greatly expand the design envelope and market applications for thermoforming. Much thicker and stronger sheets can be formed, the replication of mold surface detail is greatly improved, and it becomes possible to form relatively sharp corners and undercut features.

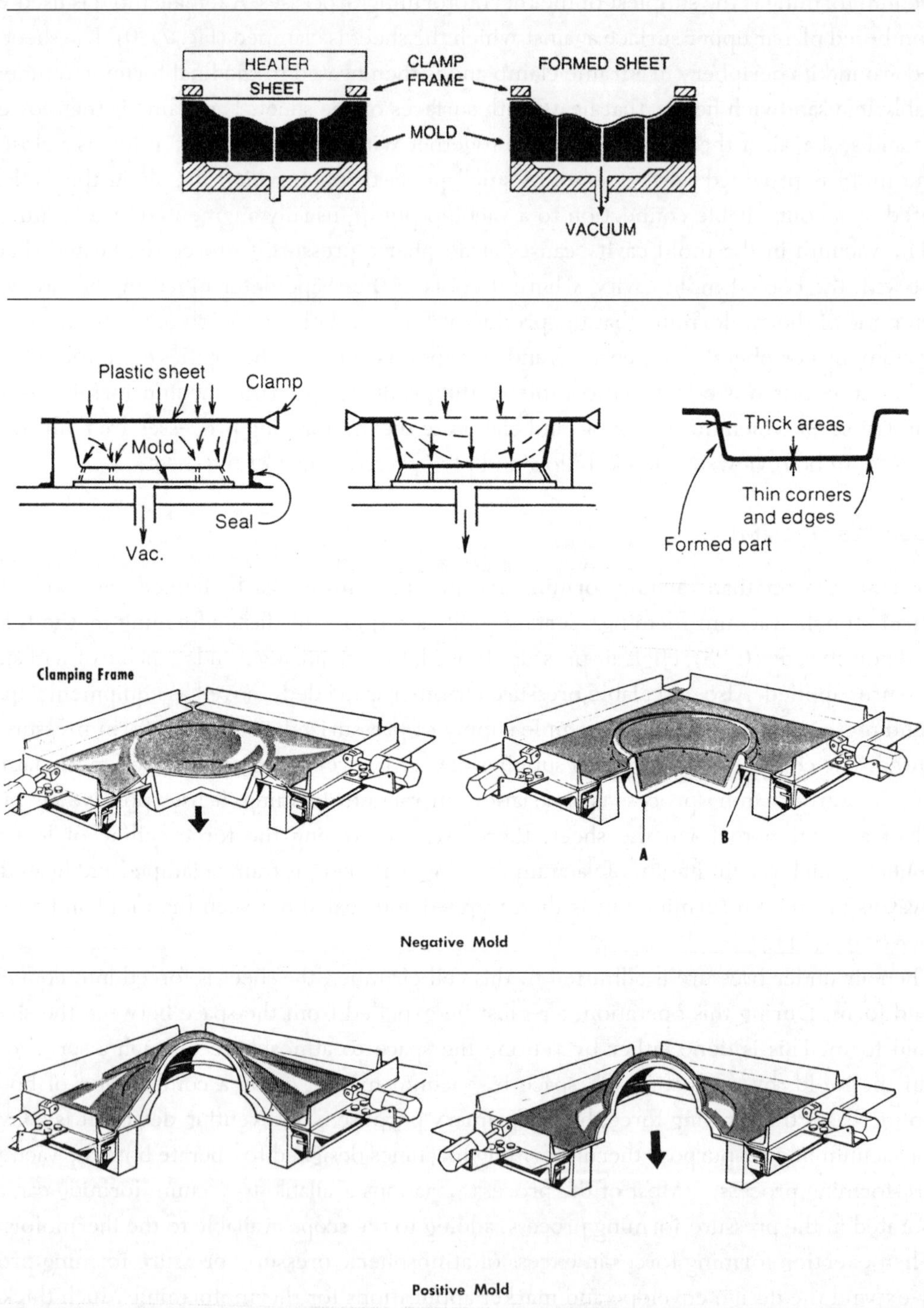

Figure 7.30 Views of vacuum thermoforming

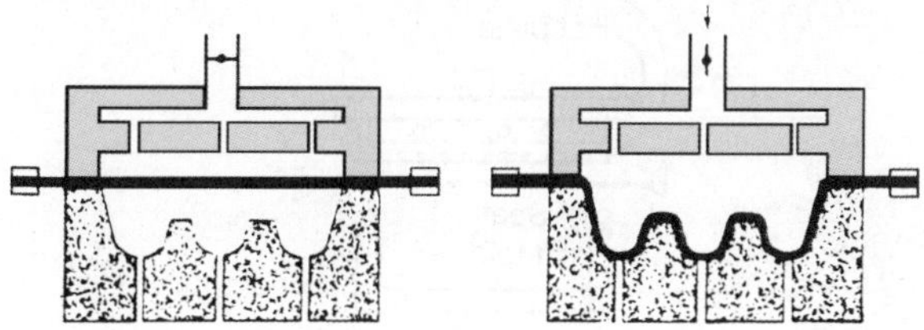

Figure 7.31 Basic pressure-forming process

VACUUM/AIR PRESSURE FORMING

There are different forming techniques that use this combination. Airflow and air pressure are used to preform a heated sheet prior to the final pulldown onto the mold, using a vacuum between the heated sheet and mold cavity (Fig. 7.32). This is a variation of pressure forming and vacuum forming. In one example, matched molds are used. The female mold is precise with respect to the finished part. The male mold need only be similar to the female in order to assist or push softened plastic sheet into the female cavity, providing a preforming action. These preforming molds are plug assists. After the sheet is forced into the female mold, vacuum and/or air pressure is applied to finalize details of the formed part in the mold. This method allows adjustments to easily be made in air pressure, vacuum, mold closing speeds, and so on, to form a better product.

Differential air pressures are produced by introducing compressed air through a chamber enclosing the top side of the mold and by vacuum evacuation of the space between the sheet and the mold. A high-volume low-vacuum pump is generally used. The pump should be capable of supplying a vacuum of 24 to 28 in of mercury and should have sufficient capacity to recover the volume of the forming unit as fast as the most rapid forming cycle.

The pump is connected to an accumulator tank approximately five times the largest volume possible in the forming unit. A tank of adequate size should prevent the vacuum from failing below 15 in of mercury. Vacuum lines and valve controls should be of sufficient size to prevent any restriction of flow for rapid evacuation of the mold. When only a slight vacuum is desired, a regulating valve is useful. A bypass on the vacuum line will allow a partial vacuum bleed-off followed by full vacuum, such as in plug-assisted forming.

BLOW FORMING

This is a method of shaping thermoplastic sheets, such as PMMAs and CAs, that uses compressed air. The process consists of securing the edges of a heated sheet to a metal backing plate and applying about 15 psi (100 kPa) of internal pressure to blow to the desired shape, such as a hemispherical building bubble or an elongated aircraft canopy bubble.

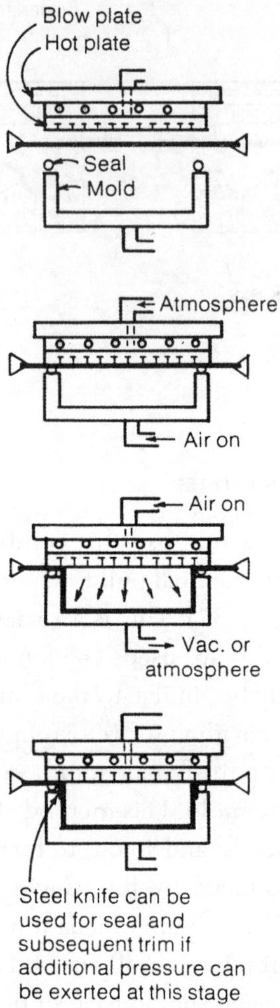

Figure 7.32 Example of pressure-vacuum thermoforming

DRAPE FORMING

The drape forming process is the simplest technique for use with a male mold. The heated clamped sheet is lowered over the mold until it seals with the mold base. This action allows the heated sheet to conform by gravity or pressure. The mold form acts as a crude sheet prestretch plug. The sheet chills rapidly on contact with the cold mold, resulting in the tendency to produce thick bases and excessively thin side walls. Techniques such as using molds that can be heated eliminate or drastically improve this potential problem.

DRAPE VACUUM FORMING

Basically, this is the forming of a hot sheet and laying it over a mold, either male or female, and allowing it to conform by vacuum (Fig. 7.33). The sheet is clamped into a movable frame, heated, and draped over high points of a male mold. The vacuum is pulled to complete the forming operation. In this technique, a male or female mold is closed into the hot sheet.

The vacuum is applied to quickly remove the air between them. Atmospheric pressure (14.7 psi [0.10 MPa]) moves the hot pliable sheet against the mold, holding it there until it cools to the heat-distortion temperature. It is very important, as with all processes that use vacuum and/or compressed air pressure, that the vacuum and/or compressed air remain on the part until it has cooled to below at least the plastic's heat-distortion temperature.

DRAPE VACUUM–ASSISTED FRAME FORMING

A frame, made of anything from thin wires to thick bars, is shaped to the peripheries of the depressed areas of the mold and suspended above the sheet to be formed. During the forming, the assist frame drops down, drawing the sheet tightly into the mold. This action prevents webbing between high areas of the mold and permits closer spacing in multiple molds.

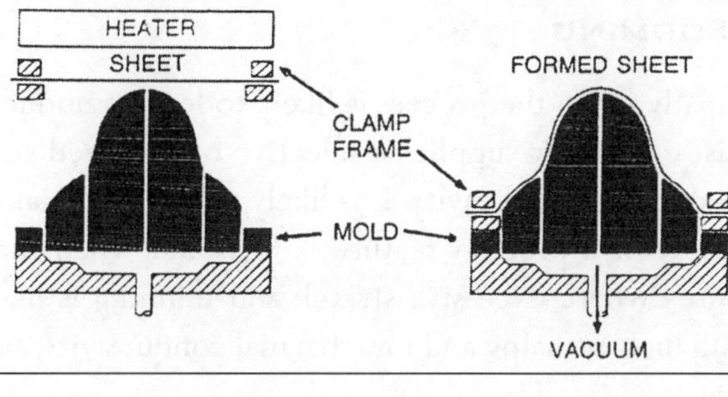

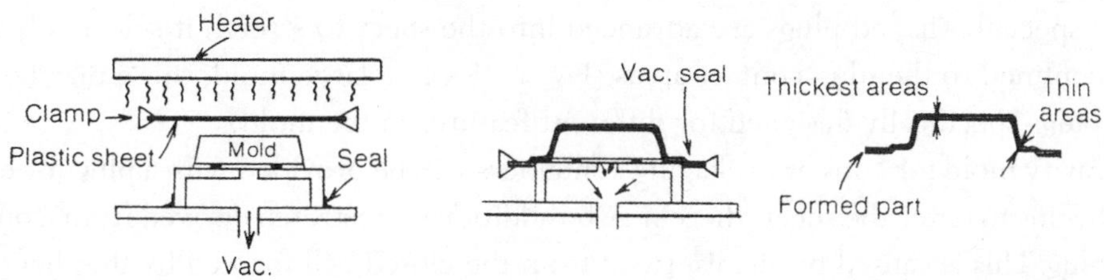

Figure 7.33 Examples of drape forming

DRAPE WITH BUBBLE STRETCHING FORMING

This is a modified system of drape forming for producing more uniform wall thickness that minimizes the dangers of tearing over the corners of large moldings because of the protective cushion of compressed air above the rising mold. The film or sheet is heated and blown into a bubble so that the sheet is prestretched before forming.

SNAP-BACK

Sheet prestretch is used when the forming force supplied by the air pressure differential is insufficient to stretch the sheet adequately or when the forming force alone is likely to lead to excessive variations in the product's wall thickness. These two techniques overlap to an extent; however, broadly speaking, the first technique is met by billow and air-slip techniques that stretch the whole sheet, while the latter is dealt with by plug-assisted means that provide selective prestretching during forming.

In the snap-back process, a vacuum (or pressure) is used both to billow prestretch and to form the sheet. The heated sheet is first clamped across a vacuum box or chamber, which is then partially evacuated, causing atmospheric pressure to billow and stretch the sheet. A male mold is then advanced into the billowed sheet, and forming is completed by drawing a vacuum (or pressure) on the mold while venting the vacuum chamber to the atmosphere (Fig. 7.34). The process is a variation of drape forming but with the advantage of sheet prestretch, which produces a much more uniform distribution of wall thickness

PLUG-ASSISTED FORMING

Plug assist is used principally when the process is likely to lead to undue variation in the product's wall thickness. Plug assist essentially supplies a selective or localized stretch that is related to the specific demands of an individual mold cavity. It is likely to be beneficial when the draw ratio (ratio of depth to projected area) of a product feature is high and when the product includes edges, corners, and other features where excessive stretch and thinning is likely to occur. Plug assist is preferred for plastics with high enthalpy and low thermal conductivity, such as polypropylene sheet formed in the solid phase.

In the plug-assisted process, the heated sheet is clamped, usually against the mold face, then one or more specially shaped plugs are advanced into the sheet to stretch it selectively in areas at least partly confined to the plug contact areas (Fig. 7.35). In a large mold, there may be a number of different plugs specifically designed for different features in the mold.

A multicavity mold for cups or packaging containers will be provided with a plug for each cavity. As the plug begins to enter the sheet, there may be additional prestretch derived from a billow effect around the plug. This is caused by air displaced from the closed cell formed by the sheet and mold and is absent if the cell is vented during the plug-assisted action. When the plug is fully advanced, the forming is made by forcing the prestretched sheet into contact with the mold by the normal vacuum means. The plug-assisted technique is also frequently used with the pressure-forming process.

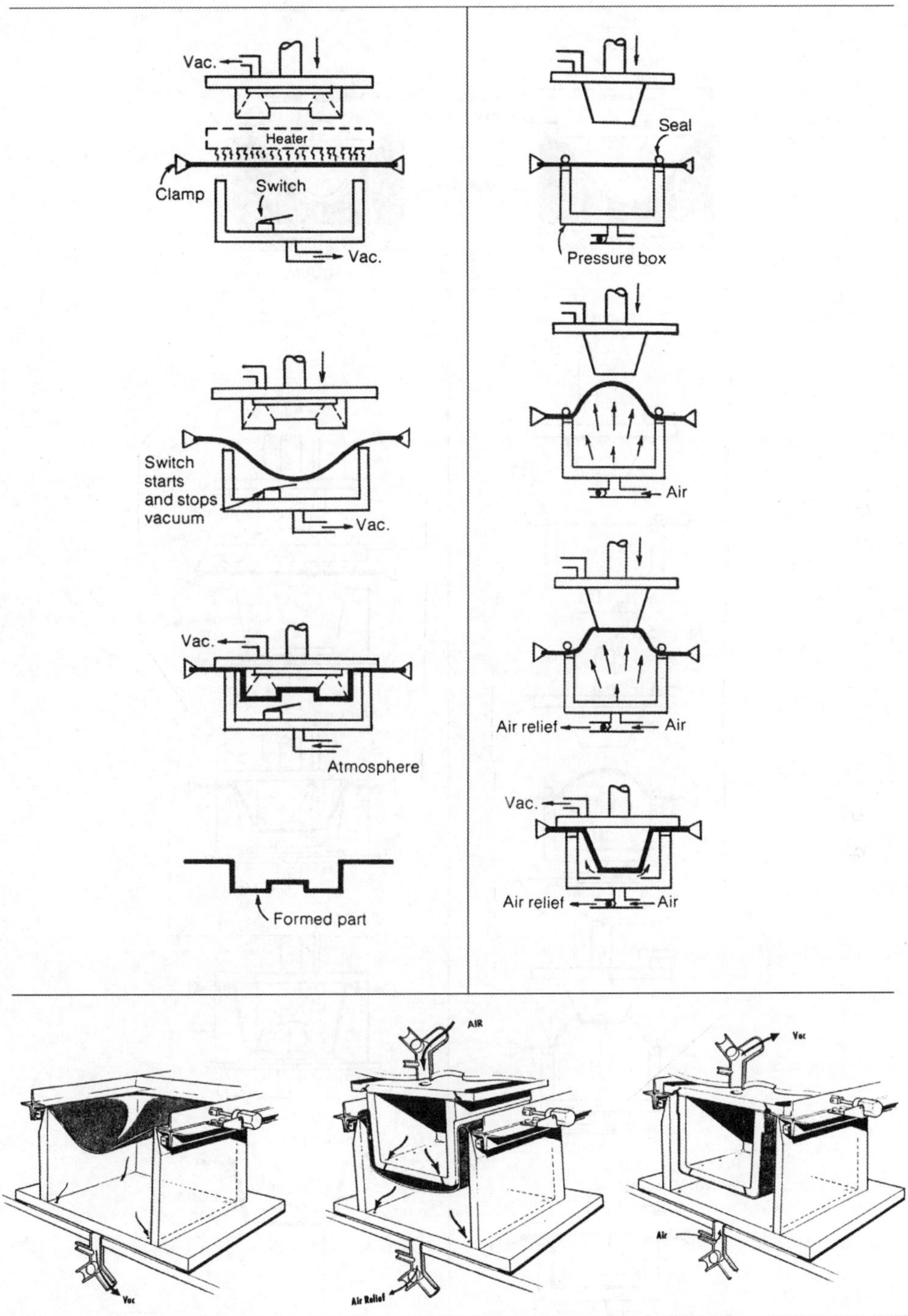

Figure 7.34 Examples of snap-back processing

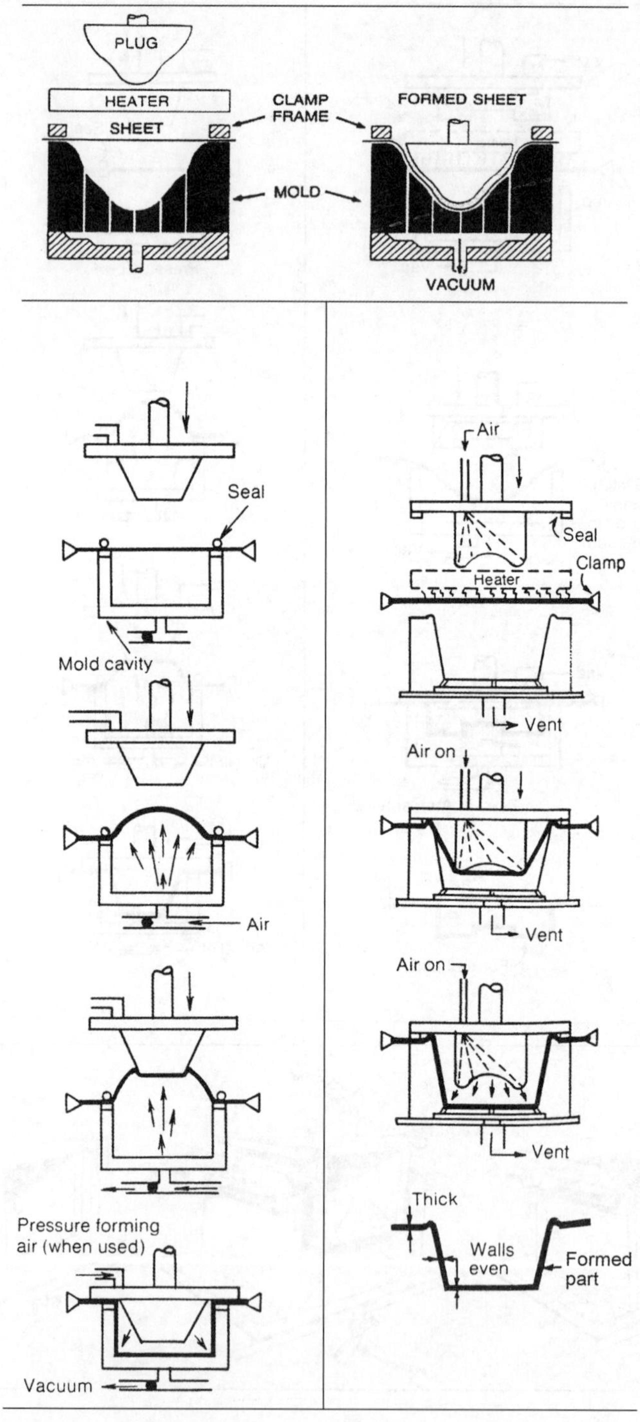

Figure 7.35 Examples of plug-assisted processes

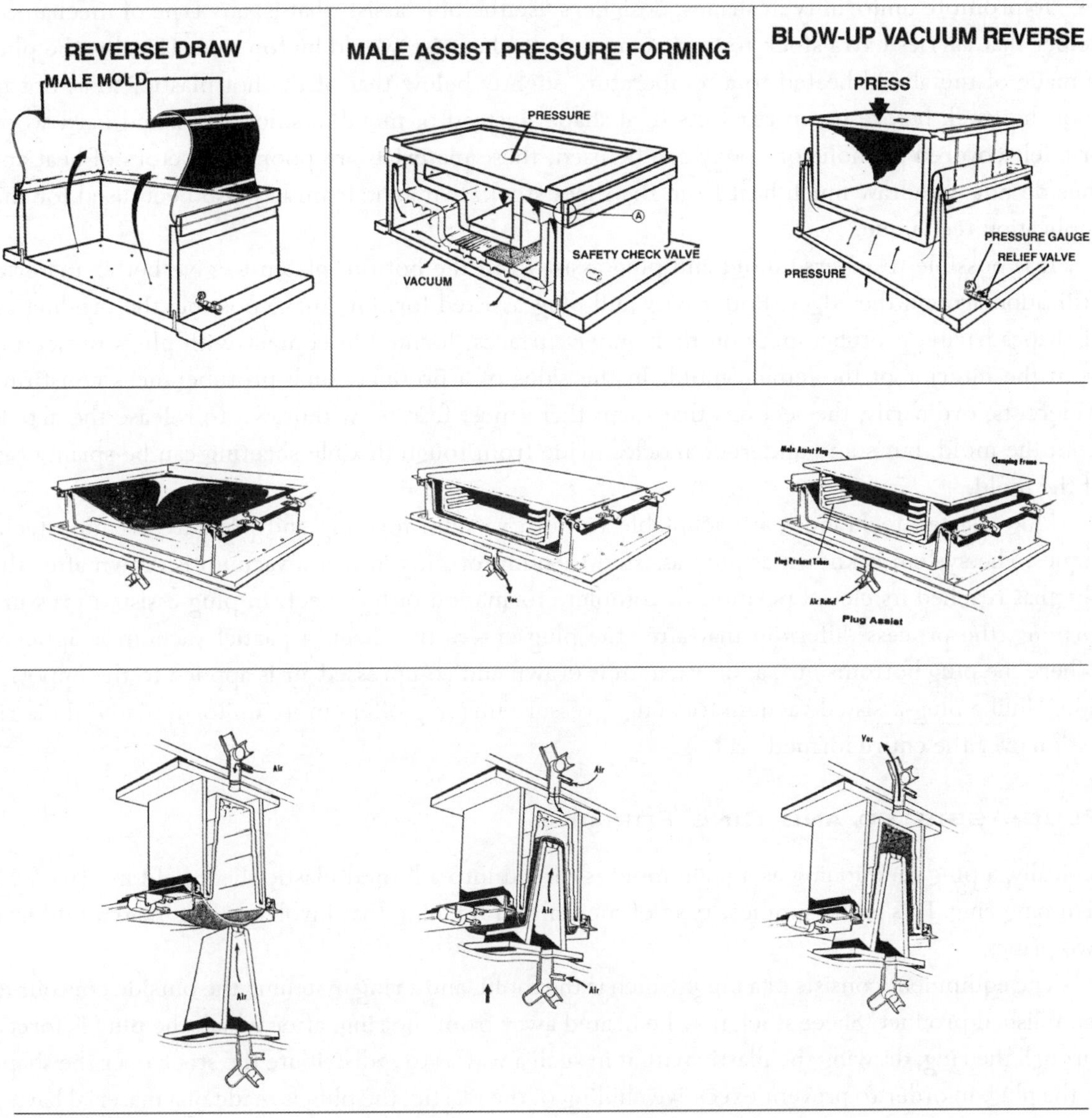

Figure 7.35 *(continued)*

To promote uniformity in shapes, designers use the plug assist, that is, any type of mechanical helper that carries extra stock toward an area that otherwise would be too thin. Usually, the plug is made of metal and heated to a temperature slightly below that of the hot plastic, so as not to cool the stock before it can reach its final shape. Instead of metal, a smooth-grained hard wood or a felt-covered phenolic or epoxy can be used; these materials are poor conductors of heat and thus do not withdraw much heat from the sheet stock. Syntactic foam has also been used for this application technique.

It is possible to reverse mold and plug assist from the bottom platen to give better material utilization around the edges. Both cavity and plug-assisted forming make possible the production of shapes having protuberances on their inner surfaces, formed in contact with plugs projecting from the interior of the female mold. In the sides of a product, such protuberances constitute undercuts; ordinarily, the sections that form them must first be withdrawn to release the article from the mold, but some undercut articles made from tough flexible sheeting can be sprung out of the mold.

Plug-assisted techniques are adaptable to both vacuum-forming and pressure-forming techniques. The system is known as plug-assisted vacuum forming in that a vacuum is drawn after the plug has reached its closed position to complete formation of the sheet. In plug-assisted pressure forming, the process differs in that after the plug enters the sheet, a partial vacuum is applied. Where the plug bottoms out, a full vacuum is drawn and compressed air is applied to the opposite side. Unlike plug-assisted vacuum forming, pressure forming offers more uniform material distribution over the entire formed part.

Plug-Assisted and Ring Forming

Basically, a plug functioning as a male mold is forced into a heated plastic sheet held in place by a clamping ring. This is the simplest type of mechanical forming that involves more than a fold into two planes.

The equipment consists of a plug, which is the mold, and a ring matching the outside contour of the finished product. Sheet stock may be heated away from the ring, after which the plug is forced through the ring, drawing the plastic with it in such a way as to redistribute the stock over the shape of the plug. In order to prevent excessive chilling of the plastic, the plug is made of a material having slow thermal conductivity; if it is made of metal, its temperature is controlled. This method has been used for the manufacture of single- and multicavity high-volume products, such as trays and packaging for candy, fruit, and so on.

Ridge Forming

This is a variation of plug-assisted and ring forming in which the plug is reduced to a skeleton frame that determines the shape of the product. By having the sheet only in contact with ridges of this frame, the intervening flat areas are free from mold surface defects or mark-off and have better surface quality than when formed against a solid mold. If a skeleton frame that surrounds a plane is used, the areas of the formed piece are plane surfaces. In other shapes with ridges that do not fall in a plane, the intervening surfaces tend to be concave. This technique works very well with plastics having high hot strength, such as cast acrylics and butyrates. Examples of products include display cases and boxes.

Billow Forming

To create different formed products with different processable plastics, different variations of this basic process are used. For example, a heated sheet is clamped over a billow box (enclosed) chamber. Air pressure in the chamber is used, causing the sheet to billow.

Billow Plug-Assisted Forming

Like the snap-back process, billow forming is a sheet prestretch process used with a male mold. In this technique, the sheet is clamped across a pressure chamber and is billow prestretched by applying a low-magnitude positive air pressure to the chamber (Fig. 7.36). The male mold is then moved into the stretched sheet until the clamp seals on the mold periphery.

Contact between the mold and the sheet can be regulated by controlling the rate at which the pressure chamber is vented to compensate for the advancing mold. Drawing a vacuum through ports in the mold completes the forming. Alternatively, forming can be effected by repressurizing the pressure chamber at a higher pressure, or vacuum and pressure can be used together. The billow process has the advantage that a greater prestretch force is available than in the snap-back process.

The female mold can be mounted on the top or bottom platen of a thermoformer. The stock is placed in the frame and heated. Then the female mold is moved into the stock close to the edge of the frame, deep enough to maintain a seal while a bubble is blown. Air is introduced into the female mold to a pressure of about 1 to 3 psi or more, sufficient to create a bubble toward the plug. The plug, in this case only a sheet-carrying assist, is heated to a temperature only slightly less than that of the hot stock sheet itself. As it is moved, it pushes the redistributed stock into the cavity in a folding or rolling action.

The plug is designed slightly undersized, at about 85%. It should move the sheet to a shape quite close to its final position in the female mold. The final shaping then is accomplished with a vacuum and/or compressed air being applied to the female mold.

By regulating the temperature of the stock, the height of the bubble, and the temperature and speed of travel of the plug assist, as well as the release of air pressure and the vacuum application, it is possible to closely control the thickness of any given section of the article. One exception is that corners never will be thin, because the folding or rolling action as the plug moves downward gathers extra stock around the comers.

Products formed into a cavity shrink away from the mold, so little draft is required in the mold. This may draw processed products with uniform wall thicknesses less than 25% of the original stock thickness, with repetitive quality, and usually without apparent impairment of the physical

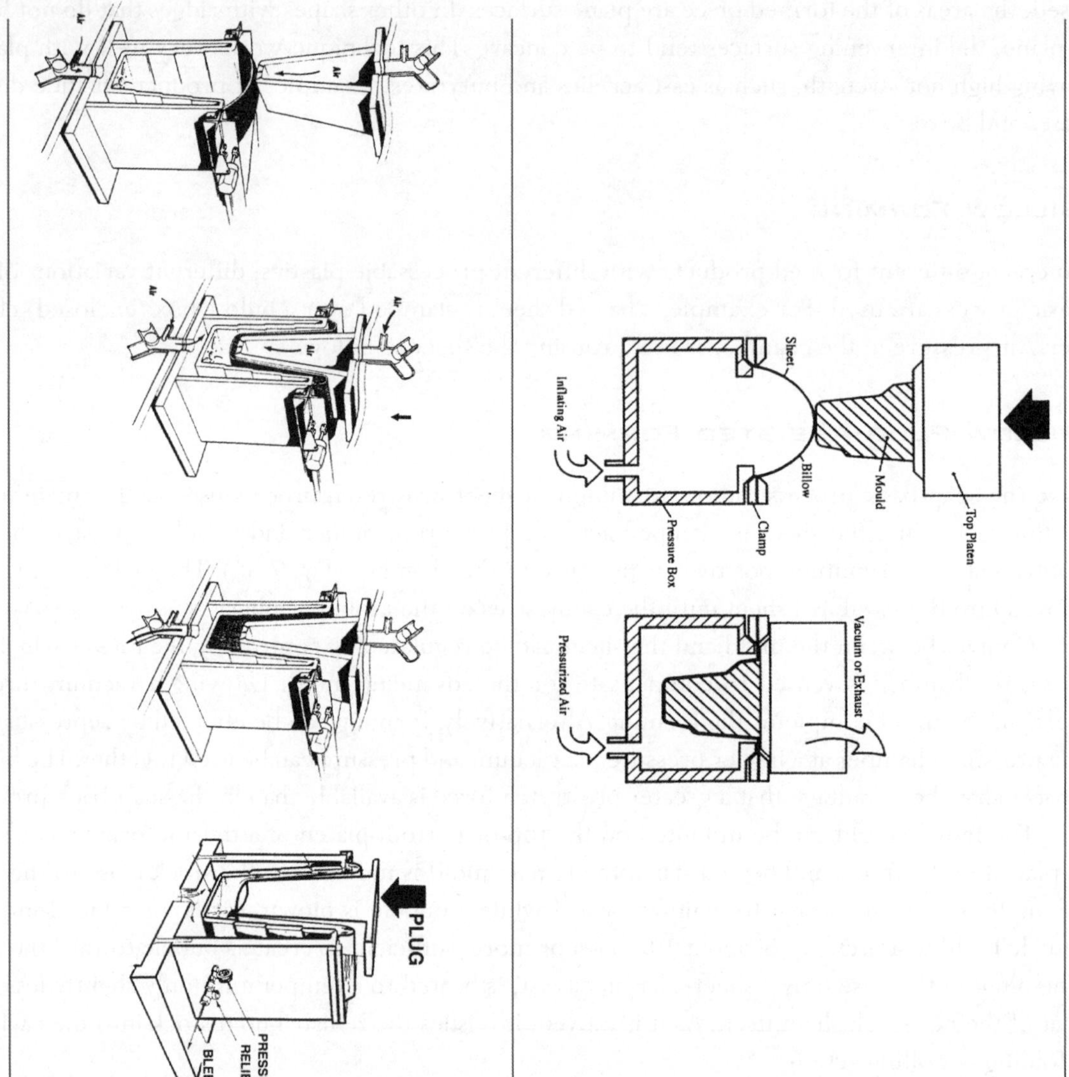

Figure 7.36 Examples of billow process

properties of the original sheet material. In some cases, a uniaxial orientation of stress may develop so that the finished article is crack sensitive parallel to the direction of stretch. This becomes more noticeable if the flanges are trimmed off.

Billow-Up Vacuum Snap-Back

This version of the billow forming method is used with grained or polished stock on a male mold to preserve the surface finish. The forming ring matches the finished product and assists in wiping the stock around the mold as the mold moves into final position. The mold should be run as warm as is compatible with efficient cooling of the product. It is usually a temperature-controlled aluminum mold.

The sheet is oversized to provide a bubble of sufficient size and shape. When the mold comes into contact with the outside of the bubble, the plastic that touches the bottom of the mold freezes there and need not be reduced further in thickness or change its shape. Corners should be cut to avoid webbing.

Billow Snap-Back Forming

Different techniques involve the use of a male or female mold and pressure chamber with or without a plug assist in the snap-back process. As an example with a male mold, instead of drawing the hot plastic into the prestretch box, it is billowed into a bubble shape. The height of the billow is usually about 65% to 70% the depth of the part during initial start-up and is then adjusted as required.

After the bubble has taken the proper shape, the mold is moved into it until the mold flange area seals against the box. Vacuum is applied and the sheet is snapped to the mold surface. This method normally produces a slightly more uniform wall thickness, but it is a little more difficult to set up. Pressure in the box needs to be controlled by relief valves as the mold is forced into it.

The bottom mold platen and a bottom moving heating panel are enclosed in a pressure chamber that is designed and constructed to withstand a minimum pressure of 10 psi (0.07 MPa). An outside top moving heater panel provides sandwich heat. With a top moving platen, great versatility in forming techniques is possible. The sheet is clamped into the top of the pressure chamber over the mold. After the sheet has been heated to forming temperature, the chamber is pressurized enough to prestretch the sheet. Usually, very low pressures are used. After the bubble is shaped, the mold is moved into it and a vacuum is applied. At the same time, the chamber is vented to the atmosphere. If a plug assist is needed, it can be actuated with the top platen.

When it is not practical to use a plug assist with a female mold, blowing a bubble and applying fast vacuum can apply a limited amount of prestretching. This is the billow-snap technique using a female mold and no plug. For best results, the female mold is mounted on the top platen. The mold is lowered into the hot sheet. After the mold seals off in the sheet, the cavity is pressurized, blowing the sheet into a uniform bubble about 60% to 80% of the depth of the cavity. A very rapid vacuum is then drawn, which adds thickness to the bottom of the product.

Air-Slip Forming

The air-slip process uses a positive pressure billow sheet prestretch that is generated by the movement of a male mold toward the heated sheet that is clamped across a pressure chamber. The mold platen acts as a piston in the chamber and drives a volume of air ahead of it that serves to billow the sheet. At the full extent of its forward travel, the mold periphery seals against the sheet frame and forming is completed by drawing a vacuum through ports in the mold (Fig. 7.37).

Conceptually, the process resembles the billow snap-back process. The billow can be regulated to an extent by venting air from the pressure chamber, but the penalty for combining the billow and mold functions is a loss of fine control over individual parameters.

This variation of snap-back forming has the mold enclosed in a box such that when the mold moves forward toward the hot plastic, air is trapped between the mold and the sheet. Upon the full travel of the sheet, a vacuum is applied to remove the air providing the sheet with a very snug fit over the mold.

Air-Slip Plug-Assisted Forming

The air-slip technique can be combined with plug assist. An example is where the forming requires both generalized and selective prestretching of the sheet. In this case, a plug is advanced into the billowed sheet when the mold is at or near the full extent of its forward travel. A partial venting of the pressure chamber compensates in a controlled manner for the volume displaced by the advancing plug. When the plug prestretch is fully extended, drawing a vacuum through ports in the mold completes the forming. The process is particularly suitable for molds that have both male and female features.

Blister Package Forming

Thin plastic film is thermoformed into simple to very complex shapes for packaging different products, such as candy, fruit, toys, and so on (Fig. 7.38). Certain packaging products can be used as molds. Machines are used chiefly for skin or contour packaging and blister packaging. Processes can involve the seating of paperboard or plastic to the formed plastic sheet, with the product to be package enclosed. Machines can be equipped to handle skin packaging followed by manual loading operations.

These machines may operate from an extruder, roll, or sheet stock, and there are machines equipped to trim the packages as they leave the machines. Rather elaborately designed machines are used for complete blister packaging. Machines will form the blisters, provide areas for manual or automatic loading, seal paperboard to the back sides (Fig. 7.39), and die-cut the finished packages. Generally, they operate with intermittent or continuous feed of roll stock and the various functions are performed simultaneously at a sequence of in-line stations.

Draw Forming

This process involves the stretching of a sheet to fit a mold and conform to the mold's cavity shape. It gradually reduces the sheet, film, or other shape of a material by stretching and/or pulling it through devices or perforations in a series of rollers and/or plates to successfully diminish its thickness. They may be cold drawn (without heat) or hot drawn (heated to their softening points).

DIP FORMING

Dip forming is frequently used for making vinyl plastisol products. The process starts by going through a dip coating operation (chapter 16). It involves dipping a mandrel into a tank that contains a

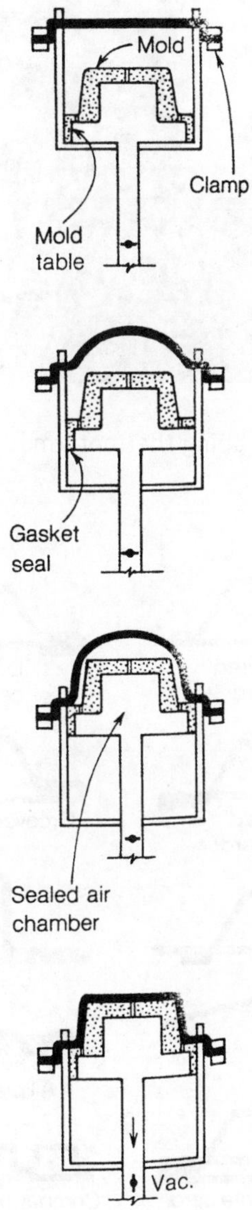

Figure 7.37 Example of air-slip process

Figure 7.38 Example of blister packages being thermoformed on a shuttle-type mold operation

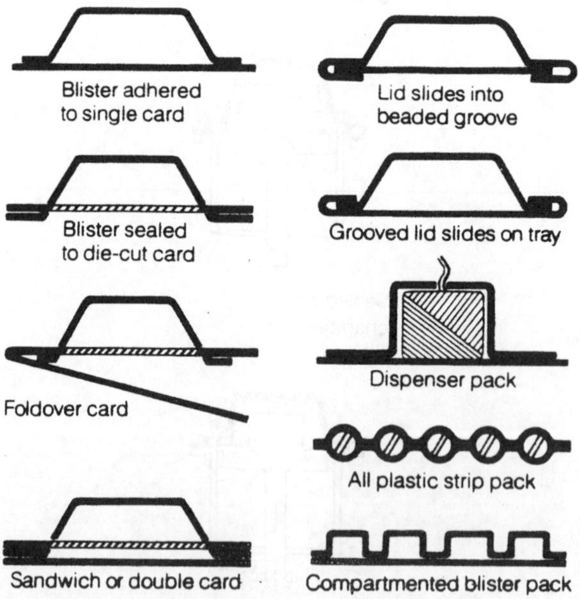

Figure 7.39 Examples of card pack blister packages

liquid plastic, such as a plastisol, and is held for a prescribed time so that the heated plastic gels against the form to the desired thickness. Shape of the mandrel simulates the shape of the finished product.

The coated form is withdrawn and usually receives a final heat to complete its fusing prior to a sufficient cooling cycle. The coating is stripped off the mandrel. After framing and heating, the stock form is mechanically stretched over a male mold to allow the framed edge to make a seal with the periphery of the mold. This stretching serves to redistribute or preform the sheet prior to application of the vacuum.

It has the disadvantage of allowing the sheet to touch prominent projections, freeze there, and perhaps rob other areas of sufficient mass to make acceptable products, particularly with improper design of a mold. Careful control of the temperature of the mold, plus selective heating of the sheets, can alleviate some of the potential difficulties that can occur. This process has the major advantages of low cost for both mold and machine as well as speed of operation.

Form, Fill, and Seal

FFS pouches are used in packaging different liquid and solid products (food, medication and medical devices, soaps, mechanical devices, etc.). They are thermoformed online with their contents automatically inserted in the pouches just prior to the sealing action.

Sheets or films are formed using a clam shaping procedure using two sheets or a single sheet that is shaped and sealed where the two edges meet to form the pouch just prior to filling. Sealing the package includes the use of heat, adhesive, or their combinations. In certain applications, a liquid product also can be used as a pressure medium to assist in forming followed with a sealing action. Because aseptic features are required for machines of this type, all operations generally are enclosed in pressurized, sterilized air chambers.

Form, Fill, and Seal vs. Preform

FFS technology generates a variety of concerns over its visual clarity. Incorporated preforms are typically thick walled and rigid; this is to protect expensive packaged items and heavy and/or sharp devices, to accommodate intricate shapes and compartments, and to prevent corner thinning. However, thickness is often accompanied by haze. FFSs are usually less thick and rigid than preforms. Although thinness improves clarity, it limits the depth of draw. Hence, FFS is well suited for packaging inexpensive high-volume products, such as syringes and gloves. They do not require protecting many physical properties (as opposed to microbiological properties). When a deeper draw is required as well as clarity, coextrusion is usually the option. There are combinations that are flexible yet tough enough for deeper draws. Nylon plastics are examples.

Form, Fill, and Seal with Zipper In-Line

In the past, this type of thermoforming technique required a secondary operation where the zipper was added after forming. Since the early 1990s, the zipper has been added in-line with the thermoforming action where the zipper is heat sealed. Speeds of these lines can match those without zippers being added. They provide sealed pouches and bags.

MULTIPLE-STEP FORMING

There are many variations in multiple-step forming, such as combining pneumatic billow and plug prestretching with zoning and pressure forming to meet final product requirements. This process forms heavy-gauge sheet where single parts are often quite complex and deep and where cost considerations make wall thickness uniformity a significant design requirement parameter. Localized nonuniform heating of the sheet, called zoning or patterning, is used to aid in material flowing in the proper directions.

Differential pressure is used to form certain filled or short fiber-reinforced plastics or to achieve sharp details in neat plastics (chapter 1). Differential pressures rarely exceed 10 atmospheres. Long fiber-reinforced and foamed plastics are frequently too stiff at the forming temperature for pressure molding; thus, matched die molding or diaphragm molding is used.

MATCHED MOLD FORMING

This method requires sufficient driving force of the equipment press platens, proper cavity air evacuation, and a realization of the depth-of-draw limitations. This process can be considered a variation of compression forming, compression molding, or reinforced plastic (RP) stamping (chapters 14 and 15).

Matched mold forming is a special case of thermoforming that differs from the mainstream in two important respects. The mold defines both surfaces of the finished part, and the forming force is not supplied by air pressure differential across the sheet but by reaction forces generated by a power press acting on the matched mold. The process amounts to compression molding using a sheet preform (chapter 14). The mold form is reproduced on both halves of the mold, offset by a dimension equal to the sheet thickness (Fig. 7.40).

Clamping and heating the sheet in the usual way then trapping it under pressure between the closing mold halves performs the process. High forming pressures allow thick sheets to be shaped with sharply defined details on both mold cavity surfaces. Wall thickness with a degree of dimensional accuracy is controlled by the clearance between the mold halves. Trapped air is allowed to escape through both mold faces. Platen press forces usually range from 5 to 50 psi (0.03 to 0.34 MPa). The result of this forming technique is that products are produced meeting excellent finished surfaces on both sides of the sheet. The surfaces of the mold cavities are reproduced on both sides of the sheet. To reduce the total forming cycle, molds can incorporate internal water-cooling systems.

Mold costs are higher than for pressure forming. Introducing moving members into the mold can permit undercuts. An exception is using flexible/elastomer thermoplastics with a degree of limitation on size and location of the undercuts so that they only require stripping the formed parts off the mold cavity.

Rubber Pad Forming This is a variation of matched metal stamping except that a metal mold half is replaced by a block of solid rubber. Processing material cannot flow to the extent that it can with

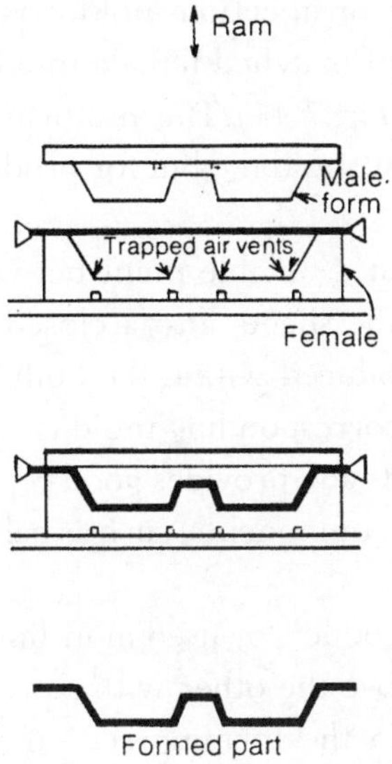

Figure 7.40 Example of matched mold process

matched metal forming. However, more uniform pressure is exerted on the material charge due to the rubber pad.

Mechanical Forming

This process does not use vacuum or compressed air to form the heated sheet. The forces necessary to shape the uniformly heated sheet are applied by mechanical or manual stretching, bending, twisting, compressing, stamping, and/or other devices. Mold components can be used, such as mechanically activated undercuts, side pulls, and plug assists.

Forging Forming

This is a production method wherein heated thermoplastic stock or billet material is formed by compression forces or by sharp hammerlike blows. Virtually all ductile materials may be forged, and in some cases preheating is not required. When forged below the melt temperature, it is called cold forging, cold forming, or solid-phase pressure forming. When worked above the melt temperature, it is said to be hot forging or hot forming.

Twin-Sheet Forming

Twin-sheet forming, also called dual-sheet forming and clamshell, is a special thermoforming technique. The process resembles blow molding (chapter 6), except that twin-heated sheets are used instead of a tubular extruded parison or injection-molded preform. Like a blow mold, the twin-sheet forming mold usually consists of matched female mold halves. Molds can include pinch-off zones at the periphery of the cavities (Fig. 7.41). Thermoforming does not provide the accuracy and tighter measurement tolerance of blow molding, but for products not requiring such performance, it provides significant cost advantages.

A blowing needle is introduced at a suitable point between the sheets, then the mold halves are closed under pressure, welding the sheets into a closed body at the pinch-off zones. At the same time, air under pressure is introduced within the body through the blowing needle, forcing each sheet into conformity with the corresponding mold cavity face. The scope for a high blowing pressure permits the use of thick sheets and provides good replication of mold detail. The process is an economical way of producing large components, such as pallets, tote trays, and tanks, particularly if in limited production.

It is used to produce individual products or used in in-line production of continuous products. Two automatically fed sheets, one above the other with a predetermined space in between, travel through the heating stations and into the forming station. These two preheated sheets or roll materials are clamped between two halves of a split mold (like a two-part mold used to form blow-molded or injection-molded products).

Each sheet is drawn into its individual mold cavities by a vacuum. In an additional step, compressed air is simultaneously injected between the sheets through a blow pin. An end contact could include an integral hinge. Where the two sheets meet with their matching perimeters at the mold

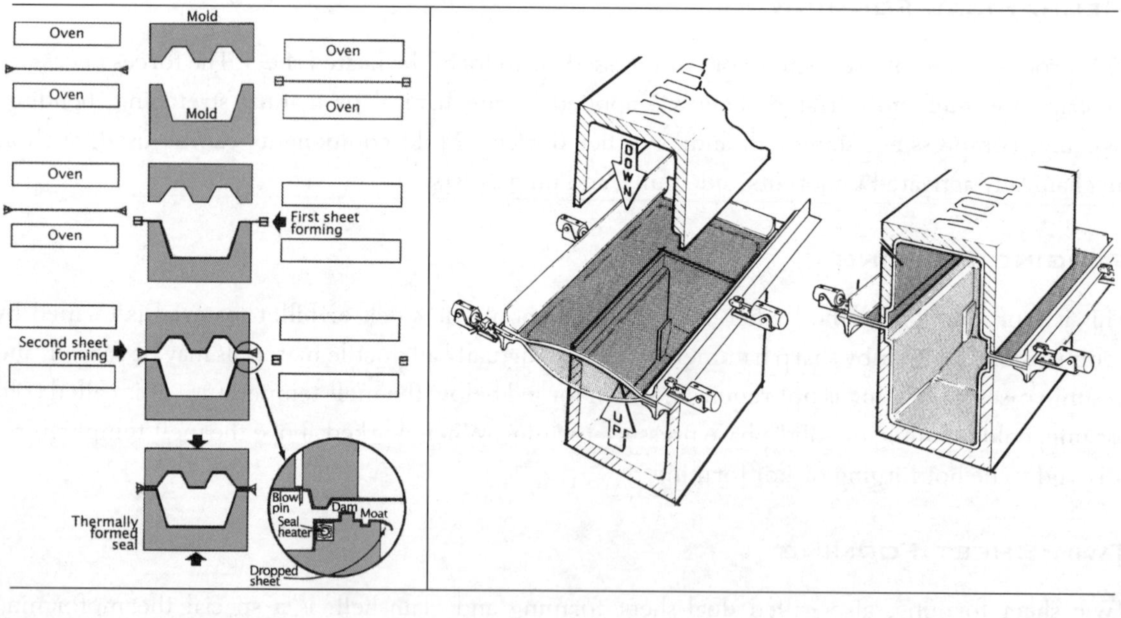

Figure 7.41 Examples of twin-sheet process

parting line, pressure applied by the two mold halves can result in a leak-proof bond. The action is similar to thermowelding with heat and pressure requiring that the surfaces of the sheets are at or above their forming temperatures.

Thermoplastic foam, such as polyurethane foam under pressure, can be introduced between the two formed and bonded sheets. The urethane bonds to the two skins, forming a rugged sandwich construction. This technique is applicable to the manufacture of polyurethane foam-filled boat hulls and other foamed core sandwich constructions. Another variation of this method uses a series of continuously moving molds (traveling on belts) that clamp onto the sheets (feeding off rolls) and travel with them as the vacuum is pulled and the sheet halves are formed.

COLD FORMING

This process forms thermoplastic sheet at room temperature. It is a stretching or orientation process used to improve properties, such as tensile strength and modulus, of thermoplastic and other materials by orientation of molecules (such as in living hinges). This process can change the shape of thermoplastic sheet, film, or billet in a solid phase through plastic (permanent) deformation with the use of pressure dies such as that shown in Figure 7.42. Split rotating pressure rolls are used to produce many different extruded-formed shapes, such as round, elliptical, square, rectangular, U-beam, I-beam, and H-beam. Through the split dies that are under a controlled pressure, a continuous plastic extruded profile is pulled.

The process can include heating that is usually well below the plastic's melt or thermoforming temperature. Thermoset (TS) plastics, such as B-stage, can be used. The main difference between metal and plastic forming is the time dependency of springback or recovery in thermoplastics. All materials exhibit some strain recovery or springback. With TPs, this process depends on temperature, time, and deformation history. For any given forming temperature, holding the part in the deformed state for a given period to allow for stress relaxation reduces the degree and rate of springback.

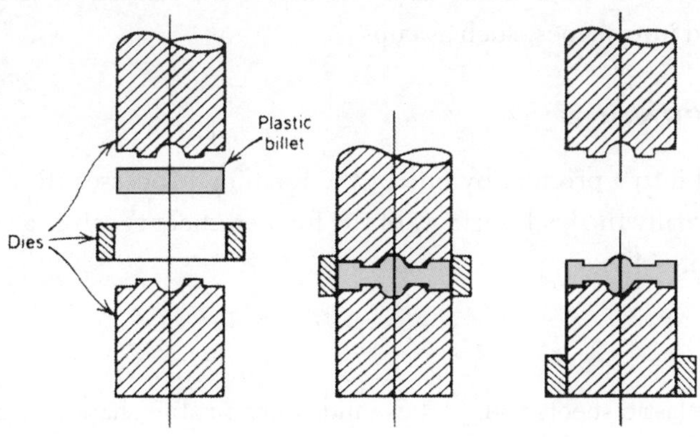

Figure 7.42 Example of compression action for the cold forming process

COMOFORM COLD FORMING

This is an extension of the cold molding process. It uses thermoformed plastic skin to improve surface and other characteristics of a cold-molded thermoplastic or a B-stage thermoset-reinforced plastic (RP). The mold is closed and the fast room-temperature curing RP system hardens it. The finished product has a smooth TP-formed sheet backed up with the cured RP.

SHRINK-WRAP FORMING

A very popular product is shrink-wrap, which forms packaging of all kinds of products. The strains developed in a plastic film are released by raising the temperature of the film, thus causing it to shrink over the package where the packaged product acts as a mold. These shrink characteristics are built into the film during its manufacture by stretching it under controlled temperatures to produce orientation of the molecules (chapter 5). Upon cooling, the film retains its stretched condition but reverts to its original dimensions when it is heated around the package. Shrink film gives good protection and excellent clarity.

SCRAPLESS FORMING

Different processes have been developed with a few in use. Their designs usually start with extruded thermoplastic sheets. Circular blanks cut from the sheet are compression molded into the desired preliminary shapes. During the compression action, the blanks can be simultaneously stretched or, being thermoplastic, can be stretched after compression molding. They generate no trim scrap when formed. They have performances and advantages such as exceptional high degrees of molecular orientation, resulting in improved part toughness, stress crack resistance, and so forth. Examples of the thermoforming processes are shown in Figures 7.43 to 7.45.

Dow's patented scrapless forming process is identified as the SFP process. An extruder produces a biaxially oriented sheet (chapter 5). In turn, the sheet is slit and cut into square blanks. These blanks are heated and pressed into circular disks with lips. Immediately or later after reheating, the disks are thermoformed into shapes, such as cups.

FORMING AND SPRAYING

Sheet is thermoformed into a product by one of the forming processes. To reinforce it or make it more rigid, one side, usually the back surface, of the formed sheet receives a spray-up of reinforced plastic (Fig. 7.46) (chapter 15).

POSTFORMING

Postforming identifies plastic sheet, rods, tubes, and other profile shapes that are formed into different shapes. Its major use is in-line with an extruder processing thermoplastics. As the plastic exits the die, it is formed into different shapes. Details on postforming thermoplastics are in chapter 5.

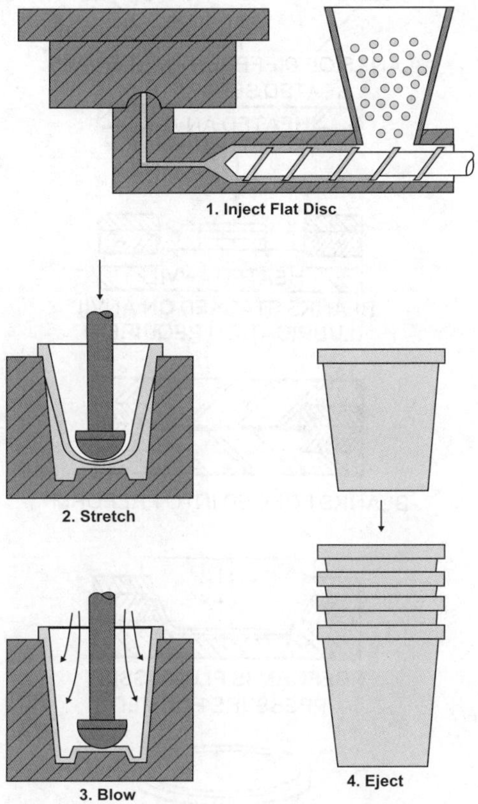

Figure 7.43 Forming occurs after a shot of melted plastic is injection molded into the forming cavity (chapter 4)

Very limited use is made of forming, bending, or shaping B-stage and certain types of fully cured C-stage reinforced thermoset plastics that have been heated to make them flexible. On cooling, the formed laminates retain their contours and shapes of the molds over which they were formed. The postformed products are subjected to final higher temperatures that cure and solidify the thermoset plastics (chapter 15). Generally, only simple shapes are produced by this technique.

Bend Forming

This postforming technique is one of the oldest thermoforming techniques used with TP and TS reinforced plastic. The bending is relatively easy to handle, particularly when it is a straight bend resulting in meeting forming requirements. If the sheet is heated only locally in the bending

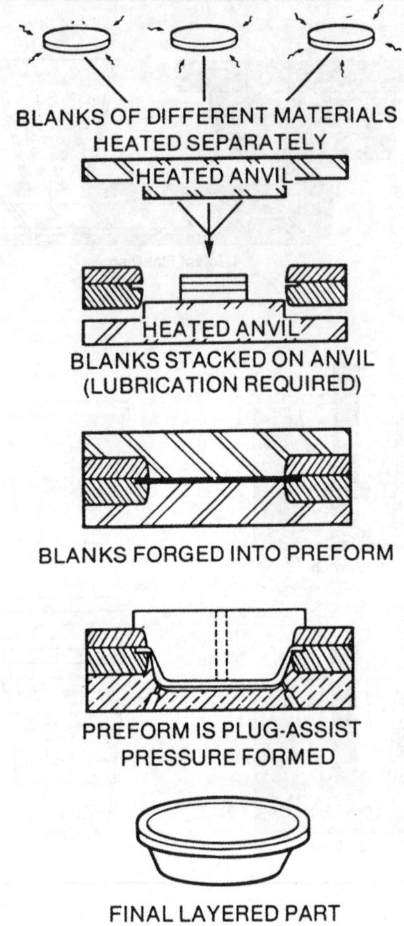

Figure 7.44 Dow's COFO process heats and forms plastic blanks

operation, no special forming tools are needed. The type of material, width of the heating zones, and thickness of the sheet determine the bending radius capability. Limitations or sags of the sheet are related to the softening point of the sheet and the intrinsic rigidity of the heated sheet. Transparent plastics, such as PMMAs and PCs, are frequently bent for use in store displays, staircases, partitions in banks, aircraft canopies, windows, and so on. RTS laminated materials are used in various markets that range from displays to kitchen countertops.

TRIMMING/SECONDARY EQUIPMENT

Because they are formed from clamped sheets, virtually all thermoformed parts must be trimmed to remove the waste/scrap that has to exist in order to clamp the sheets. About 90% of thermoformed

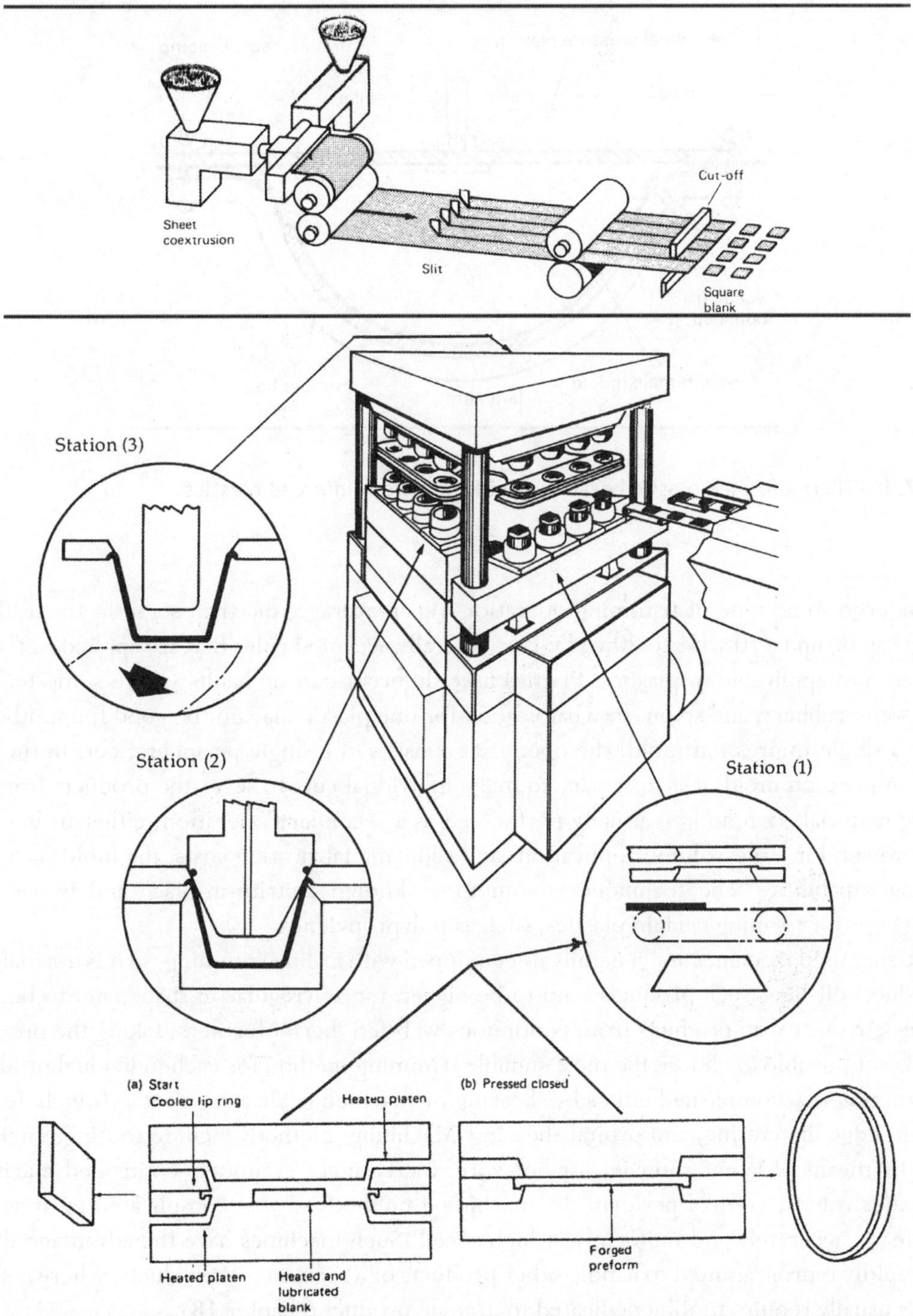

Figure 7.45 Example of Dow's SFP process going from an extruder to the formed products

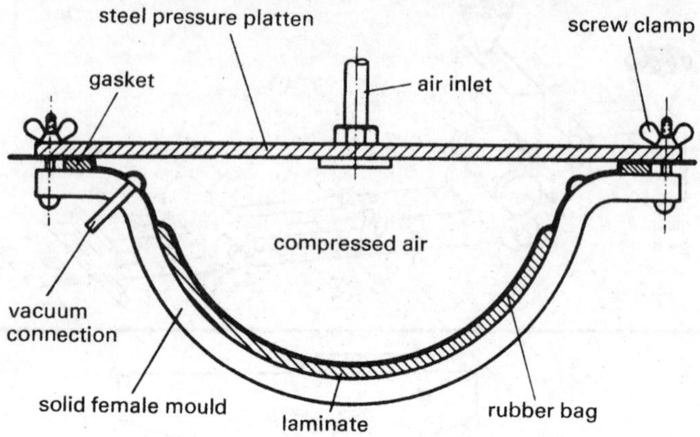

Figure 7.46 Thermoformed plastic backed up with sprayed reinforced plastics

parts undergo some type of trimming operation. As an average, the trim scrap factor in thermoforming can be up to 10 wt% to 20 wt%. The generally accepted rules that are applied to trimming metals are not applicable to plastics. Plastics have different cutting habits where some tend to be brittle, some rubbery, and so on, so what is good for one plastic may not be good for another.

For a single-impression mold, the operation consists of a single peripheral cut. In the case of a multi-impression mold, it is necessary to make individual cuts to sever the products from a web of waste material. Trimming is usually performed as a secondary operation, either in-line or off-line. However, for large-volume applications and reducing fabricating costs, the molds can include trimming capabilities. The technique is sometimes known as trim-in-place and is particularly advantageous for forming certain plastics, such as polypropylenes.

Cut-sheet-fed machines are generally not equipped with in-line trimming, so it is normal to trim the products off-line. Such products tend to be bigger, more irregular in shape, and to be formed from thicker sheet than products from continuous web-fed thermoformers. Taking the process off-line makes it possible to choose the most suitable trimming method for each individual product.

Trimming is accomplished either by shearing or machining. Shearing methods include punching, knife-edge die cutting, and orbital shearing. Machining methods include routing, sawing, and cutting by means of laser, water jet, or hot wire. Increasingly, computer-controlled machines or those under robotic control perform the machining methods so that complicated peripheral cuts can be made accurately and uniformly at high speed. Such machines have the advantage that they can be rapidly reprogrammed to handle other products or a sequence of products, whereas shearing methods usually require tooling dedicated to a single product (chapter 18).

Web-fed thermoformers using reeled or directly extruded sheet normally incorporate in-line trimming or trim-in-place tooling. The cutting method is almost always by shear punch or knife-edge means, and the machines incorporate automatic or semiautomatic waste and product-handling systems. In extruder-fed machines, the waste-handling system is usually arranged for full or partial recirculation to the sheet extruder.

With a trim-in-place tool, heat energy is extracted from the sheet and dissipated through the tool. This action requires an intensive water-cooling action of the tool. With a separation of forming and trimming, the forming tools take care of the cooling and the formed sheet is almost cooled by the time it reaches the trim press (Fig. 7.47). Potential problems can occur that influence the operation of the trim-in-place tools and costs.

With trim-in-place tools, the one tool is also doing the cutting, using punches and dies with clearances of typically 5 to 10 μm. Each tool half is fitted with individually monitored and controlled cooling circuits. During the process, the cutting parts will heat up gradually and therefore the clearance will change as the products expand. The probability is that punches and dies will warm up to different degrees.

The cooling system does not only have to extract the heat from the tool at the highest possible rate; it also needs to keep both tool halves at the same stable temperature in order to avoid severe wear or damage in the tool due to interferences as well as provide a repeatability of the trimming action on the product. The bigger the trim-in-place tool becomes, the less forgiving it is for the slightest temperature variation.

As the size of the tool bases for the large tools exceeds the maximum working sizes of many high-precision machines, some of the traditional designs need to be revised. An important factor to consider in minimizing temperature related to wear is the choice of materials used in manufacturing the tool. For the tool base, steel or aluminum is typically selected by criteria such

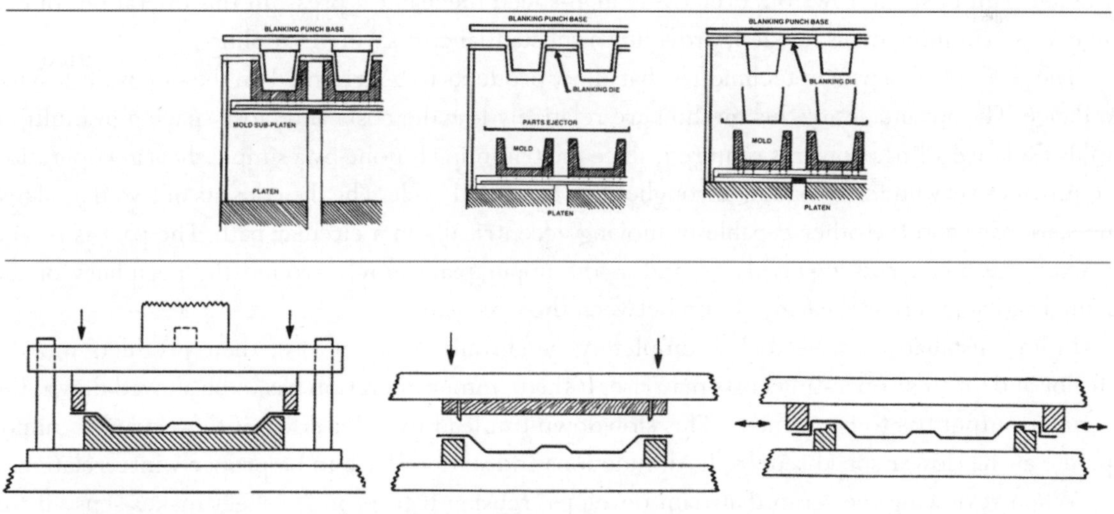

Figure 7.47 Examples of thermoforming and trimming in the same mold

as weight, corrosion resistance, and/or mechanical strength. Using both steel and aluminum in the same tool is a major potential for difficulties because of the great differences in the temperature expansion ratios of the two materials (also thermal conductivity). Sometimes, it is advisable to use steel instead of aluminum for better handling of the tool (chapter 17). Manufacturing and handling the trim-in-place tool require proper evaluation of its design and proper training of people in handling the tool.

With short runs, products may be trimmed along any line. Use can be made of band or circular saws. Operations can be manually or mechanically operated shears, shapers, routers, and so on. To provide finishing, wet or dry continuous belt sanders can be used.

Use is usually made of cutting or shearing dies mounted in a press for long production runs. Cutting dies, such as steel rule dies, are popular for trimming plastic sheet with a thickness up to 0.050 in (0.13 cm) in production runs under 100,000 strokes. These dies are mounted either in an inclined or upright mechanical clicker press. The dies usually cut against a softer material, such as hardwood or brass. Heavier machined cutting dies, referred to as high dies or clicker dies, may be used to cut heavier sheet.

Matched shearing dies with positive shearing action can be used for most materials and over a wide range of sheet thicknesses, from under 5 mils to over 200 mils. The tolerances between punch and die are much closer than are commonly used for sheet metal of the same thickness.

The mechanical presses are particularly well suited for shearing dies. Dies may be single or multiple units, depending on the size and capacity of available trimming presses and the production rates required. Normally when cut sheet is used, the trimming die will match the forming mold. The operator moves the formed sheet from the forming machine to inventory and then into the trimming press. In a continuous forming operation, it is often desirable to form several rows at one time but trim in single rows on a relatively high-speed mechanical press. In this operation, only a single row trimming die is needed, providing a substantial cost saving in tooling.

There are planetary trim techniques that allow products to be trimmed on the side walls leaving no flange. The advantages of this method are relatively low die costs and close spacing in multiple molds. Relatively little power is required, since the trimming is done by a simple shearing operation and provides very uniform loading throughout an extended cycle. The die consists of two flat plates, one stationary and the other capable of moving eccentrically in a circular path. The part is placed between the upper and lower plates, and as the upper plate moves around the periphery of the formed article, there is shearing action between the two plates.

In-line installations have to be completely synchronized; otherwise, their products may be inferior and/or their operating costs may rise. It the trimmer operation has to be slowed down, the extruder output has to be reduced. This slowdown can lead to a shutdown if the extruder cannot operate at the slower speed required. All functions and stations have to be properly interrelated.

When reviewing the scrap that trim develops, reusing it to produce sheet makes sense if the product to be formed permits its use. When the scrap cannot be reused, an approach is to consider having it extruded on the sides of the sheet (Fig. 7.48).

Figure 7.48 Example of coextruded sheet with scrap used on the sides

DESIGN

OVERVIEW

When designing a product, one should follow the inherent nature of thermoforming that basically uses thermoplastic flat panels (film or sheet) instead of the solid, enclosed, boxlike, cylindrical, rodlike, or structural shapes used in other fabricating processes. One also should be aware of and observe the material's depth-of-draw limitations, which can vary depending on the type of plastic being processed, the thickness tolerance of the material, and the degree of pinhole freedom the material enjoys. Generally, for straight vacuum forming into a female mold, the depth-to-width ratio should not exceed 0.5:1. For drape forming over a male mold, this ratio should normally not exceed 1:1. For products to be used with the plug assist, slip ring, or one of the reverse-draw methods, the ratio can exceed 1:1 and perhaps even reach 2:1 under normal circumstances. However, shallow drafts are in general more readily formed than deep ones and result in more uniform wall thickness.

Undercuts and reentrant shapes are possible in many designs. They require movable or collapsible mold members, but with small undercuts they can often be sprung from a female mold while the formed product is still warm. This type of action works best when the plastic has some flexibility, as do the TPEs, or the material is very thin. Guidelines for the maximum amounts of undercutting that can be stripped from a mold are as follows: 0.04 in (0.1 cm) for acrylics, PCs, and other rigid plastics; 0.060 in (0.15 cm) for PEs, ABSs, and PAs; 0.100 in (0.25 cm) for flexible plastics such as the PVCs.

When female tooling is split to permit the removal of products with undercuts, a parting line of the split halves becomes visible on the formed part. If this is objectionable, the designer can sometimes incorporate the parting line in the decoration of the product or at some natural line on the product.

Sharp corners generally should never be specified, since they hamper the flow of material into the mold's corners. This results in excessive thinning of the materials and causes concentrations of stress. A minimum radius of two times the stock's thickness is recommended. It is also more desirable from several standpoints to have large, flowing curves in a thermoformed product than to have squared corners or rectangular shapes. However, there are thermoforming techniques that can produce thicker corners.

The best products have smooth, natural curves and drawn sections that are spherical or nearly so in shape. Their walls will be more uniform, they will be more rigid, their surfaces will be less apt to show tool marks, and their tooling and molds will be lower in cost. Notches or square holes should be avoided when punching formed parts. Round holes are preferred to oval ones for minimizing stress buildup.

Some draft is required in side walls to facilitate the easy removal of the part from the mold. Female molds require less draft, since products tend to pull away from mold walls as they shrink during cooling. With female or male tooling, for most plastics the draft on each side wall should be at least 1°.

Metal inserts are usually not feasible, because thin walls are not sufficiently strong to hold inserts, particularly if thermal expansion and contraction of the product take place in service. It may be desirable to increase the stiffness of thermoformed products. Many are panel shaped and made of thin walls so they may lack rigidity. Corrugations, which if used are preferable in two directions or an embossed pattern, can add to their rigidity. With short production runs, it may be more economical just to use thicker sheet plastic to gain stiffness. If the function of the product permits, use curved, dished, or domed surfaces to gain stiffness.

With certain formed products, such as cups that are stacked, without controlled spacing they will jam together (as will other similarly shaped processed products), which could cause sufficient stress and products to split. To avoid jamming and control the space between parts, a stacking boss or shoulder system can be used. Within this stacking area, the plastic must be sufficiently rigid to prevent the deflection of bosses that would cause jamming. The height of the bosses is generally greater than their vertical cross-sections at the point of least taper; otherwise, the tapered walls will interfere before the stacking sections can engage. There are also other designs that can be used to eliminate jamming.

TOLERANCE

Thermoformed products lack the dimensional accuracy of injection- and compression-molded products. With its low pressure, thermoforming reduces the degree to which the sheet being formed is forced to conform to the mold. Sheet variations, mainly in thickness and degree of existing pinholes, affect the final accuracy of the product. This is particularly true because tooling is generally one sided where the product is pushed against a female or male mold. However, similar to compression molding, male and female molds are used together in a thermoforming operation.

The objective should be to use sheet that is pinhole free with tight thickness controls rather than just to determine its weight. Some fabricators purchase by the lower-cost method in which weight is the controlling factor, because they do not have to meet any tolerances. However, it is possible that using sheet thickness as a purchasing criterion could provide a significant cost saving, determined by the amount of material consumed per product. Another factor that impacts cost savings is the uniformity of heat absorption of the sheet, resulting in a quicker cycle time.

The wall-thickness distribution is a decisive quality criterion for thermoformed products. Online registration and control of the wall thickness occur during thermoforming. Pertinent measuring systems, the control strategy, and practical testing procedures are used.

The different shapes and forming techniques mean that different requirements need to be set up. Fastness, rigidity, surface properties (gloss structure, etc.), diffusion properties, shaping precision, and thermoforming stability are largely material-specific demands. However, they are considerably affected by the choice of process parameters.

Some component properties are directly influenced by the quality of wall thickness and its distribution; for most processes, poor control of thickness tolerances will lead to poor repeatability of formed shapes. The wall thickness indirectly affects the shaping precision and the thermoforming stability. Thus, the cooling behavior and the local relaxation of molecular orientation as well as the shrinkage of the material differ, depending on the thickness of the product. This is why the wall thickness and its distribution are important quality characteristics that have to be set. Important considerations are the defined tolerances and the limits on distribution variation of sheet and film to be processed, especially if the formed product must meet performance requirements and/or cut costs by using the minimum wall thickness to reduce material consumption.

Products are affected dimensionally by the differences between their forming temperatures and their service-use temperatures. Thus, a plastic's coefficient of thermal expansion and contraction has a significant effect on service conditions (chapter 2). The thermoforming pressure, time, and temperature variations that can exist will affect the final dimensions. Of these factors, evenness in heating throughout the sheet thickness before forming is usually the most important control. The type of heater has a direct effect on obtaining uniform heat. An allowance must also be made for postforming shrinkage. Molds should be designed oversized so that when shrinkage is completed the part dimensions will be correct to within the design tolerances.

The dimensional tolerances for wall thickness with the more conventional single-mold system are generally ±0.6% (±0.35% for close tolerances); with female molds they are ±0.5% (±0.3% for closer tolerances); with male molds under 3 ft they are ±0.8% (±0.4% for closer tolerances); and with male molds over 3 ft they are ± 30% (±10% for closer tolerances).

Plastics Memory

Most thermoplastics possess enough elasticity when they are hot that when stretched by mechanical or pneumatic techniques, they tend to draw tight against the force that stretches them and to stretch uniformly. These characteristics permit forming against a single mold, which is a common practice in many thermoforming methods. Reheating a formed part to the original forming temperature will activate the so-called plastic memory and cause it to relax back toward its original shape.

With this plastic memory, behavior errors in the contours of formed parts often can be corrected by reforming. It also is possible to decorate a formed product then relax it back to the original flat shape and use the distorted pattern to print subsequent sheets in distorted formats.

The printed sheets can then be formed into products in which the decorative elements will be in perfect register on the 3-D surface.

TROUBLESHOOTING

Reviews have been presented concerning different problems and solutions. Like other processes, forming is subject to many variables that influence appearance, performance, and cost. All the variables are controllable. Table 7.19 provides information on buying and selling thermoforming machines (548). Chapter 27 provides some troubleshooting guidelines.

SUMMARY

Thermoforming originated over a century ago from the early days of semisynthetic plastic processing where skived sheets of cellulose nitrate were placed in shaped steel molds, heated with steam, and quenched in water. It has become a major method of processing plastic products, consuming about 60% of extruded sheets. Growth in the thermoforming industry is now at least 8% per year in the United States, about 45% in Europe, and about 10% in the Pacific Rim area (not including when the economy is down) (549). Much of the growth is attributed to these two aspects:

1. The development of newer process technologies that allow thermoforming to compete with other process, such as injection molding and blow molding
2. A general customizing trend in design of industrial products that requires fewer total parts per design that range from small to very large products.

In addition to forming products from single-layer plastics, important multilayer or coextruded plastics are used to produce many different products, such as multicolored cups and high-barrier containers. There are also certain shaped extruded profiles that are formed usually by the postforming technique. Possibly, at least 30 wt% of all extruded products are thermoformed.

There is preprinting forming, in which a distorted pattern is printed on a plastic sheet that is then thermoformed to the desired shape, bringing the printed pattern into the proper undistorted shape.

The term *thermoforming* or *forming* is sometimes used to identify any plastic that is processed by any technique. However, the more popular definition is the forming or shaping of thermoplastic stock or billet plastics. Other fabricating techniques are basically similar to thermoforming in that a shaped form is produced, but they usually use less heat and provide for tighter shape and thickness tolerances. The terms thermoforming or forming do not include such operations as molding (injection, compression, rotational, etc.), extrusion (film, sheet, profile, etc.), casting, and others in which shapes, profiles, or parts are also "formed."

Advice for Buyers

If you're considering buying used machinery, the following guidelines can help.

- ☞ Do your homework. Know exactly what you need. What is the expected capacity or size? What brand names are preferred? How new a machine is needed? Types of controls? Type of heat? How soon will you need the equipment.
- ☞ Being prepared will help you make sure you get the right machine for the right price and on your deadline. Always have a backup machine. The market changes every day.
- ☞ Ask questions. How will the machine be shipped? Removed? Insured? Air-ride trucks and trailers should be used. Who pays for shipping and removal and how is it billed? Request blueprints, manuals, and mechanical drawings. Be certain at dismantle that everything is marked and labeled by competent personnel.
- ☞ Never hand over money without having a signed contract. The contract should include the make and model of the machine, the serial number, a delivery date, and a determination of who pays for incidentals. Be sure to get a good clean title and require a bill of sale.
- ☞ Take accurate measurements of the machine up front. How is it assembled? This will ensure no surprises when you reinstall.
- ☞ Carefully measure all facility requirements, such as load station and form station pit. This will improve your setup time and save you money.

[Ed note: After buying, notify the machine builder of your purchase. This will allow the builder to notify you of changes in federal standards such as OSHA requirements.]

Advice for Sellers

Much of the information for buyers is also relevant for sellers. Here are a few other points to consider:

- ☞ Never disconnect your machine and let a dealer warehouse it before it's sold. If a buyer can't see the equipment in operation, he or she will anticipate the worst.
- ☞ Machines well cared for and in good condition sell faster. Make the maintenance records readily available. If there is a problem, be up front about it. Be realistic in your asking price.
- ☞ If you have built your own machine, consider the potential liability risks when selling. Unknown brand machines tend to take longer to sell and usually sell at significant price reduction.
- ☞ Both parties should agree up front on what is included in the price, such as clamp frames, vacuum pumps, and temperature controls. Further where will the machine be disconnected and by whom?
- ☞ Be sure everything is in writing.

Table 7.19 Buying and selling tips for used thermoforming machines

Three-dimensional thermoformed solid, foam, and/or solid/foam plastics provide great variety in sizes, shapes, and quantities of marketable products, from millions of ounce-sized drinking cups or containers to thousands of pickup truck storage wells (each about 100 lb) and so on. The process of thermoforming is considered the fourth major fabricating process following extrusion, injection molding, and blow molding.

When compared with other processes, such as injection and blow molding, the thermoforming process has lower product cost because of lower forming and mold capital investment, less development time and cost, and predecorating of sheet before forming (Table 7.20).

This low-pressure thermoforming process permits forming large products with large surface areas at relatively low mold and equipment cost. Vacuum pressure is lower than 14.7 psi (0.1 MPa), and pressure forming usually only uses compressed air up to 300 psi (2 MPa). High-volume thin-walled products, such as drinking cups, can be produced at the lowest cost per capital investment, at production rates of 50,000 to over 200,000 units per hour. Low-volume heavy-gauge products, such as computer housings, compete favorably with injection molding in price, through lower tooling costs, and in product detail with 100 psi (0.7 MPa) forming pressures.

	Thermoforming	Injection molding
Material		
Cost	Higher (sheet costs more than resin)	Lower
Quantity	More (considerable waste, up to 40%)	Less (little waste)
Machinery cost	Lower than in injection molding	Higher than in thermoforming
Mold cost	Averages 1/10 that of an injection molding mold	Averages 10 times that of a thermoforming mold
Trimming equipment and costs	Relatively high	Negligible
Product wall characteristics		
Minimum thickness	Approximately 0.002 in. (0.05 mm)	Approximately 0.04 in. (1 mm)
Uniformity	Difficult to control	Very uniform
Rigidity	Special techniques required to obtain rigidity in larger pieces	Excellent
Finish		
Gloss	Generally not very good	Excellent
Detail	Not very sharp	Excellent
Predecorating	Most predecorated finishes can easily be thermoformed	Predecorated finishes cannot be injection molded
Production flexibility	Very high	Low
Setup time	Very short	Up to 4 times as long as in thermoforming

Table 7.20 Factors to consider when comparing thermoforming and injection molding

CHAPTER 8

FOAMING

OVERVIEW

Foamed plastics, which are expanded materials with cellular structures, have various names, such as plastic foams, cellular foams, expandable foams, structural foams, blown foams, sponges, and microcellular foams. The term originally was used to identify cellular thermoplastic products. Later, it also identified thermoset foamed products. Foamed plastics may be flexible, semirigid, or rigid. The usual process to create foamed plastics involves the introduction of a dispersed gas and subsequent cooling or curing. This expanded mass can make most plastics into foams by most plastic processing techniques. Many different products are produced, ranging from film or sheet to molded shapes. Many different properties can be obtained based on plastics used. Examples of properties are provided in Tables 8.1 through 8.6 (1, 5, 169, 212, 258, 260, 359).

A plastic foamed material is a plastic whose apparent density is decreased by the presence of numerous cells throughout its mass. It is a two-phase gas-solid system in which the solid is continuous and composed of plastic material. Their densities usually range from 1.6 kg/m^3 to over 960 kg/m^3 (0.1 lb/ft^3 to over 60 lb/ft^3). As to be reviewed in this chapter, they offer an extensive range of physical, mechanical, electrical insulating, and other properties, such as different cushioning capabilities. Their performance depends to a great extent on the type of base plastic, type of blowing system, and method of processing. Each plastic can include fillers or reinforcements to provide certain improved desirable properties (chapter 1).

There are many ways in which foams can be processed and used, such as slabs, blocks, boards, sheets, films, molded shapes, sprayed coatings, and extruded profiles. One popular technique is the foamed-in-place process in which the liquid material is poured into an existing cavity and allowed to foam. Conventional equipment can be used, such as extruders, injection-molding machines, rotational, casting, and other machines.

Property	ASTM Test	Phenolic: Foamed in Place	Phenolic: Syntactic Castable	Polyvinyl Chloride Rigid Closed Cell	Phenylene Oxide Foamable Resin	Polycarbonate	Polyethylene Medium-Density Foam	Polystyrene: Molded	Polystyrene: Extruded	Polyurethane Rigid Closed Cell
Density, lb./ft.^{3} (kg/m^3)		2–5 (32–80)	50–60 (800–960)	2–4 (32–64)	50 (800)	50 (800)	5.5–7.0 (88–112)	2.0 (32)	2–5 (32–80)	4–8 (64–130)
Tensile strength, psi (MPa)	D 1623	20–54	1000 (6.89)	1,000 (6.89)	3,300 (22.7)	5,500 (37.9)	110–210 (0.76–1.45)	42–68 (0.29–0.47)	180–200 (1.24–1.38)	90–290 (0.62–2.00)
Compression strength at 10% deflection, psi (MPa)	D 1621	22–85 (0.15–0.59)	8,000–13,000 (55.1–89.6)	——	5,500 (37.9)	7,500 (51.7)	2–18 (0.014–0.12)	25–40 (0.17–0.28)	100–180 at 3% (0.69–1.24)	70–275 (0.48–1.90)

Table 8.1 Examples of rigid plastic foams' mechanical properties

Property	ASTM Test	Phenolic: Foamed in Place	Phenolic: Syntactic Castable	Polyvinyl Chloride Rigid Closed Cell	Phenylene Oxide Foamable Resin	Polycarbonate	Polyethylene Medium-Density Foam	Polystyrene: Molded	Polystyrene: Extruded	Polyurethane Rigid Closed Cell
Maximum service temperature dry, °F (°C)		Continuous at 300 225 (149)	275 (135)	—— ——	200 (93.3)	270 (132)	180–200 (82–93)	165–175 (74–79)	165–175 (74–79)	200–250 (93–121)
Thermal conductivity BTU/in./hr.-ft.^{2}-°F (W/mK)	D 2326	0.20–0.22 (0.29–0.032)	1.0 (0.14)	2.0 (0.29)	—— —— ——	—— —— ——	0.32–0.34 (0.046–0.049)	0.23 (0.033)	0.17–0.21 (0.024–0.030)	0.15–0.21 (0.022–0.030)
Coefficient of linear expansion, 10^{-6} in./in.-°F	D 696	5	100	40–60	38	25	——	30–40	30–40	40

Table 8.2 Examples of rigid plastic foams' thermal properties

Property	ASTM test	Cellulose acetate	Epoxy	Phenolic	Polystyrene: Expanded plank			Polystyrene: Extruded plank		Polystyrene: Extruded sheet		PVC		Polyurethane: Polyether		Polyurethane: Isocyanurate: Bun	Polyurethane: Isocyanurate: Laminate	Urea-formaldehyde
Density, kg/m³		96–128	32–48	32–64	16	32	80	35	53	96	160	32	64	32–48	64–128	32	32	13–19
Mechanical properties																		
Compressive strength, kPa at 10%	D 1621	862	138–172	138–620	90–124	207–276	586–896	310	862	290	469	345	1,035	138–344	482–1,896	210	117–206	34
Tensile strength, kPa	D 1623	1,172		138–379	145–193	310–379	1,020–1,186	517		2,070–3,450	4,137–6,900	551	1,207	138–482	620–2,000	250	248–290	
Flexural strength, kPa	D 790	1,014		172–448	193–241	379–517		1,138				586	1,620	413–689	1,380–2,400			
Shear strength, kPa	C 273	965		103–207		241		241				241	793	138–207	413–896	180	117	
Compression modulus, MPa	D 1621	38–90	3.9			3.4–14		10.3				13.1	35	2.0–4.1	10.3–31			

Table 8.3 Additional mechanical properties for rigid plastic foams

Property	ASTM test	Cellulose acetate	Epoxy	Phenolic	Polystyrene: Expanded plank			Polystyrene: Extruded plank		Polystyrene: Extruded sheet		PVC	Polyurethane: Polyether		Polyurethane: Isocyanurate: Bun	Polyurethane: Isocyanurate: Laminate	Urea-formaldehyde
Thermal properties																	
Thermal conductivity, W/(m–K)	C 177	0.045–0.046	0.016–0.022	0.029–0.032	0.037	0.035	0.035	0.030		0.035	0.035	0.023	0.016–0.025	0.022–0.030	0.054	0.019	0.026–0.030
Coefficient of linear expansion, 10^{-5}/°C	D 696		1.5	0.9	5.4–7.2	5.4–7.2		6.3	6.3				5.4–7.2	7.2	7.2		
Maximum service temperature, °C		177	205–260	132	74–80	74–80	74–80	74		77–80	80		93–121	121–149	149	149	
Specific heat, kJ/(kg · K)	C 351							1.1					ca 0.9	ca 0.9	ca 0.9		
Electrical properties																	
Dielectric constant	D 1673	1.12			<1.05	1.02	1–02	1.19–1.20	<1.05	1.02	1.27	1.28		1.05	1.1	1.4	
Dissipation factor		20			<0.0004	0.0007	0.0007	0.028–0.031	<0.0004	0.0007	0.00011	0.00014			13	18	
Moisture resistance																	
Water absorption, vol. %	C 272	4.5		13–51	1–4	1–4	1–4	0.02	0.05								
Moisture vapor transmission, g/(m · s · GPa)	E 96		58		<120	35–120	23–35	35		86	56	15	35–230	50–120		230	1,610–2,000

Table 8.4 Additional thermal and other properties for rigid plastic foams

Property	ASTM test	Expanded natural rubber		Expanded SBR	Latex foam rubber		Polyethylene: Extruded plank	Polyethylene: Extruded sheet	Polyethylene: Cross-linked sheet	Polypropylene: Modified	Polypropylene: Sheet	Polyurethane: Standard cushioning	Polyurethane: High resilience type	Poly (vinyl chloride)		Silicone: Liquid	Silicone: Sheet
Density, kg/m³		56	320	72	80	130	35	43	26–28	64–96	10	24	40	56	112	272	160
Cell Structure		closed	closed	closed	open	open	closed	closed	closed	closed		open	open	closed	closed	open	open
Compression strength 25% deflection, kPa	D 3574 D 3575							48		206	4.8	5.7	4.6				
Tensile strength, kPa	D 3574		206	551	103		138	41		344		118	103	10.3	24	36 at 20%	
Tensile elongation, %	D 3574				310		60	276	276–480	1380	138–275	205	160			227	310
Rebound resilience	D 3574				73			50		75		40	62				
Tear strength (N/m) × 10²	D 3574						10.5	26				4.4	2.4				
Maximum service temperature, C°		70	70	70			82	82	79–93	135	121					350	260

Table 8.5 Properties of flexible plastic foams

Material	Specific Gravity	Elongation at Break (%)	Heat Deflection Temperature °C (°F)	Thermal Conductivity W/mK (BTU-in./hr. ft.² · °F)
Polyphenylene oxide	0.8	15.0	96 (205)	0.124 (0.860)
Polycarbonate	0.8	4.0	128 (262)	0.151 (1.05)
Epoxy resin	0.78	—	—	0.7 (4.8)
Isocyanurate	0.032	—	—	0.1 (0.69)
Polyether	0.08	—	—	—
Polystyrene	0.17	—	—	0.65 (4.5)
Polystyrene	1.04	3.5	101 (214)	—
Polyurethane	0.11	—	—	0.3 (2.1)
ABS	0.86–1.1	—	72 (162)	—
Acetal	1.130	—	153 (307)	—
Nylon 6/6	0.97	4.1	255 (491)	—
Polybutylene terephthalate	1.1	1.3	207 (405)	—
Polyimide		—	277 (531)	—
Polysulfone	0.87	3.5	177 (351)	—
Polyvinylchloride	0.6	370.0	—	—

Table 8.6 Additional properties of flexible plastic foams

The growth of plastic foams continues to be significant due to the inherent available properties and usefulness in different applications and environments. The outstanding properties of a foamed plastic are its light weight, low thermal conductivity, and high strength-to-weight ratio. Their uses range from original to replacement parts in buildings, transportation vehicles, sports equipment, boats, underwater craft, spacecraft, furniture, decorative displays, toys, and life preservers, and their uses and applications continue to advance at a rapid pace. The major plastics used as foams are the polyurethanes and polystyrenes (chapter 2).

Foamed plastic products have been a major part of and a special category within the plastics industry. The type of plastic, such as thermoplastic, thermoset, elastomer, and reinforced plastic, as well as the process used can form the basis for a classification system. The arrangement and distribution of the gas in the foam corresponds to the structure or morphology of the plastic composite system.

Hundreds of foams are on the market with many different characteristics. They include slabs, preformed logs, sheets, rods, tubes, choppings, netting tapes, and all kinds of molded shapes. The foam-in-place types are liquids, beads, molding powders, and plastisols. Foam mixtures are compounded to perform many functions, including spraying ceilings. Vinyl plastisols using chemical blowing agents can be calendered, extruded, dip coated, or slush molded (chapter 16).

In addition to the basic plastics in sold, liquid, and bead forms, there are those that require foaming agents, fillers, additives (including cell controllers and fire retardants), catalysts, surfactants, styrene monomers, and systems that vary viscosity from liquid to paste forms (chapter 1). The gas can be put directly into the plastic before the plastic solidifies. Reactant chemicals can be put in the plastic formulation that during polymerization will release a gas and produce the foam. The more popular technique used in producing foams, particularly polyurethanes, is to include in the plastic compound chemicals that will release gas during the application of heat. The chemicals decompose and produce gas.

The foaming technique can be defined as a type of molding that involves putting a mixed formulation into a cavity where simultaneous polymerization for thermoset plastics or heat for thermoplastics and gas-generating reactance occur. The mixture can be poured or sprayed into a cavity. The cavity is filled with a cellular plastic weighing from ½ to 70 lb/ft^3. Occasionally, the outside container used to restrict the foam is retained as part of the package.

Extruding expanded polyethylene, polypropylene, polyvinyl chloride, and polystyrene is very popular. Specially designed extruders can handle a mixture of plastic and a gas (such as nitrogen). The material expands as it leaves the die. Foaming will take place with a mixture of resin and blowing agents when put under pressure. Blowing agents used include methyl chloride, propylene, or butylene. A wide range of properties can be obtained in foamed vinyls by just using carbon dioxide (169). These types of foam materials find applications in containers for the liquid- and food-serving consumer markets.

Foam sheet made from expandable polystyrene beads is extruded in conventional equipment. This sheet material provides the industry with a relatively low-cost product with inherent thermal-control characteristics.

The technology of polyurethane (PUR) foams has been developing since its inception during the early 1940s in Germany, followed by its development in the United States and the rest of the world. This foam packaging material provides specific advantages. It ensures firm support and restraint for the interior item by adapting itself to the item's complex contours. This characteristic greatly reduces or eliminates risk of damage that exists in conventional foam packaging materials when handling special products.

Foamed plastics, like their solid counterparts, can be used for almost an unlimited range of products. There are different approaches to spray-foamed homes. Since about the 1950s, foamed building structures were fabricated using polystyrene foamed plastics. The initial development was by the U.S. Army. Since that time, many other foamed structures have been built worldwide. An interesting approach was designed and built in 1966. Dome-shaped buildings were built with polystyrene (PS) boards using Dow's spiral generation technique. Craftsmen heat bonded the boards in a continuous pattern to produce the dome-shaped medical clinical structures located in Lafayette, Indiana. Boards were heat bonded at the softening point of the PS in order to form a continuous pattern that produces the dome shape. Sections cut from the dome can be made into doors, connecting halls from dome to dome. These domes were self supporting, requiring no internal or external support during or after manufacture. They also provide their own insulation and other advantages (chapter 20).

Formulators are able to control versatility of foams through varying density, cell geometry and size, strength-to-weight ratio, thermal and electrical insulation, mechanical flexibility or rigidity, corrosion resistance, color, processability, degree of adhesion to other materials, and sound or microwave absorption. Parts can have multifunctional uses and properties, such as insulation and load carrying, insulation and ease of application, and buoyancy and structural rigidity. For example, urethane foamed-in-place in boat hulls or hydrofoils makes the vehicles virtually unsinkable, reduces noise level, and reduces structural vibration.

Similar to other materials, foams have limitations. No foam is fireproof, but many of them can be made flame resistant. Phenolics and silicones have excellent heat resistance but could crumble when subjected to vibrational stress. There are foams that can be affected by solvents, but fluorinated types resist them.

Basic Process

There are various combinations of plastics and blowing agents to fabricate different products. Almost all plastic fabrication processes are used to produce foamed products. Basically, during the process a blowing agent expands the plastic initiating cells that grow to produce the final foam. A gas produced at equilibrium is established between material in the gas phase and the material dissolved in the solid state. The gas dissolved in the solid state migrates from the solution into the gas phase. The cells formed are initially under higher than ambient pressure, because they must counteract the effects of the plastic's surface tension. The pressure due to surface tension depends on the reciprocal of the cell radius, so the pressure within the cell is reduced as the cell grows. Different techniques are used to control this foaming action.

Small cells tend to disappear and large cells tend to get larger. This is because the gas migrates through the substrate (plastic) or the cell walls break. After forming cells, the foam has to be stable; the gas must not diffuse out of the cell too quickly, thereby causing collapse or excessive shrinkage. The stability of the foam depends on the solubility and diffusivity of the gas in the matrix. The many processes make for many methods of cell initiation, cell growth, and cell stabilization. These methods lead to a convenient process classification.

The growth of the cell depends on the pressure difference across the cell membrane between the inside of the cell and the surrounding medium. Such pressure difference may be generated by lowering the external pressure (decompression) or by increasing the internal pressure in the cells (pressure generation). Other methods of generating the cellular structure are by dispersing gas (or solid) in the fluid state and stabilizing this foamed state or by sintering plastic particles together in a structure that contains a gas phase.

Cell Configuration

Plastic foams are composed of a cellular core structure that is created by expansion of a blowing agent. Foam structures consist of at least two phases, a plastic matrix and gaseous voids or bubbles. An open-cell or closed-cell structure is formed, with cellular walls enclosing the gaseous voids. In closed-cell foams, the gas cells are completely enclosed by cell walls, while in open-cell foams, the dispersed gas cells are unconfined and are connected by open passages. The presence of voids or discontinuities due to the cellular structure reduces material consumption. A blend of plastics can be present in the matrix in addition to thermoplastics or inorganic fillers; the plastics can be stabilized against cell rupture by cross-linking. Foams can be flexible or rigid; closed-cell foams are generally rigid, while open-cell foams are more flexible.

A basic distinction is made between closed-cell systems where spherical or roughly spherical voids (cells) are fully separated by matrix material and open-cell systems where there are interconnections between voids. The degree of interconnection can be assessed if a sample is subjected to a moderate vacuum; a liquid is then allowed to fill the interconnected spaces and the weight gain is measured.

The cell size or average cell size can be an important factor. A distinction is sometimes made between microcellular foams 0.1 to 10 μm in diameter (indistinguishable to the naked eye) and macrocellular foams of at least 250 μm (Table 8.7) (178). With microcellular foaming, products can be produced that are lightweight, high strength, and thin walled (such as 0.5 mm thick).

The cell density (number of cells per unit cross-section area or volume) is also used to characterize the coarseness or fineness of foam. Foamed products can feature a deliberately created inhomogeneous (nonuniform) morphology. Examples are when a foamed core is sandwiched between solid skins as in so-called structural foams or between elastomeric products with so-called integral skins. With cells elongated in the direction of foam rise or melt flow, the process will give anisotropic structure and properties.

<table>
<tr><td colspan="3">Extremely small closed cells from 0.1 to 10 microns in diameter can be formed in most plastics by dissolving gas in the plastic, typically supercritical CO_2, and then rapidly reducing the pressure and increasing temperature in a controlled manner to cause homogeneous and likely heterogeneous nucleation and growth of gas bubbles. The bubbles are to be smaller than naturally occurring flaws in the polymer so mechanical properties are not compromised. Particles in PS(HI) can be sites for heterogeneous nucleation.
Short diffusion paths, elevated temperatures, and gases in the supercritical state are necessary to achieve the high diffusion rate and high gas concentration needed for commercial use. Extrusion with gas injection is an efficient process to saturate polymer with gas. Manufacturing issues are determining sequence of pressure, temperature, and shaping geometries to nucleate and form cells without disruption and to shape product without distortion. Microcellular technology is covered by several patents and is offered for licensing by Axiomatics, Woburn, MA</td></tr>
<tr><th>Advantages/Disadvantages</th><th>Applications</th><th>Materials</th></tr>
<tr><td>The extremely small bubbles give weight reductions of 10-90%, no reduction in specific mechanical properties, appearance of a solid opaque surface, and foaming of thin sections. Fatigue resistance is observed to increase. Environmental advantages are use of atmospheric gases and lower material use</td><td>Siding, Pipes, Aircraft Parts, Athletic Equipment, Machine Housings, Automotive Components, Food Containers, Artificial Paper, Thermal Insulation, Fibers for Apparel and Carpets</td><td>ABS, PE, PET, PMMAPS, PS(HI), PP, PUPVC, SMC, Fluoropolymers, Poly(methylpentene)</td></tr>
</table>

Table 8.7 Microcellular plastics: formation and shaping

BLOWING AGENT

The production of foamed or cellular plastics depends on the basic materials and process. Different blowing agents (also called foaming agents) are used to produce gas and thus to generate cells or gas pockets in the plastics. They can produce rigid or flexible types. The type of blowing agents used influences all kinds of physical, mechanical, electrical, and other properties. As an example, thermal conductivity is influenced (Tables 8.8 and 8.9).

Blowing agents are liquids or solids that form gases during processing. The gas forms minute cells or bubbles within the plastic, resulting in a foamed or porous structure. The foaming process is affected by the type of blowing agent used, the gas and its solubility in the plastic, the compounding method, processing temperatures and pressures, and the melt viscosity and crystallinity of the plastic.

Foaming agents are used to decrease the density of a plastic. Density reductions of about 50% to 60% can be achieved with loading levels of about 0.5 wt% to 2.5 wt% that in turn reduce materials costs. Finishing costs can be higher due to the necessity of painting or covering the foamed material. Foamed plastics have improved electrical and thermal insulative properties and higher strength-to-weight ratios; however, mechanical properties, such as stiffness, tensile strength, and compressive strength, are decreased owing to a lower amount of plastic material in the foam compared to the solid.

Blowing Agent	Temperature		
	32°F	68°F	86°F
Air	0.168*	0.180*	---
CO_2	0.101	0.117	---
CCl_3F (CFC-11)	0.054	0.057	0.058*
$CHCl_2F_2$ (CFC-12)	0.068	0.064	0.067
CCl_2-CCl_2F_2 (CFC-113)	0.046	0.051	0.054
* BTU/°F ft^2 hr in. (To obtain SI units of W/m·K multiply by 0.144)			

Table 8.8 Thermal conductivities of blowing agents are compared to air

Blowing Agent	Mol. Wt.	Wt % to give 2 lb/ft^3 (3)	Initial K-factor(4)	Final K-factor(4)
FREON-11 (CFC-11) (CCl_3F)	137	16	0.11	0.14
Methylene Chloride(1)	85	12	0.14-0.15	0.24
n-Pentane	72	9	0.18	0.20
CO_2	44	1.7(2)	0.20-0.23	0.24
Air	29	---	0.24	0.24

(1) Properties extrapolated from blends with FREON-11 (foams with 100% methylene chloride cannot be made)
(2) As water
(3) Estimated
(4) BTU·in/hr°F ft^2 (to obtain SI units of W/m·K multiply by 0.144)

Table 8.9 Thermal conductivities of rigid polyurethane foams containing different blowing agents

The amount of blowing agent used affects the properties of the foamed plastic, and different amounts are used for particular applications; for example, about 0.1 wt% is used for elimination of sink marks in injection-molded parts, 0.2% to 0.8% for production of injection-molded structural forms, 0.3% for extruded foamed profiles, 1% to 15% for formation of vinyl foams, and 5% to 15% for compression-molded foam products. Nucleating and cell-sizing agents can be added to produce cells of more uniform sizes and to enhance the symmetrical expansion of cells during the foaming process.

Blowing agents can be classified as physical or chemical, depending on how the gas is generated. Physical blowing agents undergo a change of state during processing, while chemical blowing agents, usually solids, undergo a decomposition reaction during polymer processing, resulting in formation of a gas.

Physical Blowing Agent

Physical blowing agents (PBAs) are compressed gases or volatile liquids. Compressed gases, usually nitrogen, are injected under high pressure, such as 2,000 psi, into the plastic melt during processing. As the pressure is relieved, the gas becomes less soluble in the plastic melt and expands to form cells. Nitrogen is inert, nonflammable, and can be used at any processing temperature. No residue is left in the foamed plastic so that recycling of the plastic part is easy. When using compressed nitrogen, however, generally the result is to produce foams with coarser cell structures and poorer surface appearances than nitrogen will produce with chemical blowing agents, although nucleating agents can be added for a finer cell structures.

There are liquid physical blowing agents that are volatile and change from liquid to gaseous states when heated to the plastic processing temperatures. They are short-chain chlorinated and fluorinated aliphatic hydrocarbons (CFCs). Although they can be used over a wide temperature range and at low (atmospheric) pressures, their use has been gradually discontinued because of their role in the reduction of stratospheric ozone.

Other PBAs are reviewed in Table 8.10. Some of these PBAs are to be reviewed. The blowing efficiency listed for the PBAs is based on:

$$22{,}400/\text{molecular weight} = [\text{density @ room temperature}] \times [(273 + t)/273],$$

where t at boiling point = (C) and last column = 100°C. Vapors are treated as ideal gases.

Chemical Blowing Agent

Chemical blowing agents (CBAs) decompose at processing temperatures to form gases (Table 8.11). The most important criterion for selection of a chemical blowing agent is that the decomposition temperature matches the processing temperature of the plastic (Table 8.12). Little or no foaming will occur at processing temperatures below the decomposition temperatures. With processing temperatures too high, the result can be overblown or ruptured cells and poor surface qualities. Activators, such as alcohols, glycols, antioxidants, and metal salts, can be added to lower

Blowing agent	Molecular weight	Boiling point (C)	Blowing efficiency At boiling point	At 100C
Pentanes				
n-Pentane	72.15	36.1	216	261
2,2-Dimethylpropane	72.15	9.5	196	260
1-Pentene	70.15	30.0	227	280
Hexanes				
n-Hexane	86.17	68.7	212	232
2-Methylpentane	86.17	60.2	207	232
3-Methylpentane	86.17	63.3	211	234
2,2-Dimethylbutane	86.17	49.7	204	229
Cyclohexane	84.17	80.8	266	281
Heptanes				
n-Heptane	100.20	98.4	206	207
2,2-Dimethylpentane	100.20	79.2	193	204
2,4-Dimethylpentane	100.20	80.6	193	204
3-Ethylpentane	100.20	93.4	204	212
Toluene	92.13	110.6	294	286
Trichloromethane	119.39	61.2	342	382
Tetrachloromethane	153.84	76.7	296	316
Trichlorofluoromethane	137.38	23.8	261	329
Methanol	32.04	64.6	679	752
Isopropyl ether	102.16	67.5	198	217
Methyl ethyl ketone	72.10	79.6	324	344

Table 8.10 Blowing efficiencies for several physical blowing agents

the decomposition temperature. Other selection considerations include the type and amount of gas liberated and its effect on the final product (Table 8.13; 105).

CBAs can be classified as inorganic or organic. Their decomposition can be endothermic or exothermic. Endothermic blowing agents, usually inorganic, require the input of energy for the decomposition reaction to take place, while exothermic blowing agents, usually organic, release energy during decomposition. Exothermic CBAs commonly have higher gas yields than endothermic CBAs. The lower gas yields and pressures associated with endothermic CBAs produce foams with smaller cell structures, resulting in improved appearances and physical property performances. Endothermic and exothermic CBAs have been combined in a single product, in which the exothermic CBA provides the gas volume and pressure necessary for lower densities and the endothermic CBA produces a fine, uniform cell structure.

Chemical foaming agents are available as dry powders, liquid dispersions, and pellet concentrates. They can be incorporated by dry-blending with resin powders, tumble-blended with resin pellets, blended using a hopper blender, metered in at the feed throat, or pumped into the barrel.

Chemical Name	Decomposition Temperature (°C)	Gas Yield (cm^3/g)	Type	Typical Foamed
Azodicarbonamide (AZ)	195–215	220	Exo	EVA, HDPE, LLDPE, LDPE, PP, TPE, FPVC
4,4'-oxybisbenzene sulfonyl hydrazide (OBSH)	160	125	Exo	HDPE, FPVC
p-toluenesulfonyl semicarbizide (TSS)	228–235	140	Exo	EVA, HDPE, LLDPE, LDPE, PP, TPE, FPVC
5-phenyltetrazole (5-PT)	250–300	200	Exo	PP, PC
Sodium Bicarbonate ($NaHCO_3$)	100–140	135	Endo	LDPE, EVA, FPVC, TPE
Alkali Carbonate (Hydrocerol)	160+	100–160	Endo	LDPE, EVA, LLDPE, FPVC
Alkali Carbonate (Activex)	120	140	Endo	LDPE, EVA, FPVC
Alkali Carbonate (Safoam)	170–210	130	Endo	EVA, HDPE, LLDPE

Table 8.11 Examples of chemical blowing agents

CBA Level (% wt)	CBA Type	Oven Temperature (°C)	Oven Time (min)	Comments
1	OBSH	246	10	Good inside skin, limited foaming
1	OBSH	246	12	Good inside skin, good foam
1	OBSH	246	14	Fair inside skin, good foam
1	AZ	260	10	Good inside skin, little foam
1	AZ	260	12	Good inside skin, good foam
1	AZ	260	14	Poor inside skin, overblown with coarse cells

(OBSH = *p,p′*-oxybisbenzene sulfonyl hydrazide; AZ = azodicarbonamide)

Table 8.12 Effect of oven conditions on rotational foaming of HDPE

CAB Level (% wt)	Wall Thickness (mm)	Density (kg/m³)	Density Reduction (%)	Wall Thickness Increase (%)	Relative Stiffness (%)
None	3.5	931	None	None	100
0.2	6.0	639	32	42	132
0.5	7.8	451	52	56	88
0.8	10.8	373	60	68	76
1.0	13.0	310	68	73	53

Table 8.13 Effect of dosage of azodicarbonamide (AZ) chemical blowing agent on rotational foaming of MDPE

Inorganic Chemical Blowing Agent Typical inorganic blowing agents are sodium bicarbonate, sodium borohydride, polycarbonic acid, and citric acid, which evolve primarily carbon dioxide gas upon decomposition. Sodium bicarbonate is the most common inorganic blowing agent. It is inexpensive and it decomposes endothermically at a low temperature, over a broad temperature range of 100° to 140°C (212° to 284°F). At temperatures ≥ 142°C (287°F), decomposition becomes more rapid, facilitating its use in polyolefins. Its decomposition is less controllable than organic blowing agents, however, and it can form an open-celled foam structure. Its gas yield is 267 cc/g. Polycarbonic acid decomposes at about 160°C (287°F), with a gas yield of about 100 cc/g. It is also used as a nucleating agent in physical blowing agents.

Organic Chemical Blowing Agent Organic blowing agents evolve gas over a specific, narrow temperature range and are selected according to the processing temperature of the plastic. The most common low-temperature blowing agent is 4,4'oxybis(benzenesulfonyl hydrazide) (OBSH), with a decomposition temperature of 157° to 160°C (315° to 320°F) and a gas yield of 125 cc/g. High-temperature blowing agents, with decomposition temperatures of greater than about 230°C (450°F), include 5-phenyltetrazole, with a decomposition temperature of 240° to 250°C (460° to 480°F) and trihydrazine triazine (THT).

Azodicarbonamide (ABFA), with a decomposition temperature of 204° to 213°C (400° to 415°F), is commonly used in PP (melting temperature of 168°C [334°F]). The use of activators can reduce the decomposition temperature to 150°C (300°F). ABFA is a yellow powder that decomposes exothermically, with a gas yield of about 220 cc/g, to produce a gas mixture containing 65% nitrogen. ABFA produces a fine, uniform cell structure but can produce discoloration in the foamed part. It is nontoxic and is FDA-approved for a wide variety of applications, including those involving food contact. The high gas yield, good performance, and low cost of ABFA make it the most widely used foaming agent.

The average particle size of ABFA ranges from 2 to 20 μm; fine particle sizes decompose at lower temperatures than coarser particles. Particle size has a large effect on the final cell structure in low-pressure applications, such as casting; however, in high-pressure processing, such as injection molding or extrusion, the effect of particle size is less critical. The effect of activators is also dependent on ABFA particle size; fine particles are more easily activated than coarse particles.

Another agent used is p-toluenesulfonyl semicarbazide (TSSC), although it decomposes at intermediate to high temperatures (228° to 236°C [442° to 456°F]). Activators can be used to decrease the decomposition temperature. TSSC has a gas yield of about 140 cc/g; the gas mixture consists of nitrogen, carbon monoxide, carbon dioxide, and trace amounts of ammonia. Its white color and nonstaining residue are important in applications requiring color quality. It is flammable and burns rapidly when ignited, producing a large amount of residue.

Thermoset Plastic Foam

The overwhelming majority of foams are thermoplastics, but thermosets are also foamed with CBAs, although some of them do create problems. Popular TS foams are made from polyurethanes, polyesters, phenolics, epoxies, and rubbers. Thermal decomposition of the blowing agent with certain plastics, such as TS polyesters, cannot be applied in this system because the heat of polymerization is not high enough to induce decomposition. Chemical reactions simultaneously produce gas and free radicals; they typically involve oxidation and reduction of a hydrazine derivative and peroxide. The reactions are catalyzed by metals, which can be used repeatedly.

Most TS foams are prepared by the simultaneous occurrence of polymer formation and gas generation. The use of polyisocyanates is very popular. They are very reactive compounds and produce various polymers, such as foams, plastics, fibers, elastomers, coatings, and adhesives, by the reaction of polyaddition, polycondensation, or stepwise polymerization (chapter 1).

The first method of making isocyanate-based foams was based on the reaction of carboxyl-terminated polyester with an organic diisocyanate, such as toluene diisocyanate. The simultaneous reactions resulting in carbon dioxide generation and polyamide (nylon) formation produced cellular plastics. Hoechtlen and Dorste invented this method in 1941 (404).

Polyurethane This type of foam (often referred to as urethane foam) is prepared by the reaction of a polyisocyanate with a polyol in the presence of a blowing agent, a surfactant, and a catalyst without external beating of the foaming system (Table 8.14). The principle of preparation of urethane foams is based on the simultaneous occurrence of two reactions, that is, polyurethane formation and gas generation in the presence of a catalyst and surfactant. In flexible urethane foams, the major blowing agent is water used with an auxiliary blowing agent.

With the ban on the use of CFCs, major changes in foam formulations developed. A number of studies were carried out on the use of 100% water-blown foams for both rigid and flexible foams. These studies required modifications or improvements in raw materials, such as polyisocyanates, polyols, catalysts, and surfactants.

The first patent involving polyurethane foams was assigned to Zaunbrecher and Barth in 1942 (405). The method involved simultaneous and competitive reactions comprising polyurethane formation (reaction of an organic diisocyanate with a hydroxyl-terminated polyester) and gas generation (reaction of the diisocyanate with water to form carbon dioxide and polyurea). Both

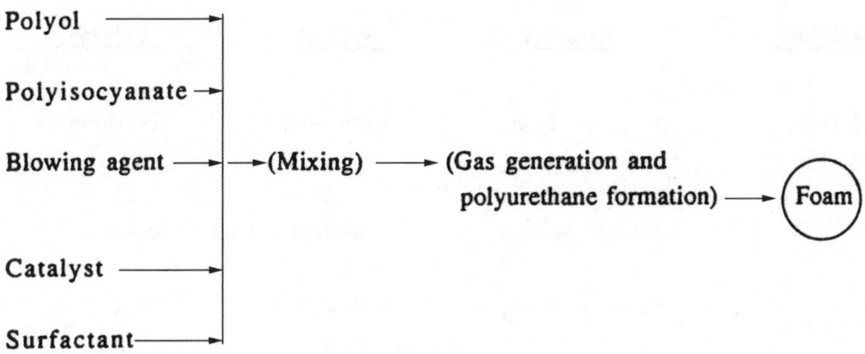

Table 8.14 Example of polyurethane formation and gas generation

methods could be recognized as a new combination of a known gas-generation and a known polymer-formation reaction. Since then, a variety of isocyanate-based foams have been developed.

Isocyanate-based foams include polyurethane, polyisocyanurate, polyurea, polycarbodiimide, polyamide, polyimide, and polyoxazolidone foams. In addition to these unmodified foams, many modified or hybrid foams have appeared in the literature, for example, urethane-modified isocyanurate foams, isocyanurate-modified urethane foams, urea-modified isocyanurate foams, and unsaturated polyester-polyurethane hybrid foams. A number of isocyanate reactions for making isocyanate-based polymers (which are expressed as model reactions) are listed in Table 8.15.

The polyisocyanate-based polymer-forming reactions can be classified into three types of reactions: addition reactions, condensation reactions, and cyclotrimerization reactions. Among the isocyanate reactions shown in Table 8.15, the addition reaction is the major isocyanate reaction in polyurethane foam preparation. The major raw materials for making isocyanate-based foams include the following compounds: polyisocyanates, polyols, catalysts, blowing agents, surfactants, epoxides, and flame retardants.

The polyisocyanates that can be used for preparing isocyanate-based foams are mainly aromatic compounds and some aliphatic or aralkyl polyisocyanates. Major polyisocyanates in the market are listed in Table 8.16. TDI is widely used for flexible foams. Pure MDI is used for elastomers and coatings. Modified TDI and modified MDI are used for high-resilience flexible foams. Polymeric isocyanates (polymeric MDI or oligomeric MDI) are mostly used for preparing rigid urethane and isocyanurate foams, and in part, for preparing flexible and semiflexible foams. Tables 8.17 and 8.18 show typical properties of TDI and MDI, respectively.

There are modified polyisocyanates to meet different polyurethane properties and processing conditions.

WATER FOAMING

Water may at first appear to be an unlikely blowing agent for plastic foams because of its low volatility and low solubility compared to chemical blowing agents. However, because manufacturers

Reaction	Material	Product	Linkage
Addition	A-NCO + B-OH	A-NH-C(=O)-O-B	Urethane
"	2 A-NCO + H_2O	A-NH-C(=O)-NH-A + CO_2	Urea
"	A-NCO + B-NH_2	A-NH-C(=O)-NH-B	Urea
"	A-NCO + B-COOH	A-NH-C(=O)-O-B + CO_2	Amide
"	A-NCO + A-NH-C(=O)-NH-B	A-N(C(=O)-NH-A)-C(=O)-NH-B	Biurette
"	A-NCO + A-NH-C(=O)-O-B	A-N(C(=O)-NH-A)-C(=O)-O-B	Allophanate
"	A-NCO + 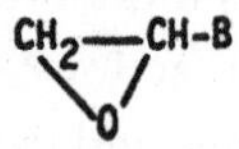	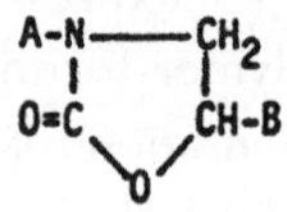	2-Oxazolidone
Cyclotrimerization			Isocyanurate
"	3 A-NCO	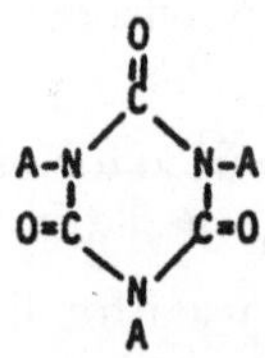	
Polycondensation	2 A-NCO	A-N=C=N-A + CO_2	Carbodiimide
"	A-NCO + 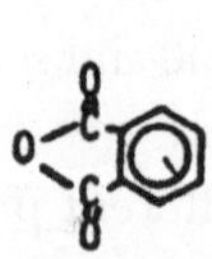	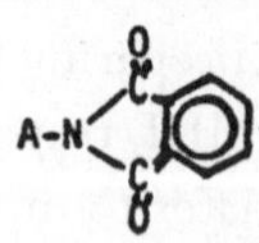 + CO_2	Imide

Table 8.15 Model reactions for foams

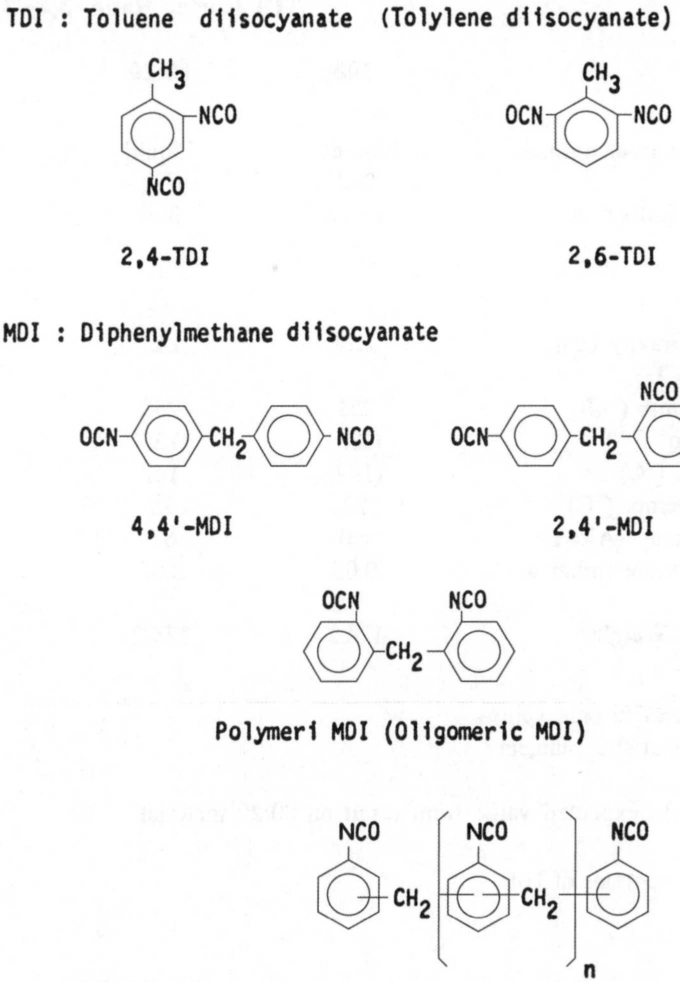

Table 8.16 Examples of polyisocyanates

have started to realize the cost, storage, handling, and environmental benefits of using water, its use as a blowing agent has increased. It offers reduction of product weight and increased production rate opportunities. In addition, the components are recyclable and exhibit excellent long-term physical properties for sealing and weathering.

Water foaming can be accomplished by modifying a standard single-screw extruder. Special requirements are focused, as with other blowing agents, on precise metering of water injection, temperature control, and mixing of the water with the plastic, such as TPE. Liquid temperature control of barrel zones and a minimum of a 30:1 L/D extruder are usually required. Water injection takes place at the 18:1 L/D position with a pump capable of at least 10 MPa with an adjustable flow rate up to 10 ml/min.

	TDI Isomer Ratio (2,4-/2,6-)		
	100	**80:20**	**65:35**
Physical state at normal temp.	Liquid/ Solid	Liquid	Liquid
Viscosity (mPa·s at 25°C)	(3–6)	3–6	(3–6)
Color		*	*
Odor		**	**
Specific Gravity (g/ml) (at 25°C)	1.21	1.21	1.21
Boiling temp. (°C)	251	251	251
Flash temp. (°C)	(135)	135	(135)
Fire temp. (°C)	(142)	142	(142)
Freezing temp. (°C)	22	15	8
Vapor density (Air=1)	6.0	6.0	6.0
Vapor pressure (mbar at 25°C)	0.03	0.03	0.03
Molecular Weight	174.2	174.2	174.2

* Colorless to pale yellow
** Characteristic pungent

NOTE: ( ) : expected value from result on 80:20 material

Table 8.17 Physical properties of TDI

Chlorofluorocarbon and Alternate

CFCs are a family of inert, nontoxic, nonflammable, and easily produced liquefied chemicals that have principally been used in refrigeration, air-conditioning, packaging, and insulation or as solvents and aerosol propellants (medical and other devices). The plastics industry, as well as other industries, has been phasing out CFCs (406). They were once widely used in producing foam products. CFC's chlorine components reportedly destroy ozone in the upper atmosphere. A targeted worldwide complete phase-out of CFCs was among the amendments to the Montreal protocol approved unanimously by ninety-three nations at a 1987 meeting in London. Participating nations also agreed to use hydrochlorofluorocarbon (HCFC) only where other alternatives were not feasible. The alternative HCFC is 98% less ozone depleting than CFCs (Tables 8.19 to 8.23).

Fully halogenated CFCs were eliminated in polystyrene foam food packaging and containers. Substitute blowing agents used are either no threat to the ozone or are a 95% improvement over fully halogenated CFCs. PS foam cups now are 100% CFC-free.

TYPE OF FOAM

Most thermoplastics and elastomers can be foamed. Examples of thermoset foams are reviewed in Table 8.24 (212). Combinations of properties vary according to the type of resins and the formulation. Not all plastic foams possess all properties in the desirable ratios. For example, those

Typical Physical Properties

	MDI (monomeric)	MDI (polymeric)
Physical State at normal temperatures	solid	liquid (oily)
Viscosity (mPa·s at 25°C)	—	100–800
Color	white to light yellow	dark brown (opaque)
Odor	—	earthy, musty (characteristic)
Specific Gravity (@ 25°C)	1.22 (43°C)	1.23
Boiling Point (°C)	171 at 1,33 mbar 200 at 6,66 mbar 230 Decomposition	polymerizes about 260°C with evolution of carbon dioxide
Flash Point (°C)	199	over 200
Fire Point (°C)	232	over 200
Freezing Point (°C)	38	below 10
Vapor Density (air = 1)	8.5	8.5
Vapor Pressure mbar at 25°C	$<10^{-5}$	$<10^{-5}$

Table 8.18 Physical properties of MDI

Designation	Formula	Mol. Wt.	B.P., °C	Vapor at 60°C W/mK	Approx. at Life Years	O.D.P.**
CFC-11	CCl_3F	138.	23.8	0.0093	60	1.0
CFC-12	CCl_2F_2	121.	-29.8	0.0114	100	0.9
CFC-113	$CCl_2F\,CClF_2$	188.	47.6	0.0044	90	0.8
CFC-114	$CClF_2\,CClF_2$	171.	3.6	0.0060	200	0.6
CFC-115	$CClF_2\,CF_3$	155.	-38.7	0.00724	400	0.3

* Chlorofluorocarbons
** Ozone depletion potential

Table 8.19 Major CFCs*

Designation	Formula	Mol. Wt.	B.P., °C	Vapor at 60°C W/mK	Approx. at Life Years	O.D.P.
HCFC-21	$CH\,Cl_2\,F$	103	8.9	-	-	<0.05
HCFC-22	$CH\,Cl\,F_2$	86.5	-40.8	0.0130	-	0.05
HCFC-31	$CH_2\,Cl\,F$	68.5	-9.1	-	-	<0.05
HCFC-123	$CH\,Cl_2\,CF_3$	153	27.9	0.0138	1.4	0.02
HCFC-124	$CH\,Cl\,F\,CF_3$	137	-11.0	0.0159	-	0.02
HCFC-132b	$CH_2\,Cl\,C\,Cl\,F_2$	99.5	46.8	-	-	<0.05
HCFC-133a	$CH_2\,Cl\,CF_3$	116	6.1	-	-	<0.05
HCFC-141b	$CCl_2\,F\,CH_3$	117	32.1	0.0138	<1	0.15
HCFC-142b	$CClF_2CH_3$	101	-9.8	0.0159	-	0.06

* Hydrochfloroflurocarbons

Table 8.20 Alternative blowing agents (HCFCs)*

Designation	Formula	Mol. Wt.	B.P., °C	Vapor at 60°C W/mK	Approx. at Life Years	O.D.P.
HFC-134a	$CF_3\,CH_2\,F$	102	-26.2	0.01717	6	0
HFC-152a	$CHF_2\,CH_3$	66.0	-24.7	0.0182	-	0
HFC-356	$C_4\,H_4\,F_6$	166	24.6	0.0095	0.4	0

*Hydrofluorocarbons

Table 8.21 Alternative blowing agents (HFCs)*

Designation	Formula	Mol. Wt.	B.P., °C	O.D.P.
PF-5040	C_4F_{10}	238	-2	0
PF-5050	C_5F_{12}	288	30	0
PF-5060	C_6F_{14}	338	56	0
PF-5070	C_7F_{16}	388	80	0

*Perfluorinated hydrocarbons.

Table 8.22 Alternative blowing agents (PFCs)*

Designation	Formula	Mol. Wt.	B.P., °C	O.D.P.
E-245	$CF_3CH_2OCF_2H$	134	26	0
E-356	CF_3CHOCH_3	150	27	0

*Hydrofluorocarbon ethers

Table 8.23 Alternative blowing agents (HFEs)*

Foam	Reaction	Property
Polyurethane	Polyaddition	Flexible and Rigid
Polyisocyanurate	Cyclotrimerization	Rigid
Polyamide	Polycondensation	Flexible and Rigid
Polyimide	Polycondensation	Semi-Rigid
Pyranyl	Radical polymerization	Rigid
Polyurea	Polyaddition	Flexible and Rigid
Epoxy	Ring-opening polymerization	Rigid
Phenolic	Polycondensation	Rigid
Urea-formaldehyde	Polycondensation	Rigid
Polycarbodiimide	Polycondensation	Rigid
Polyoxazolidone	Ring-opening polyaddition	Semi-rigid
Unsaturated polyester (or vinyl ester)	Radical polymerization	Rigid
Rubber (natural & synthetic)	Vulcanization	Flexible
Viscose	Regeneration of cellulose	Flexible
Polyvinyl alcohol	Formal formation	Flexible

Table 8.24 Classification of thermoset foams

of most interest to hospitals are urethane and vinyl foams. The latter competes with foam rubber and urethane foam as cushioning and padding material. Among the advantages of vinyl foam over other types are good resiliency, chemical resistance, and nonflammability.

Polyurethane (PUR) foams have many things in their favor. They do not have undesirable bounce-back; they do not mat, stiffen, or crumble after long use or aging; they are nonallergenic and odorless;

and they are unaffected by dampness. Moreover, since they are chemically inert, common cleaning chemicals, water, body acids, spilled foods, and liquids do not affect urethane foams.

Polyethylene, polyurethane, and polystyrene are the most common foams; however, polypropylene (PP) foams can provide favorable properties at a lower material cost. PP is stiffer than PE and performs better in load bearing and structural applications. The low glass-transition temperature of PP compared to PS provides increased flexibility and impact strength (chapter 1). Use of PP foams includes packaging, automotive, insulation, and protection of underground pipe.

It is important to know in advance what foam properties will be required for a particular application. As with other plastics or any material, it is not enough to be told that a material is plastic foam with remarkable properties. The following review on examples of different plastics used includes preliminary information on processes and applications. It starts by reviewing the subject of structural foams and reinforced plastic foams.

STRUCTURAL FOAM

Structural foam is a term originally used for cellular thermoplastic articles with integral solid skins and possessing high strength-to-weight ratios (Table 8.25). Eventually, the term covered high-density rigid cellular plastics strong enough for structural applications. As an example, thermoset foams, such as polyurethane, are frequently referred to as *structural foams*. In general, a structural foam can be made from virtually any high molecular weight thermoplastic organic polymer and will have a cellular core and an integral skin on all sides. The skin is relatively nonporous in relation to the cellular core.

Structural foam construction, when compared to an equivalent amount of a conventional foam plastic, results in a three- to fourfold increase in rigidity. (Rigidity is a function of wall thickness.) In addition to urethane and other thermoset foams, all thermoplastics can be foamed by the addition of physical or chemical blowing agents. A broad and overlapping division of thermoplastics exists between commodity and engineering groups of resins used for structural foams. The commodity group consists of the styrenics (polystyrene, styrene-acrylonitrile, etc.), olefins (polypropylene, polyethylene, etc.), and vinyl chlorides (PVCs), while the engineering group includes acetal, ABS, nylon, polycarbonate, polyester, and polyetherimide, plus various glass- or carbon-reinforced resins (chapter 1).

Structural foams have found markets where it is necessary to conserve energy and natural resources by turning to lighter and less-energy-sensitive materials. The automotive and truck industries are being forced to increase mileage relative to fuel consumption, and reduction of vehicle weight by replacing steel with structural foam is the favored way of accomplishing this objective. These plastics are targeted to solve cost, performance, and weight-reduction problems in materials handling, tanks, trucks, aircraft, and military weapons.

Most structural foam is produced in the form of complete parts by injection molding (chapter 15), rather than bulk raw stock, as in many other foams. Since a complete part is made with a skin on each outer surface, the process is ideally suited for fabrication of parts or components in

(at .250 Wall With 20% Density Reduction)											
Property	*Unit*	*Method of Testing*	*High Density Polyethylene*	*ABS*	*Modified Polyphenylene Oxide*	*Polycarbonate*	*Thermoplastic Polyester*	*Polypropylene*	*High Impact Polystyrene*	*High Impact Polystyrene w/FR*
Specific gravity	lbs./ft.^{3}	ASTM-D-792	.60	.86	.85	.90	1.2	.67	.70	.85
Deflection temperature under load	°F@66 psi	ASTM-D-792	129.6	187	205	280	405	167	189	194
	°F@264 psi		93.5	172	180	260	340	112	176	187
Coefficient of thermal expansion	in./in./°F × 10^{-5}	ASTM-D-696	12	4.9	3.8	2	4.5	5.2	9	4.5
Tensile strength	psi	ASTM-D-638	1,310	3,900	3,400	6,100	9,910	1,900	1,800	2,300
Tensile modulus	psi	ASTM-D-638		2,500,000	235,000	300,000	1,028,000	79,000	141,160	245,000
Flexural modulus	psi	ASTM-D-790	120,000	2,800,000	261,000	357,000	1,000,000	80,400	200,321	275,000
Compressive strength (10% deformation)	psi	ASTM-D-695	1,840	4,400	5,200	5,200	11,300	2,800	3,447	
Combustibility rating		UL Standard 94*		V-0	V-0/5V	V-0/5V	V-0	HB	HB	V-0

*This rating is not intended to reflect hazards presented by this or any other material under actual fire conditions.
Material properties given above are typical and vary from supplier to supplier. It is recommended that an end user contact his supplier and/or molder to obtain specific properties for use in a given application.

Table 8.25 Properties of thermoplastic structural foams

which light weight and stiffness are required. In all cases with thermoplastic structural foams, the plastic's original properties of heat and chemical resistance, as well as most electrical properties, remain the same as for the solid plastics. An example of advantages includes their dielectric constant that is improved over the solid plastic, which is the reason foamed polyethylene is used in certain electronic/electrical products, such as television cable insulation.

Reinforced Plastic Foam

The fairly dense varieties of TP and TS foams may be reinforced, usually with short glass fibers, but long fibers can also be used to provide increased performance (chapter 15). The fibers are generally introduced into the basic ingredients and are blown along with them to form part of and to reinforce the walls of the cells. These plastic foamed composites are lightweight with high strength. Tables 8.26 and 8.27 provide data on low-weight reinforced plastic foam fabricated using the reaction injection-molding process (chapter 12). Figure 8.1 compares the moduli of elasticity of plastic foams with other materials.

Acetal

This translucent crystalline polymer is one of the stiffest thermoplastics available. It provides excellent hardness and heat resistance, even in the presence of solvents and alkalies. Its low moisture

Chemistry	Polyurethane Polyisocyanurate
Dispensing Machine	Elastogran Puromat 80, typ.
Component Temperature	
A-Comp. Polyol resin	23°C
B-Comp. Isocyanates	23°C
Component Impingement Mixing Pressures	
A-Comp. polyol resin	150-170 atm.
B-Comp. isocyanate	150-170 atm.
Mold Temperature	60°C
Resin viscosity at room temperature	300-1250 cp
Isocyanate viscosity at room temperature	60-200 cp
Resin/iso ratio	100/100 to 170 typ.
Blowing agent	Typically CFC-11 or carbon dioxide generated from reaction with water
Demold time	2 minutes typ.
Cream time	20 sec. typ.
Free rise density	2.8 lb./cu./ft. typ.
Shot rate	0.5 lb./sec. typ.
Shot size	2.0 lb. typ.

Table 8.26 Properties of PUR-isotropic glass-fiber-mat-reinforced foamed composite

	Value	Test Method
Density, lb./cu. ft.	24	SAE J 315
Flex modulus, psi	150,000	ASTM D790
Flex strength, psi	4,000	ASTM D790
Moisture content, %	2.6	SAE J 315
Water absorption 2.5 hrs., %	1.0	SAE J 315
Water absorption 14 hrs., %	1.9	SAE J 315
Water swell, %	0.5	SAE J 315
Dimensional Stability, %		
Expan. in water, 70°F, 24 hrs	0.2	SAE J 315
Contrac. in water, 190°F, 24 hrs.	0.2	SAE J 315
Warpage original	0	SAE J 315
Wet	0	-
Dry	0	-
Flammability (MVSS301), inches/min.	1.5	FLTM BN24-2
Coeff. of therm. expan., in./in./°F,		ASTM D 696
- 20 to 175°F	3×10^{-6}	ASTM D 696

Table 8.27 Properties of PUR-unidirectional chopped-glass-fiber-reinforced foamed composite

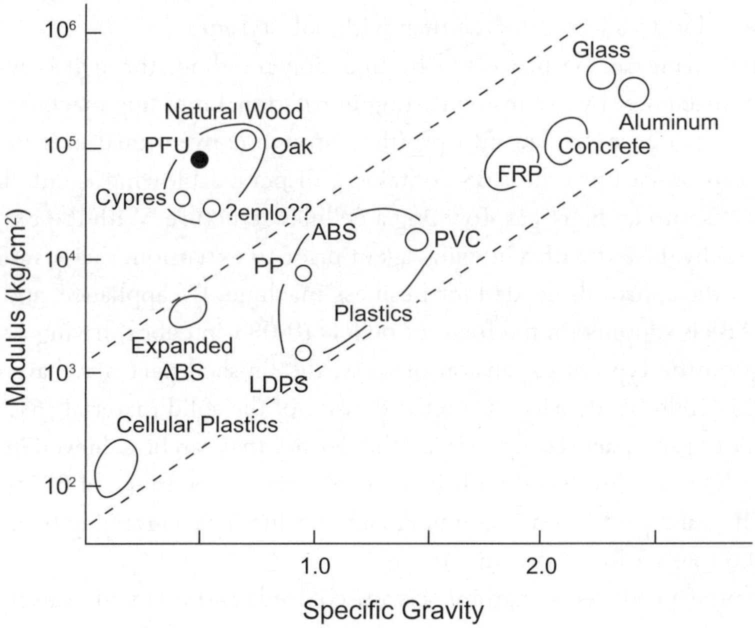

Figure 8.1 Comparison of plastic foam moduli with other materials

sensitivity and good electrical properties permit direct competition with die-cast metal in a variety of applications. In addition, acetal has extremely high creep resistance and low permeability. Acetal is also available as a copolymer (Hoechst Celanese Corp.'s Celcon®) for improved processability. The homopolymer (DuPont's Delrin®) has a very low coefficient of friction and its resistance to abrasion is second only to nylon 6/6. Acetals are frequently blended with fibers, such as glass or fluorocarbon, to enhance stiffness and friction properties. Acetal is not particularly weather resistant, but grades are available with UV stabilizers for improved outdoor performance. Acetal, whether homopolymer or copolymer, is not used to any significant degree in forming structural foams.

Acrylonitrile-Butadiene-Styrene (ABS)

ABS is an opaque, amorphous terpolymer of styrene offering impact, heat, and chemical resistance. It is available in a multitude of grades intended to emphasize key properties. Improved flammability characteristics are possible either by alloying (blending) with PVC or polycarbonate or by compounding with halogenated additives. In addition, ABS has relatively low mold shrinkage rates, good long-term dimensional stability, and platability. ABS compounds are slightly hygroscopic and should be dried prior to conventional injection molding to avoid splay marks (chapter 2). In structural foam molding, moisture can lead to excessive odor with certain chemical blowing agents.

High-flow ABS grades still display relatively stiff flow characteristics and, therefore, like all high-temperature thermoplastics, offer some resistance to foaming. ABS is susceptible to degradation and discoloration upon exposure to ultraviolet (UV) radiation. Modifying the flammability of ABS by means of halogen compounds significantly increases resin cost and decreases color stability, especially in pastels, but to a lesser degree than with polystyrenes.

ABS structural foam can be processed by injection molding, through conventional or low-pressure injection machines; by expansion casting in rotational molding machines or conveyorized oven systems; or it can be extruded into profiles through conventional extruders. For injection molding and expansion casting, the ABS contains a dispersed blowing agent that, when heat is applied, decomposes into an inert gas, forming a cellular structure. With the extrusion grade, the ABS pellets are usually dusted with a blowing agent prior to extrusion. Various ABS grades are also available that meet the approvals needed for business-machine, TV, appliance, and other markets.

Expandable ABS is supplied in the form of pellets (0.093-in cubes) having a specific gravity of 1.04. Depending on the type of expansion process, the finished part may have a specific gravity from as low as 0.3 (20 lb/ft^3 density) up to the density of the solid material (64.9 lb/ft^3). Specific conversion techniques and part size determine the density that can be achieved in a part. Injection-molded parts are normally produced with specific gravities in the range of 0.7 to 0.9 (densities of 46.7 to 60.0 lb/ft^3), and profile extrusion normally produces parts ranging from specific gravities of 0.3 to 0.7 (20.0 to 46.7 lb/ft^3 densities).

Structural foam is produced in natural (cream-colored) and in a wide variety of other colors. Because of its excellent balance of properties at reasonable cost, ABS structural foam continues to find applications in the automotive, furniture, construction, and materials-handling industries.

ABS structural foam has excellent buoyancy, a very low (desirable) stiffness-to-weight ratio, good screw and staple pullout strengths, and creep resistance superior to that of high-impact polystyrene (HIPS) and polyethylene foams. This is particularly important in load-bearing applications, such as pallets, tote boxes, furniture, and parts buried under earth loads.

IONOMER

Ionomers denote carbonyl-containing polymers or copolymers of olefins associated with monovalent or divalent cations. Although these materials are not cross-linked in the conventional sense, since they are thermoplastic and can be processed in standard thermoplastic processing equipment, they do exhibit characteristics resembling those of partially cross-linked plastics.

Ionomer foams are produced by extrusion or foam injection molding to produce tough closed-cell foam structures. The low melting point (194° to 205°F; 90° to 96°C), high melt strength of the resin, and compatibility with nucleating agents or fillers combine to produce systems that are useful in athletic products, footwear, and construction. The melt characteristics provide a tough skin in the low-density foam sheet and a better surface finish in the higher-density injection-molded products. The higher tensile strength and low melt point characteristics of an ionomer provide strong heat-seal seams for fabricated sections used in packaging applications. Products ranging from 3 to 30 lb/ft^3 are tougher and more solvent resistant than equal-density foams made from PE or PS. The ionomer-foamed sheeting may be vacuum formed, laminated, stitched, glued, and modified for flame-retardant requirements. The resilient nature of the polymer in medium-density sheeting (6 to 9 lb/ft^3) has application in floor and carpet backing, swimming pool construction, and tennis court under-liner constructions.

In foam injection-molded parts, the smooth-skinned resilient foam produces a tough energy-absorbing structure that is being used as a wood substitute in athletic products, such as lacrosse-stick heads and hockey-stick blades. Protective structures for helmets and automotive parts can make use of the foams. All are being produced in densities ranging from 0.35 to 0.7 g/cc (22 to 44 lb/ft^3). In those applications where added stiffness is desirable, the use of glass, titanate, or other fiber reinforcement with the foamed structure is very effective.

Ionomer foams are competitive with PE, PS, and some PUR structures. Where the combination of light weight, toughness, and/or direct paint adhesion are prerequisites, the foamed ionomer systems are more than competitive by reducing the overall cost through lower part weight and less-involved finishing. Ionomer foams find application in the construction and transportation industries because of their low-temperature flex strength and good thermal and moisture-barrier properties. These foams, modified with flame-retardant additives, have been used in the aircraft and packaging industries. There have been applications where they were used as protective insulation for concrete pours in construction (Libby Dam near Libby, Montana, etc.). At the dam site, the foam sheets were wired to the inside of the steel concrete forms. After the pour, the foam remained on the concrete surface for over two weeks, allowing the concrete to cure evenly without cracking, by limiting the heat transmittance from the material.

Ionomer foam is being used as pipe-insulation wrap. This soft flexible foam will not absorb moisture or collapse, and it contains no irritating material, such as glass fiber. No tools or tape are needed. The material can be secured to pipes quickly with convenient plastic ties provided with the pipe wrap. Two varieties are offered: (a) a two-layer foam lamination for indoor pipes and (b) a foil-laminated wrap suitable for outdoor use.

PHENOLIC

In the United States, a large application for phenolic foam continues to be in the decorative markets, such as the floral market. Its open-cell structure and low water absorption permits a large volume of water to remain with the flower holder. This type of foam is also used for insulation, particularly in Europe, as it has particularly good fire retardance and heat resistance (Table 8.28). Its excellent chemical resistance data is shown in Table 8.29. Properties characteristics of typical phenolic foams are given in Tables 8.30 and 8.31 and Figure 8.2. Along with its excellent performances, it also has limitations, such as low productivity, low yield rate in materials, high friability, and low strength unless reinforced (212).

Phenolic foam is divided broadly into two categories. One is the novolac type and the other is the resol type (Tables 8.32 to 8.34). In the novolac type, phenol and formaldehyde react in the presence of an acid catalyst and link the linear condensation product by a methylene bond. The basic curable cross-linking agent, such as hexamethylene tetramine, and blowing agent are added to the linear condensation product, and the resultant product is molded at elevated temperatures and high pressure.

	Acid catalyst	
	Boric-oxalic	Hydrochloric
Foam density, kg/m³	32	30
ASTM D1692-67T rating	S.E. (a)	S.E. (a)
ASTM E84-60T, Underwriter's Tunnel Test:		
Flame-spread rating	10	
Fuel contribution	23	
Smoke-density rating	0	
Bureau of Mines Flame-penetration Test:	595	
Penetration time, sec	10.3	53 (punked)
Weight loss, %		4.8
ASTM E162, Radiant-panel Test, Flame-spread rating	No ignition	

Note: (a) Self-extinguishing (no ignition).

Table 8.28 Typical flammability properties of phenolic foams

Ingredients (%)	Resol-Type Foam	Novolac-Type Foam
Sulfonic acid (50)	×	×
Sulfonic acid (20)	O	O
Nitric acid (10)	×	×
Nitric acid (5)	O	O
Hydrogen chloride (10)	O	O
Formic acid (10)	O	O
Sodium hydroxide (20)	×	×
Sodium hydroxide (1)	O	O
Ammonia water (14)	O	O
Benzene	O	O
Styrene monomer	O	O
Acetone	O	O
Acetic esters	O	O
Brine	O	O

O: OK
×: NG

Table 8.29 Typical chemical resistance after fourteen-day immersion

In the resol type, phenol and formaldehyde react in the presence of a basic catalyst and provide a liquid-type resol to which an acid catalyst and blowing agent are added and the foam is then molded. In forming the resol, a method is available for obtaining a benzylic ether-type liquid oligomer by reacting phenol and formaldehyde in a nonaqueous phase.

Materials to be used in making phenolic foam include phenol, cresol, and xylenol. Crude products of these materials can also be used as raw materials. The positions where the phenols can react with aldehydes are 2, 4, and 6 against the OH-group. Accordingly, if substitute groups (such as alkyl groups) are located at the 2, 4, and 6 positions, aldehydes can no longer react. In other words, the existence of the substituted groups governs the functionality of phenols.

Polycarbonate

This plastic has outstanding impact strength, high heat resistance (deflection temperature of 280°F [138°C] at 66 psi [0.45 MPa]), as well as very good flexural characteristics, creep resistance, and processability. PC is a good choice for structural components where load-bearing capability at elevated temperatures is a key requirement. It is an excellent alternative to metal for large components in the automotive, appliance, telecommunications, materials-handling, and business-machine industries. Foamable PC combines an unusual blend of rigidity, impact strength, and toughness with UL 94 V-0 and 5V flammability ratings.

Grades of PC are available that couple well with glass fibers to improve fatigue resistance and other properties. They offer good colorability and chemical resistance, along with very high rigidity for a resin with such good ductility. It is also one of the most moisture-sensitive thermoplastics, along with nylon. Physical properties are significantly reduced if the resin is not dried well prior to processing. Care is needed in the selection of blowing agents in foam applications, since severe reduction in physical properties (up to 30%) has been reported for certain high-temperature chemical blowing agents. Solvent resistance of PCs is poor, resulting in stress cracking in cast solid sheet.

Polybutylene Terephthalate

The outstanding characteristics of this PBT thermoplastic are its UL 94 V-0 and 5V flammability rating, heat-deflection temperature of 420°F (216°C) at 66 psi (0.45 MPa), high flexural strength and modulus, and excellent chemical and solvent resistance. PBT resins are especially suitable for applications requiring a combination of high heat endurance, stiffness, chemical resistance, and moderate creep.

They are translucent crystalline materials that are condensation products of 1,4- butanediol and terephthalic acid. They exhibit a fairly sharp melting point at approximately 440°F (227°C).

Items \ Material	Phenolic Foam				(cf) PIR Foam
	Closed Cell, Type A	Closed Cell, Type B	Open Cell, Type C	Open Cell, Type D	
Density (kg/m³)	40	40	50	25	35
Thermal Conductivity (kcal/mh ℃)	0.020	0.020	0.035	0.035	0.020
Closed-Cell Content (%)	90	90	0	0	90
Water Absorption (g/100 cm²)	4	4	12	12	3
Limiting Oxygen Index (%) (a)	50	33	33	33	26
Surface Fire Test (b)					
C_A	5	25	10	5	50
T θ (min, ℃)	50	170	50	70	80
After-glowing Time (sec)	0	0	0	0	0
Popping	None	Existing	None	None	None
Criteria	OK	NG	OK	NG (c)	OK

Notes: (a) ASTM D•2863
(b) JISA 1321 (Thickness of Test Specimen: 25 mm)
(c) NG due to large cracks

Table 8.30 Properties of typical phenolic foams

Specialty grades are available with up to 30% glass reinforcement. Among key properties are low moisture absorption and good chemical resistance. Good frictional coefficients are offered, but they are slightly inferior to that of acetal. Grades are available with high flow characteristics approaching that of the olefins, but the heat resistance of these low-melt-viscosity grades decreases rapidly as a function of improved flow.

Items \ Types	Phenolic Foam	Polyurethane Foam
Formulation	Resol: 100 Catalyst: 30 Foaming Agent: 15	Component A (MDI) 100 Component B: 100
Material Temperature (℃)	Resol: 20 Catalyst: 20 Foaming Agent: 20	Component A: 20 Component B: 20
Room Temperature (℃)	20	20
Foaming Time		
CT (seconds)	20	20
GT (seconds)	150	85
RT (seconds)	200	110
Density (kg/m³)	36	28
Maximum Temperature (℃)	96	167
Maximum Foaming Pressure (kg/cm²)	0.06	0.05

Table 8.31 Foaming characteristics of free-rise foams

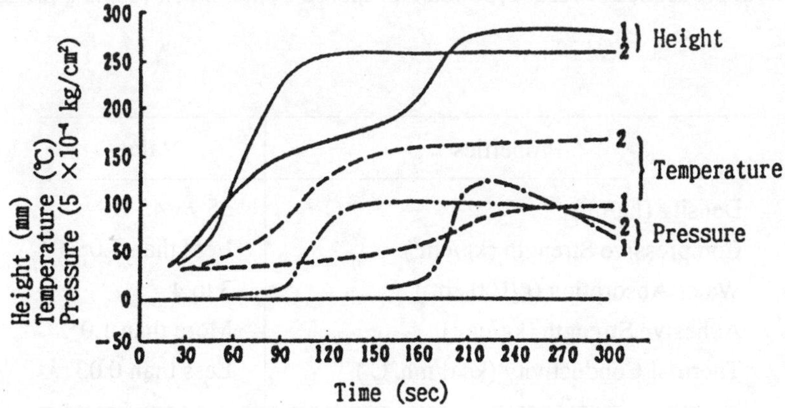

Figure 8.2 Foaming characteristics of (1) phenolic foam and (2) polyurethane foam

Properties	Values
Density (kg/m³)	40
Compressive Strength (kg/cm²)	1.8
Flexural Strength (kg/cm²)	5.9
Water Absorption (g/100 cm²)	0.5
Thermal Conductivity (kcal/mh ℃)	0.024
Specific Heat (cal/g ℃)	0.3

Table 8.32 General properties of novolac-type foam

Properties	Values	
Density (kg/m³)	60	35
Compressive Strength (kg/cm²)	2.0	1.6
Flexural Strength (kg/cm²)	8.0	5.5
Tensile Strength (kg/cm²)	1.1	1.1
Thermal Conductivity (kcal/mh ℃)	0.029	0.025
Thermal Expansion (l/ ℃)	3×10^{-5}	3×10^{-5}
Water Absorption (g/100cm²))	2	2
Limiting Oxygen Index	>40	>40
Specific Heat (cal/g ℃)	0.48	0.48

Table 8.33 General properties of resol-type foam prepared by the block-foaming process

Properties	Values
Density (kg/m³)	35 – 45
Compressive Strength (kg/cm²)	Less than 1.0
Water Absorption (g/100 cm²)	3 to 4
Adhesive Strength (kg/cm²)	More than 1.0
Thermal Conductivity (kcal/mh ℃)	Less than 0.03

Table 8.34 General properties of resol-type foam prepared by the spraying process

As with other crystalline resins, PBTs are notch sensitive. Impact resistance decreases rapidly as a function of corner radii. Thermal decomposition temperatures are only slightly higher than the melting point; therefore, close attention must be paid to residence time at high temperatures. Mold temperatures below 150°F (66°C) should be avoided to minimize excessive orientation, resulting in relatively long fabricating cycles.

POLYETHERIMIDE

PEI form provides outstanding thermal, mechanical, and electrical properties, together with exceptional flame resistance. The foamable resin has a heat-deflection temperature of 401°F (205°C) at 66 psi (0.45 MPa). Glass-reinforced grades provide even greater rigidity and dimensional stability while retaining excellent processability. The resin has excellent flammability characteristics and self-ignition temperature of 535°C and UL rating of V-0. Flame spread in the radiant-panel test is low ($1_s = 2.7$) and the LOI is high—47. NBS smoke-chamber results show a low specific optical density (Ds 4 min = 0.7, Dm = 31, flaming mode), and combustion-product-toxicity studies show results comparable to polystyrene.

POLYOLEFIN

The family of cellular polyolefins (chapter 2) that includes polyethylenes (PEs) and polypropylenes (PPs) contributes many unusual properties to the cellular plastics industry. These foams are tough, flexible, and chemical and abrasion resistant. They are known to have superior electrical and thermal insulation properties. Their mechanical properties are intermediate, between rigid and highly flexible foams. Densities are 2 lb/ft^3 and higher, approaching that of the solid plastics. The highly expanded polyolefin foams are potentially the least expensive of the cellular plastics. However, they require expensive processing techniques, and for this reason, their cost per unit volume is higher than that of low-density polystyrene and polyurethane foams.

Low-density polyolefin foams are usually considered as those in the density range of 2 to 10 lb/ft^3. They are used for producing extruded planks, rounds, tubes, and special-purpose profiles. Compression-molded items may also be produced from low-density polyolefins. High-density (10 to 40 lb/ft^3) polyolefins were used initially for electrical cable coatings.

Low-density polyolefin foams are being widely used in package cushioning. Energy absorption under continued impact provides protection for delicate electronic parts as well as heavy metal assemblies. PP foam sheeting is used widely to provide protection for polished surfaces. It is particularly easy to fabricate low-density polyolefin foams by cutting and similar methods using band saws, table saws, and routers. Slicing-type equipment or electrically heated wires can also be used to produce sawdust-free products. Cross-linked compression-molded PE foams are used commonly to produce foam packages for cameras and measuring instruments. These foams offer excellent appearance and good cushioning performance.

The production of cellular PE involves only one chemical reaction, that is, the thermal decomposition of a blowing agent at a specific temperature, which action liberates an inert gas. The choice of blowing agent for electrical service applications is critical because of several unusual requirements. The blowing agent, the gas it liberates, and the residual by-product must not absorb moisture, which would adversely affect the electrical properties of the product. It is also important that the residue left by the blowing agent be nonpolar in order to avoid losses at high frequencies.

Polyethylene Foam Electrical/Electronic Property Because cellular PE comprises roughly equal volumes of resin and gas, its properties are different from those of ordinary unfoamed PE. The cellular product has a much lower dielectric constant and therefore lower electrical losses. The composition of PE (dielectric constant 2.3) and an inert gas (dielectric constant 1.0) has a dielectric constant of 1.5. In terms of electrical insulation, the lower dielectric constant permits a reduction in space between inner and outer conductors without changing the characteristic impedance. For this reason, it is possible to reduce the attenuation by increasing the size of the inner core without increasing the overall diameter. Alternatively, decreasing the size of the inner conductor may reduce the weight.

With cellular PE being the closed-cell type, its permeability to moisture, while several times larger than that of solid polyethylene, is still relatively low. Cellular PE will provide good electrical performance in alternately wet and dry environments. Its moisture resistance is valuable in such applications as antenna lead-in wire for UHF television. This and other high-frequency applications require that the power factor and dielectric constant not be affected by changes in frequency. Variations would result in power losses, for example, a weakening of the signal brought in by a UHF antenna. Cellular PE satisfies these requirements. The dielectric constant and power factor of the foam are not seriously affected by temperature fluctuations. The power factor of cellular PE is independent of its specific gravity, while dielectric constant increases with increases in specific gravity.

Cellular PE is particularly useful in a number of electrical applications, such as in coaxial cables (CATV, military, and other) and in twin leads. The foamed material is particularly useful in modem wire-coating equipment because of the ease of handling and the economies it provides in size and weight of insulated conductors.

Low-Density Polyethylene This LDPE foam has all the advantages of the base polymer, including excellent resistance to most chemicals, both organic and inorganic. The foams are generally closed cell. They are classified as semirigid and flexible, depending on their densities and shapes. They can be made to feel as soft as low-density flexible urethane foams and as hard as rigid polystyrene foams.

There are basically two types of LDPE foam. They are the extruded and cross-linked. The extruded foam is produced in a continuous process by first blending molten polyethylene resin with a foaming or blowing agent (usually a halogenated hydrocarbon gas) under high pressure, then conveying this mixture in a temperature-controlled screw extruder though a die opening to a continuous conveyer exposed to atmospheric pressure. When the hot viscous liquid-gas solution is exposed to atmospheric pressure, the blowing gas expands to form individual cells. At the same time, the mass is cooled to solidify the molten polyethylene, thereby trapping the blowing agent in the interstitial cells. The degree of expansion, cell size, and cell orientation can be controlled by varying the flow rate, heating and cooling temperatures, gas-liquid ratio, and pressure drop through the die opening.

Cross-linked LDPE foam can be produced by batch and continuous processes. The cross-linking is accomplished by chemical or irradiation methods. Chemical cross-linking of PE foam is produced in a batch process, and because of production economics, is limited to plank products. The radiation cross-link process permits the continuous production of cross-linked PE foam. This process is limited, however, to production of relatively thin cross-sections (such as up to 3/8 in) or sheet products.

The chemical, mechanical, and thermal properties of the extruded and cross-linked LDPE foams are very similar, although the cell sizes of the cross-linked foams are generally smaller and more uniform than that of the extruded products. The cross-linked foams also feel softer than the extruded foams.

PE foam exhibits compression-deflection properties similar to flexible and semirigid foams. As would be expected, the compressive strength increases with increases in density. Compressive creep (loss of original thickness with time) is observed more in LDPE foam than in high-density (HDPE) foam. Tensile and tear strengths of PE foams are very high compared to other foams.

One of the largest uses of PE foam is in the packaging area to protect products from damage during handling, shipping, and storage. PE foams are excellent energy absorbers. Because of the wide range of dynamic cushioning properties available with different densities of foam, a wide spectrum of packaging requirements can be met with PE foams. These foams provide optimal cushioning characteristics for objects exerting static stresses ranging from approximately 0.4 to over 15 psi (0.003 to over 0.10 MPa). To date, no other flexible or semirigid foams commercially available are capable of providing optimal cushioning characteristics in this static-stress range.

PE foams are used extensively in buoyancy applications because of their excellent water-resistant properties. These basically closed-cell foams absorb less than 0.5% by volume of water after being immersed for twenty-four hours. The low density of the foams also contributes to their buoyancy. The excellent dielectric characteristics of PE are retained when it is expanded into foams.

PE foam is chemically inert and contains no water-soluble constituents. It is resistant to most chemicals and solvents at room temperature. It is unaffected by contact with fuel oil and heavier hydrocarbons, but it exhibits slight swelling when immersed for an extended period of time in gasoline. Acids and alkalies normally do not affect PE foam, but strong oxidizing agents may eventually cause degradation, especially at higher temperatures. At temperatures above 130°F (54°C), it becomes more susceptible to attack by certain solvents (chapter 2).

UV rays in sunlight cause some degradation of PE over an extended period of time. This degradation is slower in higher-density products. It is first noted as a yellowing of the foam surface. Some degradation of physical properties will be experienced with longer periods of exposure. For applications where long-term performance is required under direct sunlight, a protective coating should be used.

Ease of fabrication is one of the many advantages of PE foam. It can be skived to precise thickness, cut and shaped to form custom parts, and joined to itself or to other materials without major investment in complex equipment. It can also be vacuum formed. Expanded polyethylene (EPE) will adhere to itself by the use of heat alone. Hot air, or a plate heated to approximately 177°C (350°F), can be used to simultaneously heat the surfaces of two sections of foam to be joined. Upon softening, the two pieces are quickly joined together under moderate pressure and an excellent bond is formed, with only a short cooling period required. Release of the melted foam is aided by a coating of fluorocarbon resin or silicone dispersion on the heating surface. The foam may also be bonded to itself and to other materials by the use of solvents or commercially available adhesive.

High-Density Polyethylene HDPE foams are defined by the specific gravity (ca. 0.96 g/cc) and melt index of the basic resin. They have been used extensively worldwide, particularly for materials-handling applications, such as boxes, crates, and pallets, because, until recently, they were much cheaper than PPs.

Cross-linked Polyethylene A number of thermoplastic foams cross-linked by radiation are available. The most popular base plastic for cross-linked foam is polyethylene. These cross-linked foams offer higher stability and mechanical strength, better insulation characteristics, and improved energy-absorption properties. Most of these foams can be thermoformed, embossed, printed, laminated, or punched, using conventional equipment.

There are a number of techniques available for producing cross-linked thermoplastic foams. One such technique involves making a mixture that incorporates one or more of the monomers from the group consisting of acrylamide, methacrylamide, and acrylic acid; a foaming agent in solid or liquid state at room temperature; a cross-linking agent having more than two polymerizable double bonds in the molecule; and any monomer, if necessary, that is copolymerizable with the monomers mentioned above (chapter 1). The mixture is polymerized to a thermoplastic body containing the foaming agent homogeneously. The plastic body is then heated to a temperature above the softening point of the plastic in order to make possible, simultaneously, the foaming and the cross-linking reaction between the polymer molecules. Of the various polymerization methods, such as polymerization with catalysts or initiators, thermal polymerization, photopolymerization, and ultrasonic polymerization, the radio-induced or the catalytic polymerization followed by irradiation gives the best results.

Polypropylene PP foams supplanted HDPE foams in the structural foam field. Their use continues to increase because of the extreme range of grades and properties available, plus a favorable price advantage, compared with other thermoplastic foams. Glass-reinforced (30 wt% chopped glass fiber) PP foams are commonly used. Low-density flexible PP foam film can be extruded in the 0.7 lb/ft^3 (11.2 kg/m^3) range.

Despite its low density, the film sheeting consists of a uniform matrix of small closed-cell gas-filled bubbles. This film has outstanding toughness and strength over a wider range of temperatures and humidities. Its major characteristics, compared with other packaging films, are its light weight, resistance to tearing, chemical resistance, and moisture-barrier properties. Extrusion parameters are similar to those used for LDPE. A microcellular PP foam of this type has been used as a furniture wrap when packaging furniture in interstate commerce. The protective foam is wrapped around the item before insertion into a corrugated carton. Even with movement in the carton, the PP wrap will stay with the item it is intended to protect.

PP foam sheeting is specified in Federal Specification PP-C-1797. There are two types, Type I for general cushioning applications and Type II for electrostatic protective cushioning applications.

These foams are useful from –54° to 71°C (–65° to 160°F). The foam sheeting is intended for use as a protective cushioning wrap for low-density items. For high-density items, it can be used for protection of surfaces from abrasion. The foam is nondusting and nonlinting. Typical packaging applications are surface protection for optical lenses, equipment with critical surfaces, electrical and electronic equipment, glassware, ceramics, and magnetic-tape rolls. There is microcellular PP foam sheeting (DuPont's Microfoam® sheet) that remains flexible and useful over the temperature range of –196° to 121°C (–320° to 250°F).

Melt Strength PP tends to be more difficult to foam due to weak melt strength and low melt elasticity. Melt strength is the resistance of the melt to extension, while melt elasticity is a measure of elastic recovery. Melt strength and melt elasticity are directly related; the higher the melt elasticity, the higher the melt strength. With weak melt properties, cell walls separating gas bubbles in the foaming plastic are not strong enough to bear the extensional force as the gas expands, and they rupture. As a result, PP foam has a high open-cell foam content, which is unsatisfactory for many applications. Melt strength is commonly increased by plastic modification, such as cross-linking; other methods include the use of high injection pressures (> 3.5 MPa), two blowing agents, or two PPs with different viscosities (221).

Crystallinity/Amorphous Effect Crystalline polymers are generally more difficult to foam than amorphous polymers. The morphology and degree of crystallinity of crystalline polymers influence the solubility and diffusivity of the blowing agent and the cellular structure of the foam. The gas resulting from use of the blowing agent does not dissolve in the crystalline regions so that absorption and diffusion occur mostly in the amorphous polymer regions. As a result, bubble nucleation is nonhomogeneous and the cell structure is nonuniform. In experiments with microcellular polypropylene foams, plastics with 46% crystallinity produced nonuniform cell structures, while a crystallinity of 41% resulted in uniform cell structures.

Properties of PP foams include good heat resistance, high chemical resistance, and good insulation. Polypropylene's high melting point makes it useful in heat-sensitive applications, such as hot metal or glass processing; some foams can withstand temperatures as high as 120°C (250°F) without warpage. PP does not absorb water but is permeable to water vapor and other gases, and it is mold and mildew resistant. Several properties of commercial PP or PP-blend foams are given in Tables 8.35 to 8.37 and Figures 8.3 and 8.4.

POLYSTYRENE

This type of foam is available in two forms, extruded PS foam and expanded PS for molded foams. PS foams are light closed-cell foams with low thermal conductivities and excellent water resistance. They meet the requirements for low-temperature insulation and buoyancy media.

Extruded Polystyrene This material has been produced for over a half century and is manufactured as billets and boards. It is made by extruding (under elevated temperatures and pressure) molten PS containing a blowing agent into the atmosphere where the mass expands. The billets and boards can be used directly or can be cut into many different forms. Common tools for

Property	Unit	Microfoam
Density	kg/m^3	10
Cell Size	mm	0.5–1.0
Tensile Strength	kPa	240
Tensile Elongation	%	190[1]
Thermal Conductivity	W/(mK)	0.042
R-Factor		3.7[2]

[1] Average of two directions
[2] Rated at 1 in. thickness

Table 8.35 Properties of low-density PP closed-cell foam extruded sheet

	Permeability
Water Vapor ($g/100\ in^2/24\ hr$)	7.3[1]
Oxygen ($cc/100\ in^2/24\ hr$)	1475
Carbon Dioxide ($cc/100\ in^2/24\ hr$)	5000
Water Absorption (lbs/ft^2)	0.04

[1] Higher numbers indicate that vapors are transmitted more readily

Table 8.36 Permeability to gases and moisture of low-density PP closed-cell foam

fabrication are band saws, hot-wire cutters, planers, and routers. Boards may have cut or planed surfaces or integrally extruded skin of extremely small cells. Many sizes are available in extruded cellular PS. Billets can be as large as 10 × 24 in × 9 ft long. Boards are available 3/4 to 4 in thick by 2 to 9 ft long.

Applications include low-temperature insulation in freezers, coolers, and other types of refrigerated rooms; truck bodies and railroad cars; refrigerated pipelines; and low-temperature storage tanks for liquefied natural gas. Other applications include roof-deck insulation where the foam is placed in the last hot bitumen layer of the roof, which is then covered with gravel or stone to hold it in place. Extruded PS foam is also used in the insulation of residential housing by using the foam in place of conventional sheathing. The foam is placed over the studs and sill plates, which are sources of thermal

Part	Materials	Specific Gravity	Mold Shrinkage, Mils/in. (mm/mm)
Impact Bar 1/4 in. (6.4 mm)	100% GPP[2]	0.90	14–18 (0.014–0.018)
	50% GPP, 50% FPP[3]	0.80	23 (0.023)
	100% FPP	0.75	24 (0.024)
Plaque (Sprue Gated) 5 in. x 5 in. x ¼ in. (127 mm x 127 mm x 6.4 mm)	50% GPP, 50% FPP	0.79	23 (0.023)
	50% GPP, 50% FPP	0.75	29 (0.029)
	100% FPP	0.79	22 (0.022)
	100% FPP	0.71	23 (0.023)
	100% FPP	0.63	24 (0.024)
Plaque (Sprue Gated) 5 in. x 5 in. x 1/2 in. (127 mm x 127 mm x 12.7 mm)	50% GPP, 50% FPP	0.76	27 (0.027)
	50% GPP, 50% FPP	0.67	27 (0.027)
	50% GPP, 50% FPP	0.60	25 (0.025)
	100% FPP	0.76	24 (0.024)
	100% FPP	0.65	22 (0.022)
	100% FPP	0.47	25 (0.025)

[1] Tenite polypropylene P2635–08AA (Eastman Plastics)
[2] GPP = general purpose polypropylene
[3] FPP = Tenite polypropylene P2635–08AA

Table 8.37 Mold shrinkage of parts made with PP foam

shorts in batt-insulation buildings. This type of foam is also used in agriculture to insulate livestock buildings and low-temperature produce-storage buildings.

Extruded PS foam boards are also used as core material for structural sandwich panels, especially in the recreational vehicle and motor-home industries. The walls, floors, and roofs of many of these vehicles are constructed of PS foam core sandwich panels. White and green extruded

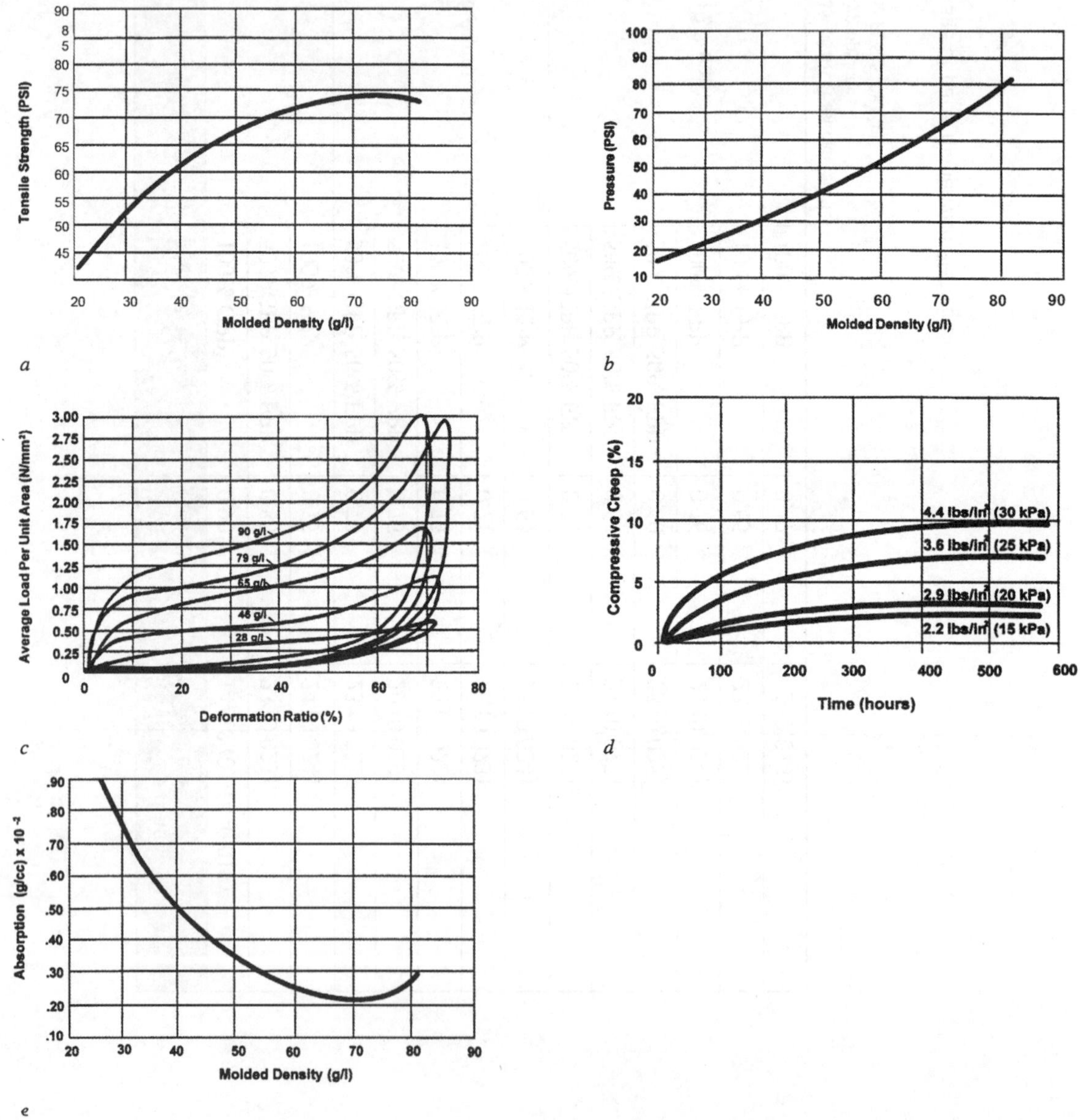

Figure 8.3 Properties of expanded PP closed-cell foam from PP and PE beads (Neopolen P, BASF)

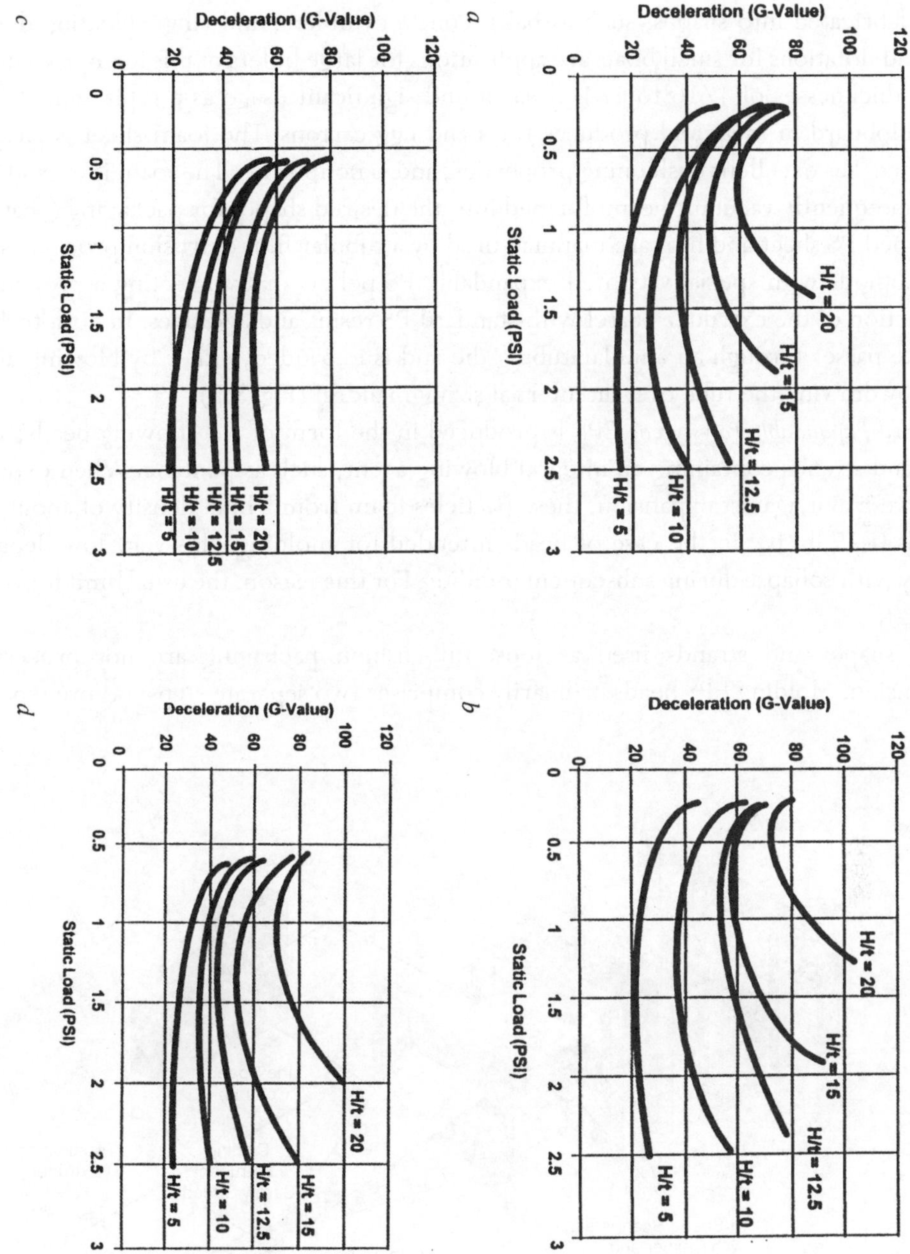

Figure 8.4 Dynamic cushioning performance of expanded PP (Neopolen P, BASF)

PS foams are used in the floral, novelty, hobby craft, and display fields, either as boards or billets, and are fabricated into shapes, such as balls, cones, cylinders, and rings. Floating docks, marker buoys, and flotations for small boats are applications for large billets of this foam. Extruded PS foam sheet in thicknesses of 1/16 to 1/4 in has found significant usage as a replacement for molded-paper pulpboard in meat and produces trays and egg cartons. The foam sheet is clean, bright in appearance, has excellent cushioning properties, and is nonporous. The foam is extruded as a sheet and is subsequently vacuum thermoformed into the desired shapes for packaging (chapter 7).

Foamed PS sheet and film are manufactured by a tubular film extrusion process using conventional methods with specially treated expandable PS pellets or by injecting a propellant directly into a section of the extruder barrel with standard PS resins and additives. In both techniques, the extrudate passes through an annular tubing die and is expanded, either by blowing air inside the tube or by drawing the tube over an internal sizing mandrel (Fig. 8.5).

Molded Expandable Polystyrene EPS is produced in the form of free-flowing beads, symmetrical shapes, and strands containing an integral blowing agent, such as pentane. When exposed to heat without restraint against expansion, these particles foam from a bulk density of about 35 lb/ft^3 to as low as 0.25 lb/ft^3. In the case of beads intended for molding, this very low density leads to difficulty with collapse during subsequent molding. For this reason, the usual limit for preexpansion is 1.0 lb/ft^3.

The shapes and strands used as loose-fill cushion packaging are not processed beyond preexpansion. Molding EPS beads ordinarily comprises two separate steps: (a) preexpansion of the

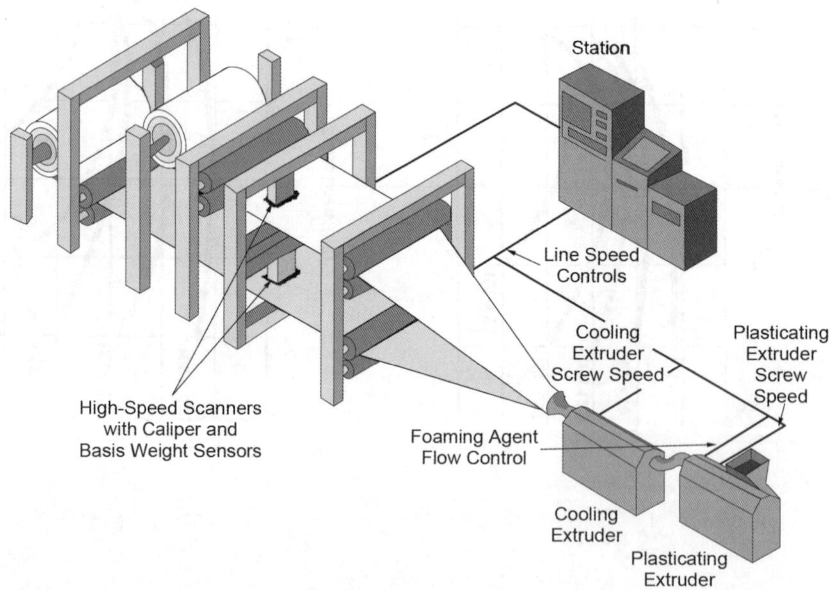

Figure 8.5 Plastic foam sheet line using dual extruders

virgin beads by heat and (b) further expansion and fusion of the preexpanded beads by heat within the shaping confines of a mold.

In unconfined preexpansion, the translucent beads grow larger and become white in color. Confined and subjected to heat, the preexpanded beads can produce smooth-skinned closed-cell foams of controlled densities, registering every detail of intricate molds. To minimize formation of density gradients and to ensure uniform expansion throughout the molded pieces, EPS beads are preexpanded to the approximate required densities by control of time and temperature, since the process of molding does not increase the density.

Continuous steam prefoaming is the most widely used method of expanding PS beads. Other methods are continuous hot-air prefoaming and batch prefoaming. Molding of EPS beads requires exposing the preexpandable or virgin beads to heat in a confined space. A number of conventional heating media are available for the fabricator to use in molding EPS. The preferred medium is steam, which is directly diffused through the preexpanded beads in the mold cavity. Other techniques involve conductive heating through the mold wall, with the heat being supplied by steam or some other energy source. Examples of the types of molding used are as follows: (a) steam-chest molding, (b) block molding, (c) foam-cup molding, (d) steam-probe molding, (e) continuous billet molding, and (f) radio-frequency molding.

In molding PS foam from beads, when heat is applied to the mold and beads, the beads are expanded enough to fill all the voids between the beads and to heat-seal all the beads together. This process permits the molder to make products with a wide range of characteristics. The density can be easily controlled. It ranges from below 1 lb/ft^3 to above 20 lb/ft^3. The density of the part is uniform throughout. Molded products can have a wide range of wall thicknesses. Most parts are molded with only very thin skin, but with an additional step in the process, it is possible to produce part with hard skin. The molded product is stress free and quickly stabilizes. Molding pressures and temperatures are low.

In the low-density range, from 0.5 to 1.0 lb/ft^3, EPS is used on boats as flotation, in packaging as an energy absorber, in building as insulation, and as a moisture barrier. In the middle-density range, from 1.0 to 4.0 lb/ft^3, the foam is used in packaging as a structural support as well as an energy absorber, in molding hot/cold drinking cups, in the construction field for such application as concrete forms, in the foundry industry as mold patterns, as insulated containers of all sizes and shapes, and in materials-handling pallets. In the high-density range, from 5.0 to 20.0 lb/ft^3, the foam exhibits almost woodlike properties. Such products as thread spools, tape cores, and furniture parts have been made from these foams.

The blowing agent most commonly used on EPS is pentane, typically about 6.3% to 7.3% by weight. To reduce pentane losses in shipping, expandable beads are ordinarily transported in 5-gallon drums constructed from fiberboard, with an inner liner of aluminum foil. In open containers at 23°C, EPS beads screened through a 10-mesh screen and retained on a 25-mesh sieve decreased their n-pentane content from 7.0% to 6.0% in six to seven days. Other blowing agents used to some extent are neopentane (tetramethylmethane) and isopentane, both solid gas-releasing blowing agents.

Recognize that PS foam has a low operating temperature of approximately 80°C (176°F).

POLYURETHANE

Flexible polyurethanes, also called urethanes, continue to dominate the market. In the furniture and mattress business, they continue to replace latex foam, previously made of natural rubber and now predominantly butadiene-styrene. The combination of providing exceptional service and favorable price has made flexible urethane the standard.

Progress in the rigid has been slower than in the flexible. The reasons include that immediate large sales for flexible existed, markets for rigid foams were more expensive to penetrate, most of the industry's time and capital were spent on the immediately profitable flexible market, and rigid has a higher cost and fewer sources for applying the foam.

Urethane foams can be classified into two principle types: flexible and rigid foams. In some cases, flexible foams can be further subdivided into flexible and semiflexible (or semirigid). The differences in physical properties of the two foams are mainly due to the differences in molecular weight per cross-link, the equivalent weight and functionality of the polyols, and the type and functionality of the isocyanates. The proper choice of equivalent weight and functionality of the polyols employed can prepare these foams. Polyisocyanates can be considered as the joining agents of the polyols. A rough classification of the three kinds of foams based on the type of polyols used is shown in Table 8.38.

The one-shot process, semi-prepolymer (quasi-prepolymer) process, or prepolymer process can prepare these foams. The one-shot process is most commonly used. The semi-prepolymer process is sometimes preferred because of the advantages of easy processing, stabilized foam rise, and lower exotherm. The prepolymer process is used only for limited purposes. Schematic diagrams of the three foaming processes are shown in Figure 8.6.

The one-shot process is used for flexible and rigid foams. In the case of slabstock foams, the ingredients are separately supplied to the mixing head. In order to adjust viscosity and mixing

Foam	Rigid Foam	Semi-Rigid Foam	Flexible Foam
Polyol:			
OH No.*	350–560	100–200	5.6–70
OH Equivalent**	160–100	560–280	10,000–800
Functionality	3.0–8.0	3.0–3.5	2.0–3.0
Elastic Modulus at 23°C			
MPa	>700	700–70	<70
lb/in²	>100,000	100,000–10,000	<10,000

* OH Number: mgKOH/g

** OH Equivalent: 56,110/OH Number

Table 8.38 Classification of polyurethane foams

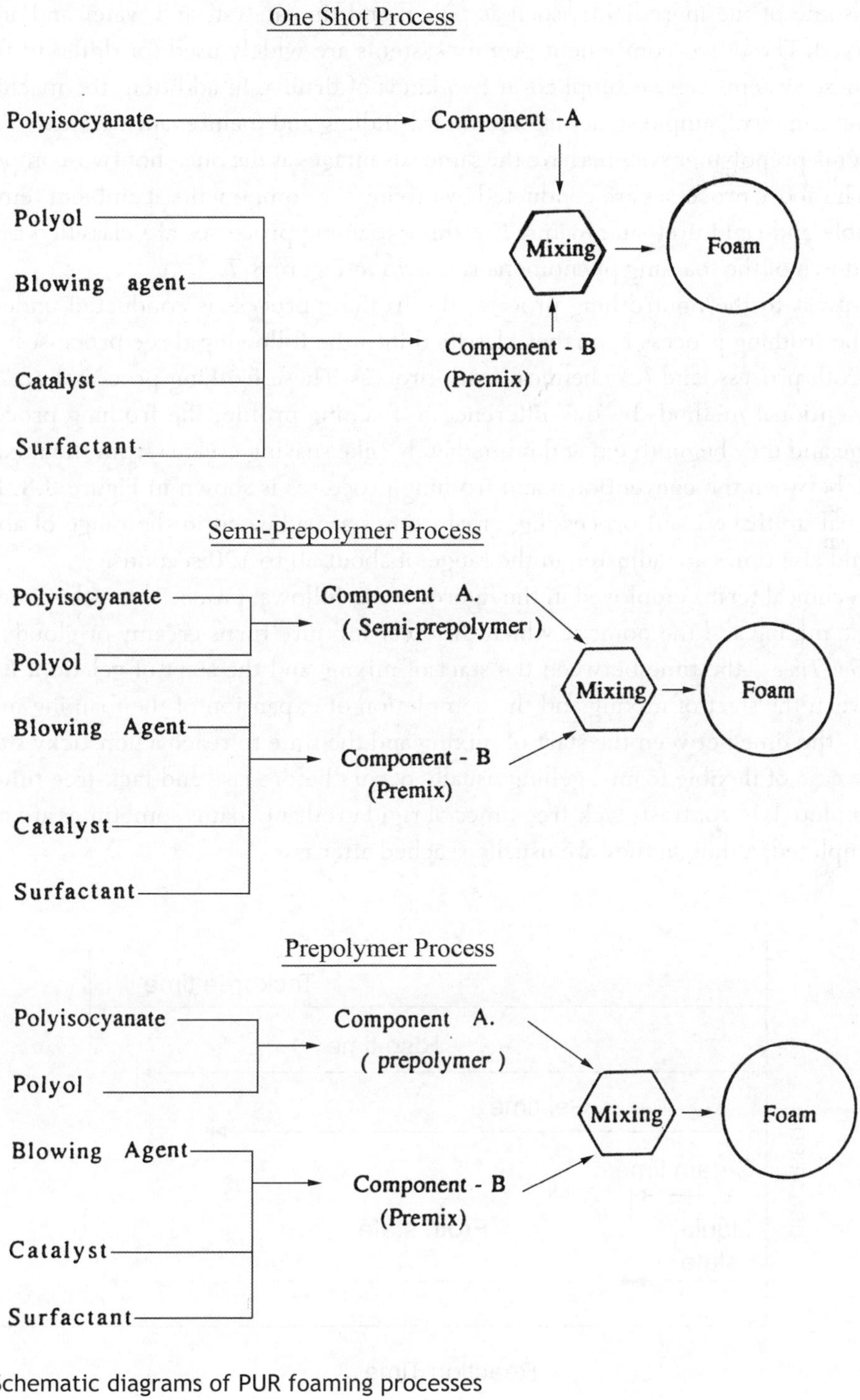

Figure 8.6 Schematic diagrams of PUR foaming processes

accuracy, some of the ingredients, such as polyol and tin catalyst, and water and amine catalyst, are premixed. These two-component premix systems are widely used for different foam systems, because these systems can be supplied in two kinds of drums. In addition, the machines for foam preparation can be of simple structure and their handling and maintenance are easy.

The semi-prepolymer systems have the same advantages as the one-shot two-component premix systems. The above processes are conducted by mixing the components at ambient temperatures for both flexible and rigid urethane foams. The three foaming processes are classified as nonfrothing; the breakdown of the foaming phenomena is shown in Figure 8.7.

In contrast to the nonfrothing process, the frothing process is conducted under pressurized mixing. The frothing process is further classified into the following three processes: (a) R-12, (b) thermal froth process, and (c) chemical froth process. These frothing processes are distinguished from conventional methods by the difference in foaming profile; the frothing processes have no cream time and they begin to expand immediately (like shaving cream) from the mixing head. The difference between the conventional and frothing processes is shown in Figure 8.8. In the case of conventional urethane foam processing, cream times are adjusted in the range of about 10 to 30 seconds and rise times are adjusted in the range of about 60 to 120 seconds.

The technical terms employed in the figures are as follows: *Cream Time*—the time between the start of the mixing and the point at which the clear mixture turns creamy or cloudy and starts to expand; *Gel Time*—the time between the start of mixing and the start of gelation; *Rise Time*—the time between the start of mixing and the completion of expansion of the foaming mass; and *Tack-Free Time*—the time between the start of mixing and the time to reach a nonsticky state.

In the case of flexible foams, gelling usually occurs before rise and tack-free times come after rise is completed. In contrast, tack-free times of rigid urethane foams sometimes are needed before rise is completed, although they are usually reached after rise.

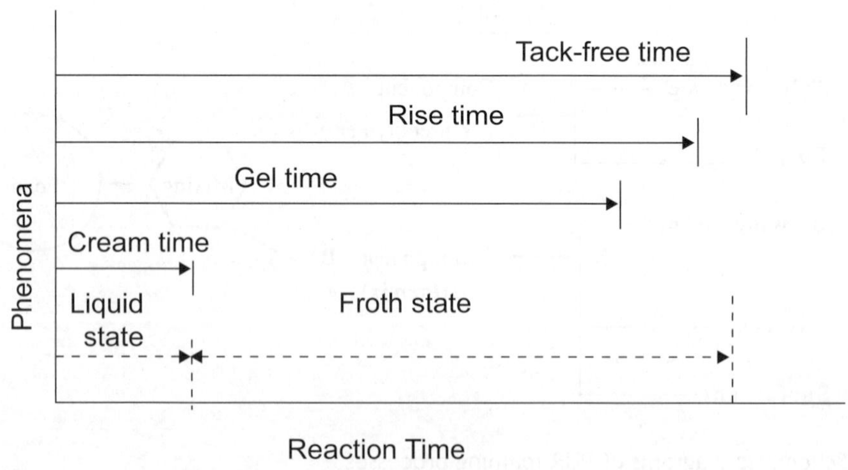

Figure 8.7 Breakdown of the foaming phenomena

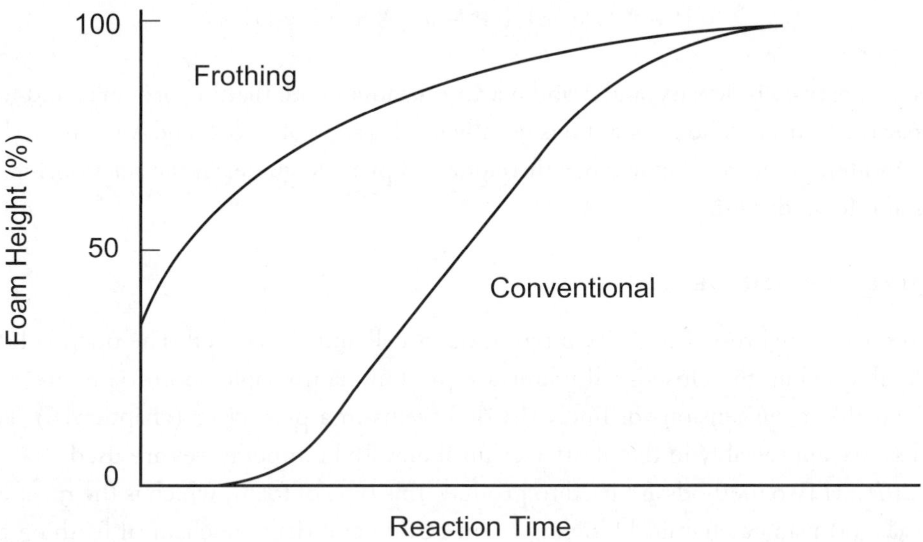

Figure 8.8 Comparison of rise time

The Du Pont Co. developed the frothing process. The process has the following advantages: isotropic physical properties and lower foaming pressure. The method is preferably used for large-panel production at in-plant production or pour-in-place foaming in field applications, for example, building panels, chemical tanks, and insulations. The R-12 frothing process is characterized by the co-use of its blowing agent (boiling point, such as 29.8°C) and occurs at ambient temperature. Elevating the temperature of the foaming systems to above the boiling point carries out the thermal frothing process. Using an accelerated gas-generation reaction by means of a special chemical reaction at ambient temperatures carries out the chemical frothing process. An example is the addition of a small amount of methanol in the foaming system.

Flexible urethane foams are defined as open-cell urethane foams having the property of complete recovery immediately after compression. They can be classified into two kinds, polyether foams and polyester foams. Polyether foams are further classified as conventional flexible foams, high-resilience flexible foams (UR foams), cold-molded foams, super-soft foams, and viscoelastic foams.

Semiflexible foams (or semirigid foams) are sometimes classified as subdivisions of flexible foams because they have higher load-bearing properties and good compression recovery. Microcellular flexible foams and integral-skin flexible foams may also be classified in this category. In some classifications, however, microcellular foams are classified as elastomers. Different foams can be prepared by the proper choice of polyols. Polyisocyanates are used as joining agents for the polyols; therefore, polyols are considered to be the major components to determine the physical and mechanical properties of the resulting foams.

Property Almost all mechanical and physical properties of rigid PUR foams depend on their foam densities, as shown by the following equation:

$$P = A\,D^n \text{ (or } \log P = \log A + n \log D\text{)},$$

where P is property, D is density, and A and n are constants. Note that log property and log density form a linear relationship. The constant n is usually in the range of 1 to 2, and A relates to both foam properties and temperatures. Figures 8.9 through 8.13 provide guides in the relationships between properties and foam densities.

Polyvinyl Chloride

Foams made from vinyl resins are of two types, open-cell and closed-cell. The open-cell foams are soft and flexible, while the closed-cell foams are predominantly rigid. Both types are made from plastisols, which are suspensions of finely divided resins in a plasticizer (chapter 16). The plastic does not dissolve appreciably in the plasticizer until elevated temperatures are used.

Open-Cell Vinyl Two methods are used to produce this type of foam, which is the most important type in vinyls: (a) using a chemical blowing agent (CBA) and (b) a mechanical frothing process in which a gas is also used as part of the blowing mechanism. In the preparation of a soft open-cell foam using a CBA, the plastisol is first chosen for the characteristics desired. The resin and plasticizer ratio in the plastisol should be one in which enough plastisol is present to assure production of a soft, flexible material. To the plastisol is added a paste made of powdered blowing agent dispersed in a plasticizer.

The CBA may be any one of a number of organic materials that, when heated, decompose to liberate a gas, usually nitrogen. One class of materials used for the large majority of vinyl foams is the azocarbonamides and other azo compounds that can be used to decompose at temperatures from about 250° to 425°F (120° to 220°C). A CBA can be chosen to accommodate practically any required resin temperature. It is extremely important that at the gas-evolution temperature the viscosity characteristics of the plastisol allow complete and even expansion and that a uniform cell structure is formed. Factors that influence these properties are the particle size of the blowing agent, the type and quantity of the resin and plasticizer, and the specific type of stabilizer and/or activator used to initiate decomposition of the blowing agent.

Soft, very flexible vinyl foams used for garment insulation, upholstery, and similar applications are made by this process. Somewhat harder foams used as underlays for rugs and flooring can also be made by this technique, but they require different resins and lower plasticizer contents. Open-cell chemically blown vinyl foams generally have densities in the range of 10 to 30 lb/ft^3.

The second method of producing open-cell vinyl foams, mechanical frothing, is used to produce a large amount of vinyl foams, particularly sheets, used for flooring underlays, wall coverings, and other applications requiring relatively close thickness tolerances. In this method, the plastisol is mixed with a given amount of air in a high-shear, temperature-controlled mixing head. The resulting product, resembling shaving cream, is cast onto a belt or fabric and knifed to a definite thickness. Passage through an oven or heating tunnel then causes fusion of the plastisol. Fusion temperatures for all the plastisol foams range from 257° to 325°F (125° to 160°C). The times required

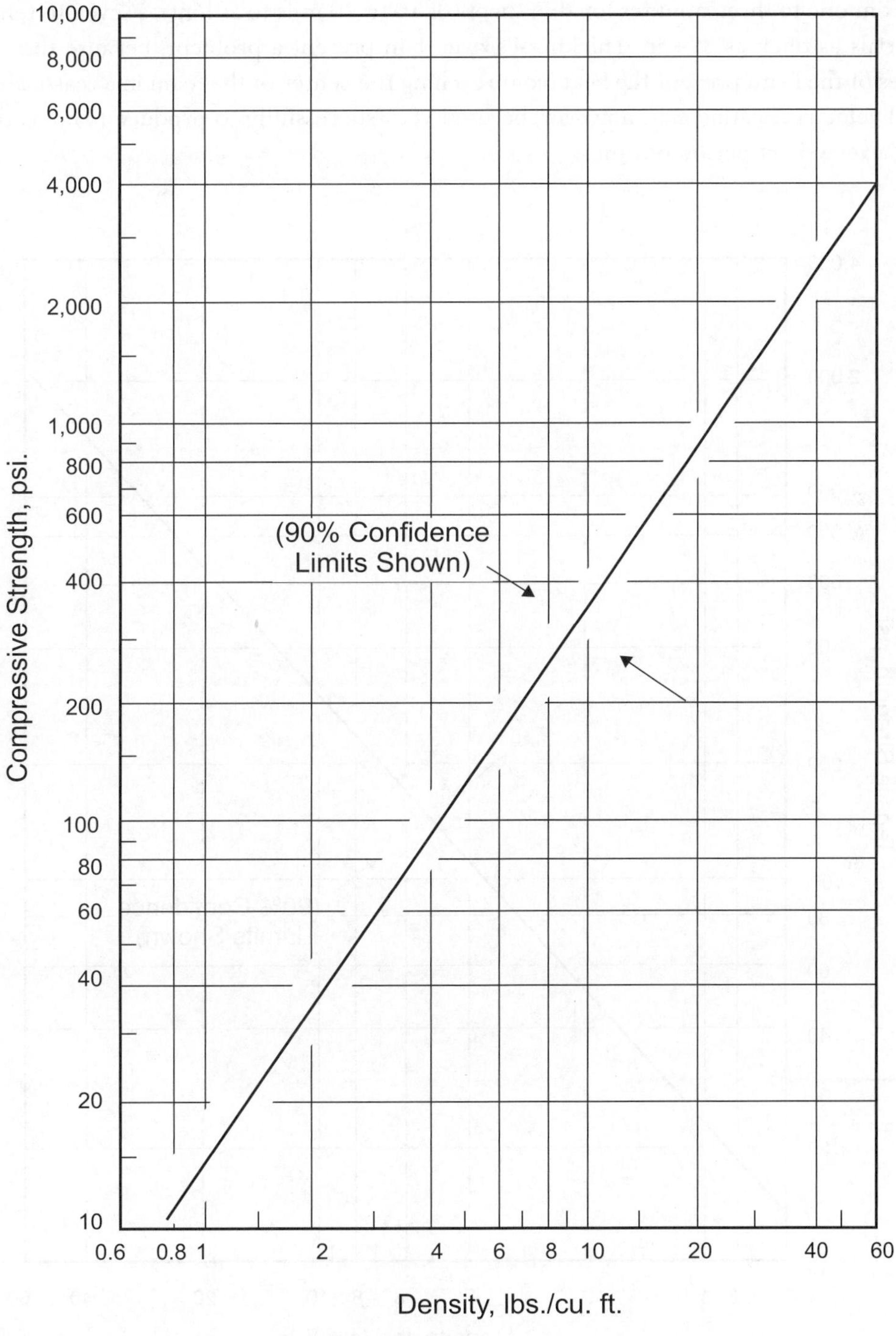

Figure 8.9 Effect of density on compressive strength of rigid polyurethane foam

may be from one to three minutes for thin sheets of 10 to 20 mils to as long as seven to ten minutes for materials as thick as 3/4 in. Thicknesses over 1 in present a problem, because the insulating properties of the foam prevent the heat from reaching the center of the foam in a reasonable length of time. Dielectric heating can, however, be used very successfully to produce foam as thick as 2 inches in a very short period of time.

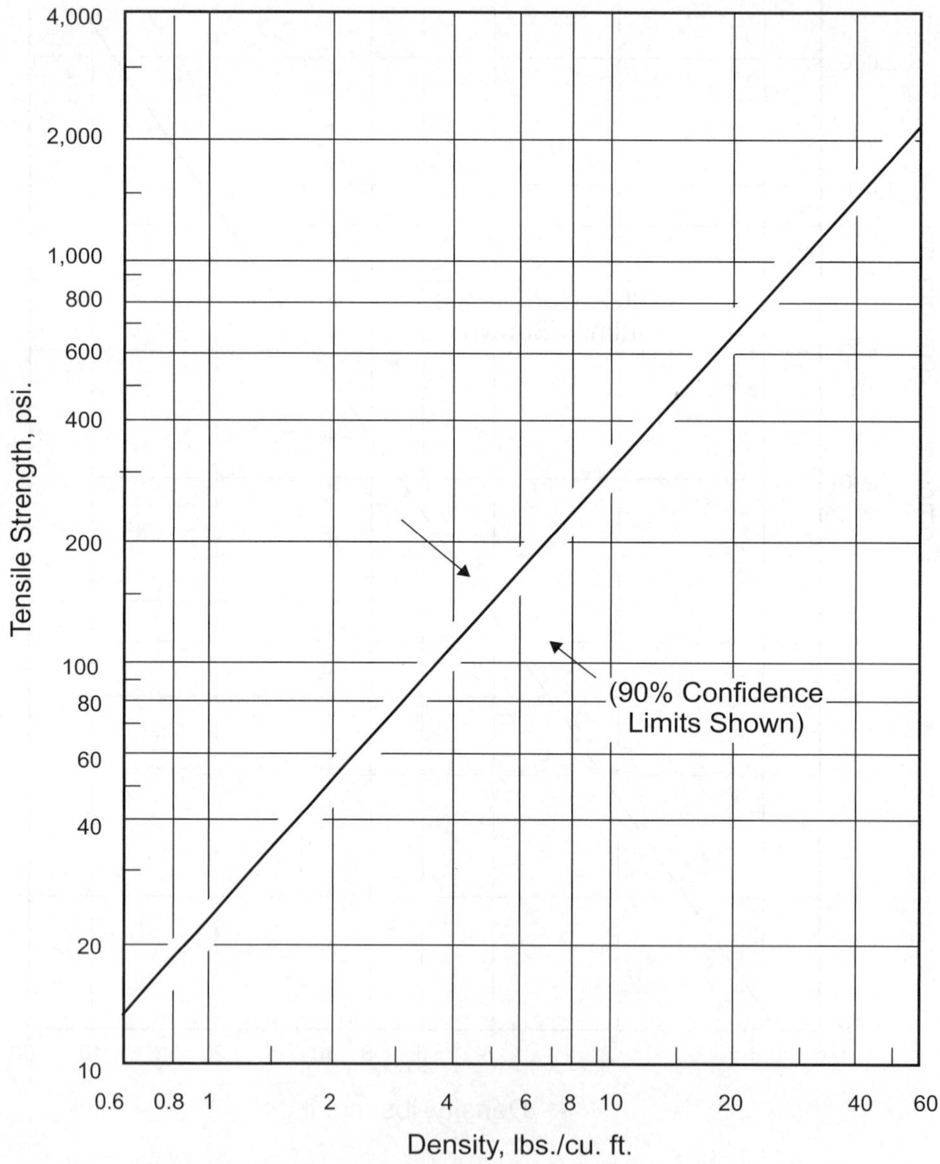

Figure 8.10 Effect of density on tensile strength of rigid polyurethane foam

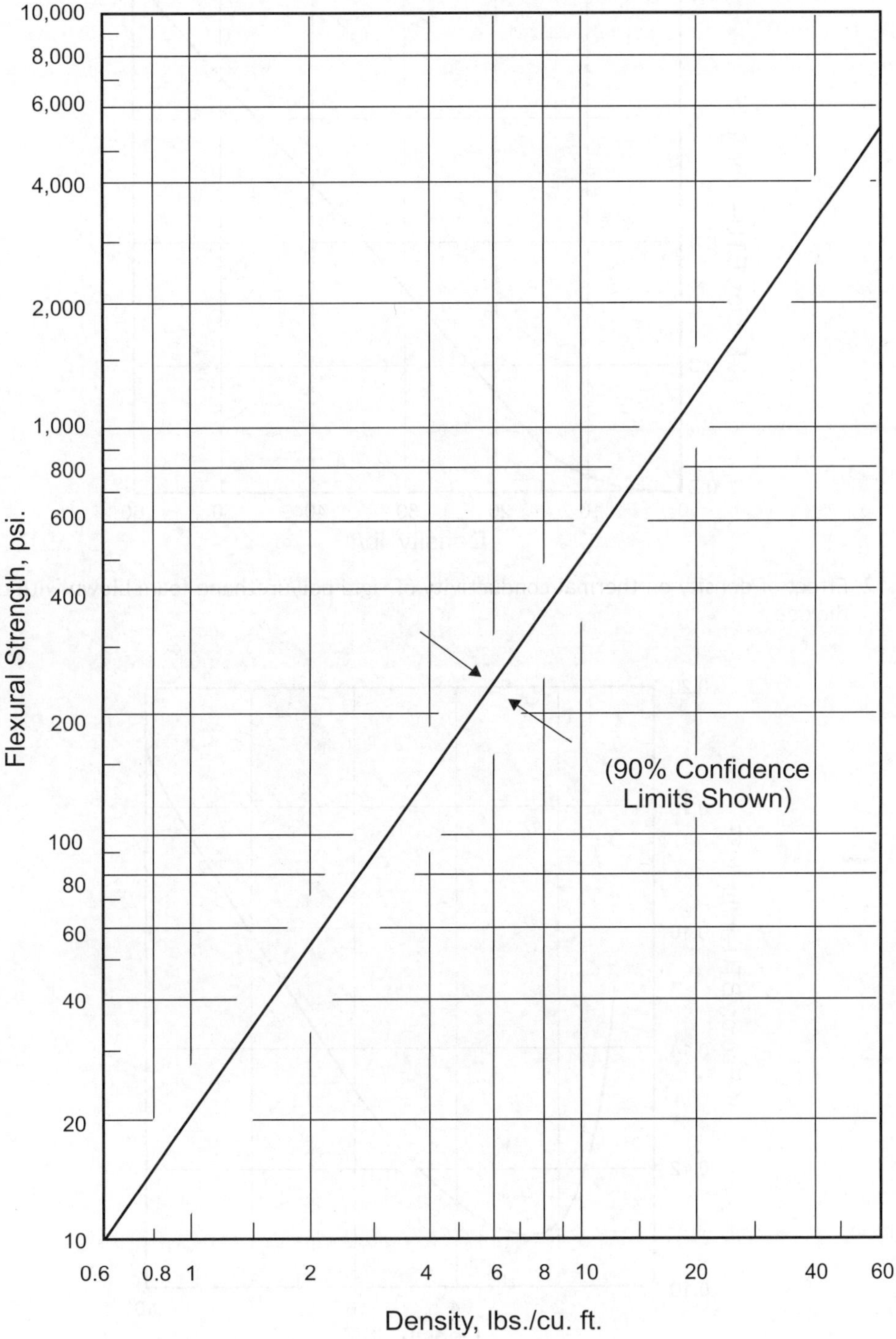

Figure 8.11 Effect of density on flexural strength of rigid polyurethane foam

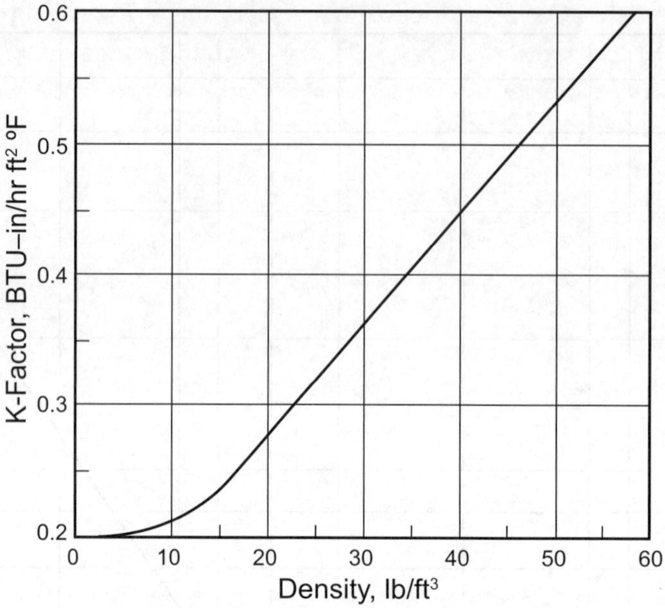

Figure 8.12 Effect of density on thermal conductivity of rigid polyurethane foam blown with carbon dioxide

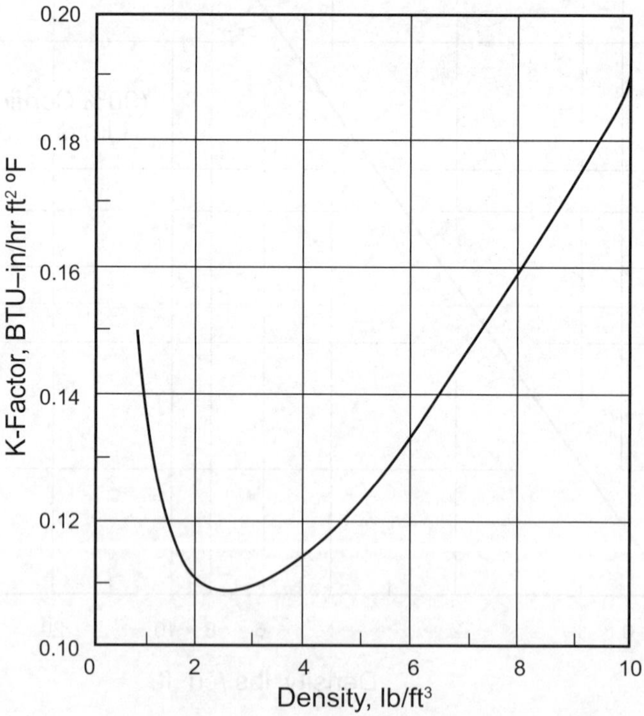

Figure 8.13 Effect of density on thermal conductivity of rigid polyurethane foam blown with CFC-11

Closed-Cell Vinyl These foams are made in a manner somewhat similar to the technique used to produce open-cell foams with CBAs, except that much higher pressures are used and the process is accomplished in two steps. First, the vinyl plastisol containing the blowing agent is placed in a mold in which very little space is left for expansion. The mold is then heated, causing decomposition of the blowing agent and, at the same time, fusion of the foam. This step raises the internal pressure in the mold to anywhere from 200 to as much as 1,600 psi (13.8 to 110 MPa). At these high pressures, the gas is dissolved in the resin in the form of microscopically small bubbles. The part is then allowed to cool and harden in the mold before being removed.

The second step consists of reheating the molded part at approximately 200° to 250°F (93° to 120°C), at which time the resin softens and the gas expands to form a closed-cell foam. With this technique, it is possible to produce foams with densities as low as 2 lb/ft^3, although the usual range is 50 to 10 lb/ft^3. Because of this two-step procedure, the process is much slower than the foaming procedure for open-cell materials. For this reason, it is used mainly for specialty items, such as marine flotation parts, special athletic mats, and similar items. Since the parts made by this process are all made at high pressures, these materials are often called "pressure sponges."

Low-density unicellular foam made from this pressure process is available in a range of densities and hardnesses in various sizes ready for fabrication into the final product. Slabstock is used to produce shock absorbing, flotation, and insulating foams. Typical densities range from 7 lb/ft^3 for shock-absorbing foams, 4 lb/ft^3 for flotation foams, and 6 lb/ft^3 for insulation foams. It is also possible to produce pressure-molded foams in cylindrical and ring-shaped molds.

Cross-linked Vinyl It is possible to produce a rigid vinyl foam by using a combination of vinyl chloride polymer and monomer plus maleic anhydride, isocyanate, and catalyst to produce a rigid, cross-linked, essentially closed-cell construction of exceptional strength. This type of foam is being laminated with metal and reinforced plastics for use in aircraft construction as a flooring material and in boat construction for hulls.

The ingredients are poured into a mold, then heated while confined under pressure. The press is brought to 100°C, and the exothermic reaction results in the maleic anhydride copolymerizing with the vinyl chloride monomer and grafting onto the PVC. The composition, then removed from the mold, is still thermoplastic, since the isocyanate has not yet reacted. The plastic is then exposed to hot water or steam, thereby causing the isocyanate to liberate CO that acts to expand the plastic mass. After expansion is completed, the water then reacts with the grafted maleic anhydride, and the resultant maleic acid reacts with the isocyanate and cross-links it.

OTHER FOAM

As reviewed, most plastics can be made into foamed products. From a practical and market-oriented view, only a few different types are used. As an example, limited use has been made of cellular cellulose acetate. The cellular cellulose acetate (CCA) was one of the first rigid foams produced and was used rather extensively during the 1940s and 1950s, mainly in aircraft sandwich constructions. The production of the foam involves extrusion of flaked cellulose acetate, employing a mixture of acetone and ethyl alcohol as the blowing agent and comminuted mineral of 100 to 325 mesh as the nucleating agent. The mixture also acts as a solvent for the cellulose acetate above 70°C and as a plasticizer during the hot extrusion. The extrusion process leads to the formation of unicellular foam by the sudden release of the superheated, pressurized molten polymer mass as it emerges from the extrusion die. The extruded foam has a denser outer skin, which may be removed by cutting off the outer edges.

CCA foam is extruded in the form of boards and rods. The foam cannot be molded, and so end-use shapes must be made by machining. Densities are in the range of 6 to 8 lb/ft^3 (96 to 118 kg/m^3). The foam has excellent strength, low water absorption, and resistance to attack by vermin and fungus. The service temperature range is broad (–57° to 177°C) and the foam will tolerate short exposures up to 194°C. It has good electrical properties, making it useful for electronics and X-ray equipment. Hydrogen bonding is responsible for CCA's insolubility in water and its inertness to most solvents and resin systems. It can also be steam formed to achieve desired shapes.

SYNTACTIC

In syntactic foams, instead of employing a blowing agent to form bubbles in the mass, preformed reinforcing bubbles of glass, ceramics, or plastics are embedded in a matrix of an unblown plastic. Microballoons or spheres range in diameter from 30 μ to 0.0004 in. The matrix is with epoxy, polyester, phenolic, or urea resin producing unique foam. This approach is to reduce weight. These high-strength 8-to-50-lb/ft^3 foams are used as void fillers in boats, cores for aircraft sandwich structures, refrigerator cores, microwave absorbers, high-frequency communication antennas, deep submergence vessels, and rocket bodies, as well as to prevent evaporation of liquids in tanks (oil, etc.) (Table 8.39).

A mixture of microspheres and the plastic can be formulated into a moldable mass that can then be shaped or pressed into cavities and molds much as molding sand and clay. The properties of the finished hardened or cured mass can then be tailored by a suitable plastic formulation. A mixture of TS polyester plastic and small hollow glass spheres (Tables 8.40 to 8.42), for example, can create synthetic wood.

Syntactic foam contains an orderly arrangement of hollow sphere fillers. They are usually glass microspheres approximately 100 μm (4 mils) in diameter that provide strong, impervious supports for otherwise weak, irregular voids. As a result, syntactic foam has attracted considerable attention both as a convenient and relatively lightweight buoyancy material and as a porous solid with excellent shock-attenuating characteristics. The latter characteristic is achieved through crushing of the spheres and filling in the voids with plastic.

These foams or porous solids are used in a variety of applications, from energy absorbers to buoyant products. Properties of these materials, such as compressive constitutive laws or equations of states, are needed to calculate the dynamic responses of the materials to suddenly applied loads. Static testing to provide such data is appealing because of its simplicity; however, the importance of rate effects cannot be determined by this one method alone. Therefore, additional but numerically limited elevated strain-rate tests must be run.

Properties	(for use at ocean depth to 20,000 ft.)	(for use at ocean depths to 10,000 ft.)	(for use at ocean depths to 1,000 ft.)	(compatible 181-glass fabric prepreg for surface finishing)	(fire-resistant syntactic prepreg primarily for surface use)
Specific gravity	0.7-0.9	0.7-0.9	0.7-0.9	1.7-1.9	0.7-0.9
Thickness, in.	0.365	0.400	0.420	0.100	0.400
Tensile properties					
Strength, p.s.i.	13,000	13,000	13,000	60,000	16,000
Modulus, 10^6 p.s.i.	1.0	1.0	1.0	3.6	1.3
Flexural properties					
Strength, p.s.i.	20,000	20,000	19,000	92,000	-
Modulus, 10^6 p.s.i.	1.0	1.0	1.0	3.7	-
Compression, flatwise					
Yield strength, p.s.i.	19,000	17,000	15,000	-	6,000
Ultimate strength, p.s.i.	22,000	20,000	16,000	-	-
Compression, edgewise					
Strength, p.s.i.	23,000	23,000	23,000	55,000	12,000
Modulus, 10^6 p.s.i.	1.0	1.0	1.0	3.8	1.1
Water absorption, %	<3[b]	-	<1[c]	0.5	-
Izod impact, ft.-lb./in.	7	7	8	16	6
Dielectric constant at 9.375 GH_z	2.3	2.3	2.3	4.4	2.5
Loss tangent at 9.375 GH_z	0.014	0.014	0.010	0.020	0.017
Thermal conductivity, 10^{-4} cal. cm/cm^2 sec. °C	2.6	2.6	2.6	-	2.6
Thermal expansion 10^{-4} in/in./°F	9.1	9.1	9.1	5	9.1

Test panels 10x10 in. were fabricated from 10 layers of syntactic foam prepreg (7804 laminated 12-ply).
The laminates were bagged with top and edge bleeder on caul sheets and cured 9 min./ply at 300°F, and 50 p.s.i.
The specimens were tested per Federal Test Methods Standard 408.
After cycling 1.500 times to a hydrostatic pressure of 13,500 p.s.i.
After cycling 1,500 to a hydrostatic pressure of 300 p.s.i.

Table 8.39 Properties of epoxy syntactic foam—molded prepregs

Name	Description	Particle Size	Bulk Density, G/ml
Kanamite grade 200	Unicellular hollow clay spheres, primarily aluminum silicate	90% 35-60 mesh 10% 20-200 mesh	0.9
Colfoam micro-ballons	Urea-formaldehyde spheres	90% 20-100 mesh 10% through 200 mesh	0.2-0.5
Microballons, BJOA-0840	Hollow phenolic spheres	10% 40-100 mesh 74% 40-200 mesh 18% through 200 mesh	0.3-0.4
Glass micro-ballons CPR 2077	Hollow glass micropheres	89% 60%-325 mesh 11% through 325 mesh	0.26

Table 8.40 Low-density hollow spheres

	Eccospheres R*	Eccospheres SI*	Eccospheres* VT
Bulk density, pcf (g/cm^2)	14 (0.22)	11 (0.18)	11 (0.18)
True particle density, pcf (g/cm^2)	26 (0.42)	17 (0.28)	17 (0.28)
Particle size, μ	30-300	30-125	30-125
Wall thickness, μ	About 2	About 2	About 2
Composition	Borosilicate glass	Over 95%	SiO, plus coating
Temperature capability, °F	1000	2500	600
Thermal conductivity, loosely packed material, Btu/(hr) (ft^2) (°F/ft)	0.04	0.03	0.03
Dielectric constant (dry),1 Mc to 8.6 Gc (approx.)	1.3	1.2	1.2
Diisipation factor (dry), 1 Mc to 8.6 Gc (approx.)	0.001	0.0005	0.0005
Bulk density, pcf	12-15		
True density (liquid displacement), pcf (g/cm^3)	21.0 (0.34)		
Particle size range, μ			
>250	3		
175-250	3		
124-175	12		
89-124	15		
61-80	19		
44-61	18		
<44	30		
Average particle size, μ	65		
Melting point, °F	1400		
Thermal conductivity, Btu/(hr)(ft^2 x °F/in)	0.4		
Oil absorption, g oil per 100 cm^3 GMBs	40		
Water absorption, % of total weight			
1 hr	0.68		
3 hr	0.68		
5 hr	0.68		
24 hr	1.40		

*Trademark, Emerson and Cuming, Inc.

Table 8.41 Properties of glass microballoons

Filler	Viscosity at 250°C, cP	Shrinkage Linear, %	Hardness, Shore D	Density at 21°C g/cc	Tensile Strength, psi	Thermal Conductivity Btu/(hr)(ft^{2}x °F/in)	Thermal Expansion 25-100°C in/(in)(°C) x 10^5	Weight Reduction Over Silica-filled compounds, %
None	13,500-19,500	0.12	80-85	1.17	8,000	2.68	8.7	26.5
Silica, 325 mesh, 100, phr	43,000-48,000	0.08	80-85	1.59	5,500	6.38	5.6	0
Phenolic spheres, 15 phr	34,000-38,500	0.14	80-84	0.86	3,300	1.91	8.2	46.0
Kanamite, 34 phr	34,000-39,000	0.06	75-80	1.01	2,000	2.47	6.7	36.5
Colfoam, 4 phr	45,000-48,000	0.17	80-85	1.01	4,050	4.15	8.6	36.5
Glass spheres, 14 phr	44,000-47,000	0.25	80-85	0.95	4,200	4.56	8.2	41.9

Electrical Properties

Filler	Dielectric constant at 25°C		Power factor at 25°C		Volume resistivity, ohm-cm			Dielectric strength at 25°C volts/mil
	1 kc	1 Mc	1 kc	1 Mc	25°C	65°C	100°C	
None	3.8	3.7	0.0035	0.015	$8.7x10^{14}$	-	$5x10^{11}$	400-500
Silica, 325 mesh, 100 phr	3.4	3.4	0.003	0.012	$1.3x10^{14}$	$6.2x10^{13}$	-	>330
Phenolic spheres, 15 phr	3.2	2.7	0.003	0.014	$1.0x10^{14}$	$5.3x10^{13}$	-	>330

Table 8.42 Physical and electrical properties of epoxy syntactic foam vs. fillers

PROCESS

The production of plastic foam materials can be carried out by mechanical, chemical, or physical means. An introduction to some of the most commonly used basic methods is as follows (212):

1. Thermal decomposition of chemical blowing agents generating either nitrogen or carbon dioxide (CO_2), or both, by application of heat or as the result of the reaction of exothermic heat during polymerization
2. Mechanical whipping of gases (frothing) into a plastic system (melt, solution, or suspension) that hardens, either by catalytic action or beat, or both, thus entrapping the gas bubbles in the plastic matrix
3. Volatilization of low-boiling liquids, such as fluorocarbons (and alternates) or methylene chloride, within the plastic mass as the result of the exothermic heat of reaction or by application of heat
4. Volatilization of gases produced by the exothermic heat of reaction during polymerization, such as occurs in the reaction of isocyanate with water to form carbon dioxide
5. Expansion of dissolved gas in a plastic mass on reduction of pressure in the system
6. Incorporation of hollow microspheres into a plastic mass (microspheres may consist of either hollow glass or hollow plastic beads)
7. Expansion of gas-fired beads by application of heat or expansion of these beads in a plastic mass by the heat of reaction; for example, expansion of polystyrene beads in a polyurethane or epoxy resin system

The production of foams can take place by many different techniques. These may include the following:

1. Continuous slabstock production by pouring or impingement, using multicomponent foam machines
2. Compression molding of foams
3. Reaction injection molding (RIM), usually by impingement
4. Foaming-in-place by pouring from a dual- or multicomponent head
5. Spraying of foams
6. Extrusion of foams using expandable beads or pellets
7. Injection molding of expandable beads or pellets
8. Rotational casting of foams
9. Frothing of foams, either by introduction of air or of a low-boiling volatile solvent
10. Lamination of foams (foam-board production)
11. Production of foam composites
12. Precipitation foam processes wherein a polymer phase is formed by polymerization or precipitation from a liquid that is later allowed to escape

Based on production output, the most important processes are (a) extrusion (using PUR, PS, PE, PVC, CA, etc.), (b) cast expandable (PS, PE, PVC, PUR, phenolic, epoxy, PF, EP, SI, etc.), (c) spray (PUR, EP, UP, etc.), (d) froth (PLTR, PVC, UF, EP, etc.), (e) injection molding (PE, PP, PS, PVC, etc.), (f) compression (PE, PVC, UP, etc.), (g) sintering (PS, PE, PTFE), and (h) leaching (PE, PVC, CA). In use are equipment that operate all-hydraulically to all-electrical and hybrids (hydraulically/electrically). Tables 8.43 and 8.44 compare conventional foam and structural foam with other processes producing solid plastic products.

As reviewed, almost every thermoplastic and thermoset resin may be produced today in cellular form by means of the mechanisms and processes cited above. The physical properties of the foams reflect in many ways those of the neat polymers, taking into account the effects of density and cell geometry.

Foaming methods vary widely. One is to whip air into suspension or a solution of the plastic, which is then hardened by heat curing. A second is to dissolve a gas in a mix, then expand it when the pressure is reduced. Another is to heat a mixture until one of its liquid components volatilizes. Similarly, water produced in an exothermic chemical reaction can be volatilized within the mass by the heat of reaction. A different technique uses a chemical reaction to produce carbon dioxide gas within solid mass. A related way is for a gas, such as nitrogen, to be liberated within a mass by thermal decomposition of a chemical blowing agent. Other techniques disperse small solid particles, tiny beads of plastic, or even glass microballoons within a plastic mix or syntactic foam.

The most common method, the dispersion process, disperses a gaseous phase throughout a fluid plastic phase and then preserves the resulting combination. The expansion process consists of the following actions: (a) creation of small discontinuities or cells in a plastic fluid phase, (b) growth of these cells to a desired volume, and (c) stabilization of the resultant cellular structure by physical or chemical means. The gas phase is usually distributed in voids or pockets called cells. They can be foamed open-cell but usually they are foamed closed-cell.

A foamed (or cellular) plastic is an expanded, spongy, resinous material having a distinctly cellular structure. Most liquid resins for plastic melts can be made into either rigid or flexible foam structures by the proper introduction of a dispersed gas and the freezing (or setting) of the expanded plastics mass by cooling. The foam-in-place types are liquids, beads, molding powders, and plastisols. A wide variety of specific methods exist for performing this operation. Foam mixtures are compounded to perform many functions, such as spraying ceilings. Vinyl plastisols using chemical blowing agents can be injection molded, extruded, calendered, dip coated, slush molded, and so on.

The processes are identified by different names with some overlap. They include bead molding, calender foaming, expandable plastic foam, expandable PS, expandable sheet stock, expandable PVC, extruded foam, injection-molded foam (low-, high-, and counter-pressure types), mechanical foaming, reaction injection molding, reticulated foam, spray foam, steam foam molding, structural foam, and syntactic foam. There are also the following, all starting with the word *foamed*: laminating, blow molding, casting, extruded film, frothing, gas counterpressure, injection molding, liquid, reservoir molding, and rubber. In the previous section entitled "Type of

Foam vs. sheet metal	Foam vs. die casting	Foam vs. sheet molding compound	Foam vs. hand lay-up fiberglass	Foam vs. injection molding
		Advantages		
1. Fabrication economy: less assembly time; tighter dimensional tolerances; increased product integrity; less final product-inspection time 2. Fewer parts required for assembly 3. Dent resistance 4. Elimination of oil canning 5. Greater design freedom 6. Better sound damping 7. Reduced damage from shipping 8. Reduced tooling costs for complex configuration	1. Much lower tooling costs 2. Longer tool life, lower maintenance 3. No trim dies required 4. Lighter weight 5. Higher impact resistance 6. Better sound damping 7. Better strength-to-weight 8. Better impact resistance	1. Uniform physical properties throughout the part 2. Warping and sink marks reduced or eliminated 3. No resin-rich areas to cause configuration problems 4. Higher impact resistance 5. Greater inherent structural capabilities 6. Lower shipping costs 7. Large parts more economical 8. Lower tooling costs 9. Better sound damping	1. More consistent part reproduction 2. Lower labor 3. Simplified assembly 4. Better dimensional stability 5. More design freedom 6. More uniform physical properties 7. Better sound damping	(Many process similarities exist) 1. Flexibility for functional engineering 2. Better low- to medium-volume economics 3. Lower tooling costs 4. Better large-part capability 5. Better sound damping 6. Lower internal stresses 7. Sink marks reduced or eliminated 8. Inherent structural strength

Table 8.43 Conventional foam process vs. other processes

Foam," reviews were included on processing methods. The following provides summation and/or additional processing information.

EXTRUDED OR CALENDERED FOAMED STOCK

This usually refers to expandable thermoplastic (TP) sheet or profile stock produced by extrusion (chapter 5) or calendering (chapter 9) in which a decomposable blowing agent is milled into a

Foam vs. sheet metal	*Foam vs. die casting*	*Foam vs. sheet molding compound*	*Foam vs. hand layup fiberglass*	*Foam vs. injection molding*
		Advangtages		
1. Fabrication economy: less assembly time; tighter dimensional tolerances; increased product integrity; less final product-inspection time 2. Fewer parts required for assembly 3. Dent resistance 4. Elimination of oil canning 5. Greater design freedom 6. Better sound damping 7. Reduced damage from shipping 8. Reduced tooling costs for complex configurations	1. Much lower tooling costs 2. Longer tool life, lower maintenance 3. No trim dies required 4. Lighter weight 5. Higher impact resistance 6. Better sound damping 7. Better strength-to-weight ratio 8. Better impact resistance	1. Uniform physical properties throughout the part 2. Warping and sink marks reduced or eliminated 3. No resin-rich areas to cause configuration problems 4. Higher impact resistance 5. Greater inherent structural capabilities 6. Lower shipping costs 7. Large parts more economical 8. Lower tooling costs 9. Better sound damping	1. More consistent part reproduction 2. Lower labor 3. Simplified assembly 4. Better dimensional stability 5. More design freedom 6. More uniform physical properties 7. Better sound damping	1. Flexibility for functional engineering 2. Better low- to medium-volume economics 3. Lower tooling costs 4. Better large-part capability 5. Better sound damping 6. Lower internal stresses 7. Sink marks reduced or eliminated 8. Inherent structural strength *Many process similarities exist*
		Limitations		
1. Smaller variety of finishes available, such as chrome or baked enamel 2. No RFI and grounding capabilities 3. Harder to retrofit to frame or skins 4. Thicker wall 5. Higher tool costs than with brakeforming	1. No heat-sink capabilities 2. No RFI and grounding capabilities 3. Fewer available finishes for cosmetic appearance 4. Higher finishing costs 5. Thicker walls 6. Possible internal voids	1. Increased finishing costs (surface swirl) 2. Heat distortion 3. Thicker wall 4. Lower physical properties 5. Possible internal voids	1. More prone to heat distortion 2. Poorer economies of part size vs. quantity 3. Thicker walls 4. Higher tooling costs	1. Poorer surface finish 2. Application of cosmetic detail for appearance parts 3. Longer cycle time 4. Thicker walls 5. Poorer high-volume economics 6. Less equipment available for various shot sizes
		Potential savings from structural foam (%)		
>50[a]	15–30	<30	>50	15–20[b]

[a] Even with limited quantities.
[b] Depending on unit volume and part size.

Table 8.44 Structural foam process vs. other processes

plastic below the decomposition temperature of the blowing agent (Figs. 8.5 and 8.14). Subsequent heating decomposes and expands the blowing agent. The expanded plastic has to be stabilized by cooling. Its uses include egg cartons, meat trays, and fast-food containers where the flat sheets have been thermoformed. Examples for foamed profiles are in packaging, insulation, and decoration.

EXTRUDING

Expandable beads (includes blowing agents with plastic) can be used, but as reviewed later in this section, there are specific expandable plastic processes that are relatively less expensive to use with more flexibility in handling. Extrusions of the more conventional thermoplastic materials containing chemical blowing agents are used. This material allows for a fairly simple operation on a normal single-screw extruder. The process is used where high densities of 0.3 to 0.6 g/cc are acceptable; however, processors use as low a density as possible to save material costs.

The most commonly used system is the direct gas extrusion process. It is possible to use the direct gas process with one extruder that does the melting of the plastic, mixing of the gas, cooling of the melt, and the extrusion through the die. Both single- and twin-screw extruders are used. Although twin-screw extruders have the advantage of requiring less energy yet melting the plastic more efficiently than the single-screw, the limiting factor is the permissible internal pressure. High pressures require very heavy bearing designs in extruders. The basic disadvantage of both systems is that they must melt, mix, and cool within a relatively short distance in an extruder with an L/D of, for example, 30:1 (chapter 3).

A tandem system specifically designed for foam processing separates these functions; the melting and mixing of the blowing agent is done in a primary extruder, while the controlled cooling is done in a secondary extruder. Although slightly more expensive than any of the above-described extrusion systems, the tandem extruder, with its better overall economics and versatility, is the most commonly used.

The basic raw materials used are general-purpose plastics in pellet forms. With regard to polystyrene, high molecular weight and low monomer content are preferred from the standpoint of foam strength and for food packaging. Certain polyethylene resins have been found to process more easily and provide lower-density foams having particular flexible qualities or higher strengths. Many different types of TPs or blends of TPs have been used to give wide ranges of properties in the finished foamed products.

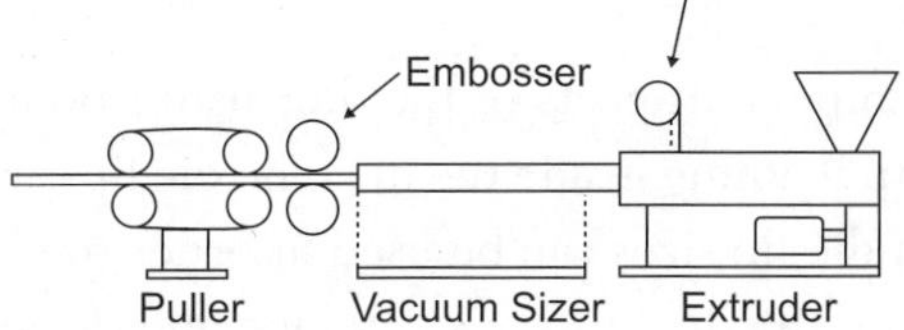

Figure 8.14 Continuous extruding of foamed profiles

A nucleating agent, usually in the form of a fine dry powder, is blended with the plastic pellets in a drum tumbler or other mixer. More sophisticated systems utilizing continuous metering and blending equipment are available. The nucleating agent controls the cell size and properties of the foam. A nucleating agent initiates the formation of bubbles by providing an imperfection in the melt where the blowing agent can come out of solution on pressure release at the die exit. Many types of nucleating agents are used depending on the resin, the fineness of cell desired, and the type of blowing agent used.

During processing, the dry blend is fed under controlled conditions to the hopper of the primary extruder. The first stage of this extruder continues the mixing of the ingredients and brings the plastic to a homogeneous melt at a pressure of about 3,000 psi and a temperature of approximately 380°F. At this point, a blowing agent is injected into the melt by means of a high-pressure metering system. The specially designed primary extruder screw has to plasticate the melt and mix it with the dry ingredients until totally dispersed, forming a homogeneous melt.

During this action, the gas has to be completely dissolved by chemical association. This is important in order to achieve consistency in output thickness. A metering system is designed to accurately meter into the extruder a stream of blowing agent at high pressures regardless of any pressure changes in the extrusion system. The gas is injected into the barrel through a specially designed valve. Common blowing agents are hydrocarbons, such as pentane (details on blowing agents are later presented). The amount of blowing agent can vary widely, depending on the resin and the type of product desired. However, generally about 7 wt% blowing agent produces a product of about 5.6 lb/ft^3, while 18% blowing agent produces a product of about 1.9 lb/ft^3. The melt is fed under pressure to the seal throat of the cooling extruder. This extruder's plasticator can contain special screws designed for uniform cooling and high-capacity heat transfer. Specially designed heater coolers are used on the barrel.

The properties of the extruded foam (sheet, film, profile) can be varied considerably. Large cells give more rigid sheet; fine cells a softer, more flexible sheet that is also less brittle. Raising density increases strength and stiffness in both transverse and machine directions. After proper cooling, the melt extrudes through the die attached to the end of the cooling barrel. It forms the extrudate into profiles, sheet, and so on.

Die design is critical because in foam extrusion the viscosity of the melt is so high that imperfections from the screw or from the die flow are easily transmitted to the output product. To prevent this, the entrance to the die is designed to have a restrictive flow passage to help heal the melt as it comes off the screw flights and through spiders and to provide for uniform flow into the die body.

The die lips are the most important parts of the die; at this point, melt is being shaped, yet at the same time is prevented from foaming inside the die. The die lips are interchangeable so that with any type of die body a range of die lip sizes can be used in order to change the blowup ratio; just as the sheet and so on emerge from the die lips the stretching and orienting take place.

Sheet usually is extruded through an annular die horizontally over a cooling mandrel that sizes the sheet. Internal and external air rings provide cooling. Also, the air supports the bubble as it moves from the die to the cooling mandrel and then helps move the sheet over the mandrel with a minimum of friction. The sheet, which is cool and self-supporting as it leaves the mandrel, is slit into one or more webs and pulled by light nip rolls or by an S-wrap puller. Sheet web is wound on large cores (usually 6 to 10 inches in diameter). Rolls are usually aged for twenty-four to forty-eight hours before going through another process, such as thermoforming (chapter 7). Profile extrusions are passed through shaping devices with air or water cooling as the product may require.

The performance of any extruder is dependent on its control systems (chapter 3). This is especially true in the case of a tandem extrusion system, since one extruder is feeding into another. Precise temperature control of the individual barrel zones is critical to prevent melt surging. Drive speeds must be stable and melt temperature and pressure are monitored at several critical locations throughout the system. Blowing agent pressure and flow rate must also be controlled to a fine degree of accuracy. Microprocessor-based control is used to monitor, record, and control many of these process variables.

Application Meat trays normally are produced from PS-foamed sheet with a thickness of approximately 0.095 in and a density of about 5 lb/ft^3. These trays are manufactured with little skin. The meat should not stick to the plastic when frozen nor should the meat juice penetrate into the tray, and the tray should not break when handled. Strength is achieved by correctly orienting the sheet, resulting in cells that are round from a planar view. The cell size, which is determined by the amount and type of nucleating agent, is kept fairly fine to give the trays soft, glossy appearances.

Where deeper draws are required, the material has to be stiffer with coarser cells. Typical are egg cartons and flats used in bulb shipping. PS sheet for trays or cartons are about 0.080 to 0.095 in thick, and the density varies between 6.9 and 8.1 lb/ft^3. On one side of the sheet, by use of the proper air ring (chapter5), a strong skin is formed, improving product appearance and allowing for better thermoforming.

Continuing with properly providing foamed products to meet their performance requirements are food trays, such as those for apples or pears. They are designed to protect the fruit under all conditions of shipping and must hold the weight of the produce. Therefore, a high degree or orientation is put into the sheet, providing for more flexible cells. Sheet thickness is approximately 0.080 in, and density is about 6.25 lb/ft^3. Trays are manufactured without skin to present soft, cushioned surfaces for the produce and to prevent cracking.

Different packaging requirements exist in the fast-food industry. Stackability, appearance, insulation, and a surface that lends itself to good printing are best seen in hamburger packs. Light skins on both sides provide glossy appearances and allow for easy printing.

Other products have their specific requirements. They include many products such as institutional feeding trays, automotive headliners based on foam cores, dielectric material for insulating the center wire of coaxial cable from the sheath, packaging to protect fragile electronic equipment, and dishware.

The most popular materials are styrenics, olefins, and engineering resins such as modified polyphenylene ether or polycarbonate (chapter 2). Fillers for enhanced physical properties, UV stabilizers, and flame retardants are common additives.

Process There is different equipment to produce the different extruded foamed products. With the equipment, special tooling is usually required (chapter 17 can provide a guide). A few of these will be reviewed in this chapter.

CASTING

Foamed casting is a simple, nonmechanical version of reaction injection molding (RIM) (chapter 12) or liquid injection molding (LIM) (chapter 16). Liquid components of the plastic with suitable additives are mixed and poured into an open mold (chapter 11). Polymerization and foaming take place in the mold cavity, which could include a matching mold cavity to enclose the foaming action. Molds are generally heated or oven cured, sometimes both.

SPRAYING

Spray techniques are used for filling molds and panels and for applying foam to plane surfaces. Spraying is particularly useful in applications where large areas are involved, such as tanks or building walls. Spraying is the simplest and least expensive way to produce thermoplastic urethane foams. In addition, spraying equipment is reasonably priced and portable. Foam can be applied without molds or jigs of any kind. In spray applications, the ingredient mixing is accomplished by atomization of the materials as they leave the nozzle of the spray gun. Resin viscosities in the range of 500 to 1,500 cP (0.5–1.5 Pas) should be maintained to ensure intimate mixing of the foam constituents. When necessary, heat may be applied to the polymer to reduce the viscosity to the desired level.

Commercial spray guns are capable of applying a 12-in (30.5-cm) circle of urethane foam when the gun is held at a distance of 3 ft (0.9 m) from the target surface. Larger deliveries and different spray configuration are possible with special guns. The spray technique may be used for filling molds (spray pouring) by confining the spray pattern as it leaves the spray gun. This is accomplished by attaching a cone to the gun head. The cone may be from 8 in (20.3 cm) to 2 ft (61cm) in length and is attached with the smaller diameter at the outlet end. This device also serves to minimize catalyst loss into the air and confines the foam discharge to a smaller area. With this method, large deliveries of foam material are possible. This technique provides interesting possibilities when filling intricate or large molds. It is relatively inexpensive and highly flexible. Adhesion of the sprayed foam to most surfaces is excellent.

Conventional spraying is usually carried out with portable metering equipment, which should be lightweight, inexpensive, and compact so that it can be carried on a small truck or trailer. Most of these units consist of air-driven dual-acting piston pumps that can be readily calibrated and used with the original 5- or 55-gal (19- or 208-l) shipping containers. Two basic types of guns are used for spray operations: the internal-mix and external-mix. The internal-mix guns use a number of different methods of mixing. The external-mix guns use high-pressure air to bring the components out of the gun and then, through a swirling action, the air mixes and atomizes the components between the gun and the substrate to be sprayed. To prevent sagging of sprayed foams, formulations with fast reaction rates are being used. The faster rates are obtained by using additional catalysts. Additional layers can be sprayed almost immediately. The nominal spray rate is about 4 to 8 lbs (1.8 to 3.6 kg) per minute. Particular attention should be given to providing adequate ventilation during the spraying operation. The development, which has been most responsible for improvement in the spray process, was the 1:1 ratio, low-viscosity, totally catalyzed one-shot system. This system brought about the development of the airless gun. Previously, atomization was achieved by the introduction of high-pressure air into a conventional mixing chamber. Such spray heads are known as air-assisted.

The rise time for airless sprayed urethane foam is about thirty seconds. It can be refoamed for additional thickness after this period. The foam can be walked on after three to four minutes and reaches its full properties in twenty-four hours. The airless spray gun is held about 30 in (76 cm) from the surface and moved steadily over it. Dispensing rates of 4 to 6 lb (1.8 to 3.6 kg) per minute are generally considered optimum for most spray applications. The surface on which the foam is sprayed must be free of loose scale or grease. The adhesion of urethane foam to steel is essentially equal to the tensile strength of the foam, provided the surface is clean. Aluminum surfaces, on the other hand, do not provide a good bond unless a primer coat, such as vinyl wash, is used prior to spray foaming.

A nonburning barrier approved by the appropriate building codes must protect spray-applied urethane foam in buildings. Without such barriers, polyurethane foam will spread fire rapidly once ignited, even with fire-retardant grades. With few exceptions, all model codes require that foam plastic insulation be covered by a thermal barrier equal in resistance to ½ in (1.27 cm) gypsum board or be used only in sprinkler buildings. The gypsum board or equivalent is supposed to prevent the foam from reaching a temperature of 325°F (163°C) for a fifteen-minute period when subjected to the ASTM E 119 time-temperature curve, which averages 1,100°F (593°C).

FROTHING

This process was introduced by Du Pont in 1961. The process operates in a two-stage expansion system. The main idea is to introduce into the formulation another more volatile liquid, such as Fluorocarbon-12 (406). With this approach, instead of delivering a liquid blend into a cavity for foaming, a preexpanded one is delivered. In this manner, a froth stream of 8 to 12 lb/ft^3 (128 to 192 kg/m^3) density is poured into a cavity where it completes its expansion to a low density of 1.5 to 2.5 lb/ft^3 (24 to 40 kg/m^3), which is the density range desired. In the froth stage, the stream is quite fluid, resembling shaving cream in appearance, and flows readily. Frothing is particularly suited for void filling or panel manufacture. The advantages of this process are the following:

1. Lower mold pressures during foaming, minimizing the need for jigs and making possible the forming of larger panels
2. Lower and more uniform foam densities

3. Ability to lay down expanding foam without causing its collapse or density change
4. Possibility of cross-section molding without excessive sacrifice in density
5. Lower densities obtainable with low-temperature molding
6. Froth may be leveled by methods similar to those used with concrete in the construction industry

Frothing of the urethane foam–component mix occurs when the volatile liquid (Fluorocarbon-12) is vaporized by a reduction in pressure as the material is discharged from the mixer. By employing a combination of fluorocarbons of different boiling points, a two-stage expansion is possible. The vaporization of the low-boiling solvent occurs in the initial stage, with the final foam density resulting from the supplemental expansion of the higher-boiling solvent (Fluorocarbon-11) due to the heat generated by the reactions between the foam components containing hydroxyl and isocyanate. With conventional molding techniques, the foam must expand thirty to forty times in the mold to reach its final volume. While in the frothing process, the final expansion in the mold is reduced to only three to six times the froth volume.

One of the major problems with the frothing process is the lack of fully perfected dispensing equipment. In addition, foam prepared by this method often has many elongated holes approximately 1/8 to 1/4 in (0.32 to 0.64 cm) in cross-sectional diameter. In molding, when using the frothing process, large vertical pours can be made in one continuous pour and the knit line or hard spots do not pose as big a problem as with a nonfroth system. The width of the cavity to be filled is limited by the diameter of the froth streams from the let-down valve. Panels and voids of 2 in (5.1 cm) width with a foam rise up to 8 ft (2.44 m) have been filled with froth foam. The main problem with the horizontal pouring of froth foam is the distribution of the froth over the mold surface. With current froth systems, conveyors and mechanical distribution are necessary with horizontal pouring.

Froth may also be sprayed. If the equipment is properly designed and adjusted, R-12 pressure reduction may be used, instead of air, to propel the material to the surface being sprayed.

Frothing techniques may be used in vertical-pouring applications of rigid urethane foam where the surface-to-volume ratio is high. Examples are found in the insulation of refrigerated railroad cars and trailer vans, large refrigerators, or simple building panels. Because of the fluidity of the product after dispensing and its quick curing, frothing is very well suited for continuous production of laminated roll stock using flexible skin materials, such as foil, corrugated paper, and cardboard. It may be necessary to use some sort of spreader to ensure even spreading of the partially expanded foam. Frothing is particularly useful in insulating railroad freight cars. This technique provides complete wraparound insulation with no through-wall connections. An average freight car can be insulated by three people in three hours.

Expandable Polystyrene

Foamed PS or expandable polystyrene (EPS) illustrates the use of plastic concentrates that include the blowing agents. Plastic beads containing the blowing agent are supplied to the molder as accurately formed solid spheres. The beads may be about 0.1 to 0.2 mm in diameter and they contain a blowing agent, usually pentane. This process is also called bead molding or steam molding, but the most popular name is expandable polystyrene (EPS). Other plastics are foamed by the same or similar methods, but EPS is produced in the largest quantities.

The process involves two major steps. As shown in Figure 8.15, the first step consists of preexpansion of the virgin beads by heat (usually steam, but also used are hot air, radiant heat, or hot water). Steam is extensively used because it is the most practical, the most economical, and has other advantages. The preexpansion step brings the beads almost to the required density within the molded product, then they are stored for six to twelve hours to allow them to reach an equilibrium.

Different preexpansion equipment is available; each type has advantages and disadvantages. Different controls exist because they are needed. The type of expander and controls will depend on the production quantity. There are continuous, single-stage, multistage, and discontinuous preexpansion systems. As an example of performance, consider the advantages of the continuous system over the other types: lower unit cost, higher throughput, easier maintenance, and greater reliability. Its principal disadvantage is the time it takes to change between materials of different bulk density. This disadvantage is avoided with the discontinuous unit.

The next step conveys these beads, usually by air through a transport tube, to a two-cavity mold. If problems exist in having the beads properly fill the cavities, a vibratory against the mold will have the beads relocated so that they fit snuggly to each other. Final expansion occurs in the mold, usually with steam heat, either by having live steam go through perforations in the mold itself (Figs. 8.16 and 8.17) or by steam probes that are withdrawn as the beads are expanding. During expansion, the beads melt together, adhering to each other and forming a relatively smooth skin, filling the cavity or cavities. Multiple cavities can be used with small parts.

The heat cycle is followed by the cooling cycle. Because EPS is an excellent thermal insulator, it takes a relatively long time to remove the heat before demolding; if the heat were to remain, the

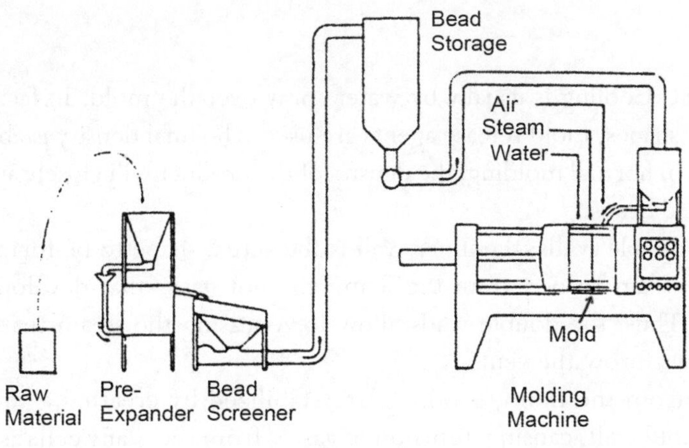

Figure 8.15 Expandable polystyrene process line starts with preexpanding the PS beads

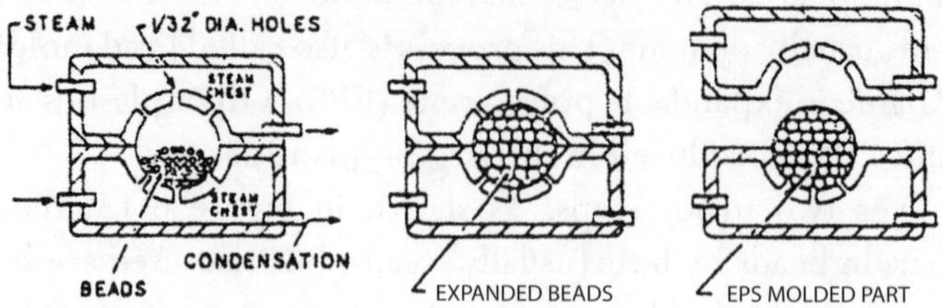

Figure 8.16 View of PS beads in a perforated mold cavity that expand when subjected to steam heat

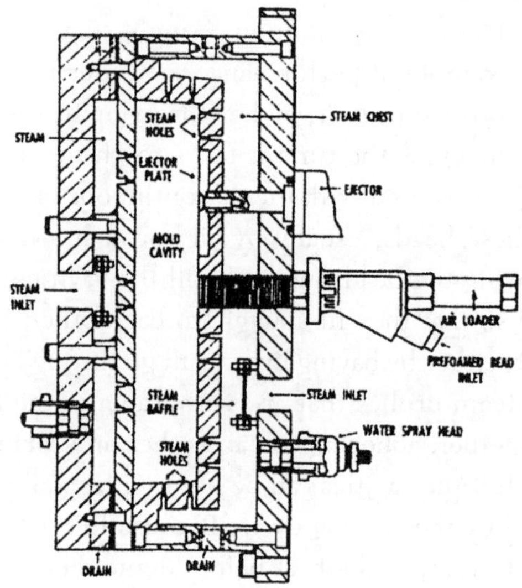

Figure 8.17 Example of an EPS steam chest mold

product would distort. Cooling is usually by water spray over the mold. To facilitate removal, particularly for complex shapes, mold release agents are used. The final density is about 0.7 to 10 lb/ft^3 (11 to 160 kg/m^3) or in normal molding, the density of the product will closely approximate the bulk density of the unheated beads.

EPS molds have double walls; the inner wall is the actual shape to be formed. It is perforated with vents to allow steam to penetrate the foam; the hot gases that develop leave the product through these vents. Thus, the double walls allow for encasing the steam that is delivered to the mold and in turn flows throw the vents.

Before removal from the mold, products are stabilized by creating a vacuum and spraying water on the inner mold wall, causing diffusion of gasses from the many cells as well as a reduction in temperature.

EPS molding generates pressures of less than 30 psi (2 kPa) in most mold applications. This low pressure allows the use of inexpensive molds such as aluminum. To process the other expandable plastic foams (EPFs), such as PEs, PPs, and PMMAs, the equipment for EPS can be used with only slight modifications.

Pentane Blowing Agent Pentane has been used as a gas-blowing agent to produce different foamed plastics or elastomers, particularly in EPS. EPS beads incorporate the pentane and in turn the beads are used during different molding processes to produce different shaped products. Pentane is used to produce certain rigid polyurethane insulation foams as an alternate to the previously used CFC blowing agents. For example, during PUR processing, it can be added separately to the mixture, bypassing on the high pressure side of the mixing head, thereby bypassing the explosive-proof mix chamber and polyol metering pump.

Because of pentane's flammability and chemical makeup, no problems exist when properly processed. It is halogen-free, nonpolar, and accepted as nontoxic. The flammability of foam products can be controlled through proper use of flame retardants. It is a hydrocarbon in the methane series occurring in petroleum.

Expandable Polyethylene

EPE is a low-density, semirigid, closed-cell, weather-stable PE homopolymer that is easier to compress than EPS but less compliant than flexible PUR. EPE foam follows a similar processing technique as that of EPS, starting with the use of PE beads. Conventional EPS molding presses can process EPE with the addition of a modified filling device, provision for higher molding pressure, and postmold oven curing. Its density range is 1.8 to 7.5 lb/ft^3 (29 to 120 kg/m^3). The most commonly used density is 1.8 lb/ft^3 (29 kg/m^3).

Expandable Polyethylene/Polystyrene

Expanded polyethylene copolymer (EPC) is a 50/50 wt% of polyethylene polystyrene. Combining the properties of both plastics widens the selection of resilient materials for packaging engineers. EPC is a material that falls between EPS and EPE in performance but exceeds both materials in toughness. The tensile and puncture resistance of EPC are superior to all of the moldable resilient foams available. It has good multiple-impact performance characteristics with better memory than EPS, but is not as good as EPE. The cushion performance of EPC parallels EPE, as well as at higher levels (7 to 12 g) even after repeated drops. EPC is especially good for reusable material-handling trays and packaging applications that require a nonabrasive, solvent-resistant, impact-absorbing material with a superior toughness that elongates, compresses, and flexes without material fatigue.

EPS molders are capable of molding this material with minor adjustments to their existing expansion equipment. EPC is a low-density, semirigid, closed-cell material that requires refrigerated storage below 40°F (4.4°C) in its raw granular form and has a shelf life of at least one month. The material expansion and conveying of the sensitive prepuff require special handling and molding within a short period of time. The molding process and equipment are similar to EPS but with slower molding cycles. Its beads contain their own blowing agents.

EXPANDABLE STYRENE-ACRYLONITRILE

ESAN is a moldable, lightweight, semirigid, closed-cell, highly resilient plastic foam. The raw materials (containing blowing agents) can be stored without refrigeration, and they have a shelf life of about six months, provided some simple precautions are observed as provided by the material supplier.

Processing is similar to EPS except that it has extended fabricating cycles owing to the higher level of blowing agent. Postmold oven curing is not necessary. A low density of more than forty times expansion can be attained during the preexpanding phase, as low as 1.0 lb/ft^3 (16 kg/rn^3) on the first pass through the expander and 0.8 lb/ft^3 (13 kg/m^3) on the second pass.

MOLDING

As to be reviewed, different processing methods are used. Basically, foam molding operations are those in which a liquid mixture of foam components is usually used. It is poured into a mold cavity to form a cellular-shaped product. The molded product is later removed after setting or curing. As reviewed in the case of expandable polystyrene beads, the preexpanded or virgin beads are poured into a mold and heated to form the desired object. In this case, liquids are not used, although the free-flowing beads might be considered as a fluid.

A number of flexible foam components are molded rather than fabricated from slabstock. For a production quantity of intricately shaped products, molding results in savings up to 15% compared to slabstock. Rigid foam can also be molded, although such applications are fewer in number. In a typical low-pressure molding operation, a mold is preheated and coated with a mold release agent. A preheated amount of liquid mix is poured into the mold, the mold is closed, and the ingredients foamed to the mold configuration. After curing and cooling, the part is stripped from the mold. The scrap rate is seldom more than 5 wt% to10 wt%. Mold temperatures are usually in the range of 150° to 160°F (66° to 71°C). The best results are usually obtained when the mold release agent is spray applied to the clean, warm mold just before each pour. Care must be taken, however, to remove all solvents from the mold release compound before the foam components are poured into the mold cavity.

Curing of the molded TS polyurethane foam product is usually carried out in two stages, a precure of fifteen to twenty minutes at about 270°F (132°C), permitting removal of the part from the mold, and a final cure of sixty minutes at the same temperature. Mold release of polyurethane foams can be difficult since uncured polyurethane has well-known adhesive properties. Two basic types of mold release agents are used. The first requires the hot molding to be stripped from the mold. Mixtures of paraffin and microcrystalline waxes are used for this technique, in which hot wax releases the part from the mold. The molds must be heated and coated with wax before each filling. There is a tendency, however, for the paraffin wax to be slowly oxidized by the repeated heating. For this reason, a release agent containing an antioxidant should be used. The breakdown products formed have no release properties, and it is important to use a thin layer of wax each time a molding procedure is carried out.

The second type of release agent, such as polyethylene waxes, is used if the mold must be stripped away when cold. In this case, the foam comes away from the release agent. Mold-release problems can be reduced by adding small amounts of dibutyl tin dilaurate, which promotes curing at the surface of the mold and thus improves mold release.

Slabstock molding requires very little curing. A bank of infrared heaters suspended above the conveyor is often sufficient to facilitate curing. In conventional molding, however, the exotherm generated is not sufficient to cure the foam and external heat must be applied. Microwave curing permits a reduction in curing time from twenty to four minutes. Plastic molds are used with a gel coat of epoxy resin containing iron powder. The plastic molds must be cooled to an even temperature before refilling, however, and this step may introduce problems. There is evidence to suggest that foams cured with microwaves have properties slightly superior to foams cured by conventional heating, especially in compression set. Dielectric heating has also been developed for use in curing.

Injection Molding

The conventional and slightly modified injection-molding machines (IMMs) are used to produce different types of foams.

Low-Pressure Low-pressure or short-shot conventional foam processing methods are the most commonly used (compared to other IM methods as reviewed in chapter 4), because they are easy, simple, and best suited to economical production, particularly of large, complex, 3-D products (178). A controlled melt mixture (plastic and blowing agent) is injected into a mold cavity from the IM plasticator (chapter 3), creating a low cavity pressure, usually 1.4 to 3.5 MPa (200 to 500 psi). As shown in Figure 8.18, two steps occur using a reciprocating IMM. First, the plastic melt with blowing agent (nitrogen, carbon dioxide, or hydrocarbon gas) is directed into an accumulator. The next step has the accumulator very quickly deliver the hot plastic mix into the mold cavity.

Along with about 0.5 wt% of CBA, this mixture can be injected directly from the barrel of a conventional IMM (with limited modifications) or via an accumulator. The mixture only partially fills the mold (short shot), and the gas bubbles, having been at higher pressure, expand immediately and fill the cavity. As the cells collapse against the mold surface, a relatively solid skin of melt is

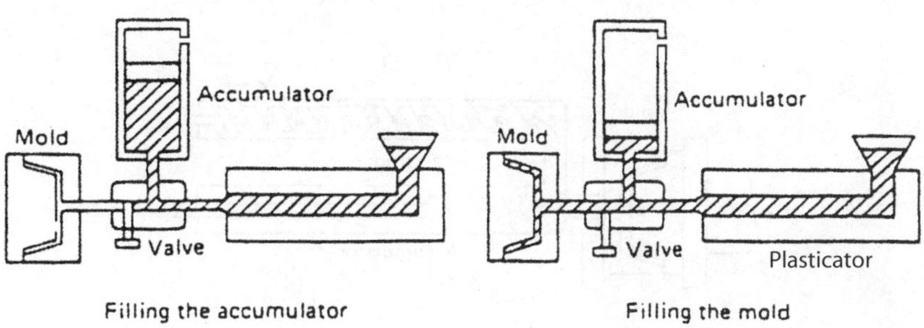

Figure 8.18 Schematic of foam reciprocating injection-molding machine for low pressure

formed over the rigid foam core. Skin thickness is controlled by the amount of melt injected, mold temperature, type and amount of blowing agent, temperature and pressure of the melt, and capabilities of the molding machine, particularly its speed of injection.

Also used are two-stage IMMs (chapter 4). As shown in Figures 8.19 and 8.20, this type of IMM has an accumulator.

Coinjection There is low pressure with coinjection. This technique involves the usual separate injection of two compatible plastics that are coinjected using two injection plasticators (chapter 4). A solid plastic is injected from one plasticator to form a solid, smooth skin against the surfaces of

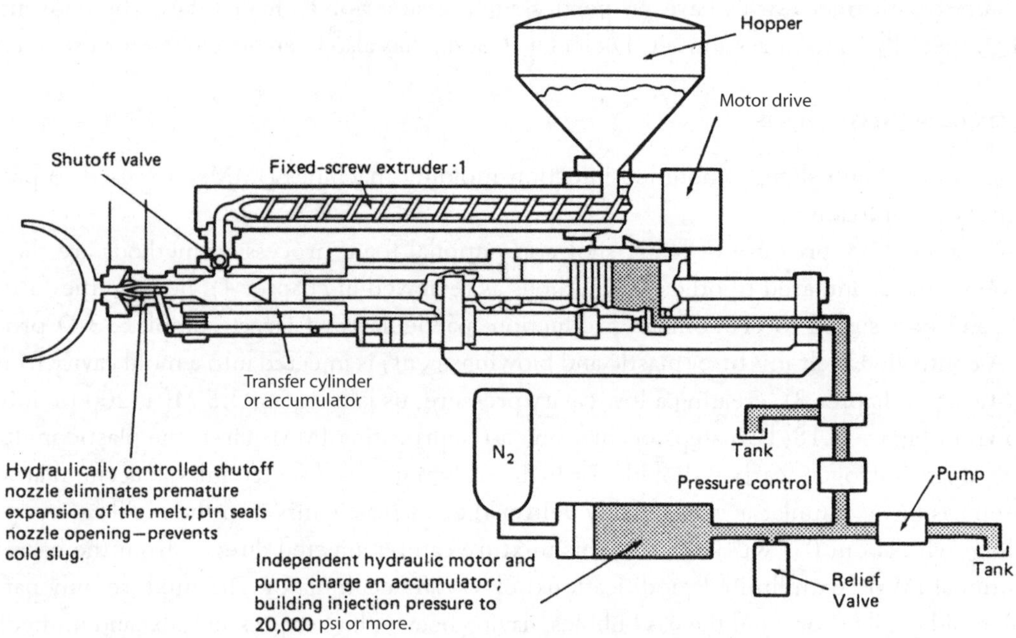

Figure 8.19 Schematic of foam two-stage injection-molding machine for low pressure with blowing agent directed into the transfer or accumulator cylinder

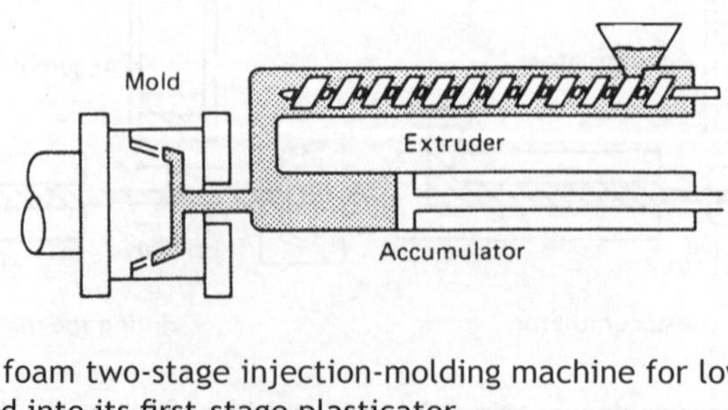

Figure 8.20 Schematic of foam two-stage injection-molding machine for low pressure with blowing agent directed into its first-stage plasticator

the mold cavity (Fig. 4.58). Simultaneously, a second material, a measured short shot containing a blowing agent, is injected to form the foamed core. This approach can also take a relatively full core shot and have the mold open (as in coining) after the skin solidifies, having the melted core expand with mold-opening action.

Surface Finish There is low pressure with surface finish in low-pressure surface-finish (LPSF) molding. In the absence of coinjection or coining, the volume of the molding cavity is always larger than the volume of the plastic in the unfoamed state. The low pressure allows microbubbles to nucleate and grow. Foam expansion occurs during filling and growing bubbles are carried to the mold surface, creating unacceptable surface irregularities and imperfections called splay or swirl patterns. The irregularities can be seen and felt; the surface roughness can be as much as 1,000 μin (25 μm). Products needing smooth, finished surfaces require secondary operations, usually sanding, filling, and painting.

There are techniques to improve surface appearance during fabrication. The principal process variables are melt and mold temperatures, injection rate, and the nature or type of blowing agent as well as its concentration. Cyclic heating and cooling of the mold surface and direct injection of the blowing agent into the melt as it is being injected into the mold are two of the methods used.

Gas Counterpressure The gas counterpressure method uses a sealed mold pressurized to 2.8 to 3.5 MPa (400 to 500 psi) with an inert gas, which is sufficient pressure to suppress foaming as the plastic mix enters the mold cavity. After the measured shot is injected, the mold pressure is released, allowing the instantaneous foaming to form the core between the already formed solid skins (Fig. 8.21). The mold action is similar to coining. Another technique is gas injection molding, used to develop similar foamed structures. Once the plastic at the mold surface has solidified, the gas pressure is released to permit the remaining melt mix to foam, creating the product's core.

There are different gas-operated patented foam IMM systems. Examples are shown in Figures 8.22 and 8.23. Figure 8.22 shows a Hoover/Carbide patent. Figure 8.23 is the patented

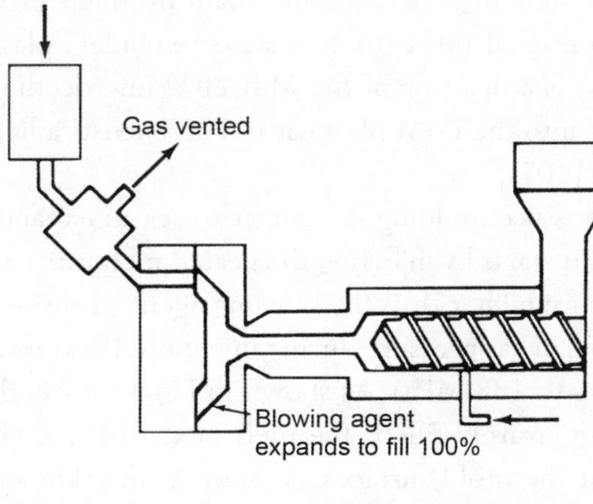

Figure 8.21 Schematic of gas counterpressure foam injection molding (Cashiers Structural Foam patent)

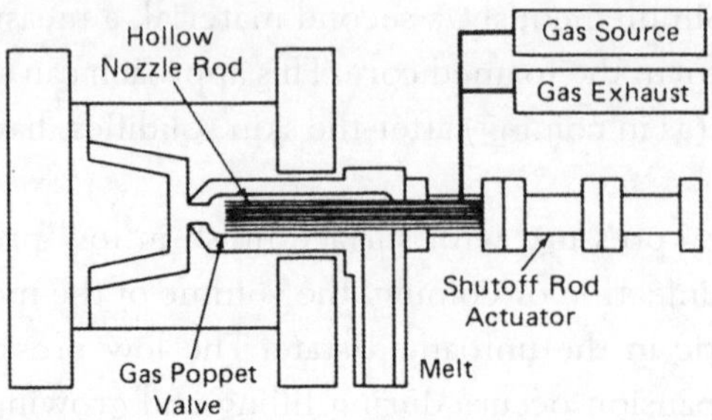

Figure 8.22 Example of an IMM-modified nozzle that handles simultaneously the melt and gas

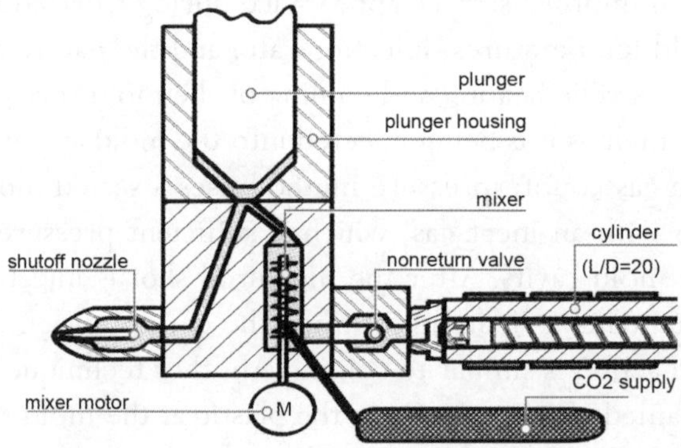

Figure 8.23 IMM microcellular foaming system directing the melt gas through its shutoff nozzle into the mold cavity

Demag Ergotech process for molding microcellular foam products (260). Liquid CO_2 enters the melt downstream of a conventional IMM (or two-stage extruder) plasticator (chapter 3). Trexel Inc.(Woburn, Massachesetts) is a licensor of the MuCell® microcellular foam IMM or extruder process. It injects CO_2 or N_2 into the IMM plasticator. There exists a licensing agreement between Trexel and Demag Ergotech (407).

High-Pressure The high-pressure molding system that uses an expandable mold is a variation of conventional IM (chapter 4). It starts by injecting the heated melt mix (with blowing agent) into the mold, creating a cavity pressure higher than the blowing agent gas pressure (usually much higher). This action is to ensure no loss in gas pressure during injection. Pressure for certain machines could be 5,000 to 20,000 psi (34.5 to 138 MPa). As shown in Figure 8.24, the next molding operation occurs. With the mold being entirely filled, the melt next to the cavity wall forms a solid skin as it starts solidifying against the mold surfaces. As soon as the skin surface hardens to a desired thickness, a second step occurs where the cavity mold opens, reducing the pressure and allowing the remaining melt to foam between the skins. The opening occurs whereby the male plug retracts but remains within the female cavity.

The mold can be modified to meet certain different shapes. The molding can be made either by withdrawing cores or by special press motions that partially open the mold halves (such as the compression molds used in coining) to provide 2-D action; 3-D mold actions are also used. The degree of foam density, wall thickness, and surface finish depend on the foam mixture (constituents and amounts), the machine controls such as the time cycle, and the mold action required.

Web Structural-web molding is a low-pressure foam molding method. It is the phrase usually used to identify the gap between structural foam (SF) molding and injection molding. Its surface does not have the usual SF characteristic swirl pattern. It can produce very large lightweight parts with smooth surfaces, like conventional injection-molded parts.

Reaction Injection Use includes fabricating flexible and rigid microcellular polyurethane foams. The process embodies high-pressure impingement mixing of the liquid components before they are injected into the mold. RIM has advantages over the standard low-pressure mechanical mixing systems in that larger parts are possible, mold cycles are shorter, there is no need for solvent-cleaning cycles, surface finishes are improved, and rapid injection into the mold is possible. Details on this process are in chapter 12.

Liquid Injection

LIM is a variation of the RIM process. The major difference is in the manner in which the liquid components are mixed. In the LIM process, the entire shot is mixed in a chamber before injection into the mold rather than being continuously mixed and injected, as in the RIM process (chapter 16).

Structural Foam

Different plastics are used, such as PSs and PVCs, to produce building trim and moldings, picture frames, and so on. The most important structural foam molding processes have been reviewed.

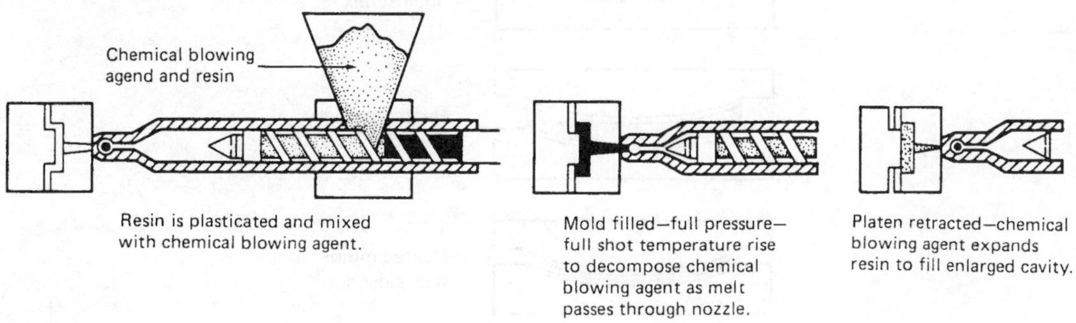

Figure 8.24 Schematic of foam injection molding for high pressure

They are the low- and high-pressure injection-molding processes. Structural foams with solid skins and cellular cores are extruded in the form of profiles, pipes, tubes, sheets, and so on using conventional extruders that include handling the blowing agents. As reviewed with IMMs, the blowing agents can be mixed with the plastic as it enters the hopper or the screw plasticator melt, or by using a mixing device to mix the melt with the blowing agent.

Foam Reservoir Molding

Foamed reservoir molding is also known as elastic reservoir molding. It has had limited use. This process creates a sandwich of plastic-impregnated, open-celled, flexible plastic foam between the face layers of fibrous reinforcements (Fig. 8.25). When this plastic composite is placed in a heated mold and squeezed, the foam is compressed, forcing the plastic and air outward and into the reinforcement. The elastic foam exerts sufficient pressure to force the plastic-impregnated reinforcement into contact with the mold surface, simultaneously removing entrapped air.

Polyurethane Process

As reviewed, when processing PURs, different processing techniques are used. This section provides more details on the extensively used products. The specific processes include free-rise, liquid (pour-in-place), froth, and spray foaming techniques. When injected, there is the common injection-molding process and others. When the PUR liquid ingredients are mixed, gases are produced that cause the mass to expand as it stiffens and hardens. The reaction is completed in a few minutes. Figure 8.26 provides chemical reactions that occur during the PUR reaction. Figure 8.27 shows different process systems in which the blowing agent can be added to either or both components A and B.

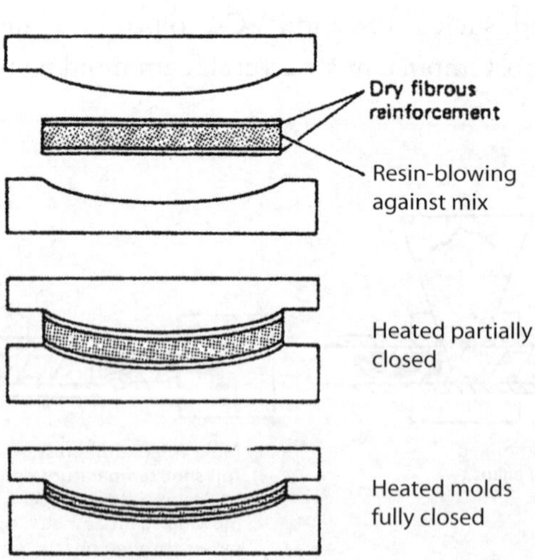

Figure 8.25 Example of stages in foamed reservoir molding

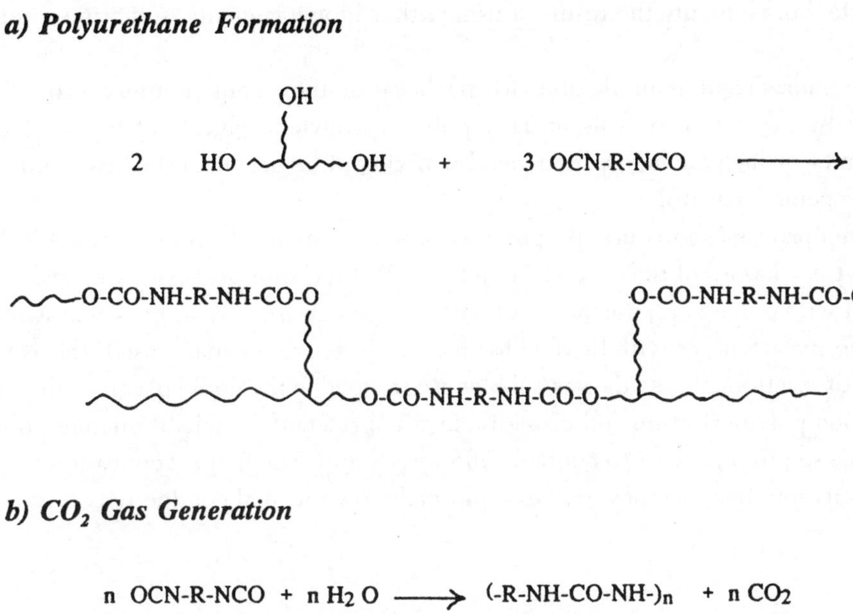

Figure 8.26 Schematics of foaming processes

In a sandwich structure, the liquid mix is poured between the cover sheets and foams between them, bonding directly to the sheets without an adhesive. As foamed-in-place materials expand, they can exert appreciable pressure, so the sandwich sheets have to be held in a rigid frame to prevent bulging until the reaction is completed. To overcome this pressure, the liquid mix may be allowed to form a froth of almost its ultimate volume prior to pouring. The result is little or no pressure and rigid foam.

Flexible PUR foam, such as that used in upholstery, is made by continuous deposition on a belt before being cut into blocks or sheets of desired shapes and sizes. Other foam materials may be handled in somewhat similar ways and may be prefoamed or foamed in place.

Water is the component-forming blowing gas in the formulation for PUR soft, flexible foam; it forms carbon dioxide with isocyanates. The evolution of heat and the change in temperature caused by this gas-generating reaction mean that other blowing agents (such as dichloromethane) have to be used, along with coolants, in order to produce low-density foam of less than 20 kg/m^3. Flexible PUR foams have open cells and are used in furniture, automotive, packaging, insulation, mechanical (sealants, sponges, etc.), and other industries (Fig. 8.28).

A rigid foamed cross-linked PUR, usually with closed cells, is formed by the reaction of a diisocyanate (sometimes containing components of a higher functionality) and often methane diisocyanate (MDI) or polymeric MDI with polyester, or more usually with a polyether polyol (Table 8.18). Foaming may result from the incorporation of water, which reacts with isocyanate groups to form carbon dioxide, but is usually the result of using other blowing agents, sometimes in combination with water.

They are more rigid than flexible foams because they contain more cross-links. This is accomplished by the use of polyols, usually polyoxypropylene glycols of low molecular weight (about 500 units), which are highly branched by mixing of higher functionality comonomers (such as sorbitols or pentaerythritols).

Prepolymer processes and quasi-prepolymer processes are used with TDI (toluene diisocyanate) to reduce the toxic hazard of this material (Table 8.17). One-shot processes are used with MDI and polymer MDI with polyethers; normally, a tertiary amine and/or organotin catalyst system is used with a silicone surfactant, as with flexible foams. Catalysts are normally used; this is to obtain the right balance of reaction rates so the gas bubbles are trapped in the liquid plastic melt as the viscosity increases during polymerization and cross-linking. A surfactant, usually a silicone-polyether block copolymer, is also often present to control cell morphology. The major application for these foams is thermal insulation, because they have exceptionally low thermal conductivity.

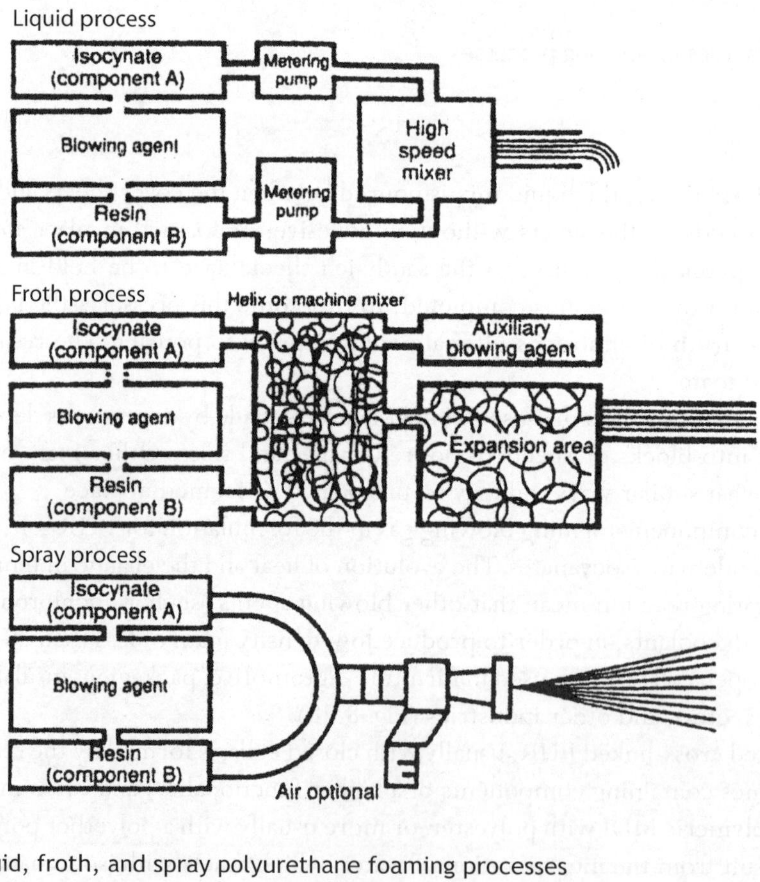

Figure 8.27 Liquid, froth, and spray polyurethane foaming processes

A variation of the PUR foam-in-place method is foam frothing. A liquid mixture of PUR chemicals is dispensed as a partially expanded froth. Frothing is achieved by using a mixture of blowing agents in the basic mix to give a two-step blowing action. The first expands the mix into froth and cools, delaying the second (and final) expansion for about thirty seconds.

With froth pour-in-place, the additional blowing agent is used in the system to create a substance more like shaving cream from the mixed liquid before it actually reacts and becomes rigid foam. A common use for froth is in large open-pour panel production, where the fixtures or jigs may be loosely constructed and the thicker liquid has no tendency to run or leak from the tool. A benefit of the process is that the material flow is slow and uniform, preventing voids in the product.

Controllable and safely operating dispensers are used to process PUR. They are designed to meet the precision mixing actions required when using these highly reactive chemicals. Ease of handling and safe pouring are required. When dispensing into open molds or large cavities, pour-in-place requires

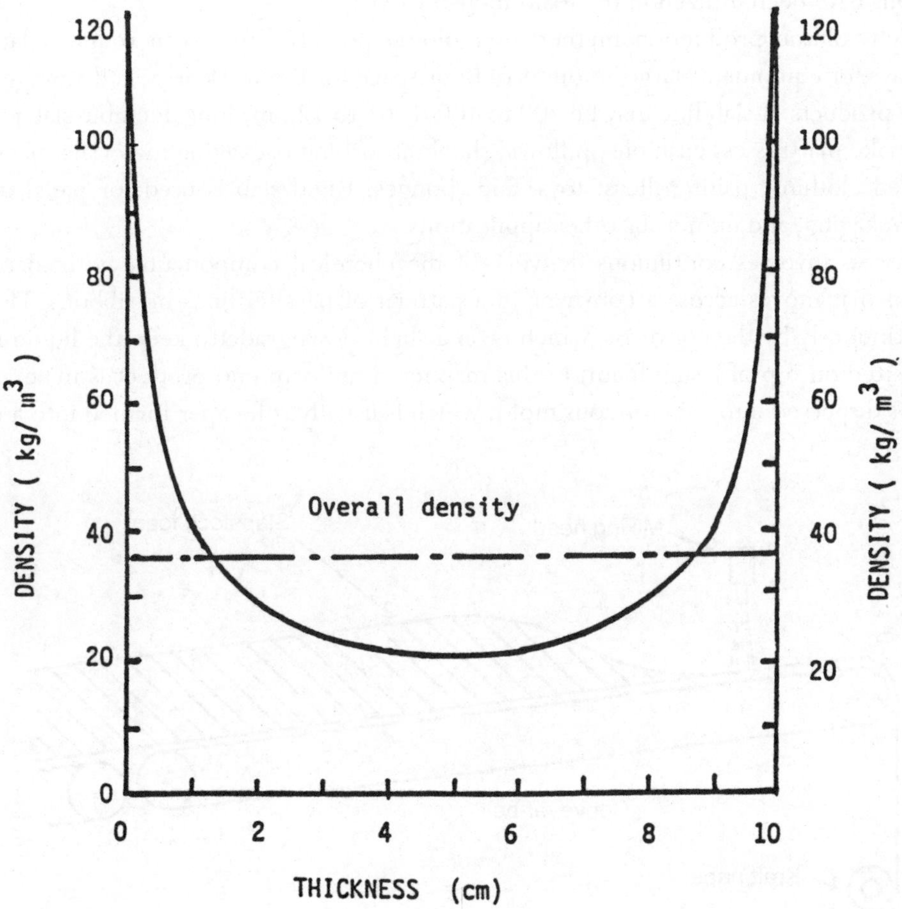

Figure 8.28 Density profile of molded flexible foam

reduced output, splatter-free mixing heads for the best results. The reaction of the mixed liquids takes place in the mold or cavity, which can be left open or closed after the shot. Open-pour is used with rigid foam systems for various molding operations and continuous panel production. Numerous flexible foam products are also produced using the open-pour method, such as foam bun stocks.

SLABSTOCK MOLDING

This is a very popular continuous free-rise foaming process providing large amounts of useful foamed products. It is called slabstock foam and a cut-off segment of the slabstock is called block foam or bun foam. The process involves the continuous pouring of liquid foaming components on a moving belt or between moving belts (Figs. 8.29 to 8.31). Different cross-sectional shapes can be formed. The cross-section of slabstock foams produced by the horizontal conveyer process is not exactly rectangular and has a slightly rounded top surface, which results in lowering the yield of fabricated foams. Accordingly, some improved processes for producing foams having rectangular cross-sections have been utilized in the foam industry (212).

This high-volume production method of molding polyurethane foam requires large initial capital outlays for equipment, large amounts of floor space for the work area, and storage space for the finished products. A slab line may be 200 to 400 ft (61 to 122 m) long. Flexible slab production is used to make mattresses, cushions, pillows, shock-absorbing packaging materials, peeled goods for laminated clothing, paint rollers, toys, and sponges. Rigid slab is used for panel insulation, flotation, packaging, and numerous other applications.

The process involves continuous delivery of the chemical components, at fixed rates, to a mixing head that moves across a conveyor in a pattern of parallel lines or ribbons. The foam is moved continuously by the conveyor, which is on a slight downgrade to keep the liquid mix from being deposited on top of rising foam. In this manner, a uniform end product is maintained. The liquid mix is dispersed into a continuous mold, which is usually roll paper formed into a trough by

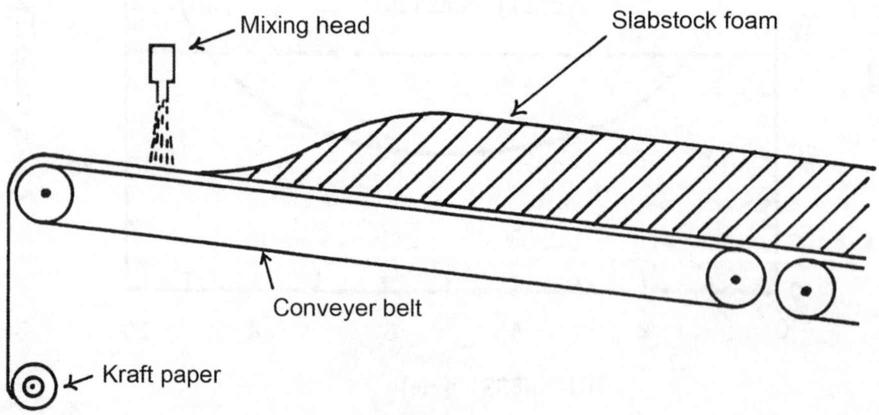

Figure 8.29 Continuous production of slabstock foam

the conveyor and adjustable sideboards. From the conveyor, the foam is moved through a curing oven. Flexible foams are usually moved through crushing rolls, although crushing is not necessary with most of the one-shot urethane formulations. However, more uniform load-bearing properties are obtained if the foam is flexed. Most one-shot foams can be flexed as soon as they are tack-free (about fifteen to thiry minutes after formation). Foams catalyzed with stannous octoate alone are slower setting and cannot be flexed for at least one hour after pouring.These foams are sometimes held as long as sixteen to twenty-four hours before flexing.

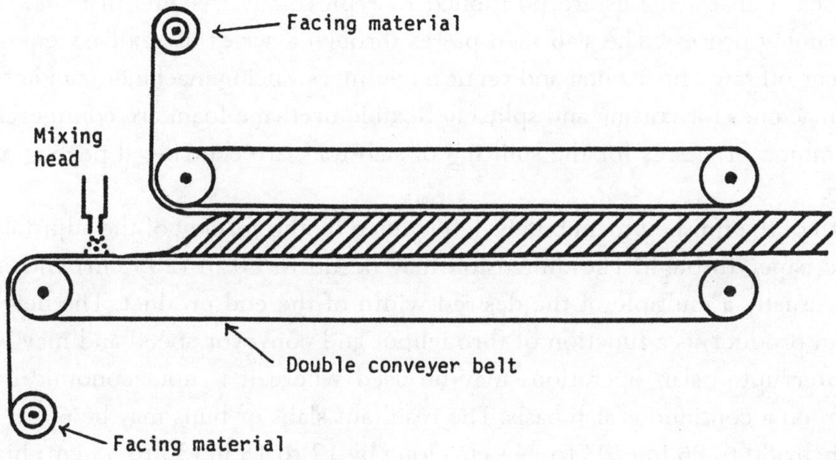

Figure 8.30 Continuous production of laminates

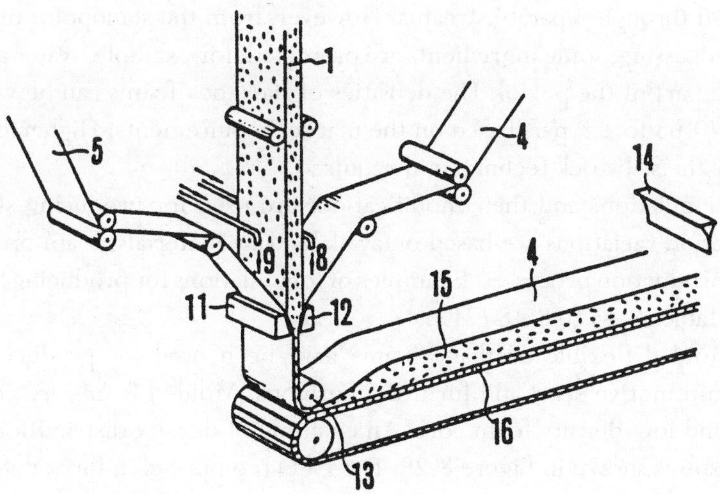

Figure 8.31 Continuous two-dimensional lamination process patented by Ashida (Japan)

Temperature control of the raw materials used in slab production is extremely important. An entire production run can be ruined if the temperature of one of the materials is allowed to change, causing a pressure change and off-ratio output. The density and other properties of slab foam are dependent upon the temperature of the foam ingredients. Foam density increases as the temperature of the foam intermediates exceeds 90°F (32°C). The temperatures of the ingredients are usually kept between 70° and 90°F (21° and 32°C).

Continuous mixers are usually used with high- or low-pressure machines to produce outputs of 100 lb (45.4 kg) or more per minute. The usual range is 150 to 200 lb (60 to 90.7 kg). After the foam has risen, it is usually passed under curing lamps, heaters, or low-pressure steam to eliminate the surface tack. Exhaust hoods are positioned to remove any irritants that may have volatized during the foaming process. The slab then passes through a series of auxiliary equipment, which may include cut-off saws, horizontal and vertical trimmers, slicing machines, and hot-wire cutters. A variety of machines for cutting and splitting flexible urethane foam are commercially available. The most common processes for the splitting of slabstock are centrifugal peeling and horizontal table splitting.

The width of the finished foam bun is determined by the spacing of the adjustable sideboards that bank the paper through. The dimension may be up to 80 in (203 cm) and, for economic reasons, it is usually a multiple of the desired width of the end product. The height of the bun (finished foam product) is a function of throughput and conveyor speed and may be up to 40 in (102 cm). Interrupted slab operations may be used where it is not economically practical to produce foam on a continuous slab basis. The resultant slabs or buns may be 60 to 72 in (152 to 183 cm) wide by 80 to 96 in (203 to 244 cm) long by 12 to 15 in (30 to 38 cm) high. While in a continuous slab operation the first few feet of material can be sacrificed, such is not the case in the interrupted slab operations.

In the case of machine foaming, a one-shot process is widely used. All ingredients are pumped into the mixing head through separate streams. However, from the standpoint of viscosity, mixing ratio, and ease of processing, some ingredients are premixed, for example, water and amine catalyst or tin catalyst and a part of the polyol. The densities of slabstock foams can be varied in the range of 10 to 30 kg/m^3 (0.63 to 1.9 pcf) based on the market requirements. Higher-density foams also can be produced by the slabstock technique, if required.

A variety of formulations and their modifications are used for producing slabstock urethane foams. The formulation variations are based on available raw materials, foam properties required, product costs, and production processes. Examples of formulations for producing free-rise slabstock foam are shown in Tables 8.45 to 8.48 (212).

Molded Foam Molded flexible urethane foams have been used for producing intricate foam products, such as automotive seats and furniture cushions. Molded foams are composed of high-density foam skin and low-density foam core. An example of density distribution of a 10cm-thick molded mattress foam is shown in Figure 8.28. The CFD (compression force deflection) as well as IFD (indentation force deflection) curves of these foams are relatively linear in comparison with slabstock foams, as shown in Figures 8.32 and 8.33.

	Weight Percent
Polyol — Glycerol-based poly(oxypropylene) poly(oxyethylene) polyol, 3500 mol.wt.	68.1
Isocyanate - TDI (80/20)	23.5
Blowing Agent - CO_2 produced by isocyanate reaction with water	1.7
Auxiliary blowing agent - Fluorocarbon	5.5
Catalyst - Organotin salt	0.2
Catalyst - Tertiary amines	0.3
Surfactant - Silicone copolymer	0.7

Physical Properties	US Unit	SI Unit
Density, lb/cu.ft. (kg/m^3)	1.80	29
ILD		
25% R, lb/50 in^2 (N/323 cm^2)	27.	120
65% R, lb/50 in^2 (N/323 cm^2)	53	236
Sag factor	1.96	1.96
Tensile strength, lb/in^2 (N/m)	12.0	83.
Elongation %	200.	200.
Tear strength, lb/in (N/m)	1.6	280
Compression set, %	7.5	7.5
Rebound, %	50.	50.

Table 8.45 Formulation of PUR slabstock without a flame retardant

In the molding process, mixed ingredients are poured through a mixing head into a preheated mold made of aluminum, steel, or epoxy resin. The mixed ingredients expand in the mold, and the resulting foams are kept at the required curing conditions, followed by demolding. This molded foam process can be classified into two types, namely, the hot-molded foam process and the cold-molded foam process. The molding temperature and oven temperature classify these two processes.

HR foam (high-resilient foam) is sometimes classified as cold-molded foam because it can be molded at low temperatures. However, HR foams have slightly different formulations than standard cold-molded foams. Cold-molded foams are prepared by using formulations based on blends of MDI and TDI, or 100% MDI; the co-use of MDI with TDI makes it possible to use low oven and mold temperatures owing to the faster reaction of MDI. The MDI can be pure MDI, modified liquid MDI, or polymeric MDI.

Some confusion may arise between cold-molded foam and high-resilience (HR) foam.

HR foams have been prepared by using almost the same molding conditions as cold-molded foams, that is, low mold temperature using the MDI-TDI blend formulation. However, in the development stage of HR foam, only TDI-based formulations were used, which required hot-mold

	Weight Percent
Polyol – Glycerol-based poly(oxypropylene) poly(oxyethylene)polyol, 3500 mol.wt.	62.9
Isocyanate – TDI (80/20)	23.5
Blowing agent – CO_2 produced by isocyanate reaction with water	1.7
Auxiliary blowing agent – Fluorocarbon	6.0
Catalyst – Organotin salt	0.2
Catalyst – Tertiary amines	0.3
Surfactant – Silicone copolymer	0.7
Flame retardant – Chlorinated phosphonate ester	4.7

Physical Properties	US Unit	SI Unit
Density, lb/ft² (kg/m³)	1.75	28
ILD		
25% R, lb/50 in² (N/323 cm³)	24	107
65% R, lb/50 in² (N/323 cm³)	49	218
Sag factor	2.04	2.04
Tensile strength, lb/in² (kPa)	12	83
Elongation, %	200	200
Tear strength, lb/in (N/m)	1.6	280.
Compression set, %	7.5	7.5
Rebound, %	53	53

Table 8.46 Formulation of PUR slabstock with a flame retardant

Component A:	
Polymeric isocyanate (NCO% : 31)	70 pbw
Component B:	
Polyether polyol (OH No, M.W. 3,000)	95.
Quadrol	5.0
Water	3.0
TMBDA	0.5
Dabco T-12	0.03
Surfactant, L-5310	2.0
Properties:	
Density, kg/m³	54.
Tensile strength, kg/cm²	1.27
Elongation, %	50.

Table 8.47 One-shot semirigid foam formulation

Foams	Conventional	Super Soft	High Load-bearing	High Resilient
Formulation (phr)				
Polyether Triol	Conventional (OHV:56) 100phr	EO-capped (OHV:50, 75% EO) (70 phr) Conventional (30 phr)	AN-grafted (OHV:31) (50 phr) Conventional (50 phr)	ST-grafteed (OHV:27) (50 phr) Conventional (50 phr)
TDI (80/20)	51.	41.	37.	37.
Water	3.9	3.5	2.8	2.7
Tertiary amine	0.1	0.2	0.25	0.18
Silicone surfactant	0.1	1.8	1.0	1.5
Stannous octoate	0.25	0.08	0.20	0.10
CFC-11	0	12.0	0	0
Properties				
Density kg/m³ *	25.	20.	30.	35.
25% Hardness, kgf *	12.5	1.5	20.5	9.0
Elongation, % **	205.	308.	165.	160.
Ball rebound, % *	42.	40.	38.	60.

* JIS-K-6401,

** JIS-K-6301. (JIS stands for Japonese Industrial Standards.)

Table 8.48 Formulations and properties of various flexible foams

conditions. For these reasons, HR foams are not exactly equal to cold-cure foams. More recently, HR foams have been produced by blends of MDI and TDI. The resulting foam systems can be molded at the same molding conditions as those for cold-molded foam systems.

In comparison with cold-molded foams, hot-molded foams have disadvantages of higher mold temperature and longer mold-retention times, but they have the following advantages: higher ratios of load-bearing/density, lower costs, and easier molding of complicated products. HR foam is characterized by high ball rebound, low hysteresis loss, and high sag factor. HR foam can be produced either as slabstock or as molded foam. They usually contain closed cells that become causes of foam shrinkage. In general, if this situation is not desired, immediately after demolding, the foams are passed through a crushing roll to break the closed-cell membranes and are then cured in an oven. The choice of proper surfactant for HR foam is also an important solution to the problem of shrunk or collapsed foams.

Specialty polyisocyanates are used to provide advantages such as high resilience, short tack-free time, and higher cross-link density. The formation of open-cell, closed-cell, or collapsed foams is reviewed in Figure 8.34. It shows that open-cell foams can be obtained only by a good balance between two reactions, that is, polyurethane formation and gas generation. When gas generation is too fast, the rising foam may collapse, like beer bubbles, because the foam is not stable.

In contrast, when polyurethane formation is too fast, the resulting foam cells are stable, cell membranes are not broken during foam rise, and closed cells are formed. The closed cells are the causes of shrunk foams because the internal gas pressure becomes lower than the atmospheric pressure.

Microcellular Elastomer Foamed urethane elastomers are also called microcellular elastomers. The densities are in the range of about 20 to 60 pcf (320 to 960 kg/m^3). Integral-skin foams having densities of 700 to 1,000 kg/m^3 are sometimes referred to as microcellular elastomers.

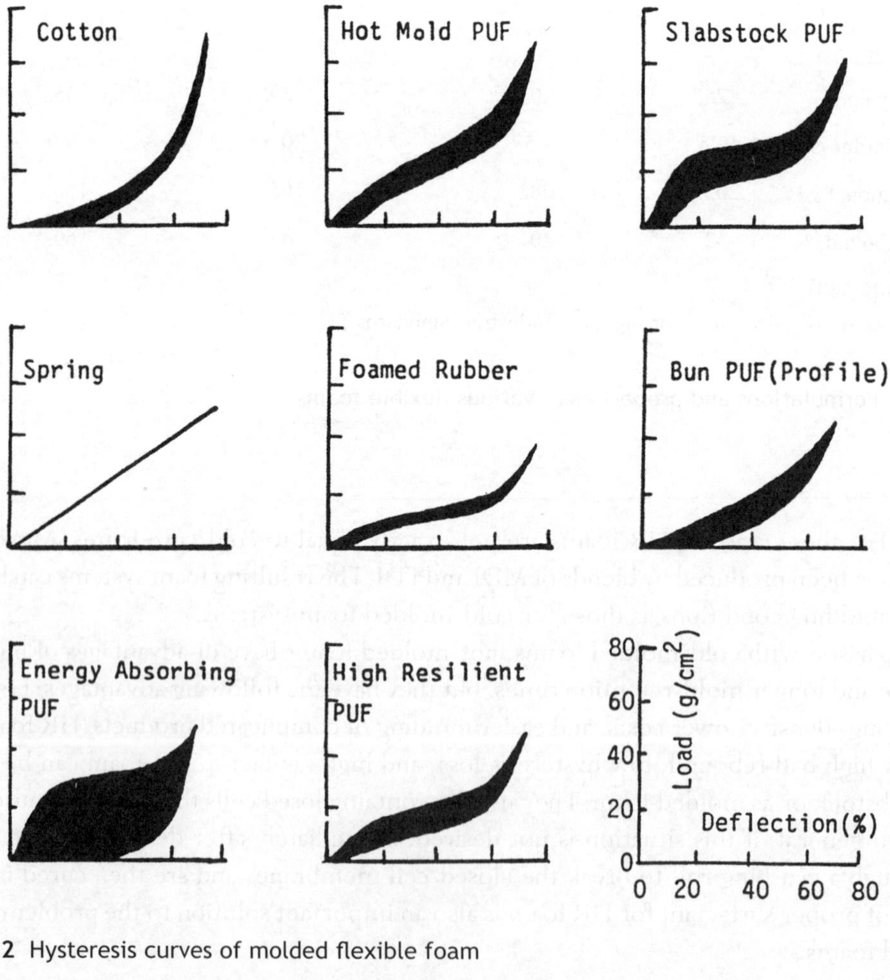

Figure 8.32 Hysteresis curves of molded flexible foam

The major polyols for microcellular elastomers include aliphatic polyester diols having a molecular weight of about 1,000 to 3,000, and poly-epsilon-caprolactones. Poly (oxytetramethylene) glycols (PTMEG) can also be used. The polyisocyanates to be used for microcellular elastomers are TDI prepolymers and liquid MDI, that is, carbodiimide-modified MDI or urethane-modified MDI. Low molecular weight active hydrogen compounds, such as chain extenders (difunctional compounds) and cross-linkers (at least three functional compounds), are used with the above-mentioned high molecular weight polyols. Examples of chain extenders are glycols (1,4-butanediol and ethylene glycol) and aromatic diamines, such as 4,4'-methylene bis (2-chloroaniline) (MBCA, formerly MOCA). MBCA results in excellent elastomers, but it is a suspected carcinogen; therefore, its use should be carefully monitored.

The blowing agent for microcellular elastomers is water. The amount of water should be accurate and its accuracy can be obtained by a water-containing solution, such as liquid sodium sulfonate (used in of vegetable oils). The catalysts to be used are those used in urethane foams, for example, tertiary amines and tin catalysts. The above ingredients are mixed and poured into a hot

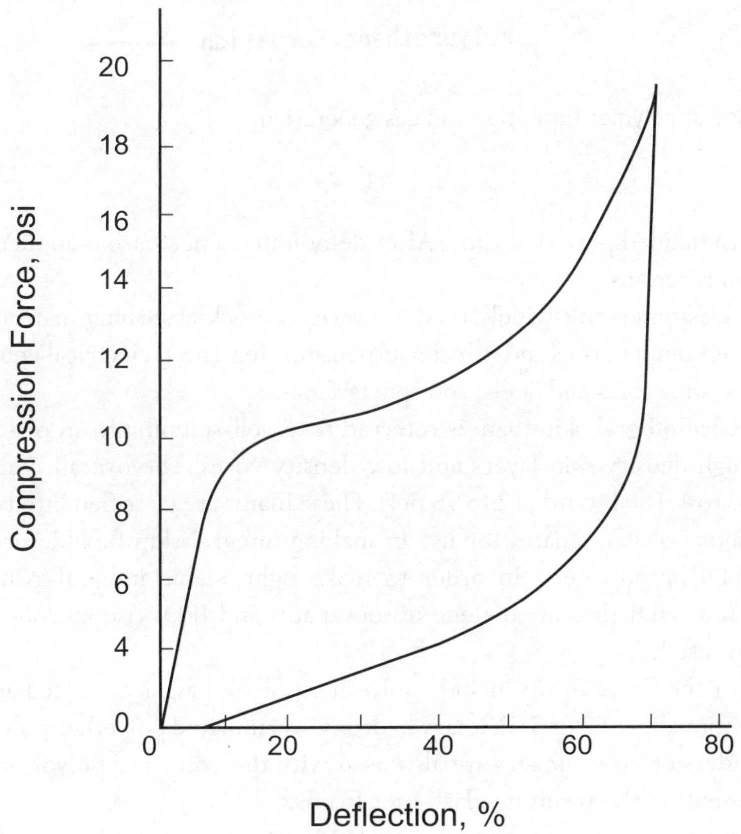

Figure 8.33 Hysteresis curves of molded semirigid foam

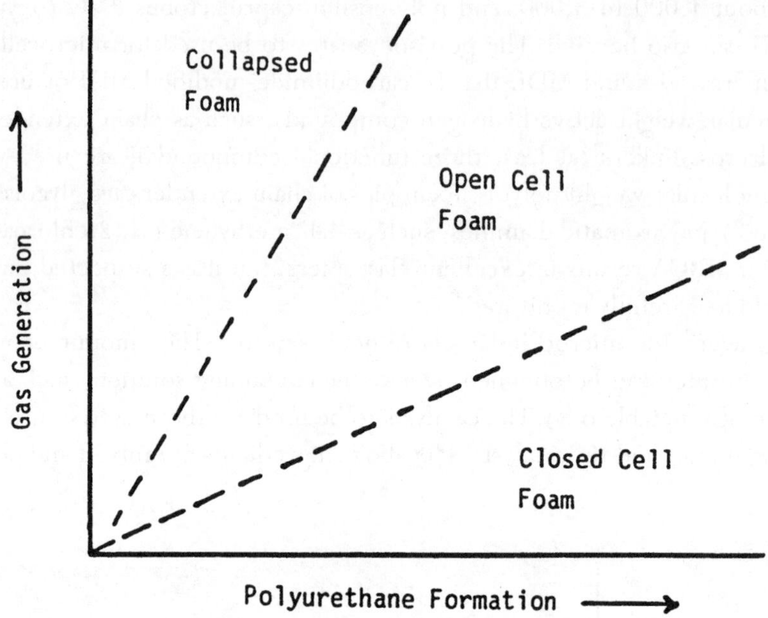

Figure 8.34 Balance of polymer formation and gas generation

mold and cured in a defined period of time. After demolding, a postcure is applied to complete the polymer formation reactions.

Microcellular elastomers are widely used for various shock-absorbing materials in automotive applications (such as bumper cores and vehicle-suspension elements), electrical equipment, cameras, precision machines, shoe soles and heels, and sports shoes.

Integral-Skin Foam Integral-skin foam is referred to as self-skinning foam or self-skinned foam. The foams have high-density skin layers and low-density cores. The overall densities vary in the range of about 200 to 1,100 kg/m^3 (12 to 70 pcf). These foams are classified into two types, flexible and rigid. The major polyisocyanates for use in making integral-skin flexible urethane foams are liquid MDI and TDI prepolymers. In order to make light, stable integral-skin foams, aliphatic diisocyanates, such as HDI (hexamethylene diisocyanate) and IPDI (isophorone diisocyanate), in modified forms are used.

Polyols used for the above foams include polyether polyols having a molecular weight of 2,000 to 8,000 and a functionality of 2 or 3. Primary hydroxyl-terminated polyether polyols are preferably used. Chain extenders or cross-linkers are also used with the polyether polyols for the purpose of forming hard segments in the resulting elastomer foams.

The major blowing agents for making integral-skin foams are nonreactive solvents having low boiling points, such as trichloromonofluoromethane (boiling point is 24°C). Such solvents do not evaporate on the cold mold surface and result in high-density-resin skin. The thickness of the skin is controlled by the difference between the mold temperature and the temperature of the foaming mixture inside the mold, as well as the foaming pressure generated by the overpacking of the foaming system into the mold.

Water is not suitable as a blowing agent for making solid skins; for this reason, to date it is used in only a few cases. Due to the ban on the use of CFCs, other methods are being studied. One method is 100% water-blown integral-skin foams. Another method is the use of inert solvents other than CFC-11, such as hydrocarbons. The typical catalysts for making integral-skin foams are tertiary amines such as triethylene diamine, organotin salts such as dibutyltin dilaurate, and mixtures thereof. Additives, such as pigments, flame retardants, and so on, can also be incorporated in the foaming systems.

The foam densities of integral-skin foams of commercial products are in the range of about 300 to 950 kg/m^3 and their Shore A hardnesses are in the range of about 90. Shore D hardnesses are about 40 to 90. Figure 8.35 provides an example of density distribution of integral-skin foams.

LAMINATING

One of the largest uses for polyester-type flexible urethane foam is in foam laminates for clothing interlinings. The first step in preparing the foam for clothing lamination is to take slabstock, cut it into roughly square logs, and peel them as one would peel wood log for veneer. Thicknesses of about 8/32 in (0.64 cm) and 3/32 in (0.24 cm) are generally used. The foam can be butt joined easily by heat to form large continuous rolls of foam.

There are basically three methods of producing foam laminates commercially. The first is the heat-fusion or flame-lamination process, which is most popular in the United States. In this process, controlled naked gas flames are allowed to impinge upon a traveling layer of foam. This produces a layer of molten urethane resin by micromelting the upper portion of the foam (about one-third the thickness). The fabric is immediately pressed against the tacky surface and adheres firmly. Good peel strength is obtained. The laminate can be rolled up immediately and prepared for use. No additional adhesive is required. At one time, only polyester foams could be utilized for flame laminations, but now polyether types can also be used. To make them susceptible to flame lamination, certain organic phosphorus compounds are added. Polyether foams are preferred in soft-clothing interlinings.

In most textile applications, a cloth lining must be sewn against the foam side. A comparatively new adaptation of flame lamination is now being used to obviate this step. In this method, two fabrics form a sandwich with a very thin foam interlining. One fabric is flame laminated to one side of the foam, followed by flame laminating of another fabric to the other side. After flame laminating, so much of the foam is used up that in reality it is only acting as an adhesive. In this way, a liner is adhered in place, rather than sewn, with little or no loss in breathability, with gain in warmth, and with practically no gain in weight. Foams as thin as 40 mils (1.0 mm) are being used. Polyester foams are preferable for this process. Polyether foams are difficult to peel in such thin layers. They also have low tear strength and a tendency toward electrostatic buildup. This results in the foam clinging to the cutting blade and bunching up. The need for antistatic bar protection is pronounced.

The second widely used method for preparing polyurethane foam laminates is the liquid adhesive or wet process. Here, special adhesives, either in the form of water solutions or as solutions in organic solvents, are applied to either the fabric or the foam. The equipment is conventional. The water or solvent is evaporated, and the bond may be set by drying or curing at elevated temperatures. One of the adhesives used for this application is based on acrylic interpolymer latices. This adhesive is used in the manufacture of thermal garments and insulated bags. Carpet underlay can also be made by adhesive laminating.

The third method, the so-called frothing technique, consists of foaming the material directly onto the base materials, such as films or impervious fabrics. The previously unsolved problem of

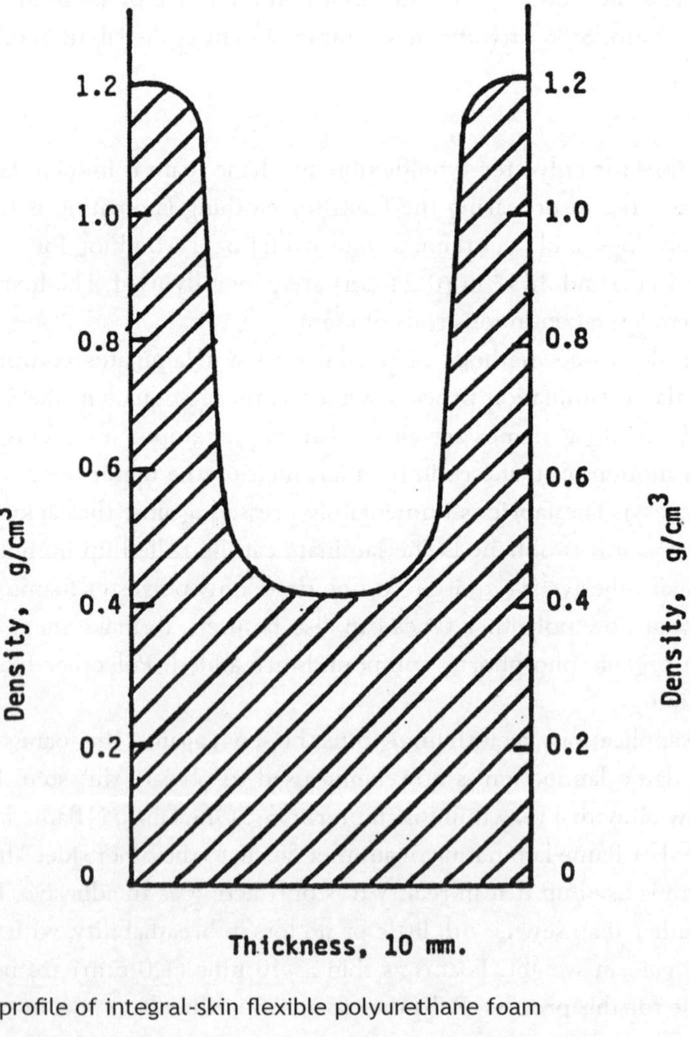

Figure 8.35 Density profile of integral-skin flexible polyurethane foam

controlling strike-through when foaming on fabrics has been solved by simply wetting the fabric with ordinary tap water. Almost any fabric can be used, although the technique is more difficult with the more porous fabrics, which may require the use of thickeners. The fabric is usually immersed in water, after which the moisture level is controlled by squeegeeing or heating. After the foam has been poured onto the textile, the entire laminate is passed through an oven. This serves the dual purpose of curing the foam and drying out the fabric.

APPLICATION

As reviewed throughout this chapter, growth continues in the use of foamed plastics in different markets using polystyrenes, polyurethanes, vinyls, and polyethylenes. Other resins, such as epoxies, phenolics, cellulose acetates, silicones, and urea-formaldehydes, are being used in limited and generally special applications. Foam can easily be made into a package so that a "snug" fit exists around its content. These materials can provide a package that can be used for shipment, storage, display, and/or functional for such parts as dishes, instruments, medicine, electronics, bowling balls, and computers. Foaming equipment is produced to simplify and/or reduce cost to foam certain products. An example is the flexible foam production line for refrigerators developed by Bayer's Hennecke Machinery Division (408).

Low-density (≈30 kg/m^3) flexible plastic foams are used in applications such as furniture and automotive seating; high-density foams are required for structural, load-bearing applications. Mechanical properties of conventional foams are lower than in solid parts because of a lower material content; however, microcellular foams with a pore or cell size of 1 to 10 μm provide good mechanical properties in addition to weight reduction.

Almost 45% of the total flexible foam produced, of which 70% is polyurethane, is used in furniture cushioning and upholstery. Another major outlet is transportation, which accounts for about 15% of flexible (one half PVC, the other half polyurethane) and 15% of rigid polyurethane. They are used mainly in automobiles for seating, crash pads, arm rests, carburetor air filters, sun visor pads, thermal and sound insulators, gaskets, and door panels.

Rigid urethane is recognized by tank car and trailer manufacturers to be the most efficient insulator. This large market permits increased storage capacity and reduced weight in refrigerated units. Use of foam insulation is in refrigerated warehouses and pipes. Pipe insulation can be made by spraying, foam-in-place, or from slabstock and shaped stock.

Other major uses for flexible foam are in mattresses (mostly polyurethane), interlines for clothing, shoes and accessories, carpet underlays, packaging, and wire and cable insulation. In addition to the rigid foam going into the transportation market, major outlets include building and construction (polystyrene and polyurethane), packaging for cups and dishes, appliances, and marine (Tables 8.49 and 8.50). Roofing insulation provides a big market for rigid (primarily polystyrene and some polyurethane) foam.

Overall, plastic foams include use in packaging, automotive applications, furniture, sporting goods, carpet underlay, textiles, gaskets, heat and sound insulation (tanks, buildings [Figs. 8.36 and

Type	Density g/cc	(lbs./ft.^{3})	Compressibilities 10^{-6} atm^{-1}	(Pa)	Useful operating depths. ft.	(m)	Net buoyancy lbs./cu. ft.	(kg/m^3)
Solids								
Polypropylene	0.90–0.92	(56–57)	20–80	(2.0–8.1)	0–38,000	(0–11,600)	7–8	(110–130)
Lithium	0.53	(33)	9	(0.91)	2,000–38,000	(610–11,600)	17–25	(270–400)
Ice	0.91–0.92	(57–57.4)	7–30	(0.7–3.0)	Surface		7–8	(110–130)
Cellular								
Wood	0.20–0.70	(12–44)	20–100	(2.0–10.1)	0–700	(0–210)	20–48	(320–770)
Syntactic foam	0.60–0.80	(37–50)	30–120	(3.0–12.2)	500–9,000	(150–2,700)	6–20	(96–320)
Plastic foam	0.10–0.80	(6.2–50)	50–1,500	(5.1–150)	0–2,000	(0–160)	12–40	(190–640)
Rigid shell								
Steel	0.15–0.80	(9–50)			0–5,500	(0–1,700)	10–50	(160–800)
Aluminum	0.15–0.80	(9–50)			0–8,000	(0–2,400)	10–55	(160–880)
Titanium	0.25–0.80	(16–50)			5,000–10,000	(1,500–3,000)	10–45	(160–720)
Reinforced plastic	0.15–0.60	(9–37)			0–25,000	(0–7,600)	20–50	(320–800)
Glass	0.15–0.40	(9–25)			0–25,000	(0–7,600)	20–50	(320–800)

Table 8.49 Syntactic foam compared to other buoyant materials

Foam density lb./ft.^{3}	Glass microballoons	Epoxy macroballoons	Uniaxial compressive yield strength. psi	Hydrostatic compressive strength, psi	Method of preparation	Resin system
24	Yes	Yes	1,700	1,800	Pack-in-place	Polyester
30	Yes	Yes	2,500	3,000	Infiltration	Polyester
35	Yes	No	10,500	14,000	Vacuum cast	Epoxy
38	Yes	No	10,000	14,000	Cast	Polyester
42	Yes	No	10,000	15,000	Cast	Polyester

Table 8.50 Syntactic foam performance in deep-water flotation

8.37], etc.), flotation products, and structural applications. They are used in concrete insulating blankets as well as overwinter protection for plants, flotation booms for oil and chemical spills, and protection for buried pipelines. Foams are used to provide cushioning and surface protection in the packaging of electronics, furniture, appliances, and agricultural and construction products. Low-density foams (> 90% voids) are generally used for sound or heat insulation; medium-density foams (50% to 90% voids) are commonly used in packaging. High-density foams (< 50% voids) are used in structural applications.

Properties useful in packaging and automotive applications include shape retention and the ability to withstand repeated impacts without breakage or a decrease in properties. Impact cushioning is especially important in the packaging of heavier products, such as appliances,

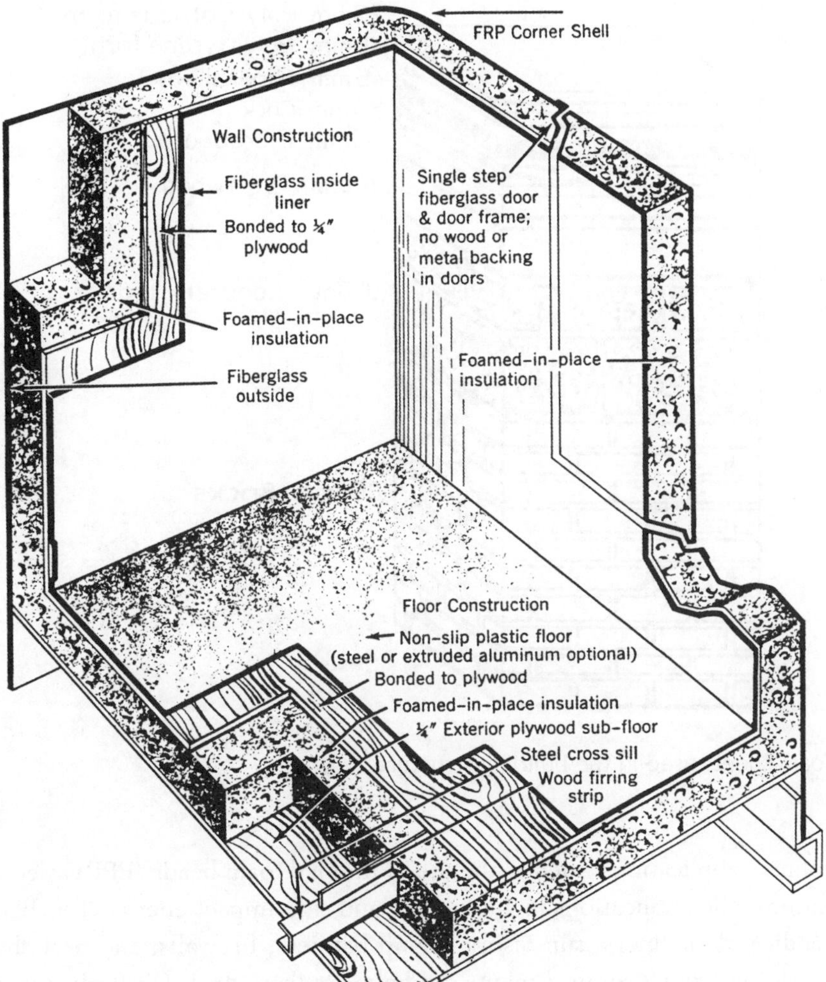

Figure 8.36 Polyurethane foamed insulated wall of a house

electronic components, and computers. As an example, the interior chassis of an auto workstation used EPP foam to provide cushioning and structural support. EPP replaced conventional metal and plastic chassis structures that required assembly, resulting in a 50% reduction in assembly time and a 30% reduction in weight, and the foam protected the computer components from impact damage during shipping.

PP foams find greater use in automotive applications than LDPE owing to the ability to withstand higher service temperatures; LDPE foams can only be used at service temperatures of up to 80°C (176°F). The use of PP foams in automotive applications continues to increase with the

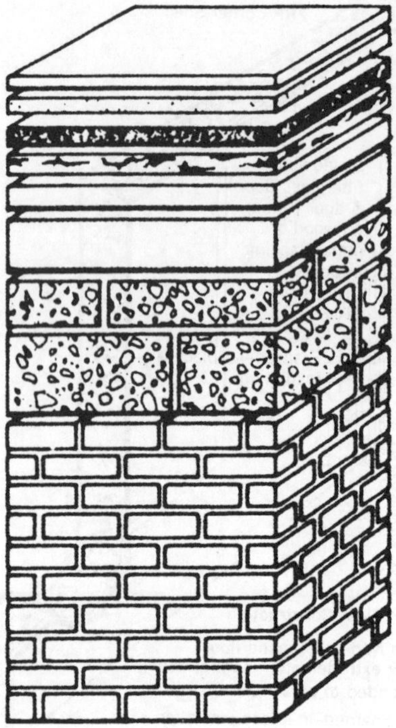

Figure 8.37 Foam sheets used in the building structure

development of new applications, such as the use of molded foam beads (EPP) as energy absorbers in bumpers. Automobile applications include head- and side-impact energy absorbers, instrument panels, tool caddies, door liners, sun visors, sound barriers, firewalls, and heat shields. Bumper cores, door panels, and glove compartments can be made from dual-density foams, in which low- and high-density foams are molded together in one operation, replacing mono-density foams that require thermoplastic inserts at high stress areas. Glove boxes use a talc-reinforced foam and a thermoplastic polyolefin skin to provide impact protection.

Automotive interior trim and other parts, such as bicycle helmets, can be produced with a solid integral skin on one surface that does not need a separate covering. The bicycle helmet has a leather-grain texture on the outer skin surface, while the usual bead-foam appearance shows on the inside and around the edges of the helmet. Other applications include hairbrush backs, shoe heels, toothbrush backs, and steering wheels. Processes and equipment range from simple operating to complex types (Figs. 8.38 to 8.40).

SHEET AND FILM

Extensive use is made of polystyrene (PS) and polyethylene (PE) foamed sheet and film. They are produced using extruders (chapter 5) and calenders (chapter 9). Their densities have a wide range, such as 3 to 12 lb/ft^3 (0.05 to 0.19 g/cm^3) with thicknesses of 15 to 150 mil (0.38 to 3.8 mm). Use of these materials can be in flat sheet form (building insulation, packaging, etc.). They are used to go through other processes, principally thermoforming (chapter 7). As reviewed, product applications include disposable packages, such as meat and produce trays and egg cartons, and containers and trays for carryout meals and disposable dinnerware. They are also used for drinking cups, bottle labels, and cushion-type packaging.

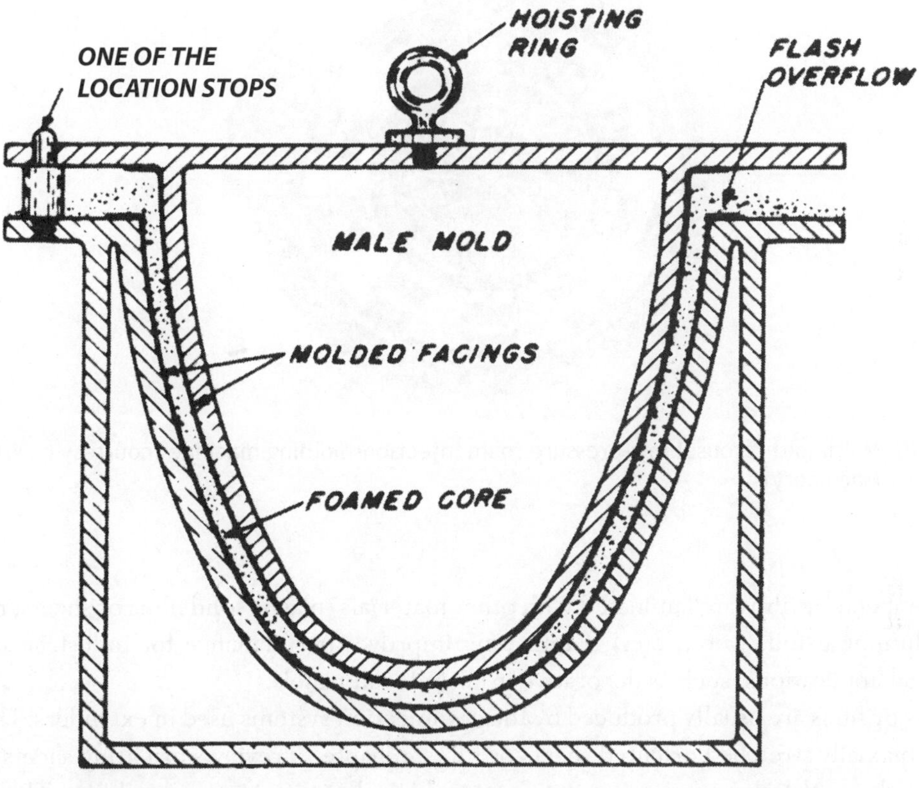

Figure 8.38 Inexpensive wood mold used for foam-in-place molding by pouring from a dual- or multicomponent mix

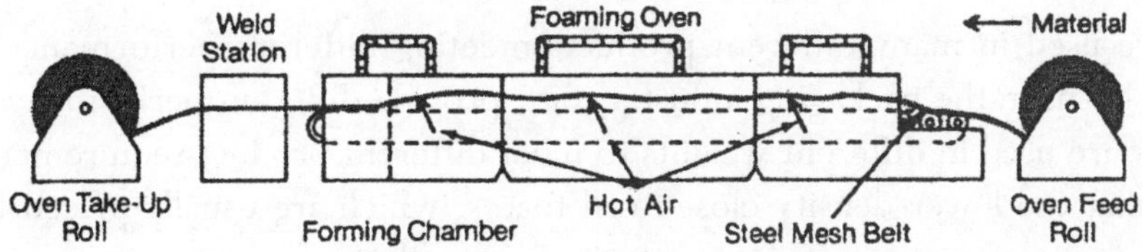

Figure 8.39 Extruded plastic blowing agent–prepared sheet is foamed going through a heating oven that can contain a thermoformer

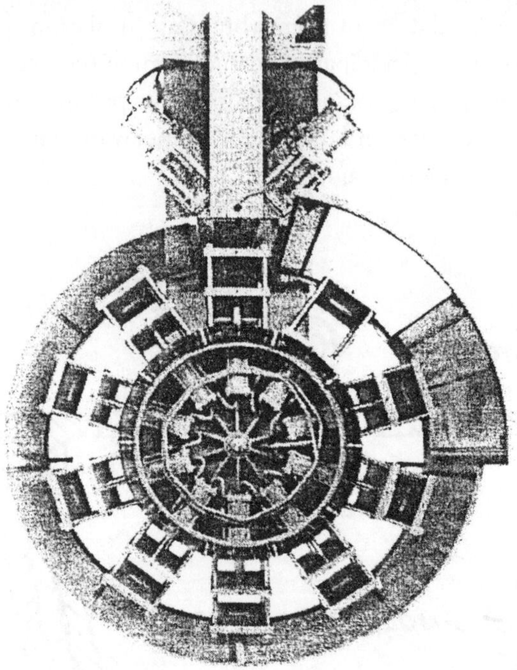

Figure 8.40 Multimold carousel low-pressure foam injection-molding machine (courtesy of Wilmington Machinery)

These sheets or films are laminated with other materials (plastic solid films or sheets, reinforced plastic, aluminum foil, wood, etc.). Laminating improves performance for industrial use and in commercial applications, such as decorative insulated wall panels.

Sheets or films are usually produced by the gas-injection systems used in extruders. The process can form biaxially stretched oriented sheets or films. (chapter 5 reviews orienting sheets or films.) Table 8.51 shows the increases in properties when biaxially stretching foamed film. This orienting processing technique has recently been used to develop new marketable products (409, 410).

Polyethylene Cushioning

Plastic foams are used in many different products meeting different performance requirements (chapters 19 and 20). In the packaging industry, they provide different performances that include cushioning. They are used in different weights to meet different product requirements. It is sometimes believed that the lower density closed-cell foams, which are usually priced lower, provide superior cushioning performance. This assumption is usually incorrect, as shown in Figure 8.41. Even though it contains less plastic, the fabricating rates, the amount and cost of the blowing agent, and the amount and cost of the base plastic all influence the final cost. As a result, very low-density

Film Properties Achieved With Biaxial Stretching Of Foamed Film

Property			Foam film (Homo PP)	Foam film* (Block-Copolymer)	Coextruded foam film (high-crystalline PP)	Test Method
Thickness		µm	65	68	80	DIN 53370
Tensile strength	MD	N/mm²	21	28	38	DIN 53455
	TD	N/mm²	29	47	48	DIN 53455
Elongation at break	MD	%	17	23	28	DIN 53455
	TD	%	12	11	18	DIN 53455
E-Modulus	MD	N/mm²	481	516	1350	DIN 53457
	TD	N/mm²	629	969	1590	DIN 53457
Gloss	T	–	87	49	20	ASTM 2457
	O		94	52	25	ASTM 2457
% Opacity	–	%	36	68	74	DIN 53146
Density	–	g/cm³	0.31	0.49	0.5	DIN 53420
Water permeation	–	g/m²/24 h	n.b.	≈13**	n.b.	DIN 53122
Heat conductivity	–	W/m-K	n.b	0.0424	n.b.	DIN 52612

Table 1
*tensile strength similar to paper
**breathable film, c. 6000 g/m²

Table 8.51 Increase in foamed film properties occurs via biaxially stretching

foams can actually be more costly to make than others. Cost of foam will not be proportional to its density.

The same type and thickness closed-cell polyethylene foams with differing densities have been subjected to dynamic static load cushioning conditions during impact. Curves relate mechanical shock experienced during an impact, where the lower the curve goes, the greater the cushioning efficiency. For curves of foam densities 6 pcf (lb/ft^3) above the maximum cushioning, efficiency of each material is not significantly different; however, a dramatic change occurs with a change in loading.

If a 40-g package were to be designed according to Figure 8.41 using a 6.0-pcf foam, the foam would measure 3 in thick, resulting in 40 g shock. If a similar package were then produced using a 2.2-pcf foam, its shock performance would not go as low as 40 g but would be 60 g, so it would endure 50% more shock or a 50% loss in shock efficiency.

To meet 40 g, the 2.2-pcf package would need to be redesigned. Greatly increasing the thickness of the pads constructed from the lower-density foam could provide adequate protection. However, the result would be increased package size, which would impair handling and shipping efficiency and possibly result in higher costs. The 6.0-pcf foam could, however, be reliably used.

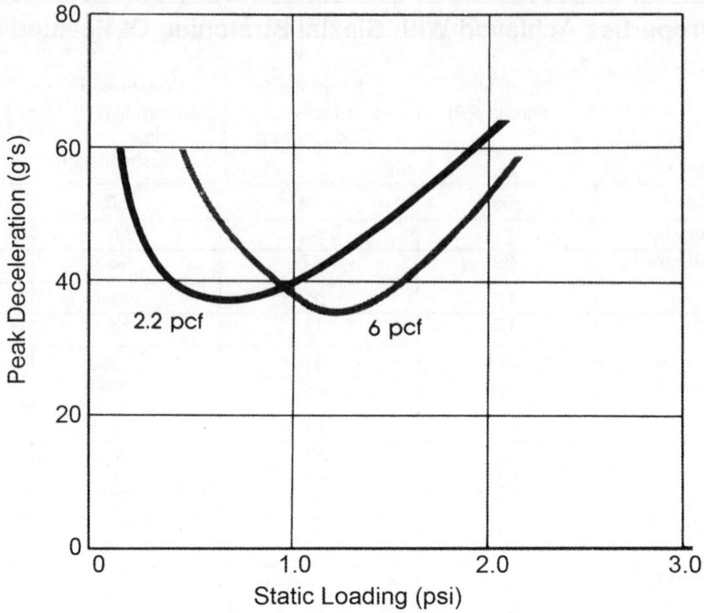

Figure 8.41 Cushioning effect of polyethylene foam density is influenced by loading

To be at the 40-g level and keep the 2.2-pcf foam thickness the same, reduce loading from 1.35 to 0.87 psi can be used. Although this approach keeps the package size the same, nearly twice as much foam must be used to meet the lower loading. Lowering the density produces a considerably higher deceleration and reduces cushioning performance. Figure 8.42 compares the effect of density. The lower-density foam costs less than half on a cost-per-unit volume basis, resulting in a cost savings. Below a density of about 2.2 pcf, the cushioning efficiency can begin to change with the density.

Significant in this figure is the narrower range of usable static loadings at the bottoms of the curves that resulted when the density was reduced. When comparing foams of different densities, it is important to consider compressive creep resistance and the ability to resist undergoing permanent thickness loss during time under loads. As the density decreases, so does the creep resistance.

PROFILE

Thermoplastic elastomer (TPE) foamed profiles currently are extensively used in industries such as automotive, appliance, and construction for sponge-sealing applications. The profiles can be coextruded with other plastics (dense and foam) as reviewed in chapter 5 (Fig. 8.43) and heat welded to create corner and ring seals (chapter 18).

STRIPPABLE

There are foam strippables, usually using vinyls that are used for protecting products, such as metal parts being packaged for shipment. Polyvinyl chloride dispersion plastic is applied in liquid form to the product. Foaming takes place during the cure cycle to produce a highly resilient spongy film. Therefore, these strippables also offer protection against denting, as well as scratching, and have taken the place of paper and corrugated wrappings at substantial savings. Types have been made available that can expand up to 300%. These can yield maximum films of about a ½ in, although a ¼ in is more commonly used.

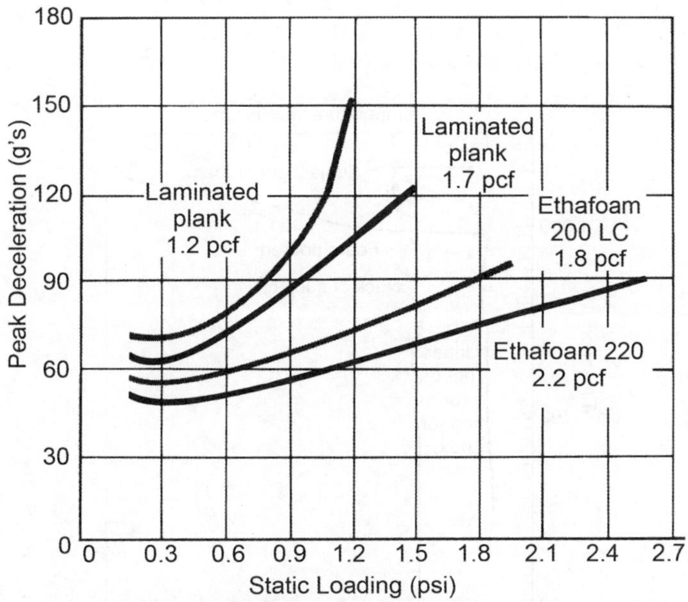

Figure 8.42 Comparison of different foam densities

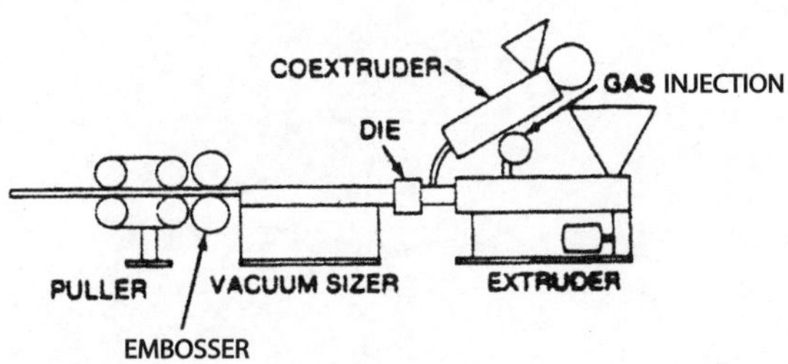

Figure 8.43 Plastic foamed profiles are coextruded to take advantage of gains over a single plastic foamed profile to meet specific increased performances

Spraying, dipping, flow, and curtain coating can apply foam-vinyl strippables over the same substrates as the nonfoam types (Fig. 8.44). Their big use is on chrome-plated automotive replacement parts, such as bumpers, headlight bezels, and decorative trim. Strippables usually boost production of wrapped parts considerably and lower their rejection rates. The output of wrapped bumpers, for example, has been increased three times.

These materials afford many other advantages. For instance, they retard corrosion by forming a tight skin around the object, which inhibits the entrance of moisture. They also help save space, since the tight fit allows more units to be stacked per cubic foot than if bulky containers were used. In addition, because one type of strippable can accommodate all sizes and shapes, there is no need to maintain a large inventory of different sized packaging materials. While auto parts packaging is one of their larger uses, other industries in which shipping metal (and even plastic) parts prevails use foam vinyl strippables.

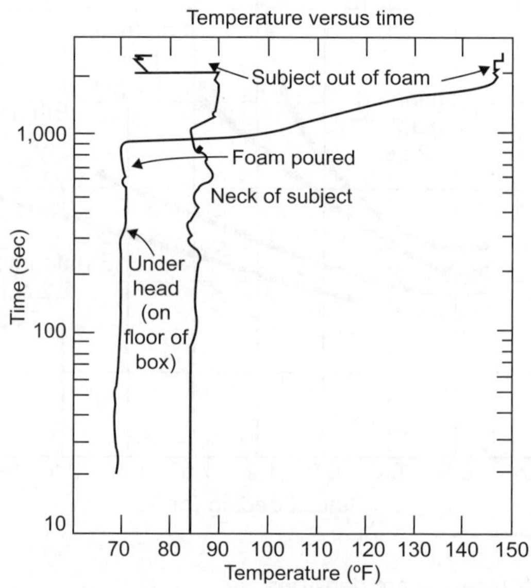

Figure 8.44 Temperature distribution in vinyl foam strippable

CHAPTER 9

CALENDERING

INTRODUCTION

Calendering dates back to the mid-eighteenth century. Figure 9.1 shows a 50-ton (with each roll weighing 3 to 4 tons) calender that operated during 1882 in an Avon Rubber Co. plant in the UK. It was installed in 1849 and operated up to about the mid-twentieth century (569). Major changes in the design of calenders occurred after 1930, following new developments in tire manufacture and, more importantly, the use of plasticized polyvinyl chloride (PVC) in the United States. However, one might say that old and new calenders have similarities, even with all the major improvements that continue to occur.

This process is used to convert thermoplastic materials into continuous sheets and films, and to apply plastic coatings to textiles, paper, and other supporting materials. When fabric, paper, or other material is feed through the rolls, plastic can be pressed into the surface of the calendered web. When operating the calendering line in this mode, it can be called a coating machine.

Calendering is an alternative to extrusion, with the usual film at three or more mils (75 microns) thick. For the production of sheet or film, plastic melt is compounded and pressed as it passes through the nips of a series of three or more heated highly polished steel rolls (Fig. 9.2). The resin "bank" is formed into a sheet in the nip between the first pair of rolls. Passing through the second and third nips further reduces the thickness. The final thickness of the sheet is determined by the gap between the last pair of rolls called the gauging rolls. Finally, a takeoff roll pulls the hot sheet around a chilled roll to cool the sheet or film (Fig. 9.3).

The revolving rollers squeeze the material between them into a sheet or film. An analogy in this case might be flattening out a pastry dough mixture with a rolling pin (Fig. 9.4). The thickness of the finished material is controlled by the space between the rolls. The surface of the plastic film or sheeting may be smooth or matted, depending on the surfacing on the rollers. When large quantities of film and sheet, particularly PVC, as well as coated fabric are to be manufactured, this process can provide lower-cost products than extrusion.

Figure 9.1 Rubber calender operating for the Avon Rubber Co., UK, during 1882

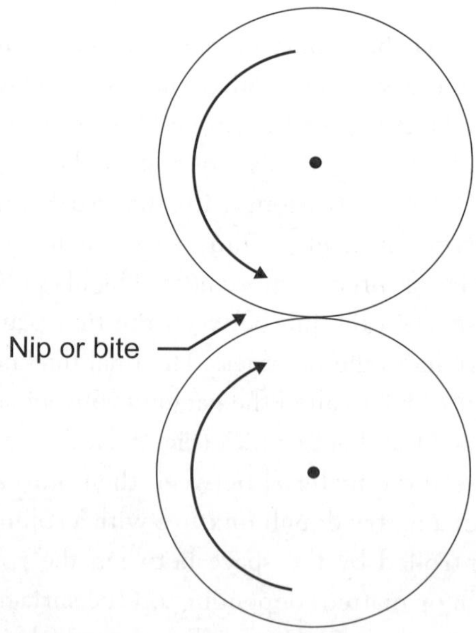

Figure 9.2 Schematic highlighting the nip section of rolls

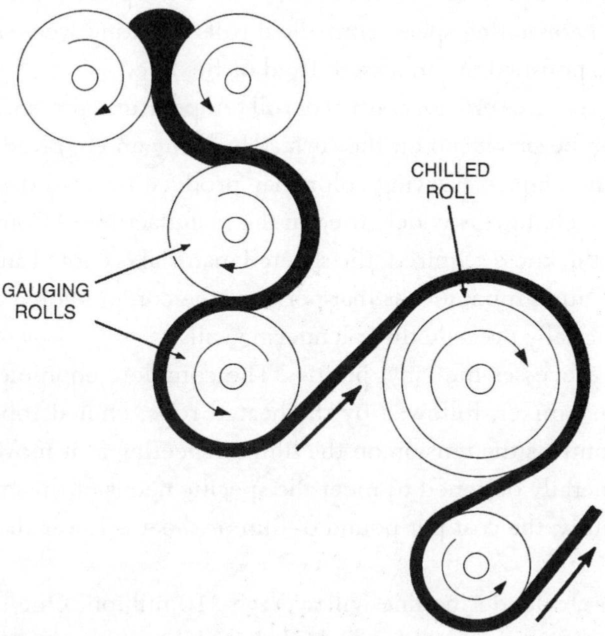

Figure 9.3 In the calendering operation, the sheet decreases in thickness while passing through a series of nip rolls

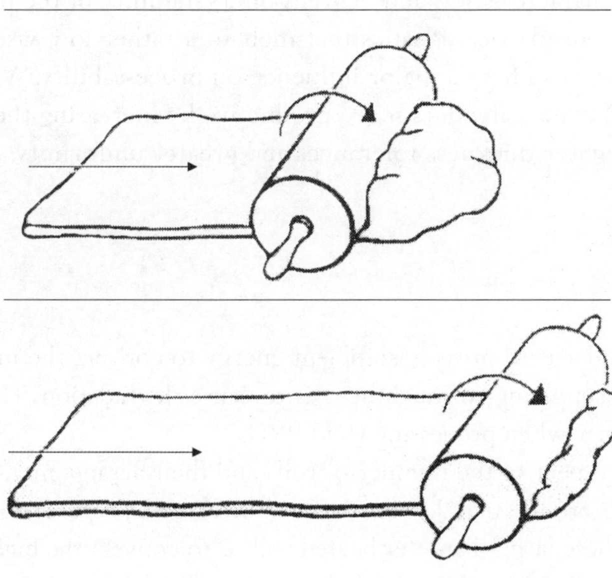

Figure 9.4 An analogy to calendering

Calendering converts plastic into a melt and then passes the pastelike melt through the roll nips of a series of heated and corotating speed-controlled rolls, creating webs of specific thickness and width. The webs may be polished or embossed, rigid or flexible.

Proper calendering requires precise control of roll temperature, pressure, and speed of rotation. An embossed design can be produced on the surface by using an engraved roll, and calendering a mixture of granular resin chips of varying color may produce unusual decorative effects, such as marbleization. This last technique is widely used in the manufacture of flooring compositions.

After rolling to the thickness required, the sheeted material is cooled and wound up in a roll for storage and shipment. Often, mixing rolls that perform the compounding operation on the plastic with other ingredients usually precede the calendering rolls.

Calendering often processes the vinyl plastics. The complete equipment usually consists of a mixer, such as a Banbury mixer, followed by the heated rolls, chilled rolls, and finally a windup roll. The windup roll controls the tension on the film or sheeting as it moves through the calender rolls. Calenders are generally designed to meet the specific needs of the customer. Once installed and operating continuously, the cost per pound of film or sheet is lower than by any other process, such as extrusion.

The capital cost for a calendering line will average $10 million. One line for processing PVC sheet, probably the largest in the world, was built by Kleinewefers Kunststoffanlagen GmbH in Munich, Germany. Cost for this five-roll L-type configuration was $40 million (2009). It has 3,500-mm roll-face widths and 770-mm diameters with an output rate of 4,000 kg/h.

Calendering is a highly developed art and theory, rather recently elucidated by combining the complexity of melt behavior with the mechanics of screw-rotating machines. Thermal properties, molecular characteristics, and degrees of crystallinity of the plastics are factors that affect the processability of plastics. Plastics that melt to a rather low viscosity are not suitable for calendering. Additives can have a major influence on processability. With this understanding comes the ability to (a) make calenders more productive by increasing their speed, (b) produce films and sheets with tighter thickness tolerances and greater uniformity, and (c) handle thicker sheets more effectively.

EQUIPMENT

The target for the calender is to provide sufficient energy to convert the mass of plastic into film or sheet form without supplying so much heat as to cause degradation. This is a very important consideration, particularly when processing rigid PVC.

Calenders vary in respect to the number of rolls and their arrangements (Fig. 9.5). Examples of the layout of the rolls are the true L, conventional inverted L, reverse-fed inverted L, I, Z, and so on. The function of these large-diameter heated rolls is to convert the high-viscosity plastic melt into film or sheet. Figure 9.6 provides nomenclature for calender parts (569). Figures 9.7 to 9.10 provide examples of calender lines.

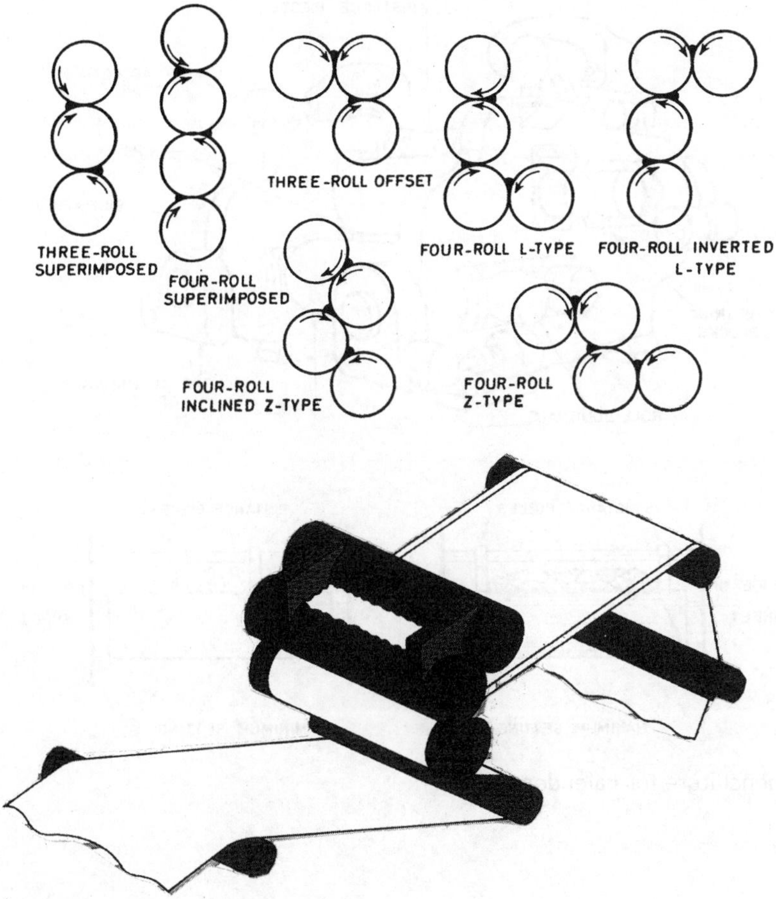

Figure 9.5 Examples of the arrangements of rolls

Roll Design

Rolls are normally numbered from the feed nip to the takeoff as shown in Figure 9.11 (569). These superimposed calender rolls were the earliest types to be used in the plastics industry in the days when a plastics calender was a modified secondhand rubber machine. These suffered from a variety of disadvantages, such as:

1. gauge variations due to spur wheel and pinion drive systems;
2. extreme difficulty in feeding horizontal nips;
3. temperature variations due to using cored rolls instead of drilled rolls;
4. no facilities for cross-axis or roll-bending adjustments.

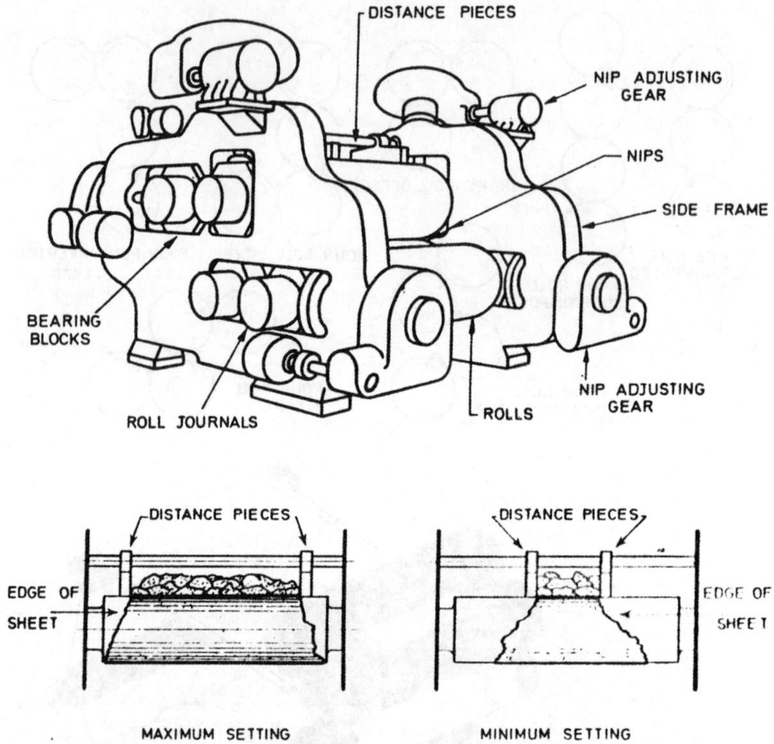

Figure 9.6 Nomenclature for calender parts

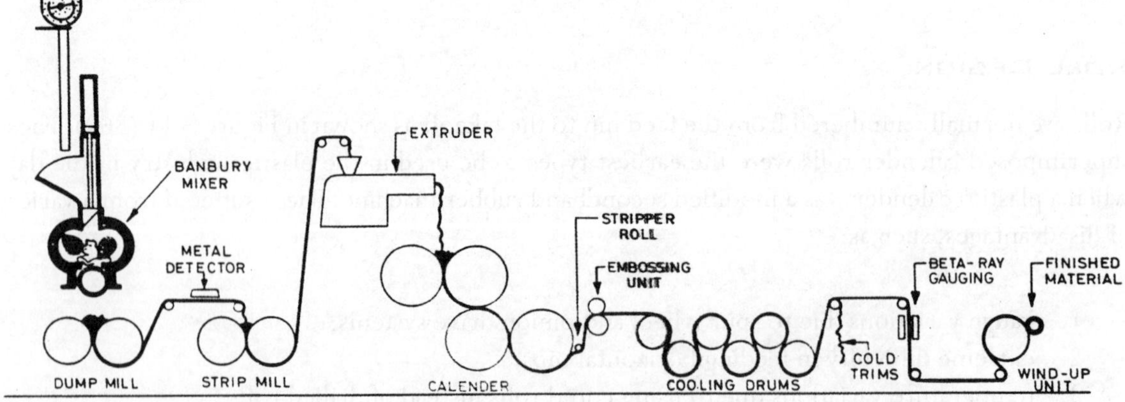

Figure 9.7 Calender layout starting with mixers

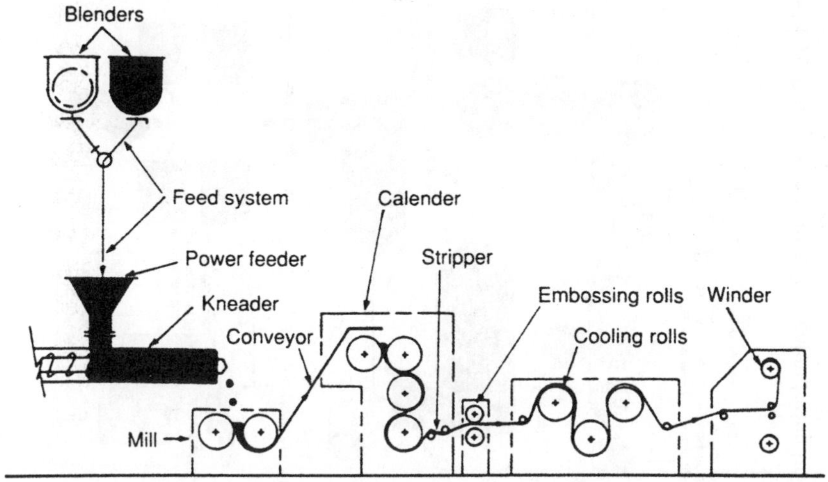

Figure 9.8 Calender layout starting with blenders and kneader

In addition to these obvious disadvantages, there was the problem of roll float. As the pressure of material in the feed nip varied, the number one roll lifted and dropped under its own weight. The number two roll floated to the full extent of the bearing clearances due to varying material pressure on either side of the roll. This position was at its worst in the case of a four-roll superimposed calender where the number one, two, and three rolls all floated. On both types of calender, only the bottom roll was stable. On this type of machine, the roll working faces were very accessible, and in the case of the four-roll type, the takeoff could be in direction A or B, depending on the requirements. It was possible to modify these machines by the addition of pullbacks, roller bearings, roll bending, and so on, but since these operations are very costly, the money might be better invested in a new machine.

Figure 9.12 shows offset calender rolls that were originally developed from superimposed machines in order to eliminate the major difficulty of the horizontal feed nip. Because the material drops by gravity into the vertical pass, the offset feed nip provides important savings in manpower and yield. Also, the pressure fluctuations of the feed to the other nips are minimized by virtue of the fact that roll number two will tend to float horizontally rather than vertically in relation to roll number three.

This degree of float still produces some gauge variation that must be overcome by fitting pull-backs to roll numbers one and two. However, roll number three remains a floating roll that may be stabilized by fitting roller bearings. Cross-axis and/or roll bending may be fitted to roll number three or four. Having the offset feed roll makes these calenders much more compact than the superimposed machines. Each roll, however, is still easily accessible for starting up and operating the machine.

Next in the development to improve the calendering process was the Z calendar (Fig. 9.13). They were developed in order to eliminate the floating number three roll on a calender fitted with bush bearings. Thus, each roll could be preloaded onto its bearings at a point that was the resultant of the material pressures and the roll weight as shown in Figure 9.14.

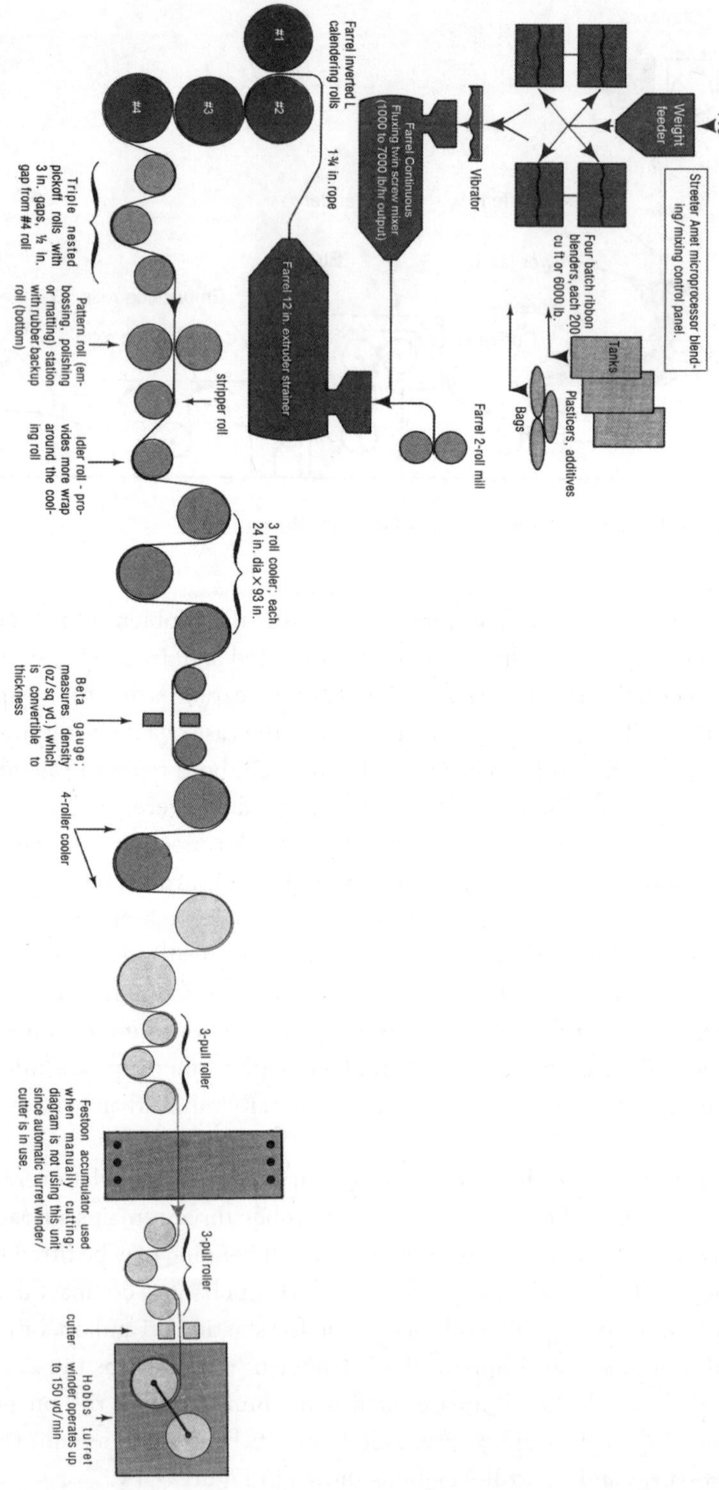

Figure 9.9 Details of a PVC calendering line

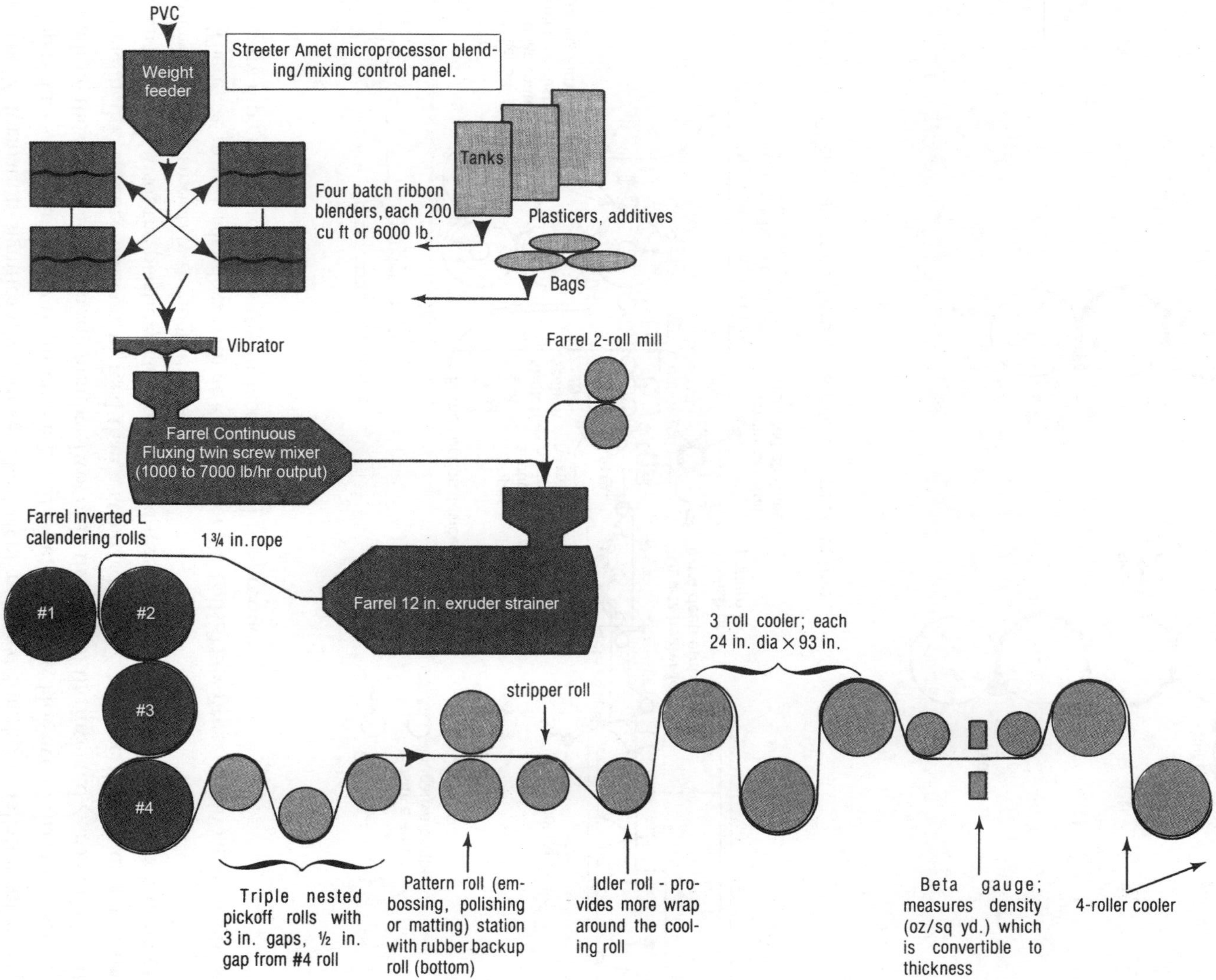

Figure 9.9 *(continued)*

The Z formation significantly reduces the height required for the installation of rolls. This action reduces the plant space requirement and the building cost. However, it has the disadvantage of limiting the ease of access to roll number two, or in the case of the inverted Z, roll number three. The inclined Z is more difficult to feed than a standard type Z because the nip does not hold as much material. Also, the plastic on top of the feed mass tends to drop back onto roll number two rather than feed smoothly through the pass.

There are in-use calenders with at least five or six rolls. They process thin rigid sheets where the extra nip greatly improves the surface finish of the product.

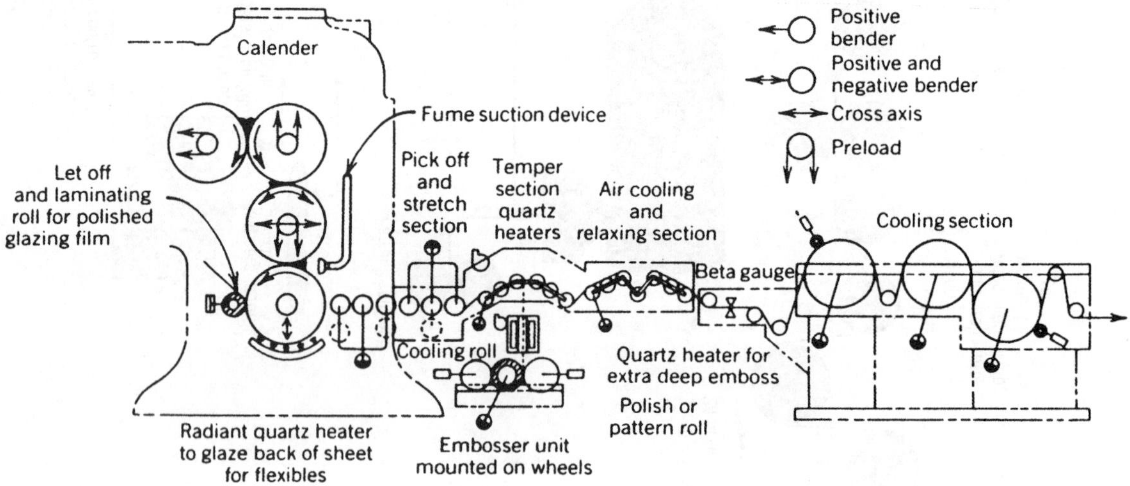

Figure 9.10 Operations going through a PVC calendering line

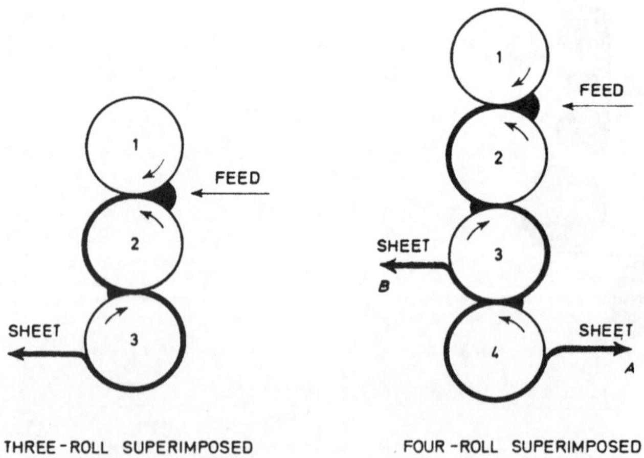

Figure 9.11 Feed and sheet plastic movement on superimposed calenders

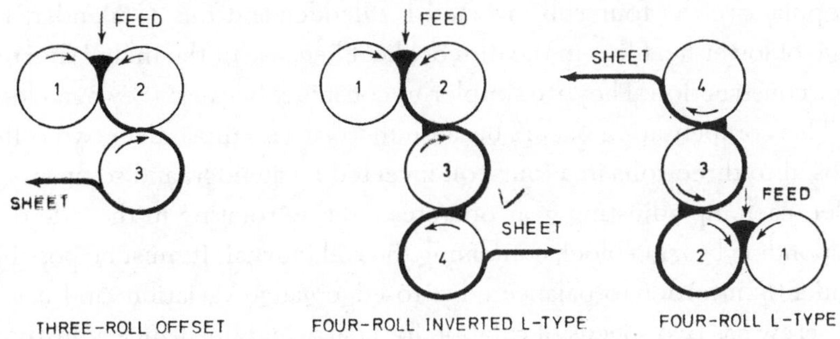

Figure 9.12 Feed and sheet plastic movement on offset calenders

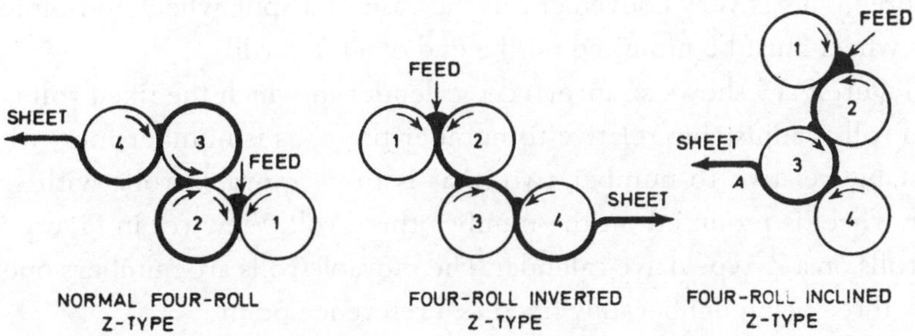

Figure 9.13 Feed and sheet plastic movement on Z calenders

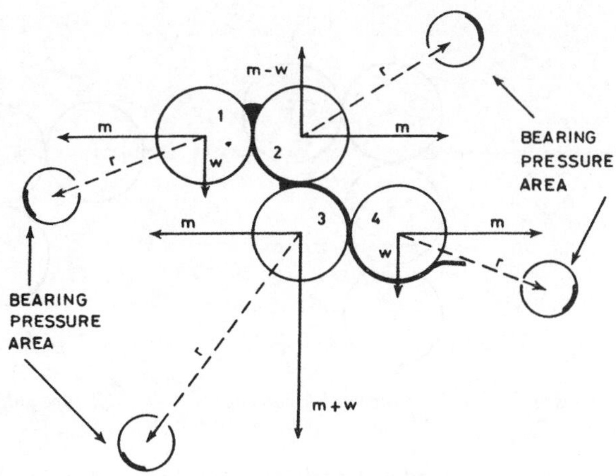

Figure 9.14 Example of preloading areas on Z calender bearings

The most popular are the four-roll inverted L calender and the Z calender. The Z calenders have the advantage of lower heat loss in the film or sheet because of the melts' shorter travel and the machines' simpler construction. They are simpler to construct because they need less compensation for roll bending. This compensation occurs because there are no more than two rolls in any vertical direction as opposed to three rolls in a four-roll inverted L calender, and so on.

On a calender, each nip adjusting gear operates a screw rotating in the side of the frame; this moves the corresponding bearing block containing the roll journal. It must be possible to move each roll end independently in order to balance edge-to-edge gauge variation, and it is preferable that the nip adjusting screw has two speeds of rotation for coarse and fine gauge adjustment. The normal type of drive for this operation gives a 10:1 reduction, that is, when the screw is rotating it is possible to alter the gauge of the sheet by 0.001 in/s (0.025 mm/s) or 0.0001 in/s (0.0025 mm/s). It is not necessary to alter the positions of all the calender rolls when adjusting gauge. On any calender, one roll is fixed. View (a) in Figure 9.15 shows a three-roll superimposed calender, the fixed roll being number two; the number one roll can be adjusted toward and away from number two, as can number three. This is very convenient in the case of a spur wheel-and-pinion drive system because the spur wheel must be mounted on the end of a fixed roll.

View (b) in Figure 9.15 shows an inverted L calender in which the fixed roll is number three. The number two roll is adjustable relative to number three, as is number four, while the number one roll is adjustable relative to number two, that is three movable rolls with six nip adjusting screws. The spur wheel is mounted on the number three roll. View (c) in Figure 9.15 shows the arrangement of rolls on a Z-type drive calender. The movable rolls are numbers one, two, and four, with roll number three being deliberately fixed as a reference point.

The speed of the calendering rolls is rarely the same. They operate at different speeds to provide the best performance of the melt, particularly the required shearing action. Typical relationships of speed settings on an inverted L can be 100% for the top in-line vertical roll, roll on its right side at 0.85%, roll just below at 1.03%, and the bottom roll at 1.06%.

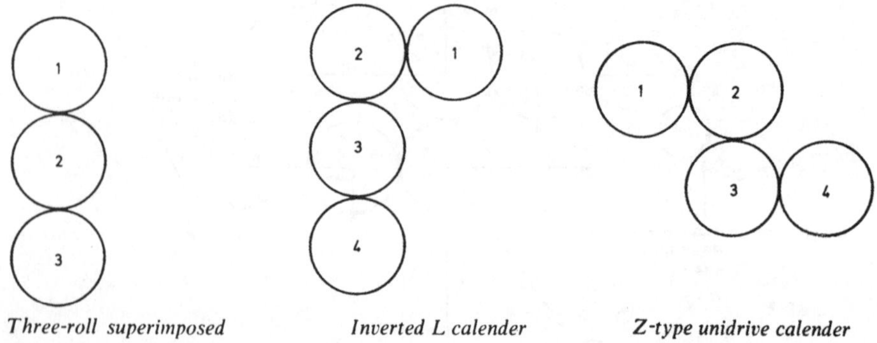

Figure 9.15 Examples of movable and fixed roll positions: (a) three-roll calender, (b) inverted L calender, and (c) Z roll calender

In roll location designs, the first roll is mounted beside the second roll. There are also designs where the first roll is next to the third roll so that four gaps are still available. Roll/film wrap length, however, is reduced so that dwell time is 40% less, resulting in the decrease of PVC heat loss and shrinkage.

PRESSURE ON ROLL

Variations in these multimillion-dollar calender lines are dictated by the very high forces exerted on the rolls to squeeze the plastic melt into thin film or sheet web constructions. High forces at least up to 6,000 psi (40 MPa) can bend or deflect the rolls. Calender bowl deflection is the distortion suffered by calender rolls resulting from the pressure of the plastic or elastomer running between them. If not corrected, the deflection produces film or sheets thicker in the middle than at the edges.

The amount of thrust exerted by the material depends on the following factors:

1. The temperature at which the plastic is being processed
2. The gauge of sheet being produced
3. The rheological properties of the plastic under the temperature and pressure conditions of the process
4. The width of the roll face being used
5. The speed of calendering
6. The method of feeding stock into the calender

To further complicate matters, a calender roll does not bend like a simple beam that is freely supported at each end and uniformly loaded along its length. A calender roll varies in thickness between the face and the journal, and because it rotates, the pressure distribution across the roll, inclusive of journals, is not exactly equal. It does not then deflect on conformation with the classical equation but in such a manner as to give a profile observed to be similar to an oxbow. This difference is a very real factor when an attempt is being made to fabricate sheet of uniform gauge that will roll out flat on a cutting table. The amount of deflection is minimized as far as possible by having stubby calender rolls where the working face width is just slightly more than twice the roll diameter. It is still essential, however, to grind the calender rolls to a special profile (crown) in order to compensate for roll deflection. The amount of crown, or in other words, the difference in roll section radius between ends and center, will vary depending on the material being processed between zero crown and as high as +0.008 in (0.2 mm) on roll radius.

The machines are set up so that the required adjustments can be made to correct this distortion. This situation can be counteracted by different methods such as the following:

1. Less deflection at high operating conditions can be achieved by the use of stiffer rolls based on higher-modulus steels or dual-metal construction.
2. Rolls are crowned resulting in having a greater diameter in the middle.

3. Crossing the rolls slightly (rather than having them truly parallel) results in increasing the nip opening at both ends of the roll.
4. Roll is bent so that the bending moment is applied to the end of each roll by having a second bearing on each roll neck, which is then loaded by a hydraulic cylinder; controls are used to perform any roll bending and crossing of the rolls.

Compensation for deflection of calender rolls can be obtained by combining the three methods of axis crossing and skewing, roll straightening (also called roll counter-bending), and roll crown (also called roll profile) where the roll crown may be ground or blocked.

Axis crossing increases the roll separation at the ends but does not change it in the middle. Roll straightening involves the application of external bending moments to both ends of the shaft of a calender roll in order to compensate for deflection (+) or excessive roll crown (–). With the usual approach, positive straightening is used to compensate for deflection caused by roll-separating forces in the nip of a calender; negative straightening is used to reduce excessive fixed crown on a roll in which it has been ground (Table 9.1).

A roll can only be contoured for one type of deflection pattern produced by a material of one flexibility, one thickness, one calendering speed, and so forth. A roll with a crown of 0.002 in (0.050 mm) may deflect as much as 0.010 in (0.25 mm), depending on the type of material being processed (569). This deflection produces sheet that is excessively thick in the center compared with the edges. To compensate for the variable deflection patterns, a device called cross-axis adjustment or screw-roll adjustment is fitted to most calenders. This operates on the principle of crossing the center lines of the two rolls in order to increase the end clearance between the rolls. As the end clearance increases, the sheet thickens at the edge relative to the center. The method of adjustment

$$F = 2vVnw(1/h_o - 1/H)$$

where: F = total separating force, MPa

v = viscosity of stock being calendered, Pa

V = peripheral speed of rolls, m/s

r = radius of rolls, M

w = width of web being calendered, m

h_o = separation between rolls at the nip, mm

H = height of melt stock at the nip entrance, mm.

Table 9.1 Example of an equation to calculate rolls' separating force

is similar to the nip adjusting gear, that is, the bearing blocks are moved in the side frames by rotating screws. Normally, the rotating screws are linked together by operating through a common shaft having a left-hand and right-hand thread. This ensures a symmetrical crossing of the rolls. The increase of end clearance can be calculated using Figure 9.16 as follows:

A. Amount of lateral end adjustment
B. Distance between centers before crossing
C. Clearance = $\sqrt{(A^2 + B^2)} - B$

It is reported that this edge clearance is theoretically equivalent to increasing the crown on the roll radius at the center. In Figure 9.17, one roll is assumed to be ground parallel as a reference. H is the height of the crown and D is the width of working face before crossing. After crossing, the working face width alters to d but H remains constant. If the curve is extended to the full working face width D, the crown has theoretically increased by the amount $(h - H)$.

The amount of cross-axis movement available is small as compared with the roll diameter of 1½ to 2 in (40 to 50 mm) on a roll diameter of approximately 30 in (760 mm). This small action has the effect of an apparent crown increase of up to 0.030 in (0.75 mm) on radius. Cross-axis roll adjustment is normally fitted either to roll number three or four on an inverted L calender or to roll number four on a Z-type calender. It is possible to fit cross-axis adjustment to more than one calender roll, for example, numbers one to four on a Z-type machine if the greater flexibility obtained is worth the extra expenditure.

Cross-axis adjustment has, however, a few disadvantages. It requires a universal coupling drive system, and the greater the degree of roll crossing, the greater the oxbow deflection effect. Also, cross-axis adjustment can only increase the apparent crown; it cannot decrease it. If the heights of the roll crowns are excessive to start with, roll crossings will only worsen the situation. Very careful roll contouring is required to compensate for these disadvantages (569).

TEMPERATURE

Calenders require high temperatures with little variations or fluctuations across the rolls during the application of the high forces or pressures. For example, during calendering thin PVC film, the melt stock temperature may be 190°C (375°F) compared with heavier gauge sheeting where 177°C (350°F) stock temperatures are usually required. For the most part, this difference is due to

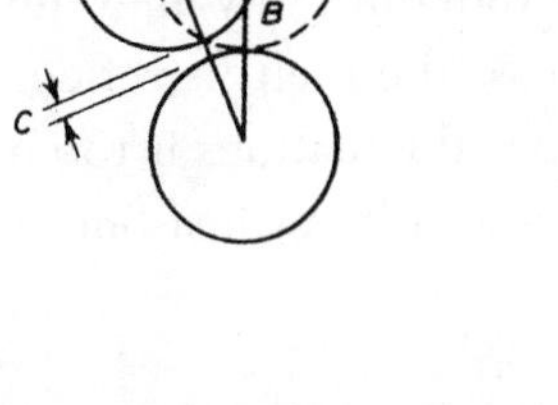

Figure 9.16 Cross-axis movement

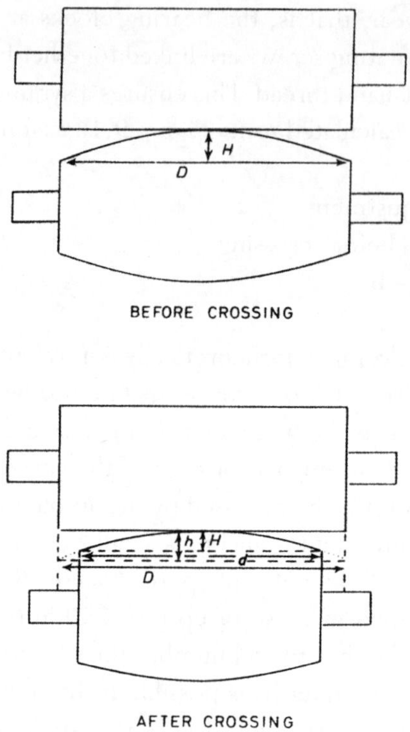

Figure 9.17 Example of effect of cross-axis adjustment to a calender roll

the higher calendering speeds, resulting in generating more frictional heat required for economical production of thin-gauge films.

The typical four-roll calender actually is a form of extrusion, with driven rotating die lips (rolls). Extrusion through the nip is dependent on friction between the stock and the roll faces, stock viscoelasticity, and hydrodynamic forces. The cooperating four rolls form three nips: the first considered a feed pass, the second metering, the final a sheet formation, gauging, and finishing pass. Material tends to track the faster and/or hotter of two rolls, and the product is directed by adjusting roll temperatures and speeds. Roll loads run 1,000 to 2,000 lb/linear in of roll face for soft sheeting, and occasionally approach 5,000 lb/linear in for thin rigid material processed cool at 330°F on larger rolls. Total connected horsepower can run from 2.5 yd/min on 24-in calenders to as much as 8 to 10 yd/min for a large 36 by 97 in machine on tough plastics.

The high forces with the temperature controls are necessary to squeeze the plastic mass into relatively thin sheets, more so for very thin film. Any unevenness in the forces along the roll, which could include uneven temperature across the melt, is reflected as variations in the product thickness. Mechanically, one cause of pressure fluctuations is too much clearance in the bearings, which can be resolved either by pulling back the rolls against one side of the bearing or by using tapered roller bearings.

Other causes of thickness changes across the calendered web include nonhomogeneous rheology of the stock (chapter 1), problems with material's lubricity, problems with pressure and temperature sensors, equipment line control malfunctioning, and use of damaged calender rolls.

Cooling of film and sheet occurs after forming through these multiple rolls of the calender. The sheet or film immediately passes through precision-surfaced cooling rolls that are kept at precisely controlled temperatures and/or a cooling tower where the web can be festooned. The temperature of the rolls is gradually reduced as the plastic travels downstream. There may be up to twenty cooling rolls. If embossing is to be applied, the embossing roll precedes these cooling rolls. After leaving the last large diameter calendering heated rolls, the film is dropped vertically into an embosser, usually with three rolls that are the embossing rolls themselves, a cooling rubber roll, and a contact cooling to the rubber roll.

These rolls are usually driven and temperature controlled in sections between two and five rolls. Temperature accuracy can be controlled within ±1°C (±2°F). In line with these rolls is usually a thickness gauge (such as a beta gauge) controlling the film or sheet thickness.

CONTROL

Proper use of all these controls and speed and stretch in the takeoff train allows the production of extremely flat sheets with a profile tolerance of ±5% across and down the web. The controls can call for adjustments on different line equipment, such as the nip openings, skew, and roll bending.

Thickness gauges are usually located within this cooling section of the production line. Continuous beta scanning equipment can be used that traverses the web constantly. This provides feedback to the upstream larger rolls that makes thickness corrections automatically. Gauge controls can also be used in the neck-down section, where the hot melt web leaves the last calendering roll and the web is transferred to the first of the cooling rolls. The amount of neck-down provides an important control in the operation of the entire production line (chapter 5).

Control features in calenders provide versatility, better quality, and higher operating rates. Complete control of friction ratio gives good tracking of stock via individual drives on all calender rolls. Close tolerances on film profiles are obtained from axis crossing, roll straightening, and gap profiling of calender rolls; greater profiling accuracy is obtained by hot grinding of calender rolls at operating temperatures. Better control over stock temperature is achieved by using drilled rolls with heat-transfer fluid circulated through them at high rates and by using accurate systems to control the fluid temperature.

Automatic web-thickness profile control can be used in calendering plastics such as PVC (chapter 5). As an example, perhaps sixty-four to seventy-six individually temperature-controlled air nozzles, arranged in a row, create a series of zones across the width of the sheet. Adjusting the coating thickness applied to each zone can control the web thickness.

With vertical stacks of rolls, the separating forces exerted by the stock at the nips may be adjusted by the operating conditions so that the middle rolls defect very little. As a first approximation, it may be assumed that the third roll on a four-roll L calender or an inverted L calender does not deflect in balanced operation. Deflection of the fourth roll primarily determines gauge variation across the width of the web. For thin plastic webs, roll-separating forces in the final nip may be as high as 6,000 psi (41 MPa), which may cause considerable deflection of the final calender roll. Unless this is compensated, gauge variation will occur across the web.

Good drawdown and pick-off control is achieved by using tightly spaced multiroll pick-offs with grouped roll-speed controls. Helical cavities in each pick-off roll allow high circulation rates of heat-transfer fluid. Quick-opening devices on all nips reduce damage to rolls from running together in the event of power failure or stock runouts. Mechanical square-cut devices with automated tension control and roll change as the roll reaches a preset diameter ensure better windup.

For many thin, unsupported film products, proper use of drawdown permits windups to be run substantially faster than the final calender roll. Calenders and takeoffs are run almost synchronously on heavy-gauge products. Films and sheets with a high gloss taken off a highly polished final calender roll tend to stick to the roll more than their matte counterparts. Very soft webs also tend to stick to the final calender roll. The fastest calender speeds are generally obtained in a median thickness range.

Very vital to this operation is the complete removal of any metal or hard surface material. This includes microscopic particles. From the start to the end of the calendering process, extreme care has to be taken to ensure there is no contamination of the equipment or plastic being processed. Preventative maintenance of these lines and the operating environments is a continuous operation. A relatively clean room has to exist.

This cleanliness is required to ensure that damage to the expensive equipment does not occur upstream or downstream. A micron-sized piece of metal will destroy the rolls, and so on. Replacing these rolls is very expensive and time consuming. Where this type of equipment may not be in the spare parts room, delay in refurbishing or replacing parts is very time consuming and expensive in lost production.

Roll Disposition

With the various calender configurations in use, the inverted L or F is the most popular in comparison with S or inclined Z calenders. The caliper error on a sheet, both theoretically and in fact, is 5% of the amount of artificial crown adjustment, whether by crossing or bending. The three-roll stack is inherently more rigid and requires less correction by crossing of the middle of the stack and with an end roll.

Sheet error problem and roll-deflection correction is complicated by the fact that the rolls are not uniformly located rotating beams because of bank variations or taper. They are not a uniform cross-sectional area (internal drillings) and there is always at least a minimal amount of thermal crown. Therefore, a stiff or high-compliance calender is desirable to reduce the amount of artificial crown.

When combined with certain forms of extruders or continuous mixers, special smaller calenders that have a reduced number of rolls but retain other essential features offer an attractive compromise of rate, cost, accuracy, and operating flexibility. A sheet head or strand head extruder directly feeds the bank of a two- or three-roll calender.

DOWNSTREAM EQUIPMENT

After the heated plastic passes through the rolls, it goes through operations of stripping, embossing, cooling, trimming, and windup. Since the heated plastic clings to the calender rolls, the sheet or film does not drop off the last roll. It has to be pulled off evenly across the width of the roll. This is accomplished by the stripper roll, which is normally positioned 3 to 6 in (75 to 150 mm) from the last roll, and at a height that gives the sheet approximately 270° lap round the roll (569).

Because here the hot plastic is in contact with a comparatively cold roller, there may be a problem of plasticizer and moisture condensing on the metal surface of the stripper roll. This condensate will mark, or in the case of condensed plasticizer, attack the sheet surface. To overcome this damaging action, the stripper roll is covered with a highly absorbent material, such as cloth.

The thinner the sheet, the greater the degree of roll cling. Thus, the speed of the stripper roll must be varied with respect to the calender speed. Once the desired speed differential is set, it is maintained. As the calender speed is altered, the stripper roll speed maintains a constant ratio with the calender speed.

If the sheet is to be embossed, this function is performed in-line with the calender. The embossing nip consists of an engraved metal roll in contact with a rubber-covered steel backup roll. The principle of embossing is to mold the hot sheet from the calender, under pressure from the backup roll, into the surface of the engraved roll that is itself water cooled, usually using refrigerated water. This sets the sheet in the pattern of the engraving. The definition of the pattern that can be obtained sharpens as the cooling of the engraved roller becomes more effective.

After cooling, the plastic is trimmed at the edges and wound. Trim material can account for up to 5% of the width, depending on the line's operating efficiency. The target is to have as little trim as possible. This operation is to cool the sheet to an ambient temperature. If warm or hot sheet is wound up, high internal strains may be caused and blocking and de-embossing problems may be introduced. Ideally, sheet should be wound up at approximately 10°C (50°F).

The usual method of cooling sheet is to run it around a series of water-cooled rolls (Figs. 9.7 to 9.10). These rolls are normally 24 in (610 mm) in diameter double-skinned drums with cooling water circulating through the double skin. Hollow drums with water sprays are not recommended because of the difficulty of statically and dynamically balancing such drums for high-speed calendering. At least two to ten cooling rolls are used, depending on the thickness of sheet and the speed of the production line. Additional cooling rolls on the line permit more gradual cooling to room temperature; this eliminates a shock cooling situation that reduces physical and mechanical properties for certain plastics, particularly rigid PVCs.

Trimming is performed either on the calender or later when the sheet is cold just prior to winding. It is economically sound to trim at the calender stage where the material, because of its temperature, can be readily conveyed back to the calender feed nip, a set of rolls, an extruder feeder for recycling, or a granulator and blended with virgin material.

At the end of the line, the windup of the sheet occurs. The two usual methods of winding into rolls are center-core winding and surface batching. Not all calendered sheet is wound up into rolls. Much of it is cut into panels by rotary cutters or automatic guillotines, which may be installed instead of windup equipment. Whichever method is used, the equipment must be capable of producing precision-cut panels. This is especially important in the manufacture of products such as wall or floor tiles.

In the case of center-core winding, one end of the mandrel is fitted into a socket, which is driven. As the roll increases in size, the moment of inertia builds up and the takeup per revolution increases. Unless the drive can compensate for this, the winding tension varies throughout the roll. Methods used to overcome this situation include inserting a slipping clutch between the mandrel socket and the drive, or more commonly, to have the drive to the mandrel transmitted by a motor drive. Such action controls the sheet tension at a predetermined value regardless of the increasing diameter as the roll winds up. To facilitate roll changing, the winding station is usually duplicated, thus allowing one roll to wind while the other is being removed. By appropriately adjusting the tensions, winding can be applied to both rigid and flexible materials.

Contact batching is a low-cost, effective method of winding flexible materials. In this case, the core is in surface contact with the windup drum. The rotation of the drum surface supplies the drive to the core that is held in position by the mandrel. The method suffers from the disadvantage of the slight tension variation in the sheet arising from the change in bearing surface pressure as the roll increases in weight. This is not serious, however, in the case of a flexible material. Contact batching is not suitable for rigid materials because the extra tension required causes slip between the drum surface and the roll. Because sheet shrinkage at ambient temperatures is controlled by the tension of winding up the cooled material, it is essential that the windup station drive be independently variable but able to be tracked with the calender speed.

Auxiliary equipment can be included in the line for other functions, such as orienting (by stretching in the machine direction and/or transverse direction using the cooling rolls or setting up bioriented stretching; chapter 5), water quenching, annealing, decorating, slitting, heat sealing, festooning, and so on.

PLASTIC STOCK

Most calendered product is PVC; flexible sheet and film represents 80% of production. Other products include rigid, expanded, polished, and clear vinyl. ABSs, PEs, PPs, PSs, natural and synthetic rubbers, fluoroplastics, and others are also processed. When calendering other plastics, there is a trade-off in economy and quality. The basic limitation of the calendering process is the need to have a sufficiently broad melt index to allow as wide a heat range as possible for the process. This behavior of the plastic stock permits the material to have a relatively high melt viscosity in the banks of the calender rolls; banks indicate where two rolls meet or at the nip of the rolls.

The proper preparation of the molding compound to be processed is important. The different plastics, each with variations in type and quantity of additives, fillers, and/or reinforcements, result in providing different processing conditions and end-product performances. The stock material prepared affects how the calender is to be operated, takeoff thickness measurements, windup system requirements, and line speed controls. These preceding factors that influence the preparation of a stock directly affect the finish (glossy, semimatte, matte, laminated to a substrate, embossed, etc.) and product (fabric, plastic film or sheet, aluminum foil, etc.) required (Fig. 9.18). Other factors could include if the web is slit in-line. With the finished product, special properties may be required, such as optical clarity and mono- or biaxial orientation.

Plastic formulations include additives and fillers to produce film and sheet that meet different performance requirements. The basic polymer is usually the largest component by weight in a formulation (chapter 1). Polymer choice is a major factor overall in determining both ease of processing and the properties of the end product.

Vinyl chloride polymers may be subdivided into homopolymers, which are usually, but not exclusively, plasticized to give flexible sheeting and copolymers more usually used in the production of rigid sheeting. There is, in addition, the possibility that the homopolymers or copolymers may have been prepared by mass, suspension, or emulsion techniques, thus providing different behaviors. The determination of molecular weight distribution (MWD) is outlined in chapter 1. Table 9.2 includes types of plasticizers used to produce flexible PVCs. Table 9.3 provides examples of plasticizers used in compounds to produce different products.

The blending or compounding of the plastic with different additives and fillers is a critical part of the process. As an example, the PVC compounds require heat stabilizers in order to be properly processed. During processing, the ideal heat stabilizer system imparts heat stability and adequate lubricating characteristics to reduce or control frictional heat.

These stabilizer systems are also very efficient for plate-out resistance. Plate-out is a condition in which the calender rolls and/or embossing rolls become coated with a deposit from the compound being processed that in turn interferes with acceptable surface finish of the film or sheet. This deposit may start out as a soft, waxy material barely visible on the metallic contact surfaces of the processing equipment. When plate-out occurs, the line is shut down and the contamination removed.

COMPOUNDING/BLENDING

The calendering process prepares a plastic material or compound. It melts the material and then passes the pastelike melt through the nips of rolls as shown in Figures 9.7 to 9.10. Upstream of the calender, the plastic mix being fed may be a simple hot melt as with PE, PP, and others. However, for PVC, it is obtained by premixing the vinyl plastic, stabilizers, plasticizers, calcium carbonate, and so on. A mixer, such as a Banbury mixer, blends the PVC raw materials that are usually in powdered form with the desired additives (plasticizers, stabilizers, etc.) and fillers (calcium carbonate, etc.). It gels for a specific time and temperature period, depending on the mix. A typical period for gelling is about five to ten minutes at 120° to 160°C (248° to 320°F).

This pastelike melt may pass through a roll mill producing a web that is fed to the calender rolls. Another approach is to have the gelled material go through an extruder/compounder to produce a

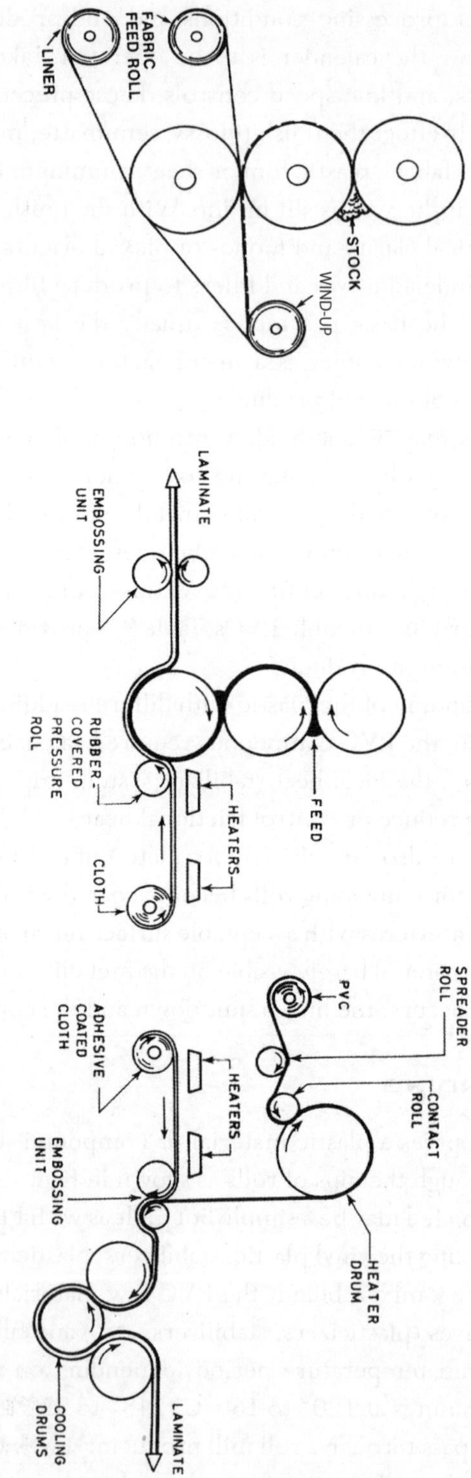

Figure 9.18 Example of contact laminating and embossing during calendering

Primary plasticisers		Secondary plasticisers		Tertiary plasticisers (Plasticiser extenders)
Di-alkyl 79 phthalate	DAP	Di-alkyl adipate	DAA	Chlorinated paraffin waxes
Di-butyl phthalate	DBP	Di-alkyl sebacate	DAS	Chlorinated paraffinic fractions
Di-capryl phthalate	DCP	Di-butyl adipate	DBA	Oil fractions
Di-(2-ethyl hexyl) phthalate (di-octyl phthalate)	DOP	Di-iso-octyl adipate	DIOA	
Di-n-octyl phthalate	DNOP	Di-iso-octyl sebacate	DIOS	
Di-iso-octyl phthalate	DIOP	Di-octyl adipate	DOA	
Di-nonyl phthalate	DNP	Di-octyl sebacate	DOS	
Di-iso-decyl phthalate	DIDP	Tri-alkyl 79 phosphate	TAP	
Tri-tolyl phosphate	TTP	Polypropylene adipate	PPA	
Tri-xylenyl phosphate	TXP	Polypropylene adipate/laurate	PPL	
		Polypropylene sebacate	PPS	
		Epoxidised Soya Bean oil		
		Epoxy stearate esters		

Table 9.2 Examples of plasticizers used to formulate flexible PVCs

Plasticiser	Parts plasticiser per 100 parts polymer	Property and use
DAP.PPA.4:1	40	Electrical adhesive tape
DAP	37	Transparent flexible windows
DAP	48	General purpose thicker sheeting
DAP. DAS. 10:1	52	General purpose for use at low temperatures (down to 0°C)
DAP	56	General purpose thin sheeting, rainwear, curtains etc.
DAP. TTP. 8:1	58	Non-flam general purpose
DIDP. PPA / DIDP. PPA / DNP : PPA — Various proportions	Quantity dependent on ratio of monomeric to polymeric but equivalent to 56 parts DAP	Thin sheeting with low soapy water extraction: baby pants etc.

Table 9.3 Examples of plasticizer blends in PVC used to produce different products

rope when exiting the machine. This rope, about 50 to 100 mm (2 to 4 in) in diameter, is fed to the calender rolls. It is directed over a conveyor belt to the nip of the top rolls. Rather than just being deposited in the center of the nip, a movable robot arm moves it from side to side, feeding the rope from almost one end of the roll to the other end. This method of delivering the rope provides for a more even distribution of the plastic.

The preparation of the material or compound is important to successful calendering. It is usually done by computer-controlled electronic weighing scales that supply precise amounts of each ingredient to a high-intensity mixer. It is designed to incorporate any liquids into the plastic particles and to secure uniform distribution of all powdered ingredients (chapter 1).

Blending is generally done for a specific time and to a specific temperature. The still-dry free-flowing blend is then charged to a feed hopper where it is screw-fed into a continuous mixer, such as an extruder and/or kneader. Under the action of a mixer's reciprocating screw in the confined volume of the mixer chamber, the blend begins to flux or masticate into the required plastic state.

The next step is to force the plastic out of the barrel of the mixing chamber through a die producing strands. The strands can exit as a continuous rope or be chopped into small baseball-sized buns. This hot plastic material may be passed through a two-roll mill and/or be directly conveyed to the top of the calender.

As reviewed in chapter 1, many different compounds are prepared to meet different processing requirements and end-item performance requirements. Preparing compounds for calendering has been an ongoing development to improve processability and performance of sheets and films. Plasticizers, heat stabilizers, and other additives and fillers are extensively used. Plasticizers, heat stabilizers, and other materials significantly improve operating performance with heat-sensitive vinyls. Examples of mixers are shown in Figure 9.19. Figure 9.20 shows examples of the action that occurs in a two-roll mill (including pressure profile) and an internal mixer widely used for solid elastomer compounding.

Coloring materials and fillers are ingredients commonly featured in compounds, including PVCs. The former are usually insoluble inert materials termed *pigments*; colors that are soluble in PVC compounds are termed *dyes*. The usual fillers are inert mineral powders and are included in formulations in order to adjust the cost or to modify the physical properties.

With a large proportion of calendered sheeting sold for decorative purposes, colors form a most important group of materials that may be added to calendering compounds. Because of the high temperatures reached in calender operations, only those pigments or dyes that will withstand sustained temperatures of over 180°C should be considered. Similarly, because calendered sheeting is subjected in use to natural weathering and the action of ultraviolet light, only colors with good resistance to attack by these agents should be used.

Most pigments suitable for coloring plastics materials are metallic salts. When choosing a pigment for use with PVC, care must be taken to avoid those pigments based on iron or zinc, because both of these metals exhibit a catalytic effect on the thermal decomposition of PVC when present in other than very small quantities. For this reason, the preferred pigments for colors other than black and white are the salts of cadmium, lead, and ultramarine. Examples of typical preferred pigments are given in Table 9.4.

In practice, it is found that colors based on lead, for example lead chromes, are less stable with respect to heat and light than similar colors based on cadmium, such as cadmium sulfide. When black sheeting is required, it is best obtained by using carbon black. If this sheeting is subsequently to be high-frequency welded or to be used as an electrical insulation, the quantity and type of carbon black must be chosen with care as the resistivity of the sheeting produced may be reduced beyond an acceptable limit. If transparent colored material is required, then dyestuffs soluble in PVC have to be used. The range of dyes available that possess adequate stability to withstand the processing conditions of PVC is small.

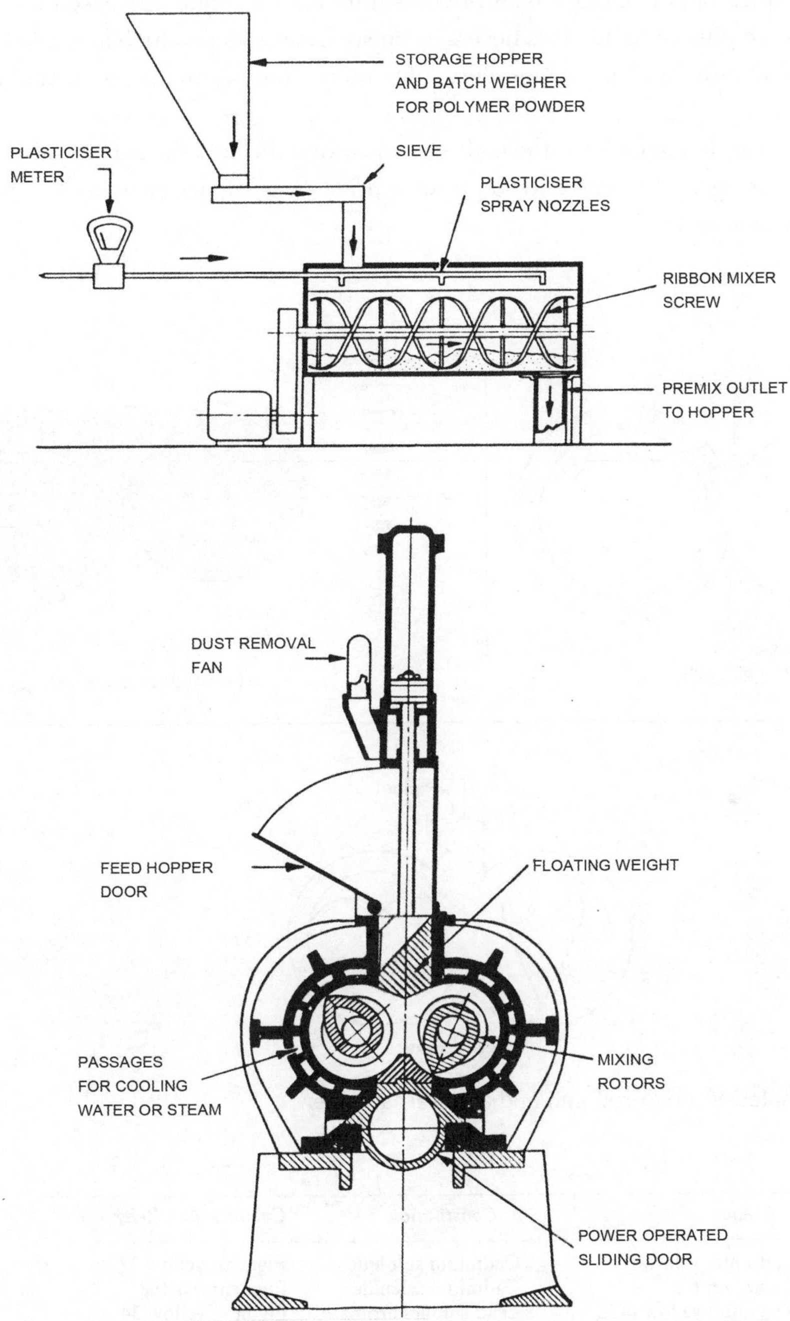

Figure 9.19 Popularly used in preparing calendering compounds are the ribbon mixer and the Banbury mixer

Where calendered sheeting is being made for use in contact with foodstuffs or for certain medical applications, the question of toxicity is of obvious importance; pigments based on compounds of lead or cadmium are unacceptable. It is necessary in such cases to produce formulations using only approved organic dyestuffs, organic pigments, phthalocyanine pigments, certain aluminum lakes, or calcium toners.

Whenever a formulation calls for the inclusion of colors, the cost should be considered carefully, because some coloring agents are extremely expensive in comparison with the other common compounding ingredients.

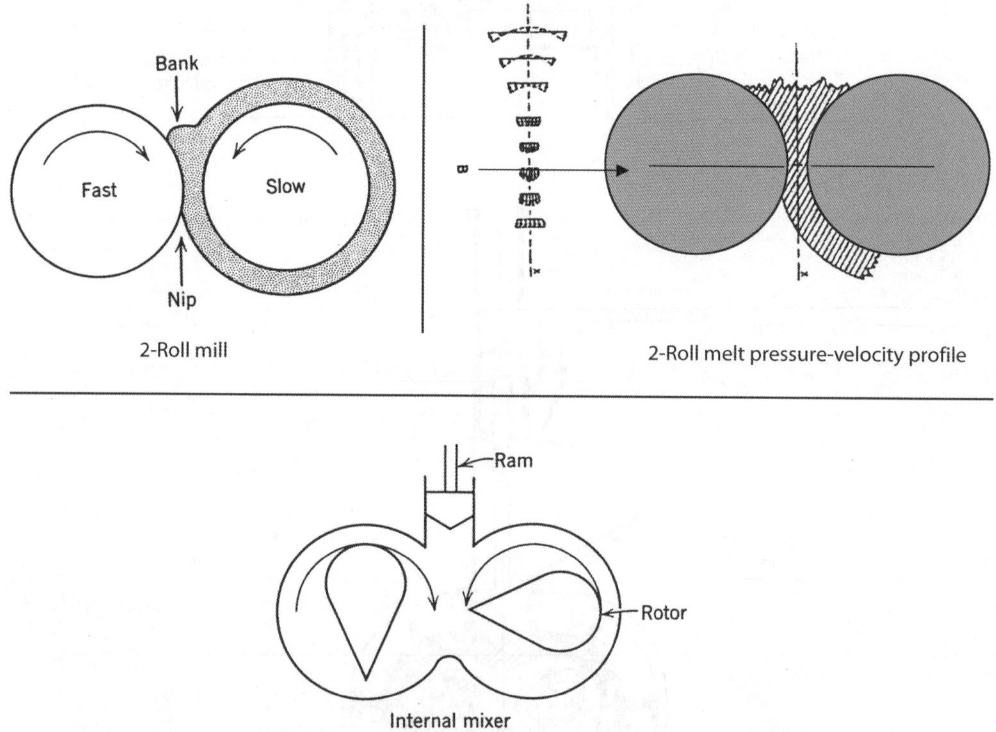

Figure 9.20 Examples of a two-roll mill and an internal mixer

Colour	*Constitution*	*Colour Index Reference*
Cadmium yellow	Cadmium sulphide	Pigment yellow 37
Cadmium red	Cadmium selenide	Pigment red 108
Vynamon yellow 6GS	Lead sulphochromate	Pigment yellow 34
Ultramarine blue	Polysulphide of sodium alumino silicate	Pigment blue 29

Table 9.4 Examples of color pigments used in PVC

Many inert inorganic substances in the form of fine powders can be added to a calendering compound with a subsequent reduction in the cost per unit weight. The most popular fillers are china clay (a hydrated aluminum silicate) and the various commercial grades of prepared chalk (calcium carbonate). Any addition of filler to a calendering compound alters the physical properties of that compound. For example, the modulus is increased, the low-temperature crack resistance is reduced, and in the case of some chalks, blooming can occur after natural aging. The greatest processing difficulty with fillers is to obtain their good dispersion throughout a compound. Some chalks are obtainable with a coating of calcium stearate or stearic acid on the particles; this helps in the dispersion of the filler, but it also limits the quantity of filler that can be included.

At processing temperatures, all vinyl plastics show some adhesion to metal surfaces. This property enables the surface finishes on calender rolls to be transferred to calendered sheeting. Adhesion to hot metal surfaces may be controlled by the addition of lubricants that are slightly incompatible waxy substances to the plastic compound.

There are two main groups of lubricants, namely, the metallic soaps that may also function as stabilizers and the waxes that are either natural or synthetic. The quantity of any particular lubricant that needs to be added to a compound in order to give a desired degree of roll release will vary from plant to plant and formulation to formulation. It is a good general rule to keep the level to a minimum consistent with the production of good quality sheeting. This level rarely exceeds two to five parts per hundred parts resin in plasticized compounds. Some lubricants, such as lead stearate and calcium stearate, are suitable only for opaque or translucent PVC sheeting. Stearic acid or acid-modified wax gives sheeting excellent clarity when incorporated at not more than five parts per hundred parts resin in either flexible or rigid formulations.

Lubricant choice and quantity will control whether the sheeting will adhere to the hotter or cooler of any two consecutive rolls. For this reason, choice of lubricant for a compound is rather specific. It depends on calendering operating conditions, calender roll surfaces, and the end-use requirements.

PROCESSING

Because the plastic is processed between the required heat and its critical heat of degradation, the time of heat becomes extremely critical and an important part of the complete process. For example, the processor will minimize the amount of melt in the bank (nip) of the rolls. The residence time of the plastic flux at high heat must be limited. PVC is especially sensitive to heat and amount of time at heat. This situation is not a problem if the machine controls are properly set and operated within set conditions. The processing variables of a PVC resin, such as flow, heat stability, and softening point, are strongly influenced by polymerization technique, MWD, and the extent of any polymerization.

As a result of the plastic's viscosity, a shear effect can be developed throughout the process. This shear is of prime importance between the calender rolls. Thus, the calender forms the web as a continuous extrusion between the rolls. Unlike when processing just through a conventional extrusion line, the plastic mass cannot be confined when being calendered. Because of the lack of confinement, the shear effect and a broad melt band are essential aspects of calendering.

Rigid PVC manufacturers usual prefer the L configuration with four to seven rolls being fed from the floor level. Since there are no disturbing vapors (condensing on rolls and web) from lower calender rolls within the pick-off area, it is preferable to have the pick-off rolls on an elevated level. Flexible PVC is commonly processed using a four-roll inverted L or an F calender. These systems enable the plasticizer-saturated vapors to escape via the usual suction hood located above the calender, where they are filtered before being released to the atmosphere.

A universal five-roll L calender is used for rigid or flexible PVC film. It provides heat stability and superior film control with good surface appearance. The major differences between this universal machine and the others are in mounting and placement of the first roll.

To improve PVC melt flow, the stock is subjected to fluxing or fusion. This is the heating of the vinyl compound to produce a homogeneous mixture. Fluxing units used in calendering lines include batch-type Banbury mixers, Farrel continuous mixers (FCMs), Buss Ko-Kneaders (BKKs), and planetary gear extruders (PGEs).

The dry blend is fed into the mixer/extruder. Both proper mixing within a short dwell time and heat-transfer control contribute to an improved product. During fluxing, each particle receives the same gentle treatment, generating less heat history and producing more uniform feed rate, color, gauge thickness, web surface, and so on. The feed can discharge onto a two-roll mill. Operating this way, it provides for a second fluxing action, mainly for working in scrap or for convenience as a buffer.

The stock delivered to the first calender nip needs to be well fused, homogeneous in composition, and relatively uniform in temperature. The optimum average temperature for good fusion depends on the formulation. A rigid PVC formulation based on a medium molecular weight plastic (intrinsic viscosity of 0.90 to 1.15) has a typical optimum temperature of 180° to 190°C (355° to 375°F) at the first calender nip. For best calendering, there should be no cold volume elements below 175°C (345°F) and no hot spots above 195°C (385°F). Flexible PVC temperature conditions are shown in Table 9.5. These typical temperature readings are necessary for close control of fusion and mixing conditions.

This interaction depends on stock temperature and in turn on the performance of PVC melts. Flexible PVC is normally calendered at temperatures of 10° to 20°C (50° to 68°F) lower than rigid PVC. In flexible PVC production, a short single-screw extruder, acting as a strainer, filters out contaminants from stock before reaching the calender. This method is not applicable to rigid PVC, because it drastically increases the head pressure, and the consequent overheating would cause the stock to decompose.

Radiation film thickness gauges are used with computer systems that process the raw data and print out detailed production reports. They also control the gap in the last calender nip by varying the opening of the rolls, the axis crossing, and the roll straightening on the final calender roll. In order for these thickness gauges (also called beta gauges) to work effectively, the web composition must be exactly known and accurately controlled. The gauges measure attenuation of

Roll no.	Sheet				Film			
	Roll temp.		Roll speed		Roll temp.		Roll speed	
	°F	°C	ft min^{-1}	m min^{-1}	°F	°C	ft min^{-1}	m min^{-1}
1	347	175	125	38	338	170	263	80
2	352	178	128	39	343	173	269	82
3	353	181	138	42	349	176	279	85
4	363	184	148	45	354	179	289	88

Table 9.5 Guide to typical four-roll temperature conditions when processing flexible PVC

radiation, which is proportional to the quantity of matter in the measured area. For known formula compositions, this can be translated into web thickness. Although good results have been reported, these gauges cannot compensate for deficiencies in older calenders or for improper control of the compounding formulations.

Bank marks are minimized by optimizing formulations, calendering speeds, and roll temperatures so as to obtain the most orderly behavior of the rolling banks of stock at the calendar nip entrances. (Bank marks are surface roughness on sheet caused by incorrect temperature or sizes of banks; bank is the quantity of plastic present in the nip formed between two rolls.) For many thin unsupported film products, proper use of drawdown permits windups to be run substantially faster than the final calender roll. Calenders and takeoffs are run almost synchronously on heavy-gauge products.

Films and sheets with a high gloss taken off a highly polished final calender roll tend to stick to the roll more than their matte counterparts. Very soft webs also tend to stick to the final calender roll. The fastest calender speeds are generally obtained using medium stiffness, moderately filled matte products in a median thickness range.

To reduce the cost of products, calendering should be done under conditions that yield the highest windup speed consistent with good quality. However, there are numerous factors that limit calender speed. At a given speed, the more highly filled and rigid stocks generate more frictional heat. Excessive speed produces degradation or sticking of many products on which the stock temperature is limited. Optimal speeds of the final calender roll for double-polished, clear melt viscosities at a given temperature degrade at about the same maximum temperatures as the rigid stocks. Air occlusion limits the speed for some heavy-gauge products, whereas the matte quality or microroughness of the film or melt fracture in the last calender nip limits the speed on others.

The amount of edge cutting influences costs. Edge cuts in excess of 3 in (76 mm) are normal to avoid undesirable edge effects. On critical products, close control of the recycle-to-virgin plastics ratio is necessary for uniformity of output. Excess recycle from critical products is used in those whose specifications permit this practice (chapters 1 and 2).

Preferably during, but also after calendering, sheet and film can be uniaxially and biaxially oriented to improve properties. Orienting methods are reviewed in chapter 5. Table 9.6 provides an example of biaxially orienting PTFE after being calendered.

	Biaxilly Oriented Unfilled Resin	Biaxilly Oriented 40% Graphite-Filled Fine Powder PTFE	35% Graphite-Filled Granular PTFE	Biaxilly Oriented 25% Glass Fiber-Filled Fine Powder PTFE	35% Glass Fiber-Filled Granular PTFE
Temperature, °C	-40	-40	-54	-40	-54
Tensile Strength, MPa	56.8	32.8	8.6	32	10.9
Break Elongation, %	85	15	6	48	8
Temperature, °C	23	23	23	23	23
Tensile Strength, MPa	35.6	16.8	6.7	18.1	7.6
Break Elongation, %	450	130	4	280	79
Temperature, °C	100	100	100	100	100
Tensile Strength, MPa	19.3	9.8	4	9.1	3.7
Break Elongation, %	380	77	5	240	144
Temperature, °C	260	260	260	260	260
Tensile Strength, MPa	9	4.9	1.8	3.2	1.5
Break Elongation, %	340	42	11	200	108

Table 9.6 Tensile properties of biaxially oriented PTFE sheeting

MARKET

Extensive use is made of calendered materials that go into different products. Products with supported flexible PVC include upholstery, luggage, water reservoirs, rainwear, loose-leaf books, and footwear. Unsupported flexible PVC is in flooring tile and material, pool liner, rainwear, packaging liquids, pressure-sensitive adhesive, shower curtains, auto interiors and trims, ditch lining, tank lining, book binders, electrical and pipe wrap tape, label tape, auto crash pads, mattress covers, crib lining, baby pants, crop covers, convertible rear windows, handbags, moisture barriers, chemical-resistant panels, and inflatable products (air beds, swim rings, children's pools, etc.). Supported rigid PVC is in window shades, wall and floor coverings, tablecloths, labels, woodgrain laminations, and book liners. Rigid PVC is in hardware and food pack, pharmaceutical pack, credit cards, lighting fixtures, ceiling tile facing, wood grain, laminate covers, signs, tank lining, and anticorrosive duct work, thin tape, strapping tape, trays, helmet liner, and printers' products.

ABS and ABS/PVC alloys go into margarine pack, luggage, panels, and trays. Chlorinated PE is used in roofing and pond liners.

A special and important application is the coating of paper, woven and nonwoven textiles, plastic films and sheets, credit cards, roll coverings, and other substrates. A calender with three rolls is usually sufficient for one-sided coating. Four rolls are used for extremely thin coatings. Double-sided coating can be applied simultaneously on both sides of substrates, such as fabric, using a four-roll calender or sequentially by two three-roll calenders. Specialized calendering equipment is used for certain products, such as floor tiles and window curtains.

The application of flexible sheet material, called roll covering, to the surface of mandrels is used in a variety of industries, including printing, paper, textiles, steel, office machinery, plastic fabricating lines, and many others (Fig. 9.21). Their uses include the following functions: squeeze, drive, emboss, convey, protect, dye, suction, treat, pickle, paint, and print.

Different programs have been set up to troubleshoot calendering. Table 9.7 is an example of a problem and a guide to the corrective action to be taken.

CALENDERING VS. EXTRUSION

Calendering and extrusion lines (chapter 5) produce film and sheet and are used for applying plastic coatings to textiles, paper, and other supporting material. Table 9.8 provides comparisons between fabricating PVC film and sheet. Compared with calendering, the extrusion process provides flexibility, including ease of changing product thicknesses, widths, and materials and providing for short production runs.

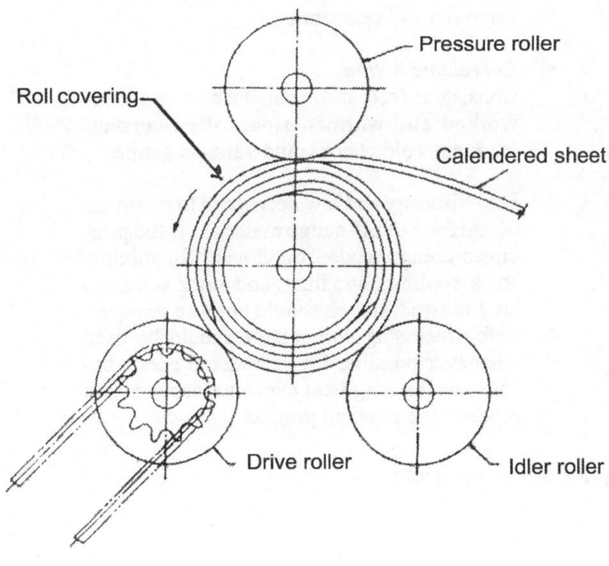

Figure 9.21 Example of roll covering

Sticking

- ***Possible Causes***
  Improper toll temperature
  Insufficient release agents
  Improper choice of oils and fillers

- ***Corrective Action***
  Suggested starting temperatures for sheet stock and friction compounds were given previously. If sticking to the top roll is experienced in calendering sheet stock, it probably means the temperature difference between the top and middle rolls is not wide enough. Heating the top roll is usually preferred over cooling the center roll because this gives smother sheets.

  with soft stocks or open-weave fabric, bottom roll temperature must be lowered on the first pass to prevent excessive strike-through. When the second side is frictioned, stick-through and penetration are less of a problem, but too cold a bottom roll may cause 'picking' of the first side onto the roll.

Feedouts*, Cold Checks**, and Variable Gauge

- ***Possible Causes***
  Cold Feed Stock
  Variable feed
  High nerve
  High viscosity
  Scorch
  Overloading the calender bank
  Improper roll temperatures

- ***Corrective Action***
  Consistent feed at the calender of uniformly worked and warmed stock often corrects feedouts, cold checks, and variable gauge.

  Low viscosity and low nerve tend to minimize occurrence of off-gauge material. If the principles recommended previously for mixing, stock cooling, handling, and stock warm-up are followed, scorch should not be a problem. Safe processing compounds should be used whenever possible for calendered goods because of the long heat exposure necessary to produce the finished product.

Table 9.7 Calendering problems/solutions

When the feed section of the calender is overloaded, the stock tends to 'pigtail' or hang out of the bank. Overloading also causes non-uniform heating of the stock which causes fluctuation in viscosity and the pressure exerted by the rolls. This invariably leads to variation in the smoothness and thickness of calendering sheets and in the penetration of friction stocks into fabric.

Variable gauge often indicates that calender temperatures are too low. This is usually accompanied by rough surfaces and/or cold checks. It may be necessary to increase the amount of release agent in the compound to permit the use of higher roll temperatures.

---

* Discontinuous sheet or coat resulting from a temporary loss of feed; also called 'holing'.

** Also called 'crows feet'; streaks caused by shrinkage of sheet in highly nervy areas after they pass through a nip.

Shrinkage

- ***Possible Causes***
  Improper roll temperatures
  Improper take-up

- ***Corrective Action***
  Shrinkage is a problem mainly in calendered unsupported sheet stock. Shrinkage can best be eliminated by calendering at high temperatures and rerolling into a fresh liner after take-up.

  Highly extended and/or low nerve compounds show little tendency to develop shrinkage problems.

Table 9.7 *(continued)*

Wrinkles

- ***Possible Cause***
  Improper take-up

- ***Corrective Action***
  A primary source of wrinkles is poorly synchronized take-up which causes buckling and foldover of the sheet. A calender set-up such as shown in Figure 1, which utilizes a spreader roll and a squeeze roll, minimizes buckling and wrinkling of the sheet. Wrinkles or splices in cloth liners invariably produce wrinkled stock. They are easily eliminated by using proper tension on the line.

Blisters

- ***Possible Causes***
  Soft Stock
  Temperature too high
  Poor choice of fillers and plasticizers
  Plies calendered too thick

- ***Corrective Action***
  Soft stock is more likely to form blisters than high viscosity material. Compounded for high viscosity will generate sufficient back pressure to force trapped air out of the stock at the calender nip. If the compound cannot be changed, running at lower temperatures or feeding the calender from the end towards the middle with a very small bank will tend to reduce the incidence of blisters.

  Sufficient squeeze roll pressure should be maintained at all times, particularly in processing stocks which are dry or highly extended.

Table 9.7 *(continued)*

The proper choice of fillers and plasticizers is important in the preparation of blister-free goods. Stocks that are too soft do not provide enough shearing action to break the trapped air bubbles. Firm stocks may also trap air, because they tend to ride in the bite rather than form a rolling bank.

Small surface blisters sometimes work out under the pressure exerted on the stock when it is rolled in a liner. Experience will reveal the size and type of blister which is likely to persist through the final curing operation. A squeeze roll should be used with stocks which tend to blister.

Picking*

- ***Possible Causes***
  Improper temperatures
  Undispersed compounding ingredients

- ***Corrective Action***
  In frictioning, when the stock picks from the center roll and adheres to the fabric being frictioned, it usually indicates that the center roll is too hot. Another cause may be non-uniform heating of the feed stock due to over-loading of the bank. Lowering the center roll temperature and maintaining consistent stock feed tends to eliminate this type of picking. When stock picks from center to top roll, the latter is probably too cold. If temperature adjustment fails to correct the picking, a small amount of release agent may have to be added to the compound. In 'second side' frictioning, picking of the first side onto the bottom roll means that the roll is too cold.

---

* The unwanted transfer of patches from one roll to another, from roll to fabric, or vice versa.

Variable Strike-Through

- ***Possible Causes***
  Variable temperature
  Variable feed
  Non-uniform fabric temperature

- ***Corrective Action***
  Variable feed and variable temperatures have been discussed previously. Both can cause variations in the viscosity of the compound

Table 9.7 *(continued)*

and thus non-uniform penetration into the fabric. Fabric which is to be frictioned should be warmed uniformly before the operation is started. The proper warming temperature depends on the type and amount of fabric used.

Ply Separation

- ***Possible Causes***
  Low temperature
  High viscosity stock
  Poor building tack
  Scorch
  Excessive liner release agent
  Low squeeze roll pressure

- ***Correction Action***
  High calender temperatures should be used to avoid ply separation. Heating the first ply as it is being fed to the calender also helps. The use of compound release agents and high calendering temperatures is preferable to cold calendering without internal lubrication.

  High viscosity, low building tack, and partial scorch are sources of ply separation. Means of eliminating these potential problems have been discussed previously.

Crushed Fabric*

- ***Possible Causes***
  Insufficient stock on middle roll
  High viscosity and high nerve
  Excessive roll pressure
  Feed stock too cold

- ***Corrective Action***
  Crushed fabric can be a problem in frictioning—usually on the second pass. This is due, primarily, to an insufficient thickness of stock on the middle roll, The coating on the middle roll should be approximately twice as thick as the fabric which is being frictioned.

  Occasionally, high viscosity or nervy stocks tend to crush or distort the fabric construction. Compounding for low viscosity and/or low nerve naturally reduces this tendency.

---

* Fabric damaged (distorted or ruptured) during frictioning

Table 9.7 *(continued)*

	Calender	Extruder-flatdie	Extruder-blown film	Extruder-flex-lip
Relative resin cost	lowest	low	higher	higher
Machine cost ($ million)	1–10	1–4	0.3–1	0.3–1
Rate and range (lb h^{-1})	800–8000	500–1500	600 ($4\frac{1}{2}$ in)	750 ($4\frac{1}{2}$ in)
Product gauge range (in)	0.002–0.050	0.002–0.005	0.001–0.003	0.001–0.125
Sheet accuracy (%)	3 (1–5)	3 (1–5)	10	10
Time to heat (h)	6	5	3	3
Time for 'on stream'	2–5 min	10 min	2 h	5 h
Gauge adjust time	seconds	seconds	5–30 min	5–30 min
Autogauging, capability	yes	yes	no	no
Color or product change time	5–30 min	10–40 min	30–60 min	30–60 min
Windup speed (vd min^{-1}) average (max.)	80 (150)	60 (80)	15 (20)	15 (30)
Limitations	High capital cost, heat time	Lower rate, versatility problem	Poor accuracy, long on stream time, low rate, degradation, reduced versatility	
Applications and advantages	Versatility, high rate, accuracy, ease and, adjustment ease at reprocess	Accuracy, gauge adjust, reduced cost	Low investment, multiplant capability, thin gauge (0.003 in and under) and heavy gauge (0.050–0.125 in)	

Table 9.8 Comparison of calendering and extrusion processes

Calendered sheet is usually less glossy than extruded material. Calendering may be preferable for certain applications requiring its higher tensile properties, product uniformity, and unusually close gauge control. Extrusion of colored films or sheets requires the extruder to be cleaned and purged when changing colors. A calender requires a minimum of cleaning between color changes. Calendering has to be used for long production runs in order to be economically profitable, producing smooth and other finishes at higher speeds.

In general, plastic materials, such as PE, PP, and PS film and sheet, are usually produced through the rather conventional extrusion lines. To produce PVC film and sheet in large quantities, calendering is almost always used since the process is less likely than extrusion to cause degradation and has dimensional and cost advantages.

The capital equipment and replacement parts in calendering lines are more expensive. A web thickness between 0.05 to 0.50 mm (0.002 to 0.020 in) is generally the kind of plasticized film and sheeting produced by calender lines. For extremely light gauges, such as those under 0.02 mm (0.001 in), calendering could become impractical or damage the equipment. For certain materials, the reasons include factors such as poor strength of the thin webs and the development of very high forces on the matting heavy-duty rolls.

For heavy/thick-gauge sheeting, such as that over 0.50 mm (0.020 in), calendering may not be the optimum method of production. The reason is that there may not be enough shearing action that can be put into the rolling banks to keep the compound at uniform temperature. In addition, the separating forces on the rolls become so low that gauge variations could become prohibitive.

In summary, the productivity of calendering the type of calendered product is significant. Heavy sheeting, the easiest product to make, can run at high speeds, depending on fluxing and feeding capacity. If the product is posttreated with laminating to a substrate, embossed, printed, or top-coated, production can be even greater since defects in the sheet can be masked.

Thin flexible film, sold straight off the calender, is difficult to make because of layflat problems, although speeds of 100 yd/min at the calender and 125 yd/min at the winder are common. Some posttreated rigid films can run at 80 yd/min, but other rigid sheets of the glossy or polished variety are limited to about 20 to 35 yd/min for top quality. Thus, the rates through a line may range from a low of 800 to a high approaching at least 9,000 lb/h.

The main disadvantages of calendering are large initial investment costs and lengthy heat-up times. The advantages that make the calender ultimately the most desirable method of all are maximum rates and speeds, accuracy of gauge, speed of gauge adjustment, processing and product range versatility, lower raw resin costs, high on-stream time factors, fast online time, and ease of accommodation of automatic gauging and control.